# The Elements

| ELEMENT | SYMBOL | ATOMIC NUMBER | ATOMIC MASS* |
|---|---|---|---|
| Actinium | Ac | 89 | (227) |
| Aluminum | Al | 13 | 26.98 |
| Americium | Am | 95 | (243) |
| Antimony | Sb | 51 | 121.8 |
| Argon | Ar | 18 | 39.95 |
| Arsenic | As | 33 | 74.92 |
| Astatine | At | 85 | (210) |
| Barium | Ba | 56 | 137.3 |
| Berkelium | Bk | 97 | (247) |
| Beryllium | Be | 4 | 9.012 |
| Bismuth | Bi | 83 | 209.0 |
| Boron | B | 5 | 10.81 |
| Bromine | Br | 35 | 79.90 |
| Cadmium | Cd | 48 | 112.4 |
| Calcium | Ca | 20 | 40.08 |
| Californium | Cf | 98 | (249) |
| Carbon | C | 6 | 12.01 |
| Cerium | Ce | 58 | 140.1 |
| Cesium | Cs | 55 | 132.9 |
| Chlorine | Cl | 17 | 35.45 |
| Chromium | Cr | 24 | 52.00 |
| Cobalt | Co | 27 | 58.93 |
| Copper | Cu | 29 | 63.55 |
| Curium | Cm | 96 | (247) |
| Dysprosium | Dy | 66 | 162.5 |
| Einsteinium | Es | 99 | (254) |
| Erbium | Er | 68 | 167.3 |
| Europium | Eu | 63 | 152.0 |
| Fermium | Fm | 100 | (253) |
| Fluorine | F | 9 | 19.00 |
| Francium | Fr | 87 | (223) |
| Gadolinium | Gd | 64 | 157.3 |
| Gallium | Ga | 31 | 69.72 |
| Germanium | Ge | 32 | 72.59 |
| Gold | Au | 79 | 197.0 |
| Hafnium | Hf | 72 | 178.5 |
| Helium | He | 2 | 4.003 |
| Holmium | Ho | 67 | 164.9 |
| Hydrogen | H | 1 | 1.008 |
| Indium | In | 49 | 114.8 |
| Iodine | I | 53 | 126.9 |
| Iridium | Ir | 77 | 192.2 |
| Iron | Fe | 26 | 55.85 |
| Krypton | Kr | 36 | 83.80 |
| Lanthanum | La | 57 | 138.9 |
| Lawrencium | Lr | 103 | (257) |
| Lead | Pb | 82 | 207.2 |
| Lithium | Li | 3 | 6.941 |
| Lutetium | Lu | 71 | 175.0 |
| Magnesium | Mg | 12 | 24.31 |
| Manganese | Mn | 25 | 54.94 |
| Mendelevium | Md | 101 | (256) |
| Mercury | Hg | 80 | 200.6 |
| Molybdenum | Mo | 42 | 95.94 |
| Neodymium | Nd | 60 | 144.2 |

| ELEMENT | SYMBOL | ATOMIC NUMBER | ATOMIC MASS* |
|---|---|---|---|
| Neon | Ne | 10 | 20.18 |
| Neptunium | Np | 93 | (244) |
| Nickel | Ni | 28 | 58.70 |
| Niobium | Nb | 41 | 92.91 |
| Nitrogen | N | 7 | 14.01 |
| Nobelium | No | 102 | (253) |
| Osmium | Os | 76 | 190.2 |
| Oxygen | O | 8 | 16.00 |
| Palladium | Pd | 46 | 106.4 |
| Phosphorus | P | 15 | 30.97 |
| Platinum | Pt | 78 | 195.1 |
| Plutonium | Pu | 94 | (242) |
| Polonium | Po | 84 | (209) |
| Potassium | K | 19 | 39.10 |
| Praseodymium | Pr | 59 | 140.9 |
| Promethium | Pm | 61 | (145) |
| Protactinium | Pa | 91 | (231) |
| Radium | Ra | 88 | (226) |
| Radon | Rn | 86 | (222) |
| Rhenium | Re | 75 | 186.2 |
| Rhodium | Rh | 45 | 102.9 |
| Rubidium | Rb | 37 | 85.47 |
| Ruthenium | Ru | 44 | 101.1 |
| Samarium | Sm | 62 | 150.4 |
| Scandium | Sc | 21 | 44.96 |
| Selenium | Se | 34 | 78.96 |
| Silicon | Si | 14 | 28.09 |
| Silver | Ag | 47 | 107.9 |
| Sodium | Na | 11 | 22.99 |
| Strontium | Sr | 38 | 87.62 |
| Sulfur | S | 16 | 32.07 |
| Tantalum | Ta | 73 | 180.9 |
| Technetium | Tc | 43 | (98) |
| Tellurium | Te | 52 | 127.6 |
| Terbium | Tb | 65 | 158.9 |
| Thallium | Tl | 81 | 204.4 |
| Thorium | Th | 90 | 232.0 |
| Thulium | Tm | 69 | 168.9 |
| Tin | Sn | 50 | 118.7 |
| Titanium | Ti | 22 | 47.90 |
| Tungsten | W | 74 | 183.9 |
| Unnilennium** | Une | 109 | (267) |
| Unnilhexium | Unh | 106 | (263) |
| Unniloctium | Uno | 108 | (265) |
| Unnilpentium | Unp | 105 | (262) |
| Unnilquadium | Unq | 104 | (261) |
| Unnilseptium | Uns | 107 | (262) |
| Uranium | U | 92 | 238.0 |
| Vanadium | V | 23 | 50.94 |
| Xenon | Xe | 54 | 131.3 |
| Ytterbium | Yb | 70 | 173.0 |
| Yttrium | Y | 39 | 88.91 |
| Zinc | Zn | 30 | 65.39 |
| Zirconium | Zr | 40 | 91.22 |

*All atomic masses have four significant figures. Values in parentheses represent the mass number of the most stable isotope.

**Although final approval has not yet been obtained, the IUPAC has recommended the following names and symbols for elements 104 through 109: 104, Dubnium (Db); 105, Joliotium (Jl); 106, Rutherfordium (Rf); 107, Bohrium (Bh); 108, Hahnium (Hn); 109, Meitnerium (Mt).

# CHEMISTRY

## The Molecular Nature of Matter and Change

# CHEMISTRY

## The Molecular Nature of Matter and Change

### Martin Silberberg

**Consultants**

L. Peter Gold
Pennsylvania State University

Charles G. Haas (emeritus)
Pennsylvania State University

Robert L. Loeschen
California State University, Long Beach

Arlan D. Norman
University of Colorado, Boulder

 Mosby

St. Louis  Baltimore  Boston  Carlsbad  Chicago  Naples  New York  Philadelphia  Portland
London  Madrid  Mexico City  Singapore  Sydney  Tokyo  Toronto  Wiesbaden

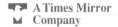

**Mosby**
Dedicated to Publishing Excellence

A Times Mirror
Company

*Editor-in-Chief:* James M. Smith
*Executive Editor:* Lloyd W. Black
*Managing Editor:* Judith Hauck
*Project Manager:* John Rogers
*Senior Production Editor:* Chris Murphy
*Manufacturing Supervisor:* Betty Richmond
*Art Developer:* Audre Newman, Martin Silberberg
*Text Development Editor:* Robin Fox
*Page Layout:* Ruth Melnick
*Special Features Designer:* Martin Silberberg, David Shaw
*Illustrations:* Michael Goodman, ArtScribe, Inc.
*Cover Design:* Michael Goodman
*Photo Research:* Donata Dettbarn

**SECOND PRINTING**

Printed in the United States of America
Composition by Graphic World, Inc.
Printing/binding by Von Hoffmann Press

Mosby–Year Book, Inc.
11830 Westline Industrial Drive
St. Louis, Missouri 63146

**International Standard Book Number 0-8151-8505-7**

96  97  98  99   00 / 9  8  7  6  5  4  3  2

*To Ruth,*
*my co-author in all things,*
*with deepest love and gratitude*

*To Daniel,*
*my sunshine, who makes me happy*
*when skies are gray*

*and*
*To the memory of Arne,*
*whose standards of excellence*
*inspire my every effort*

# Preface

No matter what your future plans in science, chemistry is one of the most exciting and useful subjects you will ever take. A chemistry course holds a three-part challenge for you while it offers a three-part reward:

- The first part is to train your mind to visualize molecular events. Everything that occurs in the observable world—from cooking an egg to digesting it, from mining aluminum ore to opening the finished soda can, from ozone depletion to curing disease—has its basis in the unobservable world of atoms and molecules. Inconceivably small objects moving at fantastic speeds populate this molecular realm, and they interact in remarkable ways to create the everyday world around and within you. To understand any material event requires the ability to picture molecular interactions.

- The second part is to develop a logical approach to solving problems—gathering necessary information, reasoning toward a conclusion, and then testing and revising it until the problem is solved, or at least clarified. This approach is the essence of science and, in fact, of most of life's endeavors. It is indispensable to performing well in this course and in any science-related career.

- The third part is to apply the principles you learn to real-world processes. Chemistry lies at the core of natural science and has essential connections to virtually every other one. The practical relevance of its concepts will help you make informed decisions on issues ranging from personal health and lifestyle to global warming and energy conservation, and it will enlighten you about how soap works, why a cave forms, what the stars are made of, and innumerable other facts. One thing is certain: a chemistry course will always give you an answer to "why is this material relevant?"

In some ways, taking this course is like traveling abroad: both provide an exciting journey, and both require a good guidebook to lead you through unfamiliar territory. The book should provide a close-up view of the customs of the land and its inhabitants; it should resolve confusion about traditions and laws by showing you how to solve any problems you may encounter; and it should help you relate new experiences to those you've already had, so you can understand both more deeply.

The challenge of the journey is yours. Here are some of the ways that this text, serving as your guidebook, can help you meet the challenge and reap the rewards of the course.

## Visualizing the Models of Chemistry

The subtitle of the text refers to the fact that matter and its changes are ultimately molecular in nature. Text discussions and illustrations work together to emphasize this theme. Models are explained to you clearly at the observable level and then from a close-up, molecular point of view, and the accompanying illustrations depict substances side-by-side at these two levels of reality. Some figures show a reaction in photos as you might see it in the lab and magnify the view to the molecular scene. Others portray a molecular view at various points on a graph so you can imagine the events that occur along the mathematical curve.

Through the latest advances in computer graphics and chemical-information software, the book includes accurate depictions of molecules and crystals, with atoms color-coded by element. A new standard of realism is depicted with watery-looking water, shiny metals, and translucent atoms. Color consistency enhances your ability to organize concepts by always showing energy as yellow, heat orange, metals blue, and so forth. The main groups of the periodic table are also colored consistently, and three-dimensional periodic tables highlight key trends among properties. In fact, the art program is so conceptual that one way to study would be to review tables, figures, and figure legends.

This interplay between words and art helps you learn to magnify scenes in your imagination, as chemists do, and to picture invisible matter in terms of colliding spheres, twisting bonds, and crystal lattices. Such mental images improve comprehension by bridging the mind-boggling gap in size between the events you see and those that cause them.

## Thinking Logically to Solve Problems

Chemistry may not be an easy "A," but if you develop a sound approach to thinking through an idea and solving problems, you *will* do well. The text provides several ways to help you develop this approach.

**Explanations that build upon previous ideas.** Many of the central concepts of chemistry form a logical sequence that build upon one another, and the text takes advantage of this hierarchy of ideas at every opportunity. Once you know that matter cannot be created or destroyed, the text applies this idea to balancing equations and calculating reaction yields. Once you know that opposite charges attract, the text shows how this fact explains why atoms bond and why gases condense to liquids. Once you understand the arrangement of electrons in an atom, the text shows how this arrangement gives rise to molecules with characteristic shapes. The discussions consistently emphasize the logical flow of ideas and interconnection of topics, as do several of the features described later. Moreover, the art includes summary diagrams that build related ideas into one overview illustration.

**A four-step method for solving problems.** Wherever an important new skill or concept is introduced, a worked-out sample problem appears.

A four-step approach is used to facilitate your reasoning, not memorizing, toward a solution: **plan, solve, check, practice.**

- *Plan.* After the problem is stated, the steps are verbally planned to show how to move from what is known to what is unknown. In early chapters or when a common type of problem is introduced, a detailed block diagram summarizes the steps in the plan.
- *Solve.* Next, the plan is executed by naming each calculation step and then carrying out the math. By planning how to solve the problem *before* you start pushing calculator buttons, you are far more likely to get the correct answer.
- *Check.* The next step is to check that the answer makes sense both chemically and mathematically. In many cases, a rough calculation appears to confirm the answer. It's easy to make an error in a complex problem, so checking is a very useful habit to form. Sometimes, a *Comment* appears about common pitfalls, alternative approaches, or interesting sidelights.
- *Practice.* A follow-up problem that requires the same concept to solve it appears immediately after the sample problem, and a brief, worked-out solution is provided at the end of the chapter.

    **A wide range of chapter problems.** The end-of-chapter problems provide a large amount of additional practice. *Concept review questions* test your general understanding of key ideas in the chapter. *Skill-building exercises* are written in pairs, with one of each pair answered at the back of the book, so you have an identical problem with which to test yourself. To build your confidence, these exercises begin simply and increase gradually in difficulty. Then, *problems in context* apply the skills you've learned to interesting scenarios and examples. These three types of problems, which are keyed by chapter section, are followed by a group of *comprehensive problems* that are presented in any order and include problems from every section and often call on concepts and skills you learned in earlier chapters.

## Applying Ideas and Skills to the Real World

No other science is as central to your everyday life as chemistry. Virtually every product of modern society undergoes at least one chemical process before it reaches you, and every natural process, both within and outside you, is governed by chemical principles. Numerous examples of this relevance are woven throughout the text discussions, but several features highlight it.

- *Boxed essays.* Two types of essays are placed throughout the text. To show you the interdisciplinary nature of chemistry and to address some of your possible career goals, Chemical Connections essays relate a principle under discussion to a major topic in another scientific field, including physiology, geology, engineering, and environmental science. To help you realize that models depend on careful measurements, Tools of the Chemistry Laboratory essays describe key instruments and techniques in the modern practice of chemistry.
- *Margin notes.* Short, interesting sidelights that relate directly to the topic you are reading appear in the margin throughout the book often accompanied by a figure. Topics range from the composition of beeswax to a biography of Albert Einstein to the atmosphere of Jupiter.
- *Galleries* are illustrated summaries that show how molecules and products in everyday life relate to chemical principles. You'll see how motor oils

and ball-point pens work, how flashlight and car batteries differ, and why gas bubbles are round and sweat cools you, as well as many other interesting applications of chemical principles.

## Aids for Study and Review

You and your professor are not alone in wanting you to learn chemistry and do well in the course. The text has been designed to help you master the concepts and skills of chemistry in every way possible. In fact, there is help wherever you look. Take a moment to examine the parts of the book and the makeup of a typical chapter:

- *The parts of the book.* Chances are you will refer to the periodic table and alphabetical list of the elements on *the inside front cover* of the book many times throughout the course. Scan the *detailed table of contents* to get a clear sense of the topics covered, and don't be concerned if your professor rearranges the order; the topic sequence is quite flexible. Next appears a complete list of *highlighted figures and tables,* by chapter and page (more about these shortly), with data tables in colored type.

    Following the body of the text are three *appendices.* One reviews some basic mathematical operations used in chemistry. The second presents a large set of important data on the elements and compounds. The third provides answers to selected problems. Finally, a *glossary* provides you with definitions of every bold-faced term and offers another way to confirm your comprehension of a chapter. Use the *index* as the surest way of finding a particular item and those associated with it. *Inside the back cover* are tables of important physical constants, some common unit-conversion factors, and a list of figures and tables, with page number, that contain frequently used data.

- *The parts of a chapter.* Every chapter opens with a *chapter outline* of the main section titles for you to get a feel for the upcoming topics, followed by a list of *concepts and skills to review* from earlier chapters that are required to understand the chapter coming up. The introductory paragraphs include a brief overview of the topic sequence to prepare you again for what's ahead and help you place it in the context of what you've already learned. As you read through the text, note the *bold-faced terms* and their nearby definitions, *italicized points to remember,* and the *numbered equations.* The *figure legends* were written to restate and reinforce concepts, so be sure to read them. Take special note of *highlighted figures and tables,* those with a partial purple border to make them stand out. Either they contain data you will need frequently or they depict a key concept you may need to review later. Follow along carefully with the steps in a *sample problem,* and try to work the *follow-up problem* right away to be sure you've understood the ideas. As soon as you finish a section, the key points are reviewed in a *section summary.* A closing *chapter perspective* orients you to the main purpose of the chapter relative to upcoming chapters. A section for *review and reference* follows that includes key (bold-faced) terms by section, key (numbered) equations and relationships by page, answers (actually brief solutions) to the follow-up problems, and a list of sample problem titles by page. The *chapter problems* follow, with reference to relevant sample problems. Green-numbered problems are answered at the back of the book. Practice by doing as many problems as you can.

# An Overview of the Chapter Content

Your professor will be especially pleased about the flexibility of chapter topics that make up the text. The main order follows a time-honored, logical sequence that covers, in four blocks of chapters, basic concepts, atomic and molecular structure, dynamic aspects of chemical change, and certain special topics. But there are innovative treatments in each chapter as well as some novel approaches to the chemistry of the elements and to two of the most popular topics in the course, organic chemistry and biochemistry. Glance at the table of contents as you look over the following rundown on the content of the book.

**Chapters 1-6: Chemical fundamentals.** The first block of chapters introduces the science of chemistry and covers some of the concepts and skills that you'll use throughout the course. You'll learn about the equation "sentences" of chemistry, the make-up of the atmosphere, why bread is less fattening than peanut butter, and the latest approaches to renewable energy. Topics include

- The origins of chemistry, scientific units and their interconversion in calculations, and the central place of chemistry in everyday life
- The historical development of understanding atomic structure, an introduction to chemical bonding in the context of the periodic table, and the chemical language of compound names and formulas
- Balancing equations and the quantitative relationships between amount and mass of a substance
- A first look at the characteristic behavior of elements and compounds in a survey of the types and essential nature of chemical reactions
- The behavior of gases and the molecular model that explains it
- The all-important relationship between heat and chemical change

If you've had a good high-school course, many of these topics may already be familiar, so you can use this block of chapters to review and consolidate these important ideas. Don't be concerned, however, if you haven't taken an earlier course: the text begins at the beginning and assumes *no* previous work in chemistry.

**Chapters 7-12: Atomic and molecular structure.** The second block of six chapters covers one of the central themes in chemistry—how physical and chemical properties of substances emerge from the properties of their component atoms and molecules. Each chapter in this block builds upon the previous one. You'll see how carbon and lead are similar and different, how your sense of smell works, why diamond is hard, and why there is nothing else in the universe with the remarkable properties of water. Topics include

- The development of quantum theory and its application to modern atomic structure
- The electronic structure of the atoms and how element properties recur throughout the periodic table
- The major types of chemical bonding, their basis in atomic properties, and their manifestation in the properties of substances
- A more extensive treatment of covalent bonding and the central importance of molecular shape
- How atomic properties and molecular shape influence the structure and physical properties of liquids and solids
- How the forces between molecules affect the properties of solutions

**A novel approach to the descriptive chemistry of the elements.** Learning how the elements and their compounds actually look and behave makes the models that predict their properties come alive. Also, learning a core of ideas several times in different contexts solidifies that knowledge. Few topics in chemistry are more important than the periodic table and the chemistry of the elements and their compounds. While these topics appear in many places throughout the text, Chapters 2, 4, 8, the Interchapter, 13, 14, 22, and 23 include sections devoted especially to the various aspects of this topic. The Interchapter and Chapters 13, 14, and 23 are unique, so they require further description.

**The Interchapter is a conceptual overview.** Through concise text and numerous summarizing illustrations, a unique conceptual overview called Interchapter: A Mid-Course Perspective on the Properties of the Elements reviews major points from Chapters 7-12 and previews their relevance to upcoming material. You or your professor may refer to the Interchapter at this point in the course or at many other times—its content is universally applicable.

**Chapter 13 applies principles to all the main-group elements.** Through a group-by-group presentation that exemplifies principles from the previous six chapters, you'll study the smoothly changing patterns of element behavior. Illustrated "Family Portraits" summarize atomic, physical, and chemical properties of each group, while accompanying essays examine key trends in behavior and focus on especially important elements. Your professor has great flexibility in how and when to cover all or part of this material.

**Chapter 14 grounds organic chemistry in atomic properties.** In a novel approach to an exciting field, you'll see the marvelously diverse chemistry of organic and biological compounds arise inevitably from the atomic nature of carbon and a handful of its bonding partners. Names, structures, and reaction patterns of organic compounds are described, with reference to similar compounds of other elements and emphasis on the enormous molecules in polymers and organisms. As in the case of Chapter 13, your professor can cover all or part of this material at many points in the course.

**Chapter 23 applies principles to the practical chemistry of the elements.** This chapter repeats the approach of Chapter 13 in applying principles from the previous block of chapters to the behavior of the elements, but with a different emphasis and organization. Chapter 23 applies principles of kinetics, equilibrium, and thermodynamics to geological, environmental, and industrial topics. You'll see how the elements became distributed in the young Earth, how they cycle through the environment, and how we extract and use them.

**Chapters 15-20: Dynamic aspects of chemical change.** The third block of six chapters covers the central theories of physical chemistry that govern all reactions and have innumerable practical applications. You'll see how enzymes function, why ozone is becoming depleted, what we can do about acid rain, and why batteries run down. Topics include

• The field of chemical kinetics, which examines the speed of reactions, the molecular pathways they follow, and the action of catalysts

• The reversibility of all reactions, which relates directly to the extent of the chemical change: the first of three chapters on the nature of equilibrium emphasizes simple gaseous systems; the second deals with acids and bases; and the third examines other aqueous systems

• The thermodynamic driving force behind all reactions and the connection between a reaction and the work we can obtain from it

**Special topics: Chapters 21-23.** A final block of three chapters covers certain applied topics. You'll learn about the use of isotopes in medical diagnosis, the chemical steps in creating a photograph, the energy advantage of recycling aluminum cans, and how the elements form during the life cycle of a star. Topics include

- Reactions of the atomic nucleus with applications to medicine, engineering, and many other fields.
- The chemistry of the transition elements—a large group of metals, some of which are very familiar—and the unique types of compounds they form
- The industrial and environmental aspects of the chemical elements

**Integration of biochemistry.** The chemistry of living things fascinates everyone. Therefore, instead of waiting until the end of the course to cover biochemistry in a final, often-skipped chapter, the text employs some of the central ideas of biochemistry to exemplify concepts throughout the chapters, in discussions, margin notes, and Chemical Connections essays. The function of the biological macromolecules and their structures as extensions of simple organic compounds occupies a major portion of Chapter 14. Other discussions explore the importance of molecular shape in our senses of smell and sight, the role of solubility in the structure of cell membranes and the action of antibiotics, the relation of equilibrium to metabolic control, the electrochemical processes in cells, and many more themes.

## A Final Word

On both the everyday and molecular levels, the world of matter and its changes is a remarkable place. By all means, let yourself be amazed by what you learn and curious about all there is still left to learn. Have a wonderful journey—and don't forget to write to let the publisher and author know how to improve the text for those coming after you!

*Martin Silberberg*

Martin Silberberg received his B.S. in Chemistry from the City University of New York in 1966 and his Ph.D. in Chemistry from the University of Oklahoma in 1971. He then accepted a research positon at the Albert Einstein College of Medicine in New York City, where, in collaboration with scientists from Rockefeller University, he studied the chemical nature of neurotransmission and Parkinson's disease.

Following a desire to teach, in 1977 Dr. Silberberg joined the faculty of Simon's Rock College of Bard (Massachusetts), a liberal arts college known for its excellence in teaching small classes of highly motivated students. As Head of the Natural Sciences Major and Director of PreMedical Studies, he taught courses in general chemistry, organic chemistry, biochemistry, and nonmajors chemistry. This close student contact afforded him insights into how students learn chemistry, where they have difficulties, and what strategies can help them succeed.

In 1983, Dr. Silberberg decided to apply these insights in a broader context and established a college-text writing and editing company. Before writing his own text, he worked on college physics, chemistry, and biochemistry texts for several major academic publishers. He resides with his wife and child in the Berkshire Mountains of western Massachusetts, where he enjoys the rich musical life of the area, bakes bread, and chases woodchucks from his vegetable garden.

L. Peter Gold grew up in Massachusetts and obtained his undergraduate and graduate education at Harvard University. He is Professor of Chemistry at The Pennsylvania State University where he has been since 1965. During that time he has taught general chemistry (to over ten thousand students) as well as physical chemistry and physical chemistry lab. His hobbies include music, both as a performer and a listener, computers, and omnivirous reading.

Charles Haas, emeritus Professor of Chemistry at the Pennsylvania State University, earned his Ph.D. at the University of Chicago under the supervision of Norman Nachtrieb. During his thirty-seven year career at Penn State he taught general chemistry, undergraduate and graduate inorganic chemistry, and courses in chemical education for public school teachers at all levels. He was honored with both the College of Science Noll Award and the University Amoco Award for teaching. His research focus is on transition metals chemistry and the stability of coordination compounds. Since his retirement he has spent much of his time reading and traveling worldwide.

Dr. Robert Loeschen earned a B.S. in Chemistry from the University of Illinois and a Ph.D. from the University of Chicago. He joined the faculty at California State University Long Beach in 1969. Trained as an organic chemist, he teaches a wide variety of courses, including general chemistry for science majors, chemistry for nursing majors (he co-authored a text for this subject), organic chemistry for nonmajors, organic chemistry for majors, and occasionally a graduate course in his research specialty, organic photochemistry. At present, he spends half his time in the chemistry department and half of his time as Associate Dean for the College of Natural Sciences and Mathematics. When he is not teaching or deaning he plays golf or putters in his woodworking shop.

Arian Norman conducted undergraduate studies at the University of North Dakota, graduate work at Indiana University, and postdoctoral research at the University of California (Berkeley); he is a Distinguished Alumnus from the University of North Dakota. Dr. Norman is Professor of Chemistry and Biochemistry at the Unviersity of Colorado (Boulder) where he has taught general and inorganic chemistry for the past 29 years, concentrating on the teaching of molecular graphics, modeling, and visualization techniques. He is a main-group element synthetic chemist, with research interests in structure/activity relationships and new materials applications of phosphorus compounds for which he has been awarded Alfred P. Sloan and University of Colorado Council of Research and Creative Work fellowships. Professor Norman is an avid cyclist and Nordic skier; most recently, he cycled across Italy.

## A Complete Package of Instructional Supplements

Several ancillary materials have been prepared to assist you and your professor in making your learning experience as complete as possible:

### For the Student

*Student Study Guide*, by Elizabeth Weberg. This extensive study guide covers the most important points in every chapter of the text. Clearly formatted and illustrated, it develops concepts and skills in a friendly, relaxed style that forestalls student confusion. The guide contains numerous worked-out examples, key points to keep in mind, and a large number of additional problems (with answers) for self-test purposes.

*Student Solutions Manual,* coauthored by Martin Silberberg. This manual contains complete worked-out solutions to all follow-up problems and half of all the chapter problems. Each chapter of solutions opens with a summary of the text-chapter content and a list of key equations needed to solve the problems.

*Laboratory Manual,* This manual includes clear descriptions of 30 experiments, including pre-lab assignments, that accommodate a wide range of equipment and stress laboratory safety. An instructor's resource guide is available to accompany the manual.

*Molecules 3D: Molecular Modeling Software,* by Mosby and Molecular Arts Corporation. This powerful and innovative, yet inexpensive, software for Windows and Macintosh incorporates an expandable library of more than 100 structures to allow construction of virtually any molecule for 3D examination. It also allows conversion of Lewis structures to molecular shapes at the click of a button.

### For the Professor

*Instructor's Resource Manual* coauthored by Martin Silberberg. This manual, also available on disc, contains discussions of chapter purpose and approach, alternative topic sequences (including specific ways to further integrate descriptive chemistry), chapter outlines, and suggestions for video lecture demonstrations and student reading.

*Instructor's Solutions Manual* by Alan J. Pribula, Frank Milio, and Thomas Berg, Towson State University. This manual contains worked-out solutions for all chapter problems in the text.

*Test Bank* by Dennis R. Flentge, Cedarville College. This resource contains 1500 multiple-choice questions organized by chapter, with every choice based on actual student responses to exam questions.

*ESATEST III Computerized Testing System* by Engineering Software Associates. This state-of-the-art test generation software, available for IBM (DOS and Windows) and Macintosh, offers an impressive array of features, including a two-track design (Easytest for the novice and Fulltest for the expert); ability to import text and graphics; and test generation by question number, topic, and formula. The Mosby version of ESATEST III contains all questions in the Test Bank plus all chapter problems from the textbook, thus offering the most complete set of digitally editable problems available.

*Transparency Acetates and Slides.* These resources contain full-color reproductions of 254 figures from the text.

*Chemical Videodisc.* Available in standard videotape format, this disc includes most of the photographs and all of the line art from the text, in addition to 40 lecture demonstrations for your course and numerous animations of chemical phenomena.

*View-Study Image Disc for General Chemistry.* Available free to adopters, and for a small fee to students, this unique new CD-ROM software for Windows and Macintosh includes a database of artwork and photographs with accompanying captions from the text. Cross-referenced by topic, concept, and figure number, this tool allows students to print items on notecard or full-size format for review and note taking, and allows instructors to show images and create transparencies and illustrated exam materials.

# Acknowledgments

**W**riting a text of this scope is both a humbling experience and a daunting task. Thankfully, numerous talented people helped, and their collective efforts have greatly improved the final result.

First and foremost, I was privileged to have worked with four professors with whom I consulted throughout the project—Peter Gold, Chuck Haas, Bob Loeschen, and Arlan Norman. Sitting together, we forged the shape of each chapter and, in the subsequent written drafts, their seemingly limitless understanding of chemical ideas and facts has informed every page. I cannot thank them enough for their contributions to the project. Knowing they were there, in steadfast support of my efforts, made the task a lot easier.

In addition to the consultants, a large number of professors and other scientific professionals played important roles in the project. Their contributions of time and experience have added immeasurably to the relevance and accuracy of the text. It is important to me that all of them know that every effort was made to incorporate their suggestions. The following is an alphabetical list of reviewers:

David L. Adams
*Bradford College*
Robert D. Allendoerfer
*State Univ. of New York, Buffalo*
Elizabeth J. Armstrong
*Skyline College*
Irwin Becker
*Villanova Univ.*
Mark Bishop
*Monterey Peninsula College*
Donna Bogner
*Wichita State Univ.*
Robert Bohn
*Univ. of Connecticut*
Kenneth L. Busch
*Georgia Institute of Technology*
Ian Butler
*McGill Univ.*
Harvey Carroll
*Kingsborough Community College*
John Clevenger
*Truckee Meadows Community College*

Derek A. Davenport
*Purdue Univ.*
William Durham
*Univ. of Arkansas*
Helmut Eckert
*Univ. of California, Santa Barbara*
Karen Eichstadt
*Ohio Univ., Athens*
John H. Forsberg
*Saint Louis Univ.*
John J. Fortman
*Wright State Univ.*
DonnaJean Fredeen
*Southern Connecticut State Univ.*
Dennis Fujita
*Santa Rosa Junior College*
Dorothy Gabel
*Indiana Univ., Bloomington*
Donald F. Gaines
*Univ. of Wisconsin, Madison*
Patrick M. Garvey
*Des Moines Area Community College*

Edward Genser
  *California State Univ., Hayward*
Charles Greenlief
  *Emporia State Univ.*
Robert W. Hamilton
  *College of Lake County*
Kenneth Hardcastle
  *California State Univ., Northridge*
Dorothea H. Hedges
  *Texas A&M Univ.*
John Hutchinson
  *Rice Univ.*
Earl Huyser
  *Univ. of Kansas*
Richard F. Jones
  *Sinclair Community College*
Edward L. King
  *Univ. of Colorado, Boulder*
Leo Kling III
  *Faulkner State Junior College*
James Krueger
  *Oregon State Univ.*
Larry Little
  *DeKalb College*
John R. Luoma
  *Cleveland State Univ.*
Leslie J. Lyons
  *Grinnell College*
Jack McKenna
  *St. Cloud State Univ.*
Jerry L. Mills
  *Texas Technical Univ.*
Steven Murov
  *Modesto Junior College*
John H. Nelson
  *Univ. of Nevada, Reno*
Alan J. Pribula
  *Towson State Univ.*
John L. Ragle
  *Univ. of Massachusetts, Amherst*

Robert R. Reeves
  *Rensselaer Polytechnic Institute*
T. W. Richardson
  *North Georgia State Univ.*
B. Ken Robertson
  *Univ. of Missouri, Rolla*
Steve Ruis
  *American River College*
Barbara Sawrey
  *Univ. of California, San Diego*
Henry Shanfield
  *Univ. of Houston, Central Campus*
C. Frank Shaw
  *Univ. of Wisconsin, Milwaukee*
Donald Showalter
  *Univ. of Wisconsin, Stevens Point*
Mary Jane Shultz
  *Tufts Univ.*
Dennis Staley
  *Southern Illinois Univ., Edwardsville*
Lee Summerlin
  *Univ. of Alabama, Birmingham*
Tamar Y. Susskind
  *Oakland Community College*
Yi-Noo Tang
  *Texas A&M Univ.*
Robert West
  *Univ. of Wisconsin, Madison*
Robert Widing
  *Univ. of Illinois, Chicago*
David Williamson
  *California Polytechnic State Univ.*
John R. Wilson
  *Shippensburg Univ.*
George Woodbury
  *Univ. of Montana*
Linda Zarzana
  *American River College*
William Zoller
  *Univ. of Washington, Seattle*

In addition to the consultants, several professors contributed chapter problems and/or provided advice for their review and editing:

Charles Baker            *Chemical engineer*
Claire Baker             *Butler University*
William Durfee           *Univ. of Colorado, Boulder*
Dorothy Gabel            *Indiana University, Bloomington*
Ronald Garber            *California State University, Long Beach*
Stephen Hawkes           *Oregon State University*
Ronald Ragsdale          *University of Utah*
Martin L. Thompson       *Lake Forest College*
Arden Zipp               *State University of New York, Cortland*

I owe a special debt of gratitude to Professor Dorothy B. Kurland of the West Virginia Institute of Technology, who reviewed earlier drafts of several chapters and then went through the entire final draft of the text, scrupulously checking every equation, calculation, and figure. Her breadth of knowledge of chemistry and meticulous attention to detail in the context of sound pedagogy are truly remarkable. Professor Steven Woeste of Scholl College provided valuable suggestions in the galleys, and Katherine Aiken, Andrea Freedman, and Jane Hoover, all experienced science proofreaders, thoroughly examined every final page.

An important group of people outside Mosby played essential roles in the project. Robin Fox, the developmental editor, helped me restructure paragraphs and clarify phrases throughout the chapters. Conceptualizing the art for the book, surely among the most remarkable in any science text, was actually fun, thanks to the exceptional creativity, boundless patience and energy, and personal warmth of Dr. Audre Newman of Artful Education, Inc. Realizing the final pieces required unstinting dedication to accuracy and aesthetics and combined the talents of three exceptional artists: Carolyn Duffy and Greg Holt of ArtScribe, Inc., and Michael Goodman. Cambridge Scientific Corporation generously supplied the Chem-3D software used to generate accurate data for Michael's superb molecular renderings.

Laboratory photographs add an indispensable visual proof of the words. I was very fortunate to have the chemical expertise of Professor Tom Gruhn of the University of San Francisco to set up the lab shots. Darcy Lanham coordinated the excellent work of Professor Gruhn and Stephen Frisch, the photographer.

I owe a special debt of gratitude to Pat Burner, a publishing professional of the first magnitude, whose tireless efforts shepherded this enormous project through much of its earlier life.

At Mosby, I found a publisher with a commitment to the highest standards of quality and the expertise to bring those standards to fruition. My admiration and gratitude go first to Jim Smith, who acquired the project. Jim is one of a noble breed of editors who actually reads the chapters because he loves the subject matter and who always puts education of the student first. I regard Jim as an inspiration and friend. Lloyd Black, the chemistry editor, came late to the project but has energized the ancillaries and provided the resolve to maintain the high standards throughout the final stages.

There is no way to express how fortunate I am to have had Judy Hauck as the managing editor of the project. With her rare combination of creativity, intelligence, and skill, Judy made numberless editorial and production decisions. She coordinated the artwork, enhanced the design, and organized the ancillaries, all the while remaining warm and humane. Her commitment to quality and her generous support of the book and its often-frazzled author have made, to the extent that such a thing is possible, the final 18 months of this project an enjoyable experience. I will be ever grateful.

Special thanks go to Donata Dettbarn, who pleasantly received my amorphous requests for images and then searched persistently, sometimes around the globe, to obtain an outstanding group of photos, and to David Shaw, who made innumerable final adjustments to the figures and was instrumental in the design and execution of the galleries.

The production team at Mosby produced an exceptionally beautiful book, and I am very appreciative of their efforts. John Rogers managed the complex production of the text; Chris Murphy, the production editor, always responded calmly and efficiently to a grinding schedule and an obsessive author; Renée Duenow produced the handsome book design; and Roger McWilliams copyedited the manuscript.

The marketing team at Mosby, led for my project by Rhonda Rogers, developed an exciting program for presenting the book to the educational community, including a handsome, informative brochure, salesperson training sessions, and a prospectus of sample chapters.

As every author of such an undertaking knows, words of thanks, no matter how effusive, become crumbs in the face of the debt one owes to all the members of one's family. Their tireless patience with my seemingly endless self-involvement, together with their unstinting belief in me, has been an invaluable gift. Daniel was born early in the first draft, and his boundless enthusiasm and cheerfulness have quite literally helped me make it through to the end. Now that he is six, I will at last have the time for guiltless play, and I can't wait!

In the beginning, there was only Ruth. Her vast experience in editing and production imprints the entire book. We thought through the overall theme and how to manifest it, the individual ideas and how to present them, the features and how to design them. Ruth was there to help with whatever needed fixing, from the clarity of an explanation to the design of a figure. She typed the problems, cropped the photos, created the page layout, and checked the page proofs. But most important of all, she still looks at me with a sparkle in her eyes.

*Martin Silberberg*

# Contents

# Detailed Contents

# Highlighted Figures and Tables

*Certain figures and tables are highlighted with a blue-purple border.*
*Page numbers are in parentheses; entries in color contain frequently used data*

# CHEMISTRY

## The Molecular Nature of Matter and Change

# CHAPTER 1

# Keys to the Study of Chemistry

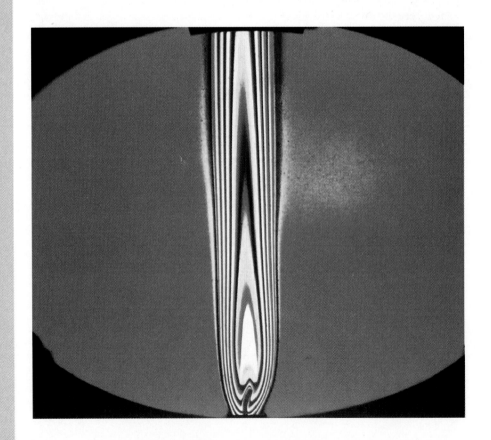

**Fire.** A symbol for change, fire is also a symbol for chemistry, the science of matter and its changes. Many themes of this chapter are evident in this view of a burning candle. When wax burns, the solid melts, the liquid vaporizes, and the gas combines with oxygen. As a result of these physical and chemical changes, other substances form and energy is given off. Insight into the process of combustion launched the science of chemistry in the 18th century. This flame is viewed through an *interferometer*, which provides quantitative data concerning the density of the combusting gases and the temperature of the flame. *Measurement*, by our senses with the aid of our instruments, is the hallmark of science.

**O**ur relationship with the physical world has undergone astonishing changes since our primitive beginnings. In the earliest reaches of human experience, people learned about the materials around them through trial and error: which stones were hard enough to chip others, which types of wood were rigid and which flexible, how to treat hides to make clothing, which plants were edible and which poisonous. They relied on knowledge acquired through everyday observation for survival. Today, the science of chemistry, with its powerful quantitative theories, helps us understand the *nature* of materials to make better use of them and even to create new ones.

As a member of a highly industrialized society, you are aware of the enormous impact that chemistry has on your life, but it might be enlightening to go through the beginning of a typical day from a chemical point of view. You wake up and throw off a thermal insulator of manufactured polymer, jump in the shower, and emulsify fatty substances on your skin and hair with chemically treated water and formulated detergents. Then, adorned in an array of processed chemicals—pleasant-smelling pigmented materials suspended in cosmetic gels, polymeric clothing, synthetic footwear, and metal-alloyed jewelry—you sit down to a bowl of nutrient-enriched, spoilage-retarded cereal and milk; a piece of fertilizer-grown, pesticide-treated fruit; and a hot cup of water-extracted, neurally stimulating alkaloid solution. You collect some books—processed cellulose and plastic, electronically printed with light- and oxygen-resistant inks—hop in your hydrocarbon-fueled, metal-vinyl-ceramic vehicle, electrically ignite a synchronized series of controlled gaseous explosions, and you're off!

Chemistry is not limited to the industrial products of modern life; its influence is much more profound. The environment—the air, natural waters, land, and organisms that thrive there—is a remarkably complex system of chemical interactions. Although modern chemical products have undeniably enhanced the quality of our lives, their manufacture and use pose increasing dangers, such as toxic waste, smog, acid rain, global warming, and ozone depletion. If our heedless application of chemical principles has led to many of these problems, our thoughtful application of the same principles will surely play a role in solving them.

Perhaps the significance of chemistry is most impressive when we ponder the chemical nature of biology. Numerous chemical events taking place within you right now allow your eyes to scan this page and your brain cells to translate fluxes of electric charge into thoughts. The most vital biological questions—How did life arise and evolve? How does an organism reproduce, grow, and age?—ultimately have chemical answers.

Comprehending the ideas of chemistry reveals a hidden level of the universe—one filled with incredibly minute particles, hurtling at fantastic speeds and colliding billions of times a second within and around you. To help you enter this world, the chapter begins with some fundamental chemical definitions and concepts. Then we examine the historical origins of the science of chemistry and the approach that chemists take to understand nature. We discuss modern systems of measurement and calculation and consider how to solve chemistry problems. Finally, we look at one example of chemistry and other sciences working together for the benefit of society.

## 1.1 Some Fundamental Definitions

Chemistry deals with an enormous subject—the entire physical universe. Perhaps the most straightforward place to begin is with the definition of a few central ideas, some of which may already be familiar to you. **Chemistry** is the study of matter and its properties, the changes that matter undergoes, and the energy associated with the changes.

### The Properties of Matter

**Matter** is the "stuff" of the universe: books, planets, trees, professors—anything that has mass and volume. (In Section 1.5, mass and volume are defined in terms of how they are measured.) Of particular interest to chemists is the **composition** of matter, the types and amounts of simpler substances that make up a sample of matter.

We investigate matter by observing its **properties,** the characteristics that give each substance a unique identity. With people, we observe such properties as height, weight, hair and eye color, fingerprints, blood type, and so on, until we obtain a unique identification. The situation is similar with a substance. Chemists distinguish between two types of properties, physical and chemical, which are closely related to two types of change that matter undergoes. **Physical properties** are those that the substance shows by itself, without interacting with another substance. Some of these are color, melting point, boiling point, electrical conductivity, and density. A **physical change** occurs when a substance alters its physical form or shape but *not* its composition. For example, when ice melts, a physical change occurs; liquid water looks different from ice because several physical properties have changed, such as hardness, density, and fluidity, but it is still water (Figure 1.1, *A*):

Physical change (same substance before and after):

Ice (solid form of water) → water (liquid form of water)

**FIGURE 1.1**

**The distinction between physical and chemical change. A,** The solid form of water undergoes a physical change to become a liquid, but its composition remains the same. **B,** A chemical change occurs when an electric current passes through water and the water breaks down into hydrogen gas *(left tube)* and oxygen gas *(right tube).*

A

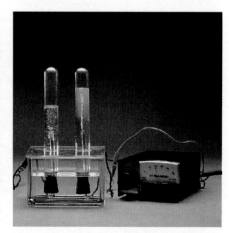

B

**TABLE 1.1 Some Characteristic Properties of Copper**

| PHYSICAL PROPERTIES | CHEMICAL PROPERTIES |
|---|---|
| Reddish brown, metallic luster | Slowly forms a green carbonate in moist air |
| Easily shaped into sheets (malleable) and wires (ductile) | |

Reacts with nitric acid and sulfuric acid

Good conductor of heat and electricity
Can be melted and mixed with zinc
   to form brass
Density = 8.95 g/cm³
Melting point = 1083°C
Boiling point = 2570°C

Slowly forms deep-blue solution in aqueous ammonia

---

**Chemical properties,** on the other hand, are those that the substance shows as it interacts with, or transforms into, other substances. Examples of chemical properties include flammability, corrosiveness, and reactivity with acids. A **chemical change,** also called a **chemical reaction,** occurs when a substance (or substances) is converted into a different substance (or substances), that is, one with different composition and therefore different properties. If you pass an electric current through water, for example, a chemical reaction occurs: the water breaks down (decomposes) into two substances, hydrogen gas and oxygen gas, each with physical and chemical properties different from each other *and* from the original water (Figure 1.1, *B*): Chemical change (different substances before and after):

$$\text{Water} \xrightarrow{\text{electric current}} \text{hydrogen gas and oxygen gas}$$

*A substance is defined by its own unique set of physical and chemical properties.* Some properties of copper appear in Table 1.1; a substance with this set of properties can be only copper.

### The Three States of Matter

Matter exists in three physical forms, or **states:** solid, liquid, and gas. Each state is defined by the way it fills a container (Figure 1.2). A **solid** has its own fixed shape, which does not necessarily conform to the container walls. Note that solids are *not* defined by rigidity or hardness: solid iron is rigid, but solid lead is flexible and solid wax is soft. A **liquid** conforms to the container shape and fills the container to the extent of its own volume; thus, a liquid forms a *surface*. A **gas** conforms to the container shape also, but it fills the *entire* container; a gas does *not* form a surface.

**FIGURE 1.2**
**The physical states of matter.** All matter exists in either the solid, liquid, or gaseous state. A solid *(left)* has its own shape regardless of the container's shape. A liquid *(center)* conforms to the container's shape but has a surface. A gas *(right)* fills the container completely. Many substances can exist in each of the three states, depending on the temperature and pressure.

◆ **The Incredible Range of Physical Change.** Scientists often study the properties of matter in remarkable settings beyond the confines of the laboratory. Far below Earth's thin crust of oceans and continents, temperatures and pressures are so great that the iron in our planet's core is molten. On Saturn's moon Titan (shown here), the spacecraft Voyager 1 measured temperatures cold enough to create blizzards of frozen methane; on Earth, methane is a component of natural gas.

Many substances can exist in all three physical states and can undergo changes in state, depending on the temperature and pressure of the surroundings. As the temperature increases, solid water melts to liquid water, and liquid water boils to gaseous water (water vapor). The reverse processes are also changes in state: with decreasing temperature, water vapor condenses to liquid water, and further cooling freezes the liquid to ice. Many other substances behave in the same way: solid iron melts to liquid (molten) iron and then at a higher temperature boils to iron gas. Cooling the iron gas changes it to liquid and then solid iron. ◆

These observations suggest that a physical change caused by a process such as heating often can be reversed by the opposite process, in this case cooling. The same is *not* true of a chemical change. Heating often causes a chemical change that cooling cannot reverse. For example, if we heat iron in moist air, a chemical reaction occurs and a brown, crumbly substance known as rust forms. Cooling does not reverse this transformation; another chemical change (or series of them) is required. When heated to a high temperature, limestone (known chemically as calcium carbonate) decomposes to lime (calcium oxide) and carbon dioxide gas, but the decomposition *cannot* be reversed by cooling: mixing lime with carbon dioxide at a low temperature will not convert them to limestone.

To recall the key distinction: a chemical change leads to a different substance (different composition), whereas a physical change merely leads to a different form of the same substance (same composition).

SAMPLE PROBLEM 1.1 _____

### Distinguishing Between Physical and Chemical Change

**Problem:** Decide whether each of the following processes is primarily a physical or a chemical change, and explain briefly:
**(a)** Frost forms as the temperature drops on a humid winter night (see photo).
**(b)** A cornstalk grows from a seed that is watered and fertilized.
**(c)** Dynamite explodes to form a mixture of gases.
**(d)** A lathe operator shapes a chair leg from a block of mahogany.
**(e)** A silver fork tarnishes in air (see photo).
**Plan:** The basic question we ask to decide whether a change is chemical or physical is, "Does the substance change composition or just change form?"
**Solution: (a)** Frost forming is a **physical change**: the temperature changes humidity (gaseous water) to ice crystals (solid water).
**(b)** A seed growing is a **chemical change**: the seed, air, fertilizer, soil, and water use energy from sunlight to undergo complex changes in composition.
**(c)** Dynamite exploding is a **chemical change**: the dynamite is converted into other substances.
**(d)** Shaping wood is a **physical change**: the lathe operator changes the wood's form, but not its composition.
**(e)** Tarnishing is a **chemical change**: silver changes to silver sulfide by reacting with sulfur-containing substances in the air.

FOLLOW-UP PROBLEM 1.1
Decide whether each of the following processes is primarily a physical or a chemical change, and explain briefly:
**(a)** Purple iodine vapor appears when solid iodine is warmed.
**(b)** Gasoline fumes are ignited by a spark in an automobile cylinder.
**(c)** A scab forms over an open cut.

Frost

Tarnish

The properties of a substance and the changes it undergoes are central to chemical investigation and are a major theme of this book. *Macroscopic* properties and behavior, such as the ones just discussed, are the results of *submicroscopic* properties and behavior. What is really happening when water boils? Why do iron and copper melt at different temperatures? What events occur in the invisible world of minute particles that cause dynamite to explode, a neon light to glow, or a nail to rust? We study *observable* changes in matter in order to understand the *unobservable* changes that cause them.

## The Importance of Energy in the Study of Matter

In general, physical and chemical changes are accompanied by energy changes. **Energy** is often defined as *the capacity to do work,* a definition that may seem vague until we look deeper. Essentially, all work involves moving something. Work is done when your arm lifts a book, when an engine moves a car's wheels, or when a falling rock moves the ground as it lands. The object doing the work (arm, engine, rock) transfers some of its energy to the object on which the work is done (book, wheels, ground).

The total energy an object possesses is the sum of its potential energy and its kinetic energy. **Potential energy** is the energy due to the *position* of the object. **Kinetic energy** is the energy due to the *motion* of the object. To illustrate the relationship between these two forms of energy, we'll examine three systems of objects: (1) a weight above the Earth, (2) two balls attached by a spring, and (3) most important to chemical systems, two electrically charged particles. A key concept illustrated by all three cases is that energy is not destroyed: it is *converted* from one form to another.

Let's first consider a weight you lift above the ground. The energy you used to move the weight against the gravitational attraction of the Earth increases the weight's potential energy (energy due to its position) (Figure 1.3, *A*). When the weight is dropped, this difference in potential energy changes to kinetic energy (energy due to its motion), and some of this kinetic energy is transferred to the ground and does work, such as driving a stake or simply moving dirt and pebbles. Note that the difference in potential energy is not destroyed: it is converted into kinetic energy. *In nature, situations of lower energy are typically favored over those of higher energy:* because the weight has less potential energy (and thus less total energy) on the ground than in the air, it will fall unless something holds it up. We can think of the situation with the weight elevated and higher in potential energy as being *less stable,* and the situation after the weight has fallen and is lower in potential energy as being *more stable.*

To bring the concept somewhat closer to chemistry, consider two balls attached by a relaxed spring (Figure 1.3, *B*). When you pull the balls apart, the energy you exert to stretch the spring increases its potential energy. This difference in potential energy is converted to kinetic energy when you release the balls and they move closer together. In the stretched position, the system of balls and spring is less stable (has more potential energy) than when the spring is relaxed. You could also increase the potential energy by pushing the balls together and compressing the spring. The balls would move apart when they are released, once again changing the potential energy into kinetic energy. Similarly, when you wind a clock, the potential energy in the mainspring increases as it is compressed. As the spring relaxes, the potential energy is slowly converted into the kinetic energy of the moving gears and hands.

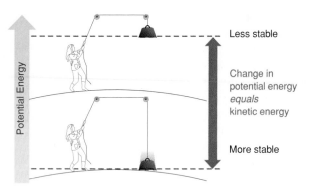

**A   A grativational system.** The $E_p$ gained when a weight is lifted is converted to $E_k$ as the weight falls.

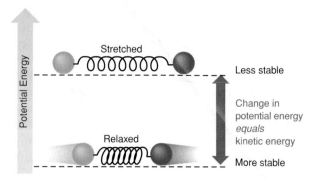

**B   A system of two balls attached by a spring.** The $E_p$ gained when the spring is stretched is converted to $E_k$ of the moving balls when it is released.

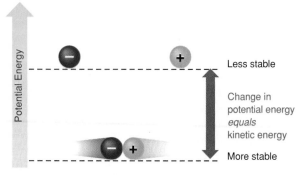

**C   A system of oppositely charged particles.** The $E_p$ gained when the charges are separated is converted to $E_k$ as the attraction pulls them together.

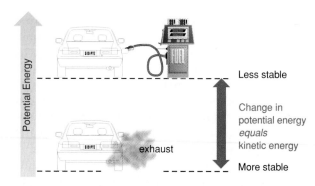

**D   A system of fuel and exhaust.** A fuel is higher in chemical $E_p$ than the exhaust. As the fuel burns, some of its $E_p$ is converted to $E_k$ of the moving car.

**FIGURE 1.3**
Potential energy *($E_p$)* is converted to kinetic energy *($E_k$).*

There are no springs in a chemical substance, of course, but a chemical system is similar in terms of energy. Much of the matter in the universe is composed of positively and negatively charged particles. There is an interaction between charged particles such that *opposite charges attract each other and like charges repel each other.* (This situation is similar to the behavior of the north and south poles of magnets.) When work is done to separate a positive particle from a negative one (like stretching the spring), the potential energy of the particles increases. That increase in potential energy is converted to kinetic energy when the particles move together again (Figure 1.3, *C*). When two positive (or two negative) particles are pushed toward each other (like compressing the spring), their potential energy also increases, and when they are allowed to move apart, that increase in potential energy is changed to kinetic energy. As with the weight above the ground and the balls connected by the stretched spring, charged particles move naturally toward a condition of lower energy, one that is more stable.

Some substances are richer than others in *chemical potential energy as a result of the relative positions and interactions among their particles.* Those used as a fuel or a food, for example, contain more potential energy than those formed as waste products. When gasoline burns in a car engine, substances with higher chemical potential energy (gasoline and air) are converted into substances with lower potential energy (exhaust gases). The difference in potential energy is converted into the kinetic energy that moves the car, heats the passenger compartment, makes the lights shine, and so forth (Figure 1.3, *D*). Similarly, the difference in potential energy between the food and air we take in and the waste products we excrete is used to move, grow, keep warm, study chemistry, and so on. The essential point to remember is that *energy is always conserved:* it is neither created nor destroyed, but rather changes from one type to another.

### Section Summary

Chemists study the composition and properties of matter and how they change. Each substance has a unique set of physical properties (attributes of the substance itself) and chemical properties (attributes of the substance as it interacts with or changes to other substances). Changes in matter can be physical (transformation to a different form of the same substance) or chemical (transformation to a different substance). A change in physical state can be reversed by reversing the process that caused it. Only other chemical changes can reverse a chemical change.

Changes in energy accompany changes in matter. An object's potential energy is due to its position, and its kinetic energy is due to its motion. Energy used to lift a weight, stretch a spring, or separate opposite charges increases the system's potential energy. In a chemical system, potential energy arises from the positions and interactions of the particles in the sample of matter. Higher energy systems are less stable than lower energy ones. A less stable system changes to a more stable one when some potential energy is converted into kinetic energy, which can do work.

## 1.2 Chemical Arts and the Origins of Modern Chemistry

Chemistry has a rich, colorful history. Even some concepts and discoveries that temporarily led theoretical developments along a confusing path contributed to the heritage of chemistry. This brief overview of early break-throughs and false directions provides some insight into how modern chemistry arose and how science progresses.

### Prechemical Traditions

Chemistry had its origins in a prescientific past that incorporated three overlapping traditions: alchemy, medicine, and technology.

**The alchemical tradition.** The occult study of nature practiced by the Greeks of northern Egypt in the 1st century AD later became known by the Arabic name **alchemy.** Its practice spread through the Near East and into Europe, where it dominated Western thinking about matter well into the 16th century. Influenced by the Greek idea that matter naturally strives toward perfection, the alchemists searched for ways to change less valued substances into precious ones. What started as a mystical search for spiritual properties in matter evolved over a thousand years into an obsession with potions to bestow eternal youth and elixirs to transmute "baser" metals, such as lead, into "purer" ones, such as gold. Greed tempted some later

**FIGURE 1.4**

**An alchemist at work.** The alchemical tradition influenced ideas about matter and its changes for 1500 years. This is a portion of a painting by the Englishman Joseph Wright entitled "The Alchymist, in search of the Philosopher's Stone, Discovers Phosphorus, and prays for the successful Conclusion of his Operation as was the custom of the Ancient Chymical Astrologers." Some suggest it portrays the German alchemist Hennig Brand in his laboratory lit by the glow of phosphorus, which he discovered in 1669. A great legacy of the alchemists was their invention of several major laboratory processes. The apparatus shown here is used for distillation, a process still commonly used to separate substances (see Chapter 2).

European alchemists into painting lead objects with a thin coating of gold to fool wealthy patrons.

Alchemy's legacy to chemistry is mixed, at best. The confusion arising from the alchemists' alternative names for the same substance and from their belief that matter could be magically altered was very difficult to eliminate. Nevertheless, through centuries of laboratory inquiry, the alchemists invented the chemical methods of distillation, percolation, and extraction, and they devised pieces of apparatus that we still use routinely today (Figure 1.4). Most important, the alchemists encouraged the widespread acceptance of observation and experimentation, which replaced the Greek approach of studying nature solely through reason.

**The medical tradition.** The alchemists greatly influenced medical practice in medieval Europe. Since the 13th century, distillates of roots, herbs, and other plant matter have been used as sources of medicines. Paracelsus (1493-1541) was an active alchemist and important physician of the time. His writings are obscure, but he seems to have considered the body to be a chemical system whose balance of substances could be restored by medical treatment. His followers introduced mineral drugs into 17th-century pharmacy. Although many of these drugs were useless and some harmful, later practitioners employed other mineral prescriptions with increasing success. Thus began an alliance between medicine and chemistry that thrives today.

**The technological tradition.** For thousands of years, people have developed technological skills to carry out changes in matter. Pottery making, dyeing, and especially metallurgy (begun about 7000 years ago) contributed greatly to our understanding of the properties of materials. During the Middle Ages and the Renaissance, technology flourished and several important treatises on metallurgy and geology were written. Books describing how to purify, assay, and coin silver and gold and how to use balances, furnaces, and crucibles were published and regularly updated. Other writings discussed making glass, pottery, dyes, and gunpowder. Some of these treatises introduced quantitative considerations, which had been lacking in the alchemical literature.

Many creations of these early artisans are unsurpassed even today. Although their working knowledge of substances was expert, their approach to understanding matter was different from ours: as in earliest times, techniques were discovered and developed by trial and error. The surviving writings indicate little interest in exploring *why* a substance changes or *how to predict* the behavior of matter.

## The Phlogiston Fiasco and the Impact of Lavoisier

Chemical investigation in the modern sense—inquiry into the causes of changes in matter—began in the late 17th century, most notably with the work of the English scientists Robert Hooke and Robert Boyle, who were friends and collaborated on some important experiments. Alchemical influences persisted, however, and understanding was hampered by an incorrect theory of **combustion,** the process of burning. Although fire had been used since prehistoric times, no satisfactory explanation existed for why some things burn and others do not, or what actually happens when a substance burns. Hooke had suggested that burning substances combine with air, but most scientists rejected this idea in favor of the **phlogiston theory,** which held sway for the next hundred years.

According to this theory, combustible materials contain *phlogiston,* an undetectable substance that is released when the material burns. A highly combustible material such as charcoal contains large amounts of phlogiston, and the ash left behind when it burns is the charcoal without phlogiston:

Charcoal → charcoal ash + phlogiston (large amount)

Slightly combustible materials such as metals contain small amounts of phlogiston. When a metal burns in air, it forms its *calx* (as metal oxides were then called), which is the metal without phlogiston:

Metal → metal calx + phlogiston (small amount)

Here is an example of how the theory might have been used to explain *smelting,* an important process in which a pure metal is obtained from its calx by heating the calx with charcoal. The phlogiston-rich charcoal burns and transfers phlogiston to the phlogiston-poor calx. This transfer converts charcoal to charcoal ash and the calx to the pure metal:

$$\overbrace{\text{Metal calx} + \text{charcoal}}^{\text{phlogiston}} \rightarrow \text{metal} + \text{charcoal ash}$$

These schemes seemed logical, but they ignored certain key observations. Why, the theory's critics asked, is air needed for combustion, and why does charcoal burn only for a short time in a closed vessel? Phlogistonists

◆ **Scientific Thinker**
**Extraordinaire.** Had Lavoisier never
performed a chemical experiment, his fame
would still be widespread. A short list of his
contributions outside chemistry: He improved
the production of French gunpowder, which
became a key factor in the success of the
American Revolution. He established on his
own farm a scientific balance between cattle,
pasture, and cultivated acreage that optimized
crop yield. He developed public assistance programs for widows and orphans. He collected
nationwide data to show the relation of fiscal
policy to agricultural production. He proposed a
national system of free public education and of
societies to foster science, politics, and the
arts. He sat on the committee that unified
weights and measures in the new metric system. Beyond this, his research into combustion
laid the foundation for the fields of thermochemistry and calorimetry (Chapter 6), and
their applications to respiration and metabolism. To support these pursuits, he had earlier
made the fateful decision to join a firm that
collected taxes for the King, and only this role
was remembered during the French Revolution.
Despite his devotion to French society, the father of modern chemistry was guillotined at the
age of 50.

responded that air is needed to "attract" the phlogiston out of the charcoal, and that burning in a vessel stops when the air is "saturated" with phlogiston. Knowing that a calx weighs more than the metal from which it forms, critics asked how the *loss* of phlogiston when a metal burns could cause a *gain* in mass. Phlogistonists either dismissed the importance of weighing or proposed that phlogiston had negative mass! Some of these responses seem ridiculous now, but they are mentioned to point out that the pursuit of science, like any other endeavor, is subject to human failings; even today it is easier to dismiss conflicting evidence than to give up an established idea.

Into this chaos of "explanations" entered the young French chemist Antoine Lavoisier (1743-1794), who demonstrated the true nature of combustion through a series of careful measurements that emphasized the importance of mass. ◆ Lavoisier heated mercury calx (mercuric oxide), decomposing it into mercury and a gas, whose combined masses equaled the starting mass of calx. The reverse experiment—heating mercury with the gas—re-formed the mercury calx, and again, *the total mass remained constant.* Lavoisier proposed that when a metal forms its calx, it does not lose phlogiston but rather combines with this gas, which must be a component of air.

To test his hypothesis, Lavoisier heated mercury in a measured volume of air until no further change occurred; mercury calx formed, and four-fifths of the air volume remained (Figure 1.5). This result confirmed that the gas was a component of air. A burning candle placed in the remaining volume of air was quickly extinguished, showing that the gas that had combined with the mercury was necessary for combustion. He named the gas *oxygen* and called metal calxes *metal oxides.*

The new theory of combustion made sense of the earlier confusion. A combustible substance such as charcoal stops burning in a closed vessel once it has combined with all the available oxygen. A metal calx weighs more than the metal because it contains the combined masses of the metal and oxygen. The true nature of smelting is that when charcoal and a metal oxide are heated, the metal, some charcoal ash, *and* carbon dioxide gas are produced; carbon, the main component of charcoal, combines with the oxygen from the metal oxide:

$$\overset{\overset{\text{oxygen}}{\frown}}{\text{Metal calx (metal oxide)} + \text{charcoal (carbon)}} \rightarrow$$

$$\text{metal} + \text{charcoal ash} + \text{carbon dioxide}$$

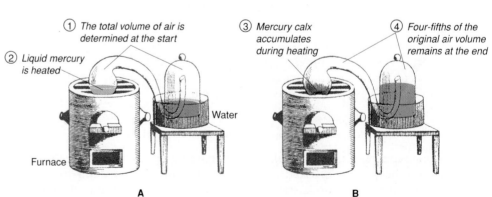

① *The total volume of air is determined at the start*
② *Liquid mercury is heated*
③ *Mercury calx accumulates during heating*
④ *Four-fifths of the original air volume remains at the end*

Water

Furnace

A

B

**FIGURE 1.5**
Lavoisier's experiment to test his hypothesis that a component of air is required for combustion.

Lavoisier's theory of combustion triumphed because it explained the observations more clearly and consistently than any other theory. ◆ Most important, the new theory relied on quantitative, reproducible measurements, not on strange properties of undetectable substances. Because this approach is at the heart of science, many propose that the *science* of chemistry began with Lavoisier.

### Section Summary

Alchemy, medicine, and technology established processes that have been important to chemists since the 17th century. These prescientific traditions placed little emphasis on objective experimentation, focusing instead on practical experience or mystical explanations. The phlogiston theory dominated thinking about combustion for almost 100 years, leading chemistry down a faulty theoretical path. In the mid-1770s, Lavoisier showed that oxygen, a component of air, is required for combustion and combines with a combustible substance as it burns.

**◆ A Great Chemist and Strict Phlogistonist.** Despite the phlogiston theory, chemists made important discoveries during the years it held sway. One of the foremost experimenters was the English scientist Joseph Priestley (1733-1804), who isolated and studied many gases, including the one obtained by heating mercury calx. In 1775, he wrote to his friend Benjamin Franklin: "Hitherto only two mice and myself have had the privilege of breathing it." Priestley also demonstrated that the gas supports burning, but he drew the wrong conclusion about combustion. He called the gas "dephlogisticated air," air devoid of phlogiston, which was thus able to attract phlogiston from a burning substance. Priestley's contributions make him one of the great chemists of the age, even though he remained a strict phlogistonist all his life, refusing to accept the new theory of combustion.

## 1.3   The Scientific Approach: Developing a Model

If we could break down a "typical" modern scientist's thinking into steps, we could organize them into an approach called the **scientific method.** It is important to realize, however, that there is no typical scientist and no single method. A scientific approach is not a checklist, but rather a process of creative thinking and testing aimed at objective, verifiable discoveries of how nature works. In general terms, the approach often includes the following parts (Figure 1.6):

• **Observations.** These are the facts that our ideas must explain. Observation is basic to scientific thinking. The most useful observations are quantitative because they can be compared and allow consistent changes, or trends, to be seen. Pieces of quantitative information are **data.** When the same observation is made by many investigators in many situations with no clear exceptions, it is summarized, often in mathematical terms, in the form of a **natural law.** The observation that mass remains constant during chemical change—made by Lavoisier and numer-

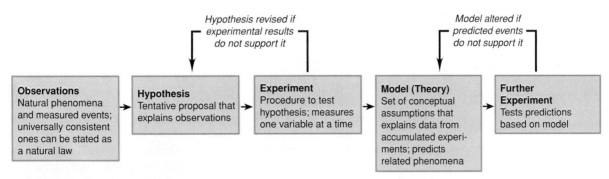

**FIGURE 1.6**
**The scientific approach to understanding nature.** Note that hypotheses and models are mental pictures that are changed to match observations and experimental results, *not* the other way around.

ous experimenters since—has been summarized as a natural law, which we discuss in Chapter 2.

- **Hypothesis.** Whether derived from actual observation or from a "spark of intuition," a hypothesis is a statement made to explain an observation. A valid hypothesis need not be correct, but it must be testable. Thus, a hypothesis is often the reason for performing an experiment. If the hypothesis turns out to be inconsistent with the experimental results, it must be either revised or discarded.

- **Experiment.** An experiment is a clear set of procedural steps that tests a hypothesis. Experimentation is the connection between our ideas, or hypotheses, about nature and nature itself. Hypothesis leads to experiment, which leads to revised hypothesis, and so forth. Hypotheses can be altered, but the results of an experiment cannot.

  An experiment always contains at least two **variables,** quantities that can have more than a single value, and is designed to show how a change in one variable affects another. A well-designed experiment is **controlled** in that it measures the effect of each variable separately by keeping the others constant. For experimental results to be accepted, they must be reproducible, not only by the person carrying out the work, but also by others. Both skill and creativity play a part in great experimental design.

- **Model.** Formulating conceptual models, or **theories,** *based on experiments* distinguishes scientific thinking from speculation. As hypotheses are altered according to experimental results, a model gradually emerges that describes how the observed phenomenon occurs. A model is not an exact representation of nature, but rather a simplified version of nature that can be used to make *predictions* about other, related phenomena. Further investigation continually refines the working model by testing its predictions and altering it to account for new facts.

  Lavoisier's overthrow of the phlogiston theory exemplifies the scientific approach. Observations of burning and smelting led some to hypothesize that combustion involved the loss of phlogiston. Experiments by others demonstrating that air is required for burning and that a metal gains mass during combustion led Lavoisier to propose a new hypothesis, which he tested repeatedly with quantitative experiments. Accumulating evidence lent support to his developing theory (model) that combustion involves *combination* with oxygen, a component of air. Innumerable predictions based on this theory have continued to support its validity. A sound model remains useful even when minor exceptions appear. An unsound one, such as the phlogiston theory, eventually falls under the weight of contrary evidence and absurd refinements.

### Section Summary

The scientific method is not a rigid sequence of steps, but rather a dynamic process designed to explain and predict real phenomena. ◆ Observations (sometimes expressed as natural laws) lead to hypotheses (conjectures) about how or why something occurs. Hypotheses are tested in controlled experiments and adjusted if necessary. If all the data collected support the hypothesis, a model (theory) can be developed to explain the observations. A good model is useful in predicting related phenomena but must be refined if conflicting data appear.

◆ **Everyday Scientific Thinking.** In an informal way, we frequently employ a scientific approach to decide on a course of action. Consider this familiar scenario. While listening to an FM broadcast on your stereo system, you notice the sound is garbled (observation) and assume it is caused by poor reception (hypothesis). To isolate this variable, you turn the selector switch to "Phono" and put on a CD (experiment): the sound is still garbled. If the problem is not poor reception, perhaps the speakers are at fault (new hypothesis). To isolate this variable, you play the CD and listen with headphones (experiment): the sound is clear. You conclude that the speakers need repair. Approaching a problem scientifically is a common practice, even if you are not aware of it.

# 1.4  Chemical Problem Solving

In many ways, learning chemistry *is* learning how to solve problems, not only those you may encounter on an examination or in a homework assignment, but also the more complex ones encountered in society and your profession. (The Chemical Connections essay at the end of this chapter provides a modern example.) With its emphasis on objective, measurable information and critical analysis of conclusions, a chemistry course offers a perfect opportunity to develop problem-solving skills.

This text has been designed to help you strengthen these skills. Almost every chapter contains sample problems that apply an important idea or skill just introduced. Unlike the problems at the end of the chapter, the sample problems are worked out in detail. In this section, we discuss a sound approach for solving them. Because many include calculations, we first consider a method for working with measured quantities.

## Units and Conversion Factors in Calculations

All measured quantities consist of a number *and* a unit; a person's height is "6 feet," not "6." Ratios of quantities have ratios of units, such as miles/hour. To minimize errors, make it a habit to include units in all calculation steps. (We discuss the most important units in chemistry in the next section.) The arithmetic operations used with measured quantities are the same as those used for pure numbers, in that units can be multiplied, divided, and canceled:

- A carpet measuring 3 feet (ft) by 4 ft has an area of

$$(3 \times 4) \ (\text{ft} \times \text{ft}) = 12 \ \text{ft}^2$$

- A car traveling 350 miles (mi) in 7 hours (h) has a speed of

$$\frac{350 \ \text{mi}}{7 \ \text{h}} = \frac{50 \ \text{mi}}{\text{h}} \quad (\text{sometimes written } 50 \ \text{mi} \cdot \text{h}^{-1})$$

- In 3 hours, the car travels

$$3 \ \cancel{\text{h}} \times \frac{50 \ \text{mi}}{\cancel{\text{h}}} = 150 \ \text{mi}$$

**Conversion factors** are used to express a measured quantity in different units. Suppose we want to know how many feet we traveled in the 150-mile car trip. To convert the distance from miles to feet, we use *equivalent quantities* to construct a conversion factor. The equivalent quantities in this case are one mile and the number of feet in one mile:

$$1 \ \text{mi} = 5280 \ \text{ft}$$

Dividing both sides of this equivalency by 5280 ft gives one conversion factor:

$$\frac{1 \ \text{mi}}{5280 \ \text{ft}} = \frac{5280 \ \cancel{\text{ft}}}{5280 \ \cancel{\text{ft}}} = 1$$

Dividing both sides by 1 mi gives another:

$$\frac{1 \ \cancel{\text{mi}}}{1 \ \cancel{\text{mi}}} = \frac{5280 \ \text{ft}}{1 \ \text{mi}} = 1$$

Each of these conversion factors is equal to one, since the numerator and denominator are equal. Therefore, *multiplying a quantity by a conversion factor* *changes the units in which it is expressed, but it does **not** change the quantity.* In our example, we want to convert the length in miles to the equivalent length in feet. Therefore, we choose the conversion factor that cancels units of miles and gives us the answer in units of feet (the conversion factor with units of feet in the numerator):

$$150 \ \cancel{mi} \times \frac{5280 \ \text{ft}}{1 \ \cancel{mi}} = 792{,}000 \ \text{ft}$$

Thinking through the relative size of the units being converted will help you check that you chose the correct conversion factor. Since a foot is smaller than a mile, the distance in feet has a larger number (792,000) than the distance in miles (150). Accordingly, the conversion factor we chose has a larger number in the numerator, so we obtained a larger number in the answer. We choose the other conversion factor to change from feet to miles. Suppose we know that Mt. Everest is 29,141 ft high and want to know its height in miles. We would write:

$$29{,}141 \ \cancel{ft} \times \frac{1 \ \text{mi}}{5280 \ \cancel{ft}} = 5.519 \ \text{mi}$$

Choosing the correct conversion factor is not a matter of chance or memorization. *Choose the factor that cancels all units except those required for the answer.* Set up the calculation so that the unit of the quantity you are converting *from* (beginning unit) is on the *opposite part of the conversion factor* (numerator or denominator) from the unit of the quantity you are converting *to* (final unit):

$$\cancel{\text{beginning unit}} \times \frac{\text{final unit}}{\cancel{\text{beginning unit}}} = \text{final unit} \qquad \text{as in} \qquad \cancel{ft} \times \frac{\text{mi}}{\cancel{ft}} = \text{mi}$$

Or, in cases that involve a ratio of units,

$$\frac{\cancel{\text{beginning unit}}}{\text{final unit}_1} \times \frac{\text{final unit}_2}{\cancel{\text{beginning unit}}} = \frac{\text{final unit}_2}{\text{final unit}_1} \qquad \text{as in} \qquad \frac{\cancel{mi}}{\text{h}} \times \frac{\text{ft}}{\cancel{mi}} = \frac{\text{ft}}{\text{h}}$$

To check your reasoning, think through the calculation to decide (1) whether the answer expressed in the new units will have a larger or smaller number, and (2) whether the conversion factor will change the size of the answer in the correct direction.

We use the same procedure to convert between systems of units, for example, between the English (or American) unit system and the International System (a revised metric system discussed fully in the next section). Suppose that instead of its height in miles, we want the height of Mt. Everest in kilometers (km). The equivalent quantities are

$$1.61 \ \text{km} = 1 \ \text{mi}$$

Since we are converting from miles to kilometers, we use the conversion factor with kilometers in the numerator, so miles cancel:

$$5.519 \ \cancel{mi} \times \frac{1.61 \ \text{km}}{1 \ \cancel{mi}} = 8.89 \ \text{km}$$

Notice that this conversion factor gave us a larger number (8.89 is larger than 5.519), consistent with a kilometer being smaller than a mile.

If we want the height of Mt. Everest in meters (m), we use the equivalent quantities 1 km = 1000 m:

$$8.89 \ \cancel{km} \times \frac{1000 \ \text{m}}{1 \ \cancel{km}} = 8890 \ \text{m}$$

In longer calculations, it is often convenient to string together several conversion factors:

$$5.519 \, \cancel{\text{mi}} \times \frac{1.61 \, \cancel{\text{km}}}{1 \, \cancel{\text{mi}}} \times \frac{1000 \, \text{m}}{1 \, \cancel{\text{km}}} = 8890 \, \text{m}$$

The use of conversion factors in calculations is known by various names, such as the factor-label method or **dimensional analysis** (because units represent a physical dimension, such as length, time, or energy). We use this method in quantitative problems throughout the text.

### How to Solve Chemistry Problems

The approach presented here is a systematic way to work through a problem. It emphasizes reasoning, not memorizing, and is based on a very simple idea: plan how to solve the problem *before* you go on to solve it, and then check your answer. Try to make a habit of applying a similar approach to homework and exams. In general, the sample problems within a chapter consist of several parts:

1. *Problem.* This part states all the information needed to solve the problem, usually framed in some interesting context.
2. *Plan.* The overall solution is broken up into two parts, plan and solution, to make a point: *think* about how to solve the problem *before* juggling numbers. If you are clear about a plan, the actual calculations will be much easier. There is often more than a single plan for a given problem. The plan will:
   - Clarify the known and unknown. (What information do we have, and what are we trying to find?)
   - Suggest the steps needed to find the solution. (What ideas, conversions, or equations link the known to the unknown?)
   - Present a "roadmap" of the solution for many quantitative problems in early chapters (and some in later ones). The roadmap is a diagrammatic summary of the planned calculation steps that shows the "stops" on the way from the known quantities to the unknown result. Each step is indicated by an arrow labeled with information about the conversion factor or operation needed to perform it.
3. *Solution.* In this part, the calculations (or operations) appear in the same general order in which they were planned.
4. *Check.* In most cases, a quick check is provided to see if the results make sense: Is the answer chemically reasonable? Are the units correct? Does the answer seem to be the right size? Did the change occur in the expected direction? We sometimes do an approximate calculation to see if the answer is "in the right ballpark." It is *always* a good idea to check your answers, especially after each step of a multipart problem, where an early error can affect later calculations.
5. *Comment.* This section is included occasionally to provide additional information, such as an application, an alternative approach, a common mistake to avoid, or an overview.
6. *Follow-up Problem.* This part consists of a problem statement only. The follow-up problem provides practice in applying the same ideas needed to work through the sample problem, so try to solve it *before* you look at the brief worked-out solution at the end of the chapter.

You should realize one additional point: you cannot learn to solve problems by reading a problem-solving approach, any more than you can learn to swim by reading about it. Practice is the key to mastery. Here are some suggestions:

- Study the sample problems with pencil, paper, and calculator.
- Do the follow-up problem as soon as you finish studying the sample problem. Check your answer against the solution at the end of the chapter.
- Read the sample problem and relevant text explanations again if you are having trouble.
- The problems at the end of the chapter review and extend the concepts and skills presented in the text and sample problems. Work on as many of them as you can. For those with colored numbers, check your answer with the one in the back of the book if necessary, but try the problem yourself first.

SAMPLE PROBLEM 1.2 _____

## Using the Sample Problem Approach to Convert Units of Length

**Problem:** What is the price of a piece of copper wire 325 centimeters (cm) long that sells for $0.15/ft?

**Plan:** We know the length of wire in centimeters and the cost in dollars per foot ($/ft). We can find the unknown price of the piece of wire by converting the length from centimeters to inches (in) and from inches to feet. Then we use the cost (1 ft = $0.15) as the factor to convert feet of wire to dollars. The roadmap starts with the known and moves through the calculation steps to the unknown.

**Solution:** Converting the known length from centimeters to inches: the equivalent quantities alongside the roadmap arrow are the ones needed to construct the conversion factor. We choose 1 in/2.54 cm, rather than the inverse, because it gives an answer in inches:

Length (cm) of wire

2.54 cm = 1 in

Length (in) of wire

12 in = 1 ft

Length (ft) of wire

1 ft = $0.15

Price ($) of wire

$$\text{Length (in)} = \text{length (cm)} \times \text{conversion factor} = 325 \ \cancel{cm} \times \frac{1 \text{ in}}{2.54 \ \cancel{cm}} = 128 \text{ in}$$

Converting the length from inches to feet:

$$\text{Length (ft)} = \text{length (in)} \times \text{conversion factor} = 128 \ \cancel{in} \times \frac{1 \text{ ft}}{12 \ \cancel{in}} = 10.7 \text{ ft}$$

Converting the length from feet to dollars:

$$\text{Price (\$)} = \text{length (ft)} \times \text{conversion factor} = 10.7 \ \cancel{ft} \times \frac{\$0.15}{1 \ \cancel{ft}} = \$1.60$$

We could also have strung the three steps together:

$$\text{Price (\$)} = 325 \ \cancel{cm} \times \frac{1 \ \cancel{in}}{2.54 \ \cancel{cm}} \times \frac{1 \ \cancel{ft}}{12 \ \cancel{in}} \times \frac{\$0.15}{1 \ \cancel{ft}} = \mathbf{\$1.60}$$

**Check:** The units are correct for each step. The conversion factors correlate with the relative sizes of the units: there are fewer inches than centimeters of wire (an inch is larger than a centimeter) and fewer feet than inches. The total price seems reasonable: a little more than 10 ft of wire at $0.15/ft *should* cost a little more than $1.50.

**Comment:** There are usually alternative sequences in unit conversion problems; for example, we would get the same answer if we had first converted the cost of wire from $/ft to $/cm and kept the wire length in cm. Try it yourself.

FOLLOW-UP PROBLEM 1.2

The speed of sound varies according to the material through which it travels. Sound travels at 5400 cm/s through rubber and at 19,700 ft/s through granite. Calculate each of these speeds in meters/second (m/s). (1 m = 100 cm = 3.28 ft)

**Section Summary**

A measured quantity consists of a number and a unit. Conversion factors are used to express a quantity in different units and are constructed as a ratio of equivalent quantities. The problem-solving approach used in this text usually has four parts: (1) devise a plan for the solution, (2) put the plan into effect in the calculations, (3) check to see if the answer makes sense, and (4) practice with similar problems.

## 1.5    Measurement in Scientific Study

Almost everything we own is manufactured with measured parts, sold in measured amounts, and paid for with measured currency. Measurement is so commonplace that we often take it for granted, but it has a long and fascinating history that is characterized by the search for *exact, invariable standards.* ◆

In 1790, the newly formed National Assembly of France, of which Lavoisier was a member, set up a committee to establish consistent standards of measurement. This effort led to the development of the metric system. In 1960, another international committee met in France to establish the International System of Units, a revised metric system now accepted by scientists throughout the world. The units of this system are called **SI units,** from the French *Système International d'Unités.*

### General Features of SI Units

The SI system is based on a set of seven **fundamental units,** or **base units,** each of which defines a standard for a physical quantity (Table 1.2). Other units, called **derived units,** are combinations of the seven base units. For example, the derived unit for speed (m/s) is the base unit for length (m) divided by the base unit for time (s). (Notice that derived units can be used as conversion factors because they contain two different units.)

For quantities that are much smaller or much larger than the base unit, we use decimal prefixes and *exponential (scientific) notation* (Table 1.3). Calculations frequently involve manipulating numbers written in exponential notation. If you are not familiar with this method of expressing numbers or just need a review, see Appendix A. Because these prefixes are based on powers of 10, SI units are easier to use in calculations than English units such as pounds and inches.

### Some Important SI Units in Chemistry

Let's discuss some of the SI units for quantities that we use early in this book: length, volume, mass, density, temperature, and time. The units for other quantities are presented at appropriate points later in the text. Table 1.4 shows some useful SI-English equivalent quantities for length, volume, and mass.

**Length.** The SI base unit of length is the **meter (m).** The standard meter is now based on two very accurately measured quantities, the speed of light in a vacuum and the second. ◆ A meter is a little longer than a yard (1 m = 1.094 yd); a centimeter ($10^{-2}$ m) is about two-fifths of an inch

◆ **How Many Barleycorns from His Majesty's Nose to His Thumb?** Systems of measurement began thousands of years ago as trade, building, and land surveying spread through the civilized world. For most of that time, however, measurement was based on inexact physical standards. For example, an inch was the length of three barleycorns placed end to end; a yard was the distance from the tip of King Henry I's nose to the tip of his thumb with his arm outstretched; and an acre was the area tilled by one man working with a pair of oxen in 1 day.

◆ **How Long Is a Meter?** The history of the meter points up the ongoing drive to find units based on unchanging standards. The scientists who set up the metric system defined the meter as 1/10,000,000 the distance from the equator to the North Pole. For convenience, the meter was later redefined as the distance between two fine lines engraved on a corrosion-resistant metal bar that was kept at the International Bureau of Weights and Measures in France. Each country received a replica of this official meter bar. Fear that the bar would be damaged by fire or war led to an exact, unchanging, universally available atomic standard: 1,650,763.73 wavelengths of orange-red light from electrically excited krypton atoms. An even more reliable standard was established in 1983: 1 meter is the distance light travels in a vacuum in 1/299,792,458 second.

**TABLE 1.2 SI Base Units**

| PHYSICAL QUANTITY (DIMENSION) | UNIT NAME | UNIT ABBREVIATION |
|---|---|---|
| Mass | kilogram | kg |
| Length | meter | m |
| Time | second | s |
| Temperature | kelvin | K |
| Electric current | ampere | A |
| Amount of substance | mole | mol |
| Luminous intensity | candela | cd |

**TABLE 1.3 Common Decimal Prefixes Used with SI Units**

| PREFIX* | PREFIX SYMBOL | MEANING NUMBER | MEANING WORD | EXPONENTIAL NOTATION |
|---|---|---|---|---|
| tera | T | 1,000,000,000,000 | trillion | $10^{12}$ |
| giga | G | 1,000,000,000 | billion | $10^{9}$ |
| **mega** | M | 1,000,000 | million | $10^{6}$ |
| **kilo** | k | 1,000 | thousand | $10^{3}$ |
| hecto | h | 100 | hundred | $10^{2}$ |
| deka | da | 10 | ten | $10^{1}$ |
| — | — | 1 | one | $10^{0}$ |
| **deci** | d | 0.1 | tenth | $10^{-1}$ |
| **centi** | c | 0.01 | hundredth | $10^{-2}$ |
| **milli** | m | 0.001 | thousandth | $10^{-3}$ |
| **micro** | μ | 0.000001 | millionth | $10^{-6}$ |
| **nano** | n | 0.000000001 | billionth | $10^{-9}$ |
| **pico** | p | 0.000000000001 | trillionth | $10^{-12}$ |
| femto | f | 0.000000000000001 | quadrillionth | $10^{-15}$ |

*The prefixes most frequently used by chemists appear in bold type.

**TABLE 1.4 Common SI-English Equivalent Quantities**

| QUANTITY | SI UNIT | SI EQUIVALENT | ENGLISH EQUIVALENT | ENGLISH TO SI EQUIVALENT |
|---|---|---|---|---|
| Length | 1 kilometer (km) | 1000 ($10^3$) meters | 0.62 mile (mi) | 1 mile = 1.61 km |
|  | 1 meter (m) | 100 ($10^2$) centimeters | 1.094 yards (yd) | 1 yard = 0.9144 m |
|  |  | 1000 millimeters (mm) | 39.37 inches (in) | 1 foot (ft) = 0.3048 m |
|  | 1 centimeter (cm) | 0.01 ($10^{-2}$) meter | 0.3937 inch | 1 inch = 2.54 cm (exactly) |
| Volume | 1 cubic meter (m³) | 1,000,000 ($10^6$) cubic centimeters | 35.3 cubic feet (ft³) | 1 cubic foot = 0.0283 m³ |
|  | 1 cubic decimeter (dm³) | 1000 cubic centimeters | 0.2642 gallon (gal) | 1 gallon = 3.785 dm³ |
|  |  |  | 1.057 quarts (qt) | 1 quart = 0.9464 dm³ |
|  | 1 cubic centimeter (cm³) | 0.001 dm³ | 0.0338 fluid ounce | 1 quart = 946.4 cm³ |
|  |  |  |  | 1 fluid ounce = 29.6 cm³ |
| Mass | 1 kilogram (kg) | 1000 grams | 2.205 pounds (lb) | 1 pound = 0.4536 kg |
|  | 1 gram (g) | 1000 milligrams (mg) | 0.03527 ounce (oz) | 1 ounce = 28.35 g |

(1 cm = 0.3937 in; 1 in = 2.54 cm). Biological cells are often measured in micrometers (1 $\mu$m = $10^{-6}$ m). On the atomic size scale, nanometers and picometers are used (1 nm = $10^{-9}$ m; 1 pm = $10^{-12}$ m); an older unit still in common use is the angstrom (1 Å = $10^{-10}$ m = 0.1 nm = 100 pm).

**Volume.** Any sample of matter has a certain **volume (V),** the amount of space it occupies. The SI derived unit of volume is the **cubic meter ($m^3$).** In chemistry, the most important volume units are the **liter (L),** a non-SI unit, and the **milliliter (mL).** (Note the uppercase L in the abbreviation.) The liter is equal to exactly one cubic decimeter ($dm^3$):

$$1 \text{ L} = 1 \text{ dm}^3 = 10^{-3} \text{ m}^3$$

As the prefix "milli-" implies, 1 mL is $^1/_{1000}$ of a liter; it is equal to exactly one cubic centimeter ($cm^3$):

$$1 \text{ mL} = 1 \text{ cm}^3 = 10^{-3} \text{ dm}^3 = 10^{-3} \text{ L} = 10^{-6} \text{ m}^3$$

A liter is slightly larger than a quart (1 L = 1.057 qt; 1 qt = 946.4 mL); one fluid ounce ($^1/_{32}$ of a quart) equals 29.6 mL (29.6 $cm^3$). Figure 1.7 shows the successive tenfold decreases in volume from the cubic meter to the cubic millimeter.

Many pieces of laboratory glassware are designed to measure liquid volume (Figure 1.8); they typically come in convenient sizes from a few milliliters to a few liters. Erlenmeyer flasks and beakers measure less accurately than graduated cylinders, pipets, and burets. Volumetric flasks have a fixed volume indicated by a mark on the neck. In careful work, liquid solutions are prepared in volumetric flasks, measured in cylinders, pipets, and burets, and then transferred to beakers or flasks for further chemical operations.

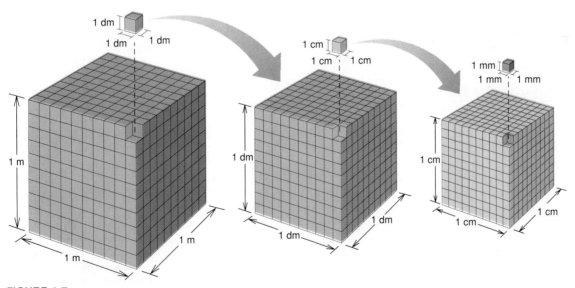

**FIGURE 1.7**
**Some volume relationships in SI.** The three cubes (not drawn to scale) demonstrate the decimal usefulness of SI units. The left cube is 1 $m^3$; each side is 1 m long and is divided into 10 dm. There are 1000 $dm^3$ in 1 $m^3$. The center cube is 1 $dm^3$; each side is 1 dm long and is divided into 10 cm; 1000 $cm^3$ = 1 $dm^3$ = 1 L. The right cube is 1 $cm^3$; each side is 1 cm long and is divided into 10 mm; 1 $cm^3$ = 1 mL. There are 1000 mL (1000 $cm^3$) in 1 L.

**FIGURE 1.8**

**Common laboratory glassware. A,** From left to right are two graduated cylinders, a pipet emptying into a beaker, a buret delivering liquid to an Erlenmeyer flask, and two volumetric flasks. **B,** In contact with glass, this liquid forms a concave meniscus (curved surface).

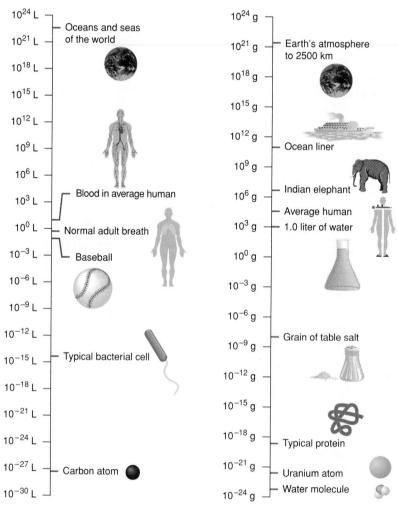

**A**

**B**

**FIGURE 1.9**

**Some interesting quantities of length (A), volume (B), and mass (C).** Note that the scales are exponential.

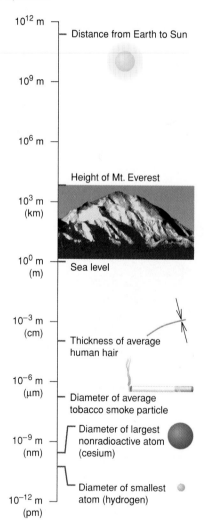

**A**

| | |
|---|---|
| $10^{12}$ m | Distance from Earth to Sun |
| $10^{9}$ m | |
| $10^{6}$ m | |
| $10^{3}$ m (km) | Height of Mt. Everest |
| $10^{0}$ m (m) | Sea level |
| $10^{-3}$ m (cm) | Thickness of average human hair |
| $10^{-6}$ m (µm) | Diameter of average tobacco smoke particle |
| $10^{-9}$ m (nm) | Diameter of largest nonradioactive atom (cesium) |
| $10^{-12}$ m (pm) | Diameter of smallest atom (hydrogen) |

**B**

| | |
|---|---|
| $10^{24}$ L | Oceans and seas of the world |
| $10^{21}$ L | |
| $10^{18}$ L | |
| $10^{15}$ L | |
| $10^{12}$ L | |
| $10^{9}$ L | |
| $10^{6}$ L | |
| $10^{3}$ L | Blood in average human |
| $10^{0}$ L | Normal adult breath |
| $10^{-3}$ L | Baseball |
| $10^{-6}$ L | |
| $10^{-9}$ L | |
| $10^{-12}$ L | Typical bacterial cell |
| $10^{-15}$ L | |
| $10^{-18}$ L | |
| $10^{-21}$ L | |
| $10^{-24}$ L | |
| $10^{-27}$ L | Carbon atom |
| $10^{-30}$ L | |

**C**

| | |
|---|---|
| $10^{24}$ g | |
| $10^{21}$ g | Earth's atmosphere to 2500 km |
| $10^{18}$ g | |
| $10^{15}$ g | |
| $10^{12}$ g | Ocean liner |
| $10^{9}$ g | Indian elephant |
| $10^{6}$ g | Average human |
| $10^{3}$ g | 1.0 liter of water |
| $10^{0}$ g | |
| $10^{-3}$ g | |
| $10^{-6}$ g | |
| $10^{-9}$ g | Grain of table salt |
| $10^{-12}$ g | |
| $10^{-15}$ g | |
| $10^{-18}$ g | Typical protein |
| $10^{-21}$ g | Uranium atom |
| $10^{-24}$ g | Water molecule |

SAMPLE PROBLEM 1.3

### Determining the Volume of a Solid by Displacement of Water

**Problem:** The volume of an irregularly shaped solid can be determined from the volume of water it displaces. A graduated cylinder contains 19.9 mL water. When a small piece of galena, an ore of lead, is submerged in the water, the volume increases to 24.5 mL. What is the volume of the piece of galena in $cm^3$ and in L?

**Plan:** The change in volume of the cylinder contents equals the volume of the added galena. We find the volume of galena in mL from the difference in the known volumes before and after the addition. Since mL and $cm^3$ represent identical volumes, the volume of the galena in mL equals the volume in $cm^3$. We convert the volume from mL to L by constructing the appropriate conversion factor. The calculation steps are shown in the roadmap.

**Solution:** Finding the volume of galena:

Volume (mL) = volume after − volume before = 24.5 mL − 19.9 mL = 4.6 mL

Converting the volume from mL to $cm^3$:

$$\text{Volume (cm}^3\text{)} = 4.6 \; \cancel{mL} \times \frac{1 \; cm^3}{1 \; \cancel{mL}} = \textbf{4.6 cm}^3$$

Converting the volume from mL to L:

$$\text{Volume (L)} = 4.6 \; \cancel{mL} \times \frac{10^{-3} \; L}{1 \; \cancel{mL}} = \textbf{4.6} \times \textbf{10}^{-3} \; \textbf{L}$$

**Check:** The units and magnitude of the answers seem correct. It makes sense that the volume expressed in mL would have a number 1000 times larger than the volume expressed in L, because a milliliter is $\frac{1}{1000}$ of a liter.

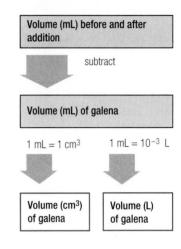

FOLLOW-UP PROBLEM 1.3

If a steel cube 15.0 mm on each side is submerged in 31.8 mL water in a graduated cylinder, what is the total volume (in L) of the cylinder contents?

---

**Mass.** The **mass** of an object refers to the quantity of matter it contains. The SI unit of mass is the **kilogram (kg),** the only base unit whose standard is a physical object—a platinum-iridium cylinder kept in France—and also the only base unit that has a prefix. ◆

The terms mass and weight have distinct meanings. Since an object's quantity of matter does not change, its *mass is constant.* The **weight** of the object, on the other hand, depends on its mass *and* the strength of the gravitational field pulling on it. Because the gravitational field varies with height above the Earth's surface, an object's *weight* also varies. For instance, you weigh slightly less on a high mountaintop than at sea level.

Does this mean that weighing an object in Miami (sea level) and in Denver (about 2 km above sea level) gives different results? No. Balances are designed to measure mass rather than weight, so this chaotic situation does not occur. (We are actually "massing" an object when we weigh it on a balance, but the term is seldom used.) Mechanical balances compare the object's unknown mass with known masses built into the balance, so the local gravitational field pulls equally on them. Electronic (analytical) balances determine mass by generating an electric field that counteracts the local gravitational field. The magnitude of the current needed to restore the pan to its zero position is then displayed as the object's mass. Figure 1.9 shows the range of some interesting lengths, volumes, and masses.

◆ **Don't Drop That Kilogram!** The U.S. copy of the master kilogram is kept in a vault at the National Bureau of Standards in Washington, D.C. Since 1889, it has been taken to France only twice for comparison with the master kilogram, which is removed from its vault only once a year for comparison with copies made from it. Two people are always present when it is moved: one to carry it with forceps, the other to catch it if the first person falls.

Optical fibers in cable.

Length (km) of fiber

1 km = 10³ m

Length (m) of fiber

1 m =1.19 × 10⁻³ lb

Mass (lb) of fiber

6 fibers = 1 cable

Mass (lb) of cable

2.205 lb = 1 kg

Mass (kg) of cable

SAMPLE PROBLEM 1.4 ⎯⎯⎯⎯⎯⎯⎯⎯⎯⎯⎯⎯⎯

**Converting Units of Mass**

**Problem:** International computer communications will soon be carried by optical fibers in cables laid along the ocean floor (see photo). If one strand of optical fiber weighs $1.19 \times 10^{-3}$ lb/m, what is the total mass (in kg) of a cable made of six strands of optical fiber, each long enough to link New York and Paris ($8.85 \times 10^3$ km)?

**Plan:** We have to find the total mass of optical cable from the mass/length of optical fiber, number of fibers per cable, and the length (distance from New York to Paris). To perform the calculation, we must express the quantities in consistent units. One approach (shown in the roadmap) is to convert the length of one fiber from km to m and then find its mass (in lb) by using the lb/m factor. The mass of the cable is six times the fiber's mass, which we convert from lb to kg.

**Solution:** Converting the fiber length from km to m:

$$\text{Length (m) of fiber} = 8.85 \times 10^3 \ \cancel{\text{km}} \times \frac{10^3 \text{ m}}{1 \ \cancel{\text{km}}} = 8.85 \times 10^6 \text{ m}$$

Converting the length of one fiber to mass (lb):

$$\text{Mass (lb) of fiber} = 8.85 \times 10^6 \ \cancel{\text{m}} \times \frac{1.19 \times 10^{-3} \text{ lb}}{1 \ \cancel{\text{m}}} = 1.05 \times 10^4 \text{ lb}$$

Finding the mass of the cable:

$$\text{Mass (lb) of cable} = \frac{1.05 \times 10^4 \text{ lb}}{1 \ \cancel{\text{fiber}}} \times \frac{6 \ \cancel{\text{fibers}}}{1 \text{ cable}} = 6.30 \times 10^4 \text{ lb/cable}$$

Converting the mass of cable from lb to kg:

$$\text{Mass (kg) of cable} = \frac{6.30 \times 10^4 \ \cancel{\text{lb}}}{1 \text{ cable}} \times \frac{1 \text{ kg}}{2.205 \ \cancel{\text{lb}}} = \textbf{2.86} \times \textbf{10}^\textbf{4} \textbf{ kg/cable}$$

**Check:** The units are correct, and the relative size of the number in each step seems reasonable. The number of m should be $10^3$ larger than the number of km. If 1 m of fiber weighs about $10^{-3}$ lb, about $10^7$ m should weigh about $10^4$ lb. The cable's mass is six times as much, or about $6 \times 10^4$ lb. Since 1 lb is about $\frac{1}{2}$ kg, the number of kg should be about half the number of lb.

**Comment:** Actually, the pound (lb) is the English unit of *weight,* not *mass.* The English unit of mass (the "slug") is rarely used.

FOLLOW-UP PROBLEM 1.4
If a raindrop weighs 65 mg on average and $5.1 \times 10^5$ raindrops fall on a lawn every minute, what mass (in kg) of rain falls in 1.5 h?

**Density.** The **density (*d*)** of an object is its mass divided by its volume:

$$\text{Density} = \frac{\text{mass}}{\text{volume}} \tag{1.1}$$

As with any variable expressed as a combination of others, each of the component variables can be isolated:

$$\text{Mass} = \text{density} \times \text{volume} \quad \text{and} \quad \text{Volume} = \frac{\text{mass}}{\text{density}}$$

Under given conditions of temperature and pressure, *density is a characteristic physical property of a substance,* but mass and volume are not. Mass and volume are examples of **extensive properties,** those dependent on the

**TABLE 1.5 Densities of Some Common Substances***

| SUBSTANCE | PHYSICAL STATE | DENSITY (g/cm³) |
|---|---|---|
| Hydrogen | Gas | 0.000089 |
| Oxygen | Gas | 0.0014 |
| Grain alcohol | Liquid | 0.789 |
| Water | Liquid | 1.0 |
| Table salt | Solid | 2.16 |
| Aluminum | Solid | 2.70 |
| Lead | Solid | 11.3 |
| Gold | Solid | 19.3 |

*At room temperature and normal atmospheric pressure.

amount of substance present. Density, on the other hand, is an **intensive property,** one that is independent of the amount of substance present. For example, the mass of a gallon of water is greater than the mass of a quart of water, but its volume is also greater; therefore, the density of the water, the ratio of its mass to its volume, is constant at a particular temperature and pressure, regardless of sample size.

The SI unit of density is the kilogram per cubic meter ($kg/m^3$), but in chemistry density is given in g/L ($g/dm^3$) or g/mL ($g/cm^3$). For example, the density of liquid water at ordinary pressure and room temperature is 1.0 g/mL. The densities of some common substances are given in Table 1.5. Note that the densities of gases are much lower than those of liquids or solids.

Lithium floating on oil on water.

**SAMPLE PROBLEM 1.5**

**Calculating Density from Mass and Length Measurements**

**Problem:** Lithium (symbol Li) is a soft, gray solid that has the lowest density of any metal (see photo). If a small slab of Li weighs $1.49 \times 10^3$ mg and has sides that measure 20.9 mm by 11.1 mm by 12.0 mm, what is the density of Li in $g/cm^3$?

**Plan:** We must find the density of Li in $g/cm^3$, so we need the mass of Li in g and the volume in $cm^3$. The mass is given in mg, so we convert mg to g. No volume data are given, but we can convert the lengths of the sides from mm to cm, and then multiply these lengths to calculate the volume in $cm^3$. Dividing the mass by the volume gives the density. The calculation steps are shown in the roadmap.

**Solution:** Converting the mass from mg to g:

$$\text{Mass (g) of Li} = 1.49 \times 10^3 \; \text{mg} \times \frac{1\,\text{g}}{10^3 \; \text{mg}} = 1.49 \; \text{g}$$

Converting side lengths from mm to cm:

$$\text{Length (cm) of one side} = 20.9 \; \text{mm} \times \frac{1 \; \text{cm}}{10 \; \text{mm}} = 2.09 \; \text{cm}$$

Similarly, the other side lengths are 1.11 cm and 1.20 cm.
Finding the volume:

$$\text{Volume (cm}^3) = 2.09 \; \text{cm} \times 1.11 \; \text{cm} \times 1.20 \; \text{cm} = 2.78 \; \text{cm}^3$$

Calculating the density:

$$\text{Density of Li} = \frac{1.49 \; \text{g}}{2.78 \; \text{cm}^3} = \textbf{0.536 g/cm}^3$$

Lengths (mm) of sides

10 mm = 1 cm

| Mass (mg) of Li | Lengths (cm) of sides |
|---|---|

$10^3$ mg = 1 g          multiply lengths

| Mass (g) of Li | Volume (cm³) |
|---|---|

divide mass by volume

Density (g/cm³) of Li

**Check:** Since 1 cm = 10 mm, the number of cm in each length should be 1/10 the number of mm. The units for density are correct and the size of the answer seems correct, since the number of grams is about half the number of cubic centimeters. Given the photo of Li floating on oil that is floating on water ($d = 1.0$ g/cm³), this answer is reasonable.

**FOLLOW-UP PROBLEM 1.5**
The piece of galena in Sample Problem 1.3 has a volume of 4.6 cm³. If the density of galena is 7.5 g/cm³, what is the mass (in kg) of that piece of galena?

---

**Temperature.** There is a common misunderstanding about the meanings of the terms heat and temperature. **Temperature (*T*)** is a measure of how hot or cold a substance is *relative to another substance.* **Heat** is energy that flows between objects that are at different temperatures. Temperature is related to the *direction* of heat flow: when two objects at different temperatures are brought into physical contact, heat flows from the one with the higher temperature to the one with the lower temperature until their temperatures are equal. When you hold an ice cube, its "cold" seems to flow *into* your hand; actually, heat flows *from* your hand into the ice. (In Chapter 6, we examine how heat is measured and how it is related to chemical and physical change.) Heat is an *extensive* property (as is volume), but temperature is an *intensive* property (as is density): a vat of boiling water contains more heat (more energy) than a cup of boiling water, but their temperatures are the same.

In the laboratory, the most common device for measuring temperature is the **thermometer,** which contains a fluid that expands when heated. When the thermometer's fluid-filled bulb is immersed in a substance hotter than itself, heat flows from the substance through the glass and into the fluid, which expands and rises within the thermometer tube. If the substance being measured is colder than the thermometer, heat flows outward from the fluid, which contracts and falls within the tube.

The three temperature scales we consider are the Celsius, formerly called centigrade (°C), Kelvin (K), and Fahrenheit (°F) scales. The SI base unit of temperature is the **kelvin (K);** note that the kelvin has no degree sign (°). The **Kelvin scale,** also known as the **absolute scale,** is preferred in all scientific work, although the **Celsius scale** is frequently used as well. In the United States, the Fahrenheit scale is sometimes used for weather reporting, body temperature, and other everyday purposes. The three scales differ in the size of the degree and/or the temperature of the zero point. Figure 1.10 shows the temperatures in the three scales at which water freezes and boils.

The Celsius scale, devised in the 18th century by the Swedish astronomer Anders Celsius, is based on changes in the physical state of water: 0°C is set at water's freezing point and 100°C at its boiling point (at normal atmospheric pressure). The Kelvin (absolute) scale was devised by the English physicist William Thomson, known as Lord Kelvin, in 1854 during his experiments on the expansion and contraction of gases. *The Kelvin scale uses the same-size degree as the Celsius scale,* that is, $\frac{1}{100}$ of the difference between the freezing and boiling points of water, but differs in its zero point. The temperature 0 K (absolute zero) is −273.15°C. Thus, water freezes at 273.15 K (0°C) and boils at 373.15 K (100°C). We can convert any temperature be-

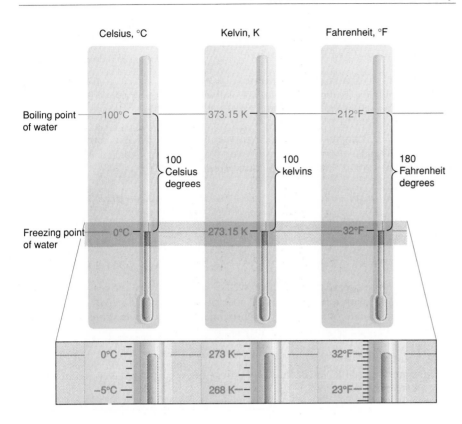

Celsius, °C          Kelvin, K          Fahrenheit, °F

Boiling point —100°C        —373.15 K        —212°F
of water

                    100                100                180
                    Celsius            kelvins            Fahrenheit
                    degrees                               degrees

Freezing point —0°C        —273.15 K        —32°F
of water

          0°C          273 K—          32°F—

          –5°C          268 K—          23°F—

**FIGURE 1.10**
**The freezing point and boiling point of water**
**in Celsius, Kelvin (absolute), and Fahrenheit**
**temperature scales.** A portion of each of the
three thermometer scales is expanded to show
the degree sizes: a Celsius degree (°C) and a
kelvin (K) are the same size, and each is 9/5 the
size of a Fahrenheit degree (°F).

tween the Celsius and Kelvin scales by taking the different zero points into account: since 0°C = 273.15 K,

$$T \text{ (in K)} = T \text{ (in °C)} + 273.15 \tag{1.2}$$

$$T \text{ (in °C)} = T \text{ (in K)} - 273.15 \tag{1.3}$$

The Fahrenheit scale differs from the other scales in its zero point *and* in the size of its degree. Water freezes at 32°F and boils at 212°F. Therefore, 180 Fahrenheit degrees (212°F − 32°F) represents the same temperature change as 100 Celsius degrees (or 100 kelvins). Since 100 Celsius degrees equal 180 Fahrenheit degrees,

1 Celsius degree = $^{180}/_{100}$ Fahrenheit degrees = $^9/_5$ Fahrenheit degrees

To convert a temperature in °C to °F, first change the degree size and then adjust the zero point:

$$T \text{ (in °F)} = {}^9/_5 \, T \text{ (in °C)} + 32 \tag{1.4}$$

To convert a temperature in °F to °C, first adjust the zero point and then change the degree size:

$$T \text{ (in °C)} = [T \text{ (in °F)} - 32] \, {}^5/_9 \tag{1.5}$$

(The only temperature with the same numerical value in the Celsius and Fahrenheit scales is −40°; that is, −40°F = −40°C.)

SAMPLE PROBLEM 1.6 _____

**Converting Units of Temperature**

**Problem:** A child has a body temperature of 38.7°C.
**(a)** If normal body temperature is 98.6°F, is the child running a fever?
**(b)** What is the child's temperature in kelvins?
**Plan: (a)** To find out if the child has a fever, we use Equation 1.4 to determine whether 38.7°C is higher than 98.6°F. **(b)** We use Equation 1.2 to convert the temperature in °C to K.
**Solution: (a)** Converting the temperature from °C to °F:

$$T \text{ (in °F)} = \tfrac{9}{5}\,T \text{ (in °C)} + 32 = \tfrac{9}{5}\,(38.7°C) + 32 = \mathbf{101.7°F}$$

**Yes,** the child is running a fever.
**(b)** Converting the temperature from °C to K:

$$T \text{ (in K)} = T \text{ (in °C)} + 273.15 = 38.7°C + 273.15 = \mathbf{311.8\ K}$$

**Check: (a)** From everyday experience, we know that 101.7°F is a reasonable number for someone with a fever. **(b)** Since a Celsius degree and a kelvin are the same size, we can estimate our answer by approximating the Celsius value to 40°C and adding 273: 40 + 273 = 313, so the answer is reasonable.

FOLLOW-UP PROBLEM 1.6
Mercury melts at 234 K, lower than any other pure metal. What is its melting point in °C and °F?

_____

**Time.** The SI base unit of time is the **second (s).** Although time was once measured by the day and year, it is now based on an atomic standard: microwave radiation from excited cesium atoms. In the laboratory, we frequently measure the time it takes for a fixed amount of substance to undergo a chemical change. A fast chemical change is over in a few nanoseconds ($10^{-9}$ s), whereas slow chemical changes, such as rusting or biological aging, take years. Some chemists are now using lasers to study changes that occur in a few picoseconds ($10^{-12}$ s) or even femtoseconds ($10^{-15}$ s). ◆

◆ **The Central Importance of Measurement in Science.** It's important to keep in mind *why* scientists measure things. Here is one person's reason: "When you can measure what you are speaking about, and express it in numbers, you know something about it; but when you cannot measure it, when you cannot express it in numbers, your knowledge is of a meager and unsatisfactory kind; it may be the beginning of knowledge, but you have scarcely, in your thoughts, advanced to the stage of science" (William Thomson, Lord Kelvin, 1824-1907).

**Section Summary**

SI units consist of seven base units and numerous derived units. Exponential notation and prefixes based on powers of 10 are used to express very small or very large numbers. The SI base unit of length is the meter (m). Length units on the atomic scale are the nanometer (nm) and picometer (pm). Volume units are derived from length units; the most important in chemistry are the cubic meter ($m^3$) and the liter (L). The mass of an object, a measure of the quantity of matter present in it, is constant. The SI unit of mass is the kilogram (kg). Density ($d$) is the ratio of mass to volume and is a characteristic physical property of a substance. Temperature ($T$) is a measure of the relative hotness of an object. Heat is energy that flows from an object at higher temperature to one at lower temperature. Temperatures are measured in scales that differ in the degree size and/or zero point. In chemistry, temperature is measured in kelvins (K) or degrees Celsius (°C). Extensive quantities, such as mass, volume, and heat, depend on an object's size. Intensive quantities, such as density and temperature, are independent of size.

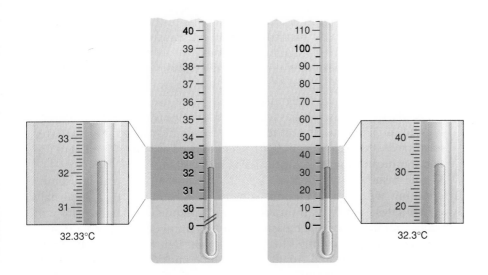

32.33°C    32.3°C

**FIGURE 1.11**
**The number of significant figures in a measurement depends on the measuring device.**
Two thermometers measuring the temperature of the same object are shown with expanded views. The thermometer on the left is graduated in 0.1°C and reads 32.33°C; the one on the right is graduated in 1°C and reads 32.3°C. Therefore, a reading with more significant figures (more certainty) can be made with the thermometer on the left.

## 1.6 Uncertainty in Measurement: Significant Figures

Because measuring devices are manufactured to limited specifications, and because we rely on our imperfect senses and skills when using them, we can never measure something exactly or know a quantity with absolute certainty. Thus, every measurement we make includes some **uncertainty.**

Our choice of a particular measuring device depends on *how much* uncertainty in a measurement is acceptable in a given situation. When you buy potatoes, a grocer's scale that measures in 0.1-kg increments is perfectly acceptable; it tells you that the mass is, for example, 2.0 ± 0.1 kg. The term "± 0.1 kg" expresses the uncertainty: the potatoes weigh between 1.9 and 2.1 kg. For a large-scale reaction, a chemist might need a balance that measures in 0.001-kg increments in order to obtain 2.036 ± 0.001 kg of a substance, that is, between 2.035 and 2.037 kg. We know the mass of the substance with more certainty than the mass of potatoes, even though the rightmost digit is still uncertain. The greater certainty is indicated by the larger number of digits in the measurement. We always *estimate the rightmost digit* when reading a measuring device such as a balance, thermometer, or buret, and this uncertainty is expressed with the "± " sign. Generally, we drop the ± sign and *assume an uncertainty of one unit in the rightmost digit.* The digits we record in a measurement, both certain and uncertain ones, are **significant figures.** There are four significant figures in 2.036 kg and two in 2.0 kg. The greater the number of significant figures, the greater is the certainty of the measurement (Figure 1.11).

### Rules for Determining Which Digits Are Significant

All digits are significant, *except zeros that are used only to position the decimal point.* Here is a set of rules for applying this generalization:
1. All nonzero digits *are* significant: 23.8 cm has three significant figures; 12.8735 g has six significant figures.
2. Zeros *between* significant digits *are* significant: 302.8 km has four significant figures; 200.05°C has five significant figures.
3. Zeros to the *left* of the first nonzero digit *are not* significant because they are used only to position the decimal point: 0.083 g has two significant figures; 0.0603 L has three significant figures.

4. A zero that ends a number and is to the *right* of the decimal point *is significant* because it was measured: 41.30 mL has four significant figures; 3.020 cm$^3$ also has four significant figures.

5. A zero that ends a number and is to the *left* of the decimal point *is significant only if it was actually measured*. For example, the quantity 5300 L may have two, three, or four significant figures; we use exponential notation to specify the number. It has four significant figures if the quantity is expressed as $5.300 \times 10^3$ L (the zeros *are* significant because they were measured), three significant figures if expressed as $5.30 \times 10^3$ L, but only two significant figures if expressed as $5.3 \times 10^3$ L.

As rule 5 shows, exponential (scientific) notation is extremely useful in determining with what certainty something was measured. A journalist who estimates that fifty thousand people attended a parade can write "50,000," but in a scientific measurement this number could imply that the people actually were counted to five significant figures (the actual number was between 49,999 and 50,001). The uncertainty in the guess is clear when the number is expressed as $5 \times 10^4$ (one significant figure) because it means that the number of people was between 40,000 and 60,000. For clarity, *in problems throughout this text, you should assume that any zero that ends a number is significant*; for example, when you see 250 mL, assume three significant figures.

### Determining the Number of Significant Figures

**Problem:** For each of the following quantities, underline the zeros that are significant figures, and determine the number of significant figures in each quantity. For **(d)** to **(f)**, express each in exponential notation first.
**(a)** 0.0030 L      **(b)** 0.1044 g    **(c)** 53,069 mL
**(d)** 0.00004715 m    **(e)** 576,004 s    **(f)** 0.0000007160 cm$^3$
**Plan:** We determine the number of significant figures using the rules just presented, paying particular attention to the position of zeros.
**Solution: (a)** 0.003<u>0</u> L has **two significant figures.**
**(b)** 0.1<u>0</u>44 g has **four significant figures.**
**(c)** 53,<u>0</u>69 mL has **five significant figures.**
**(d)** 0.00004715 m, $4.715 \times 10^{-5}$ m, has **four significant figures.**
**(e)** 576,<u>00</u>4 s, $5.76004 \times 10^5$ s, has **six significant figures.**
**(f)** 0.000000716<u>0</u> cm$^3$, $7.16\underline{0} \times 10^{-7}$ cm$^3$, has **four significant figures.**

For each of the following quantities, underline the zeros that are significant figures and determine the number of significant figures (sf) in each quantity. For (d) to (f), express each in exponential notation first.
**(a)** 31.070 mg   **(b)** 0.06060 g     **(c)** 850°C ($8.50 \times 10^2$ °C)
**(d)** 200.0 mL    **(e)** 0.0000039 m   **(f)** 0.000401 L

Certain numbers are called **exact numbers** because no uncertainty is associated with them. Some exact numbers are part of a unit definition: there are 60 minutes in 1 hour, 1000 micrograms in 1 milligram, and 2.54 centimeters in 1 inch. Other exact numbers result from actually counting individual items: there are exactly 3 quarters in my pocket, 10 fingers on my hands, and so forth. *Exact numbers have an infinite number of significant figures and, therefore, as many as a calculation requires.*

## Working with Significant Figures in Chemical Calculations

Quantities obtained from experiments often contain differing numbers of significant figures. During a calculation, we keep track of the number of significant figures in each quantity so that we do not have more significant figures (more certainty) in the answer than in the original data. If we do have too many figures, we **round off** the answer to obtain the proper number of significant figures.

Suppose you want to find the density of a new ceramic material. You measure the mass of a piece on an analytical balance and obtain 3.8056 g; you measure its volume as 2.5 mL by displacement of water in a graduated cylinder. The mass measurement has five significant figures, but the volume measurement has only two. Should you report the density as 3.8056 g/2.5 mL = 1.5222 g/mL (with five significant figures), or as 1.5 g/mL (with two significant figures)? The answer with five significant figures implies much more certainty in *all* the measurements than the answer with two. If you did not measure the volume to five significant figures, you cannot possibly know the density with that much certainty. Therefore, you report the answer as 1.5 g/mL. In other words, *the least certain measurement sets the limit on certainty for the entire calculation and determines the number of significant figures in the final answer.*

**Rules for significant figures in answers.** The following rules tell how many significant figures to show in the final answer of a calculation:

1. *For multiplication and division.* The measurement with the least certainty limits the certainty of the result. Therefore, *the answer contains the same number of significant figures as there are in the measurement with the fewest significant figures.*

   Suppose you are determining the volume of a sheet of a new graphite composite. The length (9.2 cm) and width (6.8 cm) are obtained with a meter stick and the thickness (0.3744 cm) with a set of fine calipers. The volume calculation would be

   $$9.2 \text{ cm} \times 6.8 \text{ cm} \times 0.3744 \text{ cm} = 23.4225 \text{ cm}^3 = 23 \text{ cm}^3$$

   The answer should be reported as 23 cm$^3$, with two significant figures, because the length and width measurements contain only two significant figures.

2. *For addition and subtraction.* The answer has the *same number of decimal places as there are in the measurement with the fewest decimal places.*
   Suppose you measure 83.5 mL water in a graduated cylinder and add 23.28 mL protein solution from a buret. The total volume is

   $$83.5 \text{ mL} + 23.28 \text{ mL} = 106.78 \text{ mL} = 106.8 \text{ mL}$$

   You report the volume as 106.8 mL, with one decimal place, because this is the number of decimal places in the measurement with the fewest decimal places (83.5 mL).

**Rules for rounding off.** In most calculations, you need to round off the answer to obtain the proper number of significant figures or decimal places. Notice that in calculating the volume of the graphite composite above, we removed the extra digits, but in calculating the protein solution volume, we removed the extra digit *and* increased the last digit by one. The following is a set of rules for rounding off:

1. If the digit removed is *more than 5,* the preceding number increases by 1: 5.379 rounds to 5.38 if three significant figures are retained and to 5.4 if two significant figures are retained.

2. If the digit removed is *less than 5,* the preceding number is unchanged: 0.2413 rounds to 0.241 if three significant figures are retained and to 0.24 if two significant figures are retained.

3. If the digit removed *is 5,* the preceding number increases by 1 if it is odd and remains unchanged if it is even: 17.75 rounds to 17.8, but 17.65 rounds to 17.6. If the 5 is followed only by zeros, rule 3 is followed; if the 5 is followed by nonzeros, rule 1 is followed: 17.6500 rounds to 17.6, but 17.6513 rounds to 17.7.

4. Be sure to carry two or more additional significant figures through a multistep calculation and round off only the *final* answer. (In sample problems and follow-up problems, we round off intermediate steps of a calculation to show the correct number of significant figures.)

**Significant figures and electronic calculators.** A calculator usually gives answers with more digits than the number of significant figures. For example, if your calculator displays ten digits and you divide 15.6 by 9.1, it will show 1.714285714. Obviously, most of these digits are not significant; the answer should be rounded off to 1.7 so that it has the same number of significant figures as in 9.1. A good way to see that the additional digits are not significant is to perform two calculations to obtain the highest and lowest possible answers indicated by the uncertainty in each of the last digits: for $(15.6 \pm 0.1)/(9.1 \pm 0.1)$,

$$\text{the highest answer is } \frac{15.7}{9.0} = 1.744444 \ldots$$

$$\text{the lowest answer is } \frac{15.5}{9.2} = 1.684782 \ldots$$

No matter how many digits the calculator displays, the values differ in the first decimal place, so the answer has only two significant figures and should be reported as 1.7.

**Significant figures and choice of measuring device.** As mentioned earlier, the instrument you choose determines the number of significant figures you can obtain. Suppose you are doing an experiment that requires mixing a liquid with a solid. You weigh the solid on the analytical balance and obtain a value with five significant figures. It would make sense to measure the liquid with a buret or pipet, which measures volumes to more significant figures than a graduated cylinder. If you chose the cylinder, you would have to round off more digits in the calculations, so the certainty in the mass value would be wasted. As you gain experience in the laboratory, you will learn to choose an instrument based on the number of significant figures you need in the final answer.

### Precision, Accuracy, and Instrument Calibration

Precision and accuracy are two aspects of certainty. In everyday speech, we often use these terms interchangeably, but in scientific measurements they have distinct meanings. **Precision,** or *reproducibility,* refers to how close the measurements in a series are to each other. **Accuracy** refers to how close a measurement is to the real value.

The concepts of precision and accuracy are linked with two common types of error in measurements. **Random error** produces individual values that are *both* higher and lower than the *average* value. Such error always occurs, but its size depends on the measurer's skill and the instrument's precision. Precise measurements have low random error; that is, the deviations from the average are small. **Systematic error,** on the other hand, produces values that are all *either* higher or lower than the *actual* value. Such error is part of the experimental system, perhaps caused by a faulty measuring device or by a consistent mistake in taking a reading. Accurate measurements have low systematic error and, generally, low random error as well. In some cases, when many measurements are taken that have a high random error, the *average* may still be accurate.

A simple weighing procedure illustrates these ideas. Each of four students measures 25.0 mL water in a graduated cylinder and then weighs the water in the cylinder on a balance. If the density of water is 1.00 g/mL at the temperature of the experiment, the actual mass of 25.0 mL water is 25.0 g. Each student performs the operation four times, subtracts the mass of the empty cylinder, and graphs the results (Figure 1.12). In graphs *A* and *B*, the random error is small; that is, the precision is high (the weighings are reproducible). In *A*, however, the accuracy is high as well (all the values are close to 25.0 g), whereas in *B* the accuracy is low (there is a systematic error). In graphs *C* and *D*, there is a large random error; that is, the precision is low. Large random error is often called large *scatter*. Note, however, that in *D* there is also a systematic error (all the values are high), whereas in *C* the average of the values is close to the actual value.

Systematic error can be avoided, or at least taken into account, through **calibration** of the measuring device, that is, comparing it with a known standard. The systematic error in graph *B,* for example, could be caused by a poorly manufactured cylinder that reads "25.0" when it actually contains about 27 mL. If you suspect this from a calibration procedure such as the one just described, you could adjust all volumes measured with that cylinder. Instrument calibration is an essential part of careful measurement.

### Section Summary
Because the final digit of a measurement is estimated, all measurements contain some uncertainty, which is expressed by the number of significant figures. The certainty of a result depends on the certainty of the data, so the answer has as many significant figures as in the least certain measurement. Exact numbers have as many significant figures as needed. Excess digits are rounded off in the final answer. The choice of laboratory device depends on the certainty needed.

Precision (how close values are to each other) and accuracy (how close values are to the actual value) are two aspects of certainty. Random errors are *both* higher and lower than the actual value. Systematic errors are *either* higher or lower than the actual value. Precise measurements have low random error; accurate measurements have low systematic and often low random error. The size of random errors depends on skill and instrument precision. A systematic error is usually caused by faulty equipment and can be compensated for by calibration.

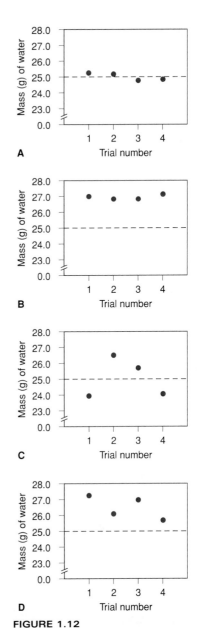

**FIGURE 1.12**

**Precision and accuracy in a laboratory calibration.** Each graph represents four measurements in a graduated cylinder that is being calibrated (see text for details). **A,** High precision, high accuracy. **B,** High precision, low accuracy (systematic error). **C,** Low precision, average value close to actual. **D,** Low precision, low accuracy.

### Chapter Perspective
*This chapter has provided several keys for you to use repeatedly in your study of chemistry: descriptions of some essential concepts; insight into how scientists think; the units of modern measurement and the mathematical skills to apply them; and a systematic approach to solving problems. You can begin using these keys in the next chapter, where we discuss the components of matter and their classification and see the winding path of scientific discovery that led to our current model of atomic structure.*

# Chemistry Problem Solving in the Real World

Working through the sample and follow-up problems and those at the end of the chapter will greatly improve your chances of doing well in this course, but that is not really the final goal. Although the vast majority of you may not become chemists, learning chemistry is essential to many scientific careers and to an understanding of many complex science-related issues in society. Therefore, before this chapter closes, let's briefly view a problem in the real world to see how it is approached by chemists interacting with scientists in related fields.

Consider the acid rain problem. Acid rain results in large part from burning high-sulfur coal, a major fuel used throughout much of North America and Europe. As the coal burns, the gaseous products, including those of its sulfur impurities, are carried away by prevailing winds. In contact with oxygen and rain, these sulfur oxides undergo chemical changes and fall as acid rain. (We discuss the chemical details in later chapters.) In the northeastern United States and adjacent parts of Canada, acid rain has killed fish, decimated forests, injured crops, and released harmful substances into the soil. Acid rain has severely damaged many forests and lakes in Germany, Sweden, Norway, and several countries in central and eastern Europe.

Chemists and other scientists are currently working together to solve this problem (Figure 1.A). As geochemists search for low-sulfur coal deposits, their engineering colleagues design better ways of removing sulfur oxides from smokestack gases. Atmospheric chemists and meteorologists track changes through the affected regions, develop computer models that can predict the changes, and coordinate their findings with those of environmental chemists at ground stations. Ecological chemists and aquatic biologists monitor the effects of acid rain on insects, birds, and fish. Agricultural chemists and agronomists study ways to protect crop yields. Biochemists and genetic engineers develop new, more acid-resistant crop species. Soil chemists measure changes in mineral content, integrating their data with those of forestry scientists to save valuable timber and recreational woodlands. Organic chemists and chemical engineers study the composition of coal in order to create cleaner fuels from it by chemical treatment. Superimposed on this intense scientific activity are economic and political pressures on business and government leaders, who rely in part on scientific information to make their decisions. With all this input, interdisciplinary understanding of the acid rain problem has increased and will continue to do so.

These professions are just a few of those involved in studying a single chemistry-related issue. Chemical principles apply to many other specialties, from medicine and pharmacology to art restoration and criminology, from genetics and space research to archaeology and oceanography. Chemistry problem solving has far-reaching relevance in much of our professional and daily lives.

**FIGURE 1.A**

**The central role of chemistry in solving real-world problems.** Researchers in many chemical specialties join with those in other sciences to investigate complex modern issues such as acid rain. **A,** An atmospheric chemist measures the components of an air sample. **B,** Ecologists sample lake water. **C,** A fisheries biologist examines specimens. **D,** An agronomist measures the acidity of soil.

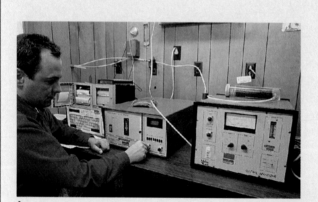

A

B

C

D

# For Review and Reference

## Key Terms

**SECTION 1.1**
chemistry
matter
composition
property
physical property
physical change
chemical property
chemical change
   (chemical reaction)
state of matter
solid
liquid
gas
energy
potential energy
kinetic energy

**SECTION 1.2**
alchemy
combustion
phlogiston theory

**SECTION 1.3**
scientific method
observation
data
natural law
hypothesis
experiment
variable
controlled experiment
model (theory)

**SECTION 1.4**
conversion factor
dimensional analysis

**SECTION 1.5**
SI unit
base (fundamental) unit
derived unit
meter (m)
volume (V)
cubic meter (m³)
liter (L)
milliliter (mL)
mass
kilogram (kg)
weight
density (d)
extensive property

intensive property
temperature (T)
heat
thermometer
Kelvin (K)
Kelvin (absolute) scale
Celsius scale
second (s)

**SECTION 1.6**
uncertainty
significant figures
exact number
rounding off
precision
accuracy
random error
systematic error
calibration

## Key Equations and Relationships

**1.1** Calculating density from mass and volume (p. 22):

$$\text{Density} = \frac{\text{mass}}{\text{volume}}$$

**1.2** Converting temperature from °C to K (p. 25):

$$T \text{ (in K)} = T \text{ (in °C)} + 273.15$$

**1.3** Converting temperature from K to °C (p. 25):

$$T \text{ (in °C)} = T \text{ (in K)} - 273.15$$

**1.4** Converting temperature from °C to °F (p. 25):

$$T \text{ (in °F)} = \tfrac{9}{5}T \text{ (in °C)} + 32$$

**1.5** Converting temperature from °F to °C (p. 25):

$$T \text{ (in °C)} = [T \text{ (in °F)} - 32]\tfrac{5}{9}$$

## Answers to Follow-up Problems

**1.1** (a) Physical. Solid iodine changes to gaseous iodine.
   (b) Chemical. Gasoline burns in air to form different substances.
   (c) Chemical. In contact with air, torn skin and blood react to form different substances.

**1.2** Rubber: Speed (m/s) $= \dfrac{5400 \text{ cm}}{1 \text{ s}} \times \dfrac{1 \text{ m}}{100 \text{ cm}} = 54$ m/s

Granite: Speed (m/s) $= \dfrac{19,700 \text{ ft}}{1 \text{ s}} \times \dfrac{1 \text{ m}}{3.28 \text{ ft}}$

$= 6006$ m/s

or $6.01 \times 10^3$ m/s to three significant figures (see Section 1.6)

**1.3** Volume (mL) of cube

$= \left(15.0 \text{ mm} \times \dfrac{1 \text{ cm}}{10 \text{ mm}}\right)^3 \times \dfrac{1 \text{ mL}}{1 \text{ cm}^3} = 3.38$ mL

Total volume (L) $= (31.8 \text{ mL} + 3.38 \text{ mL}) \times \dfrac{10^{-3} \text{ L}}{1 \text{ mL}}$
$= 3.52 \times 10^{-2}$ L

**1.4** Mass (kg) of rain

$= 1.5 \text{ h} \times \dfrac{60 \text{ min}}{1 \text{ h}} \times \dfrac{5.1 \times 10^5 \text{ drops}}{1 \text{ min}}$

$\times \dfrac{65 \text{ mg}}{1 \text{ drop}} \times \dfrac{1 \text{ g}}{10^3 \text{ mg}} \times \dfrac{1 \text{ kg}}{10^3 \text{ g}}$

$= 3.0 \times 10^3$ kg

**1.5** Mass (kg) of sample $= 4.6 \text{ cm}^3 \times \dfrac{7.5 \text{ g}}{1 \text{ cm}^3} \times \dfrac{1 \text{ kg}}{10^3 \text{ g}}$

$= 0.034$ kg

**1.6** $T$ (in °C) $= 234$ K $- 273.15 = -39°$C;
$T$ (in °F) $= \tfrac{9}{5} (-39°$C$) + 32 = -38°$F; answer contains two significant figures (see Section 1.6)

**1.7** (a) 31.070 mg, 5 sf; (b) 0.06060 g, 4 sf;
(c) $8.50 \times 10^2$ °C, 3 sf; (d) $2.000 \times 10^2$ mL, 4 sf;
(e) $3.9 \times 10^{-6}$ m, 2 sf; (f) $4.01 \times 10^{-4}$ L, 3 sf

## Sample Problem Titles

**1.1** Distinguishing Between Physical and Chemical Change (p. 4)

**1.2** Using the Sample Problem Approach to Convert Units of Length (p. 16)

**1.3** Determining the Volume of a Solid by Displacement of Water (p. 21)

**1.4** Converting Units of Mass (p. 22)

**1.5** Calculating Density From Mass and Length Measurements (p. 23)

**1.6** Converting Units of Temperature (p. 26)

**1.7** Determining the Number of Significant Figures (p. 28)

# Problems

Problems with a green number are answered at the back of the text. Most sections include three categories of problems separated by a green rule—concept review questions, *paired* skill building exercises, and problems in a relevant context.

## Some Fundamental Definitions

(Sample Problem 1.1)

**1.1** Define the following terms and give an example of each in the act of boiling an egg for breakfast: (a) matter; (b) energy; (c) chemical change; (d) physical change.

**1.2** Describe solids, liquids, and gases in terms of how they fill a container, and then use your descriptions to identify the physical state (at room temperature) of each of the following:

(a) Neon in a store window sign

(b) Mercury in a thermometer

(c) Water in a copper pipe

(d) Air in your room

(e) Aspirin tablets in a bottle

(f) Wine in a bottle

**1.3** Define physical property and chemical property. Show where each type of property appears in the following statements:

(a) Yellow-green chlorine gas attacks silvery sodium metal to form white crystals of sodium chloride (table salt).

(b) A magnet separates a mixture of iron shavings and powdered sulfur.

(c) Heating orange mercuric oxide powder produces silvery liquid mercury and gaseous oxygen.

(d) A prospector looking for gold swirls crushed ore with water in a shallow pan until the more dense gold dust settles out.

(e) The iron in discarded automobiles slowly reacts with oxygen in the air to form reddish brown, crumbly rust.

**1.4** Which of the following is a chemical change? Explain your reasoning:

(a) Acid rain destroying a marble statue

(b) Boiling soup in a pot

(c) Toasting a slice of bread

(d) Chopping a wooden log

(e) Burning a wooden log

**1.5** Which of the following changes can be reversed by changing the temperature (that is, which are physical changes)?

(a) Dew condensing on a leaf

(b) An egg turning hard when it is boiled

(c) Ice cream melting

(d) A spoonful of batter cooking on a hot griddle

**1.6** In the following description, which points relate to physical properties and which to chemical properties? Magnesium, the eighth most abundant element on Earth, is found in minerals, sea water, and all organisms. (a) It is an essential component of chlorophyll, the photosynthetic pigment of green plants. (b) Freshly cut solid magnesium metal has a silvery white luster that dulls somewhat in air as the metal reacts with oxygen. (c) Magnesium has a density of 1.738 $g/cm^3$. (d) It melts at 649°C and boils at 1105°C, a relatively low boiling temperature for a metal. (e) Care must be taken in handling magnesium, particularly in powdered form, because when heated in air, it burns rapidly with a dazzlingly bright flame. (f) Magnesium fires cannot be extinguished with water because the hot metal reacts with water to produce flammable hydrogen gas. (g) Important commercial applications of magnesium include its use in lightweight alloys for aircraft and missile bodies and in incendiary flares and fireworks.

**1.7** For each of the following pairs of objects or substances, which has higher potential energy?

(a) The fuel in your car or the products in its exhaust

(b) Wood in a fireplace or the ashes in the fireplace after the wood burns

(c) A sled resting at the top of a hill or the sled sliding down the hill

(d) Water above a dam or water that has fallen below the dam

## Chemical Arts and the Origins of Modern Chemistry

**1.8** The alchemical, medical, and technological traditions were precursors to chemistry. State a contribution that each made to the development of the science of chemistry.

**1.9** What were the alchemists trying to do? How did their efforts slow the progress of chemical science?

**1.10** How did the phlogiston theory explain combustion?

**1.11** One important observation that supporters of the phlogiston theory had trouble explaining was that the calx of a metal weighed more than the metal itself. Why was that observation important? How did the phlogistonists respond?

**1.12** Lavoisier developed a new theory of combustion that overturned the phlogiston theory. What measurements were central to his theory, and what key discovery did he make?

## The Scientific Approach: Developing a Model

**1.13** How are the key elements of scientific thinking used in the following scenario? While making your breakfast toast, you notice it fails to pop out of the toaster. Thinking the spring mechanism is stuck, you notice the bread is unchanged. Afraid that you forgot to plug in the toaster, you check and find it *is* plugged in. When you take the toaster into the dining room and plug it into a different outlet, you find the toaster works. Returning to the kitchen, you turn on the overhead light and nothing happens.

**1.14** Why is a quantitative observation more useful than a nonquantitative one? Which of the following observations are quantitative?
(a) The sun rises in the east.
(b) Ice floats on water.
(c) An astronaut weighs one-sixth as much on the moon as on Earth.
(d) In a vacuum, all objects fall at the same rate.

**1.15** Describe the essential features of a well-designed experiment.

**1.16** Describe the essential features of a scientific model.

## Chemical Problem Solving

(Sample Problem 1.2)

**1.17** When you convert feet to inches, how do you decide which portion of the conversion factor should be in the numerator and which in the denominator?

**1.18** Write the conversion factor(s) for each of the following: (a) $in^2$ to $ft^2$; (b) $km^2$ to $m^2$; (c) m/h to cm/s.

**1.19** Write the conversion factor(s) for each of the following: (a) cm/min to in/min; (b) $ft^3$ to $in^3$; (c) $m/s^2$ to $km/h^2$.

## Measurement in Scientific Study

(Sample Problems 1.3 to 1.6)

**1.20** Describe the difference between intensive and extensive properties. Which of the following properties are intensive: (a) mass; (b) density; (c) volume; (d) melting point?

**1.21** Explain the difference between mass and weight. Why is your weight on the moon one-sixth that on Earth?

**1.22** Explain the difference between heat and temperature. Does 1 L water at 65°F contain more, less, or the same amount of energy as 1 L water at 65°C?

**1.23** The radius of a gold atom is 144 pm. What is its radius in nanometers (nm)?

**1.24** The radius of a lithium atom is $1.52 \times 10^{-10}$ m. What is its radius in nanometers?

**1.25** A football field is 100 yd long. What is its length in meters (m)?

**1.26** The center on your basketball team is 7 ft 1 in tall. How tall is the player in centimeters (cm)?

**1.27** A small hole in the wing of a space shuttle requires a 23.6-$cm^2$ patch.
(a) What is the patch's area in square kilometers ($km^2$)?
(b) If the patching material costs NASA 83¢/$cm^2$, what is the cost (in $) of the patch?

**1.28** The area of a telescope lens is 5786 $mm^2$.
(a) What is its area in square meters ($m^2$)?
(b) If it takes a technician 45 s to polish 150 $mm^2$, how long does it take to polish the lens?

**1.29** Express the mass of your body in kilograms (kg).

**1.30** There are $2.60 \times 10^{15}$ short tons of oxygen in the atmosphere (1 short ton = 2000 lb). How many metric tons of oxygen are present (1 metric ton = 1000 kg)?

**1.31** The average density of the Earth is 5.52 $g/cm^3$. What is its density in (a) $kg/m^3$? (b) $lb/ft^3$?

**1.32** The speed of light in a vacuum is $2.998 \times 10^8$ m/s. What is its speed in (a) km/h? (b) mi/min?

**1.33** The volume of a certain bacterial cell is 2.25 $\mu m^3$.
(a) What is its volume in cubic millimeters ($mm^3$)?
(b) What is the volume of $10^5$ cells in liters (L)?

**1.34** (a) How many cubic meters of milk are in one quart (946.4 mL)?
(b) How many liters of milk are in 425 gallons (1 gallon = 4 quarts)?

**1.35** An empty vial weighs 31.45 g.
(a) If the vial weighs 179.56 g when filled with liquid mercury ($d = 13.53$ $g/cm^3$), what is its volume?
(b) How much would the vial weigh if it were filled with water ($d = 0.997$ $g/cm^3$ at 25°C)?

**1.36** An Erlenmeyer flask weighs 121.3 g when empty and 283.2 g when filled with water ($d = 1.00$ $g/cm^3$).
(a) What is its volume?
(b) How much does the flask weigh when filled with carbon tetrachloride ($d = 1.59$ $g/cm^3$)?

**1.37** A small cube of aluminum measures 15.6 mm on a side and weighs 10.25 g. What is the density of aluminum in $g/cm^3$?

**1.38** A steel ball–bearing with a circumference of 31.5 mm weighs 4.20 g. What is the density of the steel in $g/cm^3$? ($V$ of a sphere = $\frac{4}{3}\pi r^3$; circumference of a circle = $2\pi r$.)

**1.39** Perform each of the following conversions:
(a) 68°F (a pleasant spring day) to °C and K
(b) −164°C (the boiling point of methane, the main component of natural gas) to K and °F
(c) 0 K (absolute zero, theoretically the coldest possible temperature) to °C and °F

**1.40** Perform each of the following conversions:
(a) 106°F (the body temperature of many birds) to K and °C
(b) 3410°C (the melting point of tungsten, the highest for any element) to K and °F
(c) 6100 K (the surface temperature of the sun) to °F and °C

**1.41** Anton van Leeuwenhoek, a 17th-century pioneer in the use of microscopes, described the microorganisms he saw as "animalcules" whose length was "25 thousandths of an inch." How long were the animalcules in meters?

**1.42** The distance between two adjacent peaks on a wave is called the *wavelength*. The wavelength of visible light determines its color.
   (a) The wavelength of a beam of violet light is 423 nanometers (nm). What is its wavelength in meters?
   (b) The wavelength of a beam of red light is 690 nm. What is its wavelength in angstroms (Å)?

**1.43** It is often possible to estimate the mass of an object from its volume and density. Most common liquids have a density close to that of water (1.0 g/cm³); many common metals, including iron, copper, and silver, have densities around 9.5 g/cm³.
   (a) What is the mass of the liquid in a standard 12-oz bottle of spring water?
   (b) What is the mass of a dime? (*Hint:* a stack of five dimes has a volume of about 1 cm³.)

**1.44** To the alchemists, gold was the material representation of purity. It is a very soft metal that can be hammered into extremely thin sheets. If a 1.00-g piece of gold (*d* = 19.32 g/cm³) is hammered into a sheet whose area is 40.0 ft², what is the average thickness of the sheet in centimeters?

**1.45** Suppose your dorm room is 9 ft wide by 12 ft long by 8 ft high and has an air conditioner that exchanges 1000 L air/min. How long would it take the air conditioner to exchange the air in your room once?

**1.46** A cylindrical tank 10.0 m high and 4.0 m in diameter contains liquefied natural gas. How many liters (L) of the liquid can it hold? (*V* of a cylinder = $\pi r^2 h$.)

## Uncertainty in Measurement: Significant Figures

(Sample Problem 1.7)

**1.47** What is an exact number? Which of the following statements include exact numbers?
   (a) John Smith is 71 inches tall.
   (b) There are nine known planets in the solar system.
   (c) There are 453.59 g in 1 lb.
   (d) There are 1000 mm in 1 m.

**1.48** Which of the following include exact numbers?
   (a) The speed of light in a vacuum is a physical constant; to six significant figures, it is $2.99792 \times 10^8$ m/s.
   (b) The density of mercury at 25°C is 13.53 g/mL.
   (c) There are 3600 s in 1 h.
   (d) In 1986, 750,000 students were enrolled in general chemistry courses in the United States.

**1.49** All digits other than zero are always significant. State a rule that tells which zeros are significant. Underline the significant zeros in the following examples:
   (a) 0.39; (b) 0.039; (c) 0.0390; (d) $3.0900 \times 10^4$.

**1.50** Which of the following zeros are significant? (a) 5.08; (b) 508; (c) $5.080 \times 10^3$; (d) 0.05080? Explain.

**1.51** A newspaper reported that the attendance at Slippery Rock's home football game was 5209. How many significant figures does this number contain? Was the actual number of people counted?

**1.52** After Slippery Rock's next home game, the newspaper reported an attendance of 5000. If you assume that this number contains two significant figures, how many people could actually have been at the game?

**1.53** Round off the following numbers to the indicated number of significant figures (sf): (a) 0.0003564 (to 1 sf); (b) 21.8347 (to 4 sf); (c) 17.4555 (to 2 sf).

**1.54** Round off the following numbers to the indicated number of significant figures (sf): (a) 231.554 (to 4 sf); (b) 0.00845 (to 2 sf); (c) 144,000 (to 1 sf).

**1.55** Round off each of the following numbers to one fewer significant figure, and perform the calculation:

$$\frac{11 \times 154 \times 7.9}{2.1 \times 3.1 \times 3.9}$$

**1.56** Round off each of the following numbers to one fewer significant figure, and perform the calculation:

$$\frac{14.9 \times 6.24 \times 2.61}{25.4 \times 2.9 \times 1.48}$$

**1.57** Carry out the following calculations, making sure that your answer has the correct number of significant figures:

   (a) $\dfrac{2.795 \text{ m} \times 3.12 \text{ m}}{6.483 \text{ m}}$

   (b) $V = \frac{4}{3}\pi r^3$, where $r = 6.14$ cm
   (c) 2.110 cm + 17.2 cm + 108.2 cm + 216 cm

**1.58** Carry out the following calculations, making sure that your answer has the correct number of significant figures:

   (a) $\dfrac{2.42 \text{ g} + 15.6 \text{ g}}{5.31 \text{ g}}$   (b) $\dfrac{7.87 \text{ mL}}{16.1 \text{ mL} - 8.44 \text{ mL}}$

   (c) $V = \pi r^2 h$, where $r = 6.13$ cm and $h = 4.629$ cm

**1.59** Write the following numbers in scientific notation: (a) 131,000.0; (b) 0.00047; (c) 210,006; (d) 2160.5.

**1.60** Write the following numbers in scientific notation: (a) 281.0; (b) 0.00380; (c) 4270.8; (d) 58,200.9.

**1.61** Carry out each of the following calculations, paying special attention to significant figures, rounding, and units (J = joule, the SI unit of energy; mol = mole, the SI unit for amount of substance):

   (a) $\dfrac{(6.626 \times 10^{-34} \text{ J} \cdot \text{s}) (2.9979 \times 10^8 \text{ m/s})}{473 \times 10^{-9} \text{ m}}$

   (b) $\dfrac{(6.022 \times 10^{23} \text{ molecules/mol}) (1.04 \times 10^2 \text{ g})}{44.01 \text{ g/mol}}$

(c) $(6.022 \times 10^{23} \text{ atoms/mol}) \times (2.18 \times 10^{-18} \text{ J/atom})$

$\times (\dfrac{1}{2^2} - \dfrac{1}{3^2})$, where the numbers 2 and 3 in the last

term are exact.

**1.62** Carry out each of the following calculations, paying special attention to significant figures, rounding, and units:

(a) $\dfrac{4.336 \times 10^7 \text{ g}}{{}^{4}\!/_{3} \, (3.1416) \, (1.05 \times 10^2 \text{ cm})^3}$

(b) $\dfrac{(1.45 \times 10^2 \text{ g}) \, (44.7 \text{ m/s})^2}{2}$

(c) $\dfrac{(1.0 \times 10^{-4} \text{ mol/L})^2 \, (2.65 \times 10^{-3} \text{ mol/L})}{(8.35 \times 10^{-5} \text{ mol/L}) \, (1.48 \times 10^{-2} \text{ mol/L})^3}$

**1.63** The following dart boards illustrate the types of errors often seen in measurements. The bull's-eye represents the actual value, and the darts represent the data.
(a) Which experiments yield the same average result?
(b) Which experiment(s) display(s) high precision?
(c) Which experiment(s) display(s) high accuracy?
(d) Which experiment(s) show(s) a systematic error?

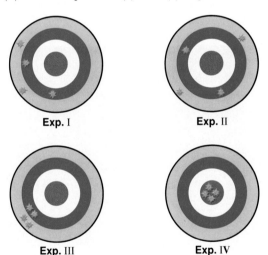

**Exp.** I      **Exp.** II

**Exp.** III      **Exp.** IV

**1.64** A laboratory instructor gives a sample of powdered metal to each of four students, I, II, III, and IV, and they weigh the samples. The results of their multiple trials follow. The true value is 6.72 g.
I: 6.71, 6.75, 6.70 g    II: 6.56, 6.76, 6.84 g
III: 6.50, 6.48, 6.52 g    IV: 6.41, 6.72, 6.55 g
(a) Calculate the average mass from each set of data, and tell which set is the most accurate.
(b) Precision is a measure of the average of the deviations of each piece of data from the true value. Which set of data is the most precise? Is this set also the most accurate?
(c) Which set of data has the best combination of accuracy and precision?

(d) Which set of data has the worst combination of accuracy and precision?

## Comprehensive Problems
Problems marked with an asterisk (*) are more challenging.

**1.65** Your sports car gets 23.4 mi/gal and holds 68.1 L gasoline.
(a) How far can you drive on a tankful?
(b) If gas costs $0.41/L, how much does the trip cost?
(c) If your average speed is 96.6 km/h, how much time does the trip take?

**1.66** An Olympic-size pool is 50.0 m long and 25.0 m wide.
(a) How many gallons of water ($d = 1.0$ g/mL) are needed to fill the pool to an average depth of 5.0 ft?
(b) What is the mass (in kg) of water in the pool?

**1.67** According to the lore of ancient Greece, Archimedes discovered the displacement method of density determination while bathing and used it to find the composition of the king's crown. If a crown weighing 4 lb 13 oz displaces 186 mL water when placed in a bathtub, is the crown made of pure gold ($d = 19.3$ g/cm$^3$)?

**1.68** Assuming the density of water is 1.00 g/cm$^3$, or 62.4 lb/ft$^3$, calculate the density of the following substances in lb/ft$^3$: (a) diamond, $d = 3.51$ g/cm$^3$; (b) copper, $d = 8.95$ g/cm$^3$; (c) carbon dioxide, $d = 1.98$ g/L.

**1.69** An empty graduated cylinder weighs 45.70 g and is filled with 40.0 mL water ($d = 1.00$ g/cm$^3$). A piece of lead submerged in the water brings the total volume to 67.4 mL and the mass of cylinder and contents to 396.4 g. What is the density of lead?

*1.70 In the buoyancy method of density determination, an object's weight in air is compared with its weight in liquid. The weight is less in the liquid because the weight of the displaced liquid acts as a force that pushes the object upward.
(a) A brass cylinder weighs 75.04 g in air and 66.11 g in water ($d = 0.998$ g/mL). What is the density of the brass?
(b) Geologists in the field use mineral oil ($d = 1.75$ g/cm$^3$) to determine the density of mineral samples. A sample of galena (an ore of lead) weighs 24.61 g in air and 18.83 g in mineral oil. What is the density of galena?

*1.71 "Copper" pennies actually contain very little copper. If a new penny is 97.3% zinc and 2.7% copper by mass, what is its density? ($d$ of copper $= 8.95$ g/cm$^3$; $d$ of zinc $= 7.14$ g/cm$^3$.)

*1.72 Temperature scales can be based on many pure liquids. If a temperature scale were based on the freezing point (5.5°C) and boiling point (80.1°C) of benzene and the temperature difference between these points was divided into 50 units (called °X), what would be the freezing and boiling points of water in °X? (See Figure 1.10 in text.)

# CHAPTER 2

# The Components of Matter

**Concepts and skills to review**

- physical and chemical change (Section 1.1)
- states of matter (Section 1.1)
- attraction and repulsion between charged particles (Section 1.1)
- meaning of a scientific model (Section 1.3)
- SI units and conversion factors (Section 1.5)
- significant figures in calculations (Section 1.6)

**The aurora borealis.** The wondrous northern lights appear when charged particles from the Sun, bent by the Earth's magnetic field, collide with gases in the atmosphere and cause them to glow. As you will learn in this chapter, almost a century ago, laboratory studies of similar phenomena led to our understanding of atomic structure and the ultimate composition of all matter.

**W**henever we look closely at a material object, we naturally ask what it is made of. This is an age-old question. Philosophers of ancient Greece believed that everything was made of one or a few elemental substances. Some believed that substance to be water, since the rivers and oceans extend everywhere. Others thought it was air, which was thinned into fire and thickened into clouds, rain, and rock. Still others believed that there were four elemental substances—whose properties accounted for taste, temperature, and all other characteristics of things.

Rather than considering specific substances, Democritus (c. 460-370 BC) focused on the *components of all substances*. He reasoned that if, for example, you cut a piece of copper smaller and smaller, you would eventually reach copper particles that could no longer be cut. Thus, he concluded that all matter is ultimately composed of indivisible particles with nothing but empty space between them; he called the particles atoms (from the Greek *atomos*, "uncuttable"): "According to convention, there is a sweet and a bitter, a hot and a cold, and according to convention, there is order. In truth, there are atoms and a void." However, Aristotle (384-322 BC), who had elaborated the idea of four elemental substances, held that it was impossible for "nothing" to exist. Because of his powerful influence on Western thinking, the concept of atoms was suppressed for 2000 years.

Finally, in 17th-century England, Robert Boyle argued that an element is composed of "simple Bodies, not made of any other Bodies, of which all mixed Bodies are compounded, and into which they are ultimately resolved," a description that is remarkably similar to our current picture of an element, in which the "simple Bodies" are atoms. Another 100 years of inquiry gave rise to laws concerning the masses of substances that react with each other. Then, as the 19th century began, John Dalton proposed an atomic model that explained these laws and led to rapid progress in chemistry. By the century's close, however, further observations exposed the need for a different atomic model. Finally, a burst of creativity in the early 20th century led to a picture of an atom with a complex internal structure of even smaller particles and eventually to our current model.

In this chapter, we discuss the properties and composition of the three types of matter—elements, compounds, and mixtures—on the macroscopic and atomic scales. We then examine the mass laws that led to Dalton's theory, analyze his theory, and see the experiments that gave rise to our current model. In the rest of the chapter, we consider how the elements are classified, how they combine to form compounds, how we derive compound names and formulas, and how mixtures are classified and separated.

## 2.1  Elements, Compounds, and Mixtures: An Atomic Overview

Matter can be broadly classified into three types of substances, two of which, elements and compounds, are considered **pure substances** because their compositions are fixed. The third type is mixtures, which are *impure* because their composition is not fixed.

An **element** is the simplest type of substance with unique physical and chemical properties. *An element consists of only one type of atom.* Therefore, it cannot be broken down into any simpler substance by physical or chemical means. Each element has a name, such as carbon, oxygen, or copper. A sample of carbon contains only carbon atoms. The properties of a piece of carbon, such as color, density, and combustibility, are different from those of a piece of copper because the properties of carbon atoms are different from those of copper atoms; thus, *each element is unique.*

Although most elements exist in nature as large populations of individual atoms (Figure 2.1, *A*), several occur in molecular form: a **molecule** is a structure consisting of two or more atoms that are chemically bound together and thus behaves as an independent unit. A molecule of an element is composed of just one type of atom (Figure 2.1, *B*). Oxygen gas, for example, occurs in air as diatomic (two-atom) molecules.

A **compound** is a substance composed of two or more elements that are chemically combined. The elements in a compound are not just mixed together; rather, their atoms have joined chemically to form the new substance (Figure 2.1, *C*). Ammonia, water, and carbon dioxide are some common compounds. One defining feature of a compound is that *the elements are present in fixed parts by mass* (fixed mass ratio). A molecule of the compound maintains this fixed mass ratio because it consists of a *fixed number ratio* of atoms of the component elements. For example, the mass of any sample of ammonia is 14 parts nitrogen and 3 parts hydrogen. One nitrogen atom has 14 times the mass of 1 hydrogen atom, so on the molecular scale, 1 molecule of ammonia consists of 1 nitrogen and 3 hydrogen atoms.

Another defining feature of a compound is that *its properties are different from those of its component elements.* For example, silvery sodium metal, which reacts violently with water, and yellow-green, poisonous chlorine gas are very different from the compound they form—white, crystalline sodium chloride, commonly known as table salt (Table 2.1). A compound *can* be broken down into its component elements. By definition, this breakdown can occur only by a *chemical change,* as when electric current decomposes sodium chloride into metallic sodium and chlorine gas.

**FIGURE 2.1**

Elements, compounds, and mixtures on the atomic scale. **A,** An element consists of identical atoms, but only some elements occur as individual atoms. **B,** Other elements occur as molecules. **C,** A molecule of a compound consists of a characteristic number of atoms of two or more elements chemically bound together. **D,** A mixture contains the individual units of two or more elements and/or compounds that are physically intermingled.

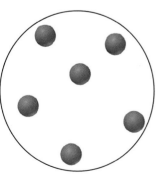

**A** Atoms of an element

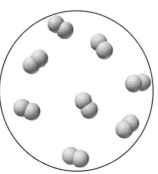

**B** Molecules of an element

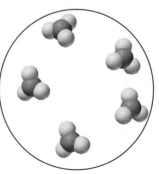

**C** Molecules of a compound

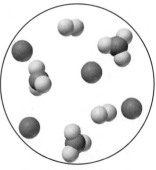

**D** Mixture of two elements and a compound

**TABLE 2.1  Some Properties of Sodium, Chlorine, and Sodium Chloride**

| PROPERTY | SODIUM + | CHLORINE → | SODIUM CHLORIDE |
|---|---|---|---|
| Melting point | 97.8°C | −101°C | 801°C |
| Boiling point | 881.4°C | −34°C | 1413°C |
| Color | Silvery | Yellow-green | Colorless (white) |
| Density | 0.97 g/cm³ | 0.0032 g/cm³ | 2.16 g/cm³ |
| Behavior in water | Reacts violently | Dissolves slightly | Dissolves freely |

A **mixture** is a group of two or more elements and/or compounds that are physically intermingled (Figure 2.1, *D*). In contrast to a compound, *the components of a mixture **can** vary widely in their parts by mass.* A mixture of sodium chloride and water, for example, can have many different proportions of salt to water. Since the components are physically mixed, not chemically combined, at the atomic scale a mixture is merely a group of the individual units that make up its component elements and compounds. Therefore, *a mixture retains many of the properties of its components.* Salt water, for instance, is colorless like water and tastes salty like sodium chloride. Unlike compounds, mixtures can be separated into their components by *physical changes;* chemical changes need not occur. Salt water can be separated into its component compounds by boiling off the water, a physical process that leaves behind the sodium chloride.

**Section Summary**
All matter exists as either elements, compounds, or mixtures. Elements are the simplest type of matter because they consist of only one type of atom. A compound forms when two or more elements combine in a chemical reaction; it exhibits different properties from its component elements. The elements of a compound occur in a fixed mass ratio because their atoms of the elements are present in a fixed number ratio. A mixture results when two or more substances are mixed together, but no chemical reaction occurs. The components can be present in any proportion, and they retain their individual properties.

## 2.2  The Observations That Led to an Atomic View of Matter

In 1808, John Dalton (1766-1844) published *A New System of Chemical Philosophy,* which contained his model of an atomic view of matter. ◆ Any model of the composition of matter had to account for two extremely important chemical observations: the *law of mass conservation* and the *law of definite (or constant) composition.* Dalton's theory explained these laws and another observation now known as the *law of multiple proportions.*

◆ **Dalton's Revival of Atomism.** John Dalton, the son of a poor weaver, had no formal education, but he established one of the most powerful concepts in science. He began teaching mathematics and science when he was only 12 years old. Later, he studied color blindness, a personal affliction still known as daltonism, and in 1787 began his life's work in meteorology, for which he recorded daily weather data until his death 57 years later. From his studies on humidity and dew point, Dalton discovered a key behavior of gases (Section 5.4), which eventually led to his atomic theory. In 1803, he stated, "I am nearly persuaded that [the mixing of gases and their solubility in water] depends upon the mass and number of the ultimate particles....An enquiry into the relative masses of [these] particles of bodies is a subject, as far as I know, entirely new: I have lately been prosecuting the enquiry with remarkable success." The atomic theory was published 5 years later.

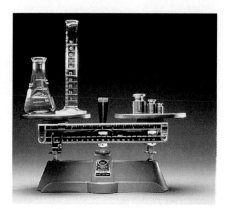
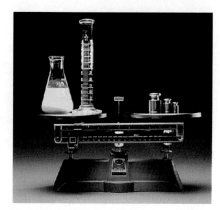

**A**                                        **B**

## FIGURE 2.2

The law of mass conservation: mass remains constant during a chemical reaction. **A,** Before the reaction, a flask with lead nitrate solution and a cylinder with sodium chromate solution are balanced by counterweights. **B,** When the solutions are mixed, a chemical reaction occurs that forms lead chromate (yellow solid) and a sodium nitrate solution. Note that no change in mass has occurred.

### ◆ Immeasurable Changes in Mass.
From the work of Albert Einstein (1879-1955), we now know that mass and energy are alternate aspects of a single entity called mass-energy. As a result, mass conservation and its counterpart, energy conservation, are combined into mass-energy conservation. Mass *does* change during a chemical change but by an extremely small amount. For example, when 100 g carbon burns in oxygen, only 0.000000036 g ($3.6 \times 10^{-8}$ g) of mass is converted to energy—a change in mass far below the limit of the best modern balance. Energy changes in chemical reactions are so small that, for all intents and purposes, mass is conserved. (As you'll see later, however, energy changes in nuclear reactions are so large that mass changes are easy to measure.)

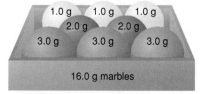

## FIGURE 2.3

The meaning of mass fraction and mass percent. The box contains three types of marbles: yellow marbles weigh 1.0 g each, purple marbles weigh 2.0 g each, and red marbles weigh 3.0 g each. Each type makes up a fraction of the total mass of marbles, 16.0 g. The *mass fraction* of the yellow (1.0-g) marbles is (3 × 1.0 g)/16.0 g = 0.19. The *mass percent* (parts per 100 parts) of the yellow marbles is 0.19 × 100 = 19% by mass. Similarly, the purple marbles have a mass fraction of 0.25 and are 25% by mass, and the red marbles have a mass fraction of 0.56, or 56% by mass. Similarly, in a compound, each element has a *fixed* mass fraction (or mass %).

## Mass Conservation: The Indestructibility of Matter During Chemical Change

The most fundamental chemical observation of the 18th century was the **law of mass conservation:** *the total mass of substances does not change during a chemical reaction.* The *number* of substances may change and their properties certainly do, but the *total amount* of matter remains constant. Lavoisier had first stated this law on the basis of his combustion experiments, in which he found the mass of oxygen plus the mass of mercury equal to the mass of mercuric oxide they formed. Figure 2.2 illustrates mass conservation in a reaction that occurs in water. Even in a complex biochemical change involving many reactions, such as the breakdown of the sugar glucose in the body, mass is conserved:

180 g glucose + 192 g oxygen gas → 264 g carbon dioxide + 108 g water

(372 g material before change) → (372 g material after change)

Mass conservation means that, based on all chemical experience, *matter cannot be created or destroyed.* ◆

## Definite Composition: Constant Mass Ratios of Elements in Compounds

Another fundamental chemical observation is summarized as the **law of definite (or constant) composition:** *no matter what its source, a particular chemical compound is composed of the same elements in the same parts (fractions) by mass.* **Percent by mass (mass percent, mass %)** is the fraction by mass expressed as a percentage. Figure 2.3 depicts the meaning of fraction by mass and percent by mass in terms of a box of marbles.

Consider the compound calcium carbonate, the major substance in a piece of chalk. The following results are obtained for the elemental mass composition of 20.0 g calcium carbonate:

| ANALYSIS BY MASS (grams/20.0 g) | MASS FRACTION (parts/1.00 part) | PERCENT BY MASS (parts/100 parts) |
|---|---|---|
| 8.0 g calcium | 0.40 calcium | 40% calcium |
| 2.4 g carbon | 0.12 carbon | 12% carbon |
| 9.6 g oxygen | 0.48 oxygen | 48% oxygen |
| 20.0 g | 1.00 part by mass | 100% by mass |

Note that the sum of the mass fractions (or percents) equals 1.00 part (or 100%) by mass. The law of definite composition tells us that if we analyze pure calcium carbonate samples obtained from Italian marble, Caribbean coral reef, Gulf Coast seashell, or any other source, we always find the same elements in the same percent by mass (Figure 2.4).

Because a given element always constitutes the same mass fraction of a given compound, we can use the mass fraction to find the mass of the element in any mass of the compound:

$$\text{Mass of element} = \text{mass of compound} \times \frac{\text{part by mass of element}}{\text{one part by mass of compound}} \qquad \textbf{(2.1)}$$

We can express the mass fraction in whatever mass units are called for in a calculation.

SAMPLE PROBLEM 2.1 _____

### Calculating the Mass of an Element in a Compound

**Problem:** Pitchblende is the most commercially important compound of uranium. Analysis shows that 84.2 g pitchblende contains 71.4 g uranium, with oxygen as the only other element. How many grams of uranium can be obtained from 102 kg pitchblende?

**Plan:** From the analysis, we know the amount of uranium in a certain sample of pitchblende and have to find the amount in a different sample. Since the mass fraction of an element in a compound applies to any sample of that compound, we find the mass fraction of uranium in 84.2 g pitchblende and then use it to find the mass (in kg) of uranium present in the 102-kg sample of pitchblende. Then we convert from kg to g.

**Solution:** Finding the mass fraction of uranium:

$$\text{Mass fraction of uranium} = \frac{\text{mass uranium}}{\text{mass pitchblende}} = \frac{71.4 \text{ g uranium}}{84.2 \text{ g pitchblende}}$$

$$= \frac{0.848 \text{ part by mass uranium}}{1.00 \text{ part by mass pitchblende}}$$

Finding the mass of uranium in 102 kg pitchblende:

$$\text{Mass (kg) of uranium} = \text{mass (kg) of pitchblende} \times \text{mass fraction of uranium}$$

$$= 102 \text{ kg pitchblende} \times \frac{0.848 \text{ kg uranium}}{1.00 \text{ kg pitchblende}}$$

$$= 86.5 \text{ kg uranium}$$

Converting the mass from kg to g:

$$\text{Mass (g) of uranium} = 86.5 \text{ kg uranium} \times \frac{1000 \text{ g}}{1 \text{ kg}} = \textbf{8.65} \times \textbf{10}^{\textbf{4}} \textbf{ g uranium}$$

**Check:** The analysis showed that most of the mass of pitchblende is due to uranium, so the large mass fraction seems correct. Round off to check the mass of uranium in pitchblende: 100 kg pitchblende × ~0.85 mass fraction uranium = 85 kg uranium.

**Comment:** Mass analysis provides part by mass, so we can use it directly with any mass unit and eliminate the need to find the mass fraction:

$$102 \text{ kg pitchblende} \times \frac{71.4 \text{ kg uranium}}{84.2 \text{ kg pitchblende}} = 86.5 \text{ kg uranium}$$

FOLLOW-UP PROBLEM 2.1
How many metric tons (t) of oxygen are combined in a sample of pitchblende that contains 2.3 t uranium? (*Hint:* Oxygen is the other element present.)

| Mass (g) of uranium from analysis |
| --- |

divide by mass (g) of pitchblende

| Mass fraction of uranium in pitchblende |
| --- |

multiply by mass (kg) of pitchblende

| Mass (kg) of uranium in sample of pitchblende |
| --- |

1 kg = 1000 g

| Mass (g) of uranium in sample of pitchblende |
| --- |

**FIGURE 2.4**

**The law of definite composition.** Calcium carbonate is found naturally in many forms, including **A,** seashells; **B,** the white cliffs of Dover, England; **C,** marble; and **D,** coral. Regardless of the compound's source, the mass percents of its component elements are the same.

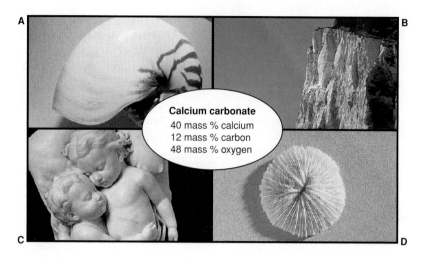

**Calcium carbonate**
40 mass % calcium
12 mass % carbon
48 mass % oxygen

## Multiple Proportions

Dalton described a phenomenon that occurs when two elements form more than one compound. His observation is now called the **law of multiple proportions:** *if elements A and B react to form two compounds, the different masses of B that combine with a fixed mass of A can be expressed as a ratio of small whole numbers.* Consider, for example, two compounds that form from the elements carbon and oxygen; for now, we can call them carbon oxides I and II. Note that the presence of more than one compound of the same elements does not violate the law of definite composition because I and II have different properties. For example, measured at the same temperature and pressure, the density of carbon oxide I is 1.25 g/L, whereas that of II is 1.98 g/L; moreover, I is poisonous and flammable, but II is not. Since they are different compounds, their compositions by mass also differ:

Carbon oxide I:  57.1 mass % oxygen and 42.9 mass % carbon
Carbon oxide II: 72.7 mass % oxygen and 27.3 mass % carbon

To see the phenomenon of multiple proportions, we use the mass percents of oxygen and of carbon in each compound to find the masses of these elements in a given mass, for example 100 g, of each compound. Then we divide the mass of oxygen by the mass of carbon in each compound to obtain the mass of oxygen that combines with a fixed mass of carbon:

|  | CARBON OXIDE I | CARBON OXIDE II |
|---|---|---|
| g oxygen/100 g compound | 57.1 | 72.7 |
| g carbon/100 g compound | 42.9 | 27.3 |
| g oxygen/g carbon | $\dfrac{57.1}{42.9} = 1.33$ | $\dfrac{72.7}{27.3} = 2.66$ |

If we then divide the grams of oxygen per gram of carbon in II by that in I, we obtain a ratio of small whole numbers:

$$\frac{2.66 \text{ g oxygen/g carbon in II}}{1.33 \text{ g oxygen/g carbon in I}} = \frac{2}{1}$$

The law of multiple proportions tells us that in two compounds of the same elements, the mass fraction of one element relative to the other changes in *whole-number increments*, rather than varying continuously; that is, II contains *2 times* as much oxygen for a given mass of carbon as I, not 1.5

times, 1.7 times, or any other intermediate amount. As you'll see next, Dalton's theory allows us to explain the composition of carbon oxides I and II on the atomic scale.

### Dalton's Atomic Theory

With almost 200 years of hindsight, it may be easy to see how the mass laws could be explained by an atomic model—matter existing in indestructible units, each with a particular mass—but it was a major breakthrough when John Dalton presented his atomic theory of matter. Dalton expressed his theory in a series of postulates, which are presented here in modern terminology and reorganized for the sake of discussion. The text in parentheses points out the key differences between Dalton's postulates and our present understanding.

1.  *All matter consists of **atoms,** tiny, indivisible particles of an element that cannot be created or destroyed.* (We now know that atoms are not indivisible, but rather are composed of smaller, subatomic particles.)
2.  *Atoms of one element cannot be converted into atoms of another element.* In chemical reactions, the original substances separate into atoms, which recombine to form different substances. (We now know that in nuclear reactions, atoms of one element often change into atoms of another, but this *never* happens in a chemical reaction.)
3.  *Atoms of an element are identical in mass and other properties and are different from atoms of any other element.* (We now know that different atoms of an element *can* differ in mass but, in general, only slightly.)
4.  *Compounds result from the chemical combination of a specific ratio of atoms of different elements.* (We now know that a few compounds can have slight variations in their atom ratios, but this postulate remains essentially unchanged.) ◆

Let's examine how Dalton's postulates explain the mass laws.

*   *Mass conservation.* Atoms cannot be created or destroyed (postulate 1) or converted into other types of atoms (postulate 2). Since each type of atom has a fixed mass (postulate 3), a chemical reaction, in which atoms are just rearranged, cannot possibly result in a mass change.
*   *Definite composition.* A compound is a combination of a *specific* ratio of different atoms (postulate 4), each of which has a particular mass (postulate 3). Therefore, each element in a compound always constitutes a fixed fraction of the total mass.
*   *Multiple proportions.* Atoms of an element have the same mass (postulate 3) and are indivisible (postulate 1). Because different numbers of B atoms combine with each A atom in the two compounds, the masses of element B that combine with a fixed mass of element A will give a small, whole-number ratio.

The *simplest* arrangement consistent with the mass data for carbon oxides I and II in our earlier example is that one atom of oxygen combines with one atom of carbon in compound I (carbon monoxide) and that two atoms of oxygen combine with one atom of carbon in compound II (carbon dioxide) (Figure 2.5).

### The Controversy over Atomic Masses

After the publication of the atomic theory, many investigators tried to determine the atomic masses of the elements from the mass ratios of the elements in a compound. Because an individual atom is so small, we can only

◆ **Building on the Ideas of Others.** Which of Dalton's postulates were original, and which came from others?
*   Postulate 1 derives from the "eternal, indestructible atoms" of Democritus more than 2000 years earlier and is consistent with mass conservation as stated by Lavoisier.
*   Postulate 2 is a statement against the alchemical belief in the magical transmutation of elements.
*   Postulate 3 contains Dalton's major new ideas, with its emphasis on unique mass and properties for all the atoms of a given element.
*   Postulate 4 follows directly from the fact of definite composition.

As with most great thinkers, Dalton incorporated the concepts of others into his own to create the new theory.

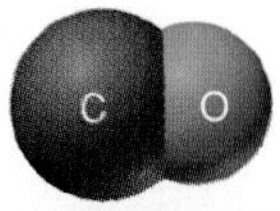

Carbon oxide I
(carbon monoxide)

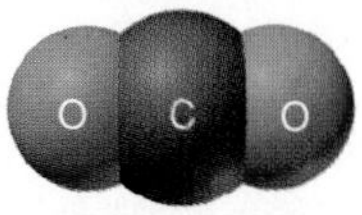

Carbon oxide II
(carbon dioxide)

**FIGURE 2.5**

**The atomic basis of the law of multiple proportions.** Carbon and oxygen can combine to form carbon oxide I (carbon monoxide) and carbon oxide II (carbon dioxide). The masses of oxygen in the two compounds relative to a fixed mass of carbon are in a ratio of small whole numbers because the molecules are composed of atoms with specific masses.

determine the mass of all the atoms of an element in a compound *relative* to the masses of the other elements. As a basis for these relative atomic masses, Dalton assigned an atomic mass of 1 to hydrogen, the lightest known substance. Lavoisier had shown earlier that water contains 8 g oxygen for every 1 g hydrogen, so Dalton assigned a relative atomic mass of 8 to oxygen. However, this relative mass would be correct *only* if there were the same number of oxygen atoms in 8 g oxygen as hydrogen atoms in 1 g hydrogen, and this would be true *only* if a water molecule consists of one oxygen atom for every hydrogen atom (symbolized HO).

At about the same time, the French chemist Joseph Gay-Lussac (1778-1850) began a series of experiments to study the atom ratio of water. Rather than measuring *masses,* however, he measured the *volumes* of hydrogen gas and oxygen gas that react to form water vapor. Gay-Lussac found that 2 L hydrogen gas combined with 1 L oxygen gas. This result implied that each water molecule consists of two hydrogen atoms and one oxygen atom (symbolized $H_2O$). However, another of Gay-Lussac's results was confusing: 2 L water vapor were produced for every 2 L hydrogen gas that reacted. Here are the expected and actual results:

If water is HO:    1 L hydrogen gas + 1 L oxygen gas → 1 L water vapor

If water is $H_2O$:    2 L hydrogen gas + 1 L oxygen gas → 1 L water vapor

Actual result:    2 L hydrogen gas + 1 L oxygen gas → 2 L water vapor

Most investigators assumed that all elements, including hydrogen and oxygen, existed naturally as individual atoms, so the observed result did not seem to make sense. Did each oxygen atom split in half in order to form two molecules of water? If so, atoms were divisible and the atomic theory was wrong. Dalton vigorously attacked Gay-Lussac's technique and results. In 1811, the Italian physicist Amadeo Avogadro (1776-1856) made two proposals that explained these confusing results:

- Each hydrogen gas particle and each oxygen gas particle is actually a molecule composed of two atoms. The molecules, not the atoms, split apart, and the separate atoms recombine to form molecules of water vapor.
- Equal volumes of a gas contain equal numbers of gas particles (under identical conditions). Therefore, 2 L hydrogen gas contain twice the number of gas particles as 1 L oxygen gas and the same number as 2 L water vapor. Figure 2.6 shows how Avogadro's hypotheses account for Gay-Lussac's results.

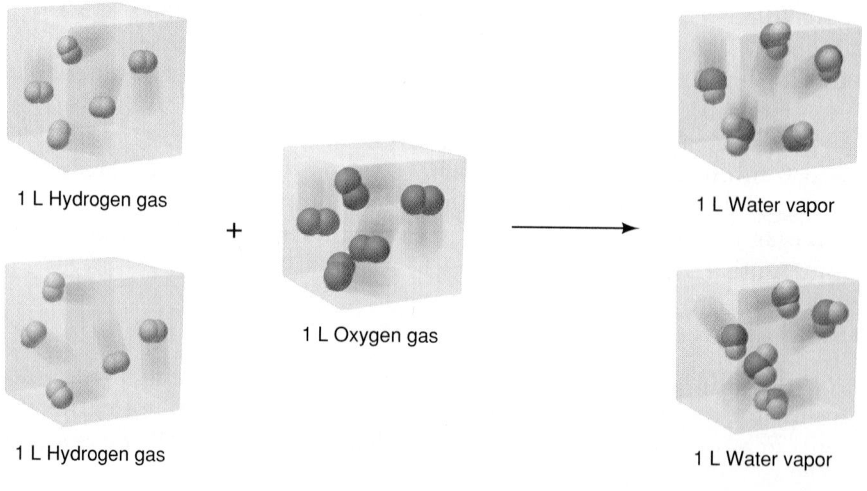

**FIGURE 2.6**

**The combining volumes of hydrogen and oxygen in the formation of water vapor.** To explain Gay-Lussac's results, Avogadro hypothesized that gas volume is proportional to the number of gas particles present. Thus, two volumes of hydrogen gas (or water vapor) contain twice the number of particles as one volume of oxygen gas. If hydrogen gas and oxygen gas occur as diatomic molecules, two hydrogen "particles" (two molecules of two atoms each) combine with one oxygen "particle" (one molecule of two atoms) to form two water "particles" (two molecules each consisting of two H atoms and one O atom).

1 L Hydrogen gas

1 L Hydrogen gas

+

1 L Oxygen gas

→

1 L Water vapor

1 L Water vapor

Even though Avogadro's explanation is correct, it was ignored because it was presented with difficult terminology and complex mathematics. Almost 50 years later, his ideas were revived and led to the determination of the correct relative atomic mass of oxygen as 16. Since the mass ratio of water is 8 g oxygen to 1 g hydrogen, the ratio of combining volumes means that 1 oxygen atom weighs 16 times as much as 1 hydrogen atom:

$$\text{(volume ratio)}\ \frac{1\ \text{oxygen (16)}}{2\ \text{hydrogen (1 each)}} \Rightarrow \frac{16}{2} = \frac{8}{1}\ \text{(mass ratio)}$$

Dalton's picture of the atom had survived its first test and has had a profound effect on the development of modern chemistry. Because his model explained the masses of reacting elements in terms of atoms, it encouraged the establishment of atomic masses and eventually of chemical formulas, such as $NH_3$ for ammonia and $H_2O$ for water. However, the model soon proved to be too limited. It could not explain *why* elements combine in *specific* ratios of atoms: why, for example, are two, and not three, hydrogen atoms combined with one oxygen atom in a water molecule? Also, Dalton's "billiard ball" view of the atom could not account for the electrically charged particles that would soon be seen in many experiments. A more complex model of the atom would be needed to understand those observations. ◆

**Section Summary**

Three widespread observations known as mass laws state that (1) the total mass remains constant during a chemical reaction; (2) any sample of a given compound has the same elements present in the same mass fractions; and (3) in different compounds of the same elements, the masses of one element that combine with a fixed mass of the other are whole-number multiples.

Dalton's atomic theory explained these laws by proposing that all matter consists of indivisible, unchangeable atoms of fixed, unique mass. Mass is constant during a reaction because atoms form new combinations; each compound has a fixed mass fraction of its elements because it is composed of a fixed number of each type of atom; and different compounds of the same elements exhibit multiple proportions because they each consist of whole atoms. Studies of the masses and volumes of elements that combine eventually led to consistent values for relative atomic masses.

◆ **Atoms? Humbug!** Rarely does a major new concept receive unanimous approval. Despite the atomic theory's impact, for another century several major scientists denied the existence of atoms. In 1877, Adolf Kolbe, an eminent organic chemist, said, "[Dalton's atoms are]…no more than stupid hallucinations…of a freshly rouged prostitute;…mere table-tapping and supernatural explanations." The influential physicist Ernst Mach believed that scientists should look at facts, not hypothetical entities such as atoms. It was not until 1908 that the famous chemist and opponent of atomism Wilhelm Ostwald wrote, "I am now convinced [by recent] experimental evidence of the discrete or grained nature of matter, which the atomic hypothesis sought in vain for hundreds and thousands of years." He was referring to the discovery of the electron, which we discuss in the next section.

## 2.3  The Observations That Led to the Nuclear Atom Model

The path of discovery is often winding and unpredictable. Basic research into the nature of electricity eventually led to the discovery of *electrons,* negatively charged particles that are part of all atoms. Soon thereafter, other experiments revealed that the atom has a *nucleus*—a tiny, central core of mass and positive charge. In this section, we examine some key experiments that led to our current model of the nuclear atom.

### Discovery of the Electron and Its Properties

Nineteenth-century investigators of electricity knew that matter and electric charge were somehow related. When amber is rubbed with fur, or glass with

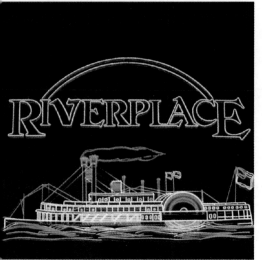

**◆ The Familiar Glow of Colliding Particles.** The electric and magnetic properties of charged particles that collide with gas particles or hit phosphor-coated screens have familiar applications. "Neon" signs glow because electrons collide with the gas particles in the tube, causing them to give off light. An aurora display occurs when the Earth's magnetic field bends streams of charged particles coming from the Sun, which collide with gases in the atmosphere (see figure at start of chapter). In a television tube or computer monitor, the cathode ray passes back and forth over the coated screen, creating a pattern that the eye sees as a picture.

silk, positive and negative charges form—the same charges that make your hair crackle and cling to your comb on a dry day. They also knew that an electric current could decompose certain compounds into their elements.

But how could electricity be studied in the *absence* of matter? The vacuum pump had recently been invented, so some investigators tried passing an electric current through sealed, evacuated tubes containing metal electrodes connected to an external source of electricity. One of those investigators was the English physicist Sir William Crookes. When he turned on the power supply in his darkened laboratory, he saw a "ray" strike the phosphor-coated end of the tube and emit a flash of light. Called **cathode rays** because they move from the negative electrode (cathode) to the positive electrode (anode), these rays travel in straight lines, unless deflected by magnetic or electric fields, and are identical, no matter what metal is used for the cathode. Figure 2.7 summarizes the investigations of cathode ray properties. It was concluded that cathode rays consist of negatively charged particles found in all matter and that the rays appear when these particles collide with the few remaining gas molecules in the evacuated tubes. ◆ Cathode ray particles were later named *electrons*.

In 1897, J. J. Thomson (1856-1940) used magnetic and electric fields to measure the ratio of the cathode ray particle's mass to its charge. By comparing this value with the mass/charge ratios for the lightest charged particles in solutions, Thomson estimated that the cathode ray particle weighed

**FIGURE 2.7**

**Experiments to determine the properties of cathode rays. A,** A cathode ray forms when high voltage is applied across the electrodes in a partially evacuated tube. A hole in the anode allows the ray to pass through and hit the coated end of the tube. In the absence of any field, the ray travels in a straight path. **B,** Because it bends in a magnetic field, the ray must consist of charged particles. **C,** Because the ray bends toward a positive electric plate, the particles' charge must be negative. **D,** Because any electrode material produces an identical ray, the particles must be part of all matter.

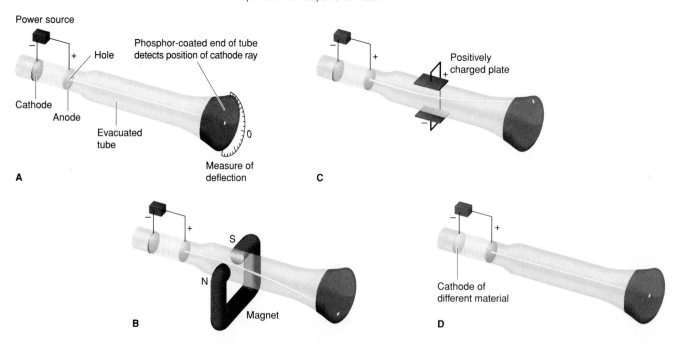

less than $\frac{1}{1000}$ as much as hydrogen, the lightest atom! He was shocked by the implication that, contrary to Dalton's atomic theory, *atoms are divisible into even smaller particles.* He concluded, "We have in the cathode rays matter in a new state, . . . in which the subdivision of matter is carried much further . . . ; this matter being the substance from which the chemical elements are built up." Fellow scientists reacted at first with disbelief; some even thought he was joking.

In 1909, the American physicist Robert Millikan measured the charge of the electron by observing the movement of tiny droplets of the "highest grade clock oil" in an apparatus that contained electrically charged plates and an x-ray source (Figure 2.8). The x-rays knocked electrons from gas molecules in the air, and as an oil droplet fell through a hole in the positive (upper) plate, the electrons stuck to the drop and gave it a negative charge. With the electric field off, Millikan measured the mass of the droplet from its rate of fall. With the field on and by varying its strength, he could make the drop fall more slowly, rise, or pause suspended. From these data, Millikan calculated the total charge of the droplet.

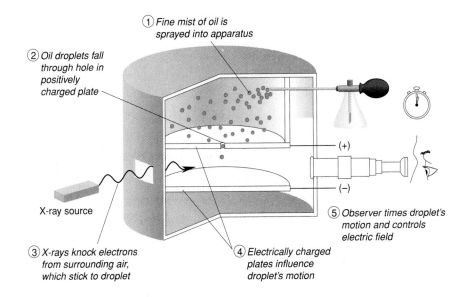

① Fine mist of oil is sprayed into apparatus

② Oil droplets fall through hole in positively charged plate

X-ray source

③ X-rays knock electrons from surrounding air, which stick to droplet

④ Electrically charged plates influence droplet's motion

(+)

(–)

⑤ Observer times droplet's motion and controls electric field

**FIGURE 2.8**
**Millikan's oil-drop experiment for measuring an electron's charge.** The motion of a given oil droplet depends on the variation in electric field and the total charge on the droplet, which depends on the number of attached electrons. Millikan reasoned that the total charge must be some whole-number multiple of the charge of the electron.

After studying many droplets, Millikan found that their various charges were always some whole-number multiple of a *minimum* charge. He reasoned that different oil droplets could pick up different numbers of electrons, so this minimum charge must be that of the electron itself. The value, which he calculated more than 85 years ago, is within 1% of the modern value of the electron's charge, $1.602 \times 10^{-19}$ C (coulomb, the SI unit of charge). With values for the electron's mass/charge ratio from the work of Thomson and others and from his own value for the electron's charge, Millikan determined the electron's *extremely* small mass:

$$\text{Mass of electron} = \frac{\text{mass}}{\text{charge}} \times \text{charge} = \left(5.686 \times 10^{-12} \frac{\text{kg}}{\cancel{\text{C}}}\right)(1.602 \times 10^{-19} \cancel{\text{C}})$$

$$= 9.109 \times 10^{-31} \text{ kg} = 9.109 \times 10^{-28} \text{ g}$$

The properties of the electron posed some problems concerning the nature of atoms. Since matter is electrically neutral, atoms must be neutral as well. If atoms contain negatively charged electrons, what positive charges balance them? And if an electron has such an incredibly tiny mass, what accounts for most of an atom's mass? To address these issues, Thomson proposed a model of a spherical atom composed of diffuse, positively charged matter, in which electrons were embedded like "raisins in a plum pudding."

### Discovery of the Atomic Nucleus

At about the turn of this century, French scientists discovered radioactivity, and a few years later, the British physicist Ernest Rutherford investigated how the tiny, dense, positively charged alpha ($\alpha$) particles that are emitted from radioactive materials interact with solid objects. Using zinc sulfide–coated screens, which showed a light flash when struck by an $\alpha$ particle, Rutherford measured the deflection (scattering) of $\alpha$ particles that were aimed at thin gold foil (Figure 2.9).

With Thomson's plum-pudding model in mind, Rutherford expected only minor deflections of the $\alpha$ particles, if any, because they should act as tiny, massive, positively charged "bullets" and go right through the atoms of gold in the foil. The embedded electrons should not deflect the $\alpha$ particles any more than a Ping-Pong ball could deflect a fast-moving baseball. Initial results confirmed this, but in his memoirs Rutherford recalled: "Then I remember two or three days later Geiger [one of his co-workers] coming to me in great excitement and saying, 'We have been able to get some of the $\alpha$ particles coming backwards. . . .' It was quite the most incredible event that has ever happened to me in my life. It was almost as incredible as if you fired a 15-inch shell at a piece of tissue paper and it came back and hit you."

The data showed that very few $\alpha$ particles were deflected at all, and that only 1 in 20,000 were deflected by more than 90° ("coming backwards"). It seemed that these few $\alpha$ particles were being repelled as they approached

**FIGURE 2.9**

Rutherford's $\alpha$-scattering experiment and discovery of the atomic nucleus. **A,** If atoms consist of electrons embedded in diffuse, positively charged matter, the speeding $\alpha$ particles should pass through the gold foil with, at most, minor deflections. **B,** In the experiment, $\alpha$ particles aimed at gold foil emit a flash of light when they pass through the gold atoms and hit a phosphor-coated screen. **C,** The actual results show occasional minor deflections and very infrequent major deflections. This could happen only if very high mass and positive charge are concentrated in a small region within the atom, the nucleus.

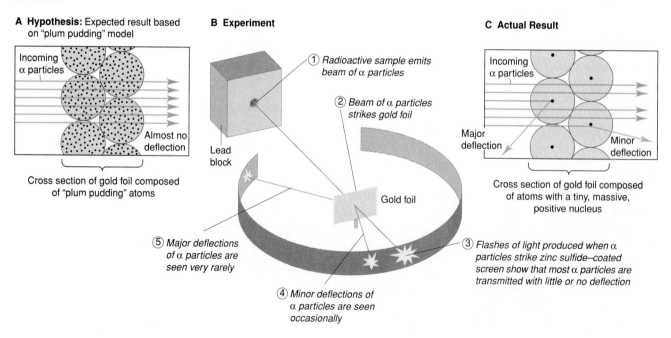

**A  Hypothesis:** Expected result based on "plum pudding" model

Incoming $\alpha$ particles

Almost no deflection

Cross section of gold foil composed of "plum pudding" atoms

**B  Experiment**

① Radioactive sample emits beam of $\alpha$ particles

② Beam of $\alpha$ particles strikes gold foil

Lead block

Gold foil

③ Flashes of light produced when $\alpha$ particles strike zinc sulfide–coated screen show that most $\alpha$ particles are transmitted with little or no deflection

④ Minor deflections of $\alpha$ particles are seen occasionally

⑤ Major deflections of $\alpha$ particles are seen very rarely

**C  Actual Result**

Incoming $\alpha$ particles

Major deflection

Minor deflection

Cross section of gold foil composed of atoms with a tiny, massive, positive nucleus

something massive and positive in the gold atoms. From the mass, charge, and velocity of the α particle, the frequency of large-angle deflections, and the properties of electrons, Rutherford calculated that an atom is mostly space occupied by electrons, but within that space lies a tiny region, which he called the **nucleus,** that contains *all the positive charge and essentially all the mass of the atom.* He proposed that positive particles lay within the nucleus and called them *protons.* Rutherford's model explained the charged nature of matter, but it could not account for all the atom's mass. Twenty years later, this issue was resolved when James Chadwick discovered the *neutron,* an uncharged massive particle that also resides in the nucleus. ◆

### Section Summary

Several major discoveries at the turn of the century led to our current model of atomic structure. Cathode rays were shown to consist of negative particles (electrons) that exist in all matter. J. J. Thomson measured their mass/charge ratio and concluded that they are much smaller and lighter than atoms. Robert Millikan determined the charge of the electron, from which he could calculate its mass. Ernest Rutherford proposed that atoms consist of a tiny, massive, positive nucleus surrounded by electrons.

## 2.4    The Atomic Theory Today

For the past 185 years, we have known that all the matter around us consists of atoms, and we have learned astonishing things about them. Dalton's hard, impenetrable spheres have given way to atoms with "fuzzy," indistinct boundaries and an elaborate internal architecture of subatomic particles. In this section, we examine our current model and begin to see how the properties of these particles affect the properties of atoms. Then we can reassess the atomic theory in the light of our present knowledge.

### Structure of the Atom

An *atom* is an electrically neutral, spherical entity composed of a positively charged central nucleus surrounded by one or more negatively charged electrons (Figure 2.10). The electrons move rapidly through the available atomic volume, held there by the attraction of the nucleus. The nucleus is incredibly dense: it contributes 99.97% of the atom's mass but occupies only about 1 ten-trillionth of its volume. A nucleus the size of a period on this page would weigh about 100 tons, as much as 50 cars! An atom's diameter ($\sim 10^{-10}$ m) is about 10,000 times the diameter of its nucleus ($\sim 10^{-14}$ m).

An atomic nucleus consists of protons and neutrons, except for the simplest hydrogen nucleus, which is a single proton. The **proton (p⁺)** has a positive charge, and the **neutron (n⁰)** has no charge; thus, the positive charge of the nucleus results from the combined charges of its protons. The *amount* of charge possessed by a proton is equal to that of an **electron (e⁻),** but the *sign* of the charge is opposite. *The number of protons in the nucleus of the atom equals the number of electrons surrounding the nucleus.* Some properties of these three subatomic particles are listed in Table 2.2.

◆ **The "Big Three" Subatomic Particles.** Of all the subatomic particles discovered (at latest count, well over 40), the electron, proton, and neutron are the most important in chemistry because they are so long-lived. While bound in the nucleus, the neutron and proton are stable for at least $10^{30}$ years. The electron is considered to be eternal.

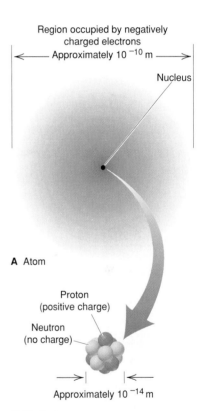

Region occupied by negatively charged electrons
←——— Approximately $10^{-10}$ m ———→

Nucleus

**A**  Atom

Proton (positive charge)

Neutron (no charge)

→| |←
Approximately $10^{-14}$ m

**B**  Nucleus

**FIGURE 2.10**

**General features of the atom. A,** A "cloud" of rapidly moving, negatively charged electrons occupies virtually all the atomic volume and surrounds the tiny, central nucleus. **B,** The nucleus contains virtually all the mass of the atom and consists of positively charged protons and uncharged neutrons. The relative sizes cannot be drawn to scale: if the nucleus were actually the size in the figure (about 1 cm across), the atom would be about 100 m across—slightly longer than a football field!

◆ **Naming an Element.** Element names have a variety of origins. Carbon came from the word for coal (Latin *carbo*, "ember"). Mercury (Hg) is named for the planet, but its symbol reveals its earlier name **h**ydra**g**yrum (Latin "liquid silver"). Some names are based on a chemical action, such as hydrogen (Greek *hydros*, "water," and *genes*, "producer"), or an obvious property, such as chlorine (Greek *chloros*, "yellow-green"). Sometimes, countries are used in the name, as in germanium for Germany, polonium for Poland, and americium for America. Ytterbium, yttrium, erbium, and terbium are all named after Ytterby, the Swedish town where these elements were discovered. Homage to a great scientist is a source of names for recently discovered elements, such as einsteinium for Albert Einstein and curium for Marie Curie.

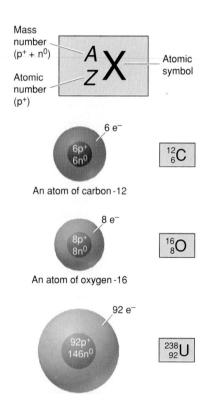

**FIGURE 2.11**
**Depicting the atom.** Atoms of carbon-12, oxygen-16, and uranium-238 are shown (nuclei not drawn to scale) with their symbolic representations. The sum of the number of protons (Z) and the number of neutrons (N) equals the mass number (A). Because an atom is neutral, the number of protons in the nucleus equals the number of electrons around the nucleus.

TABLE 2.2 **Properties of the Subatomic Particles**

| NAME (SYMBOL) | CHARGE | | MASS | | LOCATION IN ATOM |
|---|---|---|---|---|---|
| | RELATIVE | ABSOLUTE | RELATIVE | ABSOLUTE | |
| Proton ($p^+$) | 1+ | $1.602 \times 10^{-19}$ C* | 1 | $1.673 \times 10^{-24}$ g | Nucleus |
| Neutron ($n^0$) | 0 | 0 | 1 | $1.675 \times 10^{-24}$ g | Nucleus |
| Electron ($e^-$) | 1− | $-1.602 \times 10^{-19}$ C | $\frac{1}{1837}$ | $9.109 \times 10^{-28}$ g | Outside nucleus |

*The coulomb (C) is the SI unit of charge.

### Atomic Number, Mass Number, and Atomic Symbol

The **atomic number** *(Z)* of an element equals the number of protons in the nucleus of each of its atoms. *All atoms of a particular element have the same atomic number, and each element has a different atomic number from that of any other element.* All carbon atoms ($Z = 6$) have $6p^+$, all oxygen atoms ($Z = 8$) have $8p^+$, and all uranium atoms ($Z = 92$) have $92p^+$. There are currently 109 known elements, of which 90 occur in nature; the remaining 19 have been synthesized by nuclear processes.

The total number of protons and neutrons in an atom is its **mass number** *(A);* each proton and each neutron has one unit of mass. Thus, a carbon atom with 6 protons and 6 neutrons in its nucleus has a mass number of 12; a uranium atom with 92 protons and 146 neutrons in its nucleus has a mass number of 238. The mass of an electron is not included in the mass number, since it is only $\frac{1}{1837}$ of a mass unit.

Information about the nuclear mass and charge is often included with the **atomic symbol** (or element symbol). Every element has a symbol based on its English or its Latin or Greek name, such as C for carbon, O for oxygen, S for sulfur, and Na for sodium (Latin *natrium*). ◆ The atomic number *(Z)* is written as a left subscript and the mass number *(A)* as a left superscript, so element X would be symbolized $^A_Z X$. Since the mass number is the sum of protons and neutrons, the number of neutrons *(N)* equals the mass number minus the atomic number:

Number of neutrons = mass number − atomic number, or $N = A - Z$    **(2.2)**

Thus, a chlorine atom that is symbolized $^{35}_{17}Cl$ has $A = 35$, $Z = 17$, and $N = 35 - 17 = 18$. Since a given element has a given atomic number, we know the atomic number from the symbol. For example, every carbon atom has six protons, so we can write $^{12}C$ (spoken "carbon twelve") with $Z = 6$ understood, for carbon with mass number 12. Another way to write this atom is carbon-12. Figure 2.11 depicts this information for three atoms.

### Isotopes and Atomic Masses of the Elements

Not all atoms of an element are identical in mass. All carbon atoms have six protons in the nucleus ($Z = 6$), but only 98.89% of naturally occurring carbon atoms have six neutrons in the nucleus ($A = 12$). A small percentage (1.11%) have seven neutrons in the nucleus ($A = 13$), and even fewer (less than 0.01%) have eight ($A = 14$). **Isotopes** of an element are atoms that have *different numbers of neutrons* and therefore different mass numbers.

Carbon has three naturally occurring isotopes, $^{12}C$, $^{13}C$, and $^{14}C$; four other isotopes of carbon, $^{10}C$, $^{11}C$, $^{15}C$, and $^{16}C$, have been observed in the laboratory. All isotopes of carbon have six protons and six electrons. As we will note many times, the chemical properties of an element are primarily determined by the number of electrons, so *isotopes of an element have almost the same chemical behavior,* even though they have different masses.

SAMPLE PROBLEM 2.2 ⎯⎯⎯⎯⎯⎯⎯⎯⎯⎯⎯⎯⎯⎯⎯⎯⎯⎯⎯⎯⎯⎯⎯⎯⎯⎯⎯⎯⎯⎯⎯⎯⎯⎯

### Determining the Number of Subatomic Particles in the Isotopes of an Element

**Problem:** Silicon (Si) is essential to the computer industry as a major component of semiconductor chips. It has three naturally occurring isotopes: $^{28}Si$, $^{29}Si$, and $^{30}Si$. Determine the number of protons, neutrons, and electrons in each isotope.
**Plan:** The mass numbers of the three isotopes are given, so we know the sum of protons and neutrons. In the list of elements on the text's inside front cover, we find the atomic number (number of protons), which equals the number of electrons. We obtain the number of neutrons from Equation 2.2.
**Solution:** In the list of elements, the atomic number of silicon is 14. Therefore,

> **$^{28}Si$ has $14p^+$, $14e^-$, and $14n^0$ ($28 - 14$).**
> **$^{29}Si$ has $14p^+$, $14e^-$, and $15n^0$ ($29 - 14$).**
> **$^{30}Si$ has $14p^+$, $14e^-$, and $16n^0$ ($30 - 14$).**

FOLLOW-UP PROBLEM 2.2
How many protons, neutrons, and electrons are in each of the following: **(a)** $^{11}_{5}Q$; **(b)** $^{41}_{20}X$; **(c)** $^{131}_{53}Y$? What element symbols do Q, X, and Y stand for?

The mass of an atom is measured most easily *relative* to the mass of a chosen atomic standard. The modern atomic mass standard is the carbon-12 atom, whose mass is defined as *exactly* 12 atomic mass units. Thus, the **atomic mass unit (amu)** is $^1/_{12}$ the mass of a carbon-12 atom. Based on this standard, the $^1H$ atom has a mass of 1.008 amu; in other words, a $^{12}C$ atom has almost 12 times the mass of an $^1H$ atom. Although we will continue to use the term atomic mass unit in the text, the name has recently been changed to the **dalton (D);** thus, one $^{12}C$ atom has a mass of 12 daltons (12 D).

The isotopic composition of an element is determined by **mass spectrometry,** a method for measuring the relative masses of particles in a sample very precisely (see the next Tools of the Chemistry Laboratory). For example, using a mass spectrometer, we measure the mass ratio of $^{28}Si$ to $^{12}C$ as:

$$\frac{\text{Mass of }^{28}Si\text{ atom}}{\text{Mass of }^{12}C\text{ standard}} = 2.331411$$

From this mass ratio, we find the **isotopic mass** of the $^{28}Si$ atom, the mass of the isotope relative to the mass of the standard carbon-12 isotope:

$$\text{Isotopic mass of }^{28}S = \text{measured mass ratio} \times \text{mass of }^{12}C$$
$$= 2.331411 \times 12 \text{ amu} = 27.97693 \text{ amu}$$

The mass spectrometer can also measure the relative abundance (fraction) of each isotope in a sample of the element. This measurement provides a method for obtaining the **atomic mass** (also called *atomic weight*) of an element, the *average* of the masses of its naturally occurring isotopes weighted according to their abundances.

**TOOLS OF THE CHEMISTRY LABORATORY**          **Mass Spectrometry**

Mass spectrometry, the most powerful technique for measuring the mass and abundance of charged particles, is an outgrowth of electric and magnetic deflection studies on particles formed in cathode ray experiments. When a high-energy electron collides with an atom of neon-20, for example, one of the atom's electrons is knocked away and the resulting particle has one positive charge, $Ne^+$ (Figure 2.A). Therefore, its mass/charge ratio ($m/e$) equals the mass divided by $1+$. These $m/e$ values are measured to identify the masses of different isotopes of an element.

Figure 2.B is a diagram of one type of mass spectrometer and the data it provides. The sample is introduced and vaporized (if liquid or solid) and then bombarded by high-energy electrons, forming positively charged particles. These are attracted toward a series of negatively charged plates with slits in them, and some particles pass through into an evacuated tube exposed to a magnetic field. As they zoom through this region, the particles' paths bend and fan out according to their $m/e$: the path of the lightest particle is bent most and that of the heaviest particle least. At the end of the magnetic region, the particles strike a detecting device, which records their relative position and abundance. For extremely precise work, such as determining isotopic masses and abundances, the instrument is regularly calibrated with a standard substance of known amount and mass.

Mass spectrometry is also used in structural chemistry and separations science to measure the mass of virtually any atom, molecule, or fragment of a molecule. Among its many applications, mass spectrometry is employed by biochemists determining structures of proteins, by materials scientists examining the surfaces of catalysts, by organic chemists designing new drugs, and by industrial chemists investigating the components of petroleum.

**FIGURE 2.A**
Formation of a positively charged neon particle in a mass spectrometer.

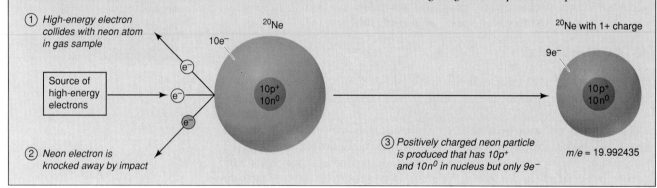

SAMPLE PROBLEM 2.3 _____

**Calculating the Atomic Mass of an Element**

**Problem:** Silver (Ag; $Z = 47$) has 16 known isotopes, but only two occur naturally, $^{107}Ag$ and $^{109}Ag$. Given the following mass spectrometric data, calculate the atomic mass of silver:

| ISOTOPE | MASS (amu) | ABUNDANCE (%) |
|---------|-----------|---------------|
| $^{107}Ag$ | 106.90509 | 51.84 |
| $^{109}Ag$ | 108.90476 | 48.16 |

**Plan:** From the mass and abundance of the two silver isotopes, we have to find the atomic mass of silver (weighted average of the isotopic masses). We multiply each isotopic mass by its fractional abundance to find the portion of the atomic mass contributed by each isotope. The sum of the isotopic portions is the atomic mass.

**Solution:** Finding the atomic mass portion of each isotope:

For $^{107}Ag$: Portion of atomic mass = isotopic mass × fractional abundance

$$= 106.90509 \text{ amu} \times 0.5184 = 55.42 \text{ amu}$$

For $^{109}Ag$: Portion of atomic mass = 108.90476 amu × 0.4816 = 52.45 amu

Mass (g) of each isotope

multiply by fractional abundance of each isotope

Portion of atomic mass from each isotope

add isotopic portions

Atomic mass

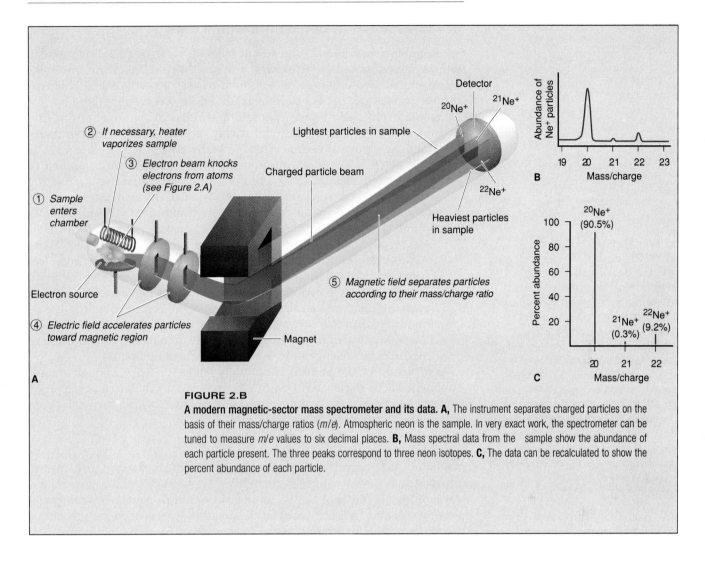

**FIGURE 2.B**

**A modern magnetic-sector mass spectrometer and its data. A,** The instrument separates charged particles on the basis of their mass/charge ratios (*m/e*). Atmospheric neon is the sample. In very exact work, the spectrometer can be tuned to measure *m/e* values to six decimal places. **B,** Mass spectral data from the sample show the abundance of each particle present. The three peaks correspond to three neon isotopes. **C,** The data can be recalculated to show the percent abundance of each particle.

Finding the atomic mass of silver:

$$\text{Atomic mass of Ag} = 55.42 \text{ amu} + 52.45 \text{ amu} = \textbf{107.87 amu}$$

**Check:** The individual portions seem right: $\sim$100 amu $\times$ 0.50 = 50 amu. The portions are almost the same because the two isotopic abundances are almost the same. We rounded each portion to four significant figures because that is the number in the abundance values. This is the correct atomic mass (to two decimal places), as shown in the list of elements.

**Comment:** It's important for you to realize that no individual silver atom in a sample of the metal has a mass of 107.87 amu; for most laboratory purposes, however, we consider the sample to consist of atoms with this average mass.

**FOLLOW-UP PROBLEM 2.3**

Boron (B; $Z = 5$) has two naturally occurring isotopes. Calculate the percent abundances of $^{10}B$ and $^{11}B$ from the following: atomic mass of B = 10.81 amu; isotopic mass of $^{10}B$ = 10.0129 amu; and isotopic mass of $^{11}B$ = 11.0093 amu. (*Hint:* The sum of the fractional abundances will be 1. If $x$ = abundance of $^{10}B$, then $1 - x$ = abundance of $^{11}B$.)

## A Modern Reassessment of the Atomic Theory

We began our discussion of the atomic basis of matter with Dalton's model, which proved inaccurate in several respects. What happens to an established model whose postulates are found by later experiment to be incorrect? No model can predict every possible future observation, but a powerful model evolves and retains its usefulness. Indeed, continual testing and revising of models are the essence of science. Let's reconstruct the postulates of the atomic theory in light of what we know now:

1. All matter is composed of atoms. Although atoms are composed of smaller particles (electrons, protons, and neutrons), the atom is the smallest body that *retains the unique identity* of the element.
2. Atoms of one element can be converted into atoms of another element *only* in a nuclear process; they are *never* transformed by a chemical reaction. ◆
3. All atoms of an element have the same number of protons and electrons, which determines the chemical behavior of the element. Isotopes of an element differ in the number of neutrons, and thus in mass number, but a sample of the element is treated as though its atoms have an *average* mass.
4. Compounds are formed by the chemical combination of two or more elements in specific ratios, as originally stated by Dalton.

Our picture of the atom is continually being revised. Although we are confident about the distribution of electrons within the atom (Chapters 7 and 8), the interactions among protons and neutrons within the nucleus are still on the frontier of discovery (Chapter 21).

### Section Summary

An atom has a central nucleus, containing positively charged protons and uncharged neutrons, which is surrounded by negatively charged electrons. An atom is neutral because the number of electrons equals the number of protons. An atom is represented by the notation $^{A}_{Z}X$, in which $Z$ is the atomic number (number of protons), $A$ the mass number (sum of protons and neutrons), and X the atomic symbol. An element occurs as a mixture of isotopes, atoms with the same number of protons but different numbers of neutrons. Each isotope has a mass relative to the $^{12}C$ mass standard. The atomic mass of an element is the average of its isotopic masses weighted according to their natural abundances and is determined by modern instruments, such as the mass spectrometer.

## 2.5   Elements: A First Look at the Periodic Table

At the end of the 18th century, Lavoisier compiled a list of the 23 elements known at that time; by 1870, 65 were known; by 1925, 88; today, there are 109 and still counting! These elements combine to form millions of compounds, so we clearly need some way to organize what we know about their behavior. By the mid-19th century, enormous amounts of information concerning reactions, properties, and atomic masses of the elements had been accumulated. Several researchers noted recurring, or *periodic*, patterns of behavior and proposed various schemes to organize the elements according to some fundamental property. In 1871, the Russian chemist Dmitri

◆ **The Heresy of Radioactive "Transmutation."** In 1902, Rutherford performed a series of experiments with radioactive elements that shocked the scientific world. When a radioactive atom of thorium ($Z = 90$) emits an $\alpha$ particle ($Z = 2$), it becomes an atom of radium ($Z = 88$), which then emits another $\alpha$ particle and becomes an atom of radon ($Z = 86$). Rutherford proposed that when an atom emits an $\alpha$ particle, it turns into a different atom—one element changes into another! Many viewed this conclusion as a return to alchemy, and, as with Thomson's discovery that atoms contain smaller particles, Rutherford's findings fell on disbelieving ears.

Mendeleev published the most successful of these organizing schemes, a table that listed the elements by increasing atomic mass and arranged so that elements with similar chemical properties would lie in the same column. The modern **periodic table of the elements,** based on Mendeleev's earlier version, is one of the great classifying schemes in science and has become an indispensable tool to chemists. Throughout your study of chemistry, the periodic table will guide you through an otherwise dizzying amount of chemical and physical behavior. The general outline of the periodic table is presented here in preparation for your close working relationship with it later.

A modern version of the periodic table appears in Figure 2.12 and inside the front cover of the text. The table consists of element boxes, each containing the atomic number, atomic symbol, and atomic mass of the element. The boxes are arranged in order of *increasing atomic number* into a grid of vertical columns **(groups)** and horizontal rows **(periods).** Each period has a number from 1 to 7. Each group has a number from 1 to 8 *and* either the letter A or B. A new system, with group numbers from 1 to 18 but no letters, appears in parentheses under the number-letter designation. (Throughout the text, we use the number-letter system, with the new numbering system in parentheses.)

The eight groups designated A (two on the left and six on the right) contain the *main-group,* or *representative, elements.* The 10 groups designated B, located between Groups 2A(2) and 3A(13), contain the *transition elements.* Two series of *inner transition elements,* the lanthanides and the actinides, fit *between* the elements in Group 3B(3) and Group 4B(4) and are usually placed below the main body of the table.

At this point in the text, the clearest distinction we can make concerning the elements is their classification as metals, nonmetals, or metalloids. Use the thick "staircase" line that runs from the top of Group 3A(13) to the bottom of Group 6A(16) as a landmark. Many main-group elements and all the transition and inner transition elements are **metals** (three shades of blue); they appear in the left and lower portion of the table. These elements are generally shiny solids at room temperature (mercury is the only liquid), conduct heat and electricity well, and can be tooled into sheets and wires. The **nonmetals** (yellow) appear in the upper right portion of the table. They are generally gases or dull, brittle solids at room temperature (bromine is the only liquid) and conduct heat and electricity poorly. Along the staircase line lie the **metalloids** (green; also called **semimetals**), elements that have properties between those of metals and nonmetals; several of them play major roles in modern electronic materials. Figure 2.13 shows some examples of these three classes of elements.

You should learn some of the traditional group names. Group 1A(1), except for hydrogen, consists of the *alkali metals,* and Group 2A(2) the *alkaline earth metals,* all highly reactive elements. The *halogens,* Group 7A(17), are highly reactive nonmetals, whereas the *noble gases,* Group 8A(18), are relatively unreactive nonmetals. Other main groups [3A(13) to 6A(16)] are often named by the first element in the group; for example, Group 6A is the oxygen family.

A point that we return to many times in the text is that *elements in a group have* **similar** *chemical properties and elements in a period have* **different** *chemical properties.* We begin applying the organizing power of the periodic table in the next section, as we discuss how elements combine to form compounds.

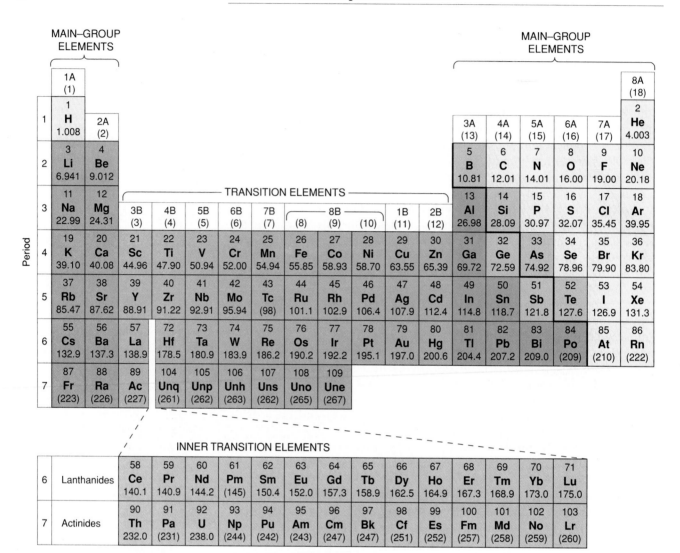

**FIGURE 2.12**

**The modern periodic table.** The table consists of element boxes arranged by *increasing* atomic number into groups (vertical columns) and periods (horizontal rows). Each box contains the atomic number, atomic symbol, and atomic mass. (A mass in parentheses is the mass number of the most stable isotope of that element.) The periods are numbered 1 to 7; the groups have a number-letter designation and a new group number in parentheses. The A groups are the main-group elements; the B groups are the transition elements. Two series of inner transition elements are placed below the main body of the table but fit between the elements indicated. The great majority of elements are metals, which lie below and to the left of the thick "staircase" line [top of 3A(13) to bottom of 6A(16)] and include main-group metals (purple-blue), transition elements (blue), and inner transition elements (gray-blue). Nonmetals (yellow) lie to the right of the line. Metalloids (green) lie along the line. Hydrogen is placed at the top of Group 1A(1) because it has similarities with other elements in that group, but it can also be placed at the top of Group 7A(17). For elements with atomic numbers 104 to 109, the interim symbols Unq, and so forth, are being replaced. At the time of this writing, the International Union of Pure and Applied Chemistry is considering, but has not yet approved, the following atomic symbols: 104, Db; 105, Jl; 106, Rf; 107, Bh; 108, Hn; 109, Mt.

Metals                    Metalloids                    Nonmetals

**FIGURE 2.13**
**Metals, metalloids, and nonmetals.** Metals (blue), from left to right: chromium, copper, cadmium, lead, and bismuth. Metalloids (green), from left to right: boron, silicon, arsenic, antimony, and tellurium. Nonmetals (yellow), from left to right: carbon (graphite), sulfur, chlorine, bromine, and iodine.

**Section Summary**

In the periodic table, the elements are arranged by atomic number into horizontal periods and vertical groups. Elements within a group have similar properties. Nonmetals appear in the right upper portion of the table, metalloids lie along a staircase line, and metals fill the rest of the table.

## 2.6 Compounds: Introduction to Bonding

Very few elements exist naturally as free elements, uncombined with others. The noble gases—helium (He), neon (Ne), argon (Ar), krypton (Kr), xenon (Xe), and radon (Rn)—occur as separate atoms. Carbon (C), oxygen (O), nitrogen (N), and sulfur (S) occur commonly as molecules, their atoms bonded to others of the same element, for example, $N_2$ and $S_8$. Some of the metals, such as copper (Cu), silver (Ag), gold (Au), and platinum (Pt), may also occur uncombined with other elements.

In contrast, *the overwhelming majority of elements occur in compounds,* that is, in chemical combination with other elements. The *electrons* of the atoms of interacting elements are primarily involved in compound formation. Elements combine in two general ways:

1. *Transferring electrons* from the atoms of one element to those of another to form **ionic compounds**
2. *Sharing electrons* between atoms of different elements to form **covalent compounds**

These processes generate **chemical bonds,** the forces that hold the atoms of elements together in the compound.

## The Formation of Ionic Compounds

Ionic compounds are composed of **ions,** charged particles that form when an atom (or small group of atoms) gains or loses one or more electrons. *Ionic compounds typically form when a metal reacts with a nonmetal.* Each metal atom loses a certain number of its electrons and becomes a **cation,** a positively charged ion; at the same time, the nonmetal atoms gain the electrons lost by the metal atoms and become **anions,** negatively charged ions. The cations and anions attract each other and form an ionic compound. A cation or anion derived from a single atom is called a **monatomic ion.** The mutual attraction of monatomic cations and anions results in a **binary ionic compound,** one composed of just two elements.

The formation of sodium chloride, common table salt, is a typical example (Figure 2.14). A sodium atom, which is neutral because it has the same number of protons as electrons, *loses* one electron and forms a sodium cation, $Na^+$. (The charge on the ion is written as a *right superscript.*) A chlorine atom *gains* the electron and becomes a chloride anion, $Cl^-$. (The name change from the nonmetal atom to the ion is discussed in the next section.)

Even the smallest visible grain of table salt is the result of an *enormous* number of sodium atoms that have transferred one electron each to an equal number of chlorine atoms. The oppositely charged ions ($Na^+$ and $Cl^-$) attract each other, whereas the similarly charged ions ($Na^+$ and $Na^+$, or $Cl^-$ and $Cl^-$) repel each other. The resulting aggregation is an ionic compound, in this case, a crystal of sodium chloride—an array of alternating $Na^+$ and $Cl^-$ ions that extends regularly in all three directions. All ionic compounds do *not* contain equal numbers of positive and negative *ions,* but they do contain equal numbers of positive and negative *charges.* Therefore, *ionic compounds are neutral; that is, they have zero net charge.*

How can we predict the number of electrons an atom will lose or gain when it forms an ion? In the formation of sodium chloride, for example, why does each sodium atom give up only 1 of its 11 electrons? Why doesn't each chlorine atom gain 2 electrons, instead of just 1? The periodic

**FIGURE 2.14**

**Electron transfer in the formation of an ionic compound. A,** The neutral sodium atom loses one electron to become a sodium cation ($Na^+$), and the chlorine atom gains one electron to become a chloride anion ($Cl^-$). (Note that when atoms lose electrons, they become smaller, and when they gain electrons, they become larger.) **B,** $Na^+$ and $Cl^-$ ions attract each other and alternate in an array extending in three directions. **C,** This cubic arrangement is reflected in the structure of crystalline NaCl, which occurs naturally as the mineral halite.

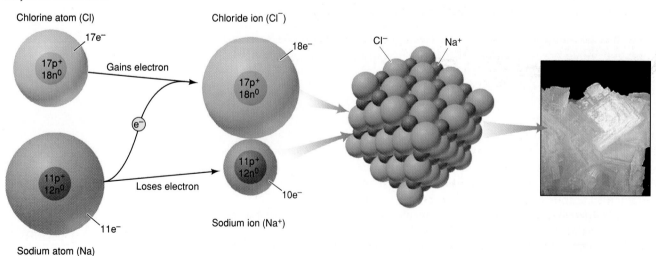

Chlorine atom (Cl)   17$e^-$   17$p^+$ 18$n^0$   Gains electron   Chloride ion ($Cl^-$)   18$e^-$   17$p^+$ 18$n^0$

Sodium atom (Na)   11$e^-$   11$p^+$ 12$n^0$   Loses electron   11$p^+$ 12$n^0$   10$e^-$   Sodium ion ($Na^+$)

$Cl^-$   $Na^+$

**A** Electron transfer          **B** Arrangement of $Na^+$ and $Cl^-$ in a crystal          **C** Sodium chloride crystal

table can provide some insight. In ionic compound formation, we generally find that metals lose electrons and nonmetals gain electrons to *form ions with the same number of electrons as in the nearest noble gas* [Group 8A(18)]. These elements have a stability (low reactivity) that is associated with their particular number (and arrangement) of electrons. A sodium atom ($11e^-$) can attain the stability of neon ($10e^-$), the nearest noble gas, by losing one electron; similarly, by gaining one electron, a chlorine atom ($17e^-$) attains the stability of argon ($18e^-$), its nearest noble gas. Thus, when an element located near a noble gas forms a monatomic ion, it gains or loses enough electrons to attain the number in that noble gas. Specifically, the elements in Group 1A(1) lose one electron, those in Group 2A(2) lose two, and aluminum in Group 3A(13) loses three; the elements in Group 7A(17) gain one electron, oxygen and sulfur in Group 6A(16) gain two, and nitrogen in Group 5A(15) gains three.

With the periodic table printed on a two-dimensional surface, it is easy to receive the false impression that the elements in Group 7A(17) are "closer" to the noble gases than the elements in Group 1A(1) (see Figure 2.12). However, both groups are only one electron away from having the same number of electrons as the nearest noble gas. A periodic table that is cut and rejoined, with the noble gases in the center, makes this point clearly (Figure 2.15). For example, fluorine (F; $Z = 9$) has one electron less and sodium (Na; $Z = 11$) has one electron more than the noble gas neon (Ne; $Z = 10$); thus they form the $F^-$ and $Na^+$ ions. Similarly, oxygen (O; $Z = 8$) gains two electrons and magnesium (Mg; $Z = 12$) loses two to form the $O^{2-}$ and $Mg^{2+}$ ions and attain the same number of electrons as neon.

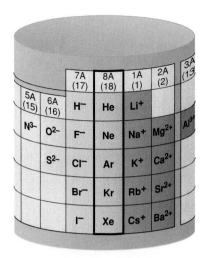

**FIGURE 2.15**
**The relationship between ions formed and the nearest noble gas.** This periodic table was redrawn (only some elements are shown) to show (1) the positions of nonmetals (yellow) and metals (blue) relative to the noble gases and (2) the ions these elements form. The ionic charges arise from the number of electrons lost or gained to attain the same number of electrons as the nearest noble gas. Species in the same row have the same number of electrons. For example, $H^-$, He, and $Li^+$ all have two electrons. [Note that H is shown here in Group 7A(17).]

SAMPLE PROBLEM 2.4 ─────────────────────────

### Predicting the Ion an Element Forms

**Problem:** What monatomic ions do the following elements form?
**(a)** Iodine ($Z = 53$)    **(b)** Calcium ($Z = 20$)    **(c)** Aluminum ($Z = 13$)
**Plan:** We consult the periodic table to see the number of electrons each element must gain or lose to attain the number of electrons in the nearest noble gas.
**Solution: (a) $I^-$** Iodine [$_{53}I$; Group 7A(17)] is a nonmetal, one of the halogens, that gains one electron to have the same number as $_{54}Xe$.
**(b) $Ca^{2+}$** Calcium [$_{20}Ca$; Group 2A(2)] is one of the alkaline earth metals and loses two electrons to attain the same number as $_{18}Ar$.
**(c) $Al^{3+}$** Aluminum [$_{13}Al$; Group 3A(13)] is a metal in the boron family and loses three electrons to attain the same number as $_{10}Ne$.

**FOLLOW-UP PROBLEM 2.4**
What monatomic ions do each of the following elements form?
**(a)** $_{36}S$       **(b)** $_{37}Rb$       **(c)** $_{56}Ba$

### The Formation of Covalent Compounds

*Covalent compounds form when elements share electrons, which usually occurs between nonmetals.* Even though relatively few nonmetals exist, they interact in many combinations to form an enormous number of covalent compounds.

**A** No interaction

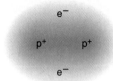

**B** Attraction begins

**C** Covalent bond

**D** Combination of forces

**FIGURE 2.16**

**Formation of a covalent bond between two H atoms. A,** The distance is too great for the atoms to affect each other. **B,** As the distance decreases, the nucleus of each atom begins to attract the electron of the other. **C,** The covalent bond forms when the two nuclei mutually attract the pair of electrons at some optimum distance. **D,** The H$_2$ molecule is more stable than the separate atoms because the attractive forces *(black arrows)* between each nucleus and the two electrons are greater than the repulsive forces *(red arrows)* between electrons and between nuclei.

Carbonate ion
$CO_3^{2-}$

**FIGURE 2.17**

**A polyatomic ion.** The covalently bonded atoms are shown as overlapping spheres.

The simplest case of electron sharing occurs not in a compound but between two hydrogen atoms (H; $Z = 1$). Imagine two separate H atoms approaching each other (Figure 2.16). As they get closer, the nucleus of each atom attracts the electron of the other atom more and more strongly. At some optimum distance between the nuclei, the two atoms form a **covalent bond,** a pair of electrons mutually attracted by the two nuclei. The result is a hydrogen molecule, in which each electron no longer "belongs" to a particular H atom: the two electrons are *shared* by the two nuclei. Repulsions between the nuclei and between the electrons also occur, but the net attraction is greater than the net repulsion.

A sample of hydrogen gas consists of many diatomic hydrogen molecules—pairs of atoms that are chemically bound and behave as an independent unit—*not* separate hydrogen atoms. Other nonmetals that exist as diatomic molecules include nitrogen, oxygen, and the halogens (fluorine, chlorine, bromine, and iodine).

In a similar way, atoms of different elements share electrons to form the molecules of a covalent compound. A sample of hydrogen chloride, for example, consists of molecules in which one hydrogen atom forms a covalent bond with one chlorine atom; ammonia consists of molecules in which one nitrogen atom forms a covalent bond with each of three hydrogen atoms. As you'll see in Chapter 9, covalent bonding provides another way for atoms to attain the same number of electrons as the nearest noble gas.

**Comparing the chemical entities in covalent and ionic compounds.** An important distinction exists between the chemical entities in a covalent substance and those in an ionic substance. Most covalent substances consist of molecules. The contents of a glass of water, for example, is a collection of individual water molecules lying next to one another. Each water molecule consists of an oxygen atom connected by covalent bonds to two hydrogen atoms.

In contrast, under ordinary conditions, no molecules exist in a sample of an ionic compound. A piece of sodium chloride, for example, is a continuous array of oppositely charged sodium and chloride ions, *not* a collection of individual "sodium chloride molecules." Likewise, a covalent bond is an actual entity—a pair of electrons shared by two atoms—but an ionic bond is not. Although the term *ionic bond* is used, the "bond" is just the force of attraction between positive and negative ions.

### Covalent Bonds in Polyatomic Ions

Covalent bonds occur within certain ionic compounds. Up to now, we've discussed monatomic ions, the type that occur in binary ionic compounds. Many other ionic compounds contain **polyatomic ions,** which consist of two or more atoms bonded covalently and having a net positive or negative charge. A polyatomic ion has its own identity and stays together as a unit during interactions with other ions. For example, the ionic compound calcium carbonate is an array of polyatomic carbonate anions and monatomic calcium cations attracted to each other. The carbonate ion consists of a carbon atom covalently bonded to three oxygen atoms, and two additional electrons give the ion its 2− charge (Figure 2.17).

The gallery on the next page previews our discussion of compound formulas in the next section by showing some of the ways chemists picture molecules.

## Picturing Molecules

One of the most exciting aspects of learning chemistry is training your imagination to view a molecular world. Molecules can be depicted in a variety of useful ways, as shown in five representations of the hydrogen molecule to the left.

H₂
Chemical formula

H:H
Electron-dot notation

H–H
Bond-line notation

Ball-and-stick model

Space-filling model

The *chemical formula* is easy to write but does not show what the molecule looks like.

*Electron-dot* and *bond-line formulas* show the bonding electrons as either a pair of dots or a line but still symbolize the atoms with letters. (*Wedge-bond formulas,* shown for some of the other molecules on this page, enhance the three-dimensional perspective.)

*Ball-and-stick models* show atoms as spheres, with relative sizes and angles based on data, but the distances between them are exaggerated.

*Space-filling models* are the most accurate because they depict scaled-up versions of actual molecules.

Dichlorodifluoromethane ($CF_2Cl_2$) an aerosol propellant, one of the molecules responsible for ozone depletion in the stratosphere

Acetic acid ($C_2H_4O_2$) a component of vinegar

Butane ($C_4H_{10}$) a fuel used in cigarette lighters and camp stoves

Glucose ($C_6H_{12}O_6$) a major sugar in energy metabolism

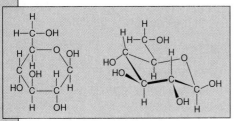

Heme ($C_{34}H_{32}FeN_4O_4$) a part of the blood protein hemoglobin that carries oxygen through the body

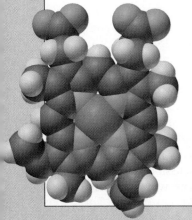

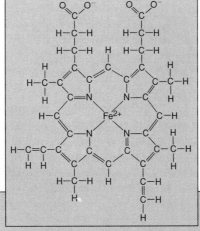

Though all molecules are minute, their relative sizes depend on the number of atoms in the molecule. A hydrogen molecule is small because it consists of only two atoms. Dichlorodifluoromethane, acetic acid, and butane are somewhat larger. The biologically important glucose and heme are larger still. Very large molecules, called *macromolecules,* can consist of thousands of atoms. The DNA molecule in one bacterial cell, if stretched out linearly, would be several centimeters long.

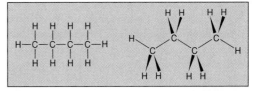

A short section of DNA, the natural macromolecule that contains genetic information

**Section Summary**

Although a few elements occur uncombined in nature, the great majority exist in compounds. Ionic compounds form when a metal transfers electrons to a nonmetal. In many cases, metal atoms lose and nonmetal atoms gain enough electrons to attain the same number of electrons as in atoms of the nearest noble gas. The resulting positive and negative ions attract each other and form a three-dimensional array. Covalent compounds form when nonmetals share electrons. Each covalent bond is an electron pair mutually attracted by two atomic nuclei. Monatomic ions are made from one atom; polyatomic ions consist of two or more covalently bonded atoms that have a net positive or negative charge due to a deficit or excess of electrons. Molecules are three-dimensional objects that range in size from the hydrogen molecule to giant biological and synthetic macromolecules.

## 2.7  Compounds: Formulas, Names, and Masses

In a **chemical formula**, element symbols and numerical subscripts show the type and number of each atom present in a unit of the substance. There are several types of chemical formulas for a compound:

1. The **empirical formula** shows the *relative* number of atoms of each element in the compound. It is the simplest type of formula and is derived from the masses of the component elements. For example, the empirical formula of hydrogen peroxide is HO; there is one H atom for every O atom.

2. The **molecular formula** shows the *actual* number of atoms of each element in a molecule of the compound. The molecular formula of hydrogen peroxide is $H_2O_2$; there are *actually* two H atoms and two O atoms in each molecule. Note that subscripts *follow* the element to which they refer.

3. A **structural formula** shows the actual number of atoms and *the bonds between them;* that is, the arrangement of atoms in the molecule. The structural formula of hydrogen peroxide is H—O—O—H; each H is bonded to an O, and the O's are bonded to each other. The bond-line representations in the Gallery (Section 2.6) are structural formulas.

In this section, we consider how to name and write formulas for ionic and simple covalent compounds and how to calculate a compound's molecular mass.

**Some Advice about Learning Names and Formulas**

The names and formulas of compounds form the vocabulary of chemistry; in this discussion, you'll begin to speak and write the language. In the future, systematic names of compounds may be applied universally. For now, however, many reference books, chemical supply catalogs, and practicing chemists still use common (trivial) names, so you will have to learn them as well. There is only one way to learn them, and that is to use them! Here are some suggestions about how to approach the task:

• Relate the main-group monatomic ions of Table 2.3 (all except $Ag^+$, $Zn^{2+}$, and $Cd^{2+}$) to their position in the periodic table (Figure 2.18). These ions

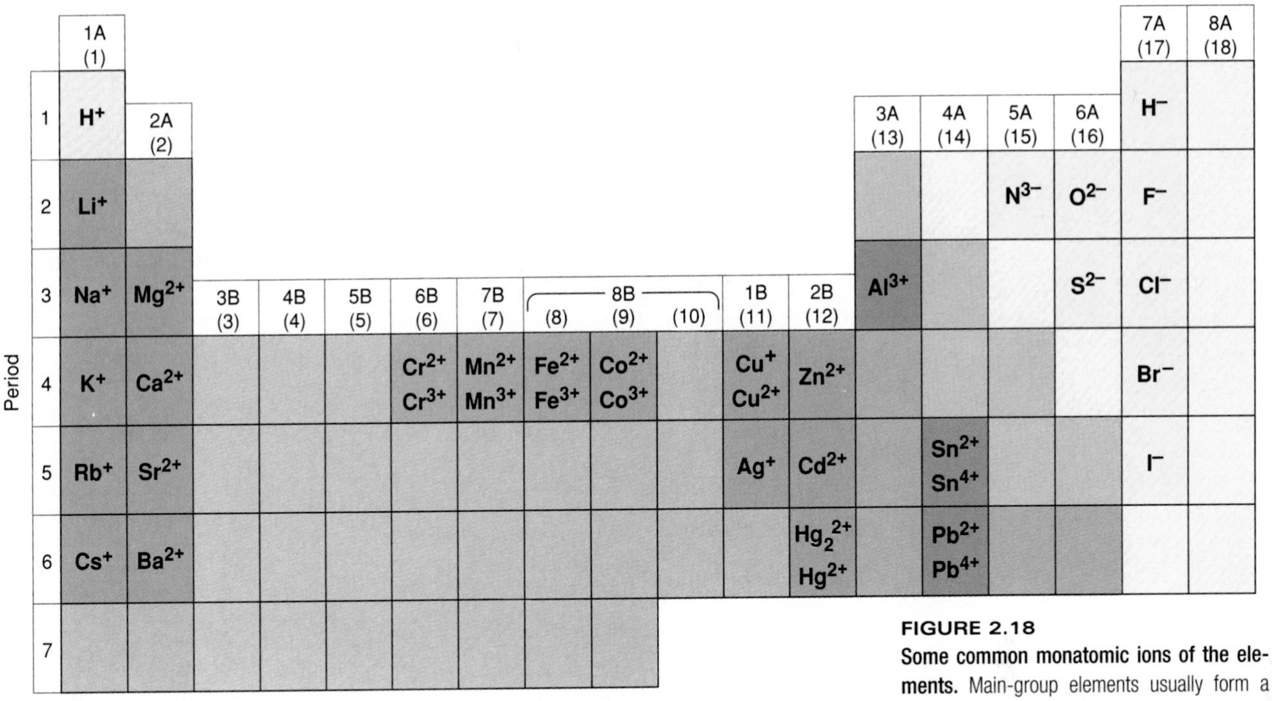

| | 1A (1) | | | | | | | | | | | | | | | | | 7A (17) | 8A (18) |
|---|---|---|---|---|---|---|---|---|---|---|---|---|---|---|---|---|---|---|---|
| 1 | **H⁺** | 2A (2) | | | | | | | | | | 3A (13) | 4A (14) | 5A (15) | 6A (16) | **H⁻** | | | |
| 2 | **Li⁺** | | | | | | | | | | | | | | $N^{3-}$ | $O^{2-}$ | **F⁻** | | |
| 3 | **Na⁺** | **Mg²⁺** | 3B (3) | 4B (4) | 5B (5) | 6B (6) | 7B (7) | (8) | 8B (9) | (10) | 1B (11) | 2B (12) | **Al³⁺** | | | $S^{2-}$ | **Cl⁻** | | |
| 4 | **K⁺** | **Ca²⁺** | | | | Cr²⁺ Cr³⁺ | Mn²⁺ Mn³⁺ | Fe²⁺ Fe³⁺ | Co²⁺ Co³⁺ | | Cu⁺ Cu²⁺ | **Zn²⁺** | | | | | **Br⁻** | | |
| 5 | **Rb⁺** | **Sr²⁺** | | | | | | | | | **Ag⁺** | Cd²⁺ | | Sn²⁺ Sn⁴⁺ | | | **I⁻** | | |
| 6 | **Cs⁺** | **Ba²⁺** | | | | | | | | | Hg₂²⁺ Hg²⁺ | | Pb²⁺ Pb⁴⁺ | | | | | | |
| 7 | | | | | | | | | | | | | | | | | | | |

Period

**FIGURE 2.18**

**Some common monatomic ions of the elements.** Main-group elements usually form a single monatomic ion. Note that members of a group have ions with the same charge. [Hydrogen is shown as both the cation $H^+$ in Group 1A(1) and the anion $H^-$ in Group 7A(17).] Many transition elements form two different monatomic ions. (Although $Hg_2^{2+}$ is a diatomic ion, it is included for comparison with $Hg^{2+}$.)

**TABLE 2.3  Common Monatomic Ions***

| CATIONS | | | ANIONS | | |
|---|---|---|---|---|---|
| CHARGE | FORMULA | NAME | CHARGE | FORMULA | NAME |
| 1+ | $H^+$ | hydrogen | 1− | $H^-$ | hydride |
| | $Li^+$ | **lithium** | | $F^-$ | **fluoride** |
| | $Na^+$ | **sodium** | | $Cl^-$ | **chloride** |
| | $K^+$ | **potassium** | | $Br^-$ | **bromide** |
| | $Cs^+$ | cesium | | $I^-$ | **iodide** |
| | $Ag^+$ | **silver** | | | |
| 2+ | $Mg^{2+}$ | **magnesium** | 2− | $O^{2-}$ | **oxide** |
| | $Ca^{2+}$ | **calcium** | | $S^{2-}$ | **sulfide** |
| | $Sr^{2+}$ | strontium | | | |
| | $Ba^{2+}$ | **barium** | | | |
| | $Zn^{2+}$ | **zinc** | | | |
| | $Cd^{2+}$ | cadmium | | | |
| 3+ | $Al^{3+}$ | **aluminum** | 3− | $N^{3-}$ | nitride |

*Listed by charge; those in **boldface** are most common.

have the same number of electrons as an atom of the nearest noble gas.

- Note that members of a periodic table group have the same ionic charge.
- Consult Table 2.4 and Figure 2.18 for some metals that form two different monatomic ions.
- Divide the tables of names and charges into smaller batches, and learn a batch each day. The most common ions are shown in **boldface** in Tables 2.3, 2.4, and 2.5, so you can focus on learning them first.

## Names and Formulas of Ionic Compounds

All ionic compound names give the positive ion (cation) first and the negative ion (anion) second.

**Compounds formed from monatomic ions.** Let's first consider binary ionic compounds, those composed of ions from two elements.

- *The name of the cation is the same as the name of the metal.* Many metal names end in "-ium."
- *The name of the anion takes the root of the nonmetal name and adds the suffix "-ide."* For example, the anion formed from brom*ine* is named brom*ide* (brom + ide). Therefore, the compound formed from the metal calcium and the nonmetal bromine is "calcium bromide."

SAMPLE PROBLEM 2.5 _____

### Naming Binary Ionic Compounds

**Problem:** Name the ionic compound that is formed from the following pairs of elements:

**(a)** Magnesium and nitrogen  **(b)** Iodine and cadmium
**(c)** Strontium and fluorine  **(d)** Sulfur and cesium

**Plan:** The key to naming a binary ionic compound is to recognize which element is the metal and which is the nonmetal. When in doubt, we can check the periodic table. We place the cation name first, add the suffix "-ide" to the nonmetal root, and place the anion name last.

**Solution: (a)** Magnesium is the metal; "nitr-" is the nonmetal root: **magnesium nitride**

**(b)** Cadmium is the metal; "iod-" is the nonmetal root: **cadmium iodide**

**(c)** Strontium is the metal; "fluor-" is the nonmetal root: **strontium fluoride** (Note the spelling is flu*o*ride, not fl*ou*ride.)

**(d)** Cesium is the metal; "sulf-" is the nonmetal root: **cesium sulfide**

FOLLOW-UP PROBLEM 2.5
Give the name and periodic table group number for the elements whose ions make up the following compounds: **(a)** zinc oxide; **(b)** silver bromide; **(c)** lithium chloride; **(d)** aluminum sulfide.

Ionic compounds are arrays of oppositely charged ions rather than separate molecular units, so we write a formula for the **formula unit,** the relative numbers of cations and anions in the compound. In other words, ionic compounds generally have only empirical formulas.* Since the compound is neutral (zero net charge), the positive charges of the cations must balance the negative charges of the anions. For example, calcium bromide is composed of $Ca^{2+}$ ions and $Br^-$ ions; therefore, two $Br^-$ balance each $Ca^{2+}$. The formula is $CaBr_2$, not $Ca_2Br$. In this and all other formulas,

- The subscript refers to the element *preceding* it.
- The *subscript 1 is understood* from the element symbol alone (that is, we do not write $Ca_1Br_2$).

_____
* Compounds of the mercury(I) ion, such as $Hg_2Cl_2$, and peroxides of the alkali metals, such as $Na_2O_2$, are the only two common exceptions.

SAMPLE PROBLEM 2.6 _____

**Determining the Formulas of Binary Ionic Compounds**

**Problem:** Write the empirical formulas for the compounds named in Sample Problem 2.5.
**Plan:** We write the empirical formula by writing the ions and finding the smallest number of each that gives the neutral compound. The number of each ion appears as a *right subscript.*
**Solution:**
(a) $Mg^{2+}$ and $N^{3-}$; three $Mg^{2+}$ ions (6+) balance two $N^{3-}$ ions (6−): **$Mg_3N_2$**
(b) $Cd^{2+}$ and $I^-$; one $Cd^{2+}$ ion (2+) balances two $I^-$ ions (2−): **$CdI_2$**
(c) $Sr^{2+}$ and $F^-$; one $Sr^{2+}$ ion (2+) balances two $F^-$ ions (2−): **$SrF_2$**
(d) $Cs^+$ and $S^{2-}$; two $Cs^+$ ions (2+) balance one $S^{2-}$ ion (2−): **$Cs_2S$**
**Comment:** Ion charges do *not* appear in the compound formula.

FOLLOW-UP PROBLEM 2.6
Write the formulas of the compounds named in Follow-up Problem 2.5.

Many metals, particularly the transition elements (B groups), can form more than one ion, each with a particular charge (Table 2.4). In naming compounds made from these elements, we indicate the metal's ionic charge by writing it in *Roman numerals within parentheses* immediately after the metal ion's name. For example, iron can form $Fe^{2+}$ and $Fe^{3+}$ ions. The two compounds that iron forms with chlorine are $FeCl_2$, named iron(II) chloride (spoken "iron two chloride"), and $FeCl_3$, named iron(III) chloride.

In common names, the Latin root of the metal is followed by the suffix "-ous" for the ion with the lower charge and "-ic" for the ion with the higher charge. Thus, iron(II) chloride is also called ferr*ous* chloride and iron(III) chloride is ferr*ic* chloride. (You can easily remember this naming relationship because there is an "*o*" in "*o*us" and "l*o*w," and an "*i*" in "*i*c" and "h*i*gh.")

**TABLE 2.4   Some Metals That Form More Than One Monatomic Ion***

| ELEMENT | ION FORMULA | SYSTEMATIC NAME | COMMON (TRIVIAL) NAME |
|---|---|---|---|
| Chromium | $Cr^{2+}$ | chromium(II) | chromous |
|  | **$Cr^{3+}$** | **chromium(III)** | chromic |
| Cobalt | $Co^{2+}$ | cobalt(II) |  |
|  | $Co^{3+}$ | cobalt(III) |  |
| Copper | **$Cu^+$** | **copper(I)** | cuprous |
|  | **$Cu^{2+}$** | **copper(II)** | cupric |
| Iron | **$Fe^{2+}$** | **iron(II)** | ferrous |
|  | **$Fe^{3+}$** | **iron(III)** | ferric |
| Lead | **$Pb^{2+}$** | **lead(II)** |  |
|  | $Pb^{4+}$ | lead(IV) |  |
| Manganese | **$Mn^{2+}$** | **manganese(II)** |  |
|  | $Mn^{3+}$ | manganese(III) |  |
| Mercury | $Hg_2^{2+}$ | mercury(I) | mercurous |
|  | **$Hg^{2+}$** | **mercury(II)** | mercuric |
| Tin | **$Sn^{2+}$** | **tin(II)** | stannous |
|  | $Sn^{4+}$ | tin(IV) | stannic |

*Listed alphabetically by metal name; those in **boldface** are most common.

## TABLE 2.5 Common Polyatomic Ions*

| FORMULA | NAME |
| --- | --- |
| **Cations** | |
| **NH$_4^+$** | **ammonium** |
| **H$_3$O$^+$** | **hydronium** |
| **Anions** | |
| **CH$_3$COO$^-$** | **acetate** |
| (or **C$_2$H$_3$O$_2^-$**) | |
| CN$^-$ | cyanide |
| **OH$^-$** | **hydroxide** |
| ClO$^-$ | hypochlorite |
| ClO$_2^-$ | chlorite |
| **ClO$_3^-$** | **chlorate** |
| **ClO$_4^-$** | **perchlorate** |
| NO$_2^-$ | nitrite |
| **NO$_3^-$** | **nitrate** |
| **MnO$_4^-$** | **permanganate** |
| **CO$_3^{2-}$** | **carbonate** |
| **HCO$_3^-$** | **hydrogen carbonate** (or **bicarbonate**) |
| CrO$_4^{2-}$ | chromate |
| **Cr$_2$O$_7^{2-}$** | **dichromate** |
| O$_2^{2-}$ | peroxide |
| **PO$_4^{3-}$** | **phosphate** |
| HPO$_4^{2-}$ | hydrogen phosphate |
| H$_2$PO$_4^-$ | dihydrogen phosphate |
| SO$_3^{2-}$ | sulfite |
| **SO$_4^{2-}$** | **sulfate** |
| HSO$_4^-$ | hydrogen sulfate or bisulfate) |

*Listed by charge and/or oxoanion family; those in **boldface** are most common.

## TABLE 2.6 Numerical Prefixes for Hydrates and Binary Covalent Compounds

| NUMBER | PREFIX |
| --- | --- |
| 1 | mono- |
| 2 | di- |
| 3 | tri- |
| 4 | tetra- |
| 5 | penta- |
| 6 | hexa- |
| 7 | hepta- |
| 8 | octa- |
| 9 | nona- |
| 10 | deca- |

SAMPLE PROBLEM 2.7

### Determining Names and Formulas of Ionic Compounds of Elements That Form More Than One Ion

**Problem:** Give the systematic names for the formulas or the formulas for the names of the following compounds:
**(a)** Tin(II) fluoride **(b)** CrI$_3$ **(c)** Ferric oxide **(d)** MnS
**Solution: (a)** Tin(II) is Sn$^{2+}$; fluoride is F$^-$. Two F$^-$ ions balance one Sn$^{2+}$ ion: tin(II) fluoride is **SnF$_2$.**
**(b)** The anion is I$^-$, iodide, and the formula shows three I$^-$. Therefore, the cation must be Cr$^{3+}$, chromium(III): CrI$_3$ is **chromium(III) iodide.**
**(c)** Ferric is the common name for iron(III), Fe$^{3+}$; oxide ion is O$^{2-}$. To balance the ionic charges, the formula of ferric oxide is **Fe$_2$O$_3$.**
**(d)** The anion is sulfide, S$^{2-}$, which requires that the cation be Mn$^{2+}$. The name is **manganese(II) sulfide.**

FOLLOW-UP PROBLEM 2.7
Give the systematic names for the formulas or the formulas for the names of the following compounds:
**(a)** lead(IV) oxide; **(b)** Cu$_2$S; **(c)** FeBr$_2$; **(d)** mercuric chloride.

**Compounds formed from polyatomic ions.** Ionic compounds in which one or both of the ions are polyatomic are very common. Table 2.5 gives the formulas and the names of some common polyatomic ions. Remember that *the polyatomic ion stays together as a charged unit.* The formula for potassium nitrate is KNO$_3$: each K$^+$ balances one NO$_3^-$. The formula for sodium carbonate is Na$_2$CO$_3$: two Na$^+$ balance one CO$_3^{2-}$. *When two or more of the same polyatomic ion are present in the formula, the ion appears in parentheses with the subscript written outside.* For example, calcium nitrate, which contains one Ca$^{2+}$ and two NO$_3^-$ ions, has the formula Ca(NO$_3$)$_2$. Parentheses and a subscript are *not* used unless *more than one* of the polyatomic ions is present; thus, sodium nitrate is NaNO$_3$, *not* Na(NO$_3$).

**Hydrated ionic compounds.** Ionic compounds called **hydrates** have a specific number of water molecules associated with each formula unit. In their formulas, the number of water molecules is shown after a centered dot. This number is indicated in the systematic name by a Greek numerical prefix (Table 2.6) before the word "hydrate." For example, Epsom salt has the formula MgSO$_4$·7H$_2$O and the name magnesium sulfate *hepta*hydrate; the mineral gypsum has the formula CaSO$_4$·2H$_2$O and the name calcium sulfate *di*hydrate.

**Families of oxoanions.** As Table 2.5 shows, most polyatomic ions are **oxoanions,** those in which an element, usually a nonmetal, is bonded to one or more oxygen atoms. In several cases, families of two or four oxoanions exist that differ only in the number of oxygen atoms. A simple naming convention is used with these ions.

With two oxoanions in the family:
- The ion with more O atoms takes the nonmetal root and the suffix "-ate."
- The ion with fewer O atoms takes the nonmetal root and the suffix "-ite."
For example, SO$_4^{2-}$ is the sulf*ate* ion; SO$_3^{2-}$ is the sulf*ite* ion.

With four oxoanions in the family (usually a halogen bonded to O):
- The ion with most O atoms has the prefix "per-," the nonmetal root, and the suffix "-ate."
- The ion with one less O atom has just the suffix "-ate."

- The ion with two less O atoms has just the suffix "-ite."
- The ion with three less O atoms has the prefix "hypo-" and the suffix "-ite."

For example, for the four chlorine oxoanions, $ClO_4^-$ is perchlorate, $ClO_3^-$ is chlorate, $ClO_2^-$ is chlorite, and $ClO^-$ is hypochlorite.

---

SAMPLE PROBLEM 2.8

### Determining Names and Formulas of Ionic Compounds Containing Polyatomic Ions

**Problem:** Give the systematic names for the formulas or the formulas for the names of the following compounds:
**(a)** $Fe(ClO_3)_2$ **(b)** Sodium sulfite **(c)** $Ba(OH)_2 \cdot 8H_2O$
**Solution:** **(a)** $ClO_3^-$ is chlorate; since it has a $1-$ charge, the cation must be $Fe^{2+}$. The name is **iron(II) chlorate.**
**(b)** Sodium is $Na^+$; sulfite is $SO_3^{2-}$. Therefore, two $Na^+$ ions balance one $SO_3^{2-}$ ion. The formula is $Na_2SO_3$.
**(c)** $Ba^{2+}$ is barium; $OH^-$ is hydroxide. There are eight (octa-) water molecules in each formula unit. The name is **barium hydroxide octahydrate.**

FOLLOW-UP PROBLEM 2.8
Give the systematic names for the formulas or the formulas for the names of the following compounds: **(a)** cupric sulfate pentahydrate; **(b)** zinc hydroxide; **(c)** LiCN.

---

SAMPLE PROBLEM 2.9

### Recognizing Incorrect Names and Formulas of Ionic Compounds

**Problem:** Something is wrong with each of the following names or formulas. Point out the mistake and correct it.
**(a)** $Ba(C_2H_3O_2)_2$ is called barium diacetate.
**(b)** Sodium sulfide has the formula $(Na)_2SO_3$.
**(c)** Iron(II) sulfate has the formula $Fe_2(SO_4)_3$.
**(d)** Cesium carbonate has the formula $Cs_2(CO_3)$.
**Solution:** **(a)** The charge of the $Ba^{2+}$ ion *must* be balanced by *two* $C_2H_3O_2^-$ ions, so the prefix "di-" is unnecessary. For ionic compounds, we do not indicate the number of ions with numerical prefixes. The correct name is **barium acetate.**
**(b)** Two mistakes occur here. The sodium ion is monatomic, so it does *not* require parentheses. The sulfide ion is $S^{2-}$, *not* $SO_3^{2-}$ (called "sulfite"). The correct formula is **$Na_2S$.**
**(c)** The Roman numeral refers to the charge of the ion, *not* the number of ions in the formula. $Fe^{2+}$ is the cation, so it requires one $SO_4^{2-}$ to balance its charge. The correct formula is **$FeSO_4$.**
**(d)** Parentheses are *not* required when only one polyatomic ion is present. The correct formula is **$Cs_2CO_3$.**

FOLLOW-UP PROBLEM 2.9
State why each of the following names or formulas is incorrect, and correct it:
**(a)** ammonium phosphate is $(NH_3)_4PO_4$; **(b)** aluminum hydroxide is $AlOH_3$;
**(c)** $Mg(HCO_3)_2$ is manganese(II) carbonate; **(d)** $Cr(NO_3)_3$ is chromic(III) nitride;
**(e)** $Ca(NO_2)_2$ is cadmium nitrate.

**Naming acids.** Acids are an important group of hydrogen-containing compounds that have been used in chemical reactions since before alchemical times. In the laboratory, acids are typically used in water solution. When naming them and writing their formulas, we can consider them as anions connected to the number of hydrogen ions ($H^+$) needed for electrical neutrality. The two common types of acid are binary acids and oxoacids:

1. *Binary acid* solutions form when certain gaseous compounds dissolve in water. For example, when gaseous hydrogen chloride (HCl) dissolves in water, it forms a solution called hydrochloric acid. The name consists of

Prefix *hydro-* + anion nonmetal *root* + suffix *-ic* + separate word *acid*.

Thus, we have hydro+chlor+ic+acid, or hydrochloric acid. This naming pattern holds for many compounds in which hydrogen combines with an anion that has an "-ide" suffix.

2. *Oxoacid* names are similar to those of the oxoanions, except for two suffix changes:

Anion "-ate" suffix becomes an "-ic" suffix in the acid. Anion "-ite" suffix becomes an "-ous" suffix in the acid.

The oxoanion prefixes "hypo-" and "per-" are retained. Thus, $BrO_4^-$ is *per*brom*ate*, and $HBrO_4$ is *per*brom*ic* acid; $IO_2^-$ is iod*ite*, and $HIO_2$ is iod*ous* acid.

SAMPLE PROBLEM 2.10 —————————————————

**Determining Names and Formulas of Anions and Acids**

**Problem:** Name the following anions and give the names and formulas of the acid solutions derived from them:
(a) $Br^-$     (b) $IO_3^-$     (c) $CN^-$     (d) $SO_4^{2-}$     (e) $NO_2^-$
**Solution:** (a) The anion is **bromide;** the acid is **hydrobromic acid, HBr.**
(b) The anion is **iodate;** the acid is **iodic acid, $HIO_3$.**
(c) The anion is **cyanide;** the acid is **hydrocyanic acid, HCN.**
(d) The anion is **sulfate;** the acid is **sulfuric acid, $H_2SO_4$.** (In this case, the suffix is added to the element name "sulfur," not to the root, "sulf-.")
(e) The anion is **nitrite;** the acid is **nitrous acid, $HNO_2$.**

FOLLOW-UP PROBLEM 2.10
Write formulas for the names or names for the formulas of the following acids:
(a) chloric acid; (b) HF; (c) acetic acid; (d) sulfurous acid; (e) HBrO.

**Names and Formulas of Binary Covalent Compounds**

Several **binary covalent compounds,** those formed by the combination of two elements, usually nonmetals, are so familiar that we use their common names—such as ammonia ($NH_3$) and water ($H_2O$)—but most are named in a systematic way according to a few basic rules:

1. The element with the lower group number in the periodic table is the first word in the name; the element with the higher group number is the second word. (*Important exception:* When the compound contains oxygen and a halogen, the halogen is named first.)
2. If both elements are in the same group, the one with the higher period number is named first.
3. The second element is named with its root and the suffix "-ide."

4. Covalent compounds have Greek numerical prefixes (see Table 2.6) to indicate the number of atoms of each element in the compound. The first word has a prefix *only* when more than one atom of the element is present; the second word *always* has a numerical prefix.

SAMPLE PROBLEM 2.11 _____

### Determining Names and Formulas of Binary Covalent Compounds

**Problem: (a)** What is the formula of carbon disulfide?
**(b)** What is the name of $PCl_5$?
**(c)** Give the name and formula of the compound formed from two N atoms and four O atoms.
**Solution: (a)** The prefix "di-" means "two." The formula is **$CS_2$.**
**(b)** P is the symbol for phosphorus; there are five chlorine atoms, which requires the prefix "penta-." The name is **phosphorus pentachloride.**
**(c)** N comes first in the name (lower group number). The compound is **dinitrogen tetraoxide, $N_2O_4$.**

FOLLOW-UP PROBLEM 2.11
Give the name or formula for **(a)** $SO_3$; **(b)** $SiO_2$; **(c)** dinitrogen monoxide; **(d)** selenium hexafluoride.

SAMPLE PROBLEM 2.12 _____

### Recognizing Incorrect Names and Formulas of Binary Covalent Compounds

**Problem:** Explain what is wrong with the name or formula of each of the following compounds, and correct it:
**(a)** $SF_4$ is monosulfur pentafluoride.
**(b)** Dichlorine heptaoxide is $Cl_2O_6$.
**(c)** $N_2O_3$ is dinitrotrioxide.
**Solution: (a)** There are two mistakes. "Mono-" is not needed if there is only one atom of the first element, and the prefix for four is "tetra-," not "penta-." The correct name is **sulfur tetrafluoride.**
**(b)** The prefix "hepta-" indicates seven, not six. The correct formula is **$Cl_2O_7$.**
**(c)** The full name of the first element is needed, and a space separates the two element names. The correct name is **dinitrogen trioxide.**

FOLLOW-UP PROBLEM 2.12
Point out the mistakes and correct the following names or formulas: **(a)** $S_2Cl_2$ is disulfurous dichloride; **(b)** nitrogen monoxide is $N_2O$; **(c)** $BrCl_3$ is trichlorine bromide.

### Molecular Masses from Chemical Formulas

In Section 2.4, we defined the atomic mass of an atom of an element as the weighted average of the element's isotopic masses in accordance with their fractional natural abundance. Using the formula and the periodic table, we calculate the **molecular mass** (also called *molecular weight*) of a formula unit of the compound as the sum of the atomic masses:

$$\text{Molecular mass} = \text{sum of atomic masses} \qquad \textbf{(2.3)}$$

The molecular mass of a water molecule (using atomic masses to four significant figures) is

$$\text{Molecular mass of } H_2O = (2 \times \text{atomic mass of H}) + (1 \times \text{atomic mass of O})$$
$$= (2 \times 1.008 \text{ amu}) + 16.00 \text{ amu} = 18.02 \text{ amu}$$

For convenience, we also use the term *molecular mass* for the mass of a formula unit of an ionic compound, even though molecules are not present. (Some texts use the term *formula mass* for ionic compounds.) For instance, the formula of barium nitrate is $Ba(NO_3)_2$; in calculating the molecular mass, note that the atomic mass of each element in the polyatomic ion is multiplied by the number of atoms of each element:

Molecular mass of $Ba(NO_3)_2$

$$= (1 \times \text{atomic mass of Ba}) + (2 \times \text{atomic mass of N}) + (6 \times \text{atomic mass of O})$$

$$= 137.3 \text{ amu} + (2 \times 14.01 \text{ amu}) + (6 \times 16.00 \text{ amu}) = 261.3 \text{ amu}$$

Note that atomic, not ionic, masses are used. Although ionic masses differ from those of the atoms by the electron masses, electron loss equals electron gain in the compound, so electron mass is balanced.

SAMPLE PROBLEM 2.13 _____

### Calculating the Molecular Mass of a Compound

**Problem:** Using the data in the periodic table, calculate the molecular mass of the following compounds:
**(a)** Tetraphosphorus trisulfide      **(b)** Ammonium nitrate
**Plan:** We first write the formula, then multiply the number of atoms (or ions) of each element by its atomic mass, and find the sum.
**Solution: (a)** The formula is $P_4S_3$.

$$\text{Molecular mass} = (4 \times \text{atomic mass of P}) + (3 \times \text{atomic mass of S})$$
$$= (4 \times 30.97 \text{ amu}) + (3 \times 32.07 \text{ amu}) = \textbf{220.09 amu}$$

**(b)** The formula is $NH_4NO_3$.

Molecular mass
$$= (2 \times \text{atomic mass of N}) + (4 \times \text{atomic mass of H}) + (3 \times \text{atomic mass of O})$$
$$= (2 \times 14.01 \text{ amu}) + (4 \times 1.008 \text{ amu}) + (3 \times 16.00 \text{ amu}) = \textbf{80.05 amu}$$

We count the total number of N atoms even though they belong to different ions.
**Check:** You can often find gross errors by rounding to the nearest 5 and adding:
**(a)** $7 \times 30 = 210 \approx 220.09$. Note that the sum has two decimal places because the atomic masses have two. **(b)** $(2 \times 15) + 5 + (3 \times 15) = 80 \approx 80.05$.

FOLLOW-UP PROBLEM 2.13
What is the formula and molecular mass of each of the following compounds: **(a)** hydrogen peroxide; **(b)** cesium chloride; **(c)** sulfuric acid; **(d)** potassium sulfate?

Each element in a compound constitutes a particular portion of the compound's mass. The molecular mass and chemical formula allow us to find the mass percent of any element X in the compound:

$$\text{Mass \% X} = \frac{\text{number of X in formula} \times \text{atomic mass of X}}{\text{molecular mass of compound}} \times 100 \qquad \textbf{(2.4)}$$

As an example, let's calculate the mass percent of nitrogen in barium nitrate. From the formula, $Ba(NO_3)_2$, there are two N atoms in every formula unit, so we have

$$\text{Mass \% N} = \frac{\text{number of N in formula} \times \text{atomic mass of N}}{\text{molecular mass of Ba(NO}_3\text{)}_2} \times 100$$

$$= \frac{2 \times 14.01 \text{ amu}}{261.3 \text{ amu}} \times 100 = 10.72 \text{ mass \%}$$

We extend this approach to determining mass percent in Chapter 3.

### Section Summary

Chemical formulas describe the simplest atom ratio (empirical formula), actual atom number (molecular formula), and atom arrangement (structural formula) of one unit of a compound. An ionic compound is named with cation first and anion last. For metals that can form more than one ion, the charge is shown with a Roman numeral after the metal name. Names of hydrates indicate the number of associated water molecules with numerical prefixes. Oxoanions have prefixes and/or suffixes attached to the element root name to indicate the number of O atoms. Acid names are based on anion names. Covalent compounds are named with the element that is leftmost or lower down in the periodic table first, with numerical prefixes showing the number of each atom. The molecular mass of a compound is the sum of the atomic masses in the formula. The mass percent of an element in a compound is the fraction of the molecular mass due to that element times 100.

## 2.8 Mixtures: Classification and Separation

Careful observation of the natural world reveals that *matter usually occurs as mixtures.* A sample of clean air, for example, consists of many elements and compounds physically mixed together, including oxygen ($O_2$), nitrogen ($N_2$), carbon dioxide ($CO_2$), the noble gases [Group 8A(18)], and water ($H_2O$). The oceans are complex mixtures of dissolved ions and covalent substances, including $Na^+$, $Mg^{2+}$, $Cl^-$, $SO_4^{2-}$, $O_2$, $CO_2$, and of course $H_2O$. Rocks and soils are mixtures of numerous compounds—calcium carbonate ($CaCO_3$), silicon dioxide ($SiO_2$), aluminum oxide ($Al_2O_3$), iron(III) oxide ($Fe_2O_3$)—perhaps a few elements (gold, silver, and carbon in the form of diamond), and petroleum and coal, which are complex mixtures themselves. Living things contain thousands of different substances: carbohydrates, lipids, proteins, and many simpler ionic and covalent compounds.

There are two broad classes of mixtures. A **heterogeneous mixture** has one or more visible boundaries between the components. Many rocks are heterogeneous, showing individual grains and flecks of different minerals. In some cases, the boundaries can be seen only with a microscope (Figure 2.19). A **homogeneous mixture** has no visible boundaries because the components are mixed as individual atoms, ions, and molecules. A mixture of sugar dissolved in water is homogeneous, for example, because the sugar molecules and water molecules are uniformly intermingled.

A homogeneous mixture is also called a **solution.** Solutions in water, called **aqueous solutions,** are extremely important in chemistry. Although we usually think of solutions as liquid, they can exist in all three physical

**A** Granite, a heterogeneous mixture

**B** Human blood, a heterogeneous mixture

**C** Copper(II) sulfate ($CuSO_4$) in water, a homogeneous mixture (solution)

**FIGURE 2.19**

**The distinction between heterogeneous and homogeneous mixtures. A,** and **B,** Microscopic examination is often required to see the physical boundaries in heterogeneous mixtures. The photomicrograph inset shows the boundaries in these mixtures. **C,** No boundaries are visible in homogeneous mixtures.

TOOLS OF THE
CHEMISTRY LABORATORY

## Basic Techniques for Separating Mixtures

Some of the most challenging and time-consuming laboratory procedures involve separating mixtures and purifying the components. Several common separation techniques are described here. Note that all these methods depend on the *physical properties* of the substances in the mixture; no chemical changes occur.

### Filtration

The process of **filtration** separates the components of a mixture on the basis of *differences in particle size*. It is used most often to separate a liquid from a solid. A familiar example is the separation of cooked pasta (larger particles) from boiling water (smaller particles) by pouring the mixture into a colander. Figure 2.C, *A* shows a simple filtration of a solid reaction product. Figure 2.C, *B* shows a more involved technique called vacuum filtration in which the pressure in the flask is lowered to speed the process by increasing the flow of the liquid through the filter.

A                                          B

**FIGURE 2.C** Filtration.

### Crystallization

The technique of **crystallization** is based on *differences in solubility* of the components of a mixture. The *solubility* of a substance is the amount that dissolves in a fixed volume of solvent at a given temperature. Many substances are more soluble in hot solvent than in cold, and this property is applied in the separation (Figure 2.D). Crystallization is often used to purify a compound by separating it from traces of impurities. Raw sugar, for example, is refined to white sugar by crystallization.

**FIGURE 2.D**
Crystallization.

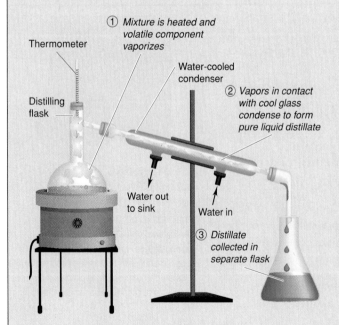

① *Mixture is heated and volatile component vaporizes*

Thermometer

Water-cooled condenser

② *Vapors in contact with cool glass condense to form pure liquid distillate*

Distilling flask

Water out to sink

Water in

③ *Distillate collected in separate flask*

### Distillation

The technique of **distillation,** which was probably invented by the alchemists, is based on *differences in volatility,* the tendency of a substance to vaporize, or become a gas. Ether, for example, is more volatile than water, which is much more volatile than sodium chloride. When a mixture boils, the more volatile component vaporizes first; it can then be condensed and collected in a separate container. The simple distillation shown in Figure 2.E is used to separate components with *large* differences in volatility, such as water from dissolved ionic compounds, which do not vaporize under these conditions. Separating components with small volatility differences requires many vaporization-condensation steps until only the more volatile component reaches the condenser (discussed in Chapter 12).

**FIGURE 2.E** Distillation.

## Extraction

Another technique based on *differences in solubility* is **extraction**. This process is usually used to remove compounds from natural materials such as plant and animal tissue. The material is ground in a blender with solvent, which extracts (dissolves) a soluble compound, or group of compounds, embedded in the less soluble natural material. (For example, coffee is a hot-water extract of coffee beans that contains many compounds.) The extracted solution is separated further by adding a second solvent that does not dissolve in the first. The two are shaken in a separatory funnel, and some components are extracted into the new solvent. Figure 2.F shows the extraction of plant pigments from water into hexane, an organic solvent.

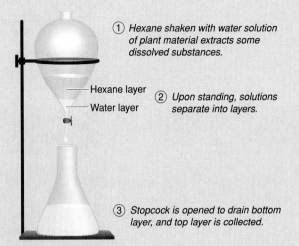

① *Hexane shaken with water solution of plant material extracts some dissolved substances.*

Hexane layer
Water layer

② *Upon standing, solutions separate into layers.*

③ *Stopcock is opened to drain bottom layer, and top layer is collected.*

**FIGURE 2.F** Extraction.

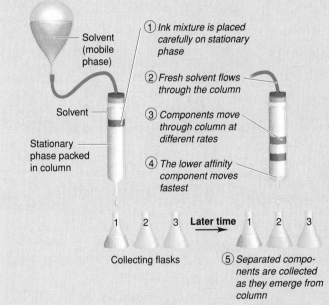

Solvent (mobile phase)

① *Ink mixture is placed carefully on stationary phase*

② *Fresh solvent flows through the column*

Solvent

③ *Components move through column at different rates*

Stationary phase packed in column

④ *The lower affinity component moves fastest*

1 2 3 **Later time** 1 2 3

Collecting flasks

⑤ *Separated components are collected as they emerge from column*

## Chromatography

The most common type of **chromatography** is also based on *differences in solubility*. The mixture is dissolved in a gas or liquid called the *mobile phase*, and the components are separated as this phase moves over a solid (or viscous liquid) surface called the *stationary phase*. A component with low solubility in the stationary phase spends less time there, thus moving faster, than a component with high solubility. Many types of chromatography are used to separate a wide variety of substances, from simple gases to biological macromolecules. Figure 2.G depicts the separation of a mixture of pigments in ink.

**FIGURE 2.G** Chromatography.

states. For example, air is a gaseous solution of mostly oxygen and nitrogen molecules, and wax is a solid solution of several fatty substances.

We have no way to tell visually whether an object is a pure substance (element or compound) or a homogeneous mixture (solution). In order to investigate pure substances, chemists have devised a variety of procedures for separating a mixture into its component elements and compounds (see Tools of the Chemistry Laboratory). Indeed, the laws and models of chemistry could never have been formulated without this ability; in fact, many of Dalton's critics thought compounds had varying composition because they were unknowingly studying mixtures.

## Section Summary

Heterogeneous mixtures have visible boundaries between the components. Homogeneous mixtures have no visible boundaries; mixing occurs at the molecular level. A solution is a homogeneous mixture and can occur in any physical state. Mixtures can be separated by physical processes, including filtration, crystallization, extraction, chromatography, and distillation. ◆

◆ **Drug Enforcement and the Separation of Mixtures.** Illicit "street" drugs are often mixed with an inactive substance, such as ascorbic acid (vitamin C). After obtaining a sample, government chemists separate the drug-vitamin mixture into its component substances. By calculating the mass of vitamin C per gram of drug sample, they can track the drug's distribution. For example, if different samples of cocaine obtained on the streets of New York, Los Angeles, and Paris all contain 1.6384 g vitamin C per gram of drug sample, they very likely come from a common source.

**2.1** Finding the mass of an element in a given mass of compound (p. 43):

Mass of element =

$$\text{mass of compound} \times \frac{\text{part by mass of element}}{\text{one part by mass of compound}}$$

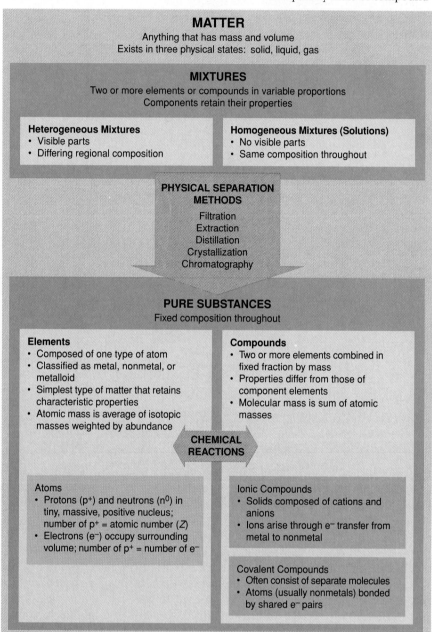

# MATTER
Anything that has mass and volume
Exists in three physical states: solid, liquid, gas

## MIXTURES
Two or more elements or compounds in variable proportions
Components retain their properties

**Heterogeneous Mixtures**
- Visible parts
- Differing regional composition

**Homogeneous Mixtures (Solutions)**
- No visible parts
- Same composition throughout

### PHYSICAL SEPARATION METHODS
Filtration
Extraction
Distillation
Crystallization
Chromatography

## PURE SUBSTANCES
Fixed composition throughout

**Elements**
- Composed of one type of atom
- Classified as metal, nonmetal, or metalloid
- Simplest type of matter that retains characteristic properties
- Atomic mass is average of isotopic masses weighted by abundance

**Compounds**
- Two or more elements combined in fixed fraction by mass
- Properties differ from those of component elements
- Molecular mass is sum of atomic masses

### CHEMICAL REACTIONS

**Atoms**
- Protons ($p^+$) and neutrons ($n^0$) in tiny, massive, positive nucleus; number of $p^+$ = atomic number ($Z$)
- Electrons ($e^-$) occupy surrounding volume; number of $p^+$ = number of $e^-$

**Ionic Compounds**
- Solids composed of cations and anions
- Ions arise through $e^-$ transfer from metal to nonmetal

**Covalent Compounds**
- Often consist of separate molecules
- Atoms (usually nonmetals) bonded by shared $e^-$ pairs

**FIGURE 2.20**

**The classification of matter from a chemical point of view.** Mixtures are separated physically into elements and compounds. Chemical reactions are required to convert elements into compounds, and vice versa.

## Chapter Perspective

*An understanding of matter at the observable and atomic levels is the essence of chemistry. In this chapter, you have learned how matter is classified in terms of its composition and how it is named in words and formulas, which are major steps toward that understanding. Figure 2.20 provides a visual review of many key terms and ideas in this chapter. In Chapter 3, we explore one of the central quantitative ideas in chemistry: how the observable amount of a substance relates to the number of atoms, molecules, or ions that make it up.*

# For Review and Reference

## Key Terms

**SECTION 2.1**
pure substance
element
molecule
compound
mixture

**SECTION 2.2**
law of mass conservation
law of definite (or
    constant) composition
percent by mass (mass
    percent, mass %)
law of multiple
    proportions
atom

**SECTION 2.3**
cathode ray
nucleus

**SECTION 2.4**
proton ($p^+$)
neutron ($n^0$)
electron ($e^-$)
atomic number *(Z)*
mass number *(A)*
atomic symbol
isotope
atomic mass unit (amu)
mass spectrometry
isotopic mass
atomic mass

**SECTION 2.5**
periodic table of the
    elements
group
period
metal

nonmetal
metalloid (semimetal)

**SECTION 2.6**
ionic compound
covalent compound
chemical bond
ion
cation
anion
monatomic ion
binary ionic compound
covalent bond
polyatomic ion

**SECTION 2.7**
chemical formula
empirical formula
molecular formula

structural formula
formula unit
hydrate
oxoanion
binary covalent
    compound
molecular mass

**SECTION 2.8**
heterogeneous mixture
homogeneous mixture
solution
aqueous solution
filtration
crystallization
extraction
chromatography
distillation
volatility

## Key Equations and Relationships

**2.1** Finding the mass of an element in a given mass of compound (p. 43):

Mass of element =

$$\text{mass of compound} \times \frac{\text{part by mass of element}}{\text{one part by mass of compound}}$$

**2.2** Calculating the number of neutrons in an atom (p. 52):

Number of neutrons = mass number − atomic number,
or $\qquad N = A - Z$

**2.3** Determining the molecular mass of a formula unit of a compound (p. 71):

Molecular mass = sum of atomic masses

**2.4** Calculating the mass percent of an element X in a compound (based on molecular mass) (p. 72):

Mass % X =

$$\frac{\text{number of X in formula} \times \text{atomic mass of X}}{\text{molecular mass of compound}} \times 100$$

## Answers to Follow-up Problems

**2.1** Mass (t) of pitchblende =

2.3 $\cancel{\text{t uranium}} \times \dfrac{1.00 \text{ t pitchblende}}{0.848 \ \cancel{\text{t uranium}}} = 2.7$ t pitchblende

Mass (t) of oxygen =

2.7 $\cancel{\text{t pitchblende}} \times \dfrac{0.152 \text{ t oxygen}}{1.00 \ \cancel{\text{t pitchblende}}} = 0.41$ t oxygen

**2.2** (a) Q = B; $5p^+$, $6n^0$, $5e^-$; (b) X = Ca; $20p^+$, $21n^0$, $20e^-$; (c) Y = I; $53p^+$, $78n^0$, $53e^-$

**2.3** $10.0129x + [11.0093(1 - x)] = 10.81$; $0.9964x = 0.1993$; $x = 0.2000$ and $1 - x = 0.8000$; % abundance of $^{10}$B = 20.00%; % abundance of $^{11}$B = 80.00%

**2.4** (a) $S^{2-}$; (b) $Rb^+$; (c) $Ba^{2+}$

**2.5** (a) Zinc [Group 2B(12)] and oxygen [Group 6A(16)]

(b) Silver [Group 1B(11)] and bromine [Group 7A(17)]

(c) Lithium [Group 1A(1)] and chlorine [Group 7A(17)]

(d) Aluminum [Group 3A(13)] and sulfur [Group 6A(16)]

**2.6** (a) ZnO; (b) AgBr; (c) LiCl; (d) $Al_2S_3$

**2.7** (a) $PbO_2$; (b) copper(I) sulfide (cuprous sulfide); (c) iron(II) bromide (ferrous bromide); (d) $HgCl_2$

**2.8** (a) $CuSO_4 \cdot 5H_2O$; (b) $Zn(OH)_2$; (c) lithium cyanide

**2.9** (a) $(NH_4)_3PO_4$; ammonium is $NH_4^+$ and phosphate is $PO_4^{3-}$.

(b) $Al(OH)_3$; parentheses are needed around the polyatomic ion $OH^-$.

(c) Magnesium hydrogen carbonate; $Mg^{2+}$ is magnesium and can have only a 2+ charge, so it does not

need (II); $HCO_3^-$ is hydrogen carbonate (or bicarbonate).

(d) Chromium(III) nitrate; the "ic" ending is not used with Roman numerals; $NO_3^-$ is nit*rate*.

(e) Calcium nitrite; $Ca^{2+}$ is calcium and $NO_2^-$ is nit*rite*.

**2.10** (a) $HClO_3$; (b) hydrofluoric acid; (c) $CH_3COOH$ (or $HC_2H_3O_2$); (d) $H_2SO_3$; (e) hypobromous acid

**2.11** (a) Sulfur trioxide; (b) silicon dioxide; (c) $N_2O$; (d) $SeF_6$

**2.12** (a) Disulfur dichloride; the "ous" suffix is not used.

(b) NO; the name indicates one nitrogen.

(c) Bromine trichloride; Br is in a higher period in Group 7A(17), so it is named first.

**2.13** (a) $H_2O_2$, 34.02 amu; (b) CsCl, 168.4 amu; (c) $H_2SO_4$, 98.09 amu; (d) $K_2SO_4$, 174.27 amu

## Sample Problem Titles

**2.1** Calculating the Mass of an Element in a Compound (p. 43)

**2.2** Determining the Number of Subatomic Particles in the Isotopes of an Element (p. 53)

**2.3** Calculating the Atomic Mass of an Element (p. 54)

**2.4** Predicting the Ion an Element Forms (p. 61)

**2.5** Naming Binary Ionic Compounds (p. 66)

**2.6** Determining the Formulas of Binary Ionic Compounds (p. 67)

**2.7** Determining Names and Formulas of Ionic Compounds of Elements That Form More Than One Ion (p. 68)

**2.8** Determining Names and Formulas of Ionic Compounds Containing Polyatomic Ions (p. 69)

**2.9** Recognizing Incorrect Names and Formulas of Ionic Compounds (p. 69)

**2.10** Determining Names and Formulas of Anions and Acids (p. 70)

**2.11** Determining Names and Formulas of Binary Covalent Compounds (p. 71)

**2.12** Recognizing Incorrect Names and Formulas of Binary Covalent Compounds (p. 71)

**2.13** Calculating the Molecular Mass of a Compound (p. 72)

## Problems

Problems with a green number are answered at the back of the text. Most sections include three categories of problems separated by a green rule—concept review questions, *paired* skill-building exercises, and problems in a relevant context.

### Elements, Compounds, and Mixtures: An Atomic Overview

**2.1** What is the key difference between an element and a compound?

**2.2** What are two differences between a compound and a mixture?

**2.3** Which of the following are pure substances? Explain your choices.

(a) Calcium chloride, which is used to melt ice on roads, consists of two elements, calcium and chlorine, in fixed mass ratios.

(b) Sulfur, which was known to the ancients, consists entirely of sulfur atoms chemically combined into molecules.

(c) Baking powder, a leavening agent, contains 26% to 30% sodium hydrogen carbonate and 30% to 35% calcium dihydrogen phosphate by mass.

(d) Water.

**2.4** Classify the substances in Problem 2.3 as an element, a compound, or a mixture, and explain your answers.

**2.5** Explain the following statement: the smallest particles that retain the identity of an element may be either atoms or molecules.

**2.6** Explain the following statement: the smallest particles that retain the identity of a compound cannot be atoms.

**2.7** Why can the relative amounts of the components of a mixture vary but not those of a compound?

**2.8** The tap water found in many areas of the United States leaves white deposits when it evaporates. Is this tap water a mixture or a compound? Explain.

**2.9** In order to determine whether a soil sample is a pure substance, a student mixes it with water and finds that much of it does not dissolve. She pours off the liquid, allows the water to evaporate, and a white deposit remains. Was the original soil sample a pure substance? Explain.

## The Observations That Led to an Atomic View of Matter

(Sample Problem 2.1)

**2.10** Why was it necessary for the techniques of chemical analysis to be developed before the laws of definite composition or multiple proportions could be formulated?

**2.11** To what classes of matter—element, compound, or mixture—do the following laws apply: (a) law of mass conservation; (b) law of definite composition; (c) law of multiple proportions?

**2.12** In view of our modern understanding of the relationship between matter and energy, is the law of mass conservation still relevant to chemical reactions? Explain.

**2.13** Identify the mass law that each of the following observations demonstrates, and explain your reasoning:
(a) A sample of potassium chloride from Chile contains the same percent by mass of potassium as one from Poland.
(b) A flashbulb contains magnesium and oxygen before being used and magnesium oxide afterward, but its mass does not change.
(c) Arsenic and oxygen form one compound that is 65.2 mass % arsenic and another that is 75.8 mass % arsenic.

**2.14** (a) Does the percent by mass of each element in a compound depend on the amount of compound? Explain.
(b) Does the mass of each element in a compound depend on the amount of compound? Explain.

**2.15** Does the percent by mass of the elements in a compound depend on the amounts of the elements used to make the compound? Explain.

**2.16** Which of Dalton's postulates about atoms have become inconsistent with later observations? Do these inconsistencies mean that Dalton was wrong? Is Dalton's model still useful?

**2.17** At a given set of conditions, 1 L hydrogen gas combines with 1 L chlorine gas to form 2 L hydrogen chloride gas.
(a) Does chlorine gas exist as single atoms or as diatomic molecules? Explain.
(b) How many chlorine atoms are present for every hydrogen atom in a hydrogen chloride molecule?
(c) Mass analysis of hydrogen chloride shows 0.0284 g hydrogen for every gram of chlorine. Suggest a relative atomic mass for chlorine.

**2.18** State the mass law(s) demonstrated by the following experimental results, and explain your reasoning:
Experiment 1: A student heats 1.00 g of a blue compound and obtains 0.64 g of a white compound and 0.36 g of a colorless gas.
Experiment 2: A second student heats 1.50 g of the same blue compound and obtains 0.96 g of a white compound and 0.54 g of a colorless gas.

**2.19** State the mass law(s) demonstrated by the following experimental results, and explain your reasoning:
Experiment 1: A student heats 1.27 g copper in 3.00 g iodine to produce 3.81 g of a white compound, and 0.46 g iodine remains.
Experiment 2: A second student heats 2.00 g copper in 3.50 g iodine to form 5.25 g of a white compound, and 0.25 g copper remains.

**2.20** Fluorite, a mineral of calcium, is a compound of the metal with fluorine. Analysis shows that a 2.76-g sample of fluorite contains 1.42 g calcium. Calculate:
(a) Mass of fluorine in the sample
(b) Mass fraction of calcium and fluorine in fluorite
(c) Mass percent of calcium and fluorine in fluorite

**2.21** Galena, a mineral of lead, is a compound of the metal with sulfur. Analysis shows that a 1.27-g sample of galena contains 1.10 g lead. Calculate:
(a) Mass of sulfur in the sample
(b) Mass fraction of lead and sulfur in galena
(c) Mass percent of lead and sulfur in galena

**2.22** Dolomite is a mineral combining magnesium carbonate and calcium carbonate. Analysis shows that 5.42 g dolomite contains 1.18 g calcium. Calculate the mass percent of calcium in dolomite. On the basis of the mass percent of calcium, and neglecting all other factors, which would be the richer source of calcium, dolomite or fluorite (see Problem 2.20)?

**2.23** Cotunnite is a mineral form of lead chloride. Analysis shows that 2.00 g cotunnite contains 1.49 g lead. On the basis of the mass percent of lead, and neglecting all other factors, which would be the better source of lead, cotunnite or galena (see Problem 2.21)?

**2.24** Magnesium oxide is 60.3% magnesium by mass.
(a) What is the mass fraction of magnesium in magnesium oxide?
(b) How many grams of magnesium can be obtained from 451 g magnesium oxide?

**2.25** Zinc sulfide is 67.1% zinc by mass.
(a) What is the mass fraction of zinc in zinc sulfide?
(b) How many grams of zinc can be obtained from 325 g zinc sulfide?

**2.26** A compound of copper and sulfur is 65.43% copper by mass. How many grams of copper can be obtained from 249 g of this compound? How many grams of sulfur remain?

**2.27** A compound of cesium and iodine is 51.16% cesium by mass. How many grams of cesium can be obtained from 124 g of this compound? How many grams of iodine remain?

**2.28** Show, with calculations, how the following data illustrate the law of multiple proportions:
Compound 1: 47.5 mass % sulfur and 52.5 mass % chlorine
Compound 2: 31.1 mass % sulfur and 68.9 mass % chlorine

**2.29** Show, with calculations, how the following data illustrate the law of multiple proportions:
Compound 1: 77.6 mass % xenon and 22.4 mass % fluorine
Compound 2: 63.3 mass % xenon and 36.7 mass % fluorine

**2.30** Hematite and magnetite are two minerals of iron from which the metal can be extracted profitably for use in making steel. If 1.64 g hematite yields 1.15 g iron, and 2.25 g magnetite yields 1.63 g iron, which mineral is a richer source of iron?

**2.31** The mass percent of sulfur in a sample of coal is a key factor in the environmental impact of the coal because the sulfur combines with oxygen and the oxide can be incorporated into acid rain after the coal is burned. Which of the following coals would have the smallest environmental impact?

| MASS (g) OF SAMPLE | | MASS (g) OF SULFUR IN SAMPLE |
|---|---|---|
| Coal A | 378 | 11.3 |
| Coal B | 286 | 11.0 |
| Coal C | 675 | 20.6 |

**The Observations That Led to the Nuclear Atom Model**

**2.32** Thomson was able to determine the mass/charge ratio of the electron from his experiments but not its mass. How did the results of Millikan's experiment make possible determination of the electron's mass?

**2.33** The following charges on individual oil droplets were obtained during an experiment similar to Millikan's. Use them to determine a charge for the electron (in C, coulomb), and explain your answer: $3.184 \times 10^{-19}$ C; $4.776 \times 10^{-19}$ C; $7.960 \times 10^{-19}$ C.

**2.34** Describe Thomson's model of the atom. How might it account for the production of cathode rays?

**2.35** How did the results obtained when Rutherford's coworkers bombarded gold foil with $\alpha$ particles compare with those they expected? What changes in Thomson's model of the atom were required to account for these results?

**The Atomic Theory Today**

(Sample Problems 2.2 and 2.3)

**2.36** Before neutrons were discovered, Rutherford suggested that electrons reside both within the nucleus *and* around it. How could such a model give an atom with mass and charge similar to that of our current model?

**2.37** Define atomic number and mass number. Which can vary without changing the identity of the element?

**2.38** Choose the correct answer. The difference between the mass number of an element and its atomic number is (a) directly related to the identity of the element; (b) the number of electrons; (c) the number of neutrons; (d) the number of naturally occurring isotopes.

**2.39** Why aren't the atomic masses of most elements whole numbers?

**2.40** The radius of a helium atom is $3.1 \times 10^{-11}$ m, and the radius of its nucleus is $2.5 \times 10^{-15}$ m. What fraction of the spherical atomic volume is occupied by the nucleus? (*V* of a sphere = $\frac{4}{3}\pi r^3$.)

**2.41** The mass of a helium-4 atom is $6.64648 \times 10^{-24}$ g, and its two electrons each have a mass of $9.10939 \times 10^{-28}$ g. What fraction of this atom's mass is contributed by its nucleus?

**2.42** Magnesium has three naturally occurring isotopes, $^{24}$Mg, $^{25}$Mg, and $^{26}$Mg. What is the mass number of each isotope? How many protons, neutrons, and electrons are present in each?

**2.43** Bromine has two naturally occurring isotopes, $^{79}$Br and $^{81}$Br. What is the mass number of each isotope? How many protons, neutrons, and electrons are present in each?

**2.44** Do both members of the following pairs have the same number of protons? Neutrons? Electrons?
(a) $^{16}_{8}$O and $^{17}_{8}$O   (b) $^{40}_{18}$Ar and $^{41}_{19}$K   (c) $^{60}_{27}$Co and $^{60}_{28}$Ni
Which pair(s) consist(s) of atoms with the same *Z* value? *N* value? *A* value?

**2.45** Do both members of the following pairs have the same number of protons? Neutrons? Electrons?
(a) $^{3}_{1}$H and $^{3}_{2}$He   (b) $^{14}_{6}$C and $^{15}_{7}$N   (c) $^{19}_{9}$F and $^{18}_{9}$F
Which pair(s) consist(s) of atoms with the same *Z* value? *N* value? *A* value?

**2.46** Gallium has two naturally occurring isotopes, $^{69}$Ga (isotopic mass 68.9256 amu, abundance 60.11%) and $^{71}$Ga (isotopic mass 70.9247 amu, abundance 39.89%). Calculate the atomic mass of gallium.

**2.47** Magnesium has three naturally occurring isotopes, $^{24}$Mg (isotopic mass 23.9850 amu, abundance 78.99%), $^{25}$Mg (isotopic mass 24.9858 amu, abundance 10.00%), and $^{26}$Mg (isotopic mass 25.9826 amu, abundance 11.01%). Calculate the atomic mass of magnesium.

**2.48** Chlorine has two naturally occurring isotopes, $^{35}$Cl (isotopic mass 34.9689 amu) and $^{37}$Cl (isotopic mass 36.9659 amu). If chlorine has an atomic mass of 35.4527 amu, what is the percent abundance of each chlorine isotope?

**2.49** Copper has two naturally occurring isotopes, $^{63}$Cu (isotopic mass 62.9396 amu) and $^{65}$Cu (isotopic mass 64.9278 amu). If copper has an atomic mass of 63.546 amu, what is the percent abundance of each copper isotope?

## Elements: A First Look at the Periodic Table

**2.50** Why is it useful to organize the elements in the form of the periodic table?

**2.51** Correct each of the following statements:
   (a) In the modern periodic table, the elements are arranged in order of increasing atomic mass.
   (b) Elements in a period have similar chemical properties.
   (c) Elements can be classified as either metalloids or nonmetals.

**2.52** What class of elements lies along the "staircase" line in the periodic table? How do their properties compare with those of metals and nonmetals?

**2.53** What are some characteristic properties of the elements that lie to the left of the "staircase"? To the right?

**2.54** The elements in Groups 1A(1) and 7A(17) are all quite reactive. What is a major difference between them?

**2.55** Give the atomic symbol for each of the following, and classify each as a metal, metalloid, or nonmetal: (a) germanium ($Z = 32$); (b) sulfur ($Z = 16$); (c) helium ($Z = 2$); (d) lithium ($Z = 3$); (e) molybdenum ($Z = 42$).

**2.56** Give the atomic symbol for each of the following, and classify each as a metal, metalloid, or nonmetal: (a) arsenic ($Z = 33$); (b) calcium ($Z = 20$); (c) bromine ($Z = 35$); (d) potassium ($Z = 19$); (e) magnesium ($Z = 12$).

**2.57** Fill in the blanks:
   (a) The symbol and atomic number of the lightest alkaline earth metal are _____ and _____.
   (b) The symbol and atomic number of the heaviest metalloid in Group 5A(15) are _____ and _____.
   (c) Group 1B(11) consists of the *coinage metals*. The symbol and atomic mass of the coinage metal whose atoms have the fewest electrons are _____ and _____.
   (d) The symbol and atomic mass of the halogen in Period 2 are _____ and _____.

**2.58** Fill in the blanks:
   (a) The symbol and atomic number of the heaviest noble gas are _____ and _____.
   (b) The symbol and group number of the transition element whose atoms have the fewest protons are _____ and _____.
   (c) The elements in Group 6A(16) are sometimes called the *chalcogens*. The symbol and atomic number of the only metallic chalcogen are _____ and _____.
   (d) The symbol and number of protons of the Period 5 alkali metal atom are _____ and _____.

## Compounds: Introduction to Bonding

(Sample Problem 2.4)

**2.59** Describe the type and nature of the bonding that occurs between active metals and nonmetals.

**2.60** Describe the type and nature of the bonding that often occurs between two nonmetals.

**2.61** How can ionic compounds be neutral if they consist of positive and negative ions?

**2.62** Are molecules present in a sample of potassium bromide (KBr)? Explain.

**2.63** Are ions present in a sample of ammonia ($NH_3$)? Explain.

**2.64** The monatomic ions of Group 1A(1) and Group 7A(17) atoms are all singly charged. In what major way do they differ? Why?

**2.65** Describe the formation of solid magnesium chloride ($MgCl_2$) from large numbers of magnesium and chlorine atoms.

**2.66** What is the difference between a monatomic and a polyatomic ion?

**2.67** Does magnesium carbonate ($MgCO_3$) incorporate ionic bonding, covalent bonding, or both? Explain.

**2.68** What monatomic ions do cesium ($Z = 55$) and bromine ($Z = 35$) form?

**2.69** What monatomic ions do strontium ($Z = 38$) and selenium ($Z = 34$) form?

**2.70** An ionic compound forms when lithium ($Z = 3$) reacts with oxygen ($Z = 8$). If a sample of the compound contains $2.8 \times 10^{20}$ lithium ions, how many oxide ions does it contain?

**2.71** An ionic compound forms when calcium ($Z = 20$) reacts with iodine ($Z = 53$). If a sample of the compound contains $1.2 \times 10^{19}$ calcium ions, how many iodide ions does it contain?

## Compounds: Formulas, Names, and Masses

(Sample Problems 2.5 to 2.13)

**2.72** What is the difference between an empirical formula and a molecular formula? Can they ever be the same?

**2.73** How is a structural formula similar to a molecular formula? How is it different?

**2.74** Consider a mixture of 10 billion $O_2$ molecules and 10 billion $H_2$ molecules. In what way is this mixture similar to a sample containing 10 billion hydrogen peroxide ($H_2O_2$) molecules? In what way is it different?

**2.75** For what type(s) of compound do we use Roman numerals in the name?

**2.76** For what type(s) of compound do we use Greek numerical prefixes in the name?

**2.77** For what type of compound are we unable to write a molecular formula?

**2.78** Write an empirical formula for each of the following:
   (a) Hydrazine, a rocket fuel, molecular formula $N_2H_4$
   (b) Glucose, a sugar, molecular formula $C_6H_{12}O_6$

**2.79** Write an empirical formula for each of the following:
(a) Ethylene glycol, a radiator antifreeze, molecular formula $C_2H_6O_2$
(b) Peroxodisulfuric acid, a compound used to make bleaching agents, molecular formula $H_2S_2O_8$

**2.80** Give the name and formula of the compound formed from the following elements: (a) lithium and nitrogen; (b) aluminum and chlorine; (c) oxygen and strontium.

**2.81** Give the name and formula of the compound formed from the following elements: (a) rubidium and bromine; (b) sulfur and barium; (c) calcium and fluorine.

**2.82** Give the systematic names for the formulas, or the formulas for the names: (a) tin(IV) chloride; (b) $FeBr_3$; (c) cuprous bromide; (d) $Mn_2O_3$; (e) $Na_2HPO_4$; (f) potassium carbonate dihydrate; (g) $NaNO_2$; (h) ammonium perchlorate.

**2.83** Give the systematic names for the formulas, or the formulas for the names: (a) CoO; (b) mercury(I) chloride; (c) $Pb(C_2H_3O_2)_2 \cdot 3H_2O$; (d) chromic oxide; (e) $Sn(SO_3)_2$; (f) potassium dichromate; (g) $FeCO_3$; (h) copper(II) nitrate.

**2.84** Correct each of the following formulas:
(a) Barium oxide is $BaO_2$.
(b) Iron(II) nitrate is $Fe(NO_3)_3$.
(c) Magnesium sulfide is $MnSO_3$.

**2.85** Correct each of the following names:
(a) CuI is cobalt(II) iodide.
(b) $Fe(HSO_4)_3$ is iron(II) sulfate.
(c) $MgCr_2O_7$ is magnesium dichromium heptaoxide.

**2.86** Give the name and formula for the acid derived from each of the following anions: (a) hydrogen sulfate; (b) $IO_3^-$; (c) cyanide; (d) $HS^-$.

**2.87** Give the name and formula for the acid derived from each of the following anions: (a) perchlorate; (b) $NO_3^-$; (c) bromite; (d) $F^-$.

**2.88** Write names for the formulas, or formulas for the names, of the following compounds: (a) $N_2O_5$; (b) chlorine trifluoride; (c) $SiS_2$.

**2.89** Write names for the formulas, or formulas for the names, of the following compounds: (a) $N_2O_3$; (b) sulfur hexafluoride; (c) $ICl_3$.

**2.90** Give the name and formula of the compound whose molecules consist of 4 phosphorus atoms and 10 sulfur atoms.

**2.91** Give the name and formula of the compound whose molecules consist of two iodine atoms and five oxygen atoms.

**2.92** Correct the names of the following compounds:
(a) CO is carbon oxide.
(b) $SO_2$ is sodium dioxide.
(c) $Cl_2O$ is chlorine dioxide.

**2.93** Correct the formulas of the following compounds:
(a) Chlorine monoxide is $ClO_1$.
(b) Disulfur decafluoride is $SF_{10}$.
(c) Phosphorus pentafluoride is $KF_5$.

**2.94** Give the number of atoms of the specified element in a formula unit of each of the following compounds, and calculate the molecular mass:
(a) Oxygen in aluminum sulfate, $Al_2(SO_4)_3$
(b) Hydrogen in ammonium hydrogen phosphate, $(NH_4)_2HPO_4$
(c) Oxygen in the mineral azurite, $Cu_3(OH)_2(CO_3)_2$

**2.95** Give the number of atoms of the specified element in a formula unit of each of the following compounds, and calculate the molecular mass:
(a) Hydrogen in ammonium benzoate, $C_6H_5COONH_4$
(b) Nitrogen in hydrazinium sulfate, $(N_2H_5)_2SO_4$
(c) Oxygen atoms in the mineral leadhillite, $Pb_4SO_4(CO_3)_2(OH)_2$

**2.96** Write the formula of each of the following compounds, and determine its molecular mass: (a) ammonium sulfate; (b) sodium dihydrogen phosphate; (c) potassium bicarbonate.

**2.97** Write the formula of each of the following compounds, and determine its molecular mass: (a) sodium dichromate; (b) ammonium perchlorate; (c) magnesium nitrite trihydrate.

**2.98** Calculate the molecular mass of each of the following compounds: (a) dinitrogen pentaoxide; (b) lead(II) nitrate; (c) calcium peroxide.

**2.99** Calculate the molecular mass of each of the following compounds: (a) iron(II) acetate tetrahydrate; (b) sulfur tetrachloride; (c) potassium permanganate.

**2.100** Calculate the mass % of each element in each compound in Problem 2.98.

**2.101** Calculate the mass % of each element in each compound in Problem 2.99.

**2.102** Aluminum cans have a long environmental "lifetime" because aluminum forms a stable oxide coating that adheres tightly to the metal underneath. What is the mass % of aluminum in aluminum oxide?

**2.103** In the process of developing black-and-white film, unexposed silver bromide is removed from the film by reaction with "hypo" (sodium thiosulfate). The resulting compound, $Na_3Ag(S_2O_3)_2$, is water soluble and washes away. What is the mass % of silver in this compound?

## Mixtures: Classification and Separation

**2.104** In what main way is separating the components of a mixture different from separating the components of a compound?

**2.105** What is the difference between a homogeneous and a heterogeneous mixture?

**2.106** Is a solution a homogeneous or a heterogeneous mixture? Give an example of an aqueous solution.

**2.107** Classify each of the following as a compound, a homogeneous mixture, or a heterogeneous mixture: (a) distilled water; (b) gasoline; (c) beach sand; (d) wine; (e) air.

**2.108** Classify each of the following as a compound, a homogeneous mixture, or a heterogeneous mixture: (a) brown sugar; (b) vegetable soup; (c) cement; (d) calcium sulfate; (e) tea.

**2.109** Name the technique(s) and briefly describe the procedure you would use to separate the following mixtures into two components: (a) table salt and pepper; (b) table sugar and sand; (c) drinking water contaminated with fuel oil; (d) vegetable oil and vinegar.

**2.110** Name the technique(s) and briefly describe the procedure you would use to separate the following mixtures into two components: (a) crushed ice and crushed glass; (b) salt dissolved in alcohol; (c) iron and sulfur; (d) two pigments (chlorophyll *a* and chlorophyll *b*) from spinach leaves.

## Comprehensive Problems

**2.111** Use the results of Lavoisier's experiment on the combustion of mercury (see Section 1.2) to decide whether air is an element, compound, or mixture.

**2.112** How can iodine ($Z = 53$) have a higher atomic number and lower atomic mass than tellurium ($Z = 52$)?

**2.113** You have a 100.0-g sample of each of the following compounds: $Na_2S_2O_3$, $K_2S_2O_8$, $Cr(NO_3)_3$, $TiO_2$, and $U_3O_8$. Which sample contains (a) the smallest mass of oxygen? (b) the largest mass of oxygen?

**2.114** You are working in the laboratory preparing sodium chloride. Consider the following results for three preparations of the compound:
Case 1: 39.34 g Na + 60.66 g $Cl_2$ → 100.00 g NaCl
Case 2: 39.34 g Na + 70.00 g $Cl_2$ →
100.00 g NaCl + 9.34 g $Cl_2$
Case 3: 50.00 g Na + 50.00 g $Cl_2$ →
82.43 g NaCl + 17.57 g Na
Explain these results in terms of the laws of conservation of mass and definite composition.

**2.115** The seven most abundant ions in sea water make up more than 99% by mass of the dissolved compounds. They are listed below in units of mg ion/kg sea water:

| ION | AMOUNT (mg/kg) |
| --- | --- |
| Chloride | 18,980 |
| Sodium | 10,560 |
| Sulfate | 2,650 |
| Magnesium | 1,270 |
| Calcium | 400 |
| Potassium | 380 |
| Hydrogen carbonate | 140 |

(a) What is the mass % of each ion in sea water?

(b) What percent of the total mass of ions does sodium ion represent?

(c) How does the total mass % of alkaline earth metal ions compare with the total mass % of alkali metal ions?

(d) Do anions or cations make up the larger mass fraction of dissolved components?

**2.116** Fluoride ion is poisonous in relatively low amounts: 0.2 g $F^-$ per 70 kg body weight can cause death. Nevertheless, in order to prevent tooth decay, $F^-$ ions are added to drinking water at a concentration of 1 mg $F^-$ ion per L water. How many liters of fluoridated drinking water would a 70-kg person have to consume in 1 day to reach this toxic level? How many kilograms of sodium fluoride would be needed to treat a $7.00 \times 10^7$ gal reservoir?

**2.117** Chromium is a component of many useful alloys, including stainless steel used in knives and kitchenware. Given the following data, calculate its atomic mass to the appropriate number of significant figures:

| ISOTOPE | NATURAL ABUNDANCE (%) | ISOTOPIC MASS (amu) |
| --- | --- | --- |
| $^{50}Cr$ | 4.35 | 49.946046 |
| $^{52}Cr$ | 83.79 | 51.940509 |
| $^{53}Cr$ | 9.50 | 52.940651 |
| $^{54}Cr$ | 2.36 | 53.938882 |

**2.118** Antimony (Latin *stibium*) was one of the elements known to the alchemists. Two antimony isotopes occur naturally: $^{121}Sb$ (isotopic mass 120.904 amu) and $^{123}Sb$ (isotopic mass 122.904 amu). If the atomic mass of antimony is 121.76 amu, what is the natural abundance of each isotope?

**2.119** The two isotopes of potassium with significant abundance in nature are $^{39}K$ (isotopic mass 38.9637 amu, 93.258%) and $^{41}K$ (isotopic mass 40.9618 amu, 6.730%). Fluorine has only one naturally occurring isotope, $^{19}F$ (isotopic mass 18.9984 amu). Calculate the molecular mass of potassium fluoride.

**2.120** A member of the family of dioxin compounds, TCDD, is one of the most toxic compounds known. It is formed as an unwanted byproduct in the manufacture of herbicides. The molecular formula of TCDD is $C_{12}H_4Cl_4O_2$. Use the empirical formula to calculate the mass % of chlorine in TCDD. Is this the same answer you would obtain by using the molecular formula? Explain.

**2.121** Give an example of the following:
(a) An element undergoing a physical change
(b) A compound undergoing a chemical change
(c) A solution that is hard and shiny
(d) A homogeneous mixture of elements that completely fills its container
(e) An element found both free and combined in nature

**2.122** What is the empirical formula of each of the following compounds: (a) $C_6H_{12}O_6$; (b) $CH_2O$; (c) $C_2H_4O_2$; (d) $C_4H_8O_4$? Explain how different compounds can have the same empirical formula.

# CHAPTER 3

## Concepts and skills to review

- isotopes and meaning of atomic mass (Section 2.3)
- names and formulas of compounds (Section 2.7)
- molecular mass of a compound (Section 2.7)
- mass percent from molecular mass and formula (Section 2.7)
- empirical and molecular formulas (Section 2.7)
- mass laws in chemical reactions (Section 2.2)

# Stoichiometry
## Mole-Mass Relationships in Chemical Systems

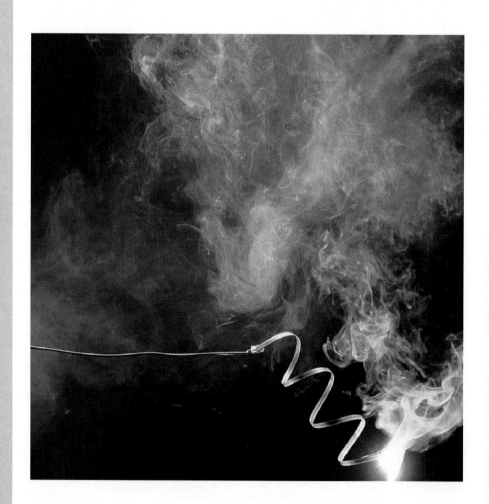

**Magnesium ribbon burning in air.** In this bright, smoky example of chemical change, as with all reactions, a chemist gains essential insight about change at the atomic level by knowing the types and amounts of substances involved at the visible level. As you'll see in this chapter, a fixed relationship exists between the *mass* of a substance and the *number* of atoms in it.

**N**owhere are the practical applications of chemistry more apparent than in determining a chemical formula or predicting the amounts of substances consumed and produced in a chemical reaction. Consider how important this information can be. A polymer chemist preparing a new plastic needs to estimate how much of the material a reaction will yield; a chemical engineer studying rocket engine thrust needs to calculate the amount of exhaust gases a fuel mixture will produce; an environmental chemist must know the quantity of air pollutants a sample of coal will release when burned; a medical researcher must establish the dosage of an experimental drug by measuring the amounts of its metabolic products. We can predict any of these quantities by examining the amounts and formulas of the substances involved in the chemical reaction.

**Stoichiometry** (Greek *stoicheion,* "element or part," + *metron,* "measure") is the study of the quantitative aspects of chemical formulas and reactions: if you know *what* is in a formula or reaction, stoichiometry tells you *how much*. In this chapter, we develop some essential quantitative skills of the chemist's trade: relating the mass of a substance to the number of chemical entities (atoms, molecules, or formula units); converting the results of analysis into a chemical formula; reading, writing, and thinking in the language of chemical equations; and applying the quantitative information held within them. All these abilities come with an understanding of the *mole,* the SI unit for amount of substance. The concept of the mole is central to chemistry.

## 3.1 The Mole

In daily life, pieces of matter are often measured either by counting them or by weighing them; the choice is determined by convenience. It is more convenient to weigh rice, for example, than to count individual grains, just as it is more convenient to count pencils than to weigh them. To measure these quantities, we use mass units (a kilogram of rice) or counting units (a dozen pencils). In the laboratory, measuring out a chemical substance to prepare a solution or "run" a reaction is a routine activity. We want to know the *number* of atoms, molecules, or formula units in a substance because these are the entities that react with each other. However, these entities are much too small to count individually, so chemists use a unit called the mole to *count them by weighing them.*

### Defining the Mole

The mole, like the dozen, represents a certain number of objects. The SI definition of the **mole** (abbreviated **mol**) is *the amount of a substance that con-*

*tains the number of entities equal to the number of atoms in exactly 12 g of carbon-12.* This number is called **Avogadro's number** in honor of the 19th-century Italian physicist, Amedeo Avogadro. As you can tell from the definition of the mole, Avogadro's number is enormous:

$$\text{1 mole contains } 6.022 \times 10^{23} \text{ entities (to four significant figures)} \qquad \textbf{(3.1)}$$

Thus, one mole of carbon-12 atoms contains $6.022 \times 10^{23}$ atoms; one mole of $H_2O$ molecules contains $6.022 \times 10^{23}$ molecules; and one mole of NaCl formula units contains $6.022 \times 10^{23}$ formula units. ◆

Unlike other counting units, *one mole of particles of a given substance has a fixed total mass.* To see why this is important, consider a simple analogy for atoms: marbles with fixed masses. Suppose you have large groups of red and of yellow marbles. Each red marble weighs 7 g, and each yellow marble weighs 4 g. Because of their fixed masses, you can count marbles by weighing them; for example, 84 g red marbles contains 12 marbles, as does 48 g yellow marbles (Figure 3.1, *A*). Moreover, any given *number* of red and of yellow marbles always has this 7:4 *mass* ratio. By the same token, any given *mass* of red and of yellow marbles always has the same 4:7 *number* ratio; for example, 280 g red marbles contains 40 marbles, whereas 280 g yellow marbles contains 70 marbles. Thus, because of their fixed masses, we can determine the number of marbles by weighing.

Similarly, because of fixed atomic masses, the mole gives us a practical way to determine the number of atoms, molecules, or formula units in a sample by weighing it. Let's focus on elements first and recall two points from Chapter 2:

- The atomic mass of an element is the weighted average of the masses of its naturally occurring isotopes.
- For purposes of weighing, all atoms of an element are considered to have this atomic mass.

That is, all sulfur atoms have an atomic mass of 32.07 amu (atomic mass units), all iron atoms have an atomic mass of 55.85 amu, and so forth.

The central relationship between the mass of one atom and the mass of one mole of those atoms is that *the atomic mass of an element expressed in **amu** is numerically the same as the mass of one mole of atoms of the element expressed in **grams.*** This means that one atom of sulfur has a mass of 32.07 amu, and one mole of sulfur atoms has a mass of 32.07 g. Similarly, one atom of iron has a mass of 55.85 amu, and one mole of iron atoms has a mass of 55.85 g. In terms of the number of atoms present, we know that 32.07 g sulfur atoms and 55.85 g iron atoms each contains $6.022 \times 10^{23}$ atoms. Moreover, as with the marbles of fixed mass in our analogy, one sulfur atom weighs

◆ **Imagine a Mole of . . .**
A mole of any ordinary object is a staggering amount: a mole of periods (.) lined up side by side would equal the radius of our galaxy; a mole of marbles stacked tightly together would cover the United States 70 miles deep. However, atoms and molecules are not ordinary objects: a mole of water molecules can be swallowed in one gulp!

**FIGURE 3.1**
Counting objects of fixed relative mass. **A,** If marbles had a fixed mass, we could count them by weighing them. Each red marble weighs 7 g, and each yellow marble weighs 4 g, so 84 g red marbles and 48 g yellow marbles each contains 12 marbles. Equal numbers of the two types of marbles always have a 7:4 mass ratio of red:yellow marbles. **B,** Since atoms of a substance have a fixed mass, we can weigh the substance to count the atoms. For example, 55.85 g Fe and 32.07 g S each contains $6.022 \times 10^{23}$ atoms (one mole of atoms). Any two samples of iron and sulfur that contain equal numbers of atoms have a 55.85:32.07 mass ratio of iron:sulfur.

A

B

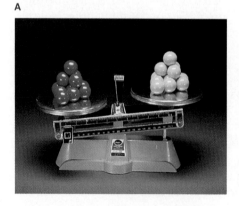

$^{32.07}/_{55.85}$ as much as one iron atom, and one mole of sulfur atoms weighs $^{32.07}/_{55.85}$ as much as one mole of iron atoms (Figure 3.1, *B*).

A similar relationship holds for compounds: *the molecular mass of a compound expressed in* **amu** *is numerically the same as the mass of one mole of the compound expressed in* **grams.** The molecular mass of a compound is the sum of the atomic masses of its elements. For instance, the molecular mass of water is

$$\text{Molecular mass of } H_2O = (2 \times \text{atomic mass of H}) + \text{atomic mass of O}$$
$$= 2(1.008 \text{ amu}) + 16.00 \text{ amu} = 18.02 \text{ amu}$$

Therefore, one mole of water ($6.022 \times 10^{23}$ molecules) has a mass of 18.02 g. The molecular mass of NaCl is the atomic mass of sodium (22.99 amu) plus the atomic mass of chlorine (35.45 amu), or 58.44 amu. Thus, one mole of NaCl ($6.022 \times 10^{23}$ formula units) has a mass of 58.44 g. As you can see, *the mole maintains the same mass relationship between macroscopic samples as exists between individual chemical entities.*

The mole is such a useful unit because it relates the *number* of chemical entities to the *mass* of a sample of those entities. A grocer cannot count a dozen eggs by weighing them because eggs vary in mass. In contrast, a chemist can obtain one mole of copper atoms ($6.022 \times 10^{23}$ atoms), or any proportion of that number, by simply weighing out 63.55 g copper, or any proportion of that mass. Figure 3.2 shows one mole of some familiar elements and compounds.

We can demonstrate this relationship by using the carbon-12 atomic mass standard to find the mass in grams of one atomic mass unit:

Mass in grams of 1 amu

$$= 1 \text{ amu} \times \frac{1 \text{ atom } ^{12}C}{12 \text{ amu}} \times \frac{1 \text{ mol } ^{12}C}{6.022 \times 10^{23} \text{ atoms } ^{12}C} \times \frac{12 \text{ g}}{1 \text{ mol } ^{12}C}$$
$$= 1.661 \times 10^{-24} \text{ g}$$

Thus, *Avogadro's number of atomic mass units has a mass of one gram:*

$$(6.022 \times 10^{23} \text{ amu})(1.661 \times 10^{-24} \text{ g/amu}) = 1.000 \text{ g}$$

### Converting Mass and Number to Moles

The **molar mass ($\mathcal{M}$)** of a substance is the mass of one mole of entities (atoms, molecules, or formula units) of the substance and has units of grams per mole (g/mol). For example, the molar mass of iron is 55.85 g/mol, and that of water is 18.02 g/mol. Table 3.1 summarizes the terms used throughout the text to express mass on the atomic and macroscopic scales.

The mole is a convenient unit for laboratory work because it allows you to calculate the mass or the number of chemical entities in a sample given the number of moles of substance in the sample. Conversely, knowing the mass or number of entities of a substance allows you to calculate the number of moles. We use the molar mass of an element or compound ($\mathcal{M}$, the number of grams per mole) to convert a given number of moles to mass:

$$\text{Mass (g)} = \text{no. of moles} \times \frac{\text{no. of grams}}{1 \text{ mole}} \qquad \textbf{(3.2)}$$

**FIGURE 3.2**
**One mole of some familiar substances.** One mole consists of $6.022 \times 10^{23}$ atoms, molecules, or formula units. From left to right: one mole of calcium carbonate (100.09 g), one mole of gaseous $O_2$ (32.00 g), one mole of copper wire (63.55 g), and one mole of liquid $H_2O$ (18.02 g).

**FIGURE 3.3**
**Masses in the periodic table.** The mass value in each element box refers to either the atomic or the molar mass. The atomic mass of S is 32.07 amu; the molar mass of S is 32.07 g.

**TABLE 3.1    Summary of Mass Terminology***

| TERM | DEFINITION | UNIT |
|------|-----------|------|
| Isotopic mass | Mass of an isotope of an element | amu |
| Atomic mass (also called atomic weight) | Average of the masses of the naturally occurring isotopes of an element weighted according to their abundance | amu |
| Molecular mass (also called molecular weight) | Sum of the atomic masses of the atoms (or ions) in a molecule (or formula unit) | amu |
| Molar mass ($\mathcal{M}$) (also called gram-molecular weight) | Mass of one mole of chemical entities (atoms, ions, molecules, formula units) | g/mol |

*All terms based on the $^{12}C$ standard: 1 atomic mass unit = $\frac{1}{12}$ mass of one $^{12}C$ atom.

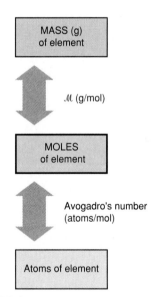

**FIGURE 3.4**
**Summary of the mass-mole-number relationships of an element.** Moles of an element are related to its mass through the molar mass ($\mathcal{M}$ in g/mol). Moles are related to the number of atoms through Avogadro's number ($6.022 \times 10^{23}$ atoms/mol). For elements that occur as molecules, Avogadro's number gives the number of *molecules* per mole. To find the number of atoms in a given mass, or vice versa, convert the information to moles first.

Alternatively, we can do the reverse with $1/\mathcal{M}$, and convert a given mass in grams to number of moles:

$$\text{No. of moles} = \text{mass (g)} \times \frac{1 \text{ mole}}{\text{no. of grams}} \quad \textbf{(3.3)}$$

Similarly, we use Avogadro's number to convert moles of substance to the number of entities:

$$\text{No. of entities} = \text{no. of moles} \times \frac{6.022 \times 10^{23} \text{ entities}}{1 \text{ mole}} \quad \textbf{(3.4)}$$

Alternatively, we can do the reverse:

$$\text{No. of moles} = \text{no. of entities} \times \frac{1 \text{ mole}}{6.022 \times 10^{23} \text{ entities}} \quad \textbf{(3.5)}$$

Notice that the mass of a sample and the number of entities it contains each relate directly to moles but *not* to each other. Therefore, *to convert between number of entities and mass, first convert to moles.*

**Moles of elements.** You can weigh out a mole of any element just by knowing the atomic mass. Suppose you want one mole of sulfur atoms. The periodic table shows that the atomic mass of sulfur is 32.07 amu (Figure 3.3), so you simply weigh out 32.07 g sulfur, its molar mass. Notice that the periodic table does not give a mass unit: the same number can refer to either the atomic mass (32.07 amu, weighted average mass of one atom) or to the molar mass (32.07 g, mass of one mole of atoms).*

When solving problems involving mass-mole-number relationships of elements, we use:
- The molar mass ($\mathcal{M}$ in g/mol) to convert between moles and mass
- Avogadro's number ($6.022 \times 10^{23}$ atoms/mol) to convert between moles and number of atoms

These relationships are summarized in Figure 3.4.

---

*The mass value in the periodic table is unitless because it is a *relative* atomic mass, given by the atomic mass (in amu) divided by 1 amu ($\frac{1}{12}$ atomic mass of one $^{12}C$ in amu):

$$\text{Relative atomic mass} = \frac{\text{atomic mass (amu)}}{\frac{1}{12} \text{ mass of } ^{12}C \text{(amu)}}$$

SAMPLE PROBLEM 3.1 ─────────────────────────────────────

## Calculating the Mass and Number of Atoms in a Given Number of Moles of Element

**Problem:** Silver (Ag) has been known since ancient times. It is still widely used in jewelry and tableware but no longer in U.S. coins.
**(a)** How many grams of Ag are in 0.0342 mol Ag?
**(b)** How many atoms of Ag are in 0.0342 mol Ag?
**Plan:** We have to find the mass and number of atoms in a certain number of moles of silver. **(a)** To convert *moles* of Ag to *grams* of Ag, we use the *molar mass* of Ag from the periodic table. **(b)** To convert *moles* to *number* of atoms, we use *Avogadro's number.*
**Solution: (a)** Converting from moles of Ag to mass:

$$\text{Mass (g) of Ag} = 0.0342 \; \text{mol Ag} \times \frac{107.9 \text{ g Ag}}{1 \text{ mol Ag}} = \textbf{3.69 g Ag}$$

**(b)** Converting from moles of Ag to number of atoms:

$$\text{No. of Ag atoms} = 0.0342 \; \text{mol Ag} \times \frac{6.022 \times 10^{23} \text{ atoms Ag}}{1 \text{ mol Ag}}$$

$$= \textbf{2.06} \times \textbf{10}^{\textbf{22}} \textbf{ atoms Ag}$$

**Check:** We rounded the answers to three significant figures because the number of moles has three. The units are correct, and the answers seem reasonable. **(a)** About 0.03 mol $\times$ 100 g/mol gives 3 g; the small mass reflects the small fraction of a mole. **(b)** About 0.03 mol $\times$ $(6 \times 10^{23}$ atoms/mol) gives $1.8 \times 10^{22}$ atoms.

**FOLLOW-UP PROBLEM 3.1**
Graphite is the crystalline form of carbon used in "lead" pencils.
**(a)** How many moles of carbon are in 315 mg graphite?
**(b)** How many carbon atoms are in this amount?

─────────────────────────────────────────────────────────────

SAMPLE PROBLEM 3.2 ─────────────────────────────────────

## Calculating the Number of Atoms in a Given Mass of Element

**Problem:** Iron (Fe) is the main component of all steel alloys and thus is the most important metal in modern society. How many iron atoms are present in a small slab of iron weighing 95.8 g?
**Plan:** We must convert the known mass of Fe to number of atoms. Since number and mass are each directly related to moles, we first use the molar mass of Fe to convert the mass of Fe to moles of Fe and then use Avogadro's number to convert moles to number of atoms:
**Solution:** Converting from mass of Fe to moles:

$$\text{Moles of Fe} = 95.8 \; \text{g Fe} \times \frac{1 \text{ mol Fe}}{55.85 \text{ g Fe}} = 1.72 \text{ mol Fe}$$

Converting from moles of Fe to number of atoms:

$$\text{No. of Fe atoms} = 1.72 \; \text{mol Fe} \times \frac{6.022 \times 10^{23} \text{ atoms Fe}}{1 \text{ mol Fe}}$$

$$= 10.4 \times 10^{23} \text{ atoms Fe} = \textbf{1.04} \times \textbf{10}^{\textbf{24}} \textbf{ atoms Fe}$$

**Check:** Approximating the mass and molar mass of Fe, we see about 2 moles present $(100/50 = 2)$, so the number of atoms should be about two times Avogadro's number: $2(6 \times 10^{23}) = 1.2 \times 10^{24}$.

Mass (g) of Fe

$\mathcal{M}$ of Fe = 55.85 g/mol

Moles of Fe

$6.022 \times 10^{23}$ atoms/mol

Number of Fe atoms

**FOLLOW-UP PROBLEM 3.2**
Manganese (Mn) is a transition element essential for the growth of strong bones. What is the mass of $3.22 \times 10^{20}$ Mn atoms, the number found in 1 kg of bone?

**Moles of compounds.** *The molar mass of a compound is the sum of the molar masses of the elements in the formula.* The formula of sulfur dioxide ($SO_2$), for example, tells us that one molecule of $SO_2$ contains one S atom and two O atoms, from which we calculate the molecular mass. It also tells us that one mole of $SO_2$ molecules contains one mole of S atoms and two moles of O atoms, from which we calculate the molar mass:

$$\text{Molar mass of } SO_2 = \text{molar mass of S} + (2 \times \text{molar mass of O})$$

$$= 32.07 \text{ g/mol} + (2 \times 16.00 \text{ g/mol}) = 64.07 \text{ g/mol}$$

One mole of $SO_2$ (64.07 g) contains $6.022 \times 10^{23}$ $SO_2$ molecules, which consist of $6.022 \times 10^{23}$ atoms of S and $2(6.022 \times 10^{23})$ atoms of O.

We use the same approach for ionic compounds, such as potassium sulfide ($K_2S$):

$$\text{Molar mass of } K_2S = (2 \times \text{molar mass of K}) + \text{molar mass of S}$$

$$= (2 \times 39.10 \text{ g/mol}) + 32.07 \text{ g/mol} = 110.27 \text{ g/mol}$$

One mole of $K_2S$ (110.27 g) contains $6.022 \times 10^{23}$ $K_2S$ formula units, which consist of $2(6.022 \times 10^{23})$ $K^+$ ions and $6.022 \times 10^{23}$ $S^{2-}$ ions. Thus, *the subscripts in a formula refer to individual atoms (or ions) as well as to moles of atoms (or ions).* Table 3.2 expresses the quantitative information in the formula of glucose ($C_6H_{12}O_6$).

When solving problems involving mass-mole-number relationships of compounds, we adopt an approach very similar to the one we used for elements. The relationships are shown diagrammatically in Figure 3.5.

**FIGURE 3.5**
**Summary of the mass-mole-number relationships of a compound.** Moles of a compound are related to its mass through the molar mass ($\mathcal{M}$ in g/mol). Moles are related to the number of molecules (or formula units) through Avogadro's number ($6.022 \times 10^{23}$ molecules/mol). With the chemical formula, you can calculate mass-mole-number information about each component element (see Figure 3.4).

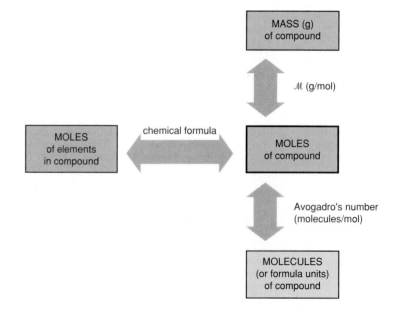

**TABLE 3.2   Information Contained in the Chemical Formula of Glucose, $C_6H_{12}O_6$ ($M$ = 180.16 g/mol)**

|                                   | CARBON (C)                              | HYDROGEN (H)                             | OXYGEN (O)                              |
|-----------------------------------|-----------------------------------------|------------------------------------------|-----------------------------------------|
| Atoms/molecule of compound        | 6 atoms                                 | 12 atoms                                 | 6 atoms                                 |
| Moles of atoms/mole of compound   | 6 moles of atoms                        | 12 moles of atoms                        | 6 moles of atoms                        |
| Atoms/mole of compound            | $6(6.022 \times 10^{23})$ atoms         | $12(6.022 \times 10^{23})$ atoms         | $6(6.022 \times 10^{23})$ atoms         |
| Mass/molecule of compound         | 6(12.01 amu) = 72.06 amu                | 12(1.008 amu) = 12.10 amu                | 6(16.00 amu) = 96.00 amu                |
| Mass/mole of compound             | 72.06 g                                 | 12.10 g                                  | 96.00 g                                 |

**SAMPLE PROBLEM 3.3** _____

**Calculating the Moles and Number of Formula Units in a Given Mass of a Compound**

**Problem:** Ammonium carbonate is a white crystalline powder that decomposes with warming. It has many uses, including being a component of baking powder, fire extinguishers, and smelling salts. How many moles and formula units are in 41.6 g ammonium carbonate?

**Plan:** We are given the mass and need to find the moles and number of formula units. Since we cannot convert directly from mass to number of formula units, we first convert mass to moles, so we need the molar mass ($M$). To find this, we determine the formula (see Table 2.5) and take the sum of the element molar masses. Then, we use the molar mass to convert the known mass to moles and use Avogadro's number to convert moles to number of formula units.

**Solution:** The formula is $(NH_4)_2CO_3$. Calculating molar mass:

$$M = (2 \times 14.01 \text{ g/mol}) + (8 \times 1.008 \text{ g/mol}) + 12.01 \text{ g/mol} + (3 \times 16.00 \text{ g/mol})$$

$$= 96.09 \text{ g/mol}$$

Converting from mass to moles:

$$\text{Moles of } (NH_4)_2CO_3 = 41.6 \text{ g } \cancel{(NH_4)_2CO_3} \times \frac{1 \text{ mol } (NH_4)_2CO_3}{96.09 \text{ g } \cancel{(NH_4)_2CO_3}}$$

$$= \textbf{0.433 mol } (NH_4)_2CO_3$$

Converting from moles to formula units:

Formula units of $(NH_4)_2CO_3$

$$= 0.433 \text{ } \cancel{\text{mol } (NH_4)_2CO_3} \times \frac{6.022 \times 10^{23} \text{ formula units } (NH_4)_2CO_3}{1 \text{ } \cancel{\text{mol } (NH_4)_2CO_3}}$$

$$= \textbf{2.61} \times \textbf{10}^{23} \textbf{ formula units } (NH_4)_2CO_3$$

**Check:** The units are correct in both cases. Since the mass is less than half the molar mass ($\sim 42/96 < 0.5$), the number of formula units should be less than half Avogadro's number ($\sim 2.6 \times 10^{23}/6.0 \times 10^{23} < 0.5$). A *common mistake* is to neglect the subscript 2 outside the parentheses in $(NH_4)_2CO_3$, thereby calculating a much lower molar mass.

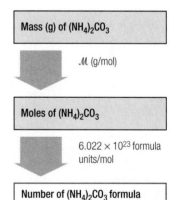

Mass (g) of $(NH_4)_2CO_3$

$M$ (g/mol)

Moles of $(NH_4)_2CO_3$

$6.022 \times 10^{23}$ formula units/mol

Number of $(NH_4)_2CO_3$ formula units

**Comment:** From here on, it is assumed you can determine the molar mass of a compound, so that calculation will not be shown.

FOLLOW-UP PROBLEM 3.3
Tetraphosphorus decaoxide reacts with water to form phosphoric acid, a major industrial acid. In the laboratory, the oxide is used as a drying agent.
**(a)** What is the mass of $4.65 \times 10^{22}$ molecules of tetraphosphorus decaoxide?
**(b)** How many P atoms are present in this sample?

---

### Mass Percent from the Chemical Formula

In Chapter 2, you saw how a compound's formula and molecular mass are used to calculate the mass percent of each element from the number of *atoms* of the element and its atomic mass (see Equation 2.4). The chemical formula also tells the number of *moles* of each element in the compound, so we can use a similar approach to calculate the mass percent of each element on a mole basis:

$$\text{Mass \% of element X} = \frac{\text{moles of X} \times \text{molar mass of X (g/mol)}}{\text{mass of one mole of compound (g)}} \times 100 \qquad \textbf{(3.6)}$$

As always, the individual mass percents of the elements in a compound add up to 100% (within rounding). An important practical use of mass percent is to determine the amount of an element in any size sample of a compound.

SAMPLE PROBLEM 3.4 ————————————————————————

### Calculating Mass Percents and Masses of Elements in a Sample of Compound

**Problem:** Glucose ($C_6H_{12}O_6$) is the most important nutrient in the cell for generating chemical potential energy.
**(a)** What is the mass percent of each element in glucose?
**(b)** How many grams of carbon are in 16.55 g glucose?

**(a) Determining the mass percent of each element**
**Plan:** We find the relative number of moles of each element (C, H, and O) in glucose from the formula. Then we convert from moles of element to grams with the element's molar mass. Dividing by the mass of one mole of glucose gives the element's mass fraction, and multiplying this fraction by 100 gives the mass percent. The calculation steps for any element X are shown in the roadmap.
**Solution:** The mass of one mole of $C_6H_{12}O_6$ is 180.16 g.
Converting moles of C to mass: There are 6 mol C/1 mol glucose, so

$$\text{Mass (g) of C} = 6\ \cancel{\text{mol C}} \times \frac{12.01 \text{ g C}}{1\ \cancel{\text{mol C}}} = 72.06 \text{ g C}$$

Finding the mass fraction of C in glucose:

$$\text{Mass fraction of C} = \frac{\text{total mass C}}{\text{mass of 1 mol glucose}} = \frac{72.06 \text{ g}}{180.16 \text{ g}} = 0.4000 \text{ g C/g glucose}$$

Finding the mass percent of C:

$$\text{Mass \% of C} = \text{mass fraction of C} \times 100 = 0.4000 \times 100$$

$$= \textbf{40.00 mass \% C}$$

---

Moles of X in one mole of compound

⬇ $\mathcal{M}$ (g/mol) of X

Mass (g) of X in one mole of compound

⬇ divide by mass (g) of one mole of compound

Mass fraction of X

⬇ multiply by 100

Mass % of X

Combining the steps for each of the other two elements:

$$\text{Mass \% of H} = \frac{\text{mol H} \times \mathcal{M} \text{ of H}}{\text{mass of 1 mol glucose}} \times 100 = \frac{12 \ \cancel{\text{mol H}} \times \dfrac{1.008 \ \text{g H}}{1 \ \cancel{\text{mol H}}}}{180.16 \ \cancel{\text{g}}} \times 100$$

$$= \textbf{6.714 mass \% H}$$

$$\text{Mass \% of O} = \frac{\text{mol O} \times \mathcal{M} \text{ of O}}{\text{mass of 1 mol glucose}} \times 100 = \frac{6 \ \cancel{\text{mol O}} \times \dfrac{16.00 \ \text{g O}}{1 \ \cancel{\text{mol O}}}}{180.16 \ \cancel{\text{g}}} \times 100$$

$$= \textbf{53.29 mass \% O}$$

**(b) Determining the mass of carbon**
**Plan:** To find the mass of C in the glucose sample, we multiply the mass of sample by the mass fraction of C from part **(a)**.
**Solution:** Finding the mass of C in a given mass of glucose (showing units for mass fraction):

$$\text{Mass (g) of C} = \text{mass of glucose} \times \text{mass fraction of C}$$

$$= 16.55 \ \cancel{\text{g glucose}} \times \frac{0.4000 \ \text{g C}}{1 \ \cancel{\text{g glucose}}} = \textbf{6.620 g C}$$

**Check:** The answers in **(a)** make sense: the mass % of O is greater than the mass % of C because, even though there are 6 moles of each in the compound, the molar mass of O is greater than the molar mass of C. The mass % of H is small because the molar mass of H is small. The total is 100.00%. The answer for **(b)** also makes sense: 16 g times less than 0.5 parts by mass should be less than 8 g.

FOLLOW-UP PROBLEM 3.4
Ammonium nitrate is used to manufacture the important dental anesthetic dinitrogen monoxide ($N_2O$, also called nitrous oxide and known informally as laughing gas). Calculate the mass percent of N in ammonium nitrate. What mass of N is present in 35.8 mg ammonium nitrate?

**Section Summary**
A mole of a substance is the amount that contains Avogadro's number ($6.022 \times 10^{23}$) of atoms, molecules, or formula units. Its mass (in grams) has the same numerical value as the mass (in amu) of the entity. Thus, the mole allows us to count entities by weighing them. Using the molar mass ($\mathcal{M}$, g/mol) of an element or a compound and Avogadro's number, we can convert among moles, mass, and number of entities. The mass fraction of an element X in a compound—the mass of the moles of X divided by the mass of one mole of the compound—can be used to calculate the mass of element X in any sample of the compound.

## 3.2 Determining the Formula of an Unknown Compound

In Sample Problem 3.4, we used the chemical formula to find the mass percent (or mass fraction) of an element in a compound and the mass of the element in a given mass of compound. In this section, we do the reverse: determine the formula of a compound from the masses of its elements.

◆ **A Rose by Any Other Name . . .** Chemists studying natural substances from animals and plants isolate compounds and determine their formulas. Geraniol ($C_{10}H_{18}O$) is the main compound that gives a rose its odor. It is used in many perfumes and cosmetics. Geraniol is also found in citronella and lemon grass and as part of a larger compound in geranium leaves, from which its name is derived.

## Empirical and Molecular Formulas

An analytical chemist investigating a compound decomposes it into simpler substances, finds the mass of each component element, and then mathematically converts these masses to moles of elements and the moles to whole-number subscripts. This procedure yields the empirical formula, the *simplest whole-number ratio* of moles of each element in the compound. ◆

Suppose, for example, that a sample of unknown compound is shown to contain 0.21 mol zinc, 0.14 mol phosphorus, and 0.56 mol oxygen, and we want to construct its empirical formula. Since the subscripts in a formula represent individual atoms or moles of atoms, we can write a preliminary formula that contains fractional subscripts based on these data: $Zn_{0.21}P_{0.14}O_{0.56}$. Then our aim is to convert these fractional subscripts to whole numbers. To do this, we employ one or two simple arithmetic steps (rounding when needed):

1. Divide each subscript by the smallest subscript:

$$Zn_{\frac{0.21}{0.14}}P_{\frac{0.14}{0.14}}O_{\frac{0.56}{0.14}} \rightarrow Zn_{1.5}P_{1.0}O_{4.0}$$

This step alone often gives integer subscripts.

2. If any of the subscripts are still not integers, multiply through by the *smallest integer* that will turn all the subscripts into integers. In this case, we multiply by 2, the smallest integer that will make the subscript 1.5 for Zn into an integer:

$$Zn_{(1.5 \times 2)}P_{(1.0 \times 2)}O_{(4.0 \times 2)} \rightarrow Zn_{3.0}P_{2.0}O_{8.0}, \text{ or } Zn_3P_2O_8$$

Notice that since we multiplied *all* the subscripts by 2, the *relative* number of moles is maintained. Always check that the subscripts obtained are the smallest set of integers that are in the same ratio as the original numbers of moles; that is, 3:2:8 are *in the same ratio* as 0.21:0.14:0.56. A more conventional way to write this formula is $Zn_3(PO_4)_2$; the compound is zinc phosphate, a dental cement.

A series of sample problems, each beginning one step farther away from the formula, helps clarify the process further. In the first of these, the elemental composition of the compound is given as masses of elements.

SAMPLE PROBLEM 3.5 _____

### Determining the Empirical Formula from Masses of Elements

**Problem:** The elemental analysis of a sample of ionic compound gave the following results: 2.82 g Na, 4.35 g Cl, and 7.83 g O. What is the empirical formula and name of the compound?

**Plan:** This problem is similar to the one just discussed, except that we are given *masses* of the elements, so we first have to calculate the *moles* of each element using its molar mass. Then we construct a preliminary formula and convert to integer subscripts.

**Solution:** Finding moles of elements:

$$\text{Moles of Na} = 2.82 \text{ g Na} \times \frac{1 \text{ mol Na}}{22.99 \text{ g Na}} = 0.123 \text{ mol Na}$$

$$\text{Moles of Cl} = 4.35 \text{ g Cl} \times \frac{1 \text{ mol Cl}}{35.45 \text{ g Cl}} = 0.123 \text{ mol Cl}$$

$$\text{Moles of O} = 7.83 \text{ g O} \times \frac{1 \text{ mol O}}{16.00 \text{ g O}} = 0.489 \text{ mol O}$$

Constructing a preliminary formula:

$$Na_{0.123}Cl_{0.123}O_{0.489}$$

Converting to integer subscripts (dividing all by smallest subscript):

$$Na_{\frac{0.123}{0.123}}Cl_{\frac{0.123}{0.123}}O_{\frac{0.489}{0.123}} \rightarrow Na_{1.00}Cl_{1.00}O_{3.98} \approx Na_1Cl_1O_4, \text{ or } NaClO_4$$

Note that we rounded the subscript of O from 3.98 to 4. The empirical formula is **$NaClO_4$**; the name is **sodium perchlorate.**
**Check:** The masses of Na and Cl are slightly more than 0.1 of their molar masses. The mass of O is greatest and its molar mass is smallest, so it should have the greatest number of moles. The ratio 1:1:4 is the same as 0.123:0.123:0.489, so the ratio of moles is maintained.

**FOLLOW-UP PROBLEM 3.5**
An unknown metal M reacts with sulfur to form a compound with formula $M_2S_3$. If 3.12 g M reacts with exactly 2.88 g S, what are the names of M and $M_2S_3$? (*Hint:* Determine moles of S and use the formula to find moles of M.)

Commercial analytical laboratories often provide composition data as mass percent of each element. With this information, we can determine the empirical formula by (1) expressing mass percent directly as mass, assuming 100 g of compound, (2) converting the mass to moles, and (3) constructing the empirical formula.

If we know the molar mass of a compound, we can use the empirical formula to obtain the molecular formula, the *actual* number of moles of each element in the smallest unit of the compound. In some cases, such as water ($H_2O$), ammonia ($NH_3$), methane ($CH_4$), and most ionic compounds, the empirical and molecular formulas are identical, but in many others the molecular formula is a *whole-number multiple* of the empirical formula. Hydrogen peroxide, for example, has the empirical formula HO and the molecular formula $H_2O_2$, so the multiple is 2. Dividing the molar mass of $H_2O_2$ (34.02 g/mol) by the empirical formula mass (17.01 g/mol) gives the whole-number multiple:

$$\text{Whole-number multiple} = \frac{\text{molar mass (g/mol)}}{\text{empirical formula mass (g/mol)}}$$

$$= \frac{34.02 \text{ g/mol}}{17.01 \text{ g/mol}} = 2$$

**SAMPLE PROBLEM 3.6**

**Determining the Molecular Formula from Elemental Composition and Molar Mass**

**Problem:** During physical activity, lactic acid ($\mathcal{M} = 90.08$ g/mol) forms in muscle tissue and is responsible for muscle soreness. Elemental analysis shows that it contains 40.0 mass % C, 6.71 mass % H, and 53.3 mass % O.
**(a)** Determine the empirical formula of lactic acid.
**(b)** Determine the molecular formula.

**(a) Determining the empirical formula**
**Plan:** The masses of elements are given as mass %, but no amount of compound is given. Since mass % is the same for any amount of compound, we can assume exactly 100 g lactic acid and thus express each mass % directly as grams. Then, we convert grams to moles and construct the empirical formula as in Sample Problem 3.5.

**Solution:** Expressing mass % as grams, assuming 100 g lactic acid:

$$\text{Mass (g) of C} = \frac{40.0 \text{ parts C by mass}}{100 \text{ parts by mass}} \times 100 \text{ g} = 40.0 \text{ g C}$$

Similarly, we have 6.71 g H and 53.3 g O.
Converting from grams of elements to moles:

$$\text{Moles of C} = \text{mass of C} \times \frac{1}{\mathcal{M} \text{ of C}} = 40.0 \text{ g C} \times \frac{1 \text{ mol C}}{12.01 \text{ g C}}$$

$$= 3.33 \text{ mol C}$$

Similarly, we have 6.66 mol H and 3.33 mol O.
Constructing the preliminary formula: $C_{3.33}H_{6.66}O_{3.33}$
Converting to integer subscripts:

$$C_{\frac{3.33}{3.33}} H_{\frac{6.66}{3.33}} O_{\frac{3.33}{3.33}} \rightarrow C_1H_2O_1; \text{ the empirical formula is } \textbf{CH}_2\textbf{O}$$

**(b) Determining the molecular formula**
**Plan:** The molecular formula subscripts are whole-number multiples of the empirical formula subscripts. To find this whole number, we divide the given molar mass by the empirical formula mass.
**Solution:** From the sum of the element molar masses in the empirical formula, the empirical formula mass is 30.03 g/mol. Finding the whole-number multiple:

$$\text{Whole-number multiple} = \frac{\mathcal{M} \text{ of lactic acid}}{\text{empirical formula mass}} = \frac{90.08 \text{ g/mol}}{30.03 \text{ g/mol}} = 3.000 = 3$$

Determining the molecular formula:

$$C_{(1 \times 3)}H_{(2 \times 3)}O_{(1 \times 3)} = \textbf{C}_3\textbf{H}_6\textbf{O}_3$$

**Check: (a)** Check as in Sample Problem 3.5. **(b)** The molecular formula has the same ratio of moles of elements (3:6:3) as the empirical formula (1:2:1). Always check that the mass of the calculated molecular formula adds up to the given molar mass:

$$\mathcal{M} \text{ of lactic acid} = (3 \times \mathcal{M} \text{ of C}) + (6 \times \mathcal{M} \text{ of H}) + (3 \times \mathcal{M} \text{ of O})$$

$$= (3 \times 12.01) + (6 \times 1.008) + (3 \times 16.00) = 90.08$$

**FOLLOW-UP PROBLEM 3.6**
One of the most widespread environmental carcinogens (cancer-causing agents) is benzo[a]pyrene ($\mathcal{M} = 252.30$ g/mol). It is found in coal dust, cigarette smoke, and even charcoal-grilled meat. Analysis of this hydrocarbon shows 95.21 mass % C and 4.79 mass % H. What is the molecular formula of benzo[a]pyrene?

Chemists have devised a variety of methods for the elemental analysis of compounds. **Combustion analysis** is used to determine the amounts of carbon and hydrogen in a sample of combustible compound. The unknown is placed in an apparatus with several chambers and burned in pure $O_2$ (Figure 3.6). All the C in the compound is converted to $CO_2$, which is absorbed in the first chamber, and all the H is converted to $H_2O$, which is absorbed in the second. By weighing the contents of the chambers before and after combustion, we find the masses of $CO_2$ and $H_2O$ and use them to calculate the masses of C and H in the compound, from which the *empirical* formula is found. Many compounds that contain carbon and hydrogen also contain oxygen, nitrogen, or a halogen. As long as the third element does not interfere with the absorption of $CO_2$ and $H_2O$ in the apparatus, its mass is calculated by subtracting the masses of C and H from the compound's original mass.

Sample of compound containing C, H, and other elements

CO₂ absorber

H₂O absorber

Other substances not absorbed

Stream of O₂

Furnace

**FIGURE 3.6**
**Combustion apparatus for determining formulas of organic compounds.** A sample of compound that contains C and H (and perhaps other elements) is burned in a stream of oxygen gas. The $CO_2$ and $H_2O$ formed from the C and H in the sample are absorbed separately, while any other element oxides are carried through by the oxygen gas stream. $CO_2$ is absorbed by NaOH on asbestos; $H_2O$ is absorbed by $Mg(ClO_4)_2$. The changes in mass of the preweighed $CO_2$ and $H_2O$ absorbers are used to calculate the moles of C and H present in the sample.

**SAMPLE PROBLEM 3.7**

## Determining a Chemical Formula from Combustion Analysis

**Problem:** Vitamin C ($\mathcal{M} = 176.12$ g/mol) is a compound of C, H, and O found in many natural sources, especially citrus fruits. A 1.000-g sample of vitamin C was placed in a combustion chamber and burned, and the following data were obtained:

Mass of $CO_2$ absorber after combustion = ¿ 5.35 g

Mass of $CO_2$ absorber before combustion = 83.85 g

Mass of $H_2O$ absorber after combustion = 37.96 g

Mass of $H_2O$ absorber before combustion = 37.55 g

What is the molecular formula of vitamin C?

**Plan:** We find the masses of $CO_2$ and $H_2O$ from the difference of the masses of the absorbers before and after the reaction. From the mass of $CO_2$, we find the mass of C, using the mass fraction of C in $CO_2$. Similarly, we find the mass of H from the mass of $H_2O$. The mass of vitamin C minus the sum of the C and H masses gives the mass of O, the third element present. Then, we proceed as in Sample Problem 3.6: calculate moles from the element molar masses and construct the empirical formula. From the given molar mass, determine the whole-number multiple and construct the molecular formula.

**Solution:** Finding the masses of combustion products:

$$\text{Mass of } CO_2 = \text{mass of } CO_2 \text{ absorber after} - \text{mass before} = 1.50 \text{ g } CO_2$$

$$\text{Mass of } H_2O = \text{mass of } H_2O \text{ absorber after} - \text{mass before} = 0.41 \text{ g } H_2O$$

Calculating mass fractions of the elements:

$$\text{Mass fraction of C in } CO_2 = \frac{\text{mol C} \times \mathcal{M} \text{ of C}}{\text{mass of 1 mol } CO_2} = \frac{1 \text{ mol C} \times \dfrac{12.01 \text{ g C}}{1 \text{ mol C}}}{44.01 \text{ g } CO_2}$$

$$= 0.2729 \text{ g C}/1 \text{ g } CO_2$$

$$\text{Mass fraction of H in } H_2O = \frac{\text{mol H} \times \mathcal{M} \text{ of H}}{\text{mass of 1 mol } H_2O} = \frac{2 \text{ mol H} \times \dfrac{1.008 \text{ g H}}{1 \text{ mol H}}}{18.02 \text{ g } H_2O}$$

$$= 0.1119 \text{ g H}/1 \text{ g } H_2O$$

Calculating masses of C and H:

$$\text{Mass of element} = \text{mass of compound} \times \text{mass fraction of element}$$

$$\text{Mass (g) of C} = 1.50 \text{ g } CO_2 \times \frac{0.2729 \text{ g C}}{1 \text{ g } CO_2} = 0.409 \text{ g C}$$

$$\text{Mass (g) of H} = 0.41 \text{ g } H_2O \times \frac{0.1119 \text{ g H}}{1 \text{ g } H_2O} = 0.046 \text{ g H}$$

Calculating the mass of O:

$$\text{Mass (g) of O} = \text{mass of vitamin C sample} - (\text{mass of C} + \text{mass of H})$$

$$= 1.000 \text{ g} - (0.409 \text{ g} + 0.046 \text{ g}) = 0.545 \text{ g O}$$

Finding the moles of elements: Dividing the element's mass by its molar mass gives 0.0341 mol C, 0.046 mol H, and 0.0341 mol O.

Constructing the preliminary formula:

$$C_{0.0341}H_{0.046}O_{0.0341}$$

Dividing through by the smallest subscript:

$$C_{\frac{0.0341}{0.0341}} H_{\frac{0.046}{0.0341}} O_{\frac{0.0341}{0.0341}} = C_{1.00}H_{1.3}O_{1.00}$$

Determining the empirical formula. By trial and error, we find that 3 is the smallest multiple that will make all subscripts into integers:

$$C_{(1.00 \times 3)}H_{(1.3 \times 3)}O_{(1.00 \times 3)} = C_{3.00}H_{3.9}O_{3.00} \approx C_3H_4O_3$$

Determining the molecular formula:

$$\text{Whole-number multiple} = \frac{\mathcal{M} \text{ of vitamin C}}{\text{empirical formula mass}} = \frac{176.12 \text{ g/mol}}{88.06 \text{ g/mol}}$$

$$= 2.000 = 2$$

$$C_{(3 \times 2)}H_{(4 \times 2)}O_{(3 \times 2)} = \mathbf{C_6H_8O_6}$$

**Check:** The element masses seem correct: C makes up slightly more than 0.25 of the mass of $CO_2$ (12 g/44 g > 0.25), as do the actual masses (0.409 g/1.50 g > 0.25). Two moles of H atoms account for slightly more than 0.10 of the mass of $H_2O$ (2 g/18 g > 0.10), as do the actual masses (0.046 g/0.41 g > 0.10). The molecular formula has the same ratio of subscripts (6:8:6) as the empirical formula (3:4:3) and adds up to the given molar mass:

$$(6 \times \mathcal{M} \text{ of C}) + (8 \times \mathcal{M} \text{ of H}) + (6 \times \mathcal{M} \text{ of O}) = \mathcal{M} \text{ of vitamin C}$$

$$(6 \times 12.01) + (8 \times 1.008) + (6 \times 16.00) = 176.12$$

**FOLLOW-UP PROBLEM 3.7**

A dry-cleaning solvent ($\mathcal{M}$ = 146.99 g/mol) that contains C, H, and Cl is suspected as a cancer-causing agent. When a 0.250-g sample was studied by combustion analysis, 0.451 g $CO_2$ and 0.0617 g $H_2O$ formed. Calculate the molecular formula.

**Chemical Formulas and the Structures of Molecules**

Contrary to the impression this chapter may give, chemistry is more than solving stoichiometry problems. Let's take a short break from calculations to recall a key point: formulas represent real three-dimensional objects. A molecular formula provides as much information as possible from mass analysis. However, only by knowing a molecule's structure—the distances and angles separating its atoms—can we begin to predict its behavior.

How much structural information is contained within a formula? The empirical formula tells only the types of atoms present and the *relative* numbers of each. However, *many different compounds can have the same empirical formula* (Table 3.3). If we know the molar mass, we can determine the molecular formula, which tells the *actual* number of each type of atom.

**TABLE 3.3 Some Compounds with Empirical Formula CH$_2$O (Composition by Mass 40.0% C, 6.71% H, 53.3% O)**

| MOLECULAR FORMULA | $\mathcal{M}$ (g/mol) | NAME | USE OR FUNCTION |
|---|---|---|---|
| CH$_2$O | 30.03 | Formaldehyde | Disinfectant; biological preservative |
| C$_2$H$_4$O$_2$ | 60.05 | Acetic acid | Acetate polymers; vinegar (5% solution) |
| C$_3$H$_6$O$_3$ | 90.08 | Lactic acid | Causes milk to sour; forms in muscle during exercise |
| C$_4$H$_8$O$_4$ | 120.10 | Erythrose | Forms during sugar metabolism |
| C$_5$H$_{10}$O$_5$ | 150.13 | Ribose | Component of many nucleic acids and vitamin B$_2$ |
| C$_6$H$_{12}$O$_6$ | 180.16 | Glucose | Major nutrient for energy in cells |

Even a molecular formula is not unique, however, because the same types and numbers of atoms can bond to each other in more than one arrangement (structural formula). Consider the empirical formula C$_2$H$_6$O. For a compound with this empirical formula and a molar mass of 46.07 g/mol, the molecular formula would also be C$_2$H$_6$O. Two *very* different compounds have this molecular formula: ethanol, the intoxicating substance present in wine and beer; and dimethyl ether, a colorless gas once used commercially in refrigeration (Table 3.4). Their radically different physical and chemical behaviors are the result of different molecular structures.

As molecular complexity increases—that is, as the number and types of atoms increase—the number of structural formulas that can be written for a given molecular formula also increases: C$_2$H$_6$O has two structural formulas, C$_3$H$_8$O three, and C$_4$H$_{10}$O seven. Imagine how many there are for C$_{16}$H$_{19}$O$_4$N$_3$S! Of the many possible structural formulas for this molecular formula, one is the antibiotic ampicillin. Whenever you write or think about a formula, try to remember that it represents a real object.

**TABLE 3.4 Two Compounds with Molecular Formula C$_2$H$_6$O**

| PROPERTY | ETHANOL | DIMETHYL ETHER |
|---|---|---|
| $\mathcal{M}$ (g/mol) | 46.07 | 46.07 |
| Color | Colorless | Colorless |
| Melting point | −117°C | −138.5°C |
| Boiling point | 78.5°C | −25°C |
| Density (at 20°C) | 0.789 g/mL | 0.00195 g/mL |
| Use | Intoxicant in alcoholic beverages | In refrigeration |
| Structural formula | 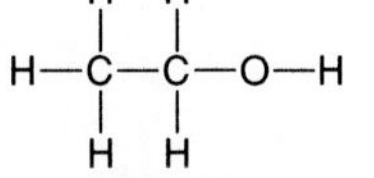 | 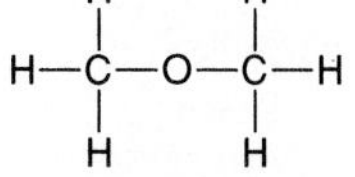 |
| Space-filling model |  |  |

**Section Summary**
From the masses of elements in an unknown compound, the relative numbers of moles of elements can be found and the empirical formula determined. If the molar mass is known, the molecular formula can also be determined. Methods such as combustion analysis provide data on masses or mass percents of the elements in a compound that can be converted to moles and used to obtain the formula of the compound. Because atoms can bond in different arrangements, a single molecular formula may correspond to more than one compound.

## 3.3 Writing and Balancing Chemical Equations

Thinking in terms of moles greatly clarifies what is happening in a reaction because it helps us view the atoms and molecules as large populations of interacting particles rather than as grams of material. Consider the formation of nitrogen monoxide gas (NO, also called nitric oxide) from $N_2$ and $O_2$, a reaction that occurs in combustion engines, blast furnaces, and lightning storms, that is, wherever air is heated strongly. If we weigh the gases, we would find that

28.02 g $N_2$ and 32.00 g $O_2$ react to form 60.02 g NO

This information tells us little except that mass is conserved. If we convert these masses to moles, however, we find that

1 mol $N_2$ and 1 mol $O_2$ react to form 2 mol NO

This information reveals that equal-size populations of $N_2$ and $O_2$ molecules combine to form a population of NO molecules twice as large. Dividing through by Avogadro's number shows us the chemical event as it occurs between individual molecules:

1 $N_2$ molecule and 1 $O_2$ molecule react to form 2 NO molecules

Notice that when we express the reaction in terms of moles, the macroscopic (molar) change corresponds to the submicroscopic (molecular) change (Figure 3.7). A balanced chemical equation, as you'll see in a moment, shows both changes.

A **chemical equation** is a statement in formulas that expresses the identities and quantities of the substances involved in a chemical or physical change. Equations are the "sentences" of chemistry, just as chemical formu-

**FIGURE 3.7**

**The formation of NO gas on the macroscopic and molecular levels.** One mole of $N_2$ (28.02 g) and one mole of $O_2$ (32.00 g) react to form two moles of NO (60.02 g). Dividing by Avogadro's number shows the change at the molecular level.

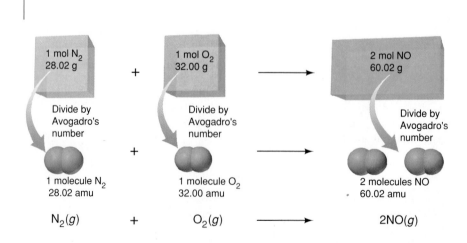

las are the "words" and atomic symbols the "letters." The left and right sides of a chemical equation represent the substances before and after the change.

For an equation to depict accurately the amounts of chemicals involved in a reaction, it must be *balanced*; that is, *the same number of each type of atom must appear on both sides of the equation*. The requirement that an equation be balanced follows directly from the mass laws and the atomic theory:

- In a chemical process, atoms cannot be created, destroyed, or changed; they can only be rearranged into different combinations.
- Compound formulas represent fixed ratios of component elements; a different ratio represents a different compound.

As an example, consider the chemical change that occurs inside a photographic flashbulb (Figure 3.8): magnesium wire and oxygen gas react to yield powdery magnesium oxide. (Light and heat are produced as well, but here we consider only the substances involved.) Several steps will convert this chemical statement into a balanced equation:

1. *Translating the statement.* We first translate the chemical statement into a "skeleton" equation: correct chemical formulas arranged in an equation format. All the substances that react in the change, called **reactants,** are placed to the left of a "yield" arrow, which points to all the substances produced, called **products:**

$$\underset{\text{reactants}}{\_\_ \text{Mg} \quad + \quad \_\_ \text{O}_2} \quad \underset{\text{yield}}{\longrightarrow} \quad \underset{\text{product}}{\_\_ \text{MgO}}$$

Magnesium and oxygen gas react to yield magnesium oxide

At the beginning of the balancing process, we put a blank in front of each substance to remind us to account for the atoms in each.

2. *Balancing the atoms.* The next step is to balance the number of each type of atom. At the end of this step, each blank will contain a **balancing (stoichiometric) coefficient,** a numerical multiplier of *all the atoms* in the formula immediately following it. Balancing is a stepwise process of matching the atoms on each side element by element. In general, it is best to begin with the most complex substance, the one with the largest number of atoms or different types of atoms. In this case, it is MgO, and we place a coefficient 1 *in front of* the compound:

$$\_\_ \text{Mg} + \_\_ \text{O}_2 \rightarrow \underline{1}\text{MgO}$$

This requires 1 Mg atom on the left to balance the Mg in MgO:

$$\underline{1}\text{Mg} + \_\_ \text{O}_2 \rightarrow \underline{1}\text{MgO}$$

The O atom on the right must be balanced by one O atom on the left. One-half an $O_2$ molecule would provide one O atom:

$$\underline{1}\text{Mg} + \tfrac{1}{2}\text{O}_2 \rightarrow \underline{1}\text{MgO}$$

In terms of number and type of atom, the equation is balanced.

3. *Adjusting the coefficients.* There are several conventions about the final form of the coefficients:

(a) In most cases, *the smallest whole-number coefficients are preferred.* Whole numbers allow us to show the actual reacting entities intact, such as $O_2$ molecules. Although one-half mole $O_2$ can exist, one-half of an $O_2$ molecule cannot, so we multiply the equation by 2:

$$2\text{Mg} + 1\text{O}_2 \rightarrow 2\text{MgO}$$

(b) We used the coefficient 1 to remind us to deal with each substance. In the final form, a coefficient of 1 is implied by the presence of the sub-

**A**

**B**

**FIGURE 3.8**
**A chemical reaction in a flashbulb. A,** Before the reaction occurs, a fine magnesium filament is surrounded by oxygen. **B,** After the reaction, white, powdery magnesium oxide coats the bulb's inner surface. By knowing the substances present before and after a reaction, we can write a balanced equation for the change.

stance and is not written:

$$2Mg + O_2 \rightarrow 2MgO$$

This convention is similar to the absence of a subscript 1 in a formula.

4. *Checking.* After balancing and adjusting the coefficients, it is always a good idea to check that the equation is balanced:

$$\text{Reactants (2 Mg, 2 O)} \rightarrow \text{products (2 Mg, 2 O)}$$

5. *Specifying the states of matter.* A balanced equation also indicates the physical state of each substance or whether it is dissolved in water. The abbreviations used for these states are solid (*s*), liquid (*l*), gas (*g*), and aqueous solution (*aq*). From the original statement, Mg "wire" is solid, $O_2$ is a gas, and "powdery" MgO is also solid:

$$2Mg(s) + O_2(g) \rightarrow 2MgO(s)$$

This is the *balanced* equation.

You must realize several other points about the balancing process:

- A coefficient operates on *all the atoms in the formula* that follows it: 2MgO means 2(MgO), or 2 Mg atoms and 2 O atoms; $2Ca(NO_3)_2$ means $2[Ca(NO_3)_2]$, or 2 Ca atoms, 4 N atoms, and 12 O atoms.

- In balancing an equation, *chemical formulas cannot be altered.* In the previous step 2, we *cannot* balance the O atoms by changing MgO to $MgO_2$ because this has a different elemental composition and is thus a different compound.

- We *cannot add other reactants or products* to balance the equation because this would represent a different chemical process. For example, we *cannot* balance the O atoms by changing $O_2$ to O or by adding one O atom to the products, because the chemical statement does not say that the reaction involves O atoms.

- A balanced equation remains balanced even if all the coefficients are multiplied by the same number. For example,

$$4Mg(s) + 2O_2(g) \rightarrow 4MgO(s)$$

is also balanced because it is just the balanced equation multiplied by 2. However, convention requires that we balance the equation with the *smallest* whole-number coefficients.

SAMPLE PROBLEM 3.8 _____

**Balancing Chemical Equations**

**Problem:** Within the cylinders of a car's engine, the hydrocarbon octane ($C_8H_{18}$), one of many components of gasoline, mixes with oxygen from the air and burns to form carbon dioxide and water vapor. Write a balanced equation for this reaction.

**Solution:**

1. *Translate* the statement into a skeleton equation (with coefficient blanks). Octane and oxygen are reactants; "oxygen from the air" implies molecular oxygen, $O_2$. Carbon dioxide and water vapor are products:

$$\_\_ C_8H_{18} + \_\_ O_2 \rightarrow \_\_ CO_2 + \_\_ H_2O$$

2. *Balance the atoms.* We start with the most complex substance, $C_8H_{18}$, and balance $O_2$ last:

$$\underline{1}C_8H_{18} + \_\_ O_2 \rightarrow \_\_ CO_2 + \_\_ H_2O$$

Each $CO_2$ contains 1 C atom, so 8 molecules of $CO_2$ are needed to balance the C atoms in each $C_8H_{18}$:

$$\underline{1}C_8H_{18} + \_\_ O_2 \rightarrow \underline{8}CO_2 + \_\_ H_2O$$

The 18 H atoms in $C_8H_{18}$ require a coefficient 9 in front of $H_2O$:

$$\underline{1}C_8H_{18} + \_\_ O_2 \rightarrow \underline{8}CO_2 + \underline{9}H_2O$$

There are 25 atoms of O (16 in $8CO_2$ and 9 in $9H_2O$) on the right, so we place the coefficient 25/2 in front of $O_2$:

$$\underline{1}C_8H_{18} + \underline{25/2}\, O_2 \rightarrow \underline{8}CO_2 + \underline{9}H_2O$$

3. *Adjust the coefficients.* Multiply through by 2 to obtain whole numbers:

$$2C_8H_{18} + 25O_2 \rightarrow 16CO_2 + 18H_2O$$

4. *Check* that the equation is balanced:

$$\text{Reactants (16 C, 36 H, 50 O)} \rightarrow \text{products (16 C, 36 H, 50 O)}$$

5. *Specify* states of matter. $C_8H_{18}$ is liquid; $O_2$, $CO_2$, and $H_2O$ vapor are gases:

$$\mathbf{2C_8H_{18}}(l) + \mathbf{25O_2}(g) \rightarrow \mathbf{16CO_2}(g) + \mathbf{18H_2O}(g)$$

**Comment:** The products in this reaction ($CO_2$, $H_2O$) are the same for any C,H-containing compound that burns in air. Compounds that contain *only* C and H are called *hydrocarbons.*

**FOLLOW-UP PROBLEM 3.8**
Write balanced equations for the following statements:
**(a)** A characteristic reaction of Group 1A(1) elements: chunks of sodium react violently with water to form hydrogen gas and sodium hydroxide solution.
**(b)** The destruction of marble statuary by acid rain: aqueous nitric acid reacts with calcium carbonate to form carbon dioxide, water, and aqueous calcium nitrate.
**(c)** Halogen compounds exchanging bonding partners: phosphorus trifluoride is prepared by the reaction of phosphorus trichloride and hydrogen fluoride; hydrogen chloride is the other product. The reaction involves gases only.
**(d)** Explosive decomposition of dynamite: liquid nitroglycerine ($C_3H_5N_3O_9$) explodes on concussion to produce a mixture of gases—carbon dioxide, water vapor, nitrogen, and oxygen.

Balancing equations is an essential skill in chemistry, and as with learning piano, tennis, or any new skill, the only way to become expert is through practice. As you develop a stepwise approach to the process, patterns will become apparent, and it will soon be a central part of your problem-solving ability.

**Section Summary**
In order to conserve mass and maintain the fixed composition of compounds, a chemical reaction must be balanced in terms of number and type of each atom. A balanced equation consists of reactant formulas on the left of a yield arrow and product formulas on the right. Coefficients are used to balance the atoms; they are integer multipliers that apply to *all* the atoms in a formula.

**TABLE 3.5    Information Contained in a Balanced Equation**

| VIEWED IN TERMS OF | REACTANTS $C_3H_8(g) + 5O_2(g)$ | $\rightarrow$ $\rightarrow$ | PRODUCTS $3CO_2(g) + 4H_2O(g)$ |
|---|---|---|---|
| Molecules | 1 molecule $C_3H_8$ + 5 molecules $O_2$ | $\rightarrow$ | 3 molecules $CO_2$ + 4 molecules $H_2O$ |
| | 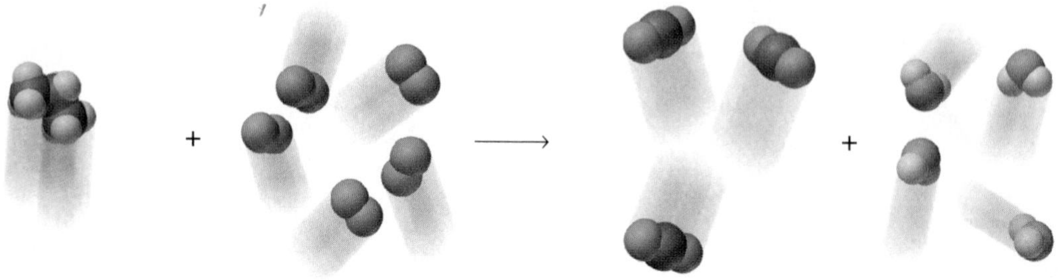 | | |
| Moles | 1 mol $C_3H_8$ + 5 mol $O_2$ | $\rightarrow$ | 3 mol $CO_2$ + 4 mol $H_2O$ |
| Mass (amu) | 44.09 amu $C_3H_8$ + 160.00 amu $O_2$ | $\rightarrow$ | 132.03 amu $CO_2$ + 72.06 amu $H_2O$ |
| Mass (g) | 44.09 g $C_3H_8$ + 160.00 g $O_2$ | $\rightarrow$ | 132.03 g $CO_2$ + 72.06 g $H_2O$ |
| Total mass (g) | 204.09 g | $\rightarrow$ | 204.09 g |

## 3.4  Calculating the Amounts of Reactant and Product

A balanced equation contains a wealth of quantitative information relating individual chemical entities (atoms, molecules, formula units), moles of substance, and masses. Consider Table 3.5, which presents various ways to view the balanced equation for the combustion of propane, an important hydrocarbon fuel used in domestic cooking and water heating:

$$C_3H_8(g) + 5O_2(g) \rightarrow 3CO_2(g) + 4H_2O(g)$$

Most important, a balanced equation is essential for all calculations involving amounts of reactants and products: if you know the number of moles of one substance, the balanced equation tells you the number of moles of all the others in the reaction.

### Using Molar Ratios in Solving Stoichiometry Problems

The quantitative, or stoichiometric, relationships in a balanced equation are expressed as *stoichiometrically equivalent molar ratios.* The term "stoichiometrically equivalent" means that a definite quantity of one substance is formed from, produces, or reacts with a definite quantity of another. In other words, *the number of moles of one substance is stoichiometrically equivalent to the number of moles of any other substance in a given reaction.* We use stoichiometrically equivalent molar ratios as conversion factors to determine how much of one substance forms from (or reacts with) another.

If, for example, we view the propane combustion reaction in terms of moles of $C_3H_8$, we see that

1 mol $C_3H_8$ reacts with 5 mol $O_2$
1 mol $C_3H_8$ produces 3 mol $CO_2$
1 mol $C_3H_8$ produces 4 mol $H_2O$

Therefore, in this reaction,

  1 mol $C_3H_8$ is stoichiometrically equivalent to 5 mol $O_2$
  1 mol $C_3H_8$ is stoichiometrically equivalent to 3 mol $CO_2$
  1 mol $C_3H_8$ is stoichiometrically equivalent to 4 mol $H_2O$

Moreover, any two of the substances are stoichiometrically equivalent to each other. Thus, 3 mol $CO_2$ is stoichiometrically equivalent to 4 mol $H_2O$, 5 mol $O_2$ is stoichiometrically equivalent to 3 mol $CO_2$, and so on.

The following problem is typical of those that chemists solve using conversion factors based on stoichiometric equivalence: In the combustion of propane, how many moles of $O_2$ are consumed when 10.0 mol $H_2O$ are produced? To solve this problem, we have to find the molar ratio between $O_2$ and $H_2O$. From the balanced equation, we see that for every 5 mol $O_2$ consumed, 4 mol $H_2O$ are formed:

$$5 \text{ mol } O_2 \text{ is stoichiometrically equivalent to } 4 \text{ mol } H_2O$$

We can construct two conversion factors from this equivalence depending on the quantity we are trying to find:

$$\frac{5 \text{ mol } O_2}{4 \text{ mol } H_2O} \quad \text{or} \quad \frac{4 \text{ mol } H_2O}{5 \text{ mol } O_2}$$

Since the moles of $O_2$ consumed are unknown and the moles of $H_2O$ produced are known, we use 5 mol $O_2$/4 mol $H_2O$, so that moles of $H_2O$ cancels:

$$\text{Moles of } O_2 \text{ consumed} = 10.0 \text{ mol } H_2O \times \frac{5 \text{ mol } O_2}{4 \text{ mol } H_2O} = 12.5 \text{ mol } O_2$$

Solving this problem depended on our knowing the molar ratios from the balanced equation. Even though a problem may not specifically ask for it, the first step in solving *any* problem that involves a chemical reaction is to *write the balanced equation.* Then do the following:

1. Convert the given mass (or number of entities) of the first substance to moles.
2. Use the appropriate molar ratio from the balanced equation to calculate the moles of the second substance.
3. Convert moles of the second substance to the desired mass (or number of entities).

This practical, quantitative approach is used frequently in the laboratory (and on chemistry exams!). It is shown diagrammatically in Figure 3.9 and demonstrated in the following sample problems.


SAMPLE PROBLEM 3.9 _____

**Using the Balanced Equation to Calculate the Amounts of Reactants and Products**

**Problem:** In a lifetime, the average American will use 1750 lb (794 kg) of copper in coins, plumbing, and wiring. Copper is obtained from sulfide ores, such as chalcocite, copper(I) sulfide, by a multistage process. After an initial grinding step, the first stage is to "roast" the ore (heat it strongly with oxygen gas) to form powdered copper(I) oxide and gaseous sulfur dioxide.
**(a)** How many moles of oxygen are required to roast 10.0 mol copper(I) sulfide?
**(b)** How many grams of sulfur dioxide are formed when 10.0 mol copper(I) sulfide is roasted?
**(c)** How many kilograms of oxygen are required to form 2.86 kg copper(I) oxide?

**FIGURE 3.9**

**Summary of the mass-mole-number relationships in a chemical reaction.** The amount of one substance in a reaction is related to that of any other. These amounts can be expressed in terms of mass, moles, or number of entities (atoms, molecules, or formula units). Start at any box in the diagram (known) and move to any other box (unknown) by using the information on the arrows as conversion factors. As an example, if you know the mass (in g) of A and want to know the number of molecules of B, the path involves three calculation steps:

1. Grams of A to moles of A, using the molar mass ($\mathcal{M}$) of A
2. Moles of A to moles of B, using the molar ratio from the balanced equation
3. Moles of B to molecules of B, using Avogadro's number

Steps 1 and 3 refer to calculations discussed in Section 3.1 (see Figure 3.5).

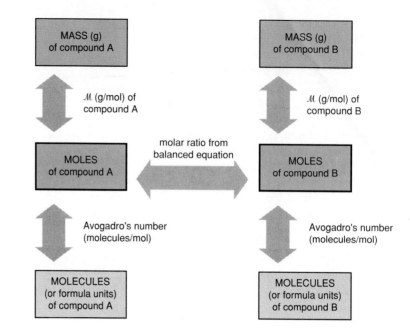

**(a)  Determining the moles of $O_2$ needed to react with 10.0 mol $Cu_2S$**

**Plan:** We *always* write the balanced equation first. The reactants are $Cu_2S$ and $O_2$, and the products are $Cu_2O$ and $SO_2$:

$$2Cu_2S(s) + 3O_2(g) \rightarrow 2Cu_2O(s) + 2SO_2(g)$$

We are given the *moles* of $Cu_2S$ and need to find the *moles* of $O_2$. The balanced equation shows that 3 mol $O_2$ is needed for every 2 mol $Cu_2S$ consumed, so the conversion factor is 3 mol $O_2$/2 mol $Cu_2S$.

**Solution:** Calculating moles of $O_2$:

$$\text{Moles of } O_2 = 10.0 \, \cancel{\text{mol } Cu_2S} \times \frac{3 \text{ mol } O_2}{2 \, \cancel{\text{mol } Cu_2S}} = \mathbf{15.0 \text{ mol } O_2}$$

**Check:** The answer is reasonable, since this $O_2$/$Cu_2S$ molar ratio (15:10) is the same as the ratio in the balanced equation (3:2).

**Comment:** A *common mistake* is to use the incorrect conversion factor; the calculation would then be

$$\text{Moles of } O_2 = 10.0 \text{ mol } Cu_2S \times \frac{2 \text{ mol } Cu_2S}{3 \text{ mol } O_2} = \frac{6.67 \text{ mol}^2 \, Cu_2S}{1 \text{ mol } O_2}$$

These strange units should signal that you made an error in setting up the conversion factor. In addition, the size of the answer, 6.67, is *less* than 10.0, whereas the balanced equation shows that *more* moles of $O_2$ than of $Cu_2S$ are needed. Be sure to *think through the calculation when setting up the conversion factor and keep track of units.*

**(b)  Determining the mass (g) of $SO_2$ formed from 10.0 mol $Cu_2S$**

**Plan:** Here we need the *mass* of product ($SO_2$) that forms from the given *moles* of reactant ($Cu_2S$). We first find the moles of $SO_2$ using the molar ratio from the balanced equation (2 mol $SO_2$/2 mol $Cu_2S$) and then convert to mass of $SO_2$ using its molar mass (64.07 g/mol). The steps appear in the roadmap.

**Solution:** Combining the two conversion steps into one calculation, we have

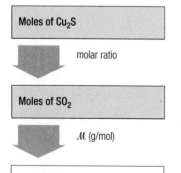

$$\text{Mass (g) of } SO_2 = 10.0 \, \cancel{\text{mol } Cu_2S} \times \frac{2 \, \cancel{\text{mol } SO_2}}{2 \, \cancel{\text{mol } Cu_2S}} \times \frac{64.07 \text{ g } SO_2}{1 \, \cancel{\text{mol } SO_2}} = \mathbf{641 \text{ g } SO_2}$$

**Check:** The answer makes sense, since the molar ratio shows that 10.0 mol $SO_2$ are formed and each weighs about 64 g. We rounded to three significant figures.

**(c) Determining the mass (kg) of $O_2$ that yields 2.86 kg $Cu_2O$**

**Plan:** Here the mass of product ($Cu_2O$) is known, and we need the mass of reactant ($O_2$) that reacts to form it. We first convert the amount of $Cu_2O$ from *mass* (kg) to *moles* (in two steps, as shown on the roadmap). Then, we use the molar ratio (3 mol $O_2$/2 mol $Cu_2O$) to find the *moles* of $O_2$ required. Finally, we convert *moles* of $O_2$ to *mass* (kg) (in two steps).

**Solution:** Converting from mass of $Cu_2O$ to moles of $Cu_2O$: Combining the mass conversion with the change to moles gives

$$\text{Moles of } Cu_2O = 2.86 \text{ kg } Cu_2O \times \frac{10^3 \text{ g}}{1 \text{ kg}} \times \frac{1 \text{ mol } Cu_2O}{143.10 \text{ g } Cu_2O}$$

$$= 20.0 \text{ mol } Cu_2O$$

Converting from moles of $Cu_2O$ to moles of $O_2$:

$$\text{Moles of } O_2 = 20.0 \text{ mol } Cu_2O \times \frac{3 \text{ mol } O_2}{2 \text{ mol } Cu_2O}$$

$$= 30.0 \text{ mol } O_2$$

Converting from moles of $O_2$ to mass of $O_2$: Combining the change to mass with the mass unit conversion gives

$$\text{Mass (kg) of } O_2 = 30.0 \text{ mol } O_2 \times \frac{32.00 \text{ g } O_2}{1 \text{ mol } O_2} \times \frac{1 \text{ kg}}{10^3 \text{ g}}$$

$$= \textbf{0.960 kg } O_2$$

**Check:** The units are correct. The answer seems OK: even though moles of $O_2$ are greater than moles of $Cu_2O$, the mass of $O_2$ is less than the mass of $Cu_2O$ because the molar mass of $O_2$ is less than the molar mass of $Cu_2O$.

**Comment:** This problem highlights a key point for solving stoichiometry problems: *convert the information given into moles*. Then, use the appropriate molar ratio and any other conversion factors to complete the problem.

**FOLLOW-UP PROBLEM 3.9**
Thermite is a mixture of iron(III) oxide and aluminum powders that was once used to weld railroad tracks. It undergoes a spectacular reaction to yield solid aluminum oxide and molten iron.
**(a)** How many grams of iron form when 135 g aluminum react?
**(b)** How many atoms of aluminum react for every 1.00 g aluminum oxide that forms?

---

## Chemical Reactions That Occur in a Sequence

In many situations, a product of one reaction becomes a reactant of the next in a sequence of reactions. For stoichiometric purposes, when a substance forms in one reaction and is used up in the next, we can write an **overall equation** that eliminates the substance altogether. The sequence of reactions is adjusted arithmetically so that the common substance "cancels out" and the reaction sequence is summed to the overall equation. In Sample Problem 3.9, for example, we looked only at the first step in the roasting of copper(I) sulfide. The second step in the process forms copper metal.

Mass (kg) of $Cu_2O$

1 kg = $10^3$ g

Mass (g) of $Cu_2O$

$\mathcal{M}$ (g/mol)

Moles of $Cu_2O$

molar ratio

Moles of $O_2$

$\mathcal{M}$ (g/mol)

Mass (g) of $O_2$

$10^3$ g = 1 kg

Mass (kg) of $O_2$

SAMPLE PROBLEM 3.10 _____

### Calculating the Amounts of Reactants and Products in a Reaction Sequence

**Problem:** Roasting chalcocite is the first step in extracting copper from this sulfide ore:

$$2Cu_2S(s) + 3O_2(g) \rightarrow 2Cu_2O(s) + 2SO_2(g) \text{ [equation 1; see Sample Problem 3.9(a)]}$$

In the next step, copper(I) oxide reacts with powdered carbon to yield copper metal and carbon monoxide gas:

$$Cu_2O(s) + C(s) \rightarrow 2Cu(s) + CO(g) \text{ [equation 2]}$$

**(a)** Write a balanced overall equation for the two-step sequence.
**(b)** How many kilograms of copper can be extracted for every 1.00 metric ton of $SO_2$ that forms? [1 metric ton (t) = 1000 kg]

### (a) Writing and balancing the overall equation
**Plan:** To obtain the overall equation, we write the individual equations in sequence, adjust coefficients to cancel the common substance (or substances), and add the equations together. In this case, only $Cu_2O$ appears as a product in one equation and as a reactant in the other, so it is the common substance.
**Solution:** Adjusting the coefficients: Since 2 mol $Cu_2O$ are produced in equation 1 but only 1 mol $Cu_2O$ reacts in equation 2, we double *all* the coefficients in equation 2. Thus, the $Cu_2O$ from equation 1 is used up in equation 2:

$$2Cu_2S(s) + 3O_2(g) \rightarrow 2Cu_2O(s) + 2SO_2(g) \text{ [equation 1]}$$

$$2Cu_2O(s) + 2C(s) \rightarrow 4Cu(s) + 2CO(g) \text{ [equation 2 doubled]}$$

Adding the two equations and canceling the common substance: We keep the reactants from both equations on the left and the products from both on the right:

$$2Cu_2S(s) + 3O_2(g) + \cancel{2Cu_2O(s)} + 2C(s) \rightarrow \cancel{2Cu_2O(s)} + 2SO_2(g) + 4Cu(s) + 2CO(g)$$

Or,  $$\mathbf{2Cu_2S(s) + 3O_2(g) + 2C(s) \rightarrow 2SO_2(g) + 4Cu(s) + 2CO(g)}$$

**Check:** Reactants (4 Cu, 2 S, 6 O, 2 C) $\rightarrow$ products (4 Cu, 2 S, 6 O, 2 C)
**Comment:** Even though $Cu_2O$ participates in the chemical change, it is not involved in the following calculation. An overall equation *may not* show which of the substances actually react; C(s) and $Cu_2S(s)$ do not interact directly here, even though both are reactants. The overall equation *does* show the relative net amounts of substances in the process. You'll see that much chemistry is "hidden under the yield arrow" when we examine stepwise reactions in Chapter 15.

### (b) Determining the mass (kg) of copper formed per metric ton of $SO_2$ formed
**Plan:** This part is similar to Sample Problem 3.9. We convert the mass of $SO_2$ to moles (in three steps), apply the molar ratio from the overall equation to find moles of Cu, and then convert moles to mass (kg) of Cu.
**Solution:** Converting from mass of $SO_2$ to moles of $SO_2$:

$$\text{Moles of } SO_2 = 1.00 \, \cancel{t \, SO_2} \times \frac{10^3 \, \cancel{kg}}{1 \, \cancel{t}} \times \frac{10^3 \, \cancel{g}}{1 \, \cancel{kg}} \times \frac{1 \, mol \, SO_2}{64.07 \, \cancel{g \, SO_2}}$$

$$= 1.56 \times 10^4 \, mol \, SO_2$$

Converting from moles of $SO_2$ to moles of Cu:

$$\text{Moles of Cu} = 1.56 \times 10^4 \, \cancel{mol \, SO_2} \times \frac{4 \, mol \, Cu}{2 \, \cancel{mol \, SO_2}} = 3.12 \times 10^4 \, mol \, Cu$$

Converting from moles of Cu to mass of Cu:

$$\text{Mass (kg) of Cu} = 3.12 \times 10^4 \, \cancel{mol \, Cu} \times \frac{63.55 \, \cancel{g} \, Cu}{1 \, \cancel{mol \, Cu}} \times \frac{1 \, kg}{10^3 \, \cancel{g}} = \mathbf{1.98 \times 10^3 \, kg \, Cu}$$

**Check:** Since there are twice as many moles of Cu as moles of $SO_2$ and the molar mass of $SO_2$ is very close to the molar mass of Cu, the mass of Cu ( ~2000 kg) should be about twice the mass of $SO_2$ (1000 kg).

**Comment:** The $SO_2$ formed in metal extraction contributes to acid rain (see Chemical Connections, Chapter 1). To help control the problem, chemists have devised microbial and electrochemical methods that avoid the roasting of sulfide ores to obtain copper and other metals.

FOLLOW-UP PROBLEM 3.10

The $SO_2$ produced in obtaining copper reacts in air with oxygen and forms sulfur trioxide. The $SO_3$, in turn, reacts with water to form a sulfuric acid solution that falls as rain or snow. Assuming all the $SO_2$ formed eventually falls as acid rain, how many grams of sulfuric acid form for every kilogram of copper that is extracted?

Multistep reaction sequences called *metabolic pathways* are common in biological systems. In a sequence that occurs in most cells, the chemical energy in glucose is released through a series of about 30 individual reactions. The product of each reaction is the reactant of the next reaction, so that all the intermediate substances cancel. The overall equation is

$$C_6H_{12}O_6(aq) + 6O_2(g) \rightarrow 6CO_2(g) + 6H_2O(l)$$

We eat food that contains the glucose, inhale the $O_2$, and excrete the $CO_2$ and $H_2O$. In our cells, these reactants and products are many steps apart; that is, $O_2$ never reacts *directly* with glucose. However, the molar ratios are the same as if the glucose were burned in a combustion apparatus filled with pure $O_2$ and formed $CO_2$ and $H_2O$ directly.

## Chemical Reactions Involving a Limiting Reactant

In previous problems, the amount of only one reactant was given, and we assumed it reacted with as much of another substance as needed for it to be completely consumed. For example, to find the amount of $SO_2$ that forms when 100 g $Cu_2S$ reacts, we convert the mass of $Cu_2S$ to moles and assume it reacts with as much $O_2$ as needed. Since all the $Cu_2S$ is used up, it determines, or limits, how much $SO_2$ can form. In this case, $Cu_2S$ is the **limiting reactant** (or *limiting reagent*) because the reaction stops once the $Cu_2S$ is gone, no matter how much $O_2$ is present. Suppose, however, that we are given the amounts of both $Cu_2S$ *and* $O_2$ and want to know how much $SO_2$ forms. We must first determine whether $Cu_2S$ or $O_2$ is the limiting reactant because *its* amount limits how much $SO_2$ can form. The other reactant is present *in excess*, and the amount that is not used in the reaction is left over.

To clarify the idea of limiting reactant, let's consider a much more appetizing situation. Suppose you have a job making sundaes in an old-fashioned ice cream parlor, and each sundae requires two scoops of ice cream, one cherry, and 50 mL of chocolate syrup:

$$2 \text{ scoops} + 1 \text{ cherry} + 50 \text{ mL syrup} \rightarrow 1 \text{ sundae}$$

A mob of 25 ravenous school kids enters; can you feed them all? You have 50 scoops of ice cream, 30 cherries, and 1 L syrup; a quick calculation shows

FIGURE 3.10

An ice cream sundae analogy for limiting reactants. **A,** The "components" combine in specific amounts to form a sundae. **B,** In this example, the number of sundaes possible is limited by the amount of syrup, the limiting "reactant." Here, only two sundaes can be made. Four scoops of ice cream and four cherries remain "in excess."

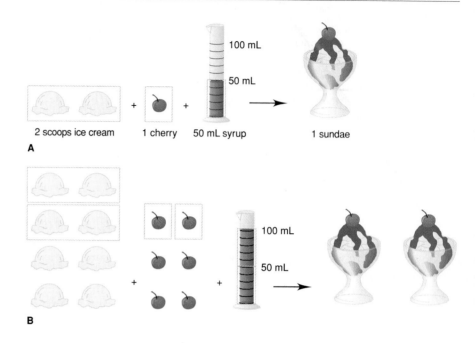

the number of sundaes you can make from each ingredient:

$$\text{Ice cream: No. of sundaes} = 50 \text{ scoops} \times \frac{1 \text{ sundae}}{2 \text{ scoops}} = 25 \text{ sundaes}$$

$$\text{Cherries: No. of sundaes} = 30 \text{ cherries} \times \frac{1 \text{ sundae}}{1 \text{ cherry}} = 30 \text{ sundaes}$$

$$\text{Syrup: No. of sundaes} = 1000 \text{ mL syrup} \times \frac{1 \text{ sundae}}{50 \text{ mL syrup}} = 20 \text{ sundaes}$$

The syrup is the limiting "reactant" here because it limits the total amount of "product" that can "form": of the three ingredients, the syrup allowed the *fewest* sundaes to be made (Figure 3.10). Some ice cream and cherries are left "unreacted" when all the syrup is gone, so they are present in excess:

50 scoops + 30 cherries + 1 L syrup → **20 sundaes** + 10 scoops + 10 cherries

In limiting-reactant problems, the amounts of two (or more) reactants are given, and we must first determine which is limiting. In practice, *we calculate the reactant that yields the least amount of product.* The simplest approach, illustrated in Sample Problem 3.11, is to work through two stoichiometry problems, each of which assumes an excess of one or the other reactant.

SAMPLE PROBLEM 3.11

### Calculating the Amounts of Reactant and Product in Reactions Involving a Limiting Reactant

**Problem:** A fuel mixture used in the early days of rocketry is composed of two liquids, hydrazine ($N_2H_4$) and dinitrogen tetraoxide ($N_2O_4$). They ignite on contact to form nitrogen gas and water vapor. How many grams of nitrogen gas form when exactly $1.00 \times 10^2$ g $N_2H_4$ and $2.00 \times 10^2$ g $N_2O_4$ are mixed?

**Plan:** As always, we first write the balanced equation. *The amounts of two reactants are given, so this is a limiting-reactant problem.* To determine which is limiting, we calculate the mass of $N_2$ formed from each reactant *assuming an excess of the other.* We convert the mass of each reactant to moles and find the moles of $N_2$ each forms from its molar ratio. Whichever yields *less* $N_2$ is the limiting reactant. Then, we convert this lower number of moles of $N_2$ to mass.

**Solution:** Writing the balanced equation:

$$2N_2H_4(l) + N_2O_4(l) \rightarrow 3N_2(g) + 4H_2O(g)$$

Finding the moles of $N_2$ assuming $N_2H_4$ is limiting:

$$\text{Moles of } N_2H_4 = 1.00 \times 10^2 \text{ g N}_2\text{H}_4 \times \frac{1 \text{ mol } N_2H_4}{32.05 \text{ g N}_2\text{H}_4} = 3.12 \text{ mol } N_2H_4$$

$$\text{Moles of } N_2 = 3.12 \text{ mol N}_2\text{H}_4 \times \frac{3 \text{ mol } N_2}{2 \text{ mol N}_2\text{H}_4} = 4.68 \text{ mol } N_2$$

Finding the moles of $N_2$ assuming $N_2O_4$ is limiting:

$$\text{Moles of } N_2O_4 = 2.00 \times 10^2 \text{ g N}_2\text{O}_4 \times \frac{1 \text{ mol } N_2O_4}{92.02 \text{ g N}_2\text{O}_4} = 2.17 \text{ mol } N_2O_4$$

$$\text{Moles of } N_2 = 2.17 \text{ mol N}_2\text{O}_4 \times \frac{3 \text{ mol } N_2}{1 \text{ mol N}_2\text{O}_4} = 6.51 \text{ mol } N_2$$

Thus, $N_2H_4$ is the limiting reactant because less $N_2$ forms when all the $N_2H_4$ reacts. Converting from moles of $N_2$ to mass:

$$\text{Mass (g) of } N_2 = 4.68 \text{ mol N}_2 \times \frac{28.02 \text{ g } N_2}{1 \text{ mol N}_2} = \textbf{131 g } \mathbf{N_2}$$

**Check:** Even though the mass of $N_2O_4$ is greater than that of $N_2H_4$, there are fewer moles because the molar mass of $N_2O_4$ is much higher. Round off to check the values; for $N_2H_4$, for example,

$$100 \text{ g} \times 1 \text{ mol}/32 \text{ g} \approx 3 \text{ mol, and } \sim 3 \text{ mol} \times \tfrac{3}{2} = 4.5 \text{ mol}$$

$$\sim 4.5 \text{ mol} \times 30 \text{ g/mol} = 135 \text{ g}$$

**Comment:** (1) A *common mistake* is to choose the *reactant* that is present in fewer moles as the limiting one (2.17 mol $N_2O_4$ vs. 3.12 mol $N_2H_4$). Instead, you use the molar ratios (3 mol $N_2$/1 mol $N_2O_4$ vs. 3 mol $N_2$/2 mol $N_2H_4$) to see which reactant gives fewer moles of *product.*

(2) A good *alternative approach* for choosing the limiting reactant is to find the number of moles of each reactant that is needed to react with the other and compare it with the given amounts. The balanced equation shows that 2 mol $N_2H_4$ react with 1 mol $N_2O_4$. Thus, the moles of $N_2O_4$ needed for the given $N_2H_4$ are

$$\text{Moles of } N_2O_4 \text{ needed} = 3.12 \text{ mol N}_2\text{H}_4 \times \frac{1 \text{ mol } N_2O_4}{2 \text{ mol N}_2\text{H}_4} = 1.56 \text{ mol } N_2O_4$$

Similarly, the moles of $N_2H_4$ needed for the given $N_2O_4$ is

$$\text{Moles of } N_2H_4 \text{ needed} = 2.17 \text{ mol N}_2\text{O}_4 \times \frac{2 \text{ mol } N_2H_4}{1 \text{ mol N}_2\text{O}_4} = 4.34 \text{ mol } N_2H_4$$

Since we have more than 1.56 mol $N_2O_4$ present but fewer than 4.34 mol $N_2H_4$, the $N_2O_4$ is in excess and the $N_2H_4$ is limiting. Then, we continue with a single calculation to find the amount of $N_2$ formed.

**FOLLOW-UP PROBLEM 3.11**
How many grams of solid aluminum sulfide can be prepared by the reaction of 10.0 g aluminum and 15.0 g sulfur? How much of the nonlimiting reactant is in excess?

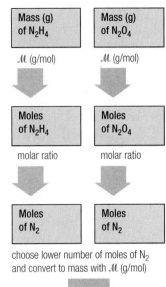

## Chemical Reactions in Practice: Theoretical, Actual, and Percent Yields

Up until now, we have adopted an optimistic attitude about the amount of product obtained from a reaction. We have assumed that 100% of the limiting reactant becomes product, that ideal separation and purification methods exist with which to isolate the product, and that we use perfect technique to obtain all the product formed. In other words, we have assumed that the **theoretical yield,** the amount indicated by the stoichiometrically equivalent molar ratio in the balanced equation, is obtainable.

It is time to face reality. The theoretical yield is *never* obtained, for reasons that are largely uncontrollable. For one thing, although the major reaction predominates, many reactant mixtures also proceed through **side reactions** that form alternative products. The amount of reactant diverted along side reactions varies with the process. In the previous rocket fuel reaction, for example, some NO might form in the following side reaction:

$$2N_2O_4(l) + N_2H_4(l) \longrightarrow 6NO(g) + 2H_2O(g)$$

This reaction would decrease the amounts of reactants available for $N_2$ production (see Problem 3.85 at the end of the chapter). Perhaps even more important, as we'll discuss in Chapter 4, many reactions seem to stop before they are complete, leaving some limiting reactant unreacted. Furthermore, even when a reaction does go to completion, some losses occur in virtually every step of a separation procedure (see Tools of the Chemistry Laboratory, Section 2.8): a tiny amount of product clings to filter paper, some distillate evaporates, a small amount of extract remains in the separatory funnel, and so forth. With careful technique, such losses can be minimized but never eliminated. The amount of product that is actually obtained is the **actual yield.** The **percent yield** (% yield) is the actual yield expressed as a percent of the theoretical yield:

$$\% \text{ yield} = \frac{\text{actual yield}}{\text{theoretical yield}} \times 100 \qquad \textbf{(3.7)}$$

Since the actual yield is always less than the theoretical yield, the percent yield is always less than 100%.

SAMPLE PROBLEM 3.12 _____

## Calculating Percent Yield

**Problem:** Silicon carbide (SiC), an important ceramic material, is manufactured by allowing sand (silicon dioxide) to react with powdered carbon at high temperature. Carbon monoxide is also formed. When 100.0 kg sand are processed, 51.4 kg SiC are recovered. What is the percent yield of SiC?

**Plan:** We are given the actual yield of SiC, so we need its theoretical yield to calculate its percent yield. After writing the balanced equation, we convert the given mass of sand ($SiO_2$) to moles, find the moles of SiC from the molar ratio, and convert moles of SiC to mass to obtain the theoretical yield [see roadmap for Sample Problem 3.9(c)]. Then, we apply Equation 3.7 to find the percent yield.

**Solution:** Writing the balanced equation:

$$SiO_2(s) + 3C(s) \longrightarrow SiC(s) + 2CO(g)$$

Converting from mass of $SiO_2$ to moles:

$$\text{Moles of } SiO_2 = 100.0 \text{ kg } SiO_2 \times \frac{1000 \text{ g}}{1 \text{ kg}} \times \frac{1 \text{ mol } SiO_2}{60.09 \text{ g } SiO_2} = 1664 \text{ mol } SiO_2$$

Converting from moles of $SiO_2$ to moles of SiC: The molar ratio is 1 mol SiC/1 mol $SiO_2$, so

$$\text{Moles of } SiO_2 = \text{moles of SiC} = 1664 \text{ mol SiC}$$

Converting from moles of SiC to mass:

$$\text{Mass (kg) of SiC} = 1664 \text{ mol SiC} \times \frac{40.10 \text{ g SiC}}{1 \text{ mol SiC}} \times \frac{1 \text{ kg}}{1000 \text{ g}} = 66.73 \text{ kg SiC}$$

Calculating the percent yield:

$$\text{\% yield SiC} = \frac{\text{actual yield}}{\text{theoretical yield}} \times 100 = \frac{51.4 \text{ kg}}{66.73 \text{ kg}} \times 100 = \textbf{77.0\%}$$

**Check:** The mass of SiC seems correct: ~1500 mol $\times$ 40 g/mol $\times$ 1 kg/1000 g = 60 kg. The molar ratio of $SiC/SiO_2$ is 1/1, and the molar mass of SiC is about two-thirds (40/60) the molar mass of $SiO_2$, so 100 kg $SiO_2$ should form about 66 kg SiC.

**FOLLOW-UP PROBLEM 3.12**
Marble (calcium carbonate) reacts with hydrochloric acid solution to form calcium chloride solution, water, and carbon dioxide. What is the percent yield of carbon dioxide if 3.65 g of the gas is collected when 10.0 g of marble reacts?

---

In multistep reaction sequences, the percent yield of each reaction step is multiplied over the whole sequence. Thus, even a series of high-yield reaction steps can result in a low actual yield for the overall reaction. Suppose a six-step sequence has a theoretical yield of 35.0 g of final product. Even if each step has a 90.0% yield, you would wind up with a much lower overall actual yield:

$$\text{Actual yield} = 35.0 \text{ g} \times 0.900 \times 0.900 \times 0.900 \times 0.900 \times 0.900 \times 0.900 = 18.6 \text{ g}$$

$$\text{\% yield} = \frac{18.6 \text{ g}}{35.0 \text{ g}} \times 100 = 53.1\%$$

Reaction sequences with many steps are common in the synthesis of organic compounds (drugs, dyes, pesticides, etc.), in which large amounts of inexpensive, simple reactants are converted to small amounts of expensive, complex products.

**Section Summary**
The substances in a balanced equation are related to each other by stoichiometrically equivalent molar ratios, which can be used as conversion factors to find the moles of one substance given the moles of another. In limiting-reactant problems, the amounts of two reactants are given, and one of them limits the amount of product that can form. In practice, side reactions, incomplete reactions, and physical losses result in less than the theoretical yield of product, the amount based solely on the molar ratio.

# 3.5   **Fundamentals of Solution Stoichiometry**

The stereotype of a chemist is a person wearing a white laboratory coat and safety glasses, surrounded by walls of glassware, carefully pouring one colored solution into another, amid raging bubbles and billowing fumes. Although most reactions in solution are not this dramatic and good technique provides for safer mixing procedures, the image is true to the extent that aqueous solution chemistry is a central part of chemical analysis. Liquid solutions are more convenient to store and mix than solids or gases, and the amounts of substances in solution can be measured very precisely. Since many environmental reactions and almost all biochemical reactions occur in solution, quantitative understanding of reactions in solution are of utmost importance in chemistry and related sciences.

One aspect of the stoichiometry of substances reacting in solution is different from what you've seen so far. We know the number of moles of pure substances by converting their masses directly into moles. When dissolved substances react, however, we must know the *concentration* of the reactant— the number of moles present in a particular volume—to calculate the volume of solution that contains a given number of moles. Several ways exist to express solution concentration, but the most important is *molarity.* (Chapter 12 covers other methods of expressing solution concentration.) First, we see how to prepare a solution of a specified molarity, and then we use solutions in stoichiometric calculations.

### Solution Concentration and the Calculation of Molarity

The simplest solution consists, in general, of a smaller amount of one substance, the **solute,** dissolved in a larger amount of another substance, the **solvent.** When a solution forms, the solute's individual chemical units become *evenly dispersed* throughout the available volume and surrounded by solvent molecules. This means that the solution **concentration,** typically *the amount of solute dissolved in a given amount of solution,* is independent of the total amount of solution: a 50-L tank of a solution has the *same concentration* (solute amount/solution amount) as a 50-mL beaker of the solution. **Molarity *(M)*** expresses the concentration in terms of *moles of solute per liter of solution:*

$$\text{Molarity} = \frac{\text{moles of solute}}{\text{liters of solution}} \quad \text{or} \quad M = \frac{\text{mol solute}}{\text{L soln}} \qquad \textbf{(3.8)}$$

---

**SAMPLE PROBLEM 3.13** _____

### Calculating the Molarity of a Solution

**Problem:** Hydrochloric acid (HCl), one of the common laboratory acids, is a solution of hydrogen chloride gas in water. Calculate the molarity of hydrochloric acid solution if 455 mL contains 1.80 mol hydrogen chloride.

**Plan:** The molarity is the number of moles of solute in each liter of solution. We are given the number of moles and the volume (in mL), so we divide moles by volume and convert the volume to liters to find the molarity.

**Solution:**

$$\text{Molarity} = \frac{1.80 \text{ mol HCl}}{455 \text{ mL soln}} \times \frac{1000 \text{ mL}}{1 \text{ L}} = \textbf{3.96 } \textit{M} \textbf{ HCl}$$

**Check:** There are almost 2 mol HCl and about 0.5 L solution, so the concentration should be about 4 mol/L, or 4 *M*.

**FOLLOW-UP PROBLEM 3.13**
How many moles of KI are in 84 mL of 0.50 *M* KI?

Moles of HCl

divide by volume (mL)

Concentration (mol/mL) of HCl

$10^3$ mL = 1 L

Molarity (mol/L) of HCl

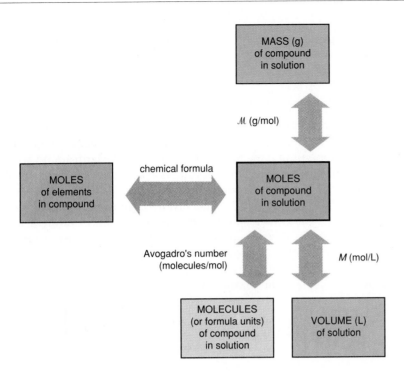

**FIGURE 3.11**
**Summary of the mass-mole-number-volume relationships in solution.** Moles of a compound in solution are related to the volume of solution in liters through the molarity (*M*) in moles per liter. The other relationships shown are identical to those in Figure 3.5, except that here they refer to the amounts *in solution*. As in previous cases, to find the amount of substance expressed in one form or another, convert the given information to moles first.

Molarity can be thought of as another conversion factor that extends the stoichiometric relationships of substances. We use it to convert between volume of solution and moles of solute, from which we find the mass or number of entities of solute (Figure 3.11).

---

SAMPLE PROBLEM 3.14

## Calculating the Mass of Solute in a Given Volume of Solution

**Problem:** How many grams of solute are in 1.75 L of 0.460 *M* sodium monohydrogen phosphate?

**Plan:** We are given the solution volume and molarity, so we can find moles of solute. Then, we convert moles to mass using the solute molar mass.

**Solution:** Calculating moles of solute in solution:

$$\text{Moles of Na}_2\text{HPO}_4 = 1.75 \; \cancel{\text{L soln}} \times \frac{0.460 \; \text{mol Na}_2\text{HPO}_4}{1 \; \cancel{\text{L soln}}}$$

$$= 0.805 \; \text{mol Na}_2\text{HPO}_4$$

Converting from moles of solute to mass:

$$\text{Mass (g) Na}_2\text{HPO}_4 = 0.805 \; \cancel{\text{mol Na}_2\text{HPO}_4} \times \frac{141.96 \; \text{g Na}_2\text{HPO}_4}{1 \; \cancel{\text{mol Na}_2\text{HPO}_4}}$$

$$= \textbf{114 g Na}_2\textbf{HPO}_4$$

**Check:** The answer seems correct: ~ 1.5 L of 0.5 mol/L should contain ~ 0.75 mol, and ~ 150 g/mol × 0.75 mol = 112 g, very close to 114 g solute.

**Comment:** We can also use molarity to find the volume of solution that contains a given mass of solute, as in the follow-up problem.

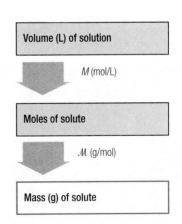

**FOLLOW-UP PROBLEM 3.14**

Concentrated solutions of sucrose (table sugar; $C_{12}H_{22}O_{11}$) are used in high-speed centrifuges to separate the parts of a cell. How many liters of 3.30 *M* sucrose contain 135 g solute?

## Laboratory Preparation of Molar Solutions

When you are preparing a solution of a specific molarity, remember that the volume term in the denominator is the *solution* volume, *not* the solvent volume. The solution volume includes contributions from both solute and solvent, so you cannot dissolve 1 mol solute in 1 L solvent and expect a 1 *M* solution to result. The added solute generally increases the solution's total volume slightly above 1 L, resulting in a lower concentration than expected. The correct preparation of a solution containing a solid solute consists of four steps. Let's go through them to prepare 0.500 L of 0.350 *M* nickel(II) nitrate hexahydrate [$Ni(NO_3)_2 \cdot 6H_2O$]:

1. *Weigh the solid on weighing paper.* First, we calculate the mass of solid needed by converting from volume to moles and from moles to mass:

$$\text{Mass (g) of solute} = 0.500 \; \text{L soln} \times \frac{0.350 \; \text{mol Ni(NO}_3)_2 \cdot 6H_2O}{1 \; \text{L soln}}$$

$$\times \frac{290.83 \; \text{g Ni(NO}_3)_2 \cdot 6H_2O}{1 \; \text{mol Ni(NO}_3)_2 \cdot 6H_2O} = 50.9 \; \text{g Ni(NO}_3)_2 \cdot 6H_2O$$

2. *Carefully transfer the solid to a volumetric flask that contains about half the final amount of solvent.* Since we need 0.500 L solution, a 500-mL volumetric flask is the best choice of glassware. We add about 250 mL distilled water and then transfer the solid. Any pieces of solid clinging to the neck are then washed down with a small amount of solvent.

3. *Dissolve the solid thoroughly by swirling.* If some solute remains undissolved, the solution will be less concentrated than expected. If necessary, wait for the solution to reach room temperature.

4. *Add solvent until the solution reaches its final volume.* We add distilled water to bring the solution exactly to the line on the flask neck, then cover and mix thoroughly again. Figure 3.12 shows the last three steps.

When lower concentrations of a solution are used frequently, it is common practice to prepare a more concentrated *stock* solution, which can be stored and diluted. When a solution is diluted, *only solvent is added,* so the solute becomes dispersed in a larger total volume. Thus, a given volume contains fewer solute particles and a lower concentration (Figure 3.13).

**FIGURE 3.12**

**Laboratory preparation of molar solutions.** After the desired amount of solid has been weighed out, the solution is prepared by **A,** carefully adding the solid to a volumetric flask about half full of solvent; **B,** swirling to dissolve the solid completely; and **C,** adding solvent to the mark on the flask neck (see *close-up, far right*). It would *not* be correct to add the solid to the entire volume of solvent, since the total volume would exceed the desired volume and result in a lower-than-calculated concentration.

A

B

C

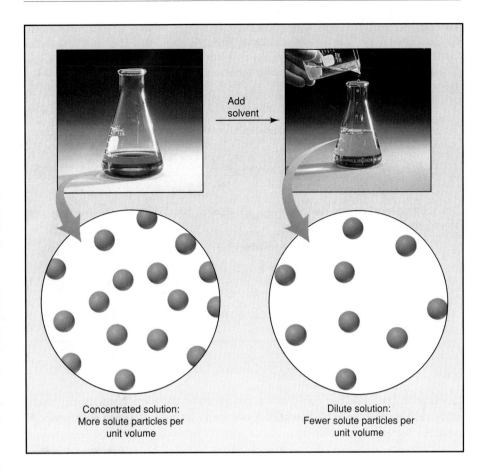

Concentrated solution:
More solute particles per
unit volume

Dilute solution:
Fewer solute particles per
unit volume

**FIGURE 3.13**
**Converting a concentrated solution to a dilute solution.** When a solution is diluted, only solvent is added**.** The solution volume increases while the number of total moles of solute remains the same. As shown in the expanded views, a unit volume of concentrated solution contains more solute particles than a unit volume of dilute solution.

SAMPLE PROBLEM 3.15 _____

## Converting a Concentrated Solution to a Dilute Solution

**Problem:** Isotonic saline is a 0.15 *M* aqueous solution of NaCl that simulates the total concentration of ions found in many cellular fluids. Its uses range from a cleansing rinse for contact lenses to a washing medium for red blood cells. How would you prepare 0.800 L isotonic saline from a 6.0 *M* stock solution?

**Plan:** To dilute a concentrated solution, we add only solvent, so the *total moles of solute is the same in both solutions.* From the given volume and molarity of dilute (dil) NaCl, we find the moles of NaCl it contains and then calculate the volume of concentrated (conc) NaCl that contains that number of moles. Then we dilute this volume with pure solvent *up to* the final volume.

**Solution:** Finding moles of solute in dilute solution:

$$\text{Moles of NaCl in dil soln} = 0.800 \ \cancel{\text{L soln}} \times \frac{0.15 \ \text{mol NaCl}}{1 \ \cancel{\text{L soln}}} = 0.12 \ \text{mol NaCl}$$

Since we add only solvent to dilute the solution,

$$\text{Moles of NaCl in dil soln} = \text{moles of NaCl in conc soln} = 0.12 \ \text{mol NaCl}$$

Finding the volume of concentrated solution that contains 0.12 mol NaCl:

$$\text{Volume (L) of conc NaCl soln} = 0.12 \ \cancel{\text{mol NaCl}} \times \frac{1 \ \text{L soln}}{6.0 \ \cancel{\text{mol NaCl}}} = 0.020 \ \text{L soln}$$

To prepare 0.800 L dilute solution, **place 0.020 L of 6.0 *M* NaCl in a 1-L flask, add distilled water ( ~780 mL) to the 0.800-L mark, and stir thoroughly.**

Volume (L) of dilute solution

*M* (mol/L) of
dilute solution

Moles of NaCl in dilute solution
= moles of NaCl in
concentrated solution

*M* (mol/L) of
concentrated solution

Volume (L) of
concentrated solution

**Check:** The answer seems reasonable because a small volume of concentrated solution is used to prepare a large volume of dilute solution. Also, the ratio of volumes (0.020 L/0.800 L) is the same as the ratio of concentrations (0.15 $M$/6.0 $M$).

**Comment:** An *alternative approach* to solving dilution problems makes use of the formula

$$M_{dil}V_{dil} = \text{moles} = M_{conc}V_{conc} \qquad \textbf{(3.9)}$$

where the $M$ and $V$ terms are the molarity and volume of the *dil*ute and *conc*entrated solutions. Solving for $V_{conc}$ gives

$$V_{conc} = \frac{M_{dil} \times V_{dil}}{M_{conc}} = \frac{0.15 \,\cancel{M} \times 0.800 \text{ L}}{6.0 \,\cancel{M}} = 0.020 \text{ L}$$

The method worked out in the solution is the same calculation broken into two parts to emphasize the thinking process:

$$V_{conc} = 0.800 \,\cancel{\text{L}} \times \frac{0.15 \,\cancel{\text{mol NaCl}}}{1 \,\cancel{\text{L}}} \times \frac{1 \text{ L}}{6.0 \,\cancel{\text{mol NaCl}}} = 0.020 \text{ L}$$

**FOLLOW-UP PROBLEM 3.15**

If 25.0 mL of 7.50 $M$ sulfuric acid is diluted to exactly 500 mL, what is the mass of sulfuric acid per milliliter?

## Chemical Reactions in Solution

Solving stoichiometry problems for reactions in solution requires the same approach as before, with the additional step of converting the solution volume to moles of reactant or product it contains: balance the equation, find the number of moles of one substance, relate it to the stoichiometrically equivalent number of moles of another substance, and convert to the desired units. In limiting-reactant problems for reactions in solution, we first determine which reactant is limiting and then determine the yield.

**SAMPLE PROBLEM 3.16** _____

### Calculating Amounts of Reactants and Products for a Reaction in Solution

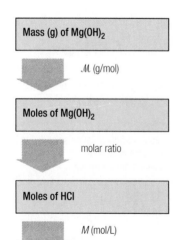

**Problem:** Specialized cells in the stomach release HCl to aid digestion. If they occasionally release too much, the excess can be neutralized with antacids. A common antacid agent is magnesium hydroxide, which reacts with the acid to form water and magnesium chloride solution. As a government chemist testing commercial antacids, you use 0.10 $M$ HCl to simulate the concentration in the stomach. How many liters of simulated stomach acid react with a tablet containing 0.10 g magnesium hydroxide?

**Plan:** We know the mass of $Mg(OH)_2$ that reacts with the acid and the acid concentration, and we must find the acid volume. After writing the balanced equation, we convert the mass of $Mg(OH)_2$ to moles, use the molar ratio to find the moles of HCl that react with this many moles of $Mg(OH)_2$, and then use the molarity of HCl to find the volume that contains this number of moles.

**Solution:** Writing the balanced equation:

$$Mg(OH)_2(s) + 2HCl(aq) \longrightarrow MgCl_2(aq) + 2H_2O(l)$$

Converting from mass of $Mg(OH)_2$ to moles:

$$\text{Moles of } Mg(OH)_2 = 0.10 \,\cancel{\text{g } Mg(OH)_2} \times \frac{1 \text{ mol } Mg(OH)_2}{58.32 \,\cancel{\text{g } Mg(OH)_2}}$$

$$= 1.7 \times 10^{-3} \text{ mol } Mg(OH)_2$$

Converting from moles of $Mg(OH)_2$ to moles of HCl:

$$\text{Moles of HCl} = 1.7 \times 10^{-3} \text{ mol Mg(OH)}_2 \times \frac{2 \text{ mol HCl}}{1 \text{ mol Mg(OH)}_2} = 3.4 \times 10^{-3} \text{ mol HCl}$$

Converting from moles of HCl to volume:

$$\text{Volume (L) of HCl} = 3.4 \times 10^{-3} \text{ mol HCl} \times \frac{1 \text{ L}}{0.10 \text{ mol HCl}} = \textbf{3.4} \times \textbf{10}^{-2} \textbf{ L}$$

**Check:** The size of the answer seems reasonable: a small volume of dilute acid (0.034 L of 0.1 $M$) reacts with a small number of moles of antacid (0.0017 mol).
**Comment:** As stated in the advertisements, antacids "neutralize" stomach acid. Neutralization reactions occur between an acid (in this case HCl) and a base [in this case $Mg(OH)_2$] to produce water and a salt (in this case $MgCl_2$). We introduce this important class of reactions in Chapter 4.

**FOLLOW-UP PROBLEM 3.16**
Another active ingredient found in some antacids is aluminum hydroxide. Which is more effective at neutralizing stomach acid, magnesium hydroxide or aluminum hydroxide? [*Hint:* Effectiveness refers to the amount of acid that reacts with a given mass of antacid. You already know the effectiveness of 0.10 g $Mg(OH)_2$].

SAMPLE PROBLEM 3.17 _____

## Solving Limiting-Reactant Problems for Reactions in Solution

**Problem:** Mercury and its compounds have many uses, from filling teeth (as a mixture with silver and tin) to the industrial production of chlorine. Because of their toxicity, soluble mercury compounds, such as mercury(II) nitrate, must be removed from industrial waste water. One method for accomplishing this is by reaction with sodium sulfide solution. The products are solid mercury(II) sulfide and sodium nitrate solution. In a laboratory simulation, 0.050 L of 0.010 $M$ mercury(II) nitrate reacts with 0.020 L of 0.10 $M$ sodium sulfide. How many grams of mercury(II) sulfide form?
**Plan:** This is a limiting-reactant problem because *the amounts of two reactants are given*. After balancing the equation, we determine the limiting reactant. We first calculate the moles of each reactant from its volume and molarity. Then, we use the molar ratio to find the moles of product formed from each reactant, *assuming the other is present in excess*. Whichever reactant forms fewer moles of product is limiting. This process is outlined in the roadmap. Once we know the limiting reactant, we convert the moles of HgS it produces to mass using the molar mass of HgS.
**Solution:** Writing the balanced equation:

$$Hg(NO_3)_2(aq) + Na_2S(aq) \rightarrow HgS(s) + 2NaNO_3(aq)$$

Finding moles of HgS assuming $Hg(NO_3)_2$ is limiting: Combining the steps gives

$$\text{Moles of HgS} = 0.050 \text{ L soln} \times \frac{0.010 \text{ mol Hg(NO}_3)_2}{1 \text{ L soln}} \times \frac{1 \text{ mol HgS}}{1 \text{ mol Hg(NO}_3)_2}$$

$$= 5.0 \times 10^{-4} \text{ mol HgS}$$

Finding moles of HgS assuming $Na_2S$ is limiting: Combining the steps gives

$$\text{Moles of HgS} = 0.020 \text{ L soln} \times \frac{0.10 \text{ mol Na}_2S}{1 \text{ L soln}} \times \frac{1 \text{ mol HgS}}{1 \text{ mol Na}_2S}$$

$$= 2.0 \times 10^{-3} \text{ mol HgS}$$

Therefore, $Hg(NO_3)_2$ is the limiting reactant.

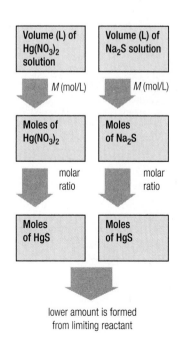

lower amount is formed
from limiting reactant

Converting from moles of HgS to mass:

$$\text{Mass (g) of HgS} = 5.0 \times 10^{-4} \, \cancel{\text{mol HgS}} \times \frac{232.7 \text{ g HgS}}{1 \, \cancel{\text{mol HgS}}} = \mathbf{0.12 \text{ g HgS}}$$

**Check:** Since the molar ratio of each reactant to product is 1/1, it makes sense that less HgS forms from $Hg(NO_3)_2$ because its molarity is one-tenth that of $Na_2S$ and fewer moles of it are present. In the final step, ($\sim$250 g/mol)($5 \times 10^{-4}$ mol) = 0.12 g, so it is correct.

**FOLLOW-UP PROBLEM 3.17**
Despite their toxicity, many compounds of lead are used as pigments.
**(a)** What volume of 1.50 $M$ lead(II) acetate contains 0.400 mol $Pb^{2+}$ ion?
**(b)** When this volume reacts with 125 mL of 3.40 $M$ sodium chloride, how many grams of solid lead(II) chloride can form? (Sodium acetate solution also forms.)

## Section Summary
When reactions occur in solution, reactant and product amounts are given in terms of their concentration and volume. Molarity is the number of moles of solute dissolved in a liter of solution. Using molarity as a conversion factor, the principles of stoichiometry apply to all aspects of reactions in solution.

**FIGURE 3.14**
An overview of the most important mole-mass stoichiometric relationships in chemical systems.

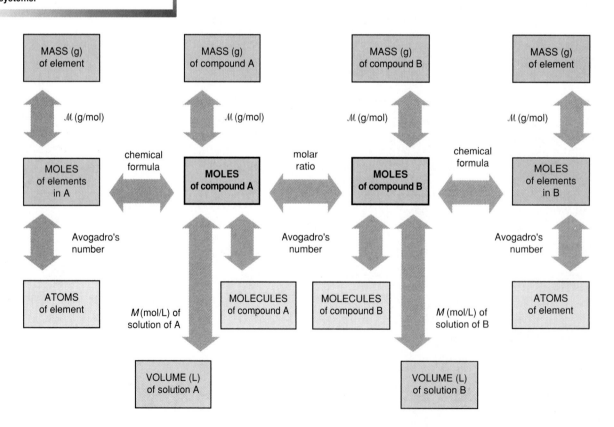

## Chapter Perspective

*Few quantitative ideas are more central to your understanding of chemistry than the mole and its relation to the amounts of substances on the macroscopic and molecular levels. You apply the mole concept every time you measure a substance or think about how much of it will react. For a general summary, Figure 3.14 combines the individual stoichiometry summary diagrams into one overall review diagram. Use it for homework, when you study for examinations, or just to obtain an overview of the various ways that the amounts involved in a reaction are interrelated.*

*We apply stoichiometry next to some of the most important types of chemical reactions (Chapter 4), to systems of reacting gases (Chapter 5), and to the heat involved in a reaction (Chapter 6). Stoichiometry appears at many places later in the text as well.*

## For Review and Reference

### Key Terms

stoichiometry

**SECTION 3.1**
mole (mol)
Avogadro's number
molar mass ($\mathcal{M}$)

**SECTION 3.2**
combustion analysis

**SECTION 3.3**
chemical equation
reactant
product
balancing (stoichiometric)
    coefficient

**SECTION 3.4**
overall equation
limiting reactant
theoretical yield
side reaction
actual yield
percent yield (% yield)

**SECTION 3.5**
solute
solvent
concentration
molarity ($M$)

### Key Equations and Relationships

**3.1** Number of entities in one mole (p. 86):

1 mole contains $6.022 \times 10^{23}$ entities (to four significant figures)

**3.2** Converting from moles to mass using $\mathcal{M}$ (p. 87):

$$\text{Mass (g)} = \text{no. of moles} \times \frac{\text{no. of grams}}{1 \text{ mole}}$$

**3.3** Converting from mass to moles using $1/\mathcal{M}$ (p. 88):

$$\text{No. of moles} = \text{mass (g)} \times \frac{1 \text{ mole}}{\text{no. of grams}}$$

**3.4** Converting from moles to number of entities (p. 88):

$$\text{No. of entities} = \text{no. of moles} \times \frac{6.022 \times 10^{23} \text{ entities}}{1 \text{ mole}}$$

**3.5** Converting from number of entities to moles (p. 88):

$$\text{No. of moles} = \text{no. of entities} \times \frac{1 \text{ mole}}{6.022 \times 10^{23} \text{ entities}}$$

**3.6** Calculating mass % (based on molar mass) (p. 92):

Mass % of element X

$$= \frac{\text{moles of X} \times \text{molar mass of X (g/mol)}}{\text{mass of one mole of compound (g)}} \times 100$$

**3.7** Calculating percent yield (p. 112):

$$\% \text{ yield} = \frac{\text{actual yield}}{\text{theoretical yield}} \times 100$$

**3.8** Defining molarity (p. 114):

$$\text{Molarity} = \frac{\text{moles of solute}}{\text{liters of solution}} \quad \text{or} \quad M = \frac{\text{mol solute}}{\text{L soln}}$$

**3.9** Diluting a concentrated solution (p. 118):

$$M_{\text{dil}} V_{\text{dil}} = \text{moles} = M_{\text{conc}} V_{\text{conc}}$$

## Answers to Follow-up Problems

**3.1** (a) Moles of C = 315 $\cancel{\text{mg C}}$ × $\dfrac{1\,\cancel{\text{g}}}{10^3\,\cancel{\text{mg}}}$ × $\dfrac{1\ \text{mol C}}{12.01\,\cancel{\text{g C}}}$

= 0.0262 mol C

(b) No. of C atoms = $2.62 \times 10^{-2}$ $\cancel{\text{mol C}}$

× $\dfrac{6.022 \times 10^{23}\ \text{C atoms}}{1\,\cancel{\text{mol C}}}$ = $1.58 \times 10^{22}$ C atoms

**3.2** Mass (g) of Mn = $3.22 \times 10^{20}$ $\cancel{\text{Mn atoms}}$

× $\dfrac{1\,\cancel{\text{mol Mn}}}{6.022 \times 10^{23}\,\cancel{\text{Mn atoms}}}$ × $\dfrac{54.94\ \text{g Mn}}{1\,\cancel{\text{mol Mn}}}$

= $2.94 \times 10^{-2}$ g Mn

**3.3** (a) Mass (g) of $P_4O_{10}$ = $4.65 \times 10^{22}$ $\cancel{\text{molecules } P_4O_{10}}$

× $\dfrac{1\,\cancel{\text{mol } P_4O_{10}}}{6.022 \times 10^{23}\,\cancel{\text{molecules } P_4O_{10}}}$ × $\dfrac{283.88\ \text{g } P_4O_{10}}{1\,\cancel{\text{mol } P_4O_{10}}}$

= 21.9 g $P_4O_{10}$

(b) No. of P atoms = $4.65 \times 10^{22}$ $\cancel{\text{molecules } P_4O_{10}}$

× $\dfrac{4\ \text{atoms P}}{1\,\cancel{\text{molecule } P_4O_{10}}}$ = $1.86 \times 10^{23}$ P atoms

**3.4** (a) Mass % of N = $\dfrac{2\,\cancel{\text{mol N}} \times \dfrac{14.01\,\cancel{\text{g N}}}{1\,\cancel{\text{mol N}}}}{80.05\,\cancel{\text{g NH}_4\text{NO}_3}}$ × 100

= 35.00 mass % N

(b) Mass (g) of N = 35.8 $\cancel{\text{mg NH}_4\text{NO}_3}$ × $\dfrac{1\,\cancel{\text{g}}}{10^3\,\cancel{\text{mg}}}$

× $\dfrac{0.3500\ \text{g N}}{1\,\cancel{\text{g NH}_4\text{NO}_3}}$ = $1.25 \times 10^{-2}$ g N

**3.5** Moles of S = 2.88 $\cancel{\text{g S}}$ × $\dfrac{1\ \text{mol S}}{32.07\,\cancel{\text{g S}}}$ = 0.0898 mol S

Moles of M = 0.0898 $\cancel{\text{mol S}}$ × $\dfrac{2\ \text{mol M}}{3\,\cancel{\text{mol S}}}$ = 0.0599 mol M

Molar mass of M = $\dfrac{3.12\ \text{g M}}{0.0599\ \text{mol M}}$ = 52.1 g/mol

M is chromium, and $M_2S_3$ is chromium(III) sulfide.

**3.6** Assuming 100 g of compound, we have 95.21 g C and 4.79 g H:

Moles of C = 95.21 $\cancel{\text{g C}}$ × $\dfrac{1\ \text{mol C}}{12.01\,\cancel{\text{g C}}}$

= 7.928 mol C; also, 4.75 mol H

Preliminary formula = $C_{7.928}H_{4.75}$ ≈ $C_{1.67}H_{1.00}$

Empirical formula = $C_5H_3$

**Whole-number multiple** = $\dfrac{252.30\,\cancel{\text{g/mol}}}{63.07\,\cancel{\text{g/mol}}}$ = 4

Molecular formula = $C_{20}H_{12}$

**3.7** Mass (g) of C = 0.451 g $CO_2$ × $\dfrac{0.2729\ \text{g C}}{\text{g } CO_2}$

= 0.123 g C; also, 0.00690 g H

Mass (g) of Cl = 0.250 g − (0.123 g + 0.00690 g)

= 0.120 g Cl

Moles of elements

= 0.0102 mol C; 0.00685 mol H; 0.00339 mol Cl

Empirical formula = $C_3H_2Cl$; Multiple = 2

Molecular formula = $C_6H_4Cl_2$

**3.8** (a) $2Na(s) + 2H_2O(l) \rightarrow H_2(g) + 2NaOH(aq)$

(b) $2HNO_3(aq) + CaCO_3(s) \rightarrow$
$Ca(NO_3)_2(aq) + H_2O(l) + CO_2(g)$

(c) $PCl_3(g) + 3HF(g) \rightarrow PF_3(g) + 3HCl(g)$

(d) $4C_3H_5N_3O_9(l) \rightarrow$
$12CO_2(g) + 10H_2O(g) + 6N_2(g) + O_2(g)$

**3.9** $Fe_2O_3(s) + 2Al(s) \rightarrow Al_2O_3(s) + 2Fe(l)$

(a) Mass (g) of Fe = 135 $\cancel{\text{g Al}}$ × $\dfrac{1\ \text{mol Al}}{26.98\,\cancel{\text{g Al}}}$

× $\dfrac{2\,\cancel{\text{mol Fe}}}{2\,\cancel{\text{mol Al}}}$ × $\dfrac{55.85\ \text{g Fe}}{1\,\cancel{\text{mol Fe}}}$ = 279 g Fe

(b) No. of Al atoms

= 1.00 $\cancel{\text{g } Al_2O_3}$ × $\dfrac{1\,\cancel{\text{mol } Al_2O_3}}{101.96\,\cancel{\text{g } Al_2O_3}}$ × $\dfrac{2\ \text{mol Al}}{1\,\cancel{\text{mol } Al_2O_3}}$

× $\dfrac{6.022 \times 10^{23}\ \text{Al atoms}}{1\,\cancel{\text{mol Al}}}$ = $1.18 \times 10^{22}$ Al atoms

**3.10** $2SO_2(g) + O_2(g) \rightarrow \cancel{2SO_3(g)}$

$\cancel{2SO_3(g)} + 2H_2O(l) \rightarrow 2H_2SO_4(aq)$

$\overline{2SO_2(g) + O_2(g) + 2H_2O(l) \rightarrow 2H_2SO_4(aq)}$

Mass (g) of $H_2SO_4$/kg Cu

= $\dfrac{1.56 \times 10^4\,\cancel{\text{mol } SO_2}}{1.98 \times 10^3\ \text{kg Cu}}$ × $\dfrac{2\,\cancel{\text{mol } H_2SO_4}}{2\,\cancel{\text{mol } SO_2}}$

× $\dfrac{98.09\ \text{g } H_2SO_4}{1\,\cancel{\text{mol } H_2SO_4}}$ = 773 g $H_2SO_4$/1 kgCu

**3.11** $2Al(s) + 3S(s) \rightarrow Al_2S_3(s)$

Mass (g) of $Al_2S_3$ formed from Al

= 10.0 $\cancel{\text{g Al}}$ × $\dfrac{1\ \text{mol Al}}{26.98\,\cancel{\text{g Al}}}$ × $\dfrac{1\ \text{mol } Al_2S_3}{2\,\cancel{\text{mol Al}}}$

× $\dfrac{150.17\ \text{g } Al_2S_3}{1\,\cancel{\text{mol } Al_2S_3}}$ = 27.8 g $Al_2S_3$

Similarly, mass (g) of $Al_2S_3$ formed from S = 23.4 g $Al_2S_3$. Therefore, S is limiting reactant and 23.4 g $Al_2S_3$ can form. Mass (g) of Al in excess

= total mass of Al − mass of Al used

$$= 10.0 \text{ g Al} - \left(15.0 \text{ g S} \times \frac{1 \text{ mol S}}{32.07 \text{ g S}} \times \frac{2 \text{ mol Al}}{3 \text{ mol S}}\right.$$

$$\left. \times \frac{26.98 \text{ g Al}}{1 \text{ mol Al}}\right) = 1.6 \text{ g Al}$$

**3.12** $2HCl(aq) + CaCO_3(s) \rightarrow CaCl_2(aq) + H_2O(l) + CO_2(g)$
Theoretical yield (g) of $CO_2$

$$= 10.0 \text{ g CaCO}_3 \times \frac{1 \text{ mol CaCO}_3}{100.09 \text{ g CaCO}_3} \times \frac{1 \text{ mol CO}_2}{1 \text{ mol CaCO}_3}$$

$$\times \frac{44.01 \text{ g CO}_2}{1 \text{ mol CO}_2} = 4.40 \text{ g CO}_2$$

$$\% \text{ yield} = \frac{3.65 \text{ g CO}_2}{4.40 \text{ g CO}_2} \times 100 = 83.0\%$$

**3.13** Moles of KI = 84 mL soln $\times \dfrac{1 \text{ L}}{10^3 \text{ mL}}$

$$\times \frac{0.50 \text{ mol KI}}{1 \text{ L soln}} = 0.042 \text{ mol KI}$$

**3.14** Volume (L) of soln

$$= 135 \text{ g sucrose} \times \frac{1 \text{ mol sucrose}}{342.30 \text{ g sucrose}}$$

$$\times \frac{1 \text{ L soln}}{3.30 \text{ mol sucrose}} = 0.120 \text{ L soln}$$

**3.15** $M_{dil}$ of $H_2SO_4 = \dfrac{7.50 \, M \times 25.0 \text{ mL}}{500 \text{ mL}}$

$$= 0.375 \, M \text{ H}_2SO_4$$

Mass (g) of $H_2SO_4$/mL soln

$$= \frac{0.375 \text{ mol H}_2SO_4}{1 \text{ L soln}} \times \frac{1 \text{ L}}{10^3 \text{ mL}}$$

$$\times \frac{98.09 \text{ g H}_2SO_4}{1 \text{ mol H}_2SO_4} = 3.68 \times 10^{-2} \text{ g/mL soln}$$

**3.16** $Al(OH)_3(s) + 3HCl(aq) \rightarrow AlCl_3(aq) + 3H_2O(l)$
Volume (L) of HCl consumed

$$= 0.10 \text{ g Al(OH)}_3 \times \frac{1 \text{ mol Al(OH)}_3}{78.00 \text{ g Al(OH)}_3} \times \frac{3 \text{ mol HCl}}{1 \text{ mol Al(OH)}_3}$$

$$\times \frac{1 \text{ L soln}}{0.10 \text{ mol HCl}} = 3.8 \times 10^{-2} \text{ L soln}$$

Therefore, $Al(OH)_3$ is more effective than $Mg(OH)_2$.

**3.17** (a) Volume (L) of soln

$$= 0.400 \text{ mol Pb}^{2+} \times \frac{1 \text{ mol Pb(C}_2H_3O_2)_2}{1 \text{ mol Pb}^{2+}}$$

$$\times \frac{1 \text{ L soln}}{1.50 \text{ mol Pb(C}_2H_3O_2)_2} = 0.267 \text{ L}$$

(b) $Pb(C_2H_3O_2)_2(aq) + 2NaCl(aq) \rightarrow$
$$PbCl_2(s) + 2NaC_2H_3O_2(aq)$$
Mass (g) of $PbCl_2$ formed from $Pb(C_2H_3O_2)_2$ soln
$$= 111 \text{ g PbCl}_2$$
Mass (g) of $PbCl_2$ formed from NaCl soln
$$= 59.1 \text{ g PbCl}_2$$
Therefore, NaCl is the limiting reactant and 59.1 g $PbCl_2$ can form.

## Sample Problem Titles

# Problems

Problems with a green number are answered at the back of the text. Most sections include three categories of problems separated by a green rule—concept review questions, *paired* skill-building exercises, and problems in a relevant context.

## The Mole

(Sample Problems 3.1 to 3.4)

**3.1** The atomic mass of Cl is 35.45 amu and that of Al is 26.98 amu. What are the masses in grams of 1 mol Al atoms and of 1 mol Cl atoms?

**3.2** (a) How many moles of C atoms are in 1 mol of sucrose ($C_{12}H_{22}O_{11}$)?
(b) How many C atoms are in 1 mol of sucrose?

**3.3** Why might the expression "a mole of oxygen" be confusing? What change would remove any uncertainty? For what other elements might a similar confusion exist? Why?

**3.4** (a) How many grams (g) are equal to one atomic mass unit (amu)?
(b) What is the ratio of the gram to the atomic mass unit?

**3.5** Why is a counting unit (the mole) used so often in chemistry instead of a mass unit?

**3.6** You need to calculate the number of $P_4$ molecules that can be formed from 2.5 g $Ca_3(PO_4)_2$. Explain how you would proceed (i.e., write a solution "plan" without doing any calculations).

---

**3.7** Calculate the molar mass of each of the following:
(a) $Mg(OH)_2$      (b) $N_2O$      (c) $K_2SO_4$
(d) $(NH_4)_3PO_4$    (e) $CH_3OH$    (f) $CuSO_4 \cdot 5H_2O$

**3.8** Calculate the molar mass of each of the following:
(a) $SnO_2$           (b) $MgF_2$      (c) $Ga_2(SO_4)_3$
(d) $N_2O_4$          (e) $C_6H_6$     (f) $CaSO_4 \cdot 2H_2O$

**3.9** Calculate each of the following quantities:
(a) Mass in grams of 0.74 mol $KMnO_4$
(b) Moles of O atoms in 9.22 g $Mg(NO_3)_2$
(c) Number of O atoms in 0.037 g $CuSO_4 \cdot 5H_2O$
(d) Mass in kilograms of $2.6 \times 10^{20}$ molecules of $SO_2$
(e) Moles of Cl atoms in 0.622 g $C_2H_4Cl_2$

**3.10** Calculate each of the following quantities:
(a) Mass in grams of 0.55 mol $CaSO_4$
(b) Moles of compound in 313 g $Fe(ClO_4)_3$
(c) Number of N atoms in 0.76 g $NH_4NO_3$
(d) Total number of ions in 14.3 g $CaBr_2$
(e) Mass in milligrams of 0.45 mol $CuCl_2 \cdot 2H_2O$

**3.11** Calculate each of the following quantities:
(a) Mass in grams of 5.36 mol silver carbonate
(b) Mass in grams of $2.78 \times 10^{21}$ molecules of dinitrogen tetraoxide
(c) Number of moles and formula units in 57.9 g potassium chlorate
(d) Number of potassium ions, chlorate ions, Cl atoms, and O atoms in (c)

**3.12** Calculate each of the following quantities:
(a) Mass in grams of 2.60 mol chromium(III) sulfate
(b) Mass in grams of $8.55 \times 10^{22}$ molecules of disulfur tetrafluoride
(c) Number of moles and formula units in 48.2 g lithium carbonate
(d) Number of lithium ions, carbonate ions, C atoms, and O atoms in (c)

**3.13** Calculate each of the following:
(a) Mass % of H in ammonium bicarbonate
(b) Mass % of O in sodium dihydrogen phosphate heptahydrate
(c) Mass % of I in strontium iodate

**3.14** Calculate each of the following:
(a) Mass fraction of K in potassium phosphate
(b) Mass fraction of O in uranyl sulfate trihydrate (the uranyl ion is $UO_2^{2+}$)
(c) Mass fraction of Cl in calcium perchlorate

---

**3.15** Oxygen is required for metabolic combustion of foods. Calculate the number of atoms in 38.0 g oxygen gas, the amount absorbed from the lungs at rest in about 15 minutes.

**3.16** Cisplatin, a drug used in the treatment of certain cancers, has the formula $Pt(NH_3)_2Cl_2$. Calculate the:
(a) Moles of compound in 136.5 g cisplatin
(b) Number of hydrogen atoms in 1.22 mol cisplatin

**3.17** Allyl sulfide [$(C_3H_5)_2S$] is the substance that gives garlic its characteristic odor. Calculate the:
(a) Mass in grams of 0.52 mol $(C_3H_5)_2S$
(b) Number of carbon atoms in 0.35 g $(C_3H_5)_2S$

**3.18** Iron reacts slowly with oxygen and water to form a compound commonly called rust ($Fe_2O_3 \cdot 4H_2O$).
(a) How many moles of compound are in 15.0 g rust?
(b) How many moles of $Fe_2O_3$ are in 15.0 g rust?
(c) How many grams of iron are in 15.0 g rust?

**3.19** Propane ($C_3H_8$) is widely used in liquid form as a fuel for barbecue grills and camp stoves.
(a) How many moles of compound are in 35.0 g propane?
(b) How many grams of carbon are in 35.0 g propane?

**3.20** Hydrogen-containing fuels have a "fuel value" based on their mass % H. Rank the following compounds in terms of their fuel value: ethane ($C_2H_6$), propane ($C_3H_8$), benzene ($C_6H_6$), ethanol ($C_2H_5OH$), cetyl palmitate (whale oil, $C_{32}H_{64}O_2$).

**3.21** The effectiveness of a nitrogen fertilizer is determined mainly by its mass % N. Rank the following fertilizers in terms of their effectiveness: potassium nitrate, ammonium nitrate, ammonium sulfate, urea $CO(NH_2)_2$.

**3.22** One useful method for recycling scrap aluminum converts some of it to common alum [potassium aluminum sulfate dodecahydrate, $KAl(SO_4)_2 \cdot 12H_2O$], an important dye fixative and food preservative.
(a) How many moles of potassium ions are in 188 g alum?
(b) What is the mass % S in alum?

(c) What is the maximum amount of alum (in metric tons) that can be prepared from 10.0 metric tons of scrap aluminum (1 metric ton = 1000 kg)?

**3.23** Plaster of Paris [$(CaSO_4)_2 \cdot H_2O$] is often used in making surgical casts.
  (a) How many moles of calcium ions are in 58.5 g plaster of Paris?
  (b) What is the mass % O in plaster of Paris?
  (c) The industrial source of calcium compounds is limestone (calcium carbonate). How many metric tons of plaster of Paris can be made from 50.0 metric tons of limestone (1 metric ton = 1000 kg)?

**3.24** Hemoglobin, a protein found in red blood cells, carries $O_2$ from the lungs to the body's cells. Iron (as ferrous ion, $Fe^{2+}$) makes up 0.33 mass % of hemoglobin. If the molar mass of hemoglobin is $6.8 \times 10^4$ g/mol, how many $Fe^{2+}$ ions are present in one molecule?

**3.25** Narceine is a narcotic in opium. It crystallizes from water solution as a hydrate that contains 10.8 mass % water. If the molar mass of narceine hydrate is 499.52 g/mol, determine $x$ in narceine$\cdot x H_2O$?

## Determining the Formula of an Unknown Compound
(Sample Problems 3.5 to 3.7)

**3.26** List three ways compositional data may be given in a problem that involves finding an empirical formula.

**3.27** Which of the following sets of information would allow you to obtain the molecular formula of a covalent compound? In each case that allows it, explain how you would proceed (write a solution "plan").
  (a) Number of moles of each type of atom in a given sample of the compound
  (b) Mass % of each element and the total number of atoms in a molecule of the compound
  (c) Mass % of each element and the number of atoms of one element in a molecule of the compound
  (d) Empirical formula and the mass % of each element in the compound
  (e) Structural formula of the compound

**3.28** Is $MgCl_2$ an empirical or a molecular formula for magnesium chloride? Explain.

**3.29** What is the empirical formula, and empirical formula mass for each of the following compounds?
  (a) $C_2H_4$  (b) $C_2H_6O_2$  (c) $N_2O_5$  (d) $Mg_3(PO_4)_2$

**3.30** What is the empirical formula and empirical formula mass for each of the following compounds?
  (a) $C_4H_8$  (b) $C_3H_6O_3$  (c) $P_4O_{10}$  (d) $Al_2(SO_4)_3$

**3.31** What is the molecular formula of each compound?
  (a) Empirical formula $CH_2$ ($\mathcal{M} = 42.08$ g/mol)
  (b) Empirical formula $NH_2$ ($\mathcal{M} = 32.05$ g/mol)
  (c) Empirical formula $NO_2$ ($\mathcal{M} = 92.02$ g/mol)

**3.32** What is the molecular formula of each compound?
  (a) Empirical formula $CH$ ($\mathcal{M} = 78.11$ g/mol)
  (b) Empirical formula $C_3H_6O_2$ ($\mathcal{M} = 74.08$ g/mol)
  (c) Empirical formula $HgCl$ ($\mathcal{M} = 472.1$ g/mol)

**3.33** Determine the empirical formula of each of the following compounds from the information given:
  (a) 0.063 mol chlorine atoms combined with 0.22 mol oxygen atoms
  (b) 2.45 g silicon combined with 12.4 g chlorine
  (c) 27.3 mass % carbon and 72.7 mass % oxygen

**3.34** Determine the empirical formula of each of the following compounds from the information given:
  (a) 0.039 mol iron atoms combined with 0.052 mol oxygen atoms
  (b) 0.903 g phosphorus combined with 6.99 g bromine
  (c) A hydrocarbon with 79.9 mass % carbon

**3.35** An oxide of nitrogen contains 26 mass % N.
  (a) What is the empirical formula of the oxide?
  (b) If the molar mass is between $105 \pm 5$ g/mol, what is the molecular formula?

**3.36** A chloride of silicon contains 79.1 mass % Cl.
  (a) What is the empirical formula of the chloride?
  (b) If the measured molar mass is 269 g/mol, what is the molecular formula?

**3.37** A sample of 0.600 mol of a metal M reacts completely with excess fluorine to form 46.8 g $MF_2$.
  (a) How many moles of F are in the sample of $MF_2$ that forms?
  (b) How many grams of M are in this sample of $MF_2$?
  (c) What element is represented by the symbol M?

**3.38** A sample of 0.370 mol of a metal oxide ($M_2O_3$) weighs 55.4 g.
  (a) How many moles of O are in the sample?
  (b) How many grams of M are in the sample?
  (c) What element is represented by the symbol M?

**3.39** Nicotine is a poisonous, addictive compound found in tobacco. A sample of nicotine contains 6.16 mmol C, 8.56 mmol H, and 1.23 mmol N [1 mmol (one millimole) = $10^{-3}$ mol]. What is the empirical formula?

**3.40** Serotonin ($\mathcal{M} = 176$ g/mol) is a compound that conducts nerve impulses in brain and muscle. It contains 68.2 mass % C, 6.86 mass % H, 15.9 mass % N, and 9.08 mass % O. What is its molecular formula?

**3.41** Cortisone ($\mathcal{M} = 360$ g/mol) is a hormone formed in the adrenal gland that is used in the treatment of rheumatoid arthritis. It has the following elemental composition by mass: 70.0% C, 7.83% H, and 22.2% O. What is its molecular formula?

**3.42** Isobutylene is a hydrocarbon used in the manufacture of synthetic rubber. When 0.847 g isobutylene was analyzed by combustion (using an apparatus similar to that of Figure 3.6), the gain in mass of the $CO_2$ absorber was 2.657 g and that of the $H_2O$ absorber was 1.089 g. What is the empirical formula of isobutylene?

**3.43** Menthol ($\mathcal{M} = 156.3$ g/mol), a strong-smelling substance used in cough drops, is a compound of carbon, hydrogen, and oxygen. When 0.1595 g menthol was subjected to combustion analysis, it produced 0.449 g $CO_2$ and 0.184 g $H_2O$. What is its molecular formula?

## Writing and Balancing Chemical Equations

(Sample Problem 3.8)

**3.44** What three types of information does a balanced chemical equation provide? How?

**3.45** How does a balanced chemical equation apply the law of conservation of mass?

**3.46** In the process of balancing the equation

$$Al + Cl_2 \rightarrow AlCl_3$$

Student I writes: $Al + Cl_2 \rightarrow AlCl_2$
Student II writes: $Al + Cl_2 + Cl \rightarrow AlCl_3$
Student III writes: $2Al + 3Cl_2 \rightarrow 2AlCl_3$
Is the approach of Student I valid? of Student II? of Student III? Explain.

**3.47** Write balanced equations for each of the following by inserting the correct coefficients in the blanks:
(a) __$Cu(s)$ + __$S_8(s)$ → __$Cu_2S(s)$
(b) __$P_4O_{10}(s)$ + __$H_2O(l)$ → __$H_3PO_4(l)$
(c) __$B_2O_3(s)$ + __$NaOH(aq)$
  → __$Na_3BO_3(aq)$ + __$H_2O(l)$
(d) __$CH_3NH_2(g)$ + __$O_2(g)$
  → __$CO_2(g)$ + __$H_2O(g)$ + __$N_2(g)$
(e) __$Cu(NO_3)_2(aq)$ + __$KOH(aq)$
  → __$Cu(OH)_2(s)$ + __$KNO_3(aq)$
(f) __$BCl_3(g)$ + __$H_2O(l)$ → __$H_3BO_3(s)$ + __$HCl(g)$
(g) __$CaSiO_3(s)$ + __$HF(g)$
  → __$SiF_4(g)$ + __$CaF_2(s)$ + __$H_2O(l)$

**3.48** Write balanced equations for each of the following by inserting the correct coefficients in the blanks:
(a) __$SO_2(g)$ + __$O_2(g)$ → __$SO_3(g)$
(b) __$Sc_2O_3(s)$ + __$H_2O(l)$ → __$Sc(OH)_3(s)$
(c) __$H_3PO_4(aq)$ + __$NaOH(aq)$
  → __$Na_2HPO_4(aq)$ + __$H_2O(l)$
(d) __$C_6H_{10}O_5(s)$ + __$O_2(g)$ → __$CO_2(g)$ + __$H_2O(g)$
(e) __$As_4S_6(s)$ + __$O_2(g)$ → __$As_4O_6(s)$ + __$SO_2(g)$
(f) __$Ca_3(PO_4)_2(s)$ + __$SiO_2(s)$ + __$C(s)$
  → __$P_4(g)$ + __$CaSiO_3(s)$ + __$CO(s)$
(g) __$Fe(s)$ + __$H_2O(g)$ → __$Fe_3O_4(s)$ + __$H_2(g)$

**3.49** Convert the following descriptions of reactions into balanced equations:
(a) When gallium metal is heated in oxygen gas, it melts and forms solid gallium(III) oxide.
(b) Liquid pentene ($C_5H_{10}$) burns in oxygen gas to form carbon dioxide and water vapor.
(c) When solutions of calcium chloride and sodium phosphate are mixed, solid calcium phosphate forms and sodium chloride remains in solution.
(d) Liquid disilicon hexachloride reacts with water to form solid silicon dioxide, hydrogen chloride gas, and hydrogen gas.
(e) When nitrogen dioxide is bubbled into water, a solution of nitric acid forms and gaseous nitrogen monoxide is released.

**3.50** Convert the following descriptions of reactions into balanced equations:
(a) In a gaseous reaction, dihydrogen sulfide burns in oxygen to form sulfur dioxide and water vapor.

(b) When crystalline potassium chlorate is heated to just above its melting point, it reacts to form two different crystalline compounds, potassium chloride and potassium perchlorate.
(c) When hydrogen gas is passed over powdered iron(III) oxide, iron metal and water vapor form.
(d) The combustion of gaseous ethane ($C_2H_6$) in air forms carbon dioxide and water vapor.
(e) Iron(II) chloride can be converted to iron(III) fluoride by treatment with chlorine trifluoride gas. Chlorine gas is also formed.

## Calculating the Amounts of Reactant and Product

(Sample Problems 3.9 to 3.12)

**3.51** What does the term *stoichiometrically equivalent molar ratio* mean, and how is it applied in solving problems?

**3.52** Reactants A and B form product C. Write a detailed "plan" to find the mass of C that forms when 5 g A reacts with excess B.

**3.53** Reactants D and E form product F. Write a detailed "plan" to find the mass of F that forms when 5 g D reacts with 8 g E.

**3.54** Percentage yields are generally calculated from mass quantities. Would the result be the same if mole quantities were used instead? Why?

**3.55** Chlorine gas can be made in the laboratory by the reaction of hydrochloric acid and manganese(IV) oxide:

$$4HCl(aq) + MnO_2(s) \rightarrow MnCl_2(aq) + 2H_2O(g) + Cl_2(g)$$

When 0.375 mol HCl reacts with excess $MnO_2$,
(a) How many moles of $Cl_2$ form?
(b) How many grams of $Cl_2$ form?

**3.56** Bismuth oxide reacts with carbon to form bismuth metal:

$$Bi_2O_3(s) + 3C(s) \rightarrow 2Bi(s) + 3CO(g)$$

When 50.7 g $Bi_2O_3$ reacts with excess carbon,
(a) How many moles of $Bi_2O_3$ react?
(b) How many moles of Bi form?

**3.57** Sodium nitrate decomposes on heating to sodium nitrite and oxygen gas:

$$2NaNO_3(s) \rightarrow 2NaNO_2(s) + O_2(g)$$

In order to produce 128 g oxygen,
(a) How many moles of $NaNO_3$ must be heated?
(b) How many grams of $NaNO_3$ must be heated?

**3.58** Chromium(III) oxide reacts with hydrogen sulfide ($H_2S$) gas to form chromium(III) sulfide and water:

$$Cr_2O_3(s) + 3H_2S(g) \rightarrow Cr_2S_3(s) + 3H_2O(l)$$

In order to produce 63.9 g $Cr_2S_3$,
(a) How many moles of $Cr_2O_3$ are required?
(b) How many grams of $Cr_2O_3$ are required?

**3.59** Calculate the mass of each product formed when 3.082 g diborane ($B_2H_6$) reacts with excess water:

$$B_2H_6(g) + H_2O(l) \longrightarrow H_3BO_3(s) + H_2(g) \text{ [unbalanced]}$$

**3.60** Calculate the mass of each product formed when 56.3 g silver sulfide reacts with excess hydrochloric acid:

$$Ag_2S(s) + HCl(aq) \longrightarrow AgCl(s) + H_2S(g) \text{ [unbalanced]}$$

**3.61** Elemental phosphorus occurs as tetratomic molecules, $P_4$. What mass of chlorine gas is needed for complete reaction with 351 g phosphorus in the formation of phosphorus pentachloride?

**3.62** Elemental sulfur occurs as octatomic molecules, $S_8$. What mass of fluorine gas is needed for complete reaction with 285 g sulfur in the formation of sulfur hexafluoride?

**3.63** Solid iodine trichloride is prepared by reaction between solid iodine and gaseous chlorine to form iodine monochloride crystals, followed by treatment with additional chlorine.
(a) Write a balanced equation for each step.
(b) Write an overall balanced equation for the formation of iodine trichloride.
(c) How many grams of iodine are needed to prepare 35.8 kg of final product?

**3.64** Lead can be prepared from galena [lead(II) sulfide] by first roasting in oxygen gas to form lead(II) oxide and sulfur dioxide. Heating the metal oxide with more galena forms the molten metal and more sulfur dioxide.
(a) Write a balanced equation for each step.
(b) Write an overall balanced equation for the process.
(c) How many metric tons of sulfur dioxide form for every metric ton of lead obtained?

**3.65** Many metals react with oxygen gas to form the metal oxide. For example, calcium reacts as follows:

$$2Ca(s) + O_2(g) \longrightarrow 2CaO(s)$$

You wish to calculate the mass of calcium oxide that can be prepared from 4.20 g Ca and 1.60 g $O_2$.
(a) How many moles of CaO can be produced from the given mass of Ca?
(b) How many moles of CaO can be produced from the given mass of $O_2$?
(c) Which is the limiting reactant?
(d) How many grams of CaO can be produced?

**3.66** Metal hydrides react with water to form hydrogen gas and the metal hydroxide. For example,

$$CaH_2(s) + 2H_2O(l) \longrightarrow Ca(OH)_2(s) + 2H_2(g)$$

You wish to calculate the mass of hydrogen gas that can be prepared from 5.00 g $CaH_2$ and 4.80 g $H_2O$.
(a) How many moles of $H_2$ can be produced from the given mass of $CaH_2$?
(b) How many moles of $H_2$ can be produced from the given mass of $H_2O$?

(c) Which is the limiting reactant?
(d) How many grams of $H_2$ can be produced?

**3.67** Calculate the maximum number of moles and mass of iodic acid ($HIO_3$) that can form when 735 g iodine trichloride reacts with 97.7 g water:

$$ICl_3 + H_2O \longrightarrow ICl + HIO_3 + HCl \text{ [unbalanced]}$$

What mass of the excess reactant remains?

**3.68** Calculate the maximum number of moles and mass of $H_2S$ that can form when 165 g aluminum sulfide reacts with 125 g water:

$$Al_2S_3 + H_2O \longrightarrow Al(OH)_3 + H_2S \text{ [unbalanced]}$$

What mass of the excess reactant remains?

**3.69** When 0.100 mol carbon is burned in a closed vessel with 8.00 g oxygen, how many grams of carbon dioxide can form? Which reactant is in excess, and how many grams of it remain after the reaction?

**3.70** A mixture of 0.0259 g hydrogen and 0.0125 mol oxygen in a closed container is sparked to initiate the reaction. How many grams of water can form? Which reactant is in excess, and how many grams of it remain after the reaction?

**3.71** Aluminum nitrite and ammonium chloride react to form aluminum chloride, nitrogen, and water. What mass of each substance is present after 33.0 g aluminum nitrite and 28.9 g ammonium chloride react completely?

**3.72** Calcium nitrate and ammonium fluoride react to form calcium fluoride, dinitrogen monoxide, and water vapor. What mass of each substance is present after 15.0 g calcium nitrate and 7.50 g ammonium fluoride react completely?

**3.73** Two successive reactions, $A \longrightarrow B$ and $B \longrightarrow C$, have yields of 82% and 65%, respectively. What is the overall percent conversion of A to C?

**3.74** Two successive reactions, $D \longrightarrow E$ and $E \longrightarrow F$, have yields of 57% and 86%, respectively. What is the overall percent conversion of D to F?

**3.75** What is the percent yield of a reaction in which 41.5 g tungsten(VI) oxide [$WO_3$] reacts with excess hydrogen gas to produce metallic tungsten and 9.50 mL water ($d = 1.00$ g/mL)?

**3.76** What is the percent yield of a reaction in which 200.0 g phosphorus trichloride reacts with excess water to form 128 g HCl and aqueous phosphorous acid ($H_3PO_3$)?

**3.77** When 153 g methane ($CH_4$) and 43.0 g chlorine gas undergo a reaction that has an 80.0% yield, what mass of chloromethane ($CH_3Cl$) forms? Hydrogen chloride is the other product.

**3.78** When 56.6 g calcium and 30.5 g nitrogen gas undergo a reaction that has a 90.0% yield, what mass of calcium nitride forms?

**3.79** The smelting of ferric oxide to form elemental iron occurs at high temperatures in a blast furnace through a reaction sequence with carbon monoxide. In the first step, ferric oxide reacts with carbon monoxide to form $Fe_3O_4$. This substance reacts with more carbon monoxide to form iron(II) oxide, which reacts with still more carbon monoxide to form molten iron. Carbon dioxide is also produced in each step.
(a) Write an overall balanced equation for the iron-smelting process.
(b) How many grams of carbon monoxide are required to form 40.0 metric tons of iron from ferric oxide?

**3.80** Cyanogen ($C_2N_2$) is used as a welding gas for heat-resistant metals and as a fumigant. In its reaction with fluorine gas, carbon tetrafluoride and nitrogen trifluoride gases are produced. What mass of carbon tetrafluoride forms when 80.0 g of each reactant are mixed?

**3.81** An intermediate step in the industrial production of nitric acid involves the reaction of ammonia with oxygen gas to form nitrogen monoxide and water. How many grams of nitrogen monoxide could form by the reaction of 466 g ammonia with 812 g oxygen?

**3.82** Gaseous butane ($C_4H_{10}$) is compressed and used as a liquid fuel in disposable cigarette lighters and lightweight camping stoves. Suppose a lighter contains 5.00 mL butane ($d = 0.579$ g/mL).
(a) How many grams of oxygen are needed to burn the butane completely?
(b) How many moles of $CO_2$ form when all the butane burns?
(c) How many total molecules of gas form when the butane burns completely?

**3.83** One of the compounds used to increase the octane rating of gasoline is toluene ($C_7H_8$). Suppose 15.0 mL toluene ($d = 0.867$ g/mL) is consumed when a sample of gasoline burns in air.
(a) How many grams of oxygen are needed for complete combustion of the toluene?
(b) How many total moles of gaseous products form?
(c) How many molecules of water vapor form?

**3.84** Diborane ($B_2H_6$), a useful reactant in organic synthesis, may be prepared by the following reaction, which occurs in a nonaqueous solvent:

$$NaBH_4(s) + BF_3(g) \rightarrow B_2H_6(g) + NaBF_4(s) \text{ [unbalanced]}$$

If the reaction has an 85.0% yield, how many grams of $NaBH_4$ are needed to make 20.0 g $B_2H_6$?

**3.85** During studies of the reaction from Sample Problem 3.11,

$$N_2O_4(l) + 2N_2H_4(l) \rightarrow 3N_2(g) + 4H_2O(g)$$

an engineer measured a less-than-expected yield of $N_2$ and discovered that the following side reaction occurs:

$$2N_2O_4(l) + N_2H_4(l) \rightarrow 6NO(g) + 2H_2O(g)$$

In one experiment, 10.0 g NO formed when 100.0 g of each reactant were mixed. What was the highest percent yield of $N_2$ that can be expected?

**Fundamentals of Solution Stoichiometry**

(Sample Problems 3.13 to 3.17)

**3.86** Given the volume and molarity of a $CaCl_2$ solution, how would you determine the number of moles and the mass of solute?

**3.87** What is wrong with the following instruction for preparing a 1.00 M solution by dilution of a 10.0 M solution: "Take 100.0 mL of the 10.0 M solution and add 900.0 mL water"?

**3.88** A mathematical equation useful for dilution calculations is $M_{dil}V_{dil} = M_{conc}V_{conc}$. What does each symbol mean, and why does the equation work?

**3.89** Calculate each of the following quantities:
(a) Grams of solute in 175.8 mL of 0.0465 M calcium acetate
(b) Molarity of a 500.0-mL solution containing 2.83 g potassium iodide
(c) Moles of solute in 3.011 L of 0.850 M sodium cyanide
(d) Volume in liters of 2.26 M potassium hydroxide that contains 8.42 g solute
(e) Number of $Cu^{2+}$ ions in 52 L of 2.3 M copper(II) chloride

**3.90** Calculate each of the following quantities:
(a) Grams of solute needed to make 475 mL of $5.62 \times 10^{-2}$ M potassium sulfate
(b) Molarity of a solution that contains 6.55 mg calcium chloride in each milliliter
(c) Number of $Mg^{2+}$ ions in each milliliter of 0.184 M magnesium bromide
(d) Molarity of the solution that results when 26.0 g silver nitrate is dissolved in enough water to give a final volume of 335 mL
(e) Volume in liters of 0.355 M manganese(II) sulfate that contains 57.0 g solute

**3.91** Calculate each of the following quantities:
(a) Molarity of a solution prepared by diluting 27.00 mL of 0.150 M potassium chloride to 150.00 mL
(b) Molarity of a solution prepared by diluting 35.71 mL of 0.0756 M ammonium sulfate to 500.00 mL
(c) Final volume of a 0.0500 M solution prepared by diluting 10.0 mL of 0.155 M lithium carbonate with water
(d) Volume of 1.050 M copper(II) nitrate that must be diluted with water to prepare 750.0 mL of 0.8543 M solution

**3.92** Calculate each of the following quantities:
(a) Volume of 18.0 M sulfuric acid that must be added to water to prepare 2.00 L of 0.309 M solution
(b) Molarity of the solution obtained by diluting 80.6 mL of 0.225 M ammonium chloride to 0.250 L
(c) Volume of water that must be added to 0.150 L of 0.0262 M sodium hydroxide to obtain a 0.0100 M solution (Assume the volumes are additive at these low concentrations.)

(d) Mass of calcium nitrate in each milliliter of a solution prepared by diluting 64.0 mL of 0.745 $M$ calcium nitrate to a final volume of 0.100 L

**3.93** Concentrated nitric acid has a density of 1.41 g/mL and contains 70% $HNO_3$ by mass.
(a) What mass of $HNO_3$ is present per liter of solution?
(b) What is the molarity of the solution?

**3.94** Concentrated sulfuric acid (18.3 $M$) has a density of 1.84 g/mL.
(a) How many moles of sulfuric acid are present per milliliter of solution?
(b) What is the mass % of $H_2SO_4$ in the solution?

**3.95** How many milliliters of 0.55 $M$ HCl are needed to react with 3.7 g $CaCO_3$?

$$2HCl(aq) + CaCO_3(s) \rightarrow CaCl_2(aq) + CO_2(g) + H_2O(l)$$

**3.96** How many grams of $NaH_2PO_4$ are needed to react with 28.74 mL of 0.225 $M$ NaOH?

$$NaH_2PO_4(s) + 2NaOH(aq) \rightarrow Na_3PO_4(aq) + 2H_2O(l)$$

**3.97** How many grams of solid barium sulfate form when 25.0 mL of 0.160 $M$ barium chloride reacts with 68.0 mL of 0.055 $M$ sodium sulfate? Aqueous sodium chloride is the other product.

**3.98** How many moles of which reactant are in excess when 350.0 mL of 0.210 $M$ sulfuric acid reacts with 0.500 L of 0.196 $M$ sodium hydroxide to form water and aqueous sodium sulfate?

---

**3.99** Ordinary household bleach is an aqueous solution of sodium hypochlorite. What is the molarity of a bleach solution that contains 20.5 g sodium hypochlorite in a total volume of 375 mL?

**3.100** Muriatic acid, an industrial grade of concentrated HCl, is used to clean masonry and etch cement for painting. Its concentration is 11.7 $M$.
(a) Write instructions for diluting the concentrated acid to make 5.0 gallons of 3.5 $M$ acid for routine use. (1 gal = 4 qt; 1 qt = 0.946 L)
(b) How many milliliters of the muriatic acid solution contain 9.55 g HCl?

**3.101** Potassium permanganate solutions are extremely useful in laboratory analysis. Their concentrations are determined by a reaction with ferrous ammonium sulfate hexahydrate, $Fe(NH_4)_2(SO_4)_2 \cdot 6H_2O$, which behaves in solution as though it were simply iron(II) sulfate:

$$2KMnO_4(aq) + 10FeSO_4(aq) + 8H_2SO_4(aq) \rightarrow$$
$$5Fe_2(SO_4)_3(aq) + 2MnSO_4(aq) + K_2SO_4(aq) + 8H_2O(l)$$

In an experiment, 2.50 g $Fe(NH_4)_2(SO_4)_2 \cdot 6H_2O$ reacted completely with 37.0 mL of a $KMnO_4$ solution. What is the molarity of the $KMnO_4$ solution?

**3.102** A sample of impure magnesium was analyzed by allowing it to react with excess HCl solution:

$$Mg(s) + 2HCl(aq) \rightarrow MgCl_2(aq) + H_2(g)$$

After 1.32 g of the impure metal was treated with 0.100 L of 0.750 $M$ HCl, 0.0125 mol HCl remained. Assuming the impurities do not react with the acid, what is the mass % Mg in the sample?

## Comprehensive Problems
Problems with an asterisk (*) are more challenging.

**3.103** Is each of the following statements true or false? Correct any that are false:
(a) A mole of one substance has the same number of atoms as a mole of any other substance.
(b) The theoretical yield for a reaction is based on the balanced chemical equation.
(c) A limiting-reactant problem can be recognized when the amount of available material is given in moles for one of the reactants.
(d) To prepare 1.00 L of 3.00 $M$ NaCl, weigh 175.5 NaCl and dissolve it in 1.00 L distilled water.
(e) The concentration of a solution is an intensive property, whereas the amount of solute in a solution is an extensive property.

**3.104** In each pair, choose the larger of the indicated quantity, or state if the samples are equal:
(a) Individual particles: 0.4 mol $O_3$ molecules or 0.4 mol O atoms
(b) Mass: 0.4 mol $O_3$ molecules or 0.4 mol O atoms
(c) Moles: 4.0 g $N_2O_4$ or 3.3 g $SO_2$
(d) Mass: 0.6 mol $C_2H_4$ or 0.6 mol $F_2$
(e) Total ions: 2.3 mol sodium chlorate or 2.2 mol magnesium chloride
(f) Molecules: 1.0 g $H_2O$ or 1.0 g $H_2O_2$
(g) $Na^+$ ions: 0.500 L of 0.500 $M$ NaBr or 0.0146 kg NaCl
(h) Mass: $6.022 \times 10^{23}$ atoms of $^{235}U$ or $6.022 \times 10^{23}$ atoms of $^{238}U$

**3.105** The number of objects in a mole is so fantastically large as to be incomprehensible from everyday experience. If a mole of ice cubes (each 1 in$^3$) were stacked neatly edge to edge over the entire continental United States, how high would the stack be? (Area of adjacent 48 states = $3.0 \times 10^6$ mi$^2$. Assume the land area is flat.)

**3.106** Oleic acid, a component of olive oil, is 76.54% C, 12.13% H, and 11.33% O by mass. An approximate determination of its molar mass gave a value of 274 ± 20 g/mol. What is the molecular formula of oleic acid?

**3.107** A balanced equation may be interpreted in several ways (see Table 3.5). Write a balanced equation for the reaction between solid tetraphosphorus trisulfide and oxygen to form solid tetraphosphorus decaoxide and gaseous sulfur trioxide. Tabulate the equation in terms of (a) molecules, (b) moles, and (c) mass (grams).

**3.108** Hydrogen gas has been suggested as a clean fuel because it produces only water vapor when it burns. If the reaction has a 98% yield, what mass of hydrogen would form 75.0 kg water?

**3.109** Assuming that the volumes are additive, what is the concentration of KBr in a solution prepared by mixing 0.200 L of 0.053 $M$ KBr with 0.500 L of 0.072 $M$ KBr?

**3.110** To 1.25 L of 0.325 $M$ HCl, you add 3.27 L of a second HCl solution of unknown concentration. The resulting solution is 0.893 $M$ HCl. Assuming the volumes are additive, calculate the molarity of the second HCl solution.

**3.111** Many reactive boron compounds are analyzed by reaction with water, which converts the boron to boric acid ($H_3BO_3$). The boric acid content is then determined with NaOH, which reacts with boric acid on a 1:1 mole basis. If 0.1352 g of a boron compound reacts with water, and the resulting boric acid requires 23.47 mL of 0.205 $M$ NaOH for reaction, what was the mass % of boron in the compound?

**3.112** Calculate each of the following quantities:
  (a) Moles of compound in 0.488 g ammonium bromide
  (b) Number of potassium ions in 65.5 g potassium nitrate
  (c) Mass in grams of 5.8 mol ethylene glycol ($C_2H_6O_2$)
  (d) Volume of 25.5 mol chloroform (CHCl$_3$; $d$ = 1.48 g/mL)
  (e) Number of sodium ions in 2.11 mol sodium carbonate
  (f) Number of atoms in 10.0 $\mu$g of the element beryllium
  (g) Number of atoms in 0.0015 mol oxygen gas

**3.113** A 0.652-g sample of a strontium halide reacts with excess sulfuric acid, and the solid strontium sulfate formed is separated, dried, and found to weigh 0.755 g. What was the formula of the original halide?

**3.114** Mixtures of hydrocarbon compounds are often used as fuels. How many grams of $CO_2(g)$ would be produced by the combustion of 200.0 g of a mixture that is 30.0% $CH_4$ and 70.0% $C_3H_8$ by mass?

**3.115** Methane ($CH_4$) and ethane ($C_2H_6$) are the two simplest hydrocarbons. What is the mass % C in a mixture that is 40.0% $CH_4$ and 60.0% $C_2H_6$ by mass?

**\*3.116** When carbon-containing compounds are burned in a limited amount of air, some CO($g$) as well as $CO_2(g)$ is produced. A gaseous product mixture is 35.0 mass % CO and 65.0 mass % $CO_2$. What is the mass % C in the product mixture?

**3.117** When ferrocene was synthesized in 1951, it was shown to be the first organo-iron compound with direct Fe—C bonds. An understanding of ferrocene's structure gave rise to new ideas about chemical bonding and led to the preparation of many useful compounds. In the combustion analysis of ferrocene, which contains only Fe, C, and H, a 0.9437-g sample produced 2.233 g $CO_2$ and 0.457 g $H_2O$. What is the empirical formula of ferrocene?

**\*3.118** When zinc metal is treated with dilute nitric acid, the reaction produces nitrogen gas, water, and aqueous zinc nitrate. Write a balanced equation for the reaction.

**\*3.119** In addition to the oxides of sulfur, various nitrogen oxides contribute to acidic rainfall through complex reaction sequences. Atmospheric nitrogen and oxygen combine to form nitrogen monoxide gas, which reacts with more oxygen to form nitrogen dioxide gas. In contact with water vapor, this forms aqueous nitric acid and more nitrogen monoxide.
  (a) Write balanced equations for these reactions.
  (b) Use the three equations to write one overall balanced equation that does *not* include nitrogen monoxide and nitrogen dioxide.
  (c) How many kilograms of nitric acid form when 1.25 kg atmospheric nitrogen is consumed?

**\*3.120** In the chemical analysis of an unknown, chemists often add an excess of a reactant, determine the amount of that reactant remaining after the reaction with the unknown, and use those amounts to calculate the amount of the unknown. For the analysis of an unknown NaOH solution, you add 50.0 mL of the solution to 0.150 L of an acid solution prepared by dissolving 0.588 g of solid oxalic acid dihydrate ($H_2C_2O_4 \cdot 2H_2O$) in water:

$$2NaOH(aq) + H_2C_2O_4(aq) \longrightarrow 2H_2O(l) + Na_2C_2O_4(aq)$$

The unreacted NaOH then reacts completely with 9.65 mL of 0.116 $M$ HCl:

$$NaOH(aq) + HCl(aq) \longrightarrow H_2O(l) + NaCl(aq)$$

What is the molarity of the oxalic acid solution?

**\*3.121** When 1.5173 g of an organic iron compound containing Fe, C, H, and O was burned in oxygen, 2.838 g $CO_2$ and 0.8122 g $H_2O$ were produced. In a separate experiment to determine the mass % of iron, 0.3355 g of the compound yielded 0.0758 g $Fe_2O_3$. What is the empirical formula of the compound?

**\*3.122** Fluorides of the relatively unreactive noble gas xenon can be formed by direct reaction of the elements at high pressure and temperature. Depending on the temperature and the reactant amounts, the product mixture may include $XeF_2$, $XeF_4$, and $XeF_6$. Under conditions that produce only $XeF_4$ and $XeF_6$, $1.85 \times 10^{-4}$ mol Xe reacted with $5.00 \times 10^{-4}$ mol $F_2$, and $9.00 \times 10^{-6}$ mol Xe was found to be in excess. What are the mass percents of $XeF_4$ and $XeF_6$ in the product

# The Major Classes of Chemical Reactions

**Concepts and skills to review**

- names and formulas of compounds (Section 2.7)
- nature of ionic and covalent bonding (Section 2.6)
- mole-mass-number conversions (Section 3.1)
- molarity and mole-volume conversions (Section 3.5)
- balancing chemical equations (Section 3.3)
- calculating the amounts of reactants and products (Section 3.4)

**Chemical change in a storm at sea.** Oceans and rivers dissolve substances from the land and air to form complex mixtures of reactants and products. A storm at sea becomes the scene of myriad natural chemical changes in the composition of matter. In fact, everywhere we look—on Earth, in outer space, in our industries, inside our bodies—we find chemical reactions in amazing variety. In this chapter, we classify chemical reactions, with special attention to those occurring in aqueous solution, as a first step toward understanding them.

The amazing variety that we see in nature is largely a consequence of the amazing variety of its chemical reactions. Rapid chemical changes occur among gas molecules as sunlight bathes the atmosphere or lightning courses through a stormy sky. The seas and oceans are gigantic reaction vessels in which aqueous solution chemistry goes on unceasingly. In every cell of your body, thousands of reactions are taking place right now.

Of the millions of chemical reactions occurring in and around you, we have examined only a tiny fraction so far; indeed, it would be impossible to examine them all. Fortunately, it isn't necessary to catalog each and every reaction, since when we survey even a small percentage of them, a few major classes emerge.

In this chapter, we consider some dominant themes in reaction chemistry and begin asking *why* reactions occur. Many common types of reactions are described, several with illustrations that allow you to view the change on the observable and atomic levels simultaneously. In the first section, we classify reactions in a traditional way, one based on the *change in numbers* of reactants and products. In the rest of the chapter, we look closer and classify reactions according to the *nature of the process* involved. As you'll see, these two ways of classifying reactions often overlap. After investigating the active solvent role of water, we discuss three important aqueous ionic reactions and then focus on oxidation-reduction reactions, one of the most universal types of chemical change. The chapter ends with an introductory look at the reversible nature of all chemical reactions.

## 4.1 Types of Chemical Reactions: An Accounting of Reactants and Products

In all chemical reactions, regardless of type, atoms or ions change their bonding environment, that is, their specific chemical attachments with one another. Elements combine into compounds, or compounds decompose into elements; large molecules break apart to form smaller ones, or vice versa. In many reactions, the atoms (or ions) in different compounds switch bonding partners, in a sort of chemical "square dance," and form a new set of compounds. When we compare the number of reactant substances with the number of product substances, noting where the bonded entities start out and wind up, three reaction types become apparent: combination, decomposition, and displacement reactions.

### Combination Reactions: Two or More Reactants Form One Product

In **combination** reactions, two or more substances combine to form one substance: X + Y → Z. In combination reactions, the reactants are typically elements or compounds and the product is a compound. Figure 4.1 presents a combination reaction between elements on the macroscopic scale, on the atomic scale, and as a balanced equation. For the sake of discussion, let's organize the most common types of combination reactions into several categories based on the kinds of reactant and product substances.

**Combination of two elements.** When two elements react, they form a binary ionic or binary covalent compound.

1. *A metal and a nonmetal combine to form a binary ionic compound.* The reaction between an alkali metal and a halogen in Figure 4.1 is an example. Another metal-nonmetal reaction occurs when a piece of steel wool (iron)

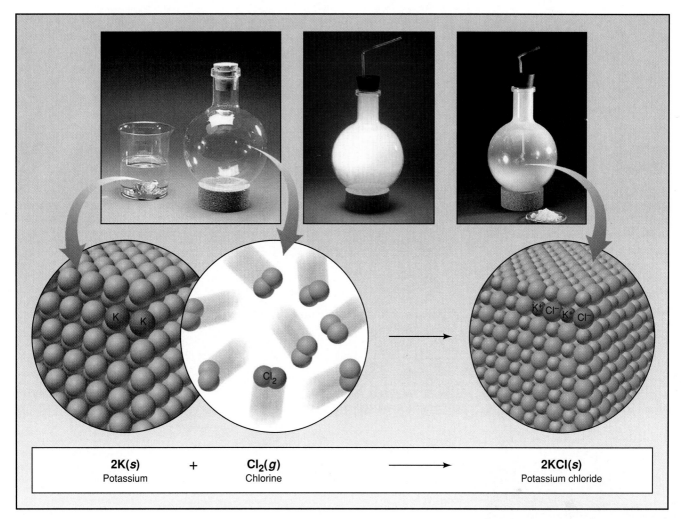

| **2K(s)** | + | **Cl₂(g)** | ⟶ | **2KCl(s)** |
|-----------|---|-----------|---|------------|
| Potassium |   | Chlorine  |   | Potassium chloride |

**FIGURE 4.1**

**Three views of a combination reaction.** In a combination reaction, two or more reactants combine to form one product. When the metal potassium and the nonmetal chlorine react, they form the solid ionic compound potassium chloride. The series of photos *(top)* presents the macroscopic view of the reaction, the view the chemist sees in the laboratory. The blowup arrows lead to an atomic-scale view *(middle)*, a representation of the chemist's mental picture of the reactants and products; the stoichiometry is indicated by the more darkly colored atoms, molecules, and ions. The balanced equation *(bottom)* is the chemist's written shorthand for the chemical change.

is ignited in air and then thrust into a container of chlorine gas: it bursts into sparks and flames as iron(III) chloride forms (Figure 4.2):

$$2Fe(s) + 3Cl_2(g) \longrightarrow 2FeCl_3(s)$$

2. *Two nonmetals combine to form a binary covalent compound.* Thousands of examples of this reaction type are known. The industrial production of ammonia from nitrogen and hydrogen gas occurs on an enormous scale:

$$N_2(g) + 3H_2(g) \longrightarrow 2NH_3(g)$$

Halogens form many compounds with other nonmetals. For example, sulfur hexafluoride, the densest gas at room temperature, forms when sulfur melts in the presence of fluorine gas:

$$S(l) + 3F_2(g) \longrightarrow SF_6(g)$$

3. *Almost every element combines with oxygen to form an oxide.* Metals form ionic oxides. As discussed in Chapter 3, magnesium and oxygen combine to form magnesium oxide. The tightly bound coating of aluminum oxide on an aluminum pan prevents the metal from reacting further with oxygen, but the brown iron(III) oxide coating on an iron pan flakes off, exposing the metal to more oxide formation (rusting). Nonmetals form covalent oxides. Chapter 3 mentions the formation of nitrogen monoxide from nitrogen and oxygen in automobile engines.

**FIGURE 4.2**
**A combination reaction between a metal and a nonmetal.** When burning steel wool (iron) reacts with chlorine gas, reddish brown iron(III) chloride forms.

**Combination of an element and a compound.** When a binary covalent compound reacts with a nonmetal, they form a larger covalent compound.

1. *Several nonmetal oxides react with additional oxygen to form "higher" oxides* (those with more oxygen atoms in each molecule):

$$2SO_2(g) + O_2(g) \rightarrow 2SO_3(g)$$

2. *Many nonmetal halides can combine with additional halogen:*

$$ClF_3(g) + F_2(g) \rightarrow ClF_5(l)$$

**Combination of two compounds.** Many combination reactions involve compounds as reactants.

1. *A metal oxide and a nonmetal oxide react to form an ionic compound with a polyatomic anion.* In this reaction, the $O^{2-}$ ion in the metal oxide combines with the nonmetal oxide to form a polyatomic oxoanion. A metal carbonate forms, for instance, when a metal oxide and carbon dioxide react. Lithium carbonate, an important antipsychotic drug, is an example:

$$Li_2O(s) + CO_2(g) \rightarrow Li_2CO_3(s)$$

Similarly, a metal oxide and silicon dioxide form a silicate in a key reaction during the industrial production of iron (Figure 4.3):

$$CaO(s) + SiO_2(s) \rightarrow CaSiO_3(l)$$

**FIGURE 4.3**
**A combination reaction between a metal oxide and a nonmetal oxide.** Calcium silicate forms in the production of iron from its ore in a blast furnace.

2. *Metal oxides react with water to form hydroxides (bases).* When lime (calcium oxide) is added to soil, it combines with the water present to form "slaked" lime (calcium hydroxide):

$$CaO(s) + H_2O(l) \rightarrow Ca(OH)_2(aq)$$

3. *Nonmetal oxides react with water to form acids.* Sulfur trioxide combines with water to form sulfuric acid, the most abundantly produced chemical in the world:

$$SO_3(g) + H_2O(l) \rightarrow H_2SO_4(aq)$$

4. *Hydrates result from the reaction of anhydrous ("without water") compounds with water.* Anhydrous calcium sulfate is used in the laboratory to remove traces of water from nonaqueous solvents or from air:

$$CaSO_4(s) + 2H_2O(g) \rightarrow CaSO_4 \cdot 2H_2O(s)$$

The rich and varied chemistry of carbon compounds displays several types of combination reactions. *Many carbon-containing compounds combine with hydrogen, halogen, or a hydrogen halide.* Some vegetable oils solidify when they combine with $H_2$; margarine and peanut butter are often prepared this way. Vinyl chloride, used in the production of polyvinyl chloride (PVC) plastic, is made when hydrogen chloride gas reacts with acetylene:

$$C_2H_2(g) + HCl(g) \rightarrow C_2H_3Cl(g)$$

In all combination reactions, *the number of reactant substances is greater than the number of product substances,* and in almost every case, the total moles of reactant are greater than the total moles of product.

### Decomposition Reactions: One Reactant Forms Two or More Products

**Decomposition** reactions are conceptually the reverse of combination reactions: in a decomposition reaction, one reactant breaks down into two or more products, $Z \rightarrow X + Y$. Typically, *the starting material is a compound, and the products are elements and/or smaller compounds.*

Decomposition reactions occur when the bonds within the reactant break and the atoms rearrange into products. This often occurs when the reactant absorbs energy, such as heat (thermal decomposition) or electricity (electrolytic decomposition).

**Thermal decomposition.** Some important types of thermal decomposition are shown next. The $\Delta$ (Greek capital letter *delta*) above the yield arrow indicates that heat is required.

1. *Many ionic compounds with oxoanions react to form a metal oxide and a non-metal oxide.* A gaseous product is often formed in these reactions. For example, carbonates yield carbon dioxide when heated. More than 200 million tons of lime are produced annually for cement, glass, and steel manufacture by heating limestone (calcium carbonate):

$$CaCO_3(s) \xrightarrow{\Delta} CaO(s) + CO_2(g)$$

Many sulfites yield sulfur dioxide:

$$FeSO_3(s) \xrightarrow{\Delta} FeO(s) + SO_2(g)$$

2. *Many metal oxides, chlorates, and perchlorates release oxygen.* As shown in Figure 4.4, the decomposition of mercury(II) oxide, used by Lavoisier and Priestley in their classic experiments, is an example.

**FIGURE 4.4**
**Three views of a decomposition reaction.** In a decomposition reaction, one reactant forms more than one product. Heating solid mercury(II) oxide decomposes it to its elements, liquid mercury and gaseous oxygen: the macroscopic (laboratory) view *(top)*; the atomic-scale view, with the more darkly colored atoms showing the stoichiometry *(middle)*; and the balanced equation *(bottom)*.

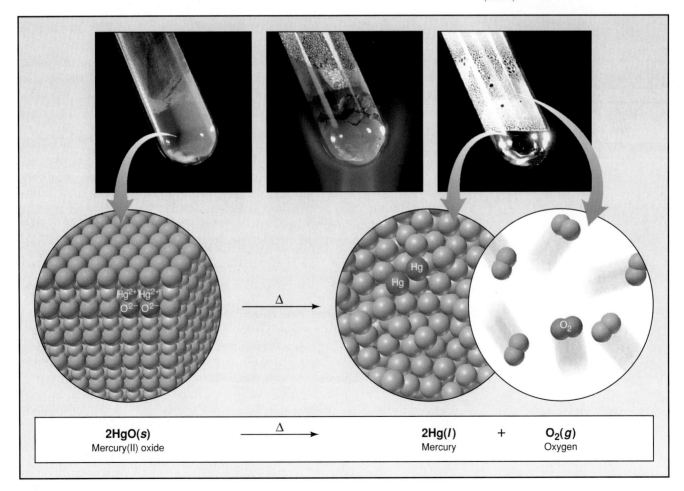

| 2HgO(s) | $\xrightarrow{\Delta}$ | 2Hg(*l*) | + | O$_2$(*g*) |
|---|---|---|---|---|
| Mercury(II) oxide | | Mercury | | Oxygen |

**FIGURE 4.5**

Decomposition of an ionic hydrate. Bright blue copper(II) sulfate pentahydrate loses water when heated.

Heating potassium chlorate is a common method for forming small amounts of oxygen in the laboratory:

$$2KClO_3(s) \xrightarrow{\Delta} 2KCl(s) + 3O_2(g)$$

3. *Hydroxides, hydrates, and some oxoacids release water:*

$$Mg(OH)_2(s) \xrightarrow{\Delta} MgO(s) + H_2O(g)$$

$$H_2SO_3(l) \xrightarrow{\Delta} SO_2(g) + H_2O(g)$$

Blue copper(II) sulfate pentahydrate loses water and becomes white, anhydrous copper(II) sulfate (Figure 4.5):

$$CuSO_4 \cdot 5H_2O(s) \xrightarrow{\Delta} CuSO_4(s) + 5H_2O(g)$$

**Electrolytic decomposition.** Electrical energy can decompose many compounds through the process of *electrolysis.*

1. *Decomposition of water.* This process, shown in Figure 1.1, *B,* was a key observation in the establishment of atomic masses:

$$2H_2O(l) \xrightarrow{\text{electricity}} 2H_2(g) + O_2(g)$$

2. *Decomposition of molten binary ionic compounds.* The electrolysis of molten halite, a mineral form of sodium chloride, produces sodium and chlorine. Magnesium, calcium, and several other metals are produced industrially by electrolysis:

$$MgCl_2(l) \xrightarrow{\text{electricity}} Mg(l) + Cl_2(g)$$

In almost every decomposition reaction, *there are more moles of products than reactant.* Notice that many types of decomposition reactions are the reverse of the types of combination reactions mentioned earlier.

### Displacement Reactions: Number of Reactants Equals Number of Products

In contrast to combination or decomposition reactions, **displacement** reactions, or *replacement* reactions, usually start and end with the same number of substances: an atom or ion in a compound is displaced by an atom or ion of another element. Displacement reactions are often classified further as single-displacement and double-displacement **(metathesis)** reactions:

$$X + YZ \rightarrow XZ + Y \text{ [single]}$$

$$WX + YZ \rightarrow WZ + YX \text{ [double; metathesis]}$$

**Single-displacement reactions.** Some important examples of single-displacement reactions are listed below.

1. *A metal displaces hydrogen from either water or an acid.* Metals vary in their reactivity, which is reflected in their ability to displace hydrogen from different sources. The most reactive metals can displace hydrogen from water. The Group 1A(1) metals and some members of Group 2A(2) (Ca, Sr, and Ba) react vigorously with water. Figure 4.6 shows the reaction of lithium. Heat is needed to speed the reaction of less reactive metals. Aluminum and zinc, for example, displace hydrogen from steam:

$$2Al(s) + 6H_2O(g) \rightarrow 2Al(OH)_3(s) + 3H_2(g)$$

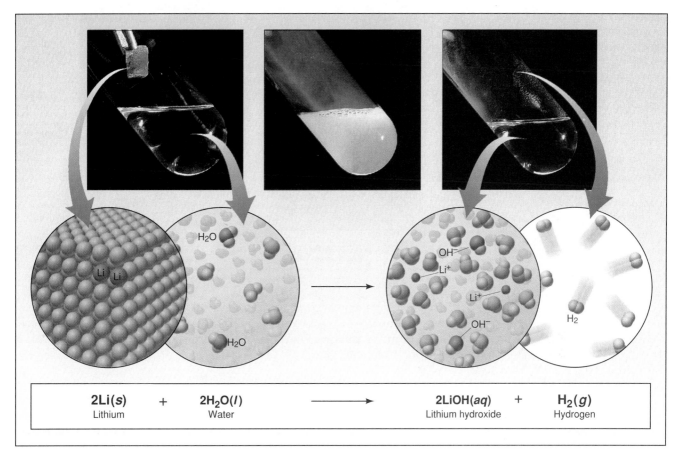

**2Li(*s*)** + **2H₂O(*l*)** ⟶ **2LiOH(*aq*)** + **H₂(*g*)**
Lithium Water Lithium hydroxide Hydrogen

**FIGURE 4.6**

**Three views of a single-displacement reaction.** In a displacement reaction, the number of reactants equals the number of products. In this single-displacement reaction, lithium displaces hydrogen from water. The vigorous reaction forms an aqueous solution of lithium hydroxide and hydrogen gas, as shown on the macroscopic scale *(top)*, on the atomic scale *(middle)*, and with the balanced equation *(bottom)*. (In the atomic-scale view, the stoichiometry is indicated by more darkly colored atoms; for clarity, the atomic view of water has been greatly simplified, and all water molecules *not* involved in the reaction are colored blue rather than red and blue.)

Still less reactive metals do not react with water but do react with acids, from which the hydrogen is displaced more easily. Nickel, for instance, reacts with hydrochloric acid:

$$Ni(s) + 2HCl(aq) \longrightarrow NiCl_2(aq) + H_2(g)$$

The least reactive metals, such as silver and gold, cannot displace hydrogen even from acids.

2. *A metal displaces another metal ion from solution.* Metals can also be ranked by their ability to displace one another from solution. Zinc metal can displace copper(II) ion from solution, so zinc is more reactive than copper:

$$Cu(NO_3)_2(aq) + Zn(s) \longrightarrow Cu(s) + Zn(NO_3)_2(aq)$$

Copper metal can, in turn, displace silver ion from solution (Figure 4.7); thus, copper is more reactive than silver.

**FIGURE 4.7**

**Copper displaces silver ions from solution.** More reactive metals can displace less reactive ones from solution. In this single-displacement reaction, copper atoms become $Cu^{2+}$ ions and leave the wire, while $Ag^+$ ions become silver atoms and coat the wire.

Copper wire

Silver nitrate solution

Copper wire coated with silver

Copper nitrate solution

$$2AgNO_3(aq) + Cu(s) \longrightarrow Cu(NO_3)_2(aq) + 2Ag(s)$$

3. *A halogen displaces another halide ion from solution.* Reactivity decreases down Group 7A(17); therefore, a halogen higher in the group can displace the anion of a halogen lower down. For example, chlorine displaces bromide ion from solution:

$$2KBr(aq) + Cl_2(aq) \rightarrow Br_2(aq) + 2KCl(aq)$$

Later in this chapter (and again in Chapter 20), we examine more closely the nature of the chemical process in these reactions.

**Double-displacement reactions.** Three common aqueous double-displacement reactions are presented here and discussed in Section 4.3. ◆

1. *In precipitation reactions, an insoluble ionic compound forms.* Silver bromide, which is used to manufacture black-and-white film, is formed in a double-displacement reaction:

$$AgNO_3(aq) + KBr(aq) \rightarrow AgBr(s) + KNO_3(aq)$$

2. *In neutralization reactions, an acid and a base form an ionic compound and water.* Some antacid tablets function through neutralization reactions:

$$3HCl(aq) + Al(OH)_3(s) \rightarrow AlCl_3(aq) + 3H_2O(l)$$

3. *A carbonate or sulfite reacts with acid to form a gas.* In a field test for carbonate-containing rocks, treating the sample with hydrochloric acid produces bubbles of gas:

$$MgCO_3(s) + 2HCl(aq) \rightarrow MgCl_2(aq) + H_2O(l) + CO_2(g)$$

(We did not show the intermediate product $H_2CO_3$ because it decomposes immediately to water and carbon dioxide; $H_2SO_3$ behaves similarly.)

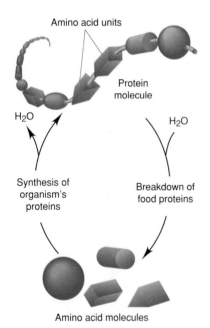

Amino acid units

Protein molecule

$H_2O$

$H_2O$

Synthesis of organism's proteins

Breakdown of food proteins

Amino acid molecules

◆ **Displacement Reactions Inside You.** The digestion of food proteins is joined to the formation of the cell's own proteins in continuous cycles of displacement reactions. A protein consists of hundreds or thousands of smaller molecules, called *amino acids*, linked in a long chain. When you eat vegetable or animal proteins, your digestive processes use $H_2O$ to displace one amino acid at a time. These are transported by the blood to your cells, where other metabolic processes link them together, while displacing $H_2O$, to make your own specific proteins.

SAMPLE PROBLEM 4.1 _____

**Identifying the Type of Chemical Reaction**

**Problem:** Classify the following as combination, decomposition, or displacement reactions, and write a balanced equation for each:
**(a)** Ammonium dichromate(s) →
                    nitrogen(g) + chromium(III) oxide(s) + water vapor
**(b)** Magnesium(s) + nitrogen(g) → magnesium nitride(s)
**(c)** $C_2H_4(COOH)_2(s)$ + sodium hydroxide(aq) → $C_2H_4(COONa)_2(aq)$ + water
**(d)** Hydrogen peroxide(l) → water + oxygen gas

**(e)** Ammonia($g$) + hydrogen chloride($g$) $\rightarrow$ ammonium chloride($s$)

**(f)** Aluminum($s$) + lead(II) nitrate($aq$) $\rightarrow$ aluminum nitrate($aq$) + lead($s$)

**Plan:** To decide on reaction type, we keep track of the number of substances on the two sides of the equation. Combination reactions produce fewer substances, decomposition reactions more substances, and displacement reactions the same number of substances.

**Solution: (a) Decomposition:** one substance forms three. This well-known but *hazardous* lecture demonstration is called the ammonium dichromate "volcano" (see photo):

$$(NH_4)_2Cr_2O_7(s) \xrightarrow{\Delta} N_2(g) + Cr_2O_3(s) + 4H_2O(g)$$

**(b) Combination:** two substances form one. This reaction occurs, along with formation of the metal oxide, when magnesium burns in air:

$$3Mg(s) + N_2(g) \rightarrow Mg_3N_2(s)$$

**(c) Displacement:** two substances form two. Succinic acid, an important biomolecule, reacts with sodium hydroxide in this neutralization reaction, which is a double displacement. The hydrogens from the two acid (COOH) groups bind to the hydroxide ions to form water:

$$C_2H_4(COOH)_2(s) + 2NaOH(aq) \rightarrow C_2H_4(COONa)_2(aq) + 2H_2O(l)$$

**(d) Decomposition:** one substance forms two. This reaction occurs within every bottle of the antiseptic hydrogen peroxide. This household medicine is stored in dark-colored bottles because the reactant is decomposed by *light:*

$$2H_2O_2(l) \rightarrow 2H_2O(l) + O_2(g)$$

**(e) Combination:** two substances form one. This reaction was once used to create a magician's "smoke screen" (see photo):

$$NH_3(g) + HCl(g) \rightarrow NH_4Cl(s)$$

**(f) Displacement:** two substances form two. This reaction (single displacement) occurs when a more reactive metal (Al) displaces a less reactive one (Pb) from solution:

$$2Al(s) + 3Pb(NO_3)_2(aq) \rightarrow 2Al(NO_3)_3(aq) + 3Pb(s)$$

**Check:** As always, be sure that the number of each type of reactant atom equals the number of that type of product atom.

Ammonium dichromate volcano.

**FOLLOW-UP PROBLEM 4.1**

Classify and balance each of the following reactions:

**(a)** $CaSO_4 \cdot \frac{1}{2}H_2O(s) + H_2O(l) \rightarrow CaSO_4 \cdot 2H_2O(s)$

**(b)** $Na_2SO_3(aq) + HCl(aq) \rightarrow NaCl(aq) + SO_2(g) + H_2O(l)$

**(c)** $NH_4NO_3(s) \xrightarrow{\Delta} N_2O(g) + H_2O(g)$

Smoke screen.

## Section Summary

Many chemical reactions can be classified by the number of reactants versus products. Combination reactions occur when two or more reactants form one product. All reactions between elements are of this type, and combination with oxygen is especially common. In decomposition reactions, one reactant forms two or more products. The reactant is decomposed by heat, light, or electricity. In displacement reactions, the number of reactants equals the number of products. In an important type of single-displacement reaction, metal reactivity is based on the ability to displace hydrogen from water or from acid. More reactive metals (or halogens) can also displace less reactive ones from solution. Precipitation and neutralization are common types of double-displacement reactions.

## 4.2  The Role of Water as a Solvent

Focusing on the number of reactant and product substances is a good starting point for classifying reactions, but it provides little insight into the underlying chemical process. Because so many reactions take place in water, our first step toward comprehending what is actually happening during a reaction is to understand the part played by water as a solvent. How a solvent participates in a reaction depends on its chemical nature. Some solvents play a passive role, dispersing the substances into individual molecules but not interacting with them much beyond that. Others take a much more active part, interacting strongly with the dispersed reactants and, in some cases, even affecting the bonds within the reactant molecules themselves. Because water is the most important example of an active solvent and is crucial to so many chemical and physical processes, we will discuss its properties frequently.

### The Solubility of Ionic Compounds

We can observe the result of the interaction between water and a soluble ionic compound with a simple apparatus that measures *electrical conductivity,* the flow of electric current (Figure 4.8). When electrodes are immersed in pure water or pushed into solid potassium bromide (KBr), no current flows. After the KBr dissolves in the water, however, a significant current flows, as the brightly lit bulb shows. The current flow in the solution implies the *movement of charged particles:* when KBr dissolves in water, the $K^+$ and $Br^-$ ions bound in the solid separate (dissociate), and each type of ion moves through the solution toward the oppositely charged electrode. A substance, such as KBr, that conducts a current when it dissolves in water is an **electrolyte.**

In the process of dissolving, each ion in KBr becomes **solvated,** that is, surrounded by solvent molecules. We express the dissociation of KBr into solvated ions as follows:

$$KBr(s) \xrightarrow{\text{H}_2\text{O}} K^+(aq) + Br^-(aq)$$

The $H_2O$ above the arrow indicates that water is required for the change to occur but is not a reactant in the usual sense. When any water-soluble ionic compound dissolves, *the oppositely charged ions separate from each other, become surrounded by water molecules, and move throughout the solution.* The chemical formula tells us the number of moles of different ions that form when an ionic compound dissolves. In the case of KBr, one mole of compound dissociates into two moles of ions.

SAMPLE PROBLEM 4.2 _____

### Determining Moles of Ions in Aqueous Solutions of Ionic Compounds

**Problem:** How many moles of each ion are in each of the following solutions?
**(a)** 5.0 mol ammonium phosphate dissolved in water
**(b)** 78.5 g cesium iodide dissolved in water
**(c)** $7.42 \times 10^{22}$ formula units copper(II) nitrate dissolved in water
**(d)** 35 mL of 0.84 $M$ zinc chloride

**A** Distilled water does not conduct a current

**B** Positive and negative ions fixed in a solid do not conduct a current

**C** In solution, positive and negative ions move and conduct a current

To (+) electrode

To (−) electrode

**FIGURE 4.8**
**The electrical conductivity of ionic solutions as a result of mobile ions. A,** When electrodes connected to a power source are placed in distilled water, no current flows and the bulb is unlit. **B,** A solid ionic compound, such as KBr, conducts no current because the ions are bound tightly together. **C,** When KBr dissolves in $H_2O$, the ions separate and move through the solution, thereby conducting a current.

**Plan:** We write the formula and then write an equation that shows the moles of ions released when *one mole* of compound dissolves. In **(a)**, we multiply the moles of ions released by 5.0. In **(b)**, we first convert mass to moles. In **(c)**, we first convert formula units to moles. In **(d)**, we first convert molarity and volume to moles.

**Solution: (a)** $(NH_4)_3PO_4(s) \xrightarrow{H_2O} 3NH_4^+(aq) + PO_4^{3-}(aq)$
Remember that *polyatomic ions remain as intact units in solution.*
Calculating moles of $NH_4^+$ ions:

$$\text{Moles of } NH_4^+ = 5.0 \ \cancel{\text{mol } (NH_4)_3PO_4} \times \frac{3 \text{ mol } NH_4^+}{1 \ \cancel{\text{mol } (NH_4)_3PO_4}} = \textbf{15 mol } NH_4^+$$

**5.0 mol $PO_4^{3-}$** are also present.

**(b)** $CsI(s) \xrightarrow{H_2O} Cs^+(aq) + I^-(aq)$
Converting from mass to moles:

$$\text{Moles of CsI} = 78.5 \ \cancel{\text{g CsI}} \times \frac{1 \text{ mol CsI}}{259.8 \ \cancel{\text{g CsI}}} = 0.302 \text{ mol CsI}$$

Thus, **0.302 mol $Cs^+$** and **0.302 mol $I^-$** are present.

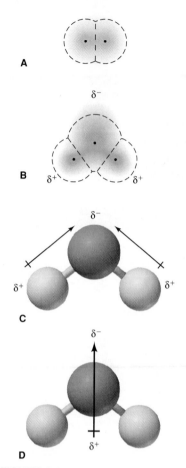

**FIGURE 4.9**

Electron distribution in molecules of $H_2$ and $H_2O$. **A,** In a pure covalent bond, such as in $H_2$, the identical nuclei attract the electrons equally, so an even distribution of negative charge exists. Note the darker region spread equally around and between the nuclei. **B,** In the polar covalent bonds of water, the O nucleus attracts the shared electrons more strongly than the H nucleus, creating an uneven charge distribution and a polar covalent bond, with the partially negative O end designated $\delta^-$ and the partially positive H end of each bond designated $\delta^+$. The dashed circles represent the outline of the space-filling model to show that the electrons are pulled toward the O atom and away from each H atom. **C,** In this ball-and-stick model of $H_2O$, a polar arrow points to the negative end of each O—H bond. **D,** The two polar O—H bonds and the bent molecular shape give rise to a polar $H_2O$ molecule, with the partially positive end located *between* the two H atoms.

**(c)** $Cu(NO_3)_2(s) \xrightarrow{H_2O} Cu^{2+}(aq) + 2NO_3^-(aq)$

Converting from formula units to moles:

$$\text{Moles of } Cu(NO_3)_2 = 7.42 \times 10^{22} \text{ formula units } Cu(NO_3)_2$$

$$\times \frac{1 \text{ mol } Cu(NO_3)_2}{6.022 \times 10^{23} \text{ formula units } Cu(NO_3)_2}$$

$$= 0.123 \text{ mol } Cu(NO_3)_2$$

$$\text{Moles of } NO_3^- = 0.123 \text{ mol } Cu(NO_3)_2 \times \frac{2 \text{ mol } NO_3^-}{1 \text{ mol } Cu(NO_3)_2}$$

$$= \mathbf{0.246 \text{ mol } NO_3^-}$$

**0.123 mol $Cu^{2+}$** is also present.

**(d)** $ZnCl_2(aq) \rightarrow Zn^{2+}(aq) + 2 Cl^-(aq)$

Converting from volume to moles:

$$\text{Moles of } ZnCl_2 = 35 \text{ mL} \times \frac{1 \text{ L}}{10^3 \text{ mL}} \times \frac{0.84 \text{ mol } ZnCl_2}{1 \text{ L}} = 2.9 \times 10^{-2} \text{ mol } ZnCl_2$$

$$\text{Moles of } Cl^- = 2.9 \times 10^{-2} \text{ mol } ZnCl_2 \times \frac{2 \text{ mol } Cl^-}{1 \text{ mol } ZnCl_2} = \mathbf{5.8 \times 10^{-2} \text{ mol } Cl^-}$$

**$2.9 \times 10^{-2}$ mol $Zn^{2+}$** is also present.

**Check:** After you round off to check the math, see if the relative moles of ions are consistent with the formula. For instance, in (a), 15 mol $NH_4^+$/5.0 mol $PO_4^{3-}$ = 3 $NH_4^+$/1 $PO_4^{3-}$, or $(NH_4)_3PO_4$. In (d), 0.029 mol $Zn^{2+}$/0.058 mol $Cl^-$ = 1 $Zn^{2+}$/ 2 $Cl^-$, or $ZnCl_2$.

**FOLLOW-UP PROBLEM 4.2**

How many moles of each ion are in each of the following solutions?

**(a)** 2 mol potassium sulfate dissolved in water
**(b)** 354 g magnesium perchlorate dissolved in water
**(c)** $1.88 \times 10^{24}$ formula units of ammonium dichromate dissolved in water
**(d)** 1.32 L of 0.55 $M$ sodium bicarbonate

## The Polar Nature of Water

Water separates ions by overcoming the *electrostatic force* of attraction between them. To see how it does this, let's closely examine the water molecule. Water's power as an ionizing solvent results from two characteristics of its molecular structure: *the distribution of its bonding electrons and its overall shape.*

Recall that the electrons in a covalent bond are shared between the bonded atoms (Section 2.6). In a *pure* covalent bond, which exists between identical atoms (as in $H_2$, $Cl_2$, and $O_2$), the sharing is equal: the shared electrons are attracted equally by the two nuclei (Figure 4.9, *A*). In covalent bonds between atoms that are *not* the same, the sharing is unequal: one atom attracts the electron pair more strongly than the other. In each of the O—H bonds of water, for example, the O atom attracts the electron pair more strongly than the H atom does, so the electrons spend more time closer to the O (Figure 4.9, *B*). In this type of bond, called a **polar covalent bond,** the electron pair's negative charge is distributed unequally, creating partially charged poles at the two ends of the bond. In the case of water, the oxygen end of each O—H bond acts as a slightly negative pole (represented as $\delta^-$), and the hydrogen end acts as a slightly positive pole (represented as $\delta^+$). The bond's polarity is indicated by a *polar arrow,* with the arrowhead pointing to the negative pole (Figure 4.9, *C*). It is important to realize that the partial charges on the atoms in a polar covalent bond are *much less* than full ionic charges. In KBr, for instance, the electron has been *transferred* from the K atom to the Br atom, forming ions. In the polar O—H bond, no ions exist; the electrons have just *shifted* their average position nearer to the O atom. (We discuss polar covalent bonds in detail in Chapter 9.)

In addition to its polar covalent bonds, the water molecule has a bent shape: the H—O—H atoms do not lie in a straight line but form an angle. The combined effect of its shape and its polar covalent bonds makes water a **polar molecule:** the O portion of the molecule is the partially negative pole, and the region midway between the H atoms is the partially positive pole (Figure 4.9, *D*).

**Ionic compounds in water.** With these bent, polar water molecules in mind, imagine a granule of an ionic compound in water (Figure 4.10). Water molecules congregate over the granule's surface; the negative ends of some are attracted to the cations, and the positive ends of others are attracted to the anions. In the electrostatic "tug of war" that ensues, the attraction between each ion and the water molecules replaces the attraction of the ions for each other. As a result, the ions are pulled apart and surrounded by solvent. This scene occurs whenever an ionic compound dissolves in water.

Although many ionic compounds dissolve in water, some do not. In these cases, the electrostatic attraction among ions in the compound is too great for the water molecules to pull them apart, and the substance remains undissolved. Actually, these so-called insoluble substances *do* dissolve to a very small extent, usually several orders of magnitude less than so-called soluble substances. Compare, for example, the dissolved amounts of NaCl (soluble) and AgCl (insoluble):

Solubility of NaCl in $H_2O$ at 20°C = 365 g/L
Solubility of AgCl in $H_2O$ at 20°C = 0.009 g/L

We examine the question of solubility quantitatively in Chapter 12; for now, the simple distinction between "soluble" and "insoluble" is useful in discussing precipitation reactions in the next section.

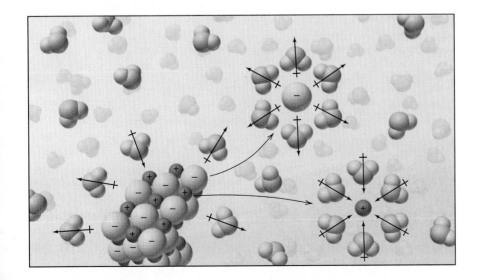

**FIGURE 4.10**

**The dissolution of an ionic compound.** When an ionic compound dissolves in water, $H_2O$ molecules separate, surround, and disperse the ions into the liquid. Note that the negative ends of the $H_2O$ molecules face the positive ions and the positive ends face the negative ions.

**Covalent compounds in water.** Water is also an excellent solvent for many covalent substances; table sugar (sucrose, $C_{12}H_{22}O_{11}$), beverage (grain) alcohol (ethanol, $C_2H_6O$), and automobile antifreeze (ethylene glycol, $C_2H_6O_2$) are some familiar water-soluble covalent compounds. All contain their own polar covalent bonds, which interact with those of water. Even though these substances dissolve, they do *not* dissociate into ions *but remain as intact molecules.* Since their aqueous solutions do not conduct an electric current, these substances are **nonelectrolytes.** Many other covalent substances, such as benzene ($C_6H_6$) and octane ($C_8H_{18}$), do not contain polar bonds and thus do not dissolve appreciably in water.

A small but very important group of hydrogen-containing, water-soluble covalent compounds interacts so strongly with water that their molecules *do dissociate into ions.* In aqueous solution they are all *acids,* as you'll see in the discussion of neutralization reactions in the next section. The molecules of these substances contain polar covalent bonds to hydrogen, such that the atom bonded to H pulls more strongly on the shared electron pair. A good example is hydrogen chloride gas. The Cl end of the HCl molecule is partially negative, and the H end is partially positive. When HCl dissolves in water, the partially charged poles of $H_2O$ molecules are attracted to the oppositely charged poles of HCl. The H—Cl bond breaks, with the H becoming the solvated cation $H^+(aq)$ and the Cl becoming the solvated anion $Cl^-(aq)$. Hydrogen bromide behaves similarly:

$$HBr(g) \xrightarrow{H_2O} H^+(aq) + Br^-(aq)$$

SAMPLE PROBLEM 4.3 _____

### Determining the Molarity of $H^+$ Ions in Aqueous Solutions of Acids

**Problem:** Nitric acid is a major chemical in the fertilizer and explosives industries. In aqueous solution, each acid molecule loses its H, which becomes a solvated $H^+$ ion. What is the molarity of $H^+(aq)$ in 1.4 $M$ nitric acid?
**Plan:** The molarity of acid is given, so we determine the formula to find the number of moles of $H^+(aq)$ present in 1 L solution.
**Solution:** One mole of $H^+(aq)$ is released per mole of nitric acid ($HNO_3$):

$$HNO_3(l) \xrightarrow{H_2O} H^+(aq) + NO_3^-(aq)$$

Therefore, 1.4 $M$ $HNO_3$ contains 1.4 mol $H^+(aq)$/L, or **1.4 $M$ $H^+(aq)$.**

FOLLOW-UP PROBLEM 4.3
How many moles of $H^+(aq)$ are present in 451 mL of 3.20 $M$ hydrobromic acid?

_____

The solvated hydrogen cation, $H^+(aq)$, is a particularly strange species. Because the $H^+$ ion is just a single proton, its positive charge is concentrated in an extremely tiny volume. Therefore, it attracts the negative pole of surrounding water molecules very strongly and becomes very tightly associated with them, forming a covalent bond to one of them. This is often indicated by writing the aqueous $H^+$ ion as $H_3O^+$ (hydronium ion); to make a point, we write it here as $(H_2O)H^+$. The hydronium ion is associated with other water molecules in a mixture of H-containing ions that includes $H_5O_2^+$ [$(H_2O)_2H^+$], $H_7O_3^+$ [$(H_2O)_3H^+$], $H_9O_4^+$ [$(H_2O)_4H^+$], and higher aggregates. These various forms of the solvated $H^+$ ion exist together in solution, so we often use $H^+(aq)$ as a general notation. Later in the text, we emphasize the role of water by showing the solvated proton as $H_3O^+(aq)$.

**Section Summary**

When an ionic compound dissolves in water, the ions dissociate and are solvated by water molecules. Because the ions are free to move, their solutions conduct electricity. Water plays an active role in dissolving ionic compounds because it consists of polar molecules that are attracted to the ions. Water also dissolves many covalent substances that have polar molecules, and it interacts with some H-containing molecules so strongly that it breaks covalent bonds and dissociates them into $H^+(aq)$ ions and anions.

## 4.3   Some Important Aqueous Ionic Reactions: Writing Ionic Equations

As you have seen, a water-soluble ionic compound dissociates in water into separated, solvated ions. When solutions of two ionic compounds are mixed, *a reaction occurs only if some ions are removed from solution in the formation of a product.* When we prepare aqueous solutions of calcium chloride and potassium nitrate, for example, their ions disperse throughout the solution:

$$CaCl_2(s) \xrightarrow{H_2O} Ca^{2+}(aq) + 2Cl^-(aq)$$

$$KNO_3(s) \xrightarrow{H_2O} K^+(aq) + NO_3^-(aq)$$

Does a reaction take place if we mix these solutions? In this case, no product forms because all the possible ion combinations—$CaCl_2$, $KNO_3$, $Ca(NO_3)_2$, and KCl—are soluble: the ions simply remain in solution. In this section, we discuss three important types of aqueous ionic reactions that occur because reactant ions *are* removed from solution. All are the double-displacement (metathesis) type: (1) precipitation reactions, (2) neutralization reactions, and (3) reactions that form a gaseous product. Note the way we depict the *actual* reaction with balanced ionic equations.

### Precipitation Reactions: Formation of an Insoluble Product

In **precipitation** reactions, two soluble ionic compounds react to form an insoluble product, a **precipitate.** For example, when solutions of silver nitrate and sodium chromate are mixed, a brick-red precipitate of $Ag_2CrO_4$ forms (Figure 4.11). We can represent this or any ionic reaction with three types of balanced equations: molecular, total ionic, and net ionic.

The **molecular equation** shows *the reactants and products as if they were intact, undissociated compounds*:

$$2AgNO_3(aq) + Na_2CrO_4(aq) \rightarrow Ag_2CrO_4(s) + 2NaNO_3(aq)$$

The molecular equation is the least revealing picture of the actual situation because the soluble compounds are not shown as ions.

We can represent the reaction more realistically with the **total ionic equation,** which shows *all the soluble ionic substances dissociated into ions.* Now the $Ag_2CrO_4(s)$ stands out as the only insoluble substance:

$$2Ag^+(aq) + 2NO_3^-(aq) + 2Na^+(aq) + CrO_4^{2-}(aq) \rightarrow$$

$$Ag_2CrO_4(s) + 2Na^+(aq) + 2NO_3^-(aq)$$

**FIGURE 4.11**

**A precipitation reaction and its three equations.** When silver nitrate and sodium chromate solutions are mixed, a precipitation reaction occurs that forms solid silver chromate and a solution of sodium nitrate. The molecular equation shows all substances intact. The total ionic equation shows all *soluble* substances as separate, solvated ions. The net ionic equation shows only the reacting species.

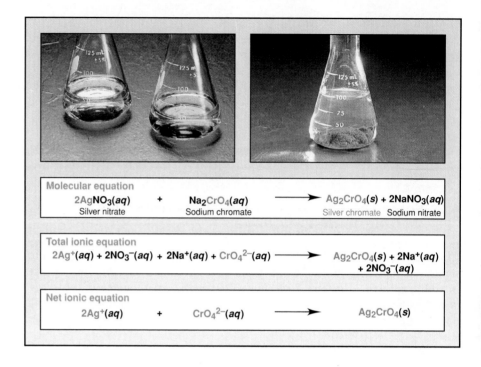

Molecular equation

$$2AgNO_3(aq) \ + \ Na_2CrO_4(aq) \longrightarrow Ag_2CrO_4(s) + 2NaNO_3(aq)$$

Silver nitrate          Sodium chromate                    Silver chromate  Sodium nitrate

Total ionic equation

$$2Ag^+(aq) + 2NO_3^-(aq) + 2Na^+(aq) + CrO_4^{2-}(aq) \longrightarrow Ag_2CrO_4(s) + 2Na^+(aq) + 2NO_3^-(aq)$$

Net ionic equation

$$2Ag^+(aq) \ + \ CrO_4^{2-}(aq) \longrightarrow Ag_2CrO_4(s)$$

Notice that the numbers of $Na^+(aq)$ and $NO_3^+(aq)$ ions are *unchanged* on both sides of the equation. They are referred to as **spectator ions** because they are not involved in the chemical change but are present as part of the reactants. In other words, we could not have added $Ag^+$ ions without also adding an anion, in this case $NO_3^-$ ion.

In the **net ionic equation,** the spectator ions are eliminated, and we see the *actual chemical change taking place*:

$$2Ag^+(aq) + CrO_4^{2-}(aq) \rightarrow Ag_2CrO_4(s)$$

The formation of solid silver chromate from silver ions and chromate ions is the only chemical reaction occurring. If we had originally mixed solutions of potassium chromate, $K_2CrO_4(aq)$, and silver acetate, $AgC_2H_3O_2(aq)$, the same reaction would have occurred, but the spectator ions would be $K^+(aq)$ and $C_2H_3O_2^-(aq)$ instead of $Na^+(aq)$ and $NO_3^-(aq)$. Writing the reaction as a net ionic equation is an excellent way to isolate mentally the key chemical event.

Precipitates form for the same reason that some ionic compounds are insoluble: the electrostatic attraction between the ions that make up the precipitate is too great for the water molecules to keep them apart. When the ions collide, they stay together and the substance "comes out of solution" as an insoluble solid. Thus, the event that "drives" a precipitation reaction to occur is *the removal of ions from solution in the form of an insoluble compound.*

How can we predict whether a precipitate will form when we mix two aqueous solutions of ionic compounds? To do so, we note the ions present in the reactants, consider the possible cation-anion combinations, and determine whether any combination is insoluble. Of the four possible ionic combinations in the case we just discussed—silver nitrate, sodium nitrate, sodium chromate, and silver chromate—only the combination of $Ag^+(aq)$ ions and $CrO_4^{2-}(aq)$ ions leads to an insoluble substance. However, there is no simple principle you can rely on to predict the solubility of a substance;

4.3    **Some Important Aqueous Ionic Reactions; Writing Ionic Equations**

**TABLE 4.1    Solubility Rules for Ionic Compounds in Water**

| SOLUBLE IONIC COMPOUNDS | INSOLUBLE IONIC COMPOUNDS |
|---|---|
| 1. All common compounds of Group 1A(1) ions ($Na^+$, $K^+$, etc.) and ammonium ion ($NH_4^+$) are soluble. <br> 2. All common nitrates ($NO_3^-$), acetates ($CH_3COO^-$), and most perchlorates ($ClO_4^-$) are soluble. <br> 3. All common chlorides ($Cl^-$), bromides ($Br^-$), and iodides ($I^-$) are soluble, *except* those of $Ag^+$, $Pb^{2+}$, $Cu^+$, and $Hg_2^{2+}$. <br> 4. All common sulfates ($SO_4^{2-}$) are soluble, *except* those of $Ca^{2+}$, $Sr^{2+}$, $Ba^{2+}$, and $Pb^{2+}$. | 1. All common metal hydroxides are insoluble, *except* those of Group 1A(1) and the larger members of Group 2A(2) (beginning with $Ca^{2+}$). <br> 2. All common carbonates ($CO_3^{2-}$) and phosphates ($PO_4^{3-}$) are insoluble, *except* those of Group 1A(1) and $NH_4^+$. <br> 3. All common sulfides are insoluble, *except* those of Group 1A(1), Group 2A(2), and $NH_4^+$. |

you need to memorize the solubility rules presented in Table 4.1. Although they do not cover every ionic compound, learning these few rules allows you to predict the outcome of many precipitation reactions.

SAMPLE PROBLEM 4.4 ———————————————————————————

### Predicting Whether a Precipitation Reaction Occurs; Writing Molecular, Total, and Net Ionic Equations

**Problem:** Predict whether a reaction occurs when each of the following pairs of solutions are mixed. If a reaction does occur, write balanced molecular, total ionic, and net ionic equations, and identify the spectator ions.
**(a)** Sodium sulfate(*aq*) + lead(II) nitrate(*aq*) →
**(b)** Ammonium perchlorate(*aq*) + sodium bromide(*aq*) →
**Plan:** For each pair of solutions, we write the cation-anion combinations and refer to Table 4.1 to see if any are insoluble. For the molecular equation, we predict the products. For the total ionic equation, we write the soluble compounds as separate ions. For the net ionic equation, we eliminate the spectator ions.
**Solution:    (a)** In addition to the reactants, the two other ion combinations are lead(II) sulfate and sodium nitrate. Table 4.1 shows that lead(II) sulfate is insoluble, so a reaction does occur. Writing the molecular equation:

$$Na_2SO_4(aq) + Pb(NO_3)_2(aq) \rightarrow PbSO_4(s) + 2NaNO_3(aq)$$

Writing the total ionic equation:

$$2Na^+(aq) + SO_4^{2-}(aq) + Pb^{2+}(aq) + 2NO_3^-(aq) \rightarrow$$
$$PbSO_4(s) + 2Na^+(aq) + 2NO_3^-(aq)$$

Writing the net ionic equation:

$$Pb^{2+}(aq) + SO_4^{2-}(aq) \rightarrow PbSO_4(s)$$

The spectator ions are **$Na^+$ and $NO_3^-$.**
**(b)** The product ion combinations are ammonium bromide and sodium perchlorate. Table 4.1 shows that all ammonium and sodium salts are soluble, and that all perchlorates are soluble and all bromides are soluble except those of $Ag^+$, $Pb^{2+}$, and $Hg_2^{2+}$. Therefore, **no reaction occurs** between these ions. The compounds remain in solution as *solvated* ions.

FOLLOW-UP PROBLEM 4.4
Predict whether a reaction occurs and write balanced total and net ionic equations:
**(a)** Iron(III) chloride(*aq*) + cesium phosphate(*aq*) →
**(b)** Potassium hydroxide(*aq*) + cadmium nitrate(*aq*) →
**(c)** Magnesium iodide(*aq*) + sodium acetate(*aq*) →
**(d)** Silver sulfate(*aq*) + barium chloride(*aq*) →

## TABLE 4.2   Common Acids and Bases

**ACIDS**

*Strong*
Hydrochloric acid, HCl
Hydrobromic acid, HBr
Hydriodic acid, HI
Nitric acid, $HNO_3$
Sulfuric acid, $H_2SO_4$
Perchloric acid, $HClO_4$

*Weak*
Hydrofluoric acid, HF
Phosphoric acid, $H_3PO_4$
Acetic acid, $CH_3COOH$
(or $HC_2H_3O_2$)

**BASES**

*Strong*
Sodium hydroxide, NaOH
Potassium hydroxide, KOH
Calcium hydroxide, $Ca(OH)_2$
Barium hydroxide, $Ba(OH)_2$

*Weak*
Ammonia, $NH_3$

## Neutralization (Acid-Base) Reactions: Formation of Water from Hydrogen and Hydroxide Ions

Another major type of aqueous ionic reaction called a neutralization (or acid-base) reaction, involves water as product as well as solvent. Reactions of this sort are the essential chemical events in many biological, industrial, and environmental processes. The biochemical synthesis of proteins, the industrial production of several fertilizers, and some of the proposed methods for revitalizing lakes damaged by acid rain are three examples.

A **neutralization** reaction occurs when an acid reacts with a base. For our purposes in this chapter, an **acid** is *a substance that produces $H^+$ ions when dissolved in water;* a **base** is *a substance that produces $OH^-$ ions when dissolved in water.* (Other definitions of acid and base are presented later in the text.) Several familiar substances contain acids and bases: most drain, window, and oven cleaners contain bases; vinegar and lemon juice contain acids.

Acids and bases are electrolytes and are often grouped in terms of "strength" by how extensively they dissociate into ions in aqueous solution. *Strong acids and bases dissociate completely* into ions when they dissolve in water; like soluble ionic compounds, they are called *strong* electrolytes. *Weak acids and bases dissociate much less;* in fact, most of the molecules remain intact. As a result, they conduct a small current and are called *weak* electrolytes. Table 4.2 lists some common laboratory acids and bases.

For our discussion here, we can say that all acids, whether strong or weak, have one or more H atoms as part of their structure. Strong bases contain either the $OH^-$ or the $O^{2-}$ ion as part of their structure. Soluble metal oxides act as strong bases in solution because when the oxide ion dissociates, it reacts with water to form hydroxide ion:

$$O^{2-}(aq) + H_2O(l) \rightarrow 2OH^-(aq)$$

The weak base ammonia does not contain $O^{2-}$ or $OH^-$ ions as part of its structure, but it produces $OH^-$ ions in a reaction with water:

$$NH_3(g) + H_2O(l) \rightarrow NH_4^+(aq) + OH^-(aq)$$

(We look more closely at the distinction between strong and weak acids and bases in the final section of this chapter and treat the topic thoroughly in Chapter 17.)

Neutralization reactions between aqueous acids and bases are double-displacement reactions, as you can see from the molecular equation for the reaction between the strong acid HCl and the strong base $Ba(OH)_2$:

$$2HCl(aq) + Ba(OH)_2(aq) \rightarrow BaCl_2(aq) + 2H_2O(l)$$

If we evaporated the water from this reaction mixture, solid barium chloride would remain. An ionic compound, such as barium chloride, that results from the reaction of an acid and a base is called a **salt.** Note that *the cation of the salt is contributed by the base and the anion by the acid.* This reaction is typical of aqueous neutralization reactions: *the reactants are an acid and a base, and the products are a salt and water.*

In fact, as the following ionic equations show, *the formation of water is the only chemical change between strong acids and bases.* Since both HCl and $Ba(OH)_2$ dissociate completely into ions, the total ionic equation is

$$2H^+(aq) + 2Cl^-(aq) + Ba^{2+}(aq) + 2OH^-(aq) \rightarrow Ba^{2+}(aq) + 2Cl^-(aq) + 2H_2O(l)$$

The $Ba^{2+}(aq)$ and $Cl^-(aq)$ are spectator ions. When we eliminate them to write the net ionic equation, we see the actual reaction:

$$2H^+(aq) + 2OH^-(aq) \rightarrow 2H_2O(l) \quad \text{or} \quad H^+(aq) + OH^-(aq) \rightarrow H_2O(l)$$

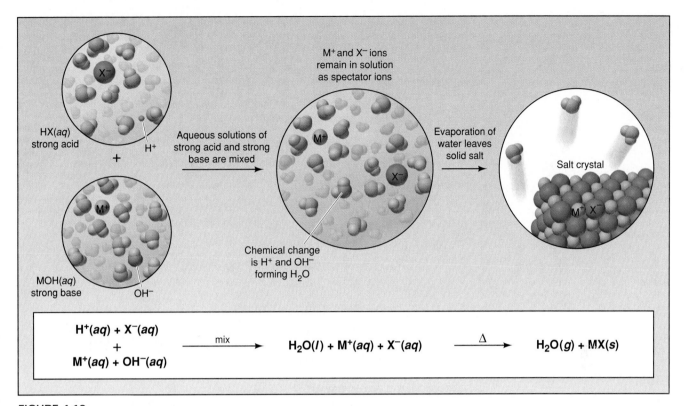

**FIGURE 4.12**

**An aqueous neutralization reaction on the atomic scale.** When solutions of a strong acid and a strong base are mixed, the $H^+$ from the acid and the $OH^-$ from the base combine to form water. Evaporation of the water leaves the spectator ions, $X^-$ and $M^+$, as a solid ionic compound called a salt.

The essential chemical change in all aqueous neutralization reactions between strong acids and bases is that *an $H^+$ ion from the acid and an $OH^-$ ion from the base form a water molecule;* only the spectator ions differ from one reaction to another (Figure 4.12).

As with precipitation reactions, neutralization reactions are driven by the electrostatic attraction of ions (in this case, $H^+$ and $OH^-$) and their removal from solution by the formation of the product ($H_2O$). The ions are removed because water consists almost entirely of undissociated molecules. We say "almost" because only one of every 555 million $H_2O$ molecules dissociates into $H^+$ and $OH^-$ ions. This tiny extent of dissociation is important, as you'll see in later chapters, but the formation of water in a neutralization reaction certainly represents an enormous net removal of ions.

SAMPLE PROBLEM 4.5

## Writing Balanced Equations for Neutralization Reactions

**Problem:** Write balanced molecular, total ionic, and net ionic equations for each of the following reactions between an acid and a base:
**(a)** Strontium hydroxide($aq$) + perchloric acid($aq$) $\rightarrow$
**(b)** Barium hydroxide($aq$) + sulfuric acid($aq$) $\rightarrow$

**Plan:** The essential reaction between these strong acids and bases (Table 4.2) is between an $H^+$ and an $OH^-$ to form water, so $H_2O$ is one of the products. The other is a salt made from the spectator ions. Note that in (b), the salt ($BaSO_4$) is insoluble (see Table 4.1), so virtually all ions are removed from solution.

**Solution:** **(a)** Writing the molecular equation:

$$Sr(OH)_2(aq) + 2HClO_4(aq) \rightarrow Sr(ClO_4)_2(aq) + 2H_2O(l)$$

Writing the total ionic equation:

$$Sr^{2+}(aq) + 2OH^-(aq) + 2H^+(aq) + 2ClO_4^-(aq) \rightarrow$$
$$Sr^{2+}(aq) + 2ClO_4^-(aq) + 2H_2O(l)$$

Writing the net ionic equation:

$$2OH^-(aq) + 2H^+(aq) \rightarrow 2H_2O(l) \quad \text{or} \quad OH^-(aq) + H^+(aq) \rightarrow H_2O(l)$$

$Sr^{2+}(aq)$ and $ClO_4^-(aq)$ are the spectator ions.

**(b)** Writing the molecular equation:

$$Ba(OH)_2(aq) + H_2SO_4(aq) \rightarrow BaSO_4(s) + 2H_2O(l)$$

Writing the total ionic equation:

$$Ba^{2+}(aq) + 2OH^-(aq) + 2H^+(aq) + SO_4^{2-}(aq) \rightarrow BaSO_4(s) + 2H_2O(l)$$

The net ionic equation is the same as the total ionic equation. This is a precipitation *and* a neutralization reaction. There are no spectator ions because all the ions are used to form the two products.

**FOLLOW-UP PROBLEM 4.5**
Write balanced molecular, total ionic, and net ionic equations for the reaction between aqueous solutions of calcium hydroxide and nitric acid.

---

**Acid-base titrations.** Chemists study neutralization reactions quantitatively in an acid-base titration. In any **titration,** *a solution of known concentration is used to determine the concentration of another solution through a monitored reaction.* In a typical acid-base titration, a *standardized* solution of base, one whose concentration has been previously determined, is slowly added to an acid solution of unknown concentration.

The laboratory procedure is straightforward but requires careful technique (Figure 4.13). A known volume of the acidic solution is placed in a flask, and a few drops of indicator solution are added. An **acid-base indicator** is a substance whose color is different in acidic and basic solutions. (In Chapter 18 we examine indicators more fully.) The standardized basic solution is added slowly from a buret clamped above the flask. As the titration nears its end, indicator molecules change color near a drop of added base due to a temporary excess of $OH^-$ ions. As soon as the solution is swirled, however, the indicator's acidic color returns. The **equivalence point** in the titration occurs when *all the moles of $H^+$ ions present in the original volume of acid solution have reacted with an equivalent number of moles of $OH^-$ ions added from the buret:*

Moles of $H^+$ (originally in flask) = moles of $OH^-$ (added from buret)

The **end point** of the titration occurs when a tiny excess of $OH^-$ ions changes the indicator to its basic color permanently. In calculations, we assume this tiny excess is insignificant, and therefore *the amount of base needed to reach the end point is the same as the amount needed to reach the equivalence point.*

**FIGURE 4.13**

A    B    C

$H^+(aq) + X^-(aq) + M^+(aq) + OH^-(aq) \longrightarrow H_2O(l) + M^+(aq) + X^-(aq)$

**An acid-base titration. A,** In this experiment, a measured volume of the unknown acid solution is placed in a flask beneath a buret containing the known (standardized) base solution. A few drops of indicator are added to the flask; the indicator used here is phenolphthalein, which is colorless in acid and pink in base. After an initial buret reading, base ($OH^-$ ions) is slowly added to the acid ($H^+$ ions). **B,** Near the end of the titration, the indicator momentarily changes to its base color but reverts to its acid color with swirling. **C,** When the end point is reached, a tiny excess of $OH^-$ is present, shown by the permanent change in color of the indicator. The difference between the final buret reading and the initial buret reading gives the volume of base used.

SAMPLE PROBLEM 4.6 _____

**Finding the Concentration of Acid from an Acid-Base Titration**

**Problem:** You perform an acid-base titration to standardize an HCl solution by placing 50.00 mL HCl in a flask with a few drops of indicator solution. You fill the buret with 0.1524 $M$ NaOH. The initial buret reading is 0.55 mL; at the end point, the buret reading is 33.87 mL. What is the concentration of the HCl solution?

**Plan:** We use the volume and molarity of base added to determine the molarity of acid present in the flask. First, we balance the equation. We find the volume of base added from the difference in buret readings and use the known molarity of base to calculate the moles of base added. Then, we use the molar ratio from the balanced equation to find moles of acid originally present and find its molarity from its original volume.

**Solution:** Writing the balanced equation:

$$NaOH(aq) + HCl(aq) \rightarrow NaCl(aq) + H_2O(l)$$

Finding volume (L) of NaOH solution added:

Volume (L) of solution

$$= (33.87 \text{ mL soln} - 0.55 \text{ mL soln}) \times \frac{1 \text{ L}}{1000 \text{ mL}} = 0.03332 \text{ L soln}$$

Finding moles of NaOH added:

$$\text{Moles of NaOH} = 0.03332 \text{ L soln} \times \frac{0.1524 \text{ mol NaOH}}{1 \text{ L soln}}$$

$$= 5.078 \times 10^{-3} \text{ mol NaOH}$$

Finding moles of HCl originally present:

$$\text{Moles of HCl} = 5.078 \times 10^{-3} \text{ mol NaOH} \times \frac{1 \text{ mol HCl}}{1 \text{ mol NaOH}} = 5.078 \times 10^{-3} \text{ mol HCl}$$

Calculating molarity of HCl:

$$\text{Molarity of HCl} = \frac{5.078 \times 10^{-3} \text{ mol HCl}}{50.00 \text{ mL}} \times \frac{1000 \text{ mL}}{1 \text{ L}} = \textbf{0.1016 } \textbf{\textit{M} HCl}$$

| Volume (L) of base (difference in buret readings) |
|---|

$M$ (mol/L) of base

| Moles of base |
|---|

molar ratio

| Moles of acid |
|---|

volume (L) of acid

| $M$ (mol/L) of acid |
|---|

**Check:** It takes ~50 mL of 0.1 $M$ H$^+$ to neutralize ~33 mL of 0.15 $M$ OH$^-$. This seems reasonable, since a larger volume of less concentrated acid neutralizes a smaller volume of more concentrated base, and the moles of H$^+$ and OH$^-$ are about equal: $50 \times 0.1 \approx 33 \times 0.15$.

**FOLLOW-UP PROBLEM 4.6**
What volume of 0.1292 $M$ Ba(OH)$_2$ would neutralize 50.00 mL of the HCl solution standardized in Sample Problem 4.6?

### Reactions That Form a Gaseous Product

The third type of aqueous ionic displacement reaction we consider is driven by *the formation of a gaseous product that removes ions from solution*. A household example of such a reaction occurs between baking soda (sodium bicarbonate) and vinegar (an aqueous 5% solution of acetic acid) (Figure 4.14). Note that because acetic acid is a weak acid, it dissociates very little, so we show it intact in the net ionic equation as one of the reactants. Therefore, only Na$^+$($aq$) is a spectator ion; CH$_3$COO$^-$($aq$) is not. The CO$_2$ leaves the mixture as bubbles of gas. Reactant ions are removed from solution here by the formation of CO$_2$ *and* H$_2$O.

Several other polyatomic ions react with H$^+$($aq$) from acids to form a gaseous product. Two examples are the loss of SO$_2$ from ionic sulfites and the loss of NO from ionic nitrites. In reaction with strong acid, the net ionic equations are

$$SO_3{}^{2-}(aq) + 2H^+(aq) \rightarrow SO_2(g) + H_2O(l)$$

$$3NO_2{}^-(aq) + 2H^+(aq) \rightarrow NO_3{}^-(aq) + 2NO(g) + H_2O(l)$$

### Section Summary

When solutions of ionic reactants are mixed, a reaction occurs only if a product forms that removes ions from the solution. A molecular equation shows reactants and products as undissociated compounds. In a total ionic equation, soluble ionic compounds appear as solvated ions. In a net ionic equation, only species that change during the reaction appear; those that remain the same on both sides of the equation are spectator ions. Precipitation reactions occur when an insoluble ionic compound forms. Neutralization reactions take place when an acid (an H$^+$-yielding compound) and a base (an OH$^-$-yielding compound) form a water molecule. In titrations, a known concentration of one reactant is used to determine the concentration of the other. Gas-forming ionic reactions occur when species in solution form a gaseous product that leaves the reaction mixture.

**FIGURE 4.14**

**The formation of a gaseous product.** Carbonates and bicarbonates react with acids to form gaseous CO$_2$ and H$_2$O. Here, sodium bicarbonate (baking soda) solution is added to dilute acetic acid solution (vinegar), and bubbles of CO$_2$ gas form.

Molecular equation
$$NaHCO_3(aq) + CH_3COOH(aq) \longrightarrow CH_3COONa(aq) + CO_2(g) + H_2O(l)$$

Total ionic equation
$$Na^+(aq) + HCO_3{}^-(aq) + CH_3COOH(aq) \longrightarrow CH_3COO^-(aq) + Na^+(aq) + CO_2(g) + H_2O(l)$$

Net ionic equation
$$HCO_3{}^-(aq) + CH_3COOH(aq) \longrightarrow CH_3COO^-(aq) + CO_2(g) + H_2O(l)$$

## 4.4   Oxidation-Reduction (Redox) Reactions

In **oxidation-reduction** (or **redox**) reactions, the key chemical event is the *net movement of electrons* from one reactant to the other. You can also think of the process as a *shift in electron charge* from one reactant to the other. Redox reactions constitute some of the most important of all chemical processes, including the formation of a compound from its elements, all combustion reactions, the reactions in batteries that generate electricity, and the production of biochemical energy.

### Redox Processes in the Formation of Ionic and Covalent Compounds

To comprehend the central idea in all redox reactions, let's reconsider the flashbulb reaction, in which an ionic compound, MgO, forms from its elements:

$$2Mg(s) + O_2(g) \rightarrow 2MgO(s)$$

During the reaction, two electrons move from each Mg atom to each O atom (Figure 4.15, *A*); that is, each Mg atom loses two electrons and each O atom gains them. This change represents a *transfer of electron charge* away from each Mg atom and toward each O atom, resulting in the formation of $Mg^{2+}$ and $O^{2-}$ ions; their mutual attraction forms the solid compound MgO.

During the formation of a covalent compound from its elements, there is a net movement of electrons also, although not enough to form ions. Consider the formation of hydrogen chloride gas:

$$H_2(g) + Cl_2(g) \rightarrow 2HCl(g)$$

To see the electron shift here, we compare the electron distributions in the reactant bonds and the product bonds (Figure 4.15, *B*). As mentioned earlier,

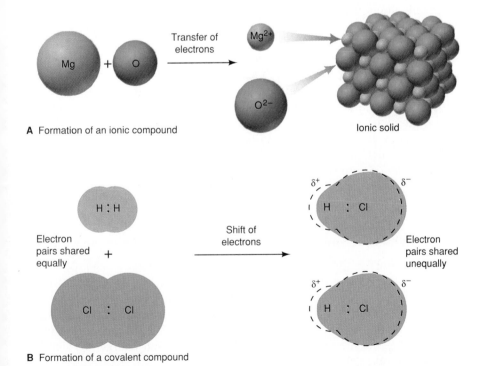

**A** Formation of an ionic compound

**B** Formation of a covalent compound

**FIGURE 4.15**

**The redox process in compound formation. A,** In forming the ionic compound MgO, each Mg atom transfers two electrons to each O atom. (Note that atoms become smaller when they lose electrons and larger when they gain them.) The resulting $Mg^{2+}$ and $O^{2-}$ ions aggregate with many other pairs to form an ionic solid. **B,** In the reactants $H_2$ and $Cl_2$, the electron pairs are shown centrally located to indicate that they are shared equally. In forming the covalent compound HCl, the Cl attracts the shared electron pair more strongly than the H does. In effect, H shifts its electron to Cl. Note the polar nature of the HCl molecule, as shown by the electron cloud not coinciding with the dashed space-filling outline.

both $H_2$ and $Cl_2$ molecules are held together by pure covalent bonds; that is, the electrons are shared equally between the bonded atoms. In the hydrogen chloride molecule, the electrons are shared unequally because the Cl atom attracts them more strongly than the H atom. Thus, some electron charge has shifted away from the H atom in HCl relative to its situation in $H_2$, and some electron charge has shifted toward the Cl atom relative to its situation in $Cl_2$. This shift is not as extreme as in MgO formation, where separate ions form, but the result is still an oxidation-reduction reaction.

Perhaps the redox process can be clarified further by showing why a neutralization reaction, for example, is *not* a redox reaction. When the $H^+$ from HCl and the $OH^-$ from NaOH react to form water, there is no net change in the relative distribution of electron charge. The greater electron charge lies nearer the O atom in H—OH just as it did in NaOH before the reaction. Similarly, the lesser electron charge lies nearer the H in H—OH just as it did in HCl before the reaction. Since there is no net shift of electron charge from one atom to another, the reaction is not a redox process.

### Some Essential Redox Terminology

Chemists use some important terminology to describe the shift in electrons that occurs in oxidation-reduction reactions.

**Oxidation** is the *loss* of electrons; **reduction** is the *gain* of electrons. (You'll see shortly why the term "reduction" is used for the act of gaining.) During the formation of magnesium oxide, Mg undergoes oxidation (electron loss) and $O_2$ undergoes reduction (electron gain). The loss and gain are simultaneous, but we can imagine them occurring in separate steps:

$$\text{Oxidation:} \quad Mg \rightarrow Mg^{2+} + 2e^- \text{ (electrons lost by Mg)}$$

$$\text{Reduction:} \quad \tfrac{1}{2}O_2 + 2e^- \rightarrow O^{2-} \text{ (electrons gained by } O_2)$$

Since $O_2$ gained the electrons that Mg lost when Mg was oxidized, we say that *$O_2$ oxidized Mg,* or that $O_2$ is the **oxidizing agent,** the species doing the oxidizing. Similarly, since Mg gave up the electrons that $O_2$ gained when $O_2$ was reduced, *Mg reduced $O_2$.* Mg is the **reducing agent,** the species doing the reducing.

Once this relationship is clear, you should note that *the oxidizing agent becomes reduced* because it removes (accepts) the electrons, whereas *the reducing agent becomes oxidized* because it gives up (donates) the electrons. In the formation of HCl, for example, $Cl_2$ oxidizes $H_2$ (H loses some electron charge and Cl gains it), which is the same as saying that $H_2$ reduces $Cl_2$. The reducing agent, $H_2$, is oxidized and the oxidizing agent, $Cl_2$, is reduced.

It is extremely important to realize that *the oxidation and the reduction occur simultaneously.* There is no such chemical change as an oxidation reaction *or* a reduction reaction; only an oxidation-reduction reaction can occur.

### Use of Oxidation Numbers to Monitor the Shift of Electron Charge

In order to follow the changes occurring in redox reactions, chemists have devised a "bookkeeping" system, a sort of positive/negative scorecard, to monitor which atom loses electron charge and which gains it. Each atom is assigned an **oxidation number (O.N., or oxidation state),** which is determined by the set of rules in Table 4.3. Note that in an oxidation number the sign precedes the number (+2), whereas in an ionic charge, the sign follows the number (2+).

**TABLE 4.3** **Rules for Assigning an Oxidation Number (O.N.)**

### General rules
1. For an atom in its elemental form (Na, $O_2$, $Cl_2$, and so forth):    O.N. = 0
2. For a monatomic ion: O.N. = ion charge
3. The sum of O.N. values for the atoms in a compound equals zero. The sum of O.N. values for the atoms in a polyatomic ion equals the ion charge.

### Rules for specific atoms or periodic table groups
1. For Group 1A(1):    O.N. = +1 in all compounds
2. For Group 2A(2):    O.N. = +2 in all compounds
3. For hydrogen:    O.N. = +1 in combination with nonmetals
     O.N. = −1 in combination with metals and boron
4. For fluorine:    O.N. = −1 in all compounds
5. For oxygen:    O.N. = −1 in peroxides
     O.N. = −2 in all other compounds (except with F)
6. For Group 7A(17):    O.N. = −1 in combination with metals, nonmetals (except O), and other halogens lower in the group

SAMPLE PROBLEM 4.7 _____

**Determining the Oxidation Number of an Element in a Compound**

**Problem:** Determine the oxidation number (O.N.) of each element in the following compounds:
**(a)** Zinc chloride            **(b)** Sulfur trioxide            **(c)** Nitric acid
**Plan:** We apply the rules in Table 4.3, always making sure that the O.N. values in a compound add up to zero, and in a polyatomic ion, to the ion's charge.
**Solution:** **(a)** $ZnCl_2$. This compound is composed of monatomic ions. The O.N. of $Zn^{2+}$ is **+2**. The O.N. of each $Cl^-$ is **−1**, for a total of −2.
**(b)** $SO_3$. The O.N. of each oxygen is **−2** for a total of −6. Since the oxidation numbers in a compound add up to zero, the O.N. of S is **+6**.
**(c)** $HNO_3$. The O.N. of H is **+1**, so the $NO_3$ group must sum to −1. The O.N. of each O is **−2** for a total of −6. Therefore, the O.N. of N is **+5**.

FOLLOW-UP PROBLEM 4.7
Determine the O.N. of each element in the following: **(a)** scandium oxide ($Sc_2O_3$); **(b)** gallium chloride ($GaCl_3$); **(c)** hydrogen phosphate ion; and **(d)** iodine trifluoride.

_____

The periodic table is a great help in learning the highest and lowest oxidation numbers of most main-group elements (Figure 4.16):
1. For most main-group elements, the A-group number (1A, 2A, and so on) is the *highest* oxidation number (always positive) of any element in the group. The exceptions are O and F (see Table 4.3).
2. For the main-group nonmetals and some metalloids, the A-group number minus 8 gives the *lowest* oxidation number (always negative) of any element in the group.

For example, the highest oxidation number of S (Group 6A) is +6, as in $SF_6$, and the lowest is (6 − 8), or −2, as in FeS and other metal sulfides.

It is important to realize that an oxidation number is just a mental tool, a conceptual tally of relative electron charge; it is *not* a real charge. Since oxidation numbers are assigned according to the atom that pulls more strongly on the electrons, they are ultimately based on atomic structure, as you'll see in Chapters 8 and 9.

When we compare the oxidation number of an atom in the reactant with that of the same atom in the product, we can determine whether the atom

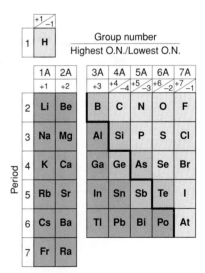

**FIGURE 4.16**
**Highest and lowest oxidation numbers of reactive main-group elements.** The A group number shows the highest possible oxidation number (O.N.) for a main-group element. (Two important exceptions are O, which never has an O.N. of +6, and F, which never has an O.N. of +7.) For nonmetals (yellow) and metalloids (green), the A group number minus 8 gives the lowest possible oxidation number.

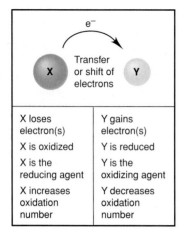

**FIGURE 4.17**
A summary of terminology for oxidation-reduction (redox) reactions.

| X loses electron(s) | Y gains electron(s) |
|---|---|
| X is oxidized | Y is reduced |
| X is the reducing agent | Y is the oxidizing agent |
| X increases oxidation number | Y decreases oxidation number |

(and therefore the species that contains it) was oxidized, reduced, or uninvolved in the redox reaction. If an atom has a higher (more positive or less negative) oxidation number in the product than in the reactant, the atom was oxidized (lost electrons) during the reaction. Thus, *oxidation is represented by an increase in oxidation number.* If an atom has a lower (more negative or less positive) oxidation number in the product than in the reactant, the atom was reduced (gained electrons) during the reaction. Thus, *the gain of electrons is represented by a decrease (a "reduction") in oxidation number.* Figure 4.17 provides a summary of redox terminology.

SAMPLE PROBLEM 4.8 _____

## Recognizing Oxidizing and Reducing Agents

**Problem:** Identify the oxidizing agent and reducing agent in each of the following:
**(a)** $2Al(s) + 3H_2SO_4(aq) \rightarrow Al_2(SO_4)_3(aq) + 3H_2(g)$
**(b)** $PbO(s) + CO(g) \rightarrow Pb(s) + CO_2(g)$
**(c)** $2H_2(g) + O_2(g) \rightarrow 2H_2O(g)$
**Plan:** We first assign an oxidation number (O.N.) to each atom (or ion) based on the rules in Table 4.3. The reactant is the reducing agent if it contains an atom that was oxidized (O.N. increased in the reaction). The reactant is the oxidizing agent if it contains an atom that was reduced (O.N. decreased).
**Solution:** **(a)** Assigning oxidation numbers:

$$\overset{0}{2Al}(s) + 3\overset{+1\ \overset{+6}{-2}}{H_2SO_4}(aq) \rightarrow \overset{+3\ \overset{+6}{-2}}{Al_2(SO_4)_3}(aq) + 3\overset{0}{H_2}(g)$$

The O.N. of Al increased from 0 to +3 (Al lost electrons), so Al was oxidized; **Al is the reducing agent.**
The O.N. of H decreased from +1 to 0 (H gained electrons), so H was reduced; **$H_2SO_4$ is the oxidizing agent.**
In terms of numbers of reactant and product substances, this redox reaction is a single-displacement reaction (Section 4.1). When metals displace hydrogen from acids or from water (see Figure 4.6), or metal ions from solution (see Figure 4.7), a redox process is occurring.
**(b)** Assigning oxidation numbers:

$$\overset{+2\ \overset{-2}{}}{PbO}(s) + \overset{+2\ \overset{-2}{}}{CO}(g) \rightarrow \overset{0}{Pb}(s) + \overset{+4\ \overset{-2}{}}{CO_2}(g)$$

Pb was reduced (O.N. decreased from +2 to 0); **PbO is the oxidizing agent.**
C was oxidized (O.N. increased from +2 to +4); **CO is the reducing agent.**
In general, when an atom (C in CO) bonds to more O atoms, it is oxidized. Similarly, when an atom (Pb in PbO) bonds to fewer O atoms, it is reduced.
**(c)** Assigning oxidation numbers:

$$\overset{0}{2H_2}(g) + \overset{0}{O_2}(g) \rightarrow \overset{+1\ -2}{2H_2O}(g)$$

O was reduced (O.N. decreased from 0 to −2); **$O_2$ is the oxidizing agent.**
H was oxidized (O.N. increased from 0 to +1); **$H_2$ is the reducing agent.**
Oxygen is always the oxidizing agent in a combustion reaction.
**Comment:** 1. Let's compare the O.N. values in (c) with those in the net ionic equation for an acid-base reaction:

$$\overset{+1}{H^+}(aq) + \overset{-2\ \overset{+1}{}}{OH^-}(aq) \rightarrow \overset{+1\ -2}{H_2O}(l)$$

In the acid-base reaction, O.N values remain the same on both sides. Therefore, as discussed earlier, the acid-base reaction is *not* a redox reaction.

2. If an elemental substance occurs as reactant or product, it must *not* have been in its elemental form on the other side of the equation, so the reaction is a redox process. Notice that elements appear in all three cases above (as well as in the combination reaction in Figure 4.1 and the decomposition reaction in Figure 4.4).

**FOLLOW-UP PROBLEM 4.8**

Identify the oxidizing agent and reducing agent in the following:

**(a)** $2Fe(s) + 3Cl_2(g) \rightarrow 2FeCl_3(s)$

**(b)** $2C_2H_6(g) + 7O_2(g) \rightarrow 4CO_2(g) + 6H_2O(g)$

## Balancing Redox Reactions

Since oxidation and reduction occur simultaneously, the transferred electrons are never free and must be accounted for in the overall process. Therefore, *the total number of electrons lost by the oxidized reactant must equal the total number of electrons gained by the reduced reactant.* By keeping track of the changes in oxidation numbers and the number of electrons transferred, we can balance redox reactions.

Two methods used to balance redox reactions are the oxidation number method and the half-reaction method. This section describes the oxidation number method in detail and the half-reaction method briefly.*

**The oxidation number method.** The **oxidation number method** consists of five steps that use the changes in oxidation numbers to generate balancing coefficients. The first two steps are identical to those in Sample Problem 4.8:

*Step 1.* Assign oxidation numbers to all elements in the equation.

*Step 2.* From the changes in oxidation numbers, identify the oxidized and reduced species.

*Step 3.* Compute the number of electrons lost in the oxidation and gained in the reduction from the oxidation number changes. Draw tie-lines between these atoms to show electron changes.

*Step 4.* Multiply one or both of these numbers by appropriate factors to make the electrons lost equal the electrons gained, and use the factors as balancing coefficients.

*Step 5.* Complete the balancing by inspection, adding states of matter.

**SAMPLE PROBLEM 4.9**

## Balancing Redox Reactions with the Oxidation Number Method

**Problem:** Use the oxidation number method to balance the following reactions:

**(a)** $Cu(s) + HNO_3(aq) \rightarrow Cu(NO_3)_2(aq) + NO_2(g) + H_2O(l)$ (see photo)

**(b)** $PbS(s) + O_2(g) \rightarrow PbO(s) + SO_2(g)$

**Solution: (a)** *Step 1.* Assign oxidation numbers to all atoms:

$$\overset{0}{Cu} + \overset{+1}{H}\overset{+5}{N}\overset{-2}{O_3} \rightarrow \overset{+2}{Cu}(\overset{+5}{N}\overset{-2}{O_3})_2 + \overset{+4}{N}\overset{-2}{O_2} + \overset{+1}{H_2}\overset{-2}{O}$$

*A thorough discussion of the half-reaction method, including reactions in acidic and basic solution, appears in Chapter 20, Section 20.1. That section, with its chapter problems, is completely transferrable to this point in Chapter 4, with no loss in continuity.

*Step 2.* Identify oxidized and reduced species. The O.N. of Cu increased from zero (in Cu metal) to +2 (in $Cu^{2+}$); Cu was oxidized. The O.N. of N decreased from +5 (in $HNO_3$) to +4 (in $NO_2$); N was reduced. Note that only some $NO_3^-$ was reduced to $NO_2$, while some acts as a spectator ion appearing unchanged in the $Cu(NO_3)_2$; this is quite common in redox reactions.

*Step 3.* Compute $e^-$ lost and $e^-$ gained and draw tie-lines between the atoms. In the oxidation, two $e^-$ were lost from each Cu. In the reduction, one $e^-$ was gained by each N:

$$Cu + HNO_3 \rightarrow Cu(NO_3)_2 + NO_2 + H_2O$$

with $-2e^-$ over Cu and $+1e^-$ under N.

*Step 4.* Multiply by factors to make $e^-$ lost equal $e^-$ gained, and use the factors as co-efficients. Cu lost two $e^-$, so the one $e^-$ gained by N should be multiplied by 2. Using 2 as a coefficient for $NO_2$ and $HNO_3$ gives

$$Cu + 2HNO_3 \rightarrow Cu(NO_3)_2 + 2NO_2 + H_2O$$

*Step 5.* Complete the balancing by inspection. Balancing N atoms requires a 4 in front of $HNO_3$ because of the additional two $NO_3^-$ ions in $Cu(NO_3)_2$:

$$Cu + 4HNO_3 \rightarrow Cu(NO_3)_2 + 2NO_2 + H_2O$$

Balancing H atoms requires a 2 in front of $H_2O$, and we add states of matter:

$$\textbf{Cu}(s) + \textbf{4HNO}_3(aq) \rightarrow \textbf{Cu(NO}_3)_2(aq) + \textbf{2NO}_2(g) + \textbf{2H}_2\textbf{O}(l)$$

**Check:** Reactants (1 Cu, 4 H, 4 N, 12 O) → products [1 Cu, 4 H, (2 + 2) N, (6 + 4 + 2) O]

**(b)** *Step 1.* Assign oxidation numbers:

$$PbS + O_2 \rightarrow PbO + SO_2$$

with +2−2, 0, +2−2, +4, −2 labels.

*Step 2.* Identify species that are oxidized and reduced. The S (O.N. = −2 in PbS) was oxidized (O.N. = +4 in $SO_2$). The O (O.N. = 0 in $O_2$) was reduced (O.N. = −2 in PbO and in $SO_2$).

*Step 3.* Compute $e^-$ lost and $e^-$ gained and draw tie-lines. The S lost $6e^-$ and each O gained $2e^-$:

$$PbS + O_2 \rightarrow PbO + SO_2$$

with $-6e^-$ and $+2e^-$ per O.

*Step 4.* Multiply by factors to make $e^-$ lost equal $e^-$ gained. The single S atom loses $6e^-$. Each O in $O_2$ gains $2e^-$, for a total of $4e^-$ gained. Thus, using $\frac{3}{2}$ as a coefficient of $O_2$ gives 3 O atoms that each gain $2e^-$, for a total of $6e^-$ gained:

$$PbS + \tfrac{3}{2}O_2 \rightarrow PbO + SO_2$$

*Step 5.* Complete the balancing by inspection. The atoms are balanced, but all coefficients must be multiplied by 2 to obtain integer coefficients, and we must add states of matter:

$$\textbf{2PbS}(s) + \textbf{3O}_2(g) \rightarrow \textbf{2PbO}(s) + \textbf{2SO}_2(g)$$

**Check:** Reactants (2 Pb, 2 S, 6 O) → products [2 Pb, 2 S, (2 + 4) O]

**FOLLOW-UP PROBLEM 4.9**
Use the oxidation number method to balance the following:
**(a)** $H_2O_2(l) \rightarrow H_2O(l) + O_2(g)$
**(b)** $K_2Cr_2O_7(aq) + HI(aq) \rightarrow KI(aq) + CrI_3(aq) + I_2(s) + H_2O(l)$

**The half-reaction method.** The other method for balancing redox reactions is called the **half-reaction method** because the overall oxidation-reduction process is viewed as the sum of an oxidation "half-reaction" and a reduction "half-reaction," each of which involves *either* the loss or the gain of electrons. In principle, any redox reaction can be viewed in this way.

Consider the reaction between zinc metal and hydrochloric acid (Figure 4.18), and look at the change in oxidation numbers:

$$\overset{0}{Zn} + \overset{+1}{H}\overset{-1}{Cl} \longrightarrow \overset{+2}{Zn}\overset{-1}{Cl_2} + \overset{0}{H_2}$$

Eliminating $Cl^-$ as a spectator ion reveals a "skeleton" equation that includes only those elements that change oxidation numbers:

$$Zn + H^+ \longrightarrow Zn^{2+} + H_2$$

We can break up this equation into *two half-reactions*. In the process of balancing them, we note that in the oxidation half-reaction, one mole of Zn metal loses *two* moles of electrons and forms one mole of $Zn^{2+}$ ions. In the reduction half-reaction, one mole of $H^+$ ions gains one mole of electrons to form one-half mole of $H_2$ gas:

$$Zn \longrightarrow Zn^{2+} + 2e^- \quad \text{[oxidation half-reaction]}$$
$$e^- + H^+ \longrightarrow \tfrac{1}{2}H_2 \quad \text{[reduction half-reaction]}$$

In *each* half-reaction, moles of atoms and amount of charge are balanced. To obtain integer coefficients *and* to make the number of electrons lost equal the number gained, we double the reduction half-reaction:

$$Zn \longrightarrow Zn^{2+} + 2e^- \quad \text{[oxidation half-reaction]}$$
$$2e^- + 2H^+ \longrightarrow H_2 \quad \text{[reduction half-reaction doubled]}$$

As you can see, the half-reaction method emphasizes the essential chemical event in redox reactions, *the loss of electrons by the reducing agent and the gain of electrons by the oxidizing agent.* Next, we add the half-reactions together, and the electrons cancel:

$$Zn + 2H^+ \longrightarrow Zn^{2+} + H_2$$

Adding the appropriate number of spectator ions and indicating states of matter gives the *balanced* total ionic equation:

$$Zn(s) + 2H^+(aq) + 2Cl^-(aq) \longrightarrow Zn^{2+}(aq) + 2Cl^-(aq) + H_2(g)$$

This particular example was chosen to show how the half-reactions reveal simultaneous electron loss and electron gain in redox reactions. You could have balanced this one by inspection and would not need to learn a new balancing method if all redox reactions were this simple. Chapter 20 reviews and extends the half-reaction method to balance more complex aqueous redox reactions.

### Redox Titrations

Just as a base is used to determine the concentration of an acid (or vice versa) in an acid-base titration, a known concentration of oxidizing agent can be used to determine the unknown concentration of a reducing agent (or vice versa) in a redox titration. This indispensable analytical tool is used

$$Zn(s) + 2HCl(aq) \longrightarrow ZnCl_2(aq) + H_2(g)$$

**FIGURE 4.18**
**The redox reaction between zinc and hydrochloric acid.** In this reaction, Zn is oxidized, so it is the reducing agent; $H^+$ is reduced, so HCl is the oxidizing agent.

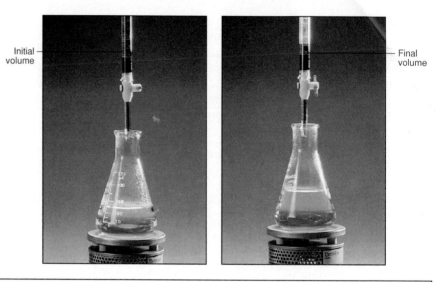

Net ionic equation

$$\overset{+7}{2MnO_4^-}(aq) + \overset{+3}{5C_2O_4^{2-}}(aq) + 16H^+(aq) \longrightarrow \overset{+2}{2Mn^{2+}}(aq) + \overset{+4}{10CO_2}(g) + 8H_2O(l)$$

**FIGURE 4.19**

**A redox titration.** The oxidizing agent in the buret, $KMnO_4$, is strongly colored, so it also serves as the indicator. When it reacts with the reducing agent $C_2O_4^{2-}$ in the flask, its color changes from deep purple to almost colorless *(left)*. When all the $C_2O_4^{2-}$ is oxidized, the next drop of $KMnO_4$ remains unreacted and turns the solution light purple *(right)*, signaling the end point of the titration.

in a wide range of situations, including measuring the iron content in drinking water and the vitamin C content in fruits and vegetables.

A common oxidizing agent in redox titrations is the permanganate ion, $MnO_4^-$. Because it is strongly colored, the ion also serves as an indicator for monitoring the reaction. In the following example, $MnO_4^-$ is used to determine the concentration of oxalate ion, $C_2O_4^{2-}$. As long as $C_2O_4^{2-}$ is present, the deep purple $MnO_4^-$ is reduced to the very faint pink (nearly colorless) $Mn^{2+}$ ion (Figure 4.19, *left*). As soon as all the available $C_2O_4^{2-}$ ions have been oxidized to $CO_2$, the next drop of $MnO_4^-$ turns the solution light purple (Figure 4.19, *right*). This color change indicates the point at which the total number of electrons lost by the oxidized species ($C_2O_4^{2-}$) equals the total number of electrons gained by the reduced species ($MnO_4^-$). We calculate the concentration of the $C_2O_4^{2-}$ solution from its known volume, the known concentration and volume of the $MnO_4^-$ solution, and the balanced redox equation.

Preparing a sample for a redox titration often requires several preliminary laboratory steps. In Sample Problem 4.10, for instance, the $Ca^{2+}$ ion concentration of blood is determined. The $Ca^{2+}$ is first precipitated as calcium oxalate ($CaC_2O_4$). Dilute $H_2SO_4$ dissolves the precipitate and releases the oxalate ion, which is then titrated with standardized $MnO_4^-$ ion as the oxidizing agent. After the $C_2O_4^{2-}$ concentration has been determined, it is used to find the $Ca^{2+}$ concentration.

SAMPLE PROBLEM 4.10 _____

## Finding an Unknown Concentration Using a Redox Titration

**Problem:** Calcium ion ($Ca^{2+}$) is required for many biological processes; it must be present for blood to clot, and an abnormal $Ca^{2+}$ concentration is indicative of several diseases. To measure the concentration of $Ca^{2+}$ ion, 1.00 mL human blood was treated with $Na_2C_2O_4$ solution, and the resulting precipitate of $CaC_2O_4$ was isolated and dissolved in dilute $H_2SO_4$ to a final volume of 10.00 mL. This solution required 2.05 mL of $4.88 \times 10^{-4}$ $M$ $KMnO_4$ to reach the titration end point.
**(a)** Calculate the molarity of $Ca^{2+}$ in the $H_2SO_4$ solution.
**(b)** Calculate the $Ca^{2+}$ ion concentration expressed in units of mg $Ca^{2+}$/100 mL blood.

### (a) Calculating the molarity of $Ca^{2+}$.
**Plan:** As always, we first write the balanced equation. All the $Ca^{2+}$ ion in the blood sample is precipitated and then dissolved in the $H_2SO_4$. By knowing the moles of $KMnO_4$ needed to reach the end point and using the molar ratio from the balanced equation, we can calculate the moles of $CaC_2O_4$ dissolved in the $H_2SO_4$. We know from the chemical formula that there is one mole of $Ca^{2+}$ ions per mole of $CaC_2O_4$. The volume of $H_2SO_4$ is given, so we can determine the molarity of $Ca^{2+}$ ion in the $H_2SO_4$ solution.
**Solution:** Writing the balanced equation:

$$2KMnO_4(aq) + 5CaC_2O_4(aq) + 8H_2SO_4(aq) \longrightarrow$$
$$2MnSO_4(aq) + K_2SO_4(aq) + 5CaSO_4(aq) + 10CO_2(g) + 8H_2O(l)$$

Converting from volume to moles of $KMnO_4$:

$$\text{Moles of } KMnO_4 = 2.05 \text{ mL soln}$$
$$\times \frac{1 \text{ L}}{1000 \text{ mL}} \times \frac{4.88 \times 10^{-4} \text{ mol } KMnO_4}{1 \text{ L soln}}$$
$$= 1.00 \times 10^{-6} \text{ mol } KMnO_4$$

Converting from moles of $KMnO_4$ to moles of $CaC_2O_4$ titrated:

$$\text{Moles of } CaC_2O_4 = 1.00 \times 10^{-6} \text{ mol } KMnO_4 \times \frac{5 \text{ mol } CaC_2O_4}{2 \text{ mol } KMnO_4}$$
$$= 2.50 \times 10^{-6} \text{ mol } CaC_2O_4$$

Finding moles of $Ca^{2+}$ in the $CaC_2O_4$ precipitate:

$$\text{Moles of } Ca^{2+} = 2.50 \times 10^{-6} \text{ mol } CaC_2O_4 \times \frac{1 \text{ mol } Ca^{2+}}{1 \text{ mol } CaC_2O_4}$$
$$= 2.50 \times 10^{-6} \text{ mol } Ca^{2+}$$

Finding the molarity of $Ca^{2+}$ in $H_2SO_4$ solution:

$$\text{Molarity of } Ca^{2+} = \frac{2.50 \times 10^{-6} \text{ mol } Ca^{2+}}{10.00 \text{ mL}} \times \frac{1000 \text{ mL}}{1 \text{ L}}$$
$$= \mathbf{2.50 \times 10^{-4} \ M \ Ca^{2+}}$$

**Check:** A very small volume of dilute $KMnO_4$ is needed, so $10^{-6}$ mol $KMnO_4$ seems reasonable. The molar ratio of $5CaC_2O_4/2KMnO_4$ gives $2.5 \times 10^{-6}$ mol $CaC_2O_4$ and thus $2.5 \times 10^{-6}$ mol $Ca^{2+}$. Finally, $2.5 \times 10^{-6}$ mol/10 mL equals $2.5 \times 10^{-4}$ mol/L.

Volume (L) of $KMnO_4$ solution

$M$ (mol/L)

Moles of $KMnO_4$

molar ratio

Moles of $CaC_2O_4$

chemical formula

Moles of $Ca^{2+}$

volume (L) of $H_2SO_4$

$M$ (mol/L) of $Ca^{2+}$

**(b) Expressing the $Ca^{2+}$ concentration as mg/100 mL blood.**

**Plan:** The amount in part (a) is the moles of $Ca^{2+}$ ion present in 1.00 mL blood. We multiply by 100 to obtain the moles of $Ca^{2+}$ ion in 100 mL blood, and then convert to milligrams of $Ca^{2+}$/100 mL blood.

**Solution:** Finding moles of $Ca^{2+}$/100 mL blood:

$$\text{Moles of } Ca^{2+}/100 \text{ mL blood} = \frac{2.50 \times 10^{-6} \text{ mol } Ca^{2+}}{1.00 \text{ mL blood}} \times 100 \text{ mL blood}$$

$$= 2.50 \times 10^{-4} \text{ mol } Ca^{2+}$$

Converting from moles of $Ca^{2+}$ to mass (mg):

Mass (mg) $Ca^{2+}$/100 mL blood

$$= 2.50 \times 10^{-4} \text{ mol } Ca^{2+} \times \frac{40.08 \text{ g } Ca^{2+}}{1 \text{ mol } Ca^{2+}} \times \frac{1000 \text{ mg } Ca^{2+}}{1 \text{ g } Ca^{2+}} = \textbf{10.0 mg } Ca^{2+}$$

**Check:** The relative amounts of $Ca^{2+}$ make sense. If there is $2.5 \times 10^{-6}$ mol/mL blood, there is $2.5 \times 10^{-4}$ mol/100 mL blood. A molar mass of about 40 g/mol for $Ca^{2+}$ gives $100 \times 10^{-4}$ g, or $10 \times 10^{-3}$ g/100 mL blood. It is easy to make an order-of-magnitude (power of 10) error in this type of calculation, so be sure to include all units.

**Comment:** 1. The normal range of $Ca^{2+}$ concentration in the adult human is 9.0 to 11.5 mg $Ca^{2+}$/100 mL blood, so the calculated value seems reasonable.

2. When blood is donated, the receiving bag contains $Na_2C_2O_4$ solution, which removes the $Ca^{2+}$ ion and prevents clotting.

3. The idea behind a redox titration is analogous to the idea behind an acid-base titration: in redox processes, electrons are lost and gained, whereas in acid-base processes, $H^+$ ions are lost and gained.

**FOLLOW-UP PROBLEM 4.10**

A 2.50-mL sample of low-fat milk was treated with sodium oxalate, and the precipitate was dissolved in $H_2SO_4$. This solution required 6.53 mL of $4.56 \times 10^{-3}$ M $KMnO_4$ to reach the end point. **(a)** Calculate the molarity of $Ca^{2+}$ in the milk. **(b)** What is the concentration of $Ca^{2+}$ in g/L? Is this value consistent with the typical value of about 1.2 g $Ca^{2+}$/L?

**Section Summary**

When there is a net movement in electron charge from one reactant to another, a redox process takes place. Electron gain (reduction) and electron loss (oxidation) occur simultaneously. The species that is oxidized (increases its oxidation number) is the reducing agent; the species that is reduced (decreases its oxidation number) is the oxidizing agent. Redox reactions can be balanced by keeping track of the changes in oxidation number or by treating the reaction as the sum of reduction and oxidation half-reactions. A redox titration is used to determine the concentration of either the oxidizing or the reducing agent from the concentration of the other.

## 4.5 Reversible Reactions: An Introduction to Chemical Equilibrium

So far, we have viewed reactions as occurring from "left to right," from reactants to products. The implication is that the reaction continues until it is complete, that is, until one of the reactants is consumed. As we noted in our

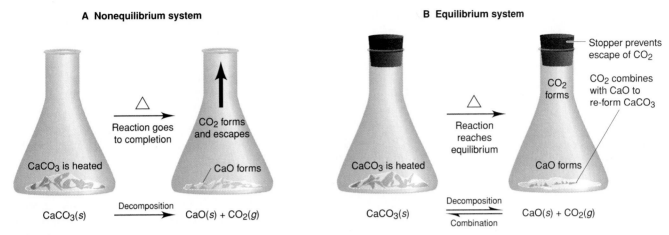

**A** Nonequilibrium system

CaCO₃ is heated → Reaction goes to completion → CO₂ forms and escapes, CaO forms

$$CaCO_3(s) \xrightarrow{\text{Decomposition}} CaO(s) + CO_2(g)$$

**B** Equilibrium system

Stopper prevents escape of CO₂

CO₂ forms

CO₂ combines with CaO to re-form CaCO₃

CaCO₃ is heated → Reaction reaches equilibrium → CaO forms

$$CaCO_3(s) \underset{\text{Combination}}{\overset{\text{Decomposition}}{\rightleftharpoons}} CaO(s) + CO_2(g)$$

**FIGURE 4.20**

**The equilibrium state. A,** In an open flask, $CaCO_3$ decomposes completely with heating because the product $CO_2$ escapes and is not present to react with the other product, CaO. **B,** When $CaCO_3$ decomposes in a closed flask, the $CO_2$ *is* present to combine with CaO and form $CaCO_3$ in a reaction that is the reverse of the decomposition. At a given temperature, no further change in the amount of products and reactants means that the reaction has reached equilibrium.

discussion of percent yield (Section 3.3), however, many reactions seem to stop before any of the reactants is totally consumed. This occurs because another reaction, the reverse of the first one, is also taking place. The forward (left-to-right) reaction has not stopped, but the reverse (right-to-left) reaction is taking place at the same speed, so *no further change occurs in the amounts of reactants or products.* This situation occurs when the reaction mixture has reached **dynamic equilibrium.** In principle, all chemical reactions are reversible and eventually reach dynamic equilibrium as long as all products remain available for the reverse reaction.

Let's examine equilibrium with a particular set of substances. Calcium carbonate can decompose to calcium oxide and carbon dioxide:

$$CaCO_3(s) \longrightarrow CaO(s) + CO_2(g) \qquad \text{[decomposition]}$$

It can also form when calcium oxide and carbon dioxide combine:

$$CaO(s) + CO_2(g) \longrightarrow CaCO_3(s) \qquad \text{[combination]}$$

Note that the combination is the reverse of the decomposition. Suppose we place 10 g $CaCO_3$ in an *open* reaction flask at 800°C (Figure 4.20, *A*). The $CaCO_3$ will begin decomposing to CaO and $CO_2$. As long as the $CO_2$ escapes from the reaction mixture, the decomposition continues because the reverse reaction cannot occur unless all the products are present.

Now suppose we perform the same experiment in a *closed* flask (Figure 4.20, *B*), so that the $CO_2$ remains in contact with the CaO. At first, very little $CaCO_3$ has decomposed, so very little $CO_2$ and CaO are available to combine. As the $CaCO_3$ continues to decompose (forward reaction), however, the amounts of $CO_2$ and CaO in the flask increase, and they interact more frequently with each other. As a result, the combination (reverse) reaction becomes faster. Eventually, the reverse reaction (combination of CaO and $CO_2$) is happening just as fast as the forward reaction (decomposition of $CaCO_3$), and the amounts of $CaCO_3$, CaO, and $CO_2$ are no longer changing:

the system has reached equilibrium. We indicate a reaction at equilibrium with a pair of arrows pointing in opposite directions:

$$CaCO_3(s) \rightleftharpoons CaO(s) + CO_2(g)$$

Bear in mind that equilibrium can be established only when all the substances involved are kept in contact with each other. The decomposition of $CaCO_3$ goes to completion if we open the flask and allow the $CO_2$ to escape. This is the reason that aqueous ionic displacement reactions that form a gaseous product go to completion in an open flask: the gas escapes, and the reverse reaction cannot take place.

The other ionic displacement reactions we saw earlier were also driven in the forward direction by removal of product. Even though all the products remained in the reaction vessel, the *ions* were unavailable for the reverse process to occur to any significant extent. In precipitation reactions, two of the four types of ions present in the reactants are tied up in the insoluble solid; in neutralization reactions, the $H^+$ and $OH^-$ ions in the reactants are effectively tied up in the water molecules.

The concept of reaction reversibility also explains why some acids and bases are weak, that is, why they dissociate only to a small extent. The dissociation quickly becomes balanced by a reassociation, such that equilibrium is reached with very few ions present. When acetic acid dissolves in water, for example, some of the molecules dissociate to $H^+$ and $CH_3COO^-$ ions. As more ions form, they interact with each other more often and re-form acetic acid:

$$CH_3COOH(aq) \rightleftharpoons H^+(aq) + CH_3COO^-(aq)$$

Weak acids and bases reach equilibrium in aqueous solution with the great majority of their molecules *undissociated.*

Dynamic equilibrium is fundamental to the functioning of many natural systems, from the cycling of water in the environment to the population balance of lion and antelope to the nuclear processes occurring in stars. We return to the applications of equilibrium in chemical and physical systems in Chapters 11, 12, and 16 to 20.

### Section Summary
Every reaction is reversible if all the substances are kept in contact with one another. As the amounts of products increase, they begin to re-form the reactants. When the reverse reaction happens as fast as the forward reaction, the amounts of the substances no longer change, and the reaction mixture has reached dynamic equilibrium. A reaction goes to completion if products are removed from the system or exist in a form that prevents them from reacting. Weak acids and bases reach equilibrium in water with a very small proportion of the molecules dissociated.

### Chapter Perspective
*Classifying facts is the first step toward understanding them, and this chapter has classified many of the most important facts of chemical reactions. We also examined the great influence that water has on reaction chemistry and introduced the dynamic equilibrium state, which is related to the central question of why physical and chemical changes occur. All these topics appear again at many places in the text.*

*In the next chapter, our focus changes to the physical behavior of gases. You'll find that your growing appreciation of events on the molecular level has become indispensable for understanding the nature of the three physical states.*

# For Review and Reference

## Key Terms

**SECTION 4.1**
combination
decomposition
displacement
metathesis

**SECTION 4.2**
electrolyte
solvated
polar covalent bond

polar molecule
nonelectrolyte

**SECTION 4.3**
precipitation
precipitate
molecular equation
total ionic equation
spectator ion
net ionic equation
neutralization

acid
base
salt
titration
acid-base indicator
equivalence point
end point

**SECTION 4.4**
oxidation-reduction
(redox)

oxidation
reduction
oxidizing agent
reducing agent
oxidation number (O.N.)
oxidation number method
half-reaction method

**SECTION 4.5**
dynamic equilibrium

## Answers to Follow-up Problems

**4.1** (a) Combination: $2(CaSO_4 \cdot \frac{1}{2}H_2O)(s) + 3H_2O(l) \rightarrow$
$$2(CaSO_4 \cdot 2H_2O)(s)$$
  (b) Displacement (double): $Na_2SO_3(aq) + 2HCl(aq) \rightarrow$
$$2NaCl(aq) + SO_2(g) + H_2O(l)$$
  (c) Decomposition: $NH_4NO_3(s) \xrightarrow{\Delta} N_2O(g) + 2H_2O(g)$

**4.2** (a) $K_2SO_4(s) \xrightarrow{H_2O} 2K^+(aq) + SO_4^{2-}(aq)$;
  4 mol $K^+$ and 2 mol $SO_4^{2-}$

  (b) $Mg(ClO_4)_2(s) \xrightarrow{H_2O} Mg^{2+}(aq) + 2\,ClO_4^{2-}(aq)$;
  1.59 mol $Mg^{2+}$ and 3.18 mol $ClO_4^-$

  (c) $(NH_4)_2Cr_2O_7(s) \xrightarrow{H_2O} 2NH_4^+(aq) + Cr_2O_7^{2-}(aq)$;
  6.24 mol $NH_4^+$ and 3.12 mol $Cr_2O_7^{2-}$

  (d) $NaHCO_3(aq) \longrightarrow Na^+(aq) + HCO_3^-(aq)$;
  0.73 mol $Na^+$ and 0.73 mol $HCO_3^-$

**4.3** Moles of $H^+$ = 451 mL $\times \dfrac{1\,L}{10^3\,mL} \times \dfrac{3.20\,mol\,HBr}{1\,L\,soln}$

$$\times \dfrac{1\,mol\,H^+}{1\,mol\,HBr} = 1.44\,mol\,H^+$$

**4.4** (a) $Fe^{3+}(aq) + 3Cl^-(aq) + 3Cs^+(aq) + PO_4^{3-}(aq) \rightarrow$
$$FePO_4(s) + 3Cl^-(aq) + 3Cs^+(aq)$$
$$Fe^{3+}(aq) + PO_4^{3-}(aq) \rightarrow FePO_4(s)$$
  (b) $2K^+(aq) + 2OH^-(aq) + Cd^{2+}(aq) + 2NO_3^-(aq) \rightarrow$
$$2K^+(aq) + 2NO_3^-(aq) + Cd(OH)_2(s)$$
$$2OH^-(aq) + Cd^{2+}(aq) \rightarrow Cd(OH)_2(s)$$
  (c) No reaction occurs
  (d) $2Ag^+(aq) + SO_4^{2-}(aq) + Ba^{2+}(aq) + 2Cl^-(aq) \rightarrow$
$$2AgCl(s) + BaSO_4(s)$$
  Total and net ionic equations are identical.

**4.5** $2HNO_3(aq) + Ca(OH)_2(aq) \rightarrow Ca(NO_3)_2(aq) + 2H_2O(l)$
$2H^+(aq) + 2NO_3^-(aq) + Ca^{2+}(aq) + 2OH^-(aq) \rightarrow$
$$Ca^{2+}(aq) + 2NO_3^-(aq) + 2H_2O(l)$$
$H^+(aq) + OH^-(aq) \rightarrow H_2O(l)$

**4.6** $2HCl(aq) + Ba(OH)_2(aq) \rightarrow BaCl_2(aq) + 2H_2O(l)$

Volume (L) of soln = 50.00 mL HCl soln $\times \dfrac{1\,L}{10^3\,mL}$

$$\times \dfrac{0.1016\,mol\,HCl}{1\,L\,soln} \times \dfrac{1\,mol\,Ba(OH)_2}{2\,mol\,HCl}$$

$$\times \dfrac{1\,L\,soln}{0.1292\,mol\,Ba(OH)_2} = 0.01966\,L$$

**4.7** (a) O.N. of Sc = +3; O.N. of O = −2
  (b) O.N. of Ga = +3; O.N. of Cl = −1
  (c) O.N. of H = +1; O.N. of P = +5; O.N. of O = −2
  (d) O.N. of I = +3; O.N. of F = −1

**4.8** (a) Fe is reducing agent; $Cl_2$ is oxidizing agent.
  (b) $C_2H_6$ is reducing agent; $O_2$ is oxidizing agent.

**4.9** (a) $2H_2O_2(l) \rightarrow 2H_2O(l) + O_2(g)$
  (b) $K_2Cr_2O_7(aq) + 14HI(aq) \rightarrow$
$$2KI(aq) + 2CrI_3(aq) + 3I_2(s) + 7H_2O(l)$$

**4.10** (a) Moles of $Ca^{2+}$ = 6.53 mL soln $\times \dfrac{1\,L}{10^3\,mL}$

$$\times \dfrac{4.56 \times 10^{-3}\,mol\,KMnO_4}{1\,L\,soln} \times \dfrac{5\,mol\,CaC_2O_4}{2\,mol\,KMnO_4}$$

$$\times \dfrac{1\,mol\,Ca^{2+}}{1\,mol\,CaC_2O_4} = 7.44 \times 10^{-5}\,mol\,Ca^{2+}$$

Molarity of $Ca^{2+}$ = $\dfrac{7.44 \times 10^{-5}\,mol\,Ca^{2+}}{2.50\,mL\,milk} \times \dfrac{10^3\,mL}{1\,L}$

$$= 2.98 \times 10^{-2}\,M\,Ca^{2+}$$

  (b) Conc. of $Ca^{2+}$ (g/L) = $\dfrac{2.98 \times 10^{-2}\,mol\,Ca^{2+}}{1\,L}$

$$\times \dfrac{40.08\,g\,Ca^{2+}}{1\,mol\,Ca^{2+}} = 1.19\,g\,Ca^{2+}/1\,L$$

## Sample Problem Titles

**4.1** Identifying the Type of Chemical Reaction (p. 138)
**4.2** Determining Moles of Ions in Aqueous Solutions of Ionic Compounds (p. 140)
**4.3** Determining the Molarity of $H^+$ Ions in Aqueous Solutions of Acids (p. 144)
**4.4** Predicting Whether a Precipitation Reaction Occurs; Writing Molecular, Total, and Net Ionic Equations (p. 147)
**4.5** Writing Balanced Equations for Neutralization Reactions (p. 149)

**4.6** Finding the Concentration of Acid from an Acid-Base Titration (p. 151)
**4.7** Determining the Oxidation Number of an Element in a Compound (p. 155)
**4.8** Recognizing Oxidizing and Reducing Agents (p. 156)
**4.9** Balancing Redox Reactions with the Oxidation Number Method (p. 157)
**4.10** Finding an Unknown Concentration Using a Redox Titration (p. 161)

# Problems

Problems with a green number are answered at the back of the text. Most sections include three categories of problems separated by a green rule: *concept review questions*, *paired skill-building exercises*, and problems in a relevant context.

## Types of Chemical Reactions: An Accounting of Reactants and Products

(Sample Problem 4.1)

**4.1** What is the name for the type of reaction that leads to the following?
(a) An increase in the number of substances
(b) A decrease in the number of substances
(c) No change in the number of substances

**4.2** Why do decomposition reactions typically have compounds as reactants, whereas combination and displacement reactions have either elements or compounds?

**4.3** Of the three types of reactions discussed in this section, which commonly produce one or more compounds?

**4.4** Balance each of the following, and classify it as a combination, decomposition, or displacement reaction:
(a) $Ca(s) + H_2O(l) \rightarrow Ca(OH)_2(aq) + H_2(g)$
(b) $NaNO_3(s) \rightarrow NaNO_2(s) + O_2(g)$
(c) $C_2H_2(g) + H_2(g) \rightarrow C_2H_6(g)$
(d) $HI(g) \rightarrow H_2(g) + I_2(g)$

**4.5** Balance each of the following, and classify it as a combination, decomposition, or displacement reaction:
(a) $Sb(s) + Cl_2(g) \rightarrow SbCl_3(s)$
(b) $AsH_3(g) \rightarrow As(s) + H_2(g)$
(c) $C_2H_5OH(l) \xrightarrow{\Delta} C_2H_4(g) + H_2O(g)$
(d) $Mg(s) + H_2O(g) \rightarrow Mg(OH)_2(s) + H_2(g)$

**4.6** Complete the following general reactions, and write a specific balanced equation to exemplify each:
(a) Metal + oxygen $\rightarrow$
(b) Nonmetal halide + halogen $\rightarrow$
(c) Nonmetal oxide + water $\rightarrow$
(d) Metal hydroxide $\xrightarrow{\Delta}$

**4.7** Complete the following general reactions, and write a specific balanced equation to exemplify each:
(a) Nonmetal + nonmetal $\rightarrow$
(b) Metal oxide + nonmetal oxide $\rightarrow$
(c) Metal carbonate $\xrightarrow{\Delta}$
(d) Acid + base $\rightarrow$

**4.8** Predict the product(s) and write a balanced equation for each of the following:
(a) $Ca(s) + Br_2(l) \rightarrow$
(b) $Ag_2O(s) \xrightarrow{\Delta}$
(c) $Ca(OH)_2(aq) + HCl(aq) \rightarrow$
(d) $LiCl(l) \xrightarrow{electricity}$

**4.9** Predict the products and write a balanced equation for each of the following:
(a) $N_2(g) + H_2(g) \rightarrow$
(b) $NaClO_3(s) \xrightarrow{\Delta}$
(c) $KOH(aq) + HClO_4(aq) \rightarrow$
(d) $S(s) + O_2(g) \rightarrow$

**4.10** Predict the products and write a balanced equation for each of the following:
(a) Cesium + iodine $\rightarrow$
(b) Aluminum + oxygen $\rightarrow$
(c) Sulfur dioxide + oxygen $\rightarrow$
(d) Carbon dioxide + barium oxide $\rightarrow$

**4.11** Predict the products and write a balanced equation for each of the following:
(a) Calcium oxide + water $\rightarrow$
(b) Phosphorus trichloride + chlorine $\rightarrow$
(c) Zinc + hydrobromic acid $\rightarrow$
(d) Aqueous potassium iodide + bromine $\rightarrow$

**4.12** How many grams of $O_2$ can be prepared from the complete decomposition of 4.27 g HgO? Name and calculate the amount of the other product.

**4.13** How many grams of CaO can be produced from the complete decomposition of 24.52 g $CaCO_3$? Name and calculate the amount of the other product.

**4.14** In a combination reaction, 1.62 g lithium is mixed with 5.00 g oxygen.
(a) Which reactant is present in excess?
(b) How many moles of product are produced?
(c) After reaction, how many grams of each reactant and product are present?

**4.15** In a combination reaction, 1.22 g magnesium is heated with 3.75 g nitrogen.
(a) Which reactant is present in excess?
(b) How many moles of product are formed?
(c) After reaction, how many grams of each reactant and product are present?

**4.16** A mixture of $KClO_3$ and KCl with a mass of 0.800 g was heated to produce $O_2$. After heating, the mass of residue was 0.700 g. Assuming all the $KClO_3$ decomposed to KCl and $O_2$, calculate the mass percent of $KClO_3$ in the original mixture.

**4.17** A mixture of $CaCO_3$ and CaO weighing 0.793 g was heated to produce $CO_2$. After heating, the remaining solid weighed 0.508 g. Assuming all the $CaCO_3$ decomposed to CaO and $CO_2$, calculate the mass percent of $CaCO_3$ in the original mixture.

**4.18** The brewing industry uses yeast microorganisms to convert glucose to ethanol for wine and beer. The baking industry uses the carbon dioxide these single-celled fungi produce to make bread rise:

$$C_6H_{12}O_6(s) \xrightarrow{yeast} 2C_2H_5OH(l) + 2CO_2(g)$$

How many grams of ethanol can be produced from 10.0 g glucose? What volume of $CO_2$ is produced? (Assume one mole of gas occupies 22.4 L under the conditions used.)

**4.19** Before arc welding was developed, a displacement reaction involving aluminum and iron(III) oxide was used to produce molten iron (the thermite process). This reaction was used, for example, to connect sections of iron railroad track. Calculate the mass of molten iron produced when 1.00 kg aluminum reacts with 2.00 mol iron(III) oxide.

**4.20** One of the first steps in the enrichment of uranium for use in nuclear power plants involves a displacement reaction between $UO_2$ and aqueous HF:

$$UO_2(s) + 4HF(aq) \rightarrow UF_4(s) + 2H_2O(l)$$

How many liters of 6.50 $M$ HF are needed to react with 1.25 kg $UO_2$?

**4.21** Calcium dihydrogen phosphate, $Ca(H_2PO_4)_2$, and sodium bicarbonate, $NaHCO_3$, are ingredients of baking powder that react with each other to produce $CO_2$, which causes dough or batter to rise:

$$Ca(H_2PO_4)_2(s) + 2NaHCO_3(s) \rightarrow$$
$$2CO_2(g) + 2H_2O(g) + CaHPO_4(s) + Na_2HPO_4(s)$$

If the baking powder contains 31% $NaHCO_3$ and 35% $Ca(H_2PO_4)_2$ by mass,
(a) How many moles of $CO_2$ are produced from 1.00 g baking powder?
(b) If one mole of $CO_2$ occupies 37.0 L at 350°F (a typical baking temperature), what volume of $CO_2$ is produced from 1.00 g baking powder?

**4.22** Ethyl chloride, a local anesthetic used by veterinarians and sports physicians, is the product of the combination of ethylene with hydrogen chloride:

$$C_2H_4(g) + HCl(g) \rightarrow C_2H_5Cl(g)$$

If 0.100 kg $C_2H_4$ and 0.100 kg HCl react,
(a) How many molecules of gas (reactants plus products) are present when the reaction is complete?
(b) How many moles of gas are present when half the product forms?

**4.23** Limestone ($CaCO_3$) is used to remove acidic pollutants from smokestack flue gases in a sequence of decomposition-combination reactions. The limestone is heated to form lime (CaO), which reacts with sulfur dioxide to form calcium sulfite. Assuming a 70% yield in the overall reaction, what mass of limestone is required to remove all the sulfur dioxide formed by the combustion of $8.5 \times 10^4$ kg coal that is 0.33 mass % sulfur?

**The Role of Water as a Solvent**

(Sample Problems 4.2 and 4.3)

**4.24** What two factors cause water to be polar?

**4.25** What types of substances are most likely to be soluble in water?

**4.26** What must be present in an aqueous solution for it to conduct an electric current? What general classes of compounds form solutions that do so?

**4.27** What occurs on the molecular level when an ionic compound dissolves in water?

**4.28** Why do some ionic compounds dissolve in water, whereas others do not?

**4.29** Why are some covalent compounds soluble in water, whereas others are not?

**4.30** Some covalent compounds dissociate into ions when they dissolve in water. What atom do these compounds have in their structures? What type of aqueous solution do they form?

**4.31** In what form(s) does the $H^+$ ion exist in water? Explain.

**4.32** State whether each of the following substances is likely to be very soluble in water. Explain.
(a) Benzene, $C_6H_6$    (b) Sodium hydroxide

**4.33** State whether each of the following substances is likely to be very soluble in water. Explain.
(a) Lithium nitrate    (b) Glycine, $H_2NCH_2COOH$

**4.34** State whether an aqueous solution of each of the following substances conducts an electric current. Explain your reasoning.
(a) Sodium iodide    (b) Hydrogen bromide

**4.35** State whether an aqueous solution of each of the following substances conducts an electric current. Explain your reasoning.
(a) Potassium hydroxide    (b) Glucose, $C_6H_{12}O_6$

**4.36** How many total moles of ions are released when each of the following samples dissolves completely in water?
(a) 0.25 mol $NH_4Cl$
(b) 26.4 g $Ba(OH)_2 \cdot 8H_2O$
(c) $1.78 \times 10^{20}$ formula units of LiCl

**4.37** How many total moles of ions are released when each of the following samples dissolves completely in water?
(a) 0.15 mol $Na_3PO_4$
(b) 47.9 g $NiBr_2 \cdot 3H_2O$
(c) $4.23 \times 10^{22}$ formula units of $FeCl_3$

**4.38** How many ions of each type are present in the following aqueous solutions?
(a) 124 mL of 2.45 $M$ aluminum chloride
(b) 2.50 L of a solution containing 3.59 g sodium sulfate
(c) 805 mL of a solution containing $2.68 \times 10^{22}$ formula units of magnesium bromide per liter

**4.39** How many ions of each type are present in the following aqueous solutions?
(a) 3.8 mL of 0.88 $M$ magnesium chloride
(b) 345 mL of a solution containing 2.22 g aluminum sulfate
(c) 1.66 L of a solution containing $6.63 \times 10^{22}$ formula units of lithium nitrate per liter

**4.40** How many moles of $H^+$ ions are present in the following aqueous solutions?
(a) 0.140 L of 2.5 $M$ perchloric acid
(b) 6.8 mL of 0.52 $M$ nitric acid
(c) 2.5 L of 0.056 $M$ hydrochloric acid

**4.41** How many moles of $H^+$ ions are present in the following aqueous solutions?
(a) 1.4 L of 0.48 $M$ hydrobromic acid
(b) 47 mL of 1.8 $M$ hydriodic acid
(c) 425 mL of 0.27 $M$ nitric acid

**4.42** In studies of ocean-dwelling specimens in the laboratory, marine biologists can use salt mixtures that simulate the ion concentrations in sea water. A 1.00-kg sample of simulated sea water is prepared by mixing 26.5 g NaCl, 2.40 g $MgCl_2$, 3.35 g $MgSO_4$, 1.20 g $CaCl_2$, 1.05 g KCl, 0.315 g $NaHCO_3$, and 0.098 g NaBr in distilled water. If the density of this solution is 1.04 $g/cm^3$, what is the molarity of each ion?

**4.43** Water "softeners" remove metal ions such as $Fe^{2+}$, $Fe^{3+}$, $Ca^{2+}$, and $Mg^{2+}$ (which make water "hard") by replacing them with enough $Na^+$ ions to maintain the same number of positive charges in the solution. If $1.0 \times 10^3$ L hard water is 0.015 $M$ $Ca^{2+}$ and 0.0010 $M$ $Fe^{3+}$, how many moles of $Na^+$ would be needed to replace these ions?

**4.44** The *salinity* of a solution is defined as the grams of total salts per kilogram of solution. An agricultural chemist uses a solution whose salinity is 35.0 g/kg to test the effect of irrigating farmland with high-salinity river water. The two solutes are NaCl and $MgSO_4$, and there are twice as many moles of NaCl as $MgSO_4$. What masses of NaCl and $MgSO_4$ are contained in 1.00 kg of the solution?

**Some Important Aqueous Ionic Reactions; Writing Ionic Equations**
(Sample Problems 4.4 to 4.6)

**4.45** When two ionic compounds dissolve, a reaction occurs only if ions are removed from solution to form product. What are three ways in which this removal can occur? What type of reaction is involved in each case?

**4.46** Why do some pairs of ions form precipitates, whereas others do not?

**4.47** Which ions do not appear in a net ionic equation? Why?

**4.48** Use Table 4.1 to determine which of the following combinations lead to reaction. How do you know the spectator ions in that reaction?
(a) Calcium nitrate($aq$) + potassium chloride($aq$) →
(b) Sodium chloride($aq$) + lead(II) nitrate($aq$) →

**4.49** Is the total ionic equation the same as the net ionic equation for the reaction between $Sr(OH)_2(aq)$ and $H_2SO_4(aq)$? Explain.

**4.50** State a general equation for a neutralization reaction.

**4.51** (a) Name three common strong acids.
(b) Name three common strong bases.
(c) What is a characteristic behavior of a strong acid or a strong base?

**4.52** (a) Name three common weak acids.
(b) Name one common weak base.
(c) What is the major difference between a weak acid and a strong acid or between a weak base and a strong base?

**4.53** What two factors drive the following reactions to completion?
(a) $MgSO_3(s) + 2HCl(aq) \rightarrow$
$$MgCl_2(aq) + SO_2(g) + H_2O(l)$$
(b) $3Ba(OH)_2(aq) + 2H_3PO_4(aq) \rightarrow$
$$Ba_3(PO_4)_2(s) + 6H_2O(l)$$

**4.54** Why do the following reactants give neutralization reactions with the same net ionic equation?
$$NaOH(aq) + HCl(aq) \rightarrow$$
$$KOH(aq) + HNO_3(aq) \rightarrow$$
$$Ba(OH)_2(aq) + 2HBr(aq) \rightarrow$$
Are the total ionic equations different? Explain.

**4.55** The net ionic equation for the aqueous neutralization between acetic acid and sodium hydroxide is different from the equation for hydrochloric acid and sodium hydroxide. Explain by writing balanced net ionic equations.

**4.56** For each of the following pairs of aqueous solutions, state whether a precipitation reaction occurs when they are mixed. Write the formulas and names of any precipitates that form.
(a) Sodium nitrate + copper(II) sulfate
(b) Ammonium iodide + silver nitrate
(c) Potassium carbonate + barium hydroxide
(d) Aluminum nitrate + sodium phosphate

**4.57** For each of the following pairs of aqueous solutions, state whether a precipitation reaction occurs when they are mixed. Write the formulas and names of any precipitates that form.
(a) Potassium chloride + iron(II) nitrate
(b) Ammonium sulfate + barium chloride
(c) Sodium sulfide + nickel(II) sulfate
(d) Lead(II) nitrate + potassium bromide

**4.58** Complete the following precipitation reactions with balanced molecular, total ionic, and net ionic equations, and identify the spectator ions:
(a) $Hg_2(NO_3)_2(aq) + KI(aq) \rightarrow$
(b) $FeSO_4(aq) + Ba(OH)_2(aq) \rightarrow$

**4.59** Complete the following precipitation reactions with balanced molecular, total ionic, and net ionic equations, and identify the spectator ions:
(a) $CaCl_2(aq) + Cs_3PO_4(aq) \rightarrow$
(b) $Na_2S(aq) + ZnSO_4(aq) \rightarrow$

**4.60** Complete the following neutralization reactions with balanced molecular, total ionic, and net ionic equations, and identify the spectator ions:
(a) Potassium hydroxide($aq$) + hydriodic acid($aq$) →
(b) Ammonia($aq$) + hydrochloric acid($aq$) →

**4.61** Complete the following neutralization reactions with balanced molecular, total ionic, and net ionic equations, and identify the spectator ions:
(a) Cesium hydroxide($aq$) + nitric acid($aq$) →
(b) Calcium hydroxide($aq$) + acetic acid($aq$) →

**4.62** Limestone (calcium carbonate) is insoluble in water but dissolves with the addition of a hydrochloric acid solution. Why? Write balanced molecular, total ionic, and net ionic equations for this reaction.

**4.63** Zinc hydroxide is insoluble in water but dissolves with the addition of a nitric acid solution. Why? Write balanced molecular, total ionic, and net ionic equations for this reaction.

**4.64** If 25.0 mL lead(II) nitrate solution reacts completely with excess sodium iodide solution to yield 0.828 g precipitate, what is the molarity of lead(II) ion in the original solution?

**4.65** If 25.0 mL silver nitrate solution reacts with excess potassium chloride solution to yield 0.642 g precipitate, what is the molarity of silver ion in the original solution?

**4.66** A standard solution of 0.1090 $M$ KOH is used to neutralize 50.00 mL $CH_3COOH$. If 17.98 mL is required to reach the end point, what is the molarity of the acid solution?

**4.67** If 36.35 mL of a standard 0.1550 $M$ NaOH solution is required to neutralize completely 25.00 mL $H_2SO_4$, what is the molarity of the acid solution?

**4.68** An auto mechanic spills 55 mL of 3.0 $M$ $H_2SO_4$ solution from an auto battery. How many milliliters of 2.0 $M$ $NaHCO_3$ must be poured on the spill in order to react completely with the sulfuric acid?

**4.69** The calcium bicarbonate impurity in a sample of phosphate rock is removed by treatment with hydrochloric acid; the products are carbon dioxide, water, and calcium chloride. When 10.0 g of the rock is ground up and treated with excess hydrochloric acid, 2.81 g carbon dioxide is formed. Calculate the mass percent of calcium bicarbonate in the rock.

**4.70** Sodium hydroxide is used extensively in acid-base titrations because it is a strong, inexpensive base. A sodium hydroxide solution was standardized by titrating 25.00 mL of 0.1628 $M$ standard hydrochloric acid. The sodium hydroxide solution was added to a buret. The initial buret reading was 1.24 mL and the final reading was 39.21 mL. What was the molarity of the base solution?

**4.71** To find the mass percent of limestone ($CaCO_3$) in a soil sample, a geochemist titrates 1.586 g of the soil with 43.56 mL of 0.2516 $M$ HCl. What is the mass percent $CaCO_3$ in the soil?

**4.72** The mass percent of $Cl^-$ in a sea water sample is determined by titrating 25.00 mL sea water with $AgNO_3$ solution, causing a precipitation reaction. An indicator is used to detect the end point, which occurs when free $Ag^+$ ion is present in solution after all the $Cl^-$ is consumed. If 43.63 mL of 0.3020 $M$ $AgNO_3$ is required to reach the end point, what is the mass percent of $Cl^-$ in the sea water? ($d$ of sea water = 1.04 g/mL.)

**4.73** Precipitation reactions are often used to prepare useful ionic compounds. For example, thousands of tons of silver bromide are prepared annually for use in making black-and-white photographic film.
(a) What mass (in kilograms) of silver bromide forms when 5.85 m³ of 1.68 $M$ potassium bromide reacts with 3.51 m³ of 2.04 $M$ silver nitrate?
(b) After the solid silver bromide is removed, what ions are present in the remaining solution? Determine the molarity of each ion. (Assume the total volume is the sum of the reactant volumes.)

**4.74** Dolomite rock, an important agricultural soil supplement, consists of $CaCO_3$ and $MgCO_3$ in a nearly 1/1 ratio by mass. When 2.44 g dolomite was treated with excess hydrochloric acid, 0.0260 mol $CO_2$ formed.
(a) What is the mass of each carbonate in the dolomite sample?
(b) How many moles of $CO_2$ would be produced if the dolomite sample were exactly $1CaCO_3/1MgCO_3$ on a mole basis?

## Oxidation-Reduction (Redox) Reactions

(Sample Problems 4.7 to 4.10)

**4.75** Describe how you would determine the oxidation number of sulfur in (a) $H_2S$ and (b) $SO_3$.

**4.76** Is the following a redox reaction? Explain.
$$NH_3(aq) + HCl(aq) \rightarrow NH_4Cl(aq)$$

**4.77** Explain why an oxidizing agent undergoes reduction.

**4.78** Why must every redox reaction involve an oxidizing agent and a reducing agent?

**4.79** Sulfuric acid functions as an oxidizing agent in (a) and as an acid in (b). How do you differentiate between these two functions?
(a) $4H^+(aq) + SO_4^{2-}(aq) + 2NaI(s) \rightarrow$
$$2Na^+(aq) + I_2(s) + SO_2(g) + 2H_2O(l)$$
(b) $BaF_2(s) + 2H^+(aq) + SO_4^{2-}(aq) \rightarrow$
$$2HF(aq) + BaSO_4(s)$$

**4.80** Identify the oxidizing agent and the reducing agent in the following reaction, and explain your answer:
$$8NH_3(g) + 6NO_2(g) \rightarrow 7N_2(g) + 12H_2O(l)$$

**4.81** Give the oxidation number of carbon in each of the following: (a) $CF_2Cl_2$; (b) $Na_2C_2O_4$; (c) $HCO_3^-$; (d) $C_2H_6$.

**4.82** Give the oxidation number of nitrogen in each of the following: (a) $NH_2OH$; (b) $N_2H_4$; (c) $NH_4^+$; (d) $HNO_2$.

**4.83** Give the oxidation number of phosphorus in each of the following: (a) $PH_3$; (b) $H_3PO_2$; (c) $H_3PO_4$; (d) $H_2P_2O_7^{2-}$; (e) $PH_4^+$.

**4.84** Give the oxidation number of manganese in each of the following: (a) $MnO_4^{2-}$; (b) $Mn_2O_3$; (c) $KMnO_4$; (d) $MnO_2^{2-}$; (e) $MnSO_4$.

**4.85** Identify the oxidizing agent and the reducing agent in each of the following balanced net ionic equations:
(a) $5H_2C_2O_4(aq) + 2MnO_4^-(aq) + 6H^+(aq) \rightarrow$
$$2Mn^{2+}(aq) + 10CO_2(g) + 8H_2O(l)$$
(b) $3Cu(s) + 8H^+(aq) + 2NO_3^-(aq) \rightarrow$
$$3Cu^{2+}(aq) + 2NO(g) + 4H_2O(l)$$
(c) $Sn(s) + 2H^+(aq) \rightarrow Sn^{2+}(aq) + H_2(g)$
(d) $2H^+(aq) + H_2O_2(aq) + 2Fe^{2+}(aq) \rightarrow$
$$2Fe^{3+}(aq) + 2H_2O(l)$$

**4.86** Identify the oxidizing agent and the reducing agent in each of the following balanced net ionic equations:
(a) $8H^+(aq) + 6Cl^-(aq) + Sn(s) + 4NO_3^-(aq) \rightarrow$
$$SnCl_6^{2-}(aq) + 4NO_2(g) + 4H_2O(l)$$
(b) $2MnO_4^-(aq) + 10Cl^-(aq) + 16H^+(aq) \rightarrow$
$$5Cl_2(g) + 2Mn^{2+}(aq) + 8H_2O(l)$$
(c) $8H^+(aq) + Cr_2O_7^{2-}(aq) + 3SO_3^{2-}(aq) \rightarrow$
$$2Cr^{3+}(aq) + 3SO_4^{2-}(aq) + 4H_2O(l)$$
(d) $NO_3^-(aq) + 4Zn(s) + 7OH^-(aq) + 6H_2O(l) \rightarrow$
$$4Zn(OH)_4^{2-}(aq) + NH_3(aq)$$

**4.87** Discuss each of the following conclusions from a study of redox reactions:
(a) The sulfide ion functions only as a reducing agent.
(b) The sulfate ion functions only as an oxidizing agent.
(c) Sulfur dioxide functions as either an oxidizing or a reducing agent.

**4.88** Discuss each of the following conclusions from a study of redox reactions:
(a) The nitride ion functions only as a reducing agent.
(b) The nitrate ion functions only as an oxidizing agent.
(c) The nitrite ion functions as either an oxidizing or a reducing agent.

**4.89** Use the oxidation number method to balance the following reactions by placing coefficients in the blanks. Identify the reducing and oxidizing agents:
(a) __$HNO_3(aq)$ + __$K_2CrO_4(aq)$ + __$Fe(NO_3)_2(aq) \rightarrow$
__$KNO_3(aq)$ + __$Fe(NO_3)_3(aq)$ + __$Cr(NO_3)_3(aq)$
+ __$H_2O(l)$
(b) __$HNO_3(aq)$ + __$C_2H_6O(l)$ + __$K_2Cr_2O_7(aq) \rightarrow$
__$KNO_3(aq)$ + __$C_2H_4O(l)$ + __$H_2O(l)$
+ __$Cr(NO_3)_3(aq)$
(c) __$HCl(aq)$ + __$NH_4Cl(aq)$ + __$K_2Cr_2O_7(aq) \rightarrow$
__$KCl(aq)$ + __$CrCl_3(aq)$ + __$N_2(g)$ + __$H_2O(l)$
(d) __$KClO_3(aq)$ + __$HBr(aq) \rightarrow$
__$Br_2(l)$ + __$H_2O(l)$ + __$KCl(aq)$

**4.90** Use the oxidation number method to balance the following reactions by placing coefficients in the blanks. Identify the reducing and oxidizing agents:
(a) __$HCl(aq)$ + __$FeCl_2(aq)$ + __$H_2O_2(aq) \rightarrow$
__$FeCl_3(aq)$ + __$H_2O(l)$

(b) __$I_2(s)$ + __$Na_2S_2O_3(aq) \rightarrow$
__$Na_2S_4O_6(aq)$ + __$NaI(aq)$
(c) __$HNO_3(aq)$ + __$KI(aq) \rightarrow$
__$NO(g)$ + __$I_2(s)$ + __$H_2O(l)$ + __$KNO_3(aq)$
(d) __$PbO(s)$ + __$NH_3(aq) \rightarrow$
__$N_2(g)$ + __$H_2O(l)$ + __$Pb(s)$

---

**4.91** The active agent in many hair bleaches is hydrogen peroxide. The amount of hydrogen peroxide in 15.8 g hair bleach was determined by titration with a standard potassium permanganate solution:

$2MnO_4^-(aq) + 5H_2O_2(aq) + 6H^+(aq) \rightarrow$
$$5O_2(g) + 2Mn^{2+}(aq) + 8H_2O(l)$$
(a) How many moles of $MnO_4^-$ were required for the titration if 43.2 mL of 0.105 $M$ $KMnO_4$ was needed to reach the end point?
(b) How many moles of $H_2O_2$ were present in the 15.8-g sample of bleach?
(c) How many grams of $H_2O_2$ were in the sample?
(d) What is the mass percent $H_2O_2$ in the sample?
(e) What is the reducing agent in the redox reaction?

**4.92** Mixtures of $CaCl_2$ and $NaCl$ are used for salting roads to prevent ice formation. A dissolved 1.9348-g sample of such a mixture was analyzed by completely precipitating the $Ca^{2+}$ as $CaC_2O_4$ with excess $Na_2C_2O_4$. The $CaC_2O_4$ was separated from the solution and then dissolved with sulfuric acid. The resulting $H_2C_2O_4$ was titrated with 37.68 mL of 0.1019 $M$ $KMnO_4$ solution.
(a) Write the balanced net ionic equation for the precipitation reaction.
(b) Write the balanced net ionic equation for the titration reaction. (See Sample Problem 4.10.)
(c) What is the oxidizing agent?
(d) What is the reducing agent?
(e) Calculate the mass percent of $CaCl_2$ in the original sample.

**4.93** A person's blood alcohol ($C_2H_5OH$) level can be determined by titrating a sample of blood plasma with a potassium dichromate solution. The balanced equation is

$16H^+(aq) + 2Cr_2O_7^{2-}(aq) + C_2H_5OH(aq) \rightarrow$
$$4Cr^{3+}(aq) + 2CO_2(g) + 11H_2O(l)$$

If 35.46 mL of 0.04961 $M$ $Cr_2O_7^{2-}$ is required to titrate 25.00 g plasma, what is the mass percent of alcohol in the blood?

**4.94** A chemical engineer determines the mass percent of iron in an ore sample by converting the Fe to $Fe^{2+}$ in acid and then titrating the $Fe^{2+}$ with $MnO_4^-$. A 1.1081-g sample was dissolved in acid and titrated with 39.32 mL of 0.03190 $M$ $KMnO_4$. The balanced equation is

$8H^+(aq) + 5Fe^{2+}(aq) + MnO_4^-(aq) \rightarrow$
$$5Fe^{3+}(aq) + Mn^{2+}(aq) + 4H_2O(l)$$
Calculate the mass percent of iron in the ore.

## Reversible Reactions: An Introduction to Chemical Equilibrium

**4.95** Why is the equilibrium state called "dynamic"?

**4.96** In a decomposition reaction involving a gaseous product, what must be done for the reaction to reach equilibrium?

**4.97** Describe what happens on the molecular level when acetic acid dissolves in water.

**4.98** When either a mixture of NO and $Br_2$ or pure nitrosyl bromide (NOBr) is placed in a reaction vessel, the product mixture contains NO, $Br_2$, and NOBr. Explain.

**4.99** Ammonia is produced in millions of tons annually for use as a fertilizer. It is commonly made from $N_2$ and $H_2$ by the Haber process. Because the reaction reaches equilibrium before going completely to product, the stoichiometric amount of ammonia is not obtained. At a particular temperature and pressure, 10.0 g $H_2$ reacts with 20.0 g $N_2$ to form ammonia. When equilibrium is reached, 15.0 g $NH_3$ has formed.
(a) Calculate the percent yield.
(b) How many moles of $N_2$ and $H_2$ are present at equilibrium?

## Comprehensive Problems

Problems with an asterisk (*) are more challenging.

**4.100** Use the oxidation number method to balance the following reactions by placing coefficients in the blanks. Identify the reducing and oxidizing agents:
(a) __$KOH(aq)$ + __$H_2O_2(aq)$ + __$Cr(OH)_3(s)$ → __$K_2CrO_4(aq)$ + __$H_2O(l)$
(b) __$MnO_4^-(aq)$ + __$ClO_2^-(aq)$ + __$H_2O(l)$ → __$MnO_2(s)$ + __$ClO_4^-(aq)$ + __$OH^-(aq)$
(c) __$KMnO_4(aq)$ + __$Na_2SO_3(aq)$ + __$H_2O(l)$ → __$MnO_2(s)$ + __$Na_2SO_4(aq)$ + __$KOH(aq)$

**4.101** Use the oxidation number method to balance the following reactions by placing coefficients in the blanks. Identify the reducing and oxidizing agents:
(a) __$CrO_4^{2-}(aq)$ + __$HSnO_2^-(aq)$ + __$H_2O(l)$ → __$CrO_2^-(aq)$ + __$HSnO_3^-(aq)$ + __$OH^-(aq)$
(b) __$KMnO_4(aq)$ + __$NaNO_2(aq)$ + __$H_2O(l)$ → __$MnO_2(s)$ + __$NaNO_3(aq)$ + __$KOH(aq)$
(c) __$I^-(aq)$ + __$O_2(g)$ + __$H_2O(l)$ → __$I_2(s)$ + __$OH^-(aq)$

**\*4.102** When $NaCl(s)$ dissolves in water, the ions separate; when the water evaporates, $NaCl(s)$ reforms. When $HCl(g)$ dissolves in water, a covalent bond breaks; when the water evaporates, $HCl(g)$ reforms. Are either or both of these chemical changes? Discuss.

**4.103** Sodium peroxide ($Na_2O_2$) is often used in self-contained breathing devices because it reacts with exhaled $CO_2$ to form $Na_2CO_3$ and $O_2$. How many liters of respired air can react with 50.0 g $Na_2O_2$ if each liter of respired air contains 0.0720 g $CO_2$?

**4.104** Magnesium is used in many lightweight alloys, including those in airplane bodies. The metal is obtained from sea water in an industrial process that includes precipitation, neutralization, evaporation, and electrolysis. How many kilograms of magnesium can be obtained from 1.00 $km^3$ sea water if the initial $Mg^{2+}$ concentration is 0.13% by mass? ($d$ of sea water = 1.04 g/mL.)

**\*4.105** A typical formulation for window glass is 75% $SiO_2$, 15% $Na_2O$, and 10% $CaO$ by mass. What masses of sand ($SiO_2$), sodium carbonate, and calcium carbonate must be combined to produce 1.00 kg glass after carbon dioxide is driven off by thermal decomposition of the carbonates?

**4.106** Carbon dioxide is removed from the atmosphere of some space capsules by reaction with a solid metal hydroxide. The products are water and the metal carbonate.
(a) Calculate the mass of $CO_2$ that can be removed by reaction with 2.50 kg lithium hydroxide.
(b) How many grams of $CO_2$ would be removed by 1.00 g of each of the following: lithium hydroxide, magnesium hydroxide, or aluminum hydroxide?

**4.107** During the process of developing black-and-white film, unexposed silver bromide is removed in a displacement reaction with sodium thiosulfate solution:
$$AgBr(s) + 2Na_2S_2O_3(aq) \rightarrow$$
$$Na_3Ag(S_2O_3)_2(aq) + NaBr(aq)$$
What volume of 0.155 $M$ $Na_2S_2O_3$ solution is needed to remove 2.66 g AgBr from a roll of film?

**\*4.108** For several years, Brazil has been engaged in a program to replace gasoline with ethanol derived from the root crop manioc (cassava).
(a) Write separate balanced equations for the complete combustion of ethanol ($C_2H_5OH$) and of gasoline (represented by the formula $C_8H_{18}$).
(b) What mass of oxygen is required to burn completely 1.00 L of a mixture that is 90.0% gasoline ($d$ = 0.742 g/mL) and 10.0% ethanol ($d$ = 0.789 g/mL) by volume?
(c) If 1.00 mol $O_2$ occupies 22.4 L, what volume of $O_2$ is needed to burn 1.00 L of the mixture?
(d) Air is 20.9% $O_2$ by volume. What volume of air is needed to burn 1.00 L of the mixture?

**4.109** One of the molecules suspected of being responsible for atmospheric ozone depletion is the refrigerant and aerosol propellant Freon 12 ($CF_2Cl_2$). Freon 12 can be prepared commercially in a sequence of two reactions:
1. A combination reaction between hydrogen and fluorine gases
2. A displacement reaction between the product of the first reaction and liquid carbon tetrachloride. Hydrogen chloride gas also forms.
(a) Write balanced equations for the two reactions.
(b) What is the maximum mass (in kg) of Freon 12 that can be produced from 0.380 kg fluorine?

# CHAPTER 5

## Concepts and skills to review

- physical states of matter (Section 1.1)
- SI unit conversions (Section 1.5)
- mass-mole-number conversions (Section 3.1)

# Gases and the Kinetic-Molecular Theory

**The atmosphere of planet Earth.** Viewed from near orbit, our gaseous planetary envelope is a picture of calm beauty, revealing nothing of the molecular chaos within. As you'll see in this chapter, however, that chaotic behavior allows us to understand gases with simple mathematical models. In fact, it might be fair to say that the young science of chemistry advanced so rapidly because gases behave so predictably under changing conditions of pressure and temperature.

A colorless, odorless mixture of gases surrounds the liquid and solid surface of our planet and the organisms thriving there. People have been observing gases, liquids, and solids throughout history—three of the four "elements" of the ancients were air (gas), water (liquid), and earth (solid)—but many questions about their behavior remain. In this chapter and its companion, Chapter 11, we examine the three physical states of matter and their interrelations. The subject of this chapter is the gaseous state, the one we understand best.

Several atmospheric gases—oxygen, nitrogen, water vapor, and carbon dioxide—are essential for life and take part in complex cycles of redox reactions as they move through the environment. ◆ Gases also have numerous roles in industry, some of which are presented in Table 5.1.

The *chemical* behavior of a gas depends on its composition, but all gases show remarkably similar *physical* behavior. For instance, although the gases involved differ, the same physical behavior is at work in the operation of a car and in the baking of bread, in the thrust of a rocket engine and in the explosion of a kernel of popcorn. The process of breathing arises from the same physical behavior as the creation of thunder.

This chapter focuses primarily on the physical behavior of gases. We begin with a brief comparison of gases, liquids, and solids and then discuss the phenomenon of gas pressure. Next, we consider several laws that each de-

◆ **The Atmosphere-Biosphere Connection.** The great diversity of organisms that make up the Earth's biosphere interact intimately with the gases of the atmosphere. Powered by solar energy, green plants reduce atmospheric $CO_2$ and incorporate the carbon atoms into their own substance. In the process, oxygen atoms in water are oxidized and released to the air as $O_2$. Certain microbes that live on plant roots reduce $N_2$ from the air and form compounds that the plant uses to make its proteins. Other microbes that feed on dead plants (and animals) oxidize the proteins and release $N_2$ again. Animals eat plants and other animals, utilizing $O_2$ to oxidize the material while returning $CO_2$ to the air.

**TABLE 5.1 Some Important Industrial Gases**

| NAME | FORMATION AND USE |
|---|---|
| Methane ($CH_4$) | Found with petroleum deposits; fuel for domestic heating; produced by bacteria living in termites, cows, sheep |
| Ammonia ($NH_3$) | Formation: $N_2(g) + 3H_2(g) \rightleftharpoons 2NH_3(g)$<br>Used in production of fertilizers and explosives |
| Chlorine ($Cl_2$) | Formed in electrolytic decomposition of ionic chloride solutions:<br>$2NaCl(aq) + 2H_2O(l) \rightarrow 2NaOH(aq) + Cl_2(g) + H_2(g)$<br>Used for bleaching paper and textiles, disinfecting domestic water |
| Carbon monoxide and hydrogen (syngas) ($CO$, $H_2$) | Formation: $C(s; coal) + H_2O(g) \rightarrow CO(g) + H_2(g)$<br>Cleaner (sulfur-free) fuel from coal; CO used in synthesis of chemicals |
| Uranium hexafluoride ($UF_6$) | Formation from uranium ore:<br>$UO_2(s) + 4HF(aq) + F_2(g) \rightarrow UF_6(g) + 2H_2O(l)$<br>Used for production of uranium fuel for nuclear energy |
| Ethylene ($C_2H_4$) | Formed by high-temperature decomposition ("cracking") of natural gas; used in plastics production |

scribe a different aspect of gas behavior; then we examine the ideal gas law, which encompasses the other laws, and apply it to reaction stoichiometry. We explain the macroscopic behavior of gases with a theory that models gases at the molecular level. Our investigation of gas behavior under extreme pressures and temperatures leads to a refinement of the simple ideal gas law. Finally, we apply these principles to the behavior of our atmosphere.

## 5.1 An Overview of the Physical States of Matter

Under appropriate conditions of pressure and temperature, most pure substances can exist in any of the three states of matter: solid, liquid, or gas. In Chapter 1 we described these physical states in terms of how a sample of matter fills a container: a solid has a fixed shape regardless of the container's shape, a liquid conforms to the container's shape but has a definite volume and a surface, and a gas expands to fill the entire container. Several other characteristics of gases distinguish them from liquids and solids:

1. *Gases are highly compressible.* When a sample of gas is confined to a container of variable volume, such as the piston-cylinder assembly of a car engine, an external force compresses the gas sample and decreases its volume. Removing the external force allows the gas volume to increase. In contrast, a liquid or solid resists being compressed.

2. *Gases are thermally expandable.* When a gas sample at constant pressure is heated, its volume increases; when it is cooled, its volume decreases. This change in volume with temperature is 50 to 100 times greater for gases than for liquids or solids. ◆

3. *Gases have low viscosity. Viscosity* is a measure of the resistance to flow. Gases flow much more freely than liquids and solids. Their low viscosity allows gases to be transported through pipes over long distances but also to leak more rapidly out of small holes.

4. *Most gases have relatively low densities* under normal conditions. Gas density is on the order of grams per *liter,* whereas liquid and solid densities are on the order of grams per *milliliter,* about 1000 times as dense (see Table 1.5). For example, at 20°C and normal atmospheric pressure, the density of $O_2(g)$ is 1.4 g/**L**, whereas the density of $H_2O(l)$ is 1.0 g/**mL** and that of NaCl(s) is 2.2 g/**mL**. When a gas is cooled, its density increases because its volume decreases: at 0°C, the density of $O_2(g)$ increases to 1.5 g/L.

5. *Gases are infinitely miscible;* that is, they mix with other gases in any proportion to form homogeneous mixtures (solutions). Clean dry air, for example, is a mixture of about 18 different gases. Liquids, on the other hand, may or may not form solutions, depending on their chemical natures: water and ethanol dissolve in each other, but water and gasoline do not. In general, solids do not form solutions with each other unless they are mixed as liquids while molten, as in the case of waxes or metal alloys, or while in solution, as in the case of mixed salts.

The change in density that occurs when a sample of gas condenses to a liquid offers a clue to the submicroscopic structure of gases compared with that of the other two states. When gaseous $N_2$ at 20°C and normal atmospheric pressure ($d = 1.25$ g/L) is cooled to below −196°C, it condenses to liquid $N_2$ ($d = 0.808$ g/mL). Note the change in units for density. The same amount of nitrogen is present, but it occupies about $1/600$ as much space. Further cooling to below −210°C freezes the liquid and forms solid $N_2$

◆ **POW! P-s-s-s-t! POP!** A jackhammer uses the force of rapidly expanding compressed air to break through rock and cement. When the nozzle on a can of spray paint is pressed, the pressurized propellant gases expand into the lower pressure of the surroundings and expel droplets of paint. The rapid expansion of heated gases results in such dissimilar phenomena as the destruction caused by a bomb, the liftoff of a rocket, and the formation of popcorn.

**FIGURE 5.1**
**A comparison of the three states of matter.** Many pure substances, such as bromine ($Br_2$), can exist under appropriate conditions as **A,** a gas; **B,** a liquid; or **C,** a solid. In **A** the liquid form of bromine has been vaporized, whereas in **C** it has been frozen. The molecular views show that molecules in a gas are much farther apart than in a liquid or solid.

**A Gas:** Molecules are far apart and fill the available space

**B Liquid:** Molecules are close together but move relative to each other

**C Solid:** Molecules are close together, packed in a regular array, and move very little relative to each other.

($d$ = 1.03 g/mL), which is only slightly more dense than the liquid. These density values show that *the molecules are much farther apart in the gas than in either the liquid or the solid.* In addition to their having much lower density, gases also lack completely the structural order that occurs to some extent in liquids and is a characteristic feature of solids (Figure 5.1).

**Section Summary**
The volume of a gas can be altered by changing the applied external force or the temperature. The corresponding volume changes for liquids and solids are much smaller. Gases flow more freely and have lower densities than liquids and solids, and they mix in any proportion to form solutions. The major reason for these differences is the greater distance between molecules in a sample of gas.

## 5.2  Measuring the Pressure of a Gas

A gas exerts pressure on the walls of its container, as you can see when you blow up a balloon or lift a car by inflating a flat tire. **Pressure (*P*)** is defined as the force exerted per unit of surface area:

$$\text{Pressure} = \text{force/area}$$

Due to the gravitational attraction of the Earth, the gases that make up the atmosphere are pulled toward its surface and exert a force on all the objects present there. The force, or weight, of these gases creates a pressure of about 14.7 pounds per square inch (lb/in$^2$; psi) of surface. ◆

◆ **The Principle of the Snowshoe.** Snowshoes allow you to walk on powdery snow without sinking because they distribute your weight over a much larger area than a boot does, thereby greatly decreasing your weight per square inch. The area of a snowshoe is typically about 10 times as large as that of a boot sole, so the snowshoe exerts only about one-tenth as much pressure as the boot. For the same reason, high-heeled shoes exert much more pressure than flat shoes.

As we'll discuss later, the molecules in a gas are moving in every direction, so the pressure is exerted uniformly on the floor, walls, ceiling, and every object in a room. You are not aware of this external pressure because it is equalized by the pressure inside your body; that is, there is no net pressure on your body's outer surface. You can see a dramatic example of what would happen if this were not the case by attaching an empty metal can to a vacuum pump (Figure 5.2). With the pump off, the can maintains its shape because the pressure on the outside of its walls is opposed by that on the inside. With the pump on, the internal pressure decreases greatly, so the much higher external pressure easily crushes the can. You may already be familiar with a vacuum-filtration flask from chemistry lab; the flask and its tubing have thick walls that can withstand the external pressure when the flask is evacuated.

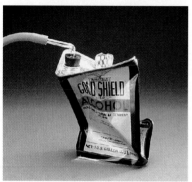

**FIGURE 5.2**

**Effect of atmospheric pressure on objects at the Earth's surface.** The atmosphere exerts a pressure of about 14.7 lb/in² on every object. A metal can before (*top*) and after (*bottom*) it is evacuated.

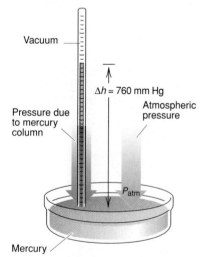

**FIGURE 5.3**

**A mercury barometer.** The pressure of the mercury is balanced by the pressure of the atmosphere. At sea level and 0°C the average atmospheric pressure can support a 760-mm column of mercury.

### Laboratory Devices for Measuring Gas Pressure

A **barometer** is a device used to measure atmospheric pressure. Invented by Evangelista Torricelli in 1643, the barometer is simply a tube, about 1 m long, that is closed at one end, filled with mercury, and inverted into a dish containing more mercury (Figure 5.3). When the tube is inverted, some mercury flows out, forming a vacuum above the mercury remaining in the tube. At sea level under ordinary atmospheric conditions, the outward flow of mercury stops when the top of the mercury column in the tube is about 760 mm above the surface of the mercury in the dish. At that point, the 760-mm column of mercury is exerting a pressure (weight/area) on the mercury surface in the dish that is equal to the pressure of the column of air that extends from the dish to the outer reaches of the atmosphere.

We can also construct a barometer by placing an evacuated tube into a dish filled with mercury. The mercury rises about 760 mm into the tube because the atmosphere exerts enough pressure to push it up to that height. Several centuries ago, people ascribed mysterious "suction" forces to a vacuum. We know now that a vacuum does not suck up mercury into the barometer tube any more than it sucks in the walls of the crushed can in Figure 5.2; only matter, in this case the atmospheric gases, can exert a force.

Notice that the diameter of the barometer tube was not specified. If the mercury in a tube with a 1-cm diameter rises to a height of 760 mm, the mercury will rise to the same height in a tube with a 2-cm diameter. The *weight* of mercury is greater in the wider tube, but the *pressure,* the *ratio* of weight to area, is the same.

Since the pressure of the mercury column is directly proportional to its height, a unit commonly used for pressure is the height of the mercury column in millimeters (mmHg). This and other units of pressure are discussed shortly. At sea level and 0°C, normal atmospheric pressure is 760 mmHg, but at the top of Mt. Everest (29,141 ft, or 8886 m), the atmospheric pressure is only about 270 mmHg. The pressure of the atmosphere decreases with altitude because the column of air above a given point is shorter and therefore weighs less.

Laboratory barometers usually contain mercury rather than some other liquid because mercury's high density allows construction of a convenient-sized instrument. For example, the pressure of the atmosphere would also

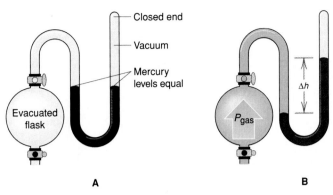

**FIGURE 5.4**
**A closed-end manometer. A,** With an evacuated flask attached, the mercury levels are equal. **B,** A gas exerts pressure on the mercury in the arm attached to the flask. The difference in heights ($\Delta h$) equals the gas pressure.

support a column of water about 10,300 mm, almost 34 ft, high. ◆ As you can see, for a given pressure, the ratio of heights ($h$) of the liquid columns is inversely related to the ratio of the densities ($d$) of the liquids:

$$\frac{h_{H_2O}}{h_{Hg}} = \frac{d_{Hg}}{d_{H_2O}}$$

**Manometers** are devices used to measure the pressure of a gas involved in a laboratory experiment. The *closed-end manometer* is a mercury-filled, U-shaped tube, closed at one end and attached to a flask at the other (Figure 5.4). When the flask is evacuated, the mercury levels in the two arms of the tube are the same because no gas exerts pressure on either mercury surface. When a gas is in the flask, it presses on the mercury, and the mercury level rises in the other arm. The *difference* in column heights ($\Delta h$) is equal to the gas pressure. If the flask is opened to the atmosphere, the closed-end manometer reads atmospheric pressure and, in effect, becomes a barometer.

The *open-end manometer* works by comparing the pressure of the gas to the pressure of the atmosphere. It also consists of a U-shaped tube filled with mercury, but one end of the tube is open to the atmosphere and the other

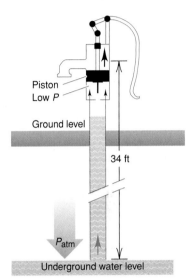

◆ **The Mystery of the Suction Pump.** When you drink through a straw, you create lower pressure above the liquid, and the atmosphere pushes the liquid up. A so-called suction pump operates by the same principle. It is a tube dipping into a water source with a piston and handle that work to lower the air pressure above the water level. The pump can raise water from a well up to 34 ft deep or up to a height of 34 ft above a river or lake. This height limit was a mystery until the 17th century, when the great Italian scientist Galileo showed that the atmosphere pushes the water up into the tube and that its pressure can support only a 34-ft column of water. Modern pumps that drill deeper for water use compressed air to increase the external pressure on the water.

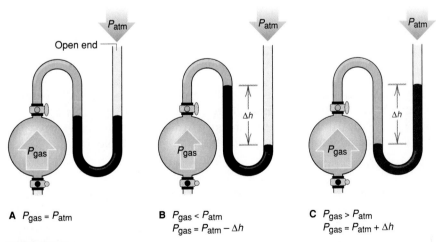

**A** $P_{gas} = P_{atm}$

**B** $P_{gas} < P_{atm}$
$P_{gas} = P_{atm} - \Delta h$

**C** $P_{gas} > P_{atm}$
$P_{gas} = P_{atm} + \Delta h$

**FIGURE 5.5**
**An open-end manometer. A,** Gas pressure is equal to atmospheric pressure. **B,** Gas pressure is lower than atmospheric pressure. **C,** Gas pressure is higher than atmospheric pressure.

**TABLE 5.2** **Common Units of Pressure**

| UNIT | ATMOSPHERIC PRESSURE | SCIENTIFIC FIELD |
|---|---|---|
| pascal (Pa); kilopascal (kPa) | $1.01325 \times 10^5$ Pa; 101.325 kPa | SI unit; physics, chemistry |
| atmosphere (atm) | 1 atm* | Chemistry |
| millimeters of mercury (mmHg) | 760 mmHg* | Chemistry, medicine, biology |
| torr | 760 torr* | Chemistry |
| pounds per square inch (psi or lb/in²) | 14.7 lb/in² | Engineering |
| bar | 1.01325 bar | Meteorology, chemistry, physics |

*This is an exact quantity; in calculations, we use as many significant figures as necessary.

is connected to the gas sample (Figure 5.5). In this type, the mercury level in the arm attached to the flask is equal to, higher than, or lower than the level in the other arm, depending on whether the gas pressure is equal to, lower than, or higher than atmospheric pressure, respectively. Since the difference between the two heights is the difference between these two pressures, a barometer is used *with* an open-end manometer to find the gas pressure.

**The Units of Pressure**

The SI unit of force is the newton (N): 1 N = 1 kg·m/s². The SI unit of pressure is the **pascal (Pa),** which equals a force of one newton exerted on an area of one square meter:

$$1 \text{ Pa} = 1 \text{ N/m}^2$$

A much larger unit in common use is the **standard atmosphere (atm),** the average atmospheric pressure measured at sea level and 0°C, which is defined in terms of the pascal:

$$1 \text{ atm} = 1.01325 \times 10^5 \text{ Pa} = 101.325 \text{ kilopascals (kPa)}$$

A commonly used pressure unit is the **millimeter of mercury (mmHg),** which is based directly on the measurement with a barometer or manometer. In honor of Torricelli, this unit has been named the **torr:** 1 mmHg = 1 torr. The torr is now defined in terms of the pascal:

$$1 \text{ torr} = \frac{1}{760} \text{ atm} = \frac{101.325}{760} \text{ kPa}$$

Despite the gradual changeover to SI units, most chemists still express pressure in units of torr and atmospheres, so they are used in this text, with reference to the equivalent pressure in pascals. These and other pressure units still used in certain scientific fields are shown in Table 5.2.

SAMPLE PROBLEM 5.1 _____

**Converting Units of Pressure**

**Problem:** A geochemist heats a limestone sample and collects the $CO_2$ released in an evacuated flask attached to a closed-end manometer (see Figure 5.4, *B*, p. 177). After the system comes to room temperature, $\Delta h$ = 291.4 mmHg. Calculate the $CO_2$ pressure in torr, atmospheres, and kilopascals.

**Plan:** The $CO_2$ pressure is given in units of mmHg, so we use appropriate conversion factors (Table 5.2) to find the pressure in the other units.

**Solution:** Converting from mmHg to torr:

$$P_{CO_2} \text{ (torr)} = 291.4 \text{ mmHg} \times \frac{1 \text{ torr}}{1 \text{ mmHg}} = \textbf{291.4 torr}$$

Converting from torr to atm:

$$P_{CO_2} \text{ (atm)} = 291.4 \text{ torr} \times \frac{1 \text{ atm}}{760 \text{ torr}} = \textbf{0.3834 atm}$$

Converting from atm to kPa:

$$P_{CO_2} \text{ (kPa)} = 0.3834 \text{ atm} \times \frac{101.325 \text{ kPa}}{1 \text{ atm}} = \textbf{38.85 kPa}$$

**Check:** There are >700 torr in 1 atm, so ~300 torr should be <0.5 atm. There are ~100 kPa in 1 atm, so <0.5 atm should be <50 kPa.

**Comment:** 1. In the conversion from torr to atm, we retained four significant figures because the number of torr in 1 atm is an *exact* number; thus, 760 torr has as many significant figures as the calculation requires.

2. From here on, except in particularly complex situations, the *canceling of units in calculations is not shown.*

**FOLLOW-UP PROBLEM 5.1**

The $CO_2$ released from another mineral sample was collected in an evacuated flask connected to an open-end manometer (see Figure 5.5, *B*, p. 177). If the barometer reading is 753.6 mmHg and $\Delta h$ is 174.0 mmHg, calculate $P_{CO_2}$ in torr, pascals, and lb/in².

**Section Summary**

Gases exert pressure (force/area) on all surfaces with which they make contact. A barometer measures atmospheric pressure by the height of the mercury column the atmosphere can support (760 mmHg at sea level and 0°C). Both closed-end and open-end manometers are used to measure the pressure of a gas sample. Chemists measure pressure in units of atmosphere (atm), torr (equivalent to mmHg), or pascal (Pa, the SI unit).

# 5.3  The Gas Laws and Their Experimental Foundations

A sample of gas can be described in terms of four variables: pressure ($P$), volume ($V$), temperature ($T$), and the number of moles of gas ($n$). The variables are interdependent: *any one of them can be determined by measuring the other three.* Furthermore, if any one of the variables is changed, the sample will change in a quantitatively predictable way. Although this predictable behavior is a direct outcome of the submicroscopic structure of gases, it was discovered entirely by macroscopic observations, most of which were made even before the publication of Dalton's atomic theory.

Three key relationships—Boyle's, Charles's, and Avogadro's laws—exist among the four gas variables. Each of these gas laws expresses the influence of one variable on another, with the remaining two variables held constant.

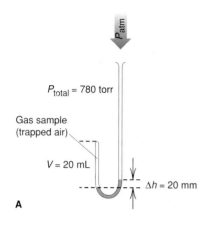

A

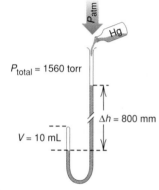

B

**FIGURE 5.6**

Boyle's apparatus for studying the relationship between the volume and pressure of a gas. **A,** A small amount of air is trapped in the short arm of a J tube; $n$ and $T$ are fixed. The total pressure on the gas ($P_{total}$) is the sum of the pressure due to the difference in heights of the mercury columns ($\Delta h$) and the pressure of the atmosphere ($P_{atm}$). If $P_{atm} = 760$ torr, $P_{total} = 780$ torr. **B,** As mercury is added, the total pressure on the gas (trapped air) increases and its volume ($V$) decreases. Note that if $P_{total}$ is doubled (1560 torr), $V$ is halved. (The mercury heights are not drawn to scale.)

**FIGURE 5.7**

The volume-pressure relationship of an ideal gas. **A,** Pressure-volume data obtained using an apparatus similar to that shown in Figure 5.6. **B,** A plot of $V$ vs. $P_{total}$ shows that $V$ is inversely proportional to $P$. **C,** A plot of $V$ vs. $1/P_{total}$ is a straight line whose slope is a constant characteristic of any gas that behaves ideally.

Since gas volume, the volume occupied by the sample of gas, is so easily measured, the laws are traditionally expressed as the effect on volume of changing the pressure, temperature, or amount (number of moles) of gas.

The three laws are special cases of an all-encompassing relationship among the variables called the *ideal gas law.* This unifying observation quantitatively describes the state of a so-called **ideal gas,** one that exhibits simple linear relationships among volume, pressure, temperature, and number of moles. Although no ideal gas actually exists, most simple gases, such as $N_2$, $O_2$, $H_2$, and the noble gases, show nearly ideal behavior under normal conditions of temperature and pressure. We discuss the ideal gas law after considering its three special cases.

### The Relationship Between Volume and Pressure: Boyle's Law

Following Torricelli's invention of the barometer, Robert Boyle, the great English chemist, performed a series of experiments that led him to conclude that at a given temperature, *the volume of a gas is inversely related to its pressure.* Boyle fashioned a J-shaped glass tube, sealed the shorter end, and poured mercury into the longer end, thereby trapping some air (Figure 5.6, *A*). From the length of the trapped air column and the diameter of the tubing, he calculated the air volume. The total pressure applied to the trapped air was the pressure of the atmosphere (measured with a barometer) plus that of the mercury column. By increasing the amount of mercury in the J tube (Figure 5.6, *B*), Boyle increased the total pressure applied on the air, and the air volume decreased. With the temperature and amount of air held constant, Boyle could directly measure the relationship between the applied pressure and the volume of air.

Some typical data Boyle might have collected are given in Figure 5.7, *A*. Note that $V$ is *inversely* proportional to $P$ (Figure 5.7, *B*). It follows that the product of corresponding $P$ and $V$ values is a constant. A plot of $V$ against $1/P$ gives a straight line whose slope has the value of this constant (Figure 5.7, *C*). This linear relationship is indicative of ideal gas behavior.

| $V$ (mL) | $P$ (torr) | | | $\frac{1}{P_{total}}$ | $PV$ (torr · mL) |
|---|---|---|---|---|---|
| | $\Delta h$ | + $P_{atm}$ | = $P_{total}$ | | |
| 20.0 | 20.0 | 760 | 780 | 0.00128 | $1.56 \times 10^4$ |
| 15.0 | 278 | 760 | 1038 | 0.000963 | $1.56 \times 10^4$ |
| 10.0 | 800 | 760 | 1560 | 0.000641 | $1.56 \times 10^4$ |
| 5.00 | 2352 | 760 | 3112 | 0.000321 | $1.56 \times 10^4$ |

A

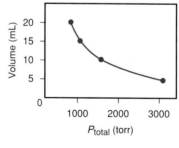

B

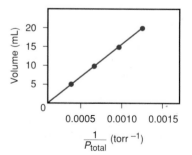

C

The generalization of Boyle's observations is known as **Boyle's law:** *at constant temperature, the volume of a fixed amount of gas is inversely proportional to the applied (external) pressure,* or

$$V \propto \frac{1}{P} \quad [T \text{ and } n \text{ fixed}]$$

This relationship can also be expressed as

$$PV = \text{constant} \quad \text{or} \quad V = \frac{\text{constant}}{P}$$

For a given quantity of gas at a given temperature, the constant is the same for the great majority of gases. Thus, tripling the external pressure reduces the gas volume to one-third its initial value; decreasing the external pressure by half doubles the gas volume; and so forth.

We have focused here on the effect of a change in *external* pressure on gas volume. In Boyle's experiment, however, the pressure exerted *on* a particular volume of gas equals the pressure exerted *by* the gas, so in measuring the applied pressure, Boyle was also determining the pressure of the gas sample. Thus, when the volume of a gas is doubled, the pressure exerted by the gas is halved: If $V_{gas}$ increases, $P_{gas}$ decreases, and vice versa.

We often use Boyle's law to solve problems that compare two sets of volume and pressure values, such as those before and after a change. Since the product of $P$ and $V$ is constant, we know that

$$P_1V_1 = P_2V_2 \quad \text{or} \quad V_2 = V_1 \times \frac{P_1}{P_2} \quad [T \text{ and } n \text{ fixed}] \qquad \textbf{(5.1)}$$

where $P_1$ and $V_1$ are initial values and $P_2$ and $V_2$ are final values.

It is very important that you approach gas law problems systematically:
1. Summarize the information given: identify known, unknown, and constant gas variables.
2. Predict the direction of the change.
3. Perform the necessary unit conversions.
4. Apply the appropriate gas law to complete the calculations.

SAMPLE PROBLEM 5.2 _____

**Applying the Volume-Pressure Relationship**

**Problem:** An apprentice of Boyle finds that the air trapped in a J tube occupies 24.8 cm³ at 1.12 atm. By adding mercury to the tube, he increases the pressure on the trapped air to 2.64 atm. Assuming constant temperature, what is the new volume of air (in L)?
**Plan:** We must find the final volume ($V_2$) in liters, given the initial volume ($V_1$), initial pressure ($P_1$), and final pressure ($P_2$). The temperature and amount of gas are fixed. We convert the units of $V_1$ from cm³ to mL and then to L and solve Boyle's law for $V_2$. We can predict the direction of the change: since the pressure increases, the volume will decrease; that is, $V_2 < V_1$. (In later problems, some unit conversions are not included on the roadmaps.)
**Solution:** Summary of gas variables:

$P_1 = 1.12 \text{ atm}$          $P_2 = 2.64 \text{ atm}$
$V_1 = 24.8 \text{ cm}^3$ (convert to L)     $V_2 = \text{unknown}$
$T$ and $n$ remain constant

Converting $V_1$ from cm³ to L:

$$V_1 = 24.8 \text{ cm}^3 \times \frac{1 \text{ mL}}{1 \text{ cm}^3} \times \frac{1 \text{ L}}{1000 \text{ mL}} = 0.0248 \text{ L}$$

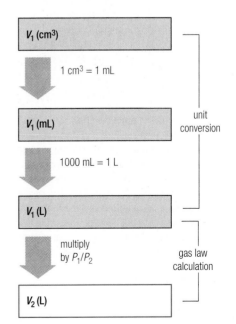

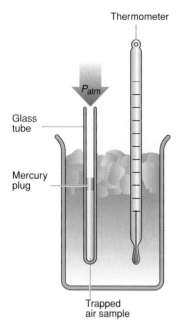

**A** Ice water bath: 0°C (273 K)

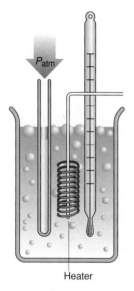

**B** Boiling water bath: 100°C (373 K)

**FIGURE 5.8**

**An experiment to study the relationship between the volume and temperature of a gas.** A sample of gas (air) is trapped under a small plug of mercury; $P$ and $n$ are fixed. The sample is shown in an ice water bath **(A)** and in a boiling water bath **(B).** As the temperature increases, the volume of the gas increases. Note that a temperature increase of about one-third (from 273 to 373 K) causes an increase in the volume of the gas of about one-third.

Solving for $V_2$:

$$V_2 = V_1 \times \frac{P_1}{P_2} = 0.0248 \text{ L} \times \frac{1.12 \text{ atm}}{2.64 \text{ atm}} = \textbf{0.0105 L}$$

**Check:** As we predicted, $V_2 < V_1$. The pressure more than doubled, so $V_2$ should be less than $\frac{1}{2}V_1$ $(0.0105/0.0248 < 1/2)$.

**Comment:** Predicting the direction of the change allows another check on the calculation: since $V_2 < V_1$, we have to multiply $V_1$ by a number less than 1. Thus, the ratio of pressures must be less than 1, so the larger pressure $(P_2)$ must be in the denominator, $P_1/P_2$.

**FOLLOW-UP PROBLEM 5.2**

A sample of argon gas occupies 105 mL at 0.871 atm. If the temperature remains constant, what is the volume (in L) at 26.3 kPa?

### The Relationship Between Volume and Temperature: Charles's Law

One of the questions raised by Boyle's work was why the pressure-volume relationship is valid only if the temperature remains constant. Experience had shown that the volume of a gas sample depends on its temperature, but it was not until the early 19th century, through the separate work of J. A. C. Charles and J. L. Gay-Lussac, that the relationship was clearly understood.

Let's examine this relationship by measuring the volume of a fixed amount of a gas under constant pressure but at different temperatures (Figure 5.8). We can use a straight tube, closed at one end, in which a fixed amount of mercury traps a fixed amount of air. The tube is immersed in a water bath that can be warmed with a heater or cooled with ice. After each change of water temperature, we measure the length of the air column, which is proportional to its volume. The pressure exerted on the gas sample is constant because the amount of mercury and the atmospheric pressure above it do not change.

Some typical data are shown for two amounts of gas in Figure 5.9. Again, we see a linear relationship between the two gas variables, but this time the variables are *directly* proportional: for a given quantity of gas at a given pressure, *volume increases as temperature increases.* The linear relationship means that the volume of a sample changes by a fixed increment for a given temperature change. For example, the volume of 0.02 mol gas at 1 atm pressure increases by 1.64 mL for each 1°C rise in temperature; for 0.04 mol gas, the volume increases by 3.28 mL for each 1°C rise; and so forth. Extrapolating the lines in the figure to lower temperatures (dashed portions), we see that the volume shrinks until the gas occupies a theoretical zero volume at −273°C (the intercept of the temperature axis). This is true regardless of the amount of gas.

One-half century after Charles's work, William Thomson (Lord Kelvin) used this linear relation between gas volume and temperature to devise the absolute temperature scale (Section 1.5). In this scale, absolute zero, 0 K or −273.15°C, is the temperature at which an ideal gas would have zero volume. (Absolute zero has never been reached, but physicists have approached it to within a small fraction of a kelvin.) Of course, a real sample of matter cannot have zero volume, and every real gas condenses to a liq-

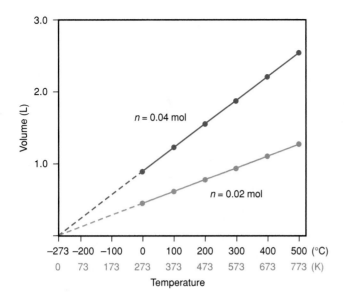

FIGURE 5.9
**The volume-temperature relationship of two samples of an ideal gas.** At constant $P$, the volume of a given amount of an ideal gas is directly proportional to the absolute temperature. Two different amounts (0.02 mol gas and 0.04 mol gas) are shown. The solid lines connect data; the dashed lines are extrapolations to lower temperatures. For any amount of an ideal gas, the predicted volume is zero at $-273.15°C$ (0 K).

uid at some temperature higher than 0 K. However, *the linear dependence of volume on absolute temperature holds for most common gases over a wide temperature range.*

The modern statement of the volume-temperature relationship is known as **Charles's law:** *at constant pressure, the volume of a fixed amount of gas is directly proportional to its absolute temperature,* or

$$V \propto T \quad [P \text{ and } n \text{ fixed}]$$

This relationship can also be expressed as

$$\frac{V}{T} = \text{constant} \quad \text{or} \quad V = \text{constant} \times T$$

If $T$ increases, $V$ increases, and vice versa. For any given $P$ and $n$, the constant is the same regardless of the gas. Once again, we can express the law in terms of two pairs of variables:

$$\frac{V_1}{T_1} = \frac{V_2}{T_2} \quad \text{or} \quad V_2 = V_1 \times \frac{T_2}{T_1} \quad [P \text{ and } n \text{ fixed}] \qquad \textbf{(5.2)}$$

Charles's law is expressed as the effect of a change on the *volume* of gas. However, volume and pressure are interdependent variables, so the effect of temperature on volume is closely related to the effect of temperature on pressure (sometimes referred to as *Amontons's law*). Measure the pressure in your automobile tires before and after a long drive, and you will find that it has risen. Frictional heating between the tire and the road increases the air temperature inside the tire; since the tire's volume cannot change appreciably, the air exerts more pressure. Thus, *at constant volume, the pressure exerted by a fixed amount of gas is directly proportional to the absolute temperature:*

$$P \propto T \quad \text{or} \quad \frac{P_1}{T_1} = \frac{P_2}{T_2} \quad [V \text{ and } n \text{ fixed}] \qquad \textbf{(5.3)}$$

SAMPLE PROBLEM 5.3 _____

### Applying the Pressure-Temperature Relationship

**Problem:** A 1-L steel tank is fitted with a safety valve that opens if the internal pressure exceeds $1.00 \times 10^3$ torr. It is filled with helium at 23°C and 0.991 atm and placed in boiling water at exactly 100°C. Will the safety valve open?

**Plan:** The question "Will the safety valve open?" translates to "Is $P_2$ greater than $1.00 \times 10^3$ torr at $T_2$?" Thus, $P_2$ is the unknown and $T_1$, $T_2$, and $P_1$ are given, with $V$ (steel container) and $n$ fixed. Both $T$ values must be converted to kelvins, $P_1$ must be converted to torr in order to compare it with the safety-limit pressure, and we solve for $P_2$. Since $T_2 > T_1$, we predict that $P_2 > P_1$.

**Solution:** Summary of gas variables:

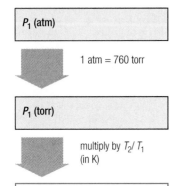

$P_1$ (atm)

1 atm = 760 torr

$P_1$ (torr)

multiply by $T_2/T_1$ (in K)

$P_2$ (torr)

| | |
|---|---|
| $P_1 = 0.991$ atm (convert to torr) | $P_2 =$ unknown |
| $T_1 = 23°C$ (convert to K) | $T_2 = 100°C$ (convert to K) |
| $V$ and $n$ remain constant | |

Converting $T$ from °C to K:

$$T_1 \text{ (K)} = 23°C + 273 = 296 \text{ K}$$

$$T_2 \text{ (K)} = 100°C + 273 = 373 \text{ K}$$

Converting $P$ from atm to torr:

$$P_1 \text{ (torr)} = 0.991 \text{ atm} \times \frac{760 \text{ torr}}{1 \text{ atm}} = 753 \text{ torr}$$

Solving for $P_2$:

$$P_2 = P_1 \times \frac{T_2}{T_1} = 753 \text{ torr} \times \frac{373 \text{ K}}{296 \text{ K}} = 949 \text{ torr}$$

$P_2$ is less than $1.00 \times 10^3$ torr, so **the valve will *not* open.**

**Check:** Our prediction is correct: since $T_2 > T_1$, $P_2 > P_1$. Thus, the temperature ratio should be >1 ($T_2$ in the numerator). The $T$ ratio is about 1.25 (373/296), so the $P$ ratio should also be about 1.25 (950/750 ≈ 1.25).

FOLLOW-UP PROBLEM 5.3

An engineer pumps air at 0°C into a newly designed piston-cylinder assembly. The volume measures 6.83 cm³. At what temperature (in K) would the volume be 9.75 cm³?

_____

### The Relationship Between Volume and Amount of Gas: Avogadro's Law

Boyle's and Charles's laws both stipulate a fixed amount of gas. Let's see why, through an experiment involving two small test tubes, A and B, each fitted with a piston-cylinder assembly (Figure 5.10). The volume of the tube is much smaller than the volume of the cylinder, so it can be neglected. A piece of dry ice (frozen $CO_2$) weighing 4.4 g (0.10 mol) is added to tube A, and a piece weighing 8.8 g (0.20 mol) is added to tube B. As the solid $CO_2$ warms, it changes directly to gaseous $CO_2$, which expands into the cylinder and pushes up the piston.

When all the solid has changed to gas and the temperature has become constant, we find that, within experimental error, volume B is twice volume A. This shows that twice the number of moles of gas occupies twice the volume. Note that for both cylinders, gas temperature equals room temperature and gas pressure equals atmospheric pressure. Thus, *at fixed temperature and pressure, the volume of gas is directly proportional to the moles of gas:*

$$V \propto n \quad [P \text{ and } T \text{ fixed}]$$

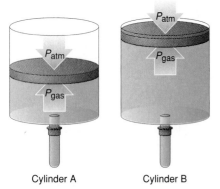

Cylinder A          Cylinder B

**FIGURE 5.10**

**An experiment to study the relationship between the volume and amount of a gas.** At a given external $P$ and $T$, twice the mass of $CO_2(s)$ is put in tube B as in tube A. After the solid has changed to gas the volume of cylinder B is twice that of cylinder A. Thus, at fixed $P$ and $T$, the volume ($V$) of a gas is directly proportional to the amount of gas ($n$).

As $n$ increases, $V$ increases, and vice versa. This relationship can also be expressed as

$$\frac{V}{n} = \text{constant} \quad \text{or} \quad V = \text{constant} \times n$$

The constant is the same for all gases at a given temperature and pressure. This mathematical relationship is another way of phrasing **Avogadro's law:** *at fixed temperature and pressure, equal volumes of any ideal gas contain equal numbers of particles (or moles).* (Recall from Chapter 2 that Avogadro proposed this idea to explain Gay-Lussac's data for the combining volumes of gases.)

As with the gas laws already discussed, we can use Avogadro's law to compare two pairs of variables at constant temperature and pressure by expressing it as

$$\frac{V_1}{n_1} = \frac{V_2}{n_2} \quad \text{or} \quad V_2 = V_1 \times \frac{n_2}{n_1} \quad [P \text{ and } T \text{ fixed}] \tag{5.4}$$

Many familiar phenomena are based on the relationships between volume, temperature, and moles of gas. For example, the operation of a car engine is based on the reaction that occurs when fewer moles of gasoline and oxygen form more moles of $CO_2$ and water vapor, which expand due to the released heat and push back the piston. Dynamite explodes because a solid decomposes rapidly to form hot gases. Dough rises in a warm room because the yeast forms $CO_2$ bubbles in the dough, which expand during baking to give the bread a still larger volume. ◆

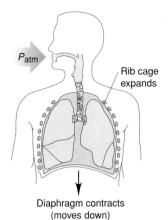

◆ **Breathing and the Gas Laws.** Taking a breath is a combined application of Boyle's, Charles's, and Avogadro's laws. When you inhale, your diaphragm muscle contracts and moves down, and your rib cage moves out (see figure). This movement increases the volume of the lung, which decreases the pressure of the air inside it, causing air to rush in. The greater amount of air expands the volume of the lung further, and the air also expands slightly as it warms to body temperature. When you exhale, the diaphragm relaxes and moves up, the rib cage moves in, and the lung volume decreases. The inside air pressure becomes greater than the outside pressure, and air rushes out. Fewer moles of air decrease the lung volume.

---

SAMPLE PROBLEM 5.4 _____

### Applying the Volume-Amount Relationship

**Problem:** Blimps are currently being considered for use as freight carriers. In an experiment, a scale model will rise off the ground when filled with helium to a volume of 55.0 dm³. If 1.10 mol He expands the blimp volume to 26.2 dm³, how many grams of He must be inside the blimp to make it rise? Assume constant $T$ and $P$.
**Plan:** We are given the initial amount of helium ($n_1$), the initial volume of the blimp ($V_1$), and the volume needed for it to rise ($V_2$), so we need to find $n_2$ and convert moles to grams. We predict that $n_2 > n_1$ because $V_2 > V_1$.
**Solution:** Summary of gas variables:

$$n_1 = 1.10 \text{ mol} \qquad\qquad n_2 = \text{unknown}$$
$$V_1 = 26.2 \text{ dm}^3 \qquad\qquad V_2 = 55.0 \text{ dm}^3$$
$$P \text{ and } T \text{ remain constant}$$

Solving for $n_2$:

$$n_2 = n_1 \times \frac{V_2}{V_1} = 1.10 \text{ mol He} \times \frac{55.0 \text{ dm}^3}{26.2 \text{ dm}^3} = 2.31 \text{ mol He}$$

Converting moles of He to grams:

$$\text{Mass (g) of He} = 2.31 \text{ mol He} \times \frac{4.003 \text{ g He}}{1 \text{ mol He}} = \textbf{9.25 g He}$$

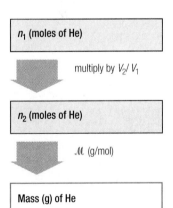

**Check:** Since $V_2$ is about twice $V_1$ ($55/26 \approx 2$), $n_2$ should be about twice $n_1$ ($2.3/1.1 \approx 2$). Since $n_2 > n_1$, it made sense to multiply $n_1$ by a number $> 1$ (that is, $V_2/V_1$). Nearly 2.5 mol $\times$ 4 g/mol = 10 g.

FOLLOW-UP PROBLEM 5.4
A rigid plastic container holds 35.0 g ethylene gas ($C_2H_4$) at a pressure of 793 torr. What is the pressure if 5.0 g ethylene is removed at constant temperature?

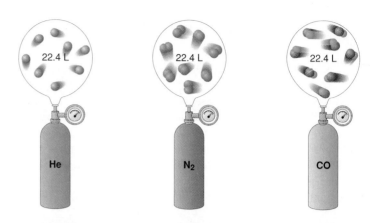

| $n$ = 1 mol | $n$ = 1 mol | $n$ = 1 mol |
|---|---|---|
| $P$ = 1 atm (760 torr) | $P$ = 1 atm (760 torr) | $P$ = 1 atm (760 torr) |
| $T$ = 0°C (273 K) | $T$ = 0°C (273 K) | $T$ = 0°C (273 K) |
| $V$ = 22.4 L | $V$ = 22.4 L | $V$ = 22.4 L |
| Number of gas particles = $6.022 \times 10^{23}$ | Number of gas particles = $6.022 \times 10^{23}$ | Number of gas particles = $6.022 \times 10^{23}$ |
| Mass = 4.003 g | Mass = 28.02 g | Mass = 28.01 g |
| $d$ = 0.179 g/L | $d$ = 1.25 g/L | $d$ = 1.25 g/L |

## Gas Behavior at Standard Conditions

To better understand the factors that influence gas behavior, a set of *standard conditions* has been chosen. Chemists designate 0°C (273.15 K) and 1 atm (760 torr) as **standard temperature and pressure (STP):**

$$\text{STP:}\quad 0°C\ (273.15\ K)\ \text{and}\ 1\ atm\ (760\ torr) \tag{5.5}$$

Under these conditions, the volume of one mole of an ideal gas is 22.414 L, which is called the **standard molar volume:**

$$\text{Standard molar volume} = 22.4\ L\quad [\text{3 significant figures}] \tag{5.6}$$

Most simple gases have a standard molar volume that, to three significant figures, equals this value (Figure 5.11). Figure 5.12 compares the volumes of familiar objects with the standard molar volume of an ideal gas.

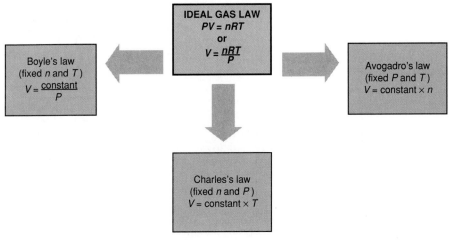

**FIGURE 5.13**
Relationship between the ideal gas law and the individual gas laws. Boyle's, Charles's, and Avogadro's laws are contained within the ideal gas law.

## The Ideal Gas Law and the Universal Gas Constant

Each of the gas laws just discussed isolates a variable and describes the effect of its variation on gas volume:

- Boyle's law isolates pressure ($V \propto 1/P$).
- Charles's law isolates temperature ($V \propto T$).
- Avogadro's law isolates amount (moles) of gas ($V \propto n$).

We can combine these individual effects into one relationship, called the **ideal gas law** (or ideal gas equation):

$$V \propto \frac{nT}{P} \quad \text{or} \quad PV \propto nT \quad \text{or} \quad \frac{PV}{nT} = R$$

where $R$ is a proportionality constant known as the **universal gas constant.** Rearranging this mathematical statement, we arrive at the most common form of the ideal gas law:

$$PV = nRT \qquad\qquad (5.7)$$

By simple rearrangements of the ideal gas law, we can see that Boyle's, Charles's, and Avogadro's laws are contained within it. In other words, the ideal gas law *becomes* each of the three individual gas laws, depending on which two of the four variables are kept constant (Figure 5.13).

We determine the value of $R$ by measuring the volume, temperature, and pressure of a given number of moles of gas and substituting the values into the ideal gas law. For example, using the standard values just presented for the gas variables, we calculate the value of $R$ in the given units:

$$R = \frac{PV}{nT} = \frac{1 \text{ atm} \times 22.4 \text{ L}}{1 \text{ mol} \times 273 \text{ K}} = 0.0821 \frac{\text{atm} \cdot \text{L}}{\text{mol} \cdot \text{K}} \quad \text{[3 significant figures]}$$

This numerical value of $R$ is used whenever the gas variables $P$, $V$, $n$, and $T$ are expressed in these units. When different units are used, $R$ has a different numerical value (Table 5.3).

The significance of $R$ extends beyond its use in the ideal gas law, and later in the text it appears in relationships that may have little to do with gases.

**TABLE 5.3  Values of $R$ (Universal Gas Constant) in Different Units**

$$R^* = 0.08206 \ \frac{\text{atm} \cdot \text{L}}{\text{mol} \cdot \text{K}}$$

$$R = 62.36 \ \frac{\text{torr} \cdot \text{L}}{\text{mol} \cdot \text{K}}$$

$$R = 8.314 \ \frac{\text{kPa} \cdot \text{dm}^3}{\text{mol} \cdot \text{K}}$$

$$R = 8.314 \ \frac{\text{J}^{**}}{\text{mol} \cdot \text{K}}$$

\* Most calculations in the text use values of $R$ to 3 significant figures.
\*\* J is the abbreviation for joule, the SI unit of energy. The joule is a derived unit composed of the base units kg·m²/s².

By expressing $P$ and $V$ in more fundamental quantities, we obtain the dimensions for energy:

$$P = \frac{\text{force}}{\text{area}} = \frac{\text{force}}{(\text{length})^2} \quad \text{and} \quad V = (\text{length})^3$$

Thus,
$$PV = \frac{\text{force}}{(\text{length})^2} \times (\text{length})^3 = \text{force} \times \text{length}$$

Energy is involved whenever an object is moved, that is, whenever a force acts over a distance. Thus, energy = force × distance (or length), so

$$PV = \text{force} \times \text{length} = \text{energy}$$

Solving for $R$ in the ideal gas law and substituting for $PV$, we obtain

$$R = \frac{PV}{nT} = \frac{\text{energy}}{\text{amount of substance} \times \text{temperature}}$$

Therefore, $R$ is the proportionality constant that relates the energy, amount of substance, and temperature of any chemical system.

   Although gas law problems are phrased in many ways, they can usually be grouped into two main types:

1. *A change in one of the four variables causes a change in another, while the two remaining variables are held constant.* In this type, the ideal gas law reduces to one of the individual gas laws, and you solve for the new value of the variable. Units must be consistent, $T$ must always be in kelvins, but $R$ is not involved. Sample Problems 5.2 to 5.4 are of this type. (A variation on this type involves simultaneous changes in two of the variables that cause a change in a third.)

2. *One variable is unknown, the other three are known, but no change occurs.* In this type, which is exemplified by Sample Problem 5.5, the ideal gas law is applied directly to find the unknown, and the units must conform to those in $R$.

SAMPLE PROBLEM 5.5 _____

### Solving for an Unknown Gas Variable

**Problem:** A tank with a fixed volume of 438 L is filled with 0.885 kg $O_2$. Calculate the pressure of the oxygen at 21°C.
**Plan:** We are given data for $V$, $T$, and $n$, so we solve for $P$, after converting $T$ to kelvins and mass of $O_2$ to moles.
**Solution:** Summary of gas variables:

$V = 438$ L                                    $T = 21°C$ (convert to K)
$n = 0.885$ kg $O_2$ (convert to moles)        $P = $ unknown

Converting $T$ from °C to K:

$$T \text{ (K)} = 21°C + 273 = 294 \text{ K}$$

Converting from mass of $O_2$ to moles:

$$n = \text{moles of } O_2 = 0.885 \text{ kg } O_2 \times \frac{1000 \text{ g}}{1 \text{ kg}} \times \frac{1 \text{ mol } O_2}{32.00 \text{ g } O_2} = 27.7 \text{ mol } O_2$$

Solving for $P$ (note the unit canceling here):

$$P = \frac{nRT}{V} = \frac{27.7 \cancel{\text{ mol}} \times 0.0821 \dfrac{\text{atm} \cdot \cancel{\text{L}}}{\cancel{\text{mol}} \cdot \cancel{\text{K}}} \times 294 \cancel{\text{K}}}{438 \cancel{\text{L}}} = \mathbf{1.53 \text{ atm}}$$

**Check:** The number of moles of $O_2$ seems correct: ~900 g/(30 g/mol) = 30 mol. To check the size of the final calculation, round off the values:

$$P = \frac{30 \text{ mol } O_2 \times 0.1 \text{ (rounded value of } R) \times 300 \text{ K}}{450 \text{ L}} = 2 \text{ atm}$$

which is close to 1.53 atm.

**FOLLOW-UP PROBLEM 5.5**
The tank in the sample problem develops a slow leak that is discovered and sealed. The new measured pressure is 1.37 atm. How many grams of $O_2$ remain?

---

### Section Summary
Four variables define the physical behavior of an ideal gas: volume ($V$), pressure ($P$), temperature ($T$), and number of moles ($n$). Most simple gases display nearly ideal behavior at ordinary temperature and pressure. Boyle's, Charles's, and Avogadro's laws relate volume to pressure, temperature, and amount of gas, respectively. The ideal gas law incorporates these individual gas laws into one equation: $PV = nRT$, where $R$ is the universal gas constant.

## 5.4 Further Applications of the Ideal Gas Law

The ideal gas law can be recast to determine other properties of gases, such as the density, molar mass, and partial pressure of each gas in a mixture.

### Gas Density

Since one mole of any gas occupies approximately the same volume at a given temperature and pressure, its density ($d = m/V$) depends on its molar mass (see Figure 5.11). For example, at STP, one mole of $O_2$ occupies a volume equal to that occupied by one mole of $N_2$, but each $O_2$ molecule has a greater mass than each $N_2$ molecule, so $O_2$ is denser.

Even though all gases are miscible, in the absence of sufficient mixing, a less dense gas will lie above a more dense one. There are several familiar examples of this phenomenon. Some types of fire extinguishers release $CO_2$ because it is denser than air and sinks onto the fire, preventing $O_2$ from reaching the flammable material. Enormous air masses of different densities moving past each other around the globe give rise to much of our weather. ◆

We can use the ideal gas law to calculate the density of a gas from its molar mass. Recall that the number of moles ($n$) is the mass ($m$) divided by the molar mass ($\mathcal{M}$), $n = m/\mathcal{M}$. Substituting for $n$ in the ideal gas law gives

$$PV = \frac{m}{\mathcal{M}}RT$$

Rearranging to isolate $m/V$ gives

$$\frac{m}{V} = d = \frac{\mathcal{M} \times P}{RT} \tag{5.8}$$

◆ **Gas Density and Human Disasters.** Many gases that are denser than air have been involved in intentional or accidental disasters. In World War I, phosgene ($COCl_2$) was used against ground troops as they lay in trenches. More recently, the unintentional release of methylisocyanate from a Union Carbide India Ltd. chemical plant in Bhopal, India, killed thousands of people as vapors spread from the outskirts into the city. In Cameroon, $CO_2$ released naturally from Lake Nyos suffocated thousands as it flowed down valleys into villages. On a far less horrific scale, the blanket of dense gases in smog over urban centers contributes to respiratory illness (photo).

◆ **Up, Up, and Away!** When the gas in a hot-air balloon is heated, its volume increases and the balloon inflates. Further heating causes some of the gas to escape. By these means, the gas density decreases and the balloon rises. Two pioneering hot-air balloonists used their knowledge of gas behavior to pursue their hobby. Jacques Charles (of Charles's law) made one of the first balloon flights in 1783. Twenty years later, Joseph Gay-Lussac (who studied the pressure-temperature relationship) set a solo altitude record that lasted for 50 years.

Using this equation, we can find the density of a known gas at any temperature and pressure near standard conditions. Two important ideas are expressed by Equation 5.8:

1. *The density of a gas is directly proportional to its molar mass* because a heavier gas occupies the same volume as a lighter gas (Avogadro's law).
2. *The density of a gas is inversely proportional to the temperature.* As the volume of a gas increases with temperature (Charles's law), the same mass occupies more space, so the density is lower.

Architectural designers and heating engineers use this principle when they place heating ducts near the floor of a room: the less dense warm air from the ducts rises and heats the room air. Safety experts recommend staying near the floor when escaping from a fire to avoid the hot, noxious gases that rise due to their lower densities. ◆

---

SAMPLE PROBLEM 5.6 _____

### Calculating Gas Density

**Problem:** Calculate the density (in g/L) of carbon dioxide and the number of molecules per liter **(a)** at STP (0°C and 1 atm) and **(b)** at ordinary room conditions (20°C exactly and 1.00 atm).

**Plan:** We must find the density ($d$) and number of molecules at two sets of $P$ and $T$. We convert $T$ to kelvins, find $\mathcal{M}$, and calculate $d$ from Equation 5.8. Then we convert the mass per liter to number of molecules per liter.

**Solution:** **(a)** Density and molecules per liter of $CO_2$ at STP. Summary of gas properties:

$$T = 0°C + 273 = 273 \text{ K} \qquad P = 1 \text{ atm} \qquad \mathcal{M} \text{ of } CO_2 = 44.01 \text{ g/mol}$$

Calculating density (note the unit canceling here):

$$d = \frac{\mathcal{M} \times P}{RT} = \frac{44.01 \text{ g/mol} \times 1 \text{ atm}}{0.0821 \frac{\text{atm} \cdot \text{L}}{\text{mol} \cdot \text{K}} \times 273 \text{ K}} = \textbf{1.96 g/L}$$

Converting from mass/L to molecules/L:

$$\text{Molecules } CO_2/L = \frac{1.96 \text{ g } CO_2}{1 \text{ L}} \times \frac{1 \text{ mol } CO_2}{44.01 \text{ g } CO_2} \times \frac{6.022 \times 10^{23} \text{ molecules } CO_2}{1 \text{ mol } CO_2}$$

$$= \textbf{2.68} \times \textbf{10}^{\textbf{22}} \textbf{ molecules } CO_2/L$$

**(b)** Density and molecules of $CO_2$ per liter at room conditions. Summary of gas properties:

$$T = 20°C + 273 = 293 \text{ K} \qquad P = 1.00 \text{ atm} \qquad \mathcal{M} \text{ of } CO_2 = 44.01 \text{ g/mol}$$

Calculating density:

$$d = \frac{\mathcal{M} \times P}{RT} = \frac{44.01 \text{ g/mol} \times 1.00 \text{ atm}}{0.0821 \frac{\text{atm} \cdot \text{L}}{\text{mol} \cdot \text{K}} \times 293 \text{ K}} = \textbf{1.83 g/L}$$

Converting from mass/L to molecules/L:

$$\text{Molecules } CO_2/L = \frac{1.83 \text{ g } CO_2}{1 \text{ L}} \times \frac{1 \text{ mol } CO_2}{44.01 \text{ g } CO_2} \times \frac{6.022 \times 10^{23} \text{ molecules } CO_2}{1 \text{ mol } CO_2}$$

$$= \textbf{2.50} \times \textbf{10}^{\textbf{22}} \textbf{ molecules } CO_2/L$$

**Check:** Round off to check the density values; for example, in **(a)**, at STP:

$$\frac{50 \text{ g/mol} \times 1 \text{ atm}}{0.1 \dfrac{\text{atm} \cdot \text{L}}{\text{mol} \cdot \text{K}} \times 250 \text{ K}} = 2 \text{ g/L} \approx 1.96 \text{ g/L}$$

At the higher temperature in **(b)**, the density should decrease, which can happen only if there are fewer molecules per liter, so the answer is reasonable.
**Comment:** An *alternative approach* for finding the density *at STP only* is to divide $\mathcal{M}$ by the standard molar volume, 22.4 L:

$$d = \frac{\mathcal{M}}{V} = \frac{44.01 \text{ g/mol}}{22.4 \text{ L/mol}} = 1.96 \text{ g/L}$$

Then, if you know the density at one temperature, you can find it at any other from $d_1/d_2 = T_2/T_1$.

**FOLLOW-UP PROBLEM 5.6**
Compare the density of $CO_2$ at 0°C and 380 torr with its density at STP.

### The Molar Mass of an Unknown Gas

Using another simple rearrangement of the ideal gas law, we can determine the molar mass of an unknown gas or a very volatile liquid (one that is easily vaporized):

$$n = \frac{m}{\mathcal{M}} = \frac{PV}{RT} \quad \text{so} \quad \mathcal{M} = \frac{mRT}{PV} \quad \text{or} \quad \mathcal{M} = \frac{dRT}{P} \qquad \textbf{(5.9)}$$

This equation is just a rearrangement of the equation (5.8) for density.

The French chemist J. B. A. Dumas (1800-1884) pioneered an ingenious method for finding the molar mass of an unknown volatile liquid. A small volume of the liquid is placed in a preweighed flask of known volume. The flask is closed with a stopper that contains a narrow tube and is immersed in a water bath whose fixed temperature exceeds the liquid's boiling point. When the liquid has completely vaporized, gas flows out the tube until the pressure of the gas remaining in the flask equals the atmospheric pressure (Figure 5.14). The flask is cooled, and the remaining gas condenses to a liquid. The mass of the liquid (obtained by reweighing the flask) equals the mass of the gas that remains in the flask. This mass of gas ($m$) occupies the flask volume ($V$) at a pressure ($P$) equal to the barometric pressure and at the temperature ($T$) of the water bath. Thus, all the variables needed to calculate molar mass have been measured directly.

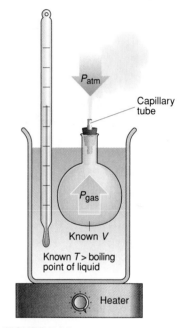

**FIGURE 5.14**
**Determining the molar mass of an unknown volatile liquid.** A small amount of unknown liquid is vaporized, and the gas fills the known flask volume at the known temperature. Excess gas escapes through the capillary tube until $P_{gas} = P_{atm}$. When the flask is cooled, the liquid is weighed, and the ideal gas law is used to calculate its molar mass (see text).

**SAMPLE PROBLEM 5.7** _____

### Finding the Molar Mass of a Gas

**Problem:** An organic chemist isolates a colorless liquid with the properties of cyclohexane ($\mathcal{M} = 84.2$ g/mol) from a petroleum sample. She uses the Dumas method and obtains the following data to determine its molar mass:

Volume ($V$) of flask = 213 mL        $T = 100.0°C$        $P = 754$ torr
Mass of flask + gas = 78.416 g        Mass of flask = 77.834 g

Is the molar mass consistent with the liquid being cyclohexane?
**Plan:** We convert $V$ to liters, $T$ to kelvins, and $P$ to atmospheres, find the mass of gas by subtracting the mass of the empty flask, and use Equation 5.9 to solve for $\mathcal{M}$.

**Solution:** Summary of gas variables:

$$m = 78.416 \text{ g} - 77.834 \text{ g} = 0.582 \text{ g}$$

$$V \text{ (L)} = 213 \text{ mL} \times \frac{1 \text{ L}}{1000 \text{ mL}} = 0.213 \text{ L}$$

$$P \text{ (atm)} = 754 \text{ torr} \times \frac{1 \text{ atm}}{760 \text{ torr}} = 0.992 \text{ atm}$$

$$T \text{ (K)} = 100.0°C + 273.15 = 373.2 \text{ K}$$

Solving for $\mathcal{M}$:

$$\mathcal{M} = \frac{mRT}{PV} = \frac{0.582 \text{ g} \times 0.0821 \dfrac{\text{atm} \cdot \text{L}}{\text{mol} \cdot \text{K}} \times 373.2 \text{ K}}{0.992 \text{ atm} \times 0.213 \text{ L}} = 84.4 \text{ g/mol}$$

Within experimental error, **the molar mass is consistent with the liquid being cyclohexane.**

**Check:** Rounding to check the arithmetic, we have

$$\frac{0.6 \text{ g} \times 0.08 \dfrac{\text{atm} \cdot \text{L}}{\text{mol} \cdot \text{K}} \times 375 \text{ K}}{1 \text{ atm} \times 0.2 \text{ L}} = 90 \text{ g/mol}$$

which is close to 84.4.

**FOLLOW-UP PROBLEM 5.7**
At 10.0°C and 102.5 kPa, the density of dry air is 1.26 g/L. What is the average "molar mass" of dry air at these conditions?

## Mixtures of Gases: Dalton's Law of Partial Pressures

All the gas behaviors discussed so far were determined from experiments with air, a complex mixture of gases. Thus, the ideal gas law is valid for virtually any gas at ordinary conditions, whether pure or a mixture. This fact holds true for two reasons:

1. Gases mix homogeneously (form a solution) over any range of proportions.
2. Each gas in a mixture behaves as if it were the only gas present (assuming no chemical interactions).

The second point was discovered by John Dalton in his studies of humidity. He observed that water vapor added to dry air increases the total air pressure by the pressure of the added water vapor:

$$P_{\text{humid air}} = P_{\text{dry air}} + P_{\text{added water vapor}}$$

In other words, each gas in the mixture exerts a **partial pressure,** a portion of the total pressure of the mixture, that is the same as the pressure it exerts when it is alone. This observation is formulated as **Dalton's law of partial pressures:** *in a mixture of unreacting gases, the total pressure is the sum of the partial pressures of the individual gases:*

$$P_{\text{total}} = P_1 + P_2 + P_3 + \cdots \tag{5.10}$$

This law allows us to see a direct relationship between pressure and moles of gas. Let's examine it with a system consisting of two chambers joined by a stopcock (Figure 5.15). The left chamber is a piston-cylinder assembly containing 500 mL $H_2$ gas at 1.0 atm. The right chamber is a tank

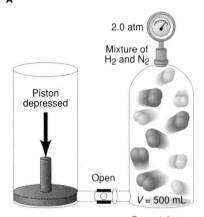

1.0 atm

$H_2$

$N_2$

Closed

$V = 500$ mL          $V = 500$ mL

$P_{H_2} = P_{\text{total}}$          $P_{N_2} = P_{\text{total}}$
$= 1.0$ atm          $= 1.0$ atm

**A**

2.0 atm

Mixture of $H_2$ and $N_2$

Piston depressed

Open

$V = 500$ mL

$P_{H_2} = 1.0$ atm
$P_{N_2} = 1.0$ atm
$P_{\text{total}} = 2.0$ atm

**B**

**FIGURE 5.15**
Demonstrating Dalton's law of partial pressures. **A,** A piston-cylinder assembly containing $H_2$ at 1.0 atm is connected to a tank of fixed volume containing $N_2$ at 1.0 atm. **B,** When the $H_2$ is forced into the tank of $N_2$, the total pressure equals the sum of the partial pressures.

with a fixed volume of 500 mL, fitted with a pressure gauge and containing $N_2$ gas at 1.0 atm. When the stopcock is opened and the piston depressed, the $H_2$ is forced into the $N_2$ chamber. Each gas spreads throughout the 500-mL tank, and the two gases form a homogeneous mixture. The tank pressure increases to 2.0 atm because each gas contributes its partial pressure to the total.

Since each gas behaves independently, we can write individual ideal gas equations for them:

$$P_{N_2} = \frac{n_{N_2}RT}{V} \quad \text{and} \quad P_{H_2} = \frac{n_{H_2}RT}{V}$$

Each gas occupies the same total volume and has the same temperature; therefore, each partial pressure is proportional to the number of moles of that gas:

$$P_{N_2} \propto n_{N_2} \quad \text{and} \quad P_{H_2} \propto n_{H_2}$$

Thus, the total pressure is given by

$$P_{\text{total}} = P_{N_2} + P_{H_2} = \frac{n_{N_2}RT}{V} + \frac{n_{H_2}RT}{V} = \frac{(n_{N_2} + n_{H_2})RT}{V} = \frac{n_{\text{total}}RT}{V}$$

where $n_{\text{total}} = n_{N_2} + n_{H_2}$.

Each component in a mixture contributes a fraction of the total number of moles in the mixture, which is the **mole fraction (X)** of that component. The sum of the mole fractions of all components in the mixture is 1. For $N_2$, the mole fraction, $X_{N_2}$, is

$$X_{N_2} = \frac{n_{N_2}}{n_{\text{total}}} = \frac{n_{N_2}}{n_{N_2} + n_{H_2}}$$

Since the total pressure is due to the total moles, the partial pressure of any gas, A, is the total pressure multiplied by the mole fraction of A:

$$P_A = X_A \times P_{\text{total}} \tag{5.11}$$

We can see this for $N_2$ if we express $P$ as $nRT/V$ and then cancel $n_{\text{total}}$:

$$\frac{n_{N_2}RT}{V} = \frac{n_{N_2}}{\cancel{n_{\text{total}}}} \times \frac{\cancel{n_{\text{total}}}RT}{V}$$

A similar relationship holds for the $H_2$ gas in the mixture. Thus,

$$P_{\text{total}} = P_{N_2} + P_{H_2}$$
$$= (X_{N_2} \times P_{\text{total}}) + (X_{H_2} \times P_{\text{total}}) = (X_{N_2} + X_{H_2})P_{\text{total}} = 1 \times P_{\text{total}}$$

SAMPLE PROBLEM 5.8 _____

### Applying Dalton's Law of Partial Pressures

**Problem:** In order to study $O_2$ uptake by muscle at high altitude, an exercise physiologist prepares an atmosphere consisting of 79% $N_2$, 17% $^{16}O_2$, and 4.0% $^{18}O_2$ by volume. (The isotope $^{18}O$ will be measured to determine how much oxygen is taken up by the muscle.) The pressure of the mixture is 0.75 atm to simulate high altitude. Calculate the mole fraction and partial pressure of $^{18}O_2$ in the mixture.

**Plan:** We are asked to find $X_{^{18}O_2}$ and $P_{^{18}O_2}$, given $P_{\text{total}}$ and the percent by volume of each gas in the mixture. Since $V$ is proportional to $n$ (Avogadro's law), the volume % equals the mole %, so dividing volume % by 100 gives $X_{^{18}O_2}$. The $P_{^{18}O_2}$ is the portion of $P_{\text{total}}$ exerted by the moles of $^{18}O$, so we apply Equation 5.11.

Volume % of $^{18}O_2$

$V \propto n$

Mole % of $^{18}O_2$

divide by 100

Mole fraction, $X_{^{18}O_2}$

multiply by $P_{\text{total}}$

Partial pressure, $P_{^{18}O_2}$

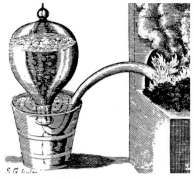

◆ **Instrumentation and Discovery.** In science, a technical break-through often leads to a burst of discovery. In the early eighteenth century, the danger of explosion from collecting gases in closed containers was stifling research. Then, in 1727, the English biologist Stephen Hales invented the pneumatic trough (see figure; the delivery tube is a bent gun barrel), and used it to collect gases safely. The Scottish chemist Joseph Black applied the method to isolate "fixed air" ($CO_2$) by heating mineral carbonates. In doing so, Black established the idea that each "air" was a unique substance, not ordinary air with different impurities. In further studies with the device, Henry Cavendish, Joseph Priestley, and others isolated inflammable air ($H_2$), nitrous air (NO), marine acid air (HCl), dephlogisticated air ($O_2$), and a host of others: $N_2$, $N_2O$, $SO_2$, $SiF_4$, $NH_3$, $H_2S$, CO, $CH_4$, $Cl_2$, and $PH_3$. From these experiments and the contributions in theory made by Lavoisier, Charles, Gay-Lussac, Dalton, and others, chemistry emerged as a science.

**Solution:** Calculating the mole fraction of $^{18}O_2$. Since volume % equals mole %, 4.0 vol % $^{18}O_2$ = 4.0 mol % $^{18}O_2$. Thus,

$$X_{^{18}O_2} = \frac{4.0 \text{ mol \% } ^{18}O_2}{100} = \textbf{0.040}$$

Solving for the partial pressure of $^{18}O_2$:

$$P_{^{18}O_2} = X_{^{18}O_2} \times P_{\text{total}} = 0.040 \times 0.75 \text{ atm} = \textbf{0.030 atm}$$

**Check:** $X_{^{18}O_2}$ is small because the mole % is small, so $P_{^{18}O_2}$ should be small also.
**Comment:** At high altitudes, specialized brain cells that are sensitive to blood $O_2$ and $CO_2$ levels trigger an increase in rate and depth of breathing for several days, until a person becomes acclimated.

**FOLLOW-UP PROBLEM 5.8**
A mixture of noble gases consisting of 5.50 g He, 15.0 g Ne, and 35.0 g Kr is placed in a piston-cylinder assembly at STP. Calculate the cylinder volume and the partial pressure of each gas.

_____

The law of partial pressures has many applications in chemical production and research. It is frequently used to determine the amount of a water-insoluble gaseous reaction product. The product is collected by being bubbled through water into an inverted container (Figure 5.16). Water evaporates into the gas until the water vapor reaches a pressure called the *vapor pressure*, which depends only on the temperature. Thus, the gas volume consists of a mixture of the gaseous product and water vapor, and the total gas pressure is the sum of the two contributing partial pressures. From a list of water vapor pressure values at various temperatures (Table 5.4), the appropriate value is found and subtracted from the total gas pressure (equalized to atmospheric pressure) to give the partial pressure of the gaseous product. With $V$ and $T$ known, the amount of gaseous product can be determined. ◆

**FIGURE 5.16**
**Determining the pressure of a water-insoluble gaseous reaction product. A,** Collecting the gaseous product. **B,** Calculating $P_{gas}$.

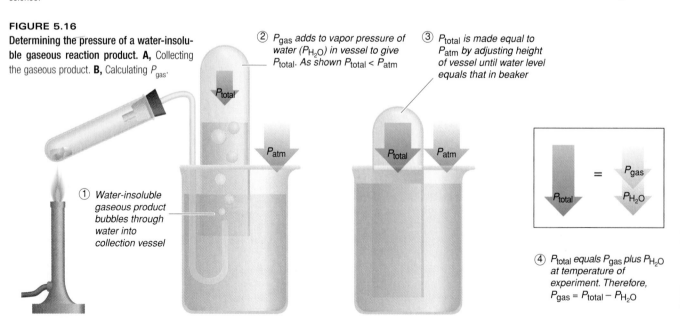

② $P_{gas}$ *adds to vapor pressure of water ($P_{H_2O}$) in vessel to give* $P_{total}$. *As shown* $P_{total} < P_{atm}$

③ $P_{total}$ *is made equal to* $P_{atm}$ *by adjusting height of vessel until water level equals that in beaker*

① *Water-insoluble gaseous product bubbles through water into collection vessel*

④ $P_{total}$ *equals* $P_{gas}$ *plus* $P_{H_2O}$ *at temperature of experiment. Therefore,* $P_{gas} = P_{total} - P_{H_2O}$

$$P_{total} = P_{gas} + P_{H_2O}$$

**A**                    **B**

SAMPLE PROBLEM 5.9

## Calculating the Amount of Gas Collected over Water

**Problem:** Acetylene ($C_2H_2$), an important fuel in welding, is produced in the laboratory when calcium carbide ($CaC_2$) reacts with water:

$$CaC_2(s) + 2H_2O(l) \rightarrow C_2H_2(g) + Ca(OH)_2(aq)$$

A sample of $C_2H_2$ was collected over water, the total gas pressure adjusted to barometric pressure (738 torr), and the volume measured as 523 mL. At the gas temperature (23°C), the vapor pressure of water is 21 torr. How many grams of acetylene were prepared?

**Plan:** We must find the mass of $C_2H_2$, so we first need $P_{C_2H_2}$. The measured $P$ is $P_{total}$, the sum of $P_{C_2H_2}$ and $P_{H_2O}$. We know $P_{total}$ (from the barometer) and are given $P_{H_2O}$, so we subtract to find $P_{C_2H_2}$. We are given $V$ and $T$, so we find moles from the ideal gas law and convert to grams.

**Solution:** Summary of gas variables:

$$P_{C_2H_2} \text{ (torr)} = P_{total} - P_{H_2O} = 738 \text{ torr} - 21 \text{ torr} = 717 \text{ torr}$$

$$P_{C_2H_2} \text{ (atm)} = 717 \text{ torr} \times \frac{1 \text{ atm}}{760 \text{ torr}} = 0.943 \text{ atm}$$

$$V \text{ (L)} = 523 \text{ mL} \times \frac{1 \text{ L}}{1000 \text{ mL}} = 0.523 \text{ L}$$

$$T \text{ (K)} = 23°C + 273 = 296 \text{ K}$$

$$n_{C_2H_2} = \text{unknown}$$

Solving for $n_{C_2H_2}$:

$$n_{C_2H_2} = \frac{PV}{RT} = \frac{0.943 \text{ atm} \times 0.523 \text{ L}}{0.0821 \frac{\text{atm} \cdot \text{L}}{\text{mol} \cdot \text{K}} \times 296 \text{ K}} = 0.0203 \text{ mol}$$

Converting moles of $C_2H_2$ to mass:

$$\text{Mass (g) of } C_2H_2 = 0.0203 \text{ mol } C_2H_2 \times \frac{26.04 \text{ g } C_2H_2}{1 \text{ mol } C_2H_2} = \textbf{0.529 g } C_2H_2$$

**Check:** We can do a quick arithmetic check for $n$:

$$n \approx \frac{1 \text{ atm} \times 0.5 \text{ L}}{0.1 \frac{\text{atm} \cdot \text{L}}{\text{mol} \cdot \text{K}} \times 300 \text{ K}} = 0.017 \text{ mol} \approx 0.0203 \text{ mol}$$

**FOLLOW-UP PROBLEM 5.9**
A small piece of zinc reacts with dilute HCl to form $H_2$, which is collected over water at 16°C into a large flask. The total pressure is adjusted to barometric pressure (752 torr), and the volume is measured as 1495 mL. Use data from Table 5.4 to help calculate the partial pressure and mass of $H_2$ collected.

## Section Summary
The ideal gas law can be rearranged to calculate the density and molar mass of a gas. In a mixture of gases, each component contributes its own partial pressure to the total (Dalton's law of partial pressures). The mole fraction of each component is the ratio of its partial pressure to the total pressure. When a gas is in contact with water, the total pressure is the sum of the gas pressure and the vapor pressure of water at the prevailing temperature.

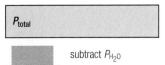

$P_{total}$

subtract $P_{H_2O}$

$P_{C_2H_2}$

$n = \frac{PV}{RT}$

Moles of $C_2H_2$

$\mathcal{M}$ (g/mol)

Mass (g) of $C_2H_2$

**TABLE 5.4 Vapor Pressure of Water ($P_{H_2O}$) at Different Temperatures**

| T (°C) | P (torr) |
|---|---|
| 0 | 4.6 |
| 5 | 6.5 |
| 10 | 9.2 |
| 11 | 9.8 |
| 12 | 10.5 |
| 13 | 11.2 |
| 14 | 12.0 |
| 15 | 12.8 |
| 16 | 13.6 |
| 18 | 15.5 |
| 20 | 17.5 |
| 22 | 19.8 |
| 24 | 22.4 |
| 26 | 25.2 |
| 28 | 28.3 |
| 30 | 31.8 |
| 35 | 42.2 |
| 40 | 55.3 |
| 45 | 71.9 |
| 50 | 92.5 |
| 55 | 118.0 |
| 60 | 149.4 |
| 65 | 187.5 |
| 70 | 233.7 |
| 75 | 289.1 |
| 80 | 355.1 |
| 85 | 433.6 |
| 90 | 525.8 |
| 95 | 633.9 |
| 100 | 760.0 |

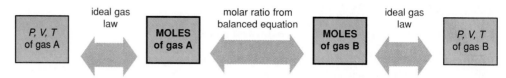

**FIGURE 5.17**
Summary of the stoichiometric relationships among moles ($n$) of gaseous reactant or product and the other gas variables ($P$, $V$, and $T$).

## 5.5  The Ideal Gas Law and Reaction Stoichiometry

In Chapters 3 and 4, we encountered many reactions that involved gases as reactants (such as combustion with $O_2$) or as products (such as decomposition of a carbonate). From the balanced equation, we used stoichiometrically equivalent molar ratios to calculate the moles of reactants and products and converted these quantities into mass, number of molecules, or solution volume (see Figures 3.11 and 3.14). Now you can expand your problem-solving repertoire by using the ideal gas law to convert gas variables ($P$, $T$, and $V$) to moles of gaseous reactants and products, and vice versa (Figure 5.17); in effect, you are combining a stoichiometry problem with a gas law problem. Moreover, this approach allows you to deal with more realistic situations involving gases because we usually measure their volume, temperature, and pressure rather than their mass or number of molecules.

SAMPLE PROBLEM 5.10 _____

### Using Gas Variables to Find Amounts of Reactants or Products

**Problem:** A laboratory-scale method for reducing a metal oxide is to heat it with $H_2$. The pure metal and water are the products (see photos). What volume of $H_2$ at 765 torr and 185°C is needed to form 35.5 g Cu from copper(II) oxide?

**Plan:** This is a combined stoichiometry and gas law problem. First we write and balance the equation. We convert mass of Cu to moles and use the molar ratio to find moles of $H_2$ required (stoichiometry portion). Then we use the ideal gas law to convert moles of $H_2$ to liters (gas law portion). A roadmap is shown this time, but you are familiar with all these steps.

**Solution:** Writing the balanced equation:

$$CuO(s) + H_2(g) \rightarrow Cu(s) + H_2O(g)$$

Calculating $n_{H_2}$:

$$n_{H_2} = 35.5 \text{ g Cu} \times \frac{1 \text{ mol Cu}}{63.55 \text{ g Cu}} \times \frac{1 \text{ mol H}_2}{1 \text{ mol Cu}} = 0.559 \text{ mol H}_2$$

Summary of gas variables:

$$V = \text{unknown} \quad P \text{ (atm)} = 765 \text{ torr} \times \frac{1 \text{ atm}}{760 \text{ torr}} = 1.01 \text{ atm}$$

$$T \text{ (K)} = 185°C + 273 = 458 \text{ K}$$

Solving for volume of $H_2$:

$$V = \frac{nRT}{P} = \frac{0.559 \text{ mol} \times 0.0821 \frac{\text{atm} \cdot \text{L}}{\text{mol} \cdot \text{K}} \times 458 \text{ K}}{1.01 \text{ atm}} = \textbf{20.8 L}$$

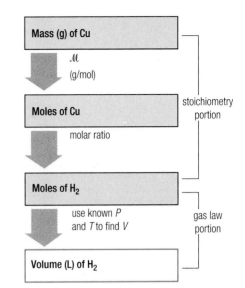

Mass (g) of Cu

$\mathcal{M}$ (g/mol)

Moles of Cu

molar ratio

Moles of H$_2$

use known $P$ and $T$ to find $V$

Volume (L) of H$_2$

stoichiometry portion

gas law portion

**Check:** Let's check the answer by comparing it with the molar volume of an ideal gas at STP (273 K and 1 atm). Since 1 mol $H_2$ at STP occupies about 22 L, a little more than 0.5 mol would occupy a little more than 11 L. Since $T$ is almost twice 273 K, $V$ should be almost twice 11 L.

**Comment:** The main point in this problem is that the stoichiometry provides one gas variable ($n$), two more are given, and the ideal gas law is used to find the fourth.

**FOLLOW-UP PROBLEM 5.10**
Sulfuric acid reacts with sodium chloride to form aqueous sodium sulfate and hydrogen chloride gas. How many milliliters of hydrogen chloride gas form at STP when 0.117 kg sodium chloride reacts with excess sulfuric acid?

---

**SAMPLE PROBLEM 5.11**

## Using the Ideal Gas Law in a Limiting-Reactant Problem

**Problem:** The alkali metals [Group 1A(1)] react with the halogens [Group 7A(17)] to form ionic metal halides. What mass of potassium chloride forms when 5.25 L chlorine gas at 0.950 atm and 293 K reacts with 17.0 g potassium?

**Plan:** The only difference between this and previous limiting-reactant problems is that here we use the ideal gas law to find the moles of gaseous reactant from the known $V$, $P$, and $T$. We then use the balanced equation to find the limiting reactant and the amount of product.

**Solution:** Writing the balanced equation:

$$2K(s) + Cl_2(g) \rightarrow 2KCl(s)$$

Summary of gas variables:

$$P = 0.950 \text{ atm} \quad V = 5.25 \text{ L} \quad T = 293 \text{ K} \quad n = \text{unknown}$$

Solving for $n_{Cl_2}$:

$$n_{Cl_2} = \frac{PV}{RT} = \frac{0.950 \text{ atm} \times 5.25 \text{ L}}{0.0821 \frac{\text{atm} \cdot \text{L}}{\text{mol} \cdot \text{K}} \times 293 \text{ K}} = 0.207 \text{ mol}$$

Converting from mass of potassium (K) to moles:

$$\text{Moles of K} = 17.0 \text{ g K} \times \frac{1 \text{ mol K}}{39.10 \text{ g K}} = 0.435 \text{ mol K}$$

Determining the limiting reactant: If $Cl_2$ is limiting,

$$\text{Moles of KCl} = 0.207 \text{ mol } Cl_2 \times \frac{2 \text{ mol KCl}}{1 \text{ mol } Cl_2}$$

$$= 0.414 \text{ mol KCl}$$

If K is limiting,

$$\text{Moles of KCl} = 0.435 \text{ mol K} \times \frac{2 \text{ mol KCl}}{2 \text{ mol K}}$$

$$= 0.435 \text{ mol KCl}$$

Therefore, $Cl_2$ is the limiting reactant.
Converting from moles of KCl to mass:

$$\text{Mass (g) of KCl} = 0.414 \text{ mol KCl} \times \frac{74.55 \text{ g KCl}}{1 \text{ mol KCl}} = \textbf{30.9 g KCl}$$

**Check:** The gas law calculation seems correct. At STP, a 5-L volume would contain about 0.25 mol $Cl_2$. Since $P$ (in numerator) is slightly lower and $T$ (in denominator) slightly higher than STP, there should be slightly less than 0.25 mol. The mass of KCl seems reasonable: less than 0.5 mol KCl gives less than $0.5 \times \mathcal{M}$ (30.9 g $< 0.5 \times$ 75 g).

**FOLLOW-UP PROBLEM 5.11**
Ammonia and hydrogen chloride gases react to form solid ammonium chloride. A 10.0-L reaction flask contains ammonia at 0.452 atm and 22°C, and 155 mL hydrogen chloride gas at 7.50 atm and 271 K is introduced. After the reaction occurs and the temperature returns to 22°C, what is the pressure inside the flask? (Neglect the volume of the solid product.)

**Section Summary**
In any reaction that involves gases, we can express the amount of gaseous reactant or product in terms of the gas variables. In this way, the ideal gas law allows us to combine stoichiometry problems with those involving gas behavior.

## 5.6    The Kinetic-Molecular Theory: A Model for Gas Behavior

So far we have discussed observations made on macroscopic samples of gas: decreasing volume in a piston-cylinder assembly, increasing pressure in a tank, and so forth. However, there is a molecular basis for macroscopic behavior, and this section presents the central model that explains gas behavior at the level of individual particles: the **kinetic-molecular theory.**

### How the Kinetic-Molecular Theory Explains the Gas Laws

Let's briefly review macroscopic gas behavior, focusing on the questions it raises at the molecular level:
1. Pressure is a measure of the force a gas exerts on a surface. What do the individual gas particles *do* that creates this force?
2. A change in gas volume in one direction is accompanied by a change in gas pressure in the other (Boyle's law). What happens to the particles

when external pressure compresses the sample volume? And why are gases compressible, whereas liquids and solids are not?

3. The pressure of a gas mixture is the sum of the pressures of the individual components (Dalton's law). Why does each gas contribute to the total interaction with a surface in proportion to its mole fraction?

4. A change in temperature is accompanied by a corresponding change in volume (Charles's law). What effect does higher temperature have on gas particles that would increase the sample volume—or increase the sample pressure if volume were fixed? This question raises a more fundamental one: what molecular phenomenon does temperature measure?

5. Gas volume (or pressure) depends on the number of moles present, not on the nature of the particular gas (Avogadro's law). But shouldn't a mole of larger molecules occupy more space than a mole of smaller ones? And why wouldn't a mole of heavier molecules exert more pressure than a mole of lighter ones?

The kinetic-molecular theory, one of the great scientific achievements of the 19th century, is expressed in highly mathematical form but is based qualitatively on three postulates.

Postulate 1: *Particle size.* The volume of an individual gas molecule is negligible compared with the space between molecules or the volume of the container. Thus, the model views gas particles as points of mass and a sample of gas as nearly empty space.

Postulate 2: *Particle motion.* The gas molecules are in constant, random, straight-line motion, until they collide with each other or with the container walls. Between collisions, the molecules do not influence each other. The collisions are *elastic*, which means that, like billiard balls, the colliding molecules exchange energy but their total kinetic energy ($E_k$) remains constant.

Postulate 3: *Particle speed, temperature, and energy.* The molecules in a sample of gas have a range of speeds ($u$), with the most probable speed near the average speed. As the temperature of a gas increases, the most probable speed increases (Figure 5.18). This increase occurs because the average kinetic energy of the molecules ($\overline{E}_k$), which incorporates the most probable speed, is proportional to the absolute temperature: $\overline{E}_k \propto T$, or $\overline{E}_k = c \times T$, where $c$ is a constant that is the same for any gas. In other words, *at a given temperature, all gases have the same average kinetic energy.*

Let's try to visualize the gas particles to see how the theory explains the macroscopic behavior of the gas sample.

1. *Origin of pressure.* When a moving object collides with a surface, it exerts a force, as when a hailstone hits a window. We conclude from postulate 2, which describes the collisions of gas particles with the container walls, that these collisions result in the observed pressure. The greater the number of molecules or the more frequently they collide with the walls per unit wall area, the greater is the pressure.

2. *Boyle's law ($V \propto 1/P$).* We picture the gas molecules as points of mass with empty space between them (postulate 1). As the pressure exerted *on* the gas increases at constant temperature, the distance between molecules decreases, so the sample volume decreases. The pressure exerted *by* the gas simultaneously increases because the smaller volume results in shorter distances between gas molecules and the walls, and therefore in more frequent collisions with the walls (Figure 5.19). Thus, gases are compressible because of the large spaces between gas molecules. The fact that liquids and solids cannot be compressed significantly implies that they contain very little free space.

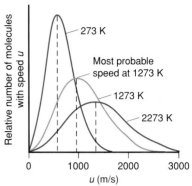

**FIGURE 5.18 The distribution of molecular speeds for N$_2$ at three temperatures.** At a given temperature, a plot of the relative number of N$_2$ molecules vs. molecular speed ($u$) results in a bell-shaped curve, with the most probable speed at the peak of each curve. Note that as the temperature *increases*, the most probable speed *increases*.

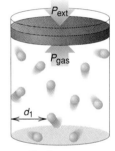

$P_{gas} = P_{ext}$

$P_{ext}$ increases, $T$ and $n$ fixed

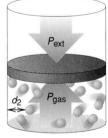

Higher $P_{ext}$ causes lower $V$, which causes more collisions until $P_{gas} = P_{ext}$

**FIGURE 5.19 A molecular description of Boyle's law ($V \propto 1/P$).** At a given $T$, gas molecules collide with the walls across an average distance ($d_1$) and give rise to a pressure ($P_{gas}$) that equals the external pressure ($P_{ext}$). If $P_{ext}$ increases, $V$ decreases, so the average distance between a molecule and the walls is shorter ($d_2 < d_1$). Molecules strike the walls more often, and $P_{gas}$ increases until it again equals $P_{ext}$. Thus, $V$ decreases when $P$ increases.

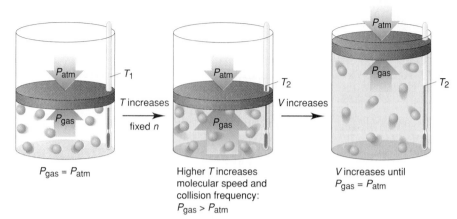

**FIGURE 5.20**

**A molecular description of Charles's law ($V \propto T$).** At a given temperature ($T_1$), $P_{gas} = P_{atm}$. When the gas is heated to $T_2$, the molecules move faster and collide with the walls more often, which increases $P_{gas}$. This increases $V$, so the molecules collide less often until $P_{gas}$ again equals $P_{atm}$. Thus, $V$ increases when $T$ increases.

3. *Dalton's law of partial pressures* ($P_{total} = P_A + P_B$). Suppose 0.6 mol gas A gives rise to a certain number of collisions per second with the walls and causes $P_A$. Assuming fixed volume and temperature, introducing 0.3 mol gas B will increase the total number of molecules, and therefore the total number of collisions per second with the wall by 50% (postulate 2). Thus, each gas exerts pressure in proportion to the number of molecules of that gas present (its mole percent or mole fraction).

4. *Charles's law* ($V \propto T$). As the temperature increases, the most probable molecular speed and kinetic energy increase (postulate 3). Thus, the molecules hit the walls more frequently and energetically. A higher frequency of collisions causes an increase in internal pressure. As a result, the walls move outward, which increases the volume and restores constant pressure (Figure 5.20).

5. *Avogadro's law* ($V \propto n$). Adding more molecules to the same container increases the total number of collisions with the walls and therefore the internal pressure. As a result, the volume expands until the number of collisions per unit wall area is the same as it was before the addition (Figure 5.21).

To explain why equal numbers of different gas molecules occupy the same volume, we must understand why heavier gas particles, such as $O_2$, *do not* bombard the container walls with more energy than lighter gas particles, such as He. The reason involves the meaning of kinetic energy, the key concept of postulate 3. The kinetic energy of an object, the energy associated with its motion, is related to its mass *and* its speed according to the equation

$$E_k = \tfrac{1}{2} \text{ mass} \times (\text{speed})^2$$

Note from the equation that if a heavy object and a light object have the same kinetic energy, the heavy object must be moving more slowly. For a large population of gas molecules, the average kinetic energy is

$$\overline{E}_k = \tfrac{1}{2} m\overline{u^2}$$

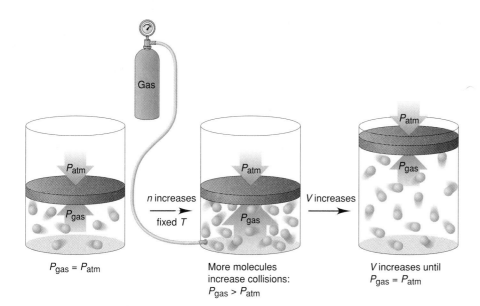

$P_{gas} = P_{atm}$

*n* increases
fixed *T*

More molecules
increase collisions:
$P_{gas} > P_{atm}$

*V* increases

*V* increases until
$P_{gas} = P_{atm}$

**FIGURE 5.21**

**A molecular description of Avogadro's law (*V* ∝ *n*).** At a given *T*, *n* moles of gas give rise to a pressure ($P_{gas}$) equal to $P_{atm}$. When more gas is added, *n* increases, so collisions with the walls are more frequent and $P_{gas}$ increases. This increases *V* until $P_{gas} = P_{atm}$ again. Thus, *V* increases when *n* increases.

where *m* is the molecular mass and $\overline{u^2}$ is the average of the squares of the molecular speeds. The square root of $\overline{u^2}$ is called the root-mean-square speed, or **rms speed ($u_{rms}$),** the speed of a molecule having the average kinetic energy. It is very close to the most probable speed, and for our purposes, we consider them essentially equal. The rms speed is related to the temperature and the molar mass:

$$u_{rms} = \sqrt{\frac{3RT}{\mathcal{M}}} \qquad \textbf{(5.12)}$$

where *R* is the gas constant (8.31 J/mol · K), *T* is the temperature in kelvins, and $\mathcal{M}$ is the molar mass expressed, in this case, in units of *kg/mol*. (To obtain a speed in units of m/s, we express the molar mass in units of kg/mol because *R* includes the joule, which has units of kg · m²/s².)

Postulate 3 states that different gases at the same temperature have the same average kinetic energy. The gas behavior expressed by Avogadro's law occurs because *molecules with a higher mass have a lower rms speed:* at the same temperature, $O_2$ molecules move more slowly, on the average, than He molecules (Figure 5.22). As a result, both types of gas particles hit the container walls with the same average kinetic energy, resulting in the same pressure.

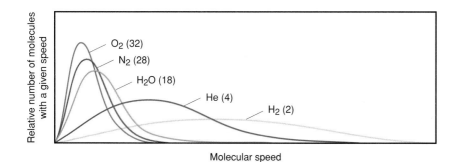

Relative number of molecules
with a given speed

$O_2$ (32)
$N_2$ (28)
$H_2O$ (18)
He (4)
$H_2$ (2)

Molecular speed

**FIGURE 5.22**

**Relationship between molar mass and molecular speed.** At a given temperature, gases with lower molar masses (numbers in parentheses) have higher rms speeds (peak of each curve).

**The meaning of temperature.** Let's look more closely at the idea that the most probable speed and the average kinetic energy of a collection of particles increase as the temperature increases. The relationship between average kinetic energy of an individual particle and absolute temperature is

$$\overline{E}_k = \frac{3}{2}\left(\frac{R}{N_A}\right)T$$

where $R$ is the gas constant and $N_A$ is Avogadro's number. This equation expresses an important point: *temperature is a measure of the energy of molecular motion.* When a beaker of water containing a thermometer is heated over a flame, the rapidly moving gas particles in the flame transfer kinetic energy to the atoms in the glass beaker, which vibrate faster and transfer kinetic energy to the molecules of water. The faster-moving water molecules collide more energetically with the atoms in the thermometer bulb, which in turn transfer kinetic energy to the atoms of mercury, causing the mercury to rise. In the macroscopic world, we see this as a temperature rise; in the molecular world, it is the transfer of kinetic energy from collections of higher energy objects to collections of lower energy objects.

### Effusion and Diffusion

One of the early triumphs of the kinetic-molecular theory was the explanation of an important observation of gas behavior. In 1846, Thomas Graham conducted a series of experiments concerning the rate of **effusion,** the process by which a gas escapes from its container through a tiny hole into an evacuated space. He concluded that the effusion rate was inversely proportional to the square root of gas density. Since density is directly proportional to molar mass, we state **Graham's law of effusion** as *the rate of effusion of a gas is inversely proportional to the square root of its molar mass:*

$$\text{Rate} \propto \frac{1}{\sqrt{\mathcal{M}}}$$

We can express rate in terms of moles (or molecules) per unit time. Argon gas (Ar), for example, effuses faster (at a higher rate) than the heavier krypton gas (Kr). The ratio of their effusion rates is

$$\frac{\text{Rate}_{Ar}}{\text{Rate}_{Kr}} = \frac{\sqrt{\mathcal{M}_{Kr}}}{\sqrt{\mathcal{M}_{Ar}}} \tag{5.13}$$

The kinetic-molecular theory explains that, at a given temperature, *the gas with the lower molar mass effuses faster because the most probable speed of its molecules is higher, so more molecules escape through the tiny hole per unit time.* ◆

Graham's law can be used to determine the molar mass of an unknown gas. The effusion rate of the unknown gas, X, is determined and compared with that of a known gas, such as He:

$$\frac{\text{Rate}_X}{\text{Rate}_{He}} = \frac{\sqrt{\mathcal{M}_{He}}}{\sqrt{\mathcal{M}_X}}$$

Squaring both sides and solving for the molar mass of X gives

$$\mathcal{M}_X = \mathcal{M}_{He} \times \left(\frac{\text{rate}_{He}}{\text{rate}_X}\right)^2$$

◆ **Preparing Nuclear Fuel.** The most important application of Graham's law is the preparation of nuclear reactor fuel: separating some nonfissionable, more abundant $^{238}U$ from fissionable $^{235}U$ to increase the proportion of $^{235}U$ in the mixture. Since the two isotopes have identical chemical properties, they are separated by differences in the effusion rates of their gaseous compounds. Uranium ore is converted to gaseous $UF_6$ (a mixture of $^{238}UF_6$ and $^{235}UF_6$), which is passed through a series of chambers with porous barriers. Molecules of $^{235}UF_6$ ($\mathcal{M} = 349.03$) move slightly faster than molecules of $^{238}UF_6$ ($\mathcal{M} = 352.04$) and effuse through each barrier at a slightly greater rate:

$$\frac{\text{Rate of } ^{235}UF_6}{\text{Rate of } ^{238}UF_6} = \frac{\sqrt{\mathcal{M} \text{ of } ^{238}UF_6}}{\sqrt{\mathcal{M} \text{ of } ^{235}UF_6}}$$

$$= \frac{\sqrt{352.04}}{\sqrt{349.03}} = 1.0043$$

Many passes are made, each increasing the fraction of $^{235}UF_6$ until a mixture is obtained that contains enough $^{235}UF_6$. This isotope-enrichment process was developed during the latter years of World War II and produced enough $^{235}U$ for the world's first three atomic bombs.

SAMPLE PROBLEM 5.12 _____

## Applying Graham's Law of Effusion

**Problem:** Calculate the ratio of the effusion rates of helium and methane ($CH_4$).
**Plan:** The effusion rate is inversely proportional to $\sqrt{\mathcal{M}}$, so we find the molar mass of each substance and take its square root. The inverse of the ratio of the square roots is the effusion rate ratio.
**Solution:**

$$\mathcal{M} \text{ of } CH_4 = 16.04 \text{ g/mol} \qquad \mathcal{M} \text{ of He} = 4.003 \text{ g/mol}$$

Calculating the effusion rate ratio:

$$\frac{\text{Rate}_{He}}{\text{Rate}_{CH_4}} = \sqrt{\frac{\mathcal{M}_{CH_4}}{\mathcal{M}_{He}}} = \sqrt{\frac{16.04 \text{ g/mol}}{4.003 \text{ g/mol}}} = \sqrt{4.007} = \mathbf{2.002}$$

**Check:** A rate ratio > 1 makes sense because the lighter He should effuse faster than the heavier $CH_4$. Because the molar mass of He is about one-fourth that of $CH_4$, He should effuse about twice as fast.

FOLLOW-UP PROBLEM 5.12
If it takes 1.25 minutes for 0.010 mol He to effuse, how long will it take for the same amount of ethane ($C_2H_6$) to effuse?

_____

A phenomenon closely related to effusion is gaseous **diffusion,** the movement of one gas through another. Comparisons of diffusion rates show that they are also described quantitatively by Graham's law:

$$\text{Rate of diffusion} \propto \frac{1}{\sqrt{\mathcal{M}}}$$

For two gases, such as $NH_3$ and HCl, moving through another gas or a mixture of gases, such as air, we find

$$\frac{\text{Rate}_{NH_3}}{\text{Rate}_{HCl}} = \sqrt{\frac{\mathcal{M}_{HCl}}{\mathcal{M}_{NH_3}}}$$

A simple experiment involving these gases demonstrates the dependence of diffusion rate on molar mass (Figure 5.23). The reason for this dependence is the same as for effusion rates: lighter molecules have higher most probable speeds than heavier molecules, so they move farther in a given amount of time.

If gas molecules move so rapidly at ordinary temperatures (see Figure 5.18), why don't you smell a perfume, for instance, the moment the bottle is opened? In the absence of air currents, a molecule must depend on diffusion to move through the available volume, but it does not move very far before it collides with another gas molecule in the air. Diffusion of gas molecules across a space is not instantaneous, therefore, because the individual molecular paths are tortuous (Figure 5.24).

Diffusion also occurs in liquids (and even to some extent in solids). However, because the distances between molecules in a liquid are much shorter than in a gas, the diffusion rates are *much* lower. Diffusion of a gas into a liquid and diffusion of a solute through a liquid are vital processes in biological systems. For example, diffusion contributes to the net movement of oxygen from lungs to blood and carbon dioxide from blood to lungs.

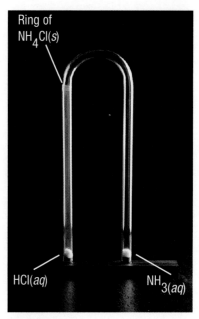

**FIGURE 5.23**
**The inverse relation between diffusion rate and molar mass.** A cotton swab with $NH_3(aq)$ and another with $HCl(aq)$ are placed at opposite ends of a U-tube, and gaseous $NH_3$ and HCl diffuse through the air in the tube. Where they meet, a white ring of $NH_4Cl$ forms:

$$NH_3(g) + HCl(g) \longrightarrow NH_4Cl(s).$$

By measuring the distance each travels, which is proportional to the diffusion rate, we can calculate the ratio of molar masses. Note that $NH_3$ ($\mathcal{M} = 17.03$ g/mol) travels farther than HCl ($\mathcal{M} = 36.46$ g/mol).

**FIGURE 5.24**
**Diffusion of a gas particle through a space filled with other particles.** In traversing a space, a gas molecule collides with many other molecules, which gives it a tortuous path. For clarity, the path of only one particle is shown (red lines).

Organisms have evolved elaborate ways to speed the diffusion of nutrients (for example, sugar and metal ions) and to slow, or even stop, the diffusion of toxins.

### The Chaotic World of Gases: Molecular Speed, Mean Free Path, and Collision Frequency

Refinements of the basic kinetic-molecular theory provide us with a view into the amazing, chaotic world of gas molecules. Let's follow the movements of an "average" $N_2$ molecule in a room at 20°C and 1 atm pressure—perhaps the room you are in now.

**Distribution of molecular speeds.** This $N_2$ molecule is hurtling through the room at a most probable speed of 0.51 km/s, or 1100 mi/h, this speed continually changing as it collides with other molecules. At any instant, it may be traveling at 2500 mi/h or standing still as it collides head on, although these extreme speeds are *much* less likely than the most probable value (see Figure 5.18).

**Mean free path.** If we know a molecule's diameter, we can use the kinetic-molecular theory to obtain the **mean free path,** the average distance the molecule travels between collisions at a given temperature and pressure. Our average $N_2$ molecule (diameter $3.7 \times 10^{-10}$ m) travels $6.6 \times 10^{-8}$ m before smashing into a fellow traveler, which corresponds to traveling an average of about 180 molecular diameters between collisions. Therefore, even though gas molecules are *not* points of mass, a gas sample *is* mostly empty space. Mean free path is a key factor in the rate of diffusion and the rate of heat flow through a gas.

**Collision frequency.** The ratio of the most probable speed (distance per second) to the mean free path (distance per collision) gives the **collision frequency,** the average number of collisions per second that a molecule undergoes. The collision frequency of our average $N_2$ molecule is

$$\text{Collision frequency} = \frac{5.1 \times 10^2 \text{ m/s}}{6.6 \times 10^{-8} \text{ m/collision}} = 7.7 \times 10^9 \text{ collisions/s} \blacklozenge$$

The ideas of speed (and kinetic energy) distribution and collision frequency are essential to understanding how fast a chemical reaction will occur, as you'll see in Chapter 15. As the upcoming Chemical Connections essay shows, many of the ideas in the kinetic-molecular theory that we just discussed apply directly to studies of our planet's atmosphere.

◆ **Danger on Molecular Highways.** To give you some idea of how astounding events are in the molecular world, we can express the collision frequency of a molecule in terms of a common experience in the macroscopic world: driving your compact car on the highway. Since a car is much larger than an $N_2$ molecule, to match the collision frequency of the $N_2$ molecule in your room, you would travel 2.8 billion mi/s and smash into another car every 700 yd!

### Section Summary

The kinetic-molecular theory postulates that gas molecules take up a negligible portion of the gas volume, move in straight-line paths between elastic collisions, and have average kinetic energies proportional to the absolute temperature. It explains the gas laws in terms of changes in distances between molecules and the container walls and changes in molecular speed. Temperature is a measure of the average kinetic energy of molecules. Effusion and diffusion rates are inversely proportional to the square root of the molar mass (Graham's law) because they are directly proportional to molecular speed. Molecular motion is characterized by a temperature-dependent most probable speed within a range of speeds, a mean free path, and a collision frequency. The atmosphere is a complex mixture of gases that exhibits variations in pressure, temperature, and composition with altitude.

Chemistry in Atmospheric Science
**Structure and Composition of the
Earth's Atmosphere**

The **atmosphere** is the envelope of gases that extends continuously from the Earth's surface outward, eventually merging with interplanetary space. Complex changes in pressure, temperature, and composition occur within this mixture of gases, and the present atmosphere is very different from the one that existed during our planet's early history.

**Variation in Pressure**

Since gases are compressible (Boyle's law), the pressure of the atmosphere decreases smoothly with distance from the surface, with a more rapid decrease at lower altitudes (Figure 5.A). Although no specific boundary delineates the outermost fringe of the atmosphere, the density and composition at around 10,000 km from the surface are identical with those of outer space. About 99% of the atmosphere's mass lies within 30 km of the surface, and 75% lies within the lowest 11 km.

**Variation in Temperature**

The atmosphere can be divided into several regions based on the pattern of temperature change (Figure 5.A). The troposphere, in which temperatures *drop* 7°C per kilometer to −55°C (218 K), includes the region from the surface to around 11 km. All our weather occurs in the troposphere, and all but a few aircraft fly there. Temperatures then *rise* through the stratosphere to about 7°C (280 K) at 50 km; we'll discuss why shortly. In the *mesosphere*, temperatures *drop* smoothly again to −93°C (180 K) at around 80 km. Within the *thermosphere*, which extends to around 500 km, temperatures *rise* again, but

**FIGURE 5.A**

Variations in pressure, temperature, and composition of the Earth's atmosphere.

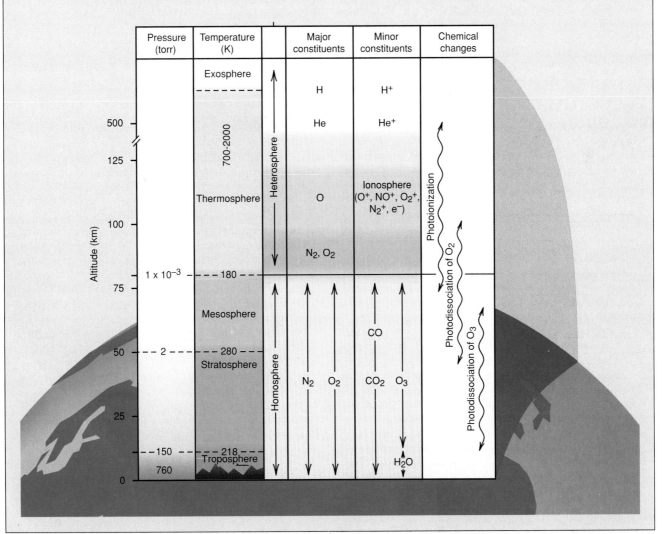

*Continued.*

Chemistry in Atmospheric Science—cont'd
# Structure and Composition of the Earth's Atmosphere

vary between 700 and 2000 K, depending on the intensity of solar radiation and sunspot activity. The *exosphere*, the outermost region, maintains these temperatures and merges with outer space.

What does a temperature of 2000 K at 500 km (300 mi) above the Earth's surface actually mean? Would a piece of iron (melting point ≈ 1700 K) glow red-hot and melt in the thermosphere? Our everyday use of the words "hot" and "cold" refers to measurements near the surface, where the density of the atmosphere is $10^6$ times greater than in the thermosphere. At an altitude of 500 km, where collision frequency is very low, a thermometer, or any other object, would experience *very little transfer of kinetic energy*. Thus, the object would not become "hot" in the usual sense; in fact, it would be very "cold." Recall that absolute temperature is a measure of average kinetic energy of the particles. The high-energy solar radiation reaching these outer regions is transferred to relatively few particles, so their average kinetic energy becomes extremely high, as indicated by the high temperature. For this reason, supersonic aircraft, such as the SST, do not reach maximum speed until they reach maximum altitude, where the air is less dense, so collisions with gas molecules are less frequent and the aircraft material becomes less hot.

## Variation in Composition

When we classify the atmosphere by chemical composition, we note two major regions called the *homosphere* and the *heterosphere*. Superimposing the regions defined by temperature on those defined by composition shows that the homosphere includes the troposphere, stratosphere, and mesosphere, and the heterosphere includes the thermosphere and exosphere (Figure 5.A).

### The Homosphere

The homosphere has a relatively constant composition, containing, by volume, approximately 78% $N_2$, 21% $O_2$, and 1% a mixture of other gases. Under homospheric conditions, the atmospheric gases behave ideally, so volume percent is equal to mole percent (Avogadro's law), and the mole fraction of a component is directly related to its partial pressure (Dalton's law). Table 5.A shows the components of a sample of clean, dry air at sea level.

The composition of the homosphere is uniform due to mixing through convection. Air directly in contact with land is warmer than the air above it. The warmer air expands (Charles's law), its density decreases, and it rises through the cooler, denser air, thereby mixing the components. The cooler air sinks, becomes warmer by contact with the land, and the convection continues. The warm air currents that rise from the ground, called *thermals*, are used by birds and glider pilots to stay aloft.

An important effect of convection is that the air above industrialized areas becomes cleaner as the rising air near

**TABLE 5.A Composition of Clean, Dry Air at Sea Level**

| COMPONENT | MOLE FRACTION |
|---|---|
| Nitrogen ($N_2$) | 0.78084 |
| Oxygen ($O_2$) | 0.20946 |
| Argon (Ar) | 0.00934 |
| Carbon dioxide ($CO_2$) | 0.00033 |
| Neon (Ne) | $1.818 \times 10^{-5}$ |
| Helium (He) | $5.24 \times 10^{-6}$ |
| Methane ($CH_4$) | $2 \times 10^{-6}$ |
| Krypton (Kr) | $1.14 \times 10^{-6}$ |
| Hydrogen ($H_2$) | $5 \times 10^{-7}$ |
| Dinitrogen monoxide ($N_2O$) | $5 \times 10^{-7}$ |
| Carbon monoxide (CO) | $1 \times 10^{-7}$ |
| Xenon (Xe) | $8 \times 10^{-8}$ |
| Ozone ($O_3$) | $2 \times 10^{-8}$ |
| Ammonia ($NH_3$) | $6 \times 10^{-9}$ |
| Nitrogen dioxide ($NO_2$) | $6 \times 10^{-9}$ |
| Nitrogen monoxide (NO) | $6 \times 10^{-10}$ |
| Sulfur dioxide ($SO_2$) | $2 \times 10^{-10}$ |
| Hydrogen sulfide ($H_2S$) | $2 \times 10^{-10}$ |

the surface carries up ground-level pollutants, which are dispersed by winds. However, under certain weather and geographical conditions, a warm air mass remains stationary over a cool one. The resulting *temperature inversion* blocks normal convection, and harmful pollutants build up, causing severe health problems.

### The Heterosphere

The heterosphere has variable composition, consisting of regions dominated by a few atomic or molecular species. Convective heating does not reach these heights, so the gas particles become layered according to molar mass: nitrogen and oxygen molecules in the lower levels, oxygen atoms (O) in the next, then helium atoms (He), and free hydrogen atoms (H).

Embedded within the lower heterosphere is the ionosphere, which contains ionic species such as $O^+$, $NO^+$, $O_2^+$, $N_2^+$, and free electrons (Figure 5.A). Ionospheric chemistry is complex, involving numerous light-induced bond breaking *(photodissociation)* and light-induced electron removal *(photoionization)* processes. One of the simpler ways that O atoms form, for instance, involves a four-step sequence:

$$N_2 \rightarrow N_2^+ + e^- \text{ [photoionization]}$$
$$N_2^+ + e^- \rightarrow N + N$$
$$N + O_2 \rightarrow NO + O$$
$$\underline{N + NO \rightarrow N_2 + O}$$
$$O_2 \rightarrow O + O \text{ [overall photodissociation]}$$

These high-energy O atoms collide with other neutral or ionic components, thereby increasing the average kinetic energy of thermospheric particles.

*The Importance of Stratospheric Ozone*
Although most high-energy radiation is absorbed by the thermosphere, a small amount reaches the stratosphere and breaks $O_2$ into O atoms. The energetic O atoms collide with more $O_2$ to form ozone ($O_3$), another molecular form of oxygen:

$$O_2(g) \xrightarrow{\text{high-energy radiation}} 2O(g)$$
$$M + O(g) + O_2(g) \rightarrow O_3(g) + M$$

where M is any particle that can remove excess energy. This reaction also gives off heat, which is the reason stratospheric temperatures increase with altitude.

Stratospheric ozone is vital to life on the surface because it absorbs ultraviolet (UV) radiation, which results in decomposition of the ozone:

$$O_3(g) \xrightarrow{\text{UV light}} O_2(g) + O(g)$$

UV radiation is extremely harmful because it can interrupt normal biological processes. Without the presence of stratospheric ozone, this radiation would reach the surface, resulting in increased mutation and cancer rates. The depletion of the ozone layer as a result of industrial gases is discussed in Chapter 15.

**Earth's Primitive Atmosphere**

The composition of the present atmosphere bears little resemblance to that covering the young Earth, but scientists disagree about what that primitive composition actually was: Did the carbon and nitrogen have low oxidation numbers, as in $CH_4$ (O.N. of C = −4) and $NH_3$ (O.N. of N = −3)? Or did these atoms have higher oxidation numbers, as in $CO_2$ (O.N. of C = +4) and $N_2$ (O.N. of N = 0)? It is generally accepted that the primitive mixture did not contain free $O_2$.

Origin-of-life models propose that about 1 billion years after the earliest organisms appeared, blue-green algae evolved. These one-celled plants used solar energy to produce glucose by photosynthesis:

$$6CO_2(g) + 6H_2O(l) \xrightarrow{\text{light}} C_6H_{12}O_6 \text{ (glucose)} + 6O_2(g)$$

As a result of this reaction, the $O_2$ content of the atmosphere increased and the $CO_2$ content decreased. More oxidation processes occurred, changing the geological and biological makeup of the early Earth. Iron(II) minerals were oxidized to iron(III) minerals, and sulfites to sulfates, and eventually organisms evolved that could oxidize other organisms to obtain energy. For these organisms to have survived, enough $O_2$ must have formed to create a protective ozone layer. Estimates indicate that the level of $O_2$ may have increased to the current level of about 20 mole % approximately 1.5 billion years ago.

## 5.7 Real Gases: Deviations from Ideal Behavior

A fundamental tenet of science is that simpler models are preferable to more complex ones—as long as they explain the data. By now, you realize the great usefulness of the ideal gas law in predicting gas behavior quantitatively. You can also appreciate how the kinetic-molecular theory explains ideal gas behavior by picturing gas molecules as infinitesimal "billiard balls" moving at speeds governed by the absolute temperature and experiencing only perfectly elastic collisions.

However, you know that molecules actually have *finite molecular volumes*, determined by the sizes of their atoms and the lengths of their bonds. You also know that molecules are composed of charged particles, which should give rise to *interactions between molecules*. (In fact, such interactions cause gases to condense to liquids.) We would expect these real properties of molecules to require adjustment of the kinetic-molecular model under some conditions, and this is indeed the case. At low temperatures and very high pressures, changes in the model are necessary in order to predict gas behavior quantitatively. ◆

### Effects of Extreme Conditions on Gas Behavior

At ordinary conditions—relatively low pressures and high temperatures—most simple gases exhibit nearly ideal behavior. Even at STP (0°C and 1

◆ **Real Gases and Industrial Processing.** Real gases may behave ideally at STP but not at the low temperatures and high pressures common in industry. Here are three of many examples. Air is cooled and eventually liquefied by first compressing it such that heat *can* flow out, and then expanding it such that heat *cannot* flow in. Distillation of the liquid separates some components as gases and the $N_2$ and $O_2$ as liquids, which have many uses, especially in steelmaking. Liquid $CO_2$ is formed by pressurizing the gas to 75 atm and then cooling it in water. It is used to freeze meat for ease in grinding and in replacing Freons as an aerosol propellant. In the production of $NH_3$ from $N_2$ and $H_2$, pressures of 200 to 300 atm are applied. The product mixture is cooled, the $NH_3$ removed as a liquid, and the remaining gases recycled to make more $NH_3$, an essential component of fertilizer and explosives.

**TABLE 5.5 Molar Volume of Some Common Gases at STP (0°C and 1 atm)**

| GAS | MOLAR VOLUME (L/mol) | CONDENSATION POINT (°C) |
|---|---|---|
| Helium (He) | 22.435 | −268.9 |
| Hydrogen ($H_2$) | 22.432 | −252.8 |
| Neon (Ne) | 22.422 | −246.1 |
| **Ideal gas** | **22.414** | — |
| Argon (Ar) | 22.397 | −185.9 |
| Nitrogen ($N_2$) | 22.396 | −195.8 |
| Oxygen ($O_2$) | 22.390 | −183.0 |
| Carbon monoxide (CO) | 22.388 | −191.5 |
| Chlorine ($Cl_2$) | 22.184 | −34.0 |
| Ammonia ($NH_3$) | 22.079 | −33.4 |

atm), however, gas behavior deviates *slightly* from the ideal. If we express the standard molar volume of an ideal gas to five significant figures, we find that the molar volumes of real gases do not quite equal the ideal value (Table 5.5). The same molecular phenomena that cause these slight deviations under standard conditions exert more and more influence as the temperature decreases toward the condensation point of the gas, the temperature at which it liquefies. For instance, the largest deviations from the standard molar volume among the gases listed are those of $Cl_2$ and $NH_3$, which have the highest condensation points. Even 0°C is close enough to their condensation points to cause noticeable deviations from ideality.

At pressures greater than 10 atm, we begin to see significant deviations from ideal behavior in many gases. Figure 5.25 shows a plot of *PV/RT* vs. *P* for one mole of several real gases and of an ideal gas. The *P* values on the *x* axis are the external pressures at which the *PV/RT* ratios are calculated; they range from normal ( ~1 atm) to very high ( ~1000 atm). For one mole of an *ideal* gas, *PV/RT* = 1, no matter what the external pressure.

The *PV/RT* curve for one mole of methane ($CH_4$) is typical of most real gases: it decreases to a minimum at moderately high pressures and then rises as pressure increases further. This curve shape is the result of two overlapping effects caused by the two characteristics of real molecules just mentioned: at high pressure, values of *PV/RT* lower than ideal (less than 1) are

**FIGURE 5.25**

**The behavior of several real gases with increasing external pressure.** The horizontal line shows the behavior of an ideal gas: *PV/RT* = 1 at all $P_{ext}$. At very high pressures, all real gases deviate significantly from such ideal behavior. Even at ordinary pressures, these deviations begin to appear (expanded region) (*inset*).

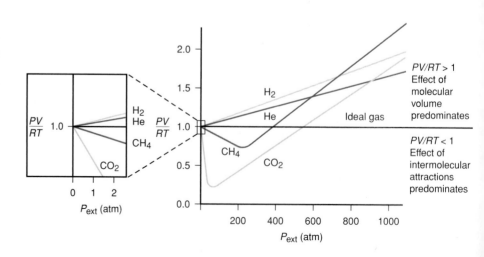

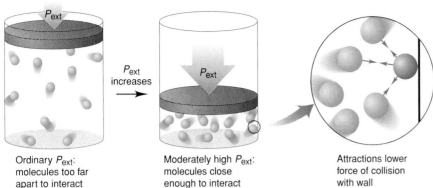

Ordinary $P_{ext}$: molecules too far apart to interact

Moderately high $P_{ext}$: molecules close enough to interact

Attractions lower force of collision with wall

**FIGURE 5.26**
**The effect of intermolecular attractions on measured gas pressure.** At ordinary pressures, the volume is large and gas molecules are too far apart to experience significant attractions. At moderately high external pressures, the volume decreases enough for the molecules to influence each other. As the close-up shows, a gas molecule approaching the container wall experiences intermolecular attractions from neighboring molecules that reduce the force of its impact. As a result, gases exert *less* pressure than that predicted from the ideal gas law.

due predominantly to *intermolecular attractions;* values of *PV/RT* greater than ideal (more than 1) are due predominantly to *molecular volume.* Let's examine these two aspects of real gases on the molecular level.

**Intermolecular attractions.** Attractive forces between molecules are *much* weaker than forces within molecules, the covalent bonds that hold a molecule together. Most intermolecular attractions are caused by slight imbalances in electron distributions and are important only over relatively short distances. At normal atmospheric pressure, the spaces between gas molecules are so large that intermolecular attractions are negligible, so the gas behaves nearly ideally. As the applied pressure rises and the volume of the sample decreases, however, the average intermolecular distance becomes smaller and intermolecular attractions exert greater influence.

At these higher pressures, a molecule approaching the container wall is attracted by nearby molecules, which lowers its speed and lessens the force of its impact (Figure 5.26). *This results in decreased gas pressure and thus a smaller numerator in PV/RT.* Lowering the temperature significantly has the same effect because it slows the molecules and allows attractive forces to be more effective. At a low enough temperature, the attractions become overwhelming, and the gas condenses to a liquid.

**Molecular volume.** At normal pressures, the space between molecules (free volume) is enormous compared with the volume of the molecules themselves, so the free volume is essentially equal to the container volume. As the applied pressure increases, however, and the free volume decreases, the volume of the molecules *themselves* takes up a greater proportion of the container volume. Thus, at very high pressures, the free volume is significantly *less* than the container volume (Figure 5.27). Since we continue to use the container volume as the "*V*" in the *PV/RT* ratio, even though the actual free volume is less, the ratio is artificially high. The molecular volume effect becomes more significant as the pressure increases, eventually outweighing the importance of the intermolecular attractions and causing *PV/RT* to rise above the ideal value.

Note that in Figure 5.25 the $H_2$ and He curves do not show the typical dip at moderate pressures. These gases consist of particles that have such small attractions among them that the influence of their molecular volumes predominates at all pressures.

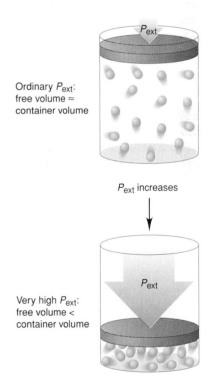

Ordinary $P_{ext}$: free volume ≈ container volume

$P_{ext}$ increases

Very high $P_{ext}$: free volume < container volume

**FIGURE 5.27**
**The effect of molecular volume on measured gas volume.** At ordinary pressures, the volume between molecules (free volume) is essentially equal to the container volume because the molecules occupy only a tiny fraction of the available space. At very high external pressures, the free volume is significantly *less* than the container volume due to the volume of the molecules themselves.

## The van der Waals Equation: The Ideal Gas Law Redesigned

To describe real gas behavior more accurately, we need an equation that adjusts the measured pressure *up* to account for decreases caused by attractive

TABLE 5.6  **Van der Waals Constants for Some Common Gases**

| GAS | $a \left( \dfrac{atm \cdot L^2}{mol^2} \right)$ | $b \left( \dfrac{L}{mol} \right)$ |
|---|---|---|
| Helium (He) | 0.034 | 0.0237 |
| Neon (Ne) | 0.211 | 0.0171 |
| Argon (Ar) | 1.35 | 0.0322 |
| Krypton (Kr) | 2.32 | 0.0398 |
| Xenon (Xe) | 4.19 | 0.0511 |
| Hydrogen ($H_2$) | 0.244 | 0.0266 |
| Nitrogen ($N_2$) | 1.39 | 0.0391 |
| Oxygen ($O_2$) | 1.36 | 0.0318 |
| Chlorine ($Cl_2$) | 6.49 | 0.0562 |
| Carbon dioxide ($CO_2$) | 3.59 | 0.0427 |
| Methane ($CH_4$) | 2.25 | 0.0428 |
| Ammonia ($NH_3$) | 4.17 | 0.0371 |
| Water vapor ($H_2O$) | 5.46 | 0.0305 |

interactions and adjusts the measured volume *down* to include only the free volume, not the entire container volume. In 1873, realizing these limitations of the ideal gas law, Johannes van der Waals proposed an equation that accounts for the behavior of real gases. The **van der Waals equation** for $n$ moles of a real gas is

$$\left( P + \frac{n^2a}{V^2} \right)(V - nb) = nRT \qquad \textbf{(5.14)}$$

$$\underset{\substack{\text{adjusts} \\ P \text{ up}}}{} \quad \underset{\substack{\text{adjusts} \\ V \text{ down}}}{}$$

where $P$ is the measured pressure, $V$ is the container volume, $n$ and $T$ have their usual meanings, and $a$ and $b$ are **van der Waals constants,** experimentally determined positive numbers that depend on the particular gas. Values of these constants for several gases are given in Table 5.6. The constant $a$ is related to the molar mass, or complexity of the molecule, and correlates with the strength of the intermolecular attractions. The constant $b$ is a measure of the actual molecular volume. ◆

◆ **Real Gases Out of This World.** Beyond the asteroid belt, in the realm of the giant planets, gases exist under conditions not to be found on Earth. Hydrogen makes up at least 90% of the atmospheres of these planets, with helium most of the remainder, together with lesser amounts of hydrocarbons ($CH_4$, $C_2H_2$, and $C_2H_6$), the smaller Group 5A(15) hydrides ($NH_3$ and $PH_3$), argon, and several others. Under titanic pressures of more than $10^6$ atm and frigid temperatures as low as 50 K, these atmospheres become compressed into layers, the impurities imparting spectacular colors, that merge into hydrogen oceans. The core of Jupiter may be an Earth-sized sphere of solid $H_2$. Downpours of He and blizzards of $NH_3$ fall on Saturn, and a crystalline haze of $CH_4$ and Ar floats perpetually through the blue-green skies of Uranus and Neptune.

SAMPLE PROBLEM 5.13

**Using the van der Waals Equation**

**Problem:** A 1.98-L vessel contains 215 g of dry ice. After standing at 26°C, the $CO_2(s)$ changes to $CO_2(g)$. The pressure calculated from the ideal gas law ($P_{IGL}$) is 60.6 atm. What is the pressure using the van der Waals equation ($P_{VDW}$)? Compare $P_{IGL}$ and $P_{VDW}$ with the actual pressure ($P_{real}$ = 44.8 atm).
**Plan:** We convert the variables to the appropriate units and substitute into Equation 5.14 to solve for $P_{VDW}$, using the constants from Table 5.6. Then we compare $P_{IGL}$ and $P_{VDW}$ with the given $P_{real}$.
**Solution:** Summary of gas variables:

$$P_{VDW} = \text{unknown} \qquad V = 1.98 \text{ L} \qquad T \text{ (K)} = 26°C + 273 = 299 \text{ K}$$

$$n = 215 \text{ g } CO_2 \times \frac{1 \text{ mol } CO_2}{44.01 \text{ g } CO_2} = 4.89 \text{ mol } CO_2$$

Solving for $P_{VDW}$: The van der Waals constants for $CO_2$ are

$$a = 3.59 \ \frac{atm \cdot L^2}{mol^2} \quad \text{and} \quad b = 0.0427 \ \frac{L}{mol}$$

Rearranging Equation 5.14 to isolate $P$ and substituting values (note canceling here):

$$\left(P + \frac{n^2 a}{V^2}\right)(V - nb) = nRT$$

$P_{VDW}$

$$= \frac{nRT}{V - nb} - \frac{n^2 a}{V^2}$$

$$= \left[\frac{4.89 \text{ mol} \times 0.0821 \dfrac{\text{atm} \cdot \text{L}}{\text{mol} \cdot \text{K}} \times 299 \text{ K}}{1.98 \text{ L} - \left(4.89 \text{ mol} \times 0.0427 \dfrac{\text{L}}{\text{mol}}\right)}\right] - \left[\frac{(4.89 \text{ mol})^2 \times 3.59 \dfrac{\text{atm} \cdot \text{L}^2}{\text{mol}^2}}{(1.98 \text{ L})^2}\right]$$

$$= \mathbf{45.9 \text{ atm}}$$

Comparing the calculated pressures with $P_{real}$: We can compare them by finding the percentage difference,

$$\text{For } P_{IGL}: \quad \frac{60.6 \text{ atm} - 44.8 \text{ atm}}{44.8 \text{ atm}} \times 100 = 35.3\%$$

$$\text{For } P_{VDW}: \quad \frac{45.9 \text{ atm} - 44.8 \text{ atm}}{44.8 \text{ atm}} \times 100 = 2.5\%$$

$P_{IGL}$ **is 35.3% greater than** $P_{real}$**, but** $P_{VDW}$ **is only 2.5% greater than** $P_{real}$**.**
**Check:** To handle such complex arithmetic, break the calculation into parts, such as $nRT$, $n^2 a/V^2$, and $V - nb$, and round within each for a quick check. All units should cancel, except $P$ in atm.

**FOLLOW-UP PROBLEM 5.13**
A slight deviation from ideal behavior exists even at normal conditions. If $Cl_2$ behaved ideally, 1 mol $Cl_2$ would occupy 22.414 L and exert 1 atm pressure at 273.15 K. Use this ideal volume and the values from Table 5.6 to calculate $P_{VDW}$ for 1.000 mol $Cl_2$ at 273.15 K. (Use $R = 0.08206$ atm · L/mol · K.)

## Section Summary
At ordinary conditions (low pressures and high temperatures), many gases behave nearly ideally, but at very high pressures or low temperatures, all gases deviate greatly from ideal behavior. As pressure increases, most real gases exhibit first a lower and then a higher $PV/RT$ ratio than the ideal value. These deviations are due to attractions between molecules, which lower the pressure (and the ratio), and to the larger fraction of the container volume occupied by the molecules, which increases the ratio. By including parameters characteristic of each gas, the van der Waals equation corrects for these deviations.

## Chapter Perspective
*As with the atomic model (Chapter 2), we have seen in this chapter how a simple molecular-scale model can explain macroscopic observations and how it often must be revised to better predict chemical behavior. Gas behavior is relatively easy to understand because gas structure is so randomized—very different from the structures of liquids and solids with their complex molecular interactions, as you'll see in Chapter 11. In Chapter 6, we return to chemical reactions, but from the standpoint of the heat involved in the change. The meaning of kinetic energy, which we discussed in this chapter, bears directly on this central topic.*

# For Review and Reference

## Key Terms

**SECTION 5.2**
pressure (P)
barometer
manometer
pascal (Pa)
standard atmosphere
(atm)
millimeter of mercury
(mmHg)
torr

**SECTION 5.3**
ideal gas
Boyle's law
Charles's law
Avogadro's law
standard temperature and
pressure (STP)
standard molar volume
ideal gas law
universal gas constant (R)

**SECTION 5.4**
partial pressure
Dalton's law of partial
pressures
mole fraction (X)

**SECTION 5.6**
kinetic-molecular theory
rms speed ($u_{rms}$)
effusion
Graham's law of effusion

diffusion
mean free path
collision frequency

**SECTION 5.7**
van der Waals equation
van der Waals constant
atmosphere
troposphere
stratosphere

## Key Equations and Relationships

**5.1** Applying the volume-pressure relationship (Boyle's law) (p. 181):

$$P_1V_1 = P_2V_2 \quad \text{or} \quad V_2 = V_1 \times \frac{P_1}{P_2} \quad [T \text{ and } n \text{ fixed}]$$

**5.2** Applying the volume-temperature relationship (Charles's law) (p. 183):

$$\frac{V_1}{T_1} = \frac{V_2}{T_2} \quad \text{or} \quad V_2 = V_1 \times \frac{T_2}{T_1} \quad [P \text{ and } n \text{ fixed}]$$

**5.3** Applying the pressure-temperature relationship (Amontons's law) (p. 183):

$$\frac{P_1}{T_1} = \frac{P_2}{T_2} \quad \text{or} \quad P_2 = P_1 \times \frac{T_2}{T_1} \quad [V \text{ and } n \text{ fixed}]$$

**5.4** Applying the volume-amount relationship (Avogadro's law) (p. 185):

$$\frac{V_1}{n_1} = \frac{V_2}{n_2} \quad \text{or} \quad V_2 = V_1 \times \frac{n_2}{n_1} \quad [P \text{ and } T \text{ fixed}]$$

**5.5** Defining standard temperature and pressure (p. 186):

STP:   0°C (273.15 K) and 1 atm (760 torr)

**5.6** Defining the volume of one mole of a gas at STP (p. 186):

Standard molar volume = 22.4 L   [3 significant figures]

**5.7** Relating volume to pressure, temperature, and amount (ideal gas law) (p. 187):

$$PV = nRT$$

**5.8** Rearranging the ideal gas law to find gas density (p. 189):

$$\frac{m}{V} = d = \frac{\mathcal{M} \times P}{RT}$$

**5.9** Rearranging the ideal gas law to find molar mass (p. 191):

$$n = \frac{m}{\mathcal{M}} = \frac{PV}{RT} \quad \text{so} \quad \mathcal{M} = \frac{mRT}{PV} \quad \text{or} \quad \mathcal{M} = \frac{dRT}{P}$$

**5.10** Relating the total pressure of a gas mixture to the partial pressures of the components (Dalton's law of partial pressures) (p. 192):

$$P_{total} = P_1 + P_2 + P_3 + \cdots$$

**5.11** Relating partial pressure to mole fraction (p. 193):

$$P_A = X_A \times P_{total}$$

**5.12** Defining rms speed as a function of molar mass and temperature (p. 201):

$$u_{rms} = \sqrt{\frac{3RT}{\mathcal{M}}}$$

**5.13** Applying Graham's laws of effusion (p. 202):

$$\frac{\text{Rate}_A}{\text{Rate}_B} = \frac{\sqrt{\mathcal{M}_B}}{\sqrt{\mathcal{M}_A}}$$

**5.14** Applying the van der Waals equation for a real gas (p. 210):

$$\left(P + \frac{n^2a}{V^2}\right)(V - nb) = nRT$$

## Answers to Follow-up Problems

**5.1** $P_{CO_2}$ (torr) = (753.6 mmHg − 174.0 mmHg)

$$\times \frac{1 \text{ torr}}{1 \text{ mmHg}} = 579.6 \text{ torr}$$

$$P_{CO_2} \text{ (Pa)} = 579.6 \text{ torr} \times \frac{1 \text{ atm}}{760 \text{ torr}}$$

$$\times \frac{1.01325 \times 10^5 \text{ Pa}}{1 \text{ atm}} = 7.727 \times 10^4 \text{ Pa}$$

$$P_{CO_2} \text{ (lb/in}^2) = 579.6 \text{ torr} \times \frac{1 \text{ atm}}{760 \text{ torr}} \times \frac{14.7 \text{ lb/in}^2}{1 \text{ atm}}$$

$$= 11.2 \text{ lb/in}^2$$

**5.2** $P_2$ (atm) = 26.3 kPa $\times \dfrac{1 \text{ atm}}{101.325 \text{ kPa}} = 0.260 \text{ atm}$

$$V_2 \text{ (L)} = 105 \text{ mL} \times \frac{1 \text{ L}}{1000 \text{ mL}} \times \frac{0.871 \text{ atm}}{0.260 \text{ atm}} = 0.352 \text{ L}$$

**5.3** $T_2$ (K) $= 273$ K $\times \dfrac{9.75 \text{ cm}^3}{6.83 \text{ cm}^3} = 390$ K

**5.4** $P_2$ (torr) $= 793$ torr $\times \dfrac{35.0 \text{ g} - 5.0 \text{ g}}{35.0 \text{ g}} = 680$ torr

(There is no need to convert moles to mass because the ratio of masses equals the ratio of moles.)

**5.5** $n = \dfrac{PV}{RT} = \dfrac{1.37 \text{ atm} \times 438 \text{ L}}{0.0821 \dfrac{\text{atm·L}}{\text{mol·K}} \times 294 \text{ K}} = 24.9$ mol $O_2$

Mass (g) of $O_2 = 24.9$ mol $O_2 \times \dfrac{32.00 \text{ g } O_2}{1 \text{ mol } O_2}$

$= 7.97 \times 10^2$ g $O_2$

**5.6** $d = \dfrac{44.01 \text{ g/mol} \times \dfrac{380 \text{ torr}}{760 \text{ torr/atm}}}{0.0821 \dfrac{\text{atm·L}}{\text{mol·K}} \times 273 \text{ K}} = 0.982$ g/L

The density is lower at the smaller $P$ because $V$ is larger.

**5.7** $\mathcal{M} = \dfrac{1.26 \text{ g} \times 0.0821 \dfrac{\text{atm·L}}{\text{mol·K}} \times 283.2 \text{ K}}{\dfrac{102.5 \text{ kPa}}{101.325 \text{ kPa/1 atm}} \times 1.00 \text{ L}} = 29.0$ g/mol

**5.8** $n_{\text{total}} = \left( 5.50 \text{ g He} \times \dfrac{1 \text{ mol He}}{4.003 \text{ g He}} \right)$

$+ \left( 15.0 \text{ g Ne} \times \dfrac{1 \text{ mol Ne}}{20.18 \text{ g Ne}} \right)$

$+ \left( 35.0 \text{ g Kr} \times \dfrac{1 \text{ mol Kr}}{83.80 \text{ g Kr}} \right)$

$= 2.53$ mol

At STP, $V = 2.53$ mol $\times \dfrac{22.4 \text{ L}}{1 \text{ mol}} = 56.7$ L

$P_{\text{He}} = \left( \dfrac{5.50 \text{ g He} \times \dfrac{1 \text{ mol He}}{4.003 \text{ g He}}}{2.53 \text{ mol}} \right) \times 1 \text{ atm} = 0.543$ atm

$P_{\text{Ne}} = 0.294$ atm; $P_{\text{Kr}} = 0.165$ atm

**5.9** $P_{H_2} = 752$ torr $- 13.6$ torr $= 738$ torr

$n_{H_2} = \left( \dfrac{\dfrac{738 \text{ torr}}{760 \text{ torr/atm}} \times 1.495 \text{ L}}{0.0821 \dfrac{\text{atm·L}}{\text{mol·K}} \times 289 \text{ K}} \right) \times \dfrac{2.016 \text{ g } H_2}{1 \text{ mol } H_2}$

$= 0.123$ g $H_2$

**5.10** $H_2SO_4(aq) + 2NaCl(s) \rightarrow Na_2SO_4(aq) + 2HCl(g)$

$n_{\text{HCl}} = 0.117 \text{ kg NaCl} \times \dfrac{10^3 \text{ g}}{1 \text{ kg}} \times \dfrac{1 \text{ mol NaCl}}{58.44 \text{ g NaCl}}$

$\times \dfrac{2 \text{ mol HCl}}{2 \text{ mol NaCl}} = 2.00$ mol HCl

At STP, $V$(mL) $= 2.00$ mol $\times \dfrac{22.4 \text{ L}}{1 \text{ mol}} \times \dfrac{10^3 \text{ mL}}{1 \text{ L}}$

$= 4.48 \times 10^4$ mL

**5.11** $NH_3(g) + HCl(g) \rightarrow NH_4Cl(s)$

$n_{NH_3} = 0.187$ mol; $n_{\text{HCl}} = 0.0522$ mol

$n_{NH_3}$ after reaction $=$

$0.187 \text{ mol } NH_3 - \left( 0.0522 \text{ mol HCl} \times \dfrac{1 \text{ mol } NH_3}{1 \text{ mol HCl}} \right)$

$= 0.135$ mol $NH_3$

$P = \dfrac{0.135 \text{ mol} \times 0.0821 \dfrac{\text{atm·L}}{\text{mol·K}} \times 295 \text{ K}}{10.0 \text{ L}} = 0.327$ atm

**5.12** $\dfrac{\text{Rate of He}}{\text{Rate of } C_2H_6} = \sqrt{\dfrac{30.07 \text{ g/mol}}{4.003 \text{ g/mol}}} = 2.741$

Time for $C_2H_6$ to effuse $= 1.25$ min $\times 2.741$

$= 3.43$ min

**5.13** $P = \left( \dfrac{1.000 \text{ mol} \times 0.08206 \dfrac{\text{atm·L}}{\text{mol·K}} \times 273.15 \text{ K}}{22.414 \text{ L} - (1.000 \text{ mol} \times 0.0562 \text{ L/1 mol})} \right)$

$- \left( \dfrac{(1.000 \text{ mol})^2 \times 6.49 \dfrac{\text{atm·L}^2}{\text{mol}^2}}{(22.414 \text{ L})^2} \right)$

$= 0.990$ atm

**Sample Problem Titles**

## Problems

Problems with a green number are answered at the back of the text. Most sections include three categories of problems separated by a green rule: concept review questions, *paired* skill-building exercises, and problems in a relevant context.

### An Overview of the Physical States of Matter

**5.1** How would a sample of gas differ in its behavior from a sample of liquid in each of the following situations?
(a) The sample is transferred from one container to a larger one.
(b) The sample is heated in an expandable container, but no change of state occurs.
(c) The sample is placed in a cylinder with a piston, and an external force is applied.

**5.2** When a substance changes from a gas to a liquid, the distances between particles decrease greatly. Use this concept to explain each of the following general observations:
(a) Gases are more compressible than liquids.
(b) Gases have lower viscosities than liquids.
(c) After thorough stirring, all gas mixtures are solutions.
(d) The density of the gas is lower than that of the liquid.

### Measuring the Pressure of a Gas

(Sample Problem 5.1)
**5.3** How does a barometer work? Is the column of mercury in a barometer shorter on a mountaintop or at sea level? Explain.
**5.4** How can a length unit such as millimeter of mercury (mmHg) be used as a unit of pressure, which has the dimensions of force per unit area?
**5.5** In a closed-end manometer, the mercury level in the arm attached to the flask can never be higher than that in the other arm, whereas in an open-end manometer, it can be higher. Explain.

---

**5.6** On a cool, rainy day, the barometric pressure was 728 mmHg. Calculate the barometric pressure in centimeters of water (cmH$_2$O). (*d* of Hg = 13.5 g/mL; *d* of H$_2$O = 1.00 g/mL.)
**5.7** A long glass tube, sealed at one end, has an inner diameter of 10.0 mm. The tube is filled with water and inverted into a pail of water. If the atmospheric pressure is 745 mmHg, how high (in mmH$_2$O) is the column of water in the tube? (*d* of Hg = 13.5 g/mL; *d* of H$_2$O = 1.00 g/mL.)
**5.8** Convert the following:
(a) 0.735 atm to mmHg  (b) 892 torr to atm
(c) 305 kPa to atm     (d) 904 mmHg to kPa
**5.9** Convert the following:
(a) 64.8 cmHg to atm   (b) 17.0 atm to kPa
(c) 3.50 atm to torr   (d) 0.967 kPa to mmHg

**5.10** If the barometer reads 748.5 torr, what is the pressure of the gas in the flask in atmospheres (Figure P5.10)?
**5.11** If the barometer reads 768.2 mmHg, what is the pressure of the gas in the flask in kilopascals (Figure P5.11)?

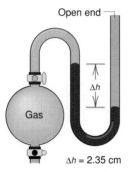

**FIGURE P5.10**

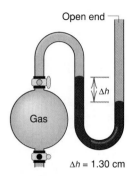

**FIGURE P5.11**

**5.12** If the sample flask is open to the air, what is the atmospheric pressure in atmospheres (Figure P5.12)?
**5.13** What is the pressure of the gas in the flask in pascals (Figure P5.13)?

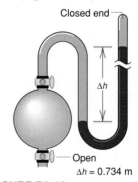

**FIGURE P5.12**

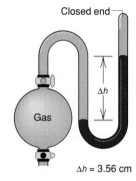

**FIGURE P5.13**

---

**5.14** An "empty" gasoline can with dimensions 20.0 cm by 40.0 cm by 12.5 cm is attached to a vacuum pump and evacuated. If the atmospheric pressure is 14.7 lb/in$^2$, what is the total force (in pounds) on the outside of the can?
**5.15** Convert the following pressures into atmospheres:
(a) At the peak of Mt. Everest, the pressure is only 2.7 × 10$^2$ mmHg.
(b) A cyclist fills her bike tires to 92 psi.
(c) The surface of Venus has an atmospheric pressure of 9.15 × 10$^6$ Pa.
(d) At 100 ft below sea level, a scuba diver experiences a pressure of 2.44 × 10$^4$ torr.

**5.16** The gravitational force exerted by an object is given by $F = mg$, where $F$ is the force in newtons, $m$ is the mass in kilograms, and $g$ is the acceleration due to gravity (9.81 m/s²).
(a) Use the definition of the pascal to calculate the mass (in kg) of the atmosphere on 1 m² of ocean.
(b) Osmium ($Z = 76$) has the highest density of any element (22.6 g/mL). If an osmium column is 1 m² in area, how high must it be for its pressure to equal atmospheric pressure? [Use the answer from part (a) in your calculation.]

**The Gas Laws and Their Experimental Foundations**

(Sample Problems 5.2 to 5.5)

**5.17** When asked to state Boyle's law, a student replies, "The volume of a gas is inversely proportional to the pressure." How is this statement incomplete? Give a correct statement of Boyle's law.

**5.18** Which quantities are variables and which are fixed in each of the following: (a) Charles's law; (b) Avogadro's law; (c) Amontons's law?

**5.19** Boyle's law relates gas volume to pressure, and Avogadro's law relates volume to number of moles. State a relationship between gas pressure and number of moles.

**5.20** What is the effect on the volume of one mole of an ideal gas when each of the following occurs?
(a) The pressure is tripled (at constant $T$).
(b) The absolute temperature is increased by a factor of 2.5 (at constant $P$).
(c) Two more moles of the gas are added (at constant $P$ and $T$).
(d) The pressure is reduced by a factor of four (at constant $T$).

**5.21** What is the effect on the pressure of one mole of an ideal gas when each of the following occurs?
(a) The temperature is decreased from 700 K to 350 K (at constant $V$).
(b) The temperature is increased from 350°C to 700°C (at constant $V$).
(c) The available volume is increased from 2 L to 8 L (at constant $T$).
(d) Half the gas escapes through a stopcock (at constant $V$ and $T$).

**5.22** A sample of sulfur hexafluoride gas occupies a volume of 5.00 L at 198°C. Assuming the pressure is constant, what will be the final temperature (in °C) if the volume is decreased to 2.50 L?

**5.23** A 93-L sample of dry air is cooled from 155°C to −32°C while the pressure is maintained at 1.85 atm. What is the final volume?

**5.24** A sample of Freon-12 ($CF_2Cl_2$) occupies 24.5 L at 298 K and 253.3 kPa. Find its volume at STP.

**5.25** Calculate the volume of carbon monoxide at −14°C and 367 torr if it occupies 3.65 L at 25°C and 745 torr.

**5.26** A sample of nitrogen gas is confined in a 6.0-L container at 228 torr and 27°C. How many moles of gas are present in the sample?

**5.27** If $1.47 \times 10^{-3}$ mol of helium occupies a 75.0-mL container at 26°C, what is the pressure (in torr)?

**5.28** You have 207 mL oxygen gas at 699 mmHg and 25°C. What is the mass (in g) of the sample?

**5.29** A 75.0-g sample of argon is confined in a 3.1-L vessel. What is the pressure (in atm) at 115°C?

**5.30** In preparation for a lecture demonstration, your professor brings a 1.5-L bottle of $SO_2$ into the lecture hall before class to allow the gas to reach room temperature. If the pressure gauge reads 85 psi and the lecture hall is 23°C, how many moles of $SO_2$ are in the bottle? (*Hint:* The gauge reads zero when 14.7 psi of gas remains.)

**5.31** Hemoglobin is the protein that transports $O_2$ through the blood from the lungs to the rest of the body. In doing so, each molecule of hemoglobin combines with four molecules of $O_2$. If 1.00 g hemoglobin combines with 1.53 mL $O_2$ at 37°C and 743 torr, what is the molar mass of hemoglobin?

**5.32** A gas-filled weather balloon with a volume of 55.0 L is released at sea-level conditions of 755 torr and 23°C. The balloon can expand to a maximum volume of 835 L. When the balloon rises to an altitude at which the temperature is −5°C and the pressure is 0.066 atm, will it reach its maximum volume?

**Further Applications of the Ideal Gas Law**

(Sample Problems 5.6 to 5.9)

**5.33** Why does moist air have a lower density than dry air?

**5.34** In order to collect a beaker of $H_2$ gas by displacing the air already in the beaker, would you hold the beaker upright or inverted? Why? How would you hold the beaker to collect $CO_2$?

**5.35** The vapors from gasoline (represented by $C_8H_{18}$) spilled at a service station can trigger a fire even if an open flame is several feet from the spill. How does an understanding of gas density help explain this behavior?

**5.36** Why can we use a gas mixture, such as air, to study the general behavior of an ideal gas under ordinary conditions?

**5.37** How does the partial pressure of gas A in a mixture compare to its mole fraction in the mixture?

**5.38** In the collection of a gas by displacement of water (see Figure 5.16), a step in the procedure states: "Raise or lower the collection tube so that the water level inside the tube equals the water level outside." What does this step accomplish, and how is it used to determine the pressure of the gas?

**5.39** What is the density of Xe gas at STP?

**5.40** What is the density of Freon-11 ($CFCl_3$) at 120°C and 1.5 atm?

**5.41** How many moles of gaseous phosphine ($PH_3$) occupy 0.0400 L at STP? What is its density?

**5.42** The density of a noble gas is 2.71 g/L at 3.00 atm and 0°C. Identify the gas.

**5.43** Calculate the molar mass of a gas at 388 torr and 45°C if 206 $\mu g$ occupies 0.206 mL.

**5.44** When an evacuated 63.8-mL glass bulb is filled with a gas at 22°C and 747 mm Hg, the bulb gains 0.103 g in mass. Is the gas $N_2$, Ne, or Ar?

**5.45** When 0.500 L $N_2$ at 1.20 atm and 227°C is mixed with 0.200 L $O_2$ at 501 torr and 127°C in a 400-mL flask at 27°C, what is the pressure in the flask?

**5.46** A 355-mL container holds 0.146 g $N_2$ and an unknown amount of Ar at 35°C and a total pressure of 626 mm Hg. Calculate the moles of Ar present.

**5.47** A sample of sodium azide ($NaN_3$), a compound used in automobile "air bags," was thermally decomposed, and 15.3 mL nitrogen gas was collected over water at 25°C and 755 torr. How many grams of nitrogen were collected?

**5.48** A sample of nickel carbonyl, $Ni(CO)_4$, was decomposed at high temperature. After the released carbon monoxide was cooled to 35°C, 0.142 L was collected by displacement of water. Barometric pressure was 769 torr. How many grams of CO were collected?

**5.49** The air in a hot-air balloon at 754 torr is heated from 27°C to 60.0°C. Assuming that the moles of air and the pressure remain constant, what is the density of the air at each temperature? (The average molar mass of air is 28.8 g/mol.)

**5.50** A sample of a liquid hydrocarbon fuel known to consist of five-carbon molecules is vaporized in a 0.204-L flask by immersion in a water bath at 101°C. The barometric pressure is 767 torr, and the remaining gas condenses to 0.482 g of liquid. What is the molecular formula of the hydrocarbon?

**5.51** A sample of air contains 78.08% nitrogen, 20.94% oxygen, 0.05% carbon dioxide, and 0.93% argon, by volume. How many molecules of each gas are present in 1.00 L of the sample at 25°C and 1.00 atm?

**5.52** An environmental chemist sampling industrial exhaust gases from a coal-burning plant collects a $CO_2$-$SO_2$-$H_2O$ mixture in a 21-L steel tank until the pressure reaches 850 torr at 35°C.
(a) How many moles of gas were collected?
(b) If the $SO_2$ concentration in the mixture is 7950 parts per million by volume (ppmv), what is its partial pressure? [*Hint:* ppmv = (volume of pure gas / total volume of gas mixture) × $10^6$.]

## The Ideal Gas Law and Reaction Stoichiometry

(Sample Problems 5.10 and 5.11)

**5.53** How many grams of phosphorus react with 41.5 L $O_2$ at STP to form tetraphosphorus decaoxide?

$$P_4(s) + 5O_2(g) \rightarrow P_4O_{10}(s)$$

**5.54** How many grams of potassium chlorate decompose to form potassium chloride and 638 mL $O_2$ at 18°C and 752 torr?

$$2KClO_3(s) \rightarrow 2KCl(s) + 3O_2(g)$$

**5.55** How many grams of phosphine ($PH_3$) can form by the reaction of 37.5 g phosphorus and 83.0 L hydrogen gas at STP?

$$P_4(s) + H_2(g) \rightarrow PH_3(g) \quad \text{[unbalanced]}$$

**5.56** When 35.6 L ammonia and 40.5 L oxygen gas at STP burn, nitrogen monoxide and water are produced. After the products return to STP, how many grams of nitrogen monoxide are present?

$$NH_3(g) + O_2(g) \rightarrow NO(g) + H_2O(l) \quad \text{[unbalanced]}$$

**5.57** Aluminum reacted with excess hydrochloric acid to form aqueous aluminum chloride and 35.8 mL hydrogen gas. How many grams of aluminum were collected over water at 27°C and 751 mmHg?

**5.58** How many liters of hydrogen gas are collected over water at 24°C and 625 mmHg when 0.054 g lithium reacts with water? Aqueous lithium hydroxide also forms.

**5.59** "Strike anywhere" matches contain the compound tetraphosphorus trisulfide, which burns to form tetraphosphorus decaoxide and sulfur dioxide gas. How many milliliters of sulfur dioxide, measured at 725 torr and 32°C, can be produced from the burning of 0.200 g tetraphosphorus trisulfide?

**5.60** Freon-12 ($CF_2Cl_2$), a refrigerant and aerosol propellant that is now known as a pollutant in the stratosphere, is prepared by reaction of gaseous carbon tetrachloride with hydrogen fluoride. Hydrogen chloride gas also forms. How many grams of carbon tetrachloride are required for the production of 16.0 L Freon-12 at 27°C and 1.20 atm?

**5.61** Ammonium nitrate is often used as an explosive in fireworks. How many liters of gas at 300°C and 1.00 atm would be formed by the explosive decomposition of 12.0 g ammonium nitrate to nitrogen, oxygen, and water vapor?

**5.62** Xenon hexafluoride was the first noble gas compound synthesized. The solid reacts rapidly with the silicon dioxide in glass or quartz containers to form liquid $XeOF_4$ and gaseous silicon tetrafluoride. What is the pressure in a 1.00-L container at 25°C after 2.00 g xenon hexafluoride react? (Assume that silicon tetrafluoride is the only gas present and that it occupies the entire volume.)

**5.63** Roasting galena [lead(II) sulfide] is an early step in the industrial isolation of lead. How many liters of sulfur dioxide, measured at STP, are produced by the reaction of 3.75 kg galena with 228 L oxygen gas at 0°C and 2.0 atm. Lead(II) oxide also forms.

**5.64** In one of his most critical studies into the nature of combustion, Lavoisier heated mercury(II) oxide and isolated elemental mercury and oxygen gas. If 40.0 g mercury(II) oxide are heated in a 502-mL vessel and 20.0% (by mass) decomposes, how many atmospheres of oxygen gas form at 25.0°C? (Assume the gas occupies the entire volume.)

**The Kinetic-Molecular Theory: A Model for Gas Behavior**

(Sample Problem 5.12)

**5.65** Use the kinetic-molecular theory to explain the change in gas pressure that results from warming a sample.

**5.66** How does the kinetic-molecular theory explain why a mole of krypton and a mole of helium have the same volume at STP?

**5.67** Is the rate of effusion of a gas higher, lower, or equal to its rate of diffusion? Explain. Is the ratio of effusion rates for two gases higher, lower, or equal to their ratio of diffusion rates? Explain.

**5.68** Consider two 1-L samples of gas, one $H_2$ and one $O_2$, both at 1 atm and 25°C. How do the samples compare in terms of (a) mass, (b) density, (c) mean free path, (d) average molecular kinetic energy, (e) average molecular speed, and (f) time for a given fraction of molecules to effuse?

**5.69** Three 5-L flasks, fixed with pressure gauges and small valves, each contain 4 g of gas at 273 K. Flask A contains $H_2$, flask B contains He, and flask C contains $CH_4$. Rank the flask contents in terms of (a) pressure, (b) average molecular kinetic energy, (c) diffusion rate after the valve is opened, (d) total kinetic energy of the molecules, (e) density, and (f) collision frequency.

**5.70** What is the ratio of effusion rates for the lightest gas, $H_2$, to the heaviest known gas, $UF_6$?

**5.71** What is the ratio of effusion rates for $O_2$ to Kr?

**5.72** At a particular pressure and temperature, it takes 4.55 min for a 1.5-L sample of He to effuse through a porous membrane. How long would it take for 1.5 L $F_2$ to effuse under the same conditions?

**5.73** A sample of an unknown gas effuses in 11.1 min. An equal volume of $H_2$ in the same apparatus at the same temperature and pressure effuses in 2.42 min. What is the molar mass of the unknown gas?

**5.74** Solid white phosphorus melts and then vaporizes at high temperature. Gaseous phosphorus effuses at a rate that is 0.404 times that of neon in the same apparatus under the same conditions. How many atoms are in a molecule of gaseous phosphorus?

**5.75** Aluminum chloride is easily vaporized above 180°C. The gas escapes through a pinhole 0.122 times as fast as helium at the same conditions of temperature and pressure in the same apparatus. What is the molecular formula of gaseous aluminum chloride?

**5.76** Helium is the lightest noble gas in air, and xenon is the heaviest.
(a) Calculate the rms speed of helium in winter (0°C) and in summer (30°C).
(b) Compare the rms speed of helium with that of xenon at 30°C.
(c) Calculate the average kinetic energy per mole of helium and of xenon at 30°C.
(d) Calculate the average kinetic energy per molecule of helium at 30°C.

**Real Gases: Deviations from Ideal Behavior**

(Sample Problem 5.13)

**5.77** Do intermolecular attractions cause negative or positive deviations from the $PV/RT$ ratio of an ideal gas? Use data from Table 5.6 to rank Kr, $CO_2$, and $N_2$ in order of increasing magnitude of these deviations.

**5.78** Does molecular size cause negative or positive deviations from the $PV/RT$ ratio of an ideal gas? Use data from Table 5.6 to rank $Cl_2$, $H_2$, and $O_2$ in order of increasing magnitude of these deviations.

**5.79** Does $N_2$ behave more ideally at 1 atm or 500 atm? Explain.

**5.80** Does $SF_6$ (boiling point 16°C at 1 atm) behave more ideally at 150°C or at 20°C? Explain.

**5.81** Chlorine is produced from concentrated sea water by the electrochemical chlor-alkali process. During the process, the chlorine is collected in a container that is isolated from the other products to prevent unwanted (and explosive) reactions. If a 15.0-L container holds 0.580 kg $Cl_2$ gas at 200°C, calculate
(a) The pressure, assuming ideal behavior
(b) The pressure, using the van der Waals equation

**5.82** Hydrogen can be produced industrially by the chlor-alkali process. When 1 mol $H_2(g)$ at 0°C is compressed until its volume is 0.4483 L, the pressure is observed to be 51.60 atm.
(a) If the hydrogen were behaving ideally, what would its calculated pressure be?
(b) What would its pressure be if it were behaving as a real gas?
(c) Compare the percentage errors in the pressures predicted by the two approaches.

**Comprehensive Problems**

Problems with an asterisk (*) are more challenging.

**5.83** Will the volume of a gas increase, decrease, or remain unchanged for each of the following sets of changes?
(a) The pressure is decreased from 2 atm to 1 atm, while the temperature is decreased from 200°C to 100°C.

(b) The pressure is increased from 1 atm to 3 atm, while the temperature is increased from 100°C to 300°C.

(c) The pressure is increased from 3 atm to 6 atm, while the temperature is increased from −73°C to 127°C.

(d) The pressure is increased from 0.2 atm to 0.4 atm, while the temperature is decreased from 300°C to 150°C.

**5.84** Nitrogen dioxide is an important gas in industry as well as a component of smog. Calculate the volume of nitrogen dioxide produced at 735 torr and 28.2°C by the reaction of 4.95 cm$^3$ copper ($d = 8.95$ g/cm$^3$) with 230.0 mL nitric acid ($d = 1.42$ g/cm$^3$, 68.0% $HNO_3$ by mass).

$$Cu(s) + 4HNO_3(aq) \rightarrow Cu(NO_3)_2(aq) + 2NO_2(g) + 2H_2O(l)$$

**5.85** In a bromine-producing plant, how many liters of gaseous elemental bromine at 300°C and 0.855 atm would be formed by the reaction of 175 g sodium bromide and 75.6 g sodium bromate in aqueous acid solution? (Assume no $Br_2$ dissolves.)

$$5NaBr(aq) + NaBrO_3(aq) + 3H_2SO_4(aq) \rightarrow$$
$$3Br_2(g) + 3Na_2SO_4(aq) + 3H_2O(g)$$

**5.86** An anesthetic gas contains 64.81% carbon, 13.60% hydrogen, and 21.59% oxygen, by mass. If 2.00 L of the gas at 25°C and 0.420 atm pressure weigh 2.57 g, what is the molecular formula of the anesthetic?

**\*5.87** (a) What is the total volume of gaseous *products*, measured at 350°C and 735 torr, when an automobile engine burns exactly 100 g $C_8H_{18}$ (a typical component of gasoline)?

(b) For part (a), the source of $O_2$ is air, which is about 80% $N_2$ and 20% $O_2$ by volume. Assuming all the $O_2$ reacts, but none of the $N_2$ does, what is the total volume of gaseous *exhaust*?

**\*5.88** An atmospheric chemist studying the reactions of the pollutant $SO_2$ places a mixture of $SO_2$ and $O_2$ in a 2.00-L container at 900 K and an initial pressure of 1.95 atm. When the reaction occurs, gaseous $SO_3$ forms, and the pressure eventually falls to 1.65 atm. How many moles of $SO_3$ form?

**5.89** A sample of liquid nitrogen trichloride was heated in a 1.50-L closed container until it decomposed completely to gaseous elements. The resulting mixture exerted a pressure of 744 mmHg at 75°C.

(a) What is the partial pressure of each gas in the container?

(b) What was the mass of the original sample?

**5.90** Analysis of a newly discovered gaseous silicon-fluorine compound shows that it contains 33.01 mass % silicon. At 27°C, 2.60 g of the compound exerts a pressure of 1.50 atm in a 0.250-L vessel. What is the molecular formula of the compound?

**5.91** A gaseous organic compound containing only carbon, hydrogen, and nitrogen is burned in oxygen gas, and the individual volume of each reactant and product is measured under the same conditions of temperature and pressure. Reaction of 4 volumes of the compound produces 4 volumes of $CO_2$, 2 volumes of $N_2$, and 10 volumes of water vapor.

(a) What volume of oxygen gas was required?

(b) What is the empirical formula of the compound?

**5.92** A piece of dry ice (solid $CO_2$, $d = 0.900$ g/mL) weighing 10.0 g is placed in a 0.800-L bottle filled with air at 0.980 atm and 550.0°C. The bottle is capped, and the dry ice changes to gas. What is the final pressure inside the bottle?

**\*5.93** By what factor would a scuba diver's lungs expand if he ascended rapidly to the surface without inhaling or exhaling from a depth of 125 ft? If an expansion factor greater than 1.5 causes lung rupture, how far could the diver safely ascend from 125 ft without breathing? Assume constant temperature. ($d$ of sea water = 1.04 g/mL; $d$ of Hg = 13.5 g/mL.)

**5.94** When 15.0 g of the mineral fluorite ($CaF_2$) react with excess sulfuric acid, hydrogen fluoride gas is collected at 744 torr and 25.5°C. Solid calcium sulfate is the other product. To what must the temperature of the gas be increased or decreased if it is to be stored in a 8.63-L container at 875 torr?

**5.95** At a height of 300 km above the Earth's surface, an astronaut finds that the atmospheric pressure is about $10^{-8}$ mmHg and the temperature 500 K. How many molecules are there per milliliter at this altitude?

**5.96** (a) What is the rms speed of $O_2$ at STP?

(b) If the mean free path of $O_2$ at STP is $6.33 \times 10^{-8}$ m, what is its collision frequency?

**5.97** A barometer tube is $1.00 \times 10^2$ cm long and has a cross-sectional area of 1.20 cm$^2$. The height of the mercury column is 74.0 cm, and the temperature is 24°C. A small amount of $N_2$ is introduced into the evacuated space above the mercury, which causes the mercury level to drop to a height of 66.0 cm. How many grams of $N_2$ were introduced?

**5.98** What is the molar concentration of the cleaning solution formed when 10.0 L ammonia gas, measured at 31°C and 735 torr, dissolves in enough water to give a final volume of 0.750 L?

**5.99** How many liters of gaseous hydrogen bromide at 27°C and 0.975 atm will a chemist need if she wishes to prepare 3.50 L of 1.20 $M$ hydrobromic acid?

**5.100** A mixture consisting of 7.0 g CO and 10.0 g $SO_2$, two atmospheric pollutants, has a pressure of 0.33 atm when placed in a sealed container. What is the partial pressure of the CO?

**\*5.101** A mixture of $CO_2$ and Kr weighs 35.0 g and exerts a pressure of 0.708 atm in its container. Since Kr is expensive, you wish to recover it from the mixture. After the $CO_2$ is completely removed by absorption with NaOH(s), the pressure in the container is 0.250 atm. How many grams of $CO_2$ were originally present? How many grams of Kr can you recover?

**5.102** When 40.3 mL of 0.200 $M$ calcium bicarbonate is treated with excess hydrochloric acid, how many milliliters of carbon dioxide will be formed at 30°C and 506 torr?

*5.103 Aqueous sulfurous acid ($H_2SO_3$) was made by dissolving 0.200 L sulfur dioxide gas at 20°C and 740 mmHg in water to yield 500.0 mL solution. The acid solution required 10.0 mL sodium hydroxide solution to reach the titration end point. What is the molarity of the sodium hydroxide solution?

5.104 A person inhales air richer in $O_2$ and exhales air richer in $CO_2$ and water vapor. During each hour of sleep, a person exhales a total of about 300 L of this $CO_2$-enriched and $H_2O$-enriched air.

(a) If the partial pressures of $CO_2$ and $H_2O$ in exhaled air are each 30.0 torr at 37.0°C, calculate the mass of $CO_2$ and of $H_2O$ exhaled in 1 hour of sleep.

(b) How many grams of body mass would the person lose in an 8-hour sleep if all the $CO_2$ and $H_2O$ exhaled came from the metabolism of glucose?

$$C_6H_{12}O_6(s) + 6O_2(g) \rightarrow 6CO_2(g) + 6H_2O(g)$$

*5.105 Given the following relationships for average kinetic energy,

$$\overline{E}_k = \tfrac{1}{2}m\overline{u^2} \quad \text{and} \quad \overline{E}_k = \frac{3}{2}\left(\frac{R}{N_A}\right)T$$

where $m$ is molecular mass, $u$ is rms speed, $R$ is the gas constant (in J/mol · K), $N_A$ is Avogadro's number, and $T$ is absolute temperature,

(a) Derive Equation 5.12.

(b) Derive Equation 5.13.

5.106 What would you observe if you tilted the barometer shown in Figure 5.3 to 30° from the vertical? Explain.

5.107 A 6.0-L flask contains a mixture of methane ($CH_4$), argon, and helium at 45°C and 1.75 atm. If the mole fractions of helium and argon are 0.25 and 0.35, respectively, how many molecules of methane are present?

*5.108 A large portion of metabolic energy arises through the biological combustion of glucose:

$$C_6H_{12}O_6(s) + 6O_2(g) \rightarrow 6CO_2(g) + 6H_2O(g)$$

(a) If this reaction were carried out in an expandable container at 35°C and 780 torr, what volume of $CO_2$ would be produced from 18.0 g glucose and excess $O_2$?

(b) If the reaction were carried out at the same conditions with the stoichiometric amount of $O_2$, what would be the partial pressure of each gas when the reaction is 50% complete (9.0 g glucose had not yet reacted)? (*Hint:* Use Table 5.4 to obtain the vapor pressure of $H_2O$ at 35°C.)

5.109 What is the average kinetic energy and rms speed of $N_2$ at STP? Compare these values with those of $H_2$ under the same conditions.

*5.110 An equimolar mixture of Ne and Xe is accidentally placed in a container that has a tiny leak. After a short while, a very small proportion of the mixture has escaped. What will be the mole fraction of Ne in the effusing gas?

*5.111 One way to prepare naturally occurring uranium (0.72% $^{235}U$ and 99.27% $^{238}U$) for use as a nuclear fuel is to enrich it (increase its $^{235}U$ content) by allowing gaseous $UF_6$ to effuse through a porous membrane (see Margin Note, p. 202). From the relative rates of effusion of $^{235}UF_6$ and $^{238}UF_6$, find the number of steps needed to produce uranium that is 3.0 mole % $^{235}U$, the enriched fuel used in many nuclear reactions.

# CHAPTER 6

## Concepts and skills to review

- energy and its interconversion (Section 1.1)
- distinction between heat and temperature (Section 1.5)
- nature of chemical bonding (Section 2.6)
- calculations of reaction stoichiometry (Section 3.4)
- properties of the gaseous state (Section 5.1)
- relation between kinetic energy and temperature (Section 5.6)

# Thermo-chemistry

# Energy Flow and Chemical Change

**Energy out and energy in.** Methane burning releases heat, and water boiling absorbs it. In this chapter, we look closely at the transfer of energy that accompanies all chemical and physical change and discover an extremely useful fact: the amount of energy involved is proportional to the amount of matter changing.

**A**ll changes in matter are accompanied by changes in its energy content. In the inferno of a forest fire, the energy content of the burning wood decreases due to changes in its composition, and much of that energy is released as heat and light. Some of the energy from a flash of lightning is absorbed when atmospheric nitrogen and oxygen react to form nitrogen monoxide. Changes in physical state also result in energy changes. Energy is absorbed when snow melts and is released when water vapor condenses to rain. The production and utilization of energy in its many forms have an enormous impact on society. Some of the largest chemical and related industries manufacture products that release, absorb, or retain energy. Fertilizers help growing crops absorb solar energy and convert it to the chemical energy of food. Common fuels—petroleum, wood, coal, and natural gas—release chemical energy either to provide heat or to power combustion engines and steam turbines. Many plastic, fiberglass, and ceramic materials serve as insulators to prevent the flow of heat. Batteries and similar devices produce electrical energy from chemical reactions.

In this chapter, we investigate the heat associated with changes in matter. After an introduction to **thermodynamics,** the study of heat (thermal energy) and its transformations, we discuss **thermochemistry,** the branch of thermodynamics that examines the heat involved in chemical reactions. We explore these changes at the molecular level to find out where the heat comes from. Then we discuss how heat is measured in order to focus on the central topic of the chapter: the amount of heat released or absorbed in a reaction. The chapter ends with an overview of current and future energy sources and the conflicts between energy demand and the quality of the environment.

## 6.1  Forms of Energy and Their Interconversion

The central idea that all energy is either potential or kinetic was presented in Chapter 1. An object has potential energy by virtue of its position and kinetic energy by virtue of its motion. The potential energy of a weight raised above the ground is converted to kinetic energy as it falls (see Figure 1.3). When the weight hits the ground, it transfers some kinetic energy to the soil and pebbles, causing some particles to move, thereby doing *work.* In addition, some of the transferred kinetic energy appears as *heat,* which warms the soil and pebbles slightly. Thus, the energy of the weight is transferred to the ground as work and as heat.

Modern atomic theory allows us to consider other forms of energy—electrical, solar, nuclear, and chemical—as examples of potential and kinetic energy on the atomic and molecular scales. The potential energy stored in the chemical bonds of foods and fuels is converted into the kinetic energy

**FIGURE 6.1**

**A chemical system and its surroundings.** Once the contents of the flask are defined as the system (red), the flask and the laboratory become defined as the surroundings.

♦ **Wherever You Look There Is a System.** In the earlier example in the text, if we define the falling weight as the system, the soil and pebbles that are moved would be the surroundings. An astronomer may define a galaxy as the system and nearby galaxies as the surroundings. An ecologist studying African wildlife could define a zebra herd as the system and other animals, plants, and water supplies as the surroundings. Thus, in general, it is the experiment and the experimenter that define the system and the surroundings.

that enables an organism or a machine to do work and generate heat, just as a falling weight does. No matter what the details of the situation, *whenever energy is transferred from one object to another, it appears as work and/or as heat.* In this section, we examine this idea in terms of the loss or gain of energy that takes place during a chemical or physical change.

Thermodynamics is a fascinating field, a rigorously logical, highly mathematical branch of science that is relevant throughout the natural world. Our discussion of thermodynamics in this chapter, and later in Chapter 19, is confined mostly to chemical applications, but both discussions attempt to convey the breadth of its ideas.

### Energy Flow to and from a System

In order to study and measure a change in energy, we must first define the **system,** that part of the universe whose change we are going to measure; everything else that is relevant to the change is defined as the **surroundings.** A typical chemical system would be the *contents* of a flask, the substances participating in the change. The flask itself, a Bunsen burner, and perhaps the rest of the laboratory, would be the surroundings (Figure 6.1). In principle, the rest of the universe is the surroundings, but in practice, only the portions of the universe relevant to the system need be considered. For example, it's not likely that a thunderstorm in central Asia or a methane blizzard on Neptune would affect the contents of the flask, but the temperature, pressure, and humidity of the laboratory could. ♦

The sum of the kinetic and potential energies of all the particles in a system is its **internal energy, $E$** (some texts use the symbol $U$). When a chemical system changes from reactants to products, the internal energy of the system changes. To determine this change, we measure the difference between the system's internal energy after the change [the *final* internal energy ($E_{final}$)] and before the change [the *initial* internal energy ($E_{initial}$)]:

$$\Delta E = E_{final} - E_{initial} = E_{products} - E_{reactants} \qquad \textbf{(6.1)}$$

where the symbol $\Delta$ (Greek *delta*) means "change (or difference) in" and refers to the *final state of the system* **minus** *the initial state.*

We can illustrate the change in a system's energy with an *energy diagram,* in which these two energy states are represented by horizontal lines placed along a vertical energy axis. The change in internal energy, $\Delta E$, is the difference between the heights of the two lines. A reacting chemical system can change its internal energy in either of two ways:

1.  By losing some energy *to* the surroundings (Figure 6.2, *A*),

$$E_{final} < E_{initial} \qquad \Delta E < 0$$

**FIGURE 6.2**

**Energy diagrams for the transfer of internal energy ($E$) between a system and its surroundings. A,** When the internal energy of a system *decreases,* the change in energy ($\Delta E$) is lost *to* the surroundings; therefore, $\Delta E$ of the system ($E_{final} - E_{initial}$) is negative. **B,** When the system's internal energy *increases,* $\Delta E$ is gained *from* the surroundings and is positive. (Note that the arrow signifying the direction of the change *always* has its tail at the initial state and its head at the final state.)

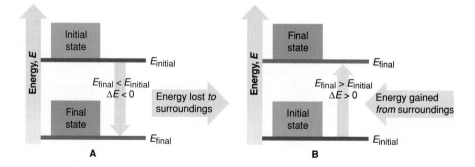

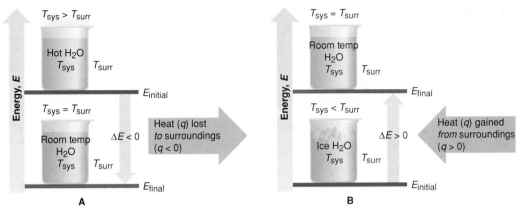

**FIGURE 6.3**
**A system transferring energy as heat only.**
**A,** The hot water transfers energy as heat ($q$) *to* the surroundings until $T_{sys} = T_{surr}$. Since $E_{initial} > E_{final}$ and $w = 0$, $\Delta E < 0$. The sign of $q$ is negative. **B,** The ice water gains energy as heat ($q$) *from* the surroundings until $T_{sys} = T_{surr}$. Since $E_{initial} < E_{final}$ and $w = 0$, $\Delta E > 0$. The sign of $q$ is positive.

2. By gaining some energy *from* the surroundings (Figure 6.2, *B*):

$$E_{final} > E_{initial} \qquad \Delta E > 0$$

As you can see, a change in the energy of the system must be accompanied by an opposite change in the energy of the surroundings: the change is always, in fact, a *transfer* of energy.

### Heat and Work: Two Forms of Energy Transfer

The energy transfer outward from the system or inward from the surroundings can appear in two forms, as heat and as work. **Heat** (symbol $q$) is the energy transferred between a system and its surroundings as a result of differences in their temperatures only. Energy is transferred from hot soup (system) to the bowl, air, and table (surroundings) in the form of heat because the surroundings are cooler. All other forms of energy transfer involve some type of **work (w),** the energy transferred when an object is moved by a force. When a football is inflated, the inside air (system) exerts a force on the inner wall of the ball (surroundings) and moves it outward; thus, energy is transferred from the air to the ball in the form of work.

Therefore the total change in a system's internal energy is

$$\Delta E = q + w \qquad \textbf{(6.2)}$$

The numerical values of $q$ and $w$ have either a positive or a negative sign, depending on the change the *system* undergoes; that is, *we define the sign of the energy transfer from the system's perspective.*

First, let's examine a system that does no work but transfers energy only as heat ($q$); since $w = 0$, from Equation 6.2 we have $\Delta E = q$. Suppose a sample of hot water is the system, with the beaker containing it and the rest of the lab the surroundings. The water transfers energy as heat to the surroundings until the temperature of the water and that of the surroundings are equal. The system's energy decreases as heat flows *out of* the system, so the final energy of the system is less than its initial energy. Heat was lost by the system, so *q is negative*, and therefore $\Delta E$ *is negative* (Figure 6.3, *A*).

On the other hand, if the system consists of ice water, it will gain energy as heat from the surroundings until the temperature of the water and that of the surroundings are equal. In this case, energy has been transferred *into* the system, so the final energy of the system is higher than its initial energy. Heat was gained by the system, so *q is positive*, and therefore $\Delta E$ *is positive* (Figure 6.3, *B*). ◆

◆ **Thermodynamics in the Kitchen.** The air in a refrigerator (surroundings) has a lower temperature than that of a newly added piece of food (system), so the food loses energy in the form of heat ($q$) to the refrigerator air. The air in a hot oven (surroundings) has a higher temperature than that of a newly added piece of food (system), so the food gains energy as heat from the oven air. In the first case, $q < 0$; in the second case, $q > 0$.

**FIGURE 6.4**

**A system losing energy as work only.** The energy of the system decreases as the reactants form products because the $H_2(g)$ does work ($w$) *on* the surroundings by pushing back the piston. The reaction vessel is insulated, so $q = 0$. Since $E_{initial} > E_{final}$, $\Delta E < 0$. The sign of $w$ is negative.

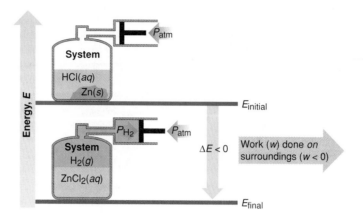

Next, let's consider a system that transfers energy only as work ($w$) being done—a chemical reaction that forms a gaseous product. In the following reaction, visualize the system as consisting of the atoms that make up the substances:

$$Zn(s) + 2HCl(aq) \longrightarrow H_2(g) + ZnCl_2(aq)$$

In the initial state, the system's energy is that of the reactants: zinc, hydrogen, and chlorine atoms in the form of metallic zinc, and hydrogen and chloride ions in solution. In the final state, the system's energy is that of the products: the same atoms in the form of hydrogen gas, and zinc and chloride ions in solution.

So that no energy is transferred as heat ($q = 0$), we place the system in an insulated reaction chamber, which is attached to a piston-cylinder assembly. As the hydrogen gas forms, some of the internal energy of the system pushes the piston and the outside air, thereby doing work *on* the surroundings. Energy was lost by the system in the form of work, so $w$ *is negative*, and since $q = 0$, $\Delta E$ *is negative* (Figure 6.4). This type of work, in which a volume change occurs against an external pressure, is called **pressure-volume work (PV work).**

Admittedly, the work done in this example is not very useful because it simply pushes back the piston and atmosphere. If the system were a ton of burning coal, however, and the surroundings were the engine of a locomotive, some of the internal energy lost when coal and oxygen react to form carbon dioxide and water vapor would appear as the work of moving a train.

The system in Figure 6.4 could also gain energy when work is done on it *by* the surroundings. Raising the external pressure would decrease the volume of the system, thereby increasing its energy by doing work on it: $w$ *is positive*, and therefore $\Delta E$ *is positive*. Table 6.1 summarizes the sign conventions of $q$ and $w$ and their effect on the sign of $\Delta E$.

**TABLE 6.1 The Sign Conventions of $q$, $w$, and $\Delta E$**

| $q$ | $+$ | $w$ | $=$ | $\Delta E$ |
|---|---|---|---|---|
| $+$ | | $+$ | | $+$ |
| $+$ | | $-$ | | Depends on *size* of $q$ and $w$ |
| $-$ | | $+$ | | Depends on *size* of $q$ and $w$ |
| $-$ | | $-$ | | $-$ |

### Units of Energy

The SI unit of energy is the **joule (J),** a derived unit composed of three base units: $1 \text{ J} = 1 \text{ kg} \cdot \text{m}^2/\text{s}^2$. Both heat and work are expressed in joules. Work is done when a *force* acts to change the velocity of a mass over a *distance.* Velocity has units of meters per second (m/s), and a change in velocity (acceleration, *a*) has units of m/s². Force, therefore, has the SI base units of mass times acceleration:

$$F = m \times a \text{ in units of } kg \times m/s^2$$

The work done on the mass is the force (*F*) times the distance (*d*) the mass travels. Therefore, work has the SI units of the joule:

$$w = F \times d \text{ in units of } (kg \cdot m/s^2) \times m = kg \cdot m^2/s^2 = J$$

Similar dimensional analyses of potential energy, kinetic energy, and *PV* work show that they are also combinations of these physical quantities and can be expressed in joules.

The **calorie (cal)** is an older unit that was defined originally as the amount of energy needed to raise the temperature of one gram of water by 1°C (from 14.5°C to 15.5°C). The calorie is now defined in terms of the joule:

$$1 \text{ cal} \equiv 4.184 \text{ J} \quad \text{or} \quad 1 \text{ J} = \frac{1}{4.184} \text{ cal} = 0.2390 \text{ cal}$$

Since the amounts of energy involved in chemical reactions are often quite large, chemists use the kilojoule (kJ), or sometimes the kilocalorie (kcal):

$$1 \text{ kJ} = 1000 \text{ J} = 0.2390 \text{ kcal} = 239.0 \text{ cal}$$

The nutritional Calorie (note the capital C), the unit used in diet charts and tables showing the energy available from food, is actually a kilocalorie. The *British thermal unit (Btu)*, a unit in engineering that you may have seen used to indicate energy output of appliances, is the amount of energy required to raise the temperature of one pound of water by 1°F and is equivalent to 1055 J. In general, the SI unit (J or kJ) is used throughout this text. Some interesting amounts of energy appear in Figure 6.5.

SAMPLE PROBLEM 6.1 _____

**Determining the Energy Change of a System**

**Problem:** When gasoline burns in an automobile engine, the heat released causes the carbon dioxide and water vapor produced to expand, which pushes the pistons outward. Excess heat is removed by the car's cooling system. Determine the change in energy (Δ*E*) in J, kJ, and kcal if the expanding gases do 451 J of work on the pistons and the system loses 325 J to the surroundings as heat.
**Plan:** First, we define the system as the reactants and products of the reaction. The surroundings consist of the pistons, the cooling system, and the rest of the car affected by the reaction. Since heat is released by the system, *q* is negative. Since the pistons are pushed outward, work is done by the system, so *w* is also negative. Δ*E* is the sum of *q* and *w*. We obtain the answer in J and then convert it to kJ and kcal.
**Solution:** Calculating Δ*E* in J:

$$q = -325 \text{ J} \qquad w = -451 \text{ J}$$

$$\Delta E = q + w = -325 \text{ J} + (-451 \text{ J}) = \textbf{-776 J}$$

Converting from J to kJ:

$$\Delta E = -776 \text{ J} \times \frac{1 \text{ kJ}}{1000 \text{ J}} = \textbf{-0.776 kJ}$$

Converting kJ to kcal:

$$\Delta E = -0.776 \text{ kJ} \times \frac{0.2390 \text{ kcal}}{1 \text{ kJ}} = \textbf{-0.185 kcal}$$

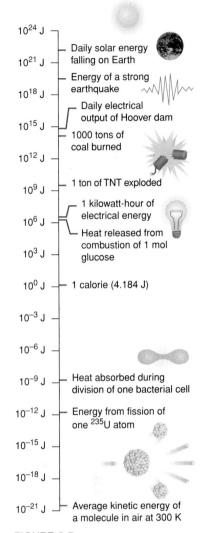

| | |
|---|---|
| $10^{24}$ J | |
| $10^{21}$ J | Daily solar energy falling on Earth |
| $10^{18}$ J | Energy of a strong earthquake |
| $10^{15}$ J | Daily electrical output of Hoover dam |
| | 1000 tons of coal burned |
| $10^{12}$ J | |
| $10^9$ J | 1 ton of TNT exploded |
| $10^6$ J | 1 kilowatt-hour of electrical energy |
| | Heat released from combustion of 1 mol glucose |
| $10^3$ J | |
| $10^0$ J | 1 calorie (4.184 J) |
| $10^{-3}$ J | |
| $10^{-6}$ J | |
| $10^{-9}$ J | Heat absorbed during division of one bacterial cell |
| $10^{-12}$ J | Energy from fission of one $^{235}$U atom |
| $10^{-15}$ J | |
| $10^{-18}$ J | |
| $10^{-21}$ J | Average kinetic energy of a molecule in air at 300 K |

**FIGURE 6.5**
**Some interesting amounts of energy.** Note that the vertical scale is exponential.

**Check:** The answer is reasonable: combustion releases energy from the system, so $E_{final} < E_{initial}$ and $\Delta E$ should be negative. Since 4 kJ ≈ 1 kcal, nearly 0.8 kJ should be nearly 0.2 kcal.

**FOLLOW-UP PROBLEM 6.1**
In a reaction, gaseous reactants form a liquid product. The heat absorbed by the surroundings is 26 kJ, and the work done on the system is 15 kJ. Calculate $\Delta E$.

## State Functions and the Path Independence of the Energy Change

The energy change of a system can involve different combinations of heat and work, but no matter what the particular combination, the same overall energy change occurs; that is, *$\Delta E$ does not depend on **how** the change takes place.*

As an example, let's define a system in its initial state as one mole of octane (a component of gasoline) together with enough oxygen to burn it. In its final state, the system is the carbon dioxide and water vapor formed from the combustion reaction:

$$C_8H_{18}(l) + {}^{25}\!/_2O_2(g) \longrightarrow 8CO_2(g) + 9H_2O(g)$$

$$\text{initial state } (E_{initial}) \qquad \text{final state } (E_{final})$$

Some of the energy in the fuel mixture is released to warm and/or do work on the surroundings, so $\Delta E$ is negative. However, the combustion of the octane and the change in internal energy can occur in a variety of ways. If we burn the octane in an open container, the energy change appears almost completely as heat (with a small amount of work done to push back the atmosphere). If we burn it in an automobile engine, a much larger portion (~30%) of the energy change appears as work that moves the car, with the rest used to heat the car, air, and exhaust gases (Figure 6.6). If we burn the octane in a lawnmower or a plane, the energy change appears as still other combinations of work and heat. Thus, the separate amounts of work and heat available from the change do depend on how the change occurs, but the *total* change in internal energy, the *sum* of the heat and work, does not: for a given change in the system, $\Delta E$ *is constant, but q and w can vary.*

The internal energy ($E$) of a system is a **state function,** a property of the system that can be determined completely by its *current* state, regardless of how it got to that state. The pressure ($P$) of an ideal gas and the volume ($V$) of water in a beaker are other examples of state functions because there are many ways that the pressure of the gas or the volume of the water could have reached its current value. Thus, changes in $E$, $P$, and $V$ *depend only on the initial and final states*. In contrast, heat ($q$) and work ($w$) are not state functions because they *do* depend on the particular path the system takes in undergoing the change in energy. (Note that symbols for state functions, such as $E$, $P$, and $V$, are capitalized.) ◆

**◆ Your Personal Financial State Function.** The *balance* in your checkbook is a state function of your personal financial system. You can open a new account with a birthday gift of $50, or you can open a new account with a deposit of a $100 paycheck and then write two $25 checks. The two paths to the balance are different, but the balance (current state) is the same.

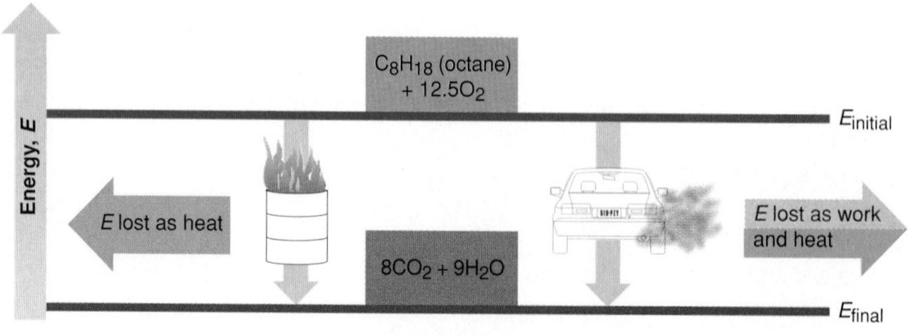

**FIGURE 6.6**
**Two different paths for the energy change of a system.** The change in energy when a given amount of octane burns in air is the same no matter how the energy is transferred. On the left, the fuel is burned in an open can, and the energy is lost almost entirely as heat. On the right, it is burned in a car engine; as a result, a portion of the energy is lost as work to move the car, and thus less is lost as heat.

$C_8H_{18}$ (octane) + 12.5$O_2$

$E_{initial}$

Energy, $E$

$E$ lost as heat

$8CO_2 + 9H_2O$

$E$ lost as work and heat

$E_{final}$

## The Law of Energy Conservation

As you have seen, whenever a system gains energy, the surroundings supply that energy and thus lose an equivalent amount; conversely, when a system loses energy, the surroundings gain an equivalent amount. Energy can be converted from one form into another as it is transferred, but it cannot simply appear or disappear— it cannot be created or destroyed. The **law of conservation of energy** states this basic observation: *the total energy of the universe is constant.* This law is also known as the **first law of thermodynamics** because it is so essential to that science. ◆

An earlier example illustrates the conservation of energy. As gasoline burns and forms exhaust gases, chemical energy is converted to both heat and work. Some of the energy transferred as work appears as mechanical energy to turn the car's wheels and belts. Some of this energy is converted into electrical energy in the generator, which powers the clock and electric windows, into the radiant energy of the headlights, and into the chemical energy of the battery. Complex biological processes also exemplify energy conservation. Through photosynthesis, green plants convert radiant energy from the sun into chemical energy when low-energy $CO_2$ and $H_2O$ are used to make high-energy carbohydrates (such as wood) and $O_2$. When the wood is burned in air, the low-energy compounds form again, with the energy difference released to the surroundings.

Thus, energy transferred from a system to the surroundings, or vice versa, can take the forms of heat and various types of work—mechanical, electrical, chemical—but *the energy of the system plus the energy of the surroundings remains constant: energy is conserved.* A mathematical expression of the first law, the energy conservation law, is

$$\Delta E_{universe} = \Delta E_{system} + \Delta E_{surroundings} = 0 \tag{6.3}$$

### Section Summary

Energy is transferred as heat ($q$) when the system and surroundings are at different temperatures; energy is transferred as work ($w$) when an object is moved by a force. Heat or work added to a system ($q > 0$; $w > 0$) increases its energy; heat or work lost by the system ($q < 0$; $w < 0$) decreases its energy. The total change in the system's energy is the sum of the heat and work: $\Delta E = q + w$. Heat and work are measured in joules (J), the SI unit of energy. Energy is a state function; therefore, the same $\Delta E$ can occur through any combination of $q$ and $w$. Energy is always conserved: it changes from one form into another, moving into or out of the system, but the total amount of energy in the universe (system *plus* surroundings) is constant.

◆ **The Tragic Life of the First Law's Discoverer.** After studying the work habits, food intake, and color of the blood of sailors in the tropics compared with those in northern Europe, the young German doctor J. R. von Mayer concluded that the amount of energy in food is used both to heat the body and to do work, and thus heat and work represent energy in different forms. When he published these ideas, they were ridiculed, and von Mayer became despondent. Soon thereafter, James Joule, an English brewer and amateur scientist, demonstrated the equivalence of heat and work experimentally but gave von Mayer no credit for the idea. The unexpected death of von Mayer's children depressed him further. He attempted suicide but was seriously injured and later sent to an insane asylum. Tainted by his conduct, his family declared him legally dead. Many years later, von Mayer was released and finally received recognition for his great insight.

## 6.2 Enthalpy: Heats of Reaction and Chemical Change

The principles of energy change pertain to many situations, from the inner workings of a clock to the movement of continents to the formation of a solar system. In chemistry, most physical and chemical changes occur under virtually constant atmospheric pressure: in an open flask, in a lake, in an organism, and so forth. By defining a thermodynamic variable called enthalpy that takes constant pressure into account, we can measure more easily the energy changes that accompany these changes in matter.

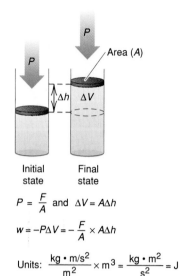

$P = \dfrac{F}{A}$ and $\Delta V = A\Delta h$

$w = -P\Delta V = -\dfrac{F}{A} \times A\Delta h$

Units: $\dfrac{kg \cdot m/s^2}{m^2} \times m^3 = \dfrac{kg \cdot m^2}{s^2} = J$

**FIGURE 6.7**

**Pressure-volume work.** When the volume ($V$) of a system increases by an amount $\Delta V$ against an external pressure ($P$), the system does $PV$ work *on* the surroundings ($w = -P\Delta V$). Note that the units of $PV$ work are the same as those of the joule, kg · m²/s².

### The Meaning of Enthalpy

To determine $\Delta E$, we must measure both heat and work. The two most important types of chemical work are electrical work, the work done by moving charged particles (Chapter 20), and $PV$ work, the work done by an expanding gas. In this chapter, we focus on $PV$ work.

We can determine the amount of $PV$ work done by multiplying the external pressure ($P$) by the change in volume of the gas ($\Delta V$, or $V_{final} - V_{initial}$). Thus, the work done by an expanding gas *on* the surroundings is

$$w = -P\Delta V \qquad (6.4)$$

The minus sign indicates that the system loses energy (Figure 6.7).

In an open flask, a reaction that produces a gas does work by pushing back the atmosphere. However, a thermodynamic variable called **enthalpy** **(H)** eliminates our need to consider $PV$ work in reactions that occur at constant pressure. Enthalpy is defined as the system's internal energy *plus* the product of its pressure and volume:

$$H = E + PV$$

The **change in enthalpy ($\Delta H$)** is the change in energy *plus* the product of the constant pressure and the change in volume:

$$\Delta H = \Delta E + P\Delta V \qquad (6.5)$$

Since $E$, $P$, and $V$ are state functions, $H$ is also a state function. Therefore, $\Delta H$ *depends only on the difference between $H_{final}$ and $H_{initial}$.* Combining Equations 6.2 and 6.4 leads to a key point about $\Delta H$:

$$\Delta E = q + w = q + (-P\Delta V) = q - P\Delta V$$

Solving for heat at constant pressure, denoted $q_P$, gives

$$q_P = \Delta E + P\Delta V$$

The right side of this equation is identical to Equation 6.5, the expression for $\Delta H$. Thus, *the change in enthalpy is equal to $q_P$, the heat gained or lost by the system at constant pressure:*

$$q_P = \Delta E + P\Delta V = \Delta H \qquad (6.6)$$

Since most chemical changes occur at constant pressure, $\Delta H$ is more relevant and easier to obtain than $\Delta E$: *to find $\Delta H$, we measure $q_P$.*

Knowing the *enthalpy* change of a system gives valuable information about its *energy* change. When the reaction involves no volume change, $P\Delta V = 0$, so $\Delta H = \Delta E$. In fact, many reactions involve little (if any) $PV$ work, so most (or all) of the energy change occurs as heat. Here are three cases:

*Case 1.* The reaction does not involve gases (neutralization, precipitation, many redox reactions). Consider the reaction

$$2KOH(aq) + H_2SO_4(aq) \rightarrow K_2SO_4(aq) + 2H_2O(l)$$

Since liquids and solids undergo *very* small volume changes, $\Delta V \approx 0$ and $P\Delta V \approx 0$, so $\Delta H \approx \Delta E$.

*Case 2.* The total moles of gaseous reactants equals the total moles of gaseous products. For example,

$$N_2(g) + O_2(g) \rightarrow 2NO(g)$$

In these cases, $V$ is constant, so $\Delta V = 0$, $P\Delta V = 0$, and $\Delta H = \Delta E$.

*Case 3.* The number of moles of gas changes during the reaction. In these cases, $P\Delta V \neq 0$. However, $q_P$ is usually *much* larger than $P\Delta V$, so most of $\Delta E$ occurs as heat and $\Delta H \approx \Delta E$. For instance, in the combustion of hydrogen, three moles of gas yield two:

$$2H_2(g) + O_2(g) \rightarrow 2H_2O(g)$$

In this reaction, $\Delta H = -483.6$ kJ and $P\Delta V = -2.5$ kJ, so $\Delta E = \Delta H - P\Delta V = -481.1$ kJ. Therefore, for many reactions, you can confidently use $\Delta H$ as either equal to or a very close estimate of $\Delta E$.

### Exothermic and Endothermic Processes

As with the energy change, the enthalpy change of a reaction *always* refers to the enthalpy of the system's final state minus the enthalpy of its initial state:

$$\Delta H = H_{final} - H_{initial} = H_{products} - H_{reactants}$$

This enthalpy change is called the **heat of reaction** (or enthalpy of reaction; often symbolized $\Delta H_{rxn}$). The sign of $\Delta H$ indicates whether heat is absorbed or released in the process. We can determine its sign by imagining the heat as a "reactant" or "product" in the change. When methane burns in air, for example, heat is released, so we show it on the right:

$$CH_4(g) + 2O_2(g) \rightarrow CO_2(g) + 2H_2O(g) + heat$$

This means that the products ($CO_2$ and $2H_2O$) have less enthalpy than the reactants ($CH_4$ and $2O_2$) by the amount of heat released, so $\Delta H$ ($H_{final} - H_{initial}$) is negative, as shown in the **enthalpy diagram** (Figure 6.8, *A*). A process that results in a decrease in the enthalpy of the system is called **exothermic** ("heat out to"); heat is released to the surroundings:

$$\text{Exothermic:} \quad H_{final} < H_{initial} \qquad \Delta H < 0$$

An **endothermic** process ("heat in from") occurs with an increase in the enthalpy of the system. When ice melts, for instance, heat flows into the ice from the surroundings, so we show the heat on the left:

$$Heat + H_2O(s) \rightarrow H_2O(l)$$

The enthalpy of the liquid water is higher than that of the solid water by the amount of heat absorbed (Figure 6.8, *B*). Therefore, $\Delta H$ ($H_{water} - H_{ice}$) is positive, as for any endothermic process:

$$\text{Endothermic:} \quad H_{final} > H_{initial} \qquad \Delta H > 0$$

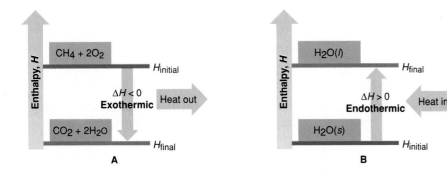

**A**

**B**

**FIGURE 6.8**

**Enthalpy diagrams for exothermic and endothermic processes. A,** The combustion of methane occurs with a decrease in enthalpy because heat *leaves* the system. Therefore, $H_{final} < H_{initial}$, and the process is exothermic: $\Delta H < 0$. **B,** The melting of ice occurs with an increase in enthalpy because heat *enters* the system. Since $H_{water} > H_{ice}$, the process is endothermic: $\Delta H > 0$.

SAMPLE PROBLEM 6.2

## Drawing Enthalpy Diagrams and Determining the Sign of $\Delta H$

**Problem:** In each of the following cases, determine the sign of $\Delta H$ and whether the reaction is exothermic or endothermic. Draw an enthalpy diagram for each reaction.

(a) $H_2(g) + \frac{1}{2}O_2(g) \rightarrow H_2O(l) + 285.8 \text{ kJ}$

(b) $40.7 \text{ kJ} + H_2O(l) \rightarrow H_2O(g)$

**Plan:** We inspect each equation to see whether heat is a "product" (exothermic; $\Delta H < 0$) or a "reactant" (endothermic; $\Delta H > 0$). For exothermic reactions, reactants are above products on the enthalpy diagram; the converse applies for endothermic reactions. The $\Delta H$ arrow *always* points from reactants to products.

**Solution: (a)** Heat is on the right (product), so $\mathbf{\Delta H < 0}$ and the reaction is **exothermic.** The enthalpy diagram appears in the margin (*top*).

**(b)** Heat is on the left (reactant), so $\mathbf{\Delta H > 0}$ and the reaction is **endothermic.** The enthalpy diagram appears in the margin (*bottom*).

**Check:** Substances on the same side of the equation as the heat have less enthalpy than substances on the other side, so we make sure they are placed on the lower line of the diagram.

**Comment:** $\Delta H$ values depend on conditions. In (b), for instance, $\Delta H = 40.7 \text{ kJ}$ at 1 atm and 100°C; at 1 atm and 25°C, $\Delta H = 44.0 \text{ kJ}$.

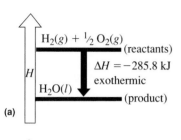

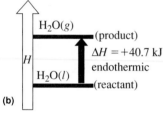

(a)

(b)

FOLLOW-UP PROBLEM 6.2

When nitroglycerine decomposes, it causes a violent explosion, releasing $5.72 \times 10^3 \text{ kJ}$ of heat for each mole of nitroglycerine:

$$C_3H_5(NO_3)_3(l) \rightarrow 3CO_2(g) + \frac{5}{2}H_2O(g) + \frac{1}{4}O_2(g) + \frac{3}{2}N_2(g)$$

Draw an enthalpy diagram for the reaction, and state whether the reaction is exothermic or endothermic.

Certain enthalpy changes represent such frequently studied processes that they are given specific names:

- When one mole of a substance combines with oxygen in a combustion reaction, the heat of reaction is the **heat of combustion ($\Delta H_{comb}$):**

$$C_4H_{10}(l) + \frac{13}{2}O_2(g) \rightarrow 4CO_2(g) + 5H_2O(g) \qquad \Delta H = \Delta H_{comb}$$

- When one mole of compound is produced from its elements, the heat of reaction is the **heat of formation ($\Delta H_f$):**

$$K(s) + \frac{1}{2}Br_2(l) \rightarrow KBr(s) \qquad \Delta H = \Delta H_f$$

- When one mole of a substance melts, the enthalpy change is the **heat of fusion ($\Delta H_{fus}$):**

$$NaCl(s) \rightarrow NaCl(l) \qquad \Delta H = \Delta H_{fus}$$

- When one mole of a substance vaporizes, the enthalpy change is the **heat of vaporization ($\Delta H_{vap}$):**

$$C_6H_6(l) \rightarrow C_6H_6(g) \qquad \Delta H = \Delta H_{vap}$$

We encounter heats of combustion and formation later in this chapter. Other special enthalpy changes are discussed in later chapters.

## Where Does the Heat Come from?

At this point, let's stop for a moment to address a central question about the heat involved in a reaction. When 2 g $H_2$ (1 mol) and 16 g $O_2$ (0.5 mol) react at 25°C, 18 g water vapor (1 mol) forms and 242 kJ of heat is released.

Where does this released heat come from? To find out, we must break down the system's internal energy into its component kinetic energy ($E_k$) and potential energy ($E_p$). Any change in the system's total energy occurs through changes in these components:

$$\Delta E_{total} = \Delta E_k + \Delta E_p$$

Let's compare the contributions to these energy components in two different systems: a wind-up clock and the substances involved in the formation of water vapor. The energy of the clock depends on the $E_p$ of the spring and the $E_k$ of its gears, hands, and so forth:

$$\Delta E_{total} = \Delta E_{p\ (spring)} + \Delta E_{k\ (gears)} + \Delta E_{k\ (hands)} + \cdots$$

When the spring is wound, the energy of the clock (initial state) is higher; when it unwinds, the energy (final state) is lower. The change in the spring's $E_p$ is slowly converted to the $E_k$ of the moving parts and then to heat from friction.

In the initial state of the chemical system (reactants $H_2$ and $O_2$ at 25°C), the energy is higher, and in the final state (product $H_2O$ at 25°C), the energy is lower. Now we can narrow our original question to "Which contribution(s) to the components of the system's internal energy is (are) responsible for this energy change?" As with the clock, the chemical system's $E_k$ and $E_p$ each has several contributions (Figure 6.9).

In general, a molecule's $E_k$ contributions are

1. The molecule moving through space, $E_{k\ (translation)}$
2. The molecule and its bound atoms rotating, $E_{k\ (rotation)}$
3. The bound atoms vibrating, $E_{k\ (vibration)}$

Thus, $E_{k\ (total)} = E_{k\ (translation)} + E_{k\ (rotation)} + E_{k\ (vibration)}$.

The major $E_p$ contributions are

1. Forces in each atom between nucleus and electrons and between electrons, $E_{p\ (atom)}$
2. Forces between the particles in each nucleus, $E_{p\ (nuclei)}$
3. Forces between atoms in each chemical bond, $E_{p\ (bond)}$

Thus, $E_{p\ (total)} = E_{p\ (atom)} + E_{p\ (nuclei)} + E_{p\ (bond)}$, and the total energy change is

$$\Delta E_{total} = \Delta E_{k\ (translation)} + \Delta E_{k\ (rotation)} + \Delta E_{k\ (vibration)}$$
$$+ \Delta E_{p\ (atom)} + \Delta E_{p\ (nuclei)} + \Delta E_{p\ (bond)}$$

The same six contributions make up the energy of the reactants and products. The $E_k$ contributions are proportional to the absolute temperature; but since reactants and products are both at 298 K (25°C), the total $E_k$ is virtually unchanged: $\Delta E_{k\ (total)} \approx 0$. Among the $E_p$ contributions, the atoms do not change, and the nuclei do not take part in chemical changes; thus, $\Delta E_{p\ (atom)} = \Delta E_{p\ (nuclei)} = 0$. Therefore, the only energy contribution that does change in a chemical reaction is $E_{p\ (bond)}$; that is, $\Delta E_{p\ (bond)} \neq 0$. *The energy released or absorbed during a chemical change is due to the difference in potential energy between the reactant bonds and the product bonds.* The answer to our opening question is that energy does not really "come from" anywhere; it exists in the different energies of the bonds of the substances. In an exothermic reaction, $E_{p\ (bond)}$ of the products is *less* than that of the reactants, so $\Delta E_{p\ (bond)} < 0$ and the system *releases* the energy difference. Conversely, in an endothermic reaction, $E_{p\ (bond)}$ of the products is *greater* than that of the reactants, so $\Delta E_{p\ (bond)} > 0$ and the system *absorbs* the energy difference.

**Breaking and forming bonds.** Based on this reasoning, we can say that the total bond energy of the reactants (1 mol of H—H bonds and 0.5 mol of O—O bonds) is greater than the total bond energy of the products

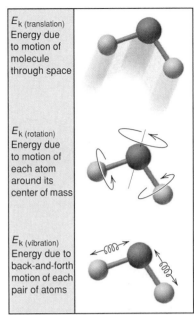

$E_{k\ (translation)}$
Energy due to motion of molecule through space

$E_{k\ (rotation)}$
Energy due to motion of each atom around its center of mass

$E_{k\ (vibration)}$
Energy due to back-and-forth motion of each pair of atoms

**A** Contributions to kinetic energy $(E_k)$

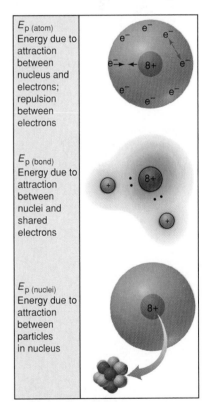

$E_{p\ (atom)}$
Energy due to attraction between nucleus and electrons; repulsion between electrons

$E_{p\ (bond)}$
Energy due to attraction between nuclei and shared electrons

$E_{p\ (nuclei)}$
Energy due to attraction between particles in nucleus

**B** Contributions to potential energy $(E_p)$

**FIGURE 6.9 Components of internal energy (E).** The total energy of a system is the sum of the $E_k$ **(A)** and $E_p$ **(B)** of its particles. The contributions to $E_k$ and $E_p$ are shown for a water molecule (or, where appropriate, just for the O atom). Only $E_{p\ (bond)}$ changes during a chemical reaction. The difference in $E_{p\ (bond)}$ between reactants and products gives rise to $\Delta H_{rxn}$.

(1 mol of H—O—H, or 2 mol of O—H bonds), and the difference in energy is released as 242 kJ of heat.

We can think of a reaction as a process in which *reactant bonds are broken and product bonds are formed.* Breaking a bond requires energy, and forming a bond releases energy. When 1 mol $H_2$ and 0.5 mol $O_2$ absorb energy, their bonds break and the atoms form the O—H bonds in 1 mol $H_2O$ in an overall process that releases energy. In other words, more energy is released when the bonds in $H_2O$ form than is absorbed when the bonds in $H_2$ and $O_2$ break, and the energy difference appears as the heat of reaction. In general, chemists speak of weak bonds as being less stable and thus requiring less energy for them to break (and releasing less energy when they form). Conversely, strong bonds are more stable and require more energy for them to break (and release more energy when they form). We discuss these ideas quantitatively in Chapter 9.

**Bond energies in fuels and foods.** The most common fuels for machines are hydrocarbon compounds and coal, whereas those for organisms are fats and carbohydrates. Both types of fuel are composed of large molecules with mostly C—C and C—H bonds. When the fuel is utilized, these bonds are broken and the atoms become bonded to O. The waste products of these fuels are carbon dioxide and water, the compounds that contain the C—O and O—H bonds. It follows from the previous discussion that the total energy of the C—C and C—H bonds in the fuel plus the energy in the bonds of $O_2$ is greater than the total energy of the C—O and O—H bonds in the waste products.

Fuels with many weaker (less stable, higher energy) bonds will yield more energy than fuels with fewer. Table 6.2 demonstrates this point for some two-carbon and one-carbon compounds. Notice that as the number

**TABLE 6.2 Heats of Combustion ($\Delta H_{comb}$) of Some Carbon Compounds**

| NAME (FORMULA) | STRUCTURAL FORMULA | SUM OF C—C AND C—H BONDS | SUM OF C—O AND O—H BONDS | $\Delta H_{comb}$ (kJ/mol) | $\Delta H_{comb}$ (kJ/g) |
|---|---|---|---|---|---|
| Two-carbon compounds | | | | | |
| Ethane ($C_2H_6$) | 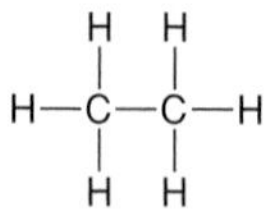 H—C—C—H (with H,H top and H,H bottom) | 7 | 0 | −1560 | −51.88 |
| Ethanol ($C_2H_5OH$) | 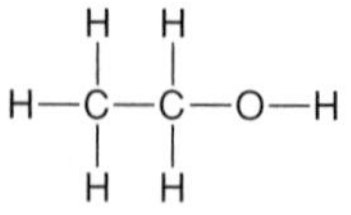 H—C—C—O—H (with H,H top and H,H bottom) | 6 | 2 | −1367 | −29.67 |
| One-carbon compounds | | | | | |
| Methane ($CH_4$) | 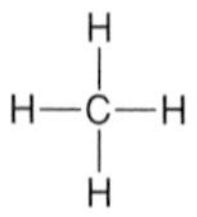 H—C—H (with H top and H bottom) | 4 | 0 | −890 | −55.5 |
| Methanol ($CH_3OH$) | 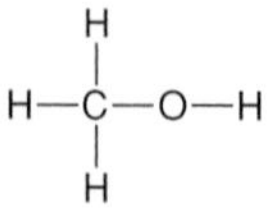 H—C—O—H (with H top and H bottom) | 3 | 2 | −727 | −22.7 |

of C—C and C—H bonds decreases and/or the number of C—O and O—H bonds increases, less energy is released on combustion; that is, $\Delta H$ is less negative. In other words, the fewer bonds to oxygen in the fuel, the more useful the fuel.

Both fats and carbohydrates are high-energy food sources. Fats consist largely of chains of carbon atoms (C—C bonds) attached to hydrogen atoms (C—H bonds). Carbohydrates have C—C bonds that attach units with the formula H—C—O—H; note the C—O and O—H bonds. As any dieter knows, fats "contain more Calories" per gram than carbohydrates. With fewer bonds to oxygen in fats, more energy is released from fat than from sugar or starch when each is converted to $CO_2$ and $H_2O$ (Table 6.3).

**Section Summary**

The change in enthalpy, $\Delta H$, is the heat lost or gained at constant pressure, $q_P$. In most cases, $\Delta H$ is identical or very close to $\Delta E$. A chemical change that releases heat is exothermic ($\Delta H < 0$); one that absorbs heat is endothermic ($\Delta H > 0$). On the molecular level, the heat lost or gained represents $\Delta E_{p\ (bond)}$ between products and reactants. A reaction involves breaking reactant bonds and forming product bonds. Product bonds have less energy than reactant bonds in an exothermic reaction and more energy than reactant bonds in an endothermic reaction. Bonds in fuels have more energy than those in waste products.

**TABLE 6.3 Heats of Combustion of Some Fats and Carbohydrates**

| SUBSTANCE | $\Delta H_{comb}$(kJ/g) |
|---|---|
| Fats | |
| Vegetable oil | 37.0 |
| Margarine | 30.1 |
| Butter | 30.0 |
| Carbohydrates | |
| Table sugar (sucrose) | 16.2 |
| Brown rice | 14.9 |
| Maple syrup | 10.4 |

## 6.3 Calorimetry: Laboratory Measurement of Heats of Reaction

As you've seen, the heat of reaction provides much useful information, including the energy value of fuels and foods and the relative strengths of chemical bonds, so it is a crucial variable to measure. Suppose we want to measure the change in enthalpy that occurs when the domestic fuel propane ($C_3H_8$) burns:

$$C_3H_8(g) + 5O_2(g) \rightarrow 3CO_2(g) + 4H_2O(g)$$

It might seem that we could simply measure the enthalpy of the reactants and subtract it from that of the products. However, the enthalpy ($H$) of a system in a particular state cannot be measured absolutely because we have no obvious starting point, no zero enthalpy. However, we *can* measure the *change* in enthalpy ($\Delta H$) as the heat released or absorbed by the system at constant pressure ($q_P$). To measure $q_P$ accurately, we must construct "surroundings" that capture and retain the heat and then measure some effect of the trapped heat, such as a temperature change on a thermometer immersed in the surroundings. Finally, we must understand how the amount of heat released or absorbed by the system is related to the temperature change we measure. This relation involves a physical property of a substance called the *specific heat capacity.*

### Specific Heat Capacity of a Substance

You know from everyday experience that the more heat an object absorbs, the higher its temperature becomes; that is, the amount of heat absorbed by the object is proportional to the temperature change:

$$\text{Heat} \propto \Delta T \quad \text{or} \quad \text{heat} = \text{constant} \times \Delta T$$

**TABLE 6.4 Specific Heat Capacities of Some Elements, Compounds, and Materials**

| SUBSTANCE | SPECIFIC HEAT CAPACITY (J/g · K)* |
|---|---|
| Elements | |
| Aluminum, Al | 0.900 |
| Graphite, C | 0.711 |
| Iron, Fe | 0.450 |
| Copper, Cu | 0.387 |
| Gold, Au | 0.129 |
| Compounds | |
| Ammonia, $NH_3(l)$ | 4.70 |
| Water, $H_2O(l)$ | 4.184 |
| Ethyl alcohol, $C_2H_5OH(l)$ | 2.46 |
| Ethyl glycol, $(CH_2OH)_2(l)$ | 2.42 |
| Carbon tetrachloride, $CCl_4(l)$ | 0.862 |
| Solid materials | |
| Wood | 1.76 |
| Cement | 0.88 |
| Glass | 0.84 |
| Granite | 0.79 |
| Steel | 0.45 |

*At 298 K (25°C)

◆ **Imagine an Earth Without Water.** Liquid water has an unusually high specific heat capacity, about 4.2 J/g · K, about six times that of rock (~0.7 J/g · K). If the Earth were devoid of oceans, the sun's energy would heat a planet composed of rock. It would take only 0.7 J of energy to increase the temperature of each gram of rock by 1 K. Daytime temperatures would soar. The oceans also limit the temperature drop when the sun sets, because the greater amount of energy absorbed during the day is released at night. If the Earth had a rocky surface, temperatures would be frigid every night.

Every object has its own particular capacity for absorbing heat, that is, its own characteristic **heat capacity,** the amount of heat required to change its temperature by one kelvin. Heat capacity is the proportionality constant in the preceding equation:

$$\text{Heat capacity} = \frac{\text{heat}}{\Delta T} \quad \text{[in J/K]}$$

A related property is **specific heat capacity (C),** the amount of heat required to change the temperature of one *gram* of a substance by one kelvin:

$$\text{Specific heat capacity } (C) = \frac{\text{heat}}{\text{grams} \times \Delta T} \quad \text{[in units of J/g · K]}$$

Specific heat capacity, like density, is a temperature-dependent physical property of a substance.* Table 6.4 lists the specific heat capacities of several common substances at 25°C. ◆

Closely related to the specific heat capacity is the **molar heat capacity,** the amount of heat required to change the temperature of one *mole* of the substance by one kelvin:

$$\text{Molar heat capacity} = \frac{\text{heat}}{\text{moles} \times \Delta T} \quad \text{[in units of J/mol · K]}$$

The specific heat capacity of liquid water is 4.184 J/g · K, so

$$\text{Molar heat capacity of } H_2O(l) = 4.184 \frac{J}{g \cdot K} \times \frac{18.02 \text{ g}}{1 \text{ mol}} = 75.40 \frac{J}{mol \cdot K}$$

If we know the specific heat capacity of a substance being heated (or cooled), we can measure its mass and temperature change and calculate the heat absorbed or released:

$$q = C \times \text{mass} \times \Delta T \qquad \textbf{(6.7)}$$

Notice that when an object becomes hotter, $\Delta T$ (that is, $T_{final} - T_{initial}$) is positive. The object gains heat, so $q > 0$. Similarly, when an object becomes cooler, $\Delta T$ is negative, and $q < 0$ because heat is lost.

SAMPLE PROBLEM 6.3

**Calculating the Amount of Heat from the Specific Heat Capacity**

**Problem:** If 125 g copper forms a welded layer on the bottom of a skillet, how much heat is needed to raise the temperature of the copper layer from 25°C to 300°C? The specific heat capacity (C) of Cu = 0.387 J/g · K.
**Plan:** We know the mass and C of Cu and can find $\Delta T$ in °C, which equals $\Delta T$ in kelvins. We use this $\Delta T$ and Equation 6.7 to solve for the heat.
**Solution:** Calculating $q$:

$$\Delta T = T_{final} - T_{initial} = 300°C - 25°C = 275°C = 275 \text{ K}$$

$$q = C \times \text{mass} \times \Delta T = 0.387 \text{ J/g · K} \times 125 \text{ g} \times 275 \text{ K} = \textbf{1.33} \times \textbf{10}^4 \textbf{ J}$$

**Check:** Heat is absorbed by the skillet, so $q$ is positive. The calculation seems reasonable: $q \approx 0.4 \times 100 \times 300 = 1.2 \times 10^4$.

FOLLOW-UP PROBLEM 6.3
Calculate the heat involved when a 5.5-g iron nail is cooled from 37°C to 25°C. (See Table 6.4 for the specific heat capacity of Fe.)

*Some texts use the term "specific heat" in place of "specific heat capacity." This usage is very common but somewhat incorrect. *Specific heat* is the ratio of the heat capacity of one gram of a substance to the heat capacity of one gram of water and therefore has no units.

### The Practice of Calorimetry

Since heat lost by the system is gained by the surroundings (and vice versa), we can design suitable surroundings and measure their temperature change to determine the heat transferred by the system. This is the principle of the **calorimeter,** a device used to measure the heat released (or absorbed) by a physical or chemical process.

A simple constant-pressure calorimeter can be made by nesting two Styrofoam coffee cups, placing a known mass of water in the inner cup, and inserting a stirring rod and thermometer through a stopper that acts as a lid (Figure 6.10). The "coffee-cup" calorimeter is useful for measuring the heat ($q_p$) of many processes that are open to the laboratory atmosphere. It is often used to determine the specific heat capacity of an unknown solid that does not react with or dissolve in water. The solid (system) is weighed, heated to some known temperature, and carefully added to the water (surroundings) of known temperature and mass in the calorimeter. The cups and lid act as thermal insulators and, with continual stirring to distribute the released heat, the final water temperature, which is also the final temperature of the solid, is measured.

The heat lost by the system ($-q$) is equal in numerical value but opposite in sign to the heat gained ($+q$) by the surroundings:

$$-q_{solid} = q_{water}$$

Substituting Equation 6.7 gives

$$-(C_{solid} \times mass_{solid} \times \Delta T_{solid}) = C_{H_2O} \times mass_{H_2O} \times \Delta T_{H_2O}$$

All the quantities are known or measured except $C_{solid}$, which is calculated:

$$C_{solid} = -\frac{C_{H_2O} \times mass_{H_2O} \times \Delta T_{H_2O}}{mass_{solid} \times \Delta T_{solid}}$$

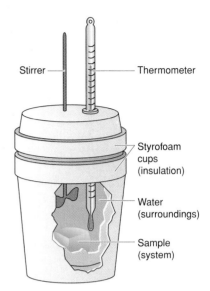

**FIGURE 6.10**
**A coffee-cup calorimeter.** This simple apparatus is used to measure the heat at constant pressure ($q_p$). It consists of a known mass of water (or solution) in an insulated container equipped with a thermometer and stirrer. The initial temperature of the water is measured, the process takes place (addition of a heated object, a soluble salt, a solution, and so on), the contents are stirred, and the final temperature of the water is measured.

---

**SAMPLE PROBLEM 6.4** _____

### Determining the Specific Heat Capacity of an Unknown Solid

**Problem:** A 25.64-g sample of an unknown solid was heated in a tube to 100.00°C in boiling water and carefully added to a coffee-cup calorimeter containing 50.00 g water. The water temperature increased from 25.10°C to 28.49°C. What is the specific heat capacity of the solid? (Assume all the heat is gained by the water.)
**Plan:** We know the masses of the water and the solid. The $\Delta T$ of the water and the solid are given; we change $\Delta T$ (°C) directly to $\Delta T$ (K). We know $C_{H_2O}$, so we find $C_{solid}$ as described in the text.
**Solution:** It is helpful to summarize the information given:

|         | MASS (g) | C (J/g · K) | $T_{initial}$ (°C) | $T_{final}$ (°C) | $\Delta T$ (K) |
|---------|----------|-------------|--------------------|------------------|----------------|
| Solid   | 25.64    | ?           | 100.00             | 28.49            | −71.51         |
| $H_2O$  | 50.00    | 4.184       | 25.10              | 28.49            | 3.39           |

Calculating $C_{solid}$:

$$C_{solid} = -\frac{C_{H_2O} \times mass_{H_2O} \times \Delta T_{H_2O}}{mass_{solid} \times \Delta T_{solid}}$$

$$= -\frac{4.184 \text{ J/g} \cdot \text{K} \times 50.00 \text{ g} \times 3.39 \text{ K}}{25.64 \text{ g} \times (-71.51 \text{ K})} = \mathbf{0.387 \text{ J/g} \cdot \text{K}}$$

**Check:** Since $-q_{solid} = q_{water}$, we can check to see if the numerical values are equal:

$$q_{water} = 4.184 \text{ J/g} \cdot \text{K} \times 50.00 \text{ g} \times 3.39 \text{ K} = 709 \text{ J}$$

$$q_{solid} = 0.387 \text{ J/g} \cdot \text{K} \times 25.64 \text{ g} \times (-71.51 \text{ K}) = -710 \text{ J}$$

The slight difference is caused by rounding.

**Comment:** A common mistake is to write the wrong sign for $\Delta T$. To avoid this error, always remember that $\Delta$ means *final − initial*.

### FOLLOW-UP PROBLEM 6.4

As a purity check for industrial diamonds, a 10.25-carat diamond is heated to 74.21°C and immersed in 26.05 g water in a constant-pressure calorimeter (1 carat = 0.2000 g). The initial temperature of the water is 27.20°C. Calculate $\Delta T$ of the water and of the diamond.($C_{diamond}$ = 0.519 J/g · K.)

In the previous calculation, we assumed that all the heat released by the object was gained by the water, but this is not true: some heat is gained by the other parts of the calorimeter (stirrer, thermometer, stopper, coffee-cup walls), and some leaks out to the laboratory. For precise work, *the heat capacity of the entire calorimeter must be known.*

The **bomb calorimeter** is designed to measure precisely the heat released in a combustion reaction (Figure 6.11). The weighed sample is placed in a metal-walled chamber (the bomb). The bomb is filled with oxygen gas and immersed in an insulated water bath fitted with a motorized stirrer and a thermometer. A heating coil connected to an electrical source ignites the sample, and the heat evolved from the combustion raises the temperature of the bomb, water, and other calorimeter parts. Knowing the mass of the sample and the heat capacity of the *entire* calorimeter, we use the measured $\Delta T$ to calculate the heat released.

### SAMPLE PROBLEM 6.5

### Calculating the Heat of Combustion with a Bomb Calorimeter

**Problem:** A manufacturer claims that its new dietetic dessert has fewer than 10 Calories per serving. An independent laboratory uses a bomb calorimeter with a heat capacity of 8.151 kJ/K to test the claim. When one serving of the dessert is burned in $O_2$, the temperature increases 4.937°C. Is the manufacturer's claim correct?
**Plan:** When the dessert burns, the heat released is gained by the calorimeter:

$$-q_{sample} = q_{calorimeter}$$

We find the heat by multiplying the heat capacity of the calorimeter by $\Delta T$.
**Solution:** Calculating the heat gained by the calorimeter:

$$q_{calorimeter} = \text{heat capacity} \times \Delta T = 8.151 \text{ kJ/K} \times 4.937 \text{ K} = 40.24 \text{ kJ}$$

One serving of the dessert has slightly less than 10 Calories (41.84 kJ), so **the manufacturer's claim is correct.**
**Check:** A quick check shows the answer is reasonable: ~8 kJ/K × 5 K = 40 kJ.

### FOLLOW-UP PROBLEM 6.5

A chemist burns 0.8650 g graphite (a form of carbon) in a new bomb calorimeter, and $CO_2$ forms. If 393.5 kJ is released per mole of graphite burned and the temperature increase is 2.613 K, what is the heat capacity of the new bomb calorimeter?

Notice that the energy released in a bomb *cannot* do *PV* work because the volume is constant: $P\Delta V = 0$. Therefore, *the energy change measured by a bomb calorimeter is the heat at constant volume ($q_V$)*, which equals $\Delta E$, not $\Delta H$:

$$\Delta E = q + w = q_V + 0 = q_V$$

Recall from Section 6.2, however, that the difference between $\Delta H$ and $\Delta E$ is usually quite small, even when the number of moles of gas changes. For ex-

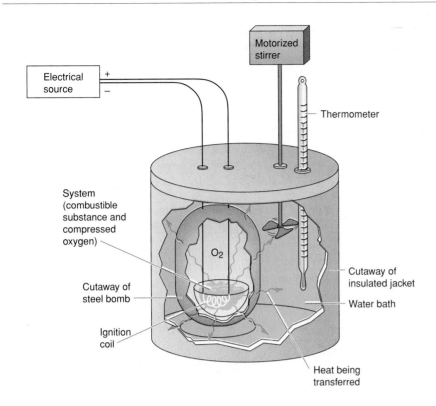

ample, $\Delta H$ is about 0.5% larger than $\Delta E$ for the combustion of $H_2$ and about 0.2% smaller for the combustion of octane (see Problem 6.91 at the end of the chapter).

### Section Summary

We measure $\Delta H$ of a process by measuring the heat at constant pressure ($q_P$). To do this, we determine $\Delta T$ of the surroundings and relate it to $q_P$ through the mass of the substance and its specific heat capacity, the amount of energy needed to raise the temperature of one gram of the substance by one kelvin. Calorimeters measure the heat released from a system either at constant pressure ($q_P = \Delta H$) or at constant volume ($q_V = \Delta E$).

## 6.4 Stoichiometry of Thermochemical Equations

The heat of reaction ($\Delta H_{rxn}$) provides us with many insights about the chemical change. In the remaining sections of the chapter, we examine some of the ways that $\Delta H_{rxn}$ values are applied. We express thermochemical changes by means of **thermochemical equations,** balanced equations that also state the heat of reaction *for the amounts of substances specified.* A heat of reaction (enthalpy change) has two key parts—a *sign* that depends on the direction of the change and a *magnitude* that depends on the amount of reacting substances:

1. Sign of $\Delta H$. The $\Delta H$ of a forward reaction is *identical in magnitude but opposite in sign* to the $\Delta H$ of the reverse reaction.

   Decomposition of 2 mol water into its elements (endothermic):

   $$2H_2O(l) \longrightarrow 2H_2(g) + O_2(g) \quad \Delta H_{rxn} = 572 \text{ kJ}$$

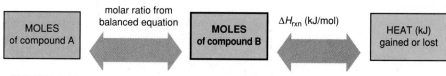

**FIGURE 6.12**
Summary of the relationship between moles of substance and the heat transferred during a reaction.

Formation of 2 mol water from its elements (exothermic):

$$2H_2(g) + O_2(g) \rightarrow 2H_2O(l) \quad \Delta H_{rxn} = -572 \text{ kJ}$$

2. **Magnitude of $\Delta H$.** The size of $\Delta H$ is *proportional to the amount of substance* in the reaction.
   Formation of 1 mol water from its elements (half the amount formed in the previous equation):

$$H_2(g) + \tfrac{1}{2}O_2(g) \rightarrow H_2O(l) \quad \Delta H_{rxn} = -286 \text{ kJ}$$

In order to specify the magnitude of $\Delta H_{rxn}$ for a *particular amount of substance,* thermochemical equations are often written with fractional coefficients. Thus, in this reaction,
   286 kJ is thermochemically equivalent to 1 mol $H_2(g)$.
   286 kJ is thermochemically equivalent to $\tfrac{1}{2}$ mol $O_2(g)$.
   286 kJ is thermochemically equivalent to 1 mol $H_2O(l)$.
Just as we use stoichiometrically equivalent molar ratios from a balanced equation to find the amounts of substances involved, we use these thermo-chemically equivalent quantities to find the heat involved when a given amount of substance reacts or is produced. Also, just as we use molar mass in g/mol, to convert moles of a substance to mass, we use the heat of reaction, in kJ/mol, to convert moles of a substance to heat (Figure 6.12).

SAMPLE PROBLEM 6.6 _____

### Calculating the Heat Involved in a Reaction

**Problem:** The major source of aluminum in the world is bauxite (mostly aluminum oxide). Its thermal decomposition can be represented by

$$Al_2O_3(s) \xrightarrow{\Delta} 2Al(s) + \tfrac{3}{2}O_2(g) \quad \Delta H_{rxn} = 1676 \text{ kJ}$$

If aluminum were produced this way (see Comment), how many grams of aluminum could form when $1.000 \times 10^3$ kJ of heat was utilized?

**Plan:** From the balanced equation and enthalpy change, we see that 2 mol Al form when 1676 kJ is absorbed. Using these equivalent quantities, we convert heat available to moles produced and then use the molar mass to convert moles to mass.

**Solution:** Converting from heat available to mass of Al:

$$\text{Mass (g) of Al} = (1.000 \times 10^3 \text{ kJ}) \times \frac{2 \text{ mol Al}}{1676 \text{ kJ}} \times \frac{26.98 \text{ g Al}}{1 \text{ mol Al}} = \textbf{32.20 g Al}$$

**Check:** The mass of Al seems correct: ~1700 kJ can form 2 mol Al (54 g), so 1000 kJ should form between 27 g and 54 g.

**Comment:** In practice, aluminum is not obtained through the application of heat alone (indeed $Al_2O_3$ does not even melt below 2000°C). Rather, it is obtained industrially by supplying electrical energy. Because $\Delta H$ is a state function, however, the total energy required for this chemical change is the same, no matter how it occurs. (We examine the industrial method in Chapter 23.)

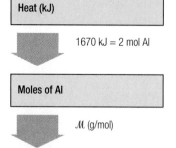

**FOLLOW-UP PROBLEM 6.6**

As Lavoisier, Priestley, and others knew, mercury(II) oxide decomposes readily with heat. If it takes 90.8 kJ to decompose one mole of oxide, how much heat is required to produce one short ton (2000 lbs; 907 kg) of Hg from decomposition of HgO?

**Section Summary**

Thermochemical equations show the balanced equation *and* its $\Delta H_{rxn}$. The sign of $\Delta H$ for a forward reaction is opposite that for the reverse reaction. The magnitude of $\Delta H$ depends on the amount of substance reacting and the $\Delta H$ per mole of substance. We use the thermochemically equivalent amounts of substance and heat from the balanced equation as conversion factors to find the amount of heat when any amount of substance reacts.

## 6.5 Hess's Law of Heat Summation

Since the enthalpy ($H$) of a substance is a state function (one that does not depend on any previous changes of the substance), we can determine the change in enthalpy ($\Delta H$) for complex reactions without concern for how the reaction occurs in reality. In fact, we can determine the $\Delta H$ of any reaction for which we can write an equation, even if we cannot carry it out in the laboratory. This state-function property of $\Delta H$ is expressed in a somewhat different way in **Hess's law of heat summation:** *the enthalpy change of an overall process is the sum of the enthalpy changes of its individual steps.* Hess's law provides a way of calculating the enthalpy changes of an enormous number of reactions because it allows us to imagine that a reaction occurs through steps for which we know the enthalpy changes.

Many chemical reactions are difficult, even impossible, to perform in one step. Some are integrated within complex biochemical processes; others may take place under environmental conditions that are difficult to simulate; and still others require a change in conditions. Consider the oxidation of sulfur to sulfur trioxide, a central process in the industrial production of sulfuric acid and in the environmental formation of acid rain. When we burn sulfur in oxygen, sulfur dioxide is obtained (equation 1; to introduce this idea, we use the formula S for 1 mol sulfur atoms, rather than the more correct $\frac{1}{8}S_8$). We must change conditions and add more oxygen to the sulfur dioxide to obtain sulfur trioxide (equation 2). Therefore, we cannot use a calorimeter to measure in one combustion step the $\Delta H$ for the overall reaction (equation 3). Nevertheless, we can calculate it with Hess's law. The three equations are

| | | |
|---|---|---|
| Equation 1: | $S(s) + O_2(g) \rightarrow SO_2(g)$ | $\Delta H_1 = -296.8 \text{ kJ}$ |
| Equation 2: | $2SO_2(g) + O_2(g) \rightarrow 2SO_3(g)$ | $\Delta H_2 = -198.4 \text{ kJ}$ |
| Equation 3: | $S(s) + \frac{3}{2}O_2(g) \rightarrow SO_3(g)$ | $\Delta H_3 = ?$ |

Hess's law tells us that if we alter equations 1 and 2 so that the substances sum to equation 3, then $\Delta H_3$ is the sum of the altered enthalpy changes of equations 1 and 2.

First, we identify equation 3 as our "target" equation, the reaction whose $\Delta H$ we are trying to find, and carefully note the moles of reactants and products. Then we see how to alter equations 1 and 2, whose $\Delta H$ values *as written* are given, to obtain equation 3. Equations 1 and 3 contain the same amount of sulfur, so we leave equation 1 unchanged. Equation 2 has twice as much $SO_3$ as we want, so we multiply it by $\frac{1}{2}$, being sure to halve $\Delta H_2$ as well. Finally, we add equation 1 to the halved equation 2 and eliminate terms that appear on both sides of the equation:

$$\text{Equation 1:} \qquad S(s) + O_2(g) \rightarrow SO_2(g) \qquad \Delta H_1 = -296.8 \text{ kJ}$$

$$\tfrac{1}{2} \text{ equation 2:} \quad SO_2(g) + \tfrac{1}{2}O_2(g) \rightarrow SO_3(g) \qquad \tfrac{1}{2}\Delta H_2 = -99.2 \text{ kJ}$$

$$S(s) + O_2(g) + \cancel{SO_2(g)} + \tfrac{1}{2}O_2(g) \rightarrow \cancel{SO_2(g)} + SO_3(g)$$

or, 
$$S(s) + \tfrac{3}{2}O_2(g) \rightarrow SO_3(g)$$

Note that equation 1 plus the halved equation 2 equals equation 3, so

$$\Delta H_3 = \Delta H_1 + \tfrac{1}{2}\Delta H_2 = -296.8 \text{ kJ} + (-99.2 \text{ kJ}) = -396.0 \text{ kJ}$$

Since enthalpy is a state function, the overall $\Delta H$ depends on only the enthalpies of the initial and final states. Therefore, whether the oxidation of sulfur occurs directly to sulfur trioxide or through the formation of sulfur dioxide first, Hess's law tells us that the difference between the enthalpy of the reactants (1 mol S + $\frac{3}{2}$ mol $O_2$) and that of the product (1 mol $SO_3$) is the same.

To summarize, calculating an unknown $\Delta H$ involves three steps:
1. Identify the target equation, the step whose $\Delta H$ is unknown, and note its moles of reactants and products.
2. Rearrange the equations of known $\Delta H$ so that the target moles of reactants and products are on the correct sides. Remember to:
   a. Change the sign of $\Delta H$ when you reverse an equation.
   b. Adjust the moles and the $\Delta H$ by the same factor.
3. Add the altered known equations to obtain the unknown $\Delta H$. All substances except those in the target equation should cancel.

SAMPLE PROBLEM 6.7 ⎯⎯⎯⎯⎯⎯⎯⎯⎯⎯⎯⎯⎯⎯⎯⎯⎯⎯⎯⎯⎯⎯⎯⎯

## Calculating an Unknown $\Delta H$ Using Hess's Law

**Problem:** Two gaseous pollutants that form in auto exhaust are CO and NO. An environmental chemist is studying ways to convert them to less harmful gases through the following equation:

$$CO(g) + NO(g) \rightarrow CO_2(g) + \tfrac{1}{2}N_2(g) \qquad \Delta H = ?$$

Given the following information, calculate the unknown $\Delta H$:

$$\text{Equation A:} \qquad CO(g) + \tfrac{1}{2}O_2(g) \rightarrow CO_2(g) \qquad \Delta H_A = -283.0 \text{ kJ}$$

$$\text{Equation B:} \qquad N_2(g) + O_2(g) \rightarrow 2NO(g) \qquad \Delta H_B = 180.6 \text{ kJ}$$

**Plan:** We note the moles of substances in the target equation, manipulate equations A *and* B *and* their $\Delta H$ values, and then add them together to obtain the target equation and the unknown $\Delta H$.

**Solution:** Noting moles of substances in the target equation: There is one mole of each reactant, one mole of product $CO_2$, and one-half mole of product $N_2$.

Manipulating the given equations: Equation A has the same number of moles of CO and $CO_2$ as the target, so we leave it as written. Equation B has twice the needed amounts of $N_2$ and NO, and they are on the wrong sides to match the target; therefore, we reverse equation B, change the sign of $\Delta H_B$, and multiply it by $\frac{1}{2}$:

$$\tfrac{1}{2}[2NO(g) \rightarrow N_2(g) + O_2(g)] \qquad \Delta H = -\tfrac{1}{2}\Delta H_B = -\tfrac{1}{2}(+180.6 \text{ kJ})$$

$$NO(g) \rightarrow \tfrac{1}{2}N_2(g) + \tfrac{1}{2}O_2(g) \qquad \Delta H = -90.3 \text{ kJ}$$

Adding the altered equations to obtain the target equation:

Equation A:  $CO(g) + \frac{1}{2}\cancel{O_2(g)} \rightarrow CO_2(g)$          $\Delta H = -283.0$ kJ

$\frac{1}{2}$ equation B reversed:    $NO(g) \rightarrow \frac{1}{2}N_2(g) + \frac{1}{2}\cancel{O_2(g)}$      $\Delta H = \phantom{-}{-90.3}$ kJ

Target:        $CO(g) + NO(g) \rightarrow CO_2(g) + \frac{1}{2}N_2(g)$      $\Delta H = -373.3$ kJ

**Check:** Obtaining the desired target equation is its own check. Be sure to remember to change the sign of any reversed equation.

**FOLLOW-UP PROBLEM 6.7**

Nitrogen oxides undergo many reactions. Calculate $\Delta H$ for the overall equation $2NO_2(g) + \frac{1}{2}O_2(g) \rightarrow N_2O_5(s)$ from the following information:

$N_2O_5(s) \rightarrow 2NO(g) + \frac{3}{2}O_2(g)$                    $\Delta H = 223.7$ kJ

$NO(g) + \frac{1}{2}O_2(g) \rightarrow NO_2(g)$                    $\Delta H = -57.1$ kJ

---

**Section Summary**

Since $H$ is a state function, $\Delta H = H_{final} - H_{initial}$ and does not depend on how the reaction takes place. Using Hess's law ($\Delta H_{total} = \Delta H_1 + \Delta H_2 + \cdots + \Delta H_n$), we can determine $\Delta H_{total}$ of any overall equation from appropriate individual reactions and their known $\Delta H$ values.

# 6.6    Standard Heats of Reaction ($\Delta H^0_{rxn}$)

Since $\Delta H$ varies somewhat with conditions, we use standardized $\Delta H$ values, which are calculated when all the substances are in their **standard state,** a set of specifications used to compare thermodynamic data. The standard states are very straightforward:

- For a *gas,* the standard state is 1 atm.*
- For a substance in *aqueous solution,* the standard state is 1 $M$ concentration (1 $M$ = 1 mol/L).
- For an *element or compound,* the standard state is the most stable form at 1 atm and the temperature of interest; in this text, the temperature of interest is usually 25°C (298 K).

A thermodynamic variable that has been determined with all substances in their standard states is indicated by a superscript zero. For example, when $\Delta H_{rxn}$ has been measured with all substances in their standard states, it is the **standard heat of reaction, $\Delta H^0_{rxn}$.**

### Formation Reactions and Their Standard Enthalpy Changes

In a **formation reaction,** one mole of a compound forms from its elements. The **standard heat of formation ($\Delta H^0_f$)** is the enthalpy change ac-

---

*The pressure at which the standard state values are obtained has been changed recently to 100 kPa, a slightly lower pressure (1 atm = 101.3 kPa). Since new tables of $\Delta H$ values that incorporate this pressure are not yet widely available, all values in this text refer to those attained at 1 atm.

**TABLE 6.5 Selected Standard Heats of Formation at 25°C (298 K)**

| FORMULA | $\Delta H_f^0$ (kJ/mol) |
|---|---|
| **Calcium** | |
| Ca(s) | 0 |
| CaO(s) | −635.1 |
| $CaCO_3(s)$ | −1206.9 |
| **Carbon** | |
| C(graphite) | 0 |
| C(diamond) | 1.9 |
| CO(g) | −110.5 |
| $CO_2(g)$ | −393.5 |
| $CH_4(g)$ | −74.9 |
| $CH_3OH(l)$ | −238.6 |
| $C_2H_5OH(l)$ | −277.6 |
| HCN(g) | 135 |
| $CS_2(l)$ | 87.9 |
| **Chlorine** | |
| $Cl_2(g)$ | 0 |
| Cl(g) | 121.0 |
| HCl(g) | −92.3 |
| **Hydrogen** | |
| H(g) | 218.0 |
| $H_2(g)$ | 0 |
| **Nitrogen** | |
| $N_2(g)$ | 0 |
| $NH_3(g)$ | −45.9 |
| NO(g) | 90.3 |
| $NO_2(g)$ | 33.2 |
| **Oxygen** | |
| $O_2(g)$ | 0 |
| $O_3(g)$ | 143 |
| $H_2O(g)$ | −241.8 |
| $H_2O(l)$ | −285.8 |
| **Silver** | |
| Ag(s) | 0 |
| AgF(s) | −203 |
| AgCl(s) | −127.0 |
| AgBr(s) | −99.5 |
| AgI(s) | −62.4 |
| **Sodium** | |
| Na(s) | 0 |
| Na(g) | 107.8 |
| NaCl(s) | −411.1 |
| $Na_2CO_3(s)$ | −1130.8 |
| **Sulfur** | |
| $S_8$(rhombic) | 0 |
| $S_8$(monoclinic) | 2 |
| $SO_2(g)$ | −296.8 |
| $SO_3(g)$ | −396.0 |

companying the formation reaction when all the substances are in their standard states. For the formation of methane ($CH_4$), we have

$$C(graphite) + 2H_2(g) \rightarrow CH_4(g) \qquad \Delta H_f^0 = -74.9 \text{ kJ}$$

Thus, the standard heat of formation of methane is −74.9 kJ/mol. Some other examples are

$$Na(s) + \tfrac{1}{2}Cl_2(g) \rightarrow NaCl(s) \qquad \Delta H_f^0 = -411.1 \text{ kJ}$$
$$2C(graphite) + 3H_2(g) + \tfrac{1}{2}O_2(g) \rightarrow C_2H_5OH(l) \qquad \Delta H_f^0 = -277.6 \text{ kJ}$$

Standard heats of formation have been tabulated for many compounds. Table 6.5 shows $\Delta H_f^0$ values for several elements and compounds; a much more extensive table appears in Appendix B.

The values in Table 6.5 were selected to make two points:

1. *An element in its standard state is assigned a $\Delta H_f^0$ of 0.* For example, note that $\Delta H_f^0 = 0$ for Na(s), but $\Delta H_f^0 = 107.8$ kJ/mol for Na(g). This means that the gaseous state is not the most stable state for sodium at 1 atm and 298 K and that heat is required to form Na(g). Similarly, $\Delta H_f^0 = 121.0$ kJ/mol for atomic chlorine because the more stable form of the element at 1 atm and 298 K is as diatomic molecules, $Cl_2$. Note also that although carbon exists as graphite and diamond, graphite is the more stable form at 1 atm and 298 K; therefore, $\Delta H_f^0 = 0$ for C(graphite). Similarly, the standard state for oxygen is $O_2$, not ozone ($O_3$), and that of sulfur is $S_8$ arranged in a rhombic crystal form.

2. *Most compounds have a negative $\Delta H_f^0$.* That is, most compounds have exothermic formation reactions under standard conditions. This means that, in most cases, *the compound is more stable than its component elements.*

**Writing Formation Reactions**

**Problem:** Write balanced equations for the formation of one mole of the following compounds from their elements and include $\Delta H_f^0$.
**(a)** Silver chloride, AgCl, a solid at standard conditions
**(b)** Calcium carbonate, $CaCO_3$, a solid at standard conditions
**(c)** Hydrogen cyanide, HCN, a gas at standard conditions
**Plan:** We write the elements as the reactants and one mole of the compound as the product, being sure all substances are in their standard states. Then we balance the atoms. $\Delta H_f^0$ values are obtained from Table 6.5.
**Solution: (a)**     $Ag(s) + \tfrac{1}{2}Cl_2(g) \rightarrow AgCl(s)$         $\Delta H_f^0 = -127.0$ kJ
**(b)**    $Ca(s) + C(graphite) + \tfrac{3}{2}O_2(g) \rightarrow CaCO_3(s)$      $\Delta H_f^0 = -1206.9$ kJ
**(c)**  $\tfrac{1}{2}H_2(g) + C(graphite) + \tfrac{1}{2}N_2(g) \rightarrow HCN(g)$        $\Delta H_f^0 = $    135 kJ

**FOLLOW-UP PROBLEM 6.8**
Write balanced equations for the formation of one mole of $CH_3OH(l)$, CaO(s), and $CS_2(l)$ from their elements in their standard states. Include $\Delta H_f^0$ for each reaction.

**Determining the Standard Heat of Reaction from Standard Heats of Formation**

By using $\Delta H_f^0$ values and applying Hess's law, we can determine $\Delta H_{rxn}^0$ for any chemical change. All we have to do is view the reaction as an imaginary two-step process:

*Step 1:* Each reactant breaks down into its elements. This step is the *reverse* of the formation reaction for the compound, so the standard enthalpy change for each reactant is $-\Delta H_f^0$.

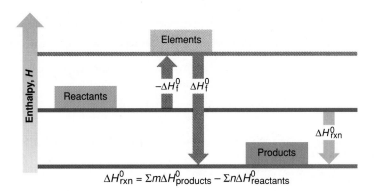

$$\Delta H^0_{rxn} = \Sigma m \Delta H^0_{products} - \Sigma n \Delta H^0_{reactants}$$

**FIGURE 6.13**

**The general process for determining $\Delta H^0_{rxn}$ from $\Delta H^0_f$ values.** For any reaction, $\Delta H^0_{rxn}$ can be considered as the sum of the enthalpy changes for the decomposition of reactants to their elements ($-\Sigma n \Delta H^0_{f\ (reactants)}$) and the formation of products from their elements ($\Sigma m \Delta H^0_{f\ (products)}$). (The coefficients $m$ and $n$ represent the moles of each substance in the balanced equation.)

*Step 2:* The appropriate elements form each product; that is, they undergo a formation reaction. The standard enthalpy change for each product is $+\Delta H^0_f$.

Hess's law states that we add the enthalpy changes for these steps to obtain the overall enthalpy change for the reaction ($\Delta H^0_{rxn}$) (Figure 6.13). Suppose we want the standard heat of reaction for

$$\text{TiCl}_4(l) + 2\text{H}_2\text{O}(g) \rightarrow \text{TiO}_2(s) + 4\text{HCl}(g)$$

Applying Hess's law, we write this equation as though it were the sum of four individual equations, one for each compound: two show the breakdown of the reactants to their elements (the *reverse* of their formation), and two show the formation of the products from their elements:

$$\text{TiCl}_4(l) \rightarrow \text{Ti}(s) + 2\text{Cl}_2(g) \qquad\qquad -\Delta H^0_f\,[\text{TiCl}_4(l)]$$

$$2\text{H}_2\text{O}(g) \rightarrow 2\text{H}_2(g) + \text{O}_2(g) \qquad\qquad -2\Delta H^0_f\,[\text{H}_2\text{O}(g)]$$

$$\text{Ti}(s) + \text{O}_2(g) \rightarrow \text{TiO}_2(s) \qquad\qquad \Delta H^0_f\,[\text{TiO}_2(s)]$$

$$2\text{H}_2(g) + 2\text{Cl}_2(g) \rightarrow 4\text{HCl}(g) \qquad\qquad 4\Delta H^0_f\,[\text{HCl}(g)]$$

$$\text{TiCl}_4(l) + 2\text{H}_2\text{O}(g) + \cancel{\text{Ti}(s)} + \cancel{\text{O}_2(g)} + \cancel{2\text{H}_2(g)} + \cancel{2\text{Cl}_2(g)} \rightarrow$$
$$\cancel{\text{Ti}(s)} + \cancel{2\text{Cl}_2(g)} + \cancel{2\text{H}_2(g)} + \cancel{\text{O}_2(g)} + \text{TiO}_2(s) + 4\text{HCl}(g)$$

Or,
$$\text{TiCl}_4(l) + 2\text{H}_2\text{O}(g) \rightarrow \text{TiO}_2(s) + 4\text{HCl}(g)$$

Note that we have viewed this change as if the reactants decomposed into their elements, which then recombined to the products, even if it did not actually occur that way. However, since $\Delta H^0_{rxn}$ is the difference between two state functions, $H^0_{products}$ minus $H^0_{reactants}$, it doesn't matter how we envision the change. We add the enthalpy changes to find the overall $\Delta H^0_{rxn}$:

$$\Delta H^0_{rxn} = \Delta H^0_f[\text{TiO}_2(s)] + 4\Delta H^0_f[\text{HCl}(g)] + \{-\Delta H^0_f[\text{TiCl}_4(l)]\} + \{-2\Delta H^0_f[\text{H}_2\text{O}(g)]\}$$

$$= \Delta H^0_f[\text{TiO}_2(s)] + 4\Delta H^0_f[\text{HCl}(g)] - \{\Delta H^0_f[\text{TiCl}_4(l)] + 2\Delta H^0_f[\text{H}_2\text{O}(g)]\}$$

By generalizing this example, we find that *the standard heat of reaction is the sum of the standard heats of formation of the products minus the sum of the standard heats of formation of the reactants:*

$$\Delta H^0_{rxn} = \Sigma m \Delta H^0_{f\ (products)} - \Sigma n \Delta H^0_{f\ (reactants)} \qquad\qquad \textbf{(6.8)}$$

where the symbol $\Sigma$ means "sum of," and $m$ and $n$ are the coefficients of the various reactants and products from the balanced equation.

SAMPLE PROBLEM 6.9

## Calculating the Heat of Reaction from Heats of Formation

**Problem:** The first step in the industrial production of nitric acid is the oxidation of ammonia according to the following equation:

$$4NH_3(g) + 5O_2(g) \rightarrow 4NO(g) + 6H_2O(g)$$

Calculate $\Delta H^0_{rxn}$ from $\Delta H^0_f$ values.

**Plan:** We find $\Delta H^0_{rxn}$ by applying Equation 6.8, using values from Table 6.5.

**Solution:** Calculating $\Delta H^0_{rxn}$:

$$\Delta H^0_{rxn} = \Sigma m\Delta H^0_{f\,(products)} - \Sigma n\Delta H^0_{f\,(reactants)}$$

$$= 4\Delta H^0_f[NO(g)] + 6\Delta H^0_f[H_2O(g)] - \{4\Delta H^0_f[NH_3(g)] + 5\Delta H^0_f[O_2(g)]\}$$

$$= 4 \text{ mol } (90.3 \text{ kJ/mol}) + 6 \text{ mol } (-241.8 \text{ kJ/mol})$$

$$- [4 \text{ mol } (-45.9 \text{ kJ/mol}) + 5 \text{ mol } (0 \text{ kJ/mol})]$$

$$= 361 \text{ kJ} - 1451 \text{ kJ} + 184 \text{ kJ} - 0 \text{ kJ} = \mathbf{-906 \text{ kJ}}$$

**Check:** One way to check is to write formation equations for the amounts of individual compounds in the correct direction and take their sum:

| | | |
|---|---|---:|
| $4NH_3(g) \rightarrow \cancel{2N_2(g)} + \cancel{6H_2(g)}$ | $-4(-45.9 \text{ kJ}) =$ | $184 \text{ kJ}$ |
| $\cancel{2N_2(g)} + 2O_2(g) \rightarrow 4NO(g)$ | $4(90.3 \text{ kJ}) =$ | $361 \text{ kJ}$ |
| $\cancel{6H_2(g)} + 3O_2(g) \rightarrow 6H_2O(g)$ | $6(-241.8 \text{ kJ}) =$ | $-1451 \text{ kJ}$ |
| $4NH_3(g) + 5O_2(g) \rightarrow 4NO(g) + 6H_2O(g)$ | | $-906 \text{ kJ}$ |

**FOLLOW-UP PROBLEM 6.9**

Given $\Delta H^0_{rxn}$, this method can also be used to calculate an unknown $\Delta H^0_f$. Use the following information to find $\Delta H^0_f$ of methanol ($CH_3OH$):

$$CH_3OH(l) + \tfrac{3}{2}O_2(g) \rightarrow CO_2(g) + 2H_2O(g) \qquad \Delta H^0_{comb} = -638.5 \text{ kJ}$$

$$\Delta H^0_f \text{ of } CO_2(g) = -393.5 \text{ kJ/mol} \qquad \Delta H^0_f \text{ of } H_2O(g) = -241.8 \text{ kJ/mol}$$

## Section Summary

Standard states are chosen conditions for substances. When one mole of a compound forms from its elements with all substances in their standard states, the enthalpy change is $\Delta H^0_f$. Hess's law allows us to picture a reaction as the breakdown of reactants to their elements, followed by the formation of products from their elements. We use tabulated $\Delta H^0_f$ values to find $\Delta H^0_{rxn}$ or use known $\Delta H^0_{rxn}$ and $\Delta H^0_f$ values to find an unknown $\Delta H^0_f$.

## Chapter Perspective

*Our investigation of energy in this chapter helps explain many chemical and physical phenomena: how gases do work, how reactions release or absorb heat, how the chemical energy in fuels and foods changes to other forms, and, as discussed in the following Chemical Connections essay, how we can utilize energy more wisely. We reexamine several of these ideas later in the text to address the question of why physical and chemical changes occur. In Chapter 7, you'll see how the energy released or absorbed by the atom created a revolutionary view of matter and energy that led to our current model of atomic energy states.*

**Chemistry in Environmental Science**
## The Future of Energy Use

Out of necessity, we are beginning to rethink our use of energy. The world's dwindling supply of fuel threatens our standard of living and has become a question of survival for hundreds of millions of people in less developed countries. Many chemical aspects of energy production and utilization provide scientists and engineers with some of the greatest challenges of our time.

Energy consumption increased enormously during the Industrial Revolution and again after World War II, and it continues to do so today. A century-long changeover has led from the use of wood to coal and now to petroleum. The **fossil fuels**—coal, petroleum, and natural gas—remain our major sources of energy, but their polluting combustion products and eventual depletion are major drawbacks to their use.

Natural processes form fossil fuels *much* more slowly than we consume them, so they are *nonrenewable:* once we use them up, they are gone. Estimates indicate that known oil reserves could be 90% depleted by the year 2020. Coal reserves are much greater, so chemical research seeks new uses for coal. The most important *renewable* combustible fuels are wood, other fuels derived from *biomass* (plant and animal matter), and hydrogen gas. The following discussion highlights some current approaches to developing more rational energy use: converting coal to cleaner fuels, producing fuels from biomass, developing a hydrogen-fueled economy, understanding the effect of carbon-based fuels on climate, utilizing noncombustible energy sources, and conserving the fuels we have.

### Chemical Approaches to Cleaner Fuels

Coal, biomass, and hydrogen are the most important fuels of the future.

*Coal.* Current U.S. reserves of coal are enormous, but coal is a highly polluting fuel because it produces $SO_2$ when it burns. Two processes are designed to reduce this noxious pollutant. Some flue-gas **desulfurization** devices called *scrubbers* use lime-water slurries [$Ca(OH)_2$] to remove $SO_2$ from the gaseous product mixture:

$$2Ca(OH)_2(aq) + 2SO_2(g) + O_2(g) \rightarrow 2CaSO_4(s) + 2H_2O(l)$$

**Coal gasification** processes alter the large molecules in coal to sulfur-free gaseous fuels. First, high temperatures (600° to 800°C) release volatile compounds that decompose coal to methane and a char:

$$\text{Coal}(s) \xrightarrow{\Delta} \text{coal volatiles}(g) \xrightarrow{\Delta} CH_4(g) + \text{char}(s)$$

The char (mostly carbon) reacts with steam in the endothermic *water-gas reaction* (or steam-carbon reaction) to form a fuel mixture of CO and $H_2$ called synthesis gas **(syngas):**

$$C(s) + H_2O(g) \rightarrow CO(g) + H_2(g) \quad \Delta H^0 = 131 \text{ kJ}$$

Syngas has a much lower fuel value than methane. For example, a mixture containing 0.5 mol CO and 0.5 mol $H_2$ (that is, 1.0 mol syngas) releases about one-third as much energy as 1.0 mol methane ($\Delta H^0_{comb} = -802$ kJ/mol):

$$\tfrac{1}{2}H_2(g) + \tfrac{1}{4}O_2(g) \rightarrow \tfrac{1}{2}H_2O(g)$$
$$\Delta H^0 = -121 \text{ kJ}$$
$$\tfrac{1}{2}CO(g) + \tfrac{1}{4}O_2(g) \rightarrow \tfrac{1}{2}CO_2(g)$$
$$\Delta H^0 = -142 \text{ kJ}$$

$$\tfrac{1}{2}H_2(g) + \tfrac{1}{2}CO(g) + \tfrac{1}{2}O_2(g) \rightarrow$$
$$\tfrac{1}{2}H_2O(g) + \tfrac{1}{2}CO_2(g) \quad \Delta H^0 = -263 \text{ kJ}$$

To upgrade its fuel value, syngas can be converted to other fuels, such as methane. In the *CO-shift reaction* (or water-gas shift reaction), some CO reacts with more steam to form $CO_2$ and additional $H_2$:

$$CO(g) + H_2O(g) \rightarrow CO_2(g) + H_2(g) \quad \Delta H^v = -41 \text{ kJ}$$

The $CO_2$ is removed, and the $H_2$-enriched mixture reacts to form methane and water vapor:

$$CO(g) + 3H_2(g) \rightarrow CH_4(g) + H_2O(g) \quad \Delta H^0 = -206 \text{ kJ}$$

Removal of the water vapor gives **synthetic natural gas (SNG).** Thus, through devolatilization and three reactions, coal is converted to methane.

*Wood and other types of biomass.* Nearly half the people in the world rely on wood for their energy needs. In principle, wood and other forms of biomass are renewable fuels, but worldwide deforestation for fuel, and for lumber and paper in richer countries, has resulted in wood being depleted much faster than it can be replenished.

A better use of vegetable (sugarcane, corn, beets) and tree waste is **biomass conversion,** which combines chemical and biological methods to convert sugar, cellulose, and other plant matter into combustible fuels. One such fuel is ethanol ($C_2H_5OH$), which is mixed with gasoline to form *gasohol.* Another is methane, which is produced in bio-gas generators through the microbial breakdown of plant and animal waste. Similar technologies produce gaseous fuels from garbage and sewage.

*Hydrogen.* Although $H_2$ ($\Delta H^0_{comb} = -242$ kJ/mol) produces only about one-third as much energy per mole as $CH_4$ ($\Delta H^0_{comb} = -802$ kJ/mol), its combustion yields nonpolluting water vapor, and it is renewable because the vapor returns as rain. Its formation from water, however, is endothermic and costly:

$$H_2O(l) \rightarrow H_2(g) + \tfrac{1}{2}O_2(g) \quad \Delta H^0 = +286 \text{ kJ}$$

*Continued.*

CHEMICAL CONNECTIONS

**Chemistry in Environmental Science—cont'd**
**The Future of Energy Use**

Alternatively, in direct applications of Hess's law, $H_2$ is obtained in reaction sequences that sum to the decomposition of water. Here is one example:

$$2CaBr_2(s) + 4H_2O(l) \rightarrow$$
$$2Ca(OH)_2(s) + 4HBr(g) \quad \Delta H^0 = 391 \text{ kJ}$$

$$2Hg(l) + 4HBr(g) \rightarrow$$
$$2HgBr_2(s) + 2H_2(g) \quad \Delta H^0 = -195 \text{ kJ}$$

$$2HgBr_2(s) + 2Ca(OH)_2(s) \rightarrow$$
$$2CaBr_2(s) + 2HgO(s) + 2H_2O(l) \quad \Delta H^0 = 193 \text{ kJ}$$

$$2HgO(s) \rightarrow 2Hg(l) + O_2(g) \quad \Delta H^0 = 182 \text{ kJ}$$

Sum: $2H_2O(l) \rightarrow 2H_2(g) + O_2(g) \quad \Delta H^0 = 571 \text{ kJ}$

How would the $H_2$ be stored, transported, and utilized? The most promising approach involves **interstitial hydrides**. Many metals absorb $H_2$, almost like a sponge, into the spaces (interstices) between their atoms. Palladium and niobium, for example, can absorb 1000 times their volume of $H_2$ gas, retain it under normal conditions, and release it at high temperatures.

### Carbon-Based Fuels and the "Greenhouse" Effect

Many natural processes produce $CO_2$, which plays a key temperature-regulating role in the atmosphere. Much of the sunlight that shines on the Earth is absorbed by the land and oceans and converted to heat (Figure 6.A). The

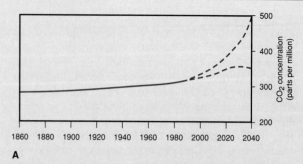

A

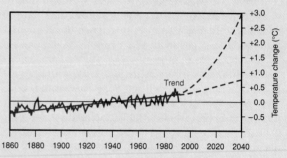

B

**FIGURE 6.B**
**Accumulating evidence of the greenhouse effect. A,** As a result of increased fossil fuel use since the onset of the Industrial Revolution in the mid-1800s, atmospheric $CO_2$ concentrations have increased. **B,** Coincident with this $CO_2$ increase, the average global temperature has risen slightly more than 0.5°C. (Zero equals the average global temperature for 1957 to 1970.) The projections in both graphs *(dashed lines)* are based on current fossil fuel consumption and deforestation continuing *(upper line)* or being curtailed *(lower line).*

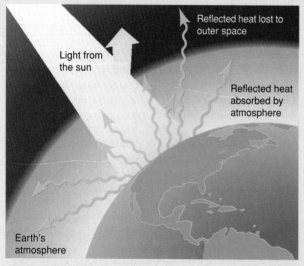

**FIGURE 6.A**
**The trapping of heat by the atmosphere.** About 25% of sunlight is reflected by the atmosphere, but most enters and is converted into heat by the air and the Earth's surface. Some of the heat emanating from the surface is trapped by atmospheric gases, mostly $CO_2$, and some is lost to outer space.

$CO_2$ does not absorb sunlight, but it does absorb heat. Thus, some of the heat radiating from the surface is trapped by the $CO_2$ and transferred to the atmosphere.

Over several billion years, the amount of $CO_2$ originally present in the Earth's atmosphere decreased to a relatively constant 0.03% by volume. However, this amount has been increasing for the past 150 years due to our burning of fossil fuels. Thus, although the same amount of solar energy passes through the atmosphere, more is being trapped as heat in a process called the "greenhouse" effect (Figure 6.B).

Based on current fossil fuel use, $CO_2$ concentrations will nearly double by 2060, but how much will the temperature of the atmosphere rise? A definitive answer is difficult to obtain. As the amounts of $CO_2$ and other "greenhouse" gases (most importantly, $CH_4$) increase from fossil fuel burning so does the amount of soot, which may block sunlight and have a cooling effect. Water vapor also traps heat; therefore, as temperatures rise, more water vapor forms, but this may thicken the cloud cover and also lead to cooling.

Despite these opposing factors, most atmospheric scientists predict a net warming of the atmosphere, and several predictions of the effects of temperature rises have been made. An average increase of 1°C in the U.S. Midwest could lower the grain yield by 20%; an increase of 3°C could alter rainfall patterns and crop yields throughout the world; a 5°C increase could melt large parts of the polar ice caps, flooding coastal regions such as the Netherlands, half of Florida, and much of India. To make matters worse, as we burn fossil fuels and wood, we simultaneously cut down the forests that naturally *absorb* $CO_2$. International conferences, such as the 1992 Earth Summit in Brazil, provide a forum for politicians and scientists to grapple with these complex issues.

### Solar Energy and Other Noncombustible Energy Sources

The amount of solar energy striking the United States annually is *600 times greater than all our energy needs.* Despite this astounding abundance, solar energy is difficult to collect, store, and convert to other forms of energy. The sun's total output is enormous but not concentrated, so vast surface areas must be devoted to collecting it.

Storage of the solar energy is necessary because intense sunlight is available over most regions for only 6 to 8 hours a day, and it is greatly reduced in rainy or cloudy weather. One storage approach makes use of ionic hydrates, such as $Na_2SO_4·10H_2O$. When warmed by sunlight to greater than 32°C, the 3 moles of ions dissolve in the 10 moles of water in an endothermic process:

$$Na_2SO_4·10H_2O(s) \xrightarrow{>32°C} Na_2SO_4(aq, \text{ from } 10H_2O)$$
$$\Delta H^0 = 354 \text{ kJ}$$

When cooled to less than 32°C after sunset, the solution recrystallizes, releasing the absorbed energy for heating:

$$Na_2SO_4(aq) \xrightarrow{<32°C} Na_2SO_4·10H_2O(s) \quad \Delta H^0 = -354 \text{ kJ}$$

**Photovoltaic cells** convert light directly into electricity, but the electrical energy produced is only about 10% of the radiant energy supplied. However, novel combinations of Group 3A(13) and 5A(15) elements, such as gallium arsenide, result in higher yields.

### Energy Conservation: More from Less

All systems in nature, whether living organisms or industrial power plants, waste some energy, which in effect is a waste of fuel. In terms of the energy used to produce it, for example, every discarded aluminum can is equivalent to 0.25 L of gasoline. Energy conservation lowers production costs, extends our supply of fossil fuels, and reduces pollution.

One example of an energy-conserving device is a high-efficiency gas-burning furnace for domestic heating. Its design channels hot waste gases through the furnace to transfer more heat to the room and to an attached domestic water system. Moreover, since the gases are cooled to less than 100°C, the water vapor condenses, releasing about 10% more heat:

$$CH_4(g) + 2O_2(g) \rightarrow CO_2(g) + 2H_2O(g) \quad \Delta H^0 = -802 \text{ kJ}$$
$$2H_2O(g) \rightarrow 2H_2O(l) \quad \Delta H^0 = -88 \text{ kJ}$$
$$CH_4(g) + 2O_2(g) \rightarrow CO_2(g) + 2H_2O(l) \quad \Delta H^0 = -890 \text{ kJ}$$

Although engineers and chemists will be on the forefront of these new energy directions, a more hopeful energy future ultimately depends on our wisdom in obtaining and conserving planetary resources.

# For Review and Reference

## Key Terms

thermodynamics
thermochemistry

**SECTION 6.1**
system
surroundings
internal energy ($E$)
heat ($q$)
work ($w$)
pressure-volume work ($PV$ work)
joule (J)
calorie (cal)
state function

law of conservation of energy (first law of thermodynamics)

**SECTION 6.2**
enthalpy ($H$)
change in enthalpy ($\Delta H$)
heat of reaction ($\Delta H_{rxn}$)
enthalpy diagram
exothermic
endothermic
heat of combustion ($\Delta H_{comb}$)
heat of formation ($\Delta H_f$)
heat of fusion ($\Delta H_{fus}$)

heat of vaporization ($\Delta H_{vap}$)

**SECTION 6.3**
heat capacity
specific heat capacity ($C$)
molar heat capacity
calorimeter
bomb calorimeter

**SECTION 6.4**
thermochemical equation

**SECTION 6.5**
Hess's law of heat summation

**SECTION 6.6**
standard state
standard heat of reaction ($\Delta H_{rxn}^0$)
formation reaction
standard heat of formation ($\Delta H_f^0$)
fossil fuel
desulfurization
coal gasification
syngas
synthetic natural gas
biomass conversion
interstitial hydride
photovoltaic cell

## Key Equations and Relationships

**6.1** Defining the change in internal energy of a system (p. 222):

$$\Delta E = E_{final} - E_{initial} = E_{products} - E_{reactants}$$

**6.2** Expressing the change in internal energy in terms of heat and work (p. 223):

$$\Delta E = q + w$$

**6.3** Stating the first law of thermodynamics (law of energy conservation) (p. 227):

$$\Delta E_{universe} = \Delta E_{system} + \Delta E_{surroundings} = 0$$

**6.4** Determining the work due to a change in volume at constant pressure (*PV* work) (p. 228):

$$w = -P\Delta V$$

**6.5** Relating the enthalpy change to the internal energy change (p. 228):

$$\Delta H = \Delta E + P\Delta V$$

**6.6** Identifying the enthalpy change with the heat at constant pressure (p. 228):

$$q_P = \Delta H$$

**6.7** Calculating the heat when a substance undergoes a temperature change (p. 234):

$$q = C \times mass \times \Delta T$$

**6.8** Calculating the standard heat of reaction for a chemical change (p. 243):

$$\Delta H^0_{rxn} = \Sigma m \Delta H^0_{f\,(products)} - \Sigma n \Delta H^0_{f\,(reactants)}$$

## Answers to Follow-up Problems

**6.1** $\Delta E = -26$ kJ $+ 15$ kJ $= -11$ kJ

**6.2** The reaction is exothermic.

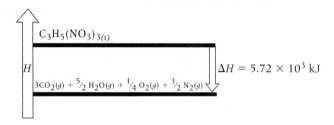

**6.3** $\Delta T = 25°C - 37°C = -12°C = -12$ K

$q = 0.450$ J/g · K $\times 5.5$ g $\times (-12$ K$) = -30$ J

**6.4** $-q_{solid} = q_{water}$

$-[(0.519$ J/g · K$) (2.050$ g$) (x - 74.21)]$

$= [(4.184$ J/g · K$) (26.05$ g$) (x - 27.20)]$; $x = 27.65$ K

$\Delta T_{diamond} = -46.56$ K and $\Delta T_{water} = 0.45$ K

**6.5** $-q_{sample} = q_{calorimeter}$

$-(0.8650$ g C$)\left(\dfrac{1 \text{ mol C}}{12.01 \text{ g C}}\right)(-393.5$ kJ/mol C$) = (2.613$K$)x$

$x = 10.85$ kJ/K

**6.6** $HgO(s) \xrightarrow{\Delta} Hg(l) + \frac{1}{2}O_2(g)$   $\Delta H = 90.8$ kJ

Heat (kJ) $= 907$ kg Hg $\times \dfrac{10^3 \text{ g}}{1 \text{ kg}} \times \dfrac{1 \text{ mol Hg}}{200.6 \text{ g Hg}} \times \dfrac{90.8 \text{ kJ}}{1 \text{ mol Hg}}$

$= 4.11 \times 10^5$ kJ

**6.7** $2NO(g) + \frac{3}{2}O_2(g) \rightarrow N_2O_5(s)$   $\Delta H = -223.7$ kJ

$2NO_2(g) \rightarrow 2NO(g) + O_2(g)$   $\Delta H = 114.2$ kJ

$\overline{2NO(g) + \frac{1}{2}\frac{3}{2}O_2(g) + 2NO_2(g) \rightarrow}$
$\quad\quad N_2O_5(s) + 2NO(g) + O_2(g)$   $\Delta H = -109.5$ kJ
$\frac{1}{2}O_2(g) + 2NO_2(g) \rightarrow N_2O_5(s)$   $\Delta H = -109.5$ kJ

**6.8** $C(graphite) + 2H_2(g) + \frac{1}{2}O_2(g) \rightarrow CH_3OH(l)$;
$\quad\quad\quad\quad\quad\quad\quad\quad\quad \Delta H^0_f = -238.6$ kJ

$Ca(s) + \frac{1}{2}O_2(g) \rightarrow CaO(s)$   $\Delta H^0_f = -635.1$ kJ

$C(graphite) + \frac{1}{4}S_8(rhombic) \rightarrow CS_2(l)$   $\Delta H^0_f = 87.9$kJ

**6.9** $\Delta H^0_f$ of $CH_3OH(l)$

$= -\Delta H^0_{comb} + 2\Delta H^0_f [H_2O(g)] + \Delta H^0_f [CO_2(g)]$

$= 638.5$ kJ $+ 2$ mol$(-241.8$ kJ/mol$)$

$\quad + 1$ mol $(-393.5$ kJ/mol$)$

$= -238.6$ kJ

## Sample Problem Titles

# Problems

Problems with a green number are answered at the back of the text. Most sections include three categories of problems separated by a green rule: concept review questions, *paired* skill-building exercises, and problems in a relevant context.

## Forms of Energy and Their Interconversion

(Sample Problem 6.1)

**6.1** Why do heat ($q$) and work ($w$) have positive values when entering a system and negative values when leaving it?

**6.2** If you feel warm after exercising, have you increased the internal energy of your body? Explain.

**6.3** For each of the following events, the system is in *italics*. State whether heat, work, or both is (are) transferred and the direction of each transfer.
(a) You pump air into an automobile *tire*.
(b) A *tree* rots in a forest.
(c) You strike a *match*.
(d) You cool juice in an *ice chest*.
(e) You cook *food* on a kitchen range.

**6.4** An *adiabatic* process is one that involves no heat transfer. What is the relationship between work and internal energy in an adiabatic process?

**6.5** State two ways that you increase the internal energy of your body and two ways that you decrease it.

**6.6** Name a common device in which each of the following energy changes occurs:
(a) Electrical energy to thermal energy
(b) Electrical energy to sound energy
(c) Electrical energy to light energy
(d) Mechanical energy to electrical energy
(e) Chemical energy to electrical energy

**6.7** In winter, an electric room heater uses a certain amount of electrical energy to heat a room to 20°C. In summer, an air conditioner uses the same amount of electrical energy to cool the room to 20°C. Is the change in internal energy of the heater larger, smaller, or the same as that of the air conditioner? Explain.

**6.8** Consider lifting your textbook into the air and dropping it on your desk top. Describe all the energy transformations (from one form to another) that have occurred, moving backward in time, from a moment after impact.

**6.9** A system receives 405 J of heat and delivers 405 J of work to its surroundings. What is the change in internal energy of the system (in J)?

**6.10** A system conducts 215 cal of heat to the surroundings while delivering 408 cal of work. What is the change in internal energy of the system (in cal)?

**6.11** What is the change in internal energy (in J) of a system that releases 575 J of thermal energy to its surroundings and has 425 cal of work done on it?

**6.12** What is the change in internal energy (in J) of a system that absorbs 0.515 kJ of heat from its surroundings and has 0.147 kcal of work done on it?

**6.13** Complete combustion of 1.0 metric ton of coal (assuming pure carbon) to gaseous carbon dioxide releases $3.3 \times 10^{10}$ J of heat. Convert this energy to (a) kilojoules; (b) kilocalories; (c) British thermal units.

**6.14** Thermal decomposition of 1.0 metric ton of limestone to lime and carbon dioxide requires $1.8 \times 10^6$ kJ of heat. Convert this energy to (a) joules; (b) calories; (c) British thermal units.

**6.15** The nutritional calorie (Calorie) is equivalent to the kilocalorie. One pound of body fat is equivalent to about $4.1 \times 10^3$ Calories. Express this energy equivalence in joules and kilojoules.

**6.16** If an athlete expends 1700 kJ/h playing tennis, how long would she have to play to work off 1.0 lb of body fat? (See Problem 6.15.)

**6.17** A typical candy bar weighs about 2 oz (1.00 oz = 28.4 g). Answer the following, assuming that a candy bar is 100% sugar and that 1.0 g sugar is equivalent to about 4.0 Calories of energy:
(a) How much energy (in kilojoules) is contained in a typical candy bar?
(b) Assuming your mass is 70 kg and you convert chemical potential energy to work with 100% efficiency, how high would you have to climb to work off this energy? (Potential energy = mass $\times$ $g \times$ height, where $g$ = 9.8 m/s$^2$.)
(c) Why is your actual conversion of potential energy to work less than 100% efficient?

## Enthalpy: Heats of Reaction and Chemical Change

(Sample Problem 6.2)

**6.18** Why is a system that undergoes an expansion against a constant external pressure assigned a negative sign for the work done?

**6.19** Why is it usually more convenient for chemists to measure $\Delta H$ than $\Delta E$?

**6.20** "Hot packs" used by skiers, climbers, and others for warmth are based on the crystallization of sodium acetate from a highly concentrated solution. What is the sign of $\Delta H$ for this crystallization reaction? Is the reaction exothermic or endothermic?

**6.21** Classify the following processes as exothermic or endothermic: (a) freezing of water; (b) boiling of water; (c) digestion of food; (d) a person running; (e) a person growing; (f) wood being chopped; (g) heating with a furnace.

**6.22** What are the two main components of the internal energy of a substance? On what are they based?

**6.23** For each of the following processes, state whether $\Delta H$ is less than (more negative), equal to, or greater than $\Delta E$ of the system. Explain.
(a) An ideal gas is cooled at constant pressure.
(b) A mixture of gases undergoes an exothermic reaction in a container of fixed volume.
(c) A solid yields a mixture of gases in an exothermic reaction that takes place in a container of variable volume.

**6.24** Draw an enthalpy diagram for a general exothermic reaction. Label axis, reactants, products, and $\Delta H$ with its sign.

**6.25** Draw an enthalpy diagram for a general endothermic reaction. Label axis, reactants, products, and $\Delta H$ with its sign.

**6.26** Write a balanced equation and draw an approximate enthalpy diagram for each of the following changes:
(a) The combustion of methane in oxygen
(b) The freezing of liquid water
(c) The formation of 1 mol sodium chloride from its elements (heat is released)

**6.27** Write a balanced equation and draw an approximate enthalpy diagram for each of the following changes:
(a) The combustion of 1 mol liquid ethanol ($C_2H_5OH$)
(b) The formation of 1 mol nitrogen dioxide from its elements (heat is absorbed)
(c) The sublimation of dry ice [conversion of $CO_2(s)$ directly to $CO_2(g)$]

**6.28** Which of the following gases would you expect to have the greater heat of combustion per mole? Why?

methane    or    formaldehyde

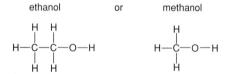

**6.29** Which of the following two fuel candidates would you expect to have the greater heat of combustion per mole? Why?

ethanol    or    methanol

## Calorimetry: Laboratory Measurement of Heats of Reaction

(Sample Problems 6.3 to 6.5)

**6.30** Why do we measure only *changes* in enthalpy, not absolute values?

**6.31** What data do you need to determine the specific heat capacity of a substance?

**6.32** Is the specific heat capacity of a substance an intensive or extensive property? Is it a state function? Explain.

**6.33** Distinguish between "specific heat capacity" and "heat capacity." Which parameter would you likely use if you were calculating heat changes in (a) a chrome-plated, brass bathroom fixture, (b) a sample of high-purity copper wire, (c) a sample of pure water? Explain.

**6.34** Both a coffee-cup calorimeter and a bomb calorimeter can be used to measure the heat involved in a reaction. Which measures $\Delta E$ and which measures $\Delta H$? Explain.

**6.35** Calculate $q$ when 10.0 g water is heated from 20°C to 100°C.

**6.36** Calculate $q$ when 0.10 g ice is cooled from 0°C to $-80$°C. ($C$ of ice = 2.087 J/g · K)

**6.37** A 195-g aluminum engine part at an initial temperature of 3.00°C absorbs 40.0 kJ of heat. What is the final temperature of the part? ($C$ of Al = 0.900 J/g · K)

**6.38** A 27.7-g sample of ethylene glycol, a car radiator coolant, loses 588 J of heat. What was the initial temperature if the final temperature is 30.5°C? ($C$ of ethylene glycol = 2.42 J/g · K)

**6.39** Two iron bolts of equal mass—one at 100°C, the other at 55°C—are placed in an insulated container. Assuming the heat capacity of the container is negligible, what is the final temperature inside the container? ($C$ of iron = 0.450 J/g · K)

**6.40** One piece of copper jewelry at 100°C has exactly twice the mass of another piece, which is at 40°C. They are placed inside a calorimeter whose heat capacity is negligible. What is the final temperature inside the calorimeter? ($C$ of copper = 0.387 J/g · K)

**6.41** When 155 mL water at 20°C is mixed with 75 mL water at 80°C, what is the final temperature? (Assume that no heat is lost to the surroundings and that the density of water is 1.00 g/mL.)

**6.42** An unknown volume of water at 18.2°C is added to 24.4 mL water at 30.0°C. If the final temperature is 21.5°C, what was the unknown volume? (Assume that no heat is lost to the surroundings and that the density of water is 1.00 g/mL.)

**6.43** A 505-g piece of copper tubing is heated to 99.9°C and placed in an insulated vessel containing 59.8 g water at 24.8°C. If the vessel has a heat capacity of 10.0 J/K, what is the final temperature of the system? ($C$ of copper = 0.387 J/g · K)

**6.44** A 30.5-g sample of an alloy at 95.0°C is placed into 50.0 g water at 25.0°C in an insulated coffee cup with a heat capacity of 9.0 J/K. If the final temperature of the system is 30.1°C, what is the specific heat capacity of the alloy?

**6.45** Two aircraft rivets, one of iron and the other of copper, are placed in a calorimeter that has an initial temperature of 20°C. The data for the metals are as follows:

|  | IRON | COPPER |
|---|---|---|
| Mass (g) | 30.0 | 20.0 |
| Initial $T$ (°C) | 0.0 | 100.0 |
| $C$ (J/g · K) | 0.450 | 0.387 |

(a) Will heat flow from Fe to Cu or from Cu to Fe?

(b) What other information is needed to correct any measurements that would be made in an actual experiment?

(c) What is the maximum attainable final temperature of the system (assuming the heat capacity of the calorimeter is negligible)?

**6.46** Pioneers and cowboys sometimes heated their coffee by placing an iron poker from the fire directly into the coffee. If a cup held 0.50 L coffee at 20°C, what would be the final temperature of the coffee when a 502-g iron poker at 800°C was placed in it? (Assume no heat is lost to the surroundings and no water vaporizes.)

**6.47** A chemical engineer studying the properties of fuels placed 1.500 g of a hydrocarbon in the bomb of a calorimeter and filled it with $O_2$ gas (see Figure 6.11). The bomb was immersed in 2.500 L water and the reaction initiated. The water temperature rose from 20.00°C to 23.55°C. If the calorimeter (excluding the water) had a heat capacity of 403 J/K, what was the heat of combustion ($q_V$) per gram of the fuel?

## Stoichiometry in Thermochemical Equations

(Sample Problem 6.6)

**6.48** Does a negative $\Delta H_{rxn}$ mean that the heat quantity appears on the reactant or product side of the equation?

**6.49** Would you expect the following thermochemical equation to have a positive or a negative $\Delta H_{rxn}$? Explain.

$$O_2(g) \rightarrow 2O(g)$$

**6.50** Is $\Delta H$ positive or negative when 1 mol water vapor condenses to liquid water? Why? How does this value compare with that from the vaporization of 2 mol liquid water to water vapor?

**6.51** Consider the following balanced thermochemical equation, sometimes used for $H_2S$ production:

$$\tfrac{1}{8}S_8(s) + H_2(g) \rightarrow H_2S(g) \quad \Delta H_{rxn} = -20.2 \text{ kJ}$$

(a) Is this an exothermic or endothermic reaction?

(b) What is $\Delta H_{rxn}$ for the reverse reaction?

(c) What is $\Delta H_{rxn}$ when 1.0 mol $S_8$ reacts?

**6.52** Consider the following balanced thermochemical equation for the decomposition of the mineral magnesite:

$$MgCO_3(s) \rightarrow MgO(s) + CO_2(g) \quad \Delta H_{rxn} = 117.3 \text{ kJ}$$

(a) Is heat absorbed or released in the reaction?

(b) What is $\Delta H_{rxn}$ for the reverse reaction?

(c) What is $\Delta H_{rxn}$ when 5.35 mol $CO_2$ reacts with excess MgO?

**6.53** When 1 mol NO(g) forms from its elements, 90.29 kJ of heat is absorbed.

(a) Write a balanced thermochemical equation for this reaction.

(b) How much heat is involved when 1.50 g NO is decomposed to its elements?

**6.54** When 1 mol KBr(s) decomposes to its elements, 394 kJ of heat is absorbed.

(a) Write a balanced thermochemical equation for this reaction.

(b) How much heat is released when 10 kg KBr forms from its elements?

**6.55** Liquid hydrogen peroxide is an oxidizing agent in many rocket fuel mixtures because it releases oxygen gas on decomposition:

$$2H_2O_2(l) \rightarrow 2H_2O(l) + O_2(g) \quad \Delta H_{rxn} = -196.1 \text{ kJ}$$

How much heat is released when 732 kg $H_2O_2$ decomposes?

**6.56** Compounds of boron and hydrogen are remarkable, not only for their unusual bonding (described in Section 13.5), but also for their reactivity. With the more reactive halogens, for example, diborane ($B_2H_6$) forms trihalides even at low temperatures:

$$B_2H_6(g) + 6Cl_2(g) \rightarrow 2BCl_3(g) + 6HCl(g)$$
$$\Delta H_{rxn} = -755.4 \text{ kJ}$$

How much heat is released per kilogram of diborane that reacts?

**6.57** Deterioration of buildings, bridges, and other structures through the rusting of iron costs millions of dollars every day. Although the actual process requires both oxygen and water, a simplified equation (with rust shown as $Fe_2O_3$) is

$$4Fe(s) + 3O_2(g) \rightarrow 2Fe_2O_3(s) \quad \Delta H_{rxn} = -1.65 \times 10^3 \text{ kJ}$$

(a) How much heat is evolved when 0.100 kg iron rusts?

(b) How much rust forms when $4.93 \times 10^3$ kJ of heat is released?

**6.58** A mercury mirror forms inside a test tube by the thermal decomposition of mercury(II) oxide:

$$2HgO(s) \rightarrow 2Hg(l) + O_2(g) \quad \Delta H_{rxn} = 181.6 \text{ kJ}$$

(a) How much heat is needed to decompose 555 g of the oxide?

(b) If 275 kJ of heat is absorbed, how many grams of mercury form?

**6.59** Ethylene ($C_2H_4$) is the starting material for the preparation of polyethylene. Although typically manufactured, it occurs naturally as a fruit-ripening hormone in plants and as a component of natural gas.

(a) If the heat of combustion of $C_2H_4$ is -1411 kJ/mol, write a balanced thermochemical equation for the combustion of $C_2H_4$.

(b) How many grams of $C_2H_4$ must be burned to give 70.0 kJ of heat?

**6.60** Sucrose ($C_{12}H_{22}O_{11}$, table sugar) is oxidized in the body by $O_2$ via a complex set of reactions that ultimately produce $CO_2(g)$ and $H_2O(g)$ and release $5.64 \times 10^3$ kJ/mol sucrose.
(a) Write a balanced thermochemical equation for this reaction.
(b) How much heat is released for every 1.00 g sucrose oxidized?

## Hess's Law of Heat Summation

(Sample Problem 6.7)

**6.61** Express in your own words the main principle underlying Hess's law.

**6.62** What is the main use of Hess's law?

**6.63** It is very difficult to burn carbon in a deficiency of $O_2$ and produce only CO; some $CO_2$ forms as well. However, carbon burns in excess $O_2$ to form only $CO_2$, and CO burns in excess $O_2$ to form only $CO_2$. Use the heats of the latter two reactions to calculate the $\Delta H_{rxn}$ for the following reaction: $C(s) + \frac{1}{2}O_2(g) \rightarrow CO(g)$

**6.64** Calculate $\Delta H_{rxn}$ for

$$Ca(s) + \frac{1}{2}O_2(g) + CO_2(g) \rightarrow CaCO_3(s)$$

given the following set of reactions:

$$Ca(s) + \frac{1}{2}O_2(g) \rightarrow CaO(s) \qquad \Delta H = -635.1 \text{ kJ}$$
$$CaCO_3(s) \rightarrow CaO(s) + CO_2(g) \quad \Delta H = 178.3 \text{ kJ}$$

**6.65** Calculate $\Delta H_{rxn}$ for

$$2NOCl(g) \rightarrow N_2(g) + O_2(g) + Cl_2(g)$$

given the following set of reactions:

$$\frac{1}{2}N_2(g) + \frac{1}{2}O_2(g) \rightarrow NO(g) \qquad \Delta H = 90.3 \text{ kJ}$$
$$NO(g) + \frac{1}{2}Cl_2(g) \rightarrow NOCl(g) \quad \Delta H = -38.6 \text{ kJ}$$

**6.66** At a specific set of conditions, 241.8 kJ is given off when 1 mol $H_2O(g)$ forms from its elements. Under the same conditions, 285.8 kJ is given off when 1 mol $H_2O(l)$ forms from its elements. Calculate the heat of vaporization of water at these conditions.

**6.67** When 1 mol $CS_2(l)$ forms from its elements at 1 atm and 25°C, 89.7 kJ is absorbed, and it takes 27.7 kJ to vaporize one mole of the liquid. How much heat is absorbed when 1 mol $CS_2(g)$ forms from its elements at these conditions?

**6.68** Phosphorus pentachloride is used in the industrial preparation of many organic phosphorus compounds. Equation I shows its preparation from $PCl_3$ and $Cl_2$:

(I) $PCl_3(l) + Cl_2(g) \rightarrow PCl_5(s)$

Use equations II and III to calculate $\Delta H_{rxn}$ of equation I:

(II) $P_4(s) + 6Cl_2(g) \rightarrow 4PCl_3(l)$ $\qquad \Delta H = -1280 \text{ kJ}$
(III) $P_4(s) + 10Cl_2(g) \rightarrow 4PCl_5(s)$ $\qquad \Delta H = -1774 \text{ kJ}$

**6.69** Diamond and graphite are two crystalline forms of carbon. At 1 atm and 25°C, diamond changes to graphite so slowly that the enthalpy change of the process must be obtained indirectly. Determine $\Delta H_{rxn}$ for

$$C(diamond) \rightarrow C(graphite)$$

by selecting equations from the following list:

(I) $C(diamond) + O_2(g) \rightarrow CO_2(g)$ $\quad \Delta H = -395.4 \text{ kJ}$
(II) $2CO_2(g) \rightarrow 2CO(g) + O_2(g)$ $\qquad \Delta H = 566.0 \text{ kJ}$
(III) $C(graphite) + O_2(g) \rightarrow CO_2(g)$ $\quad \Delta H = -393.5 \text{ kJ}$
(IV) $2CO(g) \rightarrow C(graphite) + CO_2(g)$ $\quad \Delta H = -172.5 \text{ kJ}$

## Standard Heats of Reaction ($\Delta H^0_{rxn}$)

(Sample Problems 6.8 and 6.9)

**6.70** What is the difference between the standard heat of formation and the standard heat of reaction?

**6.71** How are $\Delta H^0_f$ values used to calculate $\Delta H^0_{rxn}$?

**6.72** Make any changes needed in each of the following equations to make the $\Delta H^0_{rxn}$ equal to $\Delta H^0_f$ for the compound present:
(a) $Cl(g) + Na(s) \rightarrow NaCl(s)$
(b) $H_2O(g) \rightarrow 2H(g) + \frac{1}{2}O_2(g)$
(c) $\frac{1}{2}N_2(g) + \frac{3}{2}H_2(g) \rightarrow NH_3(g)$

**6.73** Use Table 6.5 or Appendix B to write balanced formation equations at standard conditions for each of the following compounds: (a) $CaCl_2$; (b) $NaHCO_3$; (c) $CCl_4$; (d) $HNO_3$.

**6.74** Use Table 6.5 or Appendix B to write balanced formation equations at standard conditions for each of the following compounds: (a) HI; (b) $SiF_4$; (c) $O_3$; (d) $Ca_3(PO_4)_2$.

**6.75** Calculate $\Delta H^0_{rxn}$ for each of the following:
(a) $2H_2S(g) + 3O_2(g) \rightarrow 2SO_2(g) + 2H_2O(g)$
(b) $CH_4(g) + Cl_2(g) \rightarrow CCl_4(l) + HCl(g)$ [unbalanced]

**6.76** Calculate $\Delta H^0_{rxn}$ for each of the following:
(a) $SiO_2(s) + 4HF(g) \rightarrow SiF_4(g) + 2H_2O(g)$
(b) $C_2H_6(g) + O_2(g) \rightarrow CO_2(g) + H_2O(g)$ [unbalanced]

**6.77** Copper(I) oxide is oxidized to copper(II) oxide according to the following equation:

$$Cu_2O(s) + \frac{1}{2}O_2(g) \rightarrow 2CuO(s) \qquad \Delta H^0_{rxn} = -146.0 \text{ kJ}$$

Given that $\Delta H^0_f$ of $Cu_2O(s) = -168.6$ kJ/mol, what is $\Delta H^0_f$ of $CuO(s)$?

**6.78** Acetylene burns in air according to the following equation:

$$C_2H_2(g) + \frac{5}{2}O_2(g) \rightarrow 2CO_2(g) + H_2O(g)$$
$$\Delta H^0_{rxn} = -1255.8 \text{ kJ}$$

Given that $\Delta H^0_f$ of $CO_2(g) = -393.5$ kJ/mol and $\Delta H^0_f$ of $H_2O(g) = -241.8$ kJ/mol, what is $\Delta H^0_f$ of $C_2H_2(g)$?

**6.79** Nitroglycerine, $C_3H_5(NO_3)_3(l)$, is a powerful explosive used in mining. When detonated, it produces a hot gaseous mixture of nitrogen, water, carbon dioxide, and oxygen.
  (a) Write a balanced equation for this reaction using the smallest whole-number coefficients.
  (b) If $\Delta H^0_{rxn} = -2.29 \times 10^4$ kJ for the equation as written in part (a), calculate $\Delta H^0_f$ of nitroglycerine.

**6.80** Physicians and nutritional biochemists recommend eating vegetable oils rather than animal fats to lower risks of heart disease, and olive oil is considered one of the "healthier" choices. The main fatty acid in olive oil is oleic acid ($C_{18}H_{34}O_2$), whose $\Delta H^0_{comb} = -1.11 \times 10^4$ kJ/mol. Calculate $\Delta H^0_f$ of oleic acid. [Assume $H_2O$ (g)]

**6.81** The common lead-acid car battery produces a large burst of current, even at low temperatures, and is rechargeable. The reaction that occurs while recharging a "dead" battery is

$$2PbSO_4(s) + 2H_2O(l) \rightarrow Pb(s) + PbO_2(s) + 2H_2SO_4(l)$$

  (a) Use $\Delta H^0_f$ values from Appendix B to calculate $\Delta H^0_{rxn}$.
  (b) Use the following equations to check your answer in part (a):

  (I) $Pb(s) + PbO_2(s) + 2SO_3(g) \rightarrow 2PbSO_4(s)$
  $$\Delta H^0 = -768 \text{ kJ}$$
  (II) $SO_3(g) + H_2O(l) \rightarrow H_2SO_4(l)$ $\quad \Delta H^0 = -132 \text{ kJ}$

## Comprehensive Problems
Problems with an asterisk (*) are more challenging.

**6.82** Oxidation of ClF by $F_2$ yields $ClF_3$, an important fluorinating agent formerly used to produce the uranium compounds in nuclear fuels:

$$ClF(g) + F_2(g) \rightarrow ClF_3(l)$$

Use the following thermochemical equations to calculate $\Delta H^0_{rxn}$ for the production of $ClF_3$:

  (I) $2ClF(g) + O_2(g) \rightarrow Cl_2O(g) + OF_2(g)$
  $$\Delta H^0 = 167.5 \text{ kJ}$$
  (II) $2F_2(g) + O_2(g) \rightarrow 2OF_2(g)$ $\quad \Delta H^0 = -43.5 \text{ kJ}$
  (III) $2ClF_3(l) + 2O_2(g) \rightarrow Cl_2O(g) + 3OF_2(g)$
  $$\Delta H^0 = 394.1 \text{ kJ}$$

**6.83** Iron metal is produced in a blast furnace through a complex series of reactions that involve reduction of iron(III) oxide with carbon monoxide.
  (a) Write a balanced overall equation for the process, including the other product.
  (b) Use the equations below to calculate $\Delta H^0_{rxn}$ for the overall equation:

  (I) $3Fe_2O_3(s) + CO(g) \rightarrow 2Fe_3O_4(s) + CO_2(g)$
  $$\Delta H^0 = -48.5 \text{ kJ}$$
  (II) $Fe(s) + CO_2(g) \rightarrow FeO(s) + CO(g)$
  $$\Delta H^0 = -11.0 \text{ kJ}$$
  (III) $Fe_3O_4(s) + CO(g) \rightarrow 3FeO(s) + CO_2(g)$
  $$\Delta H^0 = 22 \text{ kJ}$$

**6.84** The calorie (4.184 J) was originally defined as the amount of energy required to raise the temperature of 1.00 g liquid water 1.00°C. The British thermal unit (Btu) is defined as the amount of energy required to raise the temperature of 1.00 lb liquid water 1.00°F.
  (a) How many joules are in 1.00 British thermal unit (1 lb = 453.6 g; a change of 1.0°C = 1.8°F)?
  (b) The "therm" is a unit of energy consumption used by natural gas companies in the United States and is defined as 100,000 Btu. How many joules are in 1.00 therm?
  (c) How many moles of methane must be burned to give 1.00 therm of energy? (Assume water forms as a gas.)
  (d) If natural gas costs $0.46 per therm, what is the cost per mole of methane?
  (e) How much would it cost to warm 308 gal of water in a hot tub from 15.0°C to 40.0°C? (1 gal = 3.78 L)

**6.85** Triglycerides are the main storage form of fats in the body. During periods of starvation, a person's fat stores are utilized for energy. Tristearin ($C_{57}H_{110}O_6$) is a typical animal fat that is oxidized according to the following equation:

$$2C_{57}H_{110}O_6(s) + 163O_2(g) \rightarrow 114CO_2(g) + 110H_2O(l)$$

If $\Delta H^0_{rxn} = -7.0 \times 10^4$ kJ, how much heat is released
  (a) Per mole of $O_2$ consumed?
  (b) Per mole of $CO_2$ formed?
  (c) Per gram of tristearin oxidized?
  (d) When 325 L $O_2$ at 37°C and 755 torr is used, how many grams of tristearin can be oxidized?

**6.86** Kerosene, a common space heater fuel, is a mixture of hydrocarbons whose "average" formula is $C_{12}H_{26}$.
  (a) Write a balanced equation, using the simplest whole-number coefficients, for the complete combustion of kerosene to $CO_2$ (g) and $H_2O$ (g).
  (b) If $\Delta H^0_{comb} = -1.50 \times 10^4$ kJ for the equation as written in part (a), determine $\Delta H^0_f$ of kerosene.
  (c) Calculate the heat produced by combustion of 0.50 gal kerosene (density of kerosene = 0.749 g/mL; 1 gal = 3.78 L).
  (d) For a kerosene furnace to produce 1250 Btu, how many gallons of kerosene must be burned (1 Btu = 1.055 kJ)?

**6.87** Water has a particularly high specific heat capacity. Explain the role of water in the following two common practices:
  (a) During especially cold days, people place large tubs of water in the "root cellar" of their old country homes to prevent the stored garden vegetables from freezing.
  (b) Citrus growers in Florida spray their orchards with water when a frost is predicted.

**6.88** Consider the following hydrocarbon fuels: (I) $CH_4(g)$; (II) $C_2H_4(g)$; (III) $C_2H_6(g)$.
  (a) Rank them in terms of heat released per mole. [Assume $H_2O(g)$ forms and combustion is complete.]
  (b) Rank them in terms of heat released per gram.

**6.89** You want to determine the $\Delta H^0$ for the reaction

$$Zn(s) + 2HCl(aq) \rightarrow ZnCl_2(aq) + H_2(g)$$

(a) To do so, you first determine the heat capacity of a calorimeter using the following reaction, whose $\Delta H$ is known:

$$NaOH(aq) + HCl(aq) \rightarrow NaCl(aq) + H_2O(l)$$
$$\Delta H^0 = -57.32 \text{ kJ}$$

Calculate the heat capacity of the calorimeter from the following:
Amounts used: 50.0 mL of 2.00 $M$ HCl and 50.0 mL of 2.00 $M$ NaOH
Initial $T$ of both solutions: 16.9°C
Maximum $T$ recorded during reaction: 30.4°C
Density of resulting NaCl solution: 1.04 g/mL
$C$ of 1.00 $M$ NaCl$(aq)$ = 3.93 J/g · K

(b) Use the result from part (a) and the following data to determine $\Delta H^0_{rxn}$ for the reaction between zinc and HCl$(aq)$:
Amounts used: 100.0 mL of 1.00 $M$ HCl and 1.3078 g Zn
Initial $T$ of HCl solution and Zn: 16.8°C
Maximum $T$ recorded during reaction: 24.1°C
Density of 1.0 $M$ HCl solution = 1.015 g/mL
$C$ of resulting ZnCl$_2(aq)$ = 3.95 J/g · K

(c) Given the following values, what is the error in your experiment?

$$\Delta H^0_f \text{ of HCl}(aq) = -1.652 \times 10^2 \text{ kJ/mol}$$
$$\Delta H^0_f \text{ of ZnCl}_2(aq) = -4.822 \times 10^2 \text{ kJ/mol}$$

*6.90** One mole of nitrogen gas confined within a cylinder by a piston is heated from 0°C to 819°C at 1.00 atm.
(a) Calculate the work of expansion in joules (1 J = 9.87 × 10⁻³ atm · L). Assume all the energy is used to do work.

(b) What would be the temperature change if the gas had been heated with the same amount of energy in a container of fixed volume? (Assume the specific heat capacity of $N_2$ is 1.00 J/g · K.)

*6.91** The combustion of eight-carbon hydrocarbons $(C_8H_{18})$, such as those found in gasoline, is one of the most common redox reactions in modern society.
(a) Calculate $\Delta H^0_{comb}$ of 1 mol $C_8H_{18}$ in excess $O_2$, assuming water forms as a gas.
(b) Calculate $\Delta E$ for this reaction (1 J = 9.87 × 10⁻³ atm · L). (*Hint:* use $P\Delta V = \Delta n_{gas}RT$ to convert between $\Delta H$ and $\Delta E$.)

**6.92** High-purity benzoic acid $(C_6H_5COOH; \Delta H_{comb} = -3227 \text{ kJ/mol})$ is a combustion standard for calibrating bomb calorimeters. A 1.221-g sample burns in a calorimeter (heat capacity = 1365 J/°C) that contains exactly 1.200 kg water. What temperature change would be observed?

*6.93** Liquid methanol $(CH_3OH)$ is being used as an alternative fuel in cars and trucks, and some predict it will be the fuel of choice in the future. An industrial method for preparing methanol, sometimes called the Fischer-Tropsch process, occurs through the catalytic hydrogenation of carbon monoxide:

$$CO(g) + 2H_2(g) \xrightarrow{\text{catalyst}} CH_3OH(l)$$

How much heat is released when 15.0 L CO at 85°C and 112 kPa reacts with 18.5 L $H_2$ at 75°C and 744 torr?

*6.94** (a) How much heat is released when 25.0 g methane burns in excess $O_2$ to form gaseous products?
(b) Calculate the temperature of the product mixture if the methane with air are both at an initial temperature of 0°C. Assume a stoichiometric ratio of methane with oxygen from the air and that air is 21% $O_2$ by volume ($C$ of $CO_2$ = 57.2 J/mol · K; $C$ of $H_2O(g)$ = 36.0 J/mol · K; $C$ of $N_2$ = 30.5 J/mol · K).

# Quantum Theory and Atomic Structure

**Concepts and skills to review**

- discovery of the electron and atomic nucleus (Section 2.3)
- major features of atomic structure (Section 2.4)
- changes in energy state of a system (Section 6.1)

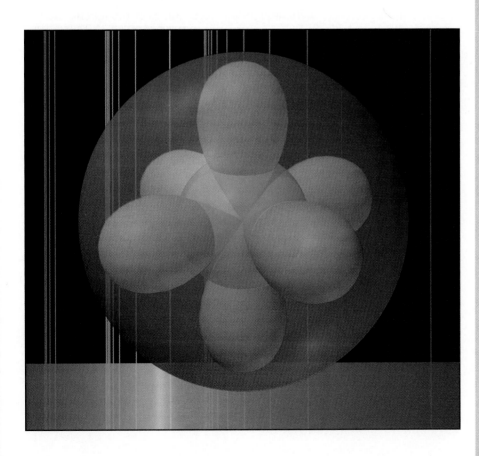

**Experiment and theory.** From the bewildering colored lines observed by 19th-century spectroscopists, 20th-century physicists developed a new view of reality and a new model of the atom. In this chapter we examine the steps in this rethinking and the remarkable entity that emerged from it.

**R**evolutions in science are not like the violent upheavals of political overthrow. Chinks appear in an established model as conflicting evidence accumulates, a startling discovery or two broadens the chinks into cracks, and the structure begins to crumble as a result of its inconsistencies. New insight verified by experiment then guides the building of a more accurate model. So it was when Lavoisier's theory of combustion overthrew the phlogiston model, when Dalton's atomic theory established the idea of individual units of matter, and when Rutherford's nuclear model substituted an atom with rich internal structure for "billiard balls" or "plum puddings." In this chapter, we see the process unfold again with the development of modern atomic theory. ◆

Almost as soon as Rutherford had proposed his nuclear model, a major problem with it arose. A nucleus and an electron attract each other, so if they are to remain apart, the energy of the electron's movement must balance the energy of attraction. However, the laws of physics had previously established that a charged particle moving in a curved path *must* give off energy. If this requirement applied to an orbiting electron, why didn't the electron continuously lose energy and spiral into the nucleus? Clearly, if electrons behaved the way classical physics predicted, all atoms would have collapsed eons ago! Subatomic matter seemed to violate common experience and accepted principles.

The breakthroughs that soon followed forced a complete rethinking of the classical picture of matter and energy. In the macroscopic world, the two are distinct: matter occurs in chunks you can hold and weigh; you can change the amount of matter in a sample piece by piece. Energy is "massless"; its amount can be changed in a continuous manner. Matter moves in specific paths, whereas light and other types of electromagnetic energy travel in diffuse waves. As soon as 20th century scientists probed the subatomic world, however, these clear distinctions between particulate matter and wavelike energy began to fade.

In this chapter and the next, we discuss *quantum mechanics*, the theory that explains our current picture of atomic structure. After considering the wave properties of electromagnetic energy, we examine the theories and experiments that led to a *quantized*, or particulate, model of light. We see why the light emitted by an excited hydrogen atom—its *atomic spectrum*—suggests an atom with distinct energy levels, and we briefly look at how atomic spectra are applied to chemical analysis today. Then we examine the central idea of quantum mechanics, the notion of wave-particle duality. The last section describes the modern model of the hydrogen atom and presents the quantum numbers we use to identify the regions of space an electron occupies in an atom. In Chapter 8, we consider atoms with more than one electron and relate electron occupancy to chemical behavior.

## 7.1 The Nature of Light

Visible light is one type of **electromagnetic (EM) radiation** (also called electromagnetic energy or radiant energy); others include x-rays, microwaves, and radio waves. All types of electromagnetic radiation travel as waves—the result of oscillating electric and magnetic fields moving simultaneously through space. This classical wave model of radiant energy is still essential for understanding everyday phenomena but not for explaining behavior on the atomic scale.

◆ **Hurrah for the Human Mind.** The first quarter of the 20th century was one of the most exciting periods in recent history for science and the arts. The invention of the car, radio, and airplane fostered a feeling of unlimited human ability. The discovery of x-rays, radioactivity, the electron, and the atomic nucleus led to the sense that the human mind would soon uncover all of nature's mysteries. Indeed, some people were convinced that few, if any, mysteries remained.

| Date | Event |
|------|-------|
| 1893 | Ford builds his first car. |
| 1895 | First pro football game is played in United States. |
| 1897 | Thomson discovers the electron. |
| 1900 | Freud proposes his theory of the unconscious mind. |
| 1900 | Planck publishes his quantum theory. |
| 1901 | Marconi invents the radio. |
| 1903 | Wright brothers fly an airplane. |
| 1905 | Rutherford develops his radioactivity theory. |
| 1905 | Einstein publishes his relativity and photon theories. |
| 1906 | Ruth St. Denis develops modern dance. |
| 1908 | Matisse and Picasso develop modern art. |
| 1909 | Schönberg and Berg develop modern music. |
| 1911 | Rutherford presents his nuclear model of the atom. |
| 1913 | Bohr proposes his atomic model. |
| 1914 | World War I begins. |
| 1923 | Compton demonstrates photon momentum. |
| 1924 | De Broglie publishes his wave theory of matter. |
| 1926 | Schrödinger develops his wave equation. |
| 1927 | Heisenberg presents his uncertainty principle. |
| 1929 | Stock market crashes. |
| 1932 | Chadwick discovers the neutron. |

## The Wave Nature of Light

The wave properties of electromagnetic radiation are described by two interdependent variables, frequency and wavelength (Figure 7.1). **Frequency** (*ν*, Greek *nu*) is the number of cycles the wave undergoes per second, expressed in units of 1/second ($s^{-1}$), or hertz (Hz). **Wavelength** (*λ*, Greek *lambda*) is the distance between any point on a wave and the corresponding point on the next wave; that is, the distance the wave travels during one cycle. Wavelength is commonly expressed in meters, but since chemists often deal with very short wavelengths, the nanometer (nm, $10^{-9}$ m), picometer (pm, $10^{-12}$ m), and angstrom (Å, $10^{-10}$ m) are also used.

The speed of the wave (distance traveled per unit time in meters per second) is the product of its frequency (cycles per second) and its wavelength (meters per cycle):

$$\text{Speed} = \frac{\text{cycles}}{\text{s}} \times \frac{\text{m}}{\text{cycle}} = \frac{\text{m}}{\text{s}}$$

In a vacuum, all electromagnetic radiation travels at $2.99792458 \times 10^{8}$ m/s ($3.00 \times 10^{8}$ m/s to 3 significant figures), a constant called the **speed of light (*c*)**:

$$c = \nu\lambda \qquad (7.1)$$

Since their product is the constant *c*, the terms *ν* and *λ* have a reciprocal relationship: when one increases, the other decreases. Thus, *radiation with a high frequency has a short wavelength, and vice versa.*

Another characteristic of a wave is its **amplitude,** the height of the crest (or depth of the trough) of each wave (Figure 7.2). The amplitude of an electromagnetic wave is a measure of the strength of its electric and magnetic fields. Thus, amplitude is related to the *intensity* of the radiation, which we perceive as brightness in the case of visible light. Light of a particular shade of red, for instance, always has the same frequency and wavelength, but it can be dim (low amplitude) or bright (high amplitude).

Visible light occupies a small portion of the continuum of radiant energy known as the **electromagnetic spectrum** (Figure 7.3). *The waves in the spectrum travel at the same speed but differ in frequency and wavelength.* Some regions of the spectrum are named after the instruments used to produce or

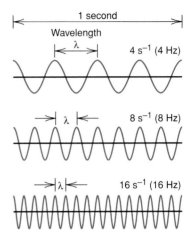

Frequency (ν) in $s^{-1}$ (Hz)

**FIGURE 7.1**
**Frequency and wavelength.** Three waves with different wavelengths (λ) and thus different frequencies (ν) are shown. Note that as the wavelength decreases, the frequency increases.

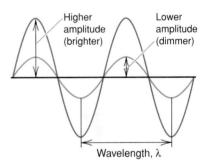

**FIGURE 7.2**
**Amplitude (intensity) of a wave.** Amplitude is represented by the height of the crest (or depth of the trough) of the wave. The two waves shown have the same wavelength (color) but different amplitudes and, therefore, different brightnesses (intensities).

**FIGURE 7.3**
**Regions of the electromagnetic spectrum.** The electromagnetic spectrum extends from the very short wavelengths (very high frequencies) of gamma rays through the very long wavelengths (very low frequencies) of radio waves. The relatively narrow visible region is expanded (and the scale made linear) to show the component colors.

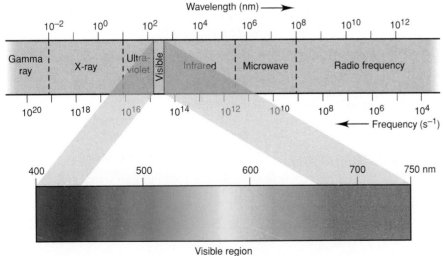

Visible region

◆ **Electromagnetic Emissions Everywhere.** We are bathed in electromagnetic radiation from the Sun. Artificial sources of radiation on Earth bombard us as well: radio and TV signals; microwave radiation from traffic radar systems and telephone relay stations; and radiation given off by light bulbs, medical x-ray equipment, and car motors. Natural sources on Earth bombard us also: lightning, radioactive decay, and even fireflies! Virtually all our knowledge of the distant universe comes from radiation entering our light, x-ray, and radio telescopes.

receive their wavelengths; thus, the long-wavelength, low-frequency portion of the spectrum comprises the microwave and radio regions. Each region overlaps the next. The **infrared (IR)** region overlaps the microwave region on one end and the visible region on the other.

We perceive different wavelengths (or frequencies) of *visible* light as different colors, from red ($\lambda \approx 750$ nm) to violet ($\lambda \approx 400$ nm). Light of a single wavelength is **monochromatic** (Greek, "one color"), whereas light of many wavelengths is **polychromatic** (Greek, "many colors"); white light is polychromatic. The region adjacent to visible light on the short-wavelength end consists of **ultraviolet (UV)** radiation (also called ultraviolet "light"). Still shorter wavelengths (higher frequencies) make up the x-ray and gamma ray ($\gamma$ ray) regions. Thus, a TV signal, a green traffic light, and a $\gamma$ ray emitted by a radioactive element differ principally in frequency and wavelength. ◆

SAMPLE PROBLEM 7.1 _____

### Interconverting Wavelength and Frequency

**Problem:** A dental hygienist uses x-rays ($\lambda = 1.00$ Å) to take a series of dental radiographs while the patient listens to an FM radio station ($\lambda = 325$ cm) and looks out the window at the blue sky ($\lambda = 473$ nm). What is the frequency (in $s^{-1}$) of the electromagnetic radiation from each source?

**Plan:** We are given the wavelengths and can find the frequencies from Equation 7.1. Since $c$ has units of m/s, we first convert all the wavelengths to meters.

**Solution:** For x-rays. Converting from angstroms to meters:

$$\lambda = 1.00 \text{ Å} \times \frac{10^{-10} \text{ m}}{1 \text{ Å}} = 1.00 \times 10^{-10} \text{ m}$$

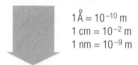

Wavelength (given units)

$1\text{Å} = 10^{-10}$ m
$1$ cm $= 10^{-2}$ m
$1$ nm $= 10^{-9}$ m

Wavelength (m)

$\nu = \frac{c}{\lambda}$

Frequency ($s^{-1}$)

Calculating the frequency:

$$\nu = \frac{c}{\lambda} = \frac{3.00 \times 10^8 \text{ m/s}}{1.00 \times 10^{-10} \text{ m}} = \mathbf{3.00 \times 10^{18} \ s^{-1}}$$

For the radio station. Combining steps to calculate the frequency:

$$\nu = \frac{c}{\lambda} = \frac{3.00 \times 10^8 \text{ m/s}}{325 \text{ cm} \times 10^{-2} \text{ m/1 cm}} = \mathbf{9.23 \times 10^7 \ s^{-1}}$$

For the sky. Combining steps to calculate the frequency:

$$\nu = \frac{c}{\lambda} = \frac{3.00 \times 10^8 \text{ m/s}}{473 \text{ nm} \times 10^{-9} \text{ m/1 nm}} = \mathbf{6.34 \times 10^{14} \ s^{-1}}$$

**Check:** The orders of magnitude are correct for the regions of the electromagnetic spectrum (see Figure 7.3): x-rays ($10^{19}$ to $10^{16} \ s^{-1}$), radio waves ($10^8$ to $10^7 \ s^{-1}$), and visible light ($7.5 \times 10^{14}$ to $4 \times 10^{14} \ s^{-1}$).

**Comment:** The radio in this problem is broadcasting at a frequency of $92.3 \times 10^6 \ s^{-1}$, or 92.3 million Hz (92.3 MHz), about midway in the FM range.

FOLLOW-UP PROBLEM 7.1
Some diamonds appear yellow because they contain nitrogen compounds that absorb purple light of frequency $7.23 \times 10^{14}$ Hz. Calculate the wavelength (in nm) of the absorbed light.

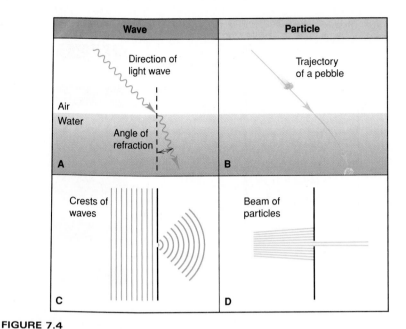

**FIGURE 7.4**

**Different behaviors of waves and particles. A,** A wave passing from air into water is *refracted* (bent at an angle). **B,** In contrast, a particle of matter (such as a pebble) slows down in a curved path under the influence of gravity. **C,** A wave is *diffracted* through a small opening, which gives rise to a circular wave on the other side. (The lines represent the wave crests of water waves as seen from above.) **D,** In contrast, particles move through an opening and continue along individual paths.

Light travels more slowly through air than through a vacuum and more slowly still through water. Similarly, it moves at particular speeds through quartz, mica, different types of glass, and any other transparent medium. Therefore, when a light wave passes through the boundary from one medium into another, for example, from air into water, the speed of the wave changes. If the light strikes the boundary at an angle other than 90°, the wave bends and continues at a different angle, a phenomenon known as **refraction** (Figure 7.4, *A*). The new angle (angle of refraction) depends on the materials on either side of the boundary and the wavelength of the light. ◆ In the process of *dispersion,* white light is separated (dispersed) into its component colors by passing it through a prism, which causes each incoming wavelength to be refracted at a slightly different angle. ◆ Refraction is a wave phenomenon, quite distinct from the behavior of particles. A pebble thrown through the air into a lake, for example, abruptly changes its speed, but not necessarily its direction, and then gradually slows down in a curved path (Figure 7.4, *B*).

When a wave strikes the edge of an object, it bends around it, a phenomenon called **diffraction.** If the wave passes through a slit about as wide as its wavelength, it bends around both edges of the slit and forms a semicircular wave on the other side of the opening (Figure 7.4, *C*). However, a stream of particles aimed at a small opening behaves quite differently: some particles hit the edge and stop, whereas those going through continue linearly in a narrower stream (Figure 7.4, *D*). If waves of light pass through two adjacent slits, the emerging circular waves interact through a process *nce* and create a diffraction pattern of brighter and darker re-

◆ **Refraction in Chemical Analysis.** The index of refraction of a material equals $c/v$, where $v$ is the speed at which light travels through the material. Since $v$ is always less than $c$, the index of refraction is always greater than 1. In the analytical technique of *refractometry*, the index of refraction is determined with light of a known wavelength passing through a fixed thickness of the substance at a known temperature. Refractometry is often used to identify a pure substance. Using yellow light ($\lambda = 589$ nm), for example, the index of refraction of water (at 20°C) = 1.33, of NaCl = 1.53, and of diamond = 2.42. Using known standards, refractometry is also applied in determining the composition of solutions from antifreeze to maple syrup.

◆ **Rainbows and Diamonds.** You can see a rainbow only when the sun is at your back. Light entering the near surface of a water droplet is dispersed and reflected off the far surface. Since red is bent least, it reaches your eye from droplets higher in the sky, while violet appears from droplets that are lower. The colors in a diamond's sparkle are due to its facets, which are cleaved at angles that disperse and reflect the incoming light, lengthening its path enough for the emerging wavelengths to separate.

**FIGURE 7.5**

The diffraction pattern caused by light passing through two slits. **A,** After passing through two closely spaced slits, emerging circular waves interfere with each other to create a diffraction (interference) pattern on the film of bright regions, where crests coincide and enhance each other (in phase), and dark regions, where crests meet troughs and cancel each other (out of phase). **B,** The interference pattern obtained on the film.

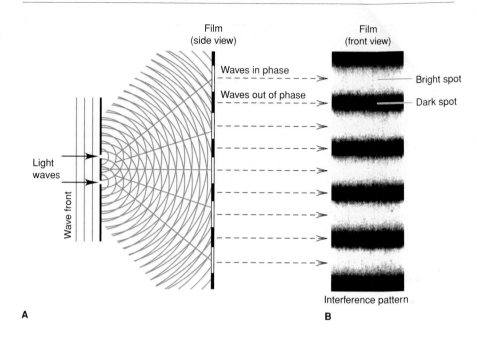

gions (Figure 7.5). At the end of the 19th century, all experience seemed to confirm these classical distinctions between the wave nature of energy and the particulate nature of matter.

## The Particulate Nature of Light

At the beginning of the 20th century, physicists were confounded by three phenomena involving matter and light: (1) the pattern of intensity and wavelength of light emitted from hot, dense objects (blackbody radiation); (2) the electric current generated when light shines on a metal plate (photoelectric effect); and (3) the individual colors emitted from electrically (or thermally) excited gases (atomic spectra). Explaining these phenomena required a radically new view of energy. We discuss the first two here and the third in the next section.

**Blackbody radiation and the quantization of energy.** When a solid object is heated to about 1000 K, it begins to emit visible light, as in the soft red glow of smoldering coal. At about 1500 K, the light is brighter and more orange, as from the heating coil of a toaster. At temperatures greater than 2000 K, the light is still brighter and whiter, as from a piece of white-hot iron or the filament of a light bulb. This phenomenon of changing intensity and wavelength of the light emitted as a dense object is heated is called **blackbody radiation,** because it approximates the light given off by a hot *blackbody,* one that absorbs all radiation falling on it. All attempts to use classical electromagnetic theory to predict the wavelengths of the emitted light had failed.

In 1900, the German physicist Max Planck made a radical assumption that led to a perfect fit with the data and, eventually, to an entirely new view of matter and energy. He proposed that the hot, glowing object could emit (or absorb) only *certain amounts of energy:*

$$E = nh\nu$$

where $E$ is the energy of the radiation, $\nu$ is its frequency, $n$ is a positive integer (1, 2, 3, and so on) called a **quantum number,** and $h$ is a protionality constant now called **Planck's constant.** With energy in joules (J) and frequency in $s^{-1}$, $h$ has units of J · s:

$$h = 6.626 \times 10^{-34} \text{ J} \cdot \text{s}$$

Later interpretations of Planck's proposal state that the hot object's radiation is emitted by the atoms contained within it. If an atom can *emit* only certain amounts of energy, it follows that *an atom can have only certain electronic energy values:* $E = 1h\nu, 2h\nu, 3h\nu$, and so forth, but not $1.424h\nu, 2.693h\nu$, and so forth.

This constraint means that the energy of an atom is *not* continuous but *quantized:* it exists only in certain fixed amounts. Each change in energy can be pictured as resulting from a "packet" of energy being gained or lost by the atom. This energy packet is called a **quantum** ("fixed amount"; plural, *quanta*) and has the energy $h\nu$: *an atom changes its energy state by emitting or absorbing a quantum of energy.* The energy of the emitted or absorbed radiation is equal to the *difference in the atom's electronic energy states:*

$$\Delta E_{\text{atom}} = E_{\text{emitted (or absorbed) radiation}} = \Delta n h\nu$$

The smallest possible energy change for an atom is from one energy state to an adjacent one ($\Delta n = 1$):

$$\Delta E = h\nu \tag{7.2}$$

**The photoelectric effect and the photon theory of light.** Despite his idea that energy is quantized, Planck and other physicists continued to picture the emitted energy as traveling in waves. The wave model, however, could not explain the **photoelectric effect.** When monochromatic light of sufficient energy shines on a metal plate, an electric current flows (Figure 7.6). This in itself was not puzzling: the light imparts energy to the electrons at the metal surface, which break free and are collected by a positive electrode. However, the photoelectric effect had certain confusing features, including the presence of a threshold frequency and the absence of a time lag:

1. *Presence of a threshold frequency.* Light shining on the metal must have a minimum frequency (which varies with the metal), or no current flows. The wave theory, however, associates the light's energy with the wave's amplitude (intensity), not its frequency (color), so it predicts that an electron should break free when it absorbs enough energy from light of *any* color.

2. *Absence of a time lag.* Current flows the moment that light of high enough frequency shines on the metal, regardless of its intensity. The wave theory, however, predicted that in dim light there should be a time lag before current flowed, while the electrons absorbed enough energy to break free.

In 1905, Albert Einstein presented his *photon theory* to explain the photoelectric effect. Carrying Planck's idea of packeted energy further, Einstein proposed that radiation is in fact particulate, occurring as quanta of electromagnetic energy, later called **photons.** ◆ Applying this idea to Planck's work, we can say that an atom changes its energy when it absorbs or emits a photon, a "piece" of light whose energy is fixed by its *frequency:*

$$E_{\text{photon}} = h\nu = \Delta E_{\text{atom}}$$

How does Einstein's photon theory explain the photoelectric effect?

~~~ce *of a threshold frequency.* According to the photon theory, a beam of
s composed of enormous numbers of photons. Light intensity

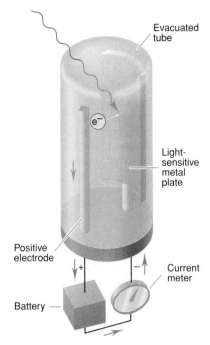

**FIGURE 7.6**
**Demonstration of the photoelectric effect.**
When monochromatic light of high enough frequency strikes the metal plate, electrons are freed from the plate and travel to the positive electrode, which creates a current.

◆ **Infinitesimal Quanta.** Why don't we see separate packets of light coming from a light bulb? The answer is that Planck's constant, $h$, $6.626 \times 10^{-34}$ J · s, is extremely small, so a single photon of visible light has very little energy. The amount of light needed to read a chemistry text, for instance, consists of so many photons that they cannot be individually perceived. Similarly, the tiny size of atoms, the quanta of mass, keeps us unaware that we are breathing individual $O_2$ and $N_2$ molecules. In addition, the quantum of electric charge, the charge of an electron, is too small for us to realize that electricity also occurs in "pieces": in a common 100-watt bulb, $6 \times 10^{18}$ quanta of charge enter and leave the filament every second!

(brightness) is related to the number of photons emitted per unit time, *not* to their energy. One electron is freed from the metal when one photon of a certain *minimum* energy is absorbed. (A photon of lower energy simply warms the metal.) Since energy depends on frequency ($h\nu$), a threshold frequency is to be expected.

2. *Absence of a time lag.* An electron is freed the moment it absorbs a photon of enough energy, not when it gradually accumulates energy from many photons of lower energy. The current will be weaker in dim light than in bright light because fewer photons of enough energy are knocking out fewer electrons per unit time, but *some* current should flow as soon as the photons reach the metal plate. ◆

SAMPLE PROBLEM 7.2 _____

## Calculating the Energy of Radiation from Its Frequency

**Problem:** A cook uses a microwave oven to heat a meal. The frequency of the radiation is $2.45 \times 10^9$ s$^{-1}$. What is the energy of one photon of this microwave radiation?

**Plan:** We know the frequency of the radiation, so we can find the energy of one quantum by using $E = h\nu$.

**Solution:** $E = h\nu = (6.626 \times 10^{-34} \text{ J} \cdot \text{s})(2.45 \times 10^9 \text{ s}^{-1}) = \mathbf{1.62 \times 10^{-24}}$ **J**

**Check:** The order of magnitude seems correct: $\sim 10^{-33} \times 10^9 = 10^{-24}$

FOLLOW-UP PROBLEM 7.2
Calculate the energies of one photon of ultraviolet ($\lambda = 1 \times 10^{-8}$ m), visible ($\lambda = 5 \times 10^{-7}$ m), and infrared ($\lambda = 1 \times 10^{-4}$ m) radiation. What do the answers indicate about the relationship between the wavelength and the energy of light?

The quantum theory and photon theory ascribed properties to radiation that had always been reserved for matter: fixed amount and discrete particles. These properties have since proved essential to explaining all interactions of matter and energy at the atomic level. But how can we match a particulate model of energy with the facts of diffraction and refraction, phenomena explained only in terms of waves? As you'll see later, the photon model did not *replace* the wave model; rather, we have had to accept *both* to understand reality. Before we discuss this astonishing notion, let's see how the new idea of quantized energy led to a key understanding about atomic behavior that became incorporated into our current model.

**Section Summary**
Electromagnetic radiation travels in waves of specific wavelength ($\lambda$) and frequency ($\nu$). All electromagnetic waves travel through a vacuum at the speed of light ($c = 3.00 \times 10^8$ m/s), which is equal to $\nu \times \lambda$. The intensity of a wave is related to its amplitude. The electromagnetic spectrum ranges from very long radio waves to very short $\gamma$ rays and includes the visible region, from approximately 750 nm (red) to 400 nm (violet). Refraction and diffraction indicate that electromagnetic radiation is wavelike, but blackbody radiation and the photoelectric effect indicate that it is particle-like. Light exists as photons (quanta), which have an energy proportional to its frequency. According to quantum theory, an atom can have only certain amounts of energy ($E = nh\nu$), which it can change only by absorbing or emitting a photon.

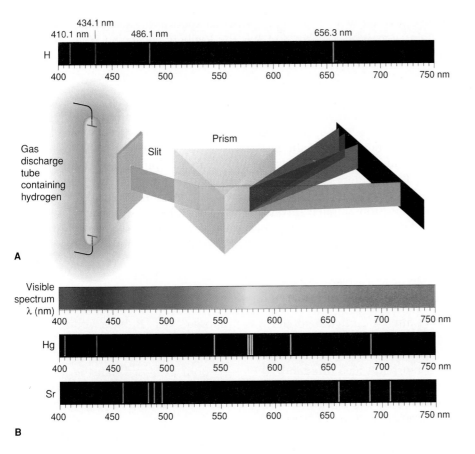

**FIGURE 7.7**

**FIGURE 7.7**

**The line spectra of several elements. A,** A gaseous sample ($H_2$ is used here) is dissociated into atoms and excited by an electric discharge. The emitted light passes through a slit and a prism, which disperses the light into individual wavelengths. The spectrum of atomic H is shown above. **B,** The continuous spectrum of white light is compared with the line spectra of mercury and strontium. Note that each line spectrum is different from the others.

## 7.2 Atomic Spectra and the Bohr Model of the Atom

The final unsolved problem in physics of the late 19th century that we discuss concerns atomic spectra. When an element is vaporized and thermally or electrically excited, it emits light. If dispersed by a prism, the light does not create a *continuous spectrum,* or rainbow, as sunlight does. Rather, it produces a **line spectrum,** a series of fine lines of individual colors separated by colorless spaces; the wavelengths at which the colored lines occur are characteristic of the element (Figure 7.7).

Spectroscopists studying the spectrum of atomic hydrogen had identified several series of spectral lines in different regions of the electromagnetic spectrum (Figure 7.8). Each series shows a similar pattern: the spaces between spectral lines become smaller at the short-wavelength (high-energy) end of the series. Equations of the following general form, called the *Rydberg equation,* were found to predict the position and wavelength of any line in a given series:

$$\frac{1}{\lambda} = R\left(\frac{1}{n_1^{\,2}} - \frac{1}{n_2^{\,2}}\right)$$ (7.3)

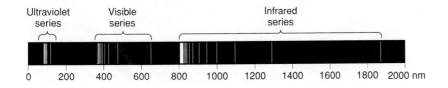

**FIGURE 7.8**

**Three series of spectral lines of atomic hydrogen.** These series appear in different regions of the electromagnetic spectrum. Note that each series shows the pattern of decreasing line spacing with shorter wavelength. The hydrogen spectrum shown in Figure 7.7, *A,* is the visible series.

where $\lambda$ is the wavelength of a particular spectral line, $n_1$ and $n_2$ are positive integers with $n_2 > n_1$, and $R$ is the Rydberg constant ($1.096776 \times 10^7$ m$^{-1}$). For the series of spectral lines in the visible range, $n_1 = 2$:

$$\frac{1}{\lambda} = R\left(\frac{1}{2^2} - \frac{1}{n_2^2}\right), \text{ with } n_2 = 3, 4, 5, \ldots$$

The Rydberg equation is empirical, that is, based on data rather than theory. No one knew *why* the spectral lines of hydrogen follow this pattern.

As was mentioned at the beginning of this chapter, classical theory predicted that an orbiting electron should emit radiation and spiral into the nucleus. Furthermore, the frequency of the radiation emitted by an orbiting charge is directly related to the time required for one revolution. Therefore, as the electron spiraled toward the nucleus, the time of each revolution would decrease, so it should emit radiation that increased smoothly in frequency—a continuous spectrum. Rutherford's nuclear model seemed totally at odds with the existence of atomic line spectra.

### The Bohr Model of the Hydrogen Atom

Soon after the nuclear model was proposed, Niels Bohr, a young Danish physicist working in Rutherford's laboratory, suggested a model for the hydrogen atom that predicted the observed line spectra. Accepting Planck's and Einstein's idea of quantized energy, Bohr proposed that the hydrogen atom had only certain allowable energy levels, which he called **stationary states.** Each of these states was associated with a fixed circular orbit of the electron around the nucleus (Figure 7.9).

Bohr's key assumption was that the atom does *not* radiate energy while in one of its stationary states, that is, while the electron is in one of the fixed orbits. Rather, he postulated that the electron can move to a different orbit (the atom can change to another stationary state) *only by absorbing or emitting a photon whose energy equals the difference in energy between the two stationary states:*

$$E_{\text{photon}} = \Delta E_{\text{stationary states}} = h\nu$$

A spectral line results from the emission of a photon of specific energy (and thus specific frequency) when the electron moves from a higher energy state to a lower one. An atomic spectrum appears as lines rather than as a continuum because *the atom's energy has only certain discrete levels, or states.*

**FIGURE 7.9**

**The Bohr model explanation of the three series of spectral lines.** According to the Bohr model, when an electron drops from an orbit farther from the nucleus to one closer, it emits a photon of a specific energy. The greater the *difference* in orbit radius, the greater is the energy of the emitted photon. For example, in the ultraviolet series, in which $n_1 = 1$, a jump from $n = 7$ to $n = 1$ emits a photon with more energy (shorter $\lambda$, higher $\nu$) than a jump from $n = 2$ to $n = 1$. Note that the three series are characterized by a particular value of $n_1$.

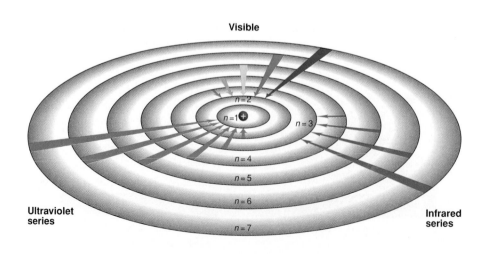

In Bohr's model, the quantum number $n$ (1, 2, 3, . . .) is associated with the radius of the electron's orbit, which is directly related to the atom's energy. *The lower the quantum number, the smaller is the radius of the orbit and the lower is the energy level of the atom.* When the electron is in the orbit closest to the nucleus ($n = 1$), the atom is in its lowest (first) energy level, called the **ground state.** By absorbing a photon whose energy equals the difference between the first and second energy levels, the electron can move to the next orbit out from the nucleus. This second energy level (second stationary state) and all higher levels are called **excited states.** The hydrogen atom in the second energy level (first excited state) can return to the ground state by emitting a photon of a particular frequency:

$$E_{photon} = h\nu = E_{first\ excited\ state} - E_{ground\ state}$$

When a sample of atomic hydrogen absorbs energy, different H atoms absorb different amounts. Even though each atom has only one electron, so many atoms are present that all the allowable energy levels (orbits) are populated by electrons. When the electrons drop from outer orbits to the $n = 3$ orbit (second excited state), the infrared series of spectral lines is produced. The visible series arises from the photons emitted when electrons drop to the $n = 2$ orbit (first excited state), and the ultraviolet series arises when these higher energy electrons drop to the $n = 1$ orbit (ground state). Since a larger orbit radius means a higher atomic energy level, the farther the electron drops, the greater is the energy (higher $\nu$, shorter $\lambda$) of the emitted photon. The spectral lines of hydrogen become closer and closer together in the short-wavelength (high-energy) region of each series because the *difference* in energy associated with the jump from $n_{initial}$ to $n_{final}$ becomes smaller and smaller with increasing distance from the nucleus (Figure 7.10).

## The Energy States of the Hydrogen Atom

From the principles of classical physics, Bohr used the known mass and charge of the electron and the proton (H nucleus) to obtain an equation for calculating the energies of the stationary states of the hydrogen atom:

$$E = -2.18 \times 10^{-18}\ \text{J}\left(\frac{1}{n^2}\right)$$

Therefore, the energy of the ground state ($n = 1$) is

$$E = -2.18 \times 10^{-18}\ \text{J}\left(\frac{1}{1^2}\right) = -2.18 \times 10^{-18}\ \text{J}$$

Don't be confused by the negative sign for the energy. It appears because we *define* zero energy as the atom's energy when *the electron is completely removed from the nucleus;* in other words, $E = 0$ when $n = \infty$, so $E < 0$ for any smaller $n$. As an analogy, consider a book resting on the floor. If you define its potential energy on the floor as zero, its potential energy is positive when it is on your desk. On the other hand, if you define the book's potential energy as zero when it is on your desk, its energy is negative when it falls to the floor; the latter case is analogous to the energy of the H atom.

Since $n$ is in the denominator of the previous energy formula, as the electron moves closer to the nucleus ($n$ decreases), the atom becomes more stable (less energetic) and its energy becomes a *larger negative number.* As the electron moves away from the nucleus ($n$ increases), the atom's energy increases (becomes a smaller negative number). The energy difference between any two energy levels is represented by

$$\Delta E = E_{final} - E_{initial} = -2.18 \times 10^{-18}\ \text{J}\left(\frac{1}{n_{final}^2} - \frac{1}{n_{initial}^2}\right) \qquad \textbf{(7.4)}$$

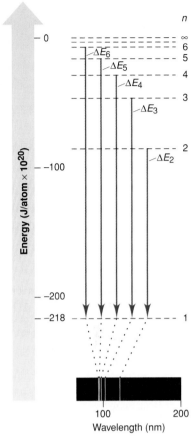

**FIGURE 7.10**

**The spacing of spectral lines as a result of emissions between atomic energy levels.** As $n$ increases, the energy difference between $n$ values becomes smaller, so the spectral lines become closer together, as shown here for the most intense lines in the H atom ultraviolet series. For example, $\Delta E_2$ differs from $\Delta E_3$ more than $\Delta E_4$ differs from $\Delta E_5$. These energy differences, depicted as downward arrows, appear as photons of a particular wavelength, each of which gives rise to one of the spectral lines shown below.

How much energy is needed to completely remove the electron from the hydrogen atom; that is, what is $\Delta E$ for the following change?

$$H(g) \rightarrow H^+(g) + e^-$$

We calculate this by substituting $n_{final} = \infty$ and $n_{initial} = 1$ into Equation 7.4:

$$\Delta E = E_{final} - E_{initial} = -2.18 \times 10^{-18} \text{ J} \left( \frac{1}{\infty^2} - \frac{1}{1^2} \right) = 2.18 \times 10^{-18} \text{ J}$$

$\Delta E$ is positive because energy is *absorbed* to remove the electron from the positive nucleus. For a mole of H atoms,

$$\Delta E = \left( 2.18 \times 10^{-18} \frac{\text{J}}{\text{atom}} \right) \left( 6.022 \times 10^{23} \frac{\text{atoms}}{\text{mol}} \right) \left( \frac{1 \text{ kJ}}{10^3 \text{ J}} \right) = 1.31 \times 10^3 \text{ kJ/mol}$$

This is the *ionization energy* of the hydrogen atom, the amount of energy required to form a mole of gaseous $H^+$ ions from a mole of gaseous H atoms; we discuss the idea thoroughly in Chapter 8.

Using the principles of circular motion and electrostatic attraction from classical physics and his idea that the H atom can have only specific values of energy, Bohr was able to predict the wavelengths of the spectral lines of the H atom and obtain a value for the Rydberg constant that differed from the spectroscopists empirical value by only 0.05%!

Despite its great success in accounting for the spectral lines of the hydrogen atom, the Bohr model failed to predict the spectrum of any other element, even helium, the next simplest. Numerous refinements of the model, such as the introduction of elliptical orbits, also failed. In essence, the Bohr model is a one-electron model and works beautifully for the hydrogen atom and some other one-electron species, but not for atoms or molecules with more than one electron. ◆ In these many-electron systems, additional nucleus-electron attractions are present, as well as electron-electron repulsions. However, a more fundamental reason exists for the model's limitations: electrons do *not* travel in circular, or even elliptical, orbits. As you'll see, their movement within the atom is far less clearly defined. As a picture of the atom, the Bohr model is incorrect, but its great contribution, which we retain in our current model, is *the existence of discrete atomic energy levels.*

Spectroscopic analysis of the H atom led to the Bohr model, the first step toward our current model of the atom. From its use by 19th century chemists as a means of identifying elements and compounds, spectrometry has become a major field unto itself and a central tool of modern chemistry (see Tools of the Chemistry Laboratory). ◆

### ◆ Predicting the Spectra of One-Electron Cations.

The Bohr model has been successfully applied to predict the energy levels and spectral lines for many other one-electron species, typically cations with all but one electron removed. A partial list is $He^+$ ($Z = 2$), $Li^{2+}$ ($Z = 3$), $Be^{3+}$ ($Z = 4$), $B^{4+}$ ($Z = 5$), $C^{5+}$ ($Z = 6$), $N^{6+}$ ($Z = 7$), and $O^{7+}$ ($Z = 8$). Some have been prepared in the laboratory; most have been seen in the spectra of stars.

### ◆ What Are Stars Made Of?

In 1868 the French astronomer Pierre Janssen noted a bright yellow line in the solar emission spectrum. After he and other scientists could not reproduce the line from any known element, they considered it due to an element unique to the Sun, which was named helium (Greek *helios*, "Sun"). Twenty years later, the British chemist William Ramsey examined the spectrum of an inert gas obtained by heating uranium-containing minerals, and it showed the same bright yellow line. Analysis of light from stars has shown many of the elements already known on Earth, but helium is the only element that was first discovered on a star. Ongoing analysis of light from Supernova 1987A is beginning to reveal the composition of that star as well.

### Section Summary

To explain the line spectrum of atomic hydrogen, Bohr proposed that the atom's energy is quantized because the electron's motion is restricted to fixed orbits. The electron can move from one orbit to another only if the atom absorbs or emits a photon whose energy equals the difference in energy levels (orbits). Line spectra are produced because these energy changes correspond to photons of specific wavelength. Bohr's model predicted the H atom spectrum but could not predict that of any other atoms because electrons do not have fixed orbits. Despite this, Bohr's idea that atoms have quantized energy levels is a cornerstone of our current model. Spectrophotometry is an instrumental technique in which emission and absorption spectra are used to identify and measure concentrations of substances.

## TOOLS OF THE CHEMISTRY LABORATORY

# Spectrophotometry in Chemical Analysis

The use of spectral data to identify and quantify substances is essential to modern chemical analysis. The terms *spectroscopy, spectrometry,* and **spectrophotometry** denote a large group of instrumental techniques that measure a substance's atomic and molecular energy levels from its spectra.

The two types of spectra most often obtained are emission and absorption spectra. An **emission spectrum,** such as the H atom line spectrum, is produced when atoms that have been excited to higher energy levels *emit* photons characteristic of the element as they return to lower energy levels. Some elements produce a very intense spectral line (or several closely spaced ones) that serves as a marker of the element's presence. Such an intense line is the basis of **flame tests,** rapid qualitative procedures performed by placing a granule of an ionic compound or a drop of its solution in a flame (Figure 7.A,

*A*). Some of the colors of fireworks and flares are due to the emissions from the same elements shown in the flame tests: red from strontium salts and blue-green from copper salts (Figure 7.A, *B*). The characteristic colors of sodium-vapor and mercury-vapor streetlamps, seen in many towns and cities, are due to one or a few prominent lines in their emission spectra.

An **absorption spectrum** is produced when atoms *absorb* photons of certain wavelengths and become excited from lower to higher energy levels. Therefore, the absorption spectrum of an element appears as dark lines against a bright background. When white light passes through sodium vapor, for example, it gives rise to a sodium absorption spectrum, the dark lines appearing at the same wavelengths that yellow lines appear in the sodium emission spectrum (Figure 7.B).

**FIGURE 7.A Flame tests and fireworks. A,** In general, the color of the flame is created by a strong emission in the line spectrum of the element and therefore is often taken as preliminary evidence of the presence of the element in a sample. Shown here are the crimson of strontium and the blue-green of copper. **B,** The same emissions from compounds that contain these metal ions often appear in the brilliant displays of fireworks.

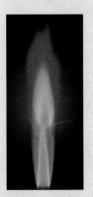

A

B

**FIGURE 7.B Emission and absorption spectra of sodium.** The wavelengths of the bright emission lines correspond to those of the dark absorption lines because both are created by the same energy change:

$$\Delta E_{\text{emission}} = -\Delta E_{\text{absorption}}$$

(Only the two most intense lines in the sodium spectra are shown here.)

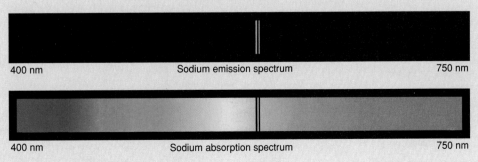

400 nm — Sodium emission spectrum — 750 nm

400 nm — Sodium absorption spectrum — 750 nm

*Continued.*

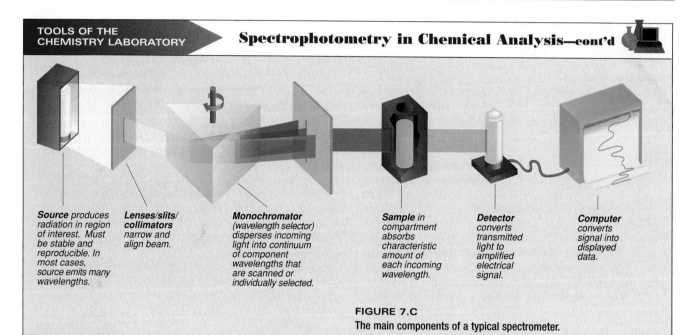

***Source*** *produces radiation in region of interest. Must be stable and reproducible. In most cases, source emits many wavelengths.*

***Lenses/slits/ collimators*** *narrow and align beam.*

***Monochromator*** *(wavelength selector) disperses incoming light into continuum of component wavelengths that are scanned or individually selected.*

***Sample*** *in compartment absorbs characteristic amount of each incoming wavelength.*

***Detector*** *converts transmitted light to amplified electrical signal.*

***Computer*** *converts signal into displayed data.*

**FIGURE 7.C**
The main components of a typical spectrometer.

Instruments based on absorption spectra are much more common than those based on emission spectra. When a solid, liquid, or dense gas is excited, it *emits* so many lines that the spectrum is a continuum (recall the continuum of colors in sunlight). Absorption is also less destructive of fragile organic and biological molecules. Despite differences that depend on the region of the electromagnetic spectrum used to illuminate the sample, all modern spectrometers have components that perform the same basic functions (Figure 7.C).

Visible light is often used to study colored substances, which absorb only some of the wavelengths from white light. A leaf looks green, for example, because its chlorophyll absorbs red and blue wavelengths strongly and green weakly, so most of the green is reflected. The absorption spectrum of chlorophyll *a* appears in Figure 7.D. The curve varies in height because chlorophyll absorbs all incoming wavelengths, but to different extents. The shape of the curve is a spectral "fingerprint" of chlorophyll *a*.

In addition to its use in identifying a substance, a spectrometer can also be used to measure its concentration, because *the absorbance, the amount of light of a given wavelength absorbed by a substance, is proportional to the number of molecules.* Suppose you want to determine the concentration of chlorophyll in a solution of leaf extract. You select a strongly absorbed wavelength from its spectrum (such as 663 nm in Figure 7.D), measure the absorbance of the leaf-extract solution, and compare it with the absorbances of a series of chlorophyll solutions of known concentration.

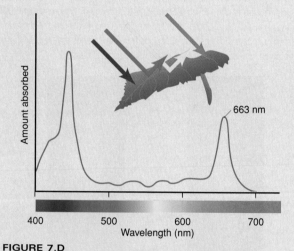

**FIGURE 7.D**
**The absorption spectrum of chlorophyll *a*.** Chlorophyll *a* is one of several leaf pigments. It absorbs red and blue wavelengths strongly but almost no green or yellow wavelengths. Thus, leaves containing large amounts of chlorophyll *a* appear green. The strong absorption at 663 nm can be used to quantify the substance in a plant extract.

# 7.3   The Wave-Particle Duality of Matter and Energy

The year 1905 was a busy one for Albert Einstein. In addition to presenting the photon theory of light and explaining the photoelectric effect, he found time to explain Brownian motion (Chapter 12), which helped establish the molecular view of matter, and to introduce a new branch of physics with his theory of relativity. One of its many startling revelations was that matter and energy are alternate forms of the same entity. This idea is revealed in his famous equation $E = mc^2$, which expresses the amount of energy that is equivalent to a given amount of mass, and vice versa. Relativity theory does not depend on quantum theory, but together they have completely blurred the sharp divisions between matter (lumpy and weighable) and energy (diffuse and massless) that we see in the macroscopic world. ◆

The early proponents of quantum theory demonstrated that *energy is particulate*. Physicists who developed the theory turned this proposition upside down and showed that *matter is wavelike*. Strange as this idea may seem, it was the key to an atomic model that would unify the diverse behavior of the elements.

### The Wave Nature of Electrons

Bohr's efforts were a perfect case of fitting theory to data: he *assumed* that an atom has only certain allowable energy levels in order to explain the observed line spectrum. However, his assumption had no basis in physical theory. In the early 1920s, a young French physics student named Louis de Broglie proposed a remarkable reason for the existence of fixed energy levels in Bohr's model. De Broglie had been thinking of other systems that display only certain allowed motions, such as the wave created by a plucked guitar string. Because the ends of the string are fixed, only certain vibrational frequencies (and wavelengths) are possible (Figure 7.11). De Broglie

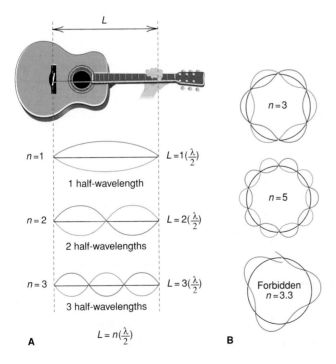

$n = 3$

$n = 1$    $L = 1(\frac{\lambda}{2})$

1 half-wavelength

$n = 2$    $L = 2(\frac{\lambda}{2})$

2 half-wavelengths

$n = 5$

$n = 3$    $L = 3(\frac{\lambda}{2})$

3 half-wavelengths

Forbidden
$n = 3.3$

$L = n(\frac{\lambda}{2})$

**A**                                        **B**

**FIGURE 7.11**

**Wave motion in restricted systems. A,** In a musical analogy to electron waves, one-half wavelength ($\lambda/2$) is the "quantum" of the guitar string's vibration. The string length $L$ is fixed, so the only allowed vibrations occur when $L$ is a whole-number multiple ($n$) of $\lambda/2$. **B,** If an electron occupies a circular orbit, only whole numbers of wavelengths are allowed ($n = 3$ and $n = 5$ are shown). A wave with a fractional number of wavelengths (such as $n = 3.3$) is "forbidden" because it rapidly dies out through overlap of crests and troughs.

reasoned that if waves of energy have some properties of particles, perhaps particles of matter have some properties of waves. Specifically, if electrons had wavelike motion and were restricted to orbits of fixed radii, they also would have only certain possible frequencies and energies.

Combining the equation for mass-energy equivalence ($E = mc^2$) with that for the energy of a photon ($E = h\nu = hc/\lambda$), de Broglie derived an equation for the wavelength of any particle of mass ($m$) moving at velocity ($u$)—planet, baseball, or electron:

$$\lambda = \frac{h}{mu} \tag{7.5}$$

where $u = c$ if the particle is a photon moving in a vacuum. According to this equation for the **de Broglie wavelength,** *matter behaves as though it were moving in a wave.* Because an object's wavelength is *inversely* proportional to its mass, however, heavy objects such as planets and baseballs have wavelengths that are *many* orders of magnitude smaller than the object itself (Table 7.1).

TABLE 7.1 **The de Broglie Wavelength of Several Objects**

| SUBSTANCE | MASS (g) | VELOCITY (m/s) | λ (m) |
|---|---|---|---|
| Slow electron | $9 \times 10^{-28}$ | 1.0 | $7 \times 10^{-4}$ |
| Fast electron | $9 \times 10^{-28}$ | $5.9 \times 10^6$ | $1 \times 10^{-10}$ |
| Alpha particle | $6.6 \times 10^{-24}$ | $1.5 \times 10^7$ | $7 \times 10^{-15}$ |
| One-gram mass | 1.0 | 0.01 | $7 \times 10^{-29}$ |
| Baseball | 142 | 25.0 | $2 \times 10^{-34}$ |
| Earth | $6.0 \times 10^{27}$ | $3.0 \times 10^4$ | $4 \times 10^{-63}$ |

SAMPLE PROBLEM 7.3

**Calculating the de Broglie Wavelength of an Electron**

**Problem:** Calculate the de Broglie wavelength of an electron with a velocity of $1.00 \times 10^6$ m/s. (Electron mass $= 9.11 \times 10^{-31}$ kg; $h = 6.626 \times 10^{-34}$ kg·m²/s.)
**Plan:** We know the velocity and mass of the electron, so we substitute these into Equation 7.5 to find λ.
**Solution:**

$$\lambda = \frac{h}{mu} = \frac{6.626 \times 10^{-34} \text{ kg} \cdot \text{m}^2/\text{s}}{(9.11 \times 10^{-31} \text{ kg})(1.00 \times 10^6 \text{ m/s})} = \textbf{7.27} \times \textbf{10}^{-10} \textbf{ m}$$

**Check:** The order of magnitude and units seem correct:

$$\lambda \approx \frac{10^{-33} \text{ kg} \cdot \text{m}^2/\text{s}}{(10^{-30} \text{ kg})(10^6 \text{ m/s})} = 10^{-9} \text{ m}$$

**Comment:** As you'll see in the upcoming discussion, such fast-moving electrons, with wavelengths in the range of atomic sizes, exhibit remarkable properties.

FOLLOW-UP PROBLEM 7.3
What is the velocity of an electron that has a de Broglie wavelength of 100 nm?

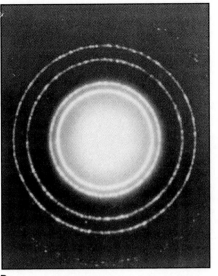

A                                              B

**FIGURE 7.12**
**Diffraction patterns of x-rays and electrons. A,** X-ray diffraction pattern of Al foil. **B,** Electron diffraction pattern of Al foil. This behavior implies that electrons travel in waves.

If de Broglie's concept is correct and all particles travel in waves, electrons should exhibit the wave properties of diffraction and interference (Section 7.1). A fast-moving electron has a wavelength of about $10^{-10}$ m, so the perfect "slit" would be the natural spacings between atoms in a crystal. Indeed, in 1927, C. Davisson and L. Germer guided a beam of electrons at a nickel crystal and obtained an electron diffraction pattern. Figure 7.12 shows the diffraction pattern obtained when x-rays or electrons impinge on aluminum foil. Apparently, electrons—particles with mass and charge—create diffraction patterns, just as electromagnetic waves do! Although electrons do not have orbits of fixed radius, as de Broglie thought, the energy levels of the atom *are* related to the wave nature of the electron. ◆

If electrons have properties of energy, do photons have properties of matter? The de Broglie equation suggests that we can calculate the momentum (*p*), the product of mass and velocity, for a photon of a given wavelength. Solving Equation 7.5 for momentum, with $u = c$, gives

$$\lambda = \frac{h}{mc} = \frac{h}{p} \quad \text{and} \quad p = \frac{h}{\lambda}$$

The inverse relationship between *p* and $\lambda$ means that shorter wavelength (higher energy) photons have greater momentum. Thus, a decrease in the momentum of a photon should appear as an increase in its wavelength. In 1923, Arthur Compton directed a beam of x-ray photons at a sample of graphite and observed that the reflected photons had a longer wavelength. This result was interpreted to mean that the photons had transferred some of their momentum to the electrons in the carbon atoms of the graphite, just as colliding billiard balls transfer momentum to one another. In this experiment, photons behave as particles with momentum.

To scientists of the early part of the 20th century, these results were very unsettling. Some experiments show that matter is particulate and energy

◆ **The Electron Microscope.**
In a transmission electron microscope a beam, focused by a lens, passes through a thin section of the specimen to a second lens. The resulting image is then magnified by a third lens to a final image. The differences between this and a light microscope are that the "beam" consists of high-speed electrons and the "lenses" are electromagnetic fields, which can be adjusted to give up to 200,000-fold magnification and 0.5 nm resolution. In a scanning electron microscope, the beam scans the specimen, knocking electrons from it, which creates a current that varies with surface irregularities. The current generates an image that looks like the object's surface [see photos: heart muscle, 90,000× colorized (*top*); tool marks on steel, 290× (*bottom*).] The great advantage of electron microscopes is that high-speed electrons have wavelengths much smaller than those of visible light and, thus, allow much greater image resolution.

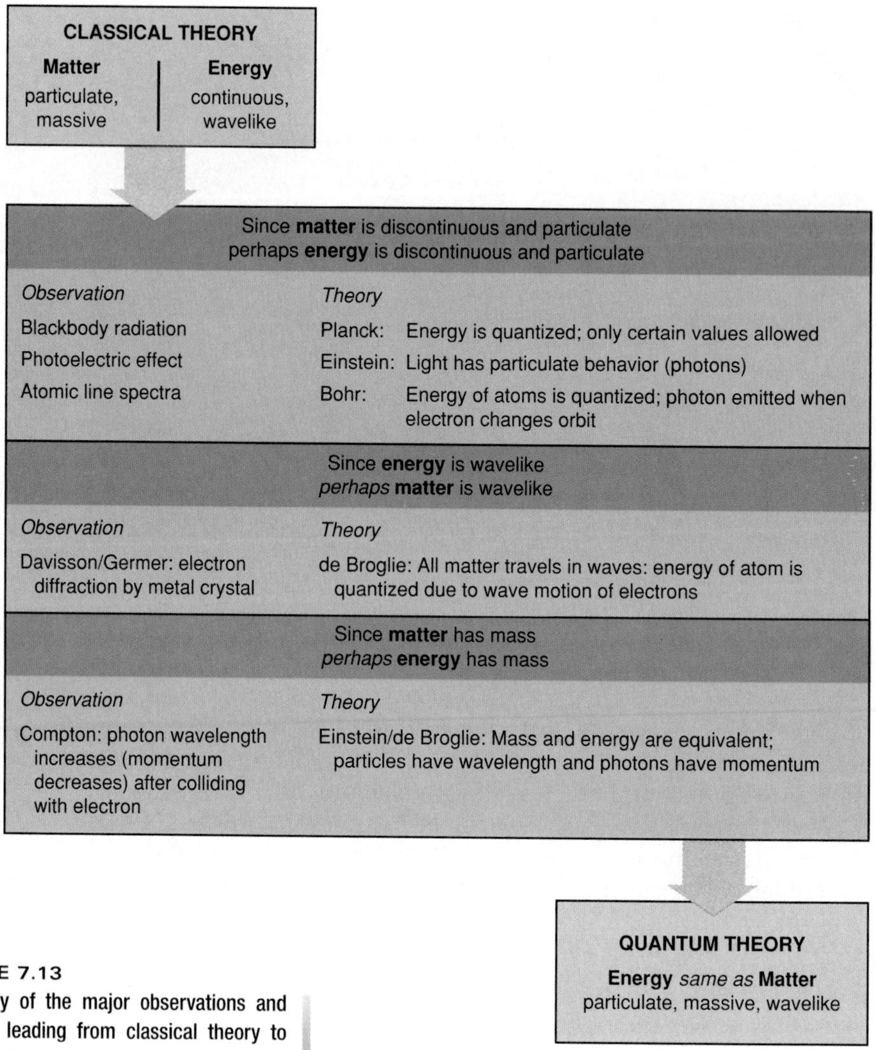

**CLASSICAL THEORY**

| Matter | Energy |
|---|---|
| particulate, massive | continuous, wavelike |

Since **matter** is discontinuous and particulate
perhaps **energy** is discontinuous and particulate

| Observation | Theory |
|---|---|
| Blackbody radiation | Planck: Energy is quantized; only certain values allowed |
| Photoelectric effect | Einstein: Light has particulate behavior (photons) |
| Atomic line spectra | Bohr: Energy of atoms is quantized; photon emitted when electron changes orbit |

Since **energy** is wavelike
*perhaps* **matter** is wavelike

| Observation | Theory |
|---|---|
| Davisson/Germer: electron diffraction by metal crystal | de Broglie: All matter travels in waves: energy of atom is quantized due to wave motion of electrons |

Since **matter** has mass
*perhaps* **energy** has mass

| Observation | Theory |
|---|---|
| Compton: photon wavelength increases (momentum decreases) after colliding with electron | Einstein/de Broglie: Mass and energy are equivalent; particles have wavelength and photons have momentum |

**QUANTUM THEORY**

**Energy** *same as* **Matter**
particulate, massive, wavelike

**FIGURE 7.13**
**Summary of the major observations and theories leading from classical theory to quantum theory.** As often happens in science, an observation (experiment) stimulates the need for an explanation (theory), and/or a theoretical insight provides the impetus for an experimental test.

wavelike, but others show just the opposite. We have come full circle in our understanding of matter and energy: every characteristic trait we had used to define one now also defines the other. Figure 7.13 summarizes the conceptual and experimental breakthroughs that led to this juncture.

The unsettling truth is that *both* matter and energy show *both* behaviors: each possesses both "faces." In some experiments, we observe one face; in other experiments, we observe the other face. The distinction between a particle and a wave is only meaningful in the macroscopic world; it disappears at the atomic level. The distinction is in our minds and our limiting definitions, not in nature. This dual character of matter and energy is known as the **wave-particle duality.**

## The Heisenberg Uncertainty Principle

In the macroscopic world, a moving particle has a definite location at any instant, whereas a wave is spread out in space. If an electron has the properties of both a particle and a wave, what can we determine about its position in the atom? In 1927, the German physicist Werner Heisenberg postulated the **uncertainty principle,** which states that it is impossible to know simultaneously the exact position *and* velocity of a particle. The principle is expressed mathematically as

$$\Delta x \cdot m\Delta u \geq \frac{h}{4\pi} \tag{7.6}$$

where $\Delta x$ is the uncertainty in position and $\Delta u$ is the uncertainty in velocity. The more accurately we know the position of the particle (smaller $\Delta x$), the less accurately we know its velocity (larger $\Delta u$), and vice versa.

By observing the position and velocity of a pitched baseball and using the classical laws of motion, we can predict its trajectory and whether it will be a strike or a ball. In this case, $\Delta x$ and $\Delta u$ are insignificant because the mass of a baseball is so large compared with $h/4\pi$. Finding the position and velocity of an electron, and from them its trajectory, is another problem entirely, however, as Sample Problem 7.4 shows.

SAMPLE PROBLEM 7.4 _____

### Applying the Uncertainty Principle

**Problem:** An electron moving near an atomic nucleus has a velocity of $6 \times 10^6 \pm$ 1% m/s. What is the uncertainty in its position?
**Plan:** The uncertainty in the velocity ($\Delta u$) is given as 1%, so we calculate its value, substitute it into Equation 7.6, and solve for the uncertainty in position ($\Delta x$).
**Solution:** Finding the uncertainty in velocity:

$$\Delta u = (6 \times 10^6 \text{ m/s}) \times 0.01 = 6 \times 10^4 \text{ m/s}$$

Calculating $\Delta x$ for the electron:

$$\Delta x \geq \frac{h}{4\pi m\Delta u} \geq \frac{6.626 \times 10^{-34} \text{ kg} \cdot \text{m}^2/\text{s}}{4\pi \, (9.11 \times 10^{-31} \text{ kg}) \, (6 \times 10^4 \text{ m/s})} \geq \mathbf{1 \times 10^{-9} \text{ m}}$$

**Check:** Be sure to check orders of magnitude of the answer.
**Comment:** The uncertainty in the electron's position is about 10 times greater than the diameter of the entire atom ($10^{-10}$ m), so we have no precise idea where in the atom it is located! In the follow-up problem, see if an umpire has any better idea where a baseball is located when calling balls and strikes.

FOLLOW-UP PROBLEM 7.4
A baseball (mass = 0.142 kg) is moving at 100.0 ± 1.00% mi/h (44.7 ± 1.00% m/s). How accurately does an umpire know its position?

_____

As the results of Sample Problem 7.4 show, the uncertainty principle has profound implications for an atomic model. It means that *we cannot prescribe exact paths for electrons,* such as the circular orbits of Bohr's model. As you'll see in the next section, the most we can ever hope to know is the *probability*—the odds—of finding an electron in a given volume of space. However, we are not *sure* it is there any more than a gambler is sure of the next roll of the dice. ◆

◆ **Uncertainty Is Unacceptable.** Einstein found the uncertainty principle difficult to accept, as reflected in his famous statement that "God does not play dice with the universe." Rutherford was also skeptical. When Niels Bohr, who had become a champion of the new physics, delivered a lecture on the uncertainty principle in Rutherford's laboratory, Rutherford said to him, "You know, Bohr, your conclusions seem to me as uncertain as the premises on which they are built." Acceptance of radical ideas does not come easily, even to fellow geniuses.

**Section Summary**

As a result of quantum theory and relativity theory, we can no longer view matter and energy as distinct entities. The de Broglie wavelength proposes that electrons (and all matter) have wavelike motion. Allowed atomic energy levels are related to allowed wavelengths of the electron's motion. Electrons exhibit diffraction patterns, as do waves of energy, and photons exhibit transfer of momentum, as do particles of mass. The wave-particle duality of matter and energy is observable only on the atomic scale. According to the uncertainty principle, we cannot know simultaneously the exact position and velocity of an electron.

## 7.4 The Quantum-Mechanical Model of the Atom

Acceptance of the dual nature of matter and energy and of the uncertainty principle culminated in the field of **quantum mechanics** (or wave mechanics), which examines the wave motion of objects on the atomic scale. In 1926, Erwin Schrödinger derived an equation that is the basis for the quantum-mechanical model of the hydrogen atom. The model describes an atom that has certain allowed amounts of energy due to the allowed wavelike motion of an electron whose exact location is impossible to know.

### The Atomic Orbital and the Probable Location of the Electron

The **Schrödinger equation** is quite complex because the electron's matter-wave occurs in three-dimensional space and is continuously but variably influenced by the nuclear charge. For our purposes, we can consider a simplified form of the equation:

$$\mathcal{H}\psi = E\psi$$

where $E$ is the energy of the atom. The symbol $\psi$ (Greek *psi*, pronounced "sigh") is called a **wave function,** a mathematical description of the motion of the electron's matter-wave as a function of time and position. The symbol $\mathcal{H}$ represents a complex set of mathematical operations that, when carried out on a particular $\psi$, yields an allowed energy state.*

Each solution to the equation is associated with a particular wave function, also called an **atomic orbital.** It is essential to realize that *an atomic orbital bears no resemblance whatsoever to an "orbit" in the Bohr model:* the *orbit* was a path supposedly followed by the electron; the *orbital* is a mathematical function with no independent physical reality.

Although the wave function (atomic orbital) has no physical meaning, the square of the wave function, $\psi^2$, does. We cannot know precisely where the electron is at any moment, but we can describe where it probably is, that

---

*The complete form of the Schrödinger equation is

$$\frac{d^2\psi}{dx^2} + \frac{d^2\psi}{dy^2} + \frac{d^2\psi}{dz^2} + \frac{8\pi^2 m_e}{h^2}[E - V(x,y,z)]\ \psi(x,y,z) = 0$$

where $\psi$ is the wave function; the first three terms describe how $\psi$ changes in space; $m_e$ is the electron's mass; $E$ is the total quantized energy of the atomic system; and $V$ is the potential energy at point $(x,y,z)$. Solving the equation for almost any practical application requires much computer time.

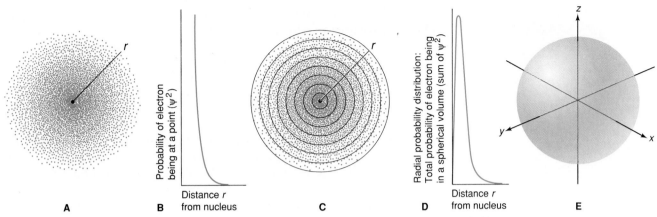

A        B        C        D        E

**FIGURE 7.14**

**Depicting electron probability in the H atom ground state. A,** An electron density diagram shows a cross section of the H atom, with dots representing the probability of the electron being at any point (or tiny volume). The probability decreases along a line emanating outward from the nucleus. **B,** A plot of the information in **A** shows that the probability ($\psi^2$) at any point along the line decreases with distance from the nucleus but does not reach zero (see text). **C,** Dividing the atom's volume into thin, concentric, spherical layers (shown in cross section) and counting the dots within each layer gives the total probability of finding the electron within that volume. **D,** A radial probability distribution plot shows total electron density in each spherical layer vs. *r*. Since electron density decreases more slowly than the volume of each concentric spherical layer increases, the plot shows a peak. **E,** A 90% contour depiction of the ground state of the H atom represents the volume in which the electron spends 90% of its time.

is, where it spends most of its time. The quantity $\psi^2$ expresses the *probability* that the electron will be found at a particular point (more precisely, in a particular very small volume) within the atom. For a given energy level, we can show this probability pictorially by means of an **electron probability density diagram,** or more simply, an **electron density diagram** (Figure 7.14, *A*). The density of the dots in each small volume of the diagram (the electron density) represents the probability of finding the electron in that particular tiny volume. That is, the higher the value of $\psi^2$ for a given volume, the greater is the density of dots.

Electron density diagrams are referred to informally as **electron cloud** representations. The idea is that if we could take a time-exposure snapshot of the electron whirling around the nucleus in wavelike motion, the picture would appear as a "cloud" of electron positions. In other words, the electron cloud is an imaginary representation of the electron rapidly changing its position over time; it does *not* mean that an electron is a diffuse cloud of charge. The electron density diagram represents the probability of finding the electron at a *particular* point at a given distance *r* along a line from the nucleus outward. Note that *the electron density decreases with distance*. We can show the same concept graphically by plotting $\psi^2$ vs. *r* (Figure 7.14, *B*). *The probability of the electron being far from the nucleus is very small, but not zero*, although the thickness of the lines in the illustration hides this fact by showing the curve touching the axis.

We may also want to know the *total* probability of finding the electron at *all* points at any distance *r* from the nucleus. To do this, we mentally divide the volume around the nucleus into thin, concentric, spherical layers, like the layers of an onion (shown in cross section in Figure 7.14, *C*) and ask in which *spherical layer* we are most likely to find the electron. This is the same as asking for the *sum of $\psi^2$ values* within each spherical layer. As we move from the innermost to the second spherical layer, the electron density (probability of finding the electron at a given point) decreases. However, the volume of the concentric layers increases faster than the electron density decreases. The second spherical layer is larger than the first, so the *total* probability of finding the electron there is higher. As we continue to move away from the nucleus, the electron density becomes lower and, even though the spherical layers become larger, the total probability of the electron being in each layer eventually becomes lower as well. Because of the opposing effects of decreasing electron density and increasing spherical layer volume, the total probability peaks in some thin, spherical layer near the nucleus but

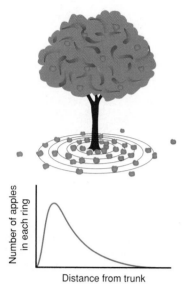

Number of apples in each ring

Distance from trunk

**◆ A Radial Probability Distribution of Apples.** An analogy might clarify why the curve in the radial probability distribution plot peaks and then falls off. Picture the fallen apples around the base of an apple tree: the density of apples is greatest near the trunk and decreases with distance. Divide the ground under the tree into foot-wide concentric rings and collect the apples within each ring. Apple density is greatest in the first ring, but the area of the second ring is larger, so there is a greater *total* number of apples. Farther out near the edge of the tree, rings have more area but lower apple density, so the total number of apples decreases. A plot of "number of apples within a ring" vs. "distance of ring from trunk" would show a peak at some close distance from the trunk, as in Figure 7.14, *D*.

beyond the first one. We show this graphically by means of a **radial probability distribution plot** (Figure 7.14, *D*). ◆

It is interesting that the peak of the radial probability distribution for the ground-state hydrogen atom appears at the same distance from the nucleus (0.529 Å, or $5.29 \times 10^{-11}$ m) as the closest Bohr orbit. Thus, at least for the ground state, the Schrödinger model predicts that the electron spends *most* of its time at the same distance that the Bohr model predicted it spent *all* of its time. The difference between "most" and "all" reflects the uncertainty of the electron's location incorporated in the Schrödinger model.

How far away from the nucleus can the electron be? This is the same as asking "How large is an atom?" As was pointed out for Figure 7.14, *B*, the probability of finding the electron at a particular point far from the nucleus is very small but not zero. Thus, we cannot assign a definite volume to an atom. To visualize atoms, however, we can use a diagram of a 90% **probability contour,** meaning that the electron spends 90% of its time within the contour limits (Figure 7.14, *E*).

## Quantum Numbers of an Atomic Orbital

So far we have discussed the electron density for the hydrogen atom ground state. When the atom absorbs energy and exists in an excited state, the wave motion of the electron is described mathematically by a different atomic orbital (wave function). As you'll see, each atomic orbital has a distinctive radial probability distribution and probability contour diagram.

*An atomic orbital is specified by three quantum numbers* that are associated, respectively, with the orbital's size, shape, and orientation in space.* The quantum numbers have a hierarchical relationship: the size-related number limits the shape-related number, which limits the orientation-related number. Let's examine this hierarchy before getting into the details of shapes and orientations.

1. The **principal quantum number ($n$)** *is a positive integer* (1, 2, 3, and so forth). It indicates the relative *size* of the orbital and therefore the relative *distance from the nucleus* of the peak in the radial probability distribution plot. The principal quantum number specifies the *energy level* of the H atom: *the higher the n value, the greater the energy.* When the electron occupies an orbital with $n = 1$, the H atom is in its ground state and has lower energy than when the electron occupies an orbital with $n = 2$ (first excited state).

2. The **azimuthal quantum number ($l$)** *is an integer from 0 to $n - 1$.* It is related to the *shape* of the orbital and is sometimes called the *orbital-shape quantum number.* Note that the principal quantum number sets a limit on the values for the azimuthal quantum number; that is, $n$ limits $l$. For an orbital with $n = 1$, $l$ can have only a value of 0. For orbitals with $n = 2$, $l$ can have a value of 0 or 1; for those with $n = 3$, $l$ can be 0, 1, or 2; and so forth. Note that the number of possible $l$ values equals the value of $n$.

3. The **magnetic quantum number ($m_l$)** *is an integer from $-l$ through 0 to $+l$.* It prescribes the *orientation* of the orbital in the three-dimensional space about the nucleus and is sometimes called the *orbital-orientation quantum number.* The possible values of an orbital's magnetic quantum

---

*For ease in discussion, we refer to the size, shape, and orientation of an "atomic orbital," although we really mean the size, shape, and orientation of an "atomic orbital's radial probability distribution." This usage is common in both introductory and advanced chemistry texts.

**TABLE 7.2 The Hierarchy of Quantum Numbers for Atomic Orbitals**

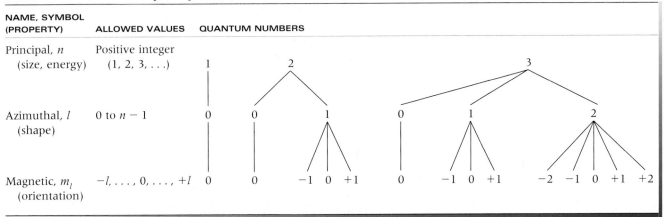

| NAME, SYMBOL (PROPERTY) | ALLOWED VALUES | QUANTUM NUMBERS |
|---|---|---|
| Principal, $n$ (size, energy) | Positive integer $(1, 2, 3, \ldots)$ | |
| Azimuthal, $l$ (shape) | 0 to $n - 1$ | |
| Magnetic, $m_l$ (orientation) | $-l, \ldots, 0, \ldots, +l$ | |

number are set by its azimuthal quantum number (that is, $l$ determines $m_l$). An orbital with $l = 0$ can have only $m_l = 0$. However, an orbital with $l = 1$ can have an $m_l$ value of $-1$, 0, or $+1$; thus, there are three possible orbitals with $l = 1$, each with its own spatial orientation. Note that the number of possible $m_l$ values, or orbitals, for a given $l$ value is $2l + 1$.

Table 7.2 summarizes the relationships among the three quantum numbers. The total number of orbitals for a given $n$ value is $n^2$.

SAMPLE PROBLEM 7.5 —————————————————————————

## Determining Quantum Numbers for a Given Energy Level

**Problem:** What values of the azimuthal ($l$) and magnetic ($m_l$) quantum numbers are allowed for a principal quantum number ($n$) of 3? How many orbitals are allowed for $n = 3$?

**Plan:** We determine the allowable quantum numbers with the rules outlined in the text: values for $l$ are integers from 0 to $n - 1$, and those for $m_l$ are integers from $-l$ to 0 to $+l$. Each $m_l$ value is assigned to a different orbital, so the number of $m_l$ values gives the number of orbitals.

**Solution:** Determining $l$ values:
   For $n = 3$, $l = $ **0, 1, 2**
Determining $m_l$ for each $l$ value:
   For $l = 0$, $m_l = $ **0**
   For $l = 1$, $m_l = $ **$-1$, 0, $+1$**
   For $l = 2$, $m_l = $ **$-2$, $-1$, 0, $+1$, $+2$**
There are nine $m_l$ values, so there are **nine orbitals with $n = 3$.**

**Check:** The total number of orbitals with a given $n$ is $n^2$. For $n = 3$, $n^2 = 9$.

FOLLOW-UP PROBLEM 7.5
Write $l$ and $m_l$ values for $n = 4$.

The quantum numbers specify the energy states of the atom:
- The atom's energy **levels,** or *shells,* are given by the $n$ value. The $n = 1$ level has lower energy and a greater probability of the electron being closer to the nucleus than the $n = 2$ level.
- The atom's **sublevels,** or *subshells,* are given by the $n$ and $l$ values. Each level contains sublevels that designate the shape of the orbital.

- The atom's *orbitals* are specified by the $n$, $l$, and $m_l$ values. Thus, the three quantum numbers that describe an orbital express its size (energy), shape, and spatial orientation.

    Each sublevel is designated by a letter:

    $l = 0$ is an $s$ sublevel.
    $l = 1$ is a $p$ sublevel.
    $l = 2$ is a $d$ sublevel.
    $l = 3$ is an $f$ sublevel.

(These letters derive from the names of spectroscopic lines: $s$, sharp; $p$, principal; $d$, diffuse; and $f$, fundamental.)

Sublevels are named by joining the $n$ value and the letter designation. For example, the sublevel (subshell) with $n = 2$ and $l = 0$ is called the $2s$ sublevel; the only orbital in this sublevel has $n = 2$, $l = 0$, and $m_l = 0$. A sublevel with $n = 3$, $l = 1$ is a $3p$ sublevel. It has three possible orbitals: one with $n = 3$, $l = 1$, and $m_l = -1$; another with $n = 3$, $l = 1$, and $m_l = 0$; and a third with $n = 3$, $l = 1$, and $m_l = +1$.

SAMPLE PROBLEM 7.6 _____

## Determining Sublevel (Subshell) Names

**Problem:** Give the name, magnetic quantum numbers, and number of orbitals for each sublevel with the following quantum numbers:

(a) $n = 3$, $l = 2$    (b) $n = 2$, $l = 0$    (c) $n = 5$, $l = 1$    (d) $n = 4$, $l = 3$

**Plan:** To name the sublevel (subshell), we combine the $n$ value and $l$ letter designation. Since we know $l$, we can find the possible $m_l$ values ($m_l = -l$ to $0$ to $+l$). We know the number of orbitals in a sublevel from the number of its $m_l$ values.

**Solution:**

|     | $n$ | $l$ | SUBLEVEL NAME | POSSIBLE $m_l$ | NO. OF ORBITALS |
|-----|-----|-----|---------------|----------------|-----------------|
| (a) | 3   | 2   | $3d$          | $-2, -1, 0, +1, +2$ | 5 |
| (b) | 2   | 0   | $2s$          | $0$ | 1 |
| (c) | 5   | 1   | $5p$          | $-1, 0, +1$ | 3 |
| (d) | 4   | 3   | $4f$          | $-3, -2, -1, 0, +1, +2, +3$ | 7 |

**Check:** Check the number of orbitals in each sublevel using

$$\text{No. of orbitals} = \text{no. of } m_l \text{ values} = 2l + 1$$

FOLLOW-UP PROBLEM 7.6
What are the $n$, $l$, and possible $m_l$ values for the $2p$ and $5f$ subshells?

SAMPLE PROBLEM 7.7 _____

## Identifying Incorrect Quantum Numbers

**Problem:** What is wrong with each of the following quantum number designations and/or sublevel names?

|     | $n$ | $l$ | $m_l$ | NAME |
|-----|-----|-----|-------|------|
| (a) | 1   | 1   | 0     | $1p$ |
| (b) | 4   | 3   | +1    | $4d$ |
| (c) | 3   | 1   | −2    | $3p$ |

**Solution:** (a) A sublevel of $n = 1$ can have **only $l = 0$,** not $l = 1$. The only possible subshell is **1s.**

(b) A sublevel with $l = 3$ is an **$f$ sublevel,** not a $d$ sublevel. The sublevel name should be **4f.**

(c) A sublevel with $l = 1$ can have **only $m_l$ of $-1$, $0$, $+1$,** not $-2$.

**Check:** Check that $l$ is always less than $n$, and $m_l$ is always $\geq -l$ or $\leq +l$.

**FOLLOW-UP PROBLEM 7.7**

Supply the missing quantum number(s) or sublevel names.

|     | $n$ | $l$ | $m_l$ | NAME |
|-----|-----|-----|-------|------|
| (a) | ?   | ?   | 0     | $4p$ |
| (b) | 2   | 1   | 0     | ?    |
| (c) | 3   | 2   | $-2$  | ?    |
| (d) | ?   | ?   | ?     | $2s$ |

## Shapes of Atomic Orbitals

Each sublevel of the hydrogen atom corresponds to orbitals with a characteristic shape. As you'll see in Chapter 8, orbitals for the other atoms have similar shapes.

**The s orbital.** An orbital with $l = 0$ is *spherically symmetrical* around the nucleus and is called an **s orbital**. The hydrogen atom's ground state, for example, has the electron in the $1s$ orbital. Figure 7.15, *A*, shows that the electron density, whether depicted graphically *(top)* or as a section of a three-

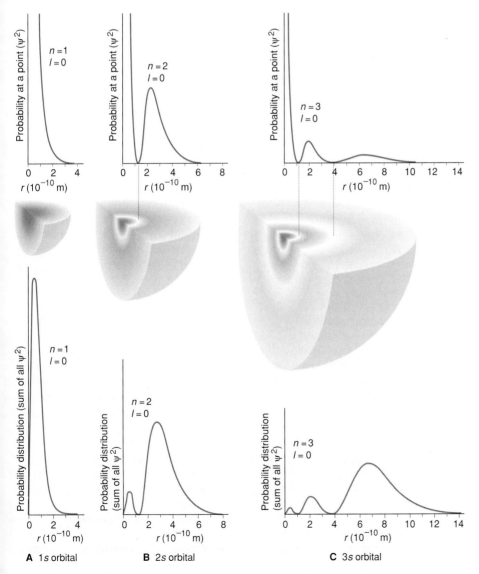

**FIGURE 7.15**

**Electron density plots, electron cloud depictions, and radial probability distribution plots for three s orbitals.** Information for each of the *s* orbitals is shown as a plot of electron density *(top)*, a cloud representation of the electron density *(middle)*, in which shading coincides with peaks in the plot above, and a radial probability distribution *(bottom)* that shows where the electron spends its time. **A,** The $1s$ orbital. **B,** The $2s$ orbital. **C,** The $3s$ orbital. The horizontal axis has units called *Bohr radii* (one Bohr radius $= 0.529 \times 10^{-10}$ m). Note the nodes in the $2s$ and $3s$ orbitals.

**A** $1s$ orbital          **B** $2s$ orbital          **C** $3s$ orbital

dimensional electron cloud picture (*middle*), is highest *at* the nucleus; on the other hand, the radial probability distribution (*bottom*) is highest *near* the nucleus. Both plots fall off smoothly with distance.

The 2*s* orbital (Figure 7.15, *B*) and 3*s* orbital (Figure 7.15, *C*) are more complex than the 1*s*. The 2*s* orbital has two regions of high electron density, with the more distant region having a *higher* radial probability distribution than the closer one. Between the two regions is a **node,** in this case a spherical node, where the probability drops to zero ($\psi^2 = 0$ at the node, analogous to zero amplitude of a wave). The 2*s* orbital is larger than the 1*s*, so when the electron occupies the 2*s* orbital, it spends more of its time *farther* from the nucleus than when it occupies the 1*s*.

The 3*s* orbital shows three regions of electron density and two nodes. Here again, the highest radial probability is at the greatest distance from the nucleus. The pattern of more nodes and higher probability with distance continues for *s* orbitals with higher *n* values. Since an *s* orbital has a spherical shape around the nucleus, it can have only one orientation and thus one value for the magnetic quantum number: for any *s* orbital, $m_l = 0$.

**The *p* orbital.** An orbital with *l* = 1 has two regions of higher probability (sometimes called "dumbbell shaped") and is called a ***p* orbital.** The two lobes of one *p* orbital lie on *either side* of the nucleus. Thus, the *nucleus lies at a nodal plane* of this orbital (Figure 7.16). Since the maximum value of *l* is *n* − 1, only levels with *n* = 2 or higher can have a *p* orbital. Therefore, the lowest energy *p* orbital (the one closest to the nucleus) is the 2*p*. Keep in mind that *one p orbital consists of both lobes* and that the electron spends equal time in both. As we would expect from the *s* orbital pattern, a 3*p* orbital is larger than a 2*p* orbital, a 4*p* is larger than a 3*p*, and so forth.

Whereas an *s* orbital is distributed spherically in space, each *p* orbital has a specific direction. The *l* = 1 value has three possible $m_l$ values: −1, 0, and +1, which refer to three *mutually perpendicular p* orbitals. We usually associate *p* orbitals with the *x*, *y*, and *z* axes and call them the $p_x$, $p_y$, and $p_z$ orbitals, but *no* direct association exists between a certain $m_l$ value and a given axis. The three *p* orbitals are equivalent in size, shape, and energy, differing only in orientation.

**The *d* orbital.** An orbital with *l* = 2 is called a ***d* orbital.** There are five possible $m_l$ values for the *l* = 2 value: −2, −1, 0, +1, and +2. Thus, a *d* orbital can have any one of five different orientations (Figure 7.17). Four of the five *d* orbitals have four lobes ("cloverleaf shaped") prescribed by two mutually perpendicular nodal planes, with the nucleus lying at the junction of the lobes. Three of these orbitals lie along the mutually perpendicular *xy*, *xz*, and *yz* planes, with their lobes *between* the axes, and are called the $d_{xy}$, $d_{xz}$, and $d_{yz}$ orbitals. A fourth, the $d_{x^2-y^2}$ orbital, also lies in the *xy* plane, but its lobes are directed *along* the axes. The fifth *d* orbital, the $d_{z^2}$, is slightly different in its shape: two major lobes lie along the *z* axis, and a "donut" of electron density girdles the center. An electron associated with a given *d* orbital has equal probability of being in any of the orbital's lobes. As before,

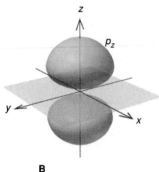

A

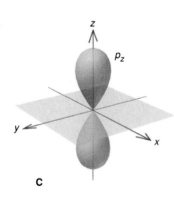

B

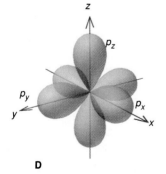

C

D

**FIGURE 7.16**

Radial probability distribution plot and contour diagrams of the three 2*p* orbitals. **A,** A 2*p* orbital shows a peak in its radial probability distribution at about the same distance as the larger peak in the 2*s* plot (see Figure 7.15, *B*). **B,** An accurate representation of the 2$p_z$ probability contour. An electron in this orbital spends 90% of its time within this volume. The $p_x$ and $p_y$ orbitals have identical shapes but lie along the *x* and *y* axes, respectively. Note the nodal plane at the nucleus. **C,** The stylized depiction of the 2*p* contour used throughout the text. **D,** In the atom, the orbitals occupy mutually perpendicular regions of space.

the axis designations are *not* associated with a given $m_l$ value. In accord with the quantum number rules, a *d* orbital must have a principal quantum number of at least $n = 3$. The 4*d* orbitals extend farther from the nucleus than the 3*d*, and the 5*d* orbitals extend still farther.

**Orbitals with higher *l* values.** Orbitals with $l = 3$ are *f* orbitals and must have a principal quantum number of at least $n = 4$. There are seven $(2l + 1 = 7)$ *f* orbitals, each with a complex, multilobed shape. Orbitals with $l = 4$ are *g* orbitals, but we will not discuss them further because they play a very minor role in chemical bonding.

## Energy Levels of the Hydrogen Atom

The energy state of the hydrogen atom depends on the principal quantum number *n*. An electron in an orbital with a higher *n* value spends its time, on the average, farther from the nucleus, so it is higher in energy. (As you'll see in Chapter 8, the energy state of an atom with more than one electron depends on both the *n and l* values of the occupied orbitals.) Thus, in the case of hydrogen *only*, all four $n = 2$ orbitals (one 2*s* and three 2*p*) have the same energy, and all nine $n = 3$ orbitals (one 3*s*, three 3*p*, and five 3*d*) have the same energy.

When the ground-state hydrogen atom $(n = 1)$ absorbs a photon whose energy equals the difference between the energies of the $n = 1$ and $n = 2$ levels, the electron jumps to one of the $n = 2$ orbitals. The excited atom can return to the ground state by emitting a photon of the same energy.

## Section Summary

The electron's wave function ($\psi$, atomic orbital) is a mathematical description of the electron's wavelike motion in an atom. Each wave function is associated with one of the atom's allowed energy states. The probability of finding the electron at a particular location is represented by $\psi^2$. An electron density diagram and a radial probability distribution plot show how the electron occupies the space near the nucleus for a particular energy level. Three features of the atomic orbital are described by quantum numbers: size (*n*), shape (*l*), and orientation ($m_l$). Orbitals are part of sublevels (defined by *n* and *l*), which are part of an energy level (defined by *n*). A sublevel with $l = 0$ has a spherical (*s*) orbital (no nodes); one with $l = 1$ has two-lobed (*p*) orbitals (one node); and one with $l = 2$ has four-lobed (*d*) orbitals (two nodes). For the H atom, the energy levels depend only on the *n* value.

## Chapter Perspective

*In this brief exploration of the origins of quantum physics, we saw the everyday distinctions between matter and energy disappear and a new picture of the hydrogen atom gradually come into focus. By the atom's very nature, however, this focus cannot be perfectly sharp. The sequel to these remarkable discoveries unfolds in Chapter 8, where we begin our discussion of how the periodic behavior of the elements is explained by the properties of the quantum-mechanical atom.*

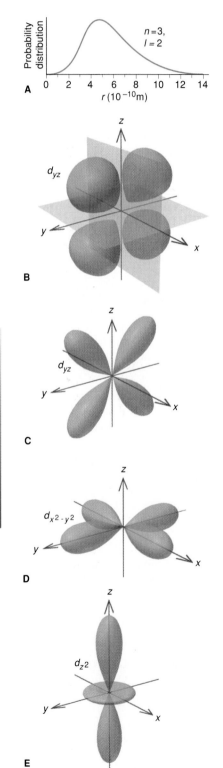

**FIGURE 7.17**

**Radial probability distribution plot and contour diagrams of the 3*d* orbitals. A,** Radial probability distribution plot. **B,** An accurate representation of the $3d_{yz}$ orbital probability contour. Note the mutually perpendicular nodal planes. The $3d_{xz}$, $3d_{xy}$, and $3d_{x^2-y^2}$ orbitals have identical shapes but different orientations. **C,** The stylized depiction of the $3d_{yz}$ orbital used through the text. **D,** The lobes of the $3d_{x^2-y^2}$ orbital lie on the *x* and *y* axes. **E,** The $d_{z^2}$ orbital has two lobes and a central donut for its probability contour.

# For Review and Reference

## Key Terms

**SECTION 7.1**
electromagnetic (EM)
  radiation
frequency ($\nu$)
wavelength ($\lambda$)
speed of light ($c$)
amplitude
electromagnetic spectrum
infrared (IR)
monochromatic
polychromatic
ultraviolet (UV)
refraction
diffraction
blackbody radiation

quantum number
Planck's constant ($h$)
quantum
photoelectric effect
photon

**SECTION 7.2**
line spectrum
stationary state
ground state
excited state
spectrophotometry
emission spectrum
flame test
absorption spectrum

**SECTION 7.3**
de Broglie wavelength
wave-particle duality
uncertainty principle

**SECTION 7.4**
quantum mechanics
Schrödinger equation
wave function
  (atomic orbital)
electron probability
  density diagram
  (electron density
  diagram)
electron cloud

radial probability
  distribution plot
probability contour
principal quantum
  number ($n$)
azimuthal quantum
  number ($l$)
magnetic quantum
  number ($m_l$)
level (shell)
sublevel (subshell)
$s$ orbital
node
$p$ orbital
$d$ orbital

## Key Equations and Relationships

**7.1** Relating the speed of light to its frequency and wavelength (p. 257):

$$c = \nu\lambda$$

**7.2** Determining the smallest change in an atom's energy (p. 261):

$$\Delta E = h\nu$$

**7.3** Calculating the wavelength of any line in the H atom spectrum (Rydberg equation) (p. 263):

$$\frac{1}{\lambda} = R\left(\frac{1}{n_1^2} - \frac{1}{n_2^2}\right)$$

**7.4** Finding the difference between two energy levels in the H atom (p. 265):

$$\Delta E = E_{final} - E_{initial} = -2.18 \times 10^{-18} \text{ J} \left(\frac{1}{n_{final}^2} - \frac{1}{n_{initial}^2}\right)$$

**7.5** Calculating the wavelength of any moving object (de Broglie wavelength) (p. 270):

$$\lambda = \frac{h}{mu}$$

**7.6** Finding the uncertainty in position or velocity of a particle (Heisenberg uncertainty principle) (p. 273):

$$\Delta x \cdot m\Delta u \geq \frac{h}{4\pi}$$

## Answers to Follow-up Problems

**7.1** $\lambda$ (nm) $= \dfrac{3.00 \times 10^8 \text{ m/s}}{7.23 \times 10^{14} \text{ s}^{-1}} \times \dfrac{10^9 \text{ nm}}{1 \text{ m}} = 415$ nm

**7.2** UV: $E = hc/\lambda$

$= \dfrac{(6.626 \times 10^{-34} \text{ J} \cdot \text{s})(3.00 \times 10^8 \text{ m/s})}{1 \times 10^{-8} \text{ m}} = 2 \times 10^{-17}$ J

Visible: $E = 4 \times 10^{-19}$ J; IR: $E = 2 \times 10^{-21}$ J
As $\lambda$ increases, $E$ decreases.

**7.3** $u = h/m\lambda$

$= \dfrac{6.626 \times 10^{-34} \text{ kg} \cdot \text{m}^2/\text{s}}{(9.11 \times 10^{-31} \text{ kg})\left(100 \text{ nm} \times \dfrac{1 \text{ m}}{10^9 \text{ nm}}\right)} = 7.27 \times 10^3$ m/s

**7.4** $\Delta x \geq \dfrac{6.626 \times 10^{-34} \text{ kg} \cdot \text{m}^2/\text{s}}{4\pi (0.142 \text{ kg})(0.447 \text{ m/s})} \geq 8.31 \times 10^{-34}$ m

**7.5** $n = 4$, so $l = 0, 1, 2, 3$. In addition to the $m_l$ values in Sample Problem 7.5, we have those for $l = 3$:

$$m_l = -3, -2, -1, 0, +1, +2, +3.$$

**7.6** For 2p: $n = 2$, $l = 1$, $m_l = -1, 0, +1$
For 5f: $n = 5$, $l = 3$, $m_l = -3, -2, -1, 0, +1, +2, +3$

**7.7** (a) $n = 4$, $l = 1$; (b) name is 2p; (c) name is 3d; (d) $n = 2$, $l = 0$, $m_l = 0$

## Sample Problem Titles

**7.1** Interconverting Wavelength and Frequency (p. 258)
**7.2** Calculating the Energy of Radiation from its Frequency (p. 262)
**7.3** Calculating the De Broglie Wavelength of an Electron (p. 270)
**7.4** Applying the Uncertainty Principle (p. 273)
**7.5** Determining Quantum Numbers for a Given Energy Level (p. 277)
**7.6** Determining Sublevel (Subshell) Names (p. 278)
**7.7** Identifying Incorrect Quantum Numbers (p. 278)

# Problems

Problems with a green number are answered at the back of the text. Most sections include three categories of problems separated by a green rule: concept review questions, *paired skill-building exercises*, and problems in a relevant context.

## The Nature of Light

(Sample Problems 7.1 and 7.2)

**7.1** In what sense are microwave and ultraviolet radiation the same? In what sense are they different?

**7.2** Consider the following types of electromagnetic radiation: (1) microwave; (2) ultraviolet; (3) radio waves; (4) infrared; (5) x-ray; (6) visible.
(a) Arrange them in order of increasing wavelength.
(b) How does this order differ from that of increasing frequency?
(c) List one way each type of radiation is used.

**7.3** Define each of the following wave phenomena, and give an example of where each occurs: (a) refraction; (b) diffraction; (c) dispersion; (d) interference.

**7.4** What new idea about energy did Planck use to explain blackbody radiation?

**7.5** What new idea about light did Einstein use to explain the photoelectric effect? Why does the photoelectric effect exhibit a threshold frequency? Why does it *not* exhibit a time lag?

---

**7.6** An AM station broadcasts rock music at "710 on your radio dial." Units for AM frequencies are given in kilohertz (kHz). Find the wavelength of these radio waves in meters (m), nanometers (nm), and angstroms (Å).

**7.7** An FM station broadcasts classical music at 103.5 MHz (megahertz, or $10^6$ Hz). Find the wavelength (in m, nm, and Å) of these radio waves.

**7.8** A radio wave has a frequency of $3.4 \times 10^{10}$ Hz. What is the energy (in J) of one photon of this radiation?

**7.9** An x-ray has a wavelength of 1.2 Å. Calculate the energy (in J) of one photon of this radiation.

**7.10** Rank the following photons in terms of increasing energy: (a) blue ($\lambda = 453$ nm); (b) red ($\lambda = 660$ nm); (c) yellow ($\lambda = 595$ nm).

**7.11** Rank the following photons in terms of decreasing energy: (a) IR ($\nu = 6.5 \times 10^{13}$ s$^{-1}$); (b) IR ($\nu = 9.8 \times 10^{12}$ s$^{-1}$); (c) UV ($\nu = 8.0 \times 10^{15}$ s$^{-1}$).

---

**7.12** In the United States, police often monitor traffic with "K-band" radar speed meters, which operate in the microwave region at 22.235 GHz (1 GHz = $10^9$ Hz). Find the wavelength (in nm and Å) of this radiation.

**7.13** Covalent bonds in a molecule absorb radiation in the IR range and vibrate at characteristic frequencies. Chemists measure these frequencies to investigate the structure of the compound.

(a) The C−O bond in an organic compound absorbs radiation of wavelength 9.6 $\mu$m. What frequency (in s$^{-1}$) corresponds to that wavelength?
(b) The H−Cl bond has a frequency of vibration of $8.652 \times 10^{13}$ Hz. What wavelength (in $\mu$m) corresponds to that frequency?
(c) One reason for the toxicity of carbon monoxide (CO) is that it binds to the blood protein hemoglobin more strongly than oxygen does. The hemoglobin−CO bond absorbs radiation of 1953 cm$^{-1}$. (The units are the reciprocal of the wavelength in centimeters.) Calculate the wavelength (in nm and Å) and the frequency (in Hz) of the absorbed radiation.

**7.14** Cobalt-60 is a radioactive isotope used to treat cancers of the brain and other tissues. A gamma ray emitted by an atom of this isotope has an energy of 1.33 MeV (million electron-volts; 1 eV = $1.602 \times 10^{-19}$ J). What is the frequency (in Hz) and wavelength (in m) of this gamma ray?

**7.15** (a) The first step in the formation of ozone in the upper atmosphere occurs when oxygen molecules absorb UV radiation of wavelengths $\leq$ 242 nm. Calculate the frequency and energy of the least energetic of these photons.
(b) Ozone absorbs light having wavelengths of 2200 to 2900 Å, thus protecting organisms on the Earth's surface from this high-energy UV radiation. What are the frequency and energy of the most energetic of these photons?

**7.16** In his explanation of the threshold frequency in the photoelectric effect, Einstein reasoned that the absorbed photon must have the minimum energy required to dislodge an electron from the metal surface. This energy is called the *work function* ($\phi$) of that metal. What is the longest wavelength of radiation (in nm) that could cause the photoelectric effect in each of the following metals?
(a) Potassium, $\phi = 3.68 \times 10^{-19}$ J
(b) Silver, $\phi = 7.59 \times 10^{-19}$ J
(c) Sodium, $\phi = 4.41 \times 10^{-19}$ J

**7.17** Refractometry is a method of chemical analysis based on the change in the speed of light as it travels through a substance compared with its speed in a vacuum (see Margin Note, p. 259).
(a) The refractive index of water is 1.33 (at 20°C). Calculate the speed of light in water.
(b) The refractive index of diamond is 2.42 (at 20°C). Calculate the speed of light in diamond.

## Atomic Spectra and the Bohr Model of the Atom

**7.18** Why does the spacing within a series of spectral lines decrease as the wavelength becomes shorter?

**7.19** How is $n_1$ in the Rydberg equation (Equation 7.3) related to the quantum number $n$ in the Bohr model?

**7.20** What was the theoretical basis for Bohr assuming that the energy of an atom has only certain values (stationary states)?

**7.21** Distinguish between an absorption spectrum and an emission spectrum. With which did Bohr work?

**7.22** Which of the following electron transitions correspond to absorption of energy and which to emission: (a) $n = 2$ to $n = 4$; (b) $n = 3$ to $n = 1$; (c) $n = 5$ to $n = 2$; (d) $n = 3$ to $n = 4$?

**7.23** Why could the Bohr model not predict line spectra for atoms other than hydrogen?

**7.24** H and $He^+$ have one electron each. Would you expect their line spectra to be identical? Explain.

**7.25** Use the Rydberg equation (Equation 7.3) to calculate the wavelength (in nm) of the photon emitted when a hydrogen atom undergoes a transition from $n = 5$ to $n = 2$.

**7.26** Use the Rydberg equation to calculate the wavelength (in Å) of the photon absorbed when a hydrogen atom undergoes a transition from $n = 1$ to $n = 3$.

**7.27** What is the wavelength (in nm) of the least energetic spectral line in the ultraviolet series of the H atom?

**7.28** What is the wavelength (in nm) of the least energetic spectral line in the visible series of the H atom?

**7.29** Calculate the energy difference ($\Delta E$) for the transition in Problem 7.25 for 1 mol H atoms.

**7.30** Calculate the energy difference ($\Delta E$) for the transition in Problem 7.26 for 1 mol H atoms.

**7.31** Arrange the following H atom electron transitions in order of *increasing* frequency of the photon absorbed or emitted: (a) $n = 2$ to $n = 4$; (b) $n = 2$ to $n = 1$; (c) $n = 2$ to $n = 5$; (d) $n = 4$ to $n = 3$.

**7.32** Arrange the following H atom electron transitions in order of *decreasing* wavelength of the photon absorbed or emitted: (a) $n = 2$ to $n = \infty$; (b) $n = 4$ to $n = 20$; (c) $n = 3$ to $n = 10$; (d) $n = 2$ to $n = 1$.

**7.33** The electron in a ground-state H atom absorbs a photon of wavelength 97.20 nm. To what energy level does the electron move?

**7.34** An electron in the $n = 5$ level of an H atom emits a photon of wavelength 1281 nm. To what energy level does the electron move?

**7.35** In addition to continuous radiation, fluorescent lamps emit sharp lines in the visible region from a mercury discharge within the tube. Much of this light has a wavelength of 436 nm. What is the energy (in J) of one photon of this light?

**7.36** The oxidizing agents used in most fireworks consist of potassium salts, such as $KClO_4$ or $KClO_3$, rather than the corresponding sodium salts. One of the problems with using sodium salts is their extremely intense yellow-orange emission at 589 nm, which obscures other colors in the display. What is the energy (in J) of one photon of this light? What is the energy (in kJ) of one einstein of this light (1 einstein = 1 mole of photons)?

**The Wave-Particle Duality of Matter and Energy**

(Sample Problems 7.3 and 7.4)

**7.37** Which of Bohr's assumptions was de Broglie attempting to generalize from other systems?

**7.38** What experimental support did de Broglie's concept receive?

**7.39** If particles have wavelike motion, why don't we observe that motion in the macroscopic world?

**7.40** Why can't we overcome the uncertainty predicted by Heisenberg's principle by building more precise measuring devices to reduce the error in measurements below the $h/4\pi$ limit?

**7.41** A 220-lb fullback runs the 40-yd dash at a speed of $19.6 \pm 0.1$ mi/h.
(a) What is his de Broglie wavelength (in meters)?
(b) What is the uncertainty in his position?

**7.42** An alpha particle (mass = $6.6 \times 10^{-24}$ g) emitted by radium travels at $3.4 \times 10^7 \pm 0.1 \times 10^7$ mi/h.
(a) What is its de Broglie wavelength (in meters)?
(b) What is the uncertainty in its position?

**7.43** In 1990, the men's singles winner of the U.S. Open tennis tournament had his serves clocked at 127 mi/h. How fast must a 56.5-g tennis ball travel to have a de Broglie wavelength equal to that of a photon of green light (5400 Å)?

**7.44** How fast must a 142-g baseball travel to have a de Broglie wavelength equal to that of an x-ray photon with $\lambda = 100$ pm?

**7.45** The yellow color of a sodium flame test is due to emissions of photons of wavelength 589 nm. What is the mass equivalence of one photon of this wavelength ($1 J = 1 kg \cdot m^2/s^2$)?

**7.46** The red color of a lithium flame test is due to emissions of photons of wavelength 671 nm. What is the mass equivalence of one mole of photons of this wavelength ($1 J = 1 kg \cdot m^2/s^2$)?

**The Quantum-Mechanical Model of the Atom**

(Sample Problems 7.5 to 7.7)

**7.47** What physical meaning is attributed to the square of the wave function, $\psi^2$?

**7.48** Explain in your own words what the "electron density" at a particular point in space means.

**7.49** Explain in your own words what it means for the peak in the radial probability distribution of the $n = 1$ level of a hydrogen atom to be at 0.529 Å. Is the probability of finding an electron at 0.529 Å from the nucleus greater for the $1s$ or the $2s$ orbital?

**7.50** What feature of an orbital is related to each of the following?
(a) Principal quantum number ($n$)
(b) Azimuthal quantum number ($l$)
(c) Magnetic quantum number ($m_l$)

**7.51** How many orbitals in an atom can have each of the following designations? (a) $1s$; (b) $3d$; (c) $4p$; (d) $n = 3$

**7.52** How many orbitals in an atom can have each of the following designations? (a) $6f$; (b) $3p$; (c) $5d$; (d) $n = 2$

**7.53** Give all possible $m_l$ values for orbitals that have each of the following: (a) $l = 2$; (b) $n = 1$; (c) $n = 4$, $l = 3$.

**7.54** Give all possible $m_l$ values for orbitals that have each of the following: (a) $l = 3$; (b) $n = 2$; (c) $n = 6$, $l = 1$.

**7.55** Draw probability contours (with axes) for each of the following orbitals: (a) $1s$; (b) $p_x$; (c) $d_{xy}$.

**7.56** Draw probability contours (with axes) for each of the following orbitals: (a) $3s$; (b) $d_{z^2}$; (c) $p_z$.

**7.57** For each of the following, give the sublevel designation, the allowable $m_l$ values, and the number of possible orbitals: (a) $n = 4$, $l = 2$; (b) $n = 5$, $l = 1$; (c) $n = 6$, $l = 3$.

**7.58** For each of the following, give the sublevel designation, the allowable $m_l$ values, and the number of possible orbitals: (a) $n = 2$, $l = 0$; (b) $n = 3$, $l = 2$; (c) $n = 5$, $l = 1$.

**7.59** For each of the following subshells, give the $n$ and $l$ values and the number of possible orbitals: (a) $5s$; (b) $3p$; (c) $4f$.

**7.60** For each of the following subshells, give the $n$ and $l$ values and the number of possible orbitals: (a) $6g$; (b) $4s$; (c) $3d$.

**7.61** Are the following quantum number combinations allowed? If not, show two ways to correct them:
(a) $n = 2$; $l = 0$; $m_l = -1$   (b) $n = 4$; $l = 3$; $m_l = -1$
(c) $n = 3$; $l = 1$; $m_l = 0$    (d) $n = 5$; $l = 2$; $m_l = +3$

**7.62** Are the following quantum number combinations allowed? If not, show two ways to correct them:
(a) $n = 1$; $l = 0$; $m_l = 0$    (b) $n = 2$; $l = 2$; $m_l = +1$
(c) $n = 7$; $l = 1$; $m_l = +2$   (d) $n = 3$; $l = 1$; $m_l = -2$

## Comprehensive Problems

Problems with an asterisk (*) are more challenging.

**7.63** Compare the wavelengths of an electron (mass = $9.11 \times 10^{-31}$ kg) and a proton (mass = $1.67 \times 10^{-27}$ kg) that each have (a) a speed of $3.0 \times 10^6$ m/s; (b) an energy of $2.5 \times 10^{-15}$ J.

**\*7.64** Five lines in the H atom spectrum have the following wavelengths (in Å): (a) 1212.7; (b) 4340.5; (c) 4861.3; (d) 6562.8; (e) 10938. Three result from transitions to $n_{final} = 2$ (visible series). The other two result from transitions in different series, one with $n_{final} = 1$ and the other with $n_{final} = 3$. Identify the $n_{initial}$ values for each line.

**7.65** The Bohr model can be used to calculate the energy levels of other one-electron species by incorporating a factor related to the charge of the nucleus ($Z$): $E = -2.18 \times 10^{-18}$ J ($Z^2/n^2$). Calculate the ionization energy (in kJ/mol) for the following species: (a) $He^+$; (b) $Li^{2+}$; (c) $C^{5+}$.

**\*7.66** In compliance with conservation of energy, Einstein explained that in the photoelectric effect, the energy of a photon ($h\nu$) absorbed by a metal is the sum of the work function ($\phi$), the minimum energy needed to dislodge an electron from the metal's surface, and the kinetic energy ($E_k$) of the electron: $h\nu = \phi + E_k$. When light of wavelength 358.1 nm falls on the surface of potassium metal, the velocity ($u$) of the dislodged electron is $6.40 \times 10^5$ m/s.
(a) What is $E_k$ ($\frac{1}{2}mu^2$) of the dislodged electron?
(b) What is $\phi$ (in J) of potassium?

**7.67** Which of the following elements can be used for a photocell that will operate with visible light? (The work function, $\phi$, of each metal is given.)
(a) Tantalum, $\phi = 6.41 \times 10^{-19}$ J
(b) Barium, $\phi = 4.3 \times 10^{-19}$ J
(c) Tungsten, $\phi = 7.16 \times 10^{-19}$ J

**\*7.68** A ground-state H atom absorbs a photon of wavelength 94.91 nm and attains a higher energy level. The atom then emits two photons: one of wavelength 1281 nm to reach an intermediate level and a second to return to the ground state.
(a) What higher level did the atom reach?
(b) What intermediate level did the atom reach?
(c) What was the wavelength of the second photon emitted?

**7.69** In the course of developing his model, Bohr arrived at a formula for the radius of the electron's orbit: $r_n = n^2h^2\epsilon_0/\pi m_e e^2$, where $m_e$ is the electron mass, $e$ is its charge, and $\epsilon_0$ is a constant related to charge attraction in a vacuum. Given $m_e = 9.109 \times 10^{-31}$ kg, $e = 1.602 \times 10^{-19}$ C, and $\epsilon_0 = 8.854 \times 10^{-12}$ $C^2/J \cdot$ m, calculate the following:
(a) The radius of the 1st ($n = 1$) orbit in the H atom
(b) The radius of the 10th ($n = 10$) orbit in the H atom

**\*7.70** (a) Calculate the Bohr radius of an electron in the $n = 3$ orbit of a hydrogen atom. (See Problem 7.69.)
(b) What is the energy (in J) of the atom in part (a)?
(c) What is the energy of a $Li^{2+}$ ion when its electron is in the $n = 3$ orbit? (See Problem 7.65.)
(d) Why are the answers to parts (b) and (c) different?

**7.71** Enormous numbers of microwave photons are needed to warm macroscopic samples of matter. A portion of soup containing 252 g water is heated in a microwave oven from 20°C to 98°C, with radiation of wavelength $1.55 \times 10^{-2}$ m. How many photons are absorbed by the water in the soup?

***7.72** The following values are the only allowable energy levels of a hypothetical one-electron atom:

$$E_6 = -2 \times 10^{-19} \text{ J}$$
$$E_5 = -7 \times 10^{-19} \text{ J}$$
$$E_4 = -11 \times 10^{-19} \text{ J}$$
$$E_3 = -15 \times 10^{-19} \text{ J}$$
$$E_2 = -17 \times 10^{-19} \text{ J}$$
$$E_1 = -20 \times 10^{-19} \text{ J}$$

(a) If the electron were in the $n = 3$ level, what would be the highest frequency (and minimum wavelength) of radiation that could be emitted?

(b) What is the ionization energy (in kJ/mol) of the atom in its ground state?

(c) If the electron were in the $n = 4$ level, what would be the shortest wavelength (in nm) of radiation that could be absorbed without ionizing?

**7.73** The following wavelengths were emitted by metal ions in a fireworks display. What are the frequency and color seen from each ion?

(a) $Ba^{2+}$, $\lambda = 551$ nm       (b) $Li^+$, $\lambda = 671$ nm
(c) $Cs^+$, $\lambda = 456$ nm       (d) $Ca^{2+}$, $\lambda = 649$ nm
(e) $Na^+$, $\lambda = 589$ nm       (f) $Sr^{2+}$, $\lambda = 661$ nm

**7.74** Electric power is typically stated in units of watts (W; 1 W = 1 J/s). About 90% of the power output of an incandescent bulb is converted to heat and 10% to light. If 10% of that light shines on your chemistry text, how many photons per second shine on the book from a 60-W bulb? (Assume the photons have a wavelength of 550 nm.)

**7.75** The use of phosphate compounds in detergents and their subsequent environmental discharge has led to serious imbalances in the natural life cycle of fresh-water lakes. A chemist studying water pollution used a spectrophotometric method to measure total phosphate and obtained the following data for known standards:

| ABSORBANCE (400 nm) | CONCENTRATION (mol/L) |
|---|---|
| 0 | 0.0 |
| 0.04 | $2.5 \times 10^{-5}$ |
| 0.16 | $3.2 \times 10^{-5}$ |
| 0.20 | $4.4 \times 10^{-5}$ |
| 0.25 | $5.6 \times 10^{-5}$ |
| 0.38 | $8.4 \times 10^{-5}$ |
| 0.48 | $10.5 \times 10^{-5}$ |
| 0.62 | $13.8 \times 10^{-5}$ |
| 0.76 | $17.0 \times 10^{-5}$ |
| 0.88 | $19.4 \times 10^{-5}$ |

(a) Draw a standard curve of absorbance at 400 nm vs. phosphate concentration.

(b) If a lake water sample has an absorbance of 0.55, what is its phosphate concentration?

## CHAPTER 8

# Electron Configuration and Chemical Periodicity

**Concepts and skills to review**

- format of the periodic table (Section 2.5)
- characteristics of metals and nonmetals (Section 2.5)
- characteristics of acids and bases (Section 4.3)
- rules for assigning quantum numbers (Section 7.4)

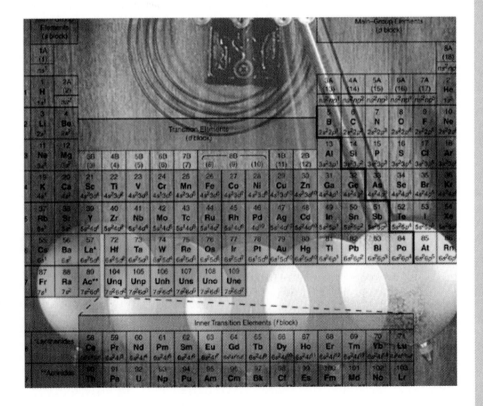

**Periodicity.** Like the movement of a pendulum, the key properties of the elements arise from the inner workings of their atoms and recur in periodic cycles. This behavior, displayed so clearly in the periodic table, is the focus of this chapter and one of the central concepts in chemistry.

In Chapter 7, you saw how the remarkable advances in our understanding of matter and energy eventually led to the quantum-mechanical model of the atom. This model is the key to one of the central questions in chemistry: why do the elements behave as they do? Or, rephrasing the question to fit this chapter: how does the distribution of electrons within the orbitals of an element's atoms—the **electron configuration**—relate to the properties of the element?

In the second half of the 19th century, before the attempts by physicists to model the atom, chemists were busy exploring the nature of solutions, establishing the kinetic-molecular theory, and developing the concepts of chemical thermodynamics. The fields of organic chemistry and biochemistry were born, as were the fertilizer, explosives, glassmaking, soapmaking, bleaching, and dyestuff industries. Also, for the first time, chemistry became a major university subject in Europe and America.

Essential to this growth of chemistry was the ability to organize the numerous facts about element behavior. One early attempt at organizing these facts placed elements with similar properties in groups of three (such as calcium, strontium, and barium); another noted similarities between every eighth element (like the similarity between every eighth note in the musical scale). As more elements were discovered, however, these early attempts lost much of their validity.

As you learned in Chapter 2, the most successful organizing scheme was that of the Russian chemist Dmitri Mendeleev, who in 1870 arranged the 65 elements known at the time into a *periodic table* and summarized his study of their behavior in the **periodic law:** when arranged by atomic mass, the elements exhibit a periodic recurrence of similar properties. The explanation of this periodic behavior more than half a century later in terms of the quantum-mechanical atom is surely one of the most satisfying achievements in science.

It is a curious quirk of history that Mendeleev and the German chemist Lothar Meyer arrived at virtually the same organization of elements simultaneously, yet independently. The greater credit has gone to Mendeleev because he was able to employ his table in very impressive ways. For example, by noting the trends in properties of neighboring elements, he *predicted* the properties of as-yet-undiscovered scandium, gallium, and germanium, for which he had left blank spaces. Table 8.1 compares Mendeleev's predictions for germanium, which he called "ekasilicon," with its actual properties.

The modern periodic table (inside front cover of text) resembles Mendeleev's table in most details, although it includes more than 40 elements that were unknown in 1870. The only substantive change is that the elements are now arranged in order of *atomic number* (number of protons) rather than atomic mass, a change made after the existence of isotopes was understood. This change was made because of the fundamental nature of atomic number and because an arrangement by atomic mass would place an element with an abundant heavy isotope, but a lower atomic number, out of order. ◆

The goal of this chapter is to show how the periodic table, condensed from countless hours of laboratory work by 19th century chemists, is perfectly consistent with the atomic model developed by 20th century physicists. We take up where we left off in Chapter 7, extending the quantum-mechanical model to *many-electron atoms,* those with more than one electron, and showing how it defines a unique set of quantum numbers for every electron in an atom. Then we consider the rules for filling orbitals of

◆ **Moseley and Atomic Number.** Soon after Rutherford's discovery of the nucleus, Niels Bohr began his studies to explain line spectra. In the process, Bohr proposed that the light emitted (usually x-rays) when an outer electron drops to an empty inner orbit had wavelengths proportional to the nuclear charge. In 1913, the 26-year-old Henry G.J. Moseley, a student of Rutherford, saw Bohr's result and used x-rays to measure the nuclear charges of several transition metals. Moseley found that the x-ray wavelengths emitted by the elements in their periodic table order correlated with the nuclear charge increasing one unit at a time. Extending this pattern, it was reasoned that nuclear charge, now called *atomic number,* was the basis for the order of the elements. Among other results, the findings confirmed the placement of Co ($Z = 28$) *before* Ni ($Z = 29$), despite its higher atomic mass, and the gap between Cl ($Z = 17$) and K ($Z = 19$) as the place for Ar ($Z = 18$). Tragically, Moseley died the next year in World War I.

**TABLE 8.1 Mendeleev's Predicted Properties of Germanium ("Ekasilicon") and Its Actual Properties**

| PROPERTY | PREDICTED PROPERTIES OF EKASILICON (E) | ACTUAL PROPERTIES OF GERMANIUM (Ge) |
|---|---|---|
| Atomic mass | 72 | 72.59 |
| Appearance | Gray metal | Gray metal |
| Density | 5.5 g/cm³ | 5.35 g/cm³ |
| Molar volume | 13 cm³/mol | 13.22 cm³/mol |
| Specific heat capacity | 0.31 J/g · K | 0.32 J/g · K |
| Oxide formula | $EO_2$ | $GeO_2$ |
| Oxide density | 4.7 g/cm³ | 4.23 g/cm³ |
| Sulfide formula and solubility | $ES_2$; insoluble in $H_2O$; soluble in aqueous $(NH_4)_2S$ | $GeS_2$; insoluble in $H_2O$; soluble in aqueous $(NH_4)_2S$ |
| Chloride formula (boiling point) | $ECl_4$ ( < 100°C) | $GeCl_4$ (84°C) |
| Chloride density | 1.9 g/cm³ | 1.844 g/cm³ |
| Element preparation | Reduction of $K_2EF_6$ with sodium | Reduction of $K_2GeF_6$ with sodium |

increasing energy with electrons and see how this filling order correlates with the placement of elements in the periodic table. All this leads to our central purpose: to understand how electron configuration and nuclear charge give rise to trends in atomic properties that explain the periodic patterns of chemical reactivity. This understanding culminates in a discussion of how electron configurations explain the behavior of metals and nonmetals and of the ions they form.

## 8.1    Characteristics of Many-Electron Atoms

The Schrödinger equation (Chapter 7) cannot be solved exactly for many-electron atoms. However, modern computers give us good approximate solutions, which show that the atomic orbitals of many-electron atoms are *hydrogen-like:* they resemble those of the hydrogen atom. This conclusion is very important because it allows us to use the same quantum numbers to describe other atoms as we used to describe hydrogen.

Nevertheless, the presence of more than one electron requires us to extend the quantum-mechanical model and consider features that were not relevant to our discussion of hydrogen. These include (1) a limit on the number of electrons allowed in an orbital, which requires the introduction of a fourth quantum number, and (2) a more complex set of orbital energy levels. Let's examine these extensions of the model so that we can determine the electron configuration for each element.

### The Electron-Spin Quantum Number and the Exclusion Principle

Recall from Chapter 7 that the three quantum numbers $n$, $l$, and $m_l$ describe the atomic orbital: its size (energy), shape, and orientation, respectively. However, an additional quantum number is needed to describe a property of the electron itself, called *spin,* which is not a property of the orbital.

**FIGURE 8.1**
**Observing the effect of electron spin.** A nonuniform magnetic field, created by magnet faces with different shapes, splits a beam of hydrogen atoms in two. The split beam results from the two possible directions of the electron's spin.

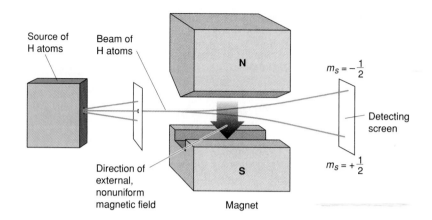

When a beam of hydrogen atoms passes through a nonuniform magnetic field, it splits into two beams that bend away from each other (Figure 8.1). The explanation of this phenomenon is that the charge of an electron generates a tiny magnetic field, as though the electron were spinning on its axis, and the single electron in each hydrogen atom can assume one of two possible spin directions. Since opposite spins produce their own oppositely directed magnetic fields, the electron in each atom is either attracted to or repelled by the external magnetic field. Half the electrons are attracted and half repelled, so the beam of H atoms splits.

The **spin quantum number ($m_s$)** indicates the direction of the electron spin and can have one of two possible values, $+\frac{1}{2}$ or $-\frac{1}{2}$. Thus, each electron in an atom is completely described by a set of *four* quantum numbers, the first three describing its orbital and the fourth describing its spin. The quantum numbers are summarized in Table 8.2.

Given the rules governing the assignment of these four numbers, we can write a set of quantum numbers for any electron in an atom. For the electron in the ground state of the hydrogen atom, the quantum numbers are $n = 1$, $l = 0$, $m_l = 0$, and $m_s = +\frac{1}{2}$. (The spin quantum number for this electron could just as well have been $-\frac{1}{2}$, but we assign $+\frac{1}{2}$ for the first electron in an orbital by convention.)

Now consider the helium atom, which has two electrons. The *first* electron in the helium ground state has the same four quantum numbers as the H atom's electron, but the *second* electron does not. Observations of the excited states of atoms led the Austrian physicist Wolfgang Pauli to formulate the **exclusion principle:** *no two electrons in the same atom can have the same*

**TABLE 8.2 Summary of Quantum Numbers of Electrons in Atoms**

| NAME | SYMBOL | PERMITTED VALUES | PROPERTY |
|---|---|---|---|
| Principal | $n$ | Positive integers (1, 2, 3, etc.) | Orbital energy (size) |
| Azimuthal | $l$ | Integers from 0 to $n - 1$ | Orbital shape (The $l$ values 0, 1, 2, and 3 correspond to $s$, $p$, $d$, and $f$ orbitals, respectively.) |
| Magnetic | $m_l$ | Integers from $-l$ to 0 to $+l$ | Orbital orientation |
| Spin | $m_s$ | $+\frac{1}{2}$ or $-\frac{1}{2}$ | Direction of $e^-$ spin |

*The combined effects of orbital shape and penetration cause an energy level to be split into energy sublevels.* Probability distribution plots show that the lower the *l* value of an orbital, the more time the electron spends penetrating near the nucleus, so the greater the nucleus-electron attraction. Therefore, *for a given n value, the lower the l value, the lower is the sublevel energy: s < p < d < f.*

The general order for energy levels and sublevels in many-electron atoms appears in Figure 8.4; note the splitting of energy levels into sublevels. In the next section, we use this order to construct a periodic table of ground-state atoms.

### Section Summary

Specifying electrons in many-electron atoms requires four quantum numbers: three ($n, l, m_l$) describe the orbitals, and a fourth ($m_s$) describes electron spin. The Pauli exclusion principle requires that each electron have a unique set of four quantum numbers; therefore, an orbital can hold no more than two electrons, and their spins must be paired (opposite). Electrostatic interactions determine orbital energies:

1. Greater nuclear charge lowers orbital energy.
2. Electron-electron repulsions raise orbital energy.
3. Electrons in outer orbitals (higher *n*) are higher in energy because inner electrons shield them from the nuclear charge.
4. Electrons that have a finite probability distribution near the nucleus (penetration) have lower energy. Thus, an energy level (shell) is split into sublevel (subshell) energies: $s < p < d < f$.

## 8.2    The Quantum-Mechanical Atom and the Periodic Table

The quantum-mechanical atom provides the theoretical foundation for the experimentally based periodic table. In this section, we fill the table and find that the electron configurations of the elements—the distributions of electrons within the orbitals of their atoms—show a recurring pattern that is the basis for recurring element properties.

### Building Up Periods 1 and 2

A useful way to determine the electron configurations of the elements is to start at the beginning of the periodic table and add one electron per element to the *lowest energy orbital available*. (Of course, a proton and one or more neutrons are also added to the nucleus.) This approach is called the **aufbau principle** (German *aufbauen*, "to build up"), and it results in *ground-state* electron configurations. Let's use this method to assign sets of quantum numbers to the electrons in the ground state of the first 10 elements, those in the first two periods.

For the hydrogen electron, as you've seen, the set is

$$\text{H } (Z = 1): \quad n = 1, l = 0, m_l = 0, m_s = +\tfrac{1}{2}$$

The first helium electron has the same set as the hydrogen electron, but the second He electron has opposing spin:

$$\text{He } (Z = 2): \quad n = 1, l = 0, m_l = 0, m_s = -\tfrac{1}{2}$$

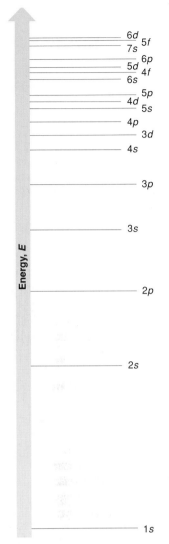

**FIGURE 8.4**

**Sublevel energies in many-electron atoms.** The relative energies of sublevels in many-electron atoms increase with principal quantum number *n* ($1 < 2 < 3$, etc.) and azimuthal quantum number *l* ($s < p < d < f$). Note that as the *n* value becomes greater, the sublevel energies become closer together. The penetration effect, together with this narrowing of energy differences, results in overlapping of some sublevels; for example, the 4*s* sublevel is slightly lower in energy than the 3*d*. (The lines have different lengths only for ease in labeling.)

**FIGURE 8.5**

**A vertical orbital diagram for the Li ground state.** Sublevel energy increases from bottom to top. Color is used to show orbital occupancy: filled (darker), half-filled (lighter), and empty (none).

When we present an element and its atomic number ($Z$) in this discussion, the quantum numbers that follow refer to the element's *last added* electron. [Of course, an atom is neutral, so the number of protons ($Z$) equals the *total* number of electrons.]

Before proceeding, we'll look at two very useful ways that will appear throughout the text to show the distribution of electrons. The first is a shorthand notation that is also called the *electron configuration*. It consists of the principal energy level ($n$ value), the letter designation of the sublevel ($l$ value), and the number of electrons (#) in the sublevel, written as a superscript: $nl^{\#}$. The electron configuration of hydrogen is $1s^1$ (spoken "one-ess-one"); that of helium is $1s^2$ (spoken "one-ess-two," *not* "one-ess-squared"). This notation does *not* indicate electron spin but assumes the reader knows that the electron spins in $1s^2$ are paired (opposite).

The other way to present this information is through an **orbital diagram,** which consists of a box (or circle, or just a line) for each orbital available in a given energy level, grouped by sublevel, with an arrow indicating the electron's presence *and* its direction of spin. (Traditionally, $\uparrow$ is $+\frac{1}{2}$ and $\downarrow$ is $-\frac{1}{2}$, but these are arbitrary, so it is necessary only to be consistent.) The orbital diagrams for the first two elements are

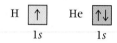

With helium, the $1s$ orbital is filled because an orbital can hold only two electrons (exclusion principle). Since the $n = 1$ level has only one orbital, it is also filled. We fill the $n = 2$ level next, beginning with the $2s$ orbital, the next lowest in energy. Thus, the first two electrons in lithium fill the $1s$ orbital, and the set of quantum numbers for the last added electron in lithium is $n = 2$, $l = 0$, $m_l = 0$, $m_s = +\frac{1}{2}$. The electron configuration for Li is $1s^2 2s^1$; note that the orbital diagram shows all the orbitals in $n = 2$, whether or not they are occupied:

Keep in mind that the energy of each sublevel (each group of boxes) increases from left to right. We could emphasize this energy order by arranging the sublevels vertically (Figure 8.5), but it is conventional to show them horizontally.

In Li, the $2s$ orbital is only half-filled, so the fourth electron of beryllium fills it with its spin paired: $n = 2$, $l = 0$, $m_l = 0$, $m_s = -\frac{1}{2}$.

With beryllium, the $2s$ sublevel is filled, and the next lowest energy sublevel is the $2p$. Since a $p$ sublevel has $l = 1$, the $m_l$ (orientation) values can be $-1$, $0$, or $+1$. The three orbitals in the $2p$ sublevel have *equal energy* (same $n$ and $l$ values), which means that the fifth electron of boron can go into *any one of the $2p$ orbitals*. For convenience, let's label the boxes from left to right, $-1$, $0$, $+1$, and assume it enters the $m_l = -1$ orbital: $n = 2$, $l = 1$, $m_l = -1$, $m_s = +\frac{1}{2}$.

For our purposes here, the designation of $m_l = -1$ for this boron electron (and the box we chose to label $-1$, in which we placed the arrow) is arbitrary. The $2p$ orbitals have equal energy and differ *only* in their orientation.

To minimize electron-electron repulsions, the last added (sixth) electron of carbon enters one of the *unoccupied* $2p$ orbitals, for example, the $m_l = 0$ orbital. Experiment shows that the spin of this electron is *parallel* to (in the same direction as) the spin of the other $2p$ electron (that is, also $+\frac{1}{2}$): $n = 2$, $l = 1$, $m_l = 0$, $m_s = +\frac{1}{2}$.

Be ($Z = 4$)     $1s^2 2s^2$

B ($Z = 5$)     $1s^2 2s^2 2p^1$
                      $-1$  $0$  $+1$

C ($Z = 6$)     $1s^2 2s^2 2p^2$

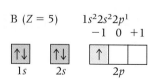

The placement of electrons shown in the ground state of carbon exemplifies a phenomenon summarized by **Hund's rule:** *when orbitals of equal energy are available, the electron configuration of lowest energy has the maximum number of unpaired electrons with parallel spins.* Based on Hund's rule, nitrogen's seventh electron enters the last empty $2p$ orbital, with its spin parallel to the two other $2p$ electrons: $n = 2$, $l = 1$, $m_l = +1$, $m_s = +\frac{1}{2}$.

The eighth electron in oxygen must enter one of these three half-filled $2p$ orbitals and "pair up" with (have opposing spin to) the electron already present. Since the $2p$ orbitals all have the same energy, we proceed as before and place it in the orbital previously designated $m_l = -1$; thus, the quantum numbers are $n = 2$, $l = 1$, $m_l = -1$, $m_s = -\frac{1}{2}$.

Fluorine's ninth electron enters either one of the two remaining half-filled $2p$ orbitals: $n = 2$, $l = 1$, $m_l = 0$, $m_s = -\frac{1}{2}$.

Only one unfilled orbital remains in the $2p$ sublevel, so the tenth electron of neon occupies it: $n = 2$, $l = 1$, $m_l = +1$, $m_s = -\frac{1}{2}$. With neon, the $n = 2$ level is filled.

SAMPLE PROBLEM 8.1 _____

**Determining Quantum Numbers from Orbital Diagrams**

**Problem:** Write a set of four quantum numbers for the third and eighth electrons in F.
**Plan:** Based on the orbital diagram, we count to the electron of interest and note its level ($n$), sublevel ($l$), orbital ($m_l$), and direction of spin ($m_s$).
**Solution:** The third electron is in the $2s$ orbital. The upward arrow indicates a spin of $+\frac{1}{2}$: $\boldsymbol{n = 2, l = 0, m_l = 0, m_s = +\frac{1}{2}}$. The eighth electron is in the first $2p$ orbital, which we designated $m_l = -1$, and has a downward arrow: $\boldsymbol{n = 2, l = 1, m_l = -1, m_s = -\frac{1}{2}}$.

**FOLLOW-UP PROBLEM 8.1**
Use the periodic table to identify the element with the electron configuration $1s^2 2s^2 2p^4$. Write its orbital diagram and give the quantum numbers of its sixth electron.

With so much attention paid to these notations for describing electrons and their configurations within atoms, it may be easy to forget that atoms

N ($Z = 7$)      $1s^2 2s^2 2p^3$

O ($Z = 8$)      $1s^2 2s^2 2p^4$

F ($Z = 9$)      $1s^2 2s^2 2p^5$

Ne ($Z = 10$)      $1s^2 2s^2 2p^6$

**FIGURE 8.6**
**The first 10 elements: H through Ne.** The first 10 elements are arranged in the periodic table format, with each box showing atomic number, atomic symbol, ground-state electron configuration, and a depiction of the atom based on the probability contours of its orbitals. Orbital occupancy is shown with shading: pale gray represents empty orbitals, half color indicates half-filled (one e$^-$) orbitals, and full shading means filled (two e$^-$) orbitals. For clarity, only the outer region of the $2s$ orbital is included.

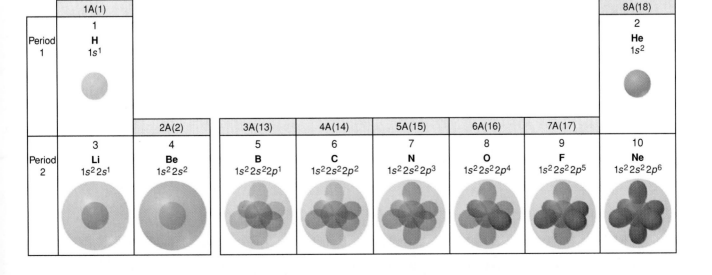

are real spherical objects and that the electrons occupy probability contours of specific shape and orientation. Figure 8.6 shows the first 10 elements arranged in the periodic table format, with orbital contours depicted. Even at this early stage of filling the table, we can see an important correlation of chemical behavior with electron configuration. Helium and neon are in the same group and have similar electron configurations in that both have filled outermost shells: $n = 1$ for He and $n = 2$ for Ne. Interestingly, neither element forms any compounds; apparently, *filled shells make these elements chemically unreactive.*

### Electron Configurations Within Groups; Building Up Period 3

The Period 3 elements, sodium through argon, lie directly under the Period 2 elements, lithium through neon, in the periodic table. The sublevels of the $n = 3$ level are filled in the order $3s$, $3p$, $3d$. Table 8.3 presents *partial* orbital diagrams ($3s$ and $3p$ sublevels only) and electron configurations (with *filled inner levels* in brackets) for the eight elements in Period 3.

In sodium (the second member of the alkali metals) and magnesium (the second member of the alkaline earth metals), electrons fill the $3s$ sublevel, which contains the single $3s$ orbital, just as they filled the $2s$ sublevel in lithium and beryllium (the first members of these groups). Then, just as with boron, carbon, and nitrogen in Period 2 above them, the last electron added to aluminum, silicon, and phosphorus half-fills each of the $3p$ orbitals with spins parallel (Hund's rule). The last electrons added to sulfur, chlorine, and argon enter the half-filled $3p$ orbitals, thereby filling the $3p$ sublevel. With argon, the next noble gas after helium and neon, we arrive at the end of Period 3. (The $3d$ orbitals are filled in Period 4.)

The rightmost column of Table 8.3 shows the *condensed electron configuration.* In this simplified notation, the electron configuration of the previous

**TABLE 8.3 Partial Orbital Diagrams and Electron Configurations for the Elements in Period 3**

| ATOMIC NUMBER | ELEMENT | ORBITAL DIAGRAM (3s AND 3p SUBLEVELS ONLY) | FULL ELECTRON CONFIGURATION | CONDENSED ELECTRON CONFIGURATION |
|---|---|---|---|---|
| 11 | Na | 3s: ↑   3p: ▢ ▢ ▢ | $[1s^22s^22p^6]\ 3s^1$ | $[Ne]\ 3s^1$ |
| 12 | Mg | 3s: ↑↓   3p: ▢ ▢ ▢ | $[1s^22s^22p^6]\ 3s^2$ | $[Ne]\ 3s^2$ |
| 13 | Al | 3s: ↑↓   3p: ↑ ▢ ▢ | $[1s^22s^22p^6]\ 3s^23p^1$ | $[Ne]\ 3s^23p^1$ |
| 14 | Si | 3s: ↑↓   3p: ↑ ↑ ▢ | $[1s^22s^22p^6]\ 3s^23p^2$ | $[Ne]\ 3s^23p^2$ |
| 15 | P | 3s: ↑↓   3p: ↑ ↑ ↑ | $[1s^22s^22p^6]\ 3s^23p^3$ | $[Ne]\ 3s^23p^3$ |
| 16 | S | 3s: ↑↓   3p: ↑↓ ↑ ↑ | $[1s^22s^22p^6]\ 3s^23p^4$ | $[Ne]\ 3s^23p^4$ |
| 17 | Cl | 3s: ↑↓   3p: ↑↓ ↑↓ ↑ | $[1s^22s^22p^6]\ 3s^23p^5$ | $[Ne]\ 3s^23p^5$ |
| 18 | Ar | 3s: ↑↓   3p: ↑↓ ↑↓ ↑↓ | $[1s^22s^22p^6]\ 3s^23p^6$ | $[Ne]\ 3s^23p^6$ |

| Period | 1A (1) | 2A (2) | 3A (13) | 4A (14) | 5A (15) | 6A (16) | 7A (17) | 8A (18) |
|---|---|---|---|---|---|---|---|---|
| 1 | 1<br>**H**<br>$1s^1$ | | | | | | | 2<br>**He**<br>$1s^2$ |
| 2 | 3<br>**Li**<br>[He] $2s^1$ | 4<br>**Be**<br>[He] $2s^2$ | 5<br>**B**<br>[He] $2s^22p^1$ | 6<br>**C**<br>[He] $2s^22p^2$ | 7<br>**N**<br>[He] $2s^22p^3$ | 8<br>**O**<br>[He] $2s^22p^4$ | 9<br>**F**<br>[He] $2s^22p^5$ | 10<br>**Ne**<br>[He] $2s^22p^6$ |
| 3 | 11<br>**Na**<br>[Ne] $3s^1$ | 12<br>**Mg**<br>[Ne] $3s^2$ | 13<br>**Al**<br>[Ne] $3s^23p^1$ | 14<br>**Si**<br>[Ne] $3s^23p^2$ | 15<br>**P**<br>[Ne] $3s^23p^3$ | 16<br>**S**<br>[Ne] $3s^23p^4$ | 17<br>**Cl**<br>[Ne] $3s^23p^5$ | 18<br>**Ar**<br>[Ne] $3s^23p^6$ |

**FIGURE 8.7**
**The first three periods.** The first 18 elements, H through Ar, are arranged in three periods containing two, eight, and eight elements. Each box shows the atomic number, atomic symbol, and condensed ground-state electron configuration. Note that *elements in a group have similar outer electron configurations.*

noble gas is represented by its element symbol in brackets, and it is followed by the electron configuration of the energy level being filled. The condensed electron configuration of oxygen, for example, is [He] $2s^22p^4$, where [He] represents $1s^2$; for sodium, it is [Ne] $3s^1$, where [Ne] stands for $1s^22s^22p^6$ (as Table 8.3 shows); and so forth.

Once again, let's arrange the elements in periodic table format, this time showing the condensed electron configuration of the first 18 elements (Figure 8.7). Note the obvious *similarities in outer electron configuration* for each group. Most importantly, *similar outer electron configurations correlate closely with similar chemical behavior.* For example, lithium and sodium [Group 1A(1)] both have the condensed electron configuration [noble gas] $ns^1$ (where $n$ is the quantum number of the outermost energy level), as do all the other alkali metals (K, Rb, Cs, Fr). All are highly reactive metals that form ionic compounds with nonmetals, having formulas such as MCl, $M_2O$, and $M_2S$ (where M represents the alkali metal), and all react with water to form $H_2$. As another example, fluorine and chlorine [Group 7A(17)] both have the condensed electron configuration [noble gas] $ns^2np^5$, as do the other halogens (Br, I, At). Little is known about the extremely rare astatine (At), but all the others are reactive nonmetals that occur as diatomic molecules, and (using X to represent a halogen) all form ionic compounds with metals (KX, $MgX_2$), covalent compounds with hydrogen (HX) that yield acidic solutions in water, and covalent compounds with carbon ($CX_4$).

This general theme appears throughout the periodic table: *within a group, similarities in chemical behavior reflect similarities in the distribution of electrons in the highest energy orbitals* (outer electron configuration). Indeed, similarities in the chemical behavior of certain elements led Mendeleev to place those elements in vertical columns. In other words—and this is one of the central points in all chemistry—*properties recur periodically because similarities in electron configurations recur periodically.* As you've seen, the recurrence of similar electron configurations is a natural consequence of *orbitals filling in order of increasing energy*.

### The First *d* Orbital Transition Series; Building Up Period 4

The $3d$ orbitals are filled in Period 4. However, due to penetration by the $4s$ orbital (Section 8.1), the $4s$ orbital is slightly *lower* in energy than the $3d$ orbitals for the first two members of the period, so the $4s$ orbital generally fills before the $3d$. We see the same sequence in each of the three periods that contains $d$ orbitals: the $5s$ orbital fills before the $4d$, and the $6s$ before the $5d$. In general, then, *the ns sublevel fills before the $(n-1)d$ sublevel.* However, because the energies of the $ns$ and $(n - 1)d$ sublevels are so close and become closer with higher values of $n$, several exceptions to this general filling pattern are observed.

Table 8.4 shows the partial orbital diagrams and ground-state electron configurations for the 18 elements in Period 4. The first two elements of the period, potassium and calcium, fill the $4s$ sublevel as the next members of the alkali and alkaline earth metals, respectively. The third element, scandium ($Z = 21$), is the first of the **transition elements,** those in which $d$ orbitals are being filled. The last electron in scandium occupies one of the five $3d$ orbitals, all of which are of equal energy, and scandium has the electron configuration [Ar] $4s^23d^1$.

The filling of the $3d$ orbitals proceeds for the most part as you would expect from filling $s$ and $p$ orbitals, but there are two exceptions: chromium ($Z = 24$) and copper ($Z = 29$). Note that vanadium ($Z = 23$), the element before chromium, has three half-filled $d$ orbitals ([Ar] $4s^23d^3$). Rather than chromium's last electron simply entering a fourth empty $d$ orbital to give the electron configuration [Ar] $4s^23d^4$, chromium has one electron in the $4s$ sublevel and five in the $3d$ sublevel. Thus, in chromium, both the $4s$ and the $3d$ sublevels are *half-filled:*

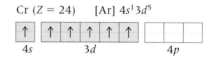

Cr ($Z = 24$)     [Ar] $4s^13d^5$

Another unexpected filling pattern occurs with copper. Following nickel ([Ar] $4s^23d^8$), we would expect copper to have the [Ar] $4s^23d^9$ configuration. Instead, the $4s$ orbital of copper is half-filled (1 electron), and the $3d$ orbitals are *filled* with 10 electrons:

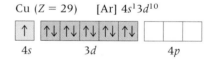

Cu ($Z = 29$)     [Ar] $4s^13d^{10}$

The unusual filling patterns in Cr and Cu lead to the conclusion that *half-filled and filled sublevels are unexpectedly stable.* These are the first two cases of a pattern seen with many other elements.

With zinc, both the $4s$ and $3d$ sublevels are filled, and the first transition series ends. As Table 8.4 shows, the $4p$ sublevel is then filled by the next six elements. Thus, Period 4 ends with krypton, the next noble gas.

### General Principles of Electron Configurations

There are 73 known elements beyond the 36 we have considered. Let's examine the ground-state electron configurations of the elements to highlight some key patterns and principles.

**Similar group outer electron configurations.** As was mentioned earlier, one of chemistry's central themes, and the key to the usefulness of the periodic table, is that *elements in a group have similar chemical properties because they have similar outer electron configurations.* Among the main-group elements (A groups), which consist of the $s$ block and $p$ block of sublevels, outer electron configurations within a group are essentially identical, as in-

# TABLE 8.4 Partial Orbital Diagrams and Electron Configurations for the Elements in Period 4

| ATOMIC NUMBER | ELEMENT | ORBITAL DIAGRAM (4s, 3d AND 4p SUBLEVELS ONLY) | | | FULL ELECTRON CONFIGURATION | CONDENSED ELECTRON CONFIGURATION |
|---|---|---|---|---|---|---|
| | | 4s | 3d | 4p | | |
| 19 | K  | ↑ | | | $[1s^22s^22p^63s^23p^6]\,4s^1$ | $[Ar]\,4s^1$ |
| 20 | Ca | ↑↓ | | | $[1s^22s^22p^63s^23p^6]\,4s^2$ | $[Ar]\,4s^2$ |
| 21 | Sc | ↑↓ | ↑ | | $[1s^22s^22p^63s^23p^6]\,4s^23d^1$ | $[Ar]\,4s^23d^1$ |
| 22 | Ti | ↑↓ | ↑ ↑ | | $[1s^22s^22p^63s^23p^6]\,4s^23d^2$ | $[Ar]\,4s^23d^2$ |
| 23 | V  | ↑↓ | ↑ ↑ ↑ | | $[1s^22s^22p^63s^23p^6]\,4s^23d^3$ | $[Ar]\,4s^23d^3$ |
| 24 | Cr | ↑ | ↑ ↑ ↑ ↑ ↑ | | $[1s^22s^22p^63s^23p^6]\,4s^13d^5$ | $[Ar]\,4s^13d^5$ |
| 25 | Mn | ↑↓ | ↑ ↑ ↑ ↑ ↑ | | $[1s^22s^22p^63s^23p^6]\,4s^23d^5$ | $[Ar]\,4s^23d^5$ |
| 26 | Fe | ↑↓ | ↑↓ ↑ ↑ ↑ ↑ | | $[1s^22s^22p^63s^23p^6]\,4s^23d^6$ | $[Ar]\,4s^23d^6$ |
| 27 | Co | ↑↓ | ↑↓ ↑↓ ↑ ↑ ↑ | | $[1s^22s^22p^63s^23p^6]\,4s^23d^7$ | $[Ar]\,4s^23d^7$ |
| 28 | Ni | ↑↓ | ↑↓ ↑↓ ↑↓ ↑ ↑ | | $[1s^22s^22p^63s^23p^6]\,4s^23d^8$ | $[Ar]\,4s^23d^8$ |
| 29 | Cu | ↑ | ↑↓ ↑↓ ↑↓ ↑↓ ↑↓ | | $[1s^22s^22p^63s^23p^6]\,4s^13d^{10}$ | $[Ar]\,4s^13d^{10}$ |
| 30 | Zn | ↑↓ | ↑↓ ↑↓ ↑↓ ↑↓ ↑↓ | | $[1s^22s^22p^63s^23p^6]\,4s^23d^{10}$ | $[Ar]\,4s^23d^{10}$ |
| 31 | Ga | ↑↓ | ↑↓ ↑↓ ↑↓ ↑↓ ↑↓ | ↑ | $[1s^22s^22p^63s^23p^6]\,4s^23d^{10}4p^1$ | $[Ar]\,4s^23d^{10}4p^1$ |
| 32 | Ge | ↑↓ | ↑↓ ↑↓ ↑↓ ↑↓ ↑↓ | ↑ ↑ | $[1s^22s^22p^63s^23p^6]\,4s^23d^{10}4p^2$ | $[Ar]\,4s^23d^{10}4p^2$ |
| 33 | As | ↑↓ | ↑↓ ↑↓ ↑↓ ↑↓ ↑↓ | ↑ ↑ ↑ | $[1s^22s^22p^63s^23p^6]\,4s^23d^{10}4p^3$ | $[Ar]\,4s^23d^{10}4p^3$ |
| 34 | Se | ↑↓ | ↑↓ ↑↓ ↑↓ ↑↓ ↑↓ | ↑↓ ↑ ↑ | $[1s^22s^22p^63s^23p^6]\,4s^23d^{10}4p^4$ | $[Ar]\,4s^23d^{10}4p^4$ |
| 35 | Br | ↑↓ | ↑↓ ↑↓ ↑↓ ↑↓ ↑↓ | ↑↓ ↑↓ ↑ | $[1s^22s^22p^63s^23p^6]\,4s^23d^{10}4p^5$ | $[Ar]\,4s^23d^{10}4p^5$ |
| 36 | Kr | ↑↓ | ↑↓ ↑↓ ↑↓ ↑↓ ↑↓ | ↑↓ ↑↓ ↑↓ | $[1s^22s^22p^63s^23p^6]\,4s^23d^{10}4p^6$ | $[Ar]\,4s^23d^{10}4p^6$ |

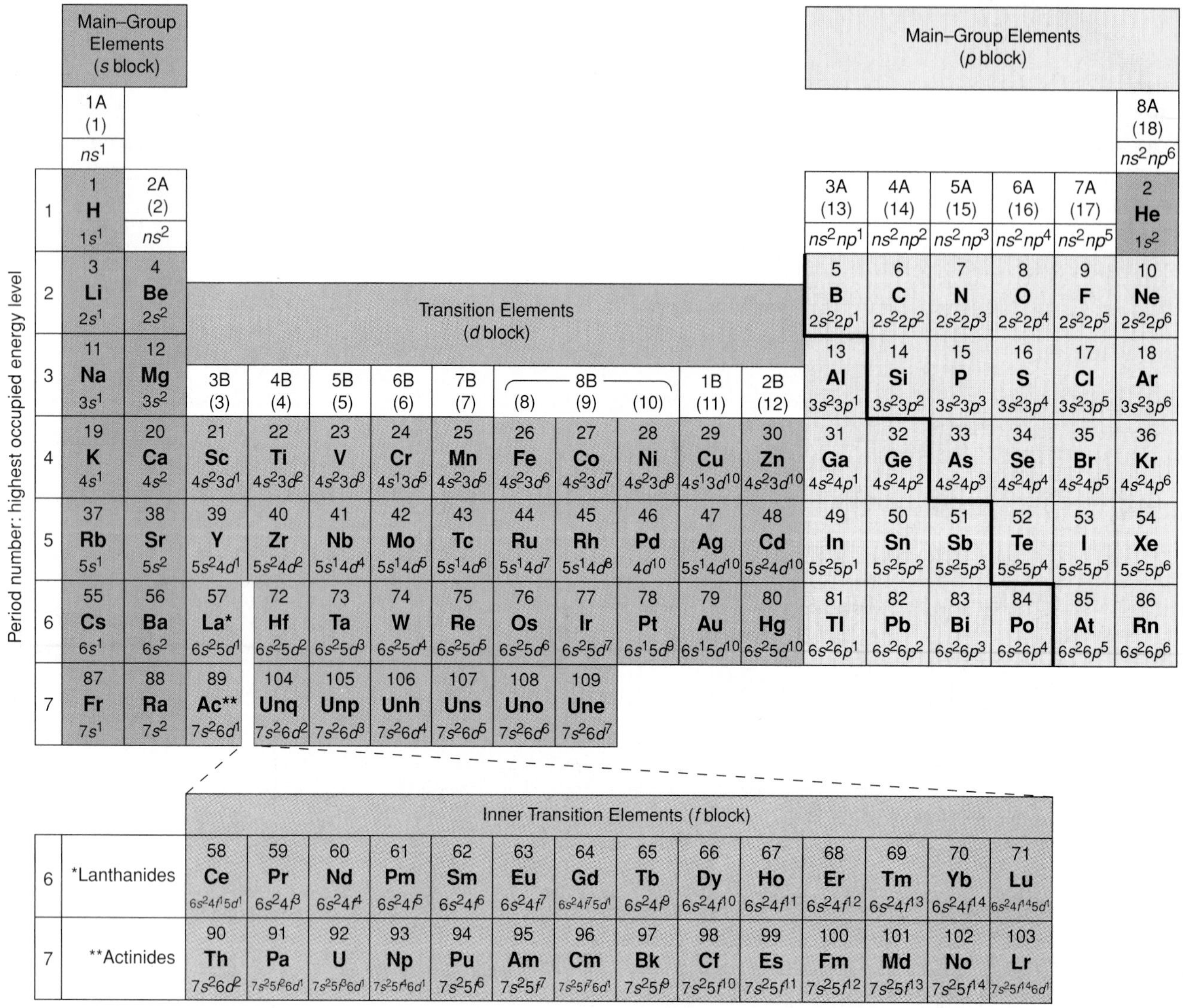

**Main–Group Elements (s block)**

**Main–Group Elements (p block)**

**Transition Elements (d block)**

Period number: highest occupied energy level

| | 1A (1) $ns^1$ | 2A (2) $ns^2$ | 3B (3) | 4B (4) | 5B (5) | 6B (6) | 7B (7) | 8B (8) | 8B (9) | 8B (10) | 1B (11) | 2B (12) | 3A (13) $ns^2np^1$ | 4A (14) $ns^2np^2$ | 5A (15) $ns^2np^3$ | 6A (16) $ns^2np^4$ | 7A (17) $ns^2np^5$ | 8A (18) $ns^2np^6$ |
|---|---|---|---|---|---|---|---|---|---|---|---|---|---|---|---|---|---|---|
| 1 | 1 H $1s^1$ | | | | | | | | | | | | | | | | | 2 He $1s^2$ |
| 2 | 3 Li $2s^1$ | 4 Be $2s^2$ | | | | | | | | | | | 5 B $2s^22p^1$ | 6 C $2s^22p^2$ | 7 N $2s^22p^3$ | 8 O $2s^22p^4$ | 9 F $2s^22p^5$ | 10 Ne $2s^22p^6$ |
| 3 | 11 Na $3s^1$ | 12 Mg $3s^2$ | | | | | | | | | | | 13 Al $3s^23p^1$ | 14 Si $3s^23p^2$ | 15 P $3s^23p^3$ | 16 S $3s^23p^4$ | 17 Cl $3s^23p^5$ | 18 Ar $3s^23p^6$ |
| 4 | 19 K $4s^1$ | 20 Ca $4s^2$ | 21 Sc $4s^23d^1$ | 22 Ti $4s^23d^2$ | 23 V $4s^23d^3$ | 24 Cr $4s^13d^5$ | 25 Mn $4s^23d^5$ | 26 Fe $4s^23d^6$ | 27 Co $4s^23d^7$ | 28 Ni $4s^23d^8$ | 29 Cu $4s^13d^{10}$ | 30 Zn $4s^23d^{10}$ | 31 Ga $4s^24p^1$ | 32 Ge $4s^24p^2$ | 33 As $4s^24p^3$ | 34 Se $4s^24p^4$ | 35 Br $4s^24p^5$ | 36 Kr $4s^24p^6$ |
| 5 | 37 Rb $5s^1$ | 38 Sr $5s^2$ | 39 Y $5s^24d^1$ | 40 Zr $5s^24d^2$ | 41 Nb $5s^14d^4$ | 42 Mo $5s^14d^5$ | 43 Tc $5s^14d^6$ | 44 Ru $5s^14d^7$ | 45 Rh $5s^14d^8$ | 46 Pd $4d^{10}$ | 47 Ag $5s^14d^{10}$ | 48 Cd $5s^24d^{10}$ | 49 In $5s^25p^1$ | 50 Sn $5s^25p^2$ | 51 Sb $5s^25p^3$ | 52 Te $5s^25p^4$ | 53 I $5s^25p^5$ | 54 Xe $5s^25p^6$ |
| 6 | 55 Cs $6s^1$ | 56 Ba $6s^2$ | 57 La* $6s^25d^1$ | 72 Hf $6s^25d^2$ | 73 Ta $6s^25d^3$ | 74 W $6s^25d^4$ | 75 Re $6s^25d^5$ | 76 Os $6s^25d^6$ | 77 Ir $6s^25d^7$ | 78 Pt $6s^15d^9$ | 79 Au $6s^15d^{10}$ | 80 Hg $6s^25d^{10}$ | 81 Tl $6s^26p^1$ | 82 Pb $6s^26p^2$ | 83 Bi $6s^26p^3$ | 84 Po $6s^26p^4$ | 85 At $6s^26p^5$ | 86 Rn $6s^26p^6$ |
| 7 | 87 Fr $7s^1$ | 88 Ra $7s^2$ | 89 Ac** $7s^26d^1$ | 104 Unq $7s^26d^2$ | 105 Unp $7s^26d^3$ | 106 Unh $7s^26d^4$ | 107 Uns $7s^26d^5$ | 108 Uno $7s^26d^6$ | 109 Une $7s^26d^7$ | | | | | | | | | |

**Inner Transition Elements (f block)**

| | | 58 Ce $6s^24f^15d^1$ | 59 Pr $6s^24f^3$ | 60 Nd $6s^24f^4$ | 61 Pm $6s^24f^5$ | 62 Sm $6s^24f^6$ | 63 Eu $6s^24f^7$ | 64 Gd $6s^24f^75d^1$ | 65 Tb $6s^24f^9$ | 66 Dy $6s^24f^{10}$ | 67 Ho $6s^24f^{11}$ | 68 Er $6s^24f^{12}$ | 69 Tm $6s^24f^{13}$ | 70 Yb $6s^24f^{14}$ | 71 Lu $6s^24f^{14}5d^1$ |
|---|---|---|---|---|---|---|---|---|---|---|---|---|---|---|---|
| 6 | *Lanthanides | | | | | | | | | | | | | | |
| 7 | **Actinides | 90 Th $7s^26d^2$ | 91 Pa $7s^25f^26d^1$ | 92 U $7s^25f^36d^1$ | 93 Np $7s^25f^46d^1$ | 94 Pu $7s^25f^6$ | 95 Am $7s^25f^7$ | 96 Cm $7s^25f^76d^1$ | 97 Bk $7s^25f^9$ | 98 Cf $7s^25f^{10}$ | 99 Es $7s^25f^{11}$ | 100 Fm $7s^25f^{12}$ | 101 Md $7s^25f^{13}$ | 102 No $7s^25f^{14}$ | 103 Lr $7s^25f^{14}6d^1$ |

**FIGURE 8.8**

**A periodic table of partial ground-state electron configurations.** These ground-state electron configurations show the electrons beyond the previous noble gas in the sublevel block being filled (excluding filled inner sublevels). Note that He is colored as an s-block element but placed with the other members of Group 8A(18). For the main-group elements, the group heading includes the general outer configuration. Unexpected electron configurations occur often among the d-block and f-block elements, the first two appearing for Cr ($Z = 24$) and Cu ($Z = 29$), as discussed in the text.

dicated by the general group headings in Figure 8.8. Some variations in the transition elements (B groups, d block) and inner transition elements (f block) occur, as you'll see.

**Orbital filling order.** When the elements are "built up" by filling their levels and sublevels in order of increasing energy, we obtain the actual sequence of elements in the periodic table. Thus, reading the table from left to right, as you do the words on a page, gives the energy order of levels and sublevels (Figure 8.9). Try to memorize the basic form and arrangement of

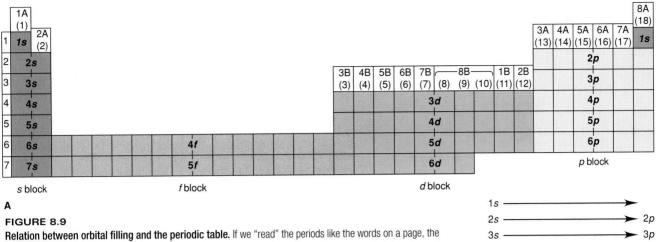

**A**

**FIGURE 8.9**

**Relation between orbital filling and the periodic table.** If we "read" the periods like the words on a page, the elements are arranged into sublevel blocks that occur in the sequence of increasing energy. **A,** A long form of the periodic table, with sublevel blocks shown. (The *f* blocks fit between the first and second elements of the *d* blocks in Periods 6 and 7.) **B,** A simple, memorizable version of the sublevel order.

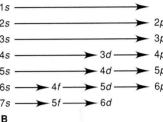

**B**

the periodic table; it is certainly the most useful way to learn the filling order of the elements. ◆

**Categories of electrons.** As we survey the periodic table, we can distinguish three categories of electrons:

1. **Inner (core) electrons** are those in the previous noble gas and in any completed transition series. They fill all the *lower energy levels* of an atom.
2. **Outer electrons** are those in the *highest energy level* (have the highest *n* value) and on average are farthest from the nucleus.
3. **Valence electrons** are those involved in forming compounds. *Among the main-group elements, the valence electrons **are** the outer electrons.* Among the transition elements, some inner *d* electrons are also often involved in bonding and are counted among the valence electrons.

**Group and period numbers.** Several key pieces of information are embedded in the structure of the periodic table:

1. Among the main-group elements (A groups), *the group number equals the number of outer electrons* (those with highest *n*). For example, chlorine (Cl; Period 3, Group **7**A) has 7 outer electrons, tellurium (Te; Period 5, Group **6**A) has 6, and so forth.
2. *The period number is the n value of the highest energy level in the period.* In Period 2, for example, the $n = 2$ level has the highest energy; in Period 5, it is the $n = 5$ level.
3. The *n* value squared ($n^2$) gives the total number of *orbitals* in that energy level. Since an orbital can hold no more than two electrons (exclusion principle), twice the square of the *n* value ($2n^2$) gives the maximum number of *electrons* in the energy level. For the $n = 3$ level, for example, the number of orbitals is $n^2 = 9$: one 3*s*, three 3*p*, and five 3*d*. The number of electrons is $2n^2$, or 18: two 3*s* and six 3*p* electrons occur in the elements of Period 3, and ten 3*d* electrons are added in the transition elements of Period 4.

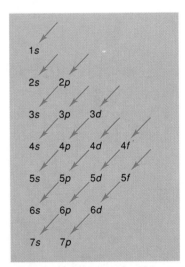

◆ **Periodic Memory Aids.** This arrangement of sublevels can serve as a reminder of the filling order. List the sublevels as shown, and then read from 1*s*, following the direction of the arrows.

### Complex Patterns: The Transition and Inner Transition Elements

Periods 4, 5, 6, and 7 incorporate the *d*-orbital transition elements. The general pattern, as you've seen, is that the $(n - 1)d$ orbitals are filled between the *ns* and *np* orbitals. Thus, Period 5 follows the same general pattern as Period 4. In Period 6, the *6s* sublevel is filled with cesium (Cs) and barium (Ba); then lanthanum (La; $Z = 57$), the first member of the *5d* transition series, occurs. At this point, a new sublevel is interspersed. Immediately following lanthanum is cerium (Ce; $Z = 58$), the first element in the first series of **inner transition elements,** those in which *f* orbitals are being filled. The *f* orbitals have $l = 3$, so the possible $m_l$ values are $-3$, $-2$, $-1$, 0, $+1$, $+2$, and $+3$; that is, there are seven *f* orbitals, for a total of 14 elements in *each* of the two inner transition series.

The Period 6 series of inner transition elements fills the *4f* orbitals and consists of the **lanthanides** (or **rare earths**) because of their placement after and similarity to lanthanum. The other inner transition series consists of the **actinides,** and it fills the *5f* orbitals that appear in Period 7 after actinium (Ac; $Z = 89$). The $(n - 2)f$ orbitals are filled in each inner transition series, after which the $(n - 1)d$ orbital filling proceeds. Period 6 ends by filling the *6p* orbitals; Period 7 is incomplete because only seven of ten possible *6d* elements are known at this time.

Several irregularities in filling pattern occur in both the *d* and *f* series. The two already mentioned occur with chromium (Cr) and copper (Cu) in Period 4. Silver (Ag) and gold (Au), the two elements under Cu in Group 1B(11), follow copper's pattern. Molybdenum (Mo) follows the pattern of Cr in Group 6B(6), but tungsten (W) does not; other unexpected configurations appear among the transition elements in Periods 5 and 6. Note, however, that even though minor variations from the expected configurations occur, the sum of *ns* electrons and $(n - 1)d$ electrons *always* equals the *new* group number within a group of transition elements. For instance, despite variations in the electron configurations in Group 6B(**6**)—Cr, Mo, W, and Unh—the sum of *ns* and $(n - 1)d$ electrons is 6; in Group 8B(**10**)—Ni, Pd, and Pt—the sum is 10.

Whenever our observations differ from our expectations, we must always remember that the phenomenon has priority over the model: the electrons don't "care" what orbitals *we* think they should occupy. As the atomic orbitals are filled with more and more electrons in these larger atoms, sublevel energies become extremely similar, and such small differences in energy result in variations from the expected pattern.

---

SAMPLE PROBLEM 8.2 _____

### Determining Electron Configurations

**Problem:** Using the periodic table on the inside cover of the text, give the (1) full and condensed electron configurations, (2) orbital diagrams for the valence electrons, and (3) number of inner electrons for the following elements:
**(a)** Potassium (K; $Z = 19$)   **(b)** Molybdenum (Mo; $Z = 42$)   **(c)** Lead (Pb; $Z = 82$)
**Plan:** We know the number of electrons in each element from its atomic number, and the periodic table shows the order for filling sublevels. In the orbital diagrams, we include all electrons after the previous noble gas except those in *filled* inner sublevels. The number of inner electrons is the sum of those in the previous noble gas and in filled *d* and *f* sublevels.

**Solution: (a)** For K ($Z = 19$), the full electron configuration is $1s^22s^22p^63s^23p^64s^1$.
The condensed electron configuration is **[Ar] $4s^1$**.
The orbital diagram for valence electrons appears in the margin.
K is a main-group element in Group 1A(1) of Period 4, so there are **18 inner electrons.**

**(b)** For Mo ($Z = 42$), we would expect the full electron configuration to be $1s^22s^22p^63s^23p^64s^23d^{10}4p^65s^24d^4$. However, Mo lies under Cr in Group 6B(6) and exhibits the same variation in filling pattern in the $ns$ and $(n - 1)d$ sublevels: $1s^22s^22p^63s^23p^64s^23d^{10}4p^65s^14d^5$.
The condensed electron configuration is **[Kr] $5s^14d^5$.**
The orbital diagram for valence electrons appears in the margin.
Mo is a transition element in Period 5, so there are **36 inner electrons.**

**(c)** For Pb ($Z = 82$), the full electron configuration is
$$1s^22s^22p^63s^23p^64s^23d^{10}4p^65s^24d^{10}5p^66s^24f^{14}5d^{10}6p^2.$$
The condensed electron configuration is **[Xe] $6s^24f^{14}5d^{10}6p^2$.**
The orbital diagram for valence electrons (no filled inner sublevels) appears in the margin.
Pb is a main-group element in Group 4A(14) of Period 6, so there are 54 (in Xe) + 14 (in $4f$ series) + 10 (in $5d$ series) = **78 inner electrons.**

**Check:** Be sure the sum of the superscripts (electrons) in the full electron configuration equals the atomic number, and that the number of *valence* electrons in the condensed configuration equals the number of electrons in the orbital diagram.

**FOLLOW-UP PROBLEM 8.2**
Write full and condensed electron configurations, orbital diagrams for valence electrons, and give the number of inner electrons for the following elements:
**(a)** Ni ($Z = 28$); **(b)** Sr ($Z = 38$); **(c)** Po ($Z = 84$).

For K

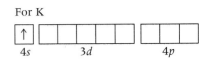

For Mo

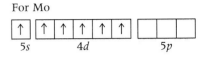

For Pb

## Section Summary

In the aufbau method of filling energy levels, we add one electron to each successive element, observing the Pauli exclusion principle (no two electrons can have the same set of quantum numbers) and Hund's rule (orbitals of equal energy become half filled, with electron spins parallel, before any pairing occurs). The elements of a group have similar outer electron configurations and similar chemical behavior. For the main-group elements, valence electrons (those involved in reactions) are in the outer (highest energy) level only. For transition elements, inner $d$ electrons are often involved in reactions also. In general, $(n - 1)d$ orbitals fill after $ns$ and before $np$ orbitals. In Periods 6 and 7, $(n - 2)f$ orbitals fill after the first $(n - 1)d$ orbital element.

# 8.3 Trends in Some Key Periodic Atomic Properties

*All physical and chemical behavior of the elements is based ultimately on the electron configurations of their atoms.* In this section, we focus on three properties of atoms that are influenced directly by the electron configuration: atomic size, ionization energy (the energy involved in removing an electron from an atom), and electron affinity (the energy involved in adding an electron to an atom). These properties are *periodic:* they generally increase and decrease in a recurring manner through the periodic table. As a result, they exhibit consistent changes, or trends, within a group or period that correlate with the behavior of the elements.

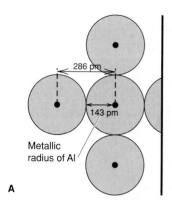

Metallic radius of Al

**A**

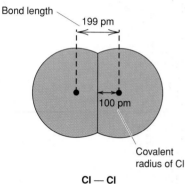

Bond length

Covalent radius of Cl

**Cl — Cl**
bond

**B**

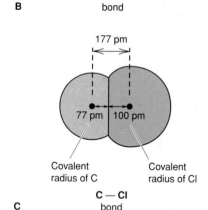

Covalent radius of C

Covalent radius of Cl

**C — Cl**
bond

**C**

**FIGURE 8.10**

**Defining metallic and covalent radii. A,** The metallic radius is one-half the distance between nuclei of adjacent atoms in a crystal of the element, as shown here for aluminum. **B,** The covalent radius is one-half the distance between bonded nuclei in a diatomic molecule of the element. In effect, it is one-half the bond length. **C,** In a covalent compound, the bond length and known covalent radii are used to determine other radii. Here the C—Cl bond length (177 pm) and the covalent radius of Cl (100 pm) are used to find a value for the covalent radius of C (177 pm − 100 pm = 77 pm).

## Trends in Atomic Size

Recall from Chapter 7 that there is a tiny, but finite, probability of an electron in an atom lying relatively far from the nucleus, so we often represent atoms with a 90% probability contour. In a sample of an element or compound, we *define the size of an atom in terms of how closely it lies to a neighboring atom.* In practice, we measure the distance between identical, adjacent atomic nuclei and divide the distance in half. (We discuss the experimental method in Chapter 11.) Keep in mind, however, that the probability contour is not a real surface, so the size will depend to some extent on the nature of the neighboring atom. In other words, the *size of a given atom typically varies somewhat from substance to substance.*

Two common definitions provide an estimate of atomic size (Figure 8.10). The **metallic radius** is one-half the distance between nuclei of adjacent atoms in a crystal of the element; we typically use this definition of atomic size for metals. For elements that commonly occur as molecules, typically nonmetals, we define atomic size by the **covalent radius,** one-half the distance between nuclei of identical covalently bonded atoms.

Atomic size greatly influences ionization energy and electron affinity, as you'll see shortly, and it is the principal factor determining the length of a covalent bond. Bond length, in turn, influences bond strength—the energy involved in breaking or forming a bond—which is crucial to a substance's reactivity. (We discuss bond characteristics and bonding in Chapter 9.) Clearly, an understanding of the trends in atomic size is essential for an appreciation of element behavior. Figure 8.11 shows the atomic radii of the main-group and transition elements. Among the main-group elements, you can see that size varies both within a group and a period. Among members of a group, this size variation is the major reason for differences in element properties. The striking differences between carbon and lead in Group 4A(14), for example, are due primarily to the different sizes of their atoms (Figure 8.12).

The observed variations in atomic size are the result of two opposing influences. As the principal quantum number ($n$) increases, the outer electrons are farther from the nucleus, so the atoms become larger. As the nuclear charge actually "felt" by an electron—the effective nuclear charge ($Z_{eff}$)—increases, however, the outer electrons are pulled closer to the nucleus, so the atoms become smaller. The overall effect of these influences on atomic size depends on the effectiveness of shielding by the inner electrons. As we move *down* a main group, each member has one additional level of *inner* electrons, which shields the *outer* electrons from the increasing nuclear charge very effectively. Calculations show $Z_{eff}$ changes little, rising only slightly for each outer level, so the atomic radius increases as we would expect from the higher $n$ value. Thus, *atomic radius generally **increases** from top to bottom in a group.* In contrast, as we move *across* a period of main-group elements, electrons are added one at a time to the *same* outer level, so the shielding by inner electrons does not change. Outer electrons shield each other much less effectively, so $Z_{eff}$ increases significantly, and the outer electrons are attracted more and more strongly by the increasing nuclear charge. Thus, *atomic radius generally **decreases** from left to right across a period.*

These size trends hold most regularly for the main-group elements; many minor variations in atomic size occur across a series or down a group of transition elements (see Figure 8.11). Remember that across a transition series, an inner *d* orbital is being filled. An initial shrinking is caused by the

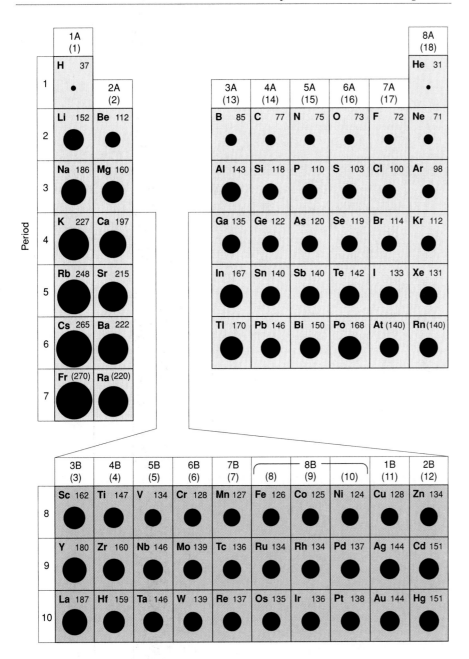

**FIGURE 8.11**

**Atomic radii of the main-group and transition elements.** The atomic radius (either metallic or covalent) in picometers is shown as circles of proportional size for the main-group elements (tan) and the transition elements (blue). Among the main-group elements, atomic radius generally increases from top to bottom and decreases from left to right. (Values in parentheses have only two significant figures.) The transition elements do not exhibit this trend in size.

**Carbon (C)**
Nonmetal
radius = 77 pm  Gaseous soluble oxides
Inert to acid

**Lead (Pb)**
Metal
Solid insoluble oxides
radius = 146 pm  Reacts with nitric acid

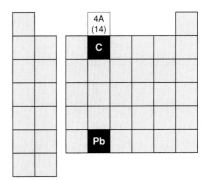

**FIGURE 8.12**

**The influence of size on the properties of carbon and lead.** Despite similar outer electron configurations, carbon, the smallest member of Group 4A(14), has properties that are very different from those of lead, the largest member.

increasing charge of the nucleus through the first two or three elements. Afterward, *the size remains relatively constant* as the shielding by inner *d* electrons counteracts the usual increase in $Z_{\text{eff}}$ across a period. For instance, vanadium (V; $Z = 23$), the third member of the Period 4 transition series, has the same radius as zinc (Zn; $Z = 30$), the last member. The same general trend appears in the Period 5 and 6 *d* transition series and through both series of inner *f* transition elements.

The filling of *d* electrons greatly affects the size of Group 3A(13) elements, the first succeeding main group. For example, in Period 4, note the large drop in size between calcium (Ca; $Z = 20$) in Group 2A(2) and gallium (Ga; $Z = 31$) in 3A(13). This size decrease between 2A and 3A main-group elements in periods *with* a transition series (Periods 4, 5, and 6) is much

**FIGURE 8.13**

**Periodicity of atomic radius.** A plot of atomic radius vs. atomic number for the elements in Periods 1 through 6 shows a periodic change: the radius generally decreases through a period to the noble gas [Group 8A(18)] and then increases suddenly to the next alkali metal [Group 1A(1)].

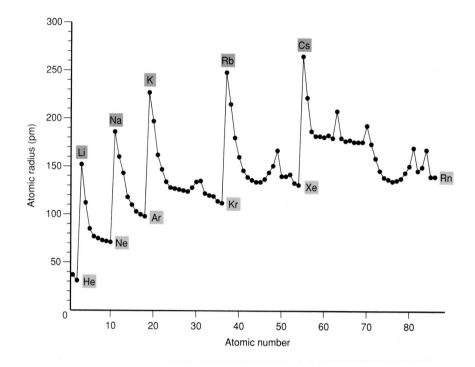

◆ **Packing 'Em In.** The increasing nuclear charge dramatically shrinks the space in which each electron can move. For example, in Group 1A(1), the atomic radius of cesium ($Z = 55$) is only 1.7 times that of lithium ($Z = 3$), so its volume is about five times as large, but cesium has 18 times as many electrons. At the opposite ends of Period 2, neon ($Z = 10$) has more than three times as many electrons as lithium ($Z = 3$), but its volume is only about $\frac{1}{8}$ as much. Thus, whether the size increases down a group or decreases across a period, a larger number of protons in the nucleus greatly crowds the electrons and compacts the atom.

greater than in periods *without* one (Periods 2 and 3). Because of penetration by the *np* orbital through the $(n - 1)d$ sublevel, the first *np* electron "feels" an increased $Z_{eff}$ from *all the protons* added during the transition series. This effect is so great that gallium is slightly *smaller* than aluminum (Al; $Z = 13$), even though it is farther down in Group 3A(13).

Figure 8.13 shows the overall variation in atomic size with increasing atomic number. Note the recurring up-and-down pattern as size decreases through a period to the noble gas and then leaps up to the alkali metal that begins the next period. ◆

SAMPLE PROBLEM 8.3 _____

**Ranking Elements According to Size**

**Problem:** Using the periodic table, but not Figure 8.11, rank each set of main-group elements in order of *decreasing* size:
**(a)** Ca, Mg, Sr  **(b)** K, Ga, Ca  **(c)** Br, Rb, Kr  **(d)** Sr, Ca, Rb
**Plan:** To rank the elements by size, we must find their relative positions in the periodic table. Since they are all main-group elements, we apply the trends of increasing size down a group and decreasing size across a period.
**Solution: (a) Sr > Ca > Mg.** These three elements are in Group 2A(2), so size decreases up the group.
**(b) K > Ca > Ga.** These three elements are in Period 4, so size decreases across a period.
**(c) Rb > Br > Kr.** Rb is largest because it has one more energy level and is farther to the left. Kr is smaller than Br because it is farther to the right in Period 4.
**(d) Rb > Sr > Ca.** Ca is smallest because it has one fewer energy level. Sr is smaller than Rb because it is farther to the right.
**Check:** Using Figure 8.11, we see that the predictions are correct.

FOLLOW-UP PROBLEM 8.3
Using only the periodic table, rank the elements in each set in order of *increasing* size:
**(a)** Se, Br, Cl; **(b)** I, Xe, Ba.

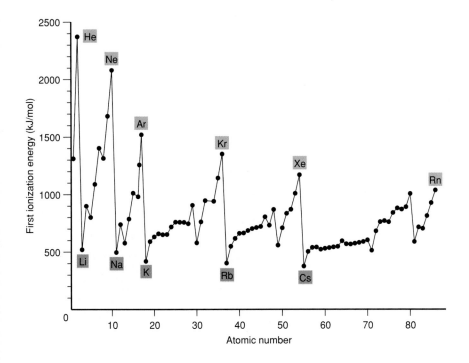

**FIGURE 8.14**
**Periodicity of first ionization energy (IE₁).** A plot of $IE_1$ (in kJ/mol) vs. atomic number for the elements in Periods 1 through 6 shows a periodic change: the lowest values occur for the alkali metals and the highest for the noble gases. This is the *inverse* of the trend in atomic size (see Figure 8.13).

## Trends in Ionization Energy

Pulling an electron away from a nucleus *requires* energy to separate the opposite charges. The **ionization energy (IE)** is the amount of energy required *to remove completely one mole of electrons from one mole of gaseous atoms or ions.* Since ionization involves putting energy *into* the system, ionization energy is always a positive number (just as $\Delta H$ of an endothermic reaction is positive).

Chapter 7 described the ionization energy of the ground-state H atom as the energy difference between $n = 1$ and $n = \infty$, the point at which the electron is completely removed. Many-electron atoms can lose more than one electron, so we number the ionization energies required to remove each electron in sequence from the ground-state atom. The first ionization energy ($IE_1$) is the energy needed to remove an electron from the *highest occupied sublevel* of the gaseous atom:

$$\text{Atom}(g) \longrightarrow \text{ion}^+(g) + e^- \qquad \Delta E = IE_1 > 0$$

The second ionization energy ($IE_2$) removes the second electron. Since the electron is being pulled away from a positively charged ion, $IE_2$ is always larger than $IE_1$:

$$\text{Ion}^+(g) \longrightarrow \text{ion}^{2+}(g) + e^- \qquad \Delta E = IE_2 \text{ (always} > IE_1)$$

The first ionization energy is a key factor in an element's chemical reactivity because, as you'll see later in this chapter, *atoms with a low $IE_1$ tend to form cations during reactions, whereas those with a high $IE_1$ (except the noble gases) often form anions.*

**Variations in first ionization energy.** The elements exhibit a periodic change in first ionization energy (Figure 8.14). By comparing Figure 8.14 with Figure 8.13, you can see a roughly *inverse* relationship between $IE_1$ and atomic size. As we move down a main group, the distance from the nucleus to the outer electron increases. Thus, the attraction between them lessens, so the electron becomes easier to remove. Therefore, in general, *the*

**FIGURE 8.15**

**First ionization energies of the main-group elements.** Values for IE$_1$ (in kJ/mol) of the main-group elements are shown as posts of varying height. Note the general increase within a period and decrease within a group. Thus, the lowest value is at the bottom of Group 1A(1), and the highest is at the top of Group 8A(18).

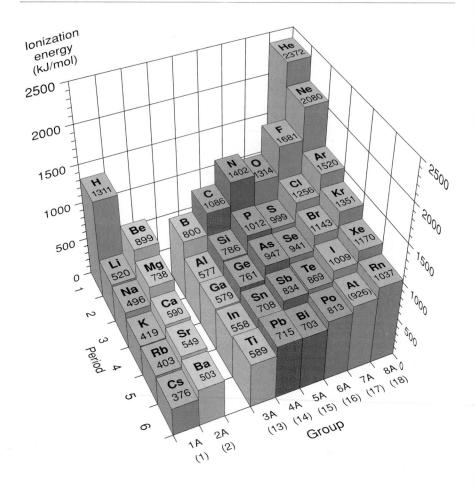

*ionization energy **decreases** down a group* (Figure 8.15). The only significant deviation from this pattern occurs in Group 3A(13). Although IE$_1$ decreases as expected from boron (B) to aluminum (Al), no decrease occurs for the rest of the group. As with the atomic size trends, filling the *d* sublevels of the intervening transition elements in Periods 4, 5, and 6 causes a greater than expected $Z_{eff}$, which holds the outer electrons tightly in these larger 3A members.

As we move across a period, $Z_{eff}$ generally increases, so atomic radii become smaller. As a result, the attraction between the nucleus and the outer electrons increases, so an electron becomes more difficult to remove. In general, *ionization energy **increases** across a period:* it is easier to remove an electron from an alkali metal than from a noble gas.

There are two small "dips" in the otherwise smooth increase in ionization energy, which occur at Groups 3A(13) and 6A(16) in Period 2 (at B and at O) and in Period 3 (at Al and at S). The first dip occurs because the *np* sublevel is higher in energy than the *ns,* so the electron is pulled off more easily, leaving a filled *ns* sublevel. The second occurs because the $np^4$ electron occupies the same orbital as another *np* electron, the first such pairing, so electron repulsions raise the orbital energy. Removing this electron relieves the repulsions and leaves a stable, half-filled *np* sublevel; thus, the fourth *p* electron is pulled off more easily.

SAMPLE PROBLEM 8.4 _____

**Ranking Elements According to First Ionization Energy**

**Problem:** Using the periodic table only, rank the elements in each of the following sets in order of *decreasing* $IE_1$:
**(a)** Kr, He, Ar   **(b)** Sb, Te, Sn   **(c)** K, Ca, Rb   **(d)** I, Xe, Cs
**Plan:** We proceed as in Sample Problem 8.3, noting the elements' positions in the periodic table and applying the general trends of decreasing $IE_1$ down a group and increasing $IE_1$ across a period.
**Solution: (a) He > Ar > Kr.** These three are all in Group 8A(18), and $IE_1$ generally decreases down a group.
**(b) Te > Sb > Sn.** These three are all in Period 5, and $IE_1$ generally increases across a period.
**(c) Ca > K > Rb.** $IE_1$ of K is larger than $IE_1$ of Rb because K is higher in Group 1A(1). $IE_1$ of Ca is larger than $IE_1$ of K because Ca is farther to the right in Period 4.
**(d) Xe > I > Cs.** $IE_1$ of I is smaller than $IE_1$ of Xe because I is farther to the left. $IE_1$ of I is larger than $IE_1$ of Cs because I is farther to the right and in the previous period.
**Check:** Since trends in $IE_1$ are generally opposite trends in size, you could rank the elements by size and check that you obtain the reverse order.

FOLLOW-UP PROBLEM 8.4
Rank the elements in each of the following sets in order of *increasing* $IE_1$: **(a)** Sb, Sn, I; **(b)** Sr, Ca, Ba.

**Variations in successive ionization energies of an element.** Successive ionization energies of a given element ($IE_1$, $IE_2$, and so on) increase because each electron is being pulled away from a more positive ion. This increase occurs for every element, but it is not a smooth one. Consider the values in Table 8.5, which shows successive ionization energies for the elements lithium through sodium (Period 2 and the first element of Period 3). As we move horizontally through the values for a given element, we reach a point that separates relatively low from relatively high IE values (shaded area to right of line). This jump appears after all outer (valence)

**TABLE 8.5 Successive Ionization Energies of the Elements Lithium Through Sodium**

| Z | ELEMENT | NUMBER OF VALENCE ELECTRONS | IONIZATION ENERGY (MJ/mol)* | | | | | | | | | |
|---|---------|------|-------|-------|-------|-------|-------|-------|-------|-------|-------|-------|
| | | | $IE_1$ | $IE_2$ | $IE_3$ | $IE_4$ | $IE_5$ | $IE_6$ | $IE_7$ | $IE_8$ | $IE_9$ | $IE_{10}$ |
| 3 | Li | 1 | 0.52 | 7.30 | 11.81 | | | | | | | |
| 4 | Be | 2 | 0.90 | 1.76 | 14.85 | 21.01 | | | Core electrons | | | |
| 5 | B | 3 | 0.80 | 2.43 | 3.66 | 25.02 | 32.82 | | | | | |
| 6 | C | 4 | 1.09 | 2.35 | 4.62 | 6.22 | 37.83 | 47.28 | | | | |
| 7 | N | 5 | 1.40 | 2.86 | 4.58 | 7.48 | 9.44 | 53.27 | 64.36 | | | |
| 8 | O | 6 | 1.31 | 3.39 | 5.30 | 7.47 | 10.98 | 13.33 | 71.33 | 84.08 | | |
| 9 | F | 7 | 1.68 | 3.37 | 6.05 | 8.41 | 11.02 | 15.16 | 17.87 | 92.04 | 106.43 | |
| 10 | Ne | 8 | 2.08 | 3.95 | 6.12 | 9.37 | 12.18 | 15.24 | 20.00 | 23.07 | 115.38 | 131.43 |
| 11 | Na | 1 | 0.50 | 4.56 | 6.91 | 9.54 | 13.35 | 16.61 | 20.11 | 25.49 | 28.93 | 141.37 |

*MJ/mol, or megajoule per mole = $10^3$ kJ/mol.

electrons have been removed and thus represents the much greater energy needed to pull an inner (core) electron away from the nucleus. For instance, with Be, $IE_1$ is lower than $IE_2$, which is *much* lower than $IE_3$, indicating that beryllium has only two electrons in the outermost level. Core electrons are *not* involved in typical chemical reactions.

SAMPLE PROBLEM 8.5

### Identifying an Element from Its Successive Ionization Energies

**Problem:** Given the following series of ionization energies (in kJ/mol) for element X in Period 3, name the element and write its electron configuration:

| $IE_1$ | $IE_2$ | $IE_3$ | $IE_4$ | $IE_5$ | $IE_6$ |
|---|---|---|---|---|---|
| 1012 | 1903 | 2910 | 4956 | 6278 | 22,230 |

**Plan:** We examine the values to find a large jump in ionization energy, which occurs after all valence electrons have been removed. Then we refer to the periodic table to find the Period 3 element with this many valence electrons and write its electron configuration.

**Solution:** The large jump in ionization energy occurs after $IE_5$, so X has five valence electrons. Thus, X is **phosphorus** (P; $Z = 15$) in Period 3, Group 5A(15). Its electron configuration is $1s^22s^22p^63s^23p^3$.

FOLLOW-UP PROBLEM 8.5

Given the following ionization energies (in kJ/mol) for element Q in Period 3, name the element and write its electron configuration.

| $IE_1$ | $IE_2$ | $IE_3$ | $IE_4$ | $IE_5$ | $IE_6$ |
|---|---|---|---|---|---|
| 577 | 1816 | 2744 | 11,576 | 14,829 | 18,375 |

### Trends in Electron Affinity

The **electron affinity (EA)** is the energy change accompanying a mole of electrons being *added* to a mole of gaseous atoms or ions. As with ionization energy, there is a first electron affinity, a second, and so forth. The *first electron affinity* ($EA_1$) accompanies the formation of a mole of 1− gaseous ions:

$$\text{Atom}(g) + e^- \rightarrow \text{ion}^-(g) \quad \Delta E = EA_1$$

In most cases, *energy is released when the first electron is added* because it is attracted to the atom's nuclear charge. Thus, $EA_1$ is usually negative (just as $\Delta H$ for an exothermic reaction is negative).* The second electron affinity ($EA_2$), on the other hand, is always positive because it *requires* energy to add another electron to an anion.

It is more difficult to measure electron affinities than ionization energies. Some values have only recently been determined, and others are being corrected with more accurate methods. Moreover, factors other than $Z_{eff}$ and atomic radius affect the values, so the trends in electron affinity are less consistent, when they appear at all (Figure 8.16). For instance, we might expect electron affinities to decrease smoothly down a group (smaller negative number) because the attracting nucleus is farther away from the electron being pulled into an outer orbital. Only Group 1A(1), however, exhibits this expected behavior. Similarly, we might expect a regular increase in electron

*Tables of first electron affinities often list them as positive values, showing the *amount* of energy released, rather than the difference between the final and initial energy values. Keep this convention in mind when researching these values in reference texts.

| 1A (1) | | | | | | | 8A (18) |
|---|---|---|---|---|---|---|---|
| **H** −72.8 | **2A** (2) | **3A** (13) | **4A** (14) | **5A** (15) | **6A** (16) | **7A** (17) | **He** (+21) |
| **Li** −59.6 | **Be** (+241) | **B** −26.7 | **C** −122 | **N** 0 | **O** −141 | **F** −328 | **Ne** (+29) |
| **Na** −52.9 | **Mg** (+230) | **Al** −42.5 | **Si** −134 | **P** −72.0 | **S** −200 | **Cl** −349 | **Ar** (+34) |
| **K** −48.4 | **Ca** (+156) | **Ga** −28.9 | **Ge** −119 | **As** −78.2 | **Se** −195 | **Br** −325 | **Kr** (+39) |
| **Rb** −46.9 | **Sr** (+167) | **In** −28.9 | **Sn** −107 | **Sb** −103 | **Te** −190 | **I** −295 | **Xe** (+40) |
| **Cs** −45.5 | **Ba** (+52) | **Tl** −19.3 | **Pb** −35.1 | **Bi** −91.3 | **Po** −183 | **At** −270 | **Rn** (+41) |

**FIGURE 8.16**
**Electron affinities of the main-group elements.** The electron affinities of the main-group elements (in kJ/mol) are shown (with estimated values in parentheses). Negative values indicate that energy is released when the anion forms. Positive values, which occur in Groups 2A(2) and 8A(18), indicate that energy is absorbed to form the anion; in fact, these anions are unstable and the values are estimated.

affinities across a period (larger negative number) because size decreases and the increasing $Z_{eff}$ would attract the electron being added more strongly. The increasing trend is there, but it certainly is not regular, with the variations arising from changes in sublevel energy and electron repulsion.

Three key points emerge from an examination of ionization energy and electron affinity values:

1. Elements in Groups 6A(16) and especially 7A(17) (halogens) have high ionization energies and highly negative (exothermic) electron affinities. These elements lose electrons with difficulty and attract electrons strongly. Therefore, *in their ionic compounds, they form negative ions.*

2. Elements in Groups 1A(1) and 2A(2) have low ionization energies and either slightly negative or positive (endothermic) electron affinities. That is, these elements lose electrons readily and do not attract them. *In their ionic compounds, they form positive ions.*

3. The noble gases, Group 8A(18), have very high ionization energies, which reflect the difficulty of removing an electron, and positive (endothermic) electron affinities, which reflect the difficulty of adding an electron. In general, *these elements neither lose nor gain electrons.*

### Section Summary

Trends in three atomic properties are summarized in Figure 8.17. Atomic size increases down a main group and decreases across a period. Across a transition series, size remains relatively constant. First ionization energy (the energy required to remove the outermost electron from a mole of gaseous atoms) is inversely related to atomic size: $IE_1$ decreases down a main group and increases across a period. Successive ionization energies show a large increase between outer and inner (core) electrons. Electron affinity (the energy involved in adding an electron to a mole of gaseous atoms) shows many variations from expected trends. Group 1A(1) and 2A(2) elements have low IEs and small EAs, whereas Group 6A(16) and 7A(17) elements have high IEs and large EAs.

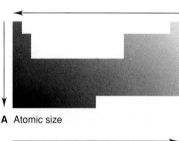

**A** Atomic size

**B** Ionization energy

**C** Electron affinity

**FIGURE 8.17**
**Trends in three atomic properties.** Periodic trends are depicted as gradations in shading on miniature periodic tables, with arrows indicating the direction of general increase in a group or period. **A,** Atomic size. **B,** Ionization energy. **C,** Electron affinity (the dashed arrow implies the numerous exceptions to clear trends).

## 8.4    The Connection Between Atomic Structure and Chemical Reactivity

In the final section of the chapter, we take our first look at how the three atomic properties we just examined affect element behavior, a topic that we consider in detail in the next few chapters. In this section, you'll see how atomic size, ionization energy, and electron affinity influence metallic behavior, how they determine the type of ion, if any, an element can form, and how electron configuration relates to magnetic properties.

### Trends in Metallic Behavior

Metals are found in the left and lower three-quarters of the periodic table and nonmetals in the upper right quarter, with the metalloids in the diagonal region between them. Metal, metalloid, and nonmetal are terms that we assign to elements based on an array of properties. *Metals* are typically shiny solids that have moderate to high melting points, are good thermal and electrical conductors, and tend to lose electrons during reaction with nonmetals. *Nonmetals* are dull substances, most of which have relatively low melting points, are poor thermal and electrical conductors, and tend to gain electrons during reaction with metals. *Metalloids* have properties between those of the other two classes.

These are, however, idealized classifications of real substances, each with its own behavior. As you've seen, there are often exceptions to our pigeonholing of the elements into neat categories. For example, carbon, in the form of graphite, is a good electrical conductor even though it is a nonmetal; iodine, another nonmetal, is a shiny solid. Gallium and cesium, both metals, melt at the temperature of your hand, and mercury is a liquid. Keeping in mind that exceptions exist, we can make several generalizations about metallic behavior throughout the periodic table (Figure 8.18).

**Relative tendency to lose electrons.** A common behavior of metals is their tendency to lose electrons during chemical reactions. Metals have low ionization energies relative to those of nonmetals. *As ionization energy decreases down a group, metallic behavior increases.* This change is particularly apparent in Groups 3A(13) through 6A(16), which contain more than one class of element. In Group 5A(15), for example, nitrogen (N) is a gaseous nonmetal and phosphorus (P) a solid nonmetal. Arsenic (As) and antimony (Sb) are metalloids, with antimony the more metallic of the two. Bismuth (Bi), the largest member, has the classic properties of a metal. Even in Group 2A(2), which consists entirely of metals, we see an increase from top to bottom in the tendency to form ions in reaction with nonmetals. Beryllium (Be), for example, forms covalent compounds with nonmetals, whereas the compounds of barium (Ba) are ionic.

As we move across a period, it becomes more difficult to lose an electron (IE increases) and easier to gain one (EA becomes more negative). Thus, with regard to monatomic ions, elements at the beginning of a period tend to form cations with the *previous* noble gas configuration, and those at the end of a period tend to form anions with the *next* noble gas configuration. In other words, *metallic behavior decreases across a period.* In Period 3, for example, the metals sodium and magnesium exist naturally as $Na^+$ and $Mg^{2+}$ ions in oceans, minerals, and organisms. Aluminum is largely metallic, forming the $Al^{3+}$ ion in some of its compounds but bonding covalently in many others. Silicon (Si) is a metalloid that does not occur as a monatomic ion. Phosphorus is a nonmetal, existing as the $P^{3-}$ ion in a few compounds. Sulfur is a nonmetal, with the sulfide ion ($S^{2-}$) appearing in many com-

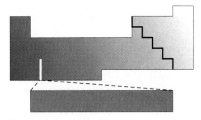

**FIGURE 8.18**
**Trend in metallic character throughout the periodic table.** The gradation in metallic character among the elements is depicted as a gradation in shading from bottom left to top right. Elements that behave as metals appear in the left and lower three-quarters of the table. (Hydrogen appears next to helium in this periodic table to emphasize that it is a nonmetal.)

pounds. Diatomic chlorine ($Cl_2$) is a gaseous nonmetal that avidly attracts electrons to form the $Cl^-$ ion.

**Acid-base behavior of the element oxides.** Metals are also distinguished from nonmetals by the acid-base behavior of their oxides. Metals transfer electrons to oxygen and form *ionic oxides that act as bases:* they produce $OH^-$ ions in aqueous solution and react with acids. Nonmetals share electrons with oxygen to form *covalent oxides that act as acids:* they produce $H^+$ ions in aqueous solution and react with bases (Figure 8.19). Some metalloids and metals form oxides that are **amphoteric:** they can act as an acid and as a base.

Let's return to the elements in Group 5A(15) and Period 3, this time to examine the acid-base behavior of their oxides (Figure 8.20). In Group 5A, a common oxide of nitrogen, $N_2O_5$, forms nitric acid, a strong acid:

$$N_2O_5(s) + H_2O(l) \rightarrow 2HNO_3(aq)$$

Tetraphosphorus decaoxide, $P_4O_{10}$, forms a weaker acid:

$$P_4O_{10}(s) + 6H_2O(l) \rightarrow 4H_3PO_4(aq)$$

The oxide of the metalloid arsenic is weakly acidic, whereas that of the metalloid antimony is weakly basic. Bismuth, the most metallic member, forms the basic bismuth oxide:

$$Bi_2O_3(s) + 3H_2O(l) \rightarrow 2Bi(OH)_3(aq)$$

Thus, as the elements become *more* metallic *down a group, their oxides become more basic.*

In Period 3, the metals sodium and magnesium form the strongly basic oxides $Na_2O$ and $MgO$. Aluminum oxide ($Al_2O_3$) is amphoteric, reacting with both acids and bases:

$$Al_2O_3(s) + 6HCl(aq) \rightarrow 2AlCl_3(aq) + 3H_2O(l)$$

$$Al_2O_3(s) + 2NaOH(aq) + 3H_2O(l) \rightarrow 2NaAl(OH)_4(aq)$$

Silicon dioxide is weakly acidic, forming a salt and water with base:

$$SiO_2(s) + 2NaOH(aq) \rightarrow Na_2SiO_3(aq) + H_2O(l)$$

The common oxides of phosphorus, sulfur, and chlorine form acids of increasing strength: $H_3PO_4$, $H_2SO_4$, and $HClO_4$. Thus, as the elements become *less* metallic *across a period, their oxides become more acidic.*

A

B

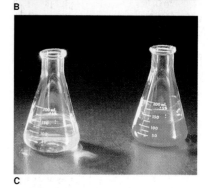

C

**FIGURE 8.19**
Acid-base behavior of a metal oxide and a nonmetal oxide. **A,** When the metal calcium reacts with $O_2$, ionic CaO forms. **B,** When the nonmetal phosphorus reacts with $O_2$, covalent $P_4O_{10}$ forms. **C,** This acid-base indicator turns yellow (acid color) in an aqueous solution of $P_4O_{10}$ *(left)* because this oxide is acidic. The indicator turns pink (base color) in an aqueous solution of CaO *(right)* because this oxide is basic.

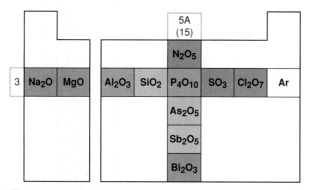

| | | | | 5A (15) | | | |
|---|---|---|---|---|---|---|---|
| | | | | $N_2O_5$ | | | |
| 3 | $Na_2O$ | $MgO$ | $Al_2O_3$ | $SiO_2$ | $P_4O_{10}$ | $SO_3$ | $Cl_2O_7$ | Ar |
| | | | | $As_2O_5$ | | | |
| | | | | $Sb_2O_5$ | | | |
| | | | | $Bi_2O_3$ | | | |

**FIGURE 8.20 The trend in acid-base behavior of element oxides.** The trend in acid-base behavior for common oxides in Group 5A(15) and Period 3 is shown as a gradation in color (red = acidic; blue = basic). Note that metals form basic oxides, and nonmetals form acidic oxides. Aluminum forms an oxide (purple) that can act as an acid and as a base. Thus, as atomic size increases, oxide basicity increases.

## Properties of Monatomic Ions

Our attention so far has focused on the reactants involved in electron loss and gain—the atoms. Now we discuss the products of these processes—the ions. In particular, we examine ionic electron configurations, why only certain elements form ions in reactions, ion magnetic properties, and ionic sizes relative to the parent atoms.

**Electron configurations of main-group ions.** In Chapter 2, we examined lists of common monatomic ions. Now we can discuss *why* a sodium ion, for instance, is $Na^+$ and not $Na^{2+}$, or *why* a fluoride ion is $F^-$ and not $F^+$, and *why* elements like boron and carbon do not exist as ions in their compounds.

The key to answering these questions lies in the very low reactivity of the noble gases. These elements have extremely high ionization energies and positive (endothermic) electron affinities. In other words, they typically do not lose or gain electrons. This chemical stability results from their *filled* outer energy level ($ns^2np^6$). *Main-group elements that readily form ions either lose or gain electrons to attain a filled outer level and thus a noble gas configuration.* Their ions are said to be **isoelectronic** (Greek *iso*, "same") with the nearest noble gas.

When an alkali metal [Group 1A(1)] loses its single valence electron, it becomes isoelectronic with the *previous* noble gas. The $Na^+$ ion, for example, is isoelectronic with neon:

$$Na\ (1s^22s^22p^63s^1) \longrightarrow e^- + Na^+\ (1s^22s^22p^6)$$
$$\text{isoelectronic with Ne } (1s^22s^22p^6)$$

When a halogen [Group 7A(17)] adds a single electron to the seven in its valence shell, it becomes isoelectronic with the *next* noble gas. Bromide ion, for example, is isoelectronic with krypton:

$$Br\ ([Ar]\ 4s^24p^5) + e^- \longrightarrow Br^-\ ([Ar]\ 4s^24p^6)$$
$$\text{isoelectronic with Kr } ([Ar]\ 4s^24p^6)$$

As you'll see in Chapter 9, the energy needed to remove one electron from Group 1A(1) metals or two electrons from Group 2A(2) metals to attain the electron configuration of the previous noble gas is supplied during their reactions with nonmetals. Removing more electrons from these atoms, however, means removing core electrons, which requires 5 to 10 times as much energy as is available in a typical reaction. This is the reason why $Na^{2+}$ or $Mg^{3+}$ ions do not form. A similar limitation exists for electrons gained. Adding more than one electron to a halogen atom requires too much energy to reach the next energy level, so this never occurs during a chemical change.

Some elements rarely, if ever, form ions for the same reason: the energy cost is much too high. Carbon, for instance, would have to lose four electrons to form $C^{4+}$ and attain the He configuration or gain four to form $C^{4-}$ and attain the Ne configuration. Both processes require much more energy than is usually available in a chemical change. (Such multivalent monatomic ions *are* observed in the spectra of stars, where temperatures exceed $10^6$ K.) As we discuss in Chapter 9, carbon and other atoms that do not form ions attain a filled shell by *sharing* electrons through covalent bonding.

How do the larger metallic members of Groups 3A(13), 4A(14), and 5A(15) form cations? They cannot possibly lose enough electrons to attain a noble gas configuration because transition elements are present. For example, for tin (Sn; $Z = 50$) to be isoelectronic with krypton (Kr; $Z = 36$), the previous noble gas, it would have to lose 14 electrons, an energetically impossible feat. Instead, it loses far fewer electrons and attains two different stable configurations. In the tin(IV) ion ($Sn^{4+}$), the element empties its outer energy level, thereby attaining the stability of empty $5s$ and $5p$ sublevels and a filled inner $4d$ sublevel. This $(n - 1)d^{10}$ configuration is called a **pseudo–noble gas configuration:**

$$Sn~([Kr]~5s^2 4d^{10} 5p^2) \rightarrow Sn^{4+}~([Kr]~4d^{10}) + 4e^-$$

On the other hand, in the more common tin(II) ion ($Sn^{2+}$), the metal loses only the $5p$ electrons and attains the stability of filled $5s$ and $4d$ sublevels:

$$Sn~([Kr]~5s^2 4d^{10} 5p^2) \rightarrow Sn^{2+}~([Kr]~5s^2 4d^{10}) + 2e^-$$

The $ns^2$ electrons retained are sometimes called an *inert pair* because they seem difficult to remove. Thallium, lead, and bismuth—the most metallic members of Groups 3A(13), 4A(14), and 5A(15), respectively—commonly exhibit this behavior of forming ions that retain the $ns^2$ pair of electrons.

SAMPLE PROBLEM 8.6 _____

## Predicting the Charge and Writing Electron Configurations of Main-Group Ions

**Problem:** Write reactions with condensed electron configurations to show the formation of the common ions of the following elements:
**(a)** Iodine ($Z = 53$)   **(b)** Potassium ($Z = 19$)   **(c)** Indium ($Z = 49$)
**Plan:** We identify the element's position in the periodic table and keep two generalizations in mind:
• Ions of elements in Groups 1A(1), 2A(2), 6A(16), and 7A(17) are typically isoelectronic with the nearest noble gas.
• Metals in Groups 3A(13) to 5A(15) can lose their $ns$ or their $ns$ and $np$ electrons.
**Solution: (a)** Iodine is in Group 7A(17), so it gains one electron to be isoelectronic with xenon:

$$\mathbf{I~([Kr]~5s^2 4d^{10} 5p^5) + e^- \rightarrow I^-~([Kr]~5s^2 4d^{10} 5p^6)}~\text{(same as Xe)}$$

**(b)** Potassium is in Group 1A(1), so it loses one electron to be isoelectronic with argon. This loss occurs so readily that K reacts vigorously even in cold water (photo):

$$\mathbf{K~([Ar]~4s^1) \rightarrow K^+~([Ar]) + e^-}$$

**(c)** Indium is a metal in Group 3A(13), so it loses one electron to form $In^+$ (inert pair) or three to form $In^{3+}$ (pseudo–noble gas):

$$\mathbf{In~([Kr]~5s^2 4d^{10} 5p^1) \rightarrow In^+~([Kr]~5s^2 4d^{10}) + e^-}$$
$$\mathbf{In~([Kr]~5s^2 4d^{10} 5p^1) \rightarrow In^{3+}~([Kr]~4d^{10}) + 3e^-}$$

**Check:** Be sure that the number of electrons in the ion electron configuration plus those gained or lost equals $Z$.

FOLLOW-UP PROBLEM 8.6
Write reactions with condensed electron configurations of the common ions formed from the following elements: **(a)** Ba ($Z = 56$); **(b)** O ($Z = 8$); **(c)** Pb ($Z = 82$).

$$2K(s) + 2H_2O(l) \rightarrow 2KOH(aq) + H_2(g)$$

_____

**Electron configurations and magnetic properties of transition metal ions.** In contrast to most main-group ions, *transition metal ions rarely attain a noble gas configuration;* once again, the energy costs are too high. The exceptions in Period 4 are scandium, which forms $Sc^{3+}$, and possibly titanium, which forms $Ti^{4+}$ in some of its compounds. The typical behavior of a transition element is to *form more than one cation by losing all of its ns and some of its (n − 1)d electrons.*

We focus here on the Period 4 transition series, but these general points hold for the other series as well. As protons are added to the nuclei and electrons to the orbitals, the $4s$ orbital in K and Ca is lower in energy than the *empty* $3d$ orbital. When we reach the transition series, however, and the $3d$ orbitals begin to fill, their electrons become increasingly attracted to the greater nuclear charge. In effect, a *changeover in orbital energy* takes place such that the energy of the $3d$ orbital becomes *lower* than that of the $4s$. Thus, although the $4s$ orbital fills before the $3d$ in the aufbau process, experiment shows that *the 4s electrons are lost first in the formation of Period 4 transition metal ions.* Indeed, except for the lanthanides, *electrons with the highest n value are always lost first.* The sequence in which electrons are removed during the formation of transition metal ions may seem unexpected, but it is important to remember that the aufbau process is a hypothetical method of *adding* an electron and a proton to form each succeeding element; it does not refer to the process of *removing* only electrons to form ions.

Since we cannot *see* electrons occupying different orbitals in an atom or ion, how do we know that a particular electron configuration is correct? Analysis of atomic spectra is the most important method used to determine configuration, but the magnetic properties of transition metals and their compounds can support or refute a configuration obtained from spectral analysis. Recall that an electron's intrinsic spin generates a tiny magnetic field, and that passing a beam of H atoms through an external nonuniform magnetic field splits the beam because the electrons spin in opposite directions (see Figure 8.1). The element used in the original 1921 split-beam experiment was silver:

$$Ag \; (Z = 47) \qquad [Kr] \; 5s^1 4d^{10}$$

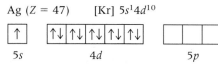

Note the unpaired $5s$ electron. Only species (atoms, ions, or molecules) with one or more *unpaired* electrons are affected by the external field. For example, a beam of cadmium atoms, the element after silver, would not be split because it has two *paired* $5s$ electrons (Cd: $[Kr] \; 5s^2 4d^{10}$).

**Paramagnetism** is the tendency of a species with unpaired electrons to be attracted by an external magnetic field. A species with all its electrons paired exhibits **diamagnetism:** it is not attracted (and in fact is slightly repelled) by a magnetic field (Figure 8.21). Many transition metals and their compounds are paramagnetic because their atoms and ions have unpaired electrons. For instance, experiment shows that compounds of the $Ti^{2+}$ ion are paramagnetic. For this to be so, the $4s$ electrons *cannot* be paired in $Ti^{2+}$. Indeed, these electrons are lost when $Ti^{2+}$ forms:

$$Ti \; ([Ar] \; 4s^2 3d^2) \rightarrow Ti^{2+} \; ([Ar] \; 3d^2) + 2e^-$$

The partial orbital diagrams are

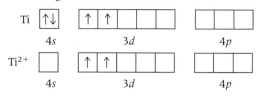

**FIGURE 8.21**

**Apparatus for measuring the magnetic behavior of a sample.** The substance is weighed on a very sensitive balance in the absence of an external magnetic field. **A,** If a substance is diamagnetic (contains no unpaired electrons), its apparent mass is unaffected (or slightly reduced) when the magnetic field is "on." **B,** If a substance is paramagnetic (contains unpaired electrons), its apparent mass increases because the balance arm feels an additional force when the field is "on." This method is used to estimate the number of unpaired electrons in transition metal compounds.

Iron, which is already paramagnetic, forms the highly paramagnetic $Fe^{3+}$ ion by losing its pair of $4s$ electrons and unpairing one of its $3d$, which relieves some electron repulsions and forms a stable half-filled sublevel:

$$Fe\ ([Ar]\ 4s^2 3d^6) \rightarrow Fe^{3+}\ ([Ar]\ 3d^5) + 3e^-$$

| | $4s$ | $3d$ | $4p$ |
|---|---|---|---|
| Fe | [↑↓] | [↑↓][↑][↑][↑][↑] | [ ][ ][ ] |
| $Fe^{3+}$ | [ ] | [↑][↑][↑][↑][↑] | [ ][ ][ ] |

The $Cu^+$ and $Zn^{2+}$ ions are diamagnetic, an observation used to support the electron configurations proposed by spectral analysis. These ions happen to be isoelectronic with each other:

$$Cu\ ([Ar]\ 4s^1 3d^{10}) \rightarrow Cu^+\ ([Ar]\ 3d^{10}) + e^-$$

$$Zn\ ([Ar]\ 4s^2 3d^{10}) \rightarrow Zn^{2+}\ ([Ar]\ 3d^{10}) + 2e^-$$

| | $4s$ | $3d$ | $4p$ |
|---|---|---|---|
| $Cu^+$ or $Zn^{2+}$ | [ ] | [↑↓][↑↓][↑↓][↑↓][↑↓] | [ ][ ][ ] |

SAMPLE PROBLEM 8.7 ——————————————————————————

**Writing Electron Configurations and Predicting Magnetic Behavior of Transition Metal Ions**

**Problem:** Write reactions with condensed electron configurations to show the formation of the following transition metal ions, and predict whether the ion is paramagnetic:
**(a)** $Mn^{2+}$ ($Z = 25$)   **(b)** $Cr^{3+}$ ($Z = 24$)   **(c)** $Hg^{2+}$ ($Z = 80$)
**Plan:** We first write the condensed electron configuration of the atom, noting the irregularity in Cr in part (b), and then remove enough electrons—$ns$ electrons first—to give the ion charge. If unpaired electrons are present, the ion is paramagnetic.
**Solution: (a)** $Mn\ ([Ar]\ 4s^2 3d^5) \rightarrow Mn^{2+}\ ([Ar]\ 3d^5) + 2e^-$
There are five unpaired $e^-$, so $Mn^{2+}$ is **paramagnetic.**
**(b)** $Cr\ ([Ar]\ 4s^1 3d^5) \rightarrow Cr^{3+}\ ([Ar]\ 3d^3) + 3e^-$
There are three unpaired $e^-$, so $Cr^{3+}$ is **paramagnetic.**
**(c)** $Hg\ ([Xe]\ 6s^2 4f^{14} 5d^{10}) \rightarrow Hg^{2+}\ ([Xe]\ 4f^{14} 5d^{10}) + 2e^-$
There are no unpaired $e^-$, so $Hg^{2+}$ is ***not* paramagnetic** (diamagnetic).
**Check:** We removed the $ns$ electrons first, and the sum of the lost electrons and those in the ion electron configuration equals $Z$.

FOLLOW-UP PROBLEM 8.7
Write the condensed electron configuration of the following transition metal ions and predict whether or not they are paramagnetic:
**(a)** $V^{3+}$ ($Z = 23$); **(b)** $Ni^{2+}$ ($Z = 28$); **(c)** $La^{3+}$ ($Z = 57$).

---

**Ionic size vs. atomic size.** The **ionic radius** is an estimate of the size of the cation or anion as measured in a crystalline ionic compound. From the relation between effective nuclear charge and atomic size, we can predict the size of an ion relative to the size of the atom from which it forms. When a cation forms, outer electrons are removed and the resulting decrease in electron-electron repulsions allows the nuclear charge to pull the remaining electrons closer. Thus, *cations are smaller than their parent atoms.* When an anion forms, electrons are added to the outer level, and the in-

**FIGURE 8.22**

**Ionic vs. atomic radii.** The atomic radius *(gray half-circle)* and ionic radius *(colored half-circle)* of some main-group elements are arranged in periodic table format (with all values in picometers). Note that metals form positive ions that are smaller than the atoms, whereas nonmetals form negative ions that are larger. The dashed outline includes ions of Period 2 nonmetals and Period 3 metals that are isoelectronic with each other. Note the size decrease from anion to cation.

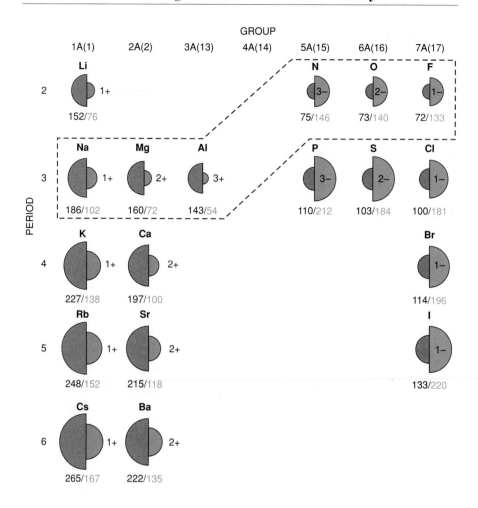

crease in repulsion causes the outer electrons to occupy more space. Thus, *anions are larger than their parent atoms.* Figure 8.22 shows the radii of some common main-group monatomic ions relative to their parent atoms.

As expected, *ionic size increases down a group* for the same reason that atomic size does: the number of electron levels increases. Across a period, however, several factors are involved, including the change from metal cations to nonmetal anions, and changes in nuclear charge and electron repulsion. As we would predict, a small decrease in ionic size occurs among the cations, followed by a tremendous increase in size when we reach the anions, and then another small decrease in size occurs among the anions.

In Period 3 (Na through Cl), for example, we see the effect of increasing nuclear charge on the ions as we progress from left to right: $Na^+$ is larger than $Mg^{2+}$, which is larger than $Al^{3+}$. Then comes the jump from cations to anions. The number of protons increases among the anions, so $Z_{eff}$ increases and size shrinks: $P^{3-}$ is larger than $S^{2-}$, which is larger than $Cl^-$. The effects of electron repulsion and nuclear charge are striking when we compare the Period 3 cations with the Period 2 anions (dashed outline in Figure 8.22). All are isoelectronic with neon, but the anions are much larger than the cations; the pattern is $3- > 2- > 1- > 1+ > 2+ > 3+$.

When an element forms more than one cation, as do the transition elements and the metals in Groups 3A(13) to 5A(15), *the greater the charge of the cation, the smaller its size.* This pattern follows directly from our previous reasoning. Consider the two iron ions, $Fe^{2+}$ and $Fe^{3+}$. The number of protons is the same, but $Fe^{3+}$ has one fewer $d$ electron. Thus, electron repulsions are reduced somewhat, the $Z_{eff}$ increases, so $Fe^{3+}$ is smaller than $Fe^{2+}$.

SAMPLE PROBLEM 8.8 _____

### Ranking Ions According to Size

**Problem:** Rank each set of ions in order of *decreasing* size, and explain your ranking:
**(a)** $Ca^{2+}$, $Sr^{2+}$, $Mg^{2+}$
**(b)** $K^+$, $S^{2-}$, $Cl^-$
**(c)** $Au^+$, $Au^{3+}$
**Plan:** We find the position of each element in the periodic table and apply the ideas presented in the text:
• Size increases down a group.
• Size decreases across a period but increases from cation to anion.
• Size decreases with increasing positive (or decreasing negative) charge in an iso-electronic series.
• Cations of the same element decrease in size as charge increases.
**Solution:** **(a)** Since $Mg^{2+}$, $Ca^{2+}$, and $Sr^{2+}$ are all from Group 2A(2), they decrease in size up the group: $\mathbf{Sr^{2+} > Ca^{2+} > Mg^{2+}}$.
**(b)** The ions $K^+$, $S^{2-}$, and $Cl^-$ are isoelectronic. $S^{2-}$ has lower $Z_{eff}$ than $Cl^-$, so it is larger. $K^+$ is a cation, and has the highest $Z_{eff}$, so it is smaller: $\mathbf{S^{2-} > Cl^- > K^+}$.
**(c)** $Au^+$ has a lower charge than $Au^{3+}$, so it is larger: $\mathbf{Au^+ > Au^{3+}}$.

FOLLOW-UP PROBLEM 8.8
Rank the ions in each set in order of *increasing* size: **(a)** $Cl^-$, $Br^-$, $F^-$; **(b)** $Na^+$, $Mg^{2+}$, $F^-$; **(c)** $Cr^{2+}$, $Cr^{3+}$.

### Section Summary

Metallic behavior correlates with large atomic size and low ionization energy. Thus, metallic behavior increases down a group and decreases across a period. Metal oxides are basic and nonmetal oxides acidic, so oxides become more acidic across a period and more basic down a group. Many main-group elements form ions that are isoelectronic with the nearest noble gas. Metals in Groups 3A(13) to 5A(15) lose either their $np$ electrons or both their $ns$ and $np$ electrons. Transition metals lose $ns$ electrons before $(n-1)d$ electrons and commonly form more than one ion. Many transition metal atoms and their ions are paramagnetic because they have unpaired electrons. Ions increase in size down a group and change from small cations to large anions across a period.

### Chapter Perspective

*This chapter is our springboard to understanding the chemistry of the elements. Now we have begun to see that recurring electron configurations lead to trends in atomic properties, which in turn lead to trends in chemical behavior. With this insight, we can go on to investigate how atoms bond (Chapter 9), how molecular shapes arise (Chapter 10), how the physical properties of liquids and solids emerge from* _____ *(Chapter 11), and how those properties influence the solution* _____ *). We will stop briefly to review these ideas and gain a perspective* _____ *(Interchapter). Then we can survey elemental behavior in greater* _____ *and see how it applies to the remarkable diversity of organic com-* _____ *4). Within this group of chapters, you will see chemical models* _____ *ical facts.*

# For Review and Reference

## Key Terms

electron configuration
periodic law

**SECTION 8.1**
spin quantum number
  $(m_s)$
Pauli exclusion principle
shielding
effective nuclear charge
  $(Z_{eff})$

penetration

**SECTION 8.2**
aufbau principle
orbital diagram
Hund's rule
transition elements
inner (core) electrons
outer electrons
valence electrons

inner transition elements
lanthanides (rare earths)
actinides

**SECTION 8.3**
metallic radius
covalent radius
ionization energy (IE)
electron affinity (EA)

**SECTION 8.4**
amphoteric
isoelectronic
pseudo–noble gas
  configuration
paramagnetism
diamagnetism
ionic radius

## Answers to Follow-up Problems

**8.1** The element has eight electrons, so $Z = 8$: oxygen.

   1s    2s      2p
$n = 2$, $l = 1$, $m_l = 0$, $m_s = +\frac{1}{2}$

**8.2** (a) For Ni, $1s^2 2s^2 2p^6 3s^2 3p^6 4s^2 3d^8$; [Ar] $4s^2 3d^8$

    4s       3d         4p

  Ni has 18 inner electrons.

  (b) For Sr, $1s^2 2s^2 2p^6 3s^2 3p^6 4s^2 3d^{10} 4p^6 5s^2$; [Kr] $5s^2$

    5s       4d         5p

  Sr has 36 inner electrons.

  (c) For Po,
    $1s^2 2s^2 2p^6 3s^2 3p^6 4s^2 3d^{10} 4p^6 5s^2 4d^{10} 5p^6 6s^2 4f^{14} 5d^{10} 6p^4$

  [Xe] $6s^2 4f^{14} 5d^{10} 6p^4$

    6s      6p

  Po has 78 inner electrons.

**8.3** (a) Cl < Br < Se; (b) Xe < I < Ba
**8.4** (a) Sn < Sb < I; (b) Ba < Sr < Ca
**8.5** Q is aluminum: $1s^2 2s^2 2p^6 3s^2 3p^1$
**8.6** (a) Ba ([Xe] $6s^2$) → $Ba^{2+}$ ([Xe]) + $2e^-$
  (b) O ([He] $2s^2 2p^4$) + $2e^- →$
                    $O^{2-}$ ([He] $2s^2 2p^6$) (same as Ne)
  (c) Pb ([Xe] $6s^2 4f^{14} 5d^{10} 6p^2$) →
                  $Pb^{2+}$ ([Xe] $6s^2 4f^{14} 5d^{10}$) + $2e^-$
    Pb ([Xe] $6s^2 4f^{14} 5d^{10} 6p^2$) →
                $Pb^{4+}$ ([Xe] $4f^{14} 5d^{10}$) + $4e^-$
**8.7** (a) $V^{3+}$: [Ar] $3d^2$; paramagnetic
  (b) $Ni^{2+}$: [Ar] $3d^8$; paramagnetic
  (c) $La^{3+}$: [Xe]; not paramagnetic (diamagnetic)
**8.8** (a) $F^- < Cl^- < Br^-$; (b) $Mg^{2+} < Na^+ < F^-$;
  (c) $Cr^{3+} < Cr^{2+}$

## Sample Problem Titles

**8.1** Determining Quantum Numbers from Orbital Diagrams (p. 295)
**8.2** Determining Electron Configurations (p. 302)
**8.3** Ranking Elements According to Size (p. 306)
**8.4** Ranking Elements According to First Ionization Energy (p. 309)

**8.5** Identifying an Element from Its Successive Ionization Energies (p. 310)
**8.6** Predicting the Charge and Writing Electron Configurations of Main-Group Ions (p. 315)
**8.7** Writing Electron Configurations and Predicting Magnetic Behavior of Transition Metal Ions (p. 317)
**8.8** Ranking Ions According to Size (p. 319)

# Problems

Problems with a green number are answered at the back of the text. Most sections include three categories of problems separated by a green rule: concept review questions, *paired* skill-building exercises, and problems in a relevant context.

## Characteristics of Many-Electron Atoms

**8.1** What would be your scientific opinion of a claim that a new element had been discovered that fit between tin (Sn) and antimony (Sb) in the periodic table?

**8.2** In his study of atomic x-ray spectra, the English physicist Henry Moseley made a discovery that replaced atomic mass as the criterion for ordering the elements. By what criterion are the elements ordered in the periodic table now? Give an example of a sequence of element order that was confirmed by Moseley's findings.

**8.3** Summarize the rules for the allowable values of the four quantum numbers of an electron in an atom.

**8.4** State the Pauli exclusion principle in your own ~~w~~ What does it imply about the number and ~~spi~~ ~~trons~~ in an atomic orbital?

**8.5** What is the key distinction between sublevel energies in one-electron species and in many-electron species? What factors lead to this distinction?

**8.6** Define shielding and effective nuclear charge. What is the connection between the two?

**8.7** What is the penetration effect? Use it to explain the difference in relative orbital energies of a $3p$ electron and a $3d$ electron in the same atom.

**8.8** How many electrons in an atom can have each of the following quantum number or sublevel designations? (a) $n = 2$, $l = 1$; (b) $3d$; (c) $4s$

**8.9** How many electrons in an atom can have each of the following quantum number or sublevel designations? (a) $4p$; (b) $n = 3$, $l = 1$, $m_l = +1$; (c) $n = 5$, $l = 3$

**The Quantum-Mechanical Atom and the Periodic Table**

(Sample Problems 8.1 and 8.2)

**8.10** State the periodic law in your own words, and explain its connection to electron configuration. (Use the relation between Na and K in your explanation.)

**8.11** State Hund's rule in your own words, and show its application in the orbital diagram of the nitrogen atom.

**8.12** How does the aufbau principle, in connection with the periodic law, lead to the format of the periodic table?

**8.13** For the main-group elements, are outer electron configurations similar or different within a group? Within a period? Explain.

**8.14** For which blocks of elements are outer electrons the same as valence electrons? For which are $d$ electrons often included among valence electrons?

**8.15** What is the general rule for the total electron capacity of an energy level? Apply it to determine the capacity of the fourth level.

**8.16** Write a full set of quantum numbers for the highest energy electron in the ground-state B atom.

**8.17** Write a full set of quantum numbers for the electron gained when an $F^-$ ion forms from an F atom.

**8.18** Write a full set of quantum numbers for the following:
(a) The outermost electron in an Rb atom
(b) The electron gained when an $S^-$ ion becomes an $S^{2-}$ ion
(c) The electron lost when an Ag atom ionizes

**8.19** Write a full set of quantum numbers for the following:
(a) The outermost electron in an Li atom
(b) The electron gained when a Br atom becomes a Br⁻ ion
(c) The electron lost when a Cs atom ionizes

Write a full ground-state electron configuration for: (a) Rb; (b) Ge; (c) Ar; (d) Br; (e) Mg.
Write a full ground-state electron configuration for: (a) Cl; (b) Si; (c) Sr; (d) Kr; (e) Cs.

**8.22** Write the symbol, group number, and period number of the element represented by each of the following condensed electron configurations: (a) [He] $2s^22p^4$; (b) [Ne] $3s^23p^3$; (c) [Kr] $5s^24d^{10}$; (d) [Ar] $4s^23d^8$.

**8.23** Write the symbol, group number, and period number of the element represented by each of the following condensed electron configurations: (a) [Ne] $3s^23p^5$; (b) [Ar] $4s^23d^{10}4p^3$; (c) [Ar] $4s^23d^5$; (d) [Kr] $5s^24d^2$.

**8.24** Draw an orbital diagram showing valence electrons, and write the condensed ground-state electron configuration for each of the following elements: (a) Ti; (b) Cl; (c) V; (d) Ba; (e) Co.

**8.25** Draw an orbital diagram showing valence electrons, and write the condensed ground-state electron configuration for each of the following elements: (a) Mn; (b) P; (c) Fe; (d) Ga; (e) Zn.

**8.26** How many inner electrons, outer electrons, and valence electrons are present in an atom of the following elements? (a) O; (b) Sn; (c) Ca; (d) Fe; (e) Se

**8.27** How many inner electrons, outer electrons, and valence electrons are present in an atom of the following elements? (a) Br; (b) Cs; (c) Cr; (d) Sr; (e) F

**8.28** Identify each element below by its condensed electron configuration, and give the symbols of the other elements in its group:
(a) [He] $2s^22p^1$     (b) [Ne] $3s^23p^4$
(c) [Xe] $6s^25d^1$     (d) [Ar] $4s^23d^{10}4p^4$
(e) [Xe] $6s^24f^{14}5d^2$   (f) [Ar] $4s^23d^5$

**8.29** Identify each element below by its condensed electron configuration, and give the symbols of the other elements in its group:
(a) [He] $2s^22p^2$     (b) [Ar] $4s^23d^3$
(c) [Ne] $3s^23p^3$     (d) [Ar] $4s^23d^{10}4p^2$
(e) [Ar] $4s^23d^7$     (f) [Kr] $5s^24d^5$

**8.30** When an atom in its ground state absorbs energy, it exists in an excited state. Spectral lines are produced when the atom returns to its ground state. The bright-yellow line in the sodium spectrum, for example, is produced by the emission of energy when excited sodium atoms return to their ground state. Write the electron configuration and the orbital diagram of the first excited state of sodium. (*Hint:* the outermost electron is excited.)

**8.31** One reason spectroscopists study excited states is to gain information about the energies of orbitals that are unoccupied in an atom's ground state. Each of the following electron configurations represents an atom in one of its excited states. Identify the element, and write its condensed ground-state electron configuration:
(a) $1s^22s^22p^63s^13p^1$
(b) $1s^22s^22p^63s^23p^44s^1$
(c) $1s^22s^22p^63s^23p^64s^23d^44p^1$
(d) $1s^22s^22p^63s^23p^64s^23d^{10}4p^65s^14d^2$
(e) $1s^22s^22p^53s^1$

## Trends in Some Key Periodic Atomic Properties

(Sample Problems 8.3 to 8.5)

**8.32** Since the exact outer limit of an isolated atom cannot be measured, what criterion can we use to determine atomic radii? What is the difference between a covalent radius and a metallic radius?

**8.33** Explain the relationship between the trends in atomic size and ionization energy (IE) within the main groups in the periodic table.

**8.34** In what region of the periodic table will you find elements with relatively high IEs? With relatively low IEs?

**8.35** Why do successive IEs of a given element always increase? When the difference between successive IEs of a given atom is exceptionally large (for example, between $IE_1$ and $IE_2$ of K), what do we learn about its electron configuration?

**8.36** In a plot of $IE_1$ for the Period 3 elements, why do the values for elements in Groups 3A(13) and 6A(16) drop slightly below the generally increasing line?

**8.37** Which group in the periodic table has elements with high $IE_1$ and very negative first electron affinities ($EA_1$)? What is the charge on the ions that these atoms form?

**8.38** The $EA_2$ of an oxygen atom is positive, even though its $EA_1$ is negative. Why does this change of sign occur? Which other elements exhibit a positive $EA_2$? Explain.

**8.39** How does $d$-electron shielding influence atomic size among the Period 4 transition elements?

---

**8.40** Arrange each set of atoms in order of *increasing* atomic size: (a) Rb, K, Cs; (b) C, O, Be; (c) Cl, K, S; (d) Mg, K, Ca.

**8.41** Arrange each set of atoms in order of *decreasing* atomic size: (a) Ge, Pb, Sn; (b) Sn, Te, Sr; (c) F, Ne, Na; (d) Be, Mg, Na.

**8.42** Arrange each set of atoms in order of *increasing* $IE_1$: (a) Sr, Ca, Ba; (b) N, B, Ne; (c) Br, Rb, Se; (d) As, Sb, Sn.

**8.43** Arrange each set of atoms in order of *decreasing* $IE_1$: (a) Na, Li, K; (b) Be, F, C; (c) Cl, Ar, Na; (d) Cl, Br, Se.

**8.44** Write the full electron configuration of the Period 2 element with the following successive IEs (in kJ/mol): $IE_1 = 801$; $IE_2 = 2427$; $IE_3 = 3659$; $IE_4 = 25,022$; $IE_5 = 32,822$.

**8.45** Write the full electron configuration of the Period 3 element with the following successive IEs (in kJ/mol): $IE_1 = 738$; $IE_2 = 1450$; $IE_3 = 7732$; $IE_4 = 10,539$; $IE_5 = 13,628$.

**8.46** Which element in the following sets would you expect to have the *highest* $IE_2$? (a) Na, Mg, Al; (b) Na, K, Fe

**8.47** Which element in the following sets would you expect to have the *lowest* $IE_3$? (a) Na, Mg, Al; (b) K, Ca, Sc

## The Connection Between Atomic Structure and Chemical Reactivity

(Sample Problems 8.6 to 8.8)

**8.48** List three ways in which metals and nonmetals differ.

**8.49** Summarize the trend in metallic character as a function of position in the periodic table. Is it the same as the trend in atomic size? Is it the same as the trend in ionization energy?

**8.50** Summarize the acid-base behavior of metal and nonmetal oxides. How does oxide acidity change down a group and across a period?

**8.51** What ions are possible for the two largest elements in Group 3A(13)? How does each arise?

**8.52** What is a pseudo–noble gas configuration? Give an example of one from Group 5A(15).

**8.53** How are measurements of paramagnetism used to support an electron configuration derived spectroscopically? Use Cu(I) and Cu(II) chlorides as examples.

**8.54** A set of isoelectronic ions have charges that vary from 3+ to 3−. Place them in order of increasing size.

---

**8.55** Which element in each of the following pairs would you expect to be *more* metallic: (a) Ca or Rb; (b) Mg or Ra; (c) Br or I; (d) S or Cl?

**8.56** Which element in each of the following pairs would you expect to be *less* metallic: (a) Sb or As; (b) Si or P; (c) Be or Na; (d) Cs or Rn?

**8.57** Does the reaction of a nonmetal oxide in water produce an acidic or a basic solution? Write a balanced equation for the reaction of a Group 6A(16) nonmetal oxide with water as an example.

**8.58** Does the reaction of a metal oxide in water produce an acidic or a basic solution? Write a balanced equation for the reaction of a Group 2A(2) oxide with water as an example.

**8.59** Write the charge and full ground-state electron configuration of the monatomic ion most likely to be formed by each of the following: (a) Cl; (b) Na; (c) Ca; (d) P; (e) Mg; (f) Se.

**8.60** Write the charge and full ground-state electron configuration of the monatomic ion most likely to be formed by each of the following: (a) Al; (b) S; (c) Sr; (d) Rb; (e) N; (f) Br.

**8.61** How many unpaired electrons are present in the ground state of an atom from each of the following groups? (a) 2A(2); (b) 5A(15); (c) 8A(18); (d) 3A(13)

**8.62** How many unpaired electrons are present in the ground state of an atom from each of the following groups? (a) 4A(14); (b) 7A(17); (c) 1A(1); (d) 6A(16)

**8.63** Which of the following atoms are paramagnetic in their ground state? (a) Ga; (b) Si; (c) Be; (d) Te

**8.64** Which of the following ions are paramagnetic in their ground state? (a) $Ti^{2+}$; (b) $Zn^{2+}$; (c) $Ca^{2+}$; (d) $Sn^2$

**8.65** Write the condensed ground-state electron configuration of the following transition metal ions, and state which are paramagnetic: (a) $V^{3+}$; (b) $Cd^{2+}$; (c) $Co^{3+}$; (d) $Ag^+$.

**8.66** Write the condensed ground-state electron configuration of the following transition metal ions, and state which are paramagnetic: (a) $Mo^{3+}$; (b) $Au^+$; (c) $Mn^{2+}$; (d) $Hf^{2+}$.

**8.67** Palladium ($Z = 46$) is diamagnetic. Draw partial orbital diagrams to show which of the following electron configurations is consistent with this fact: (a) [Kr] $5s^2 4d^8$; (b) [Kr] $4d^{10}$; (c) [Kr] $5s^1 4d^9$.

**8.68** Niobium (Nb; $Z = 41$) has an unexpected ground-state electron configuration for a Group 5B(5) element: [Kr] $5s^1 4d^4$. What is the expected electron configuration for elements in this group? Draw partial orbital diagrams to show how paramagnetic measurements could be used to support niobium's actual configuration.

**8.69** Rank the ions in each of the following sets in order of *increasing* size, and explain your ranking: (a) $Li^+$, $K^+$, $Na^+$; (b) $Se^{2-}$, $Rb^+$, $Br^-$.

**8.70** Rank the ions in each of the following sets in order of *decreasing* size, and explain your ranking: (a) $Se^{2-}$, $S^{2-}$, $O^{2-}$; (b) $Te^{2-}$, $Cs^+$, $I^-$.

## Comprehensive Problems

Problems with an asterisk (*) are more challenging.

**8.71** Some versions of the periodic table show hydrogen at the top of Group 1A(1) *and* at the top of Group 7A(17). What properties of hydrogen could justify these placements?

**8.72** Based on the general vertical and horizontal trends in periodic properties, explain why it might be difficult to rank the following pairs:
(a) K or Sr in terms of atomic size
(b) Mn or Fe in terms of $IE_1$
(c) Na or Ca in terms of metallic character
(d) P or Se in terms of oxide acidity

**8.73** Name and write a full ground-state electron configuration for each of the following elements:
(a) The smallest metal
(b) The heaviest lanthanide
(c) The lightest transition metal
(d) The Period 3 member whose 2− ion is isoelectronic with Ar
(e) The alkaline earth metal whose cation is isoelectronic with Kr
(f) The Group 5A(15) metalloid with the most acidic oxide

**\*8.74** A fundamental law of electrostatics states that the energy required to separate opposite charges of magnitude $Q_1$ and $Q_2$ that are distance $d$ apart is a function of $Q_1 Q_2 / d$. Use this law and any other factors to explain the following observations:

(a) The $IE_2$ of He ($Z = 2$) is *more* than twice the $IE_1$ of H ($Z = 1$).
(b) The $IE_1$ of He is *less* than twice the $IE_1$ of H.

**\*8.75** You are trapped in an alternate universe in which the quantum number assignments are the same as in ours, except that the quantum number $l$ is an integer from 0 to $n$. What are the atomic numbers for the two smallest noble gases in the alternate universe?

**8.76** Write the formula and name the compound formed from the following ionic interactions:
(a) The 2+ ion and the 1− ion are both isoelectronic with a chemically unreactive Period 3 element.
(b) The 2+ ion and the 2− ion are both isoelectronic with the Period 2 noble gas.
(c) The 2+ ion is the smallest with a filled $d$ subshell; the anion forms from the smallest halogen.
(d) The ions form from the largest and smallest ionizable elements in Period 3.

**8.77** Calculate the longest wavelength of electromagnetic radiation that could ionize an atom of each of the following elements:
(a) Li; $IE_1 = 520.1$ kJ/mol
(b) Au; $IE_1 = 889.9$ kJ/mol
(c) Cl; $IE_1 = 1255.7$ kJ/mol

**8.78** Gallium lies under aluminum in Group 3A(13), but it has a smaller radius. What causes this deviation from the general trend in atomic size? What other atomic property of Ga might also have a value outside the expected trend?

**8.79** In Group 4A(14), carbon (a nonmetal) and silicon (a metalloid) have high melting points (4100°C and 1420°C, respectively), whereas tin and lead are metals with low melting points (232°C and 327°C, respectively). Predict the full ground-state electron configuration and general physical properties of the as yet unknown member with atomic number 114.

**8.80** Before Mendeleev published his periodic table, J. W. Döbereiner grouped similar elements into triads in which the unknown properties of one member could be predicted by averaging known values of the others. To test this idea, predict the values of the following quantities:
(a) The atomic mass of K from the atomic masses of Na and Rb
(b) The melting point of $Br_2$ from the melting points of $Cl_2$ (−101.0°C) and $I_2$ (113.6°C) (Actual value = −7.2°C)
(c) The boiling point of HBr from the boiling points of HCl (−84.9°C) and HI (−35.4°C) (Actual value = −67.0°C)
(d) The boiling point of $AsH_3$ from the boiling points of $PH_3$ (−87.4°C), $GeH_4$ (−88.5°C), $H_2Se$ (−41.5°C), and $SbH_3$ (−17.1°C) (Actual value = −55°C)

**\*8.81** Use electron configurations to account for the stability of the lanthanide ions $Ce^{4+}$ and $Eu^{2+}$.

# CHAPTER 9

# Models of Chemical Bonding

## Concepts and skills to review

- characteristics of ionic and covalent bonding (Section 2.6)
- polar covalent bonds and the polarity of water (Section 4.2)
- Hess's law, $\Delta H^0_{rxn}$, and $\Delta H^0_f$ (Section 6.5)
- atomic and ionic electron configurations (Sections 8.2 and 8.4)
- trends in atomic properties and metallic behavior (Sections 8.3 and 8.4)

**Calcium and oxygen form calcium oxide.** This brilliant combustion reaction involves all three bonding models discussed in this chapter, as CaO (ionic) results from the reaction of $O_2$ (covalent) and Ca (metallic).

**N**early all atoms exist bound to others in elements and compounds. In Chapter 8, we examined the properties of the atoms themselves, but much of the excitement of chemistry comes in discovering how these properties influence chemical bonding. In general terms, atoms bond for one overriding reason: *bonding lowers the potential energy between positive and negative particles,* whether the particles are oppositely charged ions or the nucleus of one atom and the electrons of another. Just as the electron configuration and the strength of the nucleus-electron attraction determine the properties of an atom, the type and strength of the bonds between the atoms determine the properties of an element or compound.

This chapter begins with an overview of how atomic properties give rise to the three main ways that atoms bond. We discuss the ionic bonding model first and the energy involved in the formation of an ionic solid from its elements. Covalent bonding, which occurs in the vast majority of compounds, is treated in the next three sections. The first describes the formation and characteristics of a covalent bond. The next describes the range of bonding, from pure covalent to ionic, and explains that most bonds fall somewhere in between. The last shows how to depict molecules and polyatomic ions with electron-dot formulas. Then we focus on the relationship between the energy involved when a bond forms or breaks and the heat of reaction. The chapter ends with an introduction to metallic bonding.

## 9.1 Atomic Properties and Chemical Bonds

As a first step toward appreciating the importance of chemical bonding, let's classify the types of bonds and then develop a way to depict the bonding atoms.

### Types of Chemical Bonds

In general terms, we distinguish a metal atom from a nonmetal atom on the basis of several atomic properties that correlate with their positions across the periodic table: effective nuclear charge ($Z_{eff}$), number of valence electrons, atomic size, ionization energy (IE), and electron affinity (EA) (Figure 9.1). Metallic properties generally decrease across a period and increase down a group, so cesium [second from the bottom of Group 1A(1)] is the most reactive common metal* and fluorine [top of Group 7A(17)] the most reactive nonmetal.

---

*Based on its position in the periodic table, francium is the most reactive metal, but it is extremely rare: only 15 g francium exists in the top kilometer of the Earth's crust. Therefore, for all practical purposes, chemists refer to cesium as the most reactive metal.

| PROPERTY | METAL ATOM | NONMETAL ATOM |
|---|---|---|
| Atomic size | Larger | Smaller |
| No. valence $e^-$ | Fewer | More |
| $Z_{eff}$ | Lower | Higher |
| IE | Lower | Higher |
| EA | Lower | Higher |

**A** Relative magnitudes of atomic properties within a period

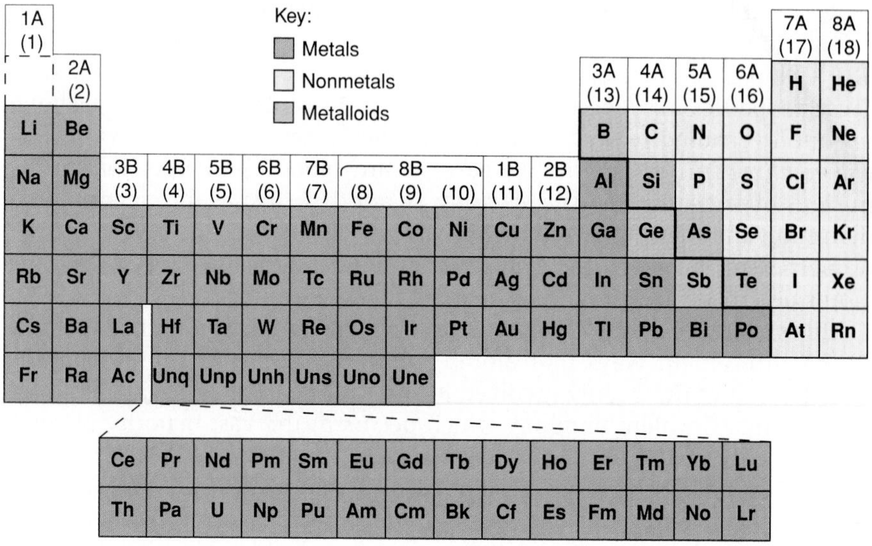

**B**

**FIGURE 9.1**

**A general comparison of metals and nonmetals. A,** The relative magnitudes of some key atomic properties change within a period and determine whether an atom is part of a metallic or a nonmetallic element. **B,** Metals *(blue)* appear in the left and lower portions of the periodic table. Nonmetals *(yellow)* appear in the upper right portion and include hydrogen also. Metalloids *(green)* are found along the stairlike line between metals and nonmetals.

These two types of atoms can combine in three possible ways: metal with nonmetal, nonmetal with nonmetal, and metal with metal:

1. *Electron transfer and ionic bonding* (Figure 9.2, *A*). We typically observe **ionic bonding** between reactive metals [Groups 1A(1) and 2A(2)] and nonmetals [Groups 7A(17) and the top of 6A(16)], those whose atoms have large differences in their tendencies to lose or gain electrons. The metal atom (low IE) loses its one or two valence electrons, whereas the nonmetal atom (large EA) gains electrons. *Electron transfer* from metal to nonmetal occurs, and each atom forms an ion with a noble gas electron configuration. The electrostatic attraction between these positive and negative ions draws them into the three-dimensional array of an ionic solid, whose chemical formula represents the cation-to-anion ratio (empirical formula).

2. *Electron sharing and covalent bonding* (Figure 9.2, *B*). When two atoms have little difference in their tendencies to lose or gain electrons, we observe *electron sharing* and **covalent bonding.** This type of bonding is most important between nonmetal atoms, although a pair of metal atoms can form a covalent bond also. Each nonmetal atom holds onto its own electrons tightly (high IE) and tends to gain other electrons as well (large EA). The attraction of each nucleus for the valence electrons of the other draws the atoms together. A shared valence electron pair is *localized* in that it spends most of its time between the two atoms, linking them in a covalent bond of specific average length and strength. Typically, separate molecules form

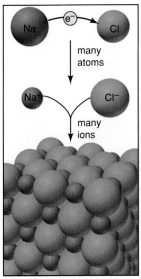

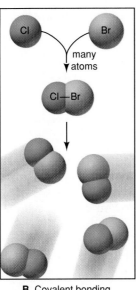

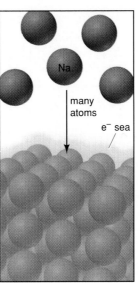

**A** Ionic bonding    **B** Covalent bonding    **C** Metallic bonding

**FIGURE 9.2**
**The three models of chemical bonding. A,** In ionic bonding, metal and nonmetal atoms form oppositely charged ions that attract each other to form a solid. **B,** In covalent bonding, two atoms share a bonding electron pair localized between their nuclei (shown as a bond-line). Separate molecules consist of two or more atoms. **C,** In metallic bonding, many metal atoms pool their valence electron(s) to form a delocalized "electron sea" that attracts all the atoms together.

when covalent bonding occurs, and the compound's formula reflects the actual number of atoms in the molecule (molecular formula).

3. *Electron pooling and metallic bonding* (Figure 9.2, *C*). In general, metal atoms are relatively large and have a small number of outer electrons that are shielded from the nuclear charge by filled inner levels. Thus, they lose outer electrons comparatively easily (low IE) but do not gain them very readily (small or positive EA). Such atomic properties lead to a third type of bonding. Unlike the covalent bonding that can occur between pairs of metal atoms, large numbers of metal atoms can share their valence electrons in another way. In the simplest model of **metallic bonding,** all the metal atoms in a sample *pool* their valence electrons into an evenly distributed "sea" of electrons that "flows" among all the metal ions and attracts them together. Unlike the localized electrons in covalent bonding, those in metallic bonding are *delocalized* and move freely throughout the piece of metal.

Keep in mind that these are idealized bonding models. In the diverse world of real substances, there are many exceptions to these simple categories. Therefore, it is not always possible to predict bond type solely from the elements' positions in the periodic table. For example, all binary ionic compounds contain a metal and a nonmetal, but all metals do not form ionic compounds with all nonmetals. When the metal beryllium [Group 2A(2)] combines with the nonmetal chlorine [Group 7A(17)], for instance, the bonding between Be and Cl fits the covalent model better than the ionic. In other words, as is emphasized in Section 9.4, just as we see a gradation in metallic behavior within groups and periods, we also see a gradation from one type of bonding to another.

### Lewis Electron-Dot Symbols: Depicting Atoms in Chemical Bonding

Because of his pioneering work in chemical bonding, the name of the great American chemist Gilbert N. Lewis (1875-1946) became associated with a simple shorthand system for depicting the electrons involved in bonding

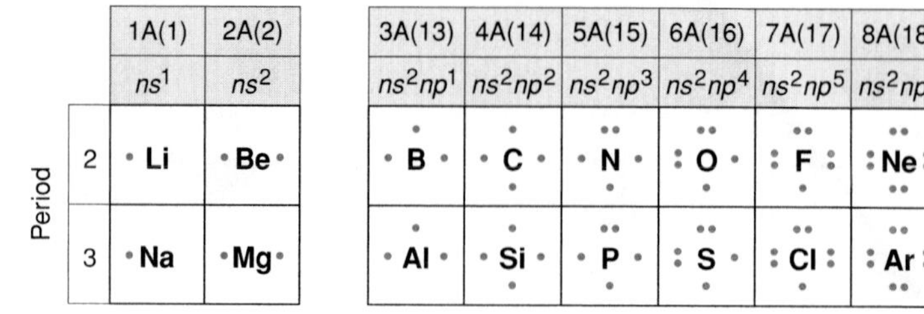

**FIGURE 9.3**

Lewis electron-dot symbols for elements in Periods 2 and 3. The element symbol represents the nucleus and inner electrons, and the dots around it represent valence electrons. (The specific positions of the paired and unpaired dots are arbitrary.) The number of unpaired dots indicates the number of electrons a metal atom loses, the number a nonmetal gains, or the number of covalent bonds a nonmetal atom usually forms.

|  | 1A(1) | 2A(2) | 3A(13) | 4A(14) | 5A(15) | 6A(16) | 7A(17) | 8A(18) |
|---|---|---|---|---|---|---|---|---|
|  | $ns^1$ | $ns^2$ | $ns^2np^1$ | $ns^2np^2$ | $ns^2np^3$ | $ns^2np^4$ | $ns^2np^5$ | $ns^2np^6$ |
| 2 | · Li | · Be · | · B̈ · | · C̈ · | · N̈ · | : Ö · | : F̈ : | : N̈e : |
| 3 | · Na | · Mg · | · Äl · | · S̈i · | · P̈ · | : S̈ · | : C̈l : | : Är : |

and the sequence of atoms in a molecule. In a **Lewis electron-dot symbol** of an atom, the element symbol represents the nucleus *and* inner electrons, and it is surrounded by a number of dots equal to the number of *valence electrons* (Figure 9.3). You can write the Lewis symbol of any main-group element from its A-group number (1A to 8A), which gives the number of valence electrons. These are placed one at a time on the four sides of the symbol and then paired up until all are used. The specific placement of dots, however, is not important. For example, the Lewis electron-dot symbol for nitrogen in Figure 9.3 could *also* have been written

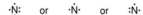

The arrangement of valence electron dots represents what we know about an element's bonding behavior. For a metal atom, the unpaired dots show the number of electrons it loses to form a cation. For a nonmetal atom, the unpaired dots indicate the number of electrons it gains to form an anion or that become paired with those of another atom in covalent bonding, that is, *the number of bonds the nonmetal usually forms.* For instance, carbon forms four bonds in its compounds, so it has four single dots rather than one pair and two single dots, as in its ground-state electron configuration: [He] $2s^2 2p^1 2p^1$. Similarly, boron forms three bonds, oxygen two, and so forth. Later, we use some other dot arrangements for the larger nonmetals.

**Section Summary**

Nearly all atoms occur bound to others. Chemical bonding allows atoms to lower their energy. Ionic bonding occurs when a metal atom transfers electrons to a nonmetal atom, and the resulting ions attract each other to form an ionic solid. Covalent bonding occurs most often between nonmetal atoms and commonly results in molecules. The bonded atoms share a pair of electrons, which remain localized between them. Metallic bonding occurs when large numbers of metal atoms pool their valence electrons in a delocalized electron sea that holds all the atoms together. The Lewis electron-dot symbol of an atom depicts the number of valence electrons for a main-group element.

## 9.2 The Ionic Bonding Model

The central theme of the ionic bonding model is the *transfer of electrons from metal to nonmetal to form ions that come together into a solid ionic compound.* Nearly every main-group monatomic ion has a filled outer level of electrons (either two or eight), the same number as in the nearest noble gas. This fact

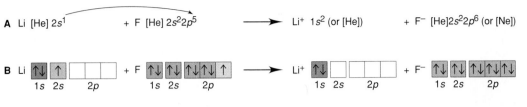

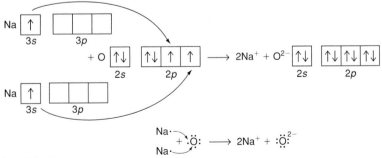

**FIGURE 9.4**

**Three ways to represent the formation of $Li^+$ and $F^-$ through electron transfer.** **A,** Electron configurations. **B,** Orbital diagrams. **C,** Lewis electron-dot symbols.

is incorporated into the **octet rule,** a useful generalization that applies to all types of bonding: when atoms bond, they lose, gain, or share electrons to attain a filled outer shell of eight electrons.

Figure 9.4 depicts the transfer of an electron from a lithium atom to a fluorine atom in several ways. Each shows that lithium loses its single outer electron and is left with a filled $n = 1$ level, and fluorine gains a single electron to fill its $n = 2$ level. Thus, the resulting $Li^+$ and $F^-$ ions attain the noble gas configurations of He and Ne, respectively. In this case, each atom is one electron away from its nearest noble gas, so the number of electrons lost by each lithium atom equals the number gained by each fluorine atom. Therefore, equal numbers of $Li^+$ and $F^-$ ions form, as the formula LiF indicates. In all cases of ionic bonding, *the total number of electrons lost by the metal atoms equals the total number of electrons gained by the nonmetal atoms.*

**SAMPLE PROBLEM 9.1**

### Depicting Ion Formation with Orbital Diagrams and Electron-Dot Symbols

**Problem:** Use partial orbital diagrams and Lewis electron-dot symbols to depict the formation of sodium and oxide ions from the atoms, and determine the formula of the compound.

**Plan:** First we draw the orbital diagrams and Lewis symbols for the Na and O atoms. To attain filled outer levels, Na loses one electron and O gains two. Therefore, to make the number of electrons lost equal the number gained, two Na atoms are needed for each O atom.

**Solution:**

The formula is **$Na_2O$.**

**FOLLOW-UP PROBLEM 9.1**

Use condensed electron configurations and Lewis electron-dot symbols to depict the formation of $Mg^{2+}$ and $Cl^-$ ions from the atoms, and write the formula of the compound.

**FIGURE 9.5**

**The reaction between sodium and bromine.**
**A,** All the Group 1A(1) metals react exothermically with all the Group 7A(17) nonmetals to form solid alkali-metal halides. **B,** The reaction is usually rapid and violent.

### Energy Considerations in Ionic Bonding: The Importance of Lattice Energy

It is important to realize that ionic compounds do *not* form as a result of the electron transfer process only; in fact, this process *requires* energy. Rather, ionic compounds form because of the tremendous decrease in potential energy that occurs when the ions aggregate in a solid.

Consider the electron transfer process for lithium fluoride. The first ionization energy ($IE_1$) of lithium is the energy required for a mole of gaseous Li atoms to lose a mole of outer electrons:

$$Li(g) \rightarrow Li^+(g) + e^- \qquad IE_1 = 520 \text{ kJ}$$

The electron affinity (EA) of fluorine is the energy released when a mole of F atoms gains a mole of electrons:

$$F(g) + e^- \rightarrow F^-(g) \qquad EA = -328 \text{ kJ}$$

Adding these values, we find that the two-step electron transfer process *by itself* requires energy:

$$Li(g) + F(g) \rightarrow Li^+(g) + F^-(g) \qquad IE_1 + EA = 192 \text{ kJ}$$

The energy cost is even more than this amount because metallic lithium and diatomic fluorine must first be converted to the separate gaseous atoms, which also requires energy. Despite this, the $\Delta H_f^0$ of solid LiF (the standard enthalpy change accompanying its formation from elemental lithium and fluorine) is $-617$ kJ/mol; that is, 617 kJ is released per mole of LiF*(s)* formed. The case of LiF is typical: despite an endothermic electron transfer, most ionic solids form readily, often vigorously (Figure 9.5).

Clearly, if the overall reaction of Li*(s)* and $F_2(g)$ to form LiF*(s)* releases energy, there must be other energy components exothermic enough to overcome the endothermic steps. These exothermic steps result from the strong *attraction between oppositely charged ions.* When a mole of $Li^+(g)$ and a mole of $F^-(g)$ form a mole of gaseous LiF molecules, heat is released:

$$Li^+(g) + F^-(g) \rightarrow LiF(g) \qquad \Delta H^0 = -755 \text{ kJ}$$

Also, since LiF does not exist as a gas under ordinary conditions, *even more energy is released when the gaseous ions coalesce into a crystalline solid,* in which each ion attracts several others of opposite charge. The **lattice energy** is the enthalpy change occurring when gaseous ions coalesce into a solid ionic compound:

$$Li^+(g) + F^-(g) \rightarrow LiF(s) \qquad \Delta H_{LiF}^0 = \text{lattice energy} = -1050 \text{ kJ}$$

The lattice energy is an important indication of the strength of ionic interactions and is a major factor influencing melting points, hardness, and solubilities of ionic compounds.

The lattice energy plays a crucial role in ionic compound formation, but it is difficult to measure directly. Nevertheless, the lattice energies of many compounds have been determined using Hess's law of heat summation (Section 6.5), which states that an overall reaction's enthalpy change is the sum of the enthalpy changes for the individual reactions that make it up: $\Delta H_{total} = \Delta H_1 + \Delta H_2 + \cdots$ . Lattice energies can be calculated through a **Born-Haber cycle,** a series of steps from elements to ionic compound for which all the enthalpies* are known except the lattice energy.

---

*Strictly speaking, ionization energy (IE) and electron affinity (EA) are internal energy changes ($\Delta E$), not enthalpy changes ($\Delta H$), but in these cases, $\Delta H = \Delta E$ because the number of moles of gas does not change.

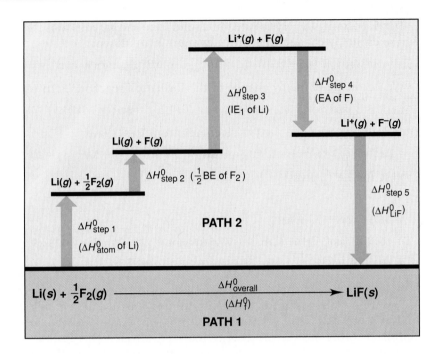

**FIGURE 9.6**
**The Born-Haber cycle for lithium fluoride.** The formation of LiF($s$) from its elements is shown as happening either in one overall reaction (Path 1) or in five steps, each with its own enthalpy change (Path 2). The overall enthalpy change for the process $(\Delta H_f^0)$ is known, as are $\Delta H_{Step\ 1}^0$ through $\Delta H_{Step\ 4}^0$. Therefore, $\Delta H_{Step\ 5}^0$, the lattice energy $(\Delta H_{LiF}^0)$, can be calculated. (The magnitudes of the $\Delta H^0$ values are not drawn to scale.)

Let's go through the Born-Haber cycle for lithium fluoride. As you'll see, *we choose steps that we can measure* to depict the energy components of ionic compound formation, from which we calculate the lattice energy; that is, *they are not necessarily the actual steps that occur* when lithium reacts with fluorine. We begin with the elements in their standard states, metallic lithium and gaseous diatomic fluorine. There are two paths to follow: either the direct combination reaction (Path 1) or the multistep cycle (Path 2), one step of which is the unknown lattice energy (Figure 9.6). From Hess's law, we know that both paths involve the same overall enthalpy change:

$$\Delta H_f^0 \text{ of LiF}(s) \text{ (Path 1)} = \text{sum of } \Delta H^0 \text{ for steps in cycle (Path 2)}$$

To preview Path 2, the elements are converted to individual gaseous atoms (Steps 1 and 2), the electron transfer steps form gaseous ions (Steps 3 and 4), and the ions form a solid (Step 5). We identify each $\Delta H^0$ by its step number:

*Step 1.* Converting solid lithium to separate gaseous lithium atoms involves breaking the metallic bonds that hold lithium atoms in the sample, so it requires energy:

$$\text{Li}(s) \longrightarrow \text{Li}(g) \qquad \Delta H_{Step\ 1}^0 = 161 \text{ kJ}$$

(This process is called *atomization*, and the enthalpy change is $\Delta H_{atom}^0$.)

*Step 2.* Converting fluorine molecules to fluorine atoms involves breaking the covalent bond in $F_2$, so it requires energy. We need 1 mol F atoms to form 1 mol LiF, so we start with $1/2$ mol $F_2$:

$$\tfrac{1}{2}F_2(g) \longrightarrow F(g) \qquad \Delta H_{Step\ 2}^0 = \tfrac{1}{2} \text{ bond energy (BE) of } F_2$$
$$= \tfrac{1}{2}(159 \text{ kJ}) = 79.5 \text{ kJ}$$

*Step 3.* Removing the $2s$ electron from Li to form $\text{Li}^+$ requires energy:

$$\text{Li}(g) \longrightarrow \text{Li}^+(g) + e^- \qquad \Delta H_{Step\ 3}^0 = IE_1 = 520 \text{ kJ}$$

*Step 4.* Adding an electron to F to form $F^-$ releases energy:

$$F(g) + e^- \longrightarrow F^-(g) \qquad \Delta H_{Step\ 4}^0 = EA = -328 \text{ kJ}$$

*Step 5.* Forming the crystalline ionic solid from the gaseous ions is the step whose enthalpy change (the lattice energy) is unknown:

$$Li^+(g) + F^-(g) \rightarrow LiF(s) \qquad \Delta H^0_{Step\ 5} = \Delta H^0_{LiF} \text{ (lattice energy)} = ?$$

We know the enthalpy change of the formation reaction (Path 1),

$$Li(s) + \tfrac{1}{2}F_2(g) \rightarrow LiF(s) \qquad \Delta H^0_{overall} = \Delta H^0_f = -617 \text{ kJ}$$

Therefore, we calculate the lattice energy from Hess's law:

$$\Delta H^0_f = -617 \text{ kJ/mol} = \Delta H^0_{Step\ 1} + \Delta H^0_{Step\ 2} + \Delta H^0_{Step\ 3} + \Delta H^0_{Step\ 4} + \Delta H^0_{LiF}$$

Solving for $\Delta H^0_{LiF}$ gives

$$\Delta H^0_{LiF}$$
$$= \Delta H^0_f - (\Delta H^0_{Step\ 1} + \Delta H^0_{Step\ 2} + \Delta H^0_{Step\ 3} + \Delta H^0_{Step\ 4})$$
$$= -617 \text{ kJ/mol} - [161 \text{ kJ/mol} + 79.5 \text{ kJ/mol} + 520 \text{ kJ/mol} + (-328 \text{ kJ/mol})]$$
$$= -1050 \text{ kJ/mol}$$

Note that *the magnitude of the lattice energy dominates the multistep process.*

The Born-Haber cycle reveals a central point about ionic bonding: *ionic solids exist only because the lattice energy drives the energetically unfavorable electron transfer.* In other words, the energy required for elements to lose or gain electrons is supplied by the electrostatic attraction between the ions they form: energy is expended to form the ions, but it is more than regained when they attract each other and form a solid.

**Periodic Trends in Lattice Energy**

Because the lattice energy is the result of electrostatic attractions among the oppositely charged ions, its magnitude depends on several factors, including ionic size, ionic charge, and the arrangement of ions in the solid. **Coulomb's law** is a fundamental relationship between charge and electrostatic force that allows us to predict the trend in lattice energy between different pairs of ions. It states that the electrostatic force associated with two charges (A and B) is directly proportional to the product of their magnitudes and inversely proportional to the square of the distance between them:

$$\text{Electrostatic force} \propto \frac{\text{charge A} \times \text{charge B}}{\text{distance}^2}$$

Since energy equals force times distance, the electrostatic energy is

$$\text{Electrostatic energy} \propto \frac{\text{charge A} \times \text{charge B}}{\text{distance}} \qquad \textbf{(9.1)}$$

In an ionic solid, the cations and anions lie as close to each other as possible, so the distance between them is the sum of their radii. Furthermore, it can be shown that the electrostatic energy is proportional to the lattice energy ($\Delta H^0_{lattice}$). Therefore,

$$\text{Electrostatic energy} \propto \frac{\text{cation charge} \times \text{anion charge}}{\text{cation radius} + \text{anion radius}} \propto \Delta H^0_{lattice} \qquad \textbf{(9.2)}$$

The effect of ionic *size* on lattice energy can be seen as we look down a group of either metals or nonmetals. Their ionic radii become larger, so the energy of attraction between metal and nonmetal ions should decrease, as should the lattice energy of their ionic compounds. This prediction is borne out in Figure 9.7, which shows the trends in lattice energy of the alkali halides, all of which have the same arrangement of ions in the solid. A

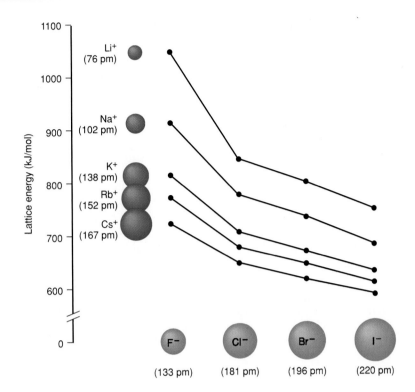

**FIGURE 9.7**
**Trends in lattice energy for the alkali halides.** The lattice energies for the alkali halides are shown, with each line representing a given 1A(1) cation *(left side)* combining with the various 7A(17) anions *(bottom)*. As ionic radii increase, the electrostatic attractions decrease, so the lattice energies of the compounds decrease. Thus, LiF (smallest ions) has the highest lattice energy and CsI (largest ions) the lowest.

smooth decrease occurs in lattice energy down a group whether we hold the cation or the anion constant. For example, the lattice energy decreases from LiF to LiI (down the halide ions with the same cation) and from LiF to CsF (down the alkali metal ions with the same anion).

The effect of ionic *charge* on lattice energy can be seen when we compare lithium fluoride with magnesium oxide. Both solids have the same structure, cations of about equal size ($Li^+$ = 76 pm and $Mg^{2+}$ = 72 pm), and anions of about equal size ($F^-$ = 133 pm and $O^{2-}$ = 140 pm). Thus, the only significant difference is the ionic charge: LiF contains the univalent ions $Li^+$ and $F^-$, whereas MgO contains the divalent ions $Mg^{2+}$ and $O^{2-}$. The difference in their lattice energies is striking:

$$\Delta H^0_{LiF} = -1050 \text{ kJ/mol} \quad \text{and} \quad \Delta H^0_{MgO} = -3923 \text{ kJ/mol}$$

The nearly fourfold increase in lattice energy reflects the fourfold increase in the product of the charges ($1 \times 1$ vs. $2 \times 2$) in the numerator of Equation 9.2. Clearly, the divalent ions attract each other much more strongly.

Given the high energy needed to form the ions, the high lattice energy of MgO explains why ionic solids with divalent ions exist at all. Forming $Mg^{2+}$ involves loss of the first *and* second electrons, so it requires the sum of the first and second ionization energies:

$$Mg(g) \rightarrow Mg^{2+}(g) + 2e^- \qquad \Delta H^0 = IE_1 + IE_2 = 738 \text{ kJ} + 1450 \text{ kJ} = 2188 \text{ kJ}$$

Adding one electron to the O atom (first electron affinity, $EA_1$) is exothermic, but adding a second electron (second electron affinity, $EA_2$) is very endothermic because an electron is being added to the negative $O^-$ ion. The overall formation of the $O^{2-}$ ion is therefore endothermic:

$$
\begin{array}{ll}
O(g) + e^- \rightarrow O^-(g) & \Delta H^0 = EA_1 = -141 \text{ kJ} \\
O^-(g) + e^- \rightarrow O^{2-}(g) & \Delta H^0 = EA_2 = 878 \text{ kJ} \\
\hline
O(g) + 2e^- \rightarrow O^{2-}(g) & \Delta H^0 = EA_1 + EA_2 = 737 \text{ kJ}
\end{array}
$$

Add to this the endothermic steps for converting Mg(*s*) to Mg(*g*) ($\Delta H^0_{atom}$ = 148 kJ/mol) and breaking $\frac{1}{2}$ mol $O_2$ molecules into 1 mol O atoms ($\Delta H^0 = \frac{1}{2}$ bond energy of $O_2 = \frac{1}{2}$ mol $\times$ 498 kJ/mol = 249 kJ). Nevertheless, as a result of the high ionic charges, solid MgO readily forms every time Mg burns in air [$\Delta H^0_f$ of MgO(*s*) = −601 kJ/mol] because the enormous lattice energy ($\Delta H^0_{MgO}$ = −3923 kJ/mol) more than compensates for these endothermic steps.

### How the Model Explains the Properties of Ionic Compounds

By magnifying our view, we can see how the ionic bonding model accounts for the properties of ionic solids, many of which are already familiar to you. A piece of rock salt (NaCl), for instance, is *hard* (does not dent), *rigid* (does not bend), and *brittle* (cracks without deforming). By contrast, a piece of metal dents or bends before breaking. These properties of ionic solids are due to the powerful attractive forces that hold the ions in place throughout the crystal. Moving the ions out of position requires overcoming these forces, so the sample resists being dented or bent. If enough pressure is applied to overcome the attractions, ions of like charge are brought close together, and their repulsions crack the sample (Figure 9.8).

**FIGURE 9.8**

**Electrostatic forces and the reason ionic compounds crack. A,** Ionic compounds are hard and crack, rather than bend, when struck with enough force. **B,** The positive and negative ions in the crystal are arranged to maximize their attraction. When an external force is applied, similar charges move near each other, and the repulsions crack the piece apart.

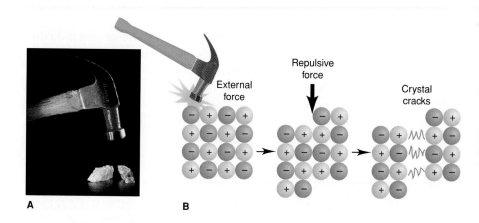

A                                B

**TABLE 9.1 Melting and Boiling Points of Some Ionic Compounds**

| COMPOUND | MP (°C) | BP (°C) |
|---|---|---|
| CsBr | 636 | 1300 |
| NaI | 661 | 1304 |
| $MgCl_2$ | 714 | 1412 |
| KBr | 734 | 1435 |
| $CaCl_2$ | 782 | >1600 |
| NaCl | 801 | 1413 |
| LiF | 845 | 1676 |
| KF | 858 | 1505 |
| MgO | 2852 | 3600 |

Most ionic compounds do not conduct electricity in the solid state but do conduct when melted or dissolved in water. (Notable exceptions to this general behavior include superionic conductors, such as AgI, and the superconducting ceramics, which have remarkable conductivity in the solid state.) According to the ionic bonding model, the solid consists of immobilized ions. When it melts or dissolves, however, the ions are free to move and carry an electric current (Figure 9.9).

The model also explains that high temperatures are needed to melt and boil an ionic compound (Table 9.1) because freeing the ions from their lattice positions (melting) requires large amounts of energy, and vaporizing them requires even more. In fact, the ionic attraction is so strong that the vapor consists of **ion pairs,** which are gaseous ionic molecules rather than individual ions (Figure 9.10). Keep in mind, however, that in their ordinary solid state, ionic compounds consist of rows of alternating ions that extend in all directions, and *no separate molecules exist.*

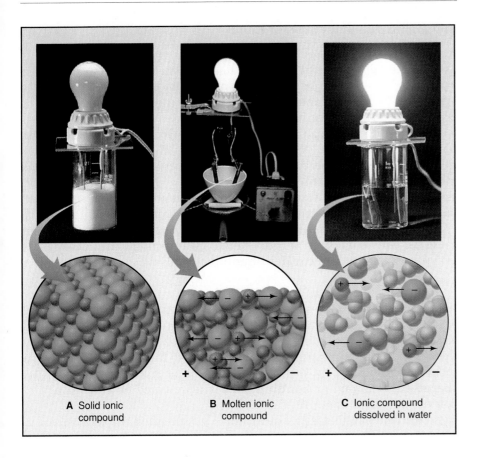

A  Solid ionic
   compound

B  Molten ionic
   compound

C  Ionic compound
   dissolved in water

**FIGURE 9.9**
**Electrical conductance of ionic compounds.**
Conduction of an electric current by an ionic compound depends on mobile ions. **A,** No current flows in the solid because ions are immobile. **B,** In the molten compound, ions flow toward the oppositely charged electrodes and carry a current. **C,** In an aqueous solution of the compound, mobile solvated ions carry a current.

### Section Summary

In ionic bonding, a metal transfers electrons to a nonmetal and the resulting ions attract each other strongly. Main-group elements often attain a filled outer level of electrons (either eight or two) by forming ions with the configuration of the nearest noble gas. Ion formation by itself requires a net input of energy. However, the lattice energy, the energy released when the gaseous ions form a solid, is large, and it is the major reason that ionic solids exist. The magnitude of the lattice energy, which depends on ionic size and charge, can be determined by applying Hess's law in a Born-Haber cycle.

The ionic bonding model pictures oppositely charged ions held rigidly in position by strong electrostatic attractions and explains why ionic solids crack rather than bend and why they conduct a current only when melted or dissolved. The ions occur as gaseous ion pairs when the compound vaporizes, which requires very high temperatures.

**FIGURE 9.10**
**Vaporizing an ionic compound.** Ionic compounds generally have very high boiling points because the ions must have high kinetic energies to break free from surrounding ions. In fact, ionic compounds usually vaporize as ion pairs.

## 9.3  The Covalent Bonding Model

The number of known ionic compounds is dwarfed by the number of known covalent compounds. Molecules held together by covalent bonds range from diatomic hydrogen to biological and synthetic macromolecules with molar masses of several million. We also find covalent bonds in the polyatomic ions of many ionic compounds. Without doubt, sharing electrons is the principal way in which atoms interact chemically.

## The Formation of a Covalent Bond

A sample of hydrogen gas consists of $H_2$ molecules. But why *do* the atoms exist bound tightly together in pairs? Imagine what would happen to two isolated H atoms as they approach each other from a distance (Figure 9.11). When the H atoms are far apart, each behaves as though the other were not present. As the distance between the two nuclei becomes shorter, each atom's nucleus starts to attract the other atom's electron, which lowers the potential energy of the system. The atoms continue to draw each other closer, and the system becomes progressively lower in energy. As attractions increase, so do repulsions between the nuclei and between the electron clouds. At some internuclear distance, the maximum attraction is achieved in the face of the increasing repulsion, and the system has its minimum energy. Any shorter internuclear distance would increase the repulsions and cause the potential energy to rise. The mutual attraction of nuclei and electron pair constitutes the **covalent bond** that holds the atoms together as a hydrogen molecule.

In covalent bonding, as in ionic bonding, each atom achieves a full outer shell of electrons, but it occurs by different means. *Each atom in a covalent bond "counts" the shared electrons as belonging entirely to itself.* Thus, one shared electron pair simultaneously fills the outer shell of *both* H atoms. [You may think of electron-pair sharing as analogous to two friends (atoms) sharing a room (electron pair) by each having full use of it simultaneously.] The **shared** electron pair, or **bonding pair,** is represented by either a pair of dots or a line, H:H or H—H.

An electron pair that is part of an atom's valence shell but *not* involved in bonding is called a **lone pair,** or **unshared** pair. The bonding pair in HF fills

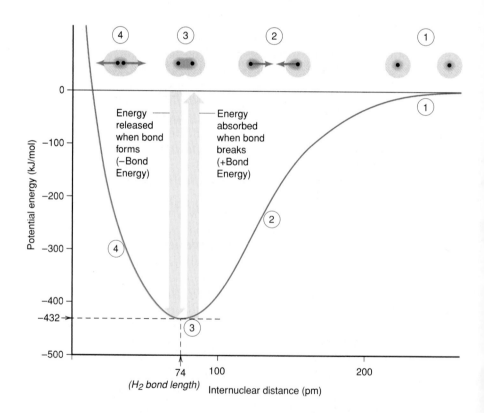

**FIGURE 9.11**

**Covalent bond formation in $H_2$.** The potential energy of a system of two H atoms plotted against the distance between the nuclei is shown, with a depiction of the atoms above. At 1, the atoms are too far apart to attract each other. At 2, each nucleus begins to attract the other atom's electron. At 3, the combination of nucleus-electron attractions and electron-electron and nucleus-nucleus repulsions gives the minimum energy of the system. If the atoms move closer (4), repulsions increase the system's energy. The energy difference between points 1 and 3 is the $H_2$ bond energy (432 kJ/mol). The internuclear distance at point 3 is the $H_2$ bond length (74 pm).

the outer shell of the H atom *and*, together with three lone pairs, fills the outer shell of the F atom as well:

$$\text{bonding pair} \quad \overset{\text{H}:\ddot{\text{F}}:}{} \quad \text{lone pairs} \qquad \text{or} \qquad \text{H—}\ddot{\text{F}}:$$

Similarly, in $F_2$, the bonding pair and three lone pairs fill the outer shells of *both* F atoms:

$$:\ddot{\text{F}}:\ddot{\text{F}}: \qquad \text{or} \qquad :\ddot{\text{F}}—\ddot{\text{F}}:$$

(This text generally shows bonding pairs as lines and lone pairs as dots.)

The covalent bond in $H_2$, HF, and $F_2$ is a **single bond,** one that consists of a single bonding pair of electrons. Many molecules (and ions) contain multiple bonds between two atoms. These occur most frequently when C, O, N, or S atoms bond to one another. A **double bond** consists of two bonding pairs, four electrons shared between two atoms. Ethylene ($C_2H_4$) is a simple hydrocarbon that contains a carbon-carbon double bond and four carbon-hydrogen single bonds:

$$\begin{array}{c} \text{H} \quad \text{H} \\ \text{C}::\text{C} \\ \text{H} \quad \text{H} \end{array} \qquad \text{or} \qquad \begin{array}{c} \text{H} \qquad \text{H} \\ \text{C}=\text{C} \\ \text{H} \qquad \text{H} \end{array}$$

*Each* carbon "counts" the four electrons in the double bond and the four in its two single bonds to hydrogen to obtain an octet. A **triple bond** consists of three bonding pairs, two atoms sharing six electrons. In the $N_2$ molecule, the atoms are held together by a triple bond, and each N also has a lone pair:

$$:\text{N}:::\text{N}: \qquad \text{or} \qquad :\text{N}\equiv\text{N}:$$

Six electrons from the triple bond and two from the lone pair give *each* N an octet.

The **bond order** is the number of electron pairs being shared between any two bonded atoms. A single bond has a bond order of one (one shared electron pair), a double bond has a bond order of two, and a triple bond has a bond order of three.

### The Properties of a Covalent Bond: Bond Energy and Bond Length

The properties of a covalent bond depend on the balance between the nucleus-electron attractions and the electron-electron and nucleus-nucleus repulsions (Figure 9.12). The strength of the bond depends on the extent to which the attractions outweigh the repulsions. The **bond energy (BE),** or bond strength, is the amount of energy required to overcome this net attraction; it is defined as the standard enthalpy change for breaking the bond in a mole of gaseous molecules. *Bond breakage is an endothermic process, so the bond energy is always positive:*

$$\text{A—B}(g) \rightarrow \text{A}(g) + \text{B}(g) \qquad \Delta H^0_{\text{bond breaking}} = \text{BE}_{\text{A—B}} \text{ (always} > 0)$$

Put another way, the bond energy is the difference in energy between the separated atoms and the bonded atoms (the energy difference between the top and bottom of the potential energy "well" in Figure 9.11). The same amount of energy that is absorbed to break the bond is released when it

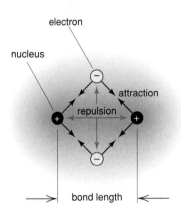

**FIGURE 9.12**
**The attractive and repulsive forces in covalent bonding.** Nucleus-electron attractions and nucleus-nucleus and electron-electron repulsions occur simultaneously. At some optimum distance (the bond length), the attractive forces are maximized in the presence of the repulsive forces. The net attraction determines the bond energy.

**TABLE 9.2  Average Bond Energies (kJ/mol)**

| BOND | ENERGY | BOND | ENERGY | BOND | ENERGY | BOND | ENERGY |
|---|---|---|---|---|---|---|---|
| | | | | **Single Bonds** | | | |
| H—H | 432 | N—H | 391 | Si—H | 323 | S—H | 347 |
| H—F | 565 | N—N | 160 | Si—Si | 226 | S—S | 266 |
| H—Cl | 427 | N—P | 209 | Si—O | 368 | S—F | 327 |
| H—Br | 363 | N—O | 201 | Si—S | 226 | S—Cl | 271 |
| H—I | 295 | N—F | 272 | Si—F | 565 | S—Br | 218 |
| | | N—Cl | 200 | Si—Cl | 381 | S—I | ~170 |
| C—H | 413 | N—Br | 243 | Si—Br | 310 | | |
| C—C | 347 | N—I | 159 | Si—I | 234 | F—F | 159 |
| C—Si | 301 | | | | | F—Cl | 193 |
| C—N | 305 | O—H | 467 | P—H | 320 | F—Br | 212 |
| C—O | 358 | O—P | 351 | P—Si | 213 | F—I | 263 |
| C—P | 264 | O—O | 204 | P—P | 200 | Cl—Cl | 243 |
| C—S | 259 | O—S | 265 | P—F | 490 | Cl—Br | 215 |
| C—F | 453 | O—F | 190 | P—Cl | 331 | Cl—I | 208 |
| C—Cl | 339 | O—Cl | 203 | P—Br | 272 | Br—Br | 193 |
| C—Br | 276 | O—Br | 234 | P—I | 184 | Br—I | 175 |
| C—I | 216 | O—I | 234 | | | I—I | 151 |
| | | | | **Multiple Bonds** | | | |
| C=C | 614 | N=N | 418 | C≡C | 839 | N≡N | 945 |
| C=N | 615 | N=O | 607 | C≡N | 891 | | |
| C=O | 745 | $O_2$ | 498 | C≡O | 1070 | | |
| | (799 in $CO_2$) | | | | | | |

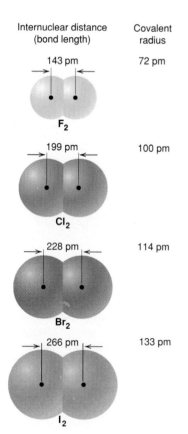

Internuclear distance (bond length)   Covalent radius

**F₂** — 143 pm — 72 pm

**Cl₂** — 199 pm — 100 pm

**Br₂** — 228 pm — 114 pm

**I₂** — 266 pm — 133 pm

forms. *Bond formation is an exothermic process, so the sign of the associated enthalpy change is negative:*

$$\text{A}(g) + \text{B}(g) \rightarrow \text{A—B}(g) \qquad \Delta H^0_{\text{bond forming}} = -\text{BE}_{\text{A—B}} \text{ (always} < 0)$$

Since bond energies depend on the bonded atoms—their electron configurations, nuclear charges, and atomic radii—each type of bond has its own bond energy (Table 9.2). Moreover, the bond energy of a given type of bond varies slightly from molecule to molecule, and even within the same molecule, so the tabulated value is an average bond energy for that type of bond.

A covalent bond has a **bond length,** the distance between the nuclei of two bonded atoms (see Figure 9.12); in Figure 9.11, it is the distance between the nuclei at their position of minimum energy. Table 9.3 shows the lengths of some covalent bonds. Here, too, the values represent average bond lengths for the given bond in different substances. Bond length is related to the sizes of the bonded atoms; in fact, most atomic radii are obtained from experimentally determined bond lengths (see Figure 8.10, *C*). Bond lengths within a series of similar bonds increase with atomic size, as shown in Figure 9.13 for the halogens.

**FIGURE 9.13**
**Bond length and covalent radius.** Within a series of similar substances, such as the diatomic halogen molecules, bond length increases as covalent radius increases.

**TABLE 9.3** **Average Bond Lengths (pm\*)**

| BOND | LENGTH | | BOND | LENGTH | BOND | LENGTH | | BOND | LENGTH |
|---|---|---|---|---|---|---|---|---|---|
| | | | | | **Single Bonds** | | | | |
| H—H | 74 | | N—H | 101 | Si—H | 148 | | S—H | 134 |
| H—F | 92 | | N—N | 146 | Si—Si | 234 | | S—P | 210 |
| H—Cl | 127 | | N—P | 177 | Si—O | 161 | | S—S | 204 |
| H—Br | 141 | | N—O | 144 | Si—S | 210 | | S—F | 158 |
| H—I | 161 | | N—S | 168 | Si—N | 172 | | S—Cl | 201 |
| | | | N—F | 139 | Si—F | 156 | | S—Br | 225 |
| | | | N—Cl | 191 | Si—Cl | 204 | | S—I | 234 |
| C—H | 109 | | N—Br | 214 | Si—Br | 216 | | | |
| C—C | 154 | | N—I | 222 | Si—I | 240 | | F—F | 143 |
| C—Si | 186 | | | | | | | F—Cl | 166 |
| C—N | 147 | | O—H | 96 | P—H | 142 | | F—Br | 178 |
| C—O | 143 | | O—P | 160 | P—Si | 227 | | F—I | 187 |
| C—P | 187 | | O—O | 148 | P—P | 221 | | Cl—Cl | 200 |
| C—S | 181 | | O—S | 151 | P—F | 156 | | Cl—Br | 214 |
| C—F | 133 | | O—F | 142 | P—Cl | 204 | | Cl—I | 243 |
| C—Cl | 177 | | O—Cl | 164 | P—Br | 222 | | Br—Br | 228 |
| C—Br | 194 | | O—Br | 172 | P—I | 243 | | Br—I | 248 |
| C—I | 213 | | O—I | 194 | | | | I—I | 266 |
| | | | | | **Multiple Bonds** | | | | |
| C=C | 134 | | N=N | 122 | C≡C | 121 | | N≡N | 110 |
| C=N | 127 | | N=O | 120 | C≡N | 115 | | N≡O | 106 |
| C=O | 123 | | $O_2$ | 121 | C≡O | 113 | | | |

A close relationship exists among bond order, bond length, and bond energy. Two nuclei are more strongly attracted to two shared electron pairs than to one: the atoms are drawn closer together *and* are more difficult to pull apart. Therefore, *for a given pair of atoms, a higher bond order results in a shorter bond length and a greater bond energy;* that is, for a given pair of atoms, a shorter bond is a stronger bond (Table 9.4). Sometimes we can extend this relationship among atomic size, bond length, and bond strength by holding one atom in the bond constant and varying the other atom within a group

**TABLE 9.4** **The Relation of Bond Order, Bond Length, and Bond Energy**

| BOND | BOND ORDER | AVERAGE BOND LENGTH (pm) | AVERAGE BOND ENERGY (kJ/mol) |
|---|---|---|---|
| C—O | 1 | 143 | 358 |
| C=O | 2 | 123 | 745 |
| C≡O | 3 | 113 | 1070 |
| C—C | 1 | 154 | 347 |
| C=C | 2 | 134 | 614 |
| C≡C | 3 | 121 | 839 |
| N—N | 1 | 146 | 160 |
| N=N | 2 | 122 | 418 |
| N≡N | 3 | 110 | 945 |

or period. For example, the trend in single bond lengths, C—N > C—O > C—F, parallels the trend in atomic size, N > O > F, and is opposite the trend in bond energy, C—F > C—O > C—N. Thus, for *single bonds*, longer bonds are usually weaker.

SAMPLE PROBLEM 9.2 _____

## Comparing Bond Length and Bond Strength

**Problem:** Using the periodic table, but not Tables 9.2 and 9.3, rank the bonds in each set in order of *decreasing* bond length and bond strength:
**(a)** P—F, P—Br, P—Cl   **(b)** C=O, C—O, C≡O
**Plan:** In part **(a)**, P is singly bonded to a halogen atom, so all members of the set have a bond order of one. The bond length increases and bond strength decreases with halogen atomic radius, which we find from the periodic table. In part **(b)**, the same two atoms are bonded, but the bond orders differ. In this case, bond strength increases and bond length decreases as bond order increases.
**Solution: (a)** Atomic size increases down a group, so F < Cl < Br.
**Bond length: P—Br > P—Cl > P—F**
**Bond strength: P—F > P—Cl > P—Br**
**(b)** By ranking the bond orders, C≡O > C=O > C—O, we obtain
**Bond length: C—O > C=O > C≡O**
**Bond strength: C≡O > C=O > C—O**
**Check:** From Tables 9.2 and 9.3, we see that the ranking is correct.
**Comment:** Remember that for different atoms, as in part **(a)**, *the relationship between length and strength holds* only *for single bonds* and not in every case, so apply it with caution.

FOLLOW-UP PROBLEM 9.2
Rank the bonds in each set in order of *increasing* bond length and bond strength:
**(a)** Si—F, Si—C, Si—O; **(b)** N=N, N—N, N≡N.

_____

## How the Model Explains the Properties of Covalent Compounds

The covalent bonding model proposes that electron sharing between pairs of atoms leads to strong, localized bonds, usually within individual molecules. At first glance, it is not at all clear how this model is consistent with the familiar physical properties of covalent substances. After all, most of these compounds are gases (such as methane and ammonia), liquids (such as benzene and water), or low-melting solids (such as sulfur and paraffin wax). If covalent bonds are strong ( ~200 to 500 kJ/mol), why do these substances melt and boil at such low temperatures?

To answer this question, we must distinguish between two completely different sets of forces: (1) the *strong covalent bonding forces* holding the atoms together within the molecule (those we have been discussing) and (2) the *weak intermolecular forces* holding the molecules near each other in the macroscopic sample. These weak forces, not the strong covalent bonds, are responsible for the physical properties of covalent substances. Consider, for example, what happens when pentane ($C_5H_{12}$) boils (boiling point at 1 atm = 36.1°C). The weak interactions *between* the pentane molecules are affected, not the strong C—C and C—H covalent bonds *within* the molecules (Figure 9.14).

Some covalent substances do not consist of separate molecules; rather, they are held together by covalent bonds *throughout* the sample. The physical properties of these substances, called *network covalent solids, do* reflect the

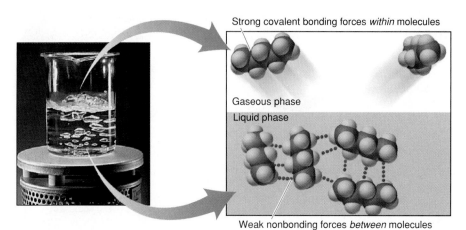

**FIGURE 9.14**
**Strong forces within molecules and weak forces between them.** When pentane boils, weak forces between molecules are overcome, and pentane molecules leave the liquid phase as separate units. The strong covalent bonds holding the atoms together within each molecule, however, remain intact.

strength of their covalent bonds (Figure 9.15). Quartz ($SiO_2$) is very hard and melts at 1550°C. It is composed of silicon and oxygen atoms connected by covalent bonds that extend throughout the sample; no separate molecules exist. Diamond consists of covalent bonds connecting all the carbon atoms in the sample. It is the hardest substance known and melts at around 3550°C. Clearly, covalent bonds *are* strong, but since most covalent substances consist of separate molecules with weak forces between them, their physical properties do not reflect this strength. (We discuss intermolecular forces in detail in Chapter 11.)

Unlike ionic compounds, most covalent substances are poor electrical conductors, even when melted or dissolved in water. Because the electrons are localized as either shared or unshared pairs, they are not free to move, and no ions are present. An electric current is carried by either mobile electrons or ions, so no entity in a covalently bonded substance can conduct a current. The essay on the following page describes an important laboratory tool for studying the properties of covalent bonds.

**Section Summary**
A shared pair of valence electrons attracts the nuclei of two atoms and holds them together in a covalent bond while making each atom's outer shell complete. The number of shared pairs between the two atoms is the bond order. For a given type of bond, the bond energy is the average energy required to completely separate the bonded atoms; the bond length is the average distance between their nuclei. For a given pair of bonded atoms, bond order is directly related to bond energy and inversely related to bond length. Substances that consist of separate molecules are generally soft and low melting due to the weak *noncovalent* forces among molecules. Those held together throughout by covalent bonds are extremely hard and high melting. Most covalent substances have low electrical conductivity because electrons are localized and ions are absent. The atoms in a covalent bond vibrate, and the energy of these vibrations can be studied with IR spectroscopy.

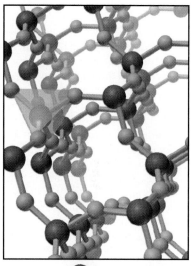

**A** Quartz   Silicon   Oxygen

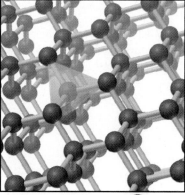

**B** Diamond   Carbon

**FIGURE 9.15**
**Covalent bond strength and the properties of network covalent solids. A,** In quartz ($SiO_2$), each Si atom is bonded covalently to four O atoms and each O atom bonds two Si atoms in a pattern that extends throughout the sample. Because no separate $SiO_2$ molecules are present, the melting point and hardness are very high. **B,** In diamond, each C atom is covalently bonded to four others throughout the crystal. Diamond is the hardest natural substance known and is extremely high melting.

| TOOLS OF THE CHEMISTRY LABORATORY | **Infrared Spectroscopy** |

**Infrared (IR) spectroscopy** is an instrumental technique used primarily for measuring the absorption of IR radiation by covalently bonded molecules. It has played a major role in our understanding of molecular structure and bonding. IR spectrometers are found in most research laboratories, particularly where organic molecules are studied, and are essential aids in identifying unknown substances. Their key components are the same as those of any spectrometer (see Tools of the Chemistry Laboratory, Section 7.2). The source emits radiation of many wavelengths, and those in the IR region are selected and directed at the sample. Certain wavelengths of light are absorbed more than others, and the IR spectrum of the compound is visually displayed.

What property of a molecule is displayed in its IR spectrum? All molecules, whether in a gas, liquid, or solid, undergo continual rotations and vibrations. Consider, for instance, a sample of ethane gas. The $H_3C$—$CH_3$ molecules zoom throughout the container, colliding with the walls and each other. If we could look closely at one molecule and disregard its motion through space, however, we would see the two $CH_3$ groups rotating relative to each other about the C—C bond. Each bond is also vibrating as though it were a flexible spring: stretching and compressing, twisting, bending, rocking, and wagging (Figure 9.A). Thus, the bond length is actually the *average* distance between nuclei, analogous to the average length of a spring when it is stretching and compressing. Each of these motions has its own natural frequency, which is based on the type of motion, the masses of the atoms, and the strengths of the bonds between them. Vibrational motions have frequencies that correspond to wave-

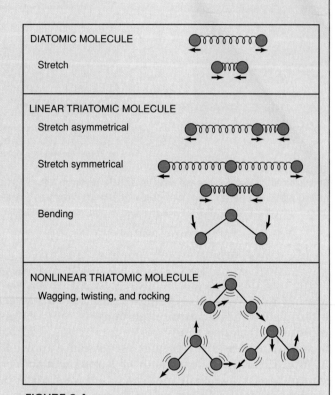

**FIGURE 9.A**
Some vibrational motions in general diatomic and triatomic molecules.

## 9.4 Between the Extremes: Electronegativity and Bond Polarity

Scientific models are idealized descriptions of reality. As we've discussed them so far in this chapter, the ionic and covalent bonding models portray compounds as being formed by either complete electron transfer *or* complete electron sharing. However, the great majority of compounds have bonds that are partly ionic and partly covalent. Now that you're familiar with the ideal models, we can explore the actual continuum in bonding that lies between these extremes.

### Electronegativity

One of the most important concepts in chemical bonding is **electronegativity (EN),** the relative ability of a bonded atom to attract the shared elec-

lengths between 2.5 and 25 μm, which make up a part of the IR region of the electromagnetic spectrum (see Figure 7.3). The energy of each of these vibrations is quantized. Just as an atom can absorb a photon whose energy corresponds to the difference between two quantized electron energy levels, a molecule can absorb an IR photon whose energy corresponds to the difference between two of its quantized vibrational energy levels.

The IR spectrum is important in compound identification because of two related factors. First, *each kind of bond has a characteristic range of absorption wavelengths*. For example, a C—C bond absorbs IR photons of different wavelength than a C=C bond, a C—H bond, or a C=O bond. Second, *the exact wavelength and amount of IR radiation absorbed by a given bond depend on the overall structure of the molecule*. This means that each compound has a characteristic spectrum that can be used to identify it, much as a fingerprint identifies a person. The spectrum appears as a series of peaks, varying in sharpness and intensity, that correspond to absorption frequencies across the IR region scanned. Figure 9.B shows the IR spectrum of acrylonitrile, a compound used to manufacture synthetic rubber and plastics; no other compound has exactly the same IR spectrum.

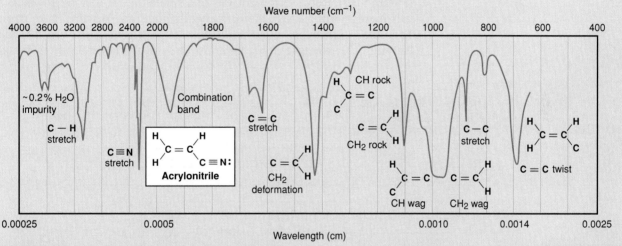

**FIGURE 9.B**

**The infrared (IR) spectrum of acrylonitrile.** The IR spectrum of acrylonitrile is typical of a molecule with several types of covalent bonds. There are many absorption bands (peaks) of differing depth and sharpness. Most peaks correspond to a particular type of vibration in a particular bonded group of atoms. Some broad peaks (for example, "combination band") represent a variety of vibrational types that are overlapping. The spectrum is reproducible and unique for acrylonitrile. (The units of the top axis are cycles/cm, or cm⁻¹. The scale expands to the right of 2000 cm⁻¹.)

trons.* More than 50 years ago, the American chemist Linus Pauling developed the most common scale of relative electronegativity values for the elements. Here is the basis of Pauling's approach. We might expect the bond energy of a molecule such as H—F to be the average of the energies of an H—H bond (432 kJ/mol) and an F—F bond (159 kJ/mol), or 296 kJ/mol. However, the actual bond energy of H—F is 565 kJ/mol, or 269 kJ/mol *higher* than the average. Pauling reasoned that this difference is due to an *electrostatic (charge) contribution* to the H—F bond energy. If F attracted the shared electron pair more strongly than H, that is, F were more electronegative than H, the electrons would spend more time closer to F. This unequal sharing of electrons would make the F end of the bond partially negative

---

*Electronegativity is not the same as electron affinity (EA), although many elements with a high EN also have a large EA. Electronegativity refers to an atom that is bound in a molecule attracting a shared electron pair; electron affinity refers to a separate atom in the gas phase gaining an electron to form a gaseous anion.

and the H end partially positive, and the attraction between these partial charges would *increase* the energy required to break the bond.

For the other hydrogen halides, the difference between the actual bond energy and the average of the bond energies of the component elements decreases down Group 7A(17) (HCl = 90, HBr = 50, and HI = 4 kJ/mol). The bond energy decreases because the electronegativity of the halogen atom, its ability to attract the bonded electron pair, decreases; thus, the relative halogen electronegativities decrease in the order F > Cl > Br > I. From many such comparisons, Pauling arrived at a *scale of relative EN values* (Figure 9.16). The values themselves are not measured quantities but are based on an assigned value of 4.0 for fluorine, the element with the highest electronegativity.

In general, an atom's electronegativity is inversely related to its size (Figure 9.17). Because the nucleus of a smaller atom is closer to the shared pair than that of a larger atom, it attracts bonding electrons more strongly. Therefore, *electronegativity generally increases up a group and across a period.* In Pauling's electronegativity scale and in any of the several others available,

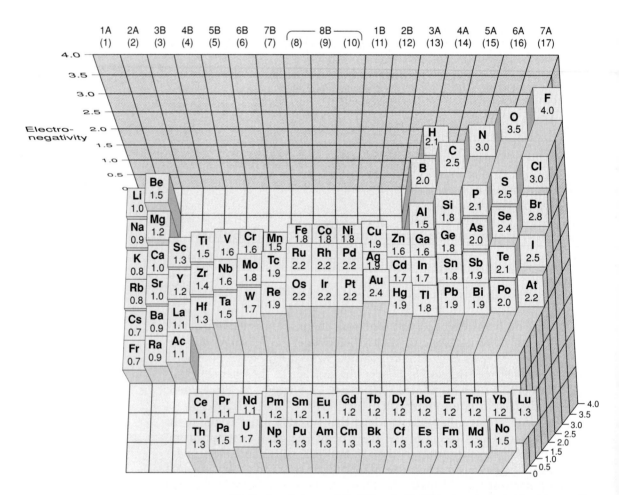

**FIGURE 9.16 The Pauling electronegativity (EN) scale.** The EN is shown by the height of the post (the value appears in the element box). In the main groups, EN generally *increases* from left to right and bottom to top. The noble gases are not shown because, except in a few cases, they do not form bonds. The transition and inner transition elements show relatively little change in EN. Hydrogen is shown near elements of similar EN to point out that its value is much higher than those of the Group 1A(1) metals.

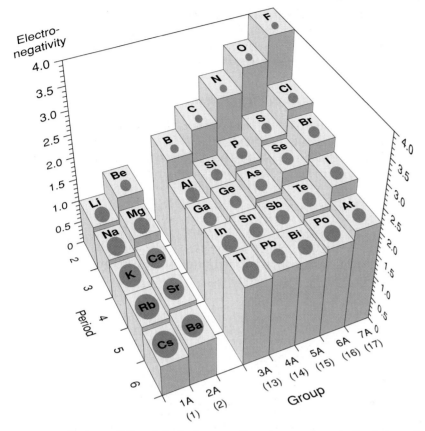

**FIGURE 9.17 Electronegativity and atomic size among the main-group elements.** The electronegativities of the main-group elements from Periods 2 to 6 (excluding noble gases) are shown as posts of different heights. On top of each post is a circle showing the relative atomic size. Note that, in general, an element with smaller atomic size has a higher electronegativity.

*nonmetals are more electronegative than metals.* The most electronegative element is fluorine, with oxygen a close second. Both elements form compounds with almost every other element. Thus, except when it bonds with fluorine, oxygen always pulls bonding electrons toward itself. The least electronegative element (also referred to as the most *electropositive*) is francium, in the lower left corner of the periodic table, but for all practical purposes, we consider cesium the most electropositive.*

One important use of electronegativity is in determining an atom's oxidation number (O.N.; see Section 4.4):

1. The more electronegative element in a bond is assigned *all* the *shared* electrons; the less electronegative element is assigned *none.*
2. Each element in a bond is assigned *all* of its *unshared* electrons.
3. The oxidation number is given by

$$\text{O.N.} = \text{no. of valence e}^- - (\text{no. of shared e}^- + \text{no. of unshared e}^-)$$

---

*In 1934, the American physicist Robert S. Mulliken developed an approach to electronegativity based solely on atomic properties:

$$\text{EN} = \frac{\text{IE} - \text{EA}}{2}$$

By this approach as well, fluorine, with a high ionization energy (IE) and a large negative electron affinity (EA), has a high EN; and cesium, with a low IE and small EA, has a low EN.

In HCl, for example, Cl is more electronegative than H. It has 7 valence electrons but is assigned 8 (2 shared + 6 unshared), so its oxidation number is $7 - 8 = -1$. The H atom has 1 valence electron and is assigned none, so its oxidation number is $1 - 0 = +1$.

## Polar Covalent Bonds and Bond Polarity

From our discussion of H—F, you can see that when atoms with different electronegativities bond, the bonding pair is not shared equally, so the bond has partially negative and positive poles. This type of bond is called a **polar covalent bond** and is depicted by a polar arrow ($+\longrightarrow$) pointing toward the negative pole, or by $\delta^+$ and $\delta^-$ symbols:

$$\overset{+\longrightarrow}{H—\ddot{\underset{\cdot\cdot}{F}}:} \quad \text{or} \quad {}^{\delta^+}H—\ddot{\underset{\cdot\cdot}{F}}:^{\delta^-}$$

In H—H or F—F, the bonded atoms are identical, so the bonding pair is shared equally, and the bond is called a **nonpolar covalent bond.** Thus, by knowing the EN values of the atoms in a bond, you can determine the direction of the bond polarity.

SAMPLE PROBLEM 9.3 _____

## Determining Bond Polarity from Electronegativity Values

**Problem: (a)** Indicate the polarity of the following bonds with a polar arrow: N—H, F—N, I—Cl.
**(b)** Rank the following bonds in order of increasing polarity: H—N, H—O, H—C.
**Plan: (a)** We use Figure 9.16 to find the EN values of the bonded atoms and point the polar arrow toward the negative end. **(b)** Each choice has H bonded to an atom from Period 2. Since EN increases across a period, the polarity is greatest for the bond whose Period 2 atom is farthest to the right.
**Solution: (a)** The EN of N = 3.0 and the EN of H = 2.1 so N is more electronegative than H: $\overset{\longleftarrow+}{\textbf{N—H}}$

The EN of F = 4.0, and the EN of N = 3.0 so F is more electronegative: $\overset{\longleftarrow+}{\textbf{F—N}}$

The EN of I = 2.5, and the EN of Cl = 3.0, so I is less electronegative: $\overset{+\longrightarrow}{\textbf{I—Cl}}$

**(b)** The order of increasing EN is C < N < O, so the order of bond polarity is **H—C < H—N < H—O**
**Comment:** As we discuss in Chapter 11, the polarity of the bonds in a molecule contributes to the overall polarity of the molecule, which is a major factor in the strength of intermolecular forces and the magnitude of physical properties.

FOLLOW-UP PROBLEM 9.3
Arrange each set of bonds in order of increasing polarity, and designate the direction of bond polarity with $\delta^+$ and $\delta^-$ symbols: **(a)** Cl—F, Br—Cl, Cl—Cl; **(b)** Si—Cl, P—Cl, S—Cl, Si—Si.

## The Partial Ionic Character of Polar Covalent Bonds

If you ask whether a bond is ionic or covalent, the answer in almost every case is "Both!" A better question is "*How* ionic or covalent is the bond?" The existence of partial charges means that a polar covalent bond behaves as if it were partially ionic. The **partial ionic character** of a bond is related directly to the **electronegativity difference (ΔEN),** the difference between the EN values of the bonded atoms: *a greater ΔEN results in larger partial charges and a higher partial ionic character.* For example, ΔEN in LiF(*g*) is 4.0 − 1.0 = 3.0; in HF(*g*), it is 4.0 − 2.1 = 1.9; in $F_2$(*g*), it is 4.0 − 4.0 = 0. Thus, the bond in LiF has more ionic character than the H—F bond, which has more than the F—F bond.

Various attempts have been made to classify the ionic character of bonds. One approach is to decide on arbitrary cutoff values of ΔEN that divide the bonds into ionic, polar covalent, and nonpolar covalent. However, a fixed cutoff value is inconsistent with the gradation of ionic character observed experimentally in a series of bonds. Based on a ΔEN range from 0 (completely nonpolar) to 3.3 (highly ionic), some approximate guidelines for this gradation appear in Figure 9.18.

Another approach to determining relative ionic or covalent character combines electronegativity difference with the behavior of a polar molecule in an electric field and allows us to calculate the *percent ionic character* of a bond. A value of 50% ionic character divides substances we recognize as "ionic" from those we recognize as "covalent." Such methods show 43% ionic character for the H—F bond and expected decreases for the other hydrogen halides: H—Cl is 19% ionic, H—Br 11%, and H—I 4%. A plot of percent ionic character vs. ΔEN for a variety of gaseous binary (two-atom) molecules is shown in Figure 9.19. The specific values are not important, but you should note that *percent ionic character generally increases with ΔEN.* Another point to note is that whereas some of the molecules, such as $Cl_2$(*g*), have 0% ionic character, none has 100% ionic character. Thus, *electron sharing occurs to some extent in every bond,* even one between an alkali metal and a halogen.

## The Continuum of Bonding Across a Period

Two reactive elements from opposite sides of the periodic table, a metal and a nonmetal, have a relatively large electronegativity difference and typically interact by electron transfer to form an ionic compound. Two reactive elements from the same side of the table, such as similar nonmetals, have a relatively small electronegativity difference and interact by electron sharing to form a covalent compound. If we combine a nonmetal such as $Cl_2$, with each of the elements in its period, we would expect to observe a gradation of bond type: from ionic through polar covalent to nonpolar covalent.

| ΔEN | IONIC CHARACTER |
|---|---|
| >1.8 | Mostly ionic |
| 0.4-1.8 | Polar covalent |
| <0.4 | Mostly covalent |
| 0 | Nonpolar covalent |

**A**

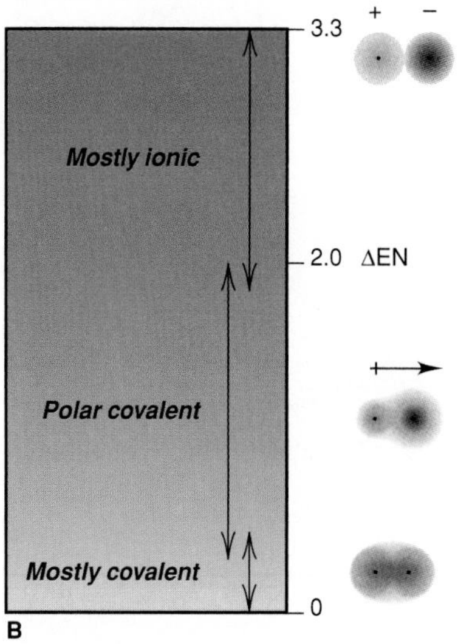

**B**

**FIGURE 9.18**
**Boundary ranges for classifying ionic character of chemical bonds. A,** The electronegativity difference (ΔEN) between bonded atoms is a general guide to a bond's relative ionic character. **B,** The gradation in ionic character is shown with shading across the entire bonding range from ionic *(green)* to covalent *(yellow).*

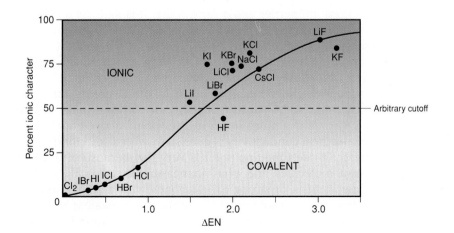

**FIGURE 9.19**
**Percent ionic character as a function of electronegativity difference (ΔEN).** The percent ionic character is plotted against ΔEN for some simple gaseous diatomic molecules. Note that, in general, ΔEN correlates with ionic character. (The arbitrary cutoff for an ionic compound is > 50% ionic character.)

| Sodium chloride NaCl | Magnesium chloride $MgCl_2$ | Aluminum trichloride $AlCl_3$ | Silicon tetrachloride $SiCl_4$ | Phosphorus trichloride $PCl_3$ | Disulfur dichloride $S_2Cl_2$ | Chlorine $Cl_2$ |
|---|---|---|---|---|---|---|

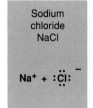  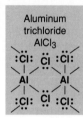 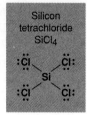   

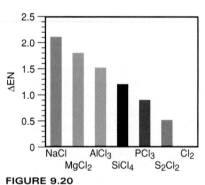

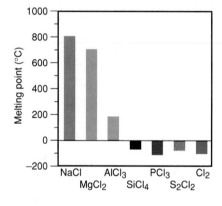

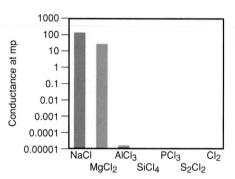

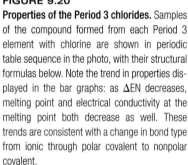

**FIGURE 9.20**

**Properties of the Period 3 chlorides.** Samples of the compound formed from each Period 3 element with chlorine are shown in periodic table sequence in the photo, with their structural formulas below. Note the trend in properties displayed in the bar graphs: as $\Delta EN$ decreases, melting point and electrical conductivity at the melting point both decrease as well. These trends are consistent with a change in bond type from ionic through polar covalent to nonpolar covalent.

Figure 9.20 shows samples of the most common Period 3 chlorides, binary compounds formed between chlorine and each of the elements in the period—NaCl, $MgCl_2$, $AlCl_3$, $SiCl_4$, $PCl_3$, $S_2Cl_2$, and $Cl_2$—along with structural formulas (discussed in the next section) and some properties. The first compound is sodium chloride, a colorless crystalline solid with typical ionic properties: high melting point and high conductivity when molten or dissolved. With these properties and a $\Delta EN$ of 2.1, it is ionic by any criterion. Magnesium chloride is mostly ionic, with a $\Delta EN$ of 1.8, but it has a lower melting point and lower conductivity.

Aluminum chloride is less ionic still. Its $\Delta EN$ value of 1.5 indicates a highly polar covalent Al—Cl bond. Rather than a lattice of $Al^{3+}$ and $Cl^-$ ions, it consists of extended layers of linked Al and Cl atoms. Strong ionic-polar interactions hold the particles together in the layer, but weak forces between layers result in a much lower melting point, above which the compound's low electrical conductivity is consistent with a scarcity of ions.

Silicon tetrachloride's melting point is very low because of weak forces *between* separate molecules, and it has no measurable electrical conductivity. Each molecule consists of strong Si—Cl bonds with a $\Delta EN$ value of 1.2, which puts these bonds near the middle of the polar covalent region of the bonding continuum. With a $\Delta EN$ value of 0.9, phosphorus trichloride continues the trend toward lower bond polarity, as evidenced by its very low melting point. The bonds in disulfur dichloride are even less polar ($\Delta$EN = 0.5). The series ends with nonpolar, diatomic chlorine, the only one of these substances that is a gas at room temperature. Thus, as the electronegativity difference becomes smaller, the bond becomes more covalent, and the properties of the Period 3 chlorides change from those of a solid consisting of ions to those of a gas consisting of individual molecules.

### Section Summary

An atom's electronegativity is related to its ability to pull bonded electrons toward it, generating partial charges at the ends of the bond. Electronegativity increases across a period and decreases down a group, the reverse of the trend in size. The greater the $\Delta$EN of the atoms in a bond, the more polar is the bond and the greater its ionic character. There is a gradation of bond type, from ionic to polar covalent to nonpolar covalent, which can be seen by combining a halogen with the other elements in its period.

## 9.5    Depicting Molecules and Ions with Lewis Structures

One of the most important ideas in chemistry is that each molecule has a particular shape, a minute architecture of atoms with specific angles and distances between them. The first step in visualizing a molecule is to write its **Lewis structure** (or **Lewis formula**), a type of structural formula consisting of electron-dot symbols that depict the sequence of atoms, the bonding pairs that hold them together, and the lone pairs that fill each atom's outer shell.* In upcoming chapters, we use Lewis structures to predict molecular shape and several physical and chemical properties. In many cases, the octet rule guides us in allotting electrons to the atoms in a Lewis structure; in many other cases, however, we set the rule aside.

### Using the Octet Rule to Write Lewis Structures

To write a Lewis structure, we decide on the sequence of the atoms in the molecule (or ion) and distribute its valence electrons among bonding and lone pairs. To begin, we examine Lewis structures for species in which each atom fills its outer level with eight electrons (or two for H).

**Lewis structures for molecules with single bonds.**  Let's first discuss how to write Lewis structures for molecules that have only singly bonded atoms, using nitrogen trifluoride, $NF_3$, as an example.

*Step 1. Decide on the sequence of atoms.* For compounds of formula $AB_n$, place the atom with lower group number in the center, the one that needs more electrons to attain an octet. In $NF_3$, the N (Group 5A) has five electrons so it needs three, whereas F (Group 7A) has seven so it needs only one; thus, N goes in the center with the three F atoms around it:

F<br>
   N<br>
F   F

We usually find this rule places the least electronegative atom in the center (EN of N = 3.0 and EN of F = 4.0). Since H can form only one bond, it is *never* a central atom.

---

*A Lewis *structure* may be more correctly called a Lewis *formula* because it provides no structural information about the actual arrangement of atoms in space. Nevertheless, use of the term Lewis "structure" is a convention we follow in this text.

*Step 2. Determine the total number of valence electrons available.* For molecules, add up the valence electrons of all the atoms (the number of valence electrons equals the A-group number). In $NF_3$, N has five valence electrons, and each F has seven:

$$[1 \times N(5e^-)] + [3 \times F(7e^-)] = 5 + 21 = 26 \text{ valence } e^-$$

For polyatomic ions, *add* one $e^-$ for each negative charge, or *subtract* one $e^-$ for each positive charge.

*Step 3. Draw a single bond from each surrounding atom to the central atom, and subtract two valence electrons for each bond.* There must be at least a single bond between bonded atoms:

$$
\begin{array}{c}
\text{F} \\
| \\
\text{F} \diagdown \overset{\text{N}}{\phantom{.}} \diagup \text{F}
\end{array}
$$

Subtract $2e^-$ for each single bond from the total number of valence electrons available (from Step 2) to find the number remaining:

$$3 \text{ N—F bonds} \times 2e^- = 6e^- \qquad 26e^- - 6e^- = 20e^- \text{ remaining}$$

*Step 4. Distribute the remaining electrons in pairs so that each atom obtains eight electrons (or two for H).* Place lone pairs on the *surrounding (more electronegative) atoms* first to give each an octet. If any electrons remain, place them around the central atom. Then check that each atom has $8e^-$:

$$
\begin{array}{c}
\text{:}\ddot{\text{F}}\text{:} \\
| \\
\text{:}\ddot{\text{F}} \diagdown \overset{\text{N}}{\underset{..}{\phantom{.}}} \diagup \ddot{\text{F}}\text{:}
\end{array}
$$

This is the Lewis structure for $NF_3$. This particular placement of F atoms around the N was chosen because it resembles the molecular shape of $NF_3$, as you will learn in Chapter 10. Lewis structures are not used to indicate shape, however, so an equally correct Lewis structure for $NF_3$ would be

$$
\begin{array}{c}
\text{:}\ddot{\text{F}}\text{:} \\
| \\
\text{:}\ddot{\text{F}}\text{—}\underset{..}{\text{N}}\text{—}\ddot{\text{F}}\text{:}
\end{array}
$$

or any other placement that retains the same *sequence* of atoms.

Using just these four steps, you can write Lewis structures for any singly bonded molecules with one of the Period 2 nonmetals C, N, or O as its central atom, in addition to several other molecules with central atoms from higher periods.

SAMPLE PROBLEM 9.4 _____

## Writing Lewis Structures for Molecules with One Central Atom

**Problem:** Write a Lewis structure for $CF_2Cl_2$, one of the compounds suspected of depleting stratospheric ozone.
**Solution:** *Step 1.* Decide on the sequence of atoms. In $CF_2Cl_2$, carbon has the lowest group number and EN, so it is the central atom. The other atoms surround it, but their specific placement is not important:

$$
\begin{array}{c}
\text{Cl} \\
\text{F} \quad \text{C} \quad \text{F} \\
\text{Cl}
\end{array}
$$

*Step 2.* Determine the total number of valence electrons:

$$[1 \times C(4e^-)] + [2 \times F(7e^-)] + [2 \times Cl(7e^-)] = 32e^-$$

*Step 3.* Draw single bonds to the central atom and subtract $2e^-$ for each bond:

$$\underset{\underset{\displaystyle Cl}{|}}{\overset{\overset{\displaystyle Cl}{|}}{F-C-F}}$$

Four single bonds use $8e^-$, so $32e^- - 8e^- = 24e^-$ remaining.
*Step 4.* Distribute the remaining valence electrons in pairs, beginning with the surrounding atoms, so that each atom has an octet:

$$\underset{\underset{\displaystyle :\ddot{C}l:}{|}}{\overset{\overset{\displaystyle :\ddot{C}l:}{|}}{:\ddot{F}-C-\ddot{F}:}}$$

**Check:** Counting the electrons shows each atom has an octet. Remember that the bonding electrons are counted as belonging to each atom in the bond. Be sure the total number of valence electrons equals $32e^-$.

**FOLLOW-UP PROBLEM 9.4**
Write Lewis structures for **(a)** $H_2S$; **(b)** $OF_2$; **(c)** $SOCl_2$.

More complex molecules have two or more central atoms bonded to each other, with the other atoms around them.

**SAMPLE PROBLEM 9.5**

**Writing Lewis Structures for Molecules with More than One Central Atom**

**Problem:** Write the Lewis structure for methanol ($CH_3OH$), an important industrial alcohol that is being used as a gasoline alternative in car engines.
**Solution:** *Step 1.* Write atom sequence. The H atoms can have only one bond, so C and O must be adjacent central atoms. In nearly all their compounds, C has four bonds and O has two, so we arrange the H atoms to show this:

$$\begin{array}{cccc} & H & & \\ H & C & O & H \\ & H & & \end{array}$$

*Step 2.* Find sum of valence electrons:

$$[1 \times C(4e^-)] + [1 \times O(6e^-)] + [4 \times H(1e^-)] = 14e^-$$

*Step 3.* Add single bonds and subtract $2e^-$ for each bond:

$$\underset{\underset{\displaystyle H}{|}}{\overset{\overset{\displaystyle H}{|}}{H-C-O-H}}$$

Five bonds use $10e^-$, so $14e^- - 10e^- = 4e^-$ remaining.
*Step 4.* Add remaining electrons in pairs:

$$\underset{\underset{\displaystyle H}{|}}{\overset{\overset{\displaystyle H}{|}}{H-C-\ddot{O}-H}}$$

Carbon already has an octet, so the four remaining valence electrons form two lone pairs on O and give the Lewis structure for methanol.
**Check:** Each H atom has $2e^-$, and the C and O each have $8e^-$. The total number of valence electrons is $14e^-$.

**FOLLOW-UP PROBLEM 9.5**
Write Lewis structures for **(a)** hydroxylamine ($NH_2OH$) and **(b)** dimethyl ether ($CH_3OCH_3$).

---

**Lewis structures for molecules with multiple bonds.** In writing some Lewis structures, you will find that after Steps 1 to 4, there are not enough electrons for the central atom (or one of the central atoms) to attain an octet. This usually means that a multiple bond must be present, and an additional step is needed:

*Step 5. Cases involving multiple bonds.* If, after Step 4, a central atom still does not have an octet, change a lone pair from one of the surrounding atoms into a bonding pair to the central atom to make a multiple bond.

**SAMPLE PROBLEM 9.6** ⎯⎯⎯⎯⎯⎯⎯⎯⎯⎯⎯⎯⎯⎯⎯⎯⎯⎯⎯⎯⎯

**Writing Lewis Structures for Molecules with Multiple Bonds**

**Problem:** Write Lewis structures for the following:
**(a)** Ethylene ($C_2H_4$), the most important reactant in polymer manufacture
**(b)** Nitrogen ($N_2$), the most abundant atmospheric gas
**Plan:** We begin the solution after going through Steps 1 to 4: arranging the atoms, counting the total valence electrons, making single bonds, and distributing the remaining valence electrons in pairs to attain octets. Then we continue with Step 5, if needed.
**Solution: (a)** For $C_2H_4$. After Steps 1 to 4, we have

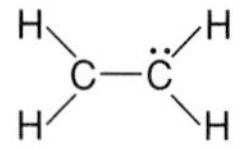

*Step 5.* If necessary, change a lone pair to a bonding pair. The C on the right has an octet, but the C on the left has only $6e^-$, so we convert the lone pair to another bonding pair between the two C atoms:

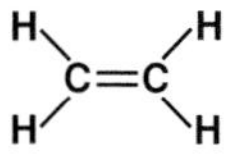

**(b)** For $N_2$. After Steps 1 to 4 we have

$$:\ddot{N}-\ddot{N}:$$

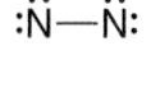

*Step 5.* Neither N has an octet, so we change a lone pair to bonding pair:

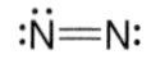

$$:\ddot{N}=N:$$

In this case, moving one lone pair to make a double bond still does not give each N an octet, so we move another lone pair to make a triple bond:

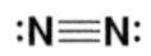

$$:N{\equiv}N:$$

**Check:** In part **(a)**, since each C counts the $4e^-$ in the double bond as part of its own octet, each now has $8e^-$. The valence electron total is $12e^-$. In part **(b)**, since each N counts the $6e^-$ in the triple bond as part of its own octet, each now has $8e^-$. The valence electron total is $10e^-$.

**FOLLOW-UP PROBLEM 9.6**
Write Lewis structures for **(a)** CO (the only common molecule in which C has three bonds); **(b)** HCN; **(c)** $CO_2$.

## Resonance: Delocalized Electron-Pair Bonding

For molecules and ions with multiple bonds adjacent to single bonds, it is often possible to write more than one Lewis structure, each with the same sequence of atoms.

Consider the case of ozone ($O_3$), a serious air pollutant at ground level but a life-sustaining absorber of harmful ultraviolet (UV) radiation in the stratosphere. Two valid Lewis structures (with lettered O atoms for clarity) are

In structure I, oxygen B has a double bond to oxygen A and a single bond to oxygen C. In structure II, the single and double bonds are reversed. These are *not* two different types of $O_3$ molecules, just two different Lewis structures for the same molecule.

The point is that *neither* Lewis structure depicts $O_3$ accurately. Bond length and bond energy measurements indicate that the two bonds in ozone are *identical*, with properties that lie *between* those of an O—O bond and an O=O bond, something like a "one-and-a-half" bond. The molecule is shown more correctly with two Lewis structures, called **resonance structures** (or **resonance forms**), with a two-headed resonance arrow (↔) between them. Resonance structures *have the same sequence of atoms but differ in the location of the bonding and lone electron pairs.* You can convert one resonance form to the other by moving lone pairs to bonding positions, and vice versa:

*Resonance structures are not real bonding depictions:* $O_3$ does *not* change back and forth between structure I this instant and structure II the next. The actual molecule is a **resonance hybrid,** something like an average of the resonance forms.

Our need for resonance structures is the result of **electron-pair delocalization.** In an isolated single, double, or triple bond, each electron pair is attracted by the two bonded atoms, and the electron density is greatest in the region between the two nuclei: each electron pair is *localized.* In the resonance hybrid for $O_3$, however, two of the electron pairs (one bonding and one nonbonding) are *delocalized:* their density is "spread" over the entire molecule. This results in two identical bonds, each consisting of a single bond (the localized electron pair) and a *partial bond* (the contribution from a delocalized electron pair). We draw the resonance hybrid with a continuous dashed line to show the delocalized pairs:

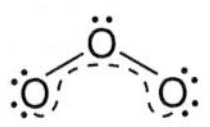

Electron delocalization stabilizes the molecule by diffusing electron density over a greater volume and reducing electron-electron repulsions. Resonance is very common, and many molecules (and ions) with a double bond next to a single bond in their Lewis structures are resonance hybrids of two or more resonance forms. Benzene ($C_6H_6$), for example, has two important resonance forms in which alternating single and double bonds have different positions. The actual molecule has six identical carbon-carbon bonds as a result of six C—C bonds and three additional electron pairs that are delocalized over all six C atoms:

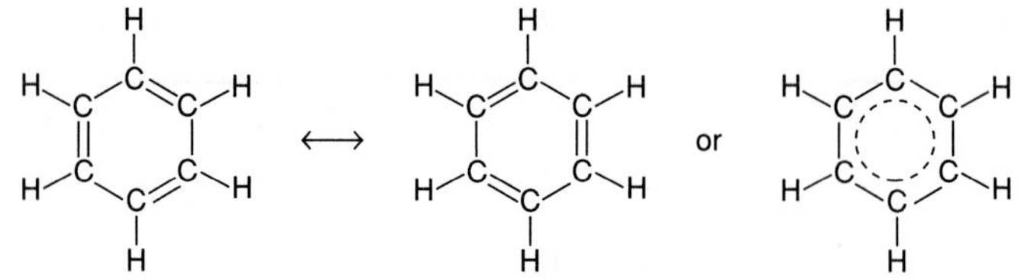

Partial bonding, such as occurs in resonance hybrids, often leads to fractional bond orders. The bond order in $O_3$ is 3 electron pairs/2 atom-to-atom linkages, or $1\frac{1}{2}$; the carbon-to-carbon bond order in benzene is 9/6, also $1\frac{1}{2}$. In the carbonate ion, $CO_3^{2-}$, three resonance structures can be drawn. Each has 4 electron pairs shared among 3 linkages, so the bond order is $1\frac{1}{3}$. One of the three resonance structures for $CO_3^{2-}$ is

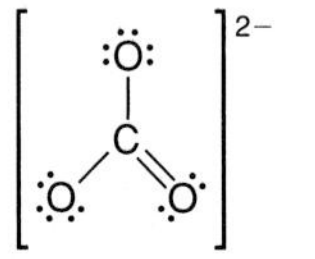

Note that *the Lewis structure of a polyatomic ion is shown in square brackets, with its charge as a right superscript outside the brackets.*

SAMPLE PROBLEM 9.7 _____

**Writing Resonance Structures**

**Problem:** Write resonance structures for the nitrate ion, $NO_3^-$.
**Plan:** We write a Lewis structure using the steps outlined earlier, remembering to add one $e^-$ to the total number of valence electrons because of the $1-$ ion charge. Then we move lone and bonding pairs to write other resonance forms and connect them with the resonance arrow.
**Solution:** After Steps 1 to 4 we have

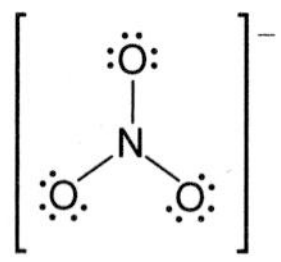

*Step 5.* We change one lone pair on an O atom to a bonding pair and form a double bond, which gives each atom an octet. All the O atoms are equivalent, however, so we can move a lone pair from each O atom and obtain three resonance structures:

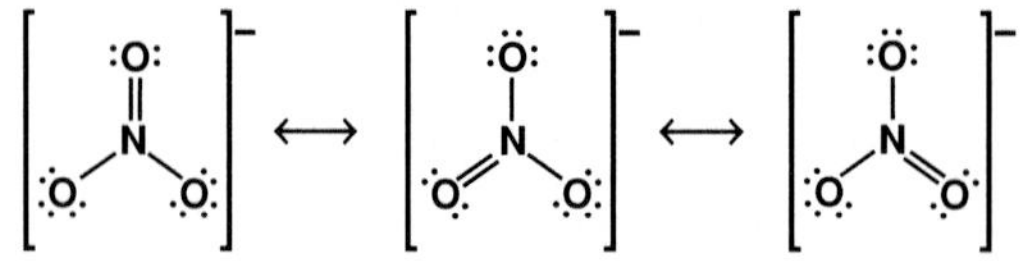

**Check:** Each structure has an octet around each atom, and each has $24e^-$, the sum of the valence electron total and $1e^-$ from the ionic charge.
**Comment:** No double bond actually occurs in the $NO_3^-$ ion. The actual ion is a resonance hybrid of these three structures with a bond order of $1\frac{1}{3}$.

FOLLOW-UP PROBLEM 9.7
Draw the two other resonance structures for $CO_3^{2-}$.

## Formal Charge: Selecting the Best Resonance Structure

In the previous examples, the resonance forms were *equally* mixed to form the resonance hybrid because the molecules (or ions) were symmetrical: the surrounding atoms were the same. When this is not the case, one resonance form may look more like the hybrid than the others. In other words, since the resonance hybrid is an "average" of the resonance forms, one form may contribute more and "weight" the average in its favor. One way to select the most important resonance form is to determine each atom's **formal charge,** the hypothetical charge it has in the molecule or ion. (We use the word "hypothetical" because formal charge is only a bookkeeping technique for assigning electrons to atoms; the charges do not really exist in the molecule.)

An atom's formal charge is the number of its valence electrons minus the number of valence electrons it is assigned. An atom is assigned *all* of its unshared valence electrons and *half* of its shared valence electrons. Thus,

Formal charge of atom

$= $ no. of valence $e^-$

$\quad - $ (no. of unshared valence $e^-$ $+ \frac{1}{2}$ no. of shared valence $e^-$) **(9.3)**

Of course, *the formal charges must sum to the actual charge on the species:* zero for a molecule and the ionic charge for an ion. In the two $O_3$ resonance forms, the formal charge of $O_A$ in structure I is

$$6 \text{ valence } e^- - 4 \text{ unshared } e^- - \tfrac{1}{2}(4 \text{ shared } e^-) = 0$$

The formal charges in the two forms are

$O_A[6 - 4 - \tfrac{1}{2}(4)] = 0$

$O_B[6 - 2 - \tfrac{1}{2}(6)] = +1$

$O_C[6 - 6 - \tfrac{1}{2}(2)] = -1$    **I**      **II**    $O_A[6 - 6 - \tfrac{1}{2}(2)] = -1$

$O_B[6 - 2 - \tfrac{1}{2}(6)] = +1$

$O_C[6 - 4 - \tfrac{1}{2}(4)] = 0$

Forms I and II are symmetrical—each has the same formal charges but on different atoms—so they contribute equally to the resonance hybrid.

Three general points govern our selection of the most important resonance structures based on formal charge: (1) smaller formal charges (whether positive or negative) are preferable to larger ones; (2) like charges on adjacent atoms are not desirable; and (3) a more negative formal charge should reside on a more electronegative atom. Consider next the cyanate ion, $NCO^-$. Three resonance forms with formal charges are

Formal charges:

Resonance forms:    **A**        **B**        **C**

We eliminate form A because it has larger formal charges than the others and a positive formal charge on the more electronegative O. Forms B and C have the same magnitude of formal charges, but form C has a $-1$ charge on O, which is more electronegative than N. Therefore, B and C are significant contributors to the resonance hybrid of the cyanate ion, with C the more important.

Formal charge (used to examine resonance structures) is *not* the same as oxidation number (used to monitor redox reactions). In determining the formal charge, the bonding electrons are assigned *equally* to the bonded atoms, so each atom receives half of them:

$$\text{Formal charge} = \text{valence } e^- - (\text{lone pair } e^- + \tfrac{1}{2} \text{ bonding } e^-)$$

In determining the oxidation number, the shared electrons are assigned *completely* to the more electronegative atom:

$$\text{Oxidation number} = \text{valence } e^- - (\text{lone pair } e^- + \text{ bonding } e^-)$$

Here are the formal charges and oxidation numbers of the three cyanate ion resonance structures:

Formal charges:

Oxidation numbers:

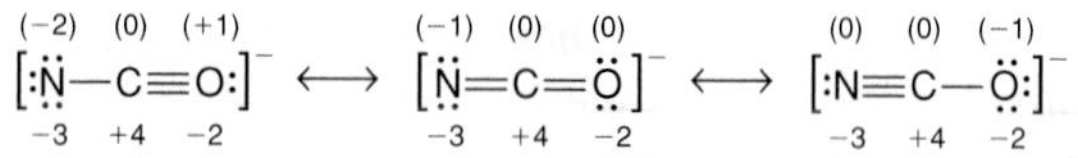

$$\underset{-3 \quad +4 \quad -2}{\overset{(-2) \quad (0) \quad (+1)}{[:\!\ddot{N}\!-\!C\!\equiv\!O\!:]^-}} \longleftrightarrow \underset{-3 \quad +4 \quad -2}{\overset{(-1) \quad (0) \quad (0)}{[\ddot{N}\!=\!C\!=\!\ddot{O}]^-}} \longleftrightarrow \underset{-3 \quad +4 \quad -2}{\overset{(0) \quad (0) \quad (-1)}{[:\!N\!\equiv\!C\!-\!\ddot{O}\!:]^-}}$$

Note that the oxidation numbers *do not* change from one resonance form to another, but the formal charges *do*.

### Lewis Structures for Exceptions to the Octet Rule

The octet rule is a useful guide for simple molecules with Period 2 central atoms, but *many compounds do not have eight electrons around each atom.* In some cases there are fewer electrons, in others more. The most significant examples of such octet rule exceptions are electron-deficient molecules, odd-electron molecules, and molecules with expanded valence shells.

**Electron-deficient molecules.** Gaseous compounds containing the element beryllium or boron as the central atom are often **electron deficient,** in that they have *fewer* than eight electrons around the Be or B atom. The Lewis structures of beryllium chloride* and boron trifluoride are

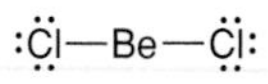

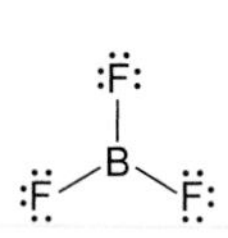

Note that there are only four electrons around beryllium and six around boron. Why don't nonbonding pairs from the surrounding halogen atoms form multiple bonds to the central atoms, thereby satisfying the octet rule? Halogen atoms are much more electronegative than beryllium or boron, and formal charges show that these are unlikely structures:

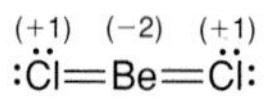

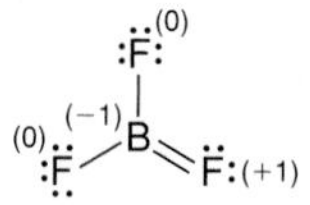

Electron-deficient molecules often attain an octet in reactions by forming additional bonds. When $BF_3$ reacts with ammonia, for instance, a compound forms in which boron attains its octet[†]:

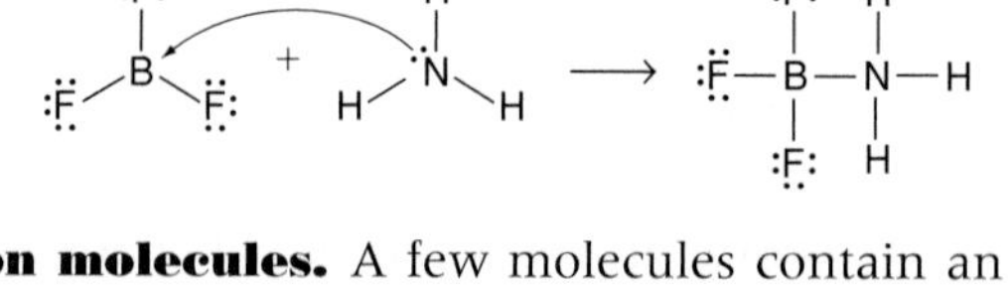

**Odd-electron molecules.** A few molecules contain an odd number of valence electrons, so they cannot possibly have all their electrons in pairs. Most examples have a central atom from an odd-numbered group, such as N [Group 5A(15)] and Cl [Group 7A(17)]. Such species, called **free radicals,** contain a lone (unpaired) electron, which makes them paramagnetic (Section 8.4) and extremely reactive. An important example is $NO_2$:

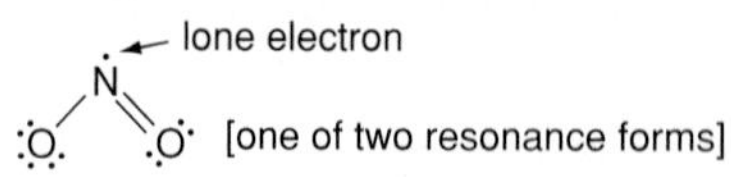

---

*Despite beryllium being a member of the alkaline earth metals [Group 2A(2)], most of its compounds have properties consistent with covalent, rather than ionic, bonding. For example, molten $BeCl_2$ does not conduct electricity, indicating the absence of ions (Chapter 13).
[†]Reactions of the sort shown here, in which one species "donates" an electron pair to another to form a covalent bond, are examples of Lewis acid-base reactions (Chapter 17).

Nitrogen dioxide is formed when the NO in auto exhaust reacts with $O_2$ in sunlight; it is a major contributor to urban smog.

Free radicals react to pair up their lone electrons. When two $NO_2$ molecules collide, for example, they form dinitrogen tetraoxide, $N_2O_4$, and each N attains an octet: ◆

◆ **Free-Radical Activity.** In the rarified outer atmosphere, free radicals, such as individual O, N, and H atoms, have significant lifetimes because collisions occur infrequently. Near the Earth's surface, where gas densities are a million times greater, such free radicals react with other substances almost immediately. In biological systems, free radicals can be extremely dangerous because they abstract H atoms from cell components to form electron-pair bonds and new free radicals, thus proliferating as they rupture genes and cell membranes.

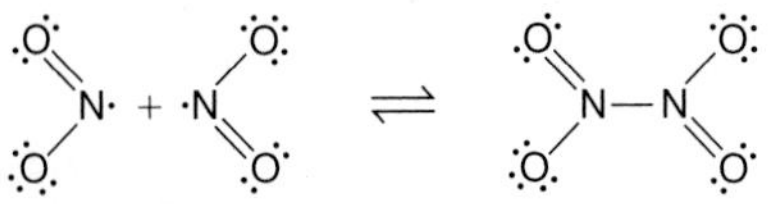

**Expanded valence shells.** Many molecules and ions have *more* than eight valence electrons around the central atom. The only way to accommodate additional pairs is for the central atom to utilize its empty *outer d* orbitals in addition to its occupied *s* and *p* orbitals. Therefore, in molecular compounds, **expanded valence shells** occur only around a *central nonmetal atom from Period 3 or higher,* those in which *d* orbitals are available.

An example is sulfur hexafluoride ($SF_6$), a remarkably inert gas used as an insulator in electrical equipment. The central sulfur is surrounded by six covalent bonds, one to each fluorine, for a total of 12 electrons:

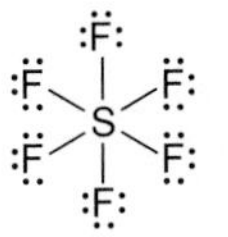

Another example is phosphorus pentachloride ($PCl_5$), a fuming yellow-white solid used in the manufacture of lacquers and films. $PCl_5$ is formed when phosphorus trichloride ($PCl_3$) reacts with chlorine gas. Note that the P in $PCl_3$ has an octet, but it uses the lone pair to form two more bonds to chlorine and expands its valence shell in $PCl_5$ to a total of 10 electrons:

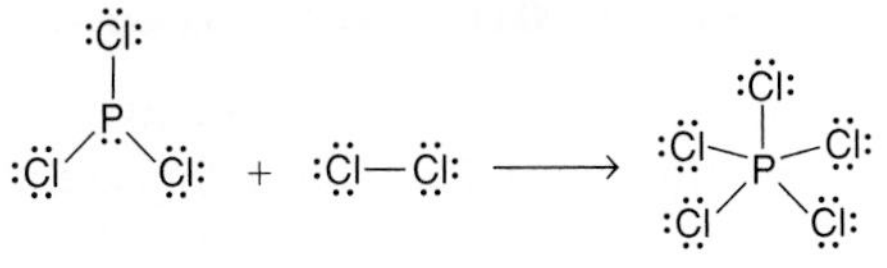

*An atom expands its valence shell to form more bonds,* a process that releases energy. Note that when $PCl_5$ forms, even though *one* Cl—Cl bond is broken (left side of the equation), *two* P—Cl bonds are formed (right side), for a net increase of one bond.

In the previous cases, the central atom expanded its valence shell to form bonds to *more than four* atoms. But there are many cases in which the central atom bonds to *four or fewer* atoms. Consider sulfuric acid, the industrial chemical produced in the greatest quantity. Two Lewis structures for $H_2SO_4$, with formal charges, are

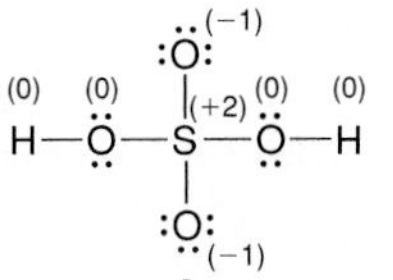

**A**

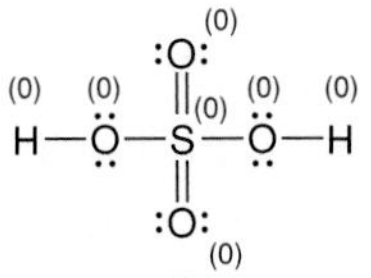

**B**

In structure B, sulfur has an expanded valence shell of 12 electrons. Structures A and B are both valid resonance forms of sulfuric acid, but B is a more significant contributor to the resonance hybrid because it has lower formal charges. Most importantly, structure B is consistent with observed bond lengths. In gaseous $H_2SO_4$, the two sulfur-oxygen bonds with H atoms attached to O are ~157 pm long, whereas the two sulfur-oxygen bonds without H atoms attached to O are ~142 pm long. Shorter bonds indicate double-bond character, which is shown in structure B.

When sulfuric acid loses two $H^+$ ions, it forms the sulfate ion, $SO_4^{2-}$. All the sulfur-oxygen bonds in $SO_4^{2-}$ are ~149 pm long, intermediate in length between the two S=O bonds (~142 pm) and the two S—O bonds (~157 pm) in the parent acid. Two of the six resonance forms consistent with these data are

Thus, the $SO_4^{2-}$ ion is a resonance hybrid with four S—O bonds and two more bonding pairs delocalized over the structure, so each sulfur-oxygen bond has a bond order of $1\frac{1}{2}$. The sulfur-oxygen bonds in $SO_2$ and $SO_3$ are all approximately the S=O bond length (142 pm), so Lewis structures for these molecules with formal charges are

Sulfur and phosphorus can accommodate a maximum of 12 electrons, and iodine as many as 14. Keep in mind that these *atoms expand their valence shells to form more bonds and minimize formal charge.*

SAMPLE PROBLEM 9.8 ⎯⎯⎯⎯⎯⎯⎯⎯⎯⎯⎯⎯⎯⎯⎯⎯⎯⎯⎯⎯⎯⎯⎯

## Writing Lewis Structures for Octet Rule Exceptions

**Problem:** Write Lewis structures for **(a)** $H_3PO_4$   **(b)** $BFCl_2$
**Plan:** We write each Lewis structure and examine it for exceptions to the octet rule. In **(a)**, the central atom is P, which is in Period 3, so it can use *d* orbitals to have more than an octet. Therefore, we can write more than one Lewis structure. We use formal charge to decide if one resonance form is more important. In **(b)**, the central atom is B, which can have fewer than an octet.
**Solution: (a)** For $H_3PO_4$, two possible Lewis structures with formal charges are

Structure II has lower formal charges, so it is a more important resonance form.
**(b)** For $BFCl_2$, the Lewis structure leaves B with only six electrons surrounding it:

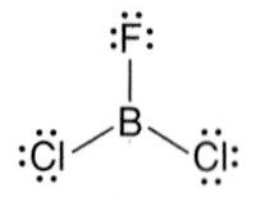

FOLLOW-UP PROBLEM 9.8
Write Lewis structures for **(a)** $POCl_3$; **(b)** $ClO_2$; **(c)** $XeF_4$.

## Section Summary

Lewis structures are two-dimensional representations of molecules (or ions) that show the sequence of atoms and distribution of valence electrons among bonding and lone pairs. When two or more Lewis structures can be drawn for the same atom sequence, the actual compound is a hybrid of those resonance forms. Formal charge can be used to determine the most important contributor to the hybrid. Electron-deficient molecules (central Be or B) and odd-electron species (free radicals) have fewer than an octet around the central atom but can attain an octet in reactions. A central non-metal from Period 3 or higher can expand its valence shell to accommodate more than eight electrons by using $d$ orbitals.

## 9.6 Using Lewis Structures and Bond Energies to Calculate Heats of Reaction

Covalent bond strength is an important factor in a molecule's chemical re-activity because, *for similar kinds of substances, one with weaker bonds is usually more reactive than one with stronger bonds.* In Chapter 6, for example, you saw that energy is released when a fuel's weaker (less stable, higher energy) bonds break apart and the waste product's stronger (more stable, lower energy) bonds form. We use Lewis structures and bond energies (bond enthalpies) to calculate the heat of reaction ($\Delta H^0_{rxn}$) by noting which reactant bonds break and which product bonds form. A simpler approach is to assume that all the reactant bonds break to give individual atoms, from which all the product bonds form (Figure 9.21). Hess's law allows us to sum the bond energies (with their appropriate signs) to arrive at the overall heat of reaction, regardless of how the actual process occurred.

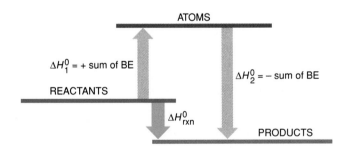

**FIGURE 9.21**
Using bond energies to calculate the enthalpy change of reaction ($\Delta H^0_{rxn}$). Any chemical reaction can be considered to occur in two steps: (1) reactant bonds break to yield separate atoms in a step that absorbs heat, and (2) the atoms combine to form product bonds in a step that releases heat. When the total bond energy of the products is greater than that of the reactants, more energy is released than is absorbed, and the reaction is exothermic (as shown); $\Delta H^0_{rxn}$ is negative. When the total bond energy of the products is less than that of the reactants, the reaction is endothermic; $\Delta H^0_{rxn}$ is positive.

Energy is required to break bonds ($\Delta H^0 > 0$) and is released when bonds form ($\Delta H^0 < 0$), and the sum of these enthalpy changes is the heat of reaction:

$$\Delta H^0_{rxn} = \Delta H^0_{\text{reactant bonds broken}} + \Delta H^0_{\text{product bonds formed}} \quad\quad (9.4)$$

Note that in an exothermic reaction, the total energy of bonds formed in the products is greater than that of bonds broken in the reactants, so the larger negative sum makes the net $\Delta H^0_{rxn}$ negative. In an endothermic reaction, on the other hand, the total energy of bonds broken in the reactants is greater, so the larger positive sum makes the net $\Delta H^0_{rxn}$ positive.

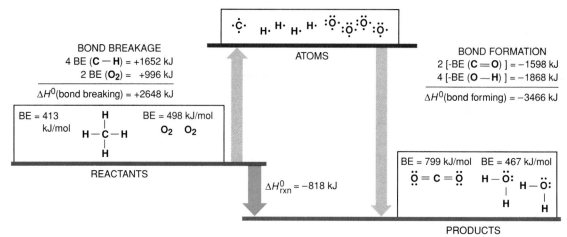

**FIGURE 9.22**

**Using bond energies to calculate $\Delta H^0_{comb}$ of methane.** By treating the combustion of methane as a two-step process (see Figure 9.21), we find that it is exothermic.

Let's apply this method to calculate $\Delta H^0_{rxn}$ for the combustion of methane and compare it with the value obtained from calorimetry:

$$CH_4(g) + 2O_2(g) \rightarrow CO_2(g) + 2H_2O(g) \quad \Delta H^0_{rxn} = -802 \text{ kJ}$$

What value do we obtain from bond energies? From the Lewis structures in Figure 9.22, we see that all the C—H bonds and the bonds in $O_2$ are broken in the reaction and all the C=O and O—H bonds are formed. We select appropriate bond energy values from Table 9.2, using a positive sign for bonds broken and a negative sign for bonds formed:

*Bonds broken*
$$4 \times C—H: 4 \text{ mol } (413 \text{ kJ/mol}) = 1652 \text{ kJ}$$
$$2 \times O_2: 2 \text{ mol } (498 \text{ kJ/mol}) = \underline{\quad 996 \text{ kJ}}$$
$$\text{Total } \Delta H^0_{\text{reactant bonds broken}} = 2648 \text{ kJ}$$

*Bonds formed*
$$2 \times C=O: 2 \text{ mol } (-799 \text{ kJ/mol}) = -1598 \text{ kJ}$$
$$4 \times O—H: 4 \text{ mol } (-467 \text{ kJ/mol}) = \underline{-1868 \text{ kJ}}$$
$$\text{Total } \Delta H^0_{\text{product bonds formed}} = -3466 \text{ kJ}$$

Adding these totals together gives

$$\Delta H^0_{rxn} = \Delta H^0_{\text{reactant bonds broken}} + \Delta H^0_{\text{product bonds formed}}$$
$$= 2648 \text{ kJ} + (-3466 \text{ kJ}) = -818 \text{ kJ}$$

Why is there a discrepancy between the bond-energy value ($-818$ kJ) and the calorimetric value ($-802$ kJ)? Variations in experimental method always introduce small discrepancies, but a more basic reason exists. As noted earlier, bond energies are *average* values obtained from many different compounds in which the bond occurs. The energy of the bond *in a particular substance* is usually close, but not equal, to this average. For example, the C—H bond energy of 413 kJ/mol is the average value of C—H bonds in many different molecules; in methane, 1660 kJ is required to break 4 mol C—H bonds, or 415 kJ/mol C—H bonds, which is slightly higher than the tabulated value of 413 kJ/mol. Thus, it isn't surprising to find discrepancies between the two $\Delta H^0_{rxn}$ values. What is surprising—and satisfying in its confirmation of bond theory—is that the values are so close.

SAMPLE PROBLEM 9.9

## Calculating Enthalpy Changes from Bond Energies

**Problem:** Use Table 9.2 to calculate $\Delta H^0_{rxn}$ for the following reaction:

$$CH_4(g) + 3Cl_2(g) \longrightarrow CHCl_3(g) + 3HCl(g)$$

**Plan:** First, we write the Lewis structures of all the substances. We assume that all reactant bonds are broken and all product bonds are formed and find their values in Table 9.2. Then we substitute the two sums with correct signs into Equation 9.4.

**Solution:** Writing the Lewis structures:

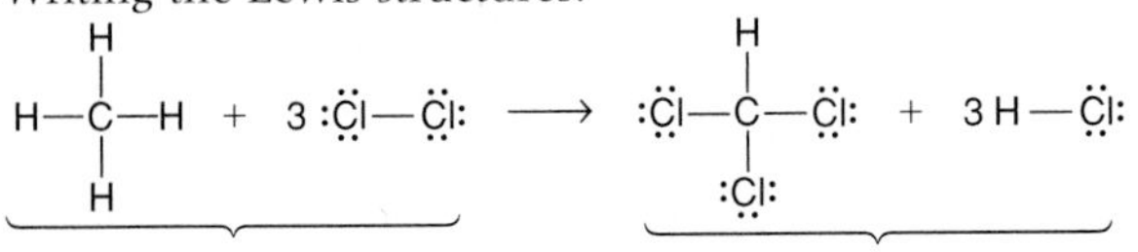

Calculating $\Delta H^0_{rxn}$: For bonds broken, the values are

$$4 \times C{-}H = 4 \text{ mol } (413 \text{ kJ/mol}) = 1652 \text{ kJ}$$
$$3 \times Cl{-}Cl = 3 \text{ mol } (243 \text{ kJ/mol}) = \underline{\phantom{0}729 \text{ kJ}}$$
$$\Delta H^0_{\text{bonds broken}} = 2381 \text{ kJ}$$

For bonds formed, the values are

$$3 \times C{-}Cl = 3 \text{ mol } (-339 \text{ kJ/mol}) = -1017 \text{ kJ}$$
$$1 \times C{-}H = 1 \text{ mol } (-413 \text{ kJ/mol}) = \phantom{0}-413 \text{ kJ}$$
$$3 \times H{-}Cl = 3 \text{ mol } (-427 \text{ kJ/mol}) = \underline{-1281 \text{ kJ}}$$
$$\Delta H^0_{\text{bonds formed}} = -2711 \text{ kJ}$$

$$\Delta H^0_{rxn} = \Delta H^0_{\text{bonds broken}} + \Delta H^0_{\text{bonds formed}} = 2381 \text{ kJ} + (-2711 \text{ kJ}) = \mathbf{-330 \text{ kJ}}$$

**Check:** The signs of the enthalpy changes are correct: $\Delta H^0_{\text{bonds broken}} > 0$ and $\Delta H^0_{\text{bonds formed}} < 0$. Since more energy is released than absorbed, $\Delta H^0_{rxn}$ is negative.

FOLLOW-UP PROBLEM 9.9
Calculate the enthalpy changes for the following reactions:
**(a)** $N_2(g) + 3H_2(g) \longrightarrow 2NH_3(g)$; **(b)** $C_2H_4(g) + HBr(g) \longrightarrow C_2H_5Br(g)$.

## Section Summary

Lewis structures of reactants and products can be used to determine the bonds broken and the bonds formed during a reaction. By applying Hess's law, we use tabulated bond energies to calculate the heat of reaction.

## 9.7 An Introduction to Metallic Bonding

The final type of bonding we consider is metallic bonding, which occurs among large numbers of metal atoms. In this section, we discuss a qualitative model; a more quantitative one is presented in Chapter 11.

### The Electron-Sea Model

In their reactions with nonmetals, reactive metals (such as sodium) transfer their outer electrons and form ionic solids (such as NaCl). Two metal atoms can also share their valence electrons in a covalent bond and form diatomic molecules, such as Na—Na. But a piece of sodium does not consist of sepa-

**TABLE 9.5** Melting and Boiling Points of Some Metals

| ELEMENT | MP (°C) | BP (°C) |
|---|---|---|
| Lithium (Li) | 180 | 1347 |
| Tin (Sn) | 232 | 2623 |
| Aluminum (Al) | 660 | 2467 |
| Barium (Ba) | 727 | 1850 |
| Silver (Ag) | 961 | 2155 |
| Copper (Cu) | 1083 | 2570 |
| Uranium (U) | 1130 | 3930 |

**FIGURE 9.23**

**The unusually low melting point of gallium.** Gallium has the widest liquid temperature range of any element. Its melting point (29.8°C) is below body temperature, indicating that metallic bonding forces holds the atoms weakly in their positions in the solid, but it boils at 2403°C, indicating that these forces are strong enough to resist separation of the atoms.

◆ **The Amazing Malleability of Gold.** All the Group 1B(11) metals—copper, silver, and gold—are soft enough to be machined easily, but gold is in a class by itself. One gram of gold would form a cube 0.37 cm on a side or a sphere the size of a small ball-bearing. Nevertheless, it can be drawn into a wire 20 μm thick and 165 m long, or hammered into a 1.0-m² sheet that is only 230 atoms (about 70 nm) thick!

rate diatomic molecules, so what holds the atoms together? The **electron-sea model** of metallic bonding proposes that all the metal atoms in the sample contribute their valence electrons to form a "sea" of electrons, which is delocalized throughout the substance. The metal ions (the nuclei with their core electrons) are submerged within this electron sea in an orderly array (see Figure 9.2, C).

In contrast to ionic bonding, the metal ions are not held in place as rigidly as in an ionic solid. In contrast to covalent bonding, no particular pair of metal atoms is bonded through any localized pair of electrons. Rather, *the valence electrons are shared among all the atoms in the substance,* which is held together by the mutual attraction of the metal cations for the mobile, highly delocalized electrons. (The extent of delocalization in a metal is much greater than that in a covalent resonance hybrid, where an electron pair is delocalized over only a few atoms.)

Although there are metallic compounds, it is more typical for metals to form **alloys,** solid mixtures with variable composition. The familiar metallic substances used for car parts, airplane bodies, coins, building girders, jewelry, and dental work are all alloys.

### How the Model Explains the Properties of Metals

Although their physical properties vary over a wide range, most metals are solids with moderate to high melting points and much higher boiling points (Table 9.5). Most metals are deformable: they bend or dent rather than crack or shatter. Many can be flattened into sheets and pulled into wires. Unlike most ionic and covalent substances, metals conduct heat and electricity well in *both* the solid and the liquid states.

Two features of the electron-sea model that account for these properties are the *regularity,* but not rigidity, of the metal-ion array and the *mobility* of the valence electrons. The melting and boiling points of metals are related to the energy of the metallic bonding. Electrostatic attractions between cations and electrons are not broken when metal atoms move from their positions during melting, so melting points are only moderately high. Boiling a metal *does* require breaking these attractions, however, so the boiling points are quite high. Gallium provides a striking example: it melts in your hand (Figure 9.23), but does not boil until the temperature reaches 2403°C.

Trends in other physical properties, such as melting point, are also consistent with the model's predictions (Figure 9.24). The increase in melting point between the alkali metals [Group 1A(1)] and the alkaline earth metals [Group 2A(2)], for example, can be explained by the 2A metals having two valence electrons available for metallic bonding, whereas the 1A metals have only one. The greater attraction between the $M^{2+}$ ions and twice the number of mobile electrons mean that higher temperatures are needed to melt the solid.

Mechanical and conducting properties are also explained by the model. When a piece of metal is struck by a hammer, the metal ions move to new lattice positions, sliding past each other through the intervening electrons. Thus, cations do not repel each other as the sample is deformed (Figure 9.25). Compare this behavior with the ionic repulsions that occur when an ionic solid is struck (see Figure 9.8). ◆

A piece of metal is easier to bend if its metal-ion array is very regular. Irregularities can make a piece of metal tougher and less workable. To see this, try bending a paper clip back and forth. At first it bends easily, but then it becomes stiffer and eventually breaks due to the movement of metal

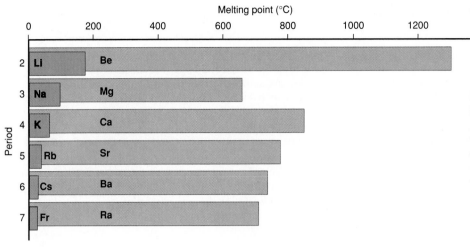

**FIGURE 9.24**

**Melting points of the Group 1A(1) and Group 2A(2) elements.** The alkaline earth metals [Group 2A(2), blue] have higher melting points than the alkali metals [Group 1A(1), brown] because of the greater attraction between $M^{2+}$ ions and the electron "sea," which contains twice as many valence electrons.

atoms from their regular positions. Irregularities are purposely introduced to toughen an alloy, as when copper is mixed with silver to make sterling silver, which is strong enough for use in jewelry and tableware.

Metals are good conductors of electricity and heat because of their mobile electrons. When a piece of metal is attached to the two terminals of a battery, electrons flow from the battery into the metal and replace electrons flowing from the metal into the battery. Irregularities in the array of metal atoms reduce this conductivity. Ordinary copper wire used to carry an electric current, for example, is greater than 99.99% pure because traces of other atoms can drastically restrict the flow of electrons.

If you place your hand on a piece of metal and a piece of wood that are both at room temperature, the metal feels colder because it conducts body heat away from your hand much faster than the wood. The mobile, delocalized electrons in the metal disperse the heat from your hand more quickly than the localized electron pairs in the covalent bonds of wood.

### Section Summary

In the electron-sea model, the valence electrons of the metal atoms in a sample are highly delocalized and attract the metal cations together. Metals have moderate melting points and high boiling points because the metal ions are not held rigidly in position and they cannot be easily separated from each other. Metals can be deformed because the electron sea prevents repulsions among the cations. Metals conduct electricity and heat because their electrons are mobile.

External force

Metal is deformed

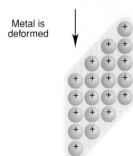

**B**

**FIGURE 9.25**

**The deformability of metals. A,** An external force applied to a piece of metal deforms the piece without breaking it. **B,** At the atomic level, the force simply moves metal ions past each other through the surrounding electron sea.

### Chapter Perspective

*Our theme throughout this chapter has been that the particular type of chemical bond—whether ionic, covalent, metallic, or some blend of these—is governed by the properties of the bonding atoms. This fundamental idea reappears many times as we investigate the forces that give rise to the properties of liquids, solids, and solutions (Chapters 11 and 12), the behavior of the main-group elements (Chapter 13), and the organic chemistry of carbon (Chapter 14). Before we can develop these topics, however, you'll see in Chapter 10 how the sequence of atoms in a molecule and the arrangement of bonding and lone pairs gives the molecule a characteristic shape, how that shape influences many of the compound's properties, and how covalent bonding theory explains the nature of the bond itself.*

# For Review and Reference

## Key Terms

**SECTION 9.1**
ionic bonding
covalent bonding
metallic bonding
Lewis electron-dot symbol

**SECTION 9.2**
octet rule
lattice energy
Born-Haber cycle
Coulomb's law
ion pair

**SECTION 9.3**
covalent bond
bonding (shared) pair
lone (unshared) pair
single bond
double bond
triple bond
bond order
bond energy (BE)
bond length
infrared (IR) spectroscopy

**SECTION 9.4**
electronegativity (EN)
polar covalent bond
nonpolar covalent bond
partial ionic character
electronegativity
   difference ($\Delta EN$)

**SECTION 9.5**
Lewis structure
   (Lewis formula)
resonance structure
   (resonance form)

resonance hybrid
electron-pair
   delocalization
formal charge
electron deficient
free radical
expanded valence shell

**SECTION 9.7**
electron-sea model
alloy

## Key Equations and Relationships

**9.1** Expressing the energy of attraction (or repulsion) between charges (p. 332):

$$\text{Electrostatic energy} \propto \frac{\text{charge A} \times \text{charge B}}{\text{distance}}$$

**9.2** Relating the energy of attraction to the lattice energy (p. 332):

$$\text{Electrostatic energy} \propto \frac{\text{cation charge} \times \text{anion charge}}{\text{cation radius} + \text{anion radius}}$$

$$\propto \Delta H^0_{\text{lattice}}$$

**9.3** Calculating the formal charge on an atom (p. 355):

Formal charge of atom

$= \text{no. of valence e}^-$
$- (\text{no. of unshared valence e}^- + \tfrac{1}{2} \text{ no. of shared valence e}^-)$

**9.4** Calculating the heat of reaction from bond energies (p. 359):

$$\Delta H^0_{\text{rxn}} = \Delta H^0_{\text{reactant bonds broken}} + \Delta H^0_{\text{product bonds formed}}$$

## Answers to Follow-up Problems

**9.1** Mg ($[\text{Ne}]3s^2$) + 2Cl ($[\text{Ne}]3s^23p^5$) →
          Mg$^{2+}$ ($[\text{Ne}]$) + 2Cl$^-$ ($[\text{Ne}] 3s^23p^6$)

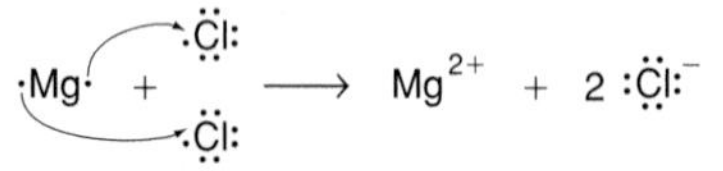

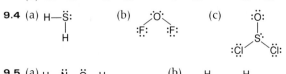

    Formula is MgCl$_2$

**9.2** (a) Bond length: Si—F < Si—O < Si—C;
       bond strength Si—C < Si—O < Si—F
   (b) Bond length: N≡N < N=N < N—N;
       bond strength: N—N < N=N < N≡N

**9.3** (a) Cl—Cl < $^{\delta+}$Br—Cl$^{\delta-}$ < $^{\delta+}$Cl—F$^{\delta-}$;
   (b) Si—Si < $^{\delta+}$S—Cl$^{\delta-}$ < $^{\delta+}$P—Cl$^{\delta-}$ < $^{\delta+}$Si—Cl$^{\delta-}$

**9.4** (a) H—S̈—   (b) :Ö:  :F̈:  :F̈:   (c) :Ö:
            |                                    S̈
            H                              :C̈l:   :C̈l:

**9.5** (a) H—N̈—Ö—H   (b)   H   H
            |               |   |
            H             H—C—Ö—C—H
                            |   |
                            H   H

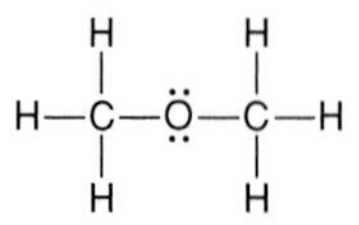

**9.6** (a) :C≡O:   (b) H—C≡N:   (c) :Ö=C=Ö:

**9.7**
$$\left[\begin{array}{c} :\ddot{O}: \\ \| \\ C \\ :\ddot{O}: \quad :\ddot{O}: \end{array}\right]^{2-} \longleftrightarrow \left[\begin{array}{c} :\ddot{O}: \\ | \\ C \\ :\ddot{O}: \quad :\ddot{O}: \end{array}\right]^{2-}$$

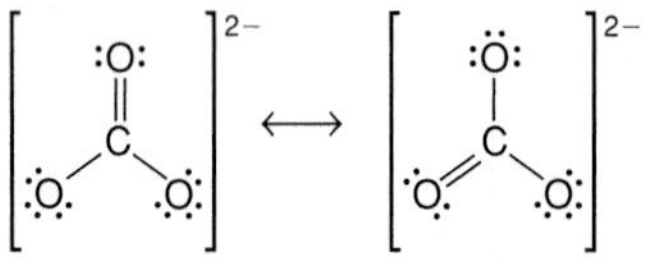

**9.8** (a)  :Ö:          (b)   :C̈l:       (c) :F̈:     :F̈:
              ‖               :Ö:   :Ö:         \\      /
              P                                 Xe
         :C̈l:  |  :C̈l:                         /      \\
              :C̈l:                         :F̈:     :F̈:

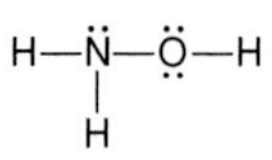

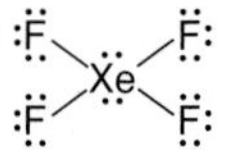

**9.9** (a)
                                         H
                                         |
   :N≡N: + 3 H—H ⟶ 2 :N—H
                                         |
                                         H

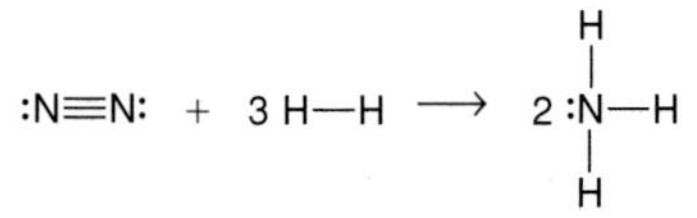

   $\Delta H^0_{\text{bonds broken}} = 1 \text{ N≡N} + 3 \text{ H—H}$
                   $= 945 \text{ kJ} + 1296 \text{ kJ} = 2241 \text{ kJ}$

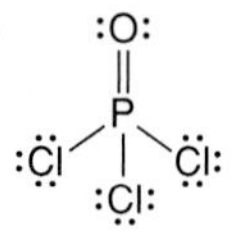

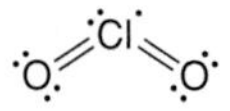

   $\Delta H^0_{\text{bonds formed}} = 6 \text{ N—H} = -2346 \text{ kJ}$

   $\Delta H^0_{\text{rxn}} = -105 \text{ kJ}$

   (b)
   H       H                          H   H
    \\     /                           |   |
     C=C     + H—B̈r:  ⟶  H—C—C—B̈r:
    /     \\                           |   |
   H       H                          H   H

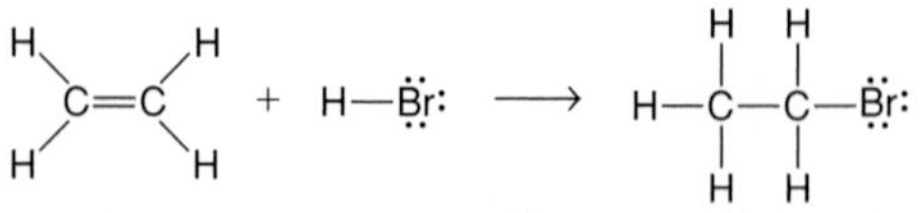

   $\Delta H^0_{\text{bonds broken}} = 1 \text{ C=C} + 4 \text{ C—H} + 1 \text{ H—Br}$
                   $= 614 \text{ kJ} + 1652 \text{ kJ} + 363 \text{ kJ}$
                   $= 2629 \text{ kJ}$

   $\Delta H^0_{\text{bonds formed}} = 1 \text{ C—C} + 5 \text{ C—H} + 1 \text{ C—Br}$
                   $= -(347 \text{ kJ} + 2065 \text{ kJ} + 276 \text{ kJ})$
                   $= -2688 \text{ kJ}$

   $\Delta H^0_{\text{rxn}} = -59 \text{ kJ}$

## Problems

Problems with a green number are answered at the back of the text. Most sections include three categories of problems separated by a green rule: concept review questions, *paired* skill-building exercises, and problems in a relevant context.

### Atomic Properties and Chemical Bonds

**9.1** In general terms, how does each of the following atomic properties influence metallic character of the main-group elements in a period: (a) ionization energy; (b) atomic radius; (c) number of outer electrons; (d) effective nuclear charge?

**9.2** Both nitrogen and bismuth are members of Group 5A(15). Which is more metallic? Explain your answer in terms of atomic properties.

**9.3** What is the relationship between the tendency of a main-group element to form a monatomic ion and its position in the periodic table? Where are the main-group elements located that typically form cations? Anions?

**9.4** Which member of each of the following pairs is *more* metallic: (a) Na or Cs; (b) Mg or Rb; (c) As or N?

**9.5** Which member of each of the following pairs is *less* metallic: (a) I or O; (b) Be or Ba; (c) Se or Ge?

**9.6** State the type of bonding—ionic, covalent, or metallic—you would expect in (a) $CsF(s)$; (b) $N_2(g)$; (c) $K(s)$; (d) $BrCl(g)$; (e) $NO(g)$; (f) $LiCl(s)$.

**9.7** State the type of bonding—ionic, covalent, or metallic—you would expect in (a) $O_3(g)$; (b) $MgCl_2(s)$; (c) $ClO_2(g)$; (d) $Mn(s)$; (e) $H_2O(s)$; (f) $CaO(s)$.

**9.8** Draw Lewis electron-dot symbols for each of the following atoms: (a) Rb; (b) Ge; (c) I; (d) Ba; (e) Xe.

**9.9** Draw Lewis electron-dot symbols for each of the following atoms: (a) Sr; (b) P; (c) S; (d) As; (e) Bi.

### The Ionic Bonding Model

(Sample Problem 9.1)

**9.10** If it requires energy to form monatomic ions from metals and nonmetals, why do ionic compounds exist?

**9.11** In general, how does the lattice energy of an ionic compound depend on the charges and sizes of the ions?

**9.12** When gaseous $Na^+$ and $Cl^-$ ions form gaseous $NaCl$ molecules, 548 kJ/mol is released. Why, then, does NaCl occur as a solid under ordinary conditions?

**9.13** It requires 214 kJ/mol to form $S^{2-}$ ions from gaseous sulfur atoms, but these ions exist in solids such as $K_2S$. Explain.

**9.14** Use condensed electron configurations and Lewis electron-dot symbols to depict the monatomic ions formed from each of the following, and predict the compound's formula: (a) Ba and Cl; (b) Sr and O; (c) Al and F.

**9.15** Use condensed electron configurations and Lewis electron-dot symbols to depict the monatomic ions formed from each of the following, and predict the compound's formula: (a) Cs and S; (b) O and Ga; (c) N and Mg.

**9.16** What is the periodic table group of X in each of the following *ionic* compound formulas: (a) $XF_2$; (b) $MgX$; (c) $X_2SO_4$?

**9.17** What is the periodic table group of X in each of the following *ionic* compound formulas: (a) $X_2O_3$; (b) $XCO_3$; (c) $Na_2X$?

**9.18** For each of the following pairs, choose the compound with the higher (more negative) lattice energy, and explain your choice: (a) BaS or CsCl; (b) LiCl or CsCl; (c) CaO or CaS.

**9.19** For each of the following pairs, choose the compound with the higher (more negative) lattice energy, and explain your choice: (a) CaS or BaS; (b) NaF or MgO; (c) NaF or NaCl.

**9.20** Use the following to calculate the lattice energy of NaCl:

$$Na(s) \rightarrow Na(g) \quad \Delta H^0 = 109 \text{ kJ}$$
$$Cl_2(g) \rightarrow 2Cl(g) \quad \Delta H^0 = 243 \text{ kJ}$$
$$Na(g) \rightarrow Na^+(g) + e^- \quad \Delta H^0 = 496 \text{ kJ}$$
$$Cl(g) + e^- \rightarrow Cl^-(g) \quad \Delta H^0 = -349 \text{ kJ}$$
$$Na(s) + \tfrac{1}{2}Cl_2(g) \rightarrow NaCl(s) \quad \Delta H_f^0 = -411 \text{ kJ}$$

Compared with the lattice energy of LiF (see text), is the value for NaCl expected? Explain.

**9.21** Use the following to calculate the lattice energy of $MgF_2$:
$$Mg(s) \rightarrow Mg(g) \quad \Delta H^0 = 148 \text{ kJ}$$
$$F_2(g) \rightarrow 2F(g) \quad \Delta H^0 = 159 \text{ kJ}$$
$$Mg(g) \rightarrow Mg^+(g) + e^- \quad \Delta H^0 = 738 \text{ kJ}$$
$$Mg^+(g) \rightarrow Mg^{2+}(g) + e^- \quad \Delta H^0 = 1450 \text{ kJ}$$
$$F(g) + e^- \rightarrow F^-(g) \quad \Delta H^0 = -328 \text{ kJ}$$
$$Mg(s) + F_2(g) \rightarrow MgF_2(s) \quad \Delta H^0_f = -1123 \text{ kJ}$$
Compared with the lattice energy of LiF (see text) or that you calculated for NaCl in Problem 9.20, does the value for $MgF_2$ surprise you? Explain.

**9.22** Aluminum oxide ($Al_2O_3$) is a widely used industrial abrasive (emery, corundum) with its specific application depending on the hardness of the crystal. What does this hardness imply about the magnitude of the lattice energy? Could you have predicted this relative magnitude from the chemical formula? Explain.

**9.23** Born-Haber cycles were used to obtain the first reliable values for electron affinity by considering it the unknown and using a theoretically calculated value for the lattice energy. Use a Born-Haber cycle for KF and the following values to calculate a value for the electron affinity of fluorine:
$$K(s) \rightarrow K(g) \quad \Delta H^0 = 90 \text{ kJ}$$
$$K(g) \rightarrow K^+(g) \quad \Delta H^0 = 419 \text{ kJ}$$
$$F_2(g) \rightarrow 2F(g) \quad \Delta H^0 = 159 \text{ kJ}$$
$$K(s) + \tfrac{1}{2}F_2(g) \rightarrow KF(s) \quad \Delta H^0_f = -569 \text{ kJ}$$
$$K^+(g) + F^-(g) \rightarrow KF(s) \quad \Delta H^0 = -821 \text{ kJ}$$

## The Covalent Bonding Model

(Sample Problem 9.2)

**9.24** Describe the interactions that occur between individual chlorine atoms as they approach each other and form $Cl_2$. What combination of forces gives rise to the energy holding the atoms together and to the internuclear distance between them?

**9.25** Define bond energy using the H—Cl bond as an example. When this bond breaks, is energy absorbed or released? Does the accompanying enthalpy change have a positive or negative sign?

**9.26** For single bonds between similar types of atoms, how does the strength of the bond relate to the sizes of the atoms? Explain.

**9.27** How does the bond energy between a given pair of atoms relate to the bond order? Why?

**9.28** When liquid benzene ($C_6H_6$) boils, does the gas consist of $C_6H_6$ molecules or separate C and H atoms? Explain.

**9.29** Using the periodic table only, arrange the members of each of the following sets in order of increasing bond *strength:* (a) Br—Br, Cl—Cl, I—I; (b) S—H, S—Br, S—Cl; (c) C=N, C—N, C≡N.

**9.30** Using the periodic table only, arrange the members of each of the following sets in order of increasing bond *length:* (a) H—F, H—I, H—Cl; (b) C—S, C=O, C—O; (c) N—H, N—S, N—O.

**9.31** Formic acid (HCOOH) is secreted by certain species of ant when they bite. It has the structural formula:

$$\overset{\displaystyle :\!O:}{\underset{}{\|}}$$
$$H-C-\ddot{O}-H$$

Rank the relative strengths of the C—O bond, the C=O bond, and the attractive force between two HCOOH molecules.

**9.32** In Figure 9.B, the peak labeled "C=C stretch" occurs at a shorter wavelength than that labeled "C—C stretch," as it would in the infrared spectrum of any substance with those types of bonds. Explain the relative positions of these peaks. In what relative position along the wavelength scale of Figure 9.B would you find a peak characteristic of a C≡C stretch? Explain.

## Between the Extremes: Electronegativity and Bond Polarity

(Sample Problem 9.3)

**9.33** Describe the vertical and horizontal trends in electronegativity (EN) among the main-group elements. According to Pauling's scale, what are the two most electronegative elements?

**9.34** What is the general relationship between $IE_1$ and EN for the elements? Why?

**9.35** Is the H—O bond in water nonpolar covalent, polar covalent, or ionic? Define each term and explain your choice.

**9.36** How does electronegativity differ from electron affinity?

**9.37** How is the partial ionic character of a diatomic molecule related to the ΔEN of the bonded atoms? Why?

**9.38** The bond energy of the C—C bond is 347 kJ/mol, and that of the Cl—Cl bond is 243 kJ/mol. Which of the following values might you expect for the C—Cl bond energy? Explain.
(a) 590 kJ/mol (sum of the values above)
(b) 104 kJ/mol (difference of the values above)
(c) 295 kJ/mol (average of the values above)
(d) 339 kJ/mol (value greater than the average of the values above)

**9.39** Using the periodic table only, arrange the elements in each of the following sets in order of *increasing* EN: (a) S, O, Si; (b) Mg, P, As; (c) I, Br, N; (d) Ca, H, F.

**9.40** Using the periodic table only, arrange the elements in each of the following sets in order of *decreasing* EN: (a) N, P, Si; (b) Ca, Ga, As; (c) Br, Cl, P; (d) I, F, O.

**9.41** Using EN values, indicate the polarity of the following bonds with *polar arrows*, and determine the more polar bond in each pair: (a) N—B or N—O; (b) C—S or S—O; (c) N—H or N—O.

**9.42** Using EN values, indicate the polarity of the following bonds with *partial charges*, and determine the more polar bond in each pair: (a) Br—Cl or F—Cl; (b) H—O or Se—H; (c) As—H or S—N.

**9.43** Are the bonds in each of the following substances ionic, nonpolar covalent, or polar covalent? Arrange the substances with polar covalent bonds in order of increasing bond polarity: (a) $S_8$; (b) RbCl; (c) $PF_3$; (d) $SCl_2$; (e) $F_2$; (f) $SF_2$.

**9.44** Are the bonds in each of the following substances ionic, nonpolar covalent, or polar covalent? Arrange the substances with polar covalent bonds in order of increasing bond polarity: (a) KCl; (b) $P_4$; (c) $BF_3$; (d) $SO_2$; (e) $Br_2$; (f) $NO_2$.

**9.45** Rank the members of each of the following sets of compounds in order of *increasing* ionic character of their bond. Use *polar arrows* to indicate the bond polarity: (a) HBr, HCl, HI; (b) $H_2O$, $CH_4$, HF; (c) $SCl_2$, $PCl_3$, $SiCl_4$.

**9.46** Rank the members of each of the following sets of compounds in order of *decreasing* ionic character of their bonds. Use *partial charges* to indicate the bond polarity: (a) $PCl_3$, $PBr_3$, $PF_3$; (b) $BF_3$, $NF_3$, $CF_4$; (c) $SeF_4$, $TeF_4$, $BrF_3$.

## Depicting Molecules and Ions with Lewis Structures

(Sample Problems 9.4 to 9.8)

**9.47** In which pattern(s) does X obey the octet rule?

(a)   (b)   (c)   (d)

(e)   (f)   (g)   (h)

**9.48** Which of the following atoms *cannot* serve as a central atom in a Lewis structure? (a) O; (b) He; (c) F; (d) H; (e) P. Explain.

**9.49** When is a resonance hybrid needed to depict adequately the bonding in a molecule? Using $NO_2$ as an example, explain how a resonance hybrid is consistent with bond length, bond strength, and bond order.

**9.50** What is the distinction between oxidation number and formal charge? Which do you think is closer to the real charges on the atoms in HF and in CO? Explain.

**9.51** What requirement is needed for an atom to expand its valence shell? Which of the following atoms can expand its valence shell: F, S, H, Al?

**9.52** Draw a Lewis structure for (a) $SiF_4$; (b) $SeCl_2$; (c) $COF_2$ (C central); (d) $PH_4^+$; (e) $C_2F_4$.

**9.53** Draw a Lewis structure for (a) $PF_3$; (b) $H_2CO_3$ (H attached to O); (c) $CS_2$; (d) $CH_4S$; (e) $S_2Cl_2$.

**9.54** Draw Lewis structures of all the important resonance forms of each of the following: (a) $NO_2$; (b) $NO_2F$ (N central); (c) $HNO_3$ ($HONO_2$).

**9.55** Draw Lewis structures of all the important resonance forms of each of the following: (a) $N_3^-$; (b) $NO_2^-$; (c) $HCO_2^-$ (H attached to C).

**9.56** Draw a Lewis structure and calculate the formal charge of each atom in the following: (a) $IF_5$; (b) $AlH_4^-$; (c) COS (C central).

**9.57** Draw a Lewis structure and calculate the formal charge of each atom in the following: (a) $CN^-$; (b) $ClO^-$; (c) $BF_4^-$.

**9.58** Draw a Lewis structure for the most important resonance form of the following, showing formal charges and oxidation numbers of the atoms: (a) $BrO_3^-$; (b) $SO_3^{2-}$

**9.59** Draw a Lewis structure for the most important resonance form of the following, showing formal charges and oxidation numbers of the atoms: (a) $AsO_4^{3-}$; (b) $ClO_2^-$

**9.60** The following species do not obey the octet rule. Draw a Lewis structure for each one and state the type of octet rule exception: (a) $BH_3$; (b) $AsF_4^-$; (c) $SeCl_4$; (d) $PF_6^-$; (e) $ClO_3$.

**9.61** The following species do not obey the octet rule. Draw a Lewis structure for each one and state the type of octet rule exception: (a) $BrF_3$; (b) $ICl_2^-$; (c) $BeF_2$; (d) $O_3^-$; (e) $XeF_2$.

**9.62** Perchlorates are powerful oxidizing agents used in fireworks, flares, and the booster rockets of the Space Shuttle. Lewis structures for the perchlorate ion ($ClO_4^-$) can be drawn with all single bonds or with one, two, or three double bonds. Draw each of these possible resonance forms, use formal charges to determine the most important form, and calculate its average bond order.

**9.63** Dinitrogen monoxide ($N_2O$), once widely used as the anesthetic laughing gas in dental surgery, supports combustion in a manner similar to oxygen, with the nitrogen atoms forming $N_2$. Draw three resonance structures for $N_2O$ (one N is central) and use formal charges to decide on the relative importance of each. What correlation can you suggest between the most important structure and the observation that $N_2O$ supports combustion?

**9.64** Molten beryllium chloride reacts with additional chloride ion to form the $BeCl_4^{2-}$ ion, in which the Be attains an octet. Use Lewis structures to depict this net ionic reaction.

**9.65** Despite many attempts, the perbromate ion ($BrO_4^-$) was not prepared in the laboratory until about 1970. (Indeed, articles were published explaining theoretically why it would never be prepared!) Draw a Lewis structure for $BrO_4^-$ that is consistent with minimal formal charges.

**9.66** Phosgene is a colorless, highly toxic gas used in warfare and as a key reactant in organic synthesis. Use formal charges to select the most important of the following resonance structures:

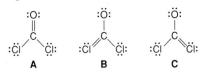

**9.67** Cryolite ($Na_3AlF_6$) is an indispensible component in the electrochemical manufacture of aluminum. Draw a Lewis structure for the $AlF_6^{3-}$ ion.

**9.68** Oxalic acid ($H_2C_2O_4$) is found in toxic concentrations in rhubarb. Its atom sequence (with single bonds) is

The acid forms two ions, $HC_2O_4^-$ and $C_2O_4^{2-}$, by the sequential loss of $H^+$. Draw Lewis structures for the three species, and comment on the relative lengths and strengths of their carbon-oxygen bonds.

**9.69** Phosphoric acid ($H_3PO_4$) is a weak acid used in the manufacture of fertilizers and foodstuffs. In basic solution, it ionizes by sequentially forming $H_2PO_4^-$, $HPO_4^{2-}$, and $PO_4^{3-}$. Draw Lewis structures for phosphoric acid and its three ions, taking into account formal charge and resonance (all H attached to O). Comment on the relative lengths of their phosphorus-oxygen bonds.

## Using Lewis Structures and Bond Energies to Calculate Heats of Reaction

(Sample Problem 9.9)

**9.70** Write a solution plan (without actual numbers) for calculating the enthalpy change of the following reaction (include the bond energies you would use and how you would combine them algebraically):

$$H_2(g) + O_2(g) \rightarrow H_2O_2(g)$$

**9.71** The text points out that, for similar types of substances, one with weaker bonds is usually more reactive than one with stronger bonds. Why is this generally true?

**9.72** Why is there often a small discrepancy between a heat of reaction obtained from $\Delta H_f^0$ values and one obtained from bond energies?

**9.73** Use bond energies from Table 9.2 to calculate the following heat of reaction:

**9.74** Use bond energies from Table 9.2 to calculate the following heat of reaction:

**9.75** Use Lewis structures and bond energies to calculate the enthalpy change of the following reaction:

$$C_2H_4(g) + H_2O(g) \rightarrow CH_3CH_2OH(l)$$

**9.76** Use Lewis structures and bond energies to calculate the enthalpy change of the following reaction:

$$HCN(g) + 2H_2(g) \rightarrow CH_3NH_2(g)$$

**9.77** Hydrazine ($N_2H_4$) is used as a rocket fuel because it reacts very exothermically with oxygen. The heat released and the increase in moles of gas provide thrust. Use Table 9.2 to calculate the heat of the following reaction:

$$N_2H_4(l) + O_2(g) \rightarrow N_2(g) + 2H_2O(g)$$

**9.78** Acetylene gas ($C_2H_2$) burns with oxygen in an oxyacetylene torch to produce carbon dioxide, water vapor, and the great heat needed for welding metals. The heat of combustion of acetylene is 1259 kJ/mol. Estimate the bond energy of the $C{\equiv}C$ bond.

## An Introduction to Metallic Bonding

**9.79** (a) List four physical characteristics of a solid metal.
(b) List two chemical characteristics that lead to an element being classified as a metal.

**9.80** Briefly account for the following relative values:
(a) The melting point of sodium is 89°C, whereas that of potassium is 63°C.
(b) The melting points of Li and Be are 180°C and 1287°C, respectively.
(c) Lithium boils more than 1100°C higher than it melts.

**9.81** Magnesium metal is easily deformed by an applied force, whereas magnesium fluoride is shattered. Why do these two solids behave so differently?

**9.82** Use electron configurations to predict the relative hardness and melting point of rubidium ($Z = 37$), vanadium ($Z = 23$), and cadmium ($Z = 48$).

## Comprehensive Problems

Problems with an asterisk (*) are more challenging.

**9.83** Use two compounds of sulfur to explain how the octet rule applies to the formation of both ionic and covalent compounds.

**9.84** Use Lewis electron-dot symbols to represent the formation of
(a) $IF_3$ from iodine and fluorine atoms
(b) $AlF_3$ from aluminum and fluorine atoms

**9.85** In addition to ammonia, nitrogen forms the hydride hydrazine ($N_2H_4$) and two thermally unstable hydrides, diazene ($N_2H_2$) and tetrazene ($N_4H_4$).
(a) Use Lewis structures to compare the strength, length, and order of nitrogen-nitrogen bonds in diazene, hydrazine, and $N_2$.
(b) Tetrazene (atom sequence $H_2NNNNH_2$) decomposes above 0°C to hydrazine and nitrogen gas. Draw a Lewis structure for tetrazene and calculate the $\Delta H_{rxn}^0$ for this decomposition.

**9.86** Draw a Lewis structure for each of the following species: (a) $PF_5$; (b) $CCl_4$; (c) $H_3O^+$; (d) $ICl_3$; (e) $BeH_2$; (f) $PH_2^-$; (g) $GeBr_4$; (h) $CH_3^-$; (i) $BCl_3$; (j) $BrF_4^+$; (k) $XeO_3$; (l) $TeF_4$.

**9.87** Nitrosyl fluoride (NOF) has an atom sequence in which all atoms have formal charges of zero. Write the Lewis structure consistent with this fact.

**\*9.88** In contrast to the cyanate ion ($NCO^-$), which is stable and found in many compounds, the fulminate ion ($CNO^-$), with its different atom sequence, is unstable and forms compounds with heavy metal ions, such as $Ag^+$ and $Hg^{2+}$, that are explosive. Like the cyanate ion, the fulminate ion has three resonance structures. Which is the most important contributor to the resonance hybrid? Suggest a reason for the instability of fulminate.

**9.89** Boron trifluoride is an important reactant in organic syntheses but is a very reactive gas at room temperature and thus difficult to handle. Boron's ability to bond another pair of electrons and attain an octet is used to prepare an easily stored liquid form: a compound of $BF_3$ and diethyl ether ($CH_3-CH_2-O-CH_2-CH_3$). Draw Lewis structures for this reaction.

**9.90** "Inert" xenon actually forms several compounds, especially with the highly electronegative elements oxygen and fluorine. The simple fluorides $XeF_2$, $XeF_4$, and $XeF_6$ are all formed by direct reaction of the elements. As one might expect from the size of the xenon atom, the Xe—F bond is not a strong one. Calculate the Xe—F bond energy in $XeF_6$, given that the heat of formation is −402 kJ/mol.

**9.91** In developing the concept of electronegativity, Pauling used the term "excess bond energy" for the difference between the actual bond energy of X—Y and the average bond energies of X—X and Y—Y (see text discussion for the case of HF). Based on the values in Figure 9.16, which of the following substances would be expected to contain bonds with *no* excess bond energy: (a) $PH_3$; (b) $CS_2$; (c) BrCl; (d) $BH_3$; (e) $Se_8$?

**9.92** Like several other bonds, carbon-oxygen bonds have lengths and strengths that depend on the bond order. Draw Lewis structures for the following species, and arrange them in order of increasing carbon-oxygen bond length and bond strength: (a) CO; (b) $CO_3^{2-}$; (c) $H_2CO$; (d) $CH_4O$; (e) $HCO_3^-$ (H attached to O).

**9.93** Chemists are employing lasers to emit light of a given energy to initiate bond breakage.
(a) What is the minimum energy and frequency of a photon that can cause the dissociation of a $Cl_2$ molecule?
(b) The first key step in the destruction of stratospheric ozone by industrial chlorofluorocarbons is thought to be photodissociation of a C—Cl bond. What is the longest wavelength of a photon that can cause this dissociation?

**9.94** Ethanol ($CH_3CH_2OH$) has been proposed as an additive or a substitute for hydrocarbon automotive fuels.

(a) Use bond energies to calculate the heat of combustion of gaseous ethanol.
(b) In its standard state at 25°C, ethanol is a liquid. The vaporization requires 40.5 kJ/mol. Correct the value from part (a) to find the heat of combustion of liquid ethanol.
(c) How does the value in part (b) compare with the value you calculate from standard heats of formation (Appendix B)?

**\*9.95** Organic compounds vary greatly in the way the C atoms bond to each other. In the following compounds, they form a single ring. Draw a Lewis structure for each, identify cases for which resonance exists, and determine the carbon-carbon bond order(s): (a) $C_3H_4$; (b) $C_3H_6$; (c) $C_4H_6$; (d) $C_4H_4$; (e) $C_6H_6$.

**9.96** An oxide of nitrogen is 25.9% N by mass, has a molar mass of 108 g/mol, and contains no nitrogen-nitrogen or oxygen-oxygen bonds. Draw its Lewis structure and name it.

**\*9.97** An experiment requires 50.0 mL of 0.040 *M* NaOH for the titration of 1.00 mmol of acid. Analysis of the acid shows 2.24% hydrogen, 26.7% carbon, and 71.1% oxygen. Draw the Lewis structure of the acid.

**\*9.98** A gaseous compound has a composition by mass of 24.8% carbon, 2.08% hydrogen, and 73.1% chlorine. At STP, the gas has a density of 4.3 g/L. Draw a Lewis structure that satisfies these facts. Would a different structure satisfy them as well? Explain.

**\*9.99** Bond energies can be combined with values for other atomic properties to obtain $\Delta H$ values that cannot be measured directly. Use bond energy, ionization energy, and electron affinity values to calculate the $\Delta H^0_{rxn}$ for the ionic dissociation of $Cl_2$ gas:
$$Cl_2(g) \rightarrow Cl^+(g) + Cl^-(g)$$

**\*9.100** An important short-lived species in flames is OH.
(a) What is unusual about the electronic structure of OH?
(b) Use the standard heat of formation of OH($g$) and bond energies to calculate the O—H bond energy in OH($g$). $\Delta H^0_f$ of OH($g$) = 39.0 kJ/mol
(c) From the average value for the O—H bond energy in Table 9.2 and your value for the O—H bond energy in OH($g$), calculate the energy needed to break the first O—H bond in water.

**9.101** Use Lewis structures to determine which *two* of the following compounds are unstable molecules: (a) $SF_2$; (b) $SF_3$; (c) $SF_4$; (d) $SF_5$; (e) $SF_6$.

**\*9.102** The HF bond length is 92 pm, 16% shorter than the sum of the covalent radii of H (37 pm) and F (72 pm). Why is the bond length less than the sum of the covalent radii in HF? Similar calculations for the other hydrogen halides show that the difference between actual and calculated bond length becomes smaller down the group from HF to HI. Explain.

# CHAPTER 10

**Concepts and skills to review**

- atomic orbital shapes (Section 7.4)
- Pauli exclusion principle (Section 8.1) and Hund's rule (Section 8.2)
- bond order, bond length, and bond energy (Section 9.3)
- polar covalent bonds and bond polarity (Section 9.4)
- writing Lewis structures (Section 9.5)
- resonance in covalent bonding (Section 9.5)

# Molecular Shape and Theories of Covalent Bonding

**Fitting Together.** In the imaginary world of M.C. Escher, shown here in a detail from *Mosaic II*, one contour fits tightly within another. The fitting together of molecular shapes initiates all major biological function. In this chapter, we explore molecular shape and the bonding theories that explain it.

**A**fter devoting so much time to writing Lewis structures, you may sometimes think of molecules as letters connected by lines and surrounded by dots, lying flat on a page. As you'll see, the reality is much more marvelous! Molecules are incredibly small conglomerations of mass and charge that swirl, twist, and vibrate incessantly. Each of their components—the atom cores (nucleus and inner electrons), the valence-electron bonds connecting the cores, and the lone electron pairs surrounding them—has its own position relative to the others, prescribed by the attractive and repulsive forces that govern all matter. With definite angles and distances separating these components, a molecule is an independent, minute architecture, extending in three dimensions throughout its tiny volume of space.

Our focus in this chapter is the shapes of molecules, the theoretical basis for them, and some effects of shape on behavior. We start by describing the VSEPR (valence-shell electron-pair repulsion) model, which allows us to convert two-dimensional Lewis structures into three-dimensional shapes. You'll see how molecular shape and bond polarity combine to create a polarity for the entire molecule and how molecular shape influences biological function. Then we consider two bonding theories based on quantum mechanics. Valence bond theory explains how the observed shape arises from the interactions of atomic orbitals. Molecular orbital theory proposes the existence of orbitals that extend over the whole molecule. Each theory complements the other and clarifies the description of certain molecular properties.

## 10.1 Valence-Shell Electron-Pair Repulsion (VSEPR) Theory and Molecular Shape

The Lewis structure of a molecule is something like the blueprint of a building: a flat drawing showing the placement of parts (sequence of atom cores), the structural connections (groups of bonding valence electrons), and the various attachments (nonbonding lone pairs of valence electrons). To construct the molecular shape from the Lewis structure, chemists employ *valence-shell electron-pair repulsion (VSEPR) theory,* whose basic principle is that the *groups of valence electrons around a central atom stay as far apart from each other as possible to minimize repulsions.* For our purposes, we define a *group* of electrons as any number of electrons that occupies a localized region around an atom. Thus, an electron group may consist of a single bond, a double bond, a triple bond, a lone pair, or in some cases even a lone electron. Each of these is a *separate* group of valence electrons that repels the other groups and occupies as much space as possible around the central atom. It is the three-dimensional arrangement of these groups that gives rise to the molecular shape.

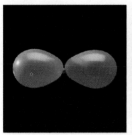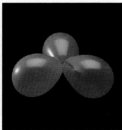

**FIGURE 10.1**

**A balloon analogy for the mutual repulsion of electron groups.** Attached balloons will move apart so that each can occupy as much space as possible. Five geometric arrangements arise by attaching two, three, four, five, or six balloons. Electron groups repel each other and become arranged in a similar way around a central atom.

**FIGURE 10.2**

**Electron-group repulsions and the five basic molecular shapes.** When a given number of electron groups attached to a central atom *(red)* repel each other, they become oriented as far apart as possible in space. If each electron group is a bonding group to a surrounding atom *(black)*, the molecular shapes and bond angles shown here are observed, and the name of the molecular shape is the same as that of the electron-group arrangement. When one or more of the electron groups is a lone pair, other molecular shapes are observed, as you'll see in upcoming figures.

## Electron-Group Arrangements and Bond Angles

When two, three, four, five, or six attached objects maximize the space each can occupy, five three-dimensional patterns result. Figure 10.1 depicts these patterns with balloons. When the objects are the valence-electron groups of a central atom, their *repulsions* give rise to the five *electron-group arrangements* of minimum energy seen in the great majority of molecules and polyatomic ions.

The electron-group arrangement includes *all the valence-electron groups,* both bonding and nonbonding, around the central atom. In contrast, the **molecular shape** is defined by the relative positions of the nuclei of the atoms, so it *depends only on the bonding groups.* Figure 10.2 shows the molecular shapes that occur when *all* the surrounding electron groups are *bonding* groups. When some are *nonbonding* groups, different molecular shapes occur. Thus, *the same electron-group arrangement can give rise to different molecular shapes:* some with all bonding groups (as in Figure 10.2) and others with bonding *and* nonbonding groups. To classify the various molecular shapes, we assign each a specific $AX_mE_n$ designation, where $m$ and $n$ are integers, A is the central atom, X is a surrounding atom, and E is a nonbonding valence-electron group.

The **bond angle** is the angle formed by two surrounding atoms with the central atom at the vertex. The angles shown in Figure 10.2 are *ideal* bond angles, those predicted by geometry. Ideal angles are observed when all the bonds around a central atom are identical and are connected to the same type of atom. When this is not the case, such as when multiple bonds, lone pairs, or different surrounding atoms are present, the bond angles deviate from the ideal angles. You'll see examples of these effects shortly.

It is important to realize that we use the VSEPR model to *explain* the molecular shapes—bond angles and bond lengths—that we *observe* through various types of spectroscopy; in almost every case, *VSEPR predictions are in accord with our observations* (see Problem 10.18 at the end of the chapter).

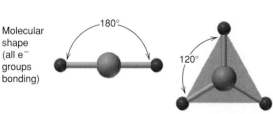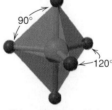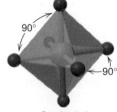

Molecular shape (all e⁻ groups bonding)

Name:    Linear    Trigonal planar    Tetrahedral    Trigonal bipyramidal    Octahedral

## Molecular Shape with Two Electron Groups (Linear Arrangement)

When two electron groups repel each other, they lie on opposite sides of the central atom in a straight line. The **linear arrangement** of electron groups results in a linear molecular shape and a bond angle of 180°. Figure 10.3 shows the general arrangement (*top*) and shape (*middle*) along with VSEPR shape class ($AX_2$) and the formulas of some linear molecules.

Gaseous beryllium chloride ($BeCl_2$) is an example of a linear molecule ($AX_2$). Gaseous Be compounds are electron deficient, with only two electron pairs around the central Be:

$$\overset{180°}{:\ddot{Cl}-Be-\ddot{Cl}:}$$

In carbon dioxide, the central carbon is surrounded by two double bonds:

$$\overset{180°}{\ddot{O}=C=\ddot{O}}$$

Each double bond acts as a separate electron group and stays 180° away from the other, so $CO_2$ is linear. Notice that the lone pairs on the O atoms of $CO_2$ or on the Cl atoms of $BeCl_2$ are not involved in the molecular shape: only the valence electron groups *around the central atom* influence the molecular shape.

## Molecular Shapes with Three Electron Groups (Trigonal Planar Arrangement)

Three electron groups around the central atom repel each other to lie at the corners of an equilateral triangle. This is the **trigonal planar arrangement,** and the ideal bond angle is 120° (Figure 10.4). Two molecular shapes are possible within this electron-group arrangement, one with three surrounding atoms and the other with two atoms and one lone pair.

When the three electron groups are bonding groups, the name of the molecular shape is also *trigonal planar* ($AX_3$). Boron trifluoride ($BF_3$), another electron-deficient molecule, is an example. It has six electrons around the central B in three single bonds to F atoms. The molecule is flat with all four atoms lying in a plane and the F atoms 120° apart:

$$:\ddot{F}: \\ | \\ :\ddot{F} \overset{B}{\underset{120°}{\diagup \diagdown}} \ddot{F}:$$

The nitrate ion ($NO_3^-$) is one of several polyatomic ions with the trigonal planar shape. This is one of three resonance forms of the nitrate ion. The resonance hybrid has three identical $1\frac{1}{3}$ bonds, so the ideal bond angle is observed:

$$\left[ \begin{array}{c} :\ddot{O}: \\ \| \\ :\ddot{O}. \underset{120°}{\overset{N}{\diagup \diagdown}} .\ddot{O}: \end{array} \right]^-$$

As we noted previously, when the surrounding atoms or the electron groups are not identical, nonideal bond angles occur. Consider formaldehyde ($CH_2O$), a substance with many uses, including the manufacture of Formica countertops, the production of methanol, and the preservation of

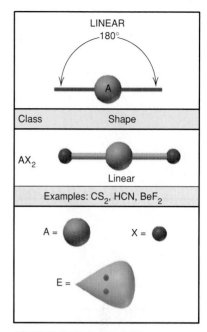

**FIGURE 10.3**
**The single molecular shape of the linear electron-group arrangement.** The key for A, X, and E (bottom) also refers to Figures 10.4, 10.5, 10.8, and 10.10.

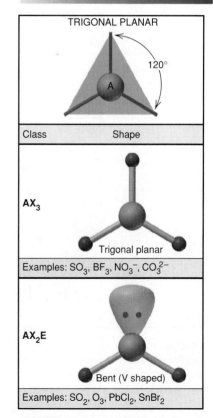

**FIGURE 10.4**
**The two molecular shapes of the trigonal planar electron-group arrangement.**

cadavers. Its trigonal planar shape is due to two types of surrounding atoms (O and H) and two types of electron groups (single and double bonds):

ideal                    actual

The actual bond angles differ from the ideal because *the four electrons in the double bond repel the two electrons in each of the two single bonds more strongly than they repel each other.*

Remember that only *atom* positions define a shape, so when one of the three electron groups is a lone pair ($AX_2E$), the molecule is **bent** or **V shaped**. The shape of gaseous tin(II) chloride has the three electron groups in a trigonal plane, with the lone pair at one of the triangle's corners. Here is the first example of the effect of a lone pair on adjacent bonding pairs. Since a lone pair is held by only one nucleus, it exerts a stronger repulsion than a bonding pair. Thus, *a lone pair repels bonding pairs more strongly than bonding pairs repel each other.* This repulsion increases the angle between lone pair and bonding pair, which decreases the angle between bonding pairs. Note the large decrease from the ideal 120° angle in $SnCl_2$:

### Molecular Shapes with Four Electron Groups (Tetrahedral Arrangement)

The shapes described so far have all been simple to depict in two dimensions, but four electron groups around a central atom utilize three dimensions to achieve maximal separation. This is a good time for you to recall that *Lewis structures do not depict shape.* The Lewis structure for methane, for example, shows the four bonds pointing to the corners of a square, which implies a 90° bond angle:

In three dimensions, the four electron groups can move farther apart than 90° and point to the corners of a tetrahedron (a polygon with four sides made of equilateral triangles), giving a bond angle of 109.5°.

*All molecules or ions with four electron groups around a central atom adopt the **tetrahedral arrangement*** (Figure 10.5). Perspective drawings, such as this one for methane, indicate depth by means of wedged lines:

The normal lines represent electron groups in the plane of the page, one wedge is the bond of a group lying toward you above the page, and the other is the bond of a group lying away from you below the page.

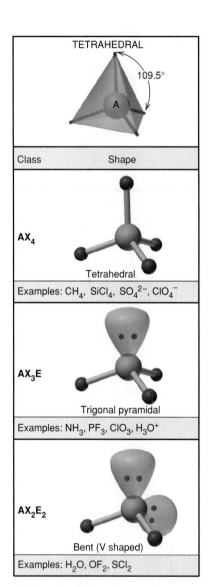

**FIGURE 10.5**

**The three molecular shapes of the tetrahedral electron-group arrangement.** The possibility of having no lone pairs ($AX_4$), one ($AX_3E$), or two ($AX_2E_2$) gives rise to the shapes in this arrangement, the most common in chemistry.

When the four electron groups are involved in bonding, the molecular shape is also called *tetrahedral* ($AX_4$), an extremely common geometry in organic molecules. In Chapter 9, we drew the Lewis structure for dichlorodifluoromethane ($CCl_2F_2$) without regard to how the halogen atoms surround the C atom. Indeed, even though it seems that we can write two Lewis structures for $CCl_2F_2$, they represent the same molecule. Figure 10.6 shows this fact clearly with a ball-and-stick model:

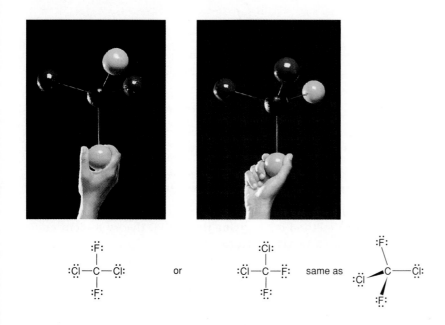

**FIGURE 10.6**

**Lewis structures and molecular shapes.** Lewis structures do not indicate geometry. For example, it may seem as if two different Lewis structures can be written for $CF_2Cl_2$, but a twist of the ball-and-stick model shows that they represent the same molecule. (Cl, *green*; F, *yellow*.)

The ammonium ion is one of many common polyatomic ions that have the tetrahedral shape. When one of the four electron groups in the tetrahedral arrangement is a lone pair, the molecular shape is that of a **trigonal pyramid** ($AX_3E$), a tetrahedron with one point "missing." As expected, the measured bond angle is slightly less than the ideal 109.5°. In ammonia ($NH_3$), for example, the lone pair forces the N—H bonding pairs together, and the H—N—H bond angle is 107.3°.

Picturing molecular shapes helps us visualize what happens during a reaction. When $NH_3$ and $H^+$ form an ammonium ion, for example, the lone pair on trigonal pyramidal $NH_3$ forms a covalent bond to $H^+$ to form tetrahedral $NH_4^+$. Note how the H—N—H bond angle expands as the repulsions due to the lone pair decrease to that of another bonding pair:

When the four electron groups around the central atom include two bonding and two nonbonding groups, the molecular shape is *bent or V shaped* ($AX_2E_2$). [Recall that one of the shapes in the trigonal planar arrangement—that with two bonding groups and one lone pair—is also called bent ($AX_2E$), but its ideal bond angle is 120°, not 109.5°.]

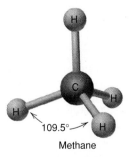

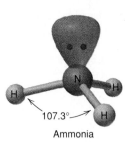

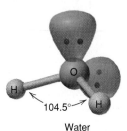

**FIGURE 10.7**
**The effect of lone-pair repulsions on bond angle.** Since lone pairs are attracted by only one atom, they exert stronger repulsions than do bonding pairs, thereby decreasing bond angles to less than the ideal values. In $CH_4$, there are no lone pairs, and the ideal bond angle of 109.5° is seen. In $NH_3$, one lone pair repels the bonding pairs, and the bond angle is decreased to 107.3°. In $H_2O$, two lone pairs cause greater repulsions, and the bond angle is 104.5°.

Water is the most important V-shaped molecule in the tetrahedral arrangement. We would expect the repulsions between its two lone pairs to have a greater effect on bond angle than the repulsions from the single lone pair in $NH_3$, and observations confirm this: two lone pairs on the central O atom compress the H—O—H bond angle to 104.5°:

Thus, for similar molecules within a given electron-group arrangement, electron-pair repulsions cause deviations from ideal bond angles in the following order:

Lone pair–lone pair > lone pair–bonding pair > bonding pair–bonding pair     **(10.1)**

The effects of electron repulsions, as they occur in the shapes of methane, ammonia, and water, are summarized in Figure 10.7.

### Using VSEPR Theory to Determine Molecular Shape

Before describing the shapes resulting from five and six electron groups, let's examine a stepwise method for applying the VSEPR theory to determine a molecular shape from a Lewis structure:

1. *Write the Lewis structure* to see the atom sequence and the number of electron groups around the central atom.
2. *Assign an electron-group arrangement* based on the number of bonding groups (single or multiple bond) *plus* nonbonding groups (lone pair or lone electron).
3. *Predict the ideal bond angle and any deviation from it* based on the electron-group arrangement and the presence of lone pairs or multiple bonds.
4. *Draw and name the molecular shape* by counting bonding groups and nonbonding groups separately.

SAMPLE PROBLEM 10.1 _____

### Predicting Molecular Shapes with Two, Three, and Four Electron Groups

**Problem:** Determine the molecular shape and ideal bond angles of **(a)** $PF_3$ and **(b)** $COCl_2$.
**Solution: (a)** For $PF_3$.
1. Write the Lewis structure:

$$:\!\ddot{F}\!-\!\ddot{P}\!-\!\ddot{F}\!:$$
$$|$$
$$:\!\ddot{F}\!:$$

2. Assign the electron-group arrangement: Four electron groups around P (three bonding and one lone pair) give the *tetrahedral arrangement.*
3. Predict the ideal bond angle and any deviation: For the tetrahedral arrangement, the **ideal angle is 109.5°.** Since there is one lone pair, the actual bond angle should be *less than 109.5°.*
4. Draw and name the molecular shape: $PF_3$ has a **trigonal pyramidal shape:**

**(b)** For $COCl_2$.

1. Write the Lewis structure:

:O:
‖
:Cl̈—C—C̈l:

2. Assign the electron-group arrangement: Three electron groups around C (two single and one double bond) give the *trigonal planar arrangement.*
3. Predict the bond angles: The **ideal angle is 120°,** but the double bond between C and O should compress the Cl—C—Cl angle.
4. Draw and name the molecular shape: The shape is **trigonal planar:**

:O:
‖ ) 124.5°
:Cl̈—C—C̈l:
111°

**Check:** We compare the answers with the general information in Figures 10.4 and 10.5.

**Comment:** Be sure the Lewis structure is correct because it determines the other steps.

**FOLLOW-UP PROBLEM 10.1**
Draw the shape and predict the ideal bond angles, and any deviations from them, for the following molecules: **(a)** $CS_2$; **(b)** $PbCl_2$; **(c)** $CBr_4$; **(d)** $SF_2$.

## Molecular Shapes with Five Electron Groups (Trigonal Bipyramidal Arrangement)

Molecules with five or six electron groups must have a central atom from Period 3 or higher with *d* orbitals available to expand its valence shell beyond eight electrons. When five electron groups maximize their separation, they form the **trigonal bipyramidal arrangement,** two trigonal pyramids sharing the same base (Figure 10.8). Note that *there are two positions for surrounding electron groups and two ideal bond angles.* Three **equatorial groups** lie in a trigonal plane that includes the central atom, and two **axial groups** lie above and below this plane. Therefore, a 120° bond angle separates equatorial groups, and a 90° angle separates axial from equatorial groups. In general, the greater the bond angle, the weaker the repulsions, so *equatorial-equatorial (120°) repulsions are weaker than axial-equatorial (90°) repulsions.* The tendency to minimize these 90° repulsions governs the four shapes within the trigonal bipyramidal arrangement.

With all five positions occupied by bonded atoms, the molecule has the *trigonal bipyramidal* shape (AX_5). Phosphorus pentachloride ($PCl_5$) is an example:

:Cl̈: 90°
:Cl̈—P—C̈l:
120° ⟋ P ⟍
:Cl̈: :Cl̈: 90°

Lone pairs exert stronger repulsive forces than bonding pairs, so to minimize repulsions, *lone pairs occupy equatorial positions.* With one lone pair present at an equatorial position, the molecule has a **seesaw shape** (AX_4E). Sulfur tetrafluoride ($SF_4$), a powerful fluorinating agent, has this shape; the seesaw is shown tipped up on an end in Figures 10.8 and 10.9. Note the ideal bond angles and how the equatorial lone pair repels all four bonding pairs to lower the actual bond angles.

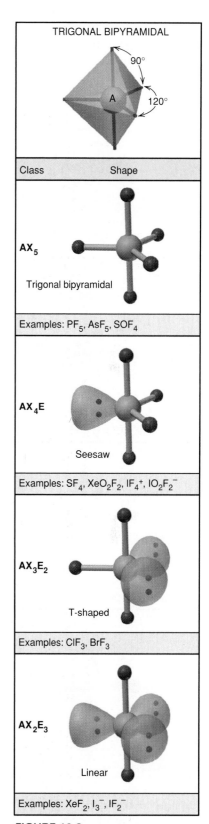

TRIGONAL BIPYRAMIDAL

90°
A
120°

| Class | Shape |
|---|---|
| AX_5 | Trigonal bipyramidal |

Examples: $PF_5$, $AsF_5$, $SOF_4$

| AX_4E | Seesaw |
|---|---|

Examples: $SF_4$, $XeO_2F_2$, $IF_4^+$, $IO_2F_2^-$

| AX_3E_2 | T-shaped |
|---|---|

Examples: $ClF_3$, $BrF_3$

| AX_2E_3 | Linear |
|---|---|

Examples: $XeF_2$, $I_3^-$, $IF_2^-$

**FIGURE 10.8**
The four molecular shapes of the trigonal bipyramidal electron-group arrangement.

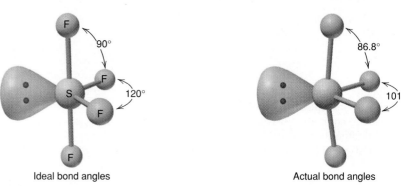

Ideal bond angles                 Actual bond angles

**FIGURE 10.9**
**Ideal and actual bond angles in SF₄.** To minimize 90° repulsions, the single lone pair in SF₄ occupies an equatorial position of the trigonal bipyramid. The lower actual bond angles result from lone-pair repulsions of the bonding pairs.

The tendency of lone pairs to occupy equatorial positions causes molecules with three bonding groups and two lone pairs to have a **T shape** ($AX_3E_2$). Bromine trifluoride ($BrF_3$), one of many compounds with fluorine bound to a larger halogen, has this shape and the predicted decrease from the ideal 90° F—Br—F bond angle:

Molecules with two bonding groups have three lone pairs in equatorial positions, so the two bonds are in axial positions, giving the molecule a *linear* shape ($AX_2E_3$) with a 180° axial atom-to-central atom-to-axial atom (X—A—X) bond angle. For example, the triiodide ion ($I_3^-$), which forms when $I_2$ dissolves in aqueous $I^-$ solution, is linear:

### Molecular Shapes with Six Electron Groups (Octahedral Arrangement)

The final electron-group arrangement we consider is the **octahedral arrangement** (Figure 10.10). Six electron groups point to the corners of an octahedron, an eight-sided regular polygon with six equidistant vertices. The six positions are equivalent, so all the groups have a 90° ideal bond angle. Three important molecular shapes are within this arrangement.

With six bonding groups, the molecular shape is *octahedral* ($AX_6$). When seesaw-shaped $SF_4$, for example, reacts with additional $F_2$, the central S atom expands its valence shell further to form sulfur hexafluoride ($SF_6$):

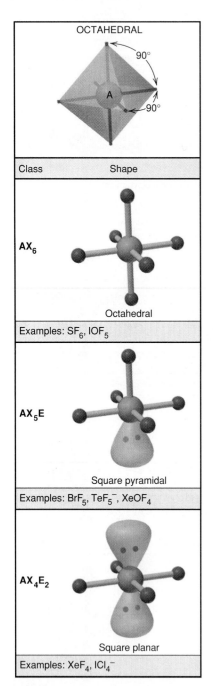

**FIGURE 10.10**
The three molecular shapes of the octahedral electron-group arrangement.

The six octahedral positions are equivalent, so it makes no difference which position one lone pair occupies. Five bonded atoms and one lone pair define the **square pyramidal shape** ($AX_5E$), as seen in iodine pentafluoride ($IF_5$).

When a molecule has two lone pairs and four bonding groups, the lone pairs always lie at *opposite vertices* to avoid the stronger lone pair–lone pair 90° repulsions. This positioning gives rise to the **square planar shape** ($AX_4E_2$). Experiment shows that xenon tetrafluoride ($XeF_4$), one of the first noble gas compounds to be prepared, has a square planar shape.

SAMPLE PROBLEM 10.2 _____

## Predicting Molecular Shapes with Five or Six Electron Groups

**Problem:** Determine the molecular shape and predict the bond angles (relative to the ideal angles) of **(a)** $SbF_5$ and **(b)** $BrF_5$.
**Plan:** We proceed as in Sample Problem 10.1, keeping in mind the need to minimize 90° repulsions.
**Solution: (a)** For $SbF_5$.
1. Lewis structure:

2. Electron-group arrangement: With five electron groups, this is the *trigonal bipyramidal* arrangement.
3. Bond angles: Since all the groups and surrounding atoms are identical, the **bond angles are ideal:** 120° between equatorial groups and 90° between axial and equatorial groups.
4. Molecular shape: Five bonding pairs give the **trigonal bipyramidal** shape:

**(b)** For $BrF_5$.
1. Lewis structure:

2. Electron-group arrangement: Six electron groups result in the *octahedral* arrangement.
3. Bond angles: The lone pair should make all the **bond angles smaller than the ideal 90°.**
4. Molecular shape: One lone pair and five bonding pairs give the **square pyramidal** shape:

FOLLOW-UP PROBLEM 10.2
Draw the molecular shapes and predict the bond angles (relative to the ideal angles) for the following molecules: **(a)** $ICl_2^-$; **(b)** $ClF_3$; **(c)** $SOF_4$.

## Molecular Shapes with More than One Central Atom

The shapes of molecules with more than one central atom are composites of molecular shapes with a single central atom. Consider ethane ($CH_3CH_3$), a component of natural gas (Figure 10.11, *A*). Four bonding groups and no lone pairs are around each of the two central carbons, so ethane is shaped like two tetrahedra that share a point.

Ethanol ($CH_3CH_2OH$), the intoxicating substance in beer and wine, has three atom centers (Figure 10.11, *B*). The $CH_3$— group is tetrahedrally shaped, and the —$CH_2$— group has four bonding groups around its central C atom, so it is also tetrahedrally shaped. The O atom has four electron groups, but the two lone pairs give it a V shape ($AX_2E_2$). For these more complex molecules, it is easiest to find the molecular shape around one central atom at a time.

SAMPLE PROBLEM 10.3 _____

### Predicting Molecular Shapes with More than One Central Atom

**Problem:** Determine the shape around each of the central atoms in acetone, $(CH_3)_2C\!=\!O$.
**Plan:** There are three central atoms, two of which are $CH_3$— groups. We determine the shape around one central atom at a time.
**Solution:**
1. Lewis structure:

$$\begin{array}{c} \text{H} \quad \text{:O:} \quad \text{H} \\ | \qquad \| \qquad | \\ \text{H}\!-\!\text{C}\!-\!\text{C}\!-\!\text{C}\!-\!\text{H} \\ | \qquad\qquad\qquad | \\ \text{H} \qquad\qquad \text{H} \end{array}$$

2. Electron-group arrangement: Each $CH_3$— group has four bonding groups around its central C, so its electron-group arrangement is *tetrahedral*. The third C has three bonding groups around it, so it has the *trigonal planar* arrangement.
3. Bond angles: The $CH_3$— groups should have close to the tetrahedral ideal angle of 109.5°. The C=O double bond should compress the C—C—C angle to less than the ideal 120°.
4. Shapes around central atoms: The $CH_3$— groups have the **tetrahedral** shape, and the three groups around the other C atom give the **trigonal planar** shape:

$$\begin{array}{c} \text{:O:} \\ \| \\ 122° \end{array}$$

(structure showing acetone with H atoms, C—C bonds, 116° and ~109.5° angles)

FOLLOW-UP PROBLEM 10.3
Determine the shape around each central atom and predict any deviations from ideal bond angles in the following: **(a)** $H_2SO_4$; **(b)** $CH_3C\!\equiv\!CH$; **(c)** $S_2F_2$.

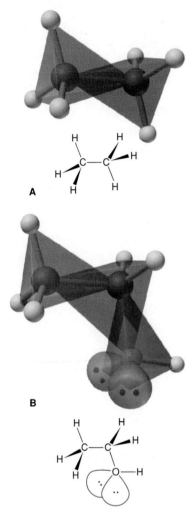

**A**

**B**

**FIGURE 10.11**
**The tetrahedral centers of ethane and of ethanol.** When a molecule has more than one central atom, the shape is a composite of the shape around each center. **A,** Ethane's shape can be viewed as two interlocking tetrahedra. **B,** Ethanol's shape can be viewed as three interlocking tetrahedral arrangements, with the shape around O bent or V shaped due to its two lone pairs.

### Section Summary

The VSEPR theory proposes that each group of electrons around the central atom (single bond, multiple bond, lone pair, or lone electron) occupies as much space as possible. Five electron-group arrangements result when two, three, four, five, and six electron groups surround a central atom. Each arrangement is associated with one or more molecular shapes. Ideal bond angles are prescribed by geometry; deviations occur when the surrounding atoms or electron groups are not identical. Lone pairs and multiple bonds exert more repulsions than single bonds. Larger molecules have shapes that are composites of the shapes around each central atom.

# 10.2 Molecular Shape and Molecular Polarity

Knowing the shape of a substance's molecules is a key to understanding many aspects of its physical and chemical behavior. One far-reaching effect of molecular shape is molecular polarity, which can influence melting and boiling points, solubility, and even reactivity.

In Chapter 9, you learned that a covalent bond is *polar* when it joins atoms of different electronegativities because the atoms share the electrons unequally. In diatomic molecules such as HF, the polar bond, in effect, causes the molecule itself to be polar. Molecules with a net imbalance of charge have a **molecular polarity** and are oriented by an electric field, with their partial charges pointing, on the average, toward the oppositely charged electric plates (Figure 10.12). In molecules with more than two atoms, both shape *and* bond polarity determine the molecular polarity. The **dipole moment ($\mu$),** which is a measure of this polarity, is the vector (directional) product of the partial charges (in coulombs, C) and the distance between them (in meters, m); it is measured in *debye* (D) units (1 D = $3.34 \times 10^{-30}$ C · m).

## Bond Polarity, Bond Angle, and Dipole Moment

Polar bonds do not *necessarily* lead to a polar molecule. In carbon dioxide, for example, the large electronegativity difference between C (EN = 2.5) and O (EN = 3.5) makes each C=O bond quite polar. However, because $CO_2$ is linear, its bonds are directed 180° from each other, so these identical bond polarities are counterbalanced and give a molecule with *no net dipole moment* ($\mu = 0$ D):

$$\overset{\longrightarrow}{\underset{\longleftarrow}{\ddot{O}=C=\ddot{O}}}$$

Water also has identical atoms bonded to the central atom, but it *does* have a significant dipole moment ($\mu = 1.85$ D) due to the effect of the lone pairs on its shape. In each O—H bond, electron density is pulled toward the more electronegative O atom, but the bond polarities do *not* counterbalance each other because the water molecule is V shaped (see Figure 4.9). Instead, the bond polarities partially reinforce each other, and the oxygen end of the molecule is considerably more negative than the other end (the region between the H atoms):

$$H \overset{\ddot{O}}{\underset{H}{\nearrow\uparrow\nwarrow}}$$

(The molecular polarity of water has some amazing effects, from determining the composition of the oceans to supporting life itself, as you'll see in Chapter 11.)

In the two previous examples, molecular shape influences polarity. When different molecules have the same shape, the nature of the atoms surrounding the central atom can have a major effect on polarity. Consider carbon tetrachloride ($CCl_4$) and chloroform ($CHCl_3$), two tetrahedral molecules with different polarities. In $CCl_4$, the surrounding atoms are all Cl atoms. Although each C—Cl bond is polar ($\Delta$EN = 0.5), the molecule is nonpolar ($\mu = 0$ D) because the individual bond polarities counterbalance each other. In $CHCl_3$, an H atom substitutes for one of the Cl atoms, disrupting the balance and giving chloroform a significant dipole moment ($\mu = 1.01$ D):

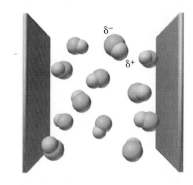

**A** Electric field off

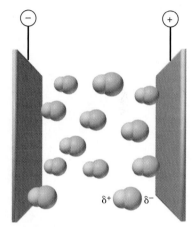

**B** Electric field on

**FIGURE 10.12**
**The orientation of polar molecules in an electric field. A,** In the absence of an external electric field, HF molecules are oriented randomly. **B,** In the presence of the field, the molecules, on average, become oriented as a result of their partial charges ($\delta^+$H—F$\delta^-$) interacting with the field.

SAMPLE PROBLEM 10.4

**Predicting the Polarity of Molecules**

**Problem:** From electronegativity (EN) values and their periodic trends (see Figure 9.16), predict whether each of the following molecules is polar and show the direction of bond dipoles and the overall molecular dipole when applicable:
**(a)** Ammonia, $NH_3$
**(b)** Boron trifluoride, $BF_3$
**(c)** Carbonyl sulfide, COS (atom sequence SCO)
**Plan:** First, we draw and name the molecular shape. Then, using relative EN values, we decide on the direction of each bond dipole. Finally, we see if the bond dipoles balance or reinforce each other in the molecule as a whole.
**Solution: (a)** For $NH_3$. The molecular shape is trigonal pyramidal. From Figure 9.16, we see that N (EN = 3.0) is more electronegative than H (EN = 2.1), so the bond dipoles point toward N. Since the bond dipoles partially reinforce each other, the molecular dipole points toward N:

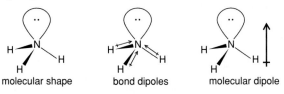

Therefore, **ammonia is polar.**
**(b)** For $BF_3$. The molecular shape is trigonal planar. Since F (EN = 4.0) is farther to the right in Period 2 than B (EN = 2.0), it is more electronegative, so each bond dipole points toward F. However, the bond angle is 120°, so the three bond dipoles counterbalance each other, and $BF_3$ has no molecular dipole:

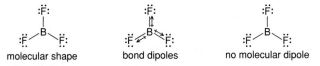

Therefore, **boron trifluoride is nonpolar.**
**(c)** For COS. The molecular shape is linear. Since C and S have the same EN, the C=S bond is nonpolar, but the C=O is quite polar (ΔEN = 1.0), so there is a net molecular dipole toward the O:

$$\ddot{S}=C=\ddot{O} \qquad \ddot{S}=C=\ddot{O} \qquad \ddot{S}=C=\ddot{O}$$
molecular shape      bond dipoles      molecular dipole

Therefore, **carbonyl sulfide is polar.**

FOLLOW-UP PROBLEM 10.4
Show the bond dipoles and molecular dipole for **(a)** dichloromethane ($CH_2Cl_2$); **(b)** iodomonoxy pentafluoride ($IOF_5$); **(c)** nitrogen tribromide ($NBr_3$).

To get a sense of the importance of molecular polarity on physical behavior, consider what effect a molecular dipole might have when many polar molecules lie near each other, as they do in a liquid. It makes sense that the partial charges on nearby polar molecules would attract each other. Thus, a molecular property such as dipole moment could affect a macroscopic property such as boiling point. A liquid boils when bubbles of gas form within it. To enter the gas phase, the molecules in the liquid phase must overcome the weak attractive forces *between* them. A molecular dipole influences the strength of these attractions. Consider the two substances *trans*-1,2-dichloroethylene and *cis*-1,2-dichloroethylene. These compounds

Chemistry in Physiology
## Molecular Shape, Biological Receptors, and the Sense of Smell

At the chemical level, a biological cell is a membrane-bound sack filled with molecular shapes interacting in an aqueous fluid. Complex processes in an organism often begin when a molecular "key" fits into a correspondingly shaped molecular "lock." The key can be a relatively small molecule circulating in a body fluid, whereas the lock is usually a large molecule, known as a biological **receptor,** that is often found embedded in a cell membrane. The surface of the receptor contains a precisely shaped cavity, or *receptor site,* that is exposed to the passing fluid. Thousands of molecules collide with this site, but when a molecule with the correct shape (that is, the molecular key) lands on it, the receptor "grabs" it through intermolecular attractions, and the biological response begins.

Let's see how this fitting together of molecular shapes operates in the sense of smell (olfaction). A substance must have certain properties to have an odor. An odorous molecule travels through the air, so it must come from a gas or a volatile liquid or solid. To reach the receptor, it must be soluble, at least to a small extent, in the thin film of aqueous solution that lines the nasal passages. Most important, the odorous molecule, or a portion of it, must have a shape that fits into one of the olfactory receptor sites that cover the nerve endings deep within the nasal passage (Figure 10.A). When this happens, the resulting nerve impulses travel from these endings to the brain, which interprets the impulses as a specific odor.

**FIGURE 10.A The location of olfactory receptors within the nose. A,** The olfactory area lies at the top of the nasal passage very close to the brain. Air containing the odorous molecules is sniffed in, warmed, moistened, and channeled toward this region. **B,** A blowup of the region shows olfactory nerve cells and their hairlike endings protruding into the liquid-coated nasal passage. **C,** A further blowup shows a receptor on one of the endings containing an odorous molecule that matches its shape. This particular molecule has a peppermint odor.

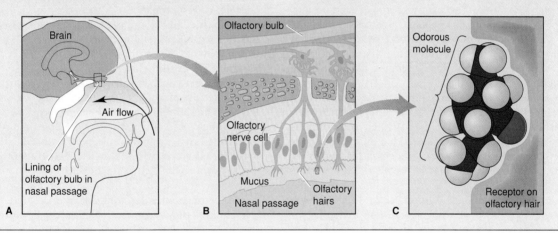

A
Brain
Air flow
Lining of olfactory bulb in nasal passage

B
Olfactory bulb
Olfactory nerve cell
Mucus
Nasal passage
Olfactory hairs

C
Odorous molecule
Receptor on olfactory hair

*Continued.*

have the same molecular formula ($C_2H_2Cl_2$) and therefore the same molar mass, but they have different physical and chemical properties; in particular, the *cis* compound boils 13°C higher than the *trans*. VSEPR theory predicts, and spectroscopic analysis shows, that both molecules are flat, with a trigonal planar shape around each C atom. The *trans* compound has no dipole moment ($\mu = 0$ D), so the C—Cl bond polarities must balance each other. In contrast, the *cis* compound is polar ($\mu = 1.90$ D), so the bond dipoles must partially reinforce each other, with the molecular dipole pointing between the Cl atoms. As a result, in liquid samples of these compounds, the polar *cis* molecules attract each other more strongly than the nonpolar *trans* molecules, so more energy is needed to overcome these stronger attractive forces. Therefore, the *cis* compound has a higher boiling point. We extend these ideas more fully to the physical properties of liquids and solids in Chapter 11. The Chemical Connections essay provides some insight into the importance of molecular shape in biological behavior.

*trans*          *cis*

Chemistry in Physiology
**Molecular Shape, Biological Receptors, and the Sense of Smell—cont'd**

Floral

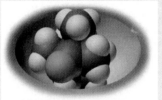

Camphor-like

Ethereal

In the 1950s, the stereochemical theory of odor (*stereo* means three-dimensional) was introduced to explain the relation between odor and molecular shape. Its basic premise is that a molecule's shape (and sometimes polarity), but *not* its chemical composition, is the primary determinant of its odor. According to this theory, there are seven primary odors, each corresponding to one of seven different types of olfactory receptor sites. The seven odors are camphor-like, musky, floral, pepperminty, ethereal, pungent, and putrid. (The last two odors depend on molecular polarity more than shape, and their receptors have partial charges opposite to those of the molecules landing there.) Figure 10.B shows the shapes of three of the seven receptor sites, each occupied by a molecule having the associated odor.

Several predictions of the theory have been verified by experiment. If two substances fit the same receptor, they should have the same odor, even if their compositions vary. The four molecules in Figure 10.C all fit the "bowl-shaped" camphor-like receptor and smell like moth repellent, despite their different formulas. If different portions of one molecule fit different receptors, the molecule should have a mixed odor. Portions of the benzaldehyde molecule fit the camphor-like, floral, and minty receptors, which gives it an almond odor; other molecules that smell like almonds also fit the same three receptors.

Despite these and other results consistent with the theory, recent evidence suggests that the original model is too simple. In many cases, a prediction of odor based on molecular shape turned out to be incorrect when the substance was actually smelled. One reason for this discrepancy is that a molecule in the gas phase may have a very different shape in solution, that is, at the receptor.

**FIGURE 10.B Shapes of some olfactory receptor sites.** Three of the seven proposed olfactory receptors are shown with a molecule having that odor occupying the site.

By the early 1990s, evidence for the presence of some 1000 receptor sites had been obtained, and it is thought that various combinations of stimulation produce the more than 10,000 odors humans can distinguish. So, although a key premise of the theory is still accepted—that odor depends on molecular shape—the nature of the dependence is complex and is an active area of research in the food, cosmetics, and insecticide industries. The sense of smell is so vital for survival that these topics are important areas of study for biologists as well.

Many other biochemical processes are controlled by one molecule fitting into a receptor site on another. Enzymes are proteins that bind cellular reactants and facilitate their reaction in this way. Nerve impulses are transmitted when small molecules released from one nerve, fit into receptors on the next. Mind-altering drugs act by chemically disrupting the molecular fit at such nerve receptors in the brain. One type of immune response is triggered when a molecule on a bacterial surface binds to the receptors on "killer" cells in the bloodstream. Hormones regulate energy production and growth by fitting into and activating key receptors. Genes function when certain nucleic acid molecules fit into specific regions of others. Indeed, *no molecular property is more crucial to living systems than shape.*

**FIGURE 10.C Different molecules with the same odor.** The proposal that odor is based on shape, not composition, is supported by these four different substances, all having a camphor-like (moth repellent) odor.

Camphor

Hexachloroethane

Thiophosphoric acid dichloride ethylamide

Cyclooctane

**Section Summary**

Bond polarity and molecular shape determine molecular polarity, which is measured as a dipole moment. When bond polarities counterbalance each other, the molecule is nonpolar; when they reinforce each other, the molecule is polar. Molecular shape and polarity can affect physical properties, such as boiling point, and they play a central role in many aspects of biological function.

## 10.3 Valence Bond (VB) Theory and Orbital Hybridization

All scientific models have limitations because they are simplifications of reality. The simple VSEPR model predicts molecular shapes by assuming that electron groups tend to minimize their repulsions. However, VSEPR theory does not reconcile the shapes of molecules with those of atomic orbitals, nor does it even address the nature of the orbitals involved in bonding. One particularly useful approach that chemists employ to explain these aspects of covalent bonding is **valence bond (VB) theory.**

Valence bond theory has limitations as well. Although it bases molecular shape on orbital characteristics, it does not adequately explain the magnetic or spectral properties of molecules, and it understates the importance of electron delocalization. To deal with these phenomena, molecular orbital theory is needed, as you'll see in the next section.

Don't be discouraged by our need for a cluster of models to explain the observations in a topic as universal as chemical bonding. What we find in every science is that one model accounts for a particular aspect of a topic better than another, and that several models are called into service to explain a broader range of phenomena.

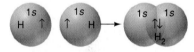

**A** Hydrogen, $H_2$

### The Central Themes of Valence Bond Theory

The basic principle of the VB model is that *a covalent bond forms when the orbitals from two atoms overlap and a pair of electrons occupies the region between the nuclei.* Three central ideas of VB theory are derived from this principle:

1. *Opposing spins of the electron pair.* As the Pauli exclusion principle (Section 8.1) prescribes, the region of space formed by the overlapping orbitals *has a maximum capacity of two electrons that must have opposite spins.* When a molecule of hydrogen forms, for instance, the $1s$ orbitals of two H atoms overlap, and the two $1s$ electrons occupy the region between the nuclei with opposite spins (Figure 10.13, A).

2. *Maximum overlap of bonding orbitals.* The bond strength depends on the attraction of the nuclei for the shared electrons, so *the greater the orbital overlap, the stronger the bond.* The extent of overlap depends on the shape and direction of the orbitals involved. The $s$ orbital is spherical, but $p$ and $d$ orbitals have particular orientations—more electron density in one direction than in another—so a bond involving $p$ or $d$ orbitals will tend to be oriented in the direction that maximizes overlap.

In the HF bond, for example, the $1s$ orbital of H overlaps the half-filled $2p$ orbital of F along the axis of that orbital (Figure 10.13, B); any other direction would result in less overlap and, thus, a weaker bond. Similarly, in the F—F bond of $F_2$, the two $2p$ orbitals interact end to end, that is, along the orbital axes, to attain maximum overlap (Figure 10.13, C).

**B** Hydrogen fluoride, HF

**C** Fluorine, $F_2$

**FIGURE 10.13**

**Orbital overlap and spin pairing in three diatomic molecules. A,** The $H_2$ molecule forms when the $1s$ orbitals of two H atoms overlap and the two electrons, with opposite spins, occupy the overlapped region. **B,** To maximize overlap in HF, half-filled H $1s$ and F $2p$ orbitals overlap along the axis of the $2p$ orbital involved in bonding. (The other two F $2p$ orbitals are not shown.) **C,** Similarly, in $F_2$, the half-filled $2p$ orbital on one F points end to end toward the other to maximize overlap.

3. *Hybridization of atomic orbitals.* To explain the bonding in simple diatomic molecules such as HF, it is sufficient to propose the direct overlap of *s* and *p* orbitals of isolated ground-state atoms, as in the previous description. But how is it possible for the overlap of spherical *s* orbitals and dumbbell-shaped, perpendicular *p* orbitals to give the linear, trigonal planar, tetrahedral, and other shapes observed in so many polyatomic molecules and ions? To cite just one example, consider a methane molecule, composed of four hydrogen atoms bonded to a central carbon. Recall that an isolated ground-state carbon atom ([He] $2s^2 2p^2$) has one filled $2s$ orbital and three $2p$ orbitals, two half filled and one empty. One might be able to see how overlap of two perpendicular *p* orbitals of carbon with the $1s$ orbitals of two hydrogens could give rise to two C—H bonds separated by a right angle, but it is not at all clear how such overlap would form the four identical, tetrahedrally oriented C—H bonds we observe in methane.

To describe the bonding in such molecules, Linus Pauling proposed that *the valence atomic orbitals in the molecule are different from those in the isolated atoms.* Quantum-mechanical calculations show that mathematical mixing of specific combinations of *nonequivalent* orbitals in a given atom (that is, *s, p, d* orbitals) gives rise to new atomic orbitals with characteristics that would lead to the best (most stable) bonds and to the observed molecular shapes. The process of orbital mixing is called **hybridization,** and the new atomic orbitals are called **hybrid orbitals.** Two key points to remember are:

1. The *number* of hybrid orbitals obtained *always equals* the number of atomic orbitals mixed.
2. The *type* of hybrid orbitals obtained *varies with* the types of atomic orbitals mixed.

You can conceive of hybridization as a process in which atomic orbitals mix to give hybrid orbitals and electrons enter them with spins parallel (Hund's rule). In truth, though, hybridization is our mathematically convenient way of explaining what occurs during bonding that gives rise to the molecules we observe.

### Types of Hybrid Orbitals

It is important to keep in mind that we postulate the presence of a certain type of hybrid orbital *after* we observe the molecular shape. As we discuss the five common types of hybridization next, notice that the spatial orientation of each type of hybrid orbital corresponds with one of the five common electron-group arrangements predicted by VSEPR theory.

*sp* **hybridization.** When two electron groups surround the central atom, we observe a linear shape, which implies a linear orientation of the orbitals involved in bonding. Valence bond theory proposes that mixing two nonequivalent orbitals of a central atom, one *s* and one *p*, gives rise to two equivalent *sp* **hybrid orbitals** that lie 180° apart (Figure 10.14, *A*). Note the shape of the hybrid orbital: with one large and one small lobe, it differs markedly from those of the atomic orbitals that were mixed. The orientations of hybrid orbitals extend electron density in the bonding direction and minimize repulsions between the electrons that occupy them. Thus, *both shape and orientation maximize overlap with the orbital of the other atom in the bond.*

In gaseous $BeCl_2$, for example, the Be atom is said to be *sp* hybridized. The other parts of Figure 10.14 depict the hybridization of Be in $BeCl_2$ in terms of a traditional orbital box diagram and one with orbital contours, then show bond formation with Cl. The $2s$ and one of the three $2p$ orbitals of Be mix and form two *sp* orbitals. These overlap $3p$ orbitals from two Cl

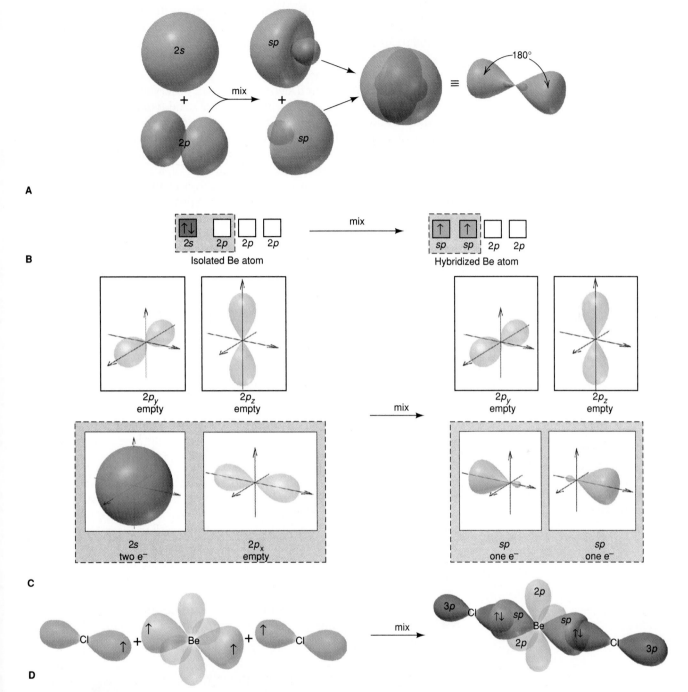

**FIGURE 10.14**

**The *sp* hybrid orbitals in gaseous BeCl₂. A,** One atomic 2*s* and one 2*p* orbital mix to form two atomic *sp* hybrid orbitals. Note the large and small lobes of the hybrid orbitals.The two *sp* orbitals lie in opposite directions. For clarity, the simplified hybrid orbitals shown at right will be used throughout the text, usually without the small lobe. **B,** The orbital box diagram for hybridization in Be shows that the 2*s* and one of the three 2*p* orbitals form two *sp* hybrid orbitals, while the two other 2*p* orbitals remain unhybridized. Electrons half fill the *sp* hybrids. **C,** The orbital box diagram is shown with orbital contours instead of arrows. **D,** BeCl₂ forms by overlap of the two *sp* hybrids with the 3*p* orbitals of two Cl atoms, while the two unhybridized Be 2*p* orbitals lie perpendicular to the *sp* hybrids. (For clarity, only the 3*p* orbital of each Cl involved in bonding is shown.)

atoms, and the four valence electrons—two from Be and one from each Cl atom—occupy the overlapped orbitals in pairs. Note the two unhybridized $2p$ orbitals of Be, which lie perpendicular to each other and to the bond axes. Thus, through hybridization, the paired $2s$ electrons in the isolated Be atom were distributed into $sp$ orbitals, which formed two Be—Cl bonds.

**$sp^2$ hybridization.** In order to explain an observed trigonal planar molecular shape, we postulate the mixing of one $s$ and two $p$ orbitals of the central atom to give three **$sp^2$ hybrid orbitals.** The superscript refers to the number of atomic orbitals of a given type that are mixed—one $s$ and two $p$ orbitals, thus $s^1p^2$, or $sp^2$—*not* to the number of electrons in the orbital. The axes of the three $sp^2$ orbitals lie in a plane 120° apart.

VB theory proposes that the central B atom in the $BF_3$ molecule is $sp^2$ hybridized. Its three $sp^2$ orbitals point to the corners of an equilateral triangle, while its third $2p$ orbital is unhybridized and lies perpendicular to this plane. Each $sp^2$ orbital overlaps the $2p$ orbital of an F atom, and the six valence electrons—three from B and one from each of the three F atoms—are in three bonding pairs (Figure 10.15).

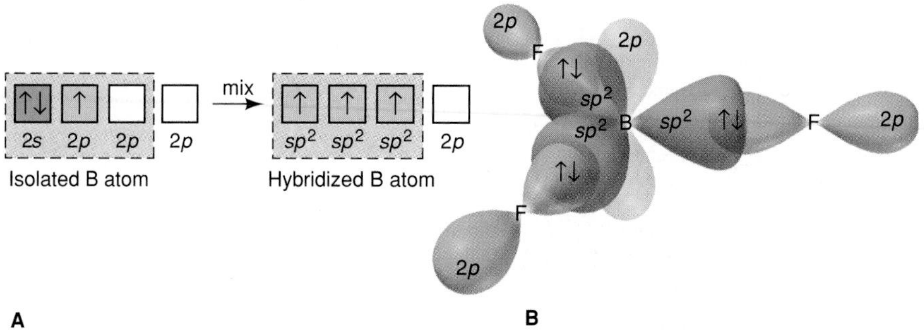

**FIGURE 10.15**
**The $sp^2$ hybrid orbitals in $BF_3$. A,** The orbital box diagram shows that the $2s$ and two of the three $2p$ orbitals of the B atom are mixed to make three $sp^2$ hybrid orbitals. Three electrons half fill the $sp^2$ hybrids. The third $2p$ orbital remains empty and unhybridized. **B,** $BF_3$ forms through overlap of $2p$ orbitals on three F atoms with the $sp^2$ hybrids. The three $sp^2$ hybrids of B lie 120° apart, and its unhybridized $2p$ orbital is perpendicular to the trigonal bonding plane.

**$sp^3$ hybridization.** To explain shapes with a tetrahedral electron-group arrangement, VB theory proposes that the one $s$ and all three $p$ orbitals of the central atom are mixed and form four **$sp^3$ hybrid orbitals,** which point toward the corners of a tetrahedron. Carbon in methane is $sp^3$ hybridized. Its four valence electrons—two $2s$ and two $2p$—half fill the four $sp^3$ hybrids, which overlap the half-filled $1s$ orbitals on four H atoms and form four C—H bonds (Figure 10.16).

Other molecular shapes within a given electron-group arrangement suggest that one or more of the hybrid orbitals are occupied by lone pairs. In $NH_3$, for example, a lone pair fills one of the four $sp^3$ orbitals, whereas in $H_2O$, lone pairs fill two of them (Figure 10.17).

**FIGURE 10.16**
**The $sp^3$ hybrid orbitals in $CH_4$. A,** The orbital box diagram shows that the $2s$ and all three $2p$ orbitals of C are mixed to form four $sp^3$ hybrids. Carbon's four valence electrons half fill the $sp^3$ hybrids. **B,** In methane, the four $sp^3$ orbitals of C point toward the corners of a tetrahedron and overlap the $1s$ orbitals of four H atoms.

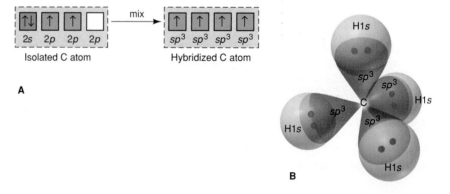

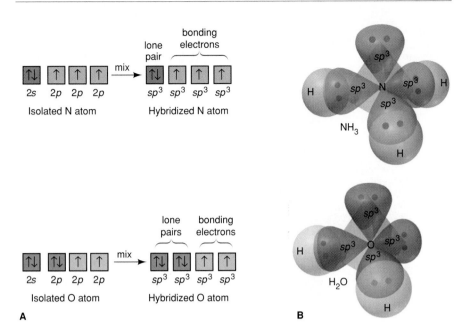

**FIGURE 10.17**
**The $sp^3$ hybrid orbitals in $NH_3$ and $H_2O$.**
**A,** The orbital box diagrams show $sp^3$ hybridization, as in $CH_4$. In $NH_3$ *(top)*, one of the $sp^3$ orbitals is filled with a lone pair. In $H_2O$, two of the $sp^3$ orbitals are filled with lone pairs. **B,** Contour diagrams show the tetrahedral orientation of the $sp^3$ orbitals and the overlap of the bonded H atoms.

**$sp^3d$ hybridization.** The shapes of molecules with trigonal bipyramidal or octahedral electron-group arrangements are explained through similar arguments. The only new point is that such molecules have central atoms from Period 3 or higher, so atomic $d$ orbitals, as well as $s$ and $p$ orbitals, are mixed to form the hybrid orbitals. To explain the trigonal bipyramidal shape of the $PCl_5$ molecule, for example, the VB model proposes that the $3s$, the three $3p$, and one of the five $3d$ orbitals of the central P atom mix and form five **$sp^3d$ hybrid orbitals,** each of which overlaps a $3p$ orbital of a Cl atom. The five valence electrons of phosphorus, together with one from each of the five Cl atoms, pair up to form five P—Cl bonds (Figure 10.18).

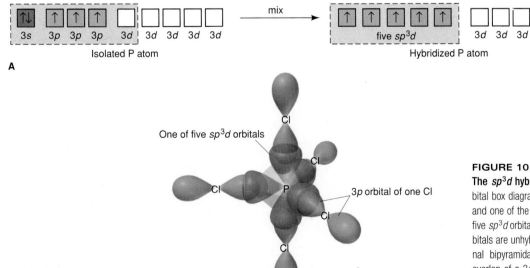

**FIGURE 10.18**
**The $sp^3d$ hybrid orbitals in $PCl_5$. A,** The orbital box diagram shows that one $3s$, three $3p$, and one of the five $3d$ orbitals of P mix to form five $sp^3d$ orbitals that are half filled. Four $3d$ orbitals are unhybridized and empty. **B,** The trigonal bipyramidal $PCl_5$ molecule forms by the overlap of a $3p$ orbital from five Cl atoms with the $sp^3d$ hybrid orbitals of P.

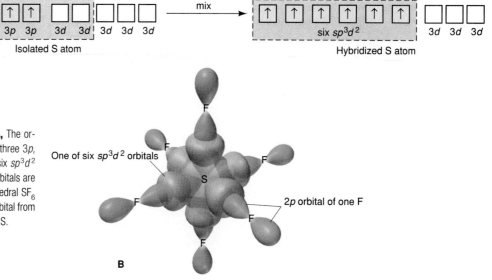

**FIGURE 10.19**
**The $sp^3d^2$ hybrid orbitals in SF$_6$. A,** The orbital box diagram shows that one 3$s$, three 3$p$, and two 3$d$ orbitals of S mix to form six $sp^3d^2$ orbitals that are half filled. Three 3$d$ orbitals are unhybridized and empty. **B,** The octahedral SF$_6$ molecule forms from overlap of a 2$p$ orbital from six F atoms with the $sp^3d^2$ orbitals of S.

**$sp^3d^2$ hybridization.** To rationalize the shape of SF$_6$, the model proposes that the one 3$s$, three 3$p$, and two of the five 3$d$ orbitals of the central S atom are mixed and form six **$sp^3d^2$ hybrid orbitals,** which point to the corners of an octahedron. Sulfur's six valence electrons half fill these hybrid orbitals, each of which overlaps the half-filled 2$p$ orbital of an F atom to produce SF$_6$ (Figure 10.19).

Table 10.1 summarizes the number and type of atomic orbitals that are mixed to obtain the five types of hybrid atomic orbitals. Note the similarities between the orientations of the hybrid orbitals proposed by VB theory and the shapes predicted by VSEPR theory (see Figure 10.2).

**TABLE 10.1 Composition and Orientation of Hybrid Orbitals**

| Atomic orbitals mixed | one $s$<br>one $p$ | one $s$<br>two $p$ | one $s$<br>three $p$ | one $s$<br>three $p$<br>one $d$ | one $s$<br>three $p$<br>two $d$ |
|---|---|---|---|---|---|
| Hybrid orbitals formed | two $sp$ | three $sp^2$ | four $sp^3$ | five $sp^3d$ | six $sp^3d^2$ |
| Shape | Linear | Trigonal planar | Tetrahedral | Trigonal bipyramidal | Octahedral |
| Orientation | | | | | |

SAMPLE PROBLEM 10.5 _____

## Postulating the Hybrid Orbitals in a Molecule

**Problem:** Describe how mixing of atomic orbitals on the central atoms leads to the hybrid orbitals in each of the following:
**(a)** Methanol, $CH_3OH$                **(b)** Sulfur tetrafluoride, $SF_4$

**Plan:** From the Lewis structure and molecular shape, we know the number and arrangement of electron groups around the central atoms, from which we postulate the type of hybrid orbitals involved. Then we write the partial orbital diagram for each central atom before and after the orbitals are hybridized.

**Solution:** **(a)** For $CH_3OH$. The shape is tetrahedral around the C and the O atoms. Therefore, each central atom is $sp^3$ hybridized. The C atom has four half-filled $sp^3$ orbitals:

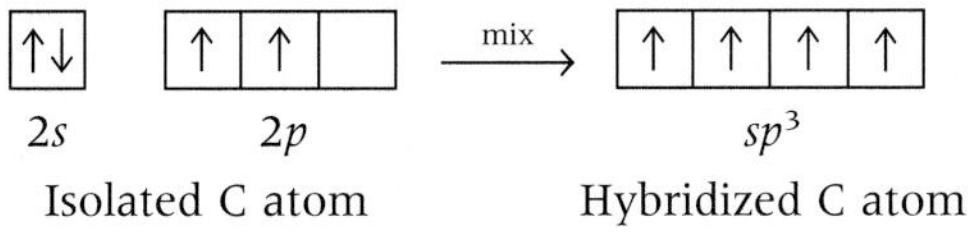

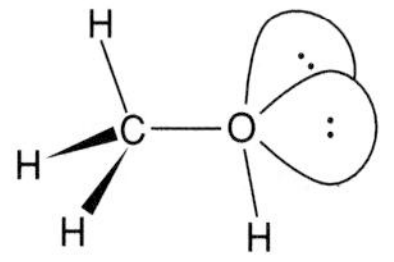

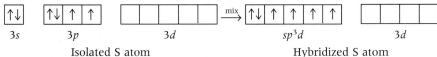

The O atom has two half-filled $sp^3$ orbitals and two filled with lone pairs:

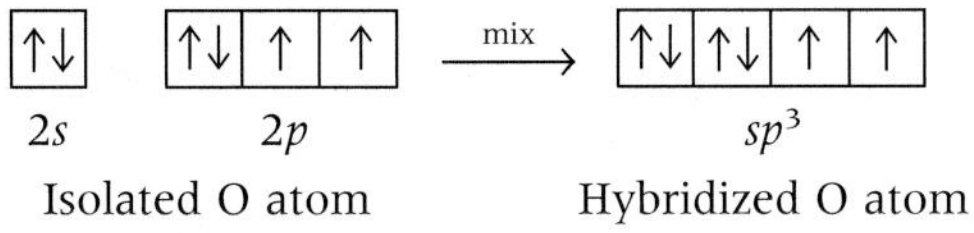

**(b)** For $SF_4$. The molecular shape is seesaw. The central S atom is surrounded by five electron groups, which implies $sp^3d$ hybridization. One $3s$, three $3p$, and one $3d$ orbital are mixed. One hybrid orbital is filled with a lone pair and four are half-filled. Four unhybridized $3d$ orbitals remain empty:

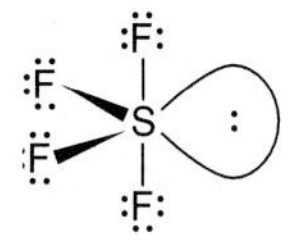

FOLLOW-UP PROBLEM 10.5
Use partial orbital diagrams to show how atomic orbitals on the central atoms lead to hybrid orbitals in the following molecules: **(a)** beryllium hydride, $BeH_2$; **(b)** silicon tetrachloride, $SiCl_4$; **(c)** xenon tetrafluoride, $XeF_4$.

_____

We employ VSEPR theory *and* VB theory to explain how a molecular shape arises from atomic orbitals, but we cannot *assume* that both theories apply in every case. Consider the Lewis structure of $H_2S$. What would we predict for its shape? Based on VSEPR theory, we would say that, as in $H_2O$, the four electron groups around $H_2S$ point to the corners of a tetrahedron, and the two lone pairs compress the H—S—H bond angle below the ideal 109.5°. Based on VB theory, we would propose that the $3s$ and $3p$ orbitals of the central S atom are mixed and form four $sp^3$ hybrid orbitals, two of which are filled with lone pairs while the other two overlap with $1s$ orbitals of two H atoms and are filled with bonding pairs. These arguments are *not* supported by observation. In fact, the $H_2S$ molecule has a bond angle of 92°, close to the 90° angle between *unhybridized p* orbitals. Similar angles also occur in the other Group 6A(16) hydrides and in the larger hydrides of Group 5A(15). As the central atom becomes larger and the bonds to H longer, crowding and electron repulsions decrease. Whatever the case, we do not need to invoke hybridization to explain the bonding in these

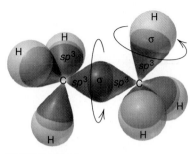

**FIGURE 10.20**
**The σ bonds in ethane ($C_2H_6$).** Both C atoms in ethane are $sp^3$ hybridized. One $sp^3$ orbital of each C overlaps to form a C—C σ bond, and the others overlap the $1s$ orbitals of six H atoms to form six C—H σ bonds. Sigma bonds allow the attached groups to rotate freely because end-to-end overlap is not affected by the rotation.

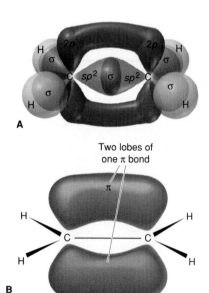

**A**

Two lobes of
one π bond

**B**

**C**

**FIGURE 10.21**
**The σ and π bonds of ethylene ($C_2H_4$).** **A,** The two C atoms are $sp^2$ hybridized and form one C—C σ bond and four C—H σ bonds. Their half-filled unhybridized $2p$ orbitals lie perpendicular to the σ-bond plane and are shown overlapping side to side. **B,** The two overlapping regions comprise one π bond. The σ bonds are shown as lines. The two lobes of one π bond lie above and below the C—C σ bond line. **C,** Bond-line drawing.

molecules. The point is that real factors—bond length, atomic size, electron repulsions, and so forth—determine molecular shape. As always, experiment is the final test of theory. We use VSEPR and VB theories to *explain* the shapes we *observe*. In the case of $H_2S$ and these other nonmetal hydrides, neither VSEPR theory nor the concept of hybridization applies.

## Types of Covalent Bonds and the Valence Bond Description of Multiple Bonding

The valence bond description of double and triple bonds is somewhat different from that of single bonds. Let's examine the bonding in the three two-carbon hydrocarbons, ethane ($C_2H_6$), ethylene ($C_2H_4$), and acetylene ($C_2H_2$).

The VSEPR model predicts, and measurements verify, different shapes for these compounds. Ethane is tetrahedrally shaped at both carbons, with bond angles of about 109.5°. Ethylene is trigonal planar at both carbons, with the double bond acting as one electron group and bond angles near the ideal 120°. Acetylene has a linear shape, with the triple bond acting as one electron group and bond angles of 180°:

ethane              ethylene              acetylene

Both carbon atoms of ethane are $sp^3$ hybridized (Figure 10.20). The C—C bond involves overlap of one $sp^3$ orbital from each C, and the six C—H bonds involve overlap of a C $sp^3$ orbital with an H $1s$ orbital. The bonds in ethane are similar to all the others described so far in this section. Each bond involves *end-to-end* overlap, which has its *highest electron density along the bond axis* (an imaginary line joining the nuclei) and is shaped like an ellipse rotated about its long axis (somewhat like a football with rounded ends). This type of bond is called a **sigma (σ) bond.**

A close look at the bonding in ethylene reveals the double nature of its carbon-carbon bond (Figure 10.21). In ethylene, each C atom is $sp^2$ hybridized. Its four valence electrons half fill the three $sp^2$ orbitals and its unhybridized $2p$ orbital, which lies perpendicular to the $sp^2$ plane. Two $sp^2$ orbitals of each C form C—H σ bonds by overlapping the $1s$ orbitals of two H atoms. The third $sp^2$ orbital forms a C—C σ bond with an $sp^2$ orbital of the other C; note that their orientation allows end-to-end overlap.

With the σ-bonded C atoms near each other, their half-filled unhybridized $2p$ orbitals are close enough to overlap *side to side* and form a **pi (π) bond.** This type of bond has *two regions of electron density,* one above and one below the σ-bond axis. One π bond holds two electrons that move through *both* regions of the bond. *A double bond always consists of one σ bond and one π bond.*

Because side-to-side overlap is not as extensive as end-to-end overlap, *a π bond is weaker than a σ bond.* Therefore, in the absence of factors such as repulsions from lone pairs on neighboring atoms, a double bond (one σ and one π) is less than twice as strong as a single bond (one σ). We can see this differing strength most clearly with C-to-C bonds, in which the C=C bond energy of 614 kJ/mol is 80 kJ/mol less than twice that of the C—C bond (347 kJ/mol).

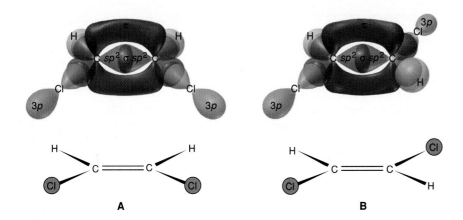

**FIGURE 10.22**
**Restricted rotation of $\pi$-bonded molecules.**
**A,** *Cis-* and **B,** *trans*-1,2-dichloroethylene occur as distinct molecules because the C-C $\pi$ bond restricts rotation and maintains two different positions of the H and Cl atoms.

Pi bonding has a major effect on the ability of atoms in a molecule to rotate relative to each other. A *$\sigma$ bond allows free rotation* of the atoms with respect to each other because the extent of overlap is not affected. That is, if you could hold one C atom of the ethane molecule, the other $CH_3$ group could spin like a pinwheel without affecting the C—C $\sigma$-bond overlap. The C—H $\sigma$ bonds also rotate freely.

However, *p* orbitals must be parallel to engage in side-to-side overlap, so *a $\pi$ bond restricts the rotation* of the atoms. Rotating one C in ethylene with respect to the other decreases the side-to-side overlap, and the $\pi$ bond breaks. This explains the existence of separate *cis* and *trans* structures in double-bonded molecules, such as dichloroethylene (Section 10.2): the $\pi$ bond allows two *different* arrangements of atoms around the two C atoms (Figure 10.22).

The C≡C bond in acetylene consists of one $\sigma$ and two $\pi$ bonds (Figure 10.23). To maximize overlap in a linear shape, one *s* and one *p* orbital in each C atom form two *sp* hybrids, and two 2*p* orbitals remain unhybridized. The four valence electrons half fill all four orbitals. Each C forms a $\sigma$ bond with an H atom using one of its *sp* orbitals, and the other *sp* orbital is used in the C—C $\sigma$ bond. Side-to-side overlap of one pair of 2*p* orbitals gives one $\pi$ bond, with electron density above and below the $\sigma$ bond, while side-to-side overlap of the other pair of 2*p* orbitals gives a second $\pi$ bond, 90° away from the first, with electron density in front of and behind the $\sigma$ bond.

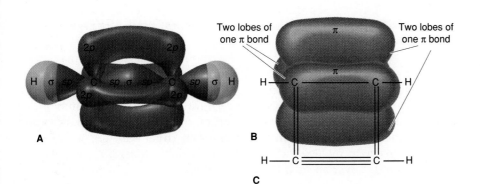

**FIGURE 10.23**
**The $\sigma$ and $\pi$ bonds of acetylene ($C_2H_2$).**
**A,** The $\sigma$ bonds of acetylene consist of the *sp* orbitals of each C overlapping to form a C—C $\sigma$ bond and overlapping H atoms to form two C—H $\sigma$ bonds. Two unhybridized 2*p* orbitals on each C remain and are shown overlapping. **B,** Two $\pi$ bonds result from this side-to-side overlap. One $\pi$ bond is above and below the C—C $\sigma$ bond, and the other is in front of and behind it, with $\sigma$ bonds shown as lines. **C,** Bond-line drawing.

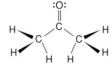

SAMPLE PROBLEM 10.6 _____

### Describing the Bonding in Molecules with Multiple Bonds

**Problem:** Describe the types of bonds and orbitals in acetone, $(CH_3)_2CO$.
**Plan:** We use the same procedure as in Sample Problem 10.5, paying special attention to the C=O bond.
**Solution:** In Sample Problem 10.3, we determined the shapes of the three central atoms of acetone to be tetrahedral around each C of the two $CH_3$ (methyl) groups and trigonal planar around the middle C atom. Thus, the middle C has three $sp^2$ orbitals and one unhybridized $p$ orbital. Each of the two methyl C atoms has four $sp^3$ orbitals. Three of them overlap the $1s$ orbitals of the H atoms to form $\sigma$ bonds; the fourth overlaps an $sp^2$ orbital of the middle C atom. Thus, two of the three $sp^2$ orbitals of the middle C form $\sigma$ bonds to the terminal C atoms.

 The O atom is $sp^2$ hybridized also. Two of its $sp^2$ orbitals hold lone pairs, and the third forms a $\sigma$ bond with the third $sp^2$ orbital of the middle C atom. The unhybridized, half-filled $p$ orbitals on C and O form a $\pi$ bond. The $\sigma$ and $\pi$ bonds constitute the C=O bond:

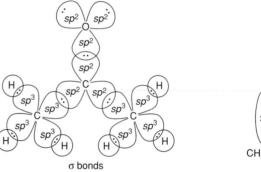

σ bonds 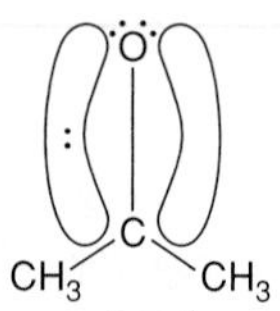            π bond

FOLLOW-UP PROBLEM 10.6
Describe the types of bonds and the orbitals in **(a)** hydrogen cyanide, HCN, and **(b)** carbon dioxide, $CO_2$.

### Section Summary
VB theory explains that a covalent bond forms when two atomic orbitals overlap. The bond holds two electrons with paired (opposite) spins. Orbital hybridization allows us to explain how orbitals on isolated atoms mix to become orbitals in molecules. We postulate the type of hybrid orbital used from the molecular shape. End-to-end orbital overlap forms a $\sigma$ bond and allows free rotation of the atoms. Side-to-side overlap forms a $\pi$ bond, which restricts rotation. A multiple bond consists of a $\sigma$ bond and either one $\pi$ bond (double bond) or two $\pi$ bonds (triple bond).

## 10.4 Molecular Orbital (MO) Theory and Electron Delocalization

Chemists choose the model that helps them answer a particular question best. If the question concerns shape, they choose the VSEPR model, perhaps followed by hybrid-orbital analysis with VB theory. If the question involves a molecule's magnetic and spectral properties, orbital energy levels, or bond energies, they choose **molecular orbital (MO) theory.**

 In VB theory, a molecule consists of individual atoms bound together through *localized* overlap of valence-shell atomic orbitals. In MO theory, a molecule is a collection of nuclei with the electron orbitals *delocalized* over the entire molecule. The MO model is a quantum-mechanical treatment for molecules similar to the one for individual atoms that we discussed in Chapter 8. Just as atoms have atomic orbitals (AOs) of a given energy and shape that are occupied by its electrons, molecules have **molecular orbitals (MOs)** of a given energy and shape that are occupied by the molecule's electrons.

Despite the great usefulness of MO theory in explaining certain molecular properties, it too has a drawback: molecular orbitals are more difficult to visualize than the easily depicted shapes of VSEPR theory or the hybrid orbitals of VB theory.

### The Central Themes of MO Theory

Several key ideas of MO theory appear in its description of the hydrogen molecule and other simple species. These ideas include the formation of MOs, their energy and shape, and how they fill with electrons.

**Formation of molecular orbitals by combination of atomic orbitals.** Because of complications arising from electron repulsions, approximations are required to solve the Schrödinger equation for many-electron atoms. Similar complications arise even with $H_2$, the simplest molecule, so we also make approximations to solve for the properties of MOs. The most common approximation used to obtain the energies and shapes of the MOs is called the *LCAO method,* which assumes that MOs are derived from a *linear* combination of *atomic orbitals.* The method mathematically *combines* (adds or subtracts) the atomic orbitals (atomic wave functions) of nearby atoms to form the molecular orbitals (molecular wave functions).

When two H nuclei lie near each other, as in an $H_2$ molecule, their AOs overlap. *Adding* their wave functions together forms a **bonding MO,** which has a *region of high electron density between the nuclei.* This additive overlap is analogous to the overlap of two light waves that reinforce each other and make the resulting wave brighter. For electron waves, the overlap *increases* the probability of the electron being between the nuclei (Figure 10.24, *A*). *Subtracting* the atomic wave functions from each other forms an **antibonding MO,** which has *a node between the nuclei, a region of zero electron density* (Figure 10.24, *B*). Similarly, when two light waves cancel each other, the waves disappear. With electron waves, subtractive overlap *decreases* the probability that the electrons will occupy that space. The two possible combinations for two hydrogen atoms $H_A$ and $H_B$ are

AO of $H_A$ + AO of $H_B$ = bonding MO of $H_2$ (more $e^-$ density between nuclei).

AO of $H_A$ − AO of $H_B$ = antibonding MO of $H_2$ (less $e^-$ density between nuclei)

Notice that *the number of AOs combined always equals the number of MOs formed:* two H atomic orbitals combine to form two $H_2$ molecular orbitals.

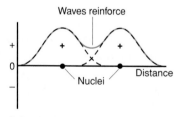

**A** Amplitudes of wave functions added

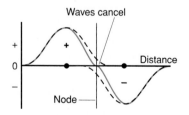

**B** Amplitudes of wave functions subtracted

**FIGURE 10.24**

**An analogy between light waves and atomic wave functions.** When light waves undergo interference, their amplitudes either add together or subtract. **A,** When the amplitudes of atomic wave functions *(dashed lines)* are added, a bonding molecular orbital (MO) results and electron density *(colored line)* increases between the nuclei. **B,** Conversely, when the amplitudes of the wave functions are subtracted, an antibonding MO results, which has a node (region of zero electron density) between the nuclei.

**FIGURE 10.25**

Contours and energies of the bonding and antibonding molecular orbitals (MOs) in $H_2$. When two H $1s$ atomic orbitals (AOs) combine, they form two $H_2$ MOs. The bonding MO ($\sigma_{1s}$) forms from addition of the AOs and is lower in energy than the AOs because most of its electron density lies *between* the nuclei. The antibonding MO ($\sigma_{1s}^*$) forms from subtraction of the AOs and is higher in energy because there is a node between the nuclei and most of the electron density lies *outside* the internuclear region.

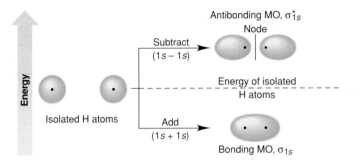

### Energy and shape of $H_2$ molecular orbitals.

The bonding MO is lower in energy and the antibonding MO higher in energy than the AOs that combined to form them. We can understand this difference in energy by examining the MO shapes (Figure 10.25).

The bonding MO in $H_2$ is spread mostly *between* the nuclei, with the nuclei attracted to the intervening electrons. An electron in this MO can delocalize its charge over a much larger volume than is possible in its separate AO. Because of this electrostatic attraction and charge delocalization, the energy of the bonding MO is lower than that of the isolated AOs. Therefore, when electrons occupy this orbital, the molecule is *more stable* than the separate atoms.

In contrast, the antibonding orbital has a node between the nuclei and most of its electron density *outside* the internuclear region. This orbital location increases nuclear repulsions and makes the antibonding MO higher in energy than the isolated AOs. Therefore, when the antibonding orbital is occupied, the molecule is *less stable* than when the orbital is empty.

Both the bonding and antibonding MOs of $H_2$ are cylindrically symmetrical about an imaginary line that runs through the nuclei; thus, they are both **sigma ($\sigma$) MOs.** The bonding MO is denoted by $\sigma_{1s}$, that is, a $\sigma$ MO formed by combination of $1s$ AOs. Antibonding orbitals are denoted with a star, so the one derived from the $1s$ AOs is $\sigma_{1s}^*$ (spoken "sigma, one ess, star").

To interact effectively and form MOs, *atomic orbitals must have similar energy and orientation.* The $1s$ orbitals on two H atoms have identical energy and orientation, so they interact strongly. This idea will be important when we consider molecules composed of atoms with many orbital energy levels.

### Filling molecular orbitals with electrons.

Electrons fill MOs under the same principles that they follow in filling AOs:

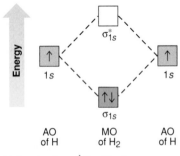

- Orbitals are filled in order of increasing energy (aufbau principle).
- An orbital has a maximum capacity of two electrons with opposite spins (Pauli exclusion principle).
- Orbitals of equal energy are half filled, with spins parallel, before any is filled (Hund's rule).

**Molecular orbital diagrams** show the relative energy and number of electrons in each MO, as well as the AOs that originally held the electrons. Figure 10.26 is the MO diagram for $H_2$.

Molecular orbital theory redefines bond order. In a Lewis structure, bond order is the number of electron pairs per linkage. The **MO bond order** is the number of electrons in bonding MOs minus the number in antibonding MOs, divided by two:

$H_2$ bond order = $\frac{1}{2}(2 - 0) = 1$

**FIGURE 10.26**

The MO diagram for $H_2$. The orbital boxes indicate the relative energies and electron occupancy of the MOs and the AOs from which they formed. Two electrons, one from each H atom, fill the lower energy $\sigma_{1s}$ MO, while the higher energy $\sigma_{1s}^*$ MO remains empty. (Orbital occupancy is also shown by full, half, or no shading.)

$$\text{Bond order} = \frac{1}{2}[(\text{no. of } e^- \text{ in bonding MO}) - (\text{no. of } e^- \text{ in antibonding MO})]$$

$$(10.2)$$

Thus, for $H_2$, the bond order is $\frac{1}{2}(2 - 0) = 1$. A bond order greater than zero indicates that the molecular species is stable relative to the separate atoms; a bond order of zero implies no net stability. In general, *the higher the bond order, the stronger is the bond.*

Another similarity between the quantum-mechanical treatments for MOs and AOs is that we can write electron configurations for a molecule. The symbol of each occupied MO is shown in parentheses, and the number of electrons in it is written outside as a superscript. Thus, the electron configuration of $H_2$ is $(\sigma_{1s})^2$.

One of the early triumphs of MO theory was its ability to *predict* the existence of $He_2^+$, which is composed of two He nuclei and three electrons. Let's use MO theory to see why $He_2^+$ exists; at the same time, we can see why $He_2$ does not. $He_2^+$ uses $1s$ atomic orbitals to form its molecular orbitals, so the MO diagram is similar to that for $H_2$ (Figure 10.27, *A*). Its three electrons are distributed such that an electron pair occupies the $\sigma_{1s}$ MO and a lone electron occupies the $\sigma_{1s}^*$ MO, so the bond order is $\frac{1}{2}(2 - 1) = \frac{1}{2}$. Thus, $He_2^+$ has a relatively weak bond, but it should exist. Indeed, this molecular ionic species has frequently been observed when He atoms collide with $He^+$ ions. Its electron configuration is $(\sigma_{1s})^2(\sigma_{1s}^*)^1$.

On the other hand, $He_2$ has four electrons to place in its $\sigma_{1s}$ and $\sigma_{1s}^*$ MOs (Figure 10.27, *B*), so both the bonding and antibonding orbitals are filled. The stabilization gained by the electron pair in the bonding MO is canceled by the destabilization from the electron pair in the antibonding MO. From its zero bond order [$\frac{1}{2}(2 - 2) = 0$], we would predict, and experiment has so far confirmed, that $He_2$ does not exist.

---

SAMPLE PROBLEM 10.7 _____

### Predicting Species Stability Using MO Diagrams

**Problem:** Use MO diagrams to predict whether $H_2^+$ and $H_2^-$ exist. Determine their bond orders and electron configurations.
**Plan:** Since these species use their $1s$ orbitals to form MOs, the MO diagrams are similar to that for $H_2$. We determine the number of electrons in each species and distribute them to the bonding and antibonding MOs in order of increasing energy. We obtain the bond order with Equation 10.2 and write the electron configuration as described in the text.
**Solution:** For $H_2^+$. Since $H_2$ has two $e^-$, $H_2^+$ has only one, as shown in the margin (*top*).
The bond order is $\frac{1}{2}(1 - 0) = \frac{1}{2}$, so we predict that **$H_2^+$ does exist.**

The electron configuration is $(\sigma_{1s})^1$.
For $H_2^-$. Since $H_2$ has two $e^-$, $H_2^-$ has three, as shown in the margin (*bottom*).
The bond order is $\frac{1}{2}(2 - 1) = \frac{1}{2}$, so we predict that **$H_2^-$ does exist.**
The electron configuration is $(\sigma_{1s})^2 (\sigma_{1s}^*)^1$.
**Check:** The number of electrons in the MOs equals the number in the AOs, as it should.
**Comment:** Both these species have been detected spectroscopically. $H_2^+$ occurs in the H-containing material around stars. $H_2^-$ has been formed in the laboratory.

FOLLOW-UP PROBLEM 10.7
Use an MO diagram to predict whether two hydride ions ($H^-$) would form $H_2^{2-}$. Calculate the bond order of $H_2^{2-}$ and write its electron configuration.

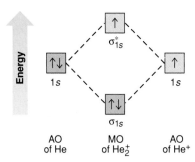

**A** $He_2^+$ bond order $= \frac{1}{2}$

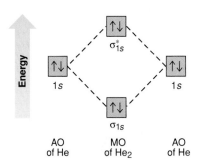

**B** $He_2$ bond order $= 0$

**FIGURE 10.27**
**MO diagrams for $He_2^+$ and $He_2$. A,** In $He_2^+$, three electrons are placed in MOs in order of increasing energy to give a filled $\sigma_{1s}$ MO and a half-filled $\sigma_{1s}^*$ MO. The bond order of $\frac{1}{2}$ implies that $He_2^+$ exists. **B,** In $He_2$, the four valence electrons fill both the $\sigma_{1s}$ and the $\sigma_{1s}^*$ MOs, so there is no net stabilization (bond order $= 0$) and $He_2$ is not known to exist.

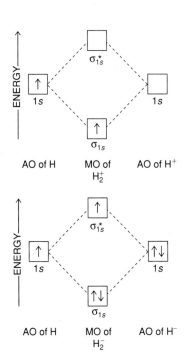

## Homonuclear Diatomic Molecules of the Period 2 Elements

**Homonuclear diatomic molecules** are those composed of identical atoms. You are already familiar with some from Period 2—$N_2$, $O_2$, and $F_2$—as the elemental forms under standard conditions. The others—$Li_2$, $Be_2$, $B_2$, $C_2$, and $Ne_2$—are observed, if at all, only in high-temperature gas-phase experiments. A molecular orbital description of these species provides some interesting tests of the model's usefulness. First, we examine those from the $s$ block, Groups 1A(1) and 2A(2), and then those from the $p$ block, Groups 3A(13) through 8A(18).

**Bonding in the s-block diatomic molecules, $Li_2$ and $Be_2$.** The first two members of this series, dilithium and diberyllium, form from atoms with a filled $1s$ level and only $2s$ electrons in their outer (valence) level. Both Li and Be occur as metals under normal conditions, but let's see what MO theory predicts for their stability as diatomic gases.

Like the MOs formed from $1s$ orbitals, those formed from atomic $2s$ orbitals are $\sigma$ orbitals; that is, they are cylindrically symmetrical around the internuclear axis. The relative energies of the MOs follow the order of the AOs that form them: a $2s$ AO is higher in energy than a $1s$, so the $\sigma_{2s}$ and $\sigma_{2s}^*$ MOs are higher in energy than the $\sigma_{1s}$ and $\sigma_{1s}^*$ MOs. The six electrons from two Li atoms enter the MOs in order of increasing energy, two electrons per orbital with opposing spins (Figure 10.28, $A$). Dilithium has four electrons in bonding MOs and two in the lower energy antibonding MO, so its bond order is $\frac{1}{2}(4-2) = 1$, and $Li_2$ has been observed. Its complete electron configuration is $(\sigma_{1s})^2(\sigma_{1s}^*)^2(\sigma_{2s})^2$, but we usually show only the configuration of the MOs formed from valence electrons, $(\sigma_{2s})^2$.

The MO diagram for $Be_2$ has filled $\sigma_{2s}$ and $\sigma_{2s}^*$ MOs (Figure 10.28, $B$). This situation is similar to that for $He_2$ discussed previously. The bond order is $\frac{1}{2}(4-4) = 0$; consistent with this, the ground state of the $Be_2$ molecule has never been observed.

The MO diagrams for $Li_2$ and $Be_2$ illustrate two important ideas. First, as stated earlier, AOs *of similar energies* interact to form MOs. Therefore, the $1s$

**FIGURE 10.28**

Bonding in $s$-block homonuclear diatomic molecules. **A,** MO diagram for $Li_2$. MOs obtained from $1s$ orbitals combining are lower in energy than those obtained from $2s$ orbitals. The six electrons from two Li atoms fill the lowest three MOs, and the $\sigma_{2s}^*$ remains empty. With a bond order of 1, $Li_2$ does form. **B,** MO diagram for $Be_2$. The eight electrons from two Be atoms fill all the available MOs to give no net stabilization. Ground-state $Be_2$ has a zero bond order and has never been observed.

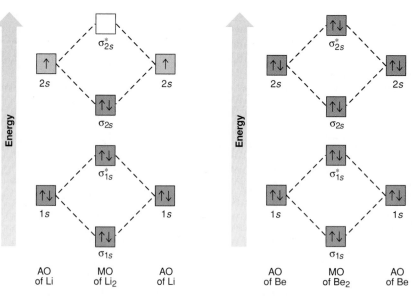

**A** $Li_2$ bond order = 1          **B** $Be_2$ bond order = 0

orbitals interact strongly with each other, as do the $2s$ orbitals, but a $1s$ orbital of the first Li interacts only very weakly with the $2s$ orbital of the second. The second point involves MOs formed by interactions of inner-electron AOs. Note that the $\sigma_{1s}$ and $\sigma_{1s}^*$ MOs are filled. Since the stabilizing effect of the filled $\sigma_{1s}$ MO equals the destabilizing effect of the $\sigma_{1s}^*$ MO, these MOs provide no *net* contribution to the bonding. From here on, the MO diagrams show only the MOs created by *combinations of the valence-electron AOs*, the $2s$ orbitals in this case.

**Molecular orbitals from atomic p–orbital combinations.** From here on, we must consider MOs that occur when atomic $2p$ orbitals combine. Recall that $p$ orbitals can interact with each other from two different directions. End-to-end combination gives a pair of $\sigma$ MOs, the $\sigma_{2p}$ and $\sigma_{2p}^*$. Side-to-side combination gives a pair of **pi ($\pi$) MOs**, $\pi_{2p}$ and $\pi_{2p}^*$ (Figure 10.29). As with MOs from $s$ orbitals, the bonding MOs from $p$-orbital combinations have their greatest electron density *between* the nuclei, whereas the antibonding MOs have a node between the nuclei and most of their electron density *outside* the internuclear region.

The order of MO energy levels, whether bonding or antibonding, is based on the order of AO energy levels *and* on the direction of the $p$-orbital combination:

1. MOs formed from $2p$ orbitals are higher in energy than MOs formed from $2s$ orbitals because $2p$ AOs are higher in energy than $2s$ AOs.
2. Bonding MOs are lower in energy than antibonding MOs, so $\sigma_{2p}$ is lower in energy than $\sigma_{2p}^*$ and $\pi_{2p}$ is lower than $\pi_{2p}^*$.
3. End-to-end overlap is greater than side-to-side overlap, so the $\sigma_{2p}$ MO is lower in energy than the $\pi_{2p}$ MO. Similarly, the destabilizing effect of the $\sigma_{2p}^*$ MO is greater than that of the $\pi_{2p}^*$ MO.

Thus, the energy order for MOs derived from $2p$ orbitals is

$$\sigma_{2p} < \pi_{2p} < \pi_{2p}^* < \sigma_{2p}^*$$

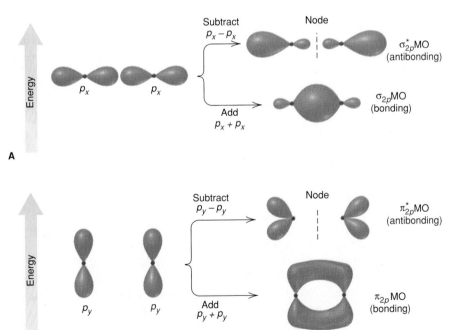

**A**

**B**

**FIGURE 10.29**

Contours and energies of $\sigma$ and $\pi$ MOs through combinations of atomic $2p$ orbitals. **A,** The $p$ orbitals lying along the line between the atoms (usually designated $p_x$) undergo end-to-end overlap and form $\sigma_{2p}$ and $\sigma_{2p}^*$ MOs. Note the greater electron density between the nuclei for the bonding orbital and the node between the nuclei for the antibonding orbital. **B,** The $p$ orbitals that lie perpendicular to the internuclear axis ($p_y$ or $p_z$) undergo side-to-side overlap to form two $\pi$ MOs. A $\pi_{2p}$ is a bonding MO with its greatest density above and below the internuclear axis; a $\pi_{2p}^*$ is an antibonding MO with a node between the nuclei and its electron density outside them.

**FIGURE 10.30**

Relative MO energy levels for Period 2 homonuclear diatomic molecules. **A,** MO energy levels for $O_2$, $F_2$, and $Ne_2$. The six atomic $2p$ orbitals of the two atoms form six MOs that are higher in energy than the two MOs formed from the two $2s$ AOs. Those forming $\pi$ orbitals yield two bonding MOs ($\pi_{2p}$) of equal energy and two antibonding MOs ($\pi^*_{2p}$) of equal energy. This sequence of energy levels arises from minimal $2s$-$2p$ orbital mixing. **B,** MO energy levels for $B_2$, $C_2$, and $N_2$. Because of significant $2s$-$2p$ orbital mixing, the energies of $\sigma$ MOs formed from $2p$ orbitals increase and those formed from $2s$ orbitals decrease. The major effect of such orbital mixing on the MO sequence is to make the $\sigma_{2p}$ higher than the $\pi_{2p}$. (For clarity, those MOs affected by mixing are shown in purple.)

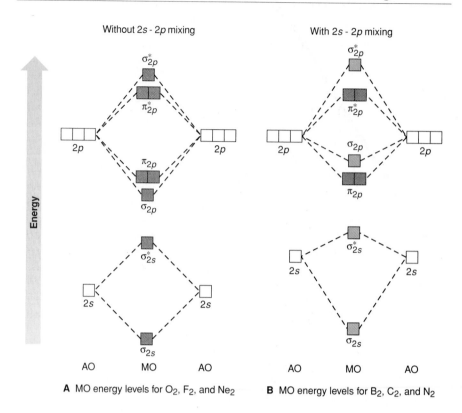

**A**  MO energy levels for $O_2$, $F_2$, and $Ne_2$    **B**  MO energy levels for $B_2$, $C_2$, and $N_2$

There are three perpendicular $2p$ orbitals on each atom, so when six $p$ orbitals combine, the two facing each other end-to-end form a $\sigma$ and a $\sigma^*$ MO, while the two pairs whose axes are parallel to each other form two $\pi$ MOs of the same energy and two $\pi^*$ MOs of the same energy. Combining these orientations with the energy order gives the *expected* MO diagram for the $p$-block Period 2 homonuclear diatomic molecules (Figure 10.30, *A*).

One other factor influences the MO energy order. Recall that only AOs of similar energy interact to form MOs. Figure 10.30, *A*, is based on the assumption that the $s$ and $p$ AOs are so different in energy that they do not interact with each other: in MO terminology, the orbitals do not *mix*. This is true for O, F, and Ne atoms, where electron repulsions that occur as the $2p$ electrons pair up raise the energy of the $2p$ orbitals sufficiently above that of the $2s$ orbitals to minimize orbital interactions (mixing). However, when the $2p$ AOs are being half filled, as in B, C, and N atoms, repulsions are small, so the $2p$ energies are much closer to the $2s$ energy. As a result, some mixing occurs between the $2s$ orbital of one atom and the end-on $2p$ orbital of the other. Such effects are enhanced by the increasing nuclear charge. This orbital mixing *lowers* the energy of the $\sigma_{2s}$ and $\sigma^*_{2s}$ MOs and *raises* the energy of the $\sigma_{2p}$ and $\sigma^*_{2p}$ MOs; the $\pi$ MOs are not affected. We use an MO diagram for $B_2$ through $N_2$ that reflects this AO mixing (Figure 10.30, *B*). The most significant change is the *reverse in order* of the $\sigma_{2p}$ and $\pi_{2p}$ MOs.

**Bonding in the *p*-block diatomic molecules, $B_2$ and $Ne_2$.** Figure 10.31 shows the sequence of MOs, their occupancy, and some properties of $B_2$ through $Ne_2$. As you examine the figure, notice how the predicted bond order compares with the measured bond energy and length. Also note how the electron populations in the MOs correlate with the observed magnetic

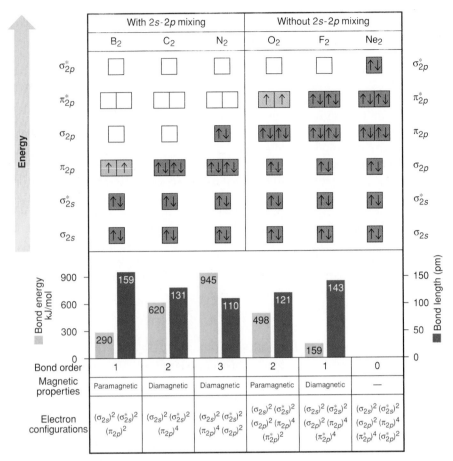

**FIGURE 10.31**

**MO occupancy and molecular properties for B₂ through Ne₂.** The sequence of MOs and their electron populations are shown for each of the homonuclear diatomic molecules in the $p$ block of Period 2 [Groups 3A(13) to 8A(18)]. The bond order, bond length, bond energy, magnetic behavior, and valence electron configuration appear below each diagram. Note the correlation between bond order and bond energy, both of which are inversely related to bond length.

behavior. Recall from Chapter 8 that the spins of unpaired electrons in an atom (or ion) cause it to be *paramagnetic,* attracted to an external magnetic field. If all the electron spins are paired, the substance is *diamagnetic,* unaffected (or weakly repelled) by the magnetic field. The same observations apply to molecules. These important molecular properties are not even addressed in VSEPR or VB theory.

The B₂ molecule has six outer electrons to place in its MOs. Four of these fill the $\sigma_{2s}$ and $\sigma_{2s}^*$ MOs. The remaining two electrons enter the two $\pi_{2p}$ MOs, one in each orbital, in keeping with Hund's rule. The four electrons in bonding MOs and two in antibonding MOs give a bond order of 1 for B₂ [$\frac{1}{2}(4-2)=1$].

From its MO diagram, we would expect B₂ to be paramagnetic, and experiment confirms this. This behavior of B₂ gives us additional confidence in the order of MOs arising from 2s-2p mixing. If the MO diagram without mixing were used (Figure 10.30, *A*), the single $\sigma_{2p}$ MO would lie *below* the

**FIGURE 10.32**

**The paramagnetic properties of $O_2$.** Liquid $O_2$ is attracted to the poles of a magnet because it is paramagnetic, as MO theory predicts. A diamagnetic substance would fall between the poles.

two $\pi_{2p}$ MOs, and the two remaining electrons in $B_2$ would fill it, incorrectly indicating that $B_2$ was diamagnetic.

The two additional electrons present in $C_2$ fill the two $\pi_{2p}$ MOs. Since $C_2$ has two more bonding electrons than $B_2$, its bond order is 2. This higher bond order corresponds to a higher bond energy and shorter bond distance. All the electrons are paired and, as the model predicts, $C_2$ is diamagnetic.

In $N_2$, the two additional electrons enter and fill the $\sigma_{2p}$ MO. The resulting bond order is 3, which is consistent with the triple bond in the Lewis structure. As the model predicts, the measured bond energy is higher, the bond length is shorter, and $N_2$ is diamagnetic.

With $O_2$, the power of MO theory over theories based on localized orbitals is most apparent. For years, it seemed impossible to reconcile bonding theories with the magnetic behavior of the $O_2$ molecule. On the one hand, the data showed a paramagnetic molecule held together with a double bond (Figure 10.32). On the other hand, the existing models indicated two possible Lewis structures: one that followed the octet rule, with a double bond and all electrons paired, and the other an octet rule exception, with a single bond and two electrons unpaired:

$$\ddot{\text{O}}{=}\ddot{\text{O}} \quad \text{or} \quad :\!\dot{\text{O}}{-}\dot{\text{O}}\!:$$

MO theory resolves this paradox beautifully. As Figure 10.31 shows, the bond order of $O_2$ is 2: eight electrons occupy bonding MOs and four occupy antibonding MOs [$\frac{1}{2}(8 - 4) = 2$]. The *two* electrons with highest energy occupy the *two* $\pi_{2p}^*$ MOs with unpaired (parallel) spins, making the molecule paramagnetic. The bond energy decreases and the bond length increases relative to $N_2$, as expected from the lower bond order.

The two additional electrons in $F_2$ fill the $\pi_{2p}^*$ orbitals, which decreases the bond order to 1, and the absence of lone electrons makes $F_2$ diamagnetic. As expected, the bond energy is lower and the bond distance longer than in $O_2$. Note that the bond energy of $F_2$ is only about half that of $B_2$, even though they have the same bond order: $F_2$ has 18 electrons in a smaller volume than $B_2$ with only 10 electrons, so greater electron repulsions in $F_2$ make its single bond easier to break.

The final member of the series, $Ne_2$, does not exist for the same reason that $He_2$ does not exist. With all the MOs filled, the stabilization from bonding electrons cancels the destabilization from antibonding electrons, and the bond order is zero.

SAMPLE PROBLEM 10.8 _____

**Using MO Theory to Explain Bond Properties**

**Problem:** Consider the following data for these homonuclear diatomic species:

|  | $N_2$ | $N_2^+$ | $O_2$ | $O_2^+$ |
|---|---|---|---|---|
| Bond energy (kJ/mol) | 945 | 841 | 498 | 623 |
| Bond length (pm) | 110 | 112 | 121 | 112 |
| No. of valence electrons | 10 | 9 | 12 | 11 |

Removing an electron from $N_2$ *decreases* the bond energy of the resulting ion, whereas removing an electron from $O_2$ *increases* the bond energy of the resulting ion. Explain these facts with diagrams that show the sequence and occupancy of MOs.

**Plan:** We first draw the sequence of MO energy levels for the four species, recalling that they differ for $N_2$ and $O_2$ (see Figure 10.31). Then we determine bond orders and compare them with the data: bond order is related directly to bond energy and inversely to bond length.

**Solution:** The MO energy levels are

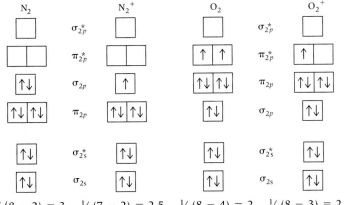

Bond orders: $\frac{1}{2}(8 - 2) = 3$   $\frac{1}{2}(7 - 2) = 2.5$   $\frac{1}{2}(8 - 4) = 2$   $\frac{1}{2}(8 - 3) = 2.5$

When an electron is removed from $N_2$ to form $N_2^+$, it comes from a bonding MO, so the bond order decreases; therefore, $N_2^+$ has a lower bond energy and longer bond length than $N_2$. When an electron is removed from $O_2$ to form $O_2^+$, it comes from an antibonding MO, so the bond order increases, and $O_2^+$ has a higher bond energy and shorter bond length than $O_2$.

**Check:** The answer makes sense in terms of the relationships among bond order, bond energy, and bond length. Check that the total number of bonding and anti-bonding electrons equals the given number of valence electrons available.

**FOLLOW-UP PROBLEM 10.8**
Determine the bond orders for the following species: $F_2^{2-}$, $F_2^-$, $F_2$, $F_2^+$, $F_2^{2+}$. List the species in order of increasing bond energy and in order of increasing bond length.

## MO Description of Some Heteronuclear Diatomic Molecules

For *heteronuclear* diatomic molecules—those composed of *different* atoms—the MO diagrams are asymmetric because the atomic orbitals of the two atoms have unequal energies. Atoms with greater effective nuclear charge ($Z_{eff}$) draw their electrons closer to the nucleus and thus have atomic orbitals of lower energy (Section 8.1).

Let's apply MO theory to the bonding in HF and NO. In HF, the much greater electronegativity of F causes *all* its orbitals to be of lower energy than the $1s$ orbital of H, such that the H $1s$ orbital interacts only with the F $2p$ orbitals. Since only one of the three $2p$ orbitals, say the $2p_x$, leads to end-on overlap, combining the H $1s$ orbital with this $2p$ orbital in F results in a $\sigma$ MO and a $\sigma^*$ MO. The two other $p$ orbitals of F ($2p_y$ and $2p_z$) are not involved in bonding. They are called **nonbonding MOs** and have the same energies as the isolated AOs (Figure 10.33).

You can see from the relative energy levels that the F orbital contributes more to the bonding MO in HF than the H orbital does. In polar covalent molecules, *bonding MOs are closer in energy to the AOs of the more electronegative atom*. In effect, fluorine's greater electronegativity lowers the energy of the bonding MO and draws the bonding electrons closer to its nucleus.

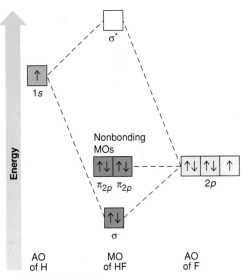

**FIGURE 10.33**
**The MO diagram for HF.** In a polar covalent molecule, the MO diagram is asymmetric because the more electronegative atom has AOs of lower energy. In HF, the bonding MO is closer in energy to the $2p$ orbital of F. Electrons not involved in bonding occupy nonbonding MOs. (The $2s$ AO of F is not shown.)

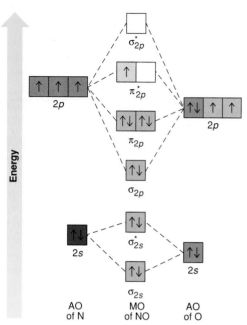

Energy

$\sigma_{2p}^*$

$\pi_{2p}^*$

2p

$\pi_{2p}$

2p

$\sigma_{2p}$

2s

$\sigma_{2s}^*$

$\sigma_{2s}$

2s

| AO | MO | AO |
| of N | of NO | of O |

**FIGURE 10.34**

**The MO diagram for NO.** Eleven electrons occupy the MOs in nitric oxide. Note that the lone electron occupies an antibonding MO whose energy is closer to that of the AO of the less electronegative N atom.

Nitrogen monoxide (nitric oxide) is a highly reactive molecule due to its lone electron. Two possible Lewis structures for NO, with formal charges (Section 9.5), are

$$\overset{0}{:\ddot{N}}=\overset{0}{\ddot{O}}: \quad \text{or} \quad \overset{-1}{:\ddot{N}}=\overset{+1}{\ddot{O}}:$$

I          II

Both structures show a double bond, but the *measured* bond energy suggests a bond order *higher* than 2. Furthermore, it is not clear where the lone electron resides, although the lower formal charges for structure I suggest that it is on the N atom.

MO theory predicts the bond order and indicates lone-electron placement with no difficulty. The MO diagram is skewed because the AOs of the more electronegative O are lower in energy (Figure 10.34). The 11 valence electrons enter MOs in order of increasing energy, leaving the lone electron in one of the $\pi_{2p}^*$ orbitals. Eight bonding electrons and three antibonding electrons give a bond order of $\frac{1}{2}(8 - 3) = 2.5$, more in keeping with experiment than either Lewis structure. The bonding electrons lie in MOs closer in energy to the AOs of the O atom. The lone electron occupies an antibonding orbital. Because this orbital receives a greater energy contribution from the 2p orbitals of N, the lone electron resides closer to the N atom.

## MO Descriptions of Ozone and Benzene

MO theory extends to molecules with more than two atoms, but the orbital shapes and MO diagrams are too complex for a detailed treatment here. However, in a final example, we briefly discuss how the model eliminates the need for resonance forms and explains the absorption of energy.

Recall that we cannot draw a single Lewis structure that adequately depicts bonding in molecules such as ozone and benzene because the adjacent single and double bonds actually have identical properties. Instead, we are forced to draw more than one structure and mentally combine them into a resonance hybrid. The VB model has the same deficiency because it also relies on *localized* electron-pair bonds.

In contrast, MO theory pictures a structure of delocalized $\sigma$ and $\pi$ bonding and antibonding MOs. Figure 10.35 shows the lowest energy $\pi$-bonding MOs in benzene and ozone. Each holds one pair of electrons. The extended electron densities allow delocalization of this $\pi$-electron pair over the entire molecule, thus eliminating the need for separate resonance depictions. In

**FIGURE 10.35**

**The lowest energy $\pi$-bonding MOs in benzene and ozone.** MO theory eliminates the need for resonance by postulating orbitals that are highly delocalized. **A,** The most stable $\pi$-bonding MO in benzene ($C_6H_6$) has hexagonal lobes of electron density above and below the $\sigma$ plane of the six C atoms. **B,** The most stable $\pi$-bonding MO in ozone ($O_3$) extends above and below the $\sigma$ plane of the three O atoms.

**A** Benzene, $C_6H_6$          **B** Ozone, $O_3$

benzene, the upper and lower hexagonal lobes of this one $\pi$-bonding MO lie above and below all six carbon nuclei. In ozone, the two lobes of this $\pi$-bonding MO extend over and under all three oxygen nuclei.

A great additional advantage of MO theory is its ability to explain excited states and spectra of molecules. It can explain, for instance, how bonding electrons in $O_3$ enter empty antibonding orbitals when the molecule absorbs ultraviolet (UV) radiation in the stratosphere, and why the UV spectrum of benzene has its characteristic absorptions.

### Section Summary
MO theory treats a molecule as a collection of nuclei with molecular orbitals delocalized over the entire structure. Atomic orbitals of comparable energy can be added and subtracted to obtain bonding and antibonding MOs, respectively. Bonding MOs have most of their electron density inside the region between the nuclei and are lower in energy than the atomic orbitals; antibonding MOs have most of their electron density outside the region between the nuclei and are higher in energy. MOs are filled in order of their energy with electrons having opposing spins. MO diagrams show the energy levels and orbital occupancy. Those for the homonuclear diatomic molecules of Period 2 explain observed bond energy, bond length, and magnetic behavior. For heteronuclear diatomic molecules, bonding MOs have a greater contribution from the more electronegative atom. MO theory eliminates the need for resonance forms in larger molecules.

### Chapter Perspective
*The covalent bond is one of the centerpieces of chemical theory. In this chapter, we examined how the location of electron pairs creates the shapes of molecules and how chemists explain molecular properties through the orbitals that electrons occupy. As you'll see in upcoming chapters, the landscape of a molecule—its geometric contours and regions of charge—give rise to all its other properties. In the next four chapters, we'll explore the physical behavior of liquids and solids, the properties of solutions, the behavior of the main-group elements, and the nature of organic molecules, all of which are based on atomic properties and their emergent molecular properties.*

# For Review and Reference

## Key Terms

**SECTION 10.1**
valence-shell electron-pair
    repulsion (VSEPR)
    theory
molecular shape
bond angle
linear arrangement
trigonal planar
    arrangement
bent (V shaped)
tetrahedral arrangement
trigonal pyramidal shape
trigonal bipyramidal
    arrangement

equatorial group
axial group
seesaw shape
T shape
octahedral arrangement
square pyramidal shape
square planar shape

**SECTION 10.2**
molecular polarity
dipole moment ($\mu$)
receptor

**SECTION 10.3**
valence bond (VB) theory
hybridization
hybrid orbital
$sp$ hybrid orbital
$sp^2$ hybrid orbital
$sp^3$ hybrid orbital
$sp^3d$ hybrid orbital
$sp^3d^2$ hybrid orbital
sigma ($\sigma$) bond
pi ($\pi$) bond

**SECTION 10.4**
molecular orbital (MO)
    theory
molecular orbital (MO)
bonding MO
antibonding MO
sigma ($\sigma$) MO
MO diagram
MO bond order
homonuclear diatomic
    molecule
pi ($\pi$) MO
nonbonding MO

## Key Equations and Relationships

**10.1** Ranking electron-pair repulsions (p. 376):

    Lone pair–lone pair > lone pair–bonding pair
                      > bonding pair–bonding pair

**10.2** Calculating the MO bond order (p. 396):

    Bond order = $\frac{1}{2}$[(number of $e^-$ in bonding MO)
                − (number of $e^-$ in antibonding MO)]

## Answers to Follow-up Problems

**10.1** (a) $\ddot{\underset{..}{S}}=C=\ddot{\underset{..}{S}}$      linear, 180°

(b) 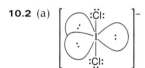      V shaped, <120°

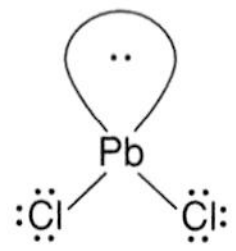

(c) 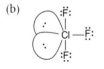      Tetrahedral, 109.5°

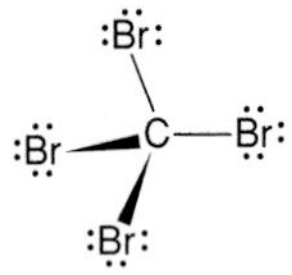

(d)       V shaped, <109.5°

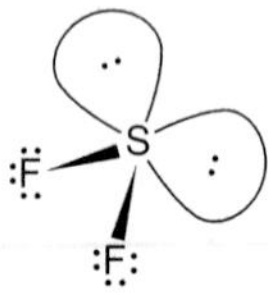

**10.2** (a) 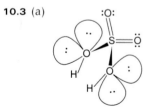      Linear, 180°

(b) 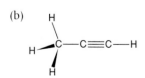      T shaped, <90°

(c)      Trigonal bipyramidal, $F_{eq}$—S—$F_{eq}$ angle <120°

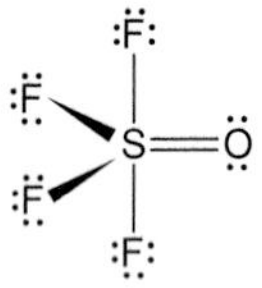

**10.3** (a)      S is tetrahedral; double bonds compress O—S—O angle to <109.5°. Each central O is V shaped; lone pairs compress H—O—S angle to <109.5°.

(b)      Methyl C tetrahedral ~109.5°; other C atoms linear, 180°.

(c) 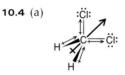      Each S is V shaped; F—S—S angle <109.5°.

**10.4** (a)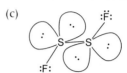

(b)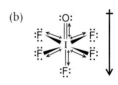

(c)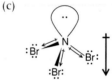

**10.5** (a)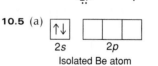

(b)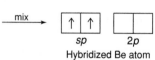

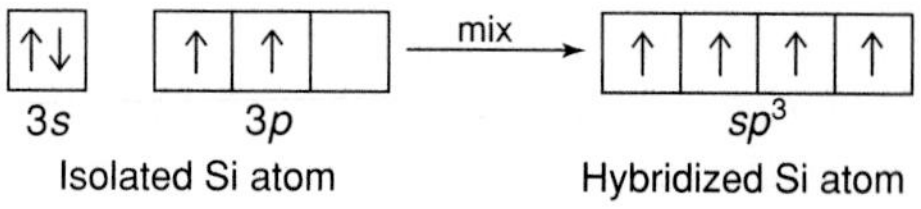

(c)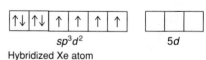

**10.6** (a) H—C≡N:      N and C are *sp* hybridized, so HCN is linear. One *sp* of C overlaps the 1*s* of H to form a σ bond. The other overlaps one *sp* of N to form a σ bond. The other *sp* of N holds a lone pair. Two unhybridized *p* orbitals of N and two of C overlap to form two π bonds.

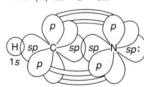

(b) $\ddot{O} = C = \ddot{O}$

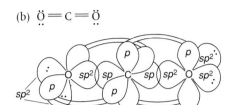

Both O atoms are $sp^2$ hybridized, and the C is $sp$ hybridized, so $CO_2$ is linear. Each $sp$ overlaps one $sp^2$ from each O to form two $\sigma$ bonds. Each of the two unhybridized $p$ orbitals of C forms a $\pi$ bond with the unhybridized $p$ of the two O atoms. Two $sp^2$ on each O hold lone pairs.

**10.7**

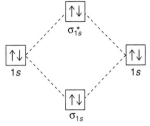

AO of H⁻    MO of $H_2^{2-}$    AO of H⁻

Not likely to exist. BO $= \frac{1}{2}(2 - 2) = 0$; $(\sigma_{1s})^2(\sigma_{1s}^*)^2$

**10.8** Bond order: $F_2^{2-} = 0$;   $F_2^- = \frac{1}{2}$;   $F_2 = 1$;   $F_2^+ = 1\frac{1}{2}$;   $F_2^{2+} = 2$

Bond energy: $F_2^{2-} < F_2^- < F_2 < F_2^+ < F_2^{2+}$

Bond length: $F_2^{2+} < F_2^+ < F_2 < F_2^-$; $F_2^{2-}$ has no bond.

## Sample Problem Titles

**10.1** Predicting Molecular Shapes With Two, Three, and Four Electron Groups (p. 376)

**10.2** Predicting Molecular Shapes With Five or Six Electron Groups (p. 379)

**10.3** Predicting Molecular Shapes With More Than One Central Atom (p. 380)

**10.4** Predicting the Polarity of Molecules (p. 382)

**10.5** Postulating the Hybrid Orbitals in a Molecule (p. 391)

**10.6** Describing the Bonding in Molecules With Multiple Bonds (p. 394)

**10.7** Predicting Species Stability Using MO Diagrams (p. 397)

**10.8** Using MO Theory to Explain Bond Properties (p. 402)

## Problems

Problems with a green number are answered at the back of the text. Most sections include three categories of problems separated by a green rule: concept review questions, *paired* skill-building exercises, and problems in a relevant context.

### Valence-Shell Electron-Pair Repulsion (VSEPR) Theory and Molecular Shape

(Sample Problems 10.1 to 10.3)

**10.1** If you know the formula of a molecule or ion, what is the first step in predicting its shape?

**10.2** In what situations is the name of the molecular shape the same as the name of the electron-group arrangement?

**10.3** Which of the following numbers of electron groups can give rise to a bent (V-shaped) molecule: two, three, four, five, six? Draw an example with the ideal bond angle in each case.

**10.4** Name all the molecular shapes that have a tetrahedral electron-group arrangement.

**10.5** Why aren't lone pairs included with surrounding bonded groups when determining the molecular shape?

**10.6** Use wedge-line structures (if necessary) to sketch the atom positions in a general molecule of formula $AX_n$ that has the following shapes: (a) bent (V shaped);

(b) trigonal planar; (c) trigonal bipyramidal; (d) T shaped; (e) trigonal pyramidal; (f) square pyramidal.

**10.7** What would you expect to be the electron-group arrangement around atom A in each of the following cases? For each arrangement, give the ideal bond angle and the direction of any expected deviation:

(a)       X       (b) X—A≡X    (c)      X
        |                                    ‖
     X—A:                                X=A—X
        |
        X

(d) X—$\ddot{A}$—X    (e) X=A=X    (f)    X
                                            X
                                       :A
                                       |    X
                                       X

**10.8** Determine the electron-group arrangement, molecular shape, and ideal bond angle(s) for each of the following: (a) $O_3$; (b) $H_3O^+$; (c) $NF_3$; (d) $SO_4^{2-}$.

**10.9** Determine the electron-group arrangement, molecular shape, and ideal bond angle(s) for each of the following: (a) $CO_3^{2-}$; (b) $SO_2$; (c) $CF_4$; (d) $SO_3$.

**10.10** Determine the molecular shape, ideal bond angle(s), and the direction of any deviations from these angles for each of the following: (a) $ClO_2^-$; (b) $PF_5$; (c) $SeF_4$; (d) $KrF_2$.

**10.11** Determine the molecular shape, ideal bond angle(s), and the direction of any deviations from these angles for each of the following: (a) $ClO_3^-$; (b) $IF_4^-$; (c) $SeOF_2$ (central Se); (d) $TeF_5^-$.

**10.12** Determine the shape around each central atom in the following molecules, and explain any deviation from ideal bond angles: (a) $CH_3OH$; (b) $N_2O_4(O_2NNO_2)$; (c) $H_3PO_4$ (no H—P bond).

**10.13** Determine the shape around each central atom in the following molecules, and explain any deviation from ideal bond angles: (a) $CH_3COOH$; (b) $H_2O_2$; (c) $H_2SO_3$ (no H—S bond).

**10.14** Arrange the following $AF_n$ species in order of *increasing* F—A—F bond angles: $BF_3$, $BeF_2$, $CF_4$, $NF_3$, $OF_2$.

**10.15** Arrange the following $ACl_n$ species in order of *decreasing* Cl—A—Cl bond angles: $SCl_2$, $OCl_2$, $PCl_3$, $SiCl_4$, $SiCl_6^{2-}$.

**10.16** State ideal values for each of the bond angles in the following molecules and where you expect deviations:

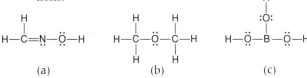

(a)          (b)          (c)

**10.17** State ideal values for each of the bond angles in the following molecules and where you expect deviations:

(a)          (b)          (c)

**10.18** The VSEPR model was developed in the 1950s, before any xenon compounds had been prepared. Thus, in the early 1960s, these compounds provided an excellent test of the model's predictive power. What would you have predicted for the shapes of $XeF_2$, $XeF_4$, and $XeF_6$?

**10.19** Because both are members of Group 4A(14), tin and carbon form structurally similar compounds. However, tin exhibits a greater variety of structures because it forms several ionic species. Predict the shapes and ideal bond angles, including any deviations, for the following molecules and ions: (a) $Sn(CH_3)_2$; (b) $SnCl_3^-$; (c) $Sn(CH_3)_4$; (d) $SnF_5^-$; (e) $SnF_6^{2-}$.

**10.20** In the gas phase, phosphorus pentachloride exists as separate molecules. In the solid phase, however, the compound is composed of alternating $PCl_4^+$ and $PCl_6^-$ ions. What change(s) in molecular shape occur(s) as $PCl_5$ solidifies? How does the Cl—P—Cl angle change?

## Molecular Shape and Molecular Polarity

(Sample Problem 10.4)

**10.21** For molecules of more than two atoms, how do you decide from its formula if a molecule is polar?

**10.22** How can a molecule with polar covalent bonds fail to be a polar molecule? Give an example.

**10.23** A molecule of formula $BY_3$ is found experimentally to be polar. Which molecular shapes are possible and which impossible for $BY_3$?

**10.24** Explain in general why the shape of a biomolecule is important to its function.

**10.25** Consider the following molecules: $SCl_2$, $F_2$, $CS_2$, $CF_4$, $BrCl$.
(a) Which has the most polar bonds?
(b) Which have a molecular dipole moment?

**10.26** Consider the following molecules: $BF_3$, $PF_3$, $BrF_3$, $SF_4$, $SF_6$.
(a) Which has the most polar bonds?
(b) Which have a molecular dipole moment?

**10.27** Which molecule in each of the following pairs has the greater dipole moment? Give the reason for your choice.
(a) $SO_2$ or $SO_3$   (b) ICl or IF   (c) $SiF_4$ or $SF_4$
(d) $H_2O$ or $H_2S$

**10.28** Which molecule in each of the following pairs has the greater dipole moment? Give the reason for your choice.
(a) $NO_2$ or $SO_2$   (b) HBr or HCl   (c) $BeCl_2$ or $SCl_2$
(d) $AsF_3$ or $AsF_5$

**10.29** Except for nitrogen, the elements of Group 5A(15) all form pentafluorides and most form pentachlorides. The chlorine atoms of $PCl_5$ can be replaced with fluorine atoms one at a time to give, successively, $PCl_4F$, $PCl_3F_2,…,PF_5$.
(a) Given the relative sizes of F and Cl, would you expect the first two F substitutions to be at axial or equatorial positions? Explain.
(b) Which of the five fluorine-containing molecules have no molecular dipole?

**10.30** Dinitrogen difluoride, $N_2F_2$, is the only stable, simple inorganic molecule with an N=N bond. The compound occurs in *cis* and *trans* forms.
(a) Draw the molecular shapes of the two forms of $N_2F_2$.
(b) Predict the polarity, if any, of each form.

**10.31** There are three different dichloroethylenes (molecular formula $C_2H_2Cl_2$), which we can designate X, Y, and Z. Compound X has no dipole moment, but compound Z does. Compounds X and Z each combine with hydrogen to give the same product:

$$C_2H_2Cl_2(X \text{ or } Z) + H_2 \longrightarrow ClCH_2—CH_2Cl$$

What are the structures of X, Y, and Z? Would you expect compound Y to have a dipole moment?

## Valence Bond (VB) Theory and Orbital Hybridization

(Sample Problems 10.5 and 10.6)

**10.32** What type of central atom orbital hybridization corresponds to each of the following electron-group arrangements? (a) trigonal planar; (b) octahedral; (c) linear; (d) tetrahedral; (e) trigonal bipyramidal

**10.33** What is the orbital hybridization of a central atom that has one lone pair and bonds to the following? (a) two other atoms; (b) three other atoms; (c) four other atoms; (d) five other atoms

**10.34** How do carbon and silicon differ with regard to the *types* of orbitals available for hybridization? Explain.

**10.35** Are the following statements true or false? Correct any that is false.
(a) Two $\sigma$ bonds comprise a double bond.
(b) A triple bond consists of one $\pi$ bond and two $\sigma$ bonds.
(c) Bonds between atomic $s$ orbitals are always $\sigma$ bonds.
(d) A $\pi$ bond restricts rotation about the $\sigma$-bond axis.
(e) A $\pi$ bond consists of two pairs of electrons.

**10.36** What is the hybridization of nitrogen in each of the following ions or molecules? (a) NO; (b) $NO_2$; (c) $NO_2^-$; (d) $NO_3^-$

**10.37** What is the hybridization of chlorine in each of the following ions or molecules? (a) $ClO_2$; (b) $ClO_3^-$; (c) $ClO_4^-$; (d) $ClF_3$

**10.38** For each of the following molecules, which types of atomic orbitals of the central atom mix to form hybrid orbitals? (a) $SiH_3Cl$; (b) $CS_2$; (c) $Cl_2O$

**10.39** For each of the following molecules, which types of atomic orbitals of the central atom mix to form hybrid orbitals? (a) $SCl_3F$; (b) $NF_3$; (c) $PF_5$

**10.40** Use partial orbital diagrams to show how the atomic orbitals of the central atom lead to hybrid orbitals in (a) $GeCl_4$; (b) $BCl_3$; (c) $CH_3^+$; (d) $BF_4^-$.

**10.41** Use partial orbital diagrams to show how the atomic orbitals of the central atom lead to hybrid orbitals in (a) $SeCl_2$; (b) $H_3O^+$; (c) $IF_4^-$; (d) $AsCl_3$.

**10.42** Describe the types of bonds and orbitals in the following: (a) $NO_3^-$; (b) $CS_2$; (c) $CH_2O$; (d) FNO; (e) $C_2F_4$.

**10.43** Describe the types of bonds and orbitals in the following: (a) $O_3$; (b) $I_3^-$; (c) $COCl_2$; (d) $BrF_3$; (e) $CH_3C\equiv CH$.

**10.44** Methyl isocyanate, $CH_3$—N=C=O, is an intermediate in the manufacture of many pesticides. It received notoriety in 1984 when a leak from a manufacturing plant resulted in the death of more than 2000 people in Bhopal, India. What is the hybridization of N and of the two C atoms in methyl isocyanate? Sketch the molecular shape.

**10.45** 2-Butene ($CH_3CH\!=\!CHCH_3$) is a starting material in the manufacture of lubricating oils and many other compounds. Draw two different forms of 2-butene, indicating the $\sigma$ and $\pi$ bonds in each structure.

## Molecular Orbital (MO) Theory and Electron Delocalization

(Sample Problems 10.7 and 10.8)

**10.46** Two $p$ orbitals from one atom and two $p$ orbitals from another atom are combined to form molecular orbitals for the joined atoms. How many MOs will result from this combination? Explain.

**10.47** How many electrons does it take to fill the following:
(a) A $\sigma$-bonding MO
(b) A $\pi$-antibonding MO
(c) The MOs formed from combination of the $1s$ orbitals of two atoms
(d) The MOs formed from combination of the $2p$ orbitals of two atoms

**10.48** How do the bonding and antibonding MOs formed from a given pair of AOs compare to each other with respect to (a) energy; (b) presence of nodal planes; (c) internuclear electron density.

**10.49** Antibonding MOs always have at least one nodal plane. Can a bonding MO have a nodal plane? If so, draw an example.

**10.50** Show the shapes of bonding and antibonding MOs formed by combination of (a) an $s$ orbital and a $p$ orbital; (b) two $p$ orbitals (end to end).

**10.51** Show the shapes of bonding and antibonding MOs formed by combination of (a) two $s$ orbitals; (b) two $p$ orbitals (side to side).

**10.52** Use MO diagrams and the bond orders you obtain from them to answer the following:
(a) Is $Be_2^+$ stable?
(b) Is $Be_2^+$ diamagnetic?

**10.53** Use MO diagrams and the bond orders you obtain from them to answer the following:
(a) Is $O_2^-$ stable?
(b) Is $O_2^-$ paramagnetic?

**10.54** Use MO diagrams to place $C_2^-$, $C_2$, and $C_2^+$ in order of (a) increasing bond energy; (b) increasing bond length.

**10.55** Use MO diagrams to place $Li_2^+$, $Li_2$, and $Li_2^-$ in order of (a) decreasing bond energy; (b) decreasing bond length.

## Comprehensive Problems

Problems with an asterisk (*) are more challenging.

**10.56** For the following ions, predict the shape, state the hybridization of the central atom, and give the ideal bond angles and any expected deviations: (a) $BrO_3^-$; (b) $AsCl_4^-$; (c) $SeO_4^{2-}$; (d) $BiF_5^{2-}$; (e) $SbF_4^+$; (f) $AlF_6^{3-}$; (g) $IF_4^+$.

**10.57** Butadiene (shown below) is a colorless gas used to make synthetic BUNA rubber and many other compounds:

$$H-C=C-C=C-H$$
(with H H H H above the carbons)

(a) How many $\sigma$ bonds and $\pi$ bonds does the molecule have?
(b) Are *cis/trans* structural arrangements about the double bonds possible? Explain.

**10.58** Epinephrine (or adrenaline) is a naturally occurring hormone that is also manufactured commercially for use as a heart stimulant, as a nasal decongestant, and for glaucoma treatment. Its Lewis structure is

(a) What is the hybridization of each C, O, and N atom?
(b) How many $\sigma$ bonds does the molecule have?
(c) How many $p$ electrons are delocalized in the ring?

**10.59** The compounds 1,1-dichloroethane ($Cl_2CH-CH_3$) and 1,2-dichloroethane ($ClCH_2-CH_2Cl$) are used as wax solvents and as fumigants. Even though both molecules contain the same types of bonds, one has a much higher dipole moment than the other. Draw structures to explain this fact.

**10.60** Isoniazid is an antibacterial agent that is particularly useful against many common strains of tuberculosis. Its structure is

(a) How many $\sigma$ bonds are in the molecule?
(b) What is the hybridization of each C and N atom?

**10.61** In each of the following equations, what hybridization change, if any, occurs at the underlined atom:
(a) $\underline{B}F_3 + NaF \rightarrow Na^+BF_4^-$
(b) $\underline{P}Cl_3 + Cl_2 \rightarrow PCl_5$
(c) $H-\underline{C}\equiv C-H + H_2 \rightarrow H_2C=CH_2$
(d) $\underline{Si}F_4 + 2F^- \rightarrow SiF_6^{2-}$
(e) $\underline{S}O_2 + \frac{1}{2}O_2 \rightarrow .SO_3$

**10.62** Which of the following molecules have a dipole moment: (a) $BI_3$; (b) $ClF_3$; (c) $XeO_3$?

**10.63** Tetrazene (atom sequence $H_2NNNNH_2$) is a rather unstable hydride of nitrogen that decomposes above 0°C to $N_2$ and hydrazine ($N_2H_4$).
(a) What is the hybridization of N in these three compounds?
(b) Rank the three types of N-to-N bonds in terms of increasing length and strength.

(c) Draw structures for the *cis* and *trans* forms of tetrazene and predict which, if any, is polar.

**10.64** There is experimental evidence that the gaseous molecules $CaF_2$, $SrCl_2$, and $BaCl_2$ have bent shapes.
(a) What structure would you predict for these molecules?
(b) What property might have been measured to provide the evidence reported?

**10.65** For the following species, give the ideal angles and any expected deviations: (a) $SO_3^{2-}$; (b) $PbCl_2$; (c) $ClF_3$.

**10.66** Use partial orbital diagrams to show how the atomic orbitals on the central atom lead to the hybrid orbitals in the following: (a) $IF_2^-$; (b) $ICl_3$; (c) $XeOF_4$; (d) $BHF_2$.

**10.67** The measured bond angle in $NO_2$ is 134.3°, and in $NO_2^-$ it is 115.4°, even though the ideal bond angle is 120° in both. Explain.

**10.68** Tryptophan is one of the amino acids typically found in proteins:

(a) What is the hybridization at the numbered C, N, and O atoms?
(b) How many $\sigma$ bonds are present in the tryptophan molecule?
(c) Predict the bond angles at points a, b, and c.

**10.69** The bond angle in $NH_3$ is 107.3°, whereas in $PH_3$ it is 93.6°. Explain these data using VB theory.

**\*10.70** The sulfate ion can be represented with four S—O bonds or with two S—O and two S=O bonds.
(a) Which representation is better from the standpoint of formal charges?
(b) What is the shape of the sulfate ion, and what hybrid orbitals are postulated for the $\sigma$ bonding?
(c) In view of the answer to part (b), what S orbitals must be used for the $\pi$ bonds? What O orbitals?
(d) Draw an overlap diagram to show how one atomic orbital from S and one from O combine to form a $\pi$ bond.

**\*10.71** The hydrocarbon allene, $H_2C=C=CH_2$, is obtained indirectly from petroleum and used as a precursor for several types of plastics. What is the hybridization at each C atom in allene? Draw a bonding picture for allene with lines for $\sigma$ bonds, and show the arrangement of the $\pi$ bonds. Be sure to represent the geometry of the molecule in three dimensions.

**10.72** Many unusual heteronuclear diatomic molecules form in hot gases. Draw an MO diagram, write the electron configuration, and predict the bond order for the following: (a) BC; (b) OF.

**10.73** Lewis structures of mescaline, a hallucinogenic compound in peyote cactus, and dopamine, a neurotransmitter in mammalian brain, appear below. Suggest a reason for mescaline's ability to disrupt nerve impulses.

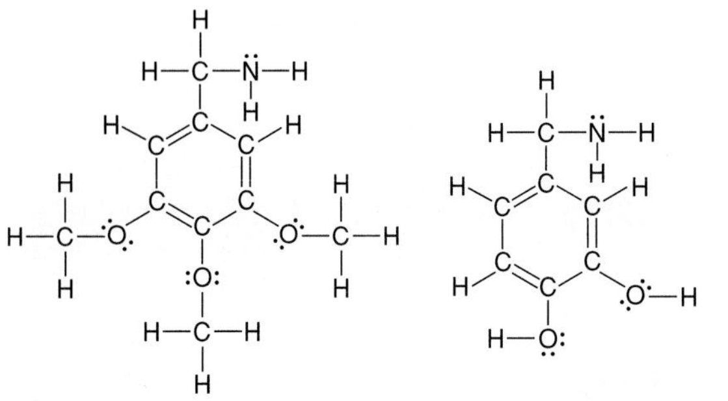

mescaline          dopamine

**\*10.74** Linoleic acid is an essential fatty acid found in many vegetable oils, such as soy, peanut, and cottonseed. A key structural feature of the molecule is the *cis* orientation around its two double bonds:

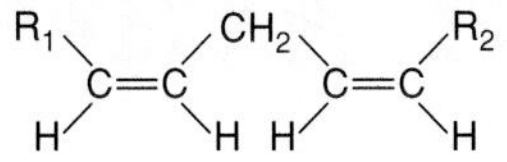

where $R_1$ and $R_2$ represent two different groups that form the rest of the molecule.

(a) How many different compounds are possible, changing only the *cis/trans* arrangements around these two double bonds?

(b) How many are possible for a similar compound with three double bonds?

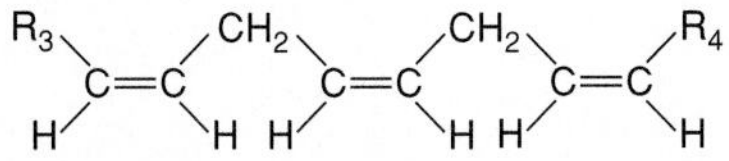

**10.75** Although LiF consists of ions in the solid state, it exists as highly polar LiF molecules in the gaseous state. Based on the relative magnitudes of properties, such as ionization energy, electron affinity, and electronegativity, explain why the molecules are polar. Draw an MO diagram for LiF, similar to that for HF, assuming that the Li $2s$ orbital interacts with the F $2p$ orbital.

**\*10.76** Simple proteins consist of amino acids linked together in a long chain, a small portion of which is

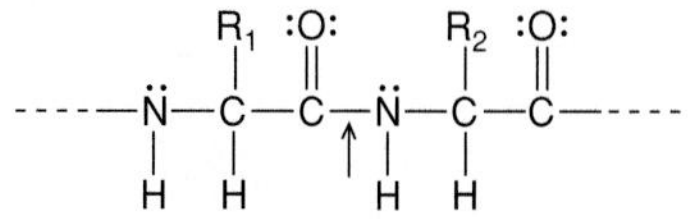

Experiment shows that rotation about the C—N bond *(arrow)* is somewhat restricted. Explain with resonance structures and show types of bonding involved.

**10.77** The compound 2,6-dimethylpyrazine gives chocolate its odor and is used extensively in flavorings. Its structure is

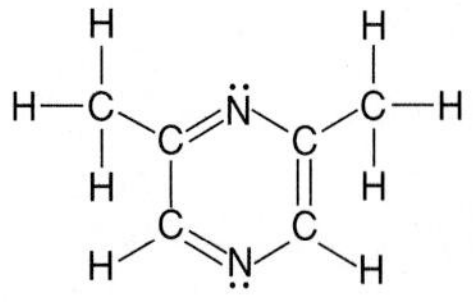

(a) Which atomic orbitals mix to form the hybrid orbitals of N?

(b) In what type of hybrid orbital do the N lone pairs reside?

(c) Is the hybridization of C in $CH_3$ the same as in the ring? Explain.

**10.78** Acetylsalicylic acid (aspirin), the most widely used medicine in the world, has the following structure:

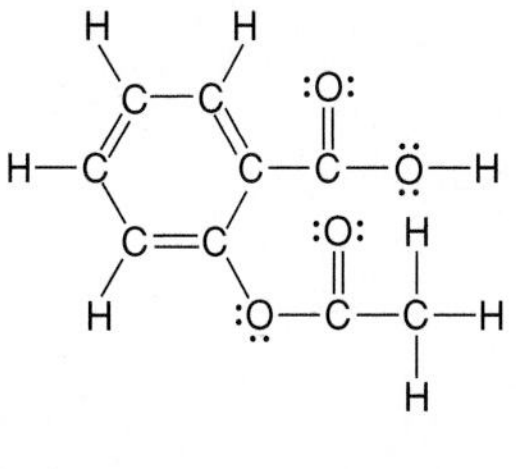

(a) What is the hybridization at each C and O atom?

(b) How many localized $\pi$ bonds are present?

(c) How many C atoms have a trigonal planar shape around them? A tetrahedral shape?

# CHAPTER 11

**Concepts and skills to review**

- properties of gases, liquids, and solids (Section 5.1)
- kinetic-molecular theory of gases (Section 5.6)
- kinetic and potential energy (Section 6.1)
- heat capacity, enthalpy change, and Hess's law (Sections 6.2, 6.3, and 6.5)
- Coulomb's law (Section 9.2)
- chemical bonding models (Sections 9.2 to 9.4, and 9.7)
- molecular polarity (Section 10.2)
- molecular orbital treatment of diatomic molecules (Section 10.4)

# Intermolecular Forces

# Liquids, Solids, and Changes of State

**Niagara Falls in winter.** Humid wind, churning water, and rigid ice display the power and beauty of the three states of water. In this chapter, we'll explore the nature of the forces between all molecules that give rise to these wondrous physical forms.

All matter occurs as either a gas, liquid, or solid. In fact, under different conditions, many pure substances can exist in all three physical states. The three states of water are most familiar: we inhale and exhale gaseous water; drink, excrete, and wash with liquid water; and slide on or cool our drinks with solid water.

After examining gases in Chapter 5, we now turn our attention to liquids and solids, which are called *condensed states* because their particles are extremely close together. The attractive and repulsive forces among the particles—molecules, atoms, or ions—in a sample of matter are known as interparticle forces or, more commonly, **intermolecular forces.** The balance between these forces and the kinetic energy of the particles give rise to the properties of each state and to **phase changes,** the changes from one state (or form) to another.

The chapter begins with a qualitative overview of the three states and their phase changes. Then we consider the various types of intermolecular forces. We discuss, in turn, the properties of liquids and solids, emphasizing the effect of the type of bonding on intermolecular forces, and how solid structures are determined experimentally. Then we focus on changes of state at the molecular level and the energy changes involved, as well as the effects of temperature and pressure. The final section combines ideas from several chapters to show how the unique properties of water arise inevitably from the electron configurations of its atoms.

## 11.1 An Overview of Physical States and Phase Changes

Two types of electrostatic forces are at work in any sample of matter. *Intra*molecular, or bonding, forces exist *within* a particle (molecule or polyatomic ion) and influence the chemical properties of the sample. *Inter*molecular forces exist *between* the particles and influence the physical properties of the sample.

If you could see one molecule from a sample of water vapor, one from liquid water, and one from ice, they would be identical bent, polar H—O—H molecules, their atoms held together by *intra*molecular covalent bonding forces. However, the samples from which each molecule comes differ greatly in their physical behavior. Whether a substance is a gas, liquid, or solid depends on the interplay between the potential energy of the intermolecular attractions, which tends to draw the molecules together, and the kinetic energy of the moving molecules, which tends to disperse them. The potential energy of the attractions depends on the charges of the particles and the distances between them (Coulomb's law, Section 9.2); the kinetic energy, which is related to their average speed, is proportional to the absolute tem-

**FIGURE 11.1**

**The three physical states of water.** Like many pure substances, water occurs in three physical states. In the gas, the molecules are far apart and moving randomly. In the liquid, they are close together but moving over each other randomly. In the solid, they are also close together but fixed in an orderly arrangement. Despite the physical differences of water on the macroscopic level, all three states consist of the same individual, covalent $H_2O$ molecules.

Water vapor: $H_2O(g)$

Water: $H_2O(l)$

Ice: $H_2O(s)$

perature. This *kinetic-molecular* view of the three states is an extension of the model we used earlier to understand gas behavior (Section 5.6).

In a gas, the energy of attraction is small relative to the energy of motion, so the particles are far apart. In a liquid, the attractions are stronger because the particles are much closer together, but their kinetic energy still allows them to tumble randomly over and around each other. In a solid, their attractions dominate their motion to the extent that the particles are organized and fixed relative to one another (Figure 11.1).

These factors directly affect three properties that define the three states: shape, compressibility, and ability to flow (Table 11.1). Since the particles in a gas are far apart, a gas moves freely throughout a container and fills it. With the particles so far apart, gases are highly compressible, and they flow and diffuse through one another easily. In a liquid, the particles are in virtual contact but still able to move freely. A liquid conforms to the shape of its container but has a surface. With very little free space between the particles, liquids resist an applied external force and thus compress only very slightly. They do flow and diffuse, but *much* more slowly than gases. Even though the particles in a solid are only slightly closer together than those in a liquid, their positions are fixed, so a solid has its own shape. Solids compress even less than liquids, and their particles do not flow to any significant extent.

**TABLE 11.1    A Macroscopic Comparison of Gases, Liquids, and Solids**

| STATE | SHAPE AND VOLUME | COMPRESSIBILITY | ABILITY TO FLOW |
|---|---|---|---|
| Gas | Conforms to shape and volume of container | High | High |
| Liquid | Conforms to shape of container; volume limited by surface | Very low | Moderate |
| Solid | Maintains its own shape and volume | Almost none | Almost none |

Phase changes are also determined by the balance between kinetic energy and intermolecular forces. As the temperature increases, so does the kinetic energy, and the faster moving particles can overcome the attractions more easily; conversely, lower temperatures allow the forces to draw the slower moving particles together. When water vapor cools, a mist appears as the particles form tiny droplets of liquid that then collect into a bulk sample with a single surface. The process of a gas changing into a liquid is called **condensation;** the opposite process, changing from a liquid into a gas, is called **vaporization.**

With further cooling, the particles move even more slowly and become fixed in position as the liquid solidifies in the process of **freezing;** the opposite change is called **melting,** or **fusion.** In common speech, freezing implies low temperature because we typically think of water, which solidifies at 0°C. However, many substances freeze at temperatures much greater than room temperature; for example, gold freezes (solidifies) at 1064°C.

As the molecules of a gas attract each other and come closer together in the liquid, and then become more organized in the solid, the system of particles loses energy, which is released as heat; thus, *condensing and freezing are exothermic changes.* On the other hand, energy must be absorbed to overcome attractive forces that restrict motion in a liquid or solid; thus, *melting and vaporizing are endothermic changes.* The Gallery on the next page displays some familiar examples of differences in the physical states and their changes.

Every pure substance has a specific enthalpy change per mole, measured at 1 atm and the temperature of the phase change, that is associated with these changes of state. For vaporization, it is the **heat of vaporization** $(\Delta H^0_{vap})$, and for fusion, it is the **heat of fusion** $(\Delta H^0_{fus})$. For water, we have

$$H_2O(l) \rightarrow H_2O(g) \quad \Delta H = \Delta H^0_{vap} = 40.7 \text{ kJ/mol (at 100°C)}$$
$$H_2O(s) \rightarrow H_2O(l) \quad \Delta H = \Delta H^0_{fus} = 6.02 \text{ kJ/mol (at 0°C)}$$

As you know from thermochemical principles, the reverse processes, condensing and freezing, have enthalpy changes that are the same magnitude but opposite in sign:

$$H_2O(g) \rightarrow H_2O(l) \quad \Delta H = -\Delta H^0_{vap} = -40.7 \text{ kJ/mol}$$
$$H_2O(l) \rightarrow H_2O(s) \quad \Delta H = -\Delta H^0_{fus} = -6.02 \text{ kJ/mol}$$

Water is typical of most pure substances in that less energy is needed to melt one mole of solid $(\Delta H^0_{fus})$ than to vaporize one mole of liquid $(\Delta H^0_{vap})$ (Figure 11.2). A change in state is essentially a change in intermolecular distance and freedom of motion. Thus, less energy is needed to overcome the forces holding the molecules in their fixed positions (to melt a solid) than to separate them completely from each other (to vaporize a liquid).

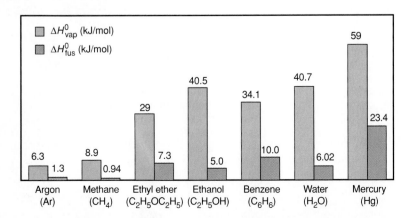

**FIGURE 11.2**

**Heats of vaporization and fusion for several common substances.** $\Delta H^0_{vap}$ is always larger than $\Delta H^0_{fus}$ because it takes more energy to separate particles completely than just to free them from their fixed positions in the solid.

## GALLERY          Physical States and Phase Changes

The properties of the condensed states and the differences in those properties when one state changes into another are manifest in so many practical and natural settings.

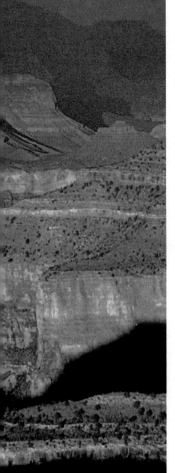

**Environmental flow.** The environment provides ample evidence of the differences in the ability of the three states to flow. In the atmosphere, components become mixed and distributed so well that the lowest 80 km have a uniform composition. There is much less mixing in the oceans, with differences in composition at various depths supporting different life forms. Rocky solids intermingle so little that adjacent strata remain separated for millions of years.

**Frozen gold.** The difference in ability to flow between liquids and solids is essential for casting applications. In the preparation of a dental crown, the niches and crevices of the tooth impression are filled with a molten metal, often a gold alloy. After cooling, the metal solidifies and withstands years of high pressure from chewing.

The casting of bronze statues, glass decorations, candles, and numerous plastic objects employs the same principle.

**Liquids in brake lines.** Low compressibility and the ability to conform to a container make liquids perfect for use in hydraulic systems, such as automobile brakes. Pressure is transferred smoothly and evenly from your foot through the brake lines, which are filled with brake fluid, to the brake pad. Gases compress too much and solids are too rigid to transfer the pressure.

**A cooling phase change.** The evaporation of sweat has a cooling effect because heat from your body is used to vaporize the water. Cats lick themselves and dogs pant to achieve the same cooling effect.

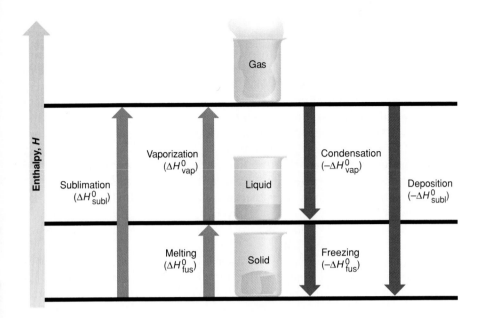

The three states of water are familiar to you because they are stable under ordinary conditions. Carbon dioxide, on the other hand, exists commonly as a gas and a solid (dry ice), but liquid $CO_2$ occurs only at pressures greater than 5 atm. Thus, at ordinary conditions, solid $CO_2$ becomes a gas without first becoming a liquid. The process of a solid changing directly into a gas is called **sublimation.** On a clear wintry day, you can dry clothes outside on a line, even though it may be too cold for ice to melt, because the ice sublimes. The opposite process, changing from gas into solid, is called **deposition**—you may have seen ice crystals form on a cold window from the deposition of water vapor. The **heat of sublimation ($\Delta H^0_{subl}$)** is the enthalpy change associated with one mole of the substance subliming. As we would expect from Hess's law, it is equal to the sum of the heats of fusion and vaporization:

$$
\begin{array}{ll}
\text{Solid} \rightarrow \text{liquid} & \Delta H^0_{fus} \\
\text{Liquid} \rightarrow \text{gas} & \Delta H^0_{vap} \\
\hline
\text{Solid} \rightarrow \text{gas} & \Delta H^0_{subl}
\end{array}
$$

Figure 11.3 summarizes the terminology of the various changes of state and the enthalpy changes associated with them.

### Section Summary

Due to the relative influence of intermolecular forces and kinetic energy, the particles in a gas are far apart, those in a liquid are in contact but mobile, and those in a solid are in contact and fixed in a rigid structure. These submicroscopic differences in the states of matter account for macroscopic differences in shape, compressibility, and ability to flow. When a solid becomes a liquid (melting, or fusion) or a liquid becomes a gas (vaporization), energy is absorbed to overcome intermolecular forces and increase the distance between particles. When particles come closer together in the opposite changes (freezing and condensation), energy is released. Sublimation is the changing of a solid directly into a gas. Each phase change is associated with a given enthalpy change under specified conditions.

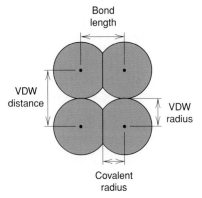

## 11.2   Types of Intermolecular Forces

Both bonding (intramolecular) forces and intermolecular forces arise from electrostatic attractions between opposite charges. Bonding forces are due to the attraction between cations and anions (ionic bonding), nuclei and electron pairs (covalent bonding), or metal cations and delocalized valence electrons (metallic bonding). Intermolecular forces, on the other hand, are due to the attraction between molecules, as a result of partial charges, or between ions and molecules. The two types of forces differ in magnitude. We know from Coulomb's law that the higher the charges and the closer they are to each other, the greater the force between them. Therefore,

- *Bonding forces are relatively strong* because they involve higher charges that are closer together.
- *Intermolecular forces are relatively weak* because they typically involve lower charges that are farther apart.

The charges that give rise to intermolecular forces are farther apart because they exist between *nonbonded* atoms in adjacent molecules. For instance, when we measure the various distances between two Cl nuclei in a sample of solid $Cl_2$, we obtain two different values (Figure 11.4). The shorter distance is between two Cl atoms *bonded to each other in the same molecule;* it is called the *bond length,* and one-half this distance is the covalent radius. The longer distance is between *two nonbonded Cl atoms in adjacent molecules;* this is called the *van der Waals distance* (named after the Dutch physicist Johannes van der Waals, who studied the effects of intermolecular forces on the behavior of real gases). This distance is the closest one $Cl_2$ molecule can approach another, the point at which intermolecular attractions balance electron-cloud repulsions. One-half this distance is the **van der Waals radius** of the Cl atom, one-half the closest distance between the nuclei of identical *nonbonded* atoms. The van der Waals radius of an atom is always larger than its covalent radius; van der Waals radii decrease across a period and increase down a group, just as covalent radii do (Figure 11.5).

There are several important types of intermolecular forces: ion-dipole, dipole-dipole, hydrogen bonding, dipole-induced dipole, and dispersion forces. Except for the ion-dipole forces, they are classified as *van der Waals forces.* Let's discuss the various intermolecular forces in decreasing order of their relative strengths. Table 11.2 compares these forces with the stronger intramolecular bonding forces.

### Ion-Dipole Forces

When an ion and a nearby polar molecule (dipole) attract each other, an **ion-dipole force** results. The most important example takes place when an ionic compound dissolves in water: the ions are separated because the attractions between the ions and the oppositely charged poles of the $H_2O$ molecules overcome the attractions between the ions themselves. Ion-dipole forces and their associated energy are discussed fully in Chapter 12.

### Dipole-Dipole Forces

When polar molecules lie near one another, as in liquids and solids, their partial charges orient them and give rise to **dipole-dipole forces:** the partially positive pole of one molecule attracts the partially negative pole of another (Figure 11.6). In Chapter 10, you saw how polar molecules are

**FIGURE 11.4**

**Covalent radii and van der Waals radii.** Two $Cl_2$ molecules approach each other as closely as their electron clouds allow. The van der Waals radius is one-half the distance between adjacent *nonbonded* atoms ($\frac{1}{2}$ × VDW distance). The covalent radius is one-half the distance between *bonded* atoms ($\frac{1}{2}$ × bond length). The covalent radius is always less than the van der Waals radius.

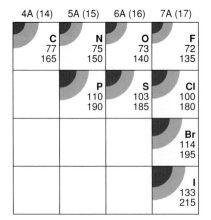

**FIGURE 11.5**

**Periodic trends in covalent and van der Waals radii of several nonmetal atoms.** Like covalent radii *(dark blue quarter circles)*, van der Waals radii *(light blue quarter circles)* increase down a group and decrease across a period.

**TABLE 11.2    Comparison of the Energies Associated with Bonding (Intramolecular) Forces and Intermolecular Forces**

| FORCE | MODEL | BASIS OF ATTRACTION | ENERGY (kJ/mol) | EXAMPLE |
|---|---|---|---|---|
| **Intramolecular** | | | | |
| Ionic | | Cation–anion | 400-4000 | NaCl |
| Covalent | | Nuclei–shared e⁻ pair | 150-1100 | H—H |
| Metallic | | Cations–delocalized electrons | 75-1000 | Fe |
| **Intermolecular** | | | | |
| Ion-dipole | | Ion charge– dipole charge | 40-600 | $Na^+ \cdots O$ H H |
| Dipole-dipole | | Dipole charges | 5-25 | I—Cl····I—Cl |
| H bond | | Polar bond to H– dipole charge (high EN of N, O, F) | 10-40 | O—H····O—H H H |
| Ion-induced dipole | | Ion charge– polarizable e⁻ cloud | 3-15 | $Fe^{2+} \cdots O_2$ |
| Dipole-induced dipole | | Dipole charge– polarizable e⁻ cloud | 2-10 | H—Cl····Cl—Cl |
| Dispersion (London) | | Polarizable e⁻ clouds | 0.05-40 | F—F····F—F |

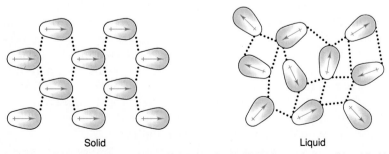

Solid                Liquid

**FIGURE 11.6 Orientation of polar molecules due to dipole-dipole forces.** In the solid and liquid states, the partial charges of polar molecules are close enough to orient the molecules. (Spaces between molecules are increased for clarity.) The arrangement is more orderly in the solid because the average kinetic energy of the particles is lower.

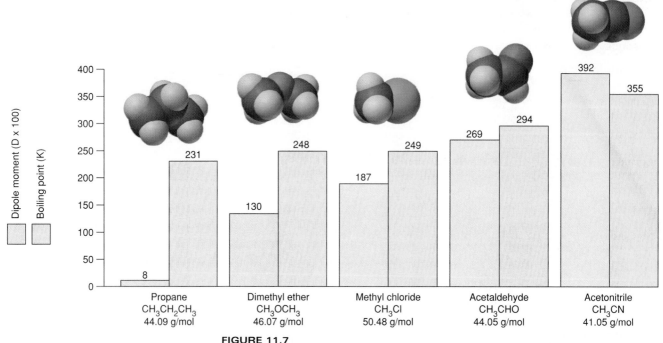

**FIGURE 11.7**

**Dipole moment and boiling point.** For compounds with similar molar masses, the boiling point increases with increasing dipole moment. The greater dipole moment creates stronger dipole-dipole forces, which require higher temperatures to overcome.

oriented in an electric field and how these forces give polar *cis*-1,2-dichloroethylene a higher boiling point than the nonpolar *trans* compound.

For substances of approximately the same size and molar mass, the greater the dipole moment, the greater is the dipole-dipole force between their molecules, so the more energy it takes to separate them—a trend that shows up in several of their physical properties. Consider the boiling points of the compounds in Figure 11.7, all of which have about the same size and molar mass. Methyl chloride, for instance, has a smaller dipole moment than acetaldehyde; therefore less energy is needed to overcome the dipole-dipole forces between its molecules and it boils at a lower temperature.

### The Hydrogen Bond

A special type of dipole-dipole force arises between molecules that have *a hydrogen atom bound to a small, highly electronegative atom with lone pairs.* The most important atoms that fit this description are oxygen, nitrogen, and fluorine. The covalent bond between hydrogen and such an atom is very polar. The partially positive H of one molecule is attracted to the partially negative lone pair on the N, O, or F of another molecule, and a **hydrogen bond (H bond)** results. To summarize, the atom sequence that leads to an H bond (dotted line) is —A:····H—B—, where *both* A and B are N, O, or F. (We do not consider the significantly weaker H-bond-type interactions that can occur with P, S, and Cl.) Three examples are

$$—\ddot{\underset{..}{F}}:····H—\ddot{\underset{..}{O}}—\qquad —\ddot{\underset{..}{O}}:····H—\overset{|}{N}—\qquad —\overset{|}{N}:····H—\ddot{\underset{..}{F}}:$$

The small size of the N, O, and F atoms is essential to hydrogen bonding because (1) these atoms are electronegative enough to make their covalently bonded H highly positive, and (2) the other atom's lone pair can come close enough to this H.

SAMPLE PROBLEM 11.1 ⎯⎯⎯⎯⎯⎯⎯⎯⎯⎯⎯⎯⎯⎯⎯⎯⎯⎯

**Drawing Hydrogen Bonds Between Molecules**

**Problem:** In which of the following substances do H bonds occur? Draw the H bonds between two molecules of the substance wherever appropriate.

(a) $C_2H_6$        (b) $CH_3OH$        (c) $CH_3\overset{\displaystyle O}{\overset{\|}{C}}-NH_2$

**Plan:** We examine each structure to see if the molecule contains N, O, or F covalently bonded to H.

**Solution:**

(a) For $C_2H_6$. No H bonds are formed:

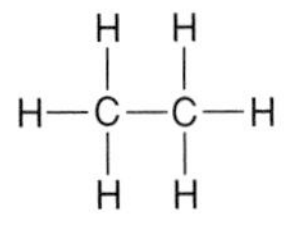

(b) For $CH_3OH$. The H covalently bonded to the O in one molecule forms an H bond to the lone pair on the O of an adjacent molecule:

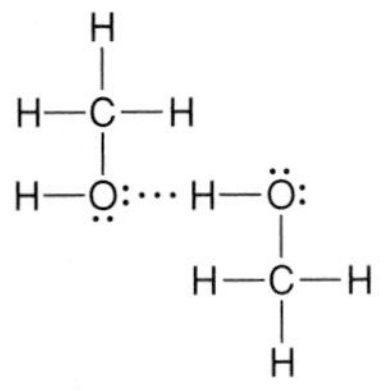

(c) For $CH_3\overset{\displaystyle O}{\overset{\|}{C}}-NH_2$. Two of these molecules can form one H bond from the H of N to O, or they can form two H bonds:

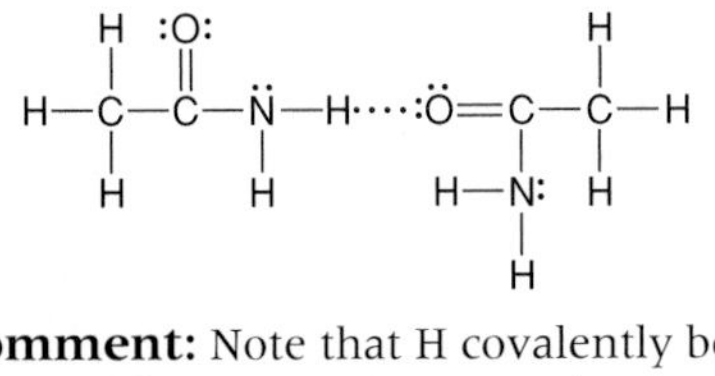

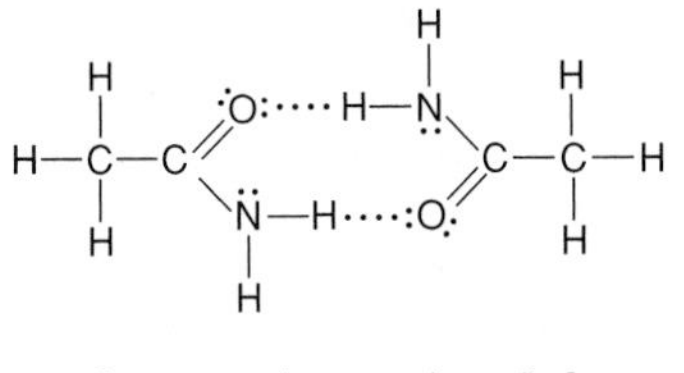

**Comment:** Note that H covalently bonded to C does not form H bonds because the C is not electronegative enough to make the C—H bond very polar.

FOLLOW-UP PROBLEM 11.1

In which of the following substances do H bonds occur? Draw the H bonds between two molecules of the substance where appropriate.

(a) $CH_3\overset{\displaystyle O}{\overset{\|}{C}}-OH$    (b) $CH_3CH_2OH$    (c) $CH_3\overset{\displaystyle O}{\overset{\|}{C}}CH_3$

The effect of H bonding on physical properties is dramatically illustrated by the boiling points of the Group 4A(14) through 7A(17) hydrides (Figure 11.8). For reasons that we'll discuss shortly, the boiling points of the 4A hydrides go up in typical fashion as the molar mass increases from $CH_4$ through $SnH_4$. In the other groups, however, the first member in each se-

**FIGURE 11.8**

**Hydrogen bonding and boiling point.** The boiling points of the binary hydrides from Groups 4A(14) to 7A(17) are plotted against period number. The H bonds in $NH_3$, $H_2O$, and HF give them much higher boiling points than would be expected from the trend based on molar mass, as seen in the values for Group 4A. In fact, if not for the strength of its H bonds, water would boil almost 200°C lower than it does *(dashed line)*.

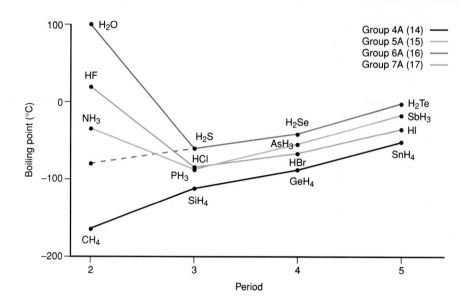

ries—$NH_3$, $H_2O$, and HF—deviates enormously from this expected increase of boiling point with molar mass because additional energy is required to break H bonds between these molecules. For example, on the basis of molar mass alone, we would expect water to boil about 200°C lower than it does! The H bonds in water have profound effects on the environment and living things, as you'll see in the final section of this chapter.

Even though the strength of one H bond is relatively small (~5% of a typical covalent bond energy), the combined strength of many H bonds can be large. Consider the H bonds in deoxyribonucleic acid (DNA), the giant molecule in all cells that acts as a genetic "blueprint," governing the function and appearance of the entire organism. DNA consists of two molecular chains wrapped around each other in a long double helix (Figure 11.9). Strong covalent bonds join the atoms in each chain, but the two intertwining chains are held together by many thousands of H bonds. The total energy of the H bonds holds the chains together during many processes, but

**FIGURE 11.9**

**Covalent bonding and H bonding in the structure of deoxyribonucleic acid (DNA).** The DNA molecule consists of two chains that are each held together by strong covalent bonds. Many thousands of H bonds link one chain tightly to the other to form a double helix.

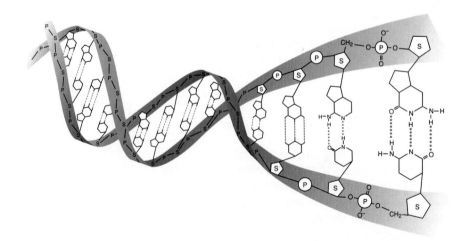

the weakness of each H bond allows a few to break at a time when the DNA separates during the processes of protein synthesis and cell reproduction. (We discuss these processes in Chapter 14.)

### Electron-Cloud Distortion and Charge-Induced Dipole Forces

Even when localized to bonding or lone pairs, electrons are in constant motion throughout their prescribed regions, and we often picture them as "clouds" of negative charge. A nearby electric field can distort an electron cloud, drawing electron density toward a positive charge or repelling it from a negative one. In effect, the field *induces* a distortion in the electron cloud. For a nonpolar molecule, this distortion results in a *temporary* dipole moment; for a polar molecule, it enhances the dipole moment already present. The source of the electric field can be the electrodes of a battery, the charge of a nearby ion, or even the partial charges of a nearby polar molecule.

The ease with which a particle's electron cloud is distorted is called its **polarizability.** Smaller atoms (or ions) are less polarizable than larger ones because their electrons are held more tightly and closer to the nucleus. *Polarizability increases down a group* of atoms (or ions) as size increases because the larger electron clouds are more easily distorted. *Polarizability decreases across a period* because the increasing effective nuclear charge holds the electrons more tightly. Among ions, cations are *less* polarizable than their parent atom, whereas anions are *more* polarizable; thus, for example, $Na^+$ is less polarizable than Na, but $F^-$ is more polarizable than F.

The two charge-induced dipole forces occur in solutions (Chapter 12), rather than pure substances, but we can preview them here. When an ion's charge distorts a nearby electron cloud, an **ion-induced dipole force** results. For example, this force arises between the $Fe^{2+}$ ion in a hemoglobin molecule and the electron cloud of an incoming $O_2$ molecule as a prelude to oxygen binding in the bloodstream. When the partial charges of a polar molecule distort a nearby electron cloud, a **dipole-induced dipole force** results. For instance, when a noble gas atom, such as Xe, is near polar $H_2O$, the electron cloud becomes distorted, and the resulting attractive force is responsible for xenon and the other noble gases dissolving (Figure 11.10). In keeping with the group trend in polarizability, xenon is more than 25 times as soluble in water at 0°C as the much smaller helium. We refer to these points again in the next chapter.

### Dispersion (London) Forces

Up to this point, we've discussed intermolecular forces that depend on an existing ionic charge or molecular dipole. But what forces cause substances like octane, carbon dioxide, and even the noble gases to condense and solidify? An attractive force must be acting between these nonpolar molecules and atoms, or they would remain gaseous under any conditions. Although bond dipoles may exert some weak attractions, the intermolecular force primarily responsible for the condensed states of nonpolar substances is the **dispersion force,** or **London force** (after Fritz London, the German physicist who first explained the quantum-mechanical basis of the attraction).

Dispersion forces are very weak and are caused by momentary oscillations of electron density (Figure 11.11). Nonpolar particles that are far apart do not influence each other. When particles are near each other, however, and especially when they collide, electron clouds are repelled, and electron density is momentarily greater in one region of the particle than in another.

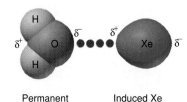

Permanent H₂O dipole    Induced Xe dipole

**FIGURE 11.10**
**Dipole-induced dipole forces and the aqueous solubility of nonpolar substances.** A polar molecule can induce a dipole in a nearby nonpolar atom or molecule. Here, water's negative end repels the polarizable electron cloud of Xe, creating partially positive and negative poles in the Xe atom. The resulting intermolecular attraction helps dissolve noble gases and other nonpolar substances.

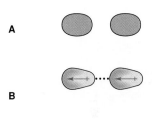

A
B

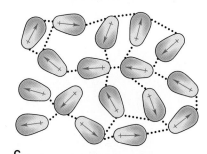

C
**FIGURE 11.11**
**Dispersion forces among nonpolar molecules.** The dispersion force is responsible for the condensed states of noble gases and nonpolar molecules. **A,** Separated H₂ molecules are nonpolar. **B,** During collisions, a dipole may form instantaneously in one molecule and induce a momentary dipole in a neighboring molecule. These momentary partial charges attract the molecules together. **C,** The process of inducing temporary dipoles takes place among molecules throughout the sample.

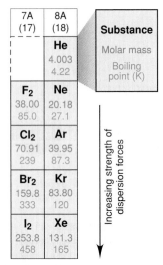

**FIGURE 11.12**

**Molar mass and boiling point.** The strength of dispersion forces increases with size, which usually correlates with molar mass. One effect of this appears in the higher boiling points down the groups of halogens and noble gases.

$CH_3-CH_2-CH_2-CH_2-CH_3$

*n*–pentane, bp = 36.1°C

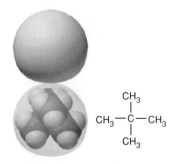

Neopentane, bp = 9.5°C

**FIGURE 11.13**

**Molecular shape and boiling point.** The spherical neopentane molecules make less contact with each other than do the cylindrical *n*-pentane molecules, so neopentane has a lower boiling point.

*Instantaneous dipoles* are induced in both particles, which induce instantaneous dipoles in adjacent particles. Thus, dispersion forces are *induced dipole–induced dipole forces*. The process occurs throughout the sample and, at low enough temperatures, keeps the particles together.

The strength of dispersion forces depends on the polarizability of the particle, which in turn depends on its size. Therefore, *dispersion forces generally increase with molar mass* because particles with greater mass have more atoms and/or larger (heavier) atoms and thus more electrons. For example, as size and molar mass increase down the halogens or noble gases, dispersion forces become greater, which is reflected in their higher boiling points (Figure 11.12).

For substances with the same molar mass, the strength of the dispersion force is often influenced by molecular shape. Shapes that allow more points of contact have more area over which electron clouds can be distorted, so stronger attractions result. Consider the hydrocarbons *n*-pentane and neopentane, which have the same formula ($C_5H_{12}$) but different shapes (Figure 11.13). The *n*-pentane molecule has a straight chain of five C atoms and is somewhat cylindrical; neopentane has two $CH_3$ branches off a three-carbon chain and, thus, a more spherical shape. As a result, two *n*-pentane molecules make more contact with each other than two neopentane molecules. Greater contact allows the dispersion forces to act at more points, so *n*-pentane has the higher boiling point.

Dispersion forces are quite weak, but they exist *between all particles*. Therefore, except for small, polar molecules with large dipole moments or those able to form H bonds, *the dispersion force is the dominant intermolecular force between identical molecules.*

SAMPLE PROBLEM 11.2

**Predicting the Type and Relative Strength of Intermolecular Forces**

**Problem:** Identify the dominant intermolecular force that is present in each of the following substances, and select the substance with the higher boiling point in each pair:
**(a)** $MgCl_2$ or $PCl_3$
**(b)** $CH_3NH_2$ or $CH_3F$
**(c)** $CH_3OH$ or $CH_3CH_2OH$
**(d)** Hexane or cyclohexane

**Plan:** By examining the formulas and picturing (or drawing) the structures, we identify a key difference between the members of each pair in order to determine the types of forces involved: Are ions present? Are the molecules polar or nonpolar? Is F, O, or N bound to H? Do the molecules have different masses or shapes? In order to rank their strengths, we must consult Table 11.2 and remember the following points:
• Bonding forces are stronger than intermolecular forces.
• Hydrogen bonding is a strong type of dipole-dipole force.
• Dispersion forces are decisive when the major difference is molar mass or molecular shape.

**Solution:**
**(a)** $MgCl_2$ consists of $Mg^{2+}$ and $Cl^-$ ions, which are held together by **ionic bonding** forces; $PCl_3$ consists of polar molecules, so intermolecular **dipole-dipole** forces are present. The forces in **$MgCl_2$** are stronger, so it has a higher boiling point.

**(b)** $CH_3NH_2$ and $CH_3F$ both contain polar molecules of about the same molar mass. $CH_3NH_2$ has N—H bonds, so it can form **H bonds** (see margin). $CH_3F$ contains a C—F bond but no H—F bond, so **dipole-dipole** forces occur but it cannot form H bonds. Therefore, **$CH_3NH_2$** has the higher boiling point.

**(c)** $CH_3OH$ and $CH_3CH_2OH$ molecules both contain an O—H bond, so they can form **H bonds**. The difference between them is the additional —$CH_2$— group in **$CH_3CH_2OH$**, so it has a larger molar mass and therefore stronger **dispersion forces**, and a higher boiling point.

**(d)** Hexane and cyclohexane are nonpolar molecules of about the same molar mass but different molecular shapes (see margin). Cylindrical hexane molecules make less intermolecular contact than disc-like cyclohexane molecules, so **cyclohexane** should have greater **dispersion forces** and a higher boiling point.

**Check:** The actual boiling points show that our predictions are correct:
**(a)** $MgCl_2$ (1412°C) and $PCl_3$ (76°C)
**(b)** $CH_3NH_2$ (−6.3°C) and $CH_3F$ (−78.4°C)
**(c)** $CH_3OH$ (64.7°C) and $CH_3CH_2OH$ (78.5°C)
**(d)** Hexane (69°C) and cyclohexane (80.7°C)
**Comment:** Dispersion forces are *always* present, but in parts **(a)** and **(b),** they are much less significant than the other forces involved.

**FOLLOW-UP PROBLEM 11.2**
In each of the following pairs, identify all the intermolecular forces present and select the substance with the higher boiling point: **(a)** $CH_3Br$ or $CH_3Cl$; **(b)** $CH_3CH_2OH$ or $CH_3OCH_3$; **(c)** $C_2H_6$ or $C_3H_8$.

Hexane

Cyclohexane

**Section Summary**

The van der Waals radius determines the shortest distance over which intermolecular forces operate; it is always larger than the covalent radius. Intermolecular forces are much weaker than bonding (intramolecular) forces and are of several types. Ion-dipole forces occur between ions and polar molecules. Dipole-dipole forces occur between oppositely charged poles on polar molecules. Hydrogen bonding, a special type of dipole-dipole force, occurs when H bound to N, O, or F is attracted to the lone pair of N, O, or F on another molecule. Electron clouds can be distorted (polarized) in an electric field. Ion- and dipole-induced dipole forces arise between a charge and the dipole it induces in another molecule. Dispersion (London) forces are instantaneous induced dipole-induced dipole forces that occur among all particles and increase with molar mass. Molecular shape determines the extent of contact between molecules and thus the strength of dispersion forces.

# 11.3 Properties of the Liquid State

Of the three states of matter, the liquid state is the least understood at the molecular level. Gases are easy to describe: due to the *randomness* of the particles, any region of a gas sample is virtually identical to any other. Also, as you'll see in the next section, different regions of a crystalline solid are identical due to the *orderliness* of the particles. Liquids, however, have a combination of these attributes that changes continually and defies simple description: a region that is orderly one moment becomes random the next, and vice versa.

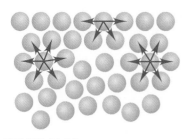

**FIGURE 11.14**

**The molecular basis of surface tension.**
Molecules in the interior of a liquid experience intermolecular attractions in all directions. Molecules at the surface experience a net attraction downward. With fewer attractions, molecules at the surface are less stable, so a liquid tends to minimize the number at the surface, which results in surface tension.

Despite this complexity at the molecular level, the macroscopic behavior of liquids is well understood. In this section, we discuss three liquid properties—surface tension, capillarity, and viscosity—in terms of the intermolecular forces that create them.

### Surface Tension

The molecules in a sample of liquid experience intermolecular forces of different strengths. Those in the interior are surrounded by molecules, whereas those at the surface have molecules only below and to the sides (Figure 11.14). As a result, molecules at the surface experience a *net attraction* pulling them downward toward the interior. For the surface area to increase, more molecules must move to the surface by breaking some attractions in the interior. Since this process makes the system less stable and requires energy, *a liquid surface tends to have the smallest possible area* and to behave like a taut skin that covers the interior.

The **surface tension** of a liquid is the energy required to increase the surface area by a given amount (in $J/m^2$). Some representative values are presented in Table 11.3. By comparing these values with those in Table 11.2, you can see that, in general, *the stronger the forces between the particles in a liquid, the greater is the surface tension.* Water has a high surface tension because its molecules form multiple H bonds. As you'll see in Chapter 12, *surfactants* (surface-active agents), such as soaps, detergents, petroleum recovery agents, and biological fat emulsifiers, decrease the surface tension of water by disrupting the H bonds.

**TABLE 11.3 Surface Tension and Forces Between Particles**

| SUBSTANCE | FORMULA | SURFACE TENSION ($J/m^2$) AT 20°C | MAJOR FORCE(S) |
|---|---|---|---|
| Diethyl ether | $CH_3CH_2OCH_2CH_3$ | $1.7 \times 10^{-2}$ | Dipole-dipole; dispersion |
| Ethanol | $CH_3CH_2OH$ | $2.3 \times 10^{-2}$ | H bonding |
| Butanol | $CH_3CH_2CH_2CH_2OH$ | $2.5 \times 10^{-2}$ | H bonding; dispersion |
| Water | $H_2O$ | $7.3 \times 10^{-2}$ | H bonding |
| Mercury | $Hg$ | $48 \times 10^{-2}$ | Metallic bonds |

### Capillarity

You may have seen a liquid rise through a narrow tube, called a capillary tube, against the pull of gravity (as when a small blood sample is taken from a pricked finger). This phenomenon is called capillary action, or **capillarity,** and it results from a competition between the intermolecular forces within the liquid (cohesive forces) and those between the liquid and the tube walls (adhesive forces).

Picture what occurs at the molecular level when you place a glass capillary tube in water. Glass is mostly silicon dioxide ($SiO_2$), so the water molecules form H bonds to the oxygen atoms that are part of the tube's inner wall. The adhesive forces (H bonding) between the water surface and the wall are stronger than the cohesive forces (H bonding) within the water, so a thin film of water creeps up the wall. At the same time, the cohesive forces that give rise to surface tension are pulling the liquid surface taut. These ad-

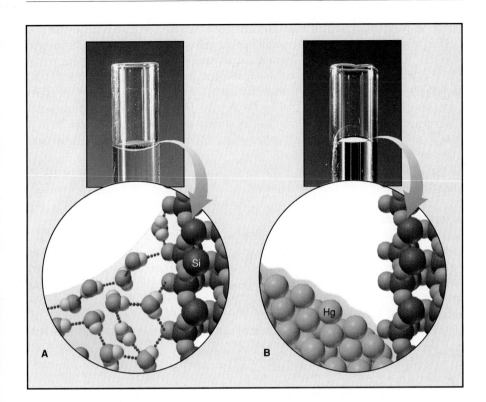

**FIGURE 11.15**
**Shape of the water or mercury meniscus in glass. A,** Water displays a concave meniscus in a glass tube because the adhesive (H bond) forces with the O—Si—O groups of the glass are *stronger* than the cohesive (H bond) forces within the water. **B,** Mercury displays a convex meniscus in a glass tube because the cohesive (metallic bonding) forces are stronger than the adhesive (dispersion) forces with the glass.

hesive and cohesive forces combine to raise the water level and produce the familiar concave meniscus (Figure 11.15, *A*).

Mercury has a higher surface tension than water (Table 11.3), which means it has stronger cohesive forces (metallic bonding). However, it has relatively weak adhesive forces (mostly dispersion) with glass. If you place a glass capillary tube in a dish of mercury, the level of the mercury in the tube drops below that in the dish. The cohesive forces among the mercury atoms are stronger than the adhesive forces between mercury and glass, so the liquid tends to pull away from the walls. At the same time, the surface atoms are being pulled toward the interior of the mercury by its high surface tension, so the level drops. These combined forces produce the convex meniscus (Figure 11.15, *B*) in a mercury barometer or manometer.

## Viscosity

When a liquid flows, the molecules slide around and past each other. The extent to which intermolecular attractions hamper this movement results in a liquid's **viscosity,** its resistance to flow. Both gases and liquids flow, but liquid viscosities are higher than those of gases because the intermolecular forces operate over shorter distances.

The *viscosity of a liquid decreases with increasing temperature* (Table 11.4). When you heat cooking oil in a pan, for example, it spreads out in a thin layer. Because the molecules move faster at higher temperatures, they overcome intermolecular forces more easily, so the resistance to flow decreases.

Molecular shape plays a role in determining a liquid's viscosity. Long molecules make more contact than spherical ones, so for the same types of forces, liquids containing longer molecules have higher viscosities. A striking example of a change in viscosity occurs during the making of syrup.

**TABLE 11.4 Viscosity of Water at Several Temperatures**

| TEMPERATURE (°C) | VISCOSITY $(N \cdot s/m^2)$* |
|---|---|
| 20 | $1.00 \times 10^{-3}$ |
| 40 | $0.65 \times 10^{-3}$ |
| 60 | $0.47 \times 10^{-3}$ |
| 80 | $0.35 \times 10^{-3}$ |

*The units of viscosity are newton-seconds per square meter.

Even at room temperature, a concentrated aqueous sugar solution has a higher viscosity than water because of H bonding among the many hydroxyl (—OH) groups on the ring-shaped sugar molecules. When the solution is slowly heated to boiling, the sugar molecules react with each other and link together, gradually forming long chains. Hydrogen bonds and dispersion forces occur at many points along the lengths of the chains, and the resulting syrup is a viscous liquid that pours slowly and clings to a spoon. When a very viscous syrup is cooled, it becomes stiff enough to be picked up and stretched—into taffy candy.

The Gallery on the facing page shows other familiar examples of these three liquid properties.

### Section Summary

Surface tension is a measure of the energy required to increase a liquid's surface area. Greater intermolecular forces within the liquid create higher surface tension. Capillary action, the rise of a liquid through a narrow space, occurs when the forces between a liquid and a solid surface (adhesive) are greater than those within the liquid itself (cohesive). Viscosity, the resistance to flow, depends on molecular shape and decreases with temperature.

## 11.4   Properties of the Solid State

Even a casual stroll through a natural history museum or school mineral collection reveals the astounding variety of solids. We categorize solids into two major groups based on their macroscopic appearance, which is based in turn on the arrangement of their component particles. **Crystalline solids** have a well-defined shape because of the orderly arrangement of their atoms, molecules, or ions (Figure 11.16); **amorphous solids** can typically be found in several shapes and lack this molecular-level order.

**FIGURE 11.16**

**The striking beauty of crystalline solids.** The symmetry and regularity of crystals are due to the orderly arrangement of their particles. Clockwise from bottom left: pyrite; franklinite; beryl; barite and calcite; aragonite.

## Properties of Liquids                                        GALLERY

Of the three states, only liquids combine the ability to flow with the strength that comes from intermolecular contact, and this combination appears in numerous applications.

**Minimizing a surface.**
In the low-gravity environment of the orbiting Space Shuttle, the tendency of a liquid to minimize its surface creates perfectly spherical droplets, unlike the elongated drops we see on Earth. For the same reason, bubbles in a soft drink are spherical because the liquid uses the minimum number of molecules to surround the gas. A water strider flits across a pond on widespread legs that do not exert enough pressure to break through the aqueous surface.

**Beaded droplets on waxy surfaces.**
The adhesive (dipole-induced dipole) forces between water and a nonpolar surface are much weaker than the cohesive (H bond) forces within water. As a result, the water pulls away from the nonpolar surface and forms beaded droplets. You have seen this effect when water beads on a flower petal or a freshly waxed car after a rainfall.

**Capillary action after a shower.**
Paper and cotton consist of fibers of cellulose, long carbon-containing molecules with many attached hydroxy (–OH) groups. A towel dries you in two ways: First, the water moves away from your body by capillary action through the spaces between the closely intertwined cellulose fibers. Second, the water molecules themselves form adhesive H bonds to the cellulose.

**How a ballpoint pen works.**
The essential parts of a ballpoint pen are the moving ball and its contact with the viscous ink. The material of the ball is chosen for its strong adhesive forces with the ink. Cohesive forces within the ink are replaced by those adhesive forces when the ink wets the ball. As the ball rolls along the paper, the adhesive forces between ball and ink are replaced by those between ink and paper. The rest of the ink stays in the pen due to its high viscosity.

Ink

Moving ball

Paper surface

**Maintaining motor oil viscosity.**
To protect engine parts during long drives or in hot weather, when an oil would ordinarily become too thin, motor oils contain additives, called "polymeric viscosity index improvers," that make the oil thicken.

As the oil heats up, the additive molecules change shape from compact spheres to spaghetti-like strands and become tangled with the hydrocarbon oil molecules. As a result, the viscosity increases and compensates for the decrease due to heating.

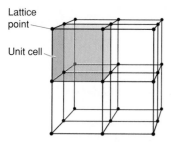

**A** Portion of 3-D lattice

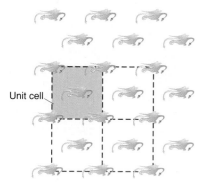

**B** Portion of 2-D lattice

**FIGURE 11.17**

**The crystal lattice and the unit cell. A,** A crystal lattice is an array of points that defines the positions of the particles in the crystal structure. It is shown here with the points connected by lines. A unit cell *(colored)* is the simplest array of points that produces the lattice, when repeated in all directions. A cubic unit cell, one of the seven types that occur in nature, is shown. **B,** A section of wallpaper is a two-dimensional analogy for a lattice.

**FIGURE 11.18** *(opposite page)*

**The three cubic unit cells. A,** Simple; **B,** body-centered; and **C,** face-centered cubic unit cells. Cubic arrangements of atoms are shown in exploded view *(top row)* and space-filling view *(second row)*. For clarity, the atoms are colored differently: corner atoms blue, body-centered atoms purple, and face-centered atoms yellow. The third row shows a unit cell shaded blue in a portion of the lattice. The coordination number is depicted by showing one particle *(dark blue in the center)* surrounded by a given number of nearest neighbors. The bottom row shows the total numbers of atoms in the actual unit cells. The simple cubic has one atom, the body-centered has two, and the face-centered has four.

## The Crystal Lattice and the Unit Cell

If you could see the particles within a crystal, you would find them packed tightly together in a three-dimensional framework. Within this framework, each particle lies centered on a specific point. The array of points forms a regular pattern that exists throughout the crystal and is called the crystal **lattice** (Figure 11.17, *A*). The **unit cell** is the *simplest* arrangement of points that, when repeated in all three directions, gives the lattice. A two-dimensional analogy for the relationship between unit cell and lattice can be seen in a checkerboard, a section of tiled floor, a piece of wallpaper, or any other pattern constructed from a repeating unit (Figure 11.17, *B*). The seven unit cells that occur in nature differ in the angles at their corners and the lengths of their sides. The **coordination number** of a particle in a crystal is the number of nearest neighbors surrounding it.

The *cubic unit cell* is the basis of the commonly occurring cubic lattice. The solid states of most of the elements, several covalent compounds, and many ionic compounds form a cubic lattice. There are three types of cubic unit cells.

1. In the **simple cubic unit cell,** eight lattice points define the corners of a cube (Figure 11.18, *A*). Interparticle attractions pull together atoms centered on these points. The atoms touch along the cube's edges but not diagonally along its faces or through its center. The coordination number of each particle is six: four in its own layer, one in the layer above, and one in the layer below.

2. A **body-centered cubic unit cell** has an atom at each corner *and* one in the center (Figure 11.18, *B*). Atoms at the corners do not touch, but they all touch the central atom. Each atom is surrounded by eight nearest neighbors, four above and four below, so the coordination number is eight.

3. The **face-centered cubic unit cell** has an atom at each corner *and* in the center of each face (Figure 11.18, *C*). Atoms at the corners touch those in the faces, but not each other. The coordination number is 12.

For particles of the same size, *the higher the coordination number, the greater is the number of particles packed into a given volume of the crystal.* Therefore, as you'll see in the following discussion, a crystal structure based on the face-centered cubic cell is packed more efficiently than one based on the body-centered cubic cell, which is packed more efficiently than one based on the simple cubic cell.

One unit cell lies directly next to another throughout the crystal, so the particles centered on a lattice point are shared by adjacent unit cells. As you can see from Figure 11.18 (third row), in the three cubic unit cells, the atom at each corner is part of eight adjacent cells, so one-eighth of each atom belongs to each cell (bottom row). Since there are eight corners in a cube, each simple cubic cell contains $8 \times \frac{1}{8}$ atom = 1 atom. Similarly, the body-centered cubic unit cell contains one atom from the eight corners and one in the center, or two atoms; and the face-centered cubic unit cell contains four atoms, one from the eight corners and three from the half-atoms in the six faces.

Since the unit cell is the smallest portion of the lattice that maintains the overall spatial arrangement, it is also the smallest portion of the substance that maintains the overall chemical composition. Thus, in ionic compounds, the unit cell has the same cation/anion ratio as the empirical formula.

**A** Simple cubic

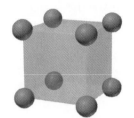

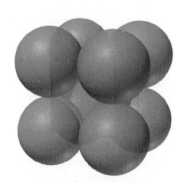

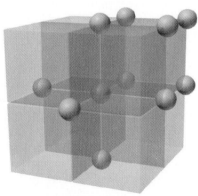

Coordination number = 6

$\frac{1}{8}$ atom
at 8 corners

Atoms /unit cell = $\frac{1}{8}$ × 8 = 1

**B** Body-centered cubic

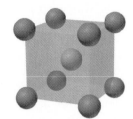

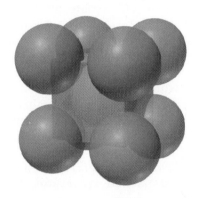

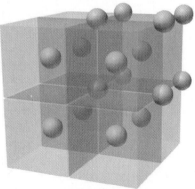

Coordination number = 8

 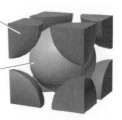

$\frac{1}{8}$ atom
at 8 corners

1 atom
in center

Atoms / unit cell = $(\frac{1}{8}$ × 8$) + 1 = 2$

**C** Face-centered cubic

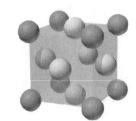

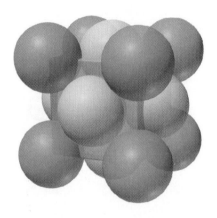

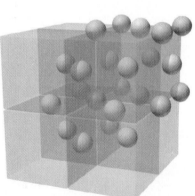

Coordination number = 12

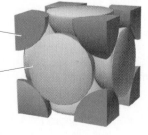

$\frac{1}{8}$ atom
at 8 corners

$\frac{1}{2}$ atom
at 6 faces

Atoms / unit cell = $(\frac{1}{8}$ × 8$) + (\frac{1}{2}$ × 6$) = 4$

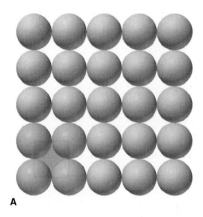

**A**

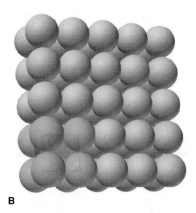

**B**

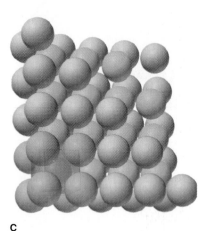

**C**

**FIGURE 11.19**

Packing of spheres in simple and body-centered cubic cells. **A,** In the first layer, each sphere is lined up next to another both horizontally and vertically; note the diamond-shaped spaces. **B,** If spheres in subsequent layers lie *directly* over those in the first, the packing arrangement is based on the simple cubic unit cell *(see lower left corner).* **C,** If spheres in subsequent layers lie in the diamond-shaped spaces of each preceding layer, the packing arrangement is based on the body-centered cubic unit cell *(see lower left corner).*

### The Packing Efficiency of Spheres

The unit cells found in nature result from the various ways atoms are packed together in crystalline solids, and these are similar to how macroscopic spheres—marbles, golf balls, fruit—are packed in shipping crates. For example, consider how oranges can be packed. Suppose you arrange the first layer as shown in Figure 11.19, *A.* If you place the oranges in subsequent layers directly above those in the first layer, you obtain an arrangement based on the simple cubic unit cell (Figure 11.19, *B*). By calculating the **packing efficiency** of this arrangement, that is, the percentage of the available volume occupied by spheres, you would find that only 52% of the available volume is occupied; 48% consists of spaces between the spheres (see Problem 11.104 at the end of the chapter). This is a very inefficient use of space, so fruit is not packed this way. Neither, it turns out, are atoms—the simple cubic unit cell is seen very rarely in nature.

Now let's try a more efficient arrangement (Figure 11.19, *C*). Rather than placing the second layer directly above the first, we place the spheres (colored differently for clarity) in the diamond-shaped hollows created by the first layer's spheres; then we can pack the third layer into the hollows of the second so that the third layer lines up vertically with the first. This arrangement produces a body-centered cubic unit cell of spheres, and its packing efficiency is 68%—much higher than that of the simple cubic structure. Several of the metallic elements, including chromium, iron, and all the Group 1A(1) elements, adopt the body-centered cubic arrangement in their crystal structures.

Spheres can be packed even more efficiently in two arrangements called **closest packed structures.** In the previous arrangements, the bottom layer has large diamond-shaped spaces between every four spheres (see Figure 11.19, *A*). Shifting the rows of the bottom layer slightly creates much smaller triangular spaces, over which we can place the spheres of the second layer. Figure 11.20, *A (top),* shows this arrangement, with the first layer labeled *a* (orange) and the second layer *b* (green).

Two different placements of the third layer are now possible. Notice that some spaces between spheres in layer *b* are orange because they lie above *spheres* in layer *a,* while others are white because they lie above *spaces* in layer *a.* If the third-layer spheres are placed over the orange spaces, they lie directly over spheres in layer *a,* and we obtain an *ababab…* layering pattern because every other layer is identically placed (Figure 11.20, *B*). This layering results in **hexagonal closest packing,** which is based on the *hexagonal unit cell.* On the other hand, if the third-layer spheres are placed over the white spaces, they lie over spaces in layer *a.* This placement differs from that in both layers *a* and *b,* so we obtain an *abcabcabc…* pattern (Figure 11.20, *C*). This results in **cubic closest packing,** which is based on the face-centered cubic cell.

The packing efficiency of both closest packed structures is 74%, and the coordination number of both is 12. There is no way to pack spheres of equal size more efficiently. Most metallic elements crystallize in either of these closest packed arrangements. Magnesium, titanium, and zinc are some elements that adopt the hexagonal structure; nickel, copper, and lead adopt the cubic structure, as do many other substances, such as frozen carbon dioxide, methane, and most noble gases. As you'll see shortly, many ionic solids utilize cubic closest packing structures, with the larger ions (usually the anion) occupying the packing sites and the smaller ions (usually the cation) lying in the holes between them.

In Sample Problem 11.3, we use the density of an element and the packing efficiency of its crystal structure to calculate its atomic radius. Variations

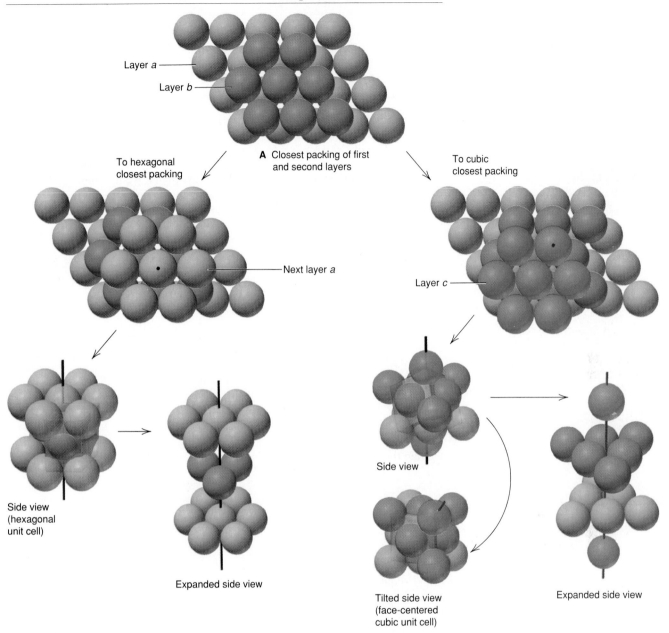

**A** Closest packing of first and second layers

Layer *a*

Layer *b*

To hexagonal closest packing

To cubic closest packing

Next layer *a*

Layer *c*

Side view (hexagonal unit cell)

Expanded side view

Side view

Tilted side view (face-centered cubic unit cell)

Expanded side view

**B**  Hexagonal closest packing (*abab...*)

**C**  Cubic closest packing (*abcabc...*)

**FIGURE 11.20**

**Packing arrangements for closest packed structures. A,** The closest possible packing of the first layer *(a; orange)* is obtained by shifting every other row in Figure 11.19, thus reducing the diamond-shaped spaces to smaller triangular spaces. The spheres of the second layer *(b, green)* are placed above these spaces; note the orange and white spaces that result. **B,** Hexagonal closest packing arises when the third layer *(a; orange)* is placed directly over the first, that is, over the orange spaces (*ababab...* pattern). Rotating the layers produces the side view, with the hexagonal unit cell shown, and the expanded side view. **C,** Cubic closest packing arises when the third layer *(c; purple)* covers the white spaces, thereby lying in a different position from the first and second layers (*abcabc...* pattern). Rotating this gives the side view, with a further tilt to show the face-centered unit cell on which this packing arrangement is based, and the expanded view.

of this approach are used to find the molar mass and as one way to determine Avogadro's number.

### Determining Atomic Radius from Crystal Structure

**Problem:** Barium, the largest nonradioactive alkaline earth metal, has a body-centered cubic unit cell and a density of 3.62 g/cm³. What is the atomic radius of barium? (Volume of a sphere: $V = \frac{4}{3}\pi r^3$.)

**Plan:** Since an atom is spherical, we can find its radius from its volume. From the density (mass/volume) and the molar mass (mass/mole), we find the volume/mole of Ba metal. Since it crystallizes in the body-centered cubic structure, 68% of this volume is occupied by the mole of Ba atoms. Dividing by Avogadro's number gives the volume of one Ba atom, from which we determine the radius.

**Solution:** Combining steps to find the volume of 1 mol Ba:

$$\text{Volume/mole of Ba metal} = \frac{1}{\text{density}} \times \mathcal{M} = \frac{1 \text{ cm}^3}{3.62 \text{ g Ba}} \times \frac{137.3 \text{ g Ba}}{1 \text{ mol Ba}}$$

$$= 37.9 \text{ cm}^3/\text{mol Ba}$$

Finding the volume of 1 mol Ba *atoms:*

$$\text{Volume/mole of Ba atoms} = \text{volume/mol Ba} \times \text{packing efficiency}$$

$$= 37.9 \text{ cm}^3/\text{mol Ba} \times 0.68 = 26 \text{ cm}^3/\text{mol Ba atoms}$$

Finding the volume of one Ba atom:

$$\text{Volume/Ba atom} = \frac{26 \text{ cm}^3}{1 \text{ mol Ba atoms}} \times \frac{1 \text{ mol Ba atoms}}{6.022 \times 10^{23} \text{ Ba atoms}}$$

$$= 4.3 \times 10^{-23} \text{ cm}^3/\text{Ba atom}$$

Finding the atomic radius of Ba from the volume of a sphere:

$$V \text{ of Ba atom} = \frac{4}{3}\pi r^3 \quad \text{and} \quad r^3 = \frac{3V}{4\pi}$$

So

$$r = \sqrt[3]{\frac{3V}{4\pi}} = \sqrt[3]{\frac{3(4.3 \times 10^{-23} \text{ cm}^3)}{4 \times 3.14}} = \mathbf{2.2 \times 10^{-8} \text{ cm}}$$

**Check:** The order of magnitude is correct for an atom ($\sim 10^{-8}$ cm $\approx 10^{-10}$ m). The actual value is $2.22 \times 10^{-8}$ cm (see Figure 8.11), so our answer seems correct.

**FOLLOW-UP PROBLEM 11.3**
Iron crystallizes in a body-centered cubic structure. The volume of one Fe atom is $8.38 \times 10^{-24}$ cm³, and the density of Fe is 7.874 g/cm³. Calculate an approximate value for Avogadro's number.

## Types and Properties of Crystalline Solids

The characteristics of the five most important types of solids—atomic, molecular, ionic, metallic, network covalent—are summarized in Table 11.5. Each type is defined by the nature of the particles making up the crystal and the interparticle forces. It might be worthwhile for you to review the bonding models we discussed in Chapter 9 because they clarify the differences in properties among the solids.

   **Atomic solids.** Individual atoms held together by dispersion forces form an **atomic solid.** The noble gases [Group 8A(18)] are the only examples of atomic solids, and their physical properties reflect the weakness of the interparticle forces. Their melting and boiling points and heats of vaporization and fusion are all very low, rising smoothly with increasing molar mass. Atomic solids crystallize in one of the two closest packing arrangements (Figure 11.21).

Density (g/cm³) of Ba

find reciprocal

Volume (cm³) per gram of Ba

$\mathcal{M}$ (g/mol)

Volume (cm³) per mole of Ba

packing efficiency

Volume (cm³) per mole of Ba atoms

Avogadro's number

Volume (cm³) per Ba atom

$V = \frac{4}{3}\pi r^3$

Radius (cm) of Ba atom

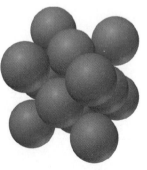

**FIGURE 11.21**
**Cubic closest packing of frozen argon.** Atomic solids are closest packed. Except for helium, the Group 8A(18) elements adopt cubic closest packing (face-centered cubic cell).

TABLE 11.5 **Characteristics of the Major Types of Crystalline Solids**

| TYPE OF SOLID | PARTICLES | FORCES BETWEEN PARTICLES | PROPERTIES OF SOLID | EXAMPLES |
|---|---|---|---|---|
| Atomic | Atoms | Dispersion | Soft, very low melting point, poor thermal and electrical conductors | Noble gases [Group 8A(18)] |
| Molecular | Molecules | Dispersion, dipole-dipole, dipole-induced dipole, hydrogen bonds | Fairly soft, low to moderately high melting point, poor thermal and electrical conductors | $CH_4$, $I_2$, table sugar ($C_{12}H_{22}O_{11}$) |
| Ionic | Positive and negative ions | Ion-ion attraction | Hard and brittle, high melting point, poor thermal and electrical conductors in solid | NaCl, Ca($NO_3$)$_2$ |
| Metallic | Atoms | Metallic bond | Soft to hard, low to very high melting point, excellent thermal and electrical conductors, malleable and ductile | Na, Fe, brass |
| Network covalent | Atoms | Covalent bond | Very hard, very high melting point, usually poor thermal and electrical conductors | Diamond (C), quartz ($SiO_2$) |

**Molecular solids.** In the many thousands of **molecular solids,** the lattice points are occupied by individual molecules. For example, iodine crystallizes in the face-centered cubic arrangement, as does methane, with the carbon of each molecule centered on the packing site (Figure 11.22). Various combinations of van der Waals forces, including dipole-dipole, dispersion, and H bonding, keep the molecules together in the crystal. As a result, molecular solids have a wide range of physical properties.

Table 11.6 shows the melting points of several types of molecular solids. In nonpolar substances, dispersion forces are primarily responsible for physical properties, and their strength increases with molar mass. Among polar molecules, dipole-dipole forces and, where possible, H bonding dominate. Except for those made of the simplest molecules, most molecular solids have much higher melting points than the atomic solids (noble gases). Nevertheless, because the intermolecular forces are still relatively weak, the melting points are much lower than those of ionic, network covalent, and metallic solids.

**Ionic solids.** In crystalline **ionic solids,** the unit cell contains two kinds of particles, cations and anions. As a result, interparticle forces are *much* stronger than the van der Waals forces in atomic or molecular solids. Despite

**FIGURE 11.22**
Cubic closest packing of frozen methane. In the solid state, methane adopts a face-centered cubic unit cell (cubic closest packing), with the C atom centered on the lattice point. Individual atoms are shown for only one $CH_4$ molecule.

TABLE 11.6 **Melting Points of Selected Molecular Solids**

| NAME (FORMULA) | MP (°C) | NAME (FORMULA) | MP (°C) |
|---|---|---|---|
| **Nonpolar** | | **Polar** | |
| *Compounds* | | | |
| | | Formaldehyde ($CH_2O$) | −92 |
| Butane ($C_4H_{10}$) | −138 | Ammonia ($NH_3$) | −77.7 |
| Boron trichloride ($BCl_3$) | −107 | Sulfur dioxide ($SO_2$) | −72.7 |
| | | Chloroform ($CHCl_3$) | −63.5 |
| Benzene ($C_6H_6$) | 5.5 | Nitric acid ($HNO_3$) | −42 |
| | | Water ($H_2O$) | 0.0 |
| *Elements* | | Acetic acid ($CH_3COOH$) | 16.6 |
| Oxygen ($O_2$) | −219 | | |
| Chlorine0 ($Cl_2$) | −101 | | |
| Phosphorus ($P_4$) | 44.1 | | |

**FIGURE 11.23**

**The sodium chloride structure. A,** In an expanded view, the sodium chloride structure can be pictured as resulting from the interpenetration of two face-centered cubic arrangements, in this case one of Na$^+$ cations and the other of Cl$^-$ anions. **B,** A space-filling view of the central overlapped portion in **A** shows the NaCl unit cell, which consists of four Cl$^-$ ions and four Na$^+$ ions.

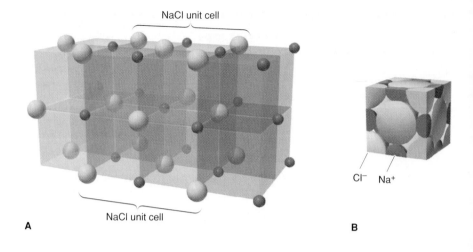

A

B

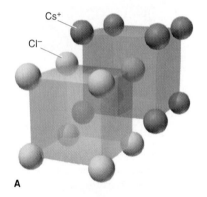

A

B

**FIGURE 11.24**

**The cesium chloride structure. A,** This structure can be viewed as consisting of interpenetrating simple cubic arrays of Cs$^+$ and Cl$^-$ ions. **B,** The space-filling unit cell shows a 1/1 ion ratio (one Cs$^+$ surrounded by 8 × $\frac{1}{8}$ Cl$^-$) and a coordination number of 8.

these differences, the structures of ionic solids are based on the same principle as in other solids—particles are arranged to maximize interparticle attractions: cations are surrounded by as many anions as possible, and vice versa. Typically, *one type of ion lies in the spaces (holes) formed by the packing of the other.*

Ionic compounds adopt several different crystal structures. Let's first consider three important crystal structures that all have a 1/1 ratio of ions. A very common arrangement is the *sodium chloride structure,* found in most of the alkali [Group 1A(1)] halides, the alkaline earth [Group 2A(2)] oxides and sulfides, several transition metal oxides, and silver chloride. Picture Cl$^-$ anions and Na$^+$ cations separately, each organized in a face-centered cubic (cubic closest packing) arrangement. The sodium chloride structure arises when these two arrays penetrate each other such that the cations end up in the holes between the anions (Figure 11.23, *A*). Thus, each cation is surrounded by six anions, and vice versa (coordination number = 6). Figure 11.23, *B*, is a space-filling depiction of the NaCl unit cell showing a face-centered cube of Cl$^-$ ions with Na$^+$ ions between them. Note the four Cl$^-$ [(8 × $\frac{1}{8}$) + (6 × $\frac{1}{2}$) = 4Cl$^-$] and four Na$^+$ [(12 × $\frac{1}{4}$) + 1 in the center = 4Na$^+$], giving a 1/1 stoichiometric ratio.

The only alkali halides that do not adopt the NaCl structure are CsCl, CsBr, and CsI, which crystallize in the *cesium chloride structure.* You can picture this structure as a result of simple cubic arrays of cations and anions penetrating each other (Figure 11.24, *A*). Each cation is surrounded by eight anions, and vice versa (coordination number = 8). Figure 11.24, *B*, shows that one cation and one anion (8 × $\frac{1}{8}$ = 1) make up the unit cell, in keeping with the empirical formula.

The very existence of the cesium chloride structure shows the importance of dispersion forces, even in the midst of more powerful ionic forces. This structure occurs when the largest alkali metal cation (Cs$^+$) combines with large halide anions (Cl$^-$, Br$^-$, and I$^-$). Since the ions are large, the interionic distance is large, so the ionic attraction is relatively weak. Nevertheless, the higher coordination number in the CsCl structure than in the NaCl structure (8 vs. 6) allows more interactions to occur. With this additional contact, the high polarizability of the large ions gives rise to strong dispersion forces, and these tip the balance in favor of the CsCl structure.

The *zinc blende (ZnS) structure* can be imagined as two face-centered arrays, one of Zn$^{2+}$ ions and the other of S$^{2-}$ ions, penetrating each other (Figure 11.25, *A*). Each ion is tetrahedrally surrounded by four ions of opposite

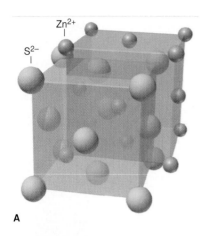

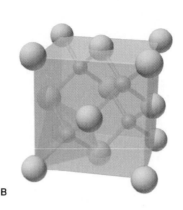

A                                    B

**FIGURE 11.25**
**The zinc blende structure. A,** Interpenetrating face-centered cubic cells of $Zn^{2+}$ and $S^{2-}$ ions adopt the zinc blende crystal structure. **B,** The translucent cube defines the unit cell, a face-centered cubic cell of four $[(8 \times \frac{1}{8}) + (6 \times \frac{1}{2})]$ $S^{2-}$ ions tetrahedrally surrounding four $Zn^{2+}$ ions inside the cell to give the 1/1 empirical formula. (The bond lines are present for clarity.)

charge (coordination number = 4), giving a ratio of 1/1 (Figure 11.25, *B*). Many other compounds, such as CuCl, BeS, and ZnO, adopt the zinc blende structure.

An important crystal structure for ionic salts with a cation/anion ratio of 1/2 is the *fluorite (CaF₂) structure.* It often appears with relatively large cations and relatively small anions. A good way to picture the $CaF_2$ unit cell is as a face-centered cubic array of $Ca^{2+}$ ions with $F^-$ ions occupying *all* eight available holes. This results in a $Ca^{2+}/F^-$ ratio of 4/8, or 1/2 (Figure 11.26). Other compounds that use the fluorite structure include $SrF_2$ and $BaCl_2$. The *antifluorite structure* is often seen in compounds with a cation/anion ratio of 2/1, in which the anion is relatively large (for example, $Li_2O$ and $K_2S$). In this structure, the cations occupy all eight holes formed by the cubic closest packing of the anions, just the opposite of the fluorite structure.

The properties of ionic solids are a direct consequence of the *fixed positions* of the ions and *very strong attractive forces* (high lattice energy). Thus, ionic solids have high melting points and low electrical conductivities. When a large amount of heat is supplied and the ions gain enough kinetic energy to break free of their positions, the solid melts and the mobile ions conduct a current. Ionic compounds are hard because only a strong external force can change the relative positions of many trillions of interacting ions (see Figure 9.8). If enough force *is* applied to move them, ions of like charge are brought next to each other, and their repulsions crack the crystal.

**Metallic solids: molecular orbital band theory.** Unlike the weak dispersion forces between the atoms in atomic solids, powerful metallic bonding forces hold individual atoms together in **metallic solids.** Most metallic elements crystallize in one of the two closest packed arrangements.

The properties of metals—high electrical and thermal conductivity, luster, and malleability—result from the nature of metallic bonding. Chapter 9 introduced a qualitative model of metallic bonding that pictures metal ions submerged in a "sea" of mobile valence electrons. Quantum mechanics offers an extension of molecular orbital (MO) theory, called **band theory,** that is more quantitative and therefore more useful. In addition to metallic properties, band theory helps explain the different characteristics of electrical conductivity in metals, metalloids, and nonmetals.

Recall that when two atoms form a diatomic molecule, their atomic orbitals combine to form an equal number of molecular orbitals (MOs). Let's consider lithium as an example (Figure 11.27). In diatomic lithium, $Li_2$, four valence orbitals (one 2s and three 2p) from each atom combine to form eight MOs, four bonding and four antibonding, that have a given range of energy and are spread over both atoms. (In Section 10.4, we showed only

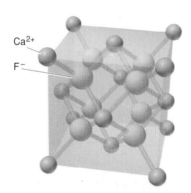

**FIGURE 11.26**
**The fluorite structure.** Calcium fluoride adopts the fluorite structure. You can picture the unit cell, which is defined by the translucent cube, as a face-centered cubic array of $Ca^{2+}$ (four $Ca^{2+}$ ions per unit cell) tetrahedrally surrounding eight $F^-$ ions.

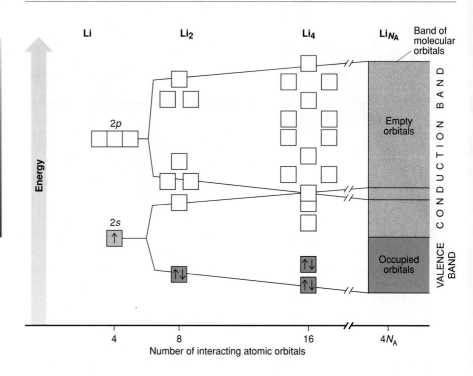

the MOs formed by $2s$ orbital overlap in Li.) If two more Li atoms combine, they form $Li_4$, a slightly larger aggregate, with 16 delocalized MOs distributed within the same energy range. As more Li atoms join the cluster, more MOs are created, their energy levels lying closer and closer together. Extending this process to a macroscopic sample of lithium metal, for example, one weighing 7 g (the molar mass), results in a mole of Li atoms ($Li_{N_A}$) combining to form an extremely large number of delocalized MOs, with energies *so closely spaced that they form a continuum, or band, of MOs.* It is almost as though the entire piece of metal were one enormous molecule of lithium.

The band model proposes that the lower energy MOs are occupied by the valence electrons and constitute the **valence band.** The empty MOs that are higher in energy constitute the **conduction band.** In lithium metal, the valence band is derived from the $2s$ orbitals in the lithium atoms, and the conduction band is derived mostly from an intermingling of the $2s$ and $2p$ orbitals. Whereas in $Li_2$ two valence electrons fill the lowest energy MO and leave the antibonding MO empty, in the piece of lithium metal a mole of valence electrons fills the valence band and leaves the conduction band empty.

*In metals, the valence and conduction bands are continuous,* which means that electrons can jump from the filled valence band to the unfilled conduction band with an infinitesimal amount of energy. In other words, the electrons are completely delocalized, *free to move throughout the piece of metal,* which explains why metals conduct electricity so well.

Metallic luster (shininess) is another effect of the continuous band of MO energy levels. With so many closely spaced levels available, electrons can absorb and release photons of many frequencies as they move between the valence and conduction bands. Malleability and thermal conductivity also result from the completely delocalized electron orbitals. Under an externally applied force, layers of positive metal ions simply move past each other, always protected from mutual repulsions by the presence of the delocalized

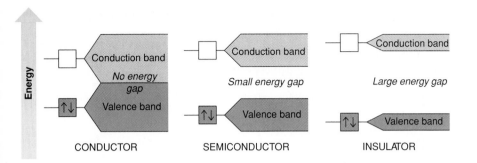

**FIGURE 11.28**
**Electrical conductivity in a conductor, a semiconductor, and an insulator.** Band theory explains differences in electrical conductivity in terms of the size of the energy gap between the valence and conduction bands. In conductors (metals), there is no gap. In semiconductors (many metalloids), electrons can jump the small gap if they are given energy, as when the sample is heated. In insulators (most nonmetals), electrons cannot jump the large energy gap.

electrons (see Figure 9.25). When a metal wire is heated, the metal ions move more rapidly and collide more frequently, transferring kinetic energy along the wire's length.

Large numbers of nonmetal or metalloid atoms can also combine to form bands of MOs. Metals conduct a current well (conductors), whereas most nonmetals do not (insulators), and the conductivity of metalloids lies somewhere in between (semiconductors). Band theory explains these differences in terms of the size of the energy gaps between the valence and conduction bands (Figure 11.28). The valence and conduction bands of a **conductor** are continuous, so electrons flow when even a tiny electrical potential difference is applied. When the temperature is raised, greater random motion of the atoms hinders free electron movement and decreases a metal's conductivity. In a **semiconductor,** a relatively small energy gap exists between the valence and conduction bands. Thermally excited electrons can cross the gap, allowing a small current to flow. Thus, in contrast to a conductor, the conductivity of a semiconductor *increases* when it is heated. In an **insulator,** the gap between the bands is too large for electrons to jump even when the substance is heated, so no current is observed.

Another type of electrical conductivity, called **superconductivity,** is currently generating great excitement. When metals conduct at ordinary temperatures, electron flow is restricted by collisions with atoms vibrating in their lattice sites. Such restricted flow appears as resistive heating and represents a loss of energy. To conduct with no energy loss—to superconduct—requires extreme cooling to minimize atom movement. This remarkable phenomenon has been observed in metals by cooling them to near absolute zero, which can be done only with liquid helium (boiling point = 4 K; price = $11/gal).

In 1986, all this changed with the synthesis of new ceramic oxides that superconduct near the boiling point of liquid nitrogen (77 K; price = $0.22/gal). Like metal conductors, superconducting oxides have no band gap. In the case of $YBa_2Cu_3O_7$ and other Cu-containing superconductors, x-ray analysis (discussed in the Tools of the Chemistry Laboratory essay on page 442) shows that the Cu ions in the oxide lattice are aligned, which may be associated with its superconducting ability. In 1989, oxides with Bi and Tl instead of Y and Ba were synthesized that could superconduct at 125 K. Engineering dreams for these materials include superconducting transmission lines that produce extremely inexpensive electrical power, electromagnets that levitate superfast railway trains (Figure 11.29), and brain-imaging methods that display remarkable clarity.

However, great difficulties must be overcome. For example, these superconducting oxides are very brittle, so they are fragile and difficult to handle.

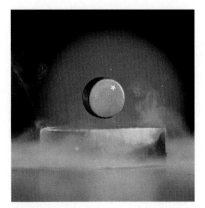

**FIGURE 11.29**
**The levitating power of a superconducting oxide.** A magnet remains suspended above a cooled high-temperature superconductor. Someday, a similar phenomenon may be used to levitate trains above their tracks for quiet, fast commuting.

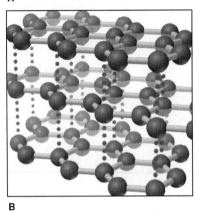

**A**

**B**

**FIGURE 11.30**

**The structures of diamond and graphite.**
**A,** In diamond, each carbon is connected to four others by localized covalent bonds in a tetrahedral arrangement. As a result, diamond is an electrical insulator and extremely hard. **B,** In graphite, the carbon atoms are covalently bound in sheets of linked hexagons. Delocalized $\pi$ electrons give high electrical conductivity in the plane of the sheets, and weak dispersion forces between the sheets make graphite soft.

◆ **A Diamond Film on Every Pot.** In the 1950s, synthetic diamonds were made slowly and expensively by exposing graphite to extreme temperatures (~1400°C) and pressures (50,000 atm). In the 1960s, a much cheaper method was discovered, that, in the 1990s, the method is being used to deposit thin films of diamond onto virtually any surface. In the process of chemical vapor deposition (CVD), a stream of methane breaks down at moderate temperatures (~600°C) and *low* pressures (~0.001 atm), and the carbon atoms deposit on the surface at rates of more than 100μm thickness per hour. Because of diamond's incredible hardness and high thermal conductivity, applications are endless: scratch proof cookware, watch crystals, hard discs, and eyeglasses; lifetime drill bits, ball bearings, and razor blades; high-temperature semiconductors. The list goes on and on.

**TABLE 11.7 Comparison of the Properties of Diamond and Graphite**

| PROPERTY | GRAPHITE | DIAMOND |
|---|---|---|
| Density (g/cm³) | 2.27 | 3.51 |
| Hardness | <1 (very soft) | 10 (hardest) |
| Melting point (K) | 4100 | 4100 |
| Color | Shiny black | Colorless transparent |
| Electrical conductivity | High (along sheet) | None |
| $\Delta H^0_{comb}$ (kJ/mol) | −393.5 | −395.4 |
| $\Delta H^0_f$ (kJ/mol) | 0 (standard state) | 1.90 |

A more fundamental problem is that when the superconducting oxide is warmed or placed in a strong magnetic field, the superconductivity often disappears and does not return again on cooling. Clearly, research into superconductivity will involve chemists, physicists, and engineers for many years to come. (We'll discuss these materials further, along with other exciting new ceramics, in Chapter 23.)

**Network covalent solids.** In the final type of crystalline solid, separate particles are not present. Strong covalent bonds link the atoms together throughout **network covalent solids.** All of these substances have extremely high melting and boiling points, but their conductivity and hardness depend on the details of their bonding. Diamond and graphite are two examples of network covalent solids. Although both consist entirely of carbon atoms, they have some strikingly different properties (Table 11.7).

Diamond crystallizes in a face-centered cubic unit cell, with each carbon atom tetrahedrally surrounded by four others in one virtually endless molecule (Figure 11.30, *A*). Strong, single bonds throughout the crystal make diamond the hardest substance known. ◆ The valence band is separated from the conduction band by a large energy gap, so diamond (like most network covalent solids) is an electrical insulator. Graphite occurs as stacked flat sheets of hexagonal carbon rings with a strong $\sigma$-bond framework and delocalized $\pi$ bonds, reminiscent of benzene (Figure 11.30, *B*). Whereas the $\pi$ electrons of benzene are delocalized over one ring, those of graphite are delocalized over the entire sheet. The mobile $\pi$ electrons allow graphite to conduct electricity, but only in the plane of the sheets. Graphite is a common electrode material and was once used for lightbulb filaments. The sheets interact via dispersion forces. Common impurities, such as $O_2$, that lodge between the sheets allow them to slide past each other easily, which explains why graphite is so soft.

By far the most important network covalent solids are the *silicates*. They exist in a variety of bonding patterns, but nearly all consist of extended arrays of covalently bonded silicon and oxygen atoms. We'll discuss silicates, which form the structure of clays, rocks, and many minerals, when we consider the chemistry of silicon in Chapter 13.

**The Importance of Crystal Defects**

The perfect regularity of a crystal is an ideal attained only when it is grown very slowly under controlled conditions. It is much more common for **crystal defects** to be present in the structure. When crystals form rapidly, planes can become misaligned, *vacancies* can be created where particles are misplaced or missing entirely, and dislocations can arise where foreign particles become lodged.

Although such defects often weaken a metal, in several important processes they are introduced intentionally to create new materials with improved properties such as greater strength. During the welding of two metals, vacancies form as surface atoms are vaporized. Atoms from lower rows rise to fill the gaps, in effect moving the vacancies deeper. The strength of the weld depends on the two types of metal atoms intermingling and filling each other's vacancies. Metal alloys have several kinds of defects, such as atoms of a second metal occupying some lattice sites of the first. Often, the alloy is harder than the pure metal; this is true of brass, an alloy of copper with zinc. One reason for increases in hardness, melting point, and similar properties is the second metal's contribution of additional valence electrons for metallic bonding.

Perhaps the most important use of crystal defects is the production of the semiconducting materials that have revolutionized modern electronics. The conductivity of semiconductors can be greatly enhanced by **doping,** adding small amounts of other elements to increase or decrease the number of valence electrons in the band of molecular orbitals. Pure silicon, which lies below carbon in Group 4A(14) and crystallizes in the diamond structure, is a poor conductor at room temperature because an energy gap separates its filled valence band from its conduction band (Figure 11.31, *A*). When the silicon crystal is doped with phosphorus [or another element from Group 5A(15)], P atoms lie at some of the lattice sites. The additional valence electron in each P atom is free to move into an empty orbital in the conduction band, bridging the energy gap and thereby increasing conductivity. This doping process creates an *n-type semiconductor,* so called because of the extra *n*egative charges (electrons) present (Figure 11.31, *B*).

When the silicon crystal is doped with gallium [or another Group 3A(13) element], Ga atoms occupy some sites (Figure 11.31, *C*). Since they have one fewer valence electron than the Si atoms, they effectively empty some of the orbitals in the valence band. Silicon electrons can migrate to these

**FIGURE 11.31**
**Crystal structures and band representations of doped semiconductors. A,** Pure silicon has the diamond crystal structure but acts as a semiconductor, the energy gap between its valence and conduction bands keeping conductivity low at room temperature. **B,** Doping silicon with phosphorus *(purple)* adds additional valence electrons, which are free to move through the crystal. They enter the lower portion of the conduction band, which is adjacent to higher energy empty orbitals, thereby increasing conductivity. **C,** Doping silicon with gallium *(orange)* removes electrons from the valence band and introduces a positive "hole." Nearby Si electrons can enter these empty orbitals, thereby increasing conductivity.

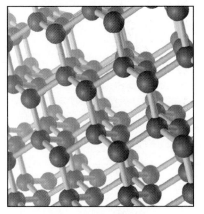

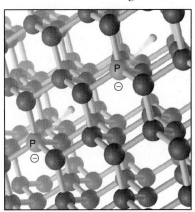

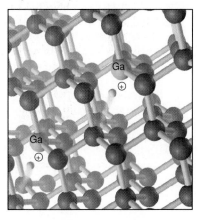

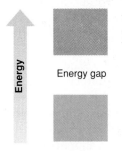
**A**  Pure silicon crystal

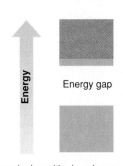
**B**  n-Type doping with phosphorus

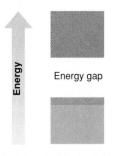
**C**  p-Type doping with gallium

| TOOLS OF THE CHEMISTRY LABORATORY | **X-Ray Diffraction Analysis and Scanning Tunneling Microscopy** |

In this chapter and in Chapter 10, we have discussed molecular shapes and crystal structures as if they have actually been seen. You may have been wondering how chemists know atomic radii, bond lengths, and bond angles when the objects are so incredibly minute. Various tools exist for peering into the molecular world and measuring its dimensions, but two of the most powerful are **x-ray diffraction analysis** and **scanning tunneling microscopy.**

### X-Ray Diffraction Analysis

This technique has been used for decades to determine crystal structures. In Chapter 7, we discussed wave diffraction to show how interference patterns of bright and dark regions appear when light passes through slits that are spaced at the distance of the light's wavelength (see Figure 7.5). In 1912, a graduate student of the Swiss physicist Max von Laue suggested that, since x-ray wavelengths are about the same size as the spaces between layers of particles in many solids, the layers might diffract x-rays. (Actually, the suggestion was made to test whether x-rays were particulate or wavelike.) X-ray diffraction was soon recognized as a powerful tool for determining structure.

Let's see how this technique is used to measure a key parameter in a crystal structure: the distance ($d$) between layers of atoms. Figure 11.A depicts a side view of two layers in a simplified lattice. Two waves impinge on the crystal at an angle $\theta$ and are diffracted at the same angle by adjacent layers. When the first wave strikes the top layer and the second strikes the next layer, the waves are "in phase" (peaks aligned with peaks and troughs with troughs). If they are still in phase after being diffracted, a bright spot appears on a nearby photographic plate. Note that this will occur only if the additional distance traveled by the second wave (DE + EF in the figure) is a whole number of wavelengths, $n\lambda$, where $n$ is an integer (1, 2, 3, and so on). From trigonometry, we find that

$$n\lambda = 2d \sin \theta$$

where $\theta$ is the known angle of incoming light, $\lambda$ is its known wavelength, and $d$ is the unknown distance between layers in the crystal. This relationship is the *Bragg equation,* named for W. H. Bragg and his son W. L. Bragg, who shared the Nobel Prize in 1915 for their work on crystal structure analysis.

Rotating the crystal changes the angle of incoming radiation and produces different sets of bright and dark spots and eventually produces a complete diffraction pattern that is used to determine the distances and angles within the lattice (Figure 11.B). The diffraction pattern is not an actual picture of the structure; the pattern must be analyzed mathematically to obtain the dimensions of the crystal. Modern x-ray diffraction equipment automatically rotates the crystal and measures thousands of dif-

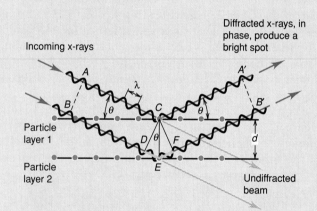

**FIGURE 11.A Diffraction of x-rays by crystal planes.** As in-phase x-ray beams *A* and *B* pass into a crystal at angle $\theta$, they are diffracted by interaction with the particles. Beam *B* travels the distance *DE + EF* farther than beam *A*. If this additional distance is equal to a whole number of wavelengths, the beams remain in phase and create a bright spot on a screen or photographic plate. From the pattern of bright spots and the Bragg equation, $n\lambda = 2d \sin \theta$, the distance *d* between layers of particles can be calculated.

empty electron spaces in the array, thereby increasing conductivity. This doping process creates a *p-type semiconductor,* so called because the empty orbitals act as *p*ositive holes. In contact with each other, an n-type and a p-type semiconductor form a *p-n junction.* When this junction is attached to a battery with the negative terminal connected to the n-type portion and the positive terminal to the p-type portion, current flows readily in the n-to-p direction, but not at all in the p-to-n direction. The ability of these minute junction devices to allow current flow in one direction only has permitted the development of electronic miniaturization, thus replacing the large vacuum tubes that performed this role previously. Our understanding of solids is based on our "seeing" the crystal structure; two techniques for doing so are described in the Tools of the Chemistry Laboratory essay.

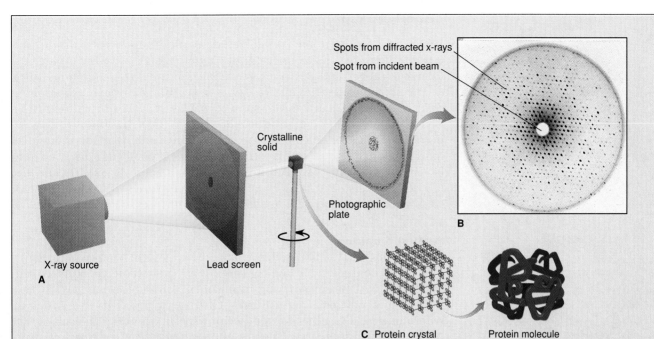

Spots from diffracted x-rays

Spot from incident beam

Crystalline solid

Photographic plate

X-ray source

Lead screen

**A**

**B**

**C** Protein crystal      Protein molecule

fractions, and a computer calculates the parameters of interest.

X-ray analysis is used to answer questions in many branches of chemistry, but its greatest impact has been in biochemistry. It has shown that DNA exists as a double helix, and it is currently helping us understand how proteins perform their vital functions.

## Scanning Tunneling Microscopy

This technique, a much more recent method than x-ray analysis, is used to observe surfaces on the atomic scale. It was invented in the early 1980s by Gerd Binnig and Heinrich Rohrer, two Swiss physicists who won the Nobel Prize for their work in 1986. The technique is based on the idea that an electron in an atom has a small probability of existing far from the nucleus, so given the right conditions, it can move ("tunnel") to end up closer to another atom.

In practice, the tunneling electrons create a current that can be used to image the atoms of an adjacent surface. An extremely sharp tungsten-tipped probe, the source of the tunneling electrons, is placed very close (about 0.5 nm) to the surface under study. A small potential is applied across this minute gap to increase the probability that the electrons will tunnel across it. The size of the gap is kept constant, by maintaining a constant tunneling current generated by the moving electrons. For this to occur, the probe must move tiny distances up and down, thus following the atomic contour of the surface. This movement is electronically monitored, and after many scans, a three-dimensional map of the surface is obtained. The method has revealed magnificent images of atoms and molecules coated on surfaces (Figure 11.C) and has already been used to study surface defects.

**FIGURE 11.B Formation of an x-ray diffraction pattern of the protein hemoglobin. A,** The crystalline sample is rotated to obtain many different angles of incoming and diffracted x-rays. **B,** A diffraction pattern of crystalline hemoglobin is obtained as a complex series of bright and dark spots. **C,** Computerized analysis relates the pattern to distances and angles within the crystal, providing data used to generate a picture of the hemoglobin molecule.

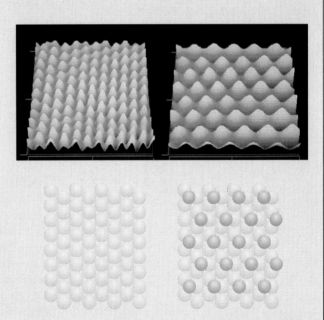

**FIGURE 11.C A scanning tunneling micrograph of metal layering.** A micrograph of the surface of gold is shown before *(left)* and after *(right)* deposition of a single layer of copper atoms. As shown in the illustrations *(below)*, the copper atoms occupy one-third of the spaces between gold atoms.

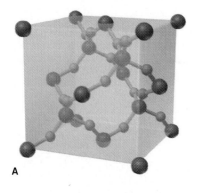

A

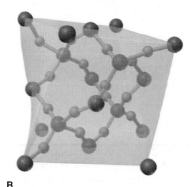

B

**FIGURE 11.32**

**Crystalline and amorphous silicon dioxide.**
**A,** Cristobalite, one of the many crystalline
forms of silica ($SiO_2$), shows the regularity of
cubic closest packing. **B,** A glass made of silica
is amorphous with a generally disordered
structure.

## Amorphous Solids

Amorphous solids are noncrystalline. Many can be described as having microcrystalline regions connected by large regions with many defects. Charcoal, rubber, and glass are some familiar examples of amorphous solids.

The process that forms quartz glass is typical of that for many amorphous solids. Crystalline quartz ($SiO_2$) has a cubic closest packed structure. After it is melted, the viscous liquid is cooled rapidly to prevent it from recrystallizing. The chains of silicon and oxygen atoms cannot orient themselves quickly enough into an orderly structure, so they solidify in a distorted jumble containing many gaps and unaligned rows (Figure 11.32). The absence of regularity in the structure confers some properties of a liquid; in fact, glasses are often referred to as supercooled liquids.

### Section Summary

The particles in crystalline solids lie at points that form a lattice of repeating unit cells. The three cubic unit cells are simple, body centered, and face centered. The most efficient packing arrangements are cubic closest packing and hexagonal closest packing. Atomic solids have a closest packed structure, with atoms held together by very weak dispersion forces. Molecular solids have molecules at the lattice points, often in the cubic closest packed structure. Their intermolecular forces (dispersion, dipole, H bond) and resulting physical properties vary greatly. Ionic solids often crystallize with one ion filling holes in a cubic closest packed structure of the other. Their high melting points, hardness, and low conductivity as a solid arise from strong ionic attractions. Most metals have a closest packed structure. Band theory proposes that orbitals in the metal atoms combine to form a continuum, or band, of molecular orbitals. Metals are electrical conductors because electrons move freely from the filled (valence band) to the empty (conduction band) portions of this energy continuum. Insulators have a large energy gap between the two portions; semiconductors have a small gap. The atoms of network covalent solids are covalently bonded throughout the crystal. Doping introduces crystal defects that increase the conductivity of semiconductors. Amorphous solids have irregular structures. Bond angles and distances in a crystal structure can be determined with x-ray diffraction analysis and scanning tunneling microscopy.

## 11.5   Quantitative Aspects of Changes in State

Some of the most spectacular, and familiar, changes in state occur in the weather. When it rains, water vapor condenses to a liquid, which changes back to a gas as puddles dry up. In the spring, snow melts and streams fill; in winter, water freezes and falls to earth. These remarkable transformations also take place whenever you make a pot of tea or a tray of ice cubes. In this section, we examine the heat absorbed or released in a phase change and the equilibrium nature of the process.

### Heat Involved in Phase Changes: A Kinetic-Molecular Approach

Let's apply the kinetic-molecular theory to describe the changes that occur as a sample of gaseous water in a closed container loses heat, condenses, and

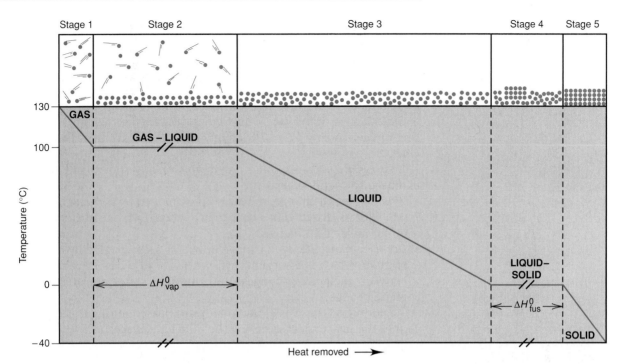

**FIGURE 11.33**

**A heating-cooling curve for the conversion of gaseous water to ice.** A plot of temperature vs. heat removed as gaseous water at 130°C changes to ice at −40°C occurs in five stages, with a molecular-level depiction shown for each stage. Stage 1: Gaseous water cools. Stage 2: Gaseous water condenses. Stage 3: Liquid water cools. Stage 4: Liquid water freezes. Stage 5: Solid water cools. The slopes of the lines in Stages 1, 3, and 5 reflect the magnitudes of the molar heat capacities of the phase. Although not drawn to scale, the line in Stage 2 is longer than the line in Stage 4 because $\Delta H^0_{vap}$ of water is greater than $\Delta H^0_{fus}$.

then freezes. The process is depicted in the **heating-cooling curve** shown in Figure 11.33, where the temperature is plotted as a function of heat released or absorbed by the system. We'll follow a 2.50-mol sample of water as it changes from gaseous water at 1 atm and 130°C to ice at −40°C. The process occurs as a smooth sequence of events, but for our purposes we can divide it into five heat-releasing (exothermic) stages corresponding to the five portions of the heating-cooling curve.

*Stage 1. Gaseous water cools.* Picture a collection of water molecules behaving like a typical gas: zooming around chaotically at a range of speeds, smashing into each other and the container walls. At a high enough temperature, the most probable speed, and thus the average kinetic energy ($E_k$), of the molecules is high enough to overcome the potential energy ($E_p$) of attraction among them. As the temperature falls, the average $E_k$ decreases and the molecules are increasingly affected by these attractions.

*Stage 2. Gaseous water condenses.* At the condensation point, the slowest of the molecules are near each other long enough for intermolecular attractions to form groups of molecules, which aggregate into microdroplets and then a bulk liquid.

While the state is changing from gas to liquid, *the temperature remains constant,* so the average $E_k$ is constant. Molecules in a gas move *farther* between collisions than those in a liquid, but their average *speed* is the same at a given temperature. Therefore, removing heat from the system must involve a decrease in the average $E_p$, as the molecules come closer together and attract each other more strongly, but not a change in the average $E_k$. In other words, at 100°C, gaseous water and liquid water have the same average $E_k$, but the liquid has lower $E_p$.

*Stage 3. Liquid water cools.* The molecules have now all condensed to the liquid state. The continued loss of heat appears as a decrease in temperature, that is, as a decrease in the most probable molecular speed and the average $E_k$. The temperature decreases as long as the sample remains liquid.

*Stage 4. Liquid water freezes.* At the freezing temperature of water, 0°C, the intermolecular attractions overcome the motion of the molecules past one another. Beginning with the slowest, the molecules lose $E_p$ and align themselves into the crystalline structure of ice. Molecular motion continues, but only about the lattice points. As they do during condensation, the temperature and average $E_k$ remain constant during freezing.

*Stage 5. Solid water cools.* With motion restricted to jiggling at each lattice site, further cooling to $-40°C$ merely reduces the average speed of this jiggling.

Two essential points stand out in this or any similar process, whether exothermic or endothermic:

- *Within a state,* a change in heat is accompanied by *a change in temperature,* which is associated with a change in average $E_k$ as the most probable *speed* of the molecules changes.
- *During a change of state,* a change in heat is accompanied by a *constant temperature,* which is associated with a change in $E_p$, as the average *distance* between molecules changes. Both physical states are present during a phase change.

According to Hess's law, the total heat released during this process is the sum of the heats released during the individual stages:

*Stage 1:* $H_2O(g)$ [130°C] $\rightarrow H_2O(g)$ [100°C]

In the first stage, the heat $(q)$ is the product of the number of moles $(n)$, the molar heat capacity of *gaseous water,* $C_{water(g)}$, and the temperature change, $\Delta T$ $(T_{final} - T_{initial})$:

$$q = n \times C_{water(g)} \times \Delta T = (2.50 \text{ mol}) (33.1 \text{ J/mol} \cdot °C) (100°C - 130°C)$$

$$= -2482 \text{ J} = -2.48 \text{ kJ}$$

The minus sign indicates that heat is released. (For purposes of canceling, the molar heat capacity includes these units, rather than J/mol·K. The magnitude of $C$ is not affected because the kelvin and Celsius degree are the same size.)

*Stage 2:* $H_2O(g)$ [100°C] $\rightarrow H_2O(l)$ [100°C]

In the second stage, temperature is constant and the phase changes. The heat released is the product of $n$ and the negative of the heat of vaporization $(-\Delta H_{vap}^0)$:

$$q = n(-\Delta H_{vap}^0) = (2.50 \text{ mol}) (-40.7 \text{ kJ/mol}) = -102 \text{ kJ}$$

*Stage 3:* $H_2O(l)$ [100°C] $\rightarrow H_2O(l)$ [0°C]

In the third stage, the heat released depends on $n$, the molar heat capacity of *liquid water,* and $\Delta T$:

$$q = n \times C_{water(l)} \times \Delta T = (2.50 \text{ mol}) (75.4 \text{ J/mol} \cdot °C) (0°C - 100°C) = -18.8 \text{ kJ}$$

*Stage 4:* $H_2O(l)$ [0°C] $\rightarrow H_2O(s)$ [0°C]

In the fourth stage, liquid $H_2O$ freezes, so $T$ is constant. The heat released is $n$ times the negative of the heat of fusion $(-\Delta H_{fus}^0)$:

$$q = n(-\Delta H_{fus}^0) = (2.50 \text{ mol}) (-6.02 \text{ kJ/mol}) = -15.0 \text{ kJ}$$

*Stage 5:* $H_2O(s)$ [0°C] $\rightarrow H_2O(s)$ [$-40°C$]

In the fifth stage, the heat released depends on $n$, the molar heat capacity of *solid water,* and $\Delta T$:

$$q = n \times C_{water(s)} \times \Delta T = (2.50 \text{ mol}) (37.6 \text{ J/mol} \cdot °C) (-40°C - 0°C) = -3.76 \text{ kJ}$$

The total heat released is the sum of $q$ values for Stages 1 to 5, or $-142$ kJ.

Two essential points stand out in this or any similar calculation:

- *Within a state*, the heat lost or gained depends on the number of moles, molar heat capacity, and change in temperature. Note that the *slopes* of the lines in Stages 1, 3, and 5 of Figure 11.33 correspond to the heat capacities of the phases.
- *During a change in state*, the heat lost or gained depends on the number of moles and the enthalpy of the phase change ($\Delta H^0_{vap}$ or $\Delta H^0_{fus}$). Temperature is constant. Note that the lines for Stages 2 and 4 in Figure 11.33 are broken (not drawn to scale), but their relative lengths indicate the greater magnitude of $\Delta H^0_{vap}$.

As in all thermochemical calculations, the sign of $q$ is positive when heat is absorbed (endothermic process) and negative when heat is released (exothermic process). (For similar calculations, see Problems 11.82, 11.83, and 11.111 at the end of the chapter.)

### The Equilibrium Nature of Phase Changes

In everyday experience, phase changes take place in open containers that lose or gain heat continuously—the outdoors, a pot on a stove, the freezer compartment of a refrigerator—so the change is not reversible. In a closed container under controlled conditions, however, *changes of state are reversible and reach equilibrium*, just as chemical changes do.

**Liquid-gas equilibria.** Picture an *open* flask of liquid at constant temperature and focus on the molecules at the surface. Within their range of molecular speeds, some are moving fast enough and in the right direction to overcome attractions and leave the surface; that is, they vaporize. Nearby molecules immediately fill the gap, and with energy supplied by the constant-temperature surroundings, the process continues until the entire liquid phase is gone.

Now picture starting with a *closed* flask at constant temperature and assume, for simplicity, that all the air above the liquid has been removed so that a vacuum exists (Figure 11.34, *A*). As before, some of the molecules at the surface have a high enough $E_k$ to vaporize. As the number of molecules in the vapor phase increases, the pressure of the vapor increases. At the same time, some of the molecules in the vapor that collide with the surface are attracted too strongly to leave the liquid; that is, they condense. Vaporization continues at a constant rate because the number of molecules that make up the surface is constant, but the rate of condensation slowly increases as the vapor becomes more populated and molecules return to the liquid more often. As condensation continues to offset vaporization, the increase in the pressure of the vapor slows. Eventually, the rate of condensation equals the rate of vaporization (Figure 11.34, *B*). At this point, and at any time thereafter, *the pressure of the vapor is constant at that temperature.* Macroscopically, the situation seems static, but at the molecular level, molecules are entering and leaving the liquid surface at equal rates. The system has reached a state of *dynamic equilibrium*:

$$\text{liquid} \rightleftharpoons \text{gas}$$

The entire process is depicted graphically in Figure 11.34, *C*.

The pressure exerted by the vapor in the flask at equilibrium is called the *equilibrium vapor pressure*, or just the **vapor pressure.** If we use a larger flask, more molecules are in the gas phase at equilibrium, but as long as both liquid and gas are present, the vapor pressure does not change. Suppose we disturb this equilibrium system by pumping out some of the va-

**A**

**B**

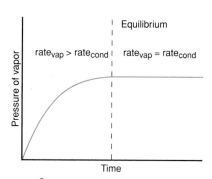

**C**

**FIGURE 11.34**

**Liquid-gas equilibrium. A,** In a closed flask at constant temperature with the air removed, the initial pressure is zero. As molecules enter the space above the liquid, the pressure of the vapor rises. **B,** With time, more molecules are in the vapor phase, so they collide with the surface and re-enter the liquid more often. At equilibrium, the same number of molecules leave as re-enter the liquid in a given time, so the pressure of the vapor reaches a constant value. **C,** Graphing pressure against time shows that the pressure of the vapor increases so long as the rate of vaporization is greater than the rate of condensation. At equilibrium, the pressure is constant because the rates are equal. The pressure at this point is the *vapor pressure* of the liquid at that temperature.

**FIGURE 11.35**

**The effect of temperature on the distribution of molecular speeds in a liquid.** With the temperature $T_1$ lower than $T_2$, the most probable molecular speed $u_1$ is less than $u_2$. (Note the similarity to Figure 5.18.) The fraction of molecules with enough energy to escape the liquid *(shaded area)* is greater at the higher temperature. The molecular views show that at the higher $T$, equilibrium is reached with more gas molecules in the same volume and thus at a higher vapor pressure.

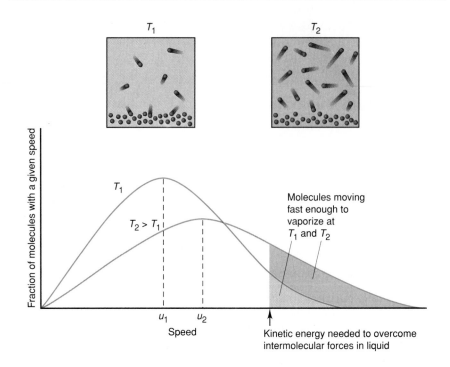

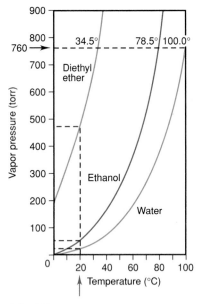

**FIGURE 11.36**

**Vapor pressure as a function of temperature.** The vapor pressure of three different liquids is plotted against temperature. At any given temperature (see the vertical line at 20°C), diethyl ether has the highest vapor pressure and water the lowest, because diethyl ether has the weakest intermolecular forces and water the strongest. The horizontal line at 760 torr shows the normal boiling points of the liquids, the temperature at which the vapor pressure equals atmospheric pressure at sea level.

por at constant temperature, thereby lowering the pressure. (If this were a cylinder with piston, we would lower the pressure by increasing the volume.) The rate of condensation temporarily falls below the rate of vaporization (the backward process occurs more slowly than the forward process). Fewer molecules enter the liquid than leave it, and the pressure rises until the condensation rate increases enough for equilibrium to be reached again. Similarly, if we disturb the system by pumping in additional vapor (or by decreasing the volume) at constant temperature, thereby raising the pressure, the rate of condensation temporarily exceeds the rate of vaporization. More molecules enter the liquid than leave it (the backward process is faster), until the pressure again reaches the equilibrium value. We conclude that *when a system at equilibrium is disturbed, it adjusts itself and eventually reattains a state of equilibrium.* This basic principle is involved in many physical and chemical changes, and you'll see it again in later chapters.

The vapor pressure of a substance depends on the temperature. Raising the temperature of a liquid *increases* the fraction of molecules moving fast enough to escape the liquid and *decreases* the fraction moving slowly enough to be recaptured (Figure 11.35). In general, *the higher the temperature, the higher is the vapor pressure.* The vapor pressure also depends on the intermolecular forces present. Since the average $E_k$ is the same for different substances at a given temperature, molecules with weaker intermolecular attractions vaporize more easily. In general, *the weaker the intermolecular forces, the higher is the vapor pressure.*

Figure 11.36 shows the vapor pressure as a function of temperature for three liquids. In each case, the curve rises more steeply as the temperature increases because the fraction of molecules with enough energy to escape the liquid increases steadily as the temperature goes up. At a given temperature, the substance with the *weakest* intermolecular forces has the *highest* vapor pressure: the dipole-dipole forces in diethyl ether are weaker than the H bonding in ethanol, which is weaker than the multiple H bonding in water.

**The relation between vapor pressure and temperature.** The non-linear relationship between vapor pressure and temperature shown in Figure 11.36 can be expressed mathematically in a linear form as

$$\ln P = \frac{-\Delta H_{vap}}{R}\left(\frac{1}{T}\right) + C$$

where $\ln P$ is the natural logarithm of the vapor pressure, $\Delta H_{vap}$ the heat of vaporization, $R$ the universal gas constant (8.31 J/mol · K), $T$ the absolute temperature, and $C$ a constant (not related to heat capacity). This is the **Clausius-Clapeyron equation,** which gives us a way of finding the heat of vaporization, the energy that must be supplied to vaporize a mole of molecules in the liquid state. A plot of $\ln P$ vs. $1/T$ has the form of a straight line, $y = mx + b$, where $y = \ln P$, $x = 1/T$, $m$ (slope)$= -\Delta H_{vap}/R$, and $b$ ($y$-axis intercept) $= C$. Figure 11.37 shows plots for two of the liquids in Figure 11.36. A two-point version of the equation allows a nongraphical determination of $\Delta H_{vap}$:

$$\ln \frac{P_2}{P_1} = \frac{-\Delta H_{vap}}{R}\left(\frac{1}{T_2} - \frac{1}{T_1}\right) \qquad \textbf{(11.1)}$$

If $\Delta H_{vap}$ is known, we can calculate the vapor pressure ($P_2$) at any other temperature ($T_2$) or the temperature at any other pressure.

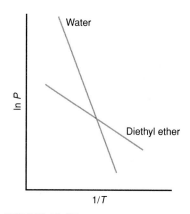

**FIGURE 11.37**
**A linear plot of the vapor pressure–temperature relationship.** The Clausius-Clapeyron equation gives a straight line when the natural logarithm of the vapor pressure (ln $P$) is plotted against the inverse of the absolute temperature (1/$T$). The slopes ($-\Delta H_{vap}/R$) allow determination of the heats of vaporization of the two liquids. Note that the slope is steeper for water because its $\Delta H_{vap}$ is greater.

SAMPLE PROBLEM 11.4 _____

### Using the Clausius-Clapeyron Equation

**Problem:** The vapor pressure of ethanol is 115 torr at 34.9°C. If $\Delta H_{vap}$ of $C_2H_5OH$ is 40.5 kJ/mol, calculate the temperature (in °C) when the vapor pressure is 760 torr.
**Plan:** We are given $\Delta H_{vap}$, $P_1$, $P_2$, and $T_1$, so we substitute in Equation 11.1 to solve for $T_2$. The value of $R$ here is 8.31 J/mol · K, so we must convert $T_1$ to K to obtain $T_2$, and then convert $T_2$ to °C.
**Solution:** Substituting the values into Equation 11.1 and solving for $T_2$:

$$\ln \frac{P_2}{P_1} = \frac{-\Delta H_{vap}}{R}\left(\frac{1}{T_2} - \frac{1}{T_1}\right)$$

$$T_1 = 34.9°C + 273.15 = 308.0 \text{ K}$$

$$\ln \frac{760 \text{ torr}}{115 \text{ torr}} = \left(-\frac{40.5 \times 10^3 \text{ J/mol}}{8.31 \text{ J/mol} \cdot \text{K}}\right)\left(\frac{1}{T_2} - \frac{1}{308.0 \text{ K}}\right)$$

$$1.89 = (-4.87 \times 10^3)\left[\frac{1}{T_2} - (3.247 \times 10^{-3})\right]$$

$$T_2 = 350 \text{ K}$$

Converting $T_2$ from K to °C:

$$T_2 = 350 \text{ K} - 273.15 = \textbf{77°C}$$

**Check:** Round off to check the math. The change is in the right direction: higher $P$ should occur at higher $T$. As we discuss next, a substance has a vapor pressure of 760 torr at its boiling point. Checking the *CRC Handbook of Chemistry and Physics* shows that the boiling point of ethanol is 78.5°C, very close to our answer.

FOLLOW-UP PROBLEM 11.4
At 34.1°C, the vapor pressure of water is 40.1 torr. What is the vapor pressure at 85.5°C? The $\Delta H_{vap}$ of water is 40.7 kJ/mol.

◆ **Superheating.** When a liquid reaches its boiling point, the bubbles need a site on which to form. If the container walls are smooth and particles of dust are absent, bubbles may not form and the liquid becomes *superheated:* its temperature rises above the boiling point. A superheated liquid can be dangerous to handle because boiling can erupt suddenly (bumping) and cause the liquid to spatter. To prevent superheating in the laboratory, a small piece of stone or ceramic, called a *boiling chip,* is added to the liquid before heating to provide a surface on which bubbles can form.

◆ **Cooking Under Low or High Pressure.** People who live or hike in mountainous regions must cook their meals under lower atmospheric pressure. Because the lower pressure results in a lower boiling point, the liquid does not become as hot, so the food takes *more* time to cook. On the other hand, in a pressure cooker, the pressure exceeds that of the atmosphere, so the temperature rises above the normal boiling point. As a result, the food becomes hotter and takes *less* time to cook.

◆ **Supercooling.** When a solid is melted and then cooled slowly, it may remain liquid at temperatures below the freezing point. The liquid becomes *supercooled* because the molecules are unable to orient themselves into the crystal structure. Like the superheated state, the supercooled state is unstable: a stir causes the liquid to solidify immediately. Supercooled thymol, a benzene derivative, is shown crystallizing in the photo.

**The relation between vapor pressure and boiling point.** In an *open* container, the atmosphere bears down on the liquid surface. As the temperature rises, molecules leave the surface more often and move more quickly throughout the entire liquid. At some temperature, the $E_k$ of the molecules is great enough for bubbles of vapor to form *in the interior,* and the liquid boils. At any lower temperature, the bubbles collapse as soon as they form because the atmospheric pressure is greater than the vapor pressure inside the bubbles. Thus, the **boiling point** is *the temperature at which the vapor pressure equals the external pressure,* in this case, that of the atmosphere. The applied heat is converted to $E_p$ as the molecules overcome attractions to enter the gas phase, so once boiling begins, the temperature remains constant until the liquid phase is gone. Boiling cannot occur in a closed container because the external pressure is the pressure exerted by the vapor on the liquid surface, which continually increases as the temperature does. Thus, *boiling is not an equilibrium phenomenon.* ◆

Of course, atmospheric pressure varies. At high elevations, a lower pressure is exerted on the liquid surface, so molecules in the interior need less kinetic energy to form bubbles. At low elevations, the opposite occurs. Thus, *the boiling point depends on the applied pressure.* The *normal boiling point* occurs at standard atmospheric pressure (760 torr, or 101.3 kPa). In Figure 11.36, the horizontal line shows the normal boiling points of the three liquids.

Since the boiling point is the temperature at which the vapor pressure equals the external pressure, we can also interpret the curves in Figure 11.36 as a plot of external pressure vs. boiling point. For instance, the $H_2O$ curve shows that water boils at 100°C at 760 torr (sea level), at 94°C at 610 torr (Denver, CO), and at about 72°C at 270 torr (top of Mt. Everest). ◆

**Solid-liquid equilibria.** If we could view a solid at the molecular level, we would see the particles continually vibrating at their sites in the crystal. As the temperature rises, the particles vibrate more violently, until some have enough kinetic energy to break free of their positions: melting has begun. As more molecules enter the liquid (molten) phase, some collide with the solid and become fixed in the structure again. Because the phases remain in contact, a dynamic equilibrium is established when the melting rate equals the freezing rate. The temperature at which this occurs is the **melting point;** it is the same temperature as the freezing point, differing only in the direction of the phase change. As with the boiling point, the temperature remains at the melting point as long as both phases are present. ◆

Because liquids and solids are nearly incompressible, a change in pressure has little effect on the rate of movement to or from the solid. Therefore, in contrast to the boiling point, the melting point is affected only slightly by pressure. A plot of pressure (*y* axis) vs. temperature (*x* axis) for the solid-liquid phase change of most substances is almost a straight vertical line, sloping slightly to the right at higher pressures: more kinetic energy (higher temperature) is needed to reach the melting point when pressure is applied because most liquids occupy slightly more space than the solid. (Water is the major exception, as you'll see.)

**Solid-gas equilibria.** Solids generally have much lower vapor pressures than liquids. Sublimation, the change of a solid directly into a gas, is much less familiar than vaporization because the conditions of pressure and temperature under which it occurs are not common ones for most substances. Some solids *do* have high enough vapor pressures to sublime at normal conditions, including dry ice (carbon dioxide), iodine (Figure 11.38), some moth repellents, and solid room deodorizers. A substance sublimes rather than melts because the combination of intermolecular attractions and pressure is not great enough to keep the particles near one another once they leave the solid state. The pressure vs. temperature plot for the solid-gas transition shows a large pressure effect; that is, it resembles the line for the liquid-gas transition in that it steadily curves upward at higher temperatures.

### Phase Diagrams: The Effect of Pressure and Temperature on Physical State

In order to describe completely the phase changes of a substance at various conditions of temperature and pressure, we construct a **phase diagram,** which combines the liquid-gas, solid-liquid, and solid-gas curves we just discussed. The shape of the phase diagram for $CO_2$ is typical for most substances (Figure 11.39, *A*). A phase diagram provides information about the physical state of a substance over a wide range of pressures and temperatures and contains the following features:

1. *Regions of the diagram.* Each region corresponds to one of the substance's phases. A particular phase is stable for any combination of pressure and temperature within its region; if any of the other phases is placed under those conditions, it will change to the stable phase. In general, the solid is stable at low temperature and high pressure, the gas at high temperature and low pressure, and the liquid at intermediate conditions.

2. *Lines between regions.* The lines separating the regions represent the phase-transition curves discussed earlier. Any point along a line shows the pressure and temperature at which the two phases exist in equilibrium.

3. *The critical point.* The liquid-gas line ends at the **critical point.** When a liquid is heated in a closed container, it expands and its density decreases as it vaporizes. At the same time, the density of the vapor above the liquid increases. At the *critical temperature* $(T_c)$, the two densities are equal and the phase boundary disappears. The average $E_k$ of the molecules at this point is

**FIGURE 11.38**
**Iodine subliming.** When solid iodine is heated at ordinary atmospheric pressure, it sublimes (changes directly from a solid into a gas). When the $I_2$ vapor comes in contact with a cold surface, it deposits $I_2$ crystals. Sublimation is a means of purification that, along with distillation, was probably discovered by the alchemists.

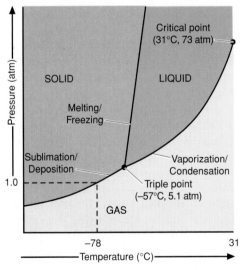

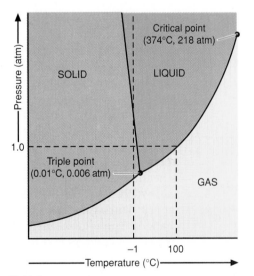

**A** $CO_2$                                                          **B** $H_2O$

**FIGURE 11.39 Phase diagrams for CO₂ and H₂O.** Each region depicts the temperatures and pressures under which the phase is stable. The curves between any two regions show the conditions at which the two phases exist in equilibrium. The critical point shows the conditions beyond which the liquid and gas phases no longer exist. At the triple point, the three phases exist in equilibrium. **A,** The phase diagram for $CO_2$ is typical of most substances in that the solid-liquid line slopes to the right with increasing pressure: the solid is *more* dense than the liquid. **B,** Water is one of the few substances whose solid-liquid line slopes to the left with increasing pressure: the solid is *less* dense than the liquid. (The slopes of the solid-liquid lines are exaggerated.)

◆ **The Remarkable Behavior of a Supercritical Fluid (SCF).** What sort of material lies beyond the familiar regions of liquid and gas? SCFs expand and contract like a gas but have the solvent properties of a liquid, properties that chemists can alter by controlling the density. Supercritical $CO_2$ has received the most attention so far from government and industry. It extracts nonpolar ingredients from complex mixtures, such as caffeine from coffee beans, nicotine from tobacco, and fats from potato and corn chips, while leaving taste and aroma ingredients behind, to produce healthier consumer products. Lower the pressure and it disperses immediately as a harmless gas. SCF $CO_2$ dissolves the fat from meat, along with pesticide and drug residues, which can then be quantified and monitored. In an unexpected finding of great potential use, SCF $H_2O$ was shown to dissolve nonpolar substances even though liquid water does not! Studies are underway for the large-scale removal of nonpolar organic toxins, such as PCBs, from industrial waste by their extraction into SCF $H_2O$; this step is followed by the introduction of $O_2$ gas, which oxidizes them to small, harmless molecules.

so high that the vapor cannot be condensed to a liquid regardless of the applied pressure. The pressure at this temperature is the *critical pressure* ($P_c$). The two most common gases in air, $N_2$ and $O_2$, have $T_c$ values far below room temperature: $O_2$ does not condense above $-119°C$, nor $N_2$ above $-147°C$, no matter what the pressure. Beyond the critical temperature, a *supercritical fluid* exists rather than the liquid and gaseous phases. ◆

4. *The triple point.* The three phase-transition curves meet at the **triple point**: the pressure and temperature at which three phases are in equilibrium. Phase diagrams for substances with several structural forms, such as sulfur, have more than one triple point. At the triple point in Figure 11.39, *A*, as strange as it sounds, $CO_2$ is subliming and depositing, melting and freezing, and vaporizing and condensing simultaneously!

A close look at the $CO_2$ phase diagram clarifies a common question: why doesn't dry ice (solid $CO_2$) melt? The answer is that it does, but not under ordinary conditions. The triple-point pressure for $CO_2$ is 5.1 atm; therefore, at normal atmospheric pressure, liquid $CO_2$ does not occur. By following the horizontal dashed line in Figure 11.39, *A*, you can see that when solid $CO_2$ is heated at 1.0 atm, it sublimes at $-78°C$ to gaseous $CO_2$ rather than melting. If our normal atmospheric pressure were, for example, 5.2 atm, liquid $CO_2$ would be commonplace.

The phase diagram for water differs somewhat from the general form, revealing an extremely important property (Figure 11.39, *B*). Unlike almost every other substance, solid water occupies more space than liquid water; that is, *water expands on freezing*. This behavior results from the unique crystal structure of ice, which we discuss in the next section. The smaller volume of the liquid means that ice *melts* under pressure. Therefore, *the solid-liquid line for water slopes to the left* rather than to the right: the higher the pressure, the lower the temperature required for water to freeze. If you follow the vertical dashed line at $-1°C$ to higher pressure, you see it cross the solid-liquid line, which means that at this temperature, ice melts due to a pressure increase only.

The triple point of water occurs at low pressure (0.006 atm). Therefore, when solid water is heated at 1.0 atm, the horizontal dashed line crosses the solid-liquid line (at 0°C, the normal melting point) and enters the liquid region. Thus, at ordinary pressures, ice melts rather than sublimes. As the temperature rises, the horizontal line crosses the liquid-gas curve (at 100°C, the normal boiling point) and enters the gas region.

### Section Summary

A heating-cooling curve depicts the change in temperature with heat loss or gain. Within a phase, temperature (and $E_k$) changes as heat is added or removed. During a phase change, temperature (and $E_k$) is constant but $E_p$ changes. The total heat change for the curve is calculated using Hess's law. In a closed container, equilibrium is established between the liquid and gas phases. Vapor pressure, the pressure of the gas at equilibrium, is related directly to temperature, and inversely to the strength of the intermolecular forces. The Clausius-Clapeyron equation allows calculation of $\Delta H_{vap}$, the vapor pressure, or the temperature, given the other two values. A liquid in an open container boils when its vapor pressure equals the external pressure. Solid-liquid equilibrium occurs at the melting point. Many solids sublime at low pressures and high temperatures. A phase diagram shows the phase that exists at a given pressure and temperature and the conditions at the critical point and the triple point of a substance.

# 11.6    The Uniqueness of Water

Water is so familiar that we take it for granted, but it has some of the most unusual properties of any substance. These properties, which are vital to our well-being and very existence, follow inevitably from the nature of the atoms that make up the water molecule.

As you know, the water molecule is composed of two types of nonmetal atoms, hydrogen and oxygen, that attain filled outer levels by sharing electrons in two single covalent bonds. With two bonding pairs and two lone pairs around the central O atom and a large electronegativity difference in each O—H bond, the $H_2O$ molecule is bent and highly polar. This arrangement is important because it allows each water molecule to engage in four H bonds with its neighbors—more than almost any other common substance (Figure 11.40). From these fundamental atomic and molecular properties emerges remarkable macroscopic behavior.

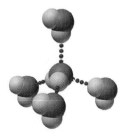

**FIGURE 11.40**
**The H-bonding ability of the water molecule.** Because it has two O—H bonds and two lone pairs, one $H_2O$ molecule can engage in as many as four H bonds, which are arranged tetrahedrally to surrounding $H_2O$ molecules.

## Solvent Properties of Water

The *great solvent power* of water is the result of its polarity and its exceptional H-bonding ability. Water dissolves ionic compounds through ion-dipole forces that separate the ions from the solid and keep them in solution. The cation becomes surrounded by water molecules with their negative ends facing the ion, whereas the anion has the positive ends facing it (see Figure 4.10). Water dissolves many polar non-ionic substances, such as ethanol ($CH_3CH_2OH$) and glucose ($C_6H_{12}O_6$), by forming H bonds to them. Water even dissolves nonpolar gases such as those in the atmosphere through dipole-induced dipole forces (see Figure 11.10). Chapter 12 highlights these solvent properties.

Because it can dissolve so many substances, water is the environmental and biological solvent, forming the complex solutions we know as oceans, rivers, lakes, and cell fluids. Aquatic animals could not survive without dissolved $O_2$, nor could aquatic plants survive without dissolved $CO_2$. The coral reefs that girdle the sea bottom in tropical latitudes are composed of carbonates made by marine animals from dissolved $CO_2$ and $HCO_3^-$ (Figure 11.41). Life itself is thought to have begun in a "primordial soup," a rich aqueous mixture of simple biomolecules from which emerged the larger molecules and their self-sustaining reactions that are characteristic of living things. From a chemical point of view, all organisms, from bacteria to humans, can be thought of as highly organized systems of membranes encompassing complex aqueous solutions.

**FIGURE 11.41**
**A coral reef.** Coral reef organisms form their carbonate-based skeletons from dissolved $CO_2$ and $HCO_3^-$. The reef is made from the skeletons of millions of these tiny animals.

## Thermal Properties of Water

Water has an exceptionally *high specific heat capacity*, higher than that of almost any other liquid. Recall from Section 6.3 that heat capacity is a measure of the energy absorbed by a substance as its temperature rises. Some of this energy increases the average molecular speed, some increases molecular vibration and rotation, and some is used to overcome intermolecular forces. This last component, in the form of strong H bonds, accounts for water's high specific heat capacity. The larger the heat capacity, the less the temperature rises when a substance absorbs a given amount of heat. Since oceans and seas cover 70% of the Earth's surface, heat from the sun causes relatively small changes in the Earth's temperature, allowing our planet to support life. On the waterless, airless moon, temperatures can range from 100°C to −150°C in the same day. Even on Earth's deserts, where a dense atmosphere tempers these extremes, day-night temperature differences of 70°F are common.

Numerous strong H bonds also give water an exceptionally *high heat of vaporization*. A quick calculation shows how essential this property is to our existence. When 1000 g water absorbs about 4 kJ of heat, its temperature rises 1°C. With a $\Delta H^0_{vap}$ of 2.3 kJ/g, however, less than 2 g water must evaporate to keep the temperature constant for the remaining 998 g. The average human adult has 40 kg body water and generates about 10,000 kJ of heat each day from metabolism. If this heat were used only to increase the average $E_k$ of body water, the temperature rise would mean immediate death. Instead, the heat is converted to $E_p$ as it breaks H bonds and evaporates sweat, resulting in a stable body temperature and minimal loss of fluid.

The sun's energy supplies the heat of vaporization for ocean water. The water vapor, formed in warm latitudes, moves through the atmosphere, and its potential energy is released as heat to warm cooler regions when the vapor condenses to rain. The enormous amount of energy involved in this cycling of water powers the storms and winds of the planet.

## Surface Properties of Water

The H bonds that give water its remarkable thermal properties are also responsible for its *high surface tension* and *high capillarity*. Except for some metals and molten salts, water has the highest surface tension of any liquid. This property is indispensable for surface aquatic life because it keeps plant debris resting on a pond surface, thereby providing shelter and nutrients for many fish, microorganisms, and insects. Water's high capillary action, a direct result of its high surface tension, is crucial to the survival of land plants. During dry periods, plant roots absorb deep ground water, which rises by capillary action through the tiny spaces between soil particles.

## The Density of Solid and Liquid Water

In the H bonds of water, each O atom becomes connected to as many as four other O atoms via four H atoms (see Figure 11.40). Continuing this pattern through many molecules in a fixed array gives ice its hexagonal, open structure (Figure 11.42, *A*). This organization explains the negative slope of the solid-liquid curve in the phase diagram for water: as pressure is applied, some H bonds break and bond angles distort. As a result, some water molecules enter the spaces, the crystal structure breaks down, and the sample liquefies. The symmetrical beauty of snowflakes (Figure 11.42, *B*) reflects this hexagonal molecular organization of water molecules in the solid state.

The large intermolecular spaces of ice give *the solid state of water a lower density than the liquid state*. When the surface of a lake freezes in winter, the ice floats on the liquid water below. If the solid were denser than the liquid, as is true for nearly every other substance, the surface of a lake would freeze and sink repeatedly until the entire lake was solid. Aquatic life would not survive from year to year.

The density of water changes in a complex way. When ice melts at 0°C, the tetrahedral arrangement around each oxygen atom becomes disordered, and the loosened molecules pack much more closely, filling spaces in the collapsing solid structure. As a result, water is most dense (1.000 g/mL) at around 4°C (3.98°C). With more heating, the density decreases through normal thermal expansion.

This change in density is vital for freshwater life. As lake water becomes colder in the fall and early winter, it becomes more dense *before* it freezes.

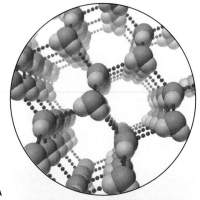

**A**

**B**

**FIGURE 11.42**

**The hexagonal structure of ice. A,** As a consequence of the geometric requirements of the H bonds in $H_2O$, ice has an open, hexagonally shaped crystal structure. Thus, when liquid water freezes, the volume increases. **B,** The delicate six-pointed beauty of a snowflake reflects the hexagonal crystal structure of ice.

Similarly, in spring, less dense ice thaws to form more dense water *before* the water expands and becomes less dense. During both these seasonal density changes, the top layer of water reaches the high-density point first and sinks. The next layer of water rises because it is slightly less dense, reaches 4°C, and likewise sinks. This convective movement distributes nutrients and dissolved oxygen.

The expansion and contraction of water also dramatically affect the land. When rain fills the crevices in rocks and boulders, and then freezes, great stress is applied, to be relieved only when the ice melts. In time, this repeated freeze-thaw stress forces the stones apart (Figure 11.43). Over eons of geologic change, this effect has helped to produce the sand and soil of the planet.

The properties of water and their far-reaching consequences illustrate the theme that runs throughout this book: the world we know is the stepwise, macroscopic outgrowth of the atomic world we seek to know. Figure 11.44 offers a fanciful summary of water's unique properties in the form of a tree whose "roots" (atomic properties) give foundation to the base of the "trunk" (molecular properties), which strengthens the next portion (polarity), from which grows another portion (H bonding). These two portions of the trunk "branch" out to the other properties of water and their global effects.

**FIGURE 11.43**
**The expansion and contraction of water.** As water freezes and thaws repeatedly, the stress can break rocks. Over geologic time, the process creates sand and soil.

**FIGURE 11.44**
**The macroscopic properties of water and their atomic and molecular "roots."**

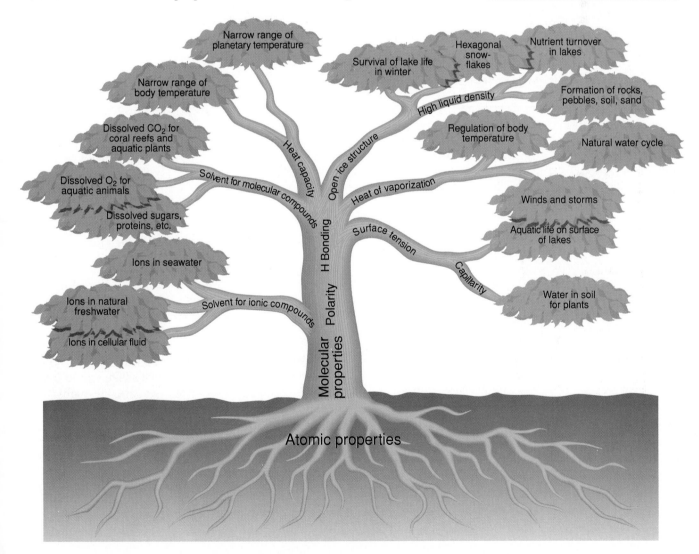

### Section Summary

The atomic properties of hydrogen and oxygen atoms result in the water molecule's bent shape, polarity, and H-bonding ability. These properties of water enable it to dissolve many ionic and polar molecular compounds. Water's H bonding results in high specific heat capacity and high heat of vaporization, which combine to give both the planet and its organisms a narrow temperature range. H bonds also confer high surface tension and capillarity, which are essential to plant and animal life. Water expands on freezing because of its H bonds, which form an open structure in ice. Seasonal density changes foster nutrient mixing in lakes. Seasonal freeze-thaw stress on rocks results eventually in soil formation.

### Chapter Perspective

*Our central focus—macroscopic behavior resulting from molecular behavior—became still clearer in this chapter, as we built on earlier ideas of bonding, molecular shape, and polarity to understand the properties of liquids and solids. In Chapter 12, we'll find many parallels to the key ideas in this one: the same intermolecular forces create solutions, temperature influences solubility as it does vapor pressure, and the equilibrium state also underlies the properties of solutions.*

## For Review and Reference

### Key Terms

intermolecular forces
phase change

**SECTION 11.1**
condensation
vaporization
freezing
melting (fusion)
heat of vaporization
 $(\Delta H^0_{vap})$
heat of fusion $(\Delta H^0_{fus})$
sublimation
deposition
heat of sublimation
 $(\Delta H^0_{subl})$

**SECTION 11.2**
van der Waals radius
ion-dipole force

dipole-dipole force
hydrogen bond (H bond)
polarizability
ion-induced dipole force
dipole-induced dipole
 force
dispersion (London) force

**SECTION 11.3**
surface tension
capillarity
viscosity

**SECTION 11.4**
crystalline solid
amorphous solid
lattice
unit cell
coordination number

simple cubic unit cell
body-centered cubic unit
 cell
face-centered cubic unit
 cell
packing efficiency
closest packed structure
hexagonal closest packing
cubic closest packing
atomic solid
molecular solid
ionic solid
metallic solid
band theory
valence band
conduction band
conductor
semiconductor
insulator

superconductivity
network covalent solid
x-ray diffraction analysis
scanning tunneling
 microscopy
crystal defect
doping

**SECTION 11.5**
heating-cooling curve
vapor pressure
Clausius-Clapeyron
 equation
boiling point
melting point
phase diagram
critical point
triple point

### Key Equations and Relationships

**11.1** Using the vapor pressure at one temperature to find the vapor pressure at another temperature (two-point form of the Clausius-Clapeyron equation) (p. 449):

$$\ln \frac{P_2}{P_1} = \frac{-\Delta H_{vap}}{R} \left( \frac{1}{T_2} - \frac{1}{T_1} \right)$$

## Answers to Follow-up Problems

**11.1** (a)

(b)

(c) No H bonds

**11.2** (a) Dipole-dipole, dispersion: $CH_3Br$
(b) H bonds, dipole-dipole, dispersion: $CH_3CH_2OH$
(c) Dispersion: $C_3H_8$

**11.3** Avogadro's no. $= \dfrac{1 \text{ cm}^3}{7.874 \text{ g Fe}} \times \dfrac{55.85 \text{ g Fe}}{1 \text{ mol Fe}} \times 0.68$

$$\times \dfrac{1 \text{ Fe atom}}{8.38 \times 10^{-24} \text{ cm}^3}$$

$$= 5.8 \times 10^{23} \text{ Fe atoms/mol Fe}$$

**11.4** $\ln \dfrac{P_2}{P_1} = \left( \dfrac{-40.7 \times 10^3 \text{ J/mol}}{8.31 \text{ J/mol} \cdot \text{K}} \right)$

$$\left( \dfrac{1}{273.15 + 85.5 \text{ K}} - \dfrac{1}{273.15 + 34.1 \text{ K}} \right)$$

$$= (-4.90 \times 10^3 \text{ K})(-4.67 \times 10^{-4} \text{ K}^{-1}) = 2.29$$

$\dfrac{P_2}{P_1} = 9.87$; thus, $P_2 = 40.1 \text{ torr} \times 9.87 = 396 \text{ torr}$

## Sample Problem Titles

**11.1** Drawing Hydrogen Bonds Between Molecules (p. 421)
**11.2** Predicting the Type and Relative Strength of Intermolecular Forces (p. 424)
**11.3** Determining Atomic Radius from Crystal Structure (p. 434)
**11.4** Using the Clausius-Clapeyron Equation (p. 449)

# Problems

Problems with a green number are answered at the back of the text. Most sections include three categories of problems separated by a green rule: concept review questions, *paired* skill-building exercises, and problems in a relevant context.

## An Overview of Physical States and Phase Changes

**11.1** How does the energy of attraction between particles compare with their energy of motion (a) in a gas; (b) in a solid?
**11.2** What types of forces, intramolecular or intermolecular,
(a) Prevent ice cubes from adopting the shape of their container?
(b) Are overcome when ice melts?
(c) Are overcome when liquid water is vaporized?
(d) Are overcome when water vapor is converted to hydrogen gas and oxygen gas?
**11.3** (a) Why are gases more easily compressed than liquids?
(b) Why do liquids have a greater ability to flow than solids?
**11.4** (a) Why is the heat of fusion ($\Delta H_{fus}$) of a substance smaller than its heat of vaporization ($\Delta H_{vap}$)?
(b) Why is the heat of sublimation ($\Delta H_{subl}$) of a substance greater than its $\Delta H_{vap}$?
(c) At a given temperature and pressure, how does the magnitude of the heat of vaporization of a substance compare with its heat of condensation?

**11.5** Which of the following forces are intramolecular and which intermolecular?
(a) Those preventing oil from evaporating at room temperature
(b) Those preventing butter from melting in a refrigerator
(c) Those allowing silver to tarnish
(d) Those preventing $O_2$ and $N_2$ in air from reacting to form NO
**11.6** Which of the following forces are intramolecular and which intermolecular?
(a) Those allowing fog to form on a cool, humid evening
(b) Those allowing water to form when $H_2$ is sparked
(c) Those causing liquid benzene to crystallize when cooled
(d) Those responsible for the low boiling point of hexane

**11.7** Name the phase change in each of these events:
(a) Dew appears on a lawn in the morning.
(b) Icicles change into liquid water.
(c) Wet clothes dry on a summer day.
**11.8** Name the phase change in each of these events:
(a) A diamond film forms in a vacuum from gaseous carbon atoms.
(b) Mothballs placed in a bureau drawer disappear over time.
(c) Molten iron from a blast furnace is cast into ingots ("pigs").

**11.9** Liquid propane, a widely used fuel, is produced by compressing gaseous propane at 20°C. During the process, approximately 15 kJ energy is released for each mole of gas liquefied. Where does this energy come from?

**11.10** In many buried, iron-containing rocks, the iron is present as greenish $Fe^{2+}$. When the rock is mined and broken into smaller pieces, the outer few millimeters of the pieces display the brown color of $Fe^{3+}$, while the center remains greenish. What property of solids prevents the change of $Fe^{2+}$ to $Fe^{3+}$?

**11.11** Many heat-sensitive and oxygen-sensitive solids, such as camphor, are purified by warming under vacuum. The solid vaporizes directly, and the vapor crystallizes on a cool surface. What phase changes are involved in the purification of camphor?

## Types of Intermolecular Forces
(Sample Problems 11.1 and 11.2)

**11.12** Why are covalent bonds typically much stronger than intermolecular forces?

**11.13** Given that dipole-dipole forces involve neutral molecules, what gives rise to the forces?

**11.14** Both oxygen and selenium are members of Group 6A(16). Water molecules form H bonds, but $H_2Se$ molecules do not. Explain.

**11.15** In solid $I_2$, is the distance between the I nuclei of one $I_2$ molecule longer or shorter than the distance between the I nuclei of adjacent $I_2$ molecules? Explain.

**11.16** Polar molecules exhibit dipole-dipole forces. Do they also exhibit dispersion forces? Explain.

**11.17** Distinguish between the terms *polarizability* and *polarity.* How does each influence intermolecular forces?

**11.18** How can one nonpolar molecule induce a dipole in a nearby nonpolar molecule?

**11.19** What is the strongest interparticle force that occurs in samples of each of the following substances? (a) $CH_3OH$; (b) $CCl_4$; (c) NaI; (d) $H_3PO_4$; (e) $SO_2$

**11.20** What is the strongest interparticle force that occurs in samples of each of the following substances? (a) $CH_3Br$; (b) $CH_3CH_3$; (c) $NH_3$; (d) Kr; (e) BrF

**11.21** Which member of each of the following pairs of compounds forms intermolecular H bonds? Draw the H-bonded structures in each case:
(a) $CH_3CHCH_3$ or $CH_3SCH_3$    (b) HF or HBr
     |
     OH

**11.22** Which member of each of the following pairs of compounds forms intermolecular H bonds? Draw the H-bonded structures in each case:
(a) $(CH_3)_2NH$ or $(CH_3)_3N$
(b) $HOCH_2CH_2OH$ or $FCH_2CH_2F.$

**11.23** Draw each of the following H-bonded structures: (a) $(HF)_2$ dimer; (b) $(CH_3OH)_4$ cyclic.

**11.24** Draw each of the following H-bonded structures: (a) $(HF)_6$ cyclic; (b) $(H_2O)_5$ tetrahedral.

**11.25** Which member of each of the following pairs would you expect to have the greater polarizability? Explain.
(a) $Br^-$ or $I^-$    (b) $CH_2{=}CH_2$ or $CH_3{-}CH_3$
(c) $H_2O$ or $H_2Se$

**11.26** Which member of each of the following pairs would you expect to have the greater polarizability? Explain.
(a) $Ca^{2+}$ or Ca    (b) $CH_3CH_3$ or $CH_3CH_2CH_3$
(c) $CCl_4$ or $CF_4$

**11.27** Which member of each of the following pairs would you predict has the *higher* boiling point? Explain.
(a) LiCl or HCl    (b) $NH_3$ or $PH_3$    (c) Xe or $I_2$
(d) $CH_3CH_2OH$ or $CH_3CH_2CH_3$    (e) NO or $N_2$

**11.28** Which member of each of the following pairs would you predict has the *lower* boiling point? Explain.
(a) $CH_3CH_2CH_2CH_3$ or $CH_2{-}CH_2$
                                                             |        |
                                                         $CH_2{-}CH_2$
(b) NaBr or $PBr_3$        (c) $H_2O$ or HBr
(d) $CH_3OH$ or $CH_3CH_3$    (e) FNO or ClNO

**11.29** For pairs of molecules in the gas phase, average H-bond dissociation energies are 17 kJ/mol for $NH_3$, 22 kJ/mol for $H_2O$, and 29 kJ/mol for HF. Explain this general increase in H-bond strength.

**11.30** Two crystalline hydrates of ammonia, $NH_3 \cdot H_2O$ and $NH_3 \cdot \frac{1}{2}H_2O$, have been isolated at low temperature. Each consists of chains of $H_2O$ molecules cross-linked by $NH_3$ molecules into a three-dimensional network. What force(s) holds the $H_2O$ chains together? What force(s) keeps the cross-linking intact?

**11.31** Why does motor oil have a high boiling point, even though dispersion forces are the only intermolecular forces present?

**11.32** Why does the antifreeze ingredient ethylene glycol ($HOCH_2CH_2OH$; $\mathcal{M} = 62.07$ g/mol) have a boiling point of 197.6°C, whereas propanol ($CH_3CH_2CH_2OH$; $\mathcal{M} = 60.09$ g/mol), a compound with a similar molar mass, has a boiling point of only 97.4°C?

## Properties of the Liquid State

**11.33** Before the phenomenon of surface tension was understood, physicists described the surface of water as being covered with a "skin." What causes this skin-like behavior?

**11.34** Why does an aqueous solution of ethanol ($CH_3CH_2OH$) have a lower surface tension than water?

**11.35** Surface tension values have units of $J/m^2$. Why are units of energy per area used?

**11.36** Does the *strength* of the intermolecular forces in a liquid change as the liquid is heated? Explain. Why does liquid viscosity decrease with rising temperature?

**11.37** Rank the following compounds in order of *increasing* surface tension at a given temperature, and explain your ranking:
(a) $CH_3CH_2CH_2OH$   (b) $HOCH_2CH(OH)CH_2OH$
(c) $HOCH_2CH_2OH$

**11.38** Rank the following compounds in order of *decreasing* surface tension at a given temperature, and explain your ranking:
(a) $CH_3OH$   (b) $CH_3CH_3$   (c) $H_2C=O$

**11.39** Rank the compounds in Problem 11.37 in order of *increasing* viscosity at a given temperature, and explain your ranking.

**11.40** Rank the compounds in Problem 11.38 in order of *decreasing* viscosity at a given temperature, and explain your ranking.

**11.41** Are the same cohesive and adhesive forces involved when a paper towel absorbs apple juice as when it absorbs cooking oil? Explain.

**11.42** Viscosity-enhancing additives increase the viscosity of a motor oil as the temperature increases. How do they work?

**11.43** Does boiling thin maple sap to make thick maple syrup violate the general observation that liquid viscosity decreases with heating? Explain.

**11.44** Pentanol ($C_5H_{11}OH$; $\mathcal{M} = 88.15$ g/mol) has nearly the same molar mass as hexane ($C_6H_{14}$; $\mathcal{M} = 86.17$ g/mol) but is more than 12 times as viscous at 20°C. Explain.

**11.45** A chemist studying the viscosity of molten ionic compounds collects data for cesium chloride (melting point = 645°C), but a computer error jumbles the values. Assuming molten CsCl behaves like other liquids, arrange the following values correctly:

| T (°C) | VISCOSITY (N·s/m²) |
|--------|--------------------|
| 650.8  | $1.432 \times 10^{-3}$ |
| 661.3  | $1.371 \times 10^{-3}$ |
| 674.9  | $1.548 \times 10^{-3}$ |
| 688.3  | $1.281 \times 10^{-3}$ |
| 711.0  | $1.494 \times 10^{-3}$ |

## Properties of the Solid State
(Sample Problem 11.3)

**11.46** What is the difference between an amorphous solid and a crystalline solid on the macroscopic and molecular levels? Give an example of each.

**11.47** How is the unit cell related to the lattice of a crystalline solid?

**11.48** How many atoms are contained in the simple, the body-centered, and the face-centered cubic unit cells? Explain how you obtained the values.

**11.49** What specific difference in the positioning of spheres gives a crystal structure based on a face-centered cubic unit cell less empty space than one based on the body-centered cubic unit cell?

**11.50** Both solid Kr and solid Cu consist of individual atoms. Why do their physical properties differ so much?

**11.51** Is it possible for a salt of formula $AB_3$ to have a face-centered cubic lattice of anions with cations in all the eight available holes? Explain.

**11.52** What property of x-rays makes them useful for investigating crystal structures?

**11.53** What is the energy gap in band theory? How does its size compare in conductors, superconductors, semiconductors, and insulators?

**11.54** Use band theory to explain why electrical conductivity increases with atomic size for elements in the same group.

**11.55** Predict the effect (if any) of an increase in temperature on the electrical conductivity of (a) a conductor; (b) a semiconductor; (c) an insulator.

**11.56** The addition of certain impurities to a crystalline semiconductor may increase its conductivity. Explain with regard to an n-type and a p-type semiconductor.

**11.57** For the metals listed, the number of atoms per unit cell is given in parentheses. What type of cubic lattice does each metal form? (a) Po (1); (b) Fe (2); (c) Ag (4)

**11.58** For the metals listed, the number of atoms per unit cell is given in parentheses. What type of cubic lattice does each metal form? (a) Ni (4); (b) Cr (2); (c) Ca (4)

**11.59** Of the five major types of crystalline solid, which would you expect each of the following to form, and why? (a) Sn; (b) Si; (c) Xe; (d) cholesterol ($C_{27}H_{45}OH$); (e) KCl; (f) BN

**11.60** Of the five major types of crystalline solid, which would you expect each of the following to form, and why? (a) Ni; (b) $F_2$; (c) $CH_3OH$; (d) SiC; (e) $Na_2SO_4$; (f) $SF_6$

**11.61** Zinc oxide adopts the zinc blende crystal structure. How many $Zn^{2+}$ ions are in the ZnO unit cell?

**11.62** Calcium sulfide adopts the rock salt (NaCl) crystal structure. How many $S^{2-}$ ions are in the CaS unit cell?

**11.63** Zinc selenide (ZnSe) crystallizes in the zinc blende structure and has a density of 5.42 g/cm³.
(a) How many Zn and Se particles are in each unit cell?
(b) What is the mass of a unit cell?
(c) What is the volume of a unit cell?
(d) What is the unit cell edge length?

**11.64** An element crystallizes in a face-centered cubic lattice and has a density of 1.45 g/cm³. The edge of its unit cell is $4.52 \times 10^{-8}$ cm.
(a) How many atoms are in each unit cell?
(b) What is the volume of a unit cell?
(c) What is the mass of a unit cell?
(d) Calculate an approximate atomic mass for the element.

**11.65** Predict the effect (if any) of an increase in temperature on the electrical conductivity of (a) antimony, Sb; (b) tellurium, Te; (c) bismuth, Bi.

**11.66** Predict the effect (if any) of a decrease in temperature on the electrical conductivity of (a) silicon, Si; (b) lead, Pb; (c) germanium, Ge.

**11.67** Classify each of the following as a conductor, insulator, n-type semiconductor, or p-type semiconductor: (a) phosphorus; (b) germanium doped with phosphorus; (c) mercury.

**11.68** Classify each of the following as a conductor, insulator, n-type semiconductor, or p-type semiconductor: (a) silicon doped with indium; (b) carbon (graphite); (c) sulfur.

**11.69** Polonium, the largest member of Group 6A(16), is a rare radioactive metal that is the only element with a crystal structure based on the simple cubic unit cell. Given a density of 9.142 g/cm³, calculate an approximate atomic radius for polonium. (*Note:* the spheres touch along the unit cell edge.)

**11.70** The coinage metals—copper, silver, and gold—crystallize in a cubic closest packed structure. Use the density of copper (8.95 g/cm³) and its molar mass (63.55 g/mol) to calculate an approximate atomic radius for copper. (*Note:* the spheres touch diagonally along a face of the unit cell.)

**11.71** Many common rocks contain large regions of silicate networks held together by ionic interactions. What properties of rocks can be attributed to this general composition?

**11.72** Semiconductors are used in many devices that measure temperature. Use band theory to explain the nature of the property that makes semiconductors suitable for this purpose.

## Quantitative Aspects of Changes in State
(Sample Problem 11.4)

**11.73** Describe the changes (if any) in potential energy and in kinetic energy among the molecules when gaseous PCl₃ condenses to a liquid at a fixed temperature.

**11.74** When benzene is at its melting point, two processes are occurring simultaneously and balancing each other. Describe these processes on the macroscopic and molecular levels.

**11.75** Liquid hexane (boiling point = 69°C) is placed in a closed container at room temperature. At first the pressure of the vapor phase increases, but after a short time it stops changing. Why?

**11.76** Explain the effect of strong intermolecular forces on each of the following parameters: (a) critical temperature; (b) boiling point; (c) vapor pressure; (d) heat of vaporization.

**11.77** At a pressure of 1.1 atm, will water boil at 100°C? Explain.

**11.78** A liquid is in equilibrium with its vapor in a closed vessel at a fixed temperature, and the vessel is connected through a stopcock to an evacuated vessel. When the stopcock is opened, will the final pressure of the vapor be different if (a) some liquid remains; (b) all the liquid is first removed? Explain.

**11.79** Substance A has a solid-liquid curve that slopes to the right with increasing pressure. Substance B has one that slopes to the left. What macroscopic property can distinguish A from B?

**11.80** Which member in each of the following pairs of liquids has the *higher* vapor pressure at a given temperature? Explain.
(a) $C_2H_6$ or $C_4H_{10}$
(b) $CH_3CH_2OH$ or $CH_3CH_2F$
(c) $NH_3$ or $PH_3$

**11.81** Which member in each of the following pairs of liquids has the *lower* vapor pressure at a given temperature? Explain.
(a) $HOCH_2CH_2OH$ or $CH_3CH_2CH_2OH$
(b) $CH_3COOH$ or $CH_3CCH_3$ ($\overset{\|}{O}$)
(c) HF or HCl

**11.82** From the following data, calculate the total heat (in J) needed to convert 12.00 g ice at −5.00°C to liquid water at 0.500°C:
Melting point at 1 atm    0.0°C
$C_{water(s)}$    2.09 J/g · °C
$C_{water(l)}$    4.21 J/g · °C
$\Delta H^0_{fus}$    6.02 kJ/mol

**11.83** From the following data, calculate the total heat (in J) needed to convert 0.333 mol ethanol gas at 300°C and 1 atm to liquid ethanol at 25.0°C and 1 atm:
Boiling point at 1 atm    78.5°C
$C_{gas}$    1.43 J/g · °C
$C_{liquid}$    2.45 J/g · °C
$\Delta H^0_{vap}$    40.5 kJ/mol

**11.84** A liquid has a heat of vaporization of 31.5 kJ/mol and a boiling point of 120°C at 1 atm. What is its vapor pressure at 110°C?

**11.85** Diethyl ether ($CH_3CH_2OCH_2CH_3$) has a heat of vaporization of 29.1 kJ/mol and a vapor pressure of 0.703 atm at 25.0°C. What is its vapor pressure at 100°C?

**11.86** What is the heat of vaporization of a liquid that has a vapor pressure of 640 torr at 85°C and a boiling point of 95°C at 1 atm?

**11.87** Methane ($CH_4$) has a boiling point of −164°C at 1 atm and a vapor pressure of 42.8 atm at −100°C. What is the heat of vaporization of $CH_4$?

**11.88** Use the following data to sketch a qualitative phase diagram for ethylene:

| | |
|---|---|
| Boiling point at 1 atm | −103.7°C |
| Melting point at 1 atm | −169.16°C |
| Critical point | 9.9°C and 50.5 atm |
| Triple point | −169.17°C and 1.20 × 10⁻³ atm |

Is solid ethylene more or less dense than liquid ethylene?

**11.89** Use the following data to sketch a qualitative phase diagram for molecular hydrogen:

| | |
|---|---|
| Melting point at 1 atm | 13.96 K |
| Boiling point at 1 atm | 20.39 K |
| Triple point | 13.95 K and 0.07 atm |
| Critical point | 33.2 K and 13.0 atm |
| Vapor pressure of solid at 10 K | 0.001 atm |

Does hydrogen sublime at 0.05 atm? Explain.

**11.90** Why does water vapor at 100°C cause more severe burns than liquid water at 100°C?

**11.91** Mercury vapor is extremely toxic and is readily absorbed through the lungs. At 20°C, mercury has a vapor pressure of $1.20 \times 10^{-3}$ torr, which is high enough to be hazardous. To reduce the danger to workers who may be exposed in processing plants, mercury is cooled to lower its vapor pressure. At what temperature would the vapor pressure of mercury be at the relatively safe level of $5.0 \times 10^{-5}$ torr ($\Delta H_{vap}$ = 59.1 kJ/mol)?

**11.92** In the process of freeze-drying, ice sublimes and the water vapor is removed by vacuum. In which region of the phase diagram for water (Figure 11.39, *B*) does sublimation take place? What is the highest temperature at which water can sublime? Why would freeze-drying become difficult (or uneconomical) at very low temperatures?

**11.93** Butane ($C_4H_{10}$) is a common fuel used in cigarette lighters and camping stoves. Normally supplied in metal containers under pressure, the fuel exists as a mixture of liquid and gas, so high temperatures may cause the container to explode. At 25.0°C, the vapor pressure of butane is 2.3 atm. What is the pressure in the container at 150°C? ($\Delta H_{vap}$ = 24.3 kJ/mol)

**The Uniqueness of Water**

**11.94** For what types of substances is water a good solvent? Explain.

**11.95** The oxygen atom of a water molecule has two covalent O—H bonds but can engage in as many as four H bonds. Explain.

**11.96** Because of their high water content, warm-blooded animals maintain a narrow range of body temperature. How do the properties of water assist in this phenomenon?

**11.97** What property of water keeps plant debris on the surface of lakes and ponds? What is the ecological significance of this?

**11.98** A drooping plant can be revived if the ground around it is watered. What property of water allows the plant to recover?

**11.99** Describe the molecular origin of the property of water responsible for the presence of ice on the surface of a lake.

**Comprehensive Problems**

Problems with an asterisk (*) are more challenging.

**11.100** Which forces are overcome when the following events occur? (a) NaCl dissolves in water; (b) krypton boils; (c) water boils; (d) $CO_2$ sublimes

**11.101** Draw one $NH_3$ molecule with H bonds to two others.

**11.102** Why are x-rays more useful than gamma rays or ultraviolet light for measuring the interatomic distances in crystals?

**11.103** Why do the densities of most liquids increase as they are cooled and solidified? How does water differ in this regard?

***11.104** The packing efficiency of spheres is the percentage of the total space (volume) of the unit cell occupied by the spheres:

Packing efficiency (%)

$$= \frac{\text{volume of spheres in unit cell}}{\text{volume of unit cell}} \times 100$$

Using spheres of radius $r$ and unit-cell edge length $A$, calculate the packing efficiency of the following (volume of a sphere = $\frac{4}{3}\pi r^3$): (a) a simple cubic unit cell; (b) a face-centered cubic unit cell.

**11.105** Consider the phase diagram for substance X:

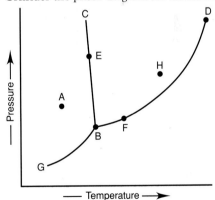

(a) What phase(s) is(are) present at point A? E? F? H? B?

(b) Which point corresponds to the critical point? Which point corresponds to the triple point?

(c) What curve corresponds to conditions at which the solid and gas are in equilibrium?

(d) Describe what happens when you start at point A and increase the temperature at constant pressure.

(e) Describe what happens when you start at point H and decrease the pressure at constant temperature.

(f) Is liquid X more or less dense than solid X?

**11.106** A capillary tube is dipped into molten paraffin wax and then allowed to drain so that the inside of the tube is coated with a thin layer of solid wax as it cools. If the tube is now dipped into a beaker of water, would you expect the water level in the tube to be higher or lower than the water in the beaker? What would you expect to be the shape of the meniscus? Explain.

**11.107** Why do uninsulated water pipes sometimes burst in cold weather?

**11.108** The density of solid gallium at its melting point is 5.9 g/cm³, whereas that of liquid gallium is 6.1 g/cm³. Is the temperature at the triple point higher or lower than the normal melting point? Which direction is the slope of the solid-liquid curve on the gallium phase diagram?

**11.109** Although the lighter halogens are insulators, it has been suggested that solid astatine ($At_2$) may be a semiconductor. Account for this prediction.

**11.110** According to band theory, application of extremely high pressures would cause a band of empty high-energy orbitals to broaden and eventually overlap a band of filled low-energy orbitals. How might this idea be used to explain the observation made by the spacecraft passing Jupiter and Saturn that they may have cores of metallic hydrogen?

**\*11.111** Substance A has the following properties.

| | |
|---|---|
| Melting point at 1 atm | −20°C |
| Boiling point at 1 atm | 85°C |
| $\Delta H_{fus}$ | 180 J/g |
| $\Delta H_{vap}$ | 500 J/g |
| $C_{solid}$ | 1.0 J/g · °C |
| $C_{liquid}$ | 2.5 J/g · °C |
| $C_{gas}$ | 0.5 J/g · °C |

At 1 atm, a 25-g sample of A is heated from −40°C to 100°C at a constant rate of 450 J/min.

(a) How many minutes does it take to heat the sample to its melting point?

(b) How many minutes does it take to melt the sample?

(c) Perform any other necessary calculations and draw a curve of temperature vs. time for the entire heating process.

**11.112** Ice is held in a cylinder by means of a piston under high pressure at −10°C. What happens to the contents of the cylinder as the piston is slowly withdrawn at a constant temperature?

**11.113** The hydrides of N, O, and F boil at higher temperatures than those of P, S, and Cl, but the hydrides of B and C boil at lower temperatures than those of Al and Si. Explain these patterns.

**\*11.114** Sodium has a crystal structure based on the body-centered cubic unit cell. What is the mass of one unit cell of sodium?

**11.115** A cubic-type unit cell contains atoms of element A at each corner and atoms of element Z on each face. What is the empirical formula of the compound?

**\*11.116** KF has the same type of crystal structure as NaCl. The unit cell of KF has an edge length of 5.39 Å. Calculate the density of KF.

**\*11.117** One of the proposed ways for storing gaseous $H_2$ fuel in the future is to pack it under high pressure into the holes of a metal's crystal structure. Palladium, which adopts a cubic closest packed structure, can absorb more $H_2$ than any other element. Although the exact Pd-H interaction is unclear, it is estimated that the density of absorbed $H_2$ approaches that of liquid hydrogen (70.8 g/L). What volume (in L) of gaseous $H_2$, measured at STP, can be packed into the spaces of 1 dm³ of palladium metal?

## CHAPTER 12

# The Properties of Mixtures

## Solutions and Colloids

### Concepts and skills to review

- classification and separation of mixtures (Section 2.8)
- calculations involving mass percent (Section 3.1) and molarity (Section 3.5)
- electrolytes; water as a solvent (Sections 4.2 and 11.6)
- mole fraction and Dalton's law of partial pressures (Section 5.4)
- types of intermolecular forces (Section 11.2)
- vapor pressure of liquids (Section 11.5)

**A mix of liquids and vapors.** To paraphrase the Bard: "All the world's a mixture." The common forms of matter within and around you are not made of pure substances but of solutions, colloids, suspensions, and so forth. The impurities make the pure substances behave differently and give them some very useful properties, as you'll see in this chapter.

**N**early all the gases, liquids, and solids that make up our world are **mixtures**—two or more substances physically mixed together but not chemically combined. Synthetic mixtures, such as glass and soap, usually contain only a few components, whereas natural mixtures, such as air, sea water, and soil, are much more complex, often containing more than 50 different substances. Living mixtures, such as trees and students, are the most complex—even a simple bacterium contains more than 5000 compounds.

Recall from Chapter 2 that any mixture has two defining characteristics: its composition is variable, and it retains some properties of its components. In this chapter, we focus on two extremely common types of mixtures: solutions and colloids.

*A solution is a homogeneous mixture,* one with no boundaries separating its components. Thus, a solution exists as one **phase,** one physically distinct portion of a system. A heterogeneous mixture has two or more phases. The small pebbles in a slab of concrete and the bubbles in a glass of champagne indicate that these are heterogeneous mixtures. A *colloid* is a heterogeneous mixture in which one component is dispersed as fine particles in another component. Smoke and milk are familiar colloids. The essential difference between these two types of mixtures is based on particle size: in a solution, the particles are separated at the molecular level; in a colloid, the particles are larger but still small enough to remain dispersed and not settle out.

The chapter opens with a survey of the types of solutions and the role of intermolecular forces in their formation. Predicting solubility is a major focus of the first section, in which we look at systems ranging from salts in water to antibiotics in bacterial cells. Next we investigate the reason a substance dissolves, in terms of the enthalpy changes involved, and we become acquainted with the concept of entropy changes in the solution process. Then we examine the equilibrium nature of solubility and see how temperature and pressure affect it. After discussing ways of expressing concentration, we use them to explore how the physical properties of solutions differ from those of pure substances. An investigation of the properties of colloids follows, and the chapter closes with the application of solution and colloid chemistry to the purification of water.

## 12.1 Types of Solutions: Intermolecular Forces and the Prediction of Solubility

We often describe solutions in terms of one substance dissolving in another: the **solute** dissolves in the **solvent.** Usually, there is much more solvent than solute. In some cases, however, the substances are **miscible,** that is, soluble in each other in any proportion, so it is not very meaningful to call one the solute and the other the solvent.

The **solubility** of a substance is the maximum amount that can dissolve in a fixed amount of a particular solvent to form a stable solution at a specified temperature. The solubility of sodium chloride in 100 mL water at 100°C is 39.12 g NaCl. The solubility of silver chloride in 100 mL water at 100°C is 0.0021 g AgCl; obviously, NaCl is much more soluble in water than is AgCl. Whereas solubility has a quantitative meaning, the words *dilute* and *concentrated* are qualitative terms that refer to the *relative amounts of solute present in a solution:* a dilute solution contains much less dissolved solute than a concentrated one.

For one substance to dissolve in another, three events must occur: (1) solute particles must separate from each other, (2) some solvent particles must separate to make room for the solute particles, and (3) solute and solvent particles must mix together. Therefore, *whether or not a solution occurs depends on the relative strengths of the solute-solute, solvent-solvent, and solute-solvent intermolecular forces.* From a knowledge of the forces present, we can predict which solutes will dissolve in which solvents.

### Intermolecular Forces in Solution

All the intermolecular forces we discussed in Chapter 11 for pure substances also occur within mixtures. Indeed, the energy of attraction created by these forces is the reason the mixtures form. Figure 12.1 summarizes these forces in order of decreasing strength.

*Ion-dipole forces* are a principal factor in the solubility of ionic compounds in water. When a salt dissolves, these forces overcome the attraction between the ions and break down the crystal structure. As each ion becomes separated, more water molecules cluster around it in **hydration shells** (Figure 12.2). The closest hydration shell is H bonded to water molecules slightly farther away, forming a less structured hydration shell, and these are H bonded to other molecules in the bulk solvent. For monatomic ions, the number of water molecules in the closest hydration shell depends on the ion's size. Four water molecules can fit tetrahedrally around small ions, such as $Li^+$; larger ions, such as $Na^+$ and $F^-$, usually have six water molecules surrounding them octahedrally.

The special type of *dipole-dipole* force known as the *H bond* is responsible for water's ability to dissolve larger O-containing and N-containing organic compounds, such as alcohols, sugars, and amines. The *ion-induced dipole force* contributes to the attractive forces in any solution that contains ions—salts in water or in less polar solvents. It is stronger than the *dipole-induced dipole force*, which operates whenever the partial charges of a dipole are present. The solubility in water of the nonpolar atmospheric gases, $O_2$, $N_2$, and the noble gases, is due in part to the attraction between polar water molecules and the temporary dipole they induce in the electron clouds of these gas particles (see Figure 11.10). Paint thinners and grease solvents also function through these forces. Ever-present *dispersion forces* contribute to the solubility of all solutes in all solvents, but they are the principal attractive force in solutions of nonpolar substances; petroleum exists as a complex homogeneous mixture due to dispersion forces.

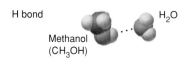

Ion-dipole

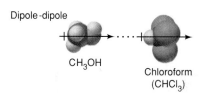

H bond

Methanol (CH₃OH)        H₂O

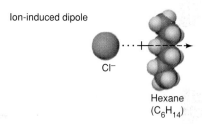

Dipole-dipole

CH₃OH        Chloroform (CHCl₃)

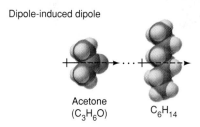

Ion-induced dipole

Cl⁻        Hexane (C₆H₁₄)

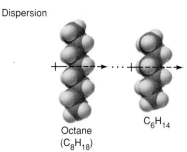

Dipole-induced dipole

Acetone (C₃H₆O)        C₆H₁₄

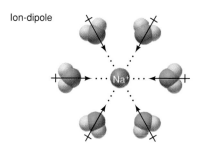

Dispersion

Octane (C₈H₁₈)        C₆H₁₄

**FIGURE 12.1**
**The major types of intermolecular forces in solutions.** Forces are listed in decreasing order of strength.

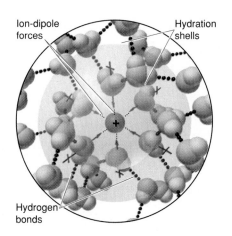

Ion-dipole forces        Hydration shells

Hydrogen bonds

**FIGURE 12.2**
**Hydration shells formed around an aqueous ion.** When an ionic compound dissolves in water, ion-dipole forces orient water molecules around the separated ions to form hydration shells. The cation shown here is octahedrally surrounded by six water molecules, which are hydrogen bonded to other water molecules in the next hydration shell, and those are bonded to molecules farther away.

Keeping these intermolecular forces in mind, let's classify types of solutions according to physical state. Solutions can be gaseous, liquid, or solid. In most cases, the physical state of the solvent determines that of the solution. The discussion highlights solutions in which the solvent is a liquid because they are by far the most common and important.

## Liquid Solutions and the Role of Molecular Polarity

From cytoplasm to tree sap, gasoline to cleaning fluid, iced tea to urine, solutions in which a liquid is the solvent appear throughout everyday life. Water is the most prominent liquid solvent because of its great ability to dissolve both ionic compounds and polar molecules and its great abundance in the environment. Nonpolar substances, such as cooking oil and car wax, will not dissolve in water but will dissolve in less polar or nonpolar solvents.

One way to predict which solutes will dissolve in which solvents is to consider the forces acting among the molecules in each case. (We consider other factors in Section 12.2.) Much experience has shown that *substances with similar types of intermolecular forces dissolve in each other.* This empirical fact is summarized in the old rule of thumb **like dissolves like,** which usually provides a good first guess at predicting solubility.

**Liquid-liquid and solid-liquid solutions.** Water dissolves many salts because the strong ion-dipole attractions that water forms with the ions are very similar to the strong attractions between the ions themselves and, therefore, can substitute for them. The same salts are insoluble in hexane ($C_6H_{14}$) because the weak ion-induced dipole forces their ions could form with this nonpolar solvent *cannot* substitute for attractions between ions. Similarly, oil does not dissolve in water because the weak dipole-induced dipole forces between them cannot substitute for the strong H bonds between water molecules. Oil does dissolve in hexane, however, because the dispersion forces in one substitute well for the dispersion forces in the other.

To examine this idea further, let's compare the solubilities of a series of solutes in water and in hexane, two solvents of widely different polarities. The solutes we'll consider are alcohols—molecules with a hydroxyl (—OH) group bound to a hydrocarbon group—with the general formula $CH_3(CH_2)_nOH$, where $n = 0$ to 5. Notice in Table 12.1 that as the hydrocarbon chain becomes longer, the solubility of the alcohol in water decreases. We can understand this by viewing the alcohol molecule as consisting of *two*

### TABLE 12.1 Solubility of a Series of Alcohols in Water and Hexane

| ALCOHOL | IN WATER AT 20°C (mol/100 g $H_2O$) | IN HEXANE AT 20°C (mol/100 g $C_6H_{14}$) |
|---|---|---|
| $CH_3OH$ (methanol) | ∞ | 0.12 |
| $CH_3CH_2OH$ (ethanol) | ∞ | ∞ |
| $CH_3CH_2CH_2OH$ (propanol) | ∞ | ∞ |
| $CH_3CH_2CH_2CH_2OH$ (butanol) | 0.11 | ∞ |
| $CH_3CH_2CH_2CH_2CH_2OH$ (pentanol) | 0.030 | ∞ |
| $CH_3CH_2CH_2CH_2CH_2CH_2OH$ (hexanol) | 0.0058 | ∞ |

portions, the polar —OH group and the nonpolar hydrocarbon chain. The hydroxyl portion forms strong H bonds with water and weak dipole-induced dipole forces with hexane. The hydrocarbon portion forms dispersion forces with hexane and dipole-induced dipole forces with water.

Let's first consider alcohol solubility in water, focusing on how the presence of the alcohol affects the solvent. In the smaller alcohols (one to three carbons), the hydroxyl group is a relatively large portion of the molecule, so the solute molecules interact with one another through H bonding, just as the solvent molecules do. When they mix, strong H bonding within solute and within solvent is replaced by strong H bonding *between* solute and solvent (Figure 12.3). Thus, the smaller alcohols are miscible with water.

When the nonpolar chain is longer, the —OH group is a smaller portion of the molecule, so the alcohol is less able to disrupt the strong H bonds in water. In effect, the longer hydrocarbon chains would have to push in among the water molecules and establish weak attractions with them that could substitute for the water H bonds. The H bonding between the alcohol —OH group and water is not enough to compensate for the necessary disruption of the H bonds in pure water. As you can see, the solubility in water decreases from three-carbon to six-carbon alcohols. Still longer-chain alcohols are virtually insoluble in water.

The opposite situation occurs when hexane is the solvent, but the change in solubility is less gradual, as Table 12.1 shows. Here the major solute-solvent *and* solvent-solvent interactions are dispersion forces. The weak forces between the hydroxyl group of methanol ($CH_3OH$) and hexane *cannot* substitute for the strong H bonding among $CH_3OH$ molecules, so the solubility of methanol in hexane is relatively low. In the larger alcohols, however, dispersion forces become increasingly more important, and they *can* substitute for dispersion forces in hexane, so solubility increases. Since no strong solvent-solvent forces must be replaced by weak solute-solvent forces, even the two-carbon chain of ethanol makes strong enough solute-solvent attractions for it to be miscible in hexane.

**SAMPLE PROBLEM 12.1** _____

**Predicting Relative Solubilities of Substances**
_____

**Problem:** Predict which solvent will dissolve more of the given solute:
**(a)** Sodium chloride in methanol ($CH_3OH$) or in propanol ($CH_3CH_2CH_2OH$)
**(b)** Ethylene glycol ($HOCH_2CH_2OH$) in water or in hexane ($CH_3CH_2CH_2CH_2CH_2CH_3$)
**(c)** Diethyl ether ($CH_3CH_2OCH_2CH_3$) in ethanol ($CH_3CH_2OH$) or in water
**Plan:** We examine the formulas of the solute and each solvent to determine the types of forces that could occur. A solute tends to be more soluble in a solvent whose intermolecular forces are similar to, and therefore can replace, those in the solute.
**Solution: (a) Methanol.** NaCl is an ionic solid that dissolves through ion-dipole forces. Both methanol and propanol contain a polar —OH group, and propanol's longer hydrocarbon chain would form only weak forces with the ions, so it would be less effective at replacing the ionic attractions in the solute.
**(b) Water.** Ethylene glycol molecules have two —OH groups, so the molecules interact with each other through H bonding. They would be more soluble in $H_2O$, whose H bonds can better replace solute H bonds than can the dispersion forces in hexane.
**(c) Ethanol.** Diethyl ether molecules interact with each other through dipole and dispersion forces and could form H bonds to both $H_2O$ and ethanol. The ether would be more soluble in ethanol because that solvent can form H bonds *and* replace the dispersion forces in the solute, whereas the H bonds in water must be partly replaced with much weaker dispersion forces.

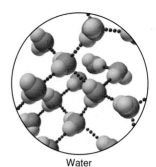

Water

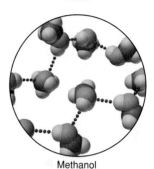

Methanol

A solution of water
and methanol

**FIGURE 12.3**
**Like dissolves like: solubility of methanol in water.** The H bonds in water and methanol are similar so they can replace one another; thus, methanol is miscible with water.

**FOLLOW-UP PROBLEM 12.1**
Predict which solute is more soluble in the given solvent:
**(a)** 1,4-Butanediol (HOCH$_2$CH$_2$CH$_2$CH$_2$OH) or butanol (CH$_3$CH$_2$CH$_2$CH$_2$OH) in water
**(b)** Chloroform (CHCl$_3$) or carbon tetrachloride (CCl$_4$) in water

---

◆ **The Hardness and Softness of Soaps.** The cation of a soap influences its properties greatly. Lithium soaps are hard and high melting and are used in car lubricants. Softer, more water-soluble sodium soaps are used as common bar soap. Potassium soaps are low melting and used as liquid soaps. The insolubility of calcium soaps explains why it is more difficult to clean clothes in hard water, which usually contains many Ca$^{2+}$ ions (Chemical Connections, Section 12.6)

The molecules of many other organic compounds have polar and nonpolar portions, and the predominance of one portion or the other determines the solubility of the substance in various solvents. In a few substances, such as soaps, the two polarities prescribe the molecule's function. A **soap** is the salt of a metal hydroxide and a fatty acid, a molecule with a long hydrocarbon chain and an organic acid group (—COOH). A typical soap molecule has a nonpolar hydrocarbon "tail" that is 15 to 19 carbons long and a polar ionic "head" consisting of the —COO$^-$ group and a metal ion; sodium stearate [CH$_3$(CH$_2$)$_{16}$COO$^-$Na$^+$] is an example (Figure 12.4, *inset*). When a grease-stained cloth is immersed in soapy water, the nonpolar tails of the soap molecules form dispersion forces with the nonpolar grease molecules, while the polar ionic heads form ion-dipole forces and H bonds with the water (Figure 12.4). ◆ Agitation creates tiny aggregates of grease surrounded by soap, which are flushed away with added water. Certain antibiotics function via dual polarities as well, as shown in the Chemical Connections essay.

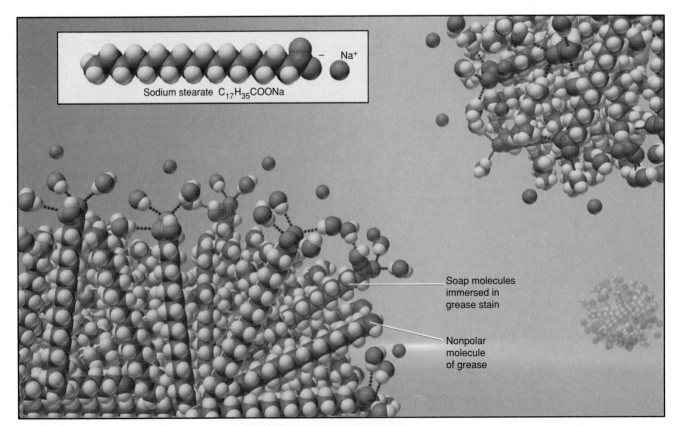

**FIGURE 12.4**
**The structure and function of a soap.** The inset shows the nonpolar tail and polar ionic head of sodium stearate, a typical soap. The tails of soap molecules interact with molecules in grease by dispersion forces, while the heads interact with water through H bonds and ion-dipole forces. Rinsing with water flushes the grease away.

**Chemistry in Pharmacology**
**How Antibiotics Work**

Antibiotics are natural substances produced by one microorganism to chemically destroy another. Many have been adopted by medical science as potent weapons against bacterial infection. Many antibiotics function by using two molecular polarities to disrupt an organism's aqueous ionic makeup. All cells have membranes that maintain their internal concentrations of ions in delicate balance with those in the external aqueous fluid; some ions, such as $Na^+$, are excluded from the cell, whereas others, such as $K^+$, are included. If this ionic balance is disrupted, the organism cannot survive. Cell membranes are thin envelopes that consist of a double layer of molecules. Each molecule combines two fatty acids, an alcohol, and a phosphate group. The nonpolar tails interact within the membrane, while the charged, polar heads interact with the watery interior and exterior fluids.

Some antibiotics disrupt ion balance by forming a channel in the bacterial membrane through which aqueous ions can flow (Figure 12.A). Gramicidin A, for example, has two polarities on one tube-shaped molecule: nonpolar chemical groups on the outside of the tube and polar groups on the inside. The nonpolar outside helps the molecule dissolve into the nonpolar membrane of the target bacteria through dispersion forces, while the polar inside allows ions to enter through H bonds and ion-dipole forces. Two gramicidin A molecules lying end to end span the bacterial membrane and remain there, providing a polar passageway for ions to diffuse in and out. The

cell's interior-exterior ionic balance is disrupted, and the organism dies.

Valinomycin is a donut-shaped antibiotic molecule that works in a different way. The polar hole of the donut binds an ion, while the nonpolar rim allows the molecule to dissolve into the bacterial membrane. Rather than remaining lodged there as gramicidin A does, valinomycin moves through to the other side and releases the ion. Millions of such transits by many valinomycin molecules soon disrupt the exterior-interior ionic balance, and the organism dies.

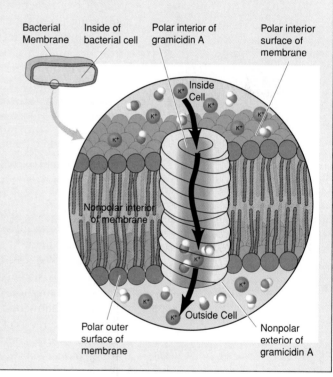

**FIGURE 12.A**

**The mode of action of the antibiotic gramicidin A.** Gramicidin A is a tube-like molecule, nonpolar on the outside and polar on the inside. After dissolving into the nonpolar interior of a bacterial cell membrane via dispersion forces, pairs of gramicidin molecules aligned end to end span the membrane's thickness. Ions diffuse through the polar interior of these molecules, which destroys the cell's ionic balance.

**Gas-liquid solutions.** Gases have low boiling points because their intermolecular forces are weak. Furthermore, gases typically are not very soluble in water for the same reason: solute-solvent intermolecular forces are weak. In fact, there is a correlation between the boiling point of a gas and its solubility in water (Table 12.2). Such weak interactions within a gas would lead us to expect that gas solubility in a liquid depends on solute-solvent and solvent-solvent forces, *not* on solute-solute forces.

In several cases, the small amount of gas that does dissolve is essential to a reaction process. The most important example is the solution of $O_2$ in water. At 25°C and 1 atm, only 0.64 mL $O_2$ dissolves in 100 mL water, but without this small amount, fish and other aquatic animal life would die.

**TABLE 12.2 Correlation Between Boiling Point and Solubility in Water**

| GAS | SOLUBILITY (*M*)* | BP (K) |
|---|---|---|
| He | $4.2 \times 10^{-4}$ | 4.2 |
| Ne | $6.6 \times 10^{-4}$ | 27.1 |
| $N_2$ | $10.4 \times 10^{-4}$ | 77.4 |
| CO | $15.6 \times 10^{-4}$ | 81.6 |
| $O_2$ | $21.8 \times 10^{-4}$ | 90.2 |
| NO | $32.7 \times 10^{-4}$ | 121.4 |

*At 273 K and 1 atm.

In some cases, the solubility of a gas may seem high because the gas is actually reacting with the solvent or with another component present. Much more $O_2$ dissolves in blood than in water because the $O_2$ continually combines with the hemoglobin molecules in red blood cells. Carbon dioxide seems very soluble in water (~81 mL $CO_2$/100 mL $H_2O$ at 25°C and 1 atm) because it is reacting in addition to simply dissolving:

$$CO_2(g) + H_2O(l) \rightleftharpoons H^+(aq) + HCO_3^-(aq)$$

### Gas Solutions and Solid Solutions

Despite the central place of liquid solutions in chemistry, gaseous solutions and solid solutions have both vital importance and numerous practical applications.

**Gas-gas solutions.** *All gases are infinitely soluble in one another.* Air is the classic example of a gaseous solution, consisting of about 18 gases in widely differing proportions. Anesthetic gas proportions are finely adjusted to the needs of the surgical procedure. The ratios of components in many industrial gas mixtures, such as $CO/H_2$ in syngas production or $N_2/H_2$ in the formation of ammonia, are controlled to optimize product yield under varying reaction conditions.

**Gas-solid solutions.** *When a gas dissolves in a solid, it occupies the spaces between the closely packed particles.* Chapter 6 mentions that hydrogen can be stored this way in some metals, most notably palladium and niobium (Figure 12.5). Almost 1000 L $H_2$ (at STP) can be pumped into spaces between the close-packed atoms in only 1 L of the metal. However, this ability of gases to penetrate a solid also has disadvantages. The electrical conductivity of copper, for example, is drastically reduced by the presence of $O_2$, which dissolves into the copper crystal structure and reacts to form copper(I) oxide. High-conductivity copper is prepared by melting and recasting the copper in an $O_2$-free atmosphere.

**Solid-solid solutions.** Since solids diffuse to such a small extent, their mixtures are usually heterogeneous, such as gravel mixed with sand. Some solid-solid solutions can be formed by melting the solids and then mixing them and allowing them to freeze. Many **alloys,** mixtures of elements that have an overall metallic character, are examples of solid-solid solutions, although several common alloys have microscopic heterogeneous regions. In *substitutional* alloys, such as sterling silver and brass, atoms of another element substitute for some atoms of the main element in the structure (Figure 12.6, *A*). In *interstitial* alloys, atoms of the main element lie at the lattice points, and other atoms (often nonmetals) fill some of the spaces, or interstices. Carbon steel, for example, is an interstitial alloy of iron with carbon (Figure 12.6, *B*).

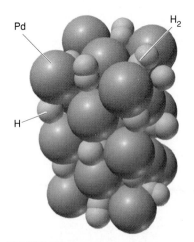

**FIGURE 12.5**
**A solution of hydrogen gas in Pd metal.** In gas-solid solutions, the gas particles lie between the particles of the solid. In this case, hydrogen molecules and atoms occupy the spaces between the cubic closest packed atoms of the palladium crystal structure.

**FIGURE 12.6**
**The arrangement of atoms in two types of alloys. A,** Brass is a substitutional alloy in which zinc atoms substitute for copper at many of its face-centered packing sites. **B,** Carbon steel is an interstitial alloy in which carbon atoms lie in the holes (interstices) of the body-centered crystal structure of iron.

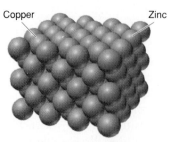

**A** Brass, a substitutional alloy

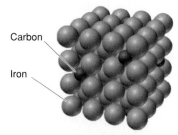

**B** Carbon steel, an interstitial alloy

Waxes are another familiar type of solid-solid solution. A *wax* is a solid of biological origin that is insoluble in water but dissolves in nonpolar solvents. Many plants have waxy coatings on their leaves and fruit to prevent evaporation of water. Most waxes are amorphous solids that sometimes contain small regions of crystalline regularity. ◆

### Section Summary

Solutions are homogeneous mixtures consisting of a solute dissolved in a solvent through the influence of intermolecular forces. The solubility of a substance in a given amount of a solvent is the maximum amount that can dissolve under specified conditions. The types of intermolecular forces in a solution are the same as in a pure substance. If similar intermolecular forces occur in solute and solvent, dissolution will often occur. The ion-dipole force enables ionic compounds to dissolve in water. The ions become surrounded by hydration shells of H-bonded water molecules. Solubility of organic molecules in water depends on the extent of their polar and nonpolar portions. Soaps and some antibiotics act through the two polarities of their molecules. Gas solubility in water is low because very weak intermolecular forces are present. Gases are miscible with one another. They dissolve in solids by fitting into spaces in the crystal structure. Solid-solid solutions, such as alloys and waxes, form when the components are mixed while molten.

◆ **Waxes for Home and Auto.** Beeswax, the remarkable structural material that bees secrete to build their hives, is a complex mixture of fatty acids and hydrocarbons in which some of the molecules contain chains more than 40 carbon atoms long. Carnauba wax from a South American palm is a mixture of compounds, each consisting of a fatty acid bound to a long-chain alcohol. It is hard and forms a thick gel in nonpolar solvents—thus its use in car waxes.

## 12.2   Energy Changes in the Solution Process

As a predictive tool, "like dissolves like" is helpful in many cases. As we would expect, this handy macroscopic rule is based on the *molecular* events that occur between solute and solvent particles. To see *why* like dissolves like, we conceptually break down the solution process into steps and examine them in terms of changes in *enthalpy* and *entropy* of the system. We discussed enthalpy in Chapter 6; the concept of entropy is introduced here and treated quantitatively in Chapter 19.

### Heats of Solution and Solution Cycles

Picture a generalized solute and solvent about to form a solution. Each consists of particles attracting each other. No matter what the nature of these attractions, energy is absorbed to overcome them, and then energy is released when the particles mix and attract one another. Because enthalpy is a state function, the actual way that dissolution occurs need not be known to find the overall enthalpy change. Let's divide the process into the following three hypothetical steps, each with its own enthalpy term:

*Step 1.* Solute separates into particles. This step involves overcoming attractions, so it must be endothermic:

$$\text{Solute (aggregated)} + heat \rightarrow \text{solute (separated)} \qquad \Delta H_{\text{solute}} > 0$$

*Step 2.* Solvent separates into particles. This step also involves overcoming intermolecular attractions and is endothermic:

$$\text{Solvent (aggregated)} + heat \rightarrow \text{solvent (separated)} \qquad \Delta H_{\text{solvent}} > 0$$

*Step 3.* Solute and solvent particles mix. The particles attract each other, so this step is exothermic:

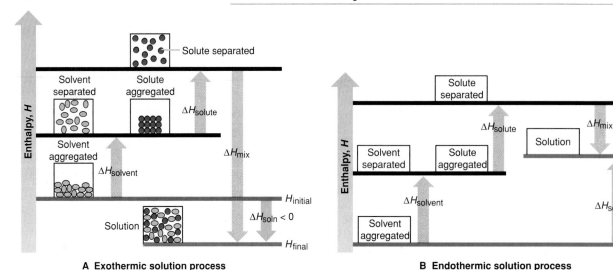

**A  Exothermic solution process**            **B  Endothermic solution process**

**FIGURE 12.7**

**Solution cycles and the enthalpy components of the heat of solution.** The $\Delta H_{soln}$ can be thought of as the sum of three enthalpy changes: $\Delta H_{solute}$ (separating the solute; always > 0), $\Delta H_{solvent}$ (separating the solvent; always > 0), and $\Delta H_{mix}$ (mixing solute and solvent; always < 0). In **A**, $\Delta H_{mix}$ is larger than $\Delta H_{solute} + \Delta H_{solvent}$, so $\Delta H_{soln}$ is negative (exothermic process). In **B**, $\Delta H_{mix}$ is smaller than the sum of the others, so $\Delta H_{soln}$ is positive (endothermic process).

$$\text{Solute (separated)} + \text{solvent (separated)} \rightarrow \text{solution} + \textit{heat} \qquad \Delta H_{mix} < 0$$

The total enthalpy change that occurs when a solution forms from solute and solvent is the **heat of solution ($\Delta H_{soln}$)**; it is found by combining the three individual enthalpy changes. The overall process is called a *thermochemical solution cycle* and is yet another application of Hess's law:

$$\Delta H_{soln} = \Delta H_{solute} + \Delta H_{solvent} + \Delta H_{mix} \qquad \textbf{(12.1)}$$

(Recall that in Section 9.2 we conceptually broke down the heat of formation of an ionic compound into component enthalpies in a similar way through the use of a Born-Haber cycle.) If the sum of the endothermic terms ($\Delta H_{solute} + \Delta H_{solvent}$) is smaller than the exothermic term ($\Delta H_{mix}$), $\Delta H_{soln}$ is exothermic, and the solution becomes warmer as it forms. Figure 12.7, *A* shows an enthalpy diagram for such a solution. If the sum of the endothermic terms is larger, $\Delta H_{soln}$ is endothermic, and the solution becomes cooler (Figure 12.7, *B*); if it is *much* larger, the solute may not dissolve in that solvent at all.

### Heats of Hydration: Ionic Solids in Water

The $\Delta H_{solvent}$ and $\Delta H_{mix}$ components of the heat of solution are especially difficult to obtain individually. These terms combined represent the enthalpy change during **solvation,** the process of surrounding a solute particle with solvent particles. Solvation in water is often called **hydration.** The enthalpy terms associated with separating the water molecules ($\Delta H_{solvent}$) and mixing the solute with them ($\Delta H_{mix}$) are combined into the **heat of hydration ($\Delta H_{hydr}$)** for the thermochemical aqueous solution cycle:

$$\Delta H_{soln} = \Delta H_{solute} + \Delta H_{hydr}$$

The heat of hydration is a crucial factor in the dissolving of an ionic solid. Breaking several H bonds in water is more than compensated for by forming several strong ion-dipole forces, so hydration of an ion is *always* exothermic. The $\Delta H_{hydr}$ of an ion is defined as the enthalpy change for the hydration of a mole of separated (gaseous) ions:

$$M^+(g) \text{ [or } X^-(g)] \xrightarrow{H_2O} M^+(aq) \text{ [or } X^-(aq)] \qquad \Delta H_{hydr \text{ of the ions}} \text{ (always} < 0)$$

Heats of hydration exhibit trends that are understandable in terms of the **charge density** of the ion, the ratio of its charge to its volume. In general, the higher the charge density, the more negative is $\Delta H_{hydr}$. According to Coulomb's law, the greater the magnitude of the charges of ion and water dipole and the closer they can approach each other, the stronger is their attraction. Thus, a 2+ ion attracts $H_2O$ molecules more strongly than a 1+ ion of similar size, and a small 1+ ion attracts $H_2O$ molecules more strongly than a large 1+ ion.

Down a group of ions, such as $Li^+$ to $Cs^+$ in Group 1A(1), the charge stays the same while the size increases, so the charge densities decrease, as do the heats of hydration. Across from Group 1A(1) to Group 2A(2), a 2A ion has a smaller radius *and* a greater charge, so its charge density and $\Delta H_{hydr}$ are greater. Figure 12.8 shows this relationship for some ionic heats of hydration.

The energy required to separate an ionic solute ($\Delta H_{solute}$) into gaseous ions is the negative of its lattice energy, so it is highly positive:

$$M^+X^-(s) \rightarrow M^+(g) + X^-(g) \qquad \Delta H_{solute} = -\Delta H_{lattice} \text{ (always} > 0)$$

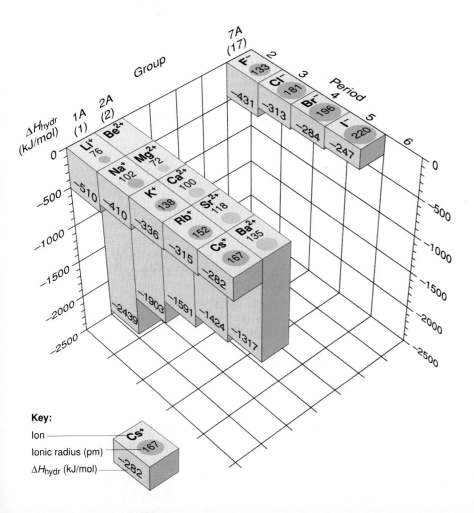

**Key:**

Ion — Cs⁺
Ionic radius (pm) — 167
$\Delta H_{hydr}$ (kJ/mol) — -282

**FIGURE 12.8**
**Trends in ionic heats of hydration.** Heats of hydration ($\Delta H_{hydr}$) are always negative because ions and water attract each other and release heat. Values for the Group 1A(1), 2A(2), and 7A(17) ions are shown as descending posts, with the ionic radius on top. The $\Delta H_{hydr}$ values depend on charge density: smaller down a group as ionic *size* increases and larger from Group 1A(1) to Group 2A(2) as ionic *charge* increases and size decreases.

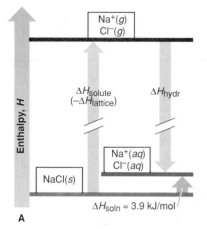

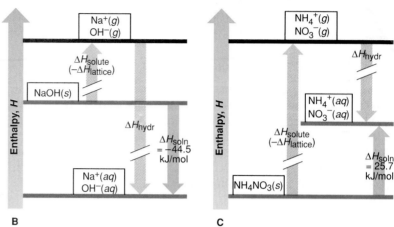

A

B

C

**FIGURE 12.9**

Enthalpy diagrams for dissolving three ionic compounds in water. The enthalpy diagram for dissolving an ionic compound in water includes the negative of the $\Delta H_{lattice}$ ($\Delta H_{solute}$; always positive) and the heat of hydration ($\Delta H_{hydr}$; always negative). **A,** For NaCl, the magnitude of $\Delta H_{lattice}$ is slightly greater than that of $\Delta H_{hydr}$, so $\Delta H_{soln}$ is small and positive. **B,** For NaOH, $\Delta H_{hydr}$ dominates, so $\Delta H_{soln}$ is large and negative. **C,** For $NH_4NO_3$, $\Delta H_{lattice}$ dominates, so $\Delta H_{soln}$ is large and positive.

♦ **Hot Packs, Cold Packs, and Self-Heating Soup.** Strain a muscle or sprain a joint and you may obtain temporary relief by applying the right heat of solution. Hot packs and cold packs consist of a heavy outer pouch containing water and a thin inner pouch containing a salt. A squeeze on the outer pouch breaks the inner pouch and the salt dissolves. Most hot packs use anhydrous $CaCl_2$ ($\Delta H_{soln} = -82.8$ kJ/mol), whereas cold packs use $NH_4NO_3$ ($\Delta H_{soln} = 25.7$ kJ/mol). The change in temperature can be quite large—a cold pack, for instance, can bring the solution from room temperature down to 0°C—but their usable time is limited to around half an hour. In Japan, double-walled cans of soup contain a salt packet and water between the walls. When you open the can, the packet breaks, and the exothermic solution process, which can quickly reach about 90°C, warms the soup.

Thus, the heat of solution for ionic compounds in water combines the previous two enthalpy terms: the negative of the lattice energy (always > 0) and the combined heats of hydration of cation and anion (always < 0),

$$\Delta H_{soln} = -\Delta H_{lattice} + \Delta H_{hydr\ of\ the\ ions} \qquad (12.2)$$

As before, the size of the individual terms determines the sign of the overall heat of solution.

Figure 12.9 shows solution enthalpy diagrams for three ionic solutes. Sodium chloride has a small endothermic heat of solution ($\Delta H_{soln} = 3.9$ kJ/mol). Its lattice energy term is only slightly greater than the combined ionic heats of hydration, so if you dissolve NaCl in water in a flask, you do not notice any temperature change. On the other hand, if you dissolve NaOH in water, the flask feels hot. The lattice energy term for NaOH is much less than the combined ionic heats of hydration, so the dissolving of NaOH is highly exothermic ($\Delta H_{soln} = -44.5$ kJ/mol). Finally, if you dissolve ammonium nitrate in water, the flask feels cold. In this case, the lattice energy term is much larger than the combined ionic heats of hydration, so the process is highly endothermic ($\Delta H_{soln} = 25.7$ kJ/mol). ♦

### The Solution Process and the Tendency Toward Disorder

The change in enthalpy of the solute-solvent system is not the only factor that determines whether a substance dissolves. Another major factor is the tendency of all systems to become more disordered over time. In thermodynamic terms, we say that the **entropy** tends to increase. *Entropy is a measure of a system's disorder,* and it increases when the system becomes more randomized. For example, stack neat piles of nickels, dimes, and quarters heads up in a closed box (the system with low entropy), give it a good shake, and look inside: the coins lie in a mixed pile, some heads up and some heads down (the system with high entropy). Chances are that if you shook the box for another century, the coins would never wind up in the stacks you started with. To get them that way, you have to expend energy sorting and restacking them to counteract the natural tendency to become mixed up. Similarly, *solutions form naturally; pure solutes and solvents do not.* Water treatment plants, oil refineries, metal foundries, and many other large industrial plants expend enormous amounts of energy reversing this natural tendency and separating mixtures into pure components.

Thus, *the solution process involves two factors—the change in heat and the change in entropy—and their relative sizes determine whether a solute dissolves in a*

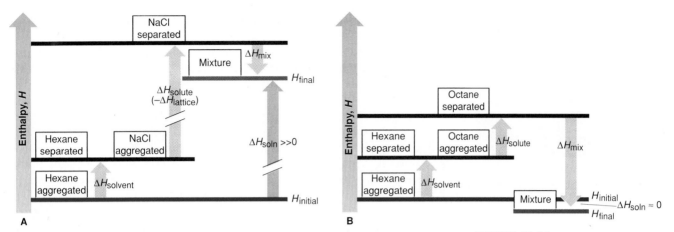

**A**                                                                                      **B**

**FIGURE 12.10**
Enthalpy diagrams for dissolving NaCl and octane in hexane. **A,** Since attractions between ions and hexane molecules are weak, $\Delta H_{mix}$ is *much* smaller than $\Delta H_{solute}$. Thus, $\Delta H_{soln}$ is so positive that NaCl does *not* dissolve in hexane. **B,** Intermolecular forces in octane and in hexane are so similar that $\Delta H_{soln}$ is very small. Octane dissolves in hexane because the solution has greater entropy (more disorder) than the pure components.

*solvent.* To see the importance of entropy in the solution process, let's consider two solute-solvent pairs, the first of which does *not* form a solution. Sodium chloride does not dissolve in hexane, as you would predict from the dissimilarity of the forces in the two compounds (Figure 12.10, *A*). Separating the solvent is relatively easy because of the weak dispersion forces, but separating the solute requires supplying the lattice energy to pull the ions apart. Mixing releases very little energy because the ion-induced dipole attractions between $Na^+$ (or $Cl^-$) ions and hexane are weak. Since the lattice energy term is so large, the sum of the endothermic terms is much larger than the small exothermic mixing term. Therefore, $\Delta H_{soln}$ is very positive: dissolution does not occur because too much energy is needed. In this case, the increase in entropy that would result from the mixing of solute and solvent does not outweigh the increase in enthalpy.

Now consider the second solute-solvent pair, octane ($C_8H_{18}$) and hexane ($C_6H_{14}$). Both consist of nonpolar molecules held together by dispersion forces of comparable strength. We would therefore predict that octane is soluble in hexane; in fact, they are infinitely soluble (miscible). At first, you might expect that a great deal of heat is released when they dissolve ($\Delta H_{soln} << 0$); actually, the heat of solution is around zero ($\Delta H_{soln} \approx 0$) (Figure 12.10, *B*). Why, then, do these substances form a solution? Octane dissolves in hexane because *the entropy of the system increases as pure substances mix together and become disordered.* In some cases, a large enough increase in entropy can cause a solution to form even when heat *is* required ($\Delta H_{soln} > 0$). As noted earlier, the solution becomes cold when $NH_4NO_3$ dissolves in water (see Figure 12.9, *C*), but the increase in entropy that occurs when the orderly crystal breaks down and the ions mix with water molecules more than compensates for the heat required.

## Section Summary

An overall heat of solution can be obtained from a thermochemical cycle as the sum of two endothermic steps, solute separation and solvent separation, and one exothermic step, solute-solvent mixing. In aqueous solutions, the combination of solvent separation and mixing is called hydration. Ionic heats of hydration are always negative because of strong ion-dipole forces. Systems have a natural tendency to increase their entropy (become disordered), and a solution has greater entropy (more disorder) than the pure solute and solvent. The combination of enthalpy and entropy changes determines whether a solution forms. A substance with a positive $\Delta H_{soln}$ dissolves *only* if the entropy increase is large enough to outweigh it.

**FIGURE 12.11**
**Sodium acetate crystallized from a supersaturated solution.** When a seed crystal of sodium acetate is added to a supersaturated solution of the compound, solute crystallizes out of solution until the solution remaining is saturated.

◆ **A Pure Liquid and Its Vapor Are Like a Saturated Solution.** Equilibrium processes in a saturated solution are very analogous to those of a pure liquid in a closed flask (Section 11.5). In the liquid, rates of vaporizing and condensing are equal; in the solution, rates of dissolving and crystallizing are equal. In the liquid, particles leave the liquid to enter the vapor, and their concentration (pressure) increases until, at equilibrium, the available space is "saturated" with vapor at a given temperature. In the solution, particles leave the solid solute to enter the solvent, and their concentration increases until, at equilibrium, the available solvent is saturated with solute at a given temperature.

**FIGURE 12.12**
**Equilibrium in a saturated solution.** In a saturated solution, equilibrium exists between the excess solute and the solution. At a particular temperature, the number of solute particles dissolving in a given time equals the number recrystallizing.

## 12.3  Solubility as an Equilibrium Process

A solution containing the maximum amount of solute that will dissolve under a given set of conditions is **saturated.** If it contains less than this amount of solute, it is **unsaturated:** add more solute, and more will dissolve until the solution becomes saturated. Add still more solute, and it will remain undissolved. In some cases, when a solution is prepared at an elevated temperature and then slowly cooled, more than the usual maximum amount of solute remains dissolved: this solution is **supersaturated.** Such a solution is unstable: the addition of a "seed" crystal of solute or any physical shock—stirring or tapping the container—causes the excess solute to crystallize immediately, leaving behind a saturated solution (Figure 12.11).

On the molecular level, *a saturated solution is in equilibrium with excess solute.* As an ionic solid dissolves, for example, ions leave the solid, so their concentration in the solution increases. Because of their increasing numbers, ions in solution collide more frequently with the undissolved solute, and some recrystallize. Eventually, their concentration becomes constant, indicating that the solution is saturated. At this point, ions from the solid are entering the solution at the same rate that dissolved ions are recrystallizing (Figure 12.12):

$$\text{Solute (crystallized)} \rightleftharpoons \text{solute (dissolved)}$$

Although dissolving and recrystallizing continue, *the net amount of dissolved solute remains constant at a given temperature.* ◆

### Effect of Temperature on Solubility

Temperature has a major effect on the solubility of most substances. You may have noticed, for example, that not only does sugar dissolve more quickly in hot tea than in iced tea, but *more* sugar dissolves: the solubility of sugar is greater at higher temperatures.

The equilibrium nature of solubility is the key to this effect of temperature. As pointed out in Chapter 11, when a system at equilibrium is disturbed, it adjusts to reduce the disturbance and reach equilibrium again. Consider a saturated solution that has a *positive* heat of solution. Since heat is needed for the solution to form, we include it in the equation for purposes of discussion:

$$\text{Solute} + \text{solvent} + heat \rightleftharpoons \text{solution}$$

If the system at equilibrium is disturbed by a temperature rise (that is, if more heat is "added"), the system absorbs some of the added heat by the forward process speeding up. More solute dissolves and equilibrium becomes reestablished with more ions in solution. *Solubility increases with temperature if the solution process is endothermic* ($\Delta H_{soln} > 0$). The solubility of most ionic compounds *increases* with temperature (Figure 12.13); that is, most have positive heats of solution, which means that the lattice energy is larger than the heat of hydration.

If a solute has a negative heat of solution, heat is given off:

$$\text{Solute} + \text{solvent} \rightleftharpoons \text{solution} + heat$$

As before, the system at equilibrium will absorb some of the added heat in response to a temperature rise. In this case, however, it does so by the reverse process speeding up. Some solute crystallizes out and equilibrium be-

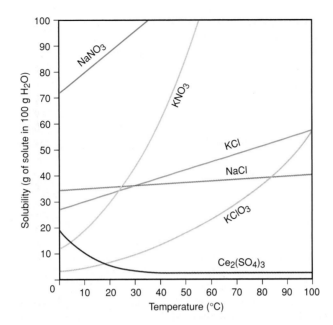

FIGURE 12.13
The relation between solubility and temperature for several ionic compounds. Most ionic compounds have a higher solubility at higher temperatures because their heats of solution are positive. Cerium sulfate is less soluble at higher temperature because its heat of solution is negative.

comes reestablished with fewer ions in solution. *Solubility decreases with temperature if the solution process is exothermic* ($\Delta H_{soln} < 0$). Potassium hydroxide and sodium sulfate are two compounds that become less soluble as the temperature rises.

SAMPLE PROBLEM 12.2

### Predicting the Effect of Temperature on Solubility

**Problem:** From the following information, predict whether the solubility of each compound increases or decreases with an increase in temperature:
(a) $\Delta H_{soln}$ NaOH($s$) = −44.5 kJ/mol
(b) When $KNO_3$ dissolves in water, the solution becomes cold.

(c) $CsCl(s) \overset{H_2O}{\rightleftharpoons} Cs^+(aq) + Cl^-(aq); \quad \Delta H_{soln} = +17.8$ kJ
**Plan:** We use the information to write a reaction that includes heat being absorbed (left) or released (right). If heat is on the left, a temperature increase shifts the system to the right, so more solute dissolves. If heat is on the right, a temperature increase shifts the system to the left, so less solute dissolves.
**Solution: (a)** A negative $\Delta H$ means the process is exothermic, so when 1 mol NaOH dissolves in water, 44.5 kJ of heat is released:

$$NaOH(s) \overset{H_2O}{\rightleftharpoons} Na^+(aq) + OH^-(aq) + heat$$

A higher temperature (more heat) **decreases** the solubility of NaOH.
**(b)** When $KNO_3$ dissolves, the solution becomes cold, so heat is absorbed:

$$KNO_3(s) + heat \overset{H_2O}{\rightleftharpoons} K^+(aq) + NO_3^-(aq)$$

A higher temperature **increases** the solubility of $KNO_3$.
**(c)** The positive $\Delta H$ means heat is absorbed when CsCl dissolves:

$$CsCl(s) + heat \overset{H_2O}{\rightleftharpoons} Cs^+(aq) + Cl^-(aq)$$

A higher temperature **increases** the solubility of CsCl.
**Check:** Be sure the sign of $\Delta H_{soln}$ corresponds with the answer. In part **(a)**, the process is exothermic, so $\Delta H_{soln}$ is negative and adding heat decreases solubility. In parts **(b)** and **(c)**, the process is endothermic, so $\Delta H_{soln}$ is positive and adding heat increases solubility.

**FOLLOW-UP PROBLEM 12.2**
Predict whether the solubility increases or decreases at lower temperature and give the sign of $\Delta H_{soln}$ for the following cases:

**(a)** When sodium cyanate (NaOCN) dissolves, the solution becomes cold.

**(b)** $KF(s) \underset{}{\overset{H_2O}{\rightleftharpoons}} \quad K^+(aq) + F^-(aq) + 17.7 \text{ kJ}$

---

The solution cycle for gases influences the effect of temperature on gas solubility. In a gas, the solute is already separated; in fact, the particles are farther apart than when the gas dissolves. Therefore, instead of energy being absorbed to separate the solute, it is released when the gas particles are brought closer together to form the solution: $\Delta H_{solute}$ is exothermic. Since the solvation step is also exothermic, all gases have a negative $\Delta H_{soln}$, which means that *gas solubility decreases with rising temperature*. This fact makes sense in terms of intermolecular forces. Gases have weak intermolecular forces and form weak attractions with the solvent. Therefore, when the temperature rises and the average kinetic energy increases, the solute molecules overcome these weak forces and enter the gas phase.

The effect of temperature on gas solubility can lead to an environmental problem known as *thermal pollution*. During many industrial processes, large amounts of water are taken from a nearby river or lake, pumped through the system to cool liquids, gases, and equipment, and then returned to the body of water at a higher temperature. Thus, less $O_2$ can dissolve in the warmer water near the plant exit. Severe $O_2$ depletion can kill fish and alter other aquatic animal populations. Farther from the plant, the water temperature returns to ambient levels, and the $O_2$ solubility increases accordingly.

## Effect of Pressure on Solubility

Since liquids and solids are almost incompressible, pressure has little effect on their solubility. In contrast, it has a major effect on gas solubility. Consider a piston-cylinder assembly with a gas above a saturated aqueous solution of the gas (Figure 12.14, *A*). At a given pressure, the system is at equilibrium; that is, the same number of gas molecules enter and leave the solution per unit time:

$$\text{Gas} + \text{solvent} \rightleftharpoons \text{solution}$$

**FIGURE 12.14**
**The effect of pressure on gas solubility. A,** A saturated solution of a gas is in equilibrium at pressure $P_1$. **B,** If the pressure is increased to $P_2$, the volume of the gas decreases, so the frequency of collisions with the surface increases and more gas is in solution when equilibrium is reestablished.

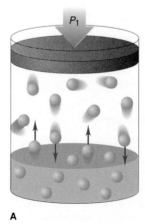
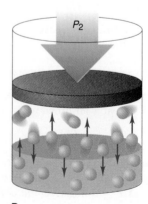

A                                    B

If you push down on the piston, the gas volume decreases, its pressure increases, and gas particles collide with the liquid more often (Figure 12.14, *B*). Therefore, more gas particles enter the solution than leave it per unit time. In terms of the equilibrium process, higher gas pressure disturbs the balance, so more gas dissolves to reduce this disturbance (the equation shifts to the right). ◆

**Henry's law** expresses the quantitative relationship between gas pressure and solubility: *the solubility of a gas ($S_{gas}$) is directly proportional to the partial pressure of the gas ($P_{gas}$) above the solution:*

$$S_{gas} = k_H \times P_{gas} \qquad (12.3)$$

where $k_H$ is the Henry's law constant and is specific for each gas-solvent combination at a given temperature. When the solubility, $S_{gas}$, is in units of mol/L and pressure is in atm, the units of $k_H$ are mol/L · atm.

SAMPLE PROBLEM 12.3 _____

### Using Henry's Law to Calculate Gas Solubility

**Problem:** In a soft-drink plant, the partial pressure of $CO_2$ gas inside a bottle of cola is adjusted to 4 atm at 25°C. What is the solubility of $CO_2$ under these conditions? Henry's law constant for $CO_2$ in water = $3 \times 10^{-2}$ mol/L · atm at 25°C.
**Plan:** We know $P_{CO_2}$ and the value of the Henry's law constant $k_H$, so we substitute them into Equation 12.3 to find $S_{CO_2}$.
**Solution:** $S_{CO_2} = k_H \times P_{CO_2} = (3 \times 10^{-2}$ mol/L · atm$) \times 4$ atm $= \mathbf{0.1\ mol/L}$
**Check:** With only one significant figure in the $k_H$ value, be sure to round the answer to one significant figure.

FOLLOW-UP PROBLEM 12.3
Henry's law constant for $N_2$ in water at 25°C is $7 \times 10^{-4}$ mol/L · atm. What is the solubility of $N_2$ in water at 25°C and 1.0 atm if air contains 78% $N_2$ by volume?

_____

### Section Summary
A stable solution that contains the maximum amount of dissolved solute at a given temperature is saturated. A state of equilibrium exists in a saturated solution, with particles entering and leaving the solution at the same rate. For an endothermic solution process ($\Delta H_{soln} > 0$), a temperature rise increases solubility; for an exothermic process ($\Delta H_{soln} < 0$), a temperature rise decreases solubility. All gases have a negative $\Delta H_{soln}$, so heating lowers gas solubility. Henry's law says that the solubility of a gas is directly proportional to its partial pressure.

◆ **Scuba Diving and Soda Pop.** Although $N_2$ is barely soluble in water and blood at sea level, high external pressures increase its solubility significantly. Scuba divers who breathe compressed air and dive below about 50 ft have considerably more $N_2$ dissolved in their blood and must ascend slowly to lower pressures. Otherwise, dissolved $N_2$ may bubble out of solution in the blood, causing the painful, sometimes fatal blockage of capillaries called "the bends," or decompression sickness. This condition can be avoided by using less soluble gases, such as He, in the breathing mixtures. In principle, the same thing happens when you open a can of a carbonated drink. In the closed can, dissolved $CO_2$ is under $CO_2$ gas at 4 atm. In the open can, it is all at once under air at 1 atm (with $P_{CO_2} = 3 \times 10^{-4}$ atm), so $CO_2$ bubbles out of solution. With time, the system again reaches equilibrium: the drink goes "flat."

## 12.4  Quantitative Ways of Expressing Concentration

Concentration is the *proportion* of a substance in a mixture, so it does *not* depend on the amount of mixture present: 1.0 L of 0.1 *M* NaCl has the same concentration as 1.0 mL of 0.1 *M* NaCl. Concentration is most often expressed as the ratio of the amount of solute to the amount of *solution,* but sometimes it is the ratio of solute to *solvent.* Since both parts of the ratio can

**TABLE 12.3 Concentration Definitions**

| CONCENTRATION TERM | RATIO |
|---|---|
| Molarity ($M$) | $\dfrac{\text{moles of solute}}{\text{liter of solution}}$ |
| Molality ($m$) | $\dfrac{\text{moles of solute}}{\text{kilogram of solvent}}$ |
| Parts by mass | $\dfrac{\text{mass of solute}}{\text{mass of solution}}$ |
| Parts by volume | $\dfrac{\text{volume of solute}}{\text{volume of solution}}$ |
| Mole fraction ($X$) | $\dfrac{\text{moles of solute}}{\text{moles of solute + moles of solvent}}$ |

be given in terms of mass, volume, or moles, chemists employ a variety of concentration terms, including molarity, molality, and various expressions of "parts of solute per parts of solution" (Table 12.3).

**Molarity and Molality**

Molarity ($M$) is defined as the *moles of solute dissolved in one liter of solution:*

$$\text{Molarity} = \frac{\text{moles of solute}}{\text{liter of solution}} \qquad \text{or} \qquad M = \frac{\text{mol solute}}{\text{L soln}} \qquad \textbf{(12.4)}$$

As you know from Chapter 3, the use of molarity allows us to convert volume of solution directly into moles of dissolved solute. Expressing concentration as molarity may have drawbacks, however. Since volume is affected by temperature, so is molarity. A solution expands slightly when heated, so each unit volume contains slightly less solute than at lower temperatures, and this can be a source of error in very precise work. More importantly, because of solute-solvent interactions that are difficult to predict, *solution volumes are not necessarily additive;* that is, adding exactly 500 mL of one solution to exactly 500 mL of another may not give exactly 1000 mL. In such cases, a solution with a desired molarity may not be easy to prepare.

One of the units that does not contain volume in its ratio of dimensions is **molality ($m$),** which is defined as *the moles of solute dissolved in 1000 g (1 kg) of solvent:*

$$\text{Molality} = \frac{\text{moles of solute}}{\text{kilogram of solvent}} \qquad \text{or} \qquad m = \frac{\text{mol solute}}{\text{kg solvent}} \qquad \textbf{(12.5)}$$

Note that molality is expressed in terms of amount of *solvent,* not solution. Molal solutions are prepared by measuring *masses* of solute and solvent, not solvent or solution volume. The mass of the resulting solution does not change with temperature, so the concentration does not change. Moreover, masses *are* additive, unlike volumes: adding exactly 500 g of one solution to exactly 500 g of another *does* give exactly 1000 g of final solution. For these reasons, molality is a preferred unit for examining properties that depend on mass (or moles) of solute and solvent, such as the effect of solute on the physical properties of a solution.

SAMPLE PROBLEM 12.4 _____

### Calculating Molality

**Problem:** What is the molality of a solution prepared by mixing 32.0 g $CaCl_2$ with 271 g water?

**Plan:** We convert mass of $CaCl_2$ to moles with the molar mass ($\mathcal{M}$) and then divide by the mass of water (in kg).

**Solution:** Converting from mass of solute to moles:

$$\text{Moles of } CaCl_2 = \frac{32.0 \text{ g } CaCl_2}{110.98 \text{ g } CaCl_2/1 \text{ mol } CaCl_2} = 0.288 \text{ mol } CaCl_2$$

Finding molality:

$$\text{Molality} = \frac{\text{mol solute}}{\text{kg solvent}} = \frac{0.288 \text{ mol } CaCl_2}{271 \text{ g } (1 \text{ kg}/10^3 \text{ g})} = \textbf{1.06 } \textit{m} \textbf{ } CaCl_2$$

**Check:** The molality seems reasonable: since the moles of $CaCl_2$ and mass (in kg) of $H_2O$ are about the same numerically, their ratio is about 1.

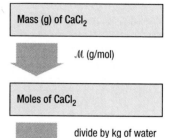

FOLLOW-UP PROBLEM 12.4
What mass of glucose ($C_6H_{12}O_6$) must be dissolved in 563 g ethanol ($C_2H_5OH$) to prepare a $2.40 \times 10^{-2} \; m$ solution?

### Parts of Solute by Parts of Solution

Several concentration terms are based on the number of solute (or solvent) parts present in a specific number of *solution* parts. The solution parts can be expressed in terms of mass, volume, or moles.

   **Parts by mass.** The most common of the parts-by-mass terms is **mass percent,** which you encountered in Chapter 3. The word "percent" means "per hundred," so mass percent of solute means the number of grams of solute in every 100 grams of solution:

$$\text{Mass percent} = \frac{\text{mass of solute}}{\text{mass of solute} + \text{mass of solvent}} \times 100$$

$$= \frac{\text{mass of solute}}{\text{mass of solution}} \times 100 \qquad \textbf{(12.6)}$$

Sometimes mass percent is written as **% (w/w),** the "w/w" indicating the percent is a ratio of weights (more accurately, masses). You may have seen mass percent values on bottles of solid chemicals to indicate the amounts of impurities present. Two very similar units are parts per million (ppm) by mass and parts per billion (ppb) by mass: grams of solute per million or per billion grams of solution. ◆

   **Parts by volume.** Solution parts measured by volume are also expressed as parts per hundred (percent), per million, or per billion. Thus, **volume percent** is the volume of solute in 100 volumes of solution:

$$\text{Volume percent} = \frac{\text{volume of solute}}{\text{volume of solution}} \times 100 \qquad \textbf{(12.7)}$$

A common symbol for this term is **% (v/v).** Commercial rubbing alcohol, for example, is an aqueous solution of isopropanol (a three-carbon alcohol) that contains 70 volumes of isopropanol per 100 volumes of solution, and the label indicates this as "70% (v/v)." Parts-by-volume concentrations are most often used for liquids and gases. Minor atmospheric components occur

◆ **Unhealthy Ultralow Concentrations.** Environmental toxicologists often measure extremely low concentrations of pollutants. The Centers for Disease Control and Prevention consider TCDD one of the most toxic substances known. It is a member of the dioxin family of chlorinated hydrocarbons, a byproduct of bleaching steps in paper manufacture. TCDD is considered unsafe at soil levels above 1 ppb. From contact with air, water, and soil, it is estimated that most Americans and Canadians already have an average of 0.01 ppb TCDD in their fatty tissue. In a recent reassessment of TCDD, the Environmental Protection Agency found that this substance appears to be unsafe at all levels measured.

◆ **Pheromones: Very Dilute Calls to Action.** Pheromones are organic compounds secreted by one member of a species to signal other members about food, danger, sexual readiness, and so forth. Many organisms, including dogs and monkeys, release pheromones, and researchers suspect that humans do as well. Most studies involve social insects. Wasp pheromones are active at only a few hundred molecules per milliliter of air, about 100 parts per quadrillion by volume!

in parts per million by volume (ppmv). For example, there are about 0.050 ppmv (50 ppbv) of carbon monoxide (CO) in clean air, 1000 times as much (about 50 ppmv of CO) in air over urban traffic, and 10 times that amount (about 500 ppmv of CO) in cigarette smoke. ◆

A symbol frequently used for aqueous solutions is % (w/v), a ratio of solute *weight* (actually mass) to solution *volume*. Thus, a 1.5% (w/v) NaCl solution contains 1.5 g NaCl per 100 mL of *solution*. You should also be aware that several handbooks, including the *Handbook of Chemistry and Physics*, cite solubility values in terms of grams of solute per 100 mL of *solvent*.

**Mole fraction.** The **mole fraction** *(X)* of solute is the ratio of solute moles to total moles (solute plus solvent), that is, parts by moles. The *mole percent* is the mole fraction expressed as a percentage:

$$\text{Mole fraction } (X) = \frac{\text{moles of solute}}{\text{moles of solute} + \text{moles of solvent}} \tag{12.8}$$

$$\text{Mole percent (mol \%)} = \text{mole fraction} \times 100$$

We discussed these terms in Chapter 5 in relation to Dalton's law of partial pressures for mixtures of gases, but they apply to liquids and solids as well. Concentrations based on mole fraction give the most direct picture of the actual number of solute (or solvent) particles present in the solution.

SAMPLE PROBLEM 12.5

### Expressing Concentration in Parts by Mass, Parts by Volume, and Mole Fraction

**Problem: (a)** Calculate the ppm by mass of calcium in a 3.50 g pill that contains 40.5 mg Ca.
**(b)** The label on a 0.750-L bottle of Italian Chianti indicates "11.5% alcohol by volume." What volume of alcohol (in L) does it contain?
**(c)** A sample of rubbing alcohol contains 142 g isopropanol ($C_3H_7OH$) and 58.0 g water. What are the mole fractions of alcohol and water?
**Plan: (a)** We convert mg Ca to g and then use mass of Ca/mass of pill and multiply by $10^6$ to obtain ppm.
**(b)** We know the vol % and the total volume, so we use Equation 12.7 to find the volume of alcohol.
**(c)** We know the mass and formula of each component, so we convert both to moles and apply Equation 12.8 to find mole fraction.
**Solution: (a)** Finding parts per million by mass. Combining the steps, we have

$$\text{ppm Ca} = \frac{\text{mass of Ca}}{\text{mass of pill}} \times 10^6 = \frac{40.5 \text{ mg Ca} \times 1 \text{ g}/10^3 \text{ mg}}{3.50 \text{ g}} \times 10^6$$

$$= \mathbf{1.16 \times 10^4 \ ppm \ Ca}$$

**(b)** Finding volume of alcohol:

$$\text{Volume (L) of alcohol} = \frac{\text{vol \%} \times \text{vol Chianti}}{100} = \frac{11.5 \times 0.750 \text{ L}}{100} = \mathbf{0.0862 \ L}$$

**(c)** Finding mole fractions. Converting from mass to moles:

$$\text{Moles of } C_3H_7OH = \frac{142 \text{ g } C_3H_7OH}{60.09 \text{ g } C_3H_7OH/1 \text{ mol } C_3H_7OH} = 2.36 \text{ mol } C_3H_7OH$$

$$\text{Moles of } H_2O = \frac{58.0 \text{ g } H_2O}{18.02 \text{ g } H_2O/1 \text{ mol } H_2O} = 3.22 \text{ mol } H_2O$$

Calculating mole fractions:

$$X_{C_3H_7OH} = \frac{\text{moles of } C_3H_7OH}{\text{total moles}} = \frac{2.36 \text{ mol}}{2.36 \text{ mol} + 3.22 \text{ mol}} = \mathbf{0.423}$$

$$X_{H_2O} = \frac{\text{moles of } H_2O}{\text{total moles}} = \frac{3.22 \text{ mol}}{2.36 \text{ mol} + 3.22 \text{ mol}} = \mathbf{0.577}$$

**Check: (a)** The mass ratio is ~$0.04/4 = 10^{-2}$, and $10^{-2} \times 10^6 = 10^4$ ppm, so it seems correct. **(b)** The vol % is slightly more than 10%, so the volume of alcohol should be slightly more than 75 mL (0.075 L). **(c)** Always check that the *mole fractions add to 1*: $0.423 + 0.577 = 1.000$.

**FOLLOW-UP PROBLEM 12.5**
An alcohol solution contains 35.0 g propanol ($C_3H_7OH$) and 150.0 g ethanol ($C_2H_5OH$). Calculate the mass percent and the mole fraction of each alcohol.

## Converting Units of Concentration

The units we just discussed merely represent different ways of expressing the concentration of a given solution, so they are interconvertible. We convert a unit based on moles to one based on mass with the molar mass. To convert a unit based on mass to one based on volume, we need the *density* of the solution. Remember that molality involves amount of *solvent*, whereas the others involve amount of *solution*.

**SAMPLE PROBLEM 12.6** _____

## Converting Concentration Units

**Problem:** Hydrogen peroxide is a powerful oxidizing agent that is used in concentrated solution in rocket fuel systems and in dilute solution as a hair bleach. An aqueous solution of $H_2O_2$ is 30.0% by mass and has a density of 1.11 g/mL. Calculate its
**(a)** Molality      **(b)** Mole fraction of $H_2O_2$      **(c)** Molarity.
**Plan:** We know the mass % and the density. **(a)** For molality, we need the moles of solute and the mass of *solvent*. Assuming 100.0 g of solution, we subtract the mass of $H_2O_2$ to obtain the mass of solvent. Then we convert mass of $H_2O_2$ to moles and divide by mass of solvent (in kg). **(b)** Converting the masses of $H_2O_2$ and $H_2O$ to moles and dividing moles of $H_2O_2$ by their sum give the mole fraction. **(c)** We use the solution density to convert its mass to volume. From the moles of $H_2O_2$ and *solution* volume, we find molarity.
**Solution: (a)** From mass % to molality. Finding mass of solvent (assuming 100.0 g of solution):

$$\text{Mass (g) of } H_2O = 100.0 \text{ g solution} - 30.0 \text{ g } H_2O_2 = 70.0 \text{ g } H_2O$$

Converting from mass of $H_2O_2$ to moles:

$$\text{Moles of } H_2O_2 = \frac{30.0 \text{ g } H_2O_2}{34.02 \text{ g } H_2O_2/1 \text{ mol } H_2O_2} = 0.882 \text{ mol } H_2O_2$$

Calculating molality:

$$\text{Molality of } H_2O_2 = \frac{0.882 \text{ mol } H_2O_2}{70.0 \text{ g } (1 \text{ kg}/10^3 \text{ g})} = \mathbf{12.6 \text{ } m \text{ } H_2O_2}$$

**(b)** From mass % of $H_2O_2$ to mole fraction:

$$\text{Moles of } H_2O_2 = 0.882 \text{ mol (from part a)}$$

$$\text{Moles of } H_2O = \frac{70.0 \text{ g } H_2O}{18.02 \text{ g } H_2O/1 \text{ mol } H_2O} = 3.88 \text{ mol } H_2O$$

$$X_{H_2O_2} = \frac{0.882 \text{ mol}}{0.882 \text{ mol} + 3.88 \text{ mol}} = \textbf{0.185}$$

**(c)** From mass % and density to molarity. Converting from solution mass to volume:

$$\text{Volume (mL) of solution} = 100.0 \text{ g} \times \frac{1 \text{ mL}}{1.11 \text{ g}} = 90.1 \text{ mL}$$

Calculating molarity:

$$\text{Molarity} = \frac{\text{mol } H_2O_2}{\text{L soln}} = \frac{0.882 \text{ mol } H_2O_2}{90.1 \text{ mL } (1 \text{ L soln}/10^3 \text{ mL})} = \textbf{9.79 } \textbf{\textit{M}} \textbf{ } \textbf{H}_2\textbf{O}_2$$

**Check:** The answers seem reasonable. **(a)** The ratio of ~0.9/0.07 is greater than 10. **(b)** Rounding gives ~1 mol $H_2O_2$/5 mol = 0.2. **(c)** The ratio is slightly less than 10. The *molarity should be lower than the molality* because the solvent is only part of the solution.

**FOLLOW-UP PROBLEM 12.6**
A sample of commercial concentrated hydrochloric acid is 11.8 *M* HCl and has a density of 1.190 g/mL. Calculate the mass % HCl, molality, and mole fraction of HCl.

**Section Summary**
The concentration of a solution is independent of the amount of solution and can be expressed by molarity (mol solute/L solution) and molality (mol solute/kg solvent), as well as parts by mass (mass solute/mass solution), parts by volume (volume solute/volume solution), and mole fraction [mol solute/(mol solute + mol solvent)]. The choice of units depends on convenience or the nature of the solution. If the solution density is known, all concentration units are interconvertible.

## 12.5  Colligative Properties of Solutions

We might expect the presence of solute particles to make the physical properties of a solution different from those of the pure solvent. However, what we might not expect is that, in the case of four important solution properties, it is the *number* of solute particles that makes the difference, *not* their chemical identity. These properties, known as **colligative properties** ("colligative" means "collective"), are vapor pressure lowering, boiling point elevation, freezing point depression, and osmotic pressure. Even though these effects are generally quite small, they have many practical applications, including some that are vital to biological systems.

   Historically, colligative properties were measured to explore how solutes exist in aqueous solution. One way to classify solutes is by their ability to conduct an electric current, which implies that moving ions are present. Chapter 4 pointed out that an aqueous solution of an **electrolyte** conducts a current because the solute separates into ions when it dissolves. Soluble

A            B            C

**FIGURE 12.15**
**The three types of electrolytes. A,** Strong electrolytes conduct a large current because they dissociate completely into ions. **B,** Weak electrolytes conduct a small current because they dissociate very little. **C,** Nonelectrolytes conduct no current because they do not dissociate.

salts, strong acids, and strong bases dissociate completely and their solutions are highly conductive, so these solutes are *strong electrolytes*. Weak acids and bases dissociate very little; they are *weak electrolytes* because their solutions conduct little current. (A slightly soluble salt is considered a strong electrolyte because the small amount that dissolves dissociates completely, even though only a weak current forms because few ions are present.) Many compounds, such as sugar and alcohol, do not dissociate when they dissolve; they are **nonelectrolytes** because their solutions do not conduct current. Figure 12.15 shows these behaviors.

To determine the moles of particles in solution, we refer to the solute formula. Each mole of dissolved nonelectrolyte yields one mole of particles. For example, 0.35 $M$ glucose contains 0.35 mol of solute particles per liter. In principle, each mole of strong electrolyte, on the other hand, dissociates into as many moles of ions as the formula unit indicates: 0.4 $M$ $Na_2SO_4$ contains 0.8 mol $Na^+$ ions and 0.4 mol $SO_4^{2-}$ ions, or 1.2 mol of particles, per liter (see Sample Problem 4.3). We discuss the dissociation of weak electrolytes in Chapters 17 and 18.

## Colligative Properties of Nonvolatile Nonelectrolytes

In this section, we focus on the simplest case, the colligative properties of dilute solutions of nonvolatile nonelectrolytes, such as sucrose (table sugar). Such solutes do not dissociate into ions and have negligible vapor pressure even at the boiling point of the solvent. Later in the section, we briefly explore the behavior of volatile nonelectrolytes and electrolytes as well.

**Vapor pressure lowering.** Experiment shows that the vapor pressure of a solution of nonvolatile nonelectrolyte is always *lower* than that of the pure solvent. A molecular view illustrates why this **vapor pressure lowering ($\Delta P$)** occurs (Figure 12.16). With the pure solvent, the surface consists entirely of molecules that can vaporize, and equilibrium occurs when the number of molecules entering the gas phase in a given time equals the number returning to the liquid; that is, when the rate of vaporization equals the rate of condensation. In the solution, some of the surface is occupied by the nonvolatile solute molecules; thus, the fraction of the surface occupied by solvent molecules is smaller, so the rate of vaporization must be lower. To establish equilibrium again at this lower vaporization rate, the molecules in the gas phase must condense at a lower rate: therefore, equilibrium occurs at a lower vapor pressure.

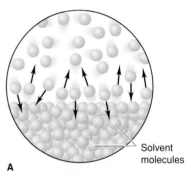

A                                    Solvent molecules

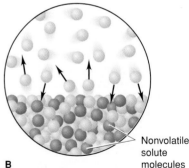

B                                    Nonvolatile solute molecules

**FIGURE 12.16**
**The effect of solute on the vapor pressure of a solution. A,** Equilibrium occurs between pure liquid and its vapor when the numbers of molecules vaporizing and condensing in a given time are equal. **B,** With a nonvolatile solute present, fewer solvent molecules vaporize in a given time, so fewer molecules need to condense to balance them. Thus, equilibrium is reached at a lower vapor pressure.

In quantitative terms, we find that the vapor pressure of solvent above the solution equals the mole fraction of the solvent in the solution (always < 1) times the vapor pressure of the pure solvent. This relationship is expressed by **Raoult's law:**

$$P_{solvent} = X_{solvent} \times P^0_{solvent} \tag{12.9}$$

where $P_{solvent}$ is the vapor pressure of the solvent above the *solution;* $X_{solvent}$ is the mole fraction of the solvent in the solution; and $P^0_{solvent}$ is the vapor pressure of the pure solvent.

An **ideal solution** follows Raoult's law, but just as most gases deviate from ideality, so do most solutions. Nevertheless, Raoult's law gives a good approximation of the behavior of *very dilute* solutions.

How does the *amount* of solute affect the *magnitude* of the vapor pressure lowering, $\Delta P$? The solution consists of solute and solvent, so the sum of their mole fractions equals one:

$$X_{solvent} + X_{solute} = 1 \qquad \text{thus} \qquad X_{solvent} = 1 - X_{solute}$$

From Raoult's law, we have

$$P_{solvent} = X_{solvent} \times P^0_{solvent} = (1 - X_{solute}) \times P^0_{solvent}$$

Multiplying through on the right side gives

$$P_{solvent} = P^0_{solvent} - X_{solute} \times P^0_{solvent}$$

Rearranging and introducing $\Delta P$ gives

$$P^0_{solvent} - P_{solvent} = \Delta P = X_{solute} \times P^0_{solvent} \tag{12.10}$$

Thus, the *magnitude* of the lowering, $\Delta P$, equals the mole fraction of the nonvolatile solute times the vapor pressure of the pure solvent.

SAMPLE PROBLEM 12.7 ———————————————

### Using Raoult's Law to Find the Vapor Pressure Lowering

**Problem:** Calculate the vapor pressure lowering when 10.0 mL glycerol ($C_3H_8O_3$) is added to 500.0 mL water at 50°C. At this temperature, the vapor pressure of pure water is 92.5 torr and its density is 0.988 g/mL. The density of glycerol is 1.26 g/mL.

**Plan:** To calculate $\Delta P$ from Equation 12.10, we use the given vapor pressure of pure water and calculate the mole fraction of glycerol. We use the given density of glycerol to convert volume to mass, and use the molar mass (obtained from the formula) to convert to moles. The same procedure gives moles of $H_2O$. From this, we calculate the mole fraction of glycerol and $\Delta P$.

**Solution:** Calculating the moles of glycerol and of water:

$$\text{Moles of glycerol} = 10.0 \text{ mL glycerol} \times \frac{1.26 \text{ g glycerol}}{1 \text{ mL glycerol}} \times \frac{1 \text{ mol glycerol}}{92.09 \text{ g glycerol}}$$

$$= 0.137 \text{ mol glycerol}$$

$$\text{Moles of } H_2O = 500.0 \text{ mL } H_2O \times \frac{0.988 \text{ g } H_2O}{1 \text{ mL } H_2O} \times \frac{1 \text{ mol } H_2O}{18.02 \text{ g } H_2O} = 27.4 \text{ mol } H_2O$$

Calculating the mole fraction of glycerol:

$$X_{glycerol} = \frac{0.137 \text{ mol}}{0.137 \text{ mol} + 27.4 \text{ mol}} = 0.00498$$

Finding the vapor pressure lowering:

$$\Delta P = X_{glycerol} \times P^0_{H_2O} = 0.00498 \times 92.5 \text{ torr} = \textbf{0.461 torr}$$

**Check:** The small $\Delta P$ is reasonable because the mole fraction of solute is small.

Volume (mL) of glycerol (or $H_2O$)

density (g/mL)

Mass (g) of glycerol (or $H_2O$)

$\mathcal{M}$ (g/mol)

Moles of glycerol (or $H_2O$)

divide by total moles

Mole fraction ($X$) of glycerol

multiply by $P^0_{H_2O}$

Vapor pressure lowering ($\Delta P$)

FOLLOW-UP PROBLEM 12.7
Calculate the vapor pressure lowering when 2.00 g aspirin ($\mathcal{M}$ = 180.16 g/mol) is dissolved in 50.0 g methanol ($CH_3OH$) at 21.2°C. Pure methanol has a vapor pressure of 101 torr at this temperature.

**Boiling point elevation.** The lowering of the solution's vapor pressure is responsible for the elevation of its boiling point above that of the pure solvent. As you saw in Chapter 11, the boiling point (boiling temperature, $T_b$) of a liquid is the temperature at which its vapor pressure equals the external pressure. A solution does not boil when it reaches the solvent's boiling point because its vapor pressure is still lower than the external pressure; a higher temperature is needed to raise its vapor pressure to equal the external pressure. We can see the **boiling point elevation ($\Delta T_b$)** that results from the presence of the nonvolatile solute by superimposing a phase diagram for the solution on one for the pure solvent (Figure 12.17). Note that the gas-liquid curve for the solution lies *below* the pure solvent curve at any temperature and to the right of the pure solvent curve at any pressure.

Since the vapor pressure lowering is proportional to the concentration of solute particles, so is the magnitude of the boiling point elevation:

$$\Delta T_b \propto m \qquad \text{or} \qquad \Delta T_b = K_b m \qquad \textbf{(12.11)}$$

where $m$ is the solution molality, $K_b$ is the *molal boiling point elevation constant,* and $\Delta T_b$ is the boiling point elevation:

$$\Delta T_b = T_{b(\text{solution})} - T_{b(\text{solvent})}$$

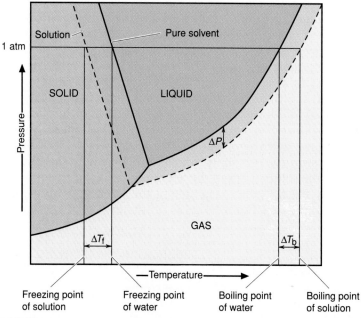

**FIGURE 12.17**
**Phase diagrams of solvent and solution.** Phase diagrams of an aqueous solution *(dashed lines)* and of pure water *(solid lines)* show that, by lowering the vapor pressure ($\Delta P$), a dissolved solute elevates the boiling point and depresses the freezing point.

**TABLE 12.4**   **Molal Boiling Point Elevation and Freezing Point Depression Constants of Several Solvents**

| SOLVENT | BOILING POINT (°C)* | $K_b$ (°C/m) | MELTING POINT (°C) | $K_f$ (°C/m) |
|---|---|---|---|---|
| Water | 100.0 | 0.512 | 0.0 | 1.86 |
| Ethanol | 78.5 | 1.22 | −117.3 | 1.99 |
| Benzene | 80.1 | 2.53 | 5.5 | 4.90 |
| Acetic acid | 117.9 | 3.07 | 16.6 | 3.90 |
| Chloroform | 61.7 | 3.63 | −63.5 | 4.70 |

*At 1 atm.

Molality is used as the concentration unit because it is directly related to mole fraction and thus to particles of solute. It also involves mass rather than volume of solvent, so it is not affected by temperature changes. The constant $K_b$ has units of degrees Celsius per molal unit (°C/m) and is specific for a given solvent; Table 12.4 lists some examples.

Notice that for water the changes in boiling point are quite small; the $K_b$ value for water is only 0.512°C/m. That is, if you dissolved 1 mol glucose (180 g; 1 mol of particles) in 1 kg water, or 0.5 mol NaCl (29.2 g; a strong electrolyte, so also 1 mol of particles) in 1 kg water, the normal boiling points of the resulting solutions would be only 100.512°C instead of 100.00°C.

**Freezing point depression.** As you just saw, only solvent molecules can vaporize from the solution, so the nonvolatile solute is left behind. Similarly, in general, *only solvent molecules can solidify*, again leaving solute molecules behind in solution. In the pure solvent at its freezing point, equilibrium occurs as molecules leave and enter the solid at the same rate. In the solution, however, some of the particles surrounding the solid are solute, and solvent molecules are somewhat blocked from entering the solid, so fewer of them can do so in a given time (Figure 12.18). Therefore, equilibrium occurs in the solution when this lower rate of solvent entering the solid is balanced by a lower rate of leaving. This occurs when solvent molecules have less kinetic energy, that is, at a lower temperature. Thus, the solution freezes at a lower temperature than the pure solvent. This **freezing point depression ($\Delta T_f$)** is shown in Figure 12.17; note that the solid-liquid curve for the solution lies to the left of the curve for the pure solvent.

As you saw for $\Delta T_b$, the magnitude of the freezing point depression is proportional to the molal concentration of solute:

$$\Delta T_f \propto m \qquad \text{or} \qquad \Delta T_f = K_f m \qquad (12.12)$$

where $\Delta T_f = T_{f(solvent)} - T_{f(solution)}$, and $K_f$ is the *molal freezing point depression constant,* which has units of °C/m (Table 12.4). Here, too, the overall effect of this colligative property in aqueous solutions is quite small because the $K_f$ value is small—only 1.86°C/m. Thus, 1 $m$ glucose, 0.5 $m$ NaCl, and 0.33 $m$ $K_2SO_4$, all solutions with 1 mole of particles per kilogram of water, freeze at −1.86°C instead of at 0°C.

Some familiar examples of freezing point depression in aqueous solutions are shown in the Gallery on the facing page.

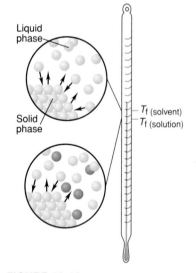

Liquid phase

Solid phase

$T_f$ (solvent)
$T_f$ (solution)

**FIGURE 12.18**

**The effect of solute on the freezing point of a solution.** In a pure solvent *(top)*, equilibrium exists at the freezing point: particles are entering and leaving the solid at the same rate. Since only the solvent freezes, the presence of solute particles in the solution lowers the rate at which solvent reenters the solid *(bottom)*, so fewer particles must leave the solid to balance this rate, and this balance occurs at a lower temperature.

# Freezing Point Depression

GALLERY

Despite its limited application in determining solute molar mass, the practical — and essential — uses of freezing point depression are easy to find.

**Plane de-icing and car antifreezing.**
Ethylene glycol ($C_2H_6O_2$) is miscible with water through extensive hydrogen bonding and has a high enough boiling point for it to be essentially nonvolatile at 100°C. It is the major ingredient in airplane "de-icers" and in automobile "year-round" anti-freeze to lower the freezing point of water in the radiator in the winter and raise its boiling point in the summer.

**Biological antifreeze.**
To survive in the Arctic and in northern winters, many fish and insects, including the common housefly, produce large amounts of glycerol ($C_3H_8O_3$) — a substance with a structure very similar to that of ethylene glycol and also miscible with water — which lowers the freezing point of their blood.

**Ice-cream making.**
When making ice cream, the temperature of the ingredients is kept below the freezing point of water with an ice bath containing large amounts of salt, which keeps the mixture at about −5°C.

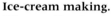

**Salts for slippery streets.**
Highway crews use salts, such as mixtures of NaCl and $CaCl_2$, to melt ice on streets. A small amount of salt dissolves in the ice by lowering its freezing point and melting it, more salt dissolves, more ice melts, and so forth. An advantage of $CaCl_2$ is that it has a highly negative $\Delta H_{soln}$, so heat is released when it dissolves, which melts more ice.

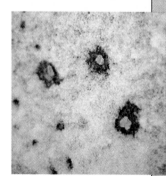

SAMPLE PROBLEM 12.8 ――――――――――――――――――――

## Determining the Boiling Point Elevation and Freezing Point Depression of a Solution

**Problem:** You add 1.00 kg of the antifreeze ethylene glycol ($C_2H_6O_2$) to your car radiator, which contains 4450 g water. What are the boiling and freezing points of the solution?

**Plan:** To find the solution boiling and freezing points, we first find the molality by converting mass of solute to moles and dividing by mass of solvent (in kg). Then we calculate $\Delta T_b$ and $\Delta T_f$ from Equations 12.11 and 12.12 (using constants from Table 12.4) and adjust the solvent boiling point up and the solvent freezing point down.

**Solution:** Calculating the molality:

$$\text{Moles of } C_2H_6O_2 = \frac{1.00 \text{ kg } C_2H_6O_2 \ (10^3 \text{ g/1 kg})}{62.07 \text{ g } C_2H_6O_2/1 \text{ mol } C_2H_6O_2} = 16.1 \text{ mol } C_2H_6O_2$$

$$\text{Molality} = \frac{\text{mol solute}}{\text{kg solvent}} = \frac{16.1 \text{ mol } C_2H_6O_2}{4450 \text{ g } H_2O \ (1 \text{ kg}/10^3 \text{ g})} = 3.62 \ m \ C_2H_6O_2$$

Finding the boiling point elevation and $T_{b(\text{solution})}$, with $K_f = 0.512°C/m$:

$$\Delta T_b = \frac{0.512°C}{m} \times 3.62 \ m = 1.85°C$$

$$T_{b(\text{solution})} = T_{b(\text{solvent})} + \Delta T_b = 100.00°C + 1.85°C = \mathbf{101.85°C}$$

Finding the freezing point depression and $T_{f(\text{solution})}$, with $K_b = 1.86°C/m$:

$$\Delta T_f = \frac{1.86°C}{m} \times 3.62 \ m = 6.73°C$$

$$T_{f(\text{solution})} = T_{f(\text{solvent})} - \Delta T_f = 0.00°C - 6.73°C = \mathbf{-6.73°C}$$

**Check:** We check to see if $\Delta T_b/\Delta T_f = K_b/K_f$: $1.85/6.73 = 0.275 = 0.512/1.86$.

**Comment:** Our answers here are only approximate because the concentration far exceeds that of a *very dilute* solution, for which Raoult's law holds.

FOLLOW-UP PROBLEM 12.8

What is the minimum concentration of ethylene glycol solution that will protect the cooling system from freezing at 0.00°F?

――――――――――――――――――――

Mass (g) of solute

$\mathcal{M}$ (g/mol)

Moles of solute

kg of solvent

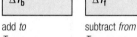

Molality *(m)*

$\Delta T_b = K_b m$ \qquad $\Delta T_f = K_f m$

$\Delta T_b$ \qquad $\Delta T_f$

add *to* \qquad subtract *from*
$T_{b(\text{solvent})}$ \qquad $T_{f(\text{solvent})}$

$T_{b(\text{solution})}$ \qquad $T_{f(\text{solution})}$

The fact that the solute remains dissolved while the solvent freezes is applied in **zone refining,** a process used to purify silicon and other semiconducting elements for electronic devices. A rod of impure silicon moves slowly through a heating coil that melts a narrow zone of the solid, forming a solution of impurities (solute) dissolved in silicon (solvent). As the next zone of impure solid melts, some silicon in the previous zone refreezes. The impurities lower the freezing point of the remaining solution, which becomes more concentrated in the dissolved impurities. This solution mixes with the newly freed impurities of the melted zone to form a still more concentrated solution (Figure 12.19, *A*). The process continues along the entire length of impure solid, each zone's impurities combining with all the previous impurities, while its silicon refreezes. The end of the rod, which contains the accumulated impurities, is removed and the process is repeated. After several passes through the coil, the rod consists of silicon that is more than 99.999999% pure (Figure 12.19, *B*).

**Osmotic pressure.** The fourth colligative property arises when two solutions of different concentrations are separated by a **semipermeable membrane,** one that allows solvent, but *not* solute, to pass through. The process by which this occurs is called **osmosis.** Many parts of organisms have semipermeable membranes that regulate internal concentrations by osmosis.

Consider a simple apparatus in which a membrane lies at the curve of a U tube and separates an aqueous sugar solution from pure water. The pores in the membrane allow water molecules to pass in *either* direction, but not the larger sugar molecules. Due to the presence of the solute, fewer water molecules lie at the membrane on the solution side, so more water molecules enter the solution in a given time than leave it (Figure 12.20, *A*). This net flow of water into the solution increases the volume of the solution and thus decreases its concentration.

As the height of the solution rises and that of the solvent falls, the resulting pressure difference pushes some water molecules *from* the solution back through the membrane. Equilibrium is reached when water molecules are pushed out of the solution at the same rate that they enter it (Figure 12.20, *B*). The pressure difference at equilibrium is the **osmotic pressure** *(π),* which is defined as the external pressure required to *prevent* the net movement of water from solvent to solution (Figure 12.20, *C*).

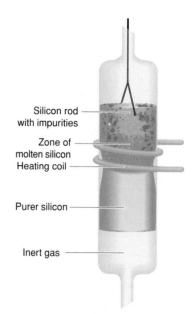

A

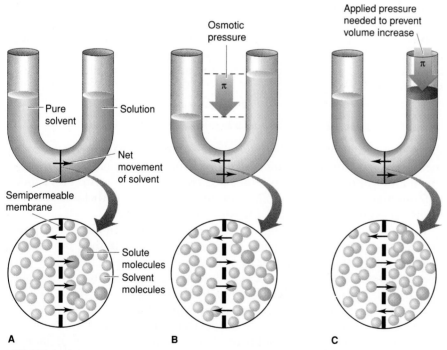

A                              B                              C

**FIGURE 12.20**

**The development of osmotic pressure. A,** In the process of osmosis, a solution and a solvent (or solutions of different concentrations) are separated by a semipermeable membrane. The molecular view *(below)* shows that more solvent molecules enter the solution than leave it. **B,** As a result, the solution volume increases and its concentration decreases. The difference in heights of liquid in the two compartments, the osmotic pressure *(π),* exerts a backward pressure that eventually equalizes the flow of solvent in both directions. **C,** The osmotic pressure is also defined as the applied pressure required to *prevent* this volume change.

B

**FIGURE 12.19**

**The process of zone refining. A,** In an inert atmosphere, a rod of impure silicon is passed through a heating coil, and a narrow zone of solid melts. The next zone melts, and the impurities from the first zone remain in solution while the purer solvent refreezes. After several passes through the coil, the rod becomes extremely pure. **B,** A zone-refined sample of silicon is ready to be sliced into thin wafers for the production of computer chips.

The osmotic pressure of a solution is proportional to the number of solute particles in a given *volume* of solution, that is, to the molarity (*M*):

$$\pi \propto \frac{n_{\text{solute}}}{V_{\text{soln}}}$$

or

$$\pi \propto M$$

The proportionality constant is the universal gas constant *R* times the absolute temperature *T*. Thus, we have

$$\pi = \frac{n_{\text{solute}}}{V_{\text{soln}}} RT = MRT \qquad \textbf{(12.13)}$$

The similarity of Equation 12.13 to the ideal gas law ($P = nRT/V$) is not surprising, since both relate the pressure of a system to its concentration and temperature. ◆ The Gallery on the facing page shows some interesting biological applications of osmotic pressure.

**The underlying nature of colligative properties.** A common thread runs through our explanations of the four colligative properties of nonvolatile solutes. Each involves the inability of solute particles to cross between two phases. They cannot enter the gas phase, which leads to vapor pressure lowering and boiling point elevation. They cannot enter the solid phase, which leads to freezing point depression. They cannot cross the semipermeable membrane, which leads to the development of osmotic pressure. Because of these inabilities, the presence of solute decreases the mole fraction of solvent, which lowers the rate at which the solvent leaves the solution, and this rate lowering requires an adjustment in rates to reach equilibrium again. The adjustment to reach the new balance results in the measured colligative property.

### Using Colligative Properties to Find Solute Molar Mass

Each of the colligative properties relates solution concentration to some measurable quantity: the number of degrees the freezing point was lowered, the amount of osmotic pressure that was created, and so forth. From these measurements, we can determine the moles of solute particles and, for a known mass of nonelectrolyte, the solute molar mass.

In principle, any of the colligative properties can be used for this purpose, but in practice, some systems provide more accurate data than others. For example, determining the molar mass of an unknown sugar from the freezing point depression of its aqueous solution is difficult because a 1 *m* solution would depress the freezing point by only 1.86°C. However, several other solvents have larger freezing point depression constants (see Table 12.4). If the sugar were soluble in acetic acid, for instance, a 1 *m* solution would depress the freezing point by 3.90°C, more than twice as large a change.

Nevertheless, of the four colligative properties, osmotic pressure results in the largest changes and therefore the most precise measurements. Biological and polymer chemists can estimate a molar mass as great as $10^5$ g/mol by measuring osmotic pressure. Because only a tiny fraction of a mole of a macromolecular solute would dissolve in any solvent, it would create only a minute change in the other colligative properties.

◆ **Sodium Ion: The Extracellular Osmoregulator.** Of the four major biological cations—$Na^+$, $K^+$, $Mg^{2+}$, and $Ca^{2+}$—$Na^+$ is essential for all animals that regulate their fluid volume (including us). The $Na^+$ ion accounts for more than 90 mol % of all cations outside the cell. A high $Na^+$ concentration draws water out of the cell by osmosis; a low concentration leaves more inside. Indeed, the primary role of $Na^+$ is to regulate the water volume of the body, and the primary role of the kidney is to regulate the concentration of $Na^+$. This regulation occurs via changes in blood pressure (volume) that activate nerves and hormones to adjust blood flow and alter kidney function. The edema (retention of fluid) that women experience before menstruation and during pregnancy is partially due to the release of estrogen hormones that trigger the kidneys to retain more $Na^+$—and the water that follows it.

# Osmotic Pressure in Biological Systems

Without question, the most important semipermeable membranes surround living cells, so applications of osmotic pressure are found throughout nature and the biosciences.

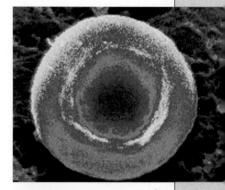

**Controlling cell shape.** The word "tonicity" refers to the tone, or firmness, of a biological cell.

An *isotonic* solution has the same concentration of particles as in the cell fluid, so water enters the cell at the same rate that it leaves, thereby maintaining the cell's normal shape.

In order to study cell contents, biochemists rupture membranes by placing the cell in a *hypotonic* solution, one that has a lower concentration of solute particles than does the cell fluid. As a result, water enters the cell faster than it leaves, causing the cell to expand and burst.

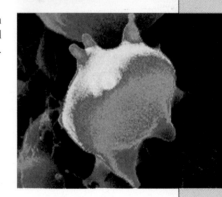

In a *hypertonic* solution, one that contains a higher concentration of solute particles, the cell shrinks from the net outward flow of water.

**Hypotonic watering of trees.** The dissolved substances in tree sap create a more concentrated solution than the surrounding ground water. Water enters membranes in the roots and rises into the tree, creating an osmotic pressure that can exceed 20 atm in the tallest trees!

**Isotonic daily care.** Contact-lens rinses consist of isotonic saline (0.15 M NaCl) to prevent any changes in the volume of corneal cells. Intravenous solutions for the delivery of nutrients or drugs are always isotonic.

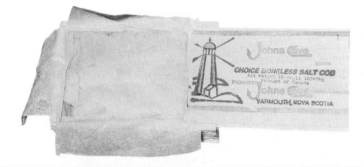

**Hypertonic meat preservation.** Before refrigeration was common, salt was used as a preservative rather than as a condiment. Food coated with salt causes microbes on the surface to shrivel and die from loss of cell water. Salt was so highly prized for this purpose that Roman soldiers were paid in salt, from which comes the word "salary."

SAMPLE PROBLEM 12.9 _____

## Determining Molar Mass from Osmotic Pressure

**Problem:** A physician studying a type of hemoglobin formed during a fatal disease dissolves 21.5 mg of the protein in water at 5.0°C to make 1.50 mL of solution in order to measure its osmotic pressure. At equilibrium, the solution has an osmotic pressure of 3.61 torr. What is the molar mass ($\mathcal{M}$) of the hemoglobin?

**Plan:** We know the osmotic pressure ($\pi$), R, and T. We convert $\pi$ from torr to atm and T from °C to K and use Equation 12.13 to solve for molarity (M). Then we calculate the moles of hemoglobin from the known volume and use the known mass to find $\mathcal{M}$.

**Solution:** Combining unit conversion steps and solving for molarity from Equation 12.13:

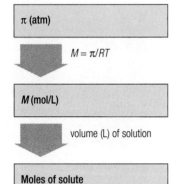

$\pi$ (atm)

$M = \pi/RT$

$M$ (mol/L)

volume (L) of solution

Moles of solute

mass (g) of solute

$\mathcal{M}$ (g/mol)

$$M = \frac{\pi}{RT} = \frac{\dfrac{3.61 \text{ torr}}{760 \text{ torr/1 atm}}}{0.0821 \dfrac{\text{atm} \cdot \text{L}}{\text{mol} \cdot \text{K}} (273.15 \text{ K} + 5.0)} = 2.08 \times 10^{-4} \ M$$

Finding moles of solute:

$$\text{Moles of solute} = M \times V = \frac{2.08 \times 10^{-4} \text{ mol}}{\text{L soln}} \times 0.00150 \text{ L soln} = 3.12 \times 10^{-7} \text{ mol}$$

Calculating molar mass of hemoglobin (after changing mg to g):

$$\mathcal{M} = \frac{0.0215 \text{ g}}{3.12 \times 10^{-7} \text{ mol}} = \mathbf{6.89 \times 10^4 \text{ g/mol}}$$

**Check:** The answers seem reasonable: The small osmotic pressure implies a very low molarity. Since hemoglobin is a protein, a biological macromolecule, we expect a small number of moles [($\sim 2 \times 10^{-4}$) ($1.5 \times 10^{-3}$) = $3 \times 10^{-7}$] and a high molar mass ($\sim 21 \times 10^{-3}/3 \times 10^{-7} = 7 \times 10^4$).

FOLLOW-UP PROBLEM 12.9
A 0.30 M solution of sucrose at 37°C has approximately the same osmotic pressure as blood. What is the osmotic pressure of blood?

## Vapor Pressure Lowering in Volatile Nonelectrolyte Solutions

What is the effect on vapor pressure when the solute *is* volatile, that is, when the vapor consists of solute *and* solvent molecules? From Raoult's law, we know that

$$P_{\text{solvent}} = X_{\text{solvent}} \times P^0_{\text{solvent}} \qquad \text{and} \qquad P_{\text{solute}} = X_{\text{solute}} \times P^0_{\text{solute}}$$

where $X_{\text{solvent}}$ and $X_{\text{solute}}$ refer to the mole fractions in the liquid phase. According to Dalton's law of partial pressures, the total vapor pressure is the sum of the partial vapor pressures:

$$P_{\text{total}} = P_{\text{solvent}} + P_{\text{solute}} = (X_{\text{solvent}} \times P^0_{\text{solvent}}) + (X_{\text{solute}} \times P^0_{\text{solute}})$$

Thus, *the presence of each substance lowers the vapor pressure of the other.*

Let's examine a solution of benzene ($C_6H_6$) and toluene ($C_7H_8$) that contains equal numbers of benzene and toluene molecules: $X_{\text{ben}} = X_{\text{tol}} = 0.500$. At 25°C, the vapor pressure of pure benzene ($P^0_{\text{ben}}$) is 95.1 torr and that of pure toluene ($P^0_{\text{tol}}$) is 28.4 torr; note that benzene is more volatile than toluene. We find their partial pressures from Raoult's law:

$$P_{\text{ben}} = X_{\text{ben}} \times P^0_{\text{ben}} = 0.500 \times 95.1 \text{ torr} = 47.6 \text{ torr}$$

$$P_{\text{tol}} = X_{\text{tol}} \times P^0_{\text{tol}} = 0.500 \times 28.4 \text{ torr} = 14.2 \text{ torr}$$

As expected, the presence of benzene lowers the vapor pressure of toluene, and vice versa.

We can calculate the mole fraction of each substance *in the vapor* by applying Dalton's law. Recall from Section 5.4 that $X_A = P_A/P_{total}$. Therefore, for benzene and toluene in the vapor,

$$X_{ben} = \frac{P_{ben}}{P_{total}} = \frac{47.6 \text{ torr}}{47.6 \text{ torr} + 14.2 \text{ torr}} = 0.770$$

$$X_{tol} = \frac{P_{tol}}{P_{total}} = \frac{14.2 \text{ torr}}{47.6 \text{ torr} + 14.2 \text{ torr}} = 0.230$$

The essential point to notice is that *the vapor has a higher mole fraction of the more volatile solution component.* The 50/50 ratio of benzene/toluene in the liquid created a 77/23 ratio of benzene/toluene in the vapor. Condense this vapor into a separate container, and the new solution has this 77/23 composition; moreover, the vapor above *it* is even further enriched in benzene.

This phenomenon of vapor enrichment is employed in **fractional distillation,** a process in which numerous vaporization-condensation steps continually enrich the vapor, until the vapor reaching the top of the fractionating column consists solely of the more volatile component (Figure 12.21, *A*). In petroleum refining, fractional distillation is used to separate the hundreds of individual compounds in crude oil into a small number of "fractions" based on boiling point range (Figure 12.21, *B*).

## Colligative Properties of Electrolyte Solutions

To conclude our discussion of colligative properties, we briefly consider the behavior of electrolyte solutions. Since colligative properties depend only on the number of particles, we would expect the boiling point elevation ($\Delta T_b$) of 0.050 *m* NaCl, for example, to be twice that of 0.050 *m* glucose because NaCl dissociates into two particles per formula unit. Thus, a multiplying factor must be included in the equations for the colligative properties of electrolyte solutions. The *van't Hoff factor (i),* named after the Dutch chemist Jacobus van't Hoff (1852-1911), relates the *measured* value of the colligative property in the electrolyte solution to the *expected* value for a nonelectrolyte solution:

$$i = \frac{\text{measured value for electrolyte solution}}{\text{expected value for nonelectrolyte solution}}$$

**FIGURE 12.21**
**The process of fractional distillation. A,** In the laboratory, a flask containing a solution of two or more volatile components is attached to a *fractionating column* that is packed with glass beads *(close up)* and connected to a condenser. As the flask is heated, the vapor mixture rises and condenses on the lowest beads, forming a liquid enriched in the more volatile component. Some of this liquid drips back to the flask, and some vaporizes again to form a more enriched vapor, which condenses on beads higher up. The process continues until the vapor reaching the condenser consists of the more volatile component only, which condenses to a pure liquid. **B,** In industry, this process is used to separate petroleum into many products. A fractionating tower can be more than 30m high and can separate components that differ by a few tenths of a degree in their boiling points. (The illustration is a highly simplified version of a multipart process.)

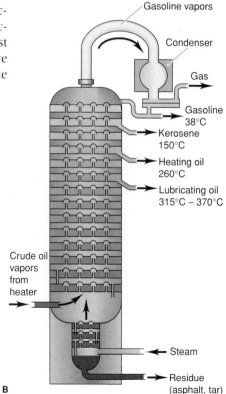

Gasoline vapors

Condenser

Gas

Gasoline
38°C

Kerosene
150°C

Heating oil
260°C

Lubricating oil
315°C – 370°C

Crude oil vapors from heater

Steam

Residue
(asphalt, tar)

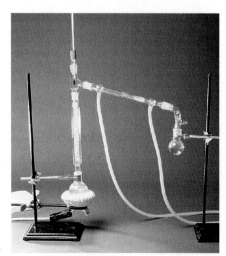

A

B

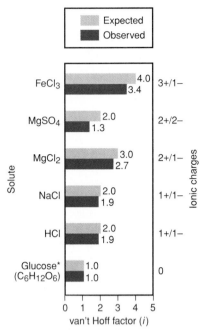

Solute

van't Hoff factor (*i*)

* Nonelectrolyte shown for comparison

**FIGURE 12.22 Nonideal behavior of electrolyte solutions.** The van't Hoff factors *(i)* for various ionic solutes in dilute (0.05 *m*) aqueous solution shows that the observed value is always lower than expected. This deviation is due to ionic interactions that, in effect, reduce the number of free ions in solution. The deviation is greatest for multivalent ions.

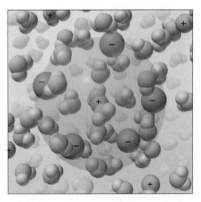

**FIGURE 12.23 An ionic atmosphere model for nonideal behavior of electrolyte solutions.** The anions and water molecules cluster near cations, and vice versa, to form ionic atmospheres of net opposite charge. Thus, the ions do not behave independently, so their concentrations are effectively *less* than expected. Such interactions cause deviations from the ideal behavior expressed by Raoult's law.

To calculate the colligative properties of electrolyte solutions, we incorporate the van't Hoff factor into the equation:

For vapor pressure lowering:    $\Delta P = i\, X_{\text{solute}}\, P^0_{\text{solvent}}$

For boiling point elevation:    $\Delta T_{\text{b}} = i\, K_{\text{b}} m$

For freezing point depression:    $\Delta T_{\text{f}} = i\, K_{\text{f}} m$

For osmotic pressure:    $\pi = i\, MRT$

If electrolyte solutions behaved ideally, the factor *i* would be 2 for NaCl, 3 for $Mg(NO_3)_2$, and so forth. Careful experiment shows, however, that *most electrolyte solutions are not ideal.* For the boiling point elevation of 0.050 *m* NaCl solution compared with 0.050 *m* glucose solution, for example, the van't Hoff factor is 1.9, not 2.0:

$$ i = \frac{\Delta T_{\text{b}} \text{ of } 0.050 \, m \text{ NaCl}}{\Delta T_{\text{b}} \text{ of } 0.050 \, m \text{ glucose}} = \frac{0.049°C}{0.026°C} = 1.9 $$

The measured value of the van't Hoff factor is typically somewhat lower than expected from the formula. This deviation implies that the ions are not behaving as independent particles. Yet we know from other evidence that soluble salts dissociate completely into ions. The fact that the deviation is greater with divalent and trivalent ions provides a clue that ionic charge is somehow involved in the behavior (Figure 12.22).

To explain such nonideal behavior, chemists picture the immediate vicinity of one ion to be occupied by other ions. Clustered near a positive ion are, on the average, more negative ions, and vice versa. That is, each ion becomes surrounded with an **ionic atmosphere** of net opposite charge (Figure 12.23). Through these electrostatic associations, the number of completely *free* ions in solution is *lower* than we would expect from the formula. The greater the ionic charge, the stronger are these associations and the larger is the effect on the colligative property.

At ordinary conditions and concentrations, nonideal behavior in solutions is much more common, and the deviations much larger, than nonideal behavior in gases because the particles are so much closer together. Nevertheless, the two systems exhibit some interesting similarities. Gases display nearly ideal behavior at low pressures, when the distances between particles are large. Similarly, van't Hoff factors approach their ideal values as the distance between ions increases, that is, as the solution becomes more dilute. With gases, attractions between particles cause deviations from the expected pressure. With solutions, attractions between particles cause deviations from the expected size of a colligative property. Finally, for both real gases and real solutions, we use empirically determined numbers (van der Waals constants or van't Hoff factors) to transform simple theories (the ideal gas law or Raoult's law) into more useful ones.

**Section Summary**

Colligative properties of solutions are due to the number of dissolved solute particles, not their chemical nature. Compared with the pure solvent, a solution of nonvolatile nonelectrolyte has a lower vapor pressure (Raoult's law), elevated boiling point, depressed freezing point, and an osmotic pressure. Colligative properties can be used to determine the solute molar mass. When the solute is volatile also, its vapor pressure is lowered by the presence of solvent, and vice versa. The vapor pressure of the more volatile component is always higher. Electrolyte solutions are nonideal because ionic interactions effectively reduce the number of free ions.

# 12.6   The Structure and Properties of Colloids

Stir a handful of fine sand in a glass of water and wait a moment. The sand particles are suspended at first but then gradually settle to the bottom. Sand in water is an example of a **suspension,** a heterogeneous mixture containing particles large enough to be seen with the naked eye and clearly distinct from the surrounding fluid. In contrast, stirring sugar in water forms a solution, a homogeneous mixture in which the particles are individual molecules distributed evenly throughout the surrounding fluid.

Between the extremes of suspensions and solutions is a large group of mixtures called colloidal dispersions, or simply **colloids,** in which a dispersed (solutelike) substance is distributed throughout a dispersing (solventlike) substance. The dispersed colloidal particles are larger than a simple molecule but small enough to remain distributed and not settle out. A colloidal particle has a diameter between 1 and 1000 nm ($10^{-9}$ to $10^{-6}$ m) and may contain many atoms, ions, or molecules. Because of their small particle size, colloids have an enormous total surface area. Consider this: a cube with 1-cm sides has a total surface area of 6 cm². If it were divided equally into $10^{12}$ cubes, they would be the size of large colloidal particles, with a total surface area of 60,000 cm², or 6 m². Such a large surface area attracts many substances through adhesive forces and gives rise to some of the practical uses of colloids.

Colloids are classified according to whether the dispersed and dispersing substances are gases, liquids, or solids (Table 12.5). A wide range of familiar commercial products and natural objects exist as colloids. For example, whipped cream is a *foam,* a gas dispersed in a liquid. Firefighting foams, such as those used in emergency airplane landings, are liquid mixtures of proteins in water made frothy with fine jets of air. Most biological fluids are aqueous *sols,* solids dispersed in water. Within a typical cell, many colloid-sized protein and nucleic acid molecules are dispersed in an aqueous solution of ions and numerous small molecules. The action of soaps and detergents that we discussed earlier occurs by the formation of an *emulsion,* a liquid (soap dissolved in grease) dispersed in another liquid (water). ◆

Most colloids are cloudy or opaque, but some are transparent to the naked eye. When light passes through a colloid, it is scattered randomly by the dispersed particles because their sizes are similar to the wavelengths of visible light (400 to 700 nm). Viewed from the side, the scattered beam is visible and broader than one passing through a solution, a phenomenon known as the **Tyndall effect** (Figure 12.24). Smoke in air displays this ef-

◆ **"Soaps" in Your Small Intestine.** In the small intestine, fats are digested by *bile salts*, soaplike molecules with a small polar and a large nonpolar portion. Secreted by the liver, stored in the gallbladder, and released in the intestine, bile salts emulsify fats just as soap emulsifies grease: fatty aggregates are broken down into colloid-sized particles and dispersed in the watery fluid. Only in this form can the fats be broken down further by enzymes and transported into the blood.

**FIGURE 12.24**
**The Tyndall effect.** When a beam of light passes through a solution *(left)*, its path remains narrow and barely visible. When it passes through a colloid *(right)*, it is easily visible, scattered and broadened by the particles.

TABLE 12.5 **Types of Colloids**

| COLLOID TYPE | DISPERSED SUBSTANCE | DISPERSING MEDIUM | EXAMPLE |
|---|---|---|---|
| Aerosol | Liquid | Gas | Fog |
| Aerosol | Solid | Gas | Smoke |
| Foam | Gas | Liquid | Whipped cream |
| Solid foam | Gas | Solid | Marshmallow |
| Emulsion | Liquid | Liquid | Milk |
| Solid emulsion | Liquid | Solid | Butter |
| Sol | Solid | Liquid | Paint; cell fluid |
| Solid sol | Solid | Solid | Ruby |

fect when sunlight shines through it, as does mist pierced by headlights at night.

Under low magnification, you can watch colloidal particles exhibit *Brownian motion,* a characteristic movement in which the particles change speed and direction erratically. This motion occurs as the colloidal particles are pushed this way and that way by molecules of the dispersing medium. These collisions are primarily responsible for keeping colloidal particles from settling, although convection helps suspend larger particles. (Einstein's explanation of Brownian motion in 1905 was a principal factor in the widespread acceptance of the molecular nature of matter.)

Why *don't* colloidal particles coagulate into larger particles and settle out? Water-dispersed colloids remain distributed because the particles have charged surfaces that interact strongly with the water. Soap molecules form spherical *micelles,* with the charged heads forming the micelle exterior and the nonpolar tails the interior. Aqueous proteins are typically spherical, with charged amino acid groups facing the water and uncharged groups buried within the molecule. Nonpolar oily particles remain dispersed in water by adsorbing ions of like charge onto their surface (Figure 12.25). Charge repulsions prevent the particles from adhering to each other when they collide.

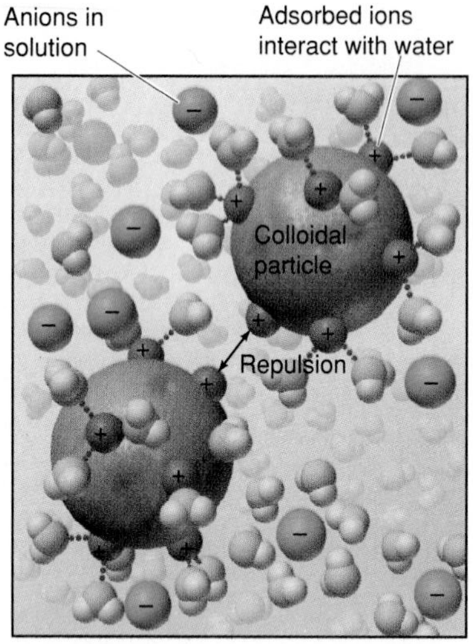

Despite these repulsions, various methods can be used to coagulate the particles and "destroy" the colloid. Heating a colloid makes the particles move faster and collide more often and strongly enough to coalesce into heavier particles that settle out. Adding an electrolyte solution introduces oppositely charged ions that neutralize the particle's surface charges, which allows them to coagulate and settle. ◆ Uncharged colloidal particles in smoke are removed by introducing ions that become adsorbed on the particles, which are then attracted to the charged plates of a Cottrell precipitator (Figure 12.26). These devices are installed in smokestacks of coal-burning plants where they collect about 90% of the colloidal particulates.

**FIGURE 12.25**
**The charged surfaces of water-dispersed, nonpolar colloids.** Nonpolar particles can be dispersed in water only if ions are adsorbed on their surfaces. Because like charges repel, the similarly charged colloidal particles repel one another during the collisions and remain dispersed.

◆ **From Colloid to Civilization.** At times, civilizations have been born where colloids are coagulated by electrolyte solutions. At the mouths of rivers, where salt concentrations increase near an ocean or sea, the clay particles dispersed in the river water come together to form muddy deltas, such as those of the Nile and the Mississippi. The ancient Egyptian empire and the city of New Orleans are the results.

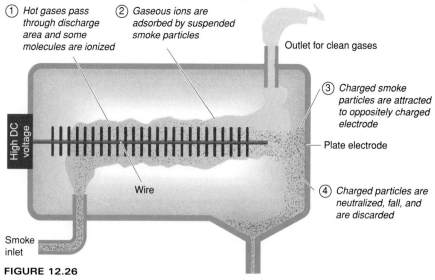

**FIGURE 12.26**
**A Cottrell precipitator for removal of particulates from industrial flue gases.**

**CHEMICAL CONNECTIONS**

**Chemistry in Sanitary Engineering**
## Solutions and Colloids in Water Purification

Clean water is a priceless and limited resource that we've begun to treasure only recently, after decades of pollution and waste. Because of the natural tendency of systems to become disordered, it requires a great deal of energy to remove dissolved, dispersed, and suspended particles from water.

Most water destined for human use comes from lakes, rivers, or reservoirs that may serve also as the final sink after the water is used. Many mineral ions, such as $NO_3^-$ and $Fe^{3+}$, may be present in high concentrations. Dissolved organic compounds, some of them toxic, may be present as well. Fine clay particles and a whole spectrum of microorganisms occur dispersed in colloidal form. Larger particles and debris of every variety may be present in suspension.

### Water Treatment Plants

As a sample of water moves from the natural source into a water treatment facility, the largest particles are physically removed at the intake site by screens (Figure 12.B, Step 1). Finer particles, including microorganisms, are removed in large settling tanks by treatment with lime (CaO) and alum [$Al_2(SO_4)_3$] (Step 2). These react to form a fluffy, gel-like precipitate of $Al(OH)_3$:

$$3CaO(s) + 3H_2O(l) + Al_2(SO_4)_3(s) \longrightarrow$$
$$2Al(OH)_3(colloidal\ gel) + 3CaSO_4(aq)$$

The fine particles are trapped within or adsorbed onto the enormous surface area of this gel, which coagulates, settles out, and is filtered through a sand bed (Step 3).

**FIGURE 12.B**

**The steps in a typical municipal water treatment plant.** Before water is sent to users, (1) it is filtered to remove large debris, (2) the finer particles are trapped in an $Al(OH)_3$ gel, (3) the gel is filtered through sand, (4) the filtrate is aerated to oxidize organic compounds, and (5) the water is disinfected with chlorine.

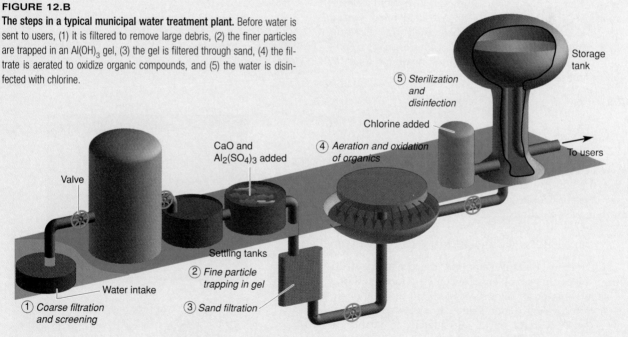

*Continued.*

### Section Summary

Colloidal particles are smaller than those in a suspension and larger than those in a solution. Colloids are classified by the physical state of the dispersed and dispersing substances and are formed from many combinations of gas, liquid, or solid. Colloids have extremely large surface areas, exhibit random (Brownian) motion, and scatter incoming light (Tyndall effect). Colloidal particles in water have charged surfaces that keep them dispersed, but they can be coagulated by heating or by the addition of ions.

CHEMICAL
CONNECTIONS

Chemistry in Sanitary Engineering
**Solutions and Colloids in Water
Purification—cont'd**

With suspended and colloidal particles removed, the water sample is aerated in large sprayers to saturate the water with oxygen, which speeds the oxidation of dissolved organic compounds (Step 4). The water is then sterilized, usually by treatment with chlorine gas (Step 5), which may give the water an unpleasant odor. Chlorine can also form toxic chlorinated hydrocarbons, but these can be removed by adsorption onto activated charcoal particles. These steps through the treatment facility dispose of debris and grit, colloidal clay, microorganisms, and much of the oxidizable organic matter, but dissolved ions remain. Many of these can be removed by water softening and reverse osmosis.

### Water Softening

Water that contains large amounts of divalent cations, such as $Ca^{2+}$, $Mg^{2+}$, and $Fe^{2+}$, is called **hard water.** These cations cause several problems. During cleaning, they combine with the fatty-acid anions in soaps to produce insoluble deposits on clothes, washing machine parts, and sinks:

$$Ca^{2+}(aq) + 2C_{17}H_{35}COONa(aq) \rightarrow$$
$$\underset{\text{soap}}{}$$

$$\underset{\text{deposit}}{Ca(C_{17}H_{35}COO)_2(s)} + 2Na^+(aq)$$

When a large amount of bicarbonate ($HCO_3^-$) is present, the hard-water cations cause *scale*, insoluble carbonate deposits within boilers and hot-water pipes that interfere with the transfer of heat and damage plumbing:

$$Ca^{2+}(aq) + 2HCO_3^-(aq) \xrightarrow{\Delta} CaCO_3(s) + CO_2(g) + H_2O(l)$$

The removal of hard-water ions, called **water softening,** solves these problems. A domestic **ion exchange** system contains an *ion exchange resin,* an insoluble polymer with covalently bound anion groups, such as $—SO_3^-$ or $—COO^-$, with $Na^+$ or $H^+$ ions attached to balance the charge (Figure 12.C). The divalent cations in hard water are attracted to the negative resin groups and displace the $Na^+$ or $H^+$ ions into the water: one type of ion is exchanged for another. The resin is replaced when all the resin sites are occupied or it is "recharged" by exchanging $Na^+$ ions for the bound $Ca^{2+}$.

### Reverse Osmosis

Another way to remove ions and other dissolved substances from water is by **reverse osmosis.** In osmosis, water moves from a dilute to a concentrated solution through a semipermeable membrane. The resulting difference in water volumes creates an osmotic pressure. In reverse osmosis, a pressure *higher* than the osmotic pressure is *applied* to the solution, forcing the water back

through the membrane and leaving the ions behind—in a sense, filtering out the ions at the molecular level.

In domestic water systems, reverse osmosis is used to remove toxic ions, such as the *heavy-metal ions,* $Pb^{2+}$, $Cd^{2+}$, and $Hg^{2+}$, present at concentrations too low for removal by ion exchange. On a much larger scale, reverse osmosis is used in **desalination** plants, which remove large amounts of ions from sea water (Figure 12.D). The sea water is pumped under high pressure into tubes containing millions of hollow fiber membranes, each the thickness of a human hair. Water molecules, but not ions, pass through the membranes into the fiber to be collected. Sea water containing about 40,000 ppm of total dissolved solids can be purified to 400 ppm (suitable for drinking) in one pass through such a system.

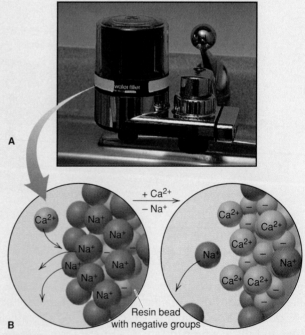

**FIGURE 12.C**

**Ion exchange for removal of hard-water ions. A,** A commercial ion-exchange column is installed in a household water system. **B,** In a typical ion-exchange resin, negatively charged groups are covalently bound to resin beads, with $Na^+$ ions present to keep the material neutral. Hard-water ions, such as $Ca^{2+}$, exchange with the $Na^+$ ions, which are displaced into the flowing water.

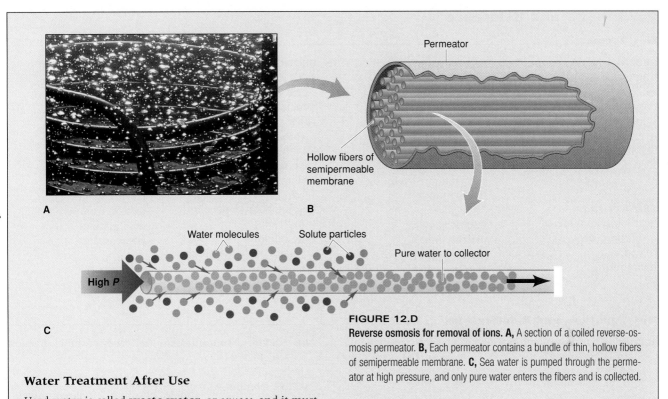

**FIGURE 12.D**
**Reverse osmosis for removal of ions. A,** A section of a coiled reverse-osmosis permeator. **B,** Each permeator contains a bundle of thin, hollow fibers of semipermeable membrane. **C,** Sea water is pumped through the permeator at high pressure, and only pure water enters the fibers and is collected.

## Water Treatment After Use

Used water is called **waste water,** or *sewage,* and it must be treated before being returned to the ground water, river, or lake. Sewage treatment is especially important for industrial waste water because it may contain toxic components. In *primary* sewage treatment, waste water undergoes the same steps as the water coming into the system. Most municipalities now also include *secondary* sewage treatment. In this stage, bacteria biologically degrade the organic compounds still present in solution or in the solids from the settling tanks. In certain cases, secondary treatment may be supplemented by *tertiary* treatment, the process being tailored to the specific pollutant involved. Heavy-metal ion contamination, for instance, can be eliminated by a precipitation step before primary and secondary treatment. Tertiary methods also exist for phosphate, nitrate, and toxic organic substances.

## Chapter Perspective

*In Chapter 11, you saw how pure liquids and solids behave, and in this chapter you've seen how their behaviors change when they are mixed together. Two features of these systems, their equilibrium nature and their tendency toward disorder, reappear in Chapters 16 through 18 and 19, respectively.*

*This chapter completes our introduction to atomic and molecular properties and their influence on the properties of matter. Immediately following is a unique Interchapter feature that reviews these properties in a pictorial format and helps preview their application in upcoming chapters. In Chapter 13, we travel through the main groups of the periodic table, applying these ideas to the chemical and physical behavior of the elements. Then, in Chapter 14, we focus on the carbon atom, comparing it with its neighbors, to see how its properties create the fascinating chemistry of the organic world.*

# For Review and Reference

## Key Terms

**SECTION 12.1**
solute
solvent
miscible
solubility
hydration shell
like-dissolves-like rule
soap
alloy

**SECTION 12.2**
heat (enthalpy) of
   solution ($\Delta H_{soln}$)
solvation
hydration

heat (enthalpy) of
   hydration ($\Delta H_{hydr}$)
charge density
entropy

**SECTION 12.3**
saturated solution
unsaturated solution
supersaturated solution
Henry's law

**SECTION 12.4**
molality ($m$)
mass percent [% (w/w)]
volume percent [% (v/v)]
mole fraction ($X$)

**SECTION 12.5**
colligative property
electrolyte
nonelectrolyte
vapor pressure lowering
   ($\Delta P$)
Raoult's law
ideal solution
boiling point elevation
   ($\Delta T_b$)
freezing point depression
   ($\Delta T_f$)
zone refining
semipermeable membrane
osmosis
osmotic pressure ($\pi$)

fractional distillation
ionic atmosphere

**SECTION 12.6**
suspension
colloid
Tyndall effect
hard water
water softening
ion exchange
reverse osmosis
desalination
waste water

## Key Equations and Relationships

**12.1** Dividing the general heat of solution into component enthalpies (p. 472):

$$\Delta H_{soln} = \Delta H_{solute} + \Delta H_{solvent} + \Delta H_{mix}$$

**12.2** Dividing the heat of solution of ionic compounds in water into component enthalpies (p. 474):

$$\Delta H_{soln} = -\Delta H_{lattice} + \Delta H_{hydr\ of\ the\ ions}$$

**12.3** Relating gas solubility to its partial pressure (Henry's law) (p. 479):

$$S_{gas} = k_H \times P_{gas}$$

**12.4** Defining concentration in terms of molarity (p. 480):

$$\text{Molarity } (M) = \frac{\text{moles of solute}}{\text{liter of solution}}$$

**12.5** Defining concentration in terms of molality (p. 480):

$$\text{Molality } (m) = \frac{\text{moles of solute}}{\text{kilogram of solvent}}$$

**12.6** Defining concentration in terms of mass percent (p. 481):

$$\text{Mass percent } (\%w/w) = \frac{\text{mass of solute}}{\text{mass of solution}} \times 100$$

**12.7** Defining concentration in terms of volume percent (p. 481):

$$\text{Volume percent } (\%v/v) = \frac{\text{volume of solute}}{\text{volume of solution}} \times 100$$

**12.8** Defining concentration in terms of mole fraction (p. 482):

$$\text{Mole fraction } (X) = \frac{\text{moles of solute}}{\text{moles of solute} + \text{moles of solvent}}$$

**12.9** Expressing the relationship between the vapor pressure of solvent above a solution and its mole fraction in the solution (Raoult's law) (p. 486):

$$P_{solvent} = X_{solvent} \times P^0_{solvent}$$

**12.10** Calculating the vapor pressure lowering due to solute (p. 486):

$$\Delta P = X_{solute} \times P^0_{solvent}$$

**12.11** Calculating the boiling point elevation of a solution (p. 487):

$$\Delta T_b = K_b m$$

**12.12** Calculating the freezing point depression of a solution (p. 488):

$$\Delta T_f = K_f m$$

**12.13** Calculating the osmotic pressure of a solution (p. 492):

$$\pi = \frac{n_{solute}}{V_{soln}} RT = MRT$$

# Answers to Follow-up Problems

**12.1** (a) 1,4-Butanediol is more soluble in water because it can form more H bonds.

(b) Chloroform is more soluble in water because it can form dipole-dipole forces.

**12.2** (a) $NaOCN(s) + heat \overset{H_2O}{\rightleftharpoons} Na^+(aq) + OCN^-(aq)$

$\Delta H_{soln} > 0$; at lower $T$, solubility decreases.

(b) $\Delta H_{soln} < 0$; at lower $T$, solubility increases.

**12.3** $S_{N_2} = (7 \times 10^{-4}$ mol/L $\cdot$ atm$)$ $(0.78$ atm$)$

$= 5 \times 10^{-4}$ mol/L

**12.4** Mass (g) of glucose

$= 563$ g ethanol $\times \dfrac{1\ kg}{10^3\ g}$

$\times \dfrac{2.40 \times 10^{-2}\ mol\ glucose}{1\ kg\ ethanol} \times \dfrac{180.16\ g\ glucose}{1\ mol\ glucose}$

$= 2.43$ g glucose

**12.5** Mass % $C_3H_7OH = \dfrac{35.0\ g}{35.0\ g + 150.0\ g} = 18.9$ mass %

Mass % $C_2H_5OH = 100.0 - 18.9 = 81.1$ mass %

$X_{C_3H_7OH}$

$= \dfrac{35.0\ g\ C_3H_7OH \times \dfrac{1\ mol\ C_3H_7OH}{60.09\ g\ C_3H_7OH}}{\left(35.0\ g\ C_3H_7OH \times \dfrac{1\ mol\ C_3H_7OH}{60.09\ g\ C_3H_7OH}\right) + \left(150.0\ g\ C_2H_5OH \times \dfrac{1\ mol\ C_2H_5OH}{46.07\ g\ C_2H_5OH}\right)}$

$= 0.152$

$X_{C_2H_5OH} = 1.000 - 0.152 = 0.848$

**12.6** Mass % HCl $= \dfrac{mass\ of\ HCl}{mass\ of\ soln} \times 100$

$= \dfrac{\dfrac{11.8\ mol\ HCl}{1\ L\ soln} \times \dfrac{36.46\ g\ HCl}{1\ mol\ HCl}}{\dfrac{1.190\ g}{1\ mL\ soln} \times \dfrac{10^3\ mL}{1\ L}} \times 100$

$= 36.2$ mass % HCl

Mass (kg) of soln $= 1$ L soln $\times \dfrac{1.190 \times 10^{-3}\ kg\ soln}{1 \times 10^{-3}\ L\ soln}$

$= 1.190$ kg soln

Mass (kg) of HCl $= 11.8$ mol HCl $\times \dfrac{36.46\ g\ HCl}{1\ mol\ HCl} \times \dfrac{1\ kg}{10^3\ g}$

$= 0.430$ kg HCl

Molality of HCl $= \dfrac{mol\ HCl}{kg\ water} = \dfrac{mol\ HCl}{kg\ soln - kg\ HCl}$

$= \dfrac{11.8\ mol\ HCl}{0.760\ kg\ H_2O} = 15.5\ m$ HCl

$X_{HCl} = \dfrac{mol\ HCl}{mol\ HCl + mol\ H_2O}$

$= \dfrac{11.8\ mol}{11.8\ mol + \left(760\ g\ H_2O \times \dfrac{1\ mol}{18.02\ g\ H_2O}\right)}$

$= 0.219$

$X_{H_2O} = 0.781$

**12.7** $\Delta P = X_{aspirin} \times P^0_{methanol}$

$= \dfrac{\dfrac{2.00\ g}{180.16\ g/mol}}{\dfrac{2.00\ g}{180.16\ g/mol} + \dfrac{50.0\ g}{32.04\ g/mol}} \times 101$ torr

$= 0.714$ torr

**12.8** $T_{f(solution)} = 0.00°F = -17.8°C$

$m = \dfrac{-17.8°C}{1.86°C/m} = 9.56\ m$

**12.9** $\pi = MRT$

$= 0.30$ mol/L $\left(0.0821 \dfrac{atm \cdot L}{mol \cdot K}\right)(273.15\ K + 37)$

$= 7.6$ atm

# Sample Problem Titles

**12.1** Predicting Relative Solubilities of Substances (p. 467)

**12.2** Predicting the Effect of Temperature on Solubility (p. 477)

**12.3** Using Henry's Law to Calculate Gas Solubility (p. 479)

**12.4** Calculating Molality (p. 481)

**12.5** Expressing Concentration in Parts by Mass, Parts by Volume, and Mole Fraction (p. 482)

**12.6** Converting Concentration Units (p. 483)

**12.7** Using Raoult's Law to Find the Vapor Pressure Lowering (p. 486)

**12.8** Determining the Boiling Point Elevation and Freezing Point Depression of a Solution (p. 490)

**12.9** Determining Molar Mass from Osmotic Pressure (p. 494)

# Problems

Problems with a green number are answered at the back of the text. Most sections include three categories of problems separated by a green rule: concept review questions, *paired* skill-building exercises, and problems in a relevant context.

## Types of Solutions: Intermolecular Forces and the Prediction of Solubility

(Sample Problem 12.1)

**12.1** The dictionary defines "homogeneous" as "uniform in composition throughout." River water is a mixture of dissolved compounds, such as calcium bicarbonate, and suspended soil particles. Is river water homogeneous? Explain.

**12.2** Use the properties of sea water to describe the two characteristics that define mixtures.

**12.3** What types of intermolecular forces give rise to hydration shells in an aqueous solution of sodium chloride?

**12.4** Acetic acid is miscible with water. Would you expect carboxylic acids, general formula $CH_3(CH_2)_nCOOH$, to become more or less water soluble as $n$ increases? Explain.

**12.5** Which is more effective as a soap, sodium acetate or sodium stearate? Explain.

**12.6** Hexane and methanol are miscible as gases but only slightly soluble in each other as liquids. Explain.

**12.7** Use the following data to state the correlation between boiling point (bp) of a gas and its solubility in water:

| SUBSTANCE | bp (°C) | SOLUBILITY (mL GAS/100 mL H$_2$O) |
|---|---|---|
| $C_3H_8$ | $-42.1$ | 0.06 |
| $C_4H_{10}$ | $-0.5$ | 0.15 |

Explain the basis for this correlation in terms of intermolecular forces.

**12.8** Hydrogen chloride (HCl) gas is much more soluble than propane gas ($C_3H_8$) in water, even though HCl has a lower boiling point. Explain.

**12.9** Which will result in the more concentrated solution, (a) $KNO_3$ in $H_2O$ or (b) $KNO_3$ in carbon tetrachloride ($CCl_4$)? Explain.

**12.10** Which will result in the more concentrated solution, (a) stearic acid [$CH_3(CH_2)_{16}COOH$] in $H_2O$ or (b) stearic acid in $CCl_4$? Explain.

**12.11** What is the strongest type of intermolecular force between solute and solvent in each of the following solutions:

(a) CsCl(s) in $H_2O(l)$    (b) $CH_3\overset{\text{O}}{\overset{\|}{C}}CH_3(l)$ in $H_2O(l)$
(c) $CH_3OH(l)$ in $CCl_4(l)$    (d) Cu(s) in Ag(s)
(e) $CH_3Cl(g)$ in $CH_3OCH_3(g)$
(f) $CH_3CH_3(g)$ in $CH_3CH_2CH_2NH_2(l)$

**12.12** What is the strongest type of intermolecular force between solute and solvent in each of the following solutions:

(a) $CH_3OCH_3(g)$ in $H_2O(l)$    (b) Ne(g) in $H_2O(l)$
(c) $N_2(g)$ in $C_4H_{10}(g)$    (d) $C_6H_{14}(l)$ in $C_8H_{18}(l)$
(e) $H_2C{=}O(g)$ in $CH_3OH(l)$    (f) $Br_2(l)$ in $CCl_4(l)$

**12.13** Which member of the following pairs is more soluble in diethylether? Why?

(a) NaCl(s) or HCl(g)    (b) $H_2O(l)$ or $CH_3\overset{\text{O}}{\overset{\|}{C}}H(l)$
(c) $MgBr_2(s)$ or $CH_3CH_2MgBr(s)$

**12.14** Which member of the following pairs is more soluble in water? Why?

(a) $CH_3CH_2OCH_2CH_3(l)$ or $CH_3CH_2OCH_3(g)$
(b) $CH_2Cl_2(l)$ or $CCl_4(l)$

(c)

cyclohexane          tetrahydropyran

---

**12.15** Pyridine (see structure) is an essential portion of many biologically active compounds, such as nicotine and vitamin B$_6$. Like ammonia, it has a lone pair on N, which makes it act as a weak base. Because it is miscible in a wide range of solvents, from water to benzene, pyridine is one of the most important bases and solvents in organic syntheses. Account for its solubility behavior in terms of intermolecular forces.

pyridine

**12.16** Gluconic acid is a derivative of glucose used in cleaners and in the dairy and brewing industries. Caproic acid is a short-chain fatty acid used in the flavoring industry. Although both are six-carbon acids (see structures), gluconic acid is soluble in water and nearly insoluble in hexane, whereas caproic acid has the opposite solubility behavior. Explain.

gluconic acid

$CH_3{-}CH_2{-}CH_2{-}CH_2{-}CH_2{-}COOH$

caproic acid

## Energy Changes in the Solution Process

**12.17** What is the relationship between solvation and hydration?

**12.18** For a general solvent, which enthalpy terms in the solution cycle would be combined to obtain $\Delta H_{solvation}$?

**12.19** (a) What is the charge density of an ion, and what two properties of an ion affect it?
(b) How do these properties affect the ionic heat of hydration, $\Delta H_{hydr}$?

**12.20** For $\Delta H_{soln}$ to be zero (or very small), what quantities must be nearly equal in magnitude? Would their signs be the same or opposite?

**12.21** Aside from shaking stacks of coins, as mentioned in the text, describe a common system undergoing an increase in entropy.

**12.22** Like $NH_4NO_3$, a flask containing solid $NH_4Cl$ in water feels cold as the salt dissolves.
(a) Is the dissolving of $NH_4Cl$ in water exothermic or endothermic?
(b) Is the magnitude of $\Delta H_{lattice}$ of $NH_4Cl$ larger or smaller than the $\Delta H_{hydr}$? Explain.
(c) Given the answer to part (a), why does $NH_4Cl$ dissolve in water?

**12.23** Sketch a qualitative enthalpy diagram for the process of dissolving $KCl(s)$ in $H_2O$ (endothermic).

**12.24** Sketch a qualitative enthalpy diagram for the process of dissolving $NaI$ in $H_2O$ (exothermic).

**12.25** Which ion has the *greater* charge density? Explain.
(a) $Na^+$ or $Cs^+$  (b) $Sr^{2+}$ or $Rb^+$  (c) $Na^+$ or $Cl^-$

**12.26** Which ion has the *lesser* charge density? Explain.
(a) $Br^-$ or $I^-$  (b) $Sc^{3+}$ or $Ca^{2+}$  (c) $Br^-$ or $K^+$

**12.27** Which ion has the *larger* $\Delta H_{hydr}$?
(a) $Mg^{2+}$ or $Ba^{2+}$  (b) $Mg^{2+}$ or $Na^+$
(c) $NO_3^-$ or $CO_3^{2-}$

**12.28** Which ion has the *smaller* $\Delta H_{hydr}$?
(a) $SO_4^{2-}$ or $ClO_4^-$  (b) $Fe^{3+}$ or $Fe^{2+}$
(c) $Ca^{2+}$ or $K^+$

**12.29** (a) Use the following data to calculate the combined heats of hydration for the ions in potassium bromate ($KBrO_3$):

$$\Delta H_{lattice} = -745 \text{ kJ/mol} \qquad \Delta H_{soln} = 41.1 \text{ kJ/mol}$$

(b) Which ion do you think contributes more to the answer to part (a)? Why?

**12.30** (a) Use the following data to calculate the combined heats of hydration for the ions in sodium acetate ($NaC_2H_3O_2$):

$$\Delta H_{lattice} = -763 \text{ kJ/mol} \quad \Delta H_{soln} = 17.3 \text{ kJ/mol}$$

(b) Which ion do you think contributes more to the answer to part (a)? Why?

**12.31** State whether the entropy of the system increases or decreases in each of the following processes:
(a) A glass vase is shattered.
(b) Gold is extracted and purified from its ore.
(c) Ethanol ($CH_3CH_2OH$) dissolves in propanol ($CH_3CH_2CH_2OH$).

**12.32** State whether the entropy of the system increases or decreases in each of the following processes:
(a) Pure gases are mixed to prepare an anesthetic.
(b) Electronic-grade silicon is prepared from sand.
(c) Dry ice (solid $CO_2$) sublimes.

**12.33** Silver nitrate is used industrially to produce silver halides for photographic film and in forensic science to accomplish a similar task. The sodium chloride left behind in the sweat of a fingerprint is treated with silver nitrate solution to form silver chloride. This precipitate is then developed to show the black-and-white fingerprint. Given that the lattice energy of silver nitrate is $-822$ kJ/mol and its heat of hydration is $-799$ kJ/mol, calculate its heat of solution.

**Solubility as an Equilibrium Process**

(Sample Problems 12.2 and 12.3)

**12.34** You are given a bottle of solid X and three aqueous solutions of X, one saturated, one unsaturated, and one supersaturated. How would you determine which solution is which?

**12.35** Potassium permanganate ($KMnO_4$) has a solubility of 6.4 g/100 g $H_2O$ at 20°C and a positive heat of solution. How would you prepare a supersaturated solution of $KMnO_4$?

**12.36** In a saturated aqueous solution of each of the following solutes at 20°C and 1 atm, will the solubility increase, decrease, or stay the same when the indicated change occurs?
(a) $O_2(g)$, increase $P$  (b) $N_2(g)$, increase $V$
(c) $KCl(s)$ ($\Delta H_{soln} > 0$), decrease $T$
(d) $He(g)$, decrease $T$
(e) $NaOH(s)$ ($\Delta H_{soln} < 0$), increase T
(f) $RbI(s)$ ($\Delta H_{soln} > 0$), increase $P$

**12.37** Why does the solubility of any gas in water decrease with rising temperature?

**12.38** The heat of solution for $NaI$ in water is $-7.5$ kJ/mol. Is $NaI$ more soluble at 20°C or at 60°C? Why?

**12.39** Cesium bromide has a heat of solution of 26.0 kJ/mol. How does its solubility in water at 50°C compare with that at 25°C? Why?

**12.40** Consider the following solubility data (in units of g solute/100 g $H_2O$):

| SOLUTE | 10°C | 20°C |
|---|---|---|
| $AgNO_3$ | 170 | 222 |
| $Li_2CO_3$ | 1.43 | 1.33 |
| $O_2$ | 0.0054 | 0.0044 |

(a) What is the sign of $\Delta H_{soln}$ for each solute?
(b) Write a reaction for each solute, with heat as reactant or product.

**12.41** Consider the following solubility data (in units of g solute/100 g $H_2O$):

| SOLUTE | 0°C | 50°C |
|---|---|---|
| $AgC_2H_3O_2$ | 0.72 | 1.6 |
| $CO_2(g)$ at 1.0 atm | 0.35 | 0.08 |
| Valine | 8.34 | 9.62 |

(a) State whether each solution process is exothermic or endothermic.
(b) Write a reaction for each solute, showing heat absorbed or released.

**12.42** The Henry's law constant ($k_H$) for $O_2$ in water at 20°C is $1.28 \times 10^{-3}$ mol/L · atm.
(a) How many grams of $O_2$ will dissolve in 2.00 L $H_2O$ that is in contact with pure $O_2$ at 1.00 atm?
(b) How many grams of $O_2$ will dissolve in 2.00 L $H_2O$ that is in contact with air, in which the partial pressure of $O_2$ is 0.209 atm?

**12.43** Argon makes up 0.93% by volume of the atmosphere. Calculate its solubility (mol/L) in water at 20°C and 1.0 atm. The Henry's law constant for Ar under these conditions is $1.5 \times 10^{-3}$ mol/L · atm.

**12.44** Caffeine is about 10 times as soluble in hot water as in cold water.
(a) Is $\Delta H_{soln}$ of caffeine in water positive or negative?
(b) A chemist puts a hot-water extract of caffeine into an ice bath, and some caffeine crystallizes. Is the remaining solution saturated, unsaturated, or supersaturated?

**12.45** A saturated $Na_2CO_3$ solution is prepared and a small excess of solid is present. A seed crystal of $Na_2{}^{14}CO_3$ is introduced ($^{14}C$ is a radioactive isotope of $^{12}C$), and the radioactivity is measured over time (see figure).

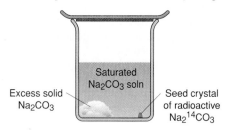

(a) Would you expect to find radioactivity in the solution? Explain.
(b) Would you expect to find radioactivity in all the solid or just in the seed crystal? Explain.

**12.46** The partial pressure of $CO_2$ gas above the liquid in a bottle of champagne at 20°C is 5.5 atm. What is the solubility of $CO_2$ in champagne? Assume that the Henry's law constant is the same for champagne as for water; at 20°C, $k_H = 2.3 \times 10^{-2}$ mol/L · atm.

**12.47** Individuals with respiratory problems are often treated with devices that deliver air with a higher partial pressure of $O_2$ than normal air. Why?

## Quantitative Ways of Expressing Concentration

(Sample Problems 12.4 to 12.6)

**12.48** Explain the difference between molarity and molality. Under what circumstances would the molality be a more accurate measure of the prepared concentration than molarity? Why?

**12.49** Which of the different ways of expressing concentration include the following: (a) volume of solution; (b) mass of solution; (c) mass of solvent?

**12.50** A solute has a solubility in water of 21 g/kg solvent. Is this value the same as 21 g/kg solution? Explain.

**12.51** Can molarity, molality, and mole fraction be interconverted *without* knowing the solution density? Explain.

**12.52** When a solution is heated, which of the different ways of expressing its concentration remain unchanged? Explain.

**12.53** Is a solution that is 50% by mass of methanol in ethanol different from one that is 50% by mass of ethanol in methanol? Explain.

**12.54** Calculate the molarity of these aqueous solutions:
(a) 42.3 g table sugar ($C_{12}H_{22}O_{11}$) in 100.0 mL solution
(b) 5.50 g $LiNO_3$ in 505 mL solution
(c) 75.0 mL of 0.250 $M$ NaOH diluted to 0.250 L with water

**12.55** Calculate the molarity of these aqueous solutions:
(a) 0.82 g ethanol ($C_2H_5OH$) in 10.5 mL solution
(b) 1.22 g gaseous $NH_3$ bubbled into water to make 33.5 mL solution
(c) 25.0 mL of 6.15 $M$ HCl diluted to 0.500 L with water

**12.56** How would you prepare these aqueous solutions:
(a) 355 mL of $8.74 \times 10^{-2}$ $M$ $KH_2PO_4$ from solid $KH_2PO_4$?
(b) 425 mL of 0.315 $M$ NaOH from 1.25 $M$ NaOH?

**12.57** How would you prepare these aqueous solutions:
(a) 1.50 L of 0.257 $M$ KBr from solid KBr?
(b) 355 mL of 0.0956 $M$ $LiNO_3$ from 0.244 $M$ $LiNO_3$?

**12.58** Calculate the molality of a solution containing 88.4 g glycine ($NH_2CH_2COOH$) dissolved in 1.250 kg $H_2O$.

**12.59** Calculate the molality of a solution containing 164 g HCl in 753 g $H_2O$.

**12.60** How would you prepare these aqueous solutions:
(a) $3.00 \times 10^2$ g of 0.115 $m$ ethylene glycol ($C_2H_6O_2$) from ethylene glycol and water?
(b) 1.00 kg of 2.00 mass % $HNO_3$ from 62.0 mass % $HNO_3$?

**12.61** How would you prepare these aqueous solutions:
(a) 1.00 kg of 0.0555 $m$ ethanol ($C_2H_5OH$) from ethanol and water?
(b) 475 g of 15.0 mass % HCl from 37.1 mass % HCl?

**12.62** A solution of isopropanol ($C_3H_7OH$) is made by dissolving 0.30 mol isopropanol in 0.80 mol water.
(a) What is the mole fraction of isopropanol?
(b) What is the mass percent of isopropanol?
(c) What is the molality of isopropanol?

**12.63** A solution is made by dissolving 0.100 mol NaCl in 8.60 mol water.
(a) What is the mole fraction of NaCl?
(b) What is the mass percent of NaCl?
(c) What is the molality of NaCl?

**12.64** What mass of cesium chloride must be added to 0.500 L water to produce a 0.400 $m$ solution? What are the mole fraction and the mass percent of CsCl? (Assume $d$ of $H_2O$ = 1.00 g/mL.)

**12.65** What are the mole fraction and the mass percent of a solution made by dissolving 0.30 g KBr in 0.400 L water ($d$ = 1.00 g/mL)?

**12.66** A 8.00 mass % aqueous solution of ammonia has a density of 0.9651 g/mL. Calculate the molality, molarity, and mole fraction of $NH_3$.

**12.67** A 28.8 mass % aqueous solution of iron(III) chloride has a density of 1.280 g/mL. Calculate the molality, molarity, and mole fraction of $FeCl_3$.

**12.68** Waste water from a cement factory contains 0.22 g $Ca^{2+}$ ion and 0.066 g $Mg^{2+}$ ion per 100.0 L solution. The solution density is 1.001 g/mL. Calculate the $Ca^{2+}$ and $Mg^{2+}$ concentrations in ppm (by mass).

**12.69** Gold occurs in sea water at an average concentration of $1.1 \times 10^{-2}$ ppb. How many liters of sea water must be processed to recover 1 troy ounce of gold, assuming 100% efficiency? (The average density of sea water is 1.025 g/mL; 1 troy ounce = 31.1 g.)

**12.70** An automobile antifreeze mixture is made by mixing equal volumes of ethylene glycol ($d$ = 1.114 g/mL; $M$ = 62.07 g/mol) and water ($d$ = 1.00 g/mL) at 20°C. The density of the mixture is 1.070 g/mL. Express the concentration of ethylene glycol in each of the following: (a) volume percent; (b) mass percent; (c) molarity; (d) molality; (e) mole fraction.

## Colligative Properties of Solutions

(Sample Problems 12.7 to 12.9)

**12.71** The composition (chemical formula) of a solute does not affect the extent of its colligative properties. What characteristic of a solute does affect these properties?

**12.72** What is a nonvolatile nonelectrolyte? Why do you think this type of solute provides the simplest case for examining colligative properties?

**12.73** In what sense is a strong electrolyte "strong"? What property of the substance makes it a strong electrolyte?

**12.74** Express Raoult's law in words. Is Raoult's law valid for a solution of a volatile solute? Explain.

**12.75** What are the most important differences between the phase diagram of a pure substance and one in which that substance is the solvent of a solution?

**12.76** Is the composition of the vapor at the top of a fractionating column different from the composition at the bottom? Explain.

**12.77** Is the boiling point of 0.01 $m$ KF higher or lower than that of 0.01 $m$ glucose? Explain.

**12.78** (a) Which electrolyte solution behaves more ideally, a dilute one or a concentrated one? Explain.
(b) Which solution has a boiling point closer to its predicted value, 0.050 $m$ NaF or 0.50 $m$ KCl?

**12.79** (a) Which electrolyte solution behaves more ideally, one with univalent ions or one with divalent ions? Explain.
(b) Which solution has a freezing point closer to its predicted value, 0.01 $m$ NaBr or 0.01 $m$ MgCl$_2$?

**12.80** The freezing point depression constants of cyclohexane and naphthalene are 20.1°C/$m$ and 6.94°C/$m$, respectively. Which would you choose to determine the molar mass by freezing point depression of a substance that is soluble in either solvent? Why?

**12.81** Classify the following substances as strong electrolytes, weak electrolytes, or nonelectrolytes: (a) hydrogen chloride (HCl); (b) potassium nitrate ($KNO_3$); (c) glucose ($C_6H_{12}O_6$); (d) ammonia ($NH_3$).

**12.82** Classify the following substances as strong electrolytes, weak electrolytes, or nonelectrolytes: (a) sodium permanganate ($NaMnO_4$); (b) acetic acid ($CH_3COOH$); (c) methanol ($CH_3OH$); (d) calcium acetate [$Ca(C_2H_3O_2)_2$].

**12.83** How many moles of solute particles are present in one liter of the following aqueous solutions: (a) 0.2 $M$ KI; (b) 0.07 $M$ HNO$_3$; (c) $10^{-4}$ $M$ K$_2$SO$_4$; (d) 0.07 $M$ ethanol ($C_2H_5OH$)?

**12.84** How many moles of solute particles are present in one milliliter of the following aqueous solutions: (a) 0.01 $M$ CuSO$_4$; (b) 0.005 $M$ Ba(OH)$_2$; (c) 0.06 $M$ pyridine ($C_5H_5N$); (d) 0.05 $M$ (NH$_4$)$_2$CO$_3$?

**12.85** Rank the following aqueous solutions—(I) 0.100 $m$ NaNO$_3$; (II) 0.200 $m$ glucose; (III) 0.100 m CaCl$_2$—in order of increasing (a) osmotic pressure; (b) boiling point; (c) freezing point; (d) vapor pressure at 50°C.

**12.86** Rank the following aqueous solutions—(I) 0.04 $m$ urea [(NH$_2$)$_2$C=O]; (II) 0.02 $m$ AgNO$_3$; (III) 0.02 $m$ CuSO$_4$—in order of decreasing (a) osmotic pressure; (b) boiling point; (c) freezing point; (d) vapor pressure at 298 K.

**12.87** Calculate the vapor pressure of a solution of 44.0 g glycerol ($C_3H_8O_3$) in 500.0 g water at 25°C. The vapor pressure of water at 25°C is 23.76 torr. (Assume ideal behavior.)

**12.88** Calculate the vapor pressure of a solution of 0.39 mol cholesterol in 5.4 mol toluene at 32°C. Pure toluene has a vapor pressure of 41 torr at this temperature. (Assume ideal behavior.)

**12.89** What is the freezing point of 0.111 $m$ urea in water?

**12.90** What is the boiling point of 0.200 $m$ lactose in water?

**12.91** The boiling point of ethanol ($C_2H_5OH$) is 78.5°C. What is the boiling point of a solution of 3.4 g vanillin ($M$ = 152.14 g/mol) in 50.0 g ethanol? ($K_b$ of ethanol = 1.22°C/$m$.)

**12.92** The freezing point of benzene is 5.5°C. What is the freezing point of a solution of 5.00 g naphthalene ($C_{10}H_8$) in 444 g benzene? ($K_f$ of benzene = 4.90°C/$m$.)

**12.93** What is the minimum mass of ethylene glycol ($C_2H_6O_2$) that must be dissolved in 14.5 kg water to prevent the solution from freezing at −10.0°F? (Assume ideal behavior.)

**12.94** What is the minimum mass of glycerol $(C_3H_8O_3)$ that must be dissolved in 11.0 mg water to prevent the solution from freezing at $-25°C$? (Assume ideal behavior.)

**12.95** Use the concept of osmotic pressure to explain why drinking sea water does not quench your thirst.

**12.96** Waste water discharged into a stream by a sugar refinery contains sucrose $(C_{12}H_{22}O_{11})$ as its main impurity. The solution contains 3.42 g sucrose/L. A government-industry project is designed to test the feasibility of removing the sugar by reverse osmosis. What pressure must be applied to the apparatus at 20°C to produce pure water?

**12.97** A biochemical engineer isolates a bacterial gene fragment and dissolves a 10.0-mg sample of the material in enough water to make 30.0 mL solution. The osmotic pressure of the solution is 0.340 torr at 25°C.
(a) What is the molar mass of the gene fragment?
(b) If the solution density is 0.997 g/mL, how large would the freezing point depression be for this solution? ($K_f$ of water = 1.86°C/$m$)

**12.98** In a study designed to prepare new gasoline-resistant coatings, a polymer chemist dissolves 6.053 g polyvinyl alcohol in enough water to make 100.0 mL solution. At 25°C, the osmotic pressure of this solution vs. pure water is 0.272 atm. What is the molar mass of the polymer sample?

**12.99** The U.S. Food and Drug Administration (FDA) lists dichloromethane $(CH_2Cl_2)$ and carbon tetrachloride $(CCl_4)$ among the many chlorinated organic compounds that are carcinogenic. What are the partial pressures of these substances in the vapor above a solution of 1.50 mol $CH_2Cl_2$ and 1.00 mol $CCl_4$ at 23.5°C? The vapor pressures of pure $CH_2Cl_2$ and $CCl_4$ at this temperature are 352 torr and 118 torr, respectively. (Assume ideal behavior.)

## The Structure and Properties of Colloids

**12.100** Is the fluid inside a bacterial cell considered a solution, a colloid, or both? Explain.

**12.101** What type of colloid is (a) milk; (b) fog; (c) shaving cream?

**12.102** What is Brownian motion, and what causes it?

**12.103** In a movie theater, you can often see the beam of projected light from below. What phenomenon does this exemplify? Why does it occur?

**12.104** Why don't soap micelles coagulate and form large globules? Is soap a more effective cleaner in fresh water or in sea water? Why?

## Comprehensive Problems

Problems with an asterisk (*) are more challenging.

**12.105** Give brief answers for each of the following:
(a) Why are lime (CaO) and alum $[Al_2(SO_4)_3]$ added during the water purification process?
(b) Why is water that contains large amounts of $Ca^{2+}$, $Mg^{2+}$, or $Fe^{2+}$ difficult to use for cleaning?

(c) What is the meaning of "reverse" in reverse osmosis?
(d) Why might a water treatment plant use ozone as the final disinfectant instead of chlorine, even though ozone is more expensive?
(e) How does passing a saturated NaCl solution through a "spent" ion-exchange resin regenerate the resin?

**12.106** An aqueous solution is 10% glucose by mass (d = 1.039 g/mL at 20°C). Calculate its freezing point, boiling point at 1 atm, and osmotic pressure.

**12.107** Gramicidin A is a common antibiotic. If the molecule were polar on the outside and nonpolar on the inside, would its function be affected? Explain.

**12.108** Solutes elevate the boiling point of a solvent but depress the freezing point. Explain in molecular terms.

**12.109** $\beta$-Pinene $(C_{10}H_{16})$ and $\alpha$-terpineol $(C_{10}H_{18}O)$ are two of the many compounds used in perfumes and cosmetics to provide a "fresh pine" scent. At 367 K, the pure substances have vapor pressures of 100.3 torr and 9.8 torr, respectively. What is the composition of the vapor (in terms of mole fractions) above a test solution containing equal masses of these compounds at 367 K? (Assume ideal behavior.)

*12.110 A pharmaceutical preparation made with ethanol $(C_2H_5OH)$ is contaminated with methanol $(CH_3OH)$. A sample of vapor above the liquid mixture is found to contain a 97/1 mass ratio of $C_2H_5OH/CH_3OH$. What is the mass ratio of these alcohols in the liquid? At the temperature of the liquid, the vapor pressures of $C_2H_5OH$ and $CH_3OH$ are 60.5 torr and 126.0 torr, respectively.

*12.111 Two beakers are placed in a closed container (left). One beaker contains water, the other a concentrated aqueous sugar solution. With time, the solution volume increases and the water volume decreases (right). Explain on the molecular level.

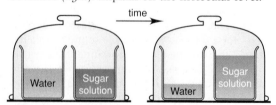

**12.112** How would you prepare 250 g of 0.150 $m$ aqueous $NaHCO_3$?

**12.113** Carbon dioxide is less soluble in dilute HCl($aq$) than in dilute NaOH($aq$). Explain.

*12.114 Derive a general equation that expresses the relationship between the molarity and the molality of a solution, and use it to explain why the numerical values of these two concentration terms would be approximately equal for very dilute solutions.

*12.115 A florist prepares a solution of nitrogen-phosphorus fertilizer by dissolving 5.66 g $NH_4NO_3$ and 4.42 g $(NH_4)_3PO_4$ in enough water to make 20.0 L solution. What is the molarity of $NH_4^+$ and $PO_4^{3-}$ in the solution?

# A Mid-Course Perspective on the Properties of the Elements

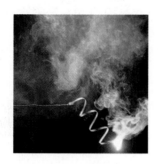

Chemistry has a central, underlying principle: *The behavior of a sample of matter is the inevitable result of the properties of its component atoms.* In the illustrated overview that follows, many ideas about these properties are summarized and organized so that you can see how they lead to the behavior of the elements in the main groups of the periodic table. A key point to keep in mind is that, despite our categories, the properties of matter, and the behaviors emerging from them, exist in gradations; clear dividing lines rarely appear in reality. This Interchapter feature reviews major points from earlier chapters and previews their application in upcoming ones.

**The molecular nature of matter and change.**
All changes in matter result from the properties of the atoms that make up the substances. These reactions are between *(clockwise from top left)* copper and nitric acid, magnesium and oxygen, phosphorus and oxygen, potassium and chlorine, zinc and hydrochloric acid, and iron and chlorine.

## TOPIC 1          The Key Atomic Properties

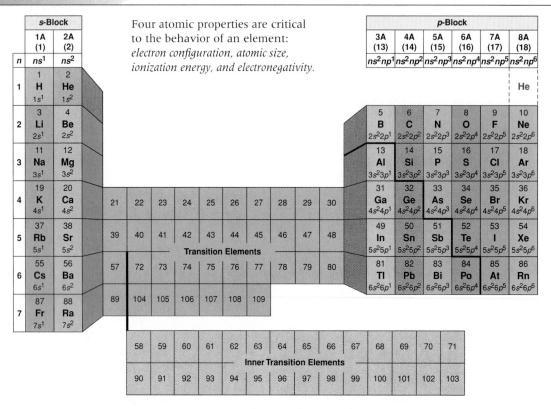

Four atomic properties are critical to the behavior of an element: *electron configuration, atomic size, ionization energy, and electronegativity.*

**Electron configuration ($nl^\#$)** is the distribution of electrons among the energy levels and sublevels of an atom *(Section 8.2)*:

- The *n* value (positive integer) indicates the *energy and relative distance* from the nucleus of orbitals in the level.
- The *l* value letter designation *(s, p, d, f)* indicates the *shape* of orbitals in the sublevel.
- The superscript (#) tells the *number* of electrons in the sublevel.

The periodic table above shows the sublevel blocks and the ground-state (lowest energy), outer (valence-level) electron configuration of the main-group elements. Note that

- The *s* block lies to the left of the transition elements and the *p* block lies to the right.
  [Although H is not in Group 1A(1) and He belongs in Group 8A(18), both are part of the *s* block.]
- Outer electron configurations are *different within a period.*
- Outer electron configurations are *similar within a group.*
- Outer electrons occupy the *ns* and *np* sublevels (*n* = period number).
- Elements in Periods 2 to 6 have four valence-level orbitals (one *ns* + three *np*).
- *Outer electrons **are** the valence electrons* of the main-group elements.
- The A-group number (**1A** to **8**A) equals the number of valence electrons.

Electrons in the same and, especially, inner levels *shield* outer electrons from the full nuclear charge, reducing the attraction to an *effective nuclear charge (Z_{eff})* *(Section 8.1)*. Within a level, electrons that *penetrate* more (spend more time near the nucleus) shield more. The probability distribution curves (right) show that

- *n* = 1 electrons shield *n* = 2 electrons very effectively.
- 2*s* electrons shield 2*p* electrons much less effectively.

$Z_{eff}$ greatly influences atomic properties. In general,

- $Z_{eff}$ increases significantly across a period.
- $Z_{eff}$ increases very little down a group.

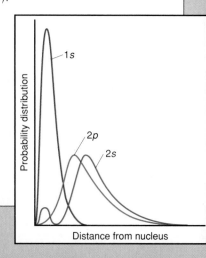

**Atomic size** is based on atomic radius, one-half the distance between nuclei of identical, bonded atoms *(Section 8.3)*. The small red periodic table shows the trends in atomic size among the main-group elements. Note that

- Atomic size generally *decreases across a period:* increasing $Z_{eff}$ pulls outer electrons closer.
- Atomic size generally *increases down a group:* outer electrons in higher periods lie farther from the nucleus.

**Ionization energy (IE)** is the energy required to remove an outer electron from a mole of gaseous atoms *(Section 8.3)*. The relative magnitude of the IE influences the types of bonds an atom forms: an element with a *low IE* is more likely to *lose* electrons, and one with a *high IE* is more likely to *share (or gain)* electrons (excluding the noble gases). From the small yellow periodic table, note that

- IE generally *increases across a period:* higher $Z_{eff}$ holds electrons tighter.
- IE generally *decreases down a group:* greater distance from the nucleus lowers the attraction for electrons.

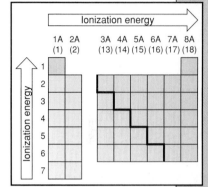

Thus, the trends in IE are *opposite* those in atomic size: it is easier to remove an electron (lower IE) if the electron is farther from the nucleus (larger atomic size).

**Electronegativity (EN)** is a number that describes the relative ability of an atom in a covalent bond to attract shared electrons *(Section 9.4)*. From the small green periodic table, note that

- EN generally *increases across a period:* higher $Z_{eff}$ and shorter distance from the nucleus strengthen the attraction for the shared pair.
- EN generally *decreases down a group:* greater distance from the nucleus weakens the attraction for the shared pair. (Group 8A is not shaded because the noble gases form few compounds.)

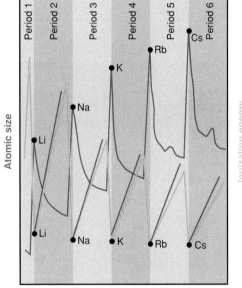

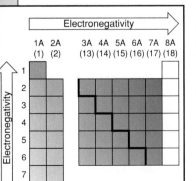

Thus, as the three periodic tables show, the trends in EN are *opposite* those in atomic size and the *same* as those in IE.

The graph plots atomic size (red), ionization energy (yellow), and electronegativity (green) versus atomic number. Note that

- Atomic size decreases gradually within a period and then increases suddenly at the beginning of the next period.
- Ionization energy and electronegativity display the opposite pattern.

The block diagram shows that the difference in electronegativity ($\Delta EN$) between the atoms in a bond greatly influences behavior.

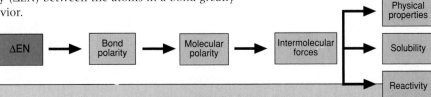

## TOPIC 2          Characteristics of Chemical Bonding

Chemical bonds are the forces that hold atoms (or ions) together in an element or compound. The type of bonding, bond properties, nature of orbital overlap, and number of bonds affect physical and chemical behavior.

### Types of Bonding
There are three idealized bonding models: *ionic, covalent,* and *metallic.*

**Covalent bonding**
results from the attraction between two nuclei and a localized electron pair. The bond arises through electron sharing between atoms with a small $\Delta EN$ (usually two nonmetals) and leads to separate molecules with specific shapes or to extended networks. (*Section 9.3*).

**Ionic bonding**
results from the attraction between positive and negative ions. The ions arise through electron transfer between atoms with a large $\Delta EN$ (from metal to nonmetal). This bonding leads to crystalline solids with ions packed tightly in regular arrays (*Section 9.2*).

**Metallic bonding**
results from the attraction between the cores of metal atoms (metal cations) and delocalized valence electrons. This bonding arises through the shared pooling of valence electrons from many atoms and leads to crystalline solids (*Sections 9.7 and 11.4*).

The actual bonding in real substances usually lies between these distinct models. The triangular diagram shows the continuum of bond types among the Period 3 elements (*Section 9.4*).
- Along the left leg of the triangle, compounds of each element with chlorine display the gradation from ionic to covalent bonding.

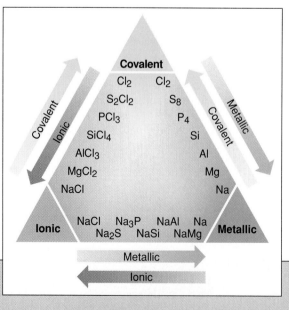

- Along the right leg, the elements themselves display the gradation from covalent to metallic bonding.
- Along the base, compounds of each element with sodium display the gradation from ionic to metallic bonding.

## Bond Properties

There are two important properties of a covalent bond (*Section 9.3*):

**Bond length** is the distance between the nuclei of bonded atoms.
**Bond energy** (bond strength) is the enthalpy change required to break a given bond in a mole of gaseous molecules.

Among similar compounds, these bond properties are related to each other and to reactivity, as shown in the graph for the carbon tetrahalides ($CX_4$). Note that

* *As bond length increases, bond energy decreases:* shorter bonds are stronger bonds.
* *As bond energy decreases, reactivity increases.*

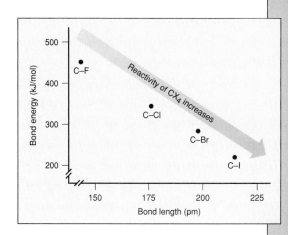

## Nature of Orbital Overlap

In a covalent bond, the shared electrons reside in the overlapping orbitals of the two atoms. The diagram depicts the bonding in ethylene ($C_2H_4$).

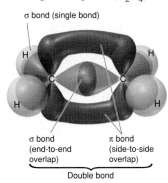

σ bond (single bond)

σ bond
(end-to-end
overlap)

π bond
(side-to-side
overlap)

Double bond

**Bond order** is one-half the number of electrons shared. Bond orders of 1 (single bond) and 2 (double bond) are common; a bond order of 3 (triple bond) is less common. Fractional orders occur in species with adjacent single and double bonds (*Section 9.3*).

Orbitals overlap in two orientations that lead to two types of bonds (*Section 10.3*).

**End-to-end overlap** (of *s*, *p*, and hybrid atomic orbitals) leads to a sigma (σ) bond, one with electron density distributed symmetrically around the bond axis.
A single bond is a σ bond.

**Side-to-side overlap** (of *p* with *p*, or sometimes *d*, orbitals) leads to a pi (π) bond, one with electron density distributed above and below the bond axis. A double bond consists of one σ bond and one π bond. Pi bonds restrict rotation around the bond axis, allowing for different arrangements of the atoms and, therefore, different molecules. Pi bonds are often sites of reactivity; for example,

$$CH_2{=}CH_2(g) + H{-}Cl(g) \rightarrow CH_3{-}CH_2{-}Cl(g)$$
$$CH_3{-}CH_3(g) + H{-}Cl(g) \rightarrow \text{no reaction}$$

## Number of Bonds

The number of bonds an atom forms determines the shapes of its molecules. The small periodic table shows that

* *The elements in Period 2 cannot form more than four bonds* because they have a total of four (one *s* and three *p*) valence orbitals. (Only carbon forms four bonds routinely). Molecular shapes (small circle) are based on linear, trigonal planar, and tetrahedral electron group arrangements (*Section 10.1*).
* *Many elements in Period 3 or higher can form more than four bonds* by using empty *d* orbitals to expand their valence shells. Shapes include those above and others based on trigonal bipyramidal and octahedral electron group arrangements (large circle).

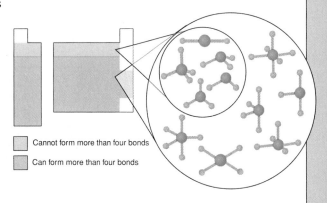

Cannot form more than four bonds

Can form more than four bonds

## TOPIC 3 — Metallic Behavior

The elements are often classified as metals, metalloids, or nonmetals. This table contrasts the *general* atomic, physical, and chemical properties of metals with those of nonmetals *(Section 8.4).*

|  | Metals | Nonmetals |
|---|---|---|
| **Atomic Properties** | Have fewer valence electrons (in period)<br>Have larger atomic size<br>Have lower ionization energies<br>Have lower electronegativities | Have more valence electrons (in period)<br>Have smaller atomic size<br>Have higher ionization energies<br>Have higher electronegativities |
| **Physical Properties** | Occur as solids at room temperature<br>Conduct electricity and heat well<br>Are malleable and ductile | Occur in all three physical states<br>Conduct electricity and heat poorly<br>Are not malleable or ductile |
| **Chemical Properties** | Lose electron(s) to become cations<br>React with nonmetals to form ionic compounds<br>Mix with other metals to form solutions (alloys) | Gain electron(s) to become anions<br>React with metals to form ionic compounds<br>React wiith other nonmetals to form covalent compounds |

From the blue, shaded periodic table, note that
- *A gradation in metallic behavior* exists among the elements.
- *Metallic behavior parallels atomic size:* Larger members of a group (bottom) or period (left) are more metallic; smaller members are less metallic.

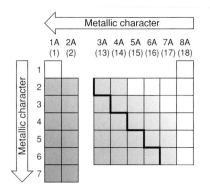

This small periodic table shows the location of metals, metalloids, and nonmetals among the main-group elements. Note that
- Metals lie in the lower left portion of the table.
- Nonmetals lie in the upper right portion of the table.
- Metalloids lie between the metals and nonmetals.

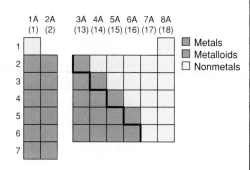

These elements have intermediate values of atomic size, IE, and EN and display an intermediate metallic character: shiny solids with low conductivity; cation-like with nonmetals (e.g., $AsF_3$) and anion-like with metals (e.g., $Na_3As$).

Metals and nonmetals typically form crystalline ionic compounds when they interact, and ionic sizes and charges determine the packing in these solids. Monatomic ions exhibit clear size trends:

- Cations are smaller and anions larger than their parent atoms.
- Ionic size *increases* down a group.
- Ionic size *decreases* across a period, but anions are larger than cations.

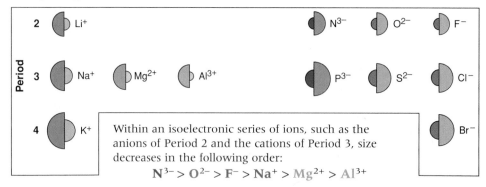

Within an isoelectronic series of ions, such as the anions of Period 2 and the cations of Period 3, size decreases in the following order:

$$N^{3-} > O^{2-} > F^- > Na^+ > Mg^{2+} > Al^{3+}$$

# Acid-Base Behavior of the Element Oxides

Oxides are known for almost every element, and the metallic behavior of an element correlates with the acid-base behavior of its oxide in water *(Section 8.4)*. Recall that

- An acid reacts with water to increase the $H^+$ concentration, and with a base to form a salt and water.
- A base reacts with water to increase the $OH^-$ concentration, and with an acid to form a salt and water.

This chart shows that the electronegativity and metallic behavior of an element (E) determine the type of bonding (E-to-O) in the oxide and, therefore, the acid-base behavior of the dissolved oxide.

| | | |
|---|---|---|
| **Basic Oxide (Ionic E-to-O)** | $+ H_2O \longrightarrow$ | more $OH^-$ |
| | $+$ acid $\longrightarrow$ | salt (E cation) $+ H_2O$ |
| **Amphoteric Oxide** | $+$ acid $\longrightarrow$ | salt (E cation) $+ H_2O$ |
| | $+$ base $\longrightarrow$ | salt (E oxoanion) $+ H_2O$ |
| **Acidic Oxide (Covalent E-to-O)** | $+$ base $\longrightarrow$ | salt (E oxoanion) $+ H_2O$ |
| | $+ H_2O \longrightarrow$ | more $H^+$ |

(Left arrows: Metallic behavior of E ↑, Electronegativity of E ↓)

Note that

- *Elements with low EN (metals) form basic oxides.* The E-to-O bonding is ionic.
  The $O^{2-}$ ion is the basic species:
  $$O^{2-}(s) + H_2O(l) \rightarrow 2OH^-(aq)$$
  For example, BaO is a basic oxide:
  Reacts with water: $BaO(s) + H_2O(l) \rightarrow Ba^{2+}(aq) + 2OH^-(aq)$
  Reacts with acid:   $BaO(s) + 2H^+(aq) \rightarrow Ba^{2+}(aq) + H_2O(l)$
- *Elements with high EN (nonmetals) form acidic oxides.* The E-to-O bonding is covalent.
  Water bonds to E to form an acid, which releases $H^+$. For example, $SO_2$ is an acidic oxide:
  Reacts with water: $SO_2(g) + H_2O(l) \rightleftharpoons [H_2SO_3(aq)] \rightleftharpoons H^+(aq) + HSO_3^-(aq)$
  Reacts with base:   $SO_2(g) + 2OH^-(aq) \rightarrow SO_3^{2-}(aq) + H_2O(l)$
- *Elements with intermediate EN (some metalloids and metals) form amphoteric oxides, which react with an acid and a base.* For example, $Al_2O_3$ is an amphoteric oxide:
  Reacts with acid:   $Al_2O_3(s) + 6H^+(aq) \rightarrow 2Al^{3+}(aq) + 3H_2O(l)$
  Reacts with base:   $Al_2O_3(s) + 2OH^-(aq) + 3H_2O(l) \rightarrow 2Al(OH)_4^-(aq)$

| | 1A (1) | 2A (2) | | 3A (13) | 4A (14) | 5A (15) | 6A (16) | 7A (17) | | 8A (18) |
|---|---|---|---|---|---|---|---|---|---|---|
| 1 | | | | | | | | | | |
| 2 | $Li_2O$ | $BeO$ | | $B_2O_3$ | $CO_2$ | $N_2O_5$ $N_2O_3$ | | | | |
| 3 | $Na_2O$ | $MgO$ | | $Al_2O_3$ | $SiO_2$ | $P_4O_{10}$ $P_4O_6$ | $SO_3$ $SO_2$ | $Cl_2O_7$ $Cl_2O$ | | |
| 4 | $K_2O$ | $CaO$ | | $Ga_2O_3$ | $GeO_2$ | $As_2O_5$ $As_4O_6$ | $SeO_3$ $SeO_2$ | $Br_2O$ | | |
| 5 | $Rb_2O$ | $SrO$ | | $In_2O_3$ $In_2O$ | $SnO_2$ $SnO$ | $Sb_2O_5$ $Sb_4O_6$ | $TeO_3$ $TeO_2$ | $I_2O_5$ | | |
| 6 | $Cs_2O$ | $BaO$ | | $Tl_2O$ | $PbO_2$ $PbO$ | $Bi_2O_3$ | $PoO_2$ $PoO$ | | | |
| 7 | $Fr_2O$ | $RaO$ | | | | | | | | |

- ■ Strongly basic
- □ Weakly basic
- ■ Amphoteric
- ■ Strongly acidic
- ■ Moderately acidic
- □ Weakly acidic

The periodic table shows the acid-base behavior of many main-group oxides in water. Note that

- *Oxide acidity increases across a period and decreases down a group*, a pattern that is *opposite* the trend in metallic behavior.
- When an element forms two oxides, *the more acidic oxide contains the element in a higher oxidation state.*

## TOPIC 5 — Redox Behavior of the Elements

The relative ability of an element to lose or gain electrons when reacting with other elements is called its redox behavior *(Section 4.4)*:

- *The oxidation state (or oxidation number, O.N.)* of an atom in an element is zero; in a compound, the O.N. is the number of electrons that have shifted away from the atom (positive O.N.) or toward it (negative O.N.). O.N. values are determined by a set of rules *(Section 4.4)*.
- *A redox reaction* occurs when the O.N. values of atoms in the reactants are different from those in the products.
- Among the countless redox reactions are all those that involve *an elemental substance as reactant or product.* These include all combustion reactions and all formation reactions, such as this one between potassium and chlorine:

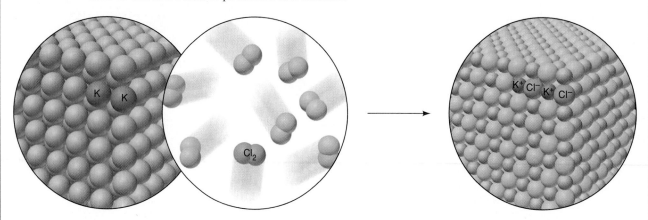

### Reducing and Oxidizing Agents

All redox reactions contain a reducing *and* an oxidizing agent. This chart shows that

- The reducing agent gives electrons to (reduces) the oxidizing agent and becomes oxidized (more positive O.N.).
- The oxidizing agent removes electrons from (oxidizes) the reducing agent and becomes reduced (more negative O.N.).

| Reducing agent | Oxidizing agent |
|---|---|
| X loses electrons | Y gains electrons |
| X is oxidized by Y | Y is reduced by X |
| X oxidation number increases | Y oxidation number decreases |

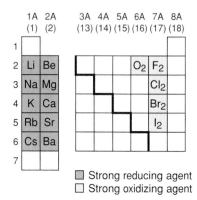

☐ Strong reducing agent
☐ Strong oxidizing agent

The periodic table shows that oxidizing and reducing ability are related to atomic properties:

- Elements with low IE and low EN [Groups 1A(1) and 2A(2)] are *strong reducing agents.*
- Elements with high IE and high EN [Group 7A(17) and oxygen in Group 6A(16)] are *strong oxidizing agents.*

## Oxidation States of the Main-Group Elements

The periodic table below shows some oxidation states (most common states in **bold** type) of the elements in their compounds. Note that

- *The highest (most positive) possible state in a group equals the A-group number.* It occurs if all the outer electrons of an atom shift toward a *more* electronegative atom.
- *Among nonmetals, the lowest (most negative) possible state equals the A-group number minus eight.* It occurs if a nonmetal atom fills its outer level with electrons that shift away from a *less* electronegative atom.
- *Nonmetals have more oxidation states than metals* within a group or period. (Oxygen and fluorine are exceptions.)
- Odd-numbered groups generally have odd-numbered states and even-numbered groups have even-numbered states. *Oxidation states differ by units of two* because electrons are lost (or gained) in pairs *(Section 13.9)*.
- For many metals and metalloids with more than one oxidation state [Groups 3A(13) to 5A(15)], *the lower state becomes more common down the group.* This state results when outer *p* electrons only are lost. *(Section 8.4)*
- An element with more than one oxidation state exhibits *greater metallic behavior in its lower state.* For example, arsenic(III) oxide is more basic, more like a metal oxide, than arsenic(V) oxide *(See Topic 4)*.

| 1A (1) | 2A (2) | | 3A (13) | 4A (14) | 5A (15) | 6A (16) | 7A (17) | 8A (18) |
|---|---|---|---|---|---|---|---|---|
| **H** −1, **+1** | | | | | | | | **He** |
| **Li** +1 | **Be** +2 | | **B** +3 | **C** −4, **+4**, +2 | **N** −3, **+5**, +4, +3, +2, +1 | **O** −1, −2 | **F** −1 | **Ne** |
| **Na** +1 | **Mg** +2 | | **Al** +3 | **Si** −4, **+4**, +2 | **P** −3, **+5**, +3 | **S** −2, **+6**, +4, +2 | **Cl** −1, **+7**, +5, +3, +1 | **Ar** |
| **K** +1 | **Ca** +2 | | **Ga** +3, +1 | **Ge** **+4**, +2 | **As** −3, **+5**, +3 | **Se** −2, **+6**, **+4**, +2 | **Br** −1, **+7**, +5, +1 | **Kr** +2 |
| **Rb** +1 | **Sr** +2 | | **In** +3, +1 | **Sn** **+4**, +2 | **Sb** −3, **+5**, +3 | **Te** −2, **+6**, **+4**, +2 | **I** −1, **+7**, +5, +1 | **Xe** **+8**, **+6**, **+4**, +2 |
| **Cs** +1 | **Ba** +2 | | **Tl** +1 | **Pb** **+4**, +2 | **Bi** +3 | **Po** **+4**, +2 | **At** −1 | **Rn** +2 |
| **Fr** +1 | **Ra** +2 | | | | | | | |

☐ Metals
☐ Metalloids
☐ Nonmetals

## TOPIC 6        Physical States and Changes of State

Physical state and the heat of a phase change reflect the relative strengths of the bonding and/or intermolecular forces between the atoms, ions, or molecules that make up an element or compound *(Sections 11.2 and 11.4)*.

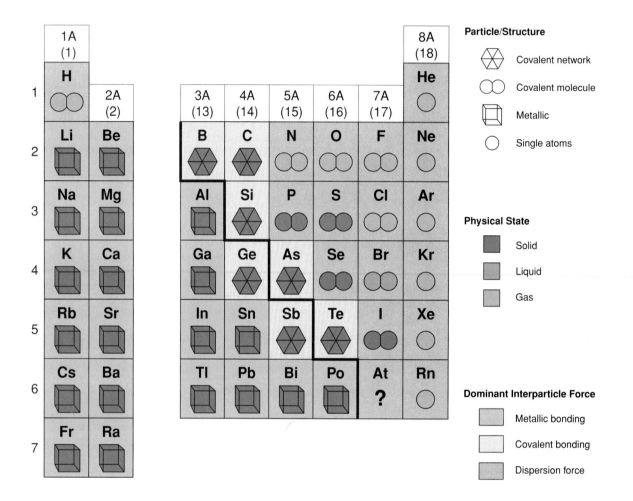

### Physical States of the Elements

The large periodic table shows the type of particle (or structure), physical state, and predominant interparticle force in the most common form of the main-group elements at room temperature. Note that

- *Metals (left and bottom) are solids:* strong metallic bonding holds the atoms in their crystal structures.
- *Metalloids (along staircase line) and carbon are solids:* strong covalent bonds hold the atoms together in extensive networks.
- *Lighter nonmetals and Group 8A(18) (right and top) are gases:* dispersion forces are weak between molecules or atoms with relatively small electron clouds. ($H_2$ is shown like other gaseous diatomic elements $N_2$, $O_2$, $F_2$, $Cl_2$.)
- *Heavier nonmetals are liquid ($Br_2$) or soft solids ($P_4$, $S_8$, and $I_2$):* dispersion forces are stronger between molecules with larger, more polarizable electron clouds.

## Phase Changes of the Elements

The small periodic table shows trends in melting point, boiling point, $\Delta H_{fus}$, and $\Delta H_{vap}$ for Groups 1A(1), 7A(17), and 8A(18):

- In Group 1A(1), these properties generally *increase up* the group: smaller atom cores attract delocalized electrons more strongly, which leads to stronger metallic bonding.
- In Groups 7A(17) and 8A(18), they generally *increase down* the group: dispersion forces become stronger with larger atoms.
- In other groups these properties reflect changes in interparticle forces down the group: lower values for molecular nonmetals, higher values for covalent networks of metalloids (and carbon), and intermediate values for metals.

Increase up
- Melting point
- Boiling point
- $\Delta H_{fus}$
- $\Delta H_{vap}$

Increase down
- Melting point
- Boiling point
- $\Delta H_{fus}$
- $\Delta H_{vap}$

## Physical Properties of Compounds

Compounds have physical properties based on the types of bonding and intermolecular forces.

**Ionic compounds,** such as sodium chloride, have very high melting points, boiling points, $\Delta H_{fus}$, and $\Delta H_{vap}$.

**Molecular compounds,** such as methane, have a physical state that depends on intermolecular forces. Most are gases, liquids, or low-melting solids at room temperature.

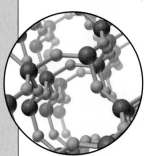

**Network covalent compounds,** such as silica, have extremely high melting points, boiling points, $\Delta H_{fus}$, and $\Delta H_{vap}$.

**Hydrogen bonding** has a major effect on physical properties. For example, despite similar molar masses, extensive H bonding in $H_2O$ (18.02 g/mol) gives it a much higher melting point, boiling point, $\Delta H_{fus}$, and $\Delta H_{vap}$ than $CH_4$ (16.04 g/mol). Water's unusually high specific heat capacity, surface tension, and viscosity also result from H bonding *(Section 11.6)*.

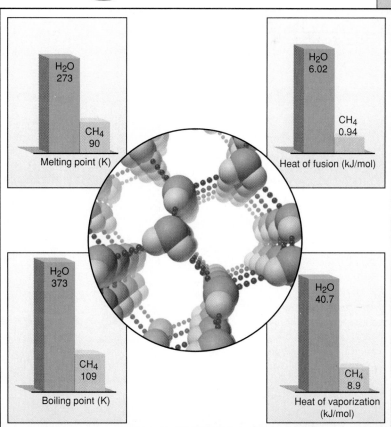

$H_2O$ 273   $CH_4$ 90
Melting point (K)

$H_2O$ 6.02   $CH_4$ 0.94
Heat of fusion (kJ/mol)

$H_2O$ 373   $CH_4$ 109
Boiling point (K)

$H_2O$ 40.7   $CH_4$ 8.9
Heat of vaporization (kJ/mol)

# CHAPTER 13

## Concepts and skills to review

• see Interchapter feature.

# Periodic Patterns in the Main-Group Elements

## Bonding, Structure, and Reactivity

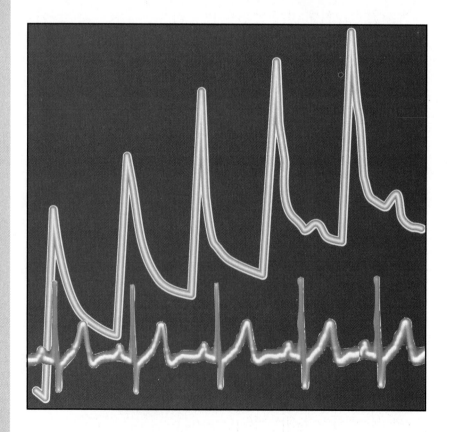

**Natural periodic patterns.** From the movement of the planets to the lifecycle of an ant and from the beat of a human heart (green) to the sizes of atoms (red), recurring patterns are the hallmark of the natural world. In this chapter, we examine the patterns in physical and chemical behavior that emerge from the recurring atomic properties of the main-group elements.

et's take stock for a moment of the ground you've covered so far. From earlier chapters, you know how to name compounds, balance equations, and calculate reaction yields. You've seen how heat is related to chemical and physical change, how electron configuration influences atomic properties, how atoms bond to form compounds, and how bond properties create molecular shapes. Most recently, you've learned how atomic and molecular properties give rise to those of gases, liquids, solids, and solutions.

With this knowledge in hand, it is time to journey through the elemental world. In this chapter, we examine the main groups of the periodic table to see how these principles help us understand the magnificent diversity of the elements—their bonding, structure, and reactivity—and how behavior correlates with position in the periodic table. We begin with hydrogen, the simplest and, in some ways, most important of all the elements, and then use Period 2 to survey the general changes *across* the periodic table. Each of the remaining sections deals with one of the eight families of main-group elements. (If you haven't already gone over the illustrated Interchapter feature preceding this chapter, now would be the perfect time to do so. It reviews the major ideas and trends seen throughout the periodic table in preparation for the upcoming discussion.)

Bear in mind that the periodic table was derived from chemical facts observed in countless hours of 18th- and 19th-century research. One of the great achievements in science is 20th-century quantum theory, which provides a theoretical foundation for the periodic table's arrangement, but a theory can rarely account for *all* the facts. Therefore, you may rightly be amazed by the predictive powers of patterns in the table, but do not be concerned if occasional exceptions to these patterns occur. After all, our models are simple, but nature is complex.

## 13.1  Hydrogen, the Simplest Atom

A hydrogen atom consists of a nucleus with a single positive charge that is surrounded by a single electron. Despite this simple structure, or perhaps because of it, hydrogen may be the most important element of all. In the sun, H nuclei combine to form He nuclei in a process that provides nearly all the energy on Earth. About 90% of all the atoms in the universe are H atoms, making it the most abundant element by far. Only tiny amounts of the free element ($H_2$) occur naturally on Earth, but hydrogen is abundant combined with oxygen in the form of water. Because of its simple structure and low molar mass, nonpolar gaseous $H_2$ is colorless and odorless, with an extremely low melting point ($-259°C$) and boiling point ($-253°C$).

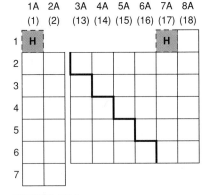

**FIGURE 13.1**

**Where does hydrogen belong?** Hydrogen's small size confers behavior that is not consistent with the members of any one periodic group. Depending on the property, hydrogen may fit better in Group 1A(1) or in 7A(17).

## In Which Periodic Group Does Hydrogen Fit?

Hydrogen has no entirely suitable position in the periodic table (Figure 13.1). With its single valence electron and +1 oxidation state, it might fit in Group 1A(1). Unlike the alkali metals, however, hydrogen shares its electron rather than losing it. Consistent with this behavior, it has a much higher ionization energy (IE = 1311 kJ/mol) than lithium (IE = 520 kJ/mol), the highest in Group 1A, and it has a higher electronegativity as well (EN of H = 2.1; EN of Li = 1.0).

On the other hand, hydrogen might fit in Group 7A(17). Like the diatomic, nonmetallic halogens, it fills its outer shell either by electron sharing or by gaining an electron from a metal to form a monatomic anion, the hydride ion ($H^-$). However, hydrogen has a lower EN than any of the halogens and lacks their three valence electron pairs. Moreover, the $H^-$ ion is rare and reactive, unlike the common and stable halide ions.

Hydrogen's unique behavior is due to its tiny size. It has a high IE because its electron is very close to the nucleus, with no core electrons to shield it from the positive charge. It has a low EN (for a nonmetal) because it has only one proton to attract bonded electrons. In this chapter, hydrogen is usually placed in Group 1A(1), but occasionally in 7A(17), depending on the property being considered.

### Highlights of Hydrogen Chemistry

In Chapters 11 and 12, we discussed *hydrogen bonding*, in which an H atom on one molecule engages in an electrostatic attraction to another, and saw its impact on melting and boiling points, heats of fusion and vaporization, specific heat capacities, and solubilities. You learned that H bonding plays a critical part in stabilizing the Earth's climate as well as the body's temperature and glimpsed at its role in the functioning of biomolecules. In this discussion, we focus on the chemical reactivity of hydrogen, since it combines with nearly every element, and examine the three types of hydrides.

**Ionic (saltlike) hydrides.** With very reactive metals, such as those in Groups 1A(1) and the larger members of 2A(2) (Ca, Sr, and Ba), hydrogen forms *saltlike hydrides:* white, crystalline solids composed of the metal cation and the hydride ion:

$$2Li(s) + H_2(g) \rightarrow 2LiH(s)$$

$$Ca(s) + H_2(g) \rightarrow CaH_2(s)$$

In water, $H^-$ is a strong base that pulls $H^+$ from surrounding $H_2O$ molecules to form $H_2$ and $OH^-$:

$$NaH(s) + H_2O(l) \rightarrow Na^+(aq) + OH^-(aq) + H_2(g)$$

The hydride ion is also a powerful reducing agent; for example, it reduces Ti(IV) to the free metal:

$$TiCl_4(l) + 4LiH(s) \rightarrow Ti(s) + 4LiCl(s) + 2H_2(g)$$

**Covalent (molecular) hydrides.** Hydrogen reacts with nonmetals to form many *covalent hydrides,* such as $CH_4$, $NH_3$, $H_2O$, and HF. Most are gases consisting of small molecules, but many hydrides of boron and carbon are liquids or solids consisting of much larger molecules. In most covalent hydrides, hydrogen has an oxidation number of +1 because the other nonmetal has a higher electronegativity.

Conditions for preparing the covalent hydrides depend on the reactivity of the other reactant. For example, with stable, triple-bonded $N_2$, hydrogen reacts only at high temperatures and pressures, and the reaction needs a catalyst to proceed at any practical speed:

$$N_2(g) + 3H_2(g) \xrightarrow{\text{Fe catalyst}} 2NH_3(g) \qquad \Delta H^0_{rxn} = -91.8 \text{ kJ}; \ T \approx 400°C; \ P \approx 250 \text{ atm}$$

Despite the extreme conditions, this reaction is used throughout the world to form ammonia for the production of fertilizers, explosives, and synthetic fibers. On the other hand, hydrogen combines rapidly with reactive, single-bonded $F_2$, even at extremely low temperatures:

$$F_2(g) + H_2(g) \rightarrow 2HF(g) \qquad \Delta H^0_{rxn} = -546 \text{ kJ}; \ T = -196°C$$

**Metallic (interstitial) hydrides.** Many of the *d* and *f* transition elements form *metallic (interstitial) hydrides,* in which $H_2$ molecules (and H atoms) occupy the holes in the metal's crystal structure (see Figure 12.5). Unlike ionic and covalent hydrides, interstitial hydrides, such as $TiH_{1.7}$, often lack a single stoichiometric formula because the metal can incorporate variable amounts of hydrogen. In fact, these substances are considered gas-solid solutions. ◆

# 13.2    Trends Across the Periodic Table: The Period 2 Elements

The behavior of the elements in Period 2, lithium through neon, changes just as we would expect from their atomic properties. Table 13.1 presents the trends in the most important atomic, physical, and chemical properties of the Period 2 elements. In general, these trends apply to the other periods as well. Note the following points:

- Electrons fill the one *ns* and the three *np* orbitals according to the Pauli exclusion principle and Hund's rule.
- Atomic size generally decreases, whereas first ionization energy and electronegativity generally increase, as a result of increasing nuclear charge and electrons being added to the same energy level (see bar graphs on p 525).
- Metallic character decreases as elements change from metals to metalloids to nonmetals.
- General reactivity is highest at the left and right ends of the period, except for the inert noble gas.
- Bonding between atoms of an element changes from metallic to network covalent to individual molecules to separate atoms. Physical properties reflect this, changing abruptly at the network/molecule boundary, which occurs between carbon (solid) and nitrogen (gas) in Period 2.
- Bonding between each element and an active nonmetal changes from ionic to polar covalent to covalent. Bonding between each element and an active metal changes from metallic to polar covalent to ionic.
- The acid-base behavior of the common oxides in water changes from basic to amphoteric to acidic as the bond between the element and O becomes more covalent.
- Reducing power decreases through the metals, and oxidizing power increases through the nonmetals. In Period 2, common oxidation numbers (O.N.s) are the A-group number for Li and Be and the A-group number minus eight for O and F. Boron has several O.N.s, Ne has none, and C and N show all possible O.N.s for their groups.

◆ **Fill 'Er Up with Hydrogen, Please!** In a possible future hydrogen economy, metallic hydrides may store hydrogen as a fuel. In addition to pure metals, such as palladium and niobium, there are also some alloys that can store large amounts of hydrogen. For example, $LaNi_5$ can absorb 7 mol H per mole of alloy and keep the hydrogen more than twice as dense as in liquid $H_2$. Imagine driving up to your local refueling station, inserting a new cylinder of H-packed alloy in your car (or repressurizing a spent one), and driving away again!

**TABLE 13.1 Trends in Atomic, Physical, and Chemical Properties of the Period 2 Elements**

| GROUP<br>ELEMENT/AT. NO. | 1A(1)<br>LITHIUM (Li) $Z = 3$ | 2A(2)<br>BERYLLIUM (Be) $Z = 4$ | 3A(13)<br>BORON (B) $Z = 5$ | 4A(14)<br>CARBON (C) $Z = 6$ |
|---|---|---|---|---|
| **Atomic Properties** | | | | |
| Condensed electron configuration; partial orbital box diagram | [He] $2s^1$ <br> 2s  2p | [He] $2s^2$ <br> 2s  2p | [He] $2s^2 2p^1$ <br> 2s  2p | [He] $2s^2 2p^2$ <br> 2s  2p |
| **Physical Properties** | | | | |
| Appearance |  |  |  |  |
| Metallic character | Metal | Metal | Metalloid | Nonmetal |
| Hardness | Soft | Hard | Very hard | Graphite: soft<br>Diamond: extremely hard |
| Melting point/ boiling point | Low mp for a metal | High mp | Extremely high mp | Extremely high mp |
| **Chemical Properties** | | | | |
| General reactivity | Reactive | Low reactivity at room temperature | Low reactivity at room temperature | Low reactivity at room temperature; graphite more reactive |
| Bonding among atoms of element | Metallic | Metallic | Network covalent | Network covalent |
| Bonding with nonmetals | Ionic | Polar covalent | Polar covalent | Covalent ($\pi$ bonds common) |
| Bonding with metals | Metallic | Metallic | Polar covalent | Polar covalent |
| Acid/base behavior of common oxide | Strongly basic | Amphoteric | Very weakly acidic | Very weakly acidic |
| Redox behavior (O.N.) | Strong reducing agent (+1) | Moderately strong reducing agent (+2) | Complex hydrides good reducing agents (+3, −3) | Every oxidation state from +4 to −4 |
| **Relevance/Uses of Element and Compounds** | | | | |
| | Li soaps for auto grease; thermonuclear bombs; high-voltage, low-weight batteries; manic-depressive treatment ($Li_2CO_3$) | Rocket nose cones; alloys for springs and gears; nuclear reactor parts; x-ray tubes | Cleaning agent (borax); eyewash, antiseptic (boric acid); armor ($B_4C$); borosilicate glass; plant nutrient | Graphite: lubricant, structural fiber; diamond: jewelry, cutting tools, protective films; limestone ($CaCO_3$); organic compounds (drugs, fuels, textiles, etc.) |

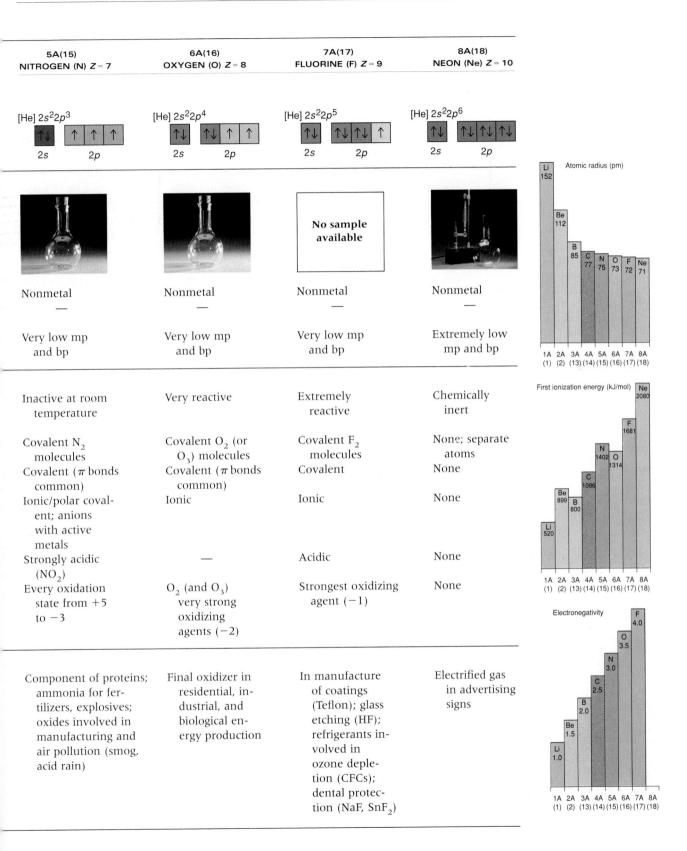

| 5A(15)<br>NITROGEN (N) $Z = 7$ | 6A(16)<br>OXYGEN (O) $Z = 8$ | 7A(17)<br>FLUORINE (F) $Z = 9$ | 8A(18)<br>NEON (Ne) $Z = 10$ |
|---|---|---|---|
| [He] $2s^2 2p^3$ | [He] $2s^2 2p^4$ | [He] $2s^2 2p^5$ | [He] $2s^2 2p^6$ |
| | | No sample available | |
| Nonmetal | Nonmetal | Nonmetal | Nonmetal |
| — | — | — | — |
| Very low mp and bp | Very low mp and bp | Very low mp and bp | Extremely low mp and bp |
| Inactive at room temperature | Very reactive | Extremely reactive | Chemically inert |
| Covalent $N_2$ molecules | Covalent $O_2$ (or $O_3$) molecules | Covalent $F_2$ molecules | None; separate atoms |
| Covalent ($\pi$ bonds common) | Covalent ($\pi$ bonds common) | Covalent | None |
| Ionic/polar covalent; anions with active metals | Ionic | Ionic | None |
| Strongly acidic ($NO_2$) | — | Acidic | None |
| Every oxidation state from $+5$ to $-3$ | $O_2$ (and $O_3$) very strong oxidizing agents ($-2$) | Strongest oxidizing agent ($-1$) | None |
| Component of proteins; ammonia for fertilizers, explosives; oxides involved in manufacturing and air pollution (smog, acid rain) | Final oxidizer in residential, industrial, and biological energy production | In manufacture of coatings (Teflon); glass etching (HF); refrigerants involved in ozone depletion (CFCs); dental protection (NaF, $SnF_2$) | Electrified gas in advertising signs |

One point we'll be noting often that is not in Table 13.1 is the anomalous behavior of the Period 2 elements within their groups: *because of their small size and limited number of available orbitals,* these elements display some behavior that is not representative of other group members.

## 13.3 Group 1A(1): The Alkali Metals

The first group in the periodic table is named for the alkaline (basic) nature of its elements' oxides and for the basic solutions the elements form in water. Group 1A(1) provides a perfect example of regular trends with no significant exceptions. All the elements in the group—lithium (Li), sodium (Na), potassium (K), rubidium (Rb), cesium (Cs), and rare, radioactive francium (Fr)—are very reactive metals. The Family Portrait of Group 1A(1) elements, on pages 528 and 529, is the first in a series that provides an overview of each of the main groups. Atomic and physical properties are summarized on the left-hand page; representative reactions and some of the many important compounds appear on the right-hand page. It will be helpful to refer to portions of these family portraits during the accompanying discussions because they present supporting information.

### Why Are the Alkali Metals Soft, Low Melting, and Lightweight?

Unlike most metals, the alkali metals are quite soft: Na has the consistency of cold butter, and K can be squeezed like clay. The alkali metals also have lower melting and boiling points than any other group of metals. Except for Li, they melt below the boiling point of water, and Cs melts a few degrees above room temperature. They also have lower densities than most metals: Li floats on light household oil.

This unusual physical behavior can be traced to the $ns^1$ electron configuration. With only one valence electron available for metallic bonding, the attraction between the delocalized electrons and the atom cores is relatively weak. This means that the alkali metal crystal structure can be easily deformed or broken down, which results in a soft consistency and low melting point. These elements also have the lowest molar masses and largest atomic radii in their respective periods, so their densities are relatively low.

### Why Are the Alkali Metals So Reactive?

The alkali metals are extremely reactive elements: powerful reducing agents that are always found in nature as 1+ cations. With water, they react vigorously (Rb and Cs explosively) to form $H_2$ and metal hydroxide solutions. In contact with $O_2$ in the air, they tarnish rapidly. In the laboratory, Na and K are usually kept under mineral oil (an unreactive liquid), and Rb and Cs are handled with gloves under an inert argon atmosphere. In reactions that release large amounts of heat, the alkali metals combine with halogens to give ionic solids.

The $ns^1$ configuration, which is the basis for their physical properties, is also the reason these metals form salts so readily. In a Born-Haber cycle of the reaction between an alkali metal and a nonmetal, for example, the solid metal separates into gaseous atoms, each of the atoms transfers its outer electron to the nonmetal, and the resulting cations attract the anions into an ionic solid. Several properties based on the $ns^1$ configuration are related to each of these crucial steps.

1. *Low heat of atomization* ($\Delta H_{atom}$). Consistent with the alkali metals' low melting and boiling points, their weak metallic bonding leads to low values for $\Delta H_{atom}$ (the energy needed to convert the solid to individual gaseous atoms), which decrease down the group:

$$M(s) \rightarrow M(g) \qquad \Delta H_{atom} \ (\text{Li} > \text{Na} > \text{K} > \text{Rb} > \text{Cs})$$

2. *Low IE and high charge density.* Each alkali metal has the largest size and lowest IE in its period. A great *decrease* in size occurs when the outer electron is lost: the volume of the $Li^+$ ion is less than 13% that of the Li atom! Thus, $M^+$ ions are small spheres with high charge density.

3. *High lattice energy.* The small cations can come close to the anions and release large amounts of energy as they crystallize in closely packed structures. Thus, the endothermic atomization and ionization steps are easily outweighed by the highly exothermic formation of the solid. For a given anion, the trend in lattice energy is the inverse of the trend in cation size: *as the cation becomes larger, the lattice energy becomes smaller (less negative).* Figure 13.2 shows the decrease in lattice energy within the Group 1A(1) and 2A(2) chlorides.

Despite the strong ionic attractions in the solid, *nearly all Group 1A salts are water soluble.* The high charge density attracts water molecules strongly to create a highly exothermic heat of hydration ($\Delta H_{hydr}$), and a large increase in disorder arises when the organized crystal becomes randomized, hydrated ions; together, these factors outweigh the lattice energy.

The magnitude of the hydration energy *decreases* as ionic size increases:

$$E^+(g) \xrightarrow{\text{H}_2\text{O}} E^+(aq) \qquad -\Delta H_{hydr} \ (\text{Li}^+ > \text{Na}^+ > \text{K}^+ > \text{Rb}^+ > \text{Cs}^+)$$

Interestingly, the *smaller* ions attract water molecules strongly enough to form *larger hydrated ions*: thus, $Li^+(aq)$ is *larger* than $Cs^+(aq)$. This size trend has a major effect on the function of nerves, kidneys, and all cell membranes because the *sizes* of $Na^+(aq)$ and $K^+(aq)$, the most common cations in cell fluids, influence their movement in and out of cells.

### The Anomalous Behavior of Lithium

As we noted in discussing Table 13.1, all the Period 2 elements display some unrepresentative behavior within their groups. Even within the regular, predictable trends of Group 1A(1), Li has some atypical chemical and solubility properties. It is the only member that forms a simple oxide and nitride, $Li_2O$ and $Li_3N$, on reaction with $O_2$ and $N_2$ in air. Only Li forms molecular compounds with hydrocarbon groups from organic halides:

$$2\text{Li}(s) + \text{CH}_3\text{CH}_2\text{Cl}(g) \rightarrow \text{CH}_3\text{CH}_2\text{Li}(s) + \text{LiCl}(s)$$

Organolithium compounds, such as $CH_3CH_2Li$, are low-melting solids or liquids that dissolve in nonpolar solvents and contain polar covalent $^{\delta-}C—Li^{\delta+}$ bonds. They are important reactants in the synthesis of organic compounds.

Many lithium salts have significant covalent bond character because the high charge density of the $Li^+$ ion deforms nearby polarizable electron clouds (Figure 13.3). For example, LiCl, LiBr, and LiI are much more soluble in polar organic solvents, such as ethanol and acetone, than the halides of Na and K. The highly positive Li end of the lithium halide dipole interacts strongly through dipole-dipole forces with these solvents. The high $Li^+$ charge density also leads to highly negative ionic lattice energies; for example, the fluoride, carbonate, hydroxide, and phosphate of Li are much less soluble in water than those of Na and K.

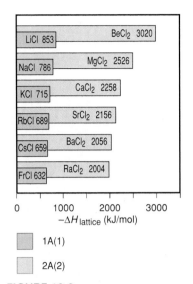

$$
\begin{array}{ll}
\text{LiCl } 853 & \text{BeCl}_2 \ 3020 \\
\text{NaCl } 786 & \text{MgCl}_2 \ 2526 \\
\text{KCl } 715 & \text{CaCl}_2 \ 2258 \\
\text{RbCl } 689 & \text{SrCl}_2 \ 2156 \\
\text{CsCl } 659 & \text{BaCl}_2 \ 2056 \\
\text{FrCl } 632 & \text{RaCl}_2 \ 2004 \\
\end{array}
$$

0        1000        2000        3000
$-\Delta H_{lattice}$ (kJ/mol)

☐ 1A(1)

☐ 2A(2)

**FIGURE 13.2**

**Lattice energies of the Group 1A(1) and 2A(2) chlorides.** Lattice energies decrease regularly in both groups of metal chlorides as the cations become larger. Lattice energies for the 2A chlorides are greater because the 2A cations have higher charge and smaller size.

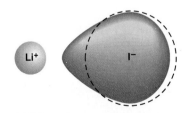

**FIGURE 13.3**

**The effect of $Li^+$ charge density on a nearby electron cloud.** The $Li^+$ cation polarizes a nearby anion, such as $I^-$, which gives some covalent character to lithium iodide, LiI.

FAMILY PORTRAIT

## Group 1A(1): The Alkali Metals
### Key Atomic and Physical Properties

**KEY**

Atomic No.
**Symbol**
Atomic mass
Valence e⁻ configuration
Common oxidation states

$ns^1$

**GROUP 1A(1)**

| 3 | |
|---|---|
| **Li** | |
| 6.941 | |
| $2s^1$ | |
| +1 | |

| 11 | |
| **Na** | |
| 22.99 | |
| $3s^1$ | |
| +1 | |

| 19 | |
| **K** | |
| 39.10 | |
| $4s^1$ | |
| +1 | |

| 37 | |
| **Rb** | |
| 85.47 | |
| $5s^1$ | |
| +1 | |

| 55 | |
| **Cs** | |
| 132.9 | |
| $6s^1$ | |
| +1 | |

| 87 | |
| **Fr** | |
| (223) | |
| $7s^1$ | |
| +1 | |

No sample available

## Atomic Properties

| Atomic radius (pm) | Ionic radius (pm) |
|---|---|
| Li 152 | Li⁺ 76 |
| Na 186 | Na⁺ 102 |
| K 227 | K⁺ 138 |
| Rb 248 | Rb⁺ 152 |
| Cs 265 | Cs⁺ 167 |
| Fr (~270) | Fr⁺ 180 |

Group electron configuration is $ns^1$. All members have the +1 oxidation state and form the E⁺ ion.

Atoms have the largest size and lowest IE and EN in their periods.

Down the group, atomic and ionic size increase, while IE and EN decrease.

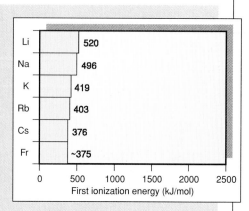

First ionization energy (kJ/mol)

| | |
|---|---|
| Li | 520 |
| Na | 496 |
| K | 419 |
| Rb | 403 |
| Cs | 376 |
| Fr | ~375 |

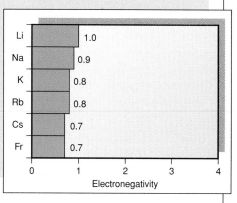

Electronegativity

| | |
|---|---|
| Li | 1.0 |
| Na | 0.9 |
| K | 0.8 |
| Rb | 0.8 |
| Cs | 0.7 |
| Fr | 0.7 |

## Physical Properties

Metallic bonding is relatively weak due to one valence electron. Therefore, these metals are soft with relatively low melting and boiling points. These values decrease down the group because larger atom cores attract delocalized electrons less strongly.

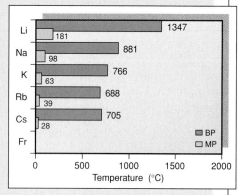

Temperature (°C)

| | BP | MP |
|---|---|---|
| Li | 1347 | 181 |
| Na | 881 | 98 |
| K | 766 | 63 |
| Rb | 688 | 39 |
| Cs | 705 | 28 |
| Fr | | |

Large atomic size and low atomic mass result in low density, which generally increases down the group because mass increases more than size.

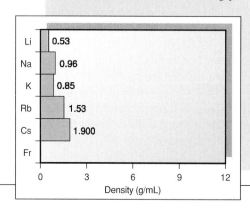

Density (g/mL)

| | |
|---|---|
| Li | 0.53 |
| Na | 0.96 |
| K | 0.85 |
| Rb | 1.53 |
| Cs | 1.900 |
| Fr | |

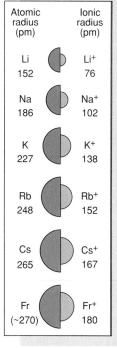

# Group 1A(1): The Alkali Metals
## Some Reactions and Compounds

## Important Reactions

The reducing power of the alkali metals (E) is shown in reactions 1 to 4. Some industrial applications of 1A(1) compounds are shown in reactions 5 to 7.

1. The alkali metals reduce H in $H_2O$ from the +1 to the zero oxidation state:

$$2E(s) + 2H_2O(l) \rightarrow 2E^+(aq) + 2OH^-(aq) + H_2(g)$$

The reaction becomes more vigorous down the group; with Rb and Cs, it is explosive.

2. The alkali metals reduce oxygen, but the product depends on the metal. Li forms the oxide, $Li_2O$; Na forms the peroxide, $Na_2O_2$; K, Rb, and Cs form the superoxide, $EO_2$:

$$4Li(s) + O_2(g) \rightarrow 2Li_2O(s)$$

$$K(s) + O_2(g) \rightarrow KO_2(s)$$

In emergency breathing units, $KO_2$ reacts with $H_2O$ and $CO_2$ to release $O_2$.

3. The alkali metals reduce hydrogen to form ionic (saltlike) hydrides:

$$2E(s) + H_2(g) \rightarrow 2EH(s)$$

NaH is an industrial base and reducing agent that is used to prepare other reducing agents, such as $NaBH_4$.

4. The alkali metals reduce halogens to form ionic halides:

$$2E(s) + X_2 \rightarrow 2EX(s) \qquad (X = F, Cl, Br, I)$$

5. Sodium chloride is the most important alkali halide.
   (a) In the Downs process for the production of sodium metal (*Section 23.3*), reaction 4 is reversed by supplying electricity to molten NaCl:

$$2NaCl(l) \xrightarrow{\text{electricity}} 2Na(l) + Cl_2(g)$$

   (b) In the chlor-alkali process (*Section 23.4*), NaCl(*aq*) is electrolyzed to form several key industrial chemicals:

$$2NaCl(aq) + 2H_2O(l) \rightarrow 2NaOH(aq) + H_2(g) + Cl_2(g)$$

   (c) In its reaction with sulfuric acid, NaCl forms two major products:

$$2\,NaCl(s) + H_2SO_4(aq) \rightarrow Na_2SO_4(aq) + 2HCl(g)$$

Sodium sulfate is important in the paper industry; HCl is essential in steel, plastics, textiles, and food production.

6. Sodium hydroxide is used in the formation of bleaching solutions:

$$2NaOH(aq) + Cl_2(g) \rightarrow NaClO(aq) + NaCl(aq) + H_2O(l)$$

7. In an ion-exchange process (*Chemical Connections, Section 12.6*), water is "softened" when $Na^+$ is displaced by "hard-water" ions ($M^{2+}$):

$$M^{2+}(aq) + Na_2Z(s) \rightarrow MZ(s) + 2Na^+(aq)$$
$$(M = Mg, Ca; Z = resin)$$

## Important Compounds

1. Lithium chloride and lithium bromide, LiCl and LiBr. Because the $Li^+$ ion is so small, Li salts have a high affinity for $H_2O$ and yet a positive heat of solution, so they are used in dehumidifiers and air-cooling units.
2. Lithium carbonate, $Li_2CO_3$. Used to make porcelain enamels and toughened glasses and as a drug in the treatment of manic-depressive disorders.
3. Sodium chloride, NaCl. Millions of tons used in the industrial production of Na, NaOH, $Na_2CO_3$/$NaHCO_3$, $Na_2SO_4$, HCl, and purified for use as table salt.
4. Sodium carbonate and sodium bicarbonate, $Na_2CO_3$ and $NaHCO_3$. Carbonate used as an industrial base and to make glass. Bicarbonate, which releases $CO_2$

at low temperatures (50° to 100°C), used in baking powder and in fire extinguishers.
5. Sodium hydroxide, NaOH. Most important industrial base; used to make bleach, sodium phosphates, and alcohols.
6. Potassium nitrate, $KNO_3$. Powerful oxidizing agent used in gunpowder and fireworks.

◆ **Versatile Magnesium.**
Magnesium, second member of the alkaline earth metals, can be rolled, forged, welded, and riveted into virtually any shape, and it forms strong, low-density alloys: some weigh 25% as much as steel but are just as strong. Alloyed with aluminum and zinc, it is used for everyday objects, such as camera bodies, luggage, and the "mag" wheels of bicycles and sports cars. Alloyed with the lanthanides (first series of inner transition elements), it is used in objects that require great strength at high temperature, such as auto engine blocks and missile parts.

◆ **Lime: The Most Useful Metal Oxide.** Calcium oxide (lime) is a slightly soluble, basic oxide that is among the five most heavily produced industrial compounds in the world; it is made by roasting limestone to high temperatures [Group 2A(2) Family Portrait, reaction 7]. It has an essential role in steelmaking, water treatment, and flue "scrubbing." Lime reacts with carbon to form calcium carbide, which is used to make acetylene, and with arsenic acid ($H_3AsO_4$) to make insecticides for treating cotton, tobacco, and potato plants. Most glass contains about 12% lime by mass. The paper industry employs lime to regenerate the NaOH needed to make pulp and to prepare the bleach [$Ca(ClO)_2$] that whitens paper. Every gardener knows that lime "sweetens" acidic soil; its alkaline nature is also employed in the dairy industry to neutralize compounds in milk to make cream, butter, and lactic acid and in the sugar industry to neutralize impurities in crude sugar.

## 13.4  Group 2A(2): The Alkaline Earth Metals

The Group 2A(2) elements are called alkaline earth metals because their oxides give basic (alkaline) solutions and melt at such high temperatures that they remained as solids ("earths") in the alchemists' fires. The group includes a fascinating collection of elements: rare beryllium (Be), common magnesium (Mg) and calcium (Ca), less familiar strontium (Sr) and barium (Ba), and radioactive radium (Ra). The Group 2A(2) Family Portrait (pp. 532 and 533) presents an overview of these elements.

### How Do the Physical Properties of the Alkaline Earth and Alkali Metals Compare?

In general terms, the elements in Groups 1A(1) and 2A(2) behave as close cousins. Whatever differences occur between the groups are those of degree, not kind, and are due to the change in electron configuration: $ns^2$ vs. $ns^1$. With two electrons available for metallic bonding and one additional positive charge in the nucleus, the attraction between delocalized electrons and atom cores is greater. Consequently, melting and boiling points are much higher than for the corresponding 1A metals; in fact, 2A melting points are close to 1A boiling points. Compared with transition metals such as iron and chromium, the 2A elements are soft and lightweight, but they are much harder and more dense than their alkali metal neighbors. ◆

### How Do the Chemical Properties of the Alkaline Earth and Alkali Metals Compare?

The alkaline earth metals display a wider range of chemical behavior than the alkali metals, largely due to the behavior of beryllium, as you'll see shortly. Since the second valence electron lies in the same sublevel as the first, it is not shielded from the additional nuclear charge very effectively. Therefore, Group 2A(2) elements have smaller atomic radii and higher ionization energies than Group 1A(1) elements. Except for Be, *the alkaline earths form ionic compounds.* So much energy is needed to remove two electrons from tiny Be that it never forms discrete $Be^{2+}$ ions, and its compounds are polar covalent.

*The alkaline earth metals are strong reducing agents,* as shown in the reactions section of the Group 2A(2) Family Portrait. The elements (E) reduce $O_2$ in air to form the oxide EO (Ba also forms the peroxide, $BaO_2$); except for Be and Mg, which form adherent oxide coatings, they reduce $H_2O$ at room temperature to form $H_2$; and, except for Be, they reduce the halogens, $N_2$, and $H_2$ to form ionic compounds. The oxides are strongly basic (except for amphoteric BeO) and react with acidic oxides to form oxosalts, such as sulfites and carbonates; for example, $EO(s) + CO_2(g) \rightarrow ECO_3(s)$. Natural carbonates, such as limestone and marble, are major structural materials and the commercial sources for most 2A compounds. ◆

The alkaline earth metals are reactive because the high lattice energies of their compounds more than compensate for the large total IE needed to form the 2+ cation (Section 9.2). Group 2A salts have much higher lattice energies than those of Group 1A (see Figure 13.2) because the 2A cations are smaller and doubly charged.

One of the main differences between the two groups is the lower solubility of 2A salts in water. The higher charge density of 2A over 1A cations increases heats of hydration, but it increases lattice energies even more. Thus, 2A fluorides, carbonates, phosphates, and sulfates are *less* soluble than the corresponding 1A compounds. Nevertheless, the ion-dipole attraction of 2+ ions for water is so strong that many soluble 2A salts crystallize as hydrates; two examples are Epsom salt, $MgSO_4 \cdot 7H_2O$, used as a soak for inflammations, and gypsum, $CaSO_4 \cdot 2H_2O$, used as the bonding material between the paper sheets in wallboard and the cement in surgical casts.

## The Anomalous Behavior of Beryllium

As the numerous exceptions just mentioned indicate, the Period 2 element Be displays many examples of anomalous behavior, far more than does Li in Group 1A(1). The high charge density of the $Be^{2+}$ ion polarizes nearby electron clouds, so that *all Be compounds exhibit covalent bonding*. Even $BeF_2$, the most ionic Be compound, has a relatively low melting point and, when melted, a low electrical conductivity.

Since Be has only two valence electrons, it does not attain an octet in its simple gaseous compounds (Section 9.5). When it bonds to an electron-rich atom, however, this electron deficiency is overcome as the gas condenses. Consider beryllium chloride ($BeCl_2$) (Figure 13.4). At temperatures greater than 900°C, it consists of linear molecules in which two *sp* hybrid orbitals hold four electrons around the central Be. As it cools, the molecules bond together, solidifying in long chains, with each Be $sp^3$ hybridized to finally attain an octet.

When beryllium bonds to hydrogen, which does not have lone pairs, Be attains an octet through an unconventional type of covalent bonding. Solid $BeH_2$ consists of long chains of connected molecular units. Three atoms, Be—H—Be, are bonded by two electrons in a *hydride bridge bond*. The hydrides of boron also display this type of bonding, so we'll examine it closely in the next section.

**FIGURE 13.4**

**Overcoming electron deficiency in beryllium chloride. A,** At high temperature, $BeCl_2$ occurs as a gaseous molecule with only four electrons around Be. **B,** In the solid state, $BeCl_2$ occurs in long chains with each Cl bridging two Be atoms, which gives each Be an octet.

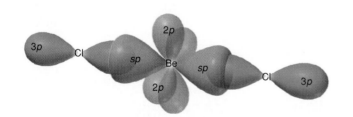

# Group 2A(2): The Alkaline Earth Metals
## Key Atomic and Physical Properties

**KEY**

Atomic No.
**Symbol**
Atomic mass
Valence e⁻ configuration
Common oxidation states

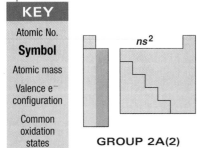

GROUP 2A(2)

$ns^2$

## Atomic Properties

| Atomic radius (pm) | Ionic radius (pm) |
|---|---|
| Be 112 | |
| Mg 160 | Mg²⁺ 72 |
| Ca 197 | Ca²⁺ 100 |
| Sr 215 | Sr²⁺ 118 |
| Ba 222 | Ba²⁺ 135 |
| Ra (~220) | Ra²⁺ 148 |

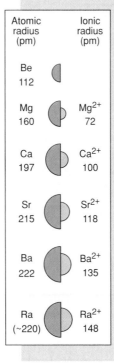

Group electron configuration is $ns^2$ (filled $ns$ sublevel). All members have the +2 oxidation state and, except for Be, form compounds with the E²⁺ ion.

Atomic and ionic sizes increase down the group but are smaller than for corresponding 1A(1) element.

IE and EN decrease down the group but are higher than for corresponding 1A(1) element.

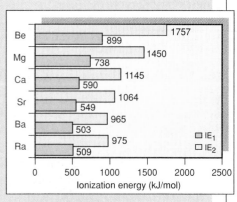

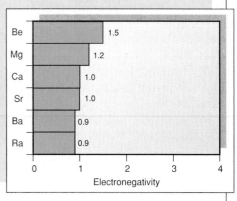

## Physical Properties

Metallic bonding involves two valence e⁻. Metals are relatively soft but are much harder than the 1A(1) metals.

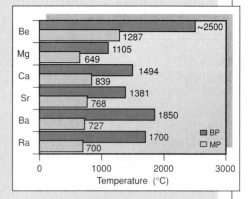

Melting and boiling points generally decrease and densities generally increase down the group. Values are much higher than for 1A(1) and the trend is not as regular.

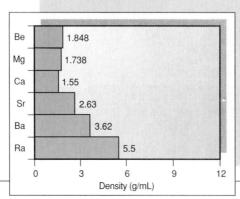

4 Be 9,012 $2s^2$ +2
12 Mg 24.30 $3s^2$ +2
20 Ca 40.08 $4s^2$ +2
38 Sr 87.62 $5s^2$ +2
56 Ba 137.3 $6s^2$ +2
88 Ra (226) $7s^2$ +2
No sample available

## Group 2A(2): The Alkaline Earth Metals
### Some Reactions and Compounds

### Important Reactions

The elements (E) act as reducing agents in reactions 1 to 5; note the similarity to reactions of Group 1A(1). Reaction 6 shows the general basicity of the oxides; reaction 7 shows the general instability of the carbonates at high temperature.

1. The metals reduce $O_2$ to form the oxide:

   $$2E(s) + O_2(g) \longrightarrow 2EO(s)$$

   Ba also forms the peroxide $BaO_2$.

Magnesium ribbon burning

2. The larger metals reduce water to form hydrogen gas:

   $$E(s) + 2H_2O(l) \longrightarrow$$
   $$E^{2+}(aq) + 2OH^-(aq) + H_2(g)$$

   (E = Ca, Sr, Ba)

   Be and Mg form an adherent oxide coating that allows only slight reaction.

3. The metals reduce halogens to form ionic halides:

   $$E(s) + X_2 \longrightarrow EX_2(s) \quad (X = F, Cl, Br, I)$$

4. Most of the elements reduce hydrogen to form ionic hydrides:

   $$E(s) + H_2(g) \longrightarrow EH_2(s)$$

   (E = all except Be)

5. Most of the elements reduce nitrogen to form ionic nitrides:

   $$3E(s) + N_2(g) \longrightarrow E_3N_2(s) \quad (E = \text{all except Be})$$

6. Except for amphoteric BeO, the oxides are basic:

   $$EO(s) + H_2O(l) \longrightarrow E^{2+}(aq) + 2OH^-(aq)$$

   $Ca(OH)_2$ is a component of cement and mortar.

7. All carbonates undergo thermal decomposition to the oxide:

   $$ECO_3(s) \xrightarrow{\Delta} EO(s) + CO_2(g)$$

   This reaction is used to produce CaO (lime) in huge amounts from naturally occurring limestone (see margin note, p. 530).

### Important Compounds

1. Beryl, $Be_3Al_2Si_6O_{18}$. Beryl occurs as a gemstone with a variety of colors (*photo*). It is chemically identical to emerald, except for the trace of $Cr^{3+}$ that gives emerald its green color. Beryl is the industrial source of Be metal.

2. Magnesium oxide, MgO. Because of its high melting point (2852°C), it is used as a refractory material for furnace brick (*photo*) and wire insulation.

Beryl

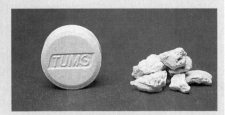
Tums, a common antacid, and seashells

Industrial kiln

3. Alkylmagnesium halides, RMgX (R = hydrocarbon group; X = halogen). These compounds, called Grignard reagents, are used to synthesize many organic compounds. Organotin agricultural fungicides are made by treating RMgX with $SnCl_4$:

   $$3RMgCl + SnCl_4 \longrightarrow 3MgCl_2 + R_3SnCl$$

4. Calcium carbonate, $CaCO_3$. Occurs as enormous natural deposits of limestone, marble, chalk, coral. Used as a building material, to make lime, and, in high purity, as toothpaste abrasive and antacid (*photo*).

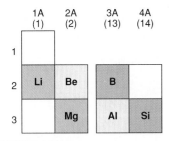

**FIGURE 13.5**
**Three diagonal relationships in the periodic table.** Certain Period 2 elements exhibit behaviors that are very similar to Period 3 elements immediately below and to the right. Three such diagonal relationships exist between Li and Mg, Be and Al, and B and Si.

### Diagonal Relationships: Lithium and Magnesium

One of the clearest ways in which atomic properties influence chemical behavior appears in the **diagonal relationships** of the periodic table, similarities between a Period 2 element and one in Period 3 diagonally down and to the right (Figure 13.5). The first of these relationships occurs between Li and Mg and reflects similarities in atomic and ionic size. Note that *one period down increases atomic (or ionic) size and one group to the right decreases it.* Thus, Li has a radius of 152 pm and Mg 160 pm; the $Li^+$ radius is 76 pm and that of $Mg^{2+}$ is 72 pm. From these similar atomic properties emerge remarkably similar chemical properties. Both elements form nitrides with $N_2$, hydroxides and carbonates that decompose easily with heat, organic compounds with polar covalent bonds from metal to a hydrocarbon group, and salts with similar solubilities. We'll discuss two other diagonal relationships—Be and Al, and B and Si—in later sections.

### Looking Backward and Forward: Groups 1A(1), 2A(2), and 3A(13)

To maintain a horizontal perspective while examining vertical groups, we stop occasionally to compare the previous, current, and upcoming groups (Figure 13.6). Little changes from 1A to 2A, the elements behaving as metals both physically and chemically. With stronger metallic bonding, 2A elements are harder, higher melting, and denser than those in 1A. Nearly all 1A and most 2A compounds are ionic. The higher ionic charge of $E^{2+}$ leads to higher lattice energies and less soluble salts. The range of behavior in 2A, wider than that in 1A because of Be, is widened much further in Group 3A from metalloid boron to metallic thallium.

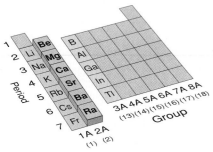

**FIGURE 13.6**
Standing in Group 2A(2), looking backward to 1A(1) and forward to 3A(13).

## 13.5  Group 3A(13): The Boron Family

The third family of main-group elements contains some unusual members and some familiar ones, some bizarre bonding, and some strange physical properties. Boron (B) heads the family, but, as you'll see, its properties certainly do not represent the other members. Metallic aluminum (Al) has properties that are more typical of the group, but its great abundance and importance contrast with the rareness of gallium (Ga), indium (In), and thallium (Tl). The atomic, physical, and chemical properties of these elements are summarized in the Group 3A(13) Family Portrait (pp. 536 and 537).

### How Do the Transition Elements Influence Group 3A(13) Properties?

Group 3A(13) is the first of the *p* block. If one looks at the main groups only, the elements of this group seem to be just one away from those of Group 2A(2). In Period 4 and higher, however, a large gap separates the two groups (see Figure 8.9). The gap holds 10 *d*-block transition elements each in Periods 4, 5, and 6 and an additional 14 *f*-block inner transition elements in Period 6. Thus, the heavier 3A members—Ga, In, and Tl—have nuclei with many more protons, but their outer (*s* and *p*) electrons are only partially shielded from the much higher positive charge because *d* and *f* electrons spend so little time near the nucleus.

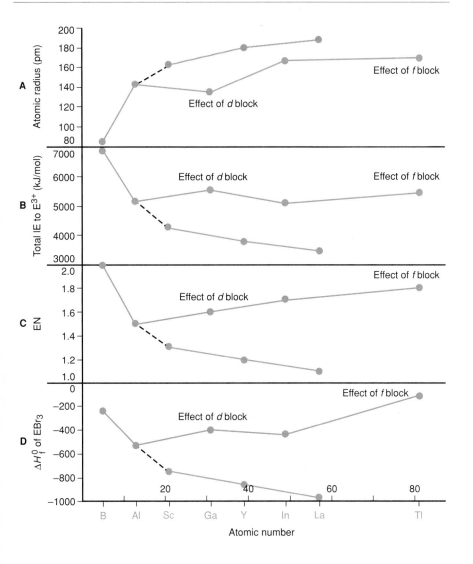

**FIGURE 13.7**
**The effect of transition elements on properties: Group 3B(3) vs. Group 3A(13).** The additional protons in the nuclei of transition elements exert an unusually strong attraction because $d$ and $f$ electrons shield the nuclear charge poorly. The greater effective charge affects the $p$-block elements in Periods 4 to 6, as can be seen by comparing properties of Group 3B(3) *(blue curves)*, the first group after the $s$ block, with 3A(13) *(orange curves)*, the first group after the $d$ block. **A,** Atomic size. In Group 3B, size increases smoothly, while in 3A, Ga and Tl are smaller than expected. **B,** Total ionization energy for $E^{3+}$. The deviations in size lead to deviations in total IE ($IE_1 + IE_2 + IE_3$). Note the regular decrease for Group 3B and the higher values for Ga and Tl in 3A. **C,** Electronegativity. The deviations in size also make Ga and Tl more electronegative than expected. **D,** Heat of formation of $EBr_3$. The deviations in IE lead to a smaller-than-expected $\Delta H_f^0$ for $GaBr_3$ and $TlBr_3$.

Many properties of these heavier 3A elements are influenced as a result. Figure 13.7 compares several properties of Group 3A(13) with those of Group 3B(3), the first group of transition elements, in which the additional protons of the $d$ block and $f$ block have *not* been added. Note the regular changes exhibited by the 3B elements, in contrast to the irregular patterns for those in 3A. The deviations for Ga reflect the $d$-block *contraction in size* caused by limited shielding of the 10 additional protons of the first transition series by the $d$ electrons, whereas the deviations for Tl reflect the $f$-block (lanthanide) contraction caused by limited shielding of the 14 additional protons of the first inner transition series by the $f$ electrons.

Physical properties are influenced by the type of bonding that occurs in the element. Boron is a network covalent metalloid—black, hard, and very high melting; the other members are metals—shiny and relatively soft and low melting. Aluminum's low density and three valence electrons make it an exceptional conductor: for a given mass, aluminum can carry twice as much current as copper. Gallium has the largest liquid temperature range of any element: it melts in your hand but does not boil until 2403°C (see Figure 9.23). The metallic bonding is too weak to keep the Ga atoms fixed when the solid is warmed, but it is strong enough to keep them from escaping the molten metal until it is very hot. ◆

◆ **Gallium Arsenide: The Next Wave of Semiconductors.** A recent use of gallium is in the production of gallium arsenide (GaAs) semiconductors. One of the factors limiting the efficiency of computer chips is the speed with which electrons can move, but they move 10 times faster through GaAs than through Si-based chips. GaAs chips also have novel optical properties: absorbing light creates a current and, conversely, supplying a current emits light. GaAs devices are already used in light-powered calculators and wristwatches and are being incorporated into solar panels. In addition, GaAs-based lasers are much smaller and more powerful than normal types.

**FAMILY PORTRAIT**

# Group 3A(13): The Boron Family
## Key Atomic and Physical Properties

### KEY

Atomic No.
**Symbol**
Atomic mass
Valence e⁻ configuration
Common oxidation states

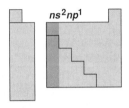

$ns^2np^1$

GROUP 3A(13)

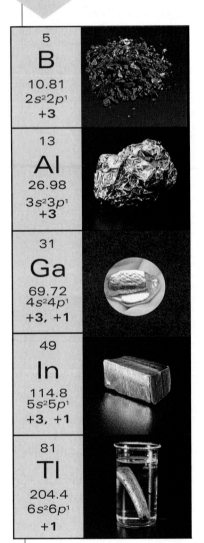

| 5 | |
|---|---|
| **B** | |
| 10.81 | |
| $2s^22p^1$ | |
| **+3** | |

| 13 | |
|---|---|
| **Al** | |
| 26.98 | |
| $3s^23p^1$ | |
| **+3** | |

| 31 | |
|---|---|
| **Ga** | |
| 69.72 | |
| $4s^24p^1$ | |
| **+3, +1** | |

| 49 | |
|---|---|
| **In** | |
| 114.8 | |
| $5s^25p^1$ | |
| **+3, +1** | |

| 81 | |
|---|---|
| **Tl** | |
| 204.4 | |
| $6s^26p^1$ | |
| **+1** | |

## Atomic Properties

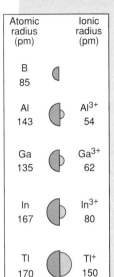

| Atomic radius (pm) | Ionic radius (pm) |
|---|---|
| B 85 | |
| Al 143 | Al³⁺ 54 |
| Ga 135 | Ga³⁺ 62 |
| In 167 | In³⁺ 80 |
| Tl 170 | Tl⁺ 150 |

Group electron configurations: $ns^2np^1$. All except Tl commonly display the +3 oxidation state. The +1 state becomes more common down the group.

Atomic size is smaller and EN is higher than 2A(2) elements, but IE is lower due to ease of removing e⁻ from higher energy $p$ sublevel.

Atomic size, IE, and EN do not change as expected down the group due to intervening transition and inner transition elements.

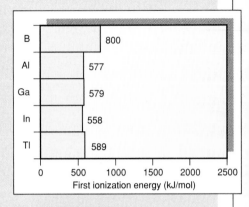

First ionization energy (kJ/mol)

| B | 800 |
| Al | 577 |
| Ga | 579 |
| In | 558 |
| Tl | 589 |

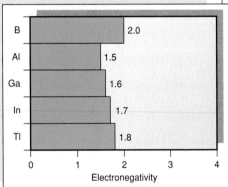

Electronegativity

| B | 2.0 |
| Al | 1.5 |
| Ga | 1.6 |
| In | 1.7 |
| Tl | 1.8 |

## Physical Properties

Bonding changes from network covalent in B to metallic in rest of group. Thus, B has a much higher melting point than others, but there is no overall trend.

Boiling points decrease.

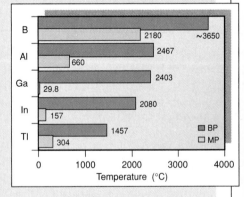

Temperature (°C)

| | BP | MP |
|---|---|---|
| B | ~3650 | 2180 |
| Al | 2467 | 660 |
| Ga | 2403 | 29.8 |
| In | 2080 | 157 |
| Tl | 1457 | 304 |

Densities increase down the group.

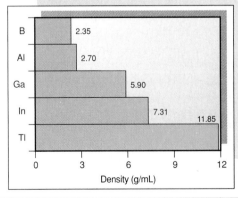

Density (g/mL)

| B | 2.35 |
| Al | 2.70 |
| Ga | 5.90 |
| In | 7.31 |
| Tl | 11.85 |

# Group 3A(13): The Boron Family
## Some Reactions and Compounds

## Important Reactions

In reactions 1 to 3, note that the elements (E) usually require higher temperatures to react than those of Groups 1A(1) and 2A(2); note the lower oxidation state of Tl. Reactions 4 to 6 show some compounds in key industrial processes.

1. The elements react sluggishly, if at all, with water:

$$2Ga(s) + 6H_2O(hot) \rightarrow 2Ga^{3+}(aq) + 6OH^-(aq) + 3H_2(g)$$

$$2Tl(s) + 2H_2O(steam) \rightarrow 2Tl^+(aq) + 2OH^-(aq) + H_2(g)$$

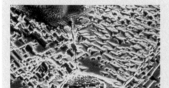

Al becomes covered with a layer of $Al_2O_3$ that prevents further reaction.

Electronmicrograph of Al

2. When strongly heated in pure $O_2$, all members form oxides:

$$4E(s) + 3O_2(g) \rightarrow 2E_2O_3(s)$$

$$(E = B, Al, Ga, In)$$

$$4Tl(s) + O_2(g) \rightarrow 2Tl_2O(s)$$

Oxide acidity decreases down the group: $B_2O_3$ (weakly acidic) $> Al_2O_3 > Ga_2O_3 > In_2O_3 > Tl_2O$ (strongly basic), and the +1 oxide is more basic than the +3 oxide.

3. All members reduce halogens ($X_2$):

$$2E(s) + 3X_2 \rightarrow 2EX_3 \quad (E = B, Al, Ga, In)$$

$$2Tl(s) + X_2 \rightarrow 2TlX(s)$$

$BX_3$ are volatile covalent molecules. Trihalides of Al, Ga, and In are (mostly) ionic solids but occur as covalent dimers in the gas phase; in this way, the 3A atom attains a filled outer level.

4. Acid treatment of $Al_2O_3$ is important in water purification:

$$Al_2O_3(s) + 3H_2SO_4(l) \rightarrow Al_2(SO_4)_3(s) + 3H_2O(l)$$

In water, $Al_2(SO_4)_3$ and CaO form a colloid that aids in removing suspended particles.

5. The overall reaction in the production of aluminum metal is a redox process:

$$2Al_2O_3(s) + 3C(s) \rightarrow 4Al(s) + 3CO_2(g)$$

This electrochemical process is carried out in the presence of cryolite ($Na_3AlF_6$), which lowers the melting point of the reactant mixture and takes part in the change *(Section 23.3)*.

6. A displacement reaction produces gallium arsenide, GaAs *(see margin note, p. 535)*:

$$(CH_3)_3Ga(g) + AsH_3(g) \rightarrow 3CH_4(g) + GaAs(s)$$

## Important Compounds

1. Boron oxide, $B_2O_3$. Used in the production of borosilicate glass *(see Highlights of Boron Chemistry)*.
2. Borax, $Na_2[B_4O_5(OH)_4] \cdot 8H_2O$. Major mineral source of boron compounds and $B_2O_3$. Used as a fireproof insulation material and as a washing powder (20-Mule Team Borax).
3. Boric acid, $H_3BO_3$ [or $B(OH)_3$]. Used as external disinfectant, eyewash, and insecticide.
4. Diborane, $B_2H_6$. A powerful reductant for possible use as a rocket fuel. Used to synthesize higher boranes, compounds that led to new theories of chemical bonding.
5. Aluminum sulfate (alum), $Al_2(SO_4)_3 \cdot 18H_2O$. Used in water purification, tanning leather, sizing paper, as a fixative for dyeing cloth, and as an antiperspirant.

6. Aluminum oxide, $Al_2O_3$. Major compound in natural source (bauxite) of Al metal. Used as abrasive in sandpaper, sanding and cutting tools, and toothpaste. Large crystals with metal ion impurities often of gemstone quality *(photo)*. Inert support for chromatography. In fibrous forms, woven into heat-resistant fabrics; also used to strengthen ceramics and metals.

Ruby

7. $Tl_2Ba_2Ca_2Cu_3O_{10}$. Becomes a high-temperature superconductor at 125 K, which is readily attained with liquid $N_2$ *(photo)*.

Magnet levitated by superconductor

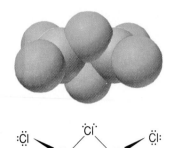

**FIGURE 13.8**

**The dimeric structure of gaseous aluminum chloride.** Despite its name, aluminum trichloride actually exists as a dimer of separate $Al_2Cl_6$ molecules in the gas phase.

## What New Features Appear in the Chemical Properties of Group 3A(13)?

Looking down Group 3A(13), we see a wide range of chemical behavior. Boron, the anomalous member from Period 2, is the first metalloid we've encountered so far and the only one in the group. It is much less reactive at room temperature than the other members and forms covalent bonds exclusively. Although aluminum acts like a metal physically, its halides exist in the gas phase as covalent dimers—molecules formed by joining two identical smaller molecules (Figure 13.8)—and its oxide is amphoteric rather than basic. Most of the other 3A compounds are ionic, but they have more covalent character than similar 2A compounds because the 3A ions are smaller and more highly charged.

Our first glimpse of multiple oxidation states occurs in this group, and it provides a chance to note some general principles. When a group exhibits more than one oxidation state, *the lower state becomes more important down the group*. In this group, for instance, all members exhibit the +3 state, but the +1 state first appears with gallium and becomes the only important state of thallium.

The larger elements in Groups 4A(14) and 5A(15) also have an important oxidation state *two lower than the A-group number*. The lower state occurs when the atoms lose their *np* electrons, but not their two *ns* electrons. This fact is often called the *inert-pair effect* (Section 8.4), but it has nothing to do with any inertness of *ns* electrons; in fact, the 6s electrons in Tl are lost *more* easily than the 4s electrons in Ga. The reason for the common appearance of the lower state involves bond energies of the compounds. Bond energy decreases as atomic size, and therefore bond length, increase. Since the Tl—Cl bond, for instance, is relatively long and weak, it takes more energy to remove the two 6s electrons from $Tl^+$ to make $Tl^{3+}$ than is released when two more Tl—Cl bonds form in $TlCl_3$. Thus, $TlCl_3$ is unstable and readily decomposes to $Cl_2$ gas and the more stable TlCl.

Another key pattern that first appears in Group 3A is that *lower state oxides are more basic than higher state oxides;* $In_2O$, for instance, is more basic than $In_2O_3$. In general, when an element has more than one oxidation state, *it acts more like a metal in its lower state:* the lower charge of the $E^+$ ion does not polarize the $O^{2-}$ ion as much as the $E^{3+}$ ion does, which makes the E-to-O bonding in $E_2O$ more ionic than in $E_2O_3$, so the $O^{2-}$ ion is more available to act as a base (see Interchapter, Topic 4).

### Highlights of Boron Chemistry

The chemical behavior of boron is strikingly different from that of the other Group 3A(13) members. As pointed out earlier, *all boron compounds are covalent*, and unlike the other 3A members, it forms network covalent compounds or large molecules with metals, H, O, N, and C. The unifying feature of many B compounds is their *electron deficiency*, but B adopts two strategies to fill its outer level: accepting a bonding pair from electron-rich atoms and forming a bridge bond with electron-poor atoms.

**Accepting a bonding pair from electron-rich atoms.** In gaseous boron trihalides ($BX_3$), the B atom is electron deficient, with only six electrons around it (Section 9.5). To attain an octet, the B atom forms a covalent bond with an electron-rich atom by accepting a lone pair:

$$BF_3(g) + :NH_3(g) \rightleftharpoons F_3B{-}NH_3(g)$$

CARBON COMPOUND

Ethane ($C_2H_6$)

Benzene ($C_6H_6$)

Graphite

Diamond   ●C

BN ANALOG

**A** Amine-borane ($BNH_6$)    **B** Borazine ($B_3N_3H_6$)    **C** Boron nitride    **D** Borazon   ●N  ●B

**FIGURE 13.9**
**Similarities between substances with C—C bonds and those with B—N bonds. A,** Ethane and its BN analog. **B,** Benzene and borazine, which is often referred to as "inorganic" benzene. **C,** Graphite and the similar extended hexagonal structure of boron nitride. **D,** Diamond and borazon have the same crystal structure and are among the hardest substances known. Note that in each case, B attains an octet of electrons by bonding with electron-rich N.

(Such reactions, in which one reactant accepts an electron pair from another to form a covalent bond, are known as Lewis acid-base reactions; we'll discuss them in Chapter 17.)

Similarly, B has only six electrons in boric acid, $B(OH)_3$. In water, the acid accepts an electron pair from the O in $H_2O$, forming a fourth bond and releasing an $H^+$ ion:

$$B(OH)_3(s) + H_2O(l) \rightleftharpoons B(OH)_4^-(aq) + H^+(aq)$$

Boron's outer shell is filled in the wide variety of borate salts, such as the mineral borax (sodium borate), $Na_2[B_4O_5(OH)_4] \cdot 8H_2O$, used for decades as a household cleaning agent. ◆

Boron also attains an octet in several nitrogen compounds whose structures are amazingly similar to elemental carbon and some of its compounds (Figure 13.9). These similarities are due to the fact that the size, IE, and EN of C are between those of B and N (see Table 13.1). Moreover, C has four valence electrons, whereas B has three and N has five, so C—C and B—N bonds have the same number of valence electrons. Thus, $H_3C$—$CH_3$ (ethane) and $H_3B$—$NH_3$ are isoelectronic, as are benzene and borazine (Figure 13.9, *A* and *B*). Boron nitride (Figure 13.9, *C*) has a multisheet, hexagonal layer structure very similar to that of graphite, although the $\pi$ electrons are *not* free to move throughout the sheet. Thus, whereas graphite is a black electrical conductor, boron nitride is a white electrical insulator. At high pressure and temperature, boron nitride forms borazon (Figure 13.9, *D*). Its diamondlike structure makes it extremely hard and useful as an abrasive.

**Forming a bridge bond with electron-poor atoms.** In elemental boron and its many hydrides (boranes), there is no electron-rich atom to supply boron with electrons. Through a combination of normal bonds and

◆ **Borates in Your Labware.**
Strong heating of boric acid (or borate salts) drives off water molecules and gives molten boron oxide: $2B(OH)_3(s) \longrightarrow B_2O_3(l) + 3H_2O(g)$— The oxide dissolves metal oxides to form borate glasses. When mixed with silica ($SiO_2$), it forms borosilicate glass. Its transparency and small change in size when heated or cooled make borosilicate glass useful in cookware and in the glassware you use in the lab.

**FIGURE 13.10**

**The two types of covalent bonding in diborane. A,** A perspective diagram of $B_2H_6$ shows the unusual B—H—B bridge bond and the tetrahedral arrangement around each B atom. **B,** A valence bond depiction shows $sp^3$-hybridized B forming normal covalent bonds at the four terminal B—H bonds and bridge bonds at the two central B—H—B groupings, in which two electrons bind three atoms.

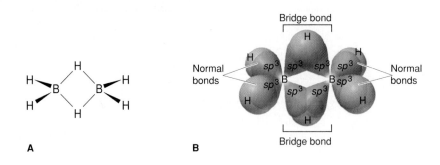

A                               B

unusual bridge bonds, however, boron attains an octet. In diborane ($B_2H_6$) and many larger boranes, for example, two types of B—H bonds exist (Figure 13.10). The first type is a typical electron-pair bond. A valence bond picture shows an $sp^3$ orbital of B overlapping the H $1s$ orbital in each of the four terminal B—H bonds, using two of the three electrons in the valence shell of each B atom.

The other type of bond is a hydride **bridge bond** (or three-center, two-electron bond), in which *each B—H—B grouping is held together by only two electrons.* An $sp^3$ orbital from *each* B overlaps an H $1s$ orbital between them. Two electrons move through this extended bonding orbital—one from one of the B atoms and the other from the H atom—and join the two B atoms via the H atom bridge. Notice that *each B atom is surrounded by eight electrons:* four from the two normal B—H bonds and four from the two B—H—B bridge bonds. In many boranes and in elemental boron, one B atom bridges two others in a three-center, two-electron B—B—B bond (Figure 13.11).

**FIGURE 13.11**

**The boron icosahedron and a larger borane. A,** The icosahedron structural unit of elemental boron uses B—B—B bridge bonds. **B,** A combination of normal covalent bonds and bridge bonds appears in $B_5H_9$, one of the many larger boranes.

**A** $B_{12}$ unit          **B** $B_5 H_9$

### Diagonal Relationships: Beryllium and Aluminum

Beryllium in Group 2A(2) and aluminum in Group 3A(13) are the next pair of diagonally related elements. Both elements form oxoanions in strong base: beryllate, $Be(OH)_4{}^{2-}$, and aluminate, $Al(OH)_4{}^-$. Both have bridge bonds in their hydrides and chlorides. Both form oxide coatings impervious to reaction with water, and both oxides are amphoteric, extremely hard, and high melting. Although their atomic and ionic sizes differ, the small, highly charged $Be^{2+}$ and $Al^{3+}$ ions strongly polarize nearby electron clouds. Therefore, some Al compounds and all Be compounds show significant covalent character in the gas phase.

# 13.6  Group 4A(14): The Carbon Family

The whole range of elemental behavior occurs within Group 4A(14): non-metallic carbon (C) leads off, followed by the metalloids silicon (Si) and germanium (Ge), with metallic tin (Sn) and lead (Pb) at the bottom of the group. Information about the compounds of C and of Si fills libraries: organic chemistry and biochemistry are based on carbon, and geochemistry is based on silicon. There are also some extremely important polymer and electronic technologies based on silicon. The Group 4A(14) Family Portrait (pp. 542 and 543) summarizes atomic, physical, and chemical properties.

### How Does the Bonding Among Atoms of an Element Affect Physical Properties?

The elements of Group 4A(14) and their neighbors in Groups 3A(13) and 5A(15) illustrate how physical properties, such as melting point and heat of fusion ($\Delta H^0_{fus}$), depend on the type of bonding in an element (Table 13.2). Within Group 4A, the large decrease in melting point between the C and Si covalent networks is due to longer, weaker bonds in the Si structure; the large decrease between Ge and Sn is due to the change from covalent network to metallic bonding. Similarly, the large horizontal increases in melting point and $\Delta H^0_{fus}$ between Al and Si and between Ga and Ge reflect the change from metallic bonding to covalent networks. Note the abrupt jumps from the values for metallic Al, Ga, and Sn to those for the network-covalent metalloids Si, Ge, and Sb, and note the abrupt drops from the covalent networks of C and Si to the individual molecules of N and P in Group 5A.

Striking changes in physical properties often appear among **allotropes,** different crystalline or molecular forms of a substance. One allotrope is usually more stable than another at a particular pressure and temperature. Carbon provides the first dramatic example of allotropism. It is difficult to imagine two substances made entirely of the same atom that are more different than graphite and diamond. One is black, the other colorless; one conducts electricity, the other insulates; one is soft and "greasy," the other extremely hard. Graphite is the standard state of carbon, the more stable form at ordinary temperature and pressure, as Figure 13.12 shows. Fortunately for jewelry owners, diamond changes to graphite at a negligible rate under normal conditions.

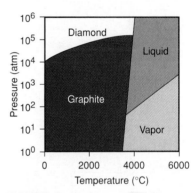

**FIGURE 13.12**
**Phase diagram of carbon.** Graphite is the more stable form at ordinary conditions (extreme lower left). Diamond is more stable at very high pressure.

**TABLE 13.2  Bond Type and Melting in Groups 3A(13) to 5A(15)**

| Group 3A(13) | | | | Group 4A(14) | | | | Group 5A(15) | | | |
|---|---|---|---|---|---|---|---|---|---|---|---|
| Element | Bond Type | Melting Point (°C) | $\Delta H_{fus}$ (kJ/mol) | Element | Bond Type | Melting Point (°C) | $\Delta H_{fus}$ (kJ/mol) | Element | Bond Type | Melting Point (°C) | $\Delta H_{fus}$ (kJ/mol) |
| B | | 2180 | 23.6 | C | | 4100 | Very high | N | | −210 | 0.7 |
| Al | | 660 | 10.5 | Si | | 1420 | 50.6 | P | | 44.1 | 2.5 |
| Ga | | 30 | 5.6 | Ge | | 945 | 36.8 | As | | 816 | 27.7 |
| In | | 157 | 3.3 | Sn | | 232 | 7.1 | Sb | | 631 | 20.0 |
| Tl | | 304 | 4.3 | Pb | | 327 | 4.8 | Bi | | 271 | 10.5 |

**Key:**
- Metallic
- Covalent network
- Covalent molecule
- Metal
- Metalloid
- Nonmetal

FAMILY PORTRAIT

# Group 4A(14): The Carbon Family
## Key Atomic and Physical Properties

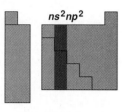

**KEY**

Atomic No.
**Symbol**
Atomic mass
Valence e⁻
configuration
Common
oxidation
states

$ns^2np^2$

**GROUP 4A(14)**

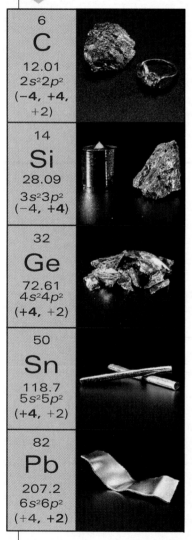

| | |
|---|---|
| 6 <br> **C** <br> 12.01 <br> $2s^2 2p^2$ <br> (−4, +4, +2) | |
| 14 <br> **Si** <br> 28.09 <br> $3s^2 3p^2$ <br> (−4, +4) | |
| 32 <br> **Ge** <br> 72.61 <br> $4s^2 4p^2$ <br> (+4, +2) | |
| 50 <br> **Sn** <br> 118.7 <br> $5s^2 5p^2$ <br> (+4, +2) | |
| 82 <br> **Pb** <br> 207.2 <br> $6s^2 6p^2$ <br> (+4, +2) | |

## Atomic Properties

| Atomic radius (pm) | Ionic radius (pm) |
|---|---|
| C <br> 77 | |
| Si <br> 118 | |
| Ge <br> 122 | |
| Sn <br> 140 | Sn²⁺ <br> 118 |
| Pb <br> 146 | Pb²⁺ <br> 119 |

Group electron configuration is $ns^2np^2$. Down the group, the number of oxidation states decreases, and the lower (+2) state becomes more common.

Down the group, size increases, as IE and EN generally decrease, but deviations occur for Ge and Pb due to the intervening transition and inner transition elements.

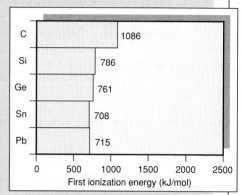

First ionization energy (kJ/mol)

| | |
|---|---|
| C | 1086 |
| Si | 786 |
| Ge | 761 |
| Sn | 708 |
| Pb | 715 |

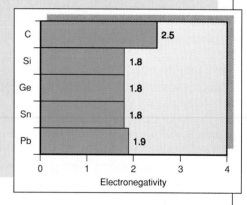

Electronegativity

| | |
|---|---|
| C | 2.5 |
| Si | 1.8 |
| Ge | 1.8 |
| Sn | 1.8 |
| Pb | 1.9 |

## Physical Properties

Trends in properties, such as decreasing hardness and melting point, are due to changes in types of bonding within the solid: covalent network in C, Si, and Ge; metallic in Sn and Pb (see text).

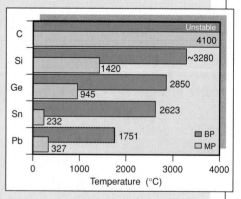

Temperature (°C) — BP / MP

| | BP | MP |
|---|---|---|
| C | 4100 (Unstable) | ~3280 |
| Si | | 1420 |
| Ge | 2850 | 945 |
| Sn | 2623 | 232 |
| Pb | 1751 | 327 |

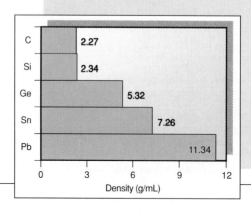

Density (g/mL)

| | |
|---|---|
| C | 2.27 |
| Si | 2.34 |
| Ge | 5.32 |
| Sn | 7.26 |
| Pb | 11.34 |

Down the group, density increases due to several factors, including differences in crystal packing.

# Group 4A(14): The Carbon Family
## Some Reactions and Compounds

## Important Reactions

Reactions 1 and 2 pertain to all the elements (E); reactions 3 to 7 concern industrial uses for compounds of C and Si.

1. The elements are oxidized by halogen:

   $$E(s) + 2X_2 \rightarrow EX_4 \quad (E = C, Si, Ge)$$

   The +2 halides are more stable for tin and lead, $SnX_2$ and $PbX_2$.

2. The elements are oxidized by $O_2$:

   $$E(s) + O_2(g) \rightarrow EO_2 \quad (E = C, Si, Ge, Sn)$$

   Pb forms the +2 oxide, PbO. Oxides become more basic down the group. The reaction of $CO_2$ and water provides the weak acidity of natural unpolluted waters:

   $$CO_2(g) + H_2O(l) \rightleftharpoons [H_2CO_3(aq)] \rightleftharpoons$$
   $$H^+(aq) + HCO_3^-(aq)$$

3. Air and steam passed over hot coke produce gaseous fuel mixtures (producer gas and water gas):

   $$C(s) + air(g) + H_2O(g) \rightarrow$$
   $$CO(g) + CO_2(g) + N_2(g) + H_2(g)$$
   [not balanced]

Computer chip

4. Hydrocarbons react with $O_2$ to form $CO_2$ and $H_2O$. The reaction for methane can be adapted to yield heat or electricity:

   $$CH_4(g) + 2O_2(g) \rightarrow$$
   $$CO_2(g) + 2H_2O(g)$$

5. Certain metal carbides produce acetylene in water:

   $$CaC_2(s) + 2H_2O(l) \rightarrow$$
   $$Ca(OH)_2(aq) + C_2H_2(g)$$

   The gas is used to make other organic compounds and as a fuel in welding (*photo*).

6. Freons (chlorofluorocarbons) are formed by fluorinating carbon tetrachloride:

   $$CCl_4(l) + HF(g) \rightarrow CFCl_3(g) + HCl(g)$$

   Production of trichlorofluoromethane (Freon-11), the major refrigerant in the world, is being eliminated because of its severe effects on the environment (*see margin note, p. 546*).

7. Silica is reduced to form elemental silicon:

   $$SiO_2(s) + 2C(s) \rightarrow Si(s) + 2CO(g)$$

   This crude silicon is made ultrapure through zone refining (*Section 12.5*) for manufacture of computer chips.

## Important Compounds

1. Carbon monoxide, CO. Used as a gaseous fuel as a precursor for one-carbon organic compounds, and as a reactant in the purification of nickel. Formed in internal combustion engines and released as a toxic air pollutant.

2. Carbon dioxide, $CO_2$. Atmospheric component used by photosynthetic plants to make carbohydrates and $O_2$. The final oxidation product of all C-based fuels; its increase in the atmosphere may lead to global warming. Used industrially as a refrigerant gas, blanketing gas in fire extinguishers, and effervescent gas in beverages. Combined with $NH_3$ to form urea for fertilizers and plastics manufacture.

3. Methane, $CH_4$. Used as a fuel and in the production of many organic compounds. Major component of natural gas. Formed by anaerobic decomposition of plants (swamp gas) and by microbes in termites and certain mammals. May contribute to global warming.

4. Silicon dioxide, $SiO_2$. Occurs in many amorphous (glassy) and crystalline forms, quartz being the most common. Used to make glass and as an inert chromatography support material.

5. Silicon carbide, SiC. Known as carborundum, a major industrial abrasive and a highly refractory ceramic for tough, high-temperature uses. Can be doped to form a high-temperature semiconductor.

6. Organotin compounds, $R_4Sn$. Used to stabilize PVC (polyvinyl chloride) plastics and to cure silicone rubbers. Agricultural biocide for insects, fungi, and weeds.

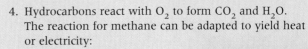

PVC pipe

7. Tetraethyl lead, $(C_2H_5)_4Pb$. Once used as a gasoline additive to improve fuel efficiency, but now removed because of its inactivation of auto catalytic converters. Major source of lead as a toxic air pollutant.

**FIGURE 13.13**
**The structure of buckminsterfullerene.** The parent compound of the fullerenes, the so-called bucky ball, is a spherical molecule of 60 carbon atoms.

In the late 1980s, interest arose among researchers about a newly discovered allotrope of carbon. Mass spectrometric evidence had been found for a soccer-ball shaped molecule of formula $C_{60}$ (Figure 13.13). This original evidence came from samples of soot, but recent findings reveal the molecule even in geological samples formed by the meteor impact that occurred around the time the dinosaurs became extinct. The molecule has been dubbed *buckminsterfullerene* (informally called a "bucky ball") after the scientist-philosopher R. Buckminster Fuller, who designed structures with similar shapes. Excitement peaked in 1990, when scientists learned how to prepare multigram samples of $C_{60}$ and related fullerenes. Structures with enclosed metal atoms, compounds with attached fluorine, hydroxyl, and sugar groups, "nanotubes" (extremely thin graphitelike tubes with fullerene ends), and fullerene films join a rapidly growing number of remarkable materials under study. With potential applications in electronics, catalysis, polymers, and medicines, fullerene chemistry will receive increasing attention well into the next century.

Tin has two allotropes. White $\beta$-tin is stable at room temperature and above, whereas gray $\alpha$-tin is the more stable form below 13°C (56°F). When white tin is kept at a low temperature, some converts to microcrystals of gray tin. The random formation and growth of these regions of gray tin, which has a different crystal structure, weaken the metal to the extent that it crumbles. In the unheated cathedrals of medieval northern Europe, the tin pipes of the magnificent organs would sometimes crumble as a result of "tin disease" caused by this allotropic transition from white to gray tin.

### How Does the Type of Bonding Change in Group 4A(14) Compounds?

With four of a possible eight valence electrons and intermediate values of EN and IE, the 4A(14) elements display a wide range of chemical behavior, from the covalent compounds of carbon to the ionic compounds of lead.

Carbon's intermediate EN of 2.5 ensures that it virtually always forms covalent bonds, but the larger members of Group 4A form bonds with increasing ionic character. With nonmetals, Si and Ge form strong polar covalent bonds, such as the Si—O bond, one of the strongest of any Period 3 element (368 kJ/mol). This bond is responsible for the physical and chemical stability of the Earth's solid surface, as we discuss later. Although individual ions rarely exist, Sn and Pb bonding to nonmetals is still more ionic.

After silicon, the shift to greater importance of the lower oxidation state, a pattern we first observed in Group 3A(13), appears here as well: Si compounds of the +4 state are more stable and common than those of the +2 state, whereas Pb compounds of the +2 state are more stable than those of the +4 state. The elements also behave more like metals in the lower state. Consider the chlorides (Figure 13.14) and oxides of Sn and Pb. The +2 chlorides $SnCl_2$ and $PbCl_2$ are white, relatively high-melting, water-soluble crystals—typical properties of a salt. In contrast, $SnCl_4$ is a volatile, benzene-soluble liquid, and $PbCl_4$ is a thermally unstable oil. The oxides SnO and PbO are more basic than $SnO_2$ and $PbO_2$ because the +2 metals are less able to polarize the $O^{2-}$ ion, so the E-to-O bonding is more ionic.

**FIGURE 13.14**
**The greater metallic character of the lower oxidation state of tin and lead.** Metals with more than one oxidation state behave more metallic in the lower state. Tin(II) (*top, left*) and lead(II) (*bottom, left*) chlorides are white, crystalline, saltlike solids. In contrast, tin(IV) (*top, right*) and lead(IV) (*bottom, right*) chlorides are volatile liquids, indicating the presence of individual molecules.

### Highlights of Carbon Chemistry

Like the other Period 2 elements, carbon is an anomaly in its group; in fact, no other element in the periodic table displays similar behavior. Carbon forms bonds with the smaller Group 1A(1) and 2A(2) metals, many transi-

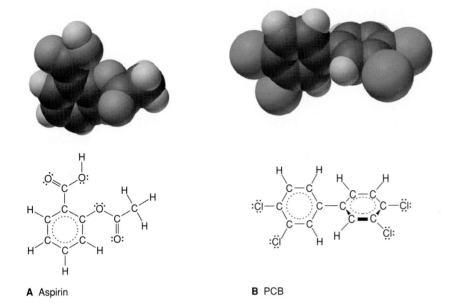

**A** Aspirin

**B** PCB

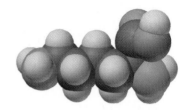

**C** Lysine

**FIGURE 13.15**

**Three organic compounds of carbon.** Three of the several million known organic compounds of carbon are **A,** aspirin (acetylsalicylic acid); **B,** one of several PCBs (polychlorinated biphenyls); and **C,** lysine (an amino acid).

tion metals, the halogens, and its neighbors: B, Si, N, O, P, and S. This versatile element exhibits every oxidation state possible for its group, from +4 in $CCl_4$ and $CO_2$ through −4 in $CH_4$.

Two features stand out in the chemistry of carbon: its ability to bond to itself *(catenation)* and its ability to form multiple bonds. As a result of its small size and its capacity for four bonds, carbon can form chains, branches, and rings that lead to myriad structures. Add a lot of H, some O and N, a bit of S, P, and halogen, and you have the whole organic world! Figure 13.15 shows three of the several million known organic compounds. Multiple bonding is common in these compounds because the C—C bond is short enough for sideways overlap of half-filled $2p$ orbitals. (In Chapter 14, we discuss in detail how the atomic properties of carbon give rise to the bonding and structure of organic compounds.)

Because the other 4A members are larger, E—E bonds become longer and weaker down the group. Thus, the longest Si chain has only eight atoms, and the longest Ge chain only five. These longer bonds also prevent sufficient overlap for $p,p$-$\pi$ bonding. Moreover, the empty $d$ orbitals of these atoms make the chains much more susceptible to chemical attack.

In contrast to its organic compounds, carbon's inorganic compounds are simple. Metal carbonates are the main mineral form. Marble, limestone, chalk, coral, and several other types are found in enormous deposits throughout the world. Many of these compounds are remnants of fossilized marine organisms. Carbonates are used in several common antacids because they react with acid, such as the HCl in stomach acid [see Group 2A(2) Family Portrait, compound 4]:

$$CaCO_3(s) + 2HCl(aq) \rightarrow CaCl_2(aq) + CO_2(g) + H_2O(l)$$

Similar reactions with sulfuric and nitric acids protect lakes bounded by limestone from destruction by acid rain.

Unlike the solid network or ionic oxides of the other 4A members, carbon forms two common gaseous oxides, $CO_2$ and $CO$. Carbon dioxide is essential to all life as the ultimate source of carbon in plants and animals through its role in photosynthesis. Its aqueous solution is the cause of acidity in natural waters. However, its atmospheric buildup from deforestation and excessive use of fossil fuels may severely affect the climate. Carbon monoxide forms when carbon or its compounds burn in an inadequate supply of $O_2$:

$$2C(s) + O_2(g) \rightarrow 2CO(g)$$

Carbon monoxide is a key component of syngas fuels (see Chemical Connections, Section 6.6) and is widely used in the production of methanol, formaldehyde, and other major industrial compounds.

CO binds strongly to many transition metals, which gives rise to its high toxicity. When inhaled from cigarette smoke or polluted air, it enters the blood and binds strongly to the Fe(II) in hemoglobin and other iron-containing proteins. The cyanide ion ($CN^-$) is *isoelectronic* with CO:

$$[:C\equiv N:]^- \quad \text{same electronic structure as} \quad :C\equiv O:$$

In fact, cyanide binds to many of the same iron-containing proteins and is also toxic.

Monocarbon halides (or halomethanes) are tetrahedral molecules whose stability to heat and light decreases as the size of the halogen atom increases and the C—X bond becomes longer (weaker) (see Interchapter, Topic 2). The chlorofluorocarbons (CFCs) have important industrial uses, but their short, strong bonds have created major environmental problems. ◆

### Highlights of Silicon Chemistry

Silicon halides are more reactive than carbon halides because Si has empty *d* orbitals that are available for chemical attack. Surprisingly, the Si halides have longer yet *stronger* bonds than the corresponding C halides (see Tables 9.2 and 9.3). One explanation for this unusual strength is that the Si—X bond has some double-bond character due to the presence of a *σ* bond *and* a different type of *π* bond, called a *p,d-π bond,* which results from sideways overlap of an Si *d* orbital and a halogen *p* orbital. A similar type of *p,d-π* bonding leads to the planar molecular shape of trisilylamine, in contrast to the trigonal pyramid of trimethylamine (Figure 13.16).

**Silicate minerals and silicone polymers.** To a great extent, the chemistry of silicon is the chemistry of the *silicon-oxygen bond.* Just as carbon forms unending C—C chains, the —Si—O— grouping repeats itself endlessly in a wide variety of **silicates,** the most important minerals on the planet, and in **silicones,** synthetic polymers with many applications.

From common sand and clay to semiprecious amethyst and carnelian, silicate minerals are the dominant form of matter in the nonliving world. Oxygen, the most abundant element on Earth, and silicon, the next most abundant, compose these minerals and thus account for four of every five atoms on the surface of the planet!

**FIGURE 13.16**
**The impact of *p,d-π* bonding on the structure of trisilylamine. A,** The molecular shape of trimethylamine is trigonal pyramidal, as in ammonia. **B,** Trisilylamine, the silicon analog, has a trigonal planar shape due to the formation of a double bond between N and Si. The ball-and-stick model shows one of three resonance forms. **C,** The lone pair in an unhybridized *p* orbital of N overlaps with an empty *d* orbital of Si to give a *p,d-π* bond. In the resonance hybrid, a *d* orbital on each Si is involved in the *π* bond.

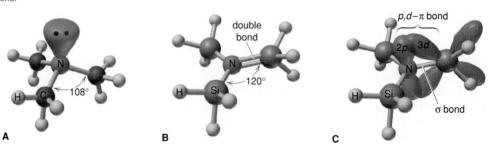

A    B    C

The silicate building unit is the *orthosilicate* grouping, —$SiO_4$—, a tetrahedral arrangement of four oxygens around a central silicon. Several well-known minerals contain $SiO_4^{4-}$ ions or small groups of them linked together. The gemstone zircon ($ZrSiO_4$) contains one unit; hemimorphite [$Zn_4(OH)_2Si_2O_7·H_2O$] contains two units linked through an oxygen corner; and beryl ($Be_3Al_2Si_6O_{18}$), the major source of beryllium, contains six units joined into a cyclic ion (Figure 13.17). As you'll see in a moment, in addition to these separate ions, $SiO_4$ units are linked more extensively to create much of the planet's mineral structure.

Unlike the naturally occurring silicates, silicone polymers are manufactured substances consisting of alternating Si and O atoms with two organic groups bound to the Si atoms. A key starting material is formed by the reaction of silicon with methyl chloride:

$$Si(s) + 2CH_3Cl(g) \xrightarrow[\sim 300°C]{Cu\ catalyst} (CH_3)_2SiCl_2(g)$$

Although the corresponding carbon compound [$(CH_3)_2CCl_2$] is inert to water, the empty *d* orbitals of Si make this compound reactive:

$$(CH_3)_2SiCl_2(l) + 2H_2O(l) \rightarrow (CH_3)_2Si(OH)_2(l) + 2HCl(g)$$

The product reacts with itself to form water and a silicone chain molecule called a *polydimethylsiloxane:*

$$n(CH_3)_2Si(OH)_2 \longrightarrow \left[\begin{array}{c} CH_3 \\ | \\ O—Si— \\ | \\ CH_3 \end{array}\right]_n + nH_2O$$

Silicones have properties of both plastics and minerals. The organic groups give them the flexibility and low intermolecular forces between chains of a plastic, while the O—Si—O backbone confers the thermal stability and nonflammability of a mineral. In the upcoming Gallery, note the structural similarities and uses of silicates and silicones.

## Diagonal Relationships: Boron and Silicon

The final diagonal relationship that we consider occurs between the metalloids B and Si. Both exhibit the electrical properties of a semiconductor. Both elements and their mineral oxoanions, borates and silicates, occur in extended covalent networks. Both boric and silicic acids are weakly acidic and occur in layers with widespread H bonding. Both elements form flammable, low-melting compounds with hydrogen—the boranes and silanes—that act as reducing agents.

## Looking Backward and Forward: Groups 3A(13), 4A(14), and 5A(15)

Standing in Group 4A(14), we look back at Group 3A(13) as the transition from the *s* block of metals to the *p* block of mostly metalloids and nonmetals (Figure 13.18). Major changes occur in physical behavior as we move across from metals to covalent networks, and in chemical behavior, as cations give way to covalent tetrahedra. Looking ahead to Group 5A(15), we find covalent bonding and, as ENs increase, the expanded valence shells of nonmetals occurring more often, as well as the first appearance of monatomic anions.

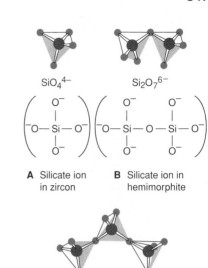

**A** Silicate ion in zircon    **B** Silicate ion in hemimorphite

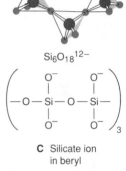

**C** Silicate ion in beryl

**FIGURE 13.17**
Structures of the silicate anions in some minerals.

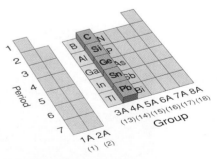

**FIGURE 13.18**
Standing in Group 4A(14), looking backward to 3A(13) and forward to 5A(15).

## GALLERY                     Silicate Minerals and Silicone Polymers

The silicates and silicones show beautifully how organization at the molecular level manifests itself in the properties of macroscopic substances. Interestingly, the major types of both materials exhibit the same three structural classes: chains, sheets, and frameworks.

### Silicate Minerals

**Chain Silicates.** The simplest structural class occurs when each $SiO_4$ unit shares two O corners and links into a chain. Two chains can link laterally into a ribbon, the most common having repeating $Si_4O_{11}^{6-}$ units. Metal ions bind the polyanionic ribbons together into neutral sheaths. With only weak intermolecular forces between sheaths, the material occurs in fibrous strands, as in the family of asbestos minerals.

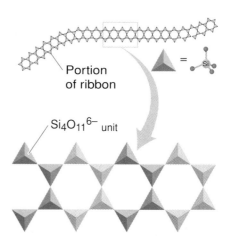

Portion of ribbon

$Si_4O_{11}^{6-}$ unit

**Sheet (Layer) Silicates.** The next structural class arises when each $SiO_4$ unit shares three of its four O corners to form a sheet; double sheets arise when the fourth O is shared with another sheet. In talc, the softest mineral, the sheets interact through weak forces, so talcum powder feels slippery. If Al substitutes for some Si, or if $Al(OH)_3$ layers interleave with silicate sheets, an aluminosilicate results, such as the clay kaolinite. Different substitutions and/or interlayering of ions give the micas. In muscovite mica, ions lie between aluminosilicate double layers (photo). Mica flakes when the ionic attractions are overcome.

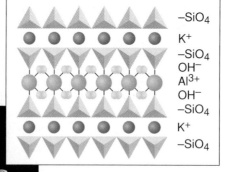

–$SiO_4$
$K^+$
–$SiO_4$
$OH^-$
$Al^{3+}$
$OH^-$
–$SiO_4$
$K^+$
–$SiO_4$

**Framework Silicates.** The final structural class arises when $SiO_4$ units share all four O corners to give a three-dimensional framework silicate, such as silica ($SiO_2$), which occurs most often as α-quartz. Some of silica's 12 crystalline forms exist as semiprecious gems. Feldspars, which comprise 60% of the Earth's crust, result when some Si is replaced by Al; granite consists of microcrystals of feldspar, mica, and quartz. Eons of weathering convert feldspars into clays. Zeolites have open frameworks of polyhedra that create minute tunnels. Synthetic zeolites are made with cavities of certain sizes to trap specific molecules; they are used to dry gas mixtures, separate hydrocarbons and prepare catalysts.

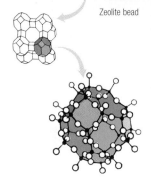

Zeolite bead

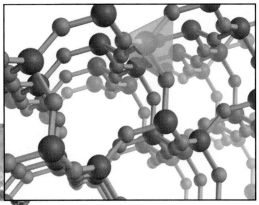

Quartz

## Silicone Polymers

**Chain Silicones:** Oils and Greases. The simplest structural class is the poly-siloxane chain, in which each $(CH_3)_2Si(OH)_2$ unit uses both OH groups to link two others (below). A chain-terminating compound with a third organic group, such as $(CH_3)_3SiOH$, is added to control chain length. Polysiloxanes are unreactive, oily liquids with high viscosity and low surface tension.

They are used as hydraulic oils and lubricants, as anti-foam agents in frying potato chips, and as components of suntan oil, car polish, digestive aids, and makeup.

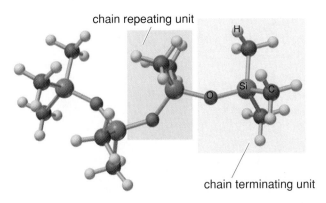

chain repeating unit

chain terminating unit

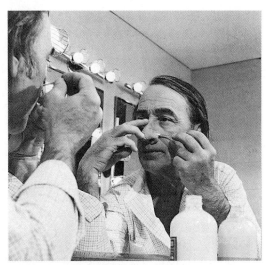

The actor Dustin Hoffman applying makeup consisting of chain and sheet silicones.

**Sheet Silicones:** Elastomers. In the next class, the third OH group of an added bridging compound, such as $CH_3Si(OH)_3$, reacts to condense chains laterally into gummy sheets that are made into elastomers (rubbers), which are flexible, elastic, and stable from $-100°C$ to $250°C$. These are used in gaskets, rollers, cable insulation, space suits, contact lenses, and dentures.

**Framework Silicones:** Resins. In the final structural class, reactions that free OH groups, while substituting larger organic groups for some $CH_3$ groups, interlink sheets to produce strong, thermally stable resins. These are used as insulating laminates on printed circuit boards, and as nonstick coatings on cookware. Sheet and framework silicones have revolutionized modern surgical practice by providing numerous parts that can be implanted permanently in a patient to replace damaged ones. Some of these (indicated by dots on the illustration) are artificial skin, bone, joints, blood vessels, and organ parts.

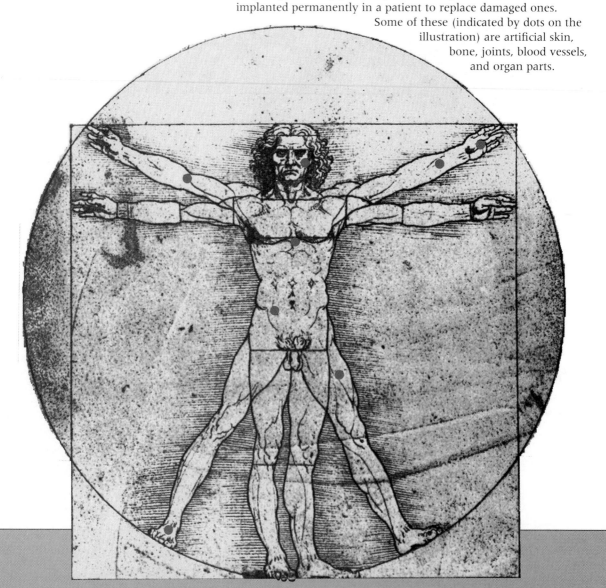

## 13.7  Group 5A(15): The Nitrogen Family

The first two elements of Group 5A(15), gaseous nonmetallic nitrogen (N) and solid nonmetallic phosphorus (P), will occupy most of our attention. The industrial and environmental significance of their compounds is matched only by their importance in the structure and function of biomolecules. Below them are two metalloids, arsenic (As) and antimony (Sb), followed by the sole metal, bismuth (Bi), the last nonradioactive element in the periodic table. The Group 5A(5) Family Portrait (pp. 552 and 553) provides an overview.

### What Accounts for the Wide Range of Physical Behavior in Group 5A(15)?

Because such large changes in bonding and intermolecular forces occur in Group 5A(15), it displays the widest range of physical behavior we've seen so far. Nitrogen occurs as a gas consisting of $N_2$ molecules, which interact through such weak dispersion forces that it *boils* over 200°C below room temperature. Elemental phosphorus exists most commonly as tetrahedral $P_4$ molecules; however, because P is heavier and more polarizable than N, stronger dispersion forces are present and the element melts about 25°C above room temperature. Arsenic also consists of $As_4$ tetrahedra, but they are linked in a covalent network, so it has the highest melting point in the group. The covalent network of Sb gives it a much higher melting point than the metallic bonding gives to Bi below it.

Phosphorus has several allotropes. The white and red forms have very different properties as a result of the way the atoms are linked. The highly reactive white form is prepared as a gas and condensed to a whitish, waxy solid under cold water to prevent it from igniting in air. It consists of tetrahedral molecules of four P atoms bonded *to each other*, with no atom in the center (Figure 13.19, *A*). Each P atom uses its $3p$ orbitals to bond to the other three. Whereas the $3p$ orbitals of an isolated P atom lie 90° apart, the bond angles in $P_4$ are 60°. With the smaller angle comes poor orbital overlap and a P—P bond strength half that of a normal bond—no wonder the white allotrope is so reactive! Heating the white form in the absence of air breaks one of the P—P bonds in each tetrahedron, and those orbitals overlap with others to form the chains of $P_4$ units that make up the red form (Figure 13.19, *B*). Individual molecules in white P make it highly reactive, low melting (44.1°C), and soluble in nonpolar solvents; the chains in red P make it much less reactive, high melting ( ~600°C), and insoluble.

**A** White phosphorus          **B** Red phosphorus

**FIGURE 13.19**
**Two allotropes of phosphorus. A,** White phosphorus exists as individual $P_4$ molecules, with the P—P bonds forming the edges of a tetrahedron. **B,** In red phosphorus, one of the P—P bonds of the white form *(bottom rear)* has broken and links the tetrahedra together. Lone pairs reside in *p* orbitals in both allotropes.

# Group 5A(15): The Nitrogen Family
## Key Atomic and Physical Properties

**KEY**

Atomic No.
**Symbol**
Atomic mass
Valence e⁻
configuration
Common
oxidation
states

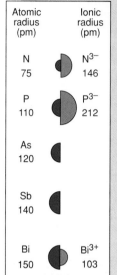

$ns^2np^3$

GROUP 5A(15)

| 7 | | |
|---|---|---|
| **N** | | |
| 14.01 | | |
| $2s^22p^3$ | | |
| (−3, +5, +4, +3, +2, +1) | | |

| 15 |
|---|
| **P** |
| 30.97 |
| $3s^23p^3$ |
| (−3, +5, +3) |

| 33 |
|---|
| **As** |
| 74.92 |
| $4s^24p^3$ |
| (−3, +5, +3) |

| 51 |
|---|
| **Sb** |
| 121.8 |
| $5s^25p^3$ |
| (−3, +5, +3) |

| 83 |
|---|
| **Bi** |
| 209.0 |
| $6s^26p^3$ |
| (+3) |

## Atomic Properties

| Atomic radius (pm) | | Ionic radius (pm) |
|---|---|---|
| N 75 | | $N^{3-}$ 146 |
| P 110 | | $P^{3-}$ 212 |
| As 120 | | |
| Sb 140 | | |
| Bi 150 | | $Bi^{3+}$ 103 |

Group electron configuration is $ns^2np^3$. The $np$ sublevel is half-filled with each $p$ orbital containing one electron. Number of oxidation states decreases down the group, and the lower (+3) state becomes more common.

Atomic properties follow generally expected trends. The large (~50%) increase in size from N to P correlates with the much lower IE and EN of P.

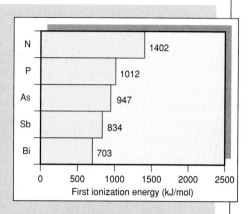

First ionization energy (kJ/mol)

| N | 1402 |
| P | 1012 |
| As | 947 |
| Sb | 834 |
| Bi | 703 |

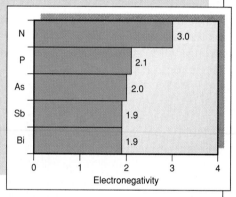

Electronegativity

| N | 3.0 |
| P | 2.1 |
| As | 2.0 |
| Sb | 1.9 |
| Bi | 1.9 |

## Physical Properties

Physical properties reflect the change from individual molecules (N, P) to covalent network (As, Sb) to metal (Bi). Thus, melting points increase and then decrease. Densities of the elements as solids increase steadily.

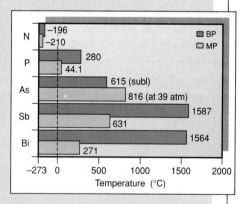

Temperature (°C)   ■ BP   □ MP

| N | −196 / −210 |
| P | 280 / 44.1 |
| As | 615 (subl) / 816 (at 39 atm) |
| Sb | 1587 / 631 |
| Bi | 1564 / 271 |

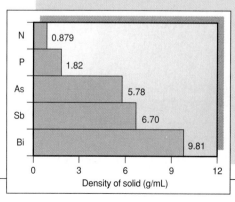

Density of solid (g/mL)

| N | 0.879 |
| P | 1.82 |
| As | 5.78 |
| Sb | 6.70 |
| Bi | 9.81 |

# Group 5A (15): The Nitrogen Family
## Some Reactions and Compounds

## Important Reactions

General group behavior is shown in reactions 1 to 3, whereas phosphorus chemistry is the theme in reactions 4 and 5.

1. Nitrogen is "fixed" industrially in the Haber process:

   $$N_2(g) + 3H_2(g) \rightleftharpoons 2NH_3(g)$$

   Further reactions convert $NH_3$ to NO, $NO_2$, and $HNO_3$ (*see Highlights of Nitrogen Chemistry*). Some other group hydrides are formed from reaction in water (or $H_3O^+$) of metal phosphide, arsenide, and so forth:

   $$Ca_3P_2(s) + 6H_2O(l) \rightarrow 2PH_3(g) + 3Ca(OH)_2(aq)$$

2. Halides are formed by direct combination of the elements:

   $$2E(s) + 3X_2 \rightarrow 2EX_3 \quad (E = \text{all except N})$$

   $$EX_3 + X_2 \rightarrow EX_5 \quad (E = \text{all except N and Bi})$$

3. Oxoacids are formed from the halide with a reaction in water that is common to many nonmetal halides:

   $$EX_3 + 3H_2O(l) \rightarrow H_3EO_3(aq) + 3HX(aq)$$

   (E = all except N)

$$EX_5 + 4H_2O(l) \rightarrow H_3EO_4(aq) + 5HX(aq)$$

(E = all except N and Bi)

Note that the oxidation number of E does *not* change.

4. Phosphate ions are dehydrated to form polyphosphates:

   $$3NaH_2PO_4(s) \rightarrow Na_3P_3O_9(s) + 3H_2O(g)$$

5. When $P_4$ reacts in basic solution, its oxidation state both decreases *and* increases:

   $$P_4(s) + 3OH^-(aq) + 3H_2O(l) \rightarrow$$
   $$PH_3(g) + 3H_2PO_2^-(aq)$$

   Analogous reactions are typical of many nonmetals, such as $S_8$ and $X_2$ (halogens).

Lake polluted by excess phosphate

## Important Compounds

1. Ammonia, $NH_3$. First substance formed when atmospheric $N_2$ is used to make N-containing compounds. Annual multimillion-ton production for use in fertilizers, explosives, rayon, and polymers such as nylon, urea-formaldehyde resins, and acrylics.
2. Hydrazine, $N_2H_4$ (*see margin note, p. 554*).

3. Nitric oxide (NO), nitrogen dioxide ($NO_2$), and nitric acid ($HNO_3$). Oxides are intermediates to $HNO_3$. Acid used in fertilizer manufacture, nylon production, metal etching, and explosives industry (*see Highlights of Nitrogen Chemistry*).
4. Amino acids, $H_3N^+$—$CH(R)$—$COO^-$ (R = one of 20 different organic groups). Occur in every organism, both free and linked together into proteins. Essential to the growth and function of all cells. Synthetic amino acids used as dietary supplements.
5. Phosphorus trichloride, $PCl_3$. Used to form many organic phosphorus compounds, including oil and fuel additives, plasticizers, flame retardants, and insecticides. Also used to make $PCl_5$, $POCl_3$, and other important P-containing compounds.
6. Tetraphosphorus decaoxide ($P_4O_{10}$) and phosphoric acid ($H_3PO_4$) (*see Highlights of Phosphorus Chemistry*).

7. Sodium tripolyphosphate, $Na_5P_3O_{10}$. As a water-softening agent (Calgon), combines with hard-water $Mg^{2+}$ and $Ca^{2+}$ ions, preventing them from reacting with soap anions, and thus improves cleaning action. Use curtailed in the United States because it pollutes lakes and streams by causing excessive algal growth.
8. Adenosine triphosphate (ATP) and other biophosphates. ATP acts to transfer chemical energy in the cell; necessary for all biological processes requiring energy. Phosphate groups occur in sugars, fats, proteins, and nucleic acids.
9. Bismuth subsalicylate, $BiO(C_7H_5O_3)$. The active ingredient in Pepto-Bismol, a widely used remedy for diarrhea and nausea. (The pink color is not due to this white compound.)

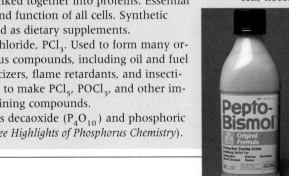

## What Patterns Appear in the Chemical Behavior of Group 5A(15)?

The same general pattern of chemical behavior that we discussed in Group 4A(14) appears again in this group, reflecting the change from nonmetallic N to metallic Bi. The overwhelming majority of 5A(15) compounds have *covalent bonds*. Whereas N can form no more than four bonds, the next three members can expand their valence shells by using empty *d* orbitals.

Forming an ion with a noble gas core means *gaining* three electrons, the last two in endothermic steps. Nevertheless, the enormous lattice energy released when such highly charged anions attract cations drives their formation, but only in the case of N in compounds with active metals, such as $Li_3N$ and $Mg_3N_2$ (and perhaps with P in $Na_3P$). Metallic Bi becomes a cation in a few compounds by *losing* three of its five outer electrons, as in $BiF_3$ and $Bi(NO_3)_3 \cdot 5H_2O$.

Again, we see familiar patterns as we move down the group. Fewer oxidation states occur, with the lower state becoming more prominent: N exhibits every state possible for a 5A element, from +5 to −3; only the +5 and +3 states are common for P, As, and Sb; and +3 is the only common state of Bi. The acid-base behavior of the oxides changes from acidic to amphoteric to basic, reflecting the increase in metallic character. In addition, the lower oxide of an element is more basic than the higher oxide, reflecting the greater ionic character of the E-to-O bonding in the lower oxide.

All the Group 5A(15) elements form gaseous hydrides of formula $EH_3$. Except for $NH_3$, they are extremely reactive and poisonous and are synthesized by reaction of a metal phosphide, arsenide, and so forth in water or aqueous acid. For example,

$$Ca_3As_2(s) + 6H_2O(l) \rightarrow 2AsH_3(g) + 3Ca(OH)_2(aq)$$

Ammonia is made industrially by direct combination of the elements at high pressure and temperature:

$$N_2(g) + 3H_2(g) \rightleftharpoons 2NH_3(g)$$

Molecular properties of the hydrides reveal some interesting bonding and structure patterns:
- Despite its much lower molar mass, $NH_3$ melts and boils at higher temperatures than the other 5A hydrides as a result of *H bonding*.
- Bond angles decrease from 107.3° for $NH_3$ to around 90° for the others, which suggests the larger atoms use unhybridized *p* orbitals.
- E—H bond lengths increase down the group, so bond strength and thermal stability decrease: $AsH_3$ decomposes at 250°C, $SbH_3$ at 20°C, and $BiH_3$ at −45°C.

We'll see these three features—H bonding for the smallest member, change in bond angles, change in bond energies—in the hydrides of Group 6A(16) as well. ◆

All Group 5A(15) elements form trihalides ($EX_3$), and some form pentahalides ($EX_5$). Nitrogen forms the trihalide only because it cannot expand its valence shell; bismuth forms the trihalide because it loses only three electrons. Most trihalides are prepared by direct combination:

$$P_4(s) + 6Cl_2(g) \rightarrow 4PCl_3(l)$$

The pentahalides form with excess halogen:

$$PCl_3(l) + Cl_2(g) \rightarrow PCl_5(s)$$

As with the hydrides, the thermal stability of the halides decreases as the E—X bond becomes longer. Among the nitrogen halides, for example, $NF_3$ is a stable, rather unreactive gas. $NCl_3$ is explosive and reacts rapidly with

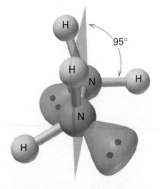

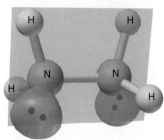

◆ **Hydrazine, Nitrogen's Other Hydride.** Aside from $NH_3$, the most important 5A(15) hydride is hydrazine, $N_2H_4$. Like $NH_3$, it is a weak base, forming $N_2H_5^+$ and $N_2H_6^{2+}$ with acids. Organic hydrazine derivatives are used in rocket and missile fuels, and hydrazine is also used to make antituberculin drugs, plant growth regulators, and fungicides. We might expect the lone pairs in the structure to lie opposite each other and give a symmetrical molecule with no dipole moment. However, repulsions between each lone pair and the bonding pairs on the other atom force the lone pairs to lie on the same side of the N—N bond, giving a large dipole moment ($\mu = 1.85$ D in the gas phase). A similar arrangement occurs in the analogous hydride of oxygen, $H_2O_2$.

water. (The chemist who first prepared it lost three fingers and an eye!) $NBr_3$ can only be made below $-87°C$. $NI_3$ has never even been prepared, but its ammonia adduct ($NI_3 \cdot NH_3$) explodes at the slightest touch.

In a reaction pattern *typical of many nonmetal halides,* the 5A halides react with water to yield the hydrogen halide and the oxoacid with E in the *same* oxidation state as in the original halide. For example, $PX_5$ (O.N. of P = +5) produces phosphoric acid (O.N. of P = +5) and HX:

$$PCl_5(s) + 4H_2O(l) \rightarrow H_3PO_4(l) + 5HCl(g)$$

### Highlights of Nitrogen Chemistry

Despite the relative inertness of $N_2$ at room temperature, it reacts at higher temperatures with $H_2$, Li, Group 2A(2), B, Al, C, Si, Ge, $O_2$, and many transition elements. Nearly every element in the periodic table forms bonds to N. Here we focus on the oxides and the oxoacids and their salts.

**Nitrogen oxides.** Nitrogen is remarkable for having six stable oxides, each with a *positive* heat of formation primarily because the N≡N bond is so strong. Their structures and some properties are shown in Table 13.3. ◆ Unlike the hydrides and halides, the oxides of nitrogen are planar. Nitrogen displays all its positive oxidation states in these compounds, and in $N_2O$ and $N_2O_3$ the two N atoms have different states.

◆ **Nitrogen Oxides in the Birth of Chemistry.** Three of the oxides—$N_2O$, NO, and $NO_2$—were among the first gaseous compounds isolated and identified; those who studied them—Joseph Priestley, Carl Scheele, Joseph Black, and Henry Cavendish—are among the greatest chemists of the 18th century. In the early 1800s, Dalton used these compounds to develop his explanation of multiple proportions. For a given mass of O, the masses of N in the three oxides are in the ratio of small whole numbers, 4/2/1.

**TABLE 13.3  Structures and Properties of the Nitrogen Oxides**

| FORMULA | NAME | SPACE-FILLING MODEL | LEWIS STRUCTURE | OXIDATION STATE OF N | $\Delta H_f^\circ$ AT 298 K (kJ/mol) | COMMENT |
|---|---|---|---|---|---|---|
| $N_2O$ | Dinitrogen monoxide (dinitrogen oxide; nitrous oxide) |  | 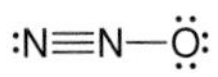 :N≡N—Ö: | +1 (0, +2) | 82.0 | Colorless gas; used as dental anesthetic ("laughing gas") and aerosol propellant |
| NO | Nitrogen monoxide (nitrogen oxide; nitric oxide) |  | 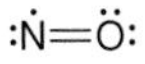 :Ṅ=Ö: | +2 | 90.3 | Colorless, paramagnetic gas; biochemical messenger; air pollutant |
| $N_2O_3$ | Dinitrogen trioxide | 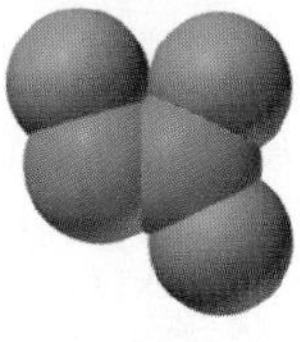 | 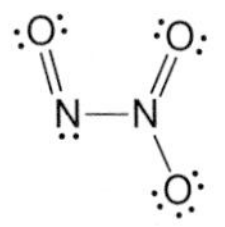 :Ö:  :Ö: \ N—N / :Ö: | +3 (+2, +4) | 83.7 | Reddish brown gas (reversibly dissociates to NO and $NO_2$) |
| $NO_2$ | Nitrogen dioxide |  | 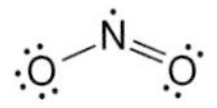 :Ö—Ṅ=Ö: | +4 | 33.2 | Orange-brown, paramagnetic gas in $HNO_3$ manufacture; poisonous air pollutant |
| $N_2O_4$ | Dinitrogen tetraoxide | 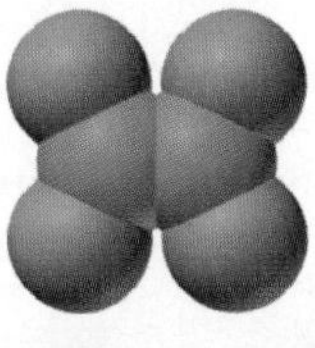 | 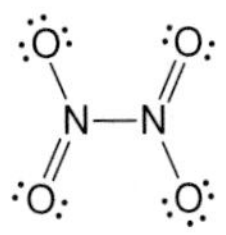 :Ö:  :Ö: N—N :Ö:  :Ö: | +4 | 9.16 | Colorless to yellow liquid (reversibly dissociates to $NO_2$) |
| $N_2O_5$ | Dinitrogen pentaoxide | 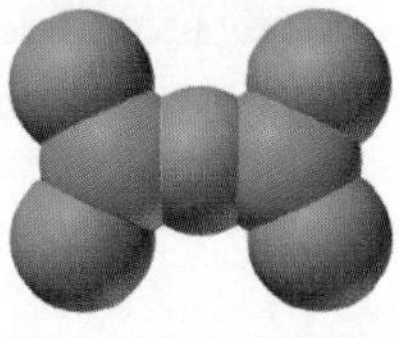 | 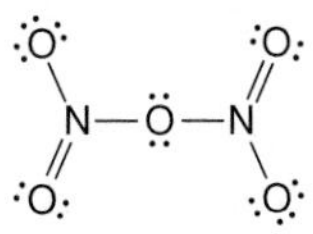 :Ö:  :Ö: N—Ö—N :Ö:  :Ö: | +5 | 11.3 | Colorless, volatile solid consisting of $NO_2^+$ and $NO_3^-$; gas consists of $N_2O_5$ molecules |

◆ **Nitric Oxide: A Biochemical Surprise.**
The tiny inorganic free radical NO has recently become a remarkable player in an enormous range of biochemical systems. NO is biosynthesized in animal species from barnacle to human. In the late 1980s and early 90s, its various roles have been shown to include neurtransmission, blood clotting, control of blood pressure, and the ability to destroy cancer cells. Recent evidence suggests that it may even be involved in memory. Living systems are full of surprises!

Dinitrogen monoxide ($N_2O$; also called dinitrogen oxide or nitrous oxide) is the dental anesthetic "laughing gas" and the propellant in canned whipped cream. It is a linear molecule with an electronic structure that is best described by three resonance forms (see Problem 9.62):

$$:N{\equiv}N{-}\ddot{\underset{..}{O}}: \leftrightarrow \ddot{N}{=}N{=}\ddot{\underset{..}{O}} \leftrightarrow :\ddot{N}{-}N{\equiv}O:$$

most important                                    negligible

Nitrogen monoxide (NO; also called nitrogen oxide or nitric oxide) is an odd-electron molecule. ◆ In Section 10.4, we used MO theory to explain its bond properties. The commercial preparation of NO occurs on the way to the production of nitric acid through the oxidation of ammonia:

$$4NH_3(g) + 5O_2(g) \rightarrow 4NO(g) + 6H_2O(g)$$

It is also produced whenever air is heated to high temperatures, as in a car engine or a lightning storm:

$$N_2(g) + O_2(g) \xrightarrow{\text{high } T} 2NO(g)$$

Heating converts NO to two other oxides:

$$3NO(g) \xrightarrow{\Delta} N_2O(g) + NO_2(g)$$

In this type of redox reaction, called a **disproportionation,** a substance with an atom in an intermediate oxidation state forms substances with the atom in both lower and higher states by reacting with itself: the oxidation state of N in NO (+2) is between that in $N_2O$ (+1) and that in $NO_2$ (+4). Thus, *NO acts as both an oxidizing and a reducing agent in the reaction.*

Nitrogen dioxide ($NO_2$), a brown poisonous gas, forms when NO reacts with additional oxygen:

$$2NO(g) + O_2(g) \rightarrow 2NO_2(g)$$

Like NO, $NO_2$ is also an odd-electron molecule, but the electron is more localized on the N atom, so $NO_2$ dimerizes reversibly to dinitrogen tetraoxide:

$$O_2N{\cdot}(g) + {\cdot}NO_2(g) \rightleftharpoons O_2N{-}NO_2(g) \text{ (or } N_2O_4)$$

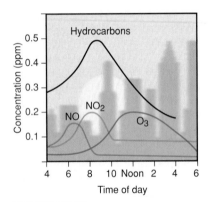

**FIGURE 13.20**
**The formation of photochemical smog.** Concentrations of the components of smog change with time of day. In early-morning traffic, NO and hydrocarbon levels increase, followed by $NO_2$, as NO reacts with air. Ozone peaks later as $NO_2$ breaks down in stronger sunlight to release O atoms that react with $O_2$. By midafternoon, all the components for production of peroxyacylnitrates (PANs) are present and result in photochemical smog.

Thunderstorms form NO and $NO_2$ and carry them down to the soil, where they act as natural fertilizers. In urban traffic, however, their formation leads to *photochemical smog* (Figure 13.20). Early in the morning, NO forms in car engines and is then oxidized to $NO_2$. As the sun climbs higher, radiant energy breaks down some $NO_2$:

$$NO_2(g) \xrightarrow{\text{sunlight}} NO(g) + O(g)$$

The O atoms collide with $O_2$ molecules and form ozone, $O_3$, a powerful oxidizing agent that damages synthetic rubber and plastic, as well as plant and animal tissue:

$$O_2(g) + O(g) \rightarrow O_3(g)$$

A complex series of reactions between $NO_2$, $O_3$, and unburned hydrocarbons in gasoline fumes forms *peroxyacylnitrates* (PANs), a group of potent nose and eye irritants. The outcome of this fascinating atmospheric chemistry is choking, brown smog.

**Nitrogen oxoacids and oxoanions.** The two common nitrogen oxoacids are nitric acid and nitrous acid (Figure 13.21). In the *Ostwald process* for the production of nitric acid, $NH_3$ is oxidized to NO, which is then oxidized to $NO_2$. The final step is a disproportionation:

$$3NO_2(g) + H_2O(l) \rightarrow 2HNO_3(l) + NO(g)$$

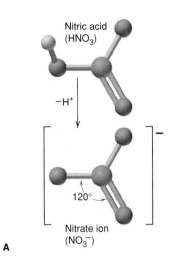

Nitric acid
(HNO$_3$)

$-H^+$

120°

Nitrate ion
(NO$_3^-$)

**A**

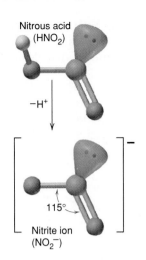

Nitrous acid
(HNO$_2$)

$-H^+$

115°

Nitrite ion
(NO$_2^-$)

**B**

**FIGURE 13.21**

**The structures of nitric and nitrous acids and their oxoanions. A,** Nitric acid loses an H$^+$ to form the trigonal planar nitrate ion (one of three resonance forms is shown). **B,** Nitrous acid, a much weaker acid, forms the planar nitrite ion. Note the effect of the lone pair in reducing the ideal 120° angle to 115° (one of two resonance forms is shown). Lone pairs are shown on N only.

In nitric acid, as in all oxoacids, *the acidic H is attached to one of the O atoms.* In the laboratory, nitric acid is used as a strong oxidizing acid. In reaction with metals, the products vary with the metal's reactivity and the HNO$_3$ concentration. With an active metal, such as Al, and dilute acid, N is reduced from the +5 state all the way to the −3 state:

$$8Al(s) + 30HNO_3(aq; <1\ M) \rightarrow 8Al(NO_3)_3(aq) + 3NH_4NO_3(aq) + 9H_2O(l)$$

With a less reactive metal, such as Cu, and more concentrated acid, N is reduced to the +2 state:

$$3Cu(s) + 8HNO_3(aq; 3\ to\ 6\ M) \rightarrow 3Cu(NO_3)_2(aq) + 4H_2O(l) + 2NO(g)$$

With more concentrated acid, N is reduced only to the +4 state:

$$Cu(s) + 4HNO_3(aq; 12\ M) \rightarrow Cu(NO_3)_2(aq) + 2H_2O(l) + 2NO_2(g)$$

Notice that in all cases, *the NO$_3^-$ ion is the oxidizing agent.* Nitrate ion that is not reduced appears as a spectator ion. Nitrates form when HNO$_3$ reacts with metals and with their hydroxides, oxides, or carbonates. *All nitrates are soluble in water.*

Nitrous acid, HNO$_2$, a much weaker acid than HNO$_3$, forms when metal nitrites are treated with a strong acid:

$$NaNO_2(aq) + HCl(aq) \rightarrow HNO_2(aq) + NaCl(aq)$$

These two acids reveal a *general pattern in relative acid strength among oxoacids:* the more O atoms bound to the central nonmetal, the stronger the acid. Thus, HNO$_3$ is stronger than HNO$_2$. The O atoms pull electron density from the N atom, which in turn pulls electron density from the O of the O—H bond, facilitating the release of the H$^+$ ion. The O atoms also act to stabilize the resulting oxoanion by delocalizing its negative charge. The same pattern occurs in the oxoacids of sulfur and the halogens; we'll discuss the pattern quantitatively in Chapter 17.

### Highlights of Phosphorus Chemistry: Oxides and Oxoacids

Phosphorus forms two important oxides, P$_4$O$_6$ and P$_4$O$_{10}$. Tetraphosphorus hexaoxide, P$_4$O$_6$, with P in its +3 oxidation state, forms when white P reacts with limited oxygen:

$$P_4(s) + 3O_2(g) \rightarrow P_4O_6(s)$$

**A** $P_4O_6$

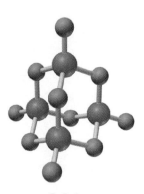

**B** $P_4O_{10}$

**FIGURE 13.22**

The structures of **(A)** $P_4O_6$ and **(B)** $P_4O_{10}$.

$P_4O_6$ has the tetrahedral orientation of the P atoms in $P_4$, with an O atom between each pair of P atoms (Figure 13.22, *A*). It reacts with water to form phosphor*ous* acid (note the spelling):

$$P_4O_6(s) + 6H_2O(l) \rightarrow 4H_3PO_3(l)$$

The formula $H_3PO_3$ is misleading because the acid has only two acidic H atoms; the third is bound to the central P and does not dissociate. Thus, a clearer formula would put the acidic H atoms first, as in $(HO)_2PHO$. It is a weak acid in water and reacts in two steps with excess strong base:

Salts of phosphorous acid contain the phosphite ion, $HPO_3^{2-}$.

In $P_4O_{10}$, P is in the +5 oxidation state. Commonly known as "phosphorus pentoxide" from the empirical formula $(P_2O_5)$ it forms when $P_4$ burns in excess $O_2$:

$$P_4(s) + 5O_2(g) \rightarrow P_4O_{10}(s)$$

Its structure can be viewed as $P_4O_6$ with another O atom bonded to each of the four corner P atoms (Figure 13.22, *B*). It is a powerful drying agent and, in a vigorous exothermic reaction with water, forms phosphoric acid $(H_3PO_4)$, one of the "Top-10" most important compounds in chemical manufacturing:

$$P_4O_{10}(s) + 6H_2O(l) \rightarrow 4H_3PO_4(l)$$

Due to the presence of many H bonds, pure $H_3PO_4$ is syrupy and more than 75 times as viscous as water. The laboratory-grade concentrated acid is an 85 mass % aqueous solution. $H_3PO_4$ is a weak triprotic acid; in water, it loses one proton in the following equilibrium reaction:

$$H_3PO_4(l) + H_2O(l) \rightleftharpoons H_2PO_4^-(aq) + H_3O^+(aq)$$

In excess strong base, however, the three protons dissociate completely in three steps to give the three phosphate oxoanions:

dihydrogen phosphate ion        hydrogen phosphate ion        phosphate ion

Phosphoric acid has a central role in fertilizer production, but it is also used as a polishing agent for aluminum car trim and as an additive in soft drinks to give a touch of tartness. Phosphates are essential in many chemical processes, and the various phosphate salts have numerous applications in industry. ◆

Polyphosphates are formed by heating hydrogen phosphates, which lose water as they form P—O—P linkages. For example, sodium diphosphate, $Na_4P_2O_7$, is prepared by heating sodium hydrogen phosphate:

$$2Na_2HPO_4(s) \xrightarrow{\Delta} Na_4P_2O_7(s) + H_2O(g)$$

◆ **The Countless Uses of Phosphates.** Phosphates have an amazing array of applications in home and industry. $Na_3PO_4$ is a paint stripper and grease remover and is still used in cleaning powders in other countries. $Na_2HPO_4$ is an emulsifier in the making of processed cheese, an additive to make ham juicier, and a starch modifier in "instant" puddings and cereals. $NaH_2PO_4$ is a laxative ingredient and is used to adjust the acidity of boiler water. The potassium salts have other uses. $K_3PO_4$ is used to stabilize latex for synthetic rubber; $K_2HPO_4$ is a radiator corrosion inhibitor. The ammonium salts are used as fertilizers and as flame-retardants on curtains and paper costumes. The various calcium phosphates are used in baking powders and toothpastes, as mineral supplements in stock feed, and (on the hundred-million-ton-scale) as fertilizers throughout the world.

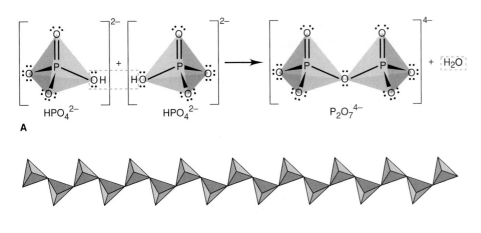

A

B

**FIGURE 13.23**

**The diphosphate ion and polyphosphates.** **A,** When two hydrogen phosphate ions undergo a dehydration-condensation reaction, they release a water molecule and join through a shared O atom to form a diphosphate ion. **B,** Polyphosphates consist of many such tetrahedral $PO_4$ units. Chain shapes depend on the orientation of the individual units. Note the similarity to silicate chains (see Gallery, Section 13.6).

The diphosphate ion, the smallest of the polyphosphates, consists of two $PO_4$ units "fused" through an oxygen corner (Figure 13.23, *A*). Polyphosphate chains consisting of tetrahedral $PO_4$ units (Figure 13.23, *B*) are structurally similar to silicate chains. This type of reaction, in which an $H_2O$ molecule is lost for every pair of OH groups that join, is called a **dehydration-condensation;** it occurs frequently in the formation of polyoxoanion chains and other polymeric structures, both synthetic and natural. As a final look at the chemical versatility of phosphorus, consider some of its numerous important compounds with sulfur and with nitrogen. ◆

## 13.8 Group 6A(16): The Oxygen Family

The first two members of this family—gaseous nonmetallic oxygen (O) and solid nonmetallic sulfur (S)—are among the most important elements in industry, the environment, and living things. Two metalloids, selenium (Se) and tellurium (Te), appear below them, and the lone metal, radioactive polonium (Po), ends the group. The Group 6A(16) Family Portrait (pages 560 and 561) displays the features of these elements.

### How Do the Oxygen and Nitrogen Families Compare Physically?

Group 6A(16) resembles Group 5A(15) in many respects, so we'll point out some common themes. The pattern of physical properties we saw in Group 5A appears again in this group. Like nitrogen, oxygen occurs as a low-boiling diatomic gas. Like phosphorus, sulfur occurs as a polyatomic molecular solid. Like arsenic, selenium commonly occurs as a gray metalloid. Like antimony, tellurium is slightly more metallic than the preceding group member but still displays network bonding. Finally, like bismuth, polonium has a metallic crystal structure. As we would expect by comparison with the 5A elements, electrical conductivities increase steadily down the group as bonding changes from individual molecules (insulators) to metalloid networks (semiconductors) to metallic solid (conductor).

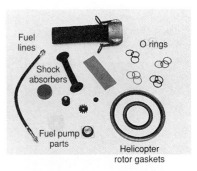

Fuel lines
Shock absorbers
Fuel pump parts
O rings
Helicopter rotor gaskets

◆ **Match Heads, Bug Sprays, and O-Rings.** Phosphorus forms many sulfides and nitrides. $P_4S_3$ is used in "strike-anywhere" match heads; $P_4S_{10}$ is used in the manufacture of organophosphorus pesticides, such as malathion. Polyphosphazenes have properties similar to those of silicones. Indeed, the —$(R_2)P$=N— unit is isoelectronic with the silicone —$(R_2)Si$—O— unit. Sheets, films, fibers, and foams are water repellent, flame and solvent resistant, and flexible at low temperature—perfect for gaskets and O-rings in spacecraft (shown above) and polar vehicles.

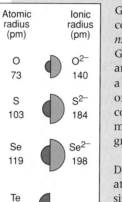

## FAMILY PORTRAIT

### Group 6A(16): The Oxygen Family
#### Key Atomic and Physical Properties

**KEY**

Atomic No.
**Symbol**
Atomic mass
Valence e⁻
configuration
Common
oxidation
states

$ns^2np^4$

**GROUP 6A(16)**

| 8 | |
|---|---|
| **O** | |
| 16.00 | |
| $2s^22p^4$ | |
| (−1, −2) | |

| 16 | |
|---|---|
| **S** | |
| 32.07 | |
| $3s^23p^4$ | |
| (−2, +6, +4, +2) | |

| 34 | |
|---|---|
| **Se** | |
| 78.96 | |
| $4s^24p^4$ | |
| (−2, +6, +4, +2) | |

| 52 | |
|---|---|
| **Te** | |
| 127.6 | |
| $5s^25p^4$ | |
| (−2, +6, +4, +2) | |

| 84 | |
|---|---|
| **Po** | |
| (210) | |
| $6s^26p^4$ | |
| (+4, +2) | |

## Atomic Properties

| Atomic radius (pm) | | Ionic radius (pm) | |
|---|---|---|---|
| O 73 | | O²⁻ 140 | |
| S 103 | | S²⁻ 184 | |
| Se 119 | | Se²⁻ 198 | |
| Te 142 | | | |
| Po 168 | | Po⁴⁺ 94 | |

Group electron configuration is $ns^2np^4$. As with Groups 3A (13) and 5A (15), a lower (+4) oxidation state becomes more common down the group.

Down the group, atomic and ionic size increase, IE and EN decrease.

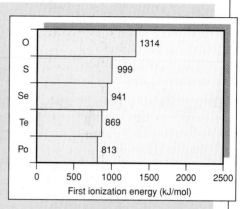

First ionization energy (kJ/mol)
- O 1314
- S 999
- Se 941
- Te 869
- Po 813

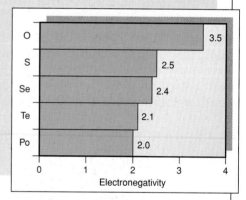

Electronegativity
- O 3.5
- S 2.5
- Se 2.4
- Te 2.1
- Po 2.0

## Physical Properties

Melting points increase through the network bonding of Te and then decrease in the metallic bonding of Po.

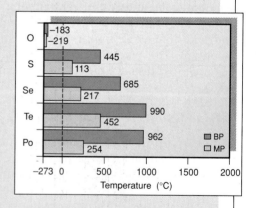

Temperature (°C) — BP / MP
- O −183 / −219
- S 445 / 113
- Se 685 / 217
- Te 990 / 452
- Po 962 / 254

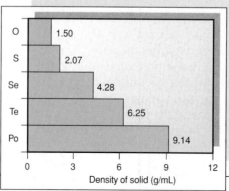

Density of solid (g/mL)
- O 1.50
- S 2.07
- Se 4.28
- Te 6.25
- Po 9.14

Densities of the elements as solids increase steadily.

# Group 6A(16): The Oxygen Family
## Some Reactions and Compounds

## Important Reactions

General halogenation and oxidation appears in reactions 1 and 2, and sulfur chemistry in reactions 3 and 4.

1. Halides are formed by direct combination:

   $$E(s) + X_2(g) \rightarrow \text{various halides}$$

   (E = S, Se, Te; X = F, Cl)

2. The other elements in the group are oxidized by $O_2$:

   $$E(s) + O_2(g) \rightarrow EO_2 \qquad (E = S, Se, Te, Po)$$

   $SO_2$ is oxidized further, and the product is used in the final step of $H_2SO_4$ manufacture (*see Highlights of Sulfur Chemistry*):

$$2SO_2(g) + O_2(g) \rightarrow 2SO_3(g)$$

3. Sulfur is recovered when hydrogen sulfide is oxidized:

   $$8H_2S(g) + 4O_2(g) \rightarrow S_8(s) + 8H_2O(g)$$

   This reaction is used to obtain sulfur when natural deposits are not available.

4. The thiosulfate ion is formed when an alkali sulfite reacts with sulfur, as in the preparation of photographer's "hypo":

   $$S_8(s) + 8Na_2SO_3(aq) \rightarrow 8Na_2S_2O_3(aq)$$

## Important Compounds

1. Water, $H_2O$. The single most important compound on Earth (*Section 11.6*).
2. Hydrogen peroxide, $H_2O_2$. Used as an oxidizing agent, disinfectant, bleach, and in the production of peroxy compounds for polymerization (*see margin note, p. 563*).
3. Hydrogen sulfide, $H_2S$. Vile-smelling toxic gas formed during anaerobic decomposition of plant and animal matter, in volcanoes, and in deep-sea thermal vents. Used as a sulfur source and in the manufacture of paper. Atmospheric traces cause silver to tarnish through formation of black $Ag_2S$.

   Tarnished spoon

4. Sulfur dioxide, $SO_2$. Colorless, choking gas formed in volcanoes or whenever an S-containing compound (coal, oil, metal sulfide ores, and so on) is burned. More than 90% of $SO_2$ produced is used to make sulfuric acid. Also used as a fumigant and preservative of fruit, syrups, and wine. As a reducing agent, removes excess $Cl_2$ from industrial waste water, removes $O_2$ from petroleum handling tanks, and prepares $ClO_2$ for bleaching paper. Major atmospheric pollutant in acid rain.
5. Sulfur trioxide ($SO_3$) and sulfuric acid ($H_2SO_4$). $SO_3$, formed from $SO_2$ over $V_2O_5$ catalysts, is then converted to sulfuric acid. The acid is the cheapest strong acid and is so widely used in industry that its production level is an indicator of a nation's economic strength. Strong dehydrating agent that

Volcano

removes water from any organic source (*see Highlights of Sulfur Chemistry*).
6. Sulfur hexafluoride, $SF_6$. Extremely inert gas used as an electrical insulator (*see Highlights of Sulfur Chemistry*).

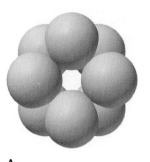

B

**FIGURE 13.24**

**The cyclo-$S_8$ molecule.** The most stable form of sulfur consists of octatomic cyclic molecules called cyclo-$S_8$. **A,** Top view of a space-filling model of the cyclo-$S_8$ molecule. **B,** Side-view ball-and-stick model of the molecule; note the crownlike shape.

A

◆ **Liquid and Plastic Sulfur.**
Sulfur undergoes remarkable changes when heated. At 112°C, the $\alpha$-$S_8$ form melts to a clear-yellow flowing liquid. By around 160°C, it has darkened to red orange and then deep red brown. $S_8$ rings break, and the —S· ends attack other rings and link into growing chains that then cross-link, causing the viscosity to rise. By about 200°C, chains of $6 \times 10^5$ S atoms occur, and the liquid barely flows. With further heating, chains break faster than they form, so that at the boiling point (445°C), the liquid flows more easily. When poured into cold water, the syrupy liquid becomes plastic sulfur, an amorphous mass of long, cross-linked S helices that impart a rubbery stretchiness.

◆ **Selenium and Xerography.** Photocopying, commonly available today, was invented in the 1930s and 40s as a rapid, inexpensive, dry means of copying documents (xerography, *Greek* "dry writing"). The process is based on the ability of selenium to photoconduct, that is, to conduct a current when illuminated. A film of amorphous Se is deposited on aluminum and electrostatically charged. Exposure to the document produces an "image" of low and high charges that correspond to the document's bright and dark lines. Black toner particles are attracted to the regions of high charge more than to those of low charge. This pattern of black particles is transferred electrostatically to print paper, and the particles are fused by heat or solvent. Excess toner is removed from the Se film, the charges are "erased" by exposure to light, and the film is ready for the page.

Allotropism is more common in Group 6A(16) than in Group 5A(15). Oxygen has two allotropes: life-giving dioxygen ($O_2$), and poisonous triatomic ozone ($O_3$). Oxygen gas is colorless, odorless, paramagnetic, and thermally stable. In contrast, ozone gas is bluish, has a pungent odor, is diamagnetic, and decomposes in heat and especially ultraviolet (UV) light:

$$2O_3(g) \xrightarrow{\text{UV}} 3O_2(g)$$

This molecular fragility in the presence of high-energy photons makes stratospheric ozone so vital to life. A thinning of the ozone layer, recently observed above the North and especially the South Poles, means that more UV light will reach the Earth's surface, with potentially hazardous effects. (We'll discuss the chemical causes of ozone depletion in Chapter 15.)

Sulfur is the allotrope "champion" of the periodic table, with more than 10 forms. The S atom's ability to bond to other S atoms creates numerous rings and chains, with S—S bond lengths that range from 180 pm to 260 pm and bond angles from 90° to 180°. At room temperature, the sulfur molecule is a crown-shaped ring of eight atoms, called *cyclo-$S_8$* (Figure 13.24). The most stable allotrope is orthorhombic $\alpha$-$S_8$, which consists entirely of these molecules; all other S allotropes eventually revert to this one. ◆

Selenium also has several allotropes, some consisting of crown-shaped $Se_8$ molecules. Gray Se is composed of layers of helical chains. Its electrical conductivity in visible light has revolutionized the photocopying industry. ◆ When molten glass, cadmium sulfide, and gray Se are mixed and heated in the absence of air, a ruby-red glass forms, which you see whenever you stop at a traffic light.

### How Do the Oxygen and Nitrogen Families Compare Chemically?

Changes in chemical behavior are also similar to those in the previous group. Even though they occur as anions much more often, nonmetallic O and S (like N and P) bond covalently with almost every other nonmetal. Covalent bonds appear in the compounds of Se and Te (as in those of As and Sb), whereas Po behaves like a metal (as does Bi) in some of its saltlike compounds. In contrast to nitrogen, oxygen has few common oxidation states, but the earlier pattern returns with the other members: the +6, +4, and −2 states occur most often, with the lower positive (+4) state becoming more common in Te and Po [as is the lower positive (+3) state in Sb and Bi].

The change in atomic properties is more extreme in this group than in 5A(15) because of oxygen's high EN (3.5) and great oxidizing strength, second only to that of fluorine. As in 5A and earlier groups, the behavior of the Period 2 element stands out. In fact, aside from a similar outer electron configuration, the larger 6A members behave very little like oxygen: they are much less electronegative, form anions much less often ($S^{2-}$ occurs with active metals), and exhibit no H bonding.

Except for O, all the 6A(16) elements form foul-smelling, poisonous, gaseous hydrides ($H_2E$) by treatment with acid of the metal sulfide, selenide, and so forth. For example,

$$FeSe(s) + 2HCl(aq) \rightarrow H_2Se(g) + FeCl_2(aq)$$

Hydrogen sulfide also forms naturally in swamps from the breakdown of organic matter. It is as toxic as HCN, and even worse, it anesthetizes your olfactory nerves, so that as its concentration increases, you smell it less! The other hydrides are about 100 times *more* toxic. (Organic compounds that contain Se are particularly dangerous, even though Se is an essential nutrient at much lower concentrations.)

In their bonding and thermal stability, these hydrides have several features in common with those of Group 5A:

- Only water can form H bonds, so it melts and boils much higher than the other $H_2E$ (see Figure 11.8). (Oxygen's other hydride is also extensively H bonded.) ◆
- Bond angles drop from the nearly tetrahedral value for $H_2O$ (104.5°) to around 90° for the larger hydrides, implying the central atom's use of unhybridized *p* orbitals.
- E—H bond length increases (bond energy decreases) down the group. Thus, $H_2Te$ decomposes above 0°C, and $H_2Po$ can be made only in extreme cold because thermal energy from the radioactive Po decomposes it. Another result of longer (weaker) bonds is that the 6A hydrides are acids in water, and their acidity increases from $H_2S$ to $H_2Po$.

Except for O, the Group 6A(16) elements form a wide range of halides, whose structure and reactivity patterns depend on the *sizes of the central atom and the surrounding halogens:*

- Sulfur forms many fluorides, a few chlorides, one bromide, but no stable iodides.
- As the central atom becomes larger, the halides become more stable. Thus, tetrachlorides and tetrabromides of Se, Te, and Po are known, as are tetraiodides of Te and Po. Hexafluorides are known only for S, Se, and Te.

The inverse relationship between bond length and bond strength that we've seen previously does not account for this pattern. Rather, it is based on the effect of electron repulsions due to crowding of lone pairs and halogen (X) atoms around the central 6A atom. With S, the larger X atoms become too crowded, which explains why sulfur iodides do not occur. With increasing size of E and therefore length of E—X bonds, however, lone pairs and X atoms do not crowd each other as much, and a greater number of stable halides form.

Two sulfur fluorides illustrate clearly how crowding and orbital availability affect reactivity. Sulfur tetrafluoride ($SF_4$) is extremely reactive. It forms $SO_2$ and HF when exposed to moisture and fluorinates many compounds:

$$3SF_4(g) + 4BCl_3(g) \rightarrow 4BF_3(g) + 3SCl_2(l) + 3Cl_2(g)$$

In contrast, sulfur hexafluoride ($SF_6$) is almost as inert as a noble gas! It is odorless, tasteless, nonflammable, nontoxic, and insoluble. Hot metals, boiling HCl, molten KOH, and high-pressure steam have no effect on it. It is used as an insulating gas in high-voltage generators, withstanding over

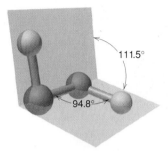

111.5°

94.8°

**◆ Hydrogen Peroxide: Hydrazine's Cousin in Group 6A(16).** Oxygen forms another hydride called hydrogen peroxide, $H_2O_2$ (HO—OH). Like hydrazine ($H_2N—NH_2$) in Group 5A(15), it has a skewed shape. It is a colorless liquid with a high density, viscosity, and boiling point due to extensive H bonding. In peroxides, O is in the −1 oxidation state, midway between that in $O_2$ (zero) and in oxides (−2); thus, $H_2O_2$ readily disproportionates:

$$H_2O_2(l) \rightarrow H_2O(l) + \tfrac{1}{2}O_2(g)$$
$$\Delta H^0 = -98 \text{ kJ}$$

Aside from its familiar use as a hair bleach and disinfectant, more than 70% of the half-million tons of $H_2O_2$ produced each year is used to bleach paper pulp, textiles, straw, and leather and to make other chemicals. $H_2O_2$ is now also used in tertiary sewage treatment (Section 12.6) to oxidize foul-smelling effluents and restore $O_2$ to waste water.

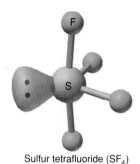

Sulfur tetrafluoride ($SF_4$)

Sulfur hexafluoride ($SF_6$)

**FIGURE 13.25**
**Structural differences between $SF_4$ and $SF_6$.**
$SF_4$ has a lone pair and empty $d$ orbitals that can become involved in bonding. $SF_6$ already has the maximum number of bonds formed by S, and the closely packed F atoms envelop it, rendering $SF_6$ chemically inert.

♦ **Acid from the Sky.** Awareness of $SO_2$ as a major air pollutant is now widespread. Even though enormous amounts of it form during volcanic and other geothermal activity, they are dwarfed by the amounts formed by human sources: coal-based power plants, petroleum refineries, and metal-ore smelters. In the atmosphere, $SO_2$ reacts with $O_2$ and water to form sulfuric acid, which rains, snows, and dusts down on animals, plants, buildings, and lakes. The destructive effects of acid precipitation are being intensively studied, and remedial action is under way (Chapter 18).

$10^6$ volts across electrodes only 50 mm apart. A look at the structures of these two fluorides provides the key to these astounding differences (Figure 13.25). $SF_4$ can form bonds to another atom by either donating its lone pair or accepting a lone pair into one of its empty $d$ orbitals. On the other hand, $SF_6$ has no central lone pair, and the six F atoms form an octahedral sheath around S that blocks chemical attack.

### Highlights of Oxygen Chemistry: Range of Oxide Properties

Oxygen is the most abundant element on the Earth's surface, occurring both as the free element and in innumerable oxides, silicates, carbonates, and phosphates, as well as in water. Virtually all free $O_2$ has a biological origin, formed by photosynthetic plants in an overall equation that looks deceptively simple:

$$n H_2O(l) + n CO_2(g) \xrightarrow{\text{light}} n O_2(g) + (CH_2O)_n \text{ (carbohydrates)}$$

The reverse process occurs during combustion and respiration. Through these $O_2$-forming and $O_2$-utilizing processes, the $1.5 \times 10^9$ km³ of water on Earth is, on average, used and remade every 2 million years!

Every element (except He, Ne, and Ar) forms at least one oxide, many by direct combination. A broad spectrum of properties characterizes these compounds. Some oxides are gases that condense at very low temperatures, such as CO (bp = $-192$°C); others are solids that melt at extremely high temperatures, such as BeO (mp = 2530°C). They cover the full range of conductivity: insulators (MgO), semiconductors (NiO), conductors ($ReO_3$), and superconductors ($YBa_2Cu_3O_7$). They may have endothermic heats of formation ($\Delta H_f^0$ of NO = $+90.3$ kJ/mol) or exothermic ones ($\Delta H_f^0$ of $CO_2 = -393.5$ kJ/mol), may be thermally stable (CaO) or unstable (HgO), and may be chemically reactive ($Li_2O$) or inert ($Fe_2O_3$).

Given this vast range of behavior, element oxides are generally classified by their acid-base properties (see Interchapter, Topic 4). The oxides of Group 6A(16) exhibit expected trends in acidity, with $SO_3$ the most acidic and $PoO_2$ the most basic.

### Highlights of Sulfur Chemistry: Oxides, Oxoacids, and Sulfides

Like phosphorus, sulfur forms two important oxides, sulfur dioxide ($SO_2$) and sulfur trioxide ($SO_3$). Sulfur is in its +4 oxidation state in $SO_2$, a colorless, choking gas that forms whenever S, $H_2S$, or a metal sulfide burns and reacts with $O_2$ in air: ♦

$$2H_2S(g) + 3O_2(g) \rightarrow 2H_2O(g) + 2SO_2(g)$$

$$4FeS_2(s) + 11O_2(g) \rightarrow 2Fe_2O_3(s) + 8SO_2(g)$$

Most $SO_2$ is used to produce sulfuric acid.

In water, sulfur dioxide forms sulfurous acid, which exists as hydrated $SO_2$ rather than stable $H_2SO_3$ molecules:

$$SO_2(aq) + H_2O(l) \rightleftharpoons [H_2SO_3] \rightleftharpoons H^+(aq) + HSO_3^-(aq)$$

(Similarly, carbonic acid occurs as hydrated $CO_2$, not $H_2CO_3$ molecules.) Sulfurous acid is weak and has two acidic protons, forming the bisulfite (hydrogen sulfite, $HSO_3^-$) and sulfite ($SO_3^{2-}$) ions with strong base. Because the S in $SO_3^{2-}$ is in the +4 state and easily oxidized to the +6 state, sulfites are good reducing agents. As such, they are used to preserve foods and wine by eliminating undesirable products of air oxidation.

En route to sulfuric acid, $SO_2$ is first oxidized to $SO_3$ (S in the +6 state) by heating in $O_2$ over a catalyst:

$$SO_2(g) + \frac{1}{2}O_2(g) \xrightleftharpoons{\text{V}_2\text{O}_5/\text{K}_2\text{O catalyst}} SO_3(g)$$

(We discuss how catalysts work in Chapter 15 and examine the energy involved in $H_2SO_4$ production in Chapter 23.) The $SO_3$ is absorbed into concentrated $H_2SO_4$ and treated with additional $H_2O$:

$$SO_3(\text{in concentrated } H_2SO_4) + H_2O(l) \rightarrow H_2SO_4(l)$$

With more than 40 million tons produced each year in the United States alone, $H_2SO_4$ ranks first among all industrial chemicals. Fertilizer production; metal, pigment, and textile processing; and soap and detergent manufacturing are just a few of the major industries that depend on sulfuric acid.

The concentrated laboratory-grade sulfuric acid is a viscous, colorless liquid that is 98% $H_2SO_4$ by mass. Like other strong acids, $H_2SO_4$ dissociates completely in water, forming the bisulfate (or hydrogen sulfate) ion, which is a much weaker acid:

$$
\begin{array}{ccc}
\text{HO}\overset{\displaystyle :O:}{\underset{\displaystyle :OH}{S}}\!\!=\!\!O & \xrightarrow{-H^+} & \left[\text{HO}\overset{\displaystyle :O:}{\underset{\displaystyle :O:}{S}}\!\!=\!\!O\right]^- \xrightleftharpoons{-H^+} \left[:O\overset{\displaystyle :O:}{\underset{\displaystyle :O:}{S}}\!\!=\!\!O\right]^{2-} \\
& & \text{bisulfate ion} \qquad\qquad \text{sulfate ion}
\end{array}
$$

There are many common bisulfates and sulfates, most are water soluble except for those of Group 2A(2) (except $MgSO_4$), $Pb^{2+}$, and $Hg_2^{2+}$.

Concentrated sulfuric acid is an excellent dehydrating agent. Its loosely held proton abstracts water in a highly exothermic conversion to hydronium ions. This process can occur even when the reacting substance contains no free water. For example, $H_2SO_4$ will dehydrate wood, natural fibers, and many other organic substances by removing the components of water from the molecular structure, leaving a carbonaceous mass (Figure 13.26).

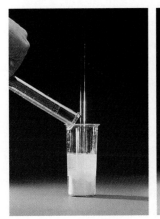

**FIGURE 13.26**

**The dehydration of carbohydrates by sulfuric acid.** The protons of concentrated $H_2SO_4$ combine exothermically with water, dehydrating many organic materials. When table sugar is treated with sulfuric acid, the components of water are removed from the carbohydrate molecules $(CH_2O)_n$. Steam forms, and the remaining carbon expands to a porous mass.

**FIGURE 13.27**
**A common sulfide mineral.** Pyrite ($FeS_2$), or fool's gold, is a beautiful but relatively cheap mineral that contains the $S_2^{2-}$ ion, but no gold.

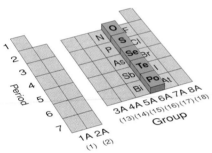

**FIGURE 13.28**
Standing in Group 6A(16), looking backward to Group 5A(15) and forward to Group 7A(17).

Thiosulfuric acid ($H_2S_2O_3$) is a structural analog of sulfuric acid in which a second S substitutes for one of the O atoms (*thio-* means S in place of O). The thiosulfate ion ($S_2O_3^{2-}$) is an important reducing agent whose largest commercial use is in the photography studio, where sodium thiosulfate pentahydrate ($Na_2S_2O_3 \cdot 5H_2O$), known as "hypo," is involved in developing the image (Section 22.3).

Many metals combine directly with S to form *metal sulfides.* Indeed, naturally occurring sulfides are ores of many metals, including copper, zinc, lead, and silver. Aside from the sulfides of Groups 1A(1) and 2A(2), most metal sulfides do not have discrete $S^{2-}$ ions. Several transition metals, such as chromium, iron, and nickel, form covalent, alloylike, nonstoichiometric compounds with S, such as $Cr_{0.88}S$ or $Fe_{0.86}S$. Some important minerals contain $S_2^{2-}$ units in the structure, such as iron pyrite, or "fool's gold" ($FeS_2$) (Figure 13.27). We discuss the metallurgy of ores in Chapter 23.

### Looking Backward and Forward: Groups 5A(15), 6A(16), and 7A(17)

Groups 5A(15) and 6A(16) are very similar in their physical and chemical trends and in the versatility of phosphorus and sulfur (Figure 13.28). Their greatest difference is the sluggish behavior of $N_2$ compared with the striking reactivity of $O_2$. In both groups, metallic character appears only in the largest members. From here on, metals and even metalloids are left behind: all the 7A(17) elements are reactive nonmetals. Anion formation, which was rare in 5A but more common in 6A, is one of the dominant features of 7A. Another feature is the number of covalent compounds with oxygen *and* with each other.

## 13.9 • Group 7A(17): The Halogens

Our last chance to view elements of great reactivity occurs in Group 7A(17). The halogens begin with fluorine (F), the strongest electron "grabber" of all. Chlorine (Cl), bromine (Br), and iodine (I) also form compounds with most elements, and even extremely rare astatine (At) is thought to be reactive. The key features of the halogens are presented in the Group 7A(17) Family Portrait (pp. 568 and 569).

### What Accounts for the Regular Change in Halogen Physical Properties?

Like the alkali metals at the other end of the periodic table, the halogens display regular trends in their physical properties. However, whereas melting and boiling points and heats of fusion and vaporization *decrease* down Group 1A(1), these properties *increase* down Group 7A(17) (see Interchapter, Topic 6). The reason for the opposite trends is the different type of bonding in the element. The alkali metals consist of atoms held together by metallic bonding, which *decreases* in strength as the atoms become larger. The halogens, on the other hand, exist as diatomic molecules that interact through dispersion forces, which *increase* in strength as the atoms become larger; thus, $F_2$ is a faint yellow gas, $Cl_2$ a yellow-green gas, $Br_2$ a brown-orange liquid, and $I_2$ a purple-black solid.

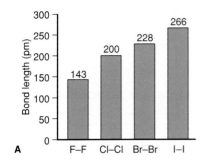

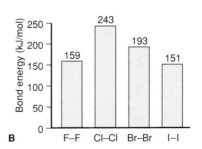

**FIGURE 13.29**

**Bond energies and bond lengths of the halogens. A,** In keeping with the increase in atomic size down the group, the bond length increases steadily. **B,** The halogens show a general decrease in bond energy as bond length increases. However, $F_2$ deviates from this trend because its small, close, electron-rich atoms repel each other, thereby lowering its bond energy.

## Why Are the Halogens So Reactive?

The Group 7A(17) elements react with most metals and nonmetals to form many ionic and covalent compounds: metal and nonmetal halides, halogen oxides, and oxoacids. The reason for halogen reactivity is the same as that for alkali metal reactivity: an electron configuration one electron away from that of a noble gas. Whereas a 1A metal atom must *lose* one electron to attain a filled outer shell, *a 7A nonmetal atom must gain one electron to fill its outer shell*. It accomplishes this filling in two ways:

1. Gaining an electron from a metal atom, thus forming a negative ion as the metal forms a positive one
2. Sharing an electron pair with a nonmetal atom, thus forming a covalent bond

Within the group, reactivity reflects the trend in electronegativity: $F_2$ is the most reactive and $I_2$ the least. The ability of atomic F to attract electrons is unsurpassed, but the exceptional reactivity of elemental $F_2$ is also related to the weakness of the F—F bond. Even though the bond is short, the small, electron-rich F atoms repel each other and make the bond easy to break (Figure 13.29). As a result of these factors, $F_2$ reacts with every element (except He, Ne, and Ar), in many cases explosively.

The halogens display the largest range in electronegativity of any group, but all are electronegative enough to behave as nonmetals. They act as *oxidizing agents* in the majority of their reactions, and halogens higher in the group can oxidize halide ions lower down:

$$F_2(g) + 2X^-(aq) \rightarrow 2F^-(aq) + X_2(aq) \qquad (X = Cl, Br, I)$$

Thus, the oxidizing ability of $X_2$ *decreases* down the group: the lower the EN, the less strongly it pulls electrons. Similarly, the reducing ability of $X^-$ *increases* down the group: the larger the ion, the more easily it gives up its electron (Figure 13.30).

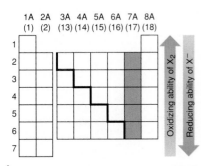

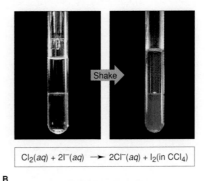

$$Cl_2(aq) + 2I^-(aq) \rightarrow 2Cl^-(aq) + I_2(in\ CCl_4)$$

**A**                                                                    **B**

**FIGURE 13.30**

**The relative oxidizing ability of the halogens. A,** Halogen redox behavior is based on atomic properties such as electron affinity and electronegativity. A halogen ($X_2$) higher in the group can oxidize a halide ion ($X^-$) lower down. **B,** As an example, when aqueous $Cl_2$ is added to a solution of $I^-$ *(top layer)*, it oxidizes the $I^-$ to $I_2$, which dissolves in the $CCl_4$ solvent *(bottom layer)* to give a purple solution.

**FAMILY PORTRAIT**

## Group 7A(17): The Halogens
### Key Atomic and Physical Properties

**KEY**

Atomic No.
**Symbol**
Atomic mass
Valence e⁻
configuration
Common
oxidation
states

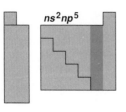

$ns^2np^5$

**GROUP 7A(17)**

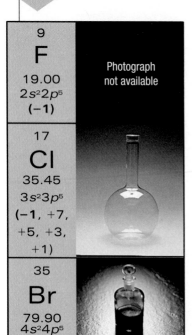

| 9 | |
|---|---|
| **F** | Photograph not available |
| 19.00 | |
| $2s^2 2p^5$ | |
| (−1) | |

| 17 | |
|---|---|
| **Cl** | |
| 35.45 | |
| $3s^2 3p^5$ | |
| (−1, +7, +5, +3, +1) | |

| 35 | |
|---|---|
| **Br** | |
| 79.90 | |
| $4s^2 4p^5$ | |
| (−1, +7, +5, +1) | |

| 53 | |
|---|---|
| **I** | |
| 126.9 | |
| $5s^2 5p^5$ | |
| (−1, +7, +5, +1) | |

| 85 | |
|---|---|
| **At** | Extremely rare, no sample available |
| (210) | |
| $6s^2 6p^5$ | |
| (−1) | |

## Atomic Properties

| Atomic radius (pm) | Ionic radius (pm) |
|---|---|
| F 72 | F⁻ 133 |
| Cl 100 | Cl⁻ 181 |
| Br 114 | Br⁻ 196 |
| I 133 | I⁻ 220 |
| At (140) | no data |

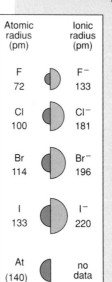

Group electron configuration is $ns^2np^5$; elements lack one e⁻ to complete their outer shell.
The -1 oxidation state is the most common for all members.
Except for F, the halogens exhibit all odd-numbered states (+7 through -1).

Down the group, atomic and ionic size increase steadily, just as IE and EN decrease.

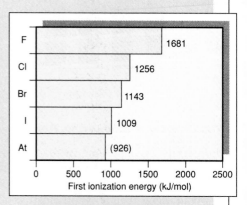

First ionization energy (kJ/mol)

F 1681
Cl 1256
Br 1143
I 1009
At (926)

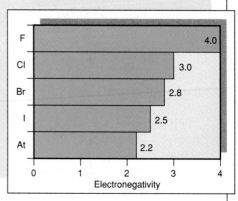

Electronegativity

F 4.0
Cl 3.0
Br 2.8
I 2.5
At 2.2

## Physical Properties

Phase-change temperatures and enthalpies exhibit smooth trends. Down the group, melting and boiling points increase as a result of stronger dispersion forces between heavier molecules.

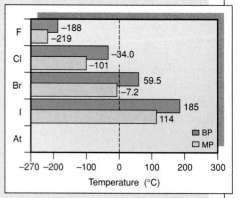

Temperature (°C)

F −188 / −219
Cl −34.0 / −101
Br 59.5 / −7.2
I 185 / 114
At

□ BP  □ MP

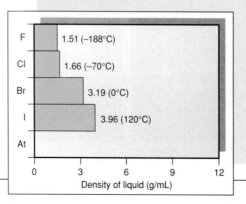

Density of liquid (g/mL)

F 1.51 (−188°C)
Cl 1.66 (−70°C)
Br 3.19 (0°C)
I 3.96 (120°C)
At

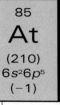

The densities of the elements as liquids (at given $T$) increase steadily with molar mass.

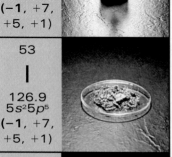

# Group 7A(17): The Halogens
## Some Reactions and Compounds

## Important reactions

Oxidizing strength and aqueous redox chemistry are shown in reactions 1 and 2 and industrial processes involving fluorine in reactions 3 and 4.

1. The halogens ($X_2$) oxidize many metals and non-metals. The reaction with hydrogen, although not used commercially for HX production (except for high-purity HCl), is characteristic of these strong oxidizing agents:

$$X_2 + H_2(g) \rightarrow 2HX(g)$$

2. The halogens disproportionate in water:

$$X_2 + H_2O(l) \rightleftharpoons HX(aq) + HXO(aq)$$

$$(X = Cl, Br, I)$$

In aqueous base, the reaction goes to completion to form hypohalites (see text) and, at higher temperatures, halates; for example:

$$3Cl_2(g) + 6OH^-(aq) \rightarrow$$
$$ClO_3^-(aq) + 5Cl^-(aq) + 3H_2O(l)$$

3. $F_2$ is produced electrolytically at moderate temperature:

$$2HF \text{ (as } KHF_2\text{, a solution of KF in HF)} \xrightarrow[90°C]{\text{electrolysis}}$$
$$H_2(g) + F_2(g)$$

A major use of $F_2$ is in the preparation of $UF_6$ for nuclear fuel.

4. Glass (amorphous silica) is etched with HF:

$$SiO_2(s) + 6HF(g) \rightarrow H_2SiF_6(aq) + 2H_2O(l)$$

## Important compounds

1. Fluorspar (fluorite), $CaF_2$. Widely distributed mineral used as a flux in steel making and in the production of HF.
2. Hydrogen fluoride, HF. Colorless, extremely toxic gas used to make $F_2$, organic fluorine compounds, and polymers. Also used in aluminum manufacture and in glass etching (*see margin note, p. 570*).

3. Hydrogen chloride, HCl. Extremely water-soluble gas that forms hydrochloric acid, which occurs naturally in stomach juice of mammals (humans produce 1.5L of 0.1 $M$ HCl daily) and in volcanic gases (from reaction of $H_2O$ on sea salt). Made by reaction of NaCl and $H_2SO_4$ and as a by-product of plastics (PVC) production. Used in the "pickling" of steel (removal of adhering oxides) and in the production of syrups, rayon, and plastic.
4. Sodium hypochlorite, NaClO, and calcium hypochlorite, $Ca(ClO)_2$. Oxidizing agents used to bleach wood pulp and textile, and disinfect swimming pools, foods, and sewage (also used to disinfect the Apollo 11 on return from the moon). Household bleach is 5.25% NaClO by mass in water.
5. Ammonium perchlorate, $NH_4ClO_4$. Strong oxidizing agent used in the Space Shuttle program.
6. Potassium iodide, KI. Most common soluble iodide. Table salt additive to prevent thyroid disease (goiter). Used in chemical analysis because it is easily oxidized to $I_2$, which forms a colored end point.
7. Polychlorinated biphenyls, PCBs. Mixture of chlorinated organic compounds used as nonflammable insulating liquids in electrical transformers. Production discontinued due to persistence in the environment, where it becomes concentrated in fish, birds, and mammals, and causes reproductive disturbances and possibly cancer.

Space Shuttle Endeavour

The halogens undergo some important aqueous redox chemistry. Fluorine is such a powerful oxidizing agent that it tears water apart, oxidizing the O to produce $O_2$, some $O_3$, and HFO (hypofluorous acid). The other halogens disproportionate:

$$\overset{0}{X}_2 + H_2O(l) \rightleftharpoons \overset{-1}{HX}(aq) + \overset{+1}{HXO}(aq) \qquad (X = Cl,\ Br,\ I)$$

At equilibrium, very little product is present unless excess $OH^-$ ion is added, which reacts with the HX and HXO and drives the reaction to completion:

$$X_2 + 2OH^-(aq) \rightarrow X^-(aq) + XO^-(aq) + H_2O(l)$$

When X is Cl, the product mixture acts as a bleach; household bleach is a dilute solution of sodium hypochlorite (NaClO). Heating causes $XO^-$ to disproportionate further, creating higher oxoanions:

$$3\overset{+1}{XO}^-(aq) \overset{\Delta}{\rightarrow} 2\overset{-1}{X}^-(aq) + \overset{+5}{XO}_3{}^-(aq)$$

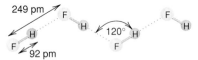

**◆ HF: Unusual Structure, Familiar Uses.** At room temperature, all the hydrogen halides except HF are diatomic gases. Because of extensive H bonding, gaseous HF becomes H bonded into short chains or rings of formula $(HF)_6$; liquid HF boils at 19.5°C, more than 50 degrees higher than even the much heavier HI; and solid HF exists as an H-bonded polymer. HF has many uses, including the synthesis of cryolite $(Na_3AlF_6)$ for aluminum production, fluorocarbons for refrigeration, and NaF for water fluoridation; nuclear fuel processing; and glass etching in the manufacture of light bulbs and TV tubes.

### Highlights of Halogen Chemistry

In this section, we examine the compounds that halogens form with hydrogen and with each other, as well as the oxides, oxoanions, and oxoacids.

**The hydrogen halides.** The halogens form gaseous hydrogen halides (HX) through direct combination with $H_2$ or through the action of a concentrated acid on the metal halide (a nonoxidizing acid is used for HBr and HI): ◆

$$CaF_2(s) + H_2SO_4(l) \rightarrow CaSO_4(s) + 2HF(g)$$
$$3NaBr(s) + H_3PO_4(l) \rightarrow Na_3PO_4(s) + 3HBr(g)$$

Most commercial HCl is formed as a byproduct in the chlorination of hydrocarbons for plastics production:

$$CH_2{=}CH_2(g) + Cl_2(g) \rightarrow ClCH_2CH_2Cl(l) \xrightarrow{500°C} \underset{\text{vinyl chloride}}{CH_2{=}CHCl(g)} + HCl(g)$$

In water, gaseous HX molecules form a *hydrohalic acid*. Only HF, with its relatively short, strong bond, forms a weak acid; the others dissociate completely to form the stoichiometric amount of hydronium ions:

$$HBr(g) + H_2O(l) \rightarrow H_3O^+(aq) + Br^-(aq)$$

(This reaction, which involves transfer of a proton from HBr to $H_2O$, is an example of a *Brønsted-Lowry acid-base reaction;* we discuss it thoroughly in Chapter 17 and examine the relation between bond length and acidity of the larger HX molecules as well.)

**Interhalogen compounds: the "halogen halides."** Halogens react exothermically with one another to form many **interhalogen compounds.** The simplest are diatomic molecules, such as ClF or BrCl. Every binary combination of the four halogens is known. The more electronegative element is in the $-1$ oxidation state, whereas the less electronegative is in the $+1$ state. Interhalogens of general formula $XY_n$ ($n = 3, 5, 7$) form when the larger members use *d* orbitals to expand their valence shells. In every case, the central atom has the *lower electronegativity* and a positive oxidation state.

The commercially useful interhalogens are powerful *fluorinating agents,* some of which react with metals, nonmetals, and oxides—even wood and asbestos:

$$Sn(s) + ClF_3(l) \rightarrow SnF_2(s) + ClF(g)$$

$$P_4(s) + 5ClF_3(l) \rightarrow 4PF_3(g) + 3ClF(g) + Cl_2(g)$$

$$2B_2O_3(s) + 4BrF_3(l) \rightarrow 4BF_3(g) + 2Br_2(l) + 3O_2(g)$$

Their reactions with water are nearly explosive and yield HF and *the oxoacid with the same oxidation state as the central halogen.* For example,

$$3H_2O(l) + \overset{+5}{Br}F_5(l) \rightarrow 5HF(g) + H\overset{+5}{Br}O_3(aq)$$

**The oddness and evenness of oxidation states.** Topic 5 of the Interchapter shows that odd-numbered groups exhibit odd-numbered oxidation states and that even-numbered groups exhibit even states. The reason for this general behavior is that almost all stable molecules have paired electrons, either bonding or lone pairs, and thus tend to be diamagnetic. Therefore, *when bonds form or break, two electrons are generally involved, so the oxidation state changes by two.*

Consider the case of the interhalogens. Four general formulas are known: XY, $XY_3$, $XY_5$, and $XY_7$. With Y in the $-1$ state, X must be in the $+1$, $+3$, $+5$, and $+7$ state, respectively. The $-1$ state arises when Y fills its valence level; the $+7$ state arises when the central halogen is completely oxidized, that is, all seven valence electrons have shifted away from it. The iodine fluorides show why the oxidation states jump by two units. When $I_2$ reacts with $F_2$, IF forms (focus on the oxidation number, O.N., of I):

$$I_2 + F_2 \rightarrow 2\overset{+1}{I}F$$

In $IF_3$, I uses *two* more valence electrons to form *two* more bonds; otherwise, an unstable lone-electron (paramagnetic) species containing two fluorines would form:

$$\overset{+1}{I}F + F_2 \rightarrow \overset{+3}{I}F_3$$

With more fluorine, another jump of two units occurs and the pentafluoride forms:

$$\overset{+3}{I}F_3 + F_2 \rightarrow \overset{+5}{I}F_5$$

With still more fluorine, the heptafluoride forms:

$$\overset{+5}{I}F_5 + F_2 \rightarrow \overset{+7}{I}F_7$$

An element in an even-numbered group, such as sulfur in Group 6A(16), shows the same tendency for its compounds to have paired electrons. Elemental sulfur (O.N. = 0) gains two electrons to complete its shell (O.N. = $-2$). It uses two electrons to react with fluorine, for example, and form $SF_2$ (O.N. = $+2$), two more electrons for $SF_4$ (O.N. = $+4$), and two more for $SF_6$ (O.N. = $+6$). Thus, an element with one even state typically has all even states, and an element with one odd state typically has all odd states. In general, *successive oxidation states differ by two units because stable molecules have electrons in pairs around their atoms.*

XY
Linear
ClF

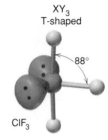
$XY_3$
T-shaped
88°
$ClF_3$

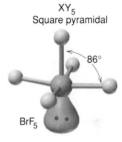
$XY_5$
Square pyramidal
86°
$BrF_5$

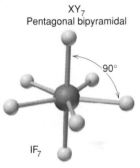
$XY_7$
Pentagonal bipyramidal
90°
$IF_7$

**FIGURE 13.31**
Molecular shapes of the main types of interhalogen compounds.

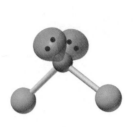

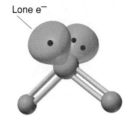

Lone e⁻

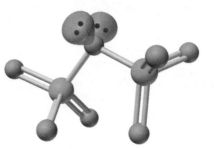

Dichlorine monoxide ($Cl_2O$)     Chlorine dioxide ($ClO_2$)     Dichlorine heptaoxide ($Cl_2O_7$)

A

B

**FIGURE 13.32**

**Chlorine oxides. A,** Dichlorine monoxide and chlorine dioxide are bent molecules; note the lone electron in $ClO_2$. Dichlorine heptaoxide can be viewed as two $ClO_4$ tetrahedra joined through an O corner. (The ball-and-stick models are drawn so each atom has its lowest formal charge. Lone pairs are shown on the central atoms only.) **B,** Several chlorine oxides are important reactants in the bleaching of paper. Here, a piece of brown paper bag is bleached.

**Halogen oxides, oxoacids, and oxoanions.** The Group 7A(17) elements form many oxides that are *powerful oxidizing agents and acids in water.* Dichlorine monoxide ($Cl_2O$) and especially chlorine dioxide ($ClO_2$) are used to bleach paper (Figure 13.32). Despite its instability to heat and shock, more than 100,000 tons of $ClO_2$ are used annually because it can be prepared on site:

$$2NaClO_3(s) + SO_2(g) + H_2SO_4(aq) \rightarrow 2ClO_2(g) + 2NaHSO_4(aq)$$

Note that the dioxide has a lone electron and Cl in the unusual +4 oxidation state.

In dichlorine heptaoxide, $Cl_2O_7$, Cl is in its highest (+7) oxidation state. It is a symmetrical molecule formed when two $HClO_4$ (HO—$ClO_3$) molecules undergo a dehydration-condensation reaction:

$$O_3Cl—O\boxed{H + HO}—ClO_3 \rightarrow O_3Cl—O—ClO_3(l) + H_2O(l)$$

The halogen oxoacids and oxoanions are produced from reaction of the halogens and their oxides with water. Most of the oxoacids are stable only in solution. Table 13.4 shows the ball-and-stick models with atoms in their lowest formal charge and formulas that emphasize that H is bonded to O. The hypohalites ($XO^-$), halites ($XO_2^-$), and halates ($XO_3^-$) are oxidizing agents formed by aqueous disproportionation reactions [see Group 7A(17) Family Portrait, reaction 2]. You may have heated solid alkali chlorates in the laboratory to form small amounts of $O_2$:

$$2MClO_3(s) \xrightarrow{\Delta} 2MCl(s) + 3O_2(g)$$

The potassium salt is the oxidizer in "safety" matches.

Several perhalates are also strong oxidizing agents. ◆ Ammonium perchlorate, prepared from sodium perchlorate, is the oxidizing agent for the aluminum powder in the solid-fuel booster rocket of the Space Shuttle; each launch uses more than 700 tons of $NH_4ClO_4$:

$$10Al(s) + 6NH_4ClO_4(s) \rightarrow 4Al_2O_3(s) + 12H_2O(g) + 3N_2(g) + 2AlCl_3(g)$$

The relative strengths of the halogen oxoacids depend on two factors:

1. *Electronegativity of the halogen.* Among oxoacids in the same oxidation state, such as the halic acids, $HXO_3$ (or $HOXO_2$), acid strength decreases as the halogen EN decreases:

$$HOClO_2 > HOBrO_2 > HOIO_2$$

The more electronegative the halogen, the more electron density it removes from the O—H bond, and the more easily the proton is lost.

◆ **Pyrotechnic Perchlorates.** Thousands of tons of perchlorates are made each year for use in explosives and fireworks. The white flash and thundering boom of a fireworks display are caused by $KClO_4$ reacting with powdered sulfur and aluminum. In rock concerts and other theatrical productions, mixtures of $KClO_4$ and Mg are often used for special effects.

**TABLE 13.4   The Known Halogen Oxoacids**

| CENTRAL ATOM | HYPOHALOUS ACID (HOX) 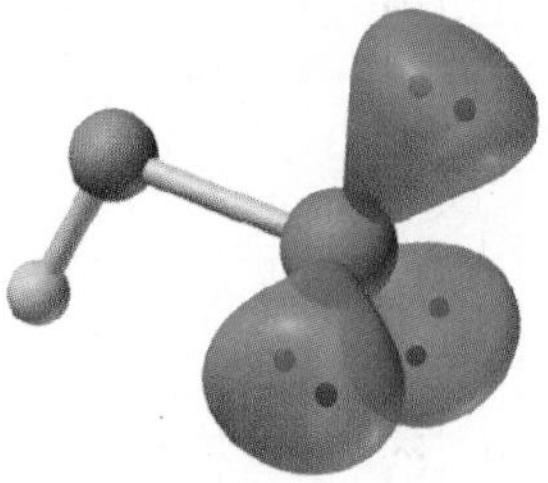 | HALOUS ACID (HOXO) 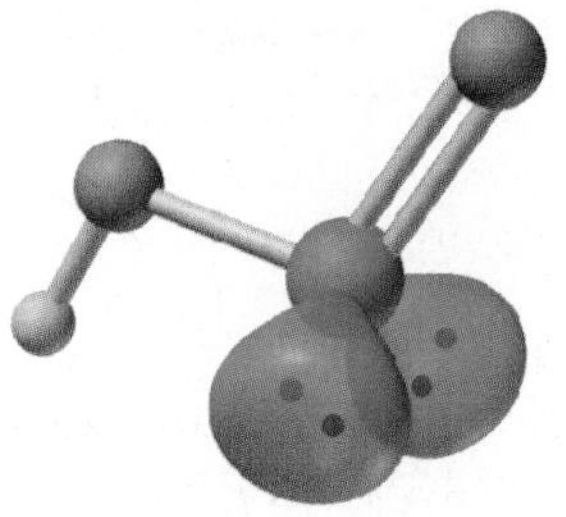 | HALIC ACID (HOXO$_2$) 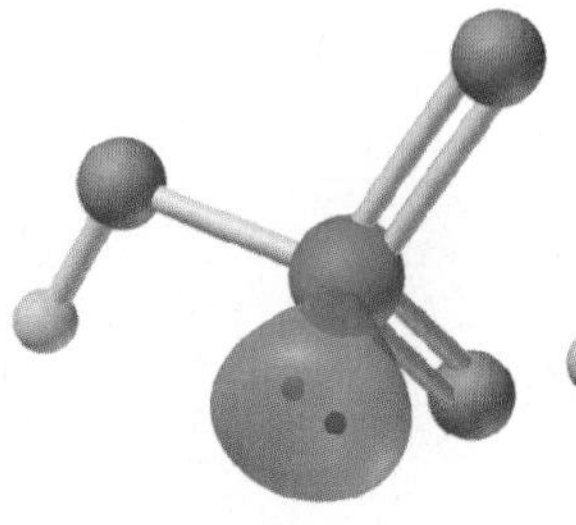 | PERHALIC ACID (HOXO$_3$) 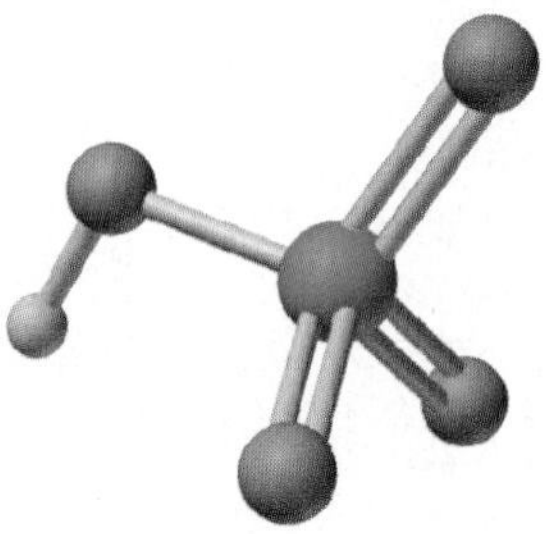 |
|---|---|---|---|---|
| Fluorine | HOF | — | — | — |
| Chlorine | HOCl | HOClO | HOClO$_2$ | HOClO$_3$ |
| Bromine | HOBr | (HOBrO)? | HOBrO$_2$ | HOBrO$_3$ |
| Iodine | HOI | — | HOIO$_2$ | HOIO$_3$, (HO)$_5$IO |
| Oxoanion | Hypohalite | Halite | Halate | Perhalate |

2. *Oxidation state of the halogen.* Among oxoacids of a given halogen, such as chlorine, acid strength decreases as the oxidation state of the halogen decreases:

$$HOClO_3 > HOClO_2 > HOClO > HOCl$$

The higher the oxidation state (number of bound O atoms) of the halogen, the more electron density it pulls from the O—H bond. Quantitative considerations of these trends in oxoacid strength appear in Chapter 17.

## 13.10   Group 8A(18): The Noble Gases

The last main group consists of individual atoms too "noble" to interact with others. The Group 8A(18) elements display regular trends in physical properties and very low, if any, reactivity. The group consists of helium (He), the second most abundant element in the universe, neon (Ne), and argon (Ar); krypton (Kr) and xenon (Xe), the only members for which compounds are known; and finally, radioactive radon (Rn). The noble gases make up about 1% by volume of the atmosphere, primarily due to the high abundance of Ar. Their properties are summarized in the Group 8A(18) Family Portrait (p. 574).

### How Can Noble Gases Form Compounds?

Lying at the far right side of the periodic table, the Group 8A(18) elements consist of individual atoms with filled outer levels and the smallest size in their periods: even Li, the smallest alkali metal (152 pm), is bigger than Rn, the largest noble gas (141 pm). The elements come as close to being ideal gases as any other substances. Only at very low temperatures do they condense and solidify. In fact, He is the only substance that does *not* solidify by a reduction in temperature alone; it requires an increase in pressure as well.

## FAMILY PORTRAIT

# Group 8A(18): The Noble Gases
### Key Atomic and Physical Properties

$ns^2np^6$

**GROUP 8A(18)**

| 2 |
| :---: |
| **He** |
| 4.003 |
| $1s^2$ |
| **(none)** |

| 10 |
| :---: |
| **Ne** |
| 20.18 |
| $2s^22p^6$ |
| **(none)** |

| 18 |
| :---: |
| **Ar** |
| 39.95 |
| $3s^23p^6$ |
| **(none)** |

| 36 |
| :---: |
| **Kr** |
| 83.80 |
| $4s^24p^6$ |
| **(+2)** |

| 54 |
| :---: |
| **Xe** |
| 131.3 |
| $5s^25p^6$ |
| **(+8, +6, +4, +2)** |

| 86 |
| :---: |
| **Rn** |
| (222) |
| $6s^26p^6$ |
| **(+2)** |

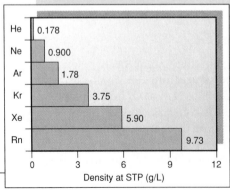
"Mass spectral peak"

## Atomic Properties

| Atomic radius (pm) |
| :---: |
| He 31 |
| Ne 71 |
| Ar 98 |
| Kr 112 |
| Xe 131 |
| Rn (140) |

Group electron configuration is $1s^2$ for He; $ns^2np^6$ for others. The valence shell is filled. Only Kr and Xe (and perhaps Rn) are known to form compounds. The more reactive Xe exhibits all even oxidation states (+2 to +6).

Group contains the smallest atoms with the highest IE in their periods. Down the group, atomic size increases and IE decreases steadily. (EN values for Kr and Xe only are given).

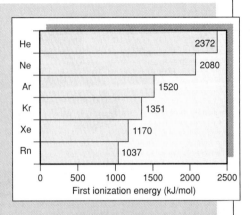

First ionization energy (kJ/mol)

| | |
| :--- | ---: |
| He | 2372 |
| Ne | 2080 |
| Ar | 1520 |
| Kr | 1351 |
| Xe | 1170 |
| Rn | 1037 |

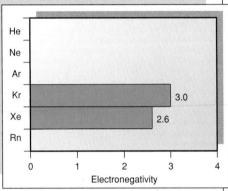

Electronegativity

| | |
| :--- | :--- |
| Kr | 3.0 |
| Xe | 2.6 |

## Physical Properties

Melting and boiling points of these gaseous elements are extremely low but increase down the group due to stronger dispersion forces. Note the extremely small liquid ranges.

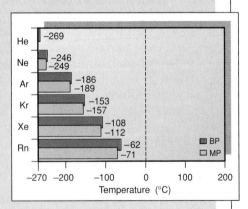

Temperature (°C)

| | BP | MP |
| :--- | ---: | ---: |
| He | –269 | |
| Ne | –246 | –249 |
| Ar | –186 | –189 |
| Kr | –153 | –157 |
| Xe | –108 | –112 |
| Rn | –62 | –71 |

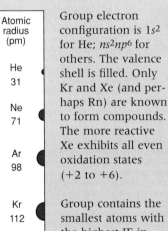

Density at STP (g/L)

| | |
| :--- | ---: |
| He | 0.178 |
| Ne | 0.900 |
| Ar | 1.78 |
| Kr | 3.75 |
| Xe | 5.90 |
| Rn | 9.73 |

Densities (at STP) increase steadily, as expected.

Helium has the lowest melting point known ($-272.2°C$), only one degree above 0 K ($-273.15°C$), and it boils only about three degrees higher. Weak dispersion forces hold these elements in condensed states, their melting and boiling points increasing with molar mass.

Ever since their discovery in the late 19th century, these elements had been considered, and even formerly named, the "inert" gases. Atomic theory and, more important, all laboratory experience had supported this idea. About 40 years ago, however, all this changed when the first noble gas compound was prepared. Nevertheless, we must ask how, with filled outer levels and extremely high ionization energies, *can* noble gases react?

The discovery of their reactivity is a classic example of clear thinking in the face of an unexpected event. In 1962, the young inorganic chemist Neil Bartlett was studying platinum fluorides, known to be strong oxidizing agents. When he accidentally exposed $PtF_6$ to air, its deep-red color lightened slightly, and analysis showed that the $PtF_6$ had oxidized $O_2$ to form ionic $[O_2]^+[PtF_6]^-$. Since the ionization energy of the oxygen molecule ($O_2 \rightarrow O_2^+ + e^-$; IE = 1175 kJ/mol) is very close to $IE_1$ of xenon (1170 kJ/mol), Bartlett reasoned that $PtF_6$ might be able to oxidize xenon. Shortly thereafter, he prepared $XePtF_6$, an orange-yellow solid. Within a few months, the white crystalline $XeF_2$ and $XeF_4$ (Figure 13.33) were also prepared. In addition to the +2 and +4 oxidation states, Xe occurs in the +6 state in several compounds, such as $XeF_6$, and in the +8 state in the unstable oxide, $XeO_4$. A few compounds of Kr have also been made. The xenon fluorides react rapidly in water to form HF and various other products, including different xenon compounds.

**FIGURE 13.33**
Crystals of xenon tetrafluoride ($XeF_4$).

## Looking Backward and Forward: Groups 7A(17), 8A(18), and 1A(1)

With metallic behavior gone, a host of 7A anions, covalent oxides, and oxoanions appear, which change oxidation states readily in a rich aqueous chemistry. The vigorous reactivity of the halogens is in stark contrast to the inertness of their 8A neighbors. Filled outer levels render these lone atoms largely unreactive, despite a limited ability to react with the most electronegative elements.

The least reactive family in the periodic table stands between the two most reactive: the halogens, which need one more electron to fill their outer level, and the alkali metals, which need one fewer (Figure 13.34). As you probably recall (Section 13.3), atomic, physical, and chemical properties change dramatically from Group 8A(18) to Group 1A(1).

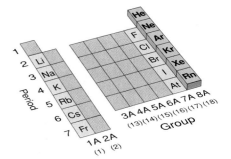

**FIGURE 13.34**
Standing in Group 8A(18), looking backward at the halogens, Group 7A(17), and ahead to the alkali metals, Group 1A(1).

## Chapter Perspective

*Our excursion through the bonding and reactivity patterns of the elements has come full circle, but we've only touched on some of their most important features. Clearly, the properties of the atoms and the resultant physical and chemical behaviors of the elements are magnificent testimony to nature's diversity. In Chapter 14, we continue with our theme of macroscopic behavior emerging from atomic properties in an investigation of the marvelously complex and essential organic compounds of carbon. In Chapter 23, we revisit the most important main-group elements to see how they occur in nature and how we isolate and use them.*

# For Review and Reference

## Key Terms

**SECTION 13.4**
diagonal relationship

**SECTION 13.5**
bridge bond

**SECTION 13.6**
allotrope
silicate
silicone

**SECTION 13.7**
disproportionation
dehydration-condensation

**SECTION 13.9**
interhalogen compound

## Problems

Problems with a green number are answered at the back of the text. Most sections include three categories of problems, separated by a green rule: concept review questions, *paired* skill-building exercises, and problems in a relevant context.

### Interchapter Review

**13.1** (a) Define the four atomic properties that are reviewed in the Interchapter.

(b) With what are the three parts of the electron configuration ($nl^{\#}$) correlated?

(c) Which part of the electron configuration is primarily associated with the size of the atom?

(d) What is the major distinction between outer electron configurations in a group compared with those in a period?

(e) What correlation, if any, exists between the group number and the number of valence shell electrons?

**13.2** (a) What trends, if any, exist for $Z_{eff}$ across a period and down a group?

(b) How does $Z_{eff}$ influence atomic size, $IE_1$, and EN of the elements across a period?

**13.3** Iodine monochloride and elemental bromine have nearly the same molar mass and liquid density but very different boiling points.

(a) What molecular property is primarily responsible for this difference in boiling point? What atomic property gives rise to it? Explain.

(b) Which substance has a higher boiling point? Why?

**13.4** Why is the trend in atomic size the inverse of the trend in ionization energy?

**13.5** How are bond energy, bond length, and reactivity related for similar compounds?

**13.6** How are covalent and metallic bonding similar? How are they different?

**13.7** If the leftmost element in a period were to combine with each of the others in the period, how would the type of bonding change from left to right? Explain in terms of atomic properties.

**13.8** Why is rotation about the bond axis possible for single-bonded atoms but not double-bonded atoms?

**13.9** Except for mercury, most metals are shiny solids at room temperature. Name two shiny solid elements that are not metals.

**13.10** Atomic size tends to decrease smoothly across a period, but ionic size exhibits a major irregularity. With what is it associated?

**13.11** Would you expect $S^{2-}$ to be larger or smaller than $Cl^-$? Would you expect $Mg^{2+}$ to be larger or smaller than $Na^+$? Explain.

**13.12** (a) How does the type of bonding in element oxides correlate with their electronegativities?

(b) How does the acid-base behavior of element oxides correlate with their electronegativities?

**13.13** (a) How does the metallic character of an element correlate with the *acidity* of its oxide?

(b) What trends, if any, exist in oxide *basicity* across a period and down a group?

**13.14** How are atomic size, $IE_1$, and EN related to redox behavior of the elements in Groups 1A(1), 2A(2), 6A(16), and 7A(17)?

**13.15** How do the physical properties of a network covalent solid and a molecular covalent solid differ? Why?

---

**13.16** Rank the following elements in order of *increasing*

(a) atomic size: Ba, Mg, Sr

(b) $IE_1$: P, Na, Al

(c) EN: Br, Cl, Se

(d) number of valence electrons: Bi, Ga, Sn

**13.17** Rank the following elements in order of *decreasing*

(a) atomic size: N, Si, P

(b) $IE_1$: Kr, K, Ar

(c) EN: In, Rb, I

(d) number of valence electrons: Sb, S, Cs

**13.18** Which of the following element pairs react to form *ionic* compounds: (a) Cl and Br; (b) Na and Br; (c) P and Se; (d) H and Ba?

**13.19** Which of the following element pairs react to form *covalent* compounds: (a) Be and C; (b) Sr and O; (c) Ca and Cl; (d) P and F?

**13.20** Draw a Lewis structure for a compound with a bond order of two throughout the molecule.

**13.21** Draw a Lewis structure for a polyatomic ion with a fractional bond order throughout the molecule.

**13.22** Rank the following compounds in order of *increasing* bond length: $SiCl_4$, $CF_4$, $GeBr_4$.

**13.23** Rank the following compounds in order of *decreasing* bond energy: $NF_3$, $NI_3$, $NCl_3$.

**13.24** Rank the following ions in order of *increasing* radius:

(a) $O^{2-}$, $F^-$, $Na^+$

(b) $S^{2-}$, $P^{3-}$, $Cl^-$

**13.25** Rank the following ions in order of *decreasing* radius:
  (a) $Ca^{2+}$, $K^+$, $Ga^{3+}$
  (b) $Br^-$, $Sr^{2+}$, $Rb^+$

**13.26** Rank $O^-$, $O^{2-}$, and O in order of *increasing* size.

**13.27** Rank $Tl^{3+}$, Tl, and $Tl^+$ in order of *decreasing* size.

**13.28** Which member of each pair of oxides will give the more *basic* solution in water? For the pair in (a), write the reaction for each oxide with water to support your answer: (a) CaO or $SO_3$; (b) BeO or BaO; (c) $CO_2$ or $NO_2$; (d) $P_4O_{10}$ or $K_2O$.

**13.29** Which member of each pair of oxides will give the more *acidic* solution in water? For the pair in (a), write the reaction for each oxide with water to support your answer: (a) $NO_2$ or SrO; (b) SnO or $SnO_2$; (c) $Cl_2O$ or $Na_2O$; (d) $SO_2$ or MgO.

**13.30** Which member of each pair will have more *covalent* character in its bonds to Cl?
  (a) LiCl or KCl; (b) $AlCl_3$ or $PCl_3$; (c) $NCl_3$ or $AsCl_3$

**13.31** Which member of each pair will have more *ionic* character in its bonds to F?
  (a) $BeF_2$ or $CaF_2$; (b) $PbF_2$ or $PbF_4$; (c) $GeF_4$ or $PF_3$

**13.32** Rank the following solids in order of *increasing*
  (a) melting point: Na, Si, Ar
  (b) $\Delta H_{fus}$: Rb, Cs, Li

**13.33** Rank the following substances in order of *decreasing*
  (a) boiling point: $O_2$, $Br_2$, As(s)
  (b) $\Delta H_{vap}$: $Cl_2$, Ar, $I_2$

## Hydrogen, the Simplest Atom

**13.34** Hydrogen has only one proton but its $IE_1$ is much greater than that of lithium, which has three protons. Explain.

**13.35** Sketch a periodic table and label the areas of elements that give rise to the three types of hydrides discussed in the text.

**13.36** $H_2$ may act as a reducing agent or an oxidizing agent, depending upon the substance reacting with it. Using sodium and chlorine as the other reactant, write balanced equations for the two reactions and characterize the role of hydrogen in each.

**13.37** Draw Lewis structures for the following compounds, and predict which member of each pair will form hydrogen bonds: (a) $NF_3$ or $NH_3$; (b) $CH_3OCH_3$ or $CH_3CH_2OH$.

**13.38** Draw Lewis structures for the following compounds, and predict which member of each pair will form hydrogen bonds: (a) $NH_3$ or $AsH_3$; (b) $CH_4$ or $H_2O$.

**13.39** Complete and balance the following reactions:
  (a) An active metal reacting with acid,

$$Al(s) + HCl(aq) \rightarrow$$

  (b) A saltlike (alkali metal) hydride reacting with water,

$$LiH(s) + H_2O(l) \rightarrow$$

**13.40** Complete and balance the following reactions:
  (a) A saltlike (alkaline earth metal) hydride reacting with water,

$$CaH_2(s) + H_2O(l) \rightarrow$$

  (b) Reduction by hydrogen, to form a metal,

$$PdCl_2(aq) + H_2(g) \rightarrow$$

**13.41** Compounds such as $NaBH_4$, $Al(BH_4)_3$, and $LiAlH_4$ are called complex hydrides and are used as reducing agents in many chemical syntheses.
  (a) Give the oxidation number of each element in these three compounds.
  (b) Write a Lewis structure for the polyatomic anion in $NaBH_4$ and predict its shape.

**13.42** Unlike the $F^-$ ion, which has an ionic radius close to 133 pm in all alkali metal fluorides, the $H^-$ ionic radius varies from 137 pm in LiH to 152 pm in CsH. Suggest an explanation for the large variability in $H^-$ but not in $F^-$.

## Trends Across the Periodic Table

**13.43** How does the maximum oxidation state vary across a period for the main groups? Is the pattern in Period 2 different?

**13.44** What correlation, if any, exists for the Period 2 elements between group number and the number of covalent bonds the element typically forms? How is the correlation different for elements in higher periods?

**13.45** Each of the chemically active Period 2 elements forms stable compounds that have bonds to fluorine.
  (a) What are the names and formulas of the substances formed?
  (b) Does $\Delta EN$ increase or decrease from left to right?
  (c) Does percent ionic character increase or decrease from left to right?
  (d) Draw Lewis structures for those compounds with predominantly covalent bonding.

**13.46** Period 6 is unusual in several ways:
  (a) It is the longest period in the table. How many elements belong to Period 6? How many metals?
  (b) It contains no metalloids. Where is the metal-nonmetal boundary in Period 6?

**13.47** An element forms an oxide $E_2O_3$ and a fluoride $EF_3$.
  (a) Of what two groups might E be a member?
  (b) How would the group in which E occurs affect the properties of the oxide and the fluoride?

**13.48** Fluorine lies between oxygen and neon in Period 2. Whereas atomic sizes and ionization energies of these three elements change smoothly, their electronegativities display a dramatic change. What is this change and how do their electron configurations explain it?

## Group 1A(1): The Alkali Metals

**13.49** Lithium salts are often much less soluble than the corresponding salts of other alkali metals. For example, at 18°C the concentration of a saturated LiF solution is $1.0 \times 10^{-2}$ M, while that of KF is 1.6 M. How would you explain this behavior?

**13.50** The alkali metals play virtually the same general chemical role in all their reactions.
(a) What is this role?
(b) How is it based on atomic properties?
(c) Using sodium, write two balanced equations that exhibit this role.

**13.51** How do atomic properties account for the low densities of the Group 1A(1) elements?

**13.52** Each of the following properties shows regular trends in Group 1A(1). Predict whether each increases or decreases *down* the group: (a) density; (b) ionic size; (c) E—E bond energy; (d) ionization energy; (e) $\Delta H_{\text{hydr}}$ of $M^+$ ion.

**13.53** Each of the following properties shows regular trends in Group 1A(1). Predict whether each increases or decreases *up* the group: (a) melting point; (b) E—E bond length; (c) hardness; (d) molar volume; (e) lattice energy of EBr.

**13.54** Write a balanced equation for the formation of sodium peroxide, an industrial bleach for fabric and paper, from its elements.

**13.55** Write a balanced equation for the formation of rubidium bromide through a neutralization reaction of a strong acid and strong base.

**13.56** Although the alkali halides can be prepared directly from the elements, the far less expensive industrial route is through treatment of the carbonate or hydroxide with aqueous hydrohalic acid (HX) followed by recrystallization. Balance the reaction between potassium carbonate and aqueous hydriodic acid.

**13.57** The main reason that alkali metal dihalides ($MX_2$) do *not* form is the high $IE_2$ of the metal.
(a) Why is the $IE_2$ so high for alkali metals?
(b) The $IE_2$ for Cs is 2255 kJ/mol, low enough for $CsF_2$ to form exothermically ($\Delta H_f^0 = -125$ kJ/mol). This compound cannot be synthesized, however, because CsF forms with a much greater release of heat ($\Delta H_f^0 = -530$ kJ/mol). Thus the breakdown of $CsF_2$ to CsF happens readily. Write the reaction for this breakdown and calculate its heat of reaction.

**13.58** The total terrestrial abundance of francium ($Z = 87$) is estimated at $2 \times 10^{-18}$ ppm, or about 15 g Fr in the top kilometer of the Earth's crust. Nevertheless, it has been predicted that if a weighable quantity of Fr could be obtained, it would be a liquid at room temperature. What is the basis of this prediction?

## Group 2A(2): The Alkaline Earth Metals

**13.59** How do Groups 1A(1) and 2A(2) compare with respect to reaction of the metals with water?

**13.60** Alkaline earth elements are involved in two key diagonal relationships in the periodic table.
(a) Give the two pairs of elements in these diagonal relationships.
(b) For each pair, cite two similarities that demonstrate the relationship.
(c) Why are the members of each pair so similar in behavior?

**13.61** The melting points of alkaline earth metals are many times higher than those of the alkali metals. Explain this difference on the basis of atomic properties. Name three other physical properties for which Group 2A(2) metals have higher values than the corresponding 1A(1) metals.

**13.62** Write a balanced equation for each of the following reactions:
(a) "Slaking" of lime (treatment with water)
(b) Combustion of calcium in air

**13.63** Write a balanced equation for each of the following reactions:
(a) Thermal decomposition of witherite (barium carbonate)
(b) Neutralization of stomach acid, HCl, by milk of magnesia (magnesium hydroxide)

**13.64** Lime (CaO) is one of the largest tonnage chemicals produced in the world. Develop balanced reactions for
(a) The preparation of lime from natural sources
(b) The use of slaked lime to remove $SO_2$ from flue gases
(c) The reaction of lime with arsenic acid ($H_3AsO_4$) to manufacture the insecticide calcium arsenate
(d) The regeneration of NaOH in the paper industry by reaction of lime with aqueous sodium carbonate

**13.65** In some reactions, Be behaves like other alkaline earth metals, and in others it does not. Complete and balance the following reactions:
(a) $BeO(s) + H_2O(l) \longrightarrow$
(b) $BeCl_2(l) + Cl^-(l; \text{from molten NaCl}) \longrightarrow$
In which reaction does Be behave like the other Group 2A(2) members?

**13.66** Grignard reagents have remarkable versatility in organic syntheses. Recently some of the solvated intermediates have been crystallized and their structure determined. Draw a Lewis structure for the species in which Mg is surrounded by $CH_3$, Br, and two molecules of ethyl ether ($CH_3CH_2$—O—$CH_2CH_3$). Predict the shape around the Mg atom. Predict the shape around the O atom in each ether group.

## Group 3A(13): The Boron Family

**13.67** How does the appearance of the transition metals in Period 4 affect the pattern of ionization energies in Group 3A(13)? How does this pattern compare with that of Group 3B(3)?

**13.68** How do the acidities of $Tl_2O$ and $Tl_2O_3$ compare with one another? Explain.

**13.69** Despite the expected decrease in atomic size, there is an unexpected drop in the first ionization energy between Groups 2A(2) and 3A(13) in Periods 2 through 4 (e.g., Be to B). Explain this pattern in terms of electron configuration and orbital energies.

**13.70** Compounds of Group 3A(13) elements have chemical behavior that reflects an electron deficiency.
 (a) What is the meaning of "electron deficiency"?
 (b) Give two examples of reactions that illustrate this behavior.

**13.71** Boron chemistry is not typical of its group.
 (a) Cite three ways in which boron and its compounds differ significantly from the other 3A(13) members.
 (b) What is the reason for these differences?

**13.72** Rank the following oxides in order of increasing *acidity*: $Ga_2O_3$, $Al_2O_3$, $In_2O_3$.

**13.73** Rank the following hydroxides in order of increasing *basicity*: $TlOH$, $B(OH)_3$, $In(OH)_3$.

**13.74** Thallium forms the compound $TlI_3$. What is the apparent oxidation state of Tl in this compound? Given that the anion is $I_3^-$, what is the actual oxidation state of Tl? Draw the shape of the anion, giving its VSEPR class and bond angles. Propose a reason that the compound does not exist as $(Tl^{3+})(I^-)_3$.

**13.75** Very stable "dihalides" of the Group 3A(13) metals are known. What is the apparent oxidation state of Ga in $GaCl_2$? Given that $GaCl_2$ consists of a $Ga^+$ cation and a $GaCl_4^-$ anion, what are the actual oxidation states of Ga? Draw the shape of the anion, giving its VSEPR class and bond angles.

**13.76** Give the name and formula of a Group 3A(13) element or compound that fits each description or use:
 (a) Component of heat-resistant (Pyrex-type) glass
 (b) Manufacture of high-speed computer chips
 (c) Largest temperature range for liquid state of an element
 (d) Elementary substance with three-center, two-electron bonds
 (e) Metal protected from oxidation by adherent oxide coat
 (f) Mild antibacterial agent (e.g., for eye infections)
 (g) Toxic metal that lies in the periodic table between two others.

**13.77** Alum (aluminum sulfate) is used as a "flocculating agent" in water purification. The "floc" is a gelatinous precipitate of aluminum hydroxide that forms on small suspended particles and bacteria, thereby carrying them down for removal by filtration.
 (a) The hydroxide is formed by the reaction of the aluminum ion with water. Write the equation for this reaction.
 (b) The acidity from this reaction is partially removed by the sulfate ion. Give an equation for this reaction.

**13.78** Just as boron nitride is isoelectronic with carbon, other compounds of Groups 3A(13) and 5A(15) are isoelectronic with elements in Group 4A(14). What element is isoelectronic with (a) aluminum phosphide; (b) gallium arsenide?

**13.79** Many compounds of Groups 3A(13) and 5A(15) are technologically important as semiconductors. They can be prepared from the elements at high pressure and temperature, but must be kept encased to prevent decomposition in moist air to the 3A hydroxide and the 5A hydride.
 (a) Write a balanced equation for the formation of indium(III) arsenide.
 (b) Write a balanced equation for the decomposition of gallium antimonide in moist air.

**13.80** Use VSEPR theory to draw structures, with ideal bond angles, for boric acid and the anion it forms in reaction with water.

**13.81** Halides of Al, Ga, and In occur as dimers in the gas and liquid phases (those of the larger halogens form dimeric solids also). Use VSEPR theory to draw the structure, with ideal bond angles, of the $GaBr_3$ dimer.

## Group 4A(14): The Carbon Family

**13.82** How does the basicity of $SnO_2$ compare with that of $CO_2$? Explain.

**13.83** Nearly every compound of silicon has the element in the +4 oxidation state. In contrast, most compounds of lead have the element in the +2 state.
 (a) What general observation do these facts illustrate?
 (b) What is the explanation for these facts in terms of atomic and molecular properties?
 (c) Give an analogous example from the chemistry of Group 3A(13).

**13.84** The sum of $IE_1$ through $IE_4$ for Group 4A(14) elements shows a decrease from C to Si, a slight increase from Si to Ge, a decrease from Ge to Sn, and an increase from Sn to Pb.
 (a) What is the expected trend for ionization energy down a group?
 (b) Explain the deviations from the expected trend.
 (c) Which group might you expect to show even greater deviations?

**13.85** Give explanations for the large drops in melting point from C to Si and from Ge to Sn.

**13.86** What is an allotrope? Name two Group 4A(14) elements that exhibit allotropy and name two of their allotropes.

**13.87** Even though EN values vary relatively little down Group 4A(14), the elements change from nonmetal to metal. Explain.

**13.88** How do atomic properties account for the enormous number of carbon compounds? Why don't other Group 4A(14) elements behave similarly?

---

**13.89** Draw a Lewis structure for
  (a) The cyclic silicate ion $Si_4O_{12}^{8-}$
  (b) The cyclic hydrocarbon $C_4H_8$

**13.90** Draw a Lewis structure for
  (a) The cyclic silicate ion $Si_6O_{18}^{12-}$
  (b) The cyclic hydrocarbon $C_6H_{12}$

**13.91** Show three units of a linear silicone polymer made from $(CH_3)_2Si(OH)_2$ with some $(CH_3)_3SiOH$ added to end the chain.

**13.92** Show two chains of three units each of a sheet silicone polymer made from $(CH_3)_2Si(OH)_2$ with $CH_3Si(OH)_3$ added to crosslink the chains.

---

**13.93** Although it is often stated that double bonding is limited to compounds of Period 2 elements, compounds of Period 3 elements also contain double bonds.
  (a) What types of atomic orbitals are used for the $\pi$ bond in C=O?
  (b) Why would the same types of orbitals not be satisfactory for an Si=O $\pi$ bond?
  (c) What type could be used for an Si=O $\pi$ bond?
  (d) Which atom(s) would supply the electrons in such a $\pi$ bond?

**13.94** Give the name and formula of a Group 4A(14) element or compound that fits each description or use:
  (a) Hardest known substance
  (b) Medicinal antacid
  (c) Atmospheric gas implicated in greenhouse effect
  (d) Waterproofing polymer based upon silicon
  (e) Synthetic abrasive composed entirely of Group 4A elements
  (f) Product formed when coke burns in a limited supply of air
  (g) Toxic metal found in gasoline, plumbing, and paints

**13.95** The diatomic species CO, $CN^-$, and $C_2^{2-}$ are isoelectronic.
  (a) Draw their Lewis structures.
  (b) Draw their MO diagrams (assume 2s-2p mixing as in $N_2$) and give the bond order and electron configuration for each.

**13.96** Producer gas is a fuel formed by passing air over red-hot coke (amorphous carbon). It consists of approximately 25% CO, 5% $CO_2$, and 70% $N_2$ by mass. What mass of producer gas can be formed from 1.75 metric tons of coke, assuming an 87% yield?

**13.97** One similarity between B and Si is the explosive combustion of their hydrides in air. Write balanced equations for the combustion of $B_2H_6$ and of $Si_4H_{10}$.

## Group 5A(15): The Nitrogen Family

**13.98** Which Group 5A(15) elements form trihalides? Pentahalides? Explain.

**13.99** Why do the melting points of Group 5A(15) elements increase and then decrease?

**13.100** (a) What is the range of oxidation states shown by the elements of Group 5A(15)?
  (b) How does this range illustrate the general rule for the range of oxidation states in groups on the right side of the periodic table?

**13.101** Solid $PCl_5$ exists not as molecules but rather as two ions, $PCl_4^+$ and $PCl_6^-$. Similarly, solid $PBr_5$ exists as ions, but in this case as $PBr_4^+$ and $Br^-$. Suggest a reason for this difference in solid state structure.

**13.102** Bismuth(V) compounds are such powerful oxidizing agents that they have not been prepared in pure form. How is this fact consistent with the location of Bi in the periodic table?

**13.103** Rank the following oxides in order of increasing acidity in water: $Sb_2O_3$, $Bi_2O_3$, $P_4O_{10}$, $Sb_2O_5$.

---

**13.104** Assuming that acid strength relates directly to electronegativity of the central atom, rank $H_3PO_4$, $HNO_3$, and $H_3AsO_4$ in order of increasing strength.

**13.105** Assuming that acid strength relates directly to number of O atoms on the central atom, rank $H_2N_2O_2$ [or $(HON)_2$]; $HNO_3$ (or $HONO_2$); $HNO_2$ (or $HONO$) in order of decreasing acid strength.

**13.106** Complete and balance the following reactions:
  (a) $As(s) + excess\ O_2(g) \rightarrow$
  (b) $Bi_2O_3(s) + HCl(aq) \rightarrow$
  (c) $Ca_3As_2(s) + H_2O(l) \rightarrow$

**13.107** Complete and balance the following reactions:
  (a) $Excess\ Sb(s) + Br_2(l) \rightarrow$
  (b) $HNO_3(aq) + MgCO_3(s) \rightarrow$
  (c) $K_2HPO_4(s) \xrightarrow{heat}$

**13.108** Complete and balance the following reactions:
  (a) $N_2(g) + Al(s) \xrightarrow{heat}$
  (b) $PF_5(g) + H_2O(l) \rightarrow$

**13.109** Complete and balance the following reactions:
  (a) $AsCl_3(l) + H_2O(l) \rightarrow$
  (b) $Sb_2O_3(s) + NaOH(aq) \rightarrow$

---

**13.110** Based on the relative sizes of F and Cl, predict the structure of $PF_2Cl_3$.

**13.111** Use the VSEPR model to predict the structure of the cyclic ion $P_3O_9^{3-}$.

---

**13.112** The white and red allotropes of phosphorus differ dramatically in toxicity. One is relatively unreactive in the body, while the other is quite poisonous. Suggest how their structures lead to a difference in toxicity and which is nontoxic.

**13.113** An important starting material for the manufacture of polyphosphazenes is the cyclic molecule $(NPCl_2)_3$. The molecule has a symmetric six-membered ring of alternating N and P atoms, with the Cl atoms bound to the P atoms. The N—P bond length is significantly less than that expected for an N—P single bond.
    (a) Draw the Lewis structure for the molecule.
    (b) How many lone pairs appear on the ring atoms?
    (c) What is the order of the N—P bond?

**13.114** Give the name and formula of a Group 5A(15) element or compound that fits each of the following descriptions or uses:
    (a) Hydride produced at multi-million ton level
    (b) Element(s) essential in plant nutrition
    (c) Hydride used in the manufacture of rocket propellants and drugs
    (d) Odd-electron molecule (two examples)
    (e) Amphoteric hydroxide with 5A element in its +3 state
    (f) P-containing water softener
    (g) Element that is an electrical conductor

**13.115** Would you expect hydrazine $(N_2H_4)$ to form H bonds? Explain with structures.

**13.116** Nitrous oxide $(N_2O)$, the "laughing gas" used as an anesthetic by dentists, is made by thermal decomposition of solid $NH_4NO_3$. Write a balanced equation for this reaction. What are the oxidation states of nitrogen in $NH_4NO_3$ and in $N_2O$?

**13.117** Write balanced equations for the thermal decomposition of potassium nitrate ($O_2$ is also formed in both cases) (a) at low temperature to the nitrite; (b) at high temperature to the oxide and nitrogen.

## Group 6A(16): The Oxygen Family

**13.118** What are the molecular formulas for the most common allotrope of oxygen and of sulfur? Why are these formulas so different?

**13.119** Rank the following in order of increasing electrical conductivity and explain your ranking: Po, S, Se.

**13.120** The oxygen and nitrogen families have some obvious similarities and differences.
    (a) State two general physical similarities between Group 5A(15) and 6A(16) elements.
    (b) State two general chemical similarities between Group 5A and 6A elements.
    (c) State two chemical similarities between P and S.
    (d) State two physical similarities between N and O.
    (e) State two chemical differences between N and O.

**13.121** As one goes down the hydrides of Group 6A(16), a molecular property changes abruptly, which has been explained by a change in the hybridization of the central atom.
    (a) Between what periods does the change occur?
    (b) What is the change in the molecular property?
    (c) What is the change in hybridization?
    (d) What other group displays a similar change?

**13.122** Complete and balance the following reactions:
    (a) $NaHSO_4(aq) + NaOH(aq) \rightarrow$
    (b) $S_8(s) + \text{excess } F_2(g) \rightarrow$
    (c) $FeS(s) + HCl(aq) \rightarrow$
    (d) $Te(s) + I_2(s) \rightarrow$

**13.123** Complete and balance the following reactions:
    (a) $H_2S(g) + O_2(g) \rightarrow$
    (b) $SO_3(g) + H_2O(l) \rightarrow$
    (c) $SF_4(g) + H_2O(l) \rightarrow$
    (d) $Al_2Se_3(s) + H_2O(l) \rightarrow$

**13.124** Predict whether each of following oxides is basic, acidic, or amphoteric in water: (a) $SeO_2$; (b) $N_2O_3$; (c) $K_2O$; (d) $BeO$; (e) $BaO$.

**13.125** Predict whether each of following oxides is basic, acidic, or amphoteric in water: (a) $MgO$; (b) $N_2O_5$; (c) $CaO$; (d) $CO_2$; (e) $TeO_2$.

**13.126** Rank the following hydrides in order of *increasing* acid strength: $H_2S$, $H_2O$, $H_2Te$.

**13.127** Rank the following species in order of *decreasing* acid strength: $H_2SO_4$, $H_2SO_3$, $HSO_3^-$.

**13.128** Describe the physical changes that are observed when solid sulfur is heated from room temperature to 440°C and then poured quickly into cold water. Explain the molecular changes that are responsible for the macroscopic changes.

**13.129** Give the name and formula of a Group 6A(16) element or compound that fits each of the following descriptions or uses:
    (a) Unreactive gas used as an electrical insulator
    (b) Unstable allotrope of oxygen
    (c) Oxide having sulfur in the same oxidation state as in sulfuric acid
    (d) Air pollutant produced by the burning of sulfur-containing coal
    (e) Powerful dehydrating agent
    (f) Compound used in solution for photographic development
    (g) Trace gas that tarnishes silver

**13.130** The best method for synthesizing $SF_4$ is the reaction of an alkali metal fluoride with $SCl_2$ in a nonaqueous solvent. Some $SCl_2$ is converted to $S_2Cl_2$ and the alkali metal fluoride is converted to the chloride.
    (a) Using NaF, develop the balanced equation for the production of $SF_4$.
    (b) Draw Lewis structures for the three sulfur-containing compounds.

**13.131** In addition to the monatomic sulfide anion $S^{2-}$, sulfur forms a variety of polyatomic sulfide anions, $S_n^{2-}$. The simplest, $S_2^{2-}$, is found in pyrite ores, such as $FeS_2$, and others through $S_6^{2-}$ are known.
    (a) Name and draw a Lewis structure for the oxygen anion that is analogous to $S_2^{2-}$.
    (b) Predict the shape and bond angle of the anion in $BaS_3$.

**13.132** Disulfur decafluoride is intermediate in reactivity between $SF_4$ and $SF_6$. It disproportionates at 150°C to these monosulfur fluorides. Write a balanced equation for this reaction and give the oxidation state of S in each compound.

**13.133** Thionyl chloride ($SOCl_2$) is a sulfur oxohalide that is used industrially to dehydrate metal halide hydrates.
(a) Write a balanced equation for its reaction with magnesium chloride hexahydrate, in which $SO_2$ and HCl form along with the metal halide.
(b) Draw the Lewis structure of thionyl chloride with the lowest formal charges.

## Group 7A(17): The Halogens

**13.134** (a) What is the physical state and color of each of the halogens at STP?
(b) Explain the change in physical state down the group in terms of molecular properties.

**13.135** (a) What are the common oxidation states of the halogens?
(b) Give an explanation based on electron configuration for the range and values of the oxidation states of chlorine.
(c) Why is fluorine an exception to the pattern of oxidation states found for the other group members?

**13.136** How many electrons does a halogen atom need to complete its octet? Give examples of the different ways that a Cl atom can do so.

**13.137** Select the stronger bond in each pair: (a) Cl—Cl or Br—Br; (b) Br—Br or I—I; (c) F—F or Cl—Cl. Why doesn't the F—F bond strength follow the group trend?

**13.138** In addition to interhalogen compounds, many polyatomic interhalogen ions exist. Would you expect interhalogen ions with a 1+ or a 1− charge to have an even or odd number of atoms? Explain.

**13.139** Perhaps surprisingly, some pure liquid interhalogen fluorides have high electrical conductivity. The explanation is that one molecule transfers a fluoride ion to another molecule. Write the equation that would explain the high electrical conductivity of bromine trifluoride.

**13.140** (a) A halogen ($X_2$) disproportionates in base in several steps to $X^-$ and $XO_3^-$. Develop the overall equation for the disproportionation of $Br_2$ to $Br^-$ and $BrO_3^-$.
(b) Write a balanced equation for the reaction of $ClF_5$ with aqueous base (by analogy with the reaction of $BrF_5$ shown in text).

**13.141** Complete and balance the following reactions. If no reaction occurs, write NR:
(a) $Rb(s) + Br_2(l) \longrightarrow$  (b) $I_2(s) + H_2O(l) \longrightarrow$
(c) $Br_2(l) + I^-(aq) \longrightarrow$  (d) $CaF_2(s) + H_2SO_4(l) \longrightarrow$

**13.142** Complete and balance the following reactions. If no reaction occurs, write NR:
(a) $H_3PO_4(l) + NaI(s) \longrightarrow$  (b) $Cl_2(g) + I^-(aq) \longrightarrow$
(c) $Br_2(l) + Cl^-(aq) \longrightarrow$  (d) $ClF(g) + F_2(g) \longrightarrow$

**13.143** Rank the following acids in order of *increasing* acid strength: HClO, $HClO_2$, HBrO, HIO.

**13.144** Rank the following acids in order of *decreasing* acid strength: $HBrO_3$, $HBrO_4$, $HIO_3$, $HClO_4$.

---

**13.145** Give the name and formula of a Group 7A(17) element or compound that fits each description or use:
(a) Used in etching glass
(b) Naturally occurring source (ore) of fluorine
(c) Oxide used in bleaching paper pulp and textiles
(d) Weakest hydrohalic acid
(e) Compound used as food additive to prevent goiter (thyroid disorder)
(f) Element that is produced in the largest quantity
(g) Organic chloride used to make plastics

**13.146** Diiodine pentaoxide ($I_2O_5$) was discovered by Gay-Lussac in 1813, but its structure was unknown until 1970! Like $Cl_2O_7$, it can be prepared by the dehydration-condensation of the corresponding oxoacid.
(a) Name the precursor oxoacid, write a reaction for formation of the oxide, and draw a likely Lewis structure.
(b) Data show that the bonds to the terminal O are shorter than the bonds to the bridging O. Why?
(c) $I_2O_5$ is one of the few chemicals that can oxidize CO rapidly and completely; elemental iodine forms in the process. Write a balanced equation for this reaction.

**13.147** An industrial chemist treats solid NaCl with concentrated $H_2SO_4$ and obtains gaseous HCl and $NaHSO_4$. When she substitutes solid NaI for NaCl, gaseous $H_2S$, solid $I_2$, and $S_8$ are obtained with no HI.
(a) What type of reaction has the $H_2SO_4$ undergone with NaI?
(b) Why does NaI, but not NaCl, cause this type of reaction?
(c) If you needed to produce HI(g) by the reaction of NaI with an acid, how would the acid have to differ from sulfuric acid?

**13.148** Rank the larger halogens in order of increasing oxidizing strength based on their products with metallic Re: $ReCl_6$, $ReBr_5$, $ReI_4$. Explain your ranking.

## Group 8A(18): The Noble Gases

**13.149** Which noble gas is the most abundant in the universe? In the atmosphere?

**13.150** What oxidation states does Xe show in its compounds?

**13.151** Why do the noble gases have such low boiling points?

**13.152** Explain why Xe, and to a limited extent Kr, form compounds, while He, Ne, and Ar do not.

**13.153** (a) Why do the stable xenon fluorides have an even number of fluorine atoms?
  (b) Why do the ionic species $XeF_3^+$ and $XeF_7^-$ have odd numbers of F atoms?
  (c) Predict the shape of $XeF_3^+$.

## Comprehensive Problems

Problems with an asterisk (*) are more challenging.

**13.154** Sodium hydride is used to prepare sodium dithionate ($Na_2S_2O_4$), used in bleaching paper pulp:

$$NaH(s) + SO_2(l) \rightarrow Na_2S_2O_4(s) + H_2(g)$$

  (a) Balance the reaction and give the O.N. of each element.
  (b) How many liters of hydrogen gas measured at 758 torr and $-45°C$ can form when exactly 1 metric ton NaH reacts with 1 metric ton $SO_2$?

**13.155** Given the following information,

$$H^+(g) + H_2O(g) \rightarrow H_3O^+(g); \Delta H = -720 \text{ kJ}$$

$$H^+(g) + H_2O(l) \xrightarrow{H_2O} H_3O^+(aq); \Delta H = -1090 \text{ kJ}$$
$$H_2O(l) \rightarrow H_2O(g); \Delta H = 40.7 \text{kJ}$$

calculate the heat of solution of the hydronium ion:

$$H_3O^+(g) \xrightarrow{H_2O} H_3O^+(aq)$$

**13.156** The electron transition in Na from $3s^1$ to $3p^1$ gives rise to the bright yellow emission at 589.2 nm. What is the energy of this transition?

**13.157** Unlike the other Group 2A(2) metals, beryllium acts like aluminum and zinc in reacting with concentrated aqueous base to release hydrogen gas and form oxoanions of formula $M(OH)_4^{n-}$. Write equations for the reaction of these metals with NaOH.

**13.158** Cyclopropane ($C_3H_6$), the smallest cyclic hydrocarbon, is reactive for the same reason that white phosphorus is. Use bond properties and valence-bond theory to explain the high reactivity of $C_3H_6$.

**13.159** Semiconductors made from Groups 3A and 5A elements are typically prepared by direct reaction of the elements at high temperature. An engineer treats 32.5g molten gallium with 20.4L white phosphorous vapor at 515K and 195kPa. If purification losses are 7.2% by mass, how many grams of gallium phosphide can be prepared?

**\*13.160** Two substances with the empirical formula HNO are hyponitrous acid ($\mathcal{M} = 62.04$ g/mol) and nitroxyl ($\mathcal{M} = 31.02$ g/mol).
  (a) What is the molecular formula of each species?
  (b) Draw a Lewis structure for each species such that formal charges are minimized. (Hint: hyponitrous acid has an $N=N$ bond.)
  (c) Predict the shape of each species around the N atoms.
  (d) When hyponitrous acid loses two protons, it forms the hyponitrite ion. Draw cis and trans forms of this ion.

**\*13.161** The Ostwald process is a series of three main reactions used for the industrial production of nitric acid from ammonia.
  (a) Write out the series of balanced equations for the Ostwald process.
  (b) If NO is not recycled, how many moles of $NH_3$ are consumed per mole of $HNO_3$ produced?
  (c) In a typical industrial unit, the process is very efficient, with a 96% yield for the first step. Assuming theoretical yields for the subsequent steps, what volume of nitric acid (60% by mass; $d = 1.37$ g/mL) can be prepared per cubic meter of air that is 10% $NH_3$ by volume at the industrial conditions of 5 atm and 850°C?

**\*13.162** Organolithium compounds such as $LiCH_3$ are important reagents in organic synthesis. The compounds are not simple substances, as the formula suggests, but exist in the solid state, and in solution with nonpolar solvents, as a tetramer [e.g., methyllithium is $Li_4(CH_3)_4$].
  (a) If the compound were simply $LiCH_3$, what violation of the octet rule would occur?
  (b) How many electrons are available for bonding among the four lithium atoms and the four methyl groups?
  (c) Considering this number of available electrons and the total number of atoms bonded in the cluster, what prediction would you make about the general nature of the bonding?

**13.163** Hypofluorous acid (HFO, or more commonly HOF) was first prepared in weighable amounts as recently as 1971. The synthesis required treatment of fluorine with ice followed by rapid removal of the oxoacid to prevent the formation of oxygen difluoride through its reaction with more $F_2$.
  (a) Write balanced equations for these reactions.
  (b) What is the O.N. of O in HOF and in $OF_2$?
  (c) Predict the shapes of HOF and $OF_2$.

**13.164** What is a disproportionation reaction, and which of the following fit the description?
  (a) $I_2(s) + KI(aq) \rightarrow KI_3(aq)$
  (b) $2ClO_2(g) + H_2O(l) \rightarrow HClO_3(aq) + HClO_2(aq)$
  (c) $Cl_2(g) + 2NaOH(aq) \rightarrow$
  $$NaCl(aq) + NaClO(aq) + H_2O(l)$$
  (d) $NH_4NO_2(s) \rightarrow N_2(g) + 2H_2O(g)$
  (e) $3MnO_4^{2-}(aq) + 2H_2O(l) \rightarrow$
  $$2MnO_4^-(aq) + MnO_2(s) + 4OH^-(aq)$$
  (f) $3AuCl(s) \rightarrow AuCl_3(s) + 2Au(s)$

**13.165** Explain the following observations:
  (a) Phosphorus forms $PCl_5$ in addition to the expected $PCl_3$, but nitrogen forms only $NCl_3$.
  (b) Carbon tetrachloride is stable toward reaction with water, but silicon tetrachloride reacts rapidly and completely. (To give what?)
  (c) The oxygen-sulfur bond in $SO_4^{2-}$ is shorter than expected for an S—O single bond.
  (d) Chlorine forms $ClF_3$ and $ClF_5$, but $ClF_4$ is unknown.

**13.166** Which group(s) of the periodic table is(are) described by the following general statements?

(a) The elements form neutral compounds of VSEPR class $AX_3E$.

(b) The free elements are strong oxidizing agents and form monatomic ions and oxoanions.

(c) The valence shell allows the atoms to form compounds by combining with two atoms that donate one electron each.

(d) The free elements are strong reducing agents, show only one nonzero oxidation state, and form mainly ionic compounds.

(e) The elements can form stable compounds with only three bonds, but the central atom can accept a pair of electrons from a fourth atom without expanding its valence shell.

(f) Only two elements in the group are known to be chemically active.

**13.167** Account for the following facts:

(a) $Ca^{2+}$ and $Na^+$ have very nearly the same radii.

(b) $CaF_2$ is insoluble, but NaF is quite soluble.

(c) Molten $BeCl_2$ is a poor electrical conductor, whereas molten $CaCl_2$ is an excellent conductor.

**13.168** Three of the hardest substances known are diamond, silicon carbide (SiC), and borazon (BN).

(a) What common structural feature do these materials possess?

(b) Explain electronically how BN fits in with the other two, whose elements come from Group 4A(14).

**13.169** $P_4$ is prepared by heating phosphate rock [principally $Ca_3(PO_4)_2$] with sand and coke:

$$Ca_3(PO_4)_2(s) + SiO_2(s) + C(s) \longrightarrow$$

$$CaSiO_3(s) + CO(g) + P_4(g) \text{ [unbalanced]}$$

How many kilograms of phosphate rock will be needed to produce 315 mol $P_4$, assuming that the conversion is 90% efficient?

**13.170** Assume you can use either $NH_4NO_3$ or $N_2H_4$ as fertilizer. Which contains more moles of N per mole of compound? Which contains the higher mass of N per gram of compound?

**13.171** From its formula, one might expect CO to be quite polar, but its dipole moment is actually low (0.11 D).

(a) Draw the Lewis structure for CO.

(b) Calculate the formal charges.

(c) Based on (a) and (b), why is the dipole moment so low?

*\***13.172** In addition to $Al_2Cl_6$, aluminum forms other species with bridging halide ions to two aluminum atoms. One such unit is the ion $Al_2Cl_7^-$. The ion is symmetrical, with a linear Al—Cl—Al bond.

(a) What orbitals does Al use to bond the Cl atoms?

(b) What is the shape around each Al?

(c) What is the hybridization of the central Cl?

(d) What do the shape and hybridization suggest about the presence of lone pairs on the central chlorine?

**13.173** The bond angles in the nitrite ion, nitrogen dioxide, and the nitronium ion ($NO_2^+$) are 115°, 134°, and 180°, respectively. Rationalize these values using Lewis structures and VSEPR theory.

**13.174** A common method for producing a gaseous hydride is to treat a salt containing the anion of the volatile hydride with a strong acid.

(a) Write an equation for each of the following examples: (1) the production of HF from $CaF_2$; (2) the production of HCl from NaCl; (3) the production of $H_2S$ from FeS.

(b) In some cases even a weak acid such as water will suffice if the salt anion has a sufficiently strong attraction for protons. An example is the production of $PH_3$ from $Ca_3P_2$ and water. Write the equation for this reaction.

(c) By analogy, predict the products and write the equation for the reaction of $Al_4C_3$ with water.

**13.175** A former use of chlorine trifluoride in the United States was the production of uranium hexafluoride for the nuclear industry:

$$U(s) + 3ClF_3(l) \longrightarrow UF_6(l) + 3ClF(g)$$

How many grams $UF_6$ can form from 1 metric ton of uranium ore that is 1.55 mass % uranium and 12.75 L chlorine trifluoride (density = 1.88 g/mL)?

*\***13.176** Chlorine is used to make household bleach solutions containing 5.2 % (by mass) NaClO. Assuming 100% yield in the reaction producing NaClO from $Cl_2$, how many liters of $Cl_2(g)$ at STP will be needed to make 1000 L bleach solution (density = 1.07 g/mL)?

**13.177** The triatomic molecule ion $H_3^+$ was first detected and characterized by J. J. Thomson using mass spectrometry. Use the bond energy of $H_2$ (432 kJ/mol) and the proton affinity of $H_2$ ($H_2 + H^+ \longrightarrow H_3^+$; $\Delta H = -337$ kJ/mol) to calculate the heat of reaction for $H + H + H^+ \longrightarrow H_3^+$.

**13.178** Carbon monoxide can be used as a fuel. Use the heats of reaction for the following processes to compute the heat of combustion of CO:

$$C(s) + O_2(g) \longrightarrow CO_2(g)$$
$$C(s) + \tfrac{1}{2}O_2(g) \longrightarrow CO(g)$$

**13.179** In aqueous HF solution, an important species is the ion $HF_2^-$, which has the bonding arrangement FHF$^-$. Draw the Lewis structure for this ion and explain how it arises.

**13.180** Which of the following oxygen ions are paramagnetic: $O^+$, $O^-$, $O^{2-}$, $O^{2+}$?

*\***13.181** Hydrogen peroxide can act as either an oxidizing agent or a reducing agent.

(a) When $H_2O_2$ is treated with aqueous KI, $I_2$ forms. In which role is $H_2O_2$ acting? What is the oxygen-containing product formed?

(b) When $H_2O_2$ is treated with aqueous $KMnO_4$, the purple color of $MnO_4^-$ disappears and a gas forms. In which role is $H_2O_2$ acting? What is the oxygen-containing product formed?

## CHAPTER 14

# Organic Compounds and the Atomic Properties of Carbon

**Concepts and skills to review**

- electronegativity difference and bond polarity (Section 9.4)
- resonance structures (Section 9.5)
- VSEPR theory of molecular shape (Section 10.1)
- σ and π bonding; bond rotation (Section 10.3)
- types of intermolecular forces (Section 11.2)
- properties of the Period 2 elements (Section 13.2)
- properties of the Group 4A(14) elements (Section 13.6)

Crystals of an organic compound. Crystalline L-DOPA (dihydroxyphenylalanine) is shown here in a color-enhanced micrograph. Like thousands of other organic substances, this one is found in living things (within brain cells) and is synthesized industrially to treat disease (parkinsonism). Fuels, pesticides, and polymers consist of organic compounds also and, as you'll see in this chapter, their physical and chemical behavior depends on the properties of the carbon atom.

A living cell is a remarkable chemical system. It oxidizes food for energy, maintains the concentrations of numerous aqueous components, interacts with its environment, synthesizes molecules for its survival and growth, and even reproduces itself. This amazing chemical machine consumes, creates, and consists largely of *organic compounds.* Except for a few inorganic salts and ever-present water, everything you put into or on your body—food, medicine, cosmetics, and clothing—consists of organic compounds. Organic fuels warm our homes, cook our meals, and power our society. Major industries are devoted to producing organic compounds, such as polymers, pharmaceuticals, and insecticides.

Dictionaries define an organic compound as "a compound of carbon," but that definition would include carbonates, cyanides, carbides, cyanates, and other carbon-containing ionic compounds usually classified as inorganic. Here is a practical, working definition: all **organic compounds** contain carbon, nearly always bonded to itself and to hydrogen, and often to other elements as well.

In the early 19th century, many prominent thinkers believed that an unobservable spiritual energy, a "vital force," existed within the compounds of living things, which made them impossible to synthesize and fundamentally different from compounds of the mineral world. This idea of *vitalism* was challenged in 1828, when the young German chemist Friedrich Wöhler heated ammonium cyanate, a "mineral-world" compound, and produced urea, a "living-world" compound:

$$NH_4OCN \xrightarrow{\Delta} H_2N{-}\overset{\overset{\displaystyle O}{\|}}{C}{-}NH_2$$

Although Wöhler did not appreciate the significance of this reaction—he was more interested by the fact that two compounds had the same molecular formula—his experiment is considered a key event in the origin of organic chemistry. Chemists soon synthesized methane, acetic acid, acetylene, and many other organic compounds from inorganic sources. ◆ Today we know that *the same chemical principles govern organic and inorganic systems* because the behavior of a compound arises from the properties of its elements, no matter how marvelous that behavior may be.

**◆ Organic Chemistry Is Enough to Drive One Mad.** Like the alchemists, phlogistonists, and antiatomists, the vitalists did not disappear or change their beliefs overnight. Indeed, organic compounds do seem strange and different from inorganic compounds because of their structural and compositional complexity. Imagine the consternation of the 19th-century chemist accustomed to studying compounds with formulas such as $CaCO_3$, $Ba(NO_3)_2$, and $CuSO_4{\cdot}5H_2O$, who isolates white crystals from a gallstone and finds them to have an empirical formula of $C_{27}H_{46}O$ (cholesterol). Even Wöhler later said, "Organic chemistry . . . is enough to drive one mad. It gives me the impression of a primeval tropical forest, full of the most remarkable things, a monstrous and boundless thicket, with no way of escape, into which one may well dread to enter."

In this brief introduction to an enormous field, we discuss how the structure and reactivity of organic molecules emerge naturally from the properties of carbon and its handful of bonding partners. First, we review the rather special atomic properties of carbon in the context of organic molecules. From there, we classify the main types of organic reactions and apply them to the major families of organic compounds. Finally, we extend these ideas to the giant molecules of commerce and life: synthetic and natural polymers.

## 14.1 The Special Nature of Carbon and the Characteristics of Organic Molecules

Although there is nothing strange or mystical about organic molecules, their central role in biology and industry leads us to ask if some extraordinary attributes give carbon a special chemical personality. Of course, each element has its own specific properties, and carbon is no more unique than sodium, hafnium, or any other. However, the atomic properties of carbon give it bonding capabilities beyond those of any other element, which in turn lead to the two most obvious characteristics of organic molecules: structural complexity and chemical diversity.

## The Structural Complexity of Organic Molecules

Most organic molecules have much more complex structures than most inorganic molecules, and a quick review of carbon's atomic properties helps explain why. The ground-state electron configuration of [He] $2s^2 2p^2$—four electrons past He and four before Ne—means that the formation of carbon ions is energetically impossible under ordinary conditions: a $C^{4+}$ cation requires the total energy of $IE_1$ through $IE_4$ to form, and a $C^{4-}$ anion requires the sum of the endothermic steps $EA_2$ through $EA_4$. Carbon's position in the periodic table (Figure 14.1) gives it an electronegativity midway between the most metallic and nonmetallic elements of Period 2: Li = 1.0, C = 2.5, F = 4.0. Therefore, *carbon shares electrons to attain a filled shell,* bonding covalently in molecules, networks, and polyatomic ions.

The *number* and *strength* of carbon's bonds lead to its outstanding ability to *catenate* (bond to itself), by which it forms a multitude of chemically and thermally stable chain, ring, and branched compounds. *Carbon forms four bonds in virtually all its compounds,* a fact explained through the concept of hybridization (Section 10.3). Thus, a central C bonds in as many as four directions at once. *Carbon forms relatively short, strong bonds* as a result of its small size, which allows close approach to another atom and thus greater orbital overlap. Moreover, the C—C bond is short enough to allow the formation of *multiple bonds,* which restrict rotation of attached groups and add more possibilities for the shapes of carbon compounds (see Figure 10.22).

Although silicon and several other elements also catenate, none can compete with carbon. Atomic and bonding properties confer three crucial differences between C and Si chains. First, as atomic size increases down Group 4A(14), bonds between identical atoms become longer and weaker. Figure 14.2 shows that a C—C bond is much stronger than an Si—Si bond. Second, note that a C—C bond is nearly as strong as a C—O bond, so relatively little energy is released when a C chain reacts and one bond replaces the other. In contrast, Si—O bonds are much stronger than Si—Si bonds, so great energy is released when an Si chain reacts similarly. Third, unlike C, Si has *d* orbitals that can be attacked by the lone pairs of incoming reactants. These factors explain why, for example, ethane ($CH_3$—$CH_3$) does not react in air unless sparked and is stable in water, whereas disilane ($SiH_3$—$SiH_3$) ignites spontaneously in air and breaks down in water.

## The Chemical Diversity of Organic Molecules

In addition to their elaborate geometries, organic compounds are noted for their sheer number and diverse behavior. Look in any compendium of known compounds, such as the *Handbook of Chemistry and Physics,* and you'll find that the number of organic compounds dwarfs that of all the other elements combined! Several million organic compounds are known, and several thousand others are discovered or synthesized each year.

This incredible variety is also founded on atomic and bonding behavior. Carbon frequently forms bonds to **heteroatoms,** atoms other than C or H, in which it appears in all possible Group 4A(14) oxidation states (−4 through +4). The most common heteroatoms are N and O, but S, P, and the halogens are common also, and organic compounds with many other ele-

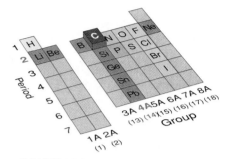

**FIGURE 14.1**

**The position of carbon in the periodic table.** Carbon lies at the center of Period 2, so it has an intermediate electronegativity (EN), and at the top of Group 4A(14), so it is relatively small. Other elements common in organic compounds are H, N, O, P, S, and the halogens.

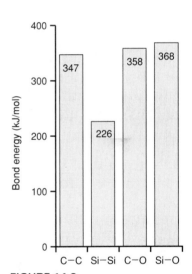

**FIGURE 14.2**

**Bond energy and the stability of carbon chains.** The C—C bond is stronger than the Si—Si bond, so it forms longer chains. Carbon chains are less easily oxidized than silicon chains because much less energy is released: C—C bond energy ≈ C—O bond energy, but Si—Si bond energy << Si—O bond energy.

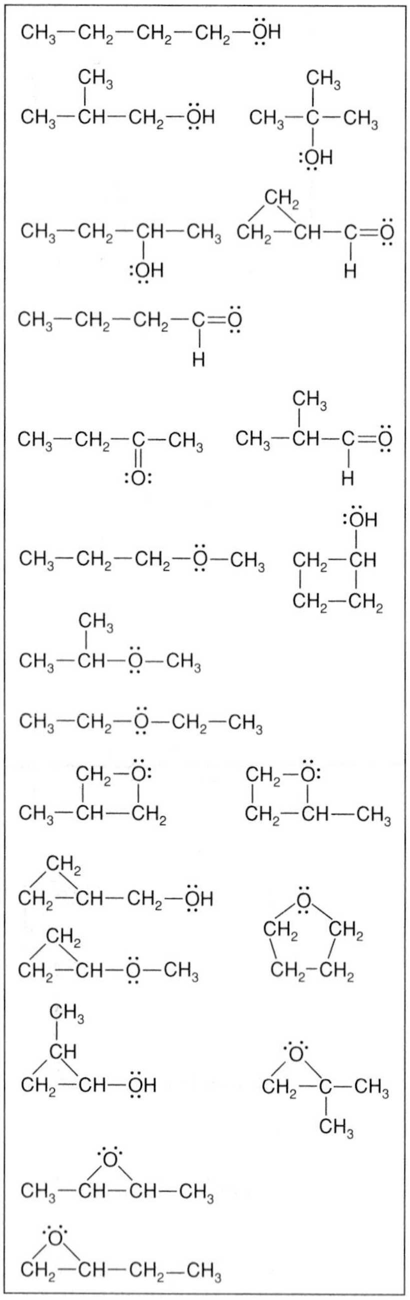

$CH_3-CH_2-CH_2-CH_2-\ddot{O}H$

$$CH_3-\underset{\underset{CH_3}{|}}{CH}-CH_2-\ddot{O}H \qquad CH_3-\underset{\underset{:\ddot{O}H}{|}}{\overset{\overset{CH_3}{|}}{C}}-CH_3$$

$$CH_3-CH_2-\underset{\underset{:\ddot{O}H}{|}}{CH}-CH_3 \qquad \underset{CH_2}{\overset{CH_2}{|}}-CH-\underset{\underset{H}{|}}{C}=\ddot{O}$$

$$CH_3-CH_2-CH_2-\underset{\underset{H}{|}}{C}=\ddot{O}$$

$$CH_3-CH_2-\underset{\underset{:O:}{|}}{\overset{\overset{CH_3}{|}}{C}}-CH_3 \qquad CH_3-\underset{\underset{H}{|}}{\overset{\overset{CH_3}{|}}{CH}}-C=\ddot{O}$$

$$CH_3-CH_2-CH_2-\ddot{O}-CH_3 \qquad \underset{CH_2-CH_2}{\overset{:\ddot{O}H}{|}}{CH_2-CH}$$

$$CH_3-\underset{\underset{CH_3}{|}}{CH}-\ddot{O}-CH_3$$

$$CH_3-CH_2-\ddot{O}-CH_2-CH_3$$

$$CH_3-\underset{\underset{CH_2}{|}}{\overset{\overset{CH_2-\ddot{O}:}{|}}{CH}} \qquad CH_2-\underset{\underset{CH_3}{|}}{CH}-CH_3$$

$$CH_2-CH-CH_2-\ddot{O}H$$

$$CH_2-CH-\ddot{O}-CH_3$$

$$CH_2-CH-\ddot{O}H$$

$$CH_3-CH-CH-CH_3$$

$$CH_2-CH-CH_2-CH_3$$

**FIGURE 14.3**
**The chemical diversity of organic compounds.** Different arrangements of chains, branches, rings, and heteroatoms give rise to many structures. There are 21 different compounds possible from just four single-bonded C atoms, one O atom, and the necessary H atoms.

ments are known as well. To get an idea of the diversity created by a heteroatom, look at Figure 14.3, which shows that 21 different compounds result from various sequences of four C atoms (no multiple bonds), the necessary number of H atoms, and just one O atom.

How does such chemical diversity lead to the wide range of organic reactivity? Most reactions start—that is, a new bond begins to form—*when a region of high electron density on one molecule meets a region of low electron density on another.* These regions may be due to the presence of a multiple bond or to the partial charges that occur in carbon-heteroatom bonds.

Consider three common bonds found in organic molecules: the C—C, C—H, and C—O bonds. In the first, C is bonded to itself, so the EN values are equal and the bond is nonpolar. In general, C—C bonds, which make up the major portion of most organic molecules, are relatively *unreactive*. The C—H bond is nearly nonpolar because it is short (109 pm) and the EN values of H (2.1) and C (2.5) are close. Thus, C—H bonds, which are plentiful in most organic molecules, are largely *unreactive* as well. In contrast, the C—O bond is highly polar ($\Delta EN = 1.0$), with the O end of the bond electron rich ($\delta^-$) and the C end electron poor ($\delta^+$). This charge imbalance makes the bond *reactive* and, given appropriate conditions, reaction will occur there. Even when a carbon-heteroatom bond has a small $\Delta EN$, as in C—Br ($\Delta EN = 0.3$), or none at all, as in C—S ($\Delta EN = 0$), the heteroatoms are large, so their bonds to carbon are long, weak, and thus reactive.

Reaction takes place at a site on the molecule called a **functional group,** a specific combination of atoms, typically a carbon multiple bond and/or carbon-heteroatom bond, that reacts in a characteristic way no matter in which molecule it occurs. A particular bond may be a functional group or part of several. For example, the C—O bond is part of four functional groups: the alcohol group ($-\overset{|}{\underset{|}{C}}-\ddot{O}-H$), the ether group ($-\overset{|}{\underset{|}{C}}-\ddot{O}-\overset{|}{\underset{|}{C}}-$), the carboxylic acid group ($-\overset{\overset{:O:}{\|}}{C}-\ddot{O}-H$), and the ester group ($-\overset{\overset{:O:}{\|}}{C}-\ddot{O}-\overset{|}{\underset{|}{C}}-$).

**Section Summary**
The structural complexity of organic compounds arises from carbon's small size, intermediate EN, four bonding orbitals, ability to form $\pi$ bonds, and nonexpandable valence shell. These factors lead to chains, branches, and rings of strong, chemically resistant bonds that extend in as many as four directions. The chemical diversity of organic compounds arises from carbon's ability to bond to itself and to many other elements and to form multiple bonds to itself and to O and N. These factors lead to compounds that contain functional groups, regions of uneven electron density that react in specific ways.

## 14.2 The Structures and Classes of Hydrocarbons

Here is a fanciful, anatomical analogy between an organic molecule and an organism. The carbon-carbon bonds form the skeleton, with the longest continual chain the backbone and any branches the limbs. Covering the skeleton is a skin of hydrogen atoms, with functional groups protruding at specific locations, like chemical fingers ready to grab an incoming reactant.

In this section, we "dissect" one group of compounds down to their skeletons and see how to name and draw them. **Hydrocarbons,** the simplest type of organic compound, are a large group of substances containing only H and C atoms. Some common fuels are hydrocarbon mixtures, such as natural gas and gasoline. Hydrocarbons are also important *feedstocks*, precursor reactants used to make other compounds. Ethylene, acetylene, and benzene, for example, are feedstocks for hundreds of other substances.

## Carbon Skeletons and Hydrogen Skins

Let's begin by examining the possible bonding arrangements of C atoms only (we'll leave off the H atoms at first) in simple skeletons without multiple bonds or rings. To distinguish different arrangements, focus on the *sequence of C atoms* and keep in mind that *single (sigma) bonds are free to rotate* (Section 10.3).

A one-, two-, or three-C chain can be arranged in only one way. Whether you draw three C atoms in a line or with a bend, the sequence is the same. Four C atoms, however, have two possible skeletons: a four-C chain or a three-C chain with a one-C branch at the central C:

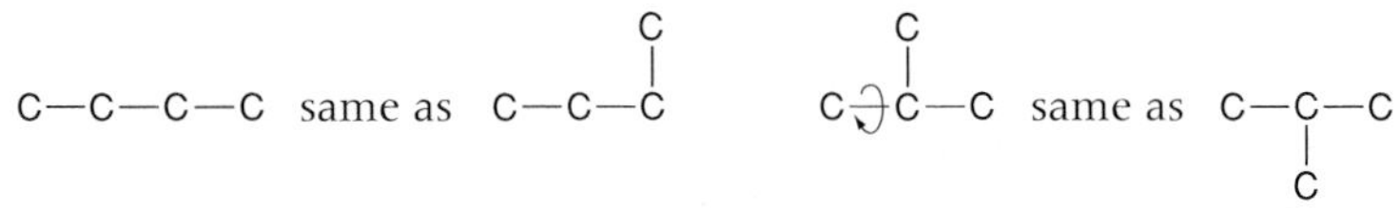

C—C—C—C  same as  C—C—C    C—C—C  same as  C—C—C

Notice that if the branch is added to either end of the three-C chain, it would simply be a bend in a four-C chain, not a different sequence. Similarly, if the branch points down instead of up, it represents the same sequence because single bonds rotate.

As the total number of C atoms increases, the number of different arrangements increases as well. Five C atoms have 3 arrangements; 6 C atoms can be arranged in 5 ways, 7 C atoms in 9 ways, 10 C atoms in 75 ways, and 20 C atoms in more than 300,000 ways! If we now consider multiple bonds and rings, the number of arrangements increases further: including one C=C bond in the 5-C skeletons creates 5 more arrangements, and including one ring creates 5 more (Figure 14.4).

When determining the number of different skeletons, remember that
* Each C can form a *maximum* of four single bonds, two single and one double bond, or one single and one triple bond.
* The *sequence* of C atoms determines the skeleton, so a straight chain and a bent chain represent the same skeleton.
* Single bonds *rotate* freely, so a branch pointing down is the same as one pointing up.

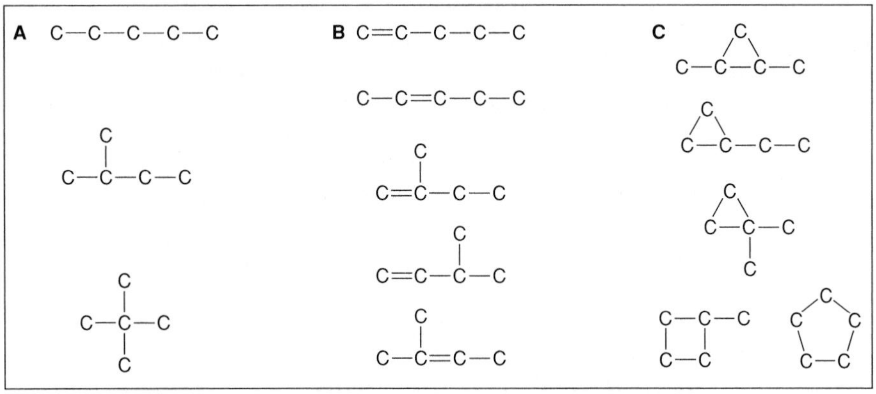

**FIGURE 14.4**
**Some five-carbon skeletons. A,** Three 5-C skeletons are possible with only single bonds. **B,** Five more skeletons are possible with one C=C bond present. **C,** Five more skeletons are possible with one ring present. Even more would be possible with a ring *and* a double bond.

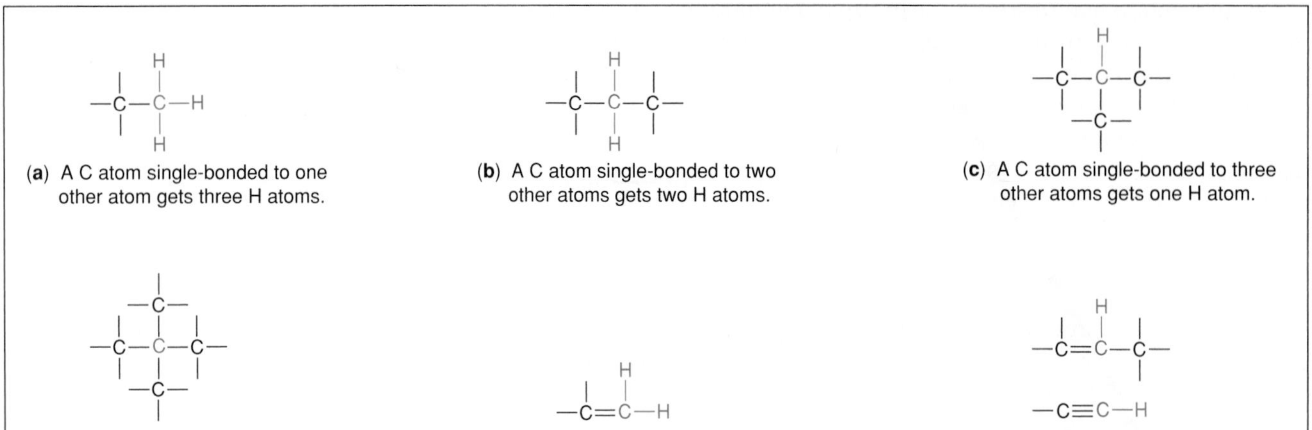

(a) A C atom single-bonded to one other atom gets three H atoms.

(b) A C atom single-bonded to two other atoms gets two H atoms.

(c) A C atom single-bonded to three other atoms gets one H atom.

(d) A C atom single-bonded to four other atoms is already fully bonded (no H atoms).

(e) A double-bonded C atom is treated as if it were bonded to two other atoms.

(f) A double- and single-bonded C atom or a triple-bonded C atom is treated as if it were bonded to three other atoms.

**FIGURE 14.5**

**Adding the H-atom skin to the C-atom skeleton.** In a hydrocarbon, each C atom bonds to as many H atoms as needed to give the C a total of four bonds.

If we put a hydrogen "skin" on a carbon skeleton, we obtain a hydrocarbon. Figure 14.5 shows that the skeleton has the correct number of H atoms when each C has four bonds. Sample Problem 14.1 provides practice in drawing hydrocarbons.

SAMPLE PROBLEM 14.1 _____

**Drawing Hydrocarbons**

**Problem:** Draw structures for hydrocarbons that have different atom sequences with
(a) Six C atoms, no multiple bonds, and no rings
(b) Four C atoms, one double bond, and no rings
(c) Four C atoms, no multiple bonds, and one ring
**Plan:** In each case, we draw the longest carbon chain and then work down to smaller chains with branches at different points along them. The process typically involves trial and error. Then we add H atoms to give each C a total of four bonds.
**Solution:**
(a) Compounds with six C atoms:

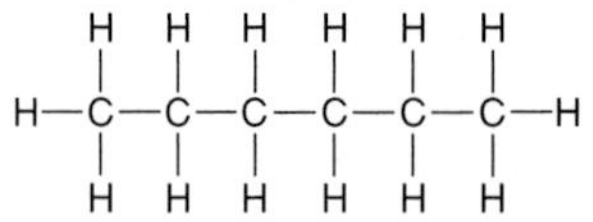

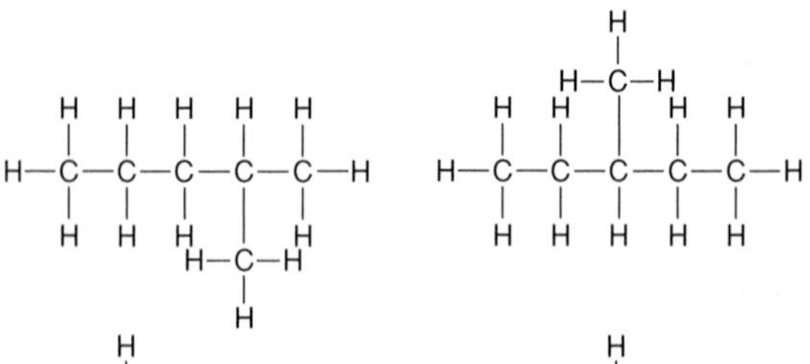

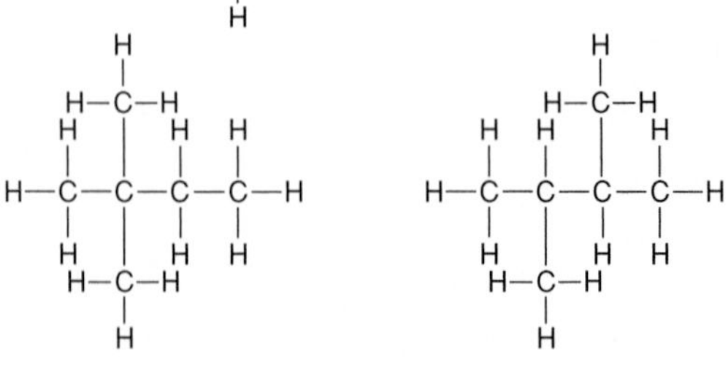

**(b)** Compounds with four C atoms and one double bond:

4-C chains:

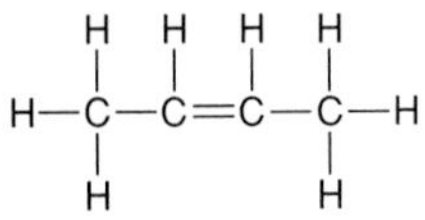

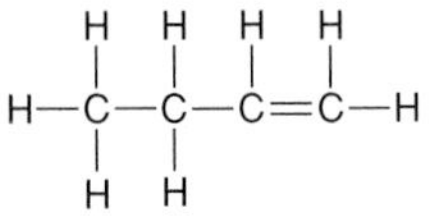

3-C chain:

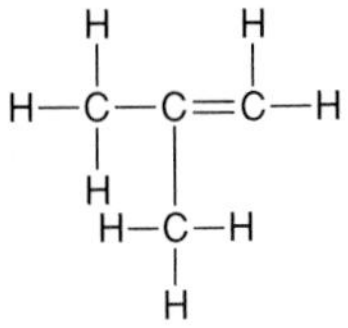

**(c)** Compounds with four C atoms and one ring:

4-C ring: 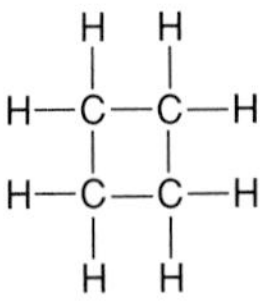                3-C ring:

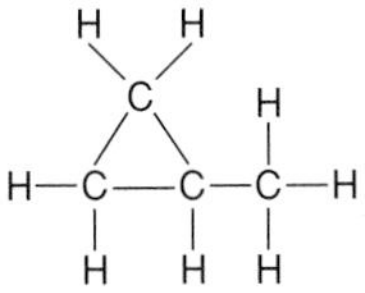

**Check:** Be sure each skeleton has the correct number of C atoms and bonds and that no sequences are repeated or omitted; remember that a double bond counts as two bonds.

**Comment:** Avoid some common mistakes:

In **(a):** C—C—C—C—C  is the same skeleton as  C—C—C—C—C

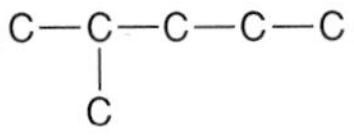

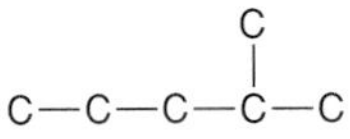

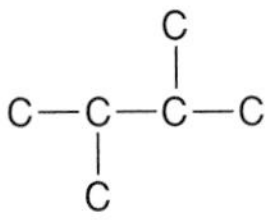

C—C—C—C  is the same skeleton as  C—C—C—C

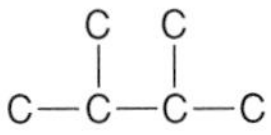

In **(b):** C—C—C=C is the same skeleton as C=C—C—C

Too many bonds in

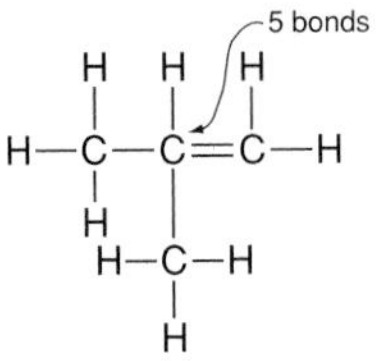

In **(c):** Too many bonds in

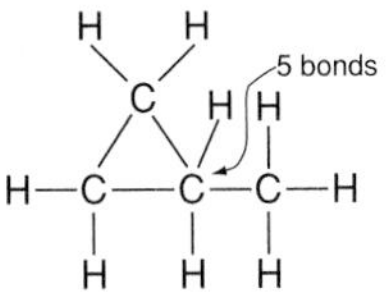

**FOLLOW-UP PROBLEM 14.1**
Draw all hydrocarbons that have different atom sequences with
**(a)** Seven C atoms, no multiple bonds, and no rings (nine arrangements)
**(b)** Five C atoms, two double bonds, and no rings (six arrangements)
**(c)** Five C atoms, one triple bond, and no rings (three arrangements)

The enormous number of hydrocarbons can be classified into four main groups. For the remainder of this section, we examine some structural features and physical properties of each group. Later, we discuss the chemical behavior of the hydrocarbons.

**TABLE 14.1 Numerical Roots for Carbon Chains and Branches**

| ROOT | NUMBER OF C ATOMS |
|------|-------------------|
| meth- | 1 |
| eth- | 2 |
| prop- | 3 |
| but- | 4 |
| pent- | 5 |
| hex- | 6 |
| hept- | 7 |
| oct- | 8 |
| non- | 9 |
| dec- | 10 |

## Alkanes: Hydrocarbons with Only Single Bonds

A hydrocarbon that contains only single bonds is an **alkane** (general formula $C_nH_{2n+2}$, where $n$ is an integer). Because each C is bonded to the maximum number of other atoms, alkanes are referred to as **saturated hydrocarbons.**

**Naming alkanes.** The key point to remember in naming any organic compound is that *each chain, branch, or ring has a name based on the **number** of C atoms.* The entire name of the compound has three portions:

- *Root:* the root tells the number of C atoms in the longest *continuous* chain in the molecule. The roots are shown in Table 14.1. There are special roots for chains with one to four C atoms; the roots of longer chains are based on Greek numbers.
- *Prefix:* placed *before* the root, the prefix tells the group(s) attached to the main chain and the carbon(s) to which they are attached.
- *Suffix:* placed *after* the root, the suffix tells the class of organic compound to which the molecule belongs.

For example, in the name 2-methylbutane, "2-methyl" is the prefix (a methyl group is attached to C-2 of the main chain), "but" is the root (the main chain has four C atoms), and "ane" is the suffix (the compound is an alkane).

The alkanes are used here to demonstrate this general approach to naming; other organic compounds are named in the same way, with additional prefixes and suffixes. In addition to the *systematic* names introduced here, many compounds are still known by *common* names, so we'll note the most important of them. To name a compound,

- First name the longest chain (root).
- Then add the compound type (suffix).
- Then name the branch (prefix).

Table 14.2 presents the rules for naming an alkane and applies them to a component of gasoline.

## TABLE 14.2 Rules for Naming an Organic Compound

1. Naming the longest chain (root)
   - (a) Find the longest *continuous* chain of C atoms.
   - (b) Select the root that corresponds to the number of C atoms in this chain.
2. Naming the compound type (suffix)
   - (a) For alkanes, add the suffix *-ane* to the chain root. (Other suffixes appear in Table 14.5 with their functional group and compound type.)
   - (b) If the chain forms a ring, the name is preceded by *cyclo-*.
3. Naming the branches (prefix)
   - (a) Each branch name consists of a subroot (number of C atoms) and the ending *-yl* to signify that it is not part of the main chain.
   - (b) Branch names precede the chain name. When two or more branches are present, name them in *alphabetical* order.
   - (c) To specify where the branch occurs along the chain, number the main-chain C atoms consecutively, starting at the end *closest* to a branch, to achieve the *lowest* numbers for the branches. Precede each branch name with the number of the chain C to which that branch is attached.
   - (d) If the compound has no branches, the name consists of the root and suffix.

5 carbons ⇒ pent-

pent- + -ane ⇒ pentane

ethylmethylpentane

3-ethyl-2-methylpentane

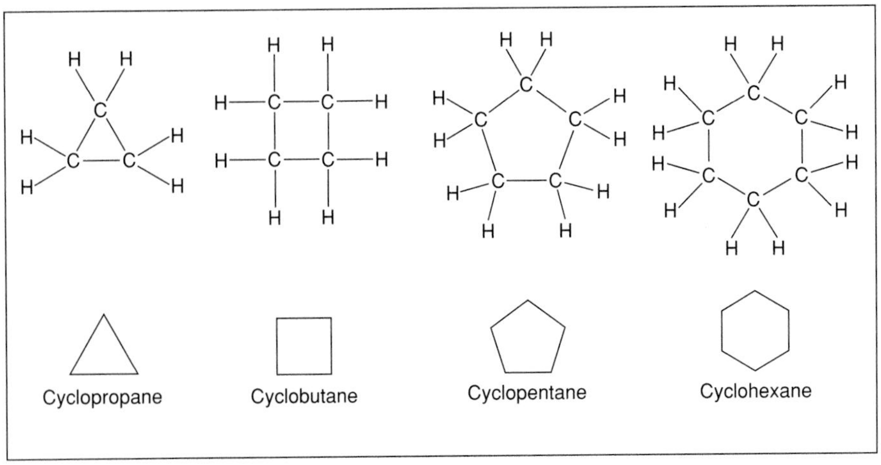

**FIGURE 14.6**
**An abbreviated way to draw cycloalkanes.**
Cycloalkanes are usually drawn as regular polygons. Each side is a C—C bond, and each corner represents one C atom with its required number of H atoms.

Cyclopropane   Cyclobutane   Cyclopentane   Cyclohexane

**Depicting alkanes.** Organic compounds are depicted with expanded and condensed structural formulas. The *expanded formula* shows each atom and bond; this type appears in Sample Problem 14.1. One common type of *condensed formula* groups the H atoms with the C atom to which they are bound. For example, the expanded and condensed formulas of the compound named in Table 14.2 are

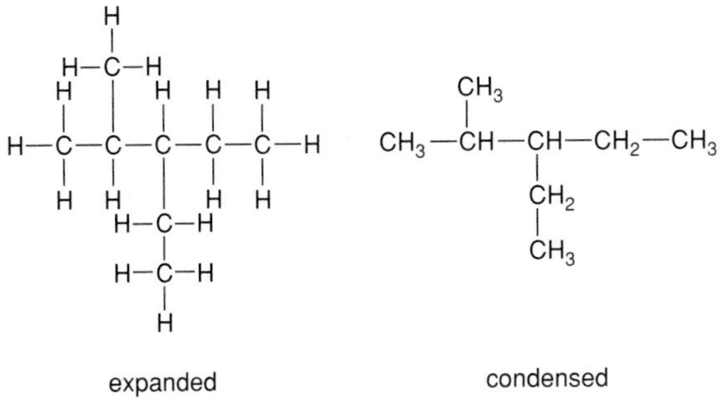

expanded                    condensed

A **cyclic hydrocarbon** contains one or more rings in its structure. Each C in the ring has two bonds already, so it can form two others; thus, *cycloalkanes* have the general formula $C_nH_{2n}$. The rings are usually drawn with *only* C—C bond lines. Each corner of the ring is *understood* to consist of a C atom and the correct number of H atoms to give a total of four bonds; structures for some cycloalkanes are shown in Figure 14.6. To minimize electron repulsions between adjacent H atoms and to maximize orbital overlap of adjacent C atoms, *cyclic molecules are nonplanar* (except for three-C rings). The most stable form of cyclohexane, for example, is called the *chair conformation* (Figure 14.7). Each C is bonded to an *axial* H atom, which points above or below the ring, and an *equatorial* H atom, which points out from the ring.

**Structural isomerism and the physical properties of alkanes.** Two or more compounds with the same molecular formula but different properties are called **isomers.** Those with *different sequences of bonded atoms* are **structural isomers;** alkanes with a given number of C atoms but different skeletons are examples. The smallest alkane to exhibit structural isomerism has four C atoms: two different compounds have the formula $C_4H_{10}$ (Table 14.3). The unbranched one is butane (common name, *n*-butane;

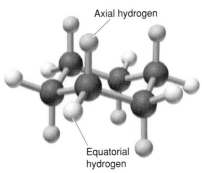

Axial hydrogen

Equatorial hydrogen

**FIGURE 14.7**
**The chair conformation of cyclohexane.**
Because of its tetrahedral bond angles, cyclohexane is not planar and usually exists in a conformation that looks like a chair. Axial hydrogens are shown blue and equatorial hydrogens gray.

**TABLE 14.3   The Structural Isomers of $C_4H_{10}$ and $C_5H_{12}$**

| SYSTEMATIC NAME (COMMON NAME) | CONDENSED FORMULA | EXPANDED FORMULA | SPACE-FILLING MODEL | DENSITY (g/mL) | BOILING POINT (°C) |
|---|---|---|---|---|---|
| Butane (*n*-butane) | $CH_3-CH_2-CH_2-CH_3$ | 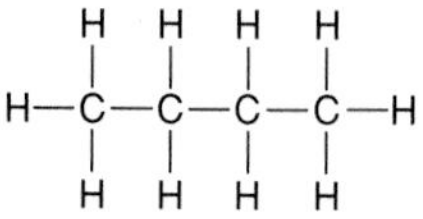 | 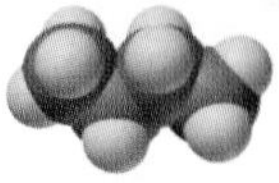 | 0.579 | −0.5 |
| 2-Methylpropane (isobutane) | $CH_3-CH-CH_3$ with $CH_3$ branch | 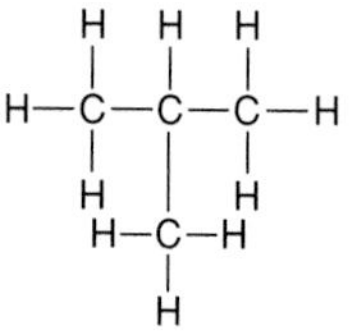 | 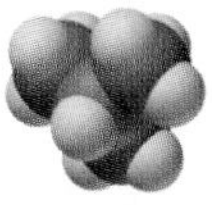 | 0.549 | −11.6 |
| Pentane (*n*-pentane) | $CH_3-CH_2-CH_2-CH_2-CH_3$ | 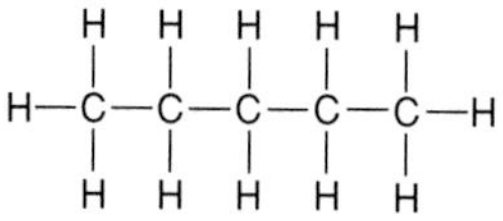 | 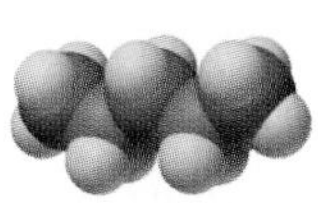 | 0.626 | 36.1 |
| 2-Methylbutane (isopentane) | $CH_3-CH-CH_2-CH_3$ with $CH_3$ branch | 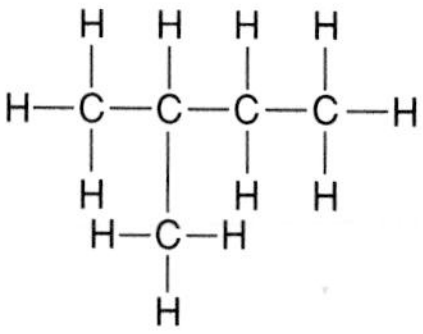 | 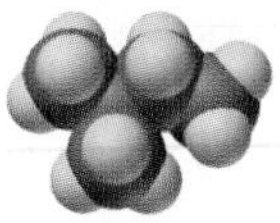 | 0.620 | 27.8 |
| 2,2-Dimethylpropane (neopentane) | $CH_3-\overset{CH_3}{\underset{CH_3}{C}}-CH_3$ | 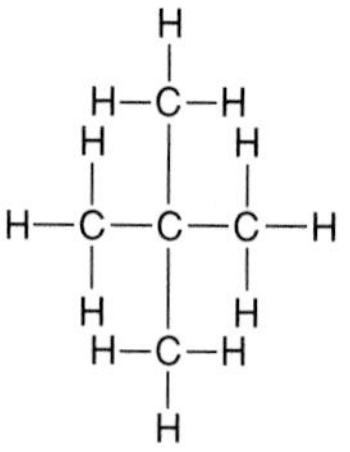 | 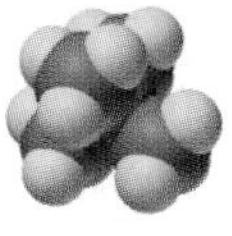 | 0.614 | 9.5 |

*n*- stands for "normal"), and the other is 2-methylpropane (common name, *iso*butane). Similarly, three compounds have the formula $C_5H_{12}$. The unbranched isomer is pentane (common name, *n*-pentane); the one with a methyl group at C-2 of a four-C chain is 2-methylbutane (common name, *iso*pentane). The third isomer has two methyl branches off C-2 of a three-C chain, so its name is 2,2-dimethylpropane (common name, *neo*pentane).

Because alkanes are nearly nonpolar, we would expect their physical properties to be determined by dispersion forces, and the boiling points in Table 14.3 certainly bear this out. The four-C alkanes boil lower than the five-C compounds. Moreover, within each group of isomers, the more spherical member (isobutane or neopentane) boils lower than the more elongated one (*n*-butane or *n*-pentane). As you saw in Chapter 11, this trend occurs because a spherical shape makes less intermolecular contact, and thus exerts a lower total dispersion force, than does an elongated shape.

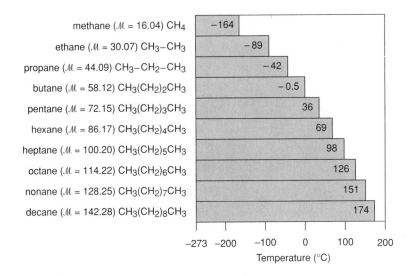

**FIGURE 14.8**
**Boiling points of the first 10 unbranched alkanes.** Boiling point increases smoothly with chain length because dispersion forces increase. Each entry includes the name, molar mass ($\mathcal{M}$, in g/mol), abbreviated formula, and normal boiling point.

A particularly clear example of the effect of dispersion forces on physical properties occurs among the unbranched alkanes (*n*-alkanes). These compounds comprise a **homologous series,** one in which each member differs from the next by a $—CH_2—$ (methylene) group. Within this series, boiling points increase steadily with chain length: the longer the chain, the greater the contact, the stronger the dispersion forces, and the higher the boiling point (Figure 14.8). Pentane (five C atoms) is the smallest *n*-alkane that exists as a liquid at room temperature. The solubility behavior of the alkanes, and of all hydrocarbons, is easy to predict from the like-dissolves-like rule (Section 12.1). Alkanes are often miscible in each other and in other nonpolar solvents, such as benzene, but nearly insoluble in water. The water solubility of pentane, for example, is only 0.36 g/L.

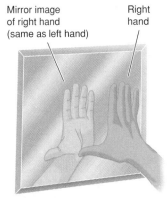

**FIGURE 14.9**
**An analogy for optical isomers.** The reflection of your right hand looks like your left hand. Because each hand is asymmetric, you cannot superimpose one on the other.

## Chiral Molecules and Optical Isomerism

Another type of isomerism that we see in some alkanes and many other organic (and inorganic) compounds is called **stereoisomerism.** It occurs in molecules with the same sequence of atoms *but different orientations of groups in space.* **Optical isomerism** is a type of stereoisomerism that arises when *an object and its mirror image cannot be superimposed on each other.* To use a familiar example, your right hand is an optical isomer of your left. Look at your right hand in a mirror and you will see that the *image* is identical to your left hand (Figure 14.9). No matter how you twist your arms around, however, your hands cannot lie on top of each other with all parts superimposed. They are not superimposable because each is *asymmetric:* there is no plane of symmetry that divides your hand into two identical parts.

An asymmetric molecule is called **chiral** (Greek *cheir,* "hand"). In most cases, *an organic molecule is chiral if it contains a C atom that is bonded to four different groups.* In 3-methylhexane, for example, C-3 is a *chiral center* (asymmetric carbon) because it is bonded to H—, $CH_3—$, $CH_3—CH_2—$, and $CH_3—CH_2—CH_2—$ (Figure 14.10, *A*). The central C atom in the amino acid alanine is also a chiral center. The two forms in Figure 14.10, *B,* are mirror images and cannot be superimposed on each other: when two of the groups are superimposed, the other two are opposite each other. Thus, the two forms are optical isomers. (Molecular models make this fact easier to see.)

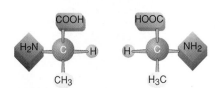

**A** 3-Methylhexane

**B** Optical isomers of alanine

**FIGURE 14.10**
**Two chiral molecules. A,** The molecule 3-methylhexane is chiral because C-3 is bonded to four different groups. **B,** The central C in the amino acid alanine is also bonded to four different groups. The two optical isomers of alanine are mirror images that cannot be superimposed on each other.

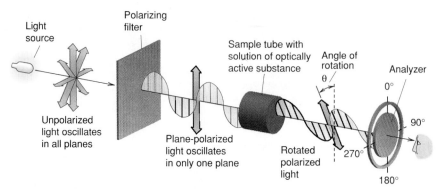

**FIGURE 14.11**

**The rotation of plane-polarized light by an optically active substance.** The source emits light moving in all planes. When the light passes through the first polarizing filter, only one plane emerges. The plane-polarized beam enters the sample compartment, which contains a known concentration of an optical isomer, and the plane rotates as it passes through the solution. The second polarizing filter (called the analyzer) is attached to a movable ring calibrated in degrees that is used to measure the angle of rotation.

Unlike structural isomers, optical isomers are identical in all but two respects:

1. In their physical properties, *optical isomers differ only in the direction that each isomer rotates the plane of polarized light.* A **polarimeter** is used to measure the angle that the plane is rotated (Figure 14.11). A beam of light consists of waves moving in all planes. A polarizing filter blocks all waves except those in one plane, so the light emerging through the filter is *plane-polarized.* An optical isomer is **optically active** because it rotates the plane of this polarized light. The *dextrorotatory* isomer (designated *d* or +) rotates the plane of light to the right; the *levorotatory* isomer (designated *l* or −) is the mirror image of the first and rotates the plane to the left. An equimolar mixture of the two isomers does not rotate the plane at all because the dextrorotation cancels the levorotation. The *specific rotation* is characteristic of the isomer at a certain temperature, concentration, and wavelength of light.

2. In their chemical properties, *optical isomers differ only in a chiral (asymmetric) chemical environment,* one that distinguishes "right-handed" and "left-handed" molecules. Typically, one isomer of an optically active reagent is added to a mixture of optical isomers of another compound. For example, the addition of *d*-lactate to a mixture of *d*-alanine and *l*-alanine is used to crystallize the *d* isomer separately from the *l* isomer.

Optical isomerism plays a vital role in living cells. Nearly all carbohydrates and amino acids are optically active, but only one of the isomers is biologically usable. For example, *d*-glucose is metabolized for energy, but *l*-glucose would be excreted unused. Similarly, *l*-alanine is incorporated naturally into proteins, but *d*-alanine is not. The organism distinguishes one optical isomer from the other through its enzymes, large molecules that speed virtually every reaction in the cell by binding to the reactants. A specific portion of the enzyme provides the asymmetric environment, so only one of the isomers can bind there and undergo reaction (Figure 14.12).

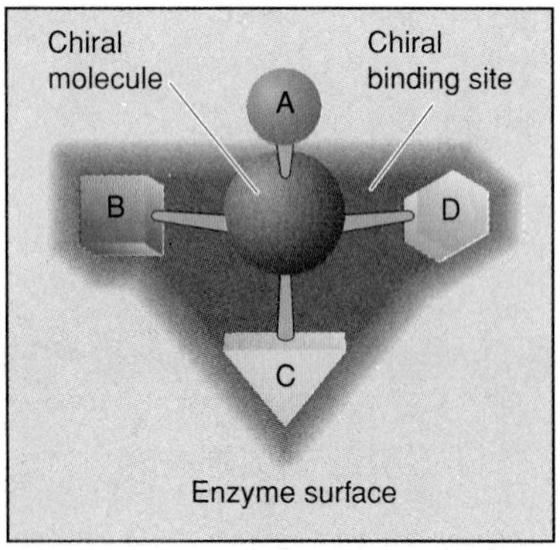

### Alkenes: Hydrocarbons with Double Bonds

A hydrocarbon that contains at least one C=C bond is called an **alkene** (general formula, $C_nH_{2n}$). Since the double-bonded C atoms bond fewer

**FIGURE 14.12**

**The binding site of an enzyme.** Organisms utilize only one of a pair of optical isomers because their enzymes have chiral (asymmetric) binding sites. The mirror image of this molecule could not bind at this enzyme site.

than the maximum of four atoms, alkenes are called **unsaturated hydro-carbons.**

Alkene names differ from those of alkanes in two respects:
1. The root chain *must* contain both C atoms of the double bond, even if it is not the longest chain. The chain is numbered from the end *closer* to the C=C bond, and the position of the bond is indicated by the number of the *first* C atom in it.
2. The suffix for alkenes is *-ene.*

For example, there are three 4-C alkenes ($C_4H_8$), two unbranched and one branched (see part b of Sample Problem 14.1). The unbranched isomer with the C=C bond between C-1 and C-2 is 1-butene, and that with the C=C bond between C-2 and C-3 is 2-butene; the branched isomer is 2-methyl-propene.

**The C=C bond and geometric isomerism.** There are two major structural differences between alkenes and alkanes. First, alkanes have a *tetrahedral* geometry ($\sim 109.5°$) around each C atom, whereas the double-bonded C atoms in alkenes are *trigonal planar* ($\sim 120°$). Second, the C—C bond *allows* rotation, so the atoms in an alkane continually change their relative positions. In contrast, the $\pi$ bond of the C=C bond *restricts* rotation, which fixes the relative positions of the bonded atoms.

This rotational restriction leads to another type of stereoisomerism: **geometric isomers** have different orientations of groups around a double bond (or similar structural feature). The alkene 2-butene, for example, has two geometric isomers (Table 14.4). *cis*-2-Butene has the $CH_3$ groups on the *same* side of the C=C bond, and *trans*-2-butene has them on *opposite* sides of the C=C bond. In general, a *cis* isomer has the *larger portions of the main chain on the same side* of the double bond. Note that each C atom in the C=C bond must have two *different* groups. (This type of geometric isomerism is also called *cis-trans isomerism,* and it occurs in many transition metal compounds as well, as we discuss in Chapter 22.) Note that the two 2-butenes are quite different in appearance *and* physical properties. The *cis* compound has a bend in the direction of the chain that the *trans* compound lacks. In Chapters 10 and 11, you saw how such a difference affects molecular polarity and physical properties. The upcoming Chemical Connections essay shows how this simple difference in geometry has profound effects in biological systems as well.

TABLE 14.4  **The Geometric Isomers of 2-Butene**

| SYSTEMATIC NAME | CONDENSED FORMULA | SPACE-FILLING MODEL | DENSITY (g/mL) | BOILING POINT (°C) |
|---|---|---|---|---|
| *cis*-2-Butene | 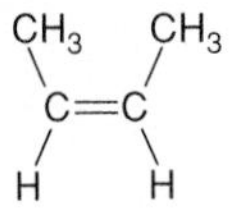 $\begin{matrix} CH_3 \quad CH_3 \\ \diagdown \quad \diagup \\ C=C \\ \diagup \quad \diagdown \\ H \qquad H \end{matrix}$ | 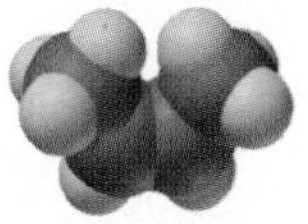 | 0.621 | 3.7 |
| *trans*-2-Butene | 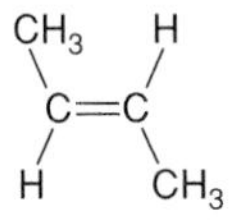 $\begin{matrix} CH_3 \quad H \\ \diagdown \quad \diagup \\ C=C \\ \diagup \quad \diagdown \\ H \qquad CH_3 \end{matrix}$ |  | 0.604 | 0.9 |

# Chemistry in Sensory Physiology:
# Isomers and the Chemistry of Vision

Of all our senses, sight provides the most information about the world. Light bouncing off objects enters the lens of the eye and is focused on the retina. From there, molecular signals are converted to mechanical signals and then to electrical signals that are transmitted to the brain. This remarkable sequence is one of the few physiological processes that we understand thoroughly at the molecular level. The first step relies on the different shapes of geometric isomers.

The molecule responsible for receiving the light energy is *retinal*. It is derived from retinol (vitamin A), which we obtain mostly from β-carotene in yellow and green vegetables. Retinal is a 20-C compound consisting of a 15-C chain and five 1-C branches (Figure 14.A). The chain includes five C=C bonds, a 6-C ring at one end, and a C=O bond at the other. The two biologically occurring isomers of retinal are the all-*trans* isomer, which has a *trans* orientation around all five double bonds, and the 11-*cis* isomer, which has a *cis* orientation around the C=C bond between C-11 and C-12.

Certain cells of the retina are densely packed with *rhodopsin*, which consists of 11-*cis*-retinal covalently bonded to a protein. The initial chemical event in vision occurs when rhodopsin absorbs a photon of visible light. The energy of visible photons (165 to 293 kJ/mol) lies in the range needed to break a C=C π bond (~250 kJ/mol). Retinal is bonded to the protein in such a way that makes the 11-*cis* π bond most susceptible to break-

age. The photon is absorbed in a few trillionths of a second, the *cis* π bond breaks, the intact σ bond rotates, and a π bond re-forms to produce all-*trans*-retinal in another few millionths of a second. In effect, light energy is converted into mechanical energy as the chain moves.

This rapid and relatively large change in the shape of retinal causes the protein portion of rhodopsin to change shape as well, a process that breaks the bond to retinal. The change in protein shape also triggers a flow of ions into the retina cells, initiating electrical impulses to the optic nerve, which leads to the brain. Meanwhile, the free all-*trans*-retinal diffuses away and is changed back to the *cis* form, which binds to the protein again.

Retinal must be ideally suited to its function because it has been selected through evolution as a photon absorber in organisms as different as purple bacteria, mollusks, insects, and vertebrates. The features that make it the perfect choice are its strong absorption in the visible region, the efficiency with which light converts the *cis* to the *trans* form, and the large structural change it undergoes as a result.

**FIGURE 14.A**

**The initial chemical event in vision.** The geometric isomer 11-*cis*-retinal is bonded to a small portion of the protein. A photon absorbed by the retinal breaks the 11-*cis* π bond. The retinal chain rotates, and the π bond forms again with a *trans* orientation. This change in retinal shape changes the protein shape, causing it to release the all-*trans*-retinal.

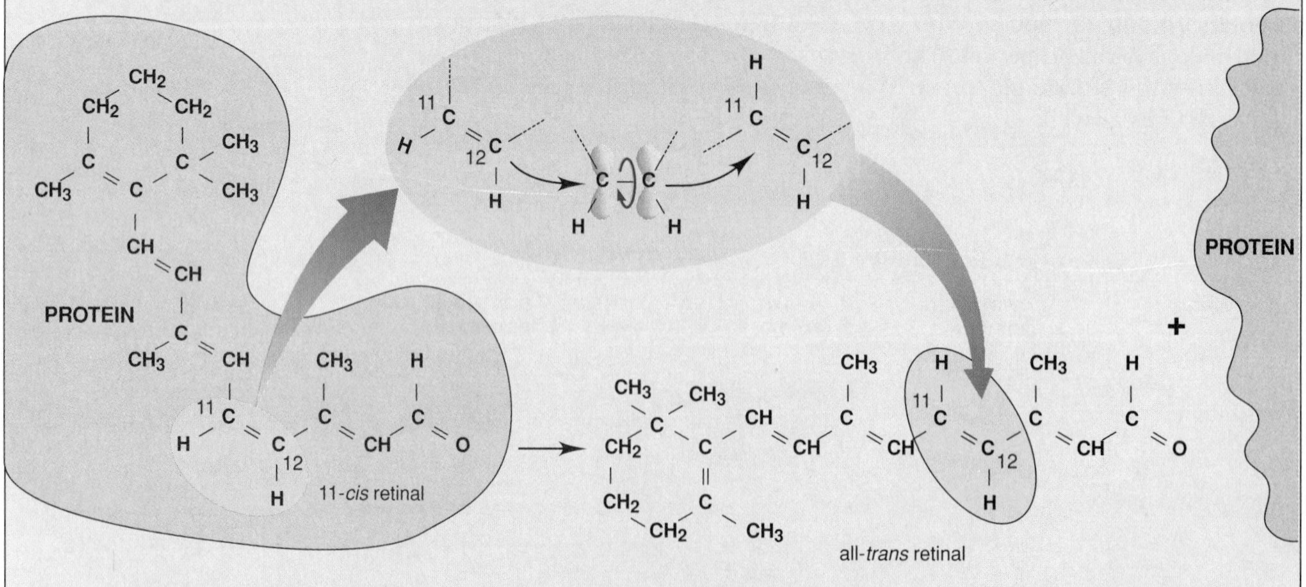

### Alkynes: Hydrocarbons with Triple Bonds

Hydrocarbons that contain at least one C≡C bond are called **alkynes** (general formula $C_nH_{2n-2}$). Since a carbon in a C≡C bond can bond to only one other atom, the geometry around each C atom is linear (180°). Alkynes are named in the same way as alkenes, except that the suffix is -*yne*. Because of their localized $\pi$ electrons, C=C and C≡C bonds are electron rich and act as functional groups. As we'll discuss in Section 14.4, alkenes and alkynes are much more reactive than alkanes, with reaction occurring at the multiple bond.

Sample Problem 14.2 brings together many ideas about hydrocarbon names and isomers.

SAMPLE PROBLEM 14.2 _____

### Naming and Drawing Alkanes, Alkenes, and Alkynes

**Problem:** Give the systematic name for each of the following, indicate the chiral center in part (d), and draw two geometric isomers for part (e):

(a) 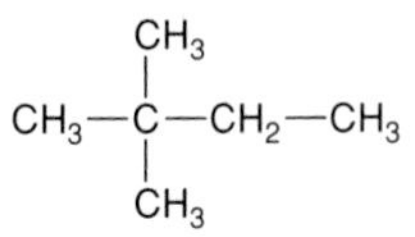 $CH_3-\overset{\overset{\displaystyle CH_3}{|}}{\underset{\underset{\displaystyle CH_3}{|}}{C}}-CH_2-CH_3$

(b) $CH_3-CH_2-\overset{\overset{\displaystyle CH_3}{|}}{CH}-CH-CH_3$ with $\overset{|}{CH_2}-CH_3$ below

(c) cyclopentane ring with $-CH_3$ and $CH_2-CH_3$ substituents

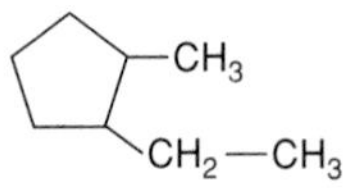

(d) 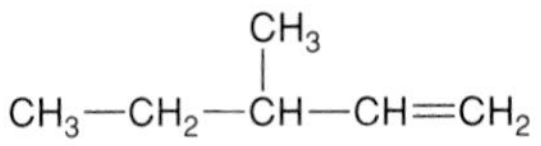 $CH_3-CH_2-\overset{\overset{\displaystyle CH_3}{|}}{CH}-CH=CH_2$

(e) 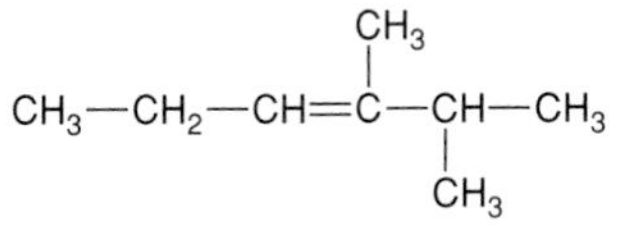 $CH_3-CH_2-CH=\overset{\overset{\displaystyle CH_3}{|}}{C}-\overset{\overset{}{}}{CH}-CH_3$ with $\overset{|}{CH_3}$ below

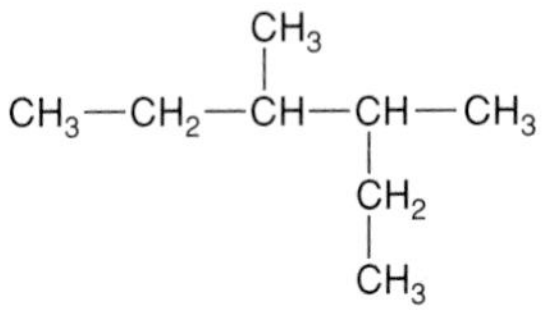

**Plan:** For **(a)** to **(c)**, we refer to Table 14.2. We first name the longest chain (root- + -ane). Then we find the *lowest* branch numbers by counting C atoms from the end *closer* to a branch. Finally, we name each branch (root- + -yl) and put them alphabetically before the chain name. For **(d)** and **(e)**, the longest chain that *includes* the multiple bond is numbered from the end closer to it. For **(d)**, the chiral center is the C atom bonded to four different groups. In **(e)**, the *cis* isomer has larger groups on the same side of the double bond, and the *trans* isomer has them on opposite sides.
**Solution:**

(a) 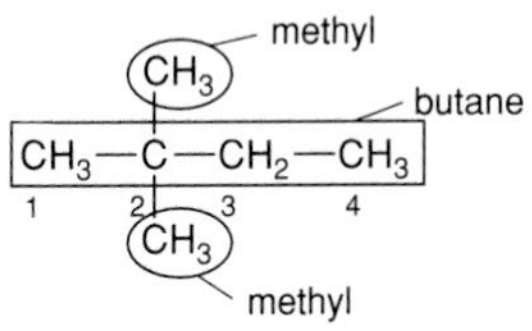 methyl, butane, methyl labels on $CH_3-\overset{}{C}-CH_2-CH_3$ (numbered 1 2 3 4) with two $CH_3$ groups

**2,2-dimethylbutane**

When a type of branch appears more than once, we group the chain numbers and indicate the number of branches with a numerical prefix, such as 2,2-*di*methyl.

(b) 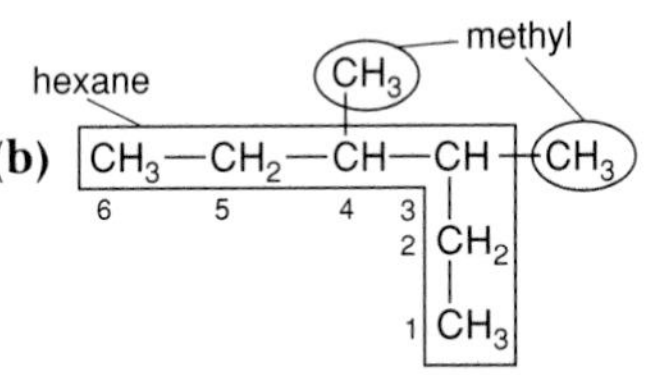 hexane, methyl labels on $CH_3-CH_2-CH-CH-CH_3$ (numbered 6 5 4 3) with $CH_2$ (2) and $CH_3$ (1)

**3,4-dimethylhexane**

In this case, we can number the chain from either end because the branches are the same and are attached to the two central C atoms.

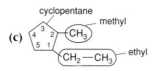

**(c)**

**1-ethyl-2-methylcyclopentane**

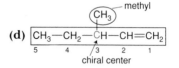

**(d)**

**3-methyl-l-pentene**

Number the ring C atoms so that
a branch is attached to C-1.

**(e)**

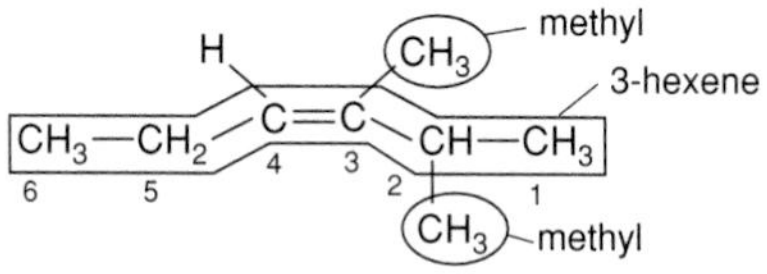

*cis*-**2,3-dimethyl-3-hexene**

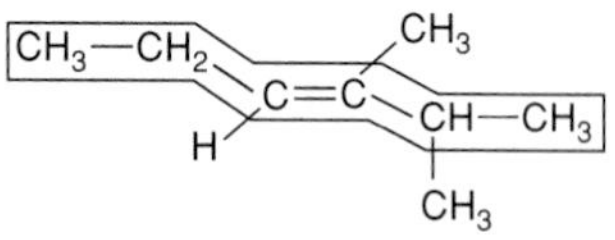

*trans*-**2,3-dimethyl-3-hexene**

**Check:** A good check (and excellent practice) is to reverse the process by drawing
structures for the names to see if you come up with the structure in the problem.
**Comment:** Note that in **(b)**, C-3 and C-4 are also chiral centers, as are C-1 and
C-2 in **(c)**.
Avoid some of these common mistakes:
In **(b)**, 2-ethyl-3-methylpentane is wrong: the longest chain is *hexane.*
In **(c)**, 1-methyl-2-ethylcyclopentane is wrong: the branches are named *alphabetically.*

**FOLLOW-UP PROBLEM 14.2**
Draw condensed formulas for the compounds with the following names: **(a)** 3-ethyl-
3-methyloctane; **(b)** 1-ethyl-3-propylcyclohexane; **(c)** 4-ethyl-2,4-dimethyldecane;
**(d)** 3,3-diethyl-l-hexyne; **(e)** *trans*-3-methyl-3-heptene; **(f)** 3-ethylcyclohexene.

### Aromatic Hydrocarbons: Cyclic Molecules with Delocalized π Electrons

**Aromatic hydrocarbons** usually have one or more rings of six C atoms
and are drawn with alternating single and double bonds. As you learned for
benzene, however, all the ring bonds are identical, with values of length and
strength *between* those of a C—C and a C=C bond. To indicate this, benzene
is often shown as a resonance hybrid, with a circle (or dashed circle) representing delocalized character (Figure 14.13, *A*). An orbital picture shows the
two lobes of the delocalized π cloud above and below the hexagonal σ-bond
plane of C atoms (Figure 14.13, *B*). ◆

The systematic naming of simple aromatic compounds is quite straightforward. Usually, benzene is the parent compound, and attached groups, or
*substituents*, are named as prefixes. However, many common names are still
in use. For example, benzene with one methyl group is systematically
named methylbenzene but is better known by its common name *toluene:*

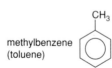

methylbenzene
(toluene)

With only one substituent present in toluene, we do not number the ring
C atoms, but when two or more groups are attached, we do so in such a way

Naphthalene

Benzo[*a*]pyrene

◆ **Aromatic Carcinogens.**
*Polycyclic aromatic compounds* have fused ring
systems, two or more rings sharing one or
more sides. The simplest is naphthalene,
which is used as a moth repellant and feedstock for dyes. Many of these compounds (and
benzene itself) have been shown to have carcinogenic (cancer-causing) activity. One of the
most potent, benzo[*a*]pyrene, is found in soot,
cigarette smoke, car exhaust, and even the
smoke from barbecue grills!

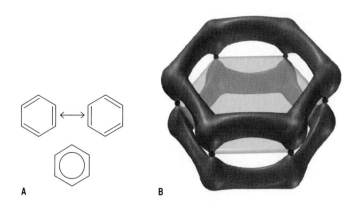

**FIGURE 14.13**
**Representations of benzene. A,** Benzene is often drawn with two resonance forms, but the molecule is more correctly depicted as a resonance hybrid. **B,** The delocalized $\pi$ cloud and the $\sigma$-bond plane of benzene.

A                                    B

that one of the groups is attached to ring C-1. Thus, the three structural isomers with two methyl groups attached are

1,2-dimethylbenzene
(*o*-xylene)
bp = 144.4°C

1,3-dimethylbenzene
(*m*-xylene)
bp = 139.1°C

1,4-dimethylbenzene
(*p*-xylene)
bp = 138.3°C

The positions of two groups in common names are indicated by *o*- (ortho) for groups on adjacent ring C atoms, *m*- (meta) for groups separated by one ring C atom, and *p*- (para) for groups on opposite ring C atoms. The dimethylbenzenes (commonly known as *xylenes*) are important solvents and feedstocks for polyester fibers and dyes.

The number of isomers increases with more than two groups. For example, there are six isomers for a compound with one methyl and three nitro ($-NO_2$) groups attached to a benzene ring; the explosive TNT is only one,

2,4,6-trinitromethylbenzene
(**trin**itro**t**oluene, TNT)

One of the most important methods for determining the structures of organic molecules is discussed in the upcoming Tools of the Chemistry Laboratory essay.

## Variations on a Theme: Catenated Inorganic Hydrides

In discussions called "Variations on a Theme," we examine similarities between organic and inorganic compounds. From this perspective, you'll see that the behavior of carbon is remarkable, but not unique, in the chemistry of the elements.

Although no element approaches carbon in the variety and complexity of its hydrides, catenation occurs frequently in the periodic table, and many ring, chain, and cage structures are known. Some of the most fascinating belong to the boron hydrides, or boranes (Section 13.5). Although their shapes

**TOOLS OF THE CHEMISTRY LABORATORY**

# Nuclear Magnetic Resonance (NMR) Spectroscopy

Chemists rely on a battery of instrumental methods to reveal the complex structures of organic molecules. Earlier we discussed mass spectrometry (Chapter 2) and infrared (IR) spectroscopy (Chapter 9). One of the most useful is **nuclear magnetic resonance (NMR) spectroscopy,** by which a chemist can learn the number of H atoms attached to each C atom in a molecule.

A proton ($^1$H), like an electron, can spin in either of two directions, each of which creates a tiny magnetic field. Generally, the fields of all the protons in a sample of compound are oriented randomly. In a strong external magnetic field ($H_0$), however, the proton fields become aligned with it (parallel) or against it (antiparallel). The parallel orientation is slightly lower in energy, with the energy difference ($\Delta E$) between the two spin states lying in the radio frequency (rf) region of the spectrum. In a process known as *resonance,* a proton in the lower spin state absorbs a photon of energy $\Delta E$ from an rf source, and its spin flips to the higher state (Figure 14.B). The

resonance gives rise to a signal, which is detected by the rf receiver. If all the protons in a sample required the same $\Delta E$ for resonance, an NMR spectrum would have only one absorption and be useless. However, the $\Delta E$ between the two states depends on the *actual* magnetic field felt by each proton, which is affected by the tiny magnetic fields of the *electrons* on atoms adjacent to that proton. Thus, the $\Delta E$ of each proton depends on the electrons of adjacent H atoms, electronegative atoms, multiple bonds, and aromatic rings—in other words, on the specific molecular environment.

The NMR spectrum of a compound is a series of peaks that represents the resonance of each proton as a function of changing magnetic field. The *chemical shift* is the position at which resonance occurs for a given proton, that is, where a peak appears. The chemical shifts are shown relative to that of an added standard, tetramethylsilane [$(CH_3)_4Si$, TMS], which has 12 protons bonded to four C atoms that are bonded to one Si atom. They are in identical environments, so they produce one peak. Figure 14.C shows the NMR spectrum of acetone. Its six protons also have identical environments—bonded to two C atoms that are bonded to the C atom in a C=O bond—so they also produce one peak, but at a different position from those in TMS. (The axes need not concern us.) The spectrum of dimethoxymethane shows *two* peaks in addition to the TMS peak (Figure 14.D). The taller one is due to the six $CH_3$ protons, and the shorter is due to the two $CH_2$ protons. The area under each peak (given here in units of chart-paper spaces) is proportional to *the number of protons in a given environment* and is automatically determined by the instrument. Note that the area ratio is $20.3/6.8 \approx 3/1$, the same as the ratio of six $CH_3$ protons/two $CH_2$ protons. Thus, from the chemical shifts and peak areas, the chemist learns the type and number of each $^1$H in the compound.

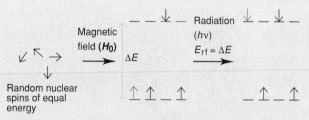

**FIGURE 14.B The basis of proton spin resonance.** The randomly oriented magnetic fields of protons in a sample become aligned in a strong external field ($H_0$), with slightly more protons in the lower spin state. In a process called resonance, the spin flips when the proton absorbs a photon with energy equal to $\Delta E$ (radio frequency region).

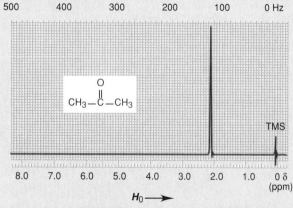

**FIGURE 14.C The $^1$H NMR spectrum of acetone.** Since the six $CH_3$ protons are in identical environments, they produce one peak, which has a different chemical shift than the 12 protons of the TMS standard.

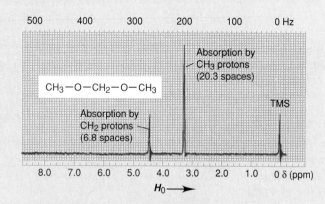

**FIGURE 14.D The $^1$H NMR spectrum of dimethoxymethane.** The two peaks represent the chemical shifts of protons in two different environments. The area under each peak is proportional to the number of protons.

Newer NMR methods measure the resonance of other nuclei and have many applications in biochemistry and medicine. $^{13}C$-NMR is used to monitor changes in protein and nucleic acid shape and function, and $^{31}P$-NMR can determine the health of various organs. For example, the extent of damage from a heart attack can be learned by using $^{31}P$-NMR to measure the concentrations of specific phosphate-containing molecules involved in energy utilization by cardiac muscle tissue. Most recently, computer-aided magnetic resonance imaging (MRI) has allowed visualization of damage to organs and greatly assisted physicians in medical diagnosis. For example, an MRI scan of the head (Figure 14.E) can show various levels of metabolic activity in different regions of the brain.

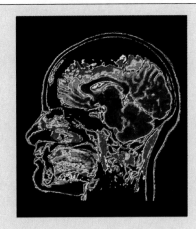

**FIGURE 14.E**
**MRI of human head.**

rival even those of the hydrocarbons, the weakness of their unusual bridge bonds renders them thermally and chemically unstable.

An obvious structural similarity exists between alkanes and the silicon hydrides, or silanes (Section 13.6). They have the analogous general formula $(Si_nH_{2n+2})$, but the longest known chain has only eight Si atoms. Branched silanes are also known, but no cyclic or unsaturated $(Si=Si)$ compounds had been prepared until recently. Unlike alkanes, silanes are unstable thermally and ignite spontaneously in air.

Sulfur's ability to catenate is second only to carbon's, and many chains and rings occur among its allotropes (Section 13.8). A large series of sulfur hydrides, or polysulfanes, is known. Since a sulfur atom has two unpaired electrons, these molecules are unbranched chains with H atoms at the ends $(H-S_n-H)$. Like the silanes, the polysulfanes are oxidized easily and decompose readily to sulfur's only stable hydride, $H_2S$, and its most stable allotrope, cyclo-$S_8$.

## Section Summary

Hydrocarbons contain only C and H atoms, so their physical properties depend on the strength of dispersion forces. The names of the organic compounds have a root for the longest chain, a prefix for the attached groups, and a suffix for the type of compound. Alkanes $(C_nH_{2n+2})$ have only single bonds. Alkenes $(C_nH_{2n})$ have at least one C=C bond. Alkynes $(C_nH_{2n-2})$ have at least one C≡C bond. Aromatic hydrocarbons have at least one ring with delocalized $\pi$ electrons.

Isomers are compounds with the same formula but different properties. Structural isomers have different atom sequences. Stereoisomers (optical and geometric) have the same sequence of atoms oriented differently in space. Optical isomers cannot be superimposed on each other because they are chiral, with four different groups bonded to the C that acts as the chiral center. They have identical physical and chemical properties except in their rotation of plane-polarized light and their reaction with chiral reactants. Geometric (*cis* and *trans*) isomers have groups oriented differently around a C=C bond, which restricts rotation. Light converts *cis*-retinal to *trans*-retinal, which initiates the visual response. NMR spectroscopy indicates the number and molecular environment of each H atom in a molecule.

## 14.3 Some Important Classes of Organic Reactions

In Chapter 4, we classified chemical reactions in terms of the number of reactants and products (combination, decomposition, and so on) and in terms
of the chemical process involved (neutralization, redox, and so on). We take
a similar approach here with organic reactions. Then, in Section 14.4, we
see how the distribution of electron density in the reactants influences the
initial step in a reaction.

From here on, we use the notation R— to signify a general organic group
attached to the atoms shown; you can usually picture R— as an **alkyl
group,** a saturated hydrocarbon chain with one bond available. Thus,
$RCH_2$—Br has an alkyl group attached to a $CH_2$ group bearing a Br atom;
$R_2C$=$CH_2$ is an alkene with two alkyl groups attached to one of the carbons in the double bond; and so forth. (Often, when more than one R group
is present, we write R, R′, R″ to indicate that these groups may be different.)

### Types of Organic Reactions

Most organic reactions are examples of three broad types that can be identified by comparing the *number of bonds to C* in the reactants and products:

1. An **addition reaction** occurs when an unsaturated reactant becomes
a saturated product:

$$R-CH{=}CH-R\ +\ X-Y \rightarrow R-\overset{\overset{\displaystyle X}{|}}{C}H-\overset{\overset{\displaystyle Y}{|}}{C}H-R$$

Note that the C atoms are bonded to *more* atoms in the product than in the
reactant.

The C=C and C≡C bonds and the C=O bond commonly undergo
addition reactions. In each case, the $\pi$ bond breaks, leaving the $\sigma$ bond intact. In the product, the C atoms (or C and O) form two additional single
bonds. Let's examine the standard heat of reaction ($\Delta H^0_{rxn}$) for a typical addition reaction to obtain some idea why these reactions occur. Consider the
reaction between ethene (common name, ethylene) and HCl:

$$CH_2{=}CH_2 + H-Cl \rightarrow H-CH_2-CH_2-Cl$$

| REACTANTS (BONDS BROKEN) | | PRODUCT (BONDS FORMED) | |
|---|---|---|---|
| 1 C=C | = 614 kJ | 1 C—C | = −347 kJ |
| 4 C—H | = 1652 kJ | 5 C—H | = −2065 kJ |
| 1 H—Cl | = 427 kJ | 1 C—Cl | = −339 kJ |
| Total | = 2693 kJ | Total | = −2751 kJ |

And

$$\Delta H^0_{rxn} = \Delta H^0_{\text{bonds broken}} + \Delta H^0_{\text{bonds formed}} = 2693\ kJ + (-2751\ kJ) = -58\ kJ$$

The reaction is exothermic. By looking at the *net* change in bonds, we see
that the driving force for many additions is the formation of two $\sigma$ bonds
(C—H and C—Cl) from one $\sigma$ bond (H—Cl) and one relatively weak $\pi$
bond. An addition reaction is the basis of a color test for the presence of
C=C bonds (Figure 14.14). Bromine is added dropwise to the unknown,
and the orange-brown color of $Br_2$ disappears if a C=C bond is present.

2. **Elimination reactions** are the opposite of addition reactions. They
occur when a saturated reactant becomes an unsaturated product:

$$R-\overset{\overset{\displaystyle X}{|}}{C}H-\overset{\overset{\displaystyle Y}{|}}{C}H_2 \longrightarrow R-CH{=}CH_2 + X-Y$$

**FIGURE 14.14**

**A color test for C=C bonds.** When $Br_2$ is
mixed with a compound that has a C=C bond
(*left* test tube), it adds to the bond and its orange-brown color disappears:

$$\overset{\diagdown}{\underset{\diagup}{C}}{=}\overset{\diagdown}{\underset{\diagup}{C} + Br_2} \longrightarrow -\overset{\overset{\displaystyle Br}{|}}{C}-\overset{\underset{\displaystyle Br}{|}}{C}-$$

The substance in the right tube has no C=C
bond, so the $Br_2$ does not react.

Note that the C atoms are bonded to *fewer* atoms in the product than in the reactant. Pairs of halogen atoms, an H and a halogen atom, or an H and an —OH group are typically eliminated, but C atoms are not. Thus, the driving force for many elimination reactions is the loss of a small, stable molecule, such as HCl($g$) or $H_2O$:

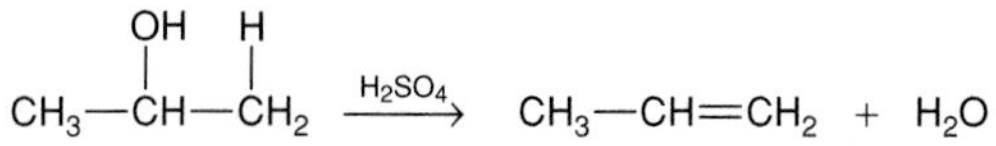

$$CH_3-\underset{\underset{OH}{|}}{C}H-\underset{\underset{H}{|}}{C}H_2 \xrightarrow{H_2SO_4} CH_3-CH{=}CH_2 + H_2O$$

3. A **substitution reaction** occurs when an atom (or group) from an added reagent substitutes for one in the organic reactant:

$$R-\overset{|}{\underset{|}{C}}-X + Y \rightarrow R-\overset{|}{\underset{|}{C}}-Y + X$$

Note that the C atom is bonded to the *same number* of atoms in the product as in the reactant. The C atom may be saturated or unsaturated, and X and Y can be many different atoms, but generally *not* C. The main flavor ingredient in banana oil, for instance, forms through a substitution reaction; note that the O substitutes for the Cl:

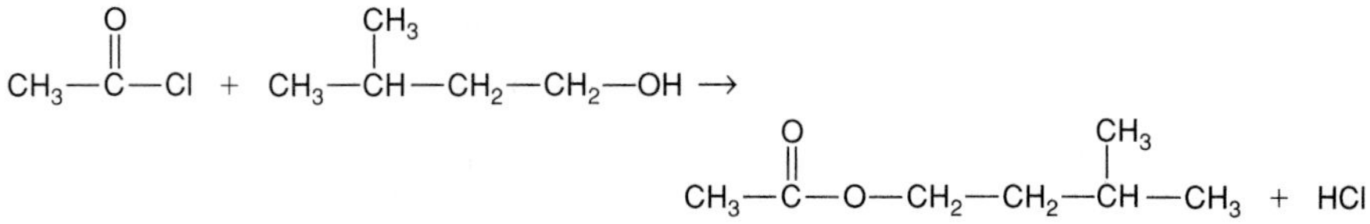

$$CH_3-\overset{\overset{O}{\|}}{C}-Cl + CH_3-\underset{\underset{CH_3}{|}}{C}H-CH_2-CH_2-OH \rightarrow$$

$$CH_3-\overset{\overset{O}{\|}}{C}-O-CH_2-CH_2-\underset{\underset{CH_3}{|}}{C}H-CH_3 + HCl$$

SAMPLE PROBLEM 14.3 ———————————————————————

## Recognizing the Type of Organic Reaction

**Problem:** State whether each of the following reactions is an addition, elimination, or substitution:

(a) $CH_3-CH_2-CH_2-Br \rightarrow CH_3-CH{=}CH_2 + HBr$

(b) 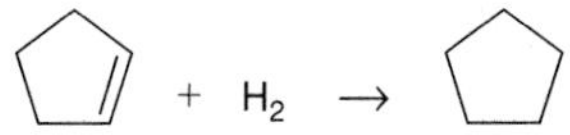 ⬠ + $H_2$ → ⬠

(c) $CH_3-CH_2-CH_2-Br + CH_3-ONa \rightarrow CH_3-CH_2-CH_2-O-CH_3 + NaBr$

**Plan:** We determine the type of reaction by examining the change in number of atoms bonded to C:
• More atoms bonded to C is an addition.
• Fewer atoms bonded to C is an elimination.
• Same number of atoms bonded to C is a substitution.
**Solution:**
(a) **Elimination:** two bonds in the reactant, C—H and C—Br, are absent in the product, so fewer atoms are bonded to C.
(b) **Addition:** two more C—H bonds form in the product, so more atoms are bonded to C.
(c) **Substitution:** the reactant C—Br bond becomes a C—O bond in the product, so the same number of atoms are bonded to C.
**Check:** Count the number of atoms bonded to the C atoms that take part in the reaction.

FOLLOW-UP PROBLEM 14.3
Write a balanced equation for the following:
(a) An addition reaction between 2-butene and $H_2O$
(b) A substitution reaction between $CH_3-CH_2-CH_2-Br$ and $OH^-$
(c) An elimination reaction between $CH_3-CHCl-CH_3$ and $CH_3-ONa$

### The Redox Process in Organic Reactions

A chemical process central to many organic reactions is *oxidation-reduction.* Because C atoms in the same molecule may have different oxidation numbers (O.N.s), let's first discuss the rules for assigning them.

Each bond to a C atom contributes to its O.N. The relative electronegativities of the C and the bonded atom determine the sign of each contribution (H is typically the only atom *less* electronegative than C). The sum of the contributions gives the O.N. of that C atom. To find the oxidation number of a particular C atom:

- Assign a zero for each bond to another C atom.
- Assign a −1 for each bond to an H atom.
- Assign a +1 for each bond to an N, O, or halogen atom.
- Add the bond contributions to obtain the O.N.

Although every redox reaction involves a reduction *and* an oxidation, we *define the change in terms of the organic compound,* by comparing the total of the O.N.s in the reactant with that in the product:

- In a reduction, the O.N. total *decreases* from reactant to product.
- In an oxidation, the O.N. total *increases* from reactant to product.

---

SAMPLE PROBLEM 14.4 ——————————————————————————

### Using Oxidation Numbers to Confirm That an Organic Reaction Is an Oxidation or a Reduction

---

**Problem:** The conversion of 2-propanol to 2-propanone is an oxidation. Use oxidation numbers to confirm this fact.

$$CH_3-\underset{\underset{OH}{|}}{CH}-CH_3 \longrightarrow CH_3-\underset{\underset{O}{\|}}{C}-CH_3$$

2-propanol         2-propanone

**Plan:** We assign O.N.s to each C atom in the compounds by taking the sum of each bond contribution. Then we find the total of the O.N.s in each compound. In an oxidation, the total is higher in the product; in a reduction, it is lower.

**Solution:** Determining the O.N.s of the C atoms.

For 2-propanol:

| | C-1 sum | C-2 sum | C-3 sum |
|---|---|---|---|
| | −1 | +1 | −1 |
| | −1 | 0 | −1 |
| | −1 | 0 | −1 |
| | 0 | −1 | 0 |
| | −3 | 0 | −3 |

For 2-propanone:

| | C-1 sum | C-2 sum | C-3 sum |
|---|---|---|---|
| | −1 | +1 | −1 |
| | −1 | +1 | −1 |
| | −1 | 0 | −1 |
| | 0 | 0 | 0 |
| | −3 | +2 | −3 |

Finding the total of the O.N.s:

For 2-propanol, $-3 + 0 + (-3) = -6$; for 2-propanone, $-3 + 2 + (-3) = -4$. The O.N. total increases (less negative), so we confirm that the change is an **oxidation;** note that the O.N. of C-2 changed from 0 (in 2-propanol) to +2 (in 2-propanone).

**Comment:** Because it is oxidized, 2-propanol is the reducing agent. One common oxidizing agent used for this reaction is sodium dichromate in sulfuric acid. As 2-propanol is oxidized, the dichromate ion, $Cr_2O_7^{2-}$ (O.N. of Cr = +6), is reduced to the chromium(III) ion, $Cr^{3+}$ (O.N. of Cr = +3):

$$Cr_2O_7^{2-} + 14H^+ + 6e^- \rightarrow 2Cr^{3+} + 7H_2O$$

**FOLLOW-UP PROBLEM 14.4**
Determine the oxidizing and reducing agents in the following addition reaction:

$$CH_2{=}CH_2 + Cl_2 \rightarrow Cl{-}CH_2{-}CH_2{-}Cl$$

The most dramatic redox reactions are combustion reactions. All organic compounds that contain C and H atoms burn in excess $O_2$ to form $CO_2$ and $H_2O$. For ethane, the reaction is

$$2CH_3{-}CH_3 + 7O_2 \rightarrow 4CO_2 + 6H_2O$$

Each C atom is oxidized from $-3$ in ethane to $+4$ in $CO_2$. Even though the reduction of $O_2$ (from 0 to $-2$) takes place also, this reaction is referred to as an *oxidation* because we focus on the O.N.s of the C atoms.

As you saw Sample Problem 14.4, however, many redox processes do not involve either breaking apart the whole molecule or such large changes in O.N. *Whenever a C atom forms more bonds to N, O, or halogen, or fewer bonds to H, it is oxidized.* Note in Sample Problem 14.4, C-2 in 2-propanone has one fewer bond to H and one more bond to O than does 2-propanol. In the following reaction, the C atom is oxidized from $-4$ to $-2$, and each Cl atom is reduced from 0 to $-1$. This chlorination reaction is an oxidation:

$$CH_4 + Cl_2 \xrightarrow{\text{light}} CH_3{-}Cl + HCl$$

On the other hand, *whenever a C atom forms fewer bonds to N, O, or halogen, or more bonds to H, it is reduced.* The addition of $H_2$ to an alkene is referred to as a *reduction:*

$$CH_2{=}CH_2 + H_2 \xrightarrow{Pd} CH_3{-}CH_3$$

Each C is reduced from $-2$ in ethene to $-3$ in ethane, and each H is oxidized from 0 to $+1$. (The palladium shown over the arrow acts as a catalyst to speed up the reaction.)

From the preceding examples, you can see that a redox reaction can be any of the three reaction types described earlier. It can be an addition, as in the reduction of ethene to ethane; an elimination, as in the oxidation of 2-propanol to 2-propanone; or a substitution, as in the light-induced oxidation (chlorination) of methane to chloromethane.

**Section Summary**
In an addition reaction, a $\pi$ bond breaks and the two C atoms bond to more atoms. In an elimination, a $\pi$ bond forms and the two C atoms bond to fewer atoms. In a substitution, one atom replaces another but the number of atoms bonded to C is not changed. In the redox process, the total of the O.N.s in the C atoms is different in reactant and product. Carbon is oxidized when it bonds to more N, O, or halogen (or fewer H) and is reduced when it bonds to more H (or fewer N, O, or halogen).

## 14.4  Properties and Reactivity of Common Functional Groups

To predict how an organic compound might react, we narrow our focus to the *functional group*. The distribution of electron density in the functional group affects the reactivity. The electron density can be high, as in the C=C and C≡C bonds, or it can be low at one end of a bond and high at the other, as in the C—Cl and C—O bonds. Such bond sites induce dipoles (or enhance existing ones) in the other reactant. The reactants attract each other and begin a sequence of bond-forming and bond-breaking steps that leads to product. Thus, *the intermolecular forces that affect physical properties and solubility also affect reactivity.* Table 14.5 lists most of the major functional groups found in organic compounds.

When we classify functional groups by bond type (single, double, and so forth), in most cases they follow certain patterns of reactivity:

• Single-bonded functional groups undergo substitution or elimination.
• Double- and triple-bonded functional groups undergo addition.
• Functional groups with single and double bonds undergo substitution.

### Functional Groups with Single Bonds

The most common single-bonded functional groups are alcohols, ethers, haloalkanes, and amines.

**Alcohols and ethers.** The **alcohol** functional group consists of carbon bonded to an —OH group, $-\overset{|}{\underset{|}{C}}-\ddot{O}-H$, and the general formula of an alcohol is R—OH. Alcohols are named by dropping the final *-e* from the parent hydrocarbon name and adding the suffix *-ol*. Thus, the two-carbon alcohol is ethanol (ethan- + -ol). The common name is the hydrocarbon *root-* + *-yl*, followed by "alcohol"; thus, the common name of ethanol is ethyl alcohol. (This substance, obtained from fermented grain, has been consumed by people as an intoxicant in beverages since ancient times; today, it is recognized as the most abused drug in the world.) Alcohols are important laboratory reagents, and the functional group occurs in many biomolecules, including carbohydrates, sterols, and some amino acids. Figure 14.15 shows the name, structure, and uses of some important compounds that contain the alcohol group.

You can think of an alcohol as a water molecule with an R group in place of one of the H atoms. In fact, alcohols react with very active metals, such as the alkali metals [Group 1A(1)], in a manner similar to water:

$$2Na + 2HOH \rightarrow 2NaOH + H_2$$

$$2Na + 2CH_3{-}CH_2{-}OH \rightarrow 2CH_3{-}CH_2{-}ONa + H_2$$

Water forms the hydroxide ion, and alcohols form the *alkoxide ion* (RO$^-$), in this case the ethoxide ion, $CH_3{-}CH_2{-}O^-$. The physical properties of the smaller alcohols are also similar to those of water. They have high melting and boiling points due to hydrogen bonding, and they dissolve polar molecules and some salts.

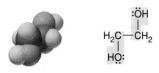

$H_3C{-}\ddot{O}H$

**Methanol (methyl alcohol)**
By-product in coal gasification; de-icing agent; gasoline substitute; precursor of organic compounds

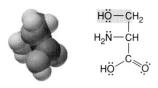

$\overset{\displaystyle :\ddot{O}H}{\underset{\displaystyle HO:}{H_2C{-}CH_2}}$

**1,2-Ethanediol (ethylene glycol)**
Main component of auto antifreeze

$\begin{array}{c} H\ddot{O}{-}CH_2 \\ H_2\ddot{N}{-}CH \\ H\ddot{O}\diagdown C\diagup\ddot{O} \end{array}$

**Serine**
Amino acid found in most proteins

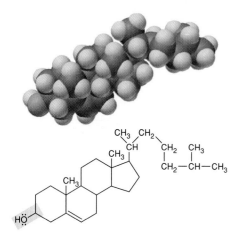

**Cholesterol**
Major sterol in animals; essential for cell membranes; precursor of steroid hormones

**FIGURE 14.15**
Some molecules with the alcohol functional group.

**TABLE 14.5  Important Functional Groups in Organic Compounds**

| FUNCTIONAL GROUP | COMPOUND TYPE | SUFFIX OR PREFIX OF NAME | EXAMPLE | SYSTEMATIC NAME (COMMON NAME) |
|---|---|---|---|---|
| 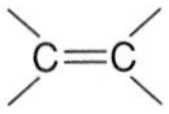 C=C | alkene | -ene | 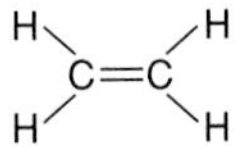 ethene (from $H_2C=CH_2$) | ethene (ethylene) |
| —C≡C— | alkyne | -yne | $H-C≡C-H$ | ethyne (acetylene) |
| 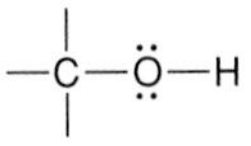 —C—Ö—H | alcohol | -ol | 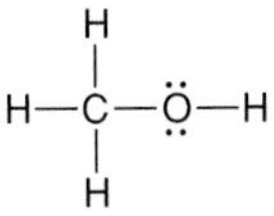 methanol (from $H_3C-OH$) | methanol (methyl alcohol) |
| 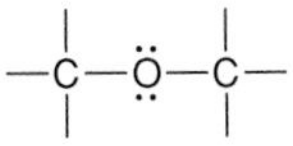 —C—Ö—C— | ether | ether | 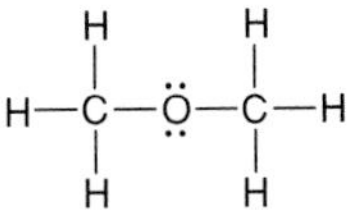 dimethyl ether (from $H_3C-O-CH_3$) | dimethyl ether |
| 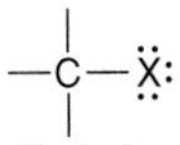 —C—Ẍ: (X=halogen) | haloalkane | halo- | 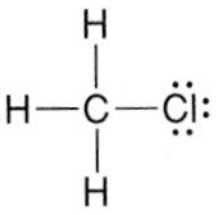 chloromethane (from $H_3C-Cl$) | chloromethane (methyl chloride) |
| 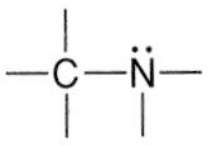 —C—N̈— | amine | -amine | 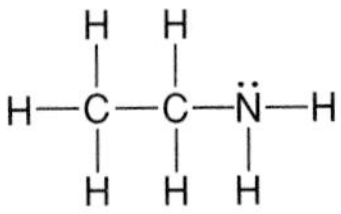 ethylamine (from $H_3C-CH_2-NH_2$) | ethylamine |
| 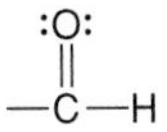 —C—H (with :O:) | aldehyde | -al | 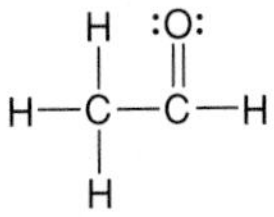 ethanal (from $H_3C-CHO$) | ethanal (acetaldehyde) |
| 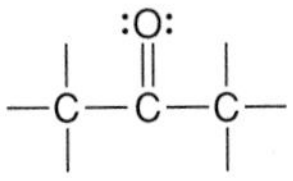 —C—C—C— (with :O: on middle C) | ketone | -one | 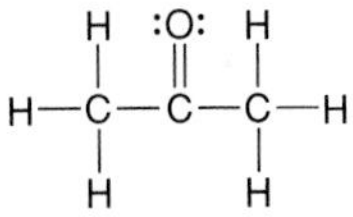 2-propanone (from $H_3C-CO-CH_3$) | 2-propanone (acetone) |
| 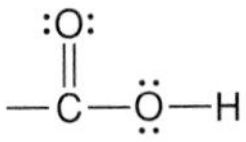 —C—Ö—H (with :O:) | carboxylic acid | -oic acid | 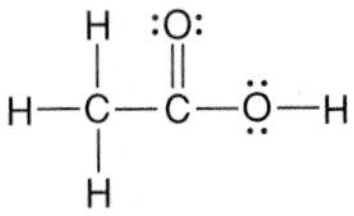 ethanoic acid (from $H_3C-COOH$) | ethanoic acid (acetic acid) |
| 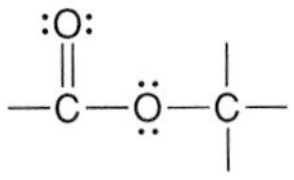 —C—Ö—C— (with :O:) | ester | -oate | 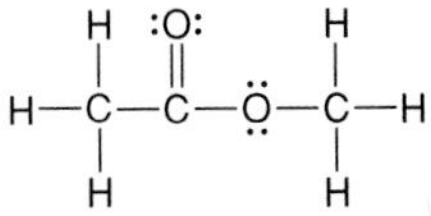 methyl ethanoate (from $H_3C-CO-O-CH_3$) | methyl ethanoate (methyl acetate) |
| 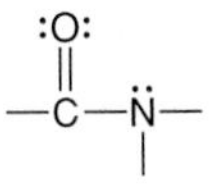 —C—N̈— (with :O:) | amide | -amide | 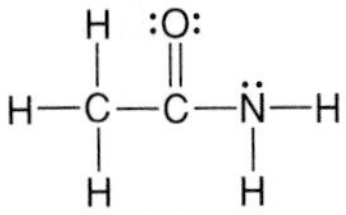 ethanamide (from $H_3C-CO-NH_2$) | ethanamide (acetamide) |
| —C≡N: | nitrile | -nitrile | 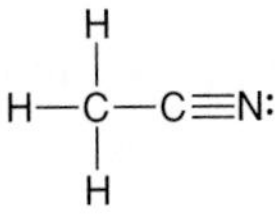 ethanenitrile (from $H_3C-C≡N$) | ethanenitrile (acetonitrile, methyl cyanide) |

Alcohols undergo elimination and substitution reactions. Dehydration, the elimination of H and OH, requires acid and forms alkenes:

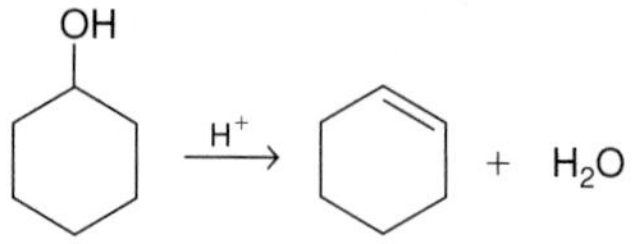

$$\text{cyclohexanol} \xrightarrow{\text{H}^+} \text{cyclohexene} + H_2O$$

Elimination of two H atoms requires inorganic oxidizing agents, such as $K_2Cr_2O_7$ in aqueous $H_2SO_4$, and produces the C$=$O group:

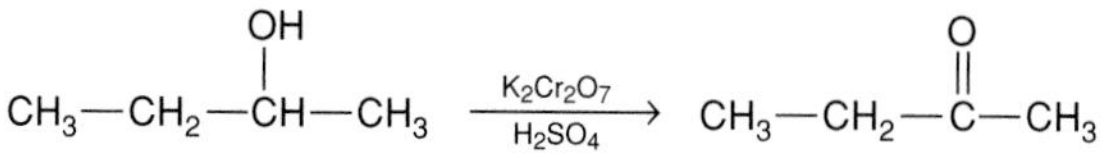

$$CH_3{-}CH_2{-}\underset{\underset{\text{2-butanol}}{}}{\overset{\overset{\displaystyle OH}{|}}{CH}}{-}CH_3 \xrightarrow[H_2SO_4]{K_2Cr_2O_7} CH_3{-}CH_2{-}\underset{\underset{\text{2-butanone}}{}}{\overset{\overset{\displaystyle O}{\|}}{C}}{-}CH_3$$

Note the change in carbon's O.N. from 0 to +2. For alcohols with an OH group at the end of the chain (R—$CH_2$—OH), another oxidation occurs, but this one is not an elimination. Wine turns sour, for example, when the ethanol is oxidized in contact with air to acetic acid:

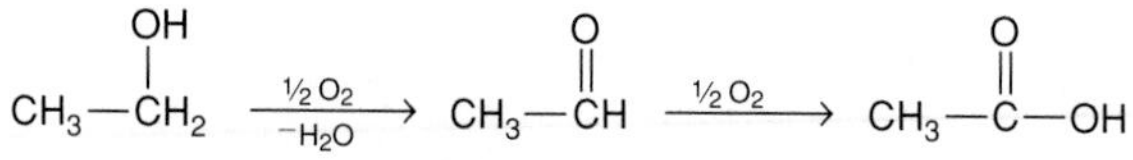

$$CH_3{-}\overset{\overset{\displaystyle OH}{|}}{CH_2} \xrightarrow[-H_2O]{\frac{1}{2}O_2} CH_3{-}\overset{\overset{\displaystyle O}{\|}}{CH} \xrightarrow{\frac{1}{2}O_2} CH_3{-}\overset{\overset{\displaystyle O}{\|}}{C}{-}OH$$

Substitution yields products with other single-bonded functional groups. With hydrohalic acids, many alcohols give haloalkanes:

$$CH_3{-}OH + HBr \rightarrow CH_3{-}Br + HOH$$

As you'll see below, *the C atom undergoing the change in a substitution is bonded to a more electronegative element,* which makes it partially positive and readily attacked by the incoming negative group, in this case, a $Br^-$ ion.

Two alcohol groups, heated in the presence of acid, lose water and form the **ether** functional group, $-\overset{\displaystyle |}{\underset{\displaystyle |}{C}}{-}\overset{\displaystyle ..}{\underset{\displaystyle ..}{O}}{-}\overset{\displaystyle |}{\underset{\displaystyle |}{C}}{-}$; the general formula for an ether is R—O—R. The anesthetic and organic solvent diethyl ether is prepared from ethanol in this way under conditions that prevent elimination to an alkene:

$$\underset{\text{ethanol}}{CH_3{-}CH_2{-}O\boxed{H}} + \underset{\text{ethanol}}{\boxed{HO}{-}CH_2{-}CH_3} \xrightarrow[\Delta]{H_2SO_4} \underset{\text{diethyl ether}}{CH_3{-}CH_2{-}O{-}CH_2{-}CH_3} + HOH$$

Ethers melt and boil at much lower temperatures than alcohols with similar numbers of C atoms because they lack the —OH group and thus the ability to form hydrogen bonds. For example, diethyl ether and 1-butanol ($CH_3$—$CH_2$—$CH_2$—$CH_2$—OH) are structural isomers, but the ether boils more than 80°C lower.

**Haloalkanes.** A *halogen* atom (X) bonded to C gives the **haloalkane** functional group, $-\overset{\displaystyle |}{\underset{\displaystyle |}{C}}{-}\overset{\displaystyle ..}{\underset{\displaystyle ..}{X}}:$; compounds have the general formula R—X.

Haloalkanes (common name, **alkyl halides**) are named with the halogen as a prefix to the parent hydrocarbon and numbering the C atom to which it is attached, as in bromomethane, 2-chloropropane, or 1,3-diiodohexane. Just as many alcohols undergo substitution to alkyl halides with halide ions in acid, the opposite change occurs in base. For example, $OH^-$ attacks the positive C end of the C—X bond and displaces $X^-$:

$$\underset{\text{1-bromobutane}}{CH_3{-}CH_2{-}CH_2{-}CH_2{-}Br} + OH^- \longrightarrow \underset{\text{1-butanol}}{CH_3{-}CH_2{-}CH_2{-}CH_2{-}OH} + Br^-$$

Substitution by groups such as —CN, —SH, —OR, and —NH$_2$ allows chemists to convert alkyl halides to a host of other compounds.

Just as the addition of HX to an alkene produces haloalkanes, the elimination of HX gives an alkene:

| 2-chloro-2-methylpropane | potassium ethoxide | 2-methylpropene |

(In this case, the CH$_3$—CH$_2$—O$^-$ does not substitute for the Cl$^-$.) As pointed out in earlier chapters, haloalkanes have important uses, but many are carcinogenic, and their stability has made them notorious environmental contaminants. ◆

**Amines.** The **amine** functional group is $-\overset{|}{\underset{|}{C}}-\overset{..}{N}:$ . Chemists classify amines as derivatives of ammonia with R groups in place of one or more H atoms on the central N; thus, primary (1°) amines are RNH$_2$, secondary (2°) amines are R$_2$NH, and tertiary (3°) amines are R$_3$N. Common names are usually used, and the suffix -*amine* follows the name of the alkyl group; thus, methylamine has one methyl group attached to N, diethylamine has two ethyl groups attached, and so forth. The amine functional group occurs in many biomolecules (Figure 14.16).

Primary and secondary amines can form H bonds, so they have higher melting and boiling points than hydrocarbons, ethers, and alkyl halides of similar molar mass. For example, dimethylamine ($\mathcal{M}$ = 45.09 g/mol) boils 30°C higher than dimethyl ether ($\mathcal{M}$ = 46.07 g/mol). Trimethylamine has a greater molar mass than dimethylamine, but it melts more than 20°C *lower* because it cannot form H bonds.

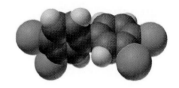

◆ **Pollutants in the Food Chain.** Until recently, halogenated aromatics, such as the *polychlorinated biphenyls (PCBs)*, were used as insulating fluids in electrical transformers and then discharged in waste water. Because of their low solubility, they accumulate for decades in river and lake sediment and are eaten by microbes and invertebrates. Fish eat the invertebrates, and birds and mammals, including humans, eat the fish. PCBs become increasingly concentrated in body fat at each stage. As a result of their health risks, PCBs in natural waters present an enormous cleanup problem.

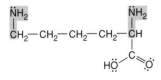

**Lysine (primary amine)**
Amino acid found in most proteins

**Adenine (primary amine)**
Component of nucleic acids

**Epinephrine (adrenaline; secondary amine)**
Neurotransmitter in brain; hormone released during stress

**Cocaine (tertiary amine)**
Brain stimulant; widely abused drug

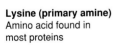

**FIGURE 14.16**
Some biomolecules with the amine functional group.

**FIGURE 14.17**

**General structures of amines.** Amines have a trigonal pyramidal shape and are classified by the number of R groups bonded to N. The N lone pair is the key to amine reactivity.

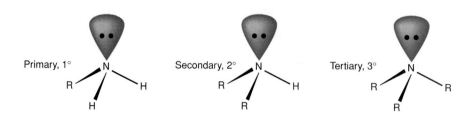

Primary, 1°    Secondary, 2°    Tertiary, 3°

Like ammonia, amines have trigonal pyramidal shapes and a lone pair of electrons on a partially negative N atom (Figure 14.17). Amines of low molar mass are fishy smelling, water soluble, and weakly basic, the reaction proceeding slightly to the right to reach equilibrium:

$$CH_3-\ddot{N}H_2 + H_2O \rightleftharpoons CH_3-\overset{+}{N}H_3 + OH^-$$

Amines undergo substitution reactions in which the N lone pair attacks the partially positive C in alkyl halides to displace $X^-$ and form a larger amine:

$$2CH_3-CH_2-\ddot{N}H_2 + CH_3-CH_2-Cl \longrightarrow CH_3-CH_2-\ddot{N}H + CH_3-CH_2-\overset{+}{N}H_3Cl^-$$
$$\underset{CH_3-CH_2}{|}$$

ethylamine          ethyl chloride          diethylamine          ethylammonium chloride

(The second molecule of ethylamine binds a released $H^+$ and prevents it from remaining on the diethylamine.)

Four R groups bonded to an N atom give an ionic compound called a *quaternary (4°) ammonium salt:*

$$4NH_3 + 4RCl \rightarrow 3NH_4Cl + R_4N^+Cl^- \text{ (a quaternary ammonium chloride)}$$

Some detergents are quaternary ammonium salts with large R groups (Figure 14.18). As with soaps (Section 12.1), the ionic portion makes these water soluble, and the R groups allow them to dissolve in grease.

**Variations on a theme: inorganic compounds with single bonds to O, X, and N.** The —OH group occurs frequently in inorganic compounds. All oxoacids contain at least one —OH usually bonded to a relatively electronegative nonmetal atom, which in most cases is bonded to other O atoms. Oxoacids are acidic because these O atoms pull electron density from the central nonmetal, which pulls electron density from the O—H bond, releasing an $H^+$ ion and stabilizing the oxoanion. Alcohols are *not* acidic because they lack the additional O atoms and the electronegative nonmetal.

Halides of nearly every nonmetal are known, and many undergo substitution reactions in base. As in the case of an alkyl halide, the process involves an attack on the partially positive central atom by $OH^-$:

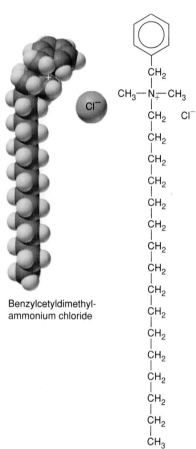

CH₃—N⁺—CH₃ with CH₂ chain: CH₃—N⁺—CH₃, CH₂, Cl⁻, CH₂, CH₂, CH₂, CH₂, CH₂, CH₂, CH₂, CH₂, CH₂, CH₂, CH₂, CH₂, CH₂, CH₃

Benzylcetyldimethyl-ammonium chloride

**FIGURE 14.18**

**Structure of a cationic detergent.** Quaternary ammonium salts are used as detergents. The charged portion dissolves in water, while the hydrocarbon portion dissolves in grease.

$$H-O^- \quad \overset{\delta^+}{C}-\overset{\delta^-}{Cl} \longrightarrow C-OH + Cl^-$$

$$H-O^- \quad \overset{\delta^+}{B}-\overset{\delta^-}{Cl} \longrightarrow B-OH + Cl^-$$

Thus, alkyl halides undergo one example of a general reaction of other nonmetal halides, such as $BCl_3$, $SiF_4$, and $PCl_5$.

The bonds from nitrogen to larger nonmetals, such as Si, P, and S, have significant double-bond character, which affects structure and reactivity. For example, trisilylamine, the Si analog of trimethylamine (see Figure 13.20), is planar, rather than trigonal pyramidal, as a result of $p,d$-$\pi$ bonding. The N lone pair is delocalized in this $\pi$ bond, so trisilylamine is not basic.

SAMPLE PROBLEM 14.5

**Predicting the Reactions of Alcohols, Alkyl Halides, and Amines**

**Problem:** Determine the reaction type and predict the product(s) in the following:
(a) $CH_3$—$CH_2$—$CH_2$—$I$ + NaOH $\longrightarrow$

(b) $CH_3$—$CH_2$—$Br$ + 2$CH_3$—$CH_2$—$CH_2$—$NH_2$ $\longrightarrow$

(c) 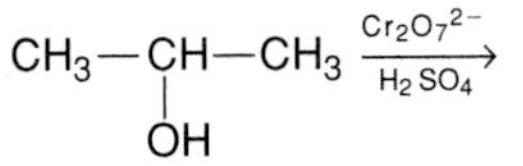 $CH_3$—$\underset{\underset{OH}{|}}{CH}$—$CH_3$ $\xrightarrow[H_2SO_4]{Cr_2O_7^{2-}}$

**Plan:** We examine the reactant(s) and any other reagent(s) to decide on the possibilities for each functional group, keeping in mind that, in general, one group changes into another.
**Solution:**
(a) **Substitution:** $CH_3$—$CH_2$—$CH_2$—$OH$ + NaI

(b) **Substitution:** $CH_3$—$CH_2$—$CH_2$—$\underset{\underset{CH_2-CH_3}{|}}{NH}$ $+ CH_3$—$CH_2$—$CH_2$—$\overset{+}{N}H_3Br^-$

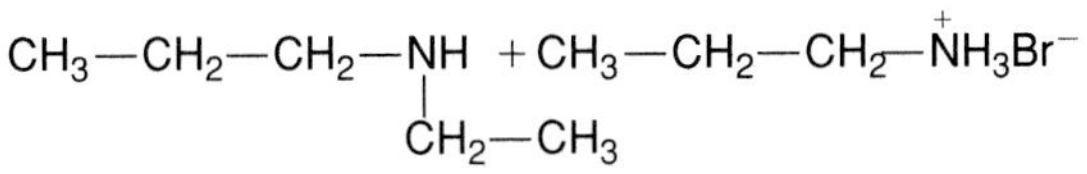

(c) **Elimination** (oxidation): $CH_3$—$\underset{\underset{O}{\|}}{C}$—$CH_3$

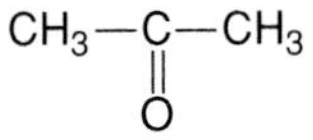

**Check:** The only changes should be at the functional group.

FOLLOW-UP PROBLEM 14.5
Fill in the blank in the following reactions. (*Hint:* examine any inorganic compounds and the organic product to determine the organic reactant.)

(a) 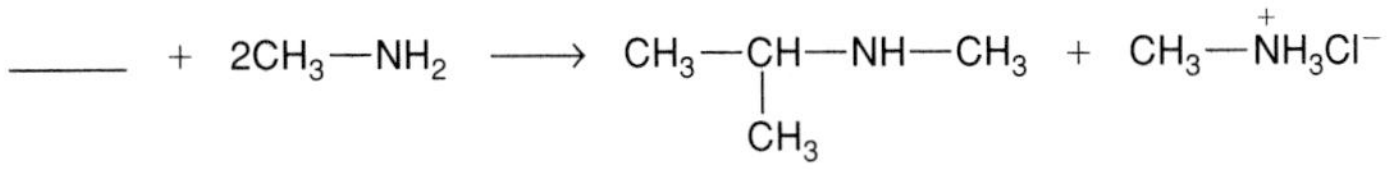 _____ + 2$CH_3$—$NH_2$ $\longrightarrow$ $CH_3$—$\underset{\underset{CH_3}{|}}{CH}$—$NH$—$CH_3$ + $CH_3$—$\overset{+}{N}H_3Cl^-$

(b) 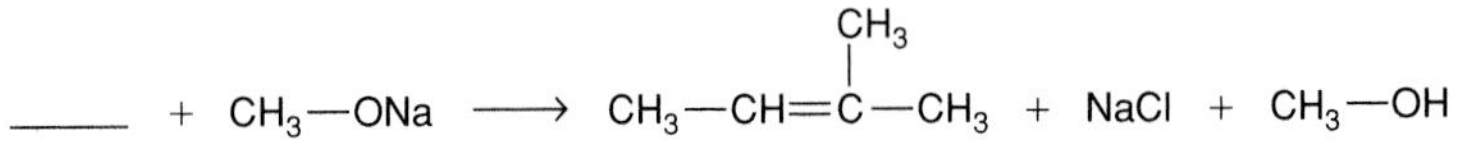 _____ + $CH_3$—$ONa$ $\longrightarrow$ $CH_3$—$CH$=$\underset{\underset{O}{\overset{\overset{CH_3}{|}}{C}}}{}$—$CH_3$ + NaCl + $CH_3$—$OH$

(c) _____ $\xrightarrow[H_2SO_4]{Cr_2O_7^{2-}}$ $CH_3$—$CH_2$—$\underset{\underset{O}{\|}}{C}$—$OH$

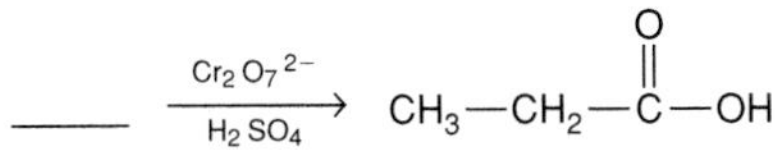

**Functional Groups with Double Bonds**

The most important double-bonded functional groups are the C==C bond of alkenes and the C==O bond of aldehydes and ketones. Both appear in many organic and biological molecules.

   **Comparing the reactivity of alkenes and aromatic compounds.** The C==C bond is the essential portion of the alkene functional group, $\underset{\diagup}{\overset{\diagdown}{C}}$==$\underset{\diagdown}{\overset{\diagup}{C}}$ . Although they can be unsaturated further to alkynes, *alkenes typically undergo addition.* The electron-rich double bond is readily attracted to the partially positive H atoms of hydronium ions and hydrohalic acids, yielding alcohols and alkyl halides, respectively:

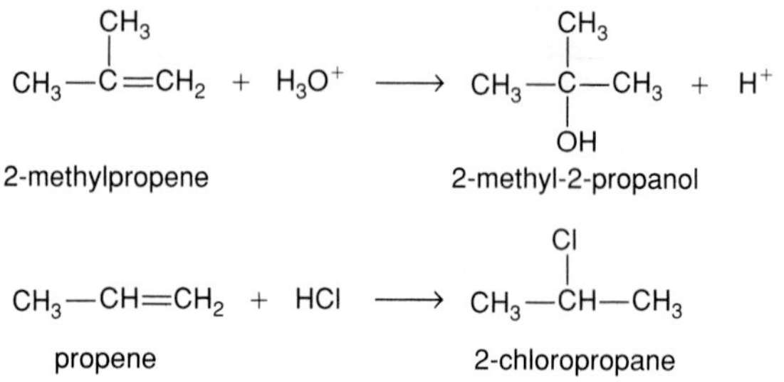

2-methylpropene     2-methyl-2-propanol

propene     2-chloropropane

The *localized* unsaturation of alkenes is very different from the *delocalized* unsaturation of benzene. Despite what its resonance forms show, benzene does *not* have double bonds and does *not* behave like an alkene. Bromine, for example, is not decolorized by benzene because there are no isolated $\pi$-electron pairs to bond with $Br_2$.

How does this electron delocalization affect reactivity? In general, aromatic rings are much *less* reactive than alkenes because delocalized $\pi$ electrons stabilize the ring, making it *lower* in energy than one with localized $\pi$ electrons. Data indicate that benzene is about 150 kJ/mol more stable than a six-C ring with three C=C bonds would be; in other words, benzene requires about 150 kJ/mol more energy to react.

An *addition* reaction with benzene, therefore, requires additional energy to break up the delocalized $\pi$ system. Benzene does undergo many *substitution* reactions, however, in which the delocalization is retained when a ring H atom is replaced by another group:

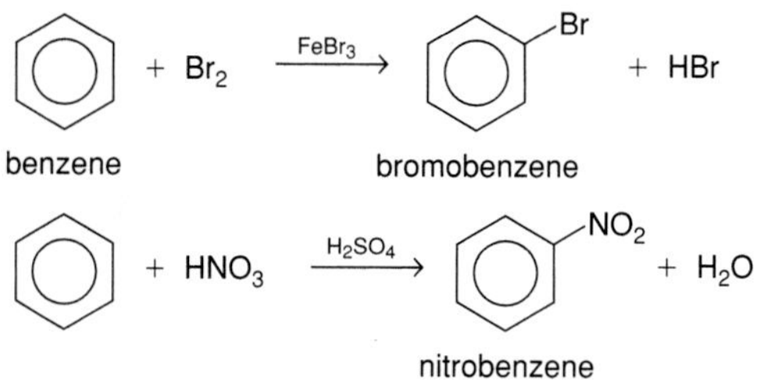

benzene     bromobenzene

nitrobenzene

**Aldehydes and ketones.** The C=O bond, or **carbonyl group,** is one of the most chemically versatile. In the **aldehyde** functional group, the carbonyl C is bonded to H (and often C), so it always occurs *at the end of a chain,* R—C=O. Aldehyde names drop the final *-e* from the parent alkane and add *-al.* For example, the three-C aldehyde is propanal. In the **ketone** functional group, the carbonyl C is bonded to two other C atoms, so it occurs *within the chain.* Ketones, R—C—R′, are named by numbering the carbonyl C, dropping the final *-e* from the alkane, and adding *-one.* For example, the unbranched, five-C ketone with the carbonyl C as C-2 in the chain is named 2-pentanone. Figure 14.19 shows some common carbonyl compounds.

Like the C=C bond, the C=O bond is *electron rich* and leads to *trigonal planar* shapes (~120°); unlike the C=C bond, it is *highly polar* ($\Delta EN = 1.0$).

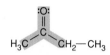

**Methanal (formaldehyde)**
Used to make resins in plywood, dishware, counter-tops; biological preservative

**Ethanal (acetaldehyde)**
Narcotic product of ethanol metabolism; used to make perfumes, flavors, plastics, other chemicals

**Benzaldehyde**
Artificial almond flavoring

**2-Propanone (acetone)**
Solvent for fat, rubber, plastic, varnish, lacquer; chemical feedstock

**2-Butanone (methyl ethyl ketone)**
Important solvent

**FIGURE 14.19**
Some common aldehydes and ketones.

Figure 14.20 emphasizes this polarity with a charged resonance form. Aldehydes and ketones are formed by the oxidation of alcohols:

$$CH_3-CH_2-OH \xrightarrow[-2H]{oxidation} CH_3-\overset{\overset{\displaystyle O}{\|}}{C}-H$$
ethanol       ethanal (common name, acetaldehyde)

$$CH_3-CH_2-\overset{\overset{\displaystyle OH}{|}}{CH}-CH_3 \xrightarrow[-2H]{oxidation} CH_3-CH_2-\overset{\overset{\displaystyle O}{\|}}{C}-CH_3$$
2-butanol       2-butanone

Conversely, they are reduced to alcohols:

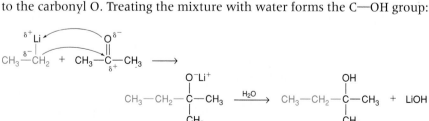

cyclobutanone       cyclobutanol

As a result of unsaturation, aldehydes and ketones can undergo *addition,* as in the reduction above. As a result of bond polarity, this type of reaction often occurs with an electron-rich group bonding to the carbonyl C and an electron-poor group bonding to the carbonyl O. In an important example, **organometallic compounds,** in which a metal has a polar covalent bond to an R group (Section 13.3), react with C=O groups to form alcohols with *different C skeletons.* For example, the electron-rich C of ethyllithium attacks the carbonyl C of 2-propanone adding its two-C portion, while the Li adds to the carbonyl O. Treating the mixture with water forms the C—OH group:

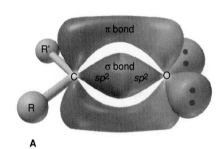

**A**

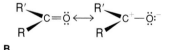

**B**

**FIGURE 14.20**
**The carbonyl group. A,** An orbital depiction of the σ bond and π bond that make up the C=O of the carbonyl group. **B,** The C=O bond is polar (ΔEN = 1.0), which is indicated by the charged resonance form.

Note that the product skeleton combines the two reactant skeletons. The field of *organic synthesis* employs such reactions to create many new compounds.

SAMPLE PROBLEM 14.6

## Predicting the Steps in an Organic Reaction Sequence

**Problem:** Fill in the blanks in the following reaction sequence:

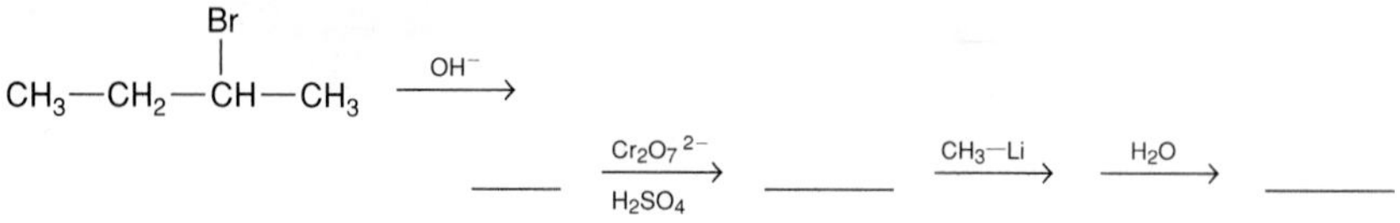

$$CH_3-CH_2-\underset{\underset{Br}{|}}{CH}-CH_3 \xrightarrow{OH^-}$$

$$\underline{\hspace{1cm}} \xrightarrow[H_2SO_4]{Cr_2O_7{}^{2-}} \underline{\hspace{1cm}} \xrightarrow{CH_3-Li} \xrightarrow{H_2O} \underline{\hspace{1cm}}$$

**Plan:** For each step, we examine the functional group of the reactant and the reagent over the yield arrow to decide on the most likely product.

**Solution:** The sequence starts with an alkyl halide reacting with $OH^-$. Substitution gives an alcohol. Oxidation of this alcohol with acidic dichromate gives a ketone. Finally, a two-step reaction of a ketone with $CH_3$—Li and then water forms an alcohol with a C skeleton that has the $CH_3$ group attached to the carbonyl C:

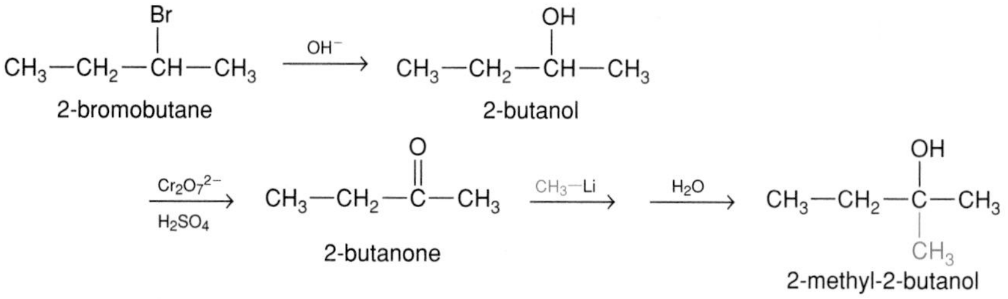

$$CH_3-CH_2-\underset{\underset{Br}{|}}{CH}-CH_3 \xrightarrow{OH^-} CH_3-CH_2-\underset{\underset{OH}{|}}{CH}-CH_3$$

2-bromobutane                              2-butanol

$$\xrightarrow[H_2SO_4]{Cr_2O_7{}^{2-}} CH_3-CH_2-\underset{\overset{O}{||}}{C}-CH_3 \xrightarrow{CH_3-Li} \xrightarrow{H_2O} CH_3-CH_2-\underset{\underset{CH_3}{|}}{\overset{\overset{OH}{|}}{C}}-CH_3$$

2-butanone                                                    2-methyl-2-butanol

**Check:** In this case, make sure that the first two reactions alter the functional group only and that the final steps change the C skeleton.

FOLLOW-UP PROBLEM 14.6

Choose reactants to obtain the following products:

(a) $\underline{\hspace{1cm}} \xrightarrow[H_2SO_4]{Cr_2O_7{}^{2-}}$

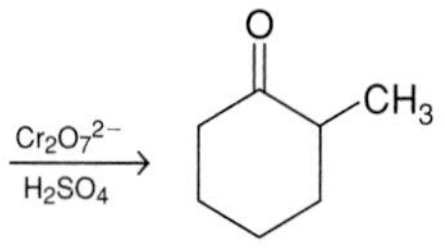

(b) $\underline{\hspace{1cm}} \xrightarrow{CH_3-CH_2-Li} \xrightarrow{H_2O} CH_3-CH_2-\underset{\underset{OH}{|}}{CH}-$

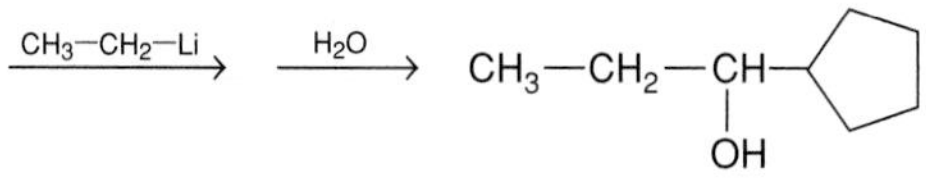

## Variations on a theme: inorganic compounds with double bonds.

Homonuclear (same atom) double bonds are rare among atoms other than C, but we've seen many double bonds to O with other nonmetals, as in the oxides of sulfur, nitrogen, and halogens. Like carbonyl compounds, these substances undergo addition reactions. In its reaction with $SO_3$, for example, the partially negative O of water attacks the partially positive S to form sulfuric acid:

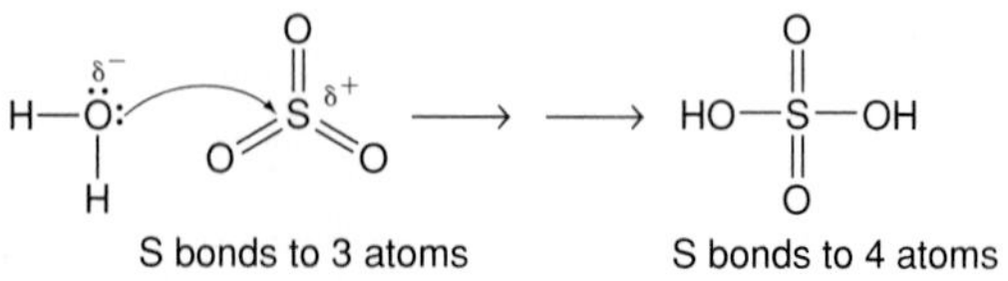

S bonds to 3 atoms          S bonds to 4 atoms

Nitrogen dioxide reacts in a similar way to form nitric acid.

## Functional Groups with Both Single and Double Bonds

A family of three functional groups contains C double bonded to O (a carbonyl group) *and* single bonded to O or N. The parent of the family is the

**carboxylic acid** group, $-\overset{\overset{\displaystyle :O:}{\|}}{C}-\ddot{O}H$, also called the *carboxyl group* and written —COOH. Substitution for the —OH by the —OR of alcohols gives

the ester group, $-\overset{\overset{\displaystyle :O:}{\|}}{C}-\ddot{O}-R$; substitution by the $-\overset{|}{\underset{}{\ddot{N}}}-$ of amines

gives the **amide** group, $-\overset{\overset{\displaystyle :O:}{\|}}{C}-\overset{|}{\underset{\cdot\cdot}{N}}-$ .

**Carboxylic acids.** Carboxylic acids, $R-\overset{\overset{\displaystyle O}{\|}}{C}-OH$, are named by dropping the *-e* from the parent alkane and adding *-oic acid;* however many common names are used. For example, the four-C acid is butanoic acid (the carboxyl C is counted in the root); its common name is butyric acid. Figure 14.21 shows some important carboxylic acids. The carboxyl C already has three bonds, so it forms only one other. In formic acid (methanoic acid), it bonds to an H, but all other carboxylic acids contain chains or rings.

Carboxylic acids are weak acids in water:

$$CH_3-\overset{\overset{\displaystyle O}{\|}}{C}-OH(l) + H_2O(l) \rightleftharpoons CH_3-\overset{\overset{\displaystyle O}{\|}}{C}-O^-(aq) + H_3O^+(aq)$$
ethanoic acid
(acetic acid)

At equilibrium, more than 99% of the acid molecules are undissociated. In base, however, they are neutralized completely to form a salt and water:

$$CH_3-\overset{\overset{\displaystyle O}{\|}}{C}-OH(l) + NaOH(aq) \longrightarrow CH_3-\overset{\overset{\displaystyle O}{\|}}{C}-O^-(aq) + Na^+(aq) + H_2O(l)$$

The anion is the *carboxylate ion,* named by dropping *-oic acid* and adding *-oate;* the sodium salt of butanoic acid, for instance, is sodium butanoate.

Carboxylic acids with long chains are **fatty acids,** an essential group of compounds found in all cells. Animal fatty acids have saturated chains, whereas many from vegetable sources are unsaturated, usually with the C=C bonds in the *cis* configuration. Nearly all fatty acid skeletons have an even number of C atoms—16 and 18 are very common—because they are

**FIGURE 14.21**
Some molecules with the carboxylic acid functional group.

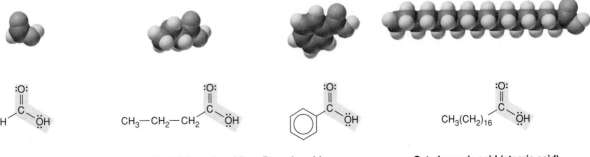

**Methanoic acid (formic acid)**
An irritating component
of ant and bee stings

**Butanoic acid (butyric acid)**
Odor of rancid butter;
suspected component of
monkey sex attractant

**Benzoic acid**
Calorimetric standard;
used in preserving food,
dyeing fabric, curing tobacco

**Octadecanoic acid (stearic acid)**
Found in animal fats; used in
making candles and soaps

made in the cell from two-C pieces. Fatty acid salts are soaps, with the cation usually from Groups 1A(1) or 2A(2) (Section 12.1).

The trigonal planar carbonyl group gives an unstable tetrahedral intermediate on addition, which immediately undergoes elimination to give a trigonal planar product. The net result of addition plus elimination is *substitution*, the most characteristic reaction type of the whole family:

$$R-\overset{\overset{\textstyle O}{\|}}{C}-X \; + \; Z-Y \;\; \underset{\text{addition}}{\rightleftharpoons} \;\; \left[ R-\overset{\overset{\textstyle O-Z}{\|}}{\underset{\underset{\textstyle Y}{|}}{C}}-X \right] \;\; \underset{\text{elimination}}{\rightleftharpoons} \;\; R-\overset{\overset{\textstyle O}{\|}}{C}-Y \; + \; Z-X$$

Strong heating of carboxylic acids forms an **acid anhydride** through a *dehydration-condensation*, in which two molecules condense into one with loss of water:

$$R-\overset{\overset{\textstyle O}{\|}}{C}-O\boxed{H \; + \; HO}-\overset{\overset{\textstyle O}{\|}}{C}-R \;\; \overset{\Delta}{\longrightarrow} \;\; R-\overset{\overset{\textstyle O}{\|}}{C}-O-\overset{\overset{\textstyle O}{\|}}{C}-R \; + \; HOH$$

**Esters.** An alcohol and a carboxylic acid form an ester; the first part of an ester name designates the alcohol portion and the second the acid portion (named in the same way as the carboxylate ion). For example, the ester formed between ethanol and ethanoic acid is ethyl ethanoate (common name, ethyl acetate), a solvent for nail polish and model glue.

The ester group occurs in **lipids,** a large group of fatty biological substances. Most dietary fats are *triglycerides,* esters composed of three fatty acids linked to the alcohol 1,2,3-trihydroxypropane (common name, glycerol), which function as energy stores. Some important lipids are shown in Figure 14.22.

**FIGURE 14.22**
Some lipid molecules with the ester functional group.

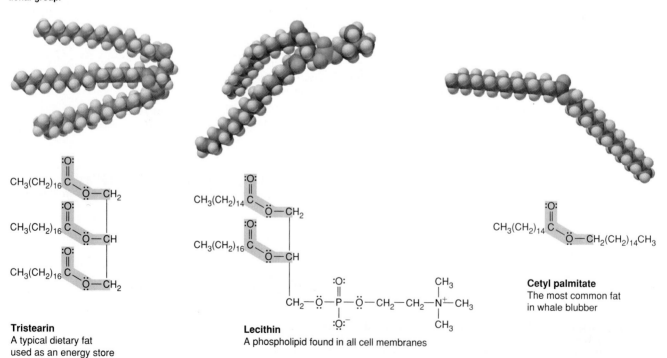

**Tristearin**
A typical dietary fat used as an energy store in animals

**Lecithin**
A phospholipid found in all cell membranes

**Cetyl palmitate**
The most common fat in whale blubber

**FIGURE 14.23**
**The contribution of groups in ester formation.** An ester forms when a carboxylic acid reacts with an alcohol. To determine which reactant supplies the ester O, the acid and alcohol were prepared by labeling with the isotope $^{18}$O. **A,** When R—$^{18}$OH reacts with the unlabeled acid, the ester contains $^{18}$O but the water does not. **B,** When R—C—$^{18}$OH reacts with the unlabeled alcohol, the water contains $^{18}$O. Thus, the alcohol supplies the —OR′ portion of the ester, and the acid supplies the RC=O portion.

Esters, like acid anhydrides, ethers, and several other organic (and inorganic) compounds, form through a dehydration-condensation:

Studies of the reaction with the $^{18}$O isotope show that the alcohol O is attracted to the carboxyl C and becomes bonded to it in the ester, and the —OH of the acid becomes part of the product water (Figure 14.23). Thus, the acid supplies the RC=O portion of the ester, and the alcohol supplies the —OR′ portion. (We discuss this multistep reaction in Chapter 15.) ◆

Note that the reaction is reversible. In the opposite direction, which is called **hydrolysis,** the O atom of water is attracted to the partially positive ester C, cleaving (lysing) the molecule into two parts. One part receives the water —OH, and the other part receives water's other H. In soap manufacture, which began in ancient times, ester bonds in animal fats are hydrolyzed with strong base in the process of *saponification* (Latin *sapon,* "soap"):

a triglyceride + 3NaOH → 3 soaps + glycerol

**Amides.** The product of a substitution between an amine (or NH$_3$) and an ester is an amide. The partially negative N is attracted to the partially positive ester C, ROH is lost, and an amide forms:

methyl ethanoate (methyl acetate) + ethylamine → N-ethylethanamide (N-ethylacetamide) + methanol

Amides are named by denoting the amine portion with *N-* and replacing *-oic acid* with *-amide.* In the amide above, the ethyl group came from the

◆ **A Pungent, Pleasant Banquet.** Many organic compounds have strong odors—fishy amines, oily smelling alkanes, ethereal alkyl halides—but none can rival the diverse odors of carboxylic acids and esters. From the pungent, vinegary odors of the one-C and two-C acids to the cheesy stench of slightly larger ones, organic acids possess some awful odors. When butanoic acid reacts with ethanol, however, its rancid-butter smell becomes the peachy pineapple scent of ethyl butanoate, or when pentanoic acid reacts with pentanol, its Limburger cheese odor becomes the fresh apple aroma of pentyl pentanoate. Naturally occurring esters are used extensively in the fragrance industry to add fruity, floral, and herbal odors to foods, cosmetics, and medicines.

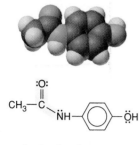

:O:

CH₃—C

NH—⟨ ⟩—ÖH

**Acetaminophen**
Active ingredient
in nonaspirin
pain relievers; used
to make dyes and
photographic chemicals

**FIGURE 14.24**
Some molecules with the amide functional
group.

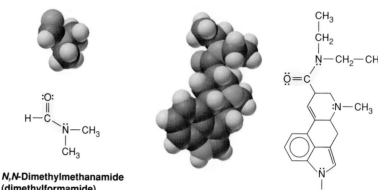

:O:

H—C

N—CH₃

CH₃

***N,N*-Dimethylmethanamide
(dimethylformamide)**
Major organic solvent;
used in production of
synthetic fibers

**Lysergic acid diethylamide (LSD-25)**
A potent hallucinogen

amine, and the acid portion came from ethanoic acid (acetic acid). Some amides are shown in Figure 14.24.

Amides are hydrolyzed in hot water (or base) to a carboxylic acid and an amine, so even though amides are not formed in the following way, they can be viewed as the result of a reversible dehydration-condensation:

$$R-\overset{\overset{\textstyle O}{\|}}{C}\boxed{-OH + H}-\overset{\overset{\textstyle H}{|}}{N}-R' \rightleftharpoons R-\overset{\overset{\textstyle O}{\|}}{C}-\overset{\overset{\textstyle H}{|}}{N}-R' + HOH$$

The most important example of the amide group is the *peptide bond*, which links amino acids in a protein, as you'll see shortly.

The carboxylic acid family undergoes reactions to other functional groups. For example, certain inorganic reducing agents convert acids and esters to alcohols, and amides to amines:

$$R-\overset{\overset{\textstyle O}{\|}}{C}-OH \text{ (or O—R')} \xrightarrow[+4H]{\text{reduction}} R-CH_2-OH + HOH \text{ (or R'—OH)}$$

$$R-\overset{\overset{\textstyle O}{\|}}{C}-\overset{\overset{\textstyle H}{|}}{N}-R' \xrightarrow[+4H]{\text{reduction}} R-CH_2-NH-R' + H_2O$$

SAMPLE PROBLEM 14.7

**Predicting the Reactions of the Carboxylic Acid Family**

**Problem:** Predict the product(s) of the following reactions:

**(a)** $CH_3-CH_2-CH_2-\overset{\overset{\textstyle O}{\|}}{C}-OH \ + \ CH_3-\overset{\overset{\textstyle OH}{|}}{CH}-CH_3 \overset{H^+}{\rightleftharpoons}$

**(b)** $CH_3-CH_2-\overset{\overset{\textstyle }{\underset{\underset{\textstyle CH_3}{|}}{CH}}}-\overset{\overset{\textstyle O}{\|}}{C}-O-CH_3 \xrightarrow[+4H]{\text{reduction}}$

**(c)** $CH_3-\overset{\overset{\textstyle CH_3}{|}}{CH}-CH_2-CH_2-\overset{\overset{\textstyle O}{\|}}{C}-NH-CH_2-CH_3 \xrightarrow[H_2O]{\text{NaOH}}$

**Plan:** We discussed substitution (including addition-elimination and dehydration-condensation), hydrolysis, and reduction. In **(a)**, a carboxylic acid and alcohol react, so it must be a substitution to form an ester and water. In **(b)**, an ester is reduced, so each portion becomes an alcohol. In **(c)**, an amide reacts with $OH^-$, so it is hydrolyzed to an amine and a sodium carboxylate.

**Solution:**

**(a)** Formation of an ester:
$$CH_3-CH_2-CH_2-\overset{\displaystyle O}{\overset{\|}{C}}-O-\overset{\displaystyle CH_3}{\overset{|}{CH}}-CH_3 \ + \ H_2O$$

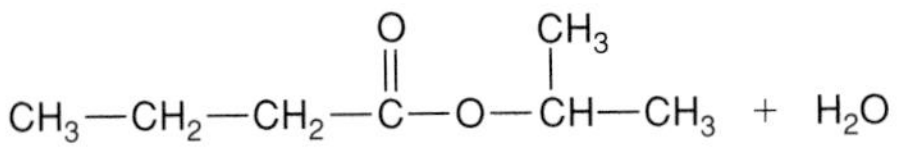

**(b)** Reduction of an ester:
$$CH_3-CH_2-\overset{}{\underset{\underset{\displaystyle CH_3}{|}}{CH}}-CH_2-OH \ + \ HO-CH_3$$

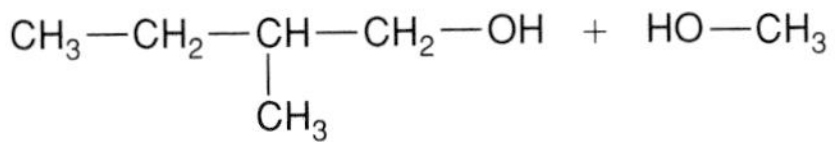

**(c)** Basic hydrolysis of an amide:
$$CH_3-\overset{}{\underset{\underset{\displaystyle CH_3}{|}}{CH}}-CH_2-CH_2-\overset{\displaystyle O}{\overset{\|}{C}}-O^- \ + \ Na^+ \ + \ H_2N-CH_2-CH_3$$

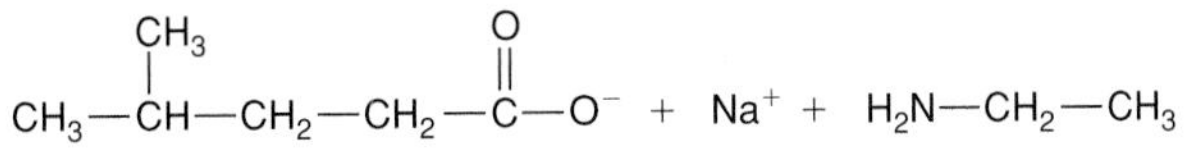

**Check:** In **(b)**, be sure that the carboxylic C reduces to a $-CH_2-$ group. In **(c)**, the carboxylate ion forms, rather than the acid, because aqueous NaOH is present.

**FOLLOW-UP PROBLEM 14.7**

Fill in the blanks in the following reactions:

**(a)** _____ + $CH_3-OH$ $\overset{H^+}{\rightleftharpoons}$
(benzene ring)$-CH_2-\overset{\displaystyle O}{\overset{\|}{C}}-O-CH_3$ + $H_2O$

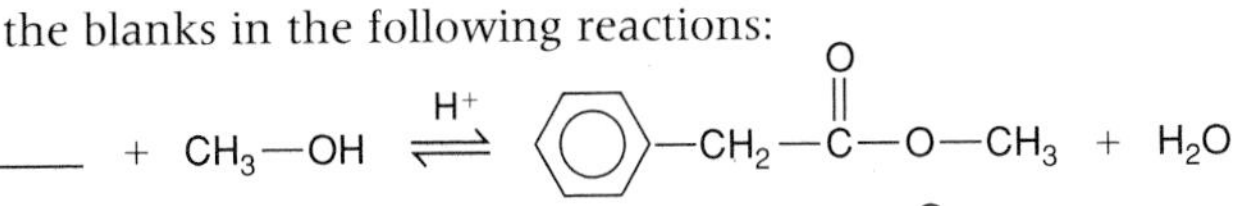

**(b)** _____ + _____ $\longrightarrow$ $CH_3-CH_2-CH_2-\overset{\displaystyle O}{\overset{\|}{C}}-NH-CH_2-CH_3$ + $CH_3-OH$

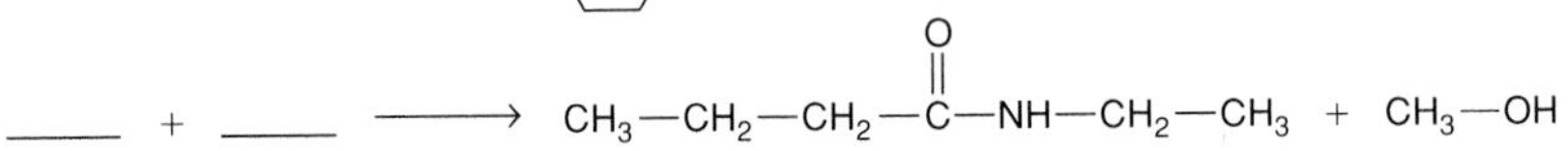

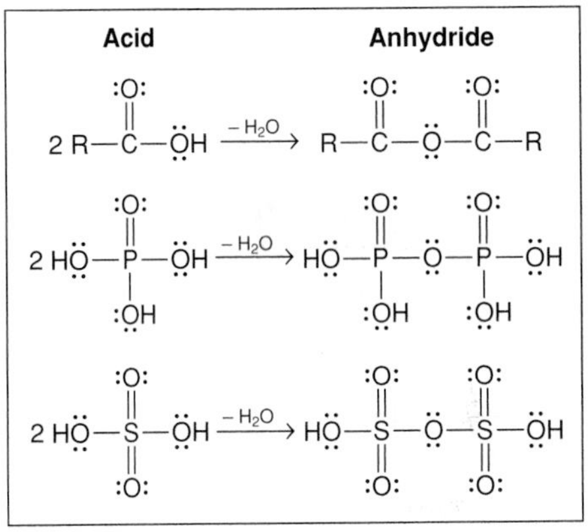

FIGURE 14.25
Formation of carboxylic, phosphoric, and sulfuric acid anhydrides.

**Variations on a theme: oxoacids, esters, and amides of other nonmetals.** A nonmetal double bonded and single bonded to O occurs in most inorganic oxoacids, such as phosphoric, sulfuric, and chlorous acids. Those with additional O atoms are stronger acids than carboxylic acids.

Diphosphoric and disulfuric acids are acid anhydrides formed in a dehydration-condensation, just as is the organic compound (Figure 14.25). Inorganic oxoacids form esters and amides that are part of many biological molecules. For example, in addition to lipids (see Figure 14.22), the first compound formed when glucose is digested is a phosphate ester (Figure 14.26, *A*), and a similar phosphate ester is a major structural feature of nucleic acids, as you'll see shortly. Amides of organic S-containing oxoacids, called sulfonamides, are potent antibiotics (Figure 14.26, *B*). More than 10,000 different sulfonamides have been synthesized.

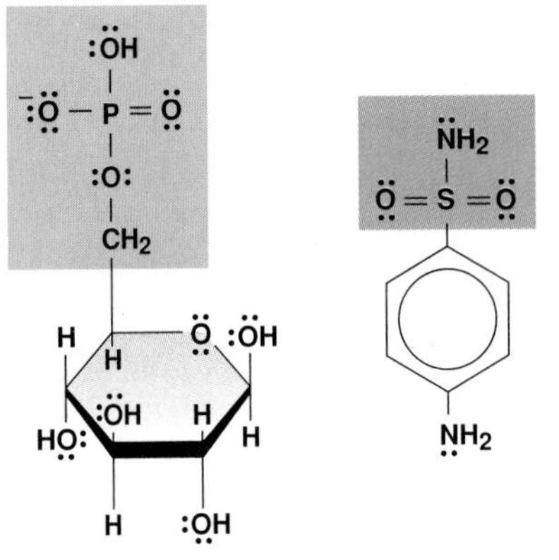

**A** Glucose-6-phosphate    **B** Sulfanilamide
FIGURE 14.26
An ester and an amide of other nonmetals.
**A,** Glucose-6-phosphate contains a phosphate ester group. **B,** Sulfanilamide contains an amide group.

## Functional Groups with Triple Bonds

There are only two important triple-bonded functional groups. *Alkynes*, with their electron-rich $-C\equiv C-$ groups, undergo addition (by $H_2O$, $H_2$, HX, $X_2$, and so forth) to form double-bonded or saturated compounds:

$$CH_3-C\equiv CH \overset{H_2}{\longrightarrow} CH_3-CH=CH_2 \overset{H_2}{\longrightarrow} CH_3-CH_2-CH_3$$
$$\text{propyne} \qquad\qquad \text{propene} \qquad\qquad \text{propane}$$

*Nitriles* ($R-C\equiv N$) contain the **nitrile** group ($-C\equiv N:$) and are made by substituting a $CN^-$ ion for $X^-$ in reaction with an alkyl halide:

$$CH_3-CH_2-Cl + NaCN \rightarrow CH_3-CH_2-C\equiv N + NaCl$$

This reaction is useful because it *increases the chain by one C atom*. Nitriles are often reduced to amines or hydrolyzed to carboxylic acids:

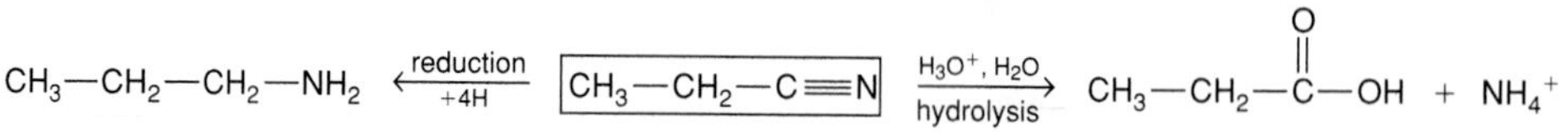

$$CH_3-CH_2-CH_2-NH_2 \xleftarrow[+4H]{reduction} \boxed{CH_3-CH_2-C \equiv N} \xrightarrow[hydrolysis]{H_3O^+, H_2O} CH_3-CH_2-\overset{\overset{\displaystyle O}{\|}}{C}-OH + NH_4^+$$

**Variations on a theme: inorganic compounds with triple bonds.** Triple bonds are scarce in the inorganic world as well as in the organic world. Carbon monoxide (:C≡O:), elemental nitrogen (:N≡N:), and the cyanide ion ([:C≡N:]⁻) are the only common examples.

---

SAMPLE PROBLEM 14.8

### Recognizing Functional Groups

---

**Problem:** Circle and name the functional groups in the following molecules:

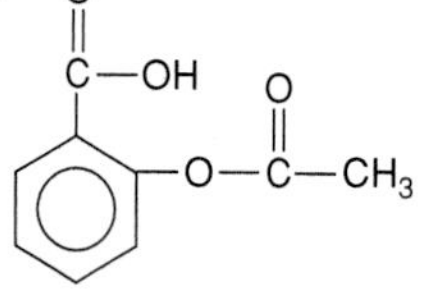

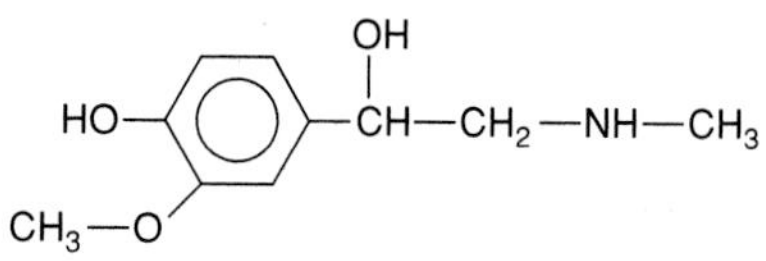

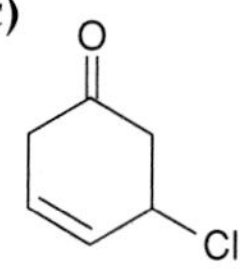

**Plan:** We use Table 14.5 to identify the various functional groups.
**Solution:**

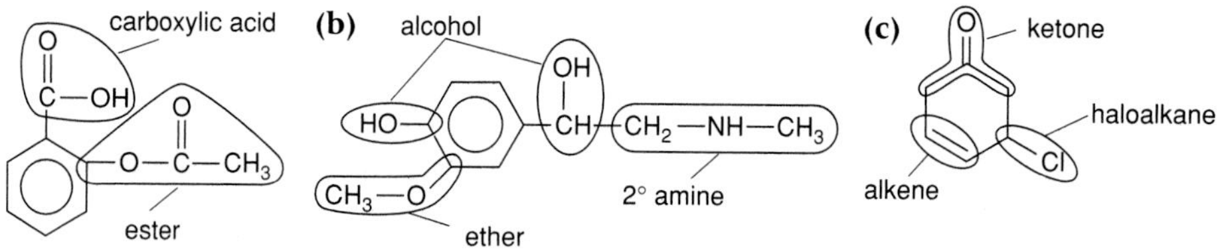

FOLLOW-UP PROBLEM 14.8

Circle and name the functional groups in the following molecules:

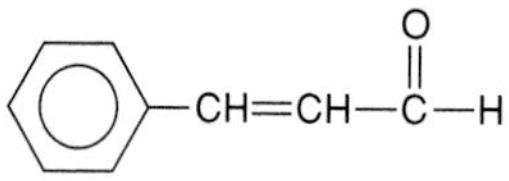

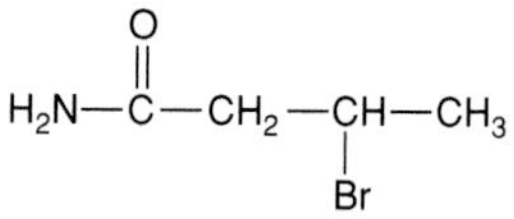

Figure 14.27 summarizes the names, structures, and interconversions of the functional groups we've discussed.

### Section Summary

Organic reactions are initiated when regions of high and low electron density on different reactant molecules attract each other. Single-bonded groups—alcohols, amines, and alkyl halides—take part in substitution and elimination reactions. Multiple-bonded groups—alkenes, aldehydes, ketones, alkynes, and nitriles—generally take part in addition reactions. Aromatic compounds typically undergo substitution, rather than addition, because electron delocalization stabilizes the ring π system. Groups with double and single bonds—carboxylic acids, esters, and amides—generally take part in substitution reactions. Many reactions change one functional group to another, but some change the C skeleton, especially those with organometallic reactants and with the cyanide ion.

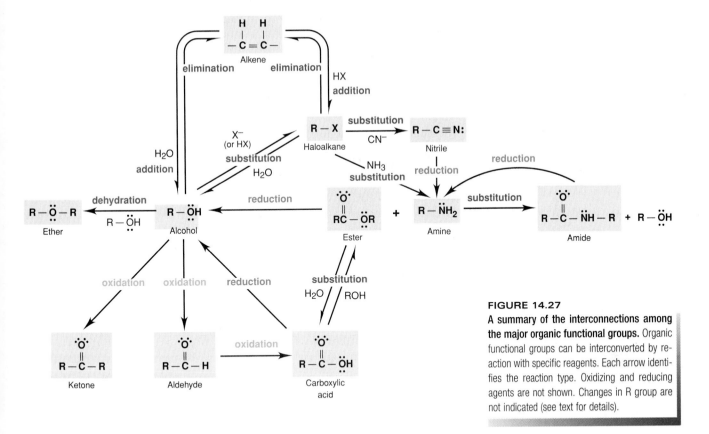

**FIGURE 14.27**
A summary of the interconnections among the major organic functional groups. Organic functional groups can be interconverted by reaction with specific reagents. Each arrow identifies the reaction type. Oxidizing and reducing agents are not shown. Changes in R group are not indicated (see text for details).

## 14.5   Giant Organic Molecules: The Monomer-Polymer Theme

**Polymers** (Greek, "many parts"), or *macromolecules,* are extremely large molecules that result from the covalent linking of many simpler molecular units called **monomers** (Greek, "one part"). *Synthetic polymers,* in the form of plastics, fibers, films, rubbers, and so forth, have revolutionized everyday life. More than half of all industrial chemists work in polymer-related fields. There are probably dozens of these materials in your home, and there will be many more in the future. Some of these products are nearly indestructible in the environment, however, so their manufacture has created a critical waste disposal problem.

*Natural polymers* are the "stuff of life"—carbohydrates, proteins, and nucleic acids. Some have structures that make wood strong, hair curly, nails hard, and wool flexible. Others speed up the myriad reactions that occur in every cell or defend us against infection. Still others possess the genetic information needed to forge other biomolecules. Remarkable as these giant molecules are, the functional groups of their monomers and the reactions that link them are identical to those of other organic molecules.

### Synthetic Polymers

The two major types of synthetic polymers are named for the reaction processes that form them—addition and condensation.

**Addition polymers.** When monomers add together, **addition polymers** form. These are also called *chain-reaction (or chain-growth) polymers* be-

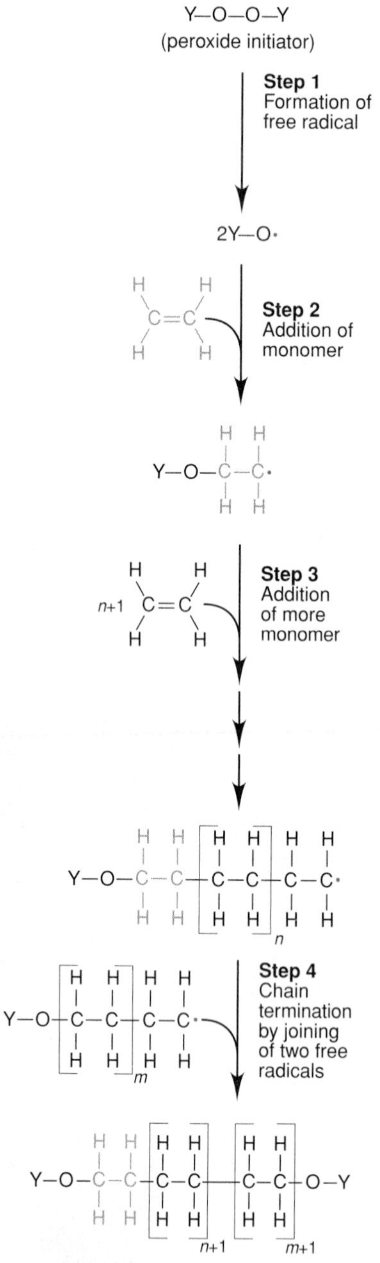

Y—O—O—Y
(peroxide initiator)

**Step 1**
Formation of
free radical

2Y—O·

**Step 2**
Addition of
monomer

**Step 3**
Addition
of more
monomer

**Step 4**
Chain
termination
by joining
of two free
radicals

**FIGURE 14.28**

**Steps in the free-radical polymerization of ethylene.** In this polymerization method, free radicals initiate, propagate, and terminate the formation of an addition polymer. An initiator (Y—O—O—Y) is split to form a free radical (Y—O·). The radical attacks the $\pi$ bond of a monomer and creates another free radical (Y—O—$CH_2$—$CH_2$·). The process proceeds and the chain grows (propagates) until an inhibitor is added (not shown) or two free radicals combine.

cause as each monomer joins the chain, it forms a new reactive site to continue the process. The common structural feature in the monomers of most addition polymers is the $\text{C}=\text{C}$ grouping.

The polymerization of ethene (ethylene, $CH_2$=$CH_2$) to polyethylene is an example of the process. In Figure 14.28, the monomer becomes a *free radical*, a species with an unpaired electron, that seeks another electron from the next monomer to form a covalent bond. The process begins when an *initiator*, usually a peroxide, generates a free radical that attacks the $\pi$ bond of an ethylene unit, forming a $\sigma$ bond with one of the $p$ electrons and leaving the other unpaired. This new free radical then attacks the $\pi$ bond of another ethylene, and the polymer chain grows. As each ethylene adds, it leaves an unpaired electron on the growing end to find an electron "mate" and make the chain one unit longer. This process, called *free-radical polymerization,* stops when two free radicals form a covalent bond or when a very stable free radical, called an *inhibitor,* is added. In a similar method, the polymerization is initiated by the formation of a cation (or anion) instead of a free radical. The cationic (or anionic) monomer attacks the $\pi$ bond of the next monomer to form another cationic (or anionic) end, and the process continues.

The most important polymerization reactions take place under relatively mild conditions through the use of catalysts that incorporate transition metals. *Ziegler-Natta catalysts,* for which Karl Ziegler and Giulio Natta received a Nobel Prize in 1963, employ an organoaluminum compound, such as $Al(C_2H_5)_3$, and tetrachlorides of either titanium or vanadium. These *stereoselective* catalysts create polymers with groups oriented spatially in a particular way.

With these catalysts, polyethylene chains are made with molar masses of $10^4$ to $10^5$ g/mol by varying conditions and reagents. One form has high density and strength as a result of extensive dispersion forces between long, unbranched chains. A lower density form has some branches that prevent the chains from packing as well and confers flexibility and transparency.

Similarly, polypropylenes, $+CH_2$—$CH+_n$ can be made that have all

the monomer $CH_3$ groups either on one side of the chain or on alternating sides. The different orientations lead to different packing efficiencies of the chains and thus differences in density, rigidity, elasticity, and so forth.

Variations on these reactions, such as using isobutylene

$$(CH_3-\underset{\overset{|}{CH_3}}{C}=CH_2)$$

or styrene ($\bigcirc$—CH=$CH_2$), give rise to many familiar products. Other substituted monomers are vinyl chloride ($CH_2$=CH—Cl) used to form PVC [poly(vinyl chloride)] and tetrafluoroethene ($CF_2$=$CF_2$) used to form Teflon. Table 14.6 shows the monomers used for the most important addition polymers. Note that the general chain is identical in all of them. Thus, the essential chemical differences between an acrylic sweater, a plastic grocery bag, and a bowling ball are the different groups attached to the main chain.

**Condensation polymers.** The monomers of **condensation polymers** must have two functional groups. Most commonly, they link when a group on one unit undergoes a dehydration-condensation with a group on another. Most condensation polymers are *copolymers* made from two or

**TABLE 14.6  Structures and Applications of Some Major Addition Polymers**

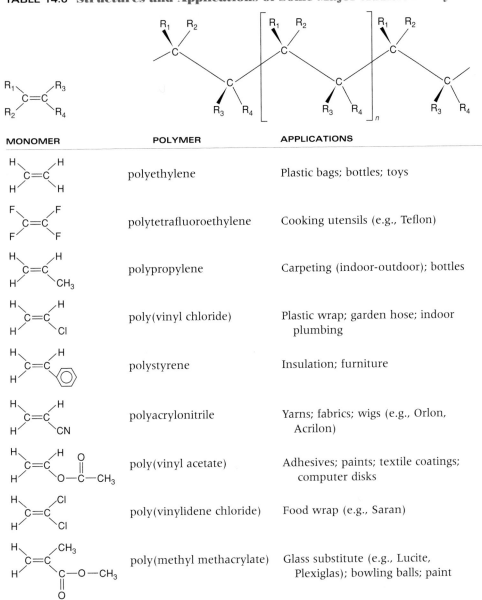

| MONOMER | POLYMER | APPLICATIONS |
|---|---|---|
| $H_2C=CH_2$ | polyethylene | Plastic bags; bottles; toys |
| $F_2C=CF_2$ | polytetrafluoroethylene | Cooking utensils (e.g., Teflon) |
| $H_2C=CHCH_3$ | polypropylene | Carpeting (indoor-outdoor); bottles |
| $H_2C=CHCl$ | poly(vinyl chloride) | Plastic wrap; garden hose; indoor plumbing |
| $H_2C=CH$(phenyl) | polystyrene | Insulation; furniture |
| $H_2C=CHCN$ | polyacrylonitrile | Yarns; fabrics; wigs (e.g., Orlon, Acrilon) |
| $H_2C=CHOC(O)CH_3$ | poly(vinyl acetate) | Adhesives; paints; textile coatings; computer disks |
| $H_2C=CCl_2$ | poly(vinylidene chloride) | Food wrap (e.g., Saran) |
| $H_2C=C(CH_3)C(O)OCH_3$ | poly(methyl methacrylate) | Glass substitute (e.g., Lucite, Plexiglas); bowling balls; paint |

more different monomers. Condensation of carboxylic acid and amine units forms *polyamides (nylons),* whereas carboxylic acid and alcohol units form *polyesters.*

One of the most common polyamides is *nylon-66,* manufactured by mixing equimolar amounts of a six-C diamine (1,6-diaminohexane) and a six-C diacid (1,6-hexanedioic acid). The basic amine reacts with the acid to form a "nylon salt." Heating drives off water and forms the amide bonds. In the laboratory, this nylon is made without heating by using a more reactive acid component (Figure 14.29). Covalent bonds in the chain give nylons great strength, while H bonds between chains give them great flexibility. About

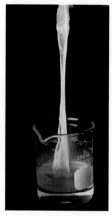

$$n\,H_2N\!-\!(CH_2)_{\overline{6}}\,NH_2 + n\,Cl\!-\!\overset{\overset{O}{\|}}{C}\!-\!(CH_2)_{\overline{4}}\,\overset{\overset{O}{\|}}{C}\!-\!Cl$$

$$\downarrow\ -(N+I)HCl$$

$$\left[-\!NH\!-\!(CH_2)_{\overline{6}}\,NH\!-\!\overset{\overset{O}{\|}}{C}\!-\!(CH_2)_{\overline{4}}\,\overset{\overset{O}{\|}}{C}\!-\right]_n$$

**FIGURE 14.29**

**The formation of nylon-66.** Nylon is formed industrially by the reaction of diamine and diacid monomers. In this laboratory demonstration, the more reactive diacid chloride monomer is used; the polyamide forms between the phases.

**♦ Polysaccharide Skeletons of Lobsters and Roaches.** The variety of polysaccharide properties that arises from simple changes in the monomers is amazing. For example, substituting an —NH₂ group for the —OH group at C-2 in glucose produces *glucosamine*. The amide of glucosamine with acetic acid (common name, *N*-acetylglucosamine) is the monomer of *chitin*, a polysaccharide that is the main component in the external skeletons of insects and crustaceans.

half of all nylons are made to reinforce automobile tires, the remainder being used for rugs, clothing, fishing line, and so forth.

Dacron, a popular polyester fiber, is woven from polymer strands formed when equimolar amounts of 1,4-benzenedicarboxylic acid and 1,2-ethanediol react. Blending these polyester fibers with various cotton products gives fabrics that are durable, easily dyed, and crease resistant. Extremely thin Mylar films, used for recording tape and food packaging, are also made from this polymer.

In keeping with our ongoing "Variations . . ." feature, you already know that some synthetic polymers are not organic. In Chapter 13, we discussed the silicones, polymers with a repeating —Si(R₂)—O— unit. Depending on the chain cross-links and the R groups, silicones range from oily liquids to elastic sheets to rigid solids and have applications that include stopcock grease, artificial limbs, and spacesuits. We also mentioned the polyphosphazenes, whose repeating —P(R₂)=N— unit exists as a flexible chain even at low temperatures.

### The Biopolymers

The monomer-polymer theme was being played out in nature eons before humans employed it to such great advantage. Biological macromolecules are nothing more than condensation polymers created by nature's reaction chemistry and improved through evolution. These remarkable molecules are the greatest proof of the versatility of carbon and its handful of atomic partners.

**Sugars and polysaccharides.** Glucose and other simple sugars are called **monosaccharides.** In addition to their roles as individual molecules, they serve as the monomer units of **polysaccharides.** In aqueous solution, the alcohol and aldehyde (or ketone) groups of the *same* sugar molecule react with each other to form a cyclic molecule (Figure 14.30, *A*). When two of these cyclic units undergo a dehydration-condensation, a **disaccharide** forms. For example, sucrose (table sugar) is a disaccharide of glucose and fructose (Figure 14.30, *B*); lactose (milk sugar) is a disaccharide of glucose and galactose; maltose, used in brewing and as a sweetener, is a disaccharide of two glucose units. A polysaccharide consists of *many* monosaccharide units linked together. The three major natural polysaccharides—cellulose, starch, and glycogen—consist entirely of glucose units, but they differ in the ring positions of the links, in bond orientation, and in the extent of crosslinking. ♦

*Cellulose* is the most abundant organic chemical on Earth. More than 50% of the carbon in plants occurs in the cellulose support structures of stems and leaves; wood is largely cellulose, and cotton is more than 90% cellulose. This polymer consists of long chains of glucose monomers linked in a particular way from C-1 in one unit to C-4 in the next. Humans lack the biochemical ability to break this link, so we cannot digest cellulose, but microorganisms in the digestive tracts of some animals, such as cows, sheep, and termites, can. The great strength of wood is due largely to the H bonds between cellulose chains.

*Starch* is a mixture of glucose polysaccharides that serves as an *energy store* in plants. When a plant needs energy, some starch is broken down by hydrolysis of the bonds between units, and the released glucose is oxidized. Starch occurs in plant cells as insoluble granules of amylose, a helical

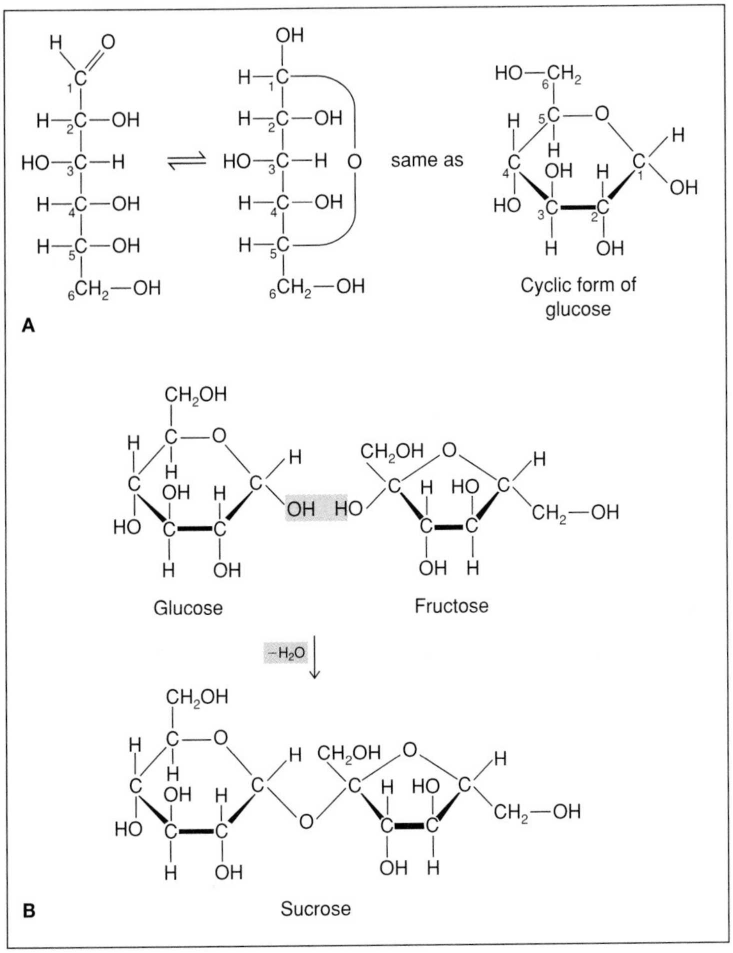

A

Cyclic form of
glucose

Glucose          Fructose

$-H_2O$

B                Sucrose

**FIGURE 14.30**
**The structure of glucose in aqueous solution and the formation of a disaccharide. A,** A molecule of glucose undergoes an internal reaction between the aldehyde group of C-1 and the alcohol group of C-5 to form a cyclic monosaccharide. **B,** Two monosaccharides undergo a dehydration-condensation to form a disaccharide. Glucose and fructose combine to form sucrose (table sugar).

molecule of several thousand glucose units, and amylopectin, a highly branched, bushlike molecule of up to a million glucose units. Most of the glucose units are linked by C-1 to C-4 bonds, as in cellulose, but a different bond orientation allows us to digest it. A C-6 to C-1 branch cross-links chains every 24 to 30 units.

*Glycogen* is the energy storage molecule in animals. It resides in liver and muscle cells as large, insoluble granules consisting of glycogen molecules made from 1000 to more than 500,000 glucose units. In glycogen the units are linked by C-1 to C-4 bonds also, but the molecule is more highly cross-linked than starch, with C-6 to C-1 cross-links every 8 to 12 units (Figure 14.31).

**Amino acids and proteins.** As you saw earlier, nylon-66 is formed from two monomers, one with a carboxyl group at each end and another with an amine group at each end. **Proteins,** the polyamides of nature, are unbranched polymers formed from about 20 types of monomers called

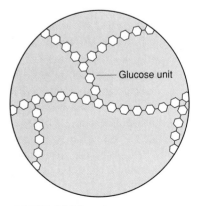

**FIGURE 14.31**
Portion of the structure of glycogen, the major storage polysaccharide in animals.

**amino acids,** which have a carboxyl and an amine group in the *same* molecule. An amino acid has these two groups attached to the *α-carbon,* the second C atom in the chain:

In the aqueous cell fluid, these groups are charged because the carboxyl group transfers its $H^+$ ion to the amine group in an acid-base reaction. An H atom is the third group bonded to the *α*-carbon, and the fourth is the R group. About 20 different R groups occur in the amino acids that make up proteins, ranging from an H atom to a polycyclic N-containing aromatic structure (Figure 14.32).

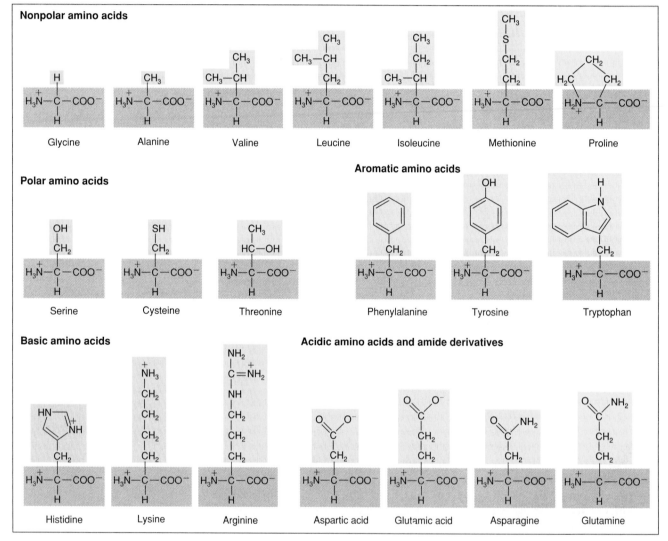

**FIGURE 14.32**

**The common amino acids.** About 20 different R groups are part of amino acids. Here the amino acids are grouped by polarity, acid-base character, and presence of an aromatic ring. The R groups play a major role in the shape and function of the protein.

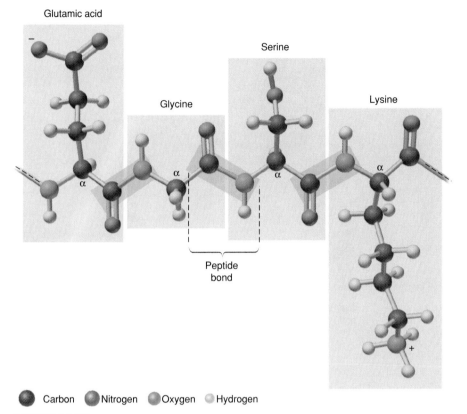

Glutamic acid

Serine

Glycine

Lysine

Peptide bond

● Carbon  ● Nitrogen  ● Oxygen  ○ Hydrogen

**FIGURE 14.33**

**Portion of a polypeptide chain.** The peptide bond holds the monomers together in a protein. Three peptide bonds occur in this portion of a polypeptide chain. The R groups dangle off the chain.

Each amino acid is linked to the next by a *peptide* (amide) bond in a dehydration-condensation by which the carboxyl group of one monomer reacts with the amine group of the other. Therefore, the repeating unit of the polypeptide chain, the "backbone" of the protein, is an $\alpha$-carbon bonded to an amide group bonded to the next $\alpha$-carbon bonded to the next amide group, and so on (Figure 14.33). The various R groups dangle from the $\alpha$-carbons on alternate sides of the chain. Proteins range in length from about 50 amino acids ($\mathcal{M} \approx 5 \times 10^3$ g/mol) to several thousand ($\mathcal{M} \approx 3 \times 10^5$ g/mol). Even for a relatively small protein of 100 amino acids, the number of possible sequences of the 20 types of amino acids is virtually limitless ($20^{100} \approx 10^{130}$). Only a tiny fraction of these possibilities occur, however; for example, in a species as complex as human beings, there are about $10^5$ different types of protein.

*Each type of protein has its own amino acid composition,* a specific number and proportion of the different amino acids. *The sequence of amino acids determines the protein's shape and function.* Shapes vary from long rods to undulating sheets, from baskets with deep crevices to Y-shaped blobs; many proteins have regions of different shapes. The forces responsible for protein shapes are *the same bonding and intermolecular forces that operate for all molecules.* Because proteins are so large, however, these forces act between regions of the same molecule.

**FIGURE 14.34**

**The forces that maintain protein structure.** A combination of covalent, ionic, and intermolecular forces is responsible for protein shape and function.

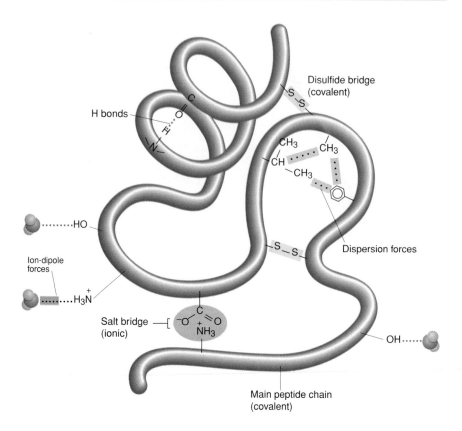

- ○ Glycine
- ● X
- ■ Proline

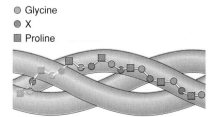

**A** Collagen

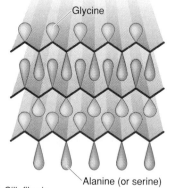

**B** Silk fibroin

**FIGURE 14.35**

**The shapes of fibrous proteins.** Fibrous proteins have largely structural roles in an organism, and their amino acid sequence is relatively simple. **A,** The triple helix of a collagen molecule has a glycine at every third amino acid along the chain. **B,** The "pleated sheet" structure of silk fibroin has a repeating sequence of glycine at every other position with alanine or serine in between. The sheet sections interact through dispersion forces.

Figure 14.34 is a schematic depiction of the forces operating in many proteins. The covalent peptide bonds create the chain. The —SH ends of two cysteine R groups often form an —S—S— bond, a covalent *disulfide bridge* that brings distant parts of the chain together. Polar and ionic R groups usually protrude into the aqueous fluid, interacting with water through ion-dipole forces and H bonds; sometimes they secure the chain's bends through an acidic (—COO⁻) R group lying near a basic (—NH₃⁺) one to form an electrostatic *salt bridge*. Helical and sheetlike segments arise from *H bonds between the C=O of one peptide bond and the N—H of another.* Other H bonds act to keep distant portions of the chain near each other. Nonpolar R groups usually congregate through dispersion forces within the protein interior.

*Fibrous proteins* have relatively simple amino acid compositions and structures shaped like extended helices or sheets. They are key components of hair, wool, skin, and connective tissue—materials that require strength and flexibility. Like synthetic polymers, these proteins have a small number of R groups in a repeating sequence. For example, collagen (Figure 14.35, *A*), the most common animal protein, is found in tendons and skin; more than 30% of the R groups come from glycine (G) and another 20% come from proline (P) to create a long, triple helix with the sequence —G—X—P—G—X—P— and so on (where X is another amino acid). In silk fibroin, more than 85% of the R groups come from glycine, serine, and alanine (Figure 14.35, *B*). Protein segments hydrogen bond to form a sheet. Stacks of sheets make fibroin flexible but not extendable—perfect for a silkworm's cocoon.

*Globular proteins* have much more complex amino acid compositions. They are fairly compact, with a wide variety of shapes and a correspondingly wide range of functions: defenders against bacterial invasion, messengers that trigger cell actions, catalysts of chemical change, membrane gatekeepers that maintain aqueous concentrations, and many others. Certain R groups occur at locations crucial to the protein's function. For example, in catalytic proteins, a few R groups form a crevice that closely matches the shapes of reactant molecules and assists in their reaction through bonding and intermolecular forces. The slightest change in one of these critical groups and function is lost. This fact supports the essential idea that the protein's amino acid sequence, its structure, and its function are inseparable: *sequence determines structure, which determines function.* As you'll see next, the sequence of amino acids in every protein of every organism is prescribed by the genetic information held within the organism's nucleic acids.

**Nucleotides and nucleic acids.** The chemical information that guides the design, construction, and function of all proteins is contained in **nucleic acids.** These unbranched polymers consist of monomers called **mononucleotides,** each of which is itself a dehydration-condensation product made from three parts: an N-containing base, a sugar, and a phosphate group (Figure 14.36, *A*). The two types of nucleic acid, *ribonucleic acid* (RNA) and *deoxyribonucleic acid* (DNA), differ in the sugar portions of their mononucleotides. RNA contains *ribose,* a five-C monosaccharide. DNA contains *deoxyribose,* in which —H substitutes for —OH on the C-2 of ribose.

Dehydration-condensation reactions create phosphate ester links between mononucleotides to give a polynucleotide chain. The repeating motif of the DNA backbone is sugar linked to phosphate linked to sugar linked to phosphate, and so on (Figure 14.36, *B*). Attached to each sugar is one of the four N-containing bases, either a pyrimidine (six-membered ring) or a purine (six- and five-membered rings sharing a side). The purines are guanine (G) and adenine (A); the pyrimidines are thymine (T) and cytosine (C); in RNA, uracil (U) substitutes for thymine. The bases dangle off the main chain, much like the R groups of proteins.

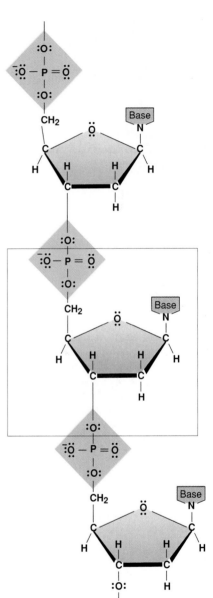

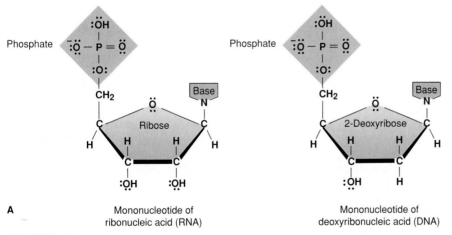

**A** Mononucleotide of ribonucleic acid (RNA)

Mononucleotide of deoxyribonucleic acid (DNA)

**B**

**FIGURE 14.36**

**The mononucleotide of RNA and DNA. A,** Each mononucleotide consists of a base, a sugar, and a phosphate group. In RNA, the sugar is ribose; in DNA, it is 2-deoxyribose. **B,** In the polynucleotide chain, one mononucleotide adds to another through a dehydration-condensation that forms a phosphate ester link with the sugar portion of the next nucleotide. Dangling off the chain are the N-containing bases.

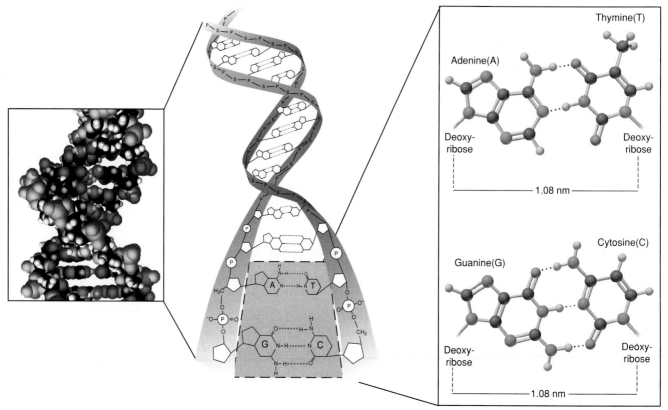

**FIGURE 14.37**

**The double helix of DNA.** In DNA, shown with a space-filling model *(left)* whose length matches the schematic drawing *(center)*, the polar sugar-phosphate backbone faces the watery outside, and the nonpolar bases form H bonds with each other in its core. A pyrimidine and a purine always form H-bonded pairs to maintain the double helix width, *(right)*, and the members of the pairs are always the same: A with T and G with C.

In the cell, DNA exists as two chains wrapped around each other in a **double helix** (Figure 14.37). The negatively charged sugar-phosphate backbones face the aqueous surroundings, and each base in one chain pairs with a base in the other by H bonds. A double-helical DNA molecule contains hundreds of thousands of H-bonded bases. Two features of these **base pairs** are crucial to the structure and function of DNA:

1. A pyrimidine and a purine are always paired, which gives the double helix a constant diameter.
2. Each base is always paired with the same partner: A with T and G with C. Thus, *the base sequence on one chain is the complement of the base sequence on the other.* For example, the sequence A—C—T on one chain is *always* paired with the sequence T—G—A on the other.

The entire DNA molecule is folded into a tangled mass that forms one of the cell's *chromosomes.* DNA is amazingly long and thin: if the largest human chromosome were stretched out, it would be 4 cm long, although it is crumpled into a structure only 5 nm long—8 million times shorter! Linear segments of the DNA molecule act as *genes,* chemical blueprints for constructing the organism's proteins.

*The information content of a gene resides in its base sequence.* In the **genetic code,** each base acts as a letter, each three-base sequence as a word, and *each word codes for a specific amino acid.* For example, the sequence C—A—C codes for the amino acid histidine, A—A—G codes for lysine, and so on. Through a complex series of interactions, greatly simplified in Figure 14.38, one amino acid at a time is positioned and linked to the next in the process

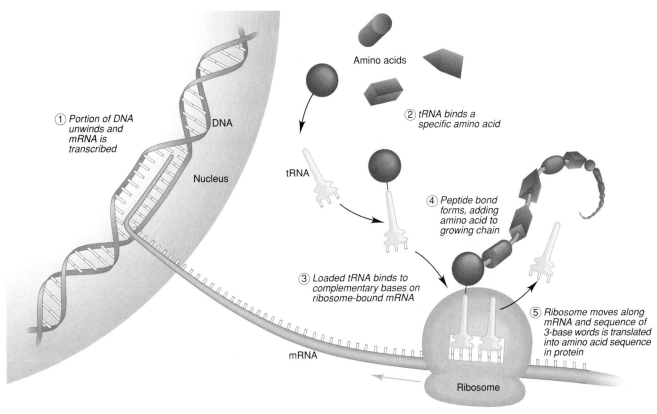

① *Portion of DNA unwinds and mRNA is transcribed*

DNA

Nucleus

Amino acids

② *tRNA binds a specific amino acid*

tRNA

④ *Peptide bond forms, adding amino acid to growing chain*

③ *Loaded tRNA binds to complementary bases on ribosome-bound mRNA*

⑤ *Ribosome moves along mRNA and sequence of 3-base words is translated into amino acid sequence in protein*

mRNA

Ribosome

**FIGURE 14.38**
The key stages in protein synthesis.

of protein synthesis. To fully appreciate just this aspect of the chemical basis of biology, try to keep in mind that *this amazingly complex process occurs largely through H bonding between base pairs.*

Here is an outline of the process of protein synthesis. DNA occurs in the cell nucleus, but the genetic message is decoded outside it, so the information must be sent to the synthesis site. RNA serves in this messenger role, as well as in several other roles. A portion of the DNA is temporarily unwound and acts as a *template* for the formation of a complementary chain of *messenger RNA* from mononucleotides; thus, the DNA words are transcribed into RNA words through base pairing. The messenger leaves the nucleus and binds, again through base pairing, to an RNA-rich particle in the cell called a *ribosome.* The words (three-base sequences) in the RNA message are then decoded by molecules of *transfer RNA,* chemical shuttles with two key portions: (1) a three-base sequence that base pairs with a word on the messenger and (2) a binding site for the amino acid coded by that word. The ribosome moves along the bound messenger, one word at a time, while transfer RNAs bind and position their amino acids near one another for peptide bond formation. In effect, the message of three-base words is translated into a sequence of amino acids that are then linked into a protein. Thus, *the base sequence in DNA determines the base sequence in RNA, which determines the amino acid sequence in the protein.*

Another complex series of interactions allows the DNA to copy itself. When a cell divides, its chromosomes are *replicated,* or reproduced, ensuring that the new cells have the same number and types of chromosomes. A

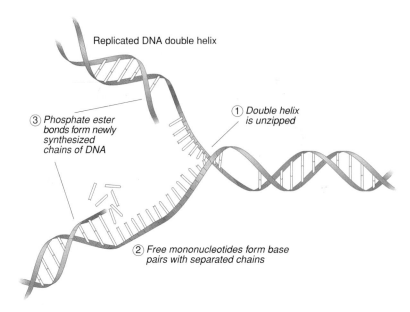

Replicated DNA double helix

③ Phosphate ester bonds form newly synthesized chains of DNA

① Double helix is unzipped

② Free mononucleotides form base pairs with separated chains

**FIGURE 14.39** The key stages in DNA replication.

small portion of the double helix is "unzipped," and each DNA chain acts as a template for the base pairing of monomers (Figure 14.39). As the next portion unzips, monomers in the first portion are covalently linked together to make a chain. Gradually, each of these new chains forms half of *two* double helices. In this way, the original double helix is copied and the genetic makeup of the cells maintained.

The biopolymers provide striking evidence for the folly of vitalism. No mystical or unknowable forces are required to explain the marvels of the living world. The same atomic properties that give rise to covalent bonds, molecular shape, and intermolecular forces provide the means for all life forms to flourish.

### Section Summary

Polymers are extremely large molecules made of many smaller monomers. Addition polymers have unsaturated units that commonly link through free-radical reactions. Most condensation polymers have two types of monomer that link through dehydration-condensation. By varying reaction conditions, catalysts, and monomer groups, polymers with different properties can be made.

The three natural polymers—polysaccharides, proteins, and nucleic acids—are formed by dehydration-condensation reactions. Polysaccharides are formed from cyclic monosaccharides, such as glucose. Cellulose, starch, and glycogen have structural or energy-storage roles. Proteins are polyamides formed from as many as 20 different amino acids. Fibrous proteins have extended shapes and structural roles. Globular proteins have compact shapes and metabolic, immunologic, and hormonal roles. The amino acid sequence of the protein determines its function. Nucleic acids (DNA and RNA) are polynucleotides formed from four different mononucleotides. The base sequence of the DNA chain determines the sequence of amino acids in an organism's proteins. Hydrogen bonding between specific base pairs is the key to protein synthesis and gene replication.

## Chapter Perspective

*The amazing diversity of organic compounds highlights the importance of atomic properties in chemical and biological behavior. While gaining an appreciation for the special properties of carbon, you've seen similar types of bonding and reactivity in organic and inorganic compounds. Upcoming chapters focus on some dynamic aspects of all reactions—their speed, extent, and direction—as we explore the kinetics, equilibrium, and thermodynamics of chemical and physical change. We'll have many opportunities to exemplify these topics with organic compounds.*

# For Review and Reference

## Key Terms

organic compound

**SECTION 14.1**
heteroatom
functional group

**SECTION 14.2**
hydrocarbon
alkane ($C_nH_{2n+2}$)
saturated hydrocarbon
cyclic hydrocarbon
isomer
structural isomer
homologous series
stereoisomer
optical isomer
chiral molecule

polarimeter
optically active
alkene ($C_nH_{2n}$)
unsaturated hydrocarbon
geometric isomer
alkyne ($C_nH_{2n-2}$)
aromatic hydrocarbon
nuclear magnetic
    resonance (NMR)
    spectroscopy

**SECTION 14.3**
alkyl group
addition reaction
elimination reaction
substitution reaction

**SECTION 14.4**
alcohol
ether
haloalkane (alkyl halide)
amine
carbonyl group
aldehyde
ketone
organometallic compound
carboxylic acid
ester
amide
fatty acid
acid anhydride
lipid
hydrolysis
nitrile

**SECTION 14.5**
polymer
monomer
addition polymer
condensation polymer
monosaccharide
polysaccharide
disaccharide
protein
amino acid
nucleic acid
mononucleotide
double helix
base pair
genetic code

## Answers to Follow-up Problems

**14.1** (a)

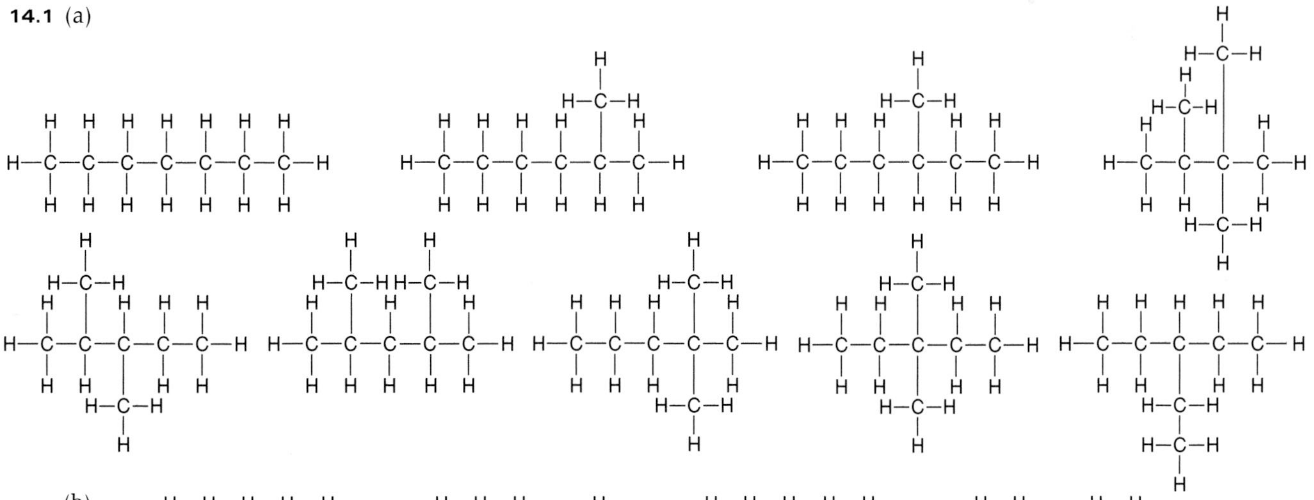

(b)

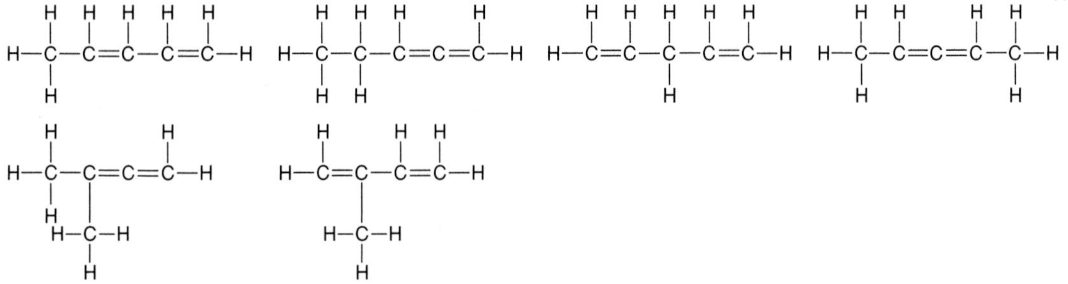

(c)

$$H-\overset{\underset{\displaystyle H}{|}}{\underset{\underset{\displaystyle H}{|}}{C}}-\overset{\underset{\displaystyle H}{|}}{\underset{\underset{\displaystyle H}{|}}{C}}-\overset{\underset{\displaystyle H}{|}}{\underset{\underset{\displaystyle H}{|}}{C}}-C\equiv C-H$$

$$H-\overset{\underset{\displaystyle H}{|}}{\underset{\underset{\displaystyle H}{|}}{C}}-\overset{\underset{\displaystyle H}{|}}{\underset{\underset{\displaystyle H}{|}}{C}}-C\equiv C-\overset{\underset{\displaystyle H}{|}}{\underset{\underset{\displaystyle H}{|}}{C}}-H$$

$$H-\overset{\underset{\displaystyle H}{|}}{\underset{\underset{\displaystyle H}{|}}{C}}-\overset{\underset{\displaystyle CH_3}{|}}{\underset{\underset{\displaystyle H}{|}}{C}}-C\equiv C-H$$

**14.2**

(a) $CH_3-CH_2-CH_2-CH_2-CH_2-\overset{\overset{\displaystyle CH_2-CH_3}{|}}{\underset{\underset{\displaystyle CH_3}{|}}{C}}-CH_2-CH_3$

(b) cyclohexane with $CH_2-CH_2-CH_3$ and $CH_2-CH_3$ substituents

(c) $CH_3-CH_2-CH_2-CH_2-CH_2-CH_2-\overset{\overset{\displaystyle CH_2-CH_3}{|}}{\underset{\underset{\displaystyle CH_3}{|}}{C}}-CH_2-\overset{\overset{\displaystyle CH_3}{|}}{CH}-CH_3$

(d) $HC\equiv C-\overset{\overset{\displaystyle CH_2-CH_3}{|}}{\underset{\underset{\displaystyle CH_2-CH_3}{|}}{C}}-CH_2-CH_2-CH_3$

(e) $\underset{CH_3}{\overset{CH_3-CH_2}{}}\!\!\diagdown C=C\diagup\!\!\underset{CH_2-CH_2-CH_3}{\overset{H}{}}$

(f) cyclohexene with $CH_2-CH_3$ substituent

**14.3** (a) $CH_3-CH=CH-CH_3 \; + \; H_2O \;\xrightarrow{H^+}\; CH_3-\overset{\overset{\displaystyle OH}{|}}{CH}-CH_2-CH_3$

(b) $CH_3-CH_2-CH_2-Br \; + \; OH^- \longrightarrow CH_3-CH_2-CH_2-OH \; + \; Br^-$

(c) $CH_3-\overset{\underset{\displaystyle Cl}{|}}{CH}-CH_3 \; + \; CH_3-ONa \longrightarrow CH_3-CH=CH_2 \; + \; NaCl \; + \; CH_3-OH$

**14.4**

$$\overset{-1}{\underset{-1}{H}}\overset{}{\underset{}{\underset{\displaystyle H}{}}}\overset{0}{C}=\overset{0}{C}\overset{-1}{\underset{-1}{H}} \; + \; Cl-Cl \longrightarrow Cl\overset{+1}{-}\overset{0}{C}\overset{-1}{-}\overset{0}{C}\overset{+1}{-}Cl$$

total O.N. $= -4$        total O.N. $= -2$

$CH_2=CH_2$ is the reducing agent; $Cl_2$ is the oxidizing agent.

**14.5** (a) $CH_3-\overset{\underset{\displaystyle Cl}{|}}{CH}-CH_3$

(b) $CH_3-CH_2-\overset{\overset{\displaystyle CH_3}{|}}{\underset{\underset{\displaystyle Cl}{|}}{C}}-CH_3$

(c) $CH_3-CH_2-CH_2-OH$

**14.6** (a) cyclohexane with OH and $CH_3$ substituents

(b) cyclopentane with $\overset{\overset{\displaystyle O}{\|}}{C}-H$

**14.7** (a) benzene ring $-CH_2-\overset{\overset{\displaystyle O}{\|}}{C}-OH$

(b) $CH_3-CH_2-CH_2-\overset{\overset{\displaystyle O}{\|}}{C}-O-CH_3 \; + \; CH_3-CH_2-NH_2$

**14.8** (a) benzene ring $-$(CH=CH)$-$($\overset{\overset{\displaystyle O}{\|}}{C}$-H)    (with labels: alkene, aldehyde)

(b) (H_2N$-\overset{\overset{\displaystyle O}{\|}}{C}$)$-CH_2-$(CH)$-CH_3$ with Br below (labels: amide, haloalkane)

## Sample Problem Titles

# Problems

Problems with a green number are answered at the back of the text. Most sections include three categories of problems separated by a green rule—*concept review questions*, *paired skill building exercises*, and problems in a relevant context.

## The Special Nature of Carbon and the Characteristics of Organic Molecules

**14.1** Give the name and formula of two carbon compounds that are organic and two that are inorganic.

**14.2** Through the first quarter of the 19th century, the idea of vitalism proposed that a key difference existed between substances isolated from animate sources and those from inanimate sources. What was the central notion of vitalism? Give an example of a finding that led to its eventual downfall.

**14.3** Explain each statement in terms of atomic properties:
(a) Carbon engages in covalent rather than ionic bonding.
(b) Carbon has four bonds in all its organic compounds.
(c) Carbon forms neither stable cations, like many metals, nor stable anions, like many nonmetals.
(d) Carbon bonds to itself more extensively than any other element.
(e) Carbon forms stable multiple bonds.

**14.4** Carbon bonds to many elements other than itself.
(a) Which elements bond to carbon most frequently in organic compounds?
(b) Which of these elements are heteroatoms?
(c) Which of these elements are more electronegative than C? Less electronegative?
(d) How does this bonding of carbon to heteroatoms increase the number of organic compounds?

**14.5** Silicon lies just below carbon in Group 4A(14) and also forms four covalent bonds. Why aren't there as many silicon compounds as carbon compounds?

**14.6** What is the range of oxidation states for carbon? Name a compound in which carbon has its highest oxidation state and one in which it has its lowest.

**14.7** Which of the following bonds to carbon would you expect to be relatively reactive: C—H, C—C, C—I, C=O, C—Li? Explain.

## The Structures and Classes of Hydrocarbons

(Sample Problems 14.1 and 14.2)

**14.8** (a) What structural feature is associated with each of the following hydrocarbons: an alkane; a cycloalkane; an alkene; an alkyne?
(b) Give the general formula for each.
(c) Which are considered "saturated"?

**14.9** Define each of the following types of isomers: (a) structural; (b) geometric; (c) optical. Which types of isomers are considered stereoisomers?

**14.10** Among alkenes, alkynes, and aromatic hydrocarbons, only alkenes exhibit *cis-trans* isomerism. Explain why the others do not.

**14.11** Which objects are asymmetric (have no plane of symmetry): (a) a circular clock face; (b) a football; (c) a dime; (d) a brick; (e) a hammer; (f) a spring?

**14.12** Explain briefly how a polarimeter works and what it measures.

**14.13** How does an aromatic hydrocarbon differ from a cycloalkane?

**14.14** Draw all possible skeletons for a 7-C compound with
(a) 1 double bond and a 6-carbon chain
(b) 1 double bond and a 5-carbon chain
(c) A 5-membered ring and no double bonds

**14.15** Draw all possible skeletons for a 6-C compound with
(a) 2 double bonds and a 5-carbon chain
(b) 1 triple bond and a 5-carbon chain
(c) A 4-membered ring and no double bonds

**14.16** Add the correct number of hydrogens to each of the skeletons in Problem 14.14.

**14.17** Add the correct number of hydrogens to each of the skeletons in Problem 14.15.

**14.18** Draw correct structures for any that are incorrect:

(a) $CH_3-CH(CH_3)-CH_2-CH_3$ (with $CH_3$ above and below central CH)

(b) $CH_3=CH-CH_2-CH_3$

(c) $CH_2\equiv C-CH_2-CH_3$ (with $CH_2-CH_3$ below)

(d) $CH_3-$⟨benzene ring⟩$-CH_3$

**14.19** Draw correct structures for any that are incorrect:

(a) $CH_3-CH=CH-CH_2-CH_3$

(b) ⟨cyclopentadiene ring with two $CH_3$ groups⟩

(c) $CH_3-C\equiv CH-CH_2-CH_3$

(d) $CH_3-CH_2-C(CH_3)-CH_2-CH_2-CH_3$

**14.20** Draw the structure or give the name of the following:
(a) 2,3-dimethyloctane
(b) 1-ethyl-3-methylcyclohexane

(c) $CH_3-CH_2-CH-CH(CH_3)-CH_2$ (with $CH_3$ above the CH and $CH_2-CH_3$ below)

(d) $CH_3-C(CH_3)-CH_2$ (with $CH_3$ above and $CH_3\ CH_3$ below)

**14.21** Draw the structure or give the name of the following:

(a) $CH_3-CH(CH_3)-CH_2-CH_3$

(b) ⟨cyclohexane ring with $H_3C$, $CH_3$, and $CH_3$ substituents⟩

(c) 1,2-diethylcyclopentane
(d) 2,4,5-trimethylnonane

**14.22** Each of the following names is wrong. Draw structures based on them and correct the names:
(a) 4-methylhexane        (b) 2-ethylpentane
(c) 2-methylcyclohexane
(d) 3,3-methyl-4-ethyloctane

**14.23** Each of the following names is wrong. Draw structures based on them and correct the names:
(a) 3,3-dimethylbutane
(b) 1,1,1-trimethylheptane
(c) 1,4-diethylcyclopentane
(d) 1-propylcyclohexane

**14.24** Each of the following compounds can exhibit optical activity. Circle the chiral center(s) in each:

(a) 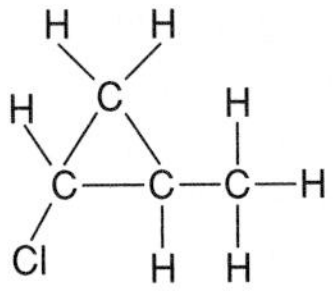          (b)

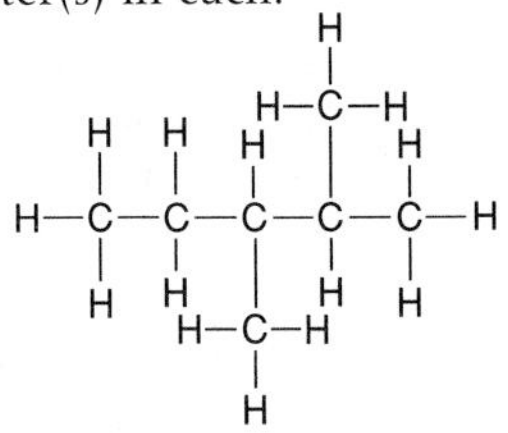

**14.25** Each of the following compounds can exhibit optical activity. Circle the chiral center(s) in each:

(a) 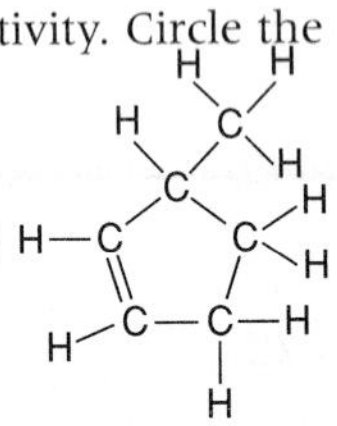          (b)

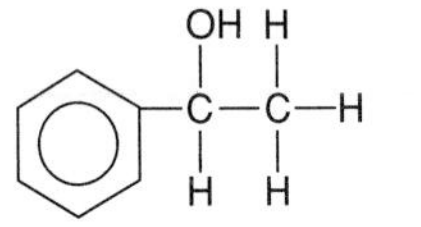

**14.26** Draw structures from the following names and determine which compounds are optically active:
(a) 3-bromohexane    (b) 3-chloro-3-methylpentane
(c) 1,2-dibromo-2-methylbutane

**14.27** Draw structures from the following names and determine which compounds are optically active:
(a) 1,3-dichloropentane
(b) 3-chloro-2,2,5-trimethylhexane
(c) 1-bromo-1-chlorobutane

**14.28** Which of the following structures exhibit geometric isomerism? Draw and name the two isomers in each case:

(a) $CH_3-CH_2-CH=CH-CH_3$

(b)

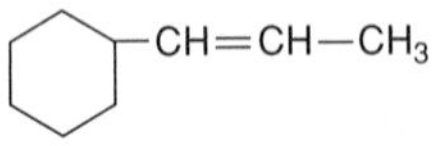

(c)

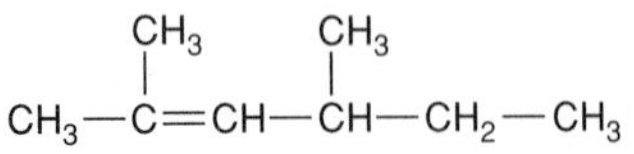

**14.29** Which of the following structures exhibit geometric isomerism? Draw and name the two isomers in each case:

(a) 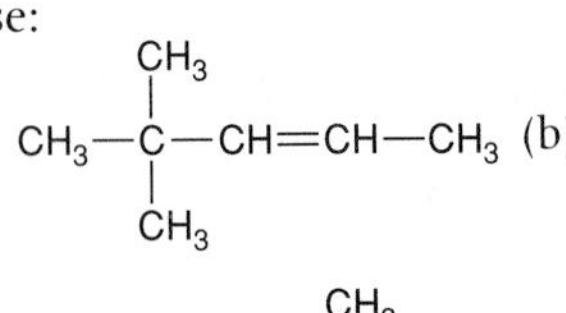 (b)

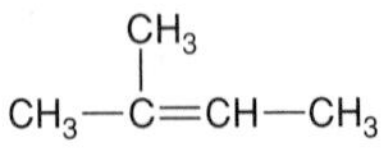

(c)

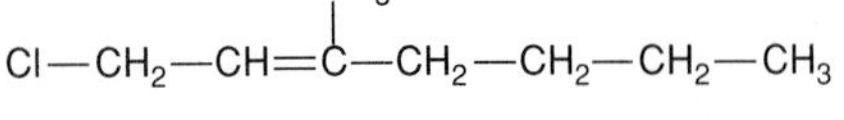

**14.30** Which of the following compounds exhibit geometric isomerism? Draw the two isomers in each case:
(a) propene              (b) 3-hexene
(c) 1,1-dichloroethene    (d) 1,2-dichloroethene

**14.31** Which of the following compounds exhibit geometric isomerism? Draw the two isomers in each case:
(a) 1-pentene            (b) 2-pentene
(c) 1-chloropropene      (d) 2-chloropropene

**14.32** Draw and name all the structural isomers of dichlorobenzene.

**14.33** Draw and name all the structural isomers of trimethylbenzene.

**14.34** The "octane number" of a gasoline sample is determined by comparing its combustion characteristics to a pure compound called "isooctane." The systematic name for isooctane is 2,2,4-trimethylpentane. Draw its structure.

**14.35** "Butylated hydroxytoluene" (BHT) is a common preservative added to cereals and other dry foods. Its systematic name is 1-hydroxy-2,6-di-*tert*-butyl-4-methylbenzene (where "*tert*-butyl" is 1,1-dimethylethyl). Draw the structure of BHT.

**14.36** There are two compounds with the name 2-methyl-3-hexene, but only one with the name 2-methyl-2-hexene. Explain with structures.

**14.37** The explosive TNT is one of six isomers of trinitrotoluene. Draw and name the other five.

**14.38** Ethanol and dimethylether are structural isomers with the molecular formula $C_2H_6O$.
(a) Draw their structures.
(b) Aside from the TMS peak, how many peaks representing different proton environments appear in the NMR spectrum of each isomer?

## Some Important Classes of Organic Reactions
(Sample Problems 14.3 and 14.4)

**14.39** In terms of numbers of reactant and product substances, which organic reaction type corresponds to (a) a combination reaction, (b) a decomposition reaction, (c) a displacement reaction?

**14.40** The same bond type is broken in an addition reaction and formed in an elimination reaction. Name the bond type.

**14.41** Can a redox reaction also be an addition, elimination, or substitution reaction? Explain with examples.

**14.42** Determine each of the following reaction types:

(a)

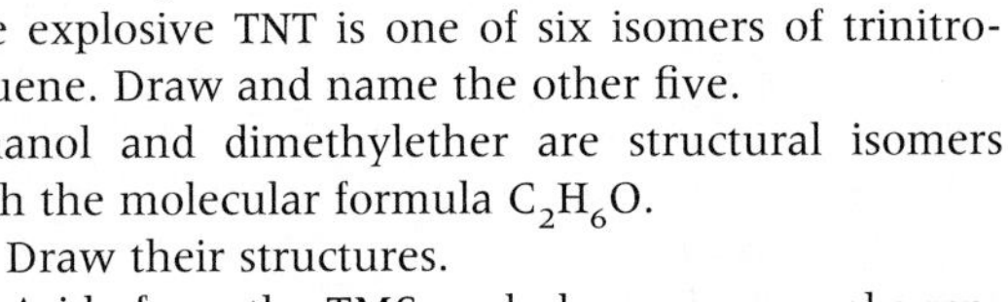

(b) $CH_3-CH=CH-CH_2-CH_3 + H_2 \xrightarrow{Pt}$
    $CH_3-CH_2-CH_2-CH_2-CH_3$

**14.43** Determine each of the following reaction types:

(a)

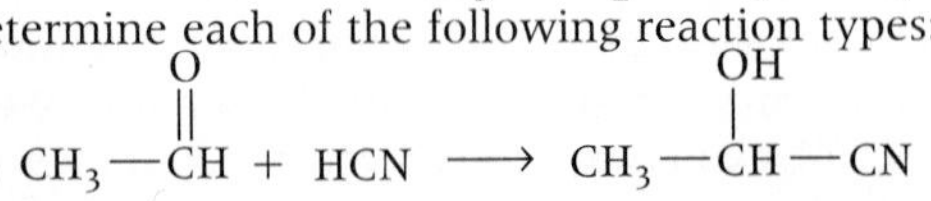

(b)

$$CH_3-\overset{\overset{\displaystyle O}{\|}}{C}-O-CH_3 + CH_3-NH_2 \xrightarrow{H^+}$$

$$CH_3-\overset{\overset{\displaystyle O}{\|}}{C}-NH-CH_3 + CH_3-OH$$

**14.44** Write equations for the following:
(a) An addition reaction between $H_2O$ and 3-hexene ($H^+$ is a catalyst)
(b) An elimination reaction between 2-bromopropane and hot $CH_3-CH_2-OK$
(c) A light-induced substitution reaction between $Cl_2$ and ethane to form 1,1-dichloroethane

**14.45** Write equations for the following:
(a) A substitution reaction between 2-bromopropane and KI
(b) An addition reaction between cyclohexene and $Cl_2$
(c) An addition reaction between 2-propanone

$$(CH_3-\overset{\overset{\displaystyle O}{\|}}{C}-CH_3) \text{ and } H_2 \text{ (Ni metal is a catalyst)}$$

**14.46** Determine the O.N. of each colored carbon:

(a) $CH_3-CH_2-O-CH_3$   (b) $CH_3-CH_2-\overset{\overset{\displaystyle O}{\|}}{C}-OH$

(c) $CH_3-CH=CH-NH_2$   (d) $CH_3-\overset{\overset{\displaystyle O}{\|}}{C}-Cl$

**14.47** Determine the O.N. of each colored carbon:

(a) $CH_3-CH_2-C\equiv N$   (b) phenyl$-\overset{\overset{\displaystyle O}{\|}}{C}H$

(c) $CH_3-C\equiv CH$   (d) $CH_3-CH_2-MgBr$

**14.48** Based on the number of bonds and the nature of the bonded atoms in each of the following changes, state whether the oxidation number of the carbon increases, decreases, or stays the same:
(a) $=CH_2$ becomes $-CH_2-Cl$
(b) $=CH-$ becomes $-CH_2-$
(c) $\equiv C-$ becomes $-CH_2-$
(d) $-CH_2-Br$ becomes $-CH_2-OH$

**14.49** Based on the number of bonds and the nature of the bonded atoms in each of the following changes, state whether the oxidation number of the carbon increases, decreases, or stays the same:

(a) $-\overset{|}{\underset{|}{C}}-OH$ becomes $-\overset{|}{C}=O$

(b) $-CH_2-OH$ becomes $=CH_2$

(c) $-\overset{\overset{\displaystyle O}{\|}}{\underset{|}{C}}-N-$ becomes $-\overset{\overset{\displaystyle O}{\|}}{C}-O-$

(d) $-\overset{|}{\underset{|}{C}}-Cl$ becomes $-\overset{|}{\underset{|}{C}}-Li$

**14.50** Is the organic reactant oxidized, reduced, or neither in each of the following reactions? Use oxidation numbers to confirm your answer.
(a) $C_5H_{12} + 8O_2 \rightarrow 5CO_2 + 6H_2O$
(b) 2-hexene $\xrightarrow[\text{cold OH}^-]{KMnO_4}$ 2,3-dihydroxyhexane
(c) cyclohexane $\xrightarrow[\text{Pd}]{\Delta}$ benzene + $3H_2$

**14.51** Is the organic reactant oxidized, reduced, or neither in each of the following reactions? Use oxidation numbers to confirm your answer.
(a) $Br-CH_2-CH_2-CH_2-Br + Zn \rightarrow \triangle + ZnBr_2$
(b) 1-butyne + $H_2 \xrightarrow{Pt}$ 1-butene
(c) benzene + $Cl_2 \xrightarrow{FeCl_3}$ chlorobenzene + HCl

**14.52** Phenylethylamine is a natural substance that resembles amphetamine structurally. It is found in sources as diverse as almond oil and human urine, where it occurs at elevated concentrations due to stress and to certain forms of schizophrenia. One method of synthesis for pharmacologic and psychiatric studies involves two steps:

phenyl$-CH_2-Cl$ + NaCN $\longrightarrow$ phenyl$-CH_2-C\equiv N \xrightarrow[\text{Pt}]{H_2}$

phenyl$-CH_2-CH_2-NH_2$

phenylethylamine

(a) Classify each step as an addition, elimination, or substitution.
(b) Is the second step a redox reaction? Explain with oxidation numbers.

## Properties and Reactivity of Common Functional Groups

(Sample Problems 14.5-14.8)

**14.53** Compounds with nearly identical molar masses often have very different physical properties. Choose the compound with the higher value for each of the following properties, and explain your choice with structures:
(a) Solubility in water: chloroethane or methylethylamine
(b) Melting point: diethylether or 1-butanol
(c) Boiling point: trimethylamine or propylamine

**14.54** Fill in the blanks with a general formula for the type of compound formed:

$$R-CH=CH_2 \underset{H^+}{\overset{H_2O, H^+}{\rightleftharpoons}} \_\_\_$$

with HBr, $CH_3O^-$, $HBr$, $OH^-$ pathways

**14.55** Of the three major types of organic reactions, which do *not* occur readily with benzene? Why?

**14.56** Why does the $C=O$ group react differently than the $C=C$ group? Show an example of the difference.

**14.57** Many substitution reactions involve an initial electrostatic attraction between reactants. Show this attraction in the formation of an amide from an amine and an ester.

**14.58** Although carboxylic acids and alcohols both contain an —OH group, one is acidic in water and the other is not. Explain.

**14.59** What reaction type is common to the formation of ethers, esters, and acid anhydrides? What is the other product?

**14.60** Alcohols and carboxylic acids both undergo substitution, but the processes are very different. Explain.

---

**14.61** Name the type of organic compound from the following description of its functional group:
  (a) Single-bonded group that cannot be at the end of a carbon chain
  (b) Triple-bonded group that is polar
  (c) Single- and double-bonded group that is acidic in water
  (d) Double-bonded group that can be only at the end of a carbon chain

**14.62** Name the type of organic compound from the following description of its functional group:
  (a) N-containing single- and double-bonded group
  (b) Double-bonded group that is not polar
  (c) Polar double-bonded group that cannot be at the end of a carbon chain
  (d) Single-bonded group that is basic in water

**14.63** In each of the following, circle and name the functional groups:

(a) $CH_3-CH=CH-CH_2-OH$

(b) $Cl-CH_2-\langle\bigcirc\rangle-\overset{\overset{\displaystyle O}{\|}}{C}-OH$

(c) $\langle\text{ring}\rangle-\overset{\overset{\displaystyle O}{\|}}{C}-NH-CH_3$

(d) $N\equiv C-CH_2-\overset{\overset{\displaystyle O}{\|}}{C}-CH_3$

(e) $\langle\triangle\rangle-\overset{\overset{\displaystyle O}{\|}}{C}-O-CH_2-CH_3$

(f) $CH_3-\overset{\overset{\displaystyle O}{\|}}{C}-CH_2-O-CH_3$

**14.64** In each of the following, circle and name the functional groups:

(a) $HO-\overset{\overset{\displaystyle O}{\|}}{C}-CH_2-\overset{\overset{\displaystyle O}{\|}}{CH}$

(b) $I-CH_2-CH_2-C\equiv CH$

(c) $CH_2=CH-CH_2-\overset{\overset{\displaystyle O}{\|}}{C}-O-CH_3$

(d) $CH_3-\overset{\overset{\displaystyle Br}{|}}{CH}-CH=CH-CH_2-NH-CH_3$

(e) $\langle\text{ring with O}\rangle$

(f) $CH_3-NH-\overset{\overset{\displaystyle O}{\|}}{C}-\overset{\overset{\displaystyle O}{\|}}{C}-O-CH_3$

**14.65** Draw and name all possible alcohols with the formula $C_5H_{12}O$.

**14.66** Draw and name all possible aldehydes and ketones with the formula $C_5H_{10}O$.

**14.67** Draw all possible amines with the formula $C_4H_{11}N$.

**14.68** Draw all possible carboxylic acids with the formula $C_5H_{10}O_2$.

**14.69** Name and draw the product resulting from mild oxidation of (a) 2-butanol; (b) 2-methylpropanal; (c) cyclopentanol.

**14.70** Name and draw the alcohol you would oxidize to produce (a) 2-methylpropanal; (b) 2-pentanone; (c) 3-methylbutanoic acid.

**14.71** Name and draw structures for the organic products formed when the following compounds react:
  (a) Acetic acid and methylamine
  (b) Butanoic acid and 2-propanol
  (c) Formic acid and 2-methyl-l-propanol

**14.72** Name and draw the products formed when the following compounds react:
  (a) Acetic acid and 1-hexanol
  (b) Propanoic acid and dimethylamine
  (c) Ethanoic acid and diethylamine

**14.73** Draw structures for the carboxylic acid and alcohol portions of the following esters:

(a) $CH_3-(CH_2)_4-\overset{\overset{\displaystyle O}{\|}}{C}-O-CH_2-CH_3$

(b) $\langle\bigcirc\rangle-\overset{\overset{\displaystyle O}{\|}}{C}-O-CH_2-CH_2-CH_3$

(c) $CH_3-CH_2-O-\overset{\overset{\displaystyle O}{\|}}{C}-CH_2-CH_2-\langle\bigcirc\rangle$

**14.74** Draw structures for the carboxylic acid and amine portions of the following amides:

(a) $H_3C-\langle\bigcirc\rangle-CH_2-\overset{\overset{\displaystyle O}{\|}}{C}-NH_2$

(b) $CH_3-\overset{\overset{\displaystyle CH_3}{|}}{CH}-\overset{\overset{\displaystyle O}{\|}}{C}-\overset{\underset{\displaystyle CH_3}{|}}{N}-CH_2-CH_3$

(c) $H\overset{\overset{\displaystyle O}{\|}}{C}-NH-\langle\bigcirc\rangle$

**14.75** Fill in the expected organic products:

(a) $CH_3-CH_2-Br \xrightarrow{OH^-} \underline{\quad} \xrightarrow[H^+]{CH_3-CH_2-\overset{\overset{\displaystyle O}{\|}}{C}-OH} \underline{\quad}$

(b) $CH_3-CH_2-\overset{\overset{\displaystyle Br}{|}}{CH}-CH_3 \xrightarrow{CN^-} \underline{\quad} \xrightarrow{H_3O^+, H_2O} \underline{\quad}$

**14.76** Fill in the expected organic products:

(a) $CH_3-CH_2-CH=CH_2 \xrightarrow{H^+, H_2O} \underline{\quad} \xrightarrow{Cr_2O_7^{2-}, H^+} \underline{\quad}$

(b) $CH_3-CH_2-\overset{\overset{\displaystyle O}{\|}}{C}-CH_3 \xrightarrow{CH_3-CH_2-Li} \xrightarrow{H_2O} \underline{\quad}$

**14.77** Supply the missing organic and/or inorganic substances:

(a) $CH_3-CH_2-OH + \underline{\quad?\quad} \xrightarrow{?}$ $\underset{\displaystyle CH_3-CH_2-O-\overset{\displaystyle O}{\overset{\|}{C}}-CH_2-CH_3}{}$

(b) $CH_3-\overset{O}{\overset{\|}{C}}-O-CH_3 \xrightarrow{?} CH_3-CH_2-NH-\overset{O}{\overset{\|}{C}}-CH_3$

**14.78** Supply the missing organic and/or inorganic substances:

(a) $CH_3-\overset{Cl}{\underset{|}{CH}}-CH_3 \xrightarrow{?} CH_3-CH=CH_2 \xrightarrow{?} CH_3-\overset{Br}{\underset{|}{CH}}-\overset{Br}{\underset{|}{CH_2}}$

(b) $CH_3-CH_2-CH_2-OH \xrightarrow{?} CH_3-CH_2-\overset{O}{\overset{\|}{C}}-OH$

$+ \underline{\quad?\quad} \xrightarrow{?} CH_3-CH_2-\overset{O}{\overset{\|}{C}}-O-CH_2-\bigcirc$

**14.79** Which of the following compounds is highly soluble in water? Explain.

(a) $CH_3-CH_2-CH=CH_2$  (b) $CH_3-CH_2-CH_2-OH$

(c) $CH_3-CH_2-CH_2-NH_2$  (d) $CH_3-CH_2-\overset{O}{\overset{\|}{C}}-OH$

(e) $CH_3-(CH_2)_6-\overset{O}{\overset{\|}{C}}-OH$  (f) $CH_3-CH_2-\overset{Br}{\underset{|}{CH}}-CH_3$

**14.80** Cadaverine (1,5-diaminopentane) and putrescine (1,4-diaminobutane) are two compounds formed by bacterial action that are responsible for the odor of rotting flesh. Draw their structures. Suggest a series of reactions to synthesize putrescine from 1,2-dibromoethane and any inorganic reagents.

**14.81** Sodium propanoate $(CH_3-CH_2-\overset{O}{\overset{\|}{C}}-O^-Na^+)$ is a common preservative found in breads, cheeses, and pies. How would you synthesize sodium propanoate from 1-propanol and any inorganic reagents?

**14.82** Ethyl formate $(H\overset{O}{\overset{\|}{C}}-O-CH_2-CH_3)$ is commonly added to foods to give them the flavoring of rum. How would you synthesize ethyl formate from ethanol, methanol, and any inorganic reagents?

## Giant Organic Molecules: The Monomer-Polymer Theme

**14.83** Name the reaction processes that lead to the two types of synthetic polymers.

**14.84** Which functional group is common to the monomers that make up addition polymers? What makes these polymers different from one another?

**14.85** What is a free radical? How is it involved in polymer formation?

**14.86** Which intermolecular force is primarily responsible for the different types of polyethylene? Explain.

**14.87** Which of the two types of synthetic polymer is more similar chemically to biopolymers? Explain.

**14.88** Which two functional groups react to form nylons? Polyesters?

**14.89** Which type of polymer is formed from each of the following monomers: (a) amino acids; (b) alkenes; (c) simple sugars; (d) mononucleotides?

**14.90** What is the key structural difference between fibrous and globular proteins? How is it related to the amino acid composition?

**14.91** Protein shape, function, and amino acid sequence are interrelated. Arrange them in order of which determines which.

**14.92** What type of functional group holds each strand of DNA together?

**14.93** What is base pairing? How does it pertain to the structure of DNA?

**14.94** RNA base sequence, protein amino acid sequence, and DNA base sequence are interrelated. Arrange them in order of which determines which in the process of protein synthesis.

**14.95** Draw an abbreviated structure for the following polymers, with brackets around the repeating unit:

(a) Poly(vinyl chloride) (PVC) from $\underset{H}{\overset{H}{\diagdown}}C=C\underset{Cl}{\overset{H}{\diagup}}$

(b) Polypropylene from $\underset{H}{\overset{H}{\diagdown}}C=C\underset{CH_3}{\overset{H}{\diagup}}$

**14.96** Draw an abbreviated structure for the following polymers, with brackets around the repeating unit:

(a) Teflon from $\underset{F}{\overset{F}{\diagdown}}C=C\underset{F}{\overset{F}{\diagup}}$

(b) Polystyrene from $\underset{H}{\overset{H}{\diagdown}}C=C\underset{\bigcirc}{\overset{H}{\diagup}}$

**14.97** Write a balanced equation for the reaction between 1,4-benzenedicarboxylic acid and 1,2-dihydroxyethane to form the polyester Dacron. Draw an abbreviated structure for the polymer, with brackets around the repeating unit.

**14.98** Write a balanced equation for the reaction of dihydroxydimethylsilane,

$HO-\underset{\underset{\displaystyle CH_3}{|}}{\overset{\overset{\displaystyle CH_3}{|}}{Si}}-OH,$

to form the condensation polymer popularly known as Silly Putty.

**14.99** Draw the structure of the R group of (a) alanine; (b) histidine; (c) methionine.

**14.100** Draw the structure of the R group of (a) glycine; (b) isoleucine; (c) tyrosine.

**14.101** Draw the structure of the following tripeptides:
(a) Aspartic acid-histidine-tryptophan
(b) Glycine-cysteine-tyrosine with the charges that exist in cell fluid

**14.102** Draw the structure of the following tripeptides:
(a) Lysine-phenylalanine-threonine
(b) Alanine-leucine-valine with the charges that exist in cell fluid

**14.103** Write the sequence of the complementary DNA strand that pairs with each of the following DNA base sequences: (a) TTAGCC; (b) AGACAT.

**14.104** Write the sequence of the complementary DNA strand that pairs with each of the following DNA base sequences: (a) GGTTAC; (b) CCCGAA.

**14.105** Write the base sequence of DNA template from which the following RNA sequence was derived: UGUUACGGA. How many amino acids are coded for in this sequence?

**14.106** Write the base sequence of DNA template from which the following RNA sequence was derived: GUAUCAAUGAACUUG. How many amino acids are coded for in this sequence?

---

**14.107** Protein shapes are maintained by a variety of forces, which arise through interaction between the amino acid R groups. Name the amino acid that possesses the R group and tell the force that could arise in each of the following interactions:
(a) $-CH_2-SH$ with $HS-CH_2-$

(b) $-(CH_2)_4-NH_3^+$ with $^-O-\overset{\displaystyle O}{\overset{\|}{C}}-CH_2-$

(c) $-CH_2-\overset{\displaystyle O}{\overset{\|}{C}}-NH_2$ with $HO-CH_2-$

(d) $-\overset{\displaystyle CH_3}{\overset{|}{CH}}-CH_3$ with ⬡$-CH_2-$

**14.108** What force is responsible for the helix and sheet structures seen in so many proteins? From which groups do they arise?

**14.109** The genetic code consists of a series of three-base words that each code for a given amino acid.
(a) Using the selections from the genetic code shown below, determine the amino acid sequence coded by the following segment of RNA:

UCCACAGCCUAUAUGGCAAACUUGAAG

AUG = methionine   CCU = proline    CAU = histidine
UGG = tryptophan   AAG = lysine     UAU = tyrosine
GCC = alanine      UUG = leucine    CGG = arginine
UGU = cysteine     AAC = asparagine   ACA = threonine
UCC = serine       GCA = alanine      UCA = serine

(b) What is the complementary DNA sequence from which this RNA sequence was made?

**Comprehensive Problems**

Problems with an asterisk (*) are more challenging.

**\*14.110** Starting with the given organic reactant and any necessary inorganic reagents, explain how you would perform each of the following syntheses:
(a) Starting with $CH_3-CH_2-CH_2-OH$,
make $CH_3-\overset{\displaystyle Br}{\overset{|}{CH}}-CH_2-Br$

(b) Starting with $CH_3-CH_2-OH$, make
$CH_3-\overset{\displaystyle O}{\overset{\|}{C}}-O-CH_2-CH_3$

**\*14.111** Methyl *tert*-butyl ether (MTBE; also named methyl 2-methyl-2-propyl ether) has been shown in many recent studies to be an excellent octane booster and clean fuel additive for gasoline. It increases the oxygen content of the fuel, which reduces CO emissions during winter months to an amount in accord with the Clean Air Act. MTBE is synthesized by reacting 2-methylpropene with methanol.
(a) Write a balanced equation for the synthesis of MTBE. (*Hint:* Alcohols add to alkenes in a manner similar to water.)
(b) If the government requires that auto fuel mixtures contain 2.7% by mass oxygen to reduce CO emissions, how many grams of MTBE must be added to each 100 g gasoline?
(c) How many liters of MTBE are present in each liter of fuel mixture? The density of both gasoline and MTBE is 0.740 g/mL.
(d) How many liters of air, which is approximately 21% $O_2$ by volume, are needed at 24°C and 1.00 atm to fully combust 1.00 L MTBE?

**14.112** Compound X has the formula $C_4H_8O$ and reacts readily with $K_2Cr_2O_7$ in $H_2SO_4$ to yield a carboxylic acid. It is synthesized from 2-methyl-1-propanol. Give a structure of compound X.

**\*14.113** Some of the most useful compounds for organic synthesis are Grignard reagents, for which Victor Grignard and Paul Sabatier were awarded the Nobel Prize in 1912. These compounds (general formula R—MgX, where X is a halogen) are made from R—X with Mg in ether solvent and used to change the carbon skeleton of the starting carbonyl reactant in a reaction similar to those with R—Li:

$$R'-\overset{\displaystyle O}{\overset{\|}{C}}-R'' + R-MgBr \longrightarrow R'-\overset{\displaystyle OMgBr}{\underset{\displaystyle R}{\overset{|}{\underset{|}{C}}}}-R'' \xrightarrow{H_2O}$$

$$R'-\overset{\displaystyle OH}{\underset{\displaystyle R}{\overset{|}{\underset{|}{C}}}}-R'' + Mg(OH)Br$$

(a) What is the product, after a final reaction step with water, of the reaction between ethanal and the Grignard reagent of bromobenzene?
(b) What is the product, after a final step with water, of the reaction between 2-butanone and the Grignard reagent of 2-bromopropane?

(c) There are often two (or more) combinations of Grignard reagent and carbonyl compound that will give the same product. Choose another pair of reactants to give the product in (a).

(d) What carbonyl compound must react with the Grignard reagent to obtain the —OH group at the *end* of the carbon chain in the product?

(e) What Grignard reagent and carbonyl compound would you use to prepare 2-methyl-2-butanol?

**14.114** A synthesis of 2-butanol was performed by treating 2-bromobutane with hot sodium hydroxide solution. The yield was 60%, indicating that a significant portion of the reactant was converted into a second product. Predict what this other product might be.

**14.115** Pyrethrins, such as jasmolin II (shown below), are a group of natural compounds that are synthesized by pyrethrum flowers to act as insecticides. (a) Circle and name the functional groups in jasmolin II. (b) What is the hybridization of the numbered carbons? (c) Which, if any, of the numbered carbons are chiral centers?

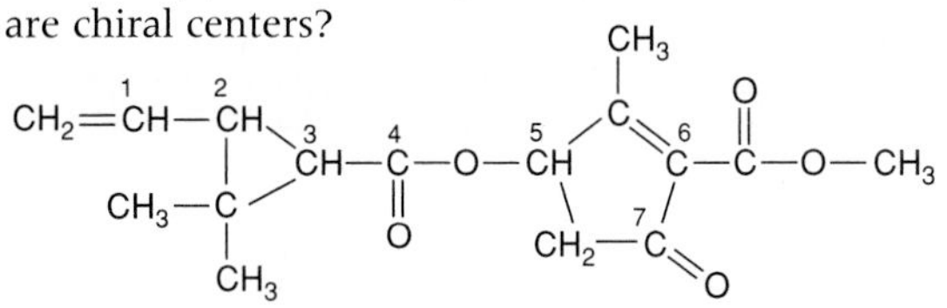

**14.116** Taste can often be affected by relatively small changes in a molecule. For example, Compound A (shown below) tastes 4000 times sweeter than glucose, whereas Compound B has no taste. If the —NH$_2$ group is attached to ring carbon 1, amino benzene is called "aniline," and CH$_3$—CH$_2$—CH$_2$—O is called a "propoxy" group, name Compounds A and B.

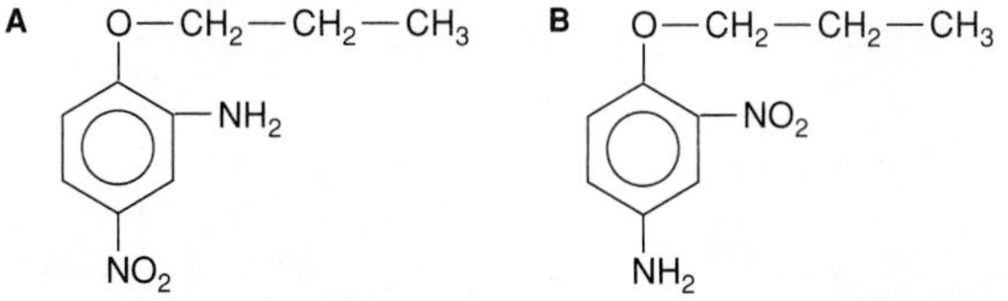

**\*14.117** Benzophenone is a ketone with two benzene rings as the R groups. It is used as a fixative for perfumes and soaps, and in the manufacture of antihistamines. Suggest a method for its synthesis, starting with benzene, benzaldehyde, and any inorganic reagents. (*Hint*: make an organometallic compound.)

**14.118** Following Wohler's synthesis of urea from an inorganic compound, many other organic substances were similarly synthesized. Explain how the reaction between calcium carbide (CaC$_2$) and water violates the notion of vitalism.

**14.119** Which features of retinal make it so useful as a photon absorber in the visual system of organisms?

**14.120** The polypeptide chain in proteins does not rotate freely due to partial double-bond character of the peptide bond. Explain this fact with resonance structures.

**14.121** Although other nonmetals form many compounds that are structurally analogous to those of carbon, the inorganic compounds are usually more reactive. Predict any missing products and write balanced equations for each of the following:
(a) The decomposition and chlorination of diborane to boron trichloride
(b) The combustion of pentaborane (B$_5$H$_9$) in O$_2$
(c) The addition of water to the double bonds of borazine (B$_3$N$_3$H$_6$, Figure 13.12). (*Hint:* The —OH group bonds to B.)
(d) The hydrolysis of trisilane (Si$_3$H$_8$) to silica (SiO$_2$) and H$_2$
(e) The complete halogenation of disilane with Cl$_2$
(f) The thermal decomposition of H$_2$S$_5$ to hydrogen sulfide and sulfur
(g) The hydrolysis of PCl$_5$

**\*14.122** Citral, the main odor constituent in oil of lemon grass, is used extensively in the flavoring and perfume industries. It occurs naturally as a mixture of the geometric isomers geranial (the *trans* isomer) and neral (the *cis* isomer). Its systematic name is 3,7-dimethyl-2,6-octadienal ("dien" refers to two C=C groups). Draw the structures of geranial and neral.

**\*14.123** Complete hydrolysis of a 100.00-g sample of a peptide gave the following amounts of individual amino acids (molar masses appear in parentheses): 3.00 g glycine (76.07); 0.90 g alanine (89.09); 3.70 g valine (117.15); 6.90 g proline (115.13); 7.30 g serine (105.09); and 86.00 g arginine (174.20).
(a) Why does the total mass of amino acids exceed the mass of peptide?
(b) What are the relative numbers of amino acids in the peptide?
(c) What is the minimum molar mass of the peptide?

**14.124** Thiols (or mercaptans) are the sulfur analogs of alcohols; that is, S appears in the molecule in place of O. Many occur naturally, and all have very disagreeable odors. They are named by adding the suffix "thiol" to the entire parent hydrocarbon name or by using "mercapto" as a prefix.
(a) Draw the structure of a compound in skunk musk, 3-methyl-1-butanethiol.
(b) Write a reaction to convert 3-mercapto-1-propene (found in garlic) to propanethiol (found in onion).

**\*14.125** 2-Butanone is reduced by hydride ion donors, such as sodium borohydride (NaBH$_4$), to the alcohol 2-butanol. Even though the alcohol has a chiral center, the product isolated from the redox reaction is not optically active. Explain.

**14.126** It has been said that "the hydrogen bond is the most important intermolecular force in nature." Give five cases to support this statement.

**14.127** Because of the behavior of their R groups in water, lysine is classified as a basic amino acid and aspartic acid as an acidic amino acid. Write balanced equations that demonstrate this behavior.

# CHAPTER 15

## Concepts and skills to review

- influence of temperature on molecular speed and collision frequency (Section 5.6)

# Kinetics

# Rates and Mechanisms of Chemical Reactions

**Temperature and biological activity.** The metabolic processes of cold-blooded animals like this magnificent hooded cobra speed up as the temperatures rise toward midday. In this chapter, you'll see how the speed of a reaction is influenced by several factors, including temperature, and how we can control them.

**C**hemical reactions are dynamic processes in which matter and energy change continuously, but until now we've taken a rather simple approach to them: reactants mix and products form. A balanced equation is a convenient shorthand for the reaction and a powerful quantitative tool for calculating product yields from reactant amounts. However, an equation tells us nothing about three crucial aspects of the reaction, which we examine in the next several chapters: How fast is the reaction proceeding at a given moment? What will the reactant and product concentrations be when it is complete? Once begun, will it proceed by itself and release energy, or will it require energy to continue?

In this chapter, we focus on *kinetics*, which deals with the speed of a reaction and its mechanism, the stepwise changes reactants undergo in their conversion to products. Chapters 16 through 18 are concerned with *equilibrium*, the dynamic balance between forward and reverse reactions and how external influences alter reactant and product concentrations. Chapters 19 and 20 present *thermodynamics* and its applications to electrochemistry, in which we investigate why a reaction occurs and how we can put it to use. These central subdisciplines of chemistry apply to all physical and chemical change. They are essential not only to our understanding of reactions but also to modern technology, the environment, and our own biology.

**Chemical kinetics** is the study of *reaction rates,* the changes in concentrations of reactants (or products) as a function of time (Figure 15.1).

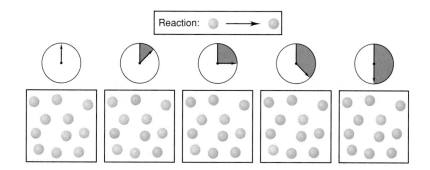

**FIGURE 15.1**
**Reaction rate: the central focus of chemical kinetics.** The rate at which reactant becomes product is the underlying theme in the study of chemical kinetics. As time elapses, reactant *(gray spheres)* decreases and product *(blue spheres)* increases.

An understanding of kinetics provides us with the means to answer our first question: How fast is a reaction proceeding at a given moment? Reactions, as you probably realize, occur at a wide range of speeds (Figure 15.2). Some, like a neutralization, a precipitation, or an explosive redox process, seem to be over as soon as the reactants make contact—in a fraction of a second. Others, such as the reactions involved in cooking or rust-

A

B

C

D

**FIGURE 15.2**
**The wide range of reaction rates.** Natural processes occur at a wide range of rates. An explosion **(A)** is much faster than the process of ripening **(B)**, which is much faster than the process of rusting **(C)**, which is much faster than the process of aging **(D)**.

ing, take a moderate amount of time, from several minutes to several months. Still others take much longer: the reactions that make up the human aging process continue for decades, and those in the formation of coal from dead plants take hundreds of millions of years.

Knowing how fast a chemical change will occur can be extremely important. How quickly a medicine acts or blood clots could make the difference between life and death. How long it takes for cement to harden, polyethylene to form, or a fabric to be dyed could make the difference between profit and loss. In general, the rates of these diverse processes depend on the same variables, most of which we can manipulate to maximize yields within a given time or to slow down an unwanted reaction.

We begin with a qualitative overview of the factors affecting reaction rate. Then we see how to express a rate quantitatively in the form of a *rate law*, and how concentration and temperature affect the rate. Next, we examine the models that explain those effects. Only then can we take apart the overall reaction, see the stages it may go through, and picture the species that exists fleetingly at the moment the reactant bonds are breaking and the product bonds forming. The chapter ends with a discussion of how catalysts increase reaction rates, highlighting the role of catalysts in the depletion of atmospheric ozone and in the reactions of a living cell.

## 15.1 A Qualitative Look at the Factors That Influence Reaction Rate

Under a given set of conditions, *each reaction has its own characteristic rate*, which is ultimately determined by the chemical nature of the reactants. At room temperature, for example, hydrogen reacts immediately with fluorine but extremely slowly with nitrogen:

$$H_2(g) + F_2(g) \rightarrow 2HF(g) \qquad \text{[very fast]}$$
$$3H_2(g) + N_2(g) \rightarrow 2NH_3(g) \qquad \text{[very slow]}$$

Magnesium reacts with water to form hydrogen gas more slowly than barium does:

$$Mg(s) + 2H_2O(l) \rightarrow Mg^{2+}(aq) + 2OH^-(aq) + H_2(g) \quad \text{[slow]}$$

$$Ba(s) + 2H_2O(l) \rightarrow Ba^{2+}(aq) + 2OH^-(aq) + H_2(g) \quad \text{[fast]}$$

The bonds in some compounds break more quickly than those in others; some elements lose or gain electrons more quickly than others. One key aspect of the diverse chemical behavior of the elements is the variation in the rates of their reactions with another substance.

For a given reaction, we can control four factors that affect its rate: the concentrations of reactants, their physical state, the temperature at which the reaction occurs, and the use of a catalyst. Let's consider the first three factors here; we'll discuss the fourth in Section 15.6.

1. *Concentration: molecules must collide in order to react.* A major factor influencing the rate of a given reaction is reactant concentration. Consider the reaction between ozone and nitric oxide that occurs in the stratosphere when nitric oxide is released in the exhaust gases of supersonic aircraft:

$$NO(g) + O_3(g) \rightarrow NO_2(g) + O_2(g)$$

Imagine what this reaction might look like at the molecular level if the reactants were confined in a reaction vessel. Nitric oxide and ozone molecules zoom every which way, crashing into each other and the vessel walls. A reaction between NO and $O_3$ can occur only when the molecules collide. The more molecules that are present in the container, the more frequently they collide, so the more often a reaction occurs. Thus, *reaction rate is proportional to the concentration of reactants:*

$$\text{Rate} \propto \text{collision frequency} \propto \text{concentration}$$

In this case, we're looking at a very simple reaction, one in which reactant molecules collide and form product molecules in one step. Even in complex reactions, however, the rate still depends on reactant concentration.

2. *Physical state: molecules must mix in order to collide.* The frequency of collisions between molecules also depends on the physical states of the reactants. When the reactants are in the same phase, as in aqueous solution, occasional stirring keeps them in contact. When they are in different phases, more vigorous mixing is needed. In these cases, the degree to which the reactant material is divided can have a significant effect on rate: the more finely divided a solid or liquid reactant, the greater its surface area per unit volume, the more contact it makes with the other reactant, and the faster the reaction. A steel nail heated in oxygen glows feebly, whereas finely divided steel wool bursts into flame (Figure 15.3). The oil-burner nozzle of a furnace produces a fine spray of oil because small droplets make greater contact with the incoming air. Likewise, you start a campfire with wood chips and thin branches, not logs.

3. *Temperature: molecules must collide with enough energy to react.* Temperature usually has a major influence on the speed of a reaction. Recall that molecules in a sample of gas have a range of speeds, with the most probable speed dependent on the temperature (Section 5.6). Thus, at higher temperature, more collisions occur in a given time. Even more important is the fact that temperature affects the kinetic energy of the molecules, and thus the *energy* of the collisions. In the jumble of molecules in the NO-$O_3$ reaction mentioned above, most collisions simply result in the molecules recoiling, like billiard balls, with no reaction taking place. However, some colli-

**A**

**B**

**FIGURE 15.3**

**The effect of surface area on reaction rate.**
**A,** A nail heated and then thrust into $O_2$ barely glows. **B,** Fine steel treated the same way bursts into flame. The steel wool has more surface area per unit volume than the nail, so more metal makes contact with $O_2$ and the reaction is faster.

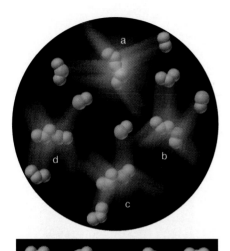

**FIGURE 15.4**
**Sufficiently energetic collisions lead to reaction.** Although many NO-$O_3$ collisions occur, relatively few have enough energy to cause reaction. At this temperature, only collision *a* is energetic enough to lead to product; the reactant molecules in collisions *b*, *c*, and *d* just bounce off each other unchanged.

sions occur with such force that the molecules react (Figure 15.4). At a higher temperature, more collisions occur with enough energy to react. Thus, *raising the temperature increases the reaction rate by increasing the number and, especially, the energy of the collisions.* Two familiar kitchen appliances employ this effect: a refrigerator slows down chemical processes that spoil food, whereas an oven speeds up other chemical processes to cook it.

Our qualitative understanding that reaction rate is influenced by the frequency and energy of reactant collisions leads us to ask several quantitative questions: How can we describe mathematically the dependence of rate on reactant concentration? Do all changes in concentration affect the rate to the same extent? Are all rates increased to the same extent by a given rise in temperature? How is the energy of collision used by reactant molecules to form product molecules, and is there a way to determine this energy? What do the reactants look like as they are turning into products? We address these questions in the following sections.

**Section Summary**
Chemical kinetics deals with reaction rates and the stepwise molecular events through which a reaction occurs. Under a given set of conditions, each reaction has its own rate. Concentration affects rate by influencing the frequency of collisions between reactant molecules. Physical state affects rate by determining the surface area per unit volume of reactants. Temperature affects rate by influencing the frequency and, more importantly, the energy of the collisions.

## 15.2 Expressing the Reaction Rate

Before we can deal quantitatively with the effects of concentration and temperature on reaction rate, we must be able to express the rate mathematically. A rate is a change in some variable per unit of time. The rate of motion of an object is the change in its position divided by the change in time. Suppose, for instance, we measure a racehorse's starting position ($x_1$) and time ($t_1$) and its final position ($x_2$) and time ($t_2$). The horse's rate of motion (speed) is given by

$$\text{Rate of motion} = \frac{\text{change in position}}{\text{change in time}} = \frac{x_2 - x_1}{t_2 - t_1} = \frac{\Delta x}{\Delta t}$$

In the case of a chemical reaction, the positions of the substances do not change over time, but their concentrations do: *reactant concentrations decrease while product concentrations increase.* At this point in the discussion, we define the **reaction rate** as the change in the concentrations of reactants or products per unit time. Consider a general reaction, A $\longrightarrow$ B. We measure the starting reactant concentration (conc $A_1$) at $t_1$, allow the reaction to proceed, and then measure the reactant concentration again (conc $A_2$) at $t_2$. The change in concentration divided by the change in time gives the rate:

$$\text{Rate of reaction} = -\frac{\text{change in concentration of A}}{\text{change in time}}$$

$$= -\frac{\text{conc } A_2 - \text{conc } A_1}{t_2 - t_1} = -\frac{\Delta \text{conc A}}{\Delta t}$$

Note the introduction of the minus sign. By convention, reaction rate is a positive number, but conc $A_2$ will always be lower than conc $A_1$, so the *change in concentration of A is always negative*. The minus sign converts the negative change in reactant concentration to a positive value for the rate. Suppose the concentration of A changes from 1.2 mol/L (conc $A_1$) to 0.75 mol/L (conc $A_2$) over a 125-s time period. The rate is

$$\text{Rate} = -\frac{0.75 \text{ mol/L} - 1.2 \text{ mol/L}}{125 \text{ s} - 0 \text{ s}} = 3.6 \times 10^{-3} \text{ mol/L} \cdot \text{s}$$

We use *square brackets [ ] to express concentration in moles per liter.* Thus, [A] is the concentration of A in mol/L, so

$$\text{Rate} = -\frac{\Delta[A]}{\Delta t} \qquad (15.1)$$

The rate has units of moles per liter per second (mol/L · s), or whatever time unit is convenient for the reaction under study (minutes, years, and so on).

If instead we measure the product to determine the reaction rate, we find its concentration *increasing* over time. That is, conc $B_2$ is always higher than conc $B_1$, so the *change* in product concentration, $\Delta[B]$, is positive and the reaction rate for A $\rightarrow$ B is

$$\text{Rate} = \frac{\Delta[B]}{\Delta t}$$

## Average, Instantaneous, and Initial Reaction Rates

Examining the rate of a real reaction reveals an important point: the rate itself varies with time. Consider the reversible gas-phase reaction between ethylene and ozone, one of many possible reactions in the formation of photochemical smog:

$$C_2H_4(g) + O_3(g) \rightleftharpoons C_2H_4O(g) + O_2(g)$$

For now, we consider only reactant concentrations. You can see from the equation coefficients that for every molecule of $C_2H_4$ that reacts, a molecule of $O_3$ reacts with it. In other words, the concentrations of both reactants decrease at the same rate in this particular reaction:

$$\text{Rate} = -\frac{\Delta[C_2H_4]}{\Delta t} = -\frac{\Delta[O_3]}{\Delta t}$$

By monitoring the concentration of either reactant, we can follow the rate of the reaction as it proceeds.

Suppose we have a known concentration of $O_3$ in a closed reaction vessel kept at 30°C (303 K). Table 15.1 shows the concentration of $O_3$ at various times during the first minute after we introduce $C_2H_4$ gas into the vessel. The rate over the entire 60 s is the total change in concentration divided by the change in time:

$$\text{Rate} = -\frac{\Delta[O_3]}{\Delta t} = -\frac{(1.10 \times 10^{-5} \text{ mol/L}) - (3.20 \times 10^{-5} \text{ mol/L})}{60.0 \text{ s} - 0.0 \text{ s}}$$

$$= 3.50 \times 10^{-7} \text{ mol/L} \cdot \text{s}$$

This calculation gives an **average rate** over that time period; that is, during the first 60 s of the reaction, ozone concentration decreases an average of $3.50 \times 10^{-7}$ mol/L each second. However, the average rate over this time does not show that the rate is changing, and it tells nothing about how fast the ozone concentration is decreasing *at any given instant*.

**TABLE 15.1 $O_3$ Concentration at Various Times in Its Reaction with $C_2H_4$ at 303 K**

| TIME (s) | CONCENTRATION OF $O_3$ (mol/L) |
|---|---|
| 0.0 | $3.20 \times 10^{-5}$ |
| 10.0 | $2.42 \times 10^{-5}$ |
| 20.0 | $1.95 \times 10^{-5}$ |
| 30.0 | $1.63 \times 10^{-5}$ |
| 40.0 | $1.40 \times 10^{-5}$ |
| 50.0 | $1.23 \times 10^{-5}$ |
| 60.0 | $1.10 \times 10^{-5}$ |

We can see that the rate changes during the reaction by calculating the average rate over two shorter time periods—one earlier and one later. Between the starting time 0.0 s and 10.0 s, the average rate is

$$\text{Rate} = -\frac{\Delta[O_3]}{\Delta t} = -\frac{(2.42 \times 10^{-5}\ \text{mol/L}) - (3.20 \times 10^{-5}\ \text{mol/L})}{10.0\ \text{s} - 0.0\ \text{s}}$$

$$= 7.8 \times 10^{-7}\ \text{mol/L} \cdot \text{s}$$

During the last 10.0 s, between 50.0 s and 60.0 s, the average rate is

$$\text{Rate} = -\frac{\Delta[O_3]}{\Delta t} = -\frac{(1.10 \times 10^{-5}\ \text{mol/L}) - (1.23 \times 10^{-5}\ \text{mol/L})}{60.0\ \text{s} - 50.0\ \text{s}}$$

$$= 1.3 \times 10^{-7}\ \text{mol/L} \cdot \text{s}$$

The earlier rate is six times as fast as the later rate. Thus, *the rate slowed during the course of the reaction.*

The change in rate can also be seen by plotting the concentrations vs. the times at which they were measured (Figure 15.5). A curve is obtained, which means that the rate changes. *The slope of the line ($\Delta y/\Delta x$, that is, $\Delta[O_3]/\Delta t$) joining any two points gives the average rate over that time period.*

The shorter the time period we choose, the closer we come to the **instantaneous rate,** the rate at a particular instant during the reaction. *The slope of a line tangent to the curve at a particular point gives the instantaneous rate at that time.* For example, the rate of the reaction 35.0 s after it began is $2.50 \times 10^{-7}$ mol/L · s, the slope of the line drawn tangent to the curve through the point at which $t = 35.0$ s. In general, we use the term "reaction rate" to mean the *instantaneous* reaction rate.

**FIGURE 15.5**

**The concentration of $O_3$ vs. time during its reaction with $C_2H_4$.** A plot of the data in Table 15.1 gives a curve because the rate changes as the reaction proceeds. The *average* rate over a given time period is the slope of a line joining any two points along the curve. The slope of line *b* is the average rate over the first 60.0 s of the reaction. The slopes of lines *c* and *e* give the average rate over the first and last 10-s intervals, respectively. Line *c* is steeper than line *e* because the average rate over the earlier time period is higher. The *instantaneous* rate at 35.0 s is the slope of line *d*, the tangent to the curve at $t = 35.0$ s. The *initial* rate is the slope of line *a*, the tangent to the curve at $t = 0.0$ s.

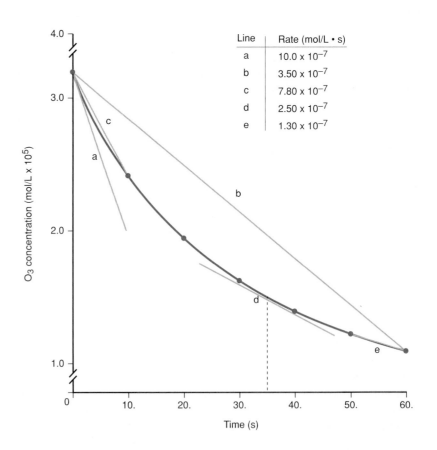

| Line | Rate (mol/L · s) |
|------|------------------|
| a | $10.0 \times 10^{-7}$ |
| b | $3.50 \times 10^{-7}$ |
| c | $7.80 \times 10^{-7}$ |
| d | $2.50 \times 10^{-7}$ |
| e | $1.30 \times 10^{-7}$ |

As a reaction continues, the product concentrations increase, so the reverse reaction becomes increasingly likely. To find the overall (net) rate, we would have to take both reactions into account and calculate the difference between the forward and reverse rates. A common way to avoid this complication for many reactions is to measure the **initial rate**, the instantaneous rate at the moment the reactants are mixed (that is, at $t = 0$). Under these conditions, product concentrations are negligible, so the reverse rate is negligible. The initial rate is measured by determining the slope of the line tangent to the curve at zero time; in Figure 15.5 (line $a$) the initial rate is $10.0 \times 10^{-7}$ mol/L·s. Unless specified otherwise, we will use initial rate data to determine other kinetic parameters.

### Expressing Rate in Terms of Reactant and Product Concentrations

So far, we've expressed the rate in terms of the decreasing concentration of one of the reactants ($O_3$). This picture is the same for the other reactant ($C_2H_4$), but it is exactly the opposite for the products: their concentrations are *increasing* as the reaction proceeds. From the balanced equation, we see that one molecule each of $C_2H_4O$ and $O_2$ appears for every molecule of $C_2H_4$ and of $O_3$ that disappears. We can express the rate in terms of any of the four substances involved:

$$\text{Rate} = -\frac{\Delta[C_2H_4]}{\Delta t} = -\frac{\Delta[O_3]}{\Delta t} = +\frac{\Delta[C_2H_4O]}{\Delta t} = +\frac{\Delta[O_2]}{\Delta t}$$

Again, note the negative values for the reactants and the positive values for the products (usually written without the plus sign). Figure 15.6 shows a plot of the simultaneous monitoring of reactants and products. Because product concentrations increase at the same rate that reactant concentrations decrease, the curves have the same shapes but are inverted.

In the reaction between ethylene and ozone, the reactants disappear and the products appear at the same rate because all the coefficients in the balanced equation equal one. Consider next the reaction between hydrogen and iodine to form hydrogen iodide:

$$H_2(g) + I_2(g) \longrightarrow 2HI(g)$$

For every molecule of $H_2$ that disappears, a molecule of $I_2$ disappears and *two* molecules of HI appear. In other words, the rate of $[H_2]$ decrease is the same as the rate of $[I_2]$ decrease but only half the rate of [HI] increase. By referring the change in $[I_2]$ and [HI] to the change in $[H_2]$, we have

$$\text{Rate} = -\frac{\Delta[H_2]}{\Delta t} = -\frac{\Delta[I_2]}{\Delta t} = \frac{1}{2}\frac{\Delta[HI]}{\Delta t}$$

If we refer the change in $[H_2]$ and $[I_2]$ to the change in [HI] instead, we obtain

$$\text{Rate} = \frac{\Delta[HI]}{\Delta t} = -2\frac{\Delta[H_2]}{\Delta t} = -2\frac{\Delta[I_2]}{\Delta t}$$

This expression is just a rearrangement of the previous one; note that it gives a numerical value for the rate that is double the previous value. Thus, the mathematical expression for the rate of a particular reaction *and* the numerical value of the rate depend on which substance serves as the reference.

We can summarize these results for any reaction:

$$a\text{A} + b\text{B} \longrightarrow c\text{C} + d\text{D}$$

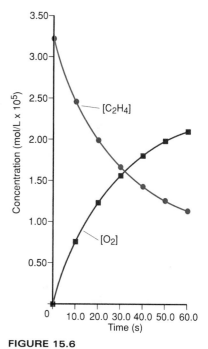

**FIGURE 15.6**
**Plots of $[C_2H_4]$ and $[O_2]$ vs. time.** Simultaneous monitoring of reactant $[C_2H_4]$ and product $[O_2]$ gives curves of identical shape changing in opposite directions. The steep upward (positive) slope of $[O_2]$ early in the reaction mirrors the steep downward (negative) slope of $[C_2H_4]$ because the faster $C_2H_4$ is used up, the faster $O_2$ is formed.

where $a$, $b$, $c$, and $d$ are coefficients of the balanced equation. In general, the rate is related to reactant or product concentrations as follows:

$$\text{Rate} = -\frac{1}{a}\frac{\Delta[A]}{\Delta t} = -\frac{1}{b}\frac{\Delta[B]}{\Delta t} = \frac{1}{c}\frac{\Delta[C]}{\Delta t} = \frac{1}{d}\frac{\Delta[D]}{\Delta t} \tag{15.2}$$

SAMPLE PROBLEM 15.1 —————————————————————————————————

**Expressing Rate in Terms of Changes in Concentration with Time**

**Problem:** One of the clean fuels that might be used in engines of the future is hydrogen gas:

$$2H_2(g) + O_2(g) \rightarrow 2H_2O(g)$$

**(a)** Express the rate of this reaction in terms of changes in $[H_2]$, $[O_2]$, and $[H_2O]$ with time.
**(b)** When $[O_2]$ is decreasing at 0.23 mol/L·s, at what rate is $[H_2O]$ increasing?
**Plan: (a)** Of the three substances in the equation, let's choose $O_2$ as the reference because its coefficient is 1. For every molecule of $O_2$ that disappears, two molecules of $H_2$ disappear, so the rate of $[O_2]$ decrease is one-half the rate of $[H_2]$ decrease. By similar reasoning, we see that the rate of $[O_2]$ decrease is one-half the rate of $[H_2O]$ increase. **(b)** Because $[O_2]$ is decreasing, the change must be negative. We substitute the negative value into the expression and solve for $\Delta[H_2O]/\Delta t$.
**Solution: (a)** Expressing the rate in terms of each component:

$$\textbf{Rate} = -\frac{1}{2}\frac{\Delta[\textbf{H}_2]}{\Delta t} = -\frac{\Delta[\textbf{O}_2]}{\Delta t} = +\frac{1}{2}\frac{\Delta[\textbf{H}_2\textbf{O}]}{\Delta t}$$

**(b)** Calculating the rate of change of $[H_2O]$:

$$\frac{1}{2}\frac{\Delta[H_2O]}{\Delta t} = -\frac{\Delta[O_2]}{\Delta t} = -(-0.23 \text{ mol/L}\cdot\text{s})$$

$$\frac{\Delta[H_2O]}{\Delta t} = 2 \times 0.23 \text{ mol/L}\cdot\text{s} = \textbf{0.46 mol/L}\cdot\textbf{s}$$

**Check: (a)** A good check is to use the rate expression to obtain the balanced equation: $[H_2]$ changes twice as fast as $[O_2]$, so two $H_2$ react for each $O_2$. $[H_2O]$ changes twice as fast as $[O_2]$, so two $H_2O$ form from each $O_2$, or $2H_2 + O_2 \rightarrow 2H_2O$. $[H_2]$ and $[O_2]$ decrease, so they take minus signs; $[H_2O]$ increases, so it takes a plus sign. Another check is to use Equation 15.2, where we have A = $H_2$, $a$ = 2; B = $O_2$, $b$ = 1; C = $H_2O$, $c$ = 2. Thus,

$$\text{Rate} = -\frac{1}{a}\frac{\Delta[A]}{\Delta t} = -\frac{1}{b}\frac{\Delta[B]}{\Delta t} = \frac{1}{c}\frac{\Delta[C]}{\Delta t}$$

or

$$\text{Rate} = -\frac{1}{2}\frac{\Delta[H_2]}{\Delta t} = -\frac{\Delta[O_2]}{\Delta t} = +\frac{1}{2}\frac{\Delta[H_2O]}{\Delta t}$$

**(b)** Given the rate expression, it makes sense for the numerical value of the rate of $[H_2O]$ increase to be twice that of $[O_2]$ decrease.
**Comment:** Thinking through this type of problem at the molecular level is the best approach, but use Equation 15.2 to confirm your answer.

FOLLOW-UP PROBLEM 15.1
**(a)** Balance the following reaction and express the rate in terms of changes in concentration with time for each substance: $NO(g) + O_2(g) \rightarrow N_2O_3(g)$.
**(b)** How fast is $[O_2]$ decreasing when $[NO]$ is decreasing at $1.60 \times 10^{-4}$ mol/L·s?

**Section Summary**
The average reaction rate is the change in reactant (or product) concentration over a change in time, $\Delta t$. The rate slows as reactants are used up. The instantaneous rate at time $t$, which occurs when $\Delta t$ is infinitely small, is obtained from the slope of the tangent to a concentration-time curve at time $t$. The initial rate, the instantaneous rate at $t = 0$, occurs before any product accumulates. The expression for a reaction rate and its value depend on which reaction component is being monitored.

## 15.3    The Rate Law and Its Components

The centerpiece of any kinetic study is the **rate law,** or *rate equation*, for the reaction in question, which expresses the rate as a function of reactant concentrations, product concentrations, and temperature. Since the rate law is based on experimental fact, any hypothesis we make about how the reaction occurs on the molecular level must conform to it.

In our discussion, we'll generally consider only reactions for which the products are not involved in the rate law. In these cases, *the reaction rate depends only on reactant concentrations and temperature.* Most of this section deals with the effect of concentration on reactions occurring at a fixed temperature; at the end, we consider the effects of temperature. For a general reaction,

$$aA + bB + \cdots \longrightarrow cC + dD + \cdots$$

the rate law has the form

$$\text{Rate} = k[A]^m[B]^n \cdots \qquad\qquad \textbf{(15.3)}$$

The proportionality constant $k$, called the **rate constant,** is specific for a given reaction at a given temperature; it does *not* change as the reaction proceeds. (As you'll see later, $k$ *does* change with temperature and therefore determines how temperature affects the rate.) The exponents $m$ and $n$, called the **reaction orders,** define how the rate is affected by the concentration of each reactant. For example, if the rate doubles when [A] doubles, the rate depends on $[A]^1$, so $m = 1$; if the rate quadruples when [B] doubles, the rate depends on $[B]^2$, so $n = 2$.

### Determining the Components of the Rate Law

A key point to keep in mind is that *the components of the rate law—rate, reaction orders, and rate constant—must be found by experiment;* they *cannot* be deduced from the reaction stoichiometry. The coefficients $a$ and $b$ in the general balanced equation are *not* necessarily related in any way to the reaction orders $m$ and $n$. We determine the initial rate and the reaction orders from concentration measurements and use them to calculate the rate constant. Once we know the rate law at a specific temperature, we can use it to predict the rate for any initial reactant concentrations.

**Determining the initial rate.** In the last section, we showed how initial rates are determined from a plot of concentration vs. time. Several experiments are carried out at *different initial concentrations* to obtain a series of *different initial rates.* Since we use the initial rate to determine the reaction orders and rate constant, an accurate experimental method for measuring concentration at various times during the reaction is the key to constructing the rate law. A few of the many techniques used for measuring changes in reactant or product concentration over time are presented in the following Tools of the Chemistry Laboratory essay.

**Measuring Reaction Rates**

Speculation about how a reaction occurs at the molecular level must be based on measurement of reaction rates. From a practical standpoint, the experimental method should measure the concentration of a single chemical species in the reaction quickly and reproducibly. Four general approaches, with specific examples, follow.

### Spectrometric Methods

These methods are used to measure the concentration of a reactant or product that absorbs (or emits) light of a narrow range of wavelengths. The reaction is typically performed in the sample compartment of a spectrometer set to measure a wavelength characteristic of one of the species (Figure 15.A). For example, in the $NO$-$O_3$ reaction, only $NO_2$ has a color:

$$NO(g, \text{ colorless}) + O_3(g, \text{ colorless}) \rightarrow$$
$$O_2(g, \text{ colorless}) + NO_2(g, \text{ brown})$$

Known amounts of reactants are injected into a gas sample tube of known volume, and the rate of $NO_2$ formation is measured by monitoring the color over time. Reactions in aqueous solution can be studied similarly.

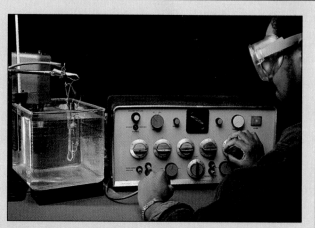

**FIGURE 15.B**
**Conductometric monitoring of a reaction.** When a reactant mixture differs in conductivity from the product mixture, the change in conductivity is proportional to the reaction rate. It is usually easier to monitor a nonionic reactant forming an ionic product.

### Conductometric Methods

When non-ionic reactants form ionic products, or vice versa, the change in conductivity of the solution over time can be used to measure the rate. Electrodes are immersed in the reaction mixture, and the increase (or decrease) in conductivity correlates with the formation of product (Figure 15.B). Consider the reaction between an organic halide, such as 2-bromo-2-methylpropane, and water:

$$(CH_3)_3CBr(l) + H_2O(l) \rightarrow (CH_3)_3COH(l) + H^+(aq) + Br^-(aq)$$

The HBr that forms is a strong acid in water, so it dissociates completely into ions. As time passes, more HBr forms and dissociates, so the conductivity of the reaction mixture increases.

### Manometric Methods

If a reaction involves a change in the number of moles of gas, the rate can be determined from the change in pressure (at constant volume and temperature) over time. In practice, a manometer is attached to a reaction vessel of known volume that is immersed in a constant-temperature bath. For example, the reaction between zinc and acetic acid can be monitored by this method:

$$Zn(s) + 2CH_3COOH(aq) \rightarrow$$
$$Zn^{2+}(aq) + 2CH_3COO^-(aq) + H_2(g)$$

As $H_2$ forms, the pressure increases in direct proportion to the reaction rate (Figure 15.C).

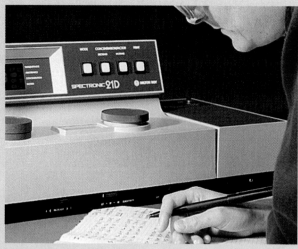

**FIGURE 15.A**
**Spectrometric monitoring of a reaction.** The investigator adds the reactant(s) to the sample tube and immediately places it in the spectrometer. In this case, the reactant is colored. Rate data can be determined from a plot of light absorbed vs. time.

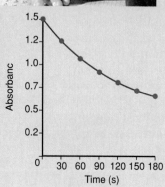

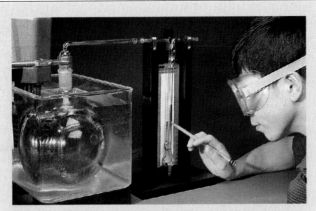

**FIGURE 15.C Manometric monitoring of a reaction.** When a reaction results in a change in the number of moles of gas, the change in pressure with time corresponds to a change in reaction rate.

### Direct Chemical Methods

Rates of slow reactions, or of those that can be easily slowed, are often studied by direct chemical methods. A small, measured portion (aliquot) of the reaction mixture is removed, and the reaction in this portion is stopped, usually by rapid cooling. The concentration of reactant or product in the aliquot is measured, while the bulk of the reaction mixture continues to react and is sampled later. For example, the previous reaction between an organic halide and water can also be studied by titration. The reaction rate in an aliquot is slowed by quickly transferring it to a chilled flask in an ice bath. HBr concentration in the aliquot is determined by titrating with standardized NaOH solution. To determine the change in HBr concentration with time, the procedure is repeated at regular intervals during the reaction.

**Determining reaction orders.** Before we see how reaction orders are determined from rate data, let's discuss the meaning of reaction order and some important terminology. We speak of the reaction as having an order "with respect to" or "in" each reactant. For example, in the reaction between nitric oxide and ozone,

$$NO(g) + O_3(g) \rightarrow NO_2(g) + O_2(g)$$

the rate equation has been experimentally determined to be

$$\text{Rate} = k[NO][O_3]$$

This reaction is first order with respect to NO (or "first order in NO"), which means that the rate depends on NO concentration raised to the first power, $[NO]^1$. It is also first order with respect to $O_3$, that is, $[O_3]^1$. The sum of the individual orders gives the *overall* reaction order; this reaction is second order overall.

Now consider a different gas-phase reaction:

$$2NO_2(g) + F_2(g) \rightarrow 2NO_2F(g)$$

for which the rate law has been determined to be

$$\text{Rate} = k[NO_2][F_2]$$

This reaction is first order in $NO_2$ and first order in $F_2$. Overall, this is also a second-order reaction.

Finally, in the hydrolysis of 2-bromo-2-methylpropane we looked at in the Tools of the Chemistry Laboratory essay,

$$(CH_3)_3CBr(l) + H_2O(l) \rightarrow (CH_3)_3COH(l) + HBr(aq)$$

the rate law has been found to be

$$\text{Rate} = k[(CH_3)_3CBr]$$

This reaction is first order in 2-bromo-2-methylpropane. Note that the concentration of $H_2O$ does not even appear in the rate law. Thus, the reaction is zero order with respect to $H_2O$ ($[H_2O]^0$). This means that the rate does not depend on the concentration of $H_2O$; we can also write the rate law for this reaction as

$$\text{Rate} = k[(CH_3)_3CBr][H_2O]^0$$

Overall, this is a first-order reaction.

These examples demonstrate a major point: *reaction orders* ***cannot be deduced from the balanced equation.*** In the reaction between $NO_2$ and $F_2$ and in the hydrolysis of 2-bromo-2-methylpropane, the reaction orders in the rate laws do *not* correspond to the coefficients of the balanced equations. Reaction orders *must* be determined from rate data.

Reaction orders are usually positive integers or zero, but they can also be fractional or negative. In the reaction

$$CHCl_3(g) + Cl_2(g) \longrightarrow CCl_4(g) + HCl(g)$$

a fractional order appears in the rate law:

$$\text{Rate} = k[CHCl_3][Cl_2]^{1/2}$$

This order means that the reaction rate depends on the square root of the $Cl_2$ concentration. If the initial $Cl_2$ concentration is increased by a factor of 4, for example, the rate increases by a factor of 2. A negative exponent means that the rate *decreases* when the concentration of that component increases. Negative orders are often seen for reactions whose rate laws include products. For example, in the atmospheric reaction

$$2O_3(g) \rightleftharpoons 3O_2(g)$$

the rate law has been shown to be

$$\text{Rate} = k[O_3]^2[O_2]^{-1} = k\frac{[O_3]^2}{[O_2]}$$

If the $O_2$ concentration doubles, the reaction proceeds half as fast.

**SAMPLE PROBLEM 15.2** _____

### Determining Reaction Order from the Rate Law

**Problem:** For each of the following reactions, determine the reaction order with respect to each reactant and the overall order from the given rate law:
**(a)** $2NO(g) + O_2(g) \longrightarrow 2NO_2(g)$; rate $= k[NO]^2[O_2]$
**(b)** $CH_3CHO(g) \longrightarrow CH_4(g) + CO(g)$; rate $= k[CH_3CHO]^{3/2}$
**(c)** $H_2O_2(aq) + 3I^-(aq) + 2H^+(aq) \longrightarrow I_3^-(aq) + 2H_2O(l)$; rate $= k[H_2O_2][I^-]$
**Plan:** We inspect the exponents in the rate law, *not* the coefficients of the balanced equation, to find the individual orders, and then take their sum to obtain the overall reaction order.
**Solution: (a)** The reaction is **second order with respect to NO, first order with respect to $O_2$,** and **third order overall.**
**(b)** The reaction is $^3/_2$ **order in $CH_3CHO$ and** $^3/_2$ **order overall.**
**(c)** The reaction is **first order in $H_2O_2$, first order in $I^-$,** and **second order overall.** The reactant $H^+$ does not appear in the rate law, so the reaction is **zero order in $H^+$.**
**Check:** Be sure the sum of the individual orders gives the overall order.

**FOLLOW-UP PROBLEM 15.2**
Experiment shows that the reaction

$$5Br^-(aq) + BrO_3^-(aq) + 6H^+(aq) \longrightarrow 3Br_2(l) + 3H_2O(l)$$

obeys the rate law: rate $= k[Br^-][BrO_3^-][H^+]^2$. What are the reaction orders in each reactant and the overall reaction order?

Next, let's see how reaction orders are determined from experimental rate data, that is, *before* the rate law is known. By comparing experiments in which the initial concentration of a single reactant is changed, we find the reaction order with respect to that reactant. Suppose we are studying the reaction between nitric oxide and oxygen, a key step in the formation of acid rain and in the industrial production of nitric acid:

$$O_2(g) + 2NO(g) \rightarrow 2NO_2(g)$$

The rate law for this reaction, expressed in general form, is

$$\text{Rate} = k[O_2]^m[NO]^n$$

To find the reaction orders, we run a series of experiments, each of which starts with a different set of reactant concentrations. From each experiment, we obtain an initial rate. As Table 15.2 shows, the experiments are designed to change one reactant concentration while keeping the other constant. If we compare the values in experiments 1 and 2, we see the effect of doubling $[O_2]$ on the rate. To do this, we take the ratio of their rate laws:

$$\frac{\text{Rate 2}}{\text{Rate 1}} = \frac{k[O_2]_2^m[NO]_2^n}{k[O_2]_1^m[NO]_1^n}$$

where $[O_2]_2$ is the $O_2$ concentration for experiment 2, $[NO]_1$ is the NO concentration for experiment 1, and so forth. Since $k$ is a constant and [NO] does not change between these two experiments, these quantities cancel:

$$\frac{\text{Rate 2}}{\text{Rate 1}} = \frac{[O_2]_2^m}{[O_2]_1^m} = \left( \frac{[O_2]_2}{[O_2]_1} \right)^m$$

Substituting the values from Table 15.2, we obtain

$$\frac{6.40 \times 10^{-3} \text{ mol/L} \cdot \text{s}}{3.21 \times 10^{-3} \text{ mol/L} \cdot \text{s}} = \left( \frac{2.20 \times 10^{-2} \text{ mol/L}}{1.10 \times 10^{-2} \text{ mol/L}} \right)^m$$

Dividing, we obtain

$$1.99 = (2.00)^m$$

Within experimental error, $1.99 \approx 2.00$, so we have

$$2.00 = (2.00)^m$$

Thus,                        $m = 1$

The reaction is first order in $O_2$: when $[O_2]$ doubles, the rate doubles.

   To find the order with respect to NO, we compare experiments 3 and 1, in which $[O_2]$ is held constant and [NO] is doubled:

$$\frac{\text{Rate 3}}{\text{Rate 1}} = \frac{k[O_2]_3^m[NO]_3^n}{k[O_2]_1^m[NO]_1^n}$$

**TABLE 15.2 Initial Rates for a Series of Experiments in the Reaction Between $O_2$ and NO**

| | INITIAL REACTANT CONCENTRATIONS (mol/L) | | INITIAL RATE |
|---|---|---|---|
| EXPERIMENT | $O_2$ | NO | (mol/L · s) |
| 1 | $1.10 \times 10^{-2}$ | $1.30 \times 10^{-2}$ | $3.21 \times 10^{-3}$ |
| 2 | $2.20 \times 10^{-2}$ | $1.30 \times 10^{-2}$ | $6.40 \times 10^{-3}$ |
| 3 | $1.10 \times 10^{-2}$ | $2.60 \times 10^{-2}$ | $12.8 \ \times 10^{-3}$ |
| 4 | $3.30 \times 10^{-2}$ | $1.30 \times 10^{-2}$ | $9.60 \times 10^{-3}$ |
| 5 | $1.10 \times 10^{-2}$ | $3.90 \times 10^{-2}$ | $28.8 \ \times 10^{-3}$ |

As before, $k$ is constant and $[O_2]$ does not change in this pair of experiments, so these quantities cancel:

$$\frac{\text{Rate 3}}{\text{Rate 1}} = \left( \frac{[NO]_3}{[NO]_1} \right)^n$$

The actual values give

$$\frac{12.8 \times 10^{-3}\ \text{mol/L} \cdot \text{s}}{3.21 \times 10^{-3}\ \text{mol/L} \cdot \text{s}} = \left( \frac{2.60 \times 10^{-2}\ \text{mol/L}}{1.30 \times 10^{-2}\ \text{mol/L}} \right)^n$$

Dividing, we obtain

$$3.99 = (2.00)^n$$

Since $3.99 \approx 4.00$, we have

$$4.00 = (2.00)^n$$

Thus,

$$n = 2$$

The reaction is second order in NO: when [NO] doubles, the rate quadruples. Thus, the rate law is

$$\text{Rate} = k[O_2][NO]^2$$

You may want to use experiment 1 in combination with experiments 4 and 5 to check this result.

SAMPLE PROBLEM 15.3 _____

**Determining Reaction Orders from Initial Rate Data**

**Problem:** Many gaseous reactions take place in a car engine and exhaust system; one of these is

$$NO_2(g) + CO(g) \longrightarrow NO(g) + CO_2(g); \qquad \text{rate} = k[NO_2]^m[CO]^n$$

Use the following data to determine the individual and overall reaction orders:

| EXPERIMENT | INITIAL RATE (mol/L · s) | INITIAL [NO$_2$] (mol/L) | INITIAL [CO] (mol/L) |
|:---:|:---:|:---:|:---:|
| 1 | 0.0050 | 0.10 | 0.10 |
| 2 | 0.080 | 0.40 | 0.10 |
| 3 | 0.0050 | 0.10 | 0.20 |

**Plan:** We need to solve the general rate law for the reaction orders $m$ and $n$. To solve for each exponent, we proceed as in text, taking the ratio of the rate laws for two experiments in which only the reactant in question changes.

**Solution:** Calculating $m$ in $[NO_2]^m$: We take the ratio of the rate laws for experiments 1 and 2, in which $[NO_2]$ varies but [CO] is constant,

$$\frac{\text{Rate 2}}{\text{Rate 1}} = \frac{k[NO_2]_2^m[CO]_2^n}{k[NO_2]_1^m[CO]_1^n} = \left( \frac{[NO_2]_2}{[NO_2]_1} \right)^m \quad \text{or} \quad \frac{0.080\ \text{mol/L} \cdot \text{s}}{0.0050\ \text{mol/L} \cdot \text{s}} = \left( \frac{0.40\ \text{mol/L}}{0.10\ \text{mol/L}} \right)^m$$

Since $16 = (4.0)^m$, $m = 2.0$. The reaction is **second order in NO$_2$.**

Calculating $n$ in $[CO]^n$: We take the ratio of the rate equations for experiments 1 and 3, in which [CO] varies but $[NO_2]$ is constant,

$$\frac{\text{Rate 3}}{\text{Rate 1}} = \frac{k[NO_2]_3^2[CO]_3^n}{k[NO_2]_1^2[CO]_1^n} = \left( \frac{[CO]_3}{[CO]_1} \right)^n \quad \text{or} \quad \frac{0.0050\ \text{mol/L} \cdot \text{s}}{0.0050\ \text{mol/L} \cdot \text{s}} = \left( \frac{0.20\ \text{mol/L}}{0.10\ \text{mol/L}} \right)^n$$

Since $1.0 = (2.0)^n$, $n = 0$. The rate does not change when [CO] varies, so the reaction is **zero order in CO.**

Therefore, the rate law is

$$\text{Rate} = k[NO_2]^2[CO]^0 = k[NO_2]^2(1) = k[NO_2]^2$$

The reaction is **second order overall.**

**Check:** Reason through the orders. If $m = 1$, quadrupling $[NO_2]$ would quadruple the rate; since the rate more than quadruples, $m > 1$. If $m = 2$, quadrupling $[NO_2]$ would increase the rate by a factor of 16 ($4^2$). The ratio of rates is $0.080/0.005 = 16$, so $m = 2$. In contrast, increasing $[CO]$ has no effect on the rate, which can happen only if $[CO]^n = 1$, so $n = 0$.

**FOLLOW-UP PROBLEM 15.3**
Find the rate law and the overall reaction order for the reaction $H_2 + I_2 \longrightarrow 2HI$ from the following data at 450°C:

| EXPERIMENT | INITIAL RATE (mol/L · s) | INITIAL $[H_2]$ (mol/L) | INITIAL $[I_2]$ (mol/L) |
|------------|--------------------------|-------------------------|-------------------------|
| 1 | $1.9 \times 10^{-23}$ | 0.0113 | 0.0011 |
| 2 | $1.1 \times 10^{-22}$ | 0.0220 | 0.0033 |
| 3 | $9.3 \times 10^{-23}$ | 0.0550 | 0.0011 |
| 4 | $1.9 \times 10^{-22}$ | 0.0220 | 0.0056 |

**Determining the rate constant.** With the rate, reactant concentrations, and reaction orders known, the sole remaining unknown in the rate law is the rate constant, $k$. The rate constant is specific for a particular reaction *at a particular temperature*. Since all the experiments in the $O_2$-NO series we are considering were conducted at the same temperature, we can use the data from any one of them to solve for $k$. From experiment 1 in Table 15.2, for instance, we obtain

$$k = \frac{\text{rate 1}}{[O_2]_1[NO]_1^2} = \frac{3.21 \times 10^{-3} \text{ mol/L} \cdot \text{s}}{(1.10 \times 10^{-2} \text{ mol/L}) (1.30 \times 10^{-2} \text{ mol/L})^2}$$

$$= \frac{3.21 \times 10^{-3} \text{ mol/L} \cdot \text{s}}{1.86 \times 10^{-6} \text{ mol}^3/\text{L}^3} = 1.73 \times 10^3 \text{ L}^2/\text{mol}^2 \cdot \text{s}$$

Always check that the values of $k$ in the same series are within experimental error. To three significant figures, the average value of $k$ for the five experiments in Table 15.2 is $1.72 \times 10^3$ L$^2$/mol$^2$ · s.

Note the units for the rate constant. Since concentrations are in mol/L and the reaction rate is in units of mol/L · time, the units for $k$ will depend on the order of the reaction and, of course, the time unit. The units for $k$ here, L$^2$/mol$^2$ · s, are required to give a rate with units of mol/L · s:

$$\frac{\text{mol}}{\text{L} \cdot \text{s}} = \frac{\text{L}^2}{\text{mol}^2 \cdot \text{s}} \times \frac{\text{mol}}{\text{L}} \times \left(\frac{\text{mol}}{\text{L}}\right)^2$$

The rate constant will *always* have these units for an overall third-order reaction with the time unit in seconds. Table 15.3 shows the units of $k$ for some common overall reaction orders.

**TABLE 15.3  Units of the Rate Constant $k$ for Several Overall Reaction Orders**

| OVERALL REACTION ORDER | UNITS OF $k$ ($t$ in seconds) |
|------------------------|-------------------------------|
| 0 | mol/L · s (or mol L$^{-1}$ s$^{-1}$) |
| 1 | 1/s (or s$^{-1}$) |
| 2 | L/mol · s (or L mol$^{-1}$ s$^{-1}$) |
| 3 | L$^2$/mol$^2$ · s (or L$^2$ mol$^{-2}$ s$^{-1}$) |

General formula:

$$\text{Units of } k = \frac{\left(\dfrac{\text{L}}{\text{mol}}\right)^{\text{order} - 1}}{\text{s}}$$

## Effect of Time on Reactant Concentration

The rate laws we have developed so far do not include time as a variable. They tell us the rate or concentration at a given instant, allowing us to answer the critical question, "How fast is the reaction proceeding at the moment when $y$ moles per liter of A are reacting with $z$ moles per liter of B?" However, by employing different forms of the rate laws, called **integrated rate laws,** we can consider the time factor and answer other questions such as "How long will it take for $x$ moles per liter of A to be used up?" or "What is the concentration of A after 15 minutes of reaction?"

**Integrated rate laws for first-order and second-order reactions.**
Consider a simple first-order reaction, A → B. The rate can be expressed as the change in the concentration of A divided by the change in time:

$$\text{Rate} = -\frac{\Delta[\text{A}]}{\Delta t}$$

It can also be expressed in terms of the rate law:

$$\text{Rate} = k[\text{A}]$$

Setting these different expressions equal to each other gives

$$-\frac{\Delta[\text{A}]}{\Delta t} = k[\text{A}]$$

Through the methods of calculus, this expression is integrated over time to obtain the integrated rate law for a first-order reaction:

$$\ln \frac{[\text{A}]_0}{[\text{A}]_t} = kt \qquad \text{(first-order reaction)} \qquad (15.4)$$

where $[\text{A}]_0$ is the concentration of A at $t = 0$ and $[\text{A}]_t$ is the concentration of A at any time $t$ during an experiment. Since $\ln \frac{a}{b} = \ln a - \ln b$, we have

$$\ln [\text{A}]_0 - \ln [\text{A}]_t = kt$$

For a general second-order rate equation, the expression including time can become slightly complex, so let's consider only the simplest case, one in which the rate law contains only one reactant:

$$\text{Rate} = -\frac{\Delta[\text{A}]}{\Delta t} = k[\text{A}]^2$$

Integrating this expression over time gives the integrated rate law for a second-order reaction involving one reactant:

$$\frac{1}{[\text{A}]_t} - \frac{1}{[\text{A}]_0} = kt \qquad \text{(second-order reaction; rate} = k[\text{A}]^2) \qquad (15.5)$$

Sample Problem 15.4 presents one way in which these integrated rate laws are applied.

---

**SAMPLE PROBLEM 15.4** _____

**Determining the Reactant Concentration at a Given Time**

**Problem:** Cyclobutane ($C_4H_8$) decomposes at 1000°C to two moles of ethylene ($C_2H_4$) with a very high first-order rate constant, 87 s$^{-1}$.
**(a)** If the initial concentration of cyclobutane is 2.00 $M$, what is the concentration after 0.010 s?
**(b)** What fraction of cyclobutane has decomposed in this time?
**Plan: (a)** We must find the concentration of cyclobutane at time $t$, $[C_4H_8]_t$. The problem tells us this is a first-order reaction, so we use the integrated first-order rate law:

$$\ln [C_4H_8]_0 - \ln [C_4H_8]_t = kt$$

We know $k$, $t$, and $[C_4H_8]_0$, so we can solve for $[C_4H_8]_t$.
**(b)** The fraction decomposed is the concentration that has decomposed divided by the initial concentration:

$$\text{Fraction decomposed} = \frac{[C_4H_8]_0 - [C_4H_8]_t}{[C_4H_8]_0}$$

**Solution:** (a) Rearranging the integrated rate expression and multiplying through by $-1$ to solve for $\ln [C_4H_8]_t$:

$$\ln [C_4H_8]_t = \ln [C_4H_8]_0 - kt$$

Substituting the data:

$$\ln [C_4H_8]_t = \ln (2.00 \text{ mol/L}) - (87 \text{ s}^{-1}) (0.010 \text{ s})$$

$$= 0.69 - 0.87 = -0.18$$

Taking the antilog of both sides:

$$[C_4H_8]_{(t = 0.010 \text{ s})} = e^{-0.18} = \textbf{0.84 mol/L } \textbf{C}_4\textbf{H}_8$$

**(b)** Finding the fraction that has decomposed after 0.010 s:

$$\frac{[C_4H_8]_0 - [C_4H_8]_t}{[C_4H_8]_0} = \frac{2.00 \text{ mol/L} - 0.84 \text{ mol/L}}{2.00 \text{ mol/L}} = \textbf{0.58}$$

**Check:** The concentration remaining after 0.010 s (0.84 mol/L) is less than the starting concentration (2.00 mol/L), which makes sense. A negative logarithm gives a number between 0 and 1, and $0 < 0.84 < 1$, which is also reasonable. Finally, the result makes sense: a high rate constant indicates a fast reaction, so it's not surprising that so much decomposes in such a short time.
**Comment:** Integrated rate laws are also used to solve for the time it takes to reach a certain reactant concentration, as in Follow-up Problem 15.4.

**FOLLOW-UP PROBLEM 15.4**
At 25°C, hydrogen iodide breaks down very slowly to hydrogen and iodine according to the following: rate $= k[HI]^2$. The rate constant at 25°C is $2.4 \times 10^{-21}$ L/mol · s. If 0.0100 mol HI(*g*) is placed in a 1.0-L container, how long will it take for the concentration of HI to reach 0.00900 mol/L?

**Determining the reaction order from the integrated rate law.** Suppose you do not know the rate law for a reaction and do not have the initial rate data needed to determine the reaction orders (as we did in Sample Problem 15.3). Another method for finding reaction orders is a graphical technique that uses concentration-time data directly.

Integrated rate laws can be rearranged into the form of an equation for a straight line, $y = mx + b$, where $m$ is the slope and $b$ is the $y$-axis intercept. For a first-order reaction, we have

$$\ln [A]_0 - \ln [A]_t = kt$$

Rearranging and changing sign gives

$$\ln [A]_t = -kt + \ln [A]_0$$
$$y \quad\quad = mx + \quad b$$

Thus, a plot of $\ln [A]_t$ vs. $t$ gives a straight line with slope $= -k$ and $y$ intercept $= \ln [A]_0$ (Figure 15.7, *A*).
For a simple second-order reaction, we have

$$\frac{1}{[A]_t} - \frac{1}{[A]_0} = kt$$

Rearranging gives

$$\frac{1}{[A]_t} = kt + \frac{1}{[A]_0}$$
$$y \quad = mx + \quad b$$

In this case, a plot of $1/[A]_t$ vs. $t$ gives a straight line with slope $= k$ and $y$ intercept $= 1/[A]_0$ (Figure 15.7, *B*).

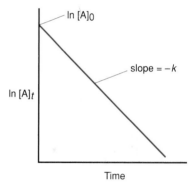

**A** First order

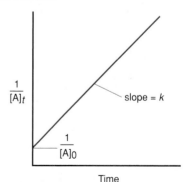

**B** Second order

**FIGURE 15.7**
**Integrated rate laws and reaction order. A,** A plot of $\ln [A]_t$ vs. $t$ gives a straight line for a reaction that is first order in A. **B,** A plot of $1/[A]_t$ vs. $t$ gives a straight line for a reaction that is second order in A.

| Time | [N₂O₅] | ln [N₂O₅] | 1/[N₂O₅] |
|------|--------|-----------|----------|
| 0    | 0.0165 | −4.104    | 60.6     |
| 10   | 0.0124 | −4.390    | 80.6     |
| 20   | 0.0093 | −4.68     | $1.1 \times 10^2$ |
| 30   | 0.0071 | −4.95     | $1.4 \times 10^2$ |
| 40   | 0.0053 | −5.24     | $1.9 \times 10^2$ |
| 50   | 0.0039 | −5.55     | $2.6 \times 10^2$ |
| 60   | 0.0029 | −5.84     | $3.4 \times 10^2$ |

**A**

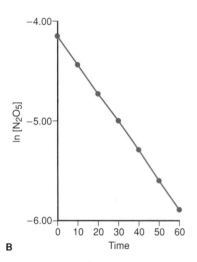

**B**

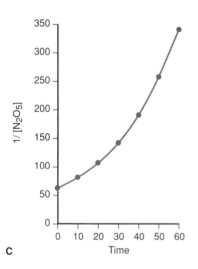

**C**

**FIGURE 15.8**

Graphical determination of the reaction order for the decomposition of $N_2O_5$. **A,** Time (min) and concentration (mol/L) data for determining reaction order. **B,** A plot of ln $[N_2O_5]$ vs. *t* gives a straight line, indicating that the reaction is first order in $N_2O_5$. **C,** A plot of 1/$[N_2O_5]$ vs. *t* is curved, indicating that the reaction is *not* second order in $N_2O_5$.

To find the reaction order from the concentration-time data, some trial-and-error graphical plotting is required:

• If you obtain a straight line when you plot ln [reactant] vs. time, the reaction is *first order* with respect to that reactant.
• If you obtain a straight line when you plot 1/[reactant] vs. time, the reaction is *second order* with respect to that reactant.

Figure 15.8 shows how this approach is used to determine that $N_2O_5$ decomposition is first order in $N_2O_5$.

**Reaction half-life.** The **half-life ($t_{1/2}$)** of a reaction is the time required to reach half the initial reactant concentration (Figure 15.9). Half-lives have time units that are reasonable for the particular reaction. The half-life for the decomposition of $N_2O_5$ at 45°C is 24.0 min, whereas that for the radioactive decay of uranium-235 is $7.1 \times 10^8$ yr. The half-life means that if we start with 0.0600 mol/L $N_2O_5$ at 45°C, after 24 min (one half-life), 0.0300 mol/L has been consumed and 0.0300 mol/L remains; after 48 min (two half-lives), 0.0150 mol/L remains; after 72 min (three half-lives), 0.00750 mol/L remains, and so forth.

At fixed conditions, *the half-life of a first-order reaction is a constant, independent of reactant concentration.* We can see this fact from the integrated first-order rate law:

$$\ln \frac{[A]_0}{[A]_t} = kt$$

After one half-life, $t = t_{1/2}$, and $[A]_t = \frac{1}{2}[A]_0$. Substituting, we obtain

$$\ln \frac{\cancel{[A]_0}}{\frac{1}{2}\cancel{[A]_0}} = kt_{1/2}$$

Thus,

$$\ln 2 = kt_{1/2}$$

Then, solving for $t_{1/2}$, we have

$$t_{1/2} = \frac{\ln 2}{k} = \frac{0.693}{k} \qquad \text{(first-order process; rate} = k[A]) \qquad \textbf{(15.6)}$$

Note that the time to reach one-half the starting concentration in a first-order reaction does not depend on what that starting concentration is. After 710 million years, a 1-kg sample of uranium-235 will contain 0.5 kg uranium-235; after the same amount of time, a 1-mg sample will contain 0.5 mg. Whether we consider a molecule or a radioactive nucleus, the decomposition of each particle in a first-order process is independent of the number of other particles present.

**SAMPLE PROBLEM 15.5**

**Determining the Half-Life of a First-Order Reaction**

**Problem:** Cyclopropane is the smallest cyclic hydrocarbon. Because its 60° bond angles allow only poor orbital overlap, its bonds are weak. As a result, it is thermally unstable and rearranges to propene at 1000°C via the following first-order reaction:

$$\underset{H_2C\!-\!CH_2}{\overset{CH_2}{\diagup}}(g) \longrightarrow CH_3\!-\!CH\!=\!CH_2(g)$$

The rate constant is 9.2 s$^{-1}$. How long does it take for the initial concentration of cyclopropane to decrease by one-half?

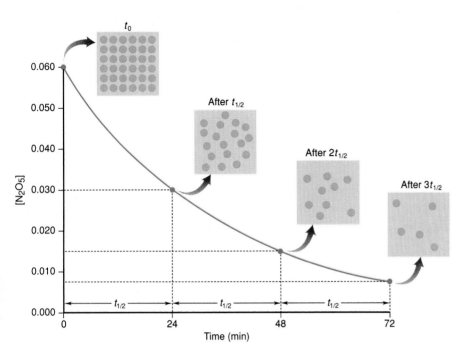

**FIGURE 15.9**
**A plot of $[N_2O_5]$ vs. time for three half-lives.** During each half-life, the concentration is halved. $T = 45°C$ and $[N_2O_5]_0 = 0.0600$ mol/L. A schematic representation, with $N_2O_5$ molecules as blue circles, shows that after three half-lives, $\frac{1}{2} \times \frac{1}{2} \times \frac{1}{2} = \frac{1}{8}$ of the original concentration remains.

**Plan:** We have to find the time it takes for one-half of the cyclopropane concentration to react, that is, this reaction's half-life. Since the reaction is first order, we use Equation 15.6, substitute for $k$, and solve for $t_{1/2}$.
**Solution:** Solving for $t_{1/2}$:

$$t_{1/2} = \frac{\ln 2}{k} = \frac{0.693}{9.2 \text{ s}^{-1}} = \textbf{0.075 s}$$

It takes 0.075 s for half the cyclopropane to form propene.
**Check:** Rounding, we have $0.7/10 = 0.07$.

**FOLLOW-UP PROBLEM 15.5**
Iodine-123 is used to study thyroid gland function. This radioactive isotope breaks down in a first-order process with a half-life of 13.1 h. What is the rate constant for the process?

In contrast to the half-life of a first-order reaction, the half-life of a second-order reaction *does* depend on reactant concentration:

$$t_{1/2} = \frac{1}{k[A]_0} \qquad \text{(second-order process; rate} = k[A]^2)$$

Note that here *the half-life is inversely proportional to the initial reactant concentration.* This result makes sense: the half-life should be shorter early in the reaction, when more molecules are present to collide with each other. A high initial concentration has a shorter half-life, whereas a low initial concentration has a longer half-life. Therefore, as the reaction proceeds, the half-life of the remaining reactants increases.

Table 15.4 summarizes the essential features of first- and second-order reactions.

TABLE 15.4 **An Overview of First-Order and Simple Second-Order Reactions**

|  | FIRST ORDER | SECOND ORDER |
|---|---|---|
| Rate law | rate = $k[A]$ | rate = $k[A]^2$ |
| Units for $k$ | $1/s$ | $L/mol \cdot s$ |
| Integrated rate law in straight-line form | $\ln [A]_t = -kt + \ln [A]_0$ | $1/[A]_t = kt + 1/[A]_0$ |
| Plot for straight line | $\ln [A]_t$ vs. $t$ | $1/[A]_t$ vs. $t$ |
| Slope, $y$ intercept | $-k, \ln [A]_0$ | $k, 1/[A]_0$ |
| Half-life | $(\ln 2)/k$ | $1/k[A]_0$ |

## The Effect of Temperature on the Rate Constant

The temperature at which a reaction occurs often has a major influence on the rate. As Figure 15.10, *A*, shows for a common organic reaction (hydrolysis, or reaction with water, of an ester), with reactant concentrations held constant the rate nearly doubles with each rise in temperature of 10 K (or 10°C). For many reactions near room temperature, an increase of 10°C causes a doubling or tripling of the rate.

How does the rate law express this effect of temperature on rate? If we collect concentration-time data for the same reaction run at *different* temperatures, and then solve each of the rate expressions for *k*, we find that *k* increases as the temperature *(T)* increases. In other words, *temperature affects the rate by affecting the rate constant*. A plot of *k* vs. *T* gives a curve that increases exponentially (Figure 15.10, *B*).

These results are consistent with studies made in 1889 by the Swedish chemist Svante Arrhenius, who found a negative exponential relationship between temperature and the rate constant. In its modern form, the **Arrhenius equation** is

$$k = A\,e^{-E_a/RT} \tag{15.7}$$

where *k* is the rate constant, *e* is the base of natural logarithms, *T* is the absolute temperature, and *R* is the universal gas constant. We'll discuss the meaning of *A*, which is related to the orientation of the colliding molecules, in the next section. The $E_a$ term is the **activation energy** of the reaction, suggested by Arrhenius as the minimum energy that the molecules must

| Expt | [Ester] | [H₂O] | T(K) | Rate (mol/L•s) | $k$ (L/mol • s) |
|---|---|---|---|---|---|
| 1 | 0.100 | 0.200 | 288 | $1.04 \times 10^{-3}$ | 0.0521 |
| 2 | 0.100 | 0.200 | 298 | $2.02 \times 10^{-3}$ | 0.101 |
| 3 | 0.100 | 0.200 | 308 | $3.68 \times 10^{-3}$ | 0.184 |
| 4 | 0.100 | 0.200 | 318 | $6.64 \times 10^{-3}$ | 0.332 |

**A**

**FIGURE 15.10**

**The dependence of the rate constant on temperature. A,** In the hydrolysis of an ester, when reactant concentrations are held constant and temperature increases, the rate and rate constant increase. Note the approximate doubling of *k* with each 10 K (10°C) temperature rise. **B,** A plot of the rate constant vs. temperature shows a smoothly increasing curve.

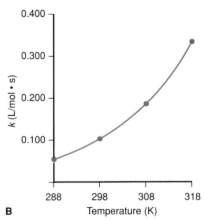

**B**

have to react; we discuss its meaning further in the next section as well. With $R$ in units of J/mol · K, $E_a$ has units of J/mol. Note that as the temperature increases, the negative exponent becomes smaller; thus the value of $k$ becomes larger, which means that the rate increases:

$$\text{Higher } T \Rightarrow \text{larger } k \Rightarrow \text{higher rate}$$

We can calculate $E_a$ from the Arrhenius equation by taking the natural logarithm of both sides and recasting the equation into the form of an equation for a straight line:

$$\ln k = \ln A - \frac{E_a}{R}\left(\frac{1}{T}\right)$$
$$\quad y \;\; = \;\; b \;\; + \;\; mx$$

A plot of $\ln k$ vs. $1/T$ gives a straight line whose slope is $-E_a/R$ and $y$ intercept is $\ln A$ (Figure 15.11). We know the constant $R$, so we can determine $E_a$ graphically from a series of $k$ values at different temperatures.

Since the relationship between $\ln k$ and $1/T$ is linear, we can use an even simpler method for finding $E_a$ if we know the rate constant at two different temperatures, $T_2$ and $T_1$:

$$\ln k_2 = \ln A - \frac{E_a}{R}\left(\frac{1}{T_2}\right) \qquad \ln k_1 = \ln A - \frac{E_a}{R}\left(\frac{1}{T_1}\right)$$

When we subtract $\ln k_1$ from $\ln k_2$, the "$\ln A$" term drops out and the other terms can be rearranged to give

$$\ln \frac{k_2}{k_1} = -\frac{E_a}{R}\left(\frac{1}{T_2} - \frac{1}{T_1}\right) \tag{15.8}$$

From this, we can solve for $E_a$.

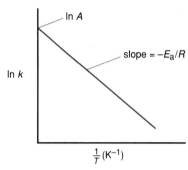

**FIGURE 15.11**
**Graphical determination of the activation energy.** A plot of $\ln k$ vs. $1/T$ gives a straight line with slope $-E_a/R$.

---

**SAMPLE PROBLEM 15.6** _____

### Determining the Energy of Activation

**Problem:** The reaction $2HI(g) \rightarrow H_2(g) + I_2(g)$ has rate constants of $9.51 \times 10^{-9}$ L/mol · s at 500 K and $1.10 \times 10^{-5}$ L/mol · s at 600 K. Find $E_a$.
**Plan:** We are given the rate constants, $k_1$ and $k_2$, at two temperatures, $T_1$ and $T_2$, so we substitute into Equation 15.8 and solve for $E_a$.
**Solution:** Rearranging Equation 15.8 to solve for $E_a$:

$$\ln \frac{k_2}{k_1} = -\frac{E_a}{R}\left(\frac{1}{T_2} - \frac{1}{T_1}\right)$$

$$E_a = -R\left(\ln \frac{k_2}{k_1}\right)\left(\frac{1}{T_2} - \frac{1}{T_1}\right)^{-1}$$

$$= -(8.31 \text{ J/mol · K})\left(\ln \frac{1.10 \times 10^{-5} \text{ L/mol · s}}{9.51 \times 10^{-9} \text{ L/mol · s}}\right)\left(\frac{1}{600 \text{ K}} - \frac{1}{500 \text{ K}}\right)^{-1}$$

$$= 1.76 \times 10^5 \text{ J/mol} = \mathbf{1.76 \times 10^2 \text{ kJ/mol}}$$

**Comment:** Be sure to retain the same number of significant figures in $1/T$ as you have in $T$, or a significant error could be introduced. Round to the correct number of significant figures only at the final answer. On most pocket calculators, the expression $(1/T_2 - 1/T_1)$ is entered as "$(T_2)\,(1/x) - (T_1)\,(1/x) = .$"

**FOLLOW-UP PROBLEM 15.6**
The reaction $2NOCl(g) \rightarrow 2NO(g) + Cl_2(g)$ has an $E_a$ of $1.00 \times 10^2$ kJ/mol and a rate constant of $0.286$ s$^{-1}$ at 500 K. What is the rate constant at 490 K?

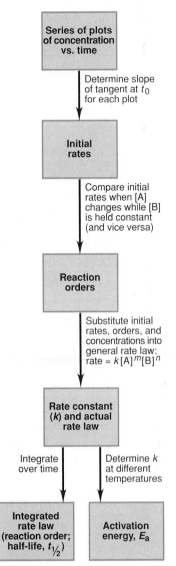

**FIGURE 15.12**

Information sequence to determine the kinetic parameters of a reaction.

In this section and the previous one, we discussed a series of experimental and mathematical methods for the study of reaction kinetics. Figure 15.12 summarizes the sequence of information in this overall approach.

### Section Summary

An experimentally determined rate law of a reaction shows how the rate depends on concentration. If we consider only initial rates, the rate law will often take the form rate = $k[A]^m[B]^n$ . . . . With an accurate method for obtaining initial rates, reaction orders are determined by comparing rates for different initial concentrations. The rate constant is calculated from these experimentally determined variables. Integrated rate laws are used to find the time needed to reach a certain concentration and to determine the reaction order graphically. The half-life is the time needed to consume half the reactant; for first-order reactions, it is independent of concentration. The Arrhenius equation shows the exponential relation between temperature and the rate constant. The activation energy, $E_a$, can be determined from $k$ and $T$ values.

## 15.4 Explaining the Effects of Concentration and Temperature on Reaction Rate

The Arrhenius equation was developed empirically from the observations of many reactions. It took more than 30 years for models to appear that could explain the effects of concentration and temperature on reaction rate. The two major models we consider here are distinct but completely compatible, each highlighting different aspects of the reaction process. **Collision theory** views the reaction rate as the result of particles colliding with a certain frequency and minimum energy. **Transition state theory** offers a close-up view of how the energy converts reactant to product.

### Collision Theory: Basis of the Rate Law

The basic tenet of collision theory is that reactant particles—atoms, molecules, and ions—must collide with each other in order to react. Therefore, the number of collisions per unit time provides an upper limit on how fast a reaction can take place. The model restricts itself to simple one-step reactions in which two particles collide and form products: A + B → products.

**Why reactant concentrations are multiplied in the rate law.** If particles must collide to react, the laws of probability tell us why the rate depends on the *product* of the reactant concentrations, not their sum. Imagine that you have only two particles of A and two of B confined in a reaction vessel. Figure 15.13 shows that four A-B collisions are possible. If you add another particle of A, there can be six A-B collisions (3 × 2), not five (3 + 2); add another particle of B and there can be nine A-B collisions (3 × 3), not six (3 + 3). Thus, the collision model is consistent with the observation that concentrations are *multiplied* in the rate law.

**How temperature affects rate.** Increasing the temperature of a reaction increases the average speed of particles and therefore their collision frequency. But, collision frequency cannot be the only factor affecting rate, or every gaseous reaction would be over instantaneously: the molecules in a milliliter of gas experience about $10^{27}$ collisions per second at 1 atm and 20°C. If every collision resulted in a reaction, our atmosphere, which is 99%

$N_2$ and $O_2$, would consist almost entirely of nitrogen oxides! *In the vast majority of collisions, the molecules rebound without reacting.*

Arrhenius proposed that every reaction has an *energy threshold* that the colliding molecules must exceed in order to react. (An analogy might be a person who must exceed the height of the bar to accomplish a high jump.) This minimum collision energy is the *activation energy ($E_a$)*, the energy required to activate the molecules into a state from which reactant bonds can change into product bonds. Recall that at any given temperature, molecules have a range of kinetic energies; thus, their collisions have a range of energies as well. According to collision theory, *only those collisions with enough energy to exceed $E_a$ can lead to reaction.*

We said earlier that many reactions near room temperature approximately double or triple their rates with a 10°C rise in temperature. However, calculations show that a 10°C rise increases the average molecular speed by only 2%. Because kinetic energy is related to the square of the speed, we would expect only a 4% increase in rate if an increase in speed were the only effect of temperature. Far more importantly, *the temperature rise enlarges the fraction of collisions with enough energy to exceed the activation energy threshold,* as shown in Figure 15.14. This second effect accounts for most of the increase in rate.

At a given temperature, the fraction $f$ of molecular collisions with energy greater than or equal to the activation energy $E_a$ is given by

$$f = e^{-E_a/RT}$$

where $e$ is the base of natural logarithms, $T$ is the absolute temperature, and $R$ is the universal gas constant. Notice that the right side of this equation is the central component in the Arrhenius equation, $k = A\,e^{-E_a/RT}$, which you saw earlier (Equation 15.7). Both $E_a$ and $T$ affect the fraction of collisions. In the top portion of Table 15.5, you can see the effect of increasing $E_a$ at 298 K on the fraction of collisions with enough energy to lead to reaction. Note how much the fraction shrinks with a 25 kJ/mol increase in activation energy. In the bottom portion you can see the effect of $T$ on the fraction for a fixed $E_a$ of 50 kJ/mol, a typical value for many reactions. Note that the fraction of collisions with sufficient energy to lead to reaction nearly doubles for a 10°C increase. This doubling of the fraction doubles the rate constant, which doubles the reaction rate.

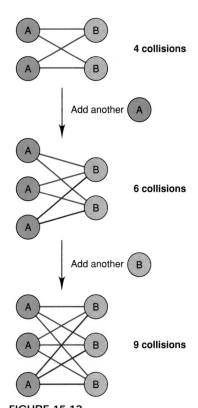

**FIGURE 15.13**
**The dependence of collision number on the product of reactant concentrations.** Concentrations are multiplied, not added, in the rate equation because the number of possible collisions is the *product*, not the sum, of the numbers of particles present.

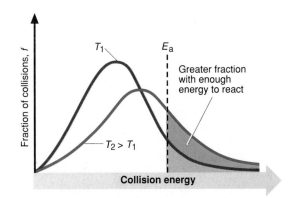

**FIGURE 15.14**
**The effect of temperature on the distribution of collision energies.** At the higher temperature, $T_2$, a larger fraction of collisions occur with enough energy to exceed $E_a$.

**TABLE 15.5 The Effect of $E_a$ and $T$ on the Fraction ($f$) of Collisions with Sufficient Energy to Allow Reaction**

| $E_a$ (kJ/mol) | $f$ (at $T$ = 298 K) |
|---|---|
| 50 | $1.70 \times 10^{-9}$ |
| 75 | $7.03 \times 10^{-14}$ |
| 100 | $2.90 \times 10^{-18}$ |

| $T$ | $f$ (at $E_a$ = 50 kJ/mol) |
|---|---|
| 25°C (298 K) | $1.70 \times 10^{-9}$ |
| 35°C (308 K) | $3.29 \times 10^{-9}$ |
| 45°C (318 K) | $6.12 \times 10^{-9}$ |

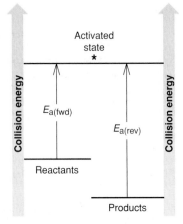

**FIGURE 15.15**

**Energy-level diagram for a reaction.** For molecules to react, they must collide with enough energy to reach an activated state. This minimum collision energy is the energy of activation, $E_a$. Because a reaction can occur in either direction, the diagram shows two activation energies. In the case shown, the forward reaction is exothermic, so $E_{a(fwd)} < E_{a(rev)}$.

A reversible reaction has two activation energies (Figure 15.15). The activation energy for the forward reaction ($E_{a(fwd)}$) is the energy difference between the activated state and the reactants; the one for the reverse reaction ($E_{a(rev)}$) is the energy difference between the activated state and the products. The figure shows an energy-level diagram for an exothermic reaction, so the products are at a lower energy than the reactants, and $E_{a(fwd)}$ is less than $E_{a(rev)}$.

By reorienting the collision-energy distribution curve of Figure 15.14 vertically and aligning it on both sides of Figure 15.15, we obtain Figure 15.16, which shows how temperature affects the fraction of collisions exceeding the activation energy for both the forward and reverse reactions. Several ideas are illustrated in this composite figure:

• In both reaction directions, a larger fraction of collisions exceeds the activation energy at the higher temperature, $T_2$; thus, higher $T$ increases reaction rate.

• For an exothermic process (forward reaction here) at any temperature, the fraction of reactant collisions with energy exceeding $E_{a(fwd)}$ is *larger* than the fraction of product collisions with energy exceeding $E_{a(rev)}$; therefore, the forward reaction is faster. Similarly, in an endothermic process (reverse reaction here), $E_{a(fwd)}$ is greater than $E_{a(rev)}$, so the fraction of product collisions with energy exceeding $E_{a(rev)}$ is larger and the reverse reaction is faster.

These conclusions are consistent with the Arrhenius equation: *the larger the $E_a$, the smaller the $k$, and the slower the reaction.*

**How molecular structure affects rate.** You've seen that the enormous number of collisions per second is greatly reduced when we count only those with enough energy to react. However, even this tiny fraction of the total collisions does not reveal the true number of **effective collisions,** those that actually lead to product. In addition to colliding with enough energy, *the molecules must collide in such a way that the reacting atoms make contact.* In other words, a collision must have enough energy *and* a particular *molecular orientation* to be an effective collision.

In the Arrhenius equation, the effect of molecular orientation is contained in the factor $A$:

$$k = A \, e^{-E_a/RT}$$

**FIGURE 15.16**

**An energy-level diagram of the fraction of collisions reaching the activated state.** A composite of Figure 15.14 aligned vertically on the left and right axes of Figure 15.15 shows that

• In either direction, the fraction of collisions exceeding $E_a$ is greater at a higher temperature.

• In an exothermic reaction at any temperature, the fraction of collisions exceeding $E_{a(fwd)}$ is greater than the fraction exceeding $E_{a(rev)}$.

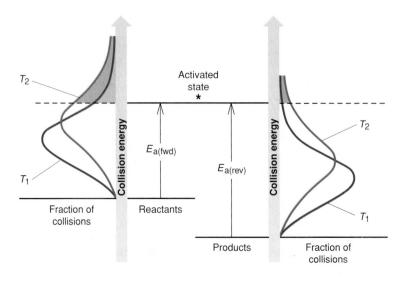

This term is called the **frequency factor,** the product of the collision frequency $Z$ and an *orientation probability factor p,* which is specific for each reaction: $A = pZ$. The factor $p$ is related to the structural complexity of the colliding particles. You can think of it as the ratio of effectively oriented collisions to all possible collisions. Figure 15.17 shows a few of the possible collision orientations for the following simple gaseous reaction:

$$NO(g) + NO_3(g) \rightarrow 2NO_2(g)$$

Of the five collisions shown, only one has an orientation in which the N of NO collides with an O of $NO_3$. Actually, the probability factor ($p$ value) for this reaction is 0.006: only 6 collisions in every thousand (1 in 167) have an orientation that leads to reaction.

Collisions between individual atoms have $p$ values near 1: almost no matter how they hit, they react. In such cases, the rate constant depends only on the frequency and energy of the collisions. At the other extreme are biochemical reactions, in which the reactants are often two small molecules that can react only when they collide with a specific tiny region of a giant molecule—a protein or nucleic acid. The orientation factor for such reactions is often less than $10^{-6}$: fewer than one in a million sufficiently energetic collisions leads to product.

### Transition State Theory: Nature of the Activated State

Collision theory is a simple model that is easy to visualize, but it provides no insight about how the activation energy is utilized and what the activated molecules look like. To understand these aspects of the process, we turn to transition state theory.

Recall from our discussions of energy changes (Chapter 6) that the internal energy of a system is the sum of its kinetic and potential energies. Two molecules that are far apart but speeding toward each other have high kinetic energy and low potential energy. As the molecules become closer, some kinetic energy is converted to potential energy as the electron clouds repel each other. At the moment of a head-on collision, the molecules stop, and their kinetic energy is converted to the potential energy of the collision. If this potential energy does not equal or exceed the activation energy, the molecules rebound, bouncing off each other like billiard balls. Repulsions decrease, speeds increase, and the molecules zoom apart without having reacted.

The tiny fraction of molecules that are oriented effectively *and* moving at the highest speed behave differently. Their kinetic energy pushes them together with enough force to overcome repulsions. Nuclei in one atom attract electrons in another; atomic orbitals overlap and electron density shifts; some bonds lengthen and weaken while others start to form. If we could watch this process in slow motion, we would see the reactant molecules gradually change their bonds and shapes as they turn into product molecules. At some point during this smooth transformation, what exists is neither reactant nor product but a transitional species with partial bonds. This species is extremely unstable (has very high potential energy) and exists only at the instant when the reacting system is highest in energy. It is called the **activated complex,** or **transition state,** of the reaction, and it forms only if the molecules collide in an effective orientation *and* the energy of the collision is equal to or greater than the activation energy. Thus, *the activation energy is used to stretch and deform bonds in order to reach the transition state.* Our knowledge of transition states comes from spectroscopic analysis because these species cannot be isolated.

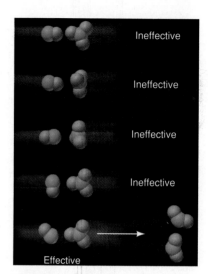

**FIGURE 15.17**
**The importance of molecular orientation in an effective collision.** Only one of the five orientations shown for the collision between NO and $NO_3$ leads to product. In the effective orientation, those atoms collide that will become bonded in the product.

**FIGURE 15.18**

**Nature of the transition state in the reaction between CH₃Br and OH⁻.** Note the partial (elongated) C—O and C—Br bonds and the trigonal bipyramidal shape of the transition state of this reaction.

**FIGURE 15.19**

Reaction energy diagram for the reaction between CH₃Br and OH⁻. A plot of energy vs. reaction progress shows the relative energy levels of reactants, products, and transition state joined by a curved line, as well as the activation energies of the forward and reverse steps. The molecular-level views depict the change at five points. Note the gradual bond forming and bond breaking as the system goes through the transition state.

Consider the reaction between methyl bromide and hydroxide ion:

$$CH_3Br + OH^- \rightarrow CH_3OH + Br^-$$

The electronegative bromine makes the carbon of methyl bromide partially positive. If the reactants are moving toward each other fast enough and oriented effectively when they collide, the negative oxygen in $OH^-$ approaches the carbon with enough energy to begin forming a C—O bond, which causes the C—Br bond to weaken. In the transition state (Figure 15.18), the carbon is surrounded by five atoms in a trigonal bipyramidal shape, which never occurs in stable carbon compounds. This high-energy species consists of three normal C—H bonds and two partial bonds, one to oxygen and the other to bromine.

Reaching the transition state is no guarantee that the reaction will proceed to products; once there, the system can change in either direction. If the C—O bond continues to shorten and strengthen, products will form; however, if the C—Br bond becomes shorter and stronger again, the transition state will decompose back to reactants.

A useful way to depict the events we just described is with a **reaction energy diagram,** which shows the potential energy of the system during the reaction as a smooth curve. Figure 15.19 shows the reaction energy diagram for the methyl bromide–hydroxide ion reaction, with a molecular view at various points during the change. The horizontal axis, labeled "reaction progress" (or reaction coordinate), implies that reactants change to products as we move from left to right. The reaction is exothermic, so the reactants are higher in energy than the products. This energy difference is due to differences in bond energy, which appear as the heat of reaction,

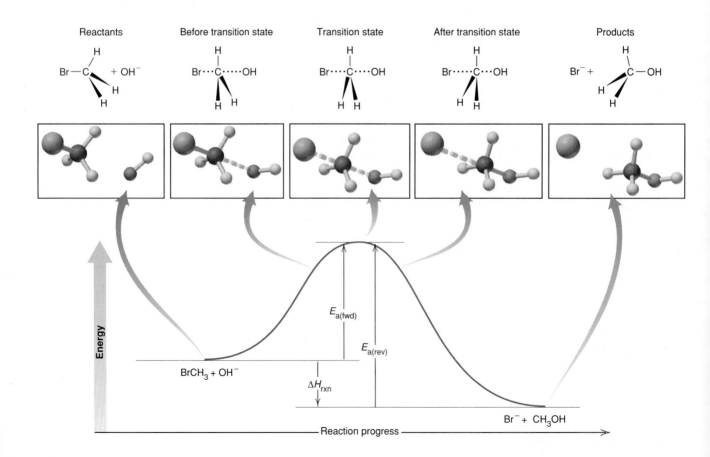

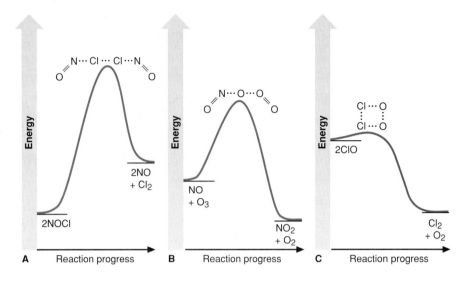

**FIGURE 15.20**
Reaction energy diagrams and possible transition states for three reactions: **A,** $2NOCl(g) \rightarrow 2NO(g) + Cl_2(g)$ (despite the formula NOCl, the atom sequence is ClNO); **B,** $NO(g) + O_3(g) \rightarrow NO_2(g) + O_2(g)$; and **C,** $2ClO(g) \rightarrow Cl_2(g) + O_2(g)$. Note that reaction **A** is endothermic, **B** is exothermic, and **C** has a very small $E_{a(fwd)}$.

$\Delta H$. The diagram also shows the activation energies for the forward and reverse reactions; in this case, $E_{a(fwd)}$ is less than $E_{a(rev)}$.

Transition state theory proposes that *every reaction goes through its own transition state*, from which point it can continue in either direction. Thus, according to this model, *all reactions are reversible*. We can suggest how the activated complex in many reactions might look by examining the reactant and product bonds that undergo change. Figure 15.20 depicts reaction energy diagrams for three simple reactions and the general shape of the postulated activated complex in each case.

---

SAMPLE PROBLEM 15.7

### Drawing Reaction Energy Diagrams

**Problem:** A key reaction in the upper atmosphere is

$$O_3(g) + O(g) \rightleftharpoons 2O_2(g)$$

The $E_{a(fwd)}$ is 19 kJ, and the $\Delta H_{rxn}$ as written is $-392$ kJ. Draw a reaction energy diagram for this reaction and calculate $E_{a(rev)}$.
**Plan:** Since the reaction is highly exothermic, the products are much lower in energy than the reactants. The small $E_{a(fwd)}$ means that the reactant energy lies slightly below that of the transition state. To reach the transition state from the products, we must supply the $\Delta H_{rxn}$ and the $E_{a(fwd)}$; thus, $E_{a(rev)} = -\Delta H_{rxn} + E_{a(fwd)}$.
**Solution:** The reaction energy diagram (not drawn to scale) is

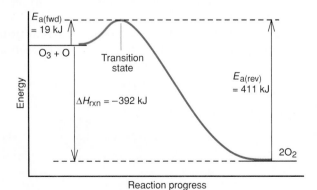

**Check:** Rounding to find $E_{a(rev)}$ gives $\sim390 + 20 = 410$.

**FOLLOW-UP PROBLEM 15.7**

The following reaction energy diagram depicts another key atmospheric reaction. Label the axes, transition state, $E_{a(fwd)}$, $E_{a(rev)}$, and $\Delta H_{rxn}$, and calculate $E_{a(rev)}$, for the reaction.

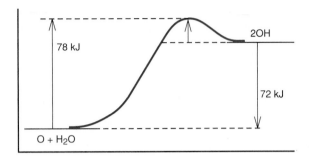

**Section Summary**

According to collision theory, reactant particles must collide to react, and the number of collisions depends on the product of the reactant concentrations. At higher temperatures, more collisions have enough energy to exceed the activation energy ($E_a$). The relative size of the $E_a$ for forward and reverse reactions depends on whether the overall reaction is exothermic or endothermic. Molecules must collide in an effective orientation for bonds to change, so structural complexity decreases rate. Transition state theory pictures the kinetic energy of the particles changing to potential energy during a collision. If the collision energy is equal to or greater than the $E_a$ and their orientation is effective, the molecules are converted to an unstable transition state, which decomposes to either reactant or product.

## 15.5 Reaction Mechanisms: Steps in the Overall Reaction

Imagine trying to figure out how a car works just by examining the body, wheels, and dashboard. It can't be done—you need to look under the hood, peer inside the engine, and see how the parts fit together and function. Similarly, if you want to know how a reaction works *at the molecular level,* examining the overall balanced equation is not much help—you need to "look under the yield arrow and peer inside the reaction" to see how reactants change into products.

You would find that most reactions occur through a **reaction mechanism,** a series of simpler reactions that sum to the overall reaction. For example, a mechanism for the overall reaction

$$2A + B \longrightarrow E + F$$

might involve these three simpler steps:

(1) $A + B \longrightarrow C$

(2) $C + A \longrightarrow D$

(3)     $D \longrightarrow E + F$

When added together, they give the overall reaction:

$$A + B + \cancel{C} + A + \cancel{D} \rightarrow \cancel{C} + \cancel{D} + E + F \quad \text{or} \quad 2A + B \rightarrow E + F$$

Chemists *propose* a reaction mechanism to explain how a given reaction might occur. This section focuses on the nature of the individual steps and how they fit together to give a rate law consistent with experiment.

### Elementary Reactions and Molecularity

The individual steps, which together make up the proposed reaction mechanism, are called **elementary reactions,** or **elementary steps.** Each describes a *single molecular event,* such as one molecule decomposing or two molecules combining. An elementary step is *not* made up of simpler steps.

An elementary step is characterized by its **molecularity,** the number of reactant particles involved in the step. Consider the mechanism for the breakdown of ozone in the stratosphere. The overall reaction is

$$2O_3(g) \rightarrow 3O_2(g)$$

A two-step mechanism has been proposed for this reaction. The first elementary step is a **unimolecular reaction,** one that involves the decomposition or rearrangement of a single particle:

(1) $O_3(g) \rightarrow O_2(g) + O(g)$

The second step is a **bimolecular reaction,** one in which two particles react:

(2) $O_3(g) + O(g) \rightarrow 2O_2(g)$

Some *termolecular* elementary steps occur, but they are rare because the probability of three particles colliding simultaneously with enough energy and with an effective orientation is quite small. Higher molecularities are not known. Unless evidence exists to the contrary, it always makes good chemical sense to propose unimolecular or bimolecular reactions as elementary steps in a mechanism.

The rate law of an elementary reaction, unlike that of an overall reaction, *can* be deduced from the reaction stoichiometry. Since an elementary reaction occurs in one step, its rate must be proportional to the product of the reactant concentrations. Therefore, we use the equation coefficients as the reaction orders in its rate law (Table 15.6). In other words, *for an elementary reaction, the reaction order equals the molecularity.* Remember that this statement holds *only* when we know that the reaction is elementary; you've already seen that for an overall reaction, the reaction orders must be determined experimentally.

**TABLE 15.6 Rate Laws for General Elementary Steps**

| ELEMENTARY STEP | MOLECULARITY | RATE LAW |
|---|---|---|
| $A \rightarrow$ product | Unimolecular | Rate = $k[A]$ |
| $2A \rightarrow$ product | Bimolecular | Rate = $k[A]^2$ |
| $A + B \rightarrow$ product | Bimolecular | Rate = $k[A][B]$ |
| $2A + B \rightarrow$ product | Termolecular | Rate = $k[A]^2[B]$ |

SAMPLE PROBLEM 15.8 _____

## Determining Molecularity and Rate Laws of Elementary Steps

**Problem:** The following two reactions are proposed as elementary steps in the mechanism for an overall reaction:

(1) $NO_2Cl(g) \rightarrow NO_2(g) + Cl(g)$

(2) $NO_2Cl(g) + Cl(g) \rightarrow NO_2(g) + Cl_2(g)$

**(a)** Write the overall balanced equation.
**(b)** What is the molecularity of each step?
**(c)** Write the rate law for each step.
**Plan:** The overall equation is the sum of the elementary steps. The molecularity of each step equals the total number of reactant particles. The rate law for each step uses the molecularities as reaction orders.
**Solution: (a)** Finding the overall equation:

(1) $\qquad\qquad NO_2Cl(g) \rightarrow NO_2(g) + Cl(g)$

(2) $\qquad\qquad NO_2Cl(g) + Cl(g) \rightarrow NO_2(g) + Cl_2(g)$

$\qquad\qquad NO_2Cl(g) + NO_2Cl(g) + \cancel{Cl(g)} \rightarrow NO_2(g) + \cancel{Cl(g)} + NO_2(g) + Cl_2(g)$

or $\qquad\qquad$ **$2NO_2Cl(g) \rightarrow 2NO_2(g) + Cl_2(g)$**

**(b)** Determining the molecularity: The first elementary step has only one reactant, $NO_2Cl$, so it is **unimolecular.** The second elementary step has two reactants, $NO_2Cl$ and $Cl$, so it is **bimolecular.**
**(c)** Writing rate laws for the elementary reactions:

(1) **Rate$_1$ = $k_1$[NO$_2$Cl]**

(2) **Rate$_2$ = $k_2$[NO$_2$Cl][Cl]**

**Check:** In **(a),** be sure the equation is balanced; in **(c),** be sure the substances in brackets are the reactants.

FOLLOW-UP PROBLEM 15.8
The following elementary steps constitute a proposed mechanism for a reaction:

(1) $\qquad 2NO(g) \rightarrow N_2O_2(g)$

(2) $\qquad 2H_2(g) \rightarrow 4H(g)$

(3) $N_2O_2(g) + H(g) \rightarrow N_2O(g) + HO(g)$

(4) $\quad HO(g) + H(g) \rightarrow H_2O(g)$

(5) $\quad H(g) + N_2O(g) \rightarrow HO(g) + N_2(g)$

(6) $\quad HO(g) + H(g) \rightarrow H_2O(g)$

**(a)** Write the balanced equation for the overall reaction.
**(b)** What is the molecularity of each step?
**(c)** Write the rate law for each step.

## The Rate-Limiting Step of a Reaction Mechanism

All the elementary reactions in a mechanism do not have the same rates. Usually, one of the steps is much slower than the others and thus limits how fast the overall reaction can proceed. This step is called the **rate-limiting step,** or **rate-determining step. ◆**

◆ **Sleeping Through the Rate-Limiting Step.** Baking provides a tasty, macroscopic example of a rate-limiting step. Five steps in making a loaf of French bread are (1) mixing the ingredients (15 min), (2) kneading the dough (8 min), (3) letting the dough rise in a refrigerator (400 min), (4) shaping the loaf (2 min), and (5) baking the loaf (25 min). Obviously, the dough-rising step limits how fast a baker can produce a loaf because it is so much slower than the other steps. Keenly aware of the baking "mechanism," bakers let the dough rise overnight, thereby sleeping through the rate-limiting step.

The importance of the slowest step in the reaction is that *the rate law for the rate-determining step represents the rate law for the overall reaction.* Consider the reaction between nitrogen dioxide and carbon monoxide:

$$NO_2(g) + CO(g) \rightarrow NO(g) + CO_2(g)$$

If the overall reaction were an elementary reaction—that is, if the mechanism consisted of only one step—we could immediately write the overall rate law as

$$Rate = k[NO_2][CO]$$

However, experiment shows that the actual rate law is

$$Rate = k[NO_2]^2$$

From this, we know immediately that the overall reaction cannot be elementary.

A proposed two-step mechanism for the reaction is

(1)  $NO_2(g) + NO_2(g) \rightarrow NO_3(g) + NO(g)$     [slow; rate-determining]

(2)   $NO_3(g) + CO(g) \rightarrow NO_2(g) + CO_2(g)$    [fast]

Rate laws for these elementary steps are

(1)  Rate = $k_1[NO_2][NO_2] = k_1[NO_2]^2$

(2)  Rate = $k_2[NO_3][CO]$

The key point here is that if $k_1 = k$, *the rate law for the rate-determining step (step 1) is identical to the experimental rate law.* The first step is so slow compared with the second that the overall reaction takes essentially as long as the first step.

It is also important to consider what happens to $NO_3$ in this mechanism. It is a product in step 1 and a reactant in step 2. Therefore, it functions here as a **reaction intermediate,** a substance that is formed and used up during the overall reaction. It does not appear in the overall balanced equation, and its existence is temporary; nevertheless, its presence is absolutely necessary for the reaction to occur. Reaction intermediates are usually unstable relative to the reactants and products, but, unlike transition states (activated complexes), they are molecules with normal bonds and are sometimes stable enough to be isolated.

### Constructing the Mechanism from the Rate Law

Conjuring up a reasonable reaction mechanism is one of the most exciting aspects of chemical kinetics and can be an excellent example of the scientific method. We use observations and data from rate experiments to hypothesize what the individual steps might be and then test our hypothesis by gathering further evidence. If the evidence supports it, we continue to use that mechanism; if not, we propose a new one. However, *we can never prove, just from the data, that a particular mechanism is the way the chemical change actually occurs.*

Regardless of the elementary steps that are proposed for the mechanism, they must obey three criteria:
1. *The elementary steps must add up to the overall equation.* We cannot wind up with more (or fewer) reactants or products than are present in the balanced equation.

2. *The elementary steps must be physically reasonable.* As we noted, most steps should involve one reactant particle (unimolecular) or two (bimolecular). Steps with three reactant particles (termolecular) are very unlikely.

3. *The mechanism must be consistent with the rate law.* A mechanism must support the experimental facts embodied in the rate law, not the other way around.

Let's see how the mechanisms of several reactions conform to these criteria and how the elementary steps fit together.

**Mechanisms with a slow initial step.** We've already seen one mechanism with a rate-determining first step in the $NO_2$-CO reaction shown previously. Another example is the reaction between nitrogen dioxide and fluorine gas:

$$2NO_2(g) + F_2(g) \longrightarrow 2NO_2F(g)$$

The experimental rate law is first order in $NO_2$ and in $F_2$:

$$Rate = k[NO_2][F_2]$$

The currently accepted mechanism for the reaction is

(1)  $NO_2(g) + F_2(g) \longrightarrow NO_2F(g) + F(g)$   [slow; rate-determining]

(2)   $NO_2(g) + F(g) \longrightarrow NO_2F(g)$            [fast]

Molecules of reactant and product appear in both elementary steps. The free fluorine atom is a reaction intermediate.

This mechanism meets the first criterion because the elementary reactions sum to the balanced equation:

$$NO_2(g) + NO_2(g) + F_2(g) + \cancel{F(g)} \longrightarrow NO_2F(g) + NO_2F(g) + \cancel{F(g)}$$

or                              $$2NO_2(g) + F_2(g) \longrightarrow 2NO_2F(g)$$

Both steps are bimolecular, so they are chemically reasonable, which satisfies the second criterion. For the mechanism to meet the third criterion, we must show that it gives the rate law of the overall equation. The rate laws for the elementary steps are

(1)  $Rate_1 = k_1[NO_2][F_2]$

(2)  $Rate_2 = k_2[NO_2][F]$

Step 1 is the rate-determining step and therefore gives the overall rate law, with $k_1 = k$. Since the second molecule of $NO_2$ appears in the step that follows the rate-determining step, it does not appear in the overall rate law. In other words, *the overall rate law includes only species up to and including those in the rate-determining step.* We saw this point in the $NO_2$-CO mechanism earlier. Carbon monoxide was absent from the overall rate law because it appeared *after* the rate-determining step.

Figure 15.21 shows a reaction energy diagram for the $NO_2$-$F_2$ reaction. Notice that *each step in the mechanism has its own transition state.* The free F atom intermediate is a reactive, unstable species (as you know from halogen chemistry), so it is higher in energy than reactants or product. The first step is slower (rate limiting), so its activation energy is *larger* than that of the second. This reaction is exothermic, so the product is lower in energy than the reactants.

**Mechanisms with a fast initial step.** If the rate-limiting step in a mechanism is not first, it acts as a bottleneck. The product of the fast initial step builds up, waiting for the slow step to remove it, and starts reverting to

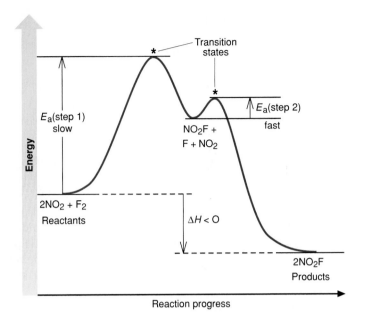

**FIGURE 15.21**
**Reaction energy diagram for the two-step $NO_2$-$F_2$ reaction.** Each step in the mechanism has its own transition state. Reactants for the second step include the intermediate F atom. Note that the first step is slower (higher $E_a$).

reactant. In other words, the *reversibility of the fast initial step becomes evident.* With time, the product of the fast step is changing back to reactant as fast as it is forming, and equilibrium is established. As you'll see, this situation allows us to fit the mechanism to the overall rate law.

Consider once again the oxidation of nitric oxide:

$$2NO(g) + O_2(g) \rightarrow 2NO_2(g)$$

The experimentally determined rate law is

$$\text{Rate} = k[NO]^2[O_2]$$

and a proposed mechanism is

(1)  $NO(g) + O_2(g) \rightleftharpoons NO_3(g)$    [fast, reversible]

(2)  $NO_3(g) + NO(g) \rightarrow 2NO_2(g)$    [slow; rate determining]

Note that, with cancellation of the intermediate $NO_3$, the sum of the steps is the overall equation, and that both steps are bimolecular. The rate laws for the elementary steps are

(1)  $\text{Rate}_{1(\text{fwd})} = k_1[NO][O_2]$

$\text{Rate}_{1(\text{rev})} = k_{-1}[NO_3]$

where $k_{-1}$ is the rate constant for the reverse reaction.

(2)  $\text{Rate}_2 = k_2[NO_3][NO]$

Now we must show that the rate law for the rate-determining step (step 2) gives the overall rate law. As written it does not, because it contains the intermediate $NO_3$, and *an overall rate law may include only substances in the overall equation.* We can eliminate $[NO_3]$ from the step 2 rate law by expressing it in terms of reactants. Step 1 reaches equilibrium when the forward and reverse rates are equal:

$$\text{Rate}_{1(\text{fwd})} = \text{rate}_{1(\text{rev})}$$

or

$$k_1[NO][O_2] = k_{-1}[NO_3]$$

Solving for [$NO_3$], we obtain

$$[NO_3] = \frac{k_1}{k_{-1}}[NO][O_2]$$

Substituting for [$NO_3$] in the rate law for step 2, we obtain

$$Rate_2 = \frac{k_2 k_1}{k_{-1}}[NO][O_2][NO] = \frac{k_2 k_1}{k_{-1}}[NO]^2[O_2]$$

This rate law is identical to the overall rate law with $k = \frac{k_2 k_1}{k_{-1}}$.

To summarize the approach for a mechanism with a fast initial, reversible step:
- Write rate laws for both directions of the fast step and for the slow step.
- Show that the slow step's rate law is equivalent to the overall rate law, by *expressing the [intermediate] in terms of [reactant]:* set the forward rate law of the fast, reversible step equal to the reverse rate law, and solve for [intermediate].
- Substitute the [intermediate] into the rate law for the slow step to obtain the overall rate law.

**Alternative mechanisms for the same reaction.** Even when a mechanism is consistent with the rate law, later experimentation may show it to be incorrect or only one of several alternatives. As an example, consider the reaction between hydrogen and iodine:

$$H_2(g) + I_2(g) \rightarrow 2HI(g); \qquad rate = k[H_2][I_2]$$

The long-accepted mechanism proposed a single bimolecular step; that is, the overall reaction was thought to be elementary. In the 1960s, however, spectroscopic evidence showed the presence of free I atoms during the reaction. This observation eliminates the one-step mechanism as a possibility, at least under those experimental conditions. Kineticists have since proposed a three-step mechanism instead, as shown in Sample Problem 15.9.

SAMPLE PROBLEM 15.9 ───────────────────────────────────

**Confirming a Reaction Mechanism with a Fast Initial Step**

**Problem:** The kinetics of the gas-phase reaction between $H_2$ and $I_2$,

$$H_2(g) + I_2(g) \rightarrow 2HI(g); \quad rate = k[H_2][I_2]$$

has been studied extensively. The accepted mechanism is

(1)           $I_2(g) \rightleftharpoons 2I(g)$    [fast, reversible]

(2)  $H_2(g) + I(g) \rightleftharpoons H_2I(g)$    [fast, reversible]

(3)  $H_2I(g) + I(g) \rightarrow 2HI(g)$    [slow; rate limiting]

Show that the mechanism is consistent with the rate law.

**Plan:** We approach the problem as just outlined for the $NO$-$O_2$ reaction, writing the rate law for the rate-limiting step and expressing intermediate concentrations in terms of reactant concentrations.

**Solution:** Writing the rate law for the rate-limiting step (step 3):

$$Rate_3 = k_3[H_2I][I]$$

Both species are intermediates.

Expressing [I] in terms of reactant concentrations: Since step 1 is reversible, we have

$$Rate_{1(fwd)} = rate_{1(rev)}$$

or
$$k_1[I_2] = k_{-1}[I]^2$$

Thus
$$[I]^2 = \frac{k_1}{k_{-1}}[I_2], \quad \text{so } [I] = \left(\frac{k_1}{k_{-1}}[I_2]\right)^{1/2}$$

Expressing $[H_2I]$ in terms of reactant concentrations: Following the same approach with step 2, we obtain

$$[H_2I] = \frac{k_2}{k_{-2}}[H_2][I]$$

Substituting the preceding relationship for $[I]$, we have

$$[H_2I] = \frac{k_2}{k_{-2}}[H_2] \times \left(\frac{k_1}{k_{-1}}[I_2]\right)^{1/2}$$

Substituting the expressions for $[H_2I]$ and $[I]$ into the rate law for the rate-limiting step (step 3):

$$\text{Rate}_3 = k_3[H_2I][I] = k_3 \times \left\{\frac{k_2}{k_{-2}}[H_2] \times \left(\frac{k_1}{k_{-1}}[I_2]\right)^{1/2}\right\} \times \left(\frac{k_1}{k_{-1}}[I_2]\right)^{1/2}$$

Grouping the constants together, we obtain

$$\textbf{Rate}_3 = k_3 \frac{k_2}{k_{-2}} \frac{k_1}{k_{-1}}[H_2][I_2]$$

This is identical to the overall rate law with $k = k_3 \dfrac{k_2}{k_{-2}} \dfrac{k_1}{k_{-1}}$

**Check:** Be sure that the fully substituted rate law for this reaction contains only *reactant* concentrations and a group of rate constants.

**FOLLOW-UP PROBLEM 15.9**
Experiment shows that the rate of formation of carbon tetrachloride from chloroform,

$$CHCl_3(g) + Cl_2(g) \rightarrow CCl_4(g) + HCl(g)$$

is first order in $CHCl_3$, $\frac{1}{2}$ order in $Cl_2$, and $\frac{3}{2}$ order overall. Is the following mechanism consistent with the overall rate law?

| | | |
|---|---|---|
| (1) | $Cl_2(g) \rightleftharpoons 2Cl(g)$ | [fast, reversible] |
| (2) | $Cl(g) + CHCl_3(g) \rightarrow HCl(g) + CCl_3(g)$ | [rate limiting] |
| (3) | $CCl_3(g) + Cl(g) \rightarrow CCl_4(g)$ | [fast] |

**Section Summary**
The mechanisms of most common reactions consist of two or more elementary steps, reactions that occur in one step and depict a single chemical change. The molecularity of an elementary step equals the number of reactant particles and is the same as the reaction order of its rate law. Unimolecular and bimolecular steps are common. The rate-limiting (slowest) step determines how fast the overall reaction occurs, and its rate law represents the overall rate law. Intermediates are species that form in one step and react in a later one. The steps in a proposed mechanism must add up to the overall reaction, be physically reasonable, and conform to the overall rate law. If a fast step precedes the slow step, the fast step reaches equilibrium, and the concentrations of intermediates in the rate law of the slow step must be expressed in terms of reactants.

## 15.6 Catalysis: Speeding Up a Chemical Reaction

There are many situations in which the rate of a reaction must be increased for it to be useful. In an industrial process, for example, a higher rate may determine whether a new product can be made economically. Sometimes a higher temperature will speed up the reaction sufficiently, but energy is costly and many substances are heat sensitive and easily decomposed. Alternatively, we can often employ a **catalyst,** a substance that increases the rate *without* being consumed in the reaction. More than 2.5 million tons of industrial catalysts are used annually in the United States alone! However, nature is the major user of catalysts: even the simplest bacterial cell employs thousands of biological catalysts, known as *enzymes,* to speed up the cellular reactions necessary for life.

Each catalyst has its own specific way of functioning, but in general, *a catalyst speeds up a reaction by making the activation energy lower,* which makes the rate constant larger and the rate higher. Two important points stand out in Figure 15.22:

1. A catalyst speeds up the forward *and* reverse reactions. *A reaction with a catalyst does not yield more product, but it yields the product more quickly.*
2. A catalyst lowers the activation energy by providing a *different mechanism* for the reaction, a new lower energy pathway.

Suppose that a general uncatalyzed reaction proceeds by a one-step mechanism involving a bimolecular collision:

$$A + B \longrightarrow product \ [slower]$$

In the catalyzed reaction, the reactants interact with the catalyst, so the mechanism might involve a two-step pathway:

$$A + catalyst \longrightarrow C \ [faster]$$

$$C + B \longrightarrow product + catalyst \ [faster]$$

**FIGURE 15.22**

**Reaction energy diagram of a catalyzed and an uncatalyzed process.** A catalyst speeds a reaction by providing a new lower energy path, in this case by replacing the one-step mechanism with a two-step mechanism. Both forward and reverse rates are increased to the same extent, so a catalyst does not affect the overall reaction yield. (Only the larger of the two activation energies is shown for each direction of the catalyzed reaction.)

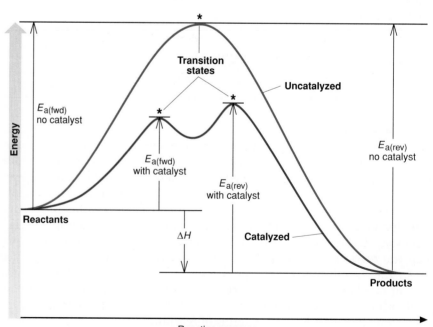

Note that the catalyst is not consumed, as its definition requires. Rather, it is regenerated, and the activation energies of both steps are lower than the activation energy of the uncatalyzed path.

There are two broad categories of catalyst—homogeneous and heterogeneous—depending on whether or not the catalyst is in the same phase as the reactant and product.

## Homogeneous Catalysis

A **homogeneous catalyst** exists in solution with the reaction mixture. Most homogeneous catalysts are gases, liquids, or soluble solids. Some industrial processes that employ these catalysts are shown in Table 15.7 (top).

A thoroughly studied example of homogeneous catalysis is the hydrolysis of an organic ester ($R—\overset{\displaystyle O}{\overset{\|}{C}}—O—R'$), a reaction we examined briefly in Section 14.4:

$$R—\overset{O}{\overset{\|}{C}}—O—R' + H_2O \rightleftharpoons R—\overset{O}{\overset{\|}{C}}—OH + R'—OH$$

Here R and R' are hydrocarbon groups, $R—\overset{O}{\overset{\|}{C}}—OH$ is an organic acid, and R'—OH is an alcohol. The reaction rate is low at room temperature but can be increased by adding a small amount of strong inorganic acid, which provides the catalyst in the reaction, $H^+$ ion.

In the first step of the catalyzed reaction (Figure 15.23), the $H^+$ ion forms a bond to the O atom that was double bonded to the C atom. This makes the C atom more positive, which increases its attraction for the partially negative O atom of water, thus enhancing their bonding, which is the rate-determining step. Several steps later, $H^+$ returns to solution. Thus, $H^+$ acts

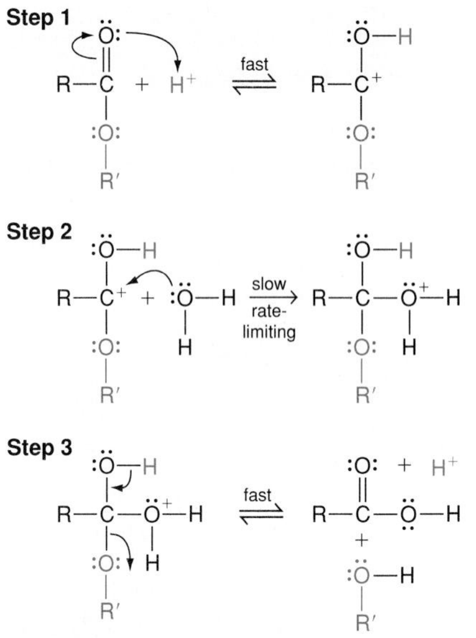

**FIGURE 15.23**
**Mechanism for the catalyzed hydrolysis of an organic ester.** In step 1, the catalytic $H^+$ ion binds to the electron-rich oxygen, making the C atom more positive than it would ordinarily be in the ester. This enhanced charge on C attracts the partially negative O of water more strongly, increasing the fraction of effective collisions and thus speeding step 2, the rate-limiting step. Loss of R'OH and release of $H^+$ occurs in a final series of fast steps.

**TABLE 15.7 Some Modern Processes Based on Catalysis**

| REACTANTS | CATALYST | PRODUCT | USE |
|---|---|---|---|
| **Homogeneous** | | | |
| Propylene, oxidizer | Mo(VI) complexes | Propylene oxide | Polyurethanes (foams); polyesters (plastics) |
| Methanol, CO | $[Rh(CO)_2I_2]^-$ | Acetic acid | Vinyl acetate (coatings); polyvinyl alcohol |
| Butadiene, HCN | Ni/P compounds | Adiponitrile | Nylon (fibers, plastics) |
| $\alpha$-Olefins, CO, $H_2$ | Rh/P compounds | Aldehydes | Plasticizers, lubricants |
| **Heterogeneous** | | | |
| Ethylene, $O_2$ | Silver, cesium chloride on alumina | Ethylene oxide | Polyesters, ethylene glycol, lubricants |
| Propylene, $NH_3$, $O_2$ | Bismuth molybdates | Acrylonitrile | Plastics, fibers, resins |
| Ethylene | Organochromium and titanium halides on silica | High-density polyethylene | Molded products |
| Propylene | Titanium chloride alkyl aluminum on magnesium chloride | Polypropylene | Plastics, fibers, films |

**Chemistry in Atmospheric Science**
# Depletion of the Earth's Ozone Layer

Homogeneous catalysis takes part in one of the most serious environmental problems of our time: the depletion of ozone from the stratosphere. The stratospheric ozone layer is vital to life on Earth because ozone absorbs short-wavelength (about $10^{-8}$ m) ultraviolet (UV) radiation from the sun:

$$O_3(g) + \text{UV photon} \rightarrow O_2(g) + O(g)$$

This radiation has enough energy to break bonds in deoxyribonucleic acid (DNA) and thereby damage genes. Were large amounts of UV radiation to reach the Earth's surface, mutations and cancers would increase dramatically, and the planet's food chain could be disrupted.

Under normal conditions, the stratospheric ozone concentration varies seasonally but remains nearly constant annually through a series of complex atmospheric reactions. Two reactions that maintain a balance in ozone concentration are

$$O_2(g) + O(g) \rightarrow O_3(g) \text{ [formation]}$$

$$O_3(g) + O(g) \rightarrow 2O_2(g) \text{ [breakdown]}$$

However, release of industrially produced chlorofluorocarbons (CFCs) has shifted this balance by catalyzing the breakdown reaction.

The CFCs are used as air-conditioner refrigerants, plastic foam reagents, and aerosol propellants. They are chemically inert and very stable in the lower atmosphere because the ozone layer shields them from high-energy, molecule-splitting UV photons. Once aloft, the CFCs reach the stratosphere, absorb UV photons, and release Cl atoms:

$$CF_2Cl_2(g) \xrightarrow{\text{UV photon}} \cdot CF_2Cl(g) + \cdot Cl(g)$$

The dots ($\cdot$) represent unpaired electrons resulting from bond breakage.

Like many species with unpaired electrons, atomic chlorine is very reactive. The Cl atoms react with stratospheric ozone to produce the intermediate chlorine monoxide ($\cdot ClO$), which then reacts with free O atoms to regenerate Cl atoms:

$$O_3(g) + \cdot Cl(g) \rightarrow \cdot ClO(g) + O_2(g)$$

$$\cdot ClO(g) + O(g) \rightarrow \cdot Cl(g) + O_2(g)$$

The sum of these two elementary reactions yields the overall reaction of ozone breakdown:

$$O_3(g) + \cdot \cancel{Cl(g)} + \cdot \cancel{ClO(g)} + O(g) \rightarrow$$

$$\cdot \cancel{ClO(g)} + O_2(g) + \cdot \cancel{Cl(g)} + O_2(g)$$

or     $$O_3(g) + O(g) \rightarrow 2O_2(g)$$

The Cl atom acts here as a typical homogeneous catalyst: it exists in the same phase as the reactants, speeds up a process via a different mechanism, and is regenerated to continue this role. Measurements of high ClO concentrations over the Antarctic are consistent with such a mechanism, although nitrogen-containing species and ice crystals in high-altitude clouds are thought to be involved as well and present a more complicated picture. The crux of the problem is that each Cl atom has a stratospheric lifetime of about 2 years, during which time it destroys about 100,000 ozone molecules.

Atmospheric scientists have documented more than 80% ozone depletion over the South Pole during certain times of the year (Figure 15.D), and there is evidence of ozone thinning over the North Pole as well.

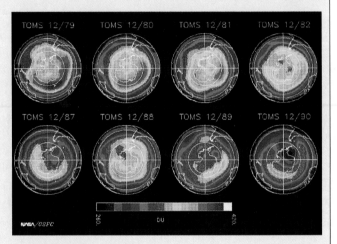

**FIGURE 15.D**
**The increasing size of the Antarctic ozone "hole."** These satellite images show changes in ozone concentration over the South Pole for several years from December 1979 through 1990. Note the enlarging "hole" in the ozone layer *(purple)*. More recent data show the hole increasing further and a similar thinning beginning to form over the North Pole.

The most recent data from NASA indicate a thinning of ozone over the temperate zones during summer months, when people are lightly clothed and outdoors. More than 1 million tons of CFCs are released from air conditioners and spray cans each year. International agreements plan to phase out CFC production before the end of the century, but this plan brings up other questions: Can we get along with less air conditioning? Are other refrigerants ($NH_3$, hydrofluorocarbons) acceptable alternatives?

Even if CFC production were to stop now, however, the amounts already released that have not yet reached the stratosphere will continue to deplete ozone for the next 100 years!

as a catalyst because it speeds up the reaction but is not itself consumed: it is used up in one step and re-formed in another. The Chemical Connections essay on the left page provides a troubling example of homogeneous catalysis in our atmosphere.

### Heterogeneous Catalysis

A **heterogeneous catalyst** speeds up a reaction that occurs in a separate phase. The catalyst is most often a solid interacting with gaseous or liquid reactants. Since reaction occurs on the solid's surface, heterogeneous catalysts usually have enormous surface areas, between 1 and 500 $m^2/g$. Table 15.7 (bottom) lists some polymer manufacturing processes that employ heterogeneous catalysts.

One of the most important examples of heterogeneous catalysis is the addition of hydrogen to the $C{=}C$ bonds of organic compounds to form $C{-}C$ bonds. The petroleum, plastics, and food industries use catalytic **hydrogenation** to change a great variety of compounds into more useful substances. The conversion of vegetable oil into margarine is one example.

The simplest hydrogenation converts ethylene to ethane:

$$H_2C{=}CH_2(g) + H_2(g) \rightarrow H_3C{-}CH_3(g)$$

In the absence of a catalyst, the reaction occurs very slowly. At high $H_2$ pressure in the presence of finely divided nickel, palladium, or platinum, the reaction becomes rapid even at ordinary temperatures. These Group 8B(10) metals catalyze by *chemically adsorbing the reactants onto their surface* (Figure 15.24). Once there, $H_2$ splits into separate H atoms chemically bound to the solid catalyst's metal atoms (catM):

$$H{-}H(g) + 2catM(s) \rightarrow 2catM{-}H \text{ (H atoms bound to metal surface)}$$

The H atoms migrate over the surface, eventually collide with a bound $C_2H_4$ molecule, and the reaction takes place. The H—H bond breakage is the rate-determining step in the overall process, and interaction with the catalyst's

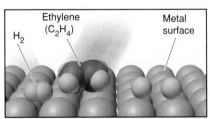
① $H_2$ and $C_2H_4$ approach and adsorb to metal surface.

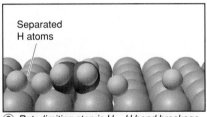
② Rate-limiting step is H—H bond breakage.

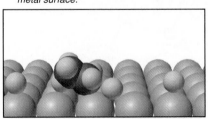
③ One H atom bonds to adsorbed $C_2H_4$.

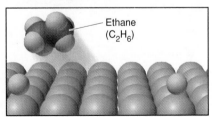
④ Another C—H bond forms and $C_2H_6$ is released.

**FIGURE 15.24**
The metal-catalyzed hydrogenation of ethylene.

# Chemistry in Enzymology
## Kinetics and Function of Biological Catalysts

Within every living cell, thousands of individual reactions occur. Many involve complex chemical changes, yet they take place in dilute solution at ordinary temperatures and pressures. The rate of each reaction responds smoothly to momentary or permanent changes in other reaction rates, various signals from other cells, and environmental stresses. Virtually every reaction in this marvelous chemical harmony is catalyzed by a specific **enzyme,** a protein whose catalytic function has been perfected through evolution.

Enzymes have complex three-dimensional shapes and molar masses ranging from about 15,000 to 1,000,000 g/mol (Section 14.5). At a specific point on an enzyme's surface is its **active site,** a molecular crevice whose shape results from those of the amino acid side chains involved in catalyzing the reaction. When the reactant molecules, called the **substrates** of the reaction, collide

with the active site, the chemical change takes place. The active site makes up only a small part of the enzyme's surface—like a tiny hollow carved into a mountainside—and may include amino acid groups from distant regions of the protein (Figure 15.E). *In most cases, substrates bind to the active site through intermolecular forces:* H bonds, dipole forces, and other weak attractions.

An enzyme has features of both a homogeneous and a heterogeneous catalyst. Most enzymes are enormous compared with their substrates, and they are often found embedded within membranes of larger cell parts. Thus, like a heterogeneous catalyst, an enzyme provides an active surface on which a reactant is immobilized temporarily, waiting for its reaction partner to land nearby. Like a homogeneous catalyst, the enzyme's amino acid groups interact actively with the substrates in multistep sequences involving intermediates.

**FIGURE 15.E**

**The widely separated amino acid groups that form an active site.** The amino acids in chymotrypsin, a digestive enzyme, are shown as linked spheres and numbered consecutively from the beginning of the chain. The R groups of the active-site amino acids 57 (histidine, His), 102 (aspartic acid, Asp), and 195 (serine, Ser) are shown because they play a crucial role in the enzyme-catalyzed reaction. Part of the substance lies within a binding pocket to anchor it during reaction.

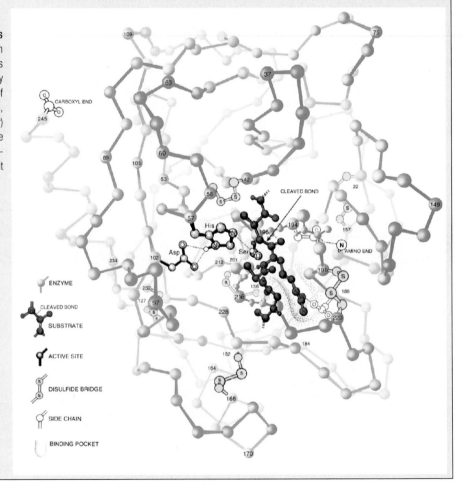

Enzymes are incredibly *efficient* catalysts. Consider the hydrolysis of urea as an example:

$$(NH_2)_2C{=}O(aq) + 2H_2O(l) + H^+(aq) \rightarrow$$

$$2NH_4^+(aq) + HCO_3^-(aq)$$

In water at room temperature, the rate constant for the uncatalyzed reaction is approximately $3 \times 10^{-10}$ s$^{-1}$. Under the same conditions in the presence of the enzyme *urease* (pronounced "*yur*-ee-ase"), the rate constant is $3 \times 10^4$ s$^{-1}$, a $10^{14}$-fold increase! Enzymes increase rates by $10^8$ to $10^{20}$ times, values that nonbiological catalysts do not even approach.

Enzymes are also extremely *specific,* in that each reaction is generally catalyzed by a particular enzyme. Urease catalyzes *only* the hydrolysis of urea, and none of the several thousand other enzymes present in the cell catalyzes that reaction. This remarkable specificity results from the particular groups that comprise the active site. Two models of enzyme action are illustrated in Figure 15.F. According to the **lock-and-key model,** when the "key" (substrate) fits the "lock" (active site), the chemical reaction occurs. In the last few decades, however, x-ray crystallography of several enzymes has shown that *the enzyme changes shape when the substrate lands at the active site.* The **induced-fit model** of enzyme action pictures the substrate inducing the active site to adopt a perfect fit. Rather than a rigidly shaped lock and key, therefore, we might picture a hand in glove, in which the "glove" (ac-

tive site) does not attain its functional shape until the "hand" (substrate) moves into place.

The kinetics of enzyme catalysis has many features in common with ordinary catalysis. In an uncatalyzed reaction, the rate is affected by the concentrations of the reactants; in a catalyzed reaction, the rate is affected by the concentration of reactant bound to catalyst. In the enzyme-catalyzed case, substrate (S) and enzyme (E) form an intermediate **enzyme-substrate complex (ES),** whose concentration determines the rate of product (P) formation (Figure 15.F). The steps common to virtually all enzyme-catalyzed reactions are

(1)  E + S $\rightleftharpoons$ ES      [fast, reversible]

(2)      ES $\rightarrow$ E + P   [slow; rate-determining]

Enzymes employ a variety of catalytic mechanisms. In some cases, the active site groups bring the reacting atoms of the bound substrates closer together. In other cases, the groups move apart slightly, stretching the substrate bond to be broken. Some enzyme groups are acidic and thus are able to provide H$^+$ ions that increase the speed of a rate-determining step. *Hydrolases* are a class of enzymes that cleave bonds by this type of acid catalysis. For example, lysozyme in tears hydrolyzes bacterial cell walls, thus protecting the eyes from microbes, and chymotrypsin in the small intestine hydrolyzes proteins into smaller molecules during digestion. No matter what their specific action, *all enzymes function by stabilizing the reaction's transition state.* For instance, in the lysozyme-catalyzed reaction, the transition state is a sugar molecule with bonds twisted and stretched into unusual lengths and angles, but it fits the lysozyme active site perfectly. By stabilizing the transition state through effective binding, lysozyme lowers the activation energy of the reaction and thus increases the rate.

**FIGURE 15.F**

**Two models of enzyme action. A,** In the lock-and-key model, the active site is thought to be an exact fit for the substrate shapes. **B,** In the induced-fit model, the active site is thought to change shape to fit those of the substrates. Most enzyme-catalyzed reactions proceed through a fast, reversible formation of an enzyme-substrate(s) complex, followed by a slow conversion to product(s) and free enzyme.

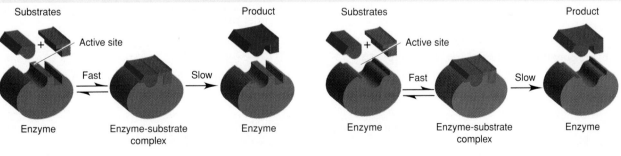

**A** Lock-and-key model

**B** Induced-fit model

◆ **Catalytically Cleaning Your Car's Exhaust.** A heterogeneous catalyst you use every day, but rarely see, is the catalytic converter in your car's exhaust system. It is designed to convert polluting exhaust gases into nontoxic ones. In a single pass through the catalyst bed, CO and unburned gasoline are oxidized to $CO_2$ and $H_2O$, while NO is reduced to $N_2$. As in the hydrogenation mechanism, the catalyst lowers the activation energy of the rate-determining step by adsorbing the molecules, thereby weakening their bonds. Mixtures of transition metals and their oxides embedded in inert supports convert as much exhaust gas as possible in the shortest period of time. It is estimated, for example, that an NO molecule is adsorbed and split into catalyst-bound N and O atoms in less than $2 \times 10^{-12}$ second!

surface provides the low–activation energy step as part of an alternative reaction mechanism. ◆ The next Chemical Connections essay introduces the remarkable abilities of the heterogeneous catalysts inside you.

**Section Summary**

A catalyst is a substance that speeds up the rate of a reaction without being consumed. It provides another reaction mechanism with lower activation energy. Homogeneous catalysts function in the same phase as the reactants, such as when chlorine atoms derived from CFC molecules catalyze the breakdown of stratospheric ozone. Heterogeneous catalysts act in a different phase from the reactants. The hydrogenation of carbon double bonds takes place on a solid catalyst, which speeds the breakage of the H—H bond in $H_2$. Enzymes are biological catalysts with high efficiency and specificity.

## Chapter Perspective

*With this introduction to chemical kinetics, we have begun to explore the dynamic inner workings of chemical change. Variations in reaction rate are observed through concentration and temperature changes, which operate on the molecular level through the energetics of particle collisions. Kinetics allows us to speculate about the molecular pathway of a reaction. Modern industry and biochemistry depend on its principles. However, speed and yield are very different aspects of a reaction. In Chapter 16, you'll see how opposing reaction rates give rise to the equilibrium state and examine how much product has formed once the net reaction has stopped.*

# For Review and Reference

## Key Terms

chemical kinetics

**SECTION 15.2**
reaction rate
average rate
instantaneous rate
initial rate

**SECTION 15.3**
rate law (rate equation)
rate constant
reaction orders

integrated rate law
half-life ($t_{1/2}$)
Arrhenius equation
activation energy ($E_a$)

**SECTION 15.4**
collision theory
transition state theory
effective collisions
frequency factor
activated complex
  (transition state)
reaction energy diagram

**SECTION 15.5**
reaction mechanism
elementary reaction
  (elementary step)
molecularity
unimolecular reaction
bimolecular reaction
rate-determining (rate-
  limiting) step
reaction intermediate

**SECTION 15.6**
catalyst
homogeneous catalyst
heterogeneous catalyst
hydrogenation
enzyme
active site
substrate
lock-and-key model
induced-fit model
enzyme-substrate
  complex (ES)

## Key Equations and Relationships

**15.1** Expressing reaction rate in terms of reactant A (p. 649):

$$\text{Rate} = -\frac{\Delta[A]}{\Delta t}$$

**15.2** Expressing the rate of a general reaction (p. 652):

$$aA + bB \longrightarrow cC + dD$$

$$\text{Rate} = -\frac{1}{a}\frac{\Delta[A]}{\Delta t} = -\frac{1}{b}\frac{\Delta[B]}{\Delta t} = \frac{1}{c}\frac{\Delta[C]}{\Delta t} = \frac{1}{d}\frac{\Delta[D]}{\Delta t}$$

**15.3** Writing a general rate law (for a case not involving products (p. 653):

$$\text{Rate} = k[A]^m[B]^n \cdots$$

**15.4** Calculating the time to reach a given [A] in a first-order reaction (rate = $k[A]$) (p. 660):

$$\ln\frac{[A]_0}{[A]_t} = kt$$

**15.5** Calculating the time to reach a given [A] in a second-order reaction (rate = $k[A]^2$) (p. 660):

$$\frac{1}{[A]_t} - \frac{1}{[A]_0} = kt$$

**15.6** Finding the half-life of a first-order process (p. 662):

$$t_{1/2} = \frac{\ln 2}{k} = \frac{0.693}{k}$$

**15.7** Relating the rate constant to the temperature (Arrhenius equation) (p. 664):

$$k = A\, e^{-E_a/RT}$$

**15.8** Calculating the activation energy (rearranged Arrhenius equation in two-point form) (p. 665):

$$\ln\frac{k_2}{k_1} = -\frac{E_a}{R}\left(\frac{1}{T_2} - \frac{1}{T_1}\right)$$

## Answers to Follow-up Problems

**15.1** (a) $4NO(g) + O_2(g) \rightarrow 2N_2O_3(g)$;

$$\text{rate} = -\frac{\Delta[O_2]}{\Delta t} = -\frac{1}{4}\frac{\Delta[NO]}{\Delta t} = \frac{1}{2}\frac{\Delta[N_2O_3]}{\Delta t}$$

(b) $-\dfrac{\Delta[O_2]}{\Delta t} = -\dfrac{1}{4}\dfrac{\Delta[NO]}{\Delta t} = -\dfrac{1}{4}$

$(-1.60 \times 10^{-4} \text{ mol/L·s}) = 4.0 \times 10^{-5} \text{ mol/L·s}$

**15.2** 1st order in $Br^-$, 1st order in $BrO_3^-$, 2nd order in $H^+$, 4th order overall

**15.3** Rate $= k[H_2]^m[I_2]^n$. From experiments 1 and 3, $m = 1$. From experiments 2 and 4, $n = 1$. Therefore, rate $= k[H_2][I_2]$; 2nd order overall.

**15.4** $1/[HI]_t - 1/[HI]_0 = kt$;
111 L/mol $-$ 100 L/mol $= 2.4 \times 10^{-21}$ L/mol · s $\times t$

$t = 4.6 \times 10^{21}$ s (or $1.5 \times 10^{14}$ yr)

**15.5** $t_{1/2} = \ln 2/k$; $k = 0.693/13.1$ h $= 5.29 \times 10^{-2}$ h$^{-1}$

**15.6** $\ln\dfrac{0.286 \text{ s}^{-1}}{k_1} = -\dfrac{1.00 \times 10^5 \text{ J/mol}}{8.31 \text{ J/mol} \cdot \text{K}}$

$\times \left(\dfrac{1}{500 \text{ K}} - \dfrac{1}{490 \text{ K}}\right)$

$= 0.491$; $k_1 = 0.175$ s$^{-1}$

**15.7**

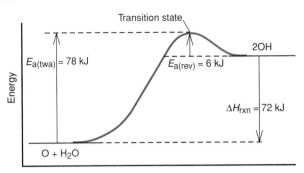

**15.8** (a) Balanced equation:
$$2NO(g) + 2H_2(g) \rightarrow N_2(g) + 2H_2O(g)$$
(b) All steps are bimolecular.
(c) Rate$_1 = k_1[NO]^2$; rate$_2 = k_2[H_2]^2$;
rate$_3 = k_3[N_2O_2]$ [H]; rate$_4 = k_4[HO][H]$;
rate$_5 = k_5[H][N_2O]$; rate$_6 = k_6[HO][H]$

**15.9** The elementary steps add to the overall reaction, and each is bimolecular. The mechanism *is* consistent with the overall rate law:

Rate$_2 = k_2[Cl][CHCl_3]$

$= k_2\left(\dfrac{k_1}{k_{-1}}\right)^{1/2}[Cl_2]^{1/2}[CHCl_3]$, with $k = k_2\left(\dfrac{k_1}{k_{-1}}\right)^{1/2}$

## Sample Problem Titles

**15.1** Expressing Rate in Terms of Changes in Concentration with Time  (p. 652)

**15.2** Determining Reaction Order from the Rate Law  (p. 656)

**15.3** Determining Reaction Orders from Initial Rate Data  (p. 658)

**15.4** Determining the Reactant Concentration at a Given Time  (p. 660)

**15.5** Determining the Half-Life of a First-Order Reaction  (p. 662)

**15.6** Determining the Energy of Activation  (p. 665)

**15.7** Drawing Reaction Energy Diagrams  (p. 671)

**15.8** Determining Molecularity and Rate Laws of Elementary Steps  (p. 674)

**15.9** Confirming a Reaction Mechanism with a Fast Initial Step  (p. 678)

## Problems

Problems with a green number are answered at the back of the text. Most sections include three categories of problems separated by a green rule—concept review questions, *paired* skill building exercises, and problems in a relevant context.

### Factors That Influence the Reaction Rate

**15.1** What variable of a chemical reaction is measured over time to obtain the reaction rate?

**15.2** How would an increase in pressure affect the rate of a gas phase reaction? Explain.

**15.3** A reaction is carried out with water as the solvent. How would the addition of more water to the reaction vessel affect the reaction rate? Explain.

**15.4** How would an increase in surface area of a solid affect the rate of its reaction with a gas? Explain.

**15.5** How would an increase in temperature affect the rate of a reaction? Explain the two factors involved.

**15.6** A pile of flour on your kitchen counter does not ignite with a match, but the flour dust in a mill or silo can react explosively. Explain.

**15.7** In a kinetic experiment, a chemist places crystals of $I_2$ in a closed reaction vessel, introduces a given quantity of $H_2$ gas, and obtains data to calculate the rate of HI formation. In a second experiment, she uses the same amounts of $I_2$ and $H_2$ but first warms the flask to 130°C, at which temperature the $I_2$ sublimes. In which experiment does the reaction proceed at a higher rate? Why?

### Expressing the Reaction Rate

(Sample Problem 15.1)

**15.8** Define reaction rate and explain why it typically changes with time. (Assume constant temperature and a closed reaction vessel.)

**15.9** (a) What is the difference between an average rate and an instantaneous rate?
(b) What is the difference between an initial rate and an instantaneous rate?

**15.10** Why are initial rates commonly used in kinetic studies?

**15.11** For the reaction $A(g) \rightarrow B(g)$, sketch two curves on the same set of axes that show
(a) The formation of product as a function of time
(b) The consumption of reactant as a function of time

**15.12** For the reaction $C(g) \rightarrow D(g)$, a plot of [C] versus time is shown below. How would you determine
(a) The average rate over the entire experiment
(b) The reaction rate at time $x$
(c) The initial reaction rate
(d) Would the values you obtained in (a), (b), and (c) be different if you plotted [D] versus time? Explain.

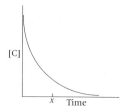

**15.13** The compound $AX_2$ decomposes according to the equation, $2AX_2(g) \rightarrow 2AX(g) + X_2(g)$. In one experiment, $[AX_2]$ was measured at various times and the following data were obtained:

| t (s) | [AX₂] |
|---|---|
| 0 | 0.0500 |
| 2.0 | 0.0458 |
| 6.0 | 0.0300 |
| 8.0 | 0.0249 |
| 10.0 | 0.0209 |
| 20.0 | 0.0088 |

(a) Find the average rate over the entire experiment.
(b) Is the initial rate higher or lower than the rate in part (a)? Use graphical methods to estimate the initial rate.

**15.14** (a) Use the data from Problem 15.13 to calculate the average rate from 8.0 to 20.0 s.
(b) Is the rate at exactly 5.0 s higher or lower than the rate in part (a)? Use graphical methods to estimate the rate at 5.0 s.

**15.15** Express the rate of reaction in terms of the change in concentration of each of the reactants and products for the reactions in (a) and (b):
(a) $2A(g) \rightarrow B(g) + C(g)$
(b) $D(g) \rightarrow \frac{3}{2}E(g) + \frac{5}{2}F(g)$
(c) For the reaction in (a), when [C] is increasing at 2 mol/L · s, how fast is [A] decreasing?
(d) For the reaction in (b), when [E] is increasing at 0.25 mol/L · s, how fast is [F] increasing?

**15.16** Express the rate of reaction in terms of the change in concentration of each of the reactants and products:

(a) $A(g) + 2B(g) \rightarrow C(g)$
(b) $2D(g) + 3E(g) + F(g) \rightarrow 2G(g) + H(g)$
(c) For the reaction in (a), when [B] is decreasing at 0.5 mol/L · s, how fast is [A] decreasing?
(d) For the reaction in (b), when [D] is decreasing at 0.1 mol/L · s, how fast is [H] increasing?

---

**15.17** The decomposition of nitrosyl bromide is followed manometrically because the number of moles of gas changes, but it cannot be followed colorimetrically because both NOBr and $Br_2$ are reddish brown:

$$2NOBr(g) \rightarrow 2NO(g) + Br_2(g)$$

Given the data below,
(a) Determine the average rate over the entire experiment.
(b) Determine the average rate between 2.00 and 4.00 s.
(c) Use graphical methods to estimate the initial reaction rate.
(d) Use graphical methods to estimate the rate at 7.00 s.

| t (s) | [NOBr] |
|---|---|
| 0.00 | 0.0100 |
| 2.00 | 0.0071 |
| 4.00 | 0.0055 |
| 6.00 | 0.0045 |
| 8.00 | 0.0038 |
| 10.00 | 0.0033 |

**15.18** The formation of ammonia is one of the most important processes in the chemical industry:

$$N_2(g) + 3H_2(g) \rightarrow 2NH_3(g)$$

Express the reaction rate in terms of changes in $[N_2]$, $[H_2]$, and $[NH_3]$.

**15.19** The chlorination of ethane is one of several synthetic routes to producing chloroethanes, a group of major industrial solvents. An example of this type of reaction is $C_2H_6(g) + 4Cl_2(g) \rightarrow C_2H_2Cl_4(l) + 4HCl(g)$. Express the reaction rate in terms of changes in concentration of each reactant and product.

**15.20** Just as the depletion of stratospheric ozone today threatens life on Earth, its accumulation was one of the crucial steps that allowed life to develop in prehistoric times: $3O_2(g) \rightarrow 2O_3(g)$.
(a) Express reaction rate in terms of $[O_2]$ and $[O_3]$.
(b) At a given instant, the rate in terms of $[O_2]$ is $2.17 \times 10^{-5}$ mol/L · s. What is it in terms of $[O_3]$?

**15.21** Nitrogen oxides undergo numerous decomposition and combination reactions both in industry and the environment. One of the most thoroughly studied is the decomposition of dinitrogen pentaoxide:

$$2N_2O_5(g) \rightarrow 4NO_2(g) + O_2(g)$$

At a given instant, $[N_2O_5]$ is decreasing at the rate of 3.15 mol/L · s. What is the rate of the reaction in terms of $[NO_2]$?

## The Rate Law and Its Components

(Sample Problems 15.2-15.6)

**15.22** The rate law for the general reaction

$$aA + bB + \cdots \rightarrow cC + dD + \cdots$$

is rate = $k\,[A]^m[B]^n \cdots$
  (a) Explain the meaning of $k$.
  (b) Explain the meanings of $m$ and $n$. Does $m = a$ and $n = b$? Explain.
  (c) If the reaction is first order in A and second order in B, and time is measured in minutes (min), what are the units for $k$?

**15.23** You are studying the reaction

$$A_2(g) + B_2(g) \rightarrow 2AB(g)$$

to determine its rate law. Assuming that you have a valid experimental procedure for obtaining $[A_2]$ and $[B_2]$ at various times, explain how you would determine (a) the initial rate; (b) the reaction orders; (c) the rate constant.

**15.24** By what factor does the rate change in each of the following cases (assume constant temperature)?
  (a) A reaction is first order with respect to reactant A and [A] is doubled.
  (b) A reaction is second order with respect to reactant B and [B] is halved.
  (c) A reaction is second order with respect to reactant C and [C] is tripled.

**15.25** How are integrated rate laws used to determine reaction order? What is the order in the reactant of the reaction if a plot of
  (a) The natural logarithm of [reactant] versus time is linear?
  (b) The inverse of [reactant] versus time is linear?
  (c) [Reactant] versus time is linear?

**15.26** Define the half-life of a reaction. Explain on the molecular level why the half-life of a first-order reaction is a constant value.

**15.27** (a) Sketch a graph to show the relationship between $k$ ($y$ axis) and $T$ ($x$ axis)
  (b) Sketch a graph to show the relationship between $\ln k$ ($y$ axis) and $1/T$ ($x$ axis). How is the activation energy determined from this graph?

**15.28** Give the individual reaction orders for each substance and the overall reaction order for each of the following rate laws:
  (a) rate = $k\,[BrO_3^-][Br^-][H^+]^2$
  (b) rate = $k\,[NH_4^+][NO_2^-]$
  (c) rate = $k\dfrac{[O_3]^2}{[O_2]}$

**15.29** Give the individual reaction orders for each substance and the overall reaction order for each rate law:
  (a) rate = $k\,[NO_2]^2[Cl_2]$
  (b) rate = $k\,[CS_2]$
  (c) rate = $k\dfrac{[HNO_2]^4}{[NO]^2}$

**15.30** For the reaction

$$4A(g) + 3B(g) \rightarrow 2C(g)$$

the following data were obtained at constant temperature:

| TRIAL | INITIAL [A] (mol/L) | INITIAL [B] (mol/L) | INITIAL RATE (mol/L · min) |
|---|---|---|---|
| 1 | 0.100 | 0.100 | 5.00 |
| 2 | 0.300 | 0.100 | 45.0 |
| 3 | 0.100 | 0.200 | 10.0 |
| 4 | 0.300 | 0.200 | 90.0 |

  (a) What is the order with respect to each reactant?
  (b) Write the rate law.
  (c) What is the value of $k$ from trial 1?

**15.31** For the reaction

$$A(g) + B(g) + C(g) \rightarrow D(g)$$

the following data were obtained at constant temperature:

| TRIAL | INITIAL [A] (mol/L) | INITIAL [B] (mol/L) | INITIAL [C] (mol/L) | INITIAL RATE (mol/L · s) |
|---|---|---|---|---|
| 1 | 0.0500 | 0.0500 | 0.0100 | $6.25 \times 10^{-3}$ |
| 2 | 0.1000 | 0.0500 | 0.0100 | $1.25 \times 10^{-2}$ |
| 3 | 0.1000 | 0.1000 | 0.0100 | $5.00 \times 10^{-2}$ |
| 4 | 0.0500 | 0.0500 | 0.0200 | $6.25 \times 10^{-3}$ |

  (a) What is the order with respect to each reactant?
  (b) Write the rate law.
  (c) What is the value of $k$ from trial 1?

**15.32** Without consulting Table 15.3, give the units of the rate constants for reactions with the following overall orders: (a) first order; (b) second order; (c) third order; (d) 5/2 order.

**15.33** Give the overall reaction order that corresponds to rate constants with the following units:
  (a) mol/L · s; (b) $yr^{-1}$; (c) $(mol/L)^{1/2} \cdot s^{-1}$;
  (d) $(mol/L)^{-5/2} \cdot min^{-1}$.

**15.34** In the simple decomposition reaction

$$AB(g) \rightarrow A(g) + B(g)$$

rate = $k[AB]^2$ and $k = 0.2$ L/mol · s. How long will it take for [AB] to reach one-third of its initial concentration of 1.50 $M$?

**15.35** For the reaction in Problem 15.34, what is [AB] after 10.0 s?

**15.36** In a first-order decomposition reaction, 50% of a compound decomposes in 10.5 min.
  (a) What is the rate constant of the reaction?
  (b) How long does it take for 75% of the compound to decompose?

**15.37** A decomposition reaction has a rate constant of 0.0012 $yr^{-1}$.
  (a) What is the half-life of the reaction?
  (b) How long will it take for [reactant] to reach 12.5% of its original value?

**15.38** The rate constant of a reaction is $3.7 \times 10^{-3}$ s$^{-1}$ at 25°C and the activation energy is 43.6 kJ/mol. What is the value of $k$ at 75°C?

**15.39** The rate constant of a reaction is $2.50 \times 10^{-5}$ L/mol·s at 190°C and $3.20 \times 10^{-3}$ L/mol·s at 250°C. What is the activation energy of the reaction?

---

**15.40** Phosgene is a toxic gas prepared by the reaction of carbon monoxide with chlorine:

$$CO(g) + Cl_2(g) \longrightarrow COCl_2(g)$$

The following data were obtained in a kinetic study of its formation:

| TRIAL | INITIAL [CO] (mol/L) | INITIAL [Cl$_2$] (mol/L) | INITIAL RATE (mol/L · s) |
|-------|-------|-------|-------|
| 1 | 1.00 | 0.100 | $1.29 \times 10^{-29}$ |
| 2 | 0.100 | 0.100 | $1.33 \times 10^{-30}$ |
| 3 | 0.100 | 1.00 | $1.30 \times 10^{-29}$ |
| 4 | 0.100 | 0.0100 | $1.32 \times 10^{-31}$ |

(a) Write the rate law for the formation of phosgene.
(b) Calculate the average value of the rate constant.

**15.41** Carbon disulfide is a poisonous, flammable liquid that is an excellent solvent for phosphorus, sulfur, and several other nonmetals. A kinetic study of its gaseous decomposition reveals the following data:

| TRIAL | INITIAL [CS$_2$] (mol/L) | INITIAL RATE (mol/L · s) |
|-------|-------|-------|
| 1 | 0.100 | $2.7 \times 10^{-7}$ |
| 2 | 0.080 | $2.2 \times 10^{-7}$ |
| 3 | 0.055 | $1.5 \times 10^{-7}$ |
| 4 | 0.044 | $1.2 \times 10^{-7}$ |

(a) Write the rate law for the decomposition of CS$_2$.
(b) Calculate the average value of the rate constant.

**15.42** For the decomposition of gaseous dinitrogen pentaoxide $2N_2O_5(g) \longrightarrow 4NO_2(g) + O_2(g)$, the rate constant is $k = 2.8 \times 10^{-3}$ s$^{-1}$ at 60°C. If the initial concentration of $N_2O_5$ is 1.58 mol/L,
(a) What is [$N_2O_5$] after 5.00 min?
(b) What fraction of the $N_2O_5$ has decomposed after 5.00 min?

**15.43** During a study of ammonia production, an industrial chemist discovers that the compound decomposes to its elements $N_2$ and $H_2$ in a first-order process. She collects the following data:

| Time (s) | 0 | 1.000 | 2.000 |
|-------|-------|-------|-------|
| [NH$_3$] (mol/L) | 4.000 | 3.986 | 3.974 |

(a) Use graphical methods to determine the rate constant.
(b) What is the half-life of ammonia decomposition?

**15.44** Understanding the characteristics of high-temperature formation and breakdown of the nitrogen oxides is essential for controlling the pollutants generated by car engines. The second-order breakdown of nitric oxide to its elements has rate constants of 0.0796 L/mol·s at 737°C and 0.0815 L/mol·s at 947°C. What is the activation energy of this reaction?

## Explaining the Effects of Concentration and Temperature on Reaction Rate

(Sample Problem 15.7)

**15.45** What is the central idea of collision theory? How does this idea explain the effect of concentration on reaction rate?

**15.46** Is collision frequency the only factor affecting reaction rate? Explain.

**15.47** Arrhenius proposed that each reaction has an energy threshold that must be reached for the particles to react. The kinetic theory of gases proposes that the average kinetic energy of the particles in a sample is proportional to the absolute temperature. Explain how these concepts relate to the effect of temperature on reaction rate.

**15.48** (a) For a reaction with a given $E_a$, how does an increase in $T$ affect the rate?
(b) For a reaction at a given $T$, how does a decrease in $E_a$ affect the rate?

**15.49** In the reaction AB + CD $\rightleftharpoons$ EF, $4 \times 10^{-5}$ mol AB collide with $4 \times 10^{-5}$ mol CD. Will $4 \times 10^{-5}$ mol EF form? Explain.

**15.50** Assuming the activation energies are equal, predict which of the following reactions will occur at a higher rate at 50°C? Explain:

$$NH_3(g) + HCl(g) \longrightarrow NH_4Cl(s)$$
$$N(CH_3)_3(g) + HCl(g) \longrightarrow (CH_3)_3NHCl(s)$$

---

**15.51** For the reaction $A(g) + B(g) \longrightarrow AB(g)$, how many unique collisions between A and B are possible if there are four particles of A and three particles of B present in the vessel?

**15.52** For the reaction $A(g) + B(g) \longrightarrow AB(g)$, how many unique collisions between A and B are possible if 1.01 mol $A(g)$ and 2.12 mol $B(g)$ are present in the vessel?

**15.53** At 25°C, what is the fraction of collisions with energy equal to or greater than an activation energy of 100 kJ/mol?

**15.54** If the temperature in Problem 15.53 is increased to 50°C, by what factor does the fraction of collisions with energy equal to or greater than the activation energy change?

---

**15.55** For the reaction $ABC + D \rightleftharpoons AB + CD$, $\Delta H^0_{rxn} = -50$ kJ/mol and $E_{a(fwd)} = 200$ kJ/mol. Assuming the reaction occurs in one step,
(a) Draw a reaction energy diagram.
(b) Calculate $E_{a(rev)}$.
(c) Sketch a possible transition state if A—B—C is V shaped.

**15.56** For the reaction $A_2 + B_2 \longrightarrow 2AB$, $E_{a(fwd)} = 125$ kJ/mol and $E_{a(rev)} = 85$ kJ/mol. Assuming the reaction occurs in one step,
(a) Draw a reaction energy diagram.
(b) Calculate $\Delta H^0_{rxn}$.
(c) Sketch a possible transition state.

**15.57** Aqua regia, a mixture of hydrochloric and nitric acids, has been used since alchemical times as a solvent for many metals, including gold. Its orange color is due to the presence of nitrosyl chloride. Consider this one-step gaseous reaction for the formation of this compound:

$$NO(g) + Cl_2(g) \rightarrow NOCl(g) + Cl(g) \qquad \Delta H^0 = 83 \text{ kJ}$$

(a) Draw a reaction energy diagram for the reaction, given that $E_{a(fwd)}$ is 86 kJ.
(b) Calculate $E_{a(rev)}$.
(c) Sketch a possible transition state for the reaction. (*Hint:* the atom sequence of nitrosyl chloride is Cl—N—O.)

**15.58** Iodide ion reacts with chloroform to displace chloride ion in a common organic substitution reaction:

$$I^- + CH_3Cl \rightarrow CH_3I + Cl^-$$

Draw the molecular shape of chloroform and indicate the direction of $I^-$ attack.

## Reaction Mechanisms: Steps in the Overall Reaction

(Sample Problems 15.8-15.9)

**15.59** Is the rate of an overall reaction lower, higher, or equal to the average rate of the individual steps? Explain.

**15.60** Explain why the coefficients of an elementary step equal the reaction orders of its rate law but those of an overall reaction do not.

**15.61** Is it possible for more than one mechanism to be consistent with the rate law of a given reaction? Explain.

**15.62** Describe the difference between a reaction intermediate and an activated complex.

**15.63** Why is a bimolecular step more reasonable physically than a termolecular step?

**15.64** If a slow step precedes a fast step in a two-step mechanism, do the substances in the fast step appear in the rate law? Explain.

**15.65** If a fast step precedes a slow step in a two-step mechanism, how is the fast step affected? How is this effect used to determine the validity of the mechanism?

**15.66** The proposed mechanism for a reaction is
(1) $A(g) + B(g) \rightleftharpoons X(g)$ [fast]
(2) $X(g) + C(g) \rightarrow Y(g)$ [slow]
(3) $Y(g) \rightarrow D(g)$ [fast]
(a) What is the overall equation?
(b) Identify the intermediates, if any.
(c) What is the molecularity and rate law of each step?
(d) Is the mechanism consistent with the actual rate law: rate = $k[A][B][C]$?
(e) Would this one-step mechanism be just as valid $A(g) + B(g) + C(g) \rightarrow D(g)$?

**15.67** Consider the following mechanism:
(1) $ClO^-(aq) + H_2O(l) \rightleftharpoons HClO(aq) + OH^-(aq)$ [fast]
(2) $I^-(aq) + HClO(aq) \rightarrow HIO(aq) + Cl^-(aq)$ [slow]
(3) $OH^-(aq) + HIO(aq) \rightarrow H_2O(l) + IO^-(aq)$ [fast]
(a) What is the overall equation?
(b) Identify the intermediates, if any.
(c) What is the molecularity and the rate law of each step?
(d) Is the mechanism consistent with the actual rate law: rate = $k[ClO^-][I^-]$?

**15.68** In a study of nitrosyl halides, a chemist proposes the following mechanism for the synthesis of nitrosyl bromide:
(1) $NO(g) + Br_2(g) \rightleftharpoons NOBr_2(g)$ [fast]
(2) $NOBr_2(g) + NO(g) \rightarrow 2NOBr(g)$ [slow]
If the rate law is rate = $k[NO]^2[Br_2]$, is the proposed mechanism valid? Explain by showing that it satisfies the three criteria for validity.

**15.69** Kinetic studies of hydrogen halide formation show that the rate law for HBr formation is different from that for HI. The rate law for the reaction

$$H_2(g) + Br_2(g) \rightarrow 2HBr(g)$$

is rate = $k[H_2][Br_2]^{1/2}$. Which of the following mechanisms is consistent with the rate law?
A. $H_2(g) + Br_2(g) \rightarrow 2HBr(g)$
B. (1) $H_2(g) \rightleftharpoons 2H(g)$ [fast]
(2) $H(g) + Br_2(g) \rightarrow HBr(g) + Br(g)$ [slow]
(3) $Br(g) + H(g) \rightarrow HBr(g)$ [fast]
C. (1) $Br_2(g) \rightleftharpoons 2Br(g)$ [fast]
(2) $Br(g) + H_2(g) \rightarrow HBr(g) + H(g)$ [slow]
(3) $H(g) + Br(g) \rightarrow HBr(g)$ [fast]

**15.70** The rate law for $2NO(g) + O_2(g) \rightarrow 2NO_2(g)$ is rate = $k[NO]^2[O_2]$. In addition to the mechanism in the text, the following ones have been proposed:
A. $2NO(g) + O_2(g) \rightarrow 2NO_2(g)$
B. (1) $2NO(g) \rightleftharpoons N_2O_2(g)$ [fast]
(2) $N_2O_2(g) + O_2(g) \rightarrow 2NO_2(g)$ [slow]
C. (1) $2NO(g) \rightleftharpoons N_2(g) + O_2(g)$ [fast]
(2) $N_2(g) + 2O_2(g) \rightarrow 2NO_2(g)$ [slow]
(a) Which of the mechanisms is consistent with the rate law?
(b) Which is most reasonable chemically? Why?

**15.71** The rate law for the reaction

$$CO(g) + NO_2(g) \rightarrow CO_2(g) + NO(g)$$

is rate = $k [NO_2]^2$, and one possible mechanism was shown on p. 675.
(a) Draw a reaction energy diagram for the in-text mechanism, given that $\Delta H^0_{overall} = -226$ kJ.
(b) An alternative mechanism has been proposed:
(1) $2NO_2(g) \rightarrow N_2(g) + 2O_2(g)$ [slow]
(2) $2CO(g) + O_2(g) \rightarrow 2CO_2(g)$ [fast]
(3) $N_2(g) + O_2(g) \rightarrow 2NO(g)$ [fast]
Is the alternative mechanism consistent with the rate law? Is one mechanism more reasonable physically? Explain.

## Catalysis: Speeding Up a Chemical Reaction

**15.72** Consider the reaction: $N_2O(g) \xrightarrow{\text{Au}} N_2(g) + \frac{1}{2}O_2(g)$.

(a) Is the gold acting as a homogeneous or a heterogeneous catalyst?

(b) On the same set of axes, sketch the reaction energy diagrams for the catalyzed and the uncatalyzed reactions.

**15.73** Does a catalyst increase reaction rate by the same means as does a rise in temperature? Explain.

**15.74** In a popular classroom demonstration, hydrogen gas and oxygen gas are mixed in a balloon. Although the mixture is stable under normal conditions, when a spark is applied to the mixture or a small amount of powdered metal dropped into it, the mixture explodes. (a) Is the spark acting as a catalyst? Explain. (b) Is the metal acting as a catalyst? Explain.

**15.75** Describe three common examples of catalyzed reactions that occur around you daily.

## Comprehensive Problems

Problems with an asterisk (*) are more challenging.

**15.76** Consider the following reaction energy diagram:

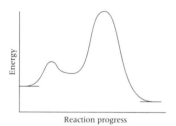

Reaction progress

(a) How many elementary steps are in the reaction mechanism?

(b) Which step is rate limiting?

(c) Is the overall reaction exothermic or endothermic?

**15.77** Reactions between certain organic (alkyl) halides and water produce alcohols. Consider the overall reaction of t-butyl bromide (2-bromo-2-methylpropane):

$$(CH_3)_3CBr(aq) + H_2O(l) \rightarrow$$
$$(CH_3)_3COH(aq) + H^+(aq) + Br^-(aq)$$

The experimental rate law is rate = $k[(CH_3)_3CBr]$. The accepted mechanism for the reaction is

(1) $(CH_3)_3C-Br(aq) \rightarrow$
$$(CH_3)_3C^+(aq) + Br^-(aq)$$  [slow]

(2) $(CH_3)_3C^+(aq) + H_2O(l) \rightarrow$
$$(CH_3)_3C-OH_2^+(aq)$$  [fast]

(3) $(CH_3)_3C-OH_2^+(aq) \rightarrow$
$$H^+(aq) + (CH_3)_3C-OH(aq)$$  [fast]

(a) Why doesn't $H_2O$ appear in the experimental rate law?

(b) Write rate laws for the elementary steps.

(c) What intermediates appear in the mechanism?

(d) Show that the mechanism is consistent with the experimental rate law.

**15.78** A student is measuring the rate of a reaction with a stopwatch but finds that the rate is too fast to get reproducible results. In each of the following cases, describe two ways to slow the reaction:
(a) The reactants are gases.
(b) The reactants are in solution.

**15.79** When studying atmospheric or other gaseous reactions, it is useful to express the rate law in terms of pressure. How would you express the general rate law in terms of pressure? If a reaction rate has units of torr/s, what are the units of $k$ for a first- and a second-order reaction?

**15.80** High-flying supersonic transports release NO into the stratosphere, which catalyzes ozone depletion by a mechanism analogous to the action of Cl. Write the reactions in the mechanism and show why NO acts as a catalyst.

**15.81** Archaeologists can determine the approximate age of artifacts made of wood or bone by measuring the concentration of the radioactive isotope carbon-14 present in the object. The amount of isotope decreases in a first-order process. If 15.5% of the original amount of carbon-14 is present in a wooden tool at the time of analysis, what is the age of the artifact? The half-life of carbon-14 is 5730 yr.

**15.82** A slightly bruised apple will rot extensively in about 4 days at room temperature (20°C). If it is kept in the refrigerator at 0°C, the same extent of rotting takes about 16 days. What is the activation energy for the rotting reaction?

**15.83** How do each of the following changes affect the reaction rate (increase, decrease, no effect):
(a) Decreasing the pressure of a gas-liquid reaction
(b) Increasing the concentration of a reactant that occurs in a step preceding the rate-limiting step
(c) Increasing the concentration of a reactant that occurs in a step following the rate-limiting step
(d) Grinding up a solid reactant in a gas-solid reaction
(e) Rapid stirring of a reaction between immiscible liquids

**15.84** Enzymes are remarkably efficient catalysts that can increase reaction rates by as many as 20 orders of magnitude.
(a) How do enzymes affect the transition state of a reaction, and how does this effect increase the reaction rate?
(b) What characteristics do enzymes have that give them this tremendous effectiveness as catalysts?

**15.85** Biochemists consider the citric acid cycle to be the central reaction sequence in metabolism. One of the key steps is an oxidation catalyzed by the enzyme isocitrate dehydrogenase and the oxidizing agent NAD. Under certain conditions, the reaction in yeast obeys 11th-order kinetics:

Rate = $k[\text{enzyme}][\text{isocitrate}]^4[\text{AMP}]^2[\text{NAD}]^m[\text{Mg}^{2+}]^2$

What is the order with respect to NAD?

**15.86** Enzymes in human liver catalyze a large number of reactions that degrade ingested toxic chemicals. By what factor would the rate of a detoxification reaction be changed if a liver enzyme lowers the activation energy by 5 kJ/mol at 37°C?

**15.87** For the reaction

$$A(g) + B(g) \rightarrow AB(g)$$

the rate is 0.20 mol/L · s, when [A] = [B] = 1.0 mol/L. If the reaction is first order in B and second order in A, what is the rate when [A] = 2.0 mol/L and [B] = 3.0 mol/L?

**\*15.88** The hydrolysis of table sugar (sucrose) occurs by the following overall reaction:

$$C_{12}H_{22}O_{11}(s) + H_2O(l) \rightarrow C_6H_{12}O_6(aq) + C_6H_{12}O_6(aq)$$

sucrose               glucose          fructose

A nutritional biochemist studies the kinetics of the process and obtains the following data:

| [SUCROSE] (mol/L) | TIME (h) |
|---|---|
| 0.501 | 0 |
| 0.451 | 0.50 |
| 0.404 | 1.00 |
| 0.363 | 1.50 |
| 0.267 | 3.00 |

(a) Use the data to determine the rate constant and the half-life of the reaction.
(b) How long does it take to hydrolyze 75% of the sucrose?
(c) Other studies have shown that this reaction is actually second order overall but appears to follow first-order kinetics (Such a reaction is termed a pseudo first-order reaction.) Suggest a reason for this apparent first-order behavior.

**15.89** Is each of the following statements true? If not, explain why.
(a) At a given temperature, all molecules possess the same kinetic energy.
(b) Halving the pressure of a gaseous reaction doubles the reaction rate.
(c) The higher the activation energy of a reaction is, the lower its rate.
(d) A temperature increase of 10°C doubles the rate of any reaction.
(e) If reactant molecules collide with greater energy than the activation energy, they change into product molecules.
(f) The activation energy of a reaction depends on the temperature.
(g) The rate of a reaction increases as the reaction proceeds.
(h) The activation energy of a reaction depends on collision frequency.
(i) A catalyst increases the rate by increasing collision frequency.
(j) Exothermic reactions have higher rates than endothermic ones.

(k) Temperature has no effect on the value of the Arrhenius factor, $A$.
(l) The activation energy of a reaction is lowered by a catalyst.
(m) For most common reactions, the enthalpy change is lowered by a catalyst.
(n) The probability factor $p$ is near unity for reactions between single atoms.
(o) The initial rate of a reaction is its maximum rate.
(p) A bimolecular reaction is generally twice as fast as a unimolecular reaction.
(q) The molecularity of an elementary reaction is proportional to the molecular complexity of the reactant(s).

**\*15.90** Ozone is one of the components of photochemical smog. It is generated in air when nitrogen dioxide, formed by the oxidation of nitric oxide from car exhaust, reacts according the following mechanism:

$$(1) \quad NO_2(g) \xrightarrow{k_1} NO(g) + O(g)$$

$$(2) \quad O(g) + O_2(g) \xrightarrow{k_2} O_3(g)$$

Assuming the rate of formation of atomic oxygen in step (1) equals the rate of its consumption in step (2), use the data below to calculate
(a) The concentration of atomic oxygen [O]
(b) The rate of ozone formation

$$k_1 = 6.0 \times 10^{-3} \text{ s}^{-1} \qquad [NO_2] = 4.0 \times 10^{-9} \text{ } M$$
$$k_2 = 1.0 \times 10^6 \text{ L/mol} \cdot \text{s} \qquad [O_2] = 1.0 \times 10^{-2} \text{ } M$$

**\*15.91** In a "clock" reaction, a dramatic color change occurs at a time determined by concentration and temperature. One of the most famous is the iodine-clock reaction. The overall equation is

$$2I^-(aq) + S_2O_8^{2-}(aq) \rightarrow I_2(aq) + 2SO_4^{2-}(aq)$$

As $I_2$ forms, it is immediately consumed by its reaction with a fixed amount of added $S_2O_3^{2-}$:

$$I_2(aq) + 2S_2O_3^{2-}(aq) \rightarrow 2I^-(aq) + S_4O_6^{2-}(aq)$$

Once the $S_2O_3^{2-}$ is consumed, the excess $I_2$ forms a blue-black product with starch solution present in the mixture: $I_2$ + starch $\rightarrow$ starch · $I_2$ (blue-black). The rate of the reaction is also influenced by the total concentration of ions, so KCl and $(NH_4)_2SO_4$ are added to maintain a constant value. Use the data below to determine
(a) The average rate for each trial
(b) The order with respect to each reactant
(c) The rate constant at the experimental temperature of 23°C
(d) The rate law of the overall equation

| | TRIAL 1 | TRIAL 2 | TRIAL 3 |
|---|---|---|---|
| 0.200 $M$ KI (mL) | 10.0 | 20.0 | 20.0 |
| 0.100 $M$ $Na_2S_2O_8$ (mL) | 20.0 | 20.0 | 10.0 |
| 0.0050 $M$ $Na_2S_2O_3$ (mL) | 10.0 | 10.0 | 10.0 |
| 0.200 $M$ KCl (mL) | 10.0 | 0 | 0 |
| 0.100 $M$ $(NH_4)_2SO_4$ (mL) | 0 | 0 | 10.0 |
| Time to color(s) | 29.0 | 14.5 | 14.5 |

# CHAPTER 16

## Concepts and skills to review

- reversibility of reactions (Section 4.5)
- equilibrium vapor pressure (Section 11.5)
- equilibrium nature of a saturated solution (Section 12.3)
- dependence of rate on concentration (Section 15.4)
- rate laws for elementary reactions (Section 15.5)
- function of a catalyst (Section 15.6)

# Equilibrium

# The Extent of Chemical Reactions

**A Plant for the Synthesis of Ammonia.** Rate and extent are critical factors in industrial production. We now examine the principles governing the second of these—how far a reaction proceeds. In a closed container at ordinary conditions, the formation of ammonia from its elements does not proceed to any great extent. Yet, as you'll learn in this chapter, this vital substance is manufactured on the multimillion-ton scale annually by applying the principles of equilibrium.

In this chapter, we turn to another of the central questions in reaction chemistry: How much product will form under a given set of starting concentrations and conditions? Whereas chemical kinetics, the topic of Chapter 15, tells us how fast a reaction is proceeding, chemical equilibrium tells us the reactant and product concentrations once it has run its course. Furthermore, studies of equilibrium provide the information we need to affect these final concentrations by altering conditions.

Just as reactions vary greatly in their speed, they also vary in their extent. These two aspects are not the same. A fast reaction may go a long way or barely at all toward products. Consider the dissociation of an acid in water, for example. With 1 $M$ HCl, virtually all the acid molecules dissociate into ions. With 1 $M$ $CH_3COOH$, fewer than 1% of the acetic acid molecules dissociate. Yet both reactions take less than a second to reach completion. Similarly, some slow reactions eventually yield a large amount of product, whereas others yield very little. After several years at ordinary temperatures, a steel water-storage tank will rust, but no matter how long you wait, the water inside will not decompose to its elements.

The point is that the principles of kinetics and of equilibrium apply to different aspects of a reaction. The *speed* of a reaction—the concentration of product that forms per unit time—depends ultimately on the *difference in energy between the reactants and the transition state*. The *extent* of a reaction—the concentration of product that is present when no further change is observed—depends ultimately on the *difference in energy between the reactants and products*.

Knowing how much product will form in a given reaction is often crucial information. How much of a new drug or industrial polymer can you obtain from a particular reaction? Would another reactant mixture that yields more product at a higher temperature be a wiser choice? If a slow reaction has a good yield, would a catalyst speed it up enough to make the reaction useful?

The chapter opens with a description of the equilibrium state at the macroscopic and molecular levels and then focuses on the equilibrium constant and its relation to the balanced equation. In the third section, we apply equilibrium concepts to a series of common quantitative problems. Then we examine how reaction conditions affect the equilibrium state. We end with a discussion of equilibrium in two areas of applied chemistry: the industrial production of ammonia and the reaction pathways in living cells.

## 16.1 The Dynamic Nature of the Equilibrium State

The observable fact for any chemical system in a state of equilibrium is that *the concentrations of reactants and products no longer change with time*. The reason for this apparent cessation of chemical activity is that all reactions are reversible. Let's examine a reacting chemical system at the macroscopic and molecular levels to see how reversibility gives rise to the equilibrium state.

The system we will examine consists of two gases, colorless dinitrogen tetraoxide and brown nitrogen dioxide:

$$N_2O_4(g; \text{colorless}) \rightleftharpoons 2NO_2(g; \text{brown})$$

If we introduce a small amount of $N_2O_4(l)$ into a sealed flask kept at 100°C, the liquid immediately vaporizes (boiling point, 21°C) and the gas begins to turn pale brown. The color slowly darkens, but after a few moments, no further change appears (Figure 16.1).

**FIGURE 16.1**

**The $N_2O_4$-$NO_2$ system reaching equilibrium.** **A,** When the experiment begins, the reaction mixture consists mostly of colorless $N_2O_4$. **B,** As $N_2O_4$ decomposes to $NO_2$, the color of the mixture becomes pale brown. **C,** When equilibrium is reached, the concentrations of $NO_2$ and $N_2O_4$ are constant, and the color reaches its final shade of brown. **D,** Because the reaction continues in the forward and reverse directions at equal rates, the concentrations remain constant, as does the color.

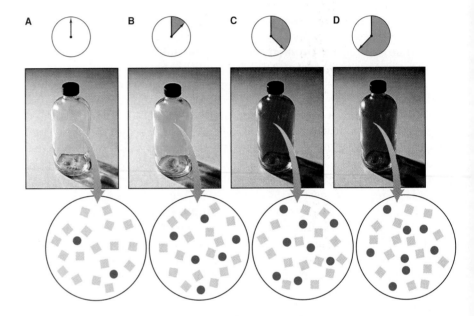

As we close in on the molecular level, a much more active scene unfolds. The $N_2O_4$ molecules fly wildly throughout the flask, some splitting into two $NO_2$ molecules. As time passes, more $N_2O_4$ molecules decompose and the concentration of $NO_2$ rises. As macroscopic observers, we see the flask contents darken because $NO_2$ is brown. As the number of $N_2O_4$ molecules decreases, $N_2O_4$ decomposition slows. At the same time, increasing numbers of $NO_2$ molecules collide and combine, so $N_2O_4$ re-formation speeds up. Eventually, $N_2O_4$ molecules decompose as fast as $NO_2$ molecules combine. The system has reached equilibrium: *reactant and product concentrations stop changing because the forward and reverse rates have become equal:*

$$\text{At equilibrium: rate}_{fwd} = \text{rate}_{rev} \qquad \textbf{(16.1)}$$

Thus, a system at equilibrium continues to be very dynamic at the molecular level, but we see no further *net* change because changes in one direction are balanced by changes in the other.

When the system reaches equilibrium at a particular temperature, the concentrations of products and reactants can be characterized by an overall numerical constant. Let's examine the rate laws for the $N_2O_4$-$NO_2$ system to see one way of deriving this constant. In this case, both the forward and reverse reactions are elementary steps (Section 15.5), so at equilibrium we have

$$\text{rate}_{\text{fwd}} = \text{rate}_{\text{rev}}$$

$$k_{\text{fwd}}[N_2O_4] = k_{\text{rev}}[NO_2]^2$$

where $k_{\text{fwd}}$ and $k_{\text{rev}}$ are the forward and reverse rate constants, respectively. By rearranging, we set the ratio of the rate constants equal to the ratio of the concentration terms:

$$\frac{k_{\text{fwd}}}{k_{\text{rev}}} = \frac{[NO_2]^2}{[N_2O_4]}$$

The ratio of constants gives rise to the new overall constant called the **equilibrium constant** ($K_{\text{eq}}$, or simply $K$):

$$K = \frac{k_{\text{fwd}}}{k_{\text{rev}}} = \frac{[NO_2]^2}{[N_2O_4]} \qquad (16.2)$$

The equilibrium constant $K$ is a *number* whose value is equal to the ratio of rate constants. In addition, and most importantly, *K is equal to a particular ratio of equilibrium product and reactant terms at a particular temperature.* We examine this idea closely in the next section.

Although the concentrations of reactants and products are constant in a system at equilibrium, this does *not* mean that the system contains 50% products and 50% reactants. It is the opposing *rates* that are equal at equilibrium, not necessarily the concentrations. Indeed, different reaction systems have a wide range of concentrations at equilibrium: from almost all reactant to almost all product, and anything in between. Here are three specific examples:

1. Small $K$. If a reaction goes very little toward product before reaching equilibrium, $K$ will be small. For example, the oxidation of nitrogen barely proceeds at 500 K (note that the equilibrium constant $K$ is always italic, and the temperature unit K always follows a number):

$$N_2(g) + O_2(g) \rightleftharpoons 2NO(g) \qquad K = 1 \times 10^{-30}$$

2. Large $K$. Conversely, if the reaction reaches equilibrium with virtually no reactant remaining, we say it "goes to completion" and has a large $K$. The oxidation of carbon monoxide goes to completion at 1000 K:

$$2CO(g) + O_2(g) \rightleftharpoons 2CO_2(g) \qquad K = 2.2 \times 10^{22}$$

3. Intermediate $K$. Significant amounts of both reactant and product are present at equilibrium of the $N_2O_4$-$NO_2$ system at 373 K, so the equilibrium constant has an intermediate value:

$$N_2O_4(g) \rightleftharpoons 2NO_2(g) \qquad K = 0.211$$

### Section Summary

Kinetics and equilibrium are distinct aspects of a reacting system; that is, rate and yield are not necessarily related. When the forward and reverse reactions occur at the same rate, the system has reached equilibrium and concentrations remain constant. The equilibrium constant ($K$) is a number based on a ratio of product and reactant terms. $K$ is large for reactions that reach equilibrium with a high concentration of products and small for reactions that reach equilibrium with a low concentration of products.

## 16.2 The Mass-Action Expression and the Equilibrium Constant

In 1864, two Norwegian chemists, Cato Guldberg and Peter Waage, observed from experiment that *at a given temperature, a chemical system reaches a state in which a particular ratio of reactant and product concentration terms has a constant value.* This is one way of stating the **law of chemical equilibrium,** or the **law of mass action.** Note that no mention of rates appears in this statement. Indeed, the law of mass action was stated several decades before the principles of kinetics had been developed. Although kinetic theory later showed that $K$ could be obtained from a ratio of rate constants, *the law of mass action does not depend on a knowledge of rates.*

Guldberg and Waage studied many reactions in which reactant and product concentrations varied widely and found that the law of mass action has a major implication: *for a particular system and temperature, the same equilibrium state is attained regardless of how the reaction is run.* This means that, for the $N_2O_4$-$NO_2$ system, we can start the reaction by introducing pure $N_2O_4$, pure $NO_2$, or any mixture of $NO_2$ and $N_2O_4$, and given enough time at 100°C, the particular ratio of concentration terms $[NO_2]^2/[N_2O_4]$ will have a fixed value (Table 16.1); note that the *ratio* is fixed even though the equilibrium *concentrations* are different in each case.

**TABLE 16.1 Initial and Equilibrium Concentrations for the $N_2O_4$-$NO_2$ System at 100°C**

| INITIAL | | EQUILIBRIUM | | RATIO |
|---|---|---|---|---|
| $[N_2O_4]$ | $[NO_2]$ | $[N_2O_4]$ | $[NO_2]$ | $[NO_2]^2/[N_2O_4]$ |
| 0.1000 | 0.0000 | 0.0491 | 0.1018 | 0.211 |
| 0.0000 | 0.1000 | 0.0185 | 0.0627 | 0.212 |
| 0.0500 | 0.0500 | 0.0332 | 0.0837 | 0.211 |
| 0.0750 | 0.0250 | 0.0411 | 0.0930 | 0.210 |

The particular ratio of terms that we write for a given reaction is called the **mass-action expression (Q)** (also called the **reaction quotient**). For the $N_2O_4$-$NO_2$ system, the mass-action expression is

$$Q = \frac{[NO_2]^2}{[N_2O_4]}$$

Although the concentration terms we use to write $Q$ remain the same, the *numerical value* of $Q$ changes as the reactant and product concentration values change during the reaction. When the system has reached equilibrium at a given temperature, the value of $Q$ equals $K$ at that temperature:

At equilibrium:   $Q = K$        (16.3)

Figure 16.2 shows the changing concentrations of $N_2O_4$ and $NO_2$, and thus the changing value of $Q$, during the course of the reaction. Once the system reaches equilibrium, the concentrations no longer change and $Q$ equals $K$. In other words, *K is a special value of Q that occurs when the reactant and product terms have their equilibrium values.*

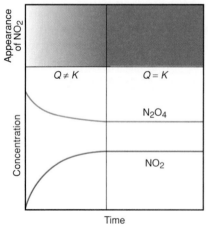

**FIGURE 16.2**

**The change in *Q* during the $N_2O_4$-$NO_2$ reaction.** This plot and the darkening brown screen above show that $[N_2O_4]$ and $[NO_2]$, and therefore the value of $Q$, change with time until equilibrium is reached (*vertical line*). After that time, $Q = K$.

## Writing the Mass-Action Expression

Unlike the rate law for a reaction, which must be determined experimentally, the mass-action expression is based directly on the balanced equation. It is a *ratio made up of product terms multiplied together divided by reactant terms multiplied together, with each term raised to the power of its stoichiometric coefficient.* The most common form of the mass-action expression shows reactant and product terms as concentrations in moles per liter ($M$), designated by square brackets, [ ]. In these cases, which include those you've seen so far, $K$ is the *equilibrium constant based on concentrations* and is designated hereafter as $K_c$. Similarly, we designate the mass-action expression based on concentrations as $Q_c$. For the general equation

$$aA + bB \rightleftharpoons cC + dD$$

where $a$, $b$, $c$, and $d$ are the stoichiometric coefficients, the mass-action expression is

$$Q_c = \frac{[C]^c\,[D]^d}{[A]^a\,[B]^b} \qquad\qquad \textbf{(16.4)}$$

(As you'll see a bit later, another form of the mass-action expression shows reactant and product terms as pressures in atm.)

For any reaction, we *write the balanced equation first and then construct the mass-action expression.* For the decomposition of $N_2O_4$, for example, the balanced equation is

$$N_2O_4(g) \rightleftharpoons 2NO_2(g)$$

and the mass-action expression is

$$Q_c = \frac{[NO_2]^2}{[N_2O_4]}$$

Similarly, for the formation of ammonia from its elements,

$$N_2(g) + 3H_2(g) \rightleftharpoons 2NH_3(g)$$

the mass-action expression is

$$Q_c = \frac{[NH_3]^2}{[N_2]\,[H_2]^3}$$

SAMPLE PROBLEM 16.1 _____

## Writing the Mass-Action Expression from the Balanced Equation

**Problem:** Write the mass-action expression for each of the following reactions:
**(a)** The decomposition of dinitrogen pentaoxide, $N_2O_5(g) \rightleftharpoons NO_2(g) + O_2(g)$
**(b)** The combustion of propane gas, $C_3H_8(g) + O_2(g) \rightleftharpoons CO_2(g) + H_2O(g)$
**Plan:** We balance the equations and then construct the mass-action expression as described by Equation 16.4.

**Solution: (a)** $2N_2O_5(g) \rightleftharpoons 4NO_2(g) + O_2(g)$ $\qquad Q_c = \dfrac{[NO_2]^4\,[O_2]}{[N_2O_5]^2}$

**(b)** $C_3H_8(g) + 5O_2(g) \rightleftharpoons 3CO_2(g) + 4H_2O(g)$ $\qquad Q_c = \dfrac{[CO_2]^3\,[H_2O]^4}{[C_3H_8]\,[O_2]^5}$

**Check:** Always be sure that the exponents in $Q$ are the same as the balancing coefficients.

**FOLLOW-UP PROBLEM 16.1**
Write a mass-action expression for each of the following unbalanced reactions:
**(a)** The first step in nitric acid production, $NH_3(g) + O_2(g) \rightleftharpoons NO(g) + H_2O(g)$
**(b)** The disproportionation of nitric oxide, $NO(g) \rightleftharpoons N_2O(g) + NO_2(g)$

## Variations in the Form of the Mass-Action Expression

As you'll see in the upcoming discussion, the mass-action expression $Q$ is a collection of terms based *exactly* on the balanced equation *as written* for a given reaction. Therefore, the value of $Q$, which varies during the reaction, and the value of $K$, the constant value of $Q$ when the system has reached equilibrium, also depend on how the balanced equation is written.

In this text (and most others), *the values of $Q$ and $K$ are shown as unitless numbers.* From a thermodynamic standpoint, each term in the mass-action expression represents the *ratio* of the measured quantity of the substance (molar concentration or pressure) to the standard state quantity of the substance (1 $M$ or 1 atm). For instance, a concentration of 1.20 $M$ becomes

$$\frac{1.20\ M}{1\ M} = 1.20;\ \text{similarly, a pressure of 0.53 atm becomes}\ \frac{0.53\ \text{atm}}{1\ \text{atm}} =$$

0.53. With the quantity terms unitless, the ratio of terms (value of $Q$ or $K$) is also unitless.

**Form of $Q$ for an overall reaction.** Notice that we've been writing mass-action expressions without knowing whether an equation represents an individual reaction step or an overall multistep reaction. We can do this because we obtain the same expression for the overall reaction as we do when we combine the expressions for the individual steps. That is, *if an overall reaction is the **sum** of two or more reactions, the overall mass-action expression (or equilibrium constant) is the **product** of the mass-action expressions (or equilibrium constants) for the steps:*

$$Q_{\text{overall}} = Q_1 \times Q_2 \times Q_3 \times \cdots$$

and

$$K_{\text{overall}} = K_1 \times K_2 \times K_3 \times \cdots \qquad \textbf{(16.5)}$$

To demonstrate this point, let's work out Sample Problem 16.2.

**SAMPLE PROBLEM 16.2** ————————————————————

### Writing the Mass-Action Expression for an Overall Reaction

**Problem:** At a typical operating temperature of an auto engine ( ~100°C), $N_2$ and $O_2$ form nitric oxide, which then combines with more $O_2$ to form nitrogen dioxide, a toxic pollutant:

(1) $N_2(g) + O_2(g) \rightleftharpoons 2NO(g)$      $K_{c1} = 4.3 \times 10^{-25}$

(2) $2NO(g) + O_2(g) \rightleftharpoons 2NO_2(g)$      $K_{c2} = 6.4 \times 10^9$

**(a)** Show that the overall $Q_c$ for this reaction sequence is the same as the product of the $Q_c$'s for the individual reactions.
**(b)** Calculate $K_c$ for the overall reaction.
**Plan:** In **(a)**, we write the overall reaction by adding the individual reactions and then write the overall $Q_c$. Since we *add* the individual steps, we *multiply* their $Q_c$'s and cancel common terms to obtain the overall expression. In **(b)**, we multiply the individual $K_c$ values to find $K_{c(\text{overall})}$.

**Solution: (a)** Writing the overall reaction and its mass-action expression:

(1)          $N_2(g) + O_2(g) \rightleftharpoons \cancel{2NO(g)}$

(2)          $\cancel{2NO(g)} + O_2(g) \rightleftharpoons 2NO_2(g)$

Overall: $N_2(g) + 2O_2(g) \rightleftharpoons 2NO_2(g)$

$$Q_{c(overall)} = \frac{[NO_2]^2}{[N_2][O_2]^2}$$

Writing the mass-action expressions for the individual steps:

For step (1),          $Q_{c1} = \dfrac{[NO]^2}{[N_2][O_2]}$

For step (2),          $Q_{c2} = \dfrac{[NO_2]^2}{[NO]^2[O_2]}$

Multiplying the individual mass-action expressions and canceling:

$$Q_{c1} \times Q_{c2} = \frac{\cancel{[NO]^2}}{[N_2][O_2]} \times \frac{[NO_2]^2}{\cancel{[NO]^2}[O_2]} = \frac{[NO_2]^2}{[N_2][O_2]^2} = Q_{c(overall)}$$

**(b)** Calculating the overall $K_c$:

$$K_{c(overall)} = K_{c1} \times K_{c2} = (4.3 \times 10^{-25})(6.4 \times 10^9) = 2.8 \times 10^{-15}$$

**Check:** Round off and check the calculation in **(b)**:

$$K_c \approx (4 \times 10^{-25})(6 \times 10^9) = 24 \times 10^{-16} = 2.4 \times 10^{-15}$$

**FOLLOW-UP PROBLEM 16.2**
The following sequence of individual steps has been proposed for the overall reaction between $H_2$ and $Br_2$ to form HBr:

(1)   $Br_2(g) \rightleftharpoons 2Br(g)$

(2)   $Br(g) + H_2(g) \rightleftharpoons HBr(g) + H(g)$

(3)   $H(g) + Br_2(g) \rightleftharpoons HBr(g) + Br(g)$

(4)   $2Br(g) \rightleftharpoons Br_2(g)$

Write the overall equation and show that the overall $Q_c$ is the product of the $Q_c$'s for the individual steps.

**Form of $Q$ for a forward and reverse reaction.** The form of the mass-action expression depends on the *direction* in which the balanced equation is written. Consider, for example, the oxidation of sulfur dioxide to sulfur trioxide:

$$2SO_2(g) + O_2(g) \rightleftharpoons 2SO_3(g)$$

The mass-action expression for this equation *as written* is

$$Q_{c(fwd)} = \frac{[SO_3]^2}{[SO_2]^2[O_2]}$$

If we had written the reverse reaction, the decomposition of sulfur trioxide,

$$2SO_3(g) \rightleftharpoons 2SO_2(g) + O_2(g)$$

the mass-action expression would be the *reciprocal* of $Q_{c(fwd)}$:

$$Q_{c(rev)} = \frac{[SO_2]^2[O_2]}{[SO_3]^2} = \frac{1}{Q_{c(fwd)}}$$

Thus, *a mass-action expression (or equilibrium constant) for a forward reaction is the **reciprocal** of the mass-action expression (or equilibrium constant) for the reverse reaction:*

$$Q_{c(\text{fwd})} = \frac{1}{Q_{c(\text{rev})}} \quad \text{and} \quad K_{c(\text{fwd})} = \frac{1}{K_{c(\text{rev})}} \tag{16.6}$$

The $K_c$ values for these reactions at 1000 K are

$$K_{c(\text{fwd})} = 261 \quad \text{and} \quad K_{c(\text{rev})} = \frac{1}{K_{c(\text{fwd})}} = 0.00383$$

These values make sense: if the forward reaction goes far to the right (high $K_c$), the reverse reaction does not (low $K_c$).

**Form of $Q$ for a reaction with coefficients multiplied by a common factor.** Multiplying all the coefficients of the equation by some factor also changes the form of $Q$. For example, multiplying all the coefficients in the previous equation for the formation of $SO_3$ by $\frac{1}{2}$ gives

$$SO_2(g) + \tfrac{1}{2}O_2(g) \rightleftharpoons SO_3(g)$$

For this equation, the mass-action expression is

$$Q'_{c(\text{fwd})} = \frac{[SO_3]}{[SO_2]\,[O_2]^{\frac{1}{2}}}$$

Notice that the mass-action expression for the halved equation is the mass-action expression for the original equation raised to the $\frac{1}{2}$ power:

$$Q'_{c(\text{fwd})} = Q_{c(\text{fwd})}{}^{\frac{1}{2}} = \left( \frac{[SO_3]^2}{[SO_2]^2\,[O_2]} \right)^{\frac{1}{2}} = \frac{[SO_3]}{[SO_2]\,[O_2]^{\frac{1}{2}}}$$

Once again, the same property holds for the equilibrium constants. Relating the halved reaction to the original, we have

$$K'_{c(\text{fwd})} = K_{c(\text{fwd})}{}^{\frac{1}{2}}$$

Thus, for the halved forward reaction, $K'_{c\,(\text{fwd})} = (261)^{\frac{1}{2}} = 16.2$.

Similarly, if you triple coefficients, the mass-action expression is the original expression raised to the third power. At first, it may seem that we have changed the extent of the reaction, as evidenced by a change in $K$, merely by changing the balancing coefficients of the equation, but this clearly cannot be true. *A particular K has meaning only in relation to a particular balanced equation.* In this case, $K_{c(\text{fwd})}$ and $K'_{c(\text{fwd})}$ relate to different equations and thus cannot be compared directly.

In general, *if all the coefficients of the balanced equation are multiplied by some factor, that factor becomes the exponent for relating the mass-action expressions and the equilibrium constants.* For a multiplying factor $n$, which we can write as, $n[aA + bB \rightleftharpoons cC + dD]$, the mass-action expression is

$$Q' = \left( \frac{[C]^c\,[D]^d}{[A]^a\,[B]^b} \right)^n \quad \text{and} \quad K' = (K)^n \tag{16.7}$$

---

SAMPLE PROBLEM 16.3 ⎯⎯⎯⎯⎯⎯⎯⎯⎯⎯⎯⎯⎯⎯⎯⎯⎯⎯⎯⎯⎯⎯

**Determining the Equilibrium Constant for an Equation Multiplied by a Common Factor**

**Problem:** For the reaction

$$N_2(g) + 3H_2(g) \rightleftharpoons 2NH_3(g)$$

the equilibrium constant $K_c$ is $2.4 \times 10^{-3}$ at 1000 K. If this is the reference (ref) equation, what are the values of $K_c$ for the following balanced equations?

**(a)** $\frac{1}{3}N_2(g) + H_2(g) \rightleftharpoons \frac{2}{3}NH_3(g)$

**(b)** $NH_3(g) \rightleftharpoons \frac{1}{2}N_2(g) + \frac{3}{2}H_2(g)$

**Plan:** We compare each equation with the reference equation to see how the direction and coefficients have changed. In **(a),** the equation is the reference equation multiplied by $\frac{1}{3}$, so $K_c$ equals the $K_{c(ref)}$ raised to the $\frac{1}{3}$ power. In **(b),** the equation is one-half the reverse of the reference equation, so $K_c$ is the reciprocal of $K_{c(ref)}$ raised to the $\frac{1}{2}$ power.

**Solution:** The mass-action expression for the reference equation is

$$Q_{c(ref)} = \frac{[NH_3]^2}{[N_2][H_2]^3}$$

**(a)**
$$Q_c = Q_{c(ref)}^{1/3} = \left( \frac{[NH_3]^2}{[N_2][H_2]^3} \right)^{1/3} = \frac{[NH_3]^{2/3}}{[N_2]^{1/3}[H_2]}$$

Thus, $K_c = K_{c(ref)}^{1/3} = (2.4 \times 10^{-3})^{1/3} = 0.13$

**(b)**
$$Q_c = \left( \frac{1}{Q_{c(ref)}} \right)^{1/2} = \left( \frac{1}{\frac{[NH_3]^2}{[N_2][H_2]^3}} \right)^{1/2} = \frac{[N_2]^{1/2}[H_2]^{3/2}}{[NH_3]}$$

Thus, $K_c = \left( \frac{1}{K_{c(ref)}} \right)^{1/2} = \left( \frac{1}{2.4 \times 10^{-3}} \right)^{1/2} = 20$

**Check:** A good check is to work the math backward. For **(a),** $(0.13)^3 = 2.2 \times 10^{-3}$, within rounding of $2.4 \times 10^{-3}$. The reaction goes in the same direction, so at equilibrium, it should be mostly reactants, as the $K_c < 1$ indicates. For **(b),** $1/(20)^2 = 2.5 \times 10^{-3}$, again within rounding. At equilibrium, the reverse reaction should contain mostly products, as the $K_c > 1$ indicates.

**FOLLOW-UP PROBLEM 16.3**
The reaction of hydrogen and chlorine to form hydrogen chloride is $H_2(g) + Cl_2(g) \rightleftharpoons 2HCl(g)$. At 1200 K, $K_c = 7.6 \times 10^8$. Calculate $K_c$ for the following reactions:

**(a)** $\frac{1}{2}H_2(g) + \frac{1}{2}Cl_2(g) \rightleftharpoons HCl(g)$

**(b)** $\frac{4}{3}HCl(g) \rightleftharpoons \frac{2}{3}H_2(g) + \frac{2}{3}Cl_2(g)$

**Form of $Q$ for a reaction involving pure liquids and solids.** Until now, we've looked at gaseous reactions only. If the components of the reaction are in the same phase, the system reaches *homogeneous equilibrium.* If they are in different phases, the system reaches *heterogeneous equilibrium.* In the latter case, one or more of the components is a liquid or solid.

Consider the decomposition of limestone to lime and carbon dioxide as an example of a system reaching heterogeneous equilibrium:

$$CaCO_3(s) \rightleftharpoons CaO(s) + CO_2(g)$$

Based on the rules for writing the mass-action expression, we have

$$Q_c = \frac{[CaO][CO_2]}{[CaCO_3]}$$

A solid, however, such as $CaCO_3$ or CaO, always has the same concentration at a given temperature: the same number of moles per liter of the solid. Moreover, since a solid's volume changes very little with temperature, its concentration also changes very little. For all intents and purposes, therefore, the concentration of a solid is constant. The same argument applies to the concentration of a pure liquid.

**FIGURE 16.3**
**The mass-action expression for a heterogeneous system.** Even though the two containers have different amounts of the two solids, as long as both solids are present, at a given temperature, the containers have the same $[CO_2]$ at equilibrium.

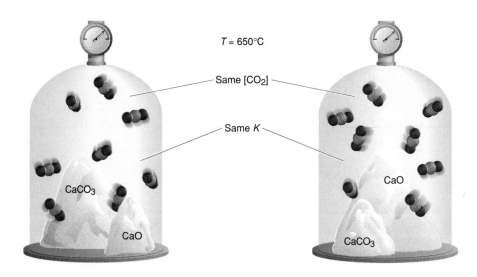

$T = 650°C$

Same $[CO_2]$

Same $K$

CaO

$CaCO_3$

$CaCO_3$

CaO

Because we are concerned only with concentrations that *change* as they approach and reach equilibrium, *we eliminate the terms for pure liquids and solids from the mass-action expression* by incorporating their constant concentrations into a rearranged mass-action expression. Thus, for the decomposition of $CaCO_3$, the only substance whose concentration can change is $CO_2$, so we obtain the new mass-action expression $Q_c'$:

$$Q_c' = Q_c \frac{[CaCO_3]}{[CaO]} = [CO_2]$$

No matter how much CaO and $CaCO_3$ are present, the mass-action expression for the reaction equals the $CO_2$ concentration (Figure 16.3).

Table 16.2 summarizes the various ways of writing the mass-action expression.

**TABLE 16.2  Ways of Expressing the Mass-Action Expression, Q**

| FORM OF CHEMICAL EQUATION | FORM OF Q |
|---|---|
| Reference reaction: A $\rightleftharpoons$ B | $Q_{(ref)} = \dfrac{[B]}{[A]}$ |
| Reverse reaction: B $\rightleftharpoons$ A | $Q = \dfrac{1}{Q_{(ref)}} = \dfrac{[A]}{[B]}$ |
| Reaction as sum of two steps:<br>(1) A $\rightleftharpoons$ C | $Q_1 = \dfrac{[C]}{[A]}; Q_2 = \dfrac{[B]}{[C]}$ |
| (2) C $\rightleftharpoons$ B | $Q_{(ref)} = Q_1 \times Q_2 = Q_{overall}$ <br> $= \dfrac{[C]}{[A]} \times \dfrac{[B]}{[C]} = \dfrac{[B]}{[A]}$ |
| Coefficients multiplied by $n$ | $Q = Q_{(ref)}^n$ |
| Reaction with pure solid or liquid component, such as A(s) | $Q' = Q_{(ref)}[A] = [B]$ |

**Equilibria Involving Gases: Relation Between $K_c$ and $K_p$**

It is usually easier to measure the pressure of a gas than its concentration and, as long as the gases behave ideally under the conditions of the experiment, the ideal gas law (Chapter 5) allows us to relate these variables to each other:

$$PV = nRT, \quad \text{so} \quad P = \frac{n}{V}RT \quad \text{or} \quad \frac{P}{RT} = \frac{n}{V}$$

where $P$ is the pressure of a gas and $n/V$ is its molar concentration. As you'll see shortly, when substances in the mass-action expression are gases, we can use partial pressures for the terms. For example, in the reaction between NO and $O_2$,

$$2NO(g) + O_2(g) \rightleftharpoons 2NO_2(g)$$

the mass-action expression based on partial pressures is

$$Q_p = \frac{P^2_{NO_2}}{P^2_{NO} \times P_{O_2}}$$

The equilibrium constant obtained when all species are present at their equilibrium partial pressures is designated $K_p$, the *equilibrium constant based on pressures*. In many cases, $K_p$ has a different value from $K_c$, but the two constants are related, so that if you know one, you can calculate the other by noting the change in number of moles of gas. Let's see their relationship by converting the terms in $Q_c$ for the NO-$O_2$ reaction to those in $Q_p$:

$$2NO(g) + O_2(g) \rightleftharpoons 2NO_2(g)$$

3 moles of gaseous reactants $\rightleftharpoons$ 2 moles of gaseous products

$\Delta n_{gas}$ = moles of gaseous product − moles of gaseous reactant = 2 − 3 = −1

Keep in mind the change in number of moles of gas, $\Delta n_{gas}$, because it appears at a crucial place in the algebraic conversion. The mass-action expression based on concentrations is

$$Q_c = \frac{[NO_2]^2}{[NO]^2\,[O_2]}$$

We write this expression with the concentration terms as $n/V$ and then convert them to partial pressures:

$$Q_c = \frac{\dfrac{n^2_{NO_2}}{V^2}}{\dfrac{n^2_{NO}}{V^2} \times \dfrac{n_{O_2}}{V}} = \frac{\dfrac{P^2_{NO_2}}{(RT)^2}}{\dfrac{P^2_{NO}}{(RT)^2} \times \dfrac{P_{O_2}}{RT}}$$

Collecting the $RT$ terms, we obtain

$$Q_c = \frac{P^2_{NO_2}}{P^2_{NO} \times P_{O_2}} \times \frac{\dfrac{1}{(RT)^2}}{\dfrac{1}{(RT)^2} \times \dfrac{1}{RT}} = \frac{P^2_{NO_2}}{P^2_{NO} \times P_{O_2}} \times RT$$

The right side of this expression is the mass-action expression based on partial pressures, which we wrote earlier, multiplied by $RT$: $Q_c = Q_p(RT)$. Thus, at equilibrium, $K_c = K_p(RT)$, or $K_p = K_c(RT)^{-1}$.

Notice that *the exponent of the RT term equals the change in the moles of gas* ($\Delta n_{gas}$) *from the balanced equation.* Thus, in general,

$$K_p = K_c(RT)^{\Delta n_{gas}} \tag{16.8}$$

The units for the partial pressure terms in $K_p$ will be atmospheres raised to some power, so the value of $R$ must be consistent with them.

SAMPLE PROBLEM 16.4

### Converting Between $K_c$ and $K_p$

**Problem:** Calculate $K_c$ for the following reaction:

$$CaCO_3(s) \rightleftharpoons CaO(s) + CO_2(g) \qquad K_p = 2.1 \times 10^{-4} \text{ (at 1000 K)}$$

**Plan:** To convert between $K_c$ and $K_p$, we determine $\Delta n_{gas}$ from the balanced equation and use Equation 16.8. $R$ is 0.0821 atm · L/mol · K.
**Solution:** Determining $\Delta n_{gas}$: There is one mole of gas on the right and none on the left, so $\Delta n_{gas} = 1$.
Calculating $K_c$:

$$K_p = K_c(RT)^1 \qquad \text{so} \qquad K_c = K_p(RT)^{-1}$$

$$K_c = (2.1 \times 10^{-4}) \left( 0.0821 \frac{\text{atm} \cdot \text{L}}{\text{mol} \cdot \text{K}} \times 1000 \text{ K} \right)^{-1} = \mathbf{2.6 \times 10^{-6}}$$

**Check:** Round off and work backward to see if you obtain the given $K_p$:
$K_p \approx (3 \times 10^{-6})(0.08 \times 1000) = 2.4 \times 10^{-4}$, within rounding of $2.1 \times 10^{-4}$.

FOLLOW-UP PROBLEM 16.4
Calculate $K_p$ for the following reaction:

$$PCl_3(g) + Cl_2(g) \rightleftharpoons PCl_5(g) \qquad K_c = 1.67 \text{ (at 500 K)}$$

### Reaction Progress: Comparing the Values of $Q$ and $K$

Suppose you are studying a reaction begun with a mixture of reactants and products for which you know the equilibrium constant at some temperature. How do you know if the reaction has reached equilibrium? If it hasn't, how do you know the direction in which it is progressing? In contrast to the value of $K$, which is constant at a given temperature, the value of $Q$ changes and can be smaller than, larger than, or, when the system attains equilibrium, equal to $K$. By comparing the value of $Q$ during the reaction with that of $K$, you can tell whether and in which direction the reaction must progress to attain equilibrium. Since product terms are in the numerator and reactant terms are in the denominator, *more product makes the ratio larger and more reactant makes the ratio smaller.*

The three possible comparisons of $Q$ and $K$ are
1. $Q < K$. If the value of $Q$ is smaller than $K$, the denominator (reactants) is too large relative to the numerator (products). For $Q$ to become equal to $K$, the denominator must decrease and the numerator increase. In other words, the reaction will progress to the right, toward products, until equilibrium is reached:

$$\text{If } Q < K, \qquad \text{reactant} \rightarrow \text{product}$$

2. $Q > K$. If $Q$ is larger than $K$, the numerator (products) will decrease and the denominator (reactants) increase until equilibrium is reached. Therefore, the reaction will progress to the left, toward reactants:

$$\text{If } Q > K, \qquad \text{reactant} \leftarrow \text{product}$$

3.  $Q = K$. This situation exists only when the reactant and product concentrations (or pressures) have attained their equilibrium values. Thus, no further net change occurs:

$$\text{If } Q = K, \qquad \text{reactant} \rightleftharpoons \text{product}$$

SAMPLE PROBLEM 16.5 ——————————————————————

**Comparing $Q$ and $K$ to Determine Reaction Progress**

**Problem:** For the reaction $N_2O_4(g) \rightleftharpoons 2NO_2(g)$, $K_c = 0.21$ at 100°C. At a point during the reaction, $[N_2O_4] = 0.12\ M$ and $[NO_2] = 0.55\ M$. Is the reaction at equilibrium? If not, in which direction is it progressing?
**Plan:** We solve for $Q_c$ by substituting the given concentrations and then compare its value with the given $K_c$.
**Solution:** Writing the mass-action expression and solving for $Q_c$:

$$Q_c = \frac{[NO_2]^2}{[N_2O_4]} = \frac{0.55^2}{0.12} = 2.5$$

Since $Q_c > K_c$, **the reaction is not at equilibrium and will proceed to the left until $Q_c = K_c$.**
**Check:** With $[NO_2] > [N_2O_4]$, we expect to obtain a value for $Q_c$ that is greater than 0.21. If $Q_c > K_c$, the denominator will increase until $Q_c = K_c$, so the reaction proceeds toward reactants.

FOLLOW-UP PROBLEM 16.5
Chloromethane forms by the reaction

$$CH_4(g) + Cl_2(g) \rightleftharpoons CH_3Cl(g) + HCl(g)$$

At 1500 K, the equilibrium constant $K_p = 1.6 \times 10^4$. In the reaction mixture, $P_{CH_4} = 0.13$ atm, $P_{Cl_2} = 0.035$ atm, $P_{CH_3Cl} = 0.24$ atm, and $P_{HCl} = 0.47$ atm. Is $CH_3Cl$ or $CH_4$ forming?

———————————————————————————————————

As we have discussed, three criteria define a system at equilibrium:
• The forward reaction rate equals the reverse reaction rate.
• Reactant and product concentrations are constant over time.
• The mass-action expression equals the equilibrium constant: $Q = K$.

**Section Summary**
The mass-action expression $Q$ is a particular ratio of product to reactant terms. Substituting experimental values into this expression gives the value of $Q$, which changes as the reaction proceeds. When the system reaches equilibrium at a particular temperature, the value of $Q = K$. If a reaction is the sum of two or more reactions, the overall $Q$ (or $K$) is the product of the individual $Q$'s (or $K$'s). The *form* of $Q$ is based directly on the balanced equation for the reaction, so it changes if the equation is reversed or multiplied by some factor, and $K$ changes accordingly. Pure liquids or solids do not appear in $Q$ because their concentrations are constant. $Q$ and $K$ can be expressed in terms of concentration ($Q_c$ and $K_c$) or, for gases, in terms of partial pressure ($Q_p$ and $K_p$). The values of $K_p$ and $K_c$ are related through the ideal gas law. We compare the values of $Q$ and $K$ to see how a reaction will progress toward equilibrium.

## 16.3  How to Solve Equilibrium Problems

Many kinds of equilibrium problems arise, in the real world as well as in chemistry courses, but we can reduce most of them to one of two types. In the first, we use concentration (or partial pressure) data to solve for the equilibrium constant; in the second, we use the equilibrium constant and initial quantities to solve for the equilibrium quantities.

### Using Concentrations to Determine the Equilibrium Constant

There are two common variations of the type of equilibrium problem in which we have to find the equilibrium constant. A straightforward case occurs when we are given the equilibrium concentrations and must calculate $K_c$.

Suppose, for example, that equal amounts of gaseous hydrogen and iodine are injected into a 1.50-L reaction flask at a fixed temperature. The following reaction occurs:

$$H_2(g) + I_2(g) \rightleftharpoons 2HI(g)$$

At equilibrium, analysis shows that the flask contains 1.80 mol $H_2$, 1.80 mol $I_2$, and 0.520 mol HI. Since these are equilibrium concentrations, we calculate $K_c$ by substituting them into the mass-action expression. From the balanced equation, we write the mass-action expression:

$$Q_c = \frac{[HI]^2}{[H_2]\,[I_2]}$$

We first have to convert the amounts given in moles to moles/liter, using the flask volume of 1.50 L:

$$[H_2] = \frac{1.80 \text{ mol}}{1.50 \text{ L}} = 1.20 \ M$$

Similarly, $[I_2] = 1.20 \ M$, and $[HI] = 0.347 \ M$. Substituting these values into the $Q_c$ expression gives $K_c$:

$$K_c = \frac{0.347^2}{1.20 \times 1.20} = 8.36 \times 10^{-2}$$

If some values are not given, we determine them first and then find $K$. In the following problem, pay close attention to a valuable technique being introduced: the reaction table.

An evacuated vessel containing a small amount of powdered graphite is heated to 1080 K, and then $CO_2$ is added to a pressure of 0.458 atm. Once the $CO_2$ is added, the system starts to produce CO. After equilibrium is reached, the total pressure inside the vessel is 0.757 atm. Calculate $K_p$.

As always, we start by writing the balanced equation and the mass-action expression:

$$CO_2(g) + C(s) \rightleftharpoons 2CO(g)$$

Since the data are given in atmospheres and we must find $K_p$, we write $Q$ in terms of partial pressures:

$$Q_p = \frac{P_{CO}^2}{P_{CO_2}}$$

We are given the initial $P_{CO_2}$ and $P_{total}$ at equilibrium. In order to find $K_p$, we must find the equilibrium pressures of the components, which requires solving a stoichiometry problem, and then substitute them into the expression for $Q_p$. To find these pressures, let's think through what happened in the vessel. An unknown amount of $CO_2$ reacted with graphite to form an unknown amount of CO. We know the *relative* amounts of $CO_2$ and CO from the balanced equation: for 1 mol $CO_2$ that reacts, 2 mol CO forms, which means that for $x$ atm of $CO_2$ that reacts, $2x$ atm CO forms:

$$x \text{ atm } CO_2 \rightarrow 2x \text{ atm } CO$$

From the information given, we already know something about the size of $K_p$. If it were very large, almost all the $CO_2$ would be converted to CO, so the final (total) pressure would be twice the initial pressure ($\sim$0.5 atm $\times$ 2 $\approx$ 1 atm). On the other hand, if $K_p$ were very small, almost no CO would form, so the final pressure would be close to the initial pressure, $\sim$0.5 atm. Since the final pressure is between these extremes, $K_p$ must have an intermediate value. The amount of $CO_2$ at equilibrium, $P_{CO_2(eq)}$, is the initial amount, $P_{CO_2(init)}$, *minus* the amount that reacts, $x$:

$$P_{CO_2(init)} - x = P_{CO_2(eq)}$$

Similarly, the amount of CO at equilibrium is the initial amount *plus* the amount that forms, $2x$:

$$P_{CO(init)} + 2x = P_{CO(eq)}$$

A good way to summarize this information is with a *reaction table* that shows the balanced equation and what we know about
- the *initial* amounts of reactants and products
- the *change* in these amounts during the reaction
- the *equilibrium* amounts

| Pressure (atm) | $CO_2(g)$ | + | $C(s)$ | $\rightleftharpoons$ | $2CO(g)$ |
|---|---|---|---|---|---|
| Initial | 0.458 | | — | | 0 |
| Change | $-x$ | | | | $+2x$ |
| Equilibrium | $0.458 - x$ | | | | $2x$ |

Note that we include data only for those substances whose concentrations will change. We use reaction tables in many of the equilibrium problems here and in later chapters.

The other piece of information we have, $P_{total}$, can now be used to find $x$, the amount of $CO_2$ that reacted. According to Dalton's law of partial pressures and using the quantities from the reaction table,

$$P_{total} = 0.757 \text{ atm} = P_{CO_2(eq)} + P_{CO(eq)} = 0.458 \text{ atm} - x + 2x$$

So $\quad\quad\quad x = 0.299$ atm

With $x$ known, we determine the equilibrium partial pressures:

$$P_{CO_2(eq)} = 0.458 \text{ atm} - x = 0.458 \text{ atm} - 0.299 \text{ atm} = 0.159 \text{ atm}$$

$$P_{CO(eq)} = 2x = 2 \times 0.299 \text{ atm} = 0.598 \text{ atm}$$

We substitute these values into the expression for $Q_p$ to find $K_p$:

$$Q_p = \frac{P^2_{CO(eq)}}{P_{CO_2(eq)}} = \frac{0.598^2}{0.159} = 2.25 = K_p$$

As we predicted, $K_p$ is neither very large nor very small.

SAMPLE PROBLEM 16.6

## Calculating $K_c$ from Concentration Data

**Problem:** In a study of hydrogen halide decomposition, a researcher fills an evacuated 2.00-L flask with 0.200 mol HI gas and allows the reaction to proceed at 453°C:

$$2HI(g) \rightleftharpoons H_2(g) + I_2(g)$$

At equilibrium, [HI] = 0.078 $M$. Calculate $K_c$.

**Plan:** To calculate $K_c$, we need the equilibrium concentrations. We can find the initial [HI] and are given [HI] at equilibrium. The balanced equation shows that when $2x$ mol HI react, $x$ mol $H_2$ and $x$ mol $I_2$ form. We set up a reaction table, solve for $x$, using the known equilibrium [HI], and substitute the concentrations into $Q_c$.

**Solution:** Calculating initial [HI]:

$$[HI] = \frac{0.200 \text{ mol}}{2.00 \text{ L}} = 0.100 \ M$$

Setting up the reaction table, with $x$ = [$H_2$] and [$I_2$] that forms:

| Concentration (M) | 2HI(g) | $\rightleftharpoons$ | $H_2$(g) | + | $I_2$(g) |
|---|---|---|---|---|---|
| Initial | 0.100 | | 0 | | 0 |
| Change | $-2x$ | | $+x$ | | $+x$ |
| Equilibrium | 0.100 − 2x | | $x$ | | $x$ |

Solving for $x$ with the known [HI] at equilibrium:

$$[HI] = 0.100 \ M - 2x = 0.078 \ M$$

$$x = 0.011 \ M$$

Therefore, the equilibrium concentrations are

$$[H_2] = [I_2] = 0.011 \ M \quad \text{and} \quad [HI] = 0.078 \ M$$

Substituting into the mass-action expression:

$$Q_c = \frac{[H_2]\,[I_2]}{[HI]^2}$$

thus

$$K_c = \frac{0.011 \times 0.011}{0.078^2} = 0.020$$

**Check:** Rounding gives ~$0.01^2/0.08^2 = 0.02$. Since the initial [HI] of 0.100 $M$ was reduced slightly to 0.078 $M$ at equilibrium, relatively little product formed, so we would expect $K_c < 1$.

FOLLOW-UP PROBLEM 16.6

The reaction between 1.000 atm NO and 1.000 atm $O_2$ to produce $NO_2$ at 184°C was studied. At equilibrium, $P_{O_2}$ = 0.506 atm. Calculate $K_p$.

## Using the Equilibrium Constant to Determine Concentrations

The other major type of equilibrium problem is very common. In one variation of this type, we are given the equilibrium constant and some of the equilibrium concentrations, and we need to find the other equilibrium concentrations. Sample Problem 16.7 is the reverse of Sample Problem 16.6, in which we were given the concentrations and had to find the equilibrium constant.

SAMPLE PROBLEM 16.7

## Determining Equilibrium Concentrations from $K_c$

**Problem:** In a study of methane conversion to other fuels, a chemical engineer mixes gaseous $CH_4$ and $H_2O$ in a 0.32-L flask at 1200 K. At equilibrium, the flask contains 0.26 mol CO, 0.091 mol $H_2$, and 0.041 mol $CH_4$. What is $[H_2O]$ at equilibrium? $K_c = 0.26$.

**Plan:** First, we write the balanced equation and mass-action expression. We can calculate the equilibrium concentrations from the given numbers of moles and the flask volume. Substituting these into $Q_c$, and setting it equal to the given $K_c$, we solve for the unknown equilibrium concentration, $[H_2O]$.

**Solution:** Writing the balanced equation and mass-action expression:

$$CH_4(g) + H_2O(g) \rightleftharpoons CO(g) + 3H_2(g) \qquad Q_c = \frac{[CO]\,[H_2]^3}{[CH_4]\,[H_2O]}$$

Determining the equilibrium concentrations:

$$[CH_4] = 0.041 \text{ mol}/0.32 \text{ L} = 0.13 \ M$$

Similarly, $[CO] = 0.81 \ M$ and $[H_2] = 0.28 \ M$.

Calculating $[H_2O]$ at equilibrium: The value of $Q_c = K_c$, so rearranging gives

$$[H_2O] = \frac{[CO]\,[H_2]^3}{[CH_4]\,K_c} = \frac{0.81 \times 0.28^3}{0.13 \times 0.26} = \textbf{0.53 } \textbf{\textit{M}}$$

**Check:** We can substitute the concentrations into $Q_c$ to confirm $K_c$:

$$Q_c = \frac{[CO]\,[H_2]^3}{[CH_4]\,[H_2O]} = \frac{0.81 \times 0.28^3}{0.13 \times 0.53} = 0.26 = K_c$$

FOLLOW-UP PROBLEM 16.7

The equilibrium between nitric oxide, oxygen, and nitrogen is described by the following equation: $2NO(g) \rightleftharpoons N_2(g) + O_2(g)$; $K_c = 2.3 \times 10^{30}$ at 298 K. In the atmosphere, $P_{O_2} = 0.209$ atm and $P_{N_2} = 0.781$ atm. What is the equilibrium partial pressure of NO in the air we breathe?

A more demanding type of problem, of which there are also many variations, involves finding the equilibrium concentrations from initial values and the equilibrium constant. This type arises whenever you mix known amounts of substances and want to know how much of each is present when the reaction reaches equilibrium. In the next sample problem, the given concentrations will allow us to apply a math shortcut, so that we can focus more easily on the conceptual approach.

SAMPLE PROBLEM 16.8

## Determining Equilibrium Concentrations from Initial Concentrations and $K_c$

**Problem:** The extent of the change from CO and $H_2O$ to $CO_2$ and $H_2$ is used to regulate the proportions in syngas fuel mixtures. If 0.250 mol CO and 0.250 mol $H_2O$ are placed in a 125-mL flask at 900 K, what is the composition of the equilibrium mixture? $K_c = 1.56$.

**Plan:** We have to find the equilibrium concentrations. As always, we write the balanced equation and the mass-action expression. We find the initial [CO] and $[H_2O]$ from the moles and volume, use the balanced equation to set up a reaction table, substitute into $Q_c$, and solve for $x$, from which we calculate the concentrations.

**Solution:** Writing the balanced equation and $Q_c$:

$$CO(g) + H_2O(g) \rightleftharpoons CO_2(g) + H_2(g) \qquad Q_c = \frac{[CO_2]\,[H_2]}{[CO]\,[H_2O]}$$

Calculating initial reactant concentrations:

$$[CO] = [H_2O] = \frac{0.250 \text{ mol}}{0.125 \text{ L}} = 2.00 \text{ } M$$

Setting up the reaction table, with $x = [CO]$ and $[H_2O]$ that reacts:

| Concentration (M) | CO(g) | + | H$_2$O(g) | $\rightleftharpoons$ | CO$_2$(g) | + | H$_2$(g) |
|---|---|---|---|---|---|---|---|
| Initial | 2.00 | | 2.00 | | 0 | | 0 |
| Change | $-x$ | | $-x$ | | $+x$ | | $+x$ |
| Equilibrium | $2.00 - x$ | | $2.00 - x$ | | $x$ | | $x$ |

Substituting into the mass-action expression and solving for $x$:

$$Q_c = \frac{[CO_2]\,[H_2]}{[CO]\,[H_2O]} = \frac{(x)(x)}{(2.00 - x)(2.00 - x)} = \frac{x^2}{(2.00 - x)^2}$$

At equilibrium, $Q_c = K_c$, so we have

$$K_c = 1.56 = \frac{x^2}{(2.00 - x)^2}$$

A math shortcut applies in this case but not in general: the right side of the equation is a perfect square, so we take the square root of both sides:

$$\sqrt{1.56} = \frac{x}{2.00 - x} = \pm 1.25$$

Although a positive number (1.56) has a positive *and* a negative square root, only the positive root has any chemical meaning, so we ignore the negative root:*

$$1.25 = \frac{x}{2.00 - x} \qquad \text{or} \qquad 2.50 - 1.25x = x$$

So $2.50 = 2.25x$; therefore, $x = 1.11$ $M$
Calculating equilibrium concentrations:

$$[CO] = [H_2O] = 2.00 \text{ M} - x = 2.00 \text{ M} - 1.11 \text{ M} = \textbf{0.89 } \textbf{\textit{M}}$$

$$[CO_2] = [H_2] = x = \textbf{1.11 } \textbf{\textit{M}}$$

**Check:** From the intermediate size of $K_c$, it makes sense that the changes in concentration are moderate. It's a good idea to check that the sign of $x$ in the reaction table makes sense—only reactants were initially present, so the change had to proceed to the right: $x$ is negative for reactants and positive for products. Also check that the concentrations give the known $K_c$: $\dfrac{1.11 \times 1.11}{0.89 \times 0.89} = 1.56$

**FOLLOW-UP PROBLEM 16.8**
The decomposition of HI at low temperature was studied by injecting 2.50 mol HI into a 10.32-L vessel at 25°C. What is $[H_2]$ at equilibrium for the reaction $2HI(g) \rightleftharpoons H_2(g) + I_2(s)$? $K_c = 1.26 \times 10^{-3}$?

---

*The negative root would give $-1.25 = \dfrac{x}{2.00 - x}$, or $-2.50 + 1.25x = x$.

So                      $-2.50 = -0.25x$, and $x = 10$ $M$

This value has no chemical meaning because we started with 2.00 $M$ of each reactant, so it is impossible for 10 $M$ to react. Moreover, the square root of an equilibrium constant is another equilibrium constant, which cannot have a negative value.

**Using the quadratic formula to solve for the unknown.** The short-cut that allowed us to simplify the math in Sample Problem 16.8 is a special case that occurs when the numerator and denominator of the mass-action expression are perfect squares. It worked previously because we started with equal reactant concentrations, but that is not ordinarily the case.

Suppose, for example, we had started the previous reaction with 2.00 $M$ CO and 1.00 $M$ $H_2O$. The reaction table is

| Concentration (M) | CO(g) | + | $H_2O(g)$ | $\rightleftharpoons$ | $CO_2(g)$ | + | $H_2(g)$ |
|---|---|---|---|---|---|---|---|
| Initial | 2.00 | | 1.00 | | 0 | | 0 |
| Change | $-x$ | | $-x$ | | $+x$ | | $+x$ |
| Equilibrium | $2.00 - x$ | | $1.00 - x$ | | $x$ | | $x$ |

Substituting these values into $Q_c$, we obtain

$$Q_c = \frac{[CO_2][H_2]}{[CO][H_2O]} = \frac{(x)(x)}{(2.00 - x)(1.00 - x)} = \frac{x^2}{x^2 - 3.00x + 2.00}$$

At equilibrium, we have

$$1.56 = \frac{x^2}{x^2 - 3.00x + 2.00}$$

To solve for $x$ in this case, we rearrange the previous expression into the form of a *quadratic equation: $ax^2 + bx + c = 0$.* Carrying out the arithmetic gives

$$0.56x^2 - 4.68x + 3.12 = 0$$

where $a = 0.56$, $b = -4.68$, and $c = 3.12$. Then we can find $x$ with the quadratic formula (Appendix A):

$$x = \frac{-b \pm \sqrt{b^2 - 4ac}}{2a}$$

We obtain two possible values for $x$, but only one will make sense chemically:

$$x = \frac{4.68 \pm \sqrt{(-4.68)^2 - 4(0.56)(3.12)}}{2(0.56)}$$

$$x = 7.6\ M \quad \text{and} \quad x = 0.73\ M$$

The larger value gives negative concentrations at equilibrium (for example, $2.00\ M - 7.6\ M = -5.6\ M$), which have no chemical meaning. Therefore, $x = 0.73\ M$ and we have

$$[CO] = 2.00\ M - x = 2.00\ M - 0.73\ M = 1.27\ M$$

$$[H_2O] = 1.00\ M - x = 0.27\ M$$

$$[CO_2] = [H_2] = x = 0.73\ M$$

Checking to see if these values give the known $K_c$, we have

$$K_c = \frac{0.73 \times 0.73}{1.27 \times 0.27} = 1.6, \text{ within rounding of 1.56}$$

**Problems involving initial mixtures of reactants and products.** In the problems we've worked so far, the direction of the reaction was obvious: since only reactants were present, the reaction had to go toward products. Thus, in the reaction tables, we knew that the change in reactant concentration had a negative sign ($-x$) and that the change in product concentration had a positive sign ($+x$). Suppose, however, we start with a mixture of reactants and products. *Whenever the reaction direction is not obvious, we first compare Q with K to find the direction in which the reaction proceeds to reach equilibrium.* This tells us whether the unknown change in concentration is positive or negative.

---

SAMPLE PROBLEM 16.9

### Predicting Reaction Direction and Calculating Equilibrium Concentrations

---

**Problem:** The research and development unit of a chemical company is studying the reaction of $CH_4$ and $H_2S$, two components of natural gas:

$$CH_4(g) + 2H_2S(g) \rightleftharpoons CS_2(g) + 4H_2(g)$$

In one experiment, 1.00 mol $CH_4$, 1.00 mol $CS_2$, 2.00 mol $H_2S$, and 2.00 mol $H_2$ are mixed in a 250-mL vessel at 960°C. At this temperature, $K_c = 0.036$.
**(a)** In which direction will the reaction proceed to reach equilibrium?
**(b)** If $[CH_4] = 5.56\ M$ at equilibrium, what are the concentrations of the other substances?
**Plan: (a)** To find the direction, we convert the given initial amounts to concentrations, calculate $Q_c$, and compare it with $K_c$. **(b)** Based on the results from (a), we determine the sign of each concentration change for the reaction table and then use the known $[CH_4]$ at equilibrium to determine $x$ and the other equilibrium concentrations.
**Solution: (a)** Calculating the initial concentrations:

$$[CH_4] = \frac{1.00\ \text{mol}}{0.250\ \text{L}} = 4.00\ M$$

Similarly, $[H_2S] = 8.00\ M$, $[CS_2] = 4.00\ M$, and $[H_2] = 8.00\ M$.
Calculating the value of $Q_c$:

$$Q_c = \frac{[CS_2]\ [H_2]^4}{[CH_4]\ [H_2S]^2} = \frac{4.00 \times 8.00^4}{4.00 \times 8.00^2} = 64.0$$

Comparing $Q_c$ and $K_c$: $Q_c > K_c$ (64.0 > 0.036), **so the reaction goes to the left.** Therefore, reactants increase and products decrease their concentrations.
**(b)** Setting up the reaction table, with $x = [CS_2]$ that reacts or $[CH_4]$ that forms:

| Concentration *(M)* | $CH_4(g)$ | + | $2H_2S(g)$ | $\rightleftharpoons$ | $CS_2(g)$ | + | $4H_2(g)$ |
|---|---|---|---|---|---|---|---|
| Initial | 4.00 | | 8.00 | | 4.00 | | 8.00 |
| Change | $+x$ | | $+2x$ | | $-x$ | | $-4x$ |
| Equilibrium | $4.00 + x$ | | $8.00 + 2x$ | | $4.00 - x$ | | $8.00 - 4x$ |

Solving for $x$:

At equilibrium,     $[CH_4] = 5.56\ M = 4.00\ M + x$

$$x = 1.56\ M$$

Thus,     $[H_2S] = 8.00\ M + 2x = 8.00\ M + 2(1.56\ M) = \mathbf{11.12\ M}$
$[CS_2] = 4.00\ M - x = \mathbf{2.44\ M}$
$[H_2] = 8.00\ M - 4x = \mathbf{1.76\ M}$

**Check:** The comparison of $Q_c$ and $K_c$ showed the reaction moving to the left, which is confirmed by the increase in $[CH_4]$ from 4.00 $M$ to 5.56 $M$ during the reaction. Check that the concentrations give the known $K_c$:

$$\frac{2.44 \times 1.76^4}{5.56 \times 11.12^2} = 0.0341 \approx 0.036$$

FOLLOW-UP PROBLEM 16.9

An inorganic chemist studying the reactions of phosphorus halides mixes 0.1050 mol $PCl_5$ with 0.0450 mol $Cl_2$ and 0.0450 mol $PCl_3$ in a 0.5000-L flask at 250°C:

$$PCl_5(g) \rightleftharpoons PCl_3(g) + Cl_2(g) \qquad K_c = 4.2 \times 10^{-2}$$

**(a)** In which direction will the reaction proceed?
**(b)** If $[PCl_5] = 0.2065\ M$ at equilibrium, what are the equilibrium concentrations of the other components?

**Simplifying assumptions for reactions with a small $K$ and high initial reactant concentration.** In many cases, we can use chemical insight to make an assumption that avoids the use of the quadratic formula to solve a problem. In general, if a reaction has a relatively small equilibrium constant *and* a relatively large initial reactant concentration, *the concentration change (x) can often be neglected* without introducing significant error. This assumption does not mean that $x = 0$, because then there would be no reaction; it means that the reactant concentration is nearly the same as its concentration at equilibrium:

$$[\text{reactant}]_{init} - x \approx [\text{reactant}]_{init} \approx [\text{reactant}]_{eq}$$

You can imagine a similar situation in everyday life. On a bathroom scale, you weigh 158 lb. Take off your wristwatch, and you still weigh 158 lb. Within the precision of the measurement, the weight of the wristwatch is so small compared with your weight that it can be neglected:

$$\text{Your weight} - \text{weight of watch} \approx \text{your weight}$$

Similarly, if the initial concentration of reactant A is, for example, 0.500 $M$ and, because of a small $K_c$, the amount that reacts is 0.002 $M$, we can assume that

$$[A]_{init} - [A]_{reacting} \approx [A]_{init} \approx [A]_{eq} \qquad \textbf{(16.9)}$$
$$0.500\ M - 0.002\ M = 0.498\ M \approx 0.500\ M$$

It's essential to check that the error introduced is not significant and thus that the assumption is justified. Here is a common criterion for doing so: *if the assumption results in a change (error) in a concentration that is less than 5%, the assumption is justified.* This approach may seem a bit vague at this point, so let's go through a sample problem that involves making this assumption, see how it simplifies the math, and then see if it is justified. We'll make a similar assumption often in Chapters 17 and 18.

SAMPLE PROBLEM 16.10

**Calculating Equilibrium Concentrations Through the Use of Simplifying Assumptions**

**Problem:** Phosgene is a potent chemical warfare agent that was outlawed by the Geneva Convention. It decomposes by the reaction

$$COCl_2(g) \rightleftharpoons CO(g) + Cl_2(g) \qquad K_c = 8.3 \times 10^{-4} \text{ (at 360°C)}$$

Calculate [CO], $[Cl_2]$, and $[COCl_2]$, when the following amounts of phosgene decompose and reach equilibrium in a 10.0-L flask:
   **(a)** 5.00 mol $COCl_2$          **(b)** 0.100 mol $COCl_2$

**Plan:** We know from the balanced equation that when $x$ mol $COCl_2$ decomposes, $x$ mol $Cl_2$ and $x$ mol CO form. We convert moles to concentration, set up the reaction table, and substitute the values into $Q_c$. Before going through the quadratic equation, we simplify the math by assuming that $x$ is negligibly small. After solving for $x$, we check the assumption and find the concentrations.

**Solution: (a)** For 5.00 mol $COCl_2$. Writing the mass-action expression for the decomposition:

$$Q_c = \frac{[CO]\,[Cl_2]}{[COCl_2]}$$

Calculating the initial $[COCl_2]$:

$$[COCl_2]_{init} = 5.00 \text{ mol}/10.0 \text{ L} = 0.500 \ M$$

Setting up the reaction table, with $x = [COCl_2]_{reacting}$:

| Concentration (M) | $COCl_2(g)$ | $\rightleftharpoons$ | $CO(g)$ | + | $Cl_2(g)$ |
|---|---|---|---|---|---|
| Initial | 0.500 | | 0 | | 0 |
| Change | $-x$ | | $+x$ | | $+x$ |
| Equilibrium | $0.500 - x$ | | $x$ | | $x$ |

If we use these values directly in $Q_c$, at equilibrium we obtain

$$Q_c = \frac{[CO]\,[Cl_2]}{[COCl_2]} = \frac{x^2}{0.500 - x} = K_c = 8.3 \times 10^{-4}$$

Since $K_c$ is small, the reaction does not proceed very far to the right, so we assume that $x$ (that is, $[COCl_2]_{reacting}$) is much smaller than 0.500 $M$:

$$0.500 \ M - x \approx 0.500 \ M$$

Using the assumption to substitute and solve for $x$:

$$K_c = \frac{x^2}{0.500} = 8.3 \times 10^{-4}$$

$$x = 2.0 \times 10^{-2}$$

Checking the assumption by finding the percent error:

$$\frac{2.0 \times 10^{-2}}{0.500} \times 100 = 4\% \text{ is less than 5\%; the assumption is justified.}$$

Solving for the equilibrium concentrations:

$$[CO] = [Cl_2] = x = \mathbf{2.0 \times 10^{-2}} \ \boldsymbol{M}$$

$$[COCl_2] = 0.500 \ M - x = \mathbf{0.480} \ \boldsymbol{M}$$

**(b)** For 0.100 mol $COCl_2$. This calculation is the same, except $[COCl_2]_{init} = 0.100$ mol/10.0 L = 0.0100 $M$. Thus, at equilibrium, we have

$$Q_c = \frac{[CO]\,[Cl_2]}{[COCl_2]} = \frac{x^2}{0.0100 - x} = K_c = 8.3 \times 10^{-4}$$

Making the assumption that $0.0100 \ M - x \approx 0.0100 \ M$ and solving for $x$:

$$K_c = \frac{x^2}{0.0100} = 8.3 \times 10^{-4}$$

$$x = 2.9 \times 10^{-3}$$

Checking the assumption:

$$\frac{2.9 \times 10^{-3}}{0.0100} \times 100 = 29\% \text{ is more than 5\%, so the assumption is } \textit{not} \text{ justified.}$$

We must solve the quadratic equation, $x^2 + (8.3 \times 10^{-4})x - (8.3 \times 10^{-6}) = 0$, for which the only meaningful root is $x = 2.5 \times 10^{-3}$.
Solving for the equilibrium concentrations:

$$[CO] = [Cl_2] = \mathbf{2.5 \times 10^{-3}} \textbf{ } \textbf{\textit{M}}$$

$$[COCl_2] = (1.00 \times 10^{-2}) \textit{ M} - x = \mathbf{7.5 \times 10^{-3}} \textbf{ } \textbf{\textit{M}}$$

**Check:** Once again, the best check is to use the calculated values to be sure you obtain the given $K_c$.
**Comment:** Note that, at the higher initial concentration, but *not* the lower, the assumption was justified.

**FOLLOW-UP PROBLEM 16.10**
In a study of halogen bond strengths, 0.50 mol $I_2$ was heated in a 2.5-L vessel, and the following reaction occurred: $I_2(g) \rightleftharpoons 2I(g)$.
**(a)** Calculate $[I_2]_{eq}$ and $[I]_{eq}$ at 600 K; $K_c = 2.94 \times 10^{-10}$.
**(b)** Calculate $[I_2]_{eq}$ and $[I]_{eq}$ at 2000 K; $K_c = 0.209$.

---

By this time, you've seen quite a few variations on the second type of equilibrium problem. Figure 16.4 summarizes the steps involved in solving these problems by grouping them into three overall parts.

### Section Summary

In most equilibrium problems, we use amounts of reactants and products to find $K$ or use $K$ to find amounts. We use a reaction table to summarize the initial amounts, how they change, and what the equilibrium amounts are. When $K$ is small and the initial amount of reactant is large, we assume the unknown amount of material that reacts ($x$) is so much smaller than the initial amount that it can be neglected. If this assumption is shown to be unjustified (error $> 5\%$), we use the quadratic formula to find the unknown amount.

## 16.4   Reaction Conditions and the Equilibrium State: Le Châtelier's Principle

Perhaps the most remarkable feature of a system at equilibrium is its ability to return to equilibrium after conditions move it away from that state. This drive to reattain equilibrium is stated in **Le Châtelier's principle:** when a chemical system in a state of equilibrium is disturbed, it reattains equilibrium by undergoing a net reaction that reduces the effect of the disturbance.

Two phrases in this statement need further explanation. First, what does it mean to "disturb" a system in a state of equilibrium? At equilibrium, $Q$ equals $K$. When a change in conditions forces the system temporarily out of equilibrium ($Q \neq K$), we say the system has been stressed, or disturbed. Three common disturbances are a change in concentration of a component, a change in pressure (volume), or a change in temperature. The "net reaction" that the system undergoes is often referred to as a shift in its *equilibrium position*, the specific concentrations (or pressures) that make up $Q$ when it equals $K$. Thus, when a disturbance occurs, the equilibrium position shifts: concentrations (or pressures) change in a way that reduces the disturbance, and the system attains a new equilibrium ($Q = K$ again).

### SOLVING EQUILIBRIUM PROBLEMS

#### PRELIMINARY SETTING UP

1. Write the balanced equation
2. Write the mass-action expression, $Q$
3. Convert all amounts into the correct units ($M$ or atm)

#### WORKING ON THE REACTION TABLE

4. When reaction direction is not known, compare $Q$ with $K$
5. Construct a reaction table

✓Check the sign of $x$, the change in amount

#### SOLVING FOR $x$ AND EQUILIBRIUM CONCENTRATIONS

6. Substitute the amounts into $Q$
7. To simplify the math, assume that $x$ is negligible ($[A]_{init} - x \approx [A]_{init}$)
8. Solve for $x$

✓Check that assumption is justified ($< 5\%$ error)

9. Solve for the equilibrium amounts

✓Check to see that calculated values give the known $K$

**FIGURE 16.4**
**How to solve a major type of equilibrium problem.** These nine steps, grouped into three tasks, present a useful approach to calculating equilibrium concentrations, given initial concentrations and $K_c$.

Le Châtelier's principle allows us to predict the direction of the shift in position and thus to create conditions that maximize yields. Let's examine each of the three disturbances (changes in conditions) to see how a system at equilibrium responds and then note the effect, if any, of a catalyst. In the following discussion, we focus on the reversible reaction between phosphorus trichloride and chlorine to produce phosphorus pentachloride:

$$PCl_3(g) + Cl_2(g) \rightleftharpoons PCl_5(g)$$

However, the principles hold for any system at equilibrium.  ◆

### The Effect of a Change in Concentration

When a system at equilibrium is disturbed by a change in a component's concentration, the system reacts in the direction that reduces the effect of the change. If the component's concentration increases, the system will shift to consume some of it; if the concentration decreases, the system will shift to produce more of it.

At 523 K, the $PCl_3$-$Cl_2$-$PCl_5$ system reaches equilibrium when $Q_c = K_c = 24.0$:

$$Q_c = \frac{[PCl_5]}{[PCl_3]\,[Cl_2]} = 24.0 = K_c$$

What happens if we now inject some $Cl_2$? The $[Cl_2]$ term increases, so the value of $Q_c$ immediately falls because the denominator becomes larger, so the system is no longer at equilibrium. As a result of this disturbance, some of the added $Cl_2$ reacts with some of the $PCl_3$ present and produces more $PCl_5$. The denominator becomes smaller, the numerator larger, and eventually $Q_c$ once again equals $K_c$. The concentrations of the components have changed, however: the concentrations of $Cl_2$ and $PCl_5$ are higher than in the original equilibrium position, and the concentration of $PCl_3$ is lower. We describe this change by saying that *the equilibrium position shifts to the right when a component on the left is added:*

$$PCl_3 + Cl_2(added) \rightarrow PCl_5$$

What happens if, instead of adding $Cl_2$, we remove some $PCl_3$? The $[PCl_3]$ term decreases, the denominator becomes smaller, and the value of $Q_c$ rises above $K_c$. As a result, some $PCl_5$ decomposes to $PCl_3$ and $Cl_2$. The numerator goes down and the denominator up until $Q_c$ equals $K_c$ again. Once again, the concentrations are different from those of the original equilibrium position. We say that *the equilibrium position shifts to the left when a component on the left is removed:*

$$PCl_3(removed) + Cl_2 \leftarrow PCl_5$$

Whenever the concentration of a component changes, *the equilibrium system compensates by using up some of the added substance or replacing some of the removed substance by means of a chemical reaction.* In this way, the system "reduces the effect of the disturbance." The effect is not completely eliminated, however, as we can show with a quantitative comparison of original and new equilibrium positions.

Consider the case in which we added $Cl_2$ to the system at equilibrium. Suppose the original equilibrium position was established with the following concentrations: $[PCl_3] = 0.200\ M$, $[Cl_2] = 0.125\ M$, and $[PCl_5] = 0.600\ M$. Thus,

$$Q_c = \frac{[PCl_5]}{[PCl_3]\,[Cl_2]} = \frac{0.600}{0.200 \times 0.125} = K_c = 24.0$$

Now we add $0.075\ M\ Cl_2$ and let the system come to a new equilibrium position. From Le Châtelier's principle, we predict that adding more of a component on the left will shift the equilibrium position to the right. Experiment shows that the new $[PCl_5]$ at equilibrium is $0.637\ M$. Table 16.3 shows a reaction table of the entire process: the original equilibrium position, the disturbance, the new initial concentrations, the size and direction of the change, and the new equilibrium position. From Table 16.3,

$$[PCl_5] = 0.600\ M + x = 0.637\ M, \quad \text{so } x = 0.037\ M$$

Also,                   $$[PCl_3] = [Cl_2] = 0.200\ M - x = 0.163\ M$$

The mass-action expression is $Q_c = \dfrac{[PCl_5]}{[PCl_3]\,[Cl_2]}$. Therefore, at equilibrium

$$K_{c(\text{original concentration})} = \frac{0.600}{0.200 \times 0.125} = 24.0$$

$$K_{c(\text{new concentration})} = \frac{0.637}{0.163 \times 0.163} = 24.0$$

There are several key points to notice about the new equilibrium concentrations:

- As we predicted, $[PCl_5]$ ($0.637\ M$) is higher than its new initial concentration ($0.600\ M$).
- $[Cl_2]$ ($0.163\ M$) is higher than its original equilibrium concentration ($0.125\ M$), but lower than its new initial concentration just after the addition ($0.200\ M$); thus, the disturbance (addition of $Cl_2$) was *reduced but not eliminated*.
- $[PCl_3]$ ($0.163\ M$), the other left-hand component, is lower than its original concentration ($0.200\ M$) because some reacted with the added $Cl_2$.
- Most importantly, although the position of equilibrium shifted to the right, $K_c$ remains the same.

It is essential to realize that the system adjusts by changing concentrations, but the value of $Q_c$ at equilibrium is constant. In other words, *at a given temperature, $K_c$ does not change with a change in concentration.*

**TABLE 16.3  The Effect of Added $Cl_2$ on the $PCl_3$-$Cl_2$-$PCl_5$ System**

| CONCENTRATION (M) | $PCl_3(g)$ | + | $Cl_2(g)$ | $\rightleftharpoons$ | $PCl_5(g)$ |
|---|---|---|---|---|---|
| Original equilibrium | 0.200 | | 0.125 | | 0.600 |
| Disturbance | | | +0.075 | | |
| New initial | 0.200 | | 0.200 | | 0.600 |
| Change | $-x$ | | $-x$ | | $+x$ |
| New equilibrium | $0.200 - x$ | | $0.200 - x$ | | $0.600 + x$ |
| | | | | | $(0.637)*$ |

*Experimentally determined value.

SAMPLE PROBLEM 16.11

## Predicting the Effect of a Change in Concentration on the Position of Equilibrium

**Problem:** Sulfur is often recovered from coal and natural gas by treating the contaminant hydrogen sulfide with oxygen:

$$2H_2S(g) + O_2(g) \rightleftharpoons 2S(s) + 2H_2O(g)$$

What would happen to
(a) $[H_2O]$ if $O_2$ is added?    (b) $[H_2S]$ if $O_2$ is added?
(c) $[O_2]$ if $H_2S$ is removed?    (d) $[H_2S]$ if sulfur is added?
**Plan:** We write the mass-action expression and apply Le Châtelier's principle to see how $Q_c$ is affected by each disturbance and in which direction the reaction will proceed for the system to reattain equilibrium.

**Solution:** Writing the mass-action expression: $Q_c = \dfrac{[H_2O]^2}{[H_2S]^2\,[O_2]}$

(a) When $O_2$ is added, the denominator of $Q_c$ increases, so $Q_c < K_c$. The reaction proceeds to the right until $Q_c = K_c$ again, so **[H$_2$O] increases.**
(b) When $O_2$ is added, $Q_c$ is less than $K_c$. Some $H_2S$ is used up in reaction with the $O_2$, so **[H$_2$S] decreases.**
(c) When $H_2S$ is removed, the denominator of $Q_c$ decreases, so $Q_c > K_c$. As the reaction proceeds to the left to re-form $H_2S$, more $O_2$ is produced as well, so **[O$_2$] increases.**
(d) Since the concentration of solid S is unchanged as long as some is present, it does not appear in the mass-action expression. Adding more S has no effect, so **[H$_2$S] is unchanged.**
**Check:** Check to see that the reaction proceeds in the direction that lowers the increased concentration or raises the decreased concentration.

FOLLOW-UP PROBLEM 16.11
In a study of the chemistry of glass etching, an inorganic chemist examines the reaction between sand ($SiO_2$) and hydrogen fluoride at high temperature:

$$SiO_2(s) + 4HF(g) \rightleftharpoons SiF_4(g) + 2H_2O(g)$$

Predict the effect on $[SiF_4]$ when (a) $H_2O(g)$ is removed; (b) some liquid water is added at 220°C; (c) HF is removed; (d) some sand is removed.

## The Effect of a Change in Pressure (Volume)

In examining the effect of pressure on a system at equilibrium, we're concerned only with systems that contain one or more gaseous components. Liquids and solids are nearly incompressible, so changing the pressure has a negligible effect on them. Pressure changes can occur by
• Changing the concentration of a gaseous component.
• Adding an inert gas (one that does not take part in the reaction).
• Changing the volume of the reaction vessel.
    We have just considered the effect of changing the concentration of a component of a system. The second means of changing the pressure, adding an inert gas to the system, has no effect on the equilibrium position. Since the volume has not changed, all reactant and product concentrations remain the same. Although the total pressure increases, the mole fractions decrease, so the partial pressures of reactant and product remain the same. Since the partial pressures are the terms we incorporate in the mass-action expression, the equilibrium position does not change.

A change in pressure due to a volume change, on the other hand, often has a large effect on the equilibrium position because a change in volume causes a change in concentration, the number of moles per liter of container volume: a decrease in container volume raises the concentration, and an increase lowers the concentration.

Suppose we let the $PCl_3$-$Cl_2$-$PCl_5$ system come to equilibrium in a cylinder-piston assembly. Recall that $Q_c = \dfrac{[PCl_5]}{[PCl_3]\,[Cl_2]}$. Now we press down on the piston and halve the volume. The concentrations double, but the denominator of $Q_c$ is the product of two concentrations, so it quadruples while the numerator only doubles. Thus, $Q_c$ becomes less than $K_c$. As a result, the reaction shifts toward formation of $PCl_5$. The denominator then becomes smaller and the numerator larger as the system moves toward a new equilibrium position. Once again, as with the change in concentration, *a change in pressure due to a change in volume does not alter $K_c$.*

Notice how the system responded to this particular disturbance: less volume moved the reaction toward the side with fewer moles of gas. Two moles were used to make one:

$$PCl_3(g) + Cl_2(g) \longrightarrow PCl_5(g)$$

The crowding created by the compression was reduced by lowering the total number of gas molecules. In this case, Le Châtelier's principle states that when a system containing gases at equilibrium undergoes a change in pressure *as a result of a change in volume*, the equilibrium position shifts to reduce the effect of the change:
- If the volume is lower (pressure is higher), the total number of gas molecules decreases.
- If the volume is higher (pressure is lower), the total number of gas molecules increases.

In many reactions, the total moles of gaseous reactants equals the total moles of gaseous products, that is, $\Delta n_{gas} = 0$. For example,

$$H_2(g) + I_2(g) \rightleftharpoons 2HI(g)$$

The mass-action expression has the same number of terms in the numerator and denominator:

$$Q_c = \frac{[HI]^2}{[H_2]\,[I_2]} = \frac{[HI]\,[HI]}{[H_2]\,[I_2]}$$

In such cases, a change in volume has the same effect on the numerator and denominator. Thus, *if $\Delta n_{gas} = 0$, there is no effect on the equilibrium position.*

---

SAMPLE PROBLEM 16.12

**Predicting the Effect of a Change in Volume (Pressure) on the Position of Equilibrium**

**Problem:** How would you change the volume of each of the following reactions to *increase* the yield of the products?
(a) $CaCO_3(s) \rightleftharpoons CaO(s) + CO_2(g)$
(b) $S(s) + 3F_2(g) \rightleftharpoons SF_6(g)$
(c) $Cl_2(g) + I_2(g) \rightleftharpoons 2ICl(g)$
**Plan:** A change in volume causes a change in concentration. If the volume decreases (pressure increases), the equilibrium position will shift to relieve the pressure by reducing the number of moles of gas. A volume increase (pressure decrease) will have the opposite effect.

**Solution: (a)** The only gas is the product $CO_2$. To make the system produce more $CO_2$, **we increase the volume (decrease the pressure).**

**(b)** With three moles of gas on the left and only one on the right, **we decrease the volume (increase the pressure)** to form more $SF_6$.

**(c)** The number of moles of gas is the same on both sides of the equation, **so a change in volume (pressure) will have no effect** on the yield of ICl.

**Check:** Let's predict the relative values of $Q_c$ and $K_c$. In **(a),** $Q_c = [CO_2]$, so increasing the volume will make $Q_c < K_c$, and the system will make more $CO_2$. In **(b),** $Q_c = [SF_6]/[F_2]^3$. Because of the difference in exponents, lowering the volume increases $[F_2]$ more than $[SF_6]$, which lowers $Q_c$. To make $Q_c = K_c$ again, $[SF_6]$ must increase. In **(c),** $Q_c = [ICl]^2/[Cl_2][I_2]$. A change in volume (pressure) affects the numerator and denominator equally, so it will have no effect.

FOLLOW-UP PROBLEM 16.12

How would you change the pressure (via a volume change) of the following reaction mixtures to *decrease* the yield of products?

**(a)** $2SO_2(g) + O_2(g) \rightleftharpoons 2SO_3(g)$     **(b)** $4NH_3(g) + 5O_2(g) \rightleftharpoons 4NO(g) + 6H_2O(g)$
**(c)** $CaC_2O_4(s) \rightleftharpoons CaCO_3(s) + CO(g)$

---

## The Effect of a Change in Temperature

Of the three types of disturbances, *only temperature changes alter K.* To see why, we must take the heat of reaction into account:

$$PCl_3(g) + Cl_2(g) \rightleftharpoons PCl_5(g) \qquad \Delta H^0_{rxn} = -111 \text{ kJ}$$

The forward reaction is exothermic, so the reverse reaction is endothermic:

$$PCl_3(g) + Cl_2(g) \rightarrow PCl_5(g) + \textbf{\textit{heat}} \text{ (exothermic)}$$

$$PCl_3(g) + Cl_2(g) \leftarrow PCl_5(g) + \textbf{\textit{heat}} \text{ (endothermic)}$$

If we consider *heat as a component of the equilibrium system,* a rise in temperature "adds" heat to the system and a drop in temperature "removes" heat from the system. As with a change in any other component, the system shifts to reduce the effect of the change. Therefore, a temperature increase favors the endothermic (heat-absorbing) direction and a temperature decrease favors the exothermic (heat-releasing) direction.

The equilibrium constant changes because the equilibrium position shifts without any *substances* being added or removed. If we start with the system at equilibrium, $Q_c$ equals $K_c$. Increase the temperature, and the system responds by absorbing the added heat as some $PCl_5$ decomposes to $PCl_3$ and $Cl_2$. Since $Q_c$ includes reactant and product terms *but no heat-related term,* the denominator becomes larger and the numerator smaller. The system reaches a new equilibrium position at a smaller ratio of concentration terms, that is, a lower $K_c$. Similarly, the system responds to a drop in temperature by releasing more heat as some $PCl_3$ and $Cl_2$ form more $PCl_5$. The numerator becomes larger, the denominator smaller, and the new equilibrium position has a higher $K_c$. Thus,

- *A temperature rise will increase $K_c$ for a system with a positive $\Delta H^0_{rxn}$.*
- *A temperature rise will decrease $K_c$ for a system with a negative $\Delta H^0_{rxn}$.*

The *van't Hoff equation* describes how the equilibrium constant is affected by changes in temperature:

$$\ln \frac{K_2}{K_1} = -\frac{\Delta H^0_{rxn}}{R} \left( \frac{1}{T_2} - \frac{1}{T_1} \right) \qquad \textbf{(16.10)}$$

where $K_1$ is the equilibrium constant at $T_1$, $K_2$ is the equilibrium constant at $T_2$, and $R$ is the universal gas constant. If we know $\Delta H^0_{rxn}$ and $K$ at one temperature, the van't Hoff equation allows us to determine $K$ at any other temperature (or to find $\Delta H^0_{rxn}$, given the $K$'s at two temperatures.) The equation confirms the qualitative prediction from Le Châtelier's principle:

$$\text{If } T_2 > T_1, \text{ then } (1/T_2 - 1/T_1) < 0$$

Therefore,
- For an endothermic reaction ($\Delta H^0_{rxn} > 0$), ln $(K_2/K_1) > 0$, so $K_2 > K_1$
- For an exothermic reaction ($\Delta H^0_{rxn} < 0$), ln $(K_2/K_1) < 0$, so $K_2 < K_1$ ◆

### SAMPLE PROBLEM 16.13

### Predicting the Effect of a Change in Temperature on the Position of Equilibrium

**Problem:** How would an *increase* in temperature affect the equilibrium concentration of the underlined substance and $K_c$ for the following reactions?
**(a)** $CaO(s) + H_2O(l) \rightleftharpoons \underline{Ca(OH)_2}(aq)$    $\Delta H^0 = -82$ kJ
**(b)** $CaCO_3(s) \rightleftharpoons \underline{CaO(s) + CO_2}(g)$    $\Delta H^0 = 178$ kJ
**(c)** $\underline{SO_2}(g) \rightleftharpoons S(s) + O_2(g)$    $\Delta H^0 = 297$ kJ
**Plan:** We write each equation to show heat as a reactant or product. Increasing the temperature adds heat, so the system shifts to absorb the heat; that is, the endothermic reaction occurs. $K_c$ will increase if the forward reaction is endothermic and decrease if it is exothermic.
**Solution: (a)** $CaO(s) + H_2O(l) \rightleftharpoons \underline{Ca(OH)_2}(aq) + \textbf{\textit{heat}}$
Adding heat shifts the system to the left: $[Ca(OH)_2]$ and $K_c$ will **decrease.**
**(b)** $CaCO_3(s) + \textbf{\textit{heat}} \rightleftharpoons \underline{CaO(s) + CO_2}(g)$
Adding heat shifts the system to the right: $[CO_2]$ and $K_c$ will **increase.**
**(c)** $\underline{SO_2}(g) + \textbf{\textit{heat}} \rightleftharpoons S(s) + O_2(g)$
Adding heat shifts the system to the right: $[SO_2]$ will **decrease** and $K_c$ will **increase.**
**Check:** You might check by going through the reasoning for a *decrease* in temperature: heat is removed and the exothermic direction is favored. All the answers should be opposite.

### FOLLOW-UP PROBLEM 16.13
How would a *decrease* in temperature affect the partial pressure of the underlined substance and $K_p$ for the following reactions?
**(a)** $C(graphite) + 2H_2(g) \rightleftharpoons \underline{CH_4}(g)$    $\Delta H^0 = -75$ kJ
**(b)** $\underline{N_2}(g) + O_2(g) \rightleftharpoons 2NO(g)$    $\Delta H^0 = 181$ kJ
**(c)** $\underline{P_4}(s) + 10Cl_2(g) \rightleftharpoons 4\underline{PCl_5}(g)$    $\Delta H^0 = -1528$ kJ

### The Lack of Effect of a Catalyst

Let's briefly consider a final external change in the reacting system: the effect of adding a catalyst. Recall from Chapter 15 that a catalyst speeds up a reaction by lowering the activation energy, thereby increasing the forward *and* reverse rates to the same extent. Thus, while it shortens the time the system takes to reach equilibrium, *a catalyst has no effect on the equilibrium position.* ◆ If, for instance, we add a catalyst to a mixture of $PCl_3$ and $Cl_2$ at 523 K, the system will attain the *same* equilibrium concentrations of $PCl_3$, $Cl_2$, and $PCl_5$ *more quickly* than it did without the catalyst. As you'll see in a moment, however, catalysts often play key roles in optimizing reaction systems.

The upcoming Chemical Connections essays highlight two areas that apply equilibrium principles: the first concerns a major industrial process, and the second examines metabolic processes in organisms.

Table 16.4 summarizes the effects of changing conditions on the position of equilibrium. Note that while all changes except the addition of a catalyst change the equilibrium position, only temperature changes affect the value of the equilibrium constant.

◆ **A Recurring Theme in Temperature-Dependent Systems.** There is a striking similarity among expressions for the temperature dependence of $K$, $k$ (rate constant), and $P$ (equilibrium vapor pressure):

$$\ln \frac{K_2}{K_1} = -\frac{\Delta H^0_{rxn}}{R}\left(\frac{1}{T_2} - \frac{1}{T_1}\right)$$

$$\ln \frac{k_2}{k_1} = -\frac{E_a}{R}\left(\frac{1}{T_2} - \frac{1}{T_1}\right)$$

$$\ln \frac{P_2}{P_1} = -\frac{\Delta H_{vap}}{R}\left(\frac{1}{T_2} - \frac{1}{T_1}\right)$$

In all three cases, a concentration-related term—$K$, $k$, or $P$—is dependent on the temperature through the energy component of the system—$\Delta H^0_{rxn}$, $E_a$, or $\Delta H_{vap}$, respectively—divided by $R$.

The similarities arise because the equations express the same relationship. $K$ equals a ratio of rate constants and $E_{a(fwd)} - E_{a(rev)} = \Delta H^0_{rxn}$. In the vapor pressure case, for the phase change, $A(l) \rightleftharpoons A(g)$, the heat of reaction is the heat of vaporization: $\Delta H_{vap} = \Delta H^0_{rxn}$. Moreover, the equilibrium constant equals the equilibrium vapor pressure: $K_p = P_A$.

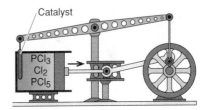
Catalyst

◆ **Impossible Catalysts and Perpetual Motion.** The engine in the figure consists of a piston attached to a flywheel, whose rocker arm holds a catalyst that moves in and out of the reaction in the cylinder. What would happen if the catalyst had the impossible property of increasing the rate of $PCl_5$ breakdown but not the rate of its formation? When in the cylinder, the catalyst speeds up the breakdown of $PCl_5$ to $PCl_3$ and $Cl_2$—one mole of gas to two—increasing gas pressure, and the piston moves out. When it is out of the cylinder, the $PCl_3$ and $Cl_2$ re-form $PCl_5$, lowering gas pressure, and the piston moves in. The process would supply power with no external input of energy. If only a catalyst *did* shift the position of equilibrium, we could design some truly amazing machines!

**TABLE 16.4    Effect of Changing Conditions on Equilibrium Position and $K$**

| CONDITION | NET DIRECTION OF REACTION | EFFECT ON VALUE OF $K$ |
|---|---|---|
| Concentration | | |
|    Increase [A] | Toward consumption of A | None |
|    Decrease [A] | Toward formation of A | None |
| Pressure (volume) | | |
|    Increase $P$ | Toward formation of fewer moles of gas | None |
|    Decrease $P$ | Toward formation of more moles of gas | None |
| Temperature | | |
|    Increase $T$ | Toward absorption of heat | Increases if $\Delta H^0_{rxn} > 0$<br>Decreases if $\Delta H^0_{rxn} < 0$ |
|    Decrease $T$ | Toward release of heat | Increases if $\Delta H^0_{rxn} < 0$<br>Decreases if $\Delta H^0_{rxn} > 0$ |
| Catalyst added | None; rates of forward and reverse reactions increase equally | None |

---

**CHEMICAL CONNECTIONS**

## Chemistry in Industrial Production:
## The Haber Process for the Synthesis of Ammonia

Nitrogen occurs in many essential natural and synthetic compounds. By far the richest source of nitrogen is the atmosphere, where four of every five molecules are $N_2$. Despite this abundance, the supply of usable nitrogen for biological and manufacturing processes is limited because of the low chemical reactivity of $N_2$. As a result of the strong triple bond holding the two N atoms together, the nitrogen atom is very difficult to "fix," that is, to combine with other atoms.

Natural nitrogen fixation is accomplished either through the activity of certain enzymes found in bacteria that live on plant roots or through the brute force of lightning storms. Nearly 13% of all the nitrogen fixation on Earth is accomplished industrially, through the **Haber process** for the formation of ammonia from its elements:

$$N_2(g) + 3H_2(g) \rightleftharpoons 2NH_3(g) \qquad \Delta H^0_{rxn} = -91.8 \text{ kJ}$$

The process was developed by the German chemist Fritz Haber and first used in 1913. From its humble beginnings in a plant with a capacity of 12,000 tons a year, current world production of ammonia has exploded to more than 110 million tons a year. On a mole basis, more ammonia is produced than any other compound. Over 80% of this ammonia finds its way into fertilizer applications. In fact, the most common form of fertilizer is compressed liquid $NH_3$ sprayed directly onto soil (Figure 16.A). Other uses include the production of explosives, via the formation of $HNO_3$, and the making of nylons and other polymers. Smaller amounts are used as re-

frigerants, rubber stabilizers, household cleaners, and in the synthesis of pharmaceuticals and other organic chemicals.

The Haber process provides an excellent setting in which to apply equilibrium principles and see the compromises needed to make an industrial process economically worthwhile. From inspection of the balanced for-

**FIGURE 16.A Liquid ammonia as fertilizer.** Most of the ammonia produced by the Haber process is used as a liquid fertilizer that is sprayed directly onto fields.

| TABLE 16.A | Effect of Temperature on $K_c$ for Ammonia Synthesis | |
| --- | --- |
| $T$ (K) | $K_c$ |
| 200 | $7.17 \times 10^{15}$ |
| 300 | $2.69 \times 10^{8}$ |
| 400 | $3.94 \times 10^{4}$ |
| 500 | $1.72 \times 10^{2}$ |
| 600 | $4.53 \times 10^{0}$ |
| 700 | $2.96 \times 10^{-1}$ |
| 800 | $3.96 \times 10^{-2}$ |

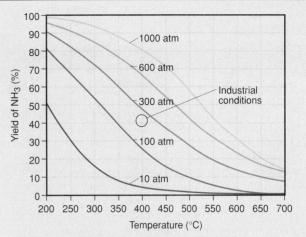

**FIGURE 16.B** Percent yield of ammonia vs. temperature (°C) at five different operating pressures. At very high pressure and low temperature *(top left)*, the yield is high, but the rate of formation is low. Industrial conditions *(circle)* are between 200 and 300 atm at about 400°C.

mation reaction, we can see three ways to maximize the yield of ammonia:

1. Decrease [$NH_3$]. Since ammonia is the product, removing it as it forms will make the system produce more in a continual drive to reattain equilibrium.
2. Decrease volume (increase pressure). Since four moles of gas react to form two, decreasing the volume will shift the equilibrium position toward fewer moles of gas, that is, toward ammonia formation.
3. Decrease temperature. Since the formation of ammonia is exothermic, decreasing the temperature (removing heat) will shift the equilibrium position in that direction, thereby increasing $K_c$ (Table 16.A).

Therefore, the ideal conditions for maximizing the yield of ammonia are continual removal of $NH_3$ as it forms, high pressure, and low temperature. Figure 16.B shows the percent yield of ammonia at various conditions of pressure and temperature. Note the almost complete conversion (98.3%) to ammonia at 473 K (200°C) and 1000 atm.

Unfortunately, a problem arises that clearly highlights the distinction between the principles of equilibrium and kinetics. Although the *yield* is favored by low temperature, the *rate* of formation is not. In fact, ammonia forms so slowly at low temperature that the process becomes uneconomical. In practice, a compromise is achieved that optimizes yield *and* rate. High pressure and continuous removal are utilized to increase yield, but the temperature is raised to a moderate level and a catalyst is employed to increase the rate. Achieving the same rate

without a catalyst would require much higher temperatures and result in a much lower yield.

Stages in the industrial production of ammonia are shown in Figure 16.C. To extend equipment life and minimize cost, modern ammonia plants operate at pressures of about 200 to 300 atm and temperatures of around 673 K (400°C). The catalyst consists of 5-mm to 10-mm chunks of iron crystals embedded in a fused mixture of MgO, $Al_2O_3$, and $SiO_2$. The mixture of compressed reactant gases ($N_2/H_2 = 1/3$ by volume) is injected into the heated, pressurized reaction chamber, where it flows over the catalyst beds. Some of the heat needed is supplied by the enthalpy change of the reaction. The emerging equilibrium mixture, which contains about 35 vol % $NH_3$, is cooled by refrigeration coils until the $NH_3$ condenses (boiling point, −33.4°C) and is removed. Since the boiling points of $N_2$ and $H_2$ are much lower, they remain gaseous and are recycled by pumps back into the reaction chamber to continue the process.

**FIGURE 16.C** Key stages in the Haber synthesis of ammonia.

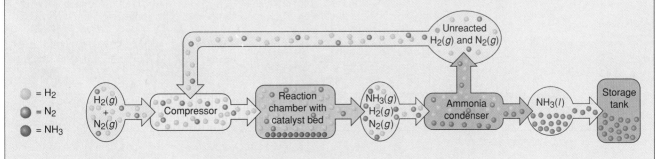

| CHEMICAL CONNECTIONS | Chemistry in Cellular Metabolism: **Design and Control of a Metabolic Pathway** |

Biological cells are microscopic wizards of chemical change. From the simplest bacterium to the most specialized nerve, every cell performs thousands of individual reactions that allow it to grow and reproduce, feed and excrete, move and communicate. Taken together, these chemical reactions constitute the cell's *metabolism.* These myriad feats of biochemical breakdown, synthesis, and energy flow are organized into reaction sequences called **metabolic pathways.** Some pathways disassemble the biopolymers in food into monomer building blocks: sugars, amino acids, and nucleotides. Others use these small molecules to extract energy and then use much of the energy to assemble molecules with other functions: hormones, neurotransmitters, defensive toxins, and the cell's own biopolymers.

In principle, *each step in a metabolic pathway is a reversible reaction catalyzed by a specific enzyme* (Section 15.6). As you'll see, however, even though the principles we discussed earlier apply, equilibrium is never reached in a pathway. As with other multistep reaction sequences, the product of the first reaction becomes the reactant of the second, the product of the second becomes the reactant of the third, and so on. Consider, for example, the five-step pathway shown in Figure 16.D, in which one amino acid, threonine, is converted into another, isoleucine, in the cells of a bread mold. Threonine, supplied from a different region of the cell, forms ketobutyrate through the catalytic action of the first enzyme (Reaction 1). The equilibrium position of Reaction 1 continuously shifts to the right because the ketobutyrate is continuously removed as the reactant in Reaction 2. Similarly, Reac-

tion 2 shifts to the right as its product is used in Reaction 3. Each subsequent reaction shifts the equilibrium position of the previous reaction in the direction of product. The final product, isoleucine, is typically removed to make proteins in another region of the cell. Thus, *the entire pathway operates in one direction.*

This continuous shift in equilibrium position results in two major features of metabolic pathways. First, *each step proceeds with nearly 100% yield.* That is, virtually every molecule of threonine that enters this region of the cell eventually changes to ketobutyrate, every molecule of ketobutyrate to the next product, and so on.

Second, *reactant and product concentrations remain constant,* or vary within extremely narrow limits, even though the system never attains equilibrium. This situation, called a *steady state,* is different in a basic way from those we have studied so far. In equilibrium systems, the rates of reactions in opposing directions give rise to constant concentrations. In steady-state systems, on the other hand, the rates of reactions in one direction—into, through, and out of the system—give rise to constant concentrations. Ketobutyrate, for example, is formed via Reaction 1 just as fast as it is used via Reaction 2, so its concentration is kept constant. (You can establish a steady-state amount of water in a bathtub by filling the tub and then adjusting the faucet and drain so that water enters the tub just as fast as it leaves.)

It is absolutely critical that a cell regulate the concentration of the final product of a pathway. Recall from Chapter 15 that an enzyme catalyzes the reaction when the substrate (reactant) occupies the active site. However, substrate concentrations are so much higher than those of the enzymes that, if no other factors are involved, all cellular reactions would occur at their maximum rates. While this might be ideal for an industrial process, it could be catastrophic for an organism. To regulate overall product formation, the rates of certain key steps are controlled by *regulatory enzymes,* which contain an *inhibitor site* on their surface in addition to an active site.

**FIGURE 16.D**

**The metabolic pathway for the biosynthesis of isoleucine from threonine.** Isoleucine is synthesized from threonine in a sequence of five enzyme-catalyzed reactions. As indicated by the unequal equilibrium arrows, each step is reversible but is shifted toward product because each product becomes a reactant of the subsequent step. When the cell's need for isoleucine is temporarily satisfied, its synthesis slows through end-product feedback inhibition: isoleucine, the end product, inhibits the catalytic activity of threonine dehydratase, the first enzyme in the pathway.

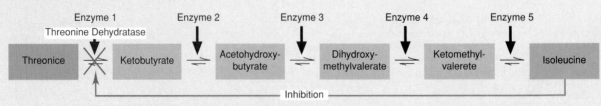

The enzyme's three-dimensional structure is such that when the inhibitor site is occupied, the shape of the active site is deformed and the reaction cannot be catalyzed (Figure 16.E).

In the simplest case of metabolic control (see Figure 16.D), *the final product of the pathway is also the inhibitor molecule, and the regulatory enzyme catalyzes the first step.* Suppose, for instance, that the cell is temporarily not making protein, so the isoleucine synthesized by the pathway is not being removed. As its concentration rises, isoleucine lands on the inhibitor site of the first enzyme in the pathway, effectively "shutting off" its own production. This process is called *end-product feedback inhibition.* More complex pathways have more elaborate regulatory schemes, but many operate through similar types of inhibitory feedback (see Problem 16.82). The regulation of metabolic pathways arises directly from the principles of chemical kinetics and equilibrium.

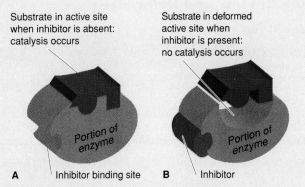

Substrate in active site when inhibitor is absent: catalysis occurs

Substrate in deformed active site when inhibitor is present: no catalysis occurs

Portion of enzyme

Portion of enzyme

**A**    Inhibitor binding site    **B**    Inhibitor

**FIGURE 16.E**

**The effect of inhibitor binding on the shape of the active site. A,** When the inhibitor site is *not* occupied, the active site has its normal shape, and the enzyme catalyzes the reaction. **B,** When the inhibitor site *is* occupied, the active site is deformed and can no longer bind the substrate, and the enzyme is nonfunctional.

## Section Summary

Le Châtelier's principle states that if an equilibrium system is disturbed, it will undergo a net reaction and reattain equilibrium. Changes in concentration cause a net reaction away from the added component or toward the removed component. If a reaction involves a change in moles of gas, a change in pressure (volume) will cause a net reaction that partially reverses the change. Although the concentration of components changes as a result of concentration and volume changes, $K$ does not change. A temperature change *does* change $K$ by causing a net reaction to replace removed heat (exothermic direction) or to absorb added heat (endothermic direction). A catalyst does not affect the equilibrium position because it speeds forward and reverse reactions equally. Ammonia is produced in a process favored by high pressure, low temperature, and continual removal of product. To make the process economical, the temperature is raised somewhat and a catalyst is used. A metabolic pathway is a cellular reaction sequence in which each step is shifted completely toward product. Its overall yield is controlled by product inhibition of certain key enzymes.

## Chapter Perspective

*The phenomenon of equilibrium is central to all natural systems. In this introduction, we discussed the nature of equilibrium on the observable and molecular levels, ways to solve relevant problems, and how conditions affect equilibrium systems. In the next two chapters, we apply these ideas to acids and bases and to other aqueous ionic systems. Then we examine the fundamental relationship between the equilibrium constant and the energy change that accompanies a reaction to understand why chemical reactions occur.*

# For Review and Reference

## Key Terms

**SECTION 16.1**
equilibrium constant ($K$)

**SECTION 16.2**
law of chemical
equilibrium (law of
mass action)

mass-action expression
(reaction quotient) ($Q$)

**SECTION 16.4**
Le Châtelier's principle
Haber process
metabolic pathway

## Key Equations and Relationships

**16.1** Defining equilibrium in terms of reaction rates (p. 696):

$$\text{At equilibrium:}\quad \text{rate}_{\text{fwd}} = \text{rate}_{\text{rev}}$$

**16.2** Defining the equilibrium constant for the reaction A $\rightleftharpoons$ 2B (p. 697):

$$K = \frac{k_{\text{fwd}}}{k_{\text{rev}}} = \frac{[\text{B}]^2}{[\text{A}]}$$

**16.3** Defining the equilibrium constant in terms of the mass-action expression (p. 698):

$$\text{At equilibrium:}\quad Q = K$$

**16.4** Expressing $Q_c$ for the reaction $a\text{A} + b\text{B} \rightleftharpoons c\text{C} + d\text{D}$ (p. 699):

$$Q_c = \frac{[\text{C}]^c\,[\text{D}]^d}{[\text{A}]^a\,[\text{B}]^b}$$

**16.5** Finding the overall $K$ for a reaction sequence (p. 700):

$$K_{\text{overall}} = K_1 \times K_2 \times K_3 \times \cdots$$

**16.6** Finding $K$ of a process from $K$ of the reverse process (p. 702):

$$K_{\text{fwd}} = \frac{1}{K_{\text{rev}}}$$

**16.7** Finding $K$ of a reaction multiplied by a factor $n$ (p. 702):

$$K' = (K)^n$$

**16.8** Relating $K$ based on pressures to $K$ based on concentrations (p. 706):

$$K_{\text{p}} = K_{\text{c}}(RT)^{\Delta n_{\text{gas}}}$$

**16.9** Assuming the concentration reacting introduces insignificant error (p. 715):

$$[\text{A}]_{\text{init}} - [\text{A}]_{\text{reacting}} \approx [\text{A}]_{\text{init}} \approx [\text{A}]_{\text{eq}}$$

**16.10** Finding $K$ at one temperature given $K$ at another (van't Hoff equation) (p. 722):

$$\ln \frac{K_2}{K_1} = -\frac{\Delta H^0_{\text{rxn}}}{R}\left(\frac{1}{T_2} - \frac{1}{T_1}\right)$$

## Answers to Follow-up Problems

**16.1** (a) $Q_c = \dfrac{[\text{NO}]^4\,[\text{H}_2\text{O}]^6}{[\text{NH}_3]^4\,[\text{O}_2]^5}$

(b) $Q_c = \dfrac{[\text{N}_2\text{O}]\,[\text{NO}_2]}{[\text{NO}]^3}$

**16.2** $\text{H}_2(g) + \text{Br}_2(g) \rightleftharpoons 2\text{HBr}(g)$;

$$Q_{c(\text{overall})} = \frac{[\text{HBr}]^2}{[\text{H}_2]\,[\text{Br}_2]}$$

$$Q_{c(\text{overall})} = Q_{c1} \times Q_{c2} \times Q_{c3} \times Q_{c4}$$

$$= \frac{[\text{Br}]^2}{[\text{Br}_2]} \times \frac{[\text{HBr}]\,[\text{H}]}{[\text{Br}]\,[\text{H}_2]} \times \frac{[\text{HBr}]\,[\text{Br}]}{[\text{H}]\,[\text{Br}_2]} \times \frac{[\text{Br}_2]}{[\text{Br}]^2}$$

$$= \frac{[\text{HBr}]^2}{[\text{H}_2]\,[\text{Br}_2]}$$

**16.3** (a) $K_c = K_{c(\text{ref})}^{\;1/2} = 2.8 \times 10^4$

(b) $K_c = \left(\dfrac{1}{K_{c(\text{ref})}}\right)^{2/3} = 1.2 \times 10^{-6}$

**16.4** $K_{\text{p}} = K_{\text{c}}(RT)^{-1} = 1.67\left(0.0821\dfrac{\text{atm} \cdot \text{L}}{\text{mol} \cdot \text{K}} \times 500\ \text{K}\right)^{-1}$

$= 4.07 \times 10^{-2}$

**16.5** $Q_{\text{p}} = \dfrac{(P_{\text{CH}_3\text{Cl}})(P_{\text{HCl}})}{(P_{\text{CH}_4})(P_{\text{Cl}_2})} = \dfrac{0.24 \times 0.47}{0.13 \times 0.035} = 25$;

$Q_{\text{p}} < K_{\text{p}}$, so $\text{CH}_3\text{Cl}$ is forming.

**16.6** From the reaction table for $2\text{NO} + \text{O}_2 \rightleftharpoons 2\text{NO}_2$,

$P_{\text{O}_2} = 1.000\ \text{atm} - x = 0.506\ \text{atm};\quad x = 0.494\ \text{atm}$

Also, $P_{\text{NO}} = 0.012\ \text{atm}$ and $P_{\text{NO}_2} = 0.988\ \text{atm}$

$$K_{\text{p}} = \frac{0.988^2}{0.012^2 \times 0.506} = 1.3 \times 10^4$$

**16.7** Since $\Delta n_{\text{gas}} = 0$,

$$K_{\text{p}} = K_{\text{c}} = 2.3 \times 10^{30} = \frac{0.781 \times 0.209}{P_{\text{NO}}^2}$$

$$P_{\text{NO}} = 2.7 \times 10^{-16}\ \text{atm}$$

**16.8** From the reaction table, $[H_2] = [I_2] = x$ and $[HI] = 0.242 - 2x$.

Therefore, $K_c = 1.26 \times 10^{-3} = \dfrac{x^2}{(0.242 - 2x)^2}$

Taking the square root of both sides, neglecting the negative root, and solving gives $x = [H_2] = 8.02 \times 10^{-3}\ M$.

**16.9** $Q_c = \dfrac{0.0900 \times 0.0900}{0.2100} = 3.86 \times 10^{-2}; Q_c < K_c,$

so reaction proceeds to the right.
From the reaction table,
$[PCl_5] = 0.2100\ M - x = 0.2065\ M; x = 0.0035\ M$
So $[Cl_2] = [PCl_3] = 0.0900\ M + x = 0.0935\ M$

**16.10** (a) Based on the reaction table and assuming $0.20\ M - x \approx 0.20\ M$,

$K_c = 2.94 \times 10^{-10} = \dfrac{4x^2}{0.20}; \quad x = 3.8 \times 10^{-6};$

error $= 1.9 \times 10^{-3}\%$, so assumption is justified; so $[I_2]_{eq} = 0.20\ M$ and $[I]_{eq} = 7.6 \times 10^{-6}\ M$

(b) Based on the same reaction table and assumption, $x = 0.10$; error is 50%, so assumption is *not* justified. Solve equation
$4x^2 + 0.209x - 0.042 = 0; \quad x = 0.079\ M.$
Therefore, $[I_2]_{eq} = 0.12\ M$ and $[I]_{eq} = 0.16\ M.$

**16.11** (a) Increases; (b) decreases; (c) decreases; (d) no effect.

**16.12** (a) Decrease $P$; (b) increase $P$; (c) increase $P$.

**16.13** (a) $P_{H_2}$ will decrease; $K_p$ will increase; (b) $P_{N_2}$ will increase; $K_p$ will decrease; (c) $P_{PCl_5}$ will increase; $K_p$ will increase.

## Sample Problem Titles

**16.1** Writing the Mass-Action Expression from the Balanced Equation (p. 699)
**16.2** Writing the Mass-Action Expression for an Overall Reaction (p. 700)
**16.3** Determining the Equilibrium Constant for an Equation Multiplied by a Common Factor (p. 702)
**16.4** Converting Between $K_c$ and $K_p$ (p. 706)
**16.5** Comparing $Q$ and $K$ to Determine Reaction Progress (p. 707)
**16.6** Calculating $K_c$ from Concentration Data (p. 710)
**16.7** Determining Equilibrium Concentrations from $K_c$ (p. 711)

**16.8** Determining Equilibrium Concentrations from Initial Concentrations and $K_c$ (p. 711)
**16.9** Predicting Reaction Direction and Calculating Equilibrium Concentrations (p. 714)
**16.10** Calculating Equilibrium Concentrations Through the Use of Simplifying Assumptions (p. 715)
**16.11** Predicting the Effect of a Change in Concentration on the Position of Equilibrium (p. 720)
**16.12** Predicting the Effect of a Change in Volume (Pressure) on the Position of Equilibrium (p. 721)
**16.13** Predicting the Effect of a Change in Temperature on the Position of Equilibrium (p. 723)

## Problems

Problems with a green number are answered at the back of the book. Most sections include three categories of problems separated by a green rule—concept review questions, *paired* skill building exercises, and problems in a relevant context.

### The Dynamic Nature of the Equilibrium State

**16.1** A change in reaction conditions increases the rate of a forward reaction more than that of the reverse reaction. What is the effect on the equilibrium constant and the concentrations of reactants and products at equilibrium?

**16.2** When a chemical company employs a new reaction to manufacture a product, the chemists consider its rate (kinetics) and yield (equilibrium). How do each of these affect the usefulness of a manufacturing process?

**16.3** Is $K$ very large or very small for a reaction that goes essentially to completion? Explain.

**16.4** Consider the following reaction energy diagram:

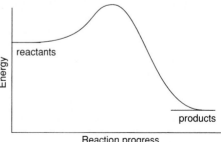

Does it describe a reaction in which $K$ is very small or very large? Explain.

## The Mass-Action Expression and the Equilibrium Constant

(Sample Problems 16.1-16.5)

**16.5** For a given reaction at a given temperature, the value of $K$ is constant. Is the value of $Q$ also constant? Explain.

**16.6** An inorganic chemist places 1 mol BrCl in container A and 0.5 mol $Br_2$ and 0.5 mol $Cl_2$ in container B. She seals the containers and heats them to 300°C. Measurement shows that, with time, the contents of both containers are identical mixtures of BrCl, $Br_2$, and $Cl_2$.
  (a) Write a balanced equation for the reaction taking place in container A.
  (b) Write the mass-action expression, $Q$, for this reaction.
  (c) How do the values of $Q$ in A compare with those in B over time?
  (d) Explain on the molecular level how it is possible for both containers to end up with identical mixtures.

**16.7** In a series of experiments on the thermal decomposition of lithium peroxide,

$$2Li_2O_2(s) \rightleftharpoons 2Li_2O(s) + O_2(g)$$

a chemist finds that, as long as enough $Li_2O_2$ is heated so that some remains at the end of the experiment, the amount of $O_2$ obtained in a given container at a given $T$ reaches the same value. Explain.

**16.8** In a study of the formation of HI from its elements,

$$H_2(g) + I_2(g) \rightleftharpoons 2HI(g)$$

equal amounts of $H_2$ and $I_2$ were placed in a container, which was then sealed and heated.
  (a) On one set of axes, sketch concentration vs. time curves for $H_2$ and HI and explain how the value of $Q$ changes as a function of time.
  (b) Would the value of $Q$ be different if $I_2$ concentration were plotted instead of $H_2$?

**16.9** Guldberg and Waage proposed the original definition of the equilibrium constant as a particular ratio of *concentrations*. What relationship allows us to use a particular ratio of *partial pressures* (for a gaseous reaction) to obtain an equilibrium constant? Explain.

**16.10** In which cases are $K_c$ and $K_p$ equal and in which cases are they not equal?

**16.11** Explain the difference between a heterogeneous and a homogeneous equilibrium reaction. Give an example of each.

**16.12** When the numerical value of $Q$ is less than $K$, in which direction does the reaction proceed to reach equilibrium? Explain.

**16.13** Balance each of the following gaseous reactions and write its mass-action expression, $Q_c$:
  (a) $NO(g) + O_2(g) \rightleftharpoons N_2O_3(g)$
  (b) $SF_6(g) + SO_3(g) \rightleftharpoons SO_2F_2(g)$
  (c) $SClF_5(g) + H_2(g) \rightleftharpoons S_2F_{10}(g) + HCl(g)$
  (d) $C_2H_6(g) + O_2(g) \rightleftharpoons CO_2(g) + H_2O(g)$

**16.14** Balance each of the following gaseous reactions and write its mass-action expression, $Q_c$:
  (a) $NO_2Cl(g) \rightleftharpoons NO_2(g) + Cl_2(g)$
  (b) $POCl_3(g) \rightleftharpoons PCl_3(g) + O_2(g)$
  (c) $NH_3(g) + O_2(g) \rightleftharpoons N_2(g) + H_2O(g)$
  (d) $O_2(g) \rightleftharpoons O_3(g)$

**16.15** The powerful chlorinating agent sulfuryl dichloride ($SO_2Cl_2$) can be prepared by the following two-step sequence:

$$H_2S(g) + O_2(g) \rightleftharpoons SO_2(g) + H_2O(g)$$
$$SO_2(g) + Cl_2(g) \rightleftharpoons SO_2Cl_2(g)$$

  (a) Balance each step and write the overall balanced equation.
  (b) Show that the overall $Q_c$ equals the product of the $Q_c$'s for the individual steps.

**16.16** The interhalogen $ClF_3$ is prepared in a two-step fluorination of chlorine gas:

$$Cl_2(g) + F_2(g) \rightleftharpoons ClF(g)$$
$$ClF(g) + F_2(g) \rightleftharpoons ClF_3(g)$$

  (a) Balance each step and write the overall balanced equation.
  (b) Show that the overall $Q_c$ equals the product of the $Q_c$'s for the individual steps.

**16.17** At a particular temperature, $K_c = 1.6 \times 10^{-2}$ for the equation

$$2H_2S(g) \rightleftharpoons 2H_2(g) + S_2(g)$$

Calculate $K_c$ for each of the following equations:
  (a) $\frac{1}{2}S_2(g) + H_2(g) \rightleftharpoons H_2S(g)$
  (b) $5H_2S(g) \rightleftharpoons 5H_2(g) + \frac{5}{2}S_2(g)$

**16.18** At a particular temperature, $K_c = 6.5 \times 10^2$ for the equation

$$2NO(g) + 2H_2(g) \rightleftharpoons N_2(g) + 2H_2O(g)$$

Calculate $K_c$ for each of the following equations:
  (a) $NO(g) + H_2(g) \rightleftharpoons \frac{1}{2}N_2(g) + H_2O(g)$
  (b) $2N_2(g) + 4H_2O(g) \rightleftharpoons 4NO(g) + 4H_2(g)$

**16.19** Balance each of the following processes and write its mass-action expression, $Q_c$:
  (a) $Na_2O_2(s) + CO_2(g) \rightleftharpoons Na_2CO_3(s) + O_2(g)$
  (b) $H_2O(l) \rightleftharpoons H_2O(g)$
  (c) $H_2SO_3(aq) \rightleftharpoons H_2O(l) + SO_2(aq)$
  (d) $H_2O(l) + SO_3(g) \rightleftharpoons H_2SO_4(aq)$
  (e) $KNO_3(s) \rightleftharpoons KNO_2(s) + O_2(g)$

**16.20** Balance each of the following processes and write its mass-action expression, $Q_c$:
  (a) $NaHCO_3(s) \rightleftharpoons Na_2CO_3(s) + CO_2(g) + H_2O(g)$
  (b) $SnO_2(s) + H_2(g) \rightleftharpoons Sn(s) + H_2O(g)$
  (c) $H_2SO_4(l) + SO_3(g) \rightleftharpoons H_2S_2O_7(l)$
  (d) $Al(s) + NaOH(aq) + H_2O(l) \rightleftharpoons$
    $$Na[Al(OH)_4](aq) + H_2(g)$$

(e) $CO_2(s) \rightleftharpoons CO_2(g)$

**16.21** Determine $\Delta n_{gas}$ for each of the following equations:
(a) $2KClO_3(s) \rightleftharpoons 2KCl(s) + 3O_2(g)$
(b) $2PbO(s) + O_2(g) \rightleftharpoons 2PbO_2(s)$
(c) $I_2(s) + 3XeF_2(s) \rightleftharpoons 2IF_3(s) + 3Xe(g)$

**16.22** Determine $\Delta n_{gas}$ for each of the following equations:
(a) $MgCO_3(s) \rightleftharpoons MgO(s) + CO_2(g)$
(b) $2H_2(g) + O_2(g) \rightleftharpoons 2H_2O(l)$
(c) $HNO_3(l) + ClF(g) \rightleftharpoons ClONO_2(g) + HF(g)$

**16.23** Calculate $K_c$ for the following equilibria:
(a) $CO(g) + Cl_2(g) \rightleftharpoons COCl_2(g)$;
$\qquad K_p = 3.9 \times 10^{-2}$ at 1000 K
(b) $S_2(g) + C(s) \rightleftharpoons CS_2(g)$; $K_p = 28.5$ at 500 K

**16.24** Calculate $K_p$ for the following equilibria:
(a) $N_2O_4(g) \rightleftharpoons 2NO_2(g)$; $K_c = 6.1 \times 10^{-3}$ at 298 K
(b) $N_2(g) + 3H_2(g) \rightleftharpoons 2NH_3(g)$;
$\qquad K_c = 2.4 \times 10^{-3}$ at 1000 K

**16.25** At 425°C, $K_p$ is $4.18 \times 10^{-9}$ for the reaction

$$2HBr(g) \rightleftharpoons H_2(g) + Br_2(g)$$

In one experiment, 0.20 atm HBr($g$), 0.010 atm H$_2$($g$) and 0.010 atm Br$_2$($g$) are introduced into a reaction vessel. Is the reaction at equilibrium? If not, in which direction will it proceed?

**16.26** At 100°C, $K_c$ is 1.98 for the reaction

$$2NOBr(g) \rightleftharpoons 2NO(g) + Br_2(g)$$

In a given experiment, 0.10 mol of each component is introduced into a 1.0-L container. Is the system at equilibrium? If not, in which direction will it proceed?

**16.27** White phosphorus, $P_4$, is produced by the reduction of phosphate rock, $Ca_3(PO_4)_2$. If exposed to oxygen, the waxy, white solid bursts into smoke and flames, and releases a large amount of heat. Does the reaction

$$P_4(g) + 5O_2(g) \rightleftharpoons P_4O_{10}(s)$$

have an equilibrium constant with a large or small value? Explain.

**16.28** The Group 8B(10) elements—Ni, Pd, and Pt—are used as some of the most important industrial catalysts. Isolation and purification of these metals involve a series of steps. In the case of nickel, for example, the sulfide ore $Ni_3S_2$ is roasted in air,

$$Ni_3S_2(s) + O_2(g) \rightleftharpoons NiO(s) + SO_2(g)$$

The oxide is reduced by the H$_2$ in water gas (CO + H$_2$) to impure Ni:

$$NiO(s) + H_2(g) \rightleftharpoons Ni(s) + H_2O(g)$$

The CO in water gas then reacts with the metal in the Mond process to form gaseous nickel carbonyl:

$$Ni(s) + CO(g) \rightleftharpoons Ni(CO)_4(g)$$

which is subsequently decomposed to the metal.
(a) Balance each of the three steps and obtain an overall balanced equation for the conversion of $Ni_3S_2$ to $Ni(CO)_4$.
(b) Show that the overall $Q_c$ is the product of the $Q_c$'s for the individual reactions.

**16.29** The water-gas shift reaction plays a central role in chemical methods for obtaining cleaner fuels from coal:

$$CO(g) + H_2O(g) \rightleftharpoons CO_2(g) + H_2(g)$$

At a given temperature, $K_p = 2.7$. If 0.13 mol CO, 0.56 mol H$_2$O, 0.62 mol CO$_2$, and 0.43 mol H$_2$ are introduced into a 2.0-L flask, in which direction must the reaction proceed to reach equilibrium?

## How to Solve Equilibrium Problems

(Sample Problems 16.6-16.10)

**16.30** The final step in the formation of CFC-11, one of the most heavily produced chlorofluorocarbons, is

$$CCl_4(g) + HF(g) \rightleftharpoons CFCl_3(g) + HCl(g)$$

If you start the reaction with equal concentrations of CCl$_4$ and HF, you obtain equal concentrations of CFCl$_3$ and HCl at equilibrium. Would the concentrations of CFCl$_3$ and HCl be equal if you start with unequal concentrations of CCl$_4$ and HF? Explain.

**16.31** In a problem involving the catalyzed reaction of methane and steam, the following reaction table was prepared:

| Pressure (atm) | $CH_4(g)$ + | $2H_2O(g)$ $\rightleftharpoons$ | $CO_2(g)$ + | $4H_2(g)$ |
|---|---|---|---|---|
| Initial | 0.30 | 0.40 | 0 | 0 |
| Change | $-x$ | $-2x$ | $+x$ | $+4x$ |
| Equilibrium | $0.30-x$ | $0.40-2x$ | $x$ | $4x$ |

Explain the entries in the "Change" and "Equilibrium" rows.

**16.32** (a) Explain the approximation made to avoid using the quadratic formula when finding an equilibrium concentration.
(b) Describe a situation in which this approximation cannot be made.

**16.33** In an experiment to study the formation of HI($g$),

$$H_2(g) + I_2(g) \rightleftharpoons 2HI(g)$$

H$_2$($g$) and I$_2$($g$) were placed in a sealed container at a certain temperature. At equilibrium, [H$_2$] = $6.50 \times 10^{-5}$ M, [I$_2$] = $1.06 \times 10^{-3}$ M, and [HI] = $1.87 \times 10^{-3}$ M. Calculate $K_c$ for the reaction at this temperature.

**16.34** Gaseous ammonia was introduced into a sealed container and heated to a certain temperature:

$$2NH_3(g) \rightleftharpoons N_2(g) + 3H_2(g)$$

At equilibrium, $[NH_3] = 0.0225\ M$, $[N_2] = 0.114\ M$, and $[H_2] = 0.342\ M$. Calculate $K_c$ for the reaction at this temperature.

**16.35** Gaseous $PCl_5$ decomposes according to the reaction

$$PCl_5(g) \rightleftharpoons PCl_3(g) + Cl_2(g)$$

In one experiment, 0.15 mol $PCl_5(g)$ was introduced into a 2.0-L container. Construct the reaction table for this process.

**16.36** Hydrogen fluoride, HF, can be made from the reaction

$$H_2(g) + F_2(g) \rightleftharpoons 2HF(g)$$

In one experiment, 0.10 mol $H_2(g)$ and 0.050 mol $F_2(g)$ are introduced into a 0.50-L container. Construct the reaction table for this process.

**16.37** When 0.15 mol $PH_3BCl_3(s)$ is introduced into a 3.0-L container at a certain temperature, $8.4 \times 10^{-3}$ mol $PH_3$ is present at equilibrium:

$$PH_3BCl_3(s) \rightleftharpoons PH_3(g) + BCl_3(g)$$

Calculate $K_c$ for the reaction at this temperature.

**16.38** When 0.0150 mol $NH_3(g)$ and 0.0150 mol $O_2(g)$ are introduced into a 1.00-L container at a certain temperature, the $N_2$ concentration at equilibrium is $1.96 \times 10^{-3}\ M$:

$$4NH_3(g) + 3O_2(g) \rightleftharpoons 2N_2(g) + 6H_2O(g)$$

Calculate $K_c$ for the reaction at this temperature.

**16.39** For the following reaction, $K_p$ is $6.5 \times 10^4$ atm$^{-1}$ at 308 K:

$$2NO(g) + Cl_2(g) \rightleftharpoons 2NOCl(g)$$

At equilibrium $P_{NO} = 0.35$ atm and $P_{Cl_2} = 0.10$ atm. What is the equilibrium partial pressure of $NOCl(g)$?

**16.40** For the following reaction, $K_p$ is 0.262 atm$^{-1}$ at 1000°C:

$$C(s) + 2H_2(g) \rightleftharpoons CH_4(g)$$

At equilibrium $P_{H_2}$ is 1.22 atm. What is the equilibrium partial pressure of $CH_4(g)$?

**16.41** Ammonium hydrogen sulfide decomposes according to the following reaction, for which $K_p = 0.11$ at 250°C:

$$NH_4HS(s) \rightleftharpoons H_2S(g) + NH_3(g)$$

If 55.0 g $NH_4HS(s)$ is in a sealed container, what is the partial pressure of $NH_3(g)$ at equilibrium?

**16.42** Hydrogen sulfide decomposes according to the following reaction, for which $K_c$ is $9.30 \times 10^{-8}$ at 700°C:

$$2H_2S(g) \rightleftharpoons 2H_2(g) + S_2(g)$$

If 0.45 mol $H_2S$ is placed in a 3.0-L container, what is the equilibrium concentration of $H_2(g)$ at 700°C?

**16.43** Even at high temperature, the formation of nitric oxide is not favored:

$$N_2(g) + O_2(g) \rightleftharpoons 2NO(g);\ K_c = 4.10 \times 10^{-4} \text{ at } 2000°C$$

What is the equilibrium concentration of $NO(g)$ when a mixture of 0.20 mol $N_2(g)$ and 0.15 mol $O_2(g)$ are allowed to come to equilibrium in a 1.0-L container at this temperature?

**16.44** Nitrogen dioxide decomposes according to the reaction:

$$2NO_2(g) \rightleftharpoons 2NO(g) + O_2(g)$$

where $K_p = 4.48 \times 10^{-13}$. A pressure of 0.75 atm of $NO_2$ is introduced into a 1.00-L container and allowed to come to equilibrium. What are the equilibrium partial pressures of $NO(g)$ and $O_2(g)$?

**16.45** Hydrogen iodide decomposes according to the reaction

$$2HI(g) \rightleftharpoons H_2(g) + I_2(g);\ K_c = 0.0184 \text{ at } 703 \text{ K}$$

A sealed 1.50-L container initially holds 0.00623 mol $H_2$, 0.00414 mol $I_2$, and 0.0244 mol HI at 703 K. When equilibrium is reached, the equilibrium concentration of $H_2(g)$ is 0.00467 $M$. What are the equilibrium concentrations of $HI(g)$ and $I_2(g)$?

**16.46** Compound A decomposes according to the reaction

$$A(g) \rightleftharpoons 2B(g) + C(g);\ K_c = 2.35 \times 10^{-8} \text{ at } 100°C$$

A sealed 1.00-L reaction vessel initially contains $1.75 \times 10^{-3}$ mol $A(g)$, $1.25 \times 10^{-3}$ mol $B(g)$, and $6.50 \times 10^{-4}$ mol $C(g)$ at 100°C. When equilibrium is reached, the concentration of $A(g)$ is $2.15 \times 10^{-3}\ M$. What are the equilibrium concentrations of $B(g)$ and $C(g)$?

**16.47** Phosgene ($COCl_2$) is an extremely toxic substance that forms readily from carbon monoxide and chlorine at elevated temperatures:

$$CO(g) + Cl_2(g) \rightleftharpoons COCl_2(g)$$

If 0.350 moles of each reactant is placed in a 0.500-L flask at 600 K, what are the concentrations of each substance at equilibrium? $K_c = 4.95$ at this temperature.

**16.48** In an analysis of interhalogen reactivity, 0.500 mol ICl was placed in a 5.00-L flask and allowed to decompose at high temperature:

$$2ICl(g) \rightleftharpoons I_2(g) + Cl_2(g)$$

Calculate the equilibrium concentrations of $I_2$, $Cl_2$, and ICl. $K_c = 0.110$ at this temperature.

**16.49** Aluminum is one of the most versatile metals. It is produced by the Hall process, in which molten cryolite, $Na_3AlF_6$, is used as a solvent for the aluminum ore. Cryolite undergoes very slight decomposition with heat to produce a tiny amount of $F_2$, which escapes into the atmosphere above the solvent. $K_c$ for the reaction

$$Na_3AlF_6(l) \rightleftharpoons 3Na(l) + Al(l) + 3F_2(g)$$

is $2 \times 10^{-104}$ at 1300 K. What is the concentration of $F_2$ over a bath of molten cryolite at this temperature?

**16.50** A United Nations toxicologist studying the properties of mustard gas $[S(CH_2CH_2Cl)_2]$, a warfare blistering agent, prepares a mixture of 0.675 $M$ $SCl_2$ and 0.973 $M$ $C_2H_4$ and allows them to react at room temperature (20°C):

$$SCl_2(g) + 2C_2H_4(g) \rightleftharpoons S(CH_2CH_2Cl)_2(g)$$

At equilibrium, 0.350 $M$ $S(CH_2CH_2Cl)_2$ is present. Calculate $K_p$.

**16.51** One of the key steps in the extraction of iron from iron ore is

$$FeO(s) + CO(g) \rightleftharpoons Fe(s) + CO_2(g); \quad K_p = 0.403 \text{ at } 1000°C$$

This step occurs in the 700-1200°C zone within a blast furnace. What are the equilibrium partial pressures of $CO(g)$ and $CO_2(g)$ when 1.00 atm $CO(g)$ and excess $FeO(s)$ react in a sealed container at 1000°C?

## Reaction Conditions and the Equilibrium State: Le Châtelier's Principle

(Sample Problems 16.11-16.13)

**16.52** In the statement of Le Châtelier's principle, what is meant by the word "disturbance"?

**16.53** What is the difference between the equilibrium position and the equilibrium constant of a reaction? Which changes as a result of a change in reactant concentration?

**16.54** What is implied by the word "constant" in the term "equilibrium constant"? Give two parameters that can be changed without changing the value of this constant.

**16.55** Le Châtelier's principle is related ultimately to the rates of the forward and reverse steps in a reaction. Use this idea to explain
(a) Why an increase in reactant concentration shifts the equilibrium position to the right but does not change the equilibrium constant.
(b) Why a decrease in volume shifts the equilibrium position toward fewer moles of gas but does not change the equilibrium constant.
(c) Why a rise in temperature shifts the equilibrium position of an exothermic reaction toward reactants and also changes the equilibrium constant.

**16.56** Le Châtelier's principle predicts that a rise in the temperature of an endothermic reaction from $T_1$ to $T_2$ would result in $K_2$ larger than $K_1$. Explain.

**16.57** Consider the equilibrium system

$$CO(g) + Fe_3O_4(s) \rightleftharpoons CO_2(g) + 3FeO(s)$$

Predict the direction the position of equilibrium will shift as a result of the following disturbances:
(a) CO is added.
(b) $CO_2$ is removed by adding solid NaOH.
(c) Additional $Fe_3O_4(s)$ is added to the system.
(d) Dry ice is added at constant $T$.

**16.58** Sodium bicarbonate undergoes thermal decomposition according to the reaction

$$2NaHCO_3(s) \rightleftharpoons Na_2CO_3(s) + CO_2(g) + H_2O(g)$$

Predict the direction the position of equilibrium will shift as a result of the following disturbances:
(a) 0.20 atm argon gas is added.
(b) $NaHCO_3(s)$ is added.
(c) Solid $Mg(ClO_4)_2$ is added as a drying agent to remove $H_2O$.
(d) Dry ice is added at constant $T$.

**16.59** Predict the effect of increasing the container volume on the *amounts* of reactants and products in each of the following:
(a) $F_2(g) \rightleftharpoons 2F(g)$
(b) $2CH_4(g) \rightleftharpoons C_2H_2(g) + 3H_2(g)$
(c) $CH_3OH(l) \rightleftharpoons CH_3OH(g)$

**16.60** Predict the effect of increasing the container volume on the *amounts* of reactants and products in each of the following:
(a) $H_2(g) + Cl_2(g) \rightleftharpoons 2HCl(g)$
(b) $2H_2(g) + O_2(g) \rightleftharpoons 2H_2O(l)$
(c) $C_3H_8(g) + 5O_2(g) \rightleftharpoons 3CO_2(g) + 4H_2O(l)$

**16.61** How would you adjust the *volume* of the reaction vessel to maximize product yield in the following reactions?
(a) $Fe_3O_4(s) + 4H_2(g) \rightleftharpoons 3Fe(s) + 4H_2O(g)$
(b) $2C(s) + O_2(g) \rightleftharpoons 2CO(g)$

**16.62** How would you adjust the *volume* of the reaction vessel to maximize product yield in the following reactions?
(a) $Na_2O_2(s) \rightleftharpoons 2Na(l) + O_2(g)$
(b) $C_2H_2(g) + 2H_2(g) \rightleftharpoons C_2H_6(g)$

**16.63** Predict the effect of *increasing* the temperature on the amounts of products in the following reactions:
(a) $CO(g) + 2H_2(g) \rightleftharpoons CH_3OH(g)$;
$$\Delta H^0_{rxn} = -90.7 \text{ kJ}$$
(b) $C(s) + H_2O(g) \rightleftharpoons CO(g) + H_2(g)$;
$$\Delta H^0_{rxn} = +131 \text{ kJ}$$
(c) $2NO_2(g) \rightleftharpoons 2NO(g) + O_2(g)$ (endothermic)
(d) $2C(s) + O_2(g) \rightleftharpoons 2CO(g)$ (exothermic)

**16.64** Predict the effect of *decreasing* the temperature on the amounts of reactants in the following reactions:
(a) $C_2H_2(g) + H_2O(g) \rightleftharpoons CH_3CHO(g)$;
$$\Delta H^0_{rxn} = -151 \text{ kJ}$$
(b) $CH_3CH_2OH(l) + O_2(g) \rightleftharpoons$
$$CH_3CO_2H(l) + H_2O(g); \Delta H^0_{rxn} = -451 \text{ kJ}$$
(c) $2C_2H_4(g) + O_2(g) \rightleftharpoons 2CH_3CHO(g)$ (exothermic)
(d) $NaCl(s) \rightleftharpoons NaCl(aq)$ (endothermic)

**16.65** Deuterium (D or $^2H$) is an isotope of hydrogen. The molecule $D_2$ undergoes an exchange reaction with ordinary $H_2$ that leads to isotopic equilibrium:

$$D_2(g) + H_2(g) \rightleftharpoons 2DH(g); \quad K_p = 1.80 \text{ at } 298 \text{ K}$$

If $\Delta H^0_{rxn}$ is 0.32 kJ/mol DH that forms, calculate $K_p$ at 500 K.

**16.66** The formation of methanol is an important industrial reaction in the processing of new fuels. At 298 K, $K_p = 2.25 \times 10^4$ for the reaction

$$CO(g) + 2H_2(g) \rightleftharpoons CH_3OH(l)$$

If $\Delta H^0_{rxn} = -128$ kJ/mol $CH_3OH$, calculate $K_p$ at 0°C.

**16.67** Lime (CaO) is used primarily in the manufacture of steel, glass, and high-quality paper. It is produced in an endothermic reaction by the thermal decomposition of limestone:

$$CaCO_3(s) \rightleftharpoons CaO(s) + CO_2(g)$$

How would you control reaction conditions to produce the maximum amount of lime?

**16.68** When isopentyl alcohol and acetic acid react, they form the pleasant-smelling compound isopentyl acetate, the essence of banana oil:

$$C_5H_{11}OH(aq) + CH_3COOH(aq) \rightleftharpoons$$
$$CH_3COOC_5H_{11}(aq) + H_2O(l)$$

A student adds a drying agent to remove $H_2O$ in an attempt to increase the yield of banana oil. Is this approach reasonable? Explain.

**16.69** The oxidation of $SO_2$ to $SO_3$ is one of the most important industrial reactions because it is the key step in sulfuric acid production:

$$SO_2(g) + \tfrac{1}{2}O_2(g) \rightleftharpoons SO_3(g); \quad \Delta H^0_{rxn} = -99.2 \text{ kJ}$$

(a) What combination of $T$ and $P$ would maximize $SO_3$ yield?

(b) How would addition of $O_2$ affect $Q$? $K$?

(c) Suggest a reason that catalysis is used for this reaction in the manufacture of $H_2SO_4$.

**16.70** A study of the water-gas shift reaction was made in which equilibrium was reached with [CO] = [$H_2O$] = [$H_2$] = 0.10 $M$ and [$CO_2$] = 0.40 $M$. After 0.60 mol $H_2$ was added to the 2.0-L container and equilibrium reestablished, what were the new concentrations of all the components?

**Comprehensive Problems**
Problems with an asterisk (*) are more challenging.

**16.71** Consider the following equilibrium system in a piston-cylinder:

$$A(g) + 3B(s) \rightleftharpoons 5D(l) + E(g); \quad \Delta H^0 < 0$$

In what direction will the equilibrium position shift upon the application of each of the following changes?
(a) Raise the temperature.
(b) Add more B.

(c) Reduce the volume of the container.
(d) Add a catalyst.
(e) Remove some E.

*16.72 Ammonium carbamate ($NH_2COONH_4$) is a salt of carbamic acid that is found in the blood and urine of mammals. At 250°C, $K_c$ is $1.58 \times 10^{-8}$ for the following equilibrium:

$$NH_2COONH_4(s) \rightleftharpoons 2NH_3(g) + CO_2(g)$$

If 7.80 g $NH_2COONH_4$ is introduced into a 0.500-L evacuated container, what is the total pressure inside the container at equilibrium?

**16.73** One of the most important industrial sources of ethanol is the exothermic reaction of steam with ethene derived from crude oil:

$$C_2H_4(g) + H_2O(g) \rightleftharpoons C_2H_5OH(g)$$

The $\Delta H^0_{rxn} = -47.8$ kJ and $K_c = 9 \times 10^3$ at 600 K.
(a) At equilibrium, $P_{C_2H_5OH} = 200$ atm and $P_{H_2O} = 400$ atm. Calculate $P_{C_2H_4}$.
(b) Would the highest yield of ethanol be obtained at high or low pressures? High or low temperatures?
(c) Calculate $K_c$ at 450 K.
(d) In the manufacture of ammonia, ammonia yield is increased by condensing it to a liquid and removing it from the vessel. Would condensing the $C_2H_5OH$ work in this process? Explain.

*16.74 An industrial chemist introduces 2.0 atm $H_2$ and 2.0 atm $CO_2$ into a 1.00-L container at 25.0°C and then raises the temperature to 700°C at which $K_c = 0.534$:

$$H_2(g) + CO_2(g) \rightleftharpoons H_2O(g) + CO(g)$$

How many grams of $H_2$ are present after equilibrium is established?

**16.75** (a) As an EPA scientist studying catalytic converters and urban smog, you want to calculate $K_c$ for the reaction

$$2NO_2(g) \rightleftharpoons N_2(g) + 2O_2(g)$$

using the following data:

$$\tfrac{1}{2}N_2(g) + \tfrac{1}{2}O_2(g) \rightleftharpoons NO(g); \quad K_c = 4.8 \times 10^{-10}$$
$$2NO_2(g) \rightleftharpoons 2NO(g) + O_2(g); \quad K_c = 1.1 \times 10^{-5}$$

(b) In combustion studies of $H_2$ as an alternative fuel, you find evidence that the hydroxyl radical (HO) is formed in flames by the reaction

$$H(g) + \tfrac{1}{2}O_2(g) \rightleftharpoons HO(g)$$

Use the following data to calculate $K_c$ for the reaction:

$$\tfrac{1}{2}H_2(g) + \tfrac{1}{2}O_2(g) \rightleftharpoons HO(g); \quad K_c = 0.58$$
$$\tfrac{1}{2}H_2(g) \rightleftharpoons H(g); \quad K_c = 1.6 \times 10^{-3}$$

**16.76** A quality control engineer examining the conversion of $SO_2$ to $SO_3$ in the manufacture of sulfuric acid determines that $K_c = 1.7 \times 10^8$ at 600 K for the reaction

$$2SO_2(g) + O_2(g) \rightleftharpoons 2SO_3(g)$$

(a) At equilibrium, $P_{SO_3} = 300$ atm and $P_{O_2} = 100$ atm. Calculate $P_{SO_2}$.
(b) She places a mixture of 0.0040 mol $SO_2(g)$ and 0.0028 mol $O_2(g)$ in a 1.0-L container and raises the temperature to 1000 K. At equilibrium, 0.0020 mol $SO_3(g)$ is present. Calculate $K_c$ and $P_{SO_2}$ for this reaction at 1000 K.

**16.77** When 0.100 mol $CaCO_3(s)$ and 0.100 mol $CaO(s)$ are introduced into an evacuated sealed 10-L container and heated to 385 K, $P_{CO_2} = 0.220$ atm after equilibrium is established:

$$CaCO_3(s) \rightleftharpoons CaO(s) + CO_2(g)$$

An additional 0.300 atm $CO_2(g)$ is then pumped into the container. How many additional grams of $CaCO_3$ will form after equilibrium is reestablished?

**16.78** Use each of the following mass-action expressions to write the balanced equation:

(a) $Q = \dfrac{[CO_2]^2[H_2O]^2}{[C_2H_4][O_2]^3}$    (b) $Q = \dfrac{[NH_3]^4[O_2]^7}{[NO_2]^4[H_2O]^6}$

**16.79** Hydrogenation of C=C bonds is a major reaction in the petroleum and food industries. The conversion of acetylene to ethylene is a simple example of the process:

$$C_2H_2(g) + H_2(g) \rightleftharpoons C_2H_4(g); \quad K_c = 2.9 \times 10^8 \text{ at } 2000 \text{ K}$$

The process is usually performed at much lower temperatures with the aid of a catalyst. Use $\Delta H$ values from Appendix B to calculate $K_c$ for this reaction at 300 K.

**\*16.80** Highly toxic disulfur decafluoride decomposes by a free-radical process:

$$S_2F_{10}(g) \rightleftharpoons SF_4(g) + SF_6(g)$$

In a study of the decomposition, $S_2F_{10}$ was placed in a 2.0-L flask and heated to 100°C and $[S_2F_{10}]$ was 0.50 $M$ at equilibrium. More $S_2F_{10}$ was added and when equilibrium was reattained, $[S_2F_{10}]$ was 2.5 $M$. How had $[SF_4]$ and $[SF_6]$ changed from the original equilibrium position to the new equilibrium position after the addition of more $S_2F_{10}$?

**\*16.81** Consider the following reaction:

$$3Fe(s) + 4H_2O(g) \rightleftharpoons Fe_3O_4(s) + 4H_2(g)$$

(a) What are the apparent oxidation states of Fe and of O in $Fe_3O_4$?
(b) $Fe_3O_4$ is a compound of iron in which Fe occurs in two oxidation states. What are the oxidation states of Fe in $Fe_3O_4$?
(c) At 900°C, $K_c$ for the reaction is 5.1. If 0.050 mol $H_2O(g)$ and 0.100 mol $Fe(s)$ are placed in a 1.0-L container at 900°C, how many grams of $Fe_3O_4$ are present when equilibrium is established?

**16.82** Many essential metabolites are products in branched pathways, such as the one shown below:

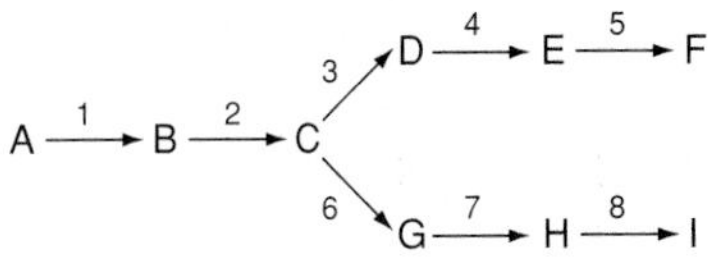

One method of control of these pathways occurs through inhibition of the first enzyme specific for a branch.
(a) Which enzyme is inhibited by F? By I?
(b) What disadvantage would there be if F inhibited Enzyme 1?

# CHAPTER 17

## Concepts and skills to review

- role of water as solvent (Section 4.2)
- acids, bases, and neutralization reactions (Section 4.3)
- writing total and net ionic equations (Section 4.3)
- properties of an equilibrium constant (Section 16.2)
- solving equilibrium problems (Section 16.3)

# Acid-Base Equilibria

**Culinary acids and bases.** The bases in fish react with the acids in lemon in this delectable neutralization reaction. We examine the equilibrium nature of these important reactions and see how acid-base definitions broaden in scope.

**A**cids and bases have been essential reagents since alchemical times. With exotic names such as oil of vitriol (sulfuric acid), aqua fortis (nitric acid), and spirits of hartshorn (ammonia), these substances found early uses in everything from fraudulent "gold-forming" recipes to the practical metallurgy of gold, silver, and copper. In the 15th century, artists frequently etched copper plates with nitric acid to create their graphic works. To make glass, sand was treated with the strongly basic potash ($K_2CO_3$) that formed among wood ashes. In modern times, acids and bases are everyday chemicals in the home and in industry (Figure 17.1) and play an indispensable role in the chemistry laboratory.

Table 17.1 shows some common acids and bases and their familiar uses. Notice that some of the acids have a sour taste. In fact, since the 17th century, sourness had been one of the defining properties of acids. An acid was any substance that had a sour taste; reacted with active metals, such as aluminum and zinc, to produce hydrogen gas; and turned certain organic compounds characteristic colors. (We discuss these compounds, known as *indicators*, later and in Chapter 18.) A base, on the other hand, was any substance that had a bitter taste and slippery feel and turned the same organic compounds different characteristic colors.* Moreover, it was commonly known that *when acids and bases react, each cancels the properties of the other in a process called neutralization.* ◆

Evolving definitions are common in science. As more phenomena are understood, an early description becomes too limited and is replaced by a broader one. Even though the early definitions of acids and bases described some distinctive features of these substances, they inevitably evolved into definitions that described acid-base behavior on the molecular level. In this

◆ **Pioneers of Acid-Base Chemistry.** Three 17th-century chemists laid the early foundations of acid-base chemistry. Johann Glauber (1604-1668) became renowned for his ability to prepare acids and their salts. Otto Tachenius (c. 1620-1690) is credited with first recognizing that a salt is the product of the reaction between an acid and a base. In addition to his studies of gas behavior, Robert Boyle (1627-1691) fostered the use of spot tests, flame colors, fume odors, and precipitates to analyze reactions. He was the first to associate the color change in syrup of violets (an organic dye) with the acidic or basic nature of the test solution.

---

*Despite what 17th-century chemists may have done, *NEVER* taste or touch chemicals in the laboratory. Many, including common laboratory acids and bases, are harmful and can cause severe burns. To satisfy your curiosity, put some lemon juice or vinegar on your next salad and note its acidic, sour taste.

**A**

**B**

**FIGURE 17.1**
**Etching with acids. A,** The insides of "frosted" bulbs are etched with hydrofluoric acid. **B,** Acid is used to remove unwanted oxides of silicon and metals during the process of manufacturing computer chips. This cross-section of a state-of-the-art microprocessor shows five layers of metal imbedded in a silicon matrix.

**TABLE 17.1  Some Common Acids and Bases and Their Household Uses**

| SUBSTANCE | FORMULA | USE | |
|---|---|---|---|
| **Acids** | | | |
| Acetic acid (vinegar) | $CH_3COOH$ (or $HC_2H_3O_2$) | Flavoring, preservative | |
| Citric acid | $H_3C_6H_5O_7$ | Flavoring | |
| Phosphoric acid | $H_3PO_4$ | Rust remover | |
| Boric acid | $B(OH)_3$ (or $H_3BO_3$) | Mild antiseptic; insecticide | |
| Aluminum salts | $Al_2(SO_4)_3$, etc. | In baking powder, with sodium bicarbonate | |
| Hydrochloric acid (muriatic acid) | HCl | Brick and ceramic tile cleaner | |
| **Bases** | | | |
| Sodium hydroxide (lye) | NaOH | Oven cleaner, unblocking plumbing | |
| Ammonia | $NH_3$ | Household cleaner | |
| Sodium carbonate | $Na_2CO_3$ | Water softener, grease remover | |
| Sodium bicarbonate | $NaHCO_3$ | Fire extinguisher, rising agent in cake mixes (baking soda), mild antacid | |
| Trisodium phosphate | $Na_3PO_4$ | Cleaner for surfaces before painting or wallpapering | |

chapter, we define acids and bases in ways that systematize their behavior and allow us to understand greater numbers of reactions. In the process, you'll see how the principles of chemical equilibrium apply to this essential group of substances.

After presenting Arrhenius acids and bases, the simplest of the more recent acid-base definitions, we examine acid dissociation as an equilibrium process to understand why acids vary in strength. Then we introduce the pH scale as a means of comparing the acidity or basicity of aqueous solutions. Next, we see how the Brønsted-Lowry definition greatly expands the meaning of "base" and, with it, the nature of acid-base reactions. We investigate the molecular structures of acids and bases to rationalize variations in strength and then see that the very designations of "acid" and "base" depend on their relative strengths and on the solvent in which the reaction occurs. Finally, we examine the Lewis acid-base definition, which expands the meaning of "acid" and broadens acid-base behavior even further.

## 17.1 Acids and Bases in Water

Although water is not an essential participant in all modern acid-base definitions, most of your laboratory work with acids and bases involves water, as do most important environmental, biological, and industrial applications. You know from our introductory discussion in Chapter 4 that *water is a product in all neutralization reactions between strong acids and strong bases:*

$$HCl(aq) + NaOH(aq) \rightarrow NaCl(aq) + H_2O(l)$$

Indeed, as the net ionic equation of this reaction shows, water is *the* product:

$$H^+(aq) + OH^-(aq) \rightarrow H_2O(l)$$

Furthermore, whenever an acid dissociates in water, solvent molecules participate in the reaction:

$$HA + H_2O(l) \rightarrow A^-(aq) + H_3O^+(aq)$$

As mentioned in that earlier discussion, water surrounds the proton to form H-bonded species with the general formula $H(H_2O)_n^+$. Because the proton is so small, its charge density is very high, so its attraction to water is especially strong. It bonds covalently to one of the lone electron pairs on a water molecule's O atom to form a **hydronium ion, $H_3O^+$,** which forms H bonds to several other water molecules (Figure 17.2). To emphasize the active role of water in aqueous acid-base chemistry and the nature of the proton-water interaction, the hydrated proton is shown generally in the text as $H_3O^+(aq)$, although in some cases this hydrated species is shown more simply as $H^+(aq)$.

### Proton/Hydroxide Release and the Classical Acid-Base Definition

The first and simplest definition of acids and bases that reflects their chemical nature on the molecular scale was suggested by Svante Arrhenius (1859-1927), whose work on the rate constant you encountered in Chapter 15. In the **classical** or **Arrhenius acid-base definition,** acids and bases are classified in terms of their formulas and their behavior *in water:*

• An *acid* is a substance that contains hydrogen and dissociates in water to yield $H_3O^+$.

• A *base* is a substance that contains the OH group and dissociates in water to yield $OH^-$.

Some typical Arrhenius acids are HCl, $HNO_3$, and HCN, and some typical bases are NaOH, KOH, and $Ba(OH)_2$. Although Arrhenius bases contain discrete $OH^-$ ions in their structures, Arrhenius acids *never* contain $H^+$ ions. On the contrary, acids contain *covalently bonded H atoms that ionize in water.*

An acid and a base undergo **neutralization** when they react. The meaning of neutralization has changed along with the definitions of acid and base, but in the Arrhenius sense, it indicates the process in which the $H^+$ ion from the acid and the $OH^-$ ion from the base combine to form $H_2O$. This description of neutralization explains an observation that puzzled many chemists of Arrhenius's time. They had observed that all neutralization reactions between what we now call strong acids and strong bases (those that dissociate completely in water) had the same heat of reaction. No matter which strong acid and base reacted, and no matter which salt formed, $\Delta H^0_{rxn}$ was about $-56$ kJ per mole of water formed. Arrhenius suggested that the heat of reaction was always the same because the actual reaction was always the same—a proton and a hydroxide ion formed water:

$$H^+(aq) + OH^-(aq) \rightarrow H_2O(l) \qquad \Delta H^0_{rxn} = -55.9 \text{ kJ}$$

The salt that formed along with the water, for example NaCl in the reaction of sodium hydroxide with hydrochloric acid,

$$Na^+(aq) + OH^-(aq) + H^+(aq) + Cl^-(aq) \rightarrow Na^+(aq) + Cl^-(aq) + H_2O(l)$$

remained present as hydrated spectator ions.

Despite its importance at the time, limitations in the classical definition soon became apparent. Arrhenius and many others realized that some substances that do *not* contain OH in their formulas *do* behave as bases. For example, $NH_3$ and $K_2CO_3$ also yield $OH^-$ in water. As you'll see shortly, broader acid-base definitions are required to include these species.

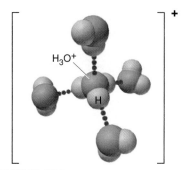

**FIGURE 17.2**

**The nature of the hydrated proton.** When an acid dissolves in water, the proton released forms a hydronium ion ($H_3O^+$) by bonding covalently to a water molecule. The $H_3O^+$ ion hydrogen bonds to other water molecules, forming species of general formula $H(H_2O)_n^+$. The $H(H_2O)_5^+$ ion is shown here.

**FIGURE 17.3**

**FIGURE 17.3**

**The extent of dissociation for strong and weak acids. A,** When a strong acid dissolves in water, it dissociates completely, yielding $H_3O^+$ and $A^-$ ions; virtually no HA molecules are present. **B,** In contrast, when a weak acid dissolves in water, it remains mostly undissociated, yielding relatively few $H_3O^+$ and $A^-$ ions.

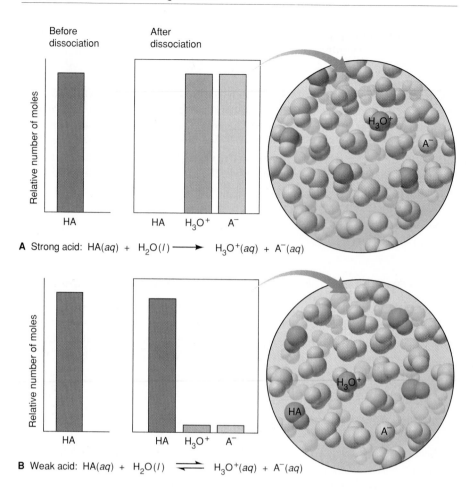

**A** Strong acid:  $HA(aq) + H_2O(l) \longrightarrow H_3O^+(aq) + A^-(aq)$

**B** Weak acid:  $HA(aq) + H_2O(l) \rightleftharpoons H_3O^+(aq) + A^-(aq)$

## Variation in Acid Strength: The Acid-Dissociation Constant ($K_a$)

Acids and bases differ greatly in their *strength* in water, that is, in the amount of $H_3O^+$ or $OH^-$ produced per mole of substance dissolved. We generally classify acids and bases as either strong or weak, according to the extent of their dissociation into ions in water. It's important to remember, however, that a *gradation* in strength exists, as we examine quantitatively in a moment. This classification of acid and base strength correlates with our earlier classification of electrolyte strength: strong electrolytes dissociate completely, and weak electrolytes dissociate partially. As you would expect, acids and bases act as electrolytes in water.

*Strong acids are completely dissociated in water* (Figure 17.3, *A*):

$$HA(aq) + H_2O(l) \rightarrow H_3O^+(aq) + A^-(aq)$$

In dilute solution, virtually none of the initial HA remains after it dissolves; that is $[H_3O^+] \approx [HA]_{init}$. In other words, $[HA]_{eq} \approx 0$, so the value of $K_c$ is extremely large:

$$Q_c = \frac{[H_3O^+][A^-]}{[HA][H_2O]} \qquad \text{At equilibrium, } Q_c = K_c \gg 1$$

Because the reaction is essentially complete, it is not useful to express it as an equilibrium process. In a dilute aqueous nitric acid solution, for example, there are virtually no undissociated nitric acid molecules:

$$HNO_3(aq) + H_2O(l) \rightarrow H_3O^+(aq) + NO_3^-(aq)$$

In contrast, *weak acids are mostly undissociated in water* (Figure 17.3, *B*):

$$HA(aq) + H_2O(l) \rightleftharpoons H_3O^+(aq) + A^-(aq)$$

Thus, $[H_3O^+] << [HA]_{init}$. Hydrocyanic acid is an example of a weak acid:

$$HCN(aq) + H_2O(l) \rightleftharpoons H_3O^+(aq) + CN^-(aq)$$

$$Q_c = \frac{[H_3O^+][CN^-]}{[HCN][H_2O]} \qquad \text{At equilibrium, } Q_c = K_c << 1$$

(From here on, square brackets without a subscript are used to mean molar concentration *at equilibrium*; that is, [X] means $[X]_{eq}$. Therefore, rather than presenting $Q$ and stating that $Q$ equals $K$ at equilibrium, $K$ will be written directly as a collection of concentration terms.)

There is a striking difference in the reaction rate when equal concentrations of a strong or a weak acid react with an active metal, such as zinc (Figure 17.4):

$$Zn(s) + 2H_3O^+(aq) \rightarrow Zn^{2+}(aq) + 2H_2O(l) + H_2(g)$$

In a strong acid such as 1 *M* HCl, zinc reacts rapidly, forming bubbles of $H_2$ vigorously. In a weak acid such as 1 *M* $CH_3COOH$, zinc reacts slowly, forming bubbles of $H_2$ sluggishly on the metal chunk. The strong acid has a high $[H_3O^+]$, so considerable $H_3O^+$ is available for reaction with the metal; the weak acid has a low $[H_3O^+]$, so little $H_3O^+$ is available for reaction.

There is a *specific* equilibrium constant for acid dissociation that highlights the substances whose concentrations are changing. The *equilibrium* expression for the dissociation of a general *weak acid, HA,* in water is

$$K_c = \frac{[H_3O^+][A^-]}{[HA][H_2O]}$$

The amount of water is so large that it changes minutely when reaction occurs, so its concentration is essentially constant. Therefore, we can simplify this expression by including the constant $[H_2O]$ term in the value of $K_c$ and define a new equilibrium constant, the **acid-dissociation constant, $K_a$** (or **acid- ionization constant**):

$$K_c[H_2O] = K_a = \frac{[H_3O^+][A^-]}{[HA]} \tag{17.1}$$

Like any equilibrium constant, $K_a$ is a number that is temperature dependent, whose magnitude tells how far to the right the reaction has proceeded to reach equilibrium. Thus, for any HA, *the higher the $K_a$, the larger is the $[H_3O^+]$ at equilibrium, and the stronger is the acid.*

Here are some benchmark $K_a$ values to give you an idea of the fraction of HA molecules that dissociates for typical weak acids:

- For a weak acid with a relatively high $K_a$ ($\sim 10^{-2}$), a 1 *M* solution will have $\sim 10\%$ of the HA molecules ionized. The $K_a$ of chlorous acid ($HClO_2$) is $1.12 \times 10^{-2}$, and 1 *M* $HClO_2$ is 10.6% dissociated.
- For a weak acid with a moderate $K_a$ ($\sim 10^{-5}$), a 1 *M* solution will have $\sim 0.3\%$ of the HA molecules dissociated. The $K_a$ of acetic acid ($CH_3COOH$) is $1.8 \times 10^{-5}$, and 1 *M* $CH_3COOH$ is 0.42% dissociated.
- For a weak acid with a relatively low $K_a$ ($\sim 10^{-10}$), a 1 *M* solution is $\sim 0.001\%$ dissociated. The $K_a$ of HCN is $6.2 \times 10^{-10}$, and 1 *M* HCN is 0.0025% dissociated.

Thus, for solutions of the same initial HA concentration, *the lower the $K_a$, the smaller is the percent dissociation of HA.*

Table 17.2 lists acid dissociation constants of some weak *monoprotic* acids (one ionizable proton) in water in order of decreasing acid strength. Strong

**A**      **B**

**FIGURE 17.4**

**Reaction of zinc with a strong and a weak acid. A,** In the reaction of zinc with a strong acid such as 1 *M* HCl, the higher concentration of $H_3O^+$ results in rapid formation of $H_2$ (bubbles). **B,** In a weak acid such as 1 *M* $CH_3COOH$, the $H_3O^+$ concentration is much lower, so the formation of $H_2$ is much slower.

**TABLE 17.2  $K_a$ Values for Some Monoprotic Acids at 25°C**

| NAME (FORMULA) | LEWIS STRUCTURE* | $K_a$ |
|---|---|---|
| Iodic acid ($HIO_3$) | 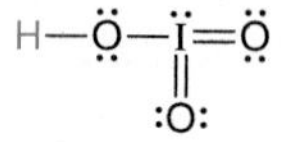 H—Ö—I̤=Ö (with :O: below, double bond) | $1.6 \times 10^{-1}$ |
| Chlorous acid ($HClO_2$) | 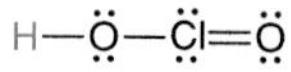 H—Ö—C̈l=Ö | $1.12 \times 10^{-2}$ |
| Nitrous acid ($HNO_2$) | 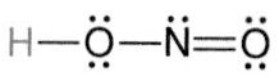 H—Ö—N̈=Ö | $7.1 \times 10^{-4}$ |
| Hydrofluoric acid (HF) | 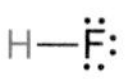 H—F̈: | $6.8 \times 10^{-4}$ |
| Formic acid (HCOOH) | 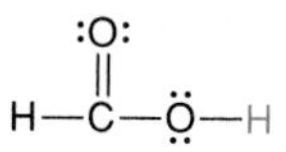 H—C(=:O:)—Ö—H | $1.8 \times 10^{-4}$ |
| Benzoic acid ($C_6H_5COOH$) | 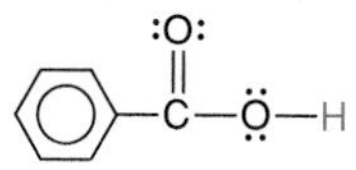 ⬡—C(=:O:)—Ö—H | $6.3 \times 10^{-5}$ |
| Acetic acid ($CH_3COOH$) | 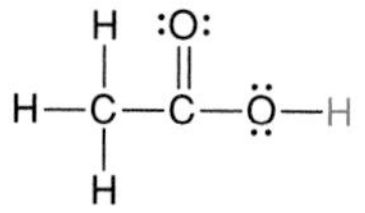 H—C(H)(H)—C(=:O:)—Ö—H | $1.8 \times 10^{-5}$ |
| Propanoic acid ($CH_3CH_2COOH$) | 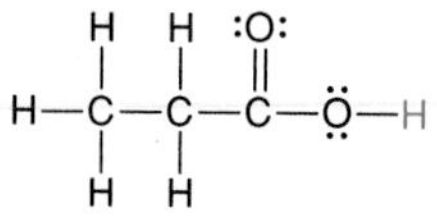 H—C(H)(H)—C(H)(H)—C(=:O:)—Ö—H | $1.3 \times 10^{-5}$ |
| Hypochlorous acid (HClO) | 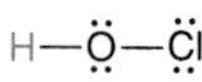 H—Ö—C̈l: | $2.9 \times 10^{-8}$ |
| Hypobromous acid (HBrO) | 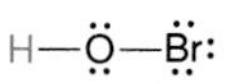 H—Ö—B̈r: | $2.3 \times 10^{-9}$ |
| Hydrocyanic acid (HCN) | 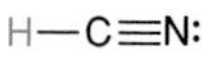 H—C≡N: | $6.2 \times 10^{-10}$ |
| Boric acid ($H_3BO_3$) | 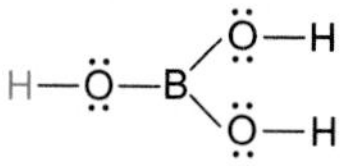 H—Ö—B(—Ö—H)(—Ö—H) | $5.8 \times 10^{-10}$ |
| Phenol ($C_6H_5OH$) | 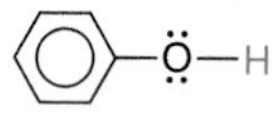 ⬡—Ö—H | $1.0 \times 10^{-10}$ |
| Hypoiodous acid (HIO) | 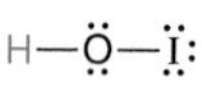 H—Ö—Ï: | $2.3 \times 10^{-11}$ |

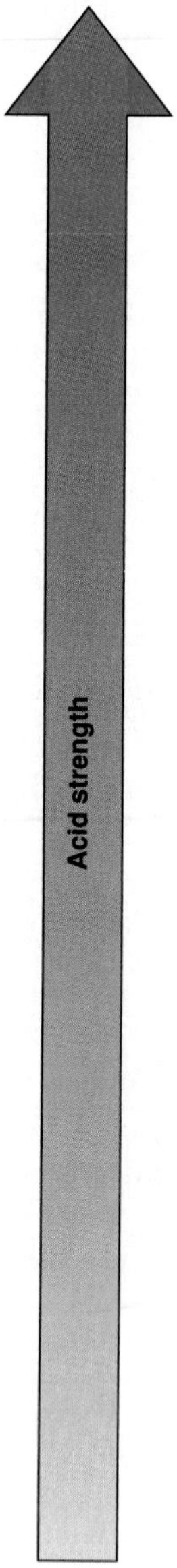

*Red type indicates the ionizable proton; structures have zero formal charge.

Acid strength ↑

acids are absent because they do not have meaningful $K_a$ values. The ionizable H atom in organic acids is attached to oxygen in the —COOH grouping; the H atoms bonded to C do *not* ionize. In Section 17.4, we discuss the behavior of polyprotic acids.

### Classifying the Relative Strengths of Acids and Bases

Using a table of acid-dissociation constants is the surest way to assess relative acid strength. Nevertheless, we can classify acids and bases in water as strong or weak qualitatively from their formulas:

*Strong acids.* The two types of strong acids, with examples that *you should memorize,* are

1. The hydrohalic acids HCl, HBr, and HI

2. Oxoacids in which the number of O atoms exceeds the number of ioniz-able H atoms by two or more, such as $HNO_3$, $H_2SO_4$, and $HClO_4$

*Weak acids.* There are many *more* weak acids than strong ones. The four types, with examples, are

1. The hydrohalic acid HF
2. Those acids in which H is not bonded to O or to halogen, such as HCN and $H_2S$
3. Oxoacids in which the number of O atoms equals or exceeds by one the number of ionizable H atoms, such as HClO, $HNO_2$, and $H_3PO_4$
4. Organic acids (general formula RCOOH), such as $CH_3COOH$ and $C_6H_5COOH$

*Strong bases.* Soluble compounds containing $O^{2-}$ or $OH^-$ ions are strong bases. The cations are usually those of the most active metals:

1. $M_2O$ or MOH, where M = Group 1A(1) metal (Li, Na, K, Rb, Cs)
2. MO or $M(OH)_2$, where M = Group 2A(2) metal (Ca, Sr, Ba)
   [MgO and $Mg(OH)_2$ are only slightly soluble, but the soluble portion dis-sociates completely to form a strongly basic solution.]

*Weak bases.* Many compounds with an electron-rich nitrogen are weak bases (none are Arrhenius bases). The common structural feature is a lone elec-tron pair on N:

1. Ammonia ($\ddot{N}H_3$)
2. Organic amines (general formula $R\ddot{N}H_2$, $R_2\ddot{N}H$, $R_3\ddot{N}$), such as $CH_3CH_2\ddot{N}H_2$, $(CH_3)_2\ddot{N}H$, $(C_3H_7)_3\ddot{N}$, and $C_5H_5\ddot{N}$

SAMPLE PROBLEM 17.1 ──────────────────────────

## Classifying Acid and Base Strength from the Chemical Formula

**Problem:** Classify each of the following compounds as a strong acid, weak acid, strong base, or weak base:

(a) $H_2CrO_4$     (b) $(CH_3)_2CHCOOH$     (c) KOH     (d) $(CH_3)_2CHNH_2$

**Plan:** Using the previous descriptions, we examine the chemical formula to deter-mine the class of acid or base in which each substance belongs.

**Solution: (a) Strong acid;** $H_2CrO_4$ is an oxoacid in which the number of O atoms exceeds the number of ionizable H atoms by two.

**(b) Weak acid;** $(CH_3)_2CHCOOH$ is an organic acid, as indicated by the —COOH group.

**(c) Strong base;** KOH is one of the Group 1A(1) hydroxides.

**(d) Weak base;** $(CH_3)_2CHNH_2$ is an organic amine.

FOLLOW-UP PROBLEM 17.1
Which of the following is the stronger acid or base: **(a)** HClO or $HClO_3$; **(b)** HCl or $CH_3COOH$; **(c)** NaOH or $CH_3NH_2$?

## Section Summary

Acids and bases are essential substances in home, industry, and the environ-ment. In aqueous solution, water combines with the proton released from an acid to form the hydrated species represented by $H_3O^+(aq)$. In the classical (Arrhenius) definition, acids contain H and yield $H_3O^+$ in water, bases con-tain OH and yield $OH^-$ in water, and a neutralization is the reaction of $H^+$ and $OH^-$ to form $H_2O$. Acid (or base) strength depends on the $[H_3O^+]$ (or $[OH^-]$) that results when a mole of acid (or base) dissolves. Strong acids and bases dissociate completely, weak acids and bases slightly. The extent of dis-sociation is expressed by the acid-dissociation constant, $K_a$. Weak acids have $K_a$ values ranging from about $10^{-1}$ to $10^{-12}$. Acids and bases can be classi-fied qualitatively as strong or weak based on their formulas.

## 17.2 Autoionization of Water and the pH Scale

Before we discuss the next major definition of acid-base behavior, let's examine a crucial property of water that enables us to quantify the $[H_3O^+]$ in any aqueous system: *water acts as a weak electrolyte*. The electrical conductivity of tap water is due almost entirely to dissolved ions, but even water that has been distilled and deionized repeatedly exhibits a tiny conductance. The reason is that water itself dissociates into ions very slightly in a process known as **autoionization** (or self-ionization):

$$H_2O(l) + H_2O(l) \rightleftharpoons H_3O^+(aq) + OH^-(aq)$$

### The Equilibrium Nature of Autoionization

Like any equilibrium system, the autoionization of water can be described by a temperature-dependent equilibrium constant:

$$K_c = \frac{[H_3O^+]\,[OH^-]}{[H_2O]^2}$$

As with the acid-dissociation constant, we can simplify this equilibrium expression, by including the constant $[H_2O]^2$ term in the value of $K_c$, and obtain a new equilibrium constant, the **ion-product constant for water, $K_w$:**

$$K_c\,[H_2O]^2 = K_w = [H_3O^+][OH^-] = 1.0 \times 10^{-14} \text{ (at 25°C)} \tag{17.2}$$

Since one $H_3O^+$ ion and one $OH^-$ ion appear for each $H_2O$ molecule that dissociates, we find that in pure water:

$$[H_3O^+] = [OH^-] = \sqrt{1.0 \times 10^{-14}} = 1.0 \times 10^{-7}\,M \text{ (at 25°C)}$$

Pure water has a concentration of about 55.5 $M$ (that is, $\frac{1000 \text{ g/L}}{18.02 \text{ g/mol}}$), so these equilibrium concentrations are attained when only one in 555 million water molecules dissociates reversibly into ions.

Autoionization has two major consequences for aqueous acid-base chemistry. First, since $K_w$ is constant at a given temperature, *a change in $[H_3O^+]$ causes an inverse change in $[OH^-]$*, and vice versa. Recall that according to Le Châtelier's principle, a change in concentration of either ion will shift the equilibrium position but *not* change the value of the equilibrium constant. Therefore, if some acid is added, $[H_3O^+]$ increases, so $[OH^-]$ must decrease; if some base is added, $[OH^-]$ increases, so $[H_3O^+]$ must decrease. In both cases, the change comes about as $H_3O^+$ and $OH^-$ combine to form $H_2O$ molecules and thus maintain the value of $K_w$. Second, *both ions are present in all aqueous systems;* thus, all acidic solutions contain some $OH^-$ ions, and all basic solutions contain some $H_3O^+$ ions.

The equilibrium nature of autoionization allows us to define "acidic" and "basic" solutions in terms of relative sizes of $[H_3O^+]$ and $[OH^-]$:

• In an *acidic* solution, $[H_3O^+] > [OH^-]$
• In a *basic* solution, $[H_3O^+] < [OH^-]$
• In a *neutral* solution, $[H_3O^+] = [OH^-]$

Moreover, if you know the value of $K_w$ at a particular temperature and the concentration of one of these ions, you can easily calculate the concentration of the other ion:

$$[H_3O^+] = \frac{K_w}{[OH^-]} \quad \text{or} \quad [OH^-] = \frac{K_w}{[H_3O^+]}$$

SAMPLE PROBLEM 17.2 _____

## Calculating [H₃O⁺] and [OH⁻] in Aqueous Solutions

**Problem:** A research chemist adds a measured amount of HCl to pure water at 25°C and obtains a solution with $[H_3O^+] = 3.0 \times 10^{-4}$ *M*. Calculate $[OH^-]$. Is the solution neutral, acidic, or basic?

**Plan:** We use the known value of $K_w$ at 25°C ($1.0 \times 10^{-14}$) and the given $[H_3O^+]$ to solve for $[OH^-]$. Then we compare $[H_3O^+]$ with $[OH^-]$ to determine whether the solution is acidic, basic, or neutral. This calculation is very common, so we present a simple roadmap.

**Solution:** Calculating $[OH^-]$:

$$[OH^-] = \frac{K_w}{[H_3O^+]} = \frac{1.0 \times 10^{-14}}{3.0 \times 10^{-4}} = \mathbf{3.3 \times 10^{-11} \ \textit{M}}$$

Since $[H_3O^+] > [OH^-]$, the solution is **acidic.**

**Check:** It makes sense that adding an acid to water would result in an acidic solution. Moreover, since $[H_3O^+]$ is greater than $10^{-7}$ *M*, $[OH^-]$ must be less than $10^{-7}$ *M* to give a constant $K_w$.

FOLLOW-UP PROBLEM 17.2
Calculate $[H_3O^+]$ in a solution at 25°C whose $[OH^-] = 6.7 \times 10^{-2}$ *M*. Is the solution neutral, acidic, or basic?

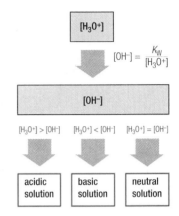

### Expressing the Hydronium Ion Concentration: The pH Scale

In aqueous solutions, $[H_3O^+]$ can vary over an enormous range: from about 10 *M* to $10^{-15}$ *M*. To handle numbers with negative exponents more conveniently in calculations, we convert them to positive numbers using a numerical system called a *p-scale:* take the common (base-10) logarithm of the number and multiply it by $-1$. Thus, **pH** is the negative logarithm of $[H^+]$ (or $[H_3O^+]$):

$$pH = -\log [H_3O^+] \qquad \text{(17.3)}$$

A $10^{-12}$ *M* $H_3O^+$ solution has a pH of 12: $-\log 10^{-12} = (-12)(-1) = 12$. A $10^{-3}$ *M* $H_3O^+$ solution has a pH of 3. A $5.4 \times 10^{-4}$ *M* solution has a pH of 3.27 [$(\log 5.4 + \log 10^{-4})(-1) = 3.27$]. (On most calculators, you would enter "5.4, EXP, 4, +/−, log, +/−".)

Note in particular that *the higher the pH, the lower the $[H_3O^+]$. Therefore, an acidic solution has a lower pH (higher $[H_3O^+]$) than a basic solution.* At 25°C in pure water, $[H_3O^+]$ is $1.0 \times 10^{-7}$ *M*, so

• pH of a neutral solution = 7.00
• pH of an acidic solution < 7.00
• pH of a basic solution    > 7.00

Since the pH scale is logarithmic, a solution of pH 1.0 has an $[H_3O^+]$ that is 10 times higher than that of a pH 2.0 solution, 100 times higher than that of a pH 3.0 solution, and so forth. To find the $[H_3O^+]$ from the pH, you perform the opposite arithmetic process: $[H_3O^+] = 10^{-pH}$; that is, you find the negative antilog of pH. ◆

As with any measurement, the number of significant figures in the pH reflects the precision with which the concentration is known. However, since it is a logarithm, the number of significant figures in the concentration equals the number of digits *to the right of the decimal in the pH*. For example, $5.4 \times 10^{-4}$ *M* has two significant figures, so its negative logarithm, 3.27, has two digits to the right of the decimal.

◆ **Logarithmic Scales in Sound and Seismology.** The p-scale is not the only logarithmic scale used in scientific measurements. The decibel scale measures the power of an acoustic signal, and the Richter scale measures the energy of ground movement.

**FIGURE 17.5**

**The relation among [H₃O⁺], pH, [OH⁻], and pOH.** Because $K_w$ is constant, $[H_3O^+]$ and $[OH^-]$ are interdependent, as are pH and pOH. These variables change in opposite directions as the acidity or basicity of the solution increases. Note that, at 25°C, the product of $[H_3O^+]$ and $[OH^-]$ is $1 \times 10^{-14}$, and the sum of pH and pOH is 14.0.

| | $[H_3O^+]$ | pH | $[OH^-]$ | pOH |
|---|---|---|---|---|
| | $1 \times 10^{-15}$ | 15.0 | $1 \times 10^{1}$ | −1.0 |
| | $1 \times 10^{-14}$ | 14.0 | $1 \times 10^{0}$ | 0.0 |
| | $1 \times 10^{-13}$ | 13.0 | $1 \times 10^{-1}$ | 1.0 |
| BASIC | $1 \times 10^{-12}$ | 12.0 | $1 \times 10^{-2}$ | 2.0 |
| | $1 \times 10^{-11}$ | 11.0 | $1 \times 10^{-3}$ | 3.0 |
| | $1 \times 10^{-10}$ | 10.0 | $1 \times 10^{-4}$ | 4.0 |
| | $1 \times 10^{-9}$ | 9.0 | $1 \times 10^{-5}$ | 5.0 |
| | $1 \times 10^{-8}$ | 8.0 | $1 \times 10^{-6}$ | 6.0 |
| NEUTRAL | $1 \times 10^{-7}$ | 7.0 | $1 \times 10^{-7}$ | 7.0 |
| | $1 \times 10^{-6}$ | 6.0 | $1 \times 10^{-8}$ | 8.0 |
| | $1 \times 10^{-5}$ | 5.0 | $1 \times 10^{-9}$ | 9.0 |
| | $1 \times 10^{-4}$ | 4.0 | $1 \times 10^{-10}$ | 10.0 |
| ACIDIC | $1 \times 10^{-3}$ | 3.0 | $1 \times 10^{-11}$ | 11.0 |
| | $1 \times 10^{-2}$ | 2.0 | $1 \times 10^{-12}$ | 12.0 |
| | $1 \times 10^{-1}$ | 1.0 | $1 \times 10^{-13}$ | 13.0 |
| | $1 \times 10^{0}$ | 0.0 | $1 \times 10^{-14}$ | 14.0 |
| | $1 \times 10^{1}$ | −1.0 | $1 \times 10^{-15}$ | 15.0 |

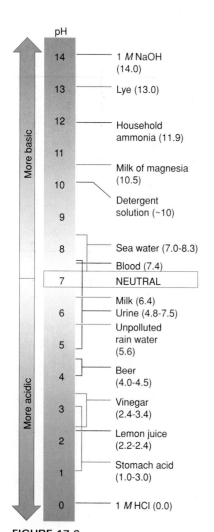

**FIGURE 17.6**

**The pH values of some familiar aqueous solutions.**

The p-scale device is used to express other quantities as well. Hydroxide ion concentration can be expressed as pOH:

$$pOH = -\log [OH^-]$$

*Acidic solutions have a higher pOH (lower $[OH^-]$) than basic solutions.* Since pH, pOH, $[H_3O^+]$, and $[OH^-]$ are interrelated through $K_w$, knowing any one of the values allows us to determine the others (Figure 17.5).

A p-scale is also used frequently for equilibrium constants; thus, $pK = -\log K$. *A low pK corresponds to a high K*: a reaction that reaches equilibrium with mostly products present (proceeds far to the right) has a low pK (high K), and one that has little product present at equilibrium has a high pK (low K). Table 17.3 shows this relationship for several acid dissociation constants.

By taking the negative log of both sides of the $K_w$ expression, we obtain a useful relationship among $pK_w$, pH, and pOH:

$$K_w = [H_3O^+][OH^-] = 1.0 \times 10^{-14} \text{ (two significant figures)}$$

$$-\log K_w = (-\log [H_3O^+]) + (-\log [OH^-]) = -\log (1.0 \times 10^{-14})$$

$$pK_w = pH + pOH = 14.00 \quad \text{(at 25°C)} \tag{17.4}$$

The sum of pH and pOH is 14.00 in any aqueous solution at 25°C (Figure 17.5). Figure 17.6 shows the pH of some familiar aqueous solutions.

**TABLE 17.3    The Relationship Between $K_a$ and $pK_a$**

| ACID NAME (FORMULA) | $K_a$ AT 25°C | $pK_a$ |
|---|---|---|
| Bisulfate ion ($HSO_4^-$) | $1.02 \times 10^{-2}$ | 1.991 |
| Nitrous acid ($HNO_2$) | $7.1 \times 10^{-4}$ | 3.15 |
| Acetic acid ($CH_3COOH$) | $1.8 \times 10^{-5}$ | 4.74 |
| Hypobromous acid (HBrO) | $2.3 \times 10^{-9}$ | 8.64 |
| Phenol ($C_6H_5OH$) | $1.0 \times 10^{-10}$ | 10.00 |

SAMPLE PROBLEM 17.3 _____

## Calculating $[H_3O^+]$, pH, $[OH^-]$, and pOH from Concentration of Acid

**Problem:** In an art restoration project, a chemist prepares etching solutions by diluting concentrated $HNO_3$ to 2.0 *M*, 0.30 *M*, and 0.0063 *M* $HNO_3$. Calculate $[H_3O^+]$, pH, $[OH^-]$, and pOH of the three solutions at 25°C.

**Plan:** We know from its formula that $HNO_3$ is a strong acid, so it dissociates completely in water; thus, $[HNO_3]_{init} = [H_3O^+]$. We use the given concentrations and the value of $K_w$ to calculate $[OH^-]$ and then use these concentrations to calculate pH and pOH.

**Solution:** Calculating the values for 2.0 *M* $HNO_3$:

$$[H_3O^+] = \mathbf{2.0\ M}$$

$$pH = -\log [H_3O^+] = -\log 2.0 = \mathbf{-0.30}$$

$$[OH^-] = \frac{K_w}{[H_3O^+]} = \frac{1.0 \times 10^{-14}}{2.0} = \mathbf{5.0 \times 10^{-15}\ M}$$

$$pOH = -\log (5.0 \times 10^{-15}) = \mathbf{14.30}$$

Calculating the values for 0.30 *M* $HNO_3$:

$$[H_3O^+] = \mathbf{0.30\ M}$$

$$pH = -\log [H_3O^+] = -\log 0.30 = \mathbf{0.52}$$

$$[OH^-] = \frac{K_w}{[H_3O^+]} = \frac{1.0 \times 10^{-14}}{0.30} = \mathbf{3.3 \times 10^{-14}\ M}$$

$$pOH = -\log (3.3 \times 10^{-14}) = \mathbf{13.48}$$

Calculating the values for 0.0063 *M* $HNO_3$:

$$[H_3O^+] = \mathbf{6.3 \times 10^{-3}\ M}$$

$$pH = -\log [H_3O^+] = -\log (6.3 \times 10^{-3}) = \mathbf{2.20}$$

$$[OH^-] = \frac{K_w}{[H_3O^+]} = \frac{1.0 \times 10^{-14}}{6.3 \times 10^{-3}} = \mathbf{1.6 \times 10^{-12}\ M}$$

$$pOH = -\log (1.6 \times 10^{-12}) = \mathbf{11.80}$$

A

**Check:** As the solution becomes more dilute, $[H_3O^+]$ decreases, so pH increases, as we would expect. Since an $[H_3O^+]$ greater than 1.0 *M*, as in 2.0 *M* $HNO_3$, gives a positive log, it results in a negative pH. The arithmetic seems correct because pH + pOH = 14.00 in each case.

FOLLOW-UP PROBLEM 17.3
A solution of NaOH has a pH of 9.52. What is pOH, $[H_3O^+]$, and $[OH^-]$ at 25°C?

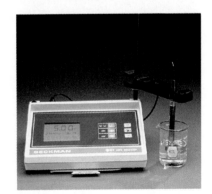

B

**FIGURE 17.7**
Methods for measuring the pH of a solution.
**A,** A few drops of the solution are placed on a strip of pH paper, and the color is compared with the color chart. **B,** The two electrodes of a pH meter immersed in the test solution measure $[H_3O^+]$. (In this equipment, the two electrodes are housed in one probe.)

As you may already know, pH values are usually obtained in the laboratory with an acid-base indicator or, more precisely, with an instrument called a pH meter. **Acid-base indicators** are organic molecules whose colors depend on the acidity or basicity of a solution. The pH of a solution is estimated quickly with *pH paper*, a paper strip impregnated with one or a mixture of indicators. A drop of test solution is placed on the paper strip, and the color of the strip is compared with a color chart (Figure 17.7, *A*).

The *pH meter* is an electrical device that measures the $[H_3O^+]$ by means of two electrodes immersed in the test solution. One electrode serves as a reference; the other has an extremely thin, conducting, glass membrane that separates a known internal $[H_3O^+]$ from the unknown external $[H_3O^+]$. Voltage difference develops across this membrane that is measured and displayed in pH units.

**Section Summary**
Pure water has a low conductivity because it autoionizes to a small extent. This process is reversible and described by the ion-product constant, $K_w$ (1.0 $\times$ $10^{-14}$ at 25°C). Thus, $[H_3O^+]$ and $[OH^-]$ are inversely related. In acidic solution, $[H_3O^+]$ is greater than $[OH^-]$, the reverse is true in basic solution, and the two are equal in neutral solution. To express small $[H_3O^+]$ more simply, we use the pH scale (pH = $-$ log $[H_3O^+]$). A high pH represents a low $[H_3O^+]$. Similarly, pOH = $-$ log $[OH^-]$, and p$K$ = $-$ log $K$. The sum of pH and pOH equals p$K_w$ (= 14.00 at 25°C). The pH of a solution is usually measured with pH paper or with a pH meter.

## 17.3 Proton Transfer and the Brønsted-Lowry Acid-Base Definition

Earlier we pointed out a major shortcoming of the Arrhenius definition: many substances yield $OH^-$ ions when they dissolve in water, but they do not contain OH in their formula. Examples include ammonia, the amines, and many salts of weak acids, such as NaF. Another limitation of the Arrhenius definition was its requirement that water be the solvent. In the early 20th century, J. N. Brønsted and T. M. Lowry suggested definitions that remove these limitations. According to the **Brønsted-Lowry acid-base definition,**

- *An acid is a **proton donor,*** *a substance that donates an $H^+$ ion.* An acid must contain H in its formula; $HNO_3$ and $H_2PO_4^-$ are two of many examples. All Arrhenius acids are Brønsted-Lowry acids.
- *A base is a **proton acceptor,*** *a substance that accepts an $H^+$ ion.* A base must contain a lone pair of electrons to bind the $H^+$ ion; a few examples are $NH_3$, $OH^-$, $F^-$, and $CO_3^{2-}$. Notice that, except for $OH^-$, Brønsted-Lowry bases are not Arrhenius bases, but all Arrhenius bases contain the Brønsted-Lowry base $OH^-$.

From the Brønsted-Lowry perspective, the only requirement for an acid-base reaction is that *one substance donate a proton and another substance accept it:* an *acid-base reaction is a proton transfer process.* Thus, acid-base reactions can occur between gases and in heterogeneous mixtures, as well as in aqueous solutions.

An acid and a base always work together in the transfer of a proton. In other words, a substance can behave as an acid only if another substance *simultaneously* behaves as a base, and vice versa. Even when an acid or a base merely dissolves in water, an acid-base reaction occurs because water acts as the other partner. Consider two typical acidic and basic solutions. When HCl dissolves in water, an $H^+$ ion (a proton) is transferred from HCl to $H_2O$, where it becomes attached to a lone pair of electrons on the O atom and forms $H_3O^+$. In effect, HCl (the acid) has *donated* the $H^+$ and $H_2O$ (the base) has *accepted* it (Figure 17.8, *A*):

$$HCl(g) + H_2\ddot{O}(l) \rightarrow Cl^-(aq) + H_3O^+(aq)$$

When ammonia dissolves in water, proton transfer also occurs. An $H^+$ from $H_2O$ attaches to the N lone pair and forms $NH_4^+$. The water molecule, with one fewer $H^+$, becomes an $OH^-$ ion:

$$H_2O(l) + \ddot{N}H_3(aq) \rightleftharpoons OH^-(aq) + NH_4^+(aq)$$

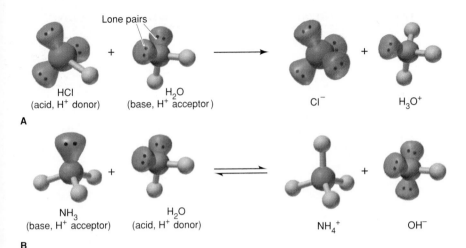

**FIGURE 17.8**

Proton transfer as the essential feature of a Brønsted-Lowry acid-base reaction. **A,** When HCl dissolves in water, it acts as an acid by donating a proton to water, which acts as the base. **B,** When $NH_3$ dissolves in water, it acts as a base by accepting a proton from water, which acts as the acid. Thus, in the Brønsted-Lowry sense, an acid-base reaction occurs in both cases.

In this case, $H_2O$ (the acid) has *donated* the $H^+$ and $NH_3$ (the base) has *accepted* it (Figure 17.8, *B*). Thus, $H_2O$ is *amphoteric:* it acts as a base in one case and as an acid in the other. Many other substances can act as either an acid or a base depending on the nature of their reaction partners.

## The Conjugate Acid-Base Pair

A new way to look at acid-base reactions emerges when we consider how the reactants *and* the products behave. For example, let's examine the reaction between hydrogen sulfide and ammonia:

$$H_2S + NH_3 \rightleftharpoons HS^- + NH_4^+$$

In the forward reaction, $H_2S$ acts as an acid by donating an $H^+$ to $NH_3$, which acts as a base. The reverse reaction involves another acid-base pair. The ammonium ion, $NH_4^+$, acts as an acid by donating a $H^+$ ion to the hydrogen sulfide ion, $HS^-$, which acts as a base. Notice that the acid, $H_2S$, becomes a base, $HS^-$, and the base, $NH_3$, becomes an acid, $NH_4^+$.

In Brønsted-Lowry terminology, $H_2S$ and $HS^-$ are a **conjugate acid-base pair:** $HS^-$ is the conjugate base of the acid $H_2S$. Similarly, $NH_3$ and $NH_4^+$ form a conjugate acid-base pair: $NH_4^+$ is the conjugate acid of the base $NH_3$. *Every acid has a conjugate base, and every base has a conjugate acid.* The conjugate base has one *fewer* H and one *more* negative charge than the acid from which it is derived. Likewise, the conjugate acid has one *more* H and one *fewer* negative charge than the base from which it is derived.

A Brønsted-Lowry neutralization occurs when *an acid and a base react to form their conjugate base and conjugate acid, respectively.* Table 17.4 shows several examples. There are three important points to notice. First, each reac-

**TABLE 17.4    The Conjugate Pairs in Some Acid-Base Reactions**

| | | CONJUGATE PAIR | | | | |
|---|---|---|---|---|---|---|
| | ACID | + | BASE | ⇌ | BASE | + | ACID |

| | ACID | + | BASE | ⇌ | BASE | + | ACID |
|---|---|---|---|---|---|---|---|
| Reaction 1 | HF | + | $H_2O$ | ⇌ | $F^-$ | + | $H_3O^+$ |
| Reaction 2 | HCOOH | + | $CN^-$ | ⇌ | $HCOO^-$ | + | HCN |
| Reaction 3 | $NH_4^+$ | + | $CO_3^{2-}$ | ⇌ | $NH_3$ | + | $HCO_3^-$ |
| Reaction 4 | $H_2PO_4^-$ | + | $OH^-$ | ⇌ | $HPO_4^{2-}$ | + | $H_2O$ |
| Reaction 5 | $H_2SO_4$ | + | $N_2H_5^+$ | ⇌ | $HSO_4^-$ | + | $N_2H_6^{2+}$ |
| Reaction 6 | $HPO_4^{2-}$ | + | $SO_3^{2-}$ | ⇌ | $PO_4^{3-}$ | + | $HSO_3^-$ |

tion has two acids and two bases: the two conjugate acid-base pairs. Second, acids and bases can be neutral, cationic, or anionic. Third, the same substance can be an acid or a base, depending on the other substance reacting. Water behaves this way in Reactions 1 and 4, and $HPO_4^{2-}$ does so in Reactions 4 and 6.

SAMPLE PROBLEM 17.4 _____

### Identifying Conjugate Acid-Base Pairs

**Problem:** The following reactions are important geochemical processes. Identify the conjugate acid-base pairs.
**(a)** $H_2PO_4^-(aq) + CO_3^{2-}(aq) \rightleftharpoons HCO_3^-(aq) + HPO_4^{2-}(aq)$
**(b)** $H_2O(l) + SO_3^{2-}(aq) \rightleftharpoons OH^-(aq) + HSO_3^-(aq)$
**Plan:** To find the conjugate pairs, we first find the species that donated an $H^+$ (the acid) and the species that accepted it (the base). The acid (or base) on the left becomes its conjugate base (or conjugate acid) on the right. Remember, the conjugate acid has one more H and a charge that is 1+ greater than that of its conjugate base.
**Solution: (a)** $H_2PO_4^-$ has one more $H^+$ than $HPO_4^{2-}$; $CO_3^{2-}$ has one fewer $H^+$ than $HCO_3^-$. Therefore, $H_2PO_4^-$ and $HCO_3^-$ are the acids, and $HPO_4^{2-}$ and $CO_3^{2-}$ are the bases. The conjugate acid-base pairs are **$H_2PO_4^-/HPO_4^{2-}$ and $HCO_3^-/CO_3^{2-}$.**
**(b)** $H_2O$ has one more $H^+$ than $OH^-$; $SO_3^{2-}$ has one fewer $H^+$ than $HSO_3^-$. The acids are $H_2O$ and $HSO_3^-$; the bases are $OH^-$ and $SO_3^{2-}$. The conjugate acid-base pairs are **$H_2O/OH^-$ and $HSO_3^-/SO_3^{2-}$.**

FOLLOW-UP PROBLEM 17.4
Identify the conjugate acid-base pairs in the following reactions:
**(a)** $CH_3COOH(aq) + H_2O(l) \rightleftharpoons CH_3COO^-(aq) + H_3O^+(aq)$
**(b)** $H_2O(l) + F^-(aq) \rightleftharpoons OH^-(aq) + HF(aq)$

### Relative Acid-Base Strength and the Direction of Reaction

The net *direction* of an acid-base reaction depends on the relative strengths of the acids and bases involved. *The reaction will proceed in the direction by which a stronger acid and stronger base form a weaker acid and weaker base.* The equilibrium position of the $H_2S$-$NH_3$ reaction lies to the right ($K_c > 1$) because $H_2S$ is a stronger acid than $NH_4^+$, the other acid present, and $NH_3$ is a stronger base than $HS^-$, the other base:

$$H_2S \quad + \quad NH_3 \quad \rightleftharpoons \quad HS^- \quad + \quad NH_4^+$$
stronger acid   +   stronger base   →   weaker base   +   weaker acid

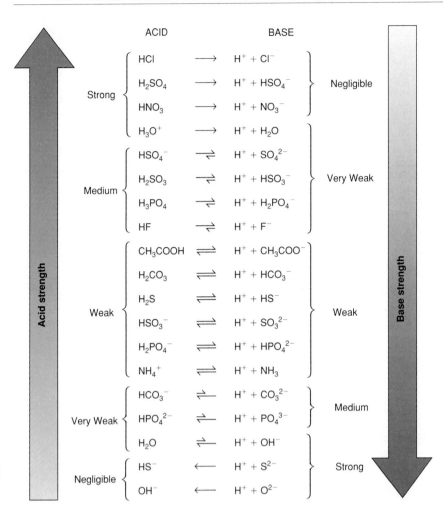

**FIGURE 17.9**
**Strengths of conjugate acid-base pairs.** The stronger the acid, the weaker is its conjugate base. The strongest acid appears on the top left and the strongest base on the bottom right. When an acid reacts with a base farther down the list, the reaction proceeds to the right ($K_c > 1$).

You might think of the process as a competition for the proton between the two bases, $NH_3$ and $HS^-$, in which $NH_3$ wins.

Similarly, the extent of acid (HA) dissociation in water depends on a competition for the proton between the two bases, $A^-$ and $H_2O$. When the acid, $HNO_3$, dissolves in water, it transfers an $H^+$ to the base, $H_2O$, forming nitrate ion, the conjugate base of $HNO_3$, and $H_3O^+$, the conjugate acid of $H_2O$:

$$HNO_3 \quad + \quad H_2O \quad \rightarrow \quad NO_3^- \quad + \quad H_3O^+$$

stronger acid + stronger base $\rightarrow$ weaker base + weaker acid

(In this case, the equilibrium position lies so far to the right that we do not show an equilibrium arrow.) $HNO_3$ is a stronger acid than $H_3O^+$, and $H_2O$ is a stronger base than $NO_3^-$. Thus, for strong acids such as $HNO_3$, the $H_2O$ wins the competition for the proton because $A^-$ ($NO_3^-$) is a much weaker base. On the other hand, for weak acids such as HF, the $A^-$ ($F^-$) wins because it is a stronger base than $H_2O$:

$$HF \quad + \quad H_2O \quad \rightleftharpoons \quad F^- \quad + \quad H_3O^+$$

weaker acid + weaker base $\leftarrow$ stronger base + stronger acid

From many such reactions, we can rank conjugate pairs in terms of the ability of the acid to transfer its proton (Figure 17.9). A major point emerges from this list of conjugate pairs: *a weaker acid has a stronger conjugate base*. This makes perfect sense: the acid gives up its proton less readily because its conjugate base holds it more strongly. We can use this list to predict the direction of a reaction between any two pairs, that is, whether the equilibrium position lies predominantly to the right ($K_c > 1$) or to the left ($K_c < 1$). A neutralization will proceed to the right if the acid reacts with a base that is lower on the list because this combination will produce a weaker conjugate base and weaker conjugate acid.

SAMPLE PROBLEM 17.5 _____

## Predicting the Direction of a Neutralization Reaction

**Problem:** Predict the net direction of each of the following reactions:
(a) $H_2PO_4^-(aq) + NH_3(aq) \rightleftharpoons NH_4^+(aq) + HPO_4^{2-}(aq)$
(b) $H_2O(l) + HS^-(aq) \rightleftharpoons OH^-(aq) + H_2S(aq)$
**Plan:** We first identify the conjugate acid-base pairs. To predict the direction, we consult Figure 17.9 to see which acid and base are stronger. The stronger acid and base form the weaker acid and base, so the reaction proceeds in that net direction.
**Solution:** **(a)** The conjugate pairs are $H_2PO_4^-/HPO_4^{2-}$ and $NH_4^+/NH_3$. Since $H_2PO_4^-$ is higher on the list of acids, it is stronger than $NH_4^+$; since $NH_3$ is lower on the list of bases, it is stronger than $HPO_4^{2-}$. Therefore,

$$H_2PO_4^-(aq) + NH_3(aq) \rightleftharpoons NH_4^+(aq) + HPO_4^{2-}(aq)$$
$$\text{stronger acid} + \text{stronger base} \longrightarrow \text{weaker acid} + \text{weaker base}$$

The equilibrium position of this reaction lies to the **right.**
**(b)** The conjugate pairs are $H_2O/OH^-$ and $H_2S/HS^-$. Since $H_2S$ is higher on the list of acids and $OH^-$ lower on the list of bases, we have

$$H_2O(l) + HS^-(aq) \rightleftharpoons OH^-(aq) + H_2S(aq)$$
$$\text{weaker acid} + \text{weaker base} \longleftarrow \text{stronger base} + \text{stronger acid}$$

The equilibrium position of this reaction lies to the **left.**

FOLLOW-UP PROBLEM 17.5
Explain with balanced equations, in which you show the net direction of the reaction, each of the following observations:
(a) You smell ammonia when $NH_3$ dissolves in water.
(b) The odor goes away when you add an excess of HCl to the solution in (a).
(c) The odor returns when you add an excess of NaOH to the solution in (b).

## Section Summary
The Brønsted-Lowry acid-base definition does not require that bases contain OH and does not require acids and bases to react in water. It defines acids as substances that donate a proton and bases as those that accept it. Acids and bases act together in proton transfer. When an acid donates its proton, it becomes a conjugate base; when a base accepts a proton, it becomes a conjugate acid. In a neutralization reaction, acids and bases react to form their conjugates. A stronger acid has a weaker conjugate base, and vice versa. Thus, the reaction proceeds in the direction in which a stronger acid and base form a weaker acid and base.

# 17.4 Solving Problems Involving Weak-Acid Equilibria

As with the equilibrium problems you've seen before, those involving weak acids and their conjugate bases are of two general types:
- Given equilibrium concentrations, find $K_a$.
- Given $K_a$ and some concentration information, find the other equilibrium concentrations.

As always, start with what is given in the problem statement and move logically toward what you want to find. Try to make a habit of applying the following steps:
1. Write the balanced equation and $K_a$ expression; these will tell you what to find.
2. Define $x$ as the unknown. Frequently, $x = [HA]_{dissoc}$, the concentration of HA that dissociates, which, through the use of certain assumptions, also equals $[H_3O^+]$ and $[A^-]$.
3. Construct a reaction table that incorporates the unknown.
4. Make assumptions that simplify the calculation.
5. Substitute the values into the $K_a$ expression and solve for $x$.
6. Check that the assumptions are justified. (We apply the 5% test that was introduced in Sample Problem 16.10, p. 715.) If they are not, use the quadratic formula to find $x$.

## Finding $K_a$ Given Concentrations

This problem type involves finding the acid-dissociation constant, $K_a$, of a weak acid:

$$HA(aq) + H_2O(l) \rightleftharpoons H_3O^+(aq) + A^-(aq)$$

$$K_a = \frac{[H_3O^+][A^-]}{[HA]}$$

A common laboratory approach to finding $K_a$ is to prepare an aqueous solution of known concentration of HA and measure its pH. We know the initial [HA], can measure $[H_3O^+]$, and determine $[A^-]$ and [HA] at equilibrium, so we can solve for $K_a$.

Note carefully the notation we use here and in the following sections. The molar concentration of each species is shown with brackets, and a subscript refers to where the species comes from or when it occurs in the reaction process. Thus, $[H_3O^+]_{from\ HA}$ is the molar concentration of $H_3O^+$ that comes from HA dissociation; $[HA]_{init}$ is the molar concentration of HA before the dissociation occurs; $[HA]_{dissoc}$ is the molar concentration of HA that dissociates; and so forth. Recall that brackets with no subscript refer to the molar concentration of the species *at equilibrium*.

Two key assumptions that simplify the arithmetic are used in these calculations:
1. The $[H_3O^+]$ from the autoionization of water is so much smaller than $[H_3O^+]$ from the HA dissociation that we can neglect it:

$$[H_3O^+] = [H_3O^+]_{from\ HA} + [H_3O^+]_{from\ H_2O}$$

Therefore,

$$[H_3O^+] \approx [H_3O^+]_{from\ HA}$$

Indeed, Le Châtelier's principle tells us that when HA dissociates, the increase in $H_3O^+$ decreases the extent of autoionization of water, so $[H_3O^+]_{\text{from } H_2O}$ in the HA solution is even less than $[H_3O^+]$ in pure water and can be neglected. Moreover, each HA that dissociates forms one $H_3O^+$ and one $A^-$, so $[A^-] = [H_3O^+]$.

2. A weak acid has a small $K_a$; therefore, it dissociates to such a small extent that we can neglect the change in its concentration to find its equilibrium concentration:

$$[HA] = [HA]_{\text{init}} - [HA]_{\text{dissoc}} \approx [HA]_{\text{init}}$$

SAMPLE PROBLEM 17.6

## Finding the $K_a$ of a Weak Acid from the pH of Its Solution

**Problem:** Phenylacetic acid ($C_6H_5CH_2COOH$, simplified here to HPAc) is one of the substances that builds up in the blood of persons with phenylketonuria, an inherited disorder that, if untreated, causes mental retardation and death. In a study of its properties, a biochemist finds that the pH of 0.12 $M$ HPAc is 2.60. What is the $K_a$ of phenylacetic acid?

**Plan:** We are given $[HPAc]_{\text{init}}$ and the pH and must find $K_a$. We first write the reaction and the expression for $K_a$ to see which values we need to find:

$$HPAc(aq) + H_2O(l) \rightleftharpoons H_3O^+(aq) + PAc^-(aq)$$

$$K_a = \frac{[H_3O^+][PAc^-]}{[HPAc]}$$

- To find $[H_3O^+]$: We know the pH, so we can find $[H_3O^+]$. Since pH 2.60 is over four pH units ($10^4$-fold) *lower* than that of pure water (pH = 7.0), we can assume that $[H_3O^+]_{\text{from HPAc}} \gg [H_3O^+]_{\text{from } H_2O}$. Thus, $[H_3O^+]_{\text{from HPAc}} + [H_3O^+]_{\text{from } H_2O} \approx [H_3O^+]_{\text{from HPAc}} \approx [H_3O^+]$.
- To find $[PAc^-]$: Since each HPAc that dissociates forms one $H_3O^+$ and one $PAc^-$, $[H_3O^+] = [PAc^-]$.
- To find $[HPAc]$: We know $[HPAc]_{\text{init}}$. Since it is a weak acid, we assume that very little dissociates, so $[HPAc]_{\text{init}} - [HPAc]_{\text{dissoc}} \approx [HPAc]_{\text{init}} \approx [HPAc]$. Now we set up a reaction table, make the assumptions, substitute the equilibrium values, solve for $K_a$, and then check the assumptions.

**Solution:** Calculating the $[H_3O^+]$:

$$[H_3O^+] = 10^{-\text{pH}} = 10^{-2.60} = 2.5 \times 10^{-3} \, M$$

Setting up the reaction table, with $x = [HPAc]_{\text{dissoc}} = [H_3O^+]_{\text{from HPAc}} = [H_3O^+] = [PAc^-]$:

| Concentration ($M$) | HPAc($aq$) | + | H$_2$O($l$) | $\rightleftharpoons$ | H$_3$O$^+$ ($aq$) | + | PAc$^-$ ($aq$) |
|---|---|---|---|---|---|---|---|
| Initial | 0.12 | | — | | $1 \times 10^{-7}$(from $H_2O$) | | 0 |
| Change | $-x$ | | — | | $+x$ | | $+x$ |
| Equilibrium | $0.12 - x$ | | — | | $<1 \times 10^{-7} + x$ | | $x$ |

Making the assumptions:

1. The given $[H_3O^+]$ ($2.5 \times 10^{-3}$ $M$) $\gg [H_3O^+]_{\text{from } H_2O}$ ($1 \times 10^{-7}$ $M$), so we assume that $[H_3O^+] \approx [H_3O^+]_{\text{from HPAc}}$, or $<1 \times 10^{-7}$ $M + x \approx x$. The $[H_3O^+]_{\text{from } H_2O}$ ($1 \times 10^{-7}$ $M$) will usually be negligible, so we will no longer show it in the reaction tables.

2. Since HPAc is weak, we assume that $[\text{HPAc}] = 0.12$   $M - x \approx 0.12$   $M$. Solving for the equilibrium concentrations:

$$x = [\text{H}_3\text{O}^+]_{\text{from HPAc}} = [\text{PAc}^-] = 2.5 \times 10^{-3}\ M$$

$$[\text{HPAc}] = 0.12\ M - (2.5 \times 10^{-3}\ M) \approx 0.12\ M \text{ (to 2 significant figures)}$$

Substituting these values into $K_a$:

$$K_a = \frac{[\text{H}_3\text{O}^+][\text{PAc}^-]}{[\text{HPAc}]} = \frac{(2.5 \times 10^{-3})\,(2.5 \times 10^{-3})}{0.12} = \textbf{5.2} \times \textbf{10}^{-5}$$

Checking the assumptions by finding the percent error in concentration:

1. For $[\text{H}_3\text{O}^+]_{\text{from H}_2\text{O}}$:   $\dfrac{<1 \times 10^{-7}\ M}{2.5 \times 10^{-3}\ M} \times 100 = <4 \times 10^{-3}\% <5\%$; assumption is justified.

2. For $[\text{HPr}]_{\text{dissoc}}$:   $\dfrac{2.5 \times 10^{-3}\ M}{0.12\ M} \times 100 = 2.1\% < 5\%$; assumption is justified.

**Check:** The $[\text{H}_3\text{O}^+]$ makes sense: pH 2.60 should give an $[\text{H}_3\text{O}^+]$ between $10^{-2}$ and $10^{-3}\ M$. The $K_a$ calculation also seems in the correct range: $(10^{-3})^2/10^{-1} = 10^{-5}$, and this value seems reasonable for a weak acid.

**FOLLOW-UP PROBLEM 17.6**
The conjugate acid of ammonia, $\text{NH}_4^+$, is a weak acid. If a 0.2 $M$ $\text{NH}_4\text{Cl}$ solution has a pH of 5.0, what is the $K_a$ of $\text{NH}_4^+$?

## Finding Concentrations Given $K_a$

The second type of problem gives some concentration data and the $K_a$ value and asks for the equilibrium concentration of some component. Such problems are very similar to those we solved in Chapter 16 in which a substance with a given initial concentration reacted to an unknown extent (Sample Problems 16.8 to 16.10).

SAMPLE PROBLEM 17.7 ————————————————————————

## Determining Equilibrium Concentrations from Known $K_a$ and [HA]

**Problem:** Propanoic acid ($\text{CH}_3\text{CH}_2\text{COOH}$, which we simplify as HPr) is an organic acid whose salts are used to retard mold growth in foods. What is the $[\text{H}_3\text{O}^+]$ of 0.10 $M$ HPr? $K_a = 1.3 \times 10^{-5}$.
**Plan:** We need to find $[\text{H}_3\text{O}^+]$. First we write the balanced equation and the expression for $K_a$:

$$\text{HPr}(aq) + \text{H}_2\text{O}(l) \rightleftharpoons \text{Pr}^-(aq) + \text{H}_3\text{O}^+(aq)$$

$$K_a = \frac{[\text{Pr}^-][\text{H}_3\text{O}^+]}{[\text{HPr}]} = 1.3 \times 10^{-5}$$

We know $[\text{HPr}]_{\text{init}}$, but not $[\text{HPr}]$. If we let $x = [\text{HPr}]_{\text{dissoc}}$, $x$ is also $[\text{H}_3\text{O}^+]_{\text{from HPr}}$ and $[\text{Pr}^-]$ because each HPr that dissociates yields one $\text{H}_3\text{O}^+$ and one $\text{Pr}^-$. With this information, we can set up a reaction table. In solving for $x$, we assume, as before, that (1) $[\text{H}_3\text{O}^+]_{\text{from H}_2\text{O}}$ is so small that $[\text{H}_3\text{O}^+]_{\text{from HPr}} \approx [\text{H}_3\text{O}^+]$, and (2) since HPr has a small $K_a$, it dissociates very little, so $[\text{HPr}]_{\text{init}} - x \approx [\text{HPr}]_{\text{init}} \approx [\text{HPr}]$. Then we find $x$ and check assumptions.

**Solution:** Setting up the reaction table, with $x = [HPr]_{dissoc} = [H_3O^+]_{from\ HPr} = [H_3O^+] = [Pr^-]$:

| Concentration (M) | HPr(aq) | + | H₂O(l) | ⇌ | H₃O⁺(aq) | + | Pr⁻(aq) |
|---|---|---|---|---|---|---|---|
| Initial | 0.10 | | — | | 0 | | 0 |
| Change | −x | | — | | +x | | +x |
| Equilibrium | 0.10 − x | | — | | x | | x |

Making the assumptions:
1. From Le Châtelier's principle, $[H_3O^+]_{from\ H_2O}$ is less than $10^{-7}$ M, so it can be neglected; note that it does not appear in the reaction table.
2. Since $K_a$ is small, $x$ is small compared with $[HPr]_{init}$, so $0.10\ M - x \approx 0.10\ M$. Substituting into the $K_a$ expression and solving for $x$:

$$K_a = \frac{[Pr^-][H_3O^+]}{[HPr]} = \frac{(x)(x)}{0.10} = 1.3 \times 10^{-5}$$

$$x = [\mathbf{H_3O^+}] = \mathbf{1.1 \times 10^{-3}}\ \boldsymbol{M}$$

Checking the assumptions:

1. For $[H_3O^+]_{from\ H_2O}$: $\dfrac{<1 \times 10^{-7}\ M}{1.1 \times 10^{-3}\ M} \times 100 = 9 \times 10^{-3}\% < 5\%$; assumption is justified.

2. For $[HPr]_{dissoc}$: $\dfrac{1.1 \times 10^{-3}\ M}{0.10\ M} \times 100 = 1.1\% < 5\%$; assumption is justified.

**Check:** The $[H_3O^+]$ seems reasonable for a dilute solution of a weak acid with a moderate $K_a$. By reversing the calculation, we can check the math: $(1.1 \times 10^{-3})^2/0.10 = 1.2 \times 10^{-5}$, which is within rounding of the given $K_a$.
**Comment:** In this problem, we assumed that since $K_a$ is relatively small, the amount that dissociates ($x$) can be neglected. However, this is true only if $[HA]_{init}$ is relatively large. A simple benchmark shows when the assumption is justified:

- If $\dfrac{[HA]_{init}}{K_a} > 400$, the assumption is justified: neglecting $[HA]_{dissoc}$ introduces less than a 5% error.

- If $\dfrac{[HA]_{init}}{K_a} < 400$, the assumption is *not* justified, and we must solve a quadratic equation to find $[HA]_{dissoc}$.

The latter situation occurs in the follow-up problem.

**FOLLOW-UP PROBLEM 17.7**
Cyanic acid (HOCN), an extremely acrid, unstable substance, is used industrially to make cyanates. What is the $[H_3O^+]$ and pH of 0.10 M HOCN? $K_a$ of cyanic acid is $3.5 \times 10^{-4}$.

## The Effect of Concentration on the Extent of Acid Dissociation

If we repeat the calculation in Sample Problem 17.7, but start with a lower [HPr], we observe a very interesting fact about the extent of weak acid dissociation. Suppose the initial concentration of HPr were 0.010 $M$ rather than 0.10 $M$. After filling in the reaction table and making the same assumptions, we would have

$$K_a = \frac{(x)(x)}{0.010} = 1.3 \times 10^{-5}$$

$$x = 3.6 \times 10^{-4} M \text{ (assumption is justified: } 3.6\% < 5\%.)$$

Now let's compare the percent of HPr molecules dissociated at the two different initial acid concentrations with the relationship

$$\text{Percent HA dissociated} = \frac{[\text{HA}]_{\text{dissoc}}}{[\text{HA}]_{\text{init}}} \times 100 \qquad \textbf{(17.5)}$$

Case 1: $[\text{HPr}]_{\text{init}} = 0.10\ M$

$$\text{Percent dissociated} = \frac{1.1 \times 10^{-3}\ M}{1.0 \times 10^{-1}\ M} \times 100 = 1.1\%$$

Case 2: $[\text{HPr}]_{\text{init}} = 0.010\ M$

$$\text{Percent dissociated} = \frac{3.6 \times 10^{-4}\ M}{1.0 \times 10^{-2}\ M} \times 100 = 3.6\%$$

*As the initial acid concentration decreases, the percent dissociation of the acid increases.* Don't confuse the $[\text{H}_3\text{O}^+]$ with the percent HA dissociated. The $[\text{H}_3\text{O}^+]$ is lower in the diluted HA solution because the actual *number* of dissociated HA molecules is less. It is the *fraction* (and thus the *percent*) of dissociated HA molecules that increases with dilution.

This phenomenon is analogous to the behavior of gases at equilibrium that undergo a change in container volume (pressure) (Section 16.4). In that case, an increase in volume shifts the equilibrium position to favor more moles of gas. In the case of HA dissociation, as the solution is diluted, there is, in effect, an increase in available volume, so the equilibrium position shifts to favor more moles of ions. Indeed, Le Châtelier's principle predicts that adding $\text{H}_2\text{O}$ shifts the equilibrium position to the right:

$$\text{H}_2\text{O(added)} + \text{HA} \xrightleftharpoons{\longrightarrow} \text{H}_3\text{O}^+ + \text{A}^-$$

### The Behavior of Polyprotic Acids

Acids with more than one ionizable proton are **polyprotic acids.** In an aqueous solution of a polyprotic acid, one proton at a time dissociates, and each step has a different $K_a$. For example, sulfurous acid is a diprotic acid and has two $K_a$ values:

$$\text{H}_2\text{SO}_3(aq) + \text{H}_2\text{O}(l) \rightleftharpoons \text{HSO}_3^-(aq) + \text{H}_3\text{O}^+(aq)$$

$$K_{a1} = \frac{[\text{HSO}_3^-][\text{H}_3\text{O}^+]}{[\text{H}_2\text{SO}_3]} = 1.4 \times 10^{-2}$$

$$\text{HSO}_3^-(aq) + \text{H}_2\text{O}(l) \rightleftharpoons \text{SO}_3^{2-}(aq) + \text{H}_3\text{O}^+(aq)$$

$$K_{a2} = \frac{[\text{SO}_3^{2-}][\text{H}_3\text{O}^+]}{[\text{HSO}_3^-]} = 6.5 \times 10^{-8}$$

As you can see from the relative $K_a$ values, $\text{H}_2\text{SO}_3$ is a much stronger acid than $\text{HSO}_3^-$. Table 17.5 lists some common polyprotic acids and their $K_a$

**TABLE 17.5  Successive $K_a$ Values for Some Polyprotic Acids at 25°C**

| NAME (FORMULA) | LEWIS STRUCTURE* | $K_{a1}$ | $K_{a2}$ | $K_{a3}$ |
|---|---|---|---|---|
| Oxalic acid ($H_2C_2O_4$) | 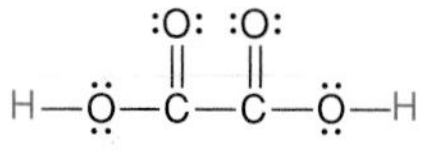 H—Ö—C—C—Ö—H (with :O: :O: double bonds) | $5.6 \times 10^{-2}$ | $5.4 \times 10^{-5}$ | |
| Phosphorous acid ($H_3PO_3$) | 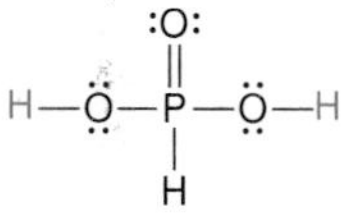 H—Ö—P—Ö—H (with :O: double bond, H below) | $3 \times 10^{-2}$ | $1.7 \times 10^{-7}$ | |
| Sulfurous acid ($H_2SO_3$) | 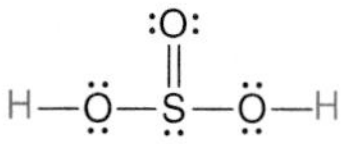 H—Ö—S—Ö—H (with :O: double bond) | $1.4 \times 10^{-2}$ | $6.5 \times 10^{-8}$ | |
| Phosphoric acid ($H_3PO_4$) | 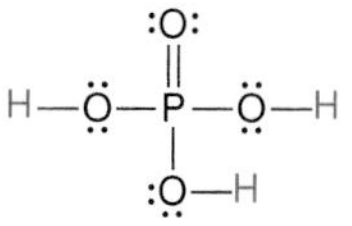 H—Ö—P—Ö—H (with :O: double bond, :O—H below) | $7.2 \times 10^{-3}$ | $6.3 \times 10^{-8}$ | $4.2 \times 10^{-13}$ |
| Arsenic acid ($H_3AsO_4$) | 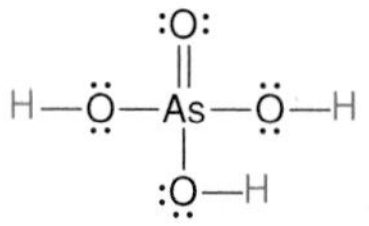 H—Ö—As—Ö—H (with :O: double bond, :O—H below) | $6 \times 10^{-3}$ | $1.1 \times 10^{-7}$ | $3 \times 10^{-12}$ |
| Citric acid ($H_3C_6H_5O_7$) | 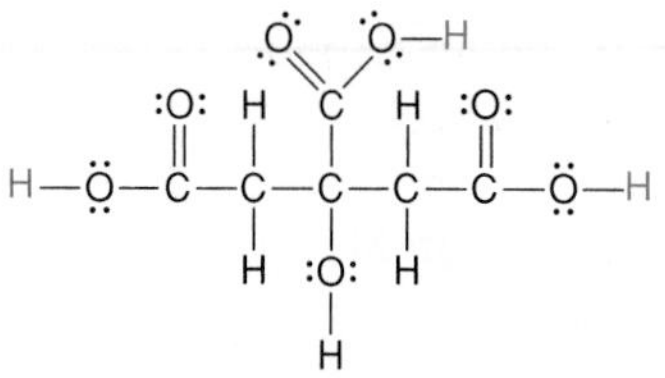 H—Ö—C—C—C—C—C—Ö—H (complex structure) | $7.5 \times 10^{-4}$ | $1.7 \times 10^{-5}$ | $4.0 \times 10^{-7}$ |
| Carbonic acid ($H_2CO_3$) | 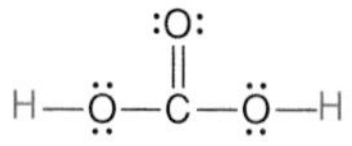 H—Ö—C—Ö—H (with :O: double bond) | $4.5 \times 10^{-7}$ | $4.7 \times 10^{-11}$ | |
| Hydrosulfuric acid ($H_2S$) | 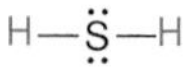 H—S̈—H | $9 \times 10^{-8}$ | $1 \times 10^{-17}$ | |

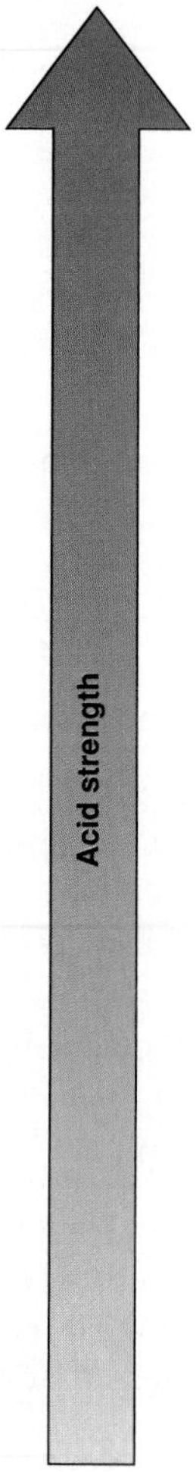

*Red type indicates the ionizable protons.

Acid strength ↑

values. Notice that in every case, the first proton comes off much more easily than the second and, where applicable, the second much more easily than the third

$$K_{a1} > K_{a2} > K_{a3}$$

This trend makes sense: it is more difficult for a $H^+$ ion to leave a negative ion (such as $HSO_3^-$) than a neutral molecule (such as $H_2SO_3$). Successive acid dissociation constants typically differ by several orders of magnitude. This fact greatly simplifies pH calculations involving polyprotic acids because *we can usually neglect the $H_3O^+$ coming from the subsequent dissociations.*

SAMPLE PROBLEM 17.8

## Calculating Equilibrium Concentrations for a Polyprotic Acid

**Problem:** Ascorbic acid ($H_2C_6H_6O_6$; $H_2Asc$ for this problem), known as vitamin C, is a diprotic acid ($K_{a1} = 1.0 \times 10^{-5}$ and $K_{a2} = 5 \times 10^{-12}$) found in citrus fruit (photo). Calculate $[H_2Asc]$, $[HAsc^-]$, $[Asc^{2-}]$, and the pH of 0.050 $M$ $H_2Asc$.
**Plan:** We must calculate the equilibrium concentrations of all species and convert $[H_3O^+]$ to pH. We first write the equations and $K_a$ expressions:

$$H_2Asc(aq) + H_2O(l) \rightleftharpoons HAsc^-(aq) + H_3O^+(aq)$$

$$K_{a1} = \frac{[HAsc^-][H_3O^+]}{[H_2Asc]} = 1.0 \times 10^{-5}$$

$$HAsc^-(aq) + H_2O(l) \rightleftharpoons Asc^{2-}(aq) + H_3O^+(aq)$$

$$K_{a2} = \frac{[Asc^{2-}][H_3O^+]}{[HAsc^-]} = 5 \times 10^{-12}$$

Then we assume that (1) since $K_{a1} \gg K_{a2} > K_w$, the first dissociation produces almost all the $H_3O^+$: $[H_3O^+]_{from\ H_2Asc} \gg [H_3O^+]_{from\ HAsc^-} > [H_3O^+]_{from\ H_2O}$, and (2) since $K_{a1}$ is small, the amount that dissociates can be neglected: $[H_2Asc]_{init} \approx [H_2Asc]$. We set up a reaction table for the first dissociation, then solve for $[H_3O^+]$ and $[HAsc^-]$. Because the second dissociation is so much less than the first, we can substitute values from the first directly to find $[Asc^{2-}]$ of the second.
**Solution:** Setting up a reaction table with $x = [H_2Asc]_{dissoc} = [H_3O^+] = [HAsc^-]$:

| Concentration ($M$) | $H_2Asc(aq)$ | $+$ | $H_2O(l)$ | $\rightleftharpoons$ | $HAsc^-(aq)$ | $+$ | $H_3O^+(aq)$ |
|---|---|---|---|---|---|---|---|
| Initial | 0.050 | | — | | 0 | | 0 |
| Change | $-x$ | | — | | $+x$ | | $+x$ |
| Equilibrium | $0.050 - x$ | | — | | $x$ | | $x$ |

Making the assumptions:
1. Since $K_{a1} \gg K_{a2} > K_w$, $[H_3O^+]_{from\ H_2Asc} \approx [H_3O^+] \gg [H_3O^+]_{from\ H_2O}$
2. Since $K_{a1}$ is small, $[H_2Asc]_{init} - x \approx [H_2Asc]_{init} = [H_2Asc]$
Thus, $[H_2Asc] = 0.050\ M - x \approx \mathbf{0.050\ M}$
Substituting into the expression for $K_{a1}$ and solving for $x$:

$$K_{a1} = \frac{[H_3O^+][HAsc^-]}{[H_2Asc]} = \frac{x^2}{0.050} = 1.0 \times 10^{-5}$$

$$x = [H_3O^+] = [HAsc^-] = \mathbf{7.1 \times 10^{-4}\ M}$$

$$pH = -\log[H_3O^+] = -\log(7.1 \times 10^{-4}) = \mathbf{3.15}$$

Checking the assumptions:

1. $\dfrac{1.0 \times 10^{-7}\ M}{7.1 \times 10^{-4}\ M} \times 100 = 0.014\% < 5\%$; assumption is justified.

2. $\dfrac{7.1 \times 10^{-4}\ M}{0.050\ M} \times 100 = 1.4\% < 5\%$; assumption is justified.

Also note that $\dfrac{[H_2Asc]_{init}}{K_{a1}} = \dfrac{0.050}{1.0 \times 10^{-5}} = 5000$, which is greater than 400.

Calculating $[Asc^{2-}]$:

$$K_{a2} = \frac{[H_3O^+][Asc^{2-}]}{[HAsc^-]} \quad \text{and} \quad [Asc^{2-}] = \frac{K_{a2} \times [HAsc^-]}{[H_3O^+]}$$

$$[Asc^{2-}] = \frac{(5 \times 10^{-12})(7.1 \times 10^{-4})}{7.1 \times 10^{-4}} = \mathbf{5 \times 10^{-12}\ M}$$

**Check:** Since $K_{a1} \gg K_{a2}$, it makes sense that $[HAsc^-] \gg [Asc^{2-}]$ because $Asc^{2-}$ is produced only in the second (much weaker) dissociation. Since both $K_a$'s are small, all concentrations except $[H_2Asc]$ should be much lower than the original 0.050 $M$.

**FOLLOW-UP PROBLEM 17.8**
Oxalic acid (HOOC—COOH, or $H_2C_2O_4$) is the simplest organic diprotic acid. Its commercial uses include bleaching straw and leather and removing rust and ink stains. Calculate $[H_2C_2O_4]$, $[HC_2O_4^-]$, $[C_2O_4^{2-}]$, and the pH of a 0.150 $M$ $H_2C_2O_4$ solution. Use $K_a$ values from Table 17.5.

**Section Summary**
Two common types of weak-acid equilibrium problems involve finding $K_a$ from concentrations and finding concentrations from $K_a$. We summarize the information in a reaction table and simplify the arithmetic by assuming (1) $[H_3O^+]_{\text{from } H_2O}$ can be neglected, and (2) weak acids dissociate so little that $[HA]_{\text{init}} \approx [HA]$ at equilibrium. The *fraction* of weak acid molecules that dissociates is greater in more dilute solution, even though the total $[H_3O^+]$ is less. Polyprotic acids have more than one ionizable proton, but we assume that the first dissociation provides virtually all the $[H_3O^+]$.

## 17.5 Weak Bases and Their Relation to Weak Acids

By focusing on where the proton comes from and goes to, the Brønsted-Lowry concept expands the definition of a base to encompass a host of substances that the Arrhenius definition excludes: a base is any substance that accepts a proton. In order to do so, *the base must have a lone electron pair.* (The lone electron pair also plays the central role in the Lewis acid-base definition, as you'll see later in this chapter.) In water solution, a base is *any substance that increases the [OH⁻] when it dissolves.* It does so by abstracting a proton from water, leaving behind an $OH^-$ ion:

$$B:(aq) + H_2O(aq) \rightleftharpoons BH^+(aq) + OH^-(aq)$$

The reversible nature of this general reaction can be shown with a mass-action expression that, at equilibrium, gives

$$K_c = \frac{[BH^+][OH^-]}{[B][H_2O]}$$

Following our earlier reasoning that $[H_2O]$ is treated as a constant in aqueous reactions, we include $[H_2O]$ in the value of $K_c$ and obtain the **base-dissociation constant, $K_b$:**

$$K_b = \frac{[BH^+][OH^-]}{[B]} \qquad (17.6)$$

Despite the name "base-dissociation constant," no base dissociates in the process. Rather, we are seeing the effect of the base abstracting a proton from water, which is acting as an acid. (Of course, the $OH^-$ ion itself can abstract a proton, so all Arrhenius bases contain the strong Brønsted-Lowry base $OH^-$.) As we've seen, water also acts as a weak base when acids dissolve in it. In aqueous solution, the two large classes of weak bases are neutral molecules, such as ammonia and the organic amines, and the anions of weak acids. ◆

◆ **Ammonia's Picturesque Past.** One of the 20th-century's key industrial chemicals has a long, vivid past. The name "ammonia" derives from an Egyptian god, whom the Romans called Ammon. Sacrificial wastes that were piled outside temples decomposed, and the remaining mineral salts were called "salts of Ammon." Later, the name "ammon" was retained for the volatile portion of these substances. In the late 15th century and for some time thereafter, ammonia was obtained from animal protein through the distillation of horns and hoofs and was called "spirits of hartshorn."

## Neutral Molecules as Bases: Ammonia and the Amines

Ammonia is the simplest member of a family of nitrogen-containing compounds that act as weak bases in water:

$$\ddot{N}H_3(aq) + H_2O(l) \rightleftharpoons NH_4^+(aq) + OH^-(aq) \qquad K_b = 1.76 \times 10^{-5} \text{ (at 25°C)}$$

As you can see from the small $K_b$ value, an aqueous solution of ammonia consists largely of unreacted $NH_3$ molecules, despite labels on reagent bottles that state "ammonium hydroxide." In a 1.0 $M$ $NH_3$ solution, for example, $[OH^-] = [NH_4^+] = 4.2 \times 10^{-3}$ $M$. Table 17.6 shows the $K_b$ values for some common neutral bases.

**TABLE 17.6** $K_b$ **Values for Some Neutral (Amine) Bases at 25°C**

| NAME (FORMULA) | LEWIS STRUCTURE* | $K_b$ |
|---|---|---|
| Diethylamine [$(CH_3CH_2)_2NH$] | 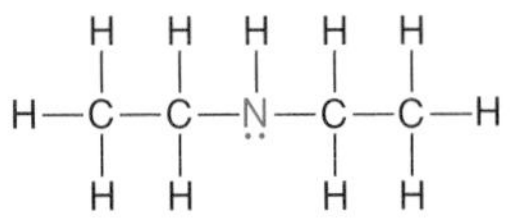 | $8.6 \times 10^{-4}$ |
| Dimethylamine [$(CH_3)_2NH$] | 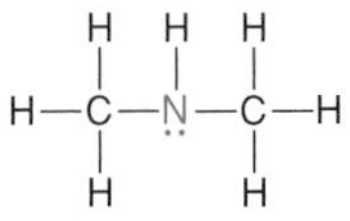 | $5.9 \times 10^{-4}$ |
| Triethylamine [$(CH_3CH_2)_3N$] | 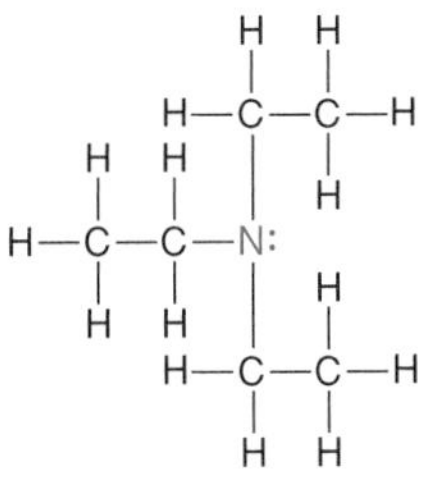 | $5.2 \times 10^{-4}$ |
| Methylamine ($CH_3NH_2$) | 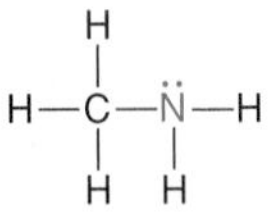 | $4.4 \times 10^{-4}$ |
| Ethanolamine ($HOCH_2CH_2NH_2$) | 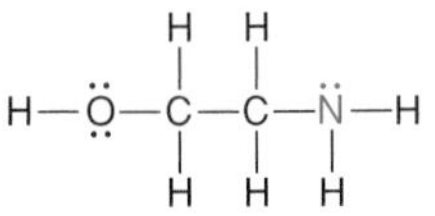 | $3.2 \times 10^{-5}$ |
| Ammonia ($NH_3$) | 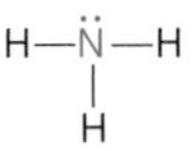 | $1.76 \times 10^{-5}$ |
| Pyridine ($C_5H_5N$) |  | $1.7 \times 10^{-9}$ |
| Aniline ($C_6H_5NH_2$) | 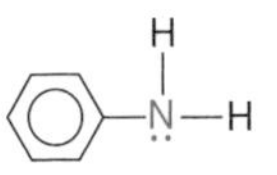 | $4.0 \times 10^{-10}$ |

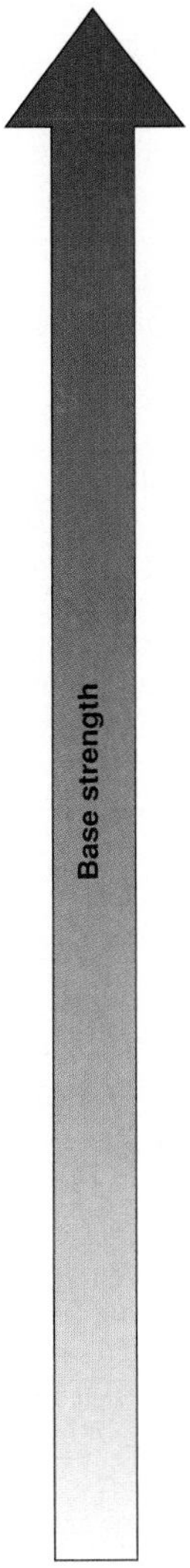

*Blue type indicates the basic nitrogen and its lone pair.

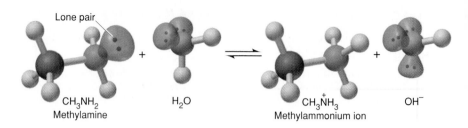

CH$_3$NH$_2$          H$_2$O                              CH$_3$NH$_3^+$                    OH$^-$
Methylamine                                      Methylammonium ion

If one or more of the H atoms in $\ddot{N}H_3$ is replaced by an organic group (designated R), an *amine* results: R$\ddot{N}H_2$, R$_2\ddot{N}H$, or R$_3\ddot{N}$ (Section 14.4). The key structural feature of these compounds, as in all Brønsted-Lowry bases, is *a lone pair of electrons that can bind the proton donated by the acid.* Figure 17.10 depicts this process for methylamine, the simplest of the amines. Solving equilibrium problems that involve neutral weak bases is very similar to solving problems involving weak acids.

SAMPLE PROBLEM 17.9

**Determining pH from $K_b$**

**Problem:** Dimethylamine, $(CH_3)_2NH$, a key intermediate in detergent manufacture, has a $K_b$ of $5.9 \times 10^{-4}$. What is the pH of 1.5 $M$ $(CH_3)_2NH$?

**Plan:** The amine forms OH$^-$ in water, so we find [OH$^-$] and then calculate [H$_3$O$^+$] and pH. The balanced equation and $K_b$ expression are

$$(CH_3)_2NH(aq) + H_2O(l) \rightleftharpoons (CH_3)_2NH_2^+(aq) + OH^-(aq)$$

$$K_b = \frac{[(CH_3)_2NH_2^+][OH^-]}{[(CH_3)_2NH]}$$

Since $K_b >> K_w$, the [OH$^-$] from the autoionization of water is negligible, so we assume that [OH$^-$]$_{\text{from base}} >>$ [OH$^-$]$_{\text{from H}_2\text{O}}$ and thus [OH$^-$]$_{\text{from base}} \approx$ [OH$^-$]. The equation tells us that [OH$^-$]$_{\text{from base}}$ also equals [$(CH_3)_2NH_2^+$]. Since $K_b$ is small, we also assume that the amount of amine reacting is small, so [$(CH_3)_2NH$]$_{\text{init}}$ − [$(CH_3)_2NH$]$_{\text{reacting}} \approx$ [$(CH_3)_2NH$]$_{\text{init}} \approx$ [$(CH_3)_2NH$]. We proceed as usual, setting up a reaction table, making the assumptions, and solving for $x$. Then we check assumptions and convert [OH$^-$] to [H$_3$O$^+$] with $K_w$ and then to pH.

**Solution:** Setting up the reaction table, with $x = [(CH_3)_2NH]_{\text{reacting}} = [OH^-] = [(CH_3)_2NH_2^+]$:

| Concentration ($M$) | $(CH_3)_2NH(aq)$ | + | $H_2O(l)$ | $\rightleftharpoons$ | $(CH_3)_2NH_2^+(aq)$ | + | $OH^-(aq)$ |
|---|---|---|---|---|---|---|---|
| Initial | 1.5 | | — | | 0 | | 0 |
| Change | $-x$ | | — | | $+x$ | | $+x$ |
| Equilibrium | $1.5 - x$ | | — | | $x$ | | $x$ |

Making the assumptions:
1. Since $K_b >> K_w$, [OH$^-$]$_{\text{from base}} >>$ [OH$^-$]$_{\text{from H}_2\text{O}}$, so [OH$^-$] = $x$ + [OH$^-$]$_{\text{from H}_2\text{O}} \approx x$
2. Since $K_b$ is small, [$(CH_3)_2NH$]$_{\text{init}} \approx$ [$(CH_3)_2NH$], so 1.5 $M$ − $x \approx$ 1.5 $M$

Substituting into the $K_b$ expression and solving for $x$:

$$K_b = \frac{[(CH_3)_2NH_2^+][OH^-]}{[(CH_3)_2NH]} = \frac{x^2}{1.5} = 5.9 \times 10^{-4}$$

$$x = [OH^-] = 3.0 \times 10^{-2} \ M$$

Checking the assumptions:

1. $\dfrac{1.0 \times 10^{-7}\, M}{3.0 \times 10^{-2}\, M} \times 100 = 3.3 \times 10^{-4}\% < 5\%$; assumption is justified.

2. $\dfrac{3.0 \times 10^{-2}\, M}{1.5\, M} \times 100 = 2.0 < 5\%$; assumption is justified.

Note that $\dfrac{[B]_{init}}{K_b} = \dfrac{1.5}{5.9 \times 10^{-4}} = 2.5 \times 10^3 > 400$

Calculating pH:

$$[H_3O^+] = \dfrac{K_w}{[OH^-]} = \dfrac{1.0 \times 10^{-14}}{3.0 \times 10^{-2}} = 3.3 \times 10^{-13}\, M$$

pH $= -\log(3.3 \times 10^{-13}) = $ **12.48**

**Check:** The value of $x$ seems reasonable: $\sqrt{(\sim 6 \times 10^{-4})(1.5)} = \sqrt{\sim 9 \times 10^{-4}} = 3 \times 10^{-2}$. Since $(CH_3)_2NH$ is a weak base, the pH should be several pH units greater than 7.

**FOLLOW-UP PROBLEM 17.9**
Pyridine ($C_5H_5\ddot{N}$), an important solvent and base in organic syntheses, has a $pK_b$ of 8.77. What is the pH of 0.10 $M$ pyridine?

## Anions of Weak Acids as Bases

The other large group of Brønsted-Lowry bases consists of the anions of weak acids:*

$$A^-(aq) + H_2O(l) \rightleftharpoons HA(aq) + OH^-(aq)$$

$$K_b = \dfrac{[HA][OH^-]}{[A^-]}$$

For example, $F^-$, the anion of the weak acid HF, acts as a weak base:

$$F^-(aq) + H_2O(l) \rightleftharpoons HF(aq) + OH^-(aq)$$

$$K_b = \dfrac{[HF][OH^-]}{[F^-]}$$

Why is a solution of HA acidic and a solution of $A^-$ basic? Let's approach the question by examining the relative concentrations of species present in 1 $M$ HF. Since HF is a weak acid, its relative concentration is large. Some HF molecules donate their protons to $H_2O$, yielding small concentrations of $H_3O^+$ and $F^-$. The equilibrium position of the system lies far to the left:

$$HF(aq) + H_2O(l) \overset{\longleftarrow}{\rightleftharpoons} H_3O^+(aq) + F^-(aq)$$

Water molecules also contribute a minute amount of $H_3O^+$ and $OH^-$, but their concentrations are extremely small:

$$2H_2O(l) \overset{\longleftarrow}{\rightleftharpoons} H_3O^+(aq) + OH^-(aq)$$

---

*This equation and equilibrium expression are sometimes referred to as a *hydrolysis reaction* and a *hydrolysis constant*, $K_h$, because water is dissociated (hydrolysed), and its parts end up in the products. Actually, it is the same as the proton abstraction process we saw with neutral bases such as ammonia, except for the charge on the base, so a new term and new equilibrium constant are unnecessary. $K_h$ is just another symbol for $K_b$, so we will simplify things and use $K_b$ throughout.

Of all the species present—HF, $H_2O$, $H_3O^+$, $F^-$, and $OH^-$—the two that can influence the acidity of the solution are $H_3O^+$, predominantly from HF, and $OH^-$ from water. Since $[H_3O^+]_{\text{from HF}} >> [OH^-]_{\text{from } H_2O'}$ the solution is acidic.

Now, consider the species present in 1 $M$ NaF. The salt dissociates completely to yield a relatively large concentration of $F^-$. Some $F^-$ reacts with water, while the $Na^+$ ion behaves as a spectator. Since $F^-$ is a weak base, it reacts with water to produce only small amounts of HF and $OH^-$:

$$F^-(aq) + H_2O(l) \overset{\longleftarrow}{\rightleftharpoons} HF(aq) + OH^-(aq)$$

As before, water dissociation contributes minute amounts of $H_3O^+$ and $OH^-$. Thus, in addition to the $Na^+$ ion, the species present are the same as in the HF solution: HF, $H_2O$, $H_3O^+$, $F^-$, and $OH^-$. The two species that affect the acidity are $OH^-$, predominantly from the $F^-$ reaction with water, and $H_3O^+$ from water. In this case, $[OH^-]_{\text{from } F^-} >> [H_3O^+]_{\text{from } H_2O'}$ so the solution is basic. *The factor that determines the relative acidity of HA and $A^-$ solutions is the relative concentrations of HA and $A^-$.* In the HF solution, $[HF] >> [F^-]$, so $[H_3O^+] >> [OH^-]$ and the solution is acidic; in the $F^-$ solution, $[F^-] >> [HF]$, so $[OH^-] >> [H_3O^+]$ and the solution is basic.

### The Relation Between $K_a$ and $K_b$ of a Conjugate Acid-Base Pair

An important relationship exists between the $K_a$ of HA and the $K_b$ of $A^-$, which we can see by treating the two dissociation reactions as a reaction sequence and adding them together:

$$\begin{array}{c} \cancel{HA} + H_2O \rightleftharpoons H_3O^+ + \cancel{A^-} \\ \underline{\cancel{A^-} + H_2O \rightleftharpoons \cancel{HA} + OH^-} \\ 2H_2O \rightleftharpoons H_3O^+ + OH^- \end{array}$$

*The sum of the two dissociation reactions is the autoionization of water.* Recall from Chapter 16 that the overall equilibrium constant for a reaction that is the *sum* of two or more reactions is the *product* of the individual equilibrium constants. Therefore, writing the expressions for each reaction gives

$$\frac{[H_3O^+]\cancel{[A^-]}}{\cancel{[HA]}} \times \frac{\cancel{[HA]}[OH^-]}{\cancel{[A^-]}} = [H_3O^+][OH^-]$$

or
$$K_a \quad \times \quad K_b \quad = \quad K_w \qquad \textbf{(17.7)}$$

This relationship allows us to find $K_a$ of the acid in a conjugate pair given $K_b$ of the base, and vice versa. Let's use this relationship to obtain a key piece of data for solving equilibrium problems. Reference tables typically have $K_a$ and $K_b$ values for neutral species only. The $K_b$ for $F^-$ ion and the $K_a$ for $CH_3NH_3^+$, for example, do not appear in standard tables, but you can calculate them simply by looking up the values of the neutral conjugate species and relating them to $K_w$. For $F^-$, for instance, we look up the $K_a$ value for HF and relate it to $K_w$:

$$K_a \text{ of HF} = 6.8 \times 10^{-4} \text{ (from Table 17.2)}$$

Since $K_a$ of HF $\times$ $K_b$ of $F^- = K_w$, we have

$$K_b \text{ of } F^- = \frac{K_w}{K_a \text{ of HF}} = \frac{1.0 \times 10^{-14}}{6.8 \times 10^{-4}} = 1.5 \times 10^{-11}$$

Then we use this calculated $K_b$ value to solve the rest of the problem.

SAMPLE PROBLEM 17.10 _____

## Determining the pH of a Solution of the Conjugate Base of a Weak Acid

**Problem:** Sodium acetate ($CH_3COONa$, or NaAc for this problem) has applications in photographic development and textile dyeing. What is the pH of 0.25 $M$ NaAc? $K_a$ of acetic acid (HAc) = $1.8 \times 10^{-5}$.
**Plan:** We have to find the pH of a solution of $Ac^-$, which acts as a base in water:

$$Ac^-(aq) + H_2O(l) \rightleftharpoons HAc(aq) + OH^-(aq)$$

$$K_b = \frac{[HAc][OH^-]}{[Ac^-]}$$

If we calculate $[OH^-]$, we can find $[H_3O^+]$ and convert to pH. To solve for $[OH^-]$, we need $K_b$ of $Ac^-$, which we obtain from the $K_a$ of HAc and $K_w$. All sodium salts are soluble, so we know that $[Ac^-] = 0.25$ $M$. Our usual assumptions mean that $[OH^-]_{from\ Ac^-} \approx [OH^-]$, and $[Ac^-]_{init} \approx [Ac^-]$.
**Solution:** Setting up the reaction table, with $x = [Ac^-]_{reacting} = [OH^-] = [HAc]$:

| Concentration ($M$) | $Ac^-(aq)$ | + | $H_2O(l)$ | $\rightleftharpoons$ | $HAc(aq)$ | + | $OH^-(aq)$ |
|---|---|---|---|---|---|---|---|
| Initial | 0.25 | | — | | 0 | | 0 |
| Change | $-x$ | | — | | $+x$ | | $+x$ |
| Equilibrium | $0.25 - x$ | | — | | $x$ | | $x$ |

Solving for $K_b$:

$$K_b = \frac{K_w}{K_a} = \frac{1.0 \times 10^{-14}}{1.8 \times 10^{-5}} = 5.6 \times 10^{-10}$$

Making the assumptions:
1. Since $K_b >> K_w$, $x + [OH^-]_{from\ H_2O} \approx x$
2. Since $K_b$ is small, 0.25 $M - x \approx 0.25$ $M$
Substituting into the expression for $K_b$ and solving for $x$:

$$K_b = \frac{[HAc][OH^-]}{[Ac^-]} = \frac{x^2}{0.25} = 5.6 \times 10^{-10}$$

$$x = 1.2 \times 10^{-5} M = [OH^-]$$

Checking the assumptions:

1. $\frac{1.0 \times 10^{-7} M}{1.2 \times 10^{-5} M} \times 100 = 0.83\% < 5\%$; assumption is justified.

2. $\frac{1.2 \times 10^{-5} M}{0.25 M} \times 100 = 4.8 \times 10^{-3}\% < 5\%$; assumption is justified.

Note that $\frac{0.25}{5.6 \times 10^{-10}} = 4.5 \times 10^8 > 400$.

Solving for pH:

$$[H_3O^+] = \frac{K_w}{[OH^-]} = \frac{1.0 \times 10^{-14}}{1.2 \times 10^{-5}} = 8.3 \times 10^{-10} M$$

$$pH = -\log(8.3 \times 10^{-10}) = \mathbf{9.08}$$

**Check:** The $K_b$ calculation seems reasonable: $\sim 10 \times 10^{-15}/2 \times 10^{-5} = 5 \times 10^{-10}$. Since $Ac^-$ is a weak base, $[OH^-] > [H_3O^+]$, so the pH > 7.

FOLLOW-UP PROBLEM 17.10
Sodium hypochlorite (NaClO) is the active ingredient in household laundry bleach. What is the pH of 0.20 $M$ NaClO?

**Section Summary**

The extent to which a weak base abstracts a proton from water to form $OH^-$ is expressed by a base dissociation constant $K_b$. Brønsted-Lowry bases include $NH_3$, the amines, and anions of weak acids. The relative concentrations of HA and $A^-$ determine the acidity of their solutions: in a solution of HA, $[HA] > [A^-]$, so $[H_3O^+] > [OH^-]$. In a solution of $A^-$, $[A^-] > [HA]$, so $[OH^-] > [H_3O^+]$. By multiplying the expressions for $K_a$ of HA and $K_b$ of $A^-$, we obtain $K_w$. This relationship allows us to calculate $K_a$ of $BH^+$, the cationic conjugate acid of a neutral weak base B, or $K_b$ of $A^-$, the anionic conjugate base of a neutral weak acid HA.

## 17.6 Molecular Properties and Acid Strength

The strength of an acid depends on its ability to donate a proton, which depends in turn on the strength of the bond to the acidic proton. In this section, we examine the trends in acid strength of binary hydrides and oxoacids and discuss the acidity of hydrated metal ions.

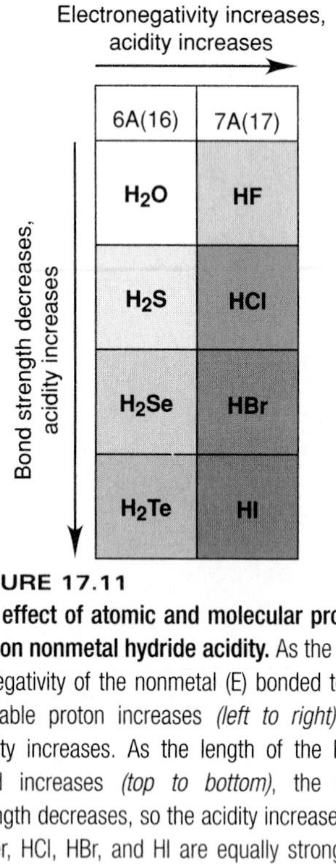

### Trends in Binary Nonmetal Hydride Acidity

Two factors determine how easily a proton is released from a binary nonmetal hydride: the electronegativity (EN) of the nonmetal (E) and the strength of the E—H bond (Figure 17.11). As E becomes more electronegative, the E—H bond becomes more polar, and the $H^+$ is more easily released. Since electronegativity increases from left to right across a period, *nonmetal hydride acidity increases across a period*. In aqueous solution, the hydrides of Groups 3A(13) to 5A(15) do not behave as acids, but the increase in acidity is seen in Groups 6A(16) and 7A(17). Thus, $H_2S$ is a weaker acid than HCl because S is less electronegative (EN = 2.5) than Cl (EN = 3.0). The same relationship holds in each period.

Down a group, E—H bond strength dominates. As E becomes larger, the E—H bond becomes longer and weaker, so $H^+$ comes off more easily. Thus, although the trend is not apparent in water, the hydrohalic acids increase in strength down the group: HF < HCl < HBr < HI.

### Trends in Oxoacid Acidity

All oxoacids have the acidic hydrogen bound to an oxygen atom, so bond strength (length) is not a factor as it is with the binary nonmetal hydrides. Rather, as you may recall from Section 13.7, two factors—the electronegativity of the central nonmetal (E) and the number of O atoms—determine oxoacid acidity. *With the same number of oxygens around E, acid strength increases with the electronegativity of E.* Consider the hypohalous acids (written here as HOE, where E is a halogen atom). The more electronegative the halogen, the more polar the O—H bond, and the more easily $H^+$ is lost. Electronegativity decreases down the group, so we predict that acid strength decreases as HOCl > HOBr > HOI. Our reasoning is confirmed by the $K_a$ values:

$K_a$ of HClO = $2.9 \times 10^{-8}$     $K_a$ of HBrO = $2.3 \times 10^{-9}$     $K_a$ of HIO = $2.3 \times 10^{-11}$

We also predict (correctly) that in Group 6A(16), $H_2SO_4$ is stronger than $H_2SeO_4$; in Group 5A(15), $H_3PO_4$ is stronger than $H_3AsO_4$, and so forth.

*With different numbers of oxygens around the same E, acid strength increases with number of O atoms.* The electronegative O atoms pull electron density away

Electronegativity increases,
acidity increases
→

Bond strength decreases,
acidity increases
↓

| 6A(16) | 7A(17) |
|--------|--------|
| $H_2O$ | HF |
| $H_2S$ | HCl |
| $H_2Se$ | HBr |
| $H_2Te$ | HI |

**FIGURE 17.11**

**The effect of atomic and molecular properties on nonmetal hydride acidity.** As the electronegativity of the nonmetal (E) bonded to the ionizable proton increases *(left to right)*, the acidity increases. As the length of the E—H bond increases *(top to bottom)*, the bond strength decreases, so the acidity increases. (In water, HCl, HBr, and HI are equally strong, for reasons discussed in Section 17.8.)

from E, which pulls electron density from the O—H bond. The more O atoms present, the greater the shift in electron density, and the more easily the $H^+$ ion comes off. Therefore, we predict that, for instance, chlorine oxoacids (written here as $HOClO_n$, with $n$ from 0 to 3) increase in strength as $HOCl < HOClO < HOClO_2 < HOClO_3$. Once again, the $K_a$ values support the prediction:

$$K_a \text{ of HOCl (hypochlorous acid)} = 2.9 \times 10^{-8}$$
$$K_a \text{ of HOClO (chlorous acid)} = 1.12 \times 10^{-2}$$
$$K_a \text{ of HOClO_2 (chloric acid)} \approx 1$$
$$K_a \text{ of HOClO_3 (perchloric acid)} = \text{very large } (>10^7)$$

It follows from this that $HNO_3$ is stronger than $HNO_2$, that $H_2SO_4$ is stronger than $H_2SO_3$, and so forth.

## Acidity of Hydrated Metal Ions

The aqueous solutions of certain metal ions are acidic because the *hydrated metal ion acts as an acid,* transferring an $H^+$ ion to water. Consider a general metal nitrate, $M(NO_3)_n$, as it dissolves in water. The ions separate and become surrounded by a certain number of tightly bound $H_2O$ molecules. The following equation shows the hydration of the cation ($M^{n+}$) with $H_2O$ molecules; hydration of the anion ($NO_3^-$) is indicated by *(aq)*:

$$M(NO_3)_n(s) + xH_2O(l) \rightarrow M(H_2O)_x^{n+}(aq) + nNO_3^-(aq)$$

If the metal ion, $M^{n+}$, is *small and highly charged,* it withdraws sufficient electron density from the O—H bonds of the bound water molecules for one of the protons to be released. Thus, the hydrated cation, $M(H_2O)_x^{n+}$, acts as a typical Brønsted-Lowry acid. In the process, the bound water molecule that releases the proton becomes a bound $OH^-$ ion:

$$M(H_2O)_x^{n+}(aq) + H_2O(l) \rightleftharpoons M(H_2O)_{x-1}OH^{(n-1)+}(aq) + H_3O^+(aq)$$

Each hydrated metal ion dissociation has a characteristic $K_a$ value. Table 17.7 shows some common examples.

Aluminum ion, for example, has the small size and high positive charge needed to produce an acidic solution. When an aluminum salt, such as $Al(NO_3)_3$, dissolves in water, the following steps occur:

$$Al(NO_3)_3(s) + 6H_2O(l) \longrightarrow Al(H_2O)_6^{3+}(aq) + 3NO_3^-(aq)$$
[dissolution and hydration]

$$Al(H_2O)_6^{3+}(aq) + H_2O(l) \rightleftharpoons Al(H_2O)_5OH^{2+}(aq) + H_3O^+(aq) \text{ [acid dissociation]}$$

Notice the formulas of the hydrated metal ions in the second step. When the $H^+$ ion is released, the number of bound $H_2O$ molecules decreases by one and the number of bound $OH^-$ ions increases by one, which reduces the ion's positive charge by one. This process is shown in Figure 17.12.

**TABLE 17.7** $K_a$ **Values of Some Hydrated Metal Ions at 25°C**

| ION | $K_a$ | |
|---|---|---|
| $Fe^{3+}(aq)$ | $6 \times 10^{-3}$ | |
| $Sn^{2+}(aq)$ | $4 \times 10^{-4}$ | |
| $Cr^{3+}(aq)$ | $1 \times 10^{-4}$ | |
| $Al^{3+}(aq)$ | $1 \times 10^{-5}$ | |
| $Be^{2+}(aq)$ | $4 \times 10^{-6}$ | |
| $Cu^{2+}(aq)$ | $3 \times 10^{-8}$ | Acid strength |
| $Pb^{2+}(aq)$ | $3 \times 10^{-8}$ | |
| $Zn^{2+}(aq)$ | $1 \times 10^{-9}$ | |
| $Co^{2+}(aq)$ | $2 \times 10^{-10}$ | |
| $Ni^{2+}(aq)$ | $1 \times 10^{-10}$ | |

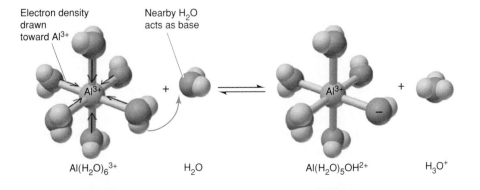

Electron density drawn toward $Al^{3+}$

Nearby $H_2O$ acts as base

$Al(H_2O)_6^{3+}$      $H_2O$      $Al(H_2O)_5OH^{2+}$      $H_3O^+$

**FIGURE 17.12**

**The acidic behavior of the hydrated $Al^{3+}$ ion.** When a metal ion enters water, it is hydrated as the solvent forms bonds to it. If the ion is small and multiply charged, as is the $Al^{3+}$ ion, it pulls sufficient electron density from the attached O—H groups of water and an $H^+$ ion is transferred to a nearby water molecule.

Through its ability to withdraw electron density from the O—H bonds of the bound water molecules, a small, highly charged central metal ion behaves like a central electronegative atom in an oxoacid. In addition to $Al^{3+}$, $Fe^{3+}$ and $Cr^{3+}$ ions are among several other cations whose salts form acidic solutions.

**Section Summary**
The strength of an acid depends on the strength of the bond to the ionizable proton. For binary nonmetal hydrides, acid strength increases across a period, with electronegativity of the nonmetal (E), and down a group, with length of E—H. For oxoacids with the same number of O atoms, acid strength increases with electronegativity of E; for oxoacids with the same E, acid strength increases with number of O atoms. Small, highly charged metal ions are acidic in water because they withdraw electron density from O—H bonds of bound $H_2O$ molecules, releasing an $H^+$ ion to the solution.

## 17.7 Acid-Base Properties of Salt Solutions

Up to now you've seen that cations of weak bases are acidic, anions of weak acids are basic, and small, highly charged metal cations are acidic. Therefore, when salts containing these ions dissolve in water, the pH of the solution is affected. You can predict the relative acidity of a salt solution from the relative acid-base strength of the cation and anion. Let's examine the types of salts that yield neutral, acidic, or basic solutions.

### Salts That Yield Neutral Solutions

*The salt of a strong acid and strong base yields a neutral solution because the ions do not react with water.* To see why the ions do not react, consider the dissociation of the parent acid and base. When a strong acid such as $HNO_3$ dissolves, complete dissociation takes place:

$$HNO_3(l) + H_2O(l) \longrightarrow NO_3^-(aq) + H_3O^+(aq)$$

$H_2O$ is a much stronger base than $NO_3^-$, so the reaction proceeds far to the right. The same argument can be made for any strong acid: *the anion of a strong acid is a much weaker base than water.* Therefore, a strong acid anion in water is hydrated, but nothing further happens.

Now consider the dissociation of a strong base, such as NaOH:

$$NaOH(s) \xrightarrow{H_2O} Na^+(aq) + OH^-(aq)$$

The $Na^+$ ion has a relatively large size and low charge and therefore does not interact strongly with the water molecules surrounding it. When $Na^+$ enters water, it becomes hydrated but undergoes no further reaction. The cations of all strong bases behave this way.

The anions of strong acids are the halide ions, except $F^-$, and those of strong oxoacids, such as $NO_3^-$, $ClO_4^-$, and $HSO_4^-$. The cations of strong bases are the Group 1A(1) and 2A(2) cations (except $Be^{2+}$). Salts containing only these ions, such as NaCl and $Ba(NO_3)_2$, yield neutral solutions because *no reaction with water takes place.*

## Salts That Yield Acidic Solutions

*The salt of a strong acid and a weak base yields an acidic solution because the cation acts as a weak acid,* and the anion does not react. For example, $NH_4Cl$ produces an acidic solution because the $NH_4^+$ ion, which forms from the weak base $NH_3$, is a weak acid, and the $Cl^-$ ion, the anion of a strong acid, does not react:

$$NH_4Cl(s) \xrightarrow{H_2O} NH_4^+(aq) + Cl^-(aq) \text{ [dissolution and hydration]}$$

$$NH_4^+(aq) + H_2O(l) \rightleftharpoons NH_3(aq) + H_3O^+(aq) \text{ [dissociation of weak acid]}$$

As we just discussed, *small, highly charged metal ions* are another group of cations that yield $H_3O^+$ in solution. For example, $Fe(NO_3)_3$ produces an acidic solution because the hydrated $Fe^{3+}$ ion acts as a weak acid, while the $NO_3^-$ ion, the anion of a strong acid, does not react:

$$Fe(NO_3)_3(s) + 6H_2O(l) \rightarrow Fe(H_2O)_6^{3+}(aq) + 3NO_3^-(aq) \text{ [dissolution and hydration]}$$

$$Fe(H_2O)_6^{3+}(aq) + H_2O(l) \rightleftharpoons Fe(H_2O)_5OH^{2+}(aq) + H_3O^+(aq)$$
$$\text{[dissociation of weak acid]}$$

## Salts That Yield Basic Solutions

*The salt of a weak acid and a strong base yields a basic solution in water because the anion acts as a weak base,* and the cation does not react. The anion of a weak acid abstracts a proton from water to yield $OH^-$ ions. Sodium acetate, for example, yields a basic solution because the $Na^+$ ion, the cation of a strong base, does not react with water, and the $CH_3COO^-$ ion, the anion of the weak acid $CH_3COOH$, acts as a weak base:

$$CH_3COONa(s) \xrightarrow{H_2O} Na^+(aq) + CH_3COO^-(aq) \text{ [dissolution and hydration]}$$

$$CH_3COO^-(aq) + H_2O(l) \rightleftharpoons CH_3COOH(aq) + OH^-(aq) \text{ [reaction of weak base]}$$

Table 17.8 shows the acid-base behavior of the three types of salts in water.

**TABLE 17.8 The Behavior of Salts in Water**

| SALT (EXAMPLE) | pH | NATURE OF IONS | PRODUCTS FORMED IN WATER | |
|---|---|---|---|---|
| Neutral (NaCl) | 7.0 | Cation of strong base: $Na^+$ | Hydrated ion: $Na^+(aq)$ | |
| | | Anion of strong acid: $Cl^-$ | Hydrated ion: $Cl^-(aq)$ | |
| Acidic ($NH_4Cl$) | <7.0 | Cation of weak base: $NH_4^+$ | Weak base + hydronium ion: $NH_3(aq) + H_3O^+(aq)$ | |
| | | Anion of strong acid: $Cl^-$ | Hydrated anion: $Cl^-(aq)$ | |
| Basic ($CH_3COONa$) | >7.0 | Cation of strong base: $Na^+$ | Hydrated cation: $Na^+(aq)$ | |
| | | Anion of weak acid: $CH_3COO^-$ | Weak acid + hydroxide ion: $CH_3COOH(aq) + OH^-(aq)$ | |

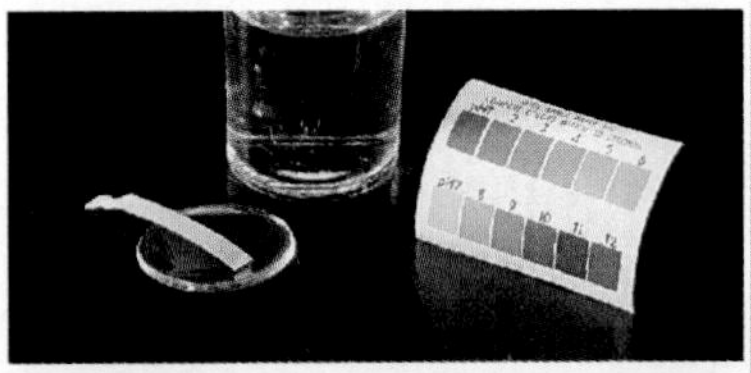

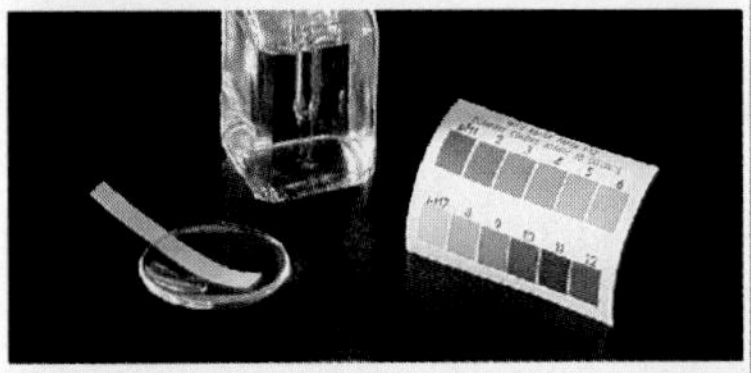

SAMPLE PROBLEM 17.11 _____

## Predicting the Relative Acidity of Salt Solutions

**Problem:** Write reactions to predict whether aqueous solutions of the following salts are acidic, basic, or neutral:
**(a)** Potassium perchlorate, $KClO_4$
**(b)** Sodium benzoate, $C_6H_5COONa$
**(c)** Chromium trichloride, $CrCl_3$
**Plan:** We examine the formulas to determine the cations and anions. Depending on the nature of these ions, the solution will be neutral (strong acid anion and strong base cation), acidic (weak base cation and strong acid anion or highly charged metal cation), or basic (weak acid anion and strong base cation).
**Solution: (a) Neutral.** The ions are $K^+$ and $ClO_4^-$. The $K^+$ ion is from the strong base KOH, and the $ClO_4^-$ anion is from the strong acid $HClO_4$. Neither ion reacts with water.
**(b) Basic.** The ions are $Na^+$ and $C_6H_5COO^-$. $Na^+$ is the cation of the strong base NaOH and does not react with water. The benzoate ion, $C_6H_5COO^-$, is from the weak acid benzoic acid, so it reacts to produce $OH^-$ ion:

$$C_6H_5COO^-(aq) + H_2O(l) \rightleftharpoons C_6H_5COOH(aq) + OH^-(aq)$$

**(c) Acidic.** The ions are $Cr^{3+}$ and $Cl^-$. $Cl^-$ is the anion of the strong acid HCl, so it does not react with water. The $Cr^{3+}$ is a small ion with a high positive charge, so the hydrated ion, which is $Cr(H_2O)_6^{3+}$, reacts with water to produce $H_3O^+$:

$$Cr(H_2O)_6^{3+}(aq) + H_2O(l) \rightleftharpoons Cr(H_2O)_5OH^{2+}(aq) + H_3O^+(aq)$$

FOLLOW-UP PROBLEM 17.11
Write equations to predict whether solutions of the following salts are acidic, basic, or neutral: **(a)** $KClO_2$; **(b)** $CH_3NH_3NO_3$; **(c)** CsI.

## Salts Composed of Weakly Acidic Cations and Weakly Basic Anions

The only salts left to consider are those consisting of a cation that acts as a weak acid and an anion that acts as a weak base. In these cases, and there are quite a few, both ions react with water. The overall acidity of the solution depends on the relative acid or base strength of the separated ions, which can be determined by comparing their equilibrium constants.

For example, will an aqueous solution of ammonium sulfide, $(NH_4)_2S$, be acidic or basic? First, we write equations for any reactions that occur between the separated ions and water. Ammonium ion is the conjugate acid of a weak base, so it acts as a weak acid:

$$NH_4^+(aq) + H_2O(l) \rightleftharpoons NH_3(aq) + H_3O^+(aq)$$

Sulfide ion is the anion of a weak acid, so it acts as a weak base:

$$S^{2-}(aq) + H_2O(l) \rightleftharpoons HS^-(aq) + OH^-(aq)$$

The reaction that goes farther to the right will have the greater influence on the pH of the solution, so we must compare the $K_a$ of $NH_4^+$ with the $K_b$ of $S^{2-}$. Recall that only neutral compounds are listed in $K_a$ and $K_b$ tables, so we have to calculate these values for the ions:

$$K_a \text{ of } NH_4^+ = \frac{K_w}{K_b \text{ of } NH_3} = \frac{1.0 \times 10^{-14}}{1.76 \times 10^{-5}} = 5.7 \times 10^{-10}$$

$$K_b \text{ of } S^{2-} = \frac{K_w}{K_{a2} \text{ of } H_2S} = \frac{1.0 \times 10^{-14}}{1 \times 10^{-17}} = 1 \times 10^3$$

The difference in magnitude of the equilibrium constants ($K_b \approx 10^{12} \times K_a$) tells us that the abstraction of a proton from $H_2O$ by $S^{2-}$ proceeds much farther than the release of a proton to $H_2O$ by $NH_4^+$. In other words, since $K_b$ of $S^{2-} >> K_a$ of $NH_4^+$, the $(NH_4)_2S$ solution is basic.

SAMPLE PROBLEM 17.12 _____

**Predicting the Relative Acidity of Salt Solutions from $K_a$ and $K_b$ of the Ions**

**Problem:** Determine whether an aqueous solution of iron(III) nitrite, $Fe(NO_2)_3$, is acidic, basic, or neutral.
**Plan:** The formula consists of the small, highly charged, and therefore weakly acidic $Fe^{3+}$ cation and the weakly basic $NO_2^-$ anion of the weak acid $HNO_2$. To determine the relative acidity of the solution, we find $K_a$ and $K_b$ of the ions to see which ion reacts with water to a greater extent.
**Solution:** Writing the reactions with water:

$$Fe(H_2O)_6^{3+}(aq) + H_2O(l) \rightleftharpoons Fe(H_2O)_5OH^{2+}(aq) + H_3O^+(aq)$$

$$NO_2^-(aq) + H_2O(l) \rightleftharpoons HNO_2(aq) + OH^-(aq)$$

Obtaining $K_a$ and $K_b$ of the ions: From Table 17.7, $K_a$ of $Fe^{3+}(aq) = 6 \times 10^{-3}$. From Table 17.2, we obtain $K_a$ of $HNO_2$ and solve for $K_b$ of $NO_2^-$:

$$K_b \text{ of } NO_2^- = \frac{K_w}{K_a \text{ of } HNO_2} = \frac{1.0 \times 10^{-14}}{7.1 \times 10^{-4}} = 1.4 \times 10^{-11}$$

Since $K_a$ of $Fe^{3+} >> K_b$ of $NO_2^-$, the solution is **acidic.**

FOLLOW-UP PROBLEM 17.12
Determine whether solutions of the following salts will be acidic, basic, or neutral:
**(a)** $Cu(CH_3COO)_2$; **(b)** $NH_4F$.

_____

**Section Summary**
If a salt consists of ions that do not react with water, its solution is neutral. Acidic salts contain a cation that releases a proton to water and an unreactive anion. Basic salts contain an anion that abstracts a proton from water and an unreactive cation. If both cation and anion react with water, the ion that reacts to the greater extent determines the overall acidity of the solution.

## 17.8 Generalizing the Brønsted-Lowry Concept: The Leveling Effect

The Brønsted-Lowry concept reveals an important principle that we can generalize to acid-base behavior in solvents other than water. Notice that, in water, all Brønsted-Lowry acids yield $H_3O^+$ and all Brønsted-Lowry bases yield $OH^-$—the ions that form when $H_2O$ autoionizes. In general, *an acid yields the cation and a base yields the anion of the autoionization of the solvent.*

Now let's extend our understanding of Brønsted-Lowry acids and bases by examining a question you may have been wondering about: Why are all strong acids and strong bases *equally* strong in water? The answer is that *the strongest acid possible in water is $H_3O^+$ and the strongest base possible is $OH^-$.* The moment we put some gaseous HCl in water, it reacts with the base $H_2O$ and forms $H_3O^+$. The same holds for $HNO_3$, $H_2SO_4$, and any strong acid. All strong acids are equally strong in water because they dissociate *completely* to form $H_3O^+$. Given that the strong acid is no longer present, we are actually observing the acid strength of $H_3O^+$.

Similarly, all strong bases yield $OH^-$ in water, even if they do not contain hydroxide ions in the solid. Suppose we dissolve a metal oxide, such as $K_2O$, in water. The oxide ion, which is a stronger base than $OH^-$, immediately reacts with water to form $OH^-$:

$$O^{2-}(aq) + H_2O(l) \rightarrow 2OH^-(aq)$$

No matter what substances we try, any acid stronger than $H_3O^+$ simply donates its proton to $H_2O$, and any base stronger than $OH^-$ abstracts a proton from $H_2O$. Thus, water acting as a base exerts a **leveling effect** on all strong acids, that is, it makes them appear equally strong, and water acting as an acid exerts a leveling effect on all strong bases.

In order to rank strong acids in terms of relative strength, we must dissolve them in a solvent that is a *weaker* base than water, one that accepts their protons less readily. For example, you saw in Figure 17.11 that the hydrohalic acids increase in strength as the halogen becomes larger, as a result of the longer, weaker H—X bond. In water, HF is weaker than the other hydrogen halides, but HCl, HBr, and HI dissociate completely, so they are equally strong. When we dissolve them in pure acetic acid, however, the acetic acid acts as a base, one that is weaker than water:

$$HCl(g) + CH_3COOH(l) \rightleftharpoons Cl^-(acet) + CH_3COOH_2^+(acet)$$
$$HBr(g) + CH_3COOH(l) \rightleftharpoons Br^-(acet) + CH_3COOH_2^+(acet)$$
$$HI(g) + CH_3COOH(l) \rightleftharpoons I^-(acet) + CH_3COOH_2^+(acet)$$

[The use of (*acet*) instead of (*aq*) indicates ions solvated by $CH_3COOH$.] Measurements show that HI protonates the solvent to a greater extent than HBr, and HBr more than HCl; that is, $K_{HI} > K_{HBr} > K_{HCl}$. Therefore, in $CH_3COOH$, HCl is a weaker acid than HBr, which is weaker than HI. Similarly, the relative strength of strong bases is determined in a solvent that is a weaker acid than $H_2O$, such as liquid $NH_3$.

### Section Summary

Strong acids (or strong bases) dissociate completely to $H_3O^+$ (or $OH^-$) in water; in effect, water equalizes (levels) their strengths. Acids that are equally strong in water show differences in strength when dissolved in a solvent that is a weaker base than water, such as acetic acid.

## 17.9 Electron-Pair Donation and the Lewis Acid-Base Definition

The final acid-base concept we will consider was developed by Gilbert N. Lewis, who also described the importance of valence electron pairs in molecular bonding (Chapter 9). While the Brønsted-Lowry concept focuses on the proton in defining a substance as an acid or a base, the Lewis concept highlights the role of the *electron pair*. The **Lewis acid-base definition** holds that
- A *base* is any substance that *donates* an electron pair.
- An *acid* is any substance that *accepts* an electron pair.

The Lewis definition, like the Brønsted-Lowry definition, requires that a base have an electron pair to donate, so it does not expand the classifications of bases. However, it greatly expands the meaning of an acid. Many substances, such as $CO_2$ and $Cu^{2+}$, that do not contain H in their formula (and thus cannot be Brønsted-Lowry acids) function as Lewis acids by accepting an electron pair in their reactions. Moreover, in the Lewis sense, *the proton itself functions as an acid* because it accepts the electron pair donated by the base:

$$H^+ + B: \rightleftharpoons BH^+$$

Thus, *all Brønsted-Lowry acids donate the Lewis acid $H^+$.*

The product of any Lewis acid-base reaction is a single species, called an **adduct,** that contains a new covalent bond:

$$A + B: \rightleftharpoons A—B \text{ (adduct)}$$

Thus, we see a radical broadening of the idea of a neutralization reaction. What to Arrhenius was the formation of $H_2O$ from $H^+$ and $OH^-$ became, to Brønsted and Lowry, the transfer of a proton from a stronger acid to a stronger base to form a weaker base and weaker acid. To Lewis, the same process became *the donation and acceptance of an electron pair to form a covalent bond in an adduct.* In order to accept an electron pair to form a new bond, *a Lewis acid must have a vacant orbital* or be able to rearrange its bonding to make one available. A variety of neutral molecules and positively charged ions satisfy this requirement.

**Neutral Molecules as Lewis Acids**

Many neutral molecules function as Lewis acids. In every case, the atom that accepts the electron pair is low in electron density and has one of two features: an electron deficiency or a multiple bond.

Some neutral Lewis acids contain a central atom that is *electron deficient,* one surrounded by fewer than eight valence electrons. The most important are compounds of the Group 3A(13) elements boron and aluminum. As noted in Chapters 9 and 13, these compounds react vigorously to complete their octet. For example, boron trifluoride accepts an electron pair from ammonia to form a covalent bond in a gaseous Lewis acid-base reaction:

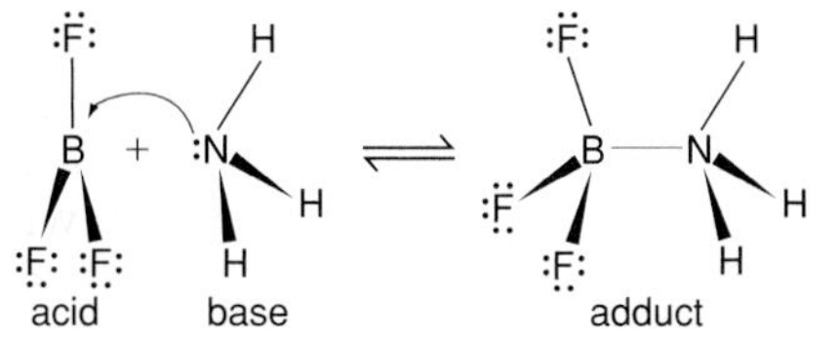

Unexpected solubility properties are sometimes due to adduct formation. Aluminum chloride, for instance, dissolves freely in ether because of a Lewis acid-base reaction, in which the ether O atom donates an electron pair to the central Al to form a covalent bond:

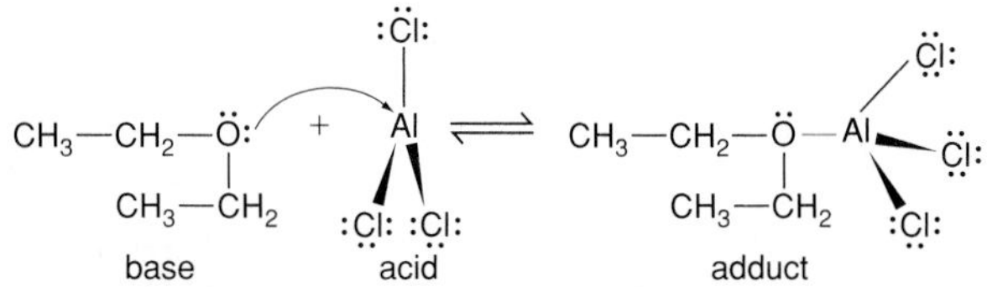

This acidic behavior of boron and aluminum halides is put to use in many industrial and laboratory organic syntheses. For example, toluene, an important solvent and organic reagent, can be made by the action of $CH_3Cl$ on benzene in the presence of $AlCl_3$. The Lewis acid $AlCl_3$ abstracts a $Cl^-$ from $CH_3Cl$ to form an adduct that contains a reactive $CH_3^+$ group, which attacks the benzene ring:

$$AlCl_3 + CH_3Cl \rightleftharpoons CH_3^+[ClAlCl_3^-]$$
$$\text{acid} \qquad \text{base} \qquad \text{adduct}$$

$$CH_3^+[ClAlCl_3^-] + C_6H_6 \rightleftharpoons C_6H_5CH_3 + AlCl_3 + HCl$$
$$\text{benzene} \qquad \text{toluene}$$

Molecules that contain a polar double bond also function as Lewis acids. The double bond becomes a single bond, as the electron pair on the base approaches and forms the new bond in the adduct. For example, consider the reaction that occurs when $SO_2$ dissolves in water. The electronegative O atoms in $SO_2$ withdraw electron density from the central S, so it is partially positive. A lone electron pair of the O atom of water is donated to the S, breaking one of the S=O $\pi$ bonds, and a proton is transferred to an oxygen. The adduct that forms is sulfurous acid:

$$\text{acid} \qquad \text{base} \qquad \text{adduct}$$

In the formation of carbonates from a metal oxide and carbon dioxide, an analogous reaction occurs in a nonaqueous heterogeneous system. The $O^{2-}$ ion, shown below from CaO, donates an electron pair to the partially positive C in $CO_2$, a $\pi$ bond breaks, and the $CO_3^{2-}$ ion forms as the adduct:

$$\text{base} \qquad \text{acid} \qquad \text{adduct}$$

### Metal Cations as Lewis Acids

Earlier we saw that certain hydrated metal ions act as Brønsted-Lowry acids. In the Lewis sense, the process of hydrating the ion is itself an acid-base reaction; thus, *any metal ion acts as a Lewis acid when it dissolves in water*. The hydrated cation is the adduct, as lone electron pairs of the O atoms of water form covalent bonds to the positively charged ion (Figure 17.13):

$$M^{2+}(aq) + 4H_2O(l) \rightleftharpoons M(H_2O)_4^{2+}(aq)$$
$$\text{acid} \qquad \text{base} \qquad \text{adduct}$$

Ammonia is a stronger Lewis base than water because it displaces $H_2O$ from a hydrated ion when aqueous $NH_3$ is added:

$$\underset{\text{hydrated adduct}}{Ni(H_2O)_6^{2+}(aq)} + \underset{\text{base}}{6NH_3(aq)} \rightleftharpoons \underset{\text{ammoniated adduct}}{Ni(NH_3)_6^{2+}(aq)}$$

We discuss the equilibrium nature of these reactions further in Chapter 18 and examine the structures of these ions in Chapter 22.

Many essential biomolecules are Lewis adducts with central metal ions. Most often, O and N atoms of organic groups, with their lone pairs, serve as the basic species. Chlorophyll is a Lewis adduct of a central $Mg^{2+}$ and the four N atoms of an organic tetrapyrrole ring system (Figure 17.14). Vitamin $B_{12}$ has a similar structure with a central $Co^{3+}$, as does heme with a central $Fe^{2+}$. Several other metal ions, such as $Zn^{2+}$, $Mo^{2+}$, and $Cu^{2+}$, are bound at the active sites of enzymes and participate in the catalytic action by virtue of their Lewis acidity.

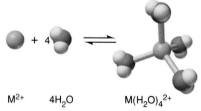

$$M^{2+} \qquad 4H_2O \qquad M(H_2O)_4^{2+}$$

**FIGURE 17.13**

**Metal ions as Lewis acids in water.** When a metal cation enters water, the $H_2O$ molecules donate electron pairs to empty orbitals on the metal ion (not shown) and form covalent bonds. Thus, the metal ion acts as a Lewis acid and the water molecule as a Lewis base.

SAMPLE PROBLEM 17.13 ──────────────────────────

### Identifying Lewis Acids and Bases

**Problem:** Identify the Lewis acids and Lewis bases in the following reactions:

**(a)** $H^+ + OH^- \rightleftharpoons H_2O$    **(b)** $Cl^- + BCl_3 \rightleftharpoons BCl_4^-$    **(c)** $K^+ + 6H_2O \rightleftharpoons K(H_2O)_6^+$

**Plan:** We examine the formulas to see which species accepts the electron pair (Lewis acid) and which donates it (Lewis base) in forming the adduct.

**Solution: (a)** The $H^+$ accepted an electron pair from the $OH^-$ in forming a bond, so it is the Lewis acid. The $OH^-$ is the Lewis base.

**(b)** The $Cl^-$ ion has four lone pairs and uses one to form a new bond to the central B. Therefore, $BCl_3$ is the acid and $Cl^-$ is the base.

**(c)** The $K^+$ ion has no valence electrons to provide, so the bond is formed when electron pairs from O atoms of water enter empty orbitals on $K^+$. Thus, $K^+$ is the Lewis acid and $H_2O$ is the Lewis base.

**Check:** The Lewis acids ($H^+$, $BCl_3$, and $K^+$) each have an unfilled valence shell that can accept an electron pair from the Lewis bases ($OH^-$, $Cl^-$, and $H_2O$).

FOLLOW-UP PROBLEM 17.13
Identify the Lewis acids and Lewis bases in the following reactions:

**(a)** $OH^- + Al(OH)_3 \rightleftharpoons Al(OH)_4^-$    **(b)** $SO_3 + H_2O \rightleftharpoons H_2SO_4$

**(c)** $Co^{3+} + 6NH_3 \rightleftharpoons Co(NH_3)_6^{3+}$

### An Overview of Acid-Base Definitions

By looking closely at the essential chemical change involved, chemists can see a common theme in reactions as diverse as the way a standardized base is used to analyze an unknown fatty acid, or the way baking soda functions in making bread, or even the way oxygen binds to hemoglobin in a blood cell. From this wider perspective, the diversity of reactions takes on a more unified overview. Now that we have discussed three definitions of acid-base behavior, let's stand back and survey the scope of each and how they fit together.

The *Arrhenius definition*, the first attempt at observing acids and bases on the molecular level, is the most limited and narrow of the three, applying only to substances that contain H or OH and release them as ions in water. Relatively few substances have these prerequisites, so Arrhenius acid-base reactions are relatively few in number.

The *Brønsted-Lowry definition* is more general, seeing acid-base reactions as proton transfer processes. Although the definition of an acid remains essentially the same, a base is defined as any substance with an electron pair available to receive the transferred proton. This definition includes a great many species that Arrhenius did not regard as bases. Furthermore, it defines the acid-base reaction in terms of conjugate acid-base pairs. The reaction system reaches an equilibrium state based on the relative strengths of the acid, the base, and their conjugates.

The *Lewis definition* has the widest scope of the three. The defining event is the donation and acceptance of an electron pair to form a new covalent bond. Any process that accomplishes this is an acid-base reaction. Bases still must have an electron pair to donate, but acids—electron-pair acceptors— can be very different from the substances encompassed by the two earlier definitions, including electron-deficient compounds, polar double bonds, metal ions, and the proton itself.

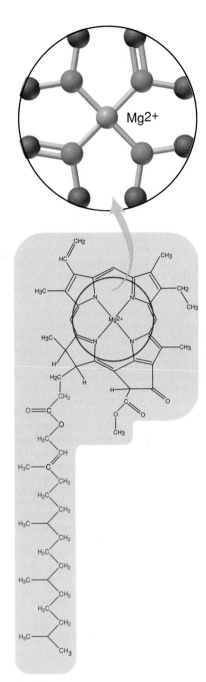

**FIGURE 17.14**
**The $Mg^{2+}$ ion as a Lewis acid in the chlorophyll molecule.** Many biomolecules contain metal ions that act as Lewis acids. In chlorophyll, $Mg^{2+}$ accepts electron pairs from surrounding N atoms that are part of the large organic portion of the molecule.

### Section Summary

The Lewis acid-base definition focuses on the donation or acceptance of an electron pair to form a new covalent bond in an adduct, the product of a neutralization reaction. Lewis bases donate the electron pair, and Lewis acids accept it. Thus, many non-H-containing substances are Lewis acids. Neutral molecules with polar double bonds act as Lewis acids, as do those with electron-deficient atoms. Metal ions act as Lewis acids when they dissolve in water, which acts as a Lewis base, to form a hydrated cation, the Lewis adduct. Many metal ions function as Lewis acids in biomolecules. The Lewis definition has the broadest scope of the three acid-base concepts.

### Chapter Perspective

*In this chapter, we extended the principles of equilibrium to acids and bases. We also investigated one way in which the science of chemistry matured, as narrow definitions of acids and bases progressively widened to encompass different species, physical states, solvent systems, and reaction types. Acids and bases, by whatever definition, are an extremely important group of substances. In Chapter 18, we continue our discussion of these systems and apply many of the ideas developed here to other aqueous equilibria.*

## For Review and Reference

### Key Terms

**SECTION 17.1**
hydronium ion, $H_3O^+$
classical (Arrhenius) acid-base definition
neutralization
acid-dissociation (acid-ionization) constant $(K_a)$

**SECTION 17.2**
autoionization of water
ion-product constant for water $(K_w)$
pH
acid-base indicator

**SECTION 17.3**
Brønsted-Lowry acid-base definition
proton donor
proton acceptor
conjugate acid-base pair

**SECTION 17.4**
polyprotic acid

**SECTION 17.5**
base-dissociation constant $(K_b)$

**SECTION 17.8**
leveling effect

**SECTION 17.9**
Lewis acid-base definition
adduct

### Key Equations and Relationships

17.1 Defining the acid-dissociation constant (p. 741):

$$K_a = \frac{[H_3O^+][A^-]}{[HA]}$$

17.2 Defining the ion-product constant for water (p. 744):

$$K_w = [H_3O^+][OH^-] = 1.0 \times 10^{-14} \text{ (at 25°C)}$$

17.3 Defining pH (p. 745):

$$pH = -\log [H_3O^+]$$

17.4 Relating $pK_w$ to pH and pOH (p. 746):

$$pK_w = pH + pOH = 14.00 \text{ (at 25°C)}$$

17.5 Finding the percent dissociation of HA (p. 757):

$$\text{Percent HA dissociated} = \frac{[HA]_{dissoc}}{[HA]_{init}} \times 100$$

17.6 Defining the base-dissociation constant (p. 760):

$$K_b = \frac{[BH^+][OH^-]}{[B]}$$

17.7 Expressing the relationship among $K_a$, $K_b$, and $K_w$ (p. 764):

$$K_a \times K_b = K_w$$

# Answers to Follow-up Problems

17.1 (a) $HClO_3$; (b) HCl; (c) NaOH

17.2 $[H_3O^+] = \dfrac{1.0 \times 10^{-14}}{6.7 \times 10^{-2}} = 1.5 \times 10^{-13}\ M$; basic

17.3 pH = 9.52; pOH = 14.00 − 9.52 = 4.48

$$[H_3O^+] = 10^{-9.52} = 3.0 \times 10^{-10}\ M$$

$$[OH^-] = \dfrac{1.0 \times 10^{-14}}{3.0 \times 10^{-10}} = 3.3 \times 10^{-5}\ M$$

17.4 (a) $CH_3COOH/CH_3COO^-$ and $H_3O^+/H_2O$
   (b) $H_2O/OH^-$ and $HF/F^-$

17.5 (a) $NH_3(g) + H_2O(l) \overset{\leftarrow}{\rightleftharpoons} NH_4^+(aq) + OH^-(aq)$
   (b) $NH_3(g) + H_3O^+(aq;$ from HCl$) \rightarrow$
   $$NH_4^+(aq) + H_2O(l)$$
   (c) $NH_4^+(aq) + OH^-(aq;$ from NaOH$) \rightarrow$
   $$NH_3(g) + H_2O(l)$$

17.6 $NH_4^+(aq) + H_2O(l) \rightleftharpoons NH_3(aq) + H_3O^+(aq)$
   $[H_3O^+] = 10^{-pH} = 10^{-5.0} = 1 \times 10^{-5}\ M = [NH_3]$

   From reaction table, $K_a = \dfrac{[NH_3][H_3O^+]}{[NH_4^+]} = \dfrac{(x)(x)}{0.2 - x}$

   $$\approx \dfrac{(1 \times 10^{-5})^2}{0.2} = 5 \times 10^{-10}$$

17.7 $K_a = \dfrac{[H_3O^+][OCN^-]}{[HOCN]} = \dfrac{(x)(x)}{0.10 - x} = 3.5 \times 10^{-4}$

   Since $\dfrac{[HOCN]_{init}}{K_a} = \dfrac{0.10}{3.5 \times 10^{-4}} = 286 < 400$, must solve

   quadratic equation: $x^2 + (3.5 \times 10^{-4})\,x - 3.5 \times 10^{-5} = 0$

   $$x = [H_3O^+] = 5.7 \times 10^{-3}\ M; \text{ pH} = 2.24$$

17.8 $K_{a1} = \dfrac{[HC_2O_4^-][H_3O^+]}{[H_2C_2O_4]} = \dfrac{x^2}{0.150 - x} = 5.6 \times 10^{-2}$

   Since $\dfrac{[H_2C_2O_4]_{init}}{K_{a1}} < 400$, must solve quadratic equation:

   $x^2 + (5.6 \times 10^{-2})\,x - (8.4 \times 10^{-3}) = 0$
   $x = [H_3O^+] = 0.068\ M$; pH = 1.17

   $x = [HC_2O_4^-] = 0.068\ M$; $[H_2C_2O_4] = 0.150\ M - x = 0.082\ M$

$[C_2O_4^{2-}] = \dfrac{K_{a2} \times [HC_2O_4^-]}{[H_3O^+]} = \dfrac{(5.4 \times 10^{-5})\,(0.068)}{0.068}$

$$= 5.4 \times 10^{-5}\ M$$

17.9 $K_b = \dfrac{[C_5H_5NH^+][OH^-]}{[C_5H_5N]} = 10^{-8.77} = 1.7 \times 10^{-9}$

Assuming $0.10\ M - x \approx 0.10\ M$, $K_b = \dfrac{(x)(x)}{0.10} = 1.7 \times 10^{-9}$;

$x = [OH^-] = 1.3 \times 10^{-5}\ M$; $[H_3O^+] = 7.7 \times 10^{-10}\ M$;
pH = 9.11

17.10 $K_b$ of $ClO^- = \dfrac{K_w}{K_a \text{ of HClO}} = \dfrac{1.0 \times 10^{-14}}{2.9 \times 10^{-8}}$

$$= 3.4 \times 10^{-7}$$

Assuming $0.20\ M - x \approx 0.20\ M$,

$$K_b = 3.4 \times 10^{-7} = \dfrac{[HClO][OH^-]}{[ClO^-]} = \dfrac{x^2}{0.20}$$

$x = [OH^-] = 2.6 \times 10^{-4}\ M$; $[H_3O^+] = 3.8 \times 10^{-11}\ M$;
pH = 10.42

17.11 (a) Basic:
   $ClO_2^-(aq) + H_2O(l) \rightleftharpoons HClO_2(aq) + OH^-(aq)$
   $K^+$ is from strong base KOH.
   (b) Acidic:
   $CH_3NH_3^+(aq) + H_2O(l) \rightleftharpoons CH_3NH_2(aq) + H_3O^+(aq)$
   $NO_3^-$ is from strong acid $HNO_3$.
   (c) Neutral: $Cs^+$ is from strong base CsOH; $I^-$ is from
   strong acid HI.

17.12 (a) $K_a$ of $Cu^{2+} = 3 \times 10^{-8}$;

   $$K_b \text{ of } CH_3COO^- = \dfrac{K_w}{K_a \text{ of } CH_3COOH} = 5.6 \times 10^{-10}$$

   Since $K_a > K_b$, $Cu(CH_3COO)_2(aq)$ is acidic.

   (b) $K_a$ of $NH_4^+ = \dfrac{K_w}{K_b \text{ of } NH_3} = 5.7 \times 10^{-10}$; also,

$K_b$ of $F^- = 1.5 \times 10^{-11}$; since $K_a > K_b$, $NH_4F(aq)$ is acidic.

17.13 (a) $OH^-$ is the Lewis base; $Al(OH)_3$ is the Lewis acid.
   (b) $H_2O$ is the Lewis base; $SO_3$ is the Lewis acid.
   (c) $NH_3$ is the Lewis base; $Co^{3+}$ is the Lewis acid.

# Sample Problem Titles

17.1 Classifying Acid and Base Strength from the Chemical Formula (p. 743)
17.2 Calculating $[H_3O^+]$ and $[OH^-]$ in Aqueous Solutions (p. 745)
17.3 Calculating $[H_3O^+]$, pH, $[OH^-]$, and pOH from Concentration of Acid (p. 747)
17.4 Identifying Conjugate Acid-Base Pairs (p. 750)
17.5 Predicting the Direction of a Neutralization Reaction (p. 752)
17.6 Finding the $K_a$ of a Weak Acid from the pH of Its Solution (p. 754)
17.7 Determining Equilibrium Concentrations from Known $K_a$ and [HA] (p. 755)
17.8 Calculating Equilibrium Concentrations for a Polyprotic Acid (p. 759)
17.9 Determining pH from $K_b$ (p. 762)
17.10 Determining the pH of a Solution of the Conjugate Base of a Weak Acid (p. 765)
17.11 Predicting the Relative Acidity of Salt Solutions (p. 770)
17.12 Predicting the Relative Acidity of Salt Solutions from $K_a$ and $K_b$ of the Ions (p. 771)
17.13 Identifying Lewis Acids and Bases (p. 775)

# Problems

Problems with a green number are answered at the back of the text. Most sections include three categories of problems separated by a green rule—concept review questions, *paired* skill building exercises, and problems in a relevant context.

## Acids and Bases in Water

(Sample Problem 17.1)

**17.1** Describe the role of water in the Arrhenius (classical) acid-base definition.

**17.2** What characteristics do all Arrhenius acids have in common? What characteristics do all Arrhenius bases have in common? Explain neutralization in terms of the Arrhenius acid-base definition. What quantitative finding led Arrhenius to propose this idea of neutralization?

**17.3** Why is the Arrhenius acid-base definition considered too limited? Give an example where the Arrhenius definition would not apply.

**17.4** Weak acids have $K_a$ values that vary over more than ten orders of magnitude. What do they have in common that classifies them as "weak?" What is meant by the words "strong" and "weak" in terms of acids and bases?

**17.5** Which of the following are Arrhenius acids?
(a) $H_2O$    (b) $Ca(OH)_2$    (c) $H_3PO_3$    (d) HI
(e) $NaHSO_4$    (f) $CH_4$      (g) NaH

**17.6** Which of the following are Arrhenius bases?
(a) $B(OH)_3$    (b) $Ba(OH)_2$    (c) HOCl    (d) KOH
(e) $CH_3COOH$    (f) HOH      (g) $CH_3OH$

**17.7** Write the $K_a$ expression for each of the following:
(a) HCN      (b) $HCO_3^-$      (c) HCOOH

**17.8** Write the $K_a$ expression for each of the following:
(a) $HNO_2$    (b) $CH_3COOH$      (c) $HBrO_2$

**17.9** Use Table 17.2 to rank the following in order of *increasing* acid strength: $HIO_3$, HI, $CH_3COOH$, HF.

**17.10** Use Table 17.2 to rank the following in order of *decreasing* acid strength: HClO, HCl, HCN, $HNO_2$.

**17.11** Classify each of the following as a strong or weak acid or base:
(a) $H_3AsO_4$    (b) $Sr(OH)_2$    (c) HIO    (d) $HClO_4$
(e) $CH_3NH_2$    (f) $K_2O$

**17.12** Classify each of the following as a strong or weak acid or base:
(a) RbOH      (b) HBr      (c) $H_2Te$    (d) $HClO_2$
(e) $HOCH_2CH_2NH_2$      (f) $H_2SeO_4$

## Autoionization of Water and the pH Scale

Unless stated otherwise, all remaining problems in this chapter refer to aqueous solutions at 298 K (25°C).
(Sample Problems 17.2-17.3)

**17.13** Explain what is meant by an autoionization reaction. Give the autoionization reactions for $H_2O$ and for $H_2SO_4$.

**17.14** What is the difference between $K_c$ and $K_w$ for the autoionization reaction of water?

**17.15** (a) What is the change in pH when the [OH$^-$] increases by a factor of ten?
(b) What is the change in [$H_3O^+$] when the pH decreases by 2 units?

**17.16** Which of the following solutions has the higher pH? Explain.
(a) A 0.1 $M$ solution of an acid with $K_a = 1 \times 10^{-4}$ or one with $K_a = 4 \times 10^{-5}$
(b) A 0.1 $M$ solution of an acid with $pK_a = 3.0$ or one with $pK_a = 3.5$
(c) A 0.1 $M$ solution of a weak acid or a 0.01 $M$ solution of the same acid
(d) A 0.1 $M$ solution of a weak acid or a 0.1 $M$ solution of a strong acid
(e) A 0.1 $M$ solution of an acid or a 0.1 $M$ solution of a base
(f) A solution of pOH = 6.0 or one of pOH = 8.0

**17.17** (a) What is the pH of 0.0111 $M$ NaOH? Is the solution neutral, acidic, or basic?
(b) What is the pOH of $1.23 \times 10^{-3}$ $M$ HCl? Is the solution neutral, acidic, or basic?

**17.18** (a) What is the pH of 0.0333 $M$ $HNO_3$? Is the solution neutral, acidic, or basic?
(b) What is the pOH of 0.0347 $M$ KOH? Is the solution neutral, acidic, or basic?

**17.19** (a) What are [$H_3O^+$], [OH$^-$], and pOH in a solution with pH 9.78?
(b) What are [$H_3O^+$], [OH$^-$], and pH in a solution with pOH 10.43?

**17.20** (a) What are [$H_3O^+$], [OH$^-$], and pOH in a solution with pH 3.47?
(b) What are [$H_3O^+$], [OH$^-$], and pH in a solution with pOH 4.33?

**17.21** A solution of HA has a pH of 3.25. How many moles of $H_3O^+$ or OH$^-$ must you add per liter of solution to adjust the pH to 3.65?

**17.22** A solution of HA has a pH of 9.33. How many moles of $H_3O^+$ or OH$^-$ must you add per liter of solution to adjust the pH to 9.07?

**17.23** Parents commonly warn their children of the danger of swimming in a pool during a lightning storm. Yet the text asserts that water is almost nonconducting. Explain this apparent contradiction.

**17.24** Seashells are mostly calcium carbonate, which reacts with $H_3O^+$ according to the equation

$$CaCO_3(s) + H_3O^+(aq) \rightleftharpoons Ca^{2+}(aq) + HCO_3^-(aq) + H_2O(l)$$

If the extent of autoionization of water increases at higher pressure, will seashells dissolve more rapidly near the surface or at great depths? Explain.

**17.25** The extent of autoionization of water also increases at higher temperature.

(a) Does $K_w$ increase or decrease at higher $T$? Explain with a reaction that includes heat as reactant or product.

(b) Bodily processes in humans maintain the pH of blood between 7.35 and 7.45. Given that the $pK_w$ of blood is 13.63 at 37°C (body temperature), what is the range in $[H_3O^+]$ and $[OH^-]$ in blood?

## Proton Transfer and the Brønsted-Lowry Acid-Base Definition

(Sample Problems 17.4-17.5)

**17.26** Pantothenic acid, $C_8H_{16}NO_3COOH$, sometimes called vitamin $B_3$, behaves like a Brønsted-Lowry acid in water. Write the reaction of pantothenic acid with water and the $K_a$ expression. Calculate $K_a/K_c$. What assumption must you make about the concentration of water so that $K_a$ does not vary with the amount of acid dissolved?

**17.27** How do the Arrhenius and Brønsted-Lowry definitions of an acid and a base differ? How are they similar?

**17.28** What is the relationship between the two species of a conjugate acid-base pair?

**17.29** A Brønsted-Lowry acid-base reaction proceeds in the direction by which a stronger acid and stronger base form a weaker acid and weaker base. Explain.

**17.30** Write balanced equations and $K_a$ expressions that describe the behavior of the following Brønsted-Lowry acids in water:
(a) $H_3PO_4$    (b) $C_6H_5COOH$    (c) $HSO_4^-$

**17.31** Write balanced equations and $K_a$ expressions that describe the behavior of the following Brønsted-Lowry acids in water:
(a) $HCOOH$    (b) $HClO_3$    (c) $H_2AsO_4^-$

**17.32** Give the formula of the conjugate base of the following acids:
(a) $HCl$    (b) $H_2CO_3$    (c) $H_2O$    (d) $HPO_4^{2-}$
(e) $NH_4^+$    (f) $HS^-$

**17.33** Give the formula of the conjugate acid of the following bases:
(a) $NH_3$    (b) $NH_2^-$    (c) nicotine, $C_{10}H_{14}N_2$
(d) $O^{2-}$    (e) $SO_4^{2-}$    (f) $H_2O$

**17.34** In each of the following equations, label the acids, the bases, and the conjugate acid-base pairs:
(a) $HCl + H_2O \rightleftharpoons Cl^- + H_3O^+$
(b) $HClO_4 + H_2SO_4 \rightleftharpoons ClO_4^- + H_3SO_4^+$
(c) $HPO_4^{2-} + H_2SO_4 \rightleftharpoons H_2PO_4^- + HSO_4^-$
(d) $NH_3 + HNO_3 \rightleftharpoons NH_4^+ + NO_3^-$
(e) $O^{2-} + H_2O \rightleftharpoons OH^- + OH^-$

**17.35** In each of the following equations, label the acids, the bases, and the conjugate acid-base pairs:
(a) $NH_3 + H_3PO_4 \rightleftharpoons NH_4^+ + H_2PO_4^-$
(b) $CH_3O^- + NH_3 \rightleftharpoons CH_3OH + NH_2^-$
(c) $HPO_4^{2-} + HSO_4^- \rightleftharpoons H_2PO_4^- + SO_4^{2-}$
(d) $NH_4^+ + CN^- \rightleftharpoons NH_3 + HCN$
(e) $H_2O + HS^- \rightleftharpoons OH^- + H_2S$

**17.36** Use the following four aqueous species to write an acid-base reaction with $K_c > 1$ and another with $K_c < 1$: $HS^-$, $Cl^-$, $HCl$, and $H_2S$.

**17.37** Use the following four aqueous species to write an acid-base reaction with $K_c > 1$ and another with $K_c < 1$: $NO_3^-$, $F^-$, $HF$, and $HNO_3$.

**17.38** Use Figure 17.9 to determine which reactions have $K_c > 1$:
(a) $HCl + NH_3 \rightleftharpoons NH_4^+ + Cl^-$
(b) $H_2S + NH_3 \rightleftharpoons HS^- + NH_4^+$
(c) $OH^- + HS^- \rightleftharpoons H_2O + S^{2-}$

**17.39** Use Figure 17.9 to determine which reactions have $K_c < 1$:
(a) $NH_4^+ + HPO_4^{2-} \rightleftharpoons NH_3 + H_2PO_4^-$
(b) $HSO_3^- + HS^- \rightleftharpoons H_2SO_3 + S^{2-}$
(c) $H_2PO_4^- + F^- \rightleftharpoons HPO_4^{2-} + HF$

## Solving Problems Involving Weak-Acid Equilibria

(Sample Problems 17.6-17.8)

**17.40** In each of the following cases, would you expect the concentration of acid before and after dissociation to be nearly the same or very different? Explain your reasoning.
(a) A concentrated solution of a strong acid
(b) A concentrated solution of a weak acid
(c) A dilute solution of a weak acid
(d) A dilute solution of a strong acid

**17.41** A sample of 0.0001 $M$ HCl has close to the same $[H_3O^+]$ as a sample of 0.1 $M$ $CH_3COOH$. Are acetic acid and hydrochloric acid equally strong in these samples? Explain.

**17.42** In which of the following solutions will $[H_3O^+]$ be approximately equal to $[CH_3COO^-]$? Explain.
(a) 0.1 $M$ $CH_3COOH$    (b) $1 \times 10^{-7}$ $M$ $CH_3COOH$
(c) 0.1 $M$ $CH_3COONa$
(d) A solution containing both 0.1 $M$ $CH_3COOH$ and 0.1 $M$ $CH_3COONa$

**17.43** Why do successive acid-dissociation constants decrease in all polyprotic acids?

**17.44** A 0.15 $M$ solution of butanoic acid, $CH_3CH_2CH_2COOH$, contains $1.51 \times 10^{-3}$ $M$ $H_3O^+$. What is $K_a$ for butanoic acid?

**17.45** A 0.035 $M$ solution of a weak acid (HA) has a pH of 4.88. What is the $K_a$ of the acid?

**17.46** Nitrous acid, $HNO_2$, has a $K_a$ of $7.1 \times 10^{-4}$. What are $[H_3O^+]$, $[NO_2^-]$, and $[OH^-]$ in 0.50 $M$ $HNO_2$?

**17.47** Hydrofluoric acid, HF, has a $K_a$ of $6.8 \times 10^{-4}$. What are $[H_3O^+]$, $[F^-]$, and $[OH^-]$ in 0.75 $M$ HF?

**17.48** Chloroacetic acid, $ClCH_2COOH$, has a $pK_a$ of 2.87. What are $[H_3O^+]$, pH, $[ClCH_2COO^-]$, and $[ClCH_2COOH]$ in 1.05 $M$ $ClCH_2COOH$?

**17.49** Hypochlorous acid, HClO, has a $pK_a$ of 7.54. What are $[H_3O^+]$, pH, $[ClO^-]$, and $[HClO]$ in 0.115 $M$ HClO?

**17.50** A 0.25 $M$ solution of a weak acid is 3.0% dissociated.
   (a) Calculate the $[H_3O^+]$, pH, $[OH^-]$, and pOH of the solution.
   (b) Calculate $K_a$ of the acid.

**17.51** A 0.735 $M$ solution of a weak acid is 12.5% dissociated.
   (a) Calculate the $[H_3O^+]$, pH, $[OH^-]$, and pOH of the solution.
   (b) Calculate $K_a$ of the acid.

**17.52** The weak acid HZ has a $K_a$ of $1.55 \times 10^{-4}$.
   (a) Calculate the pH of 0.075 $M$ HZ.
   (b) Calculate the pOH of 0.045 $M$ HZ.

**17.53** (a) Calculate the pH of 0.175 $M$ HY, if $K_a = 1.00 \times 10^{-4}$
   (b) Calculate the pOH of 0.175 $M$ HX, if $K_a = 1.00 \times 10^{-2}$

**17.54** Use Table 17.2 to calculate the percent dissociation of 0.25 $M$ benzoic acid, $C_6H_5COOH$.

**17.55** Use Table 17.2 to calculate the percent dissociation of 0.050 $M$ $CH_3COOH$.

**17.56** Use Table 17.5 to calculate $[H_2S]$, $[HS^-]$, $[S^{2-}]$, $[H_3O^+]$, pH, $[OH^-]$, and pOH in a 0.10 $M$ solution of the diprotic acid hydrosulfuric acid.

**17.57** Use Table 17.5 to calculate $[H_2C_2O_4]$, $[HC_2O_4^-]$, $[C_2O_4^{2-}]$, $[H_3O^+]$, pH, $[OH^-]$, and pOH in a 0.200 $M$ solution of the diprotic acid oxalic acid.

**17.58** Acetylsalicylic acid (aspirin), $C_9H_8O_4$, is the most widely used pain reliever and fever reducer in the world. Determine the pH of a 0.018 $M$ aqueous solution of aspirin. $K_a = 3.2 \times 10^{-4}$.

**17.59** Vinegar is a 5% (w/v) solution of acetic acid in water. For culinary use, it is flavored with apples, herbs, wine, and so forth. What is the pH of vinegar?

**17.60** Formic acid, HCOOH, the simplest carboxylic acid, has many uses in the textile and rubber industries. It is an extremely caustic liquid that is secreted as a defense by many species of ants (family *Formicidae*). Calculate the percent dissociation of 0.50 $M$ HCOOH.

## Weak Bases and Their Relation to Weak Acids

(Sample Problems 17.9-17.10)

**17.61** What is the essential structural feature of all Brønsted-Lowry bases? How does this feature function in an acid-base reaction?

**17.62** Why are almost all anions basic in $H_2O$? Give formulas of four anions that are not basic.

**17.63** Except for the $Na^+$ spectator ion, aqueous solutions of $CH_3COOH$ and $CH_3COONa$ contain the same species.
   (a) What are the species?
   (b) Why is 0.1 $M$ $CH_3COOH$ acidic and 0.1 $M$ $CH_3COONa$ basic?

**17.64** Write balanced equations and $K_b$ expressions that describe the behavior of the following Brønsted-Lowry bases in water: (a) pyridine, $C_5H_5N$; (b) $CO_3^{2-}$.

**17.65** Write balanced equations and $K_b$ expressions that describe the behavior of the following Brønsted-Lowry bases in water: (a) benzoate ion, $C_6H_5COO^-$; (b) $(CH_3)_3N$.

**17.66** What is the pH of 0.050 $M$ dimethylamine, $(CH_3)_2NH$?

**17.67** What is the pH of 0.12 $M$ diethylamine, $(CH_3CH_2)_2NH$?

**17.68** (a) What is $K_b$ of the acetate ion, $CH_3COO^-$?
   (b) What is $K_a$ of the anilinium ion, $C_6H_5NH_3^+$?

**17.69** (a) What is $K_b$ of the benzoate ion, $C_6H_5COO^-$?
   (b) What is $K_a$ of the 2-hydroxyethylammonium ion, $HOCH_2CH_2NH_3^+$? ($pK_b$ of $HOCH_2CH_2NH_2 = 7.97$)

**17.70** (a) What is the pH of 0.050 $M$ KCN?
   (b) What is the pH of 0.30 $M$ triethylammonium chloride, $(CH_3CH_2)_3NHCl$?

**17.71** (a) What is the pH of 0.100 $M$ sodium phenolate, the sodium salt of phenol, $C_6H_5ONa$?
   (b) What is the pH of 0.15 $M$ methylammonium bromide, $CH_3NH_3Br$? ($K_b$ of $CH_3NH_2 = 4.4 \times 10^{-4}$)

**17.72** Sodium hypochlorite solution is sold as "chlorine bleach" and is recognized as a potentially dangerous household solution. The dangers arise from its basicity and from the $ClO^-$, the active bleaching ingredient. What is $[OH^-]$ in an aqueous solution that is 5% NaClO by mass? What is the pH of the solution? ($d$ of solution = 1g/mL.)

**17.73** Sodium phosphate has industrial uses ranging from clarifying crude sugar to manufacturing paper. It is sold commercially as TSP and is used in solution to remove boiler scale and to wash painted brick and concrete. The label usually has a warning that TSP is harmful to the skin and eyes. What is the pH of a solution containing 33 g $Na_3PO_4$ per liter? What is the $[OH^-]$ of this solution?

**17.74** Codeine ($C_{18}H_{21}NO_3$) is a narcotic pain reliever that forms a salt with hydrochloric acid. What is the pH of 0.050 $M$ codeine hydrochloride? ($pK_b$ of codeine is 5.80.)

## Molecular Properties and Acid Strength

**17.75** (a) Explain how the electronegativity of a nonmetal affects the acidity of its binary hydride.
   (b) Explain how the atomic size of a nonmetal affects the acidity of its binary hydride.

**17.76** Is it reasonable that a strong acid has a weak bond to its acidic proton, whereas a weak acid has a strong bond to its acidic proton? Explain.

**17.77** Perchloric acid, $HClO_4$, is the strongest of the halogen oxoacids and hypoiodous acid, HIO, is the weakest. What two factors govern this difference in acid strength?

**17.78** The metallic radius of aluminum (143 pm) is similar to that of lithium (152 pm), and the ionic radius of the aluminum ion (54 pm) is similar to that of the lithium ion (76 pm). Why is an aqueous solution of aluminum nitrate acidic but one of lithium nitrate neutral?

**17.79** Choose the stronger acid in each of the following pairs:
(a) $H_2SeO_3$ or $H_2SeO_4$    (b) $H_3PO_4$ or $H_3AsO_4$
(c) $H_2S$ or $H_2Te$          (d) HBr or $H_2Se$
(e) $HClO_4$ or $H_2SO_4$     (f) $H_2SO_3$ or $H_2SO_4$

**17.80** Choose the stronger acid in each of the following pairs:
(a) $H_2Se$ or $H_3As$        (b) $B(OH)_3$ or $Al(OH)_3$
(c) $HBrO_2$ or HBrO      (d) HI or HBr
(e) $H_3AsO_4$ or $H_2SeO_4$   (f) $HNO_3$ or $HNO_2$

**17.81** Use Table 17.7 to choose the solution with the *lower* pH:
(a) 0.1 $M$ $CuSO_4$ or 0.05 $M$ $Al_2(SO_4)_3$
(b) 0.1 $M$ $ZnCl_2$ or 0.1 $M$ $PbCl_2$
(c) 0.1 $M$ $FeCl_2$ or 0.1 $M$ $AlCl_3$

**17.82** Use Table 17.7 to choose the solution with the *higher* pH:
(a) 0.1 $M$ $Ni(NO_3)_2$ or 0.1 $M$ $Co(NO_3)_2$
(b) 0.1 $M$ $Al(NO_3)_3$ or 0.1 $M$ $Co(NO_3)_2$
(c) 0.1 $M$ $NiCl_2$ or 0.1 $M$ NaCl

## Acid-Base Properties of Salt Solutions

(Sample Problems 17.11-17.12)

**17.83** What determines whether an aqueous solution of a salt will be acidic, basic, or neutral? Give an example of each type of salt.

**17.84** Why is a solution of NaF basic, whereas an NaCl solution is neutral?

**17.85** The $NH_4^+$ ion forms acidic solutions and the $C_2H_3O_2^-$ ion forms basic solutions. A solution of ammonium acetate is almost neutral. Do all ammonium salts of weak acids form neutral solutions? Explain.

**17.86** Explain with equations and, when necessary, calculations whether an aqueous solution of each of the following salts is acidic, basic, or neutral:
(a) KBr    (b) $NH_4I$    (c) KCN    (d) $Cr(NO_3)_3$
(e) NaHS   (f) $Zn(CH_3COO)_2$

**17.87** Explain with equations and, when necessary, calculations whether an aqueous solution of each of the following salts is acidic, basic, or neutral:
(a) $Na_2CO_3$     (b) $CaCl_2$      (c) $Cu(NO_3)_2$
(d) $CH_3NH_3Cl$   (e) $Mg(ClO_4)_2$   (f) $CoF_2$

**17.88** Rank the following salts in order of *increasing* pH of their aqueous solutions:
(a) $KNO_3$, $K_2SO_3$, $K_2S$, $Fe(NO_3)_2$
(b) $NH_4NO_3$, $NaHSO_4$, $NaHCO_3$, $Na_2CO_3$

**17.89** Rank the following salts in order of *decreasing* pH of their aqueous solutions:
(a) $FeCl_2$, $FeCl_3$, $MgCl_2$, $KClO_2$
(b) $NH_4Br$, $NaBrO_2$, NaBr, $NaClO_2$

## Generalizing the Brønsted-Lowry Concept: The Leveling Effect

**17.90** The methoxide ion, $CH_3O^-$, and amide ion, $NH_2^-$, are very strong bases that are "leveled" by water. What does this mean? Write the reactions that occur in the leveling process. What species do the two leveled solutions have in common?

**17.91** Explain the differing extents of dissociation of HI in $CH_3COOH$, $H_2O$, and $NH_3$.

**17.92** In $H_2O$, HF is weak and the other hydrohalic acids are equally strong. In $NH_3$, however, all the hydrohalic acids are equally strong. Explain.

## Electron-Pair Donation and the Lewis Acid-Base Definition

(Sample Problem 17.13)

**17.93** What feature must a molecule or ion possess for it to act as a Lewis base? As a Lewis acid? Explain the roles of these features.

**17.94** How do Lewis acids differ from Brønsted-Lowry acids? How are they similar?

**17.95** (a) What are the characteristics of a weak Lewis base?
(b) Would a weak Brønsted-Lowry base necessarily be a weak Lewis base? Explain with an example.
(c) Identify the Lewis bases in the following reaction:

$$Cu(H_2O)_4^{2+}(aq) + 4CN^-(aq) \rightleftharpoons Cu(CN)_4^{2-}(aq) + 4H_2O(l)$$

(d) Given that $K_c > 1$, which Lewis base is stronger?

**17.96** When iron(III) salts are dissolved in water, the solution becomes yellow and acidic due to the formation of $Fe(H_2O)_5OH^{2+}$ and $H_3O^+$. The overall process involves both Lewis and Brønsted-Lowry acid-base reactions. Write the equations for the process.

**17.97** In which of the three concepts of acid-base behavior discussed in the text can water be the product of an acid-base reaction? In which must it be?

**17.98** (a) Give an example of a substance that is a base in two of the three acid-base definitions, but not in the third.
(b) Give an example of a substance that is an acid in one of the three acid-base definitions, but not in the other two.

**17.99** Which of the following are Lewis acids and which are Lewis bases?
(a) $Be^{2+}$    (b) $Cl^-$    (c) $SnCl_2$    (d) $F_2O$
(e) $Na^+$    (f) $NH_3$

**17.100** Which of the following are Lewis acids and which are Lewis bases?
(a) $BF_3$    (b) $S^{2-}$    (c) $SO_3^{2-}$    (d) $SO_3$
(e) $Mg^{2+}$   (f) $OH^-$

**17.101** Identify the Lewis acid and the Lewis base in each of the following equations:
(a) $Na^+ + 6H_2O \rightleftharpoons Na(H_2O)_6^+$
(b) $CO_2 + H_2O \rightleftharpoons H_2CO_3$
(c) $F^- + BF_3 \rightleftharpoons BF_4^-$

**17.102** Identify the Lewis acid and the Lewis base in each of the following equations:
(a) $Fe^{3+} + 2H_2O \rightleftharpoons FeOH^{2+} + H_3O^+$
(b) $H_2O + H^- \rightleftharpoons OH^- + H_2$
(c) $CN^- + H_2O \rightleftharpoons HCN + OH^-$

**17.103** Classify the following reactions as Arrhenius, Brønsted-Lowry, and Lewis acid-base reactions. A reaction may fit all, two, one, or none of the categories:
(a) $Ag^+ + 2NH_3 \rightleftharpoons Ag(NH_3)_2^+$
(b) $H_2SO_4 + NH_3 \rightleftharpoons HSO_4^- + NH_4^+$
(c) $2HCl \rightleftharpoons H_2 + Cl_2$
(d) $AlCl_3 + Cl^- \rightleftharpoons AlCl_4^-$

**17.104** Classify the following reactions as Arrhenius, Brønsted-Lowry, and Lewis acid-base reactions. A reaction may fit all, two, one, or none of the categories:
(a) $Cu^{2+} + 4Cl^- \rightleftharpoons CuCl_4^{2-}$
(b) $Al(OH)_3 + 3HNO_3 \rightleftharpoons Al^{3+} + 3H_2O + 3NO_3^-$
(c) $N_2 + 3H_2 \rightleftharpoons 2NH_3$
(d) $CN^- + H_2O \rightleftharpoons HCN + OH^-$

---

**17.105** Esters, RCOOR′, are compounds formed by the reaction of carboxylic acids, RCOOH, and alcohols, R′OH, where R and R′ are hydrocarbon groups. Many esters are active ingredients in the odors of fruit and, thus, have important uses in the food and cosmetics industries. The mechanism of ester formation is catalyzed by $H^+$. The first two steps are Lewis acid-base reactions:

Identify the Lewis acids and Lewis bases in these two steps.

## Comprehensive Problems

Problems with an asterisk (*) are more challenging.

**17.106** At 25°C, the $pK_w$ of seawater is 13.22, lower than that of pure water as a result of the dissolved salts. If the pH of a sample of seawater is 7.54, what is pOH of the sample?

**17.107** Chlorobenzene, $C_6H_5Cl$, is a key intermediate in the manufacture of many aromatic compounds, includ-ing aniline dyes and chlorinated pesticides. It is made by the $FeCl_3$-catalyzed chlorination of bezene in the following series of steps:
Step 1: $Cl_2 + FeCl_3 \rightleftharpoons FeCl_5$ (or $Cl^+FeCl_4^-$)
Step 2: $C_6H_6 + Cl^+FeCl_4^- \rightleftharpoons C_6H_6Cl^+ + FeCl_4^-$
Step 3: $C_6H_6Cl^+ \rightleftharpoons C_6H_5Cl + H^+$
Step 4: $H^+ + FeCl_4^- \rightleftharpoons HCl + FeCl_3$
(a) Which of the steps is/are Lewis acid-base reactions?
(b) Identify the Lewis acids and Lewis bases in each of those steps.

**17.108** Hydrogen peroxide, $H_2O_2$ ($pK_a = 11.75$), is commonly used as a bleaching agent and an antiseptic. The product sold in stores is 3% $H_2O_2$ by mass and contains 0.001% phosphoric acid by mass to stabilize the solution. Which contributes more $H_3O^+$ to this commercial solution, the $H_2O_2$ or the $H_3PO_4$?

**17.109** The strengths of acids and bases are directly related to their strengths as electrolytes (Section 12.5).
(a) Why does 0.1 $M$ HCl have a much higher electrical conductivity than 0.1 $M$ $CH_3COOH$?
(b) Why do $1 \times 10^{-7}$ $M$ solutions of HCl and $CH_3COOH$ have nearly the same electrical conductivity?

**17.110** The corrosion of iron-containing objects has a major economic impact (as discussed in Chapter 20). In an investigation of corrosion of stainless steel by hydrochloric acid, some 0.010 $M$ HCl is heated in a closed stainless steel container to 200°C, which increases the pressure to $10^4$ atm. Under these conditions, $pK_w$ is 9.25. What is $[OH^-]$ in the stainless steel container?

*****17.111** Many substances undergo autoionization in a manner analogous to water. For example, the autoionization of liquid ammonia is

$$2NH_3(l) \rightleftharpoons NH_4^+(solvated) + NH_2^-(solvated)$$

(a) Write a $K_{am}$ expression for the autoionization of ammonia that is analogous to the $K_w$ expression for water.
(b) What are the strongest acid and strongest base that can exist in liquid ammonia?
(c) For water, a solution with $[OH^-] > [H_3O^+]$ is basic and one with $[OH^-] < [H_3O^+]$ is acidic. What are the analogous relationships in liquid ammonia?
(d) The acids $HNO_3$ and HCOOH are leveled in liquid $NH_3$. Explain what this means and show equations.
(e) Pure liquid sulfuric acid also undergoes autoionization. The bases hydroxide, $OH^-$, and methoxide, $CH_3O^-$, are leveled in pure liquid $H_2SO_4$. Explain what this means and show equations.

*****17.112** Autoionization (see Problem 17.111) occurs in methanol ($CH_3OH$) and in ethylenediamine ($NH_2CH_2CH_2NH_2$).

(a) The autoionization constant of methanol ($K_{met}$) is $2 \times 10^{-17}$. What is $[CH_3O^-]$ in pure $CH_3OH$?

(b) The concentration of $NH_2CH_2CH_2NH_3^+$ in pure $NH_2CH_2CH_2NH_2$ is $2 \times 10^{-8} M$. What is the autoionization constant of ethylenediamine ($K_{en}$)?

**\*17.113** Some drugs are such weak bases that titration with strong acid to determine concentration is incomplete in water. Industrial pharmacologists titrate many of these drugs using pure acetic acid as the solvent instead of water. How does this procedure aid in the titration?

**17.114** Tris(hydroxymethyl)aminomethane, commonly known as TRIS or THAM, is a water-soluble base with industrial applications in the synthesis of surfactants and pharmaceuticals, as an emulsifying agent in cosmetic creams and lotions, and as a component of various cleaning and polishing mixtures for textiles and leather. In biomedical research, its solutions are used to maintain nearly constant pH for the study of enzymes and other cellular components. Given that its $pK_b$ is 5.91, calculate the pH of 0.060 $M$ TRIS.

**17.115** Explain how you would differentiate between a strong and a weak monoprotic acid in the results of the following procedures:

(a) Equimolar solutions of each acid are titrated with 0.1 $M$ NaOH.

(b) Equal moles of each are dissolved in water.

(c) Zinc metal is added to acid solutions of equal concentration.

**\*17.116** Like any equilibrium constant, $K_w$ changes with temperature. At 50°C and 1 atm, for example, $K_w$ is $5.19 \times 10^{-14}$. Under these conditions, calculate

(a) $[H_3O^+]$ in pure water

(b) $[H_3O^+]$ in 0.010 $M$ NaOH

(c) $[OH^-]$ in 0.0010 $M$ HClO$_4$

(d) Calculate $[H_3O^+]$ in 0.0100 $M$ KOH at 100°C and 1000 atm pressure ($K_w = 1.10 \times 10^{-12}$).

(e) Calculate the pH of pure water at 100°C and 1000 atm.

**17.117** The advent of the modern supermarket has made extended shelf life of products a major concern of scientists in the food and drug industries. Calcium propionate [calcium propanoate, Ca(CH$_3$CH$_2$COO)$_2$] is one of the more common mold inhibitors in food, tobacco, and pharmaceuticals.

(a) Use balanced equations to show whether calcium propionate is an acidic, basic, or neutral salt.

(b) Use Table 17.2 to find the pH of a solution made by dissolving 7.05 g Ca(CH$_3$CH$_2$COO)$_2$ in 0.0500 L water.

**17.118** What is an acidic oxide? Which is the acidic oxide in the Lewis acid-base reaction between CaO and CO$_2$? What is required for these substances to react

in a Brønsted-Lowry acid-base reaction? Are all reactions between acidic and basic oxides Lewis acid-base reactions?

**17.119** Aqueous solutions of beryllium compounds are acidic, but the acidity of other Group 2A(2) salts is insignificant. Explain.

**17.120** How are the $H_3O^+$ and $OH^-$ ions produced in pure water? Why does the concentration of these ions remain constant? Why do the concentrations of these ions decrease when the temperature decreases? What happens to the ions that are "lost" at lower temperature?

**17.121** In pure water, $pK_a$ for benzoic acid is 4.20, but in seawater it is 3.96. In which solvent is benzoic acid a stronger acid?

**17.122** Amine bases are notorious for having foul odors. Putrescine ($NH_2CH_2CH_2CH_2CH_2NH_2$), once thought to be found only in rotting animal tissue, is now known to be a component of all cells and essential for their normal and abnormal (cancerous) growth. It also plays a key role in formation of GABA, one of the brain neurotransmitters. A 0.10 $M$ aqueous solution of putrescine has $[OH^-] = 2.1 \times 10^{-3}$. What is its $K_b$?

**\*17.123** Quinine ($C_{20}H_{24}N_2O_2$; see structure) is a natural product that was originally extracted from the bark of cinchona trees but is now synthesized by the pharmaceutical industry. Its antimalarial properties are responsible for saving thousands of lives during construction of the Panama Canal and, thus, it stands as a classic example of the untold chemical wealth of the forests of Central and South America. Both N atoms are basic, but the tertiary amine N (colored) is far more basic ($pK_b = 5.1$) than the N within the aromatic ring system ($pK_b = 9.7$).

(a) Quinine is not very soluble in water: a saturated solution is only $1.6 \times 10^{-3}$ $M$. What is the pH of this solution?

(b) Show that the aromatic N contributes negligibly to the pH of the solution.

(c) Because of its low solubility as a free base, quinine is often administered as an amine salt. Quinine hydrochloride, for example, is about 120 times more soluble in water than quinine. What is the pH of 0.53 $M$ quinine hydrochloride?

(d) What is the pH of an antimalarial solution ($d = 1.0$ g/mL) that is 1.5% quinine hydrochloride by mass, a typical medicinal concentration?

# CHAPTER 18

## Concepts and skills to review

- solubility rules for ionic compounds (Section 4.3)
- Le Châtelier's principle and the effect of concentration (Section 16.4)
- conjugate acid-base systems (Section 17.3)
- calculations of weak-acid and weak-base equilibria (Sections 17.4 and 17.5)

# Ionic Equilibria in Aqueous Systems

**The natural workings of aqueous ionic equilibria.** Remarkable formations in nature, such as those in this coral reef, arise through the interplay of weak acids and bases, slightly soluble salts, and complex ions, the three ionic equilibrium systems that we examine in this chapter.

An astronaut sees our planet from outer space as a fabulous watery world; a biologist peering through a microscope might see a cell in the same way. A chemist knows that aqueous systems predominate in both the external and internal environments and that wherever dissolved substances occur, the principles of equilibrium apply.

Consider just a few cases of aqueous equilibria. The magnificent formations in limestone caves and the vast expanses of oceanic coral reefs result from subtle shifts in carbonate solubility equilibria. Carbonates also influence soil pH and the acidification of lakes due to acid rain. Equilibria involving carbon dioxide and phosphates help organisms maintain cellular pH within narrow limits. Equilibria involving the clays found in soil control the availability of ionic nutrients for plants. In more practical areas, the principles of ionic equilibrium govern how water is softened, how substances are purified through the precipitation of unwanted ions, and even how the weak acids in wine and vinegar influence the delicate taste of a fine French sauce.

In this chapter, we explore three key aqueous ionic equilibrium systems: acid-base buffers, slightly soluble salts, and complex ions. Our discussion of buffers—solutions that minimize changes in pH—includes why they are important, how they work, and how to prepare them. We discuss the ionic species that arise during an acid-base titration and how buffers are involved in that process. Then we examine slightly soluble salts in the laboratory and in nature to see how conditions influence their solubility. Next, we investigate complex ions and how they form and change from one type to another. Finally, we see how these aqueous equilibria are employed in chemical analysis. There are not many new ideas in this chapter, but there are many opportunities to apply old ones in new ways.

## 18.1 Equilibria of Acid-Base Buffer Systems

Why do some lakes become acidic when showered by acid rain, while others remain unaffected? How does blood maintain a constant pH in contact with countless cellular acid-base reactions? How can a chemist sustain a nearly constant $[H_3O^+]$ in reactions that produce $H_3O^+$ or $OH^-$? The answer in each case is through the presence of a buffer.

In everyday language, a buffer is something that lessens the impact of an external force. An **acid-base buffer** is a solution that lessens changes in $[H_3O^+]$. When small amounts of $H_3O^+$ or $OH^-$ ions are added to an unbuffered solution, *the change in pH is much larger* than the change that results from the same additions to a buffered solution (Figure 18.1).

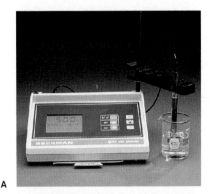

A

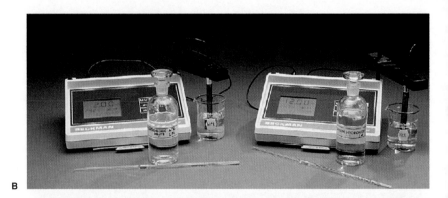

B

C

D

**FIGURE 18.1**

The effect of adding acid or base to un-buffered or buffered solutions. **A,** A 100-mL sample of an unbuffered solution of dilute HCl is adjusted to pH 5.00. **B,** After the addition of 1 mL of 1 *M* HCl *(left)* or of 1 *M* NaOH *(right)*, the pH change is large. **C,** A 100-mL sample of a buffered solution, made by mixing 1 *M* CH₃COOH with 1 *M* CH₃COONa, is adjusted to pH 5.00. **D,** After the addition of 1 mL of 1 *M* HCl *(left)* or of 1 *M* NaOH *(right)*, the pH change is negligible.

To withstand the addition of strong acid or strong base without changing its pH, a buffer solution must contain an acidic component that can react with added $OH^-$ ion *and* a basic component that can react with added $H_3O^+$ ion. They cannot be just any acid and base, however, because they must be present in the same solution without neutralizing each other: most commonly *the components of a buffer are a conjugate acid-base pair*. In Figure 18.1, *C* and *D*, for example, the buffer solution is a mixture of acetic acid ($CH_3COOH$) and acetate ion ($CH_3COO^-$). In general, *a buffer is a solution of a conjugate acid-base pair in which these components occur in similar concentrations*.

### The Common-Ion Effect and How a Buffer Works

To understand how buffers work, let's consider a phenomenon called the **common-ion effect.** An example of this phenomenon occurs in the dissociation of acetic acid in sodium acetate solution. Acetic acid dissociates slightly in water:

$$CH_3COOH(aq) + H_2O(l) \rightleftharpoons CH_3COO^-(aq) + H_3O^+(aq)$$

From Le Châtelier's principle, we know that if acetate ion is added, the extent of dissociation, and thus the $[H_3O^+]$, goes down as the equilibrium position shifts to the left:

$$CH_3COOH(aq) + H_2O(l) \underset{\longleftarrow}{\rightleftharpoons} CH_3COO^-(aq; \text{ added}) + H_3O^+(aq)$$

Similarly, if we dissolve acetic acid in a sodium acetate solution, the acetate ion already present combines with some of the released $H_3O^+$, thereby keeping the $[H_3O^+]$ lower. Thus, a smaller percentage of acetic acid dissoci-

**TABLE 18.1  The Effect of Added Acetate Ion on the Dissociation of Acetic Acid**

| [CH$_3$COOH] | [CH$_3$COO$^-$] | % DISSOCIATION* | pH |
|:---:|:---:|:---:|:---:|
| 0.10 | 0.00 | 1.3 | 2.89 |
| 0.10 | 0.050 | 0.036 | 4.44 |
| 0.10 | 0.10 | 0.018 | 4.74 |
| 0.10 | 0.15 | 0.012 | 4.92 |

$$^* \text{ \% Dissociation} = \frac{[CH_3COOH]_{dissoc}}{[CH_3COOH]_{init}} \times 100$$

ates in the presence of acetate, the *common ion.* In general, when a reactant is added to a solution that already contains one of the ions involved in the reaction (the common ion), the position of equilibrium *shifts away* from forming the common ion. Table 18.1 shows the percent dissociation and the pH of an acetic acid solution containing varying concentrations of acetate ion (supplied by adding solid sodium acetate). Note that the *common ion,* CH$_3$COO$^-$, *represses the CH$_3$COOH dissociation,* which makes the solution less acidic.

By mixing significant amounts of conjugate base (CH$_3$COO$^-$) and weak acid (CH$_3$COOH), a buffer solution is produced. How does this solution resist pH changes when H$_3$O$^+$ or OH$^-$ is added? The essential feature of a buffer is that it consists of *large reservoirs* of the acidic (HA) and basic (A$^-$) components. When H$_3$O$^+$ or OH$^-$ ions are added to the buffer, they cause *a given amount of one buffer component to convert into the other,* which changes the *relative* concentrations of the buffer-component reservoirs. As long as the amount of added H$_3$O$^+$ or OH$^-$ is much smaller than the amounts of HA and A$^-$ in the reservoirs, *the added ions have little effect on the pH because they are consumed by one or the other buffer component.*

Consider what happens to a solution containing high [CH$_3$COOH] and [CH$_3$COO$^-$] when we add small amounts of strong acid or base. The expression for the dissociation reaction in the original solution at equilibrium is given by the terms in $K_a$:

$$K_a = \frac{[CH_3COO^-][H_3O^+]}{[CH_3COOH]}$$

Solving for [H$_3$O$^+$], we obtain

$$[H_3O^+] = K_a \times \frac{[CH_3COOH]}{[CH_3COO^-]}$$

Note that since $K_a$ is constant, the [H$_3$O$^+$] of the solution depends directly on the *buffer-component ratio,* [CH$_3$COOH]/[CH$_3$COO$^-$]:
- If the ratio goes up, [H$_3$O$^+$] goes up.
- If the ratio goes down, [H$_3$O$^+$] goes down.

When we add a small amount of strong acid, the increased amount of H$_3$O$^+$ reacts with a nearly *stoichiometric amount* of acetate ion to form more acetic acid:

$$H_3O^+(aq; \text{added}) + CH_3COO^-(aq) \longrightarrow CH_3COOH(aq) + H_2O(l)$$

**FIGURE 18.2**

**How a buffer works.** A buffer consists of large reservoirs of a conjugate acid-base pair, in this case acetic acid ($CH_3COOH$) and acetate ion ($CH_3COO^-$). When $H_3O^+$ is added (*left*), a small portion of the $CH_3COO^-$ reservoir combines with it and enlarges the $CH_3COOH$ reservoir slightly. Similarly, when $OH^-$ is added (*right*), a small portion of the $CH_3COOH$ reservoir combines with it and enlarges the $CH_3COO^-$ reservoir slightly. In both cases, the relative changes are small, so the [acid]/[base] ratio, and therefore the pH, changes very little.

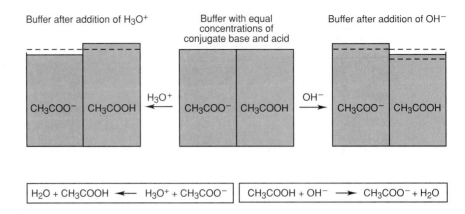

As a result, the $[CH_3COO^-]$ goes down by that amount and the $[CH_3COOH]$ goes up by that amount, which increases the buffer-component ratio, and therefore the $[H_3O^+]$, but only slightly (Figure 18.2).

Adding a small amount of strong base produces the opposite result. It supplies $OH^-$ ions, which react with a nearly *stoichiometric amount* of $CH_3COOH$, forming that much more $CH_3COO^-$:

$$CH_3COOH(aq) + OH^-(aq; \text{added}) \rightarrow CH_3COO^-(aq) + H_2O(l)$$

The buffer-component ratio decreases, as does the $[H_3O^+]$, but once again, the change is slight.

The components consume virtually all the added $H_3O^+$ or $OH^-$. As long as the amount of added $H_3O^+$ or $OH^-$ is small compared with the size of the buffer-component reservoirs, the interchange from one reservoir to the other produces a *small change in the buffer-component ratio and, consequently, a small change in $[H_3O^+]$ and in pH*. The multipart Sample Problem 18.1 demonstrates such small pH changes; note that the latter two parts combine a stoichiometry portion, like the problems in Chapter 4, and a weak-acid dissociation portion, like those in Chapter 17.

SAMPLE PROBLEM 18.1 _____

**Calculating the Effect of Added $H_3O^+$ and $OH^-$ on Buffer pH**

**Problem:** Calculate the pH of a buffer solution
**(a)** Consisting of 0.50 *M* $CH_3COOH$ and 0.50 *M* $CH_3COO^-$
**(b)** After adding 0.020 mol solid NaOH to 1.0 L of the buffer solution
**(c)** After adding 0.020 mol HCl to 1.0 L of the buffer solution.
$K_a$ of $CH_3COOH = 1.8 \times 10^{-5}$. (Assume the additions cause negligible volume changes.)
**Plan:** In each case, we know, or can find, $[CH_3COOH]_{init}$ and $[CH_3COO^-]_{init}$ and we need to find $[H_3O^+]$ at equilibrium and convert it to pH. In **(a)**, we use the given concentrations of buffer components as the initial values. As in earlier problems, we assume that $x$ ($[CH_3COOH]_{dissoc} = [H_3O^+]$) can be neglected, set up a reaction table, solve for $x$, and check the assumption. In **(b)** and **(c)**, we assume that the added $OH^-$ or $H_3O^+$ reacts completely with the buffer components to yield new $[CH_3COOH]_{init}$ and $[CH_3COO^-]_{init}$, which then dissociate to an unknown extent. We set up two reaction tables. The first summarizes the stoichiometry of adding strong base or acid. The second summarizes the dissociation of the new concentrations, so we proceed as in (a) to find the new $[H_3O^+]$.

**Solution: (a)** $[H_3O^+]$ in the original buffer.
Setting up a reaction table with $x = [CH_3COOH]_{dissoc} = [H_3O^+]$ (as in Chapter 17, we assume that the $[H_3O^+]_{from\ H_2O}$ is negligible):

| Concentration (M) | $CH_3COOH(aq)$ | $+$ | $H_2O(l)$ | $\rightleftharpoons$ | $CH_3COO^-(aq)$ | $+$ | $H_3O^+(aq)$ |
|---|---|---|---|---|---|---|---|
| Initial | 0.50 | | — | | 0.50 | | 0 |
| Change | $-x$ | | — | | $+x$ | | $+x$ |
| Equilibrium | $0.50 - x$ | | — | | $0.50 + x$ | | $x$ |

Finding the equilibrium $[CH_3COOH]$ and $[CH_3COO^-]$: Since $K_a$ is small, $x$ is small, so we assume $[CH_3COOH] = 0.50\ M - x \approx 0.50\ M$, and $[CH_3COO^-] = 0.50\ M + x \approx 0.50\ M$.
Solving for $x$ ($[H_3O^+]$ at equilibrium):

$$[H_3O^+] = K_a \times \frac{[CH_3COOH]}{[CH_3COO^-]} = (1.8 \times 10^{-5}) \times \frac{0.50}{0.50} = 1.8 \times 10^{-5}\ M$$

Checking the assumption:

$$\frac{1.8 \times 10^{-5}\ M}{0.50\ M} \times 100 = 3.6 \times 10^{-3}\% < 5\%$$

The assumption is justified, and we will use it in parts (b) and (c).
Calculating pH:

$$pH = - \log [H_3O^+] = - \log (1.8 \times 10^{-5}) = \mathbf{4.74}$$

**(b)** $[H_3O^+]$ after adding 0.020 mol NaOH to 1.0 L buffer.
Finding the $[OH^-]_{added}$:

$$[OH^-]_{added} = 0.020\ mol\ OH^-/1.0\ L\ soln = 0.020\ M\ OH^-$$

Setting up a reaction table for the stoichiometry of adding $OH^-$ to $CH_3COOH$:

| Concentration (M) | $CH_3COOH(aq)$ | $+$ | $OH^-(aq)$ | $\rightarrow$ | $CH_3COO^-(aq)$ | $+$ | $H_2O(aq)$ |
|---|---|---|---|---|---|---|---|
| Before addition | 0.50 | | — | | 0.50 | | — |
| Addition | — | | 0.020 | | — | | — |
| After addition | 0.48 | | 0 | | 0.52 | | — |

Setting up a reaction table for the acid dissociation, using these new initial concentrations. As in part (a), $x = [CH_3COOH]_{dissoc} = [H_3O^+]$:

| Concentration (M) | $CH_3COOH(aq)$ | $+$ | $H_2O(l)$ | $\rightleftharpoons$ | $CH_3COO^-(aq)$ | $+$ | $H_3O^+(aq)$ |
|---|---|---|---|---|---|---|---|
| Initial | 0.48 | | — | | 0.52 | | 0 |
| Change | $-x$ | | — | | $+x$ | | $+x$ |
| Equilibrium | $0.48 - x$ | | — | | $0.52 + x$ | | $x$ |

Making the assumption that $x$ is small, and solving for $x$:

$$[CH_3COOH] = 0.48\ M - x \approx 0.48\ M \quad \text{and} \quad [CH_3COO^-] = 0.52\ M + x \approx 0.52\ M$$

$$[H_3O^+] = K_a \times \frac{[CH_3COOH]}{[CH_3COO^-]} = (1.8 \times 10^{-5}) \times \frac{0.48}{0.52} = 1.7 \times 10^{-5}\ M$$

Calculating the pH:

$$pH = - \log [H_3O^+] = - \log (1.7 \times 10^{-5}) = \mathbf{4.77}$$

The addition of strong base increased the concentration of the basic buffer component at the expense of the acidic buffer component and *raised* the pH slightly, from 4.74 to 4.77.

**(c)** $[H_3O^+]$ after adding 0.020 mol HCl to 1.0 L of buffer.
Finding the $[H_3O^+]_{added}$:

$$[H_3O^+]_{added} = 0.020\ mol\ H_3O^+/1.0\ L\ soln = 0.020\ M\ H_3O^+$$

Proceeding as in part (b), we set up one reaction table for the stoichiometry of adding $H_3O^+$ to $CH_3COO^-$ and one for the acid dissociation, with $x = [CH_3COOH]_{dissoc} = [H_3O^+]$:

| Concentration (M) | $CH_3COO^-(aq)$ | + | $H_3O^+(aq)$ | → | $CH_3COOH(aq)$ | + | $H_2O(l)$ |
|---|---|---|---|---|---|---|---|
| Before addition | 0.50 | | — | | 0.50 | | — |
| Addition | — | | 0.020 | | — | | — |
| After addition | 0.48 | | 0 | | 0.52 | | — |

| Concentration (M) | $CH_3COOH(aq)$ | + | $H_2O(l)$ | ⇌ | $CH_3COO^-(aq)$ | + | $H_3O^+(aq)$ |
|---|---|---|---|---|---|---|---|
| Initial | 0.52 | | — | | 0.48 | | 0 |
| Change | $-x$ | | — | | $+x$ | | $+x$ |
| Equilibrium | $0.52 - x$ | | — | | $0.48 + x$ | | $x$ |

Making the assumption that $x$ is small and solving for $x$:

$$[CH_3COOH] = 0.52\ M - x \approx 0.52\ M \quad \text{and} \quad [CH_3COO^-] = 0.48\ M + x \approx 0.48\ M$$

$$[H_3O^+] = K_a \times \frac{[CH_3COOH]}{[CH_3COO^-]} = (1.8 \times 10^{-5}) \times \frac{0.52}{0.48} = 2.0 \times 10^{-5}\ M$$

Calculating the pH:

$$pH = -\log [H_3O^+] = -\log (2.0 \times 10^{-5}) = \textbf{4.70}$$

The addition of strong acid increased the concentration of the acidic buffer component at the expense of the basic component and *lowered* the pH slightly, from 4.74 to 4.70.

**Check:** The changes in $[CH_3COOH]$ and $[CH_3COO^-]$ occur in opposite directions in parts **(b)** and **(c)**, which makes sense. The additions were of equal amounts, so the pH increase in **(b)** should equal the pH decrease in **(c)**, within rounding.

**Comment:** In part **(a)**, we justified our assumption that $x$ can be neglected. Therefore, in parts **(b)** and **(c)**, we could have used the new values from the last line of the stoichiometry reaction table to obtain the ratio of buffer components and dispensed with the reaction table for the dissociation. In later problems, we follow this simplified approach.

**FOLLOW-UP PROBLEM 18.1**
Calculate the pH of a buffer consisting of 0.50 M HF and 0.45 M F⁻ **(a)** before and **(b)** after addition of 0.40 g NaOH to 1.0 L of the buffer. $K_a$ of HF = $6.8 \times 10^{-4}$.

## The Henderson-Hasselbalch Equation

For any weak acid, HA, the equation and $K_a$ expression for the acid dissociation at equilibrium is

$$HA + H_2O \rightleftharpoons H_3O^+ + A^- \qquad K_a = \frac{[H_3O^+][A^-]}{[HA]}$$

A simple mathematical rearrangement turns this expression into a more useful form for buffer calculations. As was pointed out earlier, the key variable determining $[H_3O^+]$ is the *ratio* of acid species to base species:

$$[H_3O^+] = K_a \times \frac{[HA]}{[A^-]}$$

Taking the negative logarithm (base 10) of both sides gives

$$-\log [H_3O^+] = -\log K_a - \log \left( \frac{[HA]}{[A^-]} \right)$$

from which we obtain

$$pH = pK_a + \log\left(\frac{[A^-]}{[HA]}\right)$$

(Note the inversion of the ratio when the sign of the logarithm is changed.) For any conjugate acid-base pair, regardless of formula, we have

$$pH = pK_a + \log\left(\frac{[base]}{[acid]}\right) \qquad \textbf{(18.1)}$$

This relationship is the **Henderson-Hasselbalch equation,** which allows us to solve directly for pH instead of calculating $[H_3O^+]$ first. For instance, in part (b) of Sample Problem 18.1, we could have found the pH of the buffer after addition of NaOH from

$$pH = pK_a + \log\left(\frac{[CH_3COO^-]}{[CH_3COOH]}\right) = 4.74 + \log\left(\frac{0.52}{0.48}\right) = 4.77$$

## Buffer Capacity and Buffer Range

As you've seen, buffers resist a pH change as long as the buffer-component reservoirs are *large* compared with the amount of strong acid or base added. **Buffer capacity** is a measure of the ability to resist pH change and depends on the concentration of the reservoirs: *the more concentrated the components of a buffer, the greater the buffer capacity.* You must add more $H_3O^+$ or $OH^-$ to a high-capacity (concentrated) buffer than to a low-capacity (dilute) buffer to obtain a given pH change. Conversely, adding the same amount of $H_3O^+$ or $OH^-$ to buffers of different capacities produces a smaller pH change in the higher capacity buffer, as shown in Figure 18.3. This makes sense because the more concentrated buffer exhibits a smaller change in the concentration of its two components on addition of a given amount of $H_3O^+$ or $OH^-$. Note that *the pH of a buffer is distinct from its buffer capacity.* A buffer made of equal volumes of 1.0 M $CH_3COOH$ and 1.0 M $CH_3COO^-$ has the same pH (4.74) as a buffer made of equal volumes of 0.10 M $CH_3COOH$ and 0.10 M $CH_3COO^-$, but it has a much larger capacity for resisting a pH change.

Buffer capacity is also affected by the *relative* concentrations of the buffer components. As a buffer functions, the concentration of one component increases relative to the other. Since the ratio of these concentrations determines the pH, the less the ratio changes, the less the pH changes. For a given addition of acid or base, the ratio changes less when buffer-component concentrations are similar than when they are different. Suppose we have a buffer in which $[HA] = [A^-] = 1.000$ M. When we add 0.010 mol $OH^-$ to 1.00 L of buffer, $[A^-]$ becomes 1.010 M and $[HA]$ becomes 0.990 M:

$$\frac{[A^-]_{init}}{[HA]_{init}} = \frac{1.000\ M}{1.000\ M} = 1.000 \qquad \frac{[A^-]_{final}}{[HA]_{final}} = \frac{1.010\ M}{0.990\ M} = 1.02$$

$$\text{Percent change} = \frac{1.02 - 1.000}{1.000} \times 100 = 2\%$$

Now suppose the buffer concentrations are $[HA] = 0.250$ M and $[A^-] = 1.750$ M. The same addition of 0.010 mol $OH^-$ to 1.00 L of buffer gives $[HA] = 0.240$ M and $[A^-] = 1.760$ M, so the ratios are

$$\frac{[A^-]_{init}}{[HA]_{init}} = \frac{1.750\ M}{0.250\ M} = 7.00 \qquad \frac{[A^-]_{final}}{[HA]_{final}} = \frac{1.760\ M}{0.240\ M} = 7.33$$

$$\text{Percent change} = \frac{7.33 - 7.00}{7.00} \times 100 = 4.7\%$$

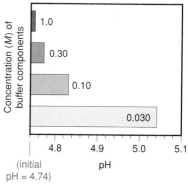

**FIGURE 18.3**

**The relation between buffer capacity and pH change.** The four bars represent acetic acid/acetate buffers with the same initial pH (4.74) but different concentrations *(labeled on each bar)*. When a given amount of strong base is added to each buffer, the pH increases *(bar length)*. Note that the greater the buffer capacity, the smaller the pH change.

Note that the change in buffer-component ratio is much larger when the component concentrations are very different from each other.

It follows that a buffer has the highest capacity when its components are present in equal concentration, that is, when $[A^-]/[HA] = 1$, which gives

$$pH = pK_a + \log\left(\frac{[A^-]}{[HA]}\right) = pK_a + \log 1 = pK_a + 0 = pK_a$$

Thus, for a given concentration, *a buffer whose pH is equal to or near the pK_a of its acid component has the highest buffer capacity.*

The **buffer range** is the pH range over which the buffer acts effectively. The further the buffer-component ratio is from 1, the less effective the buffering action (that is, the lower the buffer capacity). In practice, we find that if the $[A^-]/[HA]$ ratio is greater than 10 or less than 0.1—that is, if one component is more than 10 times as concentrated as the other—buffering action is poor. Since $\log 10 = +1$ and $\log 0.1 = -1$, we find that *buffers have a usable range within $\pm 1$ pH unit of the pK_a:*

$$pH = pK_a + \log\left(\frac{10}{1}\right) = pK_a + 1 \quad \text{and} \quad pH = pK_a + \log\left(\frac{1}{10}\right) = pK_a - 1$$

### Preparing a Buffer

Any large chemical supply-house catalog lists many common buffers available in a variety of pH values and concentrations. Does this mean that there is no need to learn how to prepare a buffer? Definitely not! When a buffer of unusual composition is required (which is often the case in medical or biological applications) or when an ordinary buffer is simply not available, even the most sophisticated and automated laboratory relies on personnel versed in wet chemical techniques and a knowledge of basic chemistry to prepare a buffer.

The first step in preparing a buffer is to *choose the conjugate acid-base pair,* which is determined to a large extent by the desired pH. Remember that a buffer is most effective when the pH $\approx$ p$K_a$ of the acid because this occurs when the component ratio is close to one. Suppose you need a buffer whose pH is 3.90: the p$K_a$ of the acid component should be as close to 3.90 as possible; or $K_a = 10^{-3.90} = 1.3 \times 10^{-4}$. Scanning a table of acid-dissociation constants (see Table 17.2) shows that formic acid ($K_a = 1.8 \times 10^{-4}$; p$K_a = 3.74$) is a good choice. Therefore, the buffer components would be formic acid, HCOOH, and formate ion, $HCOO^-$, supplied by a soluble salt, such as sodium formate, HCOONa.

Next, you have to *calculate the ratio of buffer components* that gives the desired pH. With the Henderson-Hasselbalch equation, we have

$$pH = pK_a + \log\left(\frac{[A^-]}{[HA]}\right) \quad \text{or} \quad 3.90 = 3.74 + \log\left(\frac{[HCOO^-]}{[HCOOH]}\right)$$

$$\log\left(\frac{[HCOO^-]}{[HCOOH]}\right) = 0.16 \quad \text{so} \quad \frac{[HCOO^-]}{[HCOOH]} = 10^{0.16} = 1.4$$

Thus, for every 1.0 mol HCOOH in a given volume of solution, you need 1.4 mol HCOONa.

The next step is to *determine the buffer concentration.* For most laboratory-scale applications, concentrations of about 0.50 *M* are suitable, but often the decision is based on availability of stock solutions. Suppose you have a large supply of 0.40 *M* HCOOH and you need approximately 1.0 L of final buffer.

A mole-mass calculation gives the amount of sodium formate needed:

$$\text{Moles of HCOOH} = 1.0 \text{ L soln} \times \frac{0.40 \text{ mol HCOOH}}{1.0 \text{ L soln}} = 0.40 \text{ mol HCOOH}$$

$$\text{Moles of HCOONa} = 0.40 \text{ mol HCOOH} \times \frac{1.4 \text{ mol HCOONa}}{1.0 \text{ mol HCOOH}} = 0.56 \text{ mol HCOONa}$$

$$\text{Mass (g) of HCOONa} = 0.56 \text{ mol HCOONa} \times \frac{68.01 \text{ g HCOONa}}{1 \text{ mol HCOONa}} = 38 \text{ g HCOONa}$$

The buffer is prepared by thoroughly dissolving 38 g solid sodium formate in 1.0 L of 0.40 $M$ HCOOH.

Due to the behavior of nonideal solutions (Section 12.5) and other ionic phenomena, a buffer prepared this way may vary from the desired pH by as much as several tenths of a pH unit. Therefore, the final step is to *adjust the buffer pH* to the desired value by adding strong acid or strong base, while

monitoring the solution with a pH meter. The following sample problem presents the math steps (without use of the Henderson-Hasselbalch equation).

SAMPLE PROBLEM 18.2 _____

## Calculating How to Prepare a Buffer

**Problem:** An environmental chemist needs a carbonate buffer of pH 10.00 to study the effects of soil acidification. What mass of sodium carbonate must be added to 1.5 L of 0.20 $M$ NaHCO$_3$ to make the buffer? $K_a$ of HCO$_3^-$ = $4.7 \times 10^{-11}$.
**Plan:** The conjugate pair is already chosen, HCO$_3^-$ (acid) and CO$_3^{2-}$ (base), as is the amount of the acid component, so we must find the component ratio that gives pH 10.00 and the mass of base component. We convert pH to [H$_3$O$^+$] and use the $K_a$ expression to solve for the [CO$_3^{2-}$] required. Multiplying by the volume of solution gives the moles of CO$_3^{2-}$ required, and then we use the molar mass to find the mass.
**Solution:** Calculating [H$_3$O$^+$]: [H$_3$O$^+$] = $10^{-\text{pH}}$ = $10^{-10.00}$ = $1.0 \times 10^{-10}$ $M$
Solving for the carbonate concentration in the component ratio:

$$K_a = \frac{[\text{H}_3\text{O}^+]\,[\text{CO}_3^{2-}]}{[\text{HCO}_3^-]}$$

So $\qquad [\text{CO}_3^{2-}] = \dfrac{K_a \times [\text{HCO}_3^-]}{[\text{H}_3\text{O}^+]} = \dfrac{(4.7 \times 10^{-11})\,(0.20)}{1.0 \times 10^{-10}} = 0.094$ $M$

Calculating moles of CO$_3^{2-}$ needed for the buffer volume:

$$\text{Moles of CO}_3^{2-} = 1.5 \text{ L soln} \times \frac{0.094 \text{ mol CO}_3^{2-}}{1 \text{ L soln}} = 0.14 \text{ mol CO}_3^{2-}$$

Calculating the mass of Na$_2$CO$_3$ needed:

$$\text{Mass (g) of Na}_2\text{CO}_3 = 0.14 \text{ mol Na}_2\text{CO}_3 \times \frac{105.99 \text{ g Na}_2\text{CO}_3}{1 \text{ mol Na}_2\text{CO}_3} = \textbf{15 g Na}_2\textbf{CO}_3$$

We dissolve 15 g Na$_2$CO$_3$ into 1.5 L of 0.20 $M$ NaHCO$_3$. Using a pH meter, we adjust the pH to 10.00 with strong acid or base.
**Check:** For a useful buffer range, the acidic component [HCO$_3^-$] must be within a factor of 10 of the basic component [CO$_3^{2-}$]. We have 1.5 L × 0.20 $M$ HCO$_3^-$, or

0.30 mol HCO$_3^-$, and 0.14 mol CO$_3^{2-}$; $\dfrac{0.30}{0.14}$ = 2.1, so this seems fine. Make sure

the relative amounts of components are reasonable: when pH = p$K_a$, [HCO$_3^-$] = [CO$_3^{2-}$]. The p$K_a$ of HCO$_3^-$ is 10.33. Since we want a pH lower than the p$K_a$ of HCO$_3^-$, we need more of the acidic species.

**FOLLOW-UP PROBLEM 18.2**

How would you prepare a benzoic acid/benzoate buffer with pH = 4.25, starting with 5.0 L of 0.050 $M$ sodium benzoate ($C_6H_5COONa$) solution and adding the acidic component? $K_a$ of benzoic acid ($C_6H_5COOH$) = $6.3 \times 10^{-5}$.

---

Another way to prepare a buffer is to form one of the components during the final mixing step by *partial neutralization* of the other component. For example, you can prepare an $HCOOH/HCOO^-$ buffer by mixing appropriate amounts of $HCOOH$ solution and $NaOH$ solution. As the $OH^-$ ions meet the $HCOOH$ molecules, neutralization of part of the total $HCOOH$ present produces the $HCOO^-$ needed:

$HCOOH$ (HA total) + $OH^-$ (amt added) $\longrightarrow$

$\qquad HCOOH$ (HA total $-$ $OH^-$ amt added) + $HCOO^-$ ($OH^-$ amt added) + $H_2O$

This method of preparing a buffer is based on the same chemical process that occurs when a weak acid is titrated with a strong base, as you'll see in the next section.

**Section Summary**

A buffered solution exhibits a much smaller change in pH when $H_3O^+$ or $OH^-$ is added than does an unbuffered solution. A buffer consists of large reservoirs of the components of a conjugate acid-base pair. The buffer-component ratio determines the pH, and they are related by the Henderson-Hasselbalch equation. As $H_3O^+$ or $OH^-$ is added, one buffer component is converted into the other, so their ratio, and thus the pH, changes only slightly. A concentrated buffer undergoes smaller changes in pH than a dilute one. When the buffer pH equals the $pK_a$ of the acid component, the buffer has its highest capacity. A buffer has an effective range of $pK_a \pm 1$. When preparing a buffer, you (1) choose the conjugate pair, (2) calculate the ratio of buffer components, (3) determine the buffer concentration, and (4) adjust the final buffer to the desired pH.

## 18.2   Acid-Base Titration Curves

In Chapter 4, we discussed acid-base titrations as a common analytical method. With the deeper understanding of acid-base behavior you have now, let's reexamine this method and follow the change in pH with an **acid-base titration curve,** a plot of pH vs. volume of titrant added. The behavior of an acid-base indicator and its role in the process is described first.

### Acid-Base Indicators and the Measurement of pH

The two common devices for measuring pH in the laboratory are pH meters and acid-base indicators. (We discuss the operation of pH meters in Chapter 20.) An **acid-base indicator** is a weak organic acid whose acid form (which we denote here as HIn) is a different color than its base form ($In^-$), with the color change occurring over a specific pH range. Typically, one or

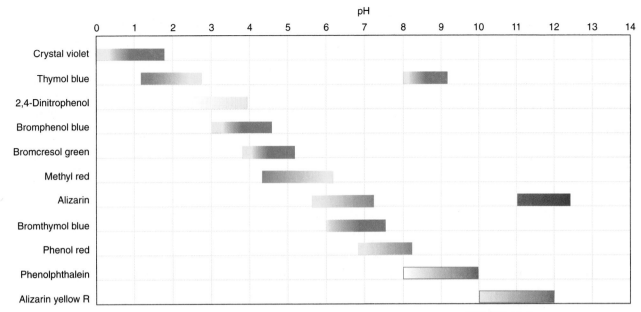

**FIGURE 18.4**
**Colors and approximate pH range of some common acid-base indicators.** Most indicators have a pH range of about two units, in keeping with a useful buffer range of about two units ($pK_a \pm 1$). (The specific pH range depends to some extent on the solvent used to prepare the indicator.)

both of the forms are intensely colored, so only a tiny amount of indicator is needed and its presence does not affect the pH of the test solution.

Indicators are used for approximate pH monitoring in acid-base titrations or in reactions. Figure 18.4 shows the colors and the pH range for the color change of some common acid-base indicators. Selecting an indicator requires that you know the approximate pH of the titration end point, which in turn requires that you know the ionic species present, as you'll see shortly in our discussion of acid-base titration curves.

Since the indicator molecule is a weak acid, the ratio of the two forms is governed by the $[H_3O^+]$ of the test solution:

$$HIn(aq) + H_2O(l) \rightleftharpoons H_3O^+(aq) + In^-(aq) \qquad K_{indicator} = \frac{[H_3O^+][In^-]}{[HIn]}$$

Therefore,

$$\frac{[HIn]}{[In^-]} = \frac{[H_3O^+]}{K_{indicator}}$$

Human color perception has a major influence on the use of indicators. The experimenter will see the HIn color if the $[HIn]/[In^-]$ ratio is 10/1 or greater and will see the $In^-$ color if the $[HIn]/[In^-]$ ratio is 1/10 or less. Between these extremes, the colors of the two forms are merged into an intermediate hue. Therefore, an indicator has a *color range* that reflects a 100-fold range in the $[HIn]/[In^-]$ ratio, which means that an *indicator changes color over a pH range of about two units*. For example, bromthymol blue is yellow below pH 6.0, blue above pH 7.6, and greenish in between (Figure 18.5).

## Strong Acid–Strong Base Titration Curves

A typical curve for the titration of a strong acid with a strong base appears in Figure 18.6, along with the data used to construct it. The pH starts out

**FIGURE 18.5**
**The color change of the indicator bromthymol blue.** Indicators are weak acids, whose conjugate acid and base have different colors. The color change takes place over a pH range of about two units. The acid form of bromthymol blue is yellow *(left)*, the basic form is blue *(right)*, and a mixture of the two forms appears green *(center)*.

**FIGURE 18.6**

Curve for a strong acid–strong base titration. **A,** Data obtained from the titration of 40.00 mL of 0.1000 $M$ HCl with 0.1000 $M$ NaOH. **B,** These data are plotted to give an acid-base titration curve. The pH increases gradually at first. When the moles of $OH^-$ added are slightly less than the moles of $H_3O^+$ originally present, a large pH change accompanies a small addition of $OH^-$. The equivalence point occurs when moles $OH^-$ added = moles $H_3O^+$ originally present. For a strong acid–strong base titration, pH = 7.00 at the equivalence point. Either methyl red or phenolphthalein would be suitable indicators in this case because they both change color on the steep portion of the curve. After this portion, added $OH^-$ causes a gradual pH increase again.

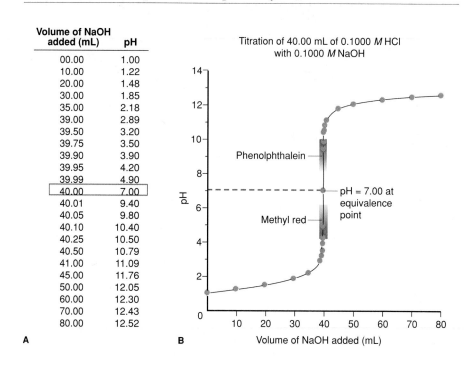

| Volume of NaOH added (mL) | pH |
|---|---|
| 00.00 | 1.00 |
| 10.00 | 1.22 |
| 20.00 | 1.48 |
| 30.00 | 1.85 |
| 35.00 | 2.18 |
| 39.00 | 2.89 |
| 39.50 | 3.20 |
| 39.75 | 3.50 |
| 39.90 | 3.90 |
| 39.95 | 4.20 |
| 39.99 | 4.90 |
| 40.00 | 7.00 |
| 40.01 | 9.40 |
| 40.05 | 9.80 |
| 40.10 | 10.40 |
| 40.25 | 10.50 |
| 40.50 | 10.79 |
| 41.00 | 11.09 |
| 45.00 | 11.76 |
| 50.00 | 12.05 |
| 60.00 | 12.30 |
| 70.00 | 12.43 |
| 80.00 | 12.52 |

A                                         B

low, reflecting the high $[H_3O^+]$ of the strong acid, and increases gradually as acid is neutralized by the added base. Suddenly, the pH rises steeply. This rise begins when the moles of $OH^-$ that have been added nearly equal the moles of $H_3O^+$ originally present in the acid. An additional drop or two of base neutralizes the final tiny excess of acid and introduces a tiny excess of base, so the pH jumps about six units. Beyond this steep portion, the curve rises slowly again because additional base causes only a gradual change in pH.

The **equivalence point** of this titration, the point at which the number of moles of added $OH^-$ equals the number of moles of $H_3O^+$ originally present, occurs within the nearly vertical portion of the curve. *At the equivalence point of a strong acid–strong base titration, the pH equals 7.00* because the solution consists of a strong-acid anion and a strong-base cation. Recall from Chapter 17 that these ions do not react with water, so the solution is neutral. The volume and concentration of base needed to reach the equivalence point allow us to calculate the amount of acid originally present (see Sample Problem 4.6).

To know when we reach the equivalence point, we add a few drops of an appropriate indicator before the titration begins. The **end point** of the titration occurs when the indicator changes color. In general, *we choose an indicator with an end point close to the equivalence point,* one that changes color in the pH range on the steep vertical portion of the curve. Figure 18.6 shows the color changes for two indicators that would be suitable for a strong acid–strong base titration. Methyl red changes from red at pH 4.2 to yellow at pH 6.3, whereas phenolphthalein changes from colorless at pH 8.3 to pink at pH 10.0. Even though neither color change occurs *at* the equivalence point (pH 7.00), both occur on the vertical portion of the curve, where a single drop of base causes a large pH change: when methyl red turns yellow, or when phenolphthalein turns pink, we know we are within a fraction of a drop from the equivalence point. For example, in going from 39.90 to 39.99 mL, one to two drops, the pH changes one whole unit.

By knowing the chemical species present during the titration, we can calculate the pH at various points along the way. In Figure 18.6, 40.00 mL of 0.1000 $M$ HCl is titrated with 0.1000 $M$ NaOH. Since a strong acid is completely dissociated, [HCl] = [$H_3O^+$] = 0.1000 $M$. Therefore, the initial pH is*

$$pH = -\log [H_3O^+] = -\log (0.1000) = 1.00$$

As soon as we start adding titrant, two changes occur that we must repeatedly incorporate into the pH calculations: (1) some acid is being neutralized, and (2) the volume is increasing. After adding 20.00 mL of 0.1000 $M$ NaOH, for example, we are halfway to the equivalence point. The pH calculation includes these two changes:

1. *Find the moles of $H_3O^+$ remaining.* Subtracting the number of moles of $H_3O^+$ reacted from the number originally present gives the number remaining. Moles of $H_3O^+$ reacted equal moles of $OH^-$ added, so

$$\begin{array}{r} \text{Initial moles of } H_3O^+ = 0.04000 \text{ L} \times 0.1000 \ M = 0.004000 \text{ mol } H_3O^+ \\ -\text{Moles of } OH^- \text{ added} = 0.02000 \text{ L} \times 0.1000 \ M = 0.002000 \text{ mol } OH^- \\ \hline \text{Moles of } H_3O^+ \text{ remaining} = 0.002000 \text{ mol } H_3O^+ \end{array}$$

2. *Calculate [$H_3O^+$] taking the total volume into account.* We add one solution to another in a titration, so *the volume increases*. To find the ion concentrations, we use the *total volume* because the water of one solution dilutes the ions of the other:

$$[H_3O^+] = \frac{\text{moles of } H_3O^+ \text{ remaining}}{\text{original volume of acid + volume of added base}}$$

$$= \frac{0.002000 \text{ mol } H_3O^+}{0.04000 \text{ L} + 0.02000 \text{ L}} = 0.03333 \ M \qquad pH = 1.48$$

Similar calculations give values up to the equivalence point. When that point is reached, all the $H_3O^+$ from the acid has been neutralized, and the solution contains $Na^+$ and $Cl^-$, neither of which reacts with water. Due to the autoionization of water, however,

$$[H_3O^+] = 1.0 \times 10^{-7} \ M \qquad pH = 7.00$$

In this example 0.004000 mol $OH^-$ reacted with 0.004000 mol $H_3O^+$ to reach the equivalence point.

From this point on, the pH calculation is based on the moles of *excess* $OH^-$ present. After the addition of 50.00 mL of NaOH, for example, the pH is found as follows:

$$\begin{array}{r} \text{Total moles of } OH^- \text{ added} = 0.05000 \text{ L} \times 0.1000 \ M = 0.005000 \text{ mol } OH^- \\ - \text{ Moles of } H_3O^+ \text{ consumed} = 0.04000 \text{ L} \times 0.1000 \ M = 0.004000 \text{ mol } H_3O^+ \\ \hline \text{Moles of excess } OH^- = 0.001000 \text{ mol } OH^- \end{array}$$

$$[OH^-] = \frac{0.001000 \text{ mol } OH^-}{0.04000 \text{ L} + 0.05000 \text{ L}} = 0.01111 \ M \qquad pOH = 1.95$$

$$pH = pK_w - pOH = 14.00 - 1.95 = 12.05$$

---

*In acid-base titrations, volumes and concentrations are usually known to four significant figures, but pH is generally reported to no more than two significant figures (two digits to the right of the decimal point).

**FIGURE 18.7**

**Curve for a weak acid–strong base titration.** The curve for the titration of 40.00 mL of 0.1000 $M$ CH$_3$CH$_2$COOH (HPr) with 0.1000 $M$ NaOH is shown and compared with that for the strong acid HCl *(dashed curve)*. The initial pH is higher than for a strong acid because HPr dissociates only slightly. At the midpoint of the gradually rising buffer region, [HPr] = [Pr$^-$], and pH = p$K_a$. The pH of the equivalence point is greater than 7 because the solution contains the weak base Pr$^-$. Phenolphthalein is a suitable indicator for this titration, but methyl red is not because its color changes over a large volume change.

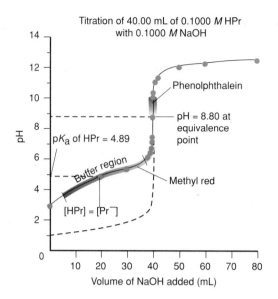

Titration of 40.00 mL of 0.1000 $M$ HPr with 0.1000 $M$ NaOH

## Weak Acid–Strong Base Titration Curves

Figure 18.7 shows the curve obtained when we use 0.1000 $M$ NaOH to titrate 40.00 mL of 0.1000 $M$ propanoic acid, a weak organic acid (CH$_3$CH$_2$COOH; $K_a = 1.3 \times 10^{-5}$). (We abbreviate the acid as HPr and the conjugate base, CH$_3$CH$_2$COO$^-$, as Pr$^-$.) When we compare this weak acid–strong base titration curve with that for a strong acid–strong base titration (dashed curve in Figure 18.7), three major differences appear:

1. *The initial pH is higher* because the weak acid (HPr) dissociates only slightly, so less H$_3$O$^+$ is present than with a strong acid of the same concentration.
2. A gradually rising portion of the curve, called the *buffer region*, occurs before the steep rise to the equivalence point because *the reaction of weak acid with strong base forms a buffer* (Section 18.1).
3. *The pH at the equivalence point is greater than 7* because the solution contains a strong-base cation (Na$^+$), which does not react with water, and a weak-acid anion (Pr$^-$), which acts as a weak base and yields OH$^-$.

Note that at the midpoint of the buffer region, *half the original HPr has been neutralized*; therefore, [HPr] = [Pr$^-$], or [Pr$^-$]/[HPr] = 1, and thus *the pH equals the p$K_a$ of the acid*:

At midpoint of buffer region:   $\text{pH} = \text{p}K_a + \log\left(\dfrac{[\text{Pr}^-]}{[\text{HPr}]}\right) = \text{p}K_a + \log 1 = \text{p}K_a$

In fact, observing the pH at the midpoint of the buffer region is one method for estimating the p$K_a$ of an unknown acid.

Our choice of indicator is more limited here than in a strong acid–strong base titration because the steep rise occurs over a smaller pH range. Phenolphthalein is suitable because its color change lies within this range (Figure 18.7). However, methyl red, our other choice for the strong acid–strong base titration, would change color earlier and slowly over a sizable volume of titrant, thereby giving a vague and false indication of the equivalence point.

The calculation procedure for the weak acid–strong base titration is different than the previous case as well because we have to consider the partial dissociation of the weak acid and the reaction with water of the conjugate base. There are four key regions of the titration curve, each of which requires a different type of calculation to find $[H_3O^+]$:

1. The initial $[H_3O^+]$ is that of a weak-acid solution. We find $[H_3O^+]$ as in Section 17.4: we set up a reaction table with $x = [HPr]_{dissoc}$, assume $[H_3O^+] = [HPr]_{dissoc} << [HPr]_{init}$, and solve for $x$. Combined into one step, the calculation is

$$[H_3O^+] = \sqrt{K_a \times [HPr]_{init}}$$

2. As soon as we add NaOH, an equivalent amount of $Pr^-$ forms, so up to the equivalence point, we have a mixture of acid and conjugate base, and a buffer solution exists over much of that interval. Therefore, we find $[H_3O^+]$ by the relationship

$$[H_3O^+] = K_a \times \frac{[HPr]}{[Pr^-]}$$

(Of course, we can find pH directly with the Henderson-Hasselbalch equation, which is just an alternative form of this relationship.) In this calculation we do *not* have to consider the new total volume because *in a ratio of concentrations, the volumes cancel.* That is, $[HPr]/[Pr^-]$ = moles of HPr/moles of $Pr^-$, so we need not calculate concentrations.

3. At the equivalence point, the original amount of HPr has been neutralized, and the flask contains a solution of $Pr^-$, a weak base that reacts with water to form $OH^-$:

$$Pr^-(aq) + H_2O(l) \rightleftharpoons HPr(aq) + OH^-(aq)$$

Therefore, the equivalence point has a pH > 7. We calculate $[H_3O^+]$ as in Section 17.5: we find $K_b$ of $Pr^-$ from $K_a$ of HPr, set up a reaction table (assume $[Pr^-] >> [Pr^-]_{reacting}$), and solve for $[OH^-]$, which we convert to $[H_3O^+]$. Since we need $[Pr^-]$ to solve for $[OH^-]$, we *do* need the total volume. Combined into one step, this final calculation is

$$[H_3O^+] = \frac{K_w}{\sqrt{K_b \times [Pr^-]}}, \quad \text{where } K_b = \frac{K_w}{K_a} \quad \text{and} \quad [Pr^-] = \frac{\text{moles of HPr}_{init}}{\text{total volume}}$$

4. Beyond the equivalence point, we are just adding excess $OH^-$ ion, so the calculation is the same as for the strong acid–strong base titration:

$$[H_3O^+] = \frac{K_w}{[OH^-]}, \quad \text{where } [OH^-] = \frac{\text{moles of excess } OH^-}{\text{total volume}}$$

Sample Problem 18.3 shows the overall approach.

SAMPLE PROBLEM 18.3 _____

### Calculating the pH During a Weak Acid–Strong Base Titration

**Problem:** Calculate the pH during the titration of 40.00 mL of 0.1000 M propanoic acid (HPr; $K_a = 1.3 \times 10^{-5}$) after adding the following volumes of 0.1000 M NaOH:
**(a)** 0.00 mL         **(b)** 30.00 mL         **(c)** 40.00 mL         **(d)** 50.00 mL
**Plan: (a)** 0.00 mL: Since no base has been added, this is a weak-acid solution, so we calculate the pH as we did in Section 17.4.
**(b)** 30.00 mL: A mixture of $Pr^-$ and HPr is present. We find the moles of each, substitute into the $K_a$ expression to solve for $[H_3O^+]$, and convert to pH.
**(c)** 40.00 mL: The moles of NaOH added equals the initial moles of HPr, so a solution of the weak base $Pr^-$ exists. We calculate the pH as we did in Section 17.5, except that we need *total* volume to find $[Pr^-]$.
**(d)** 50.00 mL: Excess NaOH is added, so we calculate the moles of excess $OH^-$ in the total volume, convert to $[H_3O^+]$ and then pH.

**Solution: (a)** 0.00 mL of 0.1000 $M$ NaOH added. Following the approach used in Sample Problem 17.7 and just described in text, we obtain

$$[H_3O^+] = \sqrt{K_a \times [HPr]_{init}} = \sqrt{(1.3 \times 10^{-5})(0.1000)} = 1.1 \times 10^{-3}\ M$$

**pH = 2.96**

**(b)** 30.00 mL of 0.1000 $M$ NaOH added. Calculating the ratio of moles of HPr to moles of Pr⁻:

$$\text{Original moles of HPr} = 0.04000\ \text{L} \times 0.1000\ M = 0.004000\ \text{mol HPr}$$

$$\text{Moles NaOH added} = 0.03000\ \text{L} \times 0.1000\ M = 0.003000\ \text{mol OH}^-$$

Since one mole of NaOH reacts to form one mole of Pr⁻, we construct the following reaction table for the stoichiometry:

| Amount (mol) | HPr($aq$) | + | OH⁻($aq$) | → | Pr⁻($aq$) | + | H₂O($l$) |
|---|---|---|---|---|---|---|---|
| Before addition | 0.004000 | | — | | 0 | | — |
| Addition | — | | 0.003000 | | — | | — |
| After addition | 0.001000 | | 0 | | 0.003000 | | — |

The last line of this table shows the new initial amounts of HPr and Pr⁻ that will react to attain a new equilibrium. However, since $x$ is very small, we assume that the [HPr]/[Pr⁻] ratio at equilibrium is essentially equal to the ratio of these new initial amounts (see Comment in Sample Problem 18.1). Thus,

$$\frac{[HPr]}{[Pr^-]} = \frac{0.001000\ \text{mol}}{0.003000\ \text{mol}} = 0.3333$$

Solving for $[H_3O^+]$:

$$[H_3O^+] = K_a \times \frac{[HPr]}{[Pr^-]} = (1.3 \times 10^{-5})(0.3333) = 4.3 \times 10^{-6}\ M \qquad \textbf{pH = 5.37}$$

**(c)** 40.00 mL of 0.1000 $M$ NaOH added. Calculating [Pr⁻] after all HPr has reacted:

$$[Pr^-] = \frac{0.004000\ \text{mol}}{0.04000\ \text{L} + 0.04000\ \text{L}} = 0.05000\ M$$

Calculating $K_b$:

$$K_b = \frac{K_w}{K_a} = \frac{1.0 \times 10^{-14}}{1.3 \times 10^{-5}} = 7.7 \times 10^{-10}$$

Solving for $[H_3O^+]$ as described in text:

$$[H_3O^+] = \frac{K_w}{\sqrt{K_b \times [Pr^-]}} = \frac{1.0 \times 10^{-14}}{\sqrt{(7.7 \times 10^{-10})(0.05000)}} = 1.6 \times 10^{-9}\ M$$

**pH = 8.80**

**(d)** 50.00 mL of 0.1000 $M$ NaOH added.

$$\text{Moles of excess OH}^- = 0.1000\ M \times (0.05000\ \text{L} - 0.04000\ \text{L}) = 0.001000\ \text{mol}$$

$$[OH^-] = \frac{\text{moles of excess OH}^-}{\text{total volume}} = \frac{0.001000\ \text{mol}}{0.09000\ \text{L}} = 0.01111\ M$$

$$[H_3O^+] = \frac{K_w}{[OH^-]} = \frac{1.0 \times 10^{-14}}{0.01111} = 9.0 \times 10^{-13}\ M \qquad \textbf{pH = 12.05}$$

**Check:** As we would expect from the continuous addition of base, the pH increases through the four stages. Be sure to round off and check the arithmetic along the way.

**FOLLOW-UP PROBLEM 18.3**
When 20.00 mL of 0.2000 $M$ HBrO is titrated with 0.1000 $M$ NaOH, what is the pH
**(a)** Before any base is added?
**(b)** When [HBrO] = [BrO$^-$]?
**(c)** At the equivalence point?
**(d)** When the moles of OH$^-$ added is twice the moles of HBrO originally present?
**(e)** Sketch the titration curve. $K_a$ of HBrO = 2.3 × 10$^{-9}$.

### Weak Base–Strong Acid Titration Curves

Up to this point, we've added bases to acids. The opposite process, adding an acid titrant to a base, is similar to that taking place when acid rain attacks marble statuary or when stomach acid reacts with an antacid tablet.

Figure 18.8 is a plot of pH vs. volume of added HCl in the titration of aqueous ammonia. Note that the shape of the curve is identical to the weak acid–strong base titration curve (Figure 18.7), but inverted. The pH starts out high and decreases gradually in the buffer region, where significant amounts of base (NH$_3$) and conjugate acid (NH$_4^+$) exist. At the midpoint of the buffer region, *the pH equals the pK$_a$ of the ammonium ion.* After the buffer region, the curve drops precipitously to the equivalence point, at which all the NH$_3$ has been neutralized and a solution of NH$_4$Cl is present. Note that the pH at the equivalence point is *below* 7 due to the acidic nature of the cation, NH$_4^+$:

$$NH_4^+(aq) + H_2O(l) \rightleftharpoons NH_3(aq) + H_3O^+(aq)$$

Beyond the sharp drop, the pH declines slowly as excess HCl is added.

Here again, we must be more careful in the choice of indicator than for a strong acid–strong base titration. Phenolphthalein would change color too soon and too slowly to indicate the equivalence point. Methyl red, on the other hand, lies on the steep portion of the curve and straddles the equivalence point, so it is a perfect choice for this titration.

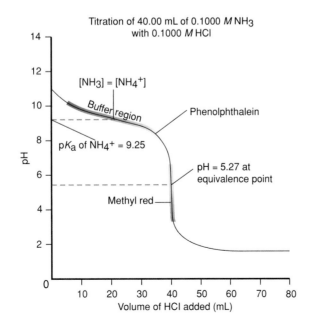

Titration of 40.00 mL of 0.1000 $M$ NH$_3$ with 0.1000 $M$ HCl

**FIGURE 18.8**
**Curve for a weak base–strong acid titration.** Titrating 40.00 mL of 0.1000 $M$ NH$_3$ with 0.1000 $M$ HCl leads to a curve whose shape is the inverse of the weak acid–strong base curve in Figure 18.7. The midpoint of the buffer region occurs when [NH$_3$] = [NH$_4^+$]; the pH at this point equals the p$K_a$ of NH$_4^+$. The pH of the equivalence point is below 7 because the solution contains the weak acid NH$_4^+$. Methyl red is a suitable indicator here, but phenolphthalein is not because its color changes over a large volume change.

### Titration Curves for Polyprotic Acids

As we discussed in Section 17.4, polyprotic acids have more than one ionizable proton. Except for sulfuric acid, the common polyprotic acids are all weak. The successive $K_a$ values for a polyprotic acid differ by several orders of magnitude, as these values for sulfurous acid show:

$$H_2SO_3(aq) + H_2O(l) \rightleftharpoons HSO_3^-(aq) + H_3O^+(aq)$$

$$K_{a1} = 1.4 \times 10^{-2} \text{ and } pK_{a1} = 1.85$$

$$HSO_3^-(aq) + H_2O(l) \rightleftharpoons SO_3^{2-}(aq) + H_3O^+(aq)$$

$$K_{a2} = 6.5 \times 10^{-8} \text{ and } pK_{a2} = 7.19$$

Two moles of $OH^-$ ion are required to remove both protons from one mole of a diprotic acid such as $H_2SO_3$. As the difference in $K_a$ values indicates, the first proton is lost much more easily than the second. As a result, each mole of protons is titrated separately; that is, all the $H_2SO_3$ molecules lose one proton before any of the $HSO_3^-$ ions lose one:

$$H_2SO_3 \xrightarrow{\text{1 mol } OH^-} HSO_3^- \xrightarrow{\text{1 mol } OH^-} SO_3^{2-}$$

The titration curve for sulfurous acid reflects this fact, with loss of each mole of protons exhibiting its own equivalence point and buffer region (Figure 18.9). As with a weak monoprotic acid, the pH at the midpoint of the buffer region is equal to the $pK_a$ of that acidic species. Note also that the same volume of added base (in this case, 40.00 mL of 0.1000 $M$ $OH^-$) is required to remove each proton.

Amino acids, the compounds that link together to form proteins (Section 14.5), have the general formula $NH_2—CH(R)—COOH$. At low pH, both the amino group ($—NH_2$) and the acid group ($—COOH$) are in their proton-

**FIGURE 18.9**

**Curve for the titration of a weak polyprotic acid.** Titrating 40.00 mL of 0.1000 $M$ $H_2SO_3$ with 0.1000 $M$ NaOH leads to a curve with two buffer regions and two equivalence points. Because the $K_a$ values are separated by several orders of magnitude, in effect the titration curve looks like two weak acid–strong base curves attached. The pH of the first equivalence point is below 7 because the solution contains $HSO_3^-$, which is a stronger acid than it is a base ($K_a$ of $HSO_3^- = 6.5 \times 10^{-8}$; $K_b$ of $HSO_3^- = 7.1 \times 10^{-13}$).

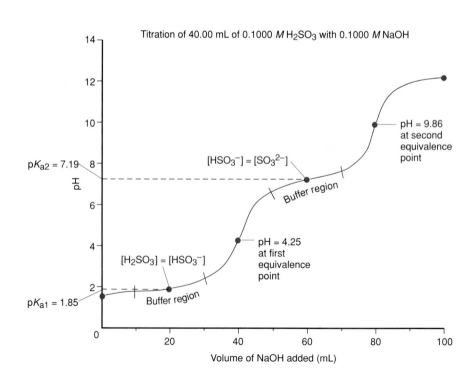

Titration of 40.00 mL of 0.1000 $M$ $H_2SO_3$ with 0.1000 $M$ NaOH

ated form, $^+NH_3$—CH(R)—COOH, which can be titrated like a polyprotic acid. In the case of glycine (R = H), the simplest amino acid, the dissociation reactions and $pK_a$ values are

$$^+NH_3CH_2COOH(aq) + H_2O(l) \rightleftharpoons {}^+NH_3CH_2COO^-(aq) + H_3O^+(aq) \quad pK_{a1} = 2.35$$

$$^+NH_3CH_2COO^-(aq) + H_2O(l) \rightleftharpoons NH_2CH_2COO^-(aq) + H_3O^+(aq) \quad pK_{a2} = 9.78$$

As with $H_2SO_3$, the large difference in $pK_a$ values indicates that virtually all the —COOH protons are removed before any —$NH_3{}^+$ protons come off:

$$^+NH_3CH_2COOH \xrightarrow{\text{1 mol OH}^-} {}^+NH_3CH_2COO^- \xrightarrow{\text{1 mol OH}^-} NH_2CH_2COO^-$$

Thus, at physiological pH (~7), *glycine exists predominantly in a dipolar form*, $^+NH_3CH_2COO^-$. Among the 20 different R groups of the amino acids found in proteins, several have *additional* —COO$^-$ or —$NH_3{}^+$ groups at pH 7. When amino acids are linked into proteins, these charged R groups give the protein its overall charge and play a major role in its function. ◆

### Section Summary

An acid-base (pH) indicator is a weak acid with differently colored acidic and basic forms that changes color over about two pH units. In a strong acid–strong base titration, the pH starts out low, rises slowly, then shoots up near the equivalence point (pH = 7). In a weak acid–strong base titration, the pH starts out higher, rises slowly in the buffer region (pH = $pK_a$ at the midpoint), then rises more quickly near the equivalence point (pH > 7). A weak base–strong acid titration is the inverse of this, its pH decreasing to the equivalence point (pH < 7). Polyprotic acids have two or more acidic protons, and each is titrated separately.

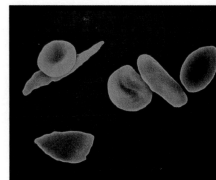

◆ **Polyprotic Proteins and Human Disease.** In a normal red blood cell, two critical amino acids in hemoglobin, the oxygen-carrying protein, contain R groups with —COO$^-$ that contribute to the molecule's functional shape (*rounded cell in photo*). In the abnormal hemoglobin that causes the hereditary disease sickle cell anemia, these charged R groups have been replaced with —$CH_3$ groups. This change in just two of hemoglobin's 574 amino acids lowers charge repulsions between hemoglobin molecules enough for them to clump together, which leads to the sickle shape of the blood cell (*elongated cells in photo*) and results in blocked capillaries. The painful course of sickle cell anemia ends in early death.

## 18.3 Equilibria of Slightly Soluble Ionic Compounds

The next type of aqueous equilibrium we explore is that of slightly soluble ionic compounds. In Chapter 12, we found that most solutes have a limited solubility in a solvent, and that in a saturated solution, equilibrium exists between the undissolved and the dissolved solute. Slightly soluble (often called "insoluble") salts have a relatively low solubility and are particularly important in chemical analysis. At this point, it would be a good idea for you to review the solubility rules in Table 4.1.

When a soluble ionic compound dissolves in water, it dissociates completely into ions. In this discussion, we assume that even though a slightly soluble salt dissolves to a very limited extent, the amount that does dissolve also dissociates completely. In reality, however, this is not the case. The equilibria that exist in solutions of slightly soluble salts, particularly those of heavy metals and transition elements, often contain other species. For example, when lead(II) chloride is shaken in water, the solution contains not only the expected $Pb^{2+}(aq)$ and $Cl^-(aq)$ ions, but also undissociated molecules of $PbCl_2(aq)$ and $PbCl^+(aq)$ ions. More advanced courses address these complexities, but we will only hint at some of them in the sample problem comments. Keep in mind, however, that our simple calculations based on complete dissociation are best treated as first approximations.

### The Solubility-Product Constant ($K_{sp}$)

Making the assumption of complete dissociation, we can state that for slightly soluble ionic compounds, *equilibrium exists between solid solute and dissolved ions*. For example, lead sulfate dissolves only to the extent of 0.04 g/L water at 25°C, but whatever dissolves exists as lead(II) and sulfate ions:

$$PbSO_4(s) \rightleftharpoons Pb^{2+}(aq) + SO_4^{2-}(aq)$$

As with all the other equilibrium systems we've looked at, the equilibrium condition for a saturated solution can be expressed by a mass-action expression. For the lead sulfate case, we have

$$Q_c = \frac{[Pb^{2+}][SO_4^{2-}]}{[PbSO_4]}$$

Combining the constant value of $[PbSO_4]$ with the value of $Q_c$, we obtain the *ion-product expression*, $Q_{sp}$:

$$Q_{sp} = Q_c[PbSO_4] = [Pb^{2+}][SO_4^{2-}]$$

When solid $PbSO_4$ attains equilibrium with $Pb^{2+}$ and $SO_4^{2-}$ ions, that is, reaches saturation, the numerical value of $Q_{sp}$ attains a constant value. This new equilibrium constant is called the **solubility-product constant, $K_{sp}$.** The value of $K_{sp}$ for $PbSO_4$ at 25°C, for example, is $1.6 \times 10^{-8}$.

As with other equilibrium constants, a particular $K_{sp}$ value depends only on the temperature, not on the individual ion concentrations. Suppose, for example, you add some lead(II) nitrate, a soluble lead salt, to increase the solution's $[Pb^{2+}]$. The equilibrium position shifts to the left and the $[SO_4^{2-}]$ goes down as more $PbSO_4$ precipitates, so the $K_{sp}$ value is maintained.

In general, for a saturated solution of a slightly soluble ionic compound, $M_pX_q$, composed of the ions $M^{n+}$ and $X^{z-}$, the equilibrium condition is

$$Q_{sp} = [M^{n+}]^p[X^{z-}]^q = K_{sp} \tag{18.2}$$

That is, the form of $Q_{sp}$ is identical to that of the other mass-action expressions we have written: *each ion concentration is raised to an exponent equal to the coefficient in the balanced equation.* Thus, the equation and $Q_{sp}$ that describes a saturated solution of $Cu(OH)_2$ is

$$Cu(OH)_2(s) \rightleftharpoons Cu^{2+}(aq) + 2OH^-(aq) \qquad Q_{sp} = [Cu^{2+}][OH^-]^2$$

The insoluble metal sulfides present a somewhat different case. Recent evidence shows that the sulfide ion, $S^{2-}$, is so basic that it never exists in water. Instead, it reacts completely with water to form the hydrogen sulfide ion and the hydroxide ion:

$$S^{2-}(aq) + H_2O(l) \rightarrow HS^-(aq) + OH^-(aq)$$

Therefore, when manganese(II) sulfide, for instance, is shaken in water, we have

$$MnS(s) + H_2O(l) \rightleftharpoons Mn^{2+}(aq) + HS^-(aq) + OH^-(aq)$$
$$Q_{sp} = [Mn^{2+}][HS^-][OH^-]$$

SAMPLE PROBLEM 18.4 _____

### Writing $Q_{sp}$ for Slightly Soluble Ionic Compounds

**Problem:** Write the ion-product expression for
**(a)** Magnesium carbonate  **(b)** Iron(II) hydroxide
**(c)** Calcium phosphate  **(d)** Silver sulfide

**Plan:** We write an equation that describes a saturated solution and then write $Q_{sp}$ according to Equation 18.2, noting the sulfide in (d).

**Solution: (a)** Magnesium carbonate:

$$MgCO_3(s) \rightleftharpoons Mg^{2+}(aq) + CO_3^{2-}(aq) \quad \boldsymbol{Q_{sp} = [Mg^{2+}][CO_3^{2-}]}$$

**(b)** Iron(II) hydroxide:

$$Fe(OH)_2(s) \rightleftharpoons Fe^{2+}(aq) + 2OH^-(aq) \quad \boldsymbol{Q_{sp} = [Fe^{2+}][OH^-]^2}$$

**(c)** Calcium phosphate:

$$Ca_3(PO_4)_2(s) \rightleftharpoons 3Ca^{2+}(aq) + 2PO_4^{3-}(aq) \quad \boldsymbol{Q_{sp} = [Ca^{2+}]^3[PO_4^{3-}]^2}$$

**(d)** Silver sulfide:

$$Ag_2S(s) + H_2O(l) \rightleftharpoons 2Ag^+(aq) + HS^-(aq) + OH^-(aq) \quad \boldsymbol{Q_{sp} = [Ag^+]^2[HS^-][OH^-]}$$

**Check:** Except for (d), you can check by reversing the process to see if you obtain the formula of the compound from $Q_{sp}$.

**FOLLOW-UP PROBLEM 18.4**
Write the ion-product expressions for **(a)** calcium sulfate; **(b)** chromium(III) carbonate; **(c)** magnesium hydroxide; **(d)** arsenic(III) sulfide.

---

Since, at saturation, the $Q_{sp}$ value is equal to $K_{sp}$, we write the ion-product expression directly as $K_{sp}$. *The value of $K_{sp}$ is a measure of how far to the right the dissolution proceeds at equilibrium (saturation).* Table 18.2 on the next page presents the $K_{sp}$ values of some slightly soluble ionic compounds. Even though the $K_{sp}$ values are all quite low, they vary over a large range.

### Calculations Involving the Solubility-Product Constant

In Chapters 16 and 17, we described two types of equilibrium problems. In one type, we use concentrations to find $K$, and in the other, we use $K$ to find concentrations. Here we encounter the same two types.

**Determining $K_{sp}$ from solubility.** The solubility of an ionic compound can be determined experimentally or found in one of the several chemical handbooks available. Most handbooks give solubility in terms of grams of solute dissolved in 100 grams of $H_2O$. Since the amount of the slightly soluble compound in solution is small, a negligible error is introduced if we assume that "100 g of water" is equal to "100 mL of solution." We convert the solubility in units of grams of solute/100 mL of solution to **molar solubility,** the number of moles of solute per liter of solution, and then use the balanced equation to find the molarity of each ion and the $K_{sp}$.

**SAMPLE PROBLEM 18.5** ———————————————————————————————

### Determining $K_{sp}$ from Solubility

**Problem: (a)** Lead(II) sulfate ($PbSO_4$) is used as a white pigment and in car batteries. Its solubility in water at 25°C is $4.25 \times 10^{-3}$ g/100 mL solution. What is the $K_{sp}$ of $PbSO_4$?
**(b)** When lead(II) fluoride ($PbF_2$) is shaken with pure water at 25°C, the solubility is found to be 0.64 g/L. Calculate $K_{sp}$ of $PbF_2$.
**Plan:** For each part, we write an equation for the dissolution of the salt to see the moles of each ion that form and write the $K_{sp}$ expression. We convert the known solubility to molar solubility, find the molarity of each ion, and substitute into the ion-product expression to find $K_{sp}$.

**TABLE 18.2** **Solubility-Product Constants ($K_{sp}$) of Some Ionic Compounds (at 25°C)**

| NAME, FORMULA | $K_{sp}$ | NAME, FORMULA | $K_{sp}$ |
|---|---|---|---|
| **Carbonates** | | Cobalt(II) hydroxide, $Co(OH)_2$ | $1.3 \times 10^{-15}$ |
| | | Copper(II) hydroxide, $Cu(OH)_2$ | $2.2 \times 10^{-20}$ |
| Barium carbonate, $BaCO_3$ | $2.0 \times 10^{-9}$ | Iron(II) hydroxide, $Fe(OH)_2$ | $4.1 \times 10^{-15}$ |
| Cadmium carbonate, $CdCO_3$ | $1.8 \times 10^{-14}$ | Iron(III) hydroxide, $Fe(OH)_3$ | $1.6 \times 10^{-39}$ |
| Calcium carbonate, $CaCO_3$ | $3.3 \times 10^{-9}$ | Magnesium hydroxide, $Mg(OH)_2$ | $6.3 \times 10^{-10}$ |
| Cobalt(II) carbonate, $CoCO_3$ | $1.0 \times 10^{-10}$ | Manganese(II) hydroxide, $Mn(OH)_2$ | $1.6 \times 10^{-13}$ |
| Copper(II) carbonate, $CuCO_3$ | $3 \times 10^{-12}$ | Nickel(II) hydroxide, $Ni(OH)_2$ | $6 \times 10^{-16}$ |
| Lead(II) carbonate, $PbCO_3$ | $7.4 \times 10^{-14}$ | Zinc hydroxide, $Zn(OH)_2$ | $3 \times 10^{-16}$ |
| Magnesium carbonate, $MgCO_3$ | $3.5 \times 10^{-8}$ | | |
| Mercury(I) carbonate, $Hg_2CO_3$ | $8.9 \times 10^{-17}$ | **Iodates** | |
| Nickel(II) carbonate, $NiCO_3$ | $1.3 \times 10^{-7}$ | | |
| Strontium carbonate, $SrCO_3$ | $5.4 \times 10^{-10}$ | Barium iodate, $Ba(IO_3)_2$ | $1.5 \times 10^{-9}$ |
| Zinc carbonate, $ZnCO_3$ | $1.0 \times 10^{-10}$ | Calcium iodate, $Ca(IO_3)_2$ | $7.1 \times 10^{-7}$ |
| | | Lead(II) iodate, $Pb(IO_3)_2$ | $2.5 \times 10^{-13}$ |
| **Chromates** | | Silver iodate, $AgIO_3$ | $3.1 \times 10^{-8}$ |
| | | Strontium iodate, $Sr(IO_3)_2$ | $3.3 \times 10^{-7}$ |
| Barium chromate, $BaCrO_4$ | $2.1 \times 10^{-10}$ | Zinc iodate, $Zn(IO_3)_2$ | $3.9 \times 10^{-6}$ |
| Calcium chromate, $CaCrO_4$ | $1 \times 10^{-8}$ | | |
| Lead(II) chromate, $PbCrO_4$ | $2.3 \times 10^{-13}$ | **Oxalates** | |
| Silver chromate, $Ag_2CrO_4$ | $2.6 \times 10^{-12}$ | Barium oxalate dihydrate, $BaC_2O_4 \cdot 2H_2O$ | $1.1 \times 10^{-7}$ |
| | | Calcium oxalate monohydrate, | $2.3 \times 10^{-9}$ |
| **Cyanides** | | $\quad CaC_2O_4 \cdot H_2O$ | |
| | | Strontium oxalate | |
| Mercury(I) cyanide, $Hg_2(CN)_2$ | $5 \times 10^{-40}$ | $\quad$ monohydrate, $SrC_2O_4 \cdot H_2O$ | $5.6 \times 10^{-8}$ |
| Silver cyanide, $AgCN$ | $2.2 \times 10^{-16}$ | | |
| | | **Phosphates** | |
| **Halides** | | | |
| *Fluorides* | | Calcium phosphate, $Ca_3(PO_4)_2$ | $1.2 \times 10^{-29}$ |
| Barium fluoride, $BaF_2$ | $1.5 \times 10^{-6}$ | Magnesium phosphate, $Mg_3(PO_4)_2$ | $5.2 \times 10^{-24}$ |
| Calcium fluoride, $CaF_2$ | $3.2 \times 10^{-11}$ | Silver phosphate, $Ag_3PO_4$ | $2.6 \times 10^{-18}$ |
| Lead(II) fluoride, $PbF_2$ | $3.6 \times 10^{-8}$ | | |
| Magnesium fluoride, $MgF_2$ | $7.4 \times 10^{-9}$ | **Sulfates** | |
| Strontium fluoride, $SrF_2$ | $2.6 \times 10^{-9}$ | | |
| | | Barium sulfate, $BaSO_4$ | $1.1 \times 10^{-10}$ |
| *Chlorides* | | Calcium sulfate, $CaSO_4$ | $2.4 \times 10^{-5}$ |
| Copper(I) chloride, $CuCl$ | $1.9 \times 10^{-7}$ | Lead(II) sulfate, $PbSO_4$ | $1.6 \times 10^{-8}$ |
| Lead(II) chloride, $PbCl_2$ | $1.7 \times 10^{-5}$ | Magnesium sulfate, $MgSO_4$ | $5.9 \times 10^{-3}$ |
| Silver chloride, $AgCl$ | $1.8 \times 10^{-10}$ | Radium sulfate, $RaSO_4$ | $2 \times 10^{-11}$ |
| | | Silver sulfate, $Ag_2SO_4$ | $1.5 \times 10^{-5}$ |
| *Bromides* | | Strontium sulfate, $SrSO_4$ | $3.2 \times 10^{-7}$ |
| Copper(I) bromide, $CuBr$ | $5 \times 10^{-9}$ | | |
| Silver bromide, $AgBr$ | $5.0 \times 10^{-13}$ | **Sulfides** | |
| | | Cadmium sulfide, $CdS$ | $1.0 \times 10^{-24}$ |
| *Iodides* | | Copper(II) sulfide, $CuS$ | $8 \times 10^{-34}$ |
| Copper(I) iodide, $CuI$ | $1 \times 10^{-12}$ | Iron(II) sulfide, $FeS$ | $8 \times 10^{-16}$ |
| Lead(II) iodide, $PbI_2$ | $7.9 \times 10^{-9}$ | Lead(II) sulfide, $PbS$ | $3 \times 10^{-25}$ |
| Mercury(I) iodide, $Hg_2I_2$ | $4.7 \times 10^{-29}$ | Manganese(II) sulfide, $MnS$ | $3 \times 10^{-11}$ |
| Silver iodide, $AgI$ | $8.3 \times 10^{-17}$ | Mercury(II) sulfide, $HgS$ | $2 \times 10^{-50}$ |
| | | Silver sulfide, $Ag_2S$ | $8 \times 10^{-48}$ |
| **Hydroxides** | | Tin(II) sulfide, $SnS$ | $1.3 \times 10^{-23}$ |
| | | Zinc sulfide, $ZnS$ | $2.0 \times 10^{-22}$ |
| Aluminum hydroxide, $Al(OH)_3$ | $3 \times 10^{-34}$ | | |
| Calcium hydroxide, $Ca(OH)_2$ | $6.5 \times 10^{-6}$ | | |

**Solution: (a)** For $PbSO_4$. Writing the equation and ion-product expression:

$$PbSO_4(s) \rightleftharpoons Pb^{2+}(aq) + SO_4^{2-}(aq) \qquad K_{sp} = [Pb^{2+}][SO_4^{2-}]$$

Converting solubility to molar solubility:

$$\text{Molar solubility of } PbSO_4 = \frac{0.00425 \text{ g } PbSO_4}{100 \text{ mL soln}} \times \frac{1000 \text{ mL}}{1 \text{ L}} \times \frac{1 \text{ mol } PbSO_4}{303.3 \text{ g } PbSO_4}$$

$$= 1.40 \times 10^{-4} \ M \ PbSO_4$$

Determining molarities of the ions: Since 1 mol $Pb^{2+}$ and 1 mol $SO_4^{2-}$ form when 1 mol $PbSO_4$ dissolves, $[Pb^{2+}] = [SO_4^{2-}] = 1.40 \times 10^{-4} \ M$.
Calculating $K_{sp}$:

$$K_{sp} = [Pb^{2+}][SO_4^{2-}] = (1.40 \times 10^{-4})^2 = \mathbf{1.96 \times 10^{-8}}$$

**(b)** For $PbF_2$. Writing the equation and $K_{sp}$ expression:

$$PbF_2(s) \rightleftharpoons Pb^{2+}(aq) + 2F^-(aq) \qquad K_{sp} = [Pb^{2+}][F^-]^2$$

Converting solubility to molar solubility:

$$\text{Molar solubility of } PbF_2 = \frac{0.64 \text{ g } PbF_2}{1 \text{ L soln}} \times \frac{1 \text{ mol } PbF_2}{245.2 \text{ g } PbF_2} = 2.6 \times 10^{-3} \ M \ PbF_2$$

Determining molarities of the ions: Since 1 mol $Pb^{2+}$ and 2 mol $F^-$ form when 1 mol $PbF_2$ dissolves,

$$[Pb^{2+}] = 2.6 \times 10^{-3} \ M \qquad \text{and} \qquad [F^-] = 2(2.6 \times 10^{-3} \ M) = 5.2 \times 10^{-3} \ M$$

Calculating $K_{sp}$:

$$K_{sp} = [Pb^{2+}][F^-]^2 = (2.6 \times 10^{-3})(5.2 \times 10^{-3})^2 = \mathbf{7.0 \times 10^{-8}}$$

**Check:** The low solubilities are consistent with $K_{sp}$ values being small. In **(a)**, the molar solubility seems about right: $\sim \dfrac{4 \times 10^{-2} \text{ g/L}}{3 \times 10^2 \text{ g/mol}} \approx 1.3 \times 10^{-4} \ M$. Squaring this number gives $1.7 \times 10^{-8}$, close to the calculated $K_{sp}$. In **(b)**, we check the final step: $\sim(3 \times 10^{-3})(5 \times 10^{-3})^2 = 7.5 \times 10^{-8}$, close to the calculated $K_{sp}$.

**Comment:** The $K_{sp}$ values for these compounds in Table 18.2 are lower than our calculated values. For $PbF_2$, for instance, the table value is $3.6 \times 10^{-8}$, but we calculated $7.0 \times 10^{-8}$ from solubility data. The discrepancy arises because we assumed that the $PbF_2$ in solution dissociates completely to $Pb^{2+}$ and $F^-$. Here is an example of the complexity pointed out earlier. Actually, about a third of the $PbF_2$ dissolves as $PbF^+(aq)$ and a small amount as undissociated $PbF_2(aq)$. The solubility (0.64 g/L) is determined experimentally and includes these other species, which we did not include in our simple calculation. In general, treat $K_{sp}$ values calculated in this way as approximations.

**FOLLOW-UP PROBLEM 18.5**
Fluorite ($CaF_2$) is a mineral source of fluorine and is used extensively in the manufacture of steel and glass. When powdered fluorite is shaken with pure water at 18°C, $1.5 \times 10^{-4}$ g dissolves for every 10.0 mL of solution. Calculate the $K_{sp}$ of $CaF_2$ at 18°C.

**Determining solubility from $K_{sp}$.** The reverse of the previous problem type involves finding the solubility of a compound based on its formula and $K_{sp}$ value. A sound approach is to define the unknown molar solubility as $S$, define the ion concentrations in terms of this unknown, and solve for $S$.

## Determining Solubility from $K_{sp}$

**Problem:** Calcium hydroxide (slaked lime) is a major component of mortar, plaster, and cement. Solutions of $Ca(OH)_2$ are used in industry as a cheap, strong base. Calculate the solubility of $Ca(OH)_2$ in water if the $K_{sp}$ is $6.5 \times 10^{-6}$.

**Plan:** We write the dissolution equation and the $K_{sp}$ expression. Letting $S$ = molar solubility, we set up a reaction table to express $[Ca^{2+}]$ and $[OH^-]$ in terms of $S$, substitute into the $K_{sp}$ expression, and solve for $S$.

**Solution:** Writing the equation and the $K_{sp}$ expression:

$$Ca(OH)_2(s) \rightleftharpoons Ca^{2+}(aq) + 2OH^-(aq) \qquad K_{sp} = [Ca^{2+}][OH^-]^2 = 6.5 \times 10^{-6}$$

Setting up a reaction table, with $S$ = molar solubility:

| Concentration ($M$) | $Ca(OH)_2(s)$ | $\rightleftharpoons$ | $Ca^{2+}(aq)$ | + | $2OH^-(aq)$ |
|---|---|---|---|---|---|
| Initial | — | | 0 | | 0 |
| Change | — | | $+S$ | | $+2S$ |
| Equilibrium | — | | $S$ | | $2S$ |

Substituting into $K_{sp}$ and solving for $S$:

$$K_{sp} = [Ca^{2+}][OH^-]^2 = S \times (2S)^2 = S \times 4S^2 = 4S^3 = 6.5 \times 10^{-6}$$

$$S = \sqrt[3]{\frac{6.5 \times 10^{-6}}{4}} = \mathbf{1.2 \times 10^{-2}\ M}$$

**Check:** We expect a low solubility from a slightly soluble salt. If we reverse the calculation, we should obtain the given $K_{sp}$: $4(1.2 \times 10^{-2})^3 = 6.9 \times 10^{-6}$, within rounding of $6.5 \times 10^{-6}$.

**Comment:** (1) Note that we did not double and *then* square $[OH^-]$. $2S$ *is* the $[OH^-]$, so we just squared it, as the $K_{sp}$ expression required.

(2) Once again, we assumed that the solid dissociates completely. Actually, the solubility is increased by the reaction $Ca(OH)_2(s) \rightleftharpoons CaOH^+(aq) + OH^-(aq)$ to about $2.0 \times 10^{-2}\ M$. Our calculated answer is only approximate because we do not take this other species into account.

A suspension of $Mg(OH)_2$ in water is sold as "milk of magnesia." It alleviates minor stomach disorders by neutralizing stomach acid, but the $[OH^-]$ is too low to harm the throat or stomach. What is the solubility of $Mg(OH)_2$ in water? $K_{sp} = 6.3 \times 10^{-10}$.

$K_{sp}$ values are a guide to relative solubility, as long as we compare compounds that produce the *same* number of ions. In such cases, *the higher the $K_{sp}$, the greater the solubility.* Table 18.3 shows this trend for several com-

### TABLE 18.3  Relationship Between $K_{sp}$ and Solubility at 25°C

| NO. OF IONS | FORMULA | CATION/ANION | $K_{sp}$ | SOLUBILITY ($M$) |
|---|---|---|---|---|
| 2 | $MgCO_3$ | 1/1 | $3.5 \times 10^{-8}$ | $1.9 \times 10^{-4}$ |
| 2 | $PbSO_4$ | 1/1 | $1.6 \times 10^{-8}$ | $1.3 \times 10^{-4}$ |
| 2 | $BaCrO_4$ | 1/1 | $2.1 \times 10^{-10}$ | $1.4 \times 10^{-5}$ |
| 3 | $Ca(OH)_2$ | 1/2 | $6.5 \times 10^{-6}$ | $1.2 \times 10^{-2}$ |
| 3 | $BaF_2$ | 1/2 | $1.5 \times 10^{-6}$ | $7.2 \times 10^{-3}$ |
| 3 | $CaF_2$ | 1/2 | $3.2 \times 10^{-11}$ | $2.0 \times 10^{-4}$ |
| 3 | $Ag_2CrO_4$ | 2/1 | $2.6 \times 10^{-12}$ | $8.7 \times 10^{-5}$ |

pounds. Note that for compounds that form three ions, the relationship holds whether the cation/anion ratio is 1/2 or 2/1.

### The Effect of a Common Ion on Solubility

As with acid-base systems, Le Châtelier's principle helps us explain the effect of a common ion on solubility. *The presence of a common ion decreases the solubility of a slightly soluble ionic compound.* This effect makes sense if we examine the equilibrium condition. Suppose we have a saturated solution of lead(II) chromate:

$$PbCrO_4(s) \rightleftharpoons Pb^{2+}(aq) + CrO_4^{2-}(aq) \qquad K_{sp} = [Pb^{2+}][CrO_4^{2-}] = 2.3 \times 10^{-13}$$

At a given temperature, $K_{sp}$ depends only on the product of the ion concentrations. If the concentration of either ion goes up, the other must go down to maintain the constant $K_{sp}$ value. Suppose we add $Na_2CrO_4$, a very soluble salt, to the saturated $PbCrO_4$ solution. The concentration of the common $CrO_4^{2-}$ ion will increase, and some of it will combine with $Pb^{2+}$ ion to form more solid $PbCrO_4$ (Figure 18.10). The overall effect will be a shift in the position of equilibrium to the left:

$$PbCrO_4(s) \overset{\longleftarrow}{\rightleftharpoons} Pb^{2+}(aq) + CrO_4^{2-}(aq;\ added)$$

The $[CrO_4^{2-}]$ is higher, but the $[Pb^{2+}]$, which represents the amount of $PbCrO_4$ dissolved, is lower; thus, less $PbCrO_4$ is in solution. The same result would be obtained if we dissolved $PbCrO_4$ in a solution of $Na_2CrO_4$.

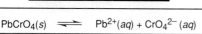

$$PbCrO_4(s) \rightleftharpoons Pb^{2+}(aq) + CrO_4^{2-}(aq) \qquad PbCrO_4(s) \overset{\longleftarrow}{\rightleftharpoons} Pb^{2+}(aq) + CrO_4^{2-}(aq;\ added)$$

**FIGURE 18.10**
**The effect of a common ion on solubility.** When a common ion is added to a saturated solution of an ionic compound, the solubility is lowered and more of the compound precipitates. **A,** Lead(II) chromate, a slightly soluble salt, forms a saturated aqueous solution. **B,** When $Na_2CrO_4$ solution is added, the amount of $PbCrO_4(s)$ increases, indicating a lower solubility in the presence of the common $CrO_4^{2-}$ ion.

SAMPLE PROBLEM 18.7 _____

### Calculating the Effect of a Common Ion on Solubility

**Problem:** In Sample Problem 18.6, we calculated the solubility of $Ca(OH)_2$ in water. What is its solubility in 0.10 *M* $Ca(NO_3)_2$? $K_{sp}$ of $Ca(OH)_2 = 6.5 \times 10^{-6}$.
**Plan:** From the equation and the ion-product expression for $Ca(OH)_2$, we see that the presence of $Ca^{2+}$ ion would lower the solubility. We set up a reaction table with $S$ equal to the $[Ca^{2+}]_{from\ Ca(OH)_2}$ and the $[Ca^{2+}]_{init}$ coming from $Ca(NO_3)_2$. This time we assume that $S$ can be neglected to simplify the math, and then we solve for $S$.

**Solution:** Writing the equation and $K_{sp}$ expression:

$$Ca(OH)_2(s) \rightleftharpoons Ca^{2+}(aq) + 2OH^-(aq) \qquad K_{sp} = [Ca^{2+}][OH^-]^2 = 6.5 \times 10^{-6}$$

Setting up the reaction table, with $S = [Ca^{2+}]_{from\ Ca(OH)_2}$:

| Concentration ($M$) | $Ca(OH)_2(s)$ | $\rightleftharpoons$ | $Ca^{2+}(aq)$ | + | $2OH^-(aq)$ |
|---|---|---|---|---|---|
| Initial | — | | 0.10 | | 0 |
| Change | — | | $+S$ | | $+2S$ |
| Equilibrium | — | | $0.10 + S$ | | $2S$ |

Making the assumption: $K_{sp}$ is small, so $S << 0.10\ M$, so $0.10\ M + S \approx 0.10\ M$. Substituting into the $K_{sp}$ expression and solving for $S$:

$$K_{sp} = [Ca^{2+}][OH^-]^2 = (0.10)(2S)^2 = 6.5 \times 10^{-6}$$

$$4S^2 = 6.5 \times 10^{-5}, \text{ so } S = \sqrt{\frac{6.5 \times 10^{-5}}{4}} = \mathbf{4.0 \times 10^{-3}\ M}$$

Checking the assumption: $\dfrac{4.0 \times 10^{-3}\ M}{0.10\ M} \times 100 = 4.0\% < 5\%$

**Check:** In Sample Problem 18.6, the solubility of $Ca(OH)_2$ was 0.012 $M$. Here, it is 0.0040 $M$, so the solubility decreased in the presence of added $Ca^{2+}$, the common ion, as we predicted.

**FOLLOW-UP PROBLEM 18.7**
To improve the quality of x-ray photos in the diagnosis of intestinal disorders, the patient drinks an aqueous suspension of $BaSO_4$. However, since $Ba^{2+}$ is toxic, its concentration is lowered by the addition of dilute $Na_2SO_4$. What is the solubility of $BaSO_4$ in **(a)** pure water and **(b)** 0.10 $M$ $Na_2SO_4$? $K_{sp}$ of $BaSO_4 = 1.1 \times 10^{-10}$.

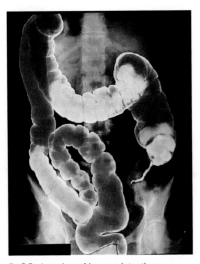

$BaSO_4$ imaging of human intestines

## The Effect of pH on Solubility

The hydronium ion can have a profound effect on the solubility of an ionic compound. *If the compound contains the anion of a weak acid, addition of $H_3O^+$ (from a strong acid) increases its solubility.* Once again, Le Châtelier's principle explains why. In a saturated solution of calcium carbonate, for example, we have

$$CaCO_3(s) \rightleftharpoons Ca^{2+}(aq) + CO_3^{2-}(aq)$$

Adding some strong acid introduces a large amount of $H_3O^+$, which immediately reacts with $CO_3^{2-}$ to form the weak acid $HCO_3^-$:

$$CO_3^{2-}(aq) + H_3O^+(aq) \rightarrow HCO_3^-(aq) + H_2O(l)$$

If enough $H_3O^+$ is added, further reaction occurs to form the unstable carbonic acid (shown as $[H_2CO_3]$), which decomposes to $H_2O$ and $CO_2$:

$$HCO_3^-(aq) + H_3O^+(aq) \rightarrow [H_2CO_3] \rightarrow CO_2(g) + 2H_2O(l) \\ + H_2O$$

**FIGURE 18.11**

**Test for the presence of a carbonate.** When a mineral that contains carbonate is treated with strong acid, the added $H_3O^+$ shifts the equilibrium position of the carbonate solubility toward more dissolution and the formation of gaseous $CO_2$.

Thus, the net effect of added $H_3O^+$ is a shift in the equilibrium position to the right, and more $CaCO_3$ dissolves:

$$CaCO_3(s) \underset{\longleftarrow}{\overset{\longrightarrow}{\rightleftharpoons}} Ca^{2+} + CO_3^{2-} \xrightarrow{H_3O^+} HCO_3^- \xrightarrow{H_3O^+} [H_2CO_3] \rightarrow CO_2(g) + H_2O$$

This particular case illustrates a qualitative field test for carbonate minerals because the $CO_2$ bubbles vigorously (Figure 18.11).

In contrast, adding $H_3O^+$ to a saturated solution of a compound with a strong-acid anion, such as silver chloride, has no effect:

$$AgCl(s) \rightleftharpoons Ag^+(aq) + Cl^-(aq)$$

The $Cl^-$ ion is the conjugate base of the strong acid HCl, and it ordinarily co-exists in solution with high $[H_3O^+]$. Since $Cl^-$ is not removed from the system, the equilibrium position is not affected.

SAMPLE PROBLEM 18.8 _____

## Predicting the Effect on Solubility of Adding Strong Acid

**Problem:** Write balanced equations to explain whether addition of the strong acid $HClO_4$ will affect the solubility of
**(a)** Lead(II) bromide          **(b)** Copper(II) hydroxide          **(c)** Iron(II) sulfide
**Plan:** We write the balanced dissolution equation and note the anion: weak-acid anions react with $H_3O^+$ and shift the equilibrium position toward more dissolution.

**Solution: (a)** $PbBr_2(s) \rightleftharpoons Pb^{2+}(aq) + 2Br^-(aq)$

**No effect**. Since $Br^-$ is the anion of HBr, a strong acid, it is a very weak base and $H_3O^+$ does not react with it.

**(b)** $Cu(OH)_2(s) \rightleftharpoons Cu^{2+}(aq) + 2OH^-(aq)$

**Increases solubility.** $OH^-$ is the anion of $H_2O$, a very weak acid, so it is a strong base, and the added $H_3O^+$ reacts with it:

$$OH^-(aq) + H_3O^+(aq) \rightleftharpoons 2H_2O(l)$$

**(c)** $FeS(s) + H_2O(l) \rightleftharpoons Fe^{2+}(aq) + HS^-(aq) + OH^-(aq)$

**Increases solubility.** We noted earlier that the $S^{2-}$ ion converts immediately to the hydrogen sulfide ion $HS^-$ in water. The added $H_3O^+$ reacts with both weak-acid anions, $HS^-$ and $OH^-$ :

$$HS^-(aq) + H_3O^+(aq) \rightarrow H_2S(aq) + H_2O(l)$$
$$OH^-(aq) + H_3O^+(aq) \rightarrow 2H_2O(l)$$

FOLLOW-UP PROBLEM 18.8
Write balanced equations to show how addition of $HNO_3$ will affect the solubility of
**(a)** calcium fluoride; **(b)** zinc sulfide; **(c)** silver iodide.

_____

Many principles of ionic equilibria are manifested in natural formations, as the Chemical Connections essay on the next page illustrates.

## Predicting the Formation of a Precipitate: $Q_{sp}$ vs. $K_{sp}$

In Chapter 16, we compared the values of $Q$ and $K$ to see if a reaction had reached equilibrium, and if not, in which net direction it would move until it did. Now we use the same approach to see if a precipitate will form and what concentrations of ions will cause it to do so.

As you know, $Q_{sp} = K_{sp}$ when the solution is saturated. If $Q_{sp}$ is greater than $K_{sp}$, the solution is momentarily supersaturated, and some solid precipitates until the remaining solution becomes saturated ($Q_{sp} = K_{sp}$). If $Q_{sp}$ is less than $K_{sp}$, the solution is unsaturated, and no precipitate forms at that temperature (more solid would dissolve). To summarize,

• $Q_{sp} = K_{sp}$: solution is saturated and no change occurs.
• $Q_{sp} > K_{sp}$: precipitate forms until solution is saturated.
• $Q_{sp} < K_{sp}$: solution is unsaturated and no precipitate forms.

**Chemistry in Geology:**
# Creation of a Limestone Cave

Caves and the detailed structures within them provide marvelous evidence of the workings of aqueous equilibria (Figure 18.A). The spires and vaults of these natural cathedrals are the products of reactions between rocks of appropriate composition and the water that has run over them for millennia. Caves are often found where the surrounding rock is rich in limestone ($CaCO_3$), a slightly soluble salt ($K_{sp} = 3.3 \times 10^{-9}$). Two facts help us understand how caves form:

1. $CO_2(g)$ is in equilibrium with $CO_2(aq)$ in natural waters:

$$CO_2(g) \overset{H_2O(l)}{\rightleftharpoons} CO_2(aq) \qquad \text{[Equation 1]}$$

The concentration of $CO_2$ in the water is proportional to the partial pressure of $CO_2(g)$ in contact with the water (Henry's law; Section 12.3):

$$[CO_2(aq)] \propto P_{CO_2}$$

Due to continual release of $CO_2$ from within the Earth (outgassing), the $P_{CO_2}$ in soil-trapped air is higher than in the atmosphere.

2. The presence of $H_3O^+(aq)$ increases the solubility of salts that contain the anion of a weak acid. Since $CO_2$ in solution produces $H_3O^+$,

$$CO_2(aq) + 2H_2O(l) \rightleftharpoons H_3O^+(aq) + HCO_3^-(aq)$$

the solubility of $CaCO_3$ increases in the presence of $CO_2(aq)$:

$$CaCO_3(s) + CO_2(aq) + H_2O(l) \rightleftharpoons$$
$$Ca^{2+}(aq) + 2HCO_3^-(aq) \qquad \text{[Equation 2]}$$

As surface water trickles through cracks in the ground, it meets soil-trapped air with a high $P_{CO_2}$, so its $[CO_2(aq)]$ increases (Equation 1 shifts to the right). When this $CO_2$-rich water contacts limestone, more $CaCO_3$ dissolves (Equation 2 shifts to the right). As a result, more water flows in, and more rock is carved out. Decades pass, and a cave slowly forms.

Eating its way through underground tunnels, some of the water [a $Ca(HCO_3)_2$ solution] passes through the ceiling of the growing cave. As it drips, it meets air with a lower $P_{CO_2}$ than that of the soil, so some $CO_2(aq)$ comes out of solution (Equation 1 shifts to the left). This causes some $CaCO_3$ to precipitate on the ceiling and on the floor

below, where the drops land (Equation 2 shifts to the left). In time, the ceiling bears an "icicle" of $CaCO_3$ called a *stalactite*, while a spike of $CaCO_3$, called a *stalagmite*, grows upward from the cave floor. Given enough time, they meet to form a column of precipitated limestone.

The same chemical process may lead to different shapes. Standing pools of $Ca(HCO_3)_2$ solution form limestone "lily pads" or "corals." Cascading solution forms delicate limestone "draperies" on a cave wall, with fabulous colors arising from trace metal ions, such as iron (reddish brown) or copper (bluish green).

**FIGURE 18.A**

**A view inside Carlsbad Caverns, NM.** The marvelous formations within this limestone cave result from subtle shifts in carbonate ionic equilibria.

SAMPLE PROBLEM 18.9 ───────────────────────────────────

## Predicting Whether a Precipitate Will Form

**Problem:** A common laboratory method for preparing a precipitate is to mix solutions of the component ions. Does a precipitate form when 0.100 L of 0.30 $M$ $Ca(NO_3)_2$ is mixed with 0.200 L of 0.060 $M$ NaF?

**Plan:** First, we must decide whether the ions present create a slightly soluble salt and, if so, look up its $K_{sp}$ value in Table 18.2. To find the initial ion concentrations, we calculate the moles of each ion and divide by the *total* volume, since one solution dilutes the other. Then we find $Q_{sp}$ and compare it with $K_{sp}$.

**Solution:** The ions present are $Ca^{2+}$, $Na^+$, $F^-$, and $NO_3^-$. All sodium and all nitrate salts are soluble, so the only possibility is $CaF_2$ ($K_{sp} = 3.2 \times 10^{-11}$). Calculating the ion concentrations:

$$\text{Moles of } Ca^{2+} = 0.30 \ M \ Ca^{2+} \times 0.100 \ L = 0.030 \ \text{mol } Ca^{2+}$$

$$[Ca^{2+}]_{init} = \frac{0.030 \ \text{mol } Ca^{2+}}{0.100 \ L + 0.200 \ L} = 0.10 \ M \ Ca^{2+}$$

$$\text{Moles of } F^- = 0.060 \ M \ F^- \times 0.200 \ L = 0.012 \ \text{mol } F^-$$

$$[F^-]_{init} = \frac{0.012 \ \text{mol } F^-}{0.100 \ L + 0.200 \ L} = 0.040 \ M \ F^-$$

Substituting into $Q_{sp}$ and comparing with $K_{sp}$:

$$Q_{sp} = [Ca^{2+}]_{init}[F^-]_{init}^2 = 0.10 \times 0.040^2 = 1.6 \times 10^{-4}$$

Since $Q_{sp} > K_{sp}$, **$CaF_2$ will precipitate** until $Q_{sp} = 3.2 \times 10^{-11}$.

**Check:** Don't forget to quickly check the math. For example, $Q_{sp} = (1 \times 10^{-1}) (4 \times 10^{-2})^2 = 1.6 \times 10^{-4}$. Since $K_{sp}$ is so low, $CaF_2$ must have a low solubility, and with the sizable concentrations being mixed, we would expect $CaF_2$ to precipitate.

FOLLOW-UP PROBLEM 18.9
Phosphate in natural waters often precipitates as insoluble salts, such as $Ca_3(PO_4)_2$. In a certain river, $[Ca^{2+}]_{init} = [PO_4^{3-}]_{init} = 1.0 \times 10^{-9} \ M$. Will $Ca_3(PO_4)_2$ precipitate? $K_{sp}$ of $Ca_3(PO_4)_2 = 1.2 \times 10^{-29}$.

As the Chemical Connections essay on the next two pages demonstrates, the principles of ionic equilibria often help us understand the chemical basis of complex environmental problems and provide ways to solve them.

## Section Summary

As a first approximation, the dissolved portion of a slightly soluble salt dissociates completely into ions. In a saturated solution, the ions are in equilibrium with the solid, and the product of the ion concentrations has a constant value ($Q_{sp} = K_{sp}$). The value of $K_{sp}$ can be obtained from the solubility, and vice versa. Adding a common ion lowers a compound's solubility. Adding $H_3O^+$ (lowering the pH) increases a compound's solubility if the anion of the compound is that of a weak acid. If $Q_{sp} > K_{sp}$ for a compound, a precipitate forms when two solutions are mixed. Limestone caves result from shifts in the $CaCO_3/CO_2$ equilibrium system. Lakes with limestone bedrock form a buffer system that prevents excess acidification.

**Chemistry in Environmental Science:**
**The Acid-Rain Problem**

The conflict between industrial society and the environment is very clear in the problem of acidic precipitation—rain, snow, or fog made unnaturally acidic by human action. Acidic precipitation has been recorded in all parts of the United States, in Canada, Mexico, and the Amazon basin, throughout Europe and Russia, and even at the North and South Poles. We've addressed several aspects of this problem in earlier chapters; now, armed with an understanding of ionic equilibria, let's examine some effects of acidic precipitation on aqueous, biological, and mineral systems. There are several troublesome chemical culprits:

1. *Sulfurous acid.* Sulfur dioxide ($SO_2$), formed primarily by the burning of high-sulfur coal, forms sulfurous acid ($H_2SO_3$) in contact with water.

2. *Sulfuric acid.* Sulfur trioxide ($SO_3$) forms through the atmospheric oxidation of $SO_2$ by ozone or by $O_2$ in a reaction catalyzed by dust:

$$SO_2(g) + O_3(g) \rightleftharpoons SO_3(g) + O_2(g)$$

$$2SO_2(g) + O_2(g) \xrightarrow{\text{dust}} 2SO_3(g)$$

Sulfur trioxide forms sulfuric acid ($H_2SO_4$) in contact with water.

3. *Nitric acid.* Nitrogen oxides (collectively known as $NO_x$) form in the reaction of $N_2$ and $O_2$. NO is formed by combustion, primarily in car engines, and it forms $NO_2$ in air. Rainwater converts $NO_x$ to $HNO_2$ and $HNO_3$. The strong acids $H_2SO_4$ and $HNO_3$ cause the greatest concern (Figure 18.B).

How does the pH of acidic precipitation compare with that of natural waters? *Normal rainwater is weakly acidic* because it contains dissolved $CO_2$ from the air:

$$CO_2(g) + 2H_2O(l) \rightleftharpoons H_3O^+(aq) + HCO_3^-(aq)$$

Based on the volume percent of $CO_2$ in air, the solubility of $CO_2$ in water, and the $K_{a1}$ of $H_2CO_3$, the pH of normal rainwater is about 5.6 (see Problem 18.113 at the end of the chapter). In stark contrast, the *average* pH of rainfall in many parts of the eastern United States was 4.2 as long ago as 1984. Worldwide, rain in Sweden and Pennsylvania share second prize with a pH of 2.7, about the same as vinegar. Rain in Wheeling, WV, wins first prize with a pH of 1.8, between that of lemon juice and stomach acid! Acidic fog in California sometimes has a pH of 1.6.

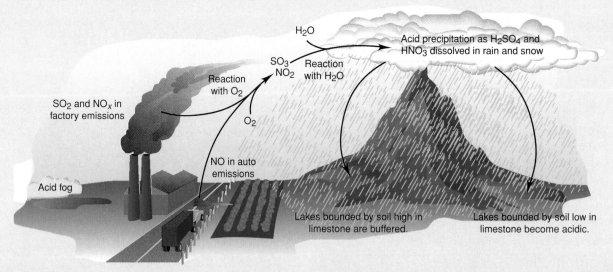

**FIGURE 18.B**
**Formation of acidic precipitation.** A complex interplay of human activities, atmospheric chemistry, and environmental distribution leads to acidic precipitation. Car exhaust and factory waste gases provide lower oxides of nitrogen and sulfur. These are oxidized by $O_2$ to higher oxides ($NO_2$, $SO_3$), which react with moisture to form acidic rain, snow, and fog. In contact with acidic precipitation, many lakes become acidified, whereas limestone-bounded lakes form a carbonate buffer that prevents acidification.

**FIGURE 18.C**
A forest damaged by acid rain.

In essence, limestone-grounded lakes become enormous $CO_3^{2-}/HCO_3^-$ buffer solutions, absorbing the additional $H_3O^+$ and maintaining a stable pH. In fact, lakes, rivers, and ground water in limestone-rich soils actually remain mildly basic.

For lakes and rivers not in contact with such soils, society's choices at present are few and expensive. A direct attack on the symptoms is the *liming* of lakes and rivers. As of 1990, Sweden had spent $25 million to neutralize slightly more than 3000 lakes by adding lime:

$$CaO(s; \text{ added}) + H_3O^+(aq; \text{ in lake}) \rightarrow$$
$$Ca^{2+}(aq) + OH^-(aq) + H_2O(l)$$

In Canada, adding lime costs more than $50 per acre, so a single large lake can cost several thousand dollars. This approach is, at best, only a stopgap because within several years the lakes are acidic again. Two longer range options aim at preventing the problem. Burning low-sulfur coal would reduce $SO_2$ formation, but such coal deposits are rare and expensive to mine. Alternatively, coal can be converted into gaseous and liquid low-sulfur fuels (see the Chemical Connections essay in Section 6.5).

The effects on soil, water, and living things of these 10-fold to 10,000-fold excesses of $[H_3O^+]$ are very destructive. Most fish and shellfish die at pH values between 4.5 and 5.0. With tens of thousands of rivers and lakes around the world becoming acidified, the loss of fish has become alarming. In addition, millions of acres of forest and crops have been harmed by the acid, which removes protective waxes from leaf surfaces (Figure 18.C).

Many principles of aqueous equilibria bear directly on the effects of acid rain. The aluminosilicates that make up most soils are extremely insoluble in water. In these materials, the $Al^{3+}$ is bonded to $OH^-$ and $O^{2-}$ ion in complex structures (see Gallery in Section 13.6). Continual contact with the $H_3O^+$ in acid rain dissolves some of the bound $Al^{3+}$ through reaction with the $OH^-$ and $O^{2-}$. The $Al^{3+}$ ion is toxic and eventually causes the death of fish. Along with the solubilized $Al^{3+}$ ions, the acid carries away nutrient ions.

Acid rain dissolves the calcium carbonate in the marble and limestone of buildings and monuments (Figure 18.D). Ironically, the same chemical system that destroys these structures is responsible for saving those lakes that lie on older soil rich in limestone. As acid rain falls, the $H_3O^+$ reacts with the dissolved carbonate ion in the lake to form bicarbonate:

$$CO_3^{2-}(aq) + H_3O^+(aq) \rightarrow HCO_3^-(aq) + H_2O(l)$$

1944

1994

**FIGURE 18.D**
**The effect of acid rain on marble statuary.** The calcium carbonate that is the major component of marble and limestone is slowly decomposed by acid rain. These photos of the same statue of George Washington, Washington Square Park, New York City, were taken 50 years apart.

## 18.4 Equilibria Involving Complex Ions

The final type of aqueous ionic equilibrium we consider involves a different type of ion than we have examined up to now. Simple ions, such as $Na^+$ or $SO_4^{2-}$, consist of one or a few atoms that have a deficit or excess of electrons. A **complex ion** consists of a central metal ion covalently bonded to two or more anions or molecules called **ligands.** Hydroxide, chloride, and cyanide ions are common ionic ligands; water and ammonia are common molecular ligands. In the complex ion $Cr(NH_3)_6^{3+}$, for example, $Cr^{3+}$ is the central metal ion and six $NH_3$ molecules are the ligands (Figure 18.12).

As we discussed in Section 17.9, *all complex ions are Lewis adducts.* The metal ion acts as a Lewis acid (accepts an electron pair) and the ligand acts as a Lewis base (donates an electron pair). The acidic hydrated metal ions that we discussed in Section 17.6 are complex ions with water molecules as ligands. In Chapter 22, we discuss the transition metals and the structures and properties of the numerous complex ions they form. Our focus here is on equilibria involving hydrated ions and other ligands besides water.

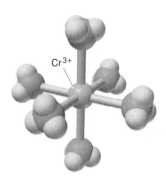

**FIGURE 18.12**

**The $Cr(NH_3)_6^{3+}$ complex ion.** A complex ion consists of a central metal ion, such as $Cr^{3+}$, covalently bonded to a specific number of ligands, such as $NH_3$.

### Formation of Complex Ions

Whenever a metal ion enters water, a complex ion forms, with water as the ligand. If this hydrated cation is treated with a solution of another ligand, the bound water molecules may exchange for the other ligand. For example, a hydrated $M^{2+}$ ion in aqueous $NH_3$ forms the $M(NH_3)_4^{2+}$ complex ion:

$$M(H_2O)_4^{2+}(aq) + 4NH_3(aq) \rightleftharpoons M(NH_3)_4^{2+}(aq) + 4H_2O(l)$$

At equilibrium, this system is expressed by a constant whose form follows that of any other equilibrium constant:

$$K_c = \frac{[M(NH_3)_4^{2+}][H_2O]^4}{[M(H_2O)_4^{2+}][NH_3]^4}$$

As with other aqueous equilibrium systems, the concentration of water is constant, so we incorporate it into $K_c$ and obtain a new equilibrium constant, the **formation constant $K_f$:**

$$K_f = \frac{K_c}{[H_2O]^4} = \frac{[M(NH_3)_4^{2+}]}{[M(H_2O)_4^{2+}][NH_3]^4}$$

At the molecular level, the actual process is stepwise, with ammonia molecules replacing water molecules one at a time to give a series of intermediate species, each with its own formation constant (Figure 18.13):

$$M(H_2O)_4^{2+}(aq) + NH_3(aq) \rightleftharpoons M(H_2O)_3(NH_3)^{2+}(aq) + H_2O(l)$$

$$K_{f1} = \frac{[M(H_2O)_3(NH_3)^{2+}]}{[M(H_2O)_4^{2+}][NH_3]}$$

$$M(H_2O)_3(NH_3)^{2+}(aq) + NH_3(aq) \rightleftharpoons M(H_2O)_2(NH_3)_2^{2+}(aq) + H_2O(l)$$

$$K_{f2} = \frac{[M(H_2O)_2(NH_3)_2^{2+}]}{[M(H_2O)_3(NH_3)^{2+}][NH_3]}$$

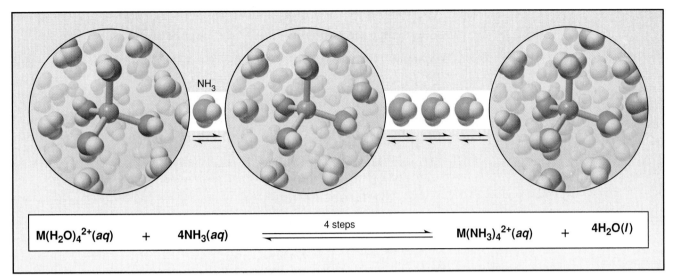

**FIGURE 18.13**

**The stepwise exchange of NH$_3$ for H$_2$O in M(H$_2$O)$_4{}^{2+}$.** The ligands of a complex ion can exchange for other ligands. When ammonia is added to a solution of a hydrated M$^{2+}$ ion, M(H$_2$O)$^{2+}$, NH$_3$ molecules replace the bound H$_2$O molecules one at a time to form the M(NH$_3$)$_4{}^{2+}$ ion. The molecular views show the first exchange and the fully ammoniated ion.

and

$$M(H_2O)_2(NH_3)_2{}^{2+}(aq) + NH_3(aq) \rightleftharpoons M(H_2O)(NH_3)_3{}^{2+}(aq) + H_2O(l)$$

$$K_{f3} = \frac{[M(H_2O)(NH_3)_3{}^{2+}]}{[M(H_2O)_2(NH_3)_2{}^{2+}][NH_3]}$$

$$M(H_2O)(NH_3)_3{}^{2+}(aq) + NH_3(aq) \rightleftharpoons M(NH_3)_4{}^{2+}(aq) + H_2O(l)$$

$$K_{f4} = \frac{[M(NH_3)_4{}^{2+}]}{[M(H_2O)(NH_3)_3{}^{2+}][NH_3]}$$

The *sum* of the equations gives the overall equation, so the *product* of the individual formation constants gives the overall formation constant:

$$K_f = K_{f1} \times K_{f2} \times K_{f3} \times K_{f4}$$

The $K_f$ for each step is much larger than one, which means that ammonia is a stronger Lewis base than water. Therefore, if we add a large excess of ammonia to the M(H$_2$O)$_4{}^{2+}$ solution, essentially all the metal ion exists as M(NH$_3$)$_4{}^{2+}$.

Table 18.4 shows the formation constants of some complex ions. Notice that the $K_f$ values are all $10^6$ or greater, which means that when these ions form they are very stable. Because of this behavior, complex-ion formation is employed to retrieve a metal from its ore or to eliminate a toxic or unwanted metal ion from a solution, or to convert it to a different form, as Sample Problem 18.10 shows for the zinc ion.

**TABLE 18.4 Formation Constants ($K_f$) of Some Complex Ions**

| COMPLEX ION | $K_f$ |
|---|---|
| Ag(CN)$_2{}^-$ | $3.0 \times 10^{20}$ |
| Fe(CN)$_6{}^{4-}$ | $3 \times 10^{35}$ |
| Fe(CN)$_6{}^{3-}$ | $4.0 \times 10^{43}$ |
| Hg(CN)$_4{}^{2-}$ | $9.3 \times 10^{38}$ |
| Zn(CN)$_4{}^{2-}$ | $4.2 \times 10^{19}$ |
| AlF$_6{}^{3-}$ | $4 \times 10^{19}$ |
| CdI$_4{}^{2-}$ | $1 \times 10^{6}$ |
| Ag(NH$_3$)$_2{}^+$ | $1.7 \times 10^{7}$ |
| Cu(NH$_3$)$_4{}^{2+}$ | $5.6 \times 10^{11}$ |
| Zn(NH$_3$)$_4{}^{2+}$ | $7.8 \times 10^{8}$ |
| Al(OH)$_4{}^-$ | $3 \times 10^{33}$ |
| Be(OH)$_4{}^{2-}$ | $4 \times 10^{18}$ |
| Co(OH)$_4{}^{2-}$ | $5 \times 10^{9}$ |
| Ni(OH)$_4{}^{2-}$ | $2 \times 10^{28}$ |
| Pb(OH)$_3{}^-$ | $8 \times 10^{13}$ |
| Sn(OH)$_3{}^-$ | $3 \times 10^{25}$ |
| Zn(OH)$_4{}^{2-}$ | $3 \times 10^{15}$ |
| Ag(S$_2$O$_3$)$_2{}^{3-}$ | $4.7 \times 10^{13}$ |

SAMPLE PROBLEM 18.10 _____

## Calculating the Concentrations of Complex Ions

**Problem:** An industrial chemist converts $Zn(H_2O)_4^{2+}$ to the more stable $Zn(NH_3)_4^{2+}$ by mixing 50.0 L of 0.0020 $M$ $Zn(H_2O)_4^{2+}$ and 25.0 L of 0.15 $M$ $NH_3$. What is the final $[Zn(H_2O)_4^{2+}]$? $K_f$ of $Zn(NH_3)_4^{2+} = 7.8 \times 10^8$.

**Plan:** We write the equation and the $K_f$ expression. As always, we must set up a reaction table to calculate the equilibrium concentrations. To do so, we first find $[Zn(H_2O)_4^{2+}]_{init}$ and $[NH_3]_{init}$ using the *total* volume. With the large excess of $NH_3$ and high $K_f$, we assume that almost all the $Zn(H_2O)_4^{2+}$ is converted to $Zn(NH_3)_4^{2+}$. Then we solve for $x$, the $[Zn(H_2O)_4^{2+}]$ at equilibrium.

**Solution:** Writing the equation and the $K_f$ expression:

$$Zn(H_2O)_4^{2+}(aq) + 4NH_3(aq) \rightleftharpoons Zn(NH_3)_4^{2+}(aq) + 4H_2O(l)$$

$$K_f = \frac{[Zn(NH_3)_4^{2+}]}{[Zn(H_2O)_4^{2+}][NH_3]^4}$$

Finding the initial reactant concentrations:

$$[Zn(H_2O)_4^{2+}]_{init} = \frac{50.0 \text{ L} \times 0.0020 \ M}{50.0 \text{ L} + 25.0 \text{ L}} = 1.3 \times 10^{-3} \ M$$

$$[NH_3]_{init} = \frac{25.0 \text{ L} \times 0.15 \ M}{50.0 \text{ L} + 25.0 \text{ L}} = 5.0 \times 10^{-2} \ M$$

Setting up a reaction table, with $x = [Zn(H_2O)_4^{2+}]$ at equilibrium: We assume that nearly all the $Zn(H_2O)_4^{2+}$ is converted to $Zn(NH_3)_4^{2+}$. So,

$$[NH_3]_{reacted} = 4(1.3 \times 10^{-3} \ M) = 5.2 \times 10^{-3} \ M$$

$$[Zn(NH_3)_4^{2+}] = 1.3 \times 10^{-3} \ M$$

| Concentration $(M)$ | $Zn(H_2O)_4^{2+}(aq)$ + | $4NH_3(aq)$ $\rightleftharpoons$ | $Zn(NH_3)_4^{2+}(aq)$ | $+4H_2O(l)$ |
|---|---|---|---|---|
| Initial | $1.3 \times 10^{-3}$ | $5.0 \times 10^{-2}$ | 0 | — |
| Change | $\sim(-1.3 \times 10^{-3})$ | $-5.2 \times 10^{-3}$ | $+1.3 \times 10^{-3}$ | — |
| Equilibrium | $x$ | $4.5 \times 10^{-2}$ | $1.3 \times 10^{-3}$ | — |

Solving for $x$, the $[Zn(H_2O)_4^{2+}]$ remaining at equilibrium:

$$K_f = \frac{[Zn(NH_3)_4^{2+}]}{[Zn(H_2O)_4^{2+}][NH_3]^4} = \frac{1.3 \times 10^{-3}}{x(4.5 \times 10^{-2})^4} = 7.8 \times 10^8$$

$$x = [Zn(H_2O)_4^{2+}] = \mathbf{4.1 \times 10^{-7} \ M}$$

**Check:** Since the $K_f$ is large, we would expect the $[Zn(H_2O)_4^{2+}]$ remaining to be very low.

FOLLOW-UP PROBLEM 18.10
Cyanide ion is toxic because it forms stable complex ions with the $Fe^{3+}$ in certain iron-containing proteins engaged in energy production. To study this effect, a biochemist mixes 25.5 mL of $3.1 \times 10^{-2} \ M$ $Fe(H_2O)_6^{3+}$ with 35.0 mL of 1.5 $M$ NaCN. What is the final $[Fe(H_2O)_6^{3+}]$? $K_f$ of $Fe(CN)_6^{3-} = 4.0 \times 10^{43}$.

_____

**Complex ions and the solubility of precipitates.** In Section 18.3, you saw that $H_3O^+$ ions increase the solubility of a slightly soluble ionic compound if its anion is that of a weak acid. Similarly, *a ligand increases the*

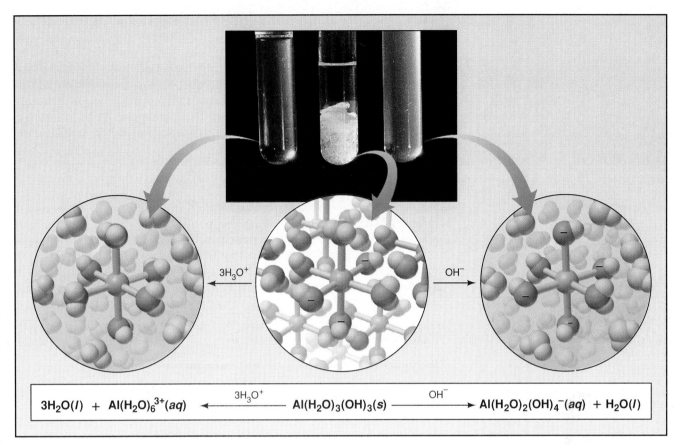

$$3H_2O(l) + Al(H_2O)_6^{3+}(aq) \xleftarrow{\phantom{xx}3H_3O^+\phantom{xx}} Al(H_2O)_3(OH)_3(s) \xrightarrow{\phantom{xx}OH^-\phantom{xx}} Al(H_2O)_2(OH)_4^-(aq) + H_2O(l)$$

**FIGURE 18.14**

**The amphoteric behavior of aluminum hydroxide.** When solid $Al(OH)_3$ is treated with $H_3O^+$ *(left)* or with $OH^-$ *(right)*, it dissolves due to the formation of soluble complex ions. The molecular views show that $Al(OH)_3$ is actually the neutral species $Al(H_2O)_3(OH)_3$. Addition of $OH^-$ forms the soluble $Al(H_2O)_2(OH)_4^-$ ion; addition of $H_3O^+$ forms the soluble $Al(H_2O)_6^{3+}$ ion.

Several other slightly soluble hydroxides, such as those of $Cr^{3+}$, $Zn^{2+}$, $Pb^{2+}$, and $Sn^{2+}$, are amphoteric due to similar reactions. In contrast, the hydroxides of $Fe^{2+}$, $Fe^{3+}$, and $Ca^{2+}$ dissolve in acid but *not* in base because the remaining bound water molecules are not acidic enough:

$$Zn(H_2O)_2(OH)_2(s) + OH^-(aq) \rightleftharpoons Zn(H_2O)(OH)_3^-(aq) + H_2O(l)$$

$$Fe(H_2O)_3(OH)_3(s) + OH^-(aq) \longrightarrow \text{no reaction}$$

The difference in solubility in base between $Al(OH)_3$ and $Fe(OH)_3$ is the key to an important separation step in the production of aluminum metal, so we'll see it again in Section 23.3. We employ it in the next section as well in analyzing a mixture of ions.

**Section Summary**

A complex ion consists of a central metal ion covalently bonded to two or more negative or neutral ligands. Its formation is described by a formation constant $K_f$. A hydrated metal ion is a complex ion with water molecules as ligands. Other ligands can displace the water in a stepwise process. In most cases, the $K_f$ value of each step is large, so the fully substituted complex ion forms almost completely in the presence of excess ligand. A ligand solution increases the solubility of an ionic precipitate if the cation forms a complex ion with the ligand. Amphoteric metal hydroxides dissolve in acid and base due to the formation of soluble complex ions.

## 18.5   Applications of Ionic Equilibria to Chemical Analysis

Many of the ideas we've discussed in this chapter are used in the analysis of the ions in a mixture. In this brief introduction to an extensive and time-honored field, we discuss how control of the precipitating ion concentration, either directly or indirectly, is used to separate one metal ion from another, and how complex mixtures of ions are separated into smaller groups and the ions identified.

### Selective Precipitation

We can often separate one ion in a solution from another by exploiting differences in the solubility of their compounds with the same precipitating ion. In the process of **selective precipitation,** we add precipitating ion until the $Q_{sp}$ of the *more soluble* compound is almost equal to its $K_{sp}$. This method ensures that the $K_{sp}$ of the *less soluble* compound will be exceeded as much as possible. As a result, the maximum amount of the less soluble compound precipitates, but none of the more soluble compound does.

SAMPLE PROBLEM 18.12 ⎯⎯⎯⎯⎯⎯⎯⎯⎯⎯⎯⎯⎯⎯⎯⎯⎯⎯⎯⎯

### Separating Ions by Selective Precipitation

**Problem:** A solution consists of 0.20 $M$ $MgCl_2$ and 0.10 $M$ $CuCl_2$. How would you separate the metal ions as their hydroxides? $K_{sp}$ of $Mg(OH)_2 = 6.3 \times 10^{-10}$; $K_{sp}$ of $Cu(OH)_2 = 2.2 \times 10^{-20}$.

**Plan:** Since the two hydroxides have the same formula type (1/2), we compare their $K_{sp}$ values and see that $Mg(OH)_2$ is about $10^{10}$ times *more* soluble than $Cu(OH)_2$, so $Cu(OH)_2$ would precipitate first. We solve for the $[OH^-]$ that will just give a saturated solution of $Mg(OH)_2$ because this $[OH^-]$ will precipitate the greatest amount of $Cu^{2+}$ ion, and then we calculate the $[Cu^{2+}]$ remaining to see if the separation was accomplished.

**Solution:** Writing the equations and $K_{sp}$ expressions:

$$Mg(OH)_2(s) \rightleftharpoons Mg^{2+}(aq) + 2OH^-(aq) \qquad K_{sp} = [Mg^{2+}][OH^-]^2$$

$$Cu(OH)_2(s) \rightleftharpoons Cu^{2+}(aq) + 2OH^-(aq) \qquad K_{sp} = [Cu^{2+}][OH^-]^2$$

Calculating the $[OH^-]$ that gives a saturated $Mg(OH)_2$ solution:

$$[OH^-] = \sqrt{\frac{K_{sp}}{[Mg^{2+}]}} = \sqrt{\frac{6.3 \times 10^{-10}}{0.20}} = 5.6 \times 10^{-5} \ M$$

This is the maximum $[OH^-]$ we can add that will *not* precipitate $Mg^{2+}$ ion. Calculating the $[Cu^{2+}]$ remaining at this $[OH^-]$:

$$[Cu^{2+}] = \frac{K_{sp}}{[OH^-]^2} = \frac{2.2 \times 10^{-20}}{(5.6 \times 10^{-5})^2} = 7.0 \times 10^{-12} \ M$$

Since the initial $[Cu^{2+}]$ was 0.10 $M$, virtually all the $Cu^{2+}$ ion would be precipitated.

**Check:** Rounding, we find that the $[OH^-]$ seems right: $\sim \sqrt{(6 \times 10^{-10})/0.2} = 5 \times 10^{-5}$. The $[Cu^{2+}]$ remaining also seems in the correct range: $200 \times 10^{-22}/(5 \times 10^{-5})^2 = 8 \times 10^{-12}$.

**Comment:** There is often more than one approach to accomplish the same analytical step. In this case, we could have added excess ammonia to make the solution basic enough to precipitate both hydroxides,

$$Mg^{2+}(aq) + 2NH_3(aq) + 2H_2O(l) \rightleftharpoons Mg(OH)_2(s) + 2NH_4^+(aq)$$

$$Cu^{2+}(aq) + 2NH_3(aq) + 2H_2O(l) \rightleftharpoons Cu(OH)_2(s) + 2NH_4^+(aq)$$

but the $Cu(OH)_2$ would then have dissolved by forming a soluble complex ion:

$$Cu(OH)_2(s) + 4NH_3(aq) \rightleftharpoons Cu(NH_3)_4^{2+}(aq) + 2OH^-(aq)$$

**FOLLOW-UP PROBLEM 18.12**

A solution of alkaline earth ions contains 0.050 $M$ $BaCl_2$ and 0.025 $M$ $CaCl_2$. What concentration of $SO_4^{2-}$ must be present to leave 99.99% of the more soluble cation in solution? $K_{sp}$ of $BaSO_4 = 1.1 \times 10^{-10}$ and $K_{sp}$ of $CaSO_4 = 2.4 \times 10^{-5}$.

Sometimes two or more types of ionic equilibria are controlled simultaneously to precipitate ions selectively. Consider the use of the hydrogen sulfide ion, $HS^-$, as a precipitating ion to separate two metal ions. In a manner similar to that used in Sample Problem 18.12, we control the $[HS^-]$ to exceed the $K_{sp}$ of one metal sulfide, but not another. We exert this control on $[HS^-]$ by controlling $H_2S$ dissociation through adjustment of the $[H_3O^+]$, because $H_2S$ is the source of the $HS^-$ ion:

$$H_2S(aq) + H_2O(l) \rightleftharpoons H_3O^+(aq) + HS^-(aq)$$

Here is the way these interactions are controlled. When the $[H_3O^+]$ is high, from addition of strong acid, the equilibrium position of the $H_2S$ dissociation shifts to the left, so $[HS^-]$ decreases. With a low $[HS^-]$, the *less* soluble sulfide precipitates. Conversely, when $[H_3O^+]$ is low, from addition of strong base, $[HS^-]$ increases and the *more* soluble sulfide precipitates. This approach utilizes **simultaneous equilibria** to separate the metal ions: we shift one equilibrium system ($H_2S$ dissociation) by adjusting a second ($H_2O$ ionization) to control a third (metal sulfide solubility).

## Qualitative Analysis: Identifying Ions in Complex Mixtures

The practice of inorganic **qualitative analysis,** the separation and identification of the ions in a mixture, was once an essential part of a chemist's skills. Today, many of these wet chemical techniques have been replaced by instrumental methods. Nevertheless, they provide an excellent means for studying ionic equilibria, including some of the very ones utilized by modern analytical instruments. In this discussion, we apply the principles of solubility and complex-ion equilibria to separate and characterize a mixture of cations; similar procedures exist for anions. There is no fixed series of steps for identifying a given set of ions; rather, the design of the procedure often depends on the analyst's ingenuity.

The general approach begins by separating the unknown mixture into *ion groups.* (Ion groups have nothing to do with periodic table groups.) The mixture is treated with a solution that precipitates a certain group of ions and leaves the others in solution. Filtration or centrifugation (the rapid spinning of the tube to collect the solid in a compact pellet at the bottom) separates the precipitated ions. The remaining dissolved ions are removed and treated with a solution that precipitates a different group of ions. These steps are repeated until the original mixture has been separated into specific ion groups. In an actual analysis, a *known solution* (one that *does* contain all the ions under study) and a *blank* of distilled water are treated in exactly the same way as the unknown solution, thereby making it much easier to judge a positive or negative result.

Figure 18.15 shows one common scheme for separating cations into ion groups. The entire mixture of soluble ions is treated with 6 $M$ HCl. Since

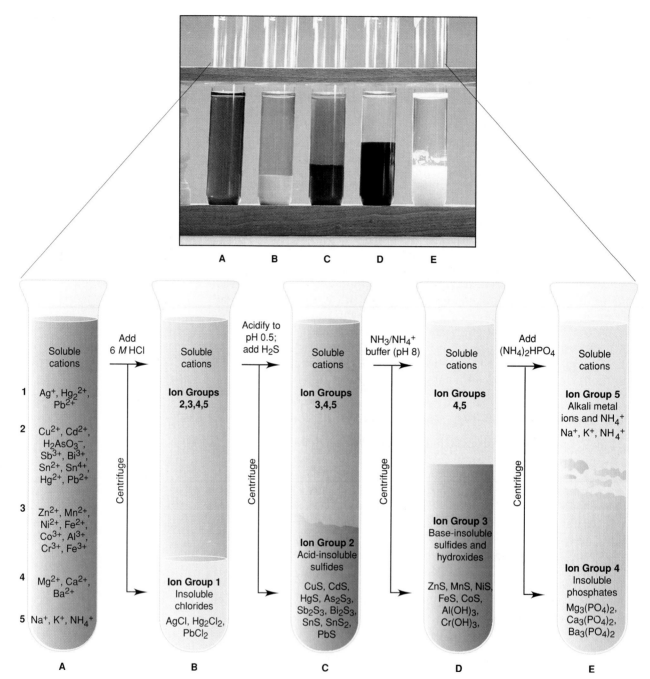

**FIGURE 18.15**

**A qualitative analysis scheme for separating cations into ion groups.** The first tube contains a solution of ions (listed according to ion group). It is treated with the first precipitating solution (6 $M$ HCl) and centrifuged to separate the precipitate (chlorides) of ion group 1 from the solution of remaining ions. This solution is decanted and treated with the next precipitating solution, and so forth.

most metal chlorides are soluble, only those few ions that form insoluble chlorides precipitate in this step. If a precipitate appears, it consists of one or more chlorides of

**Ion group 1** *(insoluble chlorides):*     $Ag^+$, $Hg_2^{2+}$, $Pb^{2+}$

If none appears, the ions from ion group 1 were not present in the mixture. If a precipitate forms, the tube is centrifuged, and the solution is carefully decanted (poured off).

The decanted solution is adjusted to pH 0.5 and treated with aqueous $H_2S$. The $H_3O^+$ present keeps the $[HS^-]$ very low, so any precipitate will contain one or more of the least soluble sulfides, which are those of

**Ion group 2** *(acid-insoluble sulfides):*
$Cu^{2+}$, $Cd^{2+}$, $Hg^{2+}$, $As^{3+}$, $Sb^{3+}$, $Bi^{3+}$, $Sn^{2+}$, $Sn^{4+}$, $Pb^{2+}$

($PbCl_2$ is slightly soluble in water, so a small amount of $Pb^{2+}$ remains in solution after addition of HCl and appears in both ion groups 1 and 2.) If no precipitate appears under these conditions, the members of ion group 2 were not present. The tube is centrifuged and the solution decanted.

Next, the decanted solution is made slightly basic with a buffer of $NH_3/NH_4^+$. The $OH^-$ present increases the $[HS^-]$, which causes precipitation of the more soluble sulfides and some hydroxides. The cations included are those of

**Ion group 3** *(base-insoluble sulfides and hydroxides):*
$Zn^{2+}$, $Mn^{2+}$, $Ni^{2+}$, $Fe^{2+}$, $Co^{2+}$ as sulfides, and $Al^{3+}$, $Cr^{3+}$ as hydroxides

($Fe^{3+}$ is reduced to $Fe^{2+}$ in this step.) If no precipitate appears, these ions were not present. Centrifuging and decanting gives the next solution.

To this slightly basic solution, $(NH_4)_2HPO_4$ is added, which precipitates any alkaline earth ions as phosphates. (Alternatively, $Na_2CO_3$ is added and the precipitate contains alkaline earth carbonates.) In either case, the ions precipitated are from

**Ion group 4** *(insoluble phosphates):*     $Mg^{2+}$, $Ca^{2+}$, $Ba^{2+}$

If no precipitate appears, these ions were not present. After centrifuging and decanting, the final solution contains ions from

**Ion group 5:**     $Na^+$, $K^+$, $NH_4^+$

With the ions separated into ion groups, the analyst then devises schemes to identify each ion in a group. For example, identification of the ions in ion group 5 is usually done through flame and color tests (Figure 18.16). In flame tests, sodium produces a characteristic yellow-orange color, and potassium gives a violet color. Moist red litmus paper turns blue when held over a solution containing $NH_4^+$ that has been made basic with NaOH because the $NH_3$ produced reacts with the $H_2O$ on the paper:

$$NH_4^+(aq) + OH^-(aq; added) \rightarrow NH_3(g) + H_2O(l)$$

$$NH_3(g) + H_2O(\text{moist red litmus}) \rightleftharpoons NH_4^+(aq) + OH^-(\text{turns litmus blue})$$

Because $NH_4^+$ ion has been added during earlier steps and $Na^+$ is a common contaminant in ammonium salts, the analyst must perform tests for these ions on the original ion mixture.

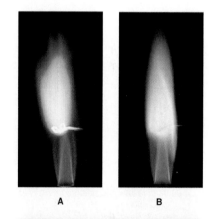

A                    B

C

**FIGURE 18.16**
**Tests to determine the presence of cations in ion group 5.** Ion group 5 is soluble through all the steps in Figure 18.15. It consists of some alkali metal ions and $NH_4^+$ ion. Flame tests give characteristic colors for $Na^+$ ion **(A)**, and $K^+$ ion **(B)**. Adding $OH^-$ to $NH_4^+$ forms gaseous $NH_3$, which turns moistened red litmus paper blue **(C)**.

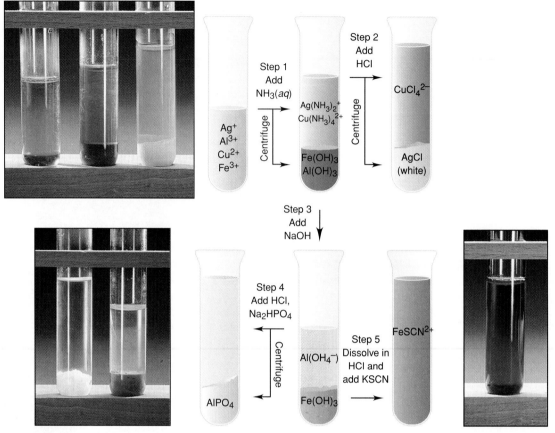

**FIGURE 18.17**
A qualitative analysis scheme for $Ag^+$, $Al^{3+}$, $Cu^{2+}$, and $Fe^{3+}$.

Qualitative analyses often include the formation of complex ions in addition to precipitates. Here is a scheme to identify four ions from several ion groups (Figure 18.17). Suppose we have a solution that contains some or all of the following ions: $Ag^+$, $Al^{3+}$, $Cu^{2+}$, and $Fe^{3+}$. We can separate and identify them by the following steps:

*Step 1.* Because $Al^{3+}$ and $Fe^{3+}$ form insoluble hydroxides, and $Ag^+$ and $Cu^{2+}$ form soluble complex ions with $NH_3$, we add aqueous $NH_3$, which provides both the $NH_3$ molecules that act as ligands and a small amount of $OH^-$ ions. A precipitate forms that may be $Fe(OH)_3$ (brown), $Al(OH)_3$ (white), or both:

$$Al^{3+}(aq) + 3OH^-(aq) \rightleftharpoons Al(OH)_3(s) \text{ [white]}$$

$$Fe^{3+}(aq) + 3OH^-(aq) \rightleftharpoons Fe(OH)_3(s) \text{ [brown]}$$

In the solution, complex ions of $Ag^+$ (colorless), $Cu^{2+}$ (blue), or both may be present:

$$Ag^+(aq) + 2NH_3(aq) \rightleftharpoons Ag(NH_3)_2^+(aq) \text{ [colorless]}$$

$$Cu^{2+}(aq) + 4NH_3(aq) \rightleftharpoons Cu(NH_3)_4^{2+}(aq) \text{ [blue]}$$

The $Cu^{2+}$ also forms an insoluble hydroxide, but the presence of high $[NH_3]$ dissolves the precipitate and forms the complex ion. We centrifuge to separate the soluble complex ions from the precipitate.

*Step 2.* To separate the $Ag^+$ and $Cu^{2+}$ that may be present as complex ions, we take advantage of the ability of $Cl^-$ to act as both a ligand and a precipitating ion. We add HCl to the solution of complex ions and centrifuge. The $H_3O^+$ ions react with the $NH_3$ in equilibrium with the complex ions to form $NH_4^+$, which does not act as a ligand:

$$Ag(NH_3)_2^+(aq) + 2H_3O^+(aq; added) \rightarrow Ag^+(aq) + 2NH_4^+(aq) + 2H_2O(l)$$

$$Cu(NH_3)_4^{2+}(aq) + 4H_3O^+(aq; added) \rightarrow Cu^{2+}(aq) + 4NH_4^+(aq) + 4H_2O(l)$$

Although $Cl^-$ ion present in the added HCl cannot displace $NH_3$ from a complex ion directly, it can react with the now hydrated cations. The $Cl^-$ ion forms a white precipitate with $Ag^+(aq)$ and a green complex ion with $Cu^{2+}(aq)$:

$$Ag^+(aq) + Cl^-(aq; added) \rightarrow AgCl(s) \text{ [white]}$$

$$Cu^{2+}(aq) + 4Cl^-(aq; added) \rightarrow CuCl_4^{2-}(aq) \text{ [green]}$$

*Step 3.* To separate the aluminum and iron compounds that may be in the precipitate of Step 1, we recall that $Al(OH)_3$ is amphoteric, but $Fe(OH)_3$ is not. We add NaOH to the solid. The amphoteric $Al(OH)_3$ dissolves, but $Fe(OH)_3$ does not:

$$Al(OH)_3(s) + OH^-(aq; added) \rightarrow Al(OH)_4^-(aq)$$

$$Fe(OH)_3(s) + OH^-(aq; added) \rightarrow \text{no reaction}$$

*Step 4.* To confirm the presence of $Al^{3+}$ in the solution of Step 3, we make the solution somewhat acidic with HCl, which first precipitates and then dissolves the $Al^{3+}$ ion. Then we add $Na_2HPO_4$, which forms a white precipitate of $AlPO_4$:

$$Al(OH)_4^-(aq) \xrightarrow{H_3O^+} Al(OH)_3(s) \xrightarrow{H_3O^+} Al^{3+}(aq)$$

$$Al^{3+}(aq) + HPO_4^{2-}(aq; added) + H_2O(l) \longrightarrow H_3O^+(aq) + AlPO_4(s) \text{ [white]}$$

*Step 5.* To confirm that the remaining solid in Step 3 is, in fact, $Fe(OH)_3$, we dissolve it in dilute HCl and add potassium thiocyanate solution (KSCN), which forms a red complex ion with $Fe^{3+}$:

$$Fe(OH)_3(s) + 3H_3O^+(aq) \rightarrow Fe^{3+}(aq) + 6H_2O(l)$$

$$Fe^{3+}(aq) + SCN^-(aq; added) \rightarrow FeSCN^{2+}(aq) \text{ [red]}$$

SAMPLE PROBLEM 18.13

### Identifying the Ions in a Mixture Through Qualitative Analysis

**Problem:** A student was given a solution containing salts of one or more of the following cations: $K^+$, $Ag^+$, $Al^{3+}$, $Cu^{2+}$, and $Fe^{3+}$. From the following laboratory results, determine the cations present:
Step 1. Added $NH_3(aq)$ to ion mixture $\Rightarrow$ brown precipitate.
Step 2. Centrifuged $\Rightarrow$ colorless solution.
Step 3. Treated precipitate with $NaOH(aq) \Rightarrow$ some solid remained, but some may have dissolved.
Step 4. Neutralized solution in Step 3 with $NH_4Cl(aq) \Rightarrow$ no precipitate.
Step 5. Added $HCl(aq)$ to solution from Step 2 $\Rightarrow$ white precipitate.
Step 6. Flame test on original mixture $\Rightarrow$ pale blue color of ordinary flame.
**Plan:** Each result provides a new or confirming clue about the presence or absence of each cation.

**Solution:**

Result 1. $Al(OH)_3$ and $Fe(OH)_3$ are both insoluble in water, suggesting that the precipitate could be brown $Fe(OH)_3$ or a mixture of $Fe(OH)_3$ and white $Al(OH)_3$. $Ag^+$ and $Cu^{2+}$ also form insoluble hydroxides, but excess $NH_3$ will form soluble ammonia complexes.

Result 2. The colorless solution suggests that $Cu^{2+}$ is absent because $Cu(NH_3)_4^{2+}$ has a deep royal-blue color.

Result 3. If $Al(OH)_3$ were present in the precipitate, it is now in the solution as $Al(OH)_4^-$.

Result 4. Neutralizing Step 3 solution with $NH_4^+$ would convert any $Al(OH)_4^-$ to solid $Al(OH)_3$. No precipitate means $Al^{3+}$ is absent.

Result 5. Adding HCl to Step 2 solution reacts with any ammonia complex and supplies $Cl^-$. The white precipitate is AgCl, so $Ag^+$ is present.

Result 6. The absence of a distinct flame color indicates that $K^+$ (red-violet flame) is absent and confirms that $Cu^{2+}$ (green flame) is absent.

Thus, **$Ag^+$ and $Fe^{3+}$ are present;** $Al^{3+}$, $Cu^{2+}$, and $K^+$ are absent.

**Comment:** It is always good practice to confirm the presence of an ion with a specific test. For example, dissolving the solid in Step 1 in dilute HCl and treating with KSCN solution should produce a red solution, thus confirming $Fe^{3+}$.

**FOLLOW-UP PROBLEM 18.13**

Another student in the class obtained the following test results on a different mixture of the same possible ions:

Step 1. Added $NH_3 \Rightarrow$ white precipitate in colored solution.

Step 2. Centrifuged $\Rightarrow$ deep-blue solution.

Step 3. Treated precipitate with NaOH $\Rightarrow$ dissolved completely. Acidified and obtained white precipitate with $Na_2HPO_4$.

Step 4. Added HCl to Step 2 solution $\Rightarrow$ turned green, and white precipitate formed.

Step 5. Centrifuged precipitate from Step 4 and washed it to remove excess HCl. Treated it with $NH_3 \Rightarrow$ dissolved to form colorless solution.

Step 6. Flame test on original mixture $\Rightarrow$ green flame with no red visible.

Which ions are present in the mixture?

## Section Summary

Ions are precipitated selectively by adding a precipitating ion, until the $K_{sp}$ of one compound is exceeded as much as possible without exceeding the $K_{sp}$ of the other. An extension of this approach is to control the equilibrium of the slightly soluble compound by controlling simultaneously an equilibrium system that contains the precipitating ion. Qualitative analysis of ion mixtures involves adding precipitating ions to separate the unknown ions into ion groups. The groups are then analyzed further through precipitation and complex ion formation.

## Chapter Perspective

*This chapter is the last of three that explored the nature and variety of equilibrium systems. In Chapter 16, we discussed the central ideas of equilibrium in the context of gaseous systems. In Chapter 17, we examined our evolving understanding of acid-base equilibria. In this chapter, we highlighted three types of aqueous ionic systems and examined their role in the laboratory and the environment. The equilibrium constant, in all its forms, is a number that provides a limit to changes in a system, whether a chemical reaction, a physical change, or a substance dissolving. You now have the skills to predict whether a change will take place and to calculate its result, but you still do not know* why *the change occurs in the first place, or* why *it stops when it does. In Chapter 19, you'll find out.*

# For Review and Reference

## Key Terms

**SECTION 18.1**
acid-base buffer
common-ion effect
Henderson-Hasselbalch
  equation
buffer capacity
buffer range

**SECTION 18.2**
acid-base titration curve
acid-base indicator
equivalence point
end point

**SECTION 18.3**
solubility-product
  constant ($K_{sp}$)
molar solubility

**SECTION 18.4**
complex ion

ligand
formation constant ($K_f$)

**SECTION 18.5**
selective precipitation
simultaneous equilibria
qualitative analysis

## Key Equations and Relationships

**18.1** Finding the pH from known concentrations of a conjugate acid-base pair (Henderson-Hasselbalch equation) (p. 791):

$$pH = pK_a + \log\left(\frac{[base]}{[acid]}\right)$$

**18.2** Defining the equilibrium condition for a slightly soluble compound, $M_pX_q$, composed of $M^{n+}$ and $X^{z-}$ ions (p. 804):

For a saturated solution, $Q_{sp} = [M^{n+}]^p[X^{z-}]^q = K_{sp}$

## Answers to Follow-up Problems

**18.1** (a) Before addition:
Assuming $x$ is small enough to be neglected,
$$[HF] = 0.50\ M \quad \text{and} \quad [F^-] = 0.45\ M$$

$$[H_3O^+] = K_a \times \frac{[HF]}{[F^-]} = (6.8 \times 10^{-4}) \times \frac{0.50}{0.45}$$

$$= 7.6 \times 10^{-4}\ M \quad pH = 3.12$$

(b) After addition of 0.40 g NaOH (0.010 mol NaOH) to 1.0 L buffer,

$$[HF] = 0.49\ M \quad \text{and} \quad [F^-] = 0.46\ M$$

$$[H_3O^+] = (6.8 \times 10^{-4}) \times \frac{0.49}{0.46}$$

$$= 7.2 \times 10^{-4}\ M \quad pH = 3.14$$

**18.2** $[H_3O^+] = 10^{-pH} = 10^{-4.25} = 5.6 \times 10^{-5}$

$$[C_6H_5COOH] = \frac{[H_3O^+][C_6H_5COO^-]}{K_a}$$

$$= \frac{(5.6 \times 10^{-5})(0.050)}{6.3 \times 10^{-5}} = 0.044\ M$$

Mass (g) of $C_6H_5COOH = 5.0$ L soln

$$\times \frac{0.044\ \text{mol } C_6H_5COOH}{1\ \text{L soln}} \times \frac{122.12\ \text{g } C_6H_5COOH}{1\ \text{mol } C_6H_5COOH}$$

$$= 27\ \text{g } C_6H_5COOH$$

Dissolve 27 g $C_6H_5COOH$ in 5.0 L 0.050 $M$ $C_6H_5COONa$. Adjust pH to 4.25 with strong acid or base.

**18.3** (a) $[H_3O^+] = \sqrt{(2.3 \times 10^{-9})(0.2000)}$

$$= 2.1 \times 10^{-5}\ M \quad pH = 4.68$$

(b) $[H_3O^+] = K_a \times \frac{[HBrO]}{[BrO^-]} = 2.3 \times 10^{-9} \times 1$

$$= 2.3 \times 10^{-9}\ M \quad pH = 8.64$$

(c) $[BrO^-] = \dfrac{\text{moles of } BrO^-}{\text{total volume}} = \dfrac{0.004000\ \text{mol}}{0.06000\ \text{L}}$

$$= 0.06667\ M$$

$$K_b \text{ of } BrO^- = \frac{K_w}{K_a \text{ of HBrO}} = 4.3 \times 10^{-6}$$

$$[H_3O^+] = \frac{K_w}{\sqrt{K_b \times [BrO^-]}}$$

$$= \frac{1.0 \times 10^{-14}}{\sqrt{(4.3 \times 10^{-6})(0.06667)}}$$

$$= 1.9 \times 10^{-11}\ M \quad pH = 10.72$$

(d) Moles of $OH^-$ added = 0.008000 mol; volume (L) of $OH^-$ soln = 0.08000 L

$$[OH^-] = \frac{\text{moles of } OH^- \text{ unreacted}}{\text{total volume}}$$

$$= \frac{0.008000\ \text{mol} - 0.004000\ \text{mol}}{(0.02000 + 0.08000)\ \text{L}} = 0.04000\ M$$

$$[H_3O^+] = \frac{K_w}{[OH^-]} = 2.5 \times 10^{-13}; pH = 12.60$$

(e)

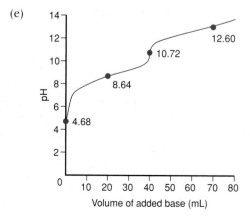

**18.4** (a) $Q_{sp} = [Ca^{2+}][SO_4^{2-}]$
    (b) $Q_{sp} = [Cr^{3+}]^2[CO_3^{2-}]^3$
    (c) $Q_{sp} = [Mg^{2+}][OH^-]^2$
    (d) $Q_{sp} = [As^{3+}]^2[HS^-]^3[OH^-]^3$

**18.5** $M$ of $CaF_2 = \dfrac{1.5 \times 10^{-4}\ g\ CaF_2}{10.0\ mL\ soln}$

$$\times \frac{1000\ mL}{1\ L} \times \frac{1\ mol\ CaF_2}{78.08\ g\ CaF_2}$$

$$= 1.9 \times 10^{-4}\ M$$

$$CaF_2(s) \rightleftharpoons Ca^{2+}(aq) + 2F^-(aq)$$

$$[Ca^{2+}] = 1.9 \times 10^{-4}\ M \quad \text{and} \quad [F^-] = 3.8 \times 10^{-4}\ M$$

$$K_{sp} = [Ca^{2+}][F^-]^2 = (1.9 \times 10^{-4})(3.8 \times 10^{-4})^2$$

$$= 2.7 \times 10^{-11}$$

**18.6** From the reaction table, $[Mg^{2+}] = S$ and $[OH^-] = 2S$

$$K_{sp} = [Mg^{2+}][OH^-]^2 = 4S^3 = 6.3 \times 10^{-10}$$

$$S = 5.4 \times 10^{-4}\ M$$

**18.7** (a) In water: $K_{sp} = [Ba^{2+}][SO_4^{2-}] = S^2 = 1.1 \times 10^{-10}$;
    $S = 1.0 \times 10^{-5}$
    (b) In $0.100\ M\ Na_2SO_4$, $[SO_4^{2-}] = 0.100\ M$
    $K_{sp} = S \times 0.100 = 1.1 \times 10^{-10}$    $S = 1.1 \times 10^{-9}\ M$
    $S$ decreases in presence of the common ion, $SO_4^{2-}$

**18.8** (a) Increases solubility. $CaF_2(s) \rightleftharpoons Ca^{2+}(aq) + 2F^-(aq)$

$$F^-(aq) + H_3O^+(aq) \rightarrow HF(aq) + H_2O(l)$$

(b) Increases solubility.

$$ZnS(s) + H_2O(l) \rightleftharpoons Zn^{2+}(aq) + HS^-(aq) + OH^-(aq)$$

$$HS^-(aq) + H_3O^+(aq) \rightarrow H_2S(aq) + H_2O(l)$$

$$OH^-(aq) + H_3O^+(aq) \rightarrow 2H_2O(l)$$

(c) No effect. $I^-(aq)$ is conjugate base of strong acid, HI.

**18.9** $Ca_3(PO_4)_2(s) \rightleftharpoons 3Ca^{2+}(aq) + 2PO_4^{3-}(aq)$

$$Q_{sp} = [Ca^{2+}]^3[PO_4^{3-}]^2 = (1.0 \times 10^{-9})^5 = 1.0 \times 10^{-45}$$

$Q_{sp} < K_{sp}$, so $Ca_3(PO_4)_2$ will not precipitate.

**18.10** $[Fe(H_2O)_6^{3+}]_{init} = \dfrac{(0.0255\ L)\ (3.1 \times 10^{-2}\ M)}{0.0255\ L + 0.0350\ L}$

$$= 1.3 \times 10^{-2}\ M$$

Similarly, $[CN^-]_{init} = 0.87\ M$. From the reaction table,

$$K_f = \frac{[Fe(CN)_6^{3-}]}{[Fe(H_2O)_6^{3+}][CN^-]^6} = \frac{1.3 \times 10^{-2}}{x(0.79)^6} = 4.0 \times 10^{43}$$

$$x = [Fe(H_2O)_6^{3+}] = 1.3 \times 10^{-45}$$

**18.11** $AgBr(s) + 2NH_3(aq) \rightleftharpoons Ag(NH_3)_2^+(aq) + Br^-(aq)$

$$K_{overall} = K_{sp}\ \text{of}\ AgBr \times K_f\ \text{of}\ Ag(NH_3)_2^+$$

$$= 8.5 \times 10^{-6}$$

From the reaction table,

$$\frac{S}{1.0 - 2S} = \sqrt{8.5 \times 10^{-6}} = 2.9 \times 10^{-3}$$

$$S = [Ag(NH_3)_2^+] = 2.9 \times 10^{-3}\ M$$

Solubility in 1 $M$ hypo is more than 150 times that in 1 $M$ NH$_3$.

**18.12** Both 1/1 salts, so $K_{sp}$ values show that $CaSO_4$ is more soluble:

$$[SO_4^{2-}] = \frac{K_{sp}}{[Ca^{2+}]} = \frac{2.4 \times 10^{-5}}{0.025 \times 0.9999}$$

$$= 9.6 \times 10^{-4}\ M$$

**18.13** Result 1. $Fe^{3+}$ absent, all others may be present.
    Result 2. $Cu^{2+}$ is present: blue $Cu(NH_3)_4^{2+}$.
    Result 3. $Al^{3+}$ is present: $Al(OH)_3$ is soluble in $OH^-$. $Al^{3+}$ is confirmed by formation of white $AlPO_4$.
    Result 4. Confirms $Cu^{2+}$ is present: green $CuCl_4^{2-}$. $Ag^+$ probably present; forms white precipitate of AgCl.
    Result 5. Confirms $Ag^+$ is present: $Ag(NH_3)_2^+$ is colorless.
    Result 6. Confirms $Cu^{2+}$ is present: green flame. $K^+$ is probably absent ($K^+$ color is very weak.)

## Sample Problem Titles

# Problems

Problems with a green number are answered at the back of the text. Most sections include three categories of problems separated by a green rule—concept review questions, *paired* skill building exercises, and problems in a relevant context.

## Equilibria of Acid-Base Buffer Systems

(Sample Problems 18.1 and 18.2)

**18.1** What is the purpose of an acid-base buffer?

**18.2** How do the acid and base components of a buffer function? Why are they typically a conjugate acid-base pair?

**18.3** What is the common-ion effect? How is it related to LeChâtelier's principle? Explain with equations that include HF and NaF.

**18.4** When $H_3O^+$ is added to a buffer, does the pH remain constant or does it change slightly? Explain.

**18.5** What is the essential difference between buffers with high or low capacities? Will the addition of 0.01 mol HCl produce a greater pH change in a buffer with a high capacity or a low one? Explain.

**18.6** Choose the factors that determine the capacity of a buffer from among the following and explain your choices:
(a) Conjugate acid-base pair
(b) pH of the buffer
(c) Concentration of buffer-component reservoirs
(d) Buffer range
(e) Buffer-component ratio
(f) $pK_a$ of the acid component

**18.7** What is the relationship between buffer range and buffer-component ratio?

**18.8** A chemist needs a pH 3.5 buffer. Should she use formic acid ($K_a = 1.8 \times 10^{-4}$) and NaOH or acetic acid ($K_a = 1.8 \times 10^{-5}$) and NaOH to prepare it? Why? What is the disadvantage of choosing the other acid? What is the role of the NaOH?

**18.9** Explain the increase or decrease in the pH and in the buffer-component ratio, [HA]/[NaA], of a buffer in each of the following cases:
(a) Add 0.1 $M$ NaOH to the buffer
(b) Add 0.1 $M$ HCl to the buffer
(c) Dissolve NaA in the buffer
(d) Dissolve HA in the buffer

**18.10** Would the pH increase or decrease, and would it do so to a large or small extent, in each of the following cases:
(a) Add 5 drops 0.1 $M$ NaOH to 100 mL of 0.5 $M$ acetate buffer
(b) Add 5 drops 0.1 $M$ HCl to 100 mL of 0.5 $M$ acetate buffer
(c) Add 5 drops 0.1 $M$ NaOH to 100 mL of 0.5 $M$ HCl
(d) Add 5 drops 0.1 $M$ NaOH to distilled water

---

**18.11** What are the $[H_3O^+]$ and the pH of a propanoate buffer that consists of 0.25 $M$ $CH_3CH_2COONa$ and 0.15 $M$ $CH_3CH_2COOH$? $K_a$ of propanoic acid $= 1.3 \times 10^{-5}$.

**18.12** What are the $[H_3O^+]$ and the pH of a benzoate buffer that consists of 0.33 $M$ $C_6H_5COOH$ and 0.28 $M$ $C_6H_5COONa$? $K_a$ of benzoic acid $= 6.3 \times 10^{-5}$.

**18.13** What is the pH of a buffer that consists of 0.55 $M$ HCOOH and 0.63 $M$ HCOONa? $pK_a$ of formic acid $= 3.74$.

**18.14** What is the pH of a buffer that consists of 0.95 $M$ HBrO and 0.68 $M$ KBrO? $pK_a$ of HBrO $= 8.64$.

**18.15** What is the pH of a buffer that consists of 0.20 $M$ $NH_3$ and 0.10 $M$ $NH_4Cl$? $pK_b$ of $NH_3 = 4.75$.

**18.16** What is the pH of a buffer that consists of 0.50 $M$ $CH_3NH_2$ and 0.60 $M$ $CH_3NH_3Cl$? $pK_b$ of methylamine is 3.35.

**18.17** A buffer consists of 0.25 $M$ $KHCO_3$ and 0.32 $M$ $K_2CO_3$. Carbonic acid is a diprotic acid with $K_{a1} = 4.5 \times 10^{-7}$ and $K_{a2} = 4.7 \times 10^{-11}$.
(a) Which $K_a$ value is more important to this buffer?
(b) What is the buffer pH?

**18.18** A buffer consists of 0.50 $M$ $NaH_2PO_4$ and 0.40 $M$ $Na_2HPO_4$. Phosphoric acid is a triprotic acid with $K_{a1} = 7.2 \times 10^{-3}$, $K_{a2} = 6.3 \times 10^{-8}$, and $K_{a3} = 4.2 \times 10^{-13}$.
(a) Which $K_a$ value is most important to this buffer?
(b) What is the buffer pH?

**18.19** What is the buffer-component ratio, [HPr]/[Pr$^-$], of a propanoate buffer that has a pH of 5.11? See Table 17.2 for $K_a$ of propanoic acid.

**18.20** What is the buffer-component ratio, [$HNO_2$]/[$NO_2^-$], of a nitrite buffer that has a pH of 2.95? See Table 17.2 for $K_a$ of nitrous acid.

**18.21** A solution that contains 0.20 $M$ of acid, HA, and 0.15 $M$ of its conjugate base, A$^-$, has a pH of 3.35. What is the pH after 0.0015 mol NaOH is added to 0.50 L of this solution?

**18.22** A solution that contains 0.40 $M$ base, B, and 0.25 $M$ of its conjugate acid, BH$^+$, has a pH of 8.88. What is the pH after 0.0020 mol HCl is added to 0.25 L of this solution?

**18.23** A buffer is prepared by mixing 184 mL of 0.442 $M$ HCl and 0.500 L of 0.400 $M$ sodium acetate. See Table 17.2.
(a) What is the pH?
(b) How many grams of KOH must be added to 0.500 L of the buffer to change the pH by 0.15 units?

**18.24** A buffer is prepared by mixing 50.0 mL of 0.050 $M$ sodium bicarbonate and 10.7 mL of 0.10 $M$ NaOH. See Table 17.5 .
(a) What is the pH?
(b) How many grams of HCl must be added to 25.0 mL of the buffer to change the pH by 0.07 units?

**18.25** Choose specific acid-base conjugate pairs from the acids or bases in Tables 17.2, 17.5, and 17.6 that are suitable for preparing the following buffers: (a) pH $\approx$ 4.0; (b) pH $\approx$ 7.0; (c) $[H_3O^+] \approx 1 \times 10^{-9}$ $M$.

**18.26** Choose specific acid-base conjugate pairs from the acids or bases in Tables 17.2, 17.5, and 17.6 that are suitable for preparing the following buffers: (a) pH ≈ 2.5; (b) pH ≈ 5.5; (c) $[OH^-]$ ≈ $1 \times 10^{-6}$ $M$.

**18.27** An industrial chemist studying hypochlorite solutions to see the effect of pH on bleaching and sterilizing properties prepares several buffers. Calculate the pH of the following solutions:
(a) 0.100 $M$ HClO and 0.100 $M$ NaClO
(b) 0.100 $M$ HClO and 0.150 $M$ NaClO
(c) 0.150 $M$ HClO and 0.100 $M$ NaClO
(d) One liter of the solution in (a) after 0.0050 mol NaOH has been added.

**18.28** Normal arterial blood has an average pH of 7.40. Phosphate ions form one of the key buffering systems in the blood. Find the buffer-component ratio of a $KH_2PO_4/Na_2HPO_4$ solution with this pH.

**18.29** As an FDA physiologist, you need 0.600 L of formate buffer with a pH of 3.74.
(a) What is the required buffer-component ratio?
(b) How would you prepare this solution from stock solutions of 1.0 $M$ HCOOH and 1.0 $M$ NaOH? (*Hint:* Mix $x$ mL of HCOOH with $600 - x$ mL of NaOH.)
(c) What is the final concentration of HCOOH in this solution?

**Acid-Base Titration Curves**

(Sample Problem 18.3)

**18.30** Acid-base indicators are weak acids. What value must you have to estimate the general range of an indicator's color change? Why do some indicators have two separate pH ranges?

**18.31** Why does the color change of an indicator take place over a range of about two pH units?

**18.32** Why doesn't the addition of an acid-base indicator affect the pH of the test solution?

**18.33** What is the difference between the end point of a titration and the equivalence point? Is the equivalence point always reached first? Explain.

**18.34** Some automatic titrators measure the slope of the titration curve to determine when the equivalence point is reached. What happens to the slope of the curve that enables the instrument to recognize this point?

**18.35** Explain how *strong acid*-strong base, *weak acid*-strong base, and strong acid-*weak base* titrations using the same concentrations differ in terms of (a) the initial pH and (b) the pH at the equivalence point. (The component in italics is in the flask.)

**18.36** What species are present in the buffer region of a weak acid-strong base titration? How are they different from the species present at the equivalence point? Are they different from the species present in the buffer region of a strong acid-weak base titration? Explain.

**18.37** Why is the center of the buffer region of a weak acid-strong base titration significant?

**18.38** How does the titration curve of a monoprotic acid differ from that of a diprotic acid?

**18.39** The indicator cresol red has $K_a = 5.0 \times 10^{-9}$. Over what approximate pH range does it change color?

**18.40** The indicator thymolphthalein has $K_a = 7.9 \times 10^{-11}$. Over what approximate pH range does it change color?

**18.41** Use Figure 18.4 to choose an indicator for the following titrations:
(a) 0.1 $M$ HCl with 0.1 $M$ NaOH
(b) 0.1 $M$ HCOOH (Table 17.2) with 0.1 $M$ NaOH
(c) 0.1 $M$ $CH_3NH_2$ (Table 17.6) with 0.1 $M$ HCl

**18.42** Use Figure 18.4 to choose an indicator for the following titrations:
(a) 0.5 $M$ $(CH_3)_2NH$ (Table 17.6) with 0.5 $M$ HBr
(b) 0.2 $M$ KOH with 0.2 $M$ $HNO_3$
(c) 0.25 $M$ $C_6H_5COOH$ (Table 17.2) with 0.25 $M$ KOH

**18.43** Calculate the pH during the titration of 50.00 mL of 0.1000 $M$ HCl with 0.1000 $M$ NaOH solution after the following additions of base: (a) 0 mL; (b) 25.00 mL; (c) 49.00 mL; (d) 49.90 mL; (e) 50.00 mL; (f) 50.10 mL; (g) 60.00 mL.

**18.44** Calculate the pH during the titration of 30.00 mL of 0.1000 $M$ KOH with 0.1000 $M$ HBr solution after the following additions of acid: (a) 0 mL; (b) 15.00 mL; (c) 29.00 mL; (d) 29.90 mL; (e) 30.00 mL; (f) 30.10 mL; (g) 40.00 mL.

**18.45** Calculate the pH during the titration of 20.00 mL of 0.1000 $M$ butanoic acid, $CH_3CH_2CH_2COOH$ ($K_a = 1.54 \times 10^{-5}$), with 0.1000 $M$ NaOH solution after the following additions of titrant: (a) 0 mL; (b) 10.00 mL; (c) 15.00 mL; (d) 19.00 mL; (e) 19.95 mL; (f) 20.00 mL; (g) 20.05 mL; (h) 25.00 mL.

**18.46** Calculate the pH during the titration of 20.00 mL of 0.1000 $M$ triethylamine, $(CH_3CH_2)_3N$ ($K_b = 5.2 \times 10^{-4}$), with 0.1000 $M$ HCl solution after the following additions of titrant: (a) 0 mL (b) 10.00 mL; (c) 15.00 mL; (d) 19.00 mL; (e) 19.95 mL; (f) 20.00 mL; (g) 20.05 mL; (h) 25.00 mL.

**18.47** Using Tables 17.2 and 17.5, find the pH and the number of milliliters of standardized 0.0372 $M$ NaOH needed to reach the following equivalence point(s):
(a) 42.2 mL of 0.0520 $M$ $CH_3COOH$
(b) 18.9 mL of 0.0890 $M$ $H_2SO_3$ (two equivalence points)

**18.48** Using Tables 17.2 and 17.5, find the pH and the number of milliliters of standardized 0.0588 $M$ KOH to reach the following equivalence point(s):
(a) 23.4 mL of 0.0390 $M$ $HNO_2$
(b) 17.3 mL of 0.130 $M$ $H_2CO_3$ (two equivalence points)

## Equilibria of Slightly Soluble Ionic Compounds

(Sample Problems 18.4-18.9)

**18.49** Why does the expression for the solubility equilibrium of an ionic solid not include the concentration of the solid?

**18.50** The molar solubility of the slightly soluble ionic compound $M_2X$ is $5 \times 10^{-5}$ $M$. What is the molarity of each of the two ions? How would you set up the calculation to find $K_{sp}$? What assumption must you make in such calculations about the dissociation of $M_2X$ into ions? Why is the calculated $K_{sp}$ higher than the actual value?

**18.51** Why does pH affect the solubility of $CaF_2$ significantly, but not that of $CaCl_2$?

**18.52** A list of solubility product constants, such as Table 18.2, can be used to compare directly the solubility of silver chloride with silver bromide but not with silver chromate. Explain.

**18.53** In gaseous equilibria, the reverse reaction predominates when $Q_c > K_c$. What occurs in an aqueous ionic solution when $Q_{sp} > K_{sp}$?

**18.54** Write the ion-product expressions for
(a) Silver carbonate, $Ag_2CO_3$
(b) Barium fluoride, $BaF_2$
(c) Copper(II) sulfide, CuS

**18.55** Write the ion-product expressions for
(a) Ferric hydroxide, $Fe(OH)_3$
(b) Barium phosphate, $Ba_3(PO_4)_2$
(c) Tin(II) sulfide, SnS

**18.56** The solubility of silver carbonate is 0.032 $M$ at 20°C. Calculate the $K_{sp}$ of silver carbonate.

**18.57** The solubility of zinc oxalate is $7.9 \times 10^{-3}$ $M$ at 18°C. Calculate the $K_{sp}$ of zinc oxalate.

**18.58** The solubility of silver dichromate at 15°C is $8.3 \times 10^{-3}$ g/100 mL solution. Calculate the $K_{sp}$ of silver dichromate.

**18.59** The solubility of calcium sulfate at 30°C is 0.209 g/100 mL solution. Calculate the $K_{sp}$ of calcium sulfate.

**18.60** Calculate the molar solubility of $SrCO_3$ ($K_{sp} = 5.4 \times 10^{-10}$) in (a) pure water and (b) 0.13 $M$ $Sr(NO_3)_2$.

**18.61** Calculate the molar solubility of $BaCrO_4$ ($K_{sp} = 2.1 \times 10^{-10}$) in (a) pure water and (b) $1.5 \times 10^{-3}$ $M$ $Na_2CrO_4$.

**18.62** Calculate the molar solubility of $Ca(IO_3)_2$ in (a) 0.060 $M$ $Ca(NO_3)_2$ and (b) 0.060 $M$ $NaIO_3$. (See Table 18.2.)

**18.63** Calculate the molar solubility of $Ag_2SO_4$ in (a) 0.22 $M$ $AgNO_3$ and (b) 0.22 $M$ $Na_2SO_4$. (See Table 18.2.)

**18.64** Use Table 18.2 to decide which of the following compounds is more soluble in water:
(a) Magnesium hydroxide or nickel(II) hydroxide
(b) Lead(II) sulfide or copper(II) sulfide
(c) Silver sulfate or magnesium fluoride

**18.65** Use Table 18.2 to decide which of the following compounds is more soluble in water:

(a) Strontium sulfate or barium chromate
(b) Calcium carbonate or copper(II) carbonate
(c) Barium iodate or silver chromate

**18.66** Write balanced equations to show which of the following compounds have a significantly different solubility with a change in pH: (a) AgCl; (b) $SrCO_3$; (c) CuBr.

**18.67** Write balanced equations to show which of the following compounds have a significantly different solubility with a change in pH: (a) $Fe(OH)_2$; (b) $Cu_2S$; (c) $PbI_2$.

**18.68** Does any solid $Cu(OH)_2$ form when 0.075 g KOH is dissolved in 1.0 L of $1.0 \times 10^{-3}$ $M$ $Cu(NO_3)_2$? (See Table 18.2.)

**18.69** Does any solid $PbCl_2$ form when 3.5 mg NaCl is dissolved in 0.250 L of 0.12 $M$ $Pb(NO_3)_2$? (See Table 18.2.)

**18.70** Tooth enamel is composed of the mineral hydroxyapatite, $Ca_5(PO_4)_3OH$ ($K_{sp} = 6.8 \times 10^{-37}$). Many water treatment plants now add fluoride to the drinking water, which reacts with $Ca_5(PO_4)_3OH$ to form the more decay-resistant fluorapatite, $Ca_5(PO_4)_3F$ ($K_{sp} = 1.0 \times 10^{-60}$). This treatment has resulted in a dramatic decrease in the number of cavities among children. Calculate the solubility of $Ca_5(PO_4)_3OH$ and of $Ca_5(PO_4)_3F$ in water.

**18.71** Calcium ion triggers the clotting of blood, so when blood is donated, the receiving bag contains sodium oxalate solution to precipitate the $Ca^{2+}$ and prevent clotting. A 104-mL sample of blood contains $9.7 \times 10^{-5}$ g $Ca^{2+}$/mL. A medical technologist treats the sample with 100.0 mL of 0.155 $M$ $Na_2C_2O_4$. Calculate the $[Ca^{2+}]$ after the treatment. (Use Table 18.2 for $K_{sp}$ of $CaC_2O_4 \cdot H_2O$.)

**18.72** The well water in an area is "hard" because it is in equilibrium with $CaCO_3$ in the surrounding rocks. What is the concentration of $Ca^{2+}$ in the well water (assuming the pH is such that the $CO_3^{2-}$ ion is not hydrolyzed by the water)? (Use Table 18.2 for $K_{sp}$ of $CaCO_3$.)

## Equilibria Involving Complex Ions

(Sample Problems 18.10 and 18.11)

**18.73** How is it possible for a positively charged metal ion to be at the center of a negatively charged complex ion?

**18.74** Write equations to show the stepwise reaction of $Cd(H_2O)_4^{2+}$ in aqueous KI to form $CdI_4^{2-}$. Show that $K_{f (overall)} = K_{f1} \times K_{f2} \times K_{f3} \times K_{f4}$.

**18.75** Consider the dissolution of PbS in water:

$$PbS(s) + H_2O(l) \rightleftharpoons Pb^{2+}(aq) + HS^-(aq) + OH^-(aq)$$

Hydroxide ion is a product, but the addition of aqueous NaOH causes more PbS to dissolve. Is this a violation of LeChâtelier's principle? Explain.

**18.76** Write a balanced equation for the reaction of $Cr(H_2O)_6^{3+}$ in aqueous KCl.

**18.77** Write a balanced equation for the reaction of $Zn(H_2O)_4^{2+}$ in aqueous NaCN.

**18.78** Potassium thiocyanate, KSCN, is often used to determine the presence of $Fe^{3+}$ ions in solution by the formation of the red $Fe(H_2O)_5SCN^{2+}$ (or more simply $FeSCN^{2+}$). What is the concentration of $Fe^{3+}$ when 0.50 L each of 0.0015 $M$ $Fe(NO_3)_3$ and 0.20 $M$ in KSCN are mixed ($K_f$ of $FeSCN^{2+}$ = 8.9 × 10²)?

**18.79** When 0.82 g $ZnCl_2$ is dissolved in 255 mL of 0.15 $M$ NaCN, what are $[Zn^{2+}]$, $[Zn(CN)_4^{2-}]$, and $[CN^-]$ [$K_f$ of $Zn(CN)_4^{2-}$ = 4.2 × 10¹⁹]?

**18.80** Calculate the solubility of AgI in 2.5 $M$ $NH_3$. $K_{sp}$ of AgI = 8.3 × 10⁻¹⁷ and $K_f$ of $Ag(NH_3)_2^+$ = 1.7 × 10⁷.

**18.81** Calculate the solubility of $Cr(OH)_3$ in a buffer of pH 13.0. $K_{sp}$ of $Cr(OH)_3$ = 6.3 × 10⁻³¹ and $K_f$ of $Cr(OH)_4^-$ = 8.0 × 10²⁹.

## Application of Ionic Equilibria to Chemical Analysis

(Sample Problems 18.12 and 18.13)

**18.82** A 50.0-mL solution of 0.50 $M$ $Fe(NO_3)_3$ is mixed with 125 mL of 0.25 $M$ in $Cd(NO_3)_2$.
(a) If aqueous NaOH is added to the solution, which ion precipitates first? (See Table 18.2 for the appropriate values.)
(b) Describe how the metal ions can be separated using NaOH.
(c) Calculate the $[OH^-]$ that will accomplish the separation.

**18.83** In a qualitative analysis procedure, a chemist adds 0.3 $M$ HCl to an unknown group of ions and then saturates the solution with $H_2S$.
(a) What is the $[HS^-]$ of the solution?
(b) Which of the following 0.01 $M$ cation solutions will form a precipitate: $Pb^{2+}$, $Mn^{2+}$, $Hg^{2+}$, $Cu^{2+}$, $Ni^{2+}$, $Fe^{2+}$, $Ag^+$, $K^+$?

**18.84** Name a reagent and describe the results that would distinguish between the following pairs of substances: (a) $NH_4Cl$ and $NiCl_2$; (b) AgCl and $PbCl_2$; (c) $Al(NO_3)_2$ and $Fe(NO_3)_2$.

**18.85** Four samples of an unknown solution are tested individually by adding HCl, $H_2S$ in acidic solution, $H_2S$ in basic solution, and $Na_2CO_3$, respectively. No precipitate forms in any of the tests. When NaOH is added to the solution, moistened litmus paper held above the solution turns blue. A drop of the original solution turns the color of the flame violet. Which ions are present in the unknown solution?

## Comprehensive Problems

Problems with an asterisk (*) are more challenging.

**18.86** The presence of cadmium ion in a solution is analyzed by precipitation as the sulfide, a yellow compound used as a pigment in everything from artists' oil colors to coloring glass and rubber. Use Table 18.2

to calculate the molar solubility of cadmium sulfide at 25°C.

**18.87** Phosphate systems form one of the most important buffers in living organisms. Use Table 17.5 to calculate the pH of a buffer made by dissolving 0.80 mol NaOH in 0.50 L of 1.0 $M$ $H_3PO_4$.

**18.88** Assuming that it is possible to detect the $NH_3$ gas over 10⁻² $M$ $NH_3$, to what pH must 0.15 $M$ $NH_4Cl$ be raised to detect $NH_3$?

**18.89** Manganese(II) sulfide is one of the compounds found in nodules on the ocean floor that may eventually be a primary source of many transition metals. The solubility of MnS is 4.7 × 10⁻⁴ g/100 mL solution. Estimate the $K_{sp}$ of MnS.

**18.90** Consider the following reaction:

$$Fe(OH)_3(s) + OH^-(aq) \rightleftharpoons Fe(OH)_4^-(aq); K_c = 4 \times 10^{-5}$$

(a) What is the solubility of $Fe(OH)_3$ in 0.010 $M$ NaOH?
(b) How is this result used in qualitative analysis?

**18.91** A biochemical engineer preparing cells for a cloning experiment bathes a small piece of rat epithelial tissue in a TRIS buffer [tris(hydroxymethyl)aminomethane, $(HOCH_2)_3CNH_2$]. The buffer is made by dissolving 43.0 g TRIS in enough 0.095 $M$ HCl to make exactly 1 L of solution. What is the molarity and pH of the buffer? ($pK_b$ of TRIS = 5.91)

**18.92** Sketch a qualitative titration curve of ethylenediamine, $H_2NCH_2CH_2NH_2$, with 0.1 $M$ HCl.

*18.93 A solution contains 0.10 $M$ $ZnCl_2$ and 0.020 $M$ $MnCl_2$. Given the following information, how would you adjust the pH to separate the ions as their sulfides?

$$MnS + H_2O \rightleftharpoons Mn^{2+} + HS^- + OH^-; K_{sp} = 3 \times 10^{-11}$$

$$ZnS + H_2O \rightleftharpoons Zn^{2+} + HS^- + OH^-; K_{sp} = 2 \times 10^{-22}$$

$$H_2S + H_2O \rightleftharpoons H_3O^+ + HS^-; K_{a1} = 9.5 \times 10^{-8}$$

[$H_2S$] in water saturated at 25°C = 0.1 $M$ $H_2S$

$$K_w = 1.0 \times 10^{-14} \text{ at } 25°C$$

*18.94 Amino acids have a general formula of $NH_2CH(R)COOH$ and can be considered polyprotic acids. Moreover, in many cases, the R group contains additional amine and carboxyl groups.
(a) An amino acid dissolved in pure water never has a protonated COOH group and an unprotonated $NH_2$ group. Use glycine, $NH_2CH_2COOH$, to explain why this is so. ($K_a$ of COOH group = 4.47 × 10⁻³; $K_b$ of $NH_2$ group = 6.03 × 10⁻⁵.)
(b) Calculate the following ratio at pH 5.5: $[^+NH_3CH_2COO^-]/[^+NH_3CH_2COOH]$
(c) The R group of lysine is $—CH_2CH_2CH_2CH_2NH_2$ ($pK_b$ = 3.47). Draw the structure of lysine at pH 1, physiologic pH (~7), and pH 13.
(d) The R group of glutamic acid is $—CH_2CH_2COOH$ ($pK_a$ = 4.07). Draw the structure of glutamic acid at pH 1, physiologic pH, and pH 13.

**18.95** The color of the acid-base indicator ethyl orange turns from red to yellow over the pH range 3.4 to 4.8. Estimate the $K_a$ of ethyl orange.

**18.96** An acid-base titration of a weak acid with a strong base has an end point at pH = 9.0. What indicator would be suitable for the titration?

**\*18.97** Use the values in Problem 18.43 to sketch the following three plots:
(a) A curve of pH vs. mL of added titrant.
(b) A curve of $[H_3O^+]$ vs. mL of added titrant
(c) A curve of $\Delta pH/\Delta mL$ vs. mL of added titrant
Are there advantages or disadvantages to viewing the results in the form of (b) and (c)?

**18.98** Can all complex ions be regarded as the result of a Lewis acid-base reaction? Write an equation and identify the acid, base, and adduct in the formation of $Ag(NH_3)_2^+$ from aqueous $Ag^+$ and $NH_3$ solutions.

**18.99** Lactic acid [$CH_3CH(OH)COOH$] is a common biomolecule. Muscle physiologists study its accumulation during exercise. Food chemists study its occurrence in sour milk products, fruit, beer, wine, and tomatoes. Industrial microbiologists study its formation by various bacterial species from carbohydrates. A biochemist prepares a lactate buffer by mixing 225 mL of 0.85 $M$ lactic acid ($K_a = 1.38 \times 10^{-4}$) with 435 mL of 0.68 $M$ sodium lactate. What is the pH of the buffer?

**18.100** Why are lakes, rivers, and groundwater in limestone-rich soils mildly basic despite the presence of acidic rainfall?

**18.101** Enormous quantities of silver halides, especially AgBr, are used to manufacture photographic film. The materials are prepared by precipitation from silver nitrate solution treated with halide solution. How many kilograms of AgBr can form when 125 $m^3$ of 0.250 $M$ $AgNO_3$ is mixed with 255 $m^3$ of 0.600 $M$ NaBr?

**18.102** The Henderson-Hasselbalch equation gives a relationship for obtaining the pH of a weak acid, HA. Derive an analogous relationship for obtaining the pOH of a weak base, B.

**18.103** Calculate the molar solubility of $Hg_2C_2O_4$ in 0.13 $M$ $Hg_2(NO_3)_2$. ($K_{sp}$ of $Hg_2C_2O_4 = 1.75 \times 10^{-13}$.)

**18.104** Human blood contains two buffer systems, one based on phosphate and one on carbonate. If the blood has a normal pH of 7.4, what are the principal phosphate and carbonate species present? What is the ratio between the two phosphate species? In the presence of the numerous dissolved ions and other species of blood, $K_{a1}$ of phosphoric acid = $1.3 \times 10^{-2}$, $K_{a2} = 2.3 \times 10^{-7}$, and $K_{a3} = 6 \times 10^{-12}$; $K_{a1}$ of carbonic acid = $8 \times 10^{-7}$ and $K_{a2} = 1.6 \times 10^{-10}$.

**\*18.105** Most qualitative analysis schemes start with the separation of ion group 1, $Ag^+$, $Hg_2^{2+}$, and $Pb^{2+}$, by formation of the white solids AgCl, $Hg_2Cl_2$, and $PbCl_2$. Develop a procedure for confirming the presence or absence of these ions from the following facts:

- $PbCl_2$ is soluble in hot water, but AgCl and $Hg_2Cl_2$ are not.
- $PbCrO_4$ is yellow solid that is insoluble in hot water.
- AgCl forms colorless $Ag(NH_3)_2^+(aq)$ in aqueous $NH_3$.
- When a solution containing $Ag(NH_3)_2^+$ is acidified with HCl, AgCl precipitates.
- $Hg_2Cl_2$ forms an insoluble mixture of $Hg(l)$ (black) and $HgNH_2Cl(s)$ (white) in aqueous $NH_3$.

**18.106** An environmental technician collects a sample of rainwater. A light on her portable pH meter indicates low battery power and the potential for inaccurate readings, so she uses indicator solutions to estimate the pH. A piece of litmus paper turns red, indicating acidity, so she divides the sample into thirds and obtains the following results: thymol blue—yellow; bromphenol blue—green; and methyl red—red. Estimate the pH of the rainwater.

**\*18.107** A 0.050 $M$ $H_2S$ solution contains 0.15 $M$ $NiCl_2$ and 0.35 $M$ $Hg(NO_3)_2$. What pH is required to precipitate the maximum amount of HgS but none of the NiS? $K_{sp}$ of NiS = $3 \times 10^{-16}$ and $K_{sp}$ of HgS = $2 \times 10^{-50}$.

**\*18.108** An ecobotanist separates the components of a tropical bark extract by liquid chromatography. She discovers a large proportion of quinidine, a dextrorotatory stereoisomer of quinine that is a cardiac muscle depressant used for control of arrhythmic heartbeat. Quinidine has two basic N groups ($K_{b1} = 4.0 \times 10^{-6}$ and $K_{b2} = 1.0 \times 10^{-10}$). To measure the concentration, she carries out a titration. Due to the low solubility of quinidine, she first protonates both N groups with excess HCl and titrates the acidified solution with standardized base. A 33.85-mg sample of quinidine ($M = 324.41$ g/mol) was acidified with 6.55 mL of 0.150 $M$ HCl.
(a) How many mL of 0.0133 $M$ NaOH are needed to titrate the excess HCl?
(b) How many additional mL of titrant are needed to reach the first equivalence point of quinidine dihydrochloride?
(c) What is the pH at the first equivalence point?

**18.109** Why was rainwater acidic in the preindustrial age? Why did its acidity not erode marble statuary? What solutes make acid rain much more acidic now? How does automobile exhaust contribute to this acidity?

**18.110** A 35.00-mL solution of 0.2500 $M$ HF is titrated with a standardized 0.1532 $M$ solution of NaOH at 25°C.
(a) What is the pH of the solution before titrant is added?
(b) How many milliliters of titrant are required to reach the equivalence point?
(c) What is the pH 0.50 mL before the equivalence point?
(d) What is the pH at the equivalence point?
(e) What is the pH 0.50 mL after the equivalence point?

*18.111 Sodium chloride is purified for use as a food condiment by adding HCl to a saturated solution of NaCl (317 g/L). When 25.5 mL of 7.85 $M$ HCl is added to 0.100 L of saturated solution, how many grams of purified NaCl precipitate?

18.112 A 35.0 mL solution of 0.075 $M$ $CaCl_2$ is mixed with 25.0 mL of 0.090 $M$ in $BaCl_2$.
   (a) If aqueous KF is added to the solution, which precipitate forms first?
   (b) Describe how the metal ions can be separated using KF to form the fluorides.
   (c) Calculate the $[F^-]$ that will accomplish the separation.

*18.113 Even before the industrial age, rainwater was slightly acidic. Use the following data to calculate pH of unpolluted rainwater at 25°C:

   Vol % in air = 0.033 vol % of $CO_2$

   Solubility in pure water at 25°C and 1 atm $CO_2$ = 88 mL $CO_2$/100 mL $H_2O$. $K_{a1}$ of $H_2CO_3$ = $4.5 \times 10^{-7}$

*18.114 Ethylenediaminetetraacetic acid ($H_4$EDTA) is a tetraprotic acid, $(HOOCCH_2)_2NCH_2CH_2N(CH_2COOH)_2$. Its salts are used to treat toxic metal poisoning through the formation of soluble complex ions that are then excreted. Because EDTA$^{4-}$ also binds essential calcium ions, it is often administered as the calcium disodium salt. For example, when $Na_2Ca(EDTA)$ is given to a patient, the $[Ca(EDTA)]^{2-}$ ion reacts with circulating $Pb^{2+}$ ions and exchanges metal ions:

$$[Ca(EDTA)]^{2-}(aq) + Pb^{2+}(aq) \rightleftharpoons$$
$$[Pb(EDTA)]^{2-}(aq) + Ca^{2+}(aq); \quad K_c = 2.5 \times 10^7$$

A child is found to have a dangerously high lead level in the blood of 120 $\mu$g/100 mL. If the child is administered 100 mL of 0.10 $M$ $Na_2Ca(EDTA)$ and assuming the exchange reaction and excretion process are 100% efficient, what is the final concentration of $Pb^{2+}$ in $\mu$g/100 mL blood? (Total blood volume = 1.5L.)

## CHAPTER 19

# Thermodynamics
## Entropy, Free Energy, and the Direction of Chemical Reactions

**Concepts and skills to
review**

- internal energy, heat, and work
  (Section 6.1)
- state functions and standard states
  (Section 6.1)
- enthalpy, $\Delta H$, and Hess's law (Sections
  6.2 and 6.5)
- entropy and disorder (Section 12.2)
- comparing $Q$ and $K$ to find reaction
  direction (Section 16.2)

**Thermodynamics and the steam engine.** In a
practical sense, the necessity to improve on the
function of the newly invented steam engine,
which required understanding of the relationship
between heat and work, led to the principles of
thermodynamics. In this chapter, we focus on
the real-world limitations to the work that can be
obtained from a chemical or physical system
and come to see why changes occur as they do.

In the last few chapters, we've posed and answered some extremely important questions about chemical and physical change. How fast does the change occur, and how is this rate affected by concentration and temperature? How much product will be present when the net change ceases, and how is this yield affected by concentration and temperature? We've explored these questions in systems ranging from the stratosphere to a limestone cave and from the cells of your body to the lakes of Sweden.

Now it's time to stand back and ask the most profound question of all: Why does a change occur in the first place? Even casual observation shows that some changes seem to have a natural direction and to happen by themselves, whereas others do not. Methane burns in oxygen with a vigorous burst of heat to yield carbon dioxide and water vapor, but these products will not remake methane and oxygen no matter how long they interact. A new garden spade left outside slowly rusts, but a rusty one will not become shiny. A cube of sugar dissolves in a cup of coffee after a few seconds of stirring, but stir for another century and the cube will not reappear.

All our investigations of nature allow us to define a **spontaneous change** as one that occurs by itself; the opposite process, the nonspontaneous change, does not occur unless we make it happen. There appears to be a driving force built into a spontaneous process that propels it in a given direction. If we want a process to occur against that natural drive, we must expend energy to accomplish it. The term *spontaneous* has nothing to do with how long a process takes to occur. Spontaneous does *not* mean instantaneous; it means that, given enough time, the process will happen by itself. Aging is a spontaneous, but happily not an instantaneous, process.

A chemical reaction proceeding toward equilibrium is an example of a spontaneous change. As you learned in Chapter 16, we can predict the net direction of the reaction—its spontaneous direction—by comparing the mass-action expression ($Q$) with the equilibrium constant ($K$). But what determines the value of the equilibrium constant? How can we explain the drive to attain equilibrium? And why do some spontaneous processes absorb energy while others release it? The principles of thermodynamics were developed in the early 19th century in order to help utilize the newly invented steam engine more efficiently. Despite that rather narrow focus, however, the principles apply, as far as we know, to every system in the universe. In this chapter, we apply them to chemical and physical change.

The chapter opens with a review of the first law of thermodynamics. As you'll see, the first law accounts for the energy change in a process but not its spontaneous direction. The enthalpy change also is not an adequate criterion for predicting reaction direction. Rather, we find this criterion in the change in *entropy*, that is, in the natural tendency of systems to become disordered. This idea leads us to the second law of thermodynamics and its quantitative application to both exothermic and endothermic changes. Then we examine the concept of *free energy*, which simplifies the criterion for spontaneous change, and see how it relates to the work a system can perform. Finally, we investigate the key relationship between the free energy of a reaction and its equilibrium constant.

## 19.1 The Second Law of Thermodynamics: Predicting Spontaneous Change

A spontaneous process, whether a chemical change, a physical change, or just a change in location, is one that occurs by itself under specified conditions, without a continuous input of energy from outside the system. The freezing of water, for example, is spontaneous at 1 atm and $-5°C$. A spontaneous process such as burning or falling may need a little "push" to get started—a spark to ignite gasoline, a shove to knock a book off your desk—but once the process begins, it continues without external aid because the system releases enough energy to keep the process going.

In contrast, for a nonspontaneous process to occur, the system must be supplied with a continuous input of energy. A book falls spontaneously, but it rises only if something else, such as a human or a hurricane-force wind, supplies energy in the form of work. Under a given set of conditions, *if a process is spontaneous in one direction, it is **not** spontaneous in the other.*

How can we tell the direction of a spontaneous change in cases that are not as familiar as burning gasoline or falling books? Can we predict the net direction of any reaction? Does the criterion for spontaneous change involve the energy of the reacting system? Since energy changes seem to be involved, let's begin addressing these questions by reviewing the idea of conservation of energy.

## Limitations of the First Law of Thermodynamics

In Chapter 6, we discussed the first law of thermodynamics (the law of conservation of energy) and its application to chemical systems. The first law states that the internal energy ($E$) of a system, the sum of the kinetic and potential energy of all its particles, changes through the addition or removal of heat ($q$) and/or work ($w$):

$$\Delta E = q + w$$

Everything that is not part of the system is the surroundings, so the system and surroundings together constitute the universe:

$$E_{universe} = E_{system} + E_{surroundings}$$

Heat or work gained by the system (sys) is lost by the surroundings (surr), and vice versa:

$$q_{sys} = -q_{surr} \quad \text{and} \quad w_{sys} = -w_{surr}$$

It follows from these ideas that *the total energy of the universe is constant, and therefore, energy cannot be created or destroyed.* (Any modern statement of conservation of energy must take into account mass-energy equivalence and the processes in stars, which convert enormous amounts of matter into energy. Therefore, we qualify the statement to be that the total *mass-energy* of the universe is constant.)

Is the first law sufficient to explain why a natural process takes place? It certainly accounts for the energy involved. When a book that was resting on your desk falls to the floor, the first law guides us through the conversion from potential energy of the resting book to the kinetic energy of the falling book to the heat dispersed in the floor near the point of contact. When gasoline burns in your car's engine, the first law explains that the potential energy difference between the chemical bonds in the fuel mixture and the exhaust gases is converted to the kinetic energy of the moving car. When an ice cube melts in your hand, the first law tells that energy from your hand was transferred to the ice to change the solid into a liquid. If you could measure the work and heat involved in each case, you would find that the energy is conserved as it is converted from one form to another.

However, the first law does not help us make sense of the *direction* of the change. Why doesn't the heat in the floor near the fallen book change to kinetic energy in the book and lift it back onto your desk? Why doesn't the heat released in the engine convert exhaust fumes back into gasoline and oxygen? Why doesn't your hand get colder and refreeze the water? None of these events would violate the first law—energy would still be conserved—but they never happen. The first law by itself tells nothing about the direction of a spontaneous change, so we must search elsewhere for a way to predict it.

### The Sign of ΔH: An Inadequate Criterion of Spontaneity

In the mid-19th century, some thermodynamicists thought that the sign of the enthalpy change ($\Delta H$), the heat added or removed at constant pressure ($q_P$), was the criterion of spontaneity. (Recall from Section 6.2 that, for most reactions, $\Delta H$ is either identical to $\Delta E$ or very close to it.) Moreover, by defining a set of *standard states*, we can compare the standard enthalpy changes ($\Delta H^0_{rxn}$) of different reactions. (The standard states, which we refer to throughout the chapter, are 1 atm for gases,* 1 $M$ for solutions, and the pure substance for solids or liquids.)

These scientists thought that exothermic processes ($\Delta H < 0$) were spontaneous and endothermic ones ($\Delta H > 0$) nonspontaneous. This hypothesis had much support; after all, many spontaneous processes *are* exothermic. All combustion reactions are spontaneous and exothermic. For example,

$$CH_4(g) + 2O_2(g) \rightarrow CO_2(g) + 2H_2O(g) \qquad \Delta H^0_{rxn} = -802 \text{ kJ}$$

Iron rusts spontaneously and exothermically:

$$2Fe(s) + \tfrac{3}{2}O_2(g) \rightarrow Fe_2O_3(s) \qquad \Delta H^0_{rxn} = -826 \text{ kJ}$$

Sodium metal and chlorine gas spontaneously combine into ionic sodium chloride with a release of heat:

$$Na(s) + \tfrac{1}{2}Cl_2(g) \rightarrow NaCl(s) \qquad \Delta H^0_{rxn} = -411 \text{ kJ}$$

In many other cases, however, the sign of $\Delta H$ is no help. An exothermic process may occur spontaneously under one set of conditions and the opposite, endothermic, process under another. At ordinary pressure, water freezes below 0°C but melts above 0°C. Both processes are spontaneous, but the first is exothermic and the second endothermic:

$$H_2O(l) \rightarrow H_2O(s) \qquad \Delta H^0_{rxn} = -6.02 \text{ kJ}$$
$$H_2O(s) \rightarrow H_2O(l) \qquad \Delta H^0_{rxn} = +6.02 \text{ kJ}$$

At ordinary pressure and room temperature, liquid water vaporizes spontaneously in dry air, another endothermic change:

$$H_2O(l) \rightarrow H_2O(g) \qquad \Delta H^0_{rxn} = +44.0 \text{ kJ}$$

In fact, all melting and vaporizing are endothermic changes that are spontaneous under appropriate conditions.

---

*The standard state for gases has been changed recently to 1 bar (100 kPa), which is only slightly different from 1 atm (101.325 kPa). Most tables of thermodynamic data, including those in this text, are based on the previous standard of 1 atm.

A

B

**FIGURE 19.1**
**A spontaneous, endothermic reaction. A,** When the crystalline, solid reactants barium hydroxide octahydrate and ammonium nitrate are mixed, a slurry soon forms as waters of hydration are released. **B,** The system absorbs heat so quickly that it cools, the beaker becomes covered with frost, and a moistened block of wood freezes to it.

There are some other broad groups of spontaneous endothermic *physical changes*. Water-soluble salts dissolve spontaneously, even though most have an endothermic $\Delta H^0_{soln}$:

$$NaCl(s) \xrightarrow{H_2O} Na^+(aq) + Cl^-(aq) \qquad \Delta H^0_{soln} = +3.9 \text{ kJ}$$

$$RbClO_3(s) \xrightarrow{H_2O} Rb^+(aq) + ClO_3^-(aq) \qquad \Delta H^0_{soln} = +47.7 \text{ kJ}$$

$$NH_4NO_3(s) \xrightarrow{H_2O} NH_4^+(aq) + NO_3^-(aq) \qquad \Delta H^0_{soln} = +25.7 \text{ kJ}$$

Some endothermic *chemical changes* are spontaneous as well:

$$N_2O_5(s) \rightarrow 2NO_2(g) + \tfrac{1}{2}O_2(g) \qquad \Delta H^0_{rxn} = +109.5 \text{ kJ}$$

$$Ba(OH)_2 \cdot 8H_2O(s) + 2NH_4NO_3(s) \rightarrow$$
$$Ba^{2+}(aq) + 2NO_3^-(aq) + 2NH_3(aq) + 10H_2O(l) \qquad \Delta H^0_{rxn} = +62.3 \text{ kJ}$$

In the second reaction, the released waters of hydration solvate the ions. The reaction mixture cannot absorb heat from the surroundings quickly enough, so it absorbs heat from itself, and the container becomes so cold that a wet block of wood freezes to it (Figure 19.1).

Is there any factor common to these spontaneous, endothermic processes that can give us a clue as to what makes them occur? The phase changes lead from a more organized to a less organized structure: solid to liquid and liquid to gas. The dissolving of salts leads from separate crystalline solid and pure liquid to dispersed ions in a solution. The chemical reactions produce gases and/or dispersed ions from crystalline solids. In all these examples, matter changes from a more ordered to a less ordered state. *This change in order is a very important factor in determining the direction of a spontaneous process.*

### The Tendency Toward Disorder

Systems become disordered spontaneously all around you. If you drop a ceramic dinner plate, it breaks into a collection of irregular pieces. Imagine what would happen to your room if you simply left everything wherever it fell, or to the college library if the staff weren't continuously organizing the books and computer files. As you know from many other familiar examples, there is a natural tendency for systems to become disordered, whereas creating order—gluing the pieces of plate together, organizing your room, reshelving library books—requires work. ◆

◆ **Vital Orderly Information.** Looking up a person's number in a phone book where the names are not arranged alphabetically or searching for an address in a city whose houses are not arranged numerically is the stuff of fitful dreams. A close relationship exists between the order of a system and the information that system conveys. One system in which the order and corresponding information is vitally important is the sequence of nucleotide bases in deoxyribonucleic acid (DNA) (Section 14.5). Disorder in the base sequence creates mutations, some of which lead to disease. Of particular relevance is the evidence showing that DNA becomes disordered through numerous natural processes, and that a host of repair mechanisms exists to do the work of keeping the sequence intact.

**FIGURE 19.2**

Order and disorder in a deck of playing cards. **A,** A new deck of cards is arranged in a specific sequence and thus is a highly ordered system. **B,** Flip the deck into the air or shuffle it and the sequence spontaneously becomes disordered because there are many more disordered sequences than ordered ones.

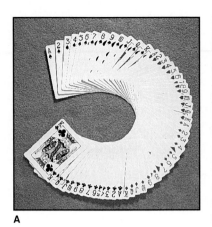

A          B

◆ **Poker Entropy.** In a game of poker, a rare combination has more worth than a common one because it has a lower probability of being dealt. In other words, the poker hand that has fewer possible arrangements wins. There are only four possible royal-flush hands (A, K, Q, J, 10 of the same suit), but 1,098,240 hands that contain one pair.

Let's analyze what we mean by *order* with a deck of playing cards (Figure 19.2). In a new deck, the cards are arranged in a specified sequence—increasing face value by suit—and thus in a highly ordered arrangement. Indeed, if the cards were arranged in any other *specified* sequence, the deck would be equally ordered. Throw the cards up in the air or shuffle them, and the deck becomes disordered, that is, randomly arranged. There are about $10^{68}$ possible arrangements of the 52 cards, and *each arrangement is equally probable.* Thus, it is extremely unlikely that *one* specific sequence would occur by chance. Shuffling the cards initiates a spontaneous process that produces disorder simply because there are *more* disordered arrangements than ordered ones. In physical and chemical systems, the change from order to disorder occurs spontaneously for the same reason: *disorder is more probable than order.* There are many more possible arrangements for the atoms, ions, or molecules in a disordered system than in an ordered one. ◆

To see how the tendency toward disorder affects reaction direction, first consider a physical system. Suppose you have an evacuated container with two bulbs of equal volume connected by a tube fitted with a stopcock (Figure 19.3). You fill the left bulb with neon to a pressure of 1 atm. When you open the stopcock, the gas eventually becomes distributed uniformly into both bulbs, as measured by a pressure of 0.5 atm in each bulb. This result is what you would expect on the basis of experience, but *why* does the gas behave this way? Why don't all the neon atoms rush to the other bulb, or just stay where they are?

The randomly moving gas particles spread throughout both bulbs because twice the original volume affords every particle twice as many places it can be. The vastly increased number of arrangements this allows makes it vastly more probable that the particles will fill the entire volume than a limited part of it. No matter how long you wait, you would never find all the neon in the left bulb again. Even for 10 Ne atoms, the odds against this happening are about 1000 to 1; for a mole of gas, the odds are about $10^{2 \times 10^{23}}$ to 1.

**FIGURE 19.3**

Spontaneous expansion of a gas. **A,** Neon gas at a pressure of 1 atm occupies one-half a two-bulb apparatus. **B,** When the stopcock is opened to the evacuated bulb, the gas expands spontaneously to fill both bulbs uniformly to a pressure of 0.5 atm.

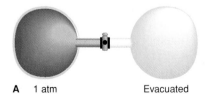

A    1 atm          Evacuated

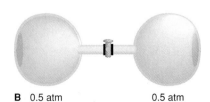

B    0.5 atm          0.5 atm

Like shuffling a new deck of cards, opening the stopcock initiates a spontaneous change from more to less order simply by providing more possible arrangements.

The redistribution of the gas particles in the container occurred with *no change in the total energy of the system,* so the various arrangements are *equivalent.* The number of ways the components of a system can be arranged without changing the system's energy is directly related to a quantity called **entropy (S).** *Entropy is a measure of the disorder of the system.* A system with relatively few equivalent ways to arrange its components, such as an ice crystal or a deck of cards in a specified sequence, has relatively small disorder and low entropy. A system with many equivalent ways to arrange its components, such as a gas or a shuffled deck of cards, has relatively large disorder and high entropy. If the entropy of a system goes up, the components of the system have become more disordered, random, or chaotic, and the system's energy has become more widely distributed: $S_{disorder} > S_{order}$.

Like internal energy and enthalpy, *entropy is a state function* of the system, one that depends only on its present state, not on how it arrived at that state. Therefore, the change in entropy of the system, $\Delta S_{sys}$, depends only on its initial and final values. Since the gas particles in the bulbs ended up with more ways to be arranged, the entropy of the system increased, so $\Delta S_{sys}$ was positive ($S_{final} > S_{initial}$). If the gas particles in both bulbs moved to one bulb, they would have fewer ways to be arranged, so the entropy of the system would decrease: $\Delta S_{sys}$ would be negative ($S_{final} < S_{initial}$). How the change in order comes about—what paths the particles take as they become rearranged—does not affect the sign or magnitude of $\Delta S_{sys}$.

## Entropy and the Second Law of Thermodynamics

Now let's return to our original question: What determines the direction of a spontaneous change? As you've just seen, natural systems tend toward disorder, toward a more random, less ordered distribution of their particles. This natural tendency is the criterion for which we've been searching, but to apply it correctly, we have to consider more than just the system. After all, some systems become less ordered spontaneously (melting of ice, dissolving a crystal) and others more ordered spontaneously (freezing of water, growing of a crystal). When we consider changes in both the system *and* its surroundings, we find that *all processes occur spontaneously in the direction that increases the total entropy of the universe* (system plus surroundings). This is one way to state the **second law of thermodynamics.**

Notice that the second law places no limitations on the entropy change of the system *or* the surroundings: either may be negative. The law does state, however, that for a spontaneous process, the *sum* of the entropy changes must be positive. When the entropy of the system decreases ($\Delta S_{sys} < 0$), the entropy of the surroundings increases even more to offset the decrease ($\Delta S_{surr} >> 0$), so the entropy of the universe (system *plus* surroundings) increases ($\Delta S_{universe} > 0$). A quantitative statement of the second law is that, for any spontaneous process,

$$\Delta S_{universe} = \Delta S_{sys} + \Delta S_{surr} > 0 \qquad \textbf{(19.1)}$$

## Determining Entropy Values in the System and Surroundings

As stated earlier, entropy ($S$) is related to the number of ways ($W$) a system can be arranged without changing its total energy. In 1877, the Austrian mathematician Ludwig Boltzmann described this relationship quantitatively:

$$S = k \ln W \tag{19.2}$$

where $k$, the *Boltzmann constant*, equals $1.38 \times 10^{-23}$ J/K. Since $W$ is unitless, you can see that *S has units of joules/kelvin (J/K).*

**Standard molar entropies, $S^0$.** Even though both are state functions, entropy values differ in a fundamental way from enthalpy values. Recall that we cannot determine absolute values of enthalpy because we have no starting point, that is, no baseline value for the enthalpy of a substance. Therefore, only enthalpy *changes* can be measured.

In contrast, we *can* determine the absolute entropy of a substance. To do so requires application of the **third law of thermodynamics,** which states that *a perfect crystal has zero entropy at absolute zero:* $S_{sys} = 0$ at 0 K. "Perfect" means that all the particles are aligned flawlessly in the crystal structure, with no defects of any kind. At absolute zero, all particles in the crystal have their minimum energy, and there is only one way they can be arranged: $W = 1$ and $S = k \ln 1 = 0$. If we warm the crystal, the total energy of the particles increases and they have a range of energies, so they can be arranged in more ways (Figure 19.4); thus, $W > 1$, $\ln W > 0$, and $S > 0$. The entropy of a substance at a given temperature is therefore an *absolute* value, equal to the entropy increase it would undergo if heated from 0 K to the temperature of interest.

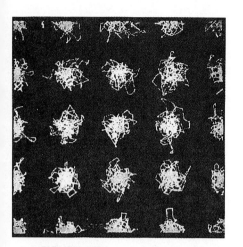

**FIGURE 19.4**

Random motion in a crystal at any temperature greater than absolute zero. This early computer simulation shows the paths of the particles in a crystalline solid at some temperature greater than 0 K. Adding thermal energy increases the total energy, and the particles can distribute the energy in more than one way, so the entropy increases.

As with other thermodynamic variables, we usually compare entropy values for substances in their standard states. Entropy is an *extensive* property, one that depends on the amount of substance, so we use the **standard molar entropy** ($S^0$) in J/mol·K ($\text{J·mol}^{-1}\text{·K}^{-1}$). A list of $S^0$ values at 298 K (25°C) for many elements, compounds, and ions appears, with the values for other thermodynamic variables, in Appendix B.

**Predicting relative $S^0$ values of the system.** Based on events at the molecular level, we can predict how the entropy of a substance is affected by temperature, physical state, dissolution, and atomic or molecular complexity. (All tabulated $S^0$ values in the following discussion have units of J/mol·K.)

1. *Temperature changes.* For a given substance, $S^0$ *increases as the temperature rises.* For example, the standard molar entropy of copper metal at 273 K (0°C) is 31.0 J/mol·K, at 295 K (22°C) it is 32.9 J/mol·K, and at 298 K (25°C) it is 33.1 J/mol·K. The temperature increase represents an increase in average kinetic energy of the particles. You know that the kinetic energies of gas particles in a sample are distributed over a range, which becomes wider as the temperature rises (Figure 5.18); the same general behavior occurs for liquids and solids. Since there are more ways to distribute the energy of the substance at the higher temperature, its entropy goes up.

2. *Physical states and phase changes.* When a more ordered phase changes to a less ordered one, the entropy change is positive. For a given substance, $S^0$ *increases as the substance changes from a solid to a liquid to a gas:*

|  | Na | $H_2O$ | C |
|---|---|---|---|
| $S^0$(s or l): | 51.4(s) | 69.9(l) | 5.7(graphite) |
| $S^0$(g): | 153.6 | 188.7 | 158.0 |

Figure 19.5 shows the behavior of a typical substance as it is heated and undergoes a change of state. Note the pattern of gradual entropy increase within a phase as the temperature rises and a large, sudden increase at the

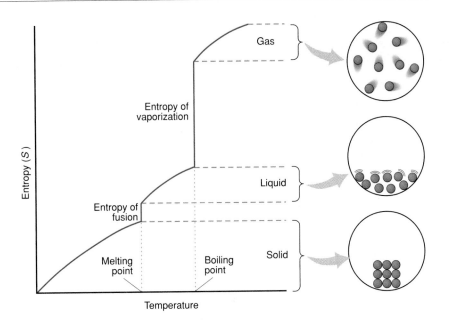

**FIGURE 19.5**
**The increase in entropy from solid to liquid to gas.** A plot of $S$ vs. temperature shows the gradual increase within a phase and the abrupt increase with a phase change. The molecular views depict the increase in randomness of the particles as the solid melts and, in particular, as the liquid vaporizes.

phase change. The solid is relatively ordered and has the lowest entropy, its particles vibrating about their positions but, on the average, remaining fixed. As the temperature rises, the entropy gradually increases as the kinetic energy increases. When the solid melts, the particles move freely between and around each other in the liquid, so there is an abrupt, large increase in entropy. Further heating increases the speed of the particles within the liquid, and the entropy increases gradually. Finally, freeing themselves from intermolecular forces, the particles move chaotically through a larger volume as a gas, with another abrupt, large entropy increase. Note that *the increase in entropy from liquid to gas is much larger than from solid to liquid.*

3. *Dissolution of a solid or liquid.* The entropy of a dissolved solid or liquid solute is *greater* than the entropy of the pure solute. In these cases, however, the type of solute *and* solvent and the nature of the solution process affects the overall entropy change:

|  | NaCl | AlCl$_3$ | CH$_3$OH |
|---|---|---|---|
| $S^0$(s or l): | 72.1(s) | 167(s) | 127(l) |
| $S^0$(aq): | 115.1 | −148 | 132 |

When an ionic solid dissolves in water, the highly ordered crystal separates into hydrated ions, which become randomly dispersed in the solvent. The entropy of the ions themselves is much greater in the solution than in the crystal. However, the entropy of the solvent must also be taken into account. Negative $S^0$ values are seen for some small, multiply charged hydrated ions because the water molecules become very highly ordered around them (Section 12.1). For example, the Al$^{3+}$(aq) ion has an $S^0$ value of −313 J/mol·K, so when AlCl$_3$ dissolves in water, the overall entropy of the solution is less than that of the solid.*

---

*An $S^0$ value for a hydrated ion can be negative because it is relative to the $S^0$ value of H$^+$(aq), which is assigned a value of 0. In other words, the Al$^{3+}$(aq) ion has a lower entropy than H$^+$(aq).

**FIGURE 19.6**
The small increase in disorder when methanol dissolves in water. Pure methanol **(A)** and pure water **(B)** have many intermolecular H bonds. **C,** When they form a solution, the molecules form H bonds to one another, so their freedom of motion does not change significantly. Thus, the entropy increase is relatively small and is due solely to random mixing.

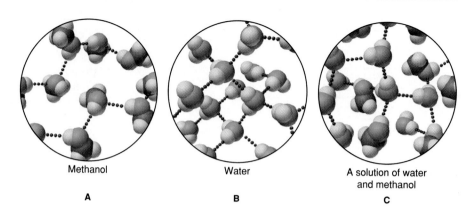

| | | |
|---|---|---|
| Methanol | Water | A solution of water and methanol |
| A | B | C |

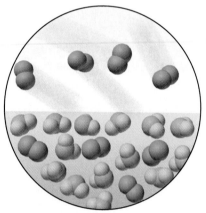

**FIGURE 19.7**
The large decrease in entropy of a gas when it dissolves in a liquid. The chaotic movement and high entropy of $O_2$ gas are reduced greatly when it dissolves in water.

With molecular solutes, the increase in entropy on dissolving is usually much smaller. For a solid, such as glucose, there is no separation into ions, and for a liquid, such as methanol, the breakdown of a crystal structure is also absent. Furthermore, in pure methanol and pure water, the molecules form many H bonds, so there is relatively little change in their freedom of movement when they are mixed. The small increase in the disorder of the dissolved methanol arises from the random mixing of the two types of molecules (Figure 19.6).

4. *Dissolution of a gas.* A gas is so disordered to begin with that it becomes *more* ordered when it dissolves in a liquid or solid. Therefore, the entropy of a solution of a gas in a liquid or solid is always *less* than the entropy of the pure gas (Figure 19.7). When $O_2$ gas [$S^0(g)$ = 205.0 J/mol·K] dissolves in water, its entropy decreases sharply [$S^0(aq)$ = 110.9 J/mol·K]. When a gas dissolves in another gas, however, the entropy increases because of the mixing of the two types of molecules.

5. *Complexity of the element or compound.* In general, differences in entropy values for substances in the same phase are based on atomic size and molecular complexity. For elements within a periodic group, those with higher molar mass have higher entropy.

| | Li | Na | K | Rb | Cs |
|---|---|---|---|---|---|
| Molar mass (g/mol): | 6.941 | 22.99 | 39.10 | 85.47 | 132.9 |
| $S^0(s)$: | 29.1 | 51.4 | 64.7 | 69.5 | 85.2 |

The same trend holds for similar compounds: $S^0$ of $CF_4(g)$ = 261.5 J/K·mol, whereas $S^0$ of $CCl_4(g)$ = 309.7 J/K·mol.

For an element that exists in different solid forms, the entropy is *less* in the form with stronger, more directional bonds. For example, carbon in the form of graphite has an $S^0$ of 5.69 J/mol·K, whereas in the form of diamond, its $S^0$ is 2.44 J/mol·K. In diamond, covalent bonds extend in three dimensions; in graphite, covalent bonds extend only within a sheet, and motion between sheets is relatively easy, so the entropy is higher. Furthermore, the delocalized $\pi$-bonding system of graphite allows more electron movement than in the localized $\sigma$ bonds of diamond.

For compounds, the chemical complexity increases as the number of atoms (or ions) in a compound increases, and so does the entropy. This trend holds for both ionic and covalent compounds, as long as we compare substances in the same physical state.

| | NaCl | AlCl$_3$ | P$_4$O$_{10}$ | NO | NO$_2$ | N$_2$O$_4$ |
|---|---|---|---|---|---|---|
| $S^0(s)$: | 72.1 | 167 | 229 | | | |
| $S^0(g)$: | | | | 211 | 240 | 304 |

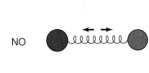

NO

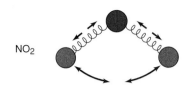

NO₂

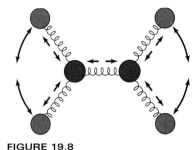

N₂O₄

**FIGURE 19.8**
**Entropy and vibrational motion.** A diatomic molecule, such as NO, can vibrate in only one way. $NO_2$ can vibrate in more ways and $N_2O_4$ in even more ways. Thus, as the number of atoms increases, a molecule can distribute its vibrational energy in more ways, so it has higher entropy.

The trend is based on the various types of movement available to the atoms (or ions) in the compound, which relates to the number of equivalent ways in which their energy can be distributed. For example, among the nitrogen oxides shown, the two atoms of NO can vibrate in only one way, toward and away from each other. The three atoms of $NO_2$ have several vibrational motions, and the six atoms of $N_2O_4$ have many more (Figure 19.8).

For larger molecules, we also consider how parts of a molecule move relative to other parts. A long hydrocarbon chain can rotate and vibrate in more ways than a short chain, so entropy increases with chain length. A ring compound, such as cyclopentane ($C_5H_{10}$), has lower entropy than a chain compound of the same general mass, such as pentane ($C_5H_{12}$), because some of the vibrations and rotations are restricted in the ring:

| | $CH_4(g)$ | $C_2H_6(g)$ | $C_3H_8(g)$ | $C_4H_{10}(g)$ | $C_5H_{12}(g)$ | $C_5H_{10}(g)$ | $C_2H_5OH(l)$ |
|---|---|---|---|---|---|---|---|
| $S^0$: | 186 | 230 | 270 | 310 | 348 | 293 | 161 |

Remember that these trends hold only for *substances in the same physical state*. Note, for instance, that gaseous methane ($CH_4$) has a greater entropy than liquid ethanol ($C_2H_5OH$), even though the ethanol molecules are more complex. When gases are compared with liquids, the effect of physical state usually dominates that of molecular complexity.

SAMPLE PROBLEM 19.1 ─────────────────────────────

### Predicting Relative Entropy Values

**Problem:** Choose the member with the higher entropy in each of the following pairs, and justify your choice:
**(a)** 1 mol $SO_2(g)$ or 1 mol $SO_3(g)$
**(b)** 1 mol $CO_2(s)$ or 1 mol $CO_2(g)$
**(c)** 3 mol oxygen gas ($O_2$) or 2 mol ozone gas ($O_3$)
**(d)** 1 mol KBr($s$) or 1 mol KBr($aq$)
**(e)** Sea water in midwinter at 2°C or in midsummer at 23°C
**(f)** 1 mol HF($g$) or 1 mol HI($g$)
**Plan:** In general, we know that less ordered systems have higher entropy than more ordered systems and that higher temperature increases entropy. We apply the general categories described in text to choose the member with the higher entropy.
**Solution: (a) 1 mol $SO_3(g)$.** For equal numbers of moles of substances with the same types of atoms in the same physical state, the more atoms in the molecule, the more types of motion available to it, so the higher its entropy.
**(b) 1 mol $CO_2(g)$.** For a given substance, entropy increases in the sequence $s < l < g$.
**(c) 3 mol $O_2(g)$.** The two samples contain the same number of oxygen atoms but different numbers of molecules. Despite the greater complexity of $O_3$, there are more ways to arrange three moles of particles than two moles.
**(d) 1 mol KBr($aq$).** The two samples have the same number of ions, but in the solid they are highly ordered, and in the solution they are randomly dispersed.
**(e) Sea water in summer.** Entropy increases with rising temperature.
**(f) 1 mol HI($g$).** For similar compounds, entropy increases with molar mass.

For each pair of substances, select the member with the higher entropy and give the reason for your choice: **(a)** $PCl_3(g)$ or $PCl_5(g)$; **(b)** $CaF_2(s)$ or $BaCl_2(s)$; **(c)** $Br_2(g)$ or $Br_2(l)$.

---

**Calculating the change in entropy of a reaction.** Based on the ideas we just discussed, when x moles of more ordered reactants yield x moles of less ordered products, the reaction will have a positive entropy change. It also follows that when the number of moles of gas decreases in the reaction, the entropy change will be negative. For example, when ammonia forms from its elements, four moles of gas produces two moles of gas, so the entropy of the products is less than that of the reactants:

$$N_2(g) + 3H_2(g) \rightleftharpoons 2NH_3(g) \qquad \Delta S^0 = S^0_{products} - S^0_{reactants} < 0$$

Recall that by applying Hess's law, we can add $\Delta H^0_f$ values to find the standard heat of reaction, $\Delta H^0_{rxn}$. Similarly, we can add standard molar entropies to find the **standard entropy of reaction, $\Delta S^0_{rxn}$:**

$$\Delta S^0_{rxn} = \Sigma m S^0_{products} - \Sigma n S^0_{reactants} \qquad \textbf{(19.3)}$$

where m and n are the coefficients of the balanced equation. For the synthesis of ammonia, we have

$$\Delta S^0_{rxn} = (2 \text{ mol } NH_3 \times S^0 \text{ of } NH_3) - [(1 \text{ mol } N_2 \times S^0 \text{ of } N_2) + (3 \text{ mol } H_2 \times S^0 \text{ of } H_2)]$$

From Appendix B, we find the appropriate $S^0$ values:

$$\Delta S^0_{rxn} = (2 \text{ mol} \times 193 \text{ J/mol·K})$$
$$- [(1 \text{ mol} \times 191.5 \text{ J/mol·K}) + (3 \text{ mol} \times 130.6 \text{ J/mol·K})]$$
$$= -197 \text{ J/K}$$

As we predicted, $\Delta S^0 < 0$.

---

**Calculating the Standard Entropy of Reaction, $\Delta S^0_{rxn}$**

**Problem:** Calculate $\Delta S^0_{rxn}$ for the combustion of 1 mol propane at 25°C:

$$C_3H_8(g) + 5O_2(g) \rightarrow 3CO_2(g) + 4H_2O(l)$$

**Plan:** To determine $\Delta S^0_{rxn}$, we apply Equation 19.3. We predict the sign of $\Delta S^0_{rxn}$ from the change in the number of moles of gas: six moles of gases yield three moles of gases, so the entropy will decrease ($\Delta S^0_{rxn} < 0$).
**Solution:** Calculating $\Delta S^0_{rxn}$. From Appendix B values,

$$\Delta S^0_{rxn} = [(3 \text{ mol } CO_2 \times S^0 \text{ of } CO_2) + (4 \text{ mol } H_2O \times S^0 \text{ of } H_2O)]$$
$$- [(1 \text{ mol } C_3H_8 \times S^0 \text{ of } C_3H_8) + (5 \text{ mol } O_2 \times S^0 \text{ of } O_2)]$$
$$= [(3 \text{ mol} \times 213.7 \text{ J/mol · K}) + (4 \text{ mol} \times 69.9 \text{ J/mol · K})]$$
$$- [(1 \text{ mol} \times 269.9 \text{ J/mol · K}) + (5 \text{ mol} \times 205.0 \text{ J/mol · K})]$$
$$= -374.2 \text{ J/K}$$

**Check:** $\Delta S^0 < 0$, so our prediction is correct. Rounding gives $[(3 \times 200) + (4 \times 70)] - [270 + (5 \times 200)] = 880 - 1270 = -390$, close to the actual value.
**Comment:** We based our prediction on the fact that $S^0$ values of gases are usually much greater than those of solids or liquids. When there is no change in moles of gas, you *cannot* confidently predict the sign of $\Delta S^0_{rxn}$.

**FOLLOW-UP PROBLEM 19.2**
Balance the following equations, predict the sign of $\Delta S^0_{rxn}$ when possible, and calculate its value at 25°C:
**(a)** $NaOH(s) + CO_2(g) \longrightarrow Na_2CO_3(s) + H_2O(l)$
**(b)** $Fe(s) + H_2O(g) \longrightarrow Fe_2O_3(s) + H_2(g)$

---

**Entropy changes in the surroundings: the other part of the total.**
As you've seen, in many spontaneous reactions, such as the synthesis of ammonia and the combustion of propane, the system becomes *more* ordered at the given conditions ($\Delta S^0_{rxn} < 0$). The second law dictates that for this to occur, entropy increases in the surroundings must compensate. Let's examine the influence of the surroundings on the *total* entropy change.

*The surroundings either add heat to the system or remove heat from it.* In essence, they function as an enormous heat source or heat sink, one so large that its temperature remains constant, even though its entropy changes through the loss or gain of heat. How do the surroundings participate in the two possible types of reaction enthalpies?

1. *Exothermic change.* Heat lost by the system is gained by the surroundings. The gain of heat increases the random motion of particles in the surroundings, so the entropy of the surroundings increases:

   For an exothermic change: $\quad q_{sys} < 0, q_{surr} > 0 \quad$ and $\quad \Delta S_{surr} > 0$

2. *Endothermic change.* Heat gained by the system is lost by the surroundings. The loss of heat reduces the disorder of the surroundings and lowers its entropy:

   For an endothermic change: $\quad q_{sys} > 0, q_{surr} < 0 \quad$ and $\quad \Delta S_{surr} < 0$

In both cases, the change in entropy of the surroundings is proportional to the change in heat of the surroundings and, most importantly, to an opposite change in heat of the system:

$$\Delta S_{surr} \propto q_{surr} \quad \text{and} \quad \Delta S_{surr} \propto -q_{sys}$$

The temperature at which heat is added to or removed from the surroundings also affects $\Delta S_{surr}$. Consider the effect of an exothermic reaction at a low and a high temperature. At a low temperature, such as 20 K, random motion in the surroundings is less, so the surroundings are relatively ordered. Therefore, adding heat has a larger effect on the degree of orderliness. At a higher temperature, such as 298 K, the surroundings are relatively disordered, so adding the same amount of heat has a smaller effect on the degree of orderliness. ◆ In other words, the change in entropy of the surroundings is greater when heat is added at a lower temperature. Thus, *the change in entropy of the surroundings is directly related to an opposite change in the heat of the system and inversely related to the temperature at which the heat is changed.* The mathematical relationships are

$$\Delta S_{surr} \propto -q_{sys} \quad \text{and} \quad \Delta S_{surr} \propto \frac{1}{T}$$

Combining them gives

$$\Delta S_{surr} = -\frac{q_{sys}}{T} \tag{19.4}$$

◆ **A Checkbook Analogy for Heating the Surroundings.** A monetary analogy may clarify the relative changes in disorder that arise from heating the surroundings at different initial temperatures. If you have $10 in your checking account, a $10 deposit represents a 100% increase in your net worth; that is, a given change to a low initial state has a large impact. If, on the other hand, you have a $1000 balance, a $10 deposit represents only a 1% increase. That is, the same change to a high initial state has a small impact.

For a process at constant pressure, the heat is $\Delta H$, so

$$\Delta S_{surr} = -\frac{\Delta H_{sys}}{T}$$

This means that we can calculate $\Delta S_{surr}$ by measuring $\Delta H_{sys}$ and the temperature at which the change takes place.

To restate the central point, if a spontaneous reaction has a negative $\Delta S_{sys}$ (the system becomes more ordered), $\Delta S_{surr}$ must be positive enough (the surroundings must become sufficiently disordered) so that $\Delta S_{universe}$ is positive (the universe becomes more disordered). Sample Problem 19.3 illustrates this situation for one of the reactions we considered earlier.

SAMPLE PROBLEM 19.3 _____

## Determining Whether a Reaction Is Spontaneous

**Problem:** At 298 K, the formation of ammonia has a negative $\Delta S_{sys}^0$:

$$N_2(g) + 3H_2(g) \rightarrow 2NH_3(g) \qquad \Delta S_{sys}^0 = -197 \text{ J/K}$$

Does the reaction occur spontaneously at this temperature?

**Plan:** For the reaction to occur spontaneously, $\Delta S_{universe}^0 > 0$, so $\Delta S_{surr}^0$ must be greater than $+197$ J/K. To find $\Delta S_{surr}^0$, we need $\Delta H_{sys}^0$, which is the $\Delta H_{rxn}^0$. We use $\Delta H_f^0$ values in Appendix B to find $\Delta H_{rxn}^0$, and use it to find $\Delta S_{surr}^0$. To find $\Delta S_{universe}^0$, we take the sum of $\Delta S_{surr}^0$ and $\Delta S_{sys}^0$.

**Solution:** Calculating $\Delta H_{sys}^0$:

$$\Delta H_{sys}^0 = \Delta H_{rxn}^0$$

$$= 2 \text{ mol NH}_3 \ (-45.9 \text{ kJ/mol}) - [3 \text{ mol H}_2 \ (0 \text{ kJ/mol}) + 1 \text{ mol N}_2 \ (0 \text{ kJ/mol})]$$

$$= -91.8 \text{ kJ}$$

Calculating $\Delta S_{surr}^0$:

$$\Delta S_{surr}^0 = -\frac{\Delta H_{sys}^0}{T} = -\frac{(-91.8 \text{ kJ} \times 1000 \text{ J/kJ})}{298 \text{ K}} = 308 \text{ J/K}$$

Determining $\Delta S_{universe}^0$:

$$\Delta S_{universe}^0 = \Delta S_{sys}^0 + \Delta S_{surr}^0 = -197 \text{ J/K} + 308 \text{ J/K} = 111 \text{ J/K}$$

$\Delta S_{universe}^0 > 0$, so **the reaction occurs spontaneously at 298 K** (see figure).

**Check:** Given the negative $\Delta H_{rxn}^0$, Le Châtelier's principle predicts that low temperature should favor $NH_3$ formation, so the answer is reasonable (see Chemical Connections essay on the Haber process, Section 16.4). Rounding to check the math, we have

$$\Delta H_f^0 \approx 2 \times (-45 \text{ kJ}) = -90 \text{ kJ}$$

$$\Delta S_{surr}^0 \approx -(-90,000 \text{ J})/300 \text{ K} = 300 \text{ J/K}$$

$$\Delta S_{universe}^0 \approx -200 \text{ J/K} + 300 \text{ J/K} = 100 \text{ J/K}$$

**Comment:** Note that $\Delta H$ has units of kJ, while $\Delta S$ has units of J/K. Don't forget to convert kJ to J, or you'll introduce a large error.

FOLLOW-UP PROBLEM 19.3
Does the oxidation of FeO(s) to $Fe_2O_3(s)$ occur spontaneously at 298 K?

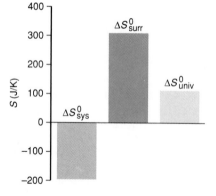

Taking the surroundings into account is critical to determining reaction spontaneity, and it resolves a confusing point about the relevance of thermodynamics to biology, as the Chemical Connections essay shows.

**CHEMICAL CONNECTIONS**

**Chemistry in Biology**
**Do Living Things Obey the Laws of Thermodynamics?**

Organisms can be thought of as chemical machines that evolved by extracting energy as efficiently as possible from the environment. Such a mechanical view holds that all processes, whether they involve living or nonliving systems, are consistent with thermodynamic principles. When applied to animals and plants, however, these ideas seem far from obvious. Let's examine the first and second laws of thermodynamics to see if they apply to living systems.

Organisms certainly comply with the first law. The chemical bond energy in food is converted into the mechanical energy of running, climbing, and countless other movements; the electrical energy of nerve conduction; the thermal energy of warming the body; and so forth. Many experiments have demonstrated that in all these energy conversions, the total energy is conserved. Some of the earliest studies of this question were performed by Lavoisier, who included animal respiration in his new theory of combustion (Figure 19.A). He was the first to show that "animal heat" was produced by a slow combustion process occurring continually in the body. In experiments with guinea pigs, he measured the intake of food and $O_2$ and the output of $CO_2$ and heat, for which he invented a calorimeter based on the melting of ice, and established the principles and methods for measuring the basal rate of metabolism. Experiments using mod-

ern room-size calorimeters designed for humans continue to confirm the conservation of energy.

It may not seem so clear, however, that an organism, or for that matter, the whole parade of life, complies with the second law. Mature humans are far more ordered and complex than the simple egg and sperm cells from which they grow. Modern organisms are far more orderly, complex systems than the one-celled ancestral specks from which they evolved. Are the growth of an organism and the evolution of life exceptions to the tendency of natural processes to become more disordered? Does biology violate the second law? Not at all, if we examine the system *and* its surroundings. In order for an organism to grow or for a species to evolve, large numbers of complex food molecules—carbohydrates, proteins, and fats—and oxygen molecules undergo combustion to form much larger numbers of gaseous $CO_2$ and $H_2O$ molecules. The formation and discharge of these waste gases represent a tremendous net increase in the entropy of the surroundings, as does the heat released from these exothermic reactions. Thus, the increase in order apparent in the spontaneous growth of organisms and their evolution occurs at the expense of far more disorder in the Earth-Sun surroundings. When system and surroundings are considered together, the entropy of the universe, as always, increases.

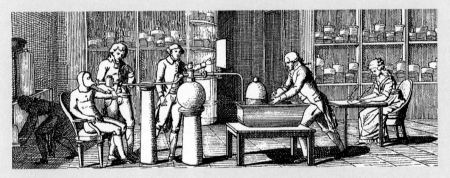

**FIGURE 19.A**
Lavoisier studying human respiration as a form of combustion. Lavoisier (standing at right) measured substances and heat to understand the chemical nature of respiration. The artist (Mme Lavoisier) shows herself taking notes (seated at right).

## The Entropy Change and the Equilibrium State

For a process at equilibrium, $\Delta S_{universe} = 0$ because any entropy change in the system is exactly balanced by an opposite entropy change in the surroundings:

$$\text{At equilibrium, } \Delta S_{sys} = -\Delta S_{surr}$$

Therefore,     $$\text{At equilibrium, } \Delta S_{universe} = \Delta S_{sys} + \Delta S_{surr} = 0$$

For example, let's calculate $\Delta S^0_{universe}$ for a phase change, such as the vaporization-condensation of 1 mol water at 100°C (373 K):

$$H_2O(l; 373 \text{ K}) \rightleftharpoons H_2O(g; 373 \text{ K})$$

First, we find $\Delta S^0_{universe}$ for the forward change (vaporization) by calculating $\Delta S^0_{sys}$:

$$\Delta S^0_{sys} = \Sigma m S^0_{products} - \Sigma n S^0_{reactants} = S^0 \text{ of } H_2O(g; 373 \text{ K}) - S^0 \text{ of } H_2O(l; 373 \text{ K})$$
$$= 195.9 \text{ J/K} - 86.8 \text{ J/K} = 109.1 \text{ J/K}$$

As we would expect, the system becomes more disordered ($\Delta S^0_{sys} > 0$) as the liquid changes to a gas.

For $\Delta S^0_{surr}$ we have

$$\Delta S^0_{surr} = -\frac{\Delta H^0_{sys}}{T}$$

where $\Delta H^0_{sys} = \Delta H^0_{vap}$ at 373 K = 40.7 kJ/mol = $40.7 \times 10^3$ J/mol. For 1 mol water, we have

$$\Delta S^0_{surr} = -\frac{\Delta H^0_{vap}}{T} = -\frac{40.7 \times 10^3 \text{ J}}{373 \text{ K}} = -109 \text{ J/K}$$

The negative sign indicates that the surroundings become more ordered because they lose heat. The two entropy changes have the same magnitude but opposite sign:

$$\Delta S^0_{universe} = 109 \text{ J/K} + (-109 \text{ J/K}) = 0$$

For the reverse change (condensation), $\Delta S^0_{universe}$ also equals zero, but $\Delta S^0_{sys}$ and $\Delta S^0_{surr}$ have signs opposite those for vaporization. This means that *when a system reaches equilibrium, neither the forward nor the reverse reaction is spontaneous* and neither proceeds any further.

### Spontaneous Exothermic and Endothermic Reactions: A Summary

We can now understand why both spontaneous exothermic and endothermic reactions occur. No matter what its *enthalpy* change, a reaction occurs because the total *entropy* of the reaction and surroundings increases.

*For an exothermic reaction* ($\Delta H_{sys} < 0$), heat is released, which increases the entropy of the surroundings ($\Delta S_{surr} > 0$). If the reacting system yields more disordered products than reactants ($\Delta S_{sys} > 0$), the total entropy change ($\Delta S_{sys} + \Delta S_{surr}$) will be positive. For example, in the oxidation of glucose, an essential reaction for all higher organisms,

$$C_6H_{12}O_6(s) + 6O_2(g) \rightarrow 6CO_2(g) + 6H_2O(g) + \textbf{\textit{heat}}$$

six moles of gas yield 12, so $\Delta S_{sys} > 0$, $\Delta S_{surr} > 0$, and $\Delta S_{universe} > 0$.

If, on the other hand, the system becomes more ordered as the reaction occurs ($\Delta S_{sys} < 0$), the surroundings must become even more disordered ($\Delta S_{surr} >> 0$) to outweigh the system's negative entropy change and make the total $\Delta S$ positive. For example, when calcium oxide and carbon dioxide gas form calcium carbonate,

$$CaO(s) + CO_2(g) \rightleftharpoons CaCO_3(s) + \textbf{\textit{heat}}$$

the system becomes more ordered because the moles of gas decrease, but the heat released disorders the surroundings even more; thus, $\Delta S_{sys} < 0$, but $\Delta S_{surr} >> 0$, so $\Delta S_{universe} > 0$.

*For an endothermic reaction* ($\Delta H_{sys} > 0$), the heat lost by the surroundings lowers its entropy ($\Delta S_{surr} < 0$). Therefore, the only way an endothermic reaction can occur spontaneously is if the system becomes so disordered ($\Delta S_{sys} >> 0$) that it outweighs the negative $\Delta S_{surr}$. The solution processes for many ionic compounds are common examples: heat is absorbed to form the solution, so the surroundings become more ordered. However, the system starts in such a highly ordered state and ends in such a disordered state that its entropy increase more than compensates for the entropy decrease in the surroundings. Thus, $\Delta S_{universe}$ is positive.

Another example is the spontaneous, endothermic reaction between barium hydroxide and ammonium nitrate (see Figure 19.1):

$$heat + Ba(OH)_2 \cdot 8H_2O(s) + 2NH_4NO_3(s) \rightarrow$$
$$Ba^{2+}(aq) + 2NO_3^-(aq) + 2NH_3(aq) + 10H_2O(l)$$

Three moles of crystalline solids absorb heat from the surroundings ($\Delta S_{surr} < 0$) and yield 15 moles of dissolved ions and molecules, thereby greatly increasing the disorder of the system ($\Delta S_{sys} >> 0$).

### Section Summary

A spontaneous change is one that occurs in a given direction under specified conditions without a continuous input of energy. Neither the first law of thermodynamics nor the sign of $\Delta H$ predicts this direction. All spontaneous processes involve a change from more to less order because a disordered arrangement is more likely than an ordered one. Entropy is a measure of disorder. The second law of thermodynamics states that, in a spontaneous process, the entropy of the universe (system plus surroundings) increases. Entropy values are absolute because perfect crystals have zero entropy at 0 K (third law). Standard molar entropy $S^0$ (J/mol · K) is affected by temperature, changes of state, dissolution, and the complexity of the substance. The standard entropy of reaction, $\Delta S^0_{rxn}$, is calculated from $S^0$ values. When the moles of gas increase in a reaction, $\Delta S^0_{rxn} > 0$. The $\Delta S^0_{surr}$ is related directly to $\Delta H^0_{sys}$ and inversely to the temperature at which the enthalpy changes. A system can become more ordered only by making the surroundings even more disordered. For a system at equilibrium, $\Delta S_{universe} = 0$.

## 19.2 Entropy, Free Energy, and Work

By making *two* separate measurements, $\Delta S_{sys}$ and $\Delta S_{surr}$, we can predict whether or not a reaction will be spontaneous at a particular temperature. It would be useful, however, to have a single criterion of spontaneity that applies only to the system. One such criterion, called the Gibbs free energy, or just **free energy (G),** is a function that combines the system's enthalpy and entropy:

$$G = H - TS$$

This function is named for Josiah Willard Gibbs, who proposed it in 1877 and laid the foundation for much of chemical thermodynamics. ◆

### The Free Energy Change and the Prediction of Reaction Spontaneity

*The free energy change ($\Delta G$) is a measure of the spontaneity of a process and of the useful energy available from it.* Let's see how the free energy change is derived from the second law. By definition, the entropy change of the universe is the sum of the entropy changes in the system and surroundings:

$$\Delta S_{universe} = \Delta S_{sys} + \Delta S_{surr}$$

At constant pressure,

$$\Delta S_{surr} = -\frac{\Delta H_{sys}}{T}$$

◆ **The Greatness and Obscurity of J. Willard Gibbs.** Even today, one of the most remarkable minds in science is barely known outside chemistry and physics. In 1878, Josiah Willard Gibbs (1839-1903), professor of mathematical physics at Yale, completed a 323-page paper that virtually established the science of chemical thermodynamics, while including major principles governing chemical equilibrium, phase-change equilibrium, and the energy changes in electrochemical cells. The European scientists James Clerk Maxwell, Wilhelm Ostwald, and Henri Le Chatelier appreciated Gibbs's significance long before his countrymen. In fact, he was elected to the Hall of Fame of Distinguished Americans only in 1950, because he had not received enough votes until then!

Substituting for $\Delta S_{surr}$, we obtain

$$\Delta S_{universe} = \Delta S_{sys} - \frac{\Delta H_{sys}}{T}$$

Multiplying both sides by $(-T)$, we obtain

$$-T\Delta S_{universe} = \Delta H_{sys} - T\Delta S_{sys}$$

Now that the entropy change of the universe is expressed in terms of the system only, we introduce the new free energy quantity to replace the enthalpy and entropy terms. Since $G = H - TS$, a change in free energy of the system ($\Delta G_{sys}$) at constant temperature and pressure is given by the *Gibbs equation:*

$$\Delta G_{sys} = \Delta H_{sys} - T\Delta S_{sys} \qquad (19.5)$$

Combining this equation with the previous relationship, we see that

$$-T\Delta S_{universe} = \Delta H_{sys} - T\Delta S_{sys} = \Delta G_{sys}$$

The sign of $\Delta G$ tells whether a reaction is spontaneous. The second law dictates that $\Delta S_{universe}$ is greater than zero for a spontaneous process, less than zero for a nonspontaneous process, and equal to zero for a process at equilibrium. Absolute temperature is always positive, so

For a spontaneous process,    $T\Delta S_{universe} > 0$  or  $-T\Delta S_{universe} < 0$

Since $\Delta G = -T\Delta S_{universe}$,

$$\Delta G < 0 \text{ for a spontaneous process}$$

and by similar reasoning,

$$\Delta G > 0 \text{ for a nonspontaneous process}$$

$$\Delta G = 0 \text{ for a process at equilibrium}$$

An important point to keep in mind is that if a process is *nonspontaneous* in one direction ($\Delta G > 0$), it is *spontaneous* in the opposite direction ($\Delta G < 0$). Note that by incorporating $\Delta G$, we have not presented any new ideas, but we have developed a way of predicting reaction spontaneity from one variable ($\Delta G_{sys}$) rather than two ($\Delta S_{sys}$ and $\Delta S_{surr}$). It is also essential to remember that the sign and magnitude of $\Delta G$ for a chemical reaction tell us nothing about its rate. The reaction between $H_2(g)$ and $O_2(g)$ at room temperature, for instance, is highly spontaneous—$\Delta G$ has a large negative value—but in the absence of a catalyst or a flame, the reaction does not occur to a measurable extent because its rate is very low.

### The Standard Free Energy Change ($\Delta G^0$) and Free Energy of Formation ($\Delta G_f^0$)

Because free energy ($G$) combines three state functions, $H$, $S$, and $T$, it is also a state function. Like enthalpy, free energy has no absolute starting point, so we can measure only free energy *changes* ($\Delta G$). As with the other thermodynamic variables, the **standard free energy change ($\Delta G^0$)** occurs when all components of the system are in their standard states. Therefore, we have

$$\Delta G_{sys}^0 = \Delta H_{sys}^0 - T\Delta S_{sys}^0 \qquad (19.6)$$

We apply this relationship in Sample Problem 19.4.

SAMPLE PROBLEM 19.4

## Calculating $\Delta G^0$ from Enthalpy and Entropy Values

**Problem:** Potassium chlorate, one of the common oxidizing agents in explosives, fireworks, and matchheads, undergoes a solid-state redox reaction when heated, in which the oxidation number of Cl in the reactant increases and decreases in the products (disproportionation):

$$4KClO_3(s) \xrightarrow{\Delta} 3KClO_4(s) + KCl(s)$$

Use $\Delta H_f^0$ and $S^0$ values to calculate $\Delta G^0$ at 25°C for this reaction.
**Plan:** To solve for $\Delta G^0$, we use $\Delta H_f^0$ values to calculate $\Delta H_{rxn}^0$ ($\Delta H_{sys}^0$), use $S^0$ values to calculate $\Delta S_{rxn}^0$ ($\Delta S_{sys}^0$), and apply Equation 19.6.
**Solution:** Calculating $\Delta H_{sys}^0$ from $\Delta H_f^0$ values (with Equation 6.8):

$$\Delta H_{sys}^0 = \Delta H_{rxn}^0 = \Sigma m\Delta H_{f(products)}^0 - \Sigma n\Delta H_{f(reactants)}^0$$

$$= [3 \text{ mol } KClO_4 \, (\Delta H_f^0 \text{ of } KClO_4) + 1 \text{ mol } KCl \, (\Delta H_f^0 \text{ of } KCl)]$$

$$- 4 \text{ mol } KClO_3 \, (\Delta H_f^0 \text{ of } KClO_3)$$

$$= [3 \text{ mol } (-432.8 \text{ kJ/mol}) + 1 \text{ mol } (-436.7 \text{ kJ/mol})]$$

$$- 4 \text{ mol } (-397.7 \text{ kJ/mol})$$

$$= -144.3 \text{ kJ}$$

Calculating $\Delta S_{sys}^0$ from $S^0$ values (with Equation 19.3):

$$\Delta S_{sys}^0 = \Delta S_{rxn}^0 = [3 \text{ mol } KClO_4 \, (S^0 \text{ of } KClO_4) + 1 \text{ mol } KCl \, (S^0 \text{ of } KCl)]$$

$$- [4 \text{ mol } KClO_3 \, (S^0 \text{ of } KClO_3)]$$

$$= [3 \text{ mol } (151.0 \text{ J/mol} \cdot \text{K}) + 1 \text{ mol } (82.6 \text{ J/mol} \cdot \text{K})]$$

$$- [4 \text{ mol } (143.1 \text{ J/mol} \cdot \text{K})]$$

$$= -36.8 \text{ J/K}$$

Calculating $\Delta G_{sys}^0$ at 298 K:

$$\Delta G_{sys}^0 = \Delta H_{sys}^0 - T\Delta S_{sys}^0$$

$$= -144.3 \text{ kJ} - [(298 \text{ K})(-36.8 \text{ J/K})(1 \text{ kJ}/1000 \text{ J})] = \mathbf{-133.3 \text{ kJ}}$$

**Check:** Rounding to check the math:

$$\Delta H^0 \approx [3(-433) + (-440)] - [4(-400)] = -1740 + 1600 = -140$$

$$\Delta S^0 \approx [3(150) + 85] - [4(145)] = 535 - 580 = -45$$

$$\Delta G^0 \approx -140 - 300(-0.04) = -140 + 12 = -128$$

All values are close to the calculated ones. Another way to check appears in Sample Problem 19.5.
**Comment:** The negative $\Delta G^0$ means the reaction is spontaneous, but the rate is extremely slow because the required O atom transfer does not occur easily in the solid. When $KClO_3$ is heated slightly above its melting point, the ions are free to move and the reaction occurs readily.

FOLLOW-UP PROBLEM 19.4
Determine the standard free energy change at 298 K for the reaction
$$2NO(g) + O_2(g) \rightarrow 2NO_2(g)$$

Another way to calculate $\Delta G_{rxn}^0$ is with values for the **standard free energy of formation, $\Delta G_f^0$,** of the components, the free energy change that occurs when one mole of a compound is made *from its elements,* with all components of the reaction in their standard states. Since free energy is a state function, we can combine $\Delta G_f^0$ values of reactants and products to calculate $\Delta G_{rxn}^0$ without regard for the reaction path:

$$\Delta G_{rxn}^0 = \Sigma m \Delta G_{f(products)}^0 - \Sigma n \Delta G_{f(reactants)}^0 \qquad \textbf{(19.7)}$$

$\Delta G_f^0$ values have properties similar to $\Delta H_f^0$ values:
- $\Delta G_f^0$ of an element in its standard state is zero.
- An equation coefficient ($m$ or $n$ above) multiplies $\Delta G_f^0$ by that number.
- Reversing a reaction changes the sign of $\Delta G_f^0$.

An extensive table of $\Delta G_f^0$ values appears with those for $\Delta H_f^0$ and $S^0$ in Appendix B.

---

SAMPLE PROBLEM 19.5 _____

### Calculating $\Delta G_{rxn}^0$ from $\Delta G_f^0$ Values

**Problem:** Use $\Delta G_f^0$ values to calculate $\Delta G_{rxn}^0$ for the reaction in Sample Problem 19.4:

$$4KClO_3(s) \longrightarrow 3KClO_4(s) + KCl(s)$$

**Plan:** We apply Equation 19.7 to calculate $\Delta G_{rxn}^0$.
**Solution:**

$$\Delta G_{rxn}^0 = \Sigma m \Delta G_{f(products)}^0 - \Sigma n \Delta G_{f(reactants)}^0$$

$$= [3 \text{ mol } KClO_4 \, (\Delta G_f^0 \text{ of } KClO_4) + 1 \text{ mol } KCl \, (\Delta G_f^0 \text{ of } KCl)]$$

$$- 4 \text{ mol } KClO_3 \, (\Delta G_f^0 \text{ of } KClO_3)$$

$$= [3 \text{ mol } (-303.2 \text{ kJ/mol}) + 1 \text{ mol } (-409.2 \text{ kJ/mol})]$$

$$- 4 \text{ mol } (-296.3 \text{ kJ/mol})$$

$$= \mathbf{-133.6 \text{ kJ}}$$

**Check:** Rounding to check the math:

$$\Delta G_{rxn}^0 \approx [3(-300) + 1(-400)] - 4(-300) = -1300 + 1200 = -100$$

**Comment:** The slight discrepancy between this value and that in Sample Problem 19.4 is within experimental error. When $\Delta G_f^0$ values are available for a reaction taking place at 25°C, this method is simpler than the calculation in Sample Problem 19.4.

FOLLOW-UP PROBLEM 19.5
Use $\Delta G_f^0$ values to calculate the free energy changes at 25°C for the following reactions:
**(a)** $2NO(g) + O_2(g) \longrightarrow 2NO_2(g)$ (from Follow-up Problem 19.4)
**(b)** $2C(graphite) + O_2(g) \longrightarrow 2 CO(g)$

---

### The Meaning of $\Delta G$: The Maximum Work a System Can Perform

So far, you've seen the usefulness of $\Delta G$ in predicting reaction spontaneity. Now let's examine the concept of free energy more closely to see what $\Delta G$ really means. One of the most practical thermodynamic relationships is that between the free energy change of a reaction and the amount of work the system can do. For a spontaneous process, $\Delta G$ is the maximum work obtainable *from* the system:

$$\Delta G = w_{max}$$

For a nonspontaneous process, $\Delta G$ is the minimum work that must be done *to* the system to make the change happen.

What do we mean by the maximum work obtainable? What determines this limit, and why can't we harness the *entire* enthalpy change of a system to do work? To answer these questions, let's return to Equation 19.5, which relates $\Delta G$ to $\Delta H$ and $\Delta S$:

$$\Delta G_{sys} = \Delta H_{sys} - T\Delta S_{sys}$$

Solving for $\Delta H_{sys}$, we obtain

$$\Delta H_{sys} = \Delta G_{sys} + T\Delta S_{sys}$$

This equation tells us that the enthalpy change of a reaction has two components: a free energy change ($\Delta G_{sys}$) and a temperature-dependent entropy change ($T\Delta S_{sys}$). Each component is a portion of the total heat transferred during the reaction ($\Delta H_{sys}$).

To see what each portion represents, consider one of the most common combustion reactions in our society, the burning of the octane in gasoline:

$$C_8H_{18}(l) + \tfrac{25}{2}O_2(g) \rightarrow 8CO_2(g) + 9H_2O(g)$$

When octane burns in an automobile engine, a large amount of energy is given off as heat ($\Delta H_{sys} < 0$), and since the moles of gas increase, the entropy of the system increases ($\Delta S_{sys} > 0$). A portion of the total released energy does work turning wheels, regenerating the battery, and so on; *the free energy change ($\Delta G_{sys}$) is that portion of the total energy change that does this work.* However, there is *always* a portion ($T\Delta S_{sys}$) of the total energy change that simply warms the engine and the outside air, which makes the motions of the particles in the universe more chaotic, in accord with the second law.

Whereas the first law considers energy in terms of its *amount*, the second law (from which we derived the relationship among $\Delta G$, $\Delta H$, and $\Delta S$) considers energy in terms of its *usability*. In the form of $\Delta G_{sys}$, energy can be used to do work; in the form of $T\Delta S_{sys}$, energy has become unusable. It has not disappeared, so the first law is satisfied, but it has lost its value because it would take even more energy to gather it up and convert it to a usable form. ◆

Thus, an overall energy change has a more and a less usable form. In the form of $T\Delta S$, it is dispersed and largely unusable; in the form of $\Delta G$, we can develop machines to harness it. However, there is a major catch: the value of $\Delta G$ represents the *maximum* amount of work we can obtain from the energy, not the *actual* amount. In reality, we can never obtain this much work: unless the free energy is released at an infinitely slow rate, some is lost as heat. *How* we release the energy determines the *efficiency* of the process, the actual amount of work done from the free energy released.

Consider a battery, which is essentially a packaged spontaneous redox reaction. When the circuit is completed between the battery's terminals, the reaction releases energy to the surroundings. If we connect the terminals to each other through a short piece of wire, the energy change takes place very quickly without performing any work: all the free energy is dissipated in heating the wire and the battery. If we connect the terminals to a motor, the energy is released much more slowly. Some of the free energy change is still converted to heat in the battery and the motor, but a large portion does the work of running the motor. The more slowly the battery discharges, the more of the free energy change does work and the less is converted to heat. However, only when the battery operates infinitely slowly can we obtain the maximum available work. This is the compromise that all engineers and machine designers must face—*in the real world, some free energy is always changed to heat and thereby unharnessed:* no process utilizes all the free energy available, so no process is 100% energy efficient. ◆

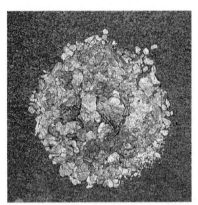

◆ **Valueless Gold.** The precious metals market provides an analogy for the *value* of the released energy in its "concentrated" usable form and its "dispersed" unusable form. Suppose you have an ounce of gold dust in a small container. At current market prices, you could sell it for about $400. Now you disperse the dust over an acre of land. Conservation of mass is satisfied because the ounce of gold is still present, but you'll receive a very negative response when you try to obtain $400 in the market. In dispersed form, the gold's value is much reduced. To extract, concentrate, and purify it would require large expenditures of energy, materials, and know-how—far more, in fact, than it is worth.

◆ **The Wide Range of Energy Efficiency.** Engineers consider the efficiency of a device as that percentage of the energy input that results in a work output. The range of efficiencies is enormous for the common "energy-conversion systems" of society. For instance, an incandescent lightbulb converts only 5% of incoming electrical energy to light, the rest being wasted as heat. At the other extreme, a large electrical generator converts 99% of the incoming mechanical energy to electricity. Although improvements are continually being made, here are some approximate values for other devices: dry cell battery, 90%; home oil furnace, 65%; handtool motor, 63%; liquid fuel rocket, 50%; car engine, <30%; fluorescent lamp, 20%; solar cell, ~10%.

To summarize the relationship between the free energy change of a reaction and the work it can accomplish:

- A spontaneous reaction ($\Delta G_{sys} < 0$) will occur *and* can do work on the surroundings.
- A nonspontaneous reaction ($\Delta G_{sys} > 0$) will not occur unless the surroundings do work on it.
- A reaction at equilibrium ($\Delta G_{sys} = 0$) can no longer do any work.

### The Effect of Temperature on Reaction Spontaneity

In most cases, the enthalpy contribution ($\Delta H$) to the free energy change ($\Delta G$) is much *larger* than the entropy contribution ($T\Delta S$). For this reason, most exothermic reactions are spontaneous: the negative $\Delta H$ helps make $\Delta G$ negative as well. However, the temperature at which a reaction occurs does influence the $T\Delta S$ term and, in many cases, makes the entropy contribution crucial to the overall spontaneity. By scrutinizing the signs of $\Delta H$ and $\Delta S$, we can predict the effect of temperature on the sign of $\Delta G$ and thus on the spontaneity of a process at any temperature. (In this discussion, we assume that $\Delta H$ and $\Delta S$ change very little with temperature, which is usually correct as long as no phase changes occur.) Let's examine the four combinations of positive and negative $\Delta H^0$ and $\Delta S^0$; two combinations do not depend on temperature and two do:

- *Temperature-independent cases.* When $\Delta H^0$ and $\Delta S^0$ have *opposite* signs, the reaction either occurs spontaneously at all temperatures ($\Delta H^0 < 0$ and $\Delta S^0 > 0$) or at none ($\Delta H^0 > 0$ and $\Delta S^0 < 0$).

1. *Reaction is spontaneous at all temperatures:* $\Delta H^0 < 0$, $\Delta S^0 > 0$. Both contributions favor the process occurring spontaneously: $\Delta H^0$ is negative and $(-T\Delta S^0)$ is negative, so $\Delta G^0$ is always negative. Many combustion reactions are in this category, such as those of glucose and octane that we saw earlier. The decomposition of hydrogen peroxide, the common disinfectant, is also spontaneous at all temperatures:

$$2H_2O_2(l) \rightarrow 2H_2O(l) + O_2(g) \qquad \Delta H^0 = -196 \text{ kJ and } \Delta S^0 = 125 \text{ J/K}$$

2. *Reaction is nonspontaneous at all temperatures:* $\Delta H^0 > 0$, $\Delta S^0 < 0$. Both contributions oppose the reaction occurring spontaneously. $\Delta H^0$ is positive and $(-T\Delta S^0)$ is positive, so $\Delta G^0$ is always positive. The formation of ozone from oxygen gas is not spontaneous at any temperature:

$$3O_2(g) \rightarrow 2O_3(g) \qquad \Delta H^0 = 286 \text{ kJ} \quad \text{and} \quad \Delta S^0 = -137 \text{ J/K}$$

This reaction occurs only if enough energy is supplied from the surroundings, as when ozone is synthesized by passing an electrical discharge through pure $O_2$.

- *Temperature-dependent cases.* When $\Delta H^0$ and $\Delta S^0$ have the *same* sign, the relative magnitudes of the $T\Delta S^0$ and $\Delta H^0$ terms determine the sign of $\Delta G^0$, so the magnitude of $T$ is crucial.

3. *Reaction is spontaneous at higher temperature:* $\Delta H^0 > 0$ *and* $\Delta S^0 > 0$. In these cases, $\Delta S^0$ favors spontaneity ($-T\Delta S^0 < 0$), but $\Delta H^0$ does not:

$$2H_2O(g) \rightarrow 2H_2(g) + O_2(g) \qquad \Delta H^0 = 483.7 \text{ kJ} \quad \text{and} \quad \Delta S^0 = 88.76 \text{ J/K}$$

With a positive $\Delta H^0$, the reaction will occur only when $-T\Delta S^0$ is large enough to make $\Delta G^0$ negative, which will happen at higher temperatures. The decomposition of water vapor occurs spontaneously at $T > 5450$ K.

**TABLE 19.1    Reaction Spontaneity and the Signs of $\Delta H^0$, $\Delta S^0$, and $\Delta G^0$**

| $\Delta H^0$ | $\Delta S^0$ | $-T\Delta S^0$ | $\Delta G^0$ | DESCRIPTION |
|---|---|---|---|---|
| − | + | − | − | Spontaneous at all $T$ |
| + | − | + | + | Nonspontaneous at all $T$ |
| − | − | + | + or − | Spontaneous at lower $T$; nonspontaneous at higher $T$ |
| + | + | − | + or − | Spontaneous at higher $T$; nonspontaneous at lower $T$ |

4. *Reaction is spontaneous at lower temperature: $\Delta H^0 < 0$ and $\Delta S^0 < 0$.* In these cases, $\Delta H^0$ favors spontaneity, but $\Delta S^0$ does not ($-T\Delta S > 0$):

$$2Na(s) + Cl_2(g) \rightarrow 2NaCl(s) \qquad \Delta H^0 = -822.2 \text{ kJ} \quad \text{and} \quad \Delta S^0 = -181.7 \text{ J/K}$$

With a negative $\Delta H^0$, the reaction will occur only if the $-T\Delta S^0$ term is smaller than the $\Delta H^0$ term, and this happens at lower temperatures. Common examples are the formation of ammonium halides from ammonia and a hydrogen halide or the formation of metal oxides, fluorides, and chlorides from their elements. The production of sodium chloride occurs spontaneously at $T < 4525$ K.

Table 19.1 summarizes these four possible combinations of $\Delta H^0$ and $\Delta S^0$.

As you saw in Sample Problem 19.4, one way to calculate $\Delta G^0$ is from enthalpy and entropy changes. Since $\Delta H^0$ and $\Delta S^0$ usually change little with temperature, we can use their values at 298 K to examine the effect of temperature on $\Delta G^0$ and thus on reaction spontaneity.

SAMPLE PROBLEM 19.6 _____

**Determining the Effect of Temperature on $\Delta G^0$**

**Problem:** An important reaction in the production of sulfuric acid is the oxidation of $SO_2(g)$ to $SO_3(g)$:

$$2SO_2(g) + O_2(g) \rightleftharpoons 2SO_3(g)$$

At 298 K, $\Delta G^0 = -141.6$ kJ; $\Delta H^0 = -198.4$ kJ; and $\Delta S^0 = -187.9$ J/K.
**(a)** Use the data to decide if the change is spontaneous at 25°C and how $\Delta G^0$ will change with increasing $T$.
**(b)** Assuming that $\Delta H^0$ and $\Delta S^0$ are constant with $T$, is the reaction spontaneous at 900°C?
**Plan: (a)** We examine the sign of $\Delta G^0$ to see if the reaction is spontaneous and the signs of $\Delta H^0$ and $\Delta S^0$ to see the effect of $T$. **(b)** We use Equation 19.6 to calculate $\Delta G^0$ from the given $\Delta H^0$, $\Delta S^0$, and the higher $T$ (in K).
**Solution: (a)** Since $\Delta G^0 < 0$, a mixture of $SO_2(g)$, $O_2(g)$, and $SO_3(g)$ in their standard states (1 atm) will spontaneously yield more $SO_3(g)$. $\Delta S^0 < 0$, so $-T\Delta S^0 > 0$ and becomes more positive at higher $T$. Therefore, **$\Delta G^0$ will be less negative, and the reaction less spontaneous, with increasing $T$.**
**(b)** Calculating $\Delta G^0$ at 900°C ($T = 273 + 900 = 1173$ K):

$$\Delta G^0 = \Delta H^0 - T\Delta S^0 = -198.4 \text{ kJ} - [(1173 \text{ K})(-187.9 \text{ J/K})(1 \text{ kJ}/1000 \text{ J})]$$

$$= 22.0 \text{ kJ}$$

Since $\Delta G^0 > 0$, the reaction is **nonspontaneous at the higher $T$.**
**Check:** The answer in (b) seems reasonable based on our prediction in (a). The arithmetic seems correct: $\Delta G^0 \approx -200 - [(1200)(-200)/1000)] = +40$, within rounding.

**FOLLOW-UP PROBLEM 19.6**

A reaction is nonspontaneous at room temperature but is spontaneous at $-40°C$. What can you say about the signs and relative magnitudes of $\Delta H^0$, $\Delta S^0$, and $T\Delta S^0$?

## The Temperature at Which a Reaction Becomes Spontaneous

As you have just seen, when the signs of $\Delta H^0$ and $\Delta S^0$ are the same, some reactions that are nonspontaneous at one temperature become spontaneous at another, and vice versa. It would certainly be useful to know the temperature at which a reaction will become spontaneous. This is the temperature at which a positive $\Delta G^0$ switches to a negative $\Delta G^0$ due to the $-T\Delta S^0$ term. We find this crossover temperature by setting $\Delta G^0$ equal to zero and solving for $T$:

$$\Delta G^0 = \Delta H^0 - T\Delta S^0 = 0$$

Therefore,

$$\Delta H^0 = T\Delta S^0 \qquad \text{and} \qquad T = \frac{\Delta H^0}{\Delta S^0} \qquad \text{(19.8)}$$

Consider the reaction of copper(I) oxide with carbon, which does not occur at lower temperature but is utilized at higher temperature in one step toward the extraction of copper metal:

$$Cu_2O(s) + C(s) \rightarrow 2Cu(s) + CO(g)$$

We would predict that this reaction has a positive entropy change due to the increase in moles of gas ($\Delta S^0 = 165$ J/K). Furthermore, since the reaction is *non*spontaneous at lower temperatures, we know that it must have a positive $\Delta H^0$ (58.1 kJ). As the $-T\Delta S^0$ term becomes larger (more negative) at higher temperatures, it will eventually overcome the positive $\Delta H^0$ term, and the reaction will occur spontaneously.

Let's calculate $\Delta G^0$ for this reaction at 25°C and then find the temperature above which the reaction is spontaneous. At 25°C (298 K),

$$\Delta G^0 = \Delta H^0 - T\Delta S^0 = 58.1 \text{ kJ} - \left(298 \text{ K} \times 165 \text{ J/K} \times \frac{1 \text{ kJ}}{1000 \text{ J}}\right) = 8.9 \text{ kJ}$$

Since $\Delta G^0$ is positive, the reaction will not proceed on its own at 25°C. At the crossover temperature, $\Delta G^0 = 0$, so

$$T = \frac{\Delta H^0}{\Delta S^0} = \frac{58.1 \text{ kJ} \times \dfrac{1000 \text{ J}}{1 \text{ kJ}}}{165 \text{ J/K}} = 352 \text{ K}$$

At any temperature above 352 K (79°C), a moderate temperature for recovering a metal from its ore, the reaction occurs spontaneously (Figure 19.9).

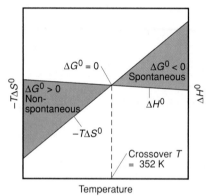

**FIGURE 19.9**
**The effect of temperature on reaction spontaneity.** A plot of the two terms that make up $\Delta G$ vs. $T$ shows a relatively constant $\Delta H^0$ and a steadily increasing $-T\Delta S^0$ for the reaction between $Cu_2O$ and C. At low $T$, the reaction is not spontaneous ($\Delta G^0 > 0$) because the positive $\Delta H^0$ term dominates. At 352 K, $\Delta H^0 = -T\Delta S^0$, so $\Delta G^0 = 0$. At any higher $T$, the reaction becomes spontaneous ($\Delta G^0 < 0$) because the $-T\Delta S^0$ term dominates.

## Coupling Reactions to Drive a Nonspontaneous Change

When we break down a spontaneous multistep reaction to its component steps, we often find that a nonspontaneous step is driven by a spontaneous step in a **coupling of reactions.** One step supplies enough free energy for the other to occur, just as the combustion of gasoline supplies enough free energy to move a car.

Consider the reaction we just discussed for the reduction of copper(I) oxide by carbon. We found that the *overall* reaction becomes spontaneous at any temperature above 352 K. When we divide the reaction into two steps,

**Chemistry in Biological Energetics**
**The Universal Role of ATP**

One of the most remarkable features of living organisms, as well as a strong indication of a common biological ancestry, is their utilization of the same few biomolecules for all the reactions of life. Despite their bewildering diversity of appearance and behavior, virtually every organism uses the same 20 amino acids to make its proteins, the same four or five nucleotides to make its nucleic acids, and the same carbohydrate (glucose) to provide energy. In addition, all organisms use the same spontaneous reaction to provide the free energy needed to drive a wide variety of nonspontaneous ones. This reaction is the hydrolysis of a high-energy molecule called **adenosine triphosphate (ATP)** to adenosine diphosphate (ADP):

$$ATP^{4-} + H_2O \rightleftharpoons ADP^{3-} + HPO_4^{2-} + H^+$$
$$\Delta G^0 = -30.5 \text{ kJ}$$

In the metabolic breakdown of glucose, for example, the initial step is the addition of a phosphate group to a glucose molecule:

$$Glucose + HPO_4^{2-} + H^+ \rightleftharpoons [glucose\ phosphate]^- + H_2O$$
$$\Delta G^0 = 13.8 \text{ kJ}$$

By coupling this nonspontaneous reaction to ATP hydrolysis, the overall reaction becomes energetically favorable. If we add the two previous reactions, $HPO_4^{2-}$, $H^+$, and $H_2O$ cancel, and we obtain

$$Glucose + ATP^{4-} \rightleftharpoons [glucose\ phosphate]^- + ADP^{3-}$$
$$\Delta G^0 = -16.7 \text{ kJ}$$

The two reactions cannot affect each other if they are physically separated because hydrolysis of ATP in water simply releases heat. The coupling is accomplished through an enzyme catalyst (Section 15.6) that simultaneously binds glucose and ATP such that the phosphate group of ATP to be transferred lies next to the particular —OH group of glucose that will accept it (Figure 19.B).

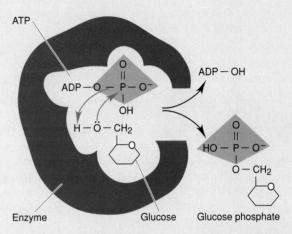

**FIGURE 19.B The coupling of a nonspontaneous reaction to the hydrolysis of ATP.** The glucose molecule (structure simplified) must lie next to the ATP molecule (written as ADP—O—PO₃H) in the enzyme's active site for the correct atoms to form bonds *(red arrows)*. ADP (written as ADP—OH) and glucose phosphate are released.

*Continued*

however, we find that even at a higher temperature, such as 375 K, copper(I) oxide does not spontaneously decompose to its elements:

$$Cu_2O(s) \rightarrow 2Cu(s) + \tfrac{1}{2}O_2(g) \qquad \Delta G^0_{375} = 140.0 \text{ kJ}$$

However, the oxidation of carbon to CO at 375 K is quite spontaneous:

$$C(s) + \tfrac{1}{2}O_2(g) \rightarrow CO(g) \qquad \Delta G^0_{375} = -143.8 \text{ kJ}$$

Coupling these reactions means having the carbon in contact with the $Cu_2O$, which allows the reaction with the larger negative $\Delta G^0$ to "drive" the one with the smaller positive $\Delta G^0$. Adding the reactions together and canceling terms gives an overall negative $\Delta G^0$:

$$Cu_2O(s) + C(s) \rightarrow 2Cu(s) + CO(g) \qquad \Delta G^0_{375} = -3.8 \text{ kJ}$$

Many essential biochemical reactions are also nonspontaneous. Key steps in the synthesis of proteins and nucleic acids, formation of fatty acids, maintenance of ion balance, and breakdown of nutrients are among numerous processes with positive $\Delta G^0$ values. Driving these energetically unfavorable steps by coupling them to a spontaneous one is a strategy common to all organisms—animals, plants, and microbes—as we discuss in the Chemical Connections essay.

**CHEMICAL CONNECTIONS**                    **The Universal Role of ATP—cont'd**

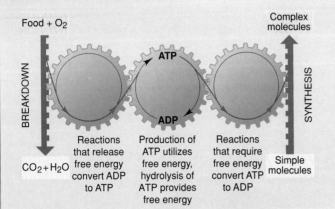

**FIGURE 19.C The cycling of metabolic free energy through ATP.** Processes that release free energy are coupled to the formation of ATP from ADP, while those that require free energy are coupled to the hydrolysis of ATP to ADP.

Enzymes play similar catalytic roles in all the reactions driven by ATP hydrolysis. The ADP that is formed in these reactions is used to regenerate ATP in other reactions. Thus, there is a continuous cycling of ATP to ADP and back to ATP again to supply energy to the cells (Figure 19.C).

What makes ATP hydrolysis such a good supplier of free energy? By examining the phosphate portions of ATP, ADP, and $HPO_4^{2-}$, we can see two reasons (Figure 19.D). The first is that the triphosphate portion of the ATP molecule at physiological pH ($\sim 7$) has an average of four negative charges grouped closely together. As a result, the molecule experiences a *built-in high charge repulsion*. In ADP, some of these repulsions are relieved.

The second reason relates to the greater delocalization of $\pi$ electrons in the hydrolysis products, which we can see using resonance structures. For example, once ATP is hydrolyzed, the $\pi$ electrons in $HPO_4^{2-}$ can be more readily delocalized, which stabilizes the ion. Thus, greater charge repulsion and less electron delocalization make ATP higher in energy than the sum of ADP and $HPO_4^{2-}$. When ATP is hydrolyzed, some of this additional energy is harnessed by the organism to drive the metabolic reactions that could not otherwise take place.

**FIGURE 19.D Why does ATP hydrolysis release free energy? A,** The high charge repulsion in the triphosphate portion of ATP is partially relieved when the molecule is hydrolyzed. **B,** Delocalization of $\pi$ electrons through resonance contributes to the stability of $HPO_4^{2-}$.

**Section Summary**

The sign of the free energy change, $\Delta G = \Delta H - T\Delta S$, is directly related to reaction spontaneity: a negative $\Delta G$ corresponds to a positive $\Delta S_{universe}$. We use the standard free energy change ($\Delta G^0$) to compare the spontaneity of reactions and the standard free energy of formation ($\Delta G_f^0$) to calculate $\Delta G_{rxn}^0$ at 25°C. A portion of $\Delta H_{rxn}$ can do work ($\Delta G$), while the other portion disorders the surroundings ($T\Delta S$). The maximum work is never obtained from a real process because some free energy is always converted to heat. The magnitude of $T$ influences the spontaneity of temperature-dependent reactions (same signs of $\Delta H$ and $\Delta S$) by affecting the size of $T\Delta S$. For such reactions, the $T$ at which the reaction becomes spontaneous can be found by setting $\Delta G = 0$. A nonspontaneous reaction ($\Delta G > 0$) can be made to occur by coupling it to a more spontaneous one ($\Delta G << 0$). In organisms, the hydrolysis of ATP drives many reactions with a positive $\Delta G$.

## 19.3    Free Energy, Equilibrium, and Reaction Direction

Observing the sign of $\Delta G$ is used to predict reaction spontaneity and thus direction, but it is not the only way to do so. In Chapter 16, we predicted reaction direction by comparing the values of the mass-action expression ($Q$) and the equilibrium constant ($K$). Recall that

- If $Q < K$ ($Q/K < 1$), the reaction as written proceeds to the right.
- If $Q > K$ ($Q/K > 1$), the reaction as written proceeds to the left.
- If $Q = K$ ($Q/K = 1$), the reaction has reached equilibrium, and there is no net reaction in either direction.

As you might expect, these two ways of predicting reaction spontaneity—the sign of $\Delta G$ and the magnitude of $Q/K$—are related mathematically. Their relationship emerges when we compare the signs of $\ln Q/K$ with $\Delta G$:

- If $Q/K < 1$, then $\ln Q/K < 0$: reaction proceeds to the right ($\Delta G < 0$).
- If $Q/K > 1$, then $\ln Q/K > 0$: reaction proceeds to the left ($\Delta G > 0$).
- If $Q/K = 1$, then $\ln Q/K = 0$: reaction is at equilibrium ($\Delta G = 0$).

Note that the signs of $\Delta G$ and $\ln Q/K$ are identical for a given reaction direction. In fact, $\Delta G$ and $\ln Q/K$ are proportional to each other through the constant $RT$:

$$\Delta G = RT \ln Q/K = RT \ln Q - RT \ln K \qquad \textbf{(19.9)}$$

What does this central relationship mean? As you know, $Q$ represents the concentrations (or pressures) of a system's components at any time during the reaction, whereas $K$ represents them when the reaction has reached equilibrium. Therefore, Equation 19.9 states that the free energy change of the system is the difference between the free energy of the system in some initial state $Q$ and the free energy of the system in its final state $K$. For a system at equilibrium, $Q$ has become equal to $K$, so $\Delta G = 0$. In other words, at equilibrium, no further free energy change occurs; in effect, the system has released all its free energy in the process of attaining equilibrium.

The size of $\Delta G$ depends on the difference in the sizes of $Q$ and $K$. By choosing standard-state values for $Q$, we obtain the standard free energy change ($\Delta G^0$), the amount of free energy a reaction releases in going from the standard-state "starting point" to its equilibrium "finishing point." When all concentrations are 1 $M$ (or all pressures 1 atm), $Q$ equals 1:

$$\Delta G^0 = RT \ln 1 - RT \ln K$$

Since $\ln 1 = 0$, the "$RT \ln Q$" term drops out, and we have

$$\Delta G^0 = -RT \ln K \qquad \textbf{(19.10)}$$

This very important relationship allows us to calculate the standard free energy change of a reaction ($\Delta G^0$) from its equilibrium constant, or vice versa. Because $\Delta G^0$ is related logarithmically to $K$, even a small change in the value of $\Delta G^0$ has a large effect on the value of $K$. Table 19.2 shows the $K$ values that correspond to a range of $\Delta G^0$ values. Note that as $\Delta G^0$ becomes more positive, the equilibrium constant becomes smaller, which means the reaction reaches equilibrium with less product and more reactant; as $\Delta G^0$ becomes more negative, the reaction reaches equilibrium with more product and less reactant. For example, if $\Delta G^0 = +10$ kJ, $K \approx 0.02$, which means

**TABLE 19.2** **The Relationship Between $\Delta G^0$ and $K$ at 298 K**

| $\Delta G^0$ (kJ) | $K$ | SIGNIFICANCE |
|---|---|---|
| 200 | $1 \times 10^{-35}$ | Essentially no forward reaction; reverse reaction goes to completion |
| 100 | $3 \times 10^{-18}$ | |
| 50 | $2 \times 10^{-9}$ | |
| 10 | $2 \times 10^{-2}$ | |
| 1 | $7 \times 10^{-1}$ | Forward and reverse reactions proceed to same extent |
| 0 | 1 | |
| −1 | 1.5 | |
| −10 | 50 | |
| −50 | $5 \times 10^{8}$ | |
| −100 | $3 \times 10^{17}$ | Forward reaction goes to completion; essentially no reverse reaction |
| −200 | $1 \times 10^{35}$ | |

that the product terms are about 50 times smaller than the reactant terms; whereas, if $\Delta G^0 = -10$ kJ, they are 50 times larger.

Of course, most reactions do not begin with all components in their standard states. By substituting the relationship between $\Delta G^0$ and $K$ (Equation 19.10) into the expression for $\Delta G$ (Equation 19.9), we obtain a relationship that applies to any starting concentrations:

$$\Delta G = \Delta G^0 + RT \ln Q \qquad \textbf{(19.11)}$$

Sample Problem 19.7 illustrates how Equations 19.10 and 19.11 are applied.

SAMPLE PROBLEM 19.7 _____

**Calculating $\Delta G$ at Nonstandard Conditions**

**Problem:** The oxidation of $SO_2$, which we considered in Sample Problem 19.6,

$$2SO_2(g) + O_2(g) \rightleftharpoons 2SO_3(g)$$

is too slow at 298 K to be useful in the manufacture of sulfuric acid, so the process is conducted at an elevated temperature.
**(a)** Calculate $K$ at 298 K and at 973 K. $\Delta G^0_{298} = -141.6$ kJ/mol of reaction as written and using $\Delta H^0$ and $\Delta S^0$ values at 973 K, $\Delta G^0_{973} = -12.12$ kJ/mol of reaction as written.
**(b)** In experiments to determine the effect of temperature on reaction spontaneity, sealed containers are filled with 0.50 atm $SO_2$, 0.010 atm $O_2$, and 0.10 atm $SO_3$ and kept at 25°C and at 700°C. In which direction, if any, will the reaction proceed to reach equilibrium at each temperature?
**(c)** Calculate $\Delta G$ for the system in part (b) at each temperature.
**Plan: (a)** We know $\Delta G^0$, $T$, and $R$, so we can calculate $K$ from Equation 19.10. **(b)** To determine if a net reaction will occur with the given pressures, we calculate $Q$ and compare it with each $K$ from part (a). **(c)** Since these are not standard-state amounts, we calculate $\Delta G$ at each $T$ from Equation 19.11 with the values of $\Delta G^0$ (given) and $Q$ [found in part (b)].

**Solution: (a)** Calculating $K$ at the two temperatures:

$$\Delta G^0 = -RT \ln K \quad \text{so} \quad K = e^{-(\Delta G^0/RT)}$$

At 298 K, the exponent is

$$-(\Delta G^0/RT) = -\left(\frac{-141.6 \text{ kJ/mol} \times 1000 \text{ J/1 kJ}}{8.31 \text{ J/mol·K} \times 298 \text{ K}}\right) = 57.2$$

So

$$K = e^{-(\Delta G^0/RT)} = e^{57.2} = \mathbf{7 \times 10^{24}}$$

At 973 K, the exponent is

$$-(\Delta G^0/RT) = -\left(\frac{-12.12 \text{ kJ/mol} \times 1000 \text{ J/1 kJ}}{8.31 \text{ J/mol·K} \times 973 \text{ K}}\right) = 1.50$$

So

$$K = e^{-(\Delta G^0/RT)} = e^{1.50} = \mathbf{4.5}$$

**(b)** Calculating the value of $Q$:

$$Q = \frac{P^2_{SO_3}}{P^2_{SO_2} \times P_{O_2}} = \frac{0.10^2}{0.50^2 \times 0.010} = 4.0$$

Since $Q < K$ at both temperatures, the denominator will decrease and the numerator increase—more $SO_3$ will form—until $Q$ equals $K$. However, at 298 K, the reaction will go **far to the right** before reaching equilibrium, whereas at 973 K, it will move only **slightly to the right.**

**(c)** Calculating $\Delta G$, the nonstandard free energy change, at 298 K:

$$\Delta G_{298} = \Delta G^0 + RT \ln Q$$

$$= -141.6 \text{ kJ/mol} + (8.31 \text{ J/mol·K} \times 1 \text{ kJ/1000 J} \times 298 \text{ K} \times \ln 4.0)$$

$$= \mathbf{-138.2 \text{ kJ/mol}}$$

Calculating $\Delta G$ at 973 K:

$$\Delta G_{973} = \Delta G^0 + RT \ln Q$$

$$= -12.12 \text{ kJ/mol} + (8.31 \text{ J/mol·K} \times 1 \text{ kJ/1000 J} \times 973 \text{ K} \times \ln 4.0)$$

$$= \mathbf{-0.91 \text{ kJ/mol}}$$

**Check:** Be sure to check the arithmetic in each step. Note that in parts (a) and (c) we made energy units in free energy changes (kJ) consistent with those in $R$ (J).
**Comment:** For these starting gas pressures at 973 K, the process is barely spontaneous ($\Delta G = -0.91$ kJ/mol), so why use a higher temperature? As in the synthesis of $NH_3$ (Section 16.4), this process is operated at a higher temperature *with a catalyst* to attain a higher *rate*, even though the *yield* is greater at a lower temperature. We discuss these details of the industrial production of sulfuric acid in Chapter 23.

**FOLLOW-UP PROBLEM 19.7**
At 298 K, hypobromous acid (HBrO) dissociates in water with a $K_a$ of $2.3 \times 10^{-9}$.
**(a)** Calculate $\Delta G^0$ for the dissociation of HBrO.
**(b)** Calculate $\Delta G$ when $[H_3O^+] = 6.0 \times 10^{-4}$ $M$, $[BrO^-] = 0.10$ $M$, and $[HBrO] = 0.20$ $M$.

At this point let's reconsider what we mean by the terms *spontaneous* and *nonspontaneous*. Consider the general reaction A $\rightleftharpoons$ B, for which $K = [B]/[A] > 1$, so the reaction goes largely from left to right (Figure 19.10, *A*). From pure A to the equilibrium point, the reaction is spontaneous ($\Delta G < 0$).

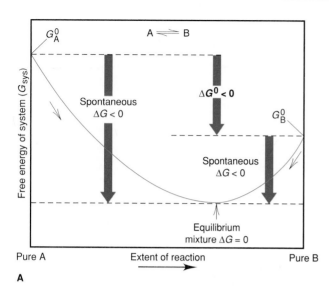

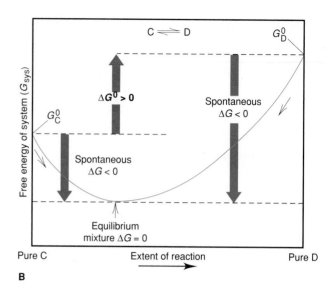

**A**                    **B**

**FIGURE 19.10**

**The relation between free energy and the extent of reaction.** The free energy of the system is plotted against the extent of reaction for two general reactions with very different $K$ values. **A,** For reaction A $\rightleftharpoons$ B, $G_A^0 > G_B^0$, so $K > 1$ and $\Delta G^0 < 0$. **B,** For reaction C $\rightleftharpoons$ D, $G_C^0 < G_D^0$, so $K < 1$ and $\Delta G^0 > 0$. Nevertheless, in both cases, the reaction proceeds spontaneously ($\Delta G < 0$), although to different extents, from the standard state of reactant *or* product to the equilibrium mixture ($\Delta G = 0$).

From there on, the reaction is nonspontaneous ($\Delta G > 0$). If we had started with pure B, the reaction would also be spontaneous until it reached equilibrium, but not thereafter. In either case, *the free energy decreases as the reaction proceeds, until it reaches a minimum at the equilibrium mixture.*

If we start with pure A, the vessel contains mostly B when the reaction stops, so we say the forward reaction A → B is spontaneous. Likewise, if we start with pure B, the reaction also stops with mostly B present. Even though a small amount of A forms to satisfy the equilibrium condition, we say the reverse reaction B → A is nonspontaneous because it proceeds very little in that direction. For the overall reaction A $\rightleftharpoons$ B (starting with all components in their standard states), $G_B^0$ is smaller than $G_A^0$, so $\Delta G^0$ is negative, which corresponds to $K > 1$ and thus a spontaneous reaction as written.

Now consider the reaction C $\rightleftharpoons$ D, where $K = $ [D]/[C] $< 1$ (Figure 19.10, *B*). Once again, starting with either pure C or pure D, the system spontaneously loses free energy until it reaches equilibrium. This time, however, it is the reverse reaction (D → C) that goes farther toward completion and is spontaneous, while the forward reaction (C → D) does not go very far and is nonspontaneous. For C $\rightleftharpoons$ D, $G_D^0$ is larger than $G_C^0$, and $\Delta G^0$ is positive, which corresponds to $K < 1$ and therefore a nonspontaneous reaction. Thus the term *spontaneous reaction* refers to one that goes predominantly, not necessarily completely, toward product.

## Section Summary

Two ways of predicting reaction direction are from the value of $\Delta G$ and from the relation of $Q$ to $K$. The signs of $\Delta G$ and of $\ln Q/K$ are the same for a given reaction direction. These two variables are related to each other by $\Delta G = RT \ln Q/K$. When $Q = K$, the system can release no more free energy. Beginning with $Q$ at the standard state, the free energy change is $\Delta G^0$, and it is related to the equilibrium constant by $\Delta G^0 = -RT \ln K$. For nonstandard conditions, $\Delta G$ has two components, $\Delta G^0$ and $RT \ln Q$. Any nonequilibrium mixture of reactants and products moves spontaneously ($\Delta G < 0$) toward the equilibrium mixture, but only when $K > 1$ is the reaction spontaneous for the components in their standard states ($\Delta G^0 < 0$).

## Chapter Perspective

*As processes move toward equilibrium, some of the released energy becomes degraded to unusable, dispersed heat, and the entropy of the universe increases. This inevitability has led some to speculate about the end of everything, when all possible processes have occurred, no free energy remains in any system, and nothing but waste heat is dispersed evenly throughout the universe—the Final Equilibrium! Even if this grim future is in store, however, it is billions of years away, so you have plenty of time to appreciate more hopeful applications of thermodynamics. In Chapter 20, you'll see how spontaneous reactions can generate electricity and how electricity supplies free energy to drive nonspontaneous ones.*

# For Review and Reference

## Key Terms

spontaneous change

**SECTION 19.1**
entropy ($S$)
second law of

third law of thermodynamics
standard molar entropy ($S^0$)
standard entropy of reaction
$(\Delta S^0_{rxn})$

**SECTION 19.2**
free energy ($G$)
standard free energy change
$(\Delta G^0)$

standard free energy of
formation ($\Delta G^0_f$)
coupling of reactions
adenosine triphosphate
(ATP)

## Key Equations and Relationships

**19.1** Stating the second law of thermodynamics (p. 843):

For a spontaneous process,

$$\Delta S_{universe} = \Delta S_{sys} + \Delta S_{surr} > 0$$

**19.2** Quantifying entropy in terms of the number of ways a system can be arranged (p. 844):

$$S = k \ln W$$

**19.3** Calculating the standard entropy of reaction from standard molar entropies of reactants and products (p. 848):

$$\Delta S^0_{rxn} = \Sigma m S^0_{products} - \Sigma n S^0_{reactants}$$

**19.4** Relating the entropy change in the surroundings to the heat of the system and the temperature (p. 849):

$$\Delta S_{surr} = -\frac{q_{sys}}{T}$$

**19.5** Expressing the free energy change of the system in terms of its component enthalpy and entropy changes (Gibbs equation) (p. 854):

$$\Delta G_{sys} = \Delta H_{sys} - T\Delta S_{sys}$$

**19.6** Calculating the standard free energy change from standard enthalpy and entropy changes (p. 854):

$$\Delta G^0_{sys} = \Delta H^0_{sys} - T\Delta S^0_{sys}$$

**19.7** Calculating the standard free energy change from standard free energies of formation (p. 856):

$$\Delta G^0_{rxn} = \Sigma m \Delta G^0_{f(products)} - \Sigma n \Delta G^0_{f(reactants)}$$

**19.8** Finding the temperature at which a reaction becomes spontaneous (p. 860):

$$T = \frac{\Delta H^0}{\Delta S^0}$$

**19.9** Expressing the free energy change as the difference between an initial state $Q$ and a final state $K$ (p. 863):

$$\Delta G = RT \ln Q/K = RT \ln Q - RT \ln K$$

**19.10** Expressing the free energy change when $Q$ is at the standard state (p. 863):

$$\Delta G^0 = RT \ln 1 - RT \ln K = -RT \ln K$$

**19.11** Expressing the free energy change for a nonstandard initial state (p. 864):

$$\Delta G = \Delta G^0 + RT \ln Q$$

## Answers to Follow-up Problems

**19.1** (a) $PCl_5(g)$: more complex molecule; (b) $BaCl_2(s)$: higher molar mass; (c) $Br_2(g)$: gases have less order than liquids.

**19.2** (a) $2NaOH(s) + CO_2(g) \rightarrow Na_2CO_3(s) + H_2O(l)$;

$\Delta n_{gas} = -1$, so $\Delta S^0_{rxn} < 0$

$\Delta S^0_{rxn} = [(1 \text{ mol } H_2O \times 69.9 \text{ J/mol} \cdot K)$
$+ (1 \text{ mol } Na_2CO_3 \times 139 \text{ J/mol} \cdot K)]$
$- [(1 \text{ mol } CO_2 \times 213.7 \text{ J/mol} \cdot K)$
$+ (2 \text{ mol } NaOH \times 64.5 \text{ J/mol} \cdot K)]$
$= -134 \text{ J/K}$

(b) $2Fe(s) + 3H_2O(g) \rightarrow Fe_2O_3(s) + 3H_2(g)$;

$\Delta n_{gas} = 0$, so cannot predict sign of $\Delta S^0_{rxn}$

$\Delta S^0_{rxn} = [(1 \text{ mol } Fe_2O_3 \times 87.4 \text{ J/mol} \cdot K)$
$+ (3 \text{ mol } H_2 \times 130.6 \text{ J/mol} \cdot K)]$
$- [(2 \text{ mol } Fe \times 27.3 \text{ J/mol} \cdot K)$
$+ (3 \text{ mol } H_2O \times 188.7 \text{ J/mol} \cdot K)]$
$= -141.5 \text{ J/K}$

**19.3** $2FeO(s) + \frac{1}{2}O_2(g) \rightarrow Fe_2O_3(s)$

$\Delta S^0_{sys} = 1 \text{ mol } Fe_2O_3 (87.4 \text{ J/mol} \cdot K)$
$- [2 \text{ mol } FeO (60.75 \text{ J/mol} \cdot K)$
$+ \frac{1}{2} \text{ mol } O_2 (205.0 \text{ J/mol} \cdot K)]$
$= -136.6 \text{ J/K}$

$\Delta H^0_{sys} = 1 \text{ mol } Fe_2O_3 (-825.5 \text{ kJ/mol})$
$- [2 \text{ mol } FeO (-272.0 \text{ kJ/mol})$
$+ \frac{1}{2} \text{ mol } O_2 (0 \text{ kJ/mol})]$
$= -281.5 \text{ kJ}$

$\Delta S^0_{surr} = -\frac{\Delta H^0_{sys}}{T} = -\frac{(-281.5 \text{ kJ} \times 1000 \text{ J/kJ})}{298 \text{ K}}$

$= +945 \text{ J/K}$

$\Delta S^0_{universe} = \Delta S^0_{sys} + \Delta S_{surr} = -136.6 \text{ J/K} + 945 \text{ J/K}$
$= 808 \text{ J/K}$; reaction is spontaneous at 298 K.

**19.4** Using $\Delta H^0_f$ and $S^0$ values from Appendix B, $\Delta H^0_{rxn} = -114.2 \text{ kJ}$ and $\Delta S^0_{rxn} = -146.5 \text{ J/K}$

$\Delta G^0_{rxn} = \Delta H^0_{rxn} - T\Delta S^0_{rxn} = -114.2 \text{ kJ}$
$- [(298 \text{ K})(-146.5 \text{ J/K})(1 \text{ kJ/1000 J})]$
$= -70.5 \text{ kJ}$

**19.5** (a) $\Delta G^0_{rxn} = 2 \text{ mol } NO_2 (51 \text{ kJ/mol})$
$- [2 \text{ mol } NO (86.60 \text{ kJ/mol})$
$+ 1 \text{ mol } O_2 (0 \text{ kJ/mol})]$
$= -71 \text{ kJ}$

(b) $\Delta G^0_{rxn} = 2 \text{ mol } CO (-137.2 \text{ kJ/mol})$
$- [2 \text{ mol } C (0 \text{ kJ/mol})$
$+ 1 \text{ mol } O_2 (0 \text{ kJ/mol})]$
$= -274.4 \text{ kJ}$

**19.6** $\Delta G^0$ becomes negative at lower $T$, so $\Delta H^0 < 0$, $\Delta S^0 < 0$, and $-T\Delta S^0 > 0$. At lower $T$, the negative $\Delta H^0$ value becomes larger than the positive $(-T\Delta S^0)$ value.

**19.7** (a) $\Delta G^0 = -RT \ln K$

$= -8.31 \text{ J/mol} \cdot K \times \frac{1 \text{ kJ}}{1000 \text{ J}} \times 298 \text{ K}$
$\times \ln (2.3 \times 10^{-9})$

$= 49 \text{ kJ/mol}$

(b) $Q = \frac{[H_3O^+][BrO^-]}{[HBrO]} = \frac{(6.0 \times 10^{-4}) \, 0.10}{0.20}$

$= 3.0 \times 10^{-4}$

$\Delta G = \Delta G^0 + RT \ln Q$

$= 49 \text{ kJ/mol} + [8.31 \text{ J/mol} \cdot K$

$\times \frac{1 \text{ kJ}}{1000 \text{ J}} \times 298 \text{ K} \times \ln (3.0 \times 10^{-4})]$

$= 29 \text{ kJ/mol}$

## Sample Problem Titles

# Problems

Problems with a green number are answered at the back of the text. Most sections include three categories of problems separated by a green rule—concept review questions, *paired* skill-building exercises, and problems in a relevant context.

## The Second Law of Thermodynamics: Predicting Spontaneous Change

(Sample Problems 19.1 to 19.3)

**19.1** Distinguish between the terms spontaneous and instantaneous. Give an example of a process that is spontaneous but very slow, and one that is very fast but not spontaneous.

**19.2** Distinguish between the terms spontaneous and nonspontaneous. Is it possible for a nonspontaneous process to occur? Explain.

**19.3** State the first law of thermodynamics in terms of (a) the energy of the universe; (b) the creation or destruction of energy; (c) the energy change of system and surroundings. Does the first law tell us the direction of spontaneous change? Explain.

**19.4** State qualitatively the relationship between entropy and probability. Use this idea to explain why you have probably never
(a) Seen a large number of coins ($>6$) flipped into the air land with all heads up.
(b) Felt suffocated because all the air near where you were sitting moved to the other side of the room.
(c) Seen half the water in your cup of tea freeze while the other half boiled.

**19.5** Banks use coin stackers to group large numbers of coins for ease in dispensing.
(a) What is the probability that three dimes will stack heads-up if you throw them in a coin stacker?
(b) Based on the answer in (a), develop a general equation for the number of ways, $W$, that $n$ identical coins can be stacked.
(c) Use this equation to find the probability that 8 dimes will stack heads-up.

**19.6** What property of entropy allows Hess's law to be used in the calculation of entropy changes for a system?

**19.7** Describe the equilibrium condition in terms of the entropy changes of a system and its surroundings. What does this description mean about the entropy change of the universe?

**19.8** Why is $\Delta S_{vaporization}$ of a substance always larger than $\Delta S_{fusion}$?

**19.9** How does the entropy of the surroundings change during the course of an exothermic reaction? An endothermic reaction? Other than those cited in text, give an example of an endothermic process that occurs spontaneously.

**19.10** (a) What is the entropy of a perfect crystal at absolute zero (0 K)?
(b) Does the entropy increase or decrease as the temperature rises?

(c) Why is $\Delta H_f^0 = 0$, but $S^0 > 0$ for an element in its standard state?
(d) Why do thermodynamic tables list $\Delta H_f^0$ values but not $\Delta S_f^0$ values?

**19.11** Consider the reaction

$$H_2O(g) + Cl_2O(g) \rightarrow 2HClO(g)$$

Given $\Delta S_{rxn}^0$ and $S^0$ of $HClO(g)$ and of $H_2O(g)$, write an expression to determine $S^0$ of $Cl_2O(g)$.

**19.12** Indicate whether each of the following statements is true or false and, if false, correct it.
(a) All spontaneous reactions occur quickly.
(b) If a reaction is spontaneous, the reverse reaction is nonspontaneous.
(c) All spontaneous processes release heat.
(d) The boiling of water at 100°C and 1 atm is a spontaneous process.
(e) If a process increases the randomness of the particles of a system, the entropy of the system decreases.
(f) The energy of the universe is constant; the entropy of the universe decreases toward a minimum.
(g) All systems become disordered spontaneously.
(h) Both $\Delta S_{sys}$ and $\Delta S_{surr}$ equal zero at equilibrium.

**19.13** Which of the following processes are spontaneous?
(a) Water evaporating from a puddle in summer
(b) A dog chasing a rabbit
(c) An unstable isotope undergoing radioactive disintegration
(d) The Earth moving around the Sun
(e) A boulder rolling up a hill

**19.14** Which of the following processes are spontaneous?
(a) Methane burning in air
(b) A teaspoonful of sugar dissolving in a cup of hot coffee
(c) A soft-boiled egg becoming raw and releasing heat
(d) A satellite falling to Earth
(e) Water decomposing to $H_2$ and $O_2$ at room temperature and pressure

**19.15** Predict the sign of $\Delta S_{sys}$ for each of the following processes:
(a) A piece of wax melting
(b) Silver chloride precipitating from solution
(c) Dew forming
(d) Gasoline vapors mixing with air in a car engine
(e) Hot air expanding

**19.16** Predict the sign of $\Delta S_{sys}$ for each of the following processes:
(a) Alcohol evaporating
(b) A solid explosive converting to a gas
(c) Perfume vapors diffusing through a room
(d) A pond freezing in winter
(e) Atmospheric $CO_2$ dissolving in the ocean

**19.17** Without referring to Appendix B, predict the sign of $\Delta S^0$ for the following processes:
(a) $2K(s) + F_2(g) \rightarrow 2KF(s)$
(b) $NH_3(g) + HBr(g) \rightarrow NH_4Br(s)$
(c) $NaClO_3(s) \rightarrow Na^+(aq) + ClO_3^-(aq)$
(d) $H_2S(g) + \frac{1}{2}O_2(g) \rightarrow \frac{1}{8}S_8(s) + H_2O(g)$
(e) $HCl(aq) + NaOH(aq) \rightarrow NaCl(aq) + H_2O(l)$

**19.18** Without referring to Appendix B, predict the sign of $\Delta S^0$ for the following processes:
(a) $CaCO_3(s) + 2HCl(aq) \rightarrow CaCl_2(aq) + H_2O(l) + CO_2(g)$
(b) $2NO(g) + O_2(g) \rightarrow 2NO_2(g)$
(c) $2KClO_3(s) \rightarrow 2KCl(s) + 3O_2(g)$
(d) $Ag^+(aq) + Cl^-(aq) \rightarrow AgCl(s)$
(e) $FeCl_3(s) \rightarrow FeCl_3(aq)$

**19.19** Predict the sign of $\Delta S$ for the following processes:
(a) $C_2H_5OH(g)$ at 350 K and 500 torr $\rightarrow$ $C_2H_5OH(g)$ at 350 K and 250 torr
(b) $N_2(g)$ at 298 K and 1 atm $\rightarrow$ $N_2(aq)$ at 298 K and 1 atm
(c) $O_2(aq)$ at 303 K and 1 atm $\rightarrow$ $O_2(g)$ at 303 K and 1 atm

**19.20** Predict the sign of $\Delta S$ for the following processes:
(a) $O_2(g)$ ($V = 1.0$ L, $P = 1$ atm) $\rightarrow$ $O_2(g)$($V = 0.10$ L, $P = 10$ atm)
(b) $Cu(s)$ at 350°C and 2.5 atm $\rightarrow$ $Cu(s)$ at 450°C and 2.5 atm
(c) $Cl_2(g)$ at 100°C and 1 atm $\rightarrow$ $Cl_2(g)$ at 10°C and 1 atm

**19.21** Predict which substance in each pair has the greater molar entropy and explain your choice:
(a) Butane, $CH_3CH_2CH_2CH_3(g)$ or 2-butene, $CH_3CH=CHCH_3(g)$
(b) $Ne(g)$ or $Xe(g)$
(c) $CH_4(g)$ or $CCl_4(l)$
(d) $NO_2(g)$ or $N_2O_4(g)$

**19.22** Predict which substance in each pair has the greater molar entropy and explain your choice:
(a) $CH_3OH(l)$ or $C_2H_5OH(l)$
(b) $KClO_3(s)$ or $KClO_3(aq)$
(c) $Na(s)$ or $K(s)$
(d) $P_4(g)$ or $P_2(g)$

**19.23** Without consulting Appendix B, arrange each of the following groups in order of *increasing* standard molar entropy, $S^0$, and explain your choice:
(a) Graphite, diamond, charcoal
(b) Ice, water vapor, liquid water
(c) $O_2$, $O_3$, O atoms
(d) Glucose ($C_6H_{12}O_6$), sucrose ($C_{12}H_{22}O_{11}$), ribose ($C_5H_{10}O_5$)
(e) $CaCO_3$, $Ca + C + \frac{3}{2}O_2$, $CaO + CO_2$
(f) $SF_6(g)$, $SF_4(g)$, $S_2F_{10}(g)$

**19.24** Without consulting Appendix B, arrange each of the following groups in order of *decreasing* standard molar entropy, $S^0$, and explain your choice:
(a) $ClO_4^-(aq)$, $ClO_2^-(aq)$, $ClO_3^-(aq)$
(b) $NO_2(g)$, $NO(g)$, $N_2(g)$

(c) $Fe_2O_3(s)$, $Al_2O_3(s)$, $Fe_3O_4(s)$
(d) Mg metal, Ca metal, Ba metal
(e) Hexane ($C_6H_{14}$), benzene ($C_6H_6$), cyclohexane ($C_6H_{12}$)
(f) $PF_2Cl_3(g)$, $PF_5(g)$, $PF_3(g)$

**19.25** For each reaction, predict the sign of $\Delta S^0$ and use Appendix B to calculate the value at 25°C:
(a) $3NO(g) \rightarrow N_2O(g) + NO_2(g)$
(b) $3H_2(g) + Fe_2O_3(s) \rightarrow 2Fe(s) + 3H_2O(g)$
(c) $P_4(s) + 5O_2(g) \rightarrow P_4O_{10}(s)$

**19.26** For each reaction, predict the sign of $\Delta S^0$ and use Appendix B to calculate the value at 25°C:
(a) $3NO_2(g) + H_2O(l) \rightarrow 2HNO_3(l) + NO(g)$
(b) $N_2(g) + 3F_2(g) \rightarrow 2NF_3(g)$
(c) $C_6H_{12}O_6(s) + 6O_2(g) \rightarrow 6CO_2(g) + 6H_2O(g)$

**19.27** Use Appendix B to calculate $\Delta S^0$ at 298 K for the combustion of ethane ($C_2H_6$) to form carbon dioxide and gaseous water. Is the answer qualitatively what you would expect?

**19.28** Use Appendix B to calculate $\Delta S^0$ at 298 K for the combustion of methane to form carbon dioxide and liquid water. Is the answer qualitatively what you would expect?

**19.29** Use Appendix B to calculate $\Delta S^0$ at 298 K for the formation of $Cu_2O(s)$ from its elements.

**19.30** Use Appendix B to calculate $\Delta S^0$ at 298 K for the formation of $HI(g)$ from its elements.

---

**19.31** Sulfur dioxide is released in the combustion of coal. Scrubbers remove much of the $SO_2$ from flue gases by reaction with lime slurries containing calcium hydroxide. Write a balanced equation for this reaction and calculate its $\Delta S^0$ at 298 K. ($S^0$ of $CaSO_3 = 101.4$ J/mol · K.)

**19.32** Oxyacetylene welding is a common method for repairing metal structures, including bridges, buildings, and even the Statue of Liberty. Calculate $\Delta S^0$ for the combustion of 1 mol acetylene ($C_2H_2$).

**Entropy, Free Energy, and Work**

(Sample Problems 19.4 to 19.6)

**19.33** What is the advantage of calculating free energy changes rather than entropy changes to determine reaction spontaneity?

**19.34** Given that $\Delta G_{sys} = -T\Delta S_{universe}$, explain how the sign of $\Delta G_{sys}$ correlates with reaction spontaneity.

**19.35** In general, is an endothermic reaction more likely to be spontaneous at higher or lower temperatures? Explain.

**19.36** With its components in their standard states, a certain reaction is spontaneous only at high temperature. What can you conclude about the signs of $\Delta H^0$ and $\Delta S^0$ for the reaction? Describe a process for which this is true.

**19.37** How can $\Delta S^0$ be relatively independent of $T$ if $S^0$ of each reactant and product increases with $T$?

**19.38** Calculate $\Delta G^0_{298}$ for the following reactions using $\Delta G^0_f$ values from Appendix B:
(a) $2Mg(s) + O_2(g) \rightarrow 2MgO(s)$
(b) $2CH_3OH(g) + 3O_2(g) \rightarrow 2CO_2(g) + 4H_2O(g)$
(c) $BaO(s) + CO_2(g) \rightarrow BaCO_3(s)$

**19.39** Calculate $\Delta G^0_{298}$ for the following reactions using $\Delta G^0_f$ values from Appendix B:
(a) $H_2(g) + I_2(s) \rightarrow 2HI(g)$
(b) $MnO_2(s) + 2CO(g) \rightarrow Mn(s) + 2CO_2(g)$
(c) $NH_4Cl(s) \rightarrow NH_3(g) + HCl(g)$

**19.40** Calculate $\Delta G^0_{298}$ for the equations in Problem 19.38 using $\Delta H^0_f$ and $S^0$ values from Appendix B.

**19.41** Calculate $\Delta G^0_{298}$ for the equations in Problem 19.39 using $\Delta H^0_f$ and $S^0$ values from Appendix B.

**19.42** For the oxidation of carbon monoxide to carbon dioxide, $CO(g) + \frac{1}{2}O_2(g) \rightarrow CO_2(g)$.
(a) Predict the signs of $\Delta S^0$ and $\Delta H^0$. Explain.
(b) Calculate $\Delta G^0$ by two different methods.

**19.43** For the combustion of butane gas,
$$C_4H_{10}(g) + \tfrac{13}{2}O_2(g) \rightarrow 4CO_2(g) + 5H_2O(g)$$
(a) Predict the signs of $\Delta S^0$ and $\Delta H^0$. Explain.
(b) Calculate $\Delta G^0$ by two different methods.

**19.44** For the gaseous reaction of xenon and fluorine to form the hexafluoride
(a) Calculate $\Delta S^0_{298}$ ($\Delta H^0_{298} = -402$ kJ/mol and $\Delta G^0_{298} = -280$ kJ/mol).
(b) Assuming $\Delta S^0$ and $\Delta H^0$ change little with temperature, calculate $\Delta G^0$ at 500 K.

**19.45** For the gaseous reaction of carbon monoxide and chlorine to form phosgene, $COCl_2$,
(a) Calculate $\Delta S^0_{298}$ ($\Delta H^0_{298} = -220$ kJ/mol and $\Delta G^0_{298} = -206$ kJ/mol).
(b) Assuming $\Delta S^0$ and $\Delta H^0$ change little with temperature, calculate $\Delta G^0$ at 450 K.

**19.46** One reaction used to produce small quantities of pure $H_2$ is $CH_3OH(g) \rightleftharpoons CO(g) + 2H_2(g)$.
(a) Calculate $\Delta H^0$ and $\Delta S^0$ for the process.
(b) Assuming these values are relatively independent of temperature, calculate $\Delta G^0$ at 38°C, 138°C, and 238°C.
(c) What do these different values of $\Delta G^0$ mean?

**19.47** Consider this chemical reaction that occurs in the internal combustion engine: $N_2(g) + O_2(g) \rightleftharpoons 2NO(g)$.
(a) Determine $\Delta H^0$ and $\Delta S^0$ for the reaction.
(b) Assuming these values are relatively independent of temperature, calculate $\Delta G^0$ at 100°C, 2560°C, and 3540°C.
(c) What do these different values of $\Delta G^0$ mean?

**19.48** The temperature at which the following process reaches equilibrium at 1 atm is the normal boiling point of bromine: $Br_2(l) \rightleftharpoons Br_2(g)$. Use $\Delta H^0$ and $\Delta S^0$ values to determine this temperature.

**19.49** The temperature at which the following process reaches equilibrium at 1 atm is the transition temper-ature for these two allotropes of crystalline sulfur: $S(\text{rhombic}) \rightleftharpoons S(\text{monoclinic})$. Use $\Delta H^0$ and $\Delta S^0$ values to determine this temperature.

**19.50** Many articles in the scientific and popular press have suggested $H_2$ as an energy source for vehicles because it produces only non-polluting $H_2O(g)$ when it burns. Moreover, when $H_2$ combines with $O_2$ in a fuel cell (Chapter 20), it provides electrical energy; such a fuel cell has been used on Space Shuttle flights.
(a) Calculate $\Delta H^0$, $\Delta S^0$, and $\Delta G^0_{298}$ per mole of $H_2$ for this process.
(b) Is the spontaneity of this reaction temperature dependent? Explain.
(c) At what temperature does the reaction become spontaneous?

**19.51** The U.S. government recently required that automobile fuels consist of a renewable component. The fermentation of glucose from corn to produce ethanol is the key process to fulfill this requirement:
$$C_6H_{12}O_6(s) \rightarrow 2C_2H_5OH(l) + 2CO_2(g)$$
Calculate $\Delta H^0$, $\Delta S^0$, and $\Delta G^0$ for the reaction at 25°C. Is the spontaneity of this reaction temperature dependent? Explain.

**19.52** Adenosine triphosphate (ATP) is essential for muscle contraction, protein building, nerve conduction, and numerous other energy-requiring processes in the body. These nonspontaneous reactions are "coupled" to the spontaneous hydrolysis of ATP to ADP (see Chemical Connections essay, Section 19.2). In a similar manner, ATP is regenerated by coupling to other energy-yielding reactions, one of which is:
creatine phosphate → creatine + phosphate
    $\Delta G^0 = -43.1$ kJ/mol
ADP + phosphate → ATP    $\Delta G^0 = +30.5$ kJ/mol
Calculate $\Delta G^0$ for the overall reaction that regenerates ATP.

**Free Energy, Equilibrium, and Reaction Direction**

(Sample Problem 19.7)

**19.53** (a) If $K \ll 1$ for a reaction, what do you know about the sign and magnitude of $\Delta G^0$?
(b) If $\Delta G^0 \ll 0$ for a reaction, what do you know about the magnitude $K$?

**19.54** How is the free energy change of a process related to the work that can be obtained from the process? Is this amount of work obtainable in practice? Explain.

**19.55** In what form is the portion of reaction enthalpy given off that cannot do work? What happens to a portion of the useful energy as it does work?

**19.56** What is the difference in the terms $\Delta G^0$ and $\Delta G$? Under what circumstances is $\Delta G = \Delta G^0$?

**19.57** Use Appendix B to calculate $K$ at 298 K for
(a) $NO(g) + \frac{1}{2}O_2(g) \rightleftharpoons NO_2(g)$
(b) $2HCl(g) \rightleftharpoons H_2(g) + Cl_2(g)$
(c) $2C(graphite) + O_2(g) \rightleftharpoons 2\,CO(g)$
(d) $MgCO_3(s) \rightleftharpoons Mg^{2+}(aq) + CO_3^{2-}(aq)$

**19.58** Use Appendix B to calculate $K$ at 298 K for
(a) $2H_2S(g) + 3O_2(g) \rightleftharpoons 2H_2O(g) + 2SO_2(g)$
(b) $H_2SO_4(l) \rightleftharpoons H_2O(l) + SO_3(g)$
(c) $HCN(aq) + NaOH(aq) \rightleftharpoons NaCN(aq) + H_2O(l)$
(d) $SrSO_4(s) \rightleftharpoons Sr^{2+}(aq) + SO_4^{2-}(aq)$

**19.59** Use Appendix B to determine the $K_{sp}$ of $Ag_2S$.

**19.60** Use Appendix B to determine the $K_{sp}$ of $CaF_2$.

**19.61** For the reaction $I_2(g) + Cl_2(g) \rightleftharpoons 2ICl(g)$, calculate $K_p$ at 25°C. ($\Delta G^0_{298}$ of $ICl(g) = -6.075$ kJ/mol).

**19.62** For the reaction $CaCO_3(s) \rightleftharpoons CaO(s) + CO_2(g)$, calculate $P_{CO_2}$ at 25°C.

**19.63** The solubility product constant for $PbCl_2$ is $1.7 \times 10^{-5}$ at 25°C. What is the value of $\Delta G^0$ at this temperature? Is it possible to prepare a solution that has $Pb^{+2}(aq)$ and $Cl^-(aq)$, each at the standard state concentration? (Neglect $Pb^{2+}$ hydrolysis.)

**19.64** The $K_{sp}$ for $ZnF_2$ is $3.0 \times 10^{-2}$ at 25°C. What is the value of $\Delta G^0$ for the reaction to which this number refers? Is it possible to prepare a solution that has $Zn^{2+}(aq)$ and $F^-(aq)$ each at the standard state concentration? (Neglect $F^-$ hydrolysis.)

**19.65** The equilibrium constant for the reaction

$$2Fe^{3+}(aq) + Hg_2^{2+}(aq) \rightleftharpoons 2Fe^{2+}(aq) + 2Hg^{2+}(aq)$$

is $K_c = 9.1 \times 10^{-6}$ at 298 K.
(a) What is $\Delta G^0$ at this temperature?
(b) If standard-state concentrations of the reactants and products were mixed, in which direction would the reaction proceed?
(c) Calculate $\Delta G$ when $[Fe^{3+}] = 0.20$ M, $[Hg_2^{2+}] = 0.010$ M, $[Fe^{2+}] = 0.010$ M, and $[Hg^{2+}] = 0.025$ M. In which direction will the reaction proceed to achieve equilibrium?

**19.66** The formation constant for the reaction

$$Ni^{2+}(aq) + 6NH_3(aq) \rightleftharpoons Ni(NH_3)_6^{2+}(aq)$$

is $K_f = 5.6 \times 10^8$ at 25°C.
(a) What is $\Delta G^0$ at this temperature?
(b) If standard-state concentrations of the reactants and products were mixed, in which direction would the reaction proceed?
(c) Determine $\Delta G$ when $[Ni(NH_3)_6^{2+}] = 0.010$ M, $[Ni^{2+}] = 0.0010$ M, and $[NH_3] = 0.0050$ M. In which direction will the reaction proceed to achieve equilibrium?

**19.67** The excessive production of ozone ($O_3$) gas in the lower atmosphere causes rubber to deteriorate, green plants to turn brown, and persons with respiratory disease to have difficulty breathing.
(a) Is the formation of $O_3$ from $O_2$ favored at all, no, high, or low temperature?

(b) Calculate $\Delta G^0$ for this reaction at 298 K.
(c) Calculate $\Delta G$ at 298 K for this reaction in urban smog where $[O_2] = 0.21$ M and $[O_3] = 5 \times 10^{-7}$ M.

**19.68** A $BaSO_4$ slurry is consumed just before an x-ray of the gastrointestinal tract because the precipitate is opaque to x-rays and thus defines the contours of the tract. The $Ba^{2+}$ ion is toxic, but the compound is nearly insoluble. If $\Delta G^0$ at 37°C (body temperature) is 59.1 kJ/mol for the process

$$BaSO_4(s) \rightleftharpoons Ba^{2+}(aq) + SO_4^{2-}(aq),$$

what is $[Ba^{2+}]$ in the intestinal tract? (Assume the only source of $SO_4^-$ is the ingested slurry.)

## Comprehensive Problems
Problems with an asterisk (*) are more challenging.

**19.69** Use $\Delta H^0$ and $S^0$ values to explain why graphite is not an easily converted, inexpensive source of diamond: $C(graphite) \rightleftharpoons C(diamond)$. Given this free energy change, suggest a possible reason why diamond owners need not worry about their jewelry.

**19.70** Supply the missing information (question marks) in each row of the following table:

| | $\Delta S_{rxn}$ | $\Delta H_{rxn}$ | $\Delta G_{rxn}$ | COMMENT |
|---|---|---|---|---|
| (a) | + | − | − | ? |
| (b) | ? | 0 | − | Spontaneous |
| (c) | − | + | ? | Not spontaneous |
| (d) | 0 | ? | − | Spontaneous |
| (e) | ? | 0 | + | ? |
| (f) | + | + | ? | $T\Delta S > \Delta H$ |

**19.71** A complex ion (Lewis adduct) is formed when a ligand (Lewis base) is bonded to a metal ion (Lewis acid) by means of a lone pair of electrons. Among the many complex ions of cobalt are the following two:

$$Co(NH_3)_6^{3+}(aq) + 3en(aq) \rightleftharpoons Co(en)_3^{3+}(aq) + 6NH_3(aq)$$

where "en" stands for ethylenediamine, $H_2NCH_2CH_2NH_2$. Since six Co—N bonds are broken and six Co—N bonds are formed in this reaction, $\Delta H^0_{rxn} \approx 0$. What are the signs of $\Delta S^0$ and $\Delta G^0$? What drives the reaction?

**\*19.72** The hemoglobin molecule carries $O_2$ in the blood from the lungs to the cells, where the $O_2$ is released for metabolic processes. The molecule can be represented as Hb in its unoxygenated form and as Hb·$O_2$ in its oxygenated form. One reason CO is toxic is that it competes with $O_2$ for binding to Hb: $Hb \cdot O_2(aq) + CO(g) \rightleftharpoons Hb \cdot CO(aq) + O_2(g)$. If $\Delta G^0 \approx -14$ kJ at 37°C (body temperature),
(a) What is the ratio of $[Hb \cdot CO]/[Hb \cdot O_2]$ at 37°C if $[O_2] = [CO]$?
(b) Use Le Châtelier's principle to suggest how to treat a victim of CO poisoning.

**19.73** The most important ore of lithium is spodumene, $LiAlSi_2O_6$. The first step in the extraction of the metal is a conversion of the $\alpha$ form of spodumene into the less dense $\beta$ form in preparation for subsequent leaching and washing steps. Use the following data to calculate the lowest temperature at which the $\alpha \rightarrow \beta$ conversion is feasible:

| | $\Delta H_f^0$ (kJ/mol) | $S^0$ (J/mol·K) |
|---|---|---|
| $\alpha$-spodumene | $-3055$ | 129.3 |
| $\beta$-spodumene | $-3027$ | 154.4 |

**19.74** Nuclear fuel is prepared by converting $U_3O_8$ to $UO_2(NO_3)_2$ through treatment with concentrated $HNO_3$. The uranyl nitrate is then converted into $UO_3$ and finally $UO_2$ ("yellow cake"). At this point, the fuel is enriched (that is, the proportion of $^{235}U$ increased) by converting the $UO_2$ into $UF_6$, a highly volatile solid, in preparation of a gaseous-diffusion separation of isotopes. The last chemical conversion occurs in two steps:

$$UO_2(s) + 4HF(g) \rightarrow UF_4(s) + 2H_2O(g)$$

$$UF_4(s) + F_2(g) \rightarrow UF_6(s)$$

Calculate $\Delta G^0$ for the overall process at 85°C:

| | $\Delta H_f^0$ (kJ/mol) | $S^0$ (J/mol·K) | $\Delta G_f^0$ (kJ/mol) |
|---|---|---|---|
| $UO_2(s)$ | $-1085$ | 77.0 | $-1032$ |
| $UF_4(s)$ | $-1921$ | 152 | $-1830$ |
| $UF_6(s)$ | $-2197$ | 225 | $-2068$ |

**19.75** Methanol, one of the most important industrial feedstocks, is made by several catalyzed reactions, one of which is $CO(g) + 2H_2(g) \rightarrow CH_3OH(l)$.
  (a) Demonstrate that this reaction is thermodynamically feasible.
  (b) Is it favored at low or at high temperature?
  (c) One concern raised about $CH_3OH$ as a fuel for cars is its partial oxidation in air to yield formaldehyde, $CH_2O(g)$, which poses a health hazard. Calculate $\Delta G^0$ at 100°C for this reaction.

**\*19.76** Calculate the equilibrium constants for decomposition of the hydrogen halides at 298 K:

$$2HX(g) \rightleftharpoons H_2(g) + X_2(g)$$

What do these values indicate about the extent of decomposition of these molecules at 298 K? Suggest a plausible reason for this trend.

**19.77** Hydrogenation is the addition of $H_2$ across double (or triple) carbon-carbon bonds. It is important in organic synthesis and the production of gasoline. Margarine, peanut butter, and most commercial baked goods include hydrogenated vegetable oils. What are the changes in enthalpy, entropy, and free energy for the hydrogenation of ethene ($C_2H_4$) to ethane ($C_2H_6$) at 25°C?

**19.78** The overall process occurring in a blast furnace during the production of iron is the reaction of $Fe_2O_3$ and carbon to Fe and $CO_2$.
  (a) Calculate $\Delta H^0$ and $\Delta S^0$. [Assume C(graphite)]

  (b) Is the reaction spontaneous at low or at high temperature? Explain.
  (c) Is the reaction spontaneous at 298 K?
  (d) At what temperature does the reaction become spontaneous?

**\*19.79** Solid dinitrogen pentaoxide ($N_2O_5$) undergoes hydrolysis to form liquid nitric acid.
  (a) Is the hydrolysis spontaneous at 25°C?
  (b) The solid decomposes to $NO_2$ and $O_2$ at 25°C. Is the decomposition spontaneous at this temperature? At what temperature does the decomposition become spontaneous?
  (c) At what temperature does the decomposition of *gaseous* $N_2O_5$ become spontaneous? Rationalize the difference between this temperature and that in (b).

**\*19.80** A key step in the metabolism of glucose for energy is the isomerization of glucose-6-phosphate (G6P) to fructose-6-phosphate (F6P): G6P $\rightleftharpoons$ F6P. At 298 K, the equilibrium constant for the isomerization is 0.510.
  (a) Calculate $\Delta G^0$ at 298 K.
  (b) Calculate $\Delta G$ when $Q$, the [F6P]/[G6P] ratio, equals 10.0.
  (c) Calculate $\Delta G$ when $Q = 0.100$.
  (d) Calculate $Q$ in the cell if $\Delta G = -2.50$ kJ/mol.

**19.81** The catalysts in an automobile exhaust system remove pollutants such as carbon monoxide and nitrogen oxides, as in

$$2CO(g) + 2NO(g) \rightarrow 2CO_2(g) + N_2(g)$$

Show that this reaction is spontaneous at standard conditions. Given this fact, what is the role of the catalyst?

**19.82** Chemists often use the phrase "thermodynamically unstable, but kinetically stable" to describe a compound such as acetylene. What do you think it means?

**19.83** The oxidation (usually unwanted) of a metal in air is called corrosion. Write equations for the corrosion of iron and aluminum. Use $\Delta G_f^0$ values to determine whether either process is spontaneous at 25°C.

**\*19.84** When heated, double-helical DNA separates into two random-coil single strands. When cooled, the random coils re-form the double helix:

double helix $\rightleftharpoons$ 2 random coils

  (a) What is the sign of $\Delta S$ for the forward process? Why?
  (b) Energy must be added to the system to overcome the H bonds between the base pairs and the dispersion forces between the stacked bases. What is the sign of $\Delta G$ for the forward process when $T\Delta S$ is smaller than $\Delta H$?
  (c) Write an expression that shows $T$ in terms of $\Delta H$ and $\Delta S$ when the reaction is at equilibrium. (This temperature is called the melting temperature of the nucleic acid.)

# CHAPTER 20

## Concepts and skills to review

- redox terminology (Section 4.4 and Interchapter Topic 5)
- balancing redox reactions (Section 4.4)
- free energy, work, and equilibrium (Sections 19.2 and 19.3)
- $Q$ vs. $K$ (Section 16.2) and $\Delta G$ vs. $\Delta G^0$ (Section 19.3)

# Electro-chemistry

# Chemical Change and Electrical Work

**The two faces of eletrochemistry.** The spontaneous face includes the wonderful convenience of batteries and the costly destruction of corrosion. The nonspontaneous face protects and beautifies as it creates some of our most useful substances. In this chapter, we look into both faces.

**D**oes thermodynamics have any everyday applications? Some are probably within your reach right now. Look around and you are bound to find examples in the form of common battery-operated devices: flashlight, radio, wristwatch, calculator. In thermodynamic terms, a battery houses a spontaneous chemical reaction that *releases* free energy to produce electricity. Modern society seems to come up with a new electrical gadget daily, but the use of electrochemical processes to provide energy is nothing new; the automotive battery, for instance, was invented more than 130 years ago. ◆

In another type of electrochemical process, a nonspontaneous chemical reaction *absorbs* free energy from an external source of electricity. Several metals are recovered from their ores this way, as are some essential compounds and nonmetals. The electroplating of surfaces for protection and beauty is accomplished through an input of electrical energy as well.

**Electrochemistry** is the study of the relationship between chemical change and electrical work. It is typically investigated through the use of **electrochemical cells,** systems that incorporate a redox reaction to produce or utilize electrical energy. Electrochemical cells that use a spontaneous reaction to provide electrical work are called *voltaic* (or *galvanic*) *cells,* whereas those in which electrical work drives a nonspontaneous reaction are called *electrolytic cells.* We examine both types and the thermodynamic principles that explain why they work.

We start with a review of redox concepts, extend a method for balancing redox equations that is useful for electrochemical cells, and provide an overview of the two cell types. Next, we focus on voltaic cells, examining the free energy change and equilibrium nature of the cell's redox reaction and how they relate to its electrical output. We describe some important types of voltaic cells: concentration cells and batteries. Then, we discuss *corrosion,* a destructive process similar in principle to the useful ones in voltaic cells. Next, we switch our focus to electrolytic cells, see how they are used to isolate elements from their compounds, how the presence of water affects the products obtained, and how the amount of current determines the amount of product. Finally, we examine the redox system that generates energy in living cells.

◆ **The Shape of Electrochemical Things to Come.** As petroleum reserves dwindle and their use threatens the atmosphere, a new generation of electrochemical devices will keep things going. Someday, and current trends show that day is not too far off, electric cars powered by banks of advanced batteries will eliminate the combustion of fossil fuels for travel, and advanced fuel cells, now used only on space missions, may be refitted for everyday applications.

## 20.1 Half-Reactions and Electrochemical Cells

Whether an electrochemical process releases or requires free energy, it always involves the movement of electrons from one chemical species to another in an oxidation-reduction, or redox, reaction. In this section, we review the redox process and elaborate on the half-reaction method of bal-

| Terminology | Example: $Zn(s) + 2H^+(aq) \longrightarrow Zn^{2+}(aq) + H_2(g)$ | |
|---|---|---|
| **OXIDATION**<br>• Electrons are lost<br>• Reducing agent is oxidized<br><br>• Oxidation number increases | Zinc **loses** electrons.<br>Zinc is the reducing agent and becomes **oxidized**.<br>The oxidation number of Zn **increases** from 0 to +2. | |
| **REDUCTION**<br>• Electrons are gained<br>• Oxidizing agent is reduced<br><br>• Oxidation number decreases | Hydrogen ion **gains** electrons.<br>Hydrogen ion is the oxidizing agent and becomes **reduced**.<br>The oxidation number of $H^+$ **decreases** from +1 to 0. | |

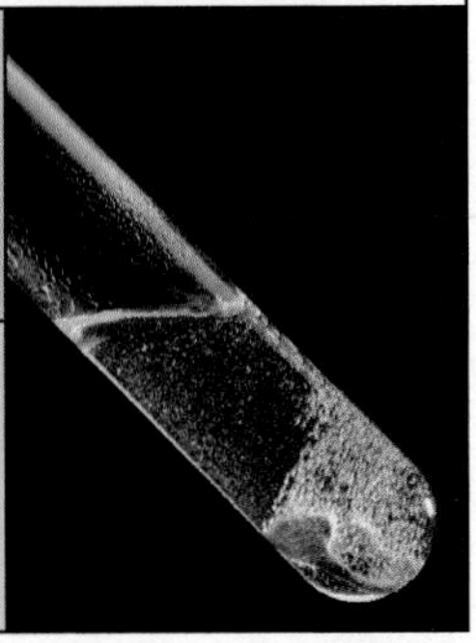

**FIGURE 20.1**

**A summary of redox terminology.** In the reaction between zinc and hydrogen ion, Zn is oxidized and $H^+$ is reduced.

ancing redox equations, which was introduced in Section 4.4. Then we take a preliminary look at how such reactions are used in the two types of electrochemical cells.

### A Quick Review of Oxidation and Reduction

In electrochemical reactions, as in any redox process, *oxidation* is the loss of electrons, and *reduction* is the gain of electrons. An *oxidizing agent* is the species that performs the oxidation, taking electrons from the substance being oxidized. A *reducing agent* performs the reduction, giving electrons to the substance being reduced. The oxidized substance ends up with a higher (more positive or less negative) oxidation number; the reduced substance with a lower (less positive or more negative) one. Thus, *the oxidizing agent is the substance being reduced, and the reducing agent is the substance being oxidized.* Keep in mind two key points:

• Oxidation (electron loss) is *always* accompanied by reduction (electron gain).
• The number of electrons lost by the oxidized reactant is *always* equal to the number gained by the reduced reactant.

Figure 20.1 presents these terms in the context of the reaction between zinc and hydrochloric acid.

The ability to identify the oxidation and reduction parts of a redox process is essential to understanding the topics in this chapter. If you still find redox reactions unclear, review the full treatment given in Chapter 4 and summarized in Topic 5 of the Interchapter.

### Half-Reaction Method for Balancing Redox Reactions

In Chapter 4, two methods for balancing redox reactions were introduced—the oxidation-number method and the half-reaction method—with emphasis on the first. The oxidation-number method assigns an oxidation number (O.N.) to each atom in the reactants and products to find those that change O.N., joins the oxidized and reduced atom (or ion) with an arrow, and uses multipliers as coefficients to make the electrons lost equal the electrons gained (see Sample Problem 4.9).

The essential difference in the half-reaction method is that *the overall redox reaction is split into half-reactions.* These are balanced separately for mass (atoms) and charge, after which it becomes clear which is the oxidation and which the reduction. Then, they are multiplied by some integer to make electrons gained equal electrons lost and recombined into the balanced redox equation. The half-reaction method offers several advantages:

- It separates the oxidation and reduction half-reactions, which reflects their actual physical separation in electrochemical cells.
- It makes it easier to balance redox reactions that take place in acidic or basic solutions, which is common in these cells.
- It (usually) does *not* require assigning O.N.s. (In cases where the half-reactions are not obvious, we assign O.N.s to determine which atoms undergo a change and write the half-reactions using the species that contain those atoms.)

In general, we begin with a "skeleton" ionic reaction, one that shows only the species that are oxidized and reduced. Thus, unless $H_2O$, $H^+$, and $OH^-$ are among the species undergoing a change, they do not appear in the skeleton reaction. The following steps are used in balancing a redox reaction by the half-reaction method:

*Step 1.* Divide the skeleton reaction into two half-reactions. (Which half-reaction is the oxidation and which the reduction becomes clear after the next step.)

*Step 2.* Balance the atoms and charges in each half-reaction. (Electrons are *added to the left in the reduction* half-reaction, because the reactant gains electrons, and are *added to the right in the oxidation* half-reaction, because the reactant loses them.)

*Step 3.* Multiply each half-reaction by some integer, if necessary, to make the $e^-$ gained in the reduction equal the $e^-$ lost in the oxidation.

*Step 4.* Add the balanced half-reactions and include states of matter.

*Step 5.* Check that the atoms and charges are balanced.

Now, let's balance a redox reaction that occurs in acidic solution and then go through Sample Problem 20.1 for balancing one in basic solution.*

**Balancing redox reactions in acidic solution.** When a redox reaction occurs in acidic solution, $H_2O$ molecules and $H^+$ ions are available for use in the balancing process. As you'll see in Sample Problem 20.1, $H_2O$ molecules and $OH^-$ ions are similarly available in basic solution. First, let's balance the redox reaction between dichromate ion and iodide ion in acidic solution (Figure 20.2) and put the balancing steps into practice. The skeleton ionic reaction shows only the oxidized and reduced species:

$$Cr_2O_7^{2-}(aq) + I^-(aq) \rightarrow Cr^{3+}(aq) + I_2(s) \qquad \text{[acidic solution]}$$

*Step 1.* Divide the reaction into half-reactions, each of which contains the oxidized and reduced forms of each species in the skeleton reaction:

$$Cr_2O_7^{2-} \rightarrow Cr^{3+}$$

$$I^- \rightarrow I_2$$

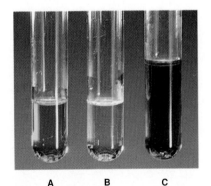

A          B          C

**FIGURE 20.2**
**The redox reaction between dichromate ion and iodide ion.** When $Cr_2O_7^{2-}$ **(A)** and $I^-$ **(B)** are mixed in acid solution, they react to form $Cr^{3+}$ and $I_2$ **(C)**.

---

*If your professor chooses to present all redox balancing earlier in the course, this discussion is completely transferrable to Chapter 4 with no loss in continuity.

*Step 2.* Balance atoms and charges in each half-reaction. We use $H_2O$ to balance O atoms, $H^+$ to balance H atoms, and $e^-$ to balance positive charges. For the $Cr_2O_7^{2-}/Cr^{3+}$ half-reaction:

a. *Balance atoms other than O and H.* We balance Cr with a coefficient 2 on the right:

$$Cr_2O_7^{2-} \rightarrow 2Cr^{3+}$$

b. *Balance O atoms by adding $H_2O$ molecules.* Each $H_2O$ has one O atom, so we add $7H_2O$ on the right to balance the O's in $Cr_2O_7^{2-}$:

$$Cr_2O_7^{2-} \rightarrow 2Cr^{3+} + 7H_2O$$

c. *Balance H atoms by adding $H^+$ ions.* Each $H_2O$ contains two H, and we added $7H_2O$, so we add $14H^+$ ions on the left:

$$14H^+ + Cr_2O_7^{2-} \rightarrow 2Cr^{3+} + 7H_2O$$

d. *Balance charge by adding electrons.* Each $H^+$ ion has a 1+ charge; $14H^+$ plus $Cr_2O_7^{2-}$ gives 12+ on the left. Two $Cr^{3+}$ give 6+ on the right. There are 6+ too many on the left, so we add $6e^-$ on the left:

$$6e^- + 14H^+ + Cr_2O_7^{2-} \rightarrow 2Cr^{3+} + 7H_2O$$

This half-reaction is balanced, and we see it is the *reduction* because electrons are added on the left: the reactant $Cr_2O_7^{2-}$ gained electrons (was reduced), so $Cr_2O_7^{2-}$ is the *oxidizing agent.* (Note that the O.N. of Cr changed from +6 to +3.)

For the $I^-/I_2$ half-reaction:

a. *Balance atoms other than O and H.* Two I atoms on the right requires a coefficient 2 on the left:

$$2I^- \rightarrow I_2$$

b. *Balance O atoms with $H_2O$.* Not needed; there are no O atoms.

c. *Balance H atoms with $H^+$.* Not needed; there are no H atoms.

d. *Balance charge with $e^-$.* We add two $e^-$ on the right:

$$2I^- \rightarrow I_2 + 2e^-$$

This balanced half-reaction is the *oxidation* because electrons are added on the right: the reactant $I^-$ lost electrons (was oxidized), so $I^-$ is the *reducing agent.* (Note that the O.N. of I changed from −1 to 0.)

*Step 3.* Multiply each half-reaction by some integer, if necessary, so that the number of $e^-$ lost in the oxidation equals the number of $e^-$ gained in the reduction. Two $e^-$ are lost in the oxidation and six $e^-$ are gained in the reduction, so we multiply the oxidation half-reaction by 3:

$$3(2I^- \rightarrow I_2 + 2e^-) \quad \text{[oxidation]}$$

$$6I^- \rightarrow 3I_2 + 6e^- \quad \text{[oxidation]}$$

*Step 4.* Add the half-reactions together, canceling substances that appear on both sides, and include states of matter. In this particular example, only the electrons cancel:

$$\cancel{6e^-} + 14H^+ + Cr_2O_7^{2-} \rightarrow 2Cr^{3+} + 7H_2O$$

$$6I^- \rightarrow 3I_2 + \cancel{6e^-}$$

$$\overline{6I^-(aq) + 14H^+(aq) + Cr_2O_7^{2-}(aq) \rightarrow 3I_2(s) + 7H_2O(l) + 2Cr^{3+}(aq)}$$

*Step 5.* Check that atoms and charges balance:

Reactants (6I, 14H, 2Cr, 7O; 6+) → products (6I, 14H, 2Cr, 7O; 6+)

**Balancing redox reactions in basic solution.** There is only one additional step needed to balance a redox equation that takes place in basic solution. It appears after both half-reactions have first been balanced *as if they took place in acidic solution* (Steps 1 and 2), the $e^-$ lost have been made equal to the $e^-$ gained (Step 3), and the half-reactions have been combined (Step 4). At this point, *we add one $OH^-$ ion* **to both sides of the equation** *for every $H^+$ ion present*. (We'll label this Step 4 Basic and include states of matter there). As a result of this step, the $H^+$ ions on one side are consumed by the $OH^-$ ions to form $H_2O$ molecules, and $OH^-$ ions appear on the other side of the equation. Excess $H_2O$ is then canceled. Finally, we check that atoms and charges balance (Step 5).

---

SAMPLE PROBLEM 20.1

### Balancing Redox Reactions by the Half-Reaction Method

**Problem:** Permanganate ion is a strong oxidizing agent frequently used in laboratory analysis. It reacts in basic solution with the oxalate ion to form carbonate ion and solid manganese dioxide. Balance the following skeleton ionic reaction that occurs between $NaMnO_4$ and $Na_2C_2O_4$ in basic solution:

$$MnO_4^-(aq) + C_2O_4^{2-}(aq) \rightarrow MnO_2(s) + CO_3^{2-}(aq) \quad \text{[basic solution]}$$

**Plan:** Although the reaction occurs in basic solution, we proceed through Step 4 as if it took place in acidic solution. Then, we add the appropriate number of $OH^-$ ions and cancel excess $H_2O$ molecules to complete the balancing.

**Solution:**

1. Divide into half-reactions

$MnO_4^- \rightarrow MnO_2$ $\qquad\qquad C_2O_4^{2-} \rightarrow CO_3^{2-}$

2. Balance.
   a. Atoms other than O and H

   Not needed $\qquad\qquad\qquad C_2O_4^{2-} \rightarrow 2CO_3^{2-}$

   b. O atoms with $H_2O$

   $MnO_4^- \rightarrow MnO_2 + 2H_2O$ $\qquad 2H_2O + C_2O_4^{2-} \rightarrow 2CO_3^{2-}$

   c. H atoms with $H^+$

   $4H^+ + MnO_4^- \rightarrow MnO_2 + 2H_2O$ $\qquad 2H_2O + C_2O_4^{2-} \rightarrow 2CO_3^{2-} + 4H^+$

   d. Charge with $e^-$

   $3e^- + 4H^+ + MnO_4^- \rightarrow MnO_2 + 2H_2O$ $\qquad 2H_2O + C_2O_4^{2-} \rightarrow 2CO_3^{2-} + 4H^+ + 2e^-$

   [reduction] $\qquad\qquad\qquad\qquad\qquad\qquad$ [oxidation]

3. Multiply each half-reaction by some integer to make $e^-$ lost equal $e^-$ gained.

   $2(3e^- + 4H^+ + MnO_4^- \rightarrow MnO_2 + 2H_2O)$ $\quad 3(2H_2O + C_2O_4^{2-} \rightarrow 2CO_3^{2-} + 4H^+ + 2e^-)$

   $6e^- + 8H^+ + 2MnO_4^- \rightarrow 2MnO_2 + 4H_2O$ $\qquad 6H_2O + 3C_2O_4^{2-} \rightarrow 6CO_3^{2-} + 12H^+ + 6e^-$

4. Add half-reactions and cancel substances appearing on both sides.

   The $e^-$ cancel, $8H^+$ cancel to leave $4H^+$ on the right, and $4H_2O$ cancel to leave $2H_2O$ on the left:

   $\cancel{6e^-} + \cancel{8}H^+ + 2MnO_4^- \rightarrow 2MnO_2 + \cancel{4H_2O}$

   $2\ \cancel{6}H_2O + 3C_2O_4^{2-} \rightarrow 6CO_3^{2-} + 4\ \cancel{12}H^+ + \cancel{6e^-}$

   ----

   $2MnO_4^- + 2H_2O + 3C_2O_4^{2-} \rightarrow 2MnO_2 + 6CO_3^{2-} + 4H^+$

4. **Basic.** Add $OH^-$ to both sides to neutralize $H^+$, and cancel $H_2O$.

   Adding $4OH^-$ to both sides forms $4H_2O$ on the right, which cancel the $2H_2O$ on the left, to leave $2H_2O$ on the right:

   $2MnO_4^- + 2H_2O + 3C_2O_4^{2-} + 4OH^- \rightarrow 2MnO_2 + 6CO_3^{2-} + [4H^+ + 4OH^-]$

   $2MnO_4^- + \cancel{2H_2O} + 3C_2O_4^{2-} + 4OH^- \rightarrow 2MnO_2 + 6CO_3^{2-} + 2\ \cancel{4}H_2O$

   Including states of matter gives the final balanced equation:
   $2MnO_4^-(aq) + 3C_2O_4^{2-}(aq) + 4OH^-(aq) \rightarrow 2MnO_2(s) + 6CO_3^{2-}(aq) + 2H_2O(l)$

5. Check that atoms and charges balance.

   (2Mn, 24 O, 6C, 4H; 12−) → (2Mn, 24 O, 6C, 4H; 12−)

**Comment:** We can obtain the balanced *molecular* equation for this reaction by including the $Na^+$ spectator ions in the starting compounds:

$$2NaMnO_4(aq) + 4NaOH(aq) + 3Na_2C_2O_4(aq) \longrightarrow$$
$$2MnO_2(s) + 6Na_2CO_3(aq) + 2H_2O(l)$$

**FOLLOW-UP PROBLEM 20.1**
Write a balanced molecular equation for the reaction between $KMnO_4$ and $KI$ in basic solution. The skeleton reaction is

$$MnO_4^-(aq) + I^-(aq) \longrightarrow MnO_4^{2-}(aq) + IO_3^-(aq) \quad \text{[basic solution]}$$

---

The half-reaction method reveals a great deal about oxidation-reduction processes. Here are the major points to remember:
• Any redox reaction can be treated as the sum of a reduction and an oxidation half-reaction.
• Mass (atoms) and charge are conserved in each half-reaction.
• Electrons lost in one half-reaction are gained in the other.
• Even though the half-reactions are treated separately, electron loss and electron gain occur simultaneously.

### An Overview of Electrochemical Cells

We distinguish two types of electrochemical cells based on the general thermodynamic nature of the reaction:

1. A **voltaic cell,** also called a *galvanic cell,* utilizes a spontaneous reaction ($\Delta G < 0$) to generate electrical energy. In the cell reaction, the difference in chemical potential energy between higher energy reactants and lower energy products is converted into electrical energy, which is used to operate the load: light bulb, radio, car starter motor, and so on. In other words, *the reacting system does work on the surroundings.* All batteries contain voltaic cells.

2. An **electrolytic cell** utilizes electrical energy to drive a nonspontaneous reaction ($\Delta G > 0$). In the cell reaction, electrical energy from an external power supply converts lower energy reactants into higher energy products. Thus, *the surroundings do work on the reacting system.* Metal plating and recovery of metals from ores involve electrolytic cells.

The two types of cell have certain design features in common (Figure 20.3). Two **electrodes,** the objects that conduct the electricity between the cell and the surroundings, are dipped into an **electrolyte,** the mixture of ions (usually in aqueous solution) that are involved in the reaction or that carry the charge. An electrode is identified as the **anode** or the **cathode,** according to the half-reaction that takes place there. *The anode is the electrode at which the oxidation half-reaction occurs.* Electrons are given up by the substance being oxidized (reducing agent) and *leave the cell* at the anode. *The cathode is the electrode at which the reduction half-reaction occurs.* Electrons *enter the cell* at the cathode and are taken up by the substance being reduced (oxidizing agent). ◆ Note in Figure 20.3 that the relative charge on the electrodes is *opposite* in the two types of cell. As you'll see in the following sections, the opposite charges result from the different phenomena that cause the electrons to flow.

**◆ Which Half-Reaction Occurs at Which Electrode?**
If you sometimes forget which half-reaction occurs at which electrode, you're not alone. Here are some memory aids to help:
1. The middle letter in anOde is "O" for "Oxidation"; therefore, the other one must be reduction at the cathode.
2. The words "anode" and "oxidation" start with vowels; the words "cathode" and "reduction" start with consonants.
3. Look at the first syllables and use your imagination: ANode OXidation; REDuction CAThode ⇒ AN OX and a RED CAT

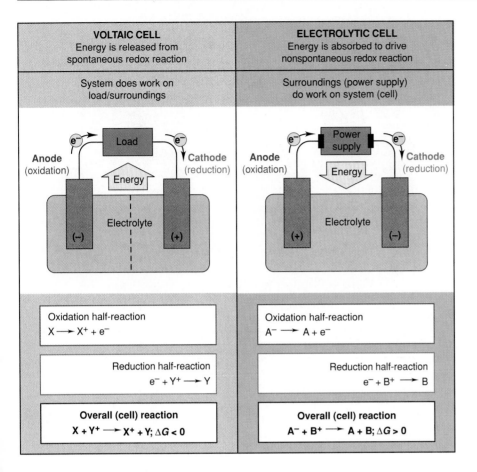

**VOLTAIC CELL**
Energy is released from spontaneous redox reaction

**ELECTROLYTIC CELL**
Energy is absorbed to drive nonspontaneous redox reaction

**FIGURE 20.3**

**General characteristics of voltaic and electrolytic cells.** A voltaic cell generates energy from a spontaneous reaction $(\Delta G < 0)$, whereas an electrolytic cell requires energy to drive a nonspontaneous reaction $(\Delta G > 0)$. In both types of cell, two electrodes dip into electrolyte solution, and an external circuit provides the means for electrons to flow. Oxidation takes place at the anode, and reduction takes place at the cathode, but the relative electrode charges are opposite in the two cells.

## Section Summary

Oxidation-reduction processes involve the transfer of electrons from a reducing agent to an oxidizing agent. The half-reaction method of balancing divides the overall reaction into half-reactions that are balanced separately and then recombined. The two types of electrochemical cells include redox reactions that generate or utilize electricity. In a voltaic cell, a spontaneous reaction does work on the surroundings (flashlight, etc.); in an electrolytic cell, the surroundings (power supply) do work on the cell and drive a nonspontaneous reaction. Both types of cell contain two electrodes that dip into the cell electrolyte and conduct electricity: oxidation occurs at the anode and reduction at the cathode.

## 20.2   Voltaic Cells: Using Spontaneous Reactions to Generate Electrical Energy

If you put a zinc metal bar in a solution of $Cu^{2+}$ ion, the blue color of the solution fades as a brown-black crust of Cu metal forms on the Zn bar (Figure 20.4). Judging from what we see, the reaction involves the reduc-

**FIGURE 20.4**

The spontaneous reaction between zinc and copper(II) ion. When a zinc metal strip is placed in a $Cu^{2+}$ solution, a redox reaction begins *(left)* in which the zinc is oxidized to $Zn^{2+}$ and the $Cu^{2+}$ is reduced to copper metal. As the reaction proceeds *(right)*, the Cu "plates out" on the Zn and falls off in chunks. (The Cu appears black because it is very finely divided.) At the molecular level, Zn atoms each lose two electrons, which are gained by the $Cu^{2+}$ ions.

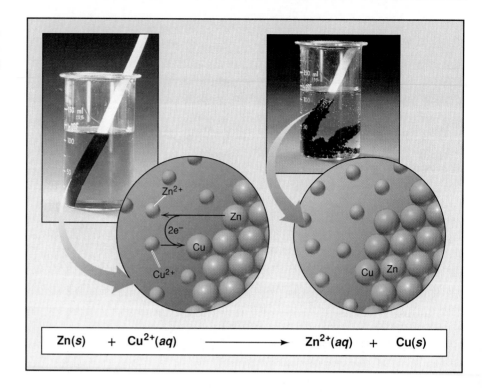

$$Zn(s) \quad + \quad Cu^{2+}(aq) \quad \longrightarrow \quad Zn^{2+}(aq) \quad + \quad Cu(s)$$

tion of $Cu^{2+}$ ion to Cu metal, which must be accompanied by the oxidation of Zn metal to $Zn^{2+}$ ion. The overall reaction consists of two half-reactions:

$$Cu^{2+}(aq) + 2e^- \rightarrow Cu(s) \qquad \text{[reduction]}$$

$$Zn(s) \rightarrow Zn^{2+}(aq) + 2e^- \qquad \text{[oxidation]}$$

$$\overline{Zn(s) + Cu^{2+}(aq) \rightarrow Zn^{2+}(aq) + Cu(s)} \qquad \text{[overall reaction]}$$

Let's examine this spontaneous reaction, which was first used in the 1830s as the basis of a voltaic (galvanic) cell.

### Construction and Functioning of a Voltaic Cell

Electrons are being transferred in the $Zn/Cu^{2+}$ reaction, but the system does not generate electrical energy because the oxidizing agent ($Cu^{2+}$) and reducing agent (Zn) are in physical contact in the same beaker. If, however, the half-reactions were physically separated and connected by an external circuit, the electrons would be transferred by traveling through the circuit and would form an electric current.

This separation of half-reactions is the essential idea behind a voltaic cell (Figure 20.5, *A*). The components of each half-reaction are placed in a separate container, or **half-cell,** which consists of an electrode dipping into an electrolyte solution. The two half-cells are joined by the circuit, which consists of a wire and a salt bridge (the inverted U tube in the figure). By convention, *the oxidation half-cell (anode compartment) is shown on the left.* In this case, it consists of a Zn metal bar, the anode, immersed in a $Zn^{2+}$ electrolyte (such as a solution of $ZnSO_4$). In addition to being an active participant in the oxidation half-reaction, the Zn bar conducts the released electrons *out* of its half-cell. The reduction half-cell in this case consists of a Cu metal bar, the cathode, immersed in a $Cu^{2+}$ electrolyte (such as a solution of $CuSO_4$).

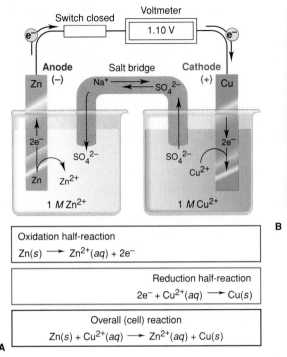

Oxidation half-reaction
$Zn(s) \longrightarrow Zn^{2+}(aq) + 2e^-$

Reduction half-reaction
$2e^- + Cu^{2+}(aq) \longrightarrow Cu(s)$

Overall (cell) reaction
$Zn(s) + Cu^{2+}(aq) \longrightarrow Zn^{2+}(aq) + Cu(s)$

A

B

**FIGURE 20.5**

**A voltaic cell based on the zinc-copper reaction. A,** The anode half-cell (oxidation) consists of a Zn electrode dipping into a $Zn^{2+}$ solution. The electrons generated in the oxidation step move through the Zn bar and the wire, and into the Cu electrode, which dips into a $Cu^{2+}$ solution in the cathode half-cell (reduction). A salt bridge contains unreactive $Na^+$ and $SO_4^{2-}$ ions that maintain neutral charge in the electrolytes. As electrons flow left to right through electrodes and wire, anions flow right to left through the salt bridge. The switch opens (breaks) or closes (completes) the circuit, and the voltmeter registers the electrical output of the cell. **B,** After the cell was running for several hours, the electrodes were removed. The Zn anode had become lighter because Zn atoms were oxidized to aqueous $Zn^{2+}$ ions. At the same time, the Cu cathode had become heavier because aqueous $Cu^{2+}$ ions were reduced to Cu metal.

In addition to participating in the reduction half-reaction, the Cu bar conducts electrons *into* its half-cell. We examine the function of the salt bridge shortly. In order to measure the voltage generated by the cell, a voltmeter is inserted in the path of the wire connecting the electrodes. A switch is included to close (complete) or open (break) the circuit.

*The charges on the electrodes are determined by the source of electrons and the direction of their flow through the circuit.* In this cell, Zn metal is oxidized at the anode to $Zn^{2+}$ ions and electrons. The $Zn^{2+}$ ions enter the solution, and the electrons enter the wire. The electrons flow left to right through the wire to the cathode, where $Cu^{2+}$ ions in that solution remove them and are reduced to Cu atoms. Electrons are continuously generated at the anode and consumed at the cathode. Therefore, the anode has an excess of electrons and a negative charge *relative* to the cathode. *In any* **voltaic** *cell, the anode is negative and the cathode is positive.*

Now let's focus on the purpose of the salt bridge. If you think about the process in each half-cell, you will notice a problem of charge imbalance. The oxidation half-cell originally contains a neutral solution of $Zn^{2+}$ and $SO_4^{2-}$ ions, but as Zn atoms in the bar lose electrons, the solution would quickly develop a net positive charge from additional $Zn^{2+}$ ions. In the reduction half-cell, the neutral solution of $Cu^{2+}$ and $SO_4^{2-}$ ions would quickly attain a net negative charge as $Cu^{2+}$ ions leave the solution to form Cu atoms. If this buildup of charge in each half-cell were to occur, the cell could not operate. To prevent this situation, the two half-cells are joined by a **salt bridge,** which acts as a "liquid wire," allowing ions to flow and complete the circuit. The salt bridge shown here is an inverted U tube containing a solution of nonreacting ions—in this case, $Na^+$ and $SO_4^{2-}$—in a gel. Thus, the solution cannot pour out, but ions can diffuse through it into and out of the half-cells. To maintain neutrality in the reduction half-cell (right; cathode compartment), as $Cu^{2+}$ ions change to Cu atoms, negative $SO_4^{2-}$ ions move from the solution into the salt bridge (and positive $Na^+$ ions move from the salt bridge into the solution). Similarly, to maintain neutrality in the oxidation half-cell (left; anode compartment), as Zn atoms change to $Zn^{2+}$ ions, negative $SO_4^{2-}$ ions move from the salt bridge into that solution. Thus, as Figure 20.5, *A*, shows, the circuit is completed as electrons move left to right through the wire and anions move right to left through the salt bridge.

The electrodes in the $Zn/Cu^{2+}$ cell are *active* ones because the metals themselves are components of the half-reactions. As the cell operates, the mass of the zinc electrode gradually decreases, and the $[Zn^{2+}]$ in the anode half-cell increases. At the same time, the mass of the copper electrode increases and the $[Cu^{2+}]$ in the cathode half-cell decreases; we say that the $Cu^{2+}$ "plates out" on the electrode. Figure 20.5, *B*, shows the electrodes removed from their half-cells after several hours of operation.

For many redox reactions, however, there are no reactants or products capable of serving as electrodes. In these cases, *inactive* electrodes are used, which conduct electrons into or out of the cell but do not take part in the half-reactions. In a voltaic cell based on the following half-reactions, for instance, the species cannot act as electrodes:

$$2I^-(aq) \rightarrow I_2(s) + 2e^- \qquad \text{[anode]}$$

$$MnO_4^-(aq) + 8H^+(aq) + 5e^- \rightarrow Mn^{2+}(aq) + 4H_2O(l) \qquad \text{[cathode]}$$

Therefore, each half-cell consists of inactive electrodes, such as graphite or platinum rods, immersed in the electrolyte solution, which contains *all the species involved in that half-reaction* (Figure 20.6). In the anode half-cell, $I^-$ ions are oxidized to solid $I_2$. The electrons released flow through the graphite anode to the wire and into the graphite cathode, where they reduce $MnO_4^-$ ions to $Mn^{2+}$ ions. (A $KNO_3$ salt bridge is used.)

As Figures 20.5, *A*, and 20.6 illustrate, there are certain consistent features in the *diagram* of a voltaic cell. The physical arrangement includes half-cell containers, electrodes, wire, and salt bridge, and the following details are shown:

- Components of the half-cells: electrode materials, ions, and other substances involved in the reaction
- Electrode name (anode or cathode) and charge. By convention, the anode compartment always appears *on the left.*
- Each half-reaction with its half-cell and the overall cell reaction
- Direction of electron flow in the external circuit
- Nature of ions and direction of ion flow in the salt bridge

You'll see how to specify these details and diagram a cell shortly.

## Notation for a Voltaic Cell

There is a useful shorthand notation (sometimes called *diagram*) for describing the components of a voltaic cell. For example, the notation for the cell based on the $Zn/Cu^{2+}$ reaction is

$$Zn(s) \,|\, Zn^{2+}(aq) \,\|\, Cu^{2+}(aq) \,|\, Cu(s)$$

The following are key parts of the notation:

- The components of the anode compartment (oxidation half-cell) are written to the left of the components of the cathode compartment (reduction half-cell).
- A double vertical line separates the half-cells and represents the wire and salt bridge.

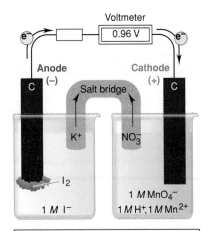

Oxidation half-reaction
$$2I^-(aq) \longrightarrow I_2(s) + 2e^-$$

Reduction half-reaction
$$MnO_4^-(aq) + 8H^+(aq) + 5e^- \longrightarrow$$
$$Mn^{2+}(aq) + 4H_2O(l)$$

Overall (cell) reaction
$$2MnO_4^-(aq) + 16H^+(aq) + 10\,I^-(aq) \longrightarrow$$
$$2Mn^{2+}(aq) + 5\,I_2(s) + 8H_2O(l)$$

**FIGURE 20.6**

**A voltaic cell using inactive electrodes.** A cell based on the reaction between $I^-$ and $MnO_4^-$ in acidic solution does not have species that can be used as electrodes, so inactive graphite (C) electrodes are used.

- Within each half-cell, a single vertical line represents a phase boundary; for example, $Zn(s)\,|\,Zn^{2+}(aq)$ indicates that the *solid* Zn is a different phase from the *aqueous* $Zn^{2+}$. A comma separates half-cell components in the same phase. For example, the notation for the voltaic cell shown in Figure 20.6 is

$$\text{Graphite}\,|\,I^-(aq)\,|\,I_2(s)\,\|\,H^+(aq),\,MnO_4^-(aq),\,Mn^{2+}(aq)\,|\,\text{graphite}$$

- Half-cell components appear in the same order as in the half-reaction, while electrodes appear at the extreme left and right of the notation.

## Diagramming Voltaic Cells

**Problem:** Diagram and write the notation for a voltaic cell that consists of one half-cell with a Cr bar in a $Cr_2(NO_3)_3$ solution, another half-cell with an Ag bar in an $AgNO_3$ solution, and a $KNO_3$ salt bridge. Measurement indicates that the Cr electrode is negative relative to the Ag electrode.

**Plan:** To diagram a cell and write its notation, we proceed as described above. The contents of the half-cells, and thus the components of the half-reactions, are given. To determine which is the anode compartment (oxidation) and which is the cathode (reduction), we must find the direction of the spontaneous redox reaction. Since electrons are released into the anode during oxidation, it has a negative charge, and we are given the relative electrode charges: Cr is negative, so it must be the anode and Ag the cathode. From this information, we proceed to diagram the cell.

**Solution:** Writing the balanced half-reactions. Since the Ag electrode is positive, the half-reaction consumes $e^-$:

$$Ag^+(aq) + e^- \longrightarrow Ag(s) \quad \text{[reduction; cathode; right]}$$

Since the Cr electrode is negative, the half-reaction releases $e^-$:

$$Cr(s) \longrightarrow Cr^{3+}(aq) + 3e^- \quad \text{[oxidation; anode; left]}$$

Writing the balanced overall cell reaction. We triple the reduction half-reaction to balance $e^-$ and combine the half-reactions to obtain the overall spontaneous reaction:

$$Cr(s) + 3Ag^+(aq) \longrightarrow Cr^{3+}(aq) + 3Ag(s)$$

Determining direction of electron and ion flow. The released $e^-$ in the Cr electrode (negative) flow through the external circuit to the Ag electrode (positive). As $Cr^{3+}$ ions enter the anode electrolyte, $NO_3^-$ ions enter from the salt bridge to maintain neutrality. As $Ag^+$ ions leave the cathode electrolyte and plate out on the Ag electrode, $NO_3^-$ ions enter the salt bridge and $K^+$ ions leave it to enter the electrolyte. The diagram of the voltaic cell is shown in the margin.
Writing the cell notation:

$$\mathbf{Cr(s)\,|\,Cr^{3+}(aq)\,\|\,Ag^+(aq)\,|\,Ag(s)}$$

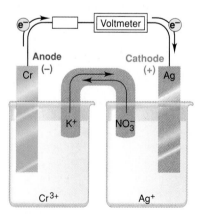

Oxidation half-reaction
$Cr(s) \longrightarrow Cr^{3+}(aq) + 3e^-$

Reduction half-reaction
$Ag^+(aq) + e^- \longrightarrow Ag(s)$

Overall (cell) reaction
$Cr(s) + 3Ag^+(aq) \longrightarrow Cr^{3+}(aq) + 3Ag(s)$

**Check:** Always be sure that the half-reactions and cell reaction are balanced, the half-cells contain *all* components of the half-reaction, and the electron and ion flow are shown. You should be able to write the half-reactions from the cell notation as a check.

**Comment:** The key to diagramming a voltaic cell is to use the spontaneous reaction to identify the oxidation (anode) and reduction (cathode) half-reactions.

**FOLLOW-UP PROBLEM 20.2**
In one compartment of a voltaic cell, a graphite rod dips into an acidic solution of $K_2Cr_2O_7$ and $Cr(NO_3)_3$; in the other, a tin bar dips into a $Sn(NO_3)_2$ solution. A $KNO_3$ salt bridge joins the half-cells. The tin electrode is negative relative to the graphite. Diagram the cell, show balanced equations, and write the cell notation.

## Why and How Long Does a Voltaic Cell Work?

By placing a light bulb in the circuit or looking at the voltmeter, we can see that the $Zn/Cu^{2+}$ cell generates electrical energy. But what principle explains how the reaction takes place, and *why* do electrons flow in the direction shown?

Let's examine what is happening when the switch is open and no reaction is occurring. In each half-cell, we can consider the metal electrode in equilibrium with the metal ions in the electrolyte, with the electrons residing in the metal:

$$Zn(s) \rightleftharpoons Zn^{2+}(aq) + 2e^- \text{ (in Zn metal)}$$

$$Cu(s) \rightleftharpoons Cu^{2+}(aq) + 2e^- \text{ (in Cu metal)}$$

From the direction of the overall spontaneous reaction, we know that Zn gives up its electrons more easily than Cu, so it is a stronger reducing agent: therefore, the equilibrium position of the Zn half-reaction lies farther to the right. You might think of the electrons in the Zn electrode as having a greater electron "pressure" than those in the Cu electrode, a greater potential energy that is ready to "push" them through the circuit the moment it is closed. This potential energy (electrical potential) difference between the electrodes is called the **voltage** of the cell.

Close the switch and electrons flow from the Zn to the Cu to equalize this electron pressure (potential energy difference). The flow disturbs the equilibrium at each electrode. The Zn half-reaction shifts to the right to restore the electrons flowing away, and the Cu half-reaction shifts to the left to remove the inflowing electrons. Thus, *the spontaneous reaction occurs as a result of the different abilities of these metals to give up their electrons and the possibility for the electrons to flow through the circuit.* ◆

◆ **Electron Flow and Water Flow.** Consider this analogy between electron "pressure" and water pressure. A U tube is separated into two arms (two half-cells) by a removable plug, and you fill the two arms with water to different heights. Remove the plug (close the switch) and the different heights (potential difference) become equal as water flows (current is generated).

## Cell Potential: Output of a Voltaic Cell

The purpose of a voltaic cell is to convert the free energy change of a spontaneous reaction into the kinetic energy of electrons moving through an external circuit (electrical energy). This electrical energy can do work and is proportional to the voltage, that is, to the difference in electrical potential between two electrodes. The voltage is also called the **cell potential ($E_{cell}$), or *electromotive force* (emf).**

As we'll discuss in the next section, a *positive* cell potential is associated with a spontaneous overall cell reaction ($\Delta G < 0$):

$$E_{cell} > 0 \text{ for a spontaneous process} \qquad \textbf{(20.1)}$$

The more positive the $E_{cell}$, the more negative the $\Delta G$, and the farther the reaction proceeds to the right. A *negative* cell potential, on the other hand, is associated with a nonspontaneous cell reaction ($\Delta G > 0$). Just as a reaction proceeds spontaneously until it reaches equilibrium, *a voltaic cell produces electricity until the reactant and product concentrations reach their equilibrium values.* At that point, the cell stops functioning because the system has no free energy to convert to electrical energy ($\Delta G = 0$).

What does the cell potential mean in terms of energy available to do work? The SI unit of electrical potential is the **volt (V),** and the SI unit of electrical charge is the **coulomb (C).** (One electron has a charge of $1.602 \times 10^{-19}$ C, and one coulomb is the charge of $6.242 \times 10^{18}$ electrons.) By definition, when two electrodes differ by one volt of potential, one joule of energy is released (that is, one joule of work can be done) for one coulomb of charge that moves between the electrodes:

$$1 \text{ V} = 1 \text{ J/C} \qquad \text{(20.2)}$$

Table 20.1 lists the voltages of some commercial and natural voltaic cells.

**Standard cell potentials.** The measured potential of a voltaic cell is affected by changes in concentration of the reactants as the reaction proceeds and by energy losses due to heating of the cell and external circuit. Therefore, in order to compare the output of different cells, we obtain a **standard cell potential ($E^0_{cell}$),** the potential measured at a specified temperature (usually 298 K) with no current flowing* and all components in their standard states: 1 atm for gases, 1 $M$ for solutions, the pure solid for electrodes. In Figure 20.5, when $[Zn^{2+}] = [Cu^{2+}] = 1$ $M$, the cell is operating at standard conditions and produces 1.10 V at 298 K:

$$Zn(s) + Cu^{2+}(aq) \longrightarrow Zn^{2+}(aq) + Cu(s) \qquad E^0_{cell} = 1.10 \text{ V}$$

**Standard electrode (half-cell) potentials.** Just as each half-reaction makes up part of the overall reaction, the potential of each half-cell makes up a part of the overall cell potential. The **standard electrode potential ($E^0_{half\text{-}cell}$)** is the potential associated with a given half-reaction (electrode compartment) when all the components are in their standard states. By convention, *a standard electrode potential always refers to the half-reaction written as a **reduction:***

$$\text{Oxidized form} + ne^- \longrightarrow \text{reduced form} \qquad E^0_{half\text{-}cell}$$

In any redox process, one species is reduced while another is oxidized, so the other half-reaction must be written as an oxidation. Because cell potentials (and half-cell potentials) are thermodynamic quantities, it is most important to remember that, as with $\Delta H$, $\Delta G$, and $\Delta S$, *reversing a reaction changes the sign of the potential:*

$$\text{Reduced form} \longrightarrow \text{oxidized form} + ne^- \qquad -E^0_{half\text{-}cell}$$

For the zinc-copper reaction, for example, the standard electrode potentials for the zinc half-reaction (anode compartment) and for the copper half-reaction (cathode compartment) refer to the processes written as reductions:

$$Zn^{2+}(aq) + 2e^- \longrightarrow Zn(s) \qquad E^0_{zinc}\ (E^0_{anode}) \qquad \text{[reduction]}$$
$$Cu^{2+}(aq) + 2e^- \longrightarrow Cu(s) \qquad E^0_{copper}\ (E^0_{cathode}) \qquad \text{[reduction]}$$

Since the overall cell reaction involves the *oxidation* of zinc at the anode, not the *reduction* of $Zn^{2+}$, we reverse the zinc half-reaction *and* change the sign of the half-cell potential:

$$Zn(s) \longrightarrow Zn^{2+}(aq) + 2e^- \qquad -E^0_{zinc}\ (-E^0_{anode}) \qquad \text{[oxidation]}$$
$$Cu^{2+}(aq) + 2e^- \longrightarrow Cu(s) \qquad E^0_{copper}\ (E^0_{cathode}) \qquad \text{[reduction]}$$

The overall redox reaction is the sum of these half-reactions, so the overall standard cell potential is the *sum* of these half-cell potentials:

$$Zn(s) + Cu^{2+}(aq) \longrightarrow Zn^{2+}(aq) + Cu(s)$$

$$E^0_{cell} = E^0_{copper} + (-E^0_{zinc}) = E^0_{copper} - E^0_{zinc}$$

---

*The current required for modern digital voltmeters to operate makes a negligible difference in the value of $E^0_{cell}$.

**TABLE 20.1 Voltages of Some Voltaic Cells**

| VOLTAIC CELL | VOLTAGE (V) |
|---|---|
| Common dry cell | 1.5 |
| Lead-acid car battery | 12 |
| Calculator battery (mercury) | 1.3 |
| Electric eel (head to tail of 6-ft eel) | 720 |
| Nerve of giant squid (across cell membrane) | 0.070 |

Generalizing this result for any voltaic cell, we see that the standard cell potential is the *difference* between the standard electrode potential of the cathode (right) half-cell and the standard electrode potential of the anode (left) half-cell:

$$E^0_{cell} = E^0_{cathode} - E^0_{anode} = E^0_{right} - E^0_{left} \qquad (20.3)$$

**Determining $E^0_{half-cell}$ with the standard hydrogen electrode.** Suppose we want to know what portion of $E^0_{cell}$ for the zinc-copper reaction is contributed by the anode half-cell (oxidation of Zn) and what portion by the cathode half-cell (reduction of $Cu^{2+}$). How can we determine half-cell potentials if we can only measure the potential of the complete cell? Chemists have solved this problem by *selecting a standard reference half-cell and defining its standard electrode potential as zero ($E^0_{reference} \equiv 0.00$ V)*. The **standard reference half-cell** is a **standard hydrogen electrode,** which consists of a specially prepared platinum electrode immersed in a 1 $M$ aqueous solution of a strong acid, $H^+(aq)$, through which $H_2$ gas at 1 atm is bubbled. Thus, the reference half-reaction is

$$2H^+(aq, 1\ M) + 2e^- \rightarrow H_2(g, 1\ atm) \qquad E^0_{reference} = 0.00\ V$$

Notice that whether we write this half-reaction as a reduction or an oxidation, that is, whether aqueous $H^+$ ions are reduced or $H_2$ gas is oxidized, the half-cell potential is zero:

$$H_2(g, 1\ atm) \rightarrow 2H^+(aq, 1\ M) + 2e^- \qquad -E^0_{reference} = 0.00\ V$$

Now we can construct a voltaic cell consisting of the reference half-cell and another half-cell whose potential is unknown. With $E^0_{reference}$ defined as zero, the overall $E^0_{cell}$ lets us find the unknown standard electrode potential. When $H_2$ is oxidized, the reference half-cell is the anode:

$$E^0_{cell} = E^0_{cathode} - E^0_{anode} = E^0_{unknown} - E^0_{reference}$$
$$= E^0_{unknown} - 0.00\ V = E^0_{unknown}$$

When $H^+$ is reduced, the reference half-cell is the cathode:

$$E^0_{cell} = E^0_{cathode} - E^0_{anode} = E^0_{reference} - E^0_{unknown}$$
$$= 0.00\ V - E^0_{unknown} = -E^0_{unknown}$$

By this approach, all other standard electrode potentials are *relative* to the standard reference electrode potential. Figure 20.7 shows a voltaic cell that has the Zn/$Zn^{2+}$ half-reaction in one compartment and the $H_2$/$H^+$ half-reaction in the other. The zinc electrode is negative relative to the hydrogen electrode, so we know the zinc is being oxidized at the anode. The measured $E^0_{cell}$ is + 0.76 V, and we use this value to find the unknown standard electrode potential, $E^0_{zinc}$:

$$2H^+(aq) + 2e^- \rightarrow H_2(g) \qquad E^0_{reference} = 0.00\ V \qquad [cathode]$$
$$Zn(s) \rightarrow Zn^{2+}(aq) + 2e^- \qquad -E^0_{zinc} = ?\ V \qquad [anode]$$

$$Zn(s) + 2H^+(aq) \rightarrow Zn^{2+}(aq) + H_2(g) \qquad E^0_{cell} = 0.76\ V$$
$$E^0_{cell} = E^0_{cathode} - E^0_{anode} = E^0_{reference} - E^0_{zinc}$$
$$E^0_{zinc} = E^0_{reference} - E^0_{cell} = 0.00\ V - 0.76\ V = -0.76\ V$$

The standard electrode potential for zinc is $-0.76$ V. That is,

$$Zn^{2+}(aq) + 2e^- \rightarrow Zn(s) \qquad E^0_{zinc} = -0.76\ V \qquad [reduction]$$

and $\qquad Zn(s) \rightarrow Zn^{2+}(aq) + 2e^- \qquad -E^0_{zinc} = +0.76\ V \qquad [oxidation]$

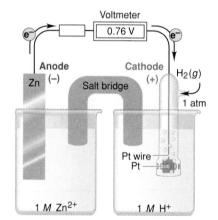

Voltmeter
0.76 V

**Anode**
(−)
Zn          Salt bridge

**Cathode**
(+)          $H_2(g)$

1 atm

Pt wire
Pt

1 $M$ $Zn^{2+}$          1 $M$ $H^+$

Oxidation half-reaction
$Zn(s) \longrightarrow Zn^{2+}(aq) + 2e^-$

Reduction half-reaction
$2H^+(aq) + 2e^- \longrightarrow H_2(g)$

Overall (cell) reaction
$Zn(s) + 2H^+(aq) \longrightarrow Zn^{2+}(aq) + H_2(g)$

**FIGURE 20.7**

**Determining an unknown $E^0$ with the standard reference (hydrogen) electrode.** When a voltaic cell is constructed with Zn/$Zn^{2+}$ in one half-cell and the hydrogen reference electrode in the other, the Zn half-cell is negative (anode), and the cell potential is 0.76 V. Since the potential of the standard reference electrode is defined as 0.00 V, the cell potential equals the anode potential, or $-E^0_{zinc}$. Therefore, $E^0_{zinc} = -0.76$ V.

Now let's return to the zinc-copper cell (Figure 20.5) and use the measured $E^0_{cell}$ (1.10 V) and the value we just found for $E^0_{zinc}$ to calculate $E^0_{copper}$:

$$E^0_{cell} = E^0_{cathode} - E^0_{anode} = E^0_{copper} - E^0_{zinc}$$

$$E^0_{copper} = E^0_{cell} + E^0_{zinc} = 1.10\ V + (-0.76\ V) = 0.34\ V$$

So the standard electrode potential for copper is 0.34 V:

$$Cu^{2+}(aq) + 2e^- \rightarrow Cu(s) \qquad E^0_{copper} = 0.34\ V$$

By continuing this process of constructing cells with one known and one unknown electrode potential, we can find many other standard electrode potentials. Let's go over these ideas once more with a sample problem.

SAMPLE PROBLEM 20.3 ⎯⎯⎯⎯⎯⎯⎯⎯⎯⎯⎯⎯⎯⎯⎯⎯⎯⎯⎯

## Calculating an Unknown $E^0_{half-cell}$ from $E^0_{cell}$

**Problem:** A voltaic cell incorporates the reaction between aqueous bromine and zinc metal:

$$Br_2(aq) + Zn(s) \rightarrow Zn^{2+}(aq) + 2Br^-(aq) \qquad E^0_{cell} = 1.83\ V$$

Calculate $E^0$ for the *oxidation* of $2Br^-(aq)$ to $Br_2(aq)$, if $E^0_{zinc} = -0.76\ V$.

**Plan:** Since $E^0_{cell}$ is positive, the reaction is spontaneous as written. By dividing the reaction into half-reactions, we see that $Br_2$ is reduced and Zn is oxidized, so the zinc half-cell is the anode. We use Equation 20.3 to find $E^0_{unknown}$, the standard electrode potential for the *reduction* of $Br_2$ to $Br^-$ ($E^0_{bromine}$). Since the problem asks for the oxidation half-reaction, we change the sign and obtain $-E^0_{bromine}$.

**Solution:** Dividing the reaction into half-reactions:

$$Br_2(aq) + 2e^- \rightarrow 2Br^-(aq) \qquad E^0_{unknown} = E^0_{bromine} = ?\ V$$

$$Zn(s) \rightarrow Zn^{2+}(aq) + 2e^- \qquad -E^0_{zinc} = +0.76\ V$$

Calculating $E^0_{bromine}$:

$$E^0_{cell} = E^0_{cathode} - E^0_{anode} = E^0_{bromine} - E^0_{zinc}$$

$$E^0_{bromine} = E^0_{cell} + E^0_{zinc} = 1.83\ V + (-0.76\ V) = 1.07\ V$$

Determining $-E^0_{bromine}$: We reverse the reaction and change the sign:

$$2Br^-(aq) \rightarrow Br_2(aq) + 2e^- \qquad -E^0_{bromine} = \mathbf{-1.07\ V}$$

**Check:** A good check is to be sure that

$$E^0_{bromine} - E^0_{zinc}\ \text{give}\ E^0_{cell}:\ 1.07\ V - (-0.76\ V) = 1.83\ V.$$

**Comment:** To avoid errors in sign, keep in mind that the standard electrode potential refers to the half-reactions written as *reductions*. Then reverse one half-reaction to the direction for which they add to give the overall cell reaction and change the sign of its $E^0_{half-cell}$.

FOLLOW-UP PROBLEM 20.3

A voltaic cell based on the reaction between aqueous $Br_2$ and vanadium(III) ions has $E^0_{cell} = 1.39\ V$:

$$Br_2(aq) + 2V^{3+}(aq) + 2H_2O(l) \rightarrow 2VO^{2+}(aq) + 4H^+(aq) + 2Br^-(aq)$$

What is the standard electrode potential for the reduction of $VO^{2+}$ to $V^{3+}$?

## Relative Strengths of Oxidizing and Reducing Agents

One of the things we learn through measurements of voltaic cells is the relative strengths of the oxidizing and reducing agents involved in the half-cells. Three oxidizing agents present in the voltaic cells just discussed are

$Cu^{2+}$, $H^+$, and $Zn^{2+}$. We can rank their relative oxidizing strengths by writing each half-reaction as a gain of electrons (reduction), with its corresponding standard electrode potential:

$$Cu^{2+}(aq) + 2e^- \rightarrow Cu(s) \qquad E^0 = +0.34 \text{ V}$$

$$2H^+(aq) + 2e^- \rightarrow H_2(g) \qquad E^0 = 0.00 \text{ V}$$

$$Zn^{2+}(aq) + 2e^- \rightarrow Zn(s) \qquad E^0 = -0.76 \text{ V}$$

*The more positive the $E^0$ value, the more the reaction (as written) tends to occur;* thus, $Cu^{2+}$ gains two electrons more easily than $H^+$, which gains them more easily than $Zn^{2+}$. In terms of oxidizing strength, therefore, $Cu^{2+} > H^+ > Zn^{2+}$. Moreover, because reversing the half-reaction changes the sign of the half-cell potential, this listing also ranks the strengths of the reducing agents: $Zn > H_2 > Cu$. The oxidizing agents (reactants) *decrease* in strength, and the reducing agents (products) *increase* in strength *from top to bottom*: $Cu^{2+}$ (top left) is the strongest oxidizing agent, and Zn (bottom right) is the strongest reducing agent.

By combining various half-cells into voltaic cells, we can create a list of reduction half-reactions and arrange them by standard electrode potential (from most positive to most negative). Such a list, called the *emf series* or table of standard electrode potentials, is presented in Table 20.2. Keep several key points in mind as you use this list:

1. All values are relative to the standard hydrogen electrode:

$$2H^+(aq, 1\ M) + 2e^- \rightleftharpoons H_2(g, 1\text{ atm}) \qquad E^0 = 0.00 \text{ V}$$

2. Since by convention, the half-reactions are written as *reductions,* all reactants are oxidizing agents and all products are reducing agents.

3. The $E^0$ value shown is for the half-reaction *as written.* The more positive the $E^0$, the greater is the tendency for the half-reaction to occur.

4. Half-reactions are shown here with an equilibrium arrow to make the point that a half-reaction can occur as a reduction or as an oxidation (that is, take place at the cathode or anode, respectively), depending on the conditions and the standard electrode potential of the other half-reaction. The standard electrode potential, $E^0$, *changes sign when the reaction is reversed.*

5. As Table 20.2 is arranged, the strength of the oxidizing agent (reactant) *increases up* (bottom to top), and the strength of the reducing agent (product) *increases down* (top to bottom).

Thus, $F_2(g)$ is the strongest oxidizing agent (has the largest positive $E^0$), so $F^-(aq)$ is the weakest reducing agent. Similarly, $Li^+(aq)$ is the weakest oxidizing agent (has the most negative $E^0$), so $Li(s)$ is the strongest reducing agent. You may have noticed an analogy to conjugate acid-base pairs: a strong acid forms a weak conjugate base, and vice versa, and a strong oxidizing agent forms a weak reducing agent, and vice versa. If you forget the ranking in the table, just rely on your chemical knowledge of the elements. You know that $F_2$ is very electronegative and typically occurs as $F^-$. Therefore, it is readily reduced (gains electrons) and must be a strong oxidizing agent (high, positive $E^0$). Similarly, Li metal has a low ionization energy and typically occurs as $Li^+$. Therefore, it is easily oxidized (loses electrons) and must be a strong reducing agent (low, negative $E^0$).

**Writing spontaneous redox reactions.** One of the ways Table 20.2 can be used is as a guide to writing spontaneous redox reactions, which can be used in laboratory analyses or to construct electrochemical cells.

# TABLE 20.2  Standard Electrode (Half-Cell) Potentials (298 K)*

| HALF-REACTION | $E^0$ (V) |
|---|---|
| $F_2(g) + 2e^- \rightleftharpoons 2F^-(aq)$ | +2.87 |
| $O_3(g) + 2H^+(aq) + 2e^- \rightleftharpoons O_2(g) + H_2O(l)$ | +2.07 |
| $Co^{3+}(aq) + e^- \rightleftharpoons Co^{2+}(aq)$ | +1.82 |
| $H_2O_2(aq) + 2H^+(aq) + 2e^- \rightleftharpoons 2H_2O(l)$ | +1.77 |
| $PbO_2(s) + 4H^+(aq) + SO_4^{2-}(aq) + 2e^- \rightleftharpoons PbSO_4(s) + 2H_2O(l)$ | +1.70 |
| $Ce^{4+}(aq) + e^- \rightleftharpoons Ce^{3+}(aq)$ | +1.61 |
| $MnO_4^-(aq) + 8H^+(aq) + 5e^- \rightleftharpoons Mn^{2+}(aq) + 4H_2O(l)$ | +1.51 |
| $Au^{3+}(aq) + 3e^- \rightleftharpoons Au(s)$ | +1.50 |
| $Cl_2(g) + 2e^- \rightleftharpoons 2Cl^-(aq)$ | +1.36 |
| $Cr_2O_7^{2-}(aq) + 14H^+(aq) + 6e^- \rightleftharpoons 2Cr^{3+}(aq) + 7H_2O(l)$ | +1.33 |
| $MnO_2(s) + 4H^+(aq) + 2e^- \rightleftharpoons Mn^{2+}(aq) + 2H_2O(l)$ | +1.23 |
| $O_2(g) + 4H^+(aq) + 4e^- \rightleftharpoons 2H_2O(l)$ | +1.23 |
| $Br_2(l) + 2e^- \rightleftharpoons 2Br^-(aq)$ | +1.07 |
| $NO_3^-(aq) + 4H^+(aq) + 3e^- \rightleftharpoons NO(g) + 2H_2O(l)$ | +0.96 |
| $2Hg^{2+}(aq) + 2e^- \rightleftharpoons Hg_2^{2+}(aq)$ | +0.92 |
| $Hg_2^{2+}(aq) + 2e^- \rightleftharpoons 2Hg(l)$ | +0.85 |
| $Ag^+(aq) + e^- \rightleftharpoons Ag(s)$ | +0.80 |
| $Fe^{3+}(aq) + e^- \rightleftharpoons Fe^{2+}(aq)$ | +0.77 |
| $O_2(g) + 2H^+(aq) + 2e^- \rightleftharpoons H_2O_2(aq)$ | +0.68 |
| $MnO_4^-(aq) + 2H_2O(l) + 3e^- \rightleftharpoons MnO_2(s) + 4OH^-(aq)$ | +0.59 |
| $I_2(s) + 2e^- \rightleftharpoons 2I^-(aq)$ | +0.53 |
| $O_2(g) + 2H_2O(l) + 4e^- \rightleftharpoons 4OH^-(aq)$ | +0.40 |
| $Cu^{2+}(aq) + 2e^- \rightleftharpoons Cu(s)$ | +0.34 |
| $AgCl(s) + e^- \rightleftharpoons Ag(s) + Cl^-(aq)$ | +0.22 |
| $SO_4^{2-}(aq) + 4H^+(aq) + 2e^- \rightleftharpoons SO_2(g) + 2H_2O(l)$ | +0.20 |
| $Cu^{2+}(aq) + e^- \rightleftharpoons Cu^+(aq)$ | +0.15 |
| $Sn^{4+}(aq) + 2e^- \rightleftharpoons Sn^{2+}(aq)$ | +0.13 |
| $2H^+(aq) + 2e^- \rightleftharpoons H_2(g)$ | 0.00 |
| $Pb^{2+}(aq) + 2e^- \rightleftharpoons Pb(s)$ | −0.13 |
| $Sn^{2+}(aq) + 2e^- \rightleftharpoons Sn(s)$ | −0.14 |
| $N_2(g) + 5H^+(aq) + 4e^- \rightleftharpoons N_2H_5^+(aq)$ | −0.23 |
| $Ni^{2+}(aq) + 2e^- \rightleftharpoons Ni(s)$ | −0.25 |
| $Co^{2+}(aq) + 2e^- \rightleftharpoons Co(s)$ | −0.28 |
| $PbSO_4(s) + 2e^- \rightleftharpoons Pb(s) + SO_4^{2-}(aq)$ | −0.31 |
| $Cd^{2+}(aq) + 2e^- \rightleftharpoons Cd(s)$ | −0.40 |
| $Fe^{2+}(aq) + 2e^- \rightleftharpoons Fe(s)$ | −0.44 |
| $Cr^{3+}(aq) + 3e^- \rightleftharpoons Cr(s)$ | −0.74 |
| $Zn^{2+}(aq) + 2e^- \rightleftharpoons Zn(s)$ | −0.76 |
| $2H_2O(l) + 2e^- \rightleftharpoons H_2(g) + 2OH^-(aq)$ | −0.83 |
| $Mn^{2+}(aq) + 2e^- \rightleftharpoons Mn(s)$ | −1.18 |
| $Al^{3+}(aq) + 3e^- \rightleftharpoons Al(s)$ | −1.66 |
| $Mg^{2+}(aq) + 2e^- \rightleftharpoons Mg(s)$ | −2.37 |
| $Na^+(aq) + e^- \rightleftharpoons Na(s)$ | −2.71 |
| $Ca^{2+}(aq) + 2e^- \rightleftharpoons Ca(s)$ | −2.87 |
| $Sr^{2+}(aq) + 2e^- \rightleftharpoons Sr(s)$ | −2.89 |
| $Ba^{2+}(aq) + 2e^- \rightleftharpoons Ba(s)$ | −2.90 |
| $K^+(aq) + e^- \rightleftharpoons K(s)$ | −2.93 |
| $Li^+(aq) + e^- \rightleftharpoons Li(s)$ | −3.05 |

*Written as reductions; $E^0$ value refers to all components in their standard states: 1 $M$ for dissolved species; 1 atm pressure for gases; the pure substance for solids and liquids.

*Every redox reaction is the sum of two half-reactions, so there is a reducing agent and an oxidizing agent on each side.* In the zinc-copper reaction, for instance, Zn and Cu are the reducing agents, and $Cu^{2+}$ and $Zn^{2+}$ are the oxidizing agents. The stronger oxidizing and reducing agents react spontaneously to form the weaker oxidizing and reducing agents:

$$\underset{\substack{\text{stronger} \\ \text{reducing agent}}}{Zn(s)} \quad + \quad \underset{\substack{\text{stronger} \\ \text{oxidizing agent}}}{Cu^{2+}(aq)} \quad \longrightarrow \quad \underset{\substack{\text{weaker} \\ \text{oxidizing agent}}}{Zn^{2+}(aq)} \quad + \quad \underset{\substack{\text{weaker} \\ \text{reducing agent}}}{Cu(s)}$$

Here, too, note the similarity to acid-base chemistry. The stronger acid and base spontaneously form the weaker base and acid, respectively. The members of a conjugate acid-base pair differ by a proton: the acid has the proton and the base does not. The members of a redox pair, or *redox couple,* such as Zn and $Zn^{2+}$, differ by one or more electrons: the reduced form (Zn) has the electrons and the oxidized form ($Zn^{2+}$) does not. In acid-base reactions, we compare acid and base strength with $K_a$ and $K_b$ values. In redox reactions, we compare oxidizing and reducing strength with $E^0$ values.

Due to the order of the $E^0$ values in Table 20.2, *a spontaneous reaction ($E^0_{cell} > 0$) will occur between an oxidizing agent (species on the left) and a reducing agent (species on the right) that lies **below** it.* For instance, $Cu^{2+}$ (left) and Zn (right) react spontaneously, and Zn lies below $Cu^{2+}$. In other words, a spontaneous reaction involves one half-reaction that proceeds as a reduction (that is, as written) and another one, listed lower in the table, that proceeds as an oxidation (that is, written in reverse). This pairing ensures that the stronger oxidizing agent (higher on the left) and stronger reducing agent (lower on the right) will be the reactants.

Let's choose a pair of half-reactions from Table 20.2 and, without referring to their relative positions in the table, arrange them into a spontaneous redox reaction:

$$Ag^+(aq) + e^- \longrightarrow Ag(s) \qquad E^0_{silver} = 0.80 \text{ V}$$
$$Sn^{2+}(aq) + 2e^- \longrightarrow Sn(s) \qquad E^0_{tin} = -0.14 \text{ V}$$

There are two steps involved:
1. Reverse one of the half-reactions into an oxidation step such that the sum of the electrode potentials gives a positive $E^0_{cell}$.
2. Add the rearranged half-reactions to obtain a balanced overall reaction.

(Don't be tempted here to add the two half-reactions as written to obtain a positive $E^0_{cell}$ because you would then have two oxidizing agents forming two reducing agents.) Trial and error quickly shows that reversing the silver half-reaction to

$$Ag(s) \longrightarrow Ag^+(aq) + e^- \qquad -E^0_{silver} = -0.80 \text{ V}$$

would give a negative $E^0_{cell}$ [$-0.80$ V + ($-0.14$ V) = $-0.94$ V], so we reverse the tin half-reaction:

$$Sn(s) \longrightarrow Sn^{2+}(aq) + 2e^- \qquad -E^0_{tin} = 0.14 \text{ V}$$

Combined with the silver half-reaction, this gives a positive $E^0_{cell}$ (0.80 V + 0.14 V = 0.94 V). [A look at Table 20.2 confirms that the reducing agent (Sn) lies *below* the oxidizing agent ($Ag^+$).]

With the half-reactions written in the correct direction, we add them to obtain a balanced equation. Be sure that the number of electrons lost in the oxidation equals the number gained in the reduction. In this case, we double the silver (reduction) half-reaction:

$$2Ag^+(aq) + 2e^- \rightarrow 2Ag(s) \qquad\qquad E^0 = 0.80 \text{ V}$$

$$\underline{Sn(s) \rightarrow Sn^{2+}(aq) + 2e^- \qquad\qquad -E^0 = 0.14 \text{ V}}$$

$$Sn(s) + 2Ag^+(aq) \rightarrow Sn^{2+}(aq) + 2Ag(s) \qquad E^0_{cell} = 0.94 \text{ V}$$

Note that $E^0$ for the doubled silver half-reaction remains 0.80 V. This example shows a very important general point: *changing the balancing coefficients of a half-reaction does **not** change the $E^0$ value.* An electrode potential is an *intensive* property, one that does *not* depend on the amount of substance present. The potential is the *ratio* of energy to charge. When we change the coefficients, thus increasing the amount of substance, the energy *and* the charge increase proportionately, so their ratio stays the same. (In a similar sense, the density of a substance, another intensive property, does not change with amount of substance because the mass *and* the volume increase proportionately.)

SAMPLE PROBLEM 20.4 _____

## Writing Spontaneous Redox Reactions and Ranking the Strengths of Oxidizing and Reducing Agents

**Problem:** Combine the half-reactions below into three spontaneous redox reactions, calculate $E^0_{cell}$ for each reaction, and rank the oxidizing and reducing agents in order of increasing strength.

(1) $NO_3^-(aq) + 4H^+(aq) + 3e^- \rightarrow NO(g) + 2H_2O(l)$ $\qquad E^0 = 0.96 \text{ V}$

(2) $\quad N_2(g) + 5H^+(aq) + 4e^- \rightarrow N_2H_5^+(aq)$ $\qquad\qquad E^0 = -0.23 \text{ V}$

(3) $MnO_2(s) + 4H^+(aq) + 2e^- \rightarrow Mn^{2+}(aq) + 2H_2O(l)$ $\quad E^0 = 1.23 \text{ V}$

**Plan:** The possible combinations of half-reactions are (1) and (2), (1) and (3), and (2) and (3). We reverse one reduction half-reaction of each pair to an oxidation such that the sum gives a positive $E^0_{cell}$. Since the stronger oxidizing and reducing agents are the reactants, we can rank relative strengths within each reaction and then compare these to obtain the overall ranking.

**Solution:** Combining half-reactions (1) and (2). If we reverse (1), $E^0_{cell} = -0.96$ V + (−0.23 V) = −1.19 V, which indicates a nonspontaneous reaction. Therefore, we must reverse (2):

(1) $NO_3^-(aq) + 4H^+(aq) + 3e^- \rightarrow NO(g) + 2H_2O(l)$ $\qquad\qquad E^0 = 0.96 \text{ V}$

(rev 2) $\qquad N_2H_5^+(aq) \rightarrow N_2(g) + 5H^+(aq) + 4e^-$ $\qquad\qquad -E^0 = 0.23 \text{ V}$

To make $e^-$ lost equal $e^-$ gained, we multiply (1) by four and the reversed (2) by three. Then we add the half-reactions and cancel excess species ($H^+$ and $e^-$) that appear on both sides:

$$4NO_3^-(aq) + 16H^+(aq) + 12e^- \rightarrow 4NO(g) + 8H_2O(l) \qquad\qquad E^0 = 0.96 \text{ V}$$

$$\underline{3N_2H_5^+(aq) \rightarrow 3N_2(g) + 15H^+(aq) + 12e^- \qquad -E^0 = 0.23 \text{ V}}$$

$$3N_2H_5^+(aq) + 4NO_3^-(aq) + H^+(aq) \rightarrow 3N_2(g) + 4NO(g) + 8H_2O(l) \qquad \mathbf{E^0_{cell} = 1.19 \text{ V}}$$

Oxidizing agents: $NO_3^- > N_2$ $\qquad$ Reducing agents: $N_2H_5^+ > NO$

Combining half-reactions (1) and (3). Reaction (1) must be reversed:

(rev 1)          $NO(g) + 2H_2O(l) \rightarrow NO_3^-(aq) + 4H^+(aq) + 3e^-$          $-E^0 = -0.96$ V

(3) $MnO_2(s) + 4H^+(aq) + 2e^- \rightarrow Mn^{2+}(aq) + 2H_2O(l)$          $E^0 = 1.23$ V

We multiply reversed (1) by two and (3) by three, then add and cancel:

$$2NO(g) + 4H_2O(l) \rightarrow 2NO_3^-(aq) + 8H^+(aq) + 6e^- \qquad -E^0 = -0.96 \text{ V}$$

$$3MnO_2(s) + 12H^+(aq) + 6e^- \rightarrow 3Mn^{2+}(aq) + 6H_2O(l) \qquad E^0 = 1.23 \text{ V}$$

$$3MnO_2(s) + 4H^+(aq) + 2NO(g) \rightarrow 3Mn^{2+}(aq) + 2H_2O(l) + 2NO_3^-(aq) \quad E^0_{cell} = \mathbf{0.27 \text{ V}}$$

Oxidizing agents: $MnO_2 > NO_3^-$      Reducing agents: $NO > Mn^{2+}$

Combining half-reactions (2) and (3). Reaction (2) must be reversed:

(rev 2)                    $N_2H_5^+(aq) \rightarrow N_2(g) + 5H^+(aq) + 4e^-$      $-E^0 = 0.23$ V

(3)      $MnO_2(s) + 4H^+(aq) + 2e^- \rightarrow Mn^{2+}(aq) + 2H_2O(l)$      $E^0 = 1.23$ V

We multiply reaction (3) by 2, add the half-reactions, and cancel:

$$N_2H_5^+(aq) \rightarrow N_2(g) + 5H^+(aq) + 4e^- \qquad -E^0 = 0.23 \text{ V}$$

$$2MnO_2(s) + 8H^+(aq) + 4e^- \rightarrow 2Mn^{2+}(aq) + 4H_2O(l) \qquad E^0 = 1.23 \text{ V}$$

$$N_2H_5^+(aq) + 2MnO_2(s) + 3H^+(aq) \rightarrow N_2(g) + 2Mn^{2+}(aq) + 4H_2O(l) \quad E^0_{cell} = \mathbf{1.46 \text{ V}}$$

Oxidizing agents: $MnO_2 > N_2$      Reducing agents: $N_2H_5^+ > Mn^{2+}$

Determining the overall ranking of oxidizing and reducing agents:

**Oxidizing agents: $MnO_2 > NO_3^- > N_2$**

**Reducing agents: $N_2H_5^+ > NO > Mn^{2+}$**

**Check:** As always check that atoms and charge balance on each of side of the equation. A good way to check the ranking and reactions is to arrange the given half-reactions in order of $E^0$ value:

$$MnO_2(s) + 4H^+(aq) + 2e^- \rightarrow Mn^{2+}(aq) + 2H_2O(l) \qquad E^0 = 1.23 \text{ V}$$

$$NO_3^-(aq) + 4H^+(aq) + 3e^- \rightarrow NO(g) + 2H_2O(l) \qquad E^0 = 0.96 \text{ V}$$

$$N_2(g) + 5H^+(aq) + 4e^- \rightarrow N_2H_5^+(aq) \qquad E^0 = -0.23 \text{ V}$$

Then the oxidizing agents decrease in strength down, so the reducing agents decrease in strength up. Moreover, the three spontaneous reactions should contain a species on the left with one lower on the right.

**FOLLOW-UP PROBLEM 20.4**
Is the following reaction spontaneous: $3Fe^{2+}(aq) \rightarrow Fe(s) + 2Fe^{3+}(aq)$? If not, write the spontaneous reaction, calculate $E^0_{cell}$, and rank the three forms of iron in order of decreasing reducing strength.

---

**Relative reactivities of metals.** Why do only some metals react with acid to produce $H_2$ gas while others do not? Why are some metals so reactive that they form $H_2$ even in water? And why can some metals displace others from aqueous solution? You can explain such variation in metal reactivity by examining standard electrode potentials.

The standard hydrogen half-reaction represents the reduction of $H^+$ from acids to $H_2$:

$$2H^+(aq) + 2e^- \rightarrow H_2(g) \qquad E^0 = 0.00 \text{ V}$$

To see which metals reduce $H^+$ (often referred to as "displacing $H_2$") from acids, choose a metal, write its half-reaction as an oxidation, combine it with the hydrogen half-reaction above, and see if $E^0_{cell}$ is positive. What you find is that the metals Li through Pb, those that lie *below* the standard hydrogen (reference) half-reaction in Table 20.2, give a positive $E^0_{cell}$ by reducing $H^+$. Iron, for example, reduces $H^+$ from an acid to $H_2$:

$$\text{Fe}(s) \rightarrow \text{Fe}^{2+}(aq) + 2e^- \qquad -E^0 = 0.44 \text{ V}$$
$$\underline{2\text{H}^+(aq) + 2e^- \rightarrow \text{H}_2(g) \qquad\qquad E^0 = 0.00 \text{ V}}$$
$$\text{Fe}(s) + 2\text{H}^+(aq) \rightarrow \text{H}_2(g) + \text{Fe}^{2+}(aq) \qquad E^0_{cell} = 0.44 \text{ V}$$

The lower the metal in the table, the more positive is its half-cell potential when the half-reaction is reversed, so the higher the $E^0_{cell}$ for its reduction of $H^+$ to $H_2$. If the $E^0_{cell}$ of metal A for the reduction of $H^+$ is more positive than the $E^0_{cell}$ of metal B, it means that metal A is a stronger reducing agent than metal B and is considered a more "active" metal.

Metals *above* the standard hydrogen (reference) half-reaction cannot reduce $H^+$ from acids because $E^0_{cell}$ is negative. For example, the coinage metals [Group 1B(11)]—Cu, Ag, and Au—are not strong enough reducing agents to reduce $H^+$ from acids:

$$\text{Ag}(s) \rightarrow \text{Ag}^+(aq) + e^- \qquad -E^0 = -0.80 \text{ V}$$
$$\underline{2\text{H}^+(aq) + 2e^- \rightarrow \text{H}_2(g) \qquad\qquad E^0 = 0.00 \text{ V}}$$
$$2\text{Ag}(s) + 2\text{H}^+(aq) \rightarrow 2\text{Ag}^+(aq) + \text{H}_2(g) \qquad E^0_{cell} = -0.80 \text{ V}$$

The higher the metal in the table, the more negative is its $E^0_{cell}$ for the reduction of $H^+$ to $H_2$, the lower its reducing strength, and the less active it is. Thus, gold is less active than silver, which is less active than copper.

Metals active enough to displace $H_2$ from water lie *below the half-reaction for the reduction of water:*

$$2\text{H}_2\text{O}(l) + 2e^- \rightarrow \text{H}_2(g) + 2\text{OH}^-(aq) \qquad E^0 = -0.83 \text{ V}$$

Consider the reaction of sodium in water (with the sodium half-reaction reversed and doubled):

$$2\text{Na}(s) \rightarrow 2\text{Na}^+(aq) + 2e^- \qquad -E^0 = 2.71 \text{ V}$$
$$\underline{2\text{H}_2\text{O}(l) + 2e^- \rightarrow \text{H}_2(g) + 2\text{OH}^-(aq) \qquad\qquad E^0 = -0.83 \text{ V}}$$
$$2\text{Na}(s) + 2\text{H}_2\text{O}(l) \rightarrow 2\text{Na}^+(aq) + \text{H}_2(g) + 2\text{OH}^-(aq) \qquad E^0_{cell} = 1.88 \text{ V}$$

The alkali metals [Group 1A(1)] and the larger alkaline earth metals [Group 2A(2)] can displace $H_2$ from $H_2O$ (Figure 20.8); Mg and especially Be form oxide coatings that prevent reaction.

Since a stronger reducing agent reacts to form a weaker reducing agent, you can also predict if one metal will reduce a different metal ion (sometimes referred to as "displacing" the metal from solution). A *metal activity series* ranks metals in order of decreasing reducing strength: a metal higher in the series can reduce a metal ion lower down (see Problem 20.113 at the end of the chapter). For example, zinc can reduce iron(II) ion:

$$\text{Zn}(s) \rightarrow \text{Zn}^{2+}(aq) + 2e^- \qquad -E^0 = 0.76 \text{ V}$$
$$\underline{\text{Fe}^{2+}(aq) + 2e^- \rightarrow \text{Fe}(s) \qquad\qquad E^0 = -0.44 \text{ V}}$$
$$\text{Zn}(s) + \text{Fe}^{2+}(aq) \rightarrow \text{Zn}^{2+}(aq) + \text{Fe}(s) \qquad E^0_{cell} = 0.32 \text{ V}$$

This reaction has tremendous economic importance in protecting iron from rusting, as you'll see shortly. The margin note points out a more personal scenario involving the reducing power of metals. ◆

Oxidation half-reaction
$$\text{Ca}(s) \longrightarrow \text{Ca}^{2+}(aq) + 2e^-$$

Reduction half-reaction
$$2\text{H}_2\text{O}(l) + 2e^- \longrightarrow 2\text{OH}^-(aq) + \text{H}_2(g)$$

Overall (cell) reaction
$$\text{Ca}(s) + 2\text{H}_2\text{O}(l) \longrightarrow \text{Ca(OH)}_2(aq) + \text{H}_2(g)$$

**FIGURE 20.8**
**The reaction of calcium in water.** Calcium is one of those metals active enough to displace $H_2$ from $H_2O$.

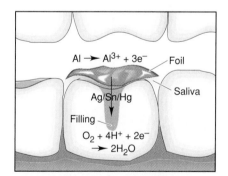

◆ **The Pain of a Dental Voltaic Cell.** Have you ever felt a jolt of pain when biting down accidentally on a scrap of aluminum foil with a filled tooth? Here's why. The foil acts as an active anode ($E^0$ of Al = $-1.66$ V), saliva as the electrolyte, and the filling (usually a silver/tin/mercury alloy) as an inactive cathode. $O_2$ is reduced to water, and the short circuit between the foil in contact with the filling creates a current that is sensed by the nerve of the tooth.

**Section Summary**

A voltaic cell consists of oxidation (anode) and reduction (cathode) half-cells, connected by a wire to conduct electrons and a salt bridge to provide ions that maintain charge neutrality as the cell operates. Electrons move from anode (left) to cathode (right), while anions move in the other direction. When all substances are in their standard states, the output of a voltaic cell is the standard cell potential ($E^0_{cell}$). For a spontaneous reaction, $E^0_{cell} > 0$. By convention, standard electrode potentials ($E^0_{half-cell}$) refer to the *reduction* half-reaction. Reversing the half-reaction to an oxidation changes the sign of $E^0_{half-cell}$. $E^0_{cell}$ equals the $E^0_{half-cell}$ of the cathode minus that of the anode. Using a standard hydrogen (reference) electrode, other $E^0_{half-cell}$ values can be measured and used to rank oxidizing (or reducing) agents (see Table 20.2). Spontaneous redox reactions combine stronger oxidizing and reducing agents to form weaker ones. A metal can reduce another species ($H^+$, $H_2O$, or metal ion) if the $E^0_{cell}$ is positive. Based on their reducing strengths, metals can be ranked in an activity series.

## 20.3  Free Energy and Electrical Work

In Chapter 19, we discussed the relationship of work, free energy, and the equilibrium state. Now we examine this central relationship in the context of electrochemical cells and see the effect of concentration on cell potential.

### Standard Cell Potential and the Equilibrium Constant

As you know, a spontaneous reaction has a *negative* change in free energy ($\Delta G < 0$). You have just seen that a spontaneous electrochemical reaction has a *positive* cell potential ($E_{cell} > 0$). Note that *the signs of $\Delta G$ and $E_{cell}$ are opposite for a spontaneous reaction.* As was mentioned earlier, these two criteria of spontaneity are related:

$$\Delta G \propto -E_{cell}$$

Let's determine the proportionality constant. The cell potential ($E_{cell}$, in volts) is the work ($w$, in joules) done *by* the system per unit of charge (in coulombs) that flows through the circuit. With no current flowing and therefore no energy lost to heating the cell components, the potential represents the *maximum* work the cell can do. Since the work is done *to* the surroundings, its sign is negative:

$$E_{cell} = \frac{-w_{max}}{charge} \qquad or \qquad w_{max} = -charge \times E_{cell}$$

The charge that flows through the cell as it functions is equal to the number of moles of electrons ($n$) transferred in the balanced redox equation times the charge of one mole of electrons (symbol $\mathscr{F}$):

$$Charge = moles\ of\ e^- \times \frac{charge}{mole\ of\ e^-} \qquad or \qquad charge = n\mathscr{F}$$

The charge of one mole of electrons is called a **faraday ($\mathscr{F}$),** in honor of Michael Faraday, the 19th-century British scientist who pioneered the study of electrochemistry:

$$1\ \mathscr{F} = 96,487\ C/mol\ e^-$$

Since $1 \text{ V} = 1 \text{ J/C}$, then $1 \text{ C} = 1 \text{ J/V}$, so

$$1 \mathcal{F} = 9.65 \times 10^4 \text{ J/V} \cdot \text{mol e}^- \quad \text{(3 significant figures)} \qquad \textbf{(20.4)}$$

Thus, substituting for charge in the equation for maximum work, we have

$$w_{\text{max}} = -\text{charge} \times E_{\text{cell}} = -n\mathcal{F}E_{\text{cell}}$$

Now we can tie together these ideas about work, $\Delta G$, and $E_{\text{cell}}$. Recall from Chapter 19 that the free energy change is the maximum work that can be obtained from a spontaneous process:

$$\Delta G = w_{\text{max}}$$

Combining these expressions for maximum work, we discover the nature of the proportionality constant in the earlier relationship and obtain the key equation in the thermodynamics of electrochemical cells:

$$\Delta G = -n\mathcal{F}E_{\text{cell}} \qquad \textbf{(20.5)}$$

When all components are in their standard states, we have

$$\Delta G^0 = -n\mathcal{F}E^0_{\text{cell}} \qquad \textbf{(20.6)}$$

Using this relationship, we can relate the standard cell potential to the equilibrium constant of the redox reaction. Recall that

$$\Delta G^0 = -RT \ln K$$

Substituting for $\Delta G^0$ in Equation 20.6, we obtain

$$-n\mathcal{F}E^0_{\text{cell}} = -RT \ln K$$

or

$$E^0_{\text{cell}} = \frac{RT}{n\mathcal{F}} \ln K \qquad \textbf{(20.7)}$$

Let's stand back to get an overview of the relationships we have just derived. Figure 20.9 summarizes the interconnections among the three reaction parameters $E^0_{\text{cell}}$, $\Delta G^0$, and $K$. Our previous procedures for determining $K$ required knowing $\Delta G^0$, either from $\Delta H^0$ and $\Delta S^0$ values or from $\Delta G^0_{\text{f}}$ values. For redox reactions, we now have a direct experimental method for determining $K$ *and* $\Delta G^0$: measure $E^0_{\text{cell}}$.

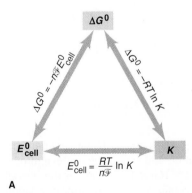

| $\Delta G^0$ | $K$ | $E^0_{\text{cell}}$ | Reaction at standard-state conditions |
|---|---|---|---|
| <0 | >1 | >0 | Spontaneous |
| 0 | 1 | 0 | At equilibrium |
| >0 | <1 | <0 | Nonspontaneous |

A          B

**FIGURE 20.9**

**The interrelationship of $\Delta G^0$, $E^0$, and $K$. A,** Any one of these three central thermodynamic variables can be used to find the other two. **B,** The signs of $\Delta G^0$, $K$, and $E^0_{\text{cell}}$ determine the reaction direction at standard-state conditions.

If we substitute values for the constants $R$ and $\mathscr{F}$ and run the cell at 25°C (298 K), we obtain a simplified form of Equation 20.7 that relates $E^0$ to $K$ when $n$ moles of $e^-$ per mole of reaction are transferred:

$$E^0_{cell} = \frac{RT}{n\mathscr{F}} \ln K = \frac{8.31 \; \frac{\cancel{J}}{\cancel{mol\;rxn} \cdot \cancel{K}} \times 298 \; \cancel{K}}{n \; \cancel{mol\;e^-} \times 96487 \; \frac{\cancel{J}}{V \cdot \cancel{mol\;e^-}}} \ln K = \frac{0.0257 \; V}{n} \ln K$$

Multiplying by 2.303 to obtain common (base-10) logarithms gives

$$E^0_{cell} = \frac{0.0592 \; V}{n} \log K \quad \text{and} \quad \log K = \frac{nE^0_{cell}}{0.0592 \; V} \qquad \text{(at 25°C)} \qquad \textbf{(20.8)}$$

SAMPLE PROBLEM 20.5 _____

### Calculating $K$ and $\Delta G^0$ from $E^0_{cell}$

**Problem:** Because lead is more active than silver, it can "displace" silver from solution. As a consequence, silver is a valuable byproduct in the extraction of lead from its ore:

$$Pb(s) + 2Ag^+(aq) \rightarrow Pb^{2+}(aq) + 2Ag(s)$$

Calculate $K$ and $\Delta G^0$ at 25°C for this reaction.

**Plan:** We use Table 20.2 to find $E^0_{cell}$ and then calculate $K$ (Equation 20.8) and $\Delta G^0$ (Equation 20.6).

**Solution:** Writing the half-reactions and their $E^0$ values:

(1)     $Ag^+(aq) + e^- \rightarrow Ag(s) \qquad E^0 = 0.80 \; V$

(2)  $Pb^{2+}(aq) + 2e^- \rightarrow Pb(s) \qquad E^0 = -0.13 \; V$

Calculating $E^0_{cell}$: We double (1), reverse (2), and add the half-reactions:

$$
\begin{array}{ll}
2Ag^+(aq) + 2e^- \rightarrow 2Ag(s) & E^0 = 0.80 \; V \\
Pb(s) \rightarrow Pb^{2+}(aq) + 2e^- & -E^0 = 0.13 \; V \\
\hline
Pb(s) + 2Ag^+(aq) \rightarrow Pb^{2+}(aq) + 2Ag(s) & E^0_{cell} = 0.93 \; V
\end{array}
$$

Calculating $K$ (Equation 20.8): The adjusted half-reactions show that 2 mol $e^-$ are transferred per mole reaction, so $n = 2$:

$$E^0_{cell} = \frac{0.0592 \; V}{n} \log K = \frac{0.0592 \; V}{2} \log K = 0.93 \; V$$

So,     $\log K = \dfrac{0.93 \; V \times 2}{0.0592 \; V} = 31.42 \qquad$ and $\qquad \textbf{K = 2.6 \times 10^{31}}$

Calculating $\Delta G^0$ (Equation 20.6):

$$\Delta G^0 = -n\mathscr{F}E^0_{cell} = -2 \; \text{mol} \; e^- \times 96.5 \; \text{kJ/V} \cdot \text{mol} \; e^- \times 0.93 \; V$$

$$= \textbf{-1.8 \times 10^2 \; kJ}$$

**Check:** The three variables are consistent with the reaction being spontaneous: $E^0_{cell} > 0$, $\Delta G^0 < 0$, and $K > 1$. Be sure to round off and check order of magnitude: In the $\Delta G^0$ calculation, for instance, $\Delta G^0 \approx -2 \times 100 \times 1 = -200$, so the overall math seems right. Another check would be to obtain $\Delta G^0$ directly from its relation with $K$:

$$\Delta G^0 = -RT \ln K = -8.31 \; \text{J/mol} \cdot K \times 298 \; K \times \ln (2.6 \times 10^{31})$$

$$= -1.8 \times 10^5 \; \text{J/mol} = -1.8 \times 10^2 \; \text{kJ/mol}$$

**FOLLOW-UP PROBLEM 20.5**

When cadmium metal reduces $Cu^{2+}$ in solution, $Cd^{2+}$ forms in addition to copper metal. If $\Delta G^0 = -143$ kJ, calculate $K$ at 25°C. What would be $E^0_{cell}$ in a voltaic cell that used this reaction?

## The Effect of Concentration on Cell Potential

So far, we have considered cells that operate with all components in their standard states, and we determined standard cell voltage ($E^0_{cell}$) from standard electrode potentials ($E^0_{half\text{-}cell}$). Even if concentrations start out at their standard states, however, they change after a few moments of cell operation. Moreover, all practical voltaic cells in batteries contain reactants at nonstandard concentrations. Clearly, we must be able to determine $E_{cell}$ (no superscript zero), the potential under nonstandard conditions.

We can derive an expression for the relation between cell potential and concentration that is based on the relation between free energy and concentration. Recall from Chapter 19 (Equation 19.11) that $\Delta G$ equals $\Delta G^0$ (the free energy change when the system moves from standard-state concentrations to equilibrium) *plus* $RT \ln Q$ (the free energy change when the system moves from nonstandard-state to standard-state concentrations):

$$\Delta G = \Delta G^0 + RT \ln Q$$

Since $\Delta G$ is related to $E_{cell}$ and $\Delta G^0$ to $E^0_{cell}$ (Equations 20.5 and 20.6), we substitute for them and obtain

$$-n\mathscr{F}E_{cell} = -n\mathscr{F}E^0_{cell} + RT \ln Q$$

Dividing both sides by $-n\mathscr{F}$, we obtain the **Nernst equation,** developed by the German chemist Walter Hermann Nernst in 1889: ◆

$$E_{cell} = E^0_{cell} - \frac{RT}{n\mathscr{F}} \ln Q \qquad \textbf{(20.9)}$$

The Nernst equation says that the cell potential under any conditions depends on its potential at standard-state concentrations and a term for the potential at nonstandard-state concentrations. From Equation 20.9, we can see that

- When $Q < 1$ and therefore [reactant] > [product], $E_{cell} > E^0_{cell}$.
- When $Q = 1$ and therefore [reactant] = [product], $E_{cell} = E^0_{cell}$.
- When $Q > 1$ and therefore [reactant] < [product], $E_{cell} < E^0_{cell}$.

As before, when we substitute known values for $R$ and $\mathscr{F}$, operate the cell at 25°C, and convert to common (base-10) logarithms, we obtain

$$E_{cell} = E^0_{cell} - \frac{0.0592\ V}{n} \log Q \quad \text{(at 25°C)} \qquad \textbf{(20.10)}$$

Remember that $Q$ contains only those species whose concentrations (or pressures) can vary; thus, solids do not appear in the mass-action expression, even though they may act as electrodes. For example, in the reaction between cadmium and silver ion,

$$Cd(s) + 2Ag^+(aq) \longrightarrow Cd^{2+}(aq) + 2Ag(s) \qquad Q = \frac{[Cd^{2+}]}{[Ag^+]^2}$$

◆ **Walther Hermann Nernst (1864-1941).** The great German physical chemist was only 25 when he developed the equation for the relationship between cell voltage and concentration. His career, which culminated in the Nobel Prize in 1920, included formulating the principle of the solubility product and the third law of thermodynamics and contributing key ideas to photochemistry and to principles underlying the Haber process. He is shown here in discussion with a few of his celebrated colleagues; from left to right, Nernst, Albert Einstein, Max Planck, Robert Millikan, and Max Von Laue.

SAMPLE PROBLEM 20.6

## Using the Nernst Equation to Calculate $E_{cell}$

**Problem:** In a test of a new reference electrode, a chemist constructs a voltaic cell consisting of a $Zn/Zn^{2+}$ electrode and the $H_2/H^+$ electrode under the following conditions:

$$[Zn^{2+}] = 0.010\ M \qquad [H^+] = 2.5\ M \qquad P_{H_2} = 0.30\ atm$$

Calculate $E_{cell}$ at 25°C.

**Plan:** To apply the Nernst equation and determine $E_{cell}$, we must know $E^0_{cell}$ and $Q$. We write the spontaneous reaction, calculate $E^0_{cell}$ from standard electrode potentials (Table 20.2), and use the given pressure and concentrations to find $Q$. Then we substitute into Equation 20.10.

**Solution:** Determining the cell reaction and $E^0_{cell}$:

$$2H^+(aq) + 2e^- \rightarrow H_2(g) \qquad\qquad E^0 = 0.00\ V$$

$$\underline{\phantom{2H^+(aq)}\ Zn(s) \rightarrow Zn^{2+}(aq) + 2e^- \qquad -E^0 = 0.76\ V}$$

$$2H^+(aq) + Zn(s) \rightarrow H_2(g) + Zn^{2+}(aq) \qquad E^0_{cell} = 0.76\ V$$

Calculating $Q$:

$$Q = \frac{P_{H_2} \times [Zn^{2+}]}{[H^+]^2} = \frac{0.30 \times 0.010}{2.5^2} = 4.8 \times 10^{-4}$$

Solving for $E_{cell}$ at 25°C (298 K), with $n = 2$:

$$E_{cell} = E^0_{cell} - \frac{0.0592\ V}{n} \log Q = 0.76\ V - \left[ \frac{0.0592\ V}{2} \log (4.8 \times 10^{-4}) \right]$$

$$= 0.76\ V - (-0.098\ V) = \mathbf{0.86\ V}$$

**Check:** After you check the arithmetic, reason through the answer: $E_{cell} > E^0_{cell}$ (0.86 > 0.76) because the log $Q$ term was negative, which means that $Q < 1$ (standard-state concentration ratio). Since the amounts of products, $P_{H_2}$ and $[Zn^{2+}]$, are lower than the standard state and that of the reactant, $[H^+]$, is higher, the reaction proceeds as written, which makes $E_{cell} > E^0_{cell}$.

**FOLLOW-UP PROBLEM 20.6**
Consider a cell based on the reaction

$$Fe(s) + Cu^{2+}(aq) \rightarrow Fe^{2+}(aq) + Cu(s)$$

If $[Cu^{2+}] = 0.30\ M$, what $[Fe^{2+}]$ is needed to increase $E_{cell}$ by 0.25 V above $E^0_{cell}$ at 25°C?

## Cell Potential and the Relation Between $Q$ and $K$

Now let's see how the potential changes as the cell operates, and we'll reconsider the zinc-copper cell:

$$Zn(s) + Cu^{2+}(aq) \rightarrow Zn^{2+}(aq) + Cu(s) \qquad Q = \frac{[Zn^{2+}]}{[Cu^{2+}]}$$

The positive $E^0_{cell}$ (1.10 V) means that this reaction proceeds spontaneously from $Q = 1$ to some $Q > 1$. Suppose we start operating the cell with

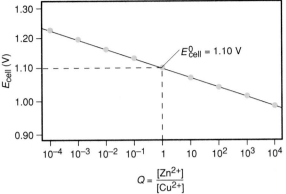

**FIGURE 20.10**
**The relation between $E_{cell}$ and log $Q$ for the zinc-copper cell.** A plot of $E_{cell}$ vs. $Q$ (on a log-arithmic scale) for the zinc-copper reaction shows a linear decrease. When $Q$ is small (left), [reactant] is relatively high, and the cell can do relatively more work. When $Q = 1$, $E_{cell} = E^0_{cell}$. When $Q > 1$, [reactant] is relatively low, and the cell can do relatively less work.

$[Zn^{2+}] = 1.0 \times 10^{-4}$ $M$ and $[Cu^{2+}] = 1.0$ $M$. Therefore, the cell potential will be *higher* than the standard cell potential because $Q = [Zn^{2+}]/[Cu^{2+}] < 1$:

$$E_{cell} = E^0_{cell} - \frac{0.0592\ V}{2} \log \frac{[Zn^{2+}]}{[Cu^{2+}]} = 1.10\ V - \left( \frac{0.0592\ V}{2} \log \frac{1.0 \times 10^{-4}}{1.0} \right)$$

$$= 1.10\ V - \left[ \frac{0.0592\ V}{2} (-4.00) \right] = 1.10\ V + 0.118\ V = 1.22\ V$$

As the cell operates, $[Zn^{2+}]$ increases (as the mass of the Zn electrode becomes lower) and $[Cu^{2+}]$ decreases (as the mass of the Cu electrode becomes higher). Therefore, $Q$ becomes larger, log $Q$ becomes less negative (more positive), so $E_{cell}$ becomes smaller. At the point when $[Zn^{2+}] = [Cu^{2+}]$, $Q = 1$, so the $[(RT/n\mathscr{F}) \log Q]$ term $= 0$ and $E_{cell} = E^0_{cell}$. As the $[Zn^{2+}]/[Cu^{2+}]$ ratio continues to increase, the $[(RT/n\mathscr{F}) \log Q]$ term eventually equals $E^0_{cell}$, so $E_{cell}$ is zero. This situation occurs when *the system has reached equilibrium*. No more free energy is released, so the cell has no more potential. At this point in the life of a battery, we say it is "dead." Figure 20.10 shows the change in $E_{cell}$ as $Q$ changes. Note that *as the cell operates, its potential decreases.* As we pointed out earlier,

- $E_{cell} > E^0_{cell}$ when $Q < 1$
- $E_{cell} = E^0_{cell}$ when $Q = 1$ (standard conditions)
- $E_{cell} < E^0_{cell}$ when $Q > 1$

At equilibrium, $E_{cell}$ is zero and $Q$ equals $K$, so Equation 20.10 becomes

$$0 = E^0_{cell} - \left( \frac{0.0592\ V}{n} \log K \right)$$

Solving for $E^0_{cell}$, we have

$$E^0_{cell} = \frac{0.0592\ V}{n} \log K$$

This is identical to Equation 20.8, which we obtained from $\Delta G^0$.

To put these conclusions another way, at any moment $E_{cell}$ is based on the $Q/K$ ratio. When $Q/K < 1$, $E_{cell}$ is positive; the smaller the ratio, the greater the value of $E_{cell}$, and the more electrical work it can do. When $Q/K$ equals 1, the cell is at equilibrium and its output is zero. If the cell begins with $Q/K > 1$, the cell will operate in reverse—the reverse redox reaction will take place—until $Q/K$ equals 1 at equilibrium.

### Concentration Cells and the Measurement of Concentration

You know that if a gas at high pressure is placed in contact with one at low pressure, they spontaneously mix and the pressures become equal. In the same way, concentrated and dilute solutions spontaneously mix and their concentrations become equal. A **concentration cell** employs this spontaneous tendency as a way to generate electrical energy. The two solutions are in separate half-cells, so they do not physically mix; rather, their concentrations become equal through the operation of the cell.

Suppose both compartments of a voltaic cell house the $Cu/Cu^{2+}$ half-reaction. The cell reaction is the sum of identical half-reactions, written in opposite directions, so the half-cell potentials, $E^0_{copper} = 0.34$ V and $-E^0_{copper} = -0.34$ V, cancel and $E^0_{cell}$ is zero. This occurs because *standard electrode potentials are based on concentrations of 1 M for each dissolved component. In a concentration cell, however, the half-reactions are the same but the concentrations are different.* As a result, even though $E^0_{cell}$ equals zero, the nonstandard voltage, $E_{cell}$, does *not* equal zero because it depends on the ratio of ion concentrations.

Consider Figure 20.11, *A*, in which a concentration cell has 0.10 *M* $Cu^{2+}$ in the anode half-cell and 1.0 *M* $Cu^{2+}$ in the cathode half-cell:

$$Cu(s) \rightarrow Cu^{2+}(aq, 0.10\ M) + 2e^- \quad \text{[anode]}$$

$$Cu^{2+}(aq, 1.0\ M) + 2e^- \rightarrow Cu(s) \quad \text{[cathode]}$$

The overall cell reaction is the sum of the half-reactions:

$$Cu^{2+}(aq, 1.0\ M) \rightarrow Cu^{2+}(aq, 0.10\ M) \qquad E_{cell} = \quad ?$$

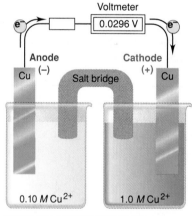

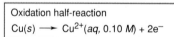

**A**

**FIGURE 20.11**

**A concentration cell based on the Cu/Cu²⁺ half-reaction. A,** Even though the half-reactions involve the same components, the cell operates because the half-cell concentrations are different. **B,** The cell operates spontaneously until the half-cell concentrations are equal. Note the change in electrodes (exaggerated here for clarity) and the equal color of solutions.

The cell potential at the initial concentrations of 0.10 $M$ (dilute) and 1.0 $M$ (concentrated), with $n = 2$, is

$$E_{cell} = E^0_{cell} - \frac{0.0592 \text{ V}}{2} \log \frac{[Cu^{2+}]_{dil}}{[Cu^{2+}]_{conc}} = 0 \text{ V} - \left( \frac{0.0592 \text{ V}}{2} \log \frac{0.10 \, M}{1.0 \, M} \right)$$

$$= 0 \text{ V} - \left[ \frac{0.0592 \text{ V}}{2} (-1.00) \right] = +0.0296 \text{ V}$$

As you can see, $E_{cell}$ for a concentration cell depends entirely on the $[(RT/n\mathscr{F}) \log Q]$ term for nonstandard conditions because $E^0_{cell}$ equals zero.

What is actually going on as this cell runs? In the half-cell with dilute electrolyte (anode), Cu atoms in the electrode give up electrons and become $Cu^{2+}$ ions, which enter the solution and make it *more* concentrated. The electrons released at the anode flow to the cathode compartment. There, $Cu^{2+}$ ions in the concentrated solution pick up the electrons and become Cu atoms, which plate out on the electrode, thereby making that solution *less* concentrated. Like any voltaic cell, this one will run until equilibrium is attained, which happens when the $Cu^{2+}$ concentrations in the half-cells are equal (Figure 20.11, *B*). We would obtain the same final concentration if the two solutions were mixed, but we would obtain no electrical work.

---

**SAMPLE PROBLEM 20.7** _____

### Calculating the Potential of a Concentration Cell

**Problem:** A concentration cell consists of two Ag/Ag$^+$ half-cells. In half-cell A, electrode A dips into 0.010 $M$ AgNO$_3$; in half-cell B, electrode B dips into $4.0 \times 10^{-4}$ $M$ AgNO$_3$. What is the cell potential at 298 K? Which electrode has a positive charge?

**Plan:** The standard half-cell reactions are identical, so $E^0_{cell}$ is zero, and we calculate $E_{cell}$ from the Nernst equation. Since half-cell A has a higher [Ag$^+$], Ag$^+$ ions will be reduced and plate out on electrode A. In half-cell B, Ag will be oxidized and Ag$^+$ ions will enter the solution. As always, the cathode is positive.

**Solution:** Writing the spontaneous reaction: The [Ag$^+$] decreases in half-cell A and increases in half-cell B, so the spontaneous reaction is

$$Ag^+(aq, 0.010 \, M) \text{ [half-cell A]} \rightarrow Ag^+(aq, 4.0 \times 10^{-4} \, M) \text{ [half-cell B]}$$

Calculating $E_{cell}$, with $n = 1$:

$$E_{cell} = E^0_{cell} - \frac{0.0592 \text{ V}}{1} \log \frac{[Ag^+]_{dil}}{[Ag^+]_{conc}}$$

$$= 0 \text{ V} - \left( 0.0592 \text{ V} \times \log \frac{4.0 \times 10^{-4}}{0.010} \right) = \mathbf{0.0829 \text{ V}}$$

Reduction occurs at the cathode, electrode A, $Ag^+(aq, 0.010 \, M) + e^- \rightarrow Ag(s)$, so it has a positive charge due to a relative electron deficiency.

**FOLLOW-UP PROBLEM 20.7**
A concentration cell is built using two Au/Au$^{3+}$ half-cells. In half-cell A, $[Au^{3+}] = 7.0 \times 10^{-4}$ $M$, and in half-cell B, $[Au^{3+}] = 2.5 \times 10^{-2}$ $M$. What is $E_{cell}$, and which electrode is negative?

---

The principle of a concentration cell has some major biological relevance. ◆ The most important commercial application is the measurement

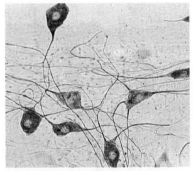

◆ **Concentration Cells in Your Nerve Cells.** The nerve cells that mediate every thought and movement function by the principle of a concentration cell. The nerve membrane, imbedded with a host of specialized protein "gates" and "pumps," separates an interior solution of high [K$^+$] and low [Na$^+$] from an exterior solution of high [Na$^+$] and low [K$^+$]. The result is a membrane more positive outside than inside. When stimulated, Na$^+$ spontaneously rushes in, such that in 0.001s the inside becomes more positive than the outside. This stage is followed by K$^+$ spontaneously rushing out, such that in another 0.001s the membrane is more positive outside again. One region of the membrane stimulates the neighboring region and the electrical impulse moves down the length of the cell.

of unknown concentrations, particularly $[H^+]$ (or pH). Suppose we construct a concentration cell based on the $H_2/H^+$ half-reaction, in which the cathode utilizes the standard hydrogen electrode and the anode has the same apparatus but an unknown $[H^+]$ in solution. The half-reactions and overall reaction are

$$H_2(g, \text{ 1 atm}) \rightarrow 2H^+(aq, \text{ unknown}) + 2e^- \quad \text{[anode]}$$

$$\underline{2H^+(aq, \text{ 1 }M) + 2e^- \rightarrow H_2(g, \text{ 1 atm}) \quad \text{[cathode]}}$$

$$2H^+(aq, \text{ 1 }M) \rightarrow 2H^+(aq, \text{ unknown}) \qquad E_{cell} = ?$$

As with the $Cu/Cu^{2+}$ concentration cell, $E^0_{cell}$ is zero, but the two half-cells differ in $[H^+]$, so $E_{cell}$ is *not* zero. Applying the Nernst equation, with $n = 2$, we obtain

$$E_{cell} = E^0_{cell} - \frac{0.0592 \text{ V}}{2} \log \frac{[H^+]^2_{unknown}}{[H^+]^2_{standard}}$$

Substituting 1 $M$ for $[H^+]_{standard}$ and zero for $E^0_{cell}$ gives

$$E_{cell} = 0 \text{ V} - \frac{0.0592 \text{ V}}{2} \log \frac{[H^+]^2_{unknown}}{1^2} = - \frac{0.0592 \text{ V}}{2} \log [H^+]^2_{unknown}$$

Since $\log x^2 = 2 \log x$, we obtain

$$E_{cell} = - \left( \frac{0.0592 \text{ V}}{2} \times 2 \log [H^+]_{unknown} \right) = - 0.0592 \text{ V} \times \log [H^+]_{unknown}$$

From $- \log [H^+] = $ pH, we obtain

$$E_{cell} = 0.0592 \text{ V} \times \text{pH}$$

Thus, by measuring $E_{cell}$, we can find the pH.

In the routine measurement of pH, a concentration cell incorporating two hydrogen electrodes would be too bulky and difficult to maintain. Instead, as was pointed out in Chapter 17, pH measurements are typically performed with a pH meter (Figure 20.12). The two electrodes of a pH meter dip into the solution being tested. One is a *glass electrode*, consisting of an Ag/AgCl half-reaction immersed in an HCl solution of fixed composition (usually 1.000 $M$) and enclosed by a thin ($\sim$0.05 mm) membrane made of a special glass that is highly sensitive to the presence of $H^+$ ions. The other electrode is often a reference *saturated calomel electrode*, consisting of a platinum wire immersed in a paste of $Hg_2Cl_2$ (calomel), liquid Hg, and saturated KCl solution. The glass electrode monitors the external $[H^+]$ in the solution

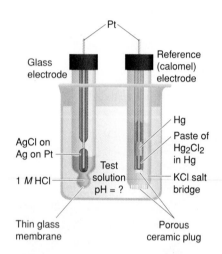

**FIGURE 20.12**

**The laboratory measurement of pH.** The glass electrode *(left)* is a self-contained Ag/AgCl half-cell immersed in an HCl solution of known concentration and enclosed by a thin glass membrane. It monitors the external $[H^+]$ in the solution relative to its fixed internal $[H^+]$. The saturated calomel electrode *(right)* acts as a reference.

**TABLE 20.3   Some Ions Measured with Ion-Specific Electrodes**

| SPECIES DETECTED | TYPICAL SAMPLE |
|---|---|
| $NH_3/NH_4^+$ | Industrial waste water, sea water |
| $CO_2/HCO_3^-$ | Blood, ground water |
| $F^-$ | Drinking water, urine, soil, industrial stack gases |
| $Br^-$ | Grain, plant tissue |
| $I^-$ | Milk, pharmaceuticals |
| $NO_3^-$ | Soil, fertilizer, drinking water |
| $K^+$ | Blood serum, soil, wine |
| $H^+$ | Laboratory solutions, soil, natural waters |

relative to its fixed internal [$H^+$], and the instrument converts the potential difference between the two electrodes into pH.

This glass electrode is one type of *ion-specific (or ion-selective) electrode*. By using glasses of specific composition and other defined membrane materials, electrodes can be designed to measure particular ion concentrations. Table 20.3 shows some of the many ions for which ion-specific electrodes are available. ◆

**Section Summary**

The two criteria for spontaneity, negative $\Delta G$ and positive $E_{cell}$ are related: $\Delta G = -n\mathscr{F}E_{cell}$. The $\Delta G$ of the cell reaction represents the maximum amount of electrical work the cell can do. Because the standard free energy change, $\Delta G^0$, is related to $E^0_{cell}$ and to $K$, we can use $E^0_{cell}$ to determine $K$. At nonstandard conditions, $E_{cell}$ depends on $E^0_{cell}$ and a correction term based on $Q$. $E_{cell}$ is high when $Q$ is small (high [reactant]), and it decreases as the cell operates. At equilibrium, $Q = K$, so $\Delta G$ and $E_{cell}$ are zero. Concentration cells use identical half-reactions, with solutions of differing concentration, that generate electrical energy as the concentrations become equal. Ion-specific electrodes, such as the pH electrode, are self-contained devices that ideally measure one species.

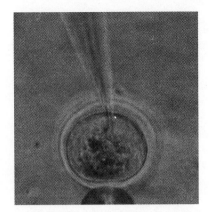

◆ **Minimicroanalysis.** Electronic miniaturization has greatly broadened the applications of electrochemistry. Environmental chemists use pocket-sized pH meters in the field to study natural waters and soil. Physiologists and biochemists employ electrodes so small that they can be surgically implanted into a single biological cell to measure ion concentrations. In the photo, a microelectrode is shown penetrating a mouse reproductive (egg) cell. A micropipette holds the cell.

## 20.4 Electrochemical Processes in Batteries

In their compactness and mobility, batteries have had a major influence on our way of life. In industrialized countries, an average of 10 batteries are used by each person every year. A **battery,** strictly speaking, is a self-contained group of voltaic cells arranged in series (plus-to-minus-to-plus, and so on), so that their individual voltages are added together. In everyday speech, however, the term may also be applied to a single voltaic cell.

Batteries are ingeniously engineered devices that operate through traditional electrochemical principles, although their half-reactions and half-cells may seem unusual. A major difference between most batteries and the voltaic cells we've seen so far is that the electrolytes are solids, pastes, or thick slurries rather than solutions. Therefore, their *concentrations are effectively constant,* as for any solid. As a result, the potential does *not* decrease gradually as the cell operates but remains relatively constant as long as all the components are present.

There are several classes of batteries. A *primary battery* cannot be recharged, so it is thrown away when the battery is "dead," that is, when the components have reached their equilibrium concentrations. In contrast, when a *secondary,* or *rechargeable, battery* runs down, it is recharged by supplying electrical energy to reverse the cell reaction and form more reactant. In other words, in this type of battery, the voltaic cells are periodically converted to electrolytic cells to restore nonequilibrium concentrations. A **fuel cell,** or flow battery, is one that is not self-contained. The reactants (usually a combustible fuel and oxygen) enter the cell and the products leave, generating electricity through the controlled oxidation of the fuel. The upcoming multipage Gallery presents some important examples of primary, secondary, and flow batteries.

## GALLERY                **Batteries and Their Applications**

### Primary (Nonrechargeable) Batteries

#### Dry Cells
Invented in the 1860s, the common dry cell, or *Leclanche cell*, has become a familiar household item. An active zinc anode in the form of a can houses a mixture of $MnO_2$ and an acidic electrolyte paste, consisting of $NH_4Cl$, $ZnCl_2$, $H_2O$, and starch. Powdered graphite improves conductivity. The inactive cathode is a graphite rod.

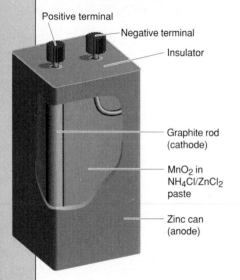

Positive terminal
Negative terminal
Insulator
Graphite rod (cathode)
$MnO_2$ in $NH_4Cl/ZnCl_2$ paste
Zinc can (anode)

*Anode (oxidation):*
$Zn(s) \rightarrow Zn^{2+}(aq) + 2e^-$
*Cathode (reduction):* The cathode half-reaction is complex and, even today, is still being studied. $MnO_2(s)$ is reduced to $Mn_2O_3(s)$ through a series of steps that may involve the presence of $Mn^{2+}$ and an acid-base reaction between $NH_4^+$ and $OH^-$ :
$2MnO_2(s) + 2NH_4^+(aq) + 2e^- \rightarrow Mn_2O_3(s) + 2NH_3(aq) + H_2O(l)$
The ammonia, some of which may be gaseous, forms a complex ion with $Zn^{2+}$ , which crystallizes in contact with $Cl^-$ ion:
$Zn^{2+}(aq) + 2NH_3(aq) + 2Cl^-(aq) \rightarrow Zn(NH_3)_2Cl_2(s)$
*Cell reaction:*
$2MnO_2(s) + 2NH_4Cl(aq) + Zn(s) \rightarrow$
$\qquad\qquad Zn(NH_3)_2Cl_2(s) + H_2O(l) + Mn_2O_3(s) \qquad E_{cell} = 1.5$ V

*Uses:* Common household items, such as portable radios, toys, flashlights.
*Advantages:* Inexpensive, safe, available in many sizes.
*Disadvantages:* At high current drain, $NH_3(g)$ builds up, causing drop in voltage; short shelf life because zinc anode reacts with the acidic $NH_4^+$ ions.

### Alkaline Battery
The alkaline battery is an improved dry cell. The half-reactions are similar, but the electrolyte is a basic KOH paste, which eliminates the buildup of gases and maintains the Zn electrode.

*Anode (oxidation):*
$Zn(s) + 2OH^-(aq) \rightarrow ZnO(s) + H_2O(l) + 2e^-$
*Cathode (reduction):*
$MnO_2(s) + 2H_2O(l) + 2e^- \rightarrow Mn(OH)_2(s) + 2OH^-(aq)$
*Cell reaction:*
$Zn(s) + MnO_2(s) + H_2O(l) \rightarrow ZnO(s) + Mn(OH)_2(s) \qquad E_{cell} = 1.5$ V

Positive button
Steel case
$MnO_2$ in KOH paste
Zn (anode)
Graphite rod (cathode)
Absorbent/separator
Negative end cap

*Uses:* Same as for dry cell.
*Advantages:* No voltage drop and longer shelf life than dry cell because of alkaline electrolyte; safe, many sizes.
*Disadvantages:* More expensive than common dry cell.

### Mercury and Silver (Button) Batteries

The mercury and silver batteries are quite similar. Like the alkaline dry cell, both use zinc in a basic medium as the anode. One employs HgO, the other $Ag_2O$, in a steel container that acts as the cathode. The solid reactants are each compressed with KOH, and moist paper acts as a salt bridge.

*Anode (oxidation):*
$Zn(s) + 2OH^-(aq) \rightarrow ZnO(s) + H_2O(l) + 2e^-$
*Cathode (reduction; mercury):*
$HgO(s) + H_2O(l) + 2e^- \rightarrow Hg(l) + 2OH^-(aq)$
*Cathode (reduction; silver):*
$Ag_2O(s) + H_2O(l) + 2e^- \rightarrow 2Ag(s) + 2OH^-(aq)$
*Cell reaction (mercury):*
$Zn(s) + HgO(s) \rightarrow ZnO(s) + Hg(l)$        $E_{cell} = 1.3$ V
*Cell reaction (silver):*
$Zn(s) + Ag_2O(s) \rightarrow ZnO(s) + 2Ag(s)$        $E_{cell} = 1.6$ V

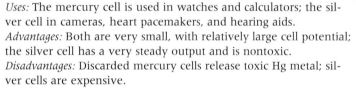

*Uses:* The mercury cell is used in watches and calculators; the silver cell in cameras, heart pacemakers, and hearing aids.
*Advantages:* Both are very small, with relatively large cell potential; the silver cell has a very steady output and is nontoxic.
*Disadvantages:* Discarded mercury cells release toxic Hg metal; silver cells are expensive.

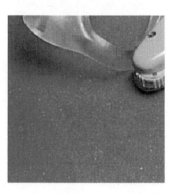

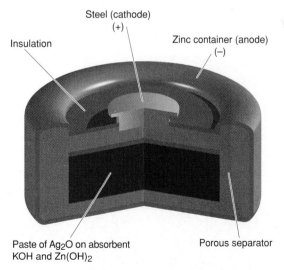

Insulation

Steel (cathode)
(+)

Zinc container (anode)
(−)

Paste of $Ag_2O$ on absorbent
KOH and $Zn(OH)_2$

Porous separator

## GALLERY            **Batteries and Their Applications**

### Secondary (Rechargeable) Batteries

### Lead-Acid Battery

A typical 12-V lead-acid car battery has six cells connected in series, each of which delivers about 2 V. Each cell contains two lead grids packed with the electrode materials: the anode is spongy Pb, and the cathode is powdered $PbO_2$. The grids are immersed in an electrolyte solution of ~4.5 $M$ $H_2SO_4$. Fiberglass sheets between the grids prevent shorting by accidental physical contact. When the cell discharges, it generates electrical energy as a voltaic cell:

*Anode (oxidation):*
$$Pb(s) + SO_4^{2-}(aq) \rightarrow PbSO_4(s) + 2e^-$$
*Cathode (reduction):*
$$PbO_2(s) + 4H^+(aq) + SO_4^{2-}(aq) + 2e^- \rightarrow PbSO_4(s) + 2H_2O(l)$$
Note that both half-reactions produce $Pb^{2+}$ ion, one through oxidation of Pb, the other through reduction of $PbO_2$. At both electrodes, the $Pb^{2+}$ reacts with $SO_4^{2-}$ to form $PbSO_4(s)$.
*Cell reaction [discharge]:*
$$PbO_2(s) + Pb(s) + 2H_2SO_4(aq) \rightarrow 2PbSO_4(s) + 2H_2O(l) \quad E_{cell} \approx 2 \text{ V}$$

When the cell recharges, it utilizes electrical energy as an electrolytic cell, and the half-cell and cell reactions are reversed (see Figure 20.18).
*Cell reaction [recharge]:*
$$2PbSO_4(s) + 2H_2O(l) \rightarrow PbO_2(s) + Pb(s) + 2H_2SO_4(aq)$$

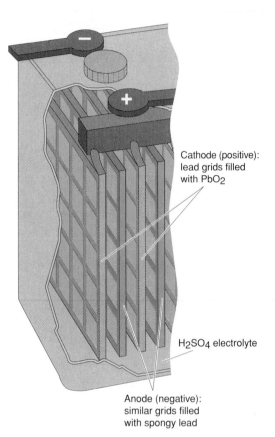

Cathode (positive): lead grids filled with PbO2

H2SO4 electrolyte

Anode (negative): similar grids filled with spongy lead

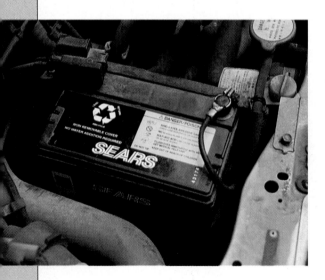

*Uses:* In automobiles and trucks.
*Advantages:* Provides a large burst of current to the engine starter motor; reliable, long life; effective at low temperature.
*Disadvantages:*
1. Loss of capacity. After discharging, $PbSO_4(s)$ coats the battery grids. $PbSO_4$ is required for recharging, but if mechanical strain and normal bumping dislodge it, battery capacity is lowered. If enough $PbSO_4$ is lost, the cell cannot be recharged.
2. Safety hazard. Older batteries had a cap on each cell to monitor electrolyte density and replace water lost during discharging. Water can be electrolyzed to $H_2$ and $O_2$ during recharging and, if sparked, the gases explode and splatter $H_2SO_4$. Modern batteries use a lead alloy that inhibits electrolysis, thereby reducing water loss, so the cells are sealed.

## Nickel-Cadmium (Nicad) Battery

The nickel-cadmium battery has an anode half-reaction that oxidizes cadmium in a basic (NaOH or KOH) electrolyte, while nickel(III) as NiO(OH) is reduced at the cathode.

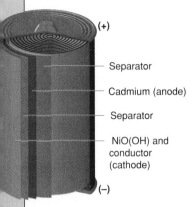

(+)

— Separator

— Cadmium (anode)

— Separator

— NiO(OH) and conductor (cathode)

(−)

*Anode (oxidation):*
$$Cd(s) + 2OH^-(aq) \rightarrow Cd(OH)_2(s) + 2e^-$$
*Cathode (reduction):*
$$2NiO(OH)(s) + 2H_2O(l) + 2e^- \rightarrow 2Ni(OH)_2(s) + 2OH^-(aq)$$
*Cell reaction:*
$$Cd(s) + 2NiO(OH)(s) + 2H_2O(l) \rightarrow$$
$$2Ni(OH)_2(s) + Cd(OH)_2(s) \qquad E_{cell} = 1.4 \text{ V}$$
The cell reaction is reversed on recharging.

*Uses:* Cordless razors, photo flash units, power tools.
*Advantages:* Lightweight.
*Disadvantages:* Disposal of toxic cadmium.

## Fuel Cells (Flow Batteries)

Fuel cells utilize combustion reactions to produce electricity. As with other batteries, the reactants undergo separated half-reactions (the fuel does not burn), and the electrons are transferred through an external circuit. One fuel cell that oxidizes $H_2$ has carbon electrodes impregnated with metal catalysts and a molten $Na_2CO_3$ electrolyte.

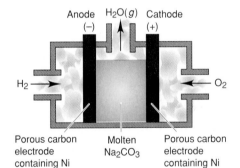

Anode   H₂O(*g*)   Cathode
(−)                (+)

H₂ →                          ← O₂

Porous carbon electrode containing Ni   Molten Na₂CO₃   Porous carbon electrode containing Ni and NiO

*Anode (oxidation):*
$$H_2(g) + CO_3{}^{2-}(l) \rightarrow H_2O(l) + CO_2(g) + 2e^-$$
*Cathode (reduction):*
$$\tfrac{1}{2}O_2(g) + CO_2(g) + 2e^- \rightarrow CO_3{}^{2-}(l)$$
*Cell reaction:*
$$H_2(g) + \tfrac{1}{2}O_2(g) \rightarrow H_2O(l) \qquad E_{cell} = 1.2 \text{ V}$$

*Uses:* Provides electricity and pure water during space flights.
*Advantages:*
1. Clean, portable sources of electricity; many fuel cells produce no pollutants.
Some other cell reactions in fuel cells are
$$2NH_3(g) + \tfrac{3}{2}O_2(g) \rightarrow N_2(g) + 3H_2O(l)$$
$$N_2H_4(g) + \tfrac{1}{2}O_2(g) \rightarrow N_2(g) + 2H_2O(l)$$
$$CH_4(g) + 2O_2(g) \rightarrow CO_2(g) + 2H_2O(l)$$
2. Fuel cells are very efficient, converting about 75% of fuel's bond energy into electricity. In contrast, an electric power plant converts about 35% to 40% of coal's bond energy into electricity, and a car engine converts about 25% of gasoline's bond energy into moving the car.

*Disadvantages:*
1. Unlike an ordinary battery, fuel cells cannot store electrical energy but operate only with a continuous flow of reactants.
2. Electrode materials are short-lived and expensive.

### Section Summary

Batteries contain several voltaic cells in series and are either primary (such as dry cell, alkaline, mercury, or silver), rechargeable (such as lead-acid, nicad), or flow (fuel cells). By supplying electricity to a rechargeable battery, the redox reaction is reversed, and more reactant is formed for further use. Fuel cells release electrons and generate a current through the controlled oxidation of combustible fuels, such as $H_2$ or $CH_4$.

## 20.5 Corrosion: A Case of Environmental Electrochemistry

By now you may be thinking that spontaneous redox reactions are always beneficial, but consider the problem of **corrosion,** the natural redox process that oxidizes metals to their oxides and sulfides. In chemical terms, corrosion is the reverse of isolating a metal from its oxide or sulfide ore. Damage from corrosion to cars, ships, buildings, and bridges runs into tens of billions of dollars annually, so it is a major problem in much of the world. Although we focus here on the corrosion of iron, many other metals, such as copper and silver, also corrode.

### The Corrosion of Iron

The most common and economically destructive form of corrosion is the rusting of iron. About 25% of the steel produced in the United States is made to replace corroded steel. Contrary to the impression our earlier discussions may have given, rust is not a direct product of the reaction between iron and oxygen but arises through a complex electrochemical process. Let's look at the facts of iron corrosion and then use the features of a voltaic cell to explain them:

1. Iron will not rust in dry air: moisture must be present.
2. Iron will not rust in air-free water: oxygen must be present.
3. Iron rusts most rapidly in ionic solutions and at low pH (high $[H^+]$).
4. The loss of iron and the depositing of rust often occur at *different* places on the *same* object.
5. Iron rusts faster in contact with a less active metal (such as Cu) and more slowly in contact with a more active metal (such as Zn).

Picture the magnified surface of a piece of iron or steel (Figure 20.13). Strains, ridges, and dents are present that, in contact with water, are usually

**FIGURE 20.13**

**The corrosion of iron. A,** Close-up view of an iron surface. Corrosion usually occurs at a surface irregularity. **B,** An enlargement of the surface, showing the steps in the corrosion process.

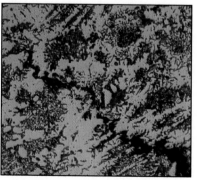

A

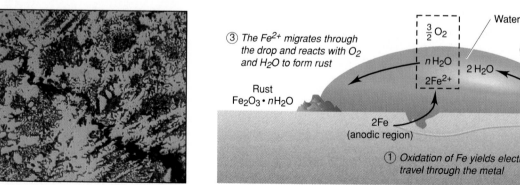

③ The $Fe^{2+}$ migrates through the drop and reacts with $O_2$ and $H_2O$ to form rust

Water droplet

$\frac{3}{2} O_2$

$n H_2O$    $2 H_2O$

$2Fe^{2+}$

② Electrons at the Fe (inactive) cathode reduce $O_2$ to $H_2O$

Rust
$Fe_2O_3 \cdot n H_2O$

2Fe
(anodic region)

$4H^+$    $O_2$

$4e^-$    (cathodic region)

① Oxidation of Fe yields electrons, which travel through the metal

B

the sites of iron loss. These sites are called *anodic regions* because of the following half-reaction that occurs there:

$$Fe(s) \rightarrow Fe^{2+}(aq) + 2e^- \qquad \text{[anodic region; oxidation]}$$

Once the iron atoms lose electrons, the damage to the object has been done: a pit forms where the iron is lost.

The freed electrons move through the external circuit—the piece of iron itself—until they reach a region of relatively high $O_2$ concentration, near the surface of a surrounding water droplet, for instance. At this *cathodic region*, the electrons released from the iron atoms reduce $O_2$ molecules:

$$O_2(g) + 4H^+(aq) + 4e^- \rightarrow 2H_2O(l) \qquad \text{[cathodic region; reduction]}$$

Notice that this overall redox process is completed without the formation of any rust:

$$2Fe(s) + O_2(g) + 4H^+(aq) \rightarrow 2H_2O(l) + 2Fe^{2+}(aq)$$

Rust forms through another redox reaction in which the reactants contact each other directly. The $Fe^{2+}$ ions formed originally at the anodic region disperse through the surrounding water and react with $O_2$. The overall reaction for this step is

$$2Fe^{2+}(aq) + \tfrac{1}{2}O_2(g) + (2 + n)H_2O(l) \rightarrow Fe_2O_3 \cdot nH_2O(s) + 4H^+(aq)$$

[The inexact coefficient $n$ for $H_2O$ in the above equation appears because rust, $Fe_2O_3 \cdot nH_2O$, is a form of iron(III) oxide with variable amounts of water of hydration.] The rust deposit is really incidental to the damage caused by loss of iron—a chemical insult added to the original injury.

Adding the previous two equations together shows the complete process for the rusting of iron:

$$2Fe(s) + \tfrac{3}{2}O_2(g) + nH_2O(l) + \cancel{4H^+(aq)} \rightarrow Fe_2O_3 \cdot nH_2O(s) + \cancel{4H^+(aq)}$$

We've shown the canceled $H^+$ ions to emphasize that they act as a catalyst, used up in one step of the overall reaction and created in another, which explains why rusting is faster at low pH (high $[H^+]$). Ionic solutions speed rusting by improving the conductivity of the aqueous medium near the anodic and cathodic regions. The effect of ions is pronounced on ocean-going vessels or on cars in cold climates where salts are used to melt ice on slippery roads (Figure 20.14).

In many ways, the components of the corrosion process resemble those of a voltaic cell. Anodic and cathodic regions are separated in space, connected via an external circuit through which the electrons travel. In the anodic region, iron behaves like an active electrode, whereas in the cathodic region, it is inactive. The moisture surrounding the pit functions somewhat like a salt bridge, a means for ions to ferry back and forth and keep the solution neutral.

**FIGURE 20.14**
**Enhanced corrosion at sea.** The high ion concentration of sea water leads to its high conductivity, which enhances corrosion of iron in ocean ship hulls.

## Protecting Against the Corrosion of Iron

A common approach to preventing or limiting corrosion is to eliminate contact with the corrosive factors. The simple act of washing off road salt removes the ionic solution from auto chassis parts. Iron objects are frequently painted to keep out $O_2$ and moisture, but if the paint layer chips, rusting proceeds. More permanent coatings include chromium on plumbing fixtures. "Blueing" of gun barrels, wood stoves, and other steel objects bonds

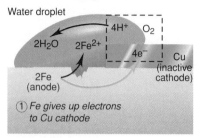

② *Electrons from Fe reduce $O_2$ to $H_2O$*

① *Fe gives up electrons to Cu cathode*

**A** Enhanced corrosion

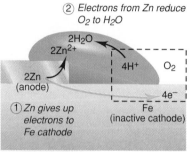

② *Electrons from Zn reduce $O_2$ to $H_2O$*

① *Zn gives up electrons to Fe cathode*

**B** Cathodic protection

**FIGURE 20.15**

**The effect of copper and zinc on the corrosion of iron. A,** When iron is in contact with a less active metal, such as copper, the iron loses electrons more readily (is more anodic), so it corrodes faster. **B,** When iron is in contact with a more active metal, such as zinc, the zinc loses electrons instead of the iron, which is cathodic, so the iron does not corrode. The process is known as *cathodic protection.*

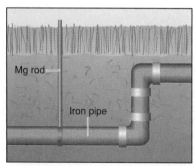

**FIGURE 20.16**

**The use of sacrificial anodes to prevent iron corrosion.** Active metals, such as magnesium and aluminum, are connected to underground iron pipes to prevent their corrosion through cathodic protection. The active metal is sacrificed instead of the iron.

an adherent coating of $Fe_3O_4$ (magnetite, called black iron, in which iron occurs in the +2 and +3 oxidation states) to the surface.

The only fact regarding corrosion that we have not yet addressed involves the relative activity of other metals in contact with iron (fact 5), and it leads to the most effective way to prevent corrosion. The essence of the idea is that *iron functions as both anode and cathode in the rusting process, but it is lost only at the anode.* Therefore, anything that makes iron behave more like the anode will increase corrosion. If iron is in contact with a *less* active metal (weaker reducing agent), such as copper, its anodic role is enhanced (Figure 20.15, *A*). When iron plumbing is connected directly to copper plumbing with no electrical insulation between them, the iron pipe corrodes rapidly. On the other hand, anything that makes iron behave more like the cathode will prevent corrosion. Application of this principle is called *cathodic protection.* For example, if the iron makes contact with a *more* active metal (stronger reducing agent), such as zinc, the now cathodic iron remains intact while the zinc acts as the anode and loses electrons (Figure 20.15, *B*). Coating steel with a "sacrificial" layer of zinc is the basis of the *galvanizing* process. In addition to blocking physical contact with $H_2O$ and $O_2$, the zinc is "sacrificed" (oxidized) instead of the iron.

Sacrificial anodes are employed to protect iron and steel structures (pipes, tanks, oil rigs, and so on) in moist underground and marine environments (Figure 20.16). The most frequently used metals for this purpose are magnesium and aluminum, elements much more active than iron, so they act as the anode while iron acts as the cathode. A further advantage of these metals is that they form adherent oxide coatings, which slows their own corrosion.

**Section Summary**

Corrosion damages metal structures through a natural electrochemical change. Iron corrosion occurs in the presence of oxygen and moisture and is increased by high $[H^+]$, high [ion], or contact with a less active metal, such as Cu. Fe is oxidized and $O_2$ is reduced in one redox reaction, while rust is formed in another reaction that often takes place at a different location. Since Fe functions as both anode and cathode in the process, the object can be protected by physically covering its surface or joining it to a more active metal (such as Zn, Mg, or Al), which acts as the anode instead of the Fe.

## 20.6   **Electrolytic Cells: Using Electrical Energy to Drive a Nonspontaneous Reaction**

Up to now, we have been considering voltaic cells, those in which a spontaneous reaction generates electrical energy. The principle of an electrolytic cell is exactly the opposite: *electrical energy from an external source drives a nonspontaneous reaction.*

**Construction and Principle of an Electrolytic Cell**

Let's examine the operation of an electrolytic cell by constructing it from a voltaic cell. Consider the tin-copper voltaic cell in Figure 20.17, *A.* The Sn

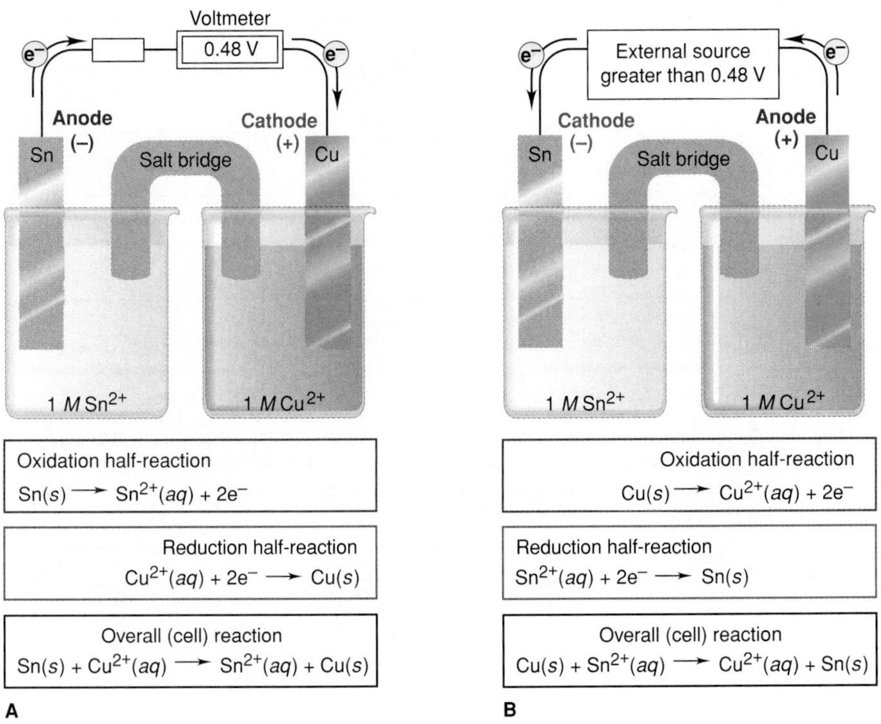

**FIGURE 20.17**
**The tin-copper reaction as the basis of a voltaic and an electrolytic cell. A,** The spontaneous reaction between Sn and $Cu^{2+}$ generates 0.48 V in a voltaic cell. **B,** If more than 0.48 V is supplied, the same apparatus is changed to an electrolytic cell, and the nonspontaneous reaction between Cu and $Sn^{2+}$ occurs. Note the change in electrode charges and direction of electron flow.

anode will gradually become oxidized to $Sn^{2+}$ ions, and the $Cu^{2+}$ ions will gradually be reduced and plate out on the Cu cathode:

$$Sn(s) \rightarrow Sn^{2+}(aq) + 2e^- \quad \text{[anode; oxidation]}$$

$$Cu^{2+}(aq) + 2e^- \rightarrow Cu(s) \quad \text{[cathode; reduction]}$$

These half-reactions occur as they do because the cell reaction is spontaneous in that direction:

$$Sn(s) + Cu^{2+}(aq) \rightarrow Sn^{2+}(aq) + Cu(s) \quad E^0_{cell} = 0.48 \text{ V} \quad \text{and} \quad \Delta G^0 = -93 \text{ kJ}$$

Therefore, the *reverse* cell reaction is *non*spontaneous, as the negative $E^0_{cell}$ and positive $\Delta G^0$ indicate:

$$Sn^{2+}(aq) + Cu(s) \rightarrow Sn(s) + Cu^{2+}(aq) \quad E^0_{cell} = -0.48 \text{ V} \quad \text{and} \quad \Delta G^0 = 93 \text{ kJ}$$

No matter how long we wait, the Cu electrode will not become oxidized to $Cu^{2+}$, and $Sn^{2+}$ ions will not plate out on the Sn electrode of their own accord. We can make that process happen, however, by supplying an electric potential greater than 0.48 V from an external source. In effect, we have converted the voltaic cell into an electrolytic cell and changed the nature of the electrodes—anode is now cathode, and vice versa:

$$Cu(s) \rightarrow Cu^{2+}(aq) + 2e^- \quad \text{[anode; oxidation]}$$

$$Sn^{2+}(aq) + 2e^- \rightarrow Sn(s) \quad \text{[cathode; reduction]}$$

Figure 20.17, *B*, illustrates this situation. Note that in an electrolytic cell, as in a voltaic cell, *oxidation still takes place at the anode and reduction still takes place at the cathode, but the signs of the electrodes are reversed.* In a voltaic cell, the electrons come from the anode, so it is negative. In an electrolytic cell, the electrons come from the external source: it *supplies* them *to* the cathode, making it negative, and *removes* them *from* the anode, making it positive.

**FIGURE 20.18**

**The processes occurring during the discharge and recharge of a lead-acid battery.** When the lead-acid battery is discharging *(top),* it behaves like a voltaic cell: the anode is negative (electrode I) and the cathode is positive (electrode II). When it is recharging *(bottom),* it behaves like an electrolytic cell: the anode is positive (electrode II) and the cathode is negative (electrode I).

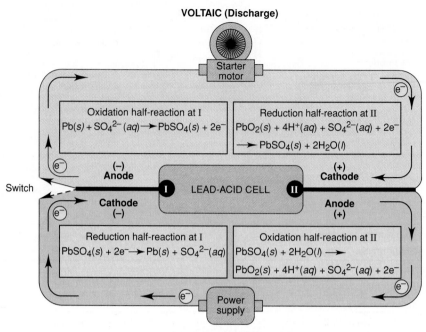

A rechargeable battery functions as a voltaic cell when it is discharging and as an electrolytic cell when it is recharging, so it provides a good way to compare these two cell types and the change in electrodes. Figure 20.18 shows these two functions in the lead-acid battery. In the discharge mode (voltaic cell), the anode is negative and occurs at one electrode (I). In the recharge mode (electrolytic cell), the anode is positive and occurs at the other electrode (II). Similarly, the cathode is positive during discharge (electrode II) and negative during recharge (electrode I). Regardless of the mode, however, oxidation occurs at the anode and reduction at the cathode. Table 20.4 summarizes the processes and signs in the two types of electrochemical cells.

## Determining the Products of Electrolysis Reactions

**Electrolysis,** the lysing (splitting) of a substance by the input of electrical energy, is often used to decompose a compound into its elements. Electrolytic cells are involved in key industrial production steps for some of the most important elements, including chlorine, aluminum, and copper.

**TABLE 20.4    Comparison of Voltaic and Electrolytic Cells**

| CELL TYPE | $\Delta G$ | $E_{CELL}$ | ELECTRODE | | |
| | | | NAME | PROCESS | SIGN |
|---|---|---|---|---|---|
| Voltaic | <0 | >0 | Anode | Oxidation | − |
| Voltaic | <0 | >0 | Cathode | Reduction | + |
| Electrolytic | >0 | <0 | Anode | Oxidation | + |
| Electrolytic | >0 | <0 | Cathode | Reduction | − |

Electrolysis of water to $H_2$ and $O_2$ was discovered in 1800, and the process is still used to produce these gases in ultrahigh purity. The electrolyte in the cell can be the pure compound (such as $H_2O$ or a molten salt), a mixture of molten salts, or an aqueous solution.

**Electrolysis of molten salts.** Many electrolytic applications involve isolating a metal or nonmetal from a molten salt. Predicting the product at each electrode is simple if the salt is pure because *the cation will be reduced and the anion oxidized*. The electrolyte is the molten salt itself, and the ions moving through the cell are attracted to the oppositely charged electrodes.

Consider the electrolysis of molten (fused) calcium chloride (melting point, 782°C). The two species present are $Ca^{2+}$ and $Cl^-$, so the $Ca^{2+}$ ion is reduced and the $Cl^-$ ion is oxidized:

$$2Cl^-(l) \rightarrow Cl_2(g) + 2e^- \qquad \text{[anode; oxidation]}$$
$$\underline{Ca^{2+}(l) + 2e^- \rightarrow Ca(s) \qquad \text{[cathode; reduction]}}$$
$$Ca^{2+}(l) + 2Cl^-(l) \rightarrow Ca(s) + Cl_2(g) \qquad \text{[overall; electrolytic cell]}$$

Metallic calcium is prepared industrially this way, as are several other active metals, such as Na, Mg, and Al, and the halogens $Cl_2$ and $Br_2$. We examine the details of these and several other electrolytic processes in Chapter 23.

More typically, the electrolyte is a mixture of molten salts, which is then electrolyzed to obtain a particular metal. When we have a choice of product, how can we tell which species will react at which electrode? The general rule for all electrolytic cells is that *the more easily oxidized species (stronger reducing agent) reacts at the anode, and the more easily reduced species (stronger oxidizing agent) reacts at the cathode.*

It is important to realize that for electrolysis of mixtures of molten salts we *cannot* use tabulated $E^0$ values to tell the relative strength of the oxidizing and reducing agents. Those values refer to the *change from aqueous ion to free element*, $M^{n+}(aq) + ne^- \rightarrow M(s)$, and there are no aqueous ions in a molten salt. Instead, we rely on our knowledge of periodic trends to predict which of the ions present gains or loses electrons more easily.

SAMPLE PROBLEM 20.8 _____

**Predicting the Electrolysis Products of a Molten Salt Mixture**

**Problem:** A chemical engineer melts a naturally occurring mixture of NaBr and $MgCl_2$ and decomposes it in an electrolytic cell. Predict the substance formed at each electrode, and write balanced half-reactions and the overall cell reaction.

**Plan:** We have to determine which metal and nonmetal will most easily form at the electrodes. We list the ions present as oxidizing or reducing agents. Whichever metal holds its electrons more tightly (has the higher ionization energy, IE) occurs as the cation that is the stronger oxidizing agent and is reduced at the cathode. Whichever nonmetal holds its electrons less tightly (has the lower electronegativity, EN) occurs as the anion that is the stronger reducing agent and is oxidized at the anode.

**Solution:** Listing the ions as oxidizing or reducing agents:

The possible oxidizing agents are $Na^+$ and $Mg^{2+}$.

The possible reducing agents are $Br^-$ and $Cl^-$.

Determining the cathode product (most easily reduced cation): Mg is to the right of Na in Period 3. IE increases from left to right, so Mg has a higher IE. It takes more energy to remove an $e^-$ from Mg than from Na, so it follows that $Mg^{2+}$ has a greater attraction for $e^-$ and thus is more easily reduced (stronger oxidizing agent):

$$Mg^{2+}(l) + 2e^- \rightarrow Mg(l) \qquad \text{[cathode; reduction]}$$

Determining the anode product (most easily oxidized anion): Br is below Cl in Group 7A(17). EN decreases down the group, so Br has a lower EN than Cl. Therefore, it follows that $Br^-$ holds its $e^-$ less tightly than $Cl^-$, so $Br^-$ is more easily oxidized (stronger reducing agent):

$$2Br^-(l) \rightarrow Br_2(g) + 2e^- \qquad \text{[anode; oxidation]}$$

Writing the overall cell reaction:

$$Mg^{2+}(l) + 2Br^-(l) \rightarrow Mg(l) + Br_2(g) \qquad \text{[overall]}$$

**Comment:** The cell temperature must be high enough to keep the salt mixture molten. In this case, the temperature is greater than the melting point of Mg, so it is a liquid, and greater than the boiling point of $Br_2$, so it is a gas.

**FOLLOW-UP PROBLEM 20.8**
A sample of $AlBr_3$ contaminated with KF is melted and electrolyzed. Determine the electrode products and the overall cell reaction.

**Electrolysis of water and nonstandard half-cell potentials.** Before we can analyze the electrolysis products of aqueous salt solutions, we must examine the electrolysis of water itself. Extremely pure water is difficult to electrolyze because very few ions are present to conduct a current. If a small amount of a nonreacting salt (such as $Na_2SO_4$) is added, however, electrolysis proceeds rapidly. A glass electrolytic cell with separated gas compartments is used to keep the $H_2$ and $O_2$ gases from mixing (Figure 20.19). At the anode, water is oxidized as the O.N. of O changes from $-2$ to zero:

$$2H_2O(l) \rightarrow O_2(g) + 4H^+(aq) + 4e^- \qquad -E = -0.82 \text{ V} \qquad \text{[anode; oxidation]}$$

At the cathode, water is reduced as the O.N. of H changes from $+1$ to zero:

$$2H_2O(l) + 2e^- \rightarrow H_2(g) + 2OH^-(aq) \qquad E = -0.42 \text{ V} \qquad \text{[cathode; reduction]}$$

After doubling the cathode half-reaction to equate $e^-$ loss and gain, adding the half-reactions, combining the $H^+$ and $OH^-$ into $H_2O$, and canceling $e^-$ and excess $H_2O$, the overall reaction is

$$2H_2O(l) \rightarrow 2H_2(g) + O_2(g) \qquad E_{cell} = -1.24 \text{ V} \qquad \text{[overall]}$$

Notice that these electrode potentials are not written with a superscript zero because they are *not* standard electrode potentials. The $[H^+]$ and $[OH^-]$ are $1.0 \times 10^{-7}$ M rather than the standard-state value of 1 M. These E values are obtained by applying the Nernst equation. For example, the anode potential (with $n = 4$) is

$$E_{cell} = E_{cell}^0 - \frac{0.0592 \text{ V}}{4} \log (P_{O_2} \times [H^+]^4)$$

The standard potential for the *oxidation* of water is $-1.23$ V (from Table 20.2) and $P_{O_2} \approx 1$ atm in the half-cell, so we have

$$E_{cell} = -1.23 \text{ V} - \{0.0148 \text{ V} \times \log [1(1.0 \times 10^{-7})^4]\} = -0.82 \text{ V}$$

In aqueous ionic solutions, $[H^+]$ and $[OH^-]$ are approximately $10^{-7}$ M also, so we use these nonstandard values to predict electrode products.

**Electrolysis of aqueous ionic solutions and the phenomenon of overvoltage.** Aqueous salt solutions are mixtures of ions *and* water, so we have to compare the various electrode potentials to predict the electrode

| Oxidation half-reaction |
| --- |
| $2H_2O(l) \longrightarrow O_2(g) + 4H^+(aq) + 4e^-$ |

| Reduction half-reaction |
| --- |
| $2H_2O(l) + 2e^- \longrightarrow H_2(g) + 2OH^-(aq)$ |

| Overall (cell) reaction |
| --- |
| $2H_2O(l) \longrightarrow 2H_2(g) + O_2(g)$ |

**FIGURE 20.19**
**The electrolysis of water.** Oxygen forms at the anode *(right)* and twice its volume of hydrogen forms at the cathode *(left)*.

products: *when two half-reactions are possible at an electrode, the one with the more positive (or less negative) electrode potential occurs.*

What happens, for instance, when a solution of KI is electrolyzed? The possible oxidizing agents are $K^+$ and $H_2O$:

$$K^+(aq) + e^- \longrightarrow K(s) \qquad\qquad E^0 = -2.93 \text{ V}$$

$$2H_2O(l) + 2e^- \longrightarrow H_2(g) + 2OH^-(aq) \qquad E = -0.42 \text{ V}$$

The less negative electrode potential for water means it is much easier to reduce than $K^+$, so $H_2$ forms at the cathode.

The possible reducing agents are $I^-$ and $H_2O$:

$$2I^-(aq) \longrightarrow I_2(s) + 2e^- \qquad\qquad -E^0 = -0.53 \text{ V}$$

$$2H_2O(l) \longrightarrow O_2(g) + 4H^+(aq) + 4e^- \qquad -E = -0.82 \text{ V}$$

Since $I^-$ is less difficult to oxidize than $H_2O$ (less negative electrode potential), $I_2$ forms at the anode.

Due to the phenomenon of **overvoltage,** however, the products predicted from this type of comparison of electrode potentials are not always the actual products. For gases such as $H_2(g)$ and $O_2(g)$ to be produced at metal electrodes an additional voltage (overvoltage) of about 0.4 to 0.6 V *more* than the electrode potential is required. The overvoltage results from kinetic factors, such as the large activation energy (Section 15.4) required for formation of gases at the electrode.

Overvoltage has major practical significance. As just one example, the industrial production of chlorine from concentrated NaCl solution is made possible by overvoltage. Water is easier to reduce than $Na^+$, so $H_2$ forms at the cathode even with an overvoltage of 0.6 V:

$$Na^+(aq) + e^- \longrightarrow Na(s) \qquad\qquad E^0 = -2.71 \text{ V}$$

$$2H_2O(l) + 2e^- \longrightarrow H_2(g) + 2OH^-(aq) \qquad E = -0.42 \text{ V } (\approx -1 \text{ V with overvoltage})$$

But $Cl_2$ forms at the anode, even though the electrode potentials suggest $O_2$ should form:

$$2H_2O(l) \longrightarrow O_2(g) + 4H^+(aq) + 4e^- \qquad -E = -0.82 \text{ V } (\approx 1.4 \text{ V with overvoltage})$$

$$2Cl^-(aq) \longrightarrow Cl_2(g) + 2e^- \qquad\qquad -E^0 = -1.36 \text{ V}$$

An overvoltage of 0.6 V makes the potential needed to form $O_2$ slightly *higher* than that for $Cl_2$. Keeping the $[Cl^-]$ high also favors $Cl_2$ formation. Thus, chlorine, one of the 10 most heavily produced industrial chemicals, can be formed from plentiful natural sources of aqueous sodium chloride. (We discuss this process in detail in Chapter 23.)

From these and other examples, we can draw some useful conclusions about which elements can be prepared electrolytically from aqueous solutions of their salts:

1. Cations of less active metals *are* reduced to the metal, including gold, silver, copper, chromium, platinum, and cadmium.
2. Cations of very active metals *are not* reduced, including those in Group 1A(1), 2A(2), and Al. Water is reduced to $H_2$ and $OH^-$ instead.
3. Anions that *are* oxidized, due to overvoltage from $O_2$ formation, include the halides ($[Cl^-]$ must be high), except for $F^-$.
4. Anions that *are not* oxidized include $F^-$ and common oxoanions, such as $SO_4^{2-}$, $CO_3^{2-}$, $NO_3^-$, and $PO_4^{3-}$, because the oxoanion nonmetal is already in its highest oxidation state. Water is oxidized to $O_2$ and $H^+$ instead.

SAMPLE PROBLEM 20.9

## Predicting the Electrolysis Products of Aqueous Ionic Solutions

**Problem:** What products form in the electrolysis of aqueous solutions of the following salts?

**(a)** KBr                **(b)** $AgNO_3$                **(c)** $MgSO_4$

**Plan:** We identify the reacting ions and compare their electrode potentials with those of water, taking the 0.4 to 0.6 V overvoltage into consideration. Whichever half-reaction has the higher electrode potential will occur at that electrode.

**Solution:**

**(a)**

$$K^+(aq) + e^- \longrightarrow K(s) \qquad\qquad E^0 = -2.93 \text{ V}$$
$$2H_2O(l) + 2e^- \longrightarrow H_2(g) + 2OH^-(aq) \qquad E = -0.42 \text{ V}$$

Despite the overvoltage, which makes $E$ for reduction of water between $-0.8$ and $-1.0$ V, $H_2O$ is still easier to reduce than $K^+$, so $H_2(g)$ **forms at the cathode.**

$$2Br^-(aq) \longrightarrow Br_2(l) + 2e^- \qquad\qquad -E^0 = -1.07 \text{ V}$$
$$2H_2O(l) \longrightarrow O_2(g) + 4H^+(aq) + 4e^- \qquad -E = -0.82 \text{ V}$$

Due to the overvoltage, which makes $E$ for oxidation of water between $-1.2$ and $-1.4$ V, $Br^-$ is easier to oxidize than water, so $Br_2(l)$ **forms at the anode** (see photo).

**(b)**

$$Ag^+(aq) + e^- \longrightarrow Ag(s) \qquad\qquad E^0 = 0.80 \text{ V}$$
$$2H_2O(l) + 2e^- \longrightarrow H_2(g) + 2OH^-(aq) \qquad E = -0.42 \text{ V}$$

As the cation of an inactive metal, $Ag^+$ is a better oxidizing agent than $H_2O$, so **Ag forms at the cathode.** $NO_3^-$ cannot be oxidized, because N is already in its highest ($+5$) oxidation state, so $O_2$ **forms at the anode:**

$$2H_2O(l) \longrightarrow O_2(g) + 4H^+(aq) + 4e^-$$

**(c)**

$$Mg^{2+}(aq) + 2e^- \longrightarrow Mg(s) \qquad E^0 = -2.37 \text{ V}$$

Like $K^+$ in part (a), $Mg^{2+}$ cannot be reduced in the presence of water, so $H_2$ **forms at the cathode.** The $SO_4^{2-}$ ion cannot be oxidized because S is in the $+6$ oxidation state, so $H_2O$ is oxidized and $O_2$ **forms at the anode.**

FOLLOW-UP PROBLEM 20.9
Write half-reactions for the products you predict will form in the electrolysis of aqueous $AuBr_3$.

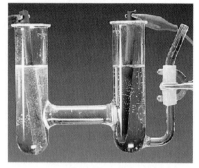

Electrolysis of KBr.

◆ **The Father of Electrochemistry and Much More.** Whereas the investigation of Michael Faraday (1791-1867) into the mass of an element that is equivalent to a given amount of charge established the science of electrochemistry, his breakthroughs in physics are even more celebrated. On Christmas Day of 1821, Faraday showed that a wire through which current flowed would move around a magnet — the precursor of the electric motor. His studies of induced currents by electric and magnetic fields eventually led to the development of the electric generator and the transformer. By the end of his career, this self-educated blacksmith's son was rated by Albert Einstein as the peer of Newton, Galileo, and Maxwell. He is shown here delivering one of his famous lectures on the "Chemical History of a Candle."

## The Stoichiometry of Electrolysis: The Relation Between Amounts of Charge and Product

As the previous discussion shows, the charge cycling through an electrolytic cell yields products at the electrodes. In the electrolysis of molten NaCl, for example, the power source supplies electrons to the cathode, where $Na^+$ ions migrate to pick them up and become Na metal. At the same time, it pulls the electrons from the anode that $Cl^-$ ions release as they become $Cl_2$ gas. It follows that the more electrons picked up by $Na^+$ ions and released by $Cl^-$ ions, the greater the masses of Na and $Cl_2$ that form. This relationship was first determined experimentally by Michael Faraday and is referred to as *Faraday's law of electrolysis: the amount of substance produced at each electrode is directly proportional to the amount of electric charge flowing through the cell.* ◆

Each balanced half-reaction shows the moles of reactant, electrons, and product involved in the change, so it contains the information we need to answer such questions as "How much material will form due to a given amount of charge?" or, conversely, "How much charge is needed to produce a given amount of material?"

To apply Faraday's law, we balance the half-reaction to find the moles of electrons needed per mole of product. Then, we use the faraday constant $(1 \ \mathscr{F} = 9.65 \times 10^4 \ C/mol \ e^-)$ to find the corresponding charge (in C) and the molar mass to find the charge needed for a given mass of product. In practice, we need some means of finding the charge flowing through the cell in order to supply the correct amount of electricity. We cannot measure charge directly, but we *can* measure current, the charge flowing per unit time. The SI unit of current is the **ampere (A),** which is defined as one coulomb flowing through a conductor in one second (s):

$$1 \ ampere = 1 \ coulomb/second \quad or \quad 1 \ A = 1 \ C/s \quad \textbf{(20.11)}$$

Thus, the current multiplied by the time gives the charge:

$$Current \times time = charge \quad or \quad A \times s = \frac{C}{s} \times s = C$$

Therefore, we know the charge by measuring the current *and* the time during which the current flows. This, in turn, relates to the moles and mass of product formed. Figure 20.20 summarizes these relationships.

Problems based on Faraday's law often ask you to calculate current, mass of material, or time. The electrode half-reaction provides the key because it is related to the mass for a certain amount of charge. Here is a typical problem in practical electrolysis: How many coulombs are needed to produce 3.0 g of $Cl_2(g)$ from the electrolysis of aqueous NaCl? From the half-reaction, we relate moles of electrons transferred to coulombs:

$$2Cl^-(aq) \rightarrow Cl_2(g) + 2e^-$$

The loss of two moles of electrons produces one mole of chlorine gas. Therefore, the change needed is

$$Charge \ (C) = 3.0 \ g \ Cl_2 \times \frac{1 \ mol \ Cl_2}{70.90 \ g \ Cl_2} \times \frac{2 \ mol \ e^-}{1 \ mol \ Cl_2} \times \frac{9.65 \times 10^4 \ C}{1 \ mol \ e^-}$$

$$= 8.2 \times 10^3 \ C$$

Let's carry the problem one step further: How long does it take to form this amount of $Cl_2$ if the power source supplies a current of 12 A?

$$Time \ (s) = \frac{charge \ (C)}{current \ (A, \ or \ C/s)} = 8.2 \times 10^3 \ C \times \frac{1 \ s}{12 \ C} = 6.8 \times 10^2 \ s \ (\sim 11 \ min)$$

Sample Problem 20.10 provides another variation of the approach.

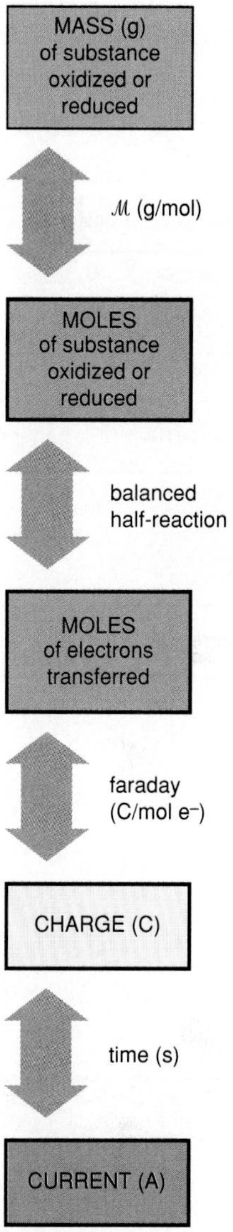

Boxes (top to bottom):

MASS (g) of substance oxidized or reduced

↕ $\mathscr{M}$ (g/mol)

MOLES of substance oxidized or reduced

↕ balanced half-reaction

MOLES of electrons transferred

↕ faraday (C/mol e⁻)

CHARGE (C)

↕ time (s)

CURRENT (A)

**FIGURE 20.20**
A summary diagram for the stoichiometry of electrolysis.

SAMPLE PROBLEM 20.10 _____

**Applying the Relationship Among Current, Time, and Amount of Substance**

**Problem:** A technician needs to plate a bathroom fixture with 0.86 g of chromium from an electrolytic bath containing aqueous $Cr_2(SO_4)_3$. If 12.5 min are allowed for the plating, what current is needed?

Mass (g) of Cr

3 mol e⁻ = 52 g Cr

Moles of e⁻ transferred

1 mol e⁻ =
$9.65 \times 10^4$ C

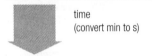

Charge (C)

time
(convert min to s)

Current (A)

**Plan:** To find the current, we divide the charge by the time, so we need the charge. First we write the half-reaction for $Cr^{3+}$ reduction. From it, we know the moles of $e^-$ required per mole of Cr, or per molar mass of Cr. As the roadmap shows, to find the charge, we convert the mass of Cr needed to moles of $e^-$ and then to charge with the faraday constant. Then we divide by the time (in s) to obtain the current.

**Solution:** Writing the half-reaction:

$$Cr^{3+}(aq) + 3e^- \longrightarrow Cr(s)$$

Determining the moles of $e^-$ transferred per molar mass of Cr plated: From the half-reaction, we have

$$\frac{3 \text{ mol } e^-}{1 \text{ mol Cr}} \times \frac{1 \text{ mol Cr}}{52.00 \text{ g Cr}} = \frac{3 \text{ mol } e^-}{52.00 \text{ g Cr}}$$

Calculating the charge:

$$\text{Charge (C)} = 0.86 \text{ g Cr} \times \frac{3 \text{ mol } e^-}{52.00 \text{ g Cr}} \times \frac{9.65 \times 10^4 \text{ C}}{1 \text{ mol } e^-} = 4.8 \times 10^3 \text{ C}$$

Calculating the current:

$$\text{Current (A)} = \frac{\text{charge (C)}}{\text{time (s)}} = \frac{4.8 \times 10^3 \text{ C}}{12.5 \text{ min}} \times \frac{1 \text{ min}}{60 \text{ s}} = 6.4 \text{ C/s} = \mathbf{6.4 \text{ A}}$$

**Check:** Rounding gives $(\sim 0.9 \text{ g})(3 \text{ mol } e^-/50 \text{ g})(10^5 \text{ C/mol } e^-) = 5 \times 10^3$ C and $(\sim 5 \times 10^3 \text{ C}/12 \text{ min})(1 \text{ min}/60 \text{ s}) = 7$ A.

**Comment:** For the sake of introducing Faraday's law, the details of the electroplating process have been simplified here. Actually, electroplating chromium is only 30% to 40% efficient and must be run at a particular temperature range for the plate to appear bright. Nearly 10,000 metric tons $(2 \times 10^8 \text{ mol})$ of chromium are used annually for electroplating.

**FOLLOW-UP PROBLEM 20.10**
Using a current of 4.75 A, how many minutes does it take to plate 1.50 g Cu onto a sculpture from a $CuSO_4$ solution?

The following Chemical Connections essay links several themes of the chapter in the setting of a living cell.

**Section Summary**
Electrolytic cells utilize electrical energy to drive a nonspontaneous reaction. Oxidation occurs at the anode and reduction at the cathode, but the charges of the electrodes are opposite those in voltaic cells. When two products can form at each electrode, the more easily oxidized substance reacts at the anode and the more easily reduced at the cathode; that is, the half-reaction with the less negative (more positive) $E^0$ occurs. The reduction or oxidation of water takes place at nonstandard conditions. Overvoltage causes an actual voltage to be unexpectedly high and can affect the electrode product that forms. The amount of product that forms depends on the amount of charge flowing through the cell, which is related to the amount of current and the time it flows. Cellular redox systems combine aspects of voltaic, concentration, and electrolytic cells to convert bond energy in food into electrochemical potential and then into the bond energy of ATP.

## Chemistry in Biological Energetics: Cellular Electrochemistry and the Production of ATP

In nearly every living cell, the principles of electrochemical cells are applied to generate energy. Bond energy in food is used to generate an electrochemical potential, which is then used to create the bond energy of the high-energy molecule adenosine triphosphate (ATP; see Chemical Connections, Section 19.2). The redox species that accomplish these steps are part of the *electron-transport chain* (ETC), which lies on the inner membranes of *mitochondria,* subcellular structures that produce the cell's energy (Figure 20.A). The ETC is a series of large molecules (mostly proteins), each of which contains a redox couple, such as $Fe^{2+}/Fe^{3+}$, that passes electrons down the chain. At three points along the chain, large potential differences are used to convert adenosine diphosphate (ADP) into ATP.

### Bond Energy to Electrochemical Potential

Cells utilize the energy in food by releasing it in controlled steps rather than all at once. The reaction that ultimately powers the ETC is the oxidation of hydrogen to form water: $H_2 + \frac{1}{2}O_2 \rightarrow H_2O$. Instead of $H_2$ gas, which does not occur in organisms, the hydrogen takes the form of two $H^+$ and two $e^-$. A biological oxidizing agent called $NAD^+$ (*n*icotinamide *a*denine *d*inucleotide) acquires them in the process of oxidizing molecules in food. The half-reaction is

$$NAD^+(aq) + 2H^+(aq) + 2e^- \rightarrow NADH(aq) + H^+(aq)$$

At the mitochondrion's inner membrane, the NADH and $H^+$ transfer the two electrons to the first redox couple of the ETC and release the two protons. The electrons are transported down the chain of redox couples, where they finally reduce $O_2$ to $H_2O$. The overall process, with standard electrode potentials,* is

$$NADH(aq) + H^+(aq) \rightarrow NAD^+(aq) + 2H^+(aq) + 2e^-$$
$$-E^{0'} = 0.32 \text{ V}$$
$$\frac{1}{2}O_2(aq) + 2e^- + 2H^+(aq) \rightarrow H_2O(l) \qquad E^{0'} = 0.82 \text{ V}$$
$$\overline{NADH(aq) + H^+(aq) + \frac{1}{2}O_2(aq) \rightarrow NAD^+(aq) + H_2O(l)}$$
$$E^{0'}_{overall} = 1.14 \text{ V}$$

Thus, for each mole of NADH that enters the ETC, the free-energy equivalent of 1.14 V is available:

$$\Delta G^{0'} = -n\mathscr{F}E^{0'}$$
$$= -2 \text{ mol } e^- \times 96.5 \text{ kJ/V} \cdot \text{mol } e^- \times 1.14 \text{ V}$$
$$= -220 \text{ kJ/mol NADH}$$

This aspect of the process functions like a voltaic cell: a spontaneous reaction, the reduction of $O_2$ to $H_2O$, is used to generate a potential. In contrast to a laboratory voltaic cell, in which the overall process occurs in one step, this

---

*In biological systems, standard potentials are designated $E^{0'}$ and the standard states include a pH of 7.0 ($[H^+] = 1 \times 10^{-7}$ M).

**FIGURE 20.A**

**The mitochondrion.** Mitochondria are subcellular particles enclosed by an outer membrane that contains a highly folded inner membrane to which the components of the electron-transport chain are attached.

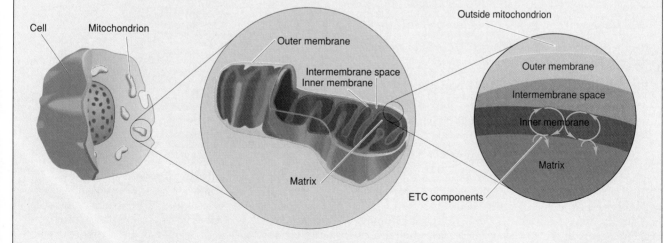

*Continued*

**CHEMICAL CONNECTIONS**

**Cellular Electrochemistry and the Production of ATP—cont'd**

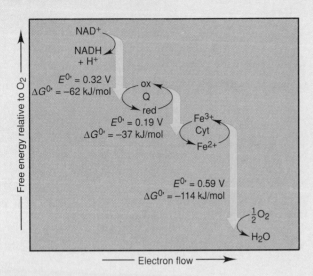

**FIGURE 20.B**

**The main components of the electron-transport chain (ETC).** The electron carriers in the ETC undergo oxidation and reduction as they pass electrons to one another along the chain. At the three points shown, the difference in potential $E^{0'}$ (or free energy, $\Delta G^{0'}$) is large enough to be used for ATP production. (Q is a large organic molecule, and Cyt is the abbreviation for cytochromes, proteins that contain a metal ion redox couple, such $Fe^{2+}/Fe^{3+}$, as the electron carrier.)

process occurs in many small steps. Figure 20.B is a greatly simplified diagram of the three key steps in the ETC that generate the high potential to produce ATP. (In the cell, each of these steps has several components, with a total molecular mass of about $1.5 \times 10^6$ amu.) Electrons are passed from one redox couple to the next down the chain, such that the reduced form of the first couple reduces the oxidized form of the second, and so forth. Most of the ETC components are iron-containing proteins, and the redox change involves the oxidation of $Fe^{2+}$ to $Fe^{3+}$ in one component by the reduction of $Fe^{3+}$ to $Fe^{2+}$ in another:

$$Fe^{2+} \text{ (in A)} + Fe^{3+} \text{ (in B)} \rightarrow Fe^{3+} \text{ (in A)} + Fe^{2+} \text{ (in B)}$$

In other words, *metal ions within the ETC proteins are the actual species undergoing the redox reaction.*

### Electrochemical Potential to Bond Energy

At the three points shown in Figure 20.B, the large potential difference is used to drive the nonspontaneous formation of ATP:

$$ADP^{3-}(aq) + HPO_4^{2-}(aq) + H^+(aq) \rightarrow$$
$$ATP^{4-}(aq) + H_2O(l) \qquad \Delta G^{0'} = +30.5 \text{ kJ/mol}$$

Note that the $\Delta G^{0'}$ values at the three ATP-producing points exceed the free energy needed.

So far, we've followed the flow of electrons through the members of the ETC, but where have the released protons gone? The answer is the key to *how* electrochemical potential is converted to bond energy in ATP. As the electrons flow and the redox couples change oxida-

tion state, free energy released at the three key steps is used to force $H^+$ ions into the intermembrane space, so that the [$H^+$] of the intermembrane space soon becomes higher than that of the matrix (Figure 20.C). In principle, this aspect of the process functions like an electrolytic cell, as the free energy supplied by the three steps *creates an $H^+$ concentration cell across the membrane*. When the difference in [$H^+$] is about 2.5-fold, it triggers the membrane to let $H^+$ ions flow back spontaneously through it (in effect, closing the switch and allowing the concentration cell to operate). The free energy released in this spontaneous process drives the nonspontaneous ATP formation. Thus, the mitochondrion uses the "electro-motive force" of redox couples on the membrane to generate a "proton-motive force" across the membrane, which converts a potential difference to bond energy.

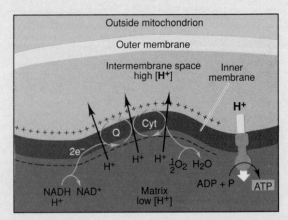

**FIGURE 20.C**

**Coupling electron transport to proton transport to ATP synthesis.** The purpose of the ETC is to convert the free energy released from food molecules into the stored free energy of ATP. It accomplishes this by transporting electrons down the chain *(colored line)*, while protons are pumped out of the mitochondrial matrix. This pumping creates an [$H^+$] difference and generates a potential across the inner membrane (concentration cell). When this potential reaches a "trigger" value, $H^+$ flows back into the inner space, and the free energy released drives the formation of ATP.

## Chapter Perspective

*The field of electrochemistry is one of the many areas in which the principles of thermodynamics lead to practical benefit. As you've seen, electrochemical cells can use a reaction to generate energy or use energy to drive a reaction. Such processes are central not only to our mobile, portable way of life, but also to our biological existence. In Chapter 23, we examine the electrochemical methods used by industry to convert raw natural resources into some of the materials modern society finds indispensable.*

# For Review and Reference

## Key Terms

electrochemistry
electrochemical cells

**SECTION 20.1**
voltaic (galvanic) cell
electrolytic cell
electrode
electrolyte
anode
cathode

**SECTION 20.2**
half-cell
salt bridge
voltage
cell potential ($E_{cell}$)
electromotive force (emf)
volt (V)
coulomb (C)
standard cell potential
($E^0_{cell}$)

standard electrode (half-cell) potential ($E^0_{half\text{-}cell}$)
standard reference half-cell (standard hydrogen electrode)

**SECTION 20.3**
faraday ($\mathscr{F}$)
Nernst equation
concentration cell

**SECTION 20.4**
battery
fuel cell

**SECTION 20.5**
corrosion

**SECTION 20.6**
electrolysis
overvoltage
ampere (A)

## Key Equations and Relationships

**20.1** Relating a spontaneous process to the sign of the cell potential (p. 886):

$$E_{cell} > 0 \text{ for a spontaneous process}$$

**20.2** Relating electric potential to energy and charge in SI units (p. 887):

$$\text{Potential} = \text{energy/charge} \quad \text{or} \quad 1 \text{ V} = 1 \text{ J/C}$$

**20.3** Relating standard cell potential to standard electrode potentials in a voltaic cell (p. 888):

$$E^0_{cell} = E^0_{cathode} - E^0_{anode} = E^0_{right} - E^0_{left}$$

**20.4** Defining the faraday constant (p. 897):

$$1 \, \mathscr{F} = 9.65 \times 10^4 \text{ J/V} \cdot \text{mol } e^- \text{ (3 significant figures)}$$

**20.5** Relating the free energy change to electrical work and cell potential (p. 897):

$$\Delta G = w_{max} = -n\mathscr{F}E_{cell}$$

**20.6** Finding the standard free energy change from the standard cell potential (p. 897):

$$\Delta G^0 = -n\mathscr{F}E^0_{cell}$$

**20.7** Finding the equilibrium constant from the standard cell potential (p. 897):

$$E^0_{cell} = \frac{RT}{n\mathscr{F}} \ln K$$

**20.8** Substituting known quantities of $R$, $\mathscr{F}$, and $T$ into Eqn 20.7 and converting to common logarithms (p. 898):

$$E^0_{cell} = \frac{0.0592 \text{ V}}{n} \log K \text{ and } \log K = \frac{nE^0_{cell}}{0.0592 \text{ V}} \quad \text{(at 25°C)}$$

**20.9** Calculating the nonstandard cell potential (Nernst equation) (p. 899):

$$E_{cell} = E^0_{cell} - \frac{RT}{n\mathscr{F}} \ln Q$$

**20.10** Substituting known quantities of $R$, $\mathscr{F}$, and $T$ into the Nernst equation and converting logarithms (p. 899):

$$E_{cell} = E^0_{cell} - \frac{0.0592 \text{ V}}{n} \log Q \quad \text{(at 25°C)}$$

**20.11** Relating current to charge and time (p. 919):

$$\text{Current} = \text{charge/time} \quad \text{or} \quad 1 \text{ A} = 1 \text{ C/s}$$

## Answers to Follow-up Problems

**20.1** $6KMnO_4(aq) + 6KOH(aq) + KI(aq) \rightarrow$
$$6K_2MnO_4(aq) + KIO_3(aq) + 3H_2O(l)$$

**20.2** $Sn(s) \rightarrow Sn^{2+}(aq) + 2e^-$    [anode; oxidation]
$6e^- + 14H^+(aq) + Cr_2O_7^{2-}(aq) \rightarrow$
$$2Cr^{3+}(aq) + 7H_2O(l)$$    [cathode; reduction]
$$\overline{3Sn(s) + Cr_2O_7^{2-}(aq) + 14H^+(aq) \rightarrow}$$
$$3Sn^{2+}(aq) + 2Cr^{3+}(aq) + 7H_2O(l)\ \text{[overall]}$$

Cell notation:
$Sn(s)\,|\,Sn^{2+}(aq)\,\|\,H^+(aq),\,Cr_2O_7^{2-}(aq),\,Cr^{3+}(aq)\,|\,graphite$

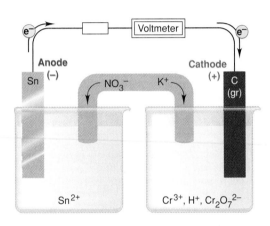

**20.3** $Br_2(aq) + 2e^- \rightarrow 2Br^-(aq)\ E^0_{bromine} = 1.07\ V$
[cathode]
$2V^{3+}(aq) + 2H_2O(l) \rightarrow 2VO^{2+}(aq) + 4H^+(aq) + 2e^-$
$-E^0_{vanadium} = ?$    [anode]
$-E^0_{vanadium} = E^0_{cell} - E^0_{bromine}$
$= 1.39\ V - 1.07\ V = 0.32\ V$
$E^0_{vanadium} = -0.32\ V$

**20.4** $Fe^{2+}(aq) + 2e^- \rightarrow Fe(s)$        $E^0 = -0.44\ V$
$2[Fe^{2+}(aq) \rightarrow Fe^{3+}(aq) + e^-]$    $-E^0 = -0.77\ V$
$$\overline{3Fe^{2+}(aq) \rightarrow 2Fe^{3+}(aq) + Fe(s)\quad E^0_{cell} = -1.21\ V}$$
The reaction is nonspontaneous. The spontaneous reaction is
$2Fe^{3+}(aq) + Fe(s) \rightarrow 3Fe^{2+}(aq);\ E^0_{cell} = +1.21\ V$
$Fe > Fe^{2+} > Fe^{3+}$

**20.5** $\Delta G^0 = -RT \ln K = -8.31\ J/mol \cdot K \times 298\ K \times \ln K$
$= -143\ kJ;\ K = 1.20 \times 10^{25}$

$$E^0_{cell} = \frac{0.0592\ V}{2} \log(1.20 \times 10^{25}) = 0.742\ V$$

**20.6**                $Fe(s) \rightarrow Fe^{2+}(aq) + 2e^-$    $-E^0 = 0.44\ V$
$Cu^{2+}(aq) + 2e^- \rightarrow Cu(s)$    $E^0 = 0.34\ V$
$$\overline{Fe(s) + Cu^{2+}(aq) \rightarrow Fe^{2+}(aq) + Cu(s)\quad E^0_{cell} = 0.78\ V}$$
so $E_{cell} = 0.78\ V + 0.25\ V = 1.03\ V$

$$1.03\ V = 0.78\ V - \frac{0.0592\ V}{2} \log \frac{[Fe^{2+}]}{[Cu^{2+}]}$$

$$\frac{[Fe^{2+}]}{[Cu^{2+}]} = 3.6 \times 10^{-9}$$

$[Fe^{2+}] = 3.6 \times 10^{-9} \times 0.30\ M = 1.1 \times 10^{-9}\ M$

**20.7** $Au^{3+}(aq,\ 2.5 \times 10^{-2}\ M)\ [B] \rightarrow$
$$Au^{3+}(aq,\ 7.0 \times 10^{-4}\ M)\ [A]$$

$$E_{cell} = 0\ V - \left(\frac{0.0592\ V}{3} \times \log \frac{7.0 \times 10^{-4}}{2.5 \times 10^{-2}}\right) = 0.031\ V$$

A is negative, so it is the anode.

**20.8** Oxidizing agents: $K^+$ and $Al^{3+}$; reducing agents: $F^-$ and $Br^-$
Al is above and to the right of K, so it has a higher IE:
$Al^{3+}(l) + 3e^- \rightarrow Al(s)$        [cathode; reduction]
Br is below F, so it has a lower EN:
$2Br^-(l) \rightarrow Br_2(g) + 2e^-$        [anode; oxidation]
$2Al^{3+}(l) + 6Br^-(l) \rightarrow 2Al(s) + 3Br_2(g)$    [overall]

**20.9** $Au^{3+}(aq) + 3e^- \rightarrow Au(s);\ E^0 = 1.50\ V$
[cathode; reduction]
Due to overvoltage, $O_2$ will not form at the anode, so $Br_2$ will form:
$2Br^-(aq) \rightarrow Br_2(l) + 2e^-;\ -E^0 = -1.07\ V$
[anode; oxidation]

**20.10** $Cu^{2+}(aq) + 2e^- \rightarrow Cu(s)$; therefore,
2 mol $e^-$ / 1 mol Cu = 2 mol $e^-$ / 63.55 g Cu

$$\text{Time (min)} = 1.50\ g\ Cu \times \frac{2\ mol\ e^-}{63.55\ g\ Cu}$$

$$\times \frac{9.65 \times 10^4\ C}{1\ mol\ e^-} \times \frac{1\ s}{4.75\ C} \times \frac{1\ min}{60\ s}$$

$$= 16.0\ min$$

## Sample Problem Titles

**20.1** Balancing Redox Reactions by the Half-Reaction Method  (p. 879)
**20.2** Diagramming Voltaic Cells  (p. 885)
**20.3** Calculating an Unknown $E^0_{half-cell}$ from $E^0_{cell}$  (p. 889)
**20.4** Writing Spontaneous Redox Reactions and Ranking the Strengths of Oxidizing and Reducing Agents  (p. 893)
**20.5** Calculating $K$ and $\Delta G^0$ from $E^0_{cell}$  (p. 898)

**20.6** Using the Nernst Equation to Calculate $E_{cell}$  (p. 900)
**20.7** Calculating the Potential of a Concentration Cell  (p. 903)
**20.8** Predicting the Electrolysis Products of a Molten Salt Mixture  (p. 915)
**20.9** Predicting the Electrolysis Products of Aqueous Ionic Solutions  (p. 918)
**20.10** Applying the Relationship Among Current, Time, and Amount of Substance  (p. 919)

# Problems

Problems with a green number are answered in the back of the text. Most sections include three categories of problems separated by a green rule—concept review questions, *paired* skill building exercises, and problems in a relevant context.

## Half-Reactions and Electrochemical Cells

(Sample Problem 20.1)

**20.1** Define oxidation and reduction in terms of electron transfer and change in oxidation number.

**20.2** Can one half-reaction in a redox process exist independently from the other? Explain.

**20.3** Why must an electrochemical process involve a redox reaction?

**20.4** Water is used to balance O atoms in the half-reaction method. Why can't $O^{2-}$ ions be used instead?

**20.5** Are spectator ions used to balance the half-reactions of a redox reaction? At what stage might spectator ions enter the process?

**20.6** How are protons removed when balancing a redox reaction in basic solution?

**20.7** During the balancing process, what step is taken to insure that $e^-$ loss equals $e^-$ gain?

**20.8** Consider the following balanced redox reaction:

$$16H^+(aq) + 2MnO_4^-(aq) + 10Cl^-(aq) \rightarrow 2Mn^{2+}(aq) + 5Cl_2(g) + 8H_2O(l)$$

(a) Which species is being oxidized?
(b) Which species is being reduced?
(c) Which species is the oxidizing agent?
(d) Which species is the reducing agent?
(e) From which species to which is electron transfer occurring?
(f) Write the balanced molecular equation, with $K^+$ and $SO_4^{2-}$ as the spectator ions.

**20.9** Which type of electrochemical cell has a $\Delta G_{sys} < 0$? Which has an increase in the free energy of the cell?

**20.10** Which statements are true? Correct any false ones.
(a) In a voltaic cell, the anode is negative relative to the cathode.
(b) Oxidation occurs at the anode of either type of cell.
(c) Electrons flow into the cathode of an electrolytic cell.
(d) In a voltaic cell, the surroundings do work on the system.
(e) If a metal is plated out of an electrolytic cell, it appears on the cathode.
(f) The electrolyte in a cell provides a solution of mobile electrons.

**20.11** Balance the following aqueous skeleton reactions and identify the oxidizing and reducing agents:
(a) $ClO_3^-(aq) + I^-(aq) \rightarrow I_2(s) + Cl^-(aq)$ [acidic]
(b) $MnO_4^-(aq) + SO_3^{2-}(aq) \rightarrow MnO_2(s) + SO_4^{2-}(aq)$ [basic]
(c) $MnO_4^-(aq) + H_2O_2(aq) \rightarrow Mn^{2+}(aq) + O_2(g)$ [acidic]

**20.12** Balance the following aqueous skeleton reactions and identify the oxidizing and reducing agents:
(a) $O_2(g) + NO(g) \rightarrow NO_3^-(aq)$ [acidic]
(b) $CrO_4^{2-}(aq) + Cu(s) \rightarrow Cr(OH)_3(s) + Cu(OH)_2(s)$ [basic]
(c) $AsO_4^{3-}(aq) + NO_2^-(aq) \rightarrow AsO_2^-(aq) + NO_3^-(aq)$ [basic]

**20.13** Balance the following aqueous skeleton reactions and identify the oxidizing and reducing agents:
(a) $Cr_2O_7^{2-}(aq) + Zn(s) \rightarrow Zn^{2+}(aq) + Cr^{3+}(aq)$ [acidic]
(b) $Fe(OH)_2(s) + MnO_4^-(aq) \rightarrow MnO_2(s) + Fe(OH)_3(s)$ [basic]
(c) $Zn(s) + NO_3^-(aq) \rightarrow Zn^{2+}(aq) + N_2(g)$ [acidic]

**20.14** Balance the following aqueous skeleton reactions and identify the oxidizing and reducing agents:
(a) $BH_4^-(aq) + ClO_3^-(aq) \rightarrow H_2BO_3^-(aq) + Cl^-(aq)$ [basic]
(b) $HCrO_4^{2-}(aq) + N_2O(g) \rightarrow Cr^{3+}(aq) + NO(g)$ [acidic]
(c) $Br_2(l) \rightarrow BrO_3^-(aq) + Br^-(aq)$ [basic]

**20.15** In many residential water systems, the aqueous $Fe^{3+}$ concentration is high enough to stain sinks and turn drinking water light brown. The analysis of iron content occurs by first reducing the $Fe^{3+}$ to $Fe^{2+}$ and then titrating with $MnO_4^-$ in acidic solution. Balance the skeleton equation of the last reaction:

$$Fe^{2+}(aq) + MnO_4^-(aq) \rightarrow Mn^{2+}(aq) + Fe^{3+}(aq)$$

**20.16** *Aqua regia*, a mixture of concentrated $HNO_3$ and HCl, was developed by the alchemists as a means to "dissolve" gold. The process is actually a redox reaction with the following simplified skeleton equation:

$$Au(s) + NO_3^-(aq) + Cl^-(aq) \rightarrow AuCl_4^-(aq) + NO_2(g)$$

(a) Balance it by the half-reaction method.
(b) What are the oxidizing and reducing agents?
(c) What is the function of the HCl?

## Voltaic Cells: Using Spontaneous Reactions to Generate Electrical Energy

(Sample Problems 20.2 to 20.4)

**20.17** Consider the following general voltaic cell:
Identify the (a) anode; (b) cathode; (c) salt bridge; (d) electrode at which $e^-$ leave the cell; (e) electrode with a positive charge; (f) electrode that gains mass as the cell operates (assuming a metal plates out).

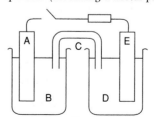

**20.18** Why does a voltaic cell not operate to any significant extent unless the two compartments are connected through an external circuit?

**20.19** What purpose does the salt bridge serve in a voltaic-cell and how does it accomplish this purpose?

**20.20** What is the difference between an active and an inactive electrode? Why are the inactive electrodes used? Name two substances commonly used for inactive electrodes.

**20.21** How does use of a standard reference electrode allow determination of unknown $E^0_{\text{half-cell}}$ values?

**20.22** What does a negative $E^0_{\text{cell}}$ indicate about a redox reaction? What does it indicate about the reverse reaction?

**20.23** How are $E^0$ values treated similarly to $\Delta H^0$, $\Delta G^0$, and $S^0$ values? How are they treated differently?

**20.24** When a piece of metal A is placed in a solution of metal B ions, metal B plates out on the piece of A.
(a) Which metal is being oxidized?
(b) Which metal is being "displaced"?
(c) Which metal would you use as the anode in a voltaic cell incorporating these two metals?
(d) If bubbles of $H_2$ form when A is placed in acid, will they form if B is placed in acid? Explain.

**20.25** Consider the following voltaic cell:

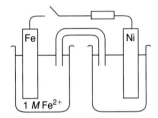

(a) In which direction do electrons flow when the switch is closed?
(b) In which half-cell does oxidation occur?
(c) In which half-cell do electrons enter the cell?
(d) At which electrode are electrons consumed?
(e) Which electrode is negatively charged?
(f) Which electrode will decrease in mass during operation of the cell?
(g) Suggest a solution for the cathode electrolyte.
(h) Suggest a pair of ions for the salt bridge.
(i) For which electrode could you use an inactive material?
(j) In which direction do cations within the salt bridge move to maintain charge neutrality?
(k) What voltage would be observed under standard conditions at 25°C?
(l) Write balanced half-reactions and an overall cell reaction.

**20.26** Consider the following voltaic cell:

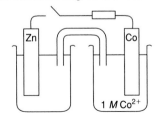

(a) In which direction do electrons flow when the switch is closed?
(b) In which half-cell does reduction occur?
(c) In which half-cell do electrons leave the cell?
(d) At which electrode are electrons generated?
(e) Which electrode is positively charged?
(f) Which electrode will increase in mass during operation of the cell?
(g) Suggest a solution for the anode electrolyte.
(h) Suggest a pair of ions for the salt bridge.
(i) For which electrode could you use an inactive material?
(j) In which net direction do cations flow within the cell?
(k) What voltage would be observed under standard conditions at 25°C?
(l) Write balanced half-reactions and an overall cell reaction.

**20.27** A voltaic cell is constructed with an Sn electrode and $Sn^{2+}$ electrolyte in one compartment and a Zn electrode and $Zn^{2+}$ electrolyte in the other compartment. Measurement shows the zinc electrode is negative.
(a) Write balanced half-reactions and the overall spontaneous reaction.
(b) Diagram the cell, labeling electrodes with their charges and showing the directions of electron and cation and anion flow.
(c) What is $E^0_{\text{cell}}$ at 25°C?

**20.28** A voltaic cell is constructed with a $Ag/Ag^+$ half-cell and a $Pb/Pb^{2+}$ half-cell. Measurement shows that the silver electrode is positive.
(a) Write balanced half-reactions and the overall spontaneous reaction.
(b) Diagram the cell, labeling electrodes with their charges and showing the directions of electron and cation and anion flow.
(c) What is $E^0_{\text{cell}}$ at 25°C?

**20.29** Write the cell notation for the voltaic cells that incorporate each of the following redox reactions:
(a) $Al(s) + Cr^{3+}(aq) \rightarrow Cr(s) + Al^{3+}(aq)$
(b) $Cu^{2+}(aq) + SO_2(g) + 2H_2O(l) \rightarrow$
$$Cu(s) + SO_4^{2-}(aq) + 4H^+(aq)$$

**20.30** Write a balanced cell reaction for each of the following cell notations:
(a) $Mn(s)|Mn^{2+}(aq)||Cd^{2+}(aq)|Cd(s)$
(b) $Fe(s)|Fe^{2+}(aq)||NO_3^-(aq)|NO(g)|Pt$

**20.31** In basic solution, selenide and sulfite ions react spontaneously:

$$2Se^{2-}(aq) + 2SO_3^{2-}(aq) + 3H_2O(l) \rightarrow$$
$$2Se(s) + 6OH^-(aq) + S_2O_3^{2-}(aq); E^0_{\text{cell}} = 0.35 \text{ V}$$

(a) Write balanced half reactions for the process.
(b) If $E^0_{\text{sulfite}}$ is $-0.57$ V, calculate $E^0_{\text{selenium}}$.

**20.32** In acidic solution, ozone and manganese(II) ion react spontaneously:

$$O_3(g) + Mn^{2+}(aq) + H_2O(l) \rightarrow$$
$$O_2(g) + MnO_2(s) + 2H^+(aq); E^0_{\text{cell}} = 0.84 \text{ V}$$

(a) Write the balanced half-reactions.
(b) Using Table 20.2 for $E^0_{ozone}$, calculate $E^0_{manganese}$.

**20.33** Use the emf series (Table 20.2) to arrange the following species
(a) In order of decreasing strength as oxidizing agents: $Fe^{3+}$, $Br_2$, $Cu^{2+}$
(b) In order of increasing strength as oxidizing agents: $Ca^{2+}$, $Cr_2O_7^{2-}$, $Ag^+$

**20.34** Use the emf series (Table 20.2) to arrange the following species
(a) In order of decreasing strength as reducing agents: $SO_2$, $PbSO_4$, $MnO_2$
(b) In order of increasing strength as reducing agents: Hg, Fe, Sn

**20.35** Balance the skeleton reactions, calculate $E^0_{cell}$, and state whether or not the reaction is spontaneous:
(a) $Co(s) + H^+(aq) \rightarrow Co^{2+}(aq) + H_2(g)$
(b) $Mn^{2+}(aq) + Br_2(l) \rightarrow MnO_4^-(aq) + Br^-(aq)$
(c) $Hg_2^{2+}(aq) \rightarrow Hg^{2+}(aq) + Hg(l)$
(d) $Cl_2(g) + Fe^{2+}(aq) \rightarrow Cl^-(aq) + Fe^{3+}(aq)$

**20.36** Balance the skeleton reactions, calculate $E^0_{cell}$, and state whether or not the reaction is spontaneous:
(a) $Ag(s) + Cu^{2+}(aq) \rightarrow Ag^+(aq) + Cu(s)$
(b) $Cd(s) + Cr_2O_7^{2-}(aq) \rightarrow Cd^{2+}(aq) + Cr^{3+}(aq)$
(c) $Ni^{2+}(aq) + Pb(s) \rightarrow Ni(s) + Pb^{2+}(aq)$
(d) $Cu^+(aq) + PbO_2(s) + SO_4^{2-}(aq) \rightarrow$ $PbSO_4(s) + Cu^{2+}(aq)$

**20.37** Use the following half-reactions to write three spontaneous reactions, calculate $E^0_{cell}$ at 25°C for each reaction, and rank the oxidizing and reducing agents:
(1) $Al^{3+}(aq) + 3e^- \rightarrow Al(s)$ $E^0 = 1.66$ V
(2) $N_2O_4(g) + 2e^- \rightarrow 2NO_2^-(aq)$ $E^0 = 0.867$ V
(3) $SO_4^{2-}(aq) + H_2O(l) + 2e^- \rightarrow$ $SO_3^{2-}(aq) + 2OH^-(aq)$ $E^0 = 0.93$ V

**20.38** Use the following half-reactions to write three spontaneous reactions, calculate $E^0_{cell}$ at 25°C for each reaction, and rank the oxidizing and reducing agents:
(1) $Au^+(aq) + e^- \rightarrow Au(s)$ $E^0 = 1.69$ V
(2) $N_2O(g) + 2H^+(aq) + 2e^- \rightarrow$ $N_2(g) + H_2O(l)$ $E^0 = 1.77$ V
(3) $Cr^{3+}(aq) + 3e^- \rightarrow Cr(s)$ $E^0 = -0.74$ V

**20.39** When metal A is placed in a solution of metal B salt, the surface of metal A changes color. When metal B is placed in acid solution, gas bubbles form on the surface of the metal. When metal A is placed in a solution of metal C, no change is observed in the solution or on the metal A surface. Would metal C cause formation of $H_2$ when placed in acid solution? Rank metals A, B, and C in order of *decreasing* reducing strength.

**20.40** Bubbles of $H_2$ form when metal D is placed in hot $H_2O$. No reaction occurs when D is placed in a solution of an E salt, but D is discolored and coated immediately when placed in a solution of an F salt. What would happen if E were placed in a solution of an F salt? Rank metals D, E, and F in order of *increasing* reducing strength.

**20.41** When a clean iron nail is placed in an aqueous solution of copper(II) sulfate, the nail immediately begins to turn a brown-black color. In a few minutes, the nail is completely coated with this color.
(a) What is the material coating on the iron?
(b) What are the oxidizing and reducing agents?
(c) Could this reaction be made into a voltaic cell?
(d) Write the balanced redox equation for the reaction
(e) Calculate $E^0_{cell}$ for the process at 25°C.

**Free Energy and Electrical Work**

(Sample Problems 20.5 to 20.7)

**20.42** (a) How do the relative magnitudes of $Q$ and $K$ relate to the sign of $\Delta G$ and $E_{cell}$? Explain.
(b) Can a cell do more work when $Q/K > 1$ or when $Q/K < 1$? Explain.

**20.43** A voltaic cell consists of a metal $A/A^+$ electrode and metal $B/B^+$ electrode, with the $A/A^+$ electrode negative. The initial $[A^+]/[B^+]$ is such that $E_{cell} > E^0_{cell}$.
(a) How do $[A^+]$ and $[B^+]$ change as the cell operates?
(b) How does $E_{cell}$ change as the cell operates?
(c) What is $[A^+]/[B^+]$ when $E_{cell} = E^0_{cell}$? Explain.
(d) Is it possible for $E_{cell}$ to be less than $E^0_{cell}$? Explain.

**20.44** Explain whether $E_{cell}$ of a voltaic cell would increase or decrease from the following changes:
(a) Decrease in cell temperature
(b) Increase in concentration of an active ion in the anode compartment
(c) Increase in concentration of an active ion in the cathode compartment
(d) Increase in pressure of a gaseous cathode half-cell component

**20.45** In a concentration cell, is the more concentrated electrolyte in the cathode or the anode compartment? Explain.

**20.46** What is the value of the equilibrium constant at 25°C for the reaction between
(a) $Ni(s)$ and $Ag^+(aq)$? (b) $Fe(s)$ and $Cr^{3+}(aq)$?

**20.47** What is the value of the equilibrium constant at 25°C for the reaction between
(a) $Al(s)$ and $Cd^{2+}(aq)$? (b) $I_2(s)$ and $Br^-(aq)$?

**20.48** Calculate $K$ and $\Delta G^0$ for each of the reactions in Problem 20.35.

**20.49** Calculate $K$ and $\Delta G^0$ for each of the reactions in Problem 20.36.

**20.50** What is $E^0_{cell}$ and $\Delta G^0$ at 25°C of a redox reaction for which (a) $n = 1$ and $K = 5.0 \times 10^3$? (b) $n = 1$ and $K = 5.0 \times 10^{-6}$?

**20.51** What is $E^0_{cell}$ and $\Delta G^0$ at 25°C of a redox reaction for which (a) $n = 2$ and $K = 75$? (b) $n = 2$ and $K = 0.075$?

**20.52** A voltaic cell consists of a standard $H_2$ electrode and a $Cu/Cu^{2+}$ electrode. Calculate $[Cu^{2+}]$ when $E_{cell}$ is 0.25 V.

**20.53** A voltaic cell consists of a $Mn/Mn^{2+}$ electrode and a $Pb/Pb^{2+}$ electrode. Calculate $[Pb^{2+}]$ when $[Mn^{2+}]$ is 1.3 $M$ and $E_{cell}$ is 0.42 V.

**20.54** A voltaic cell consists of $Ni/Ni^{2+}$ and $Co/Co^{2+}$ half-cells with the following initial concentrations: $[Ni^{2+}] = 0.80\ M$; $[Co^{2+}] = 0.20\ M$.
(a) What is the initial $E_{cell}$?
(b) What is $E_{cell}$ when $[Co^{2+}]$ reaches 0.45 $M$?
(c) What is $[Ni^{2+}]$ when $E_{cell}$ reaches 0.025 V?
(d) What are the equilibrium concentrations of the ions?

**20.55** A voltaic cell consists of $Mn/Mn^{2+}$ and $Cd/Cd^{2+}$ half-cells with the following initial concentrations: $[Mn^{2+}] = 0.090\ M$; $[Cd^{2+}] = 0.060\ M$
(a) What is the initial $E_{cell}$?
(b) What is $E_{cell}$ when $[Cd^{2+}]$ reaches 0.080 $M$?
(c) What is $[Mn^{2+}]$ when $E_{cell}$ reaches 0.055 V?
(d) What are the equilibrium concentrations of the ions?

**20.56** A concentration cell consists of two $H_2/H^+$ electrodes. Electrode A has $H_2$ at 0.90 atm bubbling into 0.10 $M$ HCl. Electrode B has $H_2$ at 0.50 atm bubbling into 2.0 $M$ HCl. Which electrode is the anode? What is the voltage of the cell?

**20.57** A concentration cell consists of two $Sn/Sn^{2+}$ electrodes. The electrolyte in compartment A is 0.13 $M$ $Sn(NO_3)_2$. The electrolyte in B is 0.87 $M$ $Sn(NO_3)_2$. Which electrode is the anode? What is the voltage of the cell?

## Electrochemical Processes in Batteries

**20.58** What is the direction of electron flow with respect to anode and cathode in a battery?

**20.59** In the everyday batteries we use for flashlights, toys, etc., no salt bridge is evident. What is used in these cells to separate the anode and cathode compartments?

**20.60** A battery operates at nearly constant voltage as long as all reactants are present. Does the Nernst equation apply to batteries? Explain.

**20.61** Many common devices require more than one battery for operation.
(a) How many alkaline batteries must be placed in series to light a flashlight with a 6.0-V bulb?
(b) What is the voltage requirement of a camera that uses six silver batteries?
(c) How many volts can a car battery deliver if two of its anode/cathode cells are shorted?

## Corrosion: A Case of Environmental Electrochemistry

**20.62** During the reconstruction of the Statue of Liberty, Teflon spacers were placed between the iron skeleton and the copper plates that cover the statue. What purpose do these spacers serve?

**20.63** Why do steel bridge-supports rust at the water line but not above or below it?

**20.64** Since the 1930s, chromium has replaced nickel for corrosion resistance and appearance on car bumpers and trim. How does the chromium protect steel from corrosion?

**20.65** Which of the following metals are suitable for use as sacrifical anodes to protect against the corrosion of underground iron pipes? If any are not suitable, explain why: (a) aluminum; (b) magnesium; (c) sodium; (d) lead; (e) nickel.

## Electrolytic Cells: Using Electrical Energy to Drive a Nonspontaneous Reaction

(Sample Problems 20.8 to 20.10)
Unless stated otherwise, assume the electrolytic cells in the following problems operate at 100% efficiency.

**20.66** Consider the following general electrolytic cell:

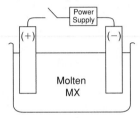

(a) At which electrode does oxidation occur?
(b) At which electrode does elemental M form?
(c) At which electrode are electrons being released by ions?
(d) At which electrode are electrons entering the cell?

**20.67** A voltaic cell consists of $Cr/Cr^{3+}$ and $Cd/Cd^{2+}$ half-cells in their standard states. After 10 minutes of operation a thin coating of cadmium metal has plated out on the cathode. Describe what would happen if you attached the negative terminal of a dry cell (1.5 V) to the cell cathode and the positive terminal to the cell anode.

**20.68** Why are $E_{half\text{-}cell}$ values for the electrolysis of water different from the $E^0_{half\text{-}cell}$ values for the oxidation and reduction of water?

**20.69** In an aqueous electrolytic cell, nitrate ions never react at the anode, but nitrite ions do. Explain.

**20.70** How does overvoltage influence the products in the electrolysis of aqueous salts?

**20.71** The commercial production of aluminum is done by the electrolysis of a bath containing $Al_2O_3$ dissolved in molten $Na_3AlF_6$. Why isn't it done by electrolysis of an aqueous $AlCl_3$ solution?

**20.72** In the electrolysis of molten NaBr,
(a) What product forms at the anode?
(b) What product forms at the cathode?

**20.73** In the electrolysis of molten $BaI_2$,
(a) What product forms at the negative electrode?
(b) What product forms at the positive electrode?

20.74 In the electrolysis of a molten mixture of KI and $MgF_2$, (a) what product forms at the anode? (b) What product forms at the cathode?

20.75 In the electrolysis of a molten mixture of CsBr and $SrCl_2$, (a) What product forms at the negative electrode? (b) What product forms at the positive electrode?

20.76 Identify those elements that can be prepared by electrolysis of their aqueous salts: copper, barium, aluminum, bromine, strontium, chlorine.

20.77 Identify those elements that can be prepared by electrolysis of their aqueous salts: lithium, iodine, zinc, silver, fluorine, cadmium.

20.78 What product would form at each electrode in the aqueous electrolysis of the following salts: (a) LiF; (b) $SnSO_4$; (c) $ZnBr_2$.

20.79 What product would form at each electrode in the aqueous electrolysis of the following salts: (a) $AgNO_3$; (b) $MnCl_2$; (c) $FeI_2$

20.80 Electrolysis of molten $MgCl_2$ is the final production step in the isolation of magnesium from sea water by the Dow process (Section 23.3). When 48.6 g Mg metal is deposited,
(a) how many faradays are required?
(b) how many coulombs are required?
(c) how many amps are required to produce this amount in 1.50 h?

20.81 Electrolysis of molten NaCl in a Downs cell is the major isolation step in the production of sodium metal (Section 23.3). When 115 g Na metal is deposited,
(a) how many faradays are required?
(b) how many coulombs are required?
(c) how many amps are required to produce this amount in 12.0 h?

20.82 How many grams of potassium can be deposited by the passage of 315 C through an electrolytic cell?

20.83 How many grams of aluminum can be deposited by the passage of 105 C through an electrolytic cell?

20.84 How long (in s) does it take to deposit 15.5 g Zn on a steel gate when 25.0 A are passed through a $ZnSO_4$ solution?

20.85 How long (in s) does it take to deposit 0.63 g Ni on a decorative drawer handle when 8.7 A are passed through a $Ni(NO_3)_2$ solution?

20.86 A professor adds $Na_2SO_4$ to facilitate the electrolysis of water in a lecture demonstration.
(a) What is the purpose of the $Na_2SO_4$?
(b) Why is the water electrolyzed instead of the salt?

20.87 Subterranean brines in parts of the United States are rich in iodides and bromides and serve as an industrial source of these elements. In one method, the brines are evaporated to dryness and then melted and electrolyzed. Which halogen is more likely to form from this treatment? Why?

20.88 Calcium is obtained industrially by electrolysis of fused $CaCl_2$ and used in aluminum alloys. How many coulombs are needed to produce 10.0 g of Ca metal? If the cell runs at 15 A, how many minutes will it take to produce the 10.0 g Ca(s)?

20.89 Zinc plating (galvanizing) is an important means of corrosion protection. Although the process is done customarily by dipping the object into molten zinc, the metal can also be electroplated from aqueous solutions. How many grams of zinc can be deposited on a steel tank from a $ZnSO_4$ solution when a 0.755-A current flows for 2.00 days?

## Comprehensive Problems
Problems with an asterisk (*) are more challenging.

20.90 Compare and contrast a voltaic cell and an electrolytic cell with respect to each of the following:
(a) Sign of the free energy change
(b) Nature of the half-reaction at the anode
(c) Nature of the half-reaction at the cathode
(d) Charge on the electrode labeled "anode"
(e) Electrode from which electrons leave the cell

20.91 A disk earring 5.00 cm in diameter is plated with gold 0.20 mm thick from an $Au^{3+}$ bath.
(a) How long would it take to deposit the gold if the current used is 0.010 A? ($d$ of gold = 19.3 g/cm³.)
(b) If the price of gold is $420 per troy ounce (31.10 g), what is the cost of the gold plating?

20.92 (a) How many minutes does it take to produce 10.0 L $O_2$ measured at 99.8 kPa and 28°C from the electrolysis of water if a current of 1.3 A passes through the electrolytic cell?
(b) What mass of $H_2$ is produced at the same time?

*20.93 Trains powered by electricity, including subways, use direct current. One conductor is the overhead wire (or "third rail" for subways) and the other is the rails upon which the wheels run. The rails are on supports in contact with the ground. In order to minimize corrosion, is the overhead wire or the rails connected to the positive terminal? Explain.

20.94 A simple laboratory test to identify the relative charge of the two leads from a DC power supply uses a piece of filter paper moistened with NaCl solution containing phenolphthalein indicator (colorless in acid, pink in base). The two wires from the power supply are touched to the paper about an inch apart. What would you expect to observe? What conclusion would you reach about the relative charges of the leads?

20.95 Electrolytic cells, as all practical apparatus, operate at less than 100% efficiency. An electrolytic cell depositing Cu from a $Cu^{2+}$ bath operates for 10 h with an average current of 5.8 A. If 53.4 g copper are deposited, at what efficiency is the cell operating?

20.96 Commercial electrolysis is performed on both molten NaCl and aqueous NaCl solutions (Chapter 23). Identify the anode product, cathode product, species reduced, and species oxidized for the (a) molten electrolysis and (b) aqueous electrolysis.

*20.97 To examine the effect of ion removal on cell voltage, a chemist constructs two voltaic cells, each with a standard hydrogen electrode in one compartment.

One cell also contains a $Pb/Pb^{2+}$ half-cell; the other also contains a $Cu/Cu^{2+}$ half-cell.
(a) What is $E^0$ of each cell at 298 K?
(b) Which electrode in each cell is negative?
(c) When $Na_2S$ solution is added to the $Pb^{2+}$ electrolyte, solid PbS forms. What happens to the cell voltage?
(d) When sufficient $Na_2S$ is added to the $Cu^{2+}$ electrolyte, CuS forms and $[Cu^{2+}]$ is lowered to $1 \times 10^{-16}$ M. What is the cell voltage?

**20.98** In the electroplating of automotive bumpers, a thin coating of copper separates the steel from a heavy coating of chromium to improve conductivity.
(a) What mass of Cu is deposited on an automobile trim piece if plating continues for 1.25 h at a current of 5.0 A?
(b) If the area of the trim piece is 50.0 cm², what is the thickness of the Cu coating? ($d$ of Cu = 8.95 g/cm³)

**20.99** Tin is used to coat "tin" cans used for food storage. If the tin is scratched and the iron of the can exposed, will the iron corrode more or less rapidly than if the tin were not present? On the inside of the can, the tin coating is itself coated with a transparent varnish. Explain.

**\*20.100** Commercial aluminum cells operate at 5 V and 100,000 A.
(a) How long does it take to produce exactly 1 metric ton (1000 kg) of aluminum?
(b) How much electrical power (in kilowatt-hours, $kW \cdot h$) is used? [One watt (W) = 1 J/s; $1 \ kW \cdot h = 3.6 \times 10^3$ kJ]
(c) Assuming electricity costs 0.90 cents per $kW \cdot h$ and cell efficiency is 90%, what is the cost of producing one pound of aluminum?

**20.101** Magnesium bars are connected electrically to underground iron pipes to serve as sacrificial anodes.
(a) Do electrons flow from the Mg to the pipe or the reverse?
(b) A 10-kg Mg bar is attached to an iron pipe. If it takes 8 yr for the Mg to be consumed, what is the average current flowing between the Mg and the Fe during this time period?

**20.102** In addition to reacting with Au, aqua regia is used to bring other precious metals into solution. Balance the skeleton equation for the reaction with Pt:

$$Pt(s) + NO_3^-(aq) + Cl^-(aq) \rightarrow PtCl_6^{2-}(aq) + NO(g)$$

**20.103** A current is applied to two electrolytic cells connected in series. In the first cell, silver is deposited. In the second cell, a zinc electrode is consumed. How much Ag is plated out if 1.2 g Zn dissolves?

**20.104** You are working in a laboratory and investigating a particular chemical reaction. State all the types of data available in standard tables that would enable you to calculate the equilibrium constant for the reaction at 298 K.

**\*20.105** The maintenance personnel at an electric-generating facility must pay careful attention to corrosion of the turbine blades. Why does iron corrode faster in steam and hot water than in cold water?

**20.106** A cell using $Cu/Cu^{2+}$ and $Sn/Sn^{2+}$ electrodes is set up in standard conditions and with each compartment having a volume of 245 mL. The cell delivers 0.15 A for 54.0 h (a) How many grams of $Cu(s)$ will have been deposited? (b) What is the $[Cu^{2+}]$ remaining?

**20.107** The $E_{half-cell}$ for the reduction of water is very different from the $E^0_{half-cell}$ value. Calculate the $E_{half-cell}$ value when $H_2$ is in its standard state.

**\*20.108** From the skeleton equations below, create a list of balanced half-reactions that has the strongest oxidizing agent on top and the weakest on the bottom:

$$U^{3+}(aq) + Cr^{3+}(aq) \rightarrow Cr^{2+}(aq) + U^{4+}(aq)$$
$$Fe(s) + Sn^{2+}(aq) \rightarrow Sn(s) + Fe^{2+}(aq)$$
$$Fe(s) + U^{4+}(aq) \rightarrow \text{no reaction}$$
$$Cr^{3+}(aq) + Fe(s) \rightarrow Cr^{2+}(aq) + Fe^{2+}(aq)$$
$$Cr^{2+}(aq) + Sn^{2+}(aq) \rightarrow Sn(s) + Cr^{3+}(aq)$$

**20.109** Use the half-reaction method to balance the reaction for the conversion of ethanol to acetic acid:

$$CH_3CH_2OH + Cr_2O_7^{2-} \rightarrow CH_3COOH + Cr^{3+} \text{ [acid solution]}$$

**\*20.110** In an experiment to determine the effect of ions on the solubility of a precipitate, a chemist designs an ion-specific probe for measuring the $[Ag^+]$ present in a NaCl solution saturated with AgCl. One half-cell is a Ag-wire electrode immersed in the unknown AgCl-saturated NaCl solution. It is connected through a salt bridge to a calomel reference electrode, a platinum wire immersed in a paste of mercury and calomel ($Hg_2Cl_2$) in a saturated KCl solution. The measured $E_{cell}$ is 0.060 V.
(a) Given the following standard half-reactions, calculate $[Ag^+]$:
Calomel: $Hg_2Cl_2(s) + 2e^- \rightarrow 2Hg(l) + 2Cl^-(aq)$
$E^0 = 0.24$ V
Silver: $Ag^+(aq) + e^- \rightarrow Ag(s)$; $E^0 = 0.80$ V
(*Hint*: Assume that $[Cl^-]$ is so high it is essentially constant.)
(b) A mining engineer sends a silver ore sample to the chemist for analysis with the $Ag^+$-selective probe. After pretreating the ore sample, the chemist measures the cell voltage as 0.57 V. What is $[Ag^+]$?

**\*20.111** Use Table 20.2 to calculate the $K_{sp}$ of AgCl.

**\*20.112** Calculate the $K_f$ of $Ag(NH_3)_2^+$ from
$$Ag^+(aq) + e^- \rightleftharpoons Ag(s) \ E^0 = 0.80 \text{ V}$$
$$Ag(NH_3)_2^+(aq) + e^- \rightleftharpoons Ag(s) + 2NH_3(aq) \ E^0 = 0.37 \text{ V}$$

**20.113** Use Table 20.2 to create an activity series of Mn, Fe, Ag, Sn, Cr, Cu, Ba, Al, Na, Hg, Ni, Li, Au, Zn, Pb. Rank them in order of decreasing reducing strength, and group them by those that can displace $H_2$ from water, those that can displace $H_2$ from acid, and those that cannot displace $H_2$.

# Nuclear Reactions and Their Applications

**Concepts and skills to review**

- discovery of the atomic nucleus (Section 2.3)
- protons, neutrons, mass number, and the $_{Z}^{A}X$ notation (Section 2.4)
- half-life and first-order reaction rate (Section 15.3)

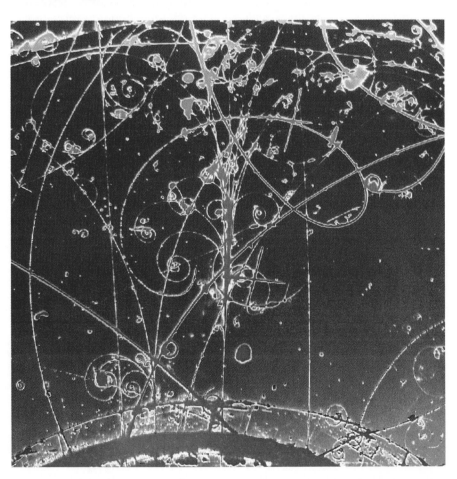

**Subatomic "fireworks."** This false-color image shows the characteristic spirals and curliques in a bubble chamber when a neutrino interacts with a proton to produce a shower of electrons and positrons. In this chapter, we meet the occupants of this alien world that lies at the core of every atom and see how they interact.

**ar** below the haze of electrons lies the atom's tiny, dense core, held together by the strongest force in the universe. For the scientists probing its structure and behavior, the atomic nucleus holds great mystery and wonder. However, society is ambivalent about the uses of nuclear research. The promise of abundant energy and the cure of disease comes hand-in-hand with the threat of nuclear waste contamination, reactor accidents, and unimaginable destruction from nuclear war. Can the power of the nucleus be harnessed for our benefit, or are the risks too great? In this chapter, we discuss the principles that can help us answer this vital question.

The changes that occur in atomic nuclei are strikingly different from chemical changes. In the reactions you've studied so far, orbital electrons are shared or transferred to form new *compounds*, while nuclei sit by passively, never changing their identities. In nuclear reactions, the roles are reversed: orbital electrons are usually bystanders as the nuclei undergo changes that, in nearly every case, form different *elements*. Nuclear reactions may be accompanied by energy changes a million times greater than those in chemical reactions, energy changes so great that changes in mass *are* detectable. Moreover, nuclear reaction yields and rates are typically *not* subject to the effects of pressure, temperature, and catalysis. Table 21.1 summarizes the differences between chemical and nuclear reactions.

We begin this brief treatment of nuclear chemistry with an investigation of why some nuclei are stable, while others are unstable, or *radioactive*. Then, we examine the detection and rate of radioactive decay. Next, we discuss how nuclei formed in particle accelerators are extending the periodic table beyond uranium, the last naturally occurring element. We consider the effects of radioactive emissions on matter, especially living matter, focusing on some major applications in science, technology, and medicine. Then, we calculate the energy released in nuclear reactions and discuss the attempts to harness this energy. Finally, we take a brief look at the nuclear processes that create chemical elements in the stars.

## 21.1 Radioactive Decay and Nuclear Stability

A stable nucleus remains intact indefinitely, but *the great majority of nuclei are unstable*. An unstable nucleus exhibits **radioactivity:** it spontaneously disintegrates, or *decays*, by emitting radiation. Each type of unstable nucleus undergoes a characteristic *rate* of radioactive decay, which ranges from a fraction of a second to several billion years. In this section, we go over important terms and notation for nuclei, discuss some key events in the discovery of radioactivity, and consider the types of radioactive decay.

### The Components of the Nucleus: Terms and Notation

Recall from Chapter 2 that the nucleus contains essentially all the atom's mass but is only about $1/10^4$ its diameter (or $1/10^{12}$ its volume). Obviously, the nucleus is incredibly dense: about $10^{14}$ g/mL. ◆ *Protons* and *neutrons*, the elementary particles that make up the nucleus, are collectively called **nucleons.** A **nuclide** is the term used for an atom with a particular nuclear composition, that is, one with specific numbers of the two types of nucleons. Most elements occur as a mixture of **isotopes,** nuclides with the characteristic number of protons of the element but different numbers of neutrons. Thus, each isotope of an element consists of a single type of nuclide.

**TABLE 21.1  Comparison of Chemical and Nuclear Reactions**

| CHEMICAL REACTIONS | NUCLEAR REACTIONS |
|---|---|
| 1. One substance is converted to another, but atoms never change identity. | 1. Atoms of one element typically change into atoms of another. |
| 2. Orbital electrons are involved as bonds break and form; nuclear particles do not take part. | 2. Protons, neutrons, and other particles are involved; orbital electrons rarely take part. |
| 3. Reactions are accompanied by relatively small changes in energy and no measurable change in mass. | 3. Reactions are accompanied by relatively large changes in energy and often measurable changes in mass. |
| 4. Reaction rates are influenced by temperature, concentration, catalysts, and the nature of the chemical substance. | 4. Reaction rates are affected by number of nuclei, but not by temperature, catalysts, or the nature of the chemical substance. |

The most abundant nuclide of oxygen, for example, has eight protons and eight neutrons in the nucleus.

The relative mass and charge of a particle—nucleon, other elementary particle, or nuclide—is described by the notation $_Z^A X$, where X is the *symbol* for the particle, $A$ is the *mass number,* or the total number of nucleons, and $Z$ is the *charge* of the particle; for nuclides, $Z$ is the same as the *number of protons* (the atomic number). Using this notation, the three subatomic elementary particles are

$$_{-1}^0 e \text{ (electron)}, \quad _1^1 p \text{ (proton)}, \text{ and } _0^1 n \text{ (neutron)}$$

(A proton is also sometimes represented as $_1^1 H$, or $_1^1 H^+$.) The number of neutrons ($N$) in a nucleus is the mass number ($A$) minus the atomic number ($Z$): $N = A - Z$. Both naturally occurring isotopes of chlorine, for example, have 17 protons ($Z = 17$), but one has 18 neutrons ($_{17}^{35}Cl$, also written $^{35}Cl$) and the other has 20 ($_{17}^{37}Cl$, or $^{37}Cl$). Nuclides are also designated with the element name followed by the mass number, for example, chlorine-35 and chlorine-37. In any *naturally occurring* sample of an element or its compounds, the isotopes of each element are present in their particular, nearly fixed proportions. Thus, in a sample of sodium chloride (or any Cl-containing substance), 75.77% of the Cl atoms are chlorine-35 and the remaining 24.23% are chlorine-37.

In this chapter, it is essential for you to be familiar with nuclear notations, so please take a moment to review Sample Problem 2.2 and Chapter Problems 2.42 to 2.45 in Chapter 2.

## The Discovery of Radioactivity

In 1896, the French physicist Antoine-Henri Becquerel discovered, quite by accident, that uranium minerals, even when wrapped in paper and stored in the dark, emit a penetrating radiation that produces bright images on a photographic plate. Becquerel also found that the radiation creates an electric discharge in air, thus providing a means for measuring its intensity. Two years later, a young doctoral student named Marie Sklodowska Curie began

◆ **Her Brilliant Career.** Marie Curie (1867-1934) is the only person to be awarded Nobel Prizes in two *different* sciences, one in physics in 1903 for her research into radioactivity and the other in chemistry in 1911 for the discovery of polonium and the discovery, isolation, and study of radium and its compounds.

ZnS-coated screen
(or photographic plate)

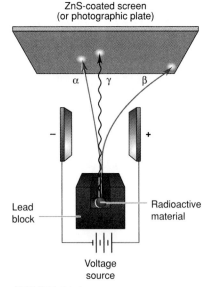

**FIGURE 21.1**

The behavior of three types of radioactive emissions in an electric field. $\alpha$ particles ($_2^4$He) are positively charged, so they bend toward the negative plate. $\beta$ particles ($_{-1}^0\beta$) are negatively charged, so they bend toward the positive plate. Note the curvature is greater for the $\beta$ particles because they have much lower mass than $\alpha$ particles. $\gamma$ rays consist of uncharged high-energy photons.

a search for other minerals that emitted radiation. She found that thorium minerals also emit radiation and showed that the intensity of the radiation is directly proportional to the *concentration* of the element in the various samples, *not* to the *nature* of the compound in which the element occurs. Curie also showed that the emissions, which she named *radioactivity*, are unaffected by temperature, pressure, or other physical and chemical conditions.

To her surprise, Curie found that certain uranium minerals were even more radioactive than pure uranium, which implied that they contained traces of one or more as yet unknown, highly radioactive elements. She and her husband, the physicist Pierre Curie, set out to isolate all the radioactive components in pitchblende, the principal ore of uranium. After months of painstaking chemical work, they isolated two extremely small, highly radioactive fractions, one that precipitated with bismuth compounds and another that precipitated with the alkaline earth compounds. Through chemical and spectroscopic analysis, Marie Curie showed that these fractions contained two new elements, which she named polonium (after her native Poland) and radium. Polonium (Po; $Z = 84$), the most metallic member of Group 6A(16), lies to the right of bismuth in Period 6. Radium (Ra; $Z = 88$) lies under barium in Group 2A(2).

Purifying radium proved to be another arduous task. Starting with several tons of pitchblende residues from which the uranium had been extracted, Curie prepared compounds of the larger Group 2A(2) elements, continually separating minuscule amounts of radium compounds from enormously larger amounts of chemically similar barium compounds. It took her 4 years to isolate 0.1 g of radium chloride, which she melted and electrolyzed to obtain pure metallic radium. ◆

During the next few years, Henri Becquerel, the Curies, and P. Villard in France and Ernest Rutherford and his co-workers in England studied the nature of radioactive emissions. When elements other than radium were observed as products of radium decay, Rutherford and Frederick Soddy proposed in 1902 that radioactive emission results in the change of one element into another. This idea, which sounded like a resurrection of alchemy, was met with disbelief and ridicule. We now know it to be true: under most circumstances, *when a nuclide of one element decays, it changes into a nuclide of a different element.*

These studies led to an understanding of the three most common types of radioactive emission. One consists of positively charged **alpha particles** (symbolized $\alpha$), which were identified as helium nuclei ($_2^4$He$^{2+}$, or just $_2^4$He). A second type consists of negatively charged **beta particles** (symbolized $\beta$, or more usually $_{-1}^0\beta$), which were identified as fast electrons. (The emission of electrons from the nucleus may seem strange, but as you'll see shortly, $\beta$ particles arise as a result of the nuclear reaction.) The third type consists of very high-energy photons called **gamma rays** (symbolized as $\gamma$, or sometimes $_0^0\gamma$). The behavior of these three emissions in an electric field is shown in Figure 21.1.

## Types of Radioactive Decay

When a nuclide decays, it forms a more stable nuclide, and the excess energy is carried off by the emitted particle. The decaying, or reactant, nuclide is called the *parent;* the product nuclide is called the *daughter.* Nuclides can

**TABLE 21.2  Modes of Radioactive Decay***

| MODE | EMISSION | DECAY PROCESS | CHANGE IN | | |
|---|---|---|---|---|---|
| | | | $A$ | $Z$ | $N$ |
| $\alpha$ Decay | $\alpha$ ($_2^4$He) | 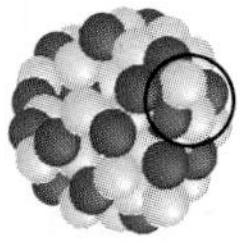 Reactant (parent) 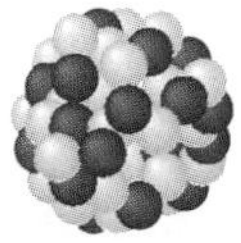     Product (daughter)      $\alpha$ expelled | $-4$ | $-2$ | $-2$ |
| $\beta$ Decay | $_{-1}^{0}\beta$ | $_0^1$n in nucleus $\longrightarrow$ $_1^1$p in nucleus $+$ $_{-1}^{0}\beta$ $\beta$ expelled | $0$ | $+1$ | $-1$ |
| Positron emission | $_1^0\beta$ | $_1^1$p in nucleus $\longrightarrow$ $_0^1$n in nucleus $+$ $_1^0\beta$ positron expelled | $0$ | $-1$ | $+1$ |
| Electron capture | x-ray photon | $_{-1}^{0}$e absorbed from low-energy orbital    $_1^1$p in nucleus $\longrightarrow$ $_0^1$n in nucleus | $0$ | $-1$ | $+1$ |
| $\gamma$ Emission | $_0^0\gamma$ | 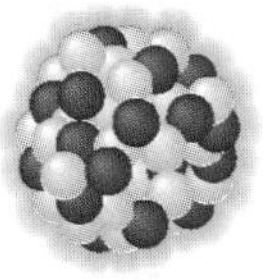 excited nucleus 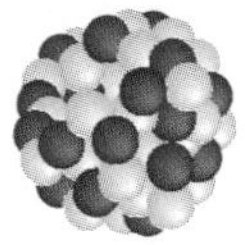 $\longrightarrow$ stable nucleus $+$ $_0^0\gamma$ $\gamma$ photon radiated | $0$ | $0$ | $0$ |

*Neutrinos ($\nu$) are involved in several of these processes but are not shown.

decay in several ways. As we discuss the major types of decay, which are summarized in Table 21.2, note carefully the principle used to balance nuclear reactions: *the total Z (charge, number of protons) and total A (sum of protons and neutrons) of the reactants equal those of the products.*

1. **Alpha decay** involves the loss of an $\alpha$ particle ($_2^4$He) from a nucleus. For each $\alpha$ particle emitted by the parent nucleus, *A decreases by 4 and Z decreases by 2*. Every element heavier than lead (Pb; $Z = 82$), as well as a few lighter ones, exhibits $\alpha$ decay. Radium undergoes $\alpha$ decay, and in Rutherford's classic gold foil experiment that established the existence of the atomic nucleus (Section 2.3), it was the source of the $\alpha$ particles used as projectiles:

$$_{88}^{226}\text{Ra} \rightarrow \,_{86}^{222}\text{Rn} + \,_2^4\text{He}$$

Note that the $A$ value for Ra equals the sum of the $A$ values for Rn and He ($226 = 222 + 4$), and that the $Z$ value for Ra equals the sum of the $Z$ values for Rn and He ($88 = 86 + 2$).

◆ **The Little Neutral One.** A neutral particle called a neutrino ($\nu$) is also emitted in many nuclear reactions, including the change of a neutron to a proton:

$$_0^1n \rightarrow {}_1^1p + {}_{-1}^0\beta + \nu$$

Theory suggests that neutrinos have masses less than $10^{-4}$ that of an electron, and that at least $10^9$ neutrinos exist in the universe for every proton. Neutrinos interact with matter so slightly that it would take a piece of lead several light-years thick to absorb them. Clearly, these particles lie outside the scope of general chemistry, so we will not mention them further.

2. **Beta decay** involves the ejection of a $\beta$ particle ($_{-1}^0\beta$) from the nucleus. This change does not involve the expulsion of a $\beta$ particle that was already present, but rather the *conversion of a neutron into a proton and a $\beta$ particle, which is expelled immediately*: ◆

$$_0^1n \rightarrow {}_1^1p + {}_{-1}^0\beta$$

As always, the totals of the $A$ and the $Z$ values in reactant and products are equal. Radioactive nickel-63 becomes stable copper-63 through $\beta$ decay:

$$_{28}^{63}Ni \rightarrow {}_{29}^{63}Cu + {}_{-1}^0\beta$$

Another example is the $\beta$ decay of carbon-14:

$$_6^{14}C \rightarrow {}_7^{14}N + {}_{-1}^0\beta$$

Note that $\beta$ *decay results in a product nuclide with the same $A$ but with $Z$ one higher (one **more** proton) than in the reactant nuclide.* In other words, an atom of the element with the next *higher* atomic number is formed.

3. **Positron decay** involves the emission of a positron from the nucleus. A key idea of modern physics is that every fundamental particle has a corresponding "antiparticle," another particle with the same mass but opposite charge. A **positron** (symbolized $_1^0\beta$) is the antiparticle of an electron. Positron decay occurs when *a proton in the nucleus is converted into a neutron and a positron,* and the positron is expelled:

$$_1^1p \rightarrow {}_0^1n + {}_1^0\beta$$

*Positron decay is the opposite of $\beta$ decay and results in a daughter nuclide with the same $A$ but with $Z$ one lower (one **fewer** proton) than the parent:* an atom of the element with the next *lower* atomic number forms. Carbon-11, an artificial isotope, decays to a stable boron isotope through emission of a positron:

$$_6^{11}C \rightarrow {}_5^{11}B + {}_1^0\beta$$

4. **Electron capture** occurs when the nucleus of an atom draws in a surrounding electron, usually one from the lowest energy level. The net effect is that *a nuclear proton is transformed into a neutron:*

$$_1^1p + {}_{-1}^0e \rightarrow {}_0^1n$$

(We use the symbol $_{-1}^0e$ to distinguish the orbital electron from a $_{-1}^0\beta$ particle.) The orbital vacancy is quickly filled by an electron moving down from a higher energy level, and that energy difference appears as an *x-ray photon.* Radioactive iron forms stable manganese through electron capture:

$$_{26}^{55}Fe + {}_{-1}^0e \rightarrow {}_{25}^{55}Mn + h\nu \text{ (x-ray)}$$

*Electron capture has the same net effect as positron decay ($Z$ lower by 1, $A$ unchanged),* even though the processes are entirely different.

5. **Gamma emission** involves the radiation of high-energy $\gamma$ photons from an excited nucleus. Recall that when an atom is in an excited *electronic* state, it reduces its energy by emitting photons, usually in the ultraviolet (UV) and visible ranges. Similarly, a nucleus in an excited state lowers its energy by emitting $\gamma$ photons, which are of much higher energy (much shorter wavelength) than UV photons. Many nuclear changes leave the nucleus in an excited state, so $\gamma$ *emission accompanies most other types of decay.* Several $\gamma$ photons ($\gamma$ rays) of different frequencies are often emitted from an excited nucleus as it relaxes, usually to the ground state. Many of Marie Curie's experiments involved the release of $\gamma$ rays, such as

$$_{92}^{238}U \rightarrow {}_{90}^{234}Th + {}_2^4He + 2{}_0^0\gamma$$

Since $\gamma$ rays have no mass or charge, *$\gamma$ emission does not change A or Z.* Gamma rays also result when a particle and an antiparticle annihilate each other, as when an emitted positron meets an orbital electron:

$$_1^0\beta\text{(from nucleus)} + \,_{-1}^0\text{e(outside nucleus)} \rightarrow 2_0^0\gamma$$

SAMPLE PROBLEM 21.1 ─────────────────────────────

## Writing Equations for Nuclear Reactions

**Problem:** Write balanced equations for the following nuclear reactions:
**(a)** Naturally occurring thorium-232 undergoes $\alpha$ decay.
**(b)** Chlorine-36 undergoes electron capture.
**Plan:** We first write a skeleton equation that includes the mass numbers, atomic numbers, and symbols of all the particles. Then we determine the unknown particle (which we symbolize $_Z^A\text{X}$) from the principle of conservation of mass numbers and charges.
**Solution:** **(a)** Writing the skeleton equation:

$$_{90}^{232}\text{Th} \rightarrow \,_Z^A\text{X} + \,_2^4\text{He}$$

Solving for $A$ and $Z$ and balancing the equation: For $A$, $232 = A + 4$, so $A = 228$. For $Z$, $90 = Z + 2$, so $Z = 88$. From the periodic table, we see that the element with $Z = 88$ is radium (Ra), so

$$_{90}^{232}\text{Th} \rightarrow \,_{88}^{228}\text{Ra} + \,_2^4\text{He}$$

**(b)** Writing the skeleton equation:

$$_{17}^{36}\text{Cl} + \,_{-1}^0\text{e} \rightarrow \,_Z^A\text{X}$$

Solving for $A$ and $Z$ and balancing the equation: For $A$, $36 + 0 = A$, so $A = 36$. For $Z$, $17 + (-1) = Z$, so $Z = 16$. The element with $Z = 16$ is sulfur (S), so

$$_{17}^{36}\text{Cl} + \,_{-1}^0\text{e} \rightarrow \,_{16}^{36}\text{S}$$

FOLLOW-UP PROBLEM 21.1
Write a balanced equation for the reaction in which a nuclide undergoes $\beta$ decay and produces cesium-133.

───────────────────────────────────────────────

A parent nuclide may undergo a series of decay steps before a stable daughter nuclide forms. The succession of steps is called a **decay series,** or **disintegration series,** and is typically depicted on a gridlike display. Figure 21.2 shows the decay series from uranium-238 (top right) to lead-206 (bottom left). Numbers of neutrons ($N$) cross numbers of protons ($Z$) to form the grid, which displays a series of $\alpha$ and $\beta$ decays. The zigzag pattern occurs because $\alpha$ decay decreases both $N$ and $Z$, while $\beta$ decay decreases $N$ and increases $Z$. This decay series is one of three that occur in nature. A second series begins with uranium-235 and ends with lead-207, and a third begins with thorium-232 and ends with lead-208. (Neptunium-237 began a fourth series, but its half-life is much shorter than the age of the Earth, so all of it has decayed.)

## Nuclear Stability and the Mode of Decay

You have now seen several ways that an unstable nuclide *might* decay, but can we predict how it *will* decay? Indeed, can we predict *whether* it will decay at all? Our knowledge of the nucleus is much less complete than our knowledge of the atom as a whole, but some patterns emerge when we examine the naturally occurring nuclides.

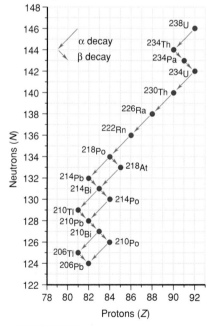

**FIGURE 21.2**
**The $^{238}$U decay series.** Uranium-238 decays through a series of $\alpha$ and $\beta$ emissions to lead-206 in 14 steps. As shown in the key *(upper left)*, each $\alpha$ emission decreases $N$ by two and $Z$ by two, while each $\beta$ emission decreases $N$ by one and increases $Z$ by one.

**Nuclear stability and the neutron/proton (N/Z) ratio.** A key factor that determines the stability of a nuclide is the ratio of the number of neutrons to the number of protons, the **N/Z ratio.** For lighter nuclides, one neutron for each proton is enough to provide stability ($N/Z \approx 1$). However, for heavier nuclides, in which the number of protons and their corresponding repulsions increases, the number of neutrons increases even more to stabilize them. If the $N/Z$ ratio is either too high or not high enough, the nuclide is unstable and decays.

Figure 21.3, *A,* is a plot of number of neutrons vs. number of protons for the stable nuclides. Note the narrow **band of stability** that gradually increases from an $N/Z$ ratio of 1, near $Z = 10$, to an $N/Z$ ratio slightly greater than 1.5, near $Z = 83$ for $^{209}$Bi. Several key points are consistent with Figure 21.3, *A:*

1. Very few stable nuclides exist with $N/Z < 1$; the only two are $^{1}_{1}$H and $^{3}_{2}$He. For lighter stable nuclides, $N/Z \approx 1$: $^{4}_{2}$He, $^{12}_{6}$C, $^{16}_{8}$O, and $^{20}_{10}$Ne are particularly stable.

2. The $N/Z$ ratio of stable nuclides gradually increases as $Z$ increases. No stable nuclide exists with $N/Z = 1$ for $Z > 20$. Thus, for $^{56}_{26}$Fe, $N/Z = 1.15$; for $^{107}_{47}$Ag, $N/Z = 1.28$; and for $^{184}_{74}$W, $N/Z = 1.49$.

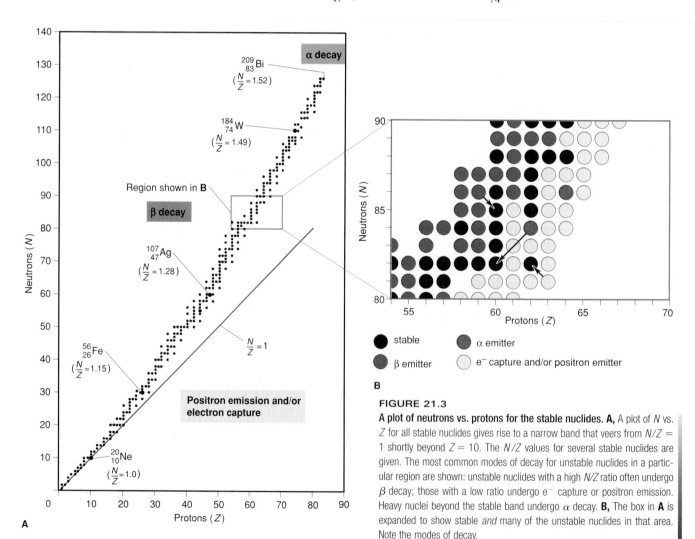

**FIGURE 21.3**

**A plot of neutrons vs. protons for the stable nuclides. A,** A plot of *N* vs. *Z* for all stable nuclides gives rise to a narrow band that veers from *N/Z* = 1 shortly beyond *Z* = 10. The *N/Z* values for several stable nuclides are given. The most common modes of decay for unstable nuclides in a particular region are shown: unstable nuclides with a high *N/Z* ratio often undergo β decay; those with a low ratio undergo e⁻ capture or positron emission. Heavy nuclei beyond the stable band undergo α decay. **B,** The box in **A** is expanded to show stable *and* many of the unstable nuclides in that area. Note the modes of decay.

3. All nuclides with $Z > 83$ are unstable. Bismuth-209 is the heaviest stable nuclide. Therefore, the largest members of Groups 1A(1), 2A(2), 6A(16), 7A(17), and 8A(18) are radioactive, as are all the actinides and known elements of the fourth transition series (Period 7).

Nuclear scientists explain these data in terms of two opposing forces. Electrostatic repulsive forces between protons would break the nucleus apart if not for the presence of an attractive force that exists between all nucleons (protons and neutrons) called the **strong force.** It is about 100 times stronger than the repulsive force but operates only over the short distances within the nucleus. Competition between the *attractive* strong force and the *repulsive* electrostatic force determines nuclear stability.

Elements with an even $Z$ (number of protons) usually have a larger number of stable nuclides than elements with an odd $Z$. Table 21.3 demonstrates this point with the elements cadmium ($Z = 48$) through xenon ($Z = 54$). Another key point emerges when we classify the known stable nuclides according to whether they have even or odd $N$ and $Z$ values (Table 21.4). Note that more than half the stable nuclides have both even $N$ and even $Z$. Of all the stable nuclides, only seven have odd $N$ and odd $Z$: $^2_1H$, $^6_3Li$, $^{10}_5B$, $^{14}_7N$, $^{50}_{23}V$, $^{138}_{57}La$, and $^{176}_{71}Lu$.

One model of nuclear structure that attempts to explain these findings postulates that protons and neutrons lie in nucleon shells, or energy levels, and that stability results from the *pairing* of like nucleons. This arrangement leads to the stability of even values of $N$ and $Z$. (The analogy to electron energy levels and the stability that arises from electron pairing is striking.)

Just as the noble gases, the elements with 2, 10, 18, 36, 54, and 84 electrons, are unusually stable due to their filled *electron* shells, nuclides with $N$ or $Z$ values of 2, 8, 20, 28, 50, 82 (and $N = 126$) are exceptionally stable as well. These so-called magic numbers are thought to correspond to the numbers of protons or neutrons in filled *nucleon* shells. A few examples are $^{50}_{22}Ti$ ($N = 28$), $^{88}_{38}Sr$ ($N = 50$), and the nine stable nuclides of tin ($Z = 50$). Some extremely stable nuclides have double magic numbers: $^4_2He$, $^{16}_8O$, $^{40}_{20}Ca$, and $^{208}_{82}Pb$ ($N = 126$).

**TABLE 21.3   Number of Stable Nuclides for Elements 48 Through 54\***

| ELEMENT | ATOMIC NUMBER ($Z$) | NUMBER OF NUCLIDES |
|---------|---------------------|--------------------|
| **Cd** | **48** | **8** |
| In | 49 | 2 |
| **Sn** | **50** | **10** |
| Sb | 51 | 2 |
| **Te** | **52** | **8** |
| I | 53 | 1 |
| **Xe** | **54** | **9** |

*Even $Z$ shown in boldface.

**TABLE 21.4   An Even-Odd Classification of the Stable Nuclides**

| $Z$ | $N$ | NUMBER OF NUCLIDES |
|-----|-----|--------------------|
| Even | Even | 157 |
| Even | Odd | 53 |
| Odd | Even | 50 |
| Odd | Odd | 7 |
| TOTAL | | 267 |

SAMPLE PROBLEM 21.2

## Predicting Nuclear Stability

**Problem:** Which of the following nuclides would you predict to be stable and which radioactive? Explain.   **(a)** $^{18}_{10}Ne$   **(b)** $^{32}_{16}S$   **(c)** $^{236}_{90}Th$   **(d)** $^{123}_{56}Ba$

**Plan:** To evaluate the stability of each nuclide, we note the $N/Z$ ratio, the value of $Z$, stable $N/Z$ ratios from Figure 21.3, *A*, and whether $Z$ and $N$ are even or odd.

**Solution: (a) Radioactive.** The $N/Z = \dfrac{18 - 10}{10} = 0.8$. Despite even $N$ and $Z$, the minimum ratio for stability is 1.0; this nuclide has too few neutrons to be stable.

**(b) Stable.** This nuclide has an $N/Z = 1.0$ and $Z < 20$, with even $N$ and $Z$, so it is most likely stable.

**(c) Radioactive.** Every nuclide with $Z > 83$ is radioactive.

**(d) Radioactive.** The $N/Z = 1.20$. For $Z$ from 55 to 60, Figure 21.3, *A* shows $N/Z \geq 1.3$, so it probably has too few neutrons to be stable.

**Check:** By consulting a table of isotopes, such as the one in the *Handbook of Chemistry and Physics*, we find that our predictions are correct.

FOLLOW-UP PROBLEM 21.2
Why is $^{31}_{15}P$ stable, but $^{30}_{15}P$ unstable?

**Predicting the mode of decay.** An unstable nuclide generally decays in a mode that shifts its $N/Z$ ratio toward the band of stability. This fact can be seen in Figure 21.3, *B*, an expansion of the boxed region in part *A* of the figure, which shows all stable *and* some of the radioactive nuclides:

1. *Neutron-rich nuclides.* Nuclides with too many neutrons for stability (a high $N/Z$) lie above the band of stability. They undergo $\beta$ *decay*, which converts a neutron into a proton, thus reducing the value of $N/Z$.

2. *Proton-rich nuclides.* Nuclides with too many protons for stability (a low $N/Z$) lie below the band. They undergo *positron decay* or *electron capture*, both of which convert a proton into a neutron, thus increasing the value of $N/Z$.

3. *Heavy nuclides.* Nuclides with $Z > 83$ are too heavy to lie within the band and undergo $\alpha$ *decay*, which reduces their $Z$ and $N$ values by two units per emission. (Several lighter nuclides also exhibit $\alpha$ decay.)

SAMPLE PROBLEM 21.3 ────────────────────────────────────

**Predicting the Mode of Nuclear Decay**

**Problem:** Predict the nature of the nuclear change(s) each of the following radionuclides is likely to undergo:

(a) $^{12}_{5}\text{B}$   (b) $^{234}_{92}\text{U}$   (c) $^{74}_{33}\text{As}$   (d) $^{127}_{57}\text{La}$

**Plan:** We use the $N/Z$ ratio to decide where the nuclide lies relative to the band of stability and how its ratio compares with others in that region of the band. Then, we predict which of the modes just discussed will yield a product nucleus that is closer to the band.

**Solution:** **(a)** This nuclide has an $N/Z$ of 1.4, which is too high for this region of the band. It will probably undergo **$\beta$ decay,** which will increase $Z$ to 6 and lower the $N/Z$ ratio to 1.

**(b)** This nuclide is heavier than those in the band of stability. It will probably undergo **$\alpha$ decay** and decrease its total mass.

**(c)** This nuclide, with $N/Z = 1.24$, lies in the band of stability, so it will probably undergo either **$\beta$ decay** or **positron emission.**

**(d)** This nuclide, with $N/Z = 1.23$, has a ratio that is too low for this region of the band, so it will decrease $Z$ by either **positron emission** or **electron capture.**

**Comment:** Both the possible modes of decay are observed for the nuclides in parts (c) and (d).

FOLLOW-UP PROBLEM 21.3
What mode of decay would you expect for **(a)** $^{61}_{26}\text{Fe}$; **(b)** $^{241}_{95}\text{Am}$?

**Section Summary**

Nuclear reactions are not affected by reaction conditions or chemical composition and release much more energy than chemical reactions. A radioactive nuclide is unstable and may emit $\alpha$ particles ($^{4}_{2}\text{He}$ nuclei), $\beta$ particles ($^{0}_{-1}\beta$; high-speed electrons), positrons ($^{0}_{1}\beta$), or $\gamma$ rays (high-energy photons), or capture an orbital electron. Radioactive decay allows a nuclide to achieve a more stable $N/Z$ ratio. Certain magic numbers of neutrons and protons are associated with very stable nuclides. By comparing a nuclide's $N/Z$ ratio with those on the band of stability, we can predict that, in general, heavy nuclides emit $\alpha$ particles, neutron-rich nuclides emit $\beta$ particles, and proton-rich nuclides undergo positron emission or electron capture.

## 21.2  The Kinetics of Nuclear Change

Chemical and nuclear systems both tend toward stability. Just as the concentrations in a chemical system change in a predictable direction to give a stable equilibrium ratio, the type and number of nucleons in an unstable nucleus change in a predictable direction to give a stable $N/Z$ ratio. As you know, however, the tendency of a chemical system to become more stable tells nothing about how long that process will take, and the same holds true for nuclear systems. In this section, we examine the kinetics of nuclear change. To begin, a Tools of the Chemistry Laboratory essay describes how radioactivity is detected and measured.

---

**TOOLS OF THE CHEMISTRY LABORATORY**           **Counters for the Detection of Radioactive Emissions**

To determine the rate of nuclear decay, we measure the radioactivity of a sample at intervals. Radioactive emissions interact with the atoms in surrounding materials. Since the effects of these interactions can be electrically amplified billions of times, it is possible to detect the decay of a single nucleus. Ionization counters and scintillation counters are two devices used to measure radioactive emissions.

An *ionization counter* detects radioactive emissions by their ionization of a gas. Ionization produces free electrons and gaseous cations, which are attracted to electrodes that conduct a current to a recording device. The most common type of ionization counter is a **Geiger-Müller counter** (Figure 21.A). It consists of a tube filled with argon gas; the tube housing acts as the cathode, and a thin wire running through the center of the tube acts as the anode. Emissions from the sample enter the tube through a thin window and strike argon atoms, producing free electrons that are accelerated toward the anode. These electrons collide with other argon atoms and free more electrons in an *avalanche* effect. The current created is amplified and converted to a meter reading or an audible click. The initial release of one electron can release $10^{10}$ electrons in a microsecond, giving the Geiger-Müller counter great sensitivity.

In a **scintillation counter**, radioactive emissions too weak to ionize surrounding atoms are detected by their ability to excite atoms and cause them to emit light. The

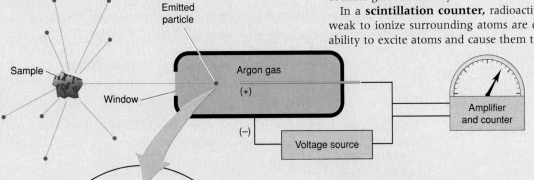

**FIGURE 21.A** Detection of radioactivity by an ionization counter.

light is emitted by a *phosphor* coated onto part of a *photomultiplier tube*, a device that increases the original electrical signal. The incoming radioactive particle strikes the phosphor, which emits a photon that, in turn, strikes a cathode, releasing an electron through the photoelectric effect (Section 7.1). This electron hits other portions of the tube that release increasing numbers of electrons, and the resulting current is recorded. Liquid scintillation counters employ organic phosphors that also function as solvent. They dissolve the sample *and* emit light when excited by the emission. These counters are often used to measure emissions from radioactive biological solutions.

## The Rate of Radioactive Decay

Radioactive nuclei decay at a characteristic rate, regardless of the chemical substance in which they occur. The *decay rate,* or **activity** (*A*), of a radioactive sample is the change in number of nuclei ($\mathcal{N}$) divided by the change in time (*t*). As we saw with chemical reaction rates, because the number of nuclei is *decreasing,* a minus sign precedes the expression:

$$\text{Decay rate (activity)} = -\frac{\Delta \mathcal{N}}{\Delta t}$$

The basic unit of radioactivity is the **curie (Ci):** one curie equals the number of nuclei disintegrating each second in 1 g radium-226:

$$1 \text{ Ci} = 3.70 \times 10^{10} \text{ disintegrations per second (dps)} \tag{21.1}$$

We often express the radioactivity of a sample in terms of *specific activity,* the decay rate per gram.

A decay rate is meaningful only when we consider the large number of nuclei in a macroscopic sample. Suppose there are $1 \times 10^{15}$ radioactive nuclei of a particular type in a macroscopic sample and they decay at a rate of 10% per hour. Although any particular nucleus in the sample might decay in a microsecond or in a million hours, the *average* of all decays results in 10% of the entire collection of nuclei disintegrating each hour. During the first hour, 10% of the *original* number, or $1 \times 10^{14}$ nuclei, will decay. During the next hour, 10% of the remaining $9 \times 10^{14}$ nuclei, or $9 \times 10^{13}$ nuclei, will decay. During the next hour, 10% of those remaining will decay, and so forth. Thus, for a large collection of radioactive nuclei, *the number decaying per unit time is proportional to the number present:*

$$\text{Decay rate} \propto \mathcal{N} \quad \text{or} \quad \text{Decay rate} = k\mathcal{N}$$

where *k* is a proportionality constant called the **decay constant** and is characteristic of the nuclide. The larger the value of *k*, the higher is the decay rate.

Combining the two rate expressions just given, we obtain

$$\text{Decay rate} = \frac{-\Delta \mathcal{N}}{\Delta t} = k\mathcal{N} \tag{21.2}$$

This expression is identical to the rate equation for a first-order process (Chapter 15). The only difference in the case of nuclear decay is that we consider the *number* of nuclei rather than their concentration. Since the decay rate depends only on $\mathcal{N}$ (and on the constant value of *k*), *radioactive decay is a first-order process.*

Decay rates are also commonly expressed in terms of the *fraction* of nuclei that decays over a given time interval. The **half-life** ($t_{1/2}$) of a nuclide is the time it takes for half the nuclei present to decay. The amount of that nuclide remaining is halved after each half-life. Thus, half-life has essentially the same meaning for a nuclear change as for a chemical change (Section 15.3). Figure 21.4 shows the decay of carbon-14, which has a half-life of 5730 years, in terms of number of $^{14}$C nuclei present. We can also consider the half-life in terms of mass of substance. After it decays to the product $^{14}$N, the $^{14}$C nucleus is no longer present. If we start with 1.0 g carbon-14, 0.50 g will be left after 5730 years, 0.25 g after another 5730 years, and so on. The decay rate depends on the number of nuclei present, so it is halved after each succeeding half-life.

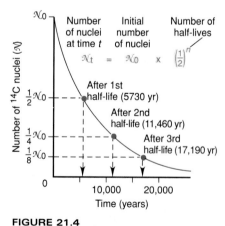

**FIGURE 21.4**

**The decrease in number of $^{14}$C nuclei over time.** A plot of number of $^{14}$C nuclei vs. time gives a smoothly decreasing curve. In each half-life (5730 years), half the $^{14}$C nuclei present undergo decay. A plot of mass of $^{14}$C vs. time would be identical because mass is directly related to number of nuclei.

We determine the half-life of a nuclear reaction from its rate constant. Rearranging Equation 21.2 and integrating over time gives

$$\ln \frac{\mathcal{N}_t}{\mathcal{N}_0} = -kt \qquad \text{or} \qquad \ln \frac{\mathcal{N}_0}{\mathcal{N}_t} = kt \qquad \textbf{(21.3)}$$

where $\mathcal{N}_0$ is the number of nuclei at $t = 0$, and $\mathcal{N}_t$ is the number of nuclei remaining at any time $t$. To calculate the half-life ($t_{1/2}$), we set $\mathcal{N}_t$ equal to $\frac{1}{2}\mathcal{N}_0$ and solve for $t_{1/2}$:

$$\ln \frac{\mathcal{N}_0}{\frac{1}{2}\mathcal{N}_0} = kt_{1/2} \qquad \text{so} \qquad t_{1/2} = \frac{\ln 2}{k} \qquad \textbf{(21.4)}$$

As you can see, *the half-life for this, or any first-order process is not dependent on the number of nuclei present and is inversely related to the decay constant:* a large $k$ means a short $t_{1/2}$, and vice versa. Decay constants and half-lives of radionuclides vary over an extremely wide range, even for isotopes of the same element (Table 21.5).

**TABLE 21.5   Decay Constants ($k$) and Half-Lives ($t_{1/2}$) of Beryllium Isotopes**

| NUCLIDE | $k$ | $t_{1/2}$ |
|---|---|---|
| $^{7}_{4}\text{Be}$ | $1.30 \times 10^{-2}$/day | 53.3 day |
| $^{8}_{4}\text{Be}$ | $1.0 \times 10^{16}$/s | $6.7 \times 10^{-17}$ s |
| $^{9}_{4}\text{Be}$ | Stable | |
| $^{10}_{4}\text{Be}$ | $4.3 \times 10^{-7}$/yr | $1.6 \times 10^{6}$ yr |
| $^{11}_{4}\text{Be}$ | $5.02 \times 10^{-2}$/s | 13.8 s |

s, second; yr, year.

**SAMPLE PROBLEM 21.4**

**Determining the Number of Radioactive Nuclei**

**Problem:** Strontium-90 is a radioactive byproduct of nuclear reactors that behaves biologically like calcium, the element above it in Group 2A(2). If ingested by mammals, it is found in their milk and eventually in the bones of those drinking the milk. If a sample of $^{90}$Sr has an activity of $1.2 \times 10^{12}$ dps, what is the activity and the fraction of nuclei that has decayed after 60 years (yr)? $t_{1/2}$ of $^{90}$Sr = 29 yr.

**Plan:** The fraction of nuclei that has decayed is the fractional decrease in number of nuclei. Since the activity of the sample ($A$) is proportional to the number of nuclei ($\mathcal{N}$), we know that

$$\text{Fraction decayed} = \frac{\mathcal{N}_0 - \mathcal{N}_t}{\mathcal{N}_0} = \frac{A_0 - A_t}{A_0}$$

We are given $A_0$, so we find $A_t$ from the integrated form of the first-order rate equation, in which $t$ is 60 yr. To solve that equation, we need $k$, which we can calculate from the given $t_{1/2}$.

**Solution:** Calculating the decay constant $k$:

$$t_{1/2} = \frac{\ln 2}{k} \qquad \text{so} \qquad k = \frac{\ln 2}{t_{1/2}} = 0.693/29 \text{ yr} = 0.024 \text{ yr}^{-1}$$

Applying Equation 21.3 to calculate the activity remaining at time $t$, $A_t$:

$$\ln \frac{\mathcal{N}_0}{\mathcal{N}_t} = \ln \frac{A_0}{A_t} = kt \quad \text{or} \quad \ln A_0 - \ln A_t = kt$$

So,   $\ln A_t = -kt + \ln A_0 = -(0.024 \text{ yr}^{-1} \times 60 \text{ yr}) + \ln (1.2 \times 10^{12} \text{ dps})$

$\ln A_t = -1.4 + 27.81 = 26.4$

$A_t = 2.9 \times 10^{11}$ dps

(Since all the data contain two significant figures, we retained two in the answer.)
Calculating the fraction decayed:

$$\text{Fraction decayed} = \frac{A_0 - A_t}{A_0} = \frac{1.2 \times 10^{12} \text{ dps} - 2.9 \times 10^{11} \text{ dps}}{1.2 \times 10^{12} \text{ dps}} = \mathbf{0.76}$$

**Check:** The answer seems reasonable: since $t$ is about 2 half-lives, $A_t$ should be about $\frac{1}{4}A_0$, or about $0.3 \times 10^{12}$, so the activity should have decreased by about $\frac{3}{4}$.
**Comment:** An *alternative approach* is to use the number of half-lives ($t/t_{1/2}$) to find the fraction of activity (or nuclei) remaining. By combining Equations 21.3 and 21.4 and substituting $\ln 2/t_{1/2}$ for $k$, we obtain

$$\ln \frac{\mathcal{N}_0}{\mathcal{N}_t} = \frac{\ln 2}{t_{1/2}}t = \frac{t}{t_{1/2}} \ln 2 = \ln 2^{t/t_{1/2}}$$

Thus,   $$\ln \frac{\mathcal{N}_t}{\mathcal{N}_0} = \ln \left(\frac{1}{2}\right)^{t/t_{1/2}}$$

Taking the antilog gives

$$\text{Fraction remaining} = \frac{\mathcal{N}_t}{\mathcal{N}_0} = \left(\frac{1}{2}\right)^{t/t_{1/2}} = \left(\frac{1}{2}\right)^{60/29} = 0.24$$

So,   Fraction decayed $= 1.00 - 0.24 = 0.76$

**FOLLOW-UP PROBLEM 21.4**
Sodium-24, which has a half-life of 15 h, is used to study blood circulation. If a patient is injected with a $^{24}$NaCl solution whose activity is $2.5 \times 10^9$ dps, how much of the activity is present in the patient's body and excreted fluids after 4 days?

## Radioisotopic Dating

The historical record dims rapidly with time and virtually disappears for events of more than a few thousand years ago. Much of our understanding of prehistory comes from a technique called **radioisotopic dating.** The method supplies data about the ages of objects in fields as diverse as art history, archeology, geology, and paleontology.

The technique of *radiocarbon dating,* for which the American chemist Willard F. Libby won the Nobel Prize in 1960, is based on measuring the amounts of $^{14}$C and $^{12}$C in materials of biological origin. The accuracy of the method falls off after about four half-lives of $^{14}$C ($t_{1/2} = 5730$ yr), so it is used to date objects up to about 24,000 years old.

Here is how the method works. High-energy neutrons in cosmic rays enter the upper atmosphere and keep the amount of $^{14}$C nearly constant through bombardment of ordinary $^{14}$N atoms:

$$^{14}_{7}\text{N} + ^{1}_{0}\text{n} \longrightarrow ^{14}_{6}\text{C} + ^{1}_{1}\text{p}$$

(Actually, cosmic ray intensity varies slightly with time, which affects the proportion of atmospheric $^{14}C$. By counting the $^{14}C$ activity in growth rings of ancient trees, we know that the amount of $^{14}C$ fell slightly around 3000 years ago to current levels. Human activity in this century, such as nuclear bomb testing and fossil fuel combustion, has also altered $^{14}C$ levels slightly.)

The $^{14}C$ atoms enter the total carbon pool as aqueous $H^{14}CO_3^-$ and atmospheric $^{14}CO_2$ and, along with ordinary $H^{12}CO_3^-$ and $^{12}CO_2$, are taken up and excreted by plants and by animals that eat the plants. Thus, the $^{12}C/^{14}C$ ratio of a living organism is constant. When an organism dies, however, it no longer takes in $^{14}C$, so the $^{12}C/^{14}C$ ratio steadily increases as the $^{14}C$ decays:

$$^{14}_6C \rightarrow \,^{14}_7N + \,^0_{-1}\beta$$

Since an organism's life span is much shorter than the half-life of $^{14}C$, the difference between the $^{12}C/^{14}C$ ratio in material from a dead organism and the ratio in living organisms reflects the time elapsed since the organism died.

As you saw in Sample Problem 21.4, the first-order rate equation can be expressed in terms of a ratio of activities:

$$\ln \frac{\mathcal{N}_0}{\mathcal{N}_t} = \ln \frac{A_0}{A_t} = kt$$

where $A_0$ is the activity in a living organism and $A_t$ is the activity in the object whose age is unknown. Solving for $t$ gives the age of the object:

$$t = \frac{1}{k} \ln \frac{A_0}{A_t} \qquad\qquad \textbf{(21.5)}$$

---

**SAMPLE PROBLEM 21.5** _____

### Applying Radiocarbon Dating

**Problem:** The charred bones of a sloth in a cave in southern Chile represent the earliest evidence of human presence in the southern tip of South America. A sample of the bone has a specific activity of 5.02 disintegrations per minute per gram of carbon (d/min · g). If the ratio of $^{12}C/^{14}C$ in living organisms results in a specific activity of 15.3 d/min · g, how old are the bones? $t_{1/2}$ of $^{14}C$ = 5730 yr.
**Plan:** We know the activities of the bones ($A_t$) and of a living organism ($A_0$), so we calculate $k$ from $t_{1/2}$ and apply Equation 21.5 to find the age ($t$) of the bones.
**Solution:** Calculating $k$ for $^{14}C$ decay:

$$k = \ln 2/t_{1/2} = 0.693/5730 \text{ yr} = 1.21 \times 10^{-4} \text{ yr}^{-1}$$

Calculating the age ($t$) of the bones:

$$t = \frac{1}{k} \ln \frac{A_0}{A_t} = \frac{1}{1.21 \times 10^{-4} \text{ yr}^{-1}} \ln \left( \frac{15.3 \text{ d/min} \cdot \text{g}}{5.02 \text{ d/min} \cdot \text{g}} \right) = 9.21 \times 10^3 \text{ yr}$$

The bones are about **9200 yr old.**
**Check:** Since the activity of the bones is between $1/2$ and $1/4$ the modern activity, the age should be between one and two half-lives (5730 to 11460 yr).

---

**FOLLOW-UP PROBLEM 21.5**
A sample of wood from an Egyptian mummy case has a specific activity of 9.41 d/min · g. How old is the wooden case?

---

**FIGURE 21.5**

**Radiocarbon dating for determining the age of artifacts.** The natural logarithms of the $^{14}$C specific activity of various artifacts are projected onto a line whose slope equals $-k$, the negative of the $^{14}$C decay constant. The unknown age of an artifact is determined from the time axis.

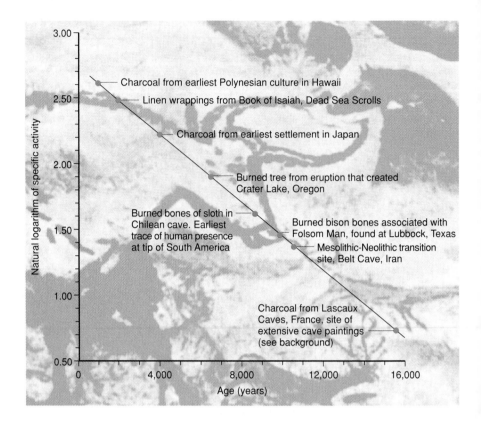

Charcoal from earliest Polynesian culture in Hawaii

Linen wrappings from Book of Isaiah, Dead Sea Scrolls

Charcoal from earliest settlement in Japan

Burned tree from eruption that created Crater Lake, Oregon

Burned bones of sloth in Chilean cave. Earliest trace of human presence at tip of South America

Burned bison bones associated with Folsom Man, found at Lubbock, Texas

Mesolithic-Neolithic transition site, Belt Cave, Iran

Charcoal from Lascaux Caves, France, site of extensive cave paintings (see background)

Natural logarithm of specific activity

Age (years)

◆ **How Old Is the Solar System?** By comparing the ratio of $^{238}$U to its final decay product, $^{206}$Pb, geochemists have found that the oldest known surface rocks—granite in western Greenland—are about 3.7 billion years old. The ratio of $^{238}$U/$^{206}$Pb in meteorites gives 4.55 billion years for the age of the solar system, and thus the Earth. These estimates have been checked with other isotope systems, such as $^{40}$K/$^{40}$Ar ($t_{1/2}$ of $^{40}$K = 1.3 × 10$^9$ yr) and $^{87}$Rb/$^{87}$Sr ($t_{1/2}$ of $^{87}$Rb = 4.9 × 10$^{10}$ yr). From rocks collected by Apollo astronauts, these methods date the moon as 4.2 billion years old and provide evidence for volcanic activity on its surface about 3.3 billion years ago, about the time that the same methods show the first organisms were evolving on Earth.

Several other results of radiocarbon dating appear in Figure 21.5. To determine the ages of more ancient objects and of objects that do not contain carbon, different radioisotopes are measured. ◆

**Section Summary**

Ionization and scintillation counters measure the number of emissions from a radioactive sample. The decay rate (activity) of a sample is proportional to the number of radioactive nuclei. Nuclear decay is a first-order process, so the half-life does not depend on number of nuclei. Radioactive dating methods, such as $^{14}$C dating, can determine the ages of objects by measuring the ratio of specific isotopes in the sample.

## 21.3   Nuclear Transmutation: Induced Changes in Nuclei

The alchemists' dream of changing base metals into gold was never realized, but in the early 20th century, scientists found that they could transmute one element into another. Research into **nuclear transmutation,** the *induced* conversion of one nucleus into another, was closely linked with research into atomic structure and led to the discovery of the neutron and to the production of artificial radioisotopes. Later, high-energy bombardment of nuclei in particle accelerators created many new isotopes and a growing number of new elements.

## Early Transmutation Experiments and Nuclear Structure

The first recognized transmutation occurred in 1919, when Ernest Rutherford showed that $\alpha$ particles emitted from radium bombarded atmospheric nitrogen to form a proton and oxygen-17:

$$^{14}_{7}N + ^{4}_{2}He \rightarrow ^{1}_{1}H + ^{17}_{8}O$$

By 1926, experimenters had found that $\alpha$ bombardment transmuted most elements with low atomic numbers to the next higher element, with ejection of a proton.

A shorthand notation commonly used for nuclear bombardment reactions shows the reactant (target) nucleus and product nucleus separated by parentheses, within which a comma separates the projectile particle from the ejected particle(s):

Reactant nucleus (particle in, particle(s) out) product nucleus

Using this notation, the previous reaction is $^{14}N\ (\alpha,p)\ ^{17}O$.

An unexpected finding in a transmutation experiment led to the discovery of the neutron. When lithium, beryllium, and boron were bombarded with $\alpha$ particles, they emitted highly penetrating radiation that could not be deflected by a magnetic or electric field. Unlike $\gamma$ radiation, these emissions were massive enough to eject protons from the substances they penetrated. In 1932, James Chadwick, a student of Rutherford, proposed that these emissions consisted of neutral particles with a mass similar to that of a proton, and he named them *neutrons*. Chadwick received the Nobel Prize in 1935 for his discovery.

In 1933, Irene and Frederic Curie-Joliot (daughter and son-in-law of Marie and Pierre Curie) created the first artificial radioisotope, phosphorus-30. When they bombarded aluminum foil with $\alpha$ particles, phosphorus-30 and neutrons were formed:

$$^{27}_{13}Al + ^{4}_{2}He \rightarrow ^{1}_{0}n + ^{30}_{15}P \qquad \text{or} \qquad ^{27}Al\ (\alpha,n)\ ^{30}P$$

Since then, other techniques for producing artificial radioisotopes have been developed. In fact, the majority of the more than 900 known radioisotopes have been produced artificially.

## Particle Accelerators and the Transuranium Elements

During the 1930s and 1940s, researchers bombarded elements with neutrons, $\alpha$ particles, protons, and **deuterons** (nuclei of the stable hydrogen isotope deuterium, $^{2}H$). Neutrons are very useful projectiles because they have no charge. Therefore, when they smash into a target nucleus, they are not repelled. The other particles are all positive, so at first it was difficult to give them enough energy to overcome their repulsion by the target nuclei. Beginning in the 1930s, however, **particle accelerators** were used to impart high kinetic energies to particles by placing them in an electric field, usually in combination with a magnetic one. In the simplest design, protons are introduced at one end of a tube and attracted to the other end by a potential difference.

A major advance occurred with the invention of the *linear accelerator,* a series of separated tubes of increasing length that change from positive to neg-

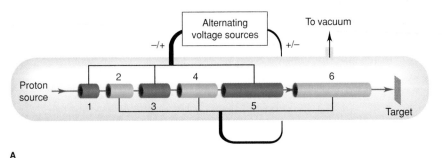

**A**

**FIGURE 21.6**

**Diagram of a linear accelerator. A,** The voltage of each tubular section is alternated, such that the positively charged particle (a proton here) is repelled from the section it is leaving and attracted to the section it is entering. As a result, the particle speed is continually increased. **B,** The linear accelerator operated by Stanford University in California.

ative in synchrony with the movement of the particle through them (Figure 21.6, *A*). A proton, for example, exits the first tube just when that tube becomes positive and the next tube negative. Repelled by the first tube and attracted by the second, the proton accelerates across the gap between them. A 40-ft linear accelerator with 46 tubes, built in California after World War II, accelerated protons to speeds several million times faster than that first accelerator. Modern designs, such as the Stanford Linear Accelerator (Figure 21.6, *B*), accelerate heavier particles, such as B, C, O, and Ne nuclei, several hundred million times faster, with correspondingly greater kinetic energies.

The *cyclotron* (Figure 21.7), invented by E. O. Lawrence in 1930, applies the principle of the linear accelerator but uses electromagnets to give the particle a spiral path, thus saving space. The magnets lie above and below two "dees," open, D-shaped electrodes within an evacuated chamber that function like the tubes in the linear design. The particle is accelerated as it passes from one dee chamber, which is momentarily positive, to the next, which is momentarily negative. Its speed and radius increase until it is deflected toward the target nucleus. The *synchrotron* uses a synchronously

**FIGURE 21.7**

**Diagram of a cyclotron accelerator.** When the positively charged particle reaches the gap between the two D-shaped electrodes ("dees"), it is repelled by one dee and attracted by the other. The particles move in a spiral path, so the cyclotron can be much smaller than a linear accelerator.

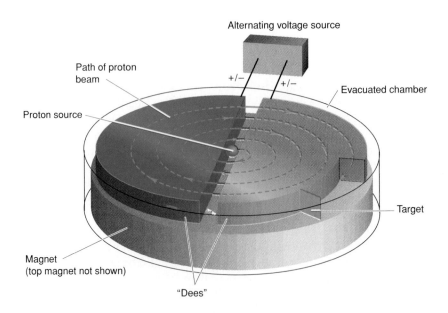

**TABLE 21.6  Formation of Some Transuranium Nuclides**

| REACTION | HALF-LIFE OF PRODUCT |
|---|---|
| $^{239}_{94}\text{Pu} + {}^{4}_{2}\text{He} \rightarrow {}^{240}_{95}\text{Am} + {}^{1}_{1}\text{H} + 2{}^{1}_{0}\text{n}$ | 50.9 h |
| $^{239}_{94}\text{Pu} + {}^{4}_{2}\text{He} \rightarrow {}^{242}_{96}\text{Cm} + {}^{1}_{0}\text{n}$ | 163 days |
| $^{244}_{96}\text{Cm} + {}^{4}_{2}\text{He} \rightarrow {}^{245}_{97}\text{Bk} + {}^{1}_{1}\text{H} + 2{}^{1}_{0}\text{n}$ | 4.94 days |
| $^{238}_{92}\text{U} + {}^{12}_{6}\text{C} \rightarrow {}^{246}_{98}\text{Cf} + 4{}^{1}_{0}\text{n}$ | 36 h |
| $^{253}_{99}\text{Es} + {}^{4}_{2}\text{He} \rightarrow {}^{256}_{101}\text{Md} + {}^{1}_{0}\text{n}$ | 76 min |
| $^{252}_{98}\text{Cf} + {}^{10}_{5}\text{B} \rightarrow {}^{256}_{103}\text{Lr} + 6{}^{1}_{0}\text{n}$ | 28 s |

increasing magnetic field to make the particle's path circular rather than spiral. ◆

Accelerators have many applications, from producing radioisotopes used in medical applications to studying the fundamental nature of matter. Perhaps their most outstanding application for chemists is the synthesis of **transuranium elements,** those with atomic numbers higher than uranium, the heaviest naturally occurring element. These elements include the remaining actinides ($Z = 93$ to $103$), which fill the $5f$ sublevel, and the known elements in the fourth transition series ($Z = 104$ to $109$), which have begun to fill the $6d$ sublevel. (Recent evidence for the formation of elements 110 and 111 has been obtained.) Some reactions that gave rise to several of the transuranium elements appear in Table 21.6. ◆

### Section Summary
One nucleus can be transmuted to another by bombarding it with high-energy particles. Accelerators increase the kinetic energy of particles in transmutation processes and are used to produce transuranium elements.

## 21.4   The Effects of Nuclear Radiation on Matter

In 1986, an accident at the Chernobyl nuclear facility in the former Soviet Union released radioactivity that may cause thousands of cancer deaths. In the same year, isotopes used in medical treatment emitted radioactivity that may prevent thousands of cancer deaths. In this section and the next, we examine the harm and benefit of radioactivity. Although a nuclear reaction occurs with little or no involvement of the atom's electrons, the emissions affect the electrons of other atoms; that is, *nuclear changes* are observed and have their effects through the *chemical changes* they cause.

### The Effects of Radioactive Emissions: Excitation and Ionization

Radioactive emissions interact with matter in ways that depend on their energies. In the process of **excitation,** a particle of relatively low energy collides with an atom of a substance, which absorbs some of the energy and then reemits it. Because electrons are not lost from the atom, the radiation that causes excitation is called **nonionizing radiation.** If the absorbed energy causes the atoms to move, vibrate, or rotate more rapidly, the material becomes hotter. Concentrated aqueous solutions of plutonium salts boil because the emissions excite the surrounding water molecules. Polonium has

◆ **The Powerful Bevatron.**
The *bevatron* includes a linear section and a synchrotron section. The instrument at the Lawrence Berkeley Laboratory in California increases the kinetic energy of the particles more than 6 billion fold. A beam of $10^{10}$ protons makes more than 4 million revolutions, a distance of 300,000 miles, in 1.8 s, attaining a final speed about 90% the speed of light! Even more powerful bevatrons are in use at the Brookhaven National Laboratory in New York and at CERN, outside Geneva, Switzerland. At CERN (photo), bending (red) and focusing (blue) magnets accelerate protons along the 4.3-mile tunnel.

◆ **Naming the Transuranium Elements.** The last naturally occurring element was named after Uranus, thought at the time to be the outermost planet, so the first two artificial elements were named after the more recently discovered Neptune and Pluto. The next few elements were named after famous scientists, such as curium, and places, such as americium. However, conflicting claims of discovery by scientists in different countries led to controversies about names for elements 104 and higher. To provide interim names until the disputes could be settled, the International Union of Pure and Applied Chemistry (IUPAC) adopted a system that uses the atomic number as the basis for a Latin name. Thus, for example, element 104 was named unnilquadium (un = 1, nil = 0, quad = 4, ium = element suffix), with the symbol Unq. At the time of this writing, the IUPAC has recommended more traditional names, but they too have proven to be very controversial and have not been accepted. The proposed names are 104, dubnium (Db); 105, joliotium (Jl); 106, rutherfordium (Rf); 107, bohrium (Bh); 108, hahnium (Hn); 109, meitnerium (Mt).

been suggested as a lightweight heat source, with no moving parts, for use on space stations. Particles of somewhat higher energy excite electrons in other atoms to higher energy levels. As the atoms return to their ground state, they emit photons, often in the blue or ultraviolet region (see scintillation counters in the previous Tools of the Chemistry Laboratory essay).

In the process of **ionization,** radiation of still higher energy collides with an atom and dislodges an electron. The free electron and positive ion that result are referred to as an *ion pair,* and the number of ion pairs produced is directly related to the energy of the incoming radiation. The high-energy radioactivity that gives rise to ion pairs is called **ionizing radiation.** The free electron of an ion pair often collides with another atom and ejects a second electron (see Geiger-Müller counters in the previous Tools essay).

### Effects of Ionizing Radiation on Living Matter

Whereas nonionizing radiation is relatively harmless, ionizing radiation has a destructive effect on living tissue.

**Units of radiation dose.** To measure the effects of ionizing radiation, we need a unit for radiation dose. The unit of radioactive decay, the curie, states the number of decay events in a given time, but not their energy or absorption by matter. The number of ion pairs produced in a given amount of living tissue is a measure of the energy absorbed by the tissue. The **rad (radiation-absorbed dose)** is the amount of radiation that results in 0.01 J of energy being absorbed per kilogram of tissue:

$$1 \text{ rad} = 1 \times 10^{-2} \text{ J/kg}$$

To account for differences in the strength of the radiation, exposure time, and the type of tissue, we multiply the number of rads by a *relative biological effectiveness* (RBE) factor, which depends on the effect of a given type of radiation on a given body part. The product is the **rem (roentgen equivalent for man),** the unit of radiation dosage for a human:

$$1 \text{ rem} = 1 \text{ rad} \times \text{RBE}$$

Doses are often expressed in millirems (1 mrem = $10^{-3}$ rem).

The radiation dose depends on the penetrating power *and* ionizing ability of the radiation. Since $\alpha$ particles are massive and highly charged, they penetrate very little. The skin easily stops $\alpha$ radiation from an external source, but if ingested, an $\alpha$ emitter, such as plutonium-239, causes grave local damage through extensive ionization. ♦ Even though it causes less ionization, a $\beta$ emitter is a more destructive external source because the particles penetrate much deeper. An external $\gamma$-ray source is the most dangerous. Because these high-energy photons are chargeless, they penetrate deepest, and when their energy *is* absorbed, it ionizes surrounding matter.

**Molecular interactions.** When ionizing radiation interacts with a molecule, it causes the loss of an electron from a bond or lone pair. The resulting pair of charged species form **free radicals,** molecular or atomic species with one or more unpaired electrons. Free radicals are very reactive and attack bonds in other molecules to form more free radicals.

When $\gamma$ radiation strikes biological tissue, for instance, the most likely molecule to absorb it is water:

$$H_2O + \gamma \rightarrow H_2O^+ + e^-$$

The $H_2O^+$ and $e^-$ collide with other water molecules to form free radicals:

$$H_2O^+ + H_2O \rightarrow H_3O^+ + \cdot OH \quad \text{and} \quad e^- + H_2O \rightarrow H\cdot + OH^-$$

♦ **A Tragic Way to Tell Time in the Dark.** In the first quarter of this century, wristwatch and clock dials were painted by hand with radium-containing paint to illuminate them in the dark. To write the numbers clearly, the young women hired to apply the paint would "tip" the fine brushes repeatedly between their lips. Small amounts of ingested $^{226}Ra^{2+}$ were incorporated into their bone, along with normal $Ca^{2+}$, which led to numerous cases of bone fracture and jaw cancer.

These free radicals go on to attack surrounding biomolecules, whose bonding and structure are delicately connected with their function.

The double bonds in membrane lipids are particularly susceptible to free-radical attack:

$$RCH\!\!=\!\!CHR' + H\!\cdot \rightarrow RCH_2\!\!-\!\!\overset{\bullet}{C}HR'$$

Damage to cell membranes causes leakage of the contents and destruction of the tissue lining around organs. Destruction of critical bonds in enzymes leads to their malfunction in metabolic reactions, and alterations in the nucleic acids and proteins that govern the rate of cell division can cause cancer. Genetic damage may occur when bonds in the DNA of sperm and egg cells are altered by free radicals.

**Assessing the risk from ionizing radiation.** We are continuously exposed to ionizing radiation from natural and artificial sources (Table 21.7). Indeed, life evolved in the presence of natural ionizing radiation, called **background radiation.** One source is cosmic radiation, which increases with altitude because of decreased absorption by the atmosphere. Thus, people in Denver absorb twice as much cosmic radiation as people in Los Angeles; even a jet flight involves significant absorption. However, the soil is the source of most background radiation due to the presence of thorium and uranium minerals. Radon, the heaviest noble gas [Group 8A(18)], is a radioactive product of uranium and thorium decay, and its concentration in the air we breathe varies with type of local soil and rocks. ◆ Moreover, about 150 g of $K^+$ ions is dissolved in the tissue water of an average adult, and 0.0118% is radioactive $^{40}K$. The presence of these substances and of atmospheric $^{14}CO_2$ means that all food, water, clothing, and building materials are slightly radioactive.

◆ **The Risk of Radon.** The most recently discovered atmospheric health threat comes from radon (Rn; Z = 86), a natural decay product that is also a gas. Once inhaled, radon decays to radioactive nuclides of Po, Pb, and Bi, along with $\alpha$, $\beta$, and $\gamma$ particles, which pose a serious potential hazard. The emitted particles damage lung tissue, and the heavy-metal atoms formed aggravate the problem. Estimates vary, but as many as 10% of lung cancer deaths have been attributed to radon. The uranium content of the local soil and rocks is a critical factor in the extent of the threat. Measurements show that houses in the northern and northwestern sections of the U.S. and adjacent areas of Canada have the highest amounts of radon.

**TABLE 21.7  Examples of Typical Radiation Doses from Natural and Artificial Sources**

| SOURCE OF RADIATION | AVERAGE ADULT EXPOSURE |
|---|---|
| Cosmic radiation | 30-50 mrem/yr |
| Radiation from the ground | |
|   From clay soil and rocks | ~25-170 mrem/yr |
|   In wooden houses | 10-20 mrem/yr |
|   In brick houses | 60-70 mrem/yr |
|   In light concrete houses | 60-160 mrem/yr |
| Radiation from the air (mainly radon) | |
|   Outdoors, average value | 20 mrem/yr |
|   In wooden houses | 70 mrem/yr |
|   In brick houses | 130 mrem/yr |
|   In light concrete houses | 260 mrem/yr |
| Internal radiation from minerals in tap water and daily intake of food ($^{40}K$, $^{14}C$, Ra) | ~20 mrem/yr |
| Diagnostic x-ray methods | |
|   Lung (local) | 0.04-0.2 rad/film |
|   Kidney (local) | 1.5-3 rad/film |
|   Dental (dose to the skin) | ≤ 1 rad/film |
| Therapeutic radiation treatment | Locally ≤ 10,000 rad |
| Other sources | |
|   Jet flight (4 h) | ~1 mrem |
|   From nuclear test | <1 mrem/yr |
| TOTAL AVERAGE VALUE | 100-200 mrem/yr |

**TABLE 21.8   Acute Effects of a Single Dose of Whole-Body Irradiation**

| rem | EFFECT | LETHAL DOSE | |
| --- | --- | --- | --- |
| | | POPULATION (%) | NO. OF DAYS |
| 5-20 | Possible late effect; possible chromosomal aberrations | — | — |
| 20-100 | Temporary reduction in white blood cells | — | — |
| 50+ | Temporary sterility in men (100+ rem = 1 yr duration) | — | — |
| 100-200 | "Mild radiation sickness": vomiting, diarrhea, tiredness in a few hours | — | — |
| | Reduction in infection resistance | | |
| | Possible bone growth retardation in children | | |
| 300+ | Permanent sterility in women | — | — |
| 300-400 | "Serious radiation sickness": marrow/intestine destruction | 50-70 | 30 |
| 400-1000 | Acute illness, early deaths | 60-95 | 30 |
| 3000+ | Acute illness, death in hours to days | 100 | 2 |

The largest artificial source of radiation, and the easiest to control, is associated with medical diagnostic techniques, especially x-rays. The radiation dosage from nuclear testing and radioactive waste disposal is extremely low for most people, but exposures for those living near test sites or disposal areas are many times higher.

How much radiation is too much? To approach the question, we must ask several others: How strong is the exposure? How long is the exposure? Which tissue is exposed? Were offspring affected? One reason we lack clear answers to these questions is that humans cannot be intentionally exposed in an experimental setting. However, accidentally exposed radiation workers and Japanese atomic-bomb survivors have been studied. Table 21.8 summarizes the immediate tissue effects on humans from an acute single dose of ionizing radiation to the whole body. The severity of the effects increases with dose; a dose of 300 rem will kill about 50% of the exposed population within 30 days.

Most data, however, come from laboratory animals, whose biological systems may differ greatly from ours. Nevertheless, studies with mice and dogs show that lesions and cancers appear after massive whole-body exposure, with rapidly dividing cells affected first. In an adult animal, these are the cells of the bone marrow, organ linings, and reproductive organs, but many other tissues are affected in an immature animal or fetus. Both animal and human studies show an increase in the incidence of cancer from either a high, single exposure or a low, chronic exposure.

Reliable data on genetic effects are few. Pioneering studies on fruit flies show a linear increase in genetic defects with both dose and exposure time. However, in the mouse, whose genetic system is obviously much more similar to ours than the fruit fly, a total dose given over a long period created

one-third as many genetic defects as the same dose given over a short period. Thus, rate of exposure is a key factor. Children of atomic bomb survivors show higher-than-normal cancer rates, which implies that their parents' reproductive systems were genetically affected. ◆

### Section Summary

Relatively low-energy emissions cause excitation of surrounding matter,
whereas high-energy emissions cause ionization. The effect of ionizing radiation on living matter depends on the amount of energy absorbed and the
extent of ionization in a given type of tissue. Dose is measured in rem.
Ionization forms free radicals, which destroy biomolecular function. All life
is exposed to varying amounts of natural ionizing radiation. Studies show
that a large acute dose or a chronic small dose are both harmful.

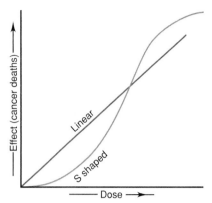

◆ **Modeling Radiation Risk.**
There are two current models of effect vs.
dose. The *linear response* model proposes that
radiation effects, such as cancer risks, accumulate over time regardless of dose and that
populations should not be exposed to any radiation above background levels. The *S-shaped
response* assumes an extremely low risk at low
doses and that only at higher doses is there
cause for concern. If the linear model is more
accurate, we should limit all excess exposure,
but this would severely restrict medical diagnosis and research, military testing, and nuclear
energy.

## 21.5 Applications of Radioisotopes

Our ability to detect minute quantities of a radioisotope makes them powerful tools for studying many processes. Such uses depend on the fact that
*isotopes of an element exhibit **very** similar chemical and physical behavior.* In other
words, except for having a less stable nucleus, a radioactive isotope has
nearly the same chemical properties as a nonradioactive isotope of that element.* In radiocarbon dating, for example, $^{14}CO_2$ is utilized by a plant in
the same way as $^{12}CO_2$.

### Radioactive Tracers: Applications of Nonionizing Radiation

Just think how useful it would be to follow a substance through the stages
of a complex process or from one region of a system to another. A tiny
amount of a radioisotope could be mixed with a large amount of the stable
isotope to act as a **tracer,** a chemical beacon emitting nonionizing radiation
that signals the presence of the substance.

   **Reaction pathways.** Tracers help us choose from among possible reaction pathways. Consider the redox reaction between periodate and iodide
ions:

$$IO_4^-(aq) + 2I^-(aq) + H_2O(l) \rightarrow I_2(s) + IO_3^-(aq) + 2OH^-(aq)$$

Is $IO_3^-$ the result of $IO_4^-$ reduction or $I^-$ oxidation? When we add "cold"
(nonradioactive) $IO_4^-$ to a solution of $I^-$ that contains some "hot" (radioactive) $^{128}I^-$, and then add a soluble $Ba^{2+}$ salt to precipitate the $IO_3^-$ as
$Ba(IO_3)_2$, we find that the $I_2$ is radioactive, not the $IO_3^-$:

$$IO_4^-(aq) + 2^{128}I^-(aq) + H_2O(l) \rightarrow {}^{128}I_2(s) + IO_3^-(aq) + 2OH^-(aq)$$

Conversely, when we add $IO_4^-$ containing some hot $^{128}IO_4^-$ to a solution
of cold $I^-$, we find that the $IO_3^-$ is radioactive, not the $I_2$:

$$^{128}IO_4^-(aq) + 2I^-(aq) + H_2O(l) \rightarrow I_2(s) + {}^{128}IO_3^-(aq) + 2OH^-(aq)$$

---

*Although this statement is generally correct, differences in isotopic mass *can* influence bond
strengths and therefore reaction rates. Such behavior is called a *kinetic isotope effect* and is particularly important for isotopes of hydrogen—$^1H$, $^2H$, and $^3H$—because their masses differ by
such large proportions. We examine an industrial application in Chapter 23.

These results show that $IO_3^-$ forms through the reduction of $IO_4^-$, and that $I_2$ forms through the oxidation of $I^-$. The tracer acts almost as a "handle" we can "hold" to follow the changing reactants.

The metabolic pathway of photosynthesis was also elucidated with radioactive tracers. The overall reaction, which uses the energy of sunlight to form the chemical bonds of glucose, looks quite simple:

$$6CO_2(g) + 6H_2O(l) \xrightarrow[\text{chlorophyll}]{\text{light}} C_6H_{12}O_6(s) + 6O_2(g)$$

However, the actual pathway is extremely complex, requiring a 13-step pathway for each molecule of $CO_2$ incorporated and six times through the pathway for each molecule of $C_6H_{12}O_6$ that forms. Using $^{14}C$ in $CO_2$ as the tracer and paper chromatography as the means of separating the products formed after different times of light exposure, Melvin Calvin and his co-workers took 7 years to determine the pathway, a remarkable achievement for which Calvin won the Nobel Prize in 1961.

**Material flow.** Tracers are used in studies of solid surfaces and the flow of materials. Metal atoms hundreds of layers deep within a solid have been shown to exchange with metal ions from the surrounding solution within a matter of minutes. Chemists and engineers use tracers to study material movement in semiconductor chips, paint, and metal plating; in detergent action; and in the process of corrosion.

Hydrologic engineers use tracers to study the volume and flow of large bodies of water. By following radioisotopes formed during atmospheric nuclear bomb tests ($^3H$ in $H_2O$, $^{90}Sr^{2+}$, and $^{137}Cs^+$), scientists have mapped the flow of water from land to lakes and streams to oceans. Surface and deep ocean currents that circulate around the globe are also studied, as are the mechanisms of hurricane formation and the mixing of the troposphere and stratosphere. Industries employ tracers to study material flow during a manufacturing process, such as the flow of ore pellets in smelting kilns, the path of wood chips and mixing of bleach in paper mills, the diffusion of fungicide into lumber, and the location of leaks in pipes and storage tanks.

**Activation analysis.** A somewhat different use of tracers occurs in *neutron activation analysis* (NAA). In this method, neutrons bombard a nonradioactive sample, converting a small fraction of its atoms to radioisotopes, which exhibit characteristic decay patterns, such as $\gamma$-ray spectra, that reveal the elements present. Unlike chemical analysis, NAA leaves the sample virtually intact, so the method can be used to determine the composition of a valuable object or a very small sample. For example, a painting thought to be a 16th-century Dutch masterpiece was shown through NAA to be a 20th-century forgery, because a microgram sample of pigment contained much less silver and antimony than the pigments used by the Dutch masters. Forensic chemists use NAA to detect trace amounts and type of ammunition on a suspect's hand or traces of arsenic in the hair of a deceased victim of poisoning.

Automotive engineers employ NAA to measure friction and wear of moving parts such as piston rings. For example, when a steel surface that has been neutron-activated to form some radioactive $^{59}Fe$ moves against a second steel surface, the amount of radioactivity on the second surface indicates the amount of material rubbing off. The radioactivity appearing in a lubricant placed between the surfaces can measure the lubricant's ability to reduce wear.

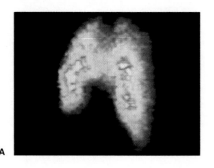

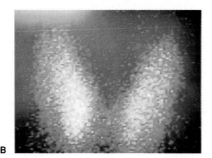

A                                        B

**FIGURE 21.8**
**The use of radioisotopes to image the thyroid gland.** Thyroid scanning is used to assess nutritional deficiencies, inflammation, tumor growth, and other thyroid-related ailments. **A,** In $^{131}$I scanning, the thyroid gland absorbs the I$^-$, and $\beta$ emissions from the $^{131}$I$^-$ expose a photographic film. **B,** A $^{99}$Tc thyroid scan.

**Medical diagnosis.** The largest use of radioisotopes is in medical science. Records show that one-quarter of U.S. hospital patients are admitted for radioisotopic diagnoses. Tracers with half-lives of a few minutes to a few days are employed to observe specific organs and body parts. For example, a healthy thyroid gland incorporates dietary I$^-$ into I-containing hormones at a known rate. To assess thyroid function, the patient drinks a solution containing a trace amount of Na$^{131}$I, and a scanning monitor follows the uptake of $^{131}$I into the thyroid. Technetium-99 ($Z = 43$) is also used for imaging the thyroid (Figure 21.8), as well as the heart, lungs, and liver. Technetium does not occur naturally, so the radioisotope is prepared just before use from radioactive molybdenum:

$$^{99}_{42}\text{Mo} \rightarrow \, ^{99}_{43}\text{Tc} + \, ^{0}_{-1}\beta$$

Tracers are also used to measure physiological processes, such as blood flow. The rate at which the heart pumps blood, for example, can be observed by injecting $^{59}$Fe, which concentrates in the hemoglobin of blood cells. Several radioisotopes used in medical diagnosis are listed in Table 21.9.

*Positron-emission tomography* (PET) is a powerful imaging method for observing brain structure and function. A positron emitter in the form of a normal biological substance is injected into a patient's bloodstream, from which it is taken up into the brain. The isotope emits positrons, each of which annihilates a nearby electron. In the process, two $\gamma$ photons are emitted simultaneously 180° apart from each other:

$$^{0}_{1}\beta + \, ^{0}_{-1}e \rightarrow 2\gamma$$

An array of detectors around the patient's head locates the sites of $\gamma$ emission, and the image is analyzed by computer. Two of the isotopes used are $^{15}$O, injected as H$_2$$^{15}$O to measure blood flow, and $^{18}$F, bonded to a glucose analog to measure glucose uptake, which is a marker for energy metabolism. Medical researchers have shown that changes in blood flow and glucose uptake accompany brain activity and malfunction (Figure 21.9).

**TABLE 21.9  Some Radionuclides Used as Medical Tracers**

| NUCLIDE | BODY PARTS OR PROCESS STUDIED |
|---|---|
| $^{18}$F | Brain |
| $^{99}$Tc | Heart, thyroid, liver, lungs |
| $^{24}$Na | Circulatory system |
| $^{131}$I | Thyroid |
| $^{32}$P | Eyes, liver, tumors |
| $^{201}$Tl | Heart muscle |
| $^{11}$C | General metabolism |
| $^{59}$Fe | Blood flow |

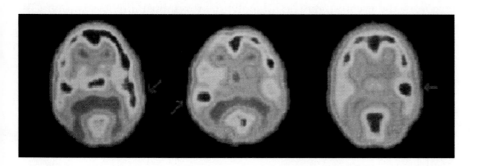

**FIGURE 21.9**
**PET and brain activity.** These scans show brain activity from musical stimulation. The left and center scans occur with the same tone in different sequences; the right scan measures different timbre. Red indicates high activity.

### Applications of Ionizing Radiation

To be used as a tracer, a radioisotope need only emit low-energy detectable radiation. Many other uses of radioisotopes, however, depend on the effects of high-energy, ionizing radiation.

The interaction between radiation and matter that causes cancer can also be used to eliminate it. Cancer cells divide more rapidly than normal cells, so radioisotopes that damage the cell-division process kill more cancer cells than normal ones. Implants of $^{198}Au$ or of a mixture of $^{90}Sr$ and $^{90}Y$ have been used to destroy pituitary and breast tumor cells, and $\gamma$ rays from $^{60}Co$ have been used to destroy brain tumors.

Irradiation of food increases shelf life by killing microorganisms that cause food to rot (Figure 21.10), but the practice is quite controversial. Advocates point to the benefits of preserving fresh foods, grains, and seeds for long periods, whereas opponents suggest that irradiation might lower the food's nutritional content or produce harmful products. The United Nations has approved irradiation for potatoes, wheat, chicken, and strawberries, and the United States allows irradiation of chicken.

Ionizing radiation has been used to control harmful insects. Captured males are sterilized by radiation and released to mate, thereby reducing the number of females who have offspring. This method has been used to control the Mediterranean fruit fly in California and disease-causing insects, such as the tsetse fly and malarial mosquito, in other parts of the world.

**FIGURE 21.10**
**The increase in shelf life by irradiating food.** Both boxes of strawberries were picked at the same time, but the fruit on the right was irradiated.

### Section Summary

Radioisotopic tracers emit nonionizing radiation and have been used to study reaction mechanisms, material flow, elemental composition, and medical conditions. Ionizing radiation has been used to destroy cancerous tissue, kill organisms that spoil food, and control insect populations.

## 21.6   Fission and Fusion: The Interconversion of Mass and Energy

Most of the nuclear reactions we've considered so far have involved radioactive decay, in which a nucleus emits one or a few small particles or photons to become a slightly lighter nucleus. In *nuclear fission,* by contrast, a heavy nucleus splits into two much lighter nuclei, emitting several small particles as well. In *nuclear fusion,* the opposite process occurs as two lighter nuclei combine to form a heavier one. Both fission and fusion release enormous quantities of energy. Of the many beneficial applications of nuclear reactions, the greatest may be the potential for almost limitless amounts of energy to power society. Let's take a quantitative look at the basis of this energy yield.

### The Mass Defect and Nuclear Binding Energy

We have known for most of this century that mass and energy are interconvertible. The traditional mass and energy conservation laws have been combined to state that *the total quantity of mass-energy in the universe is constant.*

Therefore, when *any* reaction produces or consumes energy, there must be an accompanying loss or gain in mass.

This relation between mass and energy did not concern us in earlier chapters because the energy changes in chemical reactions are so small that the mass difference is negligible. When one mole of hydrogen burns in oxygen, for example, 285.8 kJ of heat is given off:

$$H_2(g) + \tfrac{1}{2}O_2(g) \rightarrow H_2O(l) \qquad \Delta H^0_{rxn} = -285.8 \text{ kJ}$$

We calculate the mass that is equivalent to this much energy from Einstein's equation:

$$E = mc^2 \qquad \text{or} \qquad \Delta E = \Delta mc^2 \qquad \text{so} \qquad \Delta m = \frac{\Delta E}{c^2} \qquad \textbf{(21.6)}$$

where $\Delta m$ is the change in mass between reactants and products. Substituting the heat of reaction (in J/mol) for $\Delta E$ and the speed of light ($2.9979 \times 10^8$ m/s) for $c$, we obtain

$$\Delta m = \frac{-2.858 \times 10^5 \text{ J/mol}}{(2.9979 \times 10^8 \text{ m/s})^2} = -3.180 \times 10^{-12} \text{ kg/mol} = -3.180 \times 10^{-9} \text{ g/mol}$$

(Units of kg/mol are obtained because the joule includes the SI unit of mass, the kilogram: $1 \text{ J} = 1 \text{ kg} \cdot \text{m}^2/\text{s}^2$.) The mass of product (1 mol $H_2O$) is about 3 ng *less* than the combined masses of the reactants (1 mol $H_2$ plus $\tfrac{1}{2}$ mol $O_2$), a change too small to measure with even the most sophisticated equipment. Such a minute mass change in chemical reactions allows us to assume that mass is conserved.

The much larger mass change that accompanies a nuclear process reflects the enormous energy required to bind the nucleus together. Consider, for example, the change in mass and energy that occurs when a carbon-12 nucleus forms from its nucleons: six protons and six neutrons. We calculate this change in mass by combining the mass of six H *atoms* and six neutrons and then subtracting the sum from the mass of one $^{12}$C *atom*. This procedure cancels the masses of the electrons [six $e^-$ (in six $^1$H atoms) cancels six $e^-$ (in one $^{12}$C atom)]. The mass of one $^1$H atom is 1.007825 amu, and the mass of one neutron is 1.008665 amu, so we have

$$\text{Mass of six } ^1\text{H atoms} = \quad 6.046950 \text{ amu}$$
$$\underline{\text{Mass of six neutrons} = \quad 6.051990 \text{ amu}}$$
$$\text{Total mass} \qquad\quad = 12.098940 \text{ amu}$$

The mass of one $^{12}$C atom is 12 amu (exactly). The difference in mass ($\Delta m$) is the mass of the product minus the mass of the reactant, so

$$\Delta m = 12.000000 \text{ amu} - 12.098940 \text{ amu}$$
$$= -0.098940 \text{ amu}/^{12}\text{C} = -0.098940 \text{ g/mol } ^{12}\text{C}$$

Note that *the mass of the nucleus is **less** than the combined masses of its nucleons.* This mass change ($9.90 \times 10^{-2}$ g/mol) is more than 30 million times that of the previous chemical reaction ($3.18 \times 10^{-9}$ g/mol) and easily observed on any laboratory balance.

The mass decrease that occurs when nucleons are combined into a nucleus is called the **mass defect ($\Delta m$).** The energy equivalent of the mass defect for $^{12}$C is

$$\Delta E = \Delta mc^2 = (-9.8940 \times 10^{-5} \text{ kg/mol}) (2.9979 \times 10^8 \text{ m/s})^2$$
$$= -8.8921 \times 10^{12} \text{ J/mol} = -8.8921 \times 10^9 \text{ kJ/mol}$$

This is the **nuclear binding energy** for carbon-12, the energy that would be released if 1 mol $^{12}C$ nuclei were formed from their component nucleons. The nuclear binding energy is the amount of energy holding the nucleus together, and it is also the amount required to break up the nucleus into its individual nucleons:

$$Nucleons \rightarrow nucleus + binding\ energy$$

$$Nucleus + binding\ energy \rightarrow nucleons$$

Thus, the nuclear binding energy of nuclei is qualitatively similar to the heat of formation, or the sum of bond energies, of compounds. However, quantitatively, nuclear binding energies are typically several million times greater. ◆

We use joules to express the binding energy per mole of nuclei, but the joule is an impractically large unit to express the binding energy of a single nucleus. Instead, nuclear scientists use the **electron volt (eV),** the energy an electron acquires when it moves through a potential difference of one volt:

$$1\ eV = 1.602 \times 10^{-19}\ J$$

Binding energies are commonly expressed in *megaelectron volts* (MeV):

$$1\ MeV = 10^6\ eV = 1.602 \times 10^{-13}\ J$$

A particularly useful factor converts a given mass defect in atomic mass units to its energy equivalent in electron volts:

$$1\ amu = 931.5 \times 10^6\ eV = 931.5\ MeV \qquad \textbf{(21.7)}$$

Earlier we found the mass defect of the $^{12}C$ nucleus as $-0.098940$ amu. Therefore, the binding energy per $^{12}C$ nucleus is

$$\frac{Binding\ energy}{^{12}C\ nucleus} = -0.098940\ amu \times \frac{931.5\ MeV}{1\ amu} = -92.16\ MeV$$

We can compare the stability of nuclides of different elements by determining the *binding energy per nucleon*. For $^{12}C$, we have

$$\frac{Binding\ energy}{nucleon} = \frac{-92.16\ MeV}{12\ nucleons} = -7.680\ MeV/nucleon$$

◆ **The Force That Binds Us.** According to current theory, the nuclear binding energy is related to the strong force, which holds nucleons together in a nucleus. There are three other fundamental forces: (1) the weak nuclear force, which is important in $\beta$ decay, (2) the electrostatic force that we observe between charged particles, and (3) the gravitational force. Toward the end of his life, Albert Einstein tried unsuccessfully to develop a theory to explain how the four forces were really different aspects of one unified force that governed all nature. Through experiments carried out with the next generation of giant accelerators, modern particle physicists will attempt to realize Einstein's dream.

---

SAMPLE PROBLEM 21.6 _____

### Calculating the Binding Energy Per Nucleon

**Problem:** Iron-56 is an extremely stable nuclide. Compute the binding energy per nucleon for $^{56}Fe$ and compare it with that of $^{12}C$. Mass of $^{56}Fe$ atom = 55.934939 amu; mass of $^1H$ atom = 1.007825 amu; mass of neutron = 1.008665 amu.

**Plan:** Iron-56 has 26 protons and 30 neutrons in its nucleus. We calculate the mass defect ($\Delta m$) by subtracting the sum of the masses of 26 H atoms and 30 neutrons from the given mass of one $^{56}Fe$ atom. Then we multiply $\Delta m$ by the equivalent in MeV and divide by 56 (number of nucleons) to obtain the binding energy/nucleon.

**Solution:** Calculating $\Delta m$:

$$\Delta m = mass\ {}^{56}Fe\ atom - [(26 \times mass\ H\ atom) + (30 \times mass\ neutron)]$$

$$= 55.934939\ amu - [(26 \times 1.007825\ amu) + (30 \times 1.008665\ amu)]$$

$$= -0.52846\ amu$$

Calculating the binding energy per nucleon:

$$\frac{\text{Binding energy}}{\text{nucleon}} = \frac{-0.52846 \text{ amu} \times 931.5 \text{ MeV/amu}}{56 \text{ nucleons}}$$

$$= -8.790 \text{ MeV/nucleon}$$

An $^{56}$Fe nucleus would release more energy if it formed from its nucleons than would $^{12}$C ($-7.680$ MeV/nucleon), **so $^{56}$Fe is more stable.**

**Check:** The answer is consistent with the great stability of $^{56}$Fe. In view of the number of decimal places in the values, rounding to check the math is useful only to indicate a *major* error. The number of nucleons (56) is an exact number, so we retain four significant figures.

**FOLLOW-UP PROBLEM 21.6**

Uranium-235 is the essential component of the fuel in nuclear energy facilities. Calculate the binding energy per nucleon of $^{235}$U. Is it more or less stable than $^{12}$C? (Mass of $^{235}$U atom = 235.043924 amu)

Similar calculations for other nuclides show that the binding energy per nucleon varies considerably. *The greater the size of the binding energy per nucleon, the more strongly the nucleons are held together and the more stable the nuclide.* A plot of the negative of the binding energy per nucleon vs. mass number provides information about nuclide stability and the processes nuclides undergo to form more stable nuclides (Figure 21.11). Nuclides with fewer than 10 nucleons have a relatively small binding energy per nucleon. Apparently, these nuclides have too few nucleons for optimum interaction. The $^{4}$He nucleus has an exceptionally large value, however, which is why it is emitted intact as an $\alpha$ particle. Above $A = 12$, the binding energy per nucleon varies from about $-7.6$ to $-8.8$ MeV.

The most important observation is that *the binding energy per nucleon gradually becomes larger for elements up to $A \approx 60$, then slowly becomes smaller.* That is, up to around 60 nucleons, nuclides become more stable with increasing mass number, but addition of more nucleons lowers stability. The existence

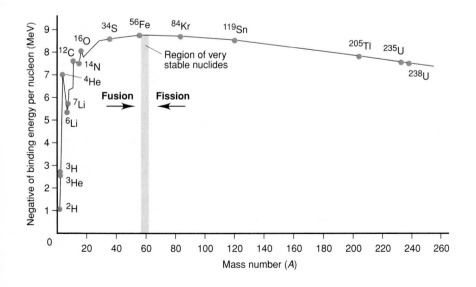

**FIGURE 21.11**

**The variation in binding energy per nucleon.** A plot of the negative of the binding energy per nucleon vs. mass number shows that nuclear stability is greatest in the region near $^{56}$Fe. Lighter nuclei may undergo fusion to become more stable; heavier ones may undergo fission. Note the exceptional stability of $^{4}$He among extremely light nuclei.

of a peak of stability suggests that there are two ways nuclides can increase
their binding energy per nucleon:

1. A heavier nucleus can split into lighter ones (closer to $A \approx 60$) in a
**fission** process. Since the product nuclei have greater binding energy per
nucleon (are more stable) than the reactant, excess energy is released.
Nuclear power plants generate energy through fission, as do atomic bombs.

2. Lighter nuclei, on the other hand, can combine to form a heavier one
(closer to $A \approx 60$) in the process of **fusion.** Once again, the product is more
stable than the reactants, and energy is released. The Sun and other stars
generate energy through fusion. In the hydrogen bomb and all current ven-
tures into developing fusion as a useful energy source, hydrogen nuclei fuse
to form the very stable helium-4 nucleus.

In the remainder of this section, we examine the processes of fission and
fusion and the facilities designed to utilize them.

### The Process of Nuclear Fission

During the mid-1930s, Enrico Fermi and co-workers bombarded uranium
($Z = 92$) with neutrons in an attempt to synthesize transuranium elements.
Many of the unstable nuclides produced were tentatively identified as having $Z > 92$, but other scientists were skeptical. Four years later, the German
chemist Otto Hahn and his associate F. Strassmann showed that one of these
unstable nuclides was an isotope of barium ($Z = 56$). The Austrian physicist
Lise Meitner, a co-worker of Hahn, and her nephew Otto Frisch proposed
that barium resulted from the *splitting* of the uranium nucleus into *smaller*
nuclei, a process they named *fission* because of its similarity to the reproductive fission of a cell. ◆

The $^{235}$U nucleus can split in many different ways, giving rise to various
product nuclei, but all routes have the same general features. Figure 21.12
depicts one of these fission patterns. Neutron bombardment results in a
highly excited $^{236}$U nucleus, which splits apart in $10^{-14}$ second. The prod-

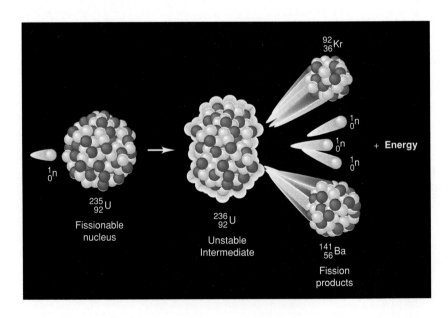

**FIGURE 21.12**

**Induced fission of $^{235}$U.** A neutron captured by
a $^{235}$U nucleus results in an extremely unstable
$^{236}$U nucleus, which becomes distorted in the
act of splitting. In this case, which shows one of
many possible splitting patterns, the products
are $^{92}$Kr and $^{141}$Ba. Three neutrons and a great
deal of energy are released also.

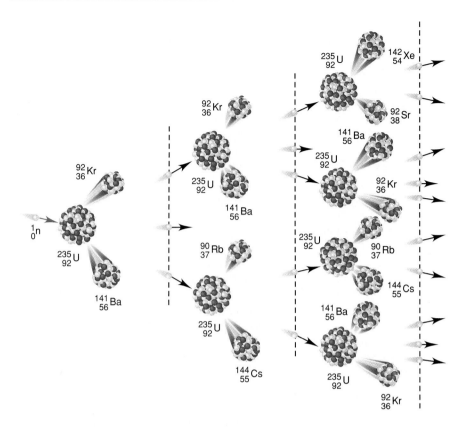

**FIGURE 21.13**

**A chain reaction of $^{235}$U.** If a sample exceeds the critical mass, neutrons produced by the first fission event collide with other nuclei, causing their fission and the production of more neutrons to continue the process. Note that various product nuclei form. The vertical dashed lines indicate each succeeding "generation" of neutrons.

ucts are two nuclei of unequal mass, two or three neutrons (average of 2.4), and a large amount of energy. A single $^{235}$U nucleus releases $3.5 \times 10^{-11}$ J when it splits; a mole of $^{235}$U (about $\frac{1}{2}$ lb) releases $2.1 \times 10^{13}$ J—a billion times as much energy as burning $\frac{1}{2}$ lb of coal (about $2 \times 10^{4}$ J)!

We harness the energy of nuclear fission, much of which appears as heat, by means of a **chain reaction** (Figure 21.13): the two to three neutrons released by fission of one nucleus collide with other fissionable nuclei and cause them to split, releasing more neutrons that collide with other nuclei, and so on, in a self-sustaining process. Thus, each fission event in a chain reaction releases two to three times as much energy as the preceding one. Whether or not a chain reaction occurs depends on the mass (and thus volume) of the fissionable sample. If the piece of uranium is large enough, product neutrons strike another fissionable nucleus *before* flying out of the sample, and a chain reaction ensues. The mass needed to achieve a chain reaction is the **critical mass.** If the sample has less than this mass *(subcritical)*, most of the product neutrons leave the sample before colliding with another $^{235}$U nucleus, and a chain reaction does not occur.

**Uncontrolled fission: the atomic bomb.** An uncontrolled chain reaction may become explosive, as the world first learned in World War II. In

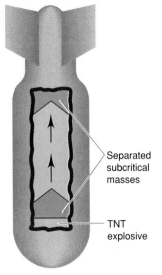

**FIGURE 21.14**
**Diagram of an atomic bomb.** Small TNT explosions bring subcritical masses together, and the chain reaction occurs.

**FIGURE 21.15**
Diagram of a light-water nuclear reactor.

August 1945, the United States detonated two atomic bombs over Japan. The horrible destructive power of these bombs was a major factor in the surrender of the Japanese a few days later.

An atomic bomb has quite a simple design (Figure 21.14). Small explosions of trinitrotoluene (TNT) bring subcritical masses of fissionable material together, the critical mass is exceeded, and the ensuing chain reaction brings about the nuclear explosion. The proliferation of nuclear power plants, which use fissionable materials to generate energy for electricity, has increased concern that more countries will have access to material for bombs. Only 1 kg of fissionable uranium was used in the first of the bombs dropped on Japan.

**Controlled fission: nuclear energy reactors.** Controlled fission can produce clean, plentiful electric power. Like a coal-fired power plant, a nuclear power plant generates heat to produce steam, which turns a turbine attached to an electric generator. In a coal plant, the heat is produced by burning coal; in a nuclear plant, by splitting uranium.

Heat generation takes place in the **reactor core** of a nuclear plant (Figure 21.15). The core contains the *fuel rods,* which consist of fuel enclosed in tubes of a corrosion-resistant zirconium alloy. The fuel is uranium(IV) oxide ($UO_2$) that has been *enriched* from 0.7% $^{235}U$, the natural abundance of this fissionable isotope, to the 3% to 4% $^{235}U$ required to sustain a chain reaction (see margin note, p. 202). Sandwiched between the fuel rods are movable *control rods* made of cadmium or boron, substances that absorb neutrons efficiently. When the control rods are lowered, the chain reaction slows as fewer neutrons are available to bombard uranium atoms; when they are raised, the chain reaction speeds up. Neutrons that leave the fuel-rod assembly collide with a *reflector,* usually made of a beryllium alloy, which absorbs very few neutrons. Reflecting the neutrons back to the fuel rods speeds the chain reaction.

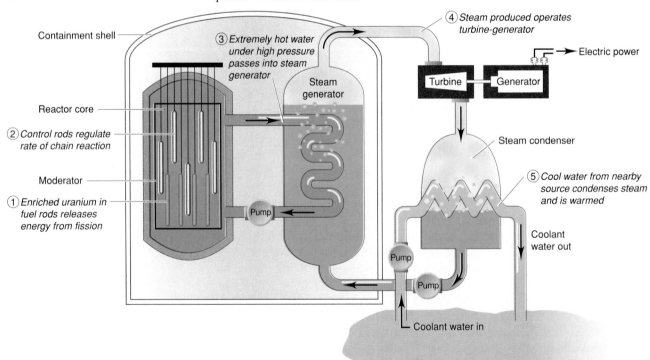

Flowing around the fuel and control rods in the reactor core is the *moderator,* a substance that slows the neutrons, making them much better at causing fission than the fast ones emerging from the fission event itself. In most modern reactors, the moderator also acts as a *coolant,* the fluid that transfers the released heat to the steam-producing region. Because $^1H$ absorbs neutrons, *light-water reactors* simply use $H_2O$ as the moderator; in heavy-water reactors, $D_2O$ is used. The advantage of $D_2O$ is that it absorbs very few neutrons, leaving more available for fission, so heavy-water reactors can use *unenriched* uranium. As the coolant flows around the encased fuel, pumps circulate it through coils that transfer its heat to the water reservoir. Steam formed in the reservoir turns the turbine that runs the generator. The steam is then condensed by cooling with water from a nearby lake or river and returned to the water reservoir. ◆

Some major accidents at nuclear plants have caused a decidedly negative public reaction. In 1979, malfunctions of coolant pumps and valves at the Three-Mile Island facility in Pennsylvania led to melting of some of the fuel, serious damage to the reactor core, and the release of radioactive gases into the atmosphere. In 1986, failure of a cooling system at the Chernobyl plant in Ukraine of the former Soviet Union caused a much greater melting of fuel and an uncontrolled reaction. High-pressure steam and ignited graphite moderator rods caused the reactor building to explode and expel radioactive debris. Carried by prevailing winds, the particles contaminated vegetables and milk in much of Europe. Thousands of people living near the accident may yet develop cancer from radiation exposure. The design of the Chernobyl plant was particularly unsafe in that the reactor was not enclosed in a massive containment building, as are the nuclear reactors in the United States.

As our need for energy grows, new reactors with enhanced safety features will be designed. However, even a smoothly operating plant has certain problems. The water used to condense the steam is several degrees warmer when returned to the river, which can result in *thermal pollution,* a condition that harms aquatic organisms (Section 12.3). A more serious problem is one of *nuclear waste disposal.* Many of the fission products formed in nuclear reactors have long half-lives, and no satisfactory plan for their permanent disposal has been devised. Proposals to place the waste in containers and bury them in deep bedrock cannot possibly be field-tested for the thousands of years the material will remain harmful. Leakage of radioactive material into ground water is always a danger, and earthquakes can occur even in geologically stable regions. It remains to be seen whether fission reactors that operate and dispose of the waste safely and inexpensively are realistic energy options.

### The Promise of Nuclear Fusion

Nuclear fusion is the ultimate source of all the energy on Earth. Nearly every nonnuclear source depends, directly or indirectly, on the energy produced by nuclear fusion in the Sun. The isotopes that provide us with energy from fission were formed, *along with all the other elements,* through fusion reactions within stars, as the upcoming Chemical Connections essay describes. Much research is now being devoted to making nuclear fusion a practical, direct source of energy.

◆ **"Breeding" Nuclear Fuel.**
Uranium-235 is not an abundant isotope. One solution to a potential fuel shortage is a *breeder reactor,* designed to consume one type of nuclear fuel as it produces another. Fuel rods are surrounded by natural $U_3O_8$, which contains 99.3% *nonfissionable* $^{238}U$ atoms. As fast neutrons, formed during $^{235}U$ fission, escape the fuel rod, they collide with $^{238}U$, which forms $^{239}Pu$, another fissionable nucleus:

$$^{238}_{92}U + ^1_0n \longrightarrow ^{239}_{92}U \; (t_{1/2} = 23.5 \text{ min})$$

$$^{239}_{92}U \longrightarrow ^{239}_{93}Np + ^{\;\;0}_{-1}\beta \; (t_{1/2} = 2.35 \text{ days})$$

$$^{239}_{93}Np \longrightarrow ^{239}_{94}Pu + ^{\;\;0}_{-1}\beta \; (t_{1/2} = 2.4 \times 10^4 \text{ yr})$$

Despite the ability of breeder reactors to make fuel as they operate, they are difficult and expensive to build, and the $^{239}Pu$ is extremely toxic and long lived. To date, a few have been built in the United States and there are several in Europe.

## Chemistry in Cosmology:
## Origin of the Elements in the Stars

How did the universe begin? Where did matter come from? How were the elements formed? Every culture has creation myths that address such questions, but only recently have astronomers, physicists, and chemists begun to offer a scientific explanation. The most accepted current model proposes that a sphere of unimaginable properties—diameter $10^{-28}$ cm, density $10^{96}$ g/mL (for a nucleus, density $\approx 10^{14}$ g/mL), and temperature $10^{32}$ K—exploded in a "Big Bang," for reasons as yet unknown, and distributed its contents through the void of space. Cosmologists consider this moment the beginning of time.

One second later, the universe was an expanding plasma of neutrons, protons, and electrons, denser than rock and hotter than an exploding H-bomb (about $10^{10}$ K). During the next few minutes, it was a gigantic fusion reactor creating the first atomic nuclei, $^2$H, $^3$He, and $^4$He. After 10 minutes, more than 25% of the mass of the universe was converted into $^4$He and one-thousandth as much into $^2$H. About 100 million years later, nearly 20 billion years ago, gravitational forces pulled this cosmic mixture into primitive, contracting stars.

This account of the origin of the universe is based on the observation of spectra from the sun, other stars, nearby galaxies, and cosmic (interstellar) dust. Spectral analysis of planets and chemical analysis of the Earth, moon, meteors, and cosmic-ray particles furnishes data about isotope abundance. From these have been developed a model for **stellar nucleogenesis,** the origin of the elements in the stars. The overall process occurs in several stages during a star's evolution, each involving a contraction of the star that produces higher temperature and heavier nuclei. Such events are forming elements in stars today (Figure 21.B). Here are the key stages in the process:

1. *Hydrogen burning produces He.* The initial contraction of a star heats its core to about $10^7$ K, at which point a fusion process called *hydrogen burning* occurs that produces helium from the abundant protons:

$$4\,^1_1\text{H} \rightarrow\,^4_2\text{He} + 2\,^0_1\beta + 2\gamma + \text{energy}$$

2. *Helium burning produces C, O, Ne, and Mg.* After several billion years of H burning, about 10% of the $^1$H is consumed, and the star contracts further. The $^4$He forms a dense core, hot enough ($2 \times 10^8$ K) to fuse $^4$He. The energy released during *helium burning* expands the remaining $^1$H into a vast envelope: the star becomes a *red giant,* more than 100 times its original diameter.

Within its core, pairs of $^4$He nuclei ($\alpha$ particles) fuse into unstable $^8$Be ($t_{1/2} = 2 \times 10^{-16}$ s), which collides with a third $^4$He to form stable $^{12}$C. Further fusion with $^4$He creates nuclei up to $^{24}$Mg:

$$^{12}\text{C} \xrightarrow{\alpha}\,^{16}\text{O} \xrightarrow{\alpha}\,^{20}\text{Ne} \xrightarrow{\alpha}\,^{24}\text{Mg}$$

3. *Elements through Fe and Ni form.* For another 10 million years, $^4$He is consumed, and the heavier nuclei form a core, which contracts and heats, expanding the star to a *supergiant.* In the hot core ($7 \times 10^8$ K), *carbon and oxygen burning* occur:

$$^{12}\text{C} +\,^{12}\text{C} \longrightarrow\,^{23}\text{Na} +\,^1\text{H}$$

$$^{12}\text{C} +\,^{16}\text{O} \longrightarrow\,^{28}\text{Si} + \gamma$$

Absorption of $\alpha$ particles forms nuclei up to calcium-40:

$$^{12}\text{C} \xrightarrow{\alpha}\,^{16}\text{O} \xrightarrow{\alpha}\,^{20}\text{Ne} \xrightarrow{\alpha}\,^{24}\text{Mg} \xrightarrow{\alpha}$$

$$^{28}\text{Si} \xrightarrow{\alpha}\,^{32}\text{S} \xrightarrow{\alpha}\,^{36}\text{Ar} \xrightarrow{\alpha}\,^{40}\text{Ca}$$

Further contraction and heating to $3 \times 10^9$ K allow reactions in which nuclei release neutrons, protons, and $\alpha$ particles and then recapture them. As a result, nuclei with lower binding energies supply nucleons to create those with higher binding energies. The process, which takes only a few minutes, stops at iron ($A = 56$) and nickel ($A = 58$), the nuclei with the highest binding energies.

4. *Heavier elements form.* For very massive stars, the next stage is the most spectacular. With all the fuel consumed, the core collapses within a second. Many Fe and Ni nuclei break down into neutrons and protons. Protons capture electrons to form neutrons, and the entire core forms an incredibly dense *neutron star.* As the core implodes, the outer layers explode in a *supernova,* which expels material throughout space. A supernova occurs every few hundred years in each galaxy; one was observed from the southern hemisphere in 1987, about 160,000 years after the event occurred (Figure 21.C). The heavier elements form during supernovas and in *second-generation stars,* those that coalesce from interstellar $^1$H and $^4$He and the debris of exploded first-generation stars.

Heavier elements form through *neutron-capture* processes. In the *s-process,* a nucleus captures a neutron and emits a $\gamma$ ray. Days, months, or even thousands of years later, the nucleus emits a $\beta$ particle to form the next element, as in this conversion of $^{68}$Zn to $^{70}$Ge:

$$^{68}\text{Zn} \xrightarrow{n}\,^{69}\text{Zn} \xrightarrow{\beta}\,^{69}\text{Ga} \xrightarrow{n}\,^{70}\text{Ga} \xrightarrow{\beta}\,^{70}\text{Ge}$$

The stable isotopes of most heavy elements form by the s-process.

Less stable isotopes and those with $A$ greater than 230 cannot form by the s-process because their half-lives are too short. These form by the *r-process* during the fury of the supernova. Multiple neutron captures, followed by multiple $\beta$ decays, occur in a second, as when $^{56}$Fe is converted to $^{79}$Br:

$$^{56}_{26}\text{Fe} + 23\,^1_0\text{n} \rightarrow\,^{79}_{26}\text{Fe} \rightarrow\,^{79}_{35}\text{Br} + 9\,^{\phantom{-}0}_{-1}\beta$$

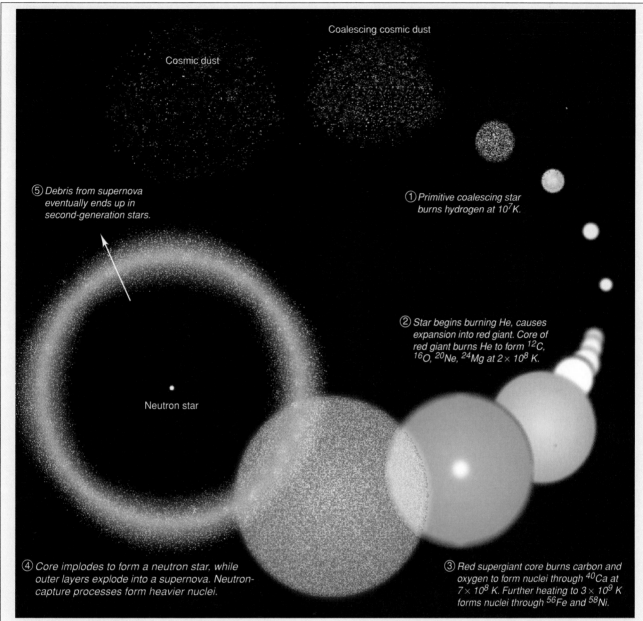

⑤ *Debris from supernova eventually ends up in second-generation stars.*

① *Primitive coalescing star burns hydrogen at $10^7$ K.*

Coalescing cosmic dust

Cosmic dust

② *Star begins burning He, causes expansion into red giant. Core of red giant burns He to form $^{12}C$, $^{16}O$, $^{20}Ne$, $^{24}Mg$ at $2 \times 10^8$ K.*

Neutron star

④ *Core implodes to form a neutron star, while outer layers explode into a supernova. Neutron-capture processes form heavier nuclei.*

③ *Red supergiant core burns carbon and oxygen to form nuclei through $^{40}Ca$ at $7 \times 10^8$ K. Further heating to $3 \times 10^9$ K forms nuclei through $^{56}Fe$ and $^{58}Ni$.*

**FIGURE 21.B** Element synthesis in the life cycle of a star.

**FIGURE 21.C** A view of Supernova 1987A.

We know from the heavy elements present in the Sun that it is at least a second-generation star presently undergoing hydrogen burning. Together with its planets, it was formed from the dust of exploded stars about $4.6 \times 10^9$ years ago. This means that many of the elements on Earth, including some within and around you, may be older than the solar system itself!

Any theory of element formation must be consistent with the element abundances we observe. Although local compositions differ, such as those between Earth and Sun, large regions of the universe have, on average, similar compositions. Thus, scientists believe that element forming reaches a dynamic equilibrium, which leads to *relatively constant amounts of the isotopes.* In Chapter 23, we examine the cosmic and terrestrial abundances of the elements.

To see the advantages of fusion, consider one of the most discussed fusion reactions, in which deuterium and tritium react:

$$^2_1H + {}^3_1H \rightarrow {}^4_2He + {}^1_0n$$

This reaction produces $1.7 \times 10^9$ kJ/mol, an enormous amount of energy, with no long-lived toxic radionuclide byproducts. Since we can obtain deuterium from water and produce tritium in accelerators, this process seems very promising, at least in principle. However, some difficult problems exist.

Fusion requires enormous energy in the form of heat to give the positively charged nuclei enough kinetic energy to force them together. The fusion of deuterium and tritium, for example, occurs at practical rates at about $10^8$ K, hotter than the Sun's core! How can such temperatures be achieved? The *hydrogen bomb* is a fusion reaction that uses an atomic bomb to provide the heat; obviously, a public energy facility cannot operate by detonating atomic bombs. A recent approach that shows promise employs focused lasers to heat the fusion reactants.

The reactor must also have a suitable core. Atoms are stripped of their electrons at high temperatures, and a gaseous *plasma* results, a neutral mixture of positive nuclei and electrons. Because of the extreme temperatures needed for fusion, no *material* can contain the plasma, so the most successful approach has been to enclose it within a magnetic field. The *tokamak* design has a donut-shaped container in which a helical magnetic field confines the plasma and prevents it from contacting the walls (Figure 21.16). Scientists at the Princeton University experimental fusion facility have recently achieved some success in generating energy from fusion; as a practical, everyday source of energy, however, fusion still seems a long way off.

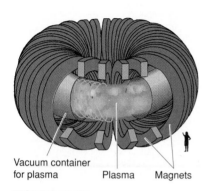

Vacuum container
for plasma          Plasma        Magnets

**FIGURE 21.16**
**The tokamak design for magnetic containment of a fusion plasma.** The donut-shaped chamber of the tokamak contains the plasma within a helical magnetic field.

### Section Summary

The mass of a nucleus is less than the sum of its nucleon masses by the mass defect. The energy equivalent to the mass defect is the nuclear binding energy, usually expressed in units of MeV. The binding energy per nucleon is a measure of nuclide stability and varies with number of nucleons. Nuclides with $A \approx 60$ are most stable. Lighter nuclides can fuse, and heavier nuclides can split to become more stable. In nuclear fission, a neutron causes a nucleus to split, releasing neutrons that split other nuclei to produce a chain reaction. A nuclear power plant controls the rate of the chain reaction to produce the heat that creates steam, which is used to generate electricity. Potential hazards, such as radiation leaks, thermal pollution, and disposal of nuclear waste, limit the current use of nuclear power. Nuclear fusion requires extremely high temperatures and is not yet a practical energy source. The elements were formed through a complex series of nuclear reactions in evolving stars.

### Chapter Perspective

*With this chapter, our earlier picture of the nucleus as a static point of positive mass at the atom's core has changed radically. Now we picture a dynamic body, capable of a host of actions that involve enormous amounts of energy. Our understanding of this minute system results from its interactions with matter, and our attempts to apply its behavior to benefit society have created some of the most fascinating and challenging fields in science today. In Chapter 22, we examine the transition elements and their compounds, but nuclear processes appear again in our examination of element abundances and separations in Chapter 23.*

# For Review and Reference

## Key Terms

**SECTION 21.1**
radioactivity
nucleon
nuclide
isotope
beta (β) particle
alpha (α) particle
gamma (γ) ray
alpha decay
beta decay
positron decay
positron
electron capture
gamma emission
decay (disintegration)
  series

*N/Z* ratio
band of stability
strong force

**SECTION 21.2**
Geiger-Müller counter
scintillation counter
activity *(A)*
curie (Ci)
decay constant
half-life *($t_{1/2}$)*
radioisotopic dating

**SECTION 21.3**
nuclear transmutation
deuteron

particle accelerator
transuranium element

**SECTION 21.4**
excitation
nonionizing radiation
ionization
ionizing radiation
rad (*radiation-absorbed dose*)
rem (*roentgen equivalent for man*)
free radical
background radiation

**SECTION 21.5**
tracer

**SECTION 21.6**
mass defect (Δ*m*)
nuclear binding energy
electron volt (eV)
fission
fusion
chain reaction
critical mass
reactor core
stellar nucleogenesis

## Key Equations and Relationships

**21.1** Defining the unit of radioactivity (curie, Ci) (p. 942):

$$1 \text{ Ci} = 3.70 \times 10^{10} \text{ disintegrations per second (dps)}$$

**21.2** Expressing the decay rate for radioactive nuclei (p. 942):

$$\text{Decay rate} = -\frac{\Delta N}{\Delta t} = kN$$

**21.3** Finding the number of nuclei remaining after a given time, $N_t$ (p. 943):

$$\ln \frac{N_0}{N_t} = kt$$

**21.4** Finding the half-life of a radioactive nuclide (p. 943):

$$t_{1/2} = \frac{\ln 2}{k}$$

**21.5** Calculating the time to reach a given specific activity (age of an object in radioisotopic dating) (p. 945):

$$t = \frac{1}{k} \ln \frac{A_0}{A_t}$$

**21.6** Using Einstein's equation and the mass defect to calculate the nuclear binding energy (p. 957):

$$\Delta E = \Delta mc^2$$

**21.7** Relating the atomic mass unit to its energy equivalent in MeV (p. 958):

$$1 \text{ amu} = 931.5 \times 10^6 \text{ eV} = 931.5 \text{ MeV}$$

## Answers to Follow-up Problems

**21.1** $^{133}_{54}\text{Xe} \rightarrow ^{133}_{55}\text{Cs} + ^{0}_{-1}\beta$

**21.2** Phosphorus-31 has a slightly higher *N/Z* ratio and an even *N* (16).

**21.3** (a) *N/Z* = 1.35; too high for this region of band: β decay

(b) Mass too high for stability: α decay

**21.4** $\ln A_t = -kt + \ln A_0$

$$= -\left(\frac{\ln 2}{15 \text{ h}} \times 4 \text{ days} \times \frac{24 \text{ h}}{1 \text{ day}}\right) + \ln (2.5 \times 10^9)$$

$$= 17.20$$

$$A_t = 3.0 \times 10^7 \text{ dps}$$

**21.5** $t = \frac{1}{k} \ln A_0/A_t = \frac{5730 \text{ yr}}{\ln 2} \ln \left(\frac{15.3 \text{ d/min} \cdot \text{g}}{9.41 \text{ d/min} \cdot \text{g}}\right)$

$$= 4.02 \times 10^3 \text{ yr}$$

The mummy case is about 4000 years old.

**21.6** $^{235}\text{U}$ has 92 $^1_1\text{p}$ and 143 $^1_0\text{n}$.

$\Delta m = 235.043924 \text{ amu}$
$\quad\quad - [(92 \times 1.007825 \text{ amu}) + (143 \times 1.008665 \text{ amu})]$
$\quad = -1.9151 \text{ amu}$

$$\frac{\text{Binding energy}}{\text{nucleon}} = \frac{-1.9151 \text{ amu} \times \dfrac{931.5 \text{ MeV}}{1 \text{ amu}}}{235 \text{ nucleons}}$$

$$= -7.591 \text{ MeV/nucleon; therefore, less stable than } ^{12}\text{C}$$

## Sample Problem Titles

**21.1** Writing Equations for Nuclear Reactions (p. 937)
**21.2** Predicting Nuclear Stability (p. 939)
**21.3** Predicting the Mode of Nuclear Decay (p. 940)
**21.4** Determining the Number of Radioactive Nuclei (p. 943)
**21.5** Applying Radiocarbon Dating (p. 945)
**21.6** Calculating the Binding Energy per Nucleon (p. 958)

# Problems

Problems with a green number are answered at the back of the text. Most sections include three categories of problems separated by a green rule: concept review questions, *paired* skill-building exercises, and problems in a relevant context.

## Radioactive Decay and Nuclear Stability

(Sample Problems 21.1 to 21.3)

**21.1** How do chemical and nuclear reactions differ in
  (a) Magnitude of the energy change?
  (b) Effect on rate of increasing temperature?
  (c) Effect on rate of higher reactant concentration?
  (d) Effect on yield of higher reactant concentration?

**21.2** Sulfur has four naturally occurring isotopes. The one with the lowest mass number is sulfur-32, which is also the most abundant (95.02%).
  (a) What percentage of the S atoms in a matchhead are $^{32}S$?
  (b) The isotopic mass of $^{32}S$ is 31.972070 amu. Is the atomic mass larger, smaller, or equal to this mass? Explain.

**21.3** What evidence led Marie Curie to conclude that
  (a) Radioactivity is a property of the element and not the compound in which it is found?
  (b) A highly radioactive element, aside from uranium, occurs in pitchblende?

**21.4** Which of the following types of radioactive decay produce an atom of a *different* element: (a) alpha; (b) beta; (c) gamma; (d) positron; (e) electron capture? Show how $Z$ and $N$ change, if at all, with each type.

**21.5** Why is $^3_2He$ stable whereas $^2_2He$ is so unstable that it has never been detected?

**21.6** How do the modes of decay differ for a neutron-rich nuclide and a proton-rich nuclide?

**21.7** Why can't you use the position of a nuclide relative to the band of stability to predict whether it is more likely to decay by positron emission or by electron capture?

---

**21.8** Write balanced nuclear equations for the following:
  (a) Alpha decay of $^{234}_{92}U$
  (b) Electron capture by neptunium-232
  (c) Positron emission by $^{12}_7N$
  (d) Beta decay of sodium-26

**21.9** Write balanced nuclear equations for the following:
  (a) Beta emission by magnesium-27
  (b) Neutron emission by $^9_3Li$
  (c) Electron capture by $^{103}_{46}Pd$
  (d) Simultaneous beta and neutron emission by helium-8

**21.10** Write balanced nuclear equations for the following:
  (a) Formation of $^{48}_{22}Ti$ through positron emission
  (b) Formation of silver-107 through electron capture
  (c) Formation of polonium-206 through $\alpha$ decay
  (d) Production of $^{241}_{95}Am$ through $\beta$ decay

**21.11** Write balanced nuclear equations for the following:
  (a) Formation of $^{186}Ir$ through electron capture
  (b) Formation of francium-221 through $\alpha$ decay
  (c) Formation of iodine-129 through $\beta$ decay
  (d) Formation of $^{52}Mn$ through positron emission

**21.12** Which nuclei would you predict to be stable: (a) $^{20}_8O$; (b) $^{59}_{27}Co$; (c) $^9_3Li$; (d) $^{146}_{60}Nd$? Why?

**21.13** Which nuclei would you predict to be stable: (a) $^{127}I$; (b) tin-106; (c) $^{68}As$; (d) $^{48}K$? Why?

**21.14** What is the most likely mode of decay for (a) $^{238}_{92}U$; (b) $^{48}_{24}Cr$; (c) $^{50}_{25}Mn$; (d) $^{61}_{26}Fe$?

**21.15** What is the most likely mode of decay for (a) $^{15}_6C$; (b) $^{120}_{54}Xe$; (c) $^{224}_{90}Th$; (d) $^{234}_{90}Th$?

**21.16** Why is $^{52}_{24}Cr$ the most stable isotope of chromium?

**21.17** Why is $^{40}_{20}Ca$ the most stable isotope of calcium?

---

**21.18** Radiochemists have proposed that a natural radioactive decay series probably began with $^{243}_{95}Am$, but all of this nuclide had decayed within the first few million years of the Earth's existence. This series would have begun by emission of an $\alpha$ particle, then a $\beta$ particle, then two more $\alpha$ particles. Write a balanced nuclear equation for each step.

**21.19** Why is helium found in deposits of uranium or thorium ores? What kind of radioactive emission produces it?

**21.20** In the natural radioactive series that starts with uranium-235, a sequence of $\alpha$ and $\beta$ emissions ends with lead-207. How many $\alpha$ and $\beta$ particles are emitted per atom of uranium-235?

## The Kinetics of Nuclear Change

(Sample Problems 21.4 and 21.5)

**21.21** What is the basis of detecting radioactive emissions in
  (a) A scintillation counter?
  (b) A Geiger-Müller counter?

**21.22** Why is radioactive decay considered a first-order process?

**21.23** After 1 minute, half the radioactive nuclei remain from an original sample of six nuclei. Is it valid to conclude that $t_{1/2}$ equals 1 minute? Would this conclusion be valid if the original sample contained $6 \times 10^{12}$ nuclei? Explain.

**21.24** Radioisotopic dating depends on the constant rate of decay and formation of various nuclides in a sample. How is the proportion of carbon-14 kept relatively constant in living organisms?

---

**21.25** What is the specific activity (in Ci/g) if 1.55 mg of an isotope emits $1.66 \times 10^6$ $\alpha$ particles per second?

**21.26** What is the specific activity (in Ci/g) if 2.6 g of an isotope emits $4.13 \times 10^8$ $\beta$ particles per hour?

**21.27** One trillionth of the atoms in a sample of radioactive isotope disintegrate in 1 h. What is the decay constant of the process?

**21.28** Measurement shows that $2.8 \times 10^{-10}\%$ of the atoms in a sample of radioactive isotope disintegrate in 1 yr. What is the decay constant of the process?

**21.29** If $1.00 \times 10^{-12}$ mol $^{135}$Cs emits $1.39 \times 10^5$ $\beta$ particles in one year, what is the decay constant?

**21.30** If $6.40 \times 10^{-9}$ mol $^{176}$W emits $1.07 \times 10^{15}$ positrons in one hour, what is the decay constant?

**21.31** A sample of bone contains enough $^{90}$Sr ($t_{1/2} = 29$ yr) to emit $8.0 \times 10^4$ $\beta$ particles per month. How long will it take for the emissions to decrease to $1.0 \times 10^4$ per month?

**21.32** What fraction of the $^{235}$U ($t_{1/2} = 7.0 \times 10^8$ yr) created when the Earth was formed would remain after $2.8 \times 10^9$ yr?

**21.33** A rock contains 270 $\mu$mol $^{238}$U ($t_{1/2} = 4.5 \times 10^9$ yr) and 110 $\mu$mol $^{206}$Pb. Assuming that all the $^{206}$Pb comes from decay of the $^{238}$U, estimate the rock's age.

**21.34** A fabric remnant from a burial site in the Near East has a $^{14}$C/$^{12}$C ratio of 0.735 of the original value. How old is the fabric?

**21.35** Plutonium-239 ($t_{1/2} = 2.41 \times 10^4$ yr) represents a serious nuclear waste disposal problem. If seven half-lives are required to reach a tolerable level of radioactivity, how long must the $^{239}$Pu be stored?

**21.36** A sample of $^{90}$Kr ($t_{1/2} = 32$ s) is to be used in a study of a patient's respiration. How soon after its creation must it be administered to the patient if the activity must be as least 90% of the original activity?

**21.37** A rock contains $2.1 \times 10^{-15}$ mol $^{232}$Th ($t_{1/2} = 1.4 \times 10^{10}$ yr). There are $9.5 \times 10^4$ fission tracks in the rock, each representing the fission of one atom of $^{232}$Th. How old does the rock appear to be?

**21.38** The approximate date of an earthquake in the San Francisco area is to be determined by measuring the $^{14}$C activity ($t_{1/2} = 5730$ yr) of parts of a tree that was uprooted during the event. The tree parts have an activity of 12.9 d/min · g C, and a living tree has 15.3 d/min · g C. What was the date of the earthquake?

**21.39** Volcanism was much more common in the distant past than today, and in some cases, specific events can be dated. A volcanic eruption melts a large area of rock and all gases are expelled. After cooling, the $^{40}_{18}$Ar accumulates from the ongoing decay of the $^{40}_{19}$K in the rock ($t_{1/2} = 1.25 \times 10^9$ yr). When a piece of the rock was analyzed, it was found to contain 1.38 mmol $^{40}$K and 1.14 mmol $^{40}$Ar. How long ago did the rock cool?

## Nuclear Transmutation: Induced Changes in Nuclei

**21.40** Irene and Frederic Curie-Joliot converted $^{27}_{13}$Al to $^{30}_{15}$P in 1933. Why was this transmutation significant?

**21.41** Early workers mistakenly thought neutron beams were gamma radiation. Why were they misled? What evidence led to the correct conclusion?

**21.42** Why must the electrical polarity of the tubes in a linear accelerator be reversed at very short time intervals?

**21.43** Why does bombardment with protons usually require higher energies than bombardment with neutrons?

**21.44** Determine the missing species in these transmutations and write a full nuclear equation from the shorthand notation: (a) $^{10}$B ($\alpha$,n)?; (b) $^{28}$Si (d,?) $^{29}$P (d= $^2$H); (c) ? ($\alpha$,2n) $^{244}$Cf.

**21.45** Determine the missing species in these transmutations and express the process in shorthand notation:
(a) Bombardment of a nuclide with a $\gamma$ photon yields a proton, a neutron, and $^{29}$Si.
(b) Bombardment of $^{252}$Cf with $^{10}$B yields 5 neutrons and a nuclide.
(c) Bombardment of $^{238}$U with a particle yields 3 neutrons and $^{239}$Pu.

**21.46** Names for elements 104, 105, and 106 have been recommended by the IUPAC as dubnium (Db), joliotium (Jl), and rutherfordium (Rf), respectively. They are synthesized from californium-249 by bombardment with carbon-12, nitrogen-15, and oxygen-18 nuclei, respectively. Four neutrons are formed in each reaction as well. (a) Write balanced nuclear equations for their formation. (b) Write the equations in shorthand notation.

## The Effects of Nuclear Radiation on Matter

**21.47** Gamma radiation and ultraviolet radiation cause different processes to occur in matter. What are these processes and how are they different?

**21.48** What is an ion pair and how does it form?

**21.49** Why is ionizing radiation more dangerous to children than to adults?

**21.50** Why is ·OH more dangerous in an organism than OH$^-$?

**21.51** A 70-kg person exposed to $^{90}$Sr receives $6.0 \times 10^5$ $\beta$ particles, each with an energy of $8.74 \times 10^{-14}$ J.
(a) How many rads does the person receive?
(b) If the RBE is 1.0, how many mrem is this?

**21.52** A laboratory rat weighs 265 g and receives $1.77 \times 10^{10}$ $\beta$ particles, each with an energy of $2.20 \times 10^{-13}$ J.
(a) How many rads does the animal receive?
(b) If the RBE is 0.75, how many mrem is this?

**21.53** If 2.50 pCi $^{239}$Pu stays in a 95-kg human for 65 h, and each disintegration has an energy of $8.25 \times 10^{-13}$ J, how many rads does the person receive? (1 pCi, picocurie = $1 \times 10^{-12}$ Ci.)

**21.54** A patient's brain is exposed for 27 min to 12.85 nCi of $^{60}$Co for treatment of a tumor. If the brain mass is 1.588 g and each $\beta$ particle emitted has an energy of $5.05 \times 10^{-14}$ J, what is the dose in rads?

## Applications of Radioisotopes

**21.55** Describe two ways that radioactive tracers are used in organisms.

**21.56** Why is neutron activation analysis (NAA) useful to art historians and criminologists?

**21.57** Positrons cannot penetrate matter more than a few atomic diameters, but positron emission of radiotracers can be monitored in medical diagnosis. Explain.

**21.58** A steel part is treated to contain some iron-59. Oil used to lubricate the part emits 298 $\beta$ particles (with the energy characteristic of $^{59}$Fe) per minute per mL of oil. What other information would you need to calculate the rate of removal of the steel from the part during use?

---

**21.59** The oxidation of methanol to formaldehyde can be accomplished by reaction with chromic acid:

$$6H^+(aq) + 3CH_3OH(aq) + 2H_2CrO_4(aq) \rightarrow$$
$$3CH_2O(aq) + 2Cr^{3+}(aq) + 8H_2O(l)$$

The atom transfer can be studied with the stable isotope tracer $^{18}$O and mass spectrometry. When a small amount of $CH_3{}^{18}OH$ is present in the alcohol reactant, $H_2C^{18}O$ forms. When a small amount of $H_2Cr^{18}O_4$ is present, $H_2{}^{18}O$ forms. Does chromic acid or methanol supply the O atom to the aldehyde? Explain.

**21.60** Neutron activation analysis bombards stable isotopes with neutrons. Depending on the isotope and the energy of the neutron, a wide variety of nuclear emissions are observed. What are the products when the following neutron-activated species decay?
(a) $^{52}_{23}V^* \rightarrow$ ———— [$\beta$ emission]
(b) $^{13}_{7}N^* \rightarrow$ ———— [positron emission]
(c) $^{28}_{13}Al^* \rightarrow$ ———— [$\beta$ emission]

## Fission and Fusion: The Interconversion of Mass and Energy

(Sample Problem 21.6)
(The following data are needed to solve some problems in this section: mass of $^1$H atom = 1.007825 amu; mass of neutron = 1.008665 amu.)

**21.61** Many scientists at first reacted skeptically to Einstein's equation, $E = mc^2$. Why?

**21.62** What is a mass defect, and how does it arise?

**21.63** When a nuclide forms from nucleons, is energy absorbed or released? Why?

**21.64** What is the binding energy per nucleon? Why is binding energy per nucleon, rather than per nuclide, used to compare nuclide stability?

**21.65** What is the minimum number of neutrons from each fission event that must be absorbed by other nuclei for a chain reaction to propagate?

**21.66** In what main way is fission different from radioactive decay? Are all fission events in a chain reaction identical? Explain.

**21.67** What is the purpose of enrichment in the preparation of fuel rods? How is it accomplished?

**21.68** Describe the nature and purpose of the following components of a nuclear reactor:
(a) control rods; (b) moderator; (c) reflector.

**21.69** State an advantage and a disadvantage of heavy-water reactors over light-water reactors.

**21.70** What are the expected advantages of fusion reactors over fission reactors?

**21.71** Account for the fact that there is a greater mass of iron in the Earth than any other element.

**21.72** Why do so many nuclides have isotopic masses that are close to multiples of 4 amu?

**21.73** What is the cosmic importance of unstable $^8$Be?

---

**21.74** A $^3$H nucleus decays with an energy of 0.01861 MeV. Convert this energy into (a) electron volts; (b) joules.

**21.75** Arsenic-84 decays with an energy of $1.57 \times 10^{-15}$ kJ per nucleus. Convert this energy into (a) electron volts; (b) MeV.

**21.76** How many joules are released when 1.0 mol $^{239}$Pu decays, if each nucleus releases 5.243 MeV?

**21.77** How many MeV are released per nucleus if $3.2 \times 10^{-3}$ mol chromium-49 releases $8.11 \times 10^5$ kJ?

**21.78** Oxygen-16 is one of the most stable nuclides. The mass of an $^{16}$O atom is 15.994915 amu. Calculate the binding energy (a) per nucleon in MeV; (b) per atom in MeV; (c) per mole in kJ.

**21.79** Lead-206 is the end product of $^{238}$U decay. One $^{206}$Pb atom has a mass of 205.974440 amu. Calculate the binding energy (a) per nucleon in MeV; (b) per atom in MeV; (c) per mole in kJ.

---

**21.80** The $^{80}$Br nuclide decays by either $\beta$ decay or electron capture. (a) What is the product of each process? (b) Which process releases more energy? (Masses of atoms: $^{80}$Br = 79.918528 amu; $^{80}$Kr = 79.916380 amu; $^{80}$Se = 79.916520 amu; neglect the mass of the electron involved.)

**21.81** The reaction that will probably power the first commercial fusion reactors is $^3_1H + ^2_1H \rightarrow ^4_2He + ^1_0n$. How much energy would be produced per mole of reaction? (Masses of atoms: $^3$H = 3.01605 amu; $^2$H = 2.0140 amu; $^4$He = 4.00260 amu)

## Comprehensive Problems

Problems with an asterisk (*) are more challenging.

**21.82** Early in the 19th century, before accurate atomic masses had been determined, a theory of atomic structure known as Prout's hypothesis held that all atoms were aggregations of H atoms. Support was found in the fact that the known atomic masses were close to integer multiples of the atomic mass of hydrogen, but it was abandoned as more accurate values were obtained. Discuss briefly whether stellar nucleogenesis provides a modern adaptation of the hypothesis.

**21.83** Plutonium "triggers" for nuclear weapons were manufactured at the Rocky Flats plant in Colorado. An 85-kg worker inhales a dust particle containing 1.00 $\mu$g $^{239}_{94}$Pu, which resides in his body for 16 h. How many rads does he receive? ($t_{1/2}$ of $^{239}$Pu = 2.41 $\times$ 10$^4$ yr; each disintegration releases 5.15 MeV.)

**\*21.84** Archeologists removed pieces of charcoal from a Native American campfire, burned them in $O_2$, and bubbled the $CO_2$ formed into $Ca(OH)_2$ solution (limewater). The $CaCO_3$ that precipitated was filtered and oven dried. A 4.38-g sample of the $CaCO_3$ had a radioactivity of 3.2 dpm. How long ago was the campfire?

**\*21.85** A 5.4-$\mu$g sample of $^{226}$RaCl$_2$ has a radioactivity of 1.5 $\times$ 10$^5$ dps. Calculate $t_{1/2}$ of $^{226}$Ra.

**21.86** How many rads does a 65-kg human receive each year from the approximately 10$^{-8}$ g of $^{14}_{6}$C naturally present in her body? ($t_{1/2}$ = 5730 yr; each disintegration releases 0.156 MeV)

**21.87** The major reaction taking place during hydrogen-burning in a young star is

$$4^1_1H \longrightarrow {}^4_2He + 2^0_1\beta + 2\gamma + energy$$

How much energy in MeV is released per He nucleus formed? Per mole of He? (Masses: $^1$H atom = 1.007825 amu; $^4$He atom = 4.00260 amu; positron = 5.48580 $\times$ 10$^{-4}$ amu)

**21.88** A precipitate of AgCl contains 175 nCi/g. The solution contains 0.338 pCi radioactive Ag$^+$/mL. What is the molar solubility of AgCl?

**21.89** Suggest a reason why the critical mass of a fissionable substance depends on its shape.

**\*21.90** Technetium-99m is a metastable nuclide used in numerous cancer diagnostic and treatment programs. It is prepared just prior to use because it decays rapidly through $\gamma$ emission: $^{99m}$Tc $\longrightarrow$ $^{99}$Tc + $\gamma$. Use the following data to determine (a) the half-life of $^{99m}$Tc; (b) the percentage of the isotope that would be lost if it takes 2.0 h to prepare and administer the dose.

| TIME (h) | $\gamma$ RAYS EMITTED/s |
|----------|-------------------------|
| 0        | 5000                    |
| 4        | 3150                    |
| 8        | 2000                    |
| 12       | 1250                    |
| 16       | 788                     |
| 20       | 495                     |

**\*21.91** How many curies are there in 1.0 mol $^{40}$K ($t_{1/2}$ = 1.25 $\times$ 10$^9$ yr)?

**21.92** Marie Curie studied elements that decay by $\alpha$, $\beta$, and $\gamma$ emissions. As a result, she proposed that the radioactivity of an element was the same in all of its compounds. Later workers learned that some elements that decay by electron capture have half-lives that vary slightly from one compound to another. Explain.

**\*21.93** The isotopic mass of $^{210}_{86}$Rn is 209.989669 amu. When it decays by electron capture it emits 2.368 MeV. What is the isotopic mass of the resulting nuclide?

**21.94** Many nuclides decay by positron emission or by electron capture. Far fewer decay by positron emission or by beta emission. Why is the first pair of decay modes much more common than the second?

**\*21.95** A metastable or excited form of $^{50}$Sc rearranges to its stable form by emitting $\gamma$ radiation with a wavelength of 8.73 pm. What is the change in mass of one mole when it undergoes this rearrangement?

**21.96** Isotopic abundances are relatively constant throughout the Earth's crust. Could the science of chemistry have developed if, for example, one sample of tin(II) oxide contained mostly $^{112}$Sn and another mostly $^{124}$Sn? Explain.

**21.97** Which isotope in each pair would you predict to be more stable? Why?

(a) $^{140}_{55}$Cs or $^{133}_{55}$Cs      (b) $^{79}_{35}$Br or $^{78}_{35}$Br

(c) $^{28}_{12}$Mg or $^{24}_{12}$Mg      (d) $^{14}_{7}$N or $^{18}_{7}$N

**21.98** Nuclear disarmament could be accomplished if weapons were not "replenished." The tritium in nuclear warheads decays with a half-life of 12.26 yr to helium, and must be periodically replaced or the weapon is useless. What fraction of the tritium is lost in 5 yr?

**21.99** A decay series starts with the synthetic isotope $^{239}_{92}$U. The first four steps are emission of a $\beta$ particle, another $\beta$, an $\alpha$ particle, and another $\alpha$. Write a balanced nuclear equation for each step. Which natural radioactive series could be started by this sequence?

**21.100** How long can a 48-lb child be exposed to 1.0 mCi of $^{222}$Rn before accumulating 1.0 mrad, if the energy of each disintegration is 5.59 MeV?

**21.101** Were organisms a billion years ago exposed to more or less ionizing radiation than similar organisms today? Explain.

**21.102** Carbon from the most recent remains of an extinct Australian marsupial, called *Diprotodon*, has an activity of 0.61 pCi/g. Modern carbon has $^{14}$C activity of 6.89 pCi/g. How long ago did the *Diprotodon* apparently become extinct?

**21.103** The reaction that allows for radiocarbon dating is the continual formation of carbon-14 in the upper atmosphere: $^{14}_{7}$N + $^1_0$n $\longrightarrow$ $^{14}_{6}$C + $^1_1$H. What is the energy change associated with this reaction in eV/reaction and in kJ/mol? (Masses of atoms: $^{14}$N = 14.003074 amu; $^{14}$C = 14.003241 amu; $^1$H = 1.007825 amu).

**21.104** What is the total nuclear binding energy of a $^7$Li nucleus in kJ/mol and in eV/nucleus? (Mass of $^7$Li atom = 7.016003 amu).

**\*21.105** Use Einstein's equation, the mass in grams of one amu, and the relation between electron volts and joules to find the energy equivalent (in MeV) of a mass defect of 1 amu.

# CHAPTER 22

## Concepts and skills to review

- properties of light (Section 7.1)
- ionic size, electron configuration, and magnetic behavior (Sections 8.2 and 8.4)
- valence-bond theory (Section 10.3)
- structural, geometric, and optical isomerism (Section 14.2)
- Lewis acid-base concepts (Section 17.9)
- complex-ion formation (Section 18.4)
- redox behavior and standard electrode potentials (Section 20.2)

# The Transition Elements and Their Coordination Compounds

**An artist's palette of transition metal compounds.** From traffic lanes to ceramic glazes, many transition metal compounds are used as pigments. These oil paints contain cadmium (three yellows, orange, and red), titanium (white), and iron and manganese (brown). In this chapter, we examine the properties of these very useful metals and their compounds.

**T**he **transition elements** (transition metals) occupy the *d* block (B groups) and *f* block of the periodic table (Figure 22.1). Many of these elements find uses in everyday objects: copper wiring, iron pipes, chromium auto parts, and gold and silver jewelry. Others, although less well known, also play important roles. Zirconium's low absorption of neutrons makes it useful in nuclear reactors. Virtually all standard light bulbs use tungsten as the filament, and tungsten carbide is the material that makes drill bits and snow-tire studs hard. Platinum is an essential component of the catalytic converters in auto exhaust systems.

In the first half of the chapter, we discuss some general properties of these elements and then focus on the chemistry of four familiar ones: chromium, manganese, silver, and mercury. The second half of the chapter is devoted to the most distinctive feature of transition metal chemistry, the formation of *coordination compounds,* substances that contain complex ions. We consider two models that explain the colors, magnetic properties, and structures of these compounds and then see some of the essential functions these substances perform in living organisms.

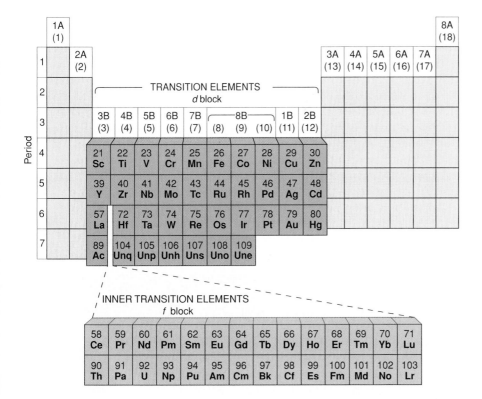

**FIGURE 22.1**

The transition elements (*d* block) and inner transition elements (*f* block) in the periodic table.

## 22.1 An Overview of Transition Element Properties

The transition elements differ considerably in their physical and chemical behavior from the main-group elements. In some ways, they are more uniform: whereas each period of main-group elements changes from metal to nonmetal, *all transition elements are metals*. In other ways, they are more diverse: whereas most main-group ionic compounds are colorless and diamagnetic, many transition metal compounds are highly colored and paramagnetic. In this section, we first discuss atomic and ionic electron configurations and then examine certain key properties of transition elements, with an occasional comparison to the main-group elements.

### Electron Configurations of the Transition Metals

As with any element, the properties of the transition elements and their compounds arise largely from the electron configurations of their atoms (Section 8.4). These are the elements in which atomic *d* or *f* orbitals are being filled. The *d*-block (B-group) elements occur in four series that lie within Periods 4 through 7 between the last *ns*-block element [Group 2A(2)] and the first *np*-block element [Group 3A(13)]. The series involve the filling of five *d* orbitals, and the first three are complete with 10 elements each; the Period 7 series is incomplete and currently contains seven elements. Interspersed between the first and second members of the series in Periods 6 and 7 are the inner transition elements, in which *f* orbitals are filled. In general, the *condensed* ground-state electron configuration for the elements in each *d*-block series is [noble gas] $ns^2(n-1)d^x$, with $x = 1$ to 10; in Periods 6 and 7, it includes the *f* sublevel: [noble gas] $ns^2(n-2)f^{14}(n-1)d^x$. The *partial* (valence-level) electron configuration for the *d*-block elements excludes the noble gas core and the filled inner *f* sublevel: $ns^2(n-1)d^x$.

The first series consists of scandium (Sc) through zinc (Zn) (Table 22.1 and Figure 22.2). Scandium has the electron configuration [Ar] $4s^2 3d^1$, and the addition of one electron at a time (along with one proton in the nucleus) first half fills, then fills, the 3d orbitals through zinc. Chromium and copper are two exceptions to this general filling pattern. The 4s and 3d orbitals in Cr are both half-filled to give [Ar] $4s^1 3d^5$, and the 4s is half-filled in Cu to give [Ar] $4s^1 3d^{10}$. The reasons for these exceptions involve the change in relative energies of the 4s and 3d orbitals as electrons are added across the series and the exceptional stability of half-filled and filled sublevels.

Transition metal ions form through *loss of the ns electrons before the (n − 1)d electrons*. Thus, the electron configuration of $Ti^{2+}$ is [Ar] $3d^2$, *not* [Ar] $4s^2$, and $Ti^{2+}$ is referred to as a $d^2$ ion. Ions of different metals with the same configuration often have similar properties. For example, both $Mn^{2+}$ and $Fe^{3+}$ are $d^5$ ions, and both have pale colors in aqueous solution and form complex ions with similar magnetic properties.

**FIGURE 22.2**

Selected Period 4 transition metals. Seven of the 10 members of the first transition series are shown: titanium (powder), chromium (chunks), manganese (flakes), iron (washers), cobalt (pellets), copper (wire), and zinc (strips).

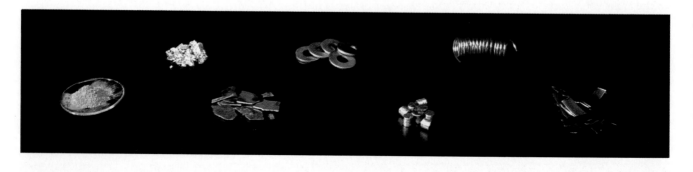

**TABLE 22.1  Orbital Occupancy of the Period 4 Transition Metals**

| ELEMENT | PARTIAL ORBITAL DIAGRAM | | | UNPAIRED ELECTRONS |
|---|---|---|---|---|
| | $4s$ | $3d$ | $4p$ | |
| Sc | [↑↓] | [↑][ ][ ][ ][ ] | [ ][ ][ ] | 1 |
| Ti | [↑↓] | [↑][↑][ ][ ][ ] | [ ][ ][ ] | 2 |
| V | [↑↓] | [↑][↑][↑][ ][ ] | [ ][ ][ ] | 3 |
| Cr | [↑] | [↑][↑][↑][↑][↑] | [ ][ ][ ] | 6 |
| Mn | [↑↓] | [↑][↑][↑][↑][↑] | [ ][ ][ ] | 5 |
| Fe | [↑↓] | [↑↓][↑][↑][↑][↑] | [ ][ ][ ] | 4 |
| Co | [↑↓] | [↑↓][↑↓][↑][↑][↑] | [ ][ ][ ] | 3 |
| Ni | [↑↓] | [↑↓][↑↓][↑↓][↑][↑] | [ ][ ][ ] | 2 |
| Cu | [↑] | [↑↓][↑↓][↑↓][↑↓][↑↓] | [ ][ ][ ] | 1 |
| Zn | [↑↓] | [↑↓][↑↓][↑↓][↑↓][↑↓] | [ ][ ][ ] | 0 |

SAMPLE PROBLEM 22.1

## Writing Electron Configurations of Transition Metal Atoms and Ions

**Problem:** Write *condensed* electron configurations for the following:
**(a)** Zr  **(b)** $V^{3+}$  **(c)** $Mo^{3+}$
(Assume that elements in higher periods behave like those in Period 4.)

**Plan:** We locate the element in the periodic table and count its position in the transition series. Since these elements are in Periods 4 and 5, the general configuration is [noble gas] $ns^2(n-1)d^x$. For the ions, we recall that $ns$ electrons are lost first.

**Solution: (a)** Zr is the second element in the $4d$ series: **[Kr] $5s^24d^2$**.

**(b)** V is the third element in the $3d$ series: [Ar] $4s^23d^3$. In forming $V^{3+}$, three electrons are lost (two $4s$ and one $3d$), so $V^{3+}$ is **[Ar] $3d^2$**.

**(c)** Mo lies below Cr in Group 6B(6), so we expect the same exception as for Cr. Thus, Mo is [Kr] $5s^14d^5$. To form the ion, Mo loses the one $5s$ and two of the $4d$ electrons, so $Mo^{3+}$ is **[Kr] $4d^3$**.

**Check:** Figure 8.8 shows that we are correct for the atoms. Be sure that the charge plus number of $d$ electrons in the ion equals the sum of the outer $s$ and $d$ electrons in the atom.

FOLLOW-UP PROBLEM 22.1
Write *partial* electron configurations for the following: **(a)** $Ag^+$; **(b)** $Cd^{2+}$; **(c)** $Ir^{3+}$.

Table 22.1 shows a general pattern in number of unpaired electrons, or half-filled orbitals, across a transition series. Note that the number increases in the first half of the series and, when pairing begins, decreases through the second half. As you'll see, it is the electron configuration of the transition

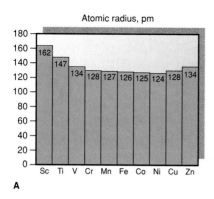

A

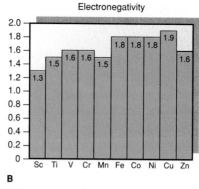

B

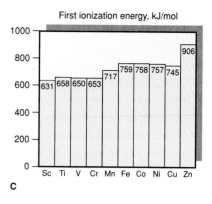

C

**FIGURE 22.3**

**Key atomic properties of the Period 4 transition metals.** Unlike a period of main-group elements, the elements in a transition series exhibit relatively small changes in atomic properties.

metal *atom* that correlates with physical properties of the *element*, such as density and magnetic behavior, while it is the electron configuration of the *ion* that determines the properties of the *compounds*.

## Atomic and Physical Properties of the Transition Metals

The atomic properties of the transition elements contrast sharply with those of a comparable set of main-group elements. Compare, for example, the variations in atomic size, electronegativity, and ionization energy of the Period 4 transition elements (Figure 22.3) with those of the Period 3 main-group elements.

Atomic size decreases steadily across Period 3 because the increasing nuclear charge is poorly shielded by the electrons added to *outer* orbitals (see Figure 8.11, p. 305). In contrast, atomic size in the transition series decreases at first but then remains fairly constant (Figure 22.3, *A*). The *d* electrons fill *inner* orbitals, so they shield outer electrons from the increasing nuclear charge much more efficiently. Consistent with this relatively small change in size is a relatively *small change in electronegativity* (Figure 22.3, *B*). Again, this trend is in marked contrast to the Period 3 elements, where a striking increase occurs between the metal Na (0.9) and the nonmetal Cl (3.0) (see Figure 9.17, p. 345). The transition elements all have intermediate electronegativity values, much like the metals of Groups 3A(13) through 5A(15).

The ionization energies of the Period 3 main-group elements rise steeply left to right (see Figure 8.15, p. 308), more than tripling from sodium (496 kJ/mol) to argon (1520 kJ/mol). Electrons become more difficult to remove because the increasing nuclear charge is poorly shielded. In the Period 4 transition metals, the *first ionization energies increase only slightly* because the inner 3*d* electrons shield effectively; thus the outer 4*s* electron experiences a relatively constant nuclear charge (Figure 22.3, *C*).

Trends within a group of transition elements are also different from those in the main groups. Atomic size increases as expected from Period 4 to 5, but there is virtually *no size increase from Period 5 to 6* (Figure 22.4, *A*). Remember that the lanthanide series, with its buried 4*f* sublevel, appears between the 4*d* (Period 5) and 5*d* (Period 6) series. Therefore, an element in Period 6 is separated from the one above it in Period 5 by 32 elements (ten 4*d*, six 5*p*, two 6*s*, and fourteen 4*f*) instead of 18. The extra shrinkage that results from the nuclear charge of these additional 14 elements is called

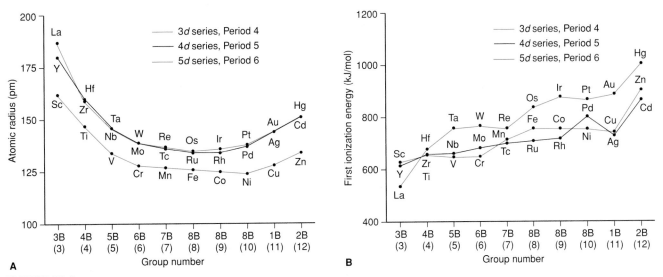

**FIGURE 22.4**

Variation in atomic size and first ionization energy in the *d* blocks of Periods 4, 5, and 6. **A,** Unlike a group of main-group elements, the second and third members of a transition metal group are nearly the same size as a result of the lanthanide contraction. **B,** With a small size increase from top to bottom of a group and a large increase in nuclear charge, first ionization energies increase.

the **lanthanide contraction.** By coincidence, this *decrease* is about equal to the normal *increase* between periods, so the Periods 5 and 6 transition elements have about the same atomic sizes.

While the atomic size increases slightly from the top to the bottom of a group of transition elements, the nuclear charge increases much more, so *the first ionization energy generally increases* (Figure 22.4, *B*). This trend runs counter to the pattern in the main-group elements, in which heavier members of a group are so much larger that their ionization energies are lower.

Atomic size bears directly on density. Since sizes, and therefore volumes, change relatively little across a series, *density generally increases as atomic mass increases*. As a result of the lanthanide contraction, some of the densest elements known are found in Period 6 (Figure 22.5): tungsten, rhenium, osmium, iridium, platinum, and gold have densities about 20 times that of water and twice that of lead.

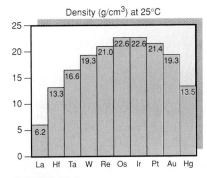

**FIGURE 22.5**

**Densities of the Period 6 *d*-block elements.** The densest elements known appear in this series.

### Chemical Properties of the First-Row Transition Metals

Next we examine the chemical properties of the transition elements and illustrate them with the Period 4 series.

**Oxidation states.** One of the most characteristic chemical properties of the transition metals is the occurrence of *multiple oxidation states*. For example, in their compounds, vanadium commonly exhibits two positive oxidation states, chromium three, and manganese three, and many others are seen less often. Since the *ns* and $(n - 1)d$ electrons are so close in energy, transition elements can involve all or most of these electrons in bonding. This behavior is markedly different from that of the main-group metals, which display one or at most two oxidation states in their compounds.

**TABLE 22.2    Oxidation States and *d*-Orbital Occupancy of the Period 4 Transition Metals\***

| OXIDATION STATE | 3B (3) Sc | 4B (4) Ti | 5B (5) V | 6B (6) Cr | 7B (7) Mn | 8B (8) Fe | 8B (9) Co | 8B (10) Ni | 1B (11) Cu | 2B (12) Zn |
|---|---|---|---|---|---|---|---|---|---|---|
| 0 | 0 $(d^1)$ | 0 $(d^2)$ | 0 $(d^3)$ | 0 $(d^5)$ | 0 $(d^5)$ | 0 $(d^6)$ | 0 $(d^7)$ | 0 $(d^8)$ | 0 $(d^{10})$ | 0 $(d^{10})$ |
| +1 | | | +1 $(d^3)$ | +1 $(d^5)$ | +1 $(d^5)$ | | +1 $(d^7)$ | +1 $(d^8)$ | +1 $(d^{10})$ | |
| +2 | | +2 $(d^2)$ | +2 $(d^3)$ | +2 $(d^4)$ | +2 $(d^5)$ | +2 $(d^6)$ | +2 $(d^7)$ | +2 $(d^8)$ | +2 $(d^9)$ | +2 $(d^{10})$ |
| +3 | +3 $(d^0)$ | +3 $(d^1)$ | +3 $(d^2)$ | +3 $(d^3)$ | +3 $(d^4)$ | +3 $(d^5)$ | +3 $(d^6)$ | +3 $(d^7)$ | +3 $(d^8)$ | |
| +4 | | +4 $(d^0)$ | +4 $(d^1)$ | +4 $(d^2)$ | +4 $(d^3)$ | +4 $(d^4)$ | +4 $(d^5)$ | +4 $(d^6)$ | | |
| +5 | | | +5 $(d^0)$ | +5 $(d^1)$ | +5 $(d^2)$ | | +5 $(d^4)$ | | | |
| +6 | | | | +6 $(d^0)$ | +6 $(d^1)$ | +6 $(d^2)$ | | | | |
| +7 | | | | | +7 $(d^0)$ | | | | | |

\*Most important in color.

**FIGURE 22.6**

**Aqueous oxoanions of V, Cr, and Mn in their highest oxidation states.** The oxidation state equals the group number in these oxoanions: $VO_4^{3-}$ *(left)*, $Cr_2O_7^{2-}$ *(middle)*, and $MnO_4^-$ *(right)*.

The highest oxidation state of elements in Groups 3B(3) through 7B(7) is equal to the group number (Table 22.2). These states are seen when the elements combine with the highly electronegative oxygen or fluorine. Thus, among their oxoanions, vanadium forms the vanadate ion ($VO_4^{3-}$), chromium forms the chromate ($CrO_4^{2-}$) and dichromate ($Cr_2O_7^{2-}$) ions, and manganese forms the permanganate ion ($MnO_4^-$) (Figure 22.6). Elements in Groups 8B(8, 9, 10) exhibit fewer oxidation states, and the highest state is less common and never equal to the group number. For instance, we never encounter Fe in the +8 state and only rarely in the +6 state. The +2 and +3 states are the most common ones for iron and cobalt, and the +2 state is most common for nickel, copper, and zinc. *The +2 oxidation state is common because ns² electrons are readily lost.*

Copper, silver, and gold in Group 1B(11) are unusual. Although copper has a fairly common oxidation state of +1, which results from loss of its single 4s electron, its most common state is +2. Silver behaves more predictably, exhibiting primarily the +1 oxidation state. Gold exhibits the +3 state and, less often, the +1 state.

**Metallic behavior.** Atomic size and oxidation state have a major effect on the nature of bonding in transition metal compounds. Like the metals in Groups 3A(13), 4A(14), and 5A(15), the transition elements in their lower oxidation states behave chemically more like metals. That is, *ionic bonding is more prevalent in the lower oxidation states, and covalent bonding is more prevalent in the higher states.* For example, at room temperature, $TiCl_2$ is an ionic solid, whereas $TiCl_4$ is a molecular liquid. In the higher oxidation states, the

FIGURE 22.7
Colors of representative compounds of the Period 4 transition metals. From left to right, the compounds are scandium oxide (white), titanium(IV) oxide (white), vanadyl sulfate dihydrate (light blue), sodium chromate (yellow), manganese(II) chloride tetrahydrate (light pink), potassium ferricyanide (red-orange), cobalt(II) chloride hexahydrate (violet), nickel(II) nitrate hexahydrate (green), copper(II) sulfate pentahydrate (blue), and zinc sulfate heptahydrate (white).

atoms have higher charge densities, so they attract nonmetal atoms more strongly and the bonding becomes more covalent. For the same reason, the oxides become less basic as the oxidation state increases: $TiO$ is weakly basic in water, whereas $TiO_2$ is amphoteric (reacts with both acid and base).

Table 22.3 shows the standard electrode potentials of the Period 4 transition metals in their +2 oxidation state in acid solution. Note that, with two exceptions, reducing strength decreases across the series. All the Period 4 transition metals, except copper, are active enough to reduce $H^+$ from aqueous acid to form hydrogen gas. In contrast to the rapid reaction at room temperature of the Group 1A(1) and 2A(2) metals with water, however, the transition metals react rapidly only with hot water or steam.

**Color and magnetism of transition metal compounds.** Most main-group ionic compounds are colorless because the metal ion has a filled outer shell (noble gas configuration). With only much higher energy orbitals available to receive an excited electron, the ion does not absorb visible light. In contrast, electrons in a partially filled $d$ sublevel can absorb visible wavelengths and move to slightly higher energy $d$ orbitals, which accounts in most cases for the striking colors of many transition metal compounds (Figure 22.7). The compounds of scandium, titanium(IV), and zinc are colorless because their metal ions have either an empty $d$ sublevel ($Sc^{3+}$ or $Ti^{4+}$: [Ar] $d^0$) or a filled one ($Zn^{2+}$: [Ar] $d^{10}$).

Magnetic properties are also related to sublevel occupancy (Section 8.4). Recall that a **paramagnetic** substance has atoms or ions with unpaired electrons, which cause it to be attracted to an external magnetic field. A **diamagnetic** substance has only paired electrons, so it is unaffected (or slightly repelled) by a magnetic field. Most main group metal ions are diamagnetic for the same reason they are colorless: all their electrons are paired. In contrast, many transition metal compounds are paramagnetic because of their unpaired $d$ electrons. For example, $MnSO_4$ is paramagnetic, but $CaSO_4$ is diamagnetic. The $Ca^{2+}$ ion has the electron configuraton of argon, whereas $Mn^{2+}$ has a $d^5$ configuration. Ions with a $d^0$ or $d^{10}$ configuration, which do not absorb visible light, are also diamagnetic.

### Chemical Behavior Within a Group of Transition Metals

The increase in reactivity seen down a group of main-group metals does not appear down a group of transition metals. Consider the chromium group [6B(6)], which shows a typical pattern (Table 22.4). As a consequence of the lanthanide contraction, the first ionization energy *increases* down the group, which makes the two heavier metals *less* reactive than the lightest one. As shown by their electrode potentials, Cr is a much stronger reducing agent than Mo and W as well.

TABLE 22.3 **Standard Electrode Potentials of Period 4 $M^{2+}$ Ions**

| HALF-REACTION | | $E^0$ (V) |
|---|---|---|
| $Ti^{2+}(aq) + 2e^-$ | $\rightleftharpoons$ $Ti(s)$ | $-1.63$ |
| $V^{2+}(aq) + 2e^-$ | $\rightleftharpoons$ $V(s)$ | $-1.19$ |
| $Cr^{2+}(aq) + 2e^-$ | $\rightleftharpoons$ $Cr(s)$ | $-0.91$ |
| $Mn^{2+}(aq) + 2e^-$ | $\rightleftharpoons$ $Mn(s)$ | $-1.18$ |
| $Fe^{2+}(aq) + 2e^-$ | $\rightleftharpoons$ $Fe(s)$ | $-0.44$ |
| $Co^{2+}(aq) + 2e^-$ | $\rightleftharpoons$ $Co(s)$ | $-0.28$ |
| $Ni^{2+}(aq) + 2e^-$ | $\rightleftharpoons$ $Ni(s)$ | $-0.25$ |
| $Cu^{2+}(aq) + 2e^-$ | $\rightleftharpoons$ $Cu(s)$ | $0.34$ |
| $Zn^{2+}(aq) + 2e^-$ | $\rightleftharpoons$ $Zn(s)$ | $-0.76$ |

TABLE 22.4 **Some Properties of Group 6B(6) Elements**

| ELEMENT | ATOMIC RADIUS (pm) | $E^0$(V) FOR $M^{3+}(aq)\|M(s)$ | $IE_1$(kJ/mol) |
|---|---|---|---|
| Cr | 128 | $-0.74$ | 653 |
| Mo | 139 | $-0.20$ | 685 |
| W | 139 | $-0.11$ | 770 |

The similarity in atomic size of Period 5 and 6 members also leads to similar chemical behavior, a fact that can have important practical consequences. Because Mo and W compounds behave similarly, for example, their ores often occur together in nature, so the elements are very difficult to separate. The same situation occurs with zirconium and hafnium in Group 4B(4) and niobium and tantalum in Group 5B(5).

**Section Summary**

All transition elements are metals. Atoms of $d$-block elements have $(n-1)d$ orbitals being filled; ions have an empty $ns$ orbital. Unlike the main-group elements, atomic size, electronegativity, and first ionization energy change little across a transition series. Because of the lanthanide contraction, atomic size changes little from Period 5 to 6 in a transition metal group; thus, first ionization energy and density *increase*. Transition metals typically have several oxidation states, with the $+2$ state most common. The elements exhibit more metallic behavior in lower states. Most Period 4 members reduce hydrogen ion from acid solution. Many transition metal compounds are colored and paramagnetic because the metal ion has unpaired $d$ electrons.

## 22.2  The Inner Transition Elements

The 14 **lanthanides** (lanthanoids), cerium (Ce; $Z = 58$) through lutetium (Lu; $Z = 71$), lie between lanthanum ($Z = 57$) and hafnium ($Z = 72$) in the third $d$ transition series. Below them are the 14 radioactive **actinides** (actinoids), thorium (Th; $Z = 90$) through lawrencium (Lr; $Z = 103$), which lie between actinium (Ac; $Z = 89$) and unnilquadium (Unq; $Z = 104$, IUPAC-recommended name dubnium, Db). The lanthanides and actinides are called **inner transition elements** because their seven inner $4f$ and $5f$ orbitals are being filled.

**The Lanthanides**

The lanthanides are still sometimes called the *rare earth* elements, a term referring to their presence in unfamiliar oxides, but they are actually not rare at all. Cerium (Ce), for instance, ranks 26th in abundance and is five times more abundant than lead. All the lanthanides are silvery, high melting (800°C to 1600°C) metals. In their chemical properties, they represent an extreme case of the small variations typical of transition elements in a period or a group, which made them very difficult to separate. ◆

The mixed natural occurrence of the lanthanides arises from their tendency to exhibit the $+3$ oxidation state, with $M^{3+}$ ions of very similar radii. The ground-state electron configuration of most lanthanides is [Xe] $6s^2 4f^x$ $5d^0$, where $x$ varies across the series. The three exceptions (Ce, Gd, and Lu) have a single electron in a $5d$ orbital, which gives the $Gd^{3+}$ and $Lu^{3+}$ ions a stable half-filled ($f^7$) or filled ($f^{14}$) sublevel; Ce ([Xe] $6s^2 4f^1 5d^1$) forms a stable 4+ ion due to its empty ($f^0$) sublevel.

Lanthanide compounds and their mixtures have many uses. Several oxides are used for tinting sunglasses and welder's goggles and for adding color to the fluorescent powder coatings in TV screens. High-quality camera lenses incorporate $La_2O_3$ because of its extremely high index of refraction. Samarium forms an alloy with cobalt, $SmCo_5$, that is used in the strongest

◆ **A Remarkable Laboratory Feat.** Discovery of the lanthanides is a testimony to the laboratory skills of 19th-century chemists. The two mineral sources, ceria and yttria, are mixtures of compounds of all 14 lanthanides; yttria also includes oxides of Sc, Y, and La. Purification of these elements was so difficult that many claims of new elements turned out to be mixtures of similar elements. To confuse matters, the periodic table of the time had space for only one element between Ba and Hf, which turned out to be La. Not until 1913, when the atomic number basis of the table was established, did chemists realize that 14 new elements could fit, and in 1918, Niels Bohr proposed that the $n = 4$ level be expanded to include the $f$ sublevel.

known permanent magnet (Sample Problem 22.2). Two industrial uses account for more than 60% of rare earth applications. In gasoline refining, the zeolite catalysts used to "crack" hydrocarbon components into different molecules (Gallery, Section 13.6) contain 5% by mass rare earth oxides. In steelmaking, a mixture of lanthanides, called misch-metal, is used to remove carbon impurities from molten iron and steel.

SAMPLE PROBLEM 22.2 _____

### Determining the Number of Unpaired Electrons

**Problem:** The alloy $SmCo_5$ forms a strong permanent magnet because both Sm and Co have unpaired electrons. How many unpaired electrons are in the Sm atom?
**Plan:** We write the condensed electron configuration of Sm and then, using Hund's rule and the aufbau principle, place the electrons in a partial orbital diagram and count the unpaired electrons.
**Solution:** Samarium (Sm; $Z = 62$) is the eighth element after Xe. Two electrons go into the $6s$ sublevel. Since, in general, the $4f$ sublevel fills before the $5d$ (only Ce, Gd, and Lu have $5d$ electrons), the remaining electrons go into the $4f$. Thus, Sm is [Xe] $6s^2 4f^6$. There are seven $f$ orbitals, so each of the six $f$ electrons enters a separate orbital:

   6s               4f                      5d                 6p

Thus, Sm has **six unpaired electrons.**
**Check:** Six $4f$ $e^-$ plus two $6s$ $e^-$ plus the 54 $e^-$ in Xe gives 62, the atomic number of Sm.

FOLLOW-UP PROBLEM 22.2
How many unpaired electrons are in the $Er^{3+}$ ion?

### The Actinides

All actinides are radioactive. Like the lanthanides, they have very similar physical and chemical properties. Thorium and uranium occur in nature, but the transuranium elements, those with $Z$ greater than 92, have been synthesized in accelerators, some in only tiny amounts (Section 21.3). In fact, macroscopic samples of mendelevium (Md), nobelium (No), and lawrencium (Lr) have never even been seen. The available metals are silvery and chemically reactive and, like the lanthanides, they form highly colored compounds. The actinides and lanthanides have similar outer-electron configurations. Although the +3 oxidation state is also common among actinide compounds, some other states commonly occur. For example, uranium exhibits the +3 through +6 states, and the +6 state is the most prevalent; thus, the most common oxide of uranium is $UO_3$.

### Section Summary

The two series of inner transition elements result from the filling of $f$ orbitals. The lanthanides ($4f$ series) have a common +3 oxidation state and exhibit very similar properties. The actinides ($5f$ series) are radioactive. All actinides have a common +3 state; several, including uranium, have common higher states as well.

## 22.3 Highlights of Selected Transition Metals

In this section, we investigate the aqueous chemistry of several transition elements, focusing on the general patterns that each reveals. We examine chromium and manganese from Period 4, silver from Period 5, and mercury from Period 6. In Chapter 23, we discuss the metallurgy of several transition elements, with particular emphasis on iron and copper.

### Chromium

Chromium is a very shiny, silvery metal, whose name (from the Greek *chroma*, color) alludes to its many colorful compounds. A solution of $Cr^{3+}$ is deep violet, for example, whereas a trace of $Cr^{3+}$ in the $Al_2O_3$ crystal structure gives the ruby its beautiful red hue. Chromium readily forms a thin, adherent, transparent coating of $Cr_2O_3$ in air, making the metal extremely useful as an attractive protective coating on easily corroded metals, such as iron. "Stainless" steels are highly resistant to corrosion and incorporate as much as 14% chromium by mass.

With six valence electrons ($[Ar]\,4s^1 3d^5$), chromium occurs in all possible positive oxidation states, but the three most important are +2, +3, and +6 (see Table 22.2). In its three common oxides, chromium exhibits the pattern seen in many elements: *oxide acidity increases with metal oxidation state*. Chromium(II) oxide (CrO) is basic and largely ionic. It forms an insoluble hydroxide in neutral or basic solution but dissolves in acidic solution to yield the $Cr^{2+}$ ion:

$$CrO(s) + 2H^+(aq) \longrightarrow Cr^{2+}(aq) + H_2O(l)$$

Chromium(III) oxide ($Cr_2O_3$) is amphoteric, dissolving in acid to yield the violet $Cr^{3+}$ ion:

$$Cr_2O_3(s) + 6H^+(aq) \longrightarrow 2Cr^{3+}(aq) + 3H_2O(l)$$

In base, it forms the green $Cr(OH)_4^-$ ion:

$$Cr_2O_3(s) + 3H_2O(l) + 2OH^-(aq) \longrightarrow 2Cr(OH)_4^-(aq)$$

Thus, chromium in its +3 state resembles the main-group metal aluminum (Section 18.4).

Deep-red chromium(VI) oxide ($CrO_3$) is covalent and acidic, forming chromic acid ($H_2CrO_4$) in water, which yields the yellow chromate ion ($CrO_4^{2-}$) in base:

$$CrO_3(s) + H_2O(l) \longrightarrow H_2CrO_4(aq)$$

$$H_2CrO_4(aq) + 2OH^-(aq) \longrightarrow CrO_4^{2-}(aq) + 2H_2O(l)$$

In acidic solution, chromate immediately forms the orange dichromate ion ($Cr_2O_7^{2-}$):

$$2CrO_4^{2-}(aq) + 2H^+(aq) \rightleftharpoons Cr_2O_7^{2-}(aq) + H_2O(l)$$

Because both ions contain chromium(VI), this is not a redox reaction; rather, it is a dehydration-condensation (Figure 22.8). Hydrogen ion concentration controls the equilibrium position: yellow $CrO_4^{2-}$ predominates at high pH and orange $Cr_2O_7^{2-}$ at low pH. The bright colors of chromium(VI) compounds lead to their wide use in pigments for artist's paints and ceramic glazes. Lead chromate (chrome yellow) is used as an oil-paint color, and also in the yellow stripes that delineate traffic lanes.

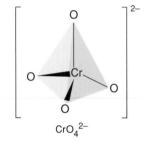

$CrO_4^{2-}$

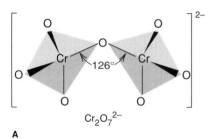

$Cr_2O_7^{2-}$

**A**

**B**

**FIGURE 22.8**

Structures of the chromate ($CrO_4^{2-}$) and dichromate ($Cr_2O_7^{2-}$) ions.

To see *why* oxide acidity increases with oxidation state, we must extend the concept of electronegativity to an element in its various states. As the oxidation state of a metal in a compound increases, the metal becomes more positively charged, which increases its attraction for electrons; in effect, its electronegativity increases. This effective electronegativity, called *valence-state electronegativity*, also has numerical values. The electronegativity of chromium metal is 1.6, close to that of aluminum (1.5), another active metal. For chromium(III), the value increases to 1.7, still characteristic of a metal. However, the electronegativity of chromium(VI) is 2.3, close to values typical of nonmetals, such as phosphorus (2.1), carbon (2.5), sulfur (2.5), and iodine (2.5). Thus, like P in $PO_4^{3-}$, it is not surprising to see chromium(VI) compounds with Cr covalently bound at the center of the oxoanion of a relatively strong acid.

Chromium metal and the $Cr^{2+}$ ion are potent reducing agents. The metal displaces hydrogen from dilute acids to form blue $Cr^{2+}(aq)$, which reduces $O_2$ in air within minutes to form the violet $Cr^{3+}$ ion:

$$4Cr^{2+}(aq) + O_2(g) + 4H^+(aq) \rightarrow 4Cr^{3+}(aq) + 2H_2O(l) \qquad E^0_{overall} = 1.64 \text{ V}$$

Chromium(VI) compounds in acid solution are strong oxidizing agents (concentrated solutions are *extremely* corrosive!), the chromium(VI) being readily reduced to chromium(III):

$$Cr_2O_7^{2-}(aq) + 14H^+(aq) + 6e^- \rightarrow 2Cr^{3+}(aq) + 7H_2O(l) \qquad E^0 = 1.33 \text{ V}$$

This reaction is often used to determine the iron content of a water or soil sample by oxidizing $Fe^{2+}$ to $Fe^{3+}$ ion. In basic solution, the $CrO_4^{2-}$ ion, which is a much weaker oxidizing agent, predominates:

$$CrO_4^{2-}(aq) + 4H_2O(l) + 3e^- \rightarrow Cr(OH)_3(s) + 5OH^-(aq) \qquad E^0 = -0.13 \text{ V}$$

## Manganese

Elemental manganese is hard and shiny and, like vanadium and chromium, is used mostly to make steel alloys. A small amount of Mn ($< 1\%$) makes steel easier to roll, forge, and weld. Steel made with 12% Mn is tough enough to be used for naval armor and bulldozer blades. Small amounts of manganese are added to aluminum beverage cans and bronze alloys to make them stiffer and tougher as well.

The chemistry of manganese resembles that of chromium in some respects. The free metal is quite reactive and readily displaces $H_2$ from acids to form the pale-pink $Mn^{2+}$ ion:

$$Mn(s) + 2H^+(aq) \rightarrow Mn^{2+}(aq) + H_2(g) \qquad E^0 = 1.18 \text{ V}$$

Like chromium, manganese can use all its valence electrons in its compounds, exhibiting every possible positive oxidation state, with the +2, +4, and +7 states most common (Table 22.5). As the oxidation state of manganese rises, its valence-state electronegativity rises and its oxides change from basic to acidic. Manganese(II) oxide (MnO) is basic, and manganese(III) oxide ($Mn_2O_3$) is amphoteric. Manganese(IV) oxide ($MnO_2$) is insoluble, so it shows no acid-base properties; it is used in dry-cell and alkaline batteries as the oxidizing agent in a redox reaction with zinc (Gallery, Section 20.4). Manganese(VII) oxide ($Mn_2O_7$), which forms by reaction of Mn with pure $O_2$, reacts with water to form permanganic acid ($HMnO_4$), which is as strong as perchloric acid ($HClO_4$).

**TABLE 22.5  Some Oxidation States of Manganese**

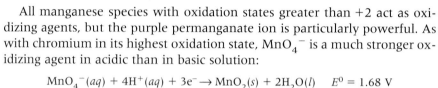

| Oxidation state* | **Mn(II)** | Mn(III) | **Mn(IV)** | Mn(VI) | **Mn(VII)** |
|---|---|---|---|---|---|
| Example | $Mn^{2+}$ | $Mn_2O_3$ | $MnO_2$ | $MnO_4^{2-}$ | $MnO_4^-$ |
| Ion configuration | $d^5$ | $d^4$ | $d^3$ | $d^1$ | $d^0$ |
| Oxide acidity | BASIC | | | | ACIDIC |

*Most common states in **boldface.**

♦ **Sharing the Ocean's Wealth.** Known reserves of most transition metals will be depleted within 50 years, so other sources must be found. One promising source is nodules strewn over large portions of the ocean floor. Varying from a few millimeters to a few meters in diameter, these chunks consist mainly of manganese and iron oxides, with other element oxides present in smaller amounts. Billions of tons are present, but mining them presents major technical and political challenges. Global agreements have designated the ocean floor as international property, so cooperation will be required to mine the nodules and share the rewards.

All manganese species with oxidation states greater than +2 act as oxidizing agents, but the purple permanganate ion is particularly powerful. As with chromium in its highest oxidation state, $MnO_4^-$ is a much stronger oxidizing agent in acidic than in basic solution:

$$MnO_4^-(aq) + 4H^+(aq) + 3e^- \rightarrow MnO_2(s) + 2H_2O(l) \quad E^0 = 1.68 \text{ V}$$
$$MnO_4^-(aq) + 2H_2O(l) + 3e^- \rightarrow MnO_2(s) + 4OH^-(aq) \quad E^0 = 0.59 \text{ V}$$

Unlike $Cr^{2+}$ and $Fe^{2+}$, the $Mn^{2+}$ ion resists oxidation in air. The $Cr^{2+}$ ion is a $d^4$ species and readily loses a $d$ electron to form the more stable $d^3$ $Cr^{3+}$ ion. The $Fe^{2+}$ ion is a $d^6$ species, and removing a $d$ electron yields the stable, half-filled $d^5$ configuration of the $Fe^{3+}$ ion. Removing an electron from the $Mn^{2+}$ ion is difficult because it disrupts the $d^5$ configuration. ♦

### Silver

Silver, the second member of the "coinage" metals [Group 1B(11)], has been admired for thousands of years and is still treasured for use in jewelry and fine flatware. Because the pure metal is too soft for these purposes, it is alloyed with copper to form the harder sterling silver. In former times, silver was used in coins, but it has been replaced almost universally by copper-nickel alloys. It has the highest electrical conductivity of any element but is not used in wiring because copper is cheaper and more plentiful. In the past, silver was found in nuggets and veins of rock, often mixed with gold, because both elements are chemically inert enough to exist uncombined. Nearly all those deposits have been mined, so most silver is now obtained as a side product during the extraction of copper from its ore (Section 23.3).

The only important oxidation state of silver is the +1 state. Its most important *soluble* compound is silver nitrate, used for electroplating and in the manufacture of the halides used for photographic film (see following discussion). Although it forms no oxide in air, silver tarnishes to black $Ag_2S$ due to traces of sulfur-containing compounds in the air. Some polishes remove the $Ag_2S$, along with some silver, by physically abrading the surface. An alternative "home remedy" removes the tarnish and restores the metal by heating the object in a conducting solution of table salt or baking soda ($NaHCO_3$) in an aluminum pan. Aluminum, a strong reducing agent, reduces the $Ag^+$ ions back to the metal:

$$2Al(s) + 3Ag_2S(s) + 6H_2O(l) \rightarrow 2Al(OH)_3(s) + 6Ag(s) + 3H_2S(g)$$

The most widespread use of silver compounds, particularly the three heavier halides, is in black-and-white photography, an art that applies tran-

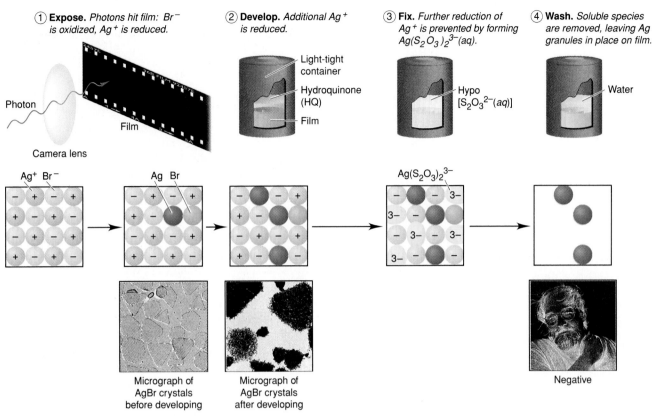

① **Expose.** *Photons hit film: Br⁻ is oxidized, Ag⁺ is reduced.*

② **Develop.** *Additional Ag⁺ is reduced.*

③ **Fix.** *Further reduction of Ag⁺ is prevented by forming* $Ag(S_2O_3)_2^{3-}(aq)$.

④ **Wash.** *Soluble species are removed, leaving Ag granules in place on film.*

Micrograph of AgBr crystals before developing

Micrograph of AgBr crystals after developing

Negative

**FIGURE 22.9**
Steps in producing a black-and-white photograph.

sition metal chemistry and solution kinetics. The steps in obtaining the final photograph are exposing the film, developing the image, fixing the image, and printing the image (Figure 22.9). The process depends on four key chemical properties of silver and its compounds:

1. Silver halides undergo a redox reaction when exposed to visible light.
2. Silver chloride, bromide, and iodide are *not* water soluble.
3. $Ag^+$ is easily reduced: $Ag^+(aq) + e^- \rightarrow Ag(s)$; $E^0 = 0.80$ V.
4. $Ag^+$ forms several stable, water-soluble complex ions.

The photographic film itself is simply a flexible plastic support for the light-sensitive emulsion, which consists of AgBr microcrystals dispersed in gelatin. Light reflected from the objects in a scene—more light from bright objects than from dark ones—enters the camera lens and strikes the film. Exposed AgBr crystals absorb photons ($h\nu$) in a very localized redox reaction, which consists of air oxidation of a $Br^-$ ion catalyzed by the photon, followed almost immediately by reduction of a nearby $Ag^+$ ion:

$$Br^- \xrightarrow{h\nu} Br + e^-$$

$$\underline{Ag^+ + e^- \rightarrow Ag}$$

$$Ag^+ + Br^- \xrightarrow{h\nu} Ag + Br$$

Wherever more light strikes a microcrystal, more Ag atoms form. The exposed crystals are called a *latent* image because the few scattered atoms of photoreduced Ag are not yet visible. Nevertheless, their presence as crystal defects within the AgBr makes the crystal highly susceptible to further reduction.

The latent image is developed into the actual image by reducing more of the silver ions in the crystal in a controlled manner. Developing is a rate-dependent step: crystals with many photoreduced Ag atoms react more quickly than those with only a few. The developer is a weak reducing agent, such as the organic substance hydroquinone (HQ):

$$Ag^+(s) + HQ_{red}(aq) \rightarrow Ag(s) + HQ_{ox}(aq) + H^+(aq)$$

The reaction rate depends on developer concentration, solution temperature, and the time the emulsion is bathed in the solution. After developing, approximately $10^6$ times as many Ag atoms are present on the film as there were in the latent image, and they form very small, black clusters of silver.

After the image is developed, it must be "fixed"; that is, the reduction of $Ag^+$ must be stopped, or the entire film will blacken on exposure to more light. The remaining $Ag^+$ is chemically removed by converting it to a soluble complex ion with sodium thiosulfate solution ("hypo"):

$$AgBr(s) + 2S_2O_3^{2-}(aq) \rightarrow Ag(S_2O_3)_2^{3-}(aq) + Br^-(aq)$$

The water-soluble ions are washed away in water. Fixation is the final step in producing a photographic *negative,* in which dark objects in the scene appear bright in the image, and vice versa.

Through the use of an enlarger, the negative is held above print paper coated with emulsion (silver halide in gelatin) and exposed to light, and the previous chemical steps are repeated to produce a "positive" of the image. The bright areas on the negative, such as rocks, allow a great deal of light to pass through and reduce many $Ag^+$ ions on the print paper. Dark areas on the negative, such as clouds, allow much less $Ag^+$ reduction. Print-paper emulsion usually contains silver chloride, which reacts more slowly than silver bromide, giving finer control of the print image. High-speed film utilizes silver iodide, the most light sensitive of the three halides.

### Mercury

Mercury has been known since ancient times because cinnabar (HgS), its principal ore, is a naturally occurring red pigment (vermilion) that readily undergoes a redox reaction in the heat of a fire. Sulfide ion, the reducing agent for the process, is already present as part of the ore:

$$HgS(s) + O_2(g) \rightarrow Hg(g) + SO_2(g)$$

The gaseous Hg condenses on cooler nearby surfaces.

The Latin name *hydrargyrum* ("liquid silver") is a good description of mercury, the only metal that is liquid at room temperature. Two factors account for this unusual property. First, due to a distorted crystal structure, each mercury atom is surrounded by 6 rather than 12 nearest neighbors. Second, a filled, tightly held *d* subshell leaves only the two 6s electrons for metallic bonding. As a result, interactions among mercury atoms are so weak that the solid breaks down at $-38.9°C$.

Many of mercury's uses arise from its unusual physical properties. Its liquid range ($-39°$ to $357°C$) encompasses most everyday temperatures, so Hg is commonly used in thermometers. As you might expect from its position in Period 6 following the lanthanide contraction, mercury is quite dense (13.5 g/mL), which makes it convenient for use in barometers. Mercury's fluidity and conductivity make it useful for "silent" switches in thermostats. At high pressures, mercury vapor can be excited electrically to emit the bright white light seen in sports stadium and highway lights.

Mercury is a good solvent for metals, and many *amalgams,* alloys of mercury, exist. In the chloralkali process for production of $Cl_2$ from sea water, which we'll discuss in Section 23.4, mercury acts as the cathode and as the solvent for the Na metal that forms. When treated with water, the resulting sodium amalgam (Na/Hg) forms two important byproducts:

$$2Na/Hg + 2H_2O(l) \rightarrow 2NaOH(aq) + H_2(g) + Hg(l)$$

The mercury is released in this step and reused in the electrolytic cell. In mercury batteries, a zinc amalgam acts as the anode and mercury(II) oxide as the cathode (Gallery, Section 20.4). In view of mercury's toxicity (see below), past contamination of waste water from the chloralkali process and the disposal of old mercury batteries are serious environmental concerns.

The chemical properties of mercury are unique within Group 2B(12). The other members, zinc and cadmium, occur in the +2 oxidation state as $d^{10}$ ions, but mercury occurs in the +1 state as well, with the condensed configuration [Xe] $6s^14f^{14}5d^{10}$. The unpaired $6s$ electron allows two Hg(I) species to form the *diatomic ion* [Hg—Hg]$^{2+}$ (written Hg$_2^{2+}$), one of the first species known with a covalent metal-metal bond. Mercury's most common oxidation state is +2. Whereas HgF$_2$ is largely ionic, many other compounds, such as HgCl$_2$, contain bonds that are predominantly covalent. Most mercury(II) compounds are insoluble in water.

The common ions $Zn^{2+}$, $Cd^{2+}$, and $Hg^{2+}$ are biopoisons. Zinc oxide is used as an external antiseptic ointment. The $Cd^{2+}$ and $Hg^{2+}$ ions are two of the so-called toxic heavy-metal ions. The cadmium in solder may be more responsible than the lead for solder's high toxicity. Mercury compounds have been used in agriculture as fungicides and pesticides and in medicine as an internal drug, but these uses have been mostly phased out. ◆

Because most mercury(II) compounds are insoluble in water, they were once thought to be harmless in the environment, but we now know otherwise. Microorganisms in sludge and river sediment convert mercury and mercury ions to the methyl mercury ion, CH$_3$—Hg$^+$, and then to organomercury compounds, such as dimethylmercury, CH$_3$—Hg—CH$_3$. These toxic compounds are nonpolar and, like chlorinated hydrocarbons, become increasingly concentrated in fatty tissues as they move up the food chain from microorganisms to fish to birds and mammals. Fish living in a mercury-polluted lake can have a mercury concentration thousands of times higher than that of the water itself.

The mechanism of toxicity of mercury and other heavy-metal ions is not fully understood. It is thought that they migrate from fatty tissue and bind strongly to sulfur-containing amino acids in proteins, thereby disrupting the proteins' structure and function. Since the brain has a high fat content, lead, cadmium, and mercury poisoning often cause devastating neurological and mental effects.

◆ **Mad as a Hatter.** The felt top hat and zany manner of the "mad hatter" in *Alice in Wonderland* is a literary reference to the toxicity of mercury compounds. Mercury(II) nitrate and chloride were used with the HNO$_3$ needed to make felt for hats from animal hair. Inhalation of the dust generated by the process led to "hatter's shakes," abnormal personality, and other neurological disorders.

## Section Summary

Chromium and manganese add corrosion resistance and hardness to steels. They are typical of transition metals in having several oxidation states. In the lower states, the element is more metallic (lower valence-state electronegativity, more ionic compounds, more basic oxides, and so on) than in the higher states. The metals produce H$_2$ in acid. Cr(VI) undergoes a pH-sensitive dehydration condensation. Both Cr(VI) and Mn(VII) are stronger oxidizing agents in acid than in base. The only important oxidation state for silver is +1. The halides are light sensitive and are used in photography. Mercury, the only liquid metal, dissolves many metals in commercially important applications. The mercury(I) ion is diatomic and has a metal-metal covalent bond. The element and its compounds are toxic and become concentrated in the food chain.

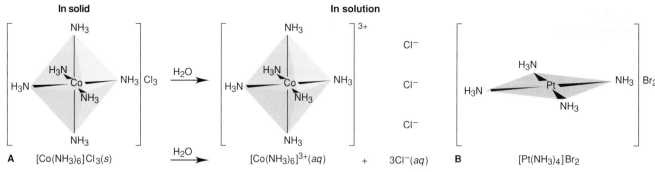

**A**    $[Co(NH_3)_6]Cl_3(s)$

$[Co(NH_3)_6]^{3+}(aq)$    +    $3Cl^-(aq)$    **B**    $[Pt(NH_3)_4]Br_2$

**FIGURE 22.10**

Components of a coordination compound. Coordination compounds consist of a complex ion (brackets) and one or more counter ions. The complex ion has a central metal ion surrounded by ligands. **A,** In solution, the counter ions and complex ion separate but the ligands remain bound. Six ligands around the metal ion adopt an octahedral geometry. **B,** Complex ions with a central $d^8$ metal ion contain four ligands that define a square plane.

## 22.4 Coordination Compounds

The most distinctive aspect of transition metal chemistry is the formation of **coordination compounds** (also called *complexes*), substances that typically contain at least one **complex ion.** As you learned in earlier chapters, a complex ion consists of a central metal cation (either transition or main-group metal) bonded to molecules and/or anions called **ligands.** To maintain electrical neutrality in the coordination compound, the complex ion is typically associated with simple ions, called **counter ions.** For example, in Figure 22.10, *A,* the coordination compound is $[Co(NH_3)_6]Cl_3$, the complex ion is $[Co(NH_3)_6]^{3+}$, the three $Cl^-$ ions are counter ions, and the six $NH_3$ molecules bonded to the central $Co^{3+}$ are ligands.

A coordination compound behaves as an electrolyte in water, the complex ions and counter ions separating from each other. *The ligands of the complex ion usually remain attached to the central metal.* Thus, 1 mol $[Co(NH_3)_6]Cl_3$ yields 1 mol $[Co(NH_3)_6]^{3+}$ ions and 3 mol $Cl^-$ ions. We discussed the Lewis acid-base properties of complex ions in Section 17.9 and their equilibria in Section 18.4. Here we examine their bonding, structure, and properties.

### Structures of Complex Ions: Coordination Numbers, Geometries, and Ligands

A complex ion is described by the metal ion and the number and types of ligands attached to it. The **coordination number** is the number of ligand atoms bonded directly to the central metal ion and is *specific* for a given metal ion in a particular oxidation state and compound. Thus, the coordination number of $[Co(NH_3)_6]^{3+}$ is 6 because six ligand atoms (from $NH_3$ molecules) are bonded to the $Co^{3+}$ ion. The coordination number of many platinum(II) complexes is 4, whereas that of platinum(IV) complexes is 6. Copper(II) may have a coordination number of 2, 4, or 6.

*The geometry of a complex ion is related to the coordination number and the metal ion.* The most common coordination number is 6, with 2 and 4 seen often as well; some higher coordination numbers are also known. The geometries associated with the coordination numbers for some complex ions are shown in Table 22.6. A complex ion with a coordination number of 2, such as $[Ag(NH_3)_2]^+$, is usually *linear.* The coordination number 4 gives rise to two geometries. The $d^8$ metal ions form *square planar* complex ions (Figure 22.10, *B*). The $d^{10}$ ions usually form *tetrahedral* complex ions. A coordination number of 6 usually results in an *octahedral* geometry, as shown by $[Co(NH_3)_6]^{3+}$ in Figure 22.10, *A.* Note the similarity with some molecular shapes in VSEPR theory (Section 10.1).

**TABLE 22.6 Coordination Numbers and Shapes of Some Complex Ions**

| COORDINATION NUMBER | SHAPE | | EXAMPLES |
|---|---|---|---|
| 2 | Linear |  | $[CuCl_2]^-$, $[Ag(NH_3)_2]^+$, $[AuCl_2]^-$ |
| 4 | Square planar |  | $[Ni(CN)_4]^{2-}$, $[PdCl_4]^{2-}$, $[Pt(NH_3)_4]^{2+}$, $Cu(NH_3)_4^{2+}$ |
| 4 | Tetrahedral | 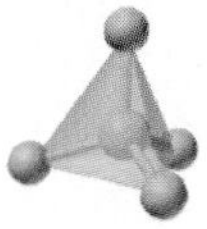 | $[Cu(CN)_4]^{3-}$, $[Zn(NH_3)_4]^{2+}$, $[CdCl_4]^{2-}$, $[MnCl_4]^{2-}$ |
| 6 | Octahedral | 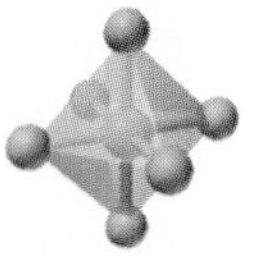 | $[Ti(H_2O)_6]^{3+}$, $[V(CN)_6]^{4-}$, $[Cr(NH_3)_4Cl_2]^+$, $[Mn(H_2O)_6]^{2+}$, $[FeCl_6]^{3-}$, $[Co(en)_3]^{3+}$ |

The ligands of complex ions are molecules and/or anions with one or more **donor atoms** that each donate a lone pair of electrons to the metal ion to form a covalent bond. Thus, *ligands are Lewis bases, the metal ion is a Lewis acid, and the complex ion is a Lewis adduct* (Section 17.9). Because they have at least one lone pair, donor atoms often come from Groups 5A(15), 6A(16), or 7A(17).

Ligands are classified in terms of the number of donor atoms, or "teeth," that each uses to bond to the central metal ion. *Unidentate* (Latin, "one-toothed") ligands, such as Cl$^-$ and NH$_3$, use a single donor atom. *Bidentate* ligands have two donor atoms, each of which bonds to the metal ion. *Polydentate* ligands have more than two donor atoms. Table 22.7 shows some common ligands in coordination compounds; note the lone pair of electrons

**TABLE 22.7 Some Common Ligands in Coordination Compounds**

| LIGAND TYPE | EXAMPLES |
|---|---|
| Unidentate | $H_2\ddot{O}:$ water    $:\ddot{F}:^-$ fluoride ion    $[:C\equiv N:]^-$ cyanide ion    $[:\ddot{O}-H]^-$ hydroxide ion <br> $:NH_3$ ammonia    $:\ddot{C}\ddot{l}:^-$ chloride ion    $[:\ddot{S}=C=\ddot{N}:]^-$ thiocyanate ion    $[:\ddot{O}-\ddot{N}=\ddot{O}:]^-$ nitrite ion |
| Bidentate | 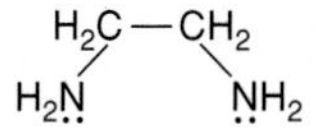 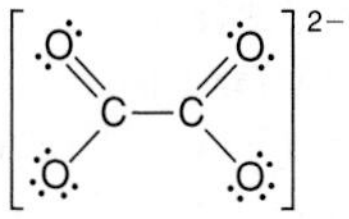 ethylenediamine (en)     oxalate ion |
| Polydentate | 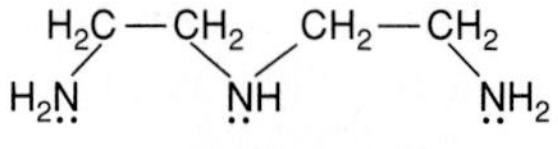 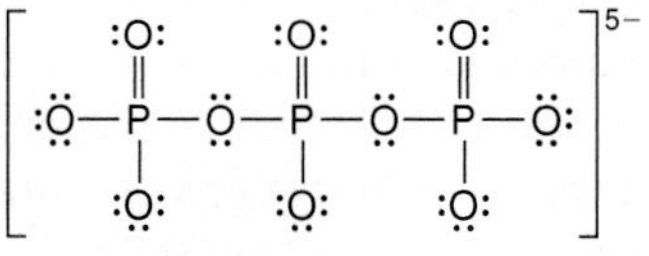 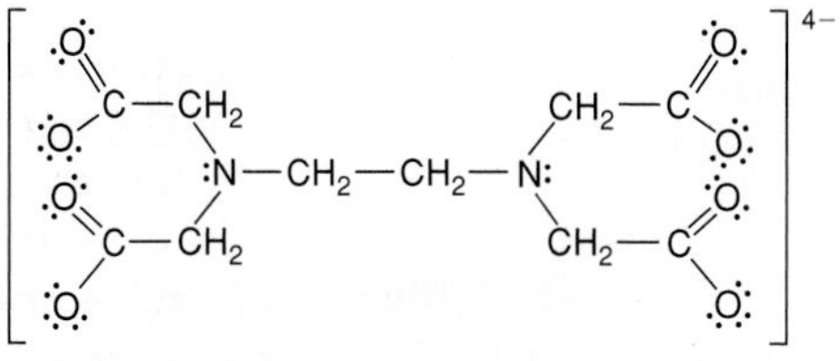 diethylenetriamine    triphosphate ion    ethylenediaminetetraacetate (EDTA) ion |

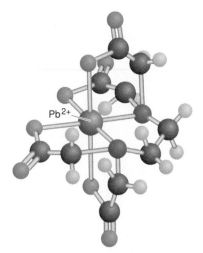

$Pb^{2+}$

◆ **Grabbing Toxic Ions.**
Because of its six donor atoms, the ethylenedi-aminetetraacetate (EDTA) ion forms very stable complexes with many metal ions. Once ingested by the patient, it acts as a scavenger to remove lead and other heavy-metal ions from the blood and other body fluids.

**TABLE 22.8  Names of Some Neutral and Anionic Ligands**

| NAME | FORMULA |
|---|---|
| **A. Neutral** | |
| Aqua | $H_2O$ |
| Ammine | $NH_3$ |
| Carbonyl | $CO$ |
| Nitrosyl | $NO$ |
| **B. Anionic** | |
| Fluoro | $F^-$ |
| Chloro | $Cl^-$ |
| Bromo | $Br^-$ |
| Iodo | $I^-$ |
| Hydroxo | $OH^-$ |
| Cyano | $CN^-$ |

on each donor atom. Bidentate and polydentate ligands give rise to rings in the complex ion. For instance, ethylenediamine (abbreviated "en") has a chain of four atoms ($\ddot{N}$—C—C—$\ddot{N}$), so it forms a five-membered ring, with the metal atom bonding to the two electron-donating N atoms. Such ligands seem to grab the metal ion like a crab's claws, so complex ions that contain them are sometimes called **chelates** (Greek *chela*, crab's claw). ◆

**Formulas and Names of Coordination Compounds**

There are three important rules for writing formulas of coordination compounds, the first two being the same for formulas of any ionic compound:
1. *The cation is written before the anion.*
2. *The charge of the cation(s) is balanced by the charge of the anion(s).*
3. *In the complex ion, neutral ligands are written before anionic ligands, and the whole ion is placed in brackets.*

Let's apply these rules as we examine the combinations of ions in coordination compounds. *A complex ion may be an anion or a cation.* A complex anion has one or more metal cations as counter ions. For example, in $K_2[Co(NH_3)_2Cl_4]$, two $K^+$ counter ions balance the charge of the complex anion $[Co(NH_3)_2Cl_4]^{2-}$, which contains two $NH_3$ molecules and four $Cl^-$ ions as ligands. The two $NH_3$ are neutral, the four $Cl^-$ have a total charge of 4−, and the entire complex ion has a charge of 2−, so the central metal ion is $Co^{2+}$. In the compound $[Co(NH_3)_4Cl_2]Cl$, the complex ion is $[Co(NH_3)_4Cl_2]^+$ and one $Cl^-$ is the counter ion. The four $NH_3$ ligands are neutral, the two $Cl^-$ ligands have a total charge of 2−, and the complex cation has a charge of 1+, so the central metal ion is $Co^{3+}$. Some coordination compounds have a complex cation *and* a complex anion, as in $[Pt(NH_3)_4][PdCl_4]$. In this compound, the complex cation is $[Pt(NH_3)_4]^{2+}$, with $Pt^{2+}$, and the complex anion is $[PdCl_4]^{2-}$ with $Pd^{2+}$.

Coordination compounds were originally named after the person who first prepared them or from their color. However, in order to name the increasing number of coordination compounds, rules for systematic naming were proposed. Here is a simplified version of these naming rules:
1. *The cation is named before the anion.* In naming $[Co(NH_3)_4Cl_2]Cl$, for example, we name the $[Co(NH_3)_4Cl_2]^+$ before the $Cl^-$ ion. The name is

tetraamminedichlorocobalt(III) chloride

The only space in the name appears between the cation and the anion.
2. *Within the complex ion, the ligands are named, in alphabetical order, **before** the metal ion.* Note that in the $[Co(NH_3)_4Cl_2]^+$ ion of the compound in Rule 1, the four $NH_3$ and two $Cl^-$ are named before the $Co^{3+}$.
3. *Neutral ligands generally have the molecule name*, but there are a few important exceptions (Table 22.8, *A*). *Anionic ligands drop the -ide and add -o after the root name*, such as fluoro for $F^-$ (Table 22.8, *B*). The two ligands in $[Co(NH_3)_4Cl_2]^+$ are ammine for $NH_3$ and chloro for $Cl^-$, with "ammine" coming before "chloro."
4. *Numerical prefixes denote the number of a particular ligand.* Thus, *tetra*ammine denotes *four* $NH_3$ and *di*chloro denotes *two* $Cl^-$. Other prefixes are tri, penta, and hexa. These prefixes do *not* affect the alphabetical order, so "tetraammine" comes before "dichloro." Some ligand names already contain a numerical prefix (such as ethylene*di*amine), so we use bis (2), tris (3), or tetrakis (4) to indicate the number of such a ligand, followed by the lig-

and name in parentheses. Thus, a complex ion with two ethylenediamine ligands would have "bis(ethylenediamine)" in the name.

5. *The oxidation state of the central metal ion is given by a Roman numeral in parentheses.* As with any ion, we follow this rule only when the metal ion has more than one oxidation state, as in the compound named in Rule 1.

6. *If the complex ion is an anion, we drop the ending of the metal name and add -ate.* Thus, the name for $K[Pt(NH_3)Cl_5]$ is

potassium amminepentachloroplatinate(IV)

(Note that there is one $K^+$ counter ion, so the complex anion has a charge of 1−. The five $Cl^-$ ligands have a total charge of 5−, so Pt must be in the +4 oxidation state.) For some metals, we use the Latin root with the "-ate" ending (Table 22.9). For example, the name for $Na_4[FeBr_6]$ is

sodium hexabromoferrate(II).

TABLE 22.9 **Names of Some Metal Ions in Complex Anions**

| METAL | NAME IN ANION |
| --- | --- |
| Iron | Ferrate |
| Copper | Cuprate |
| Lead | Plumbate |
| Silver | Argentate |
| Gold | Aurate |
| Tin | Stannate |

SAMPLE PROBLEM 22.3

## Writing Names and Formulas of Coordination Compounds

**Problem: (a)** What is the systematic name of $Na_3[AlF_6]$?
**(b)** What is the systematic name of $[Co(en)_2Cl_2]NO_3$?
**(c)** What is the formula of tetraamminebromochloroplatinum(IV) chloride?
**(d)** What is the formula of hexaamminecobalt(III) tetrachloroferrate(III)?
**Plan:** We use the rules just discussed, with reference to Tables 22.8 and 22.9.
**Solution: (a)** The complex ion is $[AlF_6]^{3-}$. There are six (hexa) $F^-$ ions (fluoro) as ligands, so we have "hexafluoro." The complex ion is an anion, so the ending of the metal ion (aluminum) must be changed to -ate: hexafluoroaluminate. Aluminum has only the +3 oxidation state, so we do *not* use a Roman numeral. The positive counter ion is named first and separated from the anion by a space:
**sodium hexafluoroaluminate.**
**(b)** Listed alphabetically, there are two $Cl^-$ (dichloro) and two en [bis(ethylenediamine)] as ligands. The complex ion is a cation, so the metal name is unchanged, but we specify its oxidation state because cobalt can have several. One $NO_3^-$ balances the cation charge; with 2− for two $Cl^-$ and 0 for two en, the metal must be cobalt(III). The word nitrate follows a space:
**dichlorobis(ethylenediamine)cobalt(III) nitrate.**
**(c)** The central metal ion is written first, followed by the neutral ligands and then (in alphabetical order) by the negative ligands. "Tetraammine" is four $NH_3$, "bromo" is one $Br^-$, "chloro" is one $Cl^-$, and "platinate(IV)" is $Pt^{4+}$, so the complex ion is $[Pt(NH_3)_4BrCl]^{2+}$. Its 2+ charge is the sum of 4+ for $Pt^{4+}$, 0 for four $NH_3$, 1− for one $Br^-$, and 1− for one $Cl^-$. To balance the 2+ charge, we need two $Cl^-$ counter ions:
**$[Pt(NH_3)_4BrCl]Cl_2$.**
**(d)** This compound consists of two complex ions. In the cation, "hexaammine" is six $NH_3$ and "cobalt(III)" is $Co^{3+}$, so the cation is $[Co(NH_3)_6]^{3+}$. The 3+ charge is the sum of 3+ for $Co^{3+}$ and 0 for six $NH_3$. In the anion, "tetrachloro" is four $Cl^-$, and "ferrate(III)" is $Fe^{3+}$, so the anion is $[FeCl_4]^-$. The 1− charge is the sum of 3+ for $Fe^{3+}$ and 4− for four $Cl^-$. In the neutral compound, one 3+ cation is balanced by three 1− anions: **$[Co(NH_3)_6][FeCl_4]_3$.**
**Check:** Reverse the process to be sure you obtain the name or formula asked for in the problem.

FOLLOW-UP PROBLEM 22.3
**(a)** What is the name of $[Cr(H_2O)_5Br]Cl_2$?
**(b)** What is the formula of barium hexacyanocobaltate(III)?

## A Historical Perspective: Alfred Werner and Coordination Theory

The substances we now call coordination compounds had been known for almost 200 years when the young Swiss chemist Alfred Werner began studying them in the 1890s. He investigated a series of compounds such as the cobalt series shown in Table 22.10, each of which contains one cobalt(III) ion, three chloride ions, and a given number of ammonia molecules. At the time, which was 30 years before the idea of atomic orbitals was proposed, no structural theory could explain how compounds with similar, even identical, formulas could have widely different properties.

**TABLE 22.10  Some Coordination Compounds Studied by Werner**

| TRADITIONAL FORMULA | WERNER'S DATA | | MODERN FORMULA | CHARGE OF COMPLEX ION |
| | TOTAL MOLES OF IONS/MOLE OF COMPOUND | MOLES OF FREE Cl⁻/MOLE OF COMPOUND | | |
| --- | --- | --- | --- | --- |
| $CoCl_3 \cdot 6NH_3$ | 4 | 3 | $[Co(NH_3)_6]Cl_3$ | $3+$ |
| $CoCl_3 \cdot 5NH_3$ | 3 | 2 | $[Co(NH_3)_5Cl]Cl_2$ | $2+$ |
| $CoCl_3 \cdot 4NH_3$ | 2 | 1 | $[Co(NH_3)_4Cl_2]Cl$ | $1+$ |
| $CoCl_3 \cdot 3NH_3$ | 0 | 0 | $[Co(NH_3)_3Cl_3]$ | — |

Werner measured the conductivity of each compound in aqueous solution to determine the total number of dissociated ions. He treated the solutions with excess $AgNO_3$ to precipitate released $Cl^-$ ions as AgCl and thus determine the number of free $Cl^-$ ions per formula unit. Previous studies had established that the $NH_3$ molecules were not free in solution. Werner's data, summarized in Table 22.10, could not be explained by the accepted, traditional formulas of the compounds. Other chemists had proposed "chain" structures, like those of organic compounds, to explain similar data. For example, a proposed structure for $[Co(NH_3)_6]Cl_3$ was

$$
\begin{array}{l}
NH_3\!-\!Cl \\
| \\
Co\!-\!NH_3\!-\!Cl \\
| \\
NH_3\!-\!NH_3\!-\!NH_3\!-\!NH_3\!-\!Cl
\end{array}
$$

However, these models proved inadequate.

Werner's novel idea was the coordination complex, a central metal ion surrounded by a *constant total number* of covalently bonded molecules and/or anions. The coordination complex could be neutral or charged; if charged, it combines with oppositely charged counter ions, in this case $Cl^-$, to form the neutral compound. Werner proposed two types of valence, or *combining ability,* for metal ions. *Primary valence,* now called oxidation state, is the positive charge on the metal ion that must be satisfied by an equivalent negative charge. In Werner's cobalt series, the primary valence is $+3$, and it is always balanced by three $Cl^-$ ions. These anions can be bonded covalently to Co as part of the complex ion and/or associated with it as counter ions.

*Secondary valence*, now called coordination number, is the constant total number of connections (anionic or neutral ligands) within the complex ion. The secondary valence in this series of cobalt compounds is 6.

As you can see from Table 22.10, Werner's data are satisfied by having the total number of ligands remain the same for each compound but varying the numbers of $Cl^-$ ions and $NH_3$ molecules in the different complex ions. For example, the first compound, $[Co(NH_3)_6]Cl_3$, has a total of four ions: one $[Co(NH_3)_6]^{3+}$ and three $Cl^-$. All three $Cl^-$ ions are free to form AgCl. The last compound, $[Co(NH_3)_3Cl_3]$, contains no separate ions. For these pioneering studies, especially his prediction of optical isomerism (discussed next), Werner received the Nobel Prize in 1913.

SAMPLE PROBLEM 22.4

## Determining the Nature of the Complex Ion from the Empirical Formula of the Compound

**Problem:** Platinum(IV) complexes, like those of cobalt(III), have a coordination number of 6. Many occur with $Cl^-$ ions and $NH_3$ molecules as ligands. Consider the following traditional formulas for coordination compounds:
**(a)** $PtCl_4·6NH_3$     **(b)** $PtCl_4·4NH_3$
For each of these compounds,
(1) Give the formula and charge of the complex ion.
(2) Predict the moles of ions formed per mole of compound dissolved and the moles of AgCl formed immediately with excess $AgNO_3$.
**Plan:** All the $NH_3$ must be bonded to the platinum(IV), so we assign $Cl^-$ ions to the complex ion to reach a total of six ligands. Any additional $Cl^-$ ions are counter ions. When dissolved, the complex ion and counter ions separate, and each mole of free $Cl^-$ forms one mole of AgCl.
**Solution: (a)** For $PtCl_4·6NH_3$
(1) The six $NH_3$ are the six ligands, so all four $Cl^-$ must be counter ions. The complex ion is **$[Pt(NH_3)_6]^{4+}$.**
(2) **Five moles of ions** are released: 1 mol $[Pt(NH_3)_6]^{4+}$ and 4 mol $Cl^-$;
   **4 mol AgCl form.**
**(b)** For $PtCl_4·4NH_3$
(1) The formula has only four $NH_3$, so two of the six ligands must be $Cl^-$, and two $Cl^-$ are counter ions. The complex ion is **$[Pt(NH_3)_4Cl_2]^{2+}$.**
(2) **Three moles of ions** are released: 1 mol $[Pt(NH_3)_4Cl_2]^{2+}$ and 2 mol $Cl^-$;
   **2 mol AgCl form.**

FOLLOW-UP PROBLEM 22.4
For the compound $PtKCl_5·NH_3$, give the formula and charge of the complex ion; and predict the moles of ions formed per mole of compound dissolved and the moles of AgCl formed immediately with excess $AgNO_3$.

## Isomerism in Coordination Compounds

**Isomers** are substances with the same chemical formula but different properties. We have already discussed many aspects of isomerism in the context of organic compounds (Section 14.2). Figure 22.11 presents an overview of the most common types of isomerism in coordination compounds.

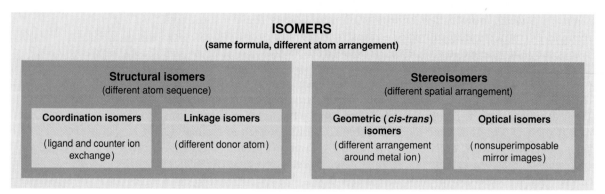

**FIGURE 22.11**
Important types of isomerism in coordination compounds.

**Structural isomers: different sequences of atoms.** Two compounds with the same formula, but with the atoms connected in different sequences, are **structural isomers.** Coordination compounds exhibit the following two types of structural isomers: one involves changes in the composition of the complex ion, the other in the donor atom of the ligand.

**Coordination isomers** occur when the complex ion changes its composition but the compound does not, as when ligand and counter ion exchange positions, as in $[Pt(NH_3)_4Cl_2](NO_2)_2$ and $[Pt(NH_3)_4(NO_2)_2]Cl_2$. In the first compound, the $Cl^-$ ions are the ligands and the $NO_2^-$ ions are counter ions; in the second, the roles are reversed. This type of isomerism also occurs in compounds consisting of two complex ions in which the two sets of ligands in one compound are reversed in the other, as in $[Cr(NH_3)_6][Co(CN)_6]$ and $[Co(NH_3)_6][Cr(CN)_6]$.

**Linkage isomers** occur when the composition of the complex ion remains the same but the ligand donor atom changes. Some ligands can bind to the metal ion through *either of two donor atoms:*

$$\left[ \begin{array}{c} \ddot{O} \\ \diagdown \\ N: \\ \diagup \\ \ddot{O} \end{array} \right]^{-} \qquad [:\ddot{O}=C=\ddot{N}:]^{-} \qquad [:\ddot{S}=C=\ddot{N}:]^{-}$$

nitrite          cyanate          thiocyanate

For example, the nitrite ion can bind through either the N atom ("nitro" $O_2N:\rightarrow$) or through one of the O atoms ("nitrito" $ONO:\rightarrow$) to give linkage isomers, as in the yellow compound pentaammine*nitro*cobalt(III) chloride $[Co(NH_3)_5(NO_2)]Cl_2$ and its red linkage isomer pentaammine*nitrito*cobalt(III) chloride $[Co(NH_3)_5(ONO)]Cl_2$. The cyanate ion can attach via the O atom ("cyanato" NCO:$\rightarrow$) or the N atom ("isocyanato" OCN:$\rightarrow$), and the thiocyanate ion behaves similarly.

**Stereoisomers: Different Spatial Arrangements of Atoms.** Stereoisomers are two compounds with the same sequence, but different spatial arrangement, of atoms. The two types we discussed for organic compounds, called geometric and optical isomers, are seen here as well. **Geometric** (or *cis-trans*) **isomers** differ in how atoms or groups of atoms are arranged in space relative to the central ion. For example, the square planar $[Pt(NH_3)_2Cl_2]$ has two arrangements, which give rise to two different compounds (Figure 22.12, *A*). The isomer with identical ligands next to

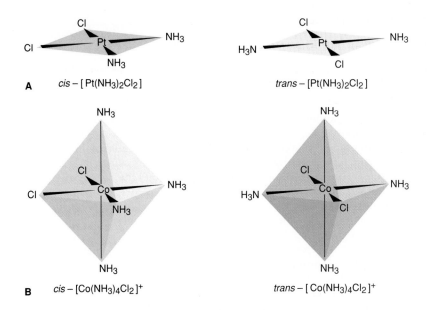

**FIGURE 22.12**
Geometric *(cis–trans)* isomerism. **A,** *Cis-* and *trans-* isomers of the square planar compound [Pt(NH$_3$)$_2$Cl$_2$]. **B,** *Cis-* and *trans-* isomers of the octahedral complex ion [Co(NH$_3$)$_4$Cl$_2$]$^+$. The colors represent the actual colors of the species.

each other is *cis*-diamminedichloroplatinum(II), and the one with the identical ligands across from each other is *trans*-diamminedichloroplatinum(II); their biological behaviors are remarkably different. ◆ Octahedral complexes also exhibit *cis-trans* isomerism (Figure 22.12, *B*). The *cis*-isomer of the [Co(NH$_3$)$_4$Cl$_2$]$^+$ ion is violet, and the *trans*-isomer is green.

**Optical isomers** occur when a molecule and its mirror image cannot be superimposed. Unlike the other types of isomers, which have distinct physical properties, optical isomers are physically identical in all ways but one: *the direction in which they rotate the plane of polarized light* (see Figure 14.11). Octahedral complexes show many examples of optical isomerism. Consider the two optical isomers of the *cis*-dichlorobis(ethylenediamine)cobalt(III) ion, [Co(en)$_2$Cl$_2$]$^+$, shown in Figure 22.13, *A*. Rotate the isomer on the left 180° around a vertical axis, and the Cl$^-$ ligands match those in the isomer on the right, but the (en) ligands do not: the two structures are not superimposable. One is designated *d*-[Co(en)$_2$Cl$_2$]$^+$ and the other is *l*-[Co(en)$_2$Cl$_2$]$^+$, depending on whether it rotates polarized light to the right (*d* for dextro-) or to the left (*l* for levo-). In contrast, *trans*-dichlorobis(ethylenediamine)cobalt(III) ion (Figure 22.13, *B*) does *not* have optical isomers: the structure on the left *is* superimposable with the one on the right.

◆ **Fighting Cancer with Geometric Isomers.** In the mid-1960s, Dr. Barnett Rosenberg and colleagues discovered that *cis*-Pt(NH$_3$)$_2$Cl$_2$ ("cisplatin") was a highly effective antitumor agent. This compound and several closely related platinum(II) complexes are still among the most effective treatments for certain types of cancer. The geometric isomer, *trans*-Pt(NH$_3$)$_2$Cl$_2$, is completely ineffective as an antitumor agent. Cisplatin may work by lying within the cancer cell's DNA double helix, such that a donor atom on each strand replaces the *cis*-Cl$^-$ ligands and binds the platinum(II) strongly, thus preventing DNA replication (Section 14.5).

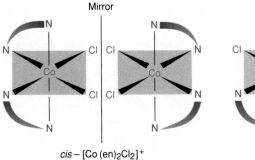

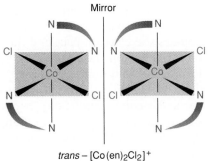

**FIGURE 22.13**
Optical isomerism in an octahedral complex ion. **A,** The *cis*-isomer of [Co(en)$_2$Cl$_2$]$^+$ and its mirror image cannot be superimposed, no matter how they are rotated, so they are optical isomers. (The curved wedge is a common representation of the bidentate ligand ethylenediamine, H$_2$N—CH$_2$—CH$_2$—NH$_2$.) **B,** The *trans*-isomer, when rotated, *can* be superimposed on its mirror image.

SAMPLE PROBLEM 22.5

### Determining the Type of Stereoisomerism

**Problem:** Which type(s) of stereoisomerism will each of the following exhibit?

**(a)** $[Pt(NH_3)_2Br_2]$ (square planar)

**(b)** $[Cr(en)_3]^{3+}$ (en = $H_2\ddot{N}CH_2CH_2\ddot{N}H_2$)

**Plan:** We first determine the geometry around each metal ion and then see if it is possible to place two ligands in different positions relative to each other. If it *is* possible, *cis-trans* isomerism occurs. Then, we see whether the mirror image of the complex is superimposable on the original. If it is *not*, optical isomerism occurs.

**Solution: (a)** The Pt(II) complex is square planar and the ligands are unidentate. Each pair of ligands can lie at adjacent or at opposite corners of the plane. Thus, ***cis-trans* isomerism** occurs. Each isomer *is* superimposable on its mirror image, so there is no optical isomerism.

**(b)** Ethylenediamine (en) is a bidentate ligand. The $Cr^{3+}$ is six-coordinate and octahedral, like $Co^{3+}$. Three bidentate ligands in an octahedral complex yield nonsuperimposable **optical isomers.**

FOLLOW-UP PROBLEM 22.5

What stereoisomers, if any, are possible for the $[Co(NH_3)_2(en)Cl_2]^+$ ion?

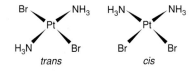

### Section Summary

Coordination compounds consist of a complex ion and charge-balancing counter ions. The complex ion has a central metal ion bonded to neutral and/or anionic ligands, which act as Lewis bases with one or more donor atoms. The most common geometry is octahedral (six ligand bonds). Formulas and names of coordination compounds follow systematic rules. Alfred Werner established the structural basis of coordination compounds. These compounds can exhibit structural isomers (coordination and linkage) and stereoisomers (*cis-trans* and optical).

## 22.5    Theoretical Basis for the Bonding and Properties of Complexes

In this section, we consider models that address, to differing extents, several key questions about complexes: how metal-ligand bonds form, why certain geometries are preferred, and why these compounds are brightly colored and often paramagnetic.

### Application of Valence-Bond Theory to Complexes

*Valence-bond (VB) theory,* which helped explain bonding and structure in main-group compounds (Section 10.3), has also been used to describe bonding in complex ions. In the formation of a complex ion, the filled ligand orbital overlaps the empty metal-ion orbital. The ligand (Lewis base) donates the electron pair, and the metal ion (Lewis acid) accepts it to form the complex ion (Lewis adduct). Such a bond is called a **coordinate covalent bond,** although, once formed, it is identical to any covalent single bond. Recall that the VB concept of hybridization proposes the mixing of particular combinations of *s*, *p*, and *d* orbitals to give hybrid-orbital sets of specific geometries. For coordination compounds, the model proposes that *the number and type of metal-ion hybrid orbitals occupied by ligand lone pairs determine the geometry of the complex ion.* The orbital combinations that lead to octahedral, square planar, and tetrahedral complexes are discussed next.

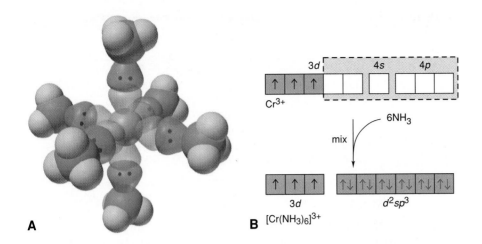

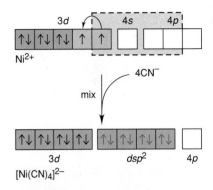

**FIGURE 22.14**
Hybrid orbitals and bonding in the octahedral $[Cr(NH_3)_6]^{3+}$ ion. **A,** The $Cr(NH_3)_6^{3+}$ ion. **B,** The partial orbital box diagrams depict the mixing of two $3d$, one $4s$, and three $4p$ orbitals in $Cr^{3+}$ and bonding with six $NH_3$ lone pairs *(red)* to form six $d^2sp^3$ hybrid orbitals.

**Octahedral complexes.** The hexaamminechromium(III) ion, $[Cr(NH_3)_6]^{3+}$, illustrates the application of VB theory to an *octahedral complex* (Figure 22.14). The six lowest energy empty orbitals of the $Cr^{3+}$ ion—two $3d$, one $4s$, and three $4p$—mix and become six equivalent $d^2sp^3$ hybrid orbitals that point toward the corners of an octahedron. Six $NH_3$ molecules donate their nitrogen lone electron pairs to form six metal-ligand bonds. The three unpaired $3d$ electrons of the central $Cr^{3+}$ ion ([Ar] $3d^3$) remain in unhybridized orbitals and make the complex ion paramagnetic.

**Square planar complexes.** The $d^8$ metal ions usually form *square planar complexes* (Figure 22.15). In the $[Ni(CN)_4]^{2-}$ ion, for example, the model proposes that one $3d$, one $4s$, and two $4p$ orbitals of $Ni^{2+}$ mix and form four $dsp^2$ hybrid orbitals, which point to the corners of a square and accept an electron pair from four $CN^-$ ligands. How can the $Ni^{2+}$ ion ([Ar] $3d^8$) offer an empty $3d$ orbital if eight $3d$ electrons lie in three filled and two half-filled orbitals? Apparently, in the $d^8$ configuration of $Ni^{2+}$, electrons in the half-filled orbitals pair and leave one $3d$ orbital empty. This reasoning is consistent with the fact that the complex is diamagnetic (no unpaired electrons). Moreover, it implies that the energy *gained* by utilizing a $3d$ orbital for bonding in the hybrid orbital is greater than the energy *required* to overcome repulsions from pairing the $3d$ electrons.

**Tetrahedral complexes.** Metal ions with a filled $d$ sublevel, such as $Zn^{2+}$ ([Ar] $d^{10}$), often form *tetrahedral complexes* (Figure 22.16). In the complex ion $[Zn(OH)_4]^{2-}$, for example, VB theory proposes that the lowest available $Zn^{2+}$ orbitals—one $4s$ and three $4p$—mix to become four $sp^3$ hybrid orbitals that point to the corners of a tetrahedron and are occupied by a lone pair from each of four $OH^-$ ligands.

## Crystal Field Theory

The VB approach is easy to picture, but it does not describe the effect of the approaching ligands on the orbital energies of the metal ion during the formation of the complex ion. Rather, it treats the orbitals as little more than empty slots for accepting electron pairs. Consequently, it imparts no insight into the colors of coordination compounds and sometimes predicts magnetic properties incorrectly.

Color and magnetism are better explained with **crystal field theory.** This model assumes little about metal-ligand bonding but does explain the

**FIGURE 22.15**
Hybrid orbitals and bonding in the square planar $[Ni(CN)_4]^{2-}$ ion. Two lone $3d$ electrons pair up and free one $3d$ orbital for hybridization with the $4s$ and two $4p$ orbitals to form four $dsp^2$ orbitals, which become occupied with lone pairs *(red)* from four $CN^-$ ligands.

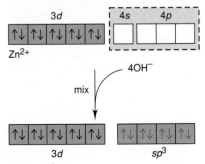

**FIGURE 22.16**
Hybrid orbitals and bonding in the tetrahedral $Zn(OH)_4^{2-}$ ion. Four $sp^3$ orbitals are available for accepting lone pairs *(red)* from $OH^-$ ligands.

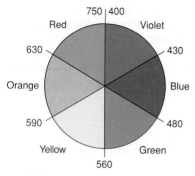

**FIGURE 22.17**

**An artist's wheel.** Primary (red, blue, and yellow) and secondary (violet, green, and orange) colors are shown, with approximate wavelength ranges. Note that two primary colors form a secondary color that is the complementary color of the third primary color.

**TABLE 22.11    Relation Between Absorbed and Observed Colors**

| ABSORBED COLOR | λ (nm) | OBSERVED COLOR | λ (nm) |
|---|---|---|---|
| Violet | 400 | Green-yellow | 560 |
| Blue | 450 | Yellow | 600 |
| Blue-green | 490 | Red | 620 |
| Yellow-green | 570 | Violet | 410 |
| Yellow | 580 | Dark blue | 430 |
| Orange | 600 | Blue | 450 |
| Red | 650 | Green | 520 |

effects on *d*-orbital energies from metal-ligand interactions. Before we discuss this theory, let's consider what causes a substance to be colored.

**What is color?** White light is electromagnetic radiation with all wavelengths (λ) in the visible range (Section 7.1). It can be dispersed into a spectrum of colors, each of which has a narrow range of wavelengths. Objects appear colored in white light because they absorb certain wavelengths and reflect or transmit others: an opaque object *reflects* light, whereas a clear one *transmits* it. The reflected or transmitted light enters the eye and is perceived as color by the brain.

There are three *primary* colors—red, yellow, and blue—and three *secondary* colors—orange, green, and violet. Mixing pairs of primary colors produces the secondary colors; for example, a red pigment mixed with a yellow one appears orange. Each primary color has a *complementary* secondary color, such that a mixture of the two absorbs all visible wavelengths and appears black. Green, for example, is the complementary color of red. Figure 22.17 shows these relationships on an artist's color wheel, a circle in which complementary colors appear as wedges opposite each other.

An object has a particular color for one of two reasons:

1.  It reflects (or transmits) light of that color.
2.  It absorbs light of the complementary color.

Thus, if an object absorbs all incoming wavelengths *except* green, the reflected (or transmitted) light enters our eyes and is interpreted as green. Alternatively, if the substance absorbs only red, the *complement* of green, our brain interprets the reflected (or transmitted) mixture of remaining wavelengths as green also. Table 22.11 lists the color absorbed and the resulting color perceived. ◆

**Splitting of *d* orbitals in an octahedral field of ligands.** The crystal field model assumes that a complex ion forms as a result of electrostatic attraction between the metal cation and the partially negative pole or full negative charge of the ligands. Figure 22.18, *A*, shows six ligands approaching a metal ion to form an octahedral complex that *minimizes* repulsions with metal *d* electrons. Let's see how the various metal *d* orbitals are affected as the complex forms. As the ligands approach, their electron pairs repel electrons in the five *d* orbitals of the metal. In the isolated metal ion, the *d* orbitals have equal energies despite their different orientations. In the electrostatic field of ligands, however, the *d* electrons are *repelled unequally because of their different orientations.*

Assume that the ligands approach the metal ion along the *x*, *y*, and *z* axes. Therefore, they approach *directly toward* the $d_{x^2-y^2}$ and $d_{z^2}$ orbitals (Figure 22.18, *B* and *C*) but *between* the $d_{xy}$, $d_{yz}$, and $d_{xz}$ orbitals (Figure 22.18, *D* to

◆ **The Colors of Spring and Fall.** In the spring and summer, leaves contain high concentrations of the photosynthetic pigment chlorophyll and lower concentrations of other pigments called xanthophylls. Chlorophyll absorbs strongly in the blue and red regions, reflecting mostly green wavelengths into your eyes. In the fall, photosynthesis slows and the leaf no longer makes chlorophyll. Gradually, the green color fades, revealing the xanthophylls that were present all along but masked by the chlorophyll. Xanthophylls absorb green and blue strongly, reflecting the bright yellows and reds of autumn.

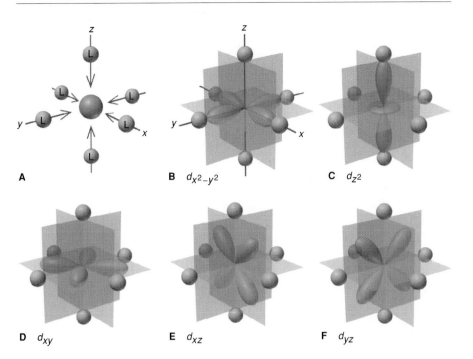

**A**  **B** $d_{x^2-y^2}$  **C** $d_{z^2}$

**D** $d_{xy}$  **E** $d_{xz}$  **F** $d_{yz}$

**FIGURE 22.18**
**The five *d* orbitals in an octahedral field of ligands.** The direction of ligand approach influences the strength of repulsions with electrons in the metal *d* orbitals. **A,** Ligands approach a metal ion along the three linear axes in an octahedral orientation. **B** and **C,** Ligands approach toward lobes of the $d_{x^2-y^2}$ and $d_{z^2}$ orbitals, so repulsions are higher. **D** to **F,** Ligands approach between lobes of the $d_{xy}$, $d_{yz}$, and $d_{xz}$ orbitals, so repulsions are lower.

*F*). As a result, electrons in the $d_{x^2-y^2}$ and $d_{z^2}$ orbitals experience *stronger* repulsions than those in the $d_{xy}$, $d_{yz}$, and $d_{xz}$ orbitals. Whereas all five *d* orbitals become higher in energy, *the orbital energies split with two d orbitals higher than the other three* (Figure 22.19). The two higher energy orbitals are called $e_g$ **orbitals,** and the other three are $t_{2g}$ **orbitals.** (These designations refer to number and symmetry properties of the orbitals that need not concern us here.) The splitting of orbital energies is called the *crystal field effect,* and the difference in energy between the $e_g$ and $t_{2g}$ sets of orbitals is the **crystal field splitting energy (Δ).**

**Colors of transition metal complexes.** The remarkably diverse colors of coordination compounds are related to the energy difference (Δ) between the $t_{2g}$ and $e_g$ orbitals in their complex ions. When the ion absorbs light in the visible range, electrons are excited ("jump") from the lower energy $t_{2g}$ level to the higher $e_g$ level. Recall that the *difference* between two

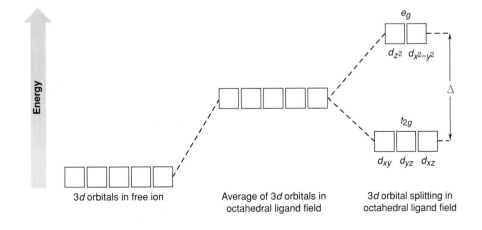

3*d* orbitals in free ion

Average of 3*d* orbitals in octahedral ligand field

3*d* orbital splitting in octahedral ligand field

$e_g$
$d_{z^2}$ $d_{x^2-y^2}$
$t_{2g}$
$d_{xy}$ $d_{yz}$ $d_{xz}$

Δ

**FIGURE 22.19**
**Splitting of *d*-orbital energies by an octahedral field of ligands.** Electrons in the *d* orbitals of the free metal ion experience an average repulsion in the negative ligand field that increases the *d*-orbital energies. Those in the $d_{xy}$, $d_{yz}$, and $d_{xz}$ orbitals, which form the $t_{2g}$ set, are repelled less than those in the $d_{x^2-y^2}$ and $d_{z^2}$ orbitals, which form the $e_g$ set. The energy difference between these two sets is the crystal field splitting energy, Δ.

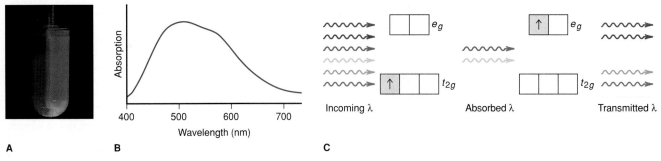

**FIGURE 22.20**
**The color of $[Ti(H_2O)_6]^{3+}$. A,** The hydrated $Ti^{3+}$ ion is light purple in aqueous solution. **B,** An absorption spectrum shows that incoming green and yellow light are absorbed, whereas other wavelengths are transmitted. **C,** An orbital diagram depicts the colors absorbed in the excitation of the $d$ electron.

electronic energy levels in the ion is equal to the energy (and inversely related to the wavelength) of the absorbed photon:

$$\Delta E_{electron} = E_{photon} = h\nu = hc/\lambda$$

Since only certain wavelengths of the incoming white light are absorbed, the substance has a color.

Consider the $[Ti(H_2O)_6]^{3+}$ ion, which appears purple in aqueous solution (Figure 22.20). Hydrated $Ti^{3+}$ is a $d^1$ ion, with the $d$ electron in one of the three lower energy $t_{2g}$ orbitals. The energy difference ($\Delta$) between the $t_{2g}$ and $e_g$ orbitals in this ion corresponds to light spanning the green and yellow range. When white light shines on the solution, these colors of light are absorbed, and the electron jumps to one of the $e_g$ orbitals. Red, blue, and violet light are transmitted, so the solution appears purple.

Different ligands split the $d$-orbital energies to different extents (Figure 22.21, *A*). **Strong-field ligands** lead to a *larger* crystal field splitting energy (larger $\Delta$); **weak-field ligands** lead to a *smaller* splitting energy (smaller $\Delta$). Therefore, the wavelengths of light absorbed by complex ions depend, to a great extent, on the ligand. Even a single ligand substitution can have a major effect on color, as Figure 22.21, *B*, shows for two $Cr^{3+}$ complex ions.

Absorption spectra show the wavelengths absorbed by a given metal ion in the presence of different ligands and by different metal ions with the same ligand. From such data, we relate the energy of the absorbed light to the $\Delta$ values, and two important observations emerge:

1. For a given ligand, *the color depends on the oxidation state of the metal ion.* A solution of $[V(H_2O)_6]^{2+}$ ion is violet, and a solution of $[V(H_2O)_6]^{3+}$ ion is green.
2. For a given metal ion, *the color depends on the ligand.*

These observations allow us to rank ligands with regard to their ability to split the $d$-orbital energies into a **spectrochemical series.** An abbreviated series moving from weak-field ligands (small splitting, small $\Delta$) to strong-

**FIGURE 22.21**
**The effect of the ligand on color. A,** Ligands interacting strongly with metal-ion $d$ orbitals produce a larger $\Delta$ than those interacting weakly. Larger $\Delta$ means photons of higher energy (shorter wavelengths) are absorbed. **B,** A change in even a single ligand can influence color. The $[Cr(NH_3)_6]^{3+}$ ion is yellow-orange *(left)*, and the $[Cr(NH_3)_5Cl]^{2+}$ ion is purple *(right)*. The weaker field $Cl^-$ ligand lowers $\Delta$, so longer $\lambda$ are absorbed.

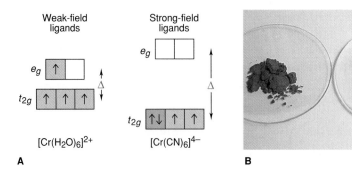

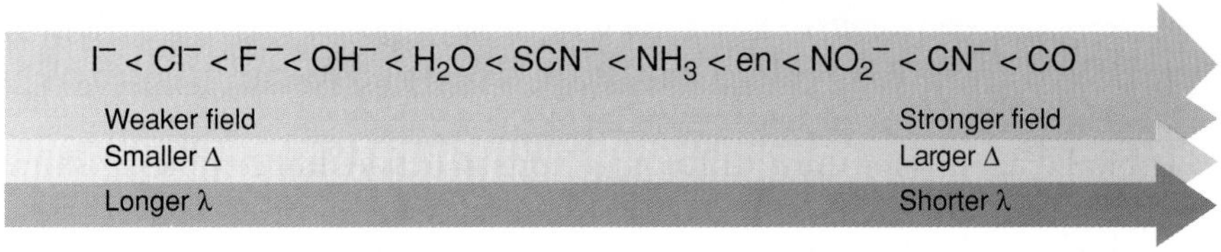

$$I^- < Cl^- < F^- < OH^- < H_2O < SCN^- < NH_3 < en < NO_2^- < CN^- < CO$$

Weaker field　　　　　　　　　　　　　　Stronger field
Smaller $\Delta$　　　　　　　　　　　　　　Larger $\Delta$
Longer $\lambda$　　　　　　　　　　　　　　Shorter $\lambda$

**FIGURE 22.22**
**The spectrochemical series.** As the crystal field strength of the ligand increases, the splitting energy ($\Delta$) increases and shorter wavelengths ($\lambda$) are absorbed. Water is usually a weak-field ligand.

field ligands (large splitting, large $\Delta$) is shown in Figure 22.22. Using this series, we can predict the *relative* size of $\Delta$ for a series of octahedral complexes of the same metal ion. Although it is difficult to predict the actual color of a given complex, we can determine whether a complex will absorb longer or shorter wavelengths than other complexes in the series.

SAMPLE PROBLEM 22.6

**Ranking Crystal Field Splitting Energies for Complex Ions of a Given Metal**

**Problem:** Rank the ions $[Ti(H_2O)_6]^{3+}$, $[Ti(NH_3)_6]^{3+}$, and $[Ti(CN)_6]^{3-}$ in terms of the value of $\Delta$ and the energy of visible light absorbed.
**Plan:** The formulas show that titanium's oxidation state is $+3$ in the three ions. From Figure 22.22, we rank the ligands in terms of their crystal field strength: the stronger the ligand, the greater the splitting, and the higher the energy of light absorbed.
**Solution:** The ligand field strength is in the order $CN^- > NH_3 > H_2O$, so the relative size of $\Delta$ and energy of light absorbed is

$$Ti(CN)_6^{3-} > Ti(NH_3)_6^{3+} > Ti(H_2O)_6^{3+}$$

FOLLOW-UP PROBLEM 22.6
Which complex ion will absorb visible light of higher energy, $[V(H_2O)_6]^{3+}$ or $[V(NH_3)_6]^{3+}$?

**Paramagnetism of transition metal complexes.** The splitting of energy levels also determines the magnetic properties of a complex by affecting *d*-orbital occupancy. Electrons occupy orbitals singly as long as empty orbitals of equal energy are available, which implies that a repulsive *pairing energy* must be overcome for two electrons to occupy the same orbital. Thus, *the relative sizes of the pairing energy ($E_{pairing}$) and the crystal field splitting energy ($\Delta$) determine the orbital occupancy.*

The isolated $Mn^{2+}$ ion ($[Ar]\,d^5$), for example, has five unpaired electrons in $3d$ orbitals of equal energy (Figure 22.23, *A*). In an octahedral field of ligands, the orbital energies split. Weak-field ligands, such as $H_2O$ in $[Mn(H_2O)_6]^{2+}$, produce such a small splitting energy that the *d* electrons remain unpaired (Figure 22.23, *B*). That is, it takes *less* energy for electrons to jump to the $e_g$ set than to pair up in the $t_{2g}$ set. With weak-field ligands, *the pairing energy is **greater** than the splitting energy ($E_{pairing} > \Delta$)*, so the number of unpaired electrons in the complex ion is the *same* as in the free ion. Indeed, a defining characteristic of a weak-field ligand is that it gives rise to **high-spin complexes,** those with the maximum number of unpaired electrons.

In contrast, strong-field ligands cause a large splitting of the *d*-orbital energies. In the $[Mn(CN)_6]^{4-}$ ion, it takes *more* energy for the electrons to jump to the $e_g$ set than to pair up in the $t_{2g}$ set (Figure 22.23, *C*). In this case,

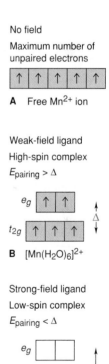

No field
Maximum number of unpaired electrons

**A**　Free $Mn^{2+}$ ion

Weak-field ligand
High-spin complex
$E_{pairing} > \Delta$

$e_g$
$t_{2g}$

**B**　$[Mn(H_2O)_6]^{2+}$

Strong-field ligand
Low-spin complex
$E_{pairing} < \Delta$

$e_g$

$t_{2g}$

**C**　$[Mn(CN)_6]^{4-}$

**FIGURE 22.23**
**High-spin and low-spin complex ions of $Mn^{2+}$. A,** Free $Mn^{2+}$ has five unpaired electrons. **B,** Bonded to weak-field ligands, such as $H_2O$, $Mn^{2+}$ still has five unpaired electrons (high spin). **C,** Bonded to strong-field ligands, such as $CN^-$, $Mn^{2+}$ has only one unpaired electron (low spin).

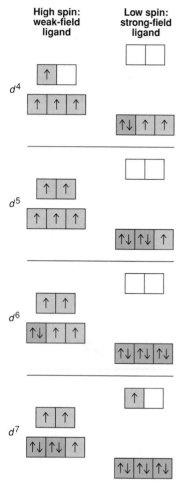

**FIGURE 22.24**
Orbital occupancy for high-spin and low-spin complexes of $d^4$ through $d^7$ metal ions.

the pairing energy is **smaller** than the splitting energy ($E_{pairing} < \Delta$ ). Strong-field ligands give rise to **low-spin complexes,** those with fewer unpaired electrons than in the free ion.

Orbital diagrams for the $d^1$ through $d^9$ ions in octahedral complexes show that high-spin/low-spin options are possible only for $d^4$, $d^5$, $d^6$, and $d^7$ ions (Figure 22.24). With three lower energy $t_{2g}$ orbitals available, the $d^1$, $d^2$, and $d^3$ ions always form high-spin complexes because there is no need to pair up. Similarly, $d^8$ and $d^9$ ions always form high-spin complexes because the $t_{2g}$ set is filled with six electrons, so the two $e_g$ orbitals *must* have either two ($d^8$) or one ($d^9$) unpaired electrons.

---

SAMPLE PROBLEM 22.7 _____

### Characterizing Complex Ions as High Spin or Low Spin

**Problem:** Iron(II) forms an essential complex in hemoglobin (see Chemical Connections, Figure 22.A). For the octahedral complex ions, $[Fe(H_2O)_6]^{2+}$ and $[Fe(CN)_6]^{4-}$, draw an orbital splitting diagram, predict the number of unpaired electrons, and characterize each as low or high spin.
**Plan:** The $Fe^{2+}$ electron configuration gives us the number of $d$ electrons, and Figure 22.22 shows the relative strengths of the two ligands. We draw the diagrams, separating the $t_{2g}$ and $e_g$ sets more for the strong-field ligand. Then we add electrons, noting that a weak-field ligand gives the maximum number of *un*paired electrons and a high-spin complex, while a strong-field ligand leads to pairing and a low-spin complex.
**Solution:** $Fe^{2+}$ has the [Ar] $d^6$ configuration. According to Figure 22.22, $H_2O$ produces smaller splitting than $CN^-$. The diagrams are

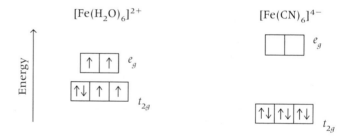

The $[Fe(H_2O)_6]^{2+}$ has **four unpaired electrons (high spin),** and the $[Fe(CN)_6]^{4-}$ has **no unpaired electrons (low spin).**

**Comment:**
1. $H_2O$ is a weak-field ligand, so it almost always forms high-spin complexes.
2. These results are correct, but in general we cannot confidently predict the spin of a complex without having values for $\Delta$ and $E_{pairing}$.
3. Cyanide ions and carbon monoxide are highly toxic because of their interaction with the iron cations in proteins.

FOLLOW-UP PROBLEM 22.7
How many unpaired electrons would you expect for $[Mn(CN)_6]^{3-}$? Is this a high-spin or low-spin complex ion?

---

**Crystal field splitting in tetrahedral and square planar complexes.** Four ligands around a metal ion also cause $d$-orbital splitting, but the magnitude and pattern of the splitting depend on whether the ligands

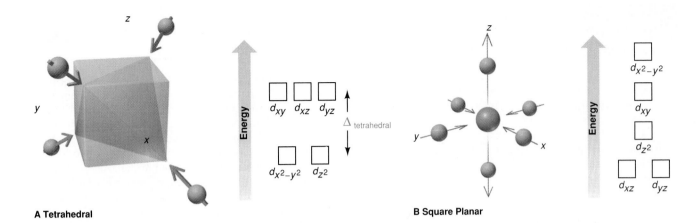

**A Tetrahedral**

**B Square Planar**

are in a tetrahedral or square planar arrangement. Let's first consider a tetrahedral complex. With the ligands approaching from the corners of a tetrahedron, none of the five $d$ orbitals is directly in their paths (Figure 22.25, A). Thus, splitting of $d$-orbital energies is less in a tetrahedral complex than in an octahedral complex of the same ligands:

$$\Delta_{\text{tetrahedral}} < \Delta_{\text{octahedral}}$$

Minimal repulsions arise if the ligands approach the $d_{xy}$, $d_{yz}$, and $d_{xz}$ orbitals closer than they approach the $d_{z^2}$ and $d_{x^2-y^2}$ orbitals. This situation is the *reverse of the octahedral case*, and the relative $d$-orbital energies are also reversed: the $d_{xy}$, $d_{yz}$, and $d_{xz}$ orbitals become *higher* in energy than the $d_{z^2}$ and $d_{x^2-y^2}$ orbitals. *Only high-spin tetrahedral complexes are known.*

In square planar complexes, the effects of the ligand field are easier to picture if we start with an octahedral geometry, and then imagine removing the two $z$-axis ligands; such a process gives the square planar geometry (Figure 22.25, B). With no interactions along the $z$ axis, the $d_{z^2}$ orbital energy decreases greatly, and the $d_{xz}$ and $d_{yz}$ orbital energies also decrease. As a result, the two $d$ orbitals in the $xy$ plane interact most strongly with the ligands, and since the $d_{x^2-y^2}$ orbital has its lobes *on* the axes, its energy is highest. As a result of this splitting pattern, square planar complexes with $d^8$ ions, such as $[PdCl_4]^{2-}$, are diamagnetic because four pairs of $d$ electrons fill the four lowest orbitals. In fact, *square planar complexes are generally low spin.*

A final word about bonding theories may be helpful. As you have seen with several other topics, no one model is satisfactory in every respect. The VB approach offers a simple picture of bond formation but does not even attempt to explain color. The crystal field model predicts color and magnetic behavior beautifully but treats the metal ion and ligands as points of opposite charge and thus offers no insight about the covalent nature of metal-ligand bonding. Despite its complexity, chemists now rely on a more refined model, called *ligand field–molecular orbital theory*, which combines aspects of the previous two models with MO theory (Section 10.4). We do not explore it here, but it is a powerful predictive tool, yielding information on bond properties resulting from the overlap of metal ion and ligand orbitals as well as the spectral and magnetic properties resulting from metal $d$-orbital splitting.

In addition to their important chemical applications, complexes of the transition elements play vital roles in living systems, as the following Chemical Connections essay describes.

**FIGURE 22.25**

Splitting of $d$-orbital energies by a tetrahedral field and a square planar field of ligands. **A,** In contrast to the octahedral case, electrons in the $d_{xy}$, $d_{yz}$, and $d_{xz}$ orbitals are closer to a tetrahedral field of ligands and experience greater repulsions than those in the $d_{x^2-y^2}$ and $d_{z^2}$ orbitals. The splitting pattern in a tetrahedral field of ligands is the reverse of the octahedral pattern. **B,** In a square planar field, the $z$-axis ligands are removed (shown moving away), so the $d_{xz}$, $d_{yz}$, and especially the $d_{z^2}$ orbitals experience weaker repulsions. Thus, their energies decrease relative to the octahedral pattern.

**Chemistry in Nutrition:**
**Transition Metals as Essential Dietary**
**Trace Elements**

Living things consist primarily of water and complex compounds of four key elements: carbon, oxygen, hydrogen, and nitrogen. All known organisms also contain seven other elements, known as *macro*nutrients because they occur in fairly high concentrations: phosphorus, sulfur, chlorine, sodium, magnesium, potassium, and calcium. In addition, organisms contain a surprisingly large number of elements in much lower concentrations, and most of these *micro*nutrients, or *trace elements*, are transition metals.

With the exception of scandium and titanium, all the Period 4 transition elements are essential for organisms, and plants require molybdenum (from Period 5) as well. Within the organism, the transition metal ion usually occurs covalently bonded to amino acid groups that act as ligands, within a bend of a surrounding protein chain. Despite the complexity of biological molecules, the principles of bonding and *d*-orbital splitting are the same as in simple inorganic systems. Table 22.A lists some of the transition metals known, or thought, to be essential in human nutrition. We focus here on iron and zinc.

Iron plays a crucial role in oxygen transport in all vertebrates. The oxygen-transporting protein hemoglobin (Figure 22.A, *A*) consists of four protein chains called *globins*, each attached to the iron-containing complex *heme*. Heme is a porphyrin, a complex derived from a metal ion and the tetradentate ring ligand known as *porphin*. Iron(II) is centered in the square plane of the ring through coordinate covalent bonds with the four N lone pairs (Figure 22.A, *B*). The complex is octahedral, with

the fifth ligand of iron(II) being an N atom from a nearby amino acid (histidine), and the sixth an O atom from either an $O_2$ or an $H_2O$ molecule.

Hemoglobin exists in two forms, depending on the nature of the sixth ligand. In the blood vessels of the lungs, where $O_2$ concentration is high, heme binds $O_2$ to form *oxyhemoglobin*, which is transported in the arteries to $O_2$-depleted tissues. The $O_2$ is released and replaced by an $H_2O$ molecule to form *deoxyhemoglobin*, which is transported in the veins back to the lungs. Since $H_2O$ is a weak-field ligand, the $d^6$ $Fe^{2+}$ ion in deoxyhemoglobin is part of a high-spin complex. Because of the relatively small *d*-orbital splitting, deoxyhemoglobin absorbs light at the red (low-energy) end of the spectrum and looks purplish blue, which accounts for the dark color of venous blood. On the other hand, $O_2$ is a strong-field ligand, so it increases the splitting energy. Oxyhemoglobin absorbs at the blue (high-energy) end of the spectrum, which accounts for the bright-red color of arterial blood.

The position of the $Fe^{2+}$ ion relative to the plane of the porphin ring also depends on the sixth ligand. Bound to $O_2$, $Fe^{2+}$ is *in* the porphin plane; bound to $H_2O$, it moves *out of* the plane slightly. This tiny ( ~70 pm = $7 \times 10^{-11}$ m) change in $Fe^{2+}$ position on release or attachment of $O_2$ influences the shape of its globin chain, which in turn alters the shape of a neighboring globin chain, triggering the release or attachment of *its* $O_2$, and so on. In this concerted action of the four globin chains, hemoglobin rapidly picks up $O_2$ from the lungs and unloads it to the tissues. Carbon monoxide is highly toxic because it binds

**TABLE 22.A**   **Some Transition Metal Trace Elements in Humans**

| ELEMENT | BIOMOLECULE CONTAINING ELEMENT | FUNCTION OF BIOMOLECULE |
|---|---|---|
| Vanadium | Protein (?) | Redox couple in fat metabolism (?) |
| Chromium | Glucose tolerance factor | Glucose utilization |
| Manganese | Isocitrate dehydrogenase | Cell respiration |
| Iron | Hemoglobin and myoglobin<br>Cytochrome *c*<br>Catalase | Oxygen transport<br>Cell respiration; ATP formation<br>Decomposition of $H_2O_2$ |
| Cobalt | Cobalamin (vitamin $B_{12}$) | Development of red blood cells |
| Copper | Ceruloplasmin<br>Cytochrome oxidase | Hemoglobin synthesis<br>Cell respiration; ATP formation |
| Zinc | Carbonic anhydrase<br>Carboxypeptidase A<br>Alcohol dehydrogenase | Elimination of $CO_2$<br>Protein digestion<br>Metabolism of ethanol |

**FIGURE 22.A**

**Hemoglobin and the octahedral complex in heme. A,** The structure of hemoglobin consists of four protein chains, each with a bound heme. **B,** In oxyhemoglobin, the octahedral complex in heme has iron(II) at the center surrounded by the four N atoms of the porphin ring, a fifth N from histidine, and an $O_2$ molecule.

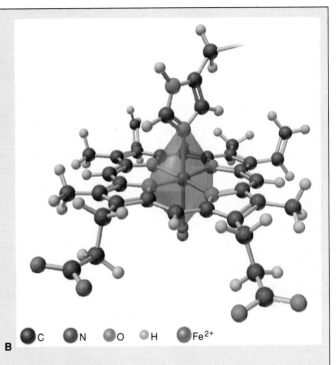

● C  ● N  ● O  ● H  ● $Fe^{2+}$

**B**

to the $Fe^{2+}$ ion in heme about 200 times more tightly than $O_2$, disrupting heme's $O_2$-carrying function.

Porphin rings are among the most common biological ligands. Chlorophyll, the photosynthetic pigment of green plants, is a porphyrin with $Mg^{2+}$ at the center of the porphin ring, and vitamin $B_{12}$ has $Co^{3+}$ at the center. Heme itself is found not only in hemoglobin, but also in several redox proteins involved in energy metabolism (see Chemical Connections, Section 20.6).

Zinc(II) occurs in many enzymes, the protein catalysts of cells (see Chemical Connections, Section 15.6). With its $d^{10}$ configuration, $Zn^{2+}$ is typically surrounded tetrahedrally by the N atoms of three amino acid groups, and the fourth position is free to interact with the molecule whose reaction is being catalyzed (Figure 22.B). In every case studied, the $Zn^{2+}$ ion acts as a Lewis acid, accepting a lone pair from the reactant as a key step in the catalytic process. Consider the enzyme carbonic anhydrase, which catalyzes the reaction between $H_2O$ and $CO_2$ during respiration:

$$CO_2(g) + H_2O(l) \rightleftharpoons H^+(aq) + HCO_3^-(aq)$$

The $Zn^{2+}$ ion at the active site binds the $H_2O$ reactant as the fourth ligand. By withdrawing electron density from the O—H bonds, the $Zn^{2+}$ makes the $H_2O$ acidic enough to lose a proton. The resulting bound $OH^-$ ion attacks the partially positive C of $CO_2$ much more vigorously than a free water molecule could, so the reaction rate is higher. One reason $Cd^{2+}$ ion is toxic is that it competes with $Zn^{2+}$ for incorporation into carbonic anhydrase.

● C  ● N  ● O  ● H

**FIGURE 22.B**

**The tetrahedral $Zn^{2+}$ complex in carbonic anhydrase.**

## Section Summary

Valence-bond theory pictures bonding in complex ions as arising from co-ordinate covalent bonding between Lewis bases (ligands) and Lewis acids (metal ions). Ligand lone pairs occupy hybridized metal-ion orbitals to form complexes with characteristic shapes. Crystal field theory explains color and magnetism of complexes. As the result of a surrounding field of negative ligands, the metal $d$-orbital energies split. The size of the split ($\Delta$) depends on the charge of the metal ion and the crystal field "strength" of the ligand. Strong-field ligands create a large split and produce low-spin complexes that absorb light of higher energy (shorter $\lambda$); the reverse is true of weak-field ligands. Several transition metals, such as iron and zinc, are essential dietary trace elements.

## Chapter Perspective

*Our study of the transition elements, a large group of metals with many essential industrial and biological roles, points up once again that macroscopic properties, such as color and magnetism, have their roots at the atomic and molecular level. In Chapter 23, we bring key ideas from this and several previous chapters to bear on a more practical side of chemistry: In what natural forms do the elements occur, and how do we obtain and make use of them?*

# For Review and Reference

## Key Terms

transition elements

### SECTION 22.1
lanthanide contraction
paramagnetism
diamagnetism

### SECTION 22.2
lanthanides
actinides
inner transition elements

### SECTION 22.4
coordination compound
complex ion
counter ion
ligand
coordination number
donor atom
chelate
isomer
structural isomer
coordination isomer

linkage isomer
stereoisomer
geometric *(cis-trans)*
  isomer
optical isomer

### SECTION 22.5
coordinate covalent bond
crystal field theory
$e_g$ orbital

$t_{2g}$ orbital
crystal field splitting
  energy ($\Delta$)
strong-field ligand
weak-field ligand
spectrochemical series
high-spin complex
low-spin complex

## Answers to Follow-up Problems

**22.1** (a) $Ag^+$: $4d^{10}$; (b) $Cd^{2+}$: $4d^{10}$; (c) $Ir^{3+}$: $5d^6$
**22.2** Three; $Er^{3+}$ is [Xe] $4f^{11}$:

| | | | | | | | | | | | | | | | | |
|--|--|--|--|--|--|--|--|--|--|--|--|--|--|--|--|--|
| | ↑↓ | ↑↓ | ↑↓ | ↑↓ | ↑ | ↑ | ↑ | | | | | | | | | |

$6s$              $4f$                              $5d$              $6p$

**22.3** (a) Pentaaquabromochromium(III) chloride;
(b) $Ba_3[Co(CN)_6]_2$
**22.4** $[Pt(NH_3)Cl_5]^-$; two moles of ions: 1 mol $K^+$ and 1 mol $[Pt(NH_3)Cl_5]^-$, but no moles of AgCl form.

**22.5** Two sets of *cis-trans* isomers, and the two *cis* isomers are optical isomers.

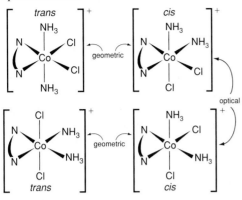

**22.6** Both metal ions are $V^{3+}$; $NH_3 > H_2O$, so $[V(NH_3)_6]^{3+}$ will absorb light of higher energy.

**22.7** The metal ion is $Mn^{3+}$: [Ar] $d^4$. $[Mn(CN)_6]^{3-}$

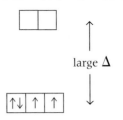

Two unpaired *d* electrons; low spin

## Sample Problem Titles

**22.1** Writing Electron Configurations of Transition Metal Atoms and Ions (p. 975)

**22.2** Determining the Number of Unpaired Electrons (p. 981)

**22.3** Writing Names and Formulas of Coordination Compounds (p. 991)

**22.4** Determining the Nature of the Complex Ion from the Empirical Formula of the Compound (p. 993)

**22.5** Determining the Type of Stereoisomerism (p. 996)

**22.6** Ranking Crystal Field Splitting Energies for Complex Ions of a Given Metal (p. 1001)

**22.7** Characterizing Complex Ions as High Spin or Low Spin (p. 1002)

## Problems

Problems with a green number are answered at the back of the text. Most sections include three categories of problems separated by a green rule: concept review questions, *paired* skill building exercises, and problems in a relevant context. (In this problem set, the term "electron configuration" refers to the condensed, ground-state electron configuration.)

### An Overview of Transition Element Properties

(Sample Problem 22.1)

**22.1** What electronic feature is characteristic of the transition elements?

**22.2** How is the *n* value of the *d* sublevel of a transition element atom related to the period number of the element?

**22.3** (a) Write the general electron configuration of a transition element in Period 5.
(b) Write the general electron configuration of a transition element in Period 6.

**22.4** What is the general rule about the order in which electrons are removed from a transition metal atom when an ion is formed? Give an example from Group 5B(5). Name two types of measurements that can be used to study ion electron configurations.

**22.5** What is the maximum number of unpaired *d* electrons that an atom or ion can possess? Give an example of an atom and an ion that have this number.

**22.6** How does the variation in atomic size across a transition series contrast with the change across the main-group elements of the same period? Why?

**22.7** (a) What is the lanthanide contraction?
(b) How does it affect atomic size down a group of transition elements?
(c) How does it influence the densities of the Period 6 transition elements?

**22.8** (a) What is the range in electronegativity values across the first (3*d*) transition series?
(b) What is the range across the fourth period of main-group elements?
(c) Explain the difference between the two ranges.

**22.9** (a) Explain the major difference between the number of oxidation states of transition elements and that of most main-group elements.
(b) Why is the +2 oxidation state so common among transition elements?

**22.10** (a) What difference in behavior distinguishes a paramagnetic substance from a diamagnetic one?
(b) Why are paramagnetic ions common among the transition elements but virtually unknown for the main-group elements?
(c) Why are colored solutions of metal ions common among the transition elements but not for the main-group elements?

**22.11** Using the periodic table to locate the element, write the electron configuration of (a) V; (b) Y; (c) Hg; (d) Ru; (e) Cu.

**22.12** Using the periodic table to locate the element, write the electron configuration of (a) Os; (b) Co; (c) Ag; (d) Zn; (e) Mn.

**22.13** Give the electron configuration and the number of unpaired electrons in each of the following ions: (a) $Sc^{3+}$; (b) $Cu^{2+}$; (c) $Fe^{3+}$; (d) $Nb^{3+}$.

**22.14** Give the electron configuration and the number of unpaired electrons in each of the following ions: (a) $Cr^{3+}$; (b) $Ti^{4+}$; (c) $Co^{3+}$; (d) $Ta^{2+}$.

**22.15** What is the highest possible oxidation state for each of the following: (a) Ta; (b) Zr; (c) Mn?

**22.16** What is the highest possible oxidation state for each of the following: (a) Nb; (b) Y; (c) Tc?

**22.17** Which transition metals have a maximum oxidation state of +6?

**22.18** Which transition metals have a maximum oxidation state of +4?

**22.19** In which compound would you expect Cr to exhibit greater metallic behavior, $CrF_2$ or $CrF_6$? Explain.

**22.20** $VF_5$ is a liquid that boils at 48°C, while $VF_3$ is a solid that melts above 800°C. How would you interpret this difference in properties?

**22.21** Would you expect it to be more difficult to oxidize Cr or Mo? Explain.

**22.22** Would you expect $MnO_4^-$ or $ReO_4^-$ to be the stronger oxidizing agent? Explain.

**22.23** Which oxide, $CrO_3$ or CrO, forms a more acidic aqueous solution? Explain.

**22.24** Which oxide, $Mn_2O_3$ or $Mn_2O_7$, displays more basic behavior? Explain.

**22.25** The green patina of the roofs of old buildings shows the result of the corrosion (oxidation) of copper in the presence of $O_2$, $H_2O$, and $CO_2$. In contrast, silver and gold—the other members of Group 1B(11)—are resistant to reaction with these atmospheric components. Corrosion of copper and silver in the presence of sulfur and its compounds leads to the familiar tarnishing of these metals. In contrast, gold is the only metal that does not react with sulfur. This pattern is markedly different from that in Group 1A(1), where ease of oxidation increases down the group. What causes the different patterns in the two groups?

**The Inner Transition Elements**

(Sample Problem 22.2)

**22.26** What atomic property of the lanthanides leads to their remarkably similar chemical properties?

**22.27** (a) What is the maximum number of unpaired electrons exhibited by an ion of a lanthanide?
(b) How does this number relate to the occupancy of the $f$ subshell?

**22.28** Which of the actinides are radioactive?

**22.29** Write the electron configurations of the following atoms and ions: (a) La; (b) $Ce^{3+}$; (c) Es; (d) $U^{4+}$.

**22.30** Write the electron configurations of the following atoms and ions: (a) Pm; (b) $Lu^{3+}$; (c) Th; (d) $Fm^{3+}$.

**22.31** Only a few of the lanthanides show any oxidation state other than the universal +3 state. Two of these, europium (Eu) and terbium (Tb), are found near the middle of the series and can be associated with a half-filled $f$ subshell.
(a) Write the electron configurations of $Eu^{2+}$, $Eu^{3+}$, and $Eu^{4+}$. Why is $Eu^{2+}$ a common ion for the element whereas $Eu^{4+}$ is unknown?
(b) Write the electron configurations of $Tb^{2+}$, $Tb^{3+}$, and $Tb^{4+}$. Would you expect terbium to show a +2 or a +4 oxidation state? Explain.

**22.32** Cerium (Ce) and ytterbium (Yb) exhibit oxidation states in addition to the universal +3 state.
(a) Write the electron configurations of $Ce^{2+}$, $Ce^{3+}$, and $Ce^{4+}$.
(b) Write the electron configurations of $Yb^{2+}$, $Yb^{3+}$, and $Yb^{4+}$.
(c) In addition to the 3+ ions, $Ce^{4+}$ and $Yb^{2+}$ are stable. Suggest a reason for this stability.

**22.33** One of the lanthanides displays the maximum possible number of unpaired electrons both for an atom and for a 3+ ion. Name the element and give the number of unpaired electrons in atom and ion.

**Highlights of Selected Transition Metals**

**22.34** What is the chemical reason that chromium is so useful for decorative plating on metals?

**22.35** What is valence-state electronegativity? Use the concept to explain the change in acidity of the oxides of Mn.

**22.36** What property does manganese confer to steel?

**22.37** What chemical property of silver leads to its use in jewelry and other decorative ware?

**22.38** How is a photographic latent image different from the image one sees on a piece of developed film?

**22.39** Mercury has an unusual physical property and an unusual 1+ ion. Explain.

**22.40** When a basic solution of $Cr(OH)_4^-$ ion is slowly acidified, solid $Cr(OH)_3$ first precipitates out and then redissolves as excess acid is added. Write two equations that represent these reactions.

**22.41** Dark green manganate salts contain the $MnO_4^{2-}$ ion. The ion is stable in basic solution but disproportionates in acid to $MnO_2(s)$ and $MnO_4^-$.
(a) What is the oxidation state of Mn in $MnO_4^{2-}$, $MnO_4^-$, and $MnO_2$?
(b) Write a balanced equation for the reaction in acidic solution.

**22.42** Use the following data to determine if $Cr^{2+}(aq)$ can be prepared by the reaction of Cr(s) with $Cr^{3+}(aq)$:

$$Cr^{3+}(aq) + e^- \rightarrow Cr^{2+}(aq) \qquad E^0 = -0.41 \text{ V}$$
$$Cr^{3+}(aq) + 3e^- \rightarrow Cr(s) \qquad E^0 = -0.74 \text{ V}$$
$$Cr^{2+}(aq) + 2e^- \rightarrow Cr(s) \qquad E^0 = -0.91 \text{ V}$$

**22.43** When solid $CrO_3$ is dissolved in water, the solution possesses an orange color rather than the yellow of $H_2CrO_4$. How does this observation indicate that $CrO_3$ is an acidic oxide?

**22.44** At one time it was common to write the formula for copper(I) chloride as $Cu_2Cl_2$, analogously to $Hg_2Cl_2$ for mercury(I) chloride instead of CuCl. Use electron configurations to explain why $Hg_2Cl_2$ is expected, but CuCl is correct.

**22.45** Solutions of $KMnO_4$ are used commonly in redox titrations. The dark purple $MnO_4^-$ serves as its own indicator, changing to the almost colorless $Mn^{2+}$ as it is reduced. The end point is taken when a pale purple color remains as the $KMnO_4$ solution is added. If a sample that has reached this end point is allowed to stand for a long period of time, the color fades and a suspension of a small amount of brown, muddy $MnO_2$ appears. Use electrode potentials to explain this result.

## Coordination Compounds

(Sample Problems 22.3 to 22.6)

**22.46** Describe the makeup of a complex ion including the nature of the ligands and their interaction with the central metal ion. Explain how a complex ion can be positive or negative and how it occurs as part of a neutral coordination compound.

**22.47** What electronic feature must a donor atom of any ligand possess?

**22.48** What is the coordination number of a complex ion? How does it differ from oxidation number?

**22.49** What structural feature is characteristic of a complex described as a "chelate"?

**22.50** What geometries are associated with coordination numbers of 2, 4, and 6?

**22.51** What are the coordination numbers for complexes of cobalt(III), platinum(II), and platinum(IV)?

**22.52** In what sense is a complex ion the adduct of a Lewis acid-base reaction?

**22.53** What does the ending "-ate" for an ion signify?

**22.54** In what order are the metal ion and ligands given in the name of a complex ion?

**22.55** Is a linkage isomer a type of structural isomer or stereoisomer? Explain.

---

**22.56** Give systematic names for the following formulas:
(a) $[Ni(H_2O)_6]Cl_2$    (b) $[Cr(en)_3](ClO_4)_3$
(c) $K_4[Mn(CN)_6]$    (d) $[Co(NH_3)_4(NO_2)_2]Cl$
(e) $[Cr(NH_3)_6][Cr(CN)_6]$

**22.57** Give systematic names for the following formulas:
(a) $K[Ag(CN)_2]$    (b) $Na_2[CdCl_4]$
(c) $[Co(NH_3)_4(H_2O)Br]Br_2$    (d) $K[Pt(NH_3)Cl_5]$
(e) $[Pt(en)(NH_3)_2][Co(en)Cl_4]$

**22.58** What is the charge of the central metal ion(s) and the coordination number of the complex ion(s) in each compound of Problem 22.56?

**22.59** What is the charge of the central metal ion(s) and the coordination number of the complex ion(s) in each compound of Problem 22.57?

**22.60** Give formulas corresponding to the following names:
(a) Tetraamminezinc sulfate
(b) Pentaamminechlorochromium(III) chloride

(c) Sodium bis(thiosulfato)argentate(I)
(d) Dibromobis(ethylenediamine)cobalt(III) sulfate
(e) Hexaamminechromium(III) tetrachlorocuprate(II)

**22.61** Give formulas corresponding to the following names:
(a) Hexaaquachromium(III) sulfate
(b) Barium tetrabromoferrate(III)
(c) Bis(ethylenediamine)platinum(II) carbonate
(d) Potassium tris(oxalato)chromate(III)
(e) Tris(ethylenediamine)cobalt(III) pentacyanoiodomanganate(II)

**22.62** What is the coordination number of the complex ion and the number of individual ions per formula unit in each of the compounds in Problem 22.60?

**22.63** What is the coordination number of the complex ion and the number of individual ions per formula unit in each of the compounds in Problem 22.61?

**22.64** Which of the following ligands can participate in linkage isomerism: (a) $NO_2^-$; (b) $SO_2$; (c) $NO_3^-$? Explain with Lewis structures.

**22.65** Which of the following ligands can participate in linkage isomerism: (a) $SCN^-$; (b) $S_2O_3^{2-}$ (thiosulfate); (c) $HS^-$? Explain with Lewis structures.

**22.66** For any of the following that can exist as isomers, state the type of isomerism and draw the structures:
(a) $[Pt(CH_3NH_2)_2Br_2]$    (b) $[Pt(NH_3)_2FCl]$
(c) $[Pt(H_2O)(NH_3)FCl]$    (d) $[Zn(en)F_2]$
(e) $[Zn(H_2O)(NH_3)FCl]$

**22.67** For any of the following that can exist as isomers, state the type of isomerism and draw the structures:
(a) $[PtCl_2Br_2]^{2-}$    (b) $[Cr(NH_3)_5(NO_2)]^{2+}$
(c) $[Pt(NH_3)_4I_2]^{2+}$    (d) $[Co(NH_3)_5Cl]Br_2$
(e) $[Pt(CH_3NH_2)_3Cl]^+$

---

**22.68** $Cr^{3+}$, like $Co^{3+}$, forms complex ions with a coordination number of 6. Compounds are known with the traditional formula $CrCl_3(NH_3)_n$, where $n = 3$ to 6. Which of the compounds has an electrical conductivity in aqueous solution similar to that of an equimolar NaCl solution?

**22.69** When $MCl_4(NH_3)_2$ is dissolved in water and treated with $AgNO_3$, two moles of AgCl precipitate immediately for each mole of the compound. What is the coordination number of the complex?

**22.70** Palladium, like its group neighbor platinum, forms four-coordinate Pd(II) and six-coordinate Pd(IV) complexes. Write correct formulas for the complexes with the following compositions:
(a) $PdK(NH_3)Cl_3$    (b) $PdCl_2(NH_3)_2$
(c) $PdK_2Cl_6$    (d) $Pd(NH_3)_4Cl_4$

**22.71** Werner prepared two compounds by heating a solution of $PtCl_2$ with triethyl phosphine, $P(C_2H_5)_3$, which is an excellent ligand for Pt. The two compounds, one white and one yellow, gave the same analysis: Pt 38.8%; Cl 14.1%; C 28.7%; P 12.4%; and H 6.02%. Write formulas, structures, and systematic names for the two isomers.

## Theoretical Basis for the Bonding and Properties of Complexes

(Sample Problems 22.7 and 22.8)

**22.72** (a) What is a coordinate covalent bond?

(b) Is it involved when $FeCl_3$ dissolves in water? Explain.

(c) Is it involved when HCl gas dissolves in water? Explain.

**22.73** According to the valence bond theory, what set of orbitals is utilized by a Period 4 metal ion in forming

(a) A square planar complex?

(b) A tetrahedral complex?

**22.74** A metal ion is described as using a $d^2sp^3$ set of orbitals in a complex. What is the coordination number and shape of the complex?

**22.75** A complex in solution absorbs green light. What is the color of the solution?

**22.76** A solution is blue. What *two* possibilities of absorption of color by the solution could give rise to the observed blue color?

**22.77** (a) What is the crystal field splitting energy, $\Delta$?

(b) How does it arise for an octahedral field of ligands?

(c) How is it different for a tetrahedral field of ligands?

**22.78** What is the distinction between a weak-field ligand and a strong-field ligand?

**22.79** What term is used for a complex having the same number of unpaired electrons as does its metal ion in the free gaseous state?

**22.80** How do the relative magnitudes of $E_{pairing}$ and $\Delta$ affect the paramagnetism of a complex?

**22.81** Why are there both high-spin and low-spin octahedral complexes but only high-spin tetrahedral complexes?

---

**22.82** Give the number of $d$ electrons ($n$ of $d^n$) for the central metal ion in each of the following species:

(a) $[TiCl_6]^{2-}$    (b) $K[AuCl_4]$    (c) $[RhCl_6]^{3-}$

(d) $[Cr(H_2O)_6](ClO_3)_2$    (e) $[Mn(CN)_6]^{2-}$

**22.83** Give the number of $d$ electrons ($n$ of $d^n$) for the central metal ion in each of the following species:

(a) $Ca[IrF_6]$    (b) $[HgI_4]^{2-}$    (c) $[Co(EDTA)]^{2-}$

(d) $[Ru(NH_3)_5Cl]SO_4$    (e) $Na_2[Os(CN)_6]$

**22.84** Sketch the orientation of the orbitals relative to the ligands in an octahedral complex to explain the splitting and the relative energies of the $d_{xy}$ and the $d_{x^2-y^2}$ orbitals.

**22.85** The two $e_g$ orbitals are identical in energy in an octahedral complex but different in a square planar complex, with the $d_{z^2}$ orbital being much lower than the $d_{x^2-y^2}$. Explain with orbital sketches.

**22.86** Which of the following metal ions *cannot* form both high-spin and low-spin octahedral complexes: (a) $Ti^{3+}$; (b) $Co^{2+}$; (c) $Fe^{2+}$; (d) $Cu^{2+}$?

**22.87** Which of the following metal ions *cannot* form both high-spin and low-spin octahedral complexes:

(a) $Mn^{3+}$; (b) $Nb^{3+}$; (c) $Ru^{3+}$; (d) $Ni^{2+}$?

**22.88** Draw orbital-energy splitting diagrams and use the spectrochemical series to show the orbital occupancy for each of the following ($H_2O$ is a weak-field ligand and $CN^-$ is a strong-field ligand): (a) $[Cr(H_2O)_6]^{3+}$; (b) $[Cu(H_2O)_4]^{2+}$; (c) $[FeF_6]^{3-}$; (d) $[Cr(CN)_6]^{3-}$; (e) $[Rh(CO)_6]^{3+}$.

**22.89** Draw orbital-energy splitting diagrams and use the spectrochemical series to show the orbital occupancy for each of the following ($H_2O$ is a weak-field ligand and $CN^-$ is a strong-field ligand): (a) $[MoCl_6]^{3-}$; (b) $[Ni(H_2O)_6]^{2+}$; (c) $[Ni(CN)_4]^{2-}$; (d) $[Fe(C_2O_4)_3]^{3-}$ ($C_2O_4^{2-}$ is weaker than $H_2O$); (e) $[Co(CN)_6]^{4-}$.

**22.90** Rank the following complex ions in order of *increasing* $\Delta$ and energy of visible light absorbed: $[Cr(NH_3)_6]^{3+}$, $[Cr(H_2O)_6]^{3+}$, $[Cr(NO_2)_6]^{3-}$.

**22.91** Rank the following complex ions in order of *decreasing* $\Delta$ and energy of visible light absorbed: $[Cr(en)_3]^{3+}$, $[Cr(CN)_6]^{3-}$, $[CrCl_6]^{3-}$.

**22.92** A complex, $ML_6^{2+}$, is violet in color. The same metal forms a complex with another ligand, Q, that creates a weaker field. What color might $MQ_6^{2+}$ be expected to show? Explain.

**22.93** $[Cr(H_2O)_6]^{2+}$ is violet. A second $CrL_6$ complex is green. Could ligand L be $CN^-$? Could it be $Cl^-$? Explain.

---

**22.94** Octahedral $[Ni(NH_3)_6]^{2+}$ is paramagnetic while planar $[Pt(NH_3)_4]^{2+}$ is diamagnetic, even though both metal ions are $d^8$. Explain.

**22.95** The hexaaqua complex $[Ni(H_2O)_6]^{2+}$ is green, whereas the corresponding ammonia complex $[Ni(NH_3)_6]^{2+}$ is violet. Explain.

**22.96** Among the complex ions formed by $Co^{3+}$ are the following: $[Co(H_2O)_6]^{3+}$, $[Co(NH_3)_6]^{3+}$, and $[CoF_6]^{3-}$. These ions have the following observed colors (listed in an arbitrary order): yellow-orange, green, and blue. Match each complex with its color. Explain.

## Comprehensive Problems

Problems with an asterisk (*) are more challenging.

**22.97** Correct each name that has an error.

(a) $Na[FeBr_4]$    sodium tetrabromoferrate(II)

(b) $[Ni(NH_3)_6]^{2+}$  nickel hexaammine ion

(c) $[Co(NH_3)_3I_3]$  triamminetriiodocobalt(III)

(d) $[V(CN)_6]^{3-}$  hexacyanovanadium(III) ion

(e) $K[FeCl_4]$    potassium tetrachloroiron(III)

**22.98** For the compound $[Co(en)_2Cl_2]Cl$, give

(a) The coordination number of the complex ion.

(b) The oxidation number of the central metal ion.

(c) The number of individual ions per formula unit.

(d) The moles of AgCl that precipitate immediately per mole of compound dissolved in water and treated with $AgNO_3$.

**22.99** (a) What are the central metal ions in chlorophyll, heme, and vitamin $B_{12}$?

(b) What similarity in structure do these compounds have?

*22.100 A powerful shortcut in looking for the presence of optical isomers is to see if the complex has a "plane of symmetry"—a plane passing through the metal atom such that every atom on one side of the plane is matched by an identical one at the same distance from the plane on the other side. Any planar complex has a plane of symmetry, since all the atoms lie in one plane. Use this approach to determine if the following exist as optical isomers:
(a) $[Zn(NH_3)_2Cl_2]$ (tetrahedral)   (b) $[Pt(en)_2]^{2+}$
(c) *trans*-$[PtBr_4Cl_2]^{2-}$   (d) *trans*-$[Co(en)_2F_2]^+$
(e) *cis*-$[Co(en)_2F_2]^+$

22.101 Hexafluorocobaltate(III) ion is a high-spin complex. Draw the orbital-energy splitting diagram for its $d$ orbitals.

22.102 Criticize and correct the following statement: Strong-field ligands always give rise to low-spin complexes.

*22.103 Two important bidentate ligands used in analytical chemistry are bipyridyl (bipy) and *ortho*-phenanthroline (*o*-phen):

bipyridyl          *o* - phenanthroline

Draw structures and discuss the possibility of isomers for the following:
(a) $[Pt(bipy)Cl_2]$     (b) $[Fe(o\text{-phen})_3]^{3+}$
(c) $[Co(bipy)_2F_2]^+$   (d) $[Co(o\text{-phen})(NH_3)_3Cl]^{2+}$

22.104 The effect of entropy on reactions is evident in the stabilities of certain complexes.
(a) Using the criterion of the number of product particles, predict which of the following will be favored in terms of $\Delta S^0_{rxn}$?

$Cu(NH_3)_4^{2+}(aq) + 4H_2O(l) \rightarrow Cu(H_2O)_4^{2+}(aq) + 4NH_3(aq)$

$Cu(H_2N—CH_2—CH_2—NH_2)_2^{2+}(aq) + 4H_2O(l) \rightarrow$
$$Cu(H_2O)_4^{2+}(aq) + 2en(aq)$$

(b) Given that the Cu—N bond strength is approximately the same in both complexes, which complex will be more stable with respect to ligand exchange in water? Explain.

22.105 For the permanganate ion, draw a Lewis structure that has the lowest formal charges.

22.106 The compound $[Pt(NH_3)_2(SCN)_2]$ displays two types of isomerism. Name the types and give names and structures for the six possible isomers.

22.107 In the sepia "toning" of a black-and-white photograph, the image is converted to a rich brownish violet by placing the finished photograph in a solution of gold(III) ions, in which metallic gold replaces the metallic silver. Use electrode potentials (Table 20.2) to explain the chemistry of this process.

22.108 In 1940, when the elements Np and Pu were prepared, a controversy arose about whether the elements starting with Ac were analogs of the transition elements (and related to Y to Mo) or analogs of the lanthanides (and related to La to Nd). At the time, the arguments hinged primarily upon comparison of the observed oxidation states to those of the earlier elements.
(a) If the actinides were analogs of the transition elements, what would you predict about the maximum oxidation state for U? For Np?
(b) If the actinides were analogs of the lanthanides, what would you predict about the maximum oxidation state for U? For Pu?

22.109 In black-and-white photography, what are the major chemical changes involved in exposing, developing, and fixing the image on the film?

22.110 Aqueous electrolysis of potassium manganate $(K_2MnO_4)$ is used for the commercial production of tens of thousands of tons of $KMnO_4$ annually. (a) Write a balanced equation for the electrolysis reaction. (Water is reduced also.) (b) How many moles of $MnO_4^-$ can be formed if 12 A flow through a tank of aqueous $MnO_4^{2-}$ for 96 h?

22.111 The actinides Pa, U, and Np form a series of complex ions , as in the compound $Na_3UF_8$, in which the central metal ion has an unusual geometry and oxidation state. In the crystal structure, the complex ion can be pictured as resulting from interpenetration of simple cubic arrays of uranium and fluoride ions. (a) What is the coordination number of the complex ion? (b) What is the oxidation state of uranium in the compound? (c) Sketch the complex ion.

# CHAPTER 23

## Concepts and skills to review

- catalysts and reaction rate (Section 15.6)
- Le Châtelier's principle (Section 16.4)
- acid-base equilibria (Sections 17.3 and 17.9)
- precipitation and complex-ion equilibria (Sections 18.3 and 18.4)
- free energy and equilibrium (Section 19.3)
- standard electrode potentials (Section 20.2)
- electrolysis of molten salts and aqueous solutions (Section 20.6)

# The Elements in Nature and Industry

**Putting ideas into practice.** From mass conservation to free energy, humans apply principles to mining copper (*photo*) or obtaining any other element they need. We examine our extraction and replacement of nature's substances in this chapter and find that our impact is enormous.

In this final chapter, we take a parting look at the periodic table, but from a new perspective. In Chapter 13, we described the atomic, physical, and chemical behavior of the main-group elements and took a similar approach toward the transition elements in the opening sections of Chapter 22. Here, we examine the elements from a practical slant to inquire how much of an element is present in nature, in what form it occurs, how organisms affect its distribution, and how we obtain it for our use.

Chemistry is, above all, a practical science, and its concepts were developed to address real-life problems: How can we isolate aluminum? How can we make iron stronger? How can we prevent excess phosphate from polluting lakes? In this chapter, we apply concepts from the chapters you've just studied to understand chemical processes—in nature and in industry—that influence our lives.

We open with a discussion of the abundances and sources of elements in the whole Earth and its crust, then consider how three essential elements cycle through the environment. After seeing where and in what form the elements are found, we focus on their isolation and utilization. We discuss general procedures for extracting an element from its ore, examine in detail the production and application of certain elements, and then quickly survey major isolation methods and uses for many of the others. The chapter ends with an application of chemical principles to the manufacture of several essential industrial products.

## 23.1 How the Elements Occur in Nature

To begin our examination of how we use the elements, let's take inventory of our elemental stock—the distribution of the elements in the Earth, especially that small portion of the planet to which we have access, and their relative amounts and mineral forms.

### Earth's Structure and the Abundance of the Elements

Any attempt to isolate an element must begin with knowledge of its **abundance,** the amount of the element in a particular region of the natural world. You saw in Chapter 21 how the elements were formed in evolving stars. Their abundance on the Earth and in its various regions is the result of the specific evolutionary history of our planet.

About 4.5 billion years ago, vast amounts of cold gases and interstellar debris from exploded older stars gradually coalesced into the Sun and planets. At first, the Earth was a cold, solid sphere of uniformly distributed elements and simple compounds. In the next billion years or so, heat from meteor impacts and radioactive decay raised the planet's temperature to

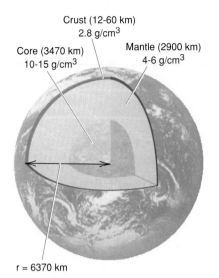

Crust (12-60 km)
2.8 g/cm$^3$

Core (3470 km)
10-15 g/cm$^3$

Mantle (2900 km)
4-6 g/cm$^3$

r = 6370 km

**FIGURE 23.1**

**The internal structure of the Earth.** The Earth consists of a large, high-density core, a thick mantle of medium density, and a very thin, low-density crust.

around $10^4$ K, sufficient to form an enormous molten mass. Any remaining gaseous elements, such as the cosmically abundant hydrogen and helium, were ejected into space.

**Layering of the Earth.** As the Earth cooled, chemical and physical processes resulted in its **differentiation,** the formation of regions of different composition and density. Differentiation gave the Earth its layered internal structure, which consists of a dense **core** (10 to 15 g/cm$^3$), a thick homogeneous **mantle** (4 to 6 g/cm$^3$), and a thin heterogeneous **crust** (average density 2.8 g/cm$^3$) (Figure 23.1). Table 23.1 compares the abundances of some key elements in the universe, the whole Earth (actually core plus mantle, which account for more than 99% of its mass), and the Earth's three regions. Because the deepest terrestrial sampling can penetrate only a few kilometers into the crust, these data represent extrapolations from meteors and from seismic studies of earthquakes. Several points stand out:

1. Cosmic and whole Earth abundances are very different, particularly for hydrogen.
2. The elements O, Si, Fe, and Mg are abundant, both cosmically and on Earth. Together, they account for more than 90% of the Earth's mass.
3. The core is particularly rich in the dense Group 8B metals: Co, Ni, and especially Fe, the most abundant element in the whole Earth.
4. Crustal abundances are very different from those of the whole Earth. The crust, which makes up only 0.4% of the Earth's mass, has the largest share of nonmetals, metalloids, and lightweight, active metals: Al, Ca, Na, and K. The mantle contains smaller proportions of these, and the core contains none. Oxygen is the most abundant element in the crust and mantle but is absent from the core.

**TABLE 23.1    Cosmic and Terrestrial Abundances of Selected Elements (Mass %)***

| ELEMENT | UNIVERSE† | EARTH | CRUST | MANTLE | CORE |
|---|---|---|---|---|---|
| O | 1.07 | 29.5 | 49.5 | 43.7 | — |
| Si | 0.06 | 15.2 | 25.7 | 21.6 | — |
| Al | — | 1.1 | 7.5 | 1.8 | — |
| Fe | 0.19 | 34.6 | 4.7 | 13.3 | 88.6 |
| Ca | 0.007 | 1.1 | 3.4 | 2.1 | — |
| Na | — | 0.6 | 2.6 | 0.8 | — |
| K | — | 0.07 | 2.4 | 0.2 | — |
| Mg | 0.06 | 12.7 | 1.9 | 16.6 | — |
| H | 73.9 | — | 0.87 | — | — |
| Ti | — | 0.05 | 0.58 | 0.18 | 0.01 |
| Cl | — | — | 0.19 | — | — |
| P | — | — | 0.12 | — | — |
| Mn | — | — | 0.09 | — | — |
| C | 0.46 | — | 0.08 | — | — |
| S | 0.04 | 1.9 | 0.06 | >2 | — |
| Ni | 0.006 | 2.4 | 0.008 | 0.3 | 8.5 |
| Co | — | 0.13 | 0.003 | 0.04 | 0.6 |

*Blank values mean either that reliable data are not available or that the value is less than 0.001 mass %.
†Helium is not abundant on Earth but accounts for 24.0 mass % of the universe.

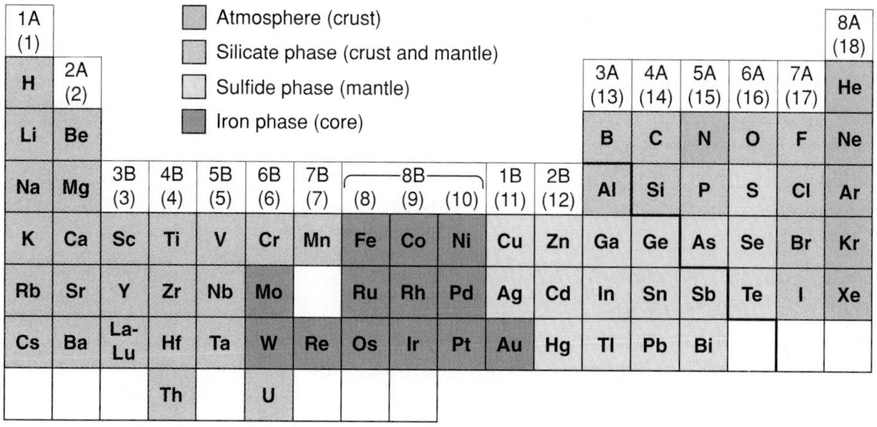

**FIGURE 23.2**
**Geochemical differentiation of the elements.** During the Earth's formation, the elements became concentrated primarily in the atmosphere or in one of three phases: silicate, sulfide, or iron.

These compositional differences in the Earth's major layers, or *phases,* arose from the effects of thermal energy. When the Earth was molten, gravity and convection led more dense materials to sink and less dense materials to rise. Most of the Fe sank to form the core, or *iron phase.* In the light outer phase, oxygen combined with Si, Al, Mg, and some Fe to form silicates, the material of rocks. This *silicate phase* later separated into the mantle and crust. The *sulfide phase,* intermediate in density and insoluble in the other two, consisted mostly of iron sulfide and mixed with parts of the silicate phase above and the iron phase below. Outgassing, the expulsion of trapped gases, produced a thin, primitive atmosphere, probably a mixture of water vapor, carbon monoxide, and nitrogen (or ammonia).

The distribution of the remaining elements was controlled by their chemical affinity for one of the three phases (Figure 23.2). In general terms, elements with low or high electronegativity, active metals (Groups 1 through 5, Cr, and Mn) and active nonmetals (O, lighter members of Groups 13 to 15, and all of Group 17), tended to congregate in the silicate phase as ionic compounds. Metals with intermediate electronegativities (many from Groups 6 to 10) dissolved in the iron core. Lower melting transition metals and many metals and metalloids in Groups 11 to 16 became concentrated in the sulfide phase.

**The impact of life on crustal abundances.** At present, the crust is the only accessible portion of our planet, so only *its* elemental abundances have practical significance. The crust is divided into solid, liquid, and gaseous portions called the **lithosphere, hydrosphere,** and **atmosphere.** For several billion years, weathering and volcanic upsurges dramatically altered the composition of the crust.

Another major influence on crustal composition has been the **biosphere,** the living systems that have inhabited the planet. When the earliest rocks were forming and the ocean basins filling with water, binary inorganic molecules in the atmosphere were reacting to form first simple, and then more complex, organic molecules. The energy for these (mostly) endothermic changes was supplied by lightning, solar radiation, geologic heating, and meteoric impact. In an amazingly short period of time, probably no more than 500 million years, the first organisms appeared. It took perhaps another billion years for these to evolve into simple algae that could derive metabolic energy from photosynthesis, converting $CO_2$ and $H_2O$ into or-

**FIGURE 23.3**
**An ancient Fe(III)-containing mineral.** With the appearance of photosynthetic organisms and the resulting oxidizing environment, much of the iron(II) in minerals was oxidized to iron(III). These ancient banded-iron formations (red beds) from Michigan contain hematite, $Fe_2O_3$.

ganic molecules and releasing $O_2$ as a byproduct. The importance of this process to crustal chemistry cannot be overstated. Over the next 300 million years, the atmosphere gradually became richer in $O_2$, and *oxidation became the major source of chemical free energy in the crust and biosphere.* Geologists mark this period by the appearance of Fe(III)-containing minerals in place of the Fe(II)-containing minerals that had predominated (Figure 23.3); paleontologists see in it an explosion of $O_2$-utilizing life forms that evolved into the organisms of today.

In addition to creating an oxidizing environment, organisms have had profound effects on specific elemental abundances. Why, for instance, is the $K^+$ concentration of the oceans so low? Forming from eons of rain and dissolved minerals, the oceans became complex ionic solutions with 30 times as much $Na^+$ as $K^+$. In addition to the fact that clays bind $K^+$ preferentially through an ion-exchange process, plants require $K^+$ for growth, absorbing dissolved $K^+$ that would otherwise wash down to the sea. As another example, consider the enormous subterranean deposits of organic carbon. Plants buried and decomposed under high pressure and temperature in the absence of free $O_2$ gradually turned into coal; animals buried under similar conditions turned into petroleum. Crustal deposits of this organic carbon provide the fuels that power society today.

Table 23.2 compares selected elemental abundances in the whole crust, the three crustal regions, and a representative portion of the biosphere—the human body. Oxygen is either the first or second most abundant element in all regions. The biosphere contains large amounts of carbon in its biomolecules and large amounts of hydrogen as water. Nitrogen is high in the atmosphere as free $N_2$ and in organisms as combined nitrogen in proteins.

**TABLE 23.2   Abundance of Selected Elements in the Crust, Its Regions, and the Biosphere (Mass %)**

| ELEMENT | CRUST | CRUSTAL REGIONS | | | HUMAN |
| | | LITHOSPHERE | HYDROSPHERE | ATMOSPHERE | |
|---|---|---|---|---|---|
| O | 49.5 | 45.5 | 85.8 | 23.0 | 65.0 |
| C | 0.08 | 0.018 | — | 0.01 | 18.0 |
| H | 0.87 | 0.15 | 10.7 | 0.02 | 10.0 |
| N | 0.03 | 0.002 | — | 75.5 | 3.0 |
| P | 0.12 | 0.11 | — | — | 1.0 |
| Mg | 1.9 | 2.76 | 0.13 | — | 0.50 |
| K | 2.4 | 1.84 | 0.04 | — | 0.34 |
| Ca | 3.4 | 4.66 | 0.05 | — | 2.4 |
| S | 0.06 | 0.034 | — | — | 0.26 |
| Na | 2.6 | 2.27 | 1.1 | — | 0.14 |
| Cl | 0.19 | 0.013 | 2.1 | — | 0.15 |
| Fe | 4.7 | 6.2 | — | — | 0.005 |
| Zn | 0.013 | 0.008 | — | — | 0.003 |
| Cr | 0.02 | 0.012 | — | — | $3 \times 10^{-6}$ |
| Co | 0.003 | 0.003 | — | — | $3 \times 10^{-6}$ |
| Cu | 0.007 | 0.007 | — | — | $4 \times 10^{-4}$ |
| Mn | 0.1 | 0.11 | — | — | $1 \times 10^{-4}$ |
| Ni | 0.008 | 0.010 | — | — | $3 \times 10^{-6}$ |
| V | 0.015 | 0.014 | — | — | $3 \times 10^{-6}$ |

One striking difference among the lithosphere, hydrosphere, and biosphere (human) is the abundances of the transition metals vanadium through zinc. Their relatively insoluble oxides and sulfides make them much scarcer in water than on land. Yet organisms evolved ways to concentrate large amounts of these elements from the aqueous environment. In every case, the accumulation is at least 100-fold, with Mn 1000-fold, and with Cu, Zn, and Fe even greater. Each of these elements performs an essential role in living systems (see Chemical Connections, Section 22.5).

### Sources of the Elements

Given an element's abundance in a region of the crust, we next determine its **occurrence,** or **source,** the form(s) in which the element exists. Practical considerations often determine the commercial source. Oxygen, for example, is abundant in all three regions, but the atmosphere is its primary industrial source because it occurs there as the free element. Nitrogen and the noble gases (except helium) are also obtained from the atmosphere. Several other elements occur uncombined, formed in large deposits by prehistoric biological action, such as sulfur in caprock salt domes and nearly pure carbon in coal. The relatively unreactive elements gold and platinum also occur in an uncombined *(native)* state.

The overwhelming majority of elements, however, occur in **ores,** natural compounds or mixtures of compounds from which an element can be extracted profitably. The phrase "extracted profitably" refers to an important point: *the financial costs of mining, isolating, and purifying an element must be considered when deciding on a process to obtain it.*

Figure 23.4 shows the most useful sources of the elements. Alkali halides are sources for both of their component groups of elements. One example of how cost influences recovery is the scarcity of usable silicate ores. Even though many elements occur as silicates, most of these sources are very stable thermodynamically, so the cost of energy required to process the ore

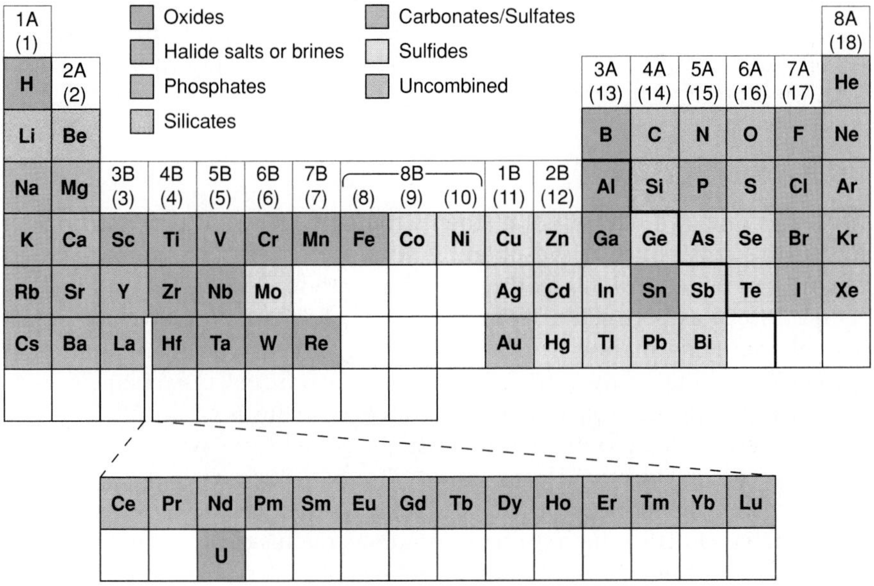

**FIGURE 23.4**
**Sources of the elements.** This periodic table shows the major sources for most of the elements. Although a few elements occur uncombined, most occur as oxides or sulfides in ores.

prohibits their use—only lithium and beryllium are obtained from their silicates. The Group 2A(2) metals occur as carbonates in the marble and limestone of mountain ranges, although magnesium's great abundance in sea water makes that the preferred source, as we discuss later.

The ores of most industrially important metals are *oxides*, which dominate the left half of the transition elements, and *sulfides*, which dominate the right half and a few of the main groups beyond. The reasons for the prominence of oxides and sulfides are complex and include processes of weathering, selective precipitation, and relative solubilities. Nevertheless, some atomic properties are relevant as well. Elements in the left half have lower electronegativities, so they tend to give up bonding electrons. The $O^{2-}$ ion is small enough to approach the cation closely, which results in a high lattice energy and stable oxide. In contrast, the elements in the right group have higher electronegativities and tend to form bonds that are more covalent, which suits the larger, more polarizable $S^{2-}$ ion.

### Section Summary

As the young Earth cooled, the elements became differentiated into a dense, metallic core, siliceous mantle, and lightweight crust. High abundances of light metals, metalloids, and nonmetals are concentrated in the crust, which consists of the lithosphere (solid), hydrosphere (liquid), atmosphere (gaseous), and biosphere (living). The biosphere has affected crustal chemistry primarily by producing free $O_2$ and thus an oxidizing environment. Some elements occur in their native state, but most are combined in ores. The most important ores of metallic elements are oxides and sulfides.

## 23.2  The Cycling of Elements Through the Environment

The distributions of many elements are in flux, changing at widely differing rates. The physical, chemical, and biological paths that the atoms of an element take on their journey within regions of the crust constitute the element's **environmental cycle.** In this section, we describe the cycles for three important elements—carbon, nitrogen, and phosphorus—and highlight the impact of humans on them.

### The Carbon Cycle

Carbon is one of a handful of elements that appear in all three portions of the Earth's crust. In the lithosphere, it occurs as elemental graphite and diamond; combined in fully oxidized form as carbonate minerals and in fully reduced form as petroleum hydrocarbons; and in complex mixtures, such as coal and living matter. In the hydrosphere, it occurs as living matter, as carbonate minerals formed by the action of coral-reef organisms, and as aqueous $CO_2$. In the atmosphere, it occurs principally as gaseous $CO_2$, a minor but essential component that exists in equilibrium with the aqueous fraction.

Figure 23.5 depicts the complex interplay of these carbon sources and the considerable effect of the biosphere on the element's cycle through the environment. It provides an estimate of the size of each source and, where reliable numbers exist, the amount of carbon moving annually between sources.

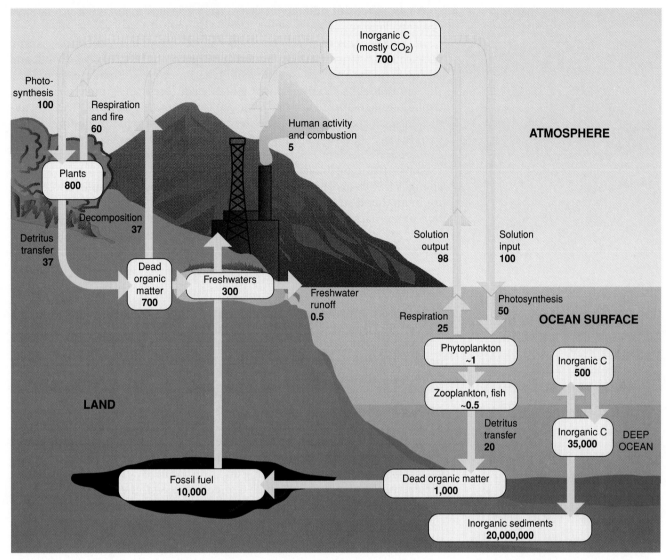

Here are some key points to note:

- The portions of the cycle are linked by the atmosphere—one between oceans and air, the other between land and air. The $2.6 \times 10^{12}$ metric tons of $CO_2$ in air cycles through the oceans and atmosphere about once every 300 years but spends a much longer time in carbonate minerals.
- Atmospheric carbon in $CO_2$ accounts for a tiny fraction (0.003%) of crustal carbon, but it is the prime mover of carbon through the other regions. Estimates indicate that a $CO_2$ molecule spends, on the average, 10 to 15 years in the atmosphere.
- The land and oceans are also in contact with the largest source of carbon, which is effectively immobilized as carbonates, coal, and oil in the rocky sediment beneath the soil.
- The biological processes of photosynthesis, respiration, and decay are major factors in the cycle. **Fixation** is the process of converting a gaseous substance into a condensed, more usable, form. Photosynthesis by marine plankton and terrestrial plants uses sunlight to fix atmospheric $CO_2$ into carbohydrates. Plants release $CO_2$ by respiration at night and by decay. Animals eat the plants and release $CO_2$ by respiration and by decay.
- Natural fires and volcanoes also release $CO_2$ to the air.

For hundreds of millions of years, the cycle has maintained a relatively constant amount of atmospheric $CO_2$. Although the size of the effect from human industry is relatively small, its recent growth has shifted this natural balance by increasing atmospheric $CO_2$. The combustion of coal, wood, and oil for fuel, the thermal decomposition of limestone to make cement, and the simultaneous clearing of vast stretches of forest and jungle for lumber, paper, and agriculture are the principal causes of this increase. These activities may be creating one of the most dire environmental situations in human history—global warming through the greenhouse effect (see Chemical Connections, Section 6.6). Higher temperatures, alterations in dry and rainy seasons, melting of polar ice, and increasing ocean acidity with associated changes in carbonate equilibria are some of the possible results. Most environmental and science policy experts are pressing for the immediate adoption of a program that combines conservation of carbon-based fuels, an end to deforestation, extensive planting of trees, and development of alternative energy sources.

### The Nitrogen Cycle

In contrast to the carbon cycle, the nitrogen cycle includes a direct interaction of land and sea (Figure 23.6). All nitrites and nitrates are soluble, so rain and runoff contribute huge amounts of nitrogen to lakes, rivers, and oceans. Human activity in the form of fertilizer production and use plays a major role in this cycle.

Like carbon dioxide, atmospheric nitrogen must be fixed to be utilized by organisms. However, whereas $CO_2$ can be incorporated by plants in either its gaseous or aqueous form, the great stability of $N_2$ prevents plants from using it directly. Fixation of $N_2$ requires a great deal of energy and occurs through atmospheric, industrial, and biological processes:

1. *Atmospheric fixation.* Lightning and fires (and, to a much smaller extent, car engines) bring about the high-temperature endothermic reaction of $N_2$ and $O_2$ to form NO, which is then oxidized exothermically to $NO_2$:

$$N_2(g) + O_2(g) \rightarrow 2NO(g) \qquad \Delta H^0 = 180.6 \text{ kJ}$$

$$2NO(g) + O_2(g) \rightarrow 2NO_2(g) \qquad \Delta H^0 = -114.2 \text{ kJ}$$

Rain converts the $NO_2$ to nitric acid, which enters both the sea and the land as $NO_3^-(aq)$ to be utilized by plants:

$$3NO_2(g) + H_2O(l) \rightarrow 2HNO_3(aq) + NO(g)$$

2. *Industrial fixation.* Nearly all industrial fixation occurs by ammonia synthesis via the Haber process (see Chemical Connections, Section 16.4), which takes place on an enormous scale: on a mole basis, $NH_3$ ranks first among compounds in amount produced. Some of this $NH_3$ is converted to $HNO_3$ in the Ostwald process (Section 13.7), but most is used as fertilizer, either directly or in the form of urea and ammonium salts (sulfate, phosphate, and nitrate), which enter the biosphere as they are taken up by plants (see below). The overuse of these fertilizers presents a serious water pollution problem in some areas. Leaching of fields by rain causes fertilizers to enter lakes and cause *eutrophication,* the depletion of $O_2$ and the death of fish from excessive plant growth and decay. Excess nitrate also spoils nearby drinkable water.

3. *Biological fixation.* The biological fixation of atmospheric $N_2$ occurs in marine blue-green algae and in nitrogen-fixing bacteria that live on the roots of leguminous plants, including peas, alfalfa, and clover. On a plane-

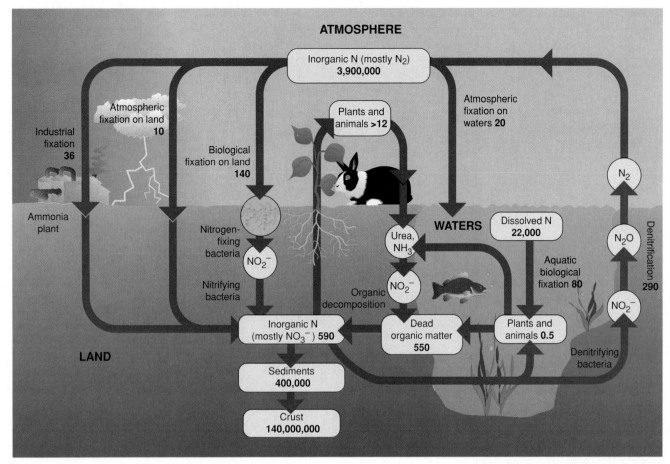

**FIGURE 23.6**

**The nitrogen cycle.** The major source of nitrogen, $N_2$, is fixed via industry, the atmosphere, and microorganisms. Various types of bacteria are indispensable throughout the cycle. Numbers in boxes are in $10^9$ metric tons of N and refer to the size of the source; numbers along arrows are in $10^6$ metric tons of N and refer to the annual movement of the element.

tary scale, these microbial processes dwarf the other two, fixing more than seven times as much nitrogen as the atmosphere and six times as much as industry. Root bacteria fix $N_2$ by reducing it to $NH_3/NH_4^+$ through the action of molybdenum-containing enzymes. Other bacteria catalyze the oxidation of the $NH_4^+$ to $NO_2^-$ and finally $NO_3^-$, which the plants reduce again to make their proteins. When the plants die, other soil bacteria oxidize the proteins to $NO_3^-$.

Animals eat the plants, use the plant proteins to make their own proteins, and excrete nitrogenous wastes, such as urea $[(H_2N)_2C{=}O]$. The nitrogen in the proteins is released when the animals die and decay, and is converted by soil bacteria to $NO_2^-$ and $NO_3^-$ again. The central inorganic nitrate pool is utilized in three ways: some enters marine and terrestrial plants, some enters the enormous sediment pool of mineral nitrates, and some is reduced by denitrifying bacteria to $NO_2^-$ and then to $N_2O$ and $N_2$, which reenter the atmosphere to complete the cycle.

## The Phosphorus Cycle

Virtually all the mineral sources of phosphorus contain the phosphate group, $PO_4^{3-}$. The most commercially important ores are **apatites,** compounds of general formula $Ca_5(PO_4)_3X$, where X is usually F, Cl, or OH. The cycling of phosphorus through the environment involves three interlocking

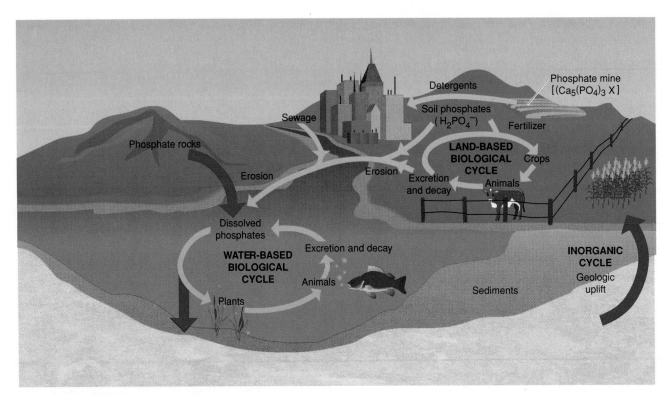

**FIGURE 23.7**

**The phosphorus cycle.** The inorganic, water-based biological, and land-based biological sub-cycles interact through erosion and through the growth and decay of organisms: the cycle has no gaseous component.

♦ **Phosphorus from Outer Space.** Although most phosphate occurs in the rocks formed by geologic activity on Earth, a sizeable amount arrived here from outer space and continues to do so. About 100 metric tons of meteorites enter our atmosphere each day. With an average P content of 0.1 mass %, they contribute 36 metric tons of P per year or, over the 4.6 billion-year lifetime of the Earth, about $10^{11}$ metric tons of P has an extraterrestrial origin!

subcycles (Figure 23.7). Two rapid biological cycles—a land-based cycle completed in a matter of years and a water-based cycle completed in weeks to years—are superimposed on an inorganic cycle that takes millions of years to complete. Unlike the carbon and nitrogen cycles, the phosphorus cycle has no gaseous component and thus does not involve the atmosphere.

**The inorganic cycle.** Most phosphate rock formed when the Earth's lithosphere solidified. ♦ Its most important components are very insoluble phosphate salts:

$$K_{sp} \text{ of } Ca_3(PO_4)_2 \approx 10^{-29}$$
$$K_{sp} \text{ of } Ca_5(PO_4)_3OH \approx 10^{-51}$$
$$K_{sp} \text{ of } Ca_5(PO_4)_3F \approx 10^{-60}$$

Weathering slowly leaches phosphates from the soil and carries the ions through rivers to the sea. Plants in the land-based biological cycle speed this process. In the ocean, some phosphate is absorbed by organisms in the water-based biological cycle, but the majority is precipitated again by $Ca^{2+}$ ion and deposited on the continental shelf. Geologic activity lifts the continental shelves, returning the phosphate to the land.

**The land-based biological cycle.** The biological cycles involve the incorporation of phosphate into organisms (biomolecules, bones and teeth, and so forth) and the release of phosphate through excretion and decay. In the land-based cycle, plants remove phosphate from the inorganic cycle. Recall that three phosphate oxoanions exist in aqueous equilibrium:

$$H_2PO_4^- \rightleftharpoons HPO_4^{2-} \rightleftharpoons PO_4^{3-}$$
$$+H^+ \quad\quad +H^+$$

In topsoil, phosphates occur as insoluble compounds of $Ca^{2+}$, $Fe^{3+}$, and $Al^{3+}$. Because plants can absorb only the soluble dihydrogen phosphates, they have evolved the ability to secrete acids near their roots to convert the insoluble salts into the soluble ion:

$$Ca_3(PO_4)_2(s; \text{ soil}) + 4H^+(aq; \text{ secreted by plants}) \longrightarrow$$

$$3Ca^{2+}(aq) + 2H_2PO_4^-(aq; \text{ absorbed by plants})$$

Animals that eat the plants excrete soluble phosphate, which is used in turn by newly growing plants. As the plants and animals excrete, die, and decay, some phosphate is washed into rivers and from there to the ocean. The majority of the 2 million metric tons of phosphate that wash from land to sea each year is from biological decay. Thus, the biosphere greatly increases the movement of phosphate from lithosphere to hydrosphere.

**The water-based biological cycle.** Various phosphorus oxoanions continually enter the aquatic environment. Tracer studies with radioactive phosphorus-32 in $H_2PO_4^-$ show that, within 1 minute, 50% of the ion is taken up by photosynthetic algae, which use it to produce their biomolecules. The overall process might be represented by

$$106CO_2(g) + 16NO_3^-(aq) + H_2PO_4^-(aq) + 122H_2O(l) + 17H^+(aq) \rightleftharpoons$$

$$C_{106}H_{263}O_{110}N_{16}P(aq; \text{ in algal cell fluid}) + 138O_2(g)$$

(The complex formula on the right represents the total composition of algal biomolecules, not some particular compound.) As on land, animals eat the plants and are eaten by other animals, and they all excrete, die, and decay (denoted by the reversible arrow in the previous reaction). Some of the released phosphate is used by other aquatic organisms, and some returns to the land when fish-eating animals (humans and birds) excrete phosphate, die, and decay. In addition, some aquatic phosphate precipitates with $Ca^{2+}$, sinks to the sea bed, and returns to the long-term mineral deposits of the inorganic cycle.

**Human effects on phosphorus movement.** From prehistoric through preindustrial times, these cycles have been balanced. Modern human activities, however, alter the movement of phosphorus considerably. Our influence on the inorganic cycle is minimal, despite our annual removal of more than 100 million metric tons of phosphate rock; its major uses appear in Figure 23.8. The end products of this removal, however, have unbalanced the two biological cycles.

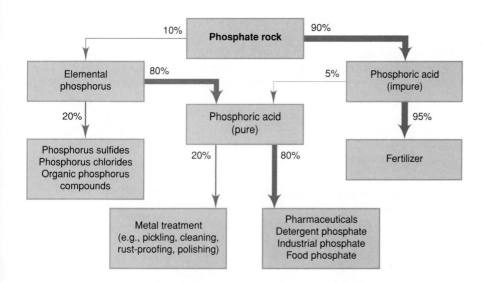

**FIGURE 23.8**
**Industrial uses of phosphorus.** Arrow thickness is proportional to amount of next product. By far the largest end product of phosphate rock (~85%) is fertilizers.

The imbalance arises from our overuse of *soluble phosphate fertilizers,* such as $Ca(H_2PO_4)_2$ and $NH_4H_2PO_4$, the end products of about 85% of phosphate rock mined. Some phosphate fertilizer finds its way into rivers, lakes, and oceans to enter the water-based cycle. Much larger pollution sources, however, are the crops grown with the fertilizer and the detergents made from phosphate rock. The great majority of the phosphate eventually arrives in urban centers as foods (crops and animals fed the crops) and consumer products, such as the tripolyphosphates in detergents. Human garbage, excrement, detergent wash water, and industrial waste water containing this phosphate return through sewers to the aquatic system. This human contribution equals the natural contribution—another 2 million metric tons of phosphate per year. The increased concentration in rivers and lakes causes eutrophication, which robs the water of $O_2$ so that it cannot support life. Such "dead" rivers and lakes are no longer usable for fishing, drinking water, or recreational use.

As early as 1912, a chemical solution to this problem was tried in Switzerland, where the enormous Lake Zurich had been choked with algae and devoid of fish due to release of human sewage. Treatment with $FeCl_3$ precipitated enough phosphate to return the lake gradually to its natural state. Soluble aluminum compounds have the same effect:

$$FeCl_3(aq) + PO_4{}^{3-}(aq) \rightarrow FePO_4(s) + 3Cl^-(aq)$$

$$KAl(SO_4)_2 \cdot 12H_2O(aq) + PO_4{}^{3-}(aq) \rightarrow AlPO_4(s) + K^+(aq) + 2SO_4{}^{2-}(aq) + 12H_2O(l)$$

After several lakes in the United States (including Lake Erie) became polluted in this way, phosphates were eliminated from detergents and stringent requirements for sewage treatment put into effect. ◆

Demeton, an insecticide

V agent, a lethal nerve gas

◆ **Phosphorus Nerve Poisons.** Because phosphorus is essential for all organisms, its compounds have many biological actions. In addition to their widespread use as fertilizer, P compounds are also used as pesticides *and* as military nerve gases. In fact, these two types of substances have similar structures (see figure) and modes of action. Both affect the target organism by inactivating an enzyme involved in muscle contraction. They bind an amino acid critical to the enzyme's function, allowing the muscle to contract but not relax; the effect on the organism—insect or human—is often fatal. Chemists have synthesized a nerve-gas antidote whose shape mimics the critical region of the enzyme. The drug binds the poison, removing it from the enzyme, and the two are excreted.

### Section Summary

The environmental distribution of many elements changes cyclically with time and is affected in major ways by organisms. Carbon occurs in all three portions of the crust, with atmospheric $CO_2$ linking the other two. Photosynthesis and decay of organisms alter the amount of carbon in the land and oceans. Human activity has increased atmospheric $CO_2$. Nitrogen is fixed by lightning, by industry, and primarily by microorganisms. When plants decay, bacteria decompose organic nitrogen and eventually return $N_2$ to the atmosphere. Through extensive use of fertilizers, humans have added excess nitrogen to fresh waters. The phosphorus cycle has no gaseous component. Inorganic phosphates leach slowly into land and water, where biological cycles interact. When plants and animals decay, they release phosphates to the natural waters, which are then absorbed by other plants and animals. Through overuse of phosphate fertilizers and detergents, humans double the amount of phosphorus entering aqueous systems.

## 23.3 Tapping the Crust: Isolation and Uses of the Elements

Once an element's abundance and sources are known, the next step is to isolate it. The approach depends on the physical and chemical properties of the source: we would isolate an element that occurs uncombined in the air differently from one dissolved in the sea or one found in a rocky ore. In this section, we discuss the general procedures for isolating metals, detail methods for recovering some elements of special interest, and then survey isolation methods and applications of all the elements in the main groups and the first transition series.

## Metallurgy: Extracting a Metal from Its Ore

**Metallurgy** is the branch of materials science concerned with the extraction and utilization of metals. The extraction process utilizes one or a combination of three types of metallurgy: *pyrometallurgy* uses heat to obtain the metal, *electrometallurgy* employs an electrochemical step, and *hydrometallurgy* relies on the metal's aqueous solution chemistry.

The extraction of a metal begins with *mining the ore.* Most ores consist of the **mineral,** that is, the compound containing the element, and **gangue,** attached debris such as sand, rock, and clay. Table 23.3 lists the common mineral sources of some metals.

**TABLE 23.3  Common Mineral Sources of Some Elements**

| ELEMENT | MINERAL NAME, FORMULA |
|---|---|
| Al | Gibbsite (in bauxite), $Al(OH)_3$ |
| Ba | Barite, $BaSO_4$ |
| Be | Beryl, $Be_3Al_2Si_6O_{18}$ |
| Ca | Limestone, $CaCO_3$ |
| Fe | Hematite, $Fe_2O_3$ |
| Hg | Cinnabar, $HgS$ |
| Na | Halite, $NaCl$ |
| Pb | Galena, $PbS$ |
| Sn | Cassiterite, $SnO_2$ |
| Zn | Sphalerite, $ZnS$ |

We have been removing metals from the Earth for thousands of years, so most concentrated known sources are long gone. In many cases, ores containing very low mass percentages of a metal are all that remain, and these require elaborate extraction methods. Nevertheless, the general procedure for extracting most metals (and many nonmetals) involves a few basic steps, described next and summarized in Figure 23.9.

1. *Pretreating the ore.* Following a crushing, grinding, or pulverizing step, which can be very expensive, pretreatment exploits some physical or chemical difference to separate mineral from gangue. For magnetic minerals, such as magnetite ($Fe_3O_4$), a magnet can remove the mineral and leave the gangue behind. Where large density differences exist, a *cyclone separator* is used to blow high-pressure air through the pulverized mixture and separate the particles. The lighter silicate-rich gangue is blown away, while the denser mineral-rich particles hit the walls and fall through the open bottom (Figure 23.10).

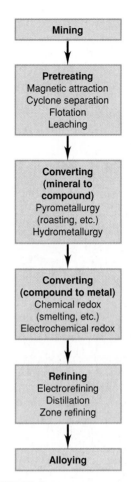

**FIGURE 23.9**
**Steps in metallurgy.**

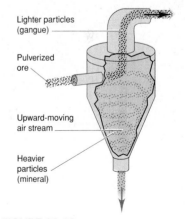

**FIGURE 23.10**
**The cyclone separator.** The pulverized ore enters at an angle, with the more dense particles (mineral) hitting the walls and spiraling downward, while the less dense particles (silicate gangue) are carried upward in a stream of air.

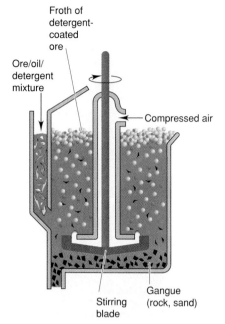

Froth of detergent-coated ore

Ore/oil/detergent mixture

Compressed air

Stirring blade

Gangue (rock, sand)

**FIGURE 23.11**

**The flotation process.** The pulverized ore is treated with an oil-detergent mixture, then stirred and aerated to create a froth that separates the soap-coated mineral from the gangue.

◆ **Panning and Fleecing for Gold.** The picture of the prospector, alone but for his faithful mule, hunched over the bank of a mountain stream, patiently panning for gold, is mostly a relic of old movies. However, the process is based on the very large density differences between gold ($d = 19.3$ g/cm$^3$) and sand ($d \approx 2.5$ g/cm$^3$). In ancient times, river sands were washed over a sheep's fleece to trap the gold grains, a practice that historians think represents the origin of the Golden Fleece of Greek mythology.

In the **flotation** process, an oil-detergent mixture is stirred with the pulverized ore in water to form a slurry (Figure 23.11). Detergents wet the mineral surface but not the silicate gangue. Rapid mixing moves air bubbles through the mixture and produces an oily, mineral froth that floats, while the silicate particles sink. Skimming, followed by solvent removal of the oil-detergent mixture, isolates the concentrated mineral fraction. Flotation is a key step in copper recovery, as you'll see later.

**Leaching** is a hydrometallurgic process that selectively extracts the metal, usually by forming a complex ion. The modern extraction of gold is a good example. ◆ The ore, which often contains as little as 25 ppm of gold, is treated with a cyanide ion solution and aerated. In the presence of $CN^-$, contact with $O_2$ oxidizes gold to gold(I), which forms the soluble dicyanoaurate(I) complex ion:

$$4Au(s) + O_2(g) + 8CN^-(aq) + 2H_2O(l) \longrightarrow 4Au(CN)_2^-(aq) + 4OH^-(aq)$$

2. *Chemically converting the mineral.* After the mineral has been freed of debris and concentrated, it is chemically converted to another compound, one that has more desirable solubility properties, is easier to reduce, or is free of a key impurity. Conversion to an oxide is common because oxides can be reduced easily. Carbonates are heated to convert them to the oxide:

$$CaCO_3(s) \xrightarrow{\Delta} CaO(s) + CO_2(g)$$

Metal sulfides, such as ZnS, are converted to oxides by **roasting** in air:

$$2ZnS(s) + 3O_2(g) \xrightarrow{\Delta} 2ZnO(s) + 2SO_2(g)$$

In most countries, hydrometallurgic methods are used now to avoid atmospheric release of $SO_2$ during roasting. One way of processing copper, for example, is by bubbling air through an acidic slurry of insoluble $Cu_2S$, which fulfills both the pretreatment and conversion steps:

$$2Cu_2S(s) + 5O_2(g) + 4H^+(aq) \longrightarrow 4Cu^{2+}(aq) + 2SO_4^{2-}(aq) + 2H_2O(l)$$

3. *Converting the compound to the metal.* The next step converts the new mineral form (usually an oxide) to the free metal by either chemical or electrochemical methods. In *chemical redox* processes, a reducing agent reacts directly with the compound. Carbon, in the form of coke (a porous residue from incomplete combustion of coal) or charcoal, is a common reductant. Heating an oxide with a reducing agent, such as coke, to obtain the metal is called **smelting.** Many metal oxides, such as zinc oxide and tin(IV) oxide, are smelted with carbon to free the metal, which may need to be condensed and solidified:

$$ZnO(s) + C(s) \longrightarrow Zn(g) + CO(g)$$
$$SnO_2(s) + 2C(s) \longrightarrow Sn(l) + 2CO(g)$$

Several nonmetals that occur in positive oxidation states in a mineral can be reduced with carbon as well. Phosphorus, for example, is produced from calcium phosphate:

$$2Ca_3(PO_4)_2(s) + 10C(s) + 6SiO_2(s) \longrightarrow 6CaSiO_3(s) + 10CO(g) + P_4(s)$$

(Metallic calcium is a much stronger reducing agent than carbon, so it is not formed.)

Low cost and ready availability explain why carbon is such a common reductant, but why is it such an effective one? Thermodynamic principles provide a reason in certain systems. Consider the standard free energy change ($\Delta G^0$) for the reduction of tin(IV) oxide:

$$SnO_2(s) + 2C(s) \rightarrow Sn(l) + 2CO(g) \qquad \Delta G^0 = 245 \text{ kJ at } 25°C$$

The magnitude and sign of $\Delta G^0$ indicate a nonspontaneous reaction at 25°C. However, because solid C becomes gaseous CO, $\Delta S^0$ (standard molar entropy change) is positive, so *the $(-T\Delta S^0)$ term of $\Delta G^0$ becomes more negative with higher temperature* (Section 19.2). Therefore, as the temperature increases, $\Delta G^0$ decreases; at 1000°C, for example, the reaction *is* spontaneous ($\Delta G^0 = -62.8$ kJ). In fact, the process is carried out at around 1250°C, so $\Delta G^0$ is even more negative.

As you know, overall reactions often hide several intermediate steps. For instance, this reduction of tin(IV) oxide may occur by first forming tin(II) oxide:

$$SnO_2(s) + C(s) \rightarrow \cancel{SnO(s)} + CO(g)$$

$$\cancel{SnO(s)} + C(s) \rightarrow Sn(l) + CO(g)$$

Also, the second step may even be composed of others, in which CO, not C, is the actual reductant:

$$SnO(s) + \cancel{CO(g)} \rightarrow Sn(l) + \cancel{CO_2(g)}$$

$$\cancel{CO_2(g)} + C(s) \rightleftharpoons 2CO(g)$$

Other pyrometallurgical processes are similarly complex. Shortly you'll see that iron is smelted according to the overall reaction

$$2Fe_2O_3(s) + 3C(s) \rightarrow 4Fe(l) + 3CO_2(g)$$

while the actual process is a multistep one with CO as the reductant.

For some metal oxides, especially some members of Groups 6B(6) and 7B(7), reduction with carbon forms metal carbides, which are difficult to convert further, so other reducing agents are used. Hydrogen gas is used for less active metals, like tungsten:

$$WO_3(s) + 3H_2(g) \rightarrow W(s) + 3H_2O(g)$$

Metals that might form undesirable hydrides are reduced by a more active metal. In the *thermite reaction,* aluminum powder reduces the metal oxide in a spectacular exothermic reaction to give the molten metal (Figure 23.12):

$$Cr_2O_3(s) + 2Al(s) \rightarrow 2Cr(l) + Al_2O_3(s)$$

Reduction by a more active metal is also employed in situations that do not involve an oxide. In the extraction of gold after pretreatment by leaching, shown before, the dicyanoaurate(I) ion is reduced with zinc:

$$2Au(CN)_2^-(aq) + Zn(s) \rightarrow 2Au(s) + Zn(CN)_4^{2-}(aq)$$

The method is also used to recover very active metals, such as rubidium, from their molten salts:

$$Ca(l) + 2RbCl(l) \rightarrow CaCl_2(l) + 2Rb(g)$$

Just as *reduction* of a mineral is used to obtain the metal, *oxidation* results in the recovery of an active nonmetal. In these cases, a stronger oxidizing agent removes electrons from the nonmetal anion to give the free nonmetal, as in the production of iodine:

$$2I^-(aq) + Cl_2(g) \rightarrow 2Cl^-(aq) + I_2(s)$$

**FIGURE 23.12**
**The thermite reaction.** This highly exothermic redox reaction uses powdered aluminum to reduce metal oxides.

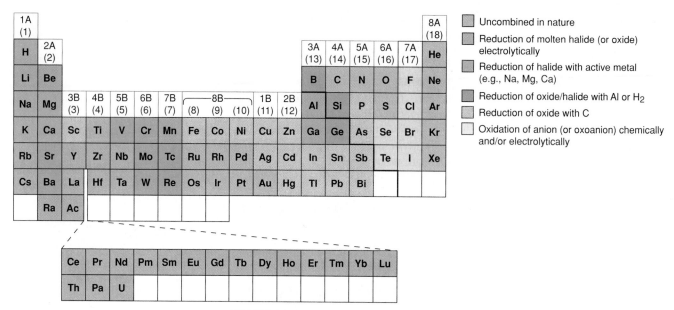

**FIGURE 23.13**

**The redox step in converting a mineral to the element.** This periodic table shows the nature of the redox step used to form the element from the mineral. Note that most of the commercially important metals are reduced from the compound with either carbon (coke or charcoal) or a more active metal (or $H_2$).

In *electrochemical redox* processes, the mineral components are converted to the elements in an electrolytic cell (Section 20.6). In some cases, the *pure* mineral, in the form of the molten halide or oxide, is used in order to prevent unwanted reactions. The cation is reduced at the cathode to the metal, and the anion is oxidized at the anode to the nonmetal:

$$BeCl_2(l) \rightarrow Be(s) + Cl_2(g)$$

High-purity hydrogen gas is prepared by electrochemical reduction:

$$2H_2O(l) \rightarrow 2H_2(g) + O_2(g)$$

Specially designed cells separate the products to prevent their recombination. Cost is a major factor in electrolysis, and an inexpensive source of electricity is essential for large-scale methods. The current and voltage requirements depend on the electrochemical potential ($E^0_{cell}$) and any overvoltage necessary to overcome electrode effects.

Figure 23.13 shows, in periodic table format, the most common step for obtaining the free element from its mineral.

4. *Refining (purifying) the element.* At this point in the isolation process, the element typically contains impurities from previous steps, so it must be refined. An electrolytic cell is employed in **electrorefining,** in which the impure element acts as the anode and a sample of the pure element acts as the cathode. (We examine this process for copper shortly.) Metals with relatively low boiling points, such as zinc and mercury, are refined by *distillation*. In the process of **zone refining** (Section 12.5), impurities are removed from a bar of the element by concentrating them in a thin molten zone, while the purified element recrystallizes. Metalloids used in electronic semiconductors, such as silicon and germanium, are zone-refined to greater than 99.999999% purity.

**TABLE 23.4 Some Familiar Alloys and Their Composition**

| NAME | COMPOSITION (MASS %) | USES |
|---|---|---|
| Stainless steel | 73-79 Fe, 14-18 Cr, 7-9 Ni | Cutlery, instruments |
| Nickel steel | 96-98 Fe, 2-4 Ni | Cables, gears |
| High-speed steels | 80-94 Fe, 14-20 W (or 6-12 Mo) | Cutting tools |
| Permalloy | 78 Ni, 22 Fe | Ocean cables |
| Bronzes | 70-95 Cu, 1-25 Zn, 1-18 Sn | Statues, castings |
| Brasses | 50-80 Cu, 20-50 Zn | Plating, ornamental objects |
| Sterling silver | 92.5 Ag, 7.5 Cu | Jewelry, tableware |
| 14-Carat gold | 58 Au, 4-28 Ag, 14-28 Cu | Jewelry |
| 18-Carat white gold | 75 Au, 12.5 Ag, 12.5 Cu | Jewelry |
| Typical tin solder | 67 Pb, 33 Sn | Electrical connections |
| Dental amalgam | 65 Ag, 21 Sn, 12 Cu, 2 Zn (dissolved in Hg) | Dental fillings |

5. *Alloying the purified element.* An **alloy** is a metal-based mixture consisting of solid phases of pure metals or solid solutions. The separate phases are often so finely divided that they can only be distinguished microscopically. Alloying a metal with other metals (and, in some cases, nonmetals) is done to enhance properties such as luster, conductivity, malleability, ductility, strength, and changes in melting point. Iron, probably the most important metal, is used only when it is alloyed. In pure form, it is soft and easily corroded. When alloyed with carbon and other metals, however, such as Mo for hardness and Cr and Ni for corrosion resistance, it forms the various steels. Copper stiffens when zinc is added to make brass. Mercury solidifies when alloyed with sodium, and vanadium becomes extremely tough with some carbon added. Table 23.4 shows the composition and uses of some common alloys.

The simplest alloys are *binary alloys,* which contain only two elements. Because of the two main ways that alloys form (see Figure 12.6, p. 470), they usually have the same lattice as the parent metal. In some cases, *the added element enters interstices,* lattice holes between the parent metal atoms (Section 11.4). In vanadium carbide (VC) alloy, for instance, carbon atoms occupy holes of a face-centered cubic vanadium lattice (Figure 23.14, *A*). Alternatively, the *added element substitutes for parent atoms* in the unit cell. β-Brass has a body-centered cubic unit cell in which a Zn atom lies at the center surrounded by Cu atoms (Figure 23.14, *B*). In one alloy of copper and gold, Cu atoms occupy the faces of a face-centered cube, and Au atoms lie at the corners (Figure 23.14, *C*).

Atomic properties, including size, electron configuration, and number of valence electrons, determine which metals can form a stable alloy. Quantitative predictions indicate that electron-poor metals (those with few *d* electrons) often form stable alloys with electron-rich metals (those with many *d* electrons). Thus, transition metals from the left half of the *d* block (Groups 3 to 5) often form alloys with metals from the right half (Groups 9 to 11). For example, electron-poor Zr, Nb, and Ta form strong alloys with the electron-rich Ir, Pt, or Au, such as the very stable $ZrPt_3$.

In the rest of this section, you'll see how these general steps in metallurgical extraction are applied to specific elements.

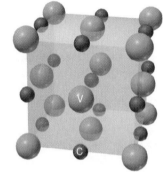

**A** Vanadium carbide

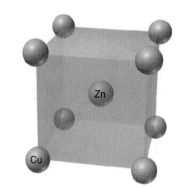

**B** β - Brass

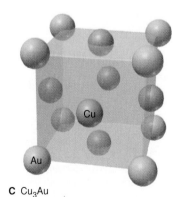

**C** $Cu_3Au$

**FIGURE 23.14**
Three binary alloys.

### Producing the Alkali Metals: Sodium and Potassium

The alkali metals are among the most reactive elements and thus are always found combined in nature. In the laboratory, they must be protected from contact with air (for example, by being stored under mineral oil) to prevent their immediate oxidation and to minimize the possibility of fires. The two most important alkali metals are sodium and potassium. Their abundant, water-soluble compounds are used throughout industry and research, and the Na⁺ and K⁺ ions are essential to life.

**Industrial production of sodium and potassium.** The sodium ore is *halite* (largely NaCl), which is obtained either by evaporation of concentrated salt solutions (brines) or by mining vast salt deposits formed from the evaporation of prehistoric seas. The Cheshire salt field in Britain, for example, is 60 km by 24 km by 400 m thick and contains more than 90% NaCl. Other large deposits occur in New Mexico, Michigan, New York, and Kansas. ◆

The brine is evaporated to dryness and the solid crushed and fused (melted) for use in an electrolytic apparatus called the **Downs cell** (Figure 23.15). To reduce heating costs, the NaCl (mp = 801°C) is mixed with $1\frac{1}{2}$ parts $CaCl_2$ to form a mixture that melts at only 580°C. Reduction of the metal ions to Na and Ca takes place at a cylindrical steel cathode, with the molten metals floating on the denser molten salt mixture. As they rise through a short collecting pipe, the liquid Na is siphoned off, while a higher melting Na/Ca alloy solidifies and falls back into the molten electrolyte. Chloride ions are oxidized to $Cl_2$ gas at a large anode within an inverted cone-shaped region. The cell design separates the metals from the $Cl_2$ to prevent their explosive recombination. The $Cl_2$ gas is collected, purified, and sold as a valuable byproduct.

Sylvite (mostly KCl) is the ore of potassium. The metal is too soluble in molten KCl to be prepared by a method similar to that for sodium. Instead, chemical reduction of K⁺ ions by liquid Na is used. An Na atom is smaller than a K atom, so it holds its electron more tightly, as these first ionization

◆ **A Plentiful Oceanic Supply of NaCl.** Will we ever run out of sodium chloride? Not likely. The rock-salt (halite) equivalent of NaCl in the world's oceans has been estimated at $1.9 \times 10^7$ km³, about 150% of the volume of North America above sea level. Or, looking at this amount another way: a row of 1 km³ cubes of NaCl would stretch from the Earth to the Moon 47 times!

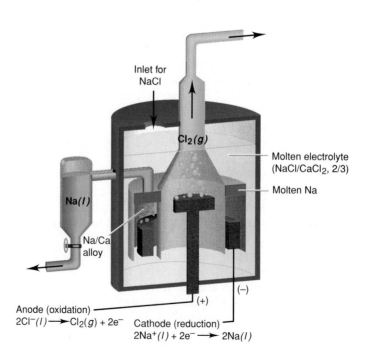

**FIGURE 23.15**

**The Downs cell for production of sodium.** The charge of solid NaCl and $CaCl_2$ forms the molten electrolyte. Sodium and calcium are formed at the cathode and float, but calcium solidifies and falls back into the bath. Chlorine gas forms at the anode.

Anode (oxidation)
$2Cl^-(l) \longrightarrow Cl_2(g) + 2e^-$     Cathode (reduction)
$2Na^+(l) + 2e^- \longrightarrow 2Na(l)$

energies indicate: $IE_1$ of Na = 496 kJ/mol; $IE_1$ of K = 419 kJ/mol. If the process were based solely on this atomic property, Na would not be very effective at reducing $K^+$, and the reaction would not move very far toward products. Applying Le Châtelier's principle overcomes this drawback. The reduction is carried out at 850°C, which is above the boiling point of K, so the equilibrium mixture contains gaseous K:

$$Na(l) + K^+(l) \rightleftharpoons Na^+(l) + K(g)$$

As the gas is removed, the equilibrium position shifts to the right, producing more K. The gas is then condensed and purified by fractional distillation. A similar approach (with Ca as the reducing agent) is used to produce rubidium and cesium.

**Uses of sodium and potassium.** The compounds of Na and K (and indeed, of all the alkali metals) have many more uses than the elements themselves. Nevertheless, there are some interesting uses of the metals that take advantage of their strong reducing power. Large amounts of Na were used as an alloy with lead to make gasoline antiknock additives, such as tetraethyllead:

$$4C_2H_5Cl(g) + 4Na(s) + Pb(s) \rightarrow (C_2H_5)_4Pb(l) + 4NaCl(s)$$

The toxic effects of environmental lead, however, have made this a declining use at the end of the 20th century.

If breeder reactors (see Margin Note, Section 21.6) were to become a practical energy generator, Na production would increase enormously. Its low melting point, viscosity, and absorption of neutrons, combined with its high thermal conductivity and heat capacity, make it the favored material for use as coolant and heat exchanger in the reactor core.

The major use of potassium at present is in an alloy with sodium for use as a heat exchanger in chemical and nuclear reactors. Another application is in the production of its superoxide, which it forms by direct contact with $O_2$:

$$K(s) + O_2(g) \rightarrow KO_2(s)$$

This material is employed as an emergency source of $O_2$ in breathing masks for miners, divers, firemen, and submarine crews:

$$4KO_2(s) + 4CO_2(g) + 2H_2O(g) \rightarrow 4KHCO_3(s) + 3O_2(g)$$

### The Indispensable Three: Steel, Copper, and Aluminum

The familiar presence, innumerable applications, and enormous production levels of three metals—iron in the form of steel, copper, and aluminum—make them stand out as indispensable materials of industrial society.

**Metallurgy of iron and steel.** Although people have practiced iron smelting for more than 3000 years, it was just slightly more than 200 years ago that iron assumed a major role in civilization. In 1773, a process to convert coal to carbon in the form of coke was discovered, and the material was used in a blast furnace. The coke process made iron smelting so much cheaper and more efficient that it led to large-scale iron production and ushered in the Industrial Revolution.

Modern society rests, quite literally, on the various alloys of iron known as **steel.** Although steel production has grown enormously since the 18th century—more than 700 million tons are produced annually—the process of recovering iron from its ores still employs the same general approach: *reduction by carbon in a blast furnace.* The most important minerals of iron are

TABLE 23.5 **Important Ores of Iron**

| MINERAL TYPE | MINERAL NAME, FORMULA |
|---|---|
| Oxide | Hematite, $Fe_2O_3$ |
|  | Magnetite, $Fe_3O_4$ |
|  | Ilmenite, $FeTiO_3$ |
| Carbonate | Siderite, $FeCO_3$ |
| Sulfide | Pyrite, $FeS_2$ |
|  | Pyrrhotite, $FeS$ |

listed in Table 23.5. All are used for steel making, except the S-containing minerals because traces of sulfur make the steel brittle. The seemingly unusual oxidation states of Fe in magnetite ($Fe_3O_4$) and pyrite ($FeS_2$) are not unusual at all. Actually, magnetite is a mixture of Fe(II) and Fe(III) oxides—1/1 $FeO/Fe_2O_3$ gives a nominal formula of $Fe_3O_4$—and pyrite contains Fe(II) combined with the disulfide ion, $S_2^{2-}$.

The process of converting iron ore to iron metal involves a series of simple redox and acid-base reactions, although the detailed chemistry is very complex and not entirely understood. A modern **blast furnace,** such as those used in Japan (Figure 23.16), is a tower, about 14 m wide by 40 m high, made of a brick material that can withstand intense heat. The *charge,* which consists of an iron ore containing the mineral (usually hematite), coke, and limestone, is fed through the top. More coke is burned in air at the bottom. The charge falls and meets a *blast* of rapidly rising hot air created by the combustion of the coke:

$$2C(s) + O_2(g) \rightarrow 2CO(g) + \textbf{\textit{heat}}$$

At the bottom of the furnace, the temperature exceeds 2000°C, while at the top, it reaches only 200°C. As a result, different stages of the process occur at different heights as the charge descends (Figure 23.16).

In the upper part of the furnace (200° to 700°C), the charge is preheated, and a *partial reduction* step occurs. The hematite is reduced to magnetite and then to iron(II) oxide (FeO) by CO, the actual reductant of the iron oxides. Because the blast passes through the entire furnace in only 10 s, the various gas-solid reactions do *not* reach equilibrium, and many intermediate prod-

**FIGURE 23.16**
**The major reactions occurring in a blast furnace.** Iron is produced from iron ore, coke, and limestone in a complex thermal process. The various stages depend on the temperature at different heights in the furnace.

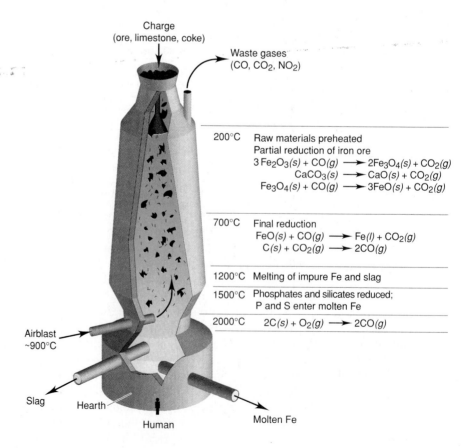

| | |
|---|---|
| 200°C | Raw materials preheated |
| | Partial reduction of iron ore |
| | $3Fe_2O_3(s) + CO(g) \longrightarrow 2Fe_3O_4(s) + CO_2(g)$ |
| | $CaCO_3(s) \longrightarrow CaO(s) + CO_2(g)$ |
| | $Fe_3O_4(s) + CO(g) \longrightarrow 3FeO(s) + CO_2(g)$ |
| 700°C | Final reduction |
| | $FeO(s) + CO(g) \longrightarrow Fe(l) + CO_2(g)$ |
| | $C(s) + CO_2(g) \longrightarrow 2CO(g)$ |
| 1200°C | Melting of impure Fe and slag |
| 1500°C | Phosphates and silicates reduced; P and S enter molten Fe |
| 2000°C | $2C(s) + O_2(g) \longrightarrow 2CO(g)$ |

ucts are formed. Limestone also decomposes to form the basic oxide CaO, which is important later in the process.

Lower down, at temperatures of 700° to 1200°C, a *final reduction* step occurs, as some of the coke reduces $CO_2$ to form more CO, which reduces the FeO to Fe. At still higher temperatures, between 1200° and 1500°C, the iron melts and drips to the bottom of the furnace. Acidic silica particles from the gangue react with basic calcium oxide to form a molten waste product called **slag:**

$$CaO(s) + SiO_2(s) \rightarrow CaSiO_3(l)$$

The slag drips down and floats on the denser iron (much as the Earth's silicate mantle and crust float on its molten iron core). Some unwanted reactions occur at this hottest stage, between 1500° and 2000°C. Any remaining phosphates and silicates are reduced to P and Si, and some Mn and traces of S dissolve into the molten iron along with carbon. The resulting impure product is called *pig iron* and contains about 3% to 4% C. A small amount of pig iron is used to make *cast iron,* but most is purified and alloyed to make various kinds of steel.

Pig iron is converted to steel in a separate furnace using the **basic-oxygen process** (Figure 23.17). High-pressure $O_2$ is blown over and through the molten iron, so that impurities (C, Si, P, Mn, S) are oxidized rapidly. The highly negative heats of formation of their oxides (such as $\Delta H_f^0$ of $CO_2$ = −394 kJ/mol and $\Delta H_f^0$ of $SiO_2$ = −911 kJ/mol) raise the temperature, which speeds the reaction. A lime (CaO) flux is added, which converts the oxides to a molten slag [primarily $CaSiO_3$ and $Ca_3(PO_4)_2$] that is decanted from the molten steel. The product is **carbon steel,** which contains 1% to 1.5% C and other impurities. It is alloyed with metals that prevent corrosion and increase its strength or flexibility.

**Isolation and electrorefining of copper.** After many centuries of mining copper ores for making bronze and brass articles, the metal has become less plentiful and thus expensive to extract. ◆ Despite this, more than 2.5 billion pounds are produced in the United States annually. The most common copper ore is chalcopyrite, $CuFeS_2$, a mixed sulfide of FeS and CuS; most remaining deposits contain less than 0.5% Cu by mass. To "win" this small amount of copper from the ore requires several metallurgical

◆ **The Dawns of Three New Ages.** Copper has been known since prehistoric times because it occurs uncombined. Its ores were probably first reduced by charcoal fires as early as 3500 BC, which marks the beginning of the Copper Age. With another 500 years of experience, people of India and Greece added molten tin to obtain a harder material, which marks the beginning of the Bronze Age. Some 1800 years later, brass, a copper-zinc alloy, was made in Palestine and used extensively by the Romans for well over 2000 years.

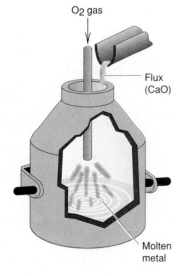

**A**            **B**

**FIGURE 23.17**
**The basic-oxygen process for making steel.** **A**, Jets of pure $O_2$ in combination with a basic flux (CaO) is used to remove impurities and lower the carbon content of pig iron in the manufacture of steel. **B**, Impure molten iron is added to a basic-oxygen furnace.

steps, including a final refining to achieve the 99.99% purity needed for electrical wiring, its most important application.

The low copper content in chalcopyrite must be enriched by removing the iron. The first step in copper extraction is pretreatment by flotation (see Figure 23.11), which concentrates the ore to around 15% Cu by mass. Next, a roasting step oxidizes FeS but leaves the CuS unaffected:

$$2FeCuS_2(s) + 3O_2(g) \rightarrow 2CuS(s) + 2FeO(s) + 2SO_2(g)$$

To remove the FeO and convert the CuS to a more convenient form, the mixture is heated to 1100°C with sand and more of the concentrated ore, which supplies additional sulfur. Several reactions occur in this step. The FeO is converted to a molten slag in a Lewis acid-base reaction with sand:

$$FeO(s) + SiO_2(s) \rightarrow FeSiO_3(l)$$

Excess sulfur in the ore reduces the CuS to $Cu_2S$.

In the final smelting step, the $Cu_2S$ is roasted in air, which converts some of it to $Cu_2O$:

$$2Cu_2S(s) + 3O_2(g) \rightarrow 2Cu_2O(s) + 2SO_2(g)$$

These two copper(I) compounds then react with each other:

$$Cu_2S(s) + 2Cu_2O(s) \rightarrow 6Cu(l) + SO_2(g)$$

As in the production of mercury (Section 22.3), sulfide ion is the reducing agent in this complex reaction.

Although usable for pipe at this stage, the copper must be purified for electrical applications, by removing unwanted impurities (Fe and Ni) as well as the valuable ones (Ag, Au, and Pt) found in copper ore. Purification is accomplished by *electrorefining*, which involves the formation of $Cu^{2+}$ ions in solution followed by the plating out of Cu metal (Figure 23.18). The impure copper obtained from smelting is cast into plates to be used as anodes, with cathodes made of already purified copper. The electrodes are immersed in acidified $CuSO_4$ solution, and a controlled voltage is applied that accomplishes two tasks simultaneously:

1. Copper and the more active impurities (Fe, Ni) are oxidized to their cations, but the less active ones (Ag, Au, Pt) are not. These unoxidized metals fall off the anode as a valuable "anode mud" and are purified separately.

**FIGURE 23.18**

**The electrorefining of copper. A,** Copper is refined electrolytically, using impure slabs of copper as anodes and sheets of pure copper as cathodes. The $Cu^{2+}$ ions released from the anode are reduced to Cu metal and plate out at the cathode. The "anode mud" contains valuable metal byproducts. **B,** An industrial facility for electrorefining copper.

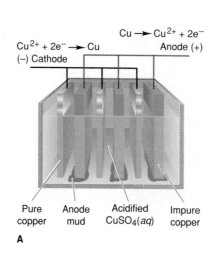

2. Because of differences in their standard electrode potentials, $Cu^{2+}$ ions are reduced at the cathode, but the $Fe^{2+}$ and $Ni^{2+}$ remain in solution:

$$Cu^{2+}(aq) + 2e^- \rightarrow Cu(s) \qquad E^0 = 0.34 \text{ V}$$

$$Ni^{2+}(aq) + 2e^- \rightarrow Ni(s) \qquad E^0 = -0.25 \text{ V}$$

$$Fe^{2+}(aq) + 2e^- \rightarrow Fe(s) \qquad E^0 = -0.44 \text{ V}$$

If not for the sale of the precious metals in the anode mud, which nearly off-sets the cost of electricity to operate the cell, Cu wire would be very expensive.

**Isolation, uses, and recycling of aluminum.** Aluminum is the most abundant metal in the Earth's crust by mass and the third most abundant element (after O and Si). It is found in numerous aluminosilicate minerals (Section 13.6), such as feldspars, micas, and clays, and in the rare gems garnet, beryl, spinel, and turquoise. Corundum, pure aluminum oxide ($Al_2O_3$), is extremely hard; mixed with traces of transition metals, it exists as ruby and sapphire. Impure $Al_2O_3$ is used in sandpaper and other abrasives.

Through eons of weathering, certain clays became *bauxite*, the major ore of aluminum. This mixed oxide-hydroxide occurs in enormous surface deposits in Mediterranean and tropical regions, but with world production approaching 100 million tons annually, it may soon be more scarce. In addition to the hydrated $Al_2O_3$ (about 75%), other components of industrial-grade bauxite that play a role in the extraction of aluminum are $Fe_2O_3$, $SiO_2$, and $TiO_2$.

The extraction of aluminum is a two-step process that combines hydro- and electrometallurgical techniques. In the first step, $Al_2O_3$ is separated from bauxite; in the second, it is converted to the metal.

1. *Isolating $Al_2O_3$ from bauxite.* After mining, bauxite is pretreated by extended boiling in 30% NaOH in the *Bayer process,* which involves acid-base, solubility, and complex-ion equilibria. The acidic $SiO_2$ and amphoteric $Al_2O_3$ dissolve, but the basic $Fe_2O_3$ and $TiO_2$ do not:

$$SiO_2(s) + 2NaOH(aq) + 2H_2O(l) \rightarrow Na_2Si(OH)_6(aq)$$

$$Al_2O_3(s) + 2NaOH(aq) + 3H_2O(l) \rightarrow 2NaAl(OH)_4(aq)$$

$$Fe_2O_3(s) + NaOH(aq) \rightarrow \text{no reaction}$$

$$TiO_2(s) + NaOH(aq) \rightarrow \text{no reaction}$$

Further heating slowly precipitates the $Na_2Si(OH)_6$ as an aluminosilicate, which is filtered out with the insoluble $Fe_2O_3$ and $TiO_2$ ("red mud").

Acidifying the filtrate precipitates $Al^{3+}$ as $Al(OH)_3$. Recall from Section 18.4 that the aluminate ion (common name), $Al(OH)_4^-(aq)$, is actually the complex ion $Al(H_2O)_2(OH)_4^-$, in which four of the six water molecules surrounding $Al^{3+}$ have each lost a proton. Weakly acidic $CO_2$ is added to produce a small amount of $H^+$ ion, which reacts with this complex ion. Cooling supersaturates the solution, and the solid forms and is filtered:

$$CO_2(g) + H_2O \rightleftharpoons H^+(aq) + HCO_3^-(aq)$$

$$Al(H_2O)_2(OH)_4^-(aq) + H^+(aq) \rightarrow Al(H_2O)_3(OH)_3(s)$$

[We usually write $Al(H_2O)_3(OH)_3$ more simply as $Al(OH)_3$.]

Drying at high temperature converts the hydroxide to the oxide:

$$2Al(H_2O)_3(OH)_3(s) \xrightarrow{\Delta} Al_2O_3(s) + 9H_2O(g)$$

**FIGURE 23.19**

The electrolytic cell in the manufacture of aluminum. Purified $Al_2O_3$ is mixed with cryolite ($Na_3AlF_6$) and melted. Reduction at the graphite furnace lining (cathode) gives molten Al. The graphite anodes must be replaced periodically due to their conversion to $CO_2$.

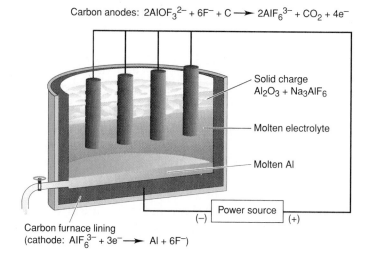

Carbon anodes: $2AlOF_3^{2-} + 6F^- + C \longrightarrow 2AlF_6^{3-} + CO_2 + 4e^-$

Solid charge
$Al_2O_3 + Na_3AlF_6$

Molten electrolyte

Molten Al

Power source
(−)    (+)

Carbon furnace lining
(cathode: $AlF_6^{3-} + 3e^- \longrightarrow Al + 6F^-$)

♦ **Energy Received and Returned.** The electron configuration of Al ([Ne] $3s^23p^1$) is the reason for the high energy needs of Al manufacture. Each $Al^{3+}$ ion needs $3e^-$ to form an Al atom, and the atomic mass of Al is so low ($\sim$ 27 g/mol) that 1 mol $e^-$ produces only 9 g Al. Compare this with the amounts of Mg or Ca (two other lightweight structural metals) that 1 mol $e^-$ would produce:

$$\tfrac{1}{3}Al^{3+} + e^- = \tfrac{1}{3}Al(s), \ \sim 9 \text{ g}$$

$$\tfrac{1}{2}Mg^{2+} + e^- = \tfrac{1}{2}Mg(s), \sim 12 \text{ g}$$

$$\tfrac{1}{2}Ca^{2+} + e^- = \tfrac{1}{2}Ca(s), \sim 20 \text{ g}$$

To turn this disadvantage around, we would need an aluminum battery. Once produced, the Al represents a concentrated form of electrical energy that can deliver 1 ℱ (96,500 C) for every 9 g of Al consumed, so its electrical output per gram of metal is high. In fact, aluminum-air batteries are now being produced (photo).

2. *Converting $Al_2O_3$ to the free metal.* Aluminum is an active metal, much too strong a reducing agent to be reduced at the cathode from aqueous solution, so the oxide itself must be electrolyzed. Because its melting point is very high (2030°C), $Al_2O_3$ is dissolved in molten *cryolite* ($Na_3AlF_6$) to give a mixture that can then be electrolyzed at $\sim$1000°C, affording a major energy savings.

The electrolytic step, called the *Hall-Heroult process,* takes place in a graphite-lined furnace, with the lining acting as the cathode. Anodes of graphite dip into the molten $Al_2O_3/Na_3AlF_6$ mixture (Figure 23.19). The cell typically operates at a moderate voltage of 4.5 V, but with an enormous current flow of between 1.0 and $2.5 \times 10^5$ A. Aluminum forms on the cathode and sinks below the less dense molten mixture. Oxygen forms at the graphite anode and reacts with it to form carbon monoxide.

Molten cryolite contains several ions, including $AlF_6^{3-}$, $AlF_4^-$, and $F^-$. The $F^-$ ion attacks $Al_2O_3$ and forms a number of fluoro-oxy ions, such as $AlOF_3^{2-}$, which dissolve in the mixture. Aluminum forms at the cathode:

$$AlF_6^{3-}(l) + 3e^- \rightarrow Al(l) + 6F^-(l)$$

The anode reaction can be written as

$$2AlOF_3^{2-}(l) + 6F^-(l) + C(s) \rightarrow 2AlF_6^{3-}(l) + CO_2(g) + 4e^-$$

The details are not well understood, but the overall cell reaction can be simplified to

$$2Al_2O_3(\text{in } Na_3AlF_6) + 3C(s) \rightarrow 4Al(l) + 3CO_2(g)$$

The Hall-Heroult process is very energy intensive: aluminum production in the United States accounts for more than 5% of our total electrical usage. ♦

Aluminum is a superb decorative, functional, and structural metal. It is lightweight, attractive, easy to work, and forms strong alloys, many of which are all around us (Figure 23.20). Although Al is very active, it does not readily corrode because of an adherent oxide layer that forms rapidly in air and prevents more $O_2$ from penetrating. Nevertheless, when it is in contact with less active metals such as Fe, Cu, and Pb, aluminum becomes the anode and deteriorates rapidly (Section 20.5). To prevent this, aluminum

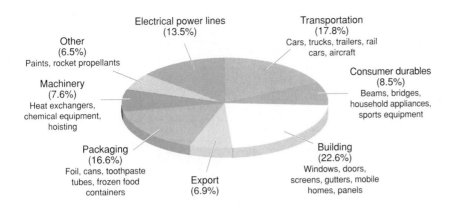

Electrical power lines (13.5%)

Other (6.5%)
Paints, rocket propellants

Machinery (7.6%)
Heat exchangers, chemical equipment, hoisting

Packaging (16.6%)
Foil, cans, toothpaste tubes, frozen food containers

Export (6.9%)

Building (22.6%)
Windows, doors, screens, gutters, mobile homes, panels

Transportation (17.8%)
Cars, trucks, trailers, rail cars, aircraft

Consumer durables (8.5%)
Beams, bridges, household appliances, sports equipment

**FIGURE 23.20**
The many familiar and essential uses of aluminum.

objects are often *anodized,* that is, made to act as the anode in an electrolysis that coats them with an oxide layer. The object is immersed in a 20% $H_2SO_4$ bath and connected to a carbon cathode. A layer of $Al_2O_3$ is deposited, whose thickness can be made from 10 to 100 $\mu$ thick, depending on the object's use.

More than 3 billion pounds of aluminum cans and packaging are discarded each year—a waste of one of the most useful materials in the world *and* the energy it represents. A quick calculation of the energy needed to prepare one mole of Al from *purified* $Al_2O_3$, compared with the energy needed for recycling, tells a clear message. The overall cell reaction in the Hall-Heroult process shown above has a $\Delta H^0$ of 2272 kJ and a $\Delta S^0$ of 635.4 J/K. Considering *only* the free energy change of the reaction at 1000°C, for one mole of Al we obtain

$$\Delta G^0 = \Delta H^0 - T\Delta S^0 = \frac{2272 \text{kJ}}{4 \text{ mol Al}} - \left(1273 \text{ K} \times \frac{0.6354 \text{ kJ/K}}{4 \text{ mol Al}}\right) = 365.8 \text{ kJ/mol Al}$$

When aluminum is recycled, the major energy input is melting the cans and foil, which has been calculated as ~26 kJ/mol Al. The ratio gives the fraction of energy needed to recycle:

$$\frac{\text{Energy to recycle 1 mol Al}}{\text{Energy to manufacture 1 mol Al}} = \frac{26 \text{ kJ}}{365.8 \text{ kJ}} = 0.071$$

Based on just these portions of the process, recycling takes about 7% as much energy as electrolysis. When mining, pretreatment, heating the electrolytic bath, and so forth are included, the amount decreases to less than 5%! The economic advantages of recycling, not to mention the environmental ones, are obvious, and recycling is now taking place in communities around the United States.

## Mining the Sea: Magnesium and Bromine

In the not too distant future, as terrestrial sources of certain elements become scarce or too costly to mine, the oceans will become an important source. Despite its abundant distribution on land, magnesium is already obtained from the sea, and its ocean-based production is a good example of the approach we might one day use for other elements. Bromine, too, is recovered from the sea as well as from inland brines.

**FIGURE 23.21**

**The production of elemental Mg from sea water.** The multistage Dow process involves obtaining $Mg^{2+}$ ion from sea water, converting the $Mg^{2+}$ to $Mg(OH)_2$ and then $MgCl_2$, which is evaporated to dryness, and then electrolytically reducing to Mg metal and $Cl_2$ gas.

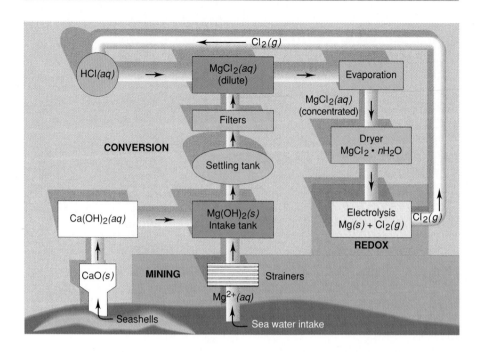

**Isolation and uses of magnesium.** The *Dow process* for the isolation of magnesium involves several steps that are similar to the procedures used for ores (Figure 23.21).

1. *Mining.* Intake of sea water and straining the debris are the "mining" steps, and no pretreatment is needed.
2. *Converting to mineral.* The dissolved $Mg^{2+}$ ion is converted to the mineral $Mg(OH)_2$, utilizing $Ca(OH)_2$. The $Ca(OH)_2$ is generated on-site (at the plant) by crushing seashells ($CaCO_3$), decomposing them with heat to CaO, and adding water to make slaked lime [$Ca(OH)_2$], which is pumped into the sea water intake tank. The dissolved $Mg^{2+}$ is precipitated as the hydroxide ($K_{sp} \approx 10^{-9}$) by reaction with $Ca(OH)_2$:

$$Ca(OH)_2(aq) + Mg^{2+}(aq) \rightarrow Mg(OH)_2(s) + Ca^{2+}(aq)$$

3. *Converting to compound.* The solid $Mg(OH)_2$ is filtered and mixed with excess HCl, which is also made on-site, to form aqueous $MgCl_2$:

$$Mg(OH)_2(s) + 2HCl(aq) \rightarrow MgCl_2(aq) + 2H_2O(l)$$

The water is evaporated in stages to give solid, hydrated $MgCl_2 \cdot nH_2O$.

4. *Electrochemical redox.* Electrolysis of molten $MgCl_2$ above 700°C gives the metal, which floats on the denser molten salt:

$$MgCl_2(l) \rightarrow Mg(l) + Cl_2(g)$$

Some of the $Cl_2$ that forms is recycled to make the HCl in Step 3, and some is sold to reduce overall plant costs.

Magnesium is the lightest structural metal available (nearly one-half the density of Al and one-fifth that of steel). Although Mg is quite reactive, it forms an extremely adherent, high-melting oxide layer of MgO, so it finds many uses in metal alloys and can be machined into any form. Magnesium alloys occur in everything from aircraft bodies to camera bodies, from luggage to auto engine blocks. The pure metal is a strong reducing agent, which makes it useful for sacrificial anodes (Section 20.5) and in the metallurgical extraction of other metals, such as Be, Ti, Zr, Hf, and U.

**Isolation and uses of bromine.** The largest source of bromine is the oceans, where it occurs as $Br^-$ at a concentration of 0.065 g/L (65 ppm). However, the added cost of concentrating such a dilute solution often shifts the choice of source, when available, to much more concentrated salt lakes and natural brines; Arkansas brines, the most important source in the United States, contain about 4.5 g $Br^-$/L (4500 ppm). The $Br^-$ is readily oxidized to $Br_2$ with aqueous $Cl_2$:

$$2Br^-(aq) + Cl_2(aq) \rightarrow Br_2(l) + 2Cl^-(aq) \qquad \Delta G^0 = -61.5 \text{ kJ}$$

The $Br_2$ is removed by passing steam through the mixture and then cooling and drying the liquid bromine.

World production of $Br_2$ is about 1% that of $Cl_2$. Its major use is in the preparation of organic chemicals, such as ethylene dibromide, which was added to leaded gasoline to prevent deposits of lead oxides from forming and depositing on engine parts. Newer products requiring bromine include the flame retardants in rugs and textiles. Bromine is needed in the synthesis of inorganic chemicals also, especially silver bromide for photographic emulsions.

## The Many Sources and Uses of Hydrogen

Although hydrogen accounts for 90% of the atoms in the universe, it makes up only 15% of the atoms in the Earth's crust, the vast majority being found in combination with oxygen in natural waters and with carbon in biomass, petroleum, and coal.

Hydrogen has three naturally occurring isotopes. Ordinary hydrogen ($^1H$), or protium, is the most abundant and has no neutrons. Deuterium ($^2H$ or D), the next in abundance, has one neutron, and tritium ($^3H$ or T) has two. All three occur as diatomic molecules, $H_2$, $D_2$ (or $^2H_2$), and $T_2$ (or $^3H_2$). The neutron/proton ($N/Z$) ratio of tritium is too high for stability, and it undergoes $\beta$ decay with a half-life of 12.4 yr (Section 21.1). Table 23.6 compares some molecular and physical properties of $H_2$, $D_2$, and $T_2$. The heavier the isotope, the higher the molar mass of the molecule, its melting and boiling points, and heats of phase change; it also effuses at a lower rate (Section 5.6).

**TABLE 23.6** **Some Molecular and Physical Properties of Diatomic Hydrogen, Deuterium, and Tritium**

| PROPERTY | $H_2$ | $D_2$ | $T_2$ |
|---|---|---|---|
| Molar mass (g/mol) | 2.016 | 4.028 | 6.032 |
| Bond length (pm) | 74.14 | 74.14 | 74.14 |
| Melting point (K) | 13.96 | 18.73 | 20.62 |
| Boiling point (K) | 20.39 | 23.67 | 25.04 |
| $\Delta H^0_{fus}$ (kJ/mol) | 0.117 | 0.197 | 0.250 |
| $\Delta H^0_{vap}$ (kJ/mol) | 0.904 | 1.226 | 1.393 |
| Bond energy (kJ/mol at 298 K) | 432 | 443 | 447 |

The mass difference also leads to different bond strengths, which affects reactivity. Thus, in contrast to the case of most elements, hydrogen isotopes differ *significantly* in chemical behavior. Because the mass of D is twice the mass of H, at a given temperature, H atoms bonded to a given atom vibrate at a higher frequency than do D atoms. As a result, the bond to D is stronger and therefore slower to break. Therefore, any reaction that includes breaking a bond to hydrogen in the rate-limiting step will occur *faster with H than with D.* This behavior is called a *kinetic isotope effect,* and we'll see an example of it shortly.

**Industrial production of hydrogen.** Hydrogen gas ($H_2$) is produced on an industrial scale worldwide: 250,000 metric tons ($3 \times 10^{12}$ L at STP) are produced annually in the United States alone. All production methods are energy intensive, so the choice is determined by energy costs. In Scandinavia, where hydroelectric power is plentiful, electrolysis is the chosen method; in the United States and Great Britain, where natural gas from oil refineries is plentiful, various thermal methods are used.

The most common thermal methods use water and a simple hydrocarbon in two steps. Modern American plants use methane, which has the highest H/C ratio of any hydrocarbon. In the first step, the reactants are heated to around 1000°C over a nickel-based catalyst in the endothermic *steam-reforming process:*

$$CH_4(g) + H_2O(g) \longrightarrow CO(g) + 3H_2(g) \qquad \Delta H = 206 \text{ kJ}$$

Heat is supplied by burning methane at the refinery. To generate more $H_2$, the product mixture (called water gas) is heated with steam at 400°C over an iron or cobalt oxide catalyst in the exothermic *water-gas shift reaction,* and the CO reacts:

$$H_2O(g) + CO(g) \rightleftharpoons CO_2(g) + H_2(g) \qquad \Delta H = -41 \text{ kJ}$$

The reaction mixture is recycled several times, which decreases the CO to around 0.2% by volume. Passing the mixture through liquid water removes the more soluble $CO_2$ (solubility = 0.034 mol/L) from the $H_2$ (solubility < 0.001 mol/L). Calcium oxide can also be used to remove $CO_2$ by conversion to $CaCO_3$. By removing $CO_2$, these process steps shift the equilibrium position to the right and produce $H_2$ that is about 98% pure. To attain greater purity (~99.9%), the gas mixture is passed through a *synthetic zeolite* (Section 13.6) that filters out nearly all molecules larger than $H_2$.

In regions where inexpensive electricity is available, very pure $H_2$ is prepared through electrolysis of water with Pt (or Ni) electrodes:

$$2H_2O(l) + 2e^- \longrightarrow H_2(g) + 2OH^-(aq) \qquad\qquad E = -0.42 \text{ V [cathode]}$$

$$\underline{H_2O(l) \longrightarrow \tfrac{1}{2}O_2(g) + 2H^+(aq) + 2e^- \qquad -E = -0.82 \text{ V [anode]}}$$

$$H_2O(l) \longrightarrow H_2(g) + \tfrac{1}{2}O_2(g) \qquad\qquad E_{\text{cell}} = -1.24 \text{ V}$$

Overvoltage increases the cell potential to ~ −2 V (Section 20.6). Thus, under typical operating conditions, it takes about 400 kJ of energy to produce 1 mol $H_2$:

$$\Delta G = -n\mathscr{F}E = -2 \text{ mol } e^- \times \frac{96.5 \text{ kJ}}{\text{V·mol } e^-} \times -2 \text{ V} = 4 \times 10^2 \text{ kJ/mol } H_2$$

High-purity $O_2$ is a valuable byproduct that offsets some of the costs.

**Laboratory production of hydrogen.** You may already have produced small amounts of $H_2$ in the lab by one of several methods. A characteristic reaction of the very active metals in Groups 1A(1) and 2A(2) is the reduction of water to $H_2$ and $OH^-$:

$$Ca(s) + 2H_2O(l) \rightarrow Ca^{2+}(aq) + 2OH^-(aq) + H_2(g)$$

Alternatively, the strongly reducing hydride ion can be used:

$$NaH(s) + H_2O(l) \rightarrow Na^+(aq) + OH^-(aq) + H_2(g)$$

Less active metals reduce $H^+$ in acids:

$$Zn(s) + 2H^+(aq) \rightarrow Zn^{2+}(aq) + H_2(g)$$

In view of hydrogen's great potential as a fuel (Chemical Connections, Section 6.6), chemists are seeking ways to decompose water by lowering the overall activation energy of the process. More than 10,000 *water-splitting* schemes have been devised; one of the more promising is

$$3FeCl_2(s) + 4H_2O(g) \rightarrow Fe_3O_4(s) + 6HCl(g) + H_2(g)$$

$$Fe_3O_4(s) + \tfrac{3}{2}Cl_2(g) + 6HCl(g) \rightarrow 3FeCl_3(s) + 3H_2O(g) + \tfrac{1}{2}O_2(g)$$

$$\underline{3FeCl_3(s) \rightarrow 3FeCl_2(s) + \tfrac{3}{2}Cl_2(g)}$$

$$H_2O(g) \rightarrow H_2(g) + \tfrac{1}{2}O_2(g)$$

All these schemes require heat ($\Delta H^0 \gg 0$), and some are now being attempted using focused sunlight.

**Industrial uses of hydrogen.** More than 95% of $H_2$ produced industrially is consumed on-site in ammonia or petrochemical facilities. In a plant that synthesizes $NH_3$ from $N_2$ and $H_2$, the reactant gases are formed through a series of reactions that involve methane, including the steam-reforming and water-gas shift reactions we discussed earlier. For this reason, the cost of $NH_3$ is closely correlated with the cost of $CH_4$. Here is how it works.

The steam-reforming reaction is performed with excess $CH_4$, which depletes the reaction mixture of $H_2O$:

$$CH_4(g; \text{excess}) + H_2O(g) \rightarrow CO(g) + 3H_2(g)$$

An excess of the product mixture ($CH_4$, CO, $H_2$) is burned in an amount of air ($N_2 + O_2$) insufficient to effect complete combustion but just enough to deplete the $O_2$, heat the mixture to 1100°C, and form additional $H_2O$:

$$4CH_4(g; \text{excess}) + 7O_2(g) \rightarrow 2CO_2(g) + 2CO(g) + 8H_2O(g)$$

$$2H_2(g; \text{excess}) + O_2(g) \rightarrow 2H_2O(g)$$

$$2CO(g; \text{excess}) + O_2(g) \rightarrow 2CO_2(g)$$

The remaining $CH_4$ reacts by the steam-reforming reaction, the remaining CO reacts by the water-gas shift reaction ($H_2O + CO \rightleftharpoons CO_2 + H_2$) to form more $H_2$, and the $CO_2$ is removed with CaO. The amounts are carefully adjusted to produce a final mixture that contains a 1/3 ratio of $N_2$ (from the added air) to $H_2$ (with traces of $CH_4$, Ar, and CO), which is used directly in the synthesis of ammonia (Chemical Connections, Section 16.4).

A second major use of $H_2$ is the *hydrogenation* of the C=C bonds in liquid oils to form the C—C bonds in solid fats and margarine. The process uses $H_2$ in contact with transition metal catalysts, such as powdered nickel. Solid fats are used not only as spreads, but also in commercial baked goods. Look at the ingredients on a package of bread, cake, or cookies, and you'll see the "partially hydrogenated vegetable oils" made by this process.

Hydrogen is also essential in the manufacture of many "bulk" chemicals, those produced in large amounts that have many further uses. One application gaining great attention is the production of methanol, in which carbon monoxide reacts with $H_2$ over a copper–zinc oxide catalyst:

$$CO(g) + 2H_2(g) \xrightarrow{\text{Cu-ZnO catalyst}} CH_3OH(l)$$

Many expect methanol to be used increasingly as a gasoline additive and, eventually, as a direct fuel alternative.

**Production and uses of deuterium and tritium.** Deuterium and its compounds are produced from $D_2O$ (heavy water), which is present as a minor component (0.016 mol % $D_2O$) in normal water and is isolated on the multiton scale by *electrolytic enrichment*. This process is based on the kinetic isotope effect for hydrogen noted earlier, specifically on the *higher rate of bond-breaking for O—H bonds than O—D bonds* and thus on the higher rate of electrolysis of $H_2O$ compared with $D_2O$. For example, using Pt electrodes, $H_2O$ electrolyzes about 14 times faster than $D_2O$. As some of the liquid decomposes to the elemental gases, the remainder becomes enriched in $D_2O$. Thus, by the time the volume of water has been reduced to 1/20,000 of its original volume, the remaining water is around 99% $D_2O$. By combining samples and continuing the process, more than 99.9% $D_2O$ is obtained.

Deuterium is produced by electrolysis of $D_2O$ or by any of the other chemical reactions that produce hydrogen from water, such as

$$2Na(s) + 2D_2O(l) \rightarrow 2Na^+(aq) + 2OD^-(aq) + D_2(g)$$

Similarly, compounds containing D (or T) are produced from reactions that give rise to the corresponding H-containing compound, such as

$$SiCl_4(l) + 2D_2O(l) \rightarrow SiO_2(s) + 4DCl(g)$$

Compounds with acidic protons undergo H/D exchange:

$$CH_3COOH(l) + D_2O(l; \text{excess}) \rightarrow CH_3COOD(l) + DHO(l; \text{small amount})$$

Notice that only the acidic H atom, the one in the COOH group, is exchanged, not those attached to carbon.

In nature, tritium forms by cosmic (neutron) irradiation of $^{14}N$,

$$^{14}_{7}N + ^{1}_{0}n \rightarrow ^{3}_{1}H + ^{12}_{6}C$$

which results in an abundance of only $10^{-7}$% $^3H$. Most tritium is produced artificially by neutron bombardment of lithium:

$$^{6}_{3}Li + ^{1}_{0}n \rightarrow ^{3}_{1}H + ^{4}_{2}He$$

Deuterium and tritium are used mostly as tracers in chemical or biochemical reaction pathways (Section 21.5).

## A Group at a Glance: Sources, Isolation, and Uses of the Elements

From a large-scale industrial perspective, certain elements may be more essential than others, but each has its own uses, in many cases, a critical one in a specialty industry. So that you can readily survey some practical aspects of the elements, the multipart Table 23.7 presents relevant facts about the main-group and Period 4 transition elements.

**TABLE 23.7A  Sources, Isolation, and Uses of Group 1A(1): The Alkali Metals**

Despite their great chemical similarity, the alkali metals rarely occur together in terrestrial minerals because of large differences in ionic size and thus crystal structure. Lithium salts occur with those of magnesium, an example of their diagonal relationship (Section 13.3). Sodium and potassium are the fifth and sixth most abundant metals in the crust. All are isolated from their molten salts, either through electrolytic or chemical reduction. Most uses rely on their low density or great reducing power.

Sea salt harvest, France

| ELEMENT | SOURCE | ISOLATION | USES |
|---|---|---|---|
| Lithium | Spodumene [$LiAl(Si_2O_6)$] | Preparation and electrolysis of molten LiCl | In strong, low-density Mg and Al alloys for armor and aerospace parts; future uses in Li batteries for electric cars |
| Sodium | NaCl in rock salt (halite); $NaNO_3$ (saltpeter) | Electrolysis of molten NaCl (Downs cell; see text) | Reducing agent for isolation of Ti, Zr, others; heat exchanger in nuclear reactors |
| Potassium | KCl (sylvite) in sea water | Na reduction of molten KCl (see text) | Reducing agent; production of $KO_2$ (see text) |
| Rubidium | Minor component of Li ores | Ca reduction of molten RbCl; byproduct of Li isolation | Reducing agent |
| Cesium | Minor component of Li ores; pollucite ($Cs_4Al_4Si_9O_{26} \cdot H_2O$) | Ca reduction of molten CsCl; byproduct of Li isolation | Reducing agent |
| Francium | Minute traces from minor branch of $^{235}U$ decay series | | |

**TABLE 23.7B  Sources, Isolation, and Uses of Group 2A(2): The Alkaline Earth Metals**

The alkaline earth metals are extracted mostly from rocky carbonates and sulfates (except for magnesium, which is also isolated from sea water). Like the alkali metals, they are strong reducing agents, so the final isolation step is either electrolytic reduction of the molten chloride or chemical reduction of the oxide with an active metal. The lighter members are used in alloys, the heavier ones as scavengers of nonmetal impurities through formation of ionic compounds.

Dolomite Mountains, Italy

| ELEMENT | SOURCE | ISOLATION | USES |
|---|---|---|---|
| Beryllium | Beryl ($Be_3Al_2Si_6O_{18}$) | Electrolysis of molten $BeCl_2$; reduction of $BeF_2$ with Mg | In high-strength alloys of Cu and Ni for aerospace engines and electronics; neutron moderator and reflector in nuclear reactors; window in x-ray tubes |
| Magnesium | Magnesite ($MgCO_3$), dolomite, ($MgCO_3 \cdot CaCO_3$), sea water | Electrolysis of molten $MgCl_2$ (see text); silicothermal method, $2(MgO \cdot CaO) + FeSi \rightarrow 2Mg + Ca_2SiO_4 + Fe$ | Lightweight alloys |
| Calcium | Limestone and aragonite ($CaCO_3$) | Electrolysis of molten $CaCl_2$ formed by HCl on $CaCO_3$ | Strengthens Al alloys; reducing agent to produce Cr, Zr, U; scavenger of traces of $O_2$, P, and S in steel |
| Strontium | Strontianite ($SrCO_3$); celestite ($SrSO_4$) | Thermal decomposition, then Al reduction of SrO | Scavenger of $O_2$ and $N_2$ in electronic devices |
| Barium | Barite ($BaSO_4$) | Al reduction of BaO | Scavenger of $O_2$ and $N_2$ in electronic devices |
| Radium | Minor (0.1 ppb) component in pitchblende (uranium ore) | Electrolysis of molten $RaCl_2$ after extensive extraction (Section 21.1) | Formerly used in cancer therapy |

**TABLE 23.7C  Sources, Isolation, and Uses of the Period 4 Transition Metals [Groups 3B(3) to 2B(12)]**

Several of the first-row transition elements—Fe, Ti, Mn—are among the most abundant metals in the crust and occur together in many ores. Extraction methods vary depending on the source, but all include a reduction step with either carbon or more active metals (Mg or Al). Alloying provides most of their uses.

Nodules of Cu, Ni, and Co

| ELEMENT | SOURCE | ISOLATION | USES |
|---|---|---|---|
| Scandium | Thortveitite (40% $Sc_2O_3$); byproduct of U extraction | C reduction of $Sc_2O_3$ | None |
| Titanium | Rutile ($TiO_2$), ilmenite ($FeTiO_3$) | Conversion to $TiCl_4$, then reduction with Mg | Very abundant (4th in crust); stronger than steel but half as dense; high-temperature, lightweight alloys for rocket and jet engines; future uses in train and car parts |
| Vanadium | Carnotite [$K(UO_2)(VO_4) \cdot 1.5\, H_2O$] | Conversion to $NaVO_3$, then reduction with Al (thermite) or FeSi | Combines with C in steel to make very strong alloy for truck springs and axles |
| Chromium | Chromite ($FeCr_2O_4$) | Conversion to $Cr_2O_3$, then reduction with Al | Nonferrous alloys; chrome plating |
| Manganese | Pyrolusite ($MnO_2$); many other ores; in future Mn "nodules" on ocean floor | Conversion to $Mn_3O_4$, then reduction with Al; conversion to $MnSO_4$, then electrolysis of aqueous Mn(II) | Scavenger of O and S in steel; high-strength steel alloys for excavators, rail crossings |
| Iron | Hematite ($Fe_2O_3$), magnetite ($Fe_3O_4$) | Reduction using C (see text) | Steel (see text) |
| Cobalt | Smaltite ($CoAs_2$), many sulfides with Ni, Cu, Pb | Roasting in $O_2$, leaching with $H_2SO_4$, precipitating $Co(OH)_3$ with $ClO^-$, heating to form CoO, and reducing with C | Cobalt blue glass and pottery; pigments for paints and inks; catalysts for organic reactions; specialty alloys with Cr and W for drill bits, lathe tools, and surgical instruments; magnetic alloys (Alnico) |
| Nickel | Pentlandite [$(Ni,Fe)_9S_8$] | Roasting in $O_2$ to NiO, reducing with C; Mond process, $Ni(CO)_4(g) \rightleftharpoons Ni(s) + 4CO(g)$ | Nickel steels for armor; stainless steel and Alnico; nonferrous alloys (tableware) with Ag; Monel with Cu for handling $F_2$; nichrome; undercoat for chrome plating; hydrogenation catalyst |
| Copper | Chalcopyrite ($CuFeS_2$) | See text | Wiring, plumbing, coins (see text) |
| Zinc | Zinc blende (ZnS); sphalerite | Roasting in $O_2$ to form ZnO, then reducing with C | Brasses (50%-80% Cu); galvanizing steel to prevent corrosion; batteries (Section 20.4) |

Steelworks

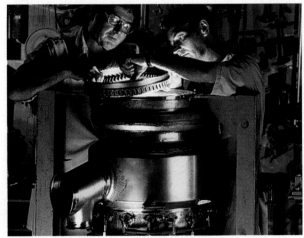

Jet engine assembly

**TABLE 23.7D** **Sources, Isolation, and Uses of Group 3A(13): The Boron Family**

Most sources are rocky oxides and sulfides. Boron is rare but concentrated in large deposits; the largest is in California's Mojave Desert (6.4 km × 1.6 km × 30 m thick). Aluminum is the most abundant crustal metal, but bauxite is its only ore. Alloys and new semiconductors are the principal end uses.

Borate deposit (tufa), California

| ELEMENT | SOURCE | ISOLATION | USES |
|---|---|---|---|
| Boron | Borax ($Na_2[B_4O_5(OH)_4]\cdot 8H_2O$); kernite ($Na_2[B_4O_5(OH)_4]\cdot 2H_2O$) | Mg reduction of $B_2O_3$; electrolysis of $KBF_4$ | $M_xB_y$ turbine blades, rocket nozzles, heat shields |
| Aluminum | Bauxite (contains gibbsite, $Al_2O_3$) | Electrolysis of $Al_2O_3$ in $Na_3AlF_6$ (see text) | Many familiar alloys; electric transmission lines (see text) |
| Gallium | Trace element in bauxite | Obtained as trace impurity in Al purification | High-speed semiconductors; future use in photovoltaic solar panels |
| Indium | Trace element in Zn/Pb sulfide ores | Recovered from sulfide roasting flue dusts | Semiconductors with P, Sb; low-melting alloys in sprinklers |
| Thallium | Trace element in Zn/Pb sulfide ores | Recovered from sulfide roasting flue dusts | Very toxic; few uses |

**TABLE 23.7E** **Sources, Isolation, and Uses of Group 4A(14): The Carbon Family**

Carbon is more abundant cosmically than on Earth; the reverse is true for silicon, the second most abundant element in the crust. Lead, the final product of most radioactive decay processes, is the crust's most abundant heavy metal. The isolation procedures vary, but those for Sn and Pb require milder conditions than those for Si and Ge. Except for Ge, each member has essential large-scale applications.

Purified silicon

| ELEMENT | SOURCE | ISOLATION | USES |
|---|---|---|---|
| Carbon | Diamond; graphite; petroleum; coal; carbonates; $CO_2$ | Used as found, or from petroleum or coal by heating in the absence of air | Graphite: composites, electrodes; control rods in nuclear reactors<br>Diamond: jewelry; abrasives; films<br>Coke, carbon black: reductant in metallurgy; rubber tire strengthener; pigment; decolorizes sugar |
| Silicon | Silica ($SiO_2$); silicate minerals | Reduction of $K_2SiF_6$ with Al; reduction of $SiO_2$ with Mg, then zone refining | Semiconductors; glass; ceramics |
| Germanium | Germanite (mixture of Cu, Fe, and Ge sulfides) | Roasting in $O_2$, then reducing $GeO_2$ with $H_2$, and zone refining | Semiconductors; infrared spectrometer windows and lenses |
| Tin | Cassiterite ($SnO_2$) | Thermal reduction of $SnO_2$ with C | Prevents corrosion in steel cans; alloys, e.g., solder, bronze, pewter |
| Lead | Galena (PbS) | Roasting in $O_2$ to PbO, then reducing with C, and electrorefining | Automotive batteries; solder; ammunition |

Pewter

Oil refinery

**TABLE 23.7F  Sources, Isolation, and Uses of Group 5A(15): The Nitrogen Family**

The high abundances of N and P contrast with the low abundances of the other members, but all occur in concentrated sources. $N_2$ is obtained from liquefied air and the others by thermal reduction of sulfide ores, usually with coke or iron. The major applications of the three larger elements are in lead alloys, with arsenic and antimony also used as dopants in new semiconductors.

$N_2$ freeze-dried rose

| ELEMENT | SOURCE | ISOLATION | USES |
|---|---|---|---|
| Nitrogen | Air | Fractional distillation of liquefied air | $N_2(g)$: inert atmosphere in metallurgical and petrochemical processing; reactant in $NH_3$ production<br>$N_2(l)$: freeze-drying food;, grinding meat for hamburger; biological preservation; future use in superconductors |
| Phosphorus | Phosphate ores, e.g., fluoroapatite $[Ca_5(PO_4)_3F]$ (see text) | Reducing phosphate rock with C | Starting material for $H_3PO_4$ (90%), $PCl_3$, $P_4S_3$ (matches), and $P_4S_{10}$ (pesticide) synthesis |
| Arsenic | Arsenopyrite (FeAsS); flue dust in Cu and Pb extraction | Heating in absence of air | Films, light-emitting (photo) diodes; lead alloys |
| Antimony | Stibnite ($Sb_2S_3$); flue dust in Cu and Pb extraction | Roasting in air to $Sb_2O_3$, then reducing with C | In lead-acid batteries (5% Sb) |
| Bismuth | Bismuthinite ($Bi_2S_3$) | Roasting to $Bi_2O_3$, then reducing with C or Fe | Alloys; medicines |

**TABLE 23.7G  Sources, Isolation, and Uses of Group 6A(16): The Oxygen Family**

Nearly one of every two atoms in the Earth's crust (and the moon's) is oxygen, but it is obtained from liquefied air. Its critical application in steel-making make its production the third highest of any chemical. Sulfur is extracted from large native deposits, with 90% destined for $H_2SO_4$ production (Section 23.4). The other elements are obtained as byproducts and, except for selenium's application in photocopying, have minor uses.

Sulfur obtained by Frasch process

| ELEMENT | SOURCE | ISOLATION | USES |
|---|---|---|---|
| Oxygen | Air | Fractional distillation of liquefied air | Oxidizing agent in steel-making (see text), sewage treatment, paper-pulp bleaching, rocket fuel; medical applications |
| Sulfur | Underground S deposits; sour natural gas or petroleum | Frasch process (see text); catalytic oxidation of $H_2S$ | Production of $H_2SO_4$ (see text); vulcanization of rubber; chemicals for pharmaceuticals, textiles, and pesticides |
| Selenium | Impurity in sulfide ores; anode muds of Cu refining | Reducing $H_2SeO_3$ with $SO_2$ | Electronics; xerography; cadmium pigments |
| Tellurium | Mixed tellurides and sulfides of Group 8 to 11 metals; anode muds of Cu refining | Oxidizing to $Na_2TeO_3$, then electrolysis | Steel-making |
| Polonium | Pitchblende; trace element formed in radium decay | Isolated in trace amounts | Future use as heat source in space satellites and lunar stations |

**TABLE 23.7H**   *Sources, Isolation, and Uses of Group 7A(17): The Halogens*

Due to their high reactivity, the halogens never occur free in nature. In rocks, their abundances decrease down the group, but chlorine dominates by its enormous abundance in the ocean (1.9 mass %). All halogens are produced by oxidation of their halides, chlorine being the oxidizing agent for bromine and iodine. Chlorine's uses, especially the formation of monomers for the plastics industry, make its production one of the 10 highest in the United States.

Fluorite

| ELEMENT | SOURCE | ISOLATION | USES |
|---|---|---|---|
| Fluorine | Fluorite; fluorspar ($CaF_2$) | Electrolysis of KF in molten anhydrous HF | Synthesis of $UF_6$ (for nuclear fuel) and $SF_6$ (electrical insulator); fluorinating agents; Teflon monomers |
| Chlorine | Halite (NaCl); sea water | Electrolysis of molten NaCl (see text); electrolysis of concentrated sea water (Section 23.4) | Oxidizing agent in bleach and disinfectant; production of polyvinyl chloride monomers; major biological anion |
| Bromine | Brine wells; sea water | Oxidation of $Br^-$ salts by $Cl_2$ (see text) | Preparation of organic bromides; AgBr in photography (see Figure 22.9) |
| Iodine | Brine wells; Chilean saltpeter ($NaIO_3$) | Oxidation of $I^-$ salts by $Cl_2$; reduction of $IO_3^-$ with $HSO_3^-$ | Essential trace element for thyroid hormones; disinfectant |
| Astatine | Extremely rare radioisotope | Obtained only in trace quantities | None |

Teflon cookware

PVC hose

**TABLE 23.7I**   *Sources, Isolation, and Uses of Group 8A(18): The Noble Gases*

Helium and the other noble gases are rare on Earth due to outgassing during the planet's formation. Helium is obtained from natural gas, a product of radioactive $\alpha$ decay. Argon's high abundance in air (~0.9 mol %) is due to radioactive decay of crustal $^{40}K$.

Neon sign

| ELEMENT | SOURCE | ISOLATION | USES |
|---|---|---|---|
| Helium | In natural gas (>0.4 mass %) | Distillation of condensed natural gas or differential diffusion of natural gas | Coolant for superconducting magnets; substitute for $N_2$ in deep-sea breathing mixture; mobile phase in gas chromatography |
| Neon | Air | Fractional distillation of liquefied air | Luminous gas in signs |
| Argon | Air | Fractional distillation of liquefied air | Arc welding |
| Krypton | Air | Fractional distillation of liquefied air | None |
| Xenon | Air | Fractional distillation of liquefied air | None |
| Radon | Air | Fractional distillation of liquefied air | None; radioactive air pollutant |

### Section Summary

Metallurgy involves mining an ore, separating it from debris, pretreating it to concentrate the mineral source, converting the mineral to a more useful form, reducing this form to the metal, purifying the metal, and in many cases, alloying the metal to obtain a more useful material. Na is isolated by electrolysis of molten NaCl in the Downs process; $Cl_2$ is a byproduct. K is produced by reduction with Na in a thermal process. Fe is produced through a multistep high-temperature process in a blast furnace. The crude pig iron is converted to carbon steel in the basic-oxygen process and then alloyed to make different steels. Cu ore is concentrated by flotation and reduced by smelting; the metal is purified by electrorefining. Al is extracted from bauxite by pretreating with concentrated base, followed by electrolysis of the $Al_2O_3$ in molten cryolite. Al alloys are used throughout home and industry. The total energy required to obtain Al is about 20 times that needed for recycling it. The $Mg^{2+}$ ion from sea water is converted to $MgCl_2$, which is electrolyzed to obtain the metal; Mg forms strong, lightweight alloys. $Br_2$ is obtained from brines by oxidizing $Br^-$ with $Cl_2$. $H_2$ is produced by electrolysis of water or in the formation of gaseous fuels from hydrocarbons. It is used in $NH_3$ production and in hydrogenation of vegetable oils. The isotopes of hydrogen differ significantly in atomic mass and thus in the rate at which their bonds to other atoms break. This difference is used obtain $D_2O$ from water.

## 23.4  Chemical Manufacturing: Three Case Studies

From laboratory-scale endeavors of the first half of the 19th century, today's chemical industries have grown to multinational corporations producing materials that define modern life: polymers, electronics, pharmaceuticals, and fuels. In this final section, we examine the interplay of theory and practice in three major industrial processes: (1) the contact process for the production of sulfuric acid, (2) the chlor-alkali process for the production of chlorine, and (3) the production of some modern ceramics.

### Sulfuric Acid, the Most Important Chemical

The manufacture of sulfuric acid began more than 400 years ago, when it was known as "oil of vitriol" and distilled from "green vitriol" ($FeSO_4 \cdot 7H_2O$). Considering its countless uses, it is not surprising that today sulfuric acid is produced throughout the world on a gigantic scale—more than 150 million tons a year. The green vitriol method and the many production methods used since have all been superseded by the modern **contact process,** which is based on the *catalyzed oxidation of* $SO_2$.

Let's begin one step before this key stage and see how the $SO_2$ is obtained. In most countries today, the production of sulfuric acid starts with the production of elemental sulfur, often by chemically separating and then oxidizing the $H_2S$ in "sour" natural gas:

$$2H_2S(g) + 2O_2(g) \xrightarrow{\text{low temperature}} \tfrac{1}{8}S_8(g) + SO_2(g) + 2H_2O(g)$$

$$2H_2S(g) + SO_2(g) \xrightarrow{\text{Fe}_2\text{O}_3 \text{ catalyst}} \tfrac{3}{8}S_8(g) + 2H_2O(g)$$

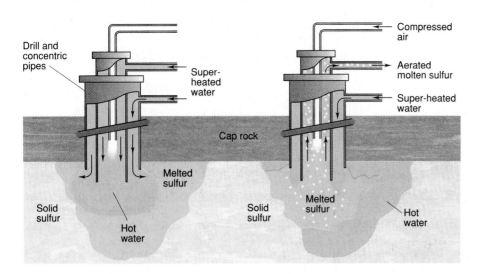

**FIGURE 23.22**
**The Frasch process for mining elemental sulfur.** This nonchemical process uses superheated water and compressed air to melt underground sulfur and bring it to the surface.

Where natural gas is not abundant, sulfur is obtained by the *Frasch process*, a nonchemical method that taps native sulfur deposits. A hole is drilled to the deposit, and superheated water (about 160°C) is pumped down two outer concentric pipes to melt the sulfur. Through a combination of hydrostatic pressure in the outermost pipe and compressed air sent through a narrow inner pipe, the sulfur is forced to the surface (Figure 23.22). The cost of drilling, pumping, and water ($5 \times 10^6$ gallons per day) are balanced somewhat by the fact that the product is very pure ($\sim$99.7% S).

Once obtained, the sulfur is burned in air to form $SO_2$, which then reacts in the contact process on its way to sulfuric acid. (Some $SO_2$ is also obtained from the roasting of metal sulfide ores.) About 90% of processed sulfur is directed to make this all-important industrial acid; indeed, sulfuric acid is central to so many chemical industries that a nation's level of sulfur production is a reliable indicator of its overall industrial capacity: the United States, Russia, Japan, and Germany are the top four sulfur producers.

The contact process oxidizes $SO_2$ with $O_2$ to $SO_3$:

$$SO_2(g) + \tfrac{1}{2}O_2(g) \rightleftharpoons SO_3(g) \qquad \Delta H^0 = -99 \text{ kJ/mol}$$

The reaction is *exothermic* and very *slow* at room temperature. From Le Châtelier's principle, we know that the yield of $SO_3$ can be increased by (1) changing the temperature, (2) increasing the pressure (more moles of gas are on the left than on the right), and (3) adjusting the concentrations (adding excess $O_2$ and removing $SO_3$) (Section 16.4).

First, let's examine the effect of temperature. Adding heat (raising the temperature) increases the frequency of $SO_2$-$O_2$ collisions and thus increases the *rate* of $SO_3$ formation. However, since the formation of $SO_3$ is exothermic, removing heat (lowering the temperature) shifts the equilibrium position to the right and thus increases the *yield* of $SO_3$. This is a classic situation that calls for use of a catalyst. By lowering the activation energy, *a catalyst allows equilibrium to be reached more quickly and at a lower temperature;* thus, rate *and* yield are optimized (Section 15.6). The catalyst in the contact process is $V_2O_5$ on inert silica and is active between 400° and 600°C.

**FIGURE 23.23**
The many indispensable applications of sulfuric acid.

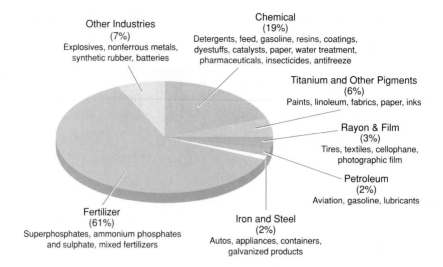

Other Industries
(7%)
Explosives, nonferrous metals,
synthetic rubber, batteries

Chemical
(19%)
Detergents, feed, gasoline, resins, coatings,
dyestuffs, catalysts, paper, water treatment,
pharmaceuticals, insecticides, antifreeze

Titanium and Other Pigments
(6%)
Paints, linoleum, fabrics, paper, inks

Rayon & Film
(3%)
Tires, textiles, cellophane,
photographic film

Petroleum
(2%)
Aviation, gasoline, lubricants

Fertilizer
(61%)
Superphosphates, ammonium phosphates
and sulphate, mixed fertilizers

Iron and Steel
(2%)
Autos, appliances, containers,
galvanized products

The pressure effect is small and not worthwhile economically. The concentration effects are controlled by providing an excess of $O_2$ in the form of a 5/1 mixture of air/$SO_2$. The mixture is passed over catalyst beds in four stages, and the $SO_3$ is removed at several points for an overall yield of 99.5%.

Sulfur trioxide is the anhydride of sulfuric acid, so a hydration step is next. However, $SO_3$ cannot be added to water because, at the operating temperature, it would first meet water vapor, which catalyzes its polymerization to $(SO_3)_x$, a smoke of solid particles that makes poor contact with water. To prevent this, previously formed $H_2SO_4$ absorbs the $SO_3$ and forms pyrosulfuric acid ($H_2S_2O_7$), which is then hydrolyzed with sufficient water:

$$SO_3(g) + H_2SO_4(l) \rightarrow H_2S_2O_7(l)$$
$$\underline{H_2S_2O_7(l) + H_2O(l) \rightarrow 2H_2SO_4(l)}$$
$$SO_3(g) + H_2O(l) \rightarrow H_2SO_4(l)$$

The uses of sulfuric acid are legion and presented in Figure 23.23.

Sulfuric acid is remarkably inexpensive (about $150/ton), largely because each step in the process is exothermic—burning S ($\Delta H^0 = -297$ kJ/mol), oxidizing $SO_2$ ($\Delta H^0 = -99$ kJ/mol), hydrating $SO_3$ ($\Delta H^0 = -132$ kJ/mol)—and the heat is a valuable byproduct. Three-quarters of the heat is sold as steam, and the rest is used to pump gases through the plant. A typical plant making 825 tons of $H_2SO_4$ per day produces enough steam to generate $7 \times 10^6$ watts of electricity.

## The Chlor-Alkali Process

Chlorine is produced and used in amounts many times greater than all the other halogens combined, ranking among the top 10 chemicals produced in the United States. All of its production methods depend on *the oxidation of $Cl^-$ ion from NaCl.*

Earlier, we discussed the Downs process for the isolation of sodium, which yields $Cl_2$ gas as the other product (Section 23.3). The **chlor-alkali process,** a much more important method that forms the basis of one of the largest inorganic chemical industries, electrolyzes concentrated aqueous NaCl to produce $Cl_2$ and several other essential chemicals. As you learned in Section 20.6, the electrolysis of aqueous NaCl does not usually yield the component elements. Chloride ion is oxidized at the anode rather than water due to the effects of overvoltage. However, $Na^+$ is not reduced at the cathode because its half-cell potential ($-2.71$ V) is much more negative than that for $H_2O$, even with the normal overvoltage ($\sim -1.0$ V). Therefore, the half reactions for electrolysis of aqueous NaCl are

$$2Cl^-(aq) \rightarrow Cl_2(g) + 2e^- \qquad\qquad -E^0 = -1.36 \text{ V [anode; oxidation]}$$

$$2H_2O(l) + 2e^- \rightarrow 2OH^-(aq) + H_2(g) \qquad E \approx -1.0 \text{ V [cathode; reduction]}$$

$$2Cl^-(aq) + 2H_2O(l) \rightarrow 2OH^-(aq) + H_2(g) + Cl_2(g) \quad E_{cell} = -2.4 \text{ V}$$

To obtain commercial quantities of $Cl_2$, however, a voltage almost twice this value and a current in excess of $3 \times 10^4$ A is used.

When we include the $Na^+$ spectator ion, the total ionic equation shows another important product made by the process:

$$2Na^+(aq) + 2Cl^-(aq) + 2H_2O(l) \rightarrow 2Na^+(aq) + 2OH^-(aq) + H_2(g) + Cl_2(g)$$

The sodium salts in the cathode compartment exist as an aqueous mixture of NaCl and NaOH, from which the NaCl is removed by fractional crystallization. Thus, in this version of the chlor-alkali process, which utilizes an asbestos diaphragm to separate the anode and cathode compartments, electrolysis of NaCl brines yields $Cl_2$, $H_2$, and industrial-grade NaOH, an important base (Figure 23.24). As with other reactive products, the $H_2$ and $Cl_2$ are kept apart to prevent explosive recombination. A slight hydrostatic pressure difference minimizes back flow of NaOH, which avoids disproportionation (self–oxidation-reduction) reactions of $Cl_2$ in the presence of $OH^-$ (Section 13.9), such as

$$Cl_2(g) + 2OH^-(aq) \rightarrow Cl^-(aq) + ClO^-(aq) + H_2O(l)$$

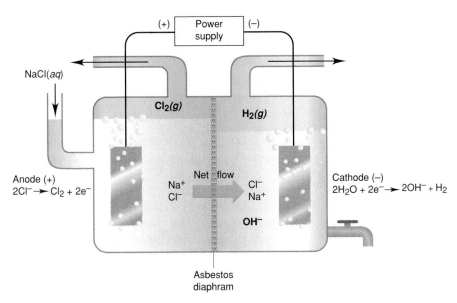

**FIGURE 23.24**

**A diaphragm chlor-alkali cell.** This process uses concentrated aqueous NaCl to make NaOH, $Cl_2$, and $H_2$ in an electrolytic cell. The height difference between compartments keeps a net movement of solution into the cathode compartment, which prevents reaction between $OH^-$ and $Cl_2$. The cathode electrolyte is concentrated and fractionally crystallized to give industrial-grade NaOH.

Where high-purity NaOH is desired, a slightly different version, called the *mercury-cell* chlor-alkali process, is employed. Mercury is used as the cathode, which creates an overvoltage for $H_2O$ reduction so large that it *does* favor reduction of $Na^+$. The sodium dissolves in the mercury to form sodium amalgam, Na(Hg). In the mercury-cell version, the half-reactions are

$$2Cl^-(aq) \rightarrow Cl_2(g) + 2e^- \quad \text{[anode; oxidation]}$$

$$2Na^+(aq) + 2e^- \xrightarrow{Hg} 2Na(Hg) \quad \text{[cathode; reduction]}$$

To obtain sodium hydroxide, the sodium amalgam is removed and treated with $H_2O$, which is reduced by the Na:

$$2Na(Hg) + 2H_2O(l) \xrightarrow{-Hg} 2Na^+(aq) + 2OH^-(aq) + H_2(g)$$

The mercury released in this step is recycled through the system. Thus, *the products are the same in both versions,* but the purity of NaOH in the latter version is much higher.

Despite the formation of higher purity NaOH, the mercury-cell method is being phased out and has been eliminated already in Japan. Cost is not the major reason for the phaseout, although the method does consume about 15% more electricity than the diaphragm version. Rather, the problem is that as the mercury is recycled, some is lost in the industrial waste water. On average, 200 g Hg are lost per ton of $Cl_2$ produced. In the 1980s, U.S. production was 2.75 million tons of $Cl_2$ annually via the mercury-cell method; thus, an effluent of 550,000 kg of this toxic heavy metal was entering U.S. waterways each year! Recently, a third method has been introduced that utilizes a polymeric membrane to separate the cell compartments. It combines the advantages of lower electricity use with the absence of Hg toxicity, so its application has been increasing steadily.

## Ceramics for Tomorrow

First developed by Stone Age people, **ceramics** are defined as nonmetallic-nonpolymeric solids, often made from clays, that are hardened by heating to high temperatures. Clay ceramics consist of silicate microcrystals suspended in a glassy, cementing medium. Ceramics such as bricks, porcelain, and glazes are so useful because they are hard and resistant to heat and chemicals. A modern kiln heats an aluminosilicate clay, such as kaolinite, to 1500°C at which temperature the bound water is driven off:

$$Si_2Al_2O_5(OH)_4(s) \rightarrow Si_2Al_2O_7(s) + 2H_2O(g)$$

In the process, the structure rearranges to an extended network of Si-centered and Al-centered tetrahedra of O atoms (Section 13.6).

In the current generation of high-technology materials, many novel ceramics with extremely useful properties are being developed (Table 23.8). In addition to the traditional ceramic characteristics of hardness, heat resistance, and chemical stability, several of the new ceramics have superior electrical and magnetic properties. As just one example, consider the unusual electrical behavior of certain zinc oxide (ZnO) composites. Ordinarily a semiconductor, ZnO can be doped to become a conductor. Imbedding particles of the doped oxide into an insulating ceramic produces a variable resistor: at low voltage, the material barely conducts, but at high voltage, it

**TABLE 23.8  Some Uses of New Ceramics and Ceramic Mixtures**

| CERAMIC | APPLICATIONS |
|---|---|
| SiC, $Si_3N_4$, $TiB_2$, $Al_2O_3$ | Whiskers (fibers) to strengthen Al or other ceramics |
| $Si_3N_4$ | Car engine parts; turbine rotors for "turbo" cars; electronic sensor units |
| $Si_3N_4$, BN, $Al_2O_3$ | Supports or layering materials (as insulators) in electronic microchips |
| SiC, $Si_3N_4$, $TiB_2$, $ZrO_2$, $Al_2O_3$, BN | Cutting tools, edge sharpeners (as coatings and whole devices), scissors, surgical tools, industrial "diamond" |
| BN, SiC | Armor-plating reinforcement fibers (as in Kevlar composites) |
| $ZrO_2$, $Al_2O_3$ | Surgical implants (hip and knee joints) |

conducts well. Best of all, the changeover voltage can be "preset" by controlling the size of ZnO particles and the thickness of the insulating medium.

Some other new ceramics being prepared are silicon carbide (SiC) and nitride ($Si_3N_4$), boron nitride (BN), and the oxide superconductors. These ceramic materials are prepared by standard chemical methods that often involve driving off a volatile component during the reaction. Silicon carbide types are prepared from compounds used in silicone polymer manufacture (Section 13.6):

$$n(CH_3)_2SiCl_2(l) + 2nNa(s) \rightarrow 2nNaCl(s) + [(CH_3)_2Si]_n(s; \text{ a polysilane})$$

The polysilane is heated to 800°C to form the ceramic:

$$[(CH_3)_2Si]_n \rightarrow nCH_4(g) + nH_2(g) + nSiC(s)$$

SiC can also be prepared by direct reaction of Si and graphite under vacuum at 1500°C:

$$Si(s) + C(graphite) \rightarrow SiC(s)$$

The nitride is prepared by reaction of the elements above 1300°C:

$$3Si(s) + 2N_2(g) \rightarrow Si_3N_4(s)$$

The formation of a BN ceramic begins with boron trichloride or boric acid reacting with ammonia:

$$B(OH)_3(s) + 3NH_3(g) \rightarrow B(NH_2)_3(s) + 3H_2O(g)$$

Heat drives off some of the bound nitrogen as $NH_3$ to yield the ceramic:

$$B(NH_2)_3(s) \rightarrow 2NH_3(g) + BN(s)$$

One of the common high-temperature superconducting ceramic oxides is made by first heating a mixture of the oxides and carbonates:

$$4BaCO_3(s) + 6CuO(s) + Y_2O_3(s) \rightarrow 2YBa_2Cu_3O_{6.5}(s) + 4CO_2(g)$$

Then, further heating in the presence of $O_2$ gives the superconductor:

$$YBa_2Cu_3O_{6.5}(s) + \frac{1}{4}O_2(g) \rightarrow YBa_2Cu_3O_7(s)$$

Structures of several ceramic materials are shown in Figure 23.25. Note the diamond-like lattice structure of silicon carbide. The network covalent

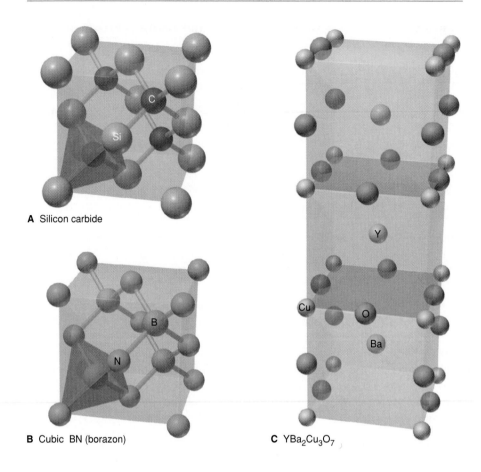

**A** Silicon carbide

**B** Cubic BN (borazon)

**C** YBa$_2$Cu$_3$O$_7$

bonding gives the material great strength. SiC is being made into thin fibers, called *whiskers,* to reinforce other ceramics in a composite structure that prevents cracking. Silicon nitride is virtually inert chemically, retains its strength and wear resistance for extended periods above 1000°C, and is dense, hard, and an excellent electrical insulator. Japanese and American automakers are testing it in high-efficiency car and truck engines because it allows an ideal combination of low weight, high operating temperature, and little need for lubrication.

The BN ceramics appear in two structures, analogous to those of the common crystalline forms of elemental carbon (Section 13.5). In the graphite-like form, BN has extraordinary properties as an electrical insulator. Under high temperature and extremely high pressure (1800°C and $8.5 \times 10^4$ atm), it converts to a diamond-like structure, which is extremely hard and durable. Both forms are virtually invisible to radar.

In Section 11.4, we mentioned some potential future uses of the superconducting oxide ceramics: storage and transmission of electricity with no loss of energy; electric power plants operating far from cities; ultrasmall microchips for ultrafast computers; high-speed trains floating frictionlessly; and inexpensive medical diagnostic equipment with high resolution.

Superconductivity, the flow of electrons with no electrical resistance, is a property acquired by certain metals around the temperature of liquid He (less than 4 K), but the new ceramics superconduct above the boiling point of $N_2$ (77 K), some as high as 125 K. All the known superconducting ceramic oxides contain copper in an unusual oxidation state. In $YBa_2Cu_3O_7$, for instance, assuming oxidation states of +3 for Y, +2 for Ba, and −2 for O, the three Cu atoms have a total oxidation state of +7. This is divided as $Cu(II)_2Cu(III)$, with Cu in the unusual +3 state. X-ray crystallography indicates that a distortion in the structure makes four of the oxide ions lean toward the $Y^{3+}$ ion, which aligns the Cu ions into chains within the crystal. Such alignment may be associated with superconduction, although the process is still poorly understood. Because of their brittleness, it has been difficult to fashion these new materials into wires, but methods for making superconducting films and ribbons have recently been developed.

Research in ceramic processing is beginning to overcome the inherent brittleness of this entire class of materials. The problem arises from the strength of the ionic-covalent bonding in many ceramics and their resulting inability to deform. Under stress, a microfine defect widens and lengthens until the material cracks. One new method forms defect-free ceramics by the controlled packing and heat-treating of extremely small, uniform oxide particles coated with organic polymers. Another method arrests the widening crack by fashioning ceramics imbedded with zirconia ($ZrO_2$), whose crystal structure expands up to 5% under the mechanical stress of a crack tip: the moment the advancing crack reaches them, the zirconia particles effectively pinch it shut.

Despite many technical difficulties, chemical ingenuity applied to the development of new ceramic materials and research into their amazing and useful properties will continue well into the next century.

## Section Summary

Sulfuric acid production starts with the extraction of sulfur, either by the oxidation of $H_2S$ or the mining of sulfur deposits. The sulfur is roasted to $SO_2$, which is oxidized to $SO_3$ by a catalytic process that optimizes the yield at lower temperatures. Absorption of the $SO_3$ into $H_2SO_4$, followed by hydration, forms sulfuric acid. In the chlor-alkali process, aqueous NaCl is electrolyzed to form $Cl_2$, $H_2$, and low-purity NaOH. The mercury-cell version produces high purity NaOH but is being phased out due to mercury pollution. Novel ceramics are being developed for structural, electronic, and superconducting applications.

## Chapter Perspective

*We began our investigation of chemistry 23 chapters ago, by seeing how chemical products influence nearly every aspect of our material existence. Now we have ended it with some modern applications of the principles you've learned. In a very real way, the end of this course is also a beginning—a chance to apply your new abilities to visualize molecular events and solve problems logically in whatever field you choose.*

*For the science of chemistry, future challenges are great and their solutions complex: What new forms of energy can satisfy our societal needs while sustaining our environment? What new products can feed, clothe, and house the world's people while maintaining precious resources? What new medicines can defend against AIDS and other dreaded diseases? The questions are many, but the science of chemistry will always be one of our most powerful means of answering them.*

# For Review and Reference

## Key Terms

**SECTION 23.1**
abundance
differentiation
core
mantle
crust
lithosphere
hydrosphere
atmosphere
biosphere

occurrence (source)
ore

**SECTION 23.2**
environmental cycle
fixation
apatites

**SECTION 23.3**
metallurgy

mineral
gangue
flotation
leaching
roasting
smelting
electrorefining
zone refining
alloy
Downs cell

steel
blast furnace
slag
basic-oxygen process
carbon steel

**SECTION 23.4**
contact process
chlor-alkali process
ceramic

## Problems

Problems with a green number are answered at the back of the book. All sections include *two* categories of problems separated by a green rule: concept review questions and problems in a relevant context. (There are no skill building exercises in this problem set.)

### How the Elements Occur in Nature

**23.1** Hydrogen is by far the most abundant element cosmically; in interstellar space, it exists mainly as $H_2$. In contrast, on Earth it exists very rarely as $H_2$ and is ninth in abundance in the crust. Why is hydrogen so abundant in the universe? Why is it so rare as a diatomic gas in the Earth's atmosphere?

**23.2** Metallic elements are recovered primarily from ores that are oxides, carbonates, halides, or sulfides. Give an example for each of these types.

**23.3** The location of elements in the regions of Earth has enormous practical importance.
(a) Define the term *differentiation* and explain which physical property of a substance is primarily responsible for this process.
(b) What are the four most abundant elements in the Earth's crust?
(c) Which element is abundant in the crust and mantle but not the core?

**23.4** (a) How does position of a metal in the periodic table relate to whether it exists in nature as an oxide or as a sulfide?
(b) How does the electronegativity of the metal bear upon this pattern?

**23.5** What material is the source for commercial production of the following elements: (a) aluminum; (b) nitrogen; (c) chlorine; (d) calcium; (e) sodium?

**23.6** Aluminum is widely distributed throughout the world in the form of aluminosilicates (see Chapter 13). What property of these readily available minerals prevents them from being a source for aluminum?

**23.7** Describe two ways in which the biosphere has influenced the composition of the Earth's crust.

### The Cycling of Elements Through the Environment

**23.8** Use atomic/molecular properties to explain why life is based on carbon chemistry and not on the chemistry of some other element such as silicon.

**23.9** Define the term *fixation*. Name two elements that undergo environmental fixation. What are the natural forms that are fixed?

**23.10** Carbon dioxide enters the atmosphere by natural processes and as a result of human activity. Why is the latter source potentially dangerous?

**23.11** Schematic cycles such as that for carbon (Figure 23.5) are simplified in order to show overall changes, or because a particular amount involved is relatively minor. For example, the production of lime from limestone is not explicitly shown. Which labeled category in the figure would include this process? What other processes would contribute to this category?

**23.12** In the nitrogen cycle (Figure 23.6), three pathways are shown for the removal of atmospheric nitrogen.
(a) Describe these three pathways
(b) What percentage of the total nitrogen removed from the atmosphere is the result of human activity?

**23.13** Why do the N-containing species shown in Figure 23.6 not include ring compounds or long-chain compounds with N—N bonds?

**23.14** (a) Which region of the Earth's crust is not involved in the phosphorus cycle?
(b) Briefly describe two roles that organisms play in the phosphorus cycle.

**23.15** An important part of the carbon cycle is the fixation of $CO_2$ by photosynthesis to produce carbohydrates and oxygen gas.

(a) Using $(CH_2O)_n$ to represent a carbohydrate, develop a general balanced equation for the photosynthetic reaction.

(b) If a moderate-sized tree fixes 45 g $CO_2$ a day, what volume of $O_2$ gas measured at 1.0 atm and 80°F does the tree produce per day?

(c) What volume of air (0.033 mol % $CO_2$) at the same conditions contains this amount of $CO_2$?

**23.16** Nitrogen fixation requires a great deal of energy because the $N_2$ bond is strong (the activation energy for its breakage is high).

(a) How do the processes of atmospheric and industrial fixation illustrate this energy requirement?

(b) How are the thermodynamics different in these two cases? (*Hint*: Examine the respective heats of formation.)

(c) In view of the mild conditions for biological fixation, what must be the source of the "great deal of energy" in this case?

(d) What would be the most obvious environmental result of a low activation energy for $N_2$ fixation?

**23.17** The following N-containing species are integral members of the nitrogen cycle: $NH_3$, $N_2O$, $NO$, $NO_2$, $NO_2^-$, $NO_3^-$. Draw a Lewis structure for each species, showing minimal formal charge, and indicate the shape, including ideal bond angles.

**23.18** The following steps are unbalanced half-reactions involved in the nitrogen cycle. Balance each half-reaction to show the number of electrons lost or gained and whether it is an oxidation or a reduction:

(a) $N_2(g) \rightarrow NO(g)$

(b) $N_2O(g) \rightarrow NO_2(g)$

(c) $NH_3(aq) \rightarrow NO_2^-(aq)$

(d) $NO_3^-(aq) \rightarrow NO_2^-(aq)$

(e) $N_2(g) \rightarrow NO_3^-(aq)$

**23.19** Why is nitric acid not produced by oxidizing atmospheric nitrogen directly in the following way?

(1) $N_2(g) + 2O_2(g) \rightarrow 2NO_2(g)$

(2) $3NO_2(g) + H_2O(l) \rightarrow 2HNO_3(aq) + NO(g)$

(3) $2NO(g) + O_2(g) \rightarrow 2NO_2(g)$

Overall: $3N_2(g) + 6O_2(g) + 2H_2O(l) \rightarrow$
$$4HNO_3(aq) + 2NO(g)$$

(*Hint*: Evaluate the thermodynamics of each step.)

**23.20** The use of silica to form slag in the production of phosphorus from phosphate rock was introduced by Robert Boyle more than 300 years ago. When fluorapatite, $[Ca_5(PO_4)_3F]$, is used in phosphorus production, most of the fluorine atoms appear in the slag but some end up in toxic and corrosive $SiF_4(g)$.

(a) If 15% by mass of the fluorine in 100 kg $Ca_5(PO_4)F$ forms $SiF_4$, what volume of this gas is collected at 1.00 atm and the industrial furnace temperature of 1450°C?

(b) In some plants, the $SiF_4$ is used to produce sodium hexafluorosilicate, $Na_2SiF_6$, which is sold for domestic water fluoridation:

$2SiF_4(g) + Na_2CO_3(s) + H_2O(l) \rightarrow$
$$Na_2SiF_6(aq) + SiO_2(s) + CO_2(g) + 2HF(aq)$$

How many cubic meters of drinking water could be fluoridated to a level of 1 ppm from the reaction of the $SiF_4$ produced in (a)?

**23.21** The compounds $(NH_4)_2HPO_4$ and $NH_4(H_2PO_4)$ are water-soluble phosphate fertilizers.

(a) Which has the higher mass % of P?

(b) Why might the one with lower mass % P be preferred?

**23.22** An impurity sometimes found in $Ca_3(PO_4)_2$ is $Fe_2O_3$. It is removed during the production of phosphorus as $Fe_2P$, a material of limited use.

(a) Why is this impurity troubling from an economic standpoint?

(b) If 50 metric tons of crude $Ca_3(PO_4)_2$ contains 2% $Fe_2O_3$ by mass and the overall yield of phosphorus is 90%, how many metric tons of $P_4$ can be isolated?

**23.23** Tetraphosphorus decaoxide, $P_4O_{10}$, a compound made from phosphate rock, is commonly used as a drying agent in the laboratory.

(a) Write a balanced equation for its reaction with water.

(b) What is the pH of a solution formed from the addition of 5.0 g $P_4O_{10}$ in sufficient water to form 0.500 L?

## Isolation and Uses of the Elements

**23.24** Define each of the following materials: (a) ore; (b) mineral; (c) gangue; (d) brine.

**23.25** Define each of the following processes: (a) roasting; (b) smelting; (c) flotation; (d) refining.

**23.26** How are each of the following involved in iron metallurgy: (a) slag; (b) pig iron; (c) steel; (d) basic-oxygen process?

**23.27** Use atomic properties to explain the reduction of a less active metal by a more active one (a) in aqueous solution; (b) in the molten state. Give a specific example of each process.

**23.28** What are the distinguishing features of the extraction processes known as pyrometallurgy, electrometallurgy, and hydrometallurgy? Explain briefly how the given types(s) of metallurgy is(are) used in the production of (a) Fe; (b) Na; (c) Au; (d) Al.

**23.29** What factors determine which reducing agent is selected for the production of a specific metal?

**23.30** What class of element is obtained by oxidation of a mineral? What class of element is obtained by reduction of a mineral?

**23.31** What property allows copper to be purified in the presence of iron and nickel impurities? Explain.

**23.32** What is the practical reason for using cryolite in the electrolysis of aluminum oxide?

**23.33** (a) What is a kinetic isotope effect?

(b) Do compounds of hydrogen exhibit a relatively large or small effect? Explain.

(c) Carbon compounds also exhibit a kinetic isotope effect. How would you expect it to compare in magnitude with that for hydrogen? Why?

**23.34** How is Hess's law involved in water-splitting schemes for the production of $H_2$?

**23.35** Select the group of element components that gives each of the following alloys: (a) brass; (b) stainless; (c) bronze; (d) sterling silver.
1. Cu, Ag    2. Cu, Sn    3. Ag, Au
4. Fe, Cr, Ni    5. Fe, V    6. Cu, Zn

**23.36** Elemental Li and Na are prepared by electrolysis of a fused salt, while K, Rb, and Cs are prepared by chemical reduction.
(a) In general terms, explain why the alkali metals cannot be prepared by electrolysis of their aqueous salt solutions.
(b) Use ionization energies to explain why calcium *should* not be able to isolate Rb from molten RbX.
(c) Use physical properties to describe how calcium *is* used to isolate Rb from molten RbX.
(d) Can calcium be used to isolate Cs from molten CsX? Explain.

**23.37** A Downs cell operates at 75.0 A and produces 30.0 kg Na metal.
(a) What volume of $Cl_2(g)$ is produced at 1 atm and 580°C?
(b) How many coulombs were passed through the cell?
(c) How long did the cell operate?

**23.38** (a) In the industrial production of iron, what is the reducing substance that is loaded into the blast furnace?
(b) In addition to furnishing the reducing power, what other function does this substance serve?
(c) What is the formula of the active reducing agent in the process?
(d) Write equations for the stepwise reduction of $Fe_2O_3$ to iron in the furnace.

**23.39** A blast furnace uses $Fe_2O_3$ to produce 8400 T Fe/day.
(a) What mass of carbon dioxide would be produced each day?
(b) Compare this amount of $CO_2$ with that produced by one million automobiles, each burning 5 gallons of gasoline a day. Assume that gasoline has the formula $C_8H_{18}$, has a density of 0.74 g/mL, and that it burns completely. (Note that the U.S. gasoline consumption is over $4 \times 10^8$ gal/day.)

**23.40** One of the substances loaded into the blast furnace is limestone, which serves to produce lime in the furnace.
(a) Give the chemical equation for the reaction forming the lime.
(b) Explain the purpose of lime in the furnace. The term *flux* is often used as a label for a substance acting as the lime does. What is the derivation of this word and how does it relate to the function of the lime?

(c) Write a chemical equation describing the action of the limestone flux.

**23.41** In the production of magnesium metal, $Mg(OH)_2$ is precipitated by using $Ca(OH)_2$, which itself is "insoluble."
(a) Use $K_{sp}$ values to show that $Mg(OH)_2$ can be precipitated from sea water in which $[Mg^{2+}]$ is initially 0.051 $M$.
(b) If the sea water is made saturated with $Ca(OH)_2$, what fraction of the $Mg^{2+}$ can be precipitated?

**23.42** The last step in the Dow process for the production of magnesium metal involves electrolysis of molten $MgCl_2$.
(a) Why isn't the electrolysis carried out in aqueous $MgCl_2$? What would be the products of aqueous electrolysis?
(b) Do the high temperatures required to melt $MgCl_2$ favor products or reactants? (*Hint*: Consider the heat of formation of $MgCl_2$.)

**23.43** (a) What is the only halogen that occurs in a positive oxidation state in nature?
(b) Is this mode of occurrence consistent with its location in the periodic table? Explain.
(c) Write a balanced equation for the production of the free halogen from this source.

**23.44** Selenium is prepared by the reaction of $H_2SeO_3$ with gaseous $SO_2$ (Table 23.7G).
(a) What redox process does the sulfur dioxide undergo? What is the oxidation state of sulfur in the product?
(b) Given that the reaction occurs in acidic aqueous solution, what is the formula of the sulfur-containing species?
(c) Write the balanced redox equation for the process.

**23.45** The halogens $F_2$ and $Cl_2$ are produced by electrolytic oxidation, whereas $Br_2$ and $I_2$ are produced by chemical oxidation of the concentrated aqueous halide (brine) by a more electronegative halogen. State two reasons why $Cl_2$ is not prepared by this method.

**23.46** Silicon is prepared by the reduction of $K_2SiF_6$ with Al (Table 23.7E). Write the equation for this reaction. (*Hint*: Can $F^-$ be changed in this reaction? Will $K^+$ be reduced?)

**23.47** What is the mass percent of iron in each of the following iron ores: $Fe_2O_3$, $Fe_3O_4$, $FeS_2$?

**23.48** Phosphorous is one of the impurities present in pig iron that are removed in the basic-oxygen process. Assuming that the phosphorus is present simply as P atoms, write equations for its oxidation and subsequent reaction in the basic slag.

**23.49** The final step in the smelting of $FeCuS_2$ is

$$Cu_2S(s) + 2Cu_2O(s) \rightarrow 6Cu(l) + SO_2(g)$$

(a) What are the oxidation states of copper in $Cu_2S$, $Cu_2O$, and Cu?

(b) What are the oxidizing and reducing agents in this reaction?

**23.50** Use balanced equations to explain how acid-base properties are used to separate iron and titanium oxides from aluminum oxide in the Bayer process.

**23.51** Graphite is used as an electrode material in voltaic and electrolytic processes.
   (a) In what direction do you think graphite conducts electricity, perpendicular or parallel to its layers? Why?
   (b) Both the Downs cell and the Hall-Héroult process use a graphite anode, but the half-reactions are entirely different. Write balanced anode half-reactions for the two processes.
   (c) In the Hall-Héroult process, what mass of $CO_2$ is released to the atmosphere per metric ton of Al produced?

**23.52** A piece of Al with a surface area of 2.3 $m^2$ is anodized to produce a film of $Al_2O_3$ 20 $\mu$ ($20 \times 10^{-6}$ m) thick.
   (a) How many coulombs flow through the cell in this process (assume the density of the $Al_2O_3$ layer is 2.8 $g/cm^3$)?
   (b) If the film grows uniformly over the course of 15 min, what current will flow through the cell?

**23.53** The production of $H_2$ gas by the electrolysis of water typically requires about 400 kJ of energy per mole.
   (a) Use the relationship between work and cell voltage (Section 20.3) to calculate the minimum work needed to form 1 mol $H_2$ gas at the reversible cell voltage of 1.24 V.
   (b) What is the energy efficiency of the commercial cell operation?
   (c) Calculate the cost of producing 500 mol $H_2$ if the cost of electric energy is exactly $0.05 per kilowatt-hour (1 watt·second = 1 joule).

**23.54** (a) What are the components of the reaction mixture following the water-gas shift reaction?
   (b) Explain how zeolites (Section 13.6) are used to purify the $H_2$ formed.

**23.55** Heavy water ($D_2O$) is used to prepare numerous deuterated chemicals.
   (a) What major species, aside from the components, would you expect to find in a solution formed by mixing $CH_3OH$ and $D_2O$?
   (b) Write equations to explain how these various species arise. (*Hint*: Consider the autoionization of both components.)

**23.56** Typically, metal sulfides are first converted to oxides by roasting in air and then reduced with carbon to produce the metal. Why are the metal sulfides not reduced directly by carbon to yield $CS_2$? Give a thermodynamic analysis of both processes for a typical case such as ZnS.

**23.57** Several transition metals are prepared by reduction of the metal halide with magnesium metal. Titanium is prepared by the Kroll method in which the ore (il-

menite) is converted to the gaseous chloride, which is then reduced to the metal by molten Mg:
   (1) $FeTiO_3(s) + Cl_2(g) + C(s) \rightarrow$
      $TiCl_4(g) + FeCl_3(l) + CO(g)$ [unbalanced]
   (2) $TiCl_4(g) + Mg(l) \rightarrow Ti(s) + MgCl_2(l)$
      [unbalanced]
Assuming yields of 84% for (1) and 93% for (2) and other reactants in excess, what mass of Ti metal can be prepared for 17.5 metric tons of ilmenite?

## Chemical Manufacturing: Three Case Studies

**23.58** Explain in detail why a catalyst is used to produce $SO_3$.

**23.59** Among the exothermic contributions in the manufacture of sulfuric acid is the process of hydrating the $SO_3$.
   (a) Write two chemical reactions that show this hydrating process.
   (b) Why is the direct reaction of $SO_3$ with water not feasible?

**23.60** What is the principal reason that commercial $H_2SO_4$ is so inexpensive?

**23.61** (a) What are the three commercial products formed in the chlor-alkali process?
   (b) State an advantage and a disadvantage of the mercury-cell version of this process.

_____

**23.62** The production of $S_8$ from the $H_2S(g)$ found in many natural gas deposits occurs through the Claus process (Section 23.4):
   (a) Use these two unbalanced steps to write an overall balanced equation for this process:
      (1) $H_2S(g) + O_2(g) \rightarrow S_8(g) + SO_2(g) + H_2O(g)$
      (2) $H_2S(g) + SO_2(g) \rightarrow S_8(g) + H_2O(g)$
   (b) Write the overall reaction with $Cl_2$ as the oxidizing agent. Use thermodynamic quantities to show whether $Cl_2(g)$ could be used to oxidize $H_2S(g)$.
   (c) Why is oxidation by $O_2$ chosen in preference to oxidation by $Cl_2$?

**23.63** (a) Calculate the standard free energy change at 25°C for the oxidation of $SO_2$ to $SO_3$ in the contact process. Is the reaction spontaneous at 25°C?
   (b) Why is the reaction not performed at 25°C?
   (c) Is the reaction spontaneous at 500°C? (Assume that $\Delta H^0$ and $\Delta S^0$ are constant with temperature.)
   (d) How does the value of the equilibrium constant at 500°C compare with that at 25°C?
   (e) What is the highest temperature at which the reaction is spontaneous?

**23.64** In the early part of the 20th century, most sulfuric acid was manufactured by a process utilizing a gaseous $NO/NO_2$ mixture as the catalyst system to convert $SO_2$ to $SO_3$.

(a) What reaction does NO undergo in contact with $O_2$?

(b) Write reactions to show how the catalyst functions and is re-formed in this process.

(c) Explain how a similar process relates to the formation of acid rain.

**23.65** If a chlor-alkali plant used an electric current of $3 \times 10^4$ A, as stated in the text, how many pounds of $Cl_2$ would be produced in a typical 8-h operating day?

**23.66** In the chlor-alkali process, the chlorine and the sodium hydroxide products are kept separate from one another.

(a) Why is this precaution necessary when $Cl_2$ is the desired product?

(b) Hypochlorite or chlorate may be formed by disproportionation of $Cl_2$ in basic solution. What condition determines which product is formed?

(c) What mole ratio of $Cl_2$ and $OH^-$ is needed to produce $ClO^-$? $ClO_3^-$?

**23.67** Small-scale production of chlorine gas is accomplished by the reaction of manganese dioxide and hydrochloric acid. What volume of $Cl_2$ gas at 755 torr and 29°C can be prepared from 253 g $MnO_2$ and 325 mL of 4.2 $M$ HCl? ($MnCl_2$ and $H_2O$ form also).

**23.68** The ceramic boron nitride (BN) has a low-pressure form with a graphite-like structure and a high-pressure form with a diamond-like structure. Which might you expect to feel slippery? Explain.

**23.69** What is the empirical formula of each of the following binary compounds used as a new ceramic: (a) calcium nitride; (b) aluminum carbide; (c) aluminum nitride; (d) gallium nitride; (e) gallium arsenide (an important semiconductor)?

**23.70** Several of the new superconducting ceramic oxides superconduct only at extremely high pressures. These pressures cause different arrangements of atoms in the lattice and result in a different phase. Would you expect the superconducting structure to have a lower or higher density than the structure at atmospheric pressure? Explain.

## Comprehensive Problems

Problems with an asterisk (*) are more challenging.

**23.71** Through the process of chelation with polydentate ligands, biological systems concentrate transition elements that occur at extremely low environmental concentrations. As a nonbiological example, consider the octahedral $[Ni(NH_3)_6]^{2+}$ and $[Ni(en)_3]^{2+}$ complex ions (where "en" is ethylenediamine, $H_2NCH_2CH_2NH_2$). The complex-ion equilibria are

$$Ni^{2+}(aq) + 6NH_3(aq) \rightleftharpoons [Ni(NH_3)_6]^{2+}(aq) \quad K_f = 4.0 \times 10^8$$

$$Ni^{2+}(aq) + 3en(aq) \rightleftharpoons [Ni(en)_3]^{2+}(aq) \quad K_f = 1.9 \times 10^{18}$$

Given that the six Ni—N bonds in both ions are of

approximately equal energy, explain the difference in $K_f$ values. (*Hint:* Recall the relationship between $K$ and $\Delta G^0$. Equal Ni—N bond strengths means that only the $\Delta H^0$ terms in $\Delta G^0$ will be similar.)

**23.72** Sodium carbonate, $Na_2CO_3$, is an industrial base that is used in the paper, soap, and detergent industries, and indispensable in glassmaking. In the U.S., it is mined as trona ($Na_2CO_3 \cdot NaHCO_3 \cdot 2H_2O$) from enormous deposits in Wyoming. In many other countries, however, it is still made by the Solvay process, which has the following *overall* reaction:

$$CaCO_3(s) + 2NaCl(aq) \rightarrow Na_2CO_3(aq) + CaCl_2(aq)$$

(a) Why would you expect the overall reaction not to proceed as written?

(b) In the industrial process, the $CaCO_3$ is heated, and the gas formed reacts with added $NH_3$ and water. The resulting aqueous bicarbonate salt then undergoes a metathesis reaction with NaCl, and the Na salt formed is heated to form the desired product. Write balanced equations for these four reactions.

(c) Sodium bicarbonate, a key intermediate in the process, has several uses as a result of its low-temperature (50-100°C) decomposition. If a fire extinguisher contains 150 g $NaHCO_3$, what volume of $CO_2$ can be formed at 100.8 kPa and 72°F?

**23.73** Several decades ago, various sets of empirical rules were devised to predict the formation of alloys. Two of the Hume-Rothery rules propose that stable alloys form if the ratios of the number of valence electrons (s plus p) to the number of atoms are 3/2 or 21/13. For example, β-brass (CuZn) is a very stable alloy: Cu has one 4s and Zn has two 4s, or 3 valence $e^-$; one Cu atom plus one Zn atom gives 2 atoms; thus, a 3 $e^-$/2 atom ratio. Which electron/atom ratio does each of the following alloys display: (a) $Cu_5Sn$; (b) $Cu_5Zn_8$; (c) $Cu_{31}Sn_8$; (d) $AgZn_3$; (e) $Ag_3Al$?

**23.74** The Ostwald process for the production of $HNO_3$ proceeds as follows:

(1) $4NH_3(g) + 5O_2(g) \xrightarrow{\text{Pt/Rh catalyst}} 4NO(g) + 6H_2O(g)$

(2) $2NO(g) + O_2(g) \rightarrow 2NO_2(g)$

(3) $3NO_2(g) + H_2O(l) \rightarrow 2HNO_3(aq) + NO(g)$

(a) Describe the nature of the redox change occuring in step 3.

(b) Write an overall equation that includes $NH_3$ and $HNO_3$ as the only N-containing species.

(c) Calculate the $\Delta H^0_{rxn}$ at 25°C for this reaction in kJ/mol N atoms

**23.75** Step (1) of the Ostwald process is

$$4NH_3(g) + 5O_2(g) \xrightarrow{\text{Pt/Rh catalyst}} 4NO(g) + 6H_2O(g)$$

An unwanted side reaction for this step is

$$4NH_3(g) + 3O_2(g) \rightarrow 2N_2(g) + 6H_2O(g)$$

(a) Calculate $K_p$ for these two $NH_3$ oxidations at 25°C.

(b) Calculate $K_p$ for these two $NH_3$ oxidations at 900°C.

(c) The Pt/Rh catalyst is one of the most efficient in industry, achieving 96% yield in 1 millisecond of contact with the reactants. However, at normal operating conditions (5 atm and 850°C), about 175 mg Pt is lost per metric ton (T) of $HNO_3$ produced. If current annual U.S. production of $HNO_3$ is $7.7 \times 10^6$ T and the current market price of platinum is \$415/troy ounce, what is the annual cost in Pt (1 kg = 32.15 troy oz)?

**23.76** Nitric oxide is a gaseous free radical with unusual molecular properties. Most surprisingly, it has recently been shown to have essential biological functions (see Margin Note p. 556)
(a) Draw a molecular orbital diagram for NO.
(b) What is the MO bond order in NO?
(c) How does the bond order change when the lone electron is lost?

**23.77** Hydrazine, $N_2H_4$, is produced commercially by the oxidation of $NH_3$ with sodium hypochlorite, NaClO.
(a) Write a balanced equation for this reaction in base. [Note: $N_2H_4$ and the intermediate $NH_2Cl$ are toxic. Attempts to produce "extra strength" cleaning solutions by mixing bleach (aqueous NaClO) and window cleaner (aqueous $NH_3$) result in hospital stays.]
(b) Why is hydrazine not produced by the reduction of $N_2$? (*Hint*: Consider thermodynamic parameters for forming $N_2H_4$ vs. $NH_3$.)

**23.78** A metal catalyst (usually nickel) is used to hydrogenate C=C bonds to form the saturated fats in commercial baked goods. What does the catalyst do to speed the reaction? Suggest a reason why other Group 8B(10) metals are not used for this purpose.

**23.79** Incomplete combustion of C-containing compounds, whether in forest fires, coal-fired power plants, or automobile engines, produces CO as well as $CO_2$. The binding of CO to transition metal atoms and ions is the reason for its toxicity as well as the basis of several industrial methods to purify metals.
(a) Draw a Lewis structure for CO that has minimal formal charges.
(b) Does CO bond to the metal through the C or the O atom? Explain.

**\*23.80** Before the development of the Downs cell, the Castner cell was used for the industrial production of Na metal. The cell was based on the electrolysis of molten NaOH.
(a) Write balanced cathode and anode half-reactions for this cell.
(b) A major problem with this cell was that the water produced at one electrode diffused to the other and reacted with the Na. If all the water produced reacted with Na, what would be the maximum efficiency of the Castner cell expressed as moles of Na produced per mole of electrons flowing through the cell?

**\*23.81** In the leaching of gold ores by $CN^-$ solutions, gold forms the complex ion, $Au(CN)_2^-$.
(a) Calculate $E_{cell}$ for the oxidation in air ($P_{O_2} = 0.21$) of gold to gold(I) in basic (pH 13.55) solution. Is this reaction spontaneous?
(b) How does the formation of the complex ion change $E^0$ so that the oxidation can be accomplished? [$E^0 = 1.68$ V for $Au^+(aq) + e^- \rightarrow Au(s)$].

**23.82** Nitric oxide is an important species in the tropospheric nitrogen cycle, but in the stratosphere it destroys ozone.
(a) Write a balanced equation for its reversible reaction with ozone.
(b) Given that the forward and reverse steps are first order in each component, write general rate laws for them.
(c) Calculate $\Delta G^0$ for this reaction at 280 K, the average temperature in the stratosphere. (Assume $\Delta H^0$ and $S^0$ do not change with temperature.)
(d) Calculate the ratio of rate constants for the reaction at 280 K.

**23.83** Some newer battery designs involve lithium or sodium and various nonmetals as electrodes.
(a) The lithium-iodine battery is dry, with electrons passing from Li metal through a crystalline LiI(s) electrolyte to an $I_2$ complex. Write the half-reactions, identify the oxidizing and reducing agents, and give the name and charge of each electrode.
(b) In the sodium-sulfur battery, Na gives up electrons, which pass through an external circuit and reduce molten $S_8$ to polysulfide anions, $S_n^{2-}$. The current is carried by $Na^+$ ions that migrate through a dry alumina electrolyte to the sulfur compartment. Write the overall cell reaction if the pentasulfide ion, $S_5^{2-}$, forms.

**\*23.84** Because of their different molar masses, $H_2$ and $D_2$ effuse at different rates (Section 5.6).
(a) If it takes 14.5 min for 0.10 mol $H_2$ to effuse, how long does it take for 0.10 mol $D_2$ in the same apparatus at the same $T$ and $P$?
(b) How many effusion steps would it take to separate an equimolar mixture of $D_2$ and $H_2$ to 99 mol % purity?

**23.85** Even though the disproportionation of carbon monoxide to graphite and carbon dioxide is thermodynamically favorable, the reaction is kinetically slow.
(a) What does this statement mean in terms of the magnitudes of the equilibrium constant $K$, the rate constant $k$, and the activation energy $E_a$?
(b) Write a balanced equation for CO disproportionation.
(c) Calculate $K_c$ at 298 K.
(d) Calculate $K_p$ at 298 K.

**23.86** Quinones are biological molecules used extensively in industrial electron-transfer processes, in which the molecules cycle between oxidized (quinone) and reduced (hydroquinone) forms. For example,

1,4-benzoquinone       1,4-benzhydroquinone

Sodium hydrosulfite, $Na_2S_2O_4$, is frequently used to reduce quinones to hydroquinones. Write a balanced equation for this reduction with 1,4-benzoquinone in basic solution. Gaseous $SO_2$ is produced in the reaction also.

*23.87 Even though many metal sulfides are very sparingly soluble in water, their solubilities differ by several orders of magnitude. This difference can sometimes be used to separate the metals in an isolation step by controlling the pH. Use the following data to find the pH at which you can separate $Cu^{2+}$ and $0.10\ M\ Ni^{2+}$:

Saturated $H_2S = 0.10\ M$
$K_{a1}$ of $H_2S = 9 \times 10^{-8}$; $K_{a2}$ of $H_2S = 1 \times 10^{-17}$
$K_{sp}$ of $NiS = 1.1 \times 10^{-18}$; $K_{sp}$ of $CuS = 8 \times 10^{-34}$

23.88 Mercury occurs in nature as the sulfide, $HgS$, cinnabar. Its metallurgy is unusual in that the mercury is freed by simply roasting in air; sulfur dioxide is formed in the process.
(a) Write the equation for the reaction.
(b) What is the reducing agent in this reaction?

*23.89 How does acid rain affect the leaching of phosphate into ground water from terrestrial phosphate rock? In order to quantify this effect, calculate the solubility of $Ca_3(PO_4)_2$ in each of the following:
(a) Pure water, pH 7.0 (Assume $PO_4^{3-}$ does not react with water.)
(b) Moderately acidic rainwater, pH 4.5. (*Hint*: Assume all the phosphate exists in the form that predominates at this pH.)

23.90 Chemosynthetic bacteria reduce $CO_2$ by "splitting" $H_2S(g)$ rather than the $H_2O(g)$ used by photosynthetic organisms. Compare the free energy change of $H_2S$ splitting with that of $H_2O$ splitting. Is there an advantage in using $H_2S$ instead of $H_2O$?

# Common Mathematical Operations in Chemistry

**I**n additon to basic arithmetic and algebra, four mathematical operations are used frequently in general chemistry: manipulating logarithms, using exponential notation, solving quadratic equations, and graphing data. Each is discussed briefly below.

## Manipulating Logarithms

### Meaning and Properties of Logarithms

A logarithm is an exponent. Specifically, if $x^n = A$, we can say that the logarithm to the base $x$ of the number $A$ is $n$, and we can denote it as

$$\log_x A = n$$

Since logarithms are exponents, they have the following properties:

$$\log_x 1 = 0$$
$$\log_x (A \times B) = \log_x A + \log_x B$$
$$\log_x \frac{A}{B} = \log_x A - \log_x B$$
$$\log_x A^y = y \log_x A$$

### Types of Logarithms

Common and natural logarithms are used frequently in chemistry and the other sciences. For common logarithms, the base ($x$ in the general examples above) is 10, but they are written without specifying the base; that is, $\log_{10} A$ is written $\log A$. Thus, for example, the common logarithm of 1000 is 3; in other words, you must raise 10 to the 3rd power to obtain 1000:

$$\log 1000 = 3 \qquad \text{or} \qquad 10^3 = 1000$$

Similarly, we have

$$\log 10 = 1 \qquad \text{or} \qquad 10^1 = 10$$
$$\log 1{,}000{,}000 = 6 \qquad \text{or} \qquad 10^6 = 1{,}000{,}000$$
$$\log 0.001 = -3 \qquad \text{or} \qquad 10^{-3} = 0.001$$
$$\log 853 = 2.931 \qquad \text{or} \qquad 10^{2.931} = 853$$

The last example shows an important point about significant figures with all logarithms: the number of significant figures in the number equals the number of digits to the right of the decimal point in the logarithm. That is, the number 853 has three significant figures, and the logarithm 2.931 has three digits to the right of the decimal point.

To find a common logarithm with an electronic calculator, you simply enter the number and press the LOG button.

For natural logarithms, the base is the number $e$, 2.71828 $\cdots$, and $\log_e A$ is written $\ln A$. The relationship between the common and natural logarithms is easily obtained:
Since

$$\log 10 = 1 \quad \text{and} \quad \ln 10 = 2.303$$

we have

$$\ln A = 2.303 \log A$$

To find a natural logarithm with an electronic calculator, you simply enter the number and press the LN button. If your calculator does not have an LN button, enter the number, press the LOG button, and multiply by 2.303.

### Antilogarithms

The antilogarithm is the number you obtain when you raise the base to the logarithm; that is,

$$\text{antilogarithm (antilog) of } n \text{ is } 10^n$$

Using two of the earlier examples, the antilog of 3 is 1000, and the antilog of 2.931 is 853. To obtain the antilog with a calculator, you enter the number and press the $10^x$ button. Similarly, to obtain the natural antilogarithm, you enter the number and press the $e^x$ button. [On some calculators, you enter the number and first press INV and then the LOG (or LN) button.]

## Using Exponential (Scientific) Notation

Many quantities in chemistry are very large or very small. For example, in the conventional way of writing numbers, the number of gold atoms in one gram of gold is

59,060,000,000,000,000,000,000 atoms (to four significant figures)

As another example, the mass in grams of one gold atom is

0.00000000000000000000003272 g (to four significant figures)

Exponential (scientific) notation provides a much more practical way of writing such numbers. In exponential notation, we express numbers in the form

$A \times 10^n$

where $A$ (the coefficient) is greater than or equal to 1 and less than 10 (that is, $1 \le A < 10$), and $n$ (the exponent) is an integer.

If the number we want to express in exponential notation is larger than 1, the exponent is positive ($n > 0$); if the number is smaller than 1, the exponent is negative ($n < 0$). The size of $n$ tells the number of places the decimal point (in conventional notation) must be moved to obtain a coefficient

exponential notation, one gram of gold contains $5.906 \times 10^{22}$ atoms, and each gold atom has a mass of $3.272 \times 10^{-22}$ g.

### Changing Between Conventional and Exponential Notation

In order to use exponential notation, you must be able to convert to it from conventional notation, and vice versa.

1. To change a number from conventional to exponential notation, move the decimal point to the left for numbers equal to or greater than 10 and to the right for numbers between 0 and 1:

    75,000,000 changes to $7.5 \times 10^7$ (decimal point 7 places to the left)
    0.006042 changes to $6.042 \times 10^{-3}$ (decimal point 3 places to the right)

2. To change a number from exponential to conventional notation, move the decimal point the number of places indicated by the exponent to the right for numbers with positive exponents and to the left for numbers with negative exponents:

    $1.38 \times 10^5$ changes to 138,000 (decimal point 5 places to the right)
    $8.41 \times 10^{-6}$ changes to 0.00000841 (decimal point 6 places to the left)

3. An exponential number with a coefficient greater than 10 or less than 1 can be changed to the standard exponential form by converting the coefficient to the standard form and adding the exponents:

    $582.3 \times 10^6$ changes to $5.823 \times 10^2 \times 10^6 = 5.823 \times 10^{(2\,+\,6)} = 5.823 \times 10^8$
    $0.0043 \times 10^{-4}$ changes to $4.3 \times 10^{-3} \times 10^{-4} = 4.3 \times 10^{[-3\,+\,(-4)]} = 4.3 \times 10^{-7}$

### Using Exponential Notation in Calculations

In calculations, you can treat the coefficient and exponents separately and apply the properties of exponents (see earlier section on logarithms).

1. To multiply exponential numbers, multiply the coefficients, add the exponents, and reconstruct the number in standard exponential notation:

    $(5.5 \times 10^3)\,(3.1 \times 10^5) = (5.5 \times 3.1) \times 10^{(3\,+\,5)} = 17 \times 10^8 = 1.7 \times 10^9$
    $(9.7 \times 10^{14})\,(4.3 \times 10^{-20}) = (9.7 \times 4.3) \times 10^{[14\,+\,(-20)]} = 42 \times 10^{-6} = 4.2 \times 10^{-5}$

2. To divide exponential numbers, divide the coefficients, subtract the exponents, and reconstruct the number in standard exponential notation:

$$\frac{2.6 \times 10^6}{5.8 \times 10^2} = \frac{2.6}{5.8} \times 10^{(6\,-\,2)} = 0.45 \times 10^4 = 4.5 \times 10^3$$

$$\frac{1.7 \times 10^{-5}}{8.2 \times 10^{-8}} = \frac{1.7}{8.2} \times 10^{[(-5)\,-\,(-8)]} = 0.21 \times 10^3 = 2.1 \times 10^2$$

3. To add or subtract exponential numbers, change all numbers to have the same exponent, then add or subtract the coefficients:

    $(1.45 \times 10^4) + (3.2 \times 10^3) = (1.45 \times 10^4) + (0.32 \times 10^4) = 1.77 \times 10^4$
    $(3.22 \times 10^5) - (9.02 \times 10^4) = (3.22 \times 10^5) - (0.902 \times 10^5) = 2.32 \times 10^5$

## Solving Quadratic Equations

A quadratic equation is one in which the highest power of $x$ is 2. The general form of a quadratic equation is

$$ax^2 + bx + c = 0$$

where $a$, $b$, and $c$ are numbers. For given values of $a$, $b$, and $c$, the values of $x$ that satisfy the equation are called solutions of the equation. We calculate $x$ with the quadratic formula:

$$x = \frac{-b \pm \sqrt{b^2 - 4ac}}{2a}$$

We commonly require the quadratic formula when solving for some concentration in an equilibrium problem. For example, if $x = [H_3O^+]$, we might have an expression that is rearranged into the quadratic equation

$$4.3x^2 + 0.65x - 8.7 = 0$$

Applying the quadratic formula, with $a = 4.3$, $b = 0.65$, and $c = -8.7$, gives

$$x = \frac{-0.65 \pm \sqrt{(0.65)^2 - 4(4.3)(-8.7)}}{2(4.3)}$$

The "$\pm$" sign indicates that there are always two possible values for $x$. In this case, they are

$$x = 1.3 \quad \text{and} \quad x = -1.5$$

In any real physical system, however, only one of the values will have any meaning. In this case, if $x$ were, for example $[H_3O^+]$, the negative value would mean a negative concentration, which has no meaning.

## Graphing Data in the Form of a Straight Line

Visualizing changes in variables by means of a graph is a very useful technique in science. In many cases, it is most useful if the data can be graphed in the form of a straight line. Any equation will appear as a straight line if it has, or can be rearranged to have, the following general form:

$$y = mx + b$$

where $y$ is the dependent variable (typically plotted along the vertical axis), $x$ is the independent variable (typically plotted along the horizontal axis), $m$ is the slope of the line, and $b$ is the intercept of the line on the $y$ axis. The intercept is the value of $y$ when $x = 0$:

$$y = m(0) + b = b$$

The slope of the line is the change in $y$ for a given change in $x$:

$$\text{Slope } (m) = \frac{y_2 - y_1}{x_2 - x_1} = \frac{\Delta y}{\Delta x}$$

The *sign* of the slope tells the *direction* of the line. If $y$ increases as $x$ increases, $m$ is positive, and the line slopes upward with higher values of $x$; if $y$ decreases as $x$ increases, $m$ is negative, and the lines slopes downward with higher values of $x$. The *magnitude* of the slope indicates the *steepness* of the line. A line with $m = 3$ is three times as steep ($y$ changes three times as much for a given change in $x$) as a line with $m = 1$.

Consider the linear equation $y = 2x + 1$. A graph of this equation is shown in Figure A.1. In practice, you can find the slope by drawing a right triangle to the line, using the line as the hypotenuse. Then, one leg gives $\Delta y$ and the other gives $\Delta x$. In the figure, $\Delta y = 8$ and $\Delta x = 4$.

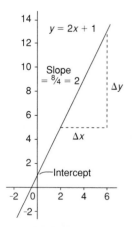

**FIGURE A.1**

At several places in the text, an equation is rearranged into the form of a straight line in order to determine information from the slope and/or the intercept. For example, in Chapter 15, we obtain the following expression:

$$\ln \frac{[A]_0}{[A]_t} = kt$$

Based on the properties of logarithms, we have

$$\ln [A]_0 - \ln [A]_t = kt$$

Rearranging into the form of an equation for a straight line gives

$$\ln [A]_t = -kt + \ln [A]_0$$
$$y = mx + b$$

Thus, a plot of $\ln [A]_t$ versus $t$ is a straight line, from which you can see that the slope is $-k$ (the negative of the rate constant) and the intercept is $\ln [A]_0$ (the natural logarithm of the initial concentration of $A$).

At many other places in the text, linear relationships occur that were not shown in graphical terms. For example, the conversion of temperatures scales in Chapter 1 can also be expressed in the form of a straight line:

$$°F = \frac{9}{5}°C + 32$$
$$y = \ mx \ + b$$

# Standard Thermodynamic Values for Selected Substances at 298 K (25°C)

| SUBSTANCE OR ION | $\Delta H_f^o$ (kJ/mol) | $\Delta G_f^o$ (kJ/mol) | $S^o$ (J/mol·K) |
|---|---|---|---|
| $e^-(g)$ | 0 | 0 | 20.87 |
| Aluminum | | | |
| $Al(s)$ | 0 | 0 | 28.3 |
| $Al^{3+}(aq)$ | −524.7 | −481.2 | −313 |
| $AlCl_3(s)$ | −704.2 | −628.9 | 110.7 |
| $Al_2O_3(s)$ | −1676 | −1582 | 50.94 |
| Barium | | | |
| $Ba(s)$ | 0 | 0 | 62.5 |
| $Ba(g)$ | 175.6 | 144.8 | 170.28 |
| $Ba^{2+}(g)$ | 1649.9 | — | — |
| $Ba^{2+}(aq)$ | −538.36 | −560.7 | 13 |
| $BaCl_2(s)$ | −806.06 | −810.9 | 126 |
| $BaCO_3(s)$ | −1219 | −1139 | 112 |
| $BaO(s)$ | −548.1 | −520.4 | 72.07 |
| $BaSO_4(s)$ | −1465 | −1353 | 132 |
| Boron | | | |
| $B(\beta$-rhombo-<br>hedral) | 0 | 0 | 5.87 |
| $BF_3(g)$ | −1137.0 | −1120.3 | 254.0 |
| $BCl_3(g)$ | −403.8 | −388.7 | 290.0 |
| $B_2H_6(g)$ | 35 | 86.6 | 232.0 |
| $B_2O_3(s)$ | −1272 | −1193 | 53.8 |
| $H_3BO_3(s)$ | −1094.3 | −969.01 | 88.83 |
| Bromine | | | |
| $Br_2(l)$ | 0 | 0 | 152.23 |
| $Br_2(g)$ | 30.91 | 3.13 | 245.38 |
| $Br(g)$ | 111.9 | 82.40 | 174.90 |
| $Br^-(g)$ | −218.9 | — | — |
| $Br^-(aq)$ | −120.9 | −102.82 | 80.71 |
| $HBr(g)$ | −36 | −53.5 | 198.59 |

| SUBSTANCE OR ION | $\Delta H_f^\circ$ (kJ/mol) | $\Delta G_f^\circ$ (kJ/mol) | $S^\circ$ (J/mol · K) |
|---|---|---|---|
| Cadmium | | | |
| $Cd(s)$ | 0 | 0 | 51.5 |
| $Cd(g)$ | 112.8 | 78.20 | 167.64 |
| $Cd^{2+}(aq)$ | −72.38 | −77.74 | −61.1 |
| $CdS(s)$ | −144 | −141 | 71 |
| Calcium | | | |
| $Ca(s)$ | 0 | 0 | 41.6 |
| $Ca(g)$ | 192.6 | 158.9 | 154.78 |
| $Ca^{2+}(g)$ | 1934.1 | — | — |
| $Ca^{2+}(aq)$ | −542.96 | −553.04 | −55.2 |
| $CaF_2(s)$ | −1215 | −1162 | 68.87 |
| $CaCl_2(s)$ | −795.0 | −750.2 | 114 |
| $CaCO_3(s)$ | −1206.9 | −1128.8 | 92.9 |
| $CaO(s)$ | −635.1 | −603.5 | 38.2 |
| $Ca(OH)_2(s)$ | −986.09 | −898.56 | 83.39 |
| $Ca_3(PO_4)_2(s)$ | −4138 | −3899 | 263 |
| $CaSO_4(s)$ | −1432.7 | −1320.3 | 107 |
| Carbon | | | |
| $C(graphite)$ | 0 | 0 | 5.686 |
| $C(diamond)$ | 1.896 | 2.866 | 2.439 |
| $C(g)$ | 715.0 | 669.6 | 158.0 |
| $CO(g)$ | −110.5 | −137.2 | 197.5 |
| $CO_2(g)$ | −393.5 | −394.4 | 213.7 |
| $CO_2(aq)$ | −412.9 | −386.2 | 121 |
| $CO_3^{2-}(aq)$ | −676.26 | −528.10 | −53.1 |
| $HCO_3^-(aq)$ | −691.11 | 587.06 | 95.0 |
| $H_2CO_3(aq)$ | −698.7 | −623.42 | 191 |
| $CH_4(g)$ | −74.87 | −50.81 | 186.1 |
| $C_2H_2(g)$ | 227 | 209 | 200.85 |
| $C_2H_4(g)$ | 52.47 | 68.36 | 219.22 |
| $C_2H_6(g)$ | −84.667 | −32.89 | 229.5 |
| $C_3H_8(g)$ | −105 | −24.5 | 269.9 |
| $C_4H_{10}(g)$ | −126 | −16.7 | 310 |
| $C_6H_6(l)$ | 49.0 | 124.5 | 172.8 |
| $CH_3OH(g)$ | −201.2 | −161.9 | 238 |
| $CH_3OH(l)$ | −238.6 | −166.2 | 127 |
| $HCHO(g)$ | −116 | −110 | 219 |
| $HCOO^-(aq)$ | −410 | −335 | 91.6 |
| $HCOOH(l)$ | −409 | −346 | 129.0 |
| $HCOOH(aq)$ | −410 | −356 | 164 |
| $C_2H_5OH(l)$ | −277.63 | −174.8 | 161 |
| $C_2H_5OH(g)$ | −235.1 | −168.6 | 282.6 |
| $CH_3CHO(g)$ | −166 | −133.7 | 266 |
| $CH_3COOH(l)$ | −487.0 | −392 | 160 |
| $C_6H_{12}O_6(s)$ | −1274.5 | −910.56 | 212.1 |
| $CN^-(aq)$ | 151 | 166 | 118 |
| $HCN(g)$ | 135 | 125 | 201.7 |
| $HCN(l)$ | 105 | 121 | 112.8 |
| $HCN(aq)$ | 105 | 112 | 129 |
| $CS_2(g)$ | 117 | 66.9 | 237.79 |
| $CS_2(l)$ | 87.9 | 63.6 | 151.0 |
| $CH_3Cl(g)$ | −83.7 | −60.2 | 234 |
| $CH_2Cl_2(l)$ | −117 | −63.2 | 179 |
| $CHCl_3(l)$ | −132 | −71.5 | 203 |
| $CCl_4(g)$ | −96.0 | −53.7 | 309.7 |

*Continued.*

| SUBSTANCE OR ION | $\Delta H_f^o$ (kJ/mol) | $\Delta G_f^o$ (kJ/mol) | $S^o$ (J/mol·K) |
|---|---|---|---|
| $CCl_4(l)$ | −139 | −68.6 | 214.4 |
| $COCl_2(g)$ | −220 | −206 | 283.74 |
| Cesium | | | |
| $Cs(s)$ | 0 | 0 | 85.15 |
| $Cs(g)$ | 76.7 | 49.7 | 175.5 |
| $Cs^+(g)$ | 458.5 | 427.1 | 169.72 |
| $Cs^+(aq)$ | −248 | −282.0 | 133 |
| $CsF(s)$ | −554.7 | −525.4 | 88 |
| $CsCl(s)$ | −442.8 | −414 | 101.18 |
| $CsBr(s)$ | −395 | −383 | 121 |
| $CsI(s)$ | −337 | −333 | 130 |
| Chlorine | | | |
| $Cl_2(g)$ | 0 | 0 | 223.0 |
| $Cl(g)$ | 121.0 | 105.0 | 165.1 |
| $Cl^-(g)$ | −234 | −240 | 153.25 |
| $Cl^-(aq)$ | −167.46 | −131.17 | 55.10 |
| $HCl(g)$ | −92.31 | −95.30 | 186.79 |
| $HCl(aq)$ | −167.46 | −131.17 | 55.06 |
| $ClO_2(g)$ | 102 | 120 | 256.7 |
| $Cl_2O(g)$ | 80.3 | 97.9 | 266.1 |
| Chromium | | | |
| $Cr(s)$ | 0 | 0 | 23.8 |
| $Cr^{3+}(aq)$ | −1971 | — | — |
| $CrO_4^{2-}(aq)$ | −863.2 | −706.3 | 38 |
| $Cr_2O_7^{2-}(aq)$ | −1461 | −1257 | 214 |
| Copper | | | |
| $Cu(s)$ | 0 | 0 | 33.1 |
| $Cu(g)$ | 341.1 | 301.4 | 166.29 |
| $Cu^+(aq)$ | 51.9 | 50.2 | −26 |
| $Cu^{2+}(aq)$ | 64.39 | 64.98 | −98.7 |
| $Cu_2O(s)$ | −168.6 | −146.0 | 93.1 |
| $CuO(s)$ | −157.3 | −130 | 42.63 |
| Fluorine | | | |
| $F_2(g)$ | 0 | 0 | 202.7 |
| $F(g)$ | 78.9 | 61.8 | 158.64 |
| $F^-(g)$ | −255.6 | −262.5 | 145.47 |
| $F^-(aq)$ | −329.1 | −276.5 | −9.6 |
| $HF(g)$ | −273 | −275 | 173.67 |
| Hydrogen | | | |
| $H_2(g)$ | 0 | 0 | 130.6 |
| $H(g)$ | 218.0 | 203.30 | 114.60 |
| $H^+(aq)$ | 0 | 0 | 0 |
| $H^+(g)$ | 1536.3 | 1517.1 | 108.83 |
| Iodine | | | |
| $I_2(s)$ | 0 | 0 | 116.14 |
| $I_2(g)$ | 62.442 | 19.38 | 260.58 |
| $I(g)$ | 106.8 | 70.21 | 180.67 |
| $I^-(g)$ | −194.7 | — | — |
| $I^-(aq)$ | −55.94 | −51.67 | 109.4 |
| $HI(g)$ | 25.9 | 1.3 | 206.33 |
| Iron | | | |
| $Fe(s)$ | 0 | 0 | 27.3 |
| $Fe^{3+}(aq)$ | −47.7 | −10.5 | −293 |
| $Fe^{2+}(aq)$ | −87.9 | −84.94 | 113 |
| $FeO(s)$ | −272.0 | −251.4 | 60.75 |

| SUBSTANCE OR ION | $\Delta H_f^0$ (kJ/mol) | $\Delta G_f^0$ (kJ/mol) | $S^0$ (J/mol·K) |
|---|---|---|---|
| $Fe_2O_3(s)$ | −825.5 | −743.6 | 87.400 |
| $Fe_3O_4(s)$ | −1121 | −1018 | 145.3 |
| Lead | | | |
| $Pb(s)$ | 0 | 0 | 64.785 |
| $Pb^{2+}(aq)$ | 1.6 | −24.3 | 21 |
| $PbCl_2(s)$ | −359 | −314 | 136 |
| $PbO(s)$ | −218 | −198 | 68.70 |
| $PbO_2(s)$ | −276.6 | −219.0 | 76.6 |
| $PbS(s)$ | −98.3 | −96.7 | 91.3 |
| $PbSO_4(s)$ | −918.39 | −811.24 | 147 |
| Lithium | | | |
| $Li(s)$ | 0 | 0 | 29.10 |
| $Li(g)$ | 161 | 128 | 138.67 |
| $Li^+(g)$ | 687.163 | 649.989 | 132.91 |
| $Li^+(aq)$ | −278.46 | −293.8 | 14 |
| $LiF(s)$ | −616.9 | −588.7 | 35.66 |
| $LiCl(s)$ | −408 | −384 | 59.30 |
| $LiBr(s)$ | −351 | −342 | 74.1 |
| $LiI(s)$ | −270 | −270 | 85.8 |
| Magnesium | | | |
| $Mg(s)$ | 0 | 0 | 32.69 |
| $Mg(g)$ | 150 | 115 | 148.55 |
| $Mg^{2+}(g)$ | 2351 | — | — |
| $Mg^{2+}(aq)$ | −461.96 | −456.01 | 118 |
| $MgCl_2(s)$ | −641.6 | −592.1 | 89.630 |
| $MgCO_3(s)$ | −1112 | −1028 | 65.86 |
| $MgO(s)$ | −601.2 | −569.0 | 26.9 |
| $Mg_3N_2(s)$ | −461 | −401 | 88 |
| Manganese | | | |
| $Mn(s, \alpha)$ | 0 | 0 | 31.8 |
| $Mn^{2+}(aq)$ | −219 | −223 | −84 |
| $MnO_2(s)$ | −520.9 | −466.1 | 53.1 |
| $MnO_4^-(aq)$ | −518.4 | −425.1 | 190 |
| Mercury | | | |
| $Hg(l)$ | 0 | 0 | 76.027 |
| $Hg(g)$ | 61.30 | 31.8 | 174.87 |
| $Hg^{2+}(aq)$ | 171 | 164.4 | −32 |
| $Hg_2^{2+}(aq)$ | 172 | 153.6 | 84.5 |
| $HgCl_2(s)$ | −230 | −184 | 144 |
| $Hg_2Cl_2(s)$ | −264.9 | −210.66 | 196 |
| $HgO(s)$ | −90.79 | −58.50 | 70.27 |
| Nitrogen | | | |
| $N_2(g)$ | 0 | 0 | 191.5 |
| $N(g)$ | 473 | 456 | 153.2 |
| $N_2O(g)$ | 82.05 | 104.2 | 219.7 |
| $NO(g)$ | 90.29 | 86.60 | 210.65 |
| $NO_2(g)$ | 33.2 | 51 | 239.9 |
| $N_2O_4(g)$ | 9.16 | 97.7 | 304.3 |
| $N_2O_5(g)$ | 11 | 118 | 346 |
| $N_2O_5(s)$ | −43.1 | 114 | 178 |
| $NH_3(g)$ | −45.9 | −16 | 193 |
| $NH_3(aq)$ | −80.83 | 26.7 | 110 |
| $N_2H_4(l)$ | 50.63 | 149.2 | 121.2 |
| $NO_3^-(aq)$ | −206.57 | −110.5 | 146 |
| $HNO_3(l)$ | −173.23 | −79.914 | 155.6 |
| $HNO_3(aq)$ | −206.57 | −110.5 | 146 |

*Continued.*

| SUBSTANCE OR ION | $\Delta H_f^0$ (kJ/mol) | $\Delta G_f^0$ (kJ/mol) | $S^0$ (J/mol·K) |
|---|---|---|---|
| $NF_3(g)$ | $-125$ | $-83.3$ | 260.6 |
| $NOCl(g)$ | 51.71 | 66.07 | 261.6 |
| $NH_4Cl(s)$ | $-314.4$ | $-203.0$ | 94.6 |
| Oxygen | | | |
| $O_2(g)$ | 0 | 0 | 205.0 |
| $O(g)$ | 249.2 | 231.7 | 160.95 |
| $O_3(g)$ | 143 | 163 | 238.82 |
| $OH^-(aq)$ | $-229.94$ | $-157.30$ | $-10.54$ |
| $H_2O(g)$ | $-241.826$ | $-228.60$ | 188.72 |
| $H_2O(l)$ | $-285.840$ | $-237.192$ | 69.940 |
| $H_2O_2(l)$ | $-187.8$ | $-120.4$ | 110 |
| $H_2O_2(aq)$ | $-191.2$ | $-134.1$ | 144 |
| Phosphorus | | | |
| $P_4(s, white)$ | 0 | 0 | 41.1 |
| $P(g)$ | 314.6 | 278.3 | 163.1 |
| $P(s, red)$ | $-17.6$ | $-12.1$ | 22.8 |
| $P_2(g)$ | 144 | 104 | 218 |
| $P_4(g)$ | 58.9 | 24.5 | 280 |
| $PCl_3(g)$ | $-287$ | $-268$ | 312 |
| $PCl_3(l)$ | $-320$ | $-272$ | 217 |
| $PCl_5(g)$ | $-402$ | $-323$ | 353 |
| $PCl_5(s)$ | $-443.5$ | — | — |
| $P_4O_{10}(s)$ | $-2984$ | $-2698$ | 229 |
| $PO_4^{3-}(aq)$ | $-1266$ | $-1013$ | $-218$ |
| $HPO_4^{2-}(aq)$ | $-1281$ | $-1082$ | $-36$ |
| $H_2PO_4^-(aq)$ | $-1285$ | $-1135$ | 89.1 |
| $H_3PO_4(aq)$ | $-1277$ | $-1019$ | 228 |
| Potassium | | | |
| $K(s)$ | 0 | 0 | 64.672 |
| $K(g)$ | 89.2 | 60.7 | 160.23 |
| $K^+(g)$ | 514.197 | 481.202 | 154.47 |
| $K^+(aq)$ | $-251.2$ | $-282.28$ | 103 |
| $KF(s)$ | $-568.6$ | $-538.9$ | 66.55 |
| $KCl(s)$ | $-436.7$ | $-409.2$ | 82.59 |
| $KBr(s)$ | $-394$ | $-380$ | 95.94 |
| $KI(s)$ | $-328$ | $-323$ | 106.39 |
| $KOH(s)$ | $-424.8$ | $-379.1$ | 78.87 |
| $KClO_3(s)$ | $-397.7$ | $-296.3$ | 143.1 |
| $KClO_4(s)$ | $-432.75$ | $-303.2$ | 151.0 |
| Rubidium | | | |
| $Rb(s)$ | 0 | 0 | 69.5 |
| $Rb(g)$ | 85.81 | 55.86 | 169.99 |
| $Rb^+(g)$ | 495.04 | — | — |
| $Rb^+(aq)$ | $-246$ | $-282.2$ | 124 |
| $RbF(s)$ | $-549.28$ | — | — |
| $RbCl(s)$ | $-435.35$ | $-407.8$ | 95.90 |
| $RbBr(s)$ | $-389.2$ | $-378.1$ | 108.3 |
| $RbI(s)$ | $-328$ | $-326$ | 118.0 |
| Silicon | | | |
| $Si(s)$ | 0 | 0 | 18.0 |
| $SiF_4(g)$ | $-1614.9$ | $-1572.7$ | 282.4 |
| $SiO_2(s)$ | $-910.9$ | $-856.5$ | 41.5 |
| Silver | | | |
| $Ag(s)$ | 0 | 0 | 42.702 |
| $Ag(g)$ | 289.2 | 250.4 | 172.892 |

| SUBSTANCE OR ION | $\Delta H_f^\circ$ (kJ/mol) | $\Delta G_f^\circ$ (kJ/mol) | $S^\circ$ (J/mol · K) |
|---|---|---|---|
| $Ag^+(aq)$ | 105.9 | 77.111 | 73.93 |
| $AgF(s)$ | −203 | −185 | 84 |
| $AgCl(s)$ | −127.03 | −109.72 | 96.11 |
| $AgBr(s)$ | −99.51 | −95.939 | 107.1 |
| $AgI(s)$ | −62.38 | −66.32 | 114 |
| $AgNO_3(s)$ | −45.06 | 19.1 | 128.2 |
| $Ag_2S(s)$ | −31.8 | −40.3 | 146 |
| Sodium | | | |
| $Na(s)$ | 0 | 0 | 51.446 |
| $Na(g)$ | 107.76 | 77.299 | 153.61 |
| $Na^+(g)$ | 609.839 | 574.877 | 147.85 |
| $Na^+(aq)$ | −239.66 | −261.87 | 60.2 |
| $NaF(s)$ | −575.4 | −545.1 | 51.21 |
| $NaCl(s)$ | −411.1 | −384.0 | 72.12 |
| $NaBr(s)$ | −361 | −349 | 86.82 |
| $NaOH(s)$ | −425.609 | −379.53 | 64.454 |
| $Na_2CO_3(s)$ | −1130.8 | −1048.1 | 139 |
| $NaHCO_3(s)$ | −947.7 | −851.9 | 102 |
| $NaI(s)$ | −288 | −285 | 98.5 |
| Strontium | | | |
| $Sr(s)$ | 0 | 0 | 54.4 |
| $Sr(g)$ | 164 | 110 | 164.54 |
| $Sr^{2+}(g)$ | 1784 | — | — |
| $Sr^{2+}(aq)$ | −545.51 | −557.3 | −39 |
| $SrCl_2(s)$ | −828.4 | −781.2 | 117 |
| $SrCO_3(s)$ | −1218 | −1138 | 97.1 |
| $SrO(s)$ | −592.0 | −562.4 | 55.5 |
| $SrSO_4(s)$ | −1445 | −1334 | 122 |
| Sulfur | | | |
| $S$(rhombic) | 0 | 0 | 31.9 |
| $S$(monoclinic) | 0.30 | 0.096 | 32.6 |
| $S(g)$ | 279 | 239 | 168 |
| $S_2(g)$ | 129 | 80.1 | 228.1 |
| $S_8(g)$ | 101 | 49.1 | 430.211 |
| $S^{2-}(aq)$ | 41.8 | 83.7 | 22 |
| $HS^-(aq)$ | −17.7 | 12.6 | 61.1 |
| $H_2S(g)$ | −20.2 | −33 | 205.6 |
| $H_2S(aq)$ | −39 | −27.4 | 122 |
| $SO_2(g)$ | −296.8 | −300.2 | 248.1 |
| $SO_3(g)$ | −396 | −371 | 256.66 |
| $SO_4^{2-}(aq)$ | −907.51 | −741.99 | 17 |
| $HSO_4^-(aq)$ | −885.75 | −752.87 | 126.9 |
| $H_2SO_4(l)$ | −813.989 | −690.059 | 156.90 |
| $H_2SO_4(aq)$ | −907.51 | −741.99 | 17 |
| Tin | | | |
| $Sn$(white) | 0 | 0 | 51.5 |
| $Sn$(gray) | 3 | 4.6 | 44.8 |
| $SnCl_4(l)$ | −545.2 | −474.0 | 259 |
| $SnO_2(s)$ | −580.7 | −519.7 | 52.3 |
| Zinc | | | |
| $Zn(s)$ | 0 | 0 | 41.6 |
| $Zn(g)$ | 130.5 | 94.93 | 160.9 |
| $Zn^{2+}(aq)$ | −152.4 | −147.21 | −106.5 |
| $ZnO(s)$ | −348.0 | −318.2 | 43.9 |
| $ZnS(s$, zinc blende$)$ | −203 | −198 | 57.7 |

# Answers to Selected Problems

## CHAPTER 1

**1.2** (a) Gas
(b) Liquid
(c) Liquid in solid
(d) Gas
(e) Solid
(f) Liquid

**1.5** (a) Reversed with temperature
(b) Not reversed with temperature
(c) Reversed with temperature
(d) Not reversed with temperature

**1.7** (a) Fuel
(b) Wood
(c) Sled at top of hill
(d) Water above the dam

**1.11** Observation was important since it helps you "keep track" of the matter undergoing the change. A gain of mass could only be explained by a loss of "phlogiston," which would have a negative mass.

**1.14** A quantitative observation contains more information than a qualitative one, can be more rigorously checked, and leads more easily to predictions (via mathematical relationships or laws) that can be verified. Observations (c) and (d) are quantitative.

**1.15** A well-designed experiment should test the relationship between at least two variables, all but one controlled, and the remaining one free to vary as the others change. It should test the relationship to see if there is any obvious cause and effect. To be meaningful, the results need to be reproducible.

**1.18** (a) $(1 \text{ ft}/12 \text{ in})^2$
(b) $(10^3 \text{ m}/1 \text{ km})^2$
(c) $(10^2 \text{ cm}/1 \text{ m}) \times (1 \text{ h}/3600 \text{ s})$

**1.22** Heat is a form of energy that flows from one object to another due to a temperature difference—specifically, from a hot object to a colder one. The cold object contains heat, and, after transfer, so does the hot object. Since 65°C is a higher temperature than 65°F, heat would flow from the 65°C object to the 65°F object. In addition, since the quantity of material is the same in both cases (1 L of water), the 65°C sample contains more heat than the 65°F object.

**1.23** 0.144 nm

**1.25** 91.4 m

**1.27** (a) $2.36 \times 10^{-9} \text{ km}^2$
(b) $20

**1.29** 77.3 kg

**1.30** $2.36 \times 10^{15}$ t

**1.31** (a) $5.52 \times 10^3$ kg/m$^3$
     (b) 344 lb/ft$^3$

**1.33** (a) $2.25 \times 10^{-9}$ mm$^3$
     (b) $2.25 \times 10^{-1}$ L/10$^5$ cells

**1.35** (a) 10.95 cm$^3$
     (b) 42.4 g

**1.37** 2.70 g/cm$^3$

**1.39** (a) 20°C; 293 K
     (b) 109 K; 263°F
     (c) −273°C; −459°F

**1.43** (a) $3.6 \times 10^2$ g
     (b) 2 g/dime

**1.45** 24 min

**1.47** An exact number is one that has no uncertainty; it re-
     sults from a count of individual objects or from a defi-
     nition. Statement (b) contains an exact number from a
     count, and (d) contains an exact number from a defi-
     nition.

**1.49** (a) 0.39   (b) 0.039   (c) 0.039$\underline{0}$   (d) $3.\underline{0900} \times 10^4$

**1.52** The uncertainty is conventionally taken to be ±1 in
     the last significant figure, so the actual attendance was
     between 4900 and 5100.

**1.53** (a) 0.0004
     (b) 21.83
     (c) 17

**1.55** $5 \times 10^2$

**1.57** (a) 1.35 m
     (b) $9.70 \times 10^2$ cm$^3$
     (c) 344 cm

**1.59** (a) $1.310000 \times 10^5$
     (b) $4.7 \times 10^{-4}$
     (c) $2.10006 \times 10^5$
     (d) $2.1605 \times 10^3$

**1.61** (a) $4.20 \times 10^{-19}$ J
     (b) $1.42 \times 10^{24}$ molecules
     (c) $1.82 \times 10^5$ J/mol

**1.62** (a) 8.94 g/cm$^3$
     (b) $1.45 \times 10^5$ g · m$^2$/s$^2$
     (c) $9.8 \times 10^{-2}$ L/mol

**1.64** (a) I: 6.72 g; II: 6.72 g; III: 6.50 g; IV: 6.56 g
     (b) Data sets I and II are equally accurate, since both
         gave the true value. Average deviations are I: 0.02
         g; II: 0.11 g; III: 0.01 g; IV: 0.11 g.
     (c) Data set I has good accuracy and precision.
     (d) Data set IV has poor accuracy and precision.

**1.65** (a) 420 mi
     (b) $28
     (c) 7.01 h

**1.67** 12 g/cm$^3$; significantly less than density of gold (19.3
     g/cm$^3$), therefore not pure gold

**1.70** (a) 8.40 g/cm$^3$
     (b) 7.45 g/cm$^3$

**1.71** 7.19 g/cm$^3$

**1.72** −3.7°X (freezing point)
     63.4X (boiling point)

**CHAPTER 2**

**2.2** A compound is pure (constant composition), but a
     mixture is impure (variable composition). A com-
     pound has distinctly different properties than its com-
     ponent elements; in a mixture, the components retain
     their individual properties.

**2.4** (a) CaCl$_2$—compound
     (b) S—element
     (c) Baking powder—mixture
     (d) Water—compound

**2.5** If the element is monoatomic (Na, Fe, Ar, etc.), its fun-
     damental particles are atoms; if it is polyatomic (Cl$_2$,
     P$_4$, S$_8$, etc.), the fundamental particles are molecules.

**2.8** The tap water must be a mixture, since it consists of
     some unknown (and almost certainly variable)
     amount of dissolved material in solution in water.

**2.11** (a) All three
     (b) Element, compound
     (c) Compound

**2.13** (a) Law of definite composition: the composition is in-
         dependent of the source.
     (b) Law of conservation of mass: total quantity of mat-
         ter does not change.
     (c) Law of multiple proportions: both materials are
         pure, but one pair of elements can combine in two
         different proportions.

**2.14** (a) No; composition is independent of amount.
     (b) Yes; more compound will contain proportionally
         more of each component element.

**2.17** (a) If chlorine were monatomic, then Avogadro's hy-
         pothesis would be violated.
     (b) Both are diatomic, therefore one atom of chlorine
         for each atom of hydrogen, since equal volumes
         react.
     (c) Assuming 1:1 atom ratio; chlorine atoms must
         weigh 1/(0.0284) = 35.2 times the mass of a hy-
         drogen atom.

**2.18** These two experiments demonstrate the law of defi-
     nite composition. The unknown blue compound de-
     composes the same way in both experiments, giving
     64% white compound and 36% colorless gas. They
     also demonstrate the law of conservation of mass,
     since in both cases, the total mass before reaction
     equals the total mass after reaction.

**2.20** (a) 1.34 g F
     (b) Ca = 0.514
         = 0.486
     (c) mass % Ca = 51.4%; mass % F = 48.6%

**2.22** 21.8%; fluorite is the richer source of calcium.

**2.24** (a) 0.603
     (b) 272 g Mg

**2.26** 163 g Cu, 86 g S

**2.28** Ratio is 2:1, supporting law of multiple proportions.

**2.30** Magnetite is the richer source of iron.

**2.32** If you know the ratio of any two quantities and the
     value of one of them, the other can always be calcu-
     lated. In this case, charge/(charge/mass) = mass.

**2.36** A helium atom would contain four protons and four electrons: the four protons, plus two electrons in the nucleus and the remaining two electrons outside the nucleus.

**2.39** The atomic mass of an element is a weighted average of the atomic masses of the element's isotopes. Usually, this will result in a fractional answer.

**2.41** 0.999726

**2.42** Mass numbers of the isotopes are 24, 25, and 26, respectively, containing 12, 13, and 14 neutrons. All the isotopes contain 12 protons and 12 electrons.

**2.44** (a) Same number of protons and electrons, different number of neutrons; same $Z$
　　　(b) Same number of neutrons, different numbers of protons and electrons; same $N$
　　　(c) Differ in all three; same $A$

**2.46** 69.72 amu

**2.48** 75.774% $^{35}Cl$; 24.226% $^{37}Cl$

**2.51** (a) Elements are arranged in order of increasing atomic number.
　　　(b) Elements in a group (or family) have similar chemical properties.
　　　(c) Elements can be classified as metals, metalloids, or nonmetals.

**2.54** Elements in group 1A(1) are metals; they lose electrons in chemical reactions. Elements in group 7A(17) are nonmetals and generally gain electrons in chemical reactions.

**2.55** (a) Ge; metalloid
　　　(b) S; nonmetal
　　　(c) He; nonmetal
　　　(d) Li; metal
　　　(e) Mo; metal

**2.57** (a) Be; 4
　　　(b) Sb; 51
　　　(c) Cu; 63.55 amu
　　　(d) F; 19.00 amu

**2.60** These atoms will form covalent bonds, in which the atoms share two or more electrons.

**2.62** There are no molecules; KBr is an ionic compound consisting of equal numbers of $K^+$ and $Br^-$ ions.

**2.67** $MgCO_3$ shows both ionic and covalent bonding; covalent bonding between C and O in $CO_3^{2-}$ and ionic between $CO_3^{2-}$ and $Mg^{2+}$.

**2.68** $Cs^+$ and $Br^-$

**2.70** Li forms $Li^+$ and O forms $O^{2-}$, so the compound would have twice as many $Li^+$ ($2.8 \times 10^{20}$) as $O^{2-}$ ($1.4 \times 10^{20}$).

**2.72** Empirical formula shows simplest ratio of atoms, while the molecular formula shows actual number of each atom in the molecule. The molecular formula is always a whole-number multiple of the empirical formula; if 1, then M.F. = E.F.

**2.74** Both contain same total number of each atom. The first is a mixture of two molecules ($H_2$, $O_2$); the second contains molecules of $H_2O_2$.

**2.78** (a) $NH_2$
　　　(b) $CH_2O$

**2.80** (a) $Li_3N$; lithium nitride
　　　(b) $AlCl_3$; aluminum chloride
　　　(c) SrO; strontium oxide

**2.82** (a) $SnCl_4$
　　　(b) Iron(III) bromide
　　　(c) CuBr
　　　(d) Manganese(III) oxide
　　　(e) Sodium hydrogen phosphate
　　　(f) $K_2CO_3 \cdot 2H_2O$
　　　(g) Sodium nitrite
　　　(h) $NH_4ClO_4$

**2.84** (a) BaO. Ba forms $Ba^{2+}$, and O forms $O^{2-}$.
　　　(b) $Fe(NO_3)_2$. Iron(II) is $Fe^{2+}$, and nitrate is $NO_3^-$.
　　　(c) MgS. Magnesium is Mg and forms $Mg^{2+}$, and sulfide is $S^{2-}$.

**2.86** (a) Sulfuric acid, $H_2SO_4$
　　　(b) Iodic acid, $HIO_3$
　　　(c) Hydrocyanic acid, HCN
　　　(d) Hydrosulfuric acid, $H_2S$

**2.88** (a) Dinitrogen pentaoxide
　　　(b) $ClF_3$
　　　(c) Silicon disulfide

**2.90** $P_4S_{10}$, tetraphosphorus decasulfide

**2.92** (a) Carbon monoxide
　　　(b) Sulfur dioxide
　　　(c) Dichlorine monoxide

**2.94** (a) 12; 342.1 amu
　　　(b) 9; 132.06 amu
　　　(c) 8; 344.6 amu

**2.96** (a) $(NH_4)_2SO_4$; 132.1 amu
　　　(b) $NaH_2PO_4$; 119.98 amu
　　　(c) $KHCO_3$; 100.12 amu

**2.98** (a) 108.0 amu
　　　(b) 331.2 amu
　　　(c) 72.08 amu

**2.100** (a) % N 25.94; % O 74.06
　　　(b) % Pb 62.56; % N 8.458; % O 28.98
　　　(c) % Ca 55.61; % O 44.39

**2.103** 26.89% Ag

**2.104** Separation of a mixture is done by physical methods only, while separation of a compound requires chemical reactions.

**2.107** (a) Compound
　　　(b) Homogeneous mixture
　　　(c) Heterogeneous mixture
　　　(d) Homogeneous mixture
　　　(e) Homogeneous mixture

**2.109** (a) Add water; salt dissolves, pepper won't. Filter and evaporate the water to recover the salt.
　　　(b) Add water; sugar dissolves, sand won't. Filter and evaporate the water to recover the sugar.
　　　(c) Heat mixture to boiling; water will evaporate first and can be condensed and collected.
　　　(d) Separatory funnel, drain off lower layer (vinegar).

**2.112** Iodine has more protons in its nucleus (higher $Z$), but iodine atoms must have, on average, fewer neutrons than Te atoms.

**2.114** Mass of reactants equals mass of products, therefore, law of conservation of mass is obeyed. In each case, the product (NaCl) contains 60.66% Cl by mass, so the law of definite composition is obeyed.

**2.115** (a) $Cl^-$: 1.8980%
   $Na^+$: 1.0560%
   $SO_4^{2-}$: 0.2650%
   $Mg^{2+}$: 0.1270%
   $Ca^{2+}$: 0.0400%
   $K^+$: 0.0380%
   $HCO_3^-$: 0.0140%
   (b) 30.715%
   (c) Alkaline-earth cations = 0.1670% (~15% of alkali metal)
      Alkali-metal cations = 1.0940%
   (d) Anions: 2.1770% > cations: 1.2610%

**2.117** 52.00 amu

**2.121** (a) Mercury freezing
   (b) Water decomposing into hydrogen and oxygen
   (c) Sterling silver (95% Ag, 5% Cu)
   (d) $N_2 + O_2$
   (e) Cu, $O_2$, etc.

**CHAPTER 3**

**3.3** "A mole of oxygen" could be interpreted as a mole of oxygen atoms or molecules. Same problem occurs with other diatomic or polyatomic molecules (e.g., $Cl_2$, $Br_2$ $N_2$, $S_8$, $P_4$).

**3.7** (a) 58.32 g/mol $Mg(OH)_2$
   (b) 44.02 g/mol $N_2O$
   (c) 174.27 g/mol $K_2SO_4$
   (d) 149.12 g/mol $(NH_4)_3PO_4$
   (e) 32.05 g/mol $CH_3OH$
   (f) 249.72 g/mol $CuSO_4 \cdot 5H_2O$

**3.9** (a) $1.2 \times 10^2$ g $KMnO_4$
   (b) 0.373 mol O atoms
   (c) $8.0 \times 10^{20}$ O atoms
   (d) $2.8 \times 10^{-5}$ kg $SO_2$
   (e) $1.26 \times 10^{-2}$ mol Cl atoms

**3.11** (a) $1.48 \times 10^3$ g $Ag_2CO_3$
   (b) 0.425 g $N_2O_4$
   (c) 0.472 mol $KClO_3$; $2.84 \times 10^{23}$ formula units $KClO_3$
   (d) $2.84 \times 10^{23}$ $K^+$ ions; $2.84 \times 10^{23}$ $ClO_3^-$ ions; $8.52 \times 10^{23}$ O atoms

**3.13** (a) 6.39% H
   (b) 71.51% O
   (c) 58.02% I

**3.16** (a) 0.4548 mol $Pt(NH_3)_2Cl_2$
   (b) $4.41 \times 10^{24}$ H atoms

**3.19** (a) 0.793 mol $C_3H_8$
   (b) 28.6 g C

**3.21** $KNO_3$      13.86% N      (4)
   $NH_4NO_3$     35.00% N      (2)
   $(NH_4)_2SO_4$  21.20% N      (3)
   $CO(NH_2)_2$   46.65% N      (1)

**3.23** (a) 0.403 mol $Ca^{2+}$
   (b) 49.60% O
   (c) 72.5 metric tons plaster of Paris

**3.24** 4 $Fe^{2+}$ ions/molecule Hb

**3.27** (b) From the mass percentages, determine the empirical formula. Add up the total number of atoms in that, and divide that into the total number of atoms in the molecule. The result is the multiplier to convert the E.F. into the M.F.
   (c) (Mass %) $\left( \dfrac{1 \text{ mol}}{\text{no. of grams}} \right)$ = moles of each element
   (e) Count the numbers of the various types of atoms in the structural formula and place into an M.F.

**3.29** (a) $C_2H_4$ = M.F., therefore $CH_2$ = E.F. (14.03 = E.F. mass)
   (b) $C_2H_6O_2$ = M.F., therefore $CH_3O$ = E.F. (31.04 = E.F. mass)
   (e) $N_2O_5$ = M.F. = E.F. (108.02 = E.F.M. = M.F.M.)
   (d) $Mg_3(PO_4)_2$ = M.F. = E.F. (262.86 = E.F.M. = M.F.M.)

**3.31** (a) M.F. = $C_3H_6$
   (b) M.F. = $N_2H_4$
   (c) M.F. = $N_2O_4$

**3.33** (a) $Cl_2O_7$
   (b) $SiCl_4$
   (c) $CO_2$

**3.35** (a) $N_2O_5$
   (b) $N_2O_5$

**3.37** (a) 1.200 mol F
   (b) 24.0 g M
   (c) Calcium

**3.40** $C_{10}H_{12}N_2O$

**3.43** $C_{10}H_{20}O$

**3.44** Amount (moles) of reactants and products
   Amount (grams) of reactants and products
   Amount (molecules) of reactants and products

**3.47** (a) $16Cu(s) + S_8(s) \rightarrow 8Cu_2S(s)$
   (b) $P_4O_{10}(s) + 6H_2O(l) \rightarrow 4H_3PO_4(l)$
   (c) $B_2O_3(s) + 6NaOH(aq) \rightarrow 2Na_3BO_3(aq) + 3H_2O(l)$
   (d) $4CH_3NH_2(g) + 9O_2(g) \rightarrow$
      $4CO_2(g) + 10H_2O(g) + 2N_2(g)$
   (e) $Cu(NO_3)_2(aq) + 2KOH(aq) \rightarrow$
      $Cu(OH)_2(s) + 2KNO_3(aq)$
   (f) $BCl_3(g) + 3H_2O(l) \rightarrow H_3BO_3(s) + 3HCl(g)$
   (g) $CaSiO_3(s) + 6HF(g) \rightarrow$
      $SiF_4(g) + CaF_2(s) + 3H_2O(l)$

**3.48** (a) $2SO_2(g) + O_2(g) \rightarrow 2SO_3(g)$
   (b) $Sc_2O_3(s) + 3H_2O(l) \rightarrow 2Sc(OH)_3(s)$
   (c) $H_3PO_4(aq) + 2NaOH(aq) \rightarrow$
      $Na_2HPO_4(aq) + 2H_2O(l)$
   (d) $C_6H_{10}O_5(s) + 6O_2(g) \rightarrow 6CO_2(g) + 5H_2O(g)$
   (e) $As_4S_6(s) + 9O_2(g) \rightarrow As_4O_6(s) + 6SO_2(g)$
   (f) $2Ca_3(PO_4)_2(s) + 6SiO_2(s) + 10C(s) \rightarrow$
      $P_4(g) + 6CaSiO_3(s) + 10CO(g)$
   (g) $3Fe(s) + 4H_2O(g) \rightarrow Fe_3O_4(s) + 4H_2(g)$

**3.49** (a) $4Ga(s) + 3O_2(g) \xrightarrow{\Delta} 2Ga_2O_3(s)$

(b) $2C_5H_{10}(l) + 15O_2(g) \xrightarrow{\Delta} 10CO_2(g) + 10H_2O(g)$

(c) $3CaCl_2(aq) + 2Na_3PO_4(aq) \rightarrow$
$$Ca_3(PO_4)_2(s) + 6NaCl(aq)$$

(d) $Si_2Cl_6(l) + 4H_2O(l) \rightarrow 2SiO_2(s) + 6HCl(g) + H_2(g)$

(e) $3NO_2(g) + H_2O(l) \rightarrow 2HNO_3(aq) + NO(g)$

**3.52** $m$ (g) B $= 5$ g A $\times \dfrac{1 \text{ mole A}}{\text{no. of grams of A}} \times \dfrac{\text{moles B}}{\text{moles A}}$

$$\times \dfrac{\text{no. of grams of B}}{1 \text{ mole of B}} = \text{g B}$$

**3.54** Yields (%) are mass ratios of actual to theoretical values. Since mass and moles are directly proportional, the % yield can be calculated by mass or mole comparisons.

**3.55** (a) $0.0938$ mol $Cl_2$

(b) $6.65$ g $Cl_2$

**3.57** (a) $8.0$ mol $NaNO_3$

(b) $680$ g $NaNO_3$

**3.59** $13.78$ g $H_3BO_3$; $1.347$ g $H_2$

**3.61** $2.01 \times 10^3$ g $Cl_2$

**3.63** (a) $I_2(s) + Cl_2(g) \rightarrow 2ICl(g)$
$$2ICl(g) + 2Cl_2(g) \rightarrow 2ICl_3(g)$$

(b) $I_2(s) + 3Cl_2(g) \rightarrow 2ICl_3(g)$

(c) $1.95 \times 10^4$ g $I_2$

**3.65** (a) $0.105$ mol $CaO$

(b) $0.100$ mol $CaO$

(c) $O_2$ is limiting reagent

(d) $5.61$ g $CaO$

**3.67** $1.58$ (mol $HIO_3$); $277$ g $HIO_3$; $12.3$ g $H_2O$ excess

**3.69** $4.40$ g $CO_2$; $4.80$ g $O_2$ in excess

**3.71** $Al(NO_2)_3$, $3.3$ g; $NH_4Cl$, none; $AlCl_3$, $24.0$ g; $N_2$, $15.1$ g; $H_2O$, $19.5$ g

**3.73** $53\%$ overall yield

**3.75** $98.3\%$ yield

**3.77** $24.5$ g $CH_3Cl$

**3.79** (a) $Fe_2O_3(s) + 3CO(g) \xrightarrow{\Delta} 2Fe(s) + 3CO_2(g)$

(b) $3.01 \times 10^7$ g $CO$

**3.80** $52.9$ g $CF_4$

**3.83** (a) $40.6$ g $O_2$

(b) $1.55$ mol of gaseous products

(c) $3.40 \times 10^{23}$ $H_2O$ molecules

**3.85** $89.8\%$ yield maximum

**3.86** $n$ (mol) $=$ molarity $\times$ liters of solution
$m$ (g) $= n$ (mol) $\times \mathcal{M}$ (g/mol of solute)

**3.89** (a) $1.29$ g $Ca(CH_3COO)_2$

(b) $0.0340$ $M$ KI

(c) $2.56$ mol NaCN

(d) $0.0664$ L

(e) $7.20 \times 10^{25}$ $Cu^{2+}$ ions

**3.91** (a) $0.0270$ $M$ KCl

(b) $0.00540$ $M$ $(NH_4)_2SO_4$

(c) $31.0$ mL

(d) $610.2$ mL

**3.93** (a) $987$ g $HNO_3$

(b) $15.7$ $M$ $HNO_3$

**3.95** $1.3 \times 10^2$ mL HCl

**3.97** $0.873$ g $BaSO_4$

**3.101** $0.0345$ $M$ $KMnO_4$

**3.102** $57.6$ mass % Mg

**3.103** (a) False; same number of units, not atoms

(b) True

(c) False; mole ratio of reactants/products is needed

(d) False; add water to make $1.00$ L

(e) True

**3.108** $8.56$ kg $H_2$

**3.110** $1.11$ $M$ HCl

**3.113** $SrCl_2$, strontium chloride

**3.115** $78.0\%$ C

**3.116** $32.7\%$ C

**3.119** (a) $N_2(g) + O_2(g) \rightarrow 2NO(g)$
$$2NO(g) + O_2(g) \rightarrow 2NO_2(g)$$
$$3NO_2(g) + H_2O(l) \rightarrow 2HNO_3(aq) + NO(aq)$$

(b) $2N_2(g) + 5O_2(g) + 2H_2O(l) \rightarrow 4HNO_3(aq)$

(c) $5.62$ kg $HNO_3$

**3.122** $13.8\%$ $XeF_4$; $86.2\%$ $XeF_6$

## CHAPTER 4

**4.3** Combination and displacement; possibly decomposition

**4.4** (a) $Ca(s) + 2H_2O(l) \rightarrow Ca(OH)_2(aq) + H_2(g)$; displacement

(b) $2NaNO_3(s) \rightarrow 2NaNO_2(s) + O_2(g)$; decomposition

(c) $C_2H_2(g) + 2H_2(g) \rightarrow C_2H_6(g)$; combination

(d) $2HI(g) \rightarrow H_2(g) + I_2(g)$; decomposition

**4.6** (a) Metal + oxygen $\rightarrow$ metal oxide
$$2Mg(s) + O_2(g) \rightarrow 2MgO(s)$$

(b) Nonmetal halide + halogen $\rightarrow$ nonmetal halide with more halogen
$$ClF_3(g) + F_2(g) \rightarrow ClF_5(g)$$

(c) Nonmetal oxide + water $\rightarrow$ oxyacid
$$SO_3(g) + H_2O(l) \rightarrow H_2SO_4(aq)$$

(d) Metal hydroxide $\xrightarrow{\Delta}$ metal oxide + water
$$Ca(OH)_2(s) \xrightarrow{\Delta} CaO(s) + H_2O(g)$$

**4.8** (a) $Ca(s) + Br_2(l) \rightarrow CaBr_2(s)$

(b) $2Ag_2O(s) \xrightarrow{\Delta} 4Ag(s) + O_2(g)$

(c) $Ca(OH)_2(aq) + 2HCl(aq) \rightarrow CaCl_2(aq) + 2H_2O(l)$

(d) $2LiCl(l) \xrightarrow{\text{electr.}} 2Li(l) + Cl_2(g)$

**4.10** (a) $2Cs(s) + I_2(s) \rightarrow 2CsI(s)$

(b) $4Al(s) + 3O_2(g) \rightarrow 2Al_2O_3(s)$

(c) $2SO_2(g) + O_2(g) \rightarrow 2SO_3(g)$

(d) $CO_2(g) + BaO(s) \rightarrow BaCO_3(s)$

**4.12** $0.315$ g $O_2$; $3.95$ g Hg

**4.14** (a) $O_2$ in excess (Li is limiting reagent)

(b) $0.117$ mol $Li_2O$

(c) $3.50$ g $Li_2O$ produced; no Li remaining; $3.13$ g $O_2$ in excess

**4.16** $31.9\%$ $KClO_3$

**4.18** $5.11$ g $C_2H_5OH$; $2.49$ L $CO_2$

**4.20** $2.85$ L HF

**4.23** $1.3 \times 10^6$ g $CaCO_3$

**4.26** Ions must be present, which could come from ionic compounds or from other electrolytes such as acids or bases.

**4.29** Depending on the structure of the molecule, if it does not interact well with water molecules, the compound will not be very soluble in water. If the covalent molecule contains polar groups, they will interact well with the polar solvent water.

**4.32** (a) Molecules of benzene are symmetrical and not much like water, so it would most likely be insoluble in water.
(b) Sodium hydroxide, an ionic compound, would be expected to be soluble in water, (and the solubility rules confirm this).

**4.34** (a) Yes; NaI is a salt and a strong electrolyte.
(b) Yes; HBr is one of the strong acids.

**4.36** (a) 0.50 mol
(b) 0.251 mol
(c) $5.92 \times 10^{-4}$ mol $Li^+$ and $Cl^-$ ions

**4.38** (a) $1.83 \times 10^{23}$ $Al^{3+}$ + $5.49 \times 10^{23}$ $Cl^-$
(b) $3.05 \times 10^{22}$ $Na^+$ + $1.52 \times 10^{22}$ $SO_4{}^{2-}$
(c) $2.16 \times 10^{22}$ $Mg^{2+}$ + $4.31 \times 10^{22}$ $Br^-$

**4.40** (a) 0.35
(b) 0.0035
(c) 0.14

**4.43** 33 mol $Na^+$

**4.44** 17.2 g NaCl, 17.8 g $MgSO_4$

**4.47** Spectator ions (i.e., those that do not change chemically in the reaction) do not appear, since they are not involved in the reaction and are only present to balance charge.

**4.52** (a) Acetic, citric, carbonic
(b) Ammonia, $NH_3$
(c) Strong acids or bases dissociate 100%; weak acids or bases dissociate less than 10% in aqueous solution.

**4.53** (a) Formation of a gas and a nonelectrolyte
(b) Formation of a precipitate and a nonelectrolyte

**4.55** $CH_3COOH(aq) + OH^-(aq) \rightarrow CH_3COO^-(aq) + H_2O(l)$
$H^+(aq) + OH^-(aq) \rightarrow H_2O(l)$
The difference is because $CH_3COOH$ is a weak acid and HCl is strong.

**4.56** (a) No reaction
(b) Silver iodide (AgI) precipitates
(c) Barium carbonate ($BaCO_3$) precipitates
(d) Aluminum phosphate ($AlPO_4$) precipitates

**4.58** (a) Molecular: $Hg_2(NO_3)_2(aq) + 2KI(aq) \rightarrow Hg_2I_2(s) + 2KNO_3(aq)$
Total ionic: $Hg_2{}^{2+}(aq) + 2NO_3{}^-(aq) + 2K^+(aq) + 2I^-(aq) \rightarrow Hg_2I_2(s) + 2K^+(aq) + 2NO_3{}^-(aq)$
Net ionic: $Hg_2{}^{2+}(aq) + 2I^-(aq) \rightarrow Hg_2I_2(s)$
Spectator ions: $K^+$ and $NO_3{}^-$
(b) Molecular: $FeSO_4(aq) + Ba(OH)_2(aq) \rightarrow Fe(OH)_2(s) + BaSO_4(s)$
Total ionic: $Fe^{2+}(aq) + SO_4{}^{2-}(aq) + Ba^{2+}(aq) + 2OH^-(aq) \rightarrow Fe(OH)_2(s) + BaSO_4(s)$
Net ionic: same as total ionic; no spectator ions

**4.60** (a) Molecular: $KOH(aq) + HI(aq) \rightarrow KI(aq) + H_2O(l)$
Total ionic: $K^+(aq) + OH^-(aq) + H^+(aq) + I^-(aq) \rightarrow K^+(aq) + I^-(aq) + H_2O(l)$
Net ionic: $OH^-(aq) + H^+(aq) \rightarrow H_2O(l)$
Spectator ions: $K^+$ and $I^-$

(b) Molecular: $NH_3(aq) + HCl(aq) \rightarrow NH_4Cl(aq)$
Total ionic: $NH_3(aq) + H^+(aq) + Cl^-(aq) \rightarrow NH_4{}^+(aq) + Cl^-(aq)$
Net ionic: $NH_3(aq) + H^+(aq) \rightarrow NH_4{}^+(aq)$
Spectator ion: $Cl^-$

**4.62** The hydrochloric acid reacts with the bound $CO_3{}^{2-}$, releasing $CO_2(g)$:
Molecular: $CaCO_3(s) + 2HCl(aq) \rightarrow CaCl_2(aq) + CO_2(g) + H_2O(l)$
Total ionic: $CaCO_3(s) + 2H^+(aq) + 2Cl^-(aq) \rightarrow Ca^{2+}(aq) + 2Cl^-(aq) + CO_2(g) + H_2O(l)$
Net ionic: $CaCO_3(s) + 2H^+(aq) \rightarrow Ca^{2+}(aq) + CO_2(g) + H_2O(l)$

**4.64** 0.0718 $M$ $Pb^{2+}$

**4.66** 0.03920 $M$ $CH_3COOH$

**4.68** 165 mL $NaHCO_3$

**4.73** (a) $1.34 \times 10^3$ kg AgBr
(b) 1.05 $M$ $K^+$; 0.285 $M$ $Br^-$; 0.765 $M$ $NO_3{}^-$

**4.76** No; the oxidation numbers of N ($-3$), H ($+1$), and Cl ($-1$) remain constant.

**4.79** (a) The sulfur is reduced (oxidation number drops from +6 to +4).
(b) The $H_2SO_4$ transfers a proton to $F^-$; the oxidation number of S is constant at +6.

**4.81** (a) +4
(b) +3
(c) +4
(d) $-3$

**4.83** (a) $-3$
(b) +1
(c) +5
(d) +5
(e) $-3$

**4.85** (a) OA = $MnO_4{}^-$; RA = $H_2C_2O_4$
(b) OA = $NO_3{}^-$; RA = Cu
(c) OA = $H^+$; RA = Sn
(d) OA = $H_2O_2$; RA = $Fe^{2+}$

**4.87** S is in group 6A(16), so its highest possible O.N. = +6 and its lowest possible O.N. = $6 - 8 = -2$.
(a) In $S^{2-}$, S is at its lowest possible O.N. and can only be oxidized.
(b) In $SO_4{}^{2-}$, S is at its highest possible O.N. and can only be reduced.
(c) In $SO_2$, O.N. (S) = +4, so it can be either oxidized or reduced.

**4.89** (a) $8HNO_3(aq) + K_2CrO_4(aq) + 3Fe(NO_3)_2(aq) \rightarrow 2KNO_3(aq) + Cr(NO_3)_3(aq) + 3Fe(NO_3)_3(aq) + 4H_2O(l)$
OA = $K_2CrO_4$; RA = $Fe(NO_3)_2$
(b) $8HNO_3(aq) + 3C_2H_6O(aq) + K_2Cr_2O_7(aq) \rightarrow 2KNO_3(aq) + 3C_2H_4O(l) + 7H_2O(l) + 2Cr(NO_3)_3(aq)$
OA = $K_2Cr_2O_7$; RA = $C_2H_6O$
(c) $6HCl(aq) + 2NH_4Cl(aq) + K_2Cr_2O_7(aq) \rightarrow 2KCl(aq) + CrCl_3(aq) + N_2(g) + 7H_2O(l)$
OA = $K_2Cr_2O_7$; RA = $NH_4Cl$
(d) $KClO_3(aq) + 6HBr(aq) \rightarrow 3Br_2(l) + 3H_2O(l) + KCl(aq)$
OA = $KClO_3$; RA = HBr

**4.91**  (a)  $4.54 \times 10^{-3}$ mol $MnO_4^-$
  (b)  0.0113 mol $H_2O_2$
  (c)  0.384 g $H_2O_2$
  (d)  2.44 % $H_2O_2$
  (e)  $H_2O_2$
**4.93**  0.1621% $C_2H_5OH$
**4.95**  Because the forward and reverse processes continue even after apparent change has ceased
**4.98**  The reaction $2NO + Br_2 \rightleftharpoons 2NOBr$ can proceed in either direction. If NO and $Br_2$ are placed in a container, they will react to form NOBr, and NOBr will decompose to form $NO + Br_2$. Eventually, the concentrations of NO, $Br_2$, and NOBr adjust so that the rates of the forward and reverse reactions become equal, and equilibrium is reached.
**4.101**  (a)  $2CrO_4^{2-}(aq) + 3HSnO_2^-(aq) + H_2O(l) \rightarrow$
           $2CrO_2^-(aq) + 3HSnO_3^-(aq) + 2OH^-(aq)$
           OA $= CrO_4^{2-}$; RA $= HSnO_2^-$
  (b)  $2KMnO_4(aq) + 3NaNO_2(aq) + H_2O(l) \rightarrow$
           $2MnO_2(s) + 3NaNO_3(aq) + 2KOH(aq)$
           OA $= KMnO_4$; RA $= NaNO_2$
  (c)  $4I^-(aq) + O_2(g) + 2H_2O(l) \rightarrow 2I_2(s) + 4OH^-(aq)$
           OA $= O_2$; RA $= I^-$
**4.105**  750 g $SiO_2$; $2.6 \times 10^2$ g $Na_2CO_3$; $1.8 \times 10^2$ g $CaCO_3$
**4.106**  (a)  $2.30 \times 10^3$ g $CO_2$; 0.919 g $CO_2$/g LiOH
  (b)  0.755 g $CO_2$/g $Mg(OH)_2$; 0.846 g $CO_2$/g $Al(OH)_3$
**4.109**  (a)  $H_2(g) + F_2(g) \rightarrow 2HF(g)$
           $2HF(g) + CCl_4(l) \rightarrow CF_2Cl_2(g) + 2HCl(g)$
  (b)  1.21 kg $CF_2Cl_2$

## CHAPTER 5

**5.1**  (a)  Volume of liquid remains constant, but the volume of gas increases to the volume of the larger container.
  (b)  Volume of the container containing the gas would increase when heated, but the volume of the liquid would remain essentially constant when heated.
  (c)  Volume of the liquid remains essentially constant, but the volume of the gas would be reduced.
**5.6**  983 cm $H_2O$
**5.8**  (a)  559 mm Hg
  (b)  1.17 atm
  (c)  3.01 atm
  (d)  121 kPa
**5.10**  0.9539 atm
**5.12**  0.966 atm
**5.15**  (a)  0.36 atm
  (b)  6.3 atm
  (c)  90.3 atm
  (d)  32.1 atm
**5.16**  (a)  $1.03 \times 10^4$ kg
  (b)  45.6 cm
**5.19**  Pressure of a gas is directly proportional to the number of moles of gas, assuming $T$ and $V$ are constant.
**5.20**  (a)  Volume is reduced to $\frac{1}{3}$ its original value.
  (b)  Volume is increased by a factor of 2.5.

  (c)  Volume is increased by a factor of 3.
  (d)  Volume is increased by a factor of 4.
**5.22**  $-37.7°C$
**5.24**  56.1 L
**5.26**  0.073 mol
**5.28**  0.249 g
**5.30**  0.42 mol
**5.31**  $6.81 \times 10^4$ g/mol
**5.33**  The molar mass of "dry air" is 29.0 g/mol. Moist air containing water ($H_2O$; M.W. 18.0 g/mol) reduced the average molar mass, and thus it would have a lower density. Molar mass of a gas is directly proportional to the density of a gas.
**5.39**  5.86 g/L (at STP)
**5.41**  1.52 g/L
**5.43**  51.1 g/mol
**5.45**  1.15 atm
**5.47**  0.0168g $N_2$
**5.49**  1.04 g/L at 60°C; 1.16 g/L at 27°C
**5.52**  (a)  0.93 mol
  (b)  6.76 torr
**5.53**  45.8 g $P_4$
**5.55**  41.2 g $PH_3$
**5.57**  0.0250 g Al
**5.59**  71.7 mL
**5.62**  0.0998 atm
**5.63**  303 L $SO_2$
**5.66**  At STP, the volume occupied by a mole of any gas will be identical. This is due to the fact that at the same temperature, all gases have the same average kinetic energy, resulting in the same pressure.
**5.69**  (a)  Pressure: A > B > C
  (b)  $E_k$: A = B = C
  (c)  Diffusion rate: A > B > C
  (d)  Total $E_k$: A > B > C
  (e)  Density: A = B = C
  (f)  Collision frequency: A > B > C
**5.70**  13.2
**5.72**  14.0 min
**5.75**  $Al_2Cl_6$
**5.77**  Intermolecular attractions cause a negative deviation from ideal behavior. Thus, $P_{real\ gases} < P_{ideal\ gases}$. Therefore, $O_2 > Kr > N_2$.
**5.81**  (a)  21.2 atm
  (b)  19.9 atm
**5.84**  35.8 L
**5.87**  (a)  788 L
  (b)  $3.11 \times 10^3$ L
**5.93**  Up 52.5 ft (to 72 ft)
**5.95**  $1.9 \times 10^8$ molecules
**5.98**  0.517 $M$
**5.101**  17.2 g $CO_2$; 17.8 g Kr
**5.104**  (a)  20.5 g $CO_2$; 8.37 g $H_2O$
  (b)  112 g body mass lost
**5.107**  $9.68 \times 10^{22}$ molecules
**5.110**  $X_{Ne} = 0.718$

**CHAPTER 6**

**6.3**  (a) No heat is transferred; work is done on the system.
(b) No work is done; heat is transferred out of the system.
(c) Heat is transferred out of the system; work is done on the system.
(d) Heat is transferred into the system; no work is done.
(e) Heat is transferred into the system; no work is done.

**6.6**  (a) Electric heater
(b) Sound amplifier
(c) Light bulb
(d) Wind turbine
(e) Battery (voltaic)

**6.9**  0 J (zero J)

**6.13**  (a) $3.3 \times 10^7$ kJ
(b) $7.9 \times 10^6$ kcal
(c) $3.1 \times 10^7$ Btu

**6.16**  8.8 h

**6.19**  $\Delta E = q_V$
$\Delta H = q_P$
Determining the heat transfer at constant pressure is more convenient.

**6.21**  (a) Exothermic
(b) Endothermic
(c) Exothermic
(d) Exothermic
(e) Endothermic
(f) Endothermic
(g) Exothermic

**6.24**

Reactants
H          $\Delta H = (-)$
Products

**6.26**  (a) $CH_4(g) + 2O_2(g) \rightarrow CO_2(g) + 2H_2O(g) + Heat$
$CH_4 + 2O_2$
H          $\Delta H = (-)$
$CO_2 + 2H_2O$

(b) $H_2O(l) \rightarrow H_2O(s) + Heat$
$H_2O_{(l)}$
H          $\Delta H = (-)$
$H_2O_{(s)}$

(c) $2Na(s) + Cl_2(g) \rightarrow 2NaCl(s) + Heat$
$2Na + Cl_2$
H          $\Delta H = (-)$
$2NaCl$

**6.28**  Both are one-carbon molecules. Since methane contains more C—H bonds, it will have the greater heat of combustion per mole.

**6.32**  Intensive property; state function since it is defined by only state functions

**6.35**  3.3 kJ

**6.37**  231°C

**6.39**  78°C

**6.41**  40°C

**6.43**  57°C

**6.45**  36.5°C

**6.47**  $-25.7 \dfrac{kJ}{g}$

**6.49**  $\Delta H_{rxn} = (+)$; energy will be required to break the O—O bond.

**6.50**  $\Delta H_{cond} = (-)$; energy is required for the change of state. It would be opposite in sign to and one-half the value for the vaporization of two moles of liquid $H_2O$ to $H_2O$ vapor.

**6.51**  (a) Exothermic
(b) $\Delta H_{rxn}$ (reverse) $= +20.2$ kJ
(c) $\Delta H_{rxn} = -162$ kJ

**6.53**  $-4.51$ kJ

**6.55**  $-2.11 \times 10^6$ kJ

**6.58**  (a) 233 kJ
(b) 608 g Hg

**6.61**  $\Delta H_{rxn}$ is independent of the number of steps on the path of the reaction.

**6.64**  $\Delta H_{rxn} = -813.4$ kJ

**6.66**  $\Delta H_{vap} = 44.0$ kJ

**6.68**  $\Delta H_{rxn} = -124$ kJ

**6.73**  (a) $Ca(s) + Cl_2(g) \rightarrow CaCl_2(s)$
(b) $Na(s) + C(graphite) + \frac{1}{2}H_2(g) + \frac{3}{2}O_2(g) \rightarrow$
$NaHCO_3(s)$
(c) $C(graphite) + 2Cl_2(g) \rightarrow CCl_4(l)$
(d) $\frac{1}{2}N_2(g) + \frac{1}{2}H_2(g) + \frac{3}{2}O_2(g) \rightarrow HNO_3(aq)$

**6.75**  (a) $\Delta H_{rxn} = -1036.8$ kJ
(b) $\Delta H_{rxn} = -433$ kJ

**6.77**  $\Delta H_f^0 = -157.3 \dfrac{kJ}{mol}$

**6.81**  (a) $\Delta H_{rxn}^0 = 503.9$ kJ
(b) $\Delta H_{rxn}^0 = 504$ kJ

**6.83**  (a) $Fe_2O_3(s) + 3CO(g) \rightarrow 2Fe(s) + 3CO_2(g)$
$\Delta H_{rxn}^0 = 21$ kJ

**6.87**  (a) and (b) in both situations; as $H_2O$ freezes, energy is released to the surroundings, thus warming the air temperature.

**6.90**  (a) $6.81 \times 10^3$ J
(b) 243°C

**CHAPTER 7**

**7.2**  (a) X-rays < ultraviolet < visible < infrared < microwave < radio waves
(b) Reverse of the above order
(c) X-rays: view bone structures in the body
Ultraviolet: tanning of the skin
Visible: vision
IR: improve night vision

Microwave: cooking food (microwave oven)
Radio waves: communications

**7.6** $\lambda = 4.23 \times 10^2$ m $= 4.23 \times 10^{11}$ nm $= 4.23 \times 10^{12}$ Å

**7.8** $2.3 \times 10^{-23}$ J

**7.10** Red < yellow < blue

**7.13** (a) $3.1 \times 10^{13}$ s$^{-1}$
(b) $3.47$ $\mu$m
(c) $\lambda = 5.120 \times 10^3$ nm $= 5.120 \times 10^4$ Å; $\nu = 5.86 \times 10^{13}$ Hz

**7.14** $\nu = 3.22 \times 10^{20}$ Hz; $\lambda = 9.32 \times 10^{-13}$ m

**7.16** (a) $5.40 \times 10^2$ nm
(b) $2.62 \times 10^2$ nm
(c) $4.51 \times 10^2$ nm

**7.20** The theoretical basis for Bohr's work was the quantum theory, developed by Max Planck and Albert Einstein, which states that energy is quantized.

**7.22** (a) Absorption
(b) Emission
(c) Emission
(d) Absorption

**7.24** The energies in a one-electron system are proportional to $Z^2$ ($Z$ = atomic number). The energy levels for He$^+$ would be four times the magnitude of those for H, so the pattern of lines would be similar, but their energies would be different.

**7.25** 434.18 nm

**7.27** 121.57 nm

**7.29** $2.76 \times 10^5$ J/mol

**7.31** (d) < (a) < (c) < (b)

**7.33** $n = 4$

**7.41** (a) $7.58 \times 10^{-37}$ m
(b) $\Delta X \geq 1.2 \times 10^{-35}$ m

**7.43** $2.2 \times 10^{-26}$ m/s

**7.45** $3.75 \times 10^{-36}$ kg/photon

**7.50** (a) Energy and distance from nucleus
(b) Shape of orbital
(c) Orientation of orbital

**7.51** (a) One
(b) Five
(c) Three
(d) Nine

**7.53** (a) $m_l = -2, -1, 0, 1, 2$
(b) $m_l = 0$
(c) $m_l = -3, -2, -1, 0, 1, 2, 3$

**7.55** (a)

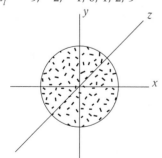

(b)

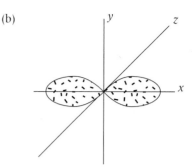

(c)

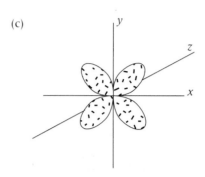

**7.57** (a) $d$; $-2, -1, 0, 1, 2$; 5
(b) $p$; $-1, 0, 1$; 3
(c) $f$; $-3, -2, -1, 0, 1, 2, 3$; 7

**7.59** (a) $n = 5$; $l = 0$; 1
(b) $n = 3$; $l = 1$; 3
(c) $n = 4$; $l = 3$; 7

**7.61** (a) No; $n = 2, l = 0, m_l = 0$ or $n = 2, l = 1, m_l = -1$
(b) OK
(c) OK
(d) No; $n = 5, l = 2, m_l = +2$ or $n = 5, l = 3, m_l = +3$

**7.63** (a) $\lambda$(electron) $= 2.42 \times 10^{-10}$ m; $\lambda$(proton) $= 1.32 \times 10^{-13}$ m
(b) $\lambda$(electron) $= 9.8 \times 10^{-12}$ m; $\lambda$(proton) $= 2.3 \times 10^{-13}$ m

**7.65** (a) $5.25 \times 10^3$ kJ/mol
(b) $1.18 \times 10^4$ kJ/mol
(c) $4.72 \times 10^4$ kJ/mol

**7.68** (a) $n = 5$
(b) $n = 3$
(c) $\lambda = 102.6$ nm

**7.71** $6.41 \times 10^{27}$

**7.73** (a) $5.44 \times 10^{14}$ Hz; yellow-green
(b) $4.47 \times 10^{14}$ Hz; red
(c) $6.58 \times 10^{14}$ Hz; blue
(d) $4.62 \times 10^{14}$ Hz; orange-red
(e) $5.09 \times 10^{14}$ Hz; yellow-orange
(f) $4.54 \times 10^{14}$ Hz; red

**7.74** $1.7 \times 10^{18}$ photons

**CHAPTER 8**

**8.2** Today, elements are listed in order of increasing atomic number. This makes a difference in the sequence of elements in only a few cases, since the larger atomic number usually has the larger atomic mass. One exception is iodine, $Z = 53$, with tellurium, $Z = 52$.

**8.4** Within an atom, no two electrons may have the identical four quantum numbers. Within a particular orbital, there can be only two electrons, and they must have paired spins.

**8.7** One orbital is said to be more penetrating if a significant portion of its electron probability density is closer to the nucleus. More penetrating orbitals are less shielded and thus at lower energy than less penetrating orbitals in the same quantum level. A $3p$ orbital (electron) is more penetrating and has a lower energy than a $3d$ orbital (electron).

**8.8** (a) 6
   (b) 10
   (c) 2

**8.11** In degenerate orbitals (those of identical energy), the filling will occur in such a manner that the spins will have the same value. N: $1s^2\ 2s^2\ 2p^3$

| ↑↓ | ↑↓ | ↑ | ↑ | ↑ | or | ↑↓ | ↑↓ | ↓ | ↓ | ↓ |
|----|----|---|---|---|----|----|----|---|---|---|
| 1s | 2s | | 2p | | | 1s | 2s | | 2p | |

**8.14** Outer electrons are the same as valence electrons for the main-group elements. The $d$ electrons are often included among the valence electrons for transition elements.

**8.16** $n = 2,\ l = 1,\ m_l = -1,\ m_s = +\frac{1}{2}$

**8.18** (a) $n = 5,\ l = 0,\ m_l = 0,\ m_s = +\frac{1}{2}$
   (b) $n = 3,\ l = 1,\ m_l = 1,\ m_s = -\frac{1}{2}$
   (c) $n = 5,\ l = 0,\ m_l = 0,\ m_s = +\frac{1}{2}$

**8.20** (a) $1s^2 2s^2 2p^6 3s^2 3p^6 4s^2 3d^{10} 4p^6 5s^1$
   (b) $1s^2 2s^2 2p^6 3s^2 3p^6 4s^2 3d^{10} 4p^2$
   (c) $1s^2 2s^2 2p^6 3s^2 3p^6$
   (d) $1s^2 2s^2 2p^6 3s^2 3p^6 4s^2 3d^{10} 4p^5$
   (e) $1s^2 2s^2 2p^6 3s^2$

**8.22** (a) O; 6A(16); 2
   (b) P; 5A(15); 3
   (c) Cd; 2B(12); 5
   (d) Ni; 8B(10); 4

**8.24** (a) $[Ar]\ 4s^2 3d^2$   $[Ar]$ ↑↓ | ↑ ↑ ☐ ☐ ☐
   (b) $[Ne]\ 3s^2 3p^5$   $[Ne]$ ↑↓ | ↑↓ ↑↓ ↑
   (c) $[Ar]\ 4s^2 3d^3$   $[Ar]$ ↑↓ | ↑ ↑ ↑ ☐ ☐
   (d) $[Xe]\ 6s^2$   $[Xe]$ ↑↓
   (e) $[Ar]\ 4s^2 3d^7$   $[Ar]$ ↑↓ | ↑↓ ↑↓ ↑ ↑ ↑

**8.26**

| | Inner | Outer | Valence |
|-----|-------|-------|---------|
| (a) | 2 | 6 | 6 |
| (b) | 46 | 4 | 4 |
| (c) | 18 | 2 | 2 |
| (d) | 18 | 2 | 8 |
| (e) | 28 | 6 | 6 |

**8.28** (a) B; Al, Ga, In, Tl
   (b) S; O, Se, Te, Po
   (c) La; Sc, Y, Ac
   (d) Se; O, S, Te, Po
   (e) Hf; Ti, Zr, Rf
   (f) Mn; Tc, Re, Uns

**8.31** (a) Mg: [Ne] $3s^2$
   (b) Cl: [Ne] $3s^2 3p^5$
   (c) Mn: [Ar] $4s^2 3d^5$
   (d) Y: [Kr] $5s^2 4d^1$
   (e) Ne: [He] $2s^2 2p^6$

**8.33** Atomic radius increases from top to bottom in a group, whereas ionization energy (IE) decreases from top to bottom. The farther away an electron is from the nucleus, the more easily it is removed.

**8.34** High IEs correspond to elements at the top right of the periodic table, whereas low IEs correspond to elements at the lower left.

**8.37** Group 7A(17) has high IE and negative first-electron affinities. These elements tend to form $-1$ ions.

**8.40** (a) K < Rb < Cs
   (b) O < C < Be
   (c) Cl < S < K
   (d) Mg < Ca < K

**8.42** (a) Ba < Sr < Ca
   (b) B < N < Ne
   (c) Rb < Se < Br
   (d) Sn < Sb < As

**8.44** B: $1s^2 2s^2 2p^1$

**8.46** (a) Na
   (b) Na

**8.49** Metallic character increases from right to left across a period and from top to bottom in a group. This trend is identical to atomic size and opposite to that of ionization energy.

**8.50** Generally, oxides of metals are basic and oxides of nonmetals are acidic. As the metallic character increases, the oxide becomes more basic. Thus, oxide basicity increases from right to left across a period and from top to bottom in a group.

**8.55** (a) Rb
   (b) Ra
   (c) I
   (d) S

**8.57** Acidic solution. $SO_3(g) + H_2O(l) \rightarrow H_2SO_4(aq)$

**8.59** (a) Cl$^-$: $1s^2 2s^2 2p^6 3s^2 3p^6$
   (b) Na$^+$: $1s^2 2s^2 2p^6$
   (c) Ca$^{2+}$: $1s^2 2s^2 2p^6 3s^2 3p^6$
   (d) P$^{3-}$: $1s^2 2s^2 2p^6 3s^2 3p^6$

(e) $Mg^{2+}$: $1s^2 2s^2 2p^6$

(f) $Se^{2-}$: $1s^2 2s^2 2p^6 3s^2 3p^6 4s^2 3d^{10} 4p^6$

**8.61** (a) 0

(b) 3

(c) 0

(d) 1

**8.63** (a), (b), and (d)

**8.65** (a) [Ar] $3d^2$           paramagnetic

(b) [Kr] $4d^{10}$        diamagnetic

(c) [Ar] $3d^6$           paramagnetic

(d) [Kr] $4d^{10}$        diamagnetic

**8.67** (a) $\boxed{\uparrow\downarrow}$   $\boxed{\uparrow\downarrow}\boxed{\uparrow\downarrow}\boxed{\uparrow\downarrow}\boxed{\uparrow}\boxed{\uparrow}$   paramagnetic

(b) $\boxed{\phantom{\uparrow}}$   $\boxed{\uparrow\downarrow}\boxed{\uparrow\downarrow}\boxed{\uparrow\downarrow}\boxed{\uparrow\downarrow}\boxed{\uparrow\downarrow}$   diamagnetic

(c) $\boxed{\uparrow}$   $\boxed{\uparrow\downarrow}\boxed{\uparrow\downarrow}\boxed{\uparrow\downarrow}\boxed{\uparrow\downarrow}\boxed{\uparrow}$   paramagnetic

**8.69** (a) $Li^+ < Na^+ < K^+$; size increases with increasing quantum level.

(b) $Rb^+ < Br^- < Se^{2-}$; size decreases with atomic number in an isoelectronic series.

**8.73** (a) Beryllium; $1s^2 2s^2$

(b) Lutetium; $1s^2 2s^2 2p^6 3s^2 3p^6 4s^2 3d^{10} 4p^6 5s^2 4d^{10} 5p^6 6s^2 4f^{14} 5d^1$

(c) Scandium; $1s^2 2s^2 2p^6 3s^2 3p^6 4s^2 3d^1$

(d) Sulfur; $1s^2 2s^2 2p^6 3s^2 3p^4$

(e) Strontium; $1s^2 2s^2 2p^6 3s^2 3p^6 4s^2 3d^{10} 4p^6 5s^2$

(f) Arsenic; $1s^2 2s^2 2p^6 3s^2 3p^6 4s^2 3d^{10} 4p^3$

**8.76** (a) $CaCl_2$; calcium chloride

(b) MgO; magnesium oxide

(c) $ZnF_2$; zinc(II) fluoride

(d) NaCl; sodium chloride

**8.79** $1s^2 2s^2 2p^6 3s^2 3p^6 4s^2 3d^{10} 4p^6 5s^2 4d^{10} 5p^6 6s^2 4f^{14} 5d^{10} 6p^6 7s^2 5f^{14} 6d^{10} 7p^2$

This would probably be a solid metal with a melting point about 400°C.

**8.81** $Ce^{4+}$ is [Xe] and $Eu^{2+}$ is [Xe] $4f^7$. $Ce^{4+}$ has a noble gas configuration, and $Eu^{2+}$ has a half-filled $f$ subshell.

## CHAPTER 9

**9.3** The tendency of the main-group elements to form cations decreases from group 1A(1) to 4A(14), and the tendency to form anions increases from 4A(14) to 7A(17). Group 1A(1) and 2A(2) elements form monovalent and divalent cations, respectively, whereas 6A(16) and 7A(17) elements form divalent and monovalent anions, respectively.

**9.4** (a) Cs

(b) Rb

(c) As

**9.6** (a) Ionic

(b) Covalent

(c) Metallic

(d) Covalent

(e) Covalent

(f) Ionic

**9.8** (a) Rb·

(b) ·G̈e·

(c) :Ï·

(d) ·Ba·

(e) :Ẍe:

**9.11** For a particular arrangement of ions, the lattice energy increases as the charges increase and as the radii decrease.

**9.14** (a) $BaCl_2$

(b) SrO

(c) $AlF_3$

**9.16** (a) 2A(2)

(b) 6A(16)

(c) 1A(1)

**9.18** (a) BaS; Ba and S have larger ion charges than Cs and Cl.

(b) LiCl; Li has a smaller radius than Cs.

(c) CaO; O has a smaller radius than S.

**9.20** Lattice energy = $-788$ kJ (less than for LiF due to the larger radii of Na and Cl)

**9.21** Lattice energy = $-2962$ kJ. Since $Mg^{2+}$ has a higher charge than $Na^+$ and there are more ions present, its lattice energy should be larger.

**9.23** $-336$ kJ

**9.25** The bond energy is the energy required for the process $HCl \rightarrow H + Cl$. Energy is absorbed; $\Delta H$ is positive.

**9.26** As the atoms become larger, the bonds become weaker due to poorer orbital overlap of larger (more diffuse) orbitals.

**9.29** (a) I—I < Br—Br < Cl—Cl

(b) S—Br < S—Cl < S—H

(c) C—N < C=N < C≡N

**9.34** $IE_1$ and EN parallel each other, since both are measures of how tightly an atom holds its electrons.

**9.38** Answer (d), since some "excess" bond energy is expected due to the difference in EN.

**9.41** (a) $\underset{\rightarrow}{N-B} > \underset{\leftarrow}{N-O}$

(b) $\underset{\rightarrow}{S-O} > \underset{none}{C-S}$

(c) $\underset{\leftarrow}{N-H} > \underset{\rightarrow}{N-O}$

**9.43** (a) Nonpolar covalent

(b) Ionic

(c) Polar covalent

(d) Polar covalent

(e) Nonpolar covalent

(f) Polar covalent

$SCl_2 < SF_2 < PF_3$

**9.45** (a) $\underset{\rightarrow}{H-I} < \underset{\rightarrow}{H-Br} < \underset{\rightarrow}{H-Cl}$

(b) $\underset{\rightarrow}{H-C} < \underset{\rightarrow}{H-O} < \underset{\rightarrow}{H-F}$

(c) $\underset{\rightarrow}{S-Cl} < \underset{\rightarrow}{P-Cl} < \underset{\rightarrow}{Si-Cl}$

**9.47** Patterns a, b, d, e, f, and h.

**9.48** (b) and (d). The outermost shell of these two cannot allow for more than one pair of electrons.

**9.50** Oxidation numbers (ONs) and formal charges are both methods of assigning an apparent charge to atoms in a molecule. They differ in that shared electrons are assigned to the more electronegative atom to determine ON, but they are shared equally in determining formal charge. In HF, the ONs are $+1$ and $-1$; the formal charges are both 0. In CO, the ONs are $+2$ and $-2$; the formal charges are $-1$ and $+1$. In general, formal charges are probably closer to the "real thing" because this takes electron sharing into account.

**9.52** (a)

$$:\!\ddot{F}\!:$$
$$|$$
$$:\!\ddot{F}\!-\!Si\!-\!\ddot{F}\!:$$
$$|$$
$$:\!\ddot{F}\!:$$

(b) $:\!\ddot{Cl}\!-\!\ddot{Se}\!-\!\ddot{Cl}\!:$

(c)
$$:\!\ddot{F}\!-\!C\!-\!\ddot{F}\!:$$
$$\|$$
$$:\!\ddot{O}\!:$$

(d)
$$\left[\begin{array}{c} H \\ | \\ H\!-\!P\!-\!H \\ | \\ H \end{array}\right]^{+}$$

(e)
$$:\!\ddot{F}\!:\ :\!\ddot{F}\!:$$
$$|\quad\ |$$
$$C\!=\!C$$
$$|\quad\ |$$
$$:\!\ddot{F}\!:\ :\!\ddot{F}\!:$$

**9.54** (a) $\ddot{O}\!=\!\dot{N}\!-\!\ddot{O}\!: \ \leftrightarrow\ :\!\ddot{O}\!-\!\dot{N}\!=\!\ddot{O}$

(b)
$$:\!\ddot{F}\!:\qquad\qquad :\!\ddot{F}\!:$$
$$|\qquad\qquad\ \ |$$
$$:\!\ddot{O}\!-\!N\!=\!\ddot{O}\ \leftrightarrow\ \ddot{O}\!=\!N\!-\!\ddot{O}\!:$$

(c)
$$:\!\ddot{O}\!-\!H\qquad\qquad :\!\ddot{O}\!-\!H$$
$$|\qquad\qquad\qquad |$$
$$:\!\ddot{O}\!-\!N\!=\!\ddot{O}\ \leftrightarrow\ \ddot{O}\!=\!N\!-\!\ddot{O}\!:$$

**9.56** (a)
$$:\!\ddot{F}\!:\ :\!\ddot{F}\!:$$
$$\diagdown\ /$$
$$:\!\ddot{F}\!\cdots\!I\!\cdots\!\ddot{F}\!:$$
$$|$$
$$:\!\ddot{F}\!:$$
FC = 0 for both

(b)
$$\left[\begin{array}{c} H \\ | \\ H\!-\!Al\!-\!H \\ | \\ H \end{array}\right]^{-}$$
FC = 0 for H, $-1$ for Al

(c) $\ddot{S}\!=\!C\!=\!\ddot{O}$    FC = 0 for all

**9.58** (a) $\left[\ddot{O}\!=\!\ddot{Br}\!=\!\ddot{O}\atop |\atop :\!\ddot{O}\!:\right]^{-}$   (b) $\left[:\!\ddot{O}\!-\!\ddot{S}\!=\!\ddot{O}\atop |\atop :\!\ddot{O}\!:\right]^{2-}$

FC $-1(-O)$          FC $-1(-O)$
$\ 0(=O)$            $\ 0(=O)$
$\ 0$ for Br         $\ 0$ for S
O.N. $+5$ for Br; $-2$ for O    O.N. $+4$ for S; $-2$ for O

**9.60** (a) $H\!-\!B\!-\!H$   electron deficient; 6 e$^-$ in B valence
$\qquad\quad |$        shell
$\qquad\quad H$

(b) $\left[:\!\ddot{F}\!-\!\ddot{As}\!-\!\ddot{F}\!:\atop :\!\ddot{F}\!:\ :\!\ddot{F}\!:\right]^{-}$   expanded As valence shell to 10 e$^-$

(c) $:\!\ddot{Cl}\diagdown_{\ddot{Se}}\diagup\ddot{Cl}\!:\atop :\!\ddot{Cl}\qquad\ddot{Cl}\!:$   expanded Se valence shell to 10 e$^-$

(d) $\left[:\!\ddot{F}\!:\ :\!\ddot{F}\!:\atop :\!\ddot{F}\!-\!P\!-\!\ddot{F}\!:\atop :\!\ddot{F}\!:\ :\!\ddot{F}\!:\right]^{-}$   expanded P valence shell to 12 e$^-$

(e) $:\!\ddot{O}\!-\!\dot{Cl}\!-\!\ddot{O}\!:\atop :\!\ddot{O}\!:$   odd electron; 7e$^-$ in Cl valence shell

**9.62**

$$\left[:\!\ddot{O}\!-\!\overset{:\ddot{O}:}{\underset{:\ddot{O}:}{Cl}}\!-\!\ddot{O}\!:\right]^{-} \ \left[:\!\ddot{O}\!-\!\overset{:O:}{\underset{:\ddot{O}:}{\overset{\|}{Cl}}}\!-\!\ddot{O}\!:\right]^{-} \ \left[:\!\ddot{O}\!-\!\overset{:O:}{\underset{:O:}{\overset{\|}{\underset{\|}{Cl}}}}\!-\!\ddot{O}\!:\right]^{-} \ \left[\ddot{O}\!=\!\overset{:O:}{\underset{:O:}{\overset{\|}{\underset{\|}{Cl}}}}\!-\!\ddot{O}\!:\right]^{-}$$

FC = Cl $+3$      Cl $+2$      Cl $+1$      Cl $0$
$\quad$ O $-1$   $-O\ -1$   $-O\ -1$   $-O\ -1$
$\qquad\qquad\quad =O\ \ 0$   $=O\ \ 0$   $=O\ \ 0$

Last form most important; average bond order = 1.75.

**9.64** $:\!\ddot{Cl}\!-\!Be\!-\!\ddot{Cl}\!: + 2 :\!\ddot{Cl}\!:^{-} \rightarrow \left[:\!\ddot{Cl}\!-\!\overset{:\ddot{Cl}:}{\underset{:\ddot{Cl}:}{Be}}\!-\!\ddot{Cl}\!:\right]^{2-}$

**9.67** $\left[:\!\ddot{F}\diagdown_{\overset{:\ddot{F}:}{Al}}\diagup\overset{\ddot{F}:}{}\atop :\!\ddot{F}\diagup\quad\diagdown\ddot{F}:\atop :\!\ddot{F}\!:\right]^{3-}$

**9.68** $H\!-\!\ddot{O}\!:\ :\!\ddot{O}\!-\!H \rightleftharpoons \left[H\!-\!\ddot{O}\!:\ :\!\ddot{O}:\atop \underset{:\ddot{O}:\ :\ddot{O}:}{\overset{|}{C}\!-\!\overset{\|}{C}} \ \longleftrightarrow\ {H\!-\!\ddot{O}\!:\ :\!\ddot{O}:\atop \underset{:\ddot{O}:\ :\ddot{O}:}{\overset{\|}{C}\!-\!\overset{|}{C}}}\right]^{-} \rightleftharpoons$

$$\left[\underset{:\ddot{O}:\ :\ddot{O}:}{\overset{:\ddot{O}:\ :\ddot{O}:}{C\!-\!C}} \ \leftrightarrow\ \underset{:\ddot{O}:\ :\ddot{O}:}{\overset{:\ddot{O}:\ :\ddot{O}:}{C\!=\!C}} \ \leftrightarrow\ \underset{:\ddot{O}:\ :\ddot{O}:}{\overset{:\ddot{O}:\ :\ddot{O}:}{C\!-\!C}} \ \leftrightarrow\ \underset{:\ddot{O}:\ :\ddot{O}:}{\overset{:\ddot{O}:\ :\ddot{O}:}{C\!-\!C}}\right]^{2-}$$

In $H_2C_2O_4$, there are two short (strong) $C\!=\!O$; two long (weak) $C\!-\!O$. In $HC_2O_4^-$, there is one short $C\!=\!O$, one long $C\!-\!O$, and two C-O bonds of intermediate length and strength. In $C_2O_4^{2-}$, all bonds are intermediate between $C\!=\!O$ and $C\!-\!O$ in length and strength.

**9.72** Bond energies are average values for a bond in a variety of compounds. Heats of formation are specific to one compound.

**9.73** $-172$ kJ

**9.75** $-37$ kJ

**9.78** 800 kJ/mol

**9.80** (a) K is a larger atom than Na, so its electrons are held more loosely and thus its metallic bond strength is weaker.

(b) Be has two valence electrons per atom compared with Li with one. The metallic bond is therefore stronger for Be.

(c) The boiling point is high due to the large amount of energy necessary to separate the metal ions from each other in the electron "sea."

**9.85** (a) H—N̈=N̈—H    H—N̈—N̈—H    :N≡N:
                       |  |
                      H  H

Bond order: $3(N_2)$, $2(N_2H_2)$, $1(N_2H_4)$
Strength: $N_2 > N_2H_2 > N_2H_4$
Length: $N_2 < N_2H_2 < N_2H_4$

(b) H—N̈—N̈=N̈—N̈—H    $\Delta H_{rxn} = -367$ kJ/mol
    |          |
    H        H

**9.88** $\left[\overset{..}{C}=N=\overset{..}{O}\right]^- \leftrightarrow \left[:C≡N—\overset{..}{O}:\right]^- \leftrightarrow \left[:\overset{..}{C}—N≡O:\right]^-$
     -2  +1  0         -1  +1  -1        -3  +1  +1

In $CNO^-$, the formal charges are larger, inappropriate based on EN, or the same sign on adjacent atoms. All these would lead to instability; center form most important.

**9.90** 144 kJ/mol

**9.94** (a) −1267 kJ/mol
(b) −1226 kJ/mol
(c) −1236 kJ/mol for $C_2H_5OH(l)$; good correlation

**9.95** (a) H—C=C—H      BO = 1, 2
        C
     H  H

(b)   H  H      BO = 1 (all)
  H—C—C—H
       C
     H  H

(c) H—C=C—H      BO = 1, 2
  H—C—C—H
     H  H

(d) H—C=C—H    H—C—C—H    BO = 1.5
           ↔
  H—C=C—H    H—C—C—H

(e)         H               H      BO = 1.5
(benzene resonance structures) ↔

**9.97** H—Ö: :Ö—H
     |     |
     C—C
     ‖    ‖
    :O: :O:

**9.99** 1150 kJ/mol

**9.101** (a) :F̈—S̈—F̈:     (d)
(b) :F̈—S̈—F̈:
       :F̈:
(c) :F̈—S̈—F̈:
    :F̈.  .F̈:
          (e) :F̈—S̈—F̈:

SF$_3$ and SF$_5$ are unstable due to their odd (unpaired) electron.

**CHAPTER 10**

**10.2** When all electron pairs are involved in bonding (i.e., there are no lone pairs)

**10.3** Three or four electron groups can give rise to a bent molecule, either $AX_2E$ (e.g., $SO_2$) or $AX_2E_2$ (e.g., $H_2O$).
:Ö=S̈ = Ö:      H—Ö—H

**10.7** (a) Tetrahedral; 109.5°; smaller
(b) Linear; 180°; none
(c) Trigonal planar; 120°; none
(d) Tetrahedral; 109.5°; smaller
(e) Linear; 180°; none
(f) Trigonal bipyramidal; 90°, 120°; smaller

**10.8** (a) Trigonal planar; bent; 120°.
(b) Tetrahedral; trigonal pyramidal; 109.5°
(c) Tetrahedral; trigonal pyramidal; 109.5°
(d) Tetrahedral; tetrahedral; 109.5°

**10.10** (a) Bent; 109.5°; smaller
(b) Trigonal pyramidal; 90°, 120°; none
(c) Seesaw; 90°, 120°; smaller
(d) Linear; 180°; none

**10.12** (a) C: tetrahedral, 109.5°
       O: bent, < 109.5°
(b) N: trigonal planar, <120°
(c) P: tetrahedral, 109.5°
       O: bent, < 109.5°
In all these cases, when an angle is less than the ideal, it is because of the pressure of lone pairs on the central atom.

**10.14** $OF_2 < NF_3 < CF_4 < BF_3 < BeF_2$

**10.16** (a) H—C—H: 120°; H—C—N: 120°; C—N—O: < 120°; N—O—H: < 109.5°
(b) H—C—H: 109.5°; H—C—O: 109.5°; C—O—H: < 109.5°
(c) H—O—B: < 109.5°; O—B—O: 120°

**10.18** $XeF_2$: linear; $XeF_4$: square planar; $XeF_6$: distorted octahedral

**10.23** The molecule could be trigonal pyramidal or T shaped but could not be trigonal planar.

**10.25** (a) $CF_4$
(b) Only $SCl_2$ and BrCl

**10.27** (a) $SO_2$, because of the lone pair on the S. $SO_3$ is planar and symmetrical.
(b) IF, because ΔEN is greater
(c) $SF_4$, because of the lone pair on the S. $SiF_4$ is symmetrical.
(d) $H_2O$, because ΔEN is greater

**10.29** (a) F would assume the axial positions first. The 120° bond angles at the equatorial positions allow more room for the larger Cl atoms.
(b) $PCl_3F_2$ and $PF_5$

**10.31** :C̈l: H     :C̈l: H     :C̈l: :C̈l:
     |  |       |  |       |  |
     C=C      C=C      C=C
     |  |       |  |       |  |
     H :C̈l:     :C̈l: H     H  H
        X            Y           Z

Compound Y would have a dipole moment.

**10.33** (a) $sp^2$
(b) $sp^3$
(c) $sp^3d$
(d) $sp^3d^2$

**10.35** (a) False; a double bond is one $\sigma$ and one $\pi$ bond.
     (b) False; a triple bond is one $\sigma$ and two $\pi$ bonds.
     (c) and (d) True
     (e) False; a $\pi$ bond consists of a second pair of electrons after a $\sigma$ bond has formed.

**10.36** (a) $sp^2$
     (b) $sp^2$
     (c) $sp^2$
     (d) $sp^2$

**10.38** (a) one $3s$ + three $3p$
     (b) one $2s$ + one $2p$
     (c) one $2s$ + three $2p$

**10.40**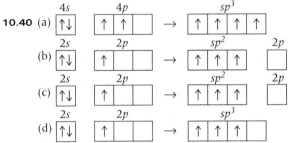

**10.42** (a) N has $sp^2$ hybridization, 3 $\sigma$ and 1 $\pi$ bonds.
     (b) C has $sp$ hybridization, 2 $\sigma$ and 2 $\pi$ bonds.
     (c) C has $sp^2$ hybridization, 3 $\sigma$ and 1 $\pi$ bonds.
     (d) N has $sp^2$ hybridization, 2 $\sigma$ and 1 $\pi$ bonds.
     (e) C has $sp^2$ hybridization, 3 $\sigma$ and 1 $\pi$ bonds.

**10.45**

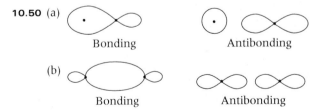

All single bonds are $\sigma$ bonds; all double bonds are 1 $\sigma$ and 1 $\pi$.

**10.47** (a) 2
     (b) 2
     (c) 4
     (d) 12

**10.48** (a) B.O. < A.O.
     (b) B.O.; none perpendicular to the bond axis
     (c) B.O. > A.O.

**10.50** (a)

Bonding      Antibonding

     (b)

Bonding      Antibonding

**10.52** (a) Yes
     (b) No

**10.54** (a) $C_2^+ < C_2 < C_2^-$
     (b) $C_2^- < C_2 < C_2^+$

**10.58** (a) C (ring): $sp^2$; C (others): $sp^3$; O, N: all $sp^3$
     (b) 26
     (c) 6 $\pi$ electrons

**10.59**

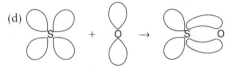

In 1,1-dichloroethane, the two C—Cl bond dipoles reinforce one another, whereas in 1,2-dichloroethane, they partially cancel, giving 1,1-dichloroethane the larger dipole moment.

**10.63** (a) $N_2$: $sp$; $N_2H_4$: $sp^3$, $N_4H_4$: $sp^3$ (end), $sp^2$ (inner)
     (b) Length: N≡N < N=N < N—N
     Energy: N—N < N=N < N≡N

**10.64** (a) Linear    (b) Dipole moment

**10.67** The lone pair in $NO_2^-$ has a greater steric effect than the bond pairs, which in turn have a greater steric effect than the single electron in $NO_2$.

**10.70** (a) The structure with the double bonds has smaller formal charges and is therefore preferred.
     (b) Tetrahedral; $sp^3$
     (c) S: $3d$; O: $2p$

     (d)

**10.75** F has a much higher ionization energy, electron affinity, and electronegativity than Li. You would therefore expect that the F end would be much more negative than the Li end. (The $2s$ orbital of F and $2p$ orbital of Li are not shown.)

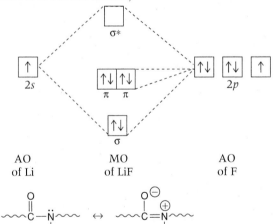

AO of Li      MO of LiF      AO of F

**10.76**

Resonance interaction gives the C—N bond partial double bond character, which hinders its rotation. The $\pi$ bond between C and O is exchanged for a $\pi$ bond between C and N.

**CHAPTER 11**

**11.1** (a) Energy of attraction is less than energy of motion.
     (b) Energy of attraction is greater than energy of motion.

**11.4** (a) Because the intermolecular forces are only partially overcome in fusion and need to be totally overcome in vaporization
     (b) Because greater intermolecular forces are found in solids as opposed to liquids

(c) $\Delta H_{cond} = -\Delta H_{vap}$

**11.5**  (a) and (b) Intermolecular

(c) and (d) Intramolecular

**11.7**  (a) Condensation

(b) Fusion

(c) Evaporation

**11.10**  The density of solids is higher, and the mobility of particles into or through (i.e., diffusion) is very slow. The oxidation of $Fe^{2+}$ to $Fe^{3+}$ requires $O_2$ to pass into the solid.

**11.13**  It is an electrostatic force, between partial charges within molecules

**11.16**  All molecules exhibit dispersion forces. Since these are so weak, however, they are generally overwhelmed if any other intermolecular forces (dipole-dipole, H-bonding) exist.

**11.17**  *Polarity* refers to a permanent imbalance in distribution of electrons in the molecule.
*Polarizability* refers to the ability of the electron distribution in a molecule to change temporarily. The polarity affects dipole-dipole interactions, while the polarizability affects dispersion forces.

**11.19**  (a) H-bonding

(b) Dispersion

(c) Ionic bonds

(d) H-bonding

(e) Dipole-dipole

**11.21**  (a) $CH_3\ CH\ CH_3$
           |
           $OH$

(b) HF

**11.23**  (a) H—F:  (b) [structure of methanol H-bonded cluster with CH₃, O, H atoms]

:F—H

**11.25**  (a) $I^-$

(b) $C_2H_4$

(c) $H_2Se$

**11.27**  (a) LiCl; ionic bonds (LiCl) vs. dipole-dipole (HCl)

(b) $NH_3$; H bonding ($NH_3$) vs. dipole-dipole ($PH_3$)

(c) $I_2$; larger molecule has larger dispersion forces.

(d) $CH_3CH_2OH$; H bonding ($CH_3CH_2OH$) vs. dispersion ($CH_3CH_2CH_3$)

(e) No; dipole-dipole (NO) vs. dispersion ($N_2$)

**11.30**  Both the intrachain and the interchain forces are H bonding.

**11.35**  Because it is defined as the amount of energy required to increase the surface area by a specific amount

**11.37**  $CH_3CH_2CH_2OH < HOCH_2CH_2OH < HOCH_2CH(OH)CH_2OH$

**11.39**  Viscosity and surface tension both increase with the strength of intermolecular forces, so the order would be the same.

**11.41**  Apple juice (primarily water) is held together by H bonding; cooking oil is held together by dispersion forces. The adhesion of the paper and water would be stronger (H bonds in paper to H bonds in water) than that between paper and cooking oil.

**11.42**  As the temperature increases, the shape of the additive molecules changes from spherical to long and stringy. The strings entangle with the oil molecules and increase the viscosity of the mixture.

**11.47**  When the unit cell is repeated infinitely in all directions, the crystal lattice is formed.

**11.50**  In atomic solids, the only interparticle forces are (weak) dispersion forces. In metallic solids, additional forces (metallic bonds) lead to different properties.

**11.52**  Only at a certain angle does the "extra" path length traveled by one of the rays equal an integral number of wavelengths of the incident x-rays.

**11.55**  (a) Conductivity decreases with temperature.

(b) Conductivity increases with temperature.

(c) Conductivity does not change with temperature.

**11.57**  (a) Simple

(b) Body centered

(c) Face centered

**11.59**  (a) Metallic, since Sn is a metal

(b) Network covalent, since Si is similar to C (diamond)

(c) Atomic, since Xe is a monatomic element

(d) Molecular, since $C_{27}H_{45}OH$ is a molecule

(e) Ionic, since KCl is an ionic compound

(f) Network covalent, since BN is isoelectronic with C (diamond)

**11.61**  4

**11.63**  (a) 4

(b) $9.66 \times 10^{-22}$ g

(c) $1.78 \times 10^{-22}$ cm³

(d) $5.63 \times 10^{-8}$ cm (5.63 Å)

**11.65**  (a) Increase (metalloid)

(b) Little effect (nonmetal)

(c) Decrease (metal)

**11.67**  (a) Insulator

(b) *n*-Type semiconductor

(c) Conductor

**11.69**  $1.69 \times 10^{-8}$ cm (1.69 Å)

**11.70**  $1.28 \times 10^{-8}$ cm (1.28 Å)

**11.73**  The gaseous $PCl_3$ molecules are moving faster than and are farther apart than the liquid molecules. As they condense, the kinetic energy is changed into potential energy stored in the dipole-dipole interactions between the molecules.

**11.76**  (a) Critical temperature increases.

(b) Boiling point increases.

(c) Vapor pressure decreases.

(d) Heat of vaporization increases.

**11.79**  If the solid is more dense than the liquid, the solid-liquid line slopes to the right; if less dense, to the left.

**11.80**  (a) $C_2H_6$; smaller molecules have smaller intermolecular forces.

(b) $CH_3CH_2F$; weaker intermolecular force (dipole-dipole vs. H bonding in $CH_3CH_2OH$)

(c) $PH_3$; weaker intermolecular force (dipole-dipole vs. H bonding in $NH_3$)

**11.82**   $4.16 \times 10^3$ J

**11.84**   591 torr

**11.86**   $1.9 \times 10^4$ J/mol (19 kJ/mol)

**11.88**   Solid is more dense.

**11.90**   Condensation of steam releases a much greater quantity of heat (40.7 kJ/mol) than does the cooling of water.

**11.91**   259 K ($-14°C$)

**11.100** (a) Ionic bonds between $Na^+$ and $Cl^-$ and H bonds between water molecules

       (b) Dispersion forces between Kr atoms

       (c) H bonds between water molecules

       (d) Dispersion forces between $CO_2$ molecules

**11.104** (a) 52.4%

       (b) 74.0%

**11.105** (a) A: solid; E: solid + liquid; F: liquid + gas; H: liquid; B: liquid + solid + gas

       (b) Critical point: D; triple point: B

       (c) BG

       (d) Substance is a solid, which melts and then boils.

       (e) Substance is a liquid, which vaporizes.

       (f) Liquid is more dense than the solid.

**11.109** At is adjacent to the "staircase" line, so it would have more metallic properties than the lighter elements in Group 7A(17). Its larger size and higher polarizibility would make its electrons more mobile and enhance any metallic (or metalloidal) properties such as conductivity.

**11.114** 45.98 amu or $7.679 \times 10^{-23}$ g

## CHAPTER 12

**12.4**   In $CH_3(CH_2)_n COOH$, as $n$ increases, the hydrophobic portion of the carboxylic acid increases and the hydrophilic part of the molecule stays the same, with a resulting decrease in water solubility.

**12.5**   Sodium stearate is more effective as a soap. The long-chain stearate ion has both a hydrophilic carboxylate and a hydrophobic hydrocarbon chain, which allows it to reduce surface tension between oil and water and to act as a soap. The acetate ion is too small to exhibit any hydrophobic qualities.

**12.8**   Hydrogen chloride (HCl) gas is actually reacting with the solvent (water) and thus shows a higher solubility than propane ($C_3H_8$) gas, which does not react, even though HCl has a lower boiling point.

**12.9**   (a) In water, $KNO_3$ ionizes completely and forms more concentrated solutions. $KNO_3$, an ionic salt, does not dissolve in or interact well with the nonpolar carbon tetrachloride.

**12.11** (a) Ion-dipole forces

       (b) Dipole-dipole forces

       (c) Dipole-induced dipole forces

       (d) Substitutional or interstitial diffusion forces

       (e) Dispersion forces

       (f) Dispersion forces

**12.13** (a) $HCl(g)$

       (b) $CH_3CHO$

       (c) $CH_3CH_2MgBr$

       Molecular interactions are greater for (a), (b), and (c).

**12.20** The solution cycle for ionic compounds in water consists of two enthalpy terms: the negative of the lattice energy and the combined heats of hydration of cation and anion:

$$\Delta H_{soln} = -\Delta H_{lattice} + \Delta H_{hydration \ of \ ions}$$

For a heat of solution to be zero (or very small):

$$\Delta H_{lattice} \approx \Delta H_{hydration \ of \ ions}$$

**12.22** (a) Endothermic

       (b) Lattice energy term is much larger than the combined ionic heats of hydration.

       (c) Increase in entropy outweighs the increase in enthalpy.

**12.23**

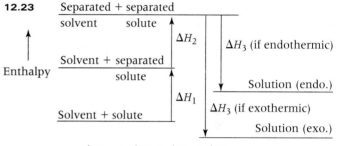

$$\Delta H_{soln} = \Delta H_1 + \Delta H_2 + \Delta H_3$$

$$\Delta H_{soln} < 0 \ (\text{exothermic})$$

$$\Delta H_{soln} > 0 \ (\text{endothermic})$$

**12.25** (a) $Na^+$ has a greater charge density than $Cs^+$. ($Na^+$ has a smaller ion volume)

       (b) $Sr^{2+}$ (has a larger ion charge than $Rb^+$)

       (c) $Na^+$ (has a smaller ion volume than $Cl^-$)

**12.27** (a) $Mg^{2+}$

       (b) $Mg^{2+}$

       (c) $CO_3^{2-}$

**12.29** (a) $\Delta H_{hyd} = -704$ kJ/mol

       (b) Probably $K^+$ due to its smaller size (larger charge density)

**12.31** (a) Entropy increases.

       (b) Entropy decreases.

       (c) Entropy increases.

**12.34** Add a pinch of solid solute to each solution. The supersaturated solution is unstable, and addition of a "seed" crystal of solute causes the excess solute to crystallize immediately, leaving behind a saturated solution. The solution in which the added solid solute dissolves is the unsaturated solution of $X$. The solution in which the added solid solute remains undissolved is the saturated solution of $X$.

**12.38** $NaI(s) + H_2O(l) \rightleftharpoons NaI(aq) + \text{heat}$

In any equilibrium, an increase in temperature favors the endothermic process. Therefore, NaI is less soluble at 60°C than at 20°C.

**12.40** (a) $\Delta H_{soln}$ for $AgNO_3$ is positive.

       $AgNO_3(s) + H_2O(l) + \text{heat} \rightleftharpoons AgNO_3(aq)$

       (b) $\Delta H_{soln}$ for $LiCO_3$ is negative.

       $Li_2CO_3(s) + H_2O(l) \rightleftharpoons Li_2CO_3(aq) + \text{heat}$

       (c) $\Delta H_{soln}$ for $O_2$ is negative.

       $O_2(g) + H_2O(l) \rightleftharpoons O_2(aq) + \text{heat}$

**12.41** (a) Endothermic
Exothermic
Endothermic
    (b) $AgC_2H_3O_2(s) + H_2O(l) + Heat \rightleftharpoons AgC_2H_3O_2(aq)$
         $CO_2(g) + H_2O(l) \rightleftharpoons CO_2(aq) + heat$
         Valine$(s) + H_2O(l) + heat \rightleftharpoons$ Valine$(aq)$

**12.42** (a) $8.19 \times 10^{-2}$ g $O_2$
    (b) $1.71 \times 10^{-2}$ g $O_2$

**12.45** (a) Yes, equilibrium is a dynamic process.
    (b) Radioactivity would be found in all the solid.

$$Na_2{}^{14}CO_3(s) + H_2O(l) \rightleftharpoons Na_2{}^{14}CO_3(aq)$$

**12.47** Solubility of gases increases with increasing partial pressure of the gas.

**12.49** (a) %w/v; molarity
    (b) %w/w; mole fraction
    (c) Molality

**12.52** %w/w; mole fraction and molality are weight-to-weight relationships that are not affected by changes in temperature.

**12.53** Both are the same because masses are additive.

**12.54** (a) 1.24 $M$
    (b) 0.158 $M$
    (c) 0.0750 $M$

**12.56** (a) Add 4.22 g $KH_2PO_4(s)$ to enough water to make 355 mL of aqueous solution.
    (b) Add 107 mL of 0.315 $M$ NaOH to enough water to make 425 mL of aqueous solution.

**12.58** 0.942 $m$

**12.60** (a) Add 2.13 g $C_2H_6O_2$ to 298 g $H_2O$.
    (b) Add 32.3 g of 62.0 mass % $HNO_3$ to 968 g $H_2O$.

**12.62** (a) 0.27
    (b) 56% w/w
    (c) 20.8 $m$

**12.64** 33.7 g CsCl; $7.15 \times 10^{-3}$ mole fraction; 6.31 mass %

**12.66** 5.11 $m$; 4.53 $M$; 0.0842 mole fraction

**12.68** $Ca^{2+}$: 2.2 ppm; $Mg^{2+}$: 0.66 ppm

**12.74** Raoult's law states that the vapor pressure of solvent above a solution equals the mole fraction of the solvent times the vapor pressure of the pure solvent. Raoult's law is not valid for a solution of a volatile solute in solution. Both solute and solvent would evaporate based on their respective vapor pressures.

**12.75** The temperature at which the solution boils is higher and the freezing point is lower for the solution compared with the pure solvent.

**12.79** (a) Univalent ions behave more ideally than divalent ions. Ionic strength (which affects "activity" concentration) is greater for divalent ions.
    (b) 0.01 $m$ NaBr has a freezing point closer to its predicted value.

**12.81** (a) Strong electrolyte
    (b) Strong electrolyte
    (c) Nonelectrolyte
    (d) Weak electrolyte

**12.83** (a) 0.4 mol ions
    (b) 0.14 mol ions
    (c) $3 \times 10^{-4}$ mol ions
    (d) 0.07 mol molecules

**12.85** (a) I = II < III
    (b) I = II < III
    (c) III < II = I
    (d) III < II = I

**12.87** 23.4 torr

**12.89** $-0.206°C$

**12.91** 79.0°C

**12.93** 11.3 kg $C_2H_6O_2$

**12.96** 0.240 atm

**12.97** (a) $1.82 \times 10^4$ g/mol
    (b) $-9.87 \times 10^{-6}$ °C

**12.102** Brownian movement is a characteristic movement in which particles change speed and direction erratically.

**12.105** (a) To shorten settling time, lime (CaO) and alum $Al_2(SO_4)_3$ are added to form a fluffy, gel-like precipitate of $Al(OH)_3$.
    (b) Water that contains large amounts of divalent cations (e.g., $Ca^{2+}$, $Mg^{2+}$, $Fe^{2+}$) is called hard water. During cleaning, these ions combine with the fatty acid anions in soap to produce insoluble deposits.

**12.109** $X_{C_{10}H_{16}} = 0.920$; $X_{C_{10}H_{18}O} = 0.080$

**12.114** $M = \dfrac{\text{moles of solute}}{\text{L of soln}}$      $m = \dfrac{\text{moles of solute}}{\text{kg of solvent}}$

$M \times$ vol of soln $= m \times$ kg of solvent

$$M = \frac{m \times \text{kg solvent}}{\text{vol of soln}} = \frac{m \times \text{kg solvent}}{\text{vol of soln} \times \text{density}}$$

As density of a solution approaches 1 g/mL, $M \approx m$. If the solvent is water, assuming a density of 1.000 g/mL, 1 kg $H_2O$ is equal to a volume of 1.0 L. In dilute solutions, the volume of the solute when added to 1 kg of water does not significantly change.

## CHAPTER 13

**13.2** (a) $Z_{eff}$ increases across a period and decreases down a group.
    (b) As you move to the right across a period, the atomic size decreases, the $IE_1$ increases, and the EN increases, all as a result of the increased $Z_{eff}$.

**13.3** (a) ICl is polar; $Br_2$ is nonpolar. The EN of Cl is greater than that of I, so the I end of ICl will be partially positively charged due to the greater "pull" of the Cl on the shared electrons.
    (b) ICl will have a higher boiling point than $Br_2$ since dipole-dipole forces in ICl are greater than dispersion forces in $Br_2$.

**13.6** They are similar in that they involve sharing of electrons between atoms. They are different in that covalent bonding includes sharing between a small number (usually two) of atoms, while metallic bonding involves essentially all the atoms in a given sample.

**13.13** (a) The metallic character of an element and the acidity of its oxide are inversely related.
    (b) In general, oxide basicity decreases as you move to the right in a period and increases as you move down a group.

**13.16** (a) Mg < Sr < Ba
(b) Na < Al < P
(c) Se < Br < Cl
(d) Ba < Sn < Bi

**13.18** (b) and (d)

**13.20** $:\ddot{O}\!\!=\!\!C\!\!=\!\!\ddot{O}:$

**13.22** $CF_4$ < $SiCl_4$ < $GeBr_4$

**13.24** (a) $Na^+$ < $F^-$ < $O^{2-}$
(b) $Cl^-$ < $S^{2-}$ < $P^{3-}$

**13.26** $O$ < $O^-$ < $O^{2-}$

**13.28** (a) $CaO(s) + H_2O(l) \rightarrow Ca(OH)_2(aq)$ (more basic)
$SO_3(g) + H_2O(l) \rightarrow H_2SO_4(aq)$
(b) BaO
(c) $CO_2$
(d) $K_2O$

**13.30** (a) LiCl
(b) $PCl_3$
(c) $NCl_3$

**13.32** (a) Ar < Na < Si
(b) Cs < Rb < Li

**13.36** As oxidizing agent: $2Na(s) + H_2(g) \rightarrow 2NaH(s)$
As reducing agent: $H_2(g) + Cl_2(g) \rightarrow 2HCl(g)$

**13.37** (a) $:\ddot{F}\!\!-\!\!\ddot{N}\!\!-\!\!\ddot{F}:$ and $H\!\!-\!\!\ddot{N}\!\!-\!\!H$
$:\ddot{F}:$                    $H$
$NH_3$ will H bond.
(b) $CH_3\!-\!\ddot{O}\!-\!CH_3$ and $CH_3\,CH_2\!-\!\ddot{O}\!-\!H$
$CH_3CH_2OH$ will H bond.

**13.39** (a) $2\,Al(s) + 6HCl(aq) \rightarrow 2AlCl_3(aq) + 3H_2(g)$
(b) $LiH(s) + H_2O(l) \rightarrow LiOH(aq) + H_2(g)$

**13.41** (a) Na = +1, B = +3, H = −1 in $NaBH_4$
Al = +3, B = +3, H = −1 in $Al(BH_4)_3$
Li = +1, Al = +3, H = −1 in $LiAlH_4$
(b)

tetrahedral

**13.43** In general, the maximum oxidation number increases as you move to the right (max O.N. = [old] group number). In the second period, the maximum oxidation number drops off below the group number in Groups 5A(15), 6A(16), and 7A(17).

**13.47** (a) Group 3B(3) or 3A(13)
(b) If in Group 3B(3), the oxide and fluoride would be more ionic.

**13.50** (a) Reducing agent
(b) Their outermost electrons are easily removed due to their large size and low ionization energy.
(c) $2Na(s) + Cl_2(g) \rightarrow 2NaCl(s)$
$2Na(s) + 2H_2O(l) \rightarrow 2NaOH(aq) + H_2(g)$

**13.52** (a) and (b) increase down the group.
(c), (d), and (e) decrease down the group.

**13.54** $2Na(s) + O_2(g) \rightarrow Na_2O_2(s)$

**13.57** (a) Because you would be "breaking into" the next lower quantum level, removing electrons that are more tightly held than the outermost level
(b) $\Delta H_{rxn} = -810$ kJ

**13.61** The Group 2A(2) metals have an additional electron available for bonding, so the bonding is stronger and the melting points increase. They are also harder, denser, and have higher boiling points.

**13.62** (a) $CaO(s) + H_2O(l) \rightarrow Ca(OH)_2(s)$
(b) $2Ca(s) + O_2(g) \rightarrow 2CaO(s)$

**13.64** (a) $CaCO_3(s) \xrightarrow{\Delta} CaO(s) + CO_2(g)$
(b) $Ca(OH)_2(s) + SO_2(g) \rightarrow CaSO_3(s) + H_2O(l)$
(c) $3CaO(s) + 2H_3AsO_4(aq) \rightarrow$
$Ca_3(AsO_4)_2(s) + 3H_2O(l)$
(d) $Na_2CO_3(s) + CaO(s) + H_2O(l) \rightarrow$
$CaCO_3(s) + 2NaOH(aq)$

**13.67** The pattern of ionization energies in Group 3B(3) is "normal"—a decrease toward the bottom of the table. The pattern in Group 3A(13) is irregular, since the appearance of the transition metals (and the 10 additional protons in their nuclei) causes a contraction of the atoms and a resulting increase in ionization energy.

**13.71** (a) Boron is a metalloid, while the other elements in the group show predominantly metallic behavior. Boron forms covalent bonds exclusively; the others at best occasionally form ions. Boron is also much less chemically reactive in general.
(b) The small size of boron

**13.72** $In_2O_3$ < $Ga_2O_3$ < $Al_2O_3$

**13.74** O.N. = +3 (apparent); = +1 (actual)
$[:\ddot{I}\!\!-\!\!\ddot{I}\!\!-\!\!\ddot{I}:]^-$ Class = $AX_2E_3$; bond angles 180°
$TlI_3$ does not exist as $(Tl^{3+})(I^-)_3$ because of the low strength of the Tl—I bond.

**13.76** (a) B     (b) Ga     (c) Ga     (d) B
(e) Al     (f) $B(OH)_3$     (g) Tl

**13.79** (a) $In(s) + As(s) \rightarrow InAs(s)$
(b) $GaSb(s) + 3H_2O(g) \rightarrow Ga(OH)_3(s) + SbH_3(g)$

**13.89** (a)
$$\left[\begin{array}{c} :\ddot{O}: \quad\quad :\ddot{O}: \\ | \quad\quad\quad | \\ :\ddot{O}\!\!-\!\!Si\!\!-\!\!\ddot{O}\!\!-\!\!Si\!\!-\!\!\ddot{O}: \\ | \quad\quad\quad | \\ :\ddot{O}: \quad\quad :\ddot{O}: \\ | \quad\quad\quad | \\ :\ddot{O}\!\!-\!\!Si\!\!-\!\!\ddot{O}\!\!-\!\!Si\!\!-\!\!\ddot{O}: \\ | \quad\quad\quad | \\ :\ddot{O}: \quad\quad :\ddot{O}: \end{array}\right]^{8-}$$

(b)
$$\begin{array}{c} H \quad H \\ | \quad\ | \\ H\!\!-\!\!C\!\!-\!\!C\!\!-\!\!H \\ | \quad\ | \\ H\!\!-\!\!C\!\!-\!\!C\!\!-\!\!H \\ | \quad\ | \\ H \quad H \end{array}$$

**13.91**
$$\begin{array}{c} CH_3 \quad\ CH_3 \quad\ CH_3 \\ | \quad\quad\ | \quad\quad\ | \\ -\ddot{O}\!\!-\!\!Si\!\!-\!\!\ddot{O}\!\!-\!\!Si\!\!-\!\!\ddot{O}\!\!-\!\!Si\!\!-\!\!CH_3 \\ | \quad\quad\ | \quad\quad\ | \\ CH_3 \quad\ CH_3 \quad\ CH_3 \end{array}$$

**13.93** (a) $2p$ from C and O
(b) Because the $2p$ orbitals on the O and the $3p$ orbitals on the Si are too different in energy
(c) A $3d$ orbital on the Si plus a $2p$ orbital on the oxygen
(d) O

**13.95** (a) $:C\equiv O:$; $[:C\equiv N:]^-$; $[:C\equiv C:]^{2-}$

(b) The three MO diagrams are qualitatively the same; there are only minor quantitative differences not relevant here.

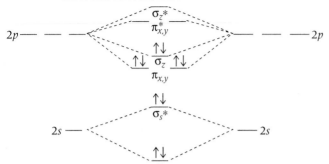

The configuration of all three is $(\sigma_s)^2(\sigma_s^*)^2(\pi_x)^2(\pi_y)^2$ $(\sigma_z)^2$, and all have a bond order of 3.

**13.96** 12 tonne

**13.101** Br is much larger than Cl, and insufficient room exists around a P atom for 6 Br to fit.

**13.104** $H_3AsO_4 < H_3PO_4 < HNO_3$

**13.106** (a) $4As(s) + 5O_2(g) \rightarrow 2As_2O_5(s)$

(b) $Bi_2O_3(s) + 6HCl(aq) \rightarrow 2BiCl_3(aq) + 3H_2O(l)$

(c) $Ca_3As_2(s) + 6H_2O(l) \rightarrow$
$$3Ca(OH)_2(aq) + 2AsH_3(g)$$

**13.108** (a) $N_2(g) + 2Al(s) \xrightarrow{\Delta} 2AlN(s)$

(b) $PF_5(g) + 4H_2O(l) \rightarrow H_3PO_4(aq) + 5HF(g)$

**13.110** A trigonal bypyramid with F axial and Cl equatorial

**13.112** The weaker bonds in the $P_4$ units in white phosphorus would make it the more reactive (and thus more toxic) form in the body.

**13.117** (a) $2KNO_3(s) \xrightarrow{\Delta} 2KNO_2(s) + O_2(g)$

(b) $4KNO_3(s) \rightarrow 2K_2O(s) + 2N_2(g) + 5O_2(g)$

**13.119** $S < Se < Po$; nonmetal < metalloid < metal

**13.122** (a) $NaHSO_4(aq) + NaOH(aq) \rightarrow Na_2SO_4(aq) +$
$H_2O(l)$

(b) $S_8(s) + 24F_2(g) \rightarrow 8SF_6(g)$

(c) $FeS(s) + 2HCl(aq) \rightarrow FeCl_2(aq) + H_2S(g)$

(d) $Te(s) + 2I_2(s) \rightarrow TeI_4(s)$

**13.124** (a) Acidic

(b) Acidic

(c) Basic

(d) Amphoteric

(e) Basic

**13.126** $H_2O < H_2S < H_2Te$

**13.129** (a) $SF_6$

(b) $O_3$

(c) $SO_3$

(d) $SO_2$

(e) $SO_3$, $H_2SO_4$

(f) $Na_2S_2O_3 \cdot 5H_2O$ (g) $H_2S$

(g) $H_2S$

**13.131** (a) Peroxide ion; $[:\ddot{O}-\ddot{O}:]^{2-}$

(b) Bent with bond angle slightly less than 109°

**13.137** (a) Cl—Cl

(b) Br—Br

(c) Cl—Cl

The F—F bond strength is weaker than expected due to the small size of the F atom. This means

that the lone electron pairs on the adjacent atoms repel, weakening the bond.

**13.139** $2BrF_3(l) \rightleftharpoons BrF_2^+(\text{solvated}) + BrF_4^-(\text{solvated})$

**13.140** (a) $3Br_2(l) + 6OH^-(aq) \rightarrow$
$$5Br^-(aq) + BrO_3^-(aq) + 3H_2O(l)$$

(b) $ClF_5(l) + 6OH^-(aq) \rightarrow$
$$5F^-(aq) + ClO_3^-(aq) + 3H_2O(l)$$

**13.141** (a) $2Rb(s) + Br_2(l) \rightarrow 2RbBr(s)$

(b) $I_2(s) + H_2O(l) \rightarrow NR$

(c) $Br_2(l) + 2I^-(aq) \rightarrow I_2(s) + 2Br^-(aq)$

(d) $CaF_2(s) + H_2SO_4(l) \rightarrow CaSO_4(s) + 2HF(g)$

**13.143** $HIO < HBrO < HClO < HClO_2$

**13.146** (a) Iodic acid;

$$2HIO_3(s) \xrightarrow{\Delta} I_2O_5(s) + H_2O(l)$$

**13.147** (a) Oxidation-reduction

(b) Because $I^-$ is more easily oxidized than $Cl^-$ due to its larger size

(c) It would have to be a nonoxidizing acid.

**13.149** Helium; argon

**13.153** (a) So that the resulting ions will have an even number of electrons

(b) Xenon fluorides containing an odd number of fluorine atoms need an odd charge to maintain an even number of electrons around xenon.

(c) $XeF_3^+$ would be T shaped.

**13.155** $\Delta H_{soln} = -411$ kJ

**13.160** (a) $H_2N_2O_2$; HNO

(b) $H-\ddot{O}-\ddot{N}=\ddot{N}-\ddot{O}-H$ ;     $H-\ddot{N}=\ddot{O}:$

(c) Both bent

(d) $\begin{bmatrix} \ddot{N}=\ddot{N} \\ :\ddot{O}. \quad .\ddot{O}: \end{bmatrix}^{2-}$ ; $\begin{bmatrix} :\ddot{O}: \\ N=\ddot{N} \\ .\ddot{O}: \end{bmatrix}^{2-}$

**13.162** (a) Li would not have a complete octet.

(b) 32

(c) Bonding will be electron deficient, since there are too few electrons for "normal" covalent bonds.

**13.163** (a) $F_2(g) + H_2O(s) \rightarrow HOF(g) + HF(g)$
$HOF(g) + F_2(g) \rightarrow OF_2(g) + HF(g)$

(b) Zero in HOF; +2 in $OF_2$

(c) Both bent at the oxygen

**13.166** (a) 5A(15)

(b) 7A(17)

(c) 6A(16)

(d) 1A(1) or 2A(2)

(e) 3A(13)

(f) 8A(18)

**13.168** (a) All covalent network solids involving an atom that can form four bonds

(b) With one Group 3A(13) and one Group 5A(15) element, there are eight valence electrons per two atoms, as in Group 4A(14).

**13.172** (a) $sp^3$ hybrid orbitals

(b) Tetrahedral

(c) $sp$

(d) There are none.

**13.175** $2.29 \times 10^4$ g $UF_6$

**13.177** $\Delta H_{rxn} = -769$ kJ

**13.181** (a) Oxidizing agent, producing $H_2O$
  (b) Reducing agent, producing $O_2$

**CHAPTER 14**

**14.4** (a) C bonds to C, H, O, N, P, S, and halogens.
  (b) Heteroatoms are all atoms other than C or H.
  (c) N, O, F, Cl, and Br are more electronegative; H and P are less electronegative; S and I have the same electronegativity as C.
  (d) Since it can bond to a wide variety of atoms, this means it can form many different compounds.

**14.7** The C—Li bond is the most reactive, since it shows the greatest difference in electronegativity. C—I and C=O would be less so, with C—H and C—C least so.

**14.8** (a) An alkane and a cycloalkane contain only single C—C bonds, while an alkene contains a C—C double bond and an alkyne contains a C—C triple bond. A cycloalkane contains a ring of C atoms.

**14.11** (a), (c), and (f)

**14.13** Aromatic hydrocarbons have C in $sp^2$ hybridization, while cycloalkanes have C in $sp^3$ hybridization.

**14.14** *(a)

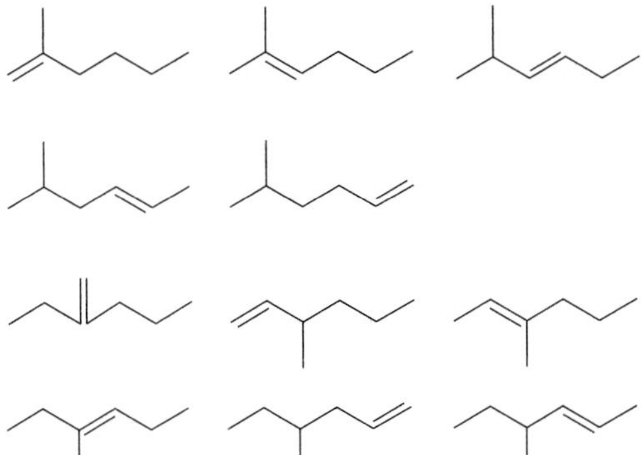

(b)

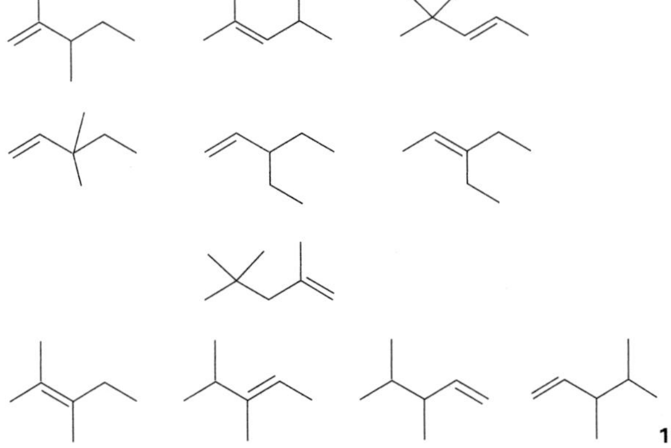

(c)

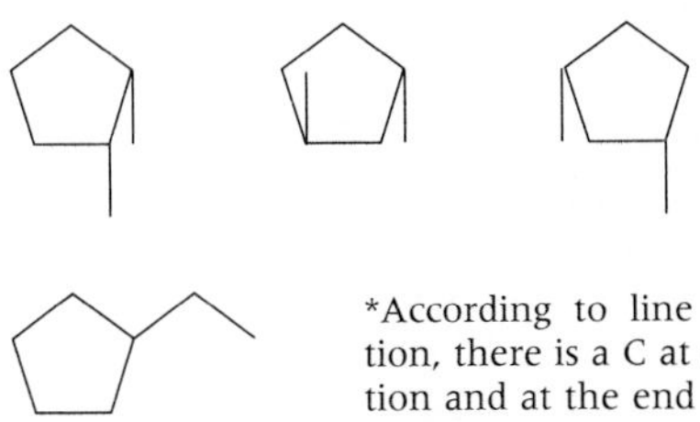

*According to line structure convention, there is a C at every line intersection and at the end of every line. Thus 14.14a is

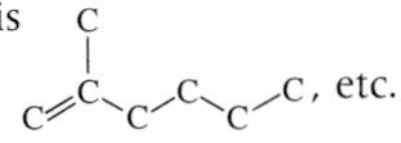

$$C=\overset{C}{C}-C-C-C-C, \text{ etc.}$$

**14.16**
(a)

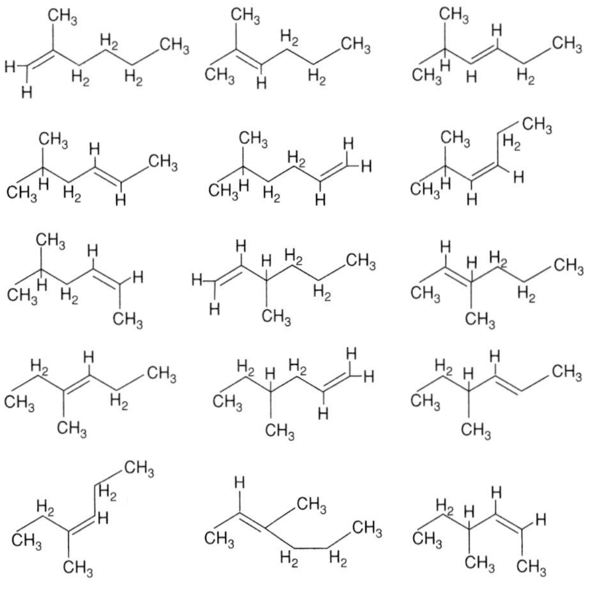

(b)

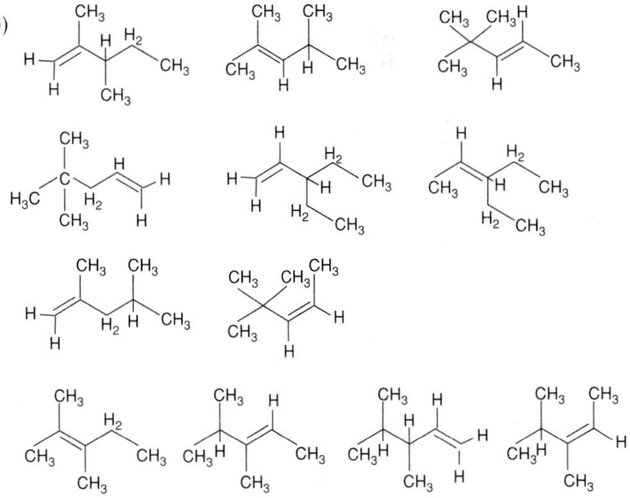

(c)

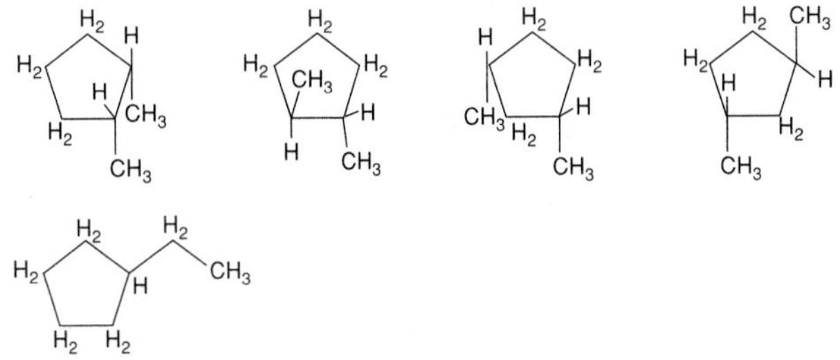

**14.18** (a)
$$CH_3-\overset{\overset{\displaystyle CH_3}{|}}{\underset{\underset{\displaystyle CH_3}{|}}{C}}-CH_2CH_3$$

(b) $CH_2{=}CH{-}CH_2CH_3$

(c) $H{-}C{\equiv}C{-}\overset{\overset{\displaystyle CH_3}{|}}{\underset{\underset{\underset{\displaystyle CH_3}{|}}{\underset{\displaystyle CH_2}{|}}}{CH}}$

(d) OK

**14.20** (a) 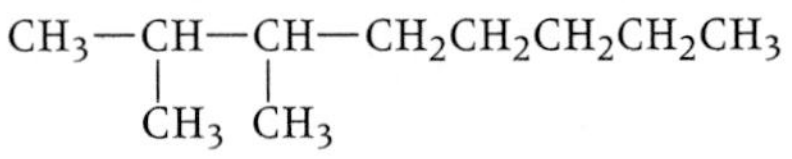 $CH_3-CH-CH-CH_2CH_2CH_2CH_2CH_3$
　　　　　　　　　　$|$　　$|$
　　　　　　　　　$CH_3$　$CH_3$

(b) 

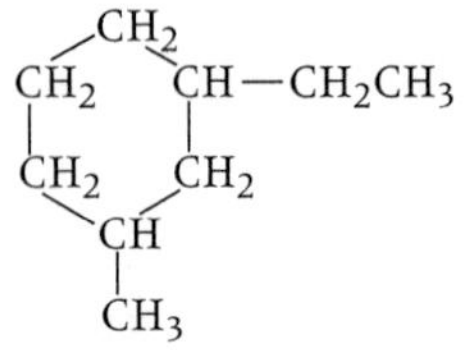

(c) 3,4-dimethylheptane

(d) 2,2-dimethylbutane

**14.22** (a) 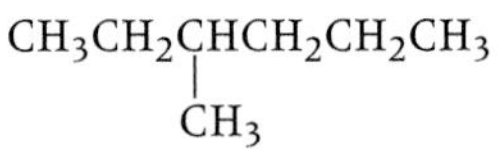 $CH_3CH_2CHCH_2CH_2CH_3$　　3-methylhexane
　　　　　　　　　$|$
　　　　　　　　$CH_3$

(b) 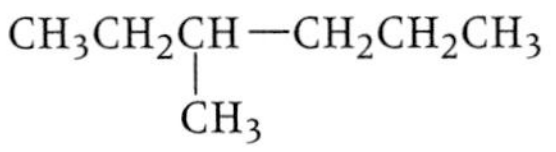 $CH_3CH_2CH-CH_2CH_2CH_3$　　3-methylhexane
　　　　　　　　　$|$
　　　　　　　　$CH_3$

(c) 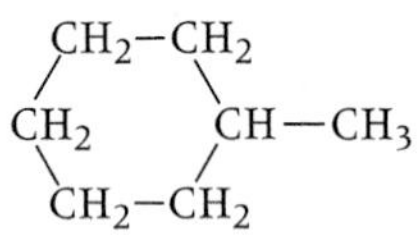 
　　methylcyclohexane

(d) 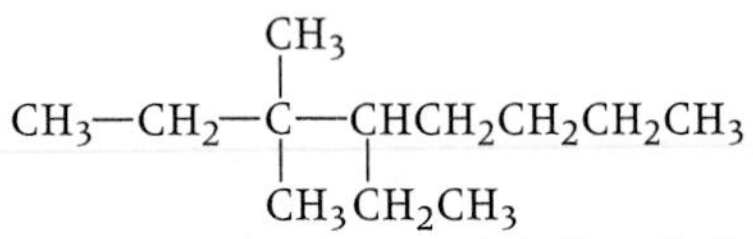 
$$CH_3-CH_2-\overset{\overset{\displaystyle CH_3}{|}}{\underset{\underset{\displaystyle CH_3CH_2CH_3}{|}}{C}}-CHCH_2CH_2CH_2CH_3$$
　　　　　　　　　　　3,3 dimethyl-4-ethyloctane

**14.24** (a) 

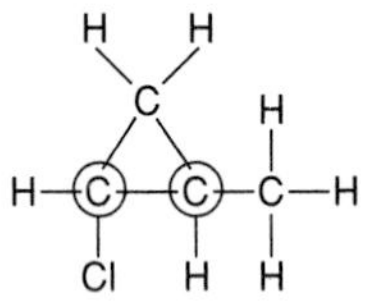

(b) 

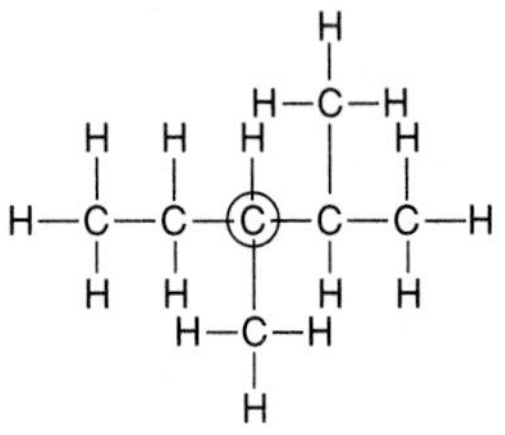

**14.26** (a) 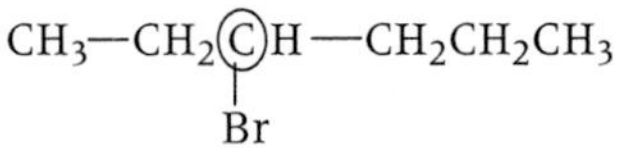 $CH_3-CH_2-\overset{\text{⟨C⟩}}{}H-CH_2CH_2CH_3$　Optically active
　　　　　　　　　　$|$
　　　　　　　　　　$Br$

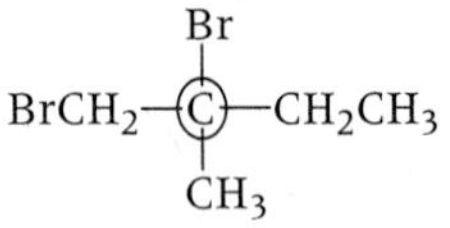

(b) 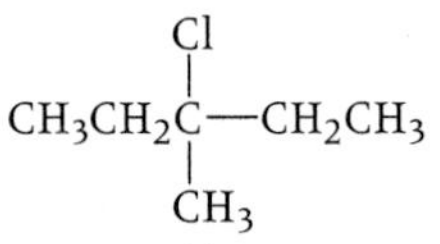 　　　　　$Cl$　　　　　　　　Inactive
　　　　　　　　$|$
　　$CH_3CH_2\overset{}{C}-CH_2CH_3$
　　　　　　　　$|$
　　　　　　　$CH_3$

(c) 　　　$Br$
　　　　　　$|$
　　$BrCH_2-\overset{}{\text{⟨C⟩}}-CH_2CH_3$　　Optically active
　　　　　　$|$
　　　　　$CH_3$

**14.28** (a) 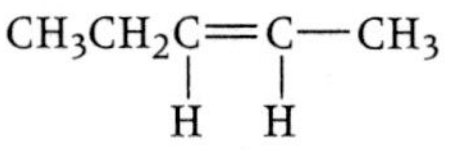 
　　　*cis*-2-pentene 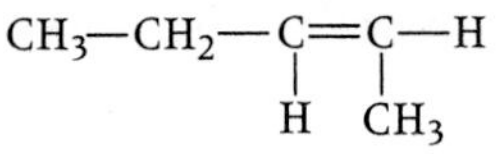　　　　　*trans*-2-pentene

(b) 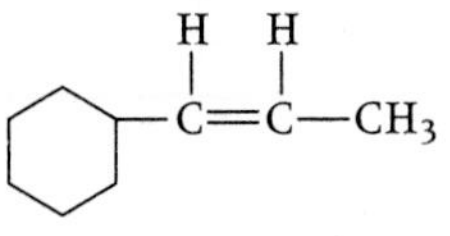 
　　*cis*-1-cyclohexylpropene 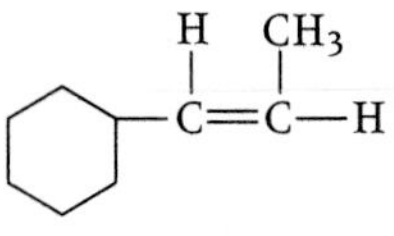　　*trans*-1-cyclohexylpropene

(c) No geometric isomers

**14.30** (a) No geometric isomers

(b) 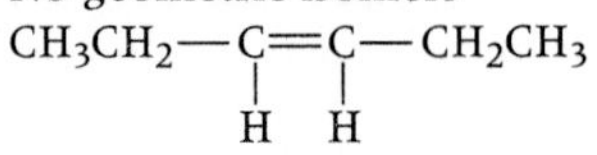 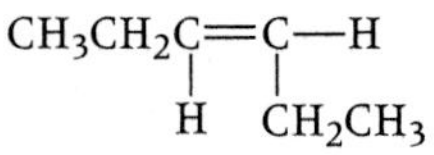 $CH_3CH_2-\overset{\overset{\displaystyle }{}}{C}=\overset{}{C}-CH_2CH_3$　　$CH_3CH_2\overset{}{C}=\overset{}{C}-H$
　　　　　　　　$|$　$|$　　　　　　　　　　　$|$　　$|$
　　　　　　　　$H$　$H$　　　　　　　　　　$H$　$CH_2CH_3$

(c) No geometric isomers

(d) 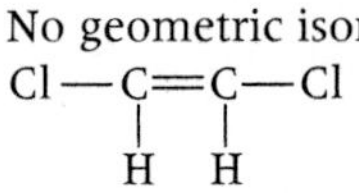 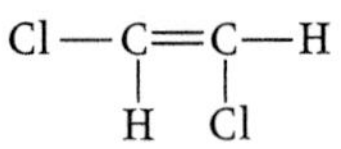 $Cl-\overset{}{C}=\overset{}{C}-Cl$　　　$Cl-\overset{}{C}=\overset{}{C}-H$
　　　　　$|$　$|$　　　　　　　　　$|$　$|$
　　　　　$H$　$H$　　　　　　　　$H$　$Cl$

**14.32**

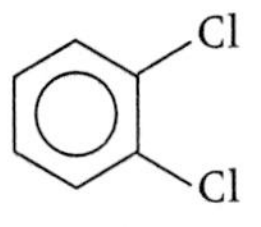

*o*-dichlorobenzene 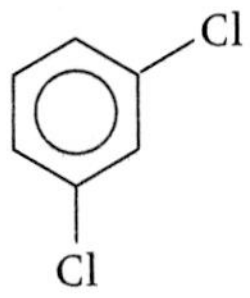　*m*-dichlorobenzene 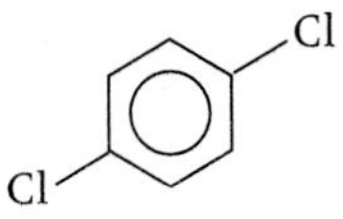　*p*-dichlorobenzene

**14.35**

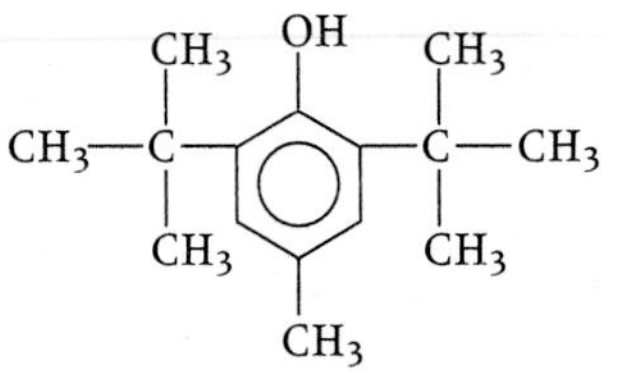

**14.38** (a) $CH_3CH_2OH$ and $CH_3OCH_3$

(b) 3 and 1, respectively

**14.40** $\pi$ bond

**14.42** (a) Elimination

(b) Addition

**14.44** (a) 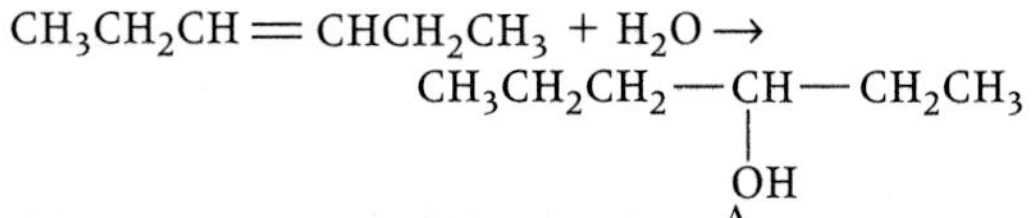 $CH_3CH_2CH=CHCH_2CH_3 + H_2O \rightarrow$
　　　　　　　　　　　$CH_3CH_2CH_2-CH-CH_2CH_3$
　　　　　　　　　　　　　　　　　　$|$
　　　　　　　　　　　　　　　　　$OH$

(b) 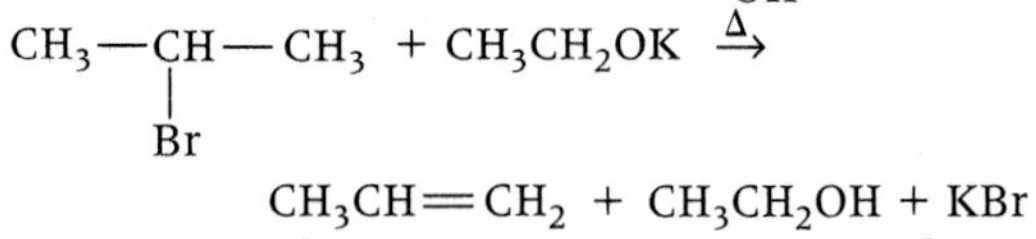 $CH_3-CH-CH_3 + CH_3CH_2OK \overset{\Delta}{\rightarrow}$
　　　　　$|$
　　　　$Br$

　　$CH_3CH=CH_2 + CH_3CH_2OH + KBr$

(c) 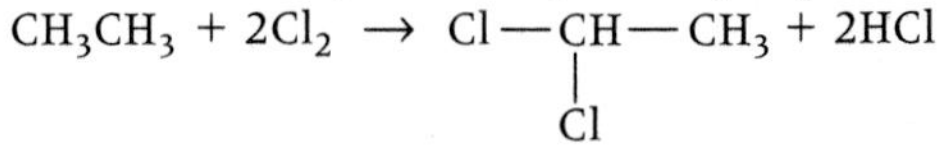 $CH_3CH_3 + 2Cl_2 \rightarrow Cl-CH-CH_3 + 2HCl$
　　　　　　　　　　　　　　　$|$
　　　　　　　　　　　　　　$Cl$

**14.46** (a) $-1$

(b) $+3$

(c) $0$

(d) $+3$

**14.48** (a) Increases

(b) Decreases

(c) Decreases

(d) No change

**14.50** (a) Oxidized; $-12/5$ to $+4$

(b) Oxidized; $-1$ to $0$

(c) Oxidized; $-2$ to $-1$

**14.54** $R-CH=CH_2 \rightleftarrows R-CH-CH_3$
$\qquad\qquad\qquad\qquad\qquad\qquad\quad |$
$\qquad\qquad\qquad\qquad\qquad\qquad\quad OH$

$R-CH-CH_3$
$\quad\quad\quad |$
$\quad\quad\quad Br$

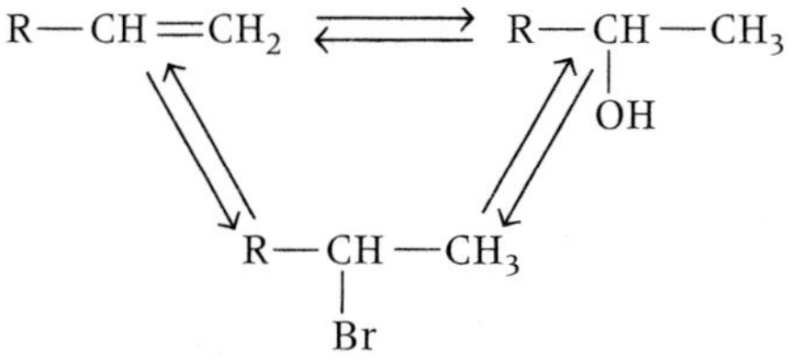

**14.55** Addition, due to the resonance stability of the ring

**14.59** Condensation; water

**14.61** (a) Ether
(b) Nitrile
(c) Carboxylic acid
(d) Aldehyde

**14.63** (a) $CH_3-\boxed{CH=CH}-CH_2-\boxed{OH}$
alkene          alcohol

(b) 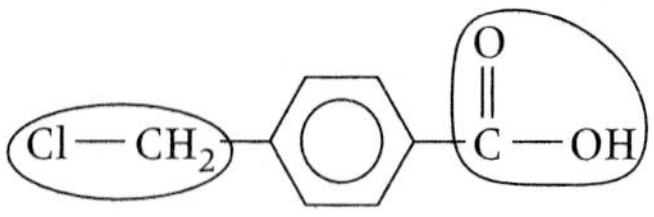 $\boxed{Cl-CH_2}-\bigcirc-\boxed{C-OH}$ (with $=O$ above C)
haloalkane  aromatic  carboxylic acid

(c) $\bigcirc$ $-\boxed{C-NH}-CH_3$ (with $=O$ above C) 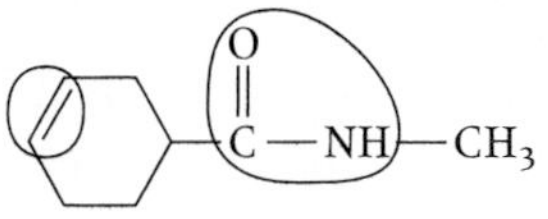
alkene       amide

(d) $\boxed{N\equiv C}-CH_2-\boxed{C}-CH_3$ (with $O$ below C)
nitrile          ketone

**14.65**

$\wedge\wedge\wedge$ OH          OH on 2nd carbon          3-pentanol structure

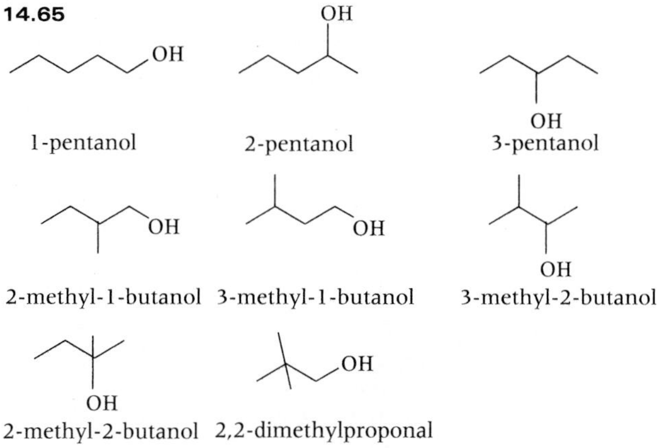

1-pentanol        2-pentanol        3-pentanol

2-methyl-1-butanol  3-methyl-1-butanol  3-methyl-2-butanol

2-methyl-2-butanol  2,2-dimethylproponal

**14.67**

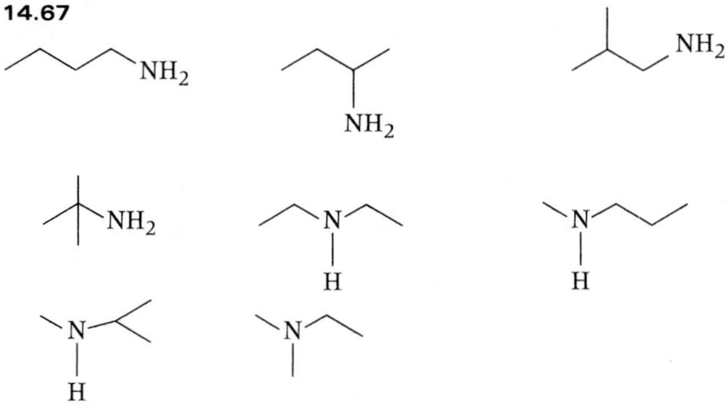

$\wedge\wedge$ NH$_2$          NH$_2$ structure          NH$_2$ structure

(various amine structures with NH$_2$, N-H)

**14.69** (a) $CH_3CCH_2CH_3$ (with $=O$)     2-butanone

(b) $CH_3-CH-C-OH$ (with $=O$ on C, $CH_3$ below)     2-methylpropanoic acid

(c) 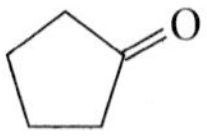 cyclopentanone (with $=O$)

**14.71** (a) $CH_3C-N-CH_3$ (with $=O$ on C, $H$ on N)     $N$-methylacetamide

(b) $CH_3CH_2CH_2C-O-CH-CH_3$ (with $=O$, $CH_3$ below)     isopropyl butanoate

(c) $H-C-O-CH_2CH-CH_3$ (with $=O$, $CH_3$ below)     isobutyl formate

**14.73** (a) $CH_3(CH_2)_4COH$ (with $=O$)    and    $CH_3CH_2OH$

(b) 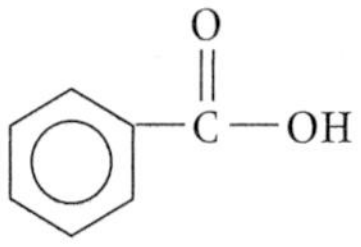 $\bigcirc-C-OH$ (with $=O$)    and    $CH_3CH_2CH_2OH$

(c) $CH_3CH_2OH$    and    $\bigcirc-CH_2CH_2-C-OH$ (with $=O$)

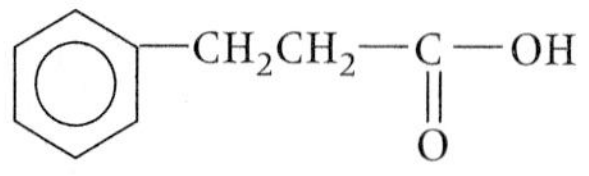

**14.75** (a) $CH_3CH_2OH$;  $CH_3CH_2C-O-CH_2CH_3$ (with $=O$)

(b) $CH_3CH_2-CH-CH_3$ (with CN);   $CH_3CH_2-CH-CH_3$ (with COOH)

**14.77** (a) $CH_3CH_2C-OH$ (with $=O$) $\xrightarrow{H^+}$
(b) $CH_3CH_2NH_2$

**14.80** $H_2NCH_2CH_2CH_2CH_2CH_2NH_2$ and
$H_2NCH_2CH_2CH_2\ CH_2NH_2$

**14.84** Double bond; substituents on the alkene

**14.87** Condensation; both are formed by the loss of water from two monomers.

**14.89** (a) Condensation
(b) Addition
(c) Condensation
(d) Condensation

**14.91** The amino acid sequence determines the shape and structure, which determine its function.

**14.94** The base sequence in DNA determines the base sequence in RNA, which determines the amino acid sequence in the protein.

**14.95** (a) $-(CH_2CH)_n-$ with Cl substituent

(b) $-(CH_2CH)_n-$ with $CH_3$ substituent

**14.97** $nHO-\overset{O}{\overset{\|}{C}}-C_6H_4-\overset{O}{\overset{\|}{C}}-OH \; + \; nHOCH_2CH_2OH \longrightarrow$

$HO-\left[CH_2CH_2O\overset{O}{\overset{\|}{C}}-C_6H_4-\overset{O}{\overset{\|}{C}}-O\right]_n H \; + \; nH_2O$

**14.99** (a) $CH_3-$    (b) imidazole ring with $CH_2-$    (c) $CH_3SCH_2CH_2-$

**14.101** (a) $H_2N-CH-\overset{O}{\overset{\|}{C}}-NH-CH-\overset{O}{\overset{\|}{C}}-NH-CH-COOH$
with side chains $CH_2$ ($CH_2$ $COOH$), $CH_2$ (imidazole H–N ring), $CH_2$ (indole ring with N–H)

(b) $H_3\overset{+}{N}-CH_2-\overset{O}{\overset{\|}{C}}-NH-CH-\overset{O}{\overset{\|}{C}}-NH-CH-CO^-$
with side chains $CH_2$ ($SH$) and $CH_2$ (phenol ring with $OH$)

**14.103** (a) AATCGG
(b) TCTGTA

**14.105** ACAATGCCT; three

**14.107** (a) Cysteine + cysteine: covalent link to form cystine
(b) Lysine + aspartic acid or glutamic acid: salt bridge
(c) Asparagine or glutamine with serine by H bonding
(d) Valine + phenylalanine: hydrophobic forces

**14.109** (a) Serine—threonine—alanine—tyrosine—methionine—alanine—leucine—lysine
(b) AGGTGTCGGATATACCGTTTGAACTTC

**14.112** $CH_3CH-\overset{O}{\overset{\|}{C}}-H$ with $CH_3$ substituent

**14.114** $CH_3CH=CH-CH_3$

**14.120** $R-\overset{\overset{\displaystyle :O:}{\|}}{C}-\overset{H}{\overset{|}{\ddot{N}}}-R \;\leftrightarrow\; R-\overset{\overset{\displaystyle :\underset{..}{O}:}{|}}{C}=\overset{H}{\overset{|}{N}}-R$

**14.122** $CH_3C=C-CH_2CH_2C=C-H$   geranial
with $CH_3$, H, and $CH_3$, $C=O$ / H substituents

$CH_3C=C-CH_2CH_2-C=C-\overset{O}{\overset{\|}{C}}-H$   neral
with $CH_3$, H and $CH_3$, H substituents

**14.123** (a) Because water was added
(b) Glycine:alanine:valine:proline:serine:arginine = 4:1:3:6:7:49
(c) 10,700 g/mol

**14.125** The reaction is not stereospecific; the $H^-$ can add from either side.

## CHAPTER 15

**15.1** Changes in concentrations of reactants (or products) as a function of time

**15.3** Additional solvent would decrease the reactant concentrations and would result in a decreased reaction rate.

**15.7** The second experiment proceeds at the higher rate. $I_2$ in the gaseous state would give more collisions and a higher rate of reaction.

**15.9** (a) The slope of the line joining any two points on the graph of concentrations *vs.* time gives the average rate over that time period. The shorter the time period, the closer the average rate will be to the instantaneous rate, the rate at a particular instant.
(b) The initial rate is the instantaneous rate at the moment the reactants are mixed (at $t = 0$).

**15.12** (a) Calculate the slope of the line connecting (0, $[uc]_0$) and ($t_f$, $[uc]_f$). The negative of this value is the average rate.
(b) Calculate the negative of the slope of the line tangent to the curve at $t = x$.
(c) Calculate the negative of the slope of the line tangent to the curve at $t = 0$.
(d) If you plotted [D] *vs.* time, you would not need to take the negative of the slopes in (a) − (c).

**15.13** (a) $2.1 \times 10^{-3}$ mol/L · s
(b) Initial rate $\approx 5.0 \times 10^{-3}$ mol/L · s; higher than (a).

**15.15** (a) Rate $= -\dfrac{1}{2}\dfrac{\Delta[A]}{\Delta t} = \dfrac{\Delta[B]}{\Delta t} = \dfrac{\Delta[C]}{\Delta t}$

(b) Rate $= -\dfrac{\Delta[D]}{\Delta t} = \dfrac{2}{3}\dfrac{\Delta[E]}{\Delta t} = \dfrac{2}{5}\dfrac{\Delta[F]}{\Delta t}$

(c) $-4$ mol/L · s
(d) $0.42$ mol/L · s

**15.18** Rate $= -\dfrac{\Delta[N_2]}{\Delta t} = -\dfrac{1}{3}\dfrac{\Delta[H_2]}{\Delta t} = \dfrac{1}{2}\dfrac{\Delta[NH_3]}{\Delta t}$

**15.21** 6.30 mol/L · s

**15.23** (a) Plot either [$A_2$] or [$B_2$] vs. time and determine the negative of the slope of the line tangent to the curve at $t = 0$.

**15.24** (a) Doubled
(b) Decreased by a factor of 4
(c) Increased by a factor of 9

**15.27** (a)

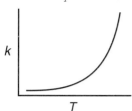

(b)

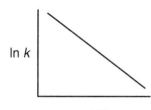

The slope of this line = $-E_a/R$.

**15.28** (a) First order in $BrO_3^-$ and $Br^-$, second order in $H^+$, fourth order overall
(b) First order in $NH_4^+$ and $NO_2^-$, second order overall
(c) Second order in $O_3$, $(-1)$-order in $O_2$, first order overall

**15.30** (a) Second order in A, first order in B
(b) Rate = $k\,[A]^2[B]$
(c) $5.00 \times 10^3$ L²/mol² · min

**15.32** (a) time⁻¹
(b) L/mol · time
(c) L²/mol² · time
(d) L$^{1/2}$/mol$^{1/2}$ · time

**15.34** 7 s

**15.36** (a) $6.60 \times 10^{-2}$ min⁻¹
(b) 21.0 min

**15.38** $4.6 \times 10^{-2}$ s⁻¹

**15.41** (a) Rate = $k\,[CS_2]$
(b) $2.7 \times 10^{-6}$ s⁻¹

**15.44** 1.15 kJ/mol

**15.46** No; for most reactions, only a small fraction of collisions actually leads to reaction. If this were not true, every gaseous reaction would be over instantaneously. Only those collisions with energy equal to or greater than the activation energy and with the proper orientation can lead to reaction.

**15.48** (a) Rate increases
(b) Rate increases

**15.50** At the same temperature, the $NH_3$ and $N(CH_3)_3$ will have the same kinetic energy. However, due to their larger mass, $N(CH_3)_3$ molecules will be moving more slowly. This will decrease the number of collisions per

unit time and reduce the rate of the second reaction. In addition, the higher molecular complexity of the $N(CH_3)_3$ would probably reduce its orientation probability factor, decreasing the rate of the second reaction still more.

**15.51** 12 unique collisions

**15.53** $2.90 \times 10^{-18}$

**15.55** (a)

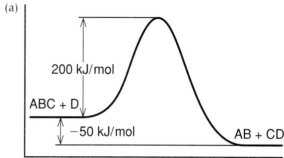

(b) 250 kJ/mol
(c) A⟍ᵦ ⟋C---D

**15.57** (a)

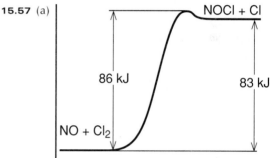

(b) 3 kJ
(c) [:Ö=N̈---:C̈l—C̈l:]

**15.59** In many cases, one of the steps is much slower than the others. If this is true, the rate equation for this slowest step represents the rate equation for the overall reaction, and the rate would be lower than the average rate of the individual steps.

**15.62** Reaction intermediates have some stability, however limited, but activated complexes are inherently unstable. Additionally, unlike activated complexes, intermediates are molecules with normal bonds.

**15.65** If the slow step is not the first one, the faster step will produce intermediates that accumulate before being consumed in the slow step. Substitution of the intermediates into the rate law for the slow step will produce the overall rate law.

**15.66** (a) $A(g) + B(g) + C(g) \rightarrow D(g)$
(b) $X(g)$ and $Y(g)$
(c) Bimolecular; rate₁ = $k_1$ [A][B]
Bimolecular; rate₂ = $k_2$ [X][C]
Unimolecular; rate₃ = $k_3$ [Y]
(d) Yes
(e) Yes

**15.69** A and B: no; C: yes

**15.71** (a)

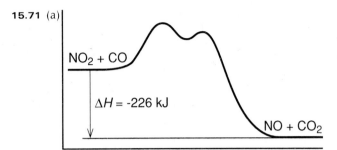

NO$_2$ + CO

$\Delta H$ = -226 kJ

NO + CO$_2$

   (b) Yes, but the in-text mechanism is more reasonable since it involves only bimolecular steps.

**15.73** No; an increase in temperature increases the reaction rate by increasing the fraction of collisions whose energy is greater than the activation energy. A catalyst increases the reaction rate by providing a different reaction mechanism with a lower activation energy.

**15.77** (a) Water is the solvent, and thus its concentration is effectively constant due to the large excess.
   (b) $Rate_1 = k_1 [(CH_3)_3CBr]$
      $Rate_2 = k_2 [(CH_3)_3C^+]$
      $Rate_3 = k_3 [(CH_3)_3COH_2{}^+]$
   (c) $(CH_3)_3C^+$ and $(CH_3)_3COH_2{}^+$
   (d) The experimental rate law matches the rate law for the slowest step.

**15.79** Rate = $k (P_A)^x (P_B)^y$
   First order: $s^{-1}$; second order: $torr^{-1} s^{-1}$

**15.82** 46 kJ/mol

**15.85** Second order

**15.88** (a) $k = 0.21 h^{-1}$; $t_{1/2} = 3.3$ h
   (b) 6.6 h
   (c) If the concentration of one reactant is much greater than the other, the change in concentration of the reactant in great excess is negligible, yielding an apparent zero-order behavior with respect to that reactant.

**15.91** (a) $Rate_1 = 3.45 \times 10^{-2} s^{-1}$
      $Rate_2 = 6.90 \times 10^{-2} s^{-1}$
      $Rate_3 = 6.90 \times 10^{-2} s^{-1}$
   (b) Zero order in $Na_2S_2O_8$; first order in KI
   (c) $0.862 s^{-1}$
   (d) Rate = $(0.862 s^{-1})[KI]$

**CHAPTER 16**

**16.3** A very large $K_{eq}$ means at equilibrium the value of product concentrations is much larger than the value of reactant concentrations, indicating that the reaction goes essentially to completion.

**16.5** No; the value of $Q$ is determined by the mass action expression with arbitrary concentrations for reactants and products.

**16.7** $K = [O_2]$. If $T$ remains constant and sufficient $Li_2O_2$ was initially present to reach equilibrium, the amount of $O_2$ obtained will be constant regardless of how much $Li_2O_2(s)$ is present.

**16.10** $K_c$ and $K_p$ are only equal when $\Delta n_{gas} = 0$.

**16.12** When $Q < K$, the concentration of products is less than its equilibrium value, so the reaction proceeds to the right to reach equilibrium.

**16.13** (a) $4NO(g) + O_2(g) \rightarrow 2N_2O_3(g)$
$$Q_c = \frac{[N_2O_3]^2}{[NO]^4[O_2]}$$
   (b) $SF_6(g) + 2SO_3(g) \rightleftharpoons 3SO_2F_2(g)$
$$Q_c = \frac{[SO_2F_2]^3}{[SF_6][SO_3]^2}$$
   (c) $2SClF_5(g) + H_2(g) \rightleftharpoons S_2F_{10}(g) + 2HCl(g)$
$$Q_c = \frac{[S_2F_{10}][HCl]^2}{[SClF_5]^2[H_2]}$$
   (d) $2C_2H_6(g) + 7O_2(g) \rightleftharpoons 4CO_2(g) + 6H_2O(g)$
$$Q_c = \frac{[CO_2]^4[H_2O]^6}{[C_2H_6]^2[O_2]^7}$$

**16.15** (a) (1) $2 H_2S(g) + 3 O_2(g) \rightleftharpoons 2 SO_2(g) + 2 H_2O(g)$
      (2) $2 SO_2(g) + 2 Cl_2(g) \rightleftharpoons 2 SO_2Cl_2(g)$
overall: $2 H_2S(g) + 3 O_2(g) + 2 Cl_2(g) \rightleftharpoons$
$2 SO_2Cl_2(g) + 2 H_2O(g)$

   (b) $Q_C$ (overall) $= \dfrac{[SO_2Cl_2]^2[H_2O]^2}{[H_2S]^2[O_2]^3[Cl_2]^2}$

$$Q_{c_1} \times Q_{c_2} = \frac{[SO_2]^2[H_2O]^2}{[H_2S]^2[O_2]^3} \times \frac{[SO_2Cl_2]^2}{[SO_2]^2[Cl_2]^2} =$$

$$\frac{[SO_2Cl_2]^2[H_2O]^2}{[H_2S]^2[O_2]^3[Cl_2]^2}$$

**16.17** (a) 7.9
   (b) $3.2 \times 10^{-5}$

**16.19** (a) $2Na_2O_2(s) + 2CO_2(g) \rightleftharpoons 2Na_2CO_3(s) + O_2(g)$
$$Q_c = \frac{[O_2]}{[CO_2]^2}$$
   (b) $H_2O(l) \rightleftharpoons H_2O(g)$
$$Q_c = [H_2O(g)]$$
   (c) $H_2SO_3(aq) \rightleftharpoons H_2O(l) + SO_2(aq)$
$$Q_c = \frac{[SO_2(aq)]}{[H_2SO_3(aq)]}$$
   (d) $H_2O(l) + SO_3(g) \rightleftharpoons H_2SO_4(aq)$
$$Q_c = \frac{[H_2SO_4(aq)]}{[SO_3(g)]}$$
   (e) $2KNO_3(s) \rightleftharpoons 2KNO_2(s) + O_2(g)$
$$Q_c = [O_2]$$

**16.21** (a) 3; (b) $-1$; (c) 3

**16.23** (a) 32; (b) 28.5

**16.25** No; to the left

**16.28** (a) $2Ni_3S_2(s) + 7O_2(g) \rightleftharpoons 6NiO(s) + 4SO_2(g)$
      $NiO(s) + H_2(g) \rightleftharpoons Ni(s) + H_2O(g)$   ($\times 6$)

$Ni(s) + 4CO(g) \rightleftharpoons Ni(CO)_4(g)$ $(\times 6)$

Overall: $2Ni_3S_2(s) + 7O_2(g) + 6H_2(g) + 24CO(g) \rightleftharpoons$
$4SO_2(g) + 6H_2O(g) + 6Ni(CO)_4(g)$

(b) $Q_1 \times Q_2 \times Q_3 = \dfrac{[SO_2]^4}{[O_2]^7} \times \dfrac{[H_2O]^6}{[H_2]^6} \times$

$\dfrac{[Ni(CO)_4]^6}{[CO]^{24}} = Q_{c(overall)}$

**16.31** When $x$ mol of $CH_4$ reacts, $2x$ mol of $H_2O$ also reacts to form $x$ mol of $CO_2$ and $4x$ mol of $H_2$. The equilibrium concentrations of reactants are the initial concentrations minus the amount that reacts. For products, it is the initial concentration plus what forms.

**16.33** 50.8

**16.35**

|  | $PCl_5(g)$ | $\rightleftharpoons$ | $PCl_3(g)$ | $+$ | $Cl_2(g)$ |
|---|---|---|---|---|---|
| Initial | 0.075 |  | 0 |  | 0 |
| Change | $-x$ |  | $+x$ |  | $+x$ |
| Equilibrium | $0.075 - x$ |  | $x$ |  | $x$ |

**16.37** $7.8 \times 10^{-6}\ M^2$

**16.39** 28 atm

**16.41** 0.33 atm

**16.43** $3.6 \times 10^{-3}\ M$

**16.45** $[HI] = 0.0153\ M$; $[I_2] = 0.00328\ M$.

**16.47** $[CO] = [Cl_2] = 0.288\ M$; $[COCl_2] = 0.412\ M$

**16.50** 0.0249 atm$^{-2}$

**16.53** The equilibrium position refers to the specific concentrations or pressures that exist when $Q = K$, the equilibrium constant. Only the position changes in response to a change in concentration.

**16.56** The rate of the forward reaction is increased more than the rate of the reverse reaction, so $K$ ($= k_f/k_r$) would increase.

**16.57** (a) Right
(b) Right
(c) No change
(d) Left

**16.59** (a) Less $F_2$; more F
(b) Less $CH_4$; more $C_2H_2$ and $H_2$
(c) Less $CH_3OH(l)$; more $CH_3OH(g)$

**16.61** (a) Amount of product is independent of pressure
(b) Lower pressure

**16.63** (a) Decrease
(b) Increase
(c) Increase
(d) Decrease

**16.65** 1.9

**16.67** Low pressure, high temperature, continual removal of $CO_2$

**16.69** (a) High pressure, low temperature
(b) $Q$ decreased, $K$ unchanged
(c) Low temperature (for increased yield) also lowers the reaction rate.

**16.71** (a) Left
(b) No change
(c) No change
(d) No change
(e) Right

**16.72** 0.203 atm

**16.73** (a) $3 \times 10^{-3}$ atm
(b) High pressure, low temperature
(c) $K_{C(450K)} = 2 \times 10^5\ M^{-1}$
(d) No; $H_2O$ would condense at a higher $T$ than $C_2H_5OH$.

**16.79** $1.9 \times 10^{34}\ M^{-1}$

**16.80** Increase by a factor of 2.2

**16.82** (a) Enzyme 5; enzyme 8
(b) Once F formed, the process would stop.

## CHAPTER 17

**17.2** All Arrhenius acids produce $H^+$ ions, and all Arrhenius bases produce $OH^-$ ions in aqueous solution. Neutralization involves the combination of $H^+$ ions and $OH^-$ ions to produce water molecules. It was this single reaction for all strong acid-base neutralizations that, according to Arrhenius, produced the identical heat of neutralization reaction value of $-56$ kJ per mole of water formed.

**17.4** Weak acids exist mainly as undissociated molecules in water. The words "strong" and "weak" refer to the extent of dissociation in water. Strong acids are essentially 100% dissociated in water.

**17.5** (a), (c), (d), and (e)

**17.7** (a) $K_a = \dfrac{[H_3O^+][CN^-]}{[HCN]}$

(b) $K_a = \dfrac{[H_3O^+][CO_3^{2-}]}{[HCO_3^-]}$

(c) $K_a = \dfrac{[H_3O^+][HCOO^-]}{[HCOOH]}$

**17.9** $CH_3COOH < HF < HIO_3 < HI$

**17.11** (a) Weak acid
(b) Strong base
(c) Weak acid
(d) Strong acid
(e) Weak base
(f) Strong base

**17.13** Autoionization reactions occur when a proton (or less frequently, another ion) is transferred from one molecule of the substance to another molecule of the same substance.

$$H_2O(l) + H_2O(l) \rightleftharpoons H_3O^+(aq) + OH^-(aq)$$

$$H_2SO_4(l) + H_2SO_4(l) \rightleftharpoons H_3SO_4^+{}_{(solvated)} + HSO_4^-{}_{(solvated)}$$

**17.15** (a) pH increases by a value of 1.
(b) $[H_3O^+]$ increases by a factor of 100.

**17.17** (a) 12.045; basic
(b) 11.090; acidic

**17.19** (a) $1.7 \times 10^{-10}\ M$; $6.0 \times 10^{-5}\ M$; 4.22
(b) $2.7 \times 10^{-4}\ M$; $3.7 \times 10^{-11}\ M$; 3.57

**17.21** $3.4 \times 10^{-4}$ mole of $OH^-$ per liter

**17.25** (a) $K_w$ increases with increase in temperature.
(b) $[H_3O^+]$: $(3.5 \times 10^{-8}) - (4.5 \times 10^{-8})\ M$
$[OH^-]$: $(5.2 \times 10^{-7}) - (6.6 \times 10^{-7})\ M$

**17.27** A Brønsted acid is a proton donor; thus, any Brønsted acid must contain at least one hydrogen atom. An Arrhenius acid donates a proton to water and thus is also an acid under the Brønsted definition. While an Arrhenius base produces $OH^-$ ions in water, a Brønsted base is a substance that accepts a proton, $H^+$.

**17.30** (a) $H_3PO_4(aq) + H_2O(l) \rightleftharpoons H_3O^+(aq) + H_2PO_4^-(aq)$

$$K_a = \frac{[H_3O^+][H_2PO_4^-]}{[H_3PO_4]}$$

(b) $C_6H_5COOH(aq) + H_2O(l) \rightleftharpoons H_3O^+(aq) + C_6H_5COO^-(aq)$

$$K_a = \frac{[H_3O^+][C_6H_5COO^-]}{[C_6H_5COOH]}$$

(c) $HSO_4^-(aq) + H_2O(l) \rightleftharpoons H_3O^+(aq) + SO_4^{2-}(aq)$

$$K_a = \frac{[H_3O^+][SO_4^{2-}]}{[HSO_4^-]}$$

**17.32** (a) $Cl^-$; (b) $HCO_3^-$; (c) $OH^-$; (d) $PO_4^{3-}$; (e) $NH_3$; (f) $S^{2-}$

**17.34** (a) $\underset{\text{acid}}{HCl} \; + \; \underset{\text{base}}{H_2O} \; \rightleftharpoons \; \underset{\text{conjugate acid}}{H_3O^+} \; + \; \underset{\text{conjugate base}}{Cl^-}$

     Conjugate pairs: $HCl/Cl^-$; $H_3O^+/H_2O$

(b) $\underset{\text{acid}}{HClO_4} \; + \; \underset{\text{base}}{H_2SO_4} \; \rightleftharpoons \; \underset{\text{conjugate acid}}{H_3SO_4^+} \; + \; \underset{\text{conjugate base}}{ClO_4^-}$

     Conjugate pairs: $HClO_4/ClO_4^-$; $H_3SO_4^+/H_2SO_4$

(c) $\underset{\text{base}}{HPO_4^{2-}} \; + \; \underset{\text{acid}}{H_2SO_4} \; \rightleftharpoons \; \underset{\text{conjugate acid}}{H_2PO_4^-} \; + \; \underset{\text{conjugate base}}{HSO_4^-}$

     Conjugate pairs: $H_2SO_4/HSO_4^-$; $H_2PO_4^-/HPO_4^{2-}$

(d) $\underset{\text{base}}{NH_3} \; + \; \underset{\text{acid}}{HNO_3} \; \rightleftharpoons \; \underset{\text{conjugate acid}}{NH_4^+} \; + \; \underset{\text{conjugate base}}{NO_3^-}$

     Conjugate pairs: $HNO_3/NO_3^-$; $NH_4^+/NH_3$

(e) $\underset{\text{base}}{O^{2-}} \; + \; \underset{\text{acid}}{H_2O} \; \rightleftharpoons \; \underset{\text{conjugate acid}}{OH^-} \; + \; \underset{\text{conjugate base}}{OH^-}$

     Conjugate pairs: $H_2O/OH^-$; $OH^-/O^{2-}$

**17.36** $K_c > 1$: $HCl + HS^- \rightleftharpoons Cl^- + H_2S$
$K_c < 1$: $Cl^- + H_2S \rightleftharpoons HCl + HS^-$

**17.38** (a) and (b)

**17.40** (a) Increased; strong acid is nearly 100% dissociated.
(b) Nearly the same; weak acid is slightly dissociated.
(c) Nearly the same; slightly greater dissociation occurs.
(d) Increased; strong acid is nearly 100% dissociated.

**17.41** No; the % dissociation of HCl is greater than that of $CH_3COOH$.

**17.44** $1.5 \times 10^{-5}$

**17.46** $[H_3O^+] = [NO_2^-] = 1.9 \times 10^{-2}\ M$
$[OH^-] = 5.3 \times 10^{-13}\ M$

**17.48** $[H_3O^+] = [ClCH_2COO^-] = 3.7 \times 10^{-2}\ M$
pH = 1.43
$[ClCH_2COOH] = 1.01\ M$

**17.50** (a) $7.5 \times 10^{-3}\ M$; 2.12; $1.3 \times 10^{-12}\ M$; 11.88
(b) $2.3 \times 10^{-4}$

**17.52** (a) 2.47
(b) 11.42

**17.54** 1.6%

**17.56** $[H_2S] = 0.10\ M$; $[HS^-] = 9.5 \times 10^{-5}$; $[S^{2-}] = 1 \times 10^{-17}\ M$; $[H_3O^+] = 9.5 \times 10^{-5}\ M$; pH = 4.02; $[OH^-] = 1.1 \times 10^{-10}\ M$; pOH = 9.96

**17.59** 2.4

**17.63** (a) $CH_3COOH(aq)$; $CH_3COO^-(aq)$; $H_3O^+(aq)$; $OH^-(aq)$
(b) $CH_3COOH$ dissociates as a weak acid. $CH_3COONa$, a salt of a weak acid and strong base, hydrolyzes in water, producing excess of $OH^-$ ions.

**17.64** (a) $C_5H_5N(aq) + H_2O(l) \rightleftharpoons C_5H_5NH^+(aq) + OH^-(aq)$

$$K_b = \frac{[C_5H_5NH^+][OH^-]}{[C_5H_5N]}$$

(b) $CO_3^{2-}(aq) + H_2O(l) \rightleftharpoons HCO_3^-(aq) + OH^-(aq)$

$$K_b = \frac{[HCO_3^-][OH^-]}{[CO_3^{2-}]}$$

**17.66** 11.72

**17.68** (a) $5.6 \times 10^{-10}$
(b) $2.5 \times 10^{-5}$

**17.70** (a) 10.96
(b) 5.62

**17.73** pH = 12.76; $[OH^-] = 5.8 \times 10^{-2}\ M$

**17.77** Greater EN and charge on Cl

**17.79** (a) $H_2SeO_4$ (b) $H_3PO_4$ (c) $H_2Te$ (d) HBr; (e) $HClO_4$ (f) $H_2SO_4$

**17.81** (a) $0.05\ M\ Al_2(SO_4)_3$
(b) $0.1\ M\ PbCl_2$
(c) $0.1\ M\ AlCl_3$

**17.84** NaF contains the $F^-$ anion of the weak acid HF, while NaCl contains the $Cl^-$ anion of the strong acid HCl.

**17.86** (a) $KBr(s) + H_2O(l) \rightarrow K^+(aq) + Br^-(aq)$   Neutral
(b) $NH_4I(s) + H_2O(l) \rightarrow NH_4^+(aq) + I^-(aq)$

$NH_4^+(aq) + H_2O(l) \rightleftharpoons NH_3(aq) + H_3O^+(aq)$   Acidic

(c) $KCN(s) + H_2O(l) \rightarrow K^+(aq) + CN^-(aq)$

$CN^-(aq) + H_2O(l) \rightleftharpoons HCN(aq) + OH^-(aq)$
Basic

(d) $Cr(NO_3)_3(s) + nH_2O(l) \rightarrow Cr(H_2O)_n^{3+}(aq) + 3NO_3^-(aq)$

$Cr(H_2O)_n^{3+} + H_2O(l) \rightleftharpoons Cr(H_2O)_{n-1}OH^{2+}(aq) + H_3O^+(aq)$
Acidic

(e) $NaHS(s) + H_2O(l) \rightarrow Na^+(aq) + HS^-(aq)$

$HS^-(aq) + H_2O(l) \rightleftharpoons H_2S(aq) + OH^-(aq)$
Basic

(f) $Zn(CH_3COO)_2(s) + nH_2O(l) \rightarrow Zn(H_2O)_n^{2+}(aq) + 2CH_3COO^-(aq)$

$Zn(H_2O)_n^{2+}(aq) + H_2O(l) \rightleftharpoons Zn(H_2O)_{n-1}OH^+(aq) + H_3O^+(aq)$

$2CH_3COO^-(aq) + -H_2O(l) \rightleftharpoons CH_3COOH(aq) + 2OH^-(aq)$

$H_3O^+ + HO^- \rightleftharpoons 2H_2O$   Close to neutral

**17.88** (a) $Fe(NO_3)_2 < KNO_3 < K_2SO_3 < K_2S$
(b) $NaHSO_4 < NH_4NO_3 < NaHCO_3 < Na_2CO_3$

**17.92**  $NH_3$ as a solvent is more basic than water and as such, weak acids such as HF act like strong ones and are 100% dissociated.

**17.93**  A Lewis base must have an electron pair to donate. A Lewis acid must have a vacant orbital or the ability to rearrange its bonding to make one available. The Lewis acid-base reaction involves the donation and acceptance of an electron pair to form a new covalent bond in an adduct.

**17.97**  All three concepts may have water as the product in an acid-base neutralization reaction. It always must be a product in an Arrhenius neutralization reaction.

**17.98**  (a) $NH_3$; Brønsted-Lowry or Lewis base
      (b) $AlCl_3$; Lewis acid only

**17.99**  (a) Acid
      (b) Base
      (c) Acid
      (d) Base
      (e) Acid
      (f) Base

**17.101**  (a) $Na^+$, acid; $H_2O$, base
       (b) $CO_2$, acid; $H_2O$, base
       (c) $BF_3$, acid; $F^-$, base

**17.103**  (a) Lewis
       (b) Brønsted-Lowry and Lewis
       (c) None
       (d) Lewis

**17.107**  (a) Steps 1, 2, and 4
       (b) Step 1: $FeCl_3$, acid; $Cl_2$, base
          Step 2: $Cl^+FeCl_4^-$, acid; $C_6H_6$, base
          Step 4: $H^+$, acid; $FeCl_4^-$, base

**17.110**  $5.6 \times 10^{-8}\ M$

**17.115**  (a) Strong-acid/strong-base solution would have lower pH.
       (b) Strong-acid solution would have lower pH.
       (c) Strong-acid solution would have lower pH.

**17.116**  (a) $2.28 \times 10^{-7}\ M$
       (b) $5.2 \times 10^{-12}\ M$
       (c) $5.2 \times 10^{-11}\ M$
       (d) $1.10 \times 10^{-10}\ M$
       (e) 5.98

**17.122**  $4.5 \times 10^{-5}$

**17.123**  (a) 10.0
       (b) $[OH^-] = 1.8 \times 10^{-7}\ M$, assuming aromatic N is unaffected by the tertiary amine
          $[OH^-] = 1.1 \times 10^{-4}\ M$, due to tertiary amine; therefore, contribution of the aromatic nitrogen is negligible.
       (c) 1.8
       (d) 2.4

### CHAPTER 18

**18.3**  The presence of an ion in common between two solutes will cause any equilibrium involving either of them to shift in accordance with LeChâtelier's principle. For example, addition of NaF to a solution of HF will cause the equilibrium to shift to the left, away from the excess of $F^-$.

$$HF(aq) + H_2O(l) \rightleftharpoons H_3O^+(aq) + F^-(aq)$$

**18.6**  Only (c) and (e) affect the buffer capacity. A high-capacity buffer will result when comparable quantities of weak acid and weak base are dissolved in relatively high concentration.

**18.9**  (a) pH will increase; ratio of weak acid/conjugate base will decrease.
      (b) pH will decrease; ratio of weak acid/conjugate base will increase.
      (c) pH will increase; ratio of weak acid/conjugate base will decrease.
      (d) pH will decrease; ratio of weak acid/conjugate base will increase.

**18.11**  $7.7 \times 10^{-6}\ M\ [H_3O^+]$; pH = 5.11

**18.13**  pH = 3.80

**18.15**  pH = 9.55

**18.17**  (a) $K_{a2} = 4.7 \times 10^{-11}$
       (b) pH = 10.44

**18.21**  pH = 3.36

**18.23**  (a) pH = 4.91
       (b) 0.67 g KOH

**18.25**  (a) $HCOOH/NaHCOO$; $C_6H_5NH_2/C_6H_5NH_3Cl$
       (b) $NaH_2PO_3/Na_2HPO_3$; $NaH_2AsO_4/Na_2HAsO_4$
       (c) $HClO/NaClO$

**18.29**  (a) $[HCOO^-]/[HCOOH] = 1$
       (b) Assuming volumes are additive, mix 400 mL of $1.0\ M$ HCOOH and 200 mL of $1.0\ M$ NaOH.
       (c) $0.333\ M$ HCOOH

**18.32**  Concentration of indicator is very small.

**18.35**  (a) Initial pH would be lower than for the weak-acid/strong-base titration.
       (b) Equivalence point pH value would be in the order strong acid/weak base (<7) less than strong acid/strong base (=7) less than weak acid/strong base (>7).

**18.37**  pH = $pK_a$

**18.39**  $pK_a \pm 1 \approx 7.3$ to 9.3

**18.41**  (a) Bromthymol blue
       (b) Thymol blue or phenolphthalein
       (c) Bromcresol purple

**18.43**  (a)  $1.00\ M$
       (b)  1.48
       (c)  3.0
       (d)  4.0
       (e)  7.00
       (f)  10.0
       (g)  11.96

**18.45**  (a)  2.91
       (b)  4.81
       (c)  5.29
       (d)  6.09
       (e)  7.41
       (f)  8.76
       (g)  10.10
       (h)  12.05

**18.47**  (a) 8.54; 59.0 mL NaOH
       (b) 4.38, 9.69; 45.2 mL, 90.4 mL

**18.51** Because $F^-$ is the anion of a weak acid and $Cl^-$ is the anion of a strong acid. The equilibrium, $F^-(aq) + H_2O(l) \rightleftharpoons HF(aq) + OH^-(aq)$, would be shifted by changing the pH, but the corresponding equilibrium is not set up for $Cl^-$.

**18.53** Solid precipitates from solution.

**18.54** (a) $Q_{sp} = [Ag^+]^2[CO_3^{2-}]$
(b) $Q_{sp} = [Ba^{2+}][F^-]^2$
(c) $Q_{sp} = [Cu^{2+}][HS^-][OH^-]$

**18.56** $K_{sp} = 1.3 \times 10^{-4}$

**18.60** (a) $2.3 \times 10^{-5}\ M$
(b) $4.2 \times 10^{-9}\ M$

**18.62** (a) $1.7 \times 10^{-3}\ M$
(b) $2.0 \times 10^{-4}\ M$

**18.64** (a) $Mg(OH)_2$
(b) $PbS$
(c) $Ag_2SO_4$

**18.66** (a) $AgCl(s) \rightleftharpoons Ag^+(aq) + Cl^-(aq)$
$Cl^-$ is the anion of a strong acid, so it is nonbasic. No change in solubility with change in pH.
(b) $SrCO_3(s) \rightleftharpoons Sr^{2+}(aq) + CO_3^{2-}(aq)$
$CO_3^{2-}$ is the anion of a weak acid.
$CO_3^{2-}(aq) + H_2O(l) \rightleftharpoons HCO_3^-(aq) + OH^-(aq)$
Change in solubility with change in pH
(c) $CuBr(s) \rightleftharpoons Cu^+(aq) + Br^-(aq)$
No change in solubility with change in pH

**18.68** Precipitate will form; $Q_{sp} > K_{sp}$.

**18.70** $2.7 \times 10^{-5}\ M$, $0.014$ g/L; $6.1 \times 10^{-8}\ M$, $3.1 \times 10^{-5}$ g/L

**18.75** This is most likely due to the formation of a complex ion by a reaction like
$$Pb^{2+}(aq) + nOH^-(aq) \rightleftharpoons Pb(OH)_n^{2-n}(aq)$$

**18.76** $Cr(H_2O)_6^{3+}(aq) + 6Cl^-(aq) \rightarrow CrCl_6^{3-}(aq) + 6H_2O\ (l)$

**18.78** $1 \times 10^{-5}\ M$

**18.80** $9.4 \times 10^{-5}\ M$

**18.83** (a) $3 \times 10^{-8}\ M$
(b) $Pb^{2+}$, $Hg^{2+}$, $Cu^{2+}$, and $Ag^+$

**18.87** pH $= 7.38$

**18.91** $0.355\ M$; $8.53$

**18.93** ZnS is less soluble, so the solution could be saturated with $H_2S$, then adjusted to a pH less than 6.60.

**18.98** Yes, with the metal ion the Lewis acid, and with the ligand(s) the Lewis base(s)
$$\underset{\text{acid}}{Ag^+(aq)} + \underset{\text{base}}{2NH_3(aq)} \rightleftharpoons \underset{\text{adduct}}{Ag(NH_3)_2^+(aq)}$$

**18.101** $5.88 \times 10^3$ kg AgBr

**18.102** $K_b = \dfrac{[BH^+][OH^-]}{[B]}$

$-\log K_b = -\log[OH^-] - \log[BH^+]/[B]$
$pOH = pK_b - \log[B]/[BH^+]$

**18.104** $[HPO_4^{2-}]/[H_2PO_4^-] = 6$

**18.106** pH $\approx 3.5$ to $4.0$

**18.108** (a) $58.2$ mL NaOH
(b) $7.84$ mL NaOH
(c) $7.70$

**18.112** (a) $CaF_2$
(b) Add KF until $[F^-]$ is such that the $CaF_2$ precipitates, but add an amount just less than that required to precipitate $BaF_2$.
(c) $[F^-] = 6.3 \times 10^{-3}\ M$

**18.113** $5.68$

**CHAPTER 19**

**19.2** Spontaneous process is capable of proceeding as written or described without need of an outside source of energy. Under a specific set of conditions, if a process is spontaneous in one direction, it is not spontaneous in the other. By changing the set of conditions, a nonspontaneous process can be made spontaneous.

**19.5** (a) $(1/2)^3$ or $1/8$
(b) $W = 2^n$
(c) $(1/2)^8$ or $1/256$

**19.8** Larger value for $\Delta S_{vap}$ results largely from the increased volume in which the molecules may be found. An increase in volume means an increase in randomness. Gases have much higher entropy than liquids or solids due to the random motion of their molecules.

**19.10** (a) Zero
(b) Increase
(c) $S^0$ must be greater than zero.
(d) Actual entropy values can be determined.

**19.13** (a), (b), (c), and (d) are spontaneous.

**19.15** (a) $\Delta S_{sys} > 0$
(b) $\Delta S_{sys} < 0$
(c) $\Delta S_{sys} < 0$
(d) $\Delta S_{sys} > 0$
(e) $\Delta S_{sys} > 0$

**19.17** (a) $\Delta S_{sys} < 0$
(b) $\Delta S_{sys} < 0$
(c) $\Delta S_{sys} > 0$
(d) $\Delta S_{sys} < 0$
(e) $\Delta S_{sys} < 0$

**19.19** (a) $\Delta S_{sys} > 0$
(b) $\Delta S_{sys} < 0$
(c) $\Delta S_{sys} > 0$

**19.21** (a) Butane (gas); 2-butene has restricted rotation about the double bond and would have less flexibility.
(b) $Xe(g)$; greater mass
(c) $CH_4(g)$; $S$(gas) is greater than $S$(liquid).
(d) $N_2O_4(g)$; greater molecular complexity

**19.23** (a) Diamond < graphite < charcoal; entropy increases from the more ordered crystalline state to the amorphous state.
(b) Ice < liquid water < water vapor; entropy increases as substance changes from a solid to a liquid to a gas.
(c) O atoms < $O_2$ < $O_3$; entropy increases with molecular complexity.
(d) Ribose < glucose < sucrose; entropy increases with molecular complexity.

(e) $CaCO_3(s) < CaO(s) + CO_2(g) < Ca(s) + C(s) + \frac{3}{2}O_2(g)$; entropy increases with moles of gas particles.

(f) $SF_4(g) < SF_6(g) < S_2F_{10}(g)$; entropy increases with molecular complexity.

**19.25** (a) $-172.4$ J/K

(b) $141.6$ J/K

(c) $-837$ J/K

**19.27** $93.1$ J/K; yes

**19.29** $-75.6$ J/K

**19.32** $-97.2$ J/K

**19.35** $\Delta G = \Delta H - T\Delta S$. Since $T\Delta S > \Delta H$ for an endothermic reaction to be spontaneous, the reaction is more likely to be spontaneous at higher temperatures.

**19.37** Entropy values increase only gradually within a particular phase. Thus, as long as no phase changes occur, the value of $\Delta S^0$ is relatively independent of temperature.

**19.38** (a) $-1138$ kJ

(b) $-1379.4$ kJ

(c) $-224$ kJ

**19.40** (a) $-1137$ kJ

(b) $-1379$ kJ

(c) $-225$ kJ

**19.42** (a) Both $\Delta S^0$ and $\Delta H^0$ are negative. A decrease in the number of moles of gas should result in a negative $\Delta S^0$ value. The combustion reaction will result in a release of energy or a negative $\Delta H^0$.

(b) $\Delta G^0 = -257.3$ kJ ($\Delta G^0 = \Delta H^0 - T\Delta S^0$)
$\Delta G^0 = -257.2$ kJ (from $\Delta G_f^0$ values)

**19.44** (a) $\Delta S^0 = -409$ J/K · mol

(b) $\Delta G^0 = -198$ kJ/mol

**19.46** (a) $\Delta H^0 = 90.7$ kJ; $\Delta S^0 = 221$ J/K

(b) $\Delta G = 22.0$ kJ, $-0.1$ kJ, and $-22.2$ kJ, respectively

(c) The reaction is unfavorable at $38°C$, favorable at $238°C$, and essentially at equilibrium at $138°C$.

**19.48** $T = 331.8$ K ($58.7°C$)

**19.51** $\Delta H_{rxn}^0 = -67.8$ kJ/mol
$\Delta S_{rxn}^0 = 537$ J/K
$\Delta G_{298}^0 = -227.8$ kJ/mol
No; a reaction with a negative value for $\Delta H$ and a positive value for $\Delta S$ is spontaneous at all temperatures.

**19.54** For a spontaneous process, $\Delta G$ is the maximum useful work from the system. In reality, the actual work is less because of energy lost as heat. The more slowly or controlled the free energy is released, the closer the actual work approaches $\Delta G$.

**19.56** Standard free energy, $\Delta G^0$, occurs when all components of the system are in their standard states. Under standard state conditions, $\Delta G = \Delta G^0$.

**19.57** (a) $K_{eq} = 2 \times 10^6$

(b) $K_{eq} = 2.4 \times 10^{33}$

(c) $K_{eq} = 1.2 \times 10^{48}$

(d) $K_{eq} = 1.9 \times 10^8$

**19.59** $K_{sp} = 5.3 \times 10^{-51}$

**19.61** $K_p = 1.3 \times 10^2$

**19.63** Since $Q > K_{sp}$, it is impossible to prepare a standard state solution of $PbCl_2$.

**19.65** (a) $\Delta G^0 = 28.7$ kJ/mol

(b) Since $\Delta G^0 > 0$, the reaction proceeds to the left.

(c) $\Delta G_{298} = 7$ kJ/mol
Since $\Delta G_{298} > 0$ and since $Q > K$, the reaction proceeds to the left to equilibrium.

**19.67** (a) Since $\Delta H_{rxn}^0 > 0$ and $\Delta S_{rxn}^0 < 0$, this reaction is not spontaneous at all temperatures.

(b) $\Delta G_{rxn}^0 = 326$ kJ

(c) $\Delta G_{298} = 266$ kJ

**19.69** Reaction as written is nonspontaneous at all temperatures. Thus, reverse reaction (conversion of diamond to graphite) is spontaneous at all temperatures. Even though the conversion to graphite is spontaneous, the rate is extremely slow.

**19.71** $\Delta S^0 = +$; $\Delta G^0 = -$; the reaction proceeds due to an increase in entropy.

**19.75** (a) $\Delta H_{rxn}^0 = -128.1$ kJ; $\Delta S^0 = -332$ J/K; $\Delta G_{298}^0 = -29.2$ kJ

(b) $\Delta H^0 < 0$; $\Delta S^0 < 0$; favored at low temperature

(c) $\Delta G_{373} = -182$ kJ

**19.78** (a) $\Delta H^0 = 470.5$ kJ; $\Delta S^0 = 558.4$ J/K

(b) At high temperature, where $T\Delta S$ is larger in magnitude than $\Delta H$

(c) No

(d) $842.6$ K

**19.79** (a) Hydrolysis is spontaneous at $25°C$.

(b) Yes, the decomposition is spontaneous at $25°C$. Decomposition of solid $N_2O_5$ becomes spontaneous at temperatures above $271$ K.

(c) Decomposition of solid $N_2O_5$ becomes spontaneous at temperatures above $230$ K. Less energy needs to be added, since the material is already vaporized.

**19.84** (a) $\Delta S^0 > 0$. Formation of a large number of components increases randomness.

(b) $\Delta G > 0$ when $T\Delta S < \Delta H$

(c) $T = \dfrac{\Delta H}{\Delta S}$

## CHAPTER 20

**20.3** An electrochemical process involves an electron flow. At least one substance must lose electron(s) and another must gain electron(s) to produce the flow.

**20.8** (a) $Cl^-$

(b) $MnO_4^-$

(c) $MnO_4^-$

(d) $Cl^-$

(e) From $Cl^-$ to $MnO_4^-$

(f) $8H_2SO_4(aq) + 2KMnO_4(aq) + 10KCl(aq) \rightarrow 2MnSO_4(aq) + 6K_2SO_4(aq) + 5Cl_2(g) + 8H_2O(l)$

**20.9** A voltaic or galvanic cell has $\Delta G_{sys} < 0$. An electrolytic cell has $\Delta G_{sys} > 0$, meaning an increase in free energy.

**20.10** (a) True

(b) True

(c) True

(d) False; in a voltaic cell, the system does work on the surroundings.

(e) True

(f) False; the electrolyte in a cell provides a solution of mobile ions to maintain charge neutrality.

**20.13** (a) $Cr_2O_7^{2-}(aq) + 3Zn(s) + 14H^+(aq) \rightarrow$
$2Cr^{3+}(aq) + 3Zn^{2+}(aq) + 7H_2O(l)$
OA $= Cr_2O_7^{2-}$; RA $= Zn$

(b) $3Fe(OH)_2(s) + MnO_4^-(aq) + 2H_2O(l) \rightarrow$
$MnO_2(s) + 3Fe(OH)_3(s) + OH^-(aq)$
OA $= MnO_4^-$    RA $= Fe(OH)_2$

(c) $5Zn(s) + 2NO_3^-(aq) + 12H^+(aq) \rightarrow$
$5Zn^{2+}(aq) + N_2(g) + 6H_2O(l)$
OA $= NO_3^-$; RA $= Zn$

**20.16** (a) $Au(s) + 3NO_3^-(aq) + 6H^+(aq) + 4Cl^-(aq) \rightarrow$
$AuCl_4^-(aq) + 3NO_2(g) + 3H_2O(l)$

(b) OA $= NO_3^-$; RA $= Au$

(c) To provide $Cl^-$ to act as a complexing agent

**20.17** (a) A    (b) E    (c) C    (d) A    (e) E    (f) E

**20.20** Active electrodes are components of the half-cell reactions. Inactive electrodes are used to conduct electrons but do not participate in the half-cell reactions. Pt and graphite

**20.23** Cell potentials change signs when the reaction is reversed. Unlike other thermodynamic quantities, the magnitude does not change when the coefficients are changed.

**20.25** (a) Left to right    (b) Left    (c) Right

(d) Ni           (e) Fe    (f) Fe

(g) $1\,M$ $NiSO_4$      (h) $K^+$, $NO_3^-$

(i) Ni           (j) Right to left      (k) $+0.19$ V

(l) Reduction: $Ni^{2+}(aq) + 2e^- \rightarrow Ni(s)$
Oxidation: $Fe(s) \rightarrow Fe^{2+}(aq) + 2e^-$
Overall: $Fe(s) + Ni^{2+}(aq) \rightarrow Ni(s) + Fe^{2+}(aq)$

**20.27** (a) Reduction: $Sn^{2+}(aq) + 2e^- \rightarrow Sn(s)$
Oxidation: $Zn(s) \rightarrow Zn^{2+}(aq) + 2e^-$
Overall: $Zn(s) + Sn^{2+}(aq) \rightarrow Sn(s) + Zn^{2+}(aq)$

(b)

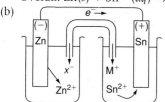

(c) 0.62 V

**20.29** (a) $Al(s)|Al^{3+}(aq)\|Cr^{3+}(aq)|Cr(s)$

(b) $Pt|SO_2(g)|SO_4^{2-}(aq), H^+(aq)\|Cu^{2+}(aq)|Cu(s)$

**20.31** (a) Reduction: $2SO_3^{2-}(aq) + 3H_2O(l) + 4e^- \rightarrow$
$S_2O_3^{2-}(aq) + 6OH^-(aq)$
Oxidation: $Se^{2-}(aq) \rightarrow Se(s) + 2e^-$

(b) $-0.92$ V

**20.33** (a) $Br_2 > Fe^{3+} > Cu^{2+}$

(b) $Ca^{2+} < Ag^+ < Cr_2O_7^{2-}$

**20.35** (a) $Co(s) + 2H^+(aq) \rightarrow Co^{2+}(aq) + H_2(g)$; $E^0 = 0.28$ V; spontaneous

(b) $2Mn^{2+}(aq) + 5Br_2(l) + 8H_2O(l) \rightarrow$
$2MnO_4^-(aq) + 10Br^-(aq) + 16H^+(aq)$;
$E^0 = -0.44$ V; not spontaneous

(c) $2Hg_2^{2+}(aq) \rightarrow 2Hg^{2+}(aq) + 2Hg(l)$; $E^0 = -0.07$ V; not spontaneous

(d) $Cl_2(g) + 2Fe^{2+}(aq) \rightarrow 2Cl^-(aq) + 2Fe^{3+}(aq)$;
$E^0 = 0.59$ V; spontaneous

**20.37** $3SO_4^{2-}(aq) + 3H_2O(l) + 2Al(s) \rightarrow 3SO_3^{2-}(aq) +$
$6OH^-(aq) + 2Al^{3+}(aq)$
$E^0 = 2.59$ V
$3N_2O_4(g) + 2Al(s) \rightarrow 6NO_2^-(aq) + 2Al^{3+}(aq)$
$E^0 = 2.53$ V
$SO_4^{2-}(aq) + H_2O(l) + 2NO_2^-(aq) \rightarrow SO_3^{2-}(aq) +$
$2OH^-(aq) + N_2O_4(g)$
$E^0 = 0.06$ V
Oxidizing strength: $SO_4^{2-} > N_2O_4 > Al^{3+}$
Reducing strength: $Al > NO_2^- > SO_3^{2-}$

**20.39** Reducing strength: $C > A > B$; C would cause formation of $H_2$.

**20.41** (a) Cu

(b) OA: $Cu^{2+}$; RA: Fe

(c) Yes

(d) $Cu^{2+}(aq) + Fe(s) \rightarrow Fe^{2+}(aq) + Cu(s)$

(e) 0.78 V

**20.43** (a) $[A^+]$ increases and $[B^+]$ decreases.

(b) Decreases

(c) $E = E^0$ when $\log ([A^+]/[B^+]) = 0$, when $[A^+]/[B^+] = 1$

(d) Yes, if $[A^+]/[B^+] > 1$

**20.44** (a) If $Q/K < 1$, $E$ decreases with decreasing $T$; if $Q/K > 1$, $E$ increases with decreasing $T$.

(b) Decrease

(c) Increase

(d) Decrease

**20.46** (a) $3 \times 10^{35}$

(b) $4 \times 10^{-31}$

**20.48** (a) $K = 3 \times 10^9$; $\Delta G^0 = -54$ kJ

(b) $K = 5 \times 10^{-75}$; $\Delta G^0 = 4.2 \times 10^2$ kJ

(c) $K = 4 \times 10^{-3}$; $\Delta G^0 = 1 \times 10^1$ kJ

(d) $K = 8 \times 10^{19}$; $\Delta G^0 = -1.1 \times 10^2$ kJ

**20.50** (a) $E^0 = 0.22$ V; $\Delta G^0 = -21$ kJ

(b) $E^0 = -0.31$ V; $\Delta G^0 = 30$ kJ

**20.52** $1 \times 10^{-3}\,M$

**20.54** (a) 0.05 V

(b) 0.03 V

(c) $0.33\,M$

(d) $[Ni^{2+}] = 0.09\,M$; $[Co^{2+}] = 0.91\,M$

**20.56** Electrode A is the anode; 0.085 V.

**20.58** From the anode to the cathode

**20.61** (a) 4

(b) 9.0 V

(c) 8 V

**20.62** Oxidation of the Cu by the atmosphere produces $Cu^{2+}$. If allowed to contact Fe, the more active Fe would react: $Fe(s) + Cu^{2+}(aq) \rightarrow Cu(s) + Fe^{2+}$.

**20.65** A more reactive metal (one preferentially oxidized) is needed. Only (a), (b), and (c) are suitable (although Na is too reactive to last very long!).

**20.68** In pure $H_2O$, $[H^+] = [OH^-] = 1 \times 10^{-7}\,M$, different from the standard value of $1\,M$

**20.69** $NO_3^-$ has N in its highest possible oxidation state, so it cannot be oxidized at the anode.

**20.72** (a) $Br_2(g)$
(b) $Na(l)$

**20.74** (a) $I_2(g)$
(b) $Mg(l)$

**20.76** Copper, bromine, and chlorine

**20.78** (a) Cathode: $H_2(g) + OH^-(aq)$; anode: $O_2(g) + H^+(aq)$
(b) Cathode: $Sn(s)$; anode: $O_2(g) + H^+(aq)$
(c) Cathode: $Zn(s)$; anode: $Br_2(l)$

**20.80** (a) $2.00 \, \mathscr{F}$
(b) $1.93 \times 10^5$ C
(c) 35.7 A

**20.82** $9.18 \times 10^{-2}$ gK

**20.84** $1.83 \times 10^3$ s

**20.89** 44.2 g

**20.90**

| | Voltaic cell | Electrolytic cell |
|---|---|---|
| (a) | $\Delta G < 0$ | $\Delta G > 0$ |
| (b) | Oxidation | Oxidation |
| (c) | Reduction | Reduction |
| (d) | $(-)$ | $(+)$ |
| (e) | Anode | Anode |

**20.92** (a) $2.0 \times 10^3$ min
(b) 1.61 g

**20.95** 77%

**20.97** (a) 0.13 V for Pb; 0.34 V for Cu
(b) $Pb(s)$; Pt (in $H_2$ cell)
(c) Increases
(d) $-0.13$ V

**20.99** Wherever the tin surface is scratched to expose the iron, the iron corrosion occurs more rapidly because iron is a better reducing agent. An electrochemical cell is set up in which iron is the anode and tin the cathode. A coating on the inside of the can separates the tin from the (normally) acidic contents, which would react with the tin and add a "metallic" taste.

**20.101** (a) From Mg to the pipe
(b) 0.3 A

**20.104** $K$ could be calculated from $\Delta G_f^0$ values (which could be calculated from $\Delta H_f^0$ and $S^0$ values) or from $E^0$ values.

**20.107** $-0.42$ V

**20.111** $1.6 \times 10^{-10}$

# CHAPTER 21

**21.1** (a) Chemical reactions are accompanied by relatively small changes in energy while nuclear reactions are accompanied by relatively large changes in energy.
(b) Rates of chemical reactions are increased by increasing temperature while nuclear reactions are not affected by temperature.
(c) Both are increased by higher reactant concentrations.
(d) Higher reaction concentrations would increase the yield of nuclear reactions but would have no change on chemical reactions.

**21.4** Reactions (a), (b), (d), and (e) all show a change in the number of protons ($Z$), and therefore a different element is formed. All show a change in the number of neutrons ($N$).

**21.7** Both kinds of decay increase the number of neutrons and decrease the number of protons. Positron emission is more common than electron capture among lighter nuclei; electron capture becomes increasingly common as nuclear charge increases.
For $Z < 20$, $\beta^+$ emission; for $Z > 80$, $e^-$ capture

**21.8** (a) $^{234}_{92}U \rightarrow \, ^4_2He + \, ^{230}_{90}Th$
(b) $^{232}_{93}Np + \, ^0_{-1}e \rightarrow \, ^{232}_{92}U$
(c) $^{12}_{7}N \rightarrow \, ^0_1\beta + \, ^{12}_6C$
(d) $^{26}_{11}Na \rightarrow \, ^0_{-1}\beta + \, ^{26}_{12}Mg$

**21.10** (a) $^{48}_{23}V \rightarrow \, ^{48}_{22}Ti + \, ^0_1\beta$
(b) $^{107}_{48}Cd + \, ^0_{-1}e \rightarrow \, ^{107}_{47}Ag$
(c) $^{210}_{86}Rn \rightarrow \, ^{206}_{84}Po + \, ^4_2He$
(d) $^{241}_{94}Pu \rightarrow \, ^{241}_{95}Am + \, ^0_{-1}\beta$

**21.12** (b) and (d) stable; $N/Z$ ratio in band of stability

**21.14** (a) Alpha decay
(b) Positron emission or electron capture
(c) Positron emission or electron capture
(d) Beta decay

**21.16** $N$ and $Z$ both even; lies in band of stability

**21.20** Net loss of 28 amu or 7 $\alpha$ and 4 $\beta$ particles.

**21.23** No; a decay rate is an average rate and is meaningful only when we consider a large number of nuclei in a macroscopic sample. The conclusion would be valid if the original sample were large.

**21.25** $2.89 \times 10^{-2}$ Ci/g

**21.27** $k = 1 \times 10^{-12}$ h$^{-1}$

**21.29** $k = 2.31 \times 10^{-7}$ yr$^{-1}$

**21.31** 87 yr

**21.33** $2.2 \times 10^9$ yr

**21.35** $1.69 \times 10^5$ yr

**21.37** $1.6 \times 10^6$ yr

**21.38** Approximately 1410 years ago

**21.41** Both gamma radiation and neutrons have no charge, but the neutron has mass approximately equal to that of a proton.

**21.43** Protons have a positive charge while neutrons are neutral.

**21.44** (a) $^{10}_{5}B + \, ^4_2He \rightarrow \, ^1_0n + \, ^{13}_7N$
(b) $^{28}_{14}Si + \, ^2_1H \rightarrow \, ^1_0n + \, ^{29}_{15}P$
(c) $^{242}_{96}Cm + \, ^4_2He \rightarrow 2^1_0n + \, ^{244}_{98}Cf$

**21.47** Relatively low-energy emission, such as ultraviolet, causes excitation of the surrounding matter, whereas high-energy emission, such as gamma, causes ionization.

**21.50** Hydroxy free radicals go on to attack surrounding biomolecules, whose bonding and structure are delicately connected to their function.

**21.51** (a) $7.5 \times 10^{-8}$ rad
(b) $7.5 \times 10^{-5}$ mrem

**21.53** $1.9 \times 10^{-8}$ rad

**21.57** In positron-emission tomography (PET), the isotope emits positrons, each of which annihilates a nearby electron. In the process, two $\gamma$ photons are emitted simultaneously 180° apart from each other. Detectors locate the sites, and the image is analyzed by computer.

**21.59** Methanol supplies oxygen to the aldehyde. The oxidation involves the loss of hydrogen atoms from the methanol to form the formaldehyde. The $H_2CrO_4$ supplies the oxygen to the water.

**21.63** Energy is released when nuclides form from their nucleons. The nuclear binding energy is the amount of energy holding the nucleus together.

**21.65** Two or three neutrons

**21.69** Light-water reactors simply use $H_2O$ as the moderator; in heavy-water reactors, $D_2O$ is used. The advantage of $D_2O$ is that it absorbs fewer neutrons, leaving more available for fission, and thus $^{235}U$ fuel does not need to be enriched. The disadvantage is the initial additional cost of $D_2O$.

**21.72** After the initial hydrogen burning process, the star's energy source changes to helium burning, in which continual heavier nuclei are formed by the fusion of helium nuclei, which have a mass of 4 amu.

**21.74** (a) $1.861 \times 10^4$ eV
(b) $2.981 \times 10^{-15}$ J

**21.76** $5.1 \times 10^{11}$ J

**21.78** (a) $-7.573 \dfrac{MeV}{nucleon}$

(b) $-121.2 \dfrac{MeV}{atom}$

(c) $-1.2313 \times 10^{10}$ kJ/mol

**21.81** $-1.054 \times 10^{25} \dfrac{MeV}{mol}$

**21.84** $7.63 \times 10^3$ yr

**21.88** $1.35 \times 10^{-5}$ $M$

**21.91** $2.85 \times 10^{-4}$ Ci

**21.95** $1.52 \times 10^{-4}$ g/mol

**21.99** $^{239}_{92}U \rightarrow {}^{0}_{-1}\beta + {}^{239}_{93}Np \rightarrow {}^{239}_{94}Pu + {}^{0}_{-1}\beta \rightarrow {}^{4}_{2}He +$
$^{235}_{92}U \rightarrow {}^{4}_{2}He + {}^{231}_{90}Th$
This could begin the $^{235}_{92}U$ decay series.

**21.101** More; number of radioactive nuclei would have been larger.

**21.103** $-6.05 \times 10^7$ kJ/mol

**CHAPTER 22**

**22.3** (a) $1s^2 2s^2 2p^6 3s^2 3p^6 4s^2 3d^{10} 4p^6 5s^2 4d^x$
(b) $1s^2 2s^2 2p^6 3s^2 3p^6 4s^2 3d^{10} 4p^6 5s^2 4d^{10} 5p^6 6s^2 4f^{14} 5d^x$

**22.5** Five; Mn ([Ar] $4s^2 3d^5$) and $Mn^{2+}$ ([Ar] $3d^5$).

**22.7** (a) The lanthanide contraction is the "shrinkage" of atoms following the filling of the $4f$ subshell (i.e., the atoms are smaller than expected).
(b) The size increases from Period 4 to Period 5 but stays fairly constant (or increases only slightly) from Period 5 to Period 6.
(c) The smaller-than-expected sizes lead to very large densities.

**22.11** (a) [Ar] $4s^2 3d^3$
(b) [Kr] $5s^2 4d^1$
(c) [Xe] $6s^2 4f^{14} 5d^{10}$
(d) [Kr] $5s^2 4d^6$

(e) [Ar] $4s^1 3d^{10}$

**22.13** (a) [Ar]; 0
(b) [Ar] $3d^9$; 1
(c) [Ar] $3d^5$; 5
(d) [Kr] $4d^2$; 2

**22.15** (a) $+5$  (b) $+4$  (c) $+7$

**22.17** Cr, Mo, and W

**22.19** In $CrF_2$, since metallic behavior is greater in lower oxidation states

**22.21** Because of the lanthanide contraction, the ionization energy of Mo is higher than that of Cr, making Mo more difficult to oxidize.

**22.23** $CrO_3$, because higher oxidation states form more acidic oxides

**22.25** The abnormally small size of Au atoms (due to the lanthanide contraction) means that its electrons are tightly held, giving it low reactivity. The Group 1A(1) elements do not show the lanthanide contraction.

**22.27** (a) 7
(b) This corresponds to a half-filled $f$ subshell.

**22.29** (a) [Xe] $6s^2 5d^1$
(b) [Xe] $4f^1$
(c) [Rn] $7s^2 5f^{11}$
(d) [Rn] $5f^2$

**22.31** (a) $Eu^{2+}$: [Xe] $4f^7$
$Eu^{3+}$: [Xe] $4f^6$
$Eu^{4+}$: [Xe] $4f^5$
The stability of the half-filled $f$ subshell makes $Eu^{2+}$ most stable.
(b) $Tb^{2+}$: [Xe] $4f^9$
$Tb^{3+}$: [Xe] $4f^8$
$Tb^{4+}$: [Xe] $4f^7$
Tb would show a $+4$ oxidation state, since that has a half-filled $f^7$ configuration.

**22.35** This is the effective electronegativity of the element in a given oxidation (valence) state. As the oxidation number increases, the atom becomes more electronegative, strengthening its M—O bonds and making the oxide more acidic. MnO is basic; $Mn_2O_7$ is acidic.

**22.37** Its lack of chemical reactivity with oxygen

**22.41** (a) $MnO_4^{2-}$: $+6$; $MnO_4^-$: $+7$; $MnO_2$: $+4$
(b) $3MnO_4^{2-}(aq) + 4H^+(aq) \rightarrow MnO_2(s) + 2MnO_4^-(aq) + 2H_2O(l)$

**22.45** $E^0 = E^0_{MnO_4^- \rightarrow MnO_2} - E^0_{MnO_2 \rightarrow Mn^{2+}} = 0.45$ V.
Since $E^0 > 0$, the reaction is favorable.

**22.46** A complex ion forms as a result of a metal ion acting as a Lewis acid by accepting one or more pairs of electrons from the ligand(s), which act(s) as a Lewis base. If the ligands are neutral (or, if negative, there are too few to neutralize the positive charge on the metal), the complex ion will be positive. If there are more than enough negative ligands to cancel the charge on the metal, the complex ion will be negative. The complex ion associates with enough ions of opposite charge to form a neutral coordination compound.

**22.50** 2: linear; 4: tetrahedral, square planar; 6: octahedral

**22.53** The ion has a negative charge.

**22.56** (a) Hexaaquanickel(II) chloride

(b) Trisethylenediaminechromium(III) perchlorate

(c) Potassium hexacyanomanganate(II)

(d) Tetraamminedinitritocobalt(III) chloride

(e) Hexaamminechromium(III) hexacyanochromate(III)

**22.58** (a) +2, 6

(b) +3, 6

(c) +2, 6

(d) +3, 6

(e) +3, 6; +3, 6

**22.60** (a) $[Zn(NH_3)_4]SO_4$

(b) $[Cr(NH_3)_5Cl]Cl_2$

(c) $Na_3[Ag(S_2O_3)_2]$

(d) $[Co(C_2H_8N_2)Br_2]_2SO_4$

(e) $[Cr(NH_3)_6]_2[CuCl_4]_3$

**22.62** (a) 4; 2

(b) 6; 3

(c) 2; 2

(d) 6; 3

(e) 6, 4; 5

**22.64** (a) $[\ddot{O}=\ddot{N}-\ddot{O}: \leftrightarrow :\ddot{O}-\ddot{N}=\ddot{O}:]^-$ can form linkage isomers, since two different atoms (O, N) have lone pairs.

(b) $\ddot{O}=\ddot{S}-\ddot{O}: \leftrightarrow :\ddot{O}-\ddot{S}=\ddot{O}$ can also form linkage isomers (O, S).

(c) $\left[:\ddot{O}-N-\ddot{O}: \leftrightarrow :\ddot{O}=N-\ddot{O}: \leftrightarrow :\ddot{O}-N=\ddot{O}:\right]^-$
$\qquad \overset{\|}{:}O: \qquad\qquad :O: \qquad\qquad :O:$

**22.66** (a) Geometric isomers

$$\begin{array}{c} Br \quad Br \\ \diagdown / \\ Pt \\ / \diagdown \\ CH_3NH_2 \quad NH_2CH_3 \end{array} \text{ and } \begin{array}{c} Br \quad NH_2CH_3 \\ \diagdown / \\ Pt \\ / \diagdown \\ CH_3NH_2 \quad Br \end{array}$$

(b) Geometric isomers

$$\begin{array}{c} H_3N \quad NH_3 \\ \diagdown / \\ Pt \\ / \diagdown \\ Cl \quad F \end{array} \text{ and } \begin{array}{c} H_3N \quad F \\ \diagdown / \\ Pt \\ / \diagdown \\ Cl \quad NH_3 \end{array}$$

(c) Geometric isomers

$$\begin{array}{c} H_2O \quad NH_3 \\ \diagdown / \\ Pt \\ / \diagdown \\ Cl \quad F \end{array} \text{ and } \begin{array}{c} H_2O \quad NH_3 \\ \diagdown / \\ Pt \\ / \diagdown \\ F \quad Cl \end{array} \text{ and } \begin{array}{c} H_2O \quad F \\ \diagdown / \\ Pt \\ / \diagdown \\ Cl \quad NH_3 \end{array}$$

(d) No isomers

(e) Optical isomers

$$\begin{array}{c} H_2O \quad NH_3 \\ \diagdown / \\ Zn \\ / \diagdown \\ Cl \quad F \end{array} \text{ and } \begin{array}{c} H_2O \quad NH_3 \\ \diagdown / \\ Zn \\ / \diagdown \\ F \quad Cl \end{array}$$

**22.68** $CrCl_3(NH_3)_4$ (really $[Cr(NH_3)_4Cl_2]Cl$)

**22.71** $Pt[P(C_2H_5)_3]_2Cl_2$

$$\begin{array}{c} P(C_2H_5)_3 \\ | \\ :\ddot{C}l-Pt-P(C_2H_5)_3 \\ | \\ :\ddot{C}l: \end{array} \text{ and } \begin{array}{c} P(C_2H_5)_3 \\ | \\ :\ddot{C}l-Pt-\ddot{C}l: \\ | \\ P(C_2H_5)_3 \end{array}$$

These are *cis*- and *trans*-dichlorobis (triethylphosphine) platinum(II), respectively.

**22.72** (a) A bond formed when both electrons came from one atom

(b) Yes; $H_2O$ molecules act as donors to $Fe^{3+}$.

(c) Yes; $H_2O$ molecules act as donors to $H^+$.

**22.75** Purple (red + violet/blue)

**22.78** A strong-field ligand causes a greater crystal field splitting (a greater $\Delta$) than a weak-field ligand.

**22.79** High spin

**22.84**

The ligands "point" directly at the electrons in the $d_{x^2-y^2}$ orbital, raising their energy. The ligands are further away from the electrons in the $d_{xy}$ orbital, so their energy is affected less.

**22.86** (a) and (d)

**22.88** (a)

(b)

(c)

(d)

(e)

**22.90** $[Cr(H_2O)_6]^{3+} < [Cr(NH_3)_6]^{3+} < [Cr(NO_2)_6]^{3-}$

**22.92** A violet complex will absorb yellow-green light. A complex of a weaker ligand will absorb lower energy yellow, orange, or red light, making the observed color blue or green.

**22.94** In an octahedral $d^8$ complex, two electrons occupy the two $e_g$ orbitals and will be unpaired. In a square-planar $d^8$ complex, the highest energy ($d_{x^2-y^2}$)

orbital is unoccupied and all other levels are full, making the complex diamagnetic.

**22.95** $NH_3$ is a stronger ligand than $H_2O$, so $[Ni(NH_3)_6]^{2+}$ will absorb higher energy light than $[Ni(H_2O)_6]^{2+}$. Being green, $[Ni(H_2O)_6]^{2+}$ is probably absorbing primarily red light, whereas violet $[Ni(NH_3)_6]^{2+}$ is probably absorbing higher energy yellow-green light.

**22.97** (a) Sodium tetrabromoferrate(III)
      (b) Hexaamminenickel(II) ion
      (c) Correct
      (d) Hexacyanovanadate(III) ion
      (e) Potassium tetrachloroferrate(III)

**22.100** (a) No
      (b) No
      (c) No
      (d) No
      (e) Yes

**22.104** (a) The first reaction would have $\Delta S \approx 0$; the second, $\Delta S < 0$. Thus, the first is more favorable (or less unfavorable).
      (b) The second $(Cu(en)_2^{2+})$ would be more stable because of the unfavorable entropy change.

**22.108** (a) $+6$; $+7$
      (b) $+6$; $+8$ (but not likely $+3$)

**22.110** (a) $2K_2MnO_4(aq) + 2H_2O(l) \rightarrow 2KMnO_4(aq) + 2KOH(aq) + H_2(g)$
      (b) 43 mol $MnO_4^-$

**22.111** (a) 8
      (b) $+5$
      (c)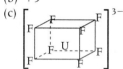

**CHAPTER 23**

**23.1** It is abundant in the universe, since its simple atoms were the first "created" after the "Big Bang." $H_2$ has the lowest density of any material, so it is only weakly held by the Earth's gravitational field.

**23.3** (a) Differentiation is the separating of the Earth's materials into different regions (layers) based primarily on their densities.
      (b) O, Si, Al, and Fe
      (c) O, Si, and all metals other than Fe and Ni

**23.5** (a) Bauxite [impure $Al(OH)_3$]
      (b) $N_2$ (air)
      (c) NaCl
      (d) $CaCO_3$
      (e) NaCl

**23.9** Fixation is the process of converting an element in a form not directly usable by animals and plants (usually in gaseous form) to a form that is usable (usually in condensed form). Nitrogen is fixed from atmospheric $N_2$, and carbon is fixed from atmosphere $CO_2$.

**23.14** (a) The atmosphere

      (b) Plants absorb phosphate in the form of $H_2PO_4^-$. Animals eat the plants and excrete phosphates, and both animals and plants produce phosphates by decay after death.

**23.17** H—N̈—H       trigonal pyramid, $109.5°$
    |
    H

   N̈=N=Ö̈  ↔  :N≡N—Ö̈:       linear, $180°$

   N̈=Ö̈       linear

   Ö̈=Ṅ—Ö̈:  ↔  :Ö̈—Ṅ=Ö̈       bent, $120°$

   $\left[\ddot{O}=\ddot{N}—\ddot{O}: \leftrightarrow :\ddot{O}—\ddot{N}=\ddot{O}\right]^-$       bent, $120°$

   $\left[:\ddot{O}—N—\ddot{O}: \leftrightarrow :\ddot{O}—N=\ddot{O} \leftrightarrow \ddot{O}=N—\ddot{O}:\right]^-$

      trigonal planar, $120°$

**23.20** (a) $1.0 \times 10^3$ L $SiF_4$
      (b) $4.2 \times 10^2$ $m^3$

**23.23** (a) $P_4O_{10}(s) + 6H_2O(l) \rightarrow 4H_3PO_4(aq)$
      (b) 1.55

**23.25** (a) Heating an ore in the presence of oxygen
      (b) Heating an ore in the presence of a reducing agent
      (c) Stirring an ore with a detergent-oil mixture to separate the mineral from the gangue
      (d) Purifying a material after its initial formation

**23.27** (a) In general, the more active metal will be the one with the lower ionization energy and/or the greater enthalpy of hydration. In this way, net energy will be given off when the more active element loses electrons and the less active one gains them, with the formation of ions of the more active element and removal from solution of the ions of the less active element. An example is $Zn(s) + Cu^{2+}(aq) \rightarrow Cu(s) + Zn^{2+}(aq)$.
      (b) This is similar to (a) except that hydration enthalpies are not involved. An example is
$$Ca(l) + 2RbCl(l) \xrightarrow{\Delta} 2Rb(g) + CaCl_2(l).$$

**23.31** Fe and Ni are more easily oxidized and less easily reduced than Cu. In the electrorefining process, all three metals are oxidized at the anode, but only Cu is reduced at the cathode.

**23.33** (a) Molecules containing different isotopes of a given element react at different rates (heavier atoms react more slowly) if a bond is broken to the atom in question.
      (b) H compounds exhibit a relatively large effect, since the mass ratio of the isotopes is the largest of any atom.
      (c) It would be smaller for C, since the mass ratio of the isotopes is smaller.

**23.36** (a) $E^0_{red} = -3.05$ V, $-2.93$ V, and $-2.71$ V for $Li^+$, $K^+$, and $Na^+$, respectively. ($Rb^+$ and $Cs^+$ would be expected to be similar.) In all of these cases, it is energetically more favorable to reduce $H_2O$ to $H_2$ than $M^+(aq)$ to M.

(b) $2RbX + Ca - CaX_2 + 2Rb$
$\Delta H = IE_1(Ca) + IE_2(Ca) - 2IE_1(Rb)$
$\quad = +941$ kJ/mol
Since $\Delta H > 0$, the reaction should be unfavorable.

(c) Rb has a lower boiling point than Ca, so the Rb produced boils out of the reaction mixture, shifting the reaction to the right.

(d) $2CsX + Ca - CaX_2 + 2Cs$
$\Delta H = IE_1(Ca) + IE_2(Ca) - 2IE_1(Cs)$
$\quad = +984$ kJ/mol
This reaction is more unfavorable than for Rb, but Cs has a lower boiling point, making the process a reasonable bet.

**23.37** (a) $4.57 \times 10^4$ LCl$_2$
(b) $1.26 \times 10^8$ C
(c) 466 h

**23.39** (a) $9.93 \times 10^3$ TCO$_2$
(b) $4.7 \times 10^4$ TCO$_2$ (greater from automobiles)

**23.43** (a) Iodine
(b) Not totally; nonmetals usually occur in negative oxidation states.
(c) $2IO_3^-(aq) + 5HSO_3^-(aq) + 2H^+(aq) \longrightarrow$
$5HSO_4^-(aq) + I_2(s) + H_2O(l)$

**23.44** (a) The $SO_2$ acts as a reducing agent and is oxidized to the +6 state.
(b) $HSO_4^-(aq)$
(c) $H_2SeO_3(aq) + 2SO_2(g) + H_2O(l) \longrightarrow Se(s) +$
$2HSO_4^-(aq) + 2H^+(aq)$

**23.47** 69.94% in $Fe_2O_3$, 72.36% in $Fe_3O_4$, and 46.55% in $FeS_2$

**23.51** (a) Parallel to the layers, since the electrons are delocalized and would be mobile in that direction
(b) Down's cell: $2Cl^- \longrightarrow Cl_2 + 2e^-$
Hall-Heroult process: $AlOF_3^{2-} + 3F^- + C \longrightarrow$
$AlF_6^{3-} + CO + 2e^-$
(c) 1.22 metric tons

**23.53** (a) $-239$ kJ
(b) 59.8%
(c) $2.78

**23.55** (a) $D^+(aq)$, $OD^-(aq)$, and traces of $CH_3O^-(aq)$ and $H^+(aq)$. Eventually, HDO, $H_2O$, and $CH_3OD$ would form.
(b) $D_2O(l) \rightleftharpoons D^+(aq) + OD^-(aq)$
$CH_3OH(aq) + OD^-(aq) \rightleftharpoons CH_3O^-(aq) + HDO(l)$
$CH_3O^-(aq) + D_2O(l) \rightleftharpoons CH_3OD(l) + OD^-(aq)$
$HDO(l) \rightleftharpoons D^+(aq) + OH^-(aq)$
$HDO(l) + OH^-(aq) \rightleftharpoons H_2O(l) + OD^-(aq)$, etc.

**23.60** All the steps in the process are exothermic, and the excess heat (in the form of steam) can be sold profitably.

**23.61** (a) $H_2$, $Cl_2$, and NaOH
(b) The product NaOH is of higher purity, but the Hg is an environmental pollutant.

**23.63** (a) $\Delta G^0 = -142$ kJ; spontaneous
(b) It is very slow at that temperature.
(c) $\Delta G_{500} = -53$ kJ; spontaneous
(d) $K_{25} = 7.8 \times 10^{24}$ atm$^{-1}$
$K_{500} = 3.8 \times 10^3$ atm$^{-1}$
(e) 781°C

**23.65** $7 \times 10^2$ lb Cl$_2$

**23.69** (a) $Ca_3N_2$
(b) $Al_4C_3$
(c) AlN
(d) GaN
(e) GaAs

**23.71** Since the $\Delta H$ terms are essentially the same, the more favorable (i.e., more negative) $\Delta G$ (as evidenced by the larger $K_{eq}$) must be due to a more favorable (or less unfavorable) $\Delta S$ term. Since fewer molecules "lose their freedom" in the second reaction, it would be entropically favored.

**23.72** (a) Because $CaCO_3$ is "insoluble" in water. In general, reactions that produce ions in solution are not favored.
(b) $CaCO_3(s) \overset{\Delta}{\longrightarrow} CaO(s) + CO_2(g)$
$CO_2(g) + NH_3(g) + H_2O(l) \longrightarrow NH_4HCO_3(aq)$
$NH_4HCO_3(aq) + NaCl(aq) \longrightarrow NaHCO_3(aq) +$
$NH_4Cl(aq)$
$2NaHCO_3(s) \overset{\Delta}{\longrightarrow} Na_2CO_3(s) + H_2O(g) + CO_2(g)$
(c) 21.7 L

**23.75** (a) $K_p(1) = 9 \times 10^{167}$; $K_p(2) = 4 \times 10^{228}$
(b) $K_p(1) = 4.5 \times 10^{49}$; $K_p(2) = 1.3 \times 10^{63}$
(c) $18 \times 10^6

**23.77** (a) $2NH_3(aq) + OCl^-(aq) \longrightarrow N_2H_4(aq) + Cl^-(aq) +$
$H_2O(l)$
(b) $\Delta G_f^0$ for $NH_3$ is $< 0$ and $\Delta G_f^0$ for $N_2H_4$ is $> 0$, so the reduction to produce $NH_3$ is favored.

**23.79** (a) $:C\equiv O:$
(b) C has a formal charge of $-1$, so it would be the more likely end for electron donation.

**23.80** (a) Cathode: $Na^+ + e^- \longrightarrow Na$
Anode: $4OH^- \longrightarrow O_2 + 2H_2O + 4e^-$
(b) $2Na(s) + 2H_2O(l) \longrightarrow 2Na^+(aq) + 2OH^-(aq) +$
$H_2(g)$
Since $^1/_2$ of the Na reacts with $H_2O$, the maximum efficiency is 50%.

**23.82** (a) $NO(g) + O_3(g) \rightleftharpoons NO_2(g) + O_2(g)$
(b) $Rate_f = k_f$ [NO][$O_3$]
$Rate_r = k_r$ [$NO_2$][$O_2$]
(c) $-199$ kJ
(d) $1.3 \times 10^{37}$

**23.84** (a) 20.5 min
(b) 14

**23.89** (a) $6.4 \times 10^{-7}$ $M$
(b) $1.1 \times 10^{-2}$ $M$

# Glossary

*The number in parentheses is the section in which the term is introduced and/or discussed.*

**absolute scale** (also *Kelvin scale*)  The preferred temperature scale in scientific work, the absolute scale takes absolute zero as the lowest temperature. (1.5) (See *kelvin*, K.)

**absorption spectrum**  The spectrum produced when atoms absorb certain wavelengths of incoming light as they become excited from lower to higher energy levels. (7.2)

**abundance**  The amount of an element in a particular region of the universe. (23.1)

**accuracy**  The closeness of a measurement to the true value. (1.6)

**acid**  In common terms, a substance that produces $H^+$ ions when dissolved in water. (4.3) (See *classical (Arrhenius), Brønsted-Lowry,* and *Lewis* acid-base definitions.)

**acid anhydride**  A compound, sometimes formed by a dehydration condensation reaction of an oxoacid, that yields two molecules of the acid when it reacts with water. (14.4)

**acid-base buffer**  (See *buffer.*)

**acid-base indicator**  A substance whose color depends on the acidity or basicity of a solution that is used to monitor the pH of a solution. (4.3, 17.2, 18.2)

**acid-base titration curve**  A plot of the pH of a solution of acid (or base) versus the volume of base (or acid) added to the solution. (18.2)

**acid dissociation constant ($K_a$)**  An equilibrium constant for the dissociation of an acid in $H_2O$ to yield the conjugate base and $H_3O^+$:

$$K_a = \frac{[H_3O^+][A^-]}{[HA]}. \quad (17.1)$$

**acid ionization constant**  (See *acid dissociation constant.*)

**actinides** (also *actinoids*)  The fourteen Period 7 elements that form the second inner transition series ($5f$ block), including thorium (Th; $Z = 90$) through lawrencium (Lr; $Z = 103$). (8.2, 22.2)

**actinoids**  (See *actinides.*)

**activated complex** (also *transition state*)  An unstable species formed in an effective collision of reactants that exists momentarily when the system is highest in energy and that can either form products or re-form reactants. (15.4)

**activation energy ($E_a$)**  The height of the energy threshold that molecules must reach in order to react. (15.3)

**active site**  The region of an enzyme formed by specific amino acid side chains at which catalysis occurs. (15.6)

**activity (A)** (or *decay rate*)  The change in number of nuclei ($\mathcal{N}$) of a radioactive sample divided by the change in time ($t$). (21.2)

**actual yield**  The amount of product actually obtained in a chemical reaction. (3.4)

**addition polymer** (also *chain-reaction polymer*)  A polymer formed when monomers (usually containing $C=C$) combine through an addition reaction. (14.4)

**addition reaction**  A reaction in which the atoms of a carbon multiple bond become bonded to more atoms. (14.3)

**adduct**  (See *Lewis adduct*)

**adenosine triphosphate (ATP)**  The most common high-energy molecule to serve as an energy store and source in organisms. (19.2)

**alchemy**  An occult study of nature that flourished for 1500 years in northern Africa and Europe and resulted in the development of several key laboratory methods. (1.2)

**alcohol**  An organic compound (general formula R—OH; ending "-ol") that contains a $-\overset{|}{\underset{|}{C}}-O-H$ functional group. (14.4)

**aldehyde**  An organic compound (general formula R—CH=O; ending "-al") that contains the carbonyl functional group ($C=O$) in which the C is also bonded to H. (14.4)

**alkane**  A hydrocarbon that contains only single bonds (general formula $C_nH_{2n+2}$). (14.2)

**alkene**  A hydrocarbon that contains at least one $C=C$ bond (general formula $C_nH_{2n}$). (14.2)

**alkyl group**  A saturated hydrocarbon chain with one bond available. (14.3)

**alkyl halide**  (See *haloalkane.*)

**alkyne**  A hydrocarbon that contains at least one $C\equiv C$ bond (general formula $C_nH_{2n-2}$). (14.2)

**allotrope**  A different crystalline or molecular form of the same element. In general, one allotrope is more stable than another at a particular pressure and temperature. (13.6)

**alloy**  A mixture of elements with an overall metallic character that consists of solid phases of pure metals or solutions. (9.7, 12.1, 23.3)

**alpha (α) decay**  A radioactive process in which an α particle is emitted from a nucleus. (21.1)

**alpha (α) particle**  A positively charged particle, identical to a helium nucleus, that is one of the common types of radioactive emissions. (21.1)

**amide**  An organic compound that contains the $-\overset{O}{\overset{||}{C}}-\overset{|}{N}-$ functional group. (14.4)

**amine**  An organic compound (general formula $R-\overset{..}{\underset{|}{N}}-$) derived structurally by replacing one or more H atoms of ammonia with alkyl groups; a weak organic base. (14.4, 17.5)

**amino acid** An organic compound [general formula $H_2N$—$CH(R)$—$COOH$] with at least one carboxyl and one amine group on the same molecule; the monomer unit of a protein. (14.4)

**amorphous solid** A solid that occurs in different shapes because it lacks a well-defined crystal structure. (11.4)

**ampere (A)** The SI unit of electric current; the current that results when 1 coulomb flows through a conductor in 1 second. (20.6)

**amphoteric** Refers to a substance that can act as an acid or a base. (8.4)

**amplitude** The height of the crest (or depth of the trough) of a wave; related to the intensity of the energy. (7.1)

**anion** A negatively charged ion. (2.6)

**anode** The electrode at which oxidation occurs. Electrons are given up by the reducing agent and leave the cell at the anode. (20.1)

**antibonding molecular orbital** A molecular orbital formed when wave functions are subtracted from each other, which decreases electron density between the nuclei and leaves a node. Electrons occupying such an orbital destabilize the molecule. (10.4)

**apatites** Compounds of general formula $Ca_5(PO_4)_3X$, where X is generally F, Cl, or OH. (23.2)

**aqueous solution** A solution in which water is the solvent. (2.8)

**aromatic hydrocarbon** A compound of C and H with one or more rings of C atoms (often drawn with alternating C—C and C=C bonds), in which there is extensive delocalization of $\pi$ electrons. (14.2)

**Arrhenius acid-base definition** (See *classical, Arrhenius, acid-base definition.*)

**Arrhenius equation** An equation, $k = A\,e^{-Ea/RT}$, expressing the negative exponential relationship between temperature and the rate constant. (15.3)

**atmosphere** The mixture of gases that extends from a planet's surface and eventually merges with outer space; the gaseous portion of the Earth's crust. (5.8, 23.1)

**atom** The smallest particle of an element that retains the chemical nature of the element. A neutral, spherical entity composed of a postively charged central nucleus surrounded by negatively charged electrons. (2.2)

**atomic mass** (also *atomic weight*) The average of the masses of the naturally occurring isotopes of an element weighted according to their abundances. (2.4)

**atomic mass unit (amu)** (also *dalton, D,*) A mass exactly equal to $1/12$ the mass of a carbon-12 atom. (2.4)

**atomic number (Z)** The unique number of protons in the nucleus of each atom of an element (equal to the number of electrons in the neutral atom). An integer that expresses the positive charge of a nucleus in multiples of the electronic charge. (2.4)

**atomic orbital** (also *wave function*) A mathematical expression that describes the wave motion of the electron in the region of the nucleus. The term is used qualitatively to mean the region of space in which there is a high probability of finding the electron. (7.4)

**atomic solid** A solid consisting of individual atoms held together by dispersion forces; the type of solid formed by the frozen noble gases. (11.4)

**atomic symbol** (also *element symbol*) A one-, two-, or three-letter notation for an element based on its English, Latin, or Greek name. (2.4)

**aufbau principle** (also *building-up principle*) A conceptual process of adding one proton and one electron at a time to build up atoms and obtain the ground-state electron configurations of the elements. (8.2)

**autoionization** (also *self-ionization*) A reaction in which two molecules of a substance react to give ions. The most important example is for water: $2H_2O(l) \rightleftharpoons H_3O^+(aq) + OH^-(aq)$. (17.2)

**average rate** The change in concentration of reactants (or products) over a finite time period. (15.2)

**Avogadro's law** The gas law stating that at fixed temperature and pressure, equal volumes of any ideal gas contain equal numbers of particles. (5.3)

**Avogadro's number** A number ($6.022 \times 10^{23}$ to four significant figures) equal to the number of atoms in exactly 12 g of carbon-12; the number of atoms or molecules in one mole of an element or compound. (3.1)

**axial group** An atom (or group) that lies above or below the trigonal plane of a trigonal bipyramidal molecule. (10.1)

**azimuthal quantum number (l)** (also *orbital-shape quantum number*) An integer from 0 to $n - 1$ related to the shape of an atomic orbital. (7.4)

**background radiation** Natural ionizing radiation, one form of which is cosmic radiation. (21.4)

**balancing coefficient** (also *stoichiometric coefficient*) A numerical multiplier of all the atoms in the formula immediately following it. (3.3)

**band of stability** The narrow band on a plot of number of neutrons versus number of protons that includes the stable nuclides. (21.1)

**band theory** An extension of molecular orbital (MO) theory that explains the properties of metals. (11.4)

**barometer** A device used to measure atmospheric pressure. (5.2)

**base** In common terms, a substance that produces $OH^-$ ions when dissolved in water. (4.3) (See *classical (Arrhenius), Brønsted-Lowry,* and *Lewis* acid-base definitions.)

**base dissociation constant ($K_b$)** The equilibrium constant for the reaction of a base with $H_2O$ to yield the conjugate acid and $OH^-$:

$$K_b = \frac{[BH^+][OH^-]}{[B]}. \quad (17.5)$$

**base pair** Two complementary bases of mononucleotides that are H bonded to each other. (14.4)

**base unit** (also *fundamental unit*) A unit that defines the standard for one of the seven physical quantities in the International System of Units (SI). (1.5)

**basic-oxygen process** The method used to convert pig iron to steel in which $O_2$ is blown over and through molten iron to oxidize impurities and decrease the amount of carbon. (23.3)

**battery** A self-contained group of voltaic cells arranged in series. (20.4)

**bent (V shaped)** A molecular shape that arises when a central atom is bonded to two other atoms and has one or two lone pairs; occurs in the trigonal planar ($AX_2E$) and tetrahedral ($AX_2E_2$) arrangements. (10.1)

**beta (β) decay** A radioactive process in which an β particle is emitted from a nucleus. (21.1)

**beta (β) particle** A negatively charged particle identified as a fast moving electron that is one of the common types of radioactive emissions. (21.1)

**bimolecular reaction** An elementary reaction involving two reactant species. (15.5)

**binary covalent compound** A compound that consists of atoms of two elements in which bonding occurs primarily through electron sharing. (2.7)

**binary ionic compound** A compound that consists of the oppositely charged ions of two elements. (2.6)

**biomass conversion** The process of applying chemical and biological methods to convert plant and/or animal matter into fuels. (6.6)

**biosphere** The totality of organisms and the regions of the Earth that support the organisms. (23.1)

**blackbody radiation** The characteristic light emitted by a dense solid as it is heated to high temperature. (7.1)

**blast furnace** A tower-shaped furnace made of brick material in which intense heat and blasts of air are used to convert iron ore and coke to iron metal and carbon dioxide. (23.3)

**body-centered cubic unit cell** A unit cell in which a particle occurs on lattice points at each corner and in the center of a cube. (11.4)

**boiling point (bp)** The temperature at which the vapor pressure of a gas equals the external (atmospheric) pressure. (11.5)

**boiling point elevation** The increase in the boiling point of a solvent due to the presence of a dissolved solute particles. (12.5)

**bomb calorimeter** A device used to measure the heat released by a combustion reaction taking place within a closed container. (6.3)

**bond angle** The angle formed by two surrounding atoms with the central atom at the vertex. (10.1)

**bond energy (BE)** (also *bond strength*) The enthalpy change required to break a given bond in a mole of gaseous molecules. (9.3)

**bond length** The distance between the nuclei of two bonded atoms. (9.3)

**bond order** The number of electron pairs being shared by any two bonded atoms. (9.3)

**bonding molecular orbital** A molecular orbital formed when wave functions are added to each other, which increases electron density between the nuclei. Electrons occupying such an orbital stabilize the molecule. (10.4)

**bonding pair** (also *shared pair*) An electron pair being shared between two nuclei and forming a covalent bond. (9.3)

**Born-Haber cycle** A cyclic series of steps from elements to ionic solid consisting of enthalpy contributions that include the lattice energy. (9.2)

**Boyle's law** The gas law stating that at constant temperature, the volume of a fixed amount of gas is inversely proportional to the applied (external) pressure. (5.3)

**bridge bond** (also *three-center, two-electron bond*) A covalent bond in which three atoms in an A—B—C sequence are held together by two electrons. (13.5)

**Brønsted-Lowry acid-base definition** A model of acid-base behavior in which acids and bases are defined, respectively, as species that donate and accept protons in an acid-base reaction. (17.3)

**buffer** A solution that resists changes in pH when small amounts of either strong acid or base are added. (18.1)

**buffer capacity** A measure of the ability of a solution to resist pH change that is related to concentration and relative proportions of buffer components. (18.1)

**buffer range** The pH range over which a buffer acts effectively. (18.1)

**calibration** The process of correcting for the accuracy of a measuring device by comparing it to a known standard. (1.6)

**calorie (cal)** A unit of energy defined as exactly 4.184 J; originally defined as the heat needed to raise the temperature of 1 g of water 1°C (from 14.5°C to 15.5°C). (6.1)

**calorimeter** A device used to measure the heat released or absorbed by a physical or chemical process taking place within it. (6.3)

**capillarity** A property that results in a liquid rising through a narrow space. (11.3)

**carbon steel** The steel produced by the basic-oxygen process that contains about 1% C and is alloyed with metals that prevent corrosion and increase strength. (23.3)

**carbonyl group** The $C{=}O$ grouping of atoms. (14.4)

**carboxylic acid** An organic compound that contains the $\overset{\displaystyle O}{\overset{\displaystyle \|}{-C}}-OH$ group. (14.4)

**catalyst** A substance that increases the rate of a reaction without being used up in the process. (15.6)

**cathode** The electrode at which reduction occurs. Electrons enter the cell and are acquired by the oxidizing agent at the cathode. (20.1)

**cathode ray** The ray emitted by the cathode (negative electrode) in a gas discharge tube. Found to travel in straight lines, unless deflected by magnetic or electric fields. (2.3)

**cation** A positively charged ion. (2.6)

**cell potential ($E_{cell}$)** (also *electromotive force, emf; cell voltage*) The potential difference between the electrodes of an electrochemical cell when no current flows. (20.2)

**Celsius scale** A temperature scale in which 100 degrees separate the freezing and boiling points of water. (1.5)

**ceramic** A nonmetallic material that is hardened by heating to high temperatures and, in most cases, consists of silicate microcrystals suspended in a glassy cementing medium. (23.4)

**chain reaction** A self-sustaining process. In nuclear fission, a process in which neutrons released by splitting of one nucleus cause other nuclei to split, which releases more neutrons, and so on. (21.6)

**change in enthalpy (ΔH)** The change in energy plus the product of the constant pressure and the change in volume: $\Delta H = \Delta E + P\Delta V$. (6.2)

**charge density** The ratio of the charge of an ion to its volume. (12.2)

**Charles's law** The gas law stating that at constant pressure, the volume of a fixed amount of gas is directly proportional to its absolute temperature. (5.3)

**chelate** A complex in which the metal ion is bonded to a polydentate ligand. (22.4)

**chemical bond** The force that holds two atoms together in a molecule. (2.6)

**chemical change** (also *chemical reaction*) A change in which a substance is converted into a substance with different composition and properties. (1.1)

**chemical equation** A statement in chemical formulas that expresses the identities and quantities of the substances involved in a chemical or physical change. (3.3)

**chemical formula** A notation of atomic symbols and numerical subscripts that shows the number of each atom present in a molecule or formula unit of a substance. (2.7)

**chemical kinetics** The study of the rates and mechanisms of reactions. (15 Intro.)

**chemical property** A characteristic of a substance shown as it interacts with, or transforms into, other substances. (1.1)

**chemical reaction** (See *chemical change*.)

**chemistry** The scientific study of matter and the changes it undergoes. (1.1)

**chiral molecule** (also *asymmetric molecule*) A molecule that is not superimposable on its mirror image; an optically active molecule. In organic compounds, a chiral molecule typically contains a C atom bonded to four different groups. (14.2)

**chlor-alkali process** An industrial method that electrolyzes concentrated aqueous NaCl and produces $Cl_2$, $H_2$, and NaOH. (23.4)

**chromatography** A technique for separating the components of a mixture in which the mixture is dissolved in a fluid (gas or liquid), and the components are separated through differences in adsorption to or solubility in a solid (or viscous liquid) surface. (2.8)

**cis-trans isomers** (See *geometric isomers*.)

**classical (Arrhenius) acid-base definition** A model of acid-base behavior in which acids and bases are defined in terms of their formula and behavior in water. An acid is a substance that contains H and produces $H^+$ in water; a base is a substance that contains OH and produces $OH^-$ in water. (17.1)

**Clausius-Clapeyron equation** An equation that expresses the relationship between vapor pressure $P$ of a liquid and temperature $T$:

$$\ln P = \left(\frac{-\Delta H_{vap}}{R}\right)\frac{1}{T} + C, \text{ where } C \text{ is a constant. (11.5)}$$

**closest packed structure** An arrangement for packing atoms, ions, or molecules in a crystal that contains the least empty space. (11.4)

**coal gasification** An industrial process for altering the large molecules in coal to sulfur-free gaseous fuels. (6.6)

**colligative property** A property of a solution that depends on the number, not the identity, of solute particles. (12.5) (See *boiling point elevation; freezing point depression; osmotic pressure, π; and vapor pressure lowering*.)

**collision frequency** The average number of collisions per second that a molecule undergoes. (5.6)

**collision theory** A model that explains reaction rate as the result of particles colliding with a certain minimum energy. (15.4)

**colloid** A suspension in which a solutelike phase is dispersed throughout a solventlike substance. (12.6)

**combination reaction** A reaction in which two or more substances combine to form one substance. (4.1)

**combustion** The process of burning in air, often with release of heat and light. (1.2)

**combustion analysis** A method of determining the formula of a compound from the amounts of its combusion products. (3.2)

**common-ion effect** The shift in the position of an ionic equilibrium away from an ion involved in the process that is caused by the addition or presence of the ion. (18.1)

**complex** (See *coordination compound*.)

**complex ion** An ion consisting of a central metal ion bonded covalently to molecules and/or anions called ligands. (18.4, 22.4)

**composition** The types and amounts of simpler substances that make up a sample of matter. (1.1)

**compound** A substance composed of two or more elements in fixed proportions that are chemically combined. (2.1)

**concentration** A measure of the amount of solute dissolved in a given amount of solution. (3.5)

**concentration cell** A voltaic cell in which both compartments contain the same components but at different concentrations. (20.3)

**condensation** The process of a gas changing into a liquid. (11.1)

**condensation polymer** A polymer formed by monomers with two functional groups that are linked together in dehydration-condensation reactions. (14.4)

**conduction band** In the band theory the empty, higher energy portion of the band of molecular orbitals through which electrons move to conduct heat and electricity. (11.4)

**conjugate acid-base pair** Two species related to each other through the gain or loss of a proton; the acid has one more proton than its conjugate base. (17.3)

**contact process**　An industrial process for the manufacture of sulfuric acid based on the catalyzed oxidation of $SO_2$. (23.4)

**controlled experiment**　An experiment that measures the effect of one variable at a time by keeping other variables constant. (1.3)

**conversion factor**　A ratio equal to one and made of equivalent quantities that is used to convert the units of a quantity. (1.4)

**coordinate covalent bond**　A covalent bond formed when one atom donates both electrons to the other atom. (17.9, 22.5)

**coordination compound**　A substance that is a neutral complex or that contains at least one complex ion. (22.4)

**coordination isomers**　Coordination compounds with the same composition in which the complex ions have different groupings of ligands. (22.4)

**coordination number**　In a crystal, the number of nearest neighbors surrounding a particle. (11.4) In a complex, the number of ligand atoms bonded to the central metal ion. (22.4)

**core**　The dense, innermost region of the Earth. (23.1)

**core electrons**　(See *inner electrons.*)

**corrosion**　The natural redox process that results in unwanted oxidation of a metal. (20.5)

**coulomb (C)**　The SI unit of electric charge. One coulomb is the charge of $6.242 \times 10^{18}$ electrons; one electron possesses a charge of $1.602 \times 10^{-19}$C. (20.2)

**Coulomb's law**　A relationship stating that the electrostatic force associated with two charges is directly proportional to the product of their magnitudes and inversely proportional to the square of the distance between them. (9.2)

**counter ion**　A simple ion associated with a complex ion in a coordination compound. (22.4)

**coupling of reactions**　The pairing of reactions in which one step supplies more than enough free energy for the other to occur. (19.2)

**covalent bond**　A type of bond in which atoms are bonded through the sharing of two electrons; the mutual attraction of the nuclei and an electron pair that holds atoms together in a molecule. (2.6, 9.3)

**covalent bonding**　The idealized bonding type that is based on localized electron-pair sharing between two atoms with little difference in their tendencies to lose or gain electrons (usually nonmetals). (9.1)

**covalent compound**　A compound that consists of atoms bonded together by shared electron pairs. (2.6)

**covalent radius**　One-half the distance between nuclei of identical covalently bonded atoms. (8.4)

**critical mass**　The mass needed to achieve a chain reaction. (21.6)

**critical point**　The point on a phase diagram above which the vapor cannot be condensed to a liquid; the end of the liquid-gas curve. (11.5)

**crust**　The thin, light, heterogeneous outer layer of the Earth. (23.1)

**crystal defect**　Any of a variety of disruptions in the regularity of a crystal structure. (11.4)

**crystal field splitting energy ($\Delta$)**　The difference in energy between two sets of metal-ion $d$ orbitals that results from electrostatic interactions with the surrounding ligands. (22.5)

**crystal field theory**　A model that explains the color and magnetism of complexes based on the effects of ligand interactions on metal-ion $d$-orbital energies. (22.5)

**crystalline solid**　Solids with a well-defined shape because of the orderly arrangement of their atoms, molecules, or ions. (11.4)

**crystallization**　A technique used to separate and purify the components of a mixture through differences in solubility in which the components comes out of solution as crystals. (2.8)

**cubic closest packing**　A crystal structure based on the face-centered cubic unit cell in which the layers have an ABCABCABC··· pattern. (11.4)

**cubic meter ($m^3$)**　The SI derived unit of volume. (1.5)

**curie (Ci)**　The unit of radioactivity, defined as the number of nuclei disintegrating each second in 1 g of radium-226: $3.70 \times 10^{10}$ dps (disintegrations per second). (21.2)

**cyclic hydrocarbon**　A hydrocarbon with one or more rings in its structure. (14.2)

**$d$ orbital**　An atomic orbital with $l = 2$. (7.4)

**dalton (D)**　(See *atomic mass unit, amu.*)

**Dalton's law of partial pressures**　A gas law stating that in a mixture of unreacting gases, the total pressure is the sum of the partial pressures of the individual gases. (5.4)

**data**　Pieces of quantitative information obtained by observation. (1.3)

**de Broglie wavelength**　The wavelength of a moving particle obtained from the de Broglie equation: $\lambda = h/mv$. (7.3)

**decay constant**　The rate constant $k$ for radioactive decay. 21.2)

**decay series**　The succession of decay steps a parent nucleus may undergo before a stable daughter nucleus forms. (21.1)

**decomposition reaction**　A reaction in which one reactant forms two or more products; the reverse of a combination reaction. (4.1)

**delocalization**　The process by which electron density is spread over several atoms rather than remaining between two. (9.5)

**density ($d$)**　An intensive physical property of a substance at a given temperature and pressure defined as the ratio of the mass of the substance to its volume. (1.5)

**deposition**　The process of changing directly from gas to solid. (11.1)

**derived unit**　Any of various combinations of the seven SI base units. (1.5)

**desalination**　A process used to remove large amounts of ions from sea water, usually by reverse osmosis. (12.6)

**desulfurization**　An industrial process in which devices called scrubbers use lime-water slurries to remove $SO_2$ from flue gases. (6.6)

**deuterons**   Nuclei of the hydrogen isotope deuterium, $^2$H. (21.3)

**diagonal relationship**   Physical and chemical similarities between a Period 2 element and one located diagonally down and to the right in Period 3. (13.5)

**diamagnetism**   The tendency of a species not to be attracted (or to be slightly repelled) by a magnetic field, which is usually due to its electrons being paired. (8.4)

**differentiation**   The formation of regions in the Earth based on differences in composition and density. (23.1)

**diffraction**   The phenomenon in which a wave striking the edge of an object bends around it. A wave passing through a slit as wide as its wavelength forms a circular wave. (7.1)

**diffusion**   The movement of one fluid through another. (5.6)

**dimensional analysis** (also *factor-label method*)   A calculation method that attains a solution through the appropriate canceling of units. (1.4)

**dipole-dipole force**   The intermolecular attraction between oppositely charged poles of nearby polar molecules. (11.2)

**dipole-induced dipole force**   The intermolecular attraction between a polar molecule and the oppositely charged pole it induces in a nearby molecule. (11.2)

**dipole moment ($\mu$)**   A measure of molecular polarity; the magnitude of the partial charges on the ends of a molecule (in coulombs) times the distance between them (in meters). (10.2)

**disaccharide**   An organic compound formed by the dehydration-condensation of two simple sugars. (14.4)

**disintegration series**   (See *decay series*.)

**dispersion force** (also *London force*)   The intermolecular attraction between all molecules as a result of instantaneous polarizations of their electron clouds; the force primarily responsible for the condensed states of nonpolar substances. (11.2)

**displacement reaction**   A reaction in which an atom or ion in one substance displaces an atom or ion in another. Classified as single-displacement and double-displacement (metathesis) reactions. (4.1)

**disproportionation**   A reaction in which a substance is both oxidized and reduced. (13.7)

**distillation**   A technique to separate the components of a mixture in which a more volatile component vaporizes and condenses separately from the less volatile components. (2.8)

**donor atom**   An atom that donates a lone pair of electrons to form a covalent bond, usually from ligand to metal ion in a complex. (22.4)

**doping**   Adding small amounts of other elements into the crystal structure of a semiconductor to increase its conductivity. (11.4)

**double bond**   A covalent bond that consists of two bonding pairs. Two atoms sharing four electrons; one $\sigma$ and one $\pi$ bond. (9.3, 10.3)

**double helix**   The two intertwined H-bonded polynucleotide strands that form the structure of DNA (deoxyribonucleic acid). (14.4)

**Downs cell**   An industrial apparatus that electrolyzes molten NaCl to produce sodium and chlorine. (23.3)

**dynamic equilibrium**   The condition at which the forward and reverse reaction are taking place at the same speed, so there is no net change in the amounts of reactants or products. (4.5)

**$e_g$ orbital set**   The set of orbitals (composed of $d_{x^2-y^2}$ and $d_{z^2}$) that results when the energies of the metal-ion $d$ orbitals are split by a ligand field. This set is higher in energy than the other ($t_{2g}$) set in an octahedral field and lower in a tetrahedral field. (22.5)

**effective collision**   A collision in which the particles meet with sufficient energy and an orientation that allows them to react. (15.4)

**effective nuclear charge ($Z_{eff}$)**   The apparent nuclear charge an electron actually experiences as a result of shielding effects by other electrons. (8.1)

**effusion**   The process by which a gas escapes from its container through a tiny hole into an evacuated space. (5.6)

**electrochemical cell**   A system that incorporates a redox reaction to produce or utilize electrical energy. (20 Intro.)

**electrochemistry**   The study of the relationship between chemical change and electrical work. (20 Intro.)

**electrodes**   The objects that conduct the electricity between the cell and the surroundings. (20.1)

**electrolysis**   The nonspontaneous lysing (splitting) of a substance, often to its component elements, by supplying electrical energy. (20.6)

**electrolyte**   A substance that conducts a current when it dissolves in water. (4.2, 12.5) A mixture of ions, in which the electrodes of an electrochemical cell are immersed, that conducts a current. (20.1)

**electrolytic cells**   An electrochemical system that utilizes electrical energy to drive a nonspontaneous reaction ($\Delta G > 0$). (20.1)

**electromagnetic (EM) radiation** (also *electromagnetic energy* or *radiant energy*)   Oscillating, perpendicular electric and magnetic fields moving simultaneously through space as waves and manifested as visible light, x-rays, microwaves, radio waves, and so on. (7.1)

**electromagnetic spectrum**   The continuum of wavelengths of radiant energy. (7.1)

**electromotive force (emf)**   (See *cell potential*.)

**electron ($e^-$)**   A subatomic particle that possesses a unit negative charge ($1.602 \times 10^{-19}$ C) and occupies the space around the atomic nucleus. (2.4)

**electron affinity (EA)**   The energy change accompanying a mole of electrons being added to a mole of gaseous atoms or ions. (8.3)

**electron capture**   The radioactive process by which a nucleus draws in an orbital electron, usually one from the lowest energy level. (21.1)

**electron cloud**   An imaginary representation of the electron rapidly changing its position around the nucleus over time. (7.4)

**electron configuration**   The distribution of electrons within the orbitals of the atoms of an element; also the notation for such a distribution. (8 Intro.)

**electron deficient**   A bonded atom that has fewer than eight valence electrons (or two for H). (9.5)

**electron density diagram**   (See *electron probability density diagram.*)

**electron-pair delocalization**   (See *delocalization.*)

**electron-sea model**   A qualitative description of metallic bonding proposing that metal atoms pool their valence electrons into a delocalized "sea" of electrons in which the metal ions are submerged in an orderly array. (9.7)

**electron volt (eV)**   The energy that an electron acquires when it moves through a potential difference of one volt: $1 \text{ eV} = 1.602 \times 10^{-19}$ J. (21.6)

**electronegativity (EN)**   The relative ability of a bonded atom to attract shared electrons. (9.4)

**electronegativity difference (ΔEN)**   The difference in electronegativities between the atoms in a bond. (9.4)

**electrorefining**   An industrial electrolytic process in which the impure metal acts as the anode and the pure metal acts as the cathode. (23.3)

**element**   The simplest type of substance with unique physical and chemical properties. An element consists of only one kind of atom, so it cannot be broken down into any simpler substances. (2.1)

**elementary reaction** (also *elementary step*)   A simple reaction that describes a single molecular event in a proposed reaction mechanism. (15.5)

**elementary step**   (See *elementary reaction.*)

**elimination reaction**   A reaction in which C atoms in the product are bonded to fewer atoms than in the reactant, leading to multiple bonding. (14.3)

**emission (atomic) spectrum**   The line spectrum produced when excited atoms emit photons characteristic of the element. (7.2)

**empirical formula**   A chemical formula that shows the lowest relative number of atoms of each element in a compound. (2.7)

**end point**   The point in a titration at which the indicator changes color. (4.3, 18.2)

**endothermic process**   A change that occurs with an absorption of heat from the surroundings ($\Delta H > 0$) and therefore an increase in the enthalpy of the system. (6.2)

**energy**   The capacity to do work, that is, to move matter. (1.1) (See *kinetic energy* [$E_k$] and *potential energy* [$E_p$].)

**enthalpy (H)**   A thermodynamic quantity that is the sum of the internal energy plus the product of the pressure and volume. (6.2)

**enthalpy diagram**   A pictorial method for showing the enthalpy change of a system. (6.2)

**enthalpy of hydration (ΔH_hydr)**   (See *heat of hydration.*)

**enthalpy of solution (ΔH_soln)**   (See *heat of solution.*)

**entropy (S)**   A thermodynamic measure of the disorder of a system that is related to the number of ways the compo-

nents can be arranged without changing the system's energy. (12.2, 19.1)

**environmental cycle**   The physical, chemical, and biological paths through which the atoms of an element move within the Earth's crust. (23.2)

**enzyme**   A biological macromolecule (usually a protein) that acts as a catalyst. (15.6)

**enzyme-substrate complex (ES)**   The intermediate in an enzyme-catalyzed reaction, whose concentration determines the rate of product formation. (15.6)

**equatorial group**   An atom (or group) that lies in a trigonal plane of a trigonal bipyramidal molecule. (10.1)

**equilibrium constant (K)**   The value obtained when equilibrium concentrations are substituted into the mass-action expression. (16.1)

**equivalence point**   The point in a titration when the number of moles of added species is stoichiometrically equivalent to the original number of moles of the other species. (4.3, 18.2)

**ester**   An organic compound (general formula RCOOR′) that contains a 

$$-\overset{\overset{\displaystyle O}{\|}}{C}-O-\overset{|}{\underset{|}{C}}-$$ group. (14.4)

**ether**   An organic compound (general formula ROR′) that contains a 

$$-\overset{|}{\underset{|}{C}}-O-\overset{|}{\underset{|}{C}}-$$ group. (14.4)

**exact number**   A quantity, usually obtained by counting or by definition, that has no uncertainty associated with it and therefore an infinite number of significant figures. (1.6)

**excitation**   The process by which a substance absorbs energy from low-energy radioactive particles and becomes warmer and/or emits light. (21.4)

**excited state**   Any energy level in an atom or molecule other than the lowest (ground) state. (7.2)

**exothermic process**   A change that occurs with a release of heat to the surroundings ($\Delta H < 0$) and therefore a decrease in the enthalpy of the system. (6.2)

**expanded valence shell**   A valence level that can accommodate more than eight electrons by utilizing available $d$ orbitals. (9.5)

**experiment**   A clear set of procedural steps that tests a hypothesis. (1.3)

**extensive property**   A property, such as mass, that depends on the amount of substance present. (1.5)

**extraction**   A method of separating the components of a mixture based on differences in their solubility in a second nonmiscible solvent. (2.8)

**face-centered cubic unit cell**   A unit cell in which a particle occurs on lattice points at each corner and in the center of each face of a cube. (11.4)

**faraday ($\mathscr{F}$)**   The charge (96,485 C) of one mole of electrons. (20.3)

**fatty acid**   A carboxylic acid with a long hydrocarbon chain. (14.4)

**filtration**   A method of separating the components of a mixture on the basis of differences in particle size. (2.8)

**first law of thermodynamics** (also *law of conservation of energy*)   A basic observation that the total energy of the universe is constant; expressed as $\Delta E_{universe} = \Delta E_{system} \leftarrow \Delta E_{surroundings} = 0$. (6.1)

**fission**   The process by which a heavier nucleus splits into lighter nuclei with the release of energy. (21.6)

**fixation**   A chemical/biochemical process that converts a gaseous substance in the environment into a condensed form. (23.2)

**flame test**   The procedure of placing a granule of a compound or a drop of its solution in a flame to observe its characteristic color. (7.2)

**flotation**   A pretreatment process in which oil and detergent are mixed with pulverized ore in water to create a slurry that separates the mineral from the gangue. (23.3)

**formal charge**   The hypothetical charge on an atom a molecule or ion. The number of valence electrons minus the sum of all the unshared and half the shared valence electrons. (9.5)

**formation constant ($K_f$)** (also *stability constant*)   An equilibrium constant for the formation of a complex ion from the hydrated metal ion and ligands. (18.4)

**formation reaction**   A reaction in which one mole of a compound forms from its elements. (6.6)

**formula unit**   The chemical unit that contains the same number of atoms or ions as in the formula of the compound. (2.7)

**fossil fuel**   A fuel derived from the products of the decay of dead organisms, including coal, petroleum, and natural gas. (6.6)

**fractional distillation**   A distillation process involving numerous vaporization-condensation steps used to separate two or more volatile components. (12.5)

**free energy ($G$)**   A thermodynamic quantity that is the difference between the enthalpy and the product of the kelvin temperature and the entropy. (19.2)

**free radical**   A molecular or atomic species with one or more unpaired electrons, which typically make it very reactive. (9.5, 21.4)

**freezing**   The process of cooling a liquid until it solidifies. (11.1)

**freezing point depression**   A lowering of the freezing point of a solvent due to the presence of dissolved solute particles. (12.5)

**frequency ($v$)**   The number of cycles a wave undergoes per second. (7.1)

**frequency factor ($A$)**   The product of the collision frequency $Z$ and an orientation probability factor $p$ specific for a reaction. (15.4)

**fuel cell** (also *flow battery*)   A battery that is not self-contained in which electricity is generated by the controlled oxidation of a combustible fuel. (20.4)

**functional group**   A specific combination of atoms, typically a carbon multiple bond and/or heteroatom bond, that reacts in a characteristic way no matter in which molecule it occurs. (14.1)

**fundamental unit**   (See *base unit*.)

**fusion**   (See *melting*.)

**fusion (nuclear)**   The process by which light nuclei combine to form a heavier nucleus and energy is released. (21.6)

**gamma ($\gamma$) emission**   The type of radioactive decay in which $\gamma$ photons are emitted from an excited nucleus. (21.1)

**gamma ($\gamma$) ray**   A very-high-energy photon. (21.1)

**gangue**   Debris associated with a mineral, such as sand, rock, and clay, that is separated during extraction of an element. (23.3)

**gas**   One of the three states of matter. A gas fills a container regardless of the shape. (1.1)

**Geiger-Müller counter**   An ionization counter that detects radioactive emissions by their ionization of gas atoms within the instrument. (21.2)

**geometric isomers** (also *cis-trans* isomers)   Compounds with the same atom sequence but different spatial arrangements of the atoms. The *cis*- isomer has similar groups on the same side of a structural feature; the *trans*- isomer has them on opposite sides. (14.2, 22.4)

**Graham's law of effusion**   A gas law stating that the rate of effusion of a gas is inversely proportional to the square root of its density (or molar mass). (5.6)

**ground state**   The lowest energy level of an atom. (7.2)

**group**   A vertical column in the periodic table. (2.5)

**Haber process**   An industrial process used to form ammonia from its elements. (16.4)

**half-cell**   A portion of an electrochemical cell in which a half-reaction takes place. (20.2)

**half-life ($t_{1/2}$)**   The time required for half the initial reactant concentration to be consumed. (15.3) The time required for half the initial number of nuclei to decay. (21.2)

**half-reaction method**   A method of balancing redox reactions by treating the oxidation and reduction half-reactions separately. (4.4, 20.1)

**haloalkane**   An organic compound (general formula R—X) containing a $-\overset{|}{\underset{|}{C}}-X$ group, where X is a halogen atom. (14.4)

**hard water**   Water that contains large amounts of the divalent cations $Ca^{2+}$ and $Mg^{2+}$. (12.6)

**heat ($q$)**   The energy transferred between objects due to differences in their temperatures only. (1.5, 6.1)

**heat capacity**   The amount of heat required to change the temperature of an object by one kelvin. (6.3)

**heat of combustion ($\Delta H_{comb}$)**   The heat of reaction when one mole of a substance combines with oxygen in a combustion reaction. (6.2)

**heat of formation ($\Delta H_f$)**   The heat of reaction when one mole of compound is produced from its elements. (6.2)

**heat of fusion ($\Delta H^0_{fus}$)**   The enthalpy change when one mole of a substance melts. (6.2, 11.1)

**heat of hydration ($\Delta H_{hydr}$)**   The enthalpy change when a gaseous species is hydrated. The sum of the enthalpies of separating water molecules and mixing the gaseous solute with them. (12.2)

**heat of reaction ($\Delta H_{rxn}$)**   The enthalpy change of a reaction. (6.2)

**heat of solution ($\Delta H_{soln}$)**    The enthalpy change when a solution forms from solute and solvent. The sum of the enthalpies from separating solute and solvent molecules and mixing them. (12.2)

**heat of sublimation ($\Delta H^0_{subl}$)**    The enthalpy change when one mole of solid changes directly to a gas. The sum of the heats of fusion and vaporization. (11.1)

**heat of vaporization ($\Delta H^0_{vap}$)**    The enthalpy change when one mole of a liquid vaporizes. (6.2, 11.1)

**heating-cooling curve**    A plot of temperature versus time for a substance when heat is released or absorbed by the system at a constant rate. (11.5)

**Heisenberg uncertainty principle**    The principle stating that it is not possible to know simultaneously the exact position and velocity of a particle. (7.3)

**Henderson-Hasselbalch equation**    An equation for calculating the pH of a buffer system. (18.1)

**Henry's law**    A law stating that the amount of a gas dissolved in a liquid is directly proportional to the partial pressure of the gas above the liquid. (12.3)

**Hess's law of heat summation**    A law stating that the enthalpy change of an overall process is the sum of the enthalpy changes of the individual steps of the process. (6.5)

**heteroatom**    Any atom in an organic compound other than C or H. (14.1)

**heterogeneous catalyst**    A catalyst that occurs in a different phase from the reactants, usually a solid interacting with gaseous or liquid reactants. (15.6)

**heterogeneous mixture**    A mixture that has one or more visible boundaries between its components. (2.8)

**hexagonal closest packing**    A crystal structure based on the hexagonal unit cell in which the layers have an ABABAB· · · pattern. (11.4)

**high-spin complex**    A complex that has the same number of unpaired electrons as in the isolated metal ion; characteristic of weak-field ligands. (22.5)

**homogeneous catalyst**    A catalyst (gas, liquid, or soluble solid) that exists in the same phase as the reactants. (15.6)

**homogeneous mixture** (also *solution*)    A mixture that has no visible boundaries among its components. (2.8)

**homologous series**    A series of organic compounds in which each member differs from the next by a —$CH_2$— (methylene) group. (14.2)

**homonuclear diatomic molecule**    A molecule composed of two identical atoms. (10.4)

**Hund's rule**    A principle stating that when orbitals of equal energy are available, the electron configuration of lowest energy has the maximum number of unpaired electrons with parallel spins. (8.2)

**hybrid orbital**    An atomic orbital postulated to form during bonding by the mathematical mixing of specific combinations of nonequivalent orbitals in a given atom. (10.3)

**hybridization**    A postulated process of orbital mixing to form hybrid orbitals. (10.3)

**hydrate**    A compound in which a specific number of water molecules is associated with each formula unit. (2.7)

**hydration**    Solvation in water. (12.2)

**hydration shell**    The oriented cluster of water molecules that surrounds an ion in aqueous solution. (12.1)

**hydrocarbon**    An organic compound that contains only H and C atoms. (14.2)

**hydrogen bond (H bond)**    A type of dipole-dipole force that arises between molecules that have an H atom bonded to a small, highly electronegative atom with lone pairs, usually N, O, or F. (11.2)

**hydrogenation**    The addition of hydrogen to the carbon multiple bonds to form C—C bonds. (15.6)

**hydrolysis**    Cleaving a molecule by reaction with water in which one part of the molecule bonds to the water —OH and the other to the water H. (14.4)

**hydronium ion ($H_3O^+$)**    A proton covalently bonded to a water molecule. (17.1)

**hydrosphere**    The liquid portion of the Earth's crust. (23.1)

**hypothesis**    A testable statement made to explain an observation. If inconsistent with experiment, a hypothesis is revised or discarded. (1.3)

**ideal gas**    A gas that would exhibit linear relationships among volume, pressure, temperature, and number of moles at all conditions. (5.3)

**ideal gas law** (also *ideal gas equation*)    An equation expressing the relationships among volume, pressure, temperature, and number of moles of an ideal gas: $PV = nRT$. (5.3)

**ideal solution**    A solution whose vapor pressure equals the mole fraction of the solvent times the vapor pressure of the pure solvent; approximated only by very dilute solutions. (12.5) (See *Raoult's law.*)

**indicator**    A substance whose color is used to monitor the equivalence point of a titration. (4.3)

**induced-fit model**    A model of enzyme action that pictures substrate binding inducing the active site to change its shape and become catalytically active. (15.6)

**infrared (IR) region**    The region of the electromagnetic spectrum between the microwave and visible regions. (7.1)

**infrared (IR) spectroscopy**    An instrumental technique for determining the types of bonds in a covalent molecule by measuring the absorption of IR radiation. (9.3)

**initial rate**    The instantaneous rate at the point at which the reactants are mixed, that is, at $t = 0$. (15.2)

**inner electrons** (also *core electrons*)    Electrons that fill all the energy levels of an atom except the valence level; those electrons in atoms of the previous noble gas and completed transition series. (8.2)

**inner transition elements**    The elements of the periodic table in which $f$ orbitals are being filled; the lanthanides and actinides. (8.2, 22.2)

**instantaneous rate**    The reaction rate at a particular time, given by the slope of a tangent to a plot of reactant concentration versus time. (15.2)

**insulator**    A substance (usually a nonmetal) that is not capable of conducting an electric current. (11.4)

**integrated rate law**    A mathematical expression for reactant concentration as a function of time. (15.3)

**intensive property**   A property, such as density, that does not depend on the amount of substance present. (1.5)

**interhalogen compound**   Compounds of two halogens. (13.9)

**intermolecular forces** (also *interparticle forces*)   The attractive and repulsive forces among the particles—molecules, atoms, or ions—in a sample of matter. (11 Intro.)

**internal energy (E)**   The sum of the kinetic and potential energies of all the particles in a system. (6.1)

**interstitial hydride**   The substance formed when metals absorb $H_2$ into the spaces between their atoms. (6.6)

**ion**   A charged particle that forms when an atom (or small group of atoms) gains or loses one or more electrons. (2.6)

**ion-dipole force**   The intermolecular attractive force between an ion and a polar molecule (dipole). (11.2)

**ion exchange**   A process of softening water by binding one type of ion (usually $Ca^{2+}$) on a special resin and displacing another (usually $Na^+$). (12.6)

**ion-induced dipole force**   The intermolecular attractive force between an ion and the dipole it induces in a nearby electron cloud. (11.2)

**ion pair**   An ionic molecule, usually formed when a salt boils. (9.2)

**ion-product constant for water ($K_w$)**   The equilibrium constant for the autoionization of water:

$$K_w = [H_3O^+][OH^-]. \ (17.2)$$

**ionic atmosphere**   A cluster of ions of net opposite charge surrounding a given ion. (12.5)

**ionic bonding**   The idealized bonding type based on the attraction of oppositely charged ions that arise through electron transfer between atoms with large differences in their tendencies to lose or gain electrons (typically metals and nonmetals). (9.1)

**ionic compound**   A compound that consists of oppositely charged ions. (2.6)

**ionic radius**   The size of an ion as measured by the distance between the centers of adjacent ions in a crystalline ionic compound. (8.4)

**ionic solid**   A solid whose unit cell contains cations and anions. (11.4)

**ionization**   The process by which a substance absorbs energy from high-energy radioactive particles and loses an electron. (21.4)

**ionization energy (IE)**   The amount of energy required to remove completely an electron from a mole of gaseous atoms or ions. (7.2, 8.3)

**ionizing radiation**   The high-energy radioactivity that can form ions in a substance. (21.4)

**isoelectronic**   Having the same number and configuration of electrons as another species. (8.4)

**isomer**   One of two or more compounds with the same molecular formula but different properties, generally as a result of different arrangements of atoms. (14.2, 22.4)

**isotope**   Atoms of an element that have different numbers of neutrons and therefore different mass numbers. (2.4, 21.1)

**isotopic mass**   The mass (in amu) of an isotope relative to the mass of carbon-12. (2.4)

**joule (J)**   The SI unit of energy: $1 \text{ J} = 1 \text{ kg} \cdot m^2/s^2$. (6.1)

**kelvin (K)**   The SI base unit of temperature. The kelvin is the same size as the Celsius degree. (1.5)

**ketone**   An organic compound (general formula $R-\overset{\overset{O}{\|}}{C}-R$; ending "-one") that contains a carbonyl group bonded to two other C atoms, $-\overset{|}{C}-\overset{\overset{O}{\|}}{C}-\overset{|}{C}-$ . (14.4)

**kilogram (kg)**   The SI base unit of mass. (1.5)

**kinetic energy ($E_k$)**   The energy an object has due to its motion. (1.1)

**kinetic-molecular theory** (also *kinetic theory*)   The model that explains gas behavior in terms of particles in random motion whose volumes and interactions are negligible. (5.6)

**lanthanides** (also *lanthanoids; rare earths*)   The Period 6 (4*f*) series of inner transition elements that includes cerium (Ce; Z = 58) through lutetium (Lu; Z = 71). (8.2, 22.2)

**lanthanide contraction**   The additional decrease in atomic and ionic size of the lanthanides over the expected trend due to the poor shielding of the increasing nuclear charge by the *f* electrons. (22.1)

**lattice**   The three-dimensional arrangement created by the particles within a crystal, each of which is centered on a point. (11.4)

**lattice energy**   The enthalpy change that occurs when gaseous ions coalesce into a solid ionic compound. (9.2)

**law** (also *natural law*)   A summary, often in mathematical form, of a universal observation. (1.3)

**law of chemical equilibrium** (also *law of mass action*)   The law stating that when a system reaches equilibrium at a given temperature, the ratio of quantities that make up the mass-action expression has a constant numerical value. (16.2)

**law of conservation of energy**   (See *first law of thermodynamics*.)

**law of definite (or constant) composition**   A mass law stating that no matter what its source, a particular chemical compound is composed of the same elements in the same parts (fractions) by mass. (2.2)

**law of mass action**   (See *law of chemical equilibrium*.)

**law of mass conservation**   A mass law stating that the total mass of substances does not change during a chemical reaction. (2.2)

**law of multiple proportions**   A mass law stating that if elements A and B react to form two compounds, the different masses of B that combine with a fixed mass of A can be expressed as a ratio of small whole numbers. (2.2)

**Le Châtelier's principle**   A principle stating that if a chemical system in a state of equilibrium is disturbed, it will undergo a chemical change that shifts its equilibrium position in a direction that reduces the effect of the disturbance. (16.4)

**leaching**   A hydrometallurgic process that selectively extracts a metal, usually through formation of a complex ion. (23.3)

**level** (also *shell*)  A specific energy state of an atom given by the principal quantum number *n*. (7.4)

**leveling effect**  The inability of a solvent to distinguish the strengths of any acid (or base) that is stronger than the conjugate acid (or conjugate base) of the solvent. (17.8)

**Lewis acid-base definition**  A model of acid-base behavior in which acids and bases are defined, respectively, as species that accept or donate an electron pair. (17.9)

**Lewis adduct**  The product of all Lewis acid-base reactions characterized by the presence of a newly formed covalent bond. (17.9)

**Lewis electron-dot symbol**  A symbol of an atom in which the element symbol represents the nucleus and inner electrons and surrounding dots represent the valence electrons. (9.1)

**Lewis structure** (also *Lewis formula*)  A structural formula consisting of electron-dot symbols, with dot pairs representing bonding and lone pairs. (9.5)

**ligand**  A molecule and/or anion bonded to a central metal ion in a complex ion. (18.4, 22.4)

**like-dissolves-like rule**  A rule of thumb stating that substances having similar kinds of intermolecular forces dissolve in each other. (12.1)

**limiting reactant** (or *limiting reagent*)  The reactant that is consumed when a reaction goes to completion and therefore determines the maximum amount of a given product that can form. (3.4)

**line spectrum**  A series of separate spectral lines whose wavelengths are characteristic of an element. (7.2) (See also *emission spectrum*.)

**linear arrangement**  The geometric arrangement obtained when two electron groups maximize their separation around a central atom. (10.1)

**linear shape**  A molecular shape formed by three atoms lying in a straight line: bond angle of 180° ($AX_2$ or $AX_2E_3$). (10.1)

**linkage isomer**  Coordination compounds in which the composition of the complex ion remains the same, but the attachment of the ligand donor atom changes. (22.4)

**lipid**  A class of biomolecules, including fats and oils, that are soluble in nonpolar solvents. (14.4)

**liquid**  One of the three states of matter. A liquid fills a container to the extent of its own volume and thus forms a surface. (1.1)

**liter (L)**  A non-SI unit of volume equivalent to 0.001 $m^3$. (1.5)

**lithosphere**  The solid portion of the Earth's crust. (23.1)

**lock-and-key model**  A model of enzyme function that pictures rigid shapes of the enzyme active site and the substrate fitting together as a lock and key, respectively. (15.6)

**London force**  (See *dispersion force*.)

**lone pair** (also *unshared pair*)  An electron pair that is part of an atom's valence shell but not involved in covalent bonding. (9.3)

**low-spin complex**  Complex ions that have fewer unpaired electrons than in the free ion due to the presence of strong-field ligands. (22.5)

**magnetic quantum number** ($m_l$) (also *orbital orientation quantum number*)  An integer from $-l$ through 0 to $+l$ that prescribes the orientation of an atomic orbital in the three-dimensional space about the nucleus. (7.4)

**manometer**  A device used to measure the pressure of a gas involved in a laboratory experiment. (5.2)

**mantle**  A thick homogeneous layer of the Earth's internal structure that lies between the core and the crust. (23.1)

**mass**  The fixed quantity of matter an object contains. Balances are designed to measure mass. (1.5)

**mass-action expression** (*Q*) (also *reaction quotient*)  A ratio of terms for a given reaction, consisting of product concentrations multiplied together divided by reactant concentrations multiplied together, each raised to the power of their balancing coefficient. (16.2)

**mass defect** ($\Delta m$)  The mass decrease that occurs when nucleons are combined into a nucleus. (21.6)

**mass number** (*A*)  The total number of protons and neutrons in an atom. (2.4)

**mass percent (mass %;** also *percent by mass*)  The proportion by mass of a component of a substance expressed as parts per hundred parts. (2.2) Solution concentration expressed as the mass in grams of solute in every 100 g of solution. (12.4)

**mass spectrometry**  An instrumental method for measuring the relative masses of particles in a sample by creating charged particles that are separated according to their mass-charge ratio. (2.4)

**matter**  Anything that possesses mass and occupies volume. (1.1)

**mean free path**  The average distance a molecule travels between collisions at a given temperature and pressure. (5.6)

**melting** (also *fusion*)  The change of a substance from a solid to a liquid. (11.1)

**melting point (mp)**  The temperature at which the solid and liquid forms of a substance are at equilibrium. (11.5)

**metabolic pathway**  A biochemical reaction sequence that flows in one direction in which each reaction is enzyme catalyzed. (16.4)

**metal**  A substance or mixture that is relatively shiny and malleable and is a good conductor of heat and electricity. In reactions, metals tend to transfer electrons to nonmetals and form ionic compounds. (2.5)

**metallic bonding**  An idealized type of bonding based on the attraction between metal ions and a delocalized "sea" of their valence electrons. (9.1)

**metallic radius** (also *crystallographic radius*)  One-half the distance between the nuclei of adjacent atoms in a crystal of an element. (8.4)

**metallic solid**  A solid whose individual atoms are held together by metallic bonding. (11.4)

**metalloid** (also *semimetal*)  An element with properties between those of metals and nonmetals. (2.5)

**metallurgy**  The branch of materials science concerned with the extraction and utilization of metals. (23.3)

**meter (m)**   The SI base unit of length. The distance light travels in a vacuum in 1/299,792,458 second. (1.5)

**milliliter (mL)**   A volume (0.001 L) equivalent to 1 $cm^3$. (1.5)

**millimeter of mercury (mmHg)**   A unit of pressure based on the difference in the heights of mercury in a barometer or manometer. Redefined as one torr. (5.2)

**mineral**   The compound in an ore that contains the element of interest. (23.3)

**miscible**   Soluble in any proportion. (12.1)

**mixture**   A group of two or more elements and/or compounds that are physically intermingled. (2.1, 12 Intro)

**model** (also *theory*)   A simplified conceptual picture based on experiment that explains how an aspect of nature occurs. (1.3)

**molality (m)**   Solution concentration expressed as the moles of solute in 1000 g (1 kg) of solvent. (12.5)

**molar heat capacity**   The amount of heat required to change the temperature of one mole of a substance by one kelvin. (6.3)

**molar mass (M)** (also *gram-molecular weight*)   The mass of one mole of entities (atoms, molecules, or formula units) of a substance, in units of g/mol. (3.1)

**molar solubility**   The number of moles of solute that will dissolve per liter of solution to form a stable solution at a given temperature. (18.4)

**molarity (M)**   Solution concentration expressed as the moles of solute in one liter of solution. (3.5)

**mole (mol)**   The SI base unit for amount of a substance. The number of objects equal to the number of atoms in exactly 12 g of carbon-12. (3.1)

**mole fraction (X)**   A concentration term expressed as the ratio of moles of one component of a mixture to the total moles present. (5.4, 12.4)

**molecular equation**   An equation that shows reactants and products as if they were intact, undissociated compounds. (4.3)

**molecular formula**   A formula that shows the actual number of atoms of each element in a molecule. (2.7)

**molecular mass**   The sum (in amu) of the atomic masses of a formula unit of a compound. (2.7)

**molecular orbital (MO)**   An orbital of given energy and shape that extends over a molecule and can be occupied by no more than two electrons. (10.4)

**molecular orbital bond order**   One-half the difference between the number of electrons in bonding and antibonding MOs. (10.4)

**molecular orbital diagram**   A depiction of the relative energy and number of electrons in each MO, as well as the atomic orbitals that originally held the electrons. (10.4)

**molecular orbital (MO) theory**   A model that describes a molecule as a collection of nuclei and electrons in which the electrons occupy orbitals that extend over the entire molecule. (10.4)

**molecular polarity**   The overall distribution of electronic charge on a molecule, determined by its shape and bond polarity. (10.2)

**molecular shape**   The three-dimensional structure defined by the relative positions of the atoms in a molecule. (10.1)

**molecular solid**   A solid composed of individual molecules that are held together by intermolecular forces. (11.4)

**molecularity**   The number of reactant particles involved in an elementary step. (15.5)

**molecule**   A structure consisting of two or more atoms that are chemically bound together and behave as an independent unit. (2.1)

**monatomic ion**   An ion derived from a single atom. (2.6)

**monochromatic**   Light of a single wavelength. (7.1)

**mononucleotide**   A monomer unit of a nucleic acid, consisting of an N-containing base, a sugar, and a phosphate group. (14.4)

**monosaccharide**   A simple sugar; a polyhydroxy ketone or aldehyde of three to nine C atoms. (14.4)

**natural law**   (See *law*.)

**Nernst equation**   An equation stating that the cell voltage under any conditions depends on the standard cell voltage and the concentrations of the cell components:

$$E_{cell} = E^0_{cell} - \frac{RT}{n\widetilde{\delta}} \ln Q. \text{ (20.3)}$$

**net ionic equation**   An equation in which spectator ions have been eliminated so that the actual chemical change is seen. (4.3)

**network covalent solid**   A solid in which all the atoms are bonded covalently. (11.4)

**neutralization**   The reaction of an acid and a base. In classical terms, the products are a salt and water (4.3); in Brønsted-Lowry terms, the products are a conjugate base and acid (17.3); in Lewis terms, the product is an adduct with a new covalent bond. (17.9)

**neutron ($n^0$)**   A subatomic particle found in the nucleus that has no charge. (2.4)

**nitrile**   An organic compound (general formula R—C≡N) containing the —C≡N group. (14.4)

**node**   A region of an orbital where the probability of finding the electron is zero. (7.4)

**nonbonding molecular orbital**   A molecular orbital that is not involved in bonding. (10.4)

**nonelectrolyte**   A substance whose aqueous solution does not conduct an electric current. (4.2, 12.5)

**nonionizing radiation**   Radiation that does not cause loss of electrons. (21.4)

**nonmetal**   An element that lacks metallic properties. In reactions, nonmetals tend to bond with each other to form covalent compounds or accept electrons from metals to form ionic compounds. (2.5)

**nonpolar covalent bond**   A covalent bond between atoms of such similar electronegativity that the bonding pair is shared equally. (9.4)

**nuclear binding energy** The energy released when a nucleus forms from its nucleons; also expressed per mole of nuclei. (21.6)

**nuclear magnetic resonance (NMR) spectroscopy** An instrumental technique used to determine the molecular environment of a given type of nucleus, most often $^1H$, in a molecule. (14.2)

**nuclear transmutation** The induced conversion of one nucleus into another by bombardment with a particle. (21.3)

**nucleic acid** An unbranched polymer consisting of mononucleotides that occurs in two types, DNA and RNA (deoxyribonucleic and ribonucleic acids), which differ chemically in the nature of the sugar. (14.4)

**nucleon** The collective name for proton and neutron. (21.1)

**nucleus** The tiny central region of the atom that contains all the positive charge and essentially all the mass. (2.3)

**nuclide** An atom with a particular number of the two types of nucleons. (21.1)

**N/Z ratio** The ratio of the number of neutrons to the number of protons, a key factor that determines the stability of a nuclide. (21.1)

**observation** A fact obtained with the senses, often with the aid of instruments. Quantitative observations can be compared objectively. (1.3)

**occurrence** (also *source*) The form(s) in which an element exists in nature. (23.1)

**octahedral arrangement** The geometric arrangement obtained when six electron groups maximize their space around a central atom. (10.1)

**octahedral shape** A molecular shape formed when six atoms surround a central atom and lie at the corners of an octahedron: bond angle 90°. (10.1)

**octet rule** The observation that when atoms bond, they lose, gain, or share electrons to attain a filled outer shell. (9.2)

**optical isomer** A stereoisomer that arises when a molecule and its mirror image cannot be superimposed on each other. (14.2, 22.4)

**optically active** A molecule that rotates the plane of polarized light. (14.2)

**orbital box diagram** A depiction of electron number and spin in an atom's orbitals by means of arrows in a series of small boxes. (8.2)

**ore** A naturally occurring compound or mixture of compounds from which an element can be profitably extracted. (23.1)

**organic compound** A compound in which carbon is nearly always bonded to itself, to hydrogen, and often to other elements. (14 Intro.)

**organometallic compound** An organic compound in which carbon is bonded covalently to a metal atom. (14.4)

**osmosis** The process in which solvent flows through a semipermeable membrane that separates solutions of different concentrations to equalize the concentrations. (12.5)

**osmotic pressure (π)** The pressure that results from the inability of solute particles to cross a semipermeable membrane. The pressure required to prevent the net movement of solvent across the membrane. (12.5)

**outer electrons** Electrons that occupy the highest energy level (highest $n$ value) and are on average farthest from the nucleus. (8.2)

**overall equation** A chemical equation that is the sum of two or more sequential equations. (3.4)

**overvoltage** The additional voltage required above the standard cell voltage to accomplish electrolysis. (20.6)

**oxidation** The loss of electrons accompanied by an increase in oxidation number. (4.4)

**oxidation number method** A method for balancing redox reactions in which the change in oxidation numbers is used to determine balancing coefficients. (4.4)

**oxidation number (O.N.)** (also *oxidation state*) A number determined by a set of rules equal to the number of charges a bonded atom would have if electrons were held completely by the atom that attracts them more strongly. (4.4)

**oxidation-reduction reaction** (also *redox reaction*) A process in which electrons are transferred from one reactant (reducing agent) to another (oxidizing agent). (4.4)

**oxidizing agent** The substance that accepts electrons in a reaction and undergoes a decrease in oxidation number. (4.4)

**oxoanion** An anion in which an element, usually a nonmetal, is bonded to one or more oxygen atoms. (2.7)

**$p$ orbital** An atomic orbital with $l = 1$. (7.4)

**packing efficiency** The percentage of the available volume occupied by atoms, ions, or molecules in a unit cell. (11.4)

**paramagnetism** The tendency of a species with unpaired electrons to be attracted by an external magnetic field. (8.4, 22.1)

**partial ionic character** An estimate of the actual charge separation in a bond due to the electronegativity difference of the bonded atoms relative to complete separation. (9.4)

**partial pressure** The portion of the total pressure contributed by a gas in a mixture of gases. (5.4)

**particle accelerator** A device used to impart high kinetic energies to nuclear particles. (21.3)

**pascal (Pa)** The SI unit of pressure: $1 \text{ Pa} = 1 \text{ N/m}^2$. (5.2)

**Pauli exclusion principle** A principle stating that no two electrons in an atom can have the same set of four quantum numbers. (8.1)

**penetration** The process by which an outer electron moves through the core electrons and spends part of its time close to the nucleus, which increases the effective nuclear charge. (8.1)

**percent by mass** (See *mass percent, mass %*.)

**percent yield (% yield)** The actual yield of a reaction expressed as a percent of the theoretical yield. (3.4)

**period** A horizontal row of the periodic table. (2.5)

**periodic law** A law stating that when the elements are arranged by atomic number, they exhibit a periodic recurrence of properties. (8 Intro.)

**periodic table of the elements**　A table in which the elements are arranged by atomic number into columns (groups) and rows (periods). (2.5)

**pH**　The negative logarithm of $[H_3O^+]$. (17.2)

**phase**　A physically distinct portion of a system. (12 Intro.)

**phase change**　A physical change from one phase to another, usually referring to a change in physical state. (11 Intro.)

**phase diagram**　A diagram used to describe the stable phase and phase change of a substance as a function of temperature and pressure. (11.5)

**phlogiston theory**　An outmoded theory of combustion proposing that a burning substance releases the undetectable material phlogiston. (1.2)

**photoelectric effect**　The electric current produced when light of sufficient energy shines on a metal. (7.1)

**photon**　A quantum of electromagnetic radiation. (7.1)

**photon theory**　A model proposed to explain the photoelectric effect, stating that radiation occurs as discrete photons. (7.1)

**photovoltaic cells**　A device capable of converting light directly into electricity. (6.6)

**physical change**　A change in which the physical form or state of a substance, but not its composition, is altered. (1.1)

**physical property**　A characteristic shown by the substance itself, without interacting with other substances. (1.1)

**pi (π) bond**　A covalent bond formed by sideways overlap of two atomic orbitals that has two regions of electron density, one above and one below the internuclear axis. (10.3)

**pi (π) molecular orbital**　A molecular orbital formed by combination of two atomic (usually $p$) orbitals whose orientation are perpendicular to the internuclear axis. (10.3)

**Planck's constant ($h$)**　A proportionality constant relating the energy and frequency of a photon: $6.626 \times 10^{-34}$ J·s. (7.1)

**polar covalent bond**　A covalent bond in which the electron pair is shared unequally, so the bond has partially negative and partially positive poles. (4.2, 9.4)

**polar molecule**　A molecule with an unequal distribution of charge as a result of its polar bonds and shape. (4.2)

**polarimeter**　A device used to measure the rotation of plane-polarized light by an optically active compound. (14.2)

**polarizability**　The ease with which the electron density around a particle can be distorted. (11.2)

**polyatomic ion**　An ion in which two or more atoms are bonded covalently. (2.6)

**polychromatic**　Light of many wavelengths. (7.1)

**polymer**　An extremely large molecule that results from the covalent linking of many simpler molecular units (monomers.) (14.4)

**polyprotic acid**　An acid with more than one ionizable proton. (17.4)

**polysaccharide**　A macromolecule composed of many simple sugars linked covalently. (14.4)

**positron ($_{1}^{0}β$)**　The antiparticle of an electron. (21.1)

**positron decay**　A type of radioactive decay in which a positron is emitted from the nucleus. (21.1)

**potential energy ($E_p$)**　The energy an object has due to its position relative to other objects or due to its composition. (1.1)

**precipitate**　(See *precipitation reaction.*)

**precipitation reaction**　A reaction in which two soluble ionic compounds form an insoluble product, a precipitate. (4.3)

**precision** (also *reproducibility*)　The closeness of a measurement to other measurements in a series. (1.6)

**pressure ($P$)**　The force exerted per unit of surface area. (5.2)

**pressure-volume work ($PV$ work)**　A type of work in which a volume change occurs against an external pressure. (6.1)

**principal quantum number ($n$)**　A positive integer that specifies the energy and relative size of an orbital. (7.4)

**probability distribution diagram** (also *electron density diagram*)　The pictorial representation for a given energy sublevel of the quantity $\psi^2$, the probability that the electron lies within a particular tiny volume, as a function of $r$, the distance from the nucleus. (7.4)

**product**　A substance formed in a chemical reaction. (3.3)

**property**　A characteristic that gives a substance its unique identity. (1.1)

**protein**　A natural, linear polymer composed of about 20 types of amino acid monomers linked together by peptide bonds. (14.4)

**proton ($p^+$)**　A subatomic particle found in the nucleus that has a unit positive charge. (2.4)

**proton acceptor**　A substance that accepts an $H^+$ ion; a Brønsted-Lowry base. (17.3)

**proton donor**　A substance that donates an $H^+$ ion; a Brønsted-Lowry acid. (17.3)

**pseudo–noble gas configuration**　The $(n-1)d^{10}$ configuration of a $p$-block metal atom that empties its outer energy level. (8.4)

**pure substance**　A type of substance whose composition is fixed; an element or compound. (2.1)

**qualitative analysis**　The separation and identification of the ions in a mixture. (18.5)

**quantum**　A packet of energy equal to $h\nu$. The smallest amount of energy that can be emitted or absorbed. (7.1)

**quantum mechanics** (also *wave mechanics*)　The branch of physics that examines the wave motion of objects on the atomic scale. (7.4)

**quantum number**　A number that specifies a property of an orbital or an electron. (7.1)

**rad**　The amount of radiation that results in 0.01 J of energy being absorbed per kilogram of tissue. (21.4)

**radial distribution plot**　The graphic depiction of the total probability distribution (sum of $\psi^2$) of an electron in the region near the nucleus. (7.4)

**radioactivity**　The emissions resulting from the spontaneous disintegration of an unstable nucleus. (21.1)

**radioisotopic dating**　A method for determining the age of an object based on the rate of decay of a particular radioactive nuclide. (21.2)

**random error** A type of error occurring in all measurements that results in values *both* higher and lower than the actual value. (1.6)

**Raoult's law** A law stating that the vapor pressure of a solution is directly proportional to the mole fraction of solvent. (12.5)

**rare earth elements** (See *lanthanides.*)

**rate constant (*k*)** The proportionality constant that relates the rate of a reaction to the concentrations of components. (15.3)

**rate-determining step** (also *rate-limiting step*) The slowest step in a reaction mechanism and therefore the step that limits the overall rate. (15.5)

**rate law** (also *rate equation*) An equation that expresses the rate as a function of reactant concentrations. (15.3)

**reactant** A starting substance in a chemical reaction. (3.3)

**reaction energy diagram** A graph that shows the potential energy of a reacting system as it changes from reactants to products. (15.4)

**reaction intermediate** A substance that is formed and used up during the overall reaction and therefore does not appear in the overall equation. (15.5)

**reaction mechanism** A series of elementary steps that sum to the overall reaction. (15.5)

**reaction order** The power to which a reactant concentration is raised that defines how the rate is affected by changes in that concentration. (15.3)

**reaction quotient** (See *mass-action expression, Q.*)

**reaction rate** The change in the concentrations of reactants or products with time. (15.2)

**reactor core** The part of a nuclear reactor that contains the fuel rods and generates heat from fission. (21.6)

**receptor** A biological macromolecule, often partially embedded in a membrane, that interacts with a small, circulating molecule and initiates a response. (10.2)

**redox reaction** (See *oxidation-reduction reaction.*)

**reducing agent** The substance that donates electrons in a reaction and undergoes an increase in oxidation number. (4.4)

**reduction** The gain of electrons accompanied by a decrease in oxidation number. (4.4)

**refraction** A phenomenon in which a wave changes its speed and therefore its angle as it passes through a phase boundary. (7.1)

**rem** The unit of radiation dosage for a human based on the product of the number of rads and a factor related to the biological tissue. (21.4)

**resonance hybrid** The average of the resonance structures of a molecule. (9.5)

**resonance structure** (also *resonance form*) One of two or more Lewis structures for a molecule that cannot be adequately described by a single structure. (9.5)

**reverse osmosis** The removal of ions from water by applying a pressure that is higher than the osmotic pressure to the solution side of the membrane. (12.6)

**rms (root-mean-square) speed (*u*$_{rms}$)** The speed of a molecule having the average kinetic energy; very close to the most probable speed. (5.6)

**roasting** A pyrometallurgical process in which metal sulfides are converted to oxides. (23.3)

**rounding off** The process of removing digits based on a series of rules to obtain an answer with the proper number of significant figures. (1.6)

***s* orbital** An atomic orbital with $l = 0$. (7.4)

**salt** An ionic compound that results from an acid-base reaction. (4.3)

**salt bridge** An inverted ⊔ tube that connects the compartments of a voltaic cell and contains a solution of nonreacting electrolyte that allows ions to flow but prevents mixing of the compartments. (20.2)

**saturated hydrocarbon** A hydrocarbon in which each C is bonded to four other atoms. (14.2)

**saturated solution** A stable solution containing the maximum amount of solute that will dissolve under a given set of conditions. (12.3)

**scanning tunneling microscopy** An instrumental technique that uses electrons moving across a minute gap to observe the topography of a surface on the atomic scale. (11.4)

**Schrödinger equation** An equation that describes how the electron matter-wave changes in space around the nucleus. Solutions of the equation provide allowable energy levels of the H atom. (7.4)

**scientific method** A process of creative thinking and testing aimed at objective, verifiable discoveries of the causes of natural events. (1.3)

**scintillation counter** A device used to measure radioactivity that causes the excitation of matter and the emission of light. (21.2)

**second (s)** The SI base unit of time. (1.5)

**second law of thermodynamics** A law stating that a process occurs spontaneously in the direction that increases the entropy of the universe. (19.1)

**seesaw shape** A molecular shape caused by the presence of an equatorial lone pair in a trigonal bipyramidal arrangement (AX$_4$E). (10.1)

**selective precipitation** The process of separating ions through differences in the solubility of their compounds with a given precipitating ion. (18.5)

**self-ionization** (See *autoionization.*)

**semiconductor** A substance whose electrical conductivity at room temperature is poor but increases significantly with temperature. (11.4)

**semimetal** (See *metalloid.*)

**semipermeable membrane** A membrane that allows solvent, but not solute, to pass through. (12.5)

**shared pair** (See *bonding pair.*)

**shell** (See *level.*)

**shielding** The ability of other electrons, especially inner ones, to lessen the nuclear charge on an outer electron. (8.1)

**SI unit** A unit composed of one or more of the base units of the Système International d'Unités, a revised metric system. (1.5)

**side reaction** A different chemical reaction that consumes some of the reactant and reduces the overall yield of the desired product. (3.4)

**sigma (σ) bond**   A type of covalent bond that arises through end-to-end orbital overlap and has most electron density along the bond axis. (10.3)

**sigma (σ) molecular orbital**   A molecular orbital that is cylindrically symmetrical about an imaginary line that runs through the nuclei of the component atoms. (10.4)

**significant figure**   A digit obtained in a measurement. The greater the number of significant figures, the greater the certainty of the measurement. (1.6)

**silicate**   A type of compound found throughout rocks and soil consisting of —Si—O bonds and, in most cases, metal cations. (13.6)

**silicone**   A type of synthetic polymer containing —Si—O chains, with organic groups and crosslinks. (13.6)

**simple cubic unit cell**   A unit cell in which a particle occurs on lattice points at each corner of a cube. (11.4)

**simultaneous equilibria**   The simultaneous adjustment of more than one equilibrium system to separate metal ions. (18.5)

**single bond**   A bond that consists of one electron pair. (9.3)

**slag**   A molten waste product formed by the reaction of acidic silica with a basic metal oxide. (23.3)

**smelting**   Heating a mineral with a reducing agent, such as coke, to obtain the metal. (23.3)

**soap**   The salt of a fatty acid and usually a Group 1A(1) or 2A(2) hydroxide. (12.1, 14.4)

**solid**   One of the three states of matter. A solid has a fixed shape that does not conform to the container shape. (1.1)

**solubility**   The maximum amount of a substance that can dissolve in a fixed amount of solvent to form a stable solution at a given temperature. (12.1)

**solubility product constant ($K_{sp}$)**   The equilibrium constant for the dissolving of a slightly soluble ionic compound in water. (18.3)

**solute**   The substance that dissolves in the solvent. (3.5, 12.1)

**solution**   A homogeneous mixture. (2.8)

**solvation**   The process of surrounding a solute particle with solvent particles. (4.2, 12.2)

**solvent**   The substance(s) in which the solute dissolves. (3.5, 12.1)

**source**   (See *occurrence*.)

***sp* hybrid orbital**   An orbital formed by the mixing of one *s* and one *p* orbital in a central atom. (10.3)

**specific heat capacity (*C*)**   The amount of heat required to change the temperature of 1 g of a substance by 1 kelvin. (6.3)

**spectator ion**   An ion that is present as part of the reactant but is not involved in the chemical change. (4.3)

**spectrochemical series**   A ranking of ligands in terms of their ability to split *d*-orbital energies. (22.5)

**spectrophotometry**   A group of instrumental techniques that measure the atomic and molecular energy levels of a substance from its electromagnetic spectra. (7.2)

**speed of light (*c*)**   A fundamental constant ($2.9979 \times 10^8$ m/s) giving the speed at which electromagnetic radiation travels in a vacuum. (7.1)

**spin quantum number ($m_s$)**   A number, either $+\frac{1}{2}$ or $-\frac{1}{2}$, that indicates the direction of electron spin. (8.1)

**spontaneous change**   A change that occurs by itself, that is, without a continuous input of energy. (19 Intro.)

***sp²* hybrid orbital**   An orbital formed by the mixing of one *s* and two *p* orbitals in a central atom. (10.3)

***sp³* hybrid orbital**   An orbital formed by the mixing of one *s* and three *p* orbitals in a central atom. (10.3)

***sp³d* hybrid orbital**   An orbital formed by the mixing of one *s*, three *p*, and one *d* orbital in a central atom. (10.3)

***sp³d²* hybrid orbital**   An orbital formed by the mixing of one *s*, three *p*, and two *d* orbitals in a central atom. (10.3)

**square planar shape**   A molecular shape caused by the presence of two axial lone pairs in an octahedral arrangement ($AX_4E_2$). (10.1)

**square pyramidal shape**   A molecular shape caused by the presence of one lone pair in an octahedral arrangement ($AX_5E$). (10.1)

**standard atmosphere (atm)**   The average atmospheric pressure measured at sea level, defined as $1.01325 \times 10^5$ Pa. (5.2)

**standard cell potential ($E^0_{cell}$)**   The potential of the cell measured with all components in their standard states and no current flowing. (20.2)

**standard electrode potential ($E^0_{half\text{-}cell}$)** (also *standard reduction potential*)   The standard potential of a half-cell, with the half-reaction written as a reduction. (20.2)

**standard entropy of reaction ($\Delta S^0_{rxn}$)**   The entropy change that occurs when all components are in their standard states. (19.1)

**standard free energy change ($\Delta G^0$)**   The free energy change that occurs when all components are in their standard states. (19.2)

**standard free energy of formation ($\Delta G^0_f$)**   The standard free energy change that occurs when one mole of a compound is made from its elements. (19.7)

**standard heat of formation ($\Delta H^0_f$)**   The enthalpy change that occurs in a reaction when one mole of a compound forms from its elements, with all substances in their standard states. (6.6)

**standard heat of reaction ($\Delta H^0_{rxn}$)**   The enthalpy change that occurs during a reaction, with all substances in their standard states. (6.6)

**standard molar entropy ($S^0$)**   The entropy of one mole of a substance in its standard state. (19.1)

**standard molar volume**   The volume of one mole of an ideal gas at standard temperature and pressure: 22.414 L. (5.3)

**standard reference half-cell** (also *standard hydrogen electrode*)   A specially prepared platinum electrode immersed in 1 *M* $H^+(aq)$ through which $H_2$ gas at 1 atm is bubbled. $E^0_{half\text{-}cell}$ is defined as 0 V. (20.2)

**standard states**   A set of specifications used to compare thermodynamic data; 1 atm for gases; 1*M* for dissolved species; the pure substance for liquids and solids. (6.6)

**standard temperature and pressure (STP)**   The reference conditions for a gas: 0°C (273.15 K) and 1 atm (760 torr). (5.3)

**state function** A property of the system determined by its current state, regardless of the path to that state. (6.1)

**state of matter** One of the three physical forms of matter: solid, liquid, or gas. (1.1)

**stationary states** In the Bohr model, one of the allowable energy levels of the atom in which it does not release or absorb energy. (7.2)

**steel** An alloy of iron with small amounts of carbon and usually other metals. (23.3)

**stellar nucleogenesis** The process in which elements are formed in the stars through nuclear fusion. (21.6)

**stereoisomer** Molecules with the same sequence of atoms but different orientations of groups in space. (14.2, 22.4)

**stoichiometric coefficient** (See *balancing coefficient.*)

**stoichiometry** The study of the mass-mole relationships of chemical formulas and reactions. (3 Intro.)

**strong-field ligand** A ligand that causes large crystal field splitting and therefore a low-spin complex. (22.5)

**strong force** An attractive force that exists between all nucleons which is about 100 times stronger than the electrostatic repulsive force.

**structural formula** A formula that shows the actual number of atoms, their sequence, and the bonds between them. (2.7)

**structural isomer** Molecules with the same molecular formula but different sequences of atoms. (14.2, 22.4)

**sublevel** (also *subshell*) The energy substates of an atom within a level. Given by the $n$ and $l$ values, the sublevel designates the size and shape of the orbitals. (7.4)

**sublimation** The process of a solid changing directly into a gas. (11.1)

**substitution reaction** A reaction that occurs when an atom (or group) from an added reactant substitutes for one in the organic reactant. (14.3)

**substrate** A reactant that binds to the active site in an enzyme-catalyzed reaction. (15.6)

**superconductivity** The ability to conduct a current with no loss of energy to resistive heating. (11.4)

**supersaturated solution** An unstable solution in which more solute is dissolved than in a saturated solution. (12.3)

**surface tension** The energy required to increase the surface area of a liquid by a given amount. (11.3)

**surroundings** All parts of the universe other than the system. (6.1)

**suspension** A heterogeneous mixture containing particles that are distinct from the surrounding medium. (12.6)

**syngas** Synthesis gas; a combustible mixture of CO and $H_2$. (6.6)

**synthetic natural gas (SNG)** A gaseous fuel mixture, mostly methane, formed from coal. (6.6)

**system** The defined part of the universe under study. (6.1)

**systematic error** A type of error producing values that are either higher or lower than the actual value and often caused by faulty equipment or a consistent fault in technique. (1.6)

**$t_{2g}$ orbital set** The set of orbitals (composed of $d_{xy}$, $d_{yz}$, and $d_{xz}$) that results when the energies of the metal-ion $d$ orbitals are split by a ligand field. This set is lower in energy than the other ($e_g$) set in an octahedral field and higher in a tetrahedral field. (22.5)

**T shape** A molecular shape caused by the presence of two equatorial lone pairs in a trigonal bipyramidal arrangement. ($AX_3E_2$) (10.1)

**temperature ($T$)** A measure of how hot or cold a substance is relative to another substance. (1.5)

**tetrahedral arrangement** The geometric arrangement formed when four electron groups maximize their separation around a central atom. (10.1)

**tetrahedral shape** A molecular shape formed when four atoms around a central atom lie at the corners of a tetrahedron: bond angle 109.5°. (10.1)

**theoretical yield** The amount of product predicted by the stoichiometrically equivalent molar ratio in the balanced equation. (3.4)

**theory** (See *model.*)

**thermochemical equation** A chemical equation that shows the heat of reaction for the amounts of substances specified. (6.4)

**thermochemistry** The branch of thermodynamics that focuses on the heat involved in chemical reactions. (6 Intro.)

**thermodynamics** The study of heat (thermal energy) and its interconversions. (6 Intro.)

**thermometer** A device for measuring temperature that contains a fluid that expands or contracts within a graduated tube. (1.5)

**third law of thermodynamics** A law stating that the entropy of a perfect crystal is 0 J/mol · K at a temperature of absolute zero. (19.1)

**titration** A method of determining the concentration of a solution by monitoring its reaction with a solution of known concentration. (4.3)

**torr** A unit of pressure identical to 1 mmHg. (5.2)

**total ionic equation** An equation that shows all the soluble ionizable substances dissociated into ions. (4.3)

**tracer** A radioisotope that signals the presence of the species of interest. (21.5)

**transition element** (also *transition metal*) An element in which $d$ orbitals are being filled; sometimes includes the inner transition elements in which $f$ orbitals are filled. (8.2, 22 Intro.)

**transition state** See *activated complex.*

**transition state theory** A model that explains how the energy of reactant collision is used to form a transitional species that can change to reactant or product. (15.4)

**transuranium element** An element with atomic number higher than uranium ($Z = 92$). (21.3)

**trigonal bipyramidal arrangement** The geometric arrangement formed when five electron groups maximize the separation around a central atom. (10.1)

**trigonal bipyramidal shape** A molecular shape formed when five atoms around a central atom lie at the corners of a trigonal bipyramid: bond angles axial-center-equatorial 90°; equatorial-center-equatorial 120° (AX$_5$). (10.1)

**trigonal planar arrangement** The geometric arrangement formed when three electron groups maximize the space around a central atom. (10.1)

**trigonal planar shape** A molecular shape formed when three atoms around a central atom lie at the corners of an equilateral triangle: bond angle 120° (AX$_3$). (10.1)

**trigonal pyramidal shape** A molecular shape caused by the presence of one lone pair in a tetrahedral arrangement (AX$_3$E). (10.1)

**triple bond** A covalent bond that consists of three bonding pairs. Two atoms share six electrons; one σ and two π bonds. (9.3, 10.3)

**triple point** The pressure and temperature at which all three phases of a substance are in equilibrium. In a phase diagram, the point at which the three phase-transition curves meet. (11.5)

**troposphere** The portion of the Earth's atmosphere that extends from the surface to an altitude of about 11 km. (5.8)

**Tyndall effect** The scattering of light by a colloidal suspension. (12.6)

**ultraviolet (UV) region** The region of the electromagnetic spectrum between the visible and the x-ray regions. (7.1)

**uncertainty** A characteristic of every measurement that results from the inexactness of the measuring device and the necessity of estimating when taking a reading. (1.6)

**unimolecular reaction** An elementary reaction that involves the decomposition or rearrangement of a single particle. (15.5)

**unit cell** The simplest arrangement of points that, when repeated in all three directions, gives the lattice. (11.4)

**universal gas constant (*R*)** A proportionality constant relating the energy, amount of substance, and temperature of a system; $R = 0.08206$ atm · L/mol · K $= 8.314$ J/mol · K. (5.3)

**unsaturated hydrocarbon** A hydrocarbon with a carbon multiple bond; one in which C is bonded to fewer than four atoms. (14.2)

**unsaturated solution** A solution in which more solute can be dissolved at a given temperature. (12.3)

**unshared pair** (See *lone pair.*)

**valence band** The lower energy portion of the band of molecular orbitals filled with valence electrons. (11.4)

**valence bond (VB) theory** A model that attempts to reconcile the shapes of molecules with those of atomic orbitals through the concepts of orbital overlap and hybridization. (10.3)

**valence electrons** The electrons involved in compound formation; in main-group elements, those in the valence (outer) level. (8.2)

**valence-shell electron-pair repulsion (VSEPR) theory** A model explaining that the shapes of molecules and ions result from minimizing electron-pair repulsions around a central atom. (10.1)

**van der Waals constants** Experimentally determined positive numbers used in the van der Waals equation to account for the molecular interactions and molecular volume of real gases. (5.7)

**van der Waals equation** An equation that accounts for the behavior of real gases. (5.7)

**van der Waals radius** One-half the closest distance between the nuclei of identical *nonbonded* atoms. (11.2)

**vapor pressure** (also *equilibrium vapor pressure*) The pressure exerted by a vapor in a closed flask at equilibrium. (11.5)

**vapor pressure lowering** The lowering of the vapor pressure of a solvent due to the presence of dissolved solute particles. (12.5)

**vaporization** The process of changing from a liquid to a gas. (11.1)

**variable** A quantity that can have more than a single value. (1.3) (See *controlled experiment.*)

**viscosity** A measure of the resistance of a liquid to flow. (11.3)

**volatility** The tendency of a substance to become a gas. (2.8)

**volt (V)** The SI unit of electrical potential: 1 V = 1 J/C. (20.2)

**voltage** (See *cell potential*, $E_{cell}$.)

**voltaic cell** (also *galvanic cell*) An electrochemical cell that utilizes a spontaneous reaction to generate electrical energy. (20.1)

**volume (*V*)** The space occupied by an object. (1.5)

**volume percent (vol %)** A concentration term defined as the volume of solute in 100 volumes of solution. (12.4)

**wastewater** (also *sewage*) Used water, usually containing industrial and/or residential waste, that is treated before being returned to the environment. (12.6)

**water softening** The process of removing the hard-water ions $Ca^{2+}$ and $Mg^{2+}$ from water. (12.6)

**wave function** (See *atomic orbital.*)

**wave-particle duality** The principle stating that both matter and energy have wavelike and particle-like properties. (7.3)

**wavelength (λ)** The distance between any point on a wave and the corresponding point on the next wave; that is, the distance a wave travels during one cycle. (7.1)

**weak-field ligand** A ligand that causes small crystal field splitting and therefore a high-spin complex. (22.5)

**weight** The force exerted by a gravitational field on an object. (1.5)

**work (*w*)** The energy transferred when an object is moved by a force. (6.1)

**x-ray diffraction analysis** An instrumental technique used to determine variables in a crystal structure by measuring the diffraction patterns caused by x-rays impinging on the crystal. (11.4)

**zone refining** A process used to purify metals and metalloids in which impurities are removed from a bar of the element by concentrating them in a thin molten zone. (12.5, 23.3)

# Photo Credits

*All unlisted laboratory shots are by Stephen Frisch*

# Index

# Fundamental Physical Constants (six significant figures)

| | | |
|---|---|---|
| Avogadro's number | $N_A$ | $= 6.02214 \times 10^{23}/\text{mol}$ |
| atomic mass unit | amu | $= 1.66054 \times 10^{-27} \text{ kg}$ |
| charge of the electron (or proton) | $e$ | $= 1.60218 \times 10^{-19} \text{ C}$ |
| Faraday constant | $\mathscr{F}$ | $= 9.64853 \times 10^4 \text{ C/mol}$ |
| mass of the electron | $m_e$ | $= 9.10939 \times 10^{-31} \text{ kg}$ |
| mass of the neutron | $m_n$ | $= 1.67493 \times 10^{-27} \text{ kg}$ |
| mass of the proton | $m_p$ | $= 1.67262 \times 10^{-27} \text{ kg}$ |
| Planck's constant | $h$ | $= 6.63608 \times 10^{-34} \text{ J·s}$ |
| speed of light in a vacuum | $c$ | $= 2.99792 \times 10^8 \text{ m/s}$ |
| standard acceleration of gravity | $g$ | $= 9.80665 \text{ m/s}^2$ |
| universal gas constant | $R$ | $= 8.31451 \text{ J/mol·K}$ |
| | | $= 8.20578 \times 10^{-2} \text{ atm·L/mol·K}$ |

# SI Unit Prefixes

| p | n | µ | m | c | d | k | M | G |
|---|---|---|---|---|---|---|---|---|
| pico- | nano- | micro- | milli- | centi- | deci- | kilo- | mega- | giga- |
| $10^{-12}$ | $10^{-9}$ | $10^{-6}$ | $10^{-3}$ | $10^{-2}$ | $10^{-1}$ | $10^3$ | $10^6$ | $10^9$ |

# Conversions and Relationships

## Length
### SI unit: meter, m

1 km $= 1000$ m
$\quad\quad = 0.62$ mile (mi)
1 inch (in) $= 2.54$ cm
1 m $= 1.094$ yards (yd)
1 pm $= 10^{-12}$ m $= 0.01$ Å

## Volume
### SI unit: cubic meter, m³

$1 \text{ dm}^3 = 10^{-3} \text{ m}^3$
$\quad\quad = 1$ liter (L)
$\quad\quad = 1.057$ quarts (qt)
$1 \text{ cm}^3 = 1$ mL
$1 \text{ m}^3 = 35.3 \text{ ft}^3$

## Mass
### SI unit: kilogram, kg

1 kg $= 10^3$ g
$\quad\quad = 2.205$ lb
1 metric ton (T) $= 10^3$ kg

## Pressure
### SI unit: pascal, Pa

$1 \text{ Pa} = 1 \text{ N/m}^2$
$\quad\quad = 1 \text{ kg/m·s}^2$
$1 \text{ atm} = 1.01325 \times 10^5 \text{ Pa}$
$\quad\quad = 760$ torr

## Energy
### SI unit: joule, J

$1 \text{ J} = 1 \text{ kg/m}^2\text{·s}^2$
$\quad\quad = 1$ coulomb·volt (1 C·V)
1 cal $= 4.184$ J
$1 \text{ eV} = 1.602 \times 10^{-19} \text{ J}$

## Temperature
### SI unit: kelvin, K

0 K $= -273.15°\text{C}$
mp of $H_2O = 0°\text{C}$ (273.15 K)
bp of $H_2O = 100°\text{C}$ (373.15 K)
K $= °\text{C} + 273.15$
$°\text{C} = (°\text{F} - 32)\tfrac{5}{9}$
$°\text{F} = \tfrac{9}{5}°\text{C} + 32$

## Math relationships

$\pi = 3.1416$
volume of sphere $= \tfrac{4}{3}\pi r^3$
volume of cylinder $= \pi r^2 h$